ESSENTIAL CALCULUS
Early Transcendentals

SECOND EDITION

JAMES STEWART

McMaster University
and
University of Toronto

BROOKS/COLE
CENGAGE Learning·

Australia · Brazil · Japan · Korea · Mexico · Singapore · Spain · United Kingdom · United States

BROOKS/COLE
CENGAGE Learning

Essential Calculus: Early Transcendentals,
Second Edition

James Stewart

Executive Editor: Liz Covello

Developmental Editor: Carolyn Crockett

Assistant Editor: Liza Neustaetter

Editorial Assistant: Jennifer Staller

Media Editor: Maureen Ross

Marketing Manager: Jennifer Jones

Marketing Coordinator: Michael Ledesma

Marketing Communications Manager:
Mary Anne Payumo

Content Project Manager: Cheryll Linthicum

Art Director: Vernon Boes

Manufacturing Planner: Becky Cross

Rights Acquisitions Specialist: Roberta Broyer

Production and Composition Service: TECH·arts

Photo Researcher: Terri Wright

Text Researcher: Terri Wright

Copy Editor: Kathi Townes

Compositor: Stephanie Kuhns

Illustrator: TECH·arts

Text Designer: Geri Davis

Cover Designer: Denise Davidson

Cover Image: Denise Davidson

For product information and technology assistance, contact us at
Cengage Learning Customer & Sales Support, 1-800-354-9706

For permission to use material from this text or product, submit all requests online at **www.cengage.com/permissions** Further permissions questions can be e-mailed to **permissionrequest@cengage.com**

Library of Congress Control Number: 2011942869

ISBN-13: 978-1-133-11228-0

ISBN-10: 1-133-11228-5

Brooks/Cole
20 Davis Drive
Belmont, CA 94002-3098
USA

Cengage Learning is a leading provider of customized learning solutions with office locations around the globe, including Singapore, the United Kingdom, Australia, Mexico, Brazil, and Japan. Locate your local office at **www.cengage.com/global**.

Cengage Learning products are represented in Canada by Nelson Education, Ltd.

To learn more about Brooks/Cole, visit
www.cengage.com/brookscole

Purchase any of our products at your local college store or at our preferred online store **www.cengagebrain.com**.

Trademarks

Derive is a registered trademark of Soft Warehouse, Inc.
Maple is a registered trademark of Waterloo Maple, Inc.
Mathematica is a registered trademark of Wolfram Research, Inc.
Tools for Enriching is a trademark used herein under license.

Printed in the United States of America
2 3 4 5 6 7 8 16 15 14 13 12

K07T12

CONTENTS

1 | FUNCTIONS AND LIMITS   1

2 | DERIVATIVES   73

3 | INVERSE FUNCTIONS: Exponential, Logarithmic, and Inverse Trigonometric Functions   145

4 | APPLICATIONS OF DIFFERENTIATION   203

5 | INTEGRALS   257

6 | TECHNIQUES OF INTEGRATION   311

10 | VECTORS AND THE GEOMETRY OF SPACE   537

11 | PARTIAL DERIVATIVES   615

12 | MULTIPLE INTEGRALS   689

PREFACE

This book is a response to those instructors who feel that calculus textbooks are too big. In writing the book I asked myself: What is *essential* for a three-semester calculus course for scientists and engineers?

The book is about two-thirds the size of my other calculus books (*Calculus,* Seventh Edition and *Calculus, Early Transcendentals,* Seventh Edition) and yet it contains almost all of the same topics. I have achieved relative brevity mainly by condensing the exposition and by putting some of the features on the website **www.stewartcalculus.com**. Here, in more detail are some of the ways I have reduced the bulk:

- I have organized topics in an efficient way and rewritten some sections with briefer exposition.

- The design saves space. In particular, chapter opening spreads and photographs have been eliminated.

- The number of examples is slightly reduced. Additional examples are provided online.

- The number of exercises is somewhat reduced, though most instructors will find that there are plenty. In addition, instructors have access to the archived problems on the website.

- Although I think projects can be a very valuable experience for students, I have removed them from the book and placed them on the website.

- A discussion of the principles of problem solving and a collection of challenging problems for each chapter have been moved to the website.

Despite the reduced size of the book, there is still a modern flavor: Conceptual understanding and technology are not neglected, though they are not as prominent as in my other books.

ALTERNATE VERSIONS

I have written several other calculus textbooks that might be preferable for some instructors. Most of them also come in single variable and multivariable versions.

- *Essential Calculus*, Second Edition, is similar to the present textbook except that the logarithm is defined as an integral and so the exponential, logarithmic, and inverse trigonometric functions are covered later than in the present book.

- *Calculus: Early Transcendentals*, Seventh Edition, has more complete coverage of calculus than the present book, with somewhat more examples and exercises.

- *Calculus: Early Transcendentals*, Seventh Edition, Hybrid Version, is similar to *Calculus: Early Transcendentals*, Seventh Edition, in content and coverage except that all of the end-of-section exercises are available only in Enhanced WebAssign. The printed text includes all end-of-chapter review material.

- *Calculus*, Seventh Edition, is similar to *Calculus: Early Transcendentals*, Seventh Edition, except that the exponential, logarithmic, and inverse trigonometric functions are covered in the second semester. It is also available in a Hybrid Version.

- *Calculus: Concepts and Contexts*, Fourth Edition, emphasizes conceptual understanding. The coverage of topics is not encyclopedic and the material on transcendental functions and on parametric equations is woven throughout the book instead of being treated in separate chapters. It is also available in a Hybrid Version.

- *Calculus: Early Vectors* introduces vectors and vector functions in the first semester and integrates them throughout the book. It is suitable for students taking Engineering and Physics courses concurrently with calculus.

- *Brief Applied Calculus* is intended for students in business, the social sciences, and the life sciences. It is also available in a Hybrid Version.

WHAT'S NEW IN THE SECOND EDITION?

The changes have resulted from talking with my colleagues and students at the University of Toronto and from reading journals, as well as suggestions from users and reviewers. Here are some of the many improvements that I've incorporated into this edition:

- At the beginning of the book there are four diagnostic tests, in Basic Algebra, Analytic Geometry, Functions, and Trigonometry. Answers are given and students who don't do well are referred to where they should seek help (Appendixes, review sections of Chapter 1, and the website).

- Section 7.5 (Area of a Surface of Revolution) is new. I had asked reviewers if there was any topic missing from the first edition that they regarded as essential. This was the only topic that was mentioned by more than one reviewer.

- Some material has been rewritten for greater clarity or for better motivation. See, for instance, the introduction to maximum and minimum values on pages 203–04 and the introduction to series on page 436.

- New examples have been added (see Example 4 on page 725 for instance). And the solutions to some of the existing examples have been amplified. A case in point: I added details to the solution of Example 1.4.9 because when I taught Section 1.4 from the first edition I realized that students need more guidance when setting up inequalities for the Squeeze Theorem.

- The data in examples and exercises have been updated to be more timely.

- Several new historical margin notes have been added.

- About 40% of the exercises are new. Here are some of my favorites: 1.6.43, 2.2.13–14, 2.5.59, 2.6.39–40, 3.2.70, 4.3.66, 5.3.44–45, 7.6.24, 8.2.29–30, 8.7.67–68, 10.1.38, 10.4.43–44

- The animations in *Tools for Enriching Calculus* (TEC) have been completely redesigned and are accessible in Enhanced WebAssign, CourseMate, and PowerLecture. Selected Visuals and Modules are available at www.stewartcalculus.com.

CONTENT

DIAGNOSTIC TESTS ▪ The book begins with four diagnostic tests, in Basic Algebra, Analytic Geometry, Functions, and Trigonometry.

CHAPTER 1 ▪ **FUNCTIONS AND LIMITS** After a brief review of the basic functions, limits and continuity are introduced, including limits of trigonometric functions, limits involving infinity, and precise definitions.

CHAPTER 2 ▪ DERIVATIVES The material on derivatives is covered in two sections in order to give students time to get used to the idea of a derivative as a function. The formulas for the derivatives of the sine and cosine functions are derived in the section on basic differentiation formulas. Exercises explore the meanings of derivatives in various contexts.

CHAPTER 3 ▪ INVERSE FUNCTIONS: EXPONENTIAL, LOGARITHMIC, AND INVERSE TRIGONO-METRIC FUNCTIONS Exponential functions are defined first and the number e is defined as a limit. Logarithms are then defined as inverse functions. Applications to exponential growth and decay follow. Inverse trigonometric functions and hyperbolic functions are also covered here. L'Hospital's Rule is included in this chapter because limits of transcendental functions so often require it.

CHAPTER 4 ▪ APPLICATIONS OF DIFFERENTIATION The basic facts concerning extreme values and shapes of curves are deduced from the Mean Value Theorem. The section on curve sketching includes a brief treatment of graphing with technology. The section on optimization problems contains a brief discussion of applications to business and economics.

CHAPTER 5 ▪ INTEGRALS The area problem and the distance problem serve to motivate the definite integral, with sigma notation introduced as needed. (Full coverage of sigma notation is provided in Appendix B.) A quite general definition of the definite integral (with unequal subintervals) is given initially before regular partitions are employed. Emphasis is placed on explaining the meanings of integrals in various contexts and on estimating their values from graphs and tables.

CHAPTER 6 ▪ TECHNIQUES OF INTEGRATION All the standard methods are covered, as well as computer algebra systems, numerical methods, and improper integrals.

CHAPTER 7 ▪ APPLICATIONS OF INTEGRATION General methods are emphasized. The goal is for students to be able to divide a quantity into small pieces, estimate with Riemann sums, and recognize the limit as an integral. The chapter concludes with an introduction to differential equations, including separable equations and direction fields.

CHAPTER 8 ▪ SERIES The convergence tests have intuitive justifications as well as formal proofs. The emphasis is on Taylor series and polynomials and their applications to physics. Error estimates include those based on Taylor's Formula (with Lagrange's form of the remainder term) and those from graphing devices.

CHAPTER 9 ▪ PARAMETRIC EQUATIONS AND POLAR COORDINATES This chapter introduces parametric and polar curves and applies the methods of calculus to them. A brief treatment of conic sections in polar coordinates prepares the way for Kepler's Laws in Chapter 10.

CHAPTER 10 ▪ VECTORS AND THE GEOMETRY OF SPACE In addition to the material on vectors, dot and cross products, lines, planes, and surfaces, this chapter covers vector-valued functions, length and curvature of space curves, and velocity and acceleration along space curves, culminating in Kepler's laws.

CHAPTER 11 ▪ PARTIAL DERIVATIVES In view of the fact that many students have difficulty forming mental pictures of the concepts of this chapter, I've placed a special emphasis on graphics to elucidate such ideas as graphs, contour maps, directional derivatives, gradients, and Lagrange multipliers.

CHAPTER 12 ▪ MULTIPLE INTEGRALS Cylindrical and spherical coordinates are introduced in the context of evaluating triple integrals.

CHAPTER 13 ▪ VECTOR CALCULUS The similarities among the Fundamental Theorem for line integrals, Green's Theorem, Stokes' Theorem, and the Divergence Theorem are emphasized.

WEBSITE

The web site **www.stewartcalulus.com** includes the following.

- Review of Algebra, Trigonometry, Analytic Geometry, and Conic Sections
- Homework Hints
- Additional Examples
- Projects
- Archived Problems (drill exercises that were in previous editions of my other books), together with their solutions
- Challenge Problems
- Lies My Calculator and Computer Told Me
- Additional Topics (complete with exercise sets): Principles of Problem Solving, Strategy for Integration, Strategy for Testing Series, Fourier Series, Linear Differential Equations, Second Order Linear Differential Equations, Nonhomogeneous Linear Equations, Applications of Second Order Differential Equations, Using Series to Solve Differential Equations, Complex Numbers, Rotation of Axes
- Links, for particular topics, to outside Web resources
- History of Mathematics, with links to the better historical websites
- TEC animations for Chapters 2 and 5

ACKNOWLEDGMENTS

I thank the following reviewers for their thoughtful comments:

SECOND EDITION REVIEWERS

Allison Arnold, *University of Georgia*

Rachel Belinsky, *Georgia State University*

Geoffrey D. Birky, *Georgetown University*

Przemyslaw Bogacki, *Old Dominion University*

Mark Brittenham, *University of Nebraska at Lincoln*

Katrina K. A. Cunningham, *Southern University* and *A&M College*

Morley Davidson, *Kent State University*

M. Hilary Davies, *University of Alaska Anchorage*

Shelby J. Kilmer, *Missouri State University*

Ilya Kofman, *College of Staten Island*, CUNY

Ramendra Krishna Bose, *University of Texas–Pan American*

Melvin Lax, *California State University Long Beach*

Derek Martinez, *Central New Mexico Community College*

Alex M. McAllister, *Centre College*

Michael McAsey, *Bradley University*

Humberto Munoz, *Southern University* and *A&M College*

Charlotte Newsom, *Tidewater Community College*

Michael Price, *University of Oregon*

Joe Rody, *Arizona State University*

Vicki Sealey, *West Virginia University*

David Shannon, *Transylvania University*

FIRST EDITION REVIEWERS

Ulrich Albrecht, *Auburn University*

Christopher Butler, *Case Western Reserve University*

Joe Fisher, *University of Cincinnati*

John Goulet, *Worchester Polytechnic Institute*

Irvin Hentzel, *Iowa State University*

Joel Irish, *University of Southern Maine*

Mary Nelson, *University of Colorado, Boulder*

Ed Slaminka, *Auburn University*

Li (Jason) Zhongshan, *Georgia State University*

I also thank Jim Propp of the University of Massachusetts—Lowell for a number of suggestions resulting from his teaching from the first edition.

In addition, I thank Kathi Townes and Stephanie Kuhns for their production services and the following Brooks/Cole staff: Cheryll Linthicum, editorial content project manager; Vernon Boes, art director; Jennifer Jones and Mary Anne Payumo, marketing team; Maureen Ross, media editor; Carolyn Crockett, development editor; Elizabeth Neustaetter, assistant editor; Jennifer Staller, editorial assistant; Roberta Broyer, rights acquisitions specialist; Becky Cross, manufacturing planner; and Denise Davidson, cover designer. They have all done an outstanding job.

The idea for this book came from my former editor Bob Pirtle, who had been hearing of the desire for a much shorter calculus text from numerous instructors. I thank my present editor Liz Covello for sustaining and supporting this idea in the second edition.

JAMES STEWART

ANCILLARIES FOR INSTRUCTORS

PowerLecture
ISBN 1-133-52566-0

This comprehensive DVD contains all art from the text in both jpeg and PowerPoint formats, complete pre-built PowerPoint lectures, an electronic version of the Instructor's Guide, Solution Builder, ExamView algorithmic testing software, Tools for Enriching Calculus, and video instruction.

Instructor's Guide
By Douglas Shaw
ISBN 1-133-52510-5

Each section of the text is discussed from several viewpoints. The Instructor's Guide contains suggested time to allot, points to stress, text discussion topics, core materials for lecture, work-show/discussion suggestions, group work exercises in a form suitable for handout, and suggested homework assignments. An electronic version of the Instructor's Guide is available on the PowerLecture DVD.

Complete Solutions Manual
ISBN 1-133-36444-6

Includes worked-out solutions to all exercises in the text.

Solution Builder
www.cengage.com/solutionbuilder

This online instructor database offers complete worked-out solutions to all exercises in the text. Solution Builder allows you to create customized, secure solution printouts (in PDF format) matched exactly to the problems you assign in class.

ExamView Algorithmic Testing

Create, deliver, and customize tests in print and online formats with ExamView, an easy-to-use assessment and tutorial software. ExamView contains hundreds of multiple-choice, numerical response, and short answer test items. ExamView algorithmic testing is available on the PowerLecture DVD.

ANCILLARIES FOR INSTRUCTORS AND STUDENTS

Stewart Website
www.stewartcalculus.com

Contents: *Review of Algebra, Trigonometry, Analytic Geometry, and Conic Sections* ▪ *Homework Hints* ▪ *Additional Examples* ▪ *Projects* ▪ *Archived Problems* ▪ *Challenge Problems* ▪ *Lies My Calculator and Computer Told Me* ▪ *Principles of Problem Solving* ▪ *Additional Topics* ▪ *Web Links* ▪ *History of Mathematics*

TEC Tools for Enriching™ Calculus

By James Stewart, Harvey Keynes, Dan Clegg, and developer Hu Hohn

Tools for Enriching Calculus (TEC) functions as both a powerful tool for instructors, as well as a tutorial environment in which students can explore and review selected topics. The Flash simulation modules in TEC include instructions, written and audio explanations, and exercises. TEC modules are assignable in Enhanced WebAssign. TEC is also available at www.stewartcalculus.com, as well as in the YouBook and CourseMate.

WebAssign Enhanced WebAssign
www.webassign.net

WebAssign's homework system lets instructors deliver, collect, and record assignments via the Web. Enhanced WebAssign for Stewart's *Essential Calculus: Early Transcendentals* now includes opportunities for students to review prerequisite skills and content both at the start of the course and at the beginning of each section. In addition, for selected problems, students can get extra help in the form of "enhanced feedback" (rejoinders) and video solutions. *Other key features include:* thousands of problems from Stewart's *Essential Calculus: Early Transcendentals,* a *QuickPrep for Calculus* review, a customizable Cengage YouBook, Just In Time Review questions, a Show My Work feature, assignable Tools for Enriching Calculus modules, quizzes, lecture videos (with associated questions), and more!

Cengage Customizable YouBook

YouBook is an eBook that is both interactive and customizable! Containing all the content from Stewart's *Essential Calculus: Early Transcendentals,* YouBook features a text edit tool that allows instructors to modify the textbook narrative as needed. With YouBook, instructors can quickly reorder entire sections and chapters or hide any content they don't teach to create an eBook that perfectly matches their syllabus. Instructors can further customize the text by adding instructor-created or YouTube video links. Additional media assets include: Tools for Enriching Calculus visuals and modules, Wolfram animations, video clips, highlighting, notes, and more! YouBook is available in Enhanced WebAssign.

CourseMate

www.cengagebrain.com

CourseMate is a perfect self-study tool for students, and requires no set-up from instructors. CourseMate brings course concepts to life with interactive learning, study, and exam preparation tools that support the printed textbook. CourseMate for Stewart's *Essential Calculus: Early Transcendentals* includes: an interactive eBook, Tools for Enriching Calculus, videos, quizzes, flashcards, and more! For instructors, CourseMate includes engagement tracker, a first-of-its kind tool that monitors student engagement.

Maple

Maple™ is an essential tool that allows you to explore, visualize, and solve even the most complex mathematical problems, reducing errors and providing greater insight into the math. Maple's world-leading computation engine offers the breadth, depth, and performance to handle every type of mathematics. With Maple, teachers can bring complex problems to life and students can focus on concepts rather than the mechanics of solutions. Maple's intuitive interface supports multiple styles of interaction, from Clickable Math™ tools to a sophisticated programming language.

CengageBrain.com

To access additional course materials and companion resources, please visit www.cengagebrain.com. At the CengageBrain.com home page, search for the ISBN of your title (from the back cover of your book) using the search box at the top of the page. This will take you to the product page where free companion resources can be found.

ANCILLARIES FOR STUDENTS

Student Solutions Manual
ISBN 1-133-49097-2

Provides completely worked-out solutions to all odd-numbered exercises in the text, giving students a chance to check their answers and ensure they took the correct steps to arrive at an answer.

CalcLabs with Maple

SINGLE VARIABLE By Robert J. Lopez and Philip B. Yasskin
ISBN 0-8400-5811-X

MULTIVARIABLE By Robert J. Lopez and Philip B. Yasskin
ISBN 0-8400-5812-8

CalcLabs with Mathematica

SINGLE VARIABLE By Selwyn Hollis
ISBN 0-8400-5814-4

MULTIVARIABLE By Selwyn Hollis
ISBN 0-8400-5813-6

Each of these comprehensive lab manuals will help students learn to use the technology tools available to them. CalcLabs contain clearly explained exercises and a variety of labs and projects.

A Companion to Calculus

By Dennis Ebersole, Doris Schattschneider, Alicia Sevilla, and Kay Somers
ISBN 0-495-01124-X

Written to improve algebra and problem-solving skills of students taking a calculus course, every chapter in this companion is keyed to a calculus topic, providing conceptual background and specific algebra techniques needed to understand and solve calculus problems related to that topic. It is designed for calculus courses that integrate the review of precalculus concepts or for individual use.

Linear Algebra for Calculus

By Konrad J. Heuvers, William P. Francis, John H. Kuisti, Deborah F. Lockhart, Daniel S. Moak, and Gene M. Ortner
ISBN 0-534-25248-6

This comprehensive book, designed to supplement the calculus course, provides an introduction to and review of the basic ideas of linear algebra.

TO THE STUDENT

Reading a calculus textbook is different from reading a newspaper or a novel, or even a physics book. Don't be discouraged if you have to read a passage more than once in order to understand it. You should have pencil and paper and calculator at hand to sketch a diagram or make a calculation.

Some students start by trying their homework problems and read the text only if they get stuck on an exercise. I suggest that a far better plan is to read and understand a section of the text before attempting the exercises. In particular, you should look at the definitions to see the exact meanings of the terms. And before you read each example, I suggest that you cover up the solution and try solving the problem yourself. You'll get a lot more from looking at the solution if you do so.

Part of the aim of this course is to train you to think logically. Learn to write the solutions of the exercises in a connected, step-by-step fashion with explanatory sentences—not just a string of disconnected equations or formulas.

The answers to the odd-numbered exercises appear at the back of the book, in Appendix E. Some exercises ask for a verbal explanation or interpretation or description. In such cases there is no single correct way of expressing the answer, so don't worry that you haven't found the definitive answer. In addition, there are often several different forms in which to express a numerical or algebraic answer, so if your answer differs from mine, don't immediately assume you're wrong. For example, if the answer given in the back of the book is $\sqrt{2} - 1$ and you obtain $1/(1 + \sqrt{2})$, then you're right and rationalizing the denominator will show that the answers are equivalent.

The icon ⊞ indicates an exercise that definitely requires the use of either a graphing calculator or a computer with graphing software. (The use of these graphing devices and some of the pitfalls that you may encounter are discussed on **stewartcalculus.com**. Go to *Additional Topics* and click on *Graphing Calculators and Computers*.) But that doesn't mean that graphing devices can't be used to check your work on the other exercises as well. The symbol CAS is reserved for problems in which the full resources of a computer algebra system (like Derive, Maple, Mathematica, or the TI-89/92) are required.

You will also encounter the symbol ⊘, which warns you against committing an error. I have placed this symbol in the margin in situations where I have observed that a large proportion of my students tend to make the same mistake.

Tools for Enriching Calculus, which is a companion to this text, is referred to by means of the symbol TEC and can be accessed in Enhanced WebAssign and CourseMate (selected Visuals and Modules are available at www.stewartcalculus.com). It directs you to modules in which you can explore aspects of calculus for which the computer is particularly useful.

There is a lot of useful information on the website **stewartcalculus.com**. There you will find a review of precalculus topics (in case your algebraic skills are rusty), as well as *Homework Hints* (see the following paragraph), *Additional Examples* (see below), *Challenge Problems*, *Projects*, *Lies My Calculator and Computer Told Me*, (explaining why calculators sometimes give the wrong answer), *Additional Topics*, and links to outside resources.

Homework Hints for representative exercises are indicated by printing the exercise number in blue: **5.** These hints can be found on stewartcalculus.com as well as Enhanced WebAssign and CourseMate. The homework hints ask you questions that allow you to make progress toward a solution without actually giving you the answer. You need to pursue each hint in an active manner with pencil and paper to work out the details. If a particular hint doesn't enable you to solve the problem, you can click to reveal the next hint.

You will see margin notes in some sections directing you to *Additional Examples* on the website. You will also see the symbol V beside two or three of the examples in every section of the text. This means that there are videos (in Enhanced WebAssign and CourseMate) of instructors explaining those examples in greater detail.

I recommend that you keep this book for reference purposes after you finish the course. Because you will likely forget some of the specific details of calculus, the book will serve as a useful reminder when you need to use calculus in subsequent courses. And, because this book contains more material than can be covered in any one course, it can also serve as a valuable resource for a working scientist or engineer.

Calculus is an exciting subject, justly considered to be one of the greatest achievements of the human intellect. I hope you will discover that it is not only useful but also intrinsically beautiful.

JAMES STEWART

DIAGNOSTIC TESTS

Success in calculus depends to a large extent on knowledge of the mathematics that precedes calculus: algebra, analytic geometry, functions, and trigonometry. The following tests are intended to diagnose weaknesses that you might have in these areas. After taking each test you can check your answers against the given answers and, if necessary, refresh your skills by referring to the review materials that are provided.

A | DIAGNOSTIC TEST: ALGEBRA

1. Evaluate each expression without using a calculator.
   (a) $(-3)^4$
   (b) -3^4
   (c) 3^{-4}
   (d) $\dfrac{5^{23}}{5^{21}}$
   (e) $\left(\dfrac{2}{3}\right)^{-2}$
   (f) $16^{-3/4}$

2. Simplify each expression. Write your answer without negative exponents.
   (a) $\sqrt{200} - \sqrt{32}$
   (b) $(3a^3b^3)(4ab^2)^2$
   (c) $\left(\dfrac{3x^{3/2}y^3}{x^2y^{-1/2}}\right)^{-2}$

3. Expand and simplify.
   (a) $3(x + 6) + 4(2x - 5)$
   (b) $(x + 3)(4x - 5)$
   (c) $(\sqrt{a} + \sqrt{b})(\sqrt{a} - \sqrt{b})$
   (d) $(2x + 3)^2$
   (e) $(x + 2)^3$

4. Factor each expression.
   (a) $4x^2 - 25$
   (b) $2x^2 + 5x - 12$
   (c) $x^3 - 3x^2 - 4x + 12$
   (d) $x^4 + 27x$
   (e) $3x^{3/2} - 9x^{1/2} + 6x^{-1/2}$
   (f) $x^3y - 4xy$

5. Simplify the rational expression.
   (a) $\dfrac{x^2 + 3x + 2}{x^2 - x - 2}$
   (b) $\dfrac{2x^2 - x - 1}{x^2 - 9} \cdot \dfrac{x + 3}{2x + 1}$
   (c) $\dfrac{x^2}{x^2 - 4} - \dfrac{x + 1}{x + 2}$
   (d) $\dfrac{\dfrac{y}{x} - \dfrac{x}{y}}{\dfrac{1}{y} - \dfrac{1}{x}}$

6. Rationalize the expression and simplify.
   (a) $\dfrac{\sqrt{10}}{\sqrt{5} - 2}$
   (b) $\dfrac{\sqrt{4 + h} - 2}{h}$

7. Rewrite by completing the square.
   (a) $x^2 + x + 1$
   (b) $2x^2 - 12x + 11$

8. Solve the equation. (Find only the real solutions.)

(a) $x + 5 = 14 - \frac{1}{2}x$

(b) $\dfrac{2x}{x+1} = \dfrac{2x-1}{x}$

(c) $x^2 - x - 12 = 0$

(d) $2x^2 + 4x + 1 = 0$

(e) $x^4 - 3x^2 + 2 = 0$

(f) $3|x - 4| = 10$

(g) $2x(4-x)^{-1/2} - 3\sqrt{4-x} = 0$

9. Solve each inequality. Write your answer using interval notation.

(a) $-4 < 5 - 3x \le 17$

(b) $x^2 < 2x + 8$

(c) $x(x-1)(x+2) > 0$

(d) $|x - 4| < 3$

(e) $\dfrac{2x-3}{x+1} \le 1$

10. State whether each equation is true or false.

(a) $(p+q)^2 = p^2 + q^2$

(b) $\sqrt{ab} = \sqrt{a}\,\sqrt{b}$

(c) $\sqrt{a^2 + b^2} = a + b$

(d) $\dfrac{1 + TC}{C} = 1 + T$

(e) $\dfrac{1}{x-y} = \dfrac{1}{x} - \dfrac{1}{y}$

(f) $\dfrac{1/x}{a/x - b/x} = \dfrac{1}{a-b}$

A | ANSWERS TO DIAGNOSTIC TEST A: ALGEBRA

1. (a) 81   (b) -81 (c) $\frac{1}{81}$
    (d) 25   (e) $\frac{9}{4}$ (f) $\frac{1}{8}$

2. (a) $6\sqrt{2}$ (b) $48a^5 b^7$ (c) $\dfrac{x}{9y^7}$

3. (a) $11x - 2$ (b) $4x^2 + 7x - 15$
    (c) $a - b$ (d) $4x^2 + 12x + 9$
    (e) $x^3 + 6x^2 + 12x + 8$

4. (a) $(2x - 5)(2x + 5)$ (b) $(2x - 3)(x + 4)$
    (c) $(x - 3)(x - 2)(x + 2)$ (d) $x(x + 3)(x^2 - 3x + 9)$
    (e) $3x^{-1/2}(x - 1)(x - 2)$ (f) $xy(x - 2)(x + 2)$

5. (a) $\dfrac{x+2}{x-2}$ (b) $\dfrac{x-1}{x-3}$
    (c) $\dfrac{1}{x-2}$ (d) $-(x + y)$

6. (a) $5\sqrt{2} + 2\sqrt{10}$ (b) $\dfrac{1}{\sqrt{4+h}+2}$

7. (a) $\left(x + \frac{1}{2}\right)^2 + \frac{3}{4}$ (b) $2(x - 3)^2 - 7$

8. (a) 6   (b) 1   (c) $-3, 4$
    (d) $-1 \pm \frac{1}{2}\sqrt{2}$ (e) $\pm 1, \pm\sqrt{2}$ (f) $\frac{2}{3}, \frac{22}{3}$
    (g) $\frac{12}{5}$

9. (a) $[-4, 3)$ (b) $(-2, 4)$
    (c) $(-2, 0) \cup (1, \infty)$ (d) $(1, 7)$
    (e) $(-1, 4]$

10. (a) False   (b) True   (c) False
    (d) False   (e) False   (f) True

If you have had difficulty with these problems, you may wish to consult the Review of Algebra on the website www.stewartcalculus.com.

B | DIAGNOSTIC TEST: ANALYTIC GEOMETRY

1. Find an equation for the line that passes through the point $(2, -5)$ and
   (a) has slope -3
   (b) is parallel to the x-axis
   (c) is parallel to the y-axis
   (d) is parallel to the line $2x - 4y = 3$

2. Find an equation for the circle that has center $(-1, 4)$ and passes through the point $(3, -2)$.

3. Find the center and radius of the circle with equation $x^2 + y^2 - 6x + 10y + 9 = 0$.

4. Let $A(-7, 4)$ and $B(5, -12)$ be points in the plane.
   (a) Find the slope of the line that contains A and B.
   (b) Find an equation of the line that passes through A and B. What are the intercepts?
   (c) Find the midpoint of the segment AB.
   (d) Find the length of the segment AB.
   (e) Find an equation of the perpendicular bisector of AB.
   (f) Find an equation of the circle for which AB is a diameter.

5. Sketch the region in the xy-plane defined by the equation or inequalities.
   (a) $-1 \leqslant y \leqslant 3$
   (b) $|x| < 4$ and $|y| < 2$
   (c) $y < 1 - \frac{1}{2}x$
   (d) $y \geqslant x^2 - 1$
   (e) $x^2 + y^2 < 4$
   (f) $9x^2 + 16y^2 = 144$

B | ANSWERS TO DIAGNOSTIC TEST B: ANALYTIC GEOMETRY

1. (a) $y = -3x + 1$
   (b) $y = -5$
   (c) $x = 2$
   (d) $y = \frac{1}{2}x - 6$

2. $(x + 1)^2 + (y - 4)^2 = 52$

3. Center $(3, -5)$, radius 5

4. (a) $-\frac{4}{3}$
   (b) $4x + 3y + 16 = 0$; x-intercept -4, y-intercept $-\frac{16}{3}$
   (c) $(-1, -4)$
   (d) 20
   (e) $3x - 4y = 13$
   (f) $(x + 1)^2 + (y + 4)^2 = 100$

5. (a)

(b)

(c)

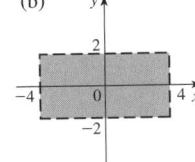

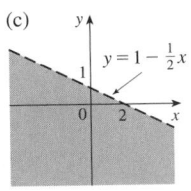

(d)

(e)

(f)

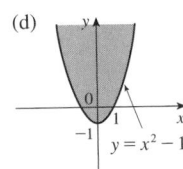

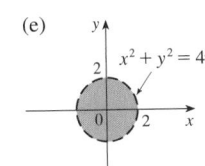

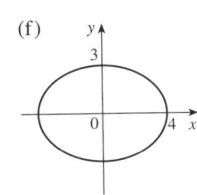

If you have had difficulty with these problems, you may wish to consult the review of analytic geometry on the website www.stewartcalculus.com.

C | DIAGNOSTIC TEST: FUNCTIONS

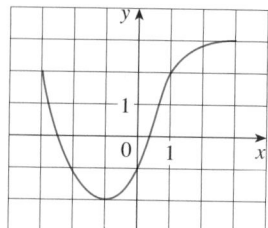

FIGURE FOR PROBLEM 1

1. The graph of a function f is given at the left.
   (a) State the value of $f(-1)$.
   (b) Estimate the value of $f(2)$.
   (c) For what values of x is $f(x) = 2$?
   (d) Estimate the values of x such that $f(x) = 0$.
   (e) State the domain and range of f.

2. If $f(x) = x^3$, evaluate the difference quotient $\dfrac{f(2 + h) - f(2)}{h}$ and simplify your answer.

3. Find the domain of the function.

   (a) $f(x) = \dfrac{2x + 1}{x^2 + x - 2}$ (b) $g(x) = \dfrac{\sqrt[3]{x}}{x^2 + 1}$ (c) $h(x) = \sqrt{4 - x} + \sqrt{x^2 - 1}$

4. How are graphs of the functions obtained from the graph of f?

   (a) $y = -f(x)$ (b) $y = 2f(x) - 1$ (c) $y = f(x - 3) + 2$

5. Without using a calculator, make a rough sketch of the graph.

   (a) $y = x^3$ (b) $y = (x + 1)^3$ (c) $y = (x - 2)^3 + 3$
   (d) $y = 4 - x^2$ (e) $y = \sqrt{x}$ (f) $y = 2\sqrt{x}$
   (g) $y = -2^x$ (h) $y = 1 + x^{-1}$

6. Let $f(x) = \begin{cases} 1 - x^2 & \text{if } x \leqslant 0 \\ 2x + 1 & \text{if } x > 0 \end{cases}$

   (a) Evaluate $f(-2)$ and $f(1)$.   (b) Sketch the graph of f.

7. If $f(x) = x^2 + 2x - 1$ and $g(x) = 2x - 3$, find each of the following functions.
   (a) $f \circ g$ (b) $g \circ f$ (c) $g \circ g \circ g$

C | ANSWERS TO DIAGNOSTIC TEST C: FUNCTIONS

1. (a) -2 (b) 2.8
(c) $-3, 1$ (d) $-2.5, 0.3$
(e) $[-3, 3], [-2, 3]$

2. $12 + 6h + h^2$

3. (a) $(-\infty, -2) \cup (-2, 1) \cup (1, \infty)$
(b) $(-\infty, \infty)$
(c) $(-\infty, -1] \cup [1, 4]$

4. (a) Reflect about the x-axis
(b) Stretch vertically by a factor of 2, then shift 1 unit downward
(c) Shift 3 units to the right and 2 units upward

(d)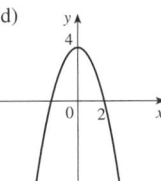

(e)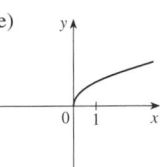

(f)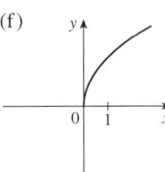

(g)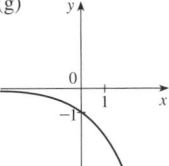

(h)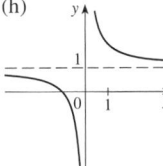

5. (a) 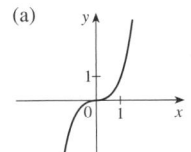   (b) 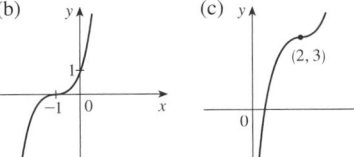   (c)

6. (a) $-3, 3$
(b)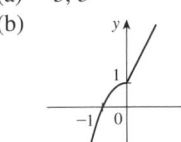

7. (a) $(f \circ g)(x) = 4x^2 - 8x + 2$
(b) $(g \circ f)(x) = 2x^2 + 4x - 5$
(c) $(g \circ g \circ g)(x) = 8x - 21$

If you have had difficulty with these problems, you should look at Sections 1.1–1.2 of this book.

D | DIAGNOSTIC TEST: TRIGONOMETRY

1. Convert from degrees to radians.
   (a) $300°$ (b) $-18°$

2. Convert from radians to degrees.
   (a) $5\pi/6$ (b) 2

3. Find the length of an arc of a circle with radius 12 cm if the arc subtends a central angle of $30°$.

4. Find the exact values.
   (a) $\tan(\pi/3)$ (b) $\sin(7\pi/6)$ (c) $\sec(5\pi/3)$

5. Express the lengths a and b in the figure in terms of θ.

6. If $\sin x = \frac{1}{3}$ and $\sec y = \frac{5}{4}$, where x and y lie between 0 and $\pi/2$, evaluate $\sin(x + y)$.

7. Prove the identities.
   (a) $\tan\theta \sin\theta + \cos\theta = \sec\theta$

   (b) $\dfrac{2 \tan x}{1 + \tan^2 x} = \sin 2x$

8. Find all values of x such that $\sin 2x = \sin x$ and $0 \leqslant x \leqslant 2\pi$.

9. Sketch the graph of the function $y = 1 + \sin 2x$ without using a calculator.

FIGURE FOR PROBLEM 5

(triangle with hypotenuse 24, angle θ, opposite side a, adjacent side b)

D | ANSWERS TO DIAGNOSTIC TEST D: TRIGONOMETRY

1. (a) $5\pi/3$ (b) $-\pi/10$

2. (a) $150°$ (b) $(360/\pi)° \approx 114.6°$

3. 2π cm

4. (a) $\sqrt{3}$ (b) $-\frac{1}{2}$ (c) 2

5. (a) $24 \sin\theta$ (b) $24 \cos\theta$

6. $\frac{1}{15}\left(4 + 6\sqrt{2}\right)$

8. $0, \pi/3, \pi, 5\pi/3, 2\pi$

9.
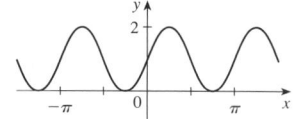

If you have had difficulty with these problems, you should look at Appendix A of this book.

1 FUNCTIONS AND LIMITS

Calculus is fundamentally different from the mathematics that you have studied previously. Calculus is less static and more dynamic. It is concerned with change and motion; it deals with quantities that approach other quantities. So in this first chapter we begin our study of calculus by investigating how the values of functions change and approach limits.

1.1 | FUNCTIONS AND THEIR REPRESENTATIONS

Functions arise whenever one quantity depends on another. Consider the following four situations.

A. The area A of a circle depends on the radius r of the circle. The rule that connects r and A is given by the equation $A = \pi r^2$. With each positive number r there is associated one value of A, and we say that A is a *function* of r.

B. The human population of the world P depends on the time t. The table gives estimates of the world population $P(t)$ at time t, for certain years. For instance,

$$P(1950) \approx 2,560,000,000$$

But for each value of the time t there is a corresponding value of P, and we say that P is a function of t.

C. The cost C of mailing an envelope depends on its weight w. Although there is no simple formula that connects w and C, the post office has a rule for determining C when w is known.

D. The vertical acceleration a of the ground as measured by a seismograph during an earthquake is a function of the elapsed time t. Figure 1 shows a graph generated by seismic activity during the Northridge earthquake that shook Los Angeles in 1994. For a given value of t, the graph provides a corresponding value of a.

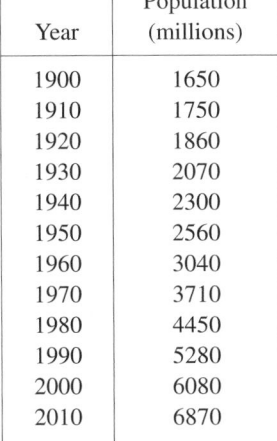

Year	Population (millions)
1900	1650
1910	1750
1920	1860
1930	2070
1940	2300
1950	2560
1960	3040
1970	3710
1980	4450
1990	5280
2000	6080
2010	6870

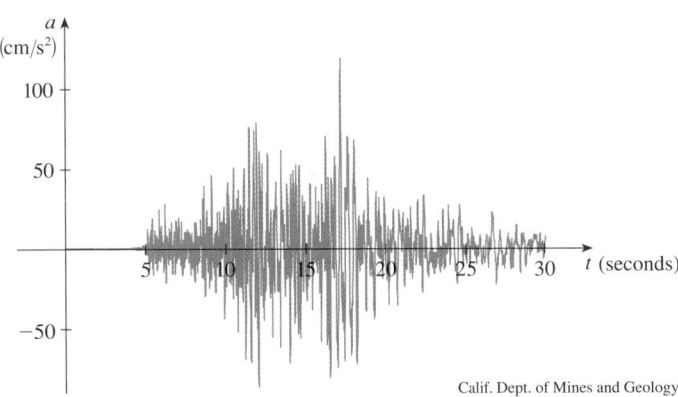

Calif. Dept. of Mines and Geology

FIGURE 1
Vertical ground acceleration during
the Northridge earthquake

Each of these examples describes a rule whereby, given a number (r, t, w, or t), another number (A, P, C, or a) is assigned. In each case we say that the second number is a function of the first number.

> A **function** f is a rule that assigns to each element x in a set D exactly one element, called $f(x)$, in a set E.

We usually consider functions for which the sets D and E are sets of real numbers. The set D is called the **domain** of the function. The number $f(x)$ is the **value of f at x** and is read "f of x." The **range** of f is the set of all possible values of $f(x)$ as x varies throughout the domain. A symbol that represents an arbitrary number in the *domain* of a function f is called an **independent variable**. A symbol that represents a number in the *range* of f is called a **dependent variable**. In Example A, for instance, r is the independent variable and A is the dependent variable.

It's helpful to think of a function as a **machine** (see Figure 2). If x is in the domain of the function f, then when x enters the machine, it's accepted as an input and the machine produces an output $f(x)$ according to the rule of the function. Thus we can think of the domain as the set of all possible inputs and the range as the set of all possible outputs.

x (input) f $f(x)$ (output)

FIGURE 2
Machine diagram for a function f

Another way to picture a function is by an **arrow diagram** as in Figure 3. Each arrow connects an element of D to an element of E. The arrow indicates that $f(x)$ is associated with x, $f(a)$ is associated with a, and so on.

The most common method for visualizing a function is its graph. If f is a function with domain D, then its **graph** is the set of ordered pairs

$$\{(x, f(x)) \mid x \in D\}$$

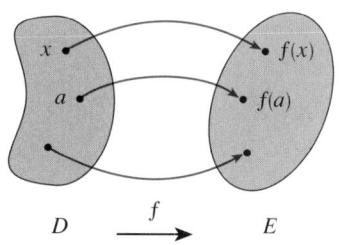

FIGURE 3
Arrow diagram for f

(Notice that these are input-output pairs.) In other words, the graph of f consists of all points (x, y) in the coordinate plane such that $y = f(x)$ and x is in the domain of f.

The graph of a function f gives us a useful picture of the behavior or "life history" of a function. Since the y-coordinate of any point (x, y) on the graph is $y = f(x)$, we can read the value of $f(x)$ from the graph as being the height of the graph above the point x. (See Figure 4.) The graph of f also allows us to picture the domain of f on the x-axis and its range on the y-axis as in Figure 5.

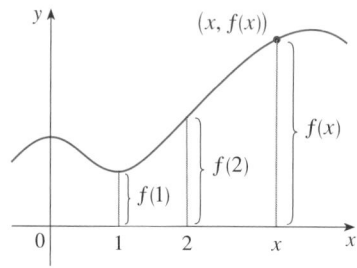

FIGURE 4

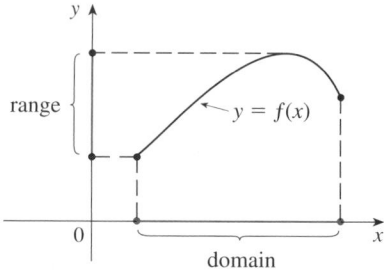

FIGURE 5

EXAMPLE 1 The graph of a function f is shown in Figure 6.
(a) Find the values of $f(1)$ and $f(5)$.
(b) What are the domain and range of f?

SOLUTION
(a) We see from Figure 6 that the point $(1, 3)$ lies on the graph of f, so the value of f at 1 is $f(1) = 3$. (In other words, the point on the graph that lies above $x = 1$ is 3 units above the x-axis.)

When $x = 5$, the graph lies about 0.7 unit below the x-axis, so we estimate that $f(5) \approx -0.7$.

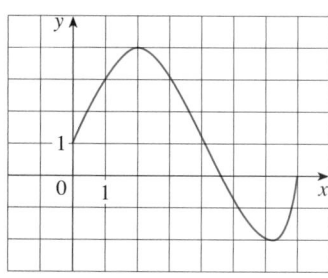

FIGURE 6

■ The notation for intervals is given on Reference Page 3. The Reference Pages are located at the back of the book.

(b) We see that $f(x)$ is defined when $0 \leqslant x \leqslant 7$, so the domain of f is the closed interval $[0, 7]$. Notice that f takes on all values from -2 to 4, so the range of f is

$$\{y \mid -2 \leqslant y \leqslant 4\} = [-2, 4] \qquad ■$$

REPRESENTATIONS OF FUNCTIONS

■ www.stewartcalculus.com

See Additional Examples A, B.

There are four possible ways to represent a function:

- ■ verbally (by a description in words)
- ■ visually (by a graph)
- ■ numerically (by a table of values)
- ■ algebraically (by an explicit formula)

If a single function can be represented in all four ways, it is often useful to go from one representation to another to gain additional insight into the function. But certain functions are described more naturally by one method than by another. With this in mind, let's reexamine the four situations that we considered at the beginning of this section.

A. The most useful representation of the area of a circle as a function of its radius is probably the algebraic formula $A(r) = \pi r^2$, though it is possible to compile a table of values or to sketch a graph (half a parabola). Because a circle has to have a positive radius, the domain is $\{r \mid r > 0\} = (0, \infty)$, and the range is also $(0, \infty)$.

t	Population (millions)
0	1650
10	1750
20	1860
30	2070
40	2300
50	2560
60	3040
70	3710
80	4450
90	5280
100	6080
110	6870

B. We are given a description of the function in words: $P(t)$ is the human population of the world at time t. Let's measure t so that $t = 0$ corresponds to the year 1900. The table of values of world population provides a convenient representation of this function. If we plot these values, we get the graph (called a *scatter plot*) in Figure 7. It too is a useful representation; the graph allows us to absorb all the data at once. What about a formula? Of course, it's impossible to devise an explicit formula that gives the exact human population $P(t)$ at any time t. But it is possible to find an expression for a function that *approximates* $P(t)$. In fact, we could use a graphing calculator with exponential regression capabilities to obtain the approximation

$$P(t) \approx f(t) = (1.43653 \times 10^9) \cdot (1.01395)^t$$

Figure 8 shows that it is a reasonably good "fit." The function f is called a *mathematical model* for population growth. In other words, it is a function with an explicit formula that approximates the behavior of our given function. We will see, however, that the ideas of calculus can be applied to a table of values; an explicit formula is not necessary.

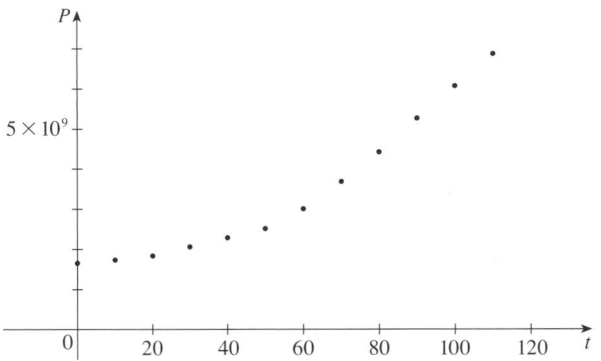

FIGURE 7 Scatter plot of data points for population growth

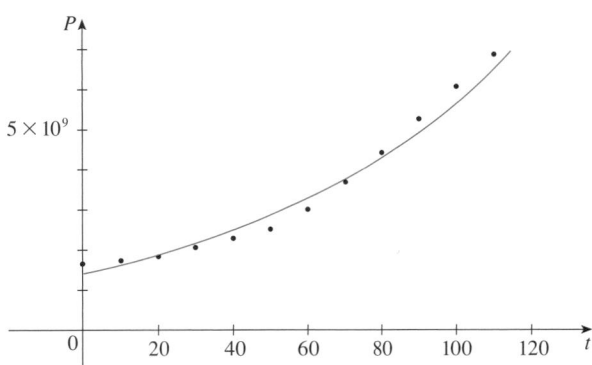

FIGURE 8 Graph of a mathematical model for population growth

▪ A function defined by a table of values is called a *tabular* function.

w (ounces)	$C(w)$ (dollars)
$0 < w \leq 1$	0.88
$1 < w \leq 2$	1.08
$2 < w \leq 3$	1.28
$3 < w \leq 4$	1.48
$4 < w \leq 5$	1.68
⋮	⋮

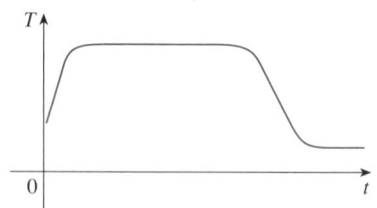

FIGURE 9

▪ If a function is given by a formula and the domain is not stated explicitly, the convention is that the domain is the set of all numbers for which the formula makes sense and defines a real number.

The function P is typical of the functions that arise whenever we attempt to apply calculus to the real world. We start with a verbal description of a function. Then we may be able to construct a table of values of the function, perhaps from instrument readings in a scientific experiment. Even though we don't have complete knowledge of the values of the function, we will see throughout the book that it is still possible to perform the operations of calculus on such a function.

C. Again the function is described in words: Let $C(w)$ be the cost of mailing a large envelope with weight w. The rule that the US Postal Service used as of 2011 is as follows: The cost is 88 cents for up to one ounce, plus 20 cents for each successive ounce (or less) up to 13 ounces. The table of values shown in the margin is the most convenient representation for this function, though it is possible to sketch a graph (see Example 6).

D. The graph shown in Figure 1 is the most natural representation of the vertical acceleration function $a(t)$. It's true that a table of values could be compiled, and it is even possible to devise an approximate formula. But everything a geologist needs to know—amplitudes and patterns—can be seen easily from the graph. (The same is true for the patterns seen in electrocardiograms of heart patients and polygraphs for lie-detection.)

In the next example we sketch the graph of a function that is defined verbally.

EXAMPLE 2 When you turn on a hot-water faucet, the temperature T of the water depends on how long the water has been running. Draw a rough graph of T as a function of the time t that has elapsed since the faucet was turned on.

SOLUTION The initial temperature of the running water is close to room temperature because the water has been sitting in the pipes. When the water from the hot-water tank starts flowing from the faucet, T increases quickly. In the next phase, T is constant at the temperature of the heated water in the tank. When the tank is drained, T decreases to the temperature of the water supply. This enables us to make the rough sketch of T as a function of t in Figure 9. ∎

EXAMPLE 3 Find the domain of each function.

(a) $f(x) = \sqrt{x + 2}$ (b) $g(x) = \dfrac{1}{x^2 - x}$

SOLUTION

(a) Because the square root of a negative number is not defined (as a real number), the domain of f consists of all values of x such that $x + 2 \geq 0$. This is equivalent to $x \geq -2$, so the domain is the interval $[-2, \infty)$.

(b) Since

$$g(x) = \frac{1}{x^2 - x} = \frac{1}{x(x - 1)}$$

and division by 0 is not allowed, we see that $g(x)$ is not defined when $x = 0$ or $x = 1$. Thus the domain of g is $\{x \mid x \neq 0, x \neq 1\}$, which could also be written in interval notation as $(-\infty, 0) \cup (0, 1) \cup (1, \infty)$. ∎

The graph of a function is a curve in the xy-plane. But the question arises: Which curves in the xy-plane are graphs of functions? This is answered by the following test.

THE VERTICAL LINE TEST A curve in the xy-plane is the graph of a function of x if and only if no vertical line intersects the curve more than once.

The reason for the truth of the Vertical Line Test can be seen in Figure 10. If each vertical line $x = a$ intersects a curve only once, at (a, b), then exactly one functional value is defined by $f(a) = b$. But if a line $x = a$ intersects the curve twice, at (a, b) and (a, c), then the curve can't represent a function because a function can't assign two different values to a.

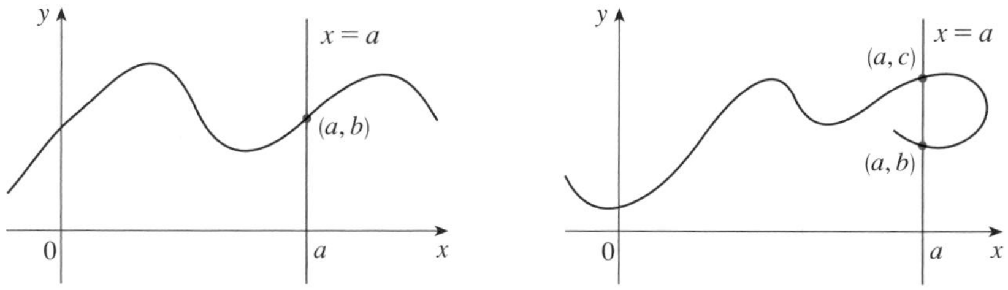

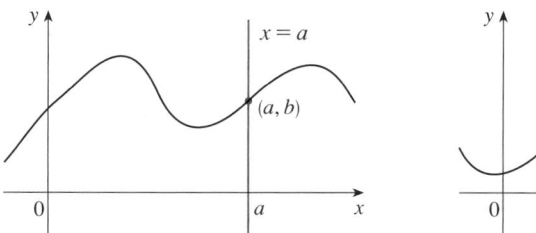

FIGURE 10

PIECEWISE DEFINED FUNCTIONS

The functions in the following three examples are defined by different formulas in different parts of their domains.

☑ **EXAMPLE 4** A function f is defined by

$$f(x) = \begin{cases} 1 - x & \text{if } x \le 1 \\ x^2 & \text{if } x > 1 \end{cases}$$

Evaluate $f(0)$, $f(1)$, and $f(2)$ and sketch the graph.

SOLUTION Remember that a function is a rule. For this particular function the rule is the following: First look at the value of the input x. If it happens that $x \le 1$, then the value of $f(x)$ is $1 - x$. On the other hand, if $x > 1$, then the value of $f(x)$ is x^2.

$$\text{Since } 0 \le 1, \text{ we have } f(0) = 1 - 0 = 1.$$

$$\text{Since } 1 \le 1, \text{ we have } f(1) = 1 - 1 = 0.$$

$$\text{Since } 2 > 1, \text{ we have } f(2) = 2^2 = 4.$$

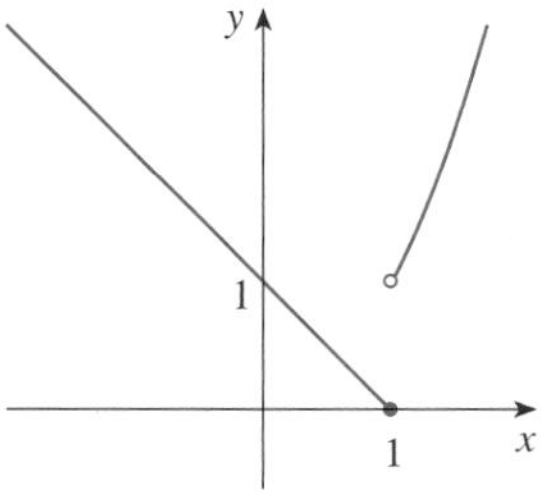

How do we draw the graph of f? We observe that if $x \le 1$, then $f(x) = 1 - x$, so the part of the graph of f that lies to the left of the vertical line $x = 1$ must coincide with the line $y = 1 - x$, which has slope -1 and y-intercept 1. If $x > 1$, then $f(x) = x^2$, so the part of the graph of f that lies to the right of the line $x = 1$ must coincide with the graph of $y = x^2$, which is a parabola. This enables us to sketch the graph in Figure 11. The solid dot indicates that the point $(1, 0)$ is included on the graph; the open dot indicates that the point $(1, 1)$ is excluded from the graph. ∎

FIGURE 11

The next example of a piecewise defined function is the absolute value function. Recall that the **absolute value** of a number a, denoted by $|a|$, is the distance from a to 0 on the real number line. Distances are always positive or 0, so we have

$$|a| \ge 0 \qquad \text{for every number } a$$

■ **www.stewartcalculus.com**
For a more extensive review of absolute values, click on *Review of Algebra*.

For example,

$$|3| = 3 \qquad |-3| = 3 \qquad |0| = 0 \qquad |\sqrt{2} - 1| = \sqrt{2} - 1 \qquad |3 - \pi| = \pi - 3$$

In general, we have

$$
\begin{array}{ll}
|a| = a & \text{if } a \geq 0 \\
|a| = -a & \text{if } a < 0
\end{array}
$$

(Remember that if a is negative, then $-a$ is positive.)

EXAMPLE 5 Sketch the graph of the absolute value function $f(x) = |x|$.

SOLUTION From the preceding discussion we know that

$$
|x| = \begin{cases} x & \text{if } x \geq 0 \\ -x & \text{if } x < 0 \end{cases}
$$

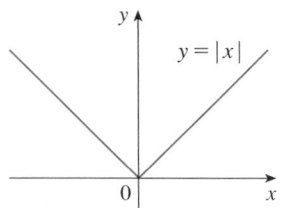

FIGURE 12

Using the same method as in Example 4, we see that the graph of f coincides with the line $y = x$ to the right of the y-axis and coincides with the line $y = -x$ to the left of the y-axis (see Figure 12). ∎

EXAMPLE 6 In Example C at the beginning of this section we considered the cost $C(w)$ of mailing a large envelope with weight w. In effect, this is a piecewise defined function because, from the table of values on page 4, we have

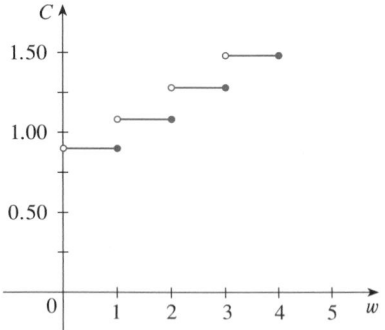

FIGURE 13

$$
C(w) = \begin{cases} 0.88 & \text{if } 0 < w \leq 1 \\ 1.08 & \text{if } 1 < w \leq 2 \\ 1.28 & \text{if } 2 < w \leq 3 \\ 1.48 & \text{if } 3 < w \leq 4 \end{cases}
$$

The graph is shown in Figure 13. You can see why functions similar to this one are called **step functions**—they jump from one value to the next. ∎

■ www.stewartcalculus.com
See Additional Examples C, D.

SYMMETRY

If a function f satisfies $f(-x) = f(x)$ for every number x in its domain, then f is called an **even function**. For instance, the function $f(x) = x^2$ is even because

$$
f(-x) = (-x)^2 = (-1)^2 x^2 = x^2 = f(x)
$$

The geometric significance of an even function is that its graph is symmetric with respect to the y-axis (see Figure 14). This means that if we have plotted the graph of

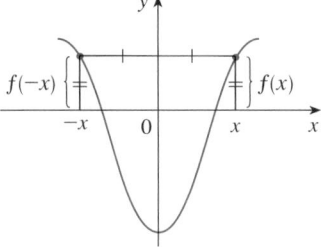

FIGURE 14 An even function

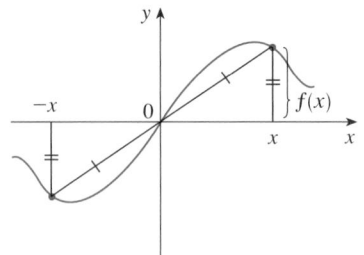

FIGURE 15 An odd function

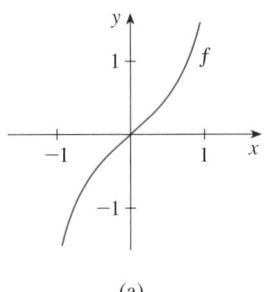

(a)

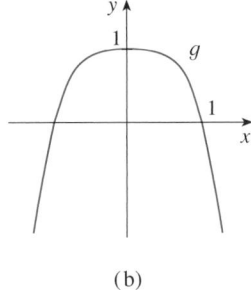

(b)

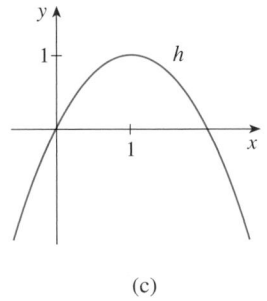

(c)

FIGURE 16

f for $x \geq 0$, we obtain the entire graph simply by reflecting this portion about the y-axis.

If f satisfies $f(-x) = -f(x)$ for every number x in its domain, then f is called an **odd function**. For example, the function $f(x) = x^3$ is odd because

$$f(-x) = (-x)^3 = (-1)^3 x^3 = -x^3 = -f(x)$$

The graph of an odd function is symmetric about the origin (see Figure 15 on page 6). If we already have the graph of f for $x \geq 0$, we can obtain the entire graph by rotating this portion through $180°$ about the origin.

▼ EXAMPLE 7 Determine whether each of the following functions is even, odd, or neither even nor odd.

(a) $f(x) = x^5 + x$ (b) $g(x) = 1 - x^4$ (c) $h(x) = 2x - x^2$

SOLUTION

(a)
$$f(-x) = (-x)^5 + (-x) = (-1)^5 x^5 + (-x)$$
$$= -x^5 - x = -(x^5 + x)$$
$$= -f(x)$$

Therefore f is an odd function.

(b)
$$g(-x) = 1 - (-x)^4 = 1 - x^4 = g(x)$$

So g is even.

(c)
$$h(-x) = 2(-x) - (-x)^2 = -2x - x^2$$

Since $h(-x) \neq h(x)$ and $h(-x) \neq -h(x)$, we conclude that h is neither even nor odd. ■

The graphs of the functions in Example 7 are shown in Figure 16. Notice that the graph of h is symmetric neither about the y-axis nor about the origin.

INCREASING AND DECREASING FUNCTIONS

The graph shown in Figure 17 rises from A to B, falls from B to C, and rises again from C to D. The function f is said to be increasing on the interval $[a, b]$, decreasing on $[b, c]$, and increasing again on $[c, d]$. Notice that if x_1 and x_2 are any two numbers between a and b with $x_1 < x_2$, then $f(x_1) < f(x_2)$. We use this as the defining property of an increasing function.

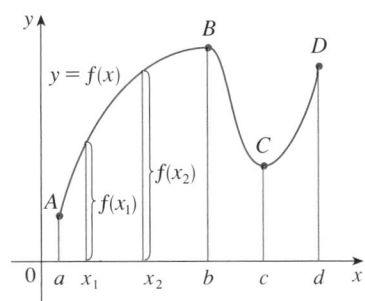

FIGURE 17

A function f is called **increasing** on an interval I if

$$f(x_1) < f(x_2) \qquad \text{whenever } x_1 < x_2 \text{ in } I$$

It is called **decreasing** on I if

$$f(x_1) > f(x_2) \qquad \text{whenever } x_1 < x_2 \text{ in } I$$

In the definition of an increasing function it is important to realize that the inequality $f(x_1) < f(x_2)$ must be satisfied for *every* pair of numbers x_1 and x_2 in I with $x_1 < x_2$.

You can see from Figure 18 that the function $f(x) = x^2$ is decreasing on the interval $(-\infty, 0]$ and increasing on the interval $[0, \infty)$.

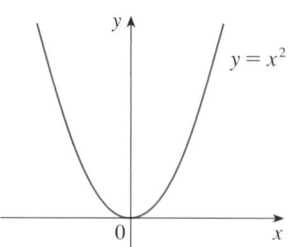

FIGURE 18

1.1 | EXERCISES

1. If $f(x) = x + \sqrt{2 - x}$ and $g(u) = u + \sqrt{2 - u}$, is it true that $f = g$?

2. If

$$f(x) = \frac{x^2 - x}{x - 1} \qquad \text{and} \qquad g(x) = x$$

is it true that $f = g$?

3. The graph of a function f is given.
   (a) State the value of $f(1)$.
   (b) Estimate the value of $f(-1)$.
   (c) For what values of x is $f(x) = 1$?
   (d) Estimate the value of x such that $f(x) = 0$.
   (e) State the domain and range of f.
   (f) On what interval is f increasing?

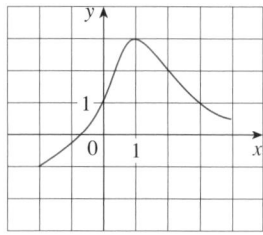

4. The graphs of f and g are given.
   (a) State the values of $f(-4)$ and $g(3)$.
   (b) For what values of x is $f(x) = g(x)$?
   (c) Estimate the solution of the equation $f(x) = -1$.
   (d) On what interval is f decreasing?
   (e) State the domain and range of f.

(f) State the domain and range of g.

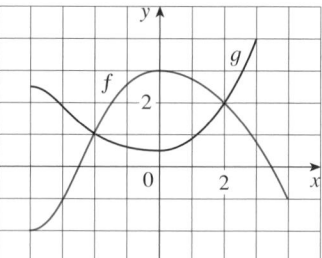

5–8 ▪ Determine whether the curve is the graph of a function of x. If it is, state the domain and range of the function.

5.
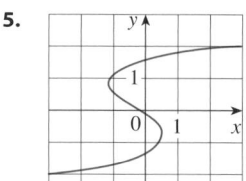

6.
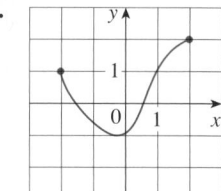

7.
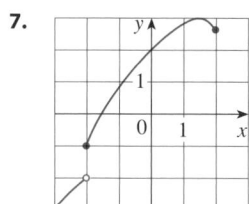

8.
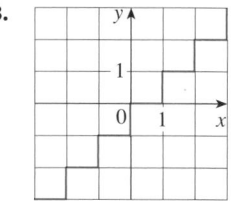

Graphing calculator or computer required   **CAS** Computer algebra system required   **1** Homework Hints at stewartcalculus.com

9. The graph shown gives the weight of a certain person as a function of age. Describe in words how this person's weight varies over time. What do you think happened when this person was 30 years old?

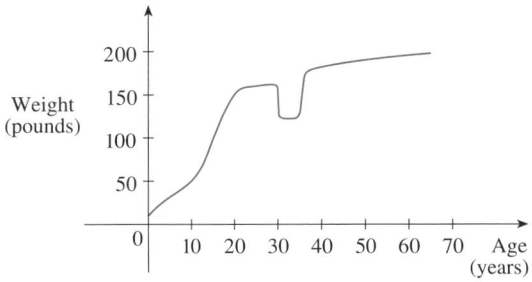

10. The graph shows the height of the water in a bathtub as a function of time. Give a verbal description of what you think happened.

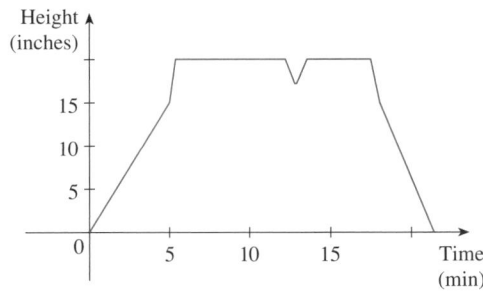

11. You put some ice cubes in a glass, fill the glass with cold water, and then let the glass sit on a table. Describe how the temperature of the water changes as time passes. Then sketch a rough graph of the temperature of the water as a function of the elapsed time.

12. Sketch a rough graph of the number of hours of daylight as a function of the time of year.

13. Sketch a rough graph of the outdoor temperature as a function of time during a typical spring day.

14. Sketch a rough graph of the market value of a new car as a function of time for a period of 20 years. Assume the car is well maintained.

15. Sketch the graph of the amount of a particular brand of coffee sold by a store as a function of the price of the coffee.

16. You place a frozen pie in an oven and bake it for an hour. Then you take it out and let it cool before eating it. Describe how the temperature of the pie changes as time passes. Then sketch a rough graph of the temperature of the pie as a function of time.

17. A homeowner mows the lawn every Wednesday afternoon. Sketch a rough graph of the height of the grass as a function of time over the course of a four-week period.

18. An airplane takes off from an airport and lands an hour later at another airport, 400 miles away. If t represents the time

in minutes since the plane has left the terminal, let $x(t)$ be the horizontal distance traveled and $y(t)$ be the altitude of the plane.
(a) Sketch a possible graph of $x(t)$.
(b) Sketch a possible graph of $y(t)$.
(c) Sketch a possible graph of the ground speed.
(d) Sketch a possible graph of the vertical velocity.

19. If $f(x) = 3x^2 - x + 2$, find $f(2)$, $f(-2)$, $f(a), f(-a)$, $f(a + 1)$, $2f(a)$, $f(2a)$, $f(a^2)$, $[f(a)]^2$, and $f(a + h)$.

20. A spherical balloon with radius r inches has volume $V(r) = \frac{4}{3}\pi r^3$. Find a function that represents the amount of air required to inflate the balloon from a radius of r inches to a radius of $r + 1$ inches.

21–24 ▪ Evaluate the difference quotient for the given function. Simplify your answer.

21. $f(x) = 4 + 3x - x^2$, $\quad \dfrac{f(3 + h) - f(3)}{h}$

22. $f(x) = x^3$, $\quad \dfrac{f(a + h) - f(a)}{h}$

23. $f(x) = \dfrac{1}{x}$, $\quad \dfrac{f(x) - f(a)}{x - a}$

24. $f(x) = \dfrac{x + 3}{x + 1}$, $\quad \dfrac{f(x) - f(1)}{x - 1}$

25–29 ▪ Find the domain of the function.

25. $f(x) = \dfrac{x + 4}{x^2 - 9}$ **26.** $f(x) = \dfrac{2x^3 - 5}{x^2 + x - 6}$

27. $F(p) = \sqrt{2 - \sqrt{p}}$

28. $g(t) = \sqrt{3 - t} - \sqrt{2 + t}$

29. $h(x) = \dfrac{1}{\sqrt[4]{x^2 - 5x}}$

30. Find the domain and range and sketch the graph of the function $h(x) = \sqrt{4 - x^2}$.

31–42 ▪ Find the domain and sketch the graph of the function.

31. $f(x) = 2 - 0.4x$ **32.** $F(x) = x^2 - 2x + 1$

33. $f(t) = 2t + t^2$ **34.** $H(t) = \dfrac{4 - t^2}{2 - t}$

35. $g(x) = \sqrt{x - 5}$ **36.** $F(x) = |2x + 1|$

37. $G(x) = \dfrac{3x + |x|}{x}$ **38.** $g(x) = |x| - x$

39. $f(x) = \begin{cases} x + 2 & \text{if } x < 0 \\ 1 - x & \text{if } x \geq 0 \end{cases}$

40. $f(x) = \begin{cases} 3 - \frac{1}{2}x & \text{if } x \leq 2 \\ 2x - 5 & \text{if } x > 2 \end{cases}$

41. $f(x) = \begin{cases} x + 2 & \text{if } x \leq -1 \\ x^2 & \text{if } x > -1 \end{cases}$

42. $f(x) = \begin{cases} -1 & \text{if } x \leq -1 \\ 3x + 2 & \text{if } |x| < 1 \\ 7 - 2x & \text{if } x \geq 1 \end{cases}$

43–46 ■ Find an expression for the function whose graph is the given curve.

43. The line segment joining the points $(1, -3)$ and $(5, 7)$

44. The line segment joining the points $(-5, 10)$ and $(7, -10)$

45. The bottom half of the parabola $x + (y - 1)^2 = 0$

46. The top half of the circle $x^2 + (y - 2)^2 = 4$

47–51 ■ Find a formula for the described function and state its domain.

47. A rectangle has perimeter 20 m. Express the area of the rectangle as a function of the length of one of its sides.

48. A rectangle has area 16 m². Express the perimeter of the rectangle as a function of the length of one of its sides.

49. Express the area of an equilateral triangle as a function of the length of a side.

50. Express the surface area of a cube as a function of its volume.

51. An open rectangular box with volume 2 m³ has a square base. Express the surface area of the box as a function of the length of a side of the base.

---

52. A cell phone plan has a basic charge of $35 a month. The plan includes 400 free minutes and charges 10 cents for each additional minute of usage. Write the monthly cost C as a function of the number x of minutes used and graph C as a function of x for $0 \leq x \leq 600$.

53. In a certain country, income tax is assessed as follows. There is no tax on income up to $10,000. Any income over $10,000 is taxed at a rate of 10%, up to an income of $20,000. Any income over $20,000 is taxed at 15%.
(a) Sketch the graph of the tax rate R as a function of the income I.
(b) How much tax is assessed on an income of $14,000? On $26,000?
(c) Sketch the graph of the total assessed tax T as a function of the income I.

54. The functions in Example 6 and Exercises 52 and 53(a) are called *step functions* because their graphs look like stairs. Give two other examples of step functions that arise in everyday life.

55–56 ■ Graphs of f and g are shown. Decide whether each function is even, odd, or neither. Explain your reasoning.

55.   **56.**

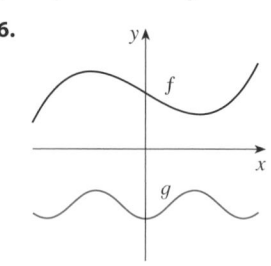

57. (a) If the point $(5, 3)$ is on the graph of an even function, what other point must also be on the graph?
(b) If the point $(5, 3)$ is on the graph of an odd function, what other point must also be on the graph?

58. A function f has domain $[-5, 5]$ and a portion of its graph is shown.
(a) Complete the graph of f if it is known that f is even.
(b) Complete the graph of f if it is known that f is odd.

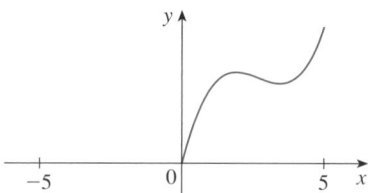

59–64 ■ Determine whether f is even, odd, or neither. If you have a graphing calculator, use it to check your answer visually.

59. $f(x) = \dfrac{x}{x^2 + 1}$

60. $f(x) = \dfrac{x^2}{x^4 + 1}$

61. $f(x) = \dfrac{x}{x + 1}$

62. $f(x) = x|x|$

63. $f(x) = 1 + 3x^2 - x^4$

64. $f(x) = 1 + 3x^3 - x^5$

---

65. If f and g are both even functions, is $f + g$ even? If f and g are both odd functions, is $f + g$ odd? What if f is even and g is odd? Justify your answers.

66. If f and g are both even functions, is the product fg even? If f and g are both odd functions, is fg odd? What if f is even and g is odd? Justify your answers.

1.2 | A CATALOG OF ESSENTIAL FUNCTIONS

In solving calculus problems you will find that it is helpful to be familiar with the graphs of some commonly occurring functions. These same basic functions are often used to model real-world phenomena, so we begin with a discussion of mathematical modeling. We also review briefly how to transform these functions by shifting, stretching, and reflecting their graphs as well as how to combine pairs of functions by the standard arithmetic operations and by composition.

MATHEMATICAL MODELING

A **mathematical model** is a mathematical description (often by means of a function or an equation) of a real-world phenomenon such as the size of a population, the demand for a product, the speed of a falling object, the concentration of a product in a chemical reaction, the life expectancy of a person at birth, or the cost of emission reductions. The purpose of the model is to understand the phenomenon and perhaps to make predictions about future behavior.

Figure 1 illustrates the process of mathematical modeling. Given a real-world problem, our first task is to formulate a mathematical model by identifying and naming the independent and dependent variables and making assumptions that simplify the phenomenon enough to make it mathematically tractable. We use our knowledge of the physical situation and our mathematical skills to obtain equations that relate the variables. In situations where there is no physical law to guide us, we may need to collect data (either from a library or the Internet or by conducting our own experiments) and examine the data in the form of a table in order to discern patterns. From this numerical representation of a function we may wish to obtain a graphical representation by plotting the data. The graph might even suggest a suitable algebraic formula in some cases.

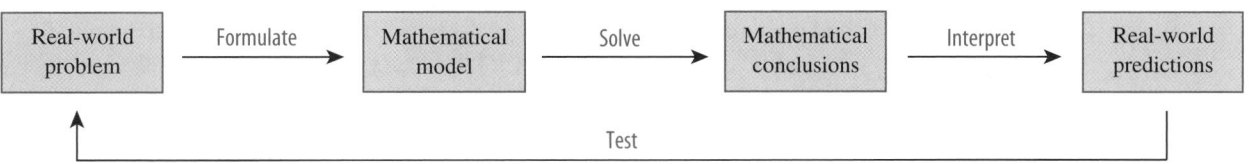

FIGURE 1   The modeling process

The second stage is to apply the mathematics that we know (such as the calculus that will be developed throughout this book) to the mathematical model that we have formulated in order to derive mathematical conclusions. Then, in the third stage, we take those mathematical conclusions and interpret them as information about the original real-world phenomenon by way of offering explanations or making predictions. The final step is to test our predictions by checking against new real data. If the predictions don't compare well with reality, we need to refine our model or to formulate a new model and start the cycle again.

A mathematical model is never a completely accurate representation of a physical situation—it is an *idealization*. A good model simplifies reality enough to permit mathematical calculations but is accurate enough to provide valuable conclusions. It is important to realize the limitations of the model. In the end, Mother Nature has the final say.

There are many different types of functions that can be used to model relationships observed in the real world. In what follows, we discuss the behavior and graphs of these functions and give examples of situations appropriately modeled by such functions.

Linear Models

▪ www.stewartcalculus.com

To review the coordinate geometry of lines, click on *Review of Analytic Geometry*.

When we say that y is a **linear function** of x, we mean that the graph of the function is a line, so we can use the slope-intercept form of the equation of a line to write a formula for the function as

$$y = f(x) = mx + b$$

where m is the slope of the line and b is the y-intercept.

A characteristic feature of linear functions is that they grow at a constant rate. For instance, Figure 2 shows a graph of the linear function $f(x) = 3x - 2$ and a table of sample values. Notice that whenever x increases by 0.1, the value of $f(x)$ increases by 0.3. So $f(x)$ increases three times as fast as x. Thus the slope of the graph $y = 3x - 2$, namely 3, can be interpreted as the rate of change of y with respect to x.

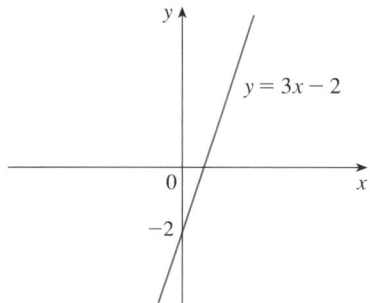

x	$f(x) = 3x - 2$
1.0	1.0
1.1	1.3
1.2	1.6
1.3	1.9
1.4	2.2
1.5	2.5

FIGURE 2

▼ EXAMPLE 1

(a) As dry air moves upward, it expands and cools. If the ground temperature is 20°C and the temperature at a height of 1 km is 10°C, express the temperature T (in °C) as a function of the height h (in kilometers), assuming that a linear model is appropriate.

(b) Draw the graph of the function in part (a). What does the slope represent?

(c) What is the temperature at a height of 2.5 km?

SOLUTION

(a) Because we are assuming that T is a linear function of h, we can write

$$T = mh + b$$

▪ www.stewartcalculus.com

See Additional Examples A, B.

We are given that $T = 20$ when $h = 0$, so

$$20 = m \cdot 0 + b = b$$

In other words, the y-intercept is $b = 20$.

We are also given that $T = 10$ when $h = 1$, so

$$10 = m \cdot 1 + 20$$

The slope of the line is therefore $m = 10 - 20 = -10$ and the required linear function is

$$T = -10h + 20$$

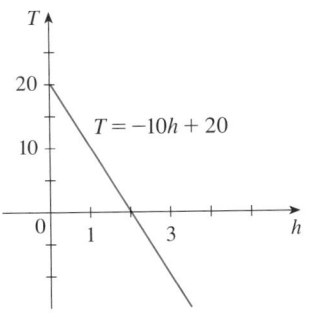

FIGURE 3

(b) The graph is sketched in Figure 3. The slope is $m = -10°C/km$, and this represents the rate of change of temperature with respect to height.

(c) At a height of $h = 2.5$ km, the temperature is

$$T = -10(2.5) + 20 = -5°C \qquad ■$$

■ Polynomials

A function P is called a **polynomial** if

$$P(x) = a_n x^n + a_{n-1} x^{n-1} + \cdots + a_2 x^2 + a_1 x + a_0$$

where n is a nonnegative integer and the numbers $a_0, a_1, a_2, \ldots, a_n$ are constants called the **coefficients** of the polynomial. The domain of any polynomial is $\mathbb{R} = (-\infty, \infty)$. If the leading coefficient $a_n \neq 0$, then the **degree** of the polynomial is n. For example, the function

$$P(x) = 2x^6 - x^4 + \tfrac{2}{5}x^3 + \sqrt{2}$$

is a polynomial of degree 6.

A polynomial of degree 1 is of the form $P(x) = mx + b$ and so it is a linear function. A polynomial of degree 2 is of the form $P(x) = ax^2 + bx + c$ and is called a **quadratic function**. Its graph is always a parabola obtained by shifting the parabola $y = ax^2$. The parabola opens upward if $a > 0$ and downward if $a < 0$. (See Figure 4.)

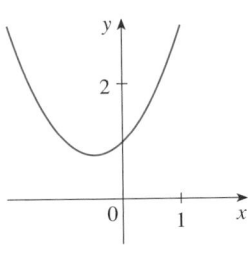

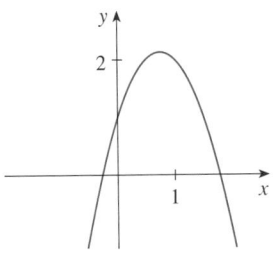

FIGURE 4

The graphs of quadratic functions are parabolas.

(a) $y = x^2 + x + 1$

(b) $y = -2x^2 + 3x + 1$

A polynomial of degree 3 is of the form

$$P(x) = ax^3 + bx^2 + cx + d \qquad a \neq 0$$

and is called a **cubic function**. Figure 5 shows the graph of a cubic function in part (a) and graphs of polynomials of degrees 4 and 5 in parts (b) and (c). We will see later why the graphs have these shapes.

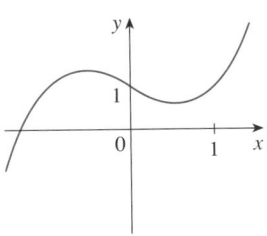

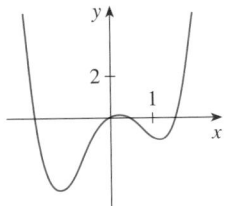

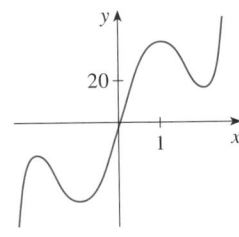

FIGURE 5

(a) $y = x^3 - x + 1$

(b) $y = x^4 - 3x^2 + x$

(c) $y = 3x^5 - 25x^3 + 60x$

Polynomials are commonly used to model various quantities that occur in the natural and social sciences. For instance, in Chapter 2 we will explain why economists often use a polynomial $P(x)$ to represent the cost of producing x units of a commodity.

■ Power Functions

A function of the form $f(x) = x^a$, where a is a constant, is called a **power function**. We consider several cases.

(i) $a = n$, where n is a positive integer

The graphs of $f(x) = x^n$ for $n = 1, 2, 3, 4,$ and 5 are shown in Figure 6. (These are polynomials with only one term.) You are familiar with the shape of the graphs of $y = x$ (a line through the origin with slope 1) and $y = x^2$ (a parabola).

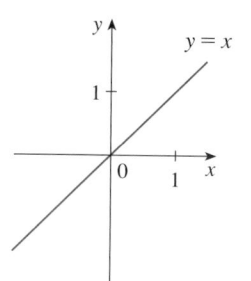
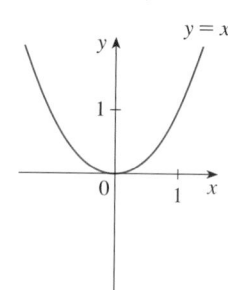
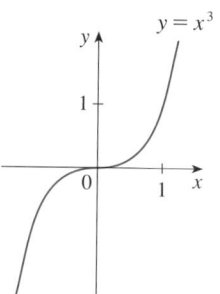
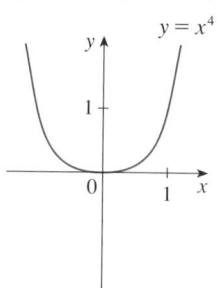
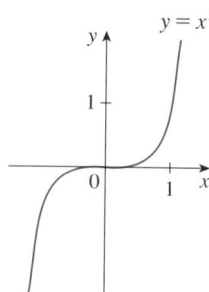

FIGURE 6   Graphs of $f(x) = x^n$ for $n = 1, 2, 3, 4, 5$

The general shape of the graph of $f(x) = x^n$ depends on whether n is even or odd. If n is even, then $f(x) = x^n$ is an even function and its graph is similar to the parabola $y = x^2$. If n is odd, then $f(x) = x^n$ is an odd function and its graph is similar to that of $y = x^3$. Notice from Figure 7, however, that as n increases, the graph of $y = x^n$ becomes flatter near 0 and steeper when $|x| \geqslant 1$. (If x is small, then x^2 is smaller, x^3 is even smaller, x^4 is smaller still, and so on.)

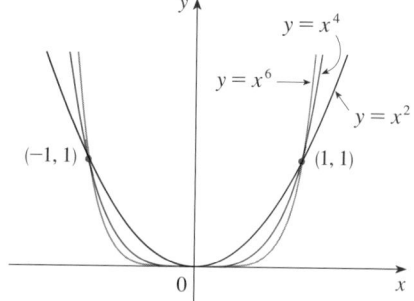
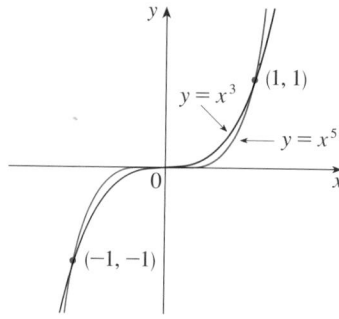

FIGURE 7
Families of power functions

(ii) $a = 1/n$, where n is a positive integer

The function $f(x) = x^{1/n} = \sqrt[n]{x}$ is a **root function**. For $n = 2$ it is the square root function $f(x) = \sqrt{x}$, whose domain is $[0, \infty)$ and whose graph is the upper half of the parabola $x = y^2$. [See Figure 8(a).] For other even values of n, the graph of $y = \sqrt[n]{x}$ is similar to that of $y = \sqrt{x}$. For $n = 3$ we have the cube root function

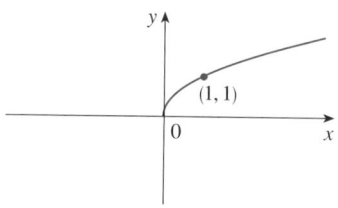
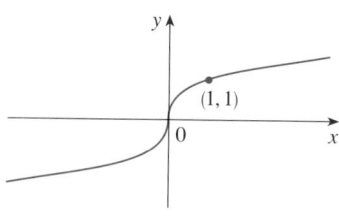

FIGURE 8
Graphs of root functions

(a) $f(x) = \sqrt{x}$

(b) $f(x) = \sqrt[3]{x}$

$f(x) = \sqrt[3]{x}$ whose domain is $\mathbb{R}$ (recall that every real number has a cube root) and whose graph is shown in Figure 8(b). The graph of $y = \sqrt[n]{x}$ for n odd ($n > 3$) is similar to that of $y = \sqrt[3]{x}$.

(iii) $a = -1$

The graph of the **reciprocal function** $f(x) = x^{-1} = 1/x$ is shown in Figure 9. Its graph has the equation $y = 1/x$, or $xy = 1$, and is a hyperbola with the coordinate axes as its asymptotes. This function arises in physics and chemistry in connection with Boyle's Law, which says that, when the temperature is constant, the volume V of a gas is inversely proportional to the pressure P:

$$V = \frac{C}{P}$$

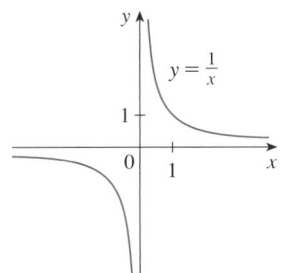

FIGURE 9
The reciprocal function

where C is a constant. Thus the graph of V as a function of P has the same general shape as the right half of Figure 9.

Power functions are also used to model species-area relationships (Exercise 16), illumination as a function of distance from a light source (Exercise 15), and the period of revolution of a planet as a function of its distance from the sun (Kepler's Second Law, page 606).

■ **Rational Functions**

A **rational function** f is a ratio of two polynomials:

$$f(x) = \frac{P(x)}{Q(x)}$$

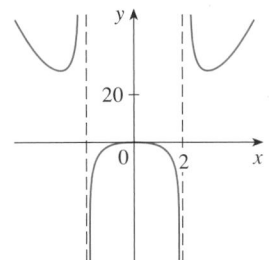

FIGURE 10
$f(x) = \dfrac{2x^4 - x^2 + 1}{x^2 - 4}$

where P and Q are polynomials. The domain consists of all values of x such that $Q(x) \neq 0$. A simple example of a rational function is the function $f(x) = 1/x$, whose domain is $\{x \mid x \neq 0\}$; this is the reciprocal function graphed in Figure 9. The function

$$f(x) = \frac{2x^4 - x^2 + 1}{x^2 - 4}$$

is a rational function with domain $\{x \mid x \neq \pm 2\}$. Its graph is shown in Figure 10.

■ **Trigonometric Functions**

Trigonometry and the trigonometric functions are reviewed on Reference Page 2 and also in Appendix A. In calculus the convention is that radian measure is always used (except when otherwise indicated). For example, when we use the function $f(x) = \sin x$, it is understood that $\sin x$ means the sine of the angle whose radian measure is x. Thus the graphs of the sine and cosine functions are as shown in Figure 11.

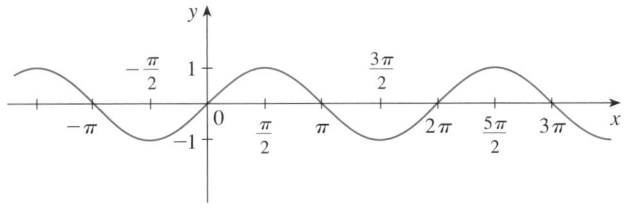

(a) $f(x) = \sin x$

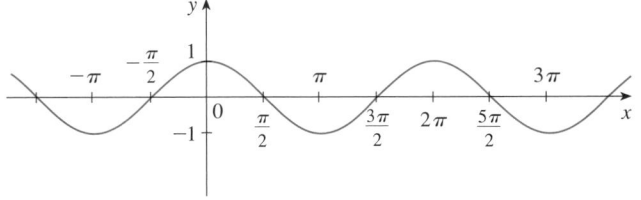

(b) $g(x) = \cos x$

FIGURE 11

Notice that for both the sine and cosine functions the domain is $(-\infty, \infty)$ and the range is the closed interval $[-1, 1]$. Thus, for all values of x, we have

$$-1 \leq \sin x \leq 1 \qquad -1 \leq \cos x \leq 1$$

or, in terms of absolute values,

$$|\sin x| \leq 1 \qquad |\cos x| \leq 1$$

Also, the zeros of the sine function occur at the integer multiples of π; that is,

$$\sin x = 0 \quad \text{when} \quad x = n\pi \quad n \text{ an integer}$$

An important property of the sine and cosine functions is that they are periodic functions and have period 2π. This means that, for all values of x,

$$\sin(x + 2\pi) = \sin x \qquad \cos(x + 2\pi) = \cos x$$

The periodic nature of these functions makes them suitable for modeling repetitive phenomena such as tides, vibrating springs, and sound waves.

The tangent function is related to the sine and cosine functions by the equation

$$\tan x = \frac{\sin x}{\cos x}$$

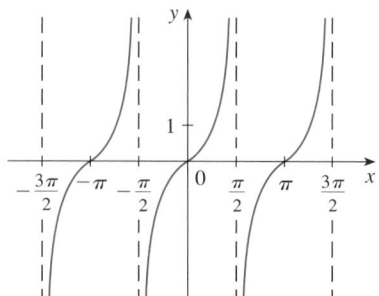

FIGURE 12
$y = \tan x$

and its graph is shown in Figure 12. It is undefined whenever $\cos x = 0$, that is, when $x = \pm\pi/2, \pm3\pi/2, \ldots$. Its range is $(-\infty, \infty)$. Notice that the tangent function has period π:

$$\tan(x + \pi) = \tan x \qquad \text{for all } x$$

The remaining three trigonometric functions (cosecant, secant, and cotangent) are the reciprocals of the sine, cosine, and tangent functions. Their graphs are shown in Appendix A.

■ Exponential Functions and Logarithms

The **exponential functions** are the functions of the form $f(x) = a^x$, where the base a is a positive constant. The graphs of $y = 2^x$ and $y = (0.5)^x$ are shown in Figure 13. In both cases the domain is $(-\infty, \infty)$ and the range is $(0, \infty)$.

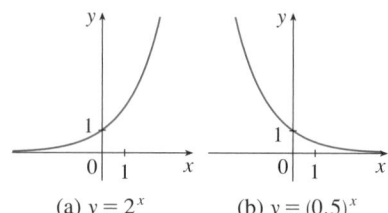

(a) $y = 2^x$ (b) $y = (0.5)^x$

FIGURE 13

Exponential functions will be studied in detail in Section 3.1, and we will see that they are useful for modeling many natural phenomena, such as population growth (if $a > 1$) and radioactive decay (if $a < 1$).

The **logarithmic functions** $f(x) = \log_a x$, where the base a is a positive constant, are the inverse functions of the exponential functions. They will be studied in Section 3.2. Figure 14 shows the graphs of four logarithmic functions with various bases. In each case the domain is $(0, \infty)$, the range is $(-\infty, \infty)$, and the function increases slowly when $x > 1$.

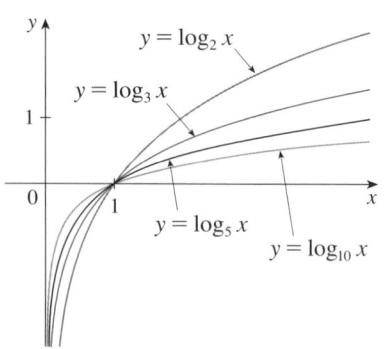

FIGURE 14

TRANSFORMATIONS OF FUNCTIONS

By applying certain transformations to the graph of a given function we can obtain the graphs of certain related functions. This will give us the ability to sketch the graphs of

■ Figure 15 illustrates these shifts by showing how the graph of $y = (x + 3)^2 + 1$ is obtained from the graph of the parabola $y = x^2$: Shift 3 units to the left and 1 unit upward.

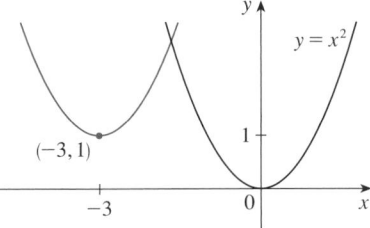

FIGURE 15
$y = (x + 3)^2 + 1$

many functions quickly by hand. It will also enable us to write equations for given graphs. Let's first consider **translations**. If c is a positive number, then the graph of $y = f(x) + c$ is just the graph of $y = f(x)$ shifted upward a distance of c units (because each y-coordinate is increased by the same number c). Likewise, if $g(x) = f(x - c)$, where $c > 0$, then the value of g at x is the same as the value of f at $x - c$ (c units to the left of x). Therefore the graph of $y = f(x - c)$ is just the graph of $y = f(x)$ shifted c units to the right.

> **VERTICAL AND HORIZONTAL SHIFTS** Suppose $c > 0$. To obtain the graph of
>
> $y = f(x) + c$, shift the graph of $y = f(x)$ a distance c units upward
>
> $y = f(x) - c$, shift the graph of $y = f(x)$ a distance c units downward
>
> $y = f(x - c)$, shift the graph of $y = f(x)$ a distance c units to the right
>
> $y = f(x + c)$, shift the graph of $y = f(x)$ a distance c units to the left

Now let's consider the **stretching** and **reflecting** transformations. If $c > 1$, then the graph of $y = cf(x)$ is the graph of $y = f(x)$ stretched by a factor of c in the vertical direction (because each y-coordinate is multiplied by the same number c). The graph of $y = -f(x)$ is the graph of $y = f(x)$ reflected about the x-axis because the point (x, y) is replaced by the point $(x, -y)$. The following chart also incorporates the results of other stretching, shrinking, and reflecting transformations.

> **VERTICAL AND HORIZONTAL STRETCHING, SHRINKING, AND REFLECTING**
> Suppose $c > 1$. To obtain the graph of
>
> $y = cf(x)$, stretch the graph of $y = f(x)$ vertically by a factor of c
>
> $y = (1/c)f(x)$, shrink the graph of $y = f(x)$ vertically by a factor of c
>
> $y = f(cx)$, shrink the graph of $y = f(x)$ horizontally by a factor of c
>
> $y = f(x/c)$, stretch the graph of $y = f(x)$ horizontally by a factor of c
>
> $y = -f(x)$, reflect the graph of $y = f(x)$ about the x-axis
>
> $y = f(-x)$, reflect the graph of $y = f(x)$ about the y-axis

Figure 16 illustrates these stretching transformations when applied to the cosine function with $c = 2$. For instance, in order to get the graph of $y = 2 \cos x$ we multiply the y-coordinate of each point on the graph of $y = \cos x$ by 2. This means that the graph of $y = \cos x$ gets stretched vertically by a factor of 2.

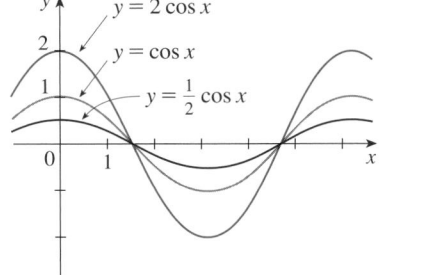

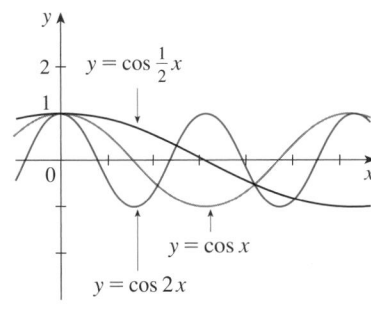

FIGURE 16

V **EXAMPLE 2** Given the graph of $y = \sqrt{x}$, use transformations to graph $y = \sqrt{x} - 2$, $y = \sqrt{x - 2}$, $y = -\sqrt{x}$, $y = 2\sqrt{x}$, and $y = \sqrt{-x}$.

SOLUTION The graph of the square root function $y = \sqrt{x}$, obtained from Figure 8(a), is shown in Figure 17(a). In the other parts of the figure we sketch $y = \sqrt{x} - 2$ by shifting 2 units downward, $y = \sqrt{x - 2}$ by shifting 2 units to the right, $y = -\sqrt{x}$ by reflecting about the x-axis, $y = 2\sqrt{x}$ by stretching vertically by a factor of 2, and $y = \sqrt{-x}$ by reflecting about the y-axis.

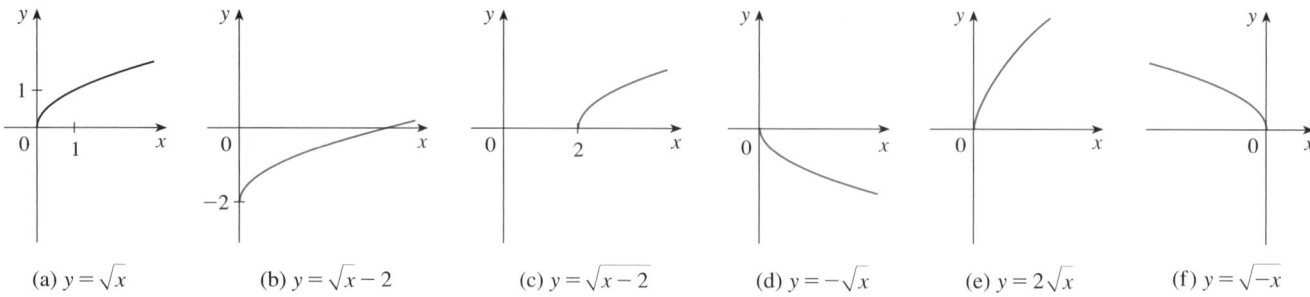

(a) $y = \sqrt{x}$ (b) $y = \sqrt{x} - 2$ (c) $y = \sqrt{x - 2}$ (d) $y = -\sqrt{x}$ (e) $y = 2\sqrt{x}$ (f) $y = \sqrt{-x}$

FIGURE 17

EXAMPLE 3 Sketch the graph of the function $y = 1 - \sin x$.

SOLUTION To obtain the graph of $y = 1 - \sin x$, we start with $y = \sin x$. We reflect about the x-axis to get the graph $y = -\sin x$ and then we shift 1 unit upward to get $y = 1 - \sin x$. (See Figure 18.)

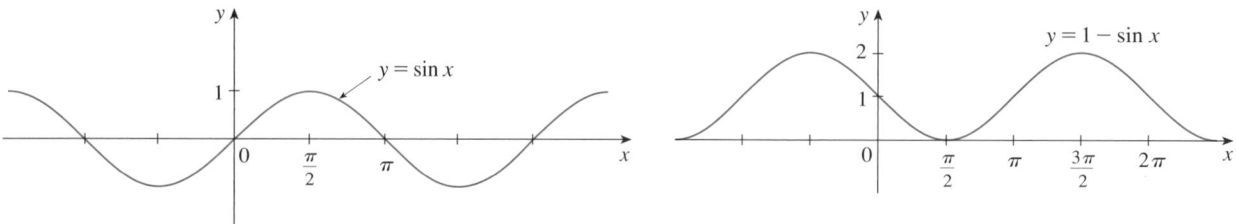

FIGURE 18

▪ **www.stewartcalculus.com**
See Additional Examples C, D, E.

COMBINATIONS OF FUNCTIONS

Two functions f and g can be combined to form new functions $f + g$, $f - g$, fg, and f/g in a manner similar to the way we add, subtract, multiply, and divide real numbers. The sum and difference functions are defined by

$$(f + g)(x) = f(x) + g(x) \qquad (f - g)(x) = f(x) - g(x)$$

If the domain of f is A and the domain of g is B, then the domain of $f + g$ is the intersection $A \cap B$ because both $f(x)$ and $g(x)$ have to be defined. For example, the domain of $f(x) = \sqrt{x}$ is $A = [0, \infty)$ and the domain of $g(x) = \sqrt{2 - x}$ is $B = (-\infty, 2]$, so the domain of $(f + g)(x) = \sqrt{x} + \sqrt{2 - x}$ is $A \cap B = [0, 2]$.

Similarly, the product and quotient functions are defined by

$$(fg)(x) = f(x)\, g(x) \qquad \left(\frac{f}{g}\right)(x) = \frac{f(x)}{g(x)}$$

The domain of fg is $A \cap B$, but we can't divide by 0 and so the domain of f/g is $\{x \in A \cap B \mid g(x) \neq 0\}$. For instance, if $f(x) = x^2$ and $g(x) = x - 1$, then the domain of the rational function $(f/g)(x) = x^2/(x - 1)$ is $\{x \mid x \neq 1\}$, or $(-\infty, 1) \cup (1, \infty)$.

There is another way of combining two functions to get a new function. For example, suppose that $y = f(u) = \sqrt{u}$ and $u = g(x) = x^2 + 1$. Since y is a function of u and u is, in turn, a function of x, it follows that y is ultimately a function of x. We compute this by substitution:

$$y = f(u) = f(g(x)) = f(x^2 + 1) = \sqrt{x^2 + 1}$$

The procedure is called *composition* because the new function is *composed* of the two given functions f and g.

In general, given any two functions f and g, we start with a number x in the domain of g and find its image $g(x)$. If this number $g(x)$ is in the domain of f, then we can calculate the value of $f(g(x))$. The result is a new function $h(x) = f(g(x))$ obtained by substituting g into f. It is called the *composition* (or *composite*) of f and g and is denoted by $f \circ g$ ("f circle g").

DEFINITION Given two functions f and g, the **composite function** $f \circ g$ (also called the **composition** of f and g) is defined by

$$(f \circ g)(x) = f(g(x))$$

The domain of $f \circ g$ is the set of all x in the domain of g such that $g(x)$ is in the domain of f. In other words, $(f \circ g)(x)$ is defined whenever both $g(x)$ and $f(g(x))$ are defined. Figure 19 shows how to picture $f \circ g$ in terms of machines.

FIGURE 19

The $f \circ g$ machine is composed of the g machine (first) and then the f machine.

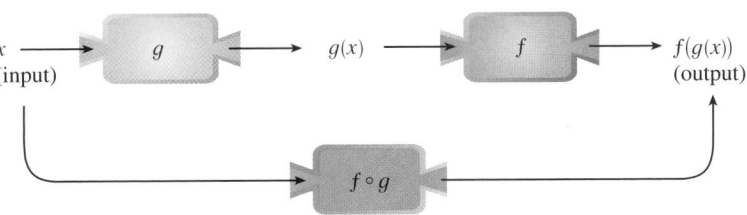

EXAMPLE 4 If $f(x) = x^2$ and $g(x) = x - 3$, find the composite functions $f \circ g$ and $g \circ f$.

SOLUTION We have

$$(f \circ g)(x) = f(g(x)) = f(x - 3) = (x - 3)^2$$

$$(g \circ f)(x) = g(f(x)) = g(x^2) = x^2 - 3$$ ■

 NOTE You can see from Example 4 that, in general, $f \circ g \neq g \circ f$. Remember, the notation $f \circ g$ means that the function g is applied first and then f is applied second. In Example 4, $f \circ g$ is the function that *first* subtracts 3 and *then* squares; $g \circ f$ is the function that *first* squares and *then* subtracts 3.

▼ EXAMPLE 5 If $f(x) = \sqrt{x}$ and $g(x) = \sqrt{2 - x}$, find each function and its domain.

(a) $f \circ g$ (b) $g \circ f$ (c) $f \circ f$ (d) $g \circ g$

SOLUTION

(a) $$(f \circ g)(x) = f(g(x)) = f\left(\sqrt{2 - x}\right) = \sqrt{\sqrt{2 - x}} = \sqrt[4]{2 - x}$$

The domain of $f \circ g$ is $\{x \mid 2 - x \geq 0\} = \{x \mid x \leq 2\} = (-\infty, 2]$.

(b) $$(g \circ f)(x) = g(f(x)) = g\left(\sqrt{x}\right) = \sqrt{2 - \sqrt{x}}$$

If $0 \leq a \leq b$, then $a^2 \leq b^2$.

For $\sqrt{x}$ to be defined we must have $x \geq 0$. For $\sqrt{2 - \sqrt{x}}$ to be defined we must have $2 - \sqrt{x} \geq 0$, that is, $\sqrt{x} \leq 2$, or $x \leq 4$. Thus we have $0 \leq x \leq 4$, so the domain of $g \circ f$ is the closed interval $[0, 4]$.

(c) $$(f \circ f)(x) = f(f(x)) = f\left(\sqrt{x}\right) = \sqrt{\sqrt{x}} = \sqrt[4]{x}$$

The domain of $f \circ f$ is $[0, \infty)$.

(d) $$(g \circ g)(x) = g(g(x)) = g\left(\sqrt{2 - x}\right) = \sqrt{2 - \sqrt{2 - x}}$$

▪ **www.stewartcalculus.com**
See Additional Examples F, G.

This expression is defined when both $2 - x \geq 0$ and $2 - \sqrt{2 - x} \geq 0$. The first inequality means $x \leq 2$, and the second is equivalent to $\sqrt{2 - x} \leq 2$, or $2 - x \leq 4$, or $x \geq -2$. Thus, $-2 \leq x \leq 2$, so the domain of $g \circ g$ is the closed interval $[-2, 2]$. ∎

It is possible to take the composition of three or more functions. For instance, the composite function $f \circ g \circ h$ is found by first applying h, then g, and then f as follows:

$$(f \circ g \circ h)(x) = f(g(h(x)))$$

So far we have used composition to build complicated functions from simpler ones. But in calculus it is often useful to be able to *decompose* a complicated function into simpler ones, as in the following example.

EXAMPLE 6 Given $F(x) = \cos^2(x + 9)$, find functions f, g, and h such that $F = f \circ g \circ h$.

SOLUTION Since $F(x) = [\cos(x + 9)]^2$, the formula for F says: First add 9, then take the cosine of the result, and finally square. So we let

$$h(x) = x + 9 \qquad g(x) = \cos x \qquad f(x) = x^2$$

Then

$$(f \circ g \circ h)(x) = f(g(h(x))) = f(g(x + 9)) = f(\cos(x + 9))$$
$$= [\cos(x + 9)]^2 = F(x)$$ ∎

1.2 | EXERCISES

1. (a) Find an equation for the family of linear functions with slope 2 and sketch several members of the family.
   (b) Find an equation for the family of linear functions such that $f(2) = 1$ and sketch several members of the family.
   (c) Which function belongs to both families?

2. What do all members of the family of linear functions $f(x) = 1 + m(x + 3)$ have in common? Sketch several members of the family.

3. What do all members of the family of linear functions $f(x) = c - x$ have in common? Sketch several members of the family.

4. Find expressions for the quadratic functions whose graphs are shown.

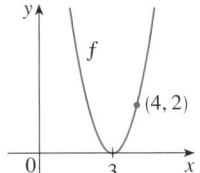
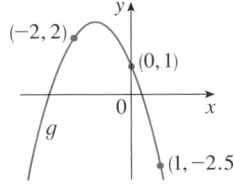

5. Find an expression for a cubic function f if $f(1) = 6$ and $f(-1) = f(0) = f(2) = 0$.

6. Some scientists believe that the average surface temperature of the world has been rising steadily. They have modeled the temperature by the linear function $T = 0.02t + 8.50$, where T is temperature in °C and t represents years since 1900.
   (a) What do the slope and T-intercept represent?
   (b) Use the equation to predict the average global surface temperature in 2100.

7. If the recommended adult dosage for a drug is D (in mg), then to determine the appropriate dosage c for a child of age a, pharmacists use the equation $c = 0.0417D(a + 1)$. Suppose the dosage for an adult is 200 mg.
   (a) Find the slope of the graph of c. What does it represent?
   (b) What is the dosage for a newborn?

8. The manager of a weekend flea market knows from past experience that if he charges x dollars for a rental space at the flea market, then the number y of spaces he can rent is given by the equation $y = 200 - 4x$.
   (a) Sketch a graph of this linear function. (Remember that the rental charge per space and the number of spaces rented can't be negative quantities.)
   (b) What do the slope, the y-intercept, and the x-intercept of the graph represent?

9. The relationship between the Fahrenheit (F) and Celsius (C) temperature scales is given by the linear function $F = \frac{9}{5}C + 32$.
   (a) Sketch a graph of this function.

   (b) What is the slope of the graph and what does it represent? What is the F-intercept and what does it represent?

10. Jason leaves Detroit at 2:00 PM and drives at a constant speed west along I-96. He passes Ann Arbor, 40 mi from Detroit, at 2:50 PM.
   (a) Express the distance traveled in terms of the time elapsed.
   (b) Draw the graph of the equation in part (a).
   (c) What is the slope of this line? What does it represent?

11. Biologists have noticed that the chirping rate of crickets of a certain species is related to temperature, and the relationship appears to be very nearly linear. A cricket produces 113 chirps per minute at 70°F and 173 chirps per minute at 80°F.
   (a) Find a linear equation that models the temperature T as a function of the number of chirps per minute N.
   (b) What is the slope of the graph? What does it represent?
   (c) If the crickets are chirping at 150 chirps per minute, estimate the temperature.

12. The manager of a furniture factory finds that it costs $2200 to manufacture 100 chairs in one day and $4800 to produce 300 chairs in one day.
   (a) Express the cost as a function of the number of chairs produced, assuming that it is linear. Then sketch the graph.
   (b) What is the slope of the graph and what does it represent?
   (c) What is the y-intercept of the graph and what does it represent?

13. At the surface of the ocean, the water pressure is the same as the air pressure above the water, 15 lb/in². Below the surface, the water pressure increases by 4.34 lb/in² for every 10 ft of descent.
   (a) Express the water pressure as a function of the depth below the ocean surface.
   (b) At what depth is the pressure 100 lb/in²?

14. The monthly cost of driving a car depends on the number of miles driven. Lynn found that in May it cost her $380 to drive 480 mi and in June it cost her $460 to drive 800 mi.
   (a) Express the monthly cost C as a function of the distance driven d, assuming that a linear relationship gives a suitable model.
   (b) Use part (a) to predict the cost of driving 1500 miles per month.
   (c) Draw the graph of the linear function. What does the slope represent?
   (d) What does the y-intercept represent?
   (e) Why does a linear function give a suitable model in this situation?

15. Many physical quantities are connected by *inverse square laws*, that is, by power functions of the form $f(x) = kx^{-2}$. In particular, the illumination of an object by a light source is inversely proportional to the square of the distance from the source. Suppose that after dark you are in a room with just one lamp and you are trying to read a book. The light is too dim and so you move halfway to the lamp. How much brighter is the light?

16. It makes sense that the larger the area of a region, the larger the number of species that inhabit the region. Many ecologists have modeled the species-area relation with a power function and, in particular, the number of species S of bats living in caves in central Mexico has been related to the surface area A of the caves by the equation $S = 0.7A^{0.3}$.
(a) The cave called *Misión Imposible* near Puebla, Mexico, has a surface area of $A = 60$ m^2. How many species of bats would you expect to find in that cave?
(b) If you discover that four species of bats live in a cave, estimate the area of the cave.

17. Suppose the graph of f is given. Write equations for the graphs that are obtained from the graph of f as follows.
(a) Shift 3 units upward.
(b) Shift 3 units downward.
(c) Shift 3 units to the right.
(d) Shift 3 units to the left.
(e) Reflect about the x-axis.
(f) Reflect about the y-axis.
(g) Stretch vertically by a factor of 3.
(h) Shrink vertically by a factor of 3.

18. Explain how each graph is obtained from the graph of $y = f(x)$.
(a) $y = f(x) + 8$　　　　　(b) $y = f(x + 8)$
(c) $y = 8f(x)$　　　　　　(d) $y = f(8x)$
(e) $y = -f(x) - 1$　　　　(f) $y = 8f(\frac{1}{8}x)$

19. The graph of $y = f(x)$ is given. Match each equation with its graph and give reasons for your choices.
(a) $y = f(x - 4)$　　　　　(b) $y = f(x) + 3$
(c) $y = \frac{1}{3}f(x)$　　　　　　(d) $y = -f(x + 4)$
(e) $y = 2f(x + 6)$

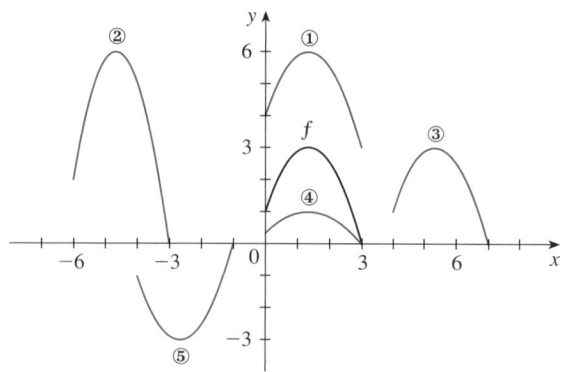

20. The graph of f is given. Draw the graphs of the following functions.
(a) $y = f(x) - 2$　　　　　(b) $y = f(x - 2)$
(c) $y = -2f(x)$　　　　　　(d) $y = f(\frac{1}{3}x) + 1$

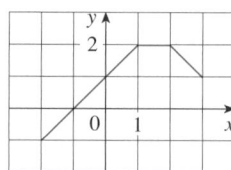

21. The graph of f is given. Use it to graph the following functions.
(a) $y = f(2x)$　　　　　　(b) $y = f(\frac{1}{2}x)$
(c) $y = f(-x)$　　　　　　(d) $y = -f(-x)$

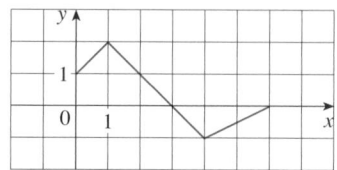

22. (a) How is the graph of $y = 2 \sin x$ related to the graph of $y = \sin x$? Use your answer and Figure 18(a) to sketch the graph of $y = 2 \sin x$.
(b) How is the graph of $y = 1 + \sqrt{x}$ related to the graph of $y = \sqrt{x}$? Use your answer and Figure 17(a) to sketch the graph of $y = 1 + \sqrt{x}$.

23–36 ▪ Graph the function by hand, not by plotting points, but by starting with the graph of one of the standard functions and then applying the appropriate transformations.

23. $y = \dfrac{1}{x + 2}$　　　　　　**24.** $y = (x - 1)^3$

25. $y = -\sqrt[3]{x}$　　　　　　　**26.** $y = x^2 + 6x + 4$

27. $y = \sqrt{x - 2} - 1$　　　　**28.** $y = 4 \sin 3x$

29. $y = \sin(\frac{1}{2}x)$　　　　　　**30.** $y = \dfrac{2}{x} - 2$

31. $y = \frac{1}{2}(1 - \cos x)$　　　　**32.** $y = 1 - 2\sqrt{x + 3}$

33. $y = 1 - 2x - x^2$　　　　**34.** $y = |x| - 2$

35. $y = \dfrac{2}{x + 1}$　　　　　　**36.** $y = \dfrac{1}{4} \tan\left(x - \dfrac{\pi}{4}\right)$

37–38 ▪ Find (a) $f + g$, (b) $f - g$, (c) fg, and (d) f/g and state their domains.

37. $f(x) = x^3 + 2x^2$, 　$g(x) = 3x^2 - 1$

38. $f(x) = \sqrt{3 - x}$, 　$g(x) = \sqrt{x^2 - 1}$

39–44 ■ Find the functions (a) $f \circ g$, (b) $g \circ f$, (c) $f \circ f$, and (d) $g \circ g$ and their domains.

39. $f(x) = x^2 - 1, \quad g(x) = 2x + 1$

40. $f(x) = x - 2, \quad g(x) = x^2 + 3x + 4$

41. $f(x) = 1 - 3x, \quad g(x) = \cos x$

42. $f(x) = \sqrt{x}, \quad g(x) = \sqrt[3]{1 - x}$

43. $f(x) = x + \dfrac{1}{x}, \quad g(x) = \dfrac{x + 1}{x + 2}$

44. $f(x) = \dfrac{x}{1 + x}, \quad g(x) = \sin 2x$

45–46 ■ Find $f \circ g \circ h$.

45. $f(x) = \sqrt{x - 3}, \quad g(x) = x^2, \quad h(x) = x^3 + 2$

46. $f(x) = \tan x, \quad g(x) = \dfrac{x}{x - 1}, \quad h(x) = \sqrt[3]{x}$

47–50 ■ Express the function in the form $f \circ g$.

47. $F(x) = (2x + x^2)^4$

48. $F(x) = \cos^2 x$

49. $v(t) = \sec(t^2) \tan(t^2)$

50. $u(t) = \dfrac{\tan t}{1 + \tan t}$

51–53 ■ Express the function in the form $f \circ g \circ h$.

51. $R(x) = \sqrt{\sqrt{x} - 1}$

52. $H(x) = \sqrt[8]{2 + |x|}$

53. $H(x) = \sec^4\!\left(\sqrt{x}\right)$

---

54. Use the table to evaluate each expression.
(a) $f(g(1))$ (b) $g(f(1))$ (c) $f(f(1))$
(d) $g(g(1))$ (e) $(g \circ f)(3)$ (f) $(f \circ g)(6)$

x	1	2	3	4	5	6
$f(x)$	3	1	4	2	2	5
$g(x)$	6	3	2	1	2	3

55. Use the given graphs of f and g to evaluate each expression, or explain why it is undefined.
(a) $f(g(2))$ (b) $g(f(0))$ (c) $(f \circ g)(0)$
(d) $(g \circ f)(6)$ (e) $(g \circ g)(-2)$ (f) $(f \circ f)(4)$

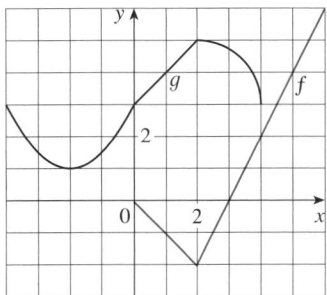

56. A spherical balloon is being inflated and the radius of the balloon is increasing at a rate of 2 cm/s.
(a) Express the radius r of the balloon as a function of the time t (in seconds).
(b) If V is the volume of the balloon as a function of the radius, find $V \circ r$ and interpret it.

57. A stone is dropped into a lake, creating a circular ripple that travels outward at a speed of 60 cm/s.
(a) Express the radius r of this circle as a function of the time t (in seconds).
(b) If A is the area of this circle as a function of the radius, find $A \circ r$ and interpret it.

58. An airplane is flying at a speed of 350 mi/h at an altitude of one mile and passes directly over a radar station at time $t = 0$.
(a) Express the horizontal distance d (in miles) that the plane has flown as a function of t.
(b) Express the distance s between the plane and the radar station as a function of d.
(c) Use composition to express s as a function of t.

59. The **Heaviside function** H is defined by

$$H(t) = \begin{cases} 0 & \text{if } t < 0 \\ 1 & \text{if } t \geq 0 \end{cases}$$

It is used in the study of electric circuits to represent the sudden surge of electric current, or voltage, when a switch is instantaneously turned on.
(a) Sketch the graph of the Heaviside function.
(b) Sketch the graph of the voltage $V(t)$ in a circuit if the switch is turned on at time $t = 0$ and 120 volts are applied instantaneously to the circuit. Write a formula for $V(t)$ in terms of $H(t)$.
(c) Sketch the graph of the voltage $V(t)$ in a circuit if the switch is turned on at time $t = 5$ seconds and 240 volts are applied instantaneously to the circuit. Write a formula for $V(t)$ in terms of $H(t)$. (Note that starting at $t = 5$ corresponds to a translation.)

60. The Heaviside function defined in Exercise 59 can also be used to define the **ramp function** $y = ctH(t)$, which represents a gradual increase in voltage or current in a circuit.
(a) Sketch the graph of the ramp function $y = tH(t)$.

(b) Sketch the graph of the voltage $V(t)$ in a circuit if the switch is turned on at time $t = 0$ and the voltage is gradually increased to 120 volts over a 60-second time interval. Write a formula for $V(t)$ in terms of $H(t)$ for $t \leq 60$.

(c) Sketch the graph of the voltage $V(t)$ in a circuit if the switch is turned on at time $t = 7$ seconds and the voltage is gradually increased to 100 volts over a period of 25 seconds. Write a formula for $V(t)$ in terms of $H(t)$ for $t \leq 32$.

61. Let f and g be linear functions with equations $f(x) = m_1 x + b_1$ and $g(x) = m_2 x + b_2$. Is $f \circ g$ also a linear function? If so, what is the slope of its graph?

62. If you invest x dollars at 4% interest compounded annually, then the amount $A(x)$ of the investment after one year is $A(x) = 1.04x$. Find $A \circ A$, $A \circ A \circ A$, and $A \circ A \circ A \circ A$.

What do these compositions represent? Find a formula for the composition of n copies of A.

63. (a) If $g(x) = 2x + 1$ and $h(x) = 4x^2 + 4x + 7$, find a function f such that $f \circ g = h$. (Think about what operations you would have to perform on the formula for g to end up with the formula for h.)

(b) If $f(x) = 3x + 5$ and $h(x) = 3x^2 + 3x + 2$, find a function g such that $f \circ g = h$.

64. If $f(x) = x + 4$ and $h(x) = 4x - 1$, find a function g such that $g \circ f = h$.

65. Suppose g is an even function and let $h = f \circ g$. Is h always an even function?

66. Suppose g is an odd function and let $h = f \circ g$. Is h always an odd function? What if f is odd? What if f is even?

1.3 | THE LIMIT OF A FUNCTION

Our aim in this section is to explore the meaning of the limit of a function. We begin by showing how the idea of a limit arises when we try to find the velocity of a falling ball.

▼ EXAMPLE 1 Suppose that a ball is dropped from the upper observation deck of the CN Tower in Toronto, 450 m above the ground. Find the velocity of the ball after 5 seconds.

SOLUTION Through experiments carried out four centuries ago, Galileo discovered that the distance fallen by any freely falling body is proportional to the square of the time it has been falling. (This model for free fall neglects air resistance.) If the distance fallen after t seconds is denoted by $s(t)$ and measured in meters, then Galileo's law is expressed by the equation

$$s(t) = 4.9t^2$$

The difficulty in finding the velocity after 5 s is that we are dealing with a single instant of time ($t = 5$), so no time interval is involved. However, we can approximate the desired quantity by computing the average velocity over the brief time interval of a tenth of a second from $t = 5$ to $t = 5.1$:

$$\text{average velocity} = \frac{\text{change in position}}{\text{time elapsed}}$$

$$= \frac{s(5.1) - s(5)}{0.1}$$

$$= \frac{4.9(5.1)^2 - 4.9(5)^2}{0.1} = 49.49 \text{ m/s}$$

Time interval	Average velocity (m/s)
$5 \leq t \leq 6$	53.9
$5 \leq t \leq 5.1$	49.49
$5 \leq t \leq 5.05$	49.245
$5 \leq t \leq 5.01$	49.049
$5 \leq t \leq 5.001$	49.0049

The table shows the results of similar calculations of the average velocity over successively smaller time periods. It appears that as we shorten the time period, the average velocity is becoming closer to 49 m/s. The **instantaneous velocity** when

$t = 5$ is defined to be the limiting value of these average velocities over shorter and shorter time periods that start at $t = 5$. Thus the (instantaneous) velocity after 5 s is

$$v = 49 \text{ m/s} \qquad \blacksquare$$

INTUITIVE DEFINITION OF A LIMIT

Let's investigate the behavior of the function f defined by $f(x) = x^2 - x + 2$ for values of x near 2. The following table gives values of $f(x)$ for values of x close to 2, but not equal to 2.

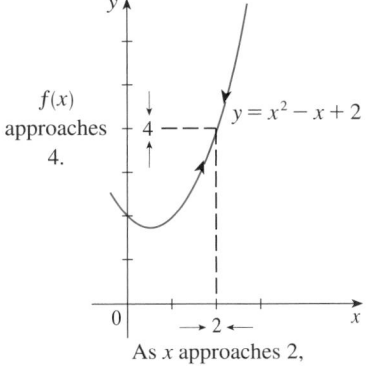

x	$f(x)$	x	$f(x)$
1.0	2.000000	3.0	8.000000
1.5	2.750000	2.5	5.750000
1.8	3.440000	2.2	4.640000
1.9	3.710000	2.1	4.310000
1.95	3.852500	2.05	4.152500
1.99	3.970100	2.01	4.030100
1.995	3.985025	2.005	4.015025
1.999	3.997001	2.001	4.003001

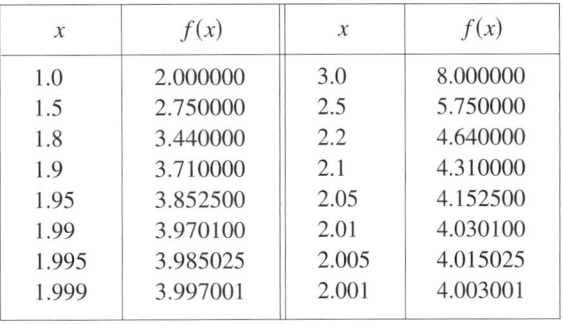

FIGURE 1

From the table and the graph of f (a parabola) shown in Figure 1 we see that when x is close to 2 (on either side of 2), $f(x)$ is close to 4. In fact, it appears that we can make the values of $f(x)$ as close as we like to 4 by taking x sufficiently close to 2. We express this by saying "the limit of the function $f(x) = x^2 - x + 2$ as x approaches 2 is equal to 4." The notation for this is

$$\lim_{x \to 2} (x^2 - x + 2) = 4$$

In general, we use the following notation.

---

1 **DEFINITION** Suppose $f(x)$ is defined when x is near the number a. (This means that f is defined on some open interval that contains a, except possibly at a itself.) Then we write

$$\lim_{x \to a} f(x) = L$$

and say   "the limit of $f(x)$, as x approaches a, equals L"

if we can make the values of $f(x)$ arbitrarily close to L (as close to L as we like) by taking x to be sufficiently close to a (on either side of a) but not equal to a.

Roughly speaking, this says that the values of $f(x)$ approach L as x approaches a. In other words, the values of $f(x)$ tend to get closer and closer to the number L as x gets closer and closer to the number a (from either side of a) but $x \neq a$.

An alternative notation for

$$\lim_{x \to a} f(x) = L$$

is $f(x) \to L \qquad \text{as} \qquad x \to a$

which is usually read "$f(x)$ approaches L as x approaches a."

Notice the phrase "but $x \neq a$" in the definition of limit. This means that in finding the limit of $f(x)$ as x approaches a, we never consider $x = a$. In fact, $f(x)$ need not even be defined when $x = a$. The only thing that matters is how f is defined *near a*.

Figure 2 shows the graphs of three functions. Note that in part (c), $f(a)$ is not defined and in part (b), $f(a) \neq L$. But in each case, regardless of what happens at a, it is true that $\lim_{x \to a} f(x) = L$.

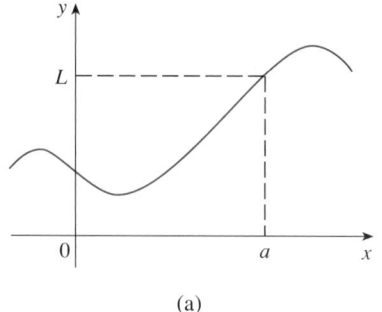
(a)

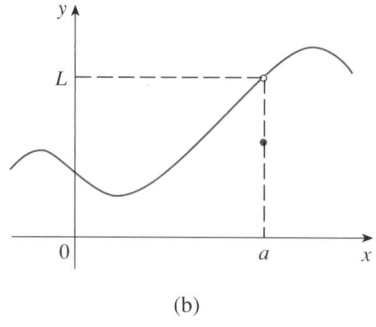
(b)

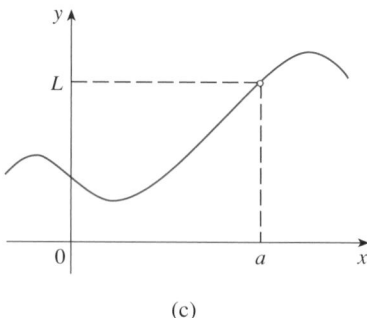
(c)

FIGURE 2

$\lim_{x \to a} f(x) = L$ in all three cases.

EXAMPLE 2 Guess the value of $\lim_{x \to 1} \dfrac{x - 1}{x^2 - 1}$.

SOLUTION Notice that the function $f(x) = (x - 1)/(x^2 - 1)$ is not defined when $x = 1$, but that doesn't matter because the definition of $\lim_{x \to a} f(x)$ says that we consider values of x that are close to a but not equal to a.

The tables at the left give values of $f(x)$ (correct to six decimal places) for values of x that approach 1 (but are not equal to 1). On the basis of the values in the tables, we make the guess that

$$\lim_{x \to 1} \frac{x - 1}{x^2 - 1} = 0.5$$ ∎

$x < 1$	$f(x)$
0.5	0.666667
0.9	0.526316
0.99	0.502513
0.999	0.500250
0.9999	0.500025

$x > 1$	$f(x)$
1.5	0.400000
1.1	0.476190
1.01	0.497512
1.001	0.499750
1.0001	0.499975

Example 2 is illustrated by the graph of f in Figure 3. Now let's change f slightly by giving it the value 2 when $x = 1$ and calling the resulting function g:

$$g(x) = \begin{cases} \dfrac{x - 1}{x^2 - 1} & \text{if } x \neq 1 \\ 2 & \text{if } x = 1 \end{cases}$$

This new function g still has the same limit as x approaches 1. (See Figure 4.)

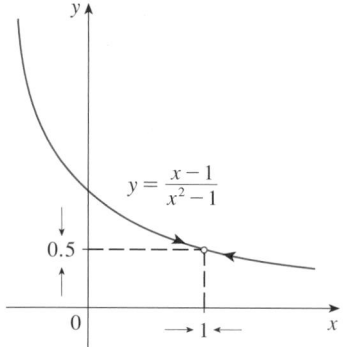

FIGURE 3

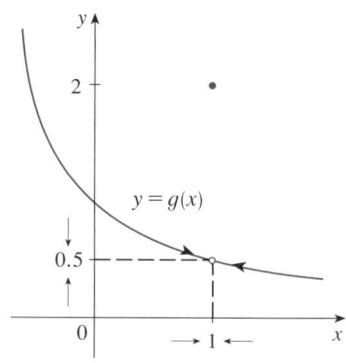

FIGURE 4

EXAMPLE 3 Estimate the value of $\lim_{t \to 0} \dfrac{\sqrt{t^2 + 9} - 3}{t^2}$.

SOLUTION The table lists values of the function for several values of t near 0.

t	$\dfrac{\sqrt{t^2 + 9} - 3}{t^2}$
± 1.0	0.16228
± 0.5	0.16553
± 0.1	0.16662
± 0.05	0.16666
± 0.01	0.16667

As t approaches 0, the values of the function seem to approach $0.1666666\ldots$ and so we guess that

$$\lim_{t \to 0} \frac{\sqrt{t^2 + 9} - 3}{t^2} = \frac{1}{6}$$
∎

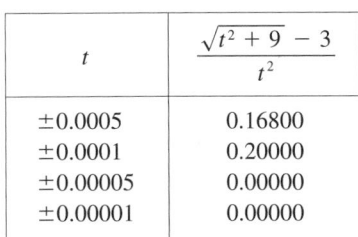

t	$\dfrac{\sqrt{t^2 + 9} - 3}{t^2}$
± 0.0005	0.16800
± 0.0001	0.20000
± 0.00005	0.00000
± 0.00001	0.00000

In Example 3 what would have happened if we had taken even smaller values of t? The table in the margin shows the results from one calculator; you can see that something strange seems to be happening.

If you try these calculations on your own calculator you might get different values, but eventually you will get the value 0 if you make t sufficiently small. Does this mean that the answer is really 0 instead of $\frac{1}{6}$? No, the value of the limit is $\frac{1}{6}$, as we will show in the next section. The problem is that the calculator gave false values because $\sqrt{t^2 + 9}$ is very close to 3 when t is small. (In fact, when t is sufficiently small, a calculator's value for $\sqrt{t^2 + 9}$ is $3.000\ldots$ to as many digits as the calculator is capable of carrying.)

Something similar happens when we try to graph the function

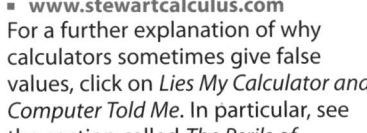

■ **www.stewartcalculus.com**
For a further explanation of why calculators sometimes give false values, click on *Lies My Calculator and Computer Told Me*. In particular, see the section called *The Perils of Subtraction*.

$$f(t) = \frac{\sqrt{t^2 + 9} - 3}{t^2}$$

of Example 3 on a graphing calculator or computer. Parts (a) and (b) of Figure 5 show quite accurate graphs of f, and when we use the trace mode (if available) we can estimate easily that the limit is about $\frac{1}{6}$. But if we zoom in too much, as in parts (c) and (d), then we get inaccurate graphs, again because of problems with subtraction.

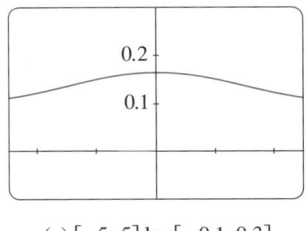

(a) $[-5, 5]$ by $[-0.1, 0.3]$

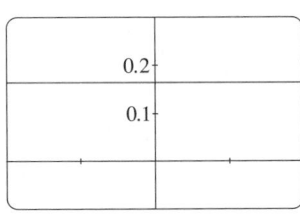

(b) $[-0.1, 0.1]$ by $[-0.1, 0.3]$

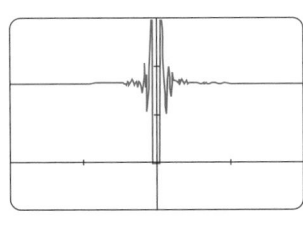

(c) $[-10^{-6}, 10^{-6}]$ by $[-0.1, 0.3]$

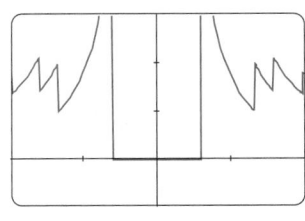

(d) $[-10^{-7}, 10^{-7}]$ by $[-0.1, 0.3]$

FIGURE 5

x	$\dfrac{\sin x}{x}$
± 1.0	0.84147098
± 0.5	0.95885108
± 0.4	0.97354586
± 0.3	0.98506736
± 0.2	0.99334665
± 0.1	0.99833417
± 0.05	0.99958339
± 0.01	0.99998333
± 0.005	0.99999583
± 0.001	0.99999983

FIGURE 6

V **EXAMPLE 4** Guess the value of $\displaystyle\lim_{x \to 0} \frac{\sin x}{x}$.

SOLUTION The function $f(x) = (\sin x)/x$ is not defined when $x = 0$. Using a calculator (and remembering that, if $x \in \mathbb{R}$, $\sin x$ means the sine of the angle whose *radian* measure is x), we construct the table of values correct to eight decimal places. From the table at the left and the graph in Figure 6 we guess that

$$\lim_{x \to 0} \frac{\sin x}{x} = 1$$

This guess is in fact correct, as will be proved in the next section using a geometric argument.

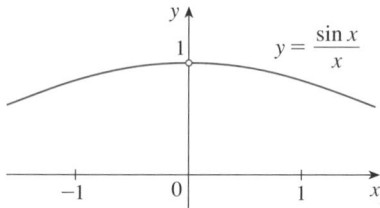

V **EXAMPLE 5** Investigate $\displaystyle\lim_{x \to 0} \sin \frac{\pi}{x}$.

■ **COMPUTER ALGEBRA SYSTEMS**

Computer algebra systems (CAS) have commands that compute limits. In order to avoid the types of pitfalls demonstrated in Examples 3 and 5, they don't find limits by numerical experimentation. Instead, they use more sophisticated techniques such as computing infinite series. If you have access to a CAS, use the limit command to compute the limits in the examples of this section and to check your answers in the exercises of this chapter.

SOLUTION Again the function $f(x) = \sin(\pi/x)$ is undefined at 0. Evaluating the function for some small values of x, we get

$$f(1) = \sin \pi = 0 \qquad\qquad f\left(\tfrac{1}{2}\right) = \sin 2\pi = 0$$

$$f\left(\tfrac{1}{3}\right) = \sin 3\pi = 0 \qquad\qquad f\left(\tfrac{1}{4}\right) = \sin 4\pi = 0$$

$$f(0.1) = \sin 10\pi = 0 \qquad f(0.01) = \sin 100\pi = 0$$

Similarly, $f(0.001) = f(0.0001) = 0$. On the basis of this information we might be tempted to guess that

$$\lim_{x \to 0} \sin \frac{\pi}{x} = 0$$

⊘ but this time our guess is wrong. Note that although $f(1/n) = \sin n\pi = 0$ for any integer n, it is also true that $f(x) = 1$ for infinitely many values of x that approach 0. The graph of f is given in Figure 7.

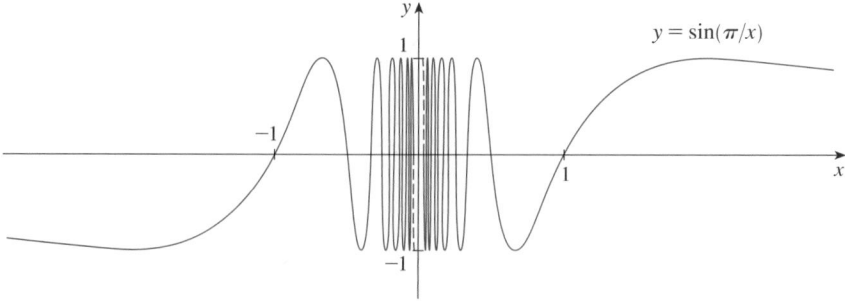

FIGURE 7

The dashed lines near the *y*-axis indicate that the values of $\sin(\pi/x)$ oscillate between 1 and -1 infinitely often as *x* approaches 0. (Use a graphing device to graph *f* and zoom in toward the origin several times. What do you observe?)

Since the values of $f(x)$ do not approach a fixed number as *x* approaches 0,

■ www.stewartcalculus.com
See Additional Example A.

$$\lim_{x \to 0} \sin \frac{\pi}{x} \quad \text{does not exist}$$ ■

⊘ Examples 3 and 5 illustrate some of the pitfalls in guessing the value of a limit. It is easy to guess the wrong value if we use inappropriate values of *x*, but it is difficult to know when to stop calculating values. And, as the discussion after Example 3 shows, sometimes calculators and computers give the wrong values. In the next section, however, we will develop foolproof methods for calculating limits.

V EXAMPLE 6 The Heaviside function *H* is defined by

$$H(t) = \begin{cases} 0 & \text{if } t < 0 \\ 1 & \text{if } t \geq 0 \end{cases}$$

[This function is named after the electrical engineer Oliver Heaviside (1850–1925) and can be used to describe an electric current that is switched on at time $t = 0$.] Its graph is shown in Figure 8.

As *t* approaches 0 from the left, $H(t)$ approaches 0. As *t* approaches 0 from the right, $H(t)$ approaches 1. There is no single number that $H(t)$ approaches as *t* approaches 0. Therefore $\lim_{t \to 0} H(t)$ does not exist. ■

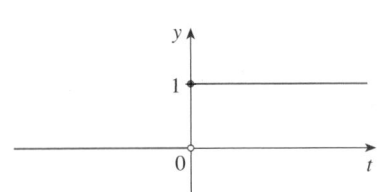

FIGURE 8

ONE-SIDED LIMITS

We noticed in Example 6 that $H(t)$ approaches 0 as *t* approaches 0 from the left and $H(t)$ approaches 1 as *t* approaches 0 from the right. We indicate this situation symbolically by writing

$$\lim_{t \to 0^-} H(t) = 0 \qquad \text{and} \qquad \lim_{t \to 0^+} H(t) = 1$$

The symbol "$t \to 0^-$" indicates that we consider only values of *t* that are less than 0. Likewise, "$t \to 0^+$" indicates that we consider only values of *t* that are greater than 0.

> **2 DEFINITION** We write
>
> $$\lim_{x \to a^-} f(x) = L$$
>
> and say the **left-hand limit of $f(x)$ as *x* approaches *a*** [or the **limit of $f(x)$ as *x* approaches *a* from the left**] is equal to *L* if we can make the values of $f(x)$ arbitrarily close to *L* by taking *x* to be sufficiently close to *a* and *x* less than *a*.

Notice that Definition 2 differs from Definition 1 only in that we require x to be less than a. Similarly, if we require that x be greater than a, we get "the **right-hand limit of $f(x)$ as x approaches a** is equal to L" and we write

$$\lim_{x \to a^+} f(x) = L$$

Thus the symbol "$x \to a^+$" means that we consider only $x > a$. These definitions are illustrated in Figure 9.

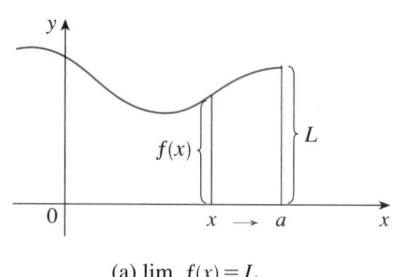

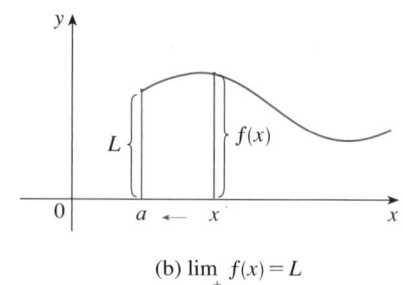

FIGURE 9
(a) $\lim\limits_{x \to a^-} f(x) = L$
(b) $\lim\limits_{x \to a^+} f(x) = L$

By comparing Definition 1 with the definitions of one-sided limits, we see that the following is true.

---

3 $\lim\limits_{x \to a} f(x) = L$ if and only if $\lim\limits_{x \to a^-} f(x) = L$ and $\lim\limits_{x \to a^+} f(x) = L$

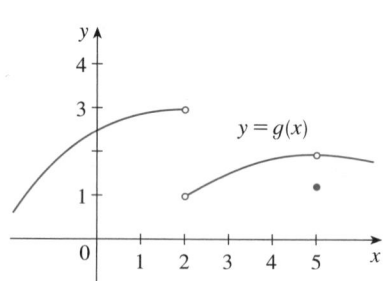

FIGURE 10

◺ EXAMPLE 7 The graph of a function g is shown in Figure 10. Use it to state the values (if they exist) of the following:

(a) $\lim\limits_{x \to 2^-} g(x)$ (b) $\lim\limits_{x \to 2^+} g(x)$ (c) $\lim\limits_{x \to 2} g(x)$

(d) $\lim\limits_{x \to 5^-} g(x)$ (e) $\lim\limits_{x \to 5^+} g(x)$ (f) $\lim\limits_{x \to 5} g(x)$

SOLUTION From the graph we see that the values of $g(x)$ approach 3 as x approaches 2 from the left, but they approach 1 as x approaches 2 from the right. Therefore

(a) $\lim\limits_{x \to 2^-} g(x) = 3$ and   (b) $\lim\limits_{x \to 2^+} g(x) = 1$

(c) Since the left and right limits are different, we conclude from ③ that $\lim_{x \to 2} g(x)$ does not exist.

The graph also shows that

(d) $\lim\limits_{x \to 5^-} g(x) = 2$ and   (e) $\lim\limits_{x \to 5^+} g(x) = 2$

(f) This time the left and right limits are the same and so, by ③, we have

$$\lim_{x \to 5} g(x) = 2$$

Despite this fact, notice that $g(5) \neq 2$.   ∎

EXAMPLE 8 Find $\lim\limits_{x \to 0} \dfrac{1}{x^2}$ if it exists.

SOLUTION As x becomes close to 0, x^2 also becomes close to 0, and $1/x^2$ becomes very large. (See the following table.) In fact, it appears from the graph of the function $f(x) = 1/x^2$ shown in Figure 11 that the values of $f(x)$ can be made arbitrarily large by taking x close enough to 0. Thus the values of $f(x)$ do not approach a number, so $\lim_{x \to 0} (1/x^2)$ does not exist.

x	$\dfrac{1}{x^2}$
± 1	1
± 0.5	4
± 0.2	25
± 0.1	100
± 0.05	400
± 0.01	10,000
± 0.001	1,000,000

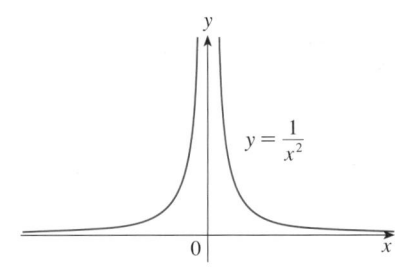

FIGURE 11

PRECISE DEFINITION OF A LIMIT

■ www.stewartcalculus.com
See Additional Example B.

Definition 1 is appropriate for an intuitive understanding of limits, but for deeper understanding and rigorous proofs we need to be more precise.

We want to express, in a quantitative manner, that $f(x)$ can be made arbitrarily close to L by taking x to be sufficiently close to a (but $x \neq a$). This means that $f(x)$ can be made to lie within any preassigned distance from L (traditionally denoted by ε, the Greek letter epsilon) by requiring that x be within a specified distance δ (the Greek letter delta) from a. That is, $|f(x) - L| < \varepsilon$ when $|x - a| < \delta$ and $x \neq a$. Notice that we can stipulate that $x \neq a$ by writing $0 < |x - a|$. The resulting precise definition of a limit is as follows.

> **4** **DEFINITION** Let f be a function defined on some open interval that contains the number a, except possibly at a itself. Then we say that the **limit of** $f(x)$ **as x approaches a is L**, and we write
>
> $$\lim_{x \to a} f(x) = L$$
>
> if for every number $\varepsilon > 0$ there is a corresponding number $\delta > 0$ such that
>
> $$\text{if} \quad 0 < |x - a| < \delta \quad \text{then} \quad |f(x) - L| < \varepsilon$$

Definition 4 is illustrated in Figures 12–14. If a number $\varepsilon > 0$ is given, then we draw the horizontal lines $y = L + \varepsilon$ and $y = L - \varepsilon$ and the graph of f. (See Figure 12.) If $\lim_{x \to a} f(x) = L$, then we can find a number $\delta > 0$ such that if we restrict x to lie in the interval $(a - \delta, a + \delta)$ and take $x \neq a$, then the curve $y = f(x)$ lies

TEC In Module 1.3/1.6 you can explore the precise definition of a limit both graphically and numerically.

between the lines $y = L - \varepsilon$ and $y = L + \varepsilon$. (See Figure 13.) You can see that if such a δ has been found, then any smaller δ will also work.

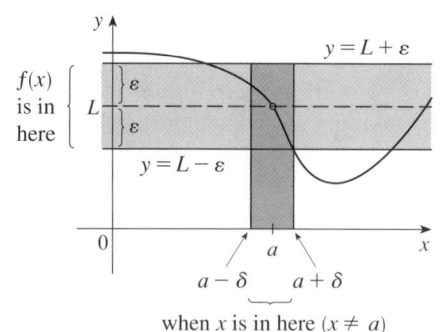

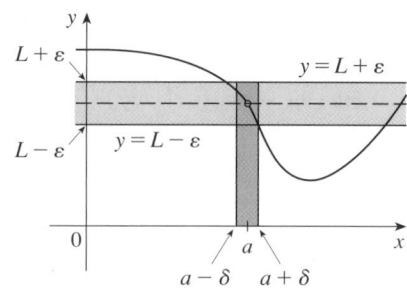

FIGURE 12

FIGURE 13

FIGURE 14

It's important to realize that the process illustrated in Figures 12 and 13 must work for *every* positive number ε, no matter how small it is chosen. Figure 14 shows that if a smaller ε is chosen, then a smaller δ may be required.

In proving limit statements it may be helpful to think of the definition of limit as a challenge. First it challenges you with a number ε. Then you must be able to produce a suitable δ. You have to be able to do this for *every* $\varepsilon > 0$, not just a particular ε.

V EXAMPLE 9 Prove that $\lim\limits_{x \to 3} (4x - 5) = 7$.

SOLUTION Let ε be a given positive number. According to Definition 4 with $a = 3$ and $L = 7$, we need to find a number δ such that

$$\text{if} \quad 0 < |x - 3| < \delta \quad \text{then} \quad |(4x - 5) - 7| < \varepsilon$$

But $|(4x - 5) - 7| = |4x - 12| = |4(x - 3)| = 4|x - 3|$. Therefore we want:

$$\text{if} \quad 0 < |x - 3| < \delta \quad \text{then} \quad 4|x - 3| < \varepsilon$$

Note that $4|x - 3| < \varepsilon \iff |x - 3| < \varepsilon/4$. So let's choose $\delta = \varepsilon/4$. We can then write the following:

$$\text{if} \quad 0 < |x - 3| < \delta \quad \text{then} \quad 4|x - 3| < \varepsilon \quad \text{so} \quad |(4x - 5) - 7| < \varepsilon$$

Therefore, by the definition of a limit,

$$\lim\limits_{x \to 3} (4x - 5) = 7$$

For a left-hand limit we restrict x so that $x < a$, so in Definition 4 we replace $0 < |x - a| < \delta$ by $a - \delta < x < a$. Similarly, for a right-hand limit we use $a < x < a + \delta$.

V EXAMPLE 10 Prove that $\lim\limits_{x \to 0^+} \sqrt{x} = 0$.

SOLUTION Let ε be a given positive number. We want to find a number δ such that

$$\text{if} \quad 0 < x < \delta \quad \text{then} \quad |\sqrt{x} - 0| < \varepsilon \quad \text{that is} \quad \sqrt{x} < \varepsilon$$

But $\sqrt{x} < \varepsilon \iff x < \varepsilon^2$. So if we choose $\delta = \varepsilon^2$ and $0 < x < \delta = \varepsilon^2$, then $\sqrt{x} < \varepsilon$. (See Figure 16.) This shows that $\sqrt{x} \to 0$ as $x \to 0^+$. ∎

■ Figure 15 shows the geometry behind Example 9.

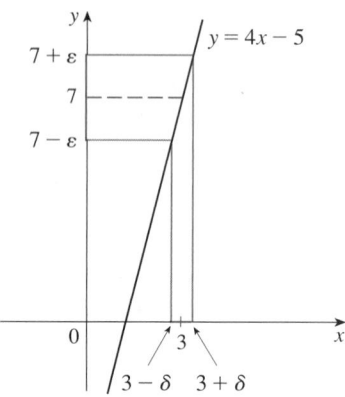

FIGURE 15

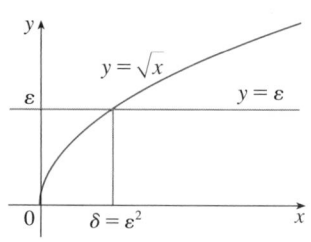

FIGURE 16

1.3 | EXERCISES

1. If a ball is thrown into the air with a velocity of 40 ft/s, its height in feet t seconds later is given by $y = 40t - 16t^2$.
   (a) Find the average velocity for the time period beginning when $t = 2$ and lasting
      (i) 0.5 second   (ii) 0.1 second
      (iii) 0.05 second   (iv) 0.01 second
   (b) Estimate the instantaneous velocity when $t = 2$.

2. If a rock is thrown upward on the planet Mars with a velocity of 10 m/s, its height in meters t seconds later is given by $y = 10t - 1.86t^2$.
   (a) Find the average velocity over the given time intervals:
      (i) [1, 2]   (ii) [1, 1.5]   (iii) [1, 1.1]
      (iv) [1, 1.01]   (v) [1, 1.001]
   (b) Estimate the instantaneous velocity when $t = 1$.

3. For the function f whose graph is given, state the value of each quantity, if it exists. If it does not exist, explain why.
   (a) $\lim\limits_{x \to 1} f(x)$ (b) $\lim\limits_{x \to 3^-} f(x)$ (c) $\lim\limits_{x \to 3^+} f(x)$
   (d) $\lim\limits_{x \to 3} f(x)$ (e) $f(3)$

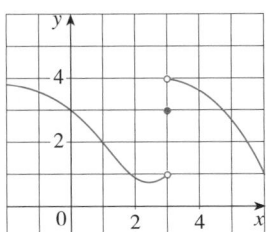

4. Use the given graph of f to state the value of each quantity, if it exists. If it does not exist, explain why.
   (a) $\lim\limits_{x \to 2^-} f(x)$ (b) $\lim\limits_{x \to 2^+} f(x)$ (c) $\lim\limits_{x \to 2} f(x)$
   (d) $f(2)$ (e) $\lim\limits_{x \to 4} f(x)$ (f) $f(4)$

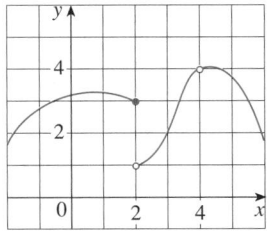

5. For the function g whose graph is given, state the value of each quantity, if it exists. If it does not exist, explain why.
   (a) $\lim\limits_{t \to 0^-} g(t)$ (b) $\lim\limits_{t \to 0^+} g(t)$ (c) $\lim\limits_{t \to 0} g(t)$
   (d) $\lim\limits_{t \to 2^-} g(t)$ (e) $\lim\limits_{t \to 2^+} g(t)$ (f) $\lim\limits_{t \to 2} g(t)$
   (g) $g(2)$ (h) $\lim\limits_{t \to 4} g(t)$

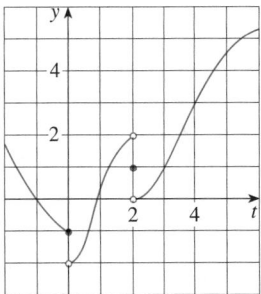

6. Sketch the graph of the following function and use it to determine the values of a for which $\lim\limits_{x \to a} f(x)$ exists:
$$f(x) = \begin{cases} 1 + \sin x & \text{if } x < 0 \\ \cos x & \text{if } 0 \le x \le \pi \\ \sin x & \text{if } x > \pi \end{cases}$$

7–10 ▪ Sketch the graph of an example of a function f that satisfies all of the given conditions.

7. $\lim\limits_{x \to 0^-} f(x) = -1$, $\lim\limits_{x \to 0^+} f(x) = 2$, $f(0) = 1$

8. $\lim\limits_{x \to 0} f(x) = 1$, $\lim\limits_{x \to 3^-} f(x) = -2$, $\lim\limits_{x \to 3^+} f(x) = 2$,
   $f(0) = -1$, $f(3) = 1$

9. $\lim\limits_{x \to 3^+} f(x) = 4$, $\lim\limits_{x \to 3^-} f(x) = 2$, $\lim\limits_{x \to -2} f(x) = 2$,
   $f(3) = 3$, $f(-2) = 1$

10. $\lim\limits_{x \to 0^-} f(x) = 2$, $\lim\limits_{x \to 0^+} f(x) = 0$, $\lim\limits_{x \to 4^-} f(x) = 3$,
    $\lim\limits_{x \to 4^+} f(x) = 0$, $f(0) = 2$, $f(4) = 1$

11–14 ▪ Guess the value of the limit (if it exists) by evaluating the function at the given numbers (correct to six decimal places).

11. $\lim\limits_{x \to 2} \dfrac{x^2 - 2x}{x^2 - x - 2}$
    $x = 2.5, 2.1, 2.05, 2.01, 2.005, 2.001,$
    $1.9, 1.95, 1.99, 1.995, 1.999$

12. $\lim\limits_{x \to -1} \dfrac{x^2 - 2x}{x^2 - x - 2}$
    $x = 0, -0.5, -0.9, -0.95, -0.99, -0.999,$
    $-2, -1.5, -1.1, -1.01, -1.001$

13. $\lim\limits_{x \to 0} \dfrac{\sin x}{x + \tan x}$
    $x = \pm 1, \pm 0.5, \pm 0.2, \pm 0.1, \pm 0.05, \pm 0.01$

14. $\lim\limits_{h \to 0} \dfrac{(2 + h)^5 - 32}{h}$

$h = \pm 0.5,\ \pm 0.1,\ \pm 0.01,\ \pm 0.001,\ \pm 0.0001$

15–18 ▪ Use a table of values to estimate the value of the limit. If you have a graphing device, use it to confirm your result graphically.

15. $\lim\limits_{x \to 0} \dfrac{\sqrt{x + 4} - 2}{x}$

16. $\lim\limits_{x \to 0} \dfrac{\tan 3x}{\tan 5x}$

17. $\lim\limits_{x \to 1} \dfrac{x^6 - 1}{x^{10} - 1}$

18. $\lim\limits_{x \to 0} \dfrac{9^x - 5^x}{x}$

---

19. (a) By graphing the function $f(x) = (\cos 2x - \cos x)/x^2$ and zooming in toward the point where the graph crosses the y-axis, estimate the value of $\lim_{x \to 0} f(x)$.
(b) Check your answer in part (a) by evaluating $f(x)$ for values of x that approach 0.

20. (a) Estimate the value of

$$\lim_{x \to 0} \frac{\sin x}{\sin \pi x}$$

by graphing the function $f(x) = (\sin x)/(\sin \pi x)$. State your answer correct to two decimal places.
(b) Check your answer in part (a) by evaluating $f(x)$ for values of x that approach 0.

21. (a) Evaluate the function $f(x) = x^2 - (2^x/1000)$ for $x = 1, 0.8, 0.6, 0.4, 0.2, 0.1,$ and 0.05, and guess the value of

$$\lim_{x \to 0} \left(x^2 - \frac{2^x}{1000} \right)$$

(b) Evaluate $f(x)$ for $x = 0.04, 0.02, 0.01, 0.005, 0.003,$ and 0.001. Guess again.

22. (a) Evaluate $h(x) = (\tan x - x)/x^3$ for $x = 1, 0.5, 0.1,$ $0.05, 0.01,$ and 0.005.
(b) Guess the value of $\lim\limits_{x \to 0} \dfrac{\tan x - x}{x^3}$.
(c) Evaluate $h(x)$ for successively smaller values of x until you finally reach 0 values for $h(x)$. Are you still confident that your guess in part (b) is correct? Explain why you eventually obtained 0 values. (In Section 3.7 a method for evaluating the limit will be explained.)
(d) Graph the function h in the viewing rectangle $[-1, 1]$ by $[0, 1]$. Then zoom in toward the point where the graph crosses the y-axis to estimate the limit of $h(x)$ as x approaches 0. Continue to zoom in until you observe distortions in the graph of h. Compare with the results of part (c).

23. Use the given graph of $f(x) = \sqrt{x}$ to find a number δ such that

$$\text{if} \quad |x - 4| < \delta \quad \text{then} \quad |\sqrt{x} - 2| < 0.4$$

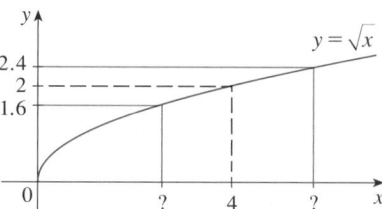

24. Use the given graph of $f(x) = x^2$ to find a number δ such that

$$\text{if} \quad |x - 1| < \delta \quad \text{then} \quad |x^2 - 1| < \tfrac{1}{2}$$

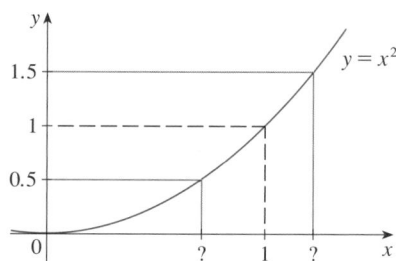

25. Use a graph to find a number δ such that

$$\text{if} \quad \left| x - \frac{\pi}{4} \right| < \delta \quad \text{then} \quad |\tan x - 1| < 0.2$$

26. Use a graph to find a number δ such that

$$\text{if} \quad |x - 1| < \delta \quad \text{then} \quad \left| \frac{2x}{x^2 + 4} - 0.4 \right| < 0.1$$

27. A machinist is required to manufacture a circular metal disk with area 1000 cm^2.
(a) What radius produces such a disk?
(b) If the machinist is allowed an error tolerance of ± 5 cm^2 in the area of the disk, how close to the ideal radius in part (a) must the machinist control the radius?
(c) In terms of the ε, δ definition of $\lim_{x \to a} f(x) = L$, what is x? What is $f(x)$? What is a? What is L? What value of ε is given? What is the corresponding value of δ?

28. A crystal growth furnace is used in research to determine how best to manufacture crystals used in electronic components for the space shuttle. For proper growth of the crystal, the temperature must be controlled accurately by adjusting the input power. Suppose the relationship is given by

$$T(w) = 0.1w^2 + 2.155w + 20$$

where T is the temperature in degrees Celsius and w is the power input in watts.
(a) How much power is needed to maintain the temperature at 200°C ?
(b) If the temperature is allowed to vary from 200°C by up to ± 1°C, what range of wattage is allowed for the input power?
(c) In terms of the ε, δ definition of $\lim_{x \to a} f(x) = L$, what is x? What is $f(x)$? What is a? What is L? What value of ε is given? What is the corresponding value of δ?

29–32 ■ Prove the statement using the ε, δ definition of a limit and illustrate with a diagram like Figure 15.

29. $\lim_{x \to 3} \left(1 + \frac{1}{3}x\right) = 2$

30. $\lim_{x \to 4} (2x - 5) = 3$

31. $\lim_{x \to -3} (1 - 4x) = 13$

32. $\lim_{x \to -2} (3x + 5) = -1$

33–44 ■ Prove the statement using the ε, δ definition of a limit.

33. $\lim_{x \to 1} \dfrac{2 + 4x}{3} = 2$

34. $\lim_{x \to 10} \left(3 - \frac{4}{5}x\right) = -5$

35. $\lim_{x \to 2} \dfrac{x^2 + x - 6}{x - 2} = 5$

36. $\lim_{x \to -1.5} \dfrac{9 - 4x^2}{3 + 2x} = 6$

37. $\lim_{x \to a} x = a$

38. $\lim_{x \to a} c = c$

39. $\lim_{x \to 0} x^2 = 0$

40. $\lim_{x \to 0} x^3 = 0$

41. $\lim_{x \to 0} |x| = 0$

42. $\lim_{x \to 9^-} \sqrt[4]{9 - x} = 0$

43. $\lim_{x \to 3} x^2 = 9$ [*Hint:* Write $|x^2 - 9| = |x + 3||x - 3|$. Show that if $|x - 3| < 1$, then $|x + 3| < 7$. If you let δ be the smaller of the numbers 1 and $\varepsilon/7$, show that this δ works.]

44. $\lim_{x \to 3} (x^2 + x - 4) = 8$ [*Hint:* If $|x - 3| < 1$, what can you say about $|x + 4|$?]

⬚ CAS **45.** (a) For the limit $\lim_{x \to 1} (x^3 + x + 1) = 3$, use a graph to find a value of δ that corresponds to $\varepsilon = 0.4$.
(b) By using a computer algebra system to solve the cubic equation $x^3 + x + 1 = 3 + \varepsilon$, find the largest possible value of δ that works for any given $\varepsilon > 0$.
(c) Put $\varepsilon = 0.4$ in your answer to part (b) and compare with your answer to part (a).

46. If H is the Heaviside function defined in Example 6, prove, using Definition 4, that $\lim_{t \to 0} H(t)$ does not exist. [*Hint:* Use an indirect proof as follows. Suppose that the limit is L. Take $\varepsilon = \frac{1}{2}$ in the definition of a limit and try to arrive at a contradiction.]

1.4 | CALCULATING LIMITS

In Section 1.3 we used calculators and graphs to guess the values of limits, but we saw that such methods don't always lead to the correct answer. In this section we use the following properties of limits, called the *Limit Laws*, to calculate limits.

> **LIMIT LAWS** Suppose that c is a constant and the limits
>
> $$\lim_{x \to a} f(x) \qquad \text{and} \qquad \lim_{x \to a} g(x)$$
>
> exist. Then
>
> **1.** $\lim_{x \to a} [f(x) + g(x)] = \lim_{x \to a} f(x) + \lim_{x \to a} g(x)$
>
> **2.** $\lim_{x \to a} [f(x) - g(x)] = \lim_{x \to a} f(x) - \lim_{x \to a} g(x)$
>
> **3.** $\lim_{x \to a} [cf(x)] = c \lim_{x \to a} f(x)$
>
> **4.** $\lim_{x \to a} [f(x) g(x)] = \lim_{x \to a} f(x) \cdot \lim_{x \to a} g(x)$
>
> **5.** $\lim_{x \to a} \dfrac{f(x)}{g(x)} = \dfrac{\lim_{x \to a} f(x)}{\lim_{x \to a} g(x)} \qquad \text{if } \lim_{x \to a} g(x) \neq 0$

These five laws can be stated verbally as follows:

Sum Law

1. The limit of a sum is the sum of the limits.

Difference Law

2. The limit of a difference is the difference of the limits.

Constant Multiple Law

3. The limit of a constant times a function is the constant times the limit of the function.

Product Law

4. The limit of a product is the product of the limits.

Quotient Law

5. The limit of a quotient is the quotient of the limits (provided that the limit of the denominator is not 0).

It is easy to believe that these properties are true. For instance, if $f(x)$ is close to L and $g(x)$ is close to M, it is reasonable to conclude that $f(x) + g(x)$ is close to $L + M$. This gives us an intuitive basis for believing that Law 1 is true. All of these laws can be proved using the precise definition of a limit. (See Appendix D.)

If we use the Product Law repeatedly with $g(x) = f(x)$, we obtain the following law.

Power Law

6. $\lim\limits_{x \to a} [f(x)]^n = \left[\lim\limits_{x \to a} f(x)\right]^n$ where n is a positive integer

In applying these six limit laws, we need to use two special limits:

7. $\lim\limits_{x \to a} c = c$ **8.** $\lim\limits_{x \to a} x = a$

These limits are obvious from an intuitive point of view (state them in words or draw graphs of $y = c$ and $y = x$), but they can be proved from the precise definition. (See Exercises 37 and 38 in Section 1.3.)

If we now put $f(x) = x$ in Law 6 and use Law 8, we get another useful special limit.

9. $\lim\limits_{x \to a} x^n = a^n$ where n is a positive integer

A similar limit holds for roots as follows.

10. $\lim\limits_{x \to a} \sqrt[n]{x} = \sqrt[n]{a}$ where n is a positive integer

(If n is even, we assume that $a > 0$.)

More generally, we have the following law.

Root Law

11. $\lim\limits_{x \to a} \sqrt[n]{f(x)} = \sqrt[n]{\lim\limits_{x \to a} f(x)}$ where n is a positive integer

$\left[\text{If } n \text{ is even, we assume that } \lim\limits_{x \to a} f(x) > 0.\right]$

Isaac Newton was born on Christmas Day in 1642, the year of Galileo's death. When he entered Cambridge University in 1661 Newton didn't know much mathematics, but he learned quickly by reading Euclid and Descartes and by attending the lectures of Isaac Barrow. Cambridge was closed because of the plague in 1665 and 1666, and Newton returned home to reflect on what he had learned. Those two years were amazingly productive for at that time he made four of his major discoveries: (1) his representation of functions as sums of infinite series, including the binomial theorem; (2) his work on differential and integral calculus; (3) his laws of motion and law of universal gravitation; and (4) his prism experiments on the nature of light and color. Because of a fear of controversy and criticism, he was reluctant to publish his discoveries and it wasn't until 1687, at the urging of the astronomer Halley, that Newton published *Principia Mathematica*. In this work, the greatest scientific treatise ever written, Newton set forth his version of calculus and used it to investigate mechanics, fluid dynamics, and wave motion, and to explain the motion of planets and comets.

The beginnings of calculus are found in the calculations of areas and volumes by ancient Greek scholars such as Eudoxus and Archimedes. Although aspects of the idea of a limit are implicit in their "method of exhaustion," Eudoxus and Archimedes never explicitly formulated the concept of a limit. Likewise, mathematicians such as Cavalieri, Fermat, and Barrow, the immediate precursors of Newton in the development of calculus, did not actually use limits. It was Isaac Newton who was the first to talk explicitly about limits. He explained that the main idea behind limits is that quantities "approach nearer than by any given difference." Newton stated that the limit was the basic concept in calculus, but it was left to later mathematicians like Cauchy to clarify his ideas about limits.

EXAMPLE 1 Evaluate the following limits and justify each step.

(a) $\lim_{x \to 5} (2x^2 - 3x + 4)$

(b) $\lim_{x \to -2} \dfrac{x^3 + 2x^2 - 1}{5 - 3x}$

SOLUTION

(a)
$$\lim_{x \to 5} (2x^2 - 3x + 4) = \lim_{x \to 5} (2x^2) - \lim_{x \to 5} (3x) + \lim_{x \to 5} 4 \quad \text{(by Laws 2 and 1)}$$

$$= 2 \lim_{x \to 5} x^2 - 3 \lim_{x \to 5} x + \lim_{x \to 5} 4 \quad \text{(by 3)}$$

$$= 2(5^2) - 3(5) + 4 \quad \text{(by 9, 8, and 7)}$$

$$= 39$$

(b) We start by using Law 5, but its use is fully justified only at the final stage when we see that the limits of the numerator and denominator exist and the limit of the denominator is not 0.

$$\lim_{x \to -2} \frac{x^3 + 2x^2 - 1}{5 - 3x} = \frac{\lim_{x \to -2} (x^3 + 2x^2 - 1)}{\lim_{x \to -2} (5 - 3x)} \quad \text{(by Law 5)}$$

$$= \frac{\lim_{x \to -2} x^3 + 2 \lim_{x \to -2} x^2 - \lim_{x \to -2} 1}{\lim_{x \to -2} 5 - 3 \lim_{x \to -2} x} \quad \text{(by 1, 2, and 3)}$$

$$= \frac{(-2)^3 + 2(-2)^2 - 1}{5 - 3(-2)} \quad \text{(by 9, 8, and 7)}$$

$$= -\frac{1}{11}$$

■

NOTE If we let $f(x) = 2x^2 - 3x + 4$, then $f(5) = 39$. In other words, we would have gotten the correct answer in Example 1(a) by substituting 5 for x. Similarly, direct substitution provides the correct answer in part (b). The functions in Example 1 are a polynomial and a rational function, respectively, and similar use of the Limit Laws proves that direct substitution always works for such functions (see Exercises 57 and 58). We state this fact as follows.

DIRECT SUBSTITUTION PROPERTY If f is a polynomial or a rational function and a is in the domain of f, then

$$\lim_{x \to a} f(x) = f(a)$$

The trigonometric functions also enjoy the Direct Substitution Property. We know from the definitions of $\sin \theta$ and $\cos \theta$ that the coordinates of the point P in Figure 1 are $(\cos \theta, \sin \theta)$. As $\theta \to 0$, we see that P approaches the point $(1, 0)$ and so $\cos \theta \to 1$ and $\sin \theta \to 0$. Thus

$$\boxed{1} \qquad \lim_{\theta \to 0} \cos \theta = 1 \qquad \lim_{\theta \to 0} \sin \theta = 0$$

Since $\cos 0 = 1$ and $\sin 0 = 0$, the equations in $\boxed{1}$ assert that the cosine and sine

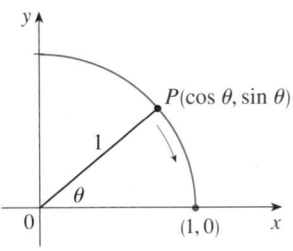

FIGURE 1

$P(\cos \theta, \sin \theta)$

$(1, 0)$

■ Another way to establish the limits in $\boxed{1}$ is to use the inequality $\sin \theta < \theta$ (for $\theta > 0$), which is proved on page 42.

functions satisfy the Direct Substitution Property at 0. The addition formulas for cosine and sine can then be used to deduce that these functions satisfy the Direct Substitution Property everywhere (see Exercises 59 and 60). In other words, for any real number a,

$$\lim_{\theta \to a} \sin \theta = \sin a \qquad \lim_{\theta \to a} \cos \theta = \cos a$$

This enables us to evaluate certain limits quite simply. For example,

$$\lim_{x \to \pi} x \cos x = \left(\lim_{x \to \pi} x\right)\left(\lim_{x \to \pi} \cos x\right) = \pi \cdot \cos \pi = -\pi$$

Functions with the Direct Substitution Property are called *continuous at a* and will be studied in Section 1.5. However, not all limits can be evaluated by direct substitution, as the following examples show.

EXAMPLE 2 Find $\displaystyle\lim_{x \to 1} \frac{x^2 - 1}{x - 1}$.

SOLUTION Let $f(x) = (x^2 - 1)/(x - 1)$. We can't find the limit by substituting $x = 1$ because $f(1)$ isn't defined. Nor can we apply the Quotient Law, because the limit of the denominator is 0. Instead, we need to do some preliminary algebra. We factor the numerator as a difference of squares:

$$\frac{x^2 - 1}{x - 1} = \frac{(x - 1)(x + 1)}{x - 1}$$

The numerator and denominator have a common factor of $x - 1$. When we take the limit as x approaches 1, we have $x \neq 1$ and so $x - 1 \neq 0$. Therefore we can cancel the common factor and compute the limit as follows:

$$\lim_{x \to 1} \frac{x^2 - 1}{x - 1} = \lim_{x \to 1} \frac{(x - 1)(x + 1)}{x - 1}$$
$$= \lim_{x \to 1} (x + 1)$$
$$= 1 + 1 = 2$$

∎

NOTE In Example 2 we were able to compute the limit by replacing the given function $f(x) = (x^2 - 1)/(x - 1)$ by a simpler function, $g(x) = x + 1$, with the same limit. This is valid because $f(x) = g(x)$ except when $x = 1$, and in computing a limit as x approaches 1 we don't consider what happens when x is actually *equal* to 1. In general, we have the following useful fact.

If $f(x) = g(x)$ when $x \neq a$, then $\displaystyle\lim_{x \to a} f(x) = \lim_{x \to a} g(x)$, provided the limits exist.

EXAMPLE 3 Find $\displaystyle\lim_{x \to 1} g(x)$ where

$$g(x) = \begin{cases} x + 1 & \text{if } x \neq 1 \\ \pi & \text{if } x = 1 \end{cases}$$

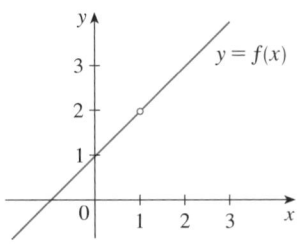

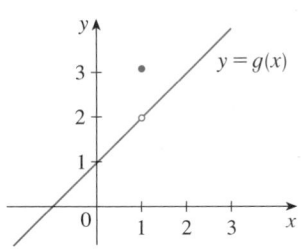

FIGURE 2

The graphs of the functions f (from Example 2) and g (from Example 3)

SOLUTION Here g is defined at $x = 1$ and $g(1) = \pi$, but the value of a limit as x approaches 1 does not depend on the value of the function at 1. Since $g(x) = x + 1$ for $x \neq 1$, we have

$$\lim_{x \to 1} g(x) = \lim_{x \to 1} (x + 1) = 2$$ ∎

Note that the values of the functions in Examples 2 and 3 are identical except when $x = 1$ (see Figure 2) and so they have the same limit as x approaches 1.

▼ EXAMPLE 4 Evaluate $\displaystyle\lim_{h \to 0} \frac{(3 + h)^2 - 9}{h}$.

SOLUTION If we define

$$F(h) = \frac{(3 + h)^2 - 9}{h}$$

then, as in Example 2, we can't compute $\lim_{h \to 0} F(h)$ by letting $h = 0$ since $F(0)$ is undefined. But if we simplify $F(h)$ algebraically, we find that

$$F(h) = \frac{(9 + 6h + h^2) - 9}{h} = \frac{6h + h^2}{h} = 6 + h$$

(Recall that we consider only $h \neq 0$ when letting h approach 0.) Thus

$$\lim_{h \to 0} \frac{(3 + h)^2 - 9}{h} = \lim_{h \to 0} (6 + h) = 6$$ ∎

EXAMPLE 5 Find $\displaystyle\lim_{t \to 0} \frac{\sqrt{t^2 + 9} - 3}{t^2}$.

SOLUTION We can't apply the Quotient Law immediately, since the limit of the denominator is 0. Here the preliminary algebra consists of rationalizing the numerator:

$$\lim_{t \to 0} \frac{\sqrt{t^2 + 9} - 3}{t^2} = \lim_{t \to 0} \frac{\sqrt{t^2 + 9} - 3}{t^2} \cdot \frac{\sqrt{t^2 + 9} + 3}{\sqrt{t^2 + 9} + 3}$$

$$= \lim_{t \to 0} \frac{(t^2 + 9) - 9}{t^2(\sqrt{t^2 + 9} + 3)} = \lim_{t \to 0} \frac{t^2}{t^2(\sqrt{t^2 + 9} + 3)}$$

$$= \lim_{t \to 0} \frac{1}{\sqrt{t^2 + 9} + 3} = \frac{1}{\sqrt{\lim_{t \to 0} (t^2 + 9)} + 3} = \frac{1}{3 + 3} = \frac{1}{6}$$

This calculation confirms the guess that we made in Example 3 in Section 1.3. ∎

Some limits are best calculated by first finding the left- and right-hand limits. The following theorem is a reminder of what we discovered in Section 1.3. It says that a two-sided limit exists if and only if both of the one-sided limits exist and are equal.

2 THEOREM $\displaystyle\lim_{x \to a} f(x) = L$ if and only if $\displaystyle\lim_{x \to a^-} f(x) = L = \lim_{x \to a^+} f(x)$

When computing one-sided limits, we use the fact that the Limit Laws also hold for one-sided limits.

EXAMPLE 6 Show that $\lim_{x \to 0} |x| = 0$.

SOLUTION Recall that

$$|x| = \begin{cases} x & \text{if } x \geq 0 \\ -x & \text{if } x < 0 \end{cases}$$

Since $|x| = x$ for $x > 0$, we have

$$\lim_{x \to 0^+} |x| = \lim_{x \to 0^+} x = 0$$

For $x < 0$ we have $|x| = -x$ and so

$$\lim_{x \to 0^-} |x| = \lim_{x \to 0^-} (-x) = 0$$

Therefore, by Theorem 2,

$$\lim_{x \to 0} |x| = 0$$

■ The result of Example 6 looks plausible from Figure 3.

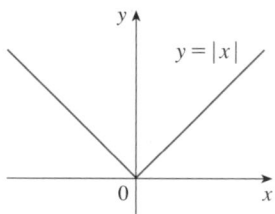

FIGURE 3

▼ EXAMPLE 7 Prove that $\lim_{x \to 0} \dfrac{|x|}{x}$ does not exist.

SOLUTION
$$\lim_{x \to 0^+} \frac{|x|}{x} = \lim_{x \to 0^+} \frac{x}{x} = \lim_{x \to 0^+} 1 = 1$$

$$\lim_{x \to 0^-} \frac{|x|}{x} = \lim_{x \to 0^-} \frac{-x}{x} = \lim_{x \to 0^-} (-1) = -1$$

Since the right- and left-hand limits are different, it follows from Theorem 2 that $\lim_{x \to 0} |x|/x$ does not exist. The graph of the function $f(x) = |x|/x$ is shown in Figure 4 and supports the one-sided limits that we found. ■

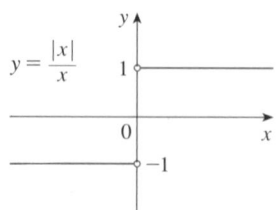

FIGURE 4

EXAMPLE 8 The **greatest integer function** is defined by $[\![x]\!] = $ the largest integer that is less than or equal to x. (For instance, $[\![4]\!] = 4$, $[\![4.8]\!] = 4$, $[\![\pi]\!] = 3$, $[\![\sqrt{2}]\!] = 1$, $[\![-\frac{1}{2}]\!] = -1$.) Show that $\lim_{x \to 3} [\![x]\!]$ does not exist.

SOLUTION The graph of the greatest integer function is shown in Figure 5. Since $[\![x]\!] = 3$ for $3 \leq x < 4$, we have

$$\lim_{x \to 3^+} [\![x]\!] = \lim_{x \to 3^+} 3 = 3$$

Since $[\![x]\!] = 2$ for $2 \leq x < 3$, we have

$$\lim_{x \to 3^-} [\![x]\!] = \lim_{x \to 3^-} 2 = 2$$

■ Other notations for $[\![x]\!]$ are $[x]$ and $\lfloor x \rfloor$. The greatest integer function is sometimes called the *floor function*.

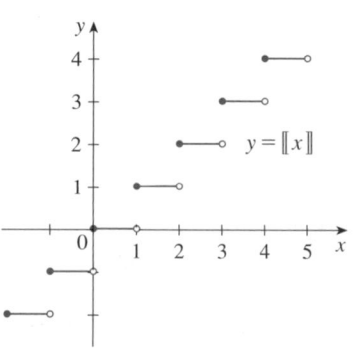

FIGURE 5
Greatest integer function

Because these one-sided limits are not equal, $\lim_{x \to 3} [\![x]\!]$ does not exist by Theorem 2. ■

■ www.stewartcalculus.com
See Additional Example A.

The next two theorems give two additional properties of limits. Their proofs can be found in Appendix D.

---

3 **THEOREM** If $f(x) \leqslant g(x)$ when x is near a (except possibly at a) and the limits of f and g both exist as x approaches a, then

$$\lim_{x \to a} f(x) \leqslant \lim_{x \to a} g(x)$$

---

4 **THE SQUEEZE THEOREM** If $f(x) \leqslant g(x) \leqslant h(x)$ when x is near a (except possibly at a) and

$$\lim_{x \to a} f(x) = \lim_{x \to a} h(x) = L$$

then

$$\lim_{x \to a} g(x) = L$$

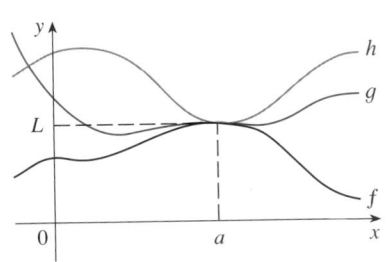

FIGURE 6

The Squeeze Theorem, which is sometimes called the Sandwich Theorem or the Pinching Theorem, is illustrated by Figure 6. It says that if $g(x)$ is squeezed between $f(x)$ and $h(x)$ near a, and if f and h have the same limit L at a, then g is forced to have the same limit L at a.

V EXAMPLE 9 Show that $\lim_{x \to 0} x^2 \sin \dfrac{1}{x} = 0$.

SOLUTION First note that we **cannot** use

$$\lim_{x \to 0} x^2 \sin \frac{1}{x} = \lim_{x \to 0} x^2 \cdot \lim_{x \to 0} \sin \frac{1}{x}$$

because $\lim_{x \to 0} \sin(1/x)$ does not exist (see Example 5 in Section 1.3).

Instead we apply the Squeeze Theorem, and so we need to find a function f smaller than $g(x) = x^2 \sin(1/x)$ and a function h bigger than g such that both $f(x)$ and $h(x)$ approach 0. To do this we use our knowledge of the sine function. Because the sine of any number lies between -1 and 1, we can write

5
$$-1 \leqslant \sin \frac{1}{x} \leqslant 1$$

Any inequality remains true when multiplied by a positive number. We know that $x^2 \geqslant 0$ for all x and so, multiplying each side of the inequalities in **5** by x^2, we get

$$-x^2 \leqslant x^2 \sin \frac{1}{x} \leqslant x^2$$

as illustrated by Figure 7. We know that

$$\lim_{x \to 0} x^2 = 0 \quad \text{and} \quad \lim_{x \to 0} (-x^2) = 0$$

Taking $f(x) = -x^2$, $g(x) = x^2 \sin(1/x)$, and $h(x) = x^2$ in the Squeeze Theorem, we obtain

$$\lim_{x \to 0} x^2 \sin \frac{1}{x} = 0 \qquad \blacksquare$$

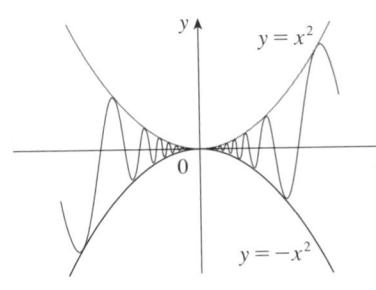

FIGURE 7
$y = x^2 \sin(1/x)$

In Example 4 in Section 1.3 we made the guess, on the basis of numerical and graphical evidence, that

6

$$\lim_{\theta \to 0} \frac{\sin \theta}{\theta} = 1$$

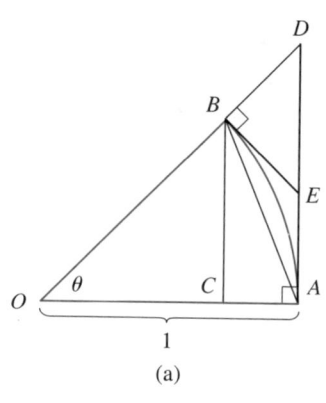

We can prove Equation 6 with help from the Squeeze Theorem. Assume first that θ lies between 0 and $\pi/2$. Figure 8(a) shows a sector of a circle with center O, central angle θ, and radius 1. BC is drawn perpendicular to OA. By the definition of radian measure, we have arc $AB = \theta$. Also, $|BC| = |OB| \sin \theta = \sin \theta$. From the diagram we see that

$$|BC| < |AB| < \text{arc } AB$$

Therefore

$$\sin \theta < \theta \qquad \text{so} \qquad \frac{\sin \theta}{\theta} < 1$$

Let the tangent lines at A and B intersect at E. You can see from Figure 8(b) that the circumference of a circle is smaller than the length of a circumscribed polygon, and so arc $AB < |AE| + |EB|$. Thus

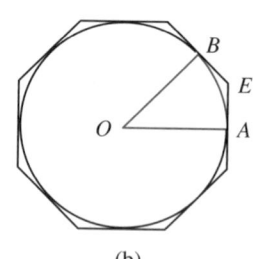

$$\theta = \text{arc } AB < |AE| + |EB| < |AE| + |ED|$$

$$= |AD| = |OA| \tan \theta = \tan \theta$$

FIGURE 8

(In Appendix D the inequality $\theta \leq \tan \theta$ is proved directly from the definition of the length of an arc without resorting to geometric intuition as we did here.) Therefore we have

$$\theta < \frac{\sin \theta}{\cos \theta} \qquad \text{and so} \qquad \cos \theta < \frac{\sin \theta}{\theta} < 1$$

We know that $\lim_{\theta \to 0} 1 = 1$ and $\lim_{\theta \to 0} \cos \theta = 1$, so by the Squeeze Theorem, we have

$$\lim_{\theta \to 0^+} \frac{\sin \theta}{\theta} = 1$$

But the function $(\sin \theta)/\theta$ is an even function, so its right and left limits must be equal. Hence we have

$$\lim_{\theta \to 0} \frac{\sin \theta}{\theta} = 1$$

so we have proved Equation 6.

EXAMPLE 10 Find $\displaystyle\lim_{x \to 0} \frac{\sin 7x}{4x}$.

SOLUTION In order to apply Equation 6, we first rewrite the function by multiplying and dividing by 7:

Note that $\sin 7x \neq 7 \sin x$.

$$\frac{\sin 7x}{4x} = \frac{7}{4}\left(\frac{\sin 7x}{7x}\right)$$

Notice that as $x \to 0$, we have $7x \to 0$, and so, by Equation 6 with $\theta = 7x$,

$$\lim_{x \to 0} \frac{\sin 7x}{7x} = \lim_{x \to 0} \frac{\sin 7x}{7x} = 1$$

Thus
$$\lim_{x \to 0} \frac{\sin 7x}{4x} = \lim_{x \to 0} \frac{7}{4}\left(\frac{\sin 7x}{7x}\right)$$

$$= \frac{7}{4} \lim_{x \to 0} \frac{\sin 7x}{7x} = \frac{7}{4} \cdot 1 = \frac{7}{4} \qquad \blacksquare$$

EXAMPLE 11 Evaluate $\displaystyle\lim_{\theta \to 0} \frac{\cos \theta - 1}{\theta}$.

SOLUTION

■ We multiply numerator and denominator by $\cos \theta + 1$ in order to put the function in a form in which we can use the limits we know.

$$\lim_{\theta \to 0} \frac{\cos \theta - 1}{\theta} = \lim_{\theta \to 0}\left(\frac{\cos \theta - 1}{\theta} \cdot \frac{\cos \theta + 1}{\cos \theta + 1}\right) = \lim_{\theta \to 0} \frac{\cos^2\theta - 1}{\theta(\cos \theta + 1)}$$

$$= \lim_{\theta \to 0} \frac{-\sin^2\theta}{\theta(\cos \theta + 1)} = -\lim_{\theta \to 0}\left(\frac{\sin \theta}{\theta} \cdot \frac{\sin \theta}{\cos \theta + 1}\right)$$

$$= -\lim_{\theta \to 0} \frac{\sin \theta}{\theta} \cdot \lim_{\theta \to 0} \frac{\sin \theta}{\cos \theta + 1}$$

$$= -1 \cdot \left(\frac{0}{1 + 1}\right) = 0 \qquad \text{(by Equation 6)} \qquad \blacksquare$$

1.4 | EXERCISES

1. Given that

$$\lim_{x \to 2} f(x) = 4 \qquad \lim_{x \to 2} g(x) = -2 \qquad \lim_{x \to 2} h(x) = 0$$

find the limits that exist. If the limit does not exist, explain why.

(a) $\displaystyle\lim_{x \to 2} [f(x) + 5g(x)]$

(b) $\displaystyle\lim_{x \to 2} [g(x)]^3$

(c) $\displaystyle\lim_{x \to 2} \sqrt{f(x)}$

(d) $\displaystyle\lim_{x \to 2} \frac{3f(x)}{g(x)}$

(e) $\displaystyle\lim_{x \to 2} \frac{g(x)}{h(x)}$

(f) $\displaystyle\lim_{x \to 2} \frac{g(x)h(x)}{f(x)}$

2. The graphs of f and g are given. Use them to evaluate each limit, if it exists. If the limit does not exist, explain why.

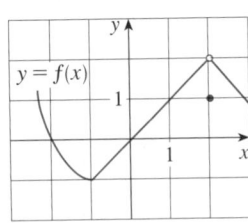
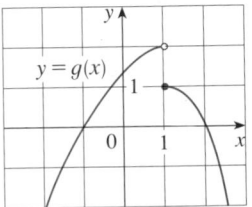

(a) $\displaystyle\lim_{x \to 2} [f(x) + g(x)]$

(b) $\displaystyle\lim_{x \to 1} [f(x) + g(x)]$

(c) $\displaystyle\lim_{x \to 0} [f(x)g(x)]$

(d) $\displaystyle\lim_{x \to -1} \frac{f(x)}{g(x)}$

(e) $\displaystyle\lim_{x \to 2} [x^3 f(x)]$

(f) $\displaystyle\lim_{x \to 1} \sqrt{3 + f(x)}$

3–9 ▪ Evaluate the limit and justify each step by indicating the appropriate Limit Law(s).

3. $\lim\limits_{x \to 3} (5x^3 - 3x^2 + x - 6)$

4. $\lim\limits_{x \to -1} (x^4 - 3x)(x^2 + 5x + 3)$

5. $\lim\limits_{t \to -2} \dfrac{t^4 - 2}{2t^2 - 3t + 2}$

6. $\lim\limits_{u \to -2} \sqrt{u^4 + 3u + 6}$

7. $\lim\limits_{x \to 8} \left(1 + \sqrt[3]{x}\right)(2 - 6x^2 + x^3)$

8. $\lim\limits_{x \to 0} \dfrac{\cos^4 x}{5 + 2x^3}$

9. $\lim\limits_{\theta \to \pi/2} \theta \sin \theta$

---

10. (a) What is wrong with the following equation?

$$\frac{x^2 + x - 6}{x - 2} = x + 3$$

(b) In view of part (a), explain why the equation

$$\lim_{x \to 2} \frac{x^2 + x - 6}{x - 2} = \lim_{x \to 2} (x + 3)$$

is correct.

11–28 ▪ Evaluate the limit, if it exists.

11. $\lim\limits_{x \to 5} \dfrac{x^2 - 6x + 5}{x - 5}$

12. $\lim\limits_{x \to 4} \dfrac{x^2 - 4x}{x^2 - 3x - 4}$

13. $\lim\limits_{x \to 5} \dfrac{x^2 - 5x + 6}{x - 5}$

14. $\lim\limits_{x \to -1} \dfrac{x^2 - 4x}{x^2 - 3x - 4}$

15. $\lim\limits_{t \to -3} \dfrac{t^2 - 9}{2t^2 + 7t + 3}$

16. $\lim\limits_{x \to -1} \dfrac{2x^2 + 3x + 1}{x^2 - 2x - 3}$

17. $\lim\limits_{h \to 0} \dfrac{(-5 + h)^2 - 25}{h}$

18. $\lim\limits_{h \to 0} \dfrac{(2 + h)^3 - 8}{h}$

19. $\lim\limits_{x \to -2} \dfrac{x + 2}{x^3 + 8}$

20. $\lim\limits_{x \to -1} \dfrac{x^2 + 2x + 1}{x^4 - 1}$

21. $\lim\limits_{h \to 0} \dfrac{\sqrt{9 + h} - 3}{h}$

22. $\lim\limits_{u \to 2} \dfrac{\sqrt{4u + 1} - 3}{u - 2}$

23. $\lim\limits_{x \to 16} \dfrac{4 - \sqrt{x}}{16x - x^2}$

24. $\lim\limits_{t \to 0} \left(\dfrac{1}{t} - \dfrac{1}{t^2 + t}\right)$

25. $\lim\limits_{x \to -4} \dfrac{\frac{1}{4} + \frac{1}{x}}{4 + x}$

26. $\lim\limits_{x \to -4} \dfrac{\sqrt{x^2 + 9} - 5}{x + 4}$

27. $\lim\limits_{h \to 0} \dfrac{(x + h)^3 - x^3}{h}$

28. $\lim\limits_{h \to 0} \dfrac{\frac{1}{(x + h)^2} - \frac{1}{x^2}}{h}$

29. (a) Estimate the value of

$$\lim_{x \to 0} \frac{x}{\sqrt{1 + 3x} - 1}$$

by graphing the function $f(x) = x/(\sqrt{1 + 3x} - 1)$.

(b) Make a table of values of $f(x)$ for x close to 0 and guess the value of the limit.

(c) Use the Limit Laws to prove that your guess is correct.

30. (a) Use a graph of

$$f(x) = \frac{\sqrt{3 + x} - \sqrt{3}}{x}$$

to estimate the value of $\lim_{x \to 0} f(x)$ to two decimal places.

(b) Use a table of values of $f(x)$ to estimate the limit to four decimal places.

(c) Use the Limit Laws to find the exact value of the limit.

31. Use the Squeeze Theorem to show that $\lim_{x \to 0} (x^2 \cos 20\pi x) = 0$. Illustrate by graphing the functions $f(x) = -x^2$, $g(x) = x^2 \cos 20\pi x$, and $h(x) = x^2$ on the same screen.

32. Use the Squeeze Theorem to show that

$$\lim_{x \to 0} \sqrt{x^3 + x^2} \, \sin \frac{\pi}{x} = 0$$

Illustrate by graphing the functions f, g, and h (in the notation of the Squeeze Theorem) on the same screen.

33. If $4x - 9 \leqslant f(x) \leqslant x^2 - 4x + 7$ for $x \geqslant 0$, find $\lim\limits_{x \to 4} f(x)$.

34. If $2x \leqslant g(x) \leqslant x^4 - x^2 + 2$ for all x, evaluate $\lim\limits_{x \to 1} g(x)$.

35. Prove that $\lim\limits_{x \to 0} x^4 \cos \dfrac{2}{x} = 0$.

36. Prove that $\lim\limits_{x \to 0^+} \sqrt{x} \left[1 + \sin^2(2\pi/x)\right] = 0$.

37–42 ▪ Find the limit, if it exists. If the limit does not exist, explain why.

37. $\lim\limits_{x \to 3} (2x + |x - 3|)$

38. $\lim\limits_{x \to -6} \dfrac{2x + 12}{|x + 6|}$

39. $\lim\limits_{x \to 0.5^-} \dfrac{2x - 1}{|2x^3 - x^2|}$

40. $\lim\limits_{x \to -2} \dfrac{2 - |x|}{2 + x}$

41. $\lim\limits_{x \to 0^-} \left(\dfrac{1}{x} - \dfrac{1}{|x|}\right)$

42. $\lim\limits_{x \to 0^+} \left(\dfrac{1}{x} - \dfrac{1}{|x|}\right)$

---

43. Let $g(x) = \dfrac{x^2 + x - 6}{|x - 2|}$.

(a) Find

   (i) $\lim\limits_{x \to 2^+} g(x)$ (ii) $\lim\limits_{x \to 2^-} g(x)$

(b) Does $\lim_{x \to 2} g(x)$ exist?

(c) Sketch the graph of g.

44. Let

$$g(x) = \begin{cases} x & \text{if } x < 1 \\ 3 & \text{if } x = 1 \\ 2 - x^2 & \text{if } 1 < x \leqslant 2 \\ x - 3 & \text{if } x > 2 \end{cases}$$

(a) Evaluate each of the following limits, if it exists.

(i) $\lim_{x \to 1^-} g(x)$ (ii) $\lim_{x \to 1} g(x)$ (iii) $g(1)$

(iv) $\lim_{x \to 2^-} g(x)$ (v) $\lim_{x \to 2^+} g(x)$ (vi) $\lim_{x \to 2} g(x)$

(b) Sketch the graph of g.

45. (a) If the symbol $[\![\ ]\!]$ denotes the greatest integer function defined in Example 8, evaluate

(i) $\lim_{x \to -2^+} [\![x]\!]$ (ii) $\lim_{x \to -2} [\![x]\!]$ (iii) $\lim_{x \to -2.4} [\![x]\!]$

(b) If n is an integer, evaluate

(i) $\lim_{x \to n^-} [\![x]\!]$ (ii) $\lim_{x \to n^+} [\![x]\!]$

(c) For what values of a does $\lim_{x \to a} [\![x]\!]$ exist?

46. Let $f(x) = [\![\cos x]\!]$, $-\pi \leqslant x \leqslant \pi$.

(a) Sketch the graph of f.

(b) Evaluate each limit, if it exists.

(i) $\lim_{x \to 0} f(x)$ (ii) $\lim_{x \to (\pi/2)^-} f(x)$

(iii) $\lim_{x \to (\pi/2)^+} f(x)$ (iv) $\lim_{x \to \pi/2} f(x)$

(c) For what values of a does $\lim_{x \to a} f(x)$ exist?

47. If $f(x) = [\![x]\!] + [\![-x]\!]$, show that $\lim_{x \to 2} f(x)$ exists but is not equal to $f(2)$.

48. In the theory of relativity, the Lorentz contraction formula

$$L = L_0 \sqrt{1 - v^2/c^2}$$

expresses the length L of an object as a function of its velocity v with respect to an observer, where L_0 is the length of the object at rest and c is the speed of light. Find $\lim_{v \to c^-} L$ and interpret the result. Why is a left-hand limit necessary?

49–56 ▪ Find the limit.

49. $\lim_{x \to 0} \dfrac{\sin 3x}{x}$

50. $\lim_{x \to 0} \dfrac{\sin 4x}{\sin 6x}$

51. $\lim_{t \to 0} \dfrac{\tan 6t}{\sin 2t}$

52. $\lim_{\theta \to 0} \dfrac{\cos \theta - 1}{\sin \theta}$

53. $\lim_{x \to 0} \dfrac{\sin 3x}{5x^3 - 4x}$

54. $\lim_{x \to 0} \dfrac{\sin 3x \sin 5x}{x^2}$

55. $\lim_{\theta \to 0} \dfrac{\sin \theta}{\theta + \tan \theta}$

56. $\lim_{x \to 0} \dfrac{\sin(x^2)}{x}$

57. If p is a polynomial, show that $\lim_{x \to a} p(x) = p(a)$.

58. If r is a rational function, use Exercise 57 to show that $\lim_{x \to a} r(x) = r(a)$ for every number a in the domain of r.

59. To prove that sine has the Direct Substitution Property we need to show that $\lim_{x \to a} \sin x = \sin a$ for every real number a. If we let $h = x - a$, then $x = a + h$ and $x \to a \iff h \to 0$. So an equivalent statement is that

$$\lim_{h \to 0} \sin(a + h) = \sin a$$

Use $\boxed{1}$ to show that this is true.

60. Prove that cosine has the Direct Substitution Property.

61. If $\lim\limits_{x \to 1} \dfrac{f(x) - 8}{x - 1} = 10$, find $\lim\limits_{x \to 1} f(x)$.

62. Prove that if $\lim_{x \to a} g(x) = 0$ and $\lim_{x \to a} f(x)$ exists and is not 0, then

$$\lim_{x \to a} \frac{f(x)}{g(x)} \quad \text{does not exist}$$

63. Show by means of an example that $\lim_{x \to a} [f(x) + g(x)]$ may exist even though neither $\lim_{x \to a} f(x)$ nor $\lim_{x \to a} g(x)$ exists.

64. Show by means of an example that $\lim_{x \to a} [f(x) g(x)]$ may exist even though neither $\lim_{x \to a} f(x)$ nor $\lim_{x \to a} g(x)$ exists.

65. Is there a number a such that

$$\lim_{x \to -2} \frac{3x^2 + ax + a + 3}{x^2 + x - 2}$$

exists? If so, find the value of a and the value of the limit.

66. The figure shows a fixed circle C_1 with equation $(x - 1)^2 + y^2 = 1$ and a shrinking circle C_2 with radius r and center the origin. P is the point $(0, r)$, Q is the upper point of intersection of the two circles, and R is the point of intersection of the line PQ and the x-axis. What happens to R as C_2 shrinks, that is, as $r \to 0^+$?

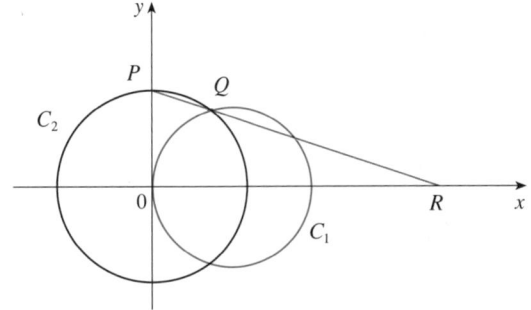

1.5 | CONTINUITY

We noticed in Section 1.4 that the limit of a function as x approaches a can often be found simply by calculating the value of the function at a. Functions with this property are called *continuous at a*. We will see that the mathematical definition of continuity corresponds closely with the meaning of the word *continuity* in everyday language. (A continuous process is one that takes place gradually, without interruption or abrupt change.)

---

1 | DEFINITION   A function f is **continuous at a number a** if

$$\lim_{x \to a} f(x) = f(a)$$

Notice that Definition 1 implicitly requires three things if f is continuous at a:

1. $f(a)$ is defined (that is, a is in the domain of f)

2. $\displaystyle\lim_{x \to a} f(x)$ exists

3. $\displaystyle\lim_{x \to a} f(x) = f(a)$

The definition says that f is continuous at a if $f(x)$ approaches $f(a)$ as x approaches a. Thus a continuous function f has the property that a small change in x produces only a small change in $f(x)$. In fact, the change in $f(x)$ can be kept as small as we please by keeping the change in x sufficiently small.

If f is defined near a (in other words, f is defined on an open interval containing a, except perhaps at a), we say that f is **discontinuous at a** (or f has a **discontinuity** at a) if f is not continuous at a.

Physical phenomena are usually continuous. For instance, the displacement or velocity of a vehicle varies continuously with time, as does a person's height. But discontinuities do occur in such situations as electric currents. [See Example 6 in Section 1.3, where the Heaviside function is discontinuous at 0 because $\lim_{t \to 0} H(t)$ does not exist.]

Geometrically, you can think of a function that is continuous at every number in an interval as a function whose graph has no break in it. The graph can be drawn without removing your pen from the paper.

EXAMPLE 1   Figure 2 shows the graph of a function f. At which numbers is f discontinuous? Why?

SOLUTION   It looks as if there is a discontinuity when $a = 1$ because the graph has a break there. The official reason that f is discontinuous at 1 is that $f(1)$ is not defined.

The graph also has a break when $a = 3$, but the reason for the discontinuity is different. Here, $f(3)$ is defined, but $\lim_{x \to 3} f(x)$ does not exist (because the left and right limits are different). So f is discontinuous at 3.

What about $a = 5$? Here, $f(5)$ is defined and $\lim_{x \to 5} f(x)$ exists (because the left and right limits are the same). But

$$\lim_{x \to 5} f(x) \neq f(5)$$

So f is discontinuous at 5. ∎

Now let's see how to detect discontinuities when a function is defined by a formula.

■ As illustrated in Figure 1, if f is continuous, then the points $(x, f(x))$ on the graph of f approach the point $(a, f(a))$ on the graph. So there is no gap in the curve.

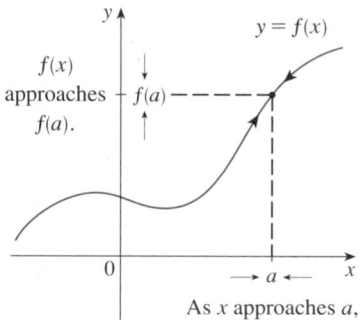

FIGURE 1

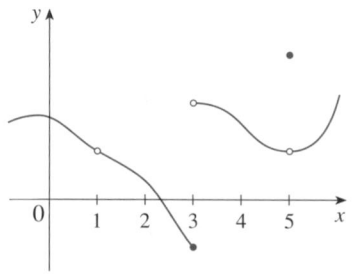

FIGURE 2

V EXAMPLE 2 Where are each of the following functions discontinuous?

(a) $f(x) = \dfrac{x^2 - x - 2}{x - 2}$

(b) $f(x) = \begin{cases} \dfrac{1}{x^2} & \text{if } x \neq 0 \\ 1 & \text{if } x = 0 \end{cases}$

(c) $f(x) = \begin{cases} \dfrac{x^2 - x - 2}{x - 2} & \text{if } x \neq 2 \\ 1 & \text{if } x = 2 \end{cases}$

(d) $f(x) = [\![x]\!]$

SOLUTION

(a) Notice that $f(2)$ is not defined, so f is discontinuous at 2. Later we'll see why f is continuous at all other numbers.

(b) Here $f(0) = 1$ is defined but

$$\lim_{x \to 0} f(x) = \lim_{x \to 0} \frac{1}{x^2}$$

does not exist. (See Example 8 in Section 1.3.) So f is discontinuous at 0.

(c) Here $f(2) = 1$ is defined and

$$\lim_{x \to 2} f(x) = \lim_{x \to 2} \frac{x^2 - x - 2}{x - 2} = \lim_{x \to 2} \frac{(x - 2)(x + 1)}{x - 2} = \lim_{x \to 2} (x + 1) = 3$$

exists. But

$$\lim_{x \to 2} f(x) \neq f(2)$$

so f is not continuous at 2.

(d) The greatest integer function $f(x) = [\![x]\!]$ has discontinuities at all of the integers because $\lim_{x \to n} [\![x]\!]$ does not exist if n is an integer. (See Example 8 and Exercise 45 in Section 1.4.) ∎

Figure 3 shows the graphs of the functions in Example 2. In each case the graph can't be drawn without lifting the pen from the paper because a hole or break or jump occurs in the graph. The kind of discontinuity illustrated in parts (a) and (c) is called **removable** because we could remove the discontinuity by redefining f at just the single number 2. [The function $g(x) = x + 1$ is continuous.] The discontinuity in part (b) is called an **infinite discontinuity**. The discontinuities in part (d) are called **jump discontinuities** because the function "jumps" from one value to another.

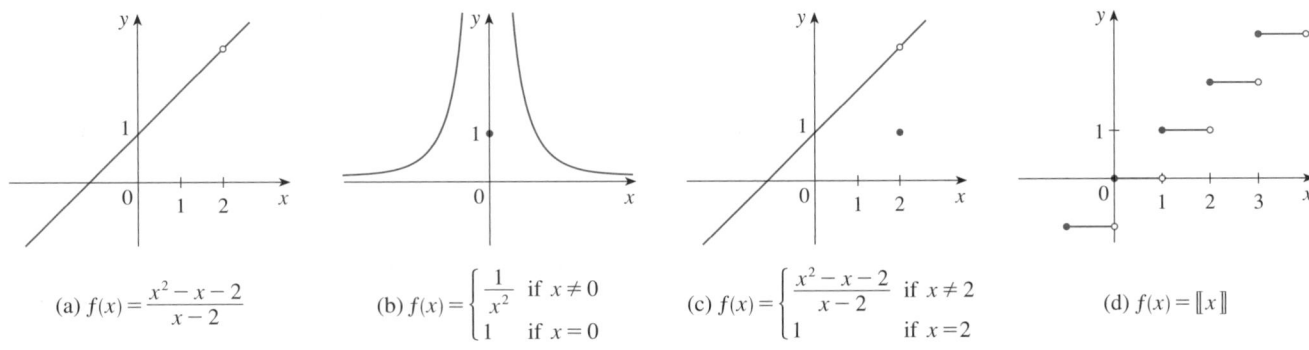

(a) $f(x) = \dfrac{x^2 - x - 2}{x - 2}$

(b) $f(x) = \begin{cases} \dfrac{1}{x^2} & \text{if } x \neq 0 \\ 1 & \text{if } x = 0 \end{cases}$

(c) $f(x) = \begin{cases} \dfrac{x^2 - x - 2}{x - 2} & \text{if } x \neq 2 \\ 1 & \text{if } x = 2 \end{cases}$

(d) $f(x) = [\![x]\!]$

FIGURE 3 Graphs of the functions in Example 2

2 **DEFINITION** A function f is **continuous from the right at a number a** if

$$\lim_{x \to a^+} f(x) = f(a)$$

and f is **continuous from the left at a** if

$$\lim_{x \to a^-} f(x) = f(a)$$

EXAMPLE 3 At each integer n, the function $f(x) = [\![x]\!]$ [see Figure 3(d)] is continuous from the right but discontinuous from the left because

$$\lim_{x \to n^+} f(x) = \lim_{x \to n^+} [\![x]\!] = n = f(n)$$

but

$$\lim_{x \to n^-} f(x) = \lim_{x \to n^-} [\![x]\!] = n - 1 \neq f(n) \qquad \blacksquare$$

3 **DEFINITION** A function f is **continuous on an interval** if it is continuous at every number in the interval. (If f is defined only on one side of an endpoint of the interval, we understand *continuous* at the endpoint to mean *continuous from the right* or *continuous from the left*.)

EXAMPLE 4 Show that the function $f(x) = 1 - \sqrt{1 - x^2}$ is continuous on the interval $[-1, 1]$.

SOLUTION If $-1 < a < 1$, then using the Limit Laws, we have

$$\lim_{x \to a} f(x) = \lim_{x \to a} \left(1 - \sqrt{1 - x^2}\right)$$

$$= 1 - \lim_{x \to a} \sqrt{1 - x^2} \qquad \text{(by Laws 2 and 7)}$$

$$= 1 - \sqrt{\lim_{x \to a} (1 - x^2)} \qquad \text{(by 11)}$$

$$= 1 - \sqrt{1 - a^2} \qquad \text{(by 2, 7, and 9)}$$

$$= f(a)$$

Thus, by Definition 1, f is continuous at a if $-1 < a < 1$. Similar calculations show that

$$\lim_{x \to -1^+} f(x) = 1 = f(-1) \qquad \text{and} \qquad \lim_{x \to 1^-} f(x) = 1 = f(1)$$

so f is continuous from the right at -1 and continuous from the left at 1. Therefore, according to Definition 3, f is continuous on $[-1, 1]$.

The graph of f is sketched in Figure 4. It is the lower half of the circle

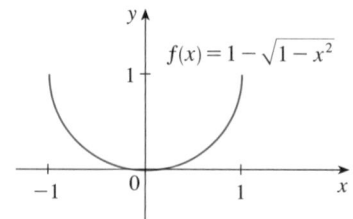

FIGURE 4

$$x^2 + (y - 1)^2 = 1 \qquad \blacksquare$$

Instead of always using Definitions 1, 2, and 3 to verify the continuity of a function as we did in Example 4, it is often convenient to use the next theorem, which shows how to build up complicated continuous functions from simple ones.

---

4 THEOREM If f and g are continuous at a and c is a constant, then the following functions are also continuous at a:

1. $f + g$ **2.** $f - g$ **3.** cf

4. fg **5.** $\dfrac{f}{g}$ if $g(a) \neq 0$

PROOF Each of the five parts of this theorem follows from the corresponding Limit Law in Section 1.4. For instance, we give the proof of part 1. Since f and g are continuous at a, we have

$$\lim_{x \to a} f(x) = f(a) \qquad \text{and} \qquad \lim_{x \to a} g(x) = g(a)$$

Therefore

$$\lim_{x \to a} (f + g)(x) = \lim_{x \to a} \left[f(x) + g(x) \right]$$

$$= \lim_{x \to a} f(x) + \lim_{x \to a} g(x) \qquad \text{(by Law 1)}$$

$$= f(a) + g(a)$$

$$= (f + g)(a)$$

This shows that $f + g$ is continuous at a.   □

It follows from Theorem 4 and Definition 3 that if f and g are continuous on an interval, then so are the functions $f + g$, $f - g$, cf, fg, and (if g is never 0) f/g. The following theorem was stated in Section 1.4 as the Direct Substitution Property.

---

5 THEOREM
(a) Any polynomial is continuous everywhere; that is, it is continuous on $\mathbb{R} = (-\infty, \infty)$.
(b) Any rational function is continuous wherever it is defined; that is, it is continuous on its domain.

PROOF
(a) A polynomial is a function of the form

$$P(x) = c_n x^n + c_{n-1} x^{n-1} + \cdots + c_1 x + c_0$$

where $c_0, c_1, \ldots, c_n$ are constants. We know that

$$\lim_{x \to a} c_0 = c_0 \qquad \text{(by Law 7)}$$

and

$$\lim_{x \to a} x^m = a^m \qquad m = 1, 2, \ldots, n \qquad \text{(by 9)}$$

This equation is precisely the statement that the function $f(x) = x^m$ is a continuous function. Thus, by part 3 of Theorem 4, the function $g(x) = cx^m$ is continuous. Since

P is a sum of functions of this form and a constant function, it follows from part 1 of Theorem 4 that P is continuous.

(b) A rational function is a function of the form

$$f(x) = \frac{P(x)}{Q(x)}$$

where P and Q are polynomials. The domain of f is $D = \{x \in \mathbb{R} \mid Q(x) \neq 0\}$. We know from part (a) that P and Q are continuous everywhere. Thus, by part 5 of Theorem 4, f is continuous at every number in D. ☐

As an illustration of Theorem 5, observe that the volume of a sphere varies continuously with its radius because the formula $V(r) = \frac{4}{3}\pi r^3$ shows that V is a polynomial function of r. Likewise, if a ball is thrown vertically into the air with a velocity of 50 ft/s, then the height of the ball in feet t seconds later is given by the formula $h = 50t - 16t^2$. Again this is a polynomial function, so the height is a continuous function of the elapsed time.

Knowledge of which functions are continuous enables us to evaluate some limits very quickly, as the following example shows. Compare it with Example 1(b) in Section 1.4.

EXAMPLE 5 Find $\lim\limits_{x \to -2} \dfrac{x^3 + 2x^2 - 1}{5 - 3x}$.

SOLUTION The function

$$f(x) = \frac{x^3 + 2x^2 - 1}{5 - 3x}$$

is rational, so by Theorem 5 it is continuous on its domain, which is $\left\{x \mid x \neq \frac{5}{3}\right\}$. Therefore

$$\lim_{x \to -2} \frac{x^3 + 2x^2 - 1}{5 - 3x} = \lim_{x \to -2} f(x) = f(-2)$$

$$= \frac{(-2)^3 + 2(-2)^2 - 1}{5 - 3(-2)} = -\frac{1}{11} \qquad ■$$

It turns out that most of the familiar functions are continuous at every number in their domains. For instance, Limit Law 10 (page 36) is exactly the statement that root functions are continuous.

From the appearance of the graphs of the sine and cosine functions (Figure 11 in Section 1.2), we would certainly guess that they are continuous. And in Section 1.4 we showed that

$$\lim_{\theta \to a} \sin \theta = \sin a \qquad \lim_{\theta \to a} \cos \theta = \cos a$$

In other words, the sine and cosine functions are continuous everywhere. It follows from part 5 of Theorem 4 that

$$\tan x = \frac{\sin x}{\cos x}$$

is continuous except where $\cos x = 0$. This happens when x is an odd integer multiple of $\pi/2$, so $y = \tan x$ has infinite discontinuities when $x = \pm\pi/2, \pm 3\pi/2, \pm 5\pi/2$, and so on (see Figure 5).

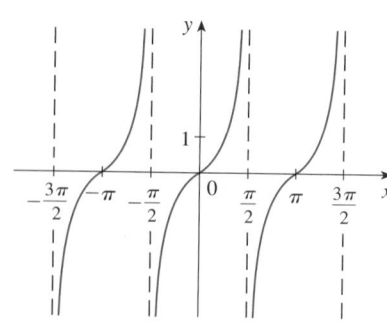

FIGURE 5
$y = \tan x$

6 **THEOREM** The following types of functions are continuous at every number in their domains: polynomials, rational functions, root functions, trigonometric functions

EXAMPLE 6 On what intervals is each function continuous?

(a) $f(x) = x^{100} - 2x^{37} + 75$ \qquad (b) $g(x) = \dfrac{x^2 + 2x + 17}{x^2 - 1}$

(c) $h(x) = \sqrt{x} + \dfrac{x + 1}{x - 1} - \dfrac{x + 1}{x^2 + 1}$

SOLUTION

(a) f is a polynomial, so it is continuous on $(-\infty, \infty)$ by Theorem 5(a).

(b) g is a rational function, so by Theorem 5(b) it is continuous on its domain, which is $D = \{x \mid x^2 - 1 \neq 0\} = \{x \mid x \neq \pm 1\}$. Thus g is continuous on the intervals $(-\infty, -1)$, $(-1, 1)$, and $(1, \infty)$.

(c) We can write $h(x) = F(x) + G(x) - H(x)$, where

$$F(x) = \sqrt{x} \qquad G(x) = \frac{x + 1}{x - 1} \qquad H(x) = \frac{x + 1}{x^2 + 1}$$

F is continuous on $[0, \infty)$ by Theorem 6. G is a rational function, so it is continuous everywhere except when $x - 1 = 0$, that is, $x = 1$. H is also a rational function, but its denominator is never 0, so H is continuous everywhere. Thus, by parts 1 and 2 of Theorem 4, h is continuous on the intervals $[0, 1)$ and $(1, \infty)$. ■

Another way of combining continuous functions f and g to get a new continuous function is to form the composite function $f \circ g$. This fact is a consequence of the following theorem.

- This theorem says that a limit symbol can be moved through a function symbol if the function is continuous and the limit exists. In other words, the order of these two symbols can be reversed.

7 **THEOREM** If f is continuous at b and $\lim_{x \to a} g(x) = b$, then $\lim_{x \to a} f(g(x)) = f(b)$. In other words,

$$\lim_{x \to a} f(g(x)) = f\left(\lim_{x \to a} g(x)\right)$$

Intuitively, Theorem 7 is reasonable because if x is close to a, then $g(x)$ is close to b, and since f is continuous at b, if $g(x)$ is close to b, then $f(g(x))$ is close to $f(b)$. A proof of Theorem 7 is given in Appendix D.

8 **THEOREM** If g is continuous at a and f is continuous at $g(a)$, then the composite function $f \circ g$ given by $(f \circ g)(x) = f(g(x))$ is continuous at a.

This theorem is often expressed informally by saying "a continuous function of a continuous function is a continuous function."

PROOF Since g is continuous at a, we have

$$\lim_{x \to a} g(x) = g(a)$$

Since f is continuous at $b = g(a)$, we can apply Theorem 7 to obtain

$$\lim_{x \to a} f(g(x)) = f(g(a))$$

which is precisely the statement that the function $h(x) = f(g(x))$ is continuous at a; that is, $f \circ g$ is continuous at a. $\square$

V EXAMPLE 7 Where are the following functions continuous?

(a) $h(x) = \sin(x^2)$

(b) $F(x) = \dfrac{1}{\sqrt{x^2 + 7} - 4}$

SOLUTION

(a) We have $h(x) = f(g(x))$, where

$$g(x) = x^2 \qquad \text{and} \qquad f(x) = \sin x$$

Now g is continuous on $\mathbb{R}$ since it is a polynomial, and f is also continuous everywhere by Theorem 6. Thus $h = f \circ g$ is continuous on $\mathbb{R}$ by Theorem 8.

(b) Notice that F can be broken up as the composition of four continuous functions:

$$F = f \circ g \circ h \circ k \qquad \text{or} \qquad F(x) = f(g(h(k(x))))$$

where $\quad f(x) = \dfrac{1}{x} \qquad g(x) = x - 4 \qquad h(x) = \sqrt{x} \qquad k(x) = x^2 + 7$

We know that each of these functions is continuous on its domain (by Theorems 5 and 6), so by Theorem 8, F is continuous on its domain, which is

$$\left\{ x \in \mathbb{R} \mid \sqrt{x^2 + 7} \neq 4 \right\} = \{ x \mid x \neq \pm 3 \} = (-\infty, -3) \cup (-3, 3) \cup (3, \infty) \quad \blacksquare$$

An important property of continuous functions is expressed by the following theorem, whose proof is found in more advanced books on calculus.

---

9 THE INTERMEDIATE VALUE THEOREM   Suppose that f is continuous on the closed interval $[a, b]$ and let N be any number between $f(a)$ and $f(b)$, where $f(a) \neq f(b)$. Then there exists a number c in (a, b) such that $f(c) = N$.

The Intermediate Value Theorem states that a continuous function takes on every intermediate value between the function values $f(a)$ and $f(b)$. It is illustrated by Figure 6. Note that the value N can be taken on once [as in part (a)] or more than once [as in part (b)].

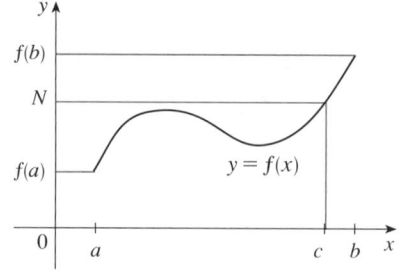

(a)

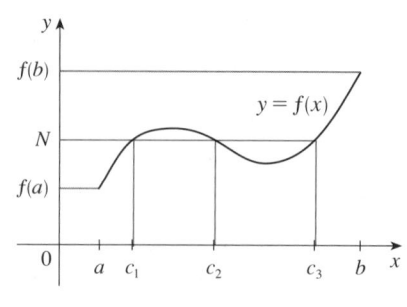

(b)

FIGURE 6

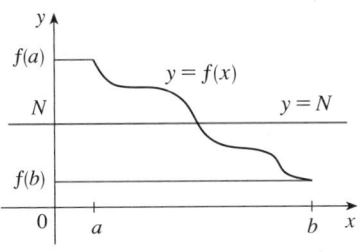

FIGURE 7

If we think of a continuous function as a function whose graph has no hole or break, then it is easy to believe that the Intermediate Value Theorem is true. In geometric terms it says that if any horizontal line $y = N$ is given between $y = f(a)$ and $y = f(b)$ as in Figure 7, then the graph of f can't jump over the line. It must intersect $y = N$ somewhere.

It is important that the function f in Theorem 9 be continuous. The Intermediate Value Theorem is not true in general for discontinuous functions (see Exercise 36).

One use of the Intermediate Value Theorem is in locating roots of equations as in the following example.

▼ EXAMPLE 8 Show that there is a root of the equation

$$4x^3 - 6x^2 + 3x - 2 = 0$$

between 1 and 2.

SOLUTION Let $f(x) = 4x^3 - 6x^2 + 3x - 2$. We are looking for a solution of the given equation, that is, a number c between 1 and 2 such that $f(c) = 0$. Therefore we take $a = 1$, $b = 2$, and $N = 0$ in Theorem 9. We have

$$f(1) = 4 - 6 + 3 - 2 = -1 < 0$$

and

$$f(2) = 32 - 24 + 6 - 2 = 12 > 0$$

Thus $f(1) < 0 < f(2)$; that is, $N = 0$ is a number between $f(1)$ and $f(2)$. Now f is continuous since it is a polynomial, so the Intermediate Value Theorem says there is a number c between 1 and 2 such that $f(c) = 0$. In other words, the equation $4x^3 - 6x^2 + 3x - 2 = 0$ has at least one root c in the interval $(1, 2)$.

In fact, we can locate a root more precisely by using the Intermediate Value Theorem again. Since

$$f(1.2) = -0.128 < 0 \qquad \text{and} \qquad f(1.3) = 0.548 > 0$$

a root must lie between 1.2 and 1.3. A calculator gives, by trial and error,

$$f(1.22) = -0.007008 < 0 \qquad \text{and} \qquad f(1.23) = 0.056068 > 0$$

so a root lies in the interval $(1.22, 1.23)$. ∎

We can use a graphing calculator or computer to illustrate the use of the Intermediate Value Theorem in Example 8. Figure 8 shows the graph of f in the viewing rectangle $[-1, 3]$ by $[-3, 3]$ and you can see that the graph crosses the x-axis between 1 and 2. Figure 9 shows the result of zooming in to the viewing rectangle $[1.2, 1.3]$ by $[-0.2, 0.2]$.

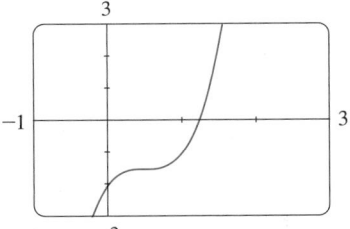

FIGURE 8

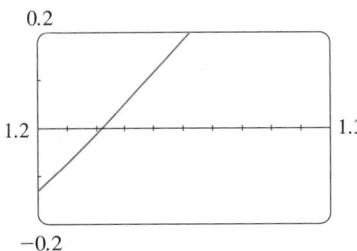

FIGURE 9

In fact, the Intermediate Value Theorem plays a role in the very way these graphing devices work. A computer calculates a finite number of points on the graph and turns on the pixels that contain these calculated points. It assumes that the function is continuous and takes on all the intermediate values between two consecutive points. The computer therefore connects the pixels by turning on the intermediate pixels.

1.5 | EXERCISES

1. Write an equation that expresses the fact that a function f is continuous at the number 4.

2. If f is continuous on $(-\infty, \infty)$, what can you say about its graph?

3. (a) From the graph of f, state the numbers at which f is discontinuous and explain why.
(b) For each of the numbers stated in part (a), determine whether f is continuous from the right, or from the left, or neither.

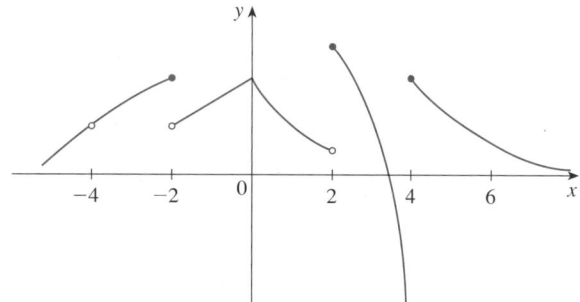

4. From the graph of g, state the intervals on which g is continuous.

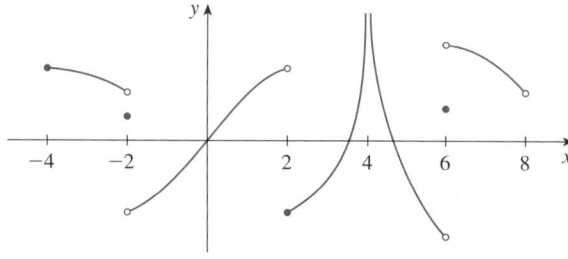

5–8 ■ Sketch the graph of a function f that is continuous except for the stated discontinuity.

5. Discontinuous, but continuous from the right, at 2

6. Discontinuities at -1 and 4, but continuous from the left at -1 and from the right at 4

7. Removable discontinuity at 3, jump discontinuity at 5

8. Neither left nor right continuous at -2, continuous only from the left at 2

9. The toll T charged for driving on a certain stretch of a toll road is \$5 except during rush hours (between 7 AM and 10 AM and between 4 PM and 7 PM) when the toll is \$7.
(a) Sketch a graph of T as a function of the time t, measured in hours past midnight.
(b) Discuss the discontinuities of this function and their significance to someone who uses the road.

10. Explain why each function is continuous or discontinuous.
(a) The temperature at a specific location as a function of time
(b) The temperature at a specific time as a function of the distance due west from New York City
(c) The altitude above sea level as a function of the distance due west from New York City
(d) The cost of a taxi ride as a function of the distance traveled
(e) The current in the circuit for the lights in a room as a function of time

11. Suppose f and g are continuous functions such that $g(2) = 6$ and $\lim_{x \to 2} [3f(x) + f(x)g(x)] = 36$. Find $f(2)$.

12–13 ■ Use the definition of continuity and the properties of limits to show that the function is continuous at the given number a.

12. $f(x) = 3x^4 - 5x + \sqrt[3]{x^2 + 4}, \quad a = 2$

13. $f(x) = (x + 2x^3)^4, \quad a = -1$

14. Use the definition of continuity and the properties of limits to show that the function $f(x) = x\sqrt{16 - x^2}$ is continuous on the interval $[-4, 4]$.

15–18 ■ Explain why the function is discontinuous at the given number a. Sketch the graph of the function.

15. $f(x) = \dfrac{1}{x + 2}$ $a = -2$

16. $f(x) = \begin{cases} \dfrac{1}{x + 2} & \text{if } x \ne -2 \\ 1 & \text{if } x = -2 \end{cases}$ $a = -2$

17. $f(x) = \begin{cases} 1 - x^2 & \text{if } x < 1 \\ 1/x & \text{if } x \ge 1 \end{cases}$ $a = 1$

18. $f(x) = \begin{cases} \dfrac{x^2 - x}{x^2 - 1} & \text{if } x \neq 1 \\ 1 & \text{if } x = 1 \end{cases}$ $\quad a = 1$

19–24 ■ Explain, using Theorems 4, 5, 6, and 8, why the function is continuous at every number in its domain. State the domain.

19. $F(x) = \dfrac{2x^2 - x - 1}{x^2 + 1}$

20. $G(x) = \dfrac{x^2 + 1}{2x^2 - x - 1}$

21. $Q(x) = \dfrac{\sqrt[3]{x - 2}}{x^3 - 2}$

22. $B(x) = \dfrac{\tan x}{\sqrt{4 - x^2}}$

23. $M(x) = \sqrt{1 + \dfrac{1}{x}}$

24. $F(x) = \sin(\cos(\sin x))$

25–26 ■ Locate the discontinuities of the function and illustrate by graphing.

25. $y = \dfrac{1}{1 + \sin x}$

26. $y = \tan \sqrt{x}$

27–28 ■ Use continuity to evaluate the limit.

27. $\displaystyle \lim_{x \to 4} \dfrac{5 + \sqrt{x}}{\sqrt{5 + x}}$

28. $\displaystyle \lim_{x \to \pi} \sin(x + \sin x)$

29–30 ■ Show that f is continuous on $(-\infty, \infty)$.

29. $f(x) = \begin{cases} x^2 & \text{if } x < 1 \\ \sqrt{x} & \text{if } x \geq 1 \end{cases}$

30. $f(x) = \begin{cases} \sin x & \text{if } x < \pi/4 \\ \cos x & \text{if } x \geq \pi/4 \end{cases}$

---

31. Find the numbers at which the function

$$f(x) = \begin{cases} x + 2 & \text{if } x < 0 \\ 2x^2 & \text{if } 0 \leq x \leq 1 \\ 2 - x & \text{if } x > 1 \end{cases}$$

is discontinuous. At which of these points is f continuous from the right, from the left, or neither? Sketch the graph of f.

32. The gravitational force exerted by the earth on a unit mass at a distance r from the center of the planet is

$$F(r) = \begin{cases} \dfrac{GMr}{R^3} & \text{if } r < R \\ \dfrac{GM}{r^2} & \text{if } r \geq R \end{cases}$$

where M is the mass of the earth, R is its radius, and G is the gravitational constant. Is F a continuous function of r?

33. For what value of the constant c is the function f continuous on $(-\infty, \infty)$?

$$f(x) = \begin{cases} cx^2 + 2x & \text{if } x < 2 \\ x^3 - cx & \text{if } x \geq 2 \end{cases}$$

34. Find the values of a and b that make f continuous everywhere.

$$f(x) = \begin{cases} \dfrac{x^2 - 4}{x - 2} & \text{if } x < 2 \\ ax^2 - bx + 3 & \text{if } 2 \leq x < 3 \\ 2x - a + b & \text{if } x \geq 3 \end{cases}$$

35. Which of the following functions f has a removable discontinuity at a? If the discontinuity is removable, find a function g that agrees with f for $x \neq a$ and is continuous at a.

(a) $f(x) = \dfrac{x^4 - 1}{x - 1}, \quad a = 1$

(b) $f(x) = \dfrac{x^3 - x^2 - 2x}{x - 2}, \quad a = 2$

(c) $f(x) = [\![\sin x]\!], \quad a = \pi$

36. Suppose that a function f is continuous on $[0, 1]$ except at 0.25 and that $f(0) = 1$ and $f(1) = 3$. Let $N = 2$. Sketch two possible graphs of f, one showing that f might not satisfy the conclusion of the Intermediate Value Theorem and one showing that f might still satisfy the conclusion of the Intermediate Value Theorem (even though it doesn't satisfy the hypothesis).

37. If $f(x) = x^2 + 10 \sin x$, show that there is a number c such that $f(c) = 1000$.

38. Suppose f is continuous on $[1, 5]$ and the only solutions of the equation $f(x) = 6$ are $x = 1$ and $x = 4$. If $f(2) = 8$, explain why $f(3) > 6$.

39–42 ■ Use the Intermediate Value Theorem to show that there is a root of the given equation in the specified interval.

39. $x^4 + x - 3 = 0, \quad (1, 2)$

40. $\sqrt[3]{x} = 1 - x, \quad (0, 1)$

41. $\cos x = x, \quad (0, 1)$

42. $\sin x = x^2 - x, \quad (1, 2)$

43–44 ▪ (a) Prove that the equation has at least one real root. (b) Use your calculator to find an interval of length 0.01 that contains a root.

43. $\cos x = x^3$

44. $x^5 - x^2 + 2x + 3 = 0$

45–46 ▪ (a) Prove that the equation has at least one real root. (b) Use your graphing device to find the root correct to three decimal places.

45. $x^5 - x^2 - 4 = 0$

46. $\sqrt{x - 5} = \dfrac{1}{x + 3}$

47. Is there a number that is exactly 1 more than its cube?

48. If a and b are positive numbers, prove that the equation

$$\frac{a}{x^3 + 2x^2 - 1} + \frac{b}{x^3 + x - 2} = 0$$

has at least one solution in the interval $(-1, 1)$.

49. Show that the function

$$f(x) = \begin{cases} x^4 \sin(1/x) & \text{if } x \neq 0 \\ 0 & \text{if } x = 0 \end{cases}$$

is continuous on $(-\infty, \infty)$.

50. (a) Show that the absolute value function $F(x) = |x|$ is continuous everywhere.
(b) Prove that if f is a continuous function on an interval, then so is $|f|$.
(c) Is the converse of the statement in part (b) also true? In other words, if $|f|$ is continuous, does it follow that f is continuous? If so, prove it. If not, find a counterexample.

51. A Tibetan monk leaves the monastery at 7:00 AM and takes his usual path to the top of the mountain, arriving at 7:00 PM. The following morning, he starts at 7:00 AM at the top and takes the same path back, arriving at the monastery at 7:00 PM. Use the Intermediate Value Theorem to show that there is a point on the path that the monk will cross at exactly the same time of day on both days.

1.6 | LIMITS INVOLVING INFINITY

In this section we investigate the global behavior of functions and, in particular, whether their graphs approach asymptotes, vertical or horizontal.

INFINITE LIMITS

In Example 8 in Section 1.3 we concluded that

$$\lim_{x \to 0} \frac{1}{x^2} \quad \text{does not exist}$$

by observing, from the table of values and the graph of $y = 1/x^2$ in Figure 1, that the values of $1/x^2$ can be made arbitrarily large by taking x close enough to 0. Thus the values of $f(x)$ do not approach a number, so $\lim_{x \to 0} (1/x^2)$ does not exist.

To indicate this kind of behavior we use the notation

$$\lim_{x \to 0} \frac{1}{x^2} = \infty$$

⊘ This does not mean that we are regarding ∞ as a number. Nor does it mean that the limit exists. It simply expresses the particular way in which the limit does not exist: $1/x^2$ can be made as large as we like by taking x close enough to 0.

In general, we write symbolically

$$\lim_{x \to a} f(x) = \infty$$

to indicate that the values of $f(x)$ become larger and larger (or "increase without bound") as x approaches a.

x	$\dfrac{1}{x^2}$
± 1	1
± 0.5	4
± 0.2	25
± 0.1	100
± 0.05	400
± 0.01	10,000
± 0.001	1,000,000

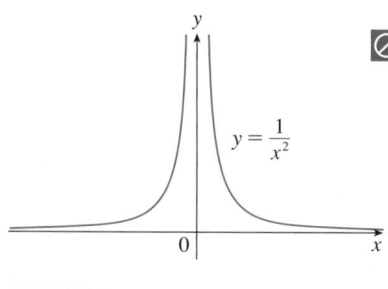

$$y = \frac{1}{x^2}$$

FIGURE 1

■ A more precise version of Definition 1 is given at the end of this section.

> **1** **DEFINITION** The notation
>
> $$\lim_{x \to a} f(x) = \infty$$
>
> means that the values of $f(x)$ can be made arbitrarily large (as large as we please) by taking x sufficiently close to a (on either side of a) but not equal to a.

Another notation for $\lim_{x \to a} f(x) = \infty$ is

$$f(x) \to \infty \qquad \text{as} \qquad x \to a$$

Again, the symbol ∞ is not a number, but the expression $\lim_{x \to a} f(x) = \infty$ is often read as

"the limit of $f(x)$, as x approaches a, is infinity"

or   "$f(x)$ becomes infinite as x approaches a"

or   "$f(x)$ increases without bound as x approaches a"

This definition is illustrated graphically in Figure 2.

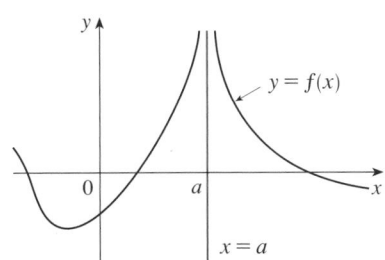

FIGURE 2

$\lim_{x \to a} f(x) = \infty$

Similarly, as shown in Figure 3,

$$\lim_{x \to a} f(x) = -\infty$$

■ When we say that a number is "large negative," we mean that it is negative but its magnitude (absolute value) is large.

means that the values of $f(x)$ are as large negative as we like for all values of x that are sufficiently close to a, but not equal to a.

The symbol $\lim_{x \to a} f(x) = -\infty$ can be read as "the limit of $f(x)$, as x approaches a, is negative infinity" or "$f(x)$ decreases without bound as x approaches a." As an example we have

$$\lim_{x \to 0} \left(-\frac{1}{x^2} \right) = -\infty$$

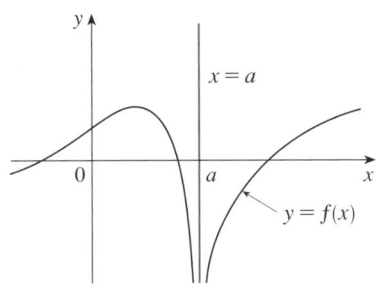

FIGURE 3

$\lim_{x \to a} f(x) = -\infty$

Similar definitions can be given for the one-sided infinite limits

$$\lim_{x \to a^-} f(x) = \infty \qquad\qquad \lim_{x \to a^+} f(x) = \infty$$

$$\lim_{x \to a^-} f(x) = -\infty \qquad\qquad \lim_{x \to a^+} f(x) = -\infty$$

remembering that "$x \to a^-$" means that we consider only values of x that are less than a, and similarly "$x \to a^+$" means that we consider only $x > a$. Illustrations of these four cases are given in Figure 4.

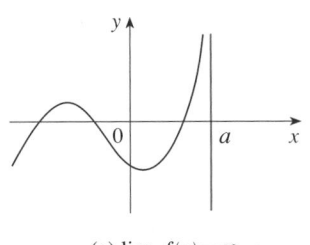 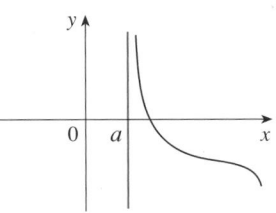 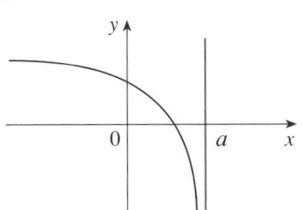 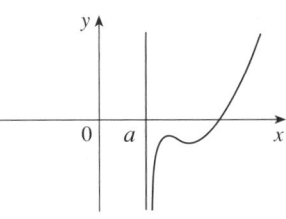

(a) $\lim_{x \to a^-} f(x) = \infty$ (b) $\lim_{x \to a^+} f(x) = \infty$ (c) $\lim_{x \to a^-} f(x) = -\infty$ (d) $\lim_{x \to a^+} f(x) = -\infty$

FIGURE 4

> **2** **DEFINITION** The line $x = a$ is called a **vertical asymptote** of the curve $y = f(x)$ if at least one of the following statements is true:
>
> $$\lim_{x \to a} f(x) = \infty \qquad \lim_{x \to a^-} f(x) = \infty \qquad \lim_{x \to a^+} f(x) = \infty$$
>
> $$\lim_{x \to a} f(x) = -\infty \qquad \lim_{x \to a^-} f(x) = -\infty \qquad \lim_{x \to a^+} f(x) = -\infty$$

For instance, the y-axis is a vertical asymptote of the curve $y = 1/x^2$ because $\lim_{x \to 0}(1/x^2) = \infty$. In Figure 4 the line $x = a$ is a vertical asymptote in each of the four cases shown.

EXAMPLE 1 Find $\displaystyle\lim_{x \to 3^+} \frac{2x}{x - 3}$ and $\displaystyle\lim_{x \to 3^-} \frac{2x}{x - 3}$.

SOLUTION If x is close to 3 but larger than 3, then the denominator $x - 3$ is a small positive number and $2x$ is close to 6. So the quotient $2x/(x - 3)$ is a large *positive* number. Thus, intuitively, we see that

$$\lim_{x \to 3^+} \frac{2x}{x - 3} = \infty$$

Likewise, if x is close to 3 but smaller than 3, then $x - 3$ is a small negative number but $2x$ is still a positive number (close to 6). So $2x/(x - 3)$ is a numerically large *negative* number. Thus

$$\lim_{x \to 3^-} \frac{2x}{x - 3} = -\infty$$

The graph of the curve $y = 2x/(x - 3)$ is given in Figure 5. The line $x = 3$ is a vertical asymptote. ∎

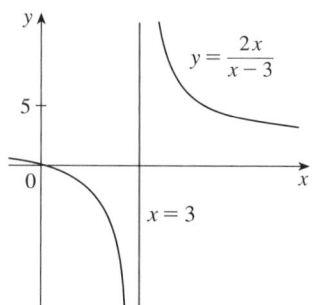

FIGURE 5

EXAMPLE 2 Find the vertical asymptotes of $f(x) = \tan x$.

SOLUTION Because

$$\tan x = \frac{\sin x}{\cos x}$$

there are potential vertical asymptotes where $\cos x = 0$. In fact, since $\cos x \to 0^+$ as $x \to (\pi/2)^-$ and $\cos x \to 0^-$ as $x \to (\pi/2)^+$, whereas $\sin x$ is positive (and not near 0) when x is near $\pi/2$, we have

$$\lim_{x \to (\pi/2)^-} \tan x = \infty \qquad \text{and} \qquad \lim_{x \to (\pi/2)^+} \tan x = -\infty$$

This shows that the line $x = \pi/2$ is a vertical asymptote. Similar reasoning shows that the lines $x = (2n + 1)\pi/2$, where n is an integer, are all vertical asymptotes of $f(x) = \tan x$. The graph in Figure 6 confirms this. ∎

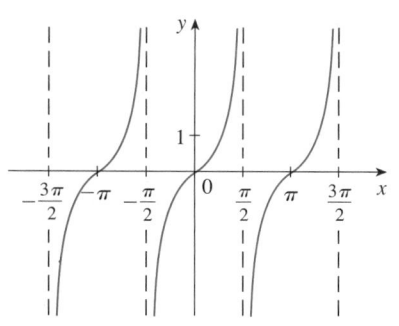

FIGURE 6
$y = \tan x$

LIMITS AT INFINITY

In computing infinite limits, we let x approach a number and the result was that the values of y became arbitrarily large (positive or negative). Here we let x become arbitrarily large (positive or negative) and see what happens to y.

x	$f(x)$
0	-1
± 1	0
± 2	0.600000
± 3	0.800000
± 4	0.882353
± 5	0.923077
± 10	0.980198
± 50	0.999200
± 100	0.999800
± 1000	0.999998

Let's begin by investigating the behavior of the function f defined by

$$f(x) = \frac{x^2 - 1}{x^2 + 1}$$

as x becomes large. The table at the left gives values of this function correct to six decimal places, and the graph of f has been drawn by a computer in Figure 7.

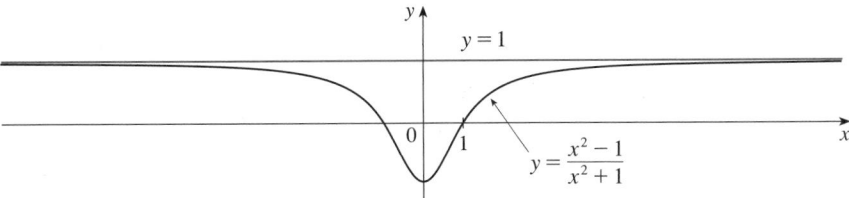

FIGURE 7

As x grows larger and larger you can see that the values of $f(x)$ get closer and closer to 1. In fact, it seems that we can make the values of $f(x)$ as close as we like to 1 by taking x sufficiently large. This situation is expressed symbolically by writing

$$\lim_{x \to \infty} \frac{x^2 - 1}{x^2 + 1} = 1$$

In general, we use the notation

$$\lim_{x \to \infty} f(x) = L$$

to indicate that the values of $f(x)$ approach L as x becomes larger and larger.

---

3   **DEFINITION**   Let f be a function defined on some interval (a, ∞). Then

$$\lim_{x \to \infty} f(x) = L$$

means that the values of $f(x)$ can be made as close to L as we like by taking x sufficiently large.

Another notation for $\lim_{x \to \infty} f(x) = L$ is

$$f(x) \to L \quad \text{as} \quad x \to \infty$$

The symbol ∞ does not represent a number. Nonetheless, the expression $\lim_{x \to \infty} f(x) = L$ is often read as

"the limit of $f(x)$, as x approaches infinity, is L"

or   "the limit of $f(x)$, as x becomes infinite, is L"

or   "the limit of $f(x)$, as x increases without bound, is L"

The meaning of such phrases is given by Definition 3. A more precise definition, similar to the ε, δ definition of Section 1.3, is given at the end of this section.

Geometric illustrations of Definition 3 are shown in Figure 8. Notice that there are many ways for the graph of f to approach the line $y = L$ (which is called a *horizontal asymptote*) as we look to the far right of each graph.

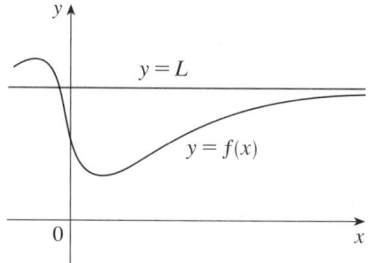
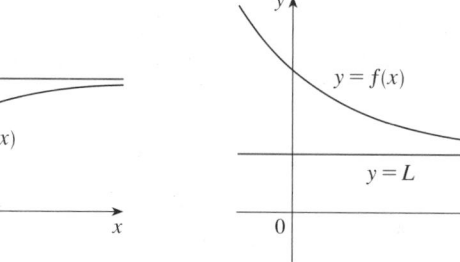
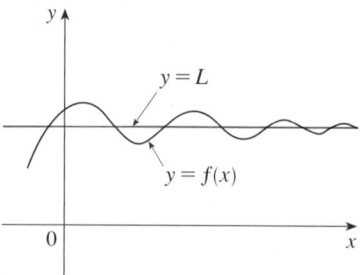

FIGURE 8

Examples illustrating $\lim_{x \to \infty} f(x) = L$

Referring back to Figure 7, we see that for numerically large negative values of x, the values of $f(x)$ are close to 1. By letting x decrease through negative values without bound, we can make $f(x)$ as close to 1 as we like. This is expressed by writing

$$\lim_{x \to -\infty} \frac{x^2 - 1}{x^2 + 1} = 1$$

In general, as shown in Figure 9, the notation

$$\lim_{x \to -\infty} f(x) = L$$

means that the values of $f(x)$ can be made arbitrarily close to L by taking x sufficiently large negative.

Again, the symbol $-\infty$ does not represent a number, but the expression $\lim_{x \to -\infty} f(x) = L$ is often read as

"the limit of $f(x)$, as x approaches negative infinity, is L"

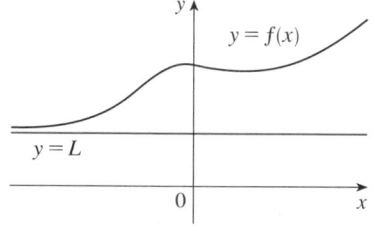

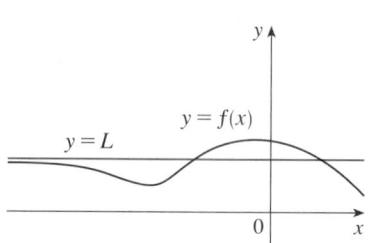

FIGURE 9

Examples illustrating $\lim_{x \to -\infty} f(x) = L$

> **4 DEFINITION** The line $y = L$ is called a **horizontal asymptote** of the curve $y = f(x)$ if either
>
> $$\lim_{x \to \infty} f(x) = L \qquad \text{or} \qquad \lim_{x \to -\infty} f(x) = L$$

For instance, the curve illustrated in Figure 7 has the line $y = 1$ as a horizontal asymptote because

$$\lim_{x \to \infty} \frac{x^2 - 1}{x^2 + 1} = 1$$

The curve $y = f(x)$ sketched in Figure 10 has both $y = -1$ and $y = 2$ as horizontal asymptotes because

$$\lim_{x \to \infty} f(x) = -1 \qquad \text{and} \qquad \lim_{x \to -\infty} f(x) = 2$$

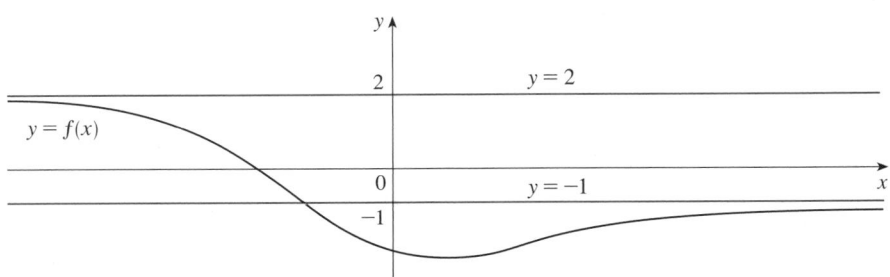

FIGURE 10

EXAMPLE 3 Find the infinite limits, limits at infinity, and asymptotes for the function f whose graph is shown in Figure 11.

SOLUTION We see that the values of $f(x)$ become large as $x \to -1$ from both sides, so

$$\lim_{x \to -1} f(x) = \infty$$

Notice that $f(x)$ becomes large negative as x approaches 2 from the left, but large positive as x approaches 2 from the right. So

$$\lim_{x \to 2^-} f(x) = -\infty \qquad \text{and} \qquad \lim_{x \to 2^+} f(x) = \infty$$

Thus both of the lines $x = -1$ and $x = 2$ are vertical asymptotes.

As x becomes large, it appears that $f(x)$ approaches 4. But as x decreases through negative values, $f(x)$ approaches 2. So

$$\lim_{x \to \infty} f(x) = 4 \qquad \text{and} \qquad \lim_{x \to -\infty} f(x) = 2$$

This means that both $y = 4$ and $y = 2$ are horizontal asymptotes. ■

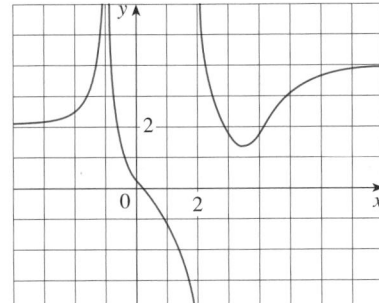

FIGURE 11

EXAMPLE 4 Find $\displaystyle\lim_{x \to \infty} \frac{1}{x}$ and $\displaystyle\lim_{x \to -\infty} \frac{1}{x}$.

SOLUTION Observe that when x is large, $1/x$ is small. For instance,

$$\frac{1}{100} = 0.01 \qquad \frac{1}{10,000} = 0.0001 \qquad \frac{1}{1,000,000} = 0.000001$$

In fact, by taking x large enough, we can make $1/x$ as close to 0 as we please. Therefore, according to Definition 3, we have

$$\lim_{x \to \infty} \frac{1}{x} = 0$$

Similar reasoning shows that when x is large negative, $1/x$ is small negative, so we also have

$$\lim_{x \to -\infty} \frac{1}{x} = 0$$

It follows that the line $y = 0$ (the x-axis) is a horizontal asymptote of the curve $y = 1/x$. (This is an equilateral hyperbola; see Figure 12.) ■

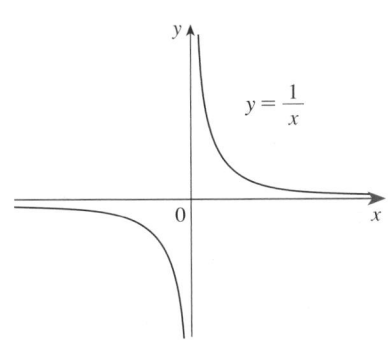

FIGURE 12

$$\lim_{x \to \infty} \frac{1}{x} = 0, \quad \lim_{x \to -\infty} \frac{1}{x} = 0$$

Most of the Limit Laws that were given in Section 1.4 also hold for limits at infinity. It can be proved that the *Limit Laws listed in Section 1.4 (with the exception of Laws 9 and 10) are also valid if "$x \to a$" is replaced by "$x \to \infty$" or "$x \to -\infty$."* In particular, if we combine Law 6 with the results of Example 4 we obtain the following important rule for calculating limits.

$\boxed{5}$ If n is a positive integer, then

$$\lim_{x \to \infty} \frac{1}{x^n} = 0 \qquad \lim_{x \to -\infty} \frac{1}{x^n} = 0$$

▼ EXAMPLE 5   Evaluate

$$\lim_{x \to \infty} \frac{3x^2 - x - 2}{5x^2 + 4x + 1}$$

SOLUTION As x becomes large, both numerator and denominator become large, so it isn't obvious what happens to their ratio. We need to do some preliminary algebra.

To evaluate the limit at infinity of any rational function, we first divide both the numerator and denominator by the highest power of x that occurs in the denominator. (We may assume that $x \neq 0$, since we are interested only in large values of x.) In this case the highest power of x is x^2, and so, using the Limit Laws, we have

$$\lim_{x \to \infty} \frac{3x^2 - x - 2}{5x^2 + 4x + 1} = \lim_{x \to \infty} \frac{\dfrac{3x^2 - x - 2}{x^2}}{\dfrac{5x^2 + 4x + 1}{x^2}} = \lim_{x \to \infty} \frac{3 - \dfrac{1}{x} - \dfrac{2}{x^2}}{5 + \dfrac{4}{x} + \dfrac{1}{x^2}}$$

$$= \frac{\displaystyle\lim_{x \to \infty} \left(3 - \frac{1}{x} - \frac{2}{x^2} \right)}{\displaystyle\lim_{x \to \infty} \left(5 + \frac{4}{x} + \frac{1}{x^2} \right)}$$

$$= \frac{\displaystyle\lim_{x \to \infty} 3 - \lim_{x \to \infty} \frac{1}{x} - 2 \lim_{x \to \infty} \frac{1}{x^2}}{\displaystyle\lim_{x \to \infty} 5 + 4 \lim_{x \to \infty} \frac{1}{x} + \lim_{x \to \infty} \frac{1}{x^2}}$$

$$= \frac{3 - 0 - 0}{5 + 0 + 0} \qquad [\text{by } \boxed{5}]$$

$$= \frac{3}{5}$$

▪ Figure 13 illustrates Example 5 by showing how the graph of the given rational function approaches the horizontal asymptote $y = \frac{3}{5}$.

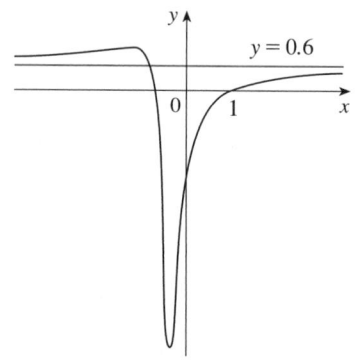

FIGURE 13

$$y = \frac{3x^2 - x - 2}{5x^2 + 4x + 1}$$

A similar calculation shows that the limit as $x \to -\infty$ is also $\frac{3}{5}$.   ▪

EXAMPLE 6   Compute $\displaystyle\lim_{x \to \infty} \left(\sqrt{x^2 + 1} - x \right)$.

SOLUTION Because both $\sqrt{x^2 + 1}$ and x are large when x is large, it's difficult to see what happens to their difference, so we use algebra to rewrite the function. We

■ We can think of the given function as having a denominator of 1.

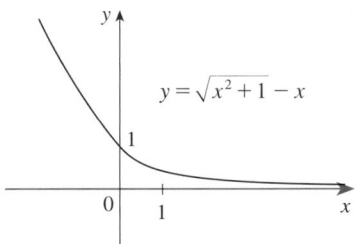

FIGURE 14

first multiply numerator and denominator by the conjugate radical:

$$\lim_{x \to \infty} \left(\sqrt{x^2 + 1} - x \right) = \lim_{x \to \infty} \left(\sqrt{x^2 + 1} - x \right) \frac{\sqrt{x^2 + 1} + x}{\sqrt{x^2 + 1} + x}$$

$$= \lim_{x \to \infty} \frac{(x^2 + 1) - x^2}{\sqrt{x^2 + 1} + x} = \lim_{x \to \infty} \frac{1}{\sqrt{x^2 + 1} + x}$$

Notice that the denominator of this last expression $\left(\sqrt{x^2 + 1} + x \right)$ becomes large as $x \to \infty$ (it's bigger than x). So

$$\lim_{x \to \infty} \left(\sqrt{x^2 + 1} - x \right) = \lim_{x \to \infty} \frac{1}{\sqrt{x^2 + 1} + x} = 0$$

Figure 14 illustrates this result. ■

EXAMPLE 7 Evaluate $\lim_{x \to \infty} \sin \dfrac{1}{x}$.

SOLUTION If we let $t = 1/x$, then $t \to 0^+$ as $x \to \infty$. Therefore

$$\lim_{x \to \infty} \sin \frac{1}{x} = \lim_{t \to 0^+} \sin t = 0$$

(See Exercise 59.) ■

EXAMPLE 8 Evaluate $\lim_{x \to \infty} \sin x$.

SOLUTION As x increases, the values of $\sin x$ oscillate between 1 and -1 infinitely often. Thus $\lim_{x \to \infty} \sin x$ does not exist. ■

INFINITE LIMITS AT INFINITY

The notation

$$\lim_{x \to \infty} f(x) = \infty$$

is used to indicate that the values of $f(x)$ become large as x becomes large. Similar meanings are attached to the following symbols:

$$\lim_{x \to -\infty} f(x) = \infty \qquad \lim_{x \to \infty} f(x) = -\infty \qquad \lim_{x \to -\infty} f(x) = -\infty$$

EXAMPLE 9 Find $\lim_{x \to \infty} x^3$ and $\lim_{x \to -\infty} x^3$.

SOLUTION When x becomes large, x^3 also becomes large. For instance,

$$10^3 = 1000 \qquad 100^3 = 1,000,000 \qquad 1000^3 = 1,000,000,000$$

In fact, we can make x^3 as big as we like by taking x large enough. Therefore we can write

$$\lim_{x \to \infty} x^3 = \infty$$

Similarly, when x is large negative, so is x^3. Thus

$$\lim_{x \to -\infty} x^3 = -\infty$$

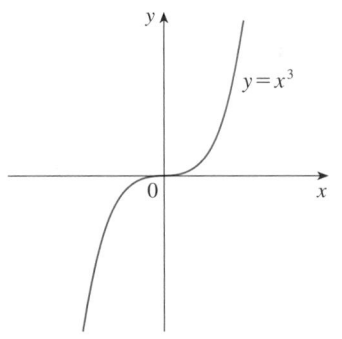

FIGURE 15

$\lim_{x \to \infty} x^3 = \infty, \quad \lim_{x \to -\infty} x^3 = -\infty$

These limit statements can also be seen from the graph of $y = x^3$ in Figure 15. ■

EXAMPLE 10 Find $\lim\limits_{x \to \infty} (x^2 - x)$.

⊘ **SOLUTION** It would be **wrong** to write

$$\lim_{x \to \infty} (x^2 - x) = \lim_{x \to \infty} x^2 - \lim_{x \to \infty} x = \infty - \infty$$

The Limit Laws can't be applied to infinite limits because ∞ is not a number ($\infty - \infty$ can't be defined). However, we can write

$$\lim_{x \to \infty} (x^2 - x) = \lim_{x \to \infty} x(x - 1) = \infty$$

because both x and $x - 1$ become arbitrarily large. ∎

EXAMPLE 11 Find $\lim\limits_{x \to \infty} \dfrac{x^2 + x}{3 - x}$.

SOLUTION We divide numerator and denominator by x (the highest power of x that occurs in the denominator):

$$\lim_{x \to \infty} \frac{x^2 + x}{3 - x} = \lim_{x \to \infty} \frac{x + 1}{\dfrac{3}{x} - 1} = -\infty$$

■ www.stewartcalculus.com
See Additional Example A.

because $x + 1 \to \infty$ and $3/x - 1 \to -1$ as $x \to \infty$. ∎

PRECISE DEFINITIONS

The following is a precise version of Definition 1.

> **6 DEFINITION** Let f be a function defined on some open interval that contains the number a, except possibly at a itself. Then
>
> $$\lim_{x \to a} f(x) = \infty$$
>
> means that for every positive number M there is a positive number δ such that
>
> $$\text{if} \qquad 0 < |x - a| < \delta \qquad \text{then} \qquad f(x) > M$$

This says that the values of $f(x)$ can be made arbitrarily large (larger than any given number M) by taking x close enough to a (within a distance δ, where δ depends on M, but with $x \neq a$). A geometric illustration is shown in Figure 16.

Given any horizontal line $y = M$, we can find a number $\delta > 0$ such that if we restrict x to lie in the interval $(a - \delta, a + \delta)$ but $x \neq a$, then the curve $y = f(x)$ lies above the line $y = M$. You can see that if a larger M is chosen, then a smaller δ may be required.

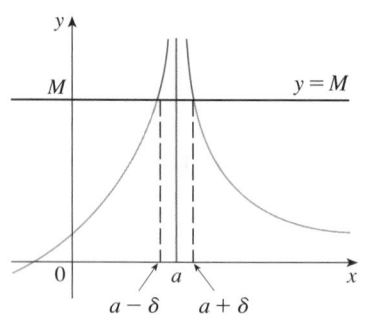

FIGURE 16

V EXAMPLE 12 Use Definition 6 to prove that $\lim\limits_{x \to 0} \dfrac{1}{x^2} = \infty$.

SOLUTION Let M be a given positive number. According to Definition 6, we need to find a number δ such that

$$\text{if} \quad 0 < |x| < \delta \quad \text{then} \quad \frac{1}{x^2} > M \quad \text{that is} \quad x^2 < \frac{1}{M}$$

But $x^2 < 1/M \iff |x| < 1/\sqrt{M}$. We can choose $\delta = 1/\sqrt{M}$ because

$$\text{if} \quad 0 < |x| < \delta = \frac{1}{\sqrt{M}} \quad \text{then} \quad \frac{1}{x^2} > \frac{1}{\delta^2} = M$$

Therefore, by Definition 6,

$$\lim_{x \to 0} \frac{1}{x^2} = \infty$$

$\blacksquare$

Similarly, $\lim_{x \to a} f(x) = -\infty$ means that for every negative number N there is a positive number δ such that if $0 < |x - a| < \delta$, then $f(x) < N$.

Definition 3 can be stated precisely as follows.

---

7 DEFINITION Let f be a function defined on some interval (a, ∞). Then

$$\lim_{x \to \infty} f(x) = L$$

means that for every $\varepsilon > 0$ there is a corresponding number N such that

$$\text{if} \quad x > N \quad \text{then} \quad |f(x) - L| < \varepsilon$$

In words, this says that the values of $f(x)$ can be made arbitrarily close to L (within a distance ε, where ε is any positive number) by taking x sufficiently large (larger than N, where N depends on ε). Graphically it says that by choosing x large enough (larger than some number N) we can make the graph of f lie between the given horizontal lines $y = L - \varepsilon$ and $y = L + \varepsilon$ as in Figure 17. This must be true no matter how small we choose ε.

TEC Module 1.3/1.6 illustrates Definition 7 graphically and numerically.

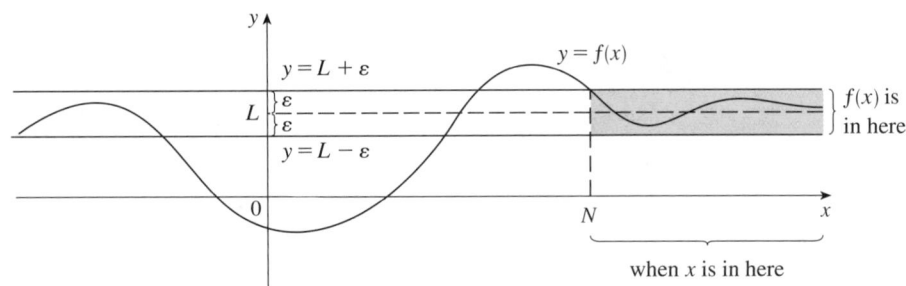

FIGURE 17
$\lim\limits_{x \to \infty} f(x) = L$

Figure 18 shows that if a smaller value of ε is chosen, then a larger value of N may be required.

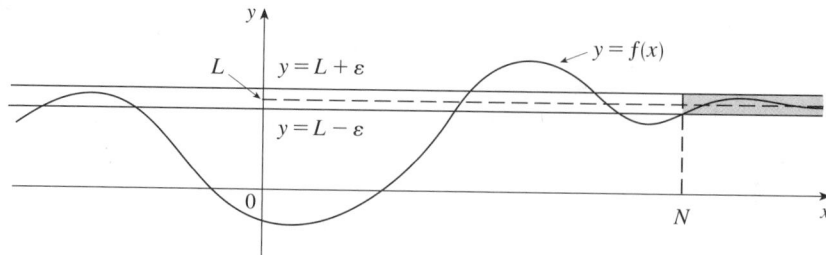

FIGURE 18

$\lim\limits_{x \to \infty} f(x) = L$

■ www.stewartcalculus.com
See Additional Example B.

Similarly, $\lim_{x \to -\infty} f(x) = L$ means that for every $\varepsilon > 0$ there is a corresponding number N such that if $x < N$, then $|f(x) - L| < \varepsilon$.

EXAMPLE 13 Use Definition 7 to prove that $\lim\limits_{x \to \infty} \dfrac{1}{x} = 0$.

SOLUTION Given $\varepsilon > 0$, we want to find N such that

$$\text{if} \quad x > N \quad \text{then} \quad \left| \frac{1}{x} - 0 \right| < \varepsilon$$

In computing the limit we may assume that $x > 0$. Then $1/x < \varepsilon \iff x > 1/\varepsilon$. Let's choose $N = 1/\varepsilon$. So

$$\text{if} \quad x > N = \frac{1}{\varepsilon} \quad \text{then} \quad \left| \frac{1}{x} - 0 \right| = \frac{1}{x} < \varepsilon$$

Therefore, by Definition 7,

$$\lim_{x \to \infty} \frac{1}{x} = 0$$

Figure 19 illustrates the proof by showing some values of ε and the corresponding values of N.

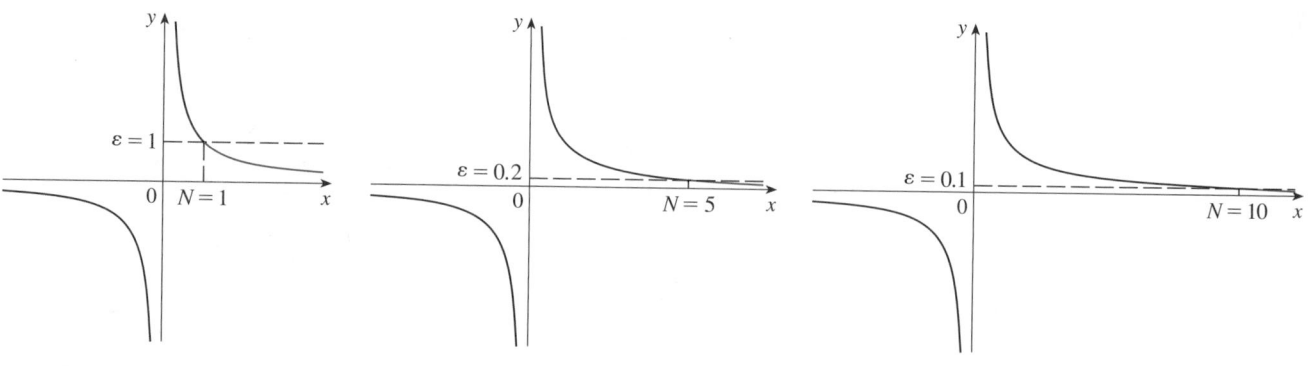

FIGURE 19

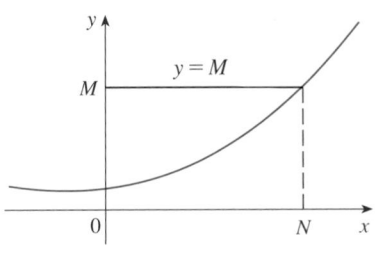

Finally we note that an infinite limit at infinity can be defined as follows. The geometric illustration is given in Figure 20.

FIGURE 20
$\lim\limits_{x \to \infty} f(x) = \infty$

---

8 **DEFINITION** Let f be a function defined on some interval (a, ∞). Then

$$\lim_{x \to \infty} f(x) = \infty$$

means that for every positive number M there is a corresponding positive number N such that

$$\text{if} \quad x > N \quad \text{then} \quad f(x) > M$$

Similar definitions apply when the symbol ∞ is replaced by $-\infty$.

1.6 | EXERCISES

1. For the function f whose graph is given, state the following.

(a) $\lim\limits_{x \to \infty} f(x)$ (b) $\lim\limits_{x \to -\infty} f(x)$

(c) $\lim\limits_{x \to 1} f(x)$ (d) $\lim\limits_{x \to 3} f(x)$

(e) The equations of the asymptotes

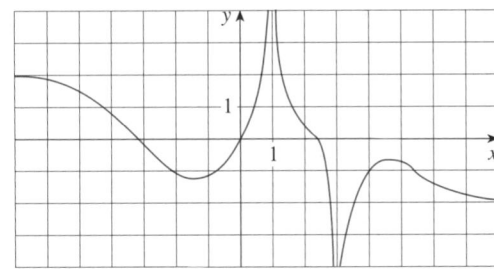

2. For the function g whose graph is given, state the following.

(a) $\lim\limits_{x \to \infty} g(x)$ (b) $\lim\limits_{x \to -\infty} g(x)$

(c) $\lim\limits_{x \to 0} g(x)$ (d) $\lim\limits_{x \to 2^-} g(x)$

(e) $\lim\limits_{x \to 2^+} g(x)$

(f) The equations of the asymptotes

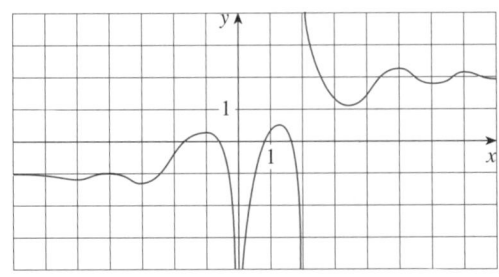

3–8 Sketch the graph of an example of a function f that satisfies all of the given conditions.

3. $\lim\limits_{x \to 0} f(x) = -\infty$, $\lim\limits_{x \to -\infty} f(x) = 5$, $\lim\limits_{x \to \infty} f(x) = -5$

4. $\lim\limits_{x \to 2} f(x) = \infty$, $\lim\limits_{x \to -2^+} f(x) = \infty$, $\lim\limits_{x \to -2^-} f(x) = -\infty$,

$\lim\limits_{x \to -\infty} f(x) = 0$, $\lim\limits_{x \to \infty} f(x) = 0$, $f(0) = 0$

5. $\lim\limits_{x \to 2} f(x) = -\infty$, $\lim\limits_{x \to \infty} f(x) = \infty$, $\lim\limits_{x \to -\infty} f(x) = 0$,

$\lim\limits_{x \to 0^+} f(x) = \infty$, $\lim\limits_{x \to 0^-} f(x) = -\infty$

6. $\lim\limits_{x \to \infty} f(x) = 3$, $\lim\limits_{x \to 2^-} f(x) = \infty$, $\lim\limits_{x \to 2^+} f(x) = -\infty$, f is odd

7. $f(0) = 3$, $\lim\limits_{x \to 0^-} f(x) = 4$, $\lim\limits_{x \to 0^+} f(x) = 2$,

$\lim\limits_{x \to -\infty} f(x) = -\infty$, $\lim\limits_{x \to 4^-} f(x) = -\infty$, $\lim\limits_{x \to 4^+} f(x) = \infty$,

$\lim\limits_{x \to \infty} f(x) = 3$

8. $\lim\limits_{x \to 3} f(x) = -\infty$, $\lim\limits_{x \to \infty} f(x) = 2$, $f(0) = 0$, f is even

---

9. Guess the value of the limit

$$\lim_{x \to \infty} \frac{x^2}{2^x}$$

by evaluating the function $f(x) = x^2/2^x$ for $x = 0, 1, 2, 3, 4, 5, 6, 7, 8, 9, 10, 20, 50,$ and 100. Then use a graph of f to support your guess.

10. Determine $\lim\limits_{x \to 1^-} \dfrac{1}{x^3 - 1}$ and $\lim\limits_{x \to 1^+} \dfrac{1}{x^3 - 1}$

(a) by evaluating $f(x) = 1/(x^3 - 1)$ for values of x that approach 1 from the left and from the right,

(b) by reasoning as in Example 1, and

 (c) from a graph of f.

11. Use a graph to estimate all the vertical and horizontal asymptotes of the curve

$$y = \frac{x^3}{x^3 - 2x + 1}$$

12. (a) Use a graph of

$$f(x) = \left(1 - \frac{2}{x}\right)^x$$

to estimate the value of $\lim_{x \to \infty} f(x)$ correct to two decimal places.

(b) Use a table of values of $f(x)$ to estimate the limit to four decimal places.

13–33 ▪ Find the limit.

13. $\displaystyle\lim_{x \to -3^+} \frac{x + 2}{x + 3}$

14. $\displaystyle\lim_{x \to -3^-} \frac{x + 2}{x + 3}$

15. $\displaystyle\lim_{x \to 1} \frac{2 - x}{(x - 1)^2}$

16. $\displaystyle\lim_{x \to \pi^-} \cot x$

17. $\displaystyle\lim_{x \to 2\pi^-} x \csc x$

18. $\displaystyle\lim_{x \to 2^-} \frac{x^2 - 2x}{x^2 - 4x + 4}$

19. $\displaystyle\lim_{x \to \infty} \frac{3x - 2}{2x + 1}$

20. $\displaystyle\lim_{x \to \infty} \frac{1 - x^2}{x^3 - x + 1}$

21. $\displaystyle\lim_{t \to \infty} \frac{\sqrt{t} + t^2}{2t - t^2}$

22. $\displaystyle\lim_{t \to \infty} \frac{t - t\sqrt{t}}{2t^{3/2} + 3t - 5}$

23. $\displaystyle\lim_{x \to \infty} \frac{(2x^2 + 1)^2}{(x - 1)^2(x^2 + x)}$

24. $\displaystyle\lim_{x \to \infty} \frac{x^2}{\sqrt{x^4 + 1}}$

25. $\displaystyle\lim_{x \to \infty} \left(\sqrt{9x^2 + x} - 3x\right)$

26. $\displaystyle\lim_{x \to \infty} \left(\sqrt{x^2 + ax} - \sqrt{x^2 + bx}\right)$

27. $\displaystyle\lim_{x \to \infty} \frac{x^4 - 3x^2 + x}{x^3 - x + 2}$

28. $\displaystyle\lim_{x \to \infty} \frac{\sin^2 x}{x^2}$

29. $\displaystyle\lim_{x \to \infty} \cos x$

30. $\displaystyle\lim_{x \to -\infty} \frac{1 + x^6}{x^4 + 1}$

31. $\displaystyle\lim_{x \to \infty} \left(x - \sqrt{x}\right)$

32. $\displaystyle\lim_{x \to \infty} \left(x^2 - x^4\right)$

33. $\displaystyle\lim_{x \to -\infty} \left(x^4 + x^5\right)$

34. (a) Graph the function

$$f(x) = \frac{\sqrt{2x^2 + 1}}{3x - 5}$$

How many horizontal and vertical asymptotes do you observe? Use the graph to estimate the values of the limits

$$\lim_{x \to \infty} \frac{\sqrt{2x^2 + 1}}{3x - 5} \quad \text{and} \quad \lim_{x \to -\infty} \frac{\sqrt{2x^2 + 1}}{3x - 5}$$

(b) By calculating values of $f(x)$, give numerical estimates of the limits in part (a).

(c) Calculate the exact values of the limits in part (a). Did you get the same value or different values for these two limits? [In view of your answer to part (a), you might have to check your calculation for the second limit.]

35–36 ▪ Find the horizontal and vertical asymptotes of each curve. Check your work by graphing the curve and estimating the asymptotes.

35. $y = \dfrac{2x^2 + x - 1}{x^2 + x - 2}$

36. $F(x) = \dfrac{x - 9}{\sqrt{4x^2 + 3x + 2}}$

37. (a) Estimate the value of

$$\lim_{x \to -\infty} \left(\sqrt{x^2 + x + 1} + x\right)$$

by graphing the function $f(x) = \sqrt{x^2 + x + 1} + x$.

(b) Use a table of values of $f(x)$ to guess the value of the limit.

(c) Prove that your guess is correct.

38. (a) Use a graph of

$$f(x) = \sqrt{3x^2 + 8x + 6} - \sqrt{3x^2 + 3x + 1}$$

to estimate the value of $\lim_{x \to \infty} f(x)$ to one decimal place.

(b) Use a table of values of $f(x)$ to estimate the limit to four decimal places.

(c) Find the exact value of the limit.

39. Estimate the horizontal asymptote of the function

$$f(x) = \frac{3x^3 + 500x^2}{x^3 + 500x^2 + 100x + 2000}$$

by graphing f for $-10 \le x \le 10$. Then calculate the equation of the asymptote by evaluating the limit. How do you explain the discrepancy?

40. Find a formula for a function that has vertical asymptotes $x = 1$ and $x = 3$ and horizontal asymptote $y = 1$.

41. Find a formula for a function f that satisfies the following conditions:

$$\lim_{x \to \pm\infty} f(x) = 0, \quad \lim_{x \to 0} f(x) = -\infty, \quad f(2) = 0,$$

$$\lim_{x \to 3^-} f(x) = \infty, \quad \lim_{x \to 3^+} f(x) = -\infty$$

42. Evaluate the limits.

(a) $\displaystyle\lim_{x \to \infty} x \sin \frac{1}{x}$

(b) $\displaystyle\lim_{x \to \infty} \sqrt{x} \, \sin \frac{1}{x}$

43. A function f is a ratio of quadratic functions and has a vertical asymptote $x = 4$ and just one x-intercept, $x = 1$. It is known that f has a removable discontinuity at $x = -1$ and $\lim_{x \to -1} f(x) = 2$. Evaluate

(a) $f(0)$

(b) $\displaystyle\lim_{x \to \infty} f(x)$

44. By the *end behavior* of a function we mean the behavior of its values as $x \to \infty$ and as $x \to -\infty$.
  (a) Describe and compare the end behavior of the functions

  $$P(x) = 3x^5 - 5x^3 + 2x \qquad Q(x) = 3x^5$$

  by graphing both functions in the viewing rectangles $[-2, 2]$ by $[-2, 2]$ and $[-10, 10]$ by $[-10,000, 10,000]$.
  (b) Two functions are said to have the *same end behavior* if their ratio approaches 1 as $x \to \infty$. Show that P and Q have the same end behavior.

45. Let P and Q be polynomials. Find

$$\lim_{x \to \infty} \frac{P(x)}{Q(x)}$$

if the degree of P is (a) less than the degree of Q and (b) greater than the degree of Q.

46. Make a rough sketch of the curve $y = x^n$ (n an integer) for the following five cases:

  (i) $n = 0$ (ii) $n > 0$, n odd
  (iii) $n > 0$, n even (iv) $n < 0$, n odd
  (v) $n < 0$, n even

  Then use these sketches to find the following limits.

  (a) $\lim_{x \to 0^+} x^n$ (b) $\lim_{x \to 0^-} x^n$
  (c) $\lim_{x \to \infty} x^n$ (d) $\lim_{x \to -\infty} x^n$

47. Find $\lim_{x \to \infty} f(x)$ if, for all $x > 5$,

$$\frac{4x - 1}{x} < f(x) < \frac{4x^2 + 3x}{x^2}$$

48. In the theory of relativity, the mass of a particle with velocity v is

$$m = \frac{m_0}{\sqrt{1 - v^2/c^2}}$$

where m_0 is the mass of the particle at rest and c is the speed of light. What happens as $v \to c^-$?

49. (a) A tank contains 5000 L of pure water. Brine that contains 30 g of salt per liter of water is pumped into the tank at a rate of 25 L/min. Show that the concentration of salt t minutes later (in grams per liter) is

$$C(t) = \frac{30t}{200 + t}$$

  (b) What happens to the concentration as $t \to \infty$?

50. (a) Show that $\lim_{x \to \infty} \dfrac{4x^2 - 5x}{2x^2 + 1} = 2$.

  (b) By graphing the function in part (a) and the line $y = 1.9$ on a common screen, find a number N such that

  if $\quad x > N \quad$ then $\quad \dfrac{4x^2 - 5x}{2x^2 + 1} > 1.9$

  What if 1.9 is replaced by 1.99?

51. How close to -3 do we have to take x so that

$$\frac{1}{(x + 3)^4} > 10{,}000$$

52. Prove, using Definition 6, that $\lim_{x \to -3} \dfrac{1}{(x + 3)^4} = \infty$.

53. Prove that $\lim_{x \to -1^-} \dfrac{5}{(x + 1)^3} = -\infty$.

54. For the limit

$$\lim_{x \to \infty} \frac{\sqrt{4x^2 + 1}}{x + 1} = 2$$

illustrate Definition 7 by finding values of N that correspond to $\varepsilon = 0.5$ and $\varepsilon = 0.1$.

55. Use a graph to find a number N such that

  if $\quad x > N \quad$ then $\quad \left| \dfrac{3x^2 + 1}{2x^2 + x + 1} - 1.5 \right| < 0.05$

56. For the limit

$$\lim_{x \to \infty} \frac{2x + 1}{\sqrt{x + 1}} = \infty$$

illustrate Definition 8 by finding a value of N that corresponds to $M = 100$.

57. (a) How large do we have to take x so that $1/x^2 < 0.0001$?
  (b) Taking $n = 2$ in $\boxed{5}$, we have the statement

$$\lim_{x \to \infty} \frac{1}{x^2} = 0$$

  Prove this directly using Definition 7.

58. Prove, using Definition 8, that $\lim_{x \to \infty} x^3 = \infty$.

59. Prove that

$$\lim_{x \to \infty} f(x) = \lim_{t \to 0^+} f(1/t)$$

and

$$\lim_{x \to -\infty} f(x) = \lim_{t \to 0^-} f(1/t)$$

if these limits exist.

CHAPTER 1   REVIEW

1. (a) What is a function? What are its domain and range?
   (b) What is the graph of a function?
   (c) How can you tell whether a given curve is the graph of a function?

2. Discuss four ways of representing a function. Illustrate your discussion with examples.

3. (a) What is an even function? How can you tell if a function is even by looking at its graph? Give three examples of an even function.
   (b) What is an odd function? How can you tell if a function is odd by looking at its graph? Give three examples of an odd function.

4. What is an increasing function?

5. What is a mathematical model?

6. Give an example of each type of function.
   (a) Linear function   (b) Power function
   (c) Exponential function   (d) Quadratic function
   (e) Polynomial of degree 5 (f) Rational function

7. Sketch by hand, on the same axes, the graphs of the following functions.
   (a) $f(x) = x$ (b) $g(x) = x^2$
   (c) $h(x) = x^3$ (d) $j(x) = x^4$

8. Draw, by hand, a rough sketch of the graph of each function.
   (a) $y = \sin x$ (b) $y = \tan x$
   (c) $y = 2^x$ (d) $y = 1/x$
   (e) $y = |x|$ (f) $y = \sqrt{x}$

9. Suppose that f has domain A and g has domain B.
   (a) What is the domain of $f + g$?
   (b) What is the domain of fg?
   (c) What is the domain of f/g?

10. How is the composite function $f \circ g$ defined? What is its domain?

11. Suppose the graph of f is given. Write an equation for each of the graphs that are obtained from the graph of f as follows.
   (a) Shift 2 units upward.

   (b) Shift 2 units downward.
   (c) Shift 2 units to the right.
   (d) Shift 2 units to the left.
   (e) Reflect about the x-axis.
   (f) Reflect about the y-axis.
   (g) Stretch vertically by a factor of 2.
   (h) Shrink vertically by a factor of 2.
   (i) Stretch horizontally by a factor of 2.
   (j) Shrink horizontally by a factor of 2.

12. Explain what each of the following means and illustrate with a sketch.
   (a) $\lim_{x \to a} f(x) = L$ (b) $\lim_{x \to a^+} f(x) = L$
   (c) $\lim_{x \to a^-} f(x) = L$ (d) $\lim_{x \to a} f(x) = \infty$
   (e) $\lim_{x \to \infty} f(x) = L$

13. Describe several ways in which a limit can fail to exist. Illustrate with sketches.

14. State the following Limit Laws.
   (a) Sum Law   (b) Difference Law
   (c) Constant Multiple Law (d) Product Law
   (e) Quotient Law   (f) Power Law
   (g) Root Law

15. What does the Squeeze Theorem say?

16. (a) What does it mean for f to be continuous at a?
   (b) What does it mean for f to be continuous on the interval $(-\infty, \infty)$? What can you say about the graph of such a function?

17. What does the Intermediate Value Theorem say?

18. (a) What does it mean to say that the line $x = a$ is a vertical asymptote of the curve $y = f(x)$? Draw curves to illustrate the various possibilities.
   (b) What does it mean to say that the line $y = L$ is a horizontal asymptote of the curve $y = f(x)$? Draw curves to illustrate the various possibilities.

Determine whether the statement is true or false. If it is true, explain why. If it is false, explain why or give an example that disproves the statement.

1. If f is a function, then $f(s + t) = f(s) + f(t)$.

2. If $f(s) = f(t)$, then $s = t$.

3. If f is a function, then $f(3x) = 3f(x)$.

4. If $x_1 < x_2$ and f is a decreasing function, then $f(x_1) > f(x_2)$.

5. A vertical line intersects the graph of a function at most once.

6. If f and g are functions, then $f \circ g = g \circ f$.

7. $\lim\limits_{x\to4}\left(\dfrac{2x}{x-4}-\dfrac{8}{x-4}\right)=\lim\limits_{x\to4}\dfrac{2x}{x-4}-\lim\limits_{x\to4}\dfrac{8}{x-4}$

8. $\lim\limits_{x\to1}\dfrac{x^2+6x-7}{x^2+5x-6}=\dfrac{\lim\limits_{x\to1}(x^2+6x-7)}{\lim\limits_{x\to1}(x^2+5x-6)}$

9. $\lim\limits_{x\to1}\dfrac{x-3}{x^2+2x-4}=\dfrac{\lim\limits_{x\to1}(x-3)}{\lim\limits_{x\to1}(x^2+2x-4)}$

10. If $\lim_{x\to5}f(x)=2$ and $\lim_{x\to5}g(x)=0$, then $\lim_{x\to5}[f(x)/g(x)]$ does not exist.

11. If $\lim_{x\to5}f(x)=0$ and $\lim_{x\to5}g(x)=0$, then $\lim_{x\to5}[f(x)/g(x)]$ does not exist.

12. If $\lim_{x\to6}[f(x)g(x)]$ exists, then the limit must be $f(6)g(6)$.

13. If p is a polynomial, then $\lim_{x\to b}p(x)=p(b)$.

14. If $\lim_{x\to0}f(x)=\infty$ and $\lim_{x\to0}g(x)=\infty$, then $\lim_{x\to0}[f(x)-g(x)]=0$.

15. A function can have two different horizontal asymptotes.

16. If f has domain $[0,\infty)$ and has no horizontal asymptote, then $\lim_{x\to\infty}f(x)=\infty$ or $\lim_{x\to\infty}f(x)=-\infty$.

17. If the line $x=1$ is a vertical asymptote of $y=f(x)$, then f is not defined at 1.

18. If f and g are polynomials and $g(2)=0$, then the rational function $h(x)=f(x)/g(x)$ has the vertical asymptote $x=2$.

19. If x is any real number, then $\sqrt{x^2}=x$.

20. If $f(1)>0$ and $f(3)<0$, then there exists a number c between 1 and 3 such that $f(c)=0$.

21. If f is continuous at 5 and $f(5)=2$ and $f(4)=3$, then $\lim_{x\to2}f(4x^2-11)=2$.

22. If f is continuous on $[-1,1]$ and $f(-1)=4$ and $f(1)=3$, then there exists a number r such that $|r|<1$ and $f(r)=\pi$.

23. Let f be a function such that $\lim_{x\to0}f(x)=6$. Then there exists a positive number δ such that if $0<|x|<\delta$, then $|f(x)-6|<1$.

24. If $f(x)>1$ for all x and $\lim_{x\to0}f(x)$ exists, then $\lim_{x\to0}f(x)>1$.

25. If f is continuous at a, so is $|f|$.

26. If $|f|$ is continuous at a, so is f.

EXERCISES

1. Let f be the function whose graph is given.
(a) Estimate the value of $f(2)$.
(b) Estimate the values of x such that $f(x)=3$.
(c) State the domain of f.
(d) State the range of f.
(e) On what interval is f increasing?
(f) Is f even, odd, or neither even nor odd? Explain.

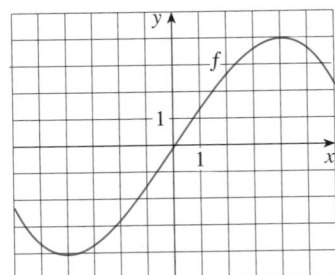

2. Determine whether each curve is the graph of a function of x. If it is, state the domain and range of the function.
(a) 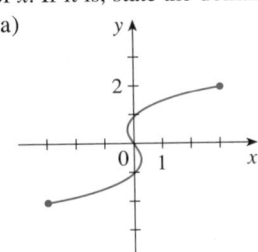 (b)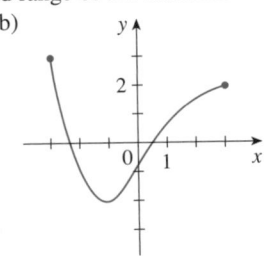

3–6 ■ Find the domain and range of the function. Write your answer in interval notation.

3. $f(x)=2/(3x-1)$

4. $g(x)=\sqrt{16-x^4}$

5. $y=1+\sin x$

6. $y=\tan 2x$

7. Suppose that the graph of f is given. Describe how the graphs of the following functions can be obtained from the graph of f.
(a) $y=f(x)+8$ (b) $y=f(x+8)$
(c) $y=1+2f(x)$ (d) $y=f(x-2)-2$
(e) $y=-f(x)$ (f) $y=3-f(x)$

8. The graph of f is given. Draw the graphs of the following functions.
(a) $y=f(x-8)$ (b) $y=-f(x)$
(c) $y=2-f(x)$ (d) $y=\frac12 f(x)-1$

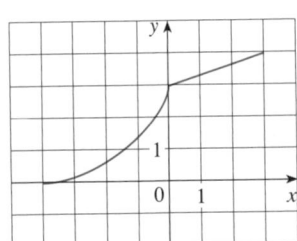

9–14 ▪ Use transformations to sketch the graph of the function.

9. $y = -\sin 2x$

10. $y = (x - 2)^2$

11. $y = 1 + \frac{1}{2}x^3$

12. $y = 2 - \sqrt{x}$

13. $f(x) = \dfrac{1}{x + 2}$

14. $f(x) = \begin{cases} 1 + x & \text{if } x < 0 \\ 1 + x^2 & \text{if } x \geq 0 \end{cases}$

15. Determine whether f is even, odd, or neither even nor odd.
(a) $f(x) = 2x^5 - 3x^2 + 2$ (b) $f(x) = x^3 - x^7$
(c) $f(x) = \cos(x^2)$ (d) $f(x) = 1 + \sin x$

16. Find an expression for the function whose graph consists of the line segment from the point $(-2, 2)$ to the point $(-1, 0)$ together with the top half of the circle with center the origin and radius 1.

17. If $f(x) = \sqrt{x}$ and $g(x) = \sin x$, find the functions (a) $f \circ g$, (b) $g \circ f$, (c) $f \circ f$, (d) $g \circ g$, and their domains.

18. Express the function $F(x) = 1/\sqrt{x + \sqrt{x}}$ as a composition of three functions.

19. The graph of f is given.
(a) Find each limit, or explain why it does not exist.

(i) $\lim\limits_{x \to 2^+} f(x)$ (ii) $\lim\limits_{x \to -3^+} f(x)$ (iii) $\lim\limits_{x \to -3} f(x)$

(iv) $\lim\limits_{x \to 4} f(x)$ (v) $\lim\limits_{x \to 0} f(x)$ (vi) $\lim\limits_{x \to 2^-} f(x)$

(vii) $\lim\limits_{x \to \infty} f(x)$ (viii) $\lim\limits_{x \to -\infty} f(x)$

(b) State the equations of the horizontal asymptotes.
(c) State the equations of the vertical asymptotes.
(d) At what numbers is f discontinuous? Explain.

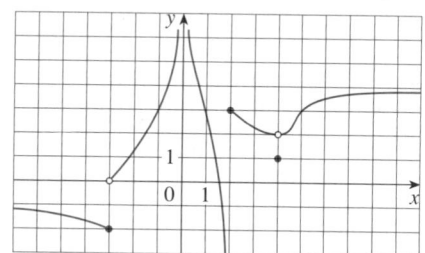

20. Sketch the graph of an example of a function f that satisfies all of the following conditions:

$\lim\limits_{x \to -\infty} f(x) = -2$, $\lim\limits_{x \to \infty} f(x) = 0$, $\lim\limits_{x \to -3} f(x) = \infty$,

$\lim\limits_{x \to 3^-} f(x) = -\infty$, $\lim\limits_{x \to 3^+} f(x) = 2$,

f is continuous from the right at 3

21–36 ▪ Find the limit.

21. $\lim\limits_{x \to 0} \cos(x + \sin x)$

22. $\lim\limits_{x \to 3} \dfrac{x^2 - 9}{x^2 + 2x - 3}$

23. $\lim\limits_{x \to -3} \dfrac{x^2 - 9}{x^2 + 2x - 3}$

24. $\lim\limits_{x \to 1^+} \dfrac{x^2 - 9}{x^2 + 2x - 3}$

25. $\lim\limits_{h \to 0} \dfrac{(h - 1)^3 + 1}{h}$

26. $\lim\limits_{t \to 2} \dfrac{t^2 - 4}{t^3 - 8}$

27. $\lim\limits_{r \to 9} \dfrac{\sqrt{r}}{(r - 9)^4}$

28. $\lim\limits_{v \to 4^+} \dfrac{4 - v}{|4 - v|}$

29. $\lim\limits_{s \to 16} \dfrac{4 - \sqrt{s}}{s - 16}$

30. $\lim\limits_{v \to 2} \dfrac{v^2 + 2v - 8}{v^4 - 16}$

31. $\lim\limits_{x \to \infty} \dfrac{1 + 2x - x^2}{1 - x + 2x^2}$

32. $\lim\limits_{x \to -\infty} \dfrac{1 - 2x^2 - x^4}{5 + x - 3x^4}$

33. $\lim\limits_{x \to \infty} \left(\sqrt{x^2 + 4x + 1} - x\right)$

34. $\lim\limits_{x \to 1} \left(\dfrac{1}{x - 1} + \dfrac{1}{x^2 - 3x + 2}\right)$

35. $\lim\limits_{x \to 0} \dfrac{\cot 2x}{\csc x}$

36. $\lim\limits_{t \to 0} \dfrac{t^3}{\tan^3 2t}$

37–38 ▪ Use graphs to discover the asymptotes of the curve. Then prove what you have discovered.

37. $y = \dfrac{\cos^2 x}{x^2}$

38. $y = \sqrt{x^2 + x + 1} - \sqrt{x^2 - x}$

39. If $2x - 1 \leq f(x) \leq x^2$ for $0 < x < 3$, find $\lim\limits_{x \to 1} f(x)$.

40. Prove that $\lim\limits_{x \to 0} x^2 \cos(1/x^2) = 0$.

41–44 ▪ Prove the statement using the precise definition of a limit.

41. $\lim\limits_{x \to 2} (14 - 5x) = 4$

42. $\lim\limits_{x \to 0} \sqrt[3]{x} = 0$

43. $\lim\limits_{x \to \infty} \dfrac{1}{x^4} = 0$

44. $\lim\limits_{x \to 4^+} \dfrac{2}{\sqrt{x - 4}} = \infty$

45. Let
$$f(x) = \begin{cases} \sqrt{-x} & \text{if } x < 0 \\ 3 - x & \text{if } 0 \leq x < 3 \\ (x - 3)^2 & \text{if } x > 3 \end{cases}$$

(a) Evaluate each limit, if it exists.

(i) $\lim\limits_{x \to 0^+} f(x)$ (ii) $\lim\limits_{x \to 0^-} f(x)$ (iii) $\lim\limits_{x \to 0} f(x)$

(iv) $\lim\limits_{x \to 3^-} f(x)$ (v) $\lim\limits_{x \to 3^+} f(x)$ (vi) $\lim\limits_{x \to 3} f(x)$

(b) Where is f discontinuous?
(c) Sketch the graph of f.

46. Show that each function is continuous on its domain. State the domain.

(a) $g(x) = \dfrac{\sqrt{x^2 - 9}}{x^2 - 2}$

(b) $h(x) = \sqrt[4]{x} + x^3 \cos x$

47–48 ▪ Use the Intermediate Value Theorem to show that there is a root of the equation in the given interval.

47. $x^5 - x^3 + 3x - 5 = 0$, $(1, 2)$

48. $2 \sin x = 3 - 2x$, $(0, 1)$

2 DERIVATIVES

In this chapter we study a special type of limit, called a derivative, that occurs when we want to find a slope of a tangent line, or a velocity, or any instantaneous rate of change.

2.1 | DERIVATIVES AND RATES OF CHANGE

The problem of finding the tangent line to a curve and the problem of finding the velocity of an object involve finding the same type of limit, which we call a *derivative*.

THE TANGENT PROBLEM

The word *tangent* is derived from the Latin word *tangens*, which means "touching." Thus a tangent to a curve is a line that touches the curve. In other words, a tangent line should have the same direction as the curve at the point of contact. How can this idea be made precise?

For a circle we could simply follow Euclid and say that a tangent is a line that intersects the circle once and only once as in Figure 1(a). For more complicated curves this definition is inadequate. Figure l(b) shows two lines L and T passing through a point P on a curve C. The line L intersects C only once, but it certainly does not look like what we think of as a tangent. The line T, on the other hand, looks like a tangent but it intersects C twice.

■ www.stewartcalculus.com
See Additional Example A.

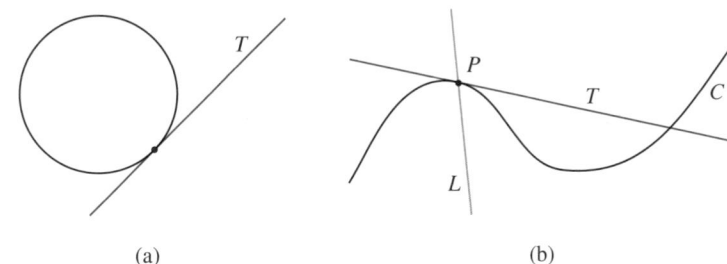

FIGURE 1   (a)   (b)

To be specific, let's look at the problem of trying to find a tangent line T to the parabola $y = x^2$ in the following example.

▶ EXAMPLE 1 Find an equation of the tangent line to the parabola $y = x^2$ at the point $P(1, 1)$.

SOLUTION We will be able to find an equation of the tangent line T as soon as we know its slope m. The difficulty is that we know only one point, P, on T, whereas we need two points to compute the slope. But observe that we can compute an approximation to m by choosing a nearby point $Q(x, x^2)$ on the parabola (as in Figure 2) and computing the slope m_{PQ} of the secant line PQ. [A **secant line**, from the Latin word *secans*, meaning cutting, is a line that cuts (intersects) a curve more than once.]

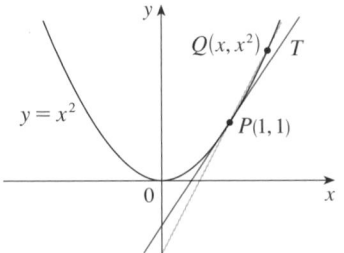

FIGURE 2

We choose $x \neq 1$ so that $Q \neq P$. Then

$$m_{PQ} = \frac{x^2 - 1}{x - 1}$$

What happens as x approaches 1? From Figure 3 we see that Q approaches P along the parabola and the secant lines PQ rotate about P and approach the tangent line T.

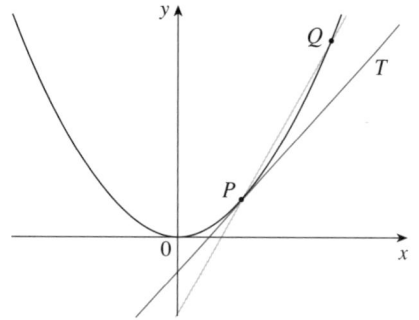 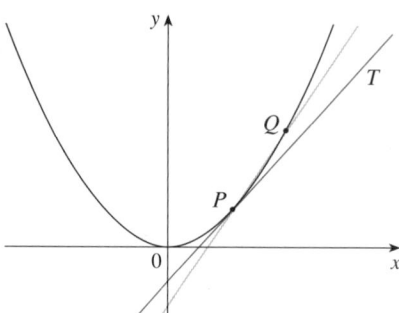 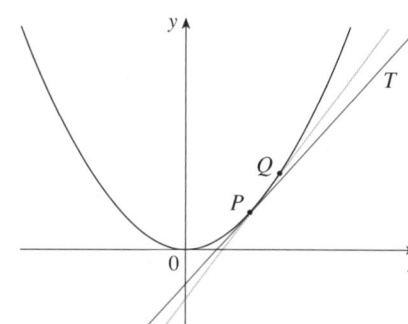

Q approaches P from the right

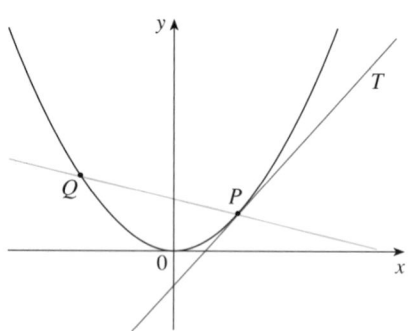 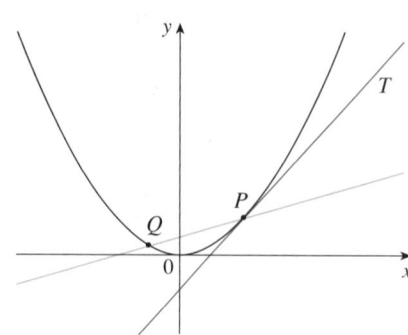 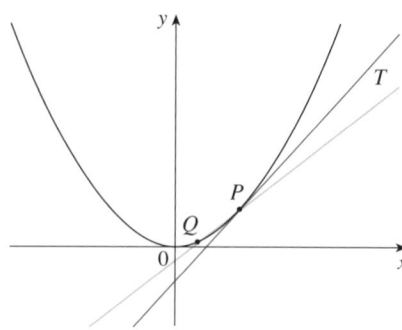

Q approaches P from the left

FIGURE 3

It appears that the slope m of the tangent line is the limit of the slopes of the secant lines as x approaches 1:

TEC In Visual 2.1A you can see how the process in Figure 3 works for additional functions.

$$m = \lim_{x \to 1} \frac{x^2 - 1}{x - 1} = \lim_{x \to 1} \frac{(x - 1)(x + 1)}{x - 1}$$

$$= \lim_{x \to 1} (x + 1) = 1 + 1 = 2$$

■ Point-slope form for a line through the point (x_1, y_1) with slope m:

$$y - y_1 = m(x - x_1)$$

Using the point-slope form of the equation of a line, we find that an equation of the tangent line at $(1, 1)$ is

$$y - 1 = 2(x - 1) \qquad \text{or} \qquad y = 2x - 1 \qquad \blacksquare$$

We sometimes refer to the slope of the tangent line to a curve at a point as the **slope of the curve** at the point. The idea is that if we zoom in far enough toward the point, the curve looks almost like a straight line. Figure 4 illustrates this procedure for the curve $y = x^2$ in Example 1. The more we zoom in, the more the parabola looks like a line. In other words, the curve becomes almost indistinguishable from its tangent line.

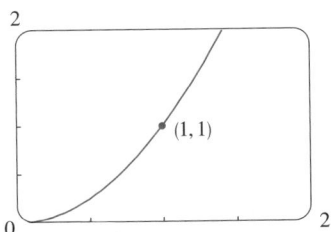
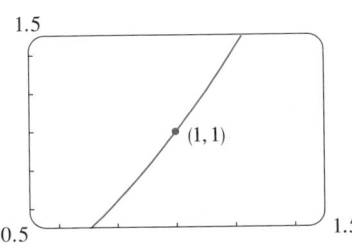
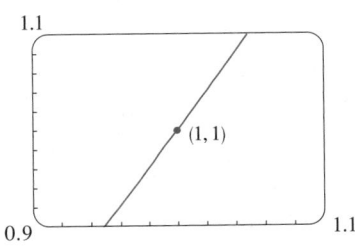

FIGURE 4 Zooming in toward the point $(1, 1)$ on the parabola $y = x^2$

TEC Visual 2.1B shows an animation of Figure 4.

In general, if a curve C has equation $y = f(x)$ and we want to find the tangent line to C at the point $P(a, f(a))$, then we consider a nearby point $Q(x, f(x))$, where $x \neq a$, and compute the slope of the secant line PQ:

$$m_{PQ} = \frac{f(x) - f(a)}{x - a}$$

Then we let Q approach P along the curve C by letting x approach a. If m_{PQ} approaches a number m, then we define the *tangent* T to be the line through P with slope m. (This amounts to saying that the tangent line is the limiting position of the secant line PQ as Q approaches P. See Figure 5.)

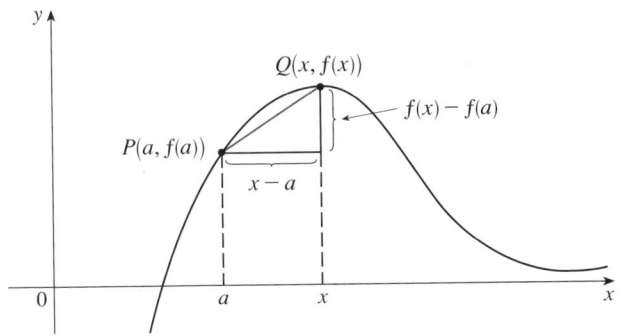
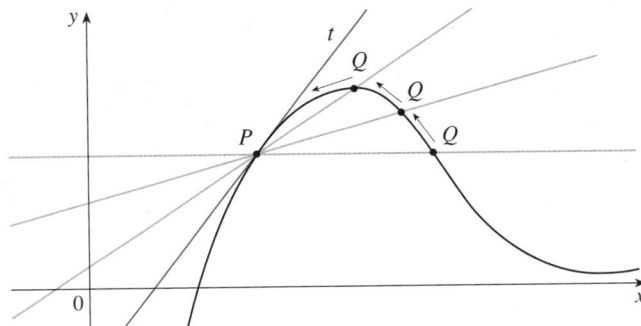

FIGURE 5

---

1 DEFINITION The **tangent line** to the curve $y = f(x)$ at the point $P(a, f(a))$ is the line through P with slope

$$m = \lim_{x \to a} \frac{f(x) - f(a)}{x - a}$$

provided that this limit exists.

There is another expression for the slope of a tangent line that is sometimes easier to use. If $h = x - a$, then $x = a + h$ and so the slope of the secant line PQ is

$$m_{PQ} = \frac{f(a + h) - f(a)}{h}$$

(See Figure 6 where the case $h > 0$ is illustrated and Q is to the right of P. If it happened that $h < 0$, however, Q would be to the left of P.) Notice that as x approaches a, h approaches 0 (because $h = x - a$) and so the expression for the slope of the

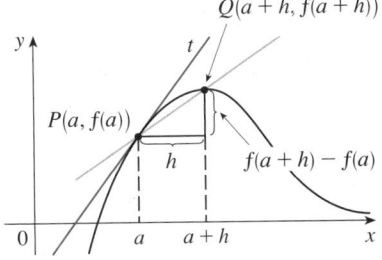

FIGURE 6

tangent line in Definition 1 becomes

$$
\boxed{2} \qquad m = \lim_{h \to 0} \frac{f(a + h) - f(a)}{h}
$$

EXAMPLE 2 Find an equation of the tangent line to the hyperbola $y = 3/x$ at the point $(3, 1)$.

SOLUTION Let $f(x) = 3/x$. Then the slope of the tangent at $(3, 1)$ is

$$
m = \lim_{h \to 0} \frac{f(3 + h) - f(3)}{h}
$$

$$
= \lim_{h \to 0} \frac{\dfrac{3}{3 + h} - 1}{h} = \lim_{h \to 0} \frac{\dfrac{3 - (3 + h)}{3 + h}}{h}
$$

$$
= \lim_{h \to 0} \frac{-h}{h(3 + h)} = \lim_{h \to 0} -\frac{1}{3 + h} = -\frac{1}{3}
$$

Therefore an equation of the tangent at the point $(3, 1)$ is

$$
y - 1 = -\tfrac{1}{3}(x - 3)
$$

which simplifies to

$$
x + 3y - 6 = 0
$$

The hyperbola and its tangent are shown in Figure 7. ∎

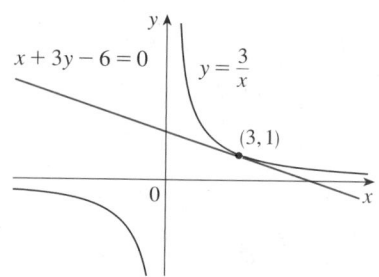

FIGURE 7

THE VELOCITY PROBLEM

In Section 1.3 we investigated the motion of a ball dropped from the CN Tower and defined its velocity to be the limiting value of average velocities over shorter and shorter time periods.

In general, suppose an object moves along a straight line according to an equation of motion $s = f(t)$, where s is the displacement (directed distance) of the object from the origin at time t. The function f that describes the motion is called the **position function** of the object. In the time interval from $t = a$ to $t = a + h$ the change in position is $f(a + h) - f(a)$. (See Figure 8.) The average velocity over this time interval is

$$
\text{average velocity} = \frac{\text{displacement}}{\text{time}} = \frac{f(a + h) - f(a)}{h}
$$

which is the same as the slope of the secant line PQ in Figure 9.

Now suppose we compute the average velocities over shorter and shorter time intervals $[a, a + h]$. In other words, we let h approach 0. As in the example of the falling ball, we define the **velocity** (or **instantaneous velocity**) $v(a)$ at time $t = a$ to be the limit of these average velocities:

$$
\boxed{3} \qquad v(a) = \lim_{h \to 0} \frac{f(a + h) - f(a)}{h}
$$

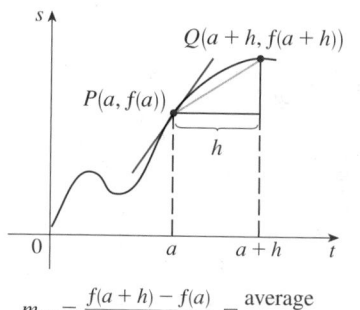

FIGURE 8

FIGURE 9

This means that the velocity at time $t = a$ is equal to the slope of the tangent line at P (compare Equations 2 and 3).

Now that we know how to compute limits, let's reconsider the problem of the falling ball.

■ **EXAMPLE 3** Suppose that a ball is dropped from the upper observation deck of the CN Tower, 450 m above the ground.
(a) What is the velocity of the ball after 5 seconds?
(b) How fast is the ball traveling when it hits the ground?

■ Recall from Section 1.3: The distance (in meters) fallen after t seconds is $4.9t^2$.

SOLUTION We first use the equation of motion $s = f(t) = 4.9t^2$ to find the velocity $v(a)$ after a seconds:

$$v(a) = \lim_{h \to 0} \frac{f(a + h) - f(a)}{h} = \lim_{h \to 0} \frac{4.9(a + h)^2 - 4.9a^2}{h}$$

$$= \lim_{h \to 0} \frac{4.9(a^2 + 2ah + h^2 - a^2)}{h} = \lim_{h \to 0} \frac{4.9(2ah + h^2)}{h}$$

$$= \lim_{h \to 0} 4.9(2a + h) = 9.8a$$

(a) The velocity after 5 s is $v(5) = (9.8)(5) = 49$ m/s.

(b) Since the observation deck is 450 m above the ground, the ball will hit the ground at the time t_1 when $s(t_1) = 450$, that is,

$$4.9t_1^2 = 450$$

■ **www.stewartcalculus.com**
See Additional Example B.

This gives

$$t_1^2 = \frac{450}{4.9} \qquad \text{and} \qquad t_1 = \sqrt{\frac{450}{4.9}} \approx 9.6 \text{ s}$$

The velocity of the ball as it hits the ground is therefore

$$v(t_1) = 9.8t_1 = 9.8\sqrt{\frac{450}{4.9}} \approx 94 \text{ m/s} \qquad\blacksquare$$

DERIVATIVES

We have seen that the same type of limit arises in finding the slope of a tangent line (Equation 2) or the velocity of an object (Equation 3). In fact, limits of the form

$$\lim_{h \to 0} \frac{f(a + h) - f(a)}{h}$$

arise whenever we calculate a rate of change in any of the sciences or in engineering, such as a rate of reaction in chemistry or a marginal cost in economics. Since this type of limit occurs so widely, it is given a special name and notation.

> **4 DEFINITION** The **derivative of a function** f **at a number** a, denoted by $f'(a)$, is
>
> ■ $f'(a)$ is read "f prime of a."
>
> $$f'(a) = \lim_{h \to 0} \frac{f(a + h) - f(a)}{h}$$
>
> if this limit exists.

If we write $x = a + h$, then $h = x - a$ and h approaches 0 if and only if x approaches a. Therefore an equivalent way of stating the definition of the derivative, as we saw in finding tangent lines, is

5
$$f'(a) = \lim_{x \to a} \frac{f(x) - f(a)}{x - a}$$

☑ EXAMPLE 4 Find the derivative of the function $f(x) = x^2 - 8x + 9$ at the number a.

SOLUTION From Definition 4 we have

$$f'(a) = \lim_{h \to 0} \frac{f(a + h) - f(a)}{h}$$

$$= \lim_{h \to 0} \frac{[(a + h)^2 - 8(a + h) + 9] - [a^2 - 8a + 9]}{h}$$

$$= \lim_{h \to 0} \frac{a^2 + 2ah + h^2 - 8a - 8h + 9 - a^2 + 8a - 9}{h}$$

$$= \lim_{h \to 0} \frac{2ah + h^2 - 8h}{h} = \lim_{h \to 0} (2a + h - 8)$$

$$= 2a - 8 \qquad \blacksquare$$

We defined the tangent line to the curve $y = f(x)$ at the point $P(a, f(a))$ to be the line that passes through P and has slope m given by Equation 1 or 2. Since, by Definition 4, this is the same as the derivative $f'(a)$, we can now say the following.

> The tangent line to $y = f(x)$ at $(a, f(a))$ is the line through $(a, f(a))$ whose slope is equal to $f'(a)$, the derivative of f at a.

If we use the point-slope form of the equation of a line, we can write an equation of the tangent line to the curve $y = f(x)$ at the point $(a, f(a))$:

$$y - f(a) = f'(a)(x - a)$$

☑ EXAMPLE 5 Find an equation of the tangent line to the parabola $y = x^2 - 8x + 9$ at the point $(3, -6)$.

SOLUTION From Example 4 we know that the derivative of $f(x) = x^2 - 8x + 9$ at the number a is $f'(a) = 2a - 8$. Therefore the slope of the tangent line at $(3, -6)$ is $f'(3) = 2(3) - 8 = -2$. Thus an equation of the tangent line, shown in Figure 10, is

$$y - (-6) = (-2)(x - 3) \qquad \text{or} \qquad y = -2x \qquad \blacksquare$$

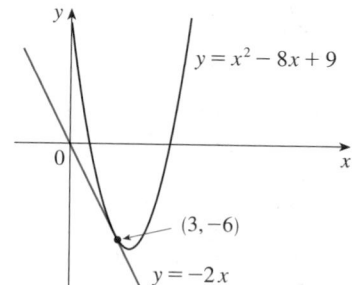

$y = x^2 - 8x + 9$

$(3, -6)$

$y = -2x$

FIGURE 10

RATES OF CHANGE

Suppose y is a quantity that depends on another quantity x. Thus y is a function of x and we write $y = f(x)$. If x changes from x_1 to x_2, then the change in x (also called the **increment** of x) is

$$\Delta x = x_2 - x_1$$

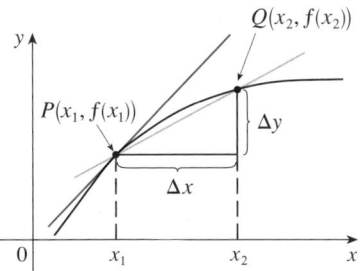

average rate of change $= m_{PQ}$

instantaneous rate of change =
                    slope of tangent at P

FIGURE 11

and the corresponding change in y is

$$\Delta y = f(x_2) - f(x_1)$$

The difference quotient

$$\frac{\Delta y}{\Delta x} = \frac{f(x_2) - f(x_1)}{x_2 - x_1}$$

is called the **average rate of change of y with respect to x** over the interval $[x_1, x_2]$ and can be interpreted as the slope of the secant line PQ in Figure 11.

By analogy with velocity, we consider the average rate of change over smaller and smaller intervals by letting x_2 approach x_1 and therefore letting Δx approach 0. The limit of these average rates of change is called the (**instantaneous**) **rate of change of y with respect to x** at $x = x_1$, which is interpreted as the slope of the tangent to the curve $y = f(x)$ at $P(x_1, f(x_1))$:

$$\boxed{6} \qquad \text{instantaneous rate of change} = \lim_{\Delta x \to 0} \frac{\Delta y}{\Delta x} = \lim_{x_2 \to x_1} \frac{f(x_2) - f(x_1)}{x_2 - x_1}$$

We recognize this limit as being the derivative $f'(x_1)$.

We know that one interpretation of the derivative $f'(a)$ is as the slope of the tangent line to the curve $y = f(x)$ when $x = a$. We now have a second interpretation:

> The derivative $f'(a)$ is the instantaneous rate of change of $y = f(x)$ with respect to x when $x = a$.

The connection with the first interpretation is that if we sketch the curve $y = f(x)$, then the instantaneous rate of change is the slope of the tangent to this curve at the point where $x = a$. This means that when the derivative is large (and therefore the curve is steep, as at the point P in Figure 12), the y-values change rapidly. When the derivative is small, the curve is relatively flat and the y-values change slowly.

In particular, if $s = f(t)$ is the position function of a particle that moves along a straight line, then $f'(a)$ is the rate of change of the displacement s with respect to the time t. In other words, $f'(a)$ *is the velocity of the particle at time $t = a$.* The **speed** of the particle is the absolute value of the velocity, that is, $|f'(a)|$.

In the following example we estimate the rate of change of the national debt with respect to time. Here the function is defined not by a formula but by a table of values.

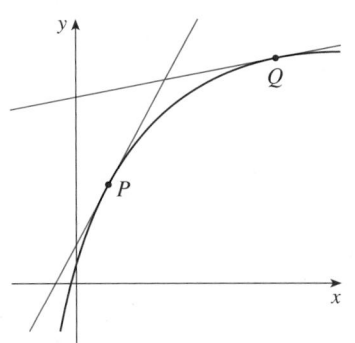

FIGURE 12

The y-values are changing rapidly at P and slowly at Q.

t	$D(t)$
1990	3,233.3
1995	4,974.0
2000	5,674.2
2005	7,932.7
2010	13,050.8

▼ **EXAMPLE 6** Let $D(t)$ be the US national debt at time t. The table in the margin gives approximate values of this function by providing end of year estimates, in billions of dollars, from 1990 to 2010. Interpret and estimate the value of $D'(2000)$.

SOLUTION The derivative $D'(2000)$ means the rate of change of D with respect to t when $t = 2000$, that is, the rate of increase of the national debt in 2000.

According to Equation 5,

$$D'(2000) = \lim_{t \to 2000} \frac{D(t) - D(2000)}{t - 2000}$$

So we compute and tabulate values of the difference quotient (the average rates of change) as follows.

t	$\dfrac{D(t) - D(2000)}{t - 2000}$
1990	244.09
1995	144.04
2005	451.70
2010	736.66

From this table we see that $D'(2000)$ lies somewhere between 140.04 and 451.70 billion dollars per year. [Here we are making the reasonable assumption that the debt didn't fluctuate wildly between 1995 and 2005.] We estimate that the rate of increase of the national debt of the United States in 2000 was the average of these two numbers, namely

$$D'(2000) \approx 296 \text{ billion dollars per year}$$

Another method would be to plot the debt function and estimate the slope of the tangent line when $t = 2000$. ■

The rate of change of the debt with respect to time in Example 6 is just one example of a rate of change. Here are a few of the many others:

The velocity of a particle is the rate of change of displacement with respect to time. Physicists are interested in other rates of change as well—for instance, the rate of change of work with respect to time (which is called *power*). Chemists who study a chemical reaction are interested in the rate of change in the concentration of a reactant with respect to time (called the *rate of reaction*). A steel manufacturer is interested in the rate of change of the cost of producing x tons of steel per day with respect to x (called the *marginal cost*). A biologist is interested in the rate of change of the population of a colony of bacteria with respect to time. In fact, the computation of rates of change is important in all of the natural sciences, in engineering, and even in the social sciences.

All these rates of change can be interpreted as slopes of tangents. This gives added significance to the solution of the tangent problem. Whenever we solve a problem involving tangent lines, we are not just solving a problem in geometry. We are also implicitly solving a great variety of problems involving rates of change in science and engineering.

■ **www.stewartcalculus.com**
See Additional Examples C, D.

2.1 | EXERCISES

1. (a) Find the slope of the tangent line to the parabola
$y = 4x - x^2$ at the point $(1, 3)$
  (i) using Definition 1   (ii) using Equation 2
(b) Find an equation of the tangent line in part (a).
(c) Graph the parabola and the tangent line. As a check on your work, zoom in toward the point $(1, 3)$ until the parabola and the tangent line are indistinguishable.

2. (a) Find the slope of the tangent line to the curve
$y = x - x^3$ at the point $(1, 0)$
  (i) using Definition 1   (ii) using Equation 2
(b) Find an equation of the tangent line in part (a).
(c) Graph the curve and the tangent line in successively smaller viewing rectangles centered at $(1, 0)$ until the curve and the line appear to coincide.

Graphing calculator or computer required   **CAS** Computer algebra system required   **1** Homework Hints at stewartcalculus.com

3–6 ▪ Find an equation of the tangent line to the curve at the given point.

3. $y = 4x - 3x^2$, $(2, -4)$ **4.** $y = x^3 - 3x + 1$, $(2, 3)$

5. $y = \sqrt{x}$, $(1, 1)$ **6.** $y = \dfrac{2x + 1}{x + 2}$, $(1, 1)$

7. (a) Find the slope of the tangent to the curve
$y = 3 + 4x^2 - 2x^3$ at the point where $x = a$.
  (b) Find equations of the tangent lines at the points $(1, 5)$ and $(2, 3)$.
  (c) Graph the curve and both tangents on a common screen.

8. (a) Find the slope of the tangent to the curve $y = 1/\sqrt{x}$ at the point where $x = a$.
  (b) Find equations of the tangent lines at the points $(1, 1)$ and $\left(4, \frac{1}{2}\right)$.
  (c) Graph the curve and both tangents on a common screen.

9. The graph shows the position function of a car. Use the shape of the graph to explain your answers to the following questions.
  (a) What was the initial velocity of the car?
  (b) Was the car going faster at B or at C?
  (c) Was the car slowing down or speeding up at A, B, and C?
  (d) What happened between D and E?

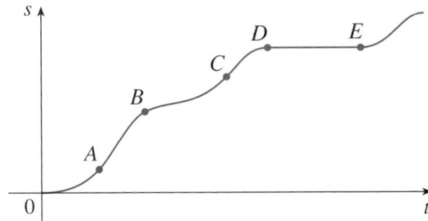

10. Shown are graphs of the position functions of two runners, A and B, who run a 100-m race and finish in a tie.

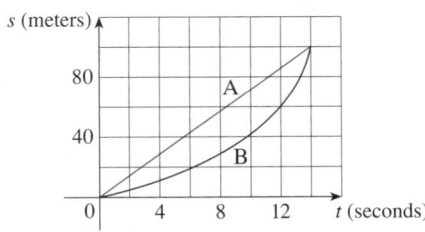

  (a) Describe and compare how the runners run the race.
  (b) At what time is the distance between the runners the greatest?
  (c) At what time do they have the same velocity?

11. If a ball is thrown into the air with a velocity of 40 ft/s, its height (in feet) after t seconds is given by $y = 40t - 16t^2$. Find the velocity when $t = 2$.

12. If an arrow is shot upward on the moon with a velocity of 58 m/s, its height (in meters) after t seconds is given by $H = 58t - 0.83t^2$.
  (a) Find the velocity of the arrow after one second.
  (b) Find the velocity of the arrow when $t = a$.
  (c) When will the arrow hit the moon?
  (d) With what velocity will the arrow hit the moon?

13. The displacement (in meters) of a particle moving in a straight line is given by the equation of motion $s = 1/t^2$, where t is measured in seconds. Find the velocity of the particle at times $t = a$, $t = 1$, $t = 2$, and $t = 3$.

14. The displacement (in meters) of a particle moving in a straight line is given by $s = t^2 - 8t + 18$, where t is measured in seconds.
  (a) Find the average velocity over each time interval:
    (i) $[3, 4]$ (ii) $[3.5, 4]$
    (iii) $[4, 5]$ (iv) $[4, 4.5]$
  (b) Find the instantaneous velocity when $t = 4$.
  (c) Draw the graph of s as a function of t and draw the secant lines whose slopes are the average velocities in part (a) and the tangent line whose slope is the instantaneous velocity in part (b).

15. For the function g whose graph is given, arrange the following numbers in increasing order and explain your reasoning:

$$0 \quad g'(-2) \quad g'(0) \quad g'(2) \quad g'(4)$$

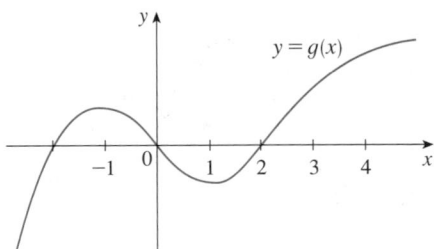

16. Find an equation of the tangent line to the graph of $y = g(x)$ at $x = 5$ if $g(5) = -3$ and $g'(5) = 4$.

17. If an equation of the tangent line to the curve $y = f(x)$ at the point where $a = 2$ is $y = 4x - 5$, find $f(2)$ and $f'(2)$.

18. If the tangent line to $y = f(x)$ at $(4, 3)$ passes through the point $(0, 2)$, find $f(4)$ and $f'(4)$.

19. Sketch the graph of a function f for which $f(0) = 0$, $f'(0) = 3$, $f'(1) = 0$, and $f'(2) = -1$.

20. Sketch the graph of a function g for which $g(0) = g(2) = g(4) = 0$, $g'(1) = g'(3) = 0$, $g'(0) = g'(4) = 1$, $g'(2) = -1$, $\lim_{x \to \infty} g(x) = \infty$, and $\lim_{x \to -\infty} g(x) = -\infty$.

21. If $f(x) = 3x^2 - x^3$, find $f'(1)$ and use it to find an equation of the tangent line to the curve $y = 3x^2 - x^3$ at the point $(1, 2)$.

22. If $g(x) = x^4 - 2$, find $g'(1)$ and use it to find an equation of the tangent line to the curve $y = x^4 - 2$ at the point $(1, -1)$.

23. (a) If $F(x) = 5x/(1 + x^2)$, find $F'(2)$ and use it to find an equation of the tangent line to the curve

$$y = \frac{5x}{1 + x^2}$$

at the point $(2, 2)$.

 (b) Illustrate part (a) by graphing the curve and the tangent line on the same screen.

24. (a) If $G(x) = 4x^2 - x^3$, find $G'(a)$ and use it to find equations of the tangent lines to the curve $y = 4x^2 - x^3$ at the points $(2, 8)$ and $(3, 9)$.

 (b) Illustrate part (a) by graphing the curve and the tangent lines on the same screen.

25–30 ▪ Find $f'(a)$.

25. $f(x) = 3x^2 - 4x + 1$

26. $f(t) = 2t^3 + t$

27. $f(t) = \dfrac{2t + 1}{t + 3}$

28. $f(x) = x^{-2}$

29. $f(x) = \sqrt{1 - 2x}$

30. $f(x) = \dfrac{4}{\sqrt{1 - x}}$

31–36 ▪ Each limit represents the derivative of some function f at some number a. State such an f and a in each case.

31. $\displaystyle\lim_{h \to 0} \frac{(1 + h)^{10} - 1}{h}$

32. $\displaystyle\lim_{h \to 0} \frac{\sqrt[4]{16 + h} - 2}{h}$

33. $\displaystyle\lim_{x \to 5} \frac{2^x - 32}{x - 5}$

34. $\displaystyle\lim_{x \to \pi/4} \frac{\tan x - 1}{x - \pi/4}$

35. $\displaystyle\lim_{h \to 0} \frac{\cos(\pi + h) + 1}{h}$

36. $\displaystyle\lim_{t \to 1} \frac{t^4 + t - 2}{t - 1}$

37. A warm can of soda is placed in a cold refrigerator. Sketch the graph of the temperature of the soda as a function of time. Is the initial rate of change of temperature greater or less than the rate of change after an hour?

38. A roast turkey is taken from an oven when its temperature has reached 185°F and is placed on a table in a room where the temperature is 75°F. The graph shows how the temperature of the turkey decreases and eventually approaches room temperature. By measuring the slope of the tangent, estimate the rate of change of the temperature after an hour.

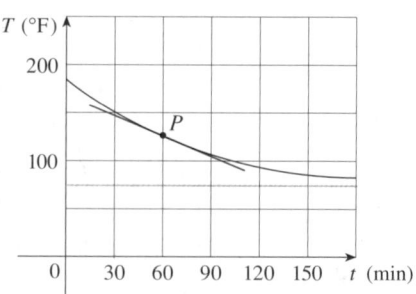

39. The number N of US cellular phone subscribers (in millions) is shown in the table. (Midyear estimates are given.)

t	1996	1998	2000	2002	2004	2006
N	44	69	109	141	182	233

(a) Find the average rate of cell phone growth
  (i) from 2002 to 2006   (ii) from 2002 to 2004
  (iii) from 2000 to 2002
  In each case, include the units.

(b) Estimate the instantaneous rate of growth in 2002 by taking the average of two average rates of change. What are its units?

(c) Estimate the instantaneous rate of growth in 2002 by measuring the slope of a tangent.

40. The number N of locations of a popular coffeehouse chain is given in the table. (The numbers of locations as of October 1 are given.)

Year	2004	2005	2006	2007	2008
N	8569	10,241	12,440	15,011	16,680

(a) Find the average rate of growth
  (i) from 2006 to 2008   (ii) from 2006 to 2007
  (iii) from 2005 to 2006
  In each case, include the units.

(b) Estimate the instantaneous rate of growth in 2006 by taking the average of two average rates of change. What are its units?

(c) Estimate the instantaneous rate of growth in 2006 by measuring the slope of a tangent.

(d) Estimate the intantaneous rate of growth in 2007 and compare it with the growth rate in 2006. What do you conclude?

41. The cost (in dollars) of producing x units of a certain commodity is $C(x) = 5000 + 10x + 0.05x^2$.

(a) Find the average rate of change of C with respect to x when the production level is changed
  (i) from $x = 100$ to $x = 105$
  (ii) from $x = 100$ to $x = 101$

(b) Find the instantaneous rate of change of C with respect to x when $x = 100$. (This is called the *marginal cost*. Its significance will be explained in Section 2.3.)

42. If a cylindrical tank holds 100,000 gallons of water, which can be drained from the bottom of the tank in an hour, then Torricelli's Law gives the volume V of water remaining in the tank after t minutes as

$$V(t) = 100,000\left(1 - \tfrac{1}{60}t\right)^2 \qquad 0 \le t \le 60$$

Find the rate at which the water is flowing out of the tank (the instantaneous rate of change of V with respect to t) as a function of t. What are its units? For times $t = 0, 10, 20, 30, 40, 50,$ and 60 min, find the flow rate and the amount of water remaining in the tank. Summarize your findings in a sentence or two. At what time is the flow rate the greatest? The least?

43. The cost of producing x ounces of gold from a new gold mine is $C = f(x)$ dollars.
   (a) What is the meaning of the derivative $f'(x)$? What are its units?
   (b) What does the statement $f'(800) = 17$ mean?
   (c) Do you think the values of $f'(x)$ will increase or decrease in the short term? What about the long term? Explain.

44. The number of bacteria after t hours in a controlled laboratory experiment is $n = f(t)$.
   (a) What is the meaning of the derivative $f'(5)$? What are its units?
   (b) Suppose there is an unlimited amount of space and nutrients for the bacteria. Which do you think is larger, $f'(5)$ or $f'(10)$? If the supply of nutrients is limited, would that affect your conclusion? Explain.

45. Let $T(t)$ be the temperature (in °F) in Phoenix t hours after midnight on September 10, 2008. The table shows values of this function recorded every two hours. What is the meaning of $T'(8)$? Estimate its value.

t	0	2	4	6	8	10	12	14
T	82	75	74	75	84	90	93	94

46. The quantity (in pounds) of a gourmet ground coffee that is sold by a coffee company at a price of p dollars per pound is $Q = f(p)$.
   (a) What is the meaning of the derivative $f'(8)$? What are its units?
   (b) Is $f'(8)$ positive or negative? Explain.

47. The quantity of oxygen that can dissolve in water depends on the temperature of the water. (So thermal pollution influences the oxygen content of water.) The graph shows how oxygen solubility S varies as a function of the water temperature T.
   (a) What is the meaning of the derivative $S'(T)$? What are its units?
   (b) Estimate the value of $S'(16)$ and interpret it.

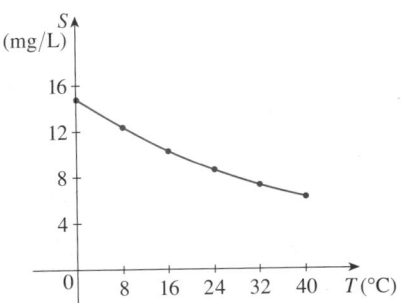

Adapted from Kupchella & Hyland, *Environmental Science: Living Within the System of Nature*, 2d ed.; © 1989. Printed and electronically reproduced by permission of Pearson Education, Inc., Upper Saddle River, NJ.

48. The graph shows the influence of the temperature T on the maximum sustainable swimming speed S of Coho salmon.
   (a) What is the meaning of the derivative $S'(T)$? What are its units?
   (b) Estimate the values of $S'(15)$ and $S'(25)$ and interpret them.

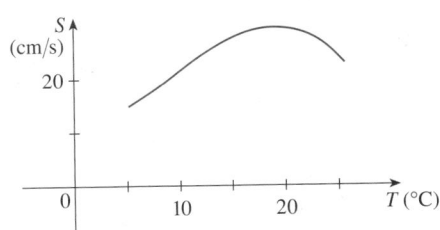

49–50 ▪ Determine whether $f'(0)$ exists.

49. $f(x) = \begin{cases} x \sin \dfrac{1}{x} & \text{if } x \ne 0 \\ 0 & \text{if } x = 0 \end{cases}$

50. $f(x) = \begin{cases} x^2 \sin \dfrac{1}{x} & \text{if } x \ne 0 \\ 0 & \text{if } x = 0 \end{cases}$

2.2 | THE DERIVATIVE AS A FUNCTION

In Section 2.1 we considered the derivative of a function f at a fixed number a:

$$\boxed{1} \qquad f'(a) = \lim_{h \to 0} \frac{f(a + h) - f(a)}{h}$$

Here we change our point of view and let the number a vary. If we replace a in Equation 1 by a variable x, we obtain

$$\boxed{2} \qquad f'(x) = \lim_{h \to 0} \frac{f(x + h) - f(x)}{h}$$

Given any number x for which this limit exists, we assign to x the number $f'(x)$. So we can regard f' as a new function, called the **derivative of f** and defined by Equation 2. We know that the value of f' at x, $f'(x)$, can be interpreted geometrically as the slope of the tangent line to the graph of f at the point $(x, f(x))$.

The function f' is called the derivative of f because it has been "derived" from f by the limiting operation in Equation 2. The domain of f' is the set $\{x \mid f'(x) \text{ exists}\}$ and may be smaller than the domain of f.

V EXAMPLE 1 The graph of a function f is given in Figure 1. Use it to sketch the graph of the derivative f'.

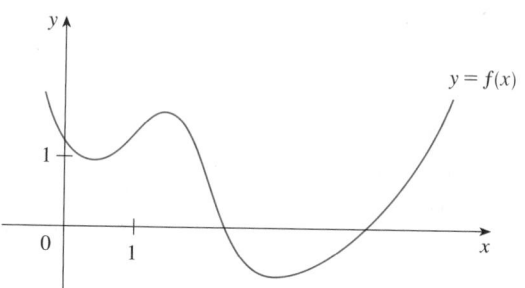

FIGURE 1

SOLUTION We can estimate the value of the derivative at any value of x by drawing the tangent at the point $(x, f(x))$ and estimating its slope. For instance, for $x = 5$ we draw the tangent at P in Figure 2(a) and estimate its slope to be about $\frac{3}{2}$, so $f'(5) \approx 1.5$. This allows us to plot the point $P'(5, 1.5)$ on the graph of f' directly beneath P. Repeating this procedure at several points, we get the graph shown in Figure 2(b). Notice that the tangents at A, B, and C are horizontal, so the derivative is 0 there and the graph of f' crosses the x-axis at the points A', B', and C', directly beneath A, B, and C. Between A and B the tangents have positive slope, so $f'(x)$ is positive there. But between B and C the tangents have negative slope, so $f'(x)$ is negative there.

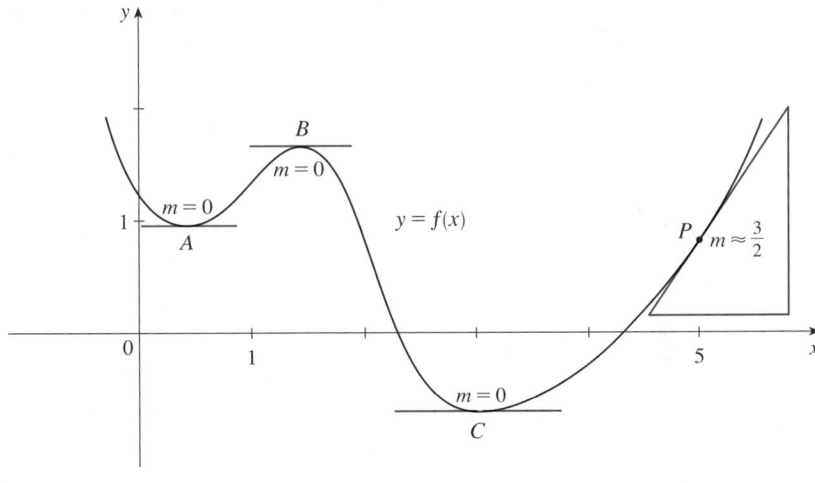

(a)

TEC Visual 2.2 shows an animation of Figure 2 for several functions.

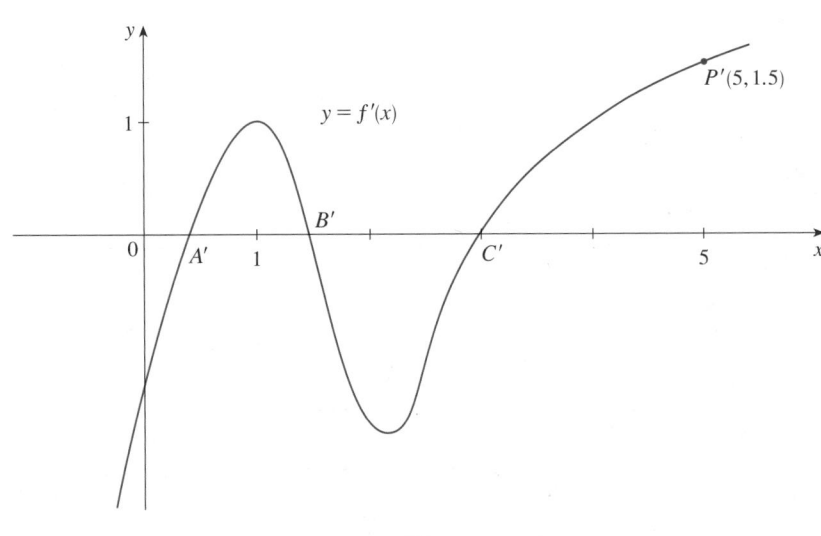

FIGURE 2   (b)   ∎

▼ EXAMPLE 2

(a) If $f(x) = x^3 - x$, find a formula for $f'(x)$.

(b) Illustrate by comparing the graphs of f and f'.

SOLUTION

(a) When using Equation 2 to compute a derivative, we must remember that the variable is h and that x is temporarily regarded as a constant during the calculation of the limit.

$$f'(x) = \lim_{h \to 0} \frac{f(x + h) - f(x)}{h} = \lim_{h \to 0} \frac{[(x + h)^3 - (x + h)] - [x^3 - x]}{h}$$

$$= \lim_{h \to 0} \frac{x^3 + 3x^2h + 3xh^2 + h^3 - x - h - x^3 + x}{h}$$

$$= \lim_{h \to 0} \frac{3x^2h + 3xh^2 + h^3 - h}{h} = \lim_{h \to 0} (3x^2 + 3xh + h^2 - 1) = 3x^2 - 1$$

(b) We use a graphing device to graph f and f' in Figure 3. Notice that $f'(x) = 0$ when f has horizontal tangents and $f'(x)$ is positive when the tangents have positive slopes. So these graphs serve as a check on our work in part (a).

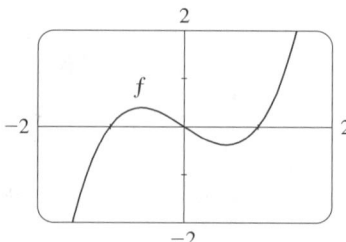 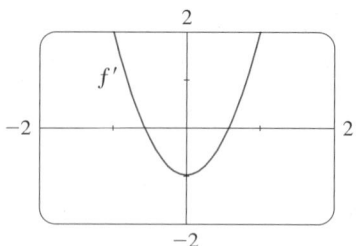

FIGURE 3

EXAMPLE 3 If $f(x) = \sqrt{x}$, find the derivative of f. State the domain of f'.

SOLUTION

$$f'(x) = \lim_{h \to 0} \frac{f(x+h) - f(x)}{h} = \lim_{h \to 0} \frac{\sqrt{x+h} - \sqrt{x}}{h}$$

Here we rationalize the numerator.

$$= \lim_{h \to 0} \left(\frac{\sqrt{x+h} - \sqrt{x}}{h} \cdot \frac{\sqrt{x+h} + \sqrt{x}}{\sqrt{x+h} + \sqrt{x}} \right)$$

$$= \lim_{h \to 0} \frac{(x+h) - x}{h(\sqrt{x+h} + \sqrt{x})} = \lim_{h \to 0} \frac{1}{\sqrt{x+h} + \sqrt{x}}$$

$$= \frac{1}{\sqrt{x} + \sqrt{x}} = \frac{1}{2\sqrt{x}}$$

We see that $f'(x)$ exists if $x > 0$, so the domain of f' is $(0, \infty)$. This is smaller than the domain of f, which is $[0, \infty)$.

Let's check to see that the result of Example 3 is reasonable by looking at the graphs of f and f' in Figure 4. When x is close to 0, $\sqrt{x}$ is also close to 0, so $f'(x) = 1/(2\sqrt{x})$ is very large and this corresponds to the steep tangent lines near $(0, 0)$ in Figure 4(a) and the large values of $f'(x)$ just to the right of 0 in Figure 4(b). When x is large, $f'(x)$ is very small and this corresponds to the flatter tangent lines at the far right of the graph of f and the horizontal asymptote of the graph of f'.

EXAMPLE 4 Find f' if $f(x) = \dfrac{1 - x}{2 + x}$.

SOLUTION

$$f'(x) = \lim_{h \to 0} \frac{f(x+h) - f(x)}{h} = \lim_{h \to 0} \frac{\dfrac{1 - (x+h)}{2 + (x+h)} - \dfrac{1 - x}{2 + x}}{h}$$

$$= \lim_{h \to 0} \frac{(1 - x - h)(2 + x) - (1 - x)(2 + x + h)}{h(2 + x + h)(2 + x)}$$

$$= \lim_{h \to 0} \frac{(2 - x - 2h - x^2 - xh) - (2 - x + h - x^2 - xh)}{h(2 + x + h)(2 + x)}$$

$$= \lim_{h \to 0} \frac{-3h}{h(2 + x + h)(2 + x)} = \lim_{h \to 0} \frac{-3}{(2 + x + h)(2 + x)} = -\frac{3}{(2 + x)^2} \quad \blacksquare$$

(a) $f(x) = \sqrt{x}$

(b) $f'(x) = \dfrac{1}{2\sqrt{x}}$

FIGURE 4

$$\frac{\dfrac{a}{b} - \dfrac{c}{d}}{e} = \frac{ad - bc}{bd} \cdot \frac{1}{e}$$

OTHER NOTATIONS

If we use the traditional notation $y = f(x)$ to indicate that the independent variable is x and the dependent variable is y, then some common alternative notations for the derivative are as follows:

$$f'(x) = y' = \frac{dy}{dx} = \frac{df}{dx} = \frac{d}{dx} f(x) = Df(x) = D_x f(x)$$

The symbols D and d/dx are called **differentiation operators** because they indicate the operation of **differentiation**, which is the process of calculating a derivative.

The symbol dy/dx, which was introduced by Leibniz, should not be regarded as a ratio (for the time being); it is simply a synonym for $f'(x)$. Nonetheless, it is a very useful and suggestive notation, especially when used in conjunction with increment notation. Referring to Equation 2.1.6, we can rewrite the definition of derivative in Leibniz notation in the form

$$\frac{dy}{dx} = \lim_{\Delta x \to 0} \frac{\Delta y}{\Delta x}$$

If we want to indicate the value of a derivative dy/dx in Leibniz notation at a specific number a, we use the notation

$$\frac{dy}{dx}\bigg|_{x=a} \qquad \text{or} \qquad \frac{dy}{dx}\bigg]_{x=a}$$

which is a synonym for $f'(a)$.

DIFFERENTIABLE FUNCTIONS

> **3 DEFINITION** A function f is **differentiable at a** if $f'(a)$ exists. It is **differentiable on an open interval** (a, b) [or (a, ∞) or $(-\infty, a)$ or $(-\infty, \infty)$] if it is differentiable at every number in the interval.

☑ **EXAMPLE 5** Where is the function $f(x) = |x|$ differentiable?

SOLUTION If $x > 0$, then $|x| = x$ and we can choose h small enough that $x + h > 0$ and hence $|x + h| = x + h$. Therefore, for $x > 0$, we have

$$f'(x) = \lim_{h \to 0} \frac{|x + h| - |x|}{h}$$

$$= \lim_{h \to 0} \frac{(x + h) - x}{h} = \lim_{h \to 0} \frac{h}{h} = \lim_{h \to 0} 1 = 1$$

and so f is differentiable for any $x > 0$.

Similarly, for $x < 0$ we have $|x| = -x$ and h can be chosen small enough that $x + h < 0$ and so $|x + h| = -(x + h)$. Therefore, for $x < 0$,

$$f'(x) = \lim_{h \to 0} \frac{|x + h| - |x|}{h}$$

$$= \lim_{h \to 0} \frac{-(x + h) - (-x)}{h} = \lim_{h \to 0} \frac{-h}{h} = \lim_{h \to 0} (-1) = -1$$

and so f is differentiable for any $x < 0$.

■ **LEIBNIZ**

Gottfried Wilhelm Leibniz was born in Leipzig in 1646 and studied law, theology, philosophy, and mathematics at the university there, graduating with a bachelor's degree at age 17. After earning his doctorate in law at age 20, Leibniz entered the diplomatic service and spent most of his life traveling to the capitals of Europe on political missions. In particular, he worked to avert a French military threat against Germany and attempted to reconcile the Catholic and Protestant churches.

His serious study of mathematics did not begin until 1672 while he was on a diplomatic mission in Paris. There he built a calculating machine and met scientists, like Huygens, who directed his attention to the latest developments in mathematics and science. Leibniz sought to develop a symbolic logic and system of notation that would simplify logical reasoning. In particular, the version of calculus that he published in 1684 established the notation and the rules for finding derivatives that we use today.

Unfortunately, a dreadful priority dispute arose in the 1690s between the followers of Newton and those of Leibniz as to who had invented calculus first. Leibniz was even accused of plagiarism by members of the Royal Society in England. The truth is that each man invented calculus independently. Newton arrived at his version of calculus first but, because of his fear of controversy, did not publish it immediately. So Leibniz's 1684 account of calculus was the first to be

For $x = 0$ we have to investigate

$$f'(0) = \lim_{h \to 0} \frac{f(0 + h) - f(0)}{h}$$

$$= \lim_{h \to 0} \frac{|0 + h| - |0|}{h} \qquad \text{(if it exists)}$$

Let's compute the left and right limits separately:

$$\lim_{h \to 0^+} \frac{|0 + h| - |0|}{h} = \lim_{h \to 0^+} \frac{|h|}{h} = \lim_{h \to 0^+} \frac{h}{h} = \lim_{h \to 0^+} 1 = 1$$

and

$$\lim_{h \to 0^-} \frac{|0 + h| - |0|}{h} = \lim_{h \to 0^-} \frac{|h|}{h} = \lim_{h \to 0^-} \frac{-h}{h} = \lim_{h \to 0^-} (-1) = -1$$

Since these limits are different, $f'(0)$ does not exist. Thus f is differentiable at all x except 0.

A formula for f' is given by

$$f'(x) = \begin{cases} 1 & \text{if } x > 0 \\ -1 & \text{if } x < 0 \end{cases}$$

and its graph is shown in Figure 5(b). The fact that $f'(0)$ does not exist is reflected geometrically in the fact that the curve $y = |x|$ does not have a tangent line at $(0, 0)$. [See Figure 5(a).] ∎

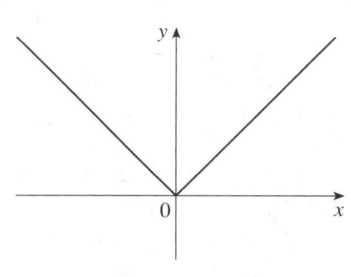

(a) $y = f(x) = |x|$

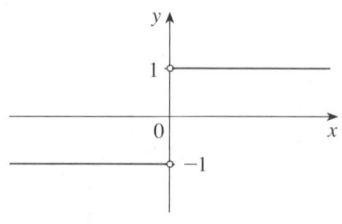

(b) $y = f'(x)$

FIGURE 5

Both continuity and differentiability are desirable properties for a function to have. The following theorem shows how these properties are related.

---

4 **THEOREM** If f is differentiable at a, then f is continuous at a.

PROOF To prove that f is continuous at a, we have to show that $\lim_{x \to a} f(x) = f(a)$. We do this by showing that the difference $f(x) - f(a)$ approaches 0.

The given information is that f is differentiable at a, that is,

$$f'(a) = \lim_{x \to a} \frac{f(x) - f(a)}{x - a}$$

exists (see Equation 2.1.5). To connect the given and the unknown, we divide and multiply $f(x) - f(a)$ by $x - a$ (which we can do when $x \neq a$):

$$f(x) - f(a) = \frac{f(x) - f(a)}{x - a} (x - a)$$

Thus, using the Product Law and (2.1.5), we can write

$$\lim_{x \to a} [f(x) - f(a)] = \lim_{x \to a} \frac{f(x) - f(a)}{x - a} (x - a)$$

$$= \lim_{x \to a} \frac{f(x) - f(a)}{x - a} \cdot \lim_{x \to a} (x - a)$$

$$= f'(a) \cdot 0 = 0$$

To use what we have just proved, we start with $f(x)$ and add and subtract $f(a)$:

$$\lim_{x \to a} f(x) = \lim_{x \to a} \left[f(a) + (f(x) - f(a)) \right]$$

$$= \lim_{x \to a} f(a) + \lim_{x \to a} \left[f(x) - f(a) \right]$$

$$= f(a) + 0 = f(a)$$

Therefore f is continuous at a. $\qquad\qquad\qquad\qquad\qquad\qquad\qquad\quad \square$

⊘   **NOTE** The converse of Theorem 4 is false; that is, there are functions that are continuous but not differentiable. For instance, the function $f(x) = |x|$ is continuous at 0 because

$$\lim_{x \to 0} f(x) = \lim_{x \to 0} |x| = 0 = f(0)$$

(See Example 6 in Section 1.4.) But in Example 5 we showed that f is not differentiable at 0.

HOW CAN A FUNCTION FAIL TO BE DIFFERENTIABLE?

We saw that the function $y = |x|$ in Example 5 is not differentiable at 0 and Figure 5(a) shows that its graph changes direction abruptly when $x = 0$. In general, if the graph of a function f has a "corner" or "kink" in it, then the graph of f has no tangent at this point and f is not differentiable there. [In trying to compute $f'(a)$, we find that the left and right limits are different.]

Theorem 4 gives another way for a function not to have a derivative. It says that if f is not continuous at a, then f is not differentiable at a. So at any discontinuity (for instance, a jump discontinuity) f fails to be differentiable.

A third possibility is that the curve has a **vertical tangent line** when $x = a$; that is, f is continuous at a and

$$\lim_{x \to a} |f'(x)| = \infty$$

This means that the tangent lines become steeper and steeper as $x \to a$. Figure 6 shows one way that this can happen; Figure 7(c) shows another. Figure 7 illustrates the three possibilities that we have discussed.

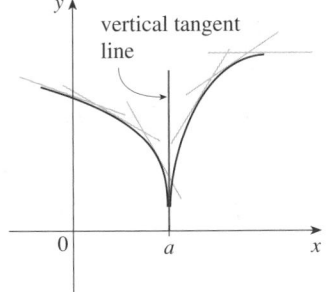

FIGURE 6

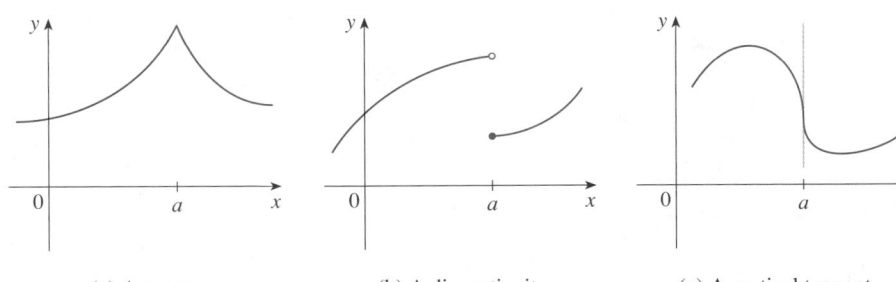

FIGURE 7
Three ways for f not to be
differentiable at a

(a) A corner   (b) A discontinuity   (c) A vertical tangent

A graphing calculator or computer provides another way of looking at differentiability. If f is differentiable at a, then when we zoom in toward the point $(a, f(a))$ the graph straightens out and appears more and more like a line. (See Figure 8. We saw a specific example of this in Figure 4 in Section 2.1.) But no matter how much we zoom in toward a point like the ones in Figures 6 and 7(a), we can't eliminate the sharp point or corner (see Figure 9).

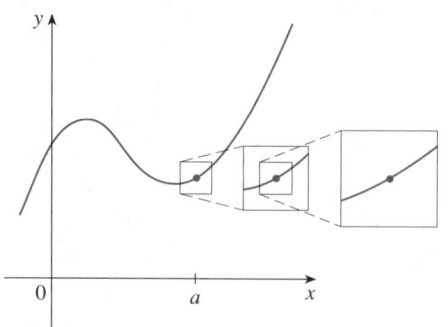

FIGURE 8
f is differentiable at a.

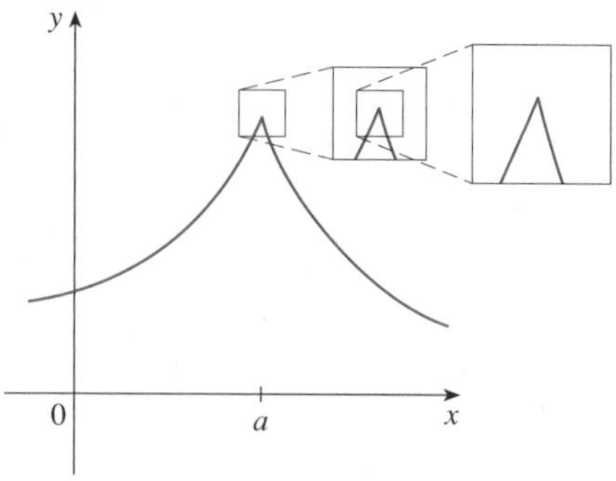

FIGURE 9
f is not differentiable at a.

HIGHER DERIVATIVES

If f is a differentiable function, then its derivative f' is also a function, so f' may have a derivative of its own, denoted by $(f')' = f''$. This new function f'' is called the **second derivative** of f because it is the derivative of the derivative of f. Using Leibniz notation, we write the second derivative of $y = f(x)$ as

$$\frac{d}{dx}\left(\frac{dy}{dx}\right) = \frac{d^2y}{dx^2}$$

EXAMPLE 6 If $f(x) = x^3 - x$, find and interpret $f''(x)$.

SOLUTION In Example 2 we found that the first derivative is $f'(x) = 3x^2 - 1$. So the second derivative is

$$f''(x) = \lim_{h \to 0} \frac{f'(x + h) - f'(x)}{h}$$

$$= \lim_{h \to 0} \frac{[3(x + h)^2 - 1] - [3x^2 - 1]}{h}$$

$$= \lim_{h \to 0} \frac{3x^2 + 6xh + 3h^2 - 1 - 3x^2 + 1}{h}$$

$$= \lim_{h \to 0} (6x + 3h) = 6x$$

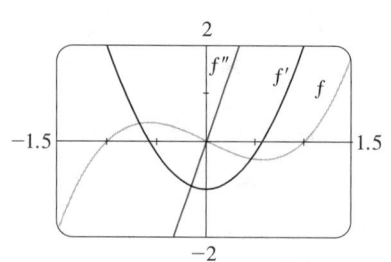

FIGURE 10

TEC In Module 2.2 you can see how changing the coefficients of a polynomial f affects the appearance of the graphs of f, f', and f''.

The graphs of f, f', f'' are shown in Figure 10.

We can interpret $f''(x)$ as the slope of the curve $y = f'(x)$ at the point $(x, f'(x))$. In other words, it is the rate of change of the slope of the original curve $y = f(x)$.

Notice from Figure 10 that $f''(x)$ is negative when $y = f'(x)$ has negative slope and positive when $y = f'(x)$ has positive slope. So the graphs serve as a check on our calculations. ∎

In general, we can interpret a second derivative as a rate of change of a rate of change. The most familiar example of this is *acceleration,* which we define as follows.

If $s = s(t)$ is the position function of an object that moves in a straight line, we know that its first derivative represents the velocity $v(t)$ of the object as a function of time:

$$v(t) = s'(t) = \frac{ds}{dt}$$

The instantaneous rate of change of velocity with respect to time is called the **acceleration** $a(t)$ of the object. Thus the acceleration function is the derivative of the velocity function and is therefore the second derivative of the position function:

$$a(t) = v'(t) = s''(t)$$

▪ www.stewartcalculus.com
See Additional Example A.

or, in Leibniz notation,

$$a = \frac{dv}{dt} = \frac{d^2s}{dt^2}$$

The **third derivative** f''' is the derivative of the second derivative: $f''' = (f'')'$. So $f'''(x)$ can be interpreted as the slope of the curve $y = f''(x)$ or as the rate of change of $f''(x)$. If $y = f(x)$, then alternative notations for the third derivative are

$$y''' = f'''(x) = \frac{d}{dx}\left(\frac{d^2y}{dx^2}\right) = \frac{d^3y}{dx^3}$$

The process can be continued. The fourth derivative f'''' is usually denoted by $f^{(4)}$. In general, the nth derivative of f is denoted by $f^{(n)}$ and is obtained from f by differentiating n times. If $y = f(x)$, we write

$$y^{(n)} = f^{(n)}(x) = \frac{d^n y}{dx^n}$$

EXAMPLE 7 If $f(x) = x^3 - x$, find $f'''(x)$ and $f^{(4)}(x)$.

SOLUTION In Example 6 we found that $f''(x) = 6x$. The graph of the second derivative has equation $y = 6x$ and so it is a straight line with slope 6. Since the derivative $f'''(x)$ is the slope of $f''(x)$, we have

$$f'''(x) = 6$$

for all values of x. So f''' is a constant function and its graph is a horizontal line. Therefore, for all values of x,

$$f^{(4)}(x) = 0 \qquad \blacksquare$$

We have seen that one application of second derivatives occurs in analyzing the motion of objects using acceleration. We will investigate another application of second derivatives in Section 4.3, where we show how knowledge of f'' gives us information about the shape of the graph of f. In Section 8.7 we will see how second and higher derivatives enable us to represent functions as sums of infinite series.

2.2 | EXERCISES

1–2 ■ Use the given graph to estimate the value of each derivative. Then sketch the graph of f'.

1. (a) $f'(-3)$
(b) $f'(-2)$
(c) $f'(-1)$
(d) $f'(0)$
(e) $f'(1)$
(f) $f'(2)$
(g) $f'(3)$

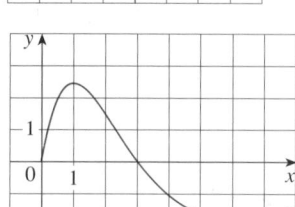

2. (a) $f'(0)$
(b) $f'(1)$
(c) $f'(2)$
(d) $f'(3)$
(e) $f'(4)$
(f) $f'(5)$
(g) $f'(6)$
(h) $f'(7)$

3. Match the graph of each function in (a)–(d) with the graph
of its derivative in I–IV. Give reasons for your choices.

(a)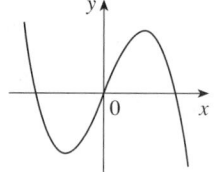

(b)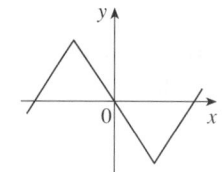

(c)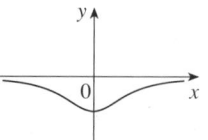

(d)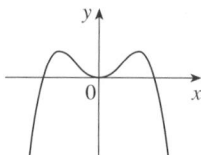

I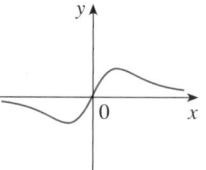

II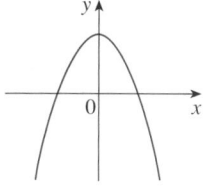

III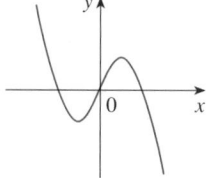

IV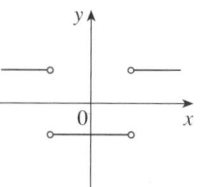

4–11 ■ Trace or copy the graph of the given function f.
(Assume that the axes have equal scales.) Then use the method
of Example 1 to sketch the graph of f' below it.

4.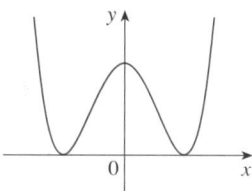

5.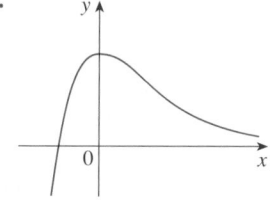

6.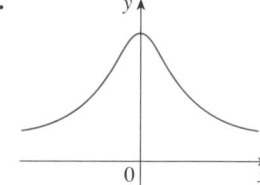

7.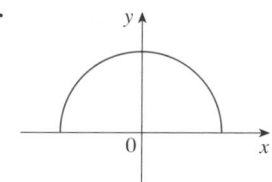

8.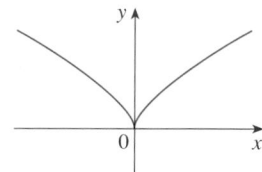

9.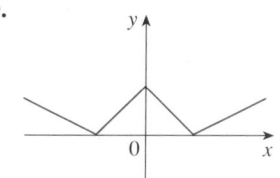

10.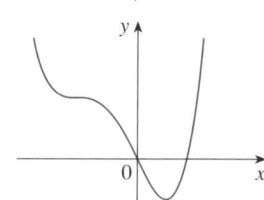

11.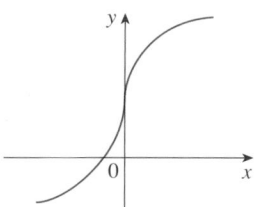

12. Shown is the graph of the population function $P(t)$ for yeast
cells in a laboratory culture. Use the method of Example 1 to

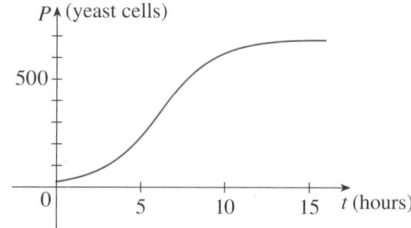

graph the derivative $P'(t)$. What does the graph of P' tell us about the yeast population?

13. A rechargeable battery is plugged into a charger. The graph shows $C(t)$, the percentage of full capacity that the battery reaches as a function of time t elapsed (in hours).
 (a) What is the meaning of the derivative $C'(t)$?
 (b) Sketch the graph of $C'(t)$. What does the graph tell you?

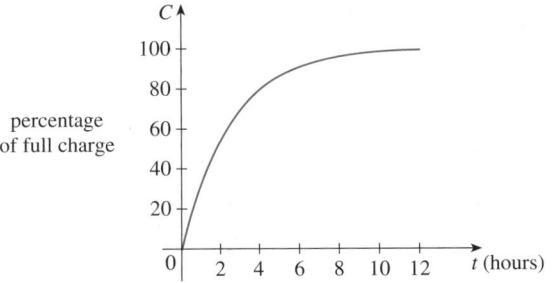

14. The graph (from the US Department of Energy) shows how driving speed affects gas mileage. Fuel economy F is measured in miles per gallon and speed v is measured in miles per hour.
 (a) What is the meaning of the derivative $F'(v)$?
 (b) Sketch the graph of $F'(v)$.
 (c) At what speed should you drive if you want to save on gas?

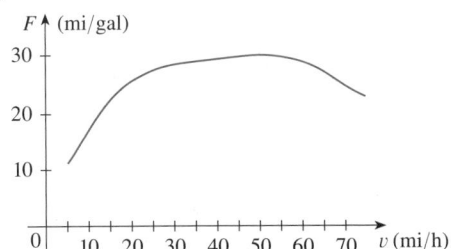

15. The graph shows how the average age of first marriage of Japanese men varied in the last half of the 20th century. Sketch the graph of the derivative function $M'(t)$. During which years was the derivative negative?

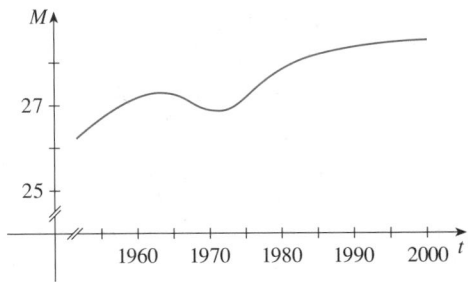

16. Make a careful sketch of the graph of the sine function and below it sketch the graph of its derivative in the same man-

ner as in Exercises 4–11. Can you guess what the derivative of the sine function is from its graph?

17. Let $f(x) = x^2$.
 (a) Estimate the values of $f'(0)$, $f'\left(\frac{1}{2}\right)$, $f'(1)$, and $f'(2)$ by using a graphing device to zoom in on the graph of f.
 (b) Use symmetry to deduce the values of $f'\left(-\frac{1}{2}\right)$, $f'(-1)$, and $f'(-2)$.
 (c) Use the results from parts (a) and (b) to guess a formula for $f'(x)$.
 (d) Use the definition of a derivative to prove that your guess in part (c) is correct.

18. Let $f(x) = x^3$.
 (a) Estimate the values of $f'(0)$, $f'\left(\frac{1}{2}\right)$, $f'(1)$, $f'(2)$, and $f'(3)$ by using a graphing device to zoom in on the graph of f.
 (b) Use symmetry to deduce the values of $f'\left(-\frac{1}{2}\right)$, $f'(-1)$, $f'(-2)$, and $f'(-3)$.
 (c) Use the values from parts (a) and (b) to graph f'.
 (d) Guess a formula for $f'(x)$.
 (e) Use the definition of a derivative to prove that your guess in part (d) is correct.

19–27 ■ Find the derivative of the function using the definition of derivative. State the domain of the function and the domain of its derivative.

19. $f(x) = \frac{1}{2}x - \frac{1}{3}$

20. $f(x) = 1.5x^2 - x + 3.7$

21. $f(x) = x^2 - 2x^3$

22. $g(t) = \dfrac{1}{\sqrt{t}}$

23. $g(x) = \sqrt{9 - x}$

24. $f(x) = \dfrac{x^2 - 1}{2x - 3}$

25. $G(t) = \dfrac{1 - 2t}{3 + t}$

26. $f(x) = x^{3/2}$

27. $f(x) = x^4$

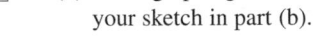

28. (a) Sketch the graph of $f(x) = \sqrt{6 - x}$ by starting with the graph of $y = \sqrt{x}$ and using the transformations of Section 1.2.
 (b) Use the graph from part (a) to sketch the graph of f'.
 (c) Use the definition of a derivative to find $f'(x)$. What are the domains of f and f'?
 (d) Use a graphing device to graph f' and compare with your sketch in part (b).

29. (a) If $f(x) = x^4 + 2x$, find $f'(x)$.
 (b) Check to see that your answer to part (a) is reasonable by comparing the graphs of f and f'.

30. (a) If $f(x) = x + 1/x$, find $f'(x)$.
 (b) Check to see that your answer to part (a) is reasonable by comparing the graphs of f and f'.

31. The unemployment rate $U(t)$ varies with time. The table (from the Bureau of Labor Statistics) gives the percentage of unemployed in the US labor force from 1999 to 2008.

t	$U(t)$	t	$U(t)$
1999	4.2	2004	5.5
2000	4.0	2005	5.1
2001	4.7	2006	4.6
2002	5.8	2007	4.6
2003	6.0	2008	5.8

(a) What is the meaning of $U'(t)$? What are its units?
(b) Construct a table of estimated values for $U'(t)$.

32. Let $P(t)$ be the percentage of Americans under the age of 18 at time t. The table gives values of this function in census years from 1950 to 2000.

t	$P(t)$	t	$P(t)$
1950	31.1	1980	28.0
1960	35.7	1990	25.7
1970	34.0	2000	25.7

(a) What is the meaning of $P'(t)$? What are its units?
(b) Construct a table of estimated values for $P'(t)$.
(c) Graph P and P'.
(d) How would it be possible to get more accurate values for $P'(t)$?

33–36 ▪ The graph of f is given. State, with reasons, the numbers at which f is not differentiable.

33.
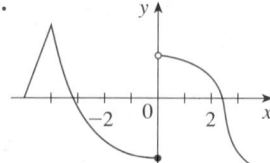

34.
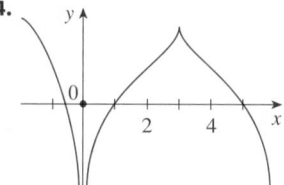

35.
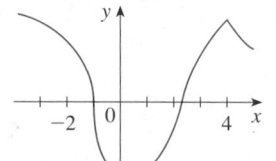

36.
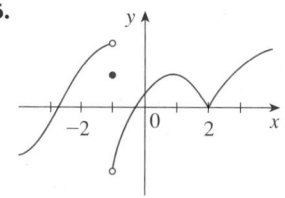

37. Graph the function $f(x) = x + \sqrt{|x|}$. Zoom in repeatedly, first toward the point $(-1, 0)$ and then toward the origin. What is different about the behavior of f in the vicinity of these two points? What do you conclude about the differentiability of f?

38. Zoom in toward the points $(1, 0)$, $(0, 1)$, and $(-1, 0)$ on the graph of the function $g(x) = (x^2 - 1)^{2/3}$. What do you

notice? Account for what you see in terms of the differentiability of g.

39. The figure shows the graphs of f, f', and f''. Identify each curve, and explain your choices.

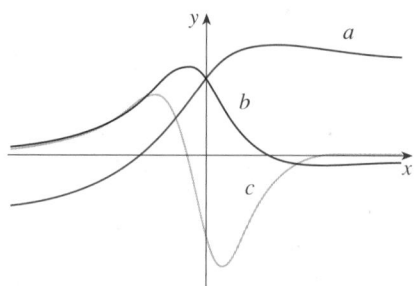

40. The figure shows graphs of f, f', f'', and f'''. Identify each curve, and explain your choices.

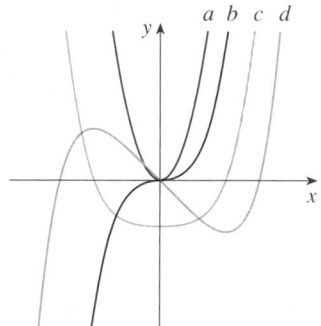

41. The figure shows the graphs of three functions. One is the position function of a car, one is the velocity of the car, and one is its acceleration. Identify each curve, and explain your choices.

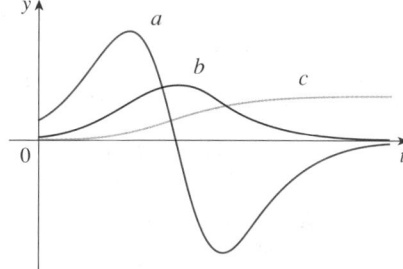

42–43 ▪ Use the definition of a derivative to find $f'(x)$ and $f''(x)$. Then graph f, f', and f'' on a common screen and check to see if your answers are reasonable.

42. $f(x) = x^3 - 3x$

43. $f(x) = 3x^2 + 2x + 1$

44. If $f(x) = 2x^2 - x^3$, find $f'(x)$, $f''(x)$, $f'''(x)$, and $f^{(4)}(x)$. Graph f, f', f'', and f''' on a common screen. Are the graphs consistent with the geometric interpretations of these derivatives?

45. Let $f(x) = \sqrt[3]{x}$.
  (a) If $a \neq 0$, use Equation 2.1.5 to find $f'(a)$.
  (b) Show that $f'(0)$ does not exist.
  (c) Show that $y = \sqrt[3]{x}$ has a vertical tangent line at $(0, 0)$. (Recall the shape of the graph of f. See Figure 8 in Section 1.2.)

46. (a) If $g(x) = x^{2/3}$, show that $g'(0)$ does not exist.
  (b) If $a \neq 0$, find $g'(a)$.
  (c) Show that $y = x^{2/3}$ has a vertical tangent line at $(0, 0)$.
  (d) Illustrate part (c) by graphing $y = x^{2/3}$.

47. Show that the function $f(x) = |x - 6|$ is not differentiable at 6. Find a formula for f' and sketch its graph.

48. Where is the greatest integer function $f(x) = [\![x]\!]$ not differentiable? Find a formula for f' and sketch its graph.

49. Recall that a function f is called *even* if $f(-x) = f(x)$ for all x in its domain and *odd* if $f(-x) = -f(x)$ for all such x. Prove each of the following.
  (a) The derivative of an even function is an odd function.
  (b) The derivative of an odd function is an even function.

50. When you turn on a hot-water faucet, the temperature T of the water depends on how long the water has been running.
  (a) Sketch a possible graph of T as a function of the time t that has elapsed since the faucet was turned on.
  (b) Describe how the rate of change of T with respect to t varies as t increases.
  (c) Sketch a graph of the derivative of T.

51. Let ℓ be the tangent line to the parabola $y = x^2$ at the point $(1, 1)$. The *angle of inclination* of ℓ is the angle ϕ that ℓ makes with the positive direction of the x-axis. Calculate ϕ correct to the nearest degree.

2.3 | BASIC DIFFERENTIATION FORMULAS

If it were always necessary to compute derivatives directly from the definition, as we did in the preceding section, such computations would be tedious and the evaluation of some limits would require ingenuity. Fortunately, several rules have been developed for finding derivatives without having to use the definition directly. These formulas greatly simplify the task of differentiation.

In this section we learn how to differentiate constant functions, power functions, polynomials, and the sine and cosine functions. Then we use this knowledge to compute rates of change.

Let's start with the simplest of all functions, the *constant function* $f(x) = c$. The graph of this function is the horizontal line $y = c$, which has slope 0, so we must have $f'(x) = 0$. (See Figure 1.) A formal proof, from the definition of a derivative, is also easy:

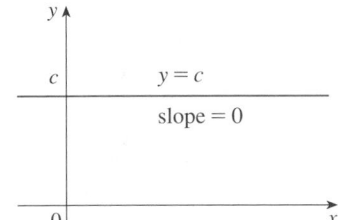

$$f'(x) = \lim_{h \to 0} \frac{f(x + h) - f(x)}{h} = \lim_{h \to 0} \frac{c - c}{h}$$

$$= \lim_{h \to 0} 0 = 0$$

FIGURE 1
The graph of $f(x) = c$ is the line $y = c$, so $f'(x) = 0$.

In Leibniz notation, we write this rule as follows.

DERIVATIVE OF A CONSTANT FUNCTION

$$\frac{d}{dx}(c) = 0$$

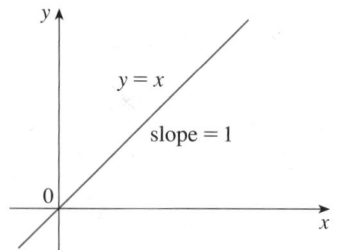

FIGURE 2
The graph of $f(x) = x$ is the
line $y = x$, so $f'(x) = 1$.

POWER FUNCTIONS

We next look at the functions $f(x) = x^n$, where n is a positive integer. If $n = 1$, the graph of $f(x) = x$ is the line $y = x$, which has slope 1. (See Figure 2.) So

$$\boxed{1} \qquad \boxed{\frac{d}{dx}(x) = 1}$$

(You can also verify Equation 1 from the definition of a derivative.) We have already investigated the cases $n = 2$ and $n = 3$. In fact, in Section 2.2 (Exercises 17 and 18) we found that

$$\boxed{2} \qquad \frac{d}{dx}(x^2) = 2x \qquad \frac{d}{dx}(x^3) = 3x^2$$

For $n = 4$ we find the derivative of $f(x) = x^4$ as follows:

$$f'(x) = \lim_{h \to 0} \frac{f(x+h) - f(x)}{h} = \lim_{h \to 0} \frac{(x+h)^4 - x^4}{h}$$

$$= \lim_{h \to 0} \frac{x^4 + 4x^3h + 6x^2h^2 + 4xh^3 + h^4 - x^4}{h}$$

$$= \lim_{h \to 0} \frac{4x^3h + 6x^2h^2 + 4xh^3 + h^4}{h}$$

$$= \lim_{h \to 0} (4x^3 + 6x^2h + 4xh^2 + h^3) = 4x^3$$

Thus

$$\boxed{3} \qquad \frac{d}{dx}(x^4) = 4x^3$$

Comparing the equations in (1), (2), and (3), we see a pattern emerging. It seems to be a reasonable guess that, when n is a positive integer, $(d/dx)(x^n) = nx^{n-1}$. This turns out to be true.

THE POWER RULE   If n is a positive integer, then

$$\frac{d}{dx}(x^n) = nx^{n-1}$$

PROOF   If $f(x) = x^n$, then

$$f'(x) = \lim_{h \to 0} \frac{f(x+h) - f(x)}{h} = \lim_{h \to 0} \frac{(x+h)^n - x^n}{h}$$

■ The Binomial Theorem is given on Reference Page 1.

In finding the derivative of x^4 we had to expand $(x + h)^4$. Here we need to expand $(x + h)^n$ and we use the Binomial Theorem to do so:

$$f'(x) = \lim_{h \to 0} \frac{\left[x^n + nx^{n-1}h + \dfrac{n(n-1)}{2}x^{n-2}h^2 + \cdots + nxh^{n-1} + h^n \right] - x^n}{h}$$

$$= \lim_{h \to 0} \frac{nx^{n-1}h + \dfrac{n(n-1)}{2}x^{n-2}h^2 + \cdots + nxh^{n-1} + h^n}{h}$$

$$= \lim_{h \to 0} \left[nx^{n-1} + \frac{n(n-1)}{2}x^{n-2}h + \cdots + nxh^{n-2} + h^{n-1} \right]$$

$$= nx^{n-1}$$

because every term except the first has h as a factor and therefore approaches 0.   □

We illustrate the Power Rule using various notations in Example 1.

EXAMPLE 1

(a) If $f(x) = x^6$, then $f'(x) = 6x^5$.   (b) If $y = x^{1000}$, then $y' = 1000x^{999}$.

(c) If $y = t^4$, then $\dfrac{dy}{dt} = 4t^3$.   (d) $\dfrac{d}{dr}(r^3) = 3r^2$ ■

What about power functions with negative integer exponents? In Exercise 57 we ask you to verify from the definition of a derivative that

$$\frac{d}{dx}\left(\frac{1}{x}\right) = -\frac{1}{x^2}$$

We can rewrite this equation as

$$\frac{d}{dx}(x^{-1}) = (-1)x^{-2}$$

and so the Power Rule is true when $n = -1$. In fact, we will show in the next section [Exercise 57(c)] that it holds for all negative integers.

What if the exponent is a fraction? In Example 3 in Section 2.2 we found that

$$\frac{d}{dx}\sqrt{x} = \frac{1}{2\sqrt{x}}$$

which can be written as

$$\frac{d}{dx}(x^{1/2}) = \tfrac{1}{2}x^{-1/2}$$

This shows that the Power Rule is true even when $n = \tfrac{1}{2}$. In fact, we will show in Section 3.3 that it is true for all real numbers n.

THE POWER RULE (GENERAL VERSION)   If n is any real number, then

$$\frac{d}{dx}(x^n) = nx^{n-1}$$

EXAMPLE 2 Differentiate:

(a) $f(x) = \dfrac{1}{x^2}$ 　　　　　　　　　　　　　(b) $y = \sqrt[3]{x^2}$

SOLUTION In each case we rewrite the function as a power of x.

(a) Since $f(x) = x^{-2}$, we use the Power Rule with $n = -2$:

$$f'(x) = \frac{d}{dx}(x^{-2}) = -2x^{-2-1} = -2x^{-3} = -\frac{2}{x^3}$$

(b)
$$\frac{dy}{dx} = \frac{d}{dx}\left(\sqrt[3]{x^2}\right) = \frac{d}{dx}(x^{2/3}) = \tfrac{2}{3}x^{(2/3)-1} = \tfrac{2}{3}x^{-1/3}\quad\blacksquare$$

The Power Rule enables us to find tangent lines without having to resort to the definition of a derivative. It also enables us to find *normal lines*. The **normal line** to a curve C at a point P is the line through P that is perpendicular to the tangent line at P. (In the study of optics, one needs to consider the angle between a light ray and the normal line to a lens.)

▼ EXAMPLE 3 Find equations of the tangent line and normal line to the curve $y = x\sqrt{x}$ at the point $(1, 1)$. Illustrate by graphing the curve and these lines.

SOLUTION The derivative of $f(x) = x\sqrt{x} = xx^{1/2} = x^{3/2}$ is

$$f'(x) = \tfrac{3}{2}x^{(3/2)-1} = \tfrac{3}{2}x^{1/2} = \tfrac{3}{2}\sqrt{x}$$

So the slope of the tangent line at $(1, 1)$ is $f'(1) = \tfrac{3}{2}$. Therefore an equation of the tangent line is

$$y - 1 = \tfrac{3}{2}(x - 1) \qquad \text{or} \qquad y = \tfrac{3}{2}x - \tfrac{1}{2}$$

The normal line is perpendicular to the tangent line, so its slope is the negative reciprocal of $\tfrac{3}{2}$, that is, $-\tfrac{2}{3}$. Thus an equation of the normal line is

$$y - 1 = -\tfrac{2}{3}(x - 1) \qquad \text{or} \qquad y = -\tfrac{2}{3}x + \tfrac{5}{3}$$

We graph the curve and its tangent line and normal line in Figure 4.　　　　■

■ Figure 3 shows the function y in Example 2(b) and its derivative y'. Notice that y is not differentiable at 0 (y' is not defined there). Observe that y' is positive when y increases and is negative when y decreases.

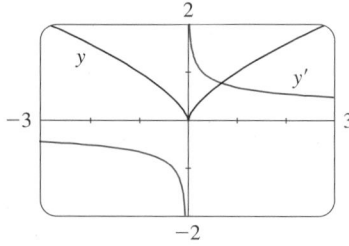

FIGURE 3
$y = \sqrt[3]{x^2}$

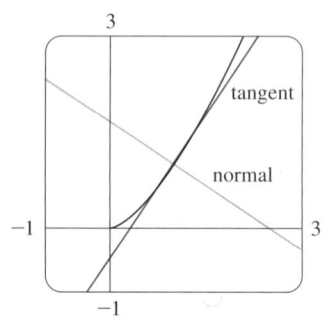

FIGURE 4

NEW DERIVATIVES FROM OLD

When new functions are formed from old functions by addition, subtraction, or multiplication by a constant, their derivatives can be calculated in terms of derivatives of the old functions. In particular, the following formula says that *the derivative of a constant times a function is the constant times the derivative of the function.*

> **THE CONSTANT MULTIPLE RULE**   If c is a constant and f is a differentiable function, then
> $$\frac{d}{dx}\,[cf(x)] = c\,\frac{d}{dx}\,f(x)$$

- **GEOMETRIC INTERPRETATION OF THE CONSTANT MULTIPLE RULE**

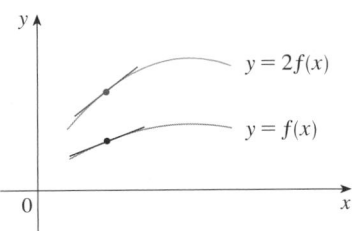

Multiplying by $c = 2$ stretches the graph vertically by a factor of 2. All the rises have been doubled but the runs stay the same. So the slopes are doubled too.

PROOF   Let $g(x) = cf(x)$. Then

$$g'(x) = \lim_{h\to 0}\frac{g(x+h) - g(x)}{h} = \lim_{h\to 0}\frac{cf(x+h) - cf(x)}{h}$$

$$= \lim_{h\to 0} c\left[\frac{f(x+h) - f(x)}{h}\right]$$

$$= c \lim_{h\to 0}\frac{f(x+h) - f(x)}{h} \qquad \text{(by Law 3 of limits)}$$

$$= cf'(x) \qquad \square$$

EXAMPLE 4

(a) $\dfrac{d}{dx}\,(3x^4) = 3\,\dfrac{d}{dx}\,(x^4) = 3(4x^3) = 12x^3$

(b) $\dfrac{d}{dx}\,(-x) = \dfrac{d}{dx}\,[(-1)x] = (-1)\,\dfrac{d}{dx}\,(x) = -1(1) = -1$ ∎

The next rule tells us that *the derivative of a sum of functions is the sum of the derivatives.*

> **THE SUM RULE**   If f and g are both differentiable, then
> $$\frac{d}{dx}\,[f(x) + g(x)] = \frac{d}{dx}\,f(x) + \frac{d}{dx}\,g(x)$$

- Using prime notation, we can write the Sum Rule as
$$(f + g)' = f' + g'$$

PROOF   Let $F(x) = f(x) + g(x)$. Then

$$F'(x) = \lim_{h\to 0}\frac{F(x+h) - F(x)}{h}$$

$$= \lim_{h\to 0}\frac{[f(x+h) + g(x+h)] - [f(x) + g(x)]}{h}$$

$$= \lim_{h\to 0}\left[\frac{f(x+h) - f(x)}{h} + \frac{g(x+h) - g(x)}{h}\right]$$

$$= \lim_{h\to 0}\frac{f(x+h) - f(x)}{h} + \lim_{h\to 0}\frac{g(x+h) - g(x)}{h} \qquad \text{(by Law 1)}$$

$$= f'(x) + g'(x) \qquad \square$$

The Sum Rule can be extended to the sum of any number of functions. For instance, using this theorem twice, we get

$$(f + g + h)' = [(f + g) + h]' = (f + g)' + h' = f' + g' + h'$$

By writing $f - g$ as $f + (-1)g$ and applying the Sum Rule and the Constant Multiple Rule, we get the following formula.

THE DIFFERENCE RULE If f and g are both differentiable, then

$$\frac{d}{dx}[f(x) - g(x)] = \frac{d}{dx}f(x) - \frac{d}{dx}g(x)$$

The Constant Multiple Rule, the Sum Rule, and the Difference Rule can be combined with the Power Rule to differentiate any polynomial, as the following examples demonstrate.

EXAMPLE 5

■ www.stewartcalculus.com
See Additional Example A.

$$\frac{d}{dx}(x^8 + 12x^5 - 4x^4 + 10x^3 - 6x + 5)$$

$$= \frac{d}{dx}(x^8) + 12\frac{d}{dx}(x^5) - 4\frac{d}{dx}(x^4) + 10\frac{d}{dx}(x^3) - 6\frac{d}{dx}(x) + \frac{d}{dx}(5)$$

$$= 8x^7 + 12(5x^4) - 4(4x^3) + 10(3x^2) - 6(1) + 0$$

$$= 8x^7 + 60x^4 - 16x^3 + 30x^2 - 6 \qquad \blacksquare$$

V EXAMPLE 6 Find the points on the curve $y = x^4 - 6x^2 + 4$ where the tangent line is horizontal.

SOLUTION Horizontal tangents occur where the derivative is zero. We have

$$\frac{dy}{dx} = \frac{d}{dx}(x^4) - 6\frac{d}{dx}(x^2) + \frac{d}{dx}(4)$$

$$= 4x^3 - 12x + 0 = 4x(x^2 - 3)$$

Thus $dy/dx = 0$ if $x = 0$ or $x^2 - 3 = 0$, that is, $x = \pm\sqrt{3}$. So the given curve has horizontal tangents when $x = 0$, $\sqrt{3}$, and $-\sqrt{3}$. The corresponding points are $(0, 4)$, $(\sqrt{3}, -5)$, and $(-\sqrt{3}, -5)$. (See Figure 5.)

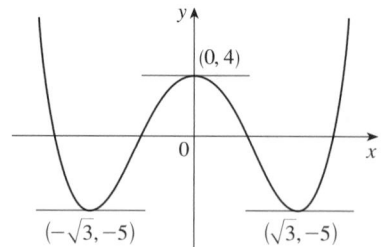

FIGURE 5
The curve $y = x^4 - 6x^2 + 4$ and
its horizontal tangents

$\blacksquare$

THE SINE AND COSINE FUNCTIONS

If we sketch the graph of the function $f(x) = \sin x$ and use the interpretation of $f'(x)$ as the slope of the tangent to the sine curve in order to sketch the graph of f' (see Exercise 16 in Section 2.2), then it looks as if the graph of f' may be the same as the cosine curve (see Figure 6).

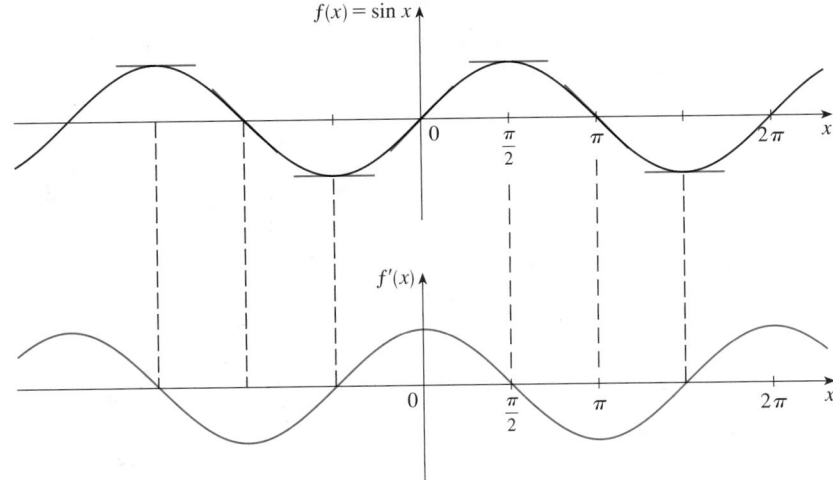

TEC Visual 2.3 shows an animation of Figure 6.

FIGURE 6

To prove that this is true we need to use two limits from Section 1.4 (see Equation 6 and Example 11 in that section):

$$\lim_{\theta \to 0} \frac{\sin \theta}{\theta} = 1 \qquad \lim_{\theta \to 0} \frac{\cos \theta - 1}{\theta} = 0$$

$\boxed{4}$

$$\boxed{\frac{d}{dx}(\sin x) = \cos x}$$

PROOF If $f(x) = \sin x$, then

$$f'(x) = \lim_{h \to 0} \frac{f(x + h) - f(x)}{h} = \lim_{h \to 0} \frac{\sin(x + h) - \sin x}{h}$$

- We have used the addition formula for sine. See Appendix A.

$$= \lim_{h \to 0} \frac{\sin x \cos h + \cos x \sin h - \sin x}{h}$$

$$= \lim_{h \to 0} \left[\frac{\sin x \cos h - \sin x}{h} + \frac{\cos x \sin h}{h} \right]$$

$$= \lim_{h \to 0} \left[\sin x \left(\frac{\cos h - 1}{h} \right) + \cos x \left(\frac{\sin h}{h} \right) \right]$$

- Note that we regard x as a constant when computing a limit as $h \to 0$, so $\sin x$ and $\cos x$ are also constants.

$$= \lim_{h \to 0} \sin x \cdot \lim_{h \to 0} \frac{\cos h - 1}{h} + \lim_{h \to 0} \cos x \cdot \lim_{h \to 0} \frac{\sin h}{h}$$

$$= (\sin x) \cdot 0 + (\cos x) \cdot 1 = \cos x$$

Using the same methods as in the proof of Formula 4, one can prove (see Exercise 58) that

5

$$\frac{d}{dx}(\cos x) = -\sin x$$

EXAMPLE 7 Differentiate $y = 3 \sin \theta + 4 \cos \theta$.

SOLUTION

$$\frac{dy}{d\theta} = 3\frac{d}{d\theta}(\sin \theta) + 4\frac{d}{d\theta}(\cos \theta) = 3 \cos \theta - 4 \sin \theta \qquad \blacksquare$$

EXAMPLE 8 Find the 27th derivative of $\cos x$.

SOLUTION The first few derivatives of $f(x) = \cos x$ are as follows:

$$f'(x) = -\sin x$$
$$f''(x) = -\cos x$$
$$f'''(x) = \sin x$$
$$f^{(4)}(x) = \cos x$$
$$f^{(5)}(x) = -\sin x$$

Looking for a pattern, we see that the successive derivatives occur in a cycle of length 4 and, in particular, $f^{(n)}(x) = \cos x$ whenever n is a multiple of 4. Therefore

$$f^{(24)}(x) = \cos x$$

and, differentiating three more times, we have

$$f^{(27)}(x) = \sin x \qquad \blacksquare$$

APPLICATIONS TO RATES OF CHANGE

We discussed velocity and other rates of change in Section 2.1, but now that we know some differentiation formulas we can solve problems involving rates of change more easily.

V EXAMPLE 9 The position of a particle is given by the equation

$$s = f(t) = t^3 - 6t^2 + 9t$$

where t is measured in seconds and s in meters.
(a) Find the velocity at time t.
(b) What is the velocity after 2 s? After 4 s?
(c) When is the particle at rest?
(d) When is the particle moving forward (that is, in the positive direction)?
(e) Draw a diagram to represent the motion of the particle.
(f) Find the total distance traveled by the particle during the first five seconds.
(g) Find the acceleration at time t and after 4 s.
(h) Graph the position, velocity, and acceleration functions for $0 \le t \le 5$.

SOLUTION

(a) The velocity function is the derivative of the position function.

$$s = f(t) = t^3 - 6t^2 + 9t$$

$$v(t) = \frac{ds}{dt} = 3t^2 - 12t + 9$$

(b) The velocity after 2 s means the instantaneous velocity when $t = 2$, that is,

$$v(2) = \frac{ds}{dt}\bigg|_{t=2} = 3(2)^2 - 12(2) + 9 = -3 \text{ m/s}$$

The velocity after 4 s is

$$v(4) = 3(4)^2 - 12(4) + 9 = 9 \text{ m/s}$$

(c) The particle is at rest when $v(t) = 0$, that is,

$$3t^2 - 12t + 9 = 3(t^2 - 4t + 3) = 3(t - 1)(t - 3) = 0$$

and this is true when $t = 1$ or $t = 3$. Thus the particle is at rest after 1 s and after 3 s.

(d) The particle moves in the positive direction when $v(t) > 0$, that is,

$$3t^2 - 12t + 9 = 3(t - 1)(t - 3) > 0$$

This inequality is true when both factors are positive ($t > 3$) or when both factors are negative ($t < 1$). Thus the particle moves in the positive direction in the time intervals $t < 1$ and $t > 3$. It moves backward (in the negative direction) when $1 < t < 3$.

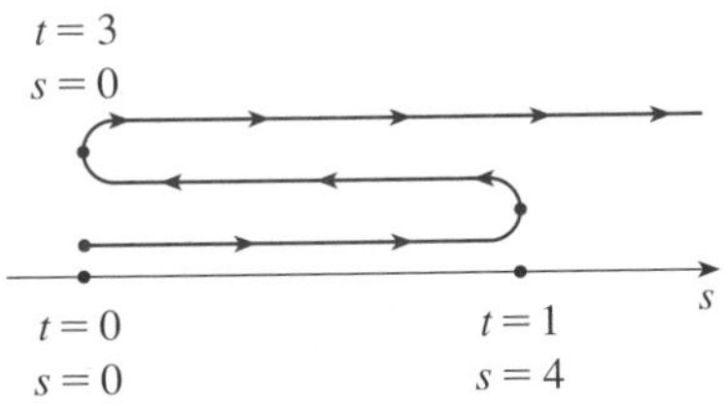

(e) Using the information from part (d) we make a schematic sketch in Figure 7 of the motion of the particle back and forth along a line (the s-axis).

(f) Because of what we learned in parts (d) and (e), we need to calculate the distances traveled during the time intervals [0, 1], [1, 3], and [3, 5] separately.

The distance traveled in the first second is

$$|f(1) - f(0)| = |4 - 0| = 4 \text{ m}$$

From $t = 1$ to $t = 3$ the distance traveled is

$$|f(3) - f(1)| = |0 - 4| = 4 \text{ m}$$

From $t = 3$ to $t = 5$ the distance traveled is

$$|f(5) - f(3)| = |20 - 0| = 20 \text{ m}$$

The total distance is $4 + 4 + 20 = 28$ m.

(g) The acceleration is the derivative of the velocity function:

$$a(t) = \frac{d^2s}{dt^2} = \frac{dv}{dt} = 6t - 12$$

$$a(4) = 6(4) - 12 = 12 \text{ m/s}^2$$

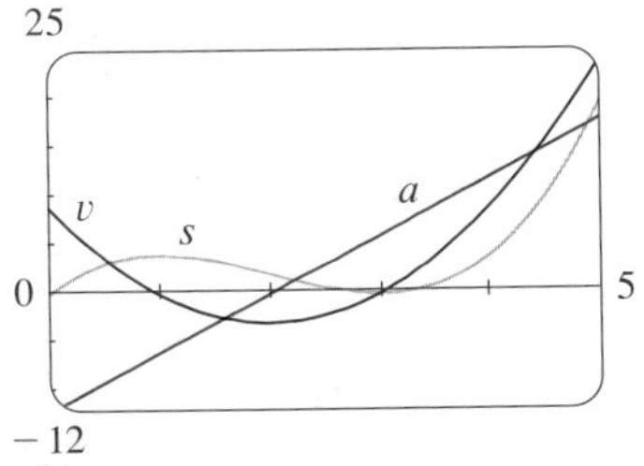

(h) Figure 8 shows the graphs of s, v, and a.

$t = 3$
$s = 0$

$t = 0$ $t = 1$
$s = 0$ $s = 4$

FIGURE 7

25

v s a

0   5

-12

FIGURE 8

V EXAMPLE 10 Suppose $C(x)$ is the total cost that a company incurs in producing x units of a certain commodity. The function C is called a **cost function**. If the number of items produced is increased from x_1 to x_2, then the additional cost is $\Delta C = C(x_2) - C(x_1)$, and the average rate of change of the cost is

$$\frac{\Delta C}{\Delta x} = \frac{C(x_2) - C(x_1)}{x_2 - x_1} = \frac{C(x_1 + \Delta x) - C(x_1)}{\Delta x}$$

The limit of this quantity as $\Delta x \to 0$, that is, the instantaneous rate of change of cost with respect to the number of items produced, is called the **marginal cost** by economists:

■ www.stewartcalculus.com
See Additional Examples B–F.

$$\text{marginal cost} = \lim_{\Delta x \to 0} \frac{\Delta C}{\Delta x} = \frac{dC}{dx}$$

[Since x often takes on only integer values, it may not make literal sense to let Δx approach 0, but we can always replace $C(x)$ by a smooth approximating function.]
   Taking $\Delta x = 1$ and n large (so that Δx is small compared to n), we have

$$C'(n) \approx C(n + 1) - C(n)$$

Thus the marginal cost of producing n units is approximately equal to the cost of producing one more unit [the $(n + 1)$st unit].
   It is often appropriate to represent a total cost function by a polynomial

$$C(x) = a + bx + cx^2 + dx^3$$

where a represents the overhead cost (rent, heat, maintenance) and the other terms represent the cost of raw materials, labor, and so on. (The cost of raw materials may be proportional to x, but labor costs might depend partly on higher powers of x because of overtime costs and inefficiencies involved in large-scale operations.)
   For instance, suppose a company has estimated that the cost (in dollars) of producing x items is

$$C(x) = 10{,}000 + 5x + 0.01x^2$$

Then the marginal cost function is

$$C'(x) = 5 + 0.02x$$

The marginal cost at the production level of 500 items is

$$C'(500) = 5 + 0.02(500) = \$15/\text{item}$$

This gives the rate at which costs are increasing with respect to the production level when $x = 500$ and predicts the cost of the 501st item.
   The actual cost of producing the 501st item is

$$C(501) - C(500) = [10{,}000 + 5(501) + 0.01(501)^2]$$
$$- [10{,}000 + 5(500) + 0.01(500)^2]$$
$$= \$15.01$$

Notice that $C'(500) \approx C(501) - C(500)$.   ■

2.3 | EXERCISES

1–26 ■ Differentiate the function.

1. $f(x) = 2^{40}$

2. $f(x) = \pi^2$

3. $f(t) = 2 - \frac{2}{3}t$

4. $F(x) = \frac{3}{4}x^8$

5. $f(x) = x^3 - 4x + 6$

6. $f(t) = 1.4t^5 - 2.5t^2 + 6.7$

7. $f(x) = 3x^2 - 2\cos x$

8. $y = \sin t + \pi \cos t$

9. $g(x) = x^2(1 - 2x)$

10. $h(x) = (x - 2)(2x + 3)$

11. $g(t) = 2t^{-3/4}$

12. $B(y) = cy^{-6}$

13. $A(s) = -\dfrac{12}{s^5}$

14. $y = x^{5/3} - x^{2/3}$

15. $R(a) = (3a + 1)^2$

16. $y = \sqrt{x}\,(x - 1)$

17. $S(p) = \sqrt{p} - p$

18. $S(R) = 4\pi R^2$

19. $y = \dfrac{x^2 + 4x + 3}{\sqrt{x}}$

20. $g(u) = \sqrt{2}\,u + \sqrt{3u}$

21. $v = t^2 - \dfrac{1}{\sqrt[4]{t^3}}$

22. $y = \dfrac{\sqrt{x} + x}{x^2}$

23. $z = \dfrac{A}{y^{10}} + B \cos y$

24. $y = \dfrac{\sin\theta}{2} + \dfrac{c}{\theta}$

25. $H(x) = (x + x^{-1})^3$

26. $u = \sqrt[3]{t^2} + 2\sqrt{t^3}$

27–28 ■ Find equations of the tangent line and normal line to the curve at the given point.

27. $y = 6\cos x$, $(\pi/3, 3)$

28. $y = x^2 - x^4$, $(1, 0)$

29–30 ■ Find an equation of the tangent line to the curve at the given point. Illustrate by graphing the curve and the tangent line on the same screen.

29. $y = 3x^2 - x^3$, $(1, 2)$

30. $y = x - \sqrt{x}$, $(1, 0)$

31–34 ■ Find the first and second derivatives of the function.

31. $f(x) = x^4 - 3x^3 + 16x$

32. $G(r) = \sqrt{r} + \sqrt[3]{r}$

33. $g(t) = 2\cos t - 3\sin t$

34. $h(t) = \sqrt{t} + 5\sin t$

35. Find $\dfrac{d^{99}}{dx^{99}}(\sin x)$.

36. Find the nth derivative of each function by calculating the first few derivatives and observing the pattern that occurs.
(a) $f(x) = x^n$
(b) $f(x) = 1/x$

37. For what values of x does the graph of $f(x) = x + 2\sin x$ have a horizontal tangent?

38. Find the points on the curve $y = 2x^3 + 3x^2 - 12x + 1$ where the tangent is horizontal.

39. Show that the curve $y = 6x^3 + 5x - 3$ has no tangent line with slope 4.

40. Find an equation of the tangent line to the curve $y = x\sqrt{x}$ that is parallel to the line $y = 1 + 3x$.

41. Find an equation of the normal line to the parabola $y = x^2 - 5x + 4$ that is parallel to the line $x - 3y = 5$.

42. Where does the normal line to the parabola $y = x - x^2$ at the point $(1, 0)$ intersect the parabola a second time? Illustrate with a sketch.

43. The equation of motion of a particle is $s = t^3 - 3t$, where s is in meters and t is in seconds. Find
(a) the velocity and acceleration as functions of t,
(b) the acceleration after 2 s, and
(c) the acceleration when the velocity is 0.

44. The equation of motion of a particle is $s = 2t^3 - 7t^2 + 4t + 1$, where s is in meters and t is in seconds.
(a) Find the velocity and acceleration as functions of t.
(b) Find the acceleration after 1 s.
(c) Graph the position, velocity, and acceleration functions on the same screen.

45–46 ■ A particle moves according to a law of motion $s = f(t)$, $t \geq 0$, where t is measured in seconds and s in feet.
(a) Find the velocity at time t.
(b) What is the velocity after 3 s?
(c) When is the particle at rest?
(d) When is the particle moving in the positive direction?
(e) Find the total distance traveled during the first 8 s.
(f) Draw a diagram like Figure 2 to illustrate the motion of the particle.
(g) Find the acceleration at time t and after 3 s.
(h) Graph the position, velocity, and acceleration functions for $0 \leq t \leq 8$.

45. $f(t) = t^3 - 12t^2 + 36t$

46. $f(t) = 0.01t^4 - 0.04t^3$

47. The position function of a particle is given by $s = t^3 - 4.5t^2 - 7t$, $t \geq 0$.
(a) When does the particle reach a velocity of 5 m/s?
(b) When is the acceleration 0? What is the significance of this value of t?

48. If a ball is given a push so that it has an initial velocity of 5 m/s down a certain inclined plane, then the distance it has rolled after t seconds is $s = 5t + 3t^2$.
(a) Find the velocity after 2 s.
(b) How long does it take for the velocity to reach 35 m/s?

49. If a rock is thrown vertically upward from the surface of Mars with velocity 15 m/s, its height after t seconds is $h = 15t - 1.86t^2$.
(a) What is the velocity of the rock after 2 s?
(b) What is the velocity of the rock when its height is 25 m on its way up? On its way down?

50. If a ball is thrown vertically upward with a velocity of 80 ft/s, then its height after t seconds is $s = 80t - 16t^2$.
(a) What is the maximum height reached by the ball?
(b) What is the velocity of the ball when it is 96 ft above the ground on its way up? On its way down?

51. The cost, in dollars, of producing x yards of a certain fabric is

$$C(x) = 1200 + 12x - 0.1x^2 + 0.0005x^3$$

(a) Find the marginal cost function.
(b) Find $C'(200)$ and explain its meaning. What does it predict?
(c) Compare $C'(200)$ with the cost of manufacturing the 201st yard of fabric.

52. The cost function for production of a commodity is

$$C(x) = 339 + 25x - 0.09x^2 + 0.0004x^3$$

(a) Find and interpret $C'(100)$.
(b) Compare $C'(100)$ with the cost of producing the 101st item.

53. A spherical balloon is being inflated. Find the rate of increase of the surface area ($S = 4\pi r^2$) with respect to the radius r when r is (a) 1 ft, (b) 2 ft, and (c) 3 ft. What conclusion can you make?

54. If a tank holds 5000 gallons of water, which drains from the bottom of the tank in 40 minutes, then Torricelli's Law gives the volume V of water remaining in the tank after t minutes as

$$V = 5000\left(1 - \tfrac{1}{40}t\right)^2 \qquad 0 \le t \le 40$$

Find the rate at which water is draining from the tank after (a) 5 min, (b) 10 min, (c) 20 min, and (d) 40 min. At what time is the water flowing out the fastest? The slowest? Summarize your findings.

55. Boyle's Law states that when a sample of gas is compressed at a constant pressure, the pressure P of the gas is inversely proportional to the volume V of the gas.
(a) Suppose that the pressure of a sample of air that occupies 0.106 m^3 at 25°C is 50 kPa. Write V as a function of P.

(b) Calculate dV/dP when $P = 50$ kPa. What is the meaning of the derivative? What are its units?

56. Newton's Law of Gravitation says that the magnitude F of the force exerted by a body of mass m on a body of mass M is

$$F = \frac{GmM}{r^2}$$

where G is the gravitational constant and r is the distance between the bodies.
(a) Find dF/dr and explain its meaning. What does the minus sign indicate?
(b) Suppose it is known that the earth attracts an object with a force that decreases at the rate of 2 N/km when $r = 20{,}000$ km. How fast does this force change when $r = 10{,}000$ km?

57. Use the definition of a derivative to show that if $f(x) = 1/x$, then $f'(x) = -1/x^2$. (This proves the Power Rule for the case $n = -1$.)

58. Prove, using the definition of derivative, that if $f(x) = \cos x$, then $f'(x) = -\sin x$.

59. The equation $y'' + y' - 2y = \sin x$ is called a **differential equation** because it involves an unknown function y and its derivatives y' and y''. Find constants A and B such that the function $y = A \sin x + B \cos x$ satisfies this equation. (Differential equations will be studied in detail in Section 7.7.)

60. Find constants A, B, and C such that the function $y = Ax^2 + Bx + C$ satisfies the differential equation $y'' + y' - 2y = x^2$.

61. Draw a diagram to show that there are two tangent lines to the parabola $y = x^2$ that pass through the point $(0, -4)$. Find the coordinates of the points where these tangent lines intersect the parabola.

62. (a) Find equations of both lines through the point $(2, -3)$ that are tangent to the parabola $y = x^2 + x$.
(b) Show that there is no line through the point $(2, 7)$ that is tangent to the parabola. Then draw a diagram to see why.

63. For what values of a and b is the line $2x + y = b$ tangent to the parabola $y = ax^2$ when $x = 2$?

64. Find a parabola with equation $y = ax^2 + bx + c$ that has slope 4 at $x = 1$, slope -8 at $x = -1$, and passes through the point $(2, 15)$.

65. Find a cubic function $y = ax^3 + bx^2 + cx + d$ whose graph has horizontal tangents at the points $(-2, 6)$ and $(2, 0)$.

66. Suppose the curve $y = x^4 + ax^3 + bx^2 + cx + d$ has a tangent line when $x = 0$ with equation $y = 2x + 1$ and a tangent line when $x = 1$ with equation $y = 2 - 3x$. Find the values of a, b, c, and d.

67. Find the parabola with equation $y = ax^2 + bx$ whose tangent line at $(1, 1)$ has equation $y = 3x - 2$.

68. A tangent line is drawn to the hyperbola $xy = c$ at a point P.
  (a) Show that the midpoint of the line segment cut from this tangent line by the coordinate axes is P.
  (b) Show that the triangle formed by the tangent line and the coordinate axes always has the same area, no matter where P is located on the hyperbola.

69. Evaluate $\displaystyle\lim_{x \to 1} \frac{x^{1000} - 1}{x - 1}$.

70. Draw a diagram showing two perpendicular lines that intersect on the y-axis and are both tangent to the parabola $y = x^2$. Where do these lines intersect?

71. If $c > \frac{1}{2}$, how many lines through the point $(0, c)$ are normal lines to the parabola $y = x^2$? What if $c \leq \frac{1}{2}$?

72. Sketch the parabolas $y = x^2$ and $y = x^2 - 2x + 2$. Do you think there is a line that is tangent to both curves? If so, find its equation. If not, why not?

2.4 | THE PRODUCT AND QUOTIENT RULES

The formulas of this section enable us to differentiate new functions formed from old functions by multiplication or division.

THE PRODUCT RULE

⊘ By analogy with the Sum and Difference Rules, one might be tempted to guess, as Leibniz did three centuries ago, that the derivative of a product is the product of the derivatives. We can see, however, that this guess is wrong by looking at a particular example. Let $f(x) = x$ and $g(x) = x^2$. Then the Power Rule gives $f'(x) = 1$ and $g'(x) = 2x$. But $(fg)(x) = x^3$, so $(fg)'(x) = 3x^2$. Thus $(fg)' \neq f'g'$. The correct formula was discovered by Leibniz (soon after his false start) and is called the Product Rule.

■ We can write the Product Rule in prime notation as
$$(fg)' = fg' + gf'$$

> **THE PRODUCT RULE** If f and g are both differentiable, then
> $$\frac{d}{dx}[f(x)g(x)] = f(x)\frac{d}{dx}[g(x)] + g(x)\frac{d}{dx}[f(x)]$$

PROOF Let $F(x) = f(x)g(x)$. Then

$$F'(x) = \lim_{h \to 0} \frac{F(x + h) - F(x)}{h} = \lim_{h \to 0} \frac{f(x + h)g(x + h) - f(x)g(x)}{h}$$

In order to evaluate this limit, we would like to separate the functions f and g as in the proof of the Sum Rule. We can achieve this separation by subtracting and adding the term $f(x + h)g(x)$ in the numerator:

$$F'(x) = \lim_{h \to 0} \frac{f(x + h)g(x + h) - f(x + h)g(x) + f(x + h)g(x) - f(x)g(x)}{h}$$

$$= \lim_{h \to 0}\left[f(x + h)\frac{g(x + h) - g(x)}{h} + g(x)\frac{f(x + h) - f(x)}{h}\right]$$

$$= \lim_{h \to 0} f(x + h) \cdot \lim_{h \to 0}\frac{g(x + h) - g(x)}{h} + \lim_{h \to 0} g(x) \cdot \lim_{h \to 0}\frac{f(x + h) - f(x)}{h}$$

$$= f(x)g'(x) + g(x)f'(x)$$

Note that $\lim_{h \to 0} g(x) = g(x)$ because $g(x)$ is a constant with respect to the variable h. Also, since f is differentiable at x, it is continuous at x by Theorem 2.2.4, and so $\lim_{h \to 0} f(x + h) = f(x)$.   □

In words, the Product Rule says that *the derivative of a product of two functions is the first function times the derivative of the second function plus the second function times the derivative of the first function.*

■ Figure 1 shows the graphs of the function of Example 1 and its derivative. Notice that $y' = 0$ whenever y has a horizontal tangent.

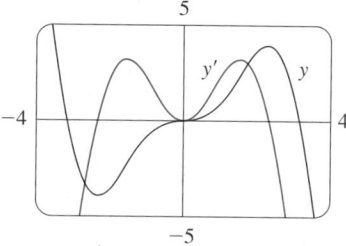

FIGURE 1

Ⅴ EXAMPLE 1 Differentiate $y = x^2 \sin x$.

SOLUTION Using the Product Rule, we have

$$\frac{dy}{dx} = x^2 \frac{d}{dx} (\sin x) + \sin x \frac{d}{dx} (x^2)$$

$$= x^2 \cos x + 2x \sin x$$ ■

EXAMPLE 2 Differentiate the function $f(t) = \sqrt{t}\,(a + bt)$.

■ In Example 2, a and b are constants. It is customary in mathematics to use letters near the beginning of the alphabet to represent constants and letters near the end of the alphabet to represent variables.

SOLUTION 1 Using the Product Rule, we have

$$f'(t) = \sqrt{t}\ \frac{d}{dt} (a + bt) + (a + bt) \frac{d}{dt} \left(\sqrt{t} \right)$$

$$= \sqrt{t} \cdot b + (a + bt) \cdot \tfrac{1}{2} t^{-1/2}$$

$$= b\sqrt{t} + \frac{a + bt}{2\sqrt{t}} = \frac{a + 3bt}{2\sqrt{t}}$$

SOLUTION 2 If we first use the laws of exponents to rewrite $f(t)$, then we can proceed directly without using the Product Rule.

$$f(t) = a\sqrt{t} + bt\sqrt{t} = at^{1/2} + bt^{3/2}$$

$$f'(t) = \tfrac{1}{2} at^{-1/2} + \tfrac{3}{2} bt^{1/2}$$

which is equivalent to the answer given in Solution 1.   ■

Example 2 shows that it is sometimes easier to simplify a product of functions than to use the Product Rule. In Example 1, however, the Product Rule is the only possible method.

EXAMPLE 3 If $h(x) = xg(x)$ and it is known that $g(3) = 5$ and $g'(3) = 2$, find $h'(3)$.

■ **www.stewartcalculus.com**
See Additional Examples A, B.

SOLUTION Applying the Product Rule, we get

$$h'(x) = \frac{d}{dx} [xg(x)] = x \frac{d}{dx} [g(x)] + g(x) \frac{d}{dx} [x]$$

$$= xg'(x) + g(x)$$

Therefore $h'(3) = 3g'(3) + g(3) = 3 \cdot 2 + 5 = 11$ ■

THE QUOTIENT RULE

The following rule enables us to differentiate the quotient of two differentiable functions.

- In prime notation we can write the Quotient Rule as

$$\left(\frac{f}{g}\right)' = \frac{gf' - fg'}{g^2}$$

> **THE QUOTIENT RULE**   If f and g are differentiable, then
>
> $$\frac{d}{dx}\left[\frac{f(x)}{g(x)}\right] = \frac{g(x)\dfrac{d}{dx}[f(x)] - f(x)\dfrac{d}{dx}[g(x)]}{[g(x)]^2}$$

PROOF   Let $F(x) = f(x)/g(x)$. Then

$$F'(x) = \lim_{h \to 0} \frac{F(x+h) - F(x)}{h} = \lim_{h \to 0} \frac{\dfrac{f(x+h)}{g(x+h)} - \dfrac{f(x)}{g(x)}}{h}$$

$$= \lim_{h \to 0} \frac{f(x+h)g(x) - f(x)g(x+h)}{hg(x+h)g(x)}$$

We can separate f and g in this expression by subtracting and adding the term $f(x)\,g(x)$ in the numerator:

$$F'(x) = \lim_{h \to 0} \frac{f(x+h)g(x) - f(x)g(x) + f(x)g(x) - f(x)g(x+h)}{hg(x+h)g(x)}$$

$$= \lim_{h \to 0} \frac{g(x)\dfrac{f(x+h) - f(x)}{h} - f(x)\dfrac{g(x+h) - g(x)}{h}}{g(x+h)g(x)}$$

$$= \frac{\displaystyle\lim_{h \to 0} g(x) \cdot \lim_{h \to 0} \frac{f(x+h) - f(x)}{h} - \lim_{h \to 0} f(x) \cdot \lim_{h \to 0} \frac{g(x+h) - g(x)}{h}}{\displaystyle\lim_{h \to 0} g(x+h) \cdot \lim_{h \to 0} g(x)}$$

$$= \frac{g(x)f'(x) - f(x)g'(x)}{[g(x)]^2}$$

Again g is continuous by Theorem 2.2.4, so $\lim_{h \to 0} g(x+h) = g(x)$.   □

In words, the Quotient Rule says that the *derivative of a quotient is the denominator times the derivative of the numerator minus the numerator times the derivative of the denominator, all divided by the square of the denominator.*

The Quotient Rule and the other differentiation formulas enable us to compute the derivative of any rational function, as the next example illustrates.

■ We can use a graphing device to check that the answer to Example 4 is plausible. Figure 2 shows the graphs of the function of Example 4 and its derivative. Notice that when y grows rapidly (near -2), y' is large. And when y grows slowly, y' is near 0.

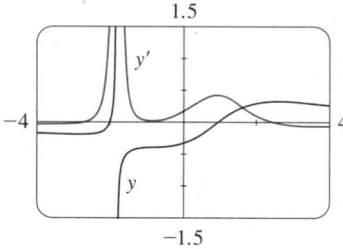

FIGURE 2

EXAMPLE 4 Let $y = \dfrac{x^2 + x - 2}{x^3 + 6}$.

Then

$$y' = \frac{(x^3 + 6)\dfrac{d}{dx}(x^2 + x - 2) - (x^2 + x - 2)\dfrac{d}{dx}(x^3 + 6)}{(x^3 + 6)^2}$$

$$= \frac{(x^3 + 6)(2x + 1) - (x^2 + x - 2)(3x^2)}{(x^3 + 6)^2}$$

$$= \frac{(2x^4 + x^3 + 12x + 6) - (3x^4 + 3x^3 - 6x^2)}{(x^3 + 6)^2}$$

$$= \frac{-x^4 - 2x^3 + 6x^2 + 12x + 6}{(x^3 + 6)^2}$$
■

EXAMPLE 5 Find an equation of the tangent line to the curve $y = \sqrt{x}/(1 + x^2)$ at the point $\left(1, \tfrac{1}{2}\right)$.

SOLUTION According to the Quotient Rule, we have

$$\frac{dy}{dx} = \frac{(1 + x^2)\dfrac{d}{dx}\left(\sqrt{x}\right) - \sqrt{x}\,\dfrac{d}{dx}(1 + x^2)}{(1 + x^2)^2}$$

$$= \frac{(1 + x^2)\dfrac{1}{2\sqrt{x}} - \sqrt{x}\,(2x)}{(1 + x^2)^2}$$

$$= \frac{(1 + x^2) - 4x^2}{2\sqrt{x}\,(1 + x^2)^2} = \frac{1 - 3x^2}{2\sqrt{x}\,(1 + x^2)^2}$$

So the slope of the tangent line at $\left(1, \tfrac{1}{2}\right)$ is

$$\left.\frac{dy}{dx}\right|_{x=1} = \frac{1 - 3 \cdot 1^2}{2\sqrt{1}\,(1 + 1^2)^2} = -\frac{1}{4}$$

We use the point-slope form to write an equation of the tangent line at $\left(1, \tfrac{1}{2}\right)$:

$$y - \tfrac{1}{2} = -\tfrac{1}{4}(x - 1) \qquad \text{or} \qquad y = -\tfrac{1}{4}x + \tfrac{3}{4}$$

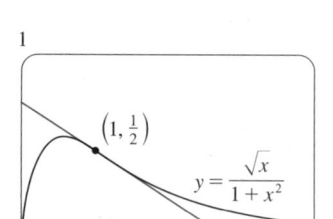

FIGURE 3

The curve and its tangent line are graphed in Figure 3. ■

NOTE Don't use the Quotient Rule *every* time you see a quotient. Sometimes it's easier to rewrite a quotient first to put it in a form that is simpler for the purpose of differentiation. For instance, although it is possible to differentiate the function

$$F(x) = \frac{3x^2 + 2\sqrt{x}}{x}$$

using the Quotient Rule, it is much easier to perform the division first and write the function as

$$F(x) = 3x + 2x^{-1/2}$$

before differentiating.

TRIGONOMETRIC FUNCTIONS

Knowing the derivatives of the sine and cosine functions, we can use the Quotient Rule to find the derivative of the tangent function:

$$\frac{d}{dx} (\tan x) = \frac{d}{dx} \left(\frac{\sin x}{\cos x} \right)$$

$$= \frac{\cos x \dfrac{d}{dx} (\sin x) - \sin x \dfrac{d}{dx} (\cos x)}{\cos^2 x}$$

$$= \frac{\cos x \cdot \cos x - \sin x (-\sin x)}{\cos^2 x}$$

$$= \frac{\cos^2 x + \sin^2 x}{\cos^2 x}$$

$$= \frac{1}{\cos^2 x} = \sec^2 x$$

$$\boxed{\frac{d}{dx} (\tan x) = \sec^2 x}$$

The derivatives of the remaining trigonometric functions, csc, sec, and cot, can also be found easily using the Quotient Rule (see Exercises 37–39). We collect all the differentiation formulas for trigonometric functions in the following table. Remember that they are valid only when x is measured in radians.

■ When you memorize this table, it is helpful to notice that the minus signs go with the derivatives of the "cofunctions," that is, cosine, cosecant, and cotangent.

DERIVATIVES OF TRIGONOMETRIC FUNCTIONS

$\dfrac{d}{dx} (\sin x) = \cos x$	$\dfrac{d}{dx} (\csc x) = -\csc x \cot x$
$\dfrac{d}{dx} (\cos x) = -\sin x$	$\dfrac{d}{dx} (\sec x) = \sec x \tan x$
$\dfrac{d}{dx} (\tan x) = \sec^2 x$	$\dfrac{d}{dx} (\cot x) = -\csc^2 x$

EXAMPLE 6 Differentiate $f(x) = \dfrac{\sec x}{1 + \tan x}$. For what values of x does the graph of f have a horizontal tangent?

SOLUTION The Quotient Rule gives

$$f'(x) = \frac{(1 + \tan x)\dfrac{d}{dx}(\sec x) - \sec x \dfrac{d}{dx}(1 + \tan x)}{(1 + \tan x)^2}$$

$$= \frac{(1 + \tan x)\sec x \tan x - \sec x \cdot \sec^2 x}{(1 + \tan x)^2}$$

$$= \frac{\sec x (\tan x + \tan^2 x - \sec^2 x)}{(1 + \tan x)^2}$$

$$= \frac{\sec x (\tan x - 1)}{(1 + \tan x)^2}$$

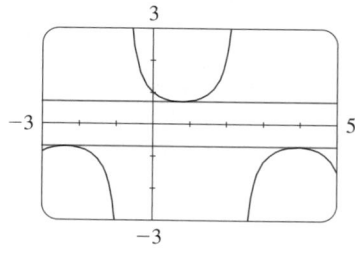

FIGURE 4
The horizontal tangents in Example 6

In simplifying the answer we have used the identity $\tan^2 x + 1 = \sec^2 x$.

Since $\sec x$ is never 0, we see that $f'(x) = 0$ when $\tan x = 1$, and this occurs when $x = n\pi + \pi/4$, where n is an integer (see Figure 4).   ■

2.4 | EXERCISES

1. Find the derivative of $f(x) = (1 + 2x^2)(x - x^2)$ in two ways: by using the Product Rule and by performing the multiplication first. Do your answers agree?

2. Find the derivative of the function

$$F(x) = \frac{x^4 - 5x^3 + \sqrt{x}}{x^2}$$

in two ways: by using the Quotient Rule and by simplifying first. Show that your answers are equivalent. Which method do you prefer?

3–26 ■ Differentiate.

3. $g(t) = t^3 \cos t$

4. $f(x) = \sqrt{x}\, \sin x$

5. $F(y) = \left(\dfrac{1}{y^2} - \dfrac{3}{y^4}\right)(y + 5y^3)$

6. $J(v) = (v^3 - 2v)(v^{-4} + v^{-2})$

7. $f(x) = \sin x + \frac{1}{2}\cot x$

8. $y = 2 \sec x - \csc x$

9. $h(\theta) = \theta \csc \theta - \cot \theta$

10. $y = \sin \theta \cos \theta$

11. $g(x) = \dfrac{1 + 2x}{3 - 4x}$

12. $G(x) = \dfrac{x^2 - 2}{2x + 1}$

13. $y = \dfrac{x^3}{1 - x^2}$

14. $y = \dfrac{x + 1}{x^3 + x - 2}$

15. $y = \dfrac{v^3 - 2v\sqrt{v}}{v}$

16. $g(t) = \dfrac{t - \sqrt{t}}{t^{1/3}}$

17. $f(t) = \dfrac{2t}{2 + \sqrt{t}}$

18. $y = \dfrac{\sqrt{x} - 1}{\sqrt{x} + 1}$

19. $y = \dfrac{x}{2 - \tan x}$

20. $y = \dfrac{\cos x}{1 - \sin x}$

21. $f(\theta) = \dfrac{\sec \theta}{1 + \sec \theta}$

22. $y = \dfrac{1 - \sec x}{\tan x}$

23. $y = \dfrac{t \sin t}{1 + t}$

24. $y = \dfrac{t}{(t - 1)^2}$

25. $f(x) = \dfrac{x}{x + \dfrac{c}{x}}$

26. $y = x^2 \sin x \tan x$

27–30 ■ Find an equation of the tangent line to the curve at the given point.

27. $y = \dfrac{x^2 - 1}{x^2 + x + 1}$, $(1, 0)$

28. $y = \dfrac{\sqrt{x}}{x + 1}$, $(4, 0.4)$

29. $y = \cos x - \sin x$, $(\pi, -1)$

30. $y = x + \tan x$, (π, π)

31. (a) The curve $y = 1/(1 + x^2)$ is called a **witch of Maria Agnesi**. Find an equation of the tangent line to this curve at the point $\left(-1, \frac{1}{2}\right)$.

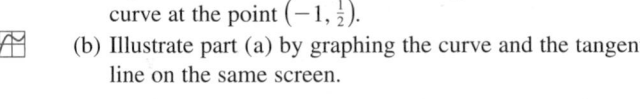

 (b) Illustrate part (a) by graphing the curve and the tangent line on the same screen.

32. (a) The curve $y = x/(1 + x^2)$ is called a **serpentine**. Find an equation of the tangent line to this curve at the point $(3, 0.3)$.

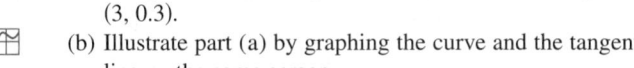

 (b) Illustrate part (a) by graphing the curve and the tangent line on the same screen.

33. If $f(x) = x^2/(1 + x)$, find $f''(1)$.

34. If $f(x) = \sec x$, find $f''(\pi/4)$.

35. If $H(\theta) = \theta \sin \theta$, find $H'(\theta)$ and $H''(\theta)$.

36. Find $\dfrac{d^{35}}{dx^{35}}(x \sin x)$.

37. Prove that $\dfrac{d}{dx}(\csc x) = -\csc x \cot x$.

38. Prove that $\dfrac{d}{dx}(\sec x) = \sec x \tan x$.

39. Prove that $\dfrac{d}{dx}(\cot x) = -\csc^2 x$.

40. Suppose $f(\pi/3) = 4$ and $f'(\pi/3) = -2$, and let $g(x) = f(x) \sin x$ and $h(x) = (\cos x)/f(x)$. Find
 (a) $g'(\pi/3)$ (b) $h'(\pi/3)$

41. Suppose that $f(5) = 1$, $f'(5) = 6$, $g(5) = -3$, and $g'(5) = 2$. Find the following values.
 (a) $(fg)'(5)$ (b) $(f/g)'(5)$ (c) $(g/f)'(5)$

42. If $f(3) = 4$, $g(3) = 2$, $f'(3) = -6$, and $g'(3) = 5$, find the following numbers.
 (a) $(f + g)'(3)$ (b) $(fg)'(3)$ (c) $(f/g)'(3)$

43. If f and g are the functions whose graphs are shown, let $u(x) = f(x) g(x)$ and $v(x) = f(x)/g(x)$.
 (a) Find $u'(1)$.   (b) Find $v'(5)$.

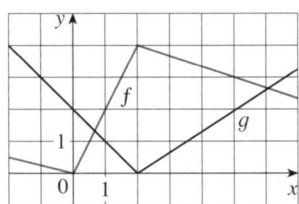

44. Let $P(x) = F(x) G(x)$ and $Q(x) = F(x)/G(x)$, where F and G are the functions whose graphs are shown.
 (a) Find $P'(2)$.   (b) Find $Q'(7)$.

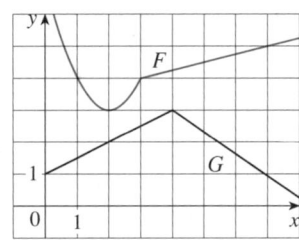

45. If g is a differentiable function, find an expression for the derivative of each of the following functions.
 (a) $y = xg(x)$ (b) $y = \dfrac{x}{g(x)}$ (c) $y = \dfrac{g(x)}{x}$

46. If f is a differentiable function, find an expression for the derivative of each of the following functions.
 (a) $y = x^2 f(x)$ (b) $y = \dfrac{f(x)}{x^2}$
 (c) $y = \dfrac{x^2}{f(x)}$ (d) $y = \dfrac{1 + xf(x)}{\sqrt{x}}$

47. How many tangent lines to the curve $y = x/(x + 1)$ pass through the point $(1, 2)$? At which points do these tangent lines touch the curve?

48. Find equations of the tangent lines to the curve
$$y = \frac{x - 1}{x + 1}$$
that are parallel to the line $x - 2y = 2$.

49. Find $R'(0)$, where
$$R(x) = \frac{x - 3x^3 + 5x^5}{1 + 3x^3 + 6x^6 + 9x^9}$$
Hint: Instead of finding $R'(x)$ first, let $f(x)$ be the numerator and $g(x)$ the denominator of $R(x)$ and compute $R'(0)$ from $f(0)$, $f'(0)$, $g(0)$, and $g'(0)$.

50. A manufacturer produces bolts of a fabric with a fixed width. The quantity q of this fabric (measured in yards) that is sold is a function of the selling price p (in dollars per yard), so we can write $q = f(p)$. Then the total revenue earned with selling price p is $R(p) = pf(p)$.
 (a) What does it mean to say that $f(20) = 10{,}000$ and $f'(20) = -350$?
 (b) Assuming the values in part (a), find $R'(20)$ and interpret your answer.

51. A mass on a spring vibrates horizontally on a smooth level surface (see the figure). Its equation of motion is $x(t) = 8 \sin t$, where t is in seconds and x in centimeters.
 (a) Find the velocity and acceleration at time t.

(b) Find the position, velocity, and acceleration of the mass at time $t = 2\pi/3$. In what direction is it moving at that time? Is it speeding up or slowing down?

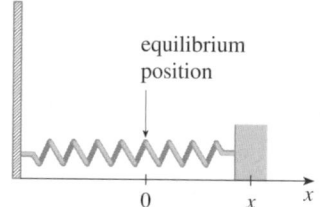

52. An object with weight W is dragged along a horizontal plane by a force acting along a rope attached to the object. If the rope makes an angle θ with the plane, then the magnitude of the force is

$$F = \frac{\mu W}{\mu \sin \theta + \cos \theta}$$

where μ is a constant called the *coefficient of friction*.
(a) Find the rate of change of F with respect to θ.
(b) When is this rate of change equal to 0?
(c) If $W = 50$ lb and $\mu = 0.6$, draw the graph of F as a function of θ and use it to locate the value of θ for which $dF/d\theta = 0$. Is the value consistent with your answer to part (b)?

53. The gas law for an ideal gas at absolute temperature T (in kelvins), pressure P (in atmospheres), and volume V (in liters) is $PV = nRT$, where n is the number of moles of the gas and $R = 0.0821$ is the gas constant. Suppose that, at a certain instant, $P = 8.0$ atm and is increasing at a rate of 0.10 atm/min and $V = 10$ L and is decreasing at a rate of 0.15 L/min. Find the rate of change of T with respect to time at that instant if $n = 10$ mol.

54. If R denotes the reaction of the body to some stimulus of strength x, the *sensitivity* S is defined to be the rate of change of the reaction with respect to x. A particular example is that when the brightness x of a light source is increased, the eye reacts by decreasing the area R of the pupil. The experimental formula

$$R = \frac{40 + 24x^{0.4}}{1 + 4x^{0.4}}$$

has been used to model the dependence of R on x when R is measured in square millimeters and x is measured in appropriate units of brightness.
(a) Find the sensitivity.
(b) Illustrate part (a) by graphing both R and S as functions of x. Comment on the values of R and S at low levels of brightness. Is this what you would expect?

55. (a) Use the Product Rule twice to prove that if f, g, and h are differentiable, then $(fgh)' = f'gh + fg'h + fgh'$.
(b) Use part (a) to differentiate $y = x \sin x \cos x$.

56. (a) If $F(x) = f(x)g(x)$, where f and g have derivatives of all orders, show that $F'' = f''g + 2f'g' + fg''$.
(b) Find similar formulas for F''' and $F^{(4)}$.
(c) Guess a formula for $F^{(n)}$.

57. (a) If g is differentiable, the **Reciprocal Rule** says that

$$\frac{d}{dx}\left[\frac{1}{g(x)}\right] = -\frac{g'(x)}{[g(x)]^2}$$

Use the Quotient Rule to prove the Reciprocal Rule.
(b) Use the Reciprocal Rule to differentiate the function $y = 1/(x^4 + x^2 + 1)$.
(c) Use the Reciprocal Rule to verify that the Power Rule is valid for negative integers, that is,

$$\frac{d}{dx}(x^{-n}) = -nx^{-n-1}$$

for all positive integers n.

2.5 | THE CHAIN RULE

Suppose you are asked to differentiate the function

$$F(x) = \sqrt{x^2 + 1}$$

The differentiation formulas you learned in the previous sections of this chapter do not enable you to calculate $F'(x)$.

Observe that F is a composite function. In fact, if we let $y = f(u) = \sqrt{u}$ and let $u = g(x) = x^2 + 1$, then we can write $y = F(x) = f(g(x))$, that is, $F = f \circ g$. We know how to differentiate both f and g, so it would be useful to have a rule that tells us how to find the derivative of $F = f \circ g$ in terms of the derivatives of f and g.

It turns out that the derivative of the composite function $f \circ g$ is the product of the derivatives of f and g. This fact is one of the most important of the differentiation rules

■ See Section 1.2 for a review of composite functions.

and is called the *Chain Rule*. It seems plausible if we interpret derivatives as rates of change. Regard du/dx as the rate of change of u with respect to x, dy/du as the rate of change of y with respect to u, and dy/dx as the rate of change of y with respect to x. If u changes twice as fast as x and y changes three times as fast as u, then it seems reasonable that y changes six times as fast as x, and so we expect that

$$\frac{dy}{dx} = \frac{dy}{du}\,\frac{du}{dx}$$

THE CHAIN RULE   If f and g are both differentiable and $F = f \circ g$ is the composite function defined by $F(x) = f(g(x))$, then F is differentiable and F' is given by the product

$$F'(x) = f'(g(x)) \cdot g'(x)$$

In Leibniz notation, if $y = f(u)$ and $u = g(x)$ are both differentiable functions, then

$$\frac{dy}{dx} = \frac{dy}{du}\,\frac{du}{dx}$$

COMMENTS ON THE PROOF OF THE CHAIN RULE   Let Δu be the change in u corresponding to a change of Δx in x, that is,

$$\Delta u = g(x + \Delta x) - g(x)$$

Then the corresponding change in y is

$$\Delta y = f(u + \Delta u) - f(u)$$

It is tempting to write

$$\frac{dy}{dx} = \lim_{\Delta x \to 0} \frac{\Delta y}{\Delta x}$$

$$\boxed{1} \qquad = \lim_{\Delta x \to 0} \frac{\Delta y}{\Delta u} \cdot \frac{\Delta u}{\Delta x}$$

$$= \lim_{\Delta x \to 0} \frac{\Delta y}{\Delta u} \cdot \lim_{\Delta x \to 0} \frac{\Delta u}{\Delta x}$$

$$= \lim_{\Delta u \to 0} \frac{\Delta y}{\Delta u} \cdot \lim_{\Delta x \to 0} \frac{\Delta u}{\Delta x} \qquad \text{(Note that } \Delta u \to 0 \text{ as } \Delta x \to 0 \\ \text{since } g \text{ is continuous.)}$$

$$= \frac{dy}{du}\,\frac{du}{dx}$$

The only flaw in this reasoning is that in $\boxed{1}$ it might happen that $\Delta u = 0$ (even when $\Delta x \neq 0$) and, of course, we can't divide by 0.

Nonetheless, this reasoning does at least *suggest* that the Chain Rule is true. A full proof of the Chain Rule is given at the end of this section.

The Chain Rule can be written either in the prime notation

$$\boxed{2} \qquad (f \circ g)'(x) = f'(g(x)) \cdot g'(x)$$

or, if $y = f(u)$ and $u = g(x)$, in Leibniz notation:

$$\boxed{3} \qquad \frac{dy}{dx} = \frac{dy}{du}\,\frac{du}{dx}$$

Equation 3 is easy to remember because if dy/du and du/dx were quotients, then we could cancel du. Remember, however, that du has not been defined and du/dx should not be thought of as an actual quotient.

EXAMPLE 1   Find $F'(x)$ if $F(x) = \sqrt{x^2 + 1}$.

SOLUTION 1 (using Equation 2):   At the beginning of this section we expressed F as $F(x) = (f \circ g)(x) = f(g(x))$ where $f(u) = \sqrt{u}$ and $g(x) = x^2 + 1$. Since

$$f'(u) = \tfrac{1}{2}u^{-1/2} = \frac{1}{2\sqrt{u}} \qquad \text{and} \qquad g'(x) = 2x$$

we have
$$F'(x) = f'(g(x)) \cdot g'(x)$$

$$= \frac{1}{2\sqrt{x^2 + 1}} \cdot 2x = \frac{x}{\sqrt{x^2 + 1}}$$

SOLUTION 2 (using Equation 3):   If we let $u = x^2 + 1$ and $y = \sqrt{u}$, then

$$F'(x) = \frac{dy}{du}\,\frac{du}{dx} = \frac{1}{2\sqrt{u}}\,(2x)$$

$$= \frac{1}{2\sqrt{x^2 + 1}}(2x) = \frac{x}{\sqrt{x^2 + 1}} \qquad \blacksquare$$

When using Formula 3 we should bear in mind that dy/dx refers to the derivative of y when y is considered as a function of x (called the *derivative of y with respect to x*), whereas dy/du refers to the derivative of y when considered as a function of u (the derivative of y with respect to u). For instance, in Example 1, y can be considered as a function of x $\left(y = \sqrt{x^2 + 1}\right)$ and also as a function of u $\left(y = \sqrt{u}\,\right)$. Note that

$$\frac{dy}{dx} = F'(x) = \frac{x}{\sqrt{x^2 + 1}} \qquad \text{whereas} \qquad \frac{dy}{du} = f'(u) = \frac{1}{2\sqrt{u}}$$

NOTE   In using the Chain Rule we work from the outside to the inside. Formula 2 says that *we differentiate the outer function f [at the inner function $g(x)$] and then we multiply by the derivative of the inner function.*

$$\frac{d}{dx} \quad \underbrace{f}_{\substack{\text{outer} \\ \text{function}}} \quad \underbrace{(g(x))}_{\substack{\text{evaluated} \\ \text{at inner} \\ \text{function}}} \quad = \quad \underbrace{f'}_{\substack{\text{derivative} \\ \text{of outer} \\ \text{function}}} \quad \underbrace{(g(x))}_{\substack{\text{evaluated} \\ \text{at inner} \\ \text{function}}} \quad \cdot \quad \underbrace{g'(x)}_{\substack{\text{derivative} \\ \text{of inner} \\ \text{function}}}$$

▼ EXAMPLE 2 Differentiate (a) $y = \sin(x^2)$ and (b) $y = \sin^2 x$.

SOLUTION

(a) If $y = \sin(x^2)$, then the outer function is the sine function and the inner function is the squaring function, so the Chain Rule gives

$$\frac{dy}{dx} = \frac{d}{dx} \underbrace{\sin}_{\substack{\text{outer} \\ \text{function}}} \underbrace{(x^2)}_{\substack{\text{evaluated} \\ \text{at inner} \\ \text{function}}} = \underbrace{\cos}_{\substack{\text{derivative} \\ \text{of outer} \\ \text{function}}} \underbrace{(x^2)}_{\substack{\text{evaluated} \\ \text{at inner} \\ \text{function}}} \cdot \underbrace{2x}_{\substack{\text{derivative} \\ \text{of inner} \\ \text{function}}}$$

$$= 2x \cos(x^2)$$

(b) Note that $\sin^2 x = (\sin x)^2$. Here the outer function is the squaring function and the inner function is the sine function. So

$$\frac{dy}{dx} = \frac{d}{dx} \underbrace{(\sin x)^2}_{\substack{\text{inner} \\ \text{function}}} = \underbrace{2}_{\substack{\text{derivative} \\ \text{of outer} \\ \text{function}}} \cdot \underbrace{(\sin x)}_{\substack{\text{evaluated} \\ \text{at inner} \\ \text{function}}} \cdot \underbrace{\cos x}_{\substack{\text{derivative} \\ \text{of inner} \\ \text{function}}}$$

The answer can be left as $2 \sin x \, \cos x$ or written as $\sin 2x$ (by a trigonometric identity known as the double-angle formula). ∎

■ See Reference Page 2 or Appendix A.

In Example 2(a) we combined the Chain Rule with the rule for differentiating the sine function. In general, if $y = \sin u$, where u is a differentiable function of x, then, by the Chain Rule,

$$\frac{dy}{dx} = \frac{dy}{du}\frac{du}{dx} = \cos u \frac{du}{dx}$$

Thus

$$\frac{d}{dx}(\sin u) = \cos u \frac{du}{dx}$$

In a similar fashion, all of the formulas for differentiating trigonometric functions can be combined with the Chain Rule.

Let's make explicit the special case of the Chain Rule where the outer function f is a power function. If $y = [g(x)]^n$, then we can write $y = f(u) = u^n$ where $u = g(x)$. By using the Chain Rule and then the Power Rule, we get

$$\frac{dy}{dx} = \frac{dy}{du}\frac{du}{dx} = nu^{n-1}\frac{du}{dx} = n[g(x)]^{n-1}g'(x)$$

---

4 **THE POWER RULE COMBINED WITH THE CHAIN RULE** If n is any real number and $u = g(x)$ is differentiable, then

$$\frac{d}{dx}(u^n) = nu^{n-1}\frac{du}{dx}$$

Alternatively,

$$\frac{d}{dx}[g(x)]^n = n[g(x)]^{n-1} \cdot g'(x)$$

Notice that the derivative in Example 1 could be calculated by taking $n = \frac{1}{2}$ in Rule 4.

EXAMPLE 3 Differentiate $y = (x^3 - 1)^{100}$.

SOLUTION Taking $u = g(x) = x^3 - 1$ and $n = 100$ in $\boxed{4}$, we have

$$\frac{dy}{dx} = \frac{d}{dx}(x^3 - 1)^{100} = 100(x^3 - 1)^{99} \frac{d}{dx}(x^3 - 1)$$

$$= 100(x^3 - 1)^{99} \cdot 3x^2 = 300x^2(x^3 - 1)^{99} \qquad \blacksquare$$

$\boxed{\text{V}}$ EXAMPLE 4 Find $f'(x)$ if $f(x) = \dfrac{1}{\sqrt[3]{x^2 + x + 1}}$.

SOLUTION First rewrite f: $\quad f(x) = (x^2 + x + 1)^{-1/3}$. Thus

$$f'(x) = -\tfrac{1}{3}(x^2 + x + 1)^{-4/3} \frac{d}{dx}(x^2 + x + 1)$$

$$= -\tfrac{1}{3}(x^2 + x + 1)^{-4/3}(2x + 1) \qquad \blacksquare$$

EXAMPLE 5 Find the derivative of the function

$$g(t) = \left(\frac{t - 2}{2t + 1}\right)^9$$

SOLUTION Combining the Power Rule, Chain Rule, and Quotient Rule, we get

$$g'(t) = 9\left(\frac{t - 2}{2t + 1}\right)^8 \frac{d}{dt}\left(\frac{t - 2}{2t + 1}\right)$$

$$= 9\left(\frac{t - 2}{2t + 1}\right)^8 \frac{(2t + 1) \cdot 1 - 2(t - 2)}{(2t + 1)^2} = \frac{45(t - 2)^8}{(2t + 1)^{10}} \qquad \blacksquare$$

EXAMPLE 6 Differentiate $y = (2x + 1)^5(x^3 - x + 1)^4$.

SOLUTION In this example we must use the Product Rule before using the Chain Rule:

$$\frac{dy}{dx} = (2x + 1)^5 \frac{d}{dx}(x^3 - x + 1)^4 + (x^3 - x + 1)^4 \frac{d}{dx}(2x + 1)^5$$

$$= (2x + 1)^5 \cdot 4(x^3 - x + 1)^3 \frac{d}{dx}(x^3 - x + 1)$$

$$\quad + (x^3 - x + 1)^4 \cdot 5(2x + 1)^4 \frac{d}{dx}(2x + 1)$$

$$= 4(2x + 1)^5(x^3 - x + 1)^3(3x^2 - 1) + 5(x^3 - x + 1)^4(2x + 1)^4 \cdot 2$$

■ The graphs of the functions y and y' in Example 6 are shown in Figure 1. Notice that y' is large when y increases rapidly and $y' = 0$ when y has a horizontal tangent. So our answer appears to be reasonable.

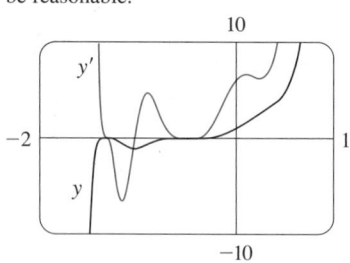

FIGURE 1

Noticing that each term has the common factor $2(2x + 1)^4(x^3 - x + 1)^3$, we could factor it out and write the answer as

$$\frac{dy}{dx} = 2(2x + 1)^4(x^3 - x + 1)^3(17x^3 + 6x^2 - 9x + 3) \qquad \blacksquare$$

The reason for the name "Chain Rule" becomes clear when we make a longer chain by adding another link. Suppose that $y = f(u)$, $u = g(x)$, and $x = h(t)$, where f, g, and *h* are differentiable functions. Then, to compute the derivative of *y* with respect to *t*, we use the Chain Rule twice:

$$\frac{dy}{dt} = \frac{dy}{dx}\frac{dx}{dt} = \frac{dy}{du}\frac{du}{dx}\frac{dx}{dt}$$

▼ EXAMPLE 7 If $f(x) = \sin(\cos(\tan x))$, then

$$f'(x) = \cos(\cos(\tan x))\,\frac{d}{dx}\cos(\tan x)$$

$$= \cos(\cos(\tan x))[-\sin(\tan x)]\frac{d}{dx}(\tan x)$$

$$= -\cos(\cos(\tan x))\,\sin(\tan x)\,\sec^2 x$$

Notice that we used the Chain Rule twice. ∎

EXAMPLE 8 Differentiate $y = \sqrt{\sec x^3}$.

SOLUTION Here the outer function is the square root function, the middle function is the secant function, and the inner function is the cubing function. So we have

$$\frac{dy}{dx} = \frac{1}{2\sqrt{\sec x^3}}\,\frac{d}{dx}(\sec x^3)$$

$$= \frac{1}{2\sqrt{\sec x^3}}\,\sec x^3 \tan x^3 \frac{d}{dx}(x^3)$$

$$= \frac{3x^2 \sec x^3 \tan x^3}{2\sqrt{\sec x^3}}$$

∎

HOW TO PROVE THE CHAIN RULE

Recall that if $y = f(x)$ and *x* changes from *a* to $a + \Delta x$, we defined the increment of *y* as

$$\Delta y = f(a + \Delta x) - f(a)$$

According to the definition of a derivative, we have

$$\lim_{\Delta x \to 0} \frac{\Delta y}{\Delta x} = f'(a)$$

So if we denote by ε the difference between the difference quotient and the derivative, we obtain

$$\lim_{\Delta x \to 0}\varepsilon = \lim_{\Delta x \to 0}\left(\frac{\Delta y}{\Delta x} - f'(a)\right) = f'(a) - f'(a) = 0$$

But $\quad\quad \varepsilon = \dfrac{\Delta y}{\Delta x} - f'(a) \quad\Rightarrow\quad \Delta y = f'(a)\,\Delta x + \varepsilon\,\Delta x$

If we define ε to be 0 when $\Delta x = 0$, then ε becomes a continuous function of Δx.

Thus, for a differentiable function f, we can write

$$\boxed{5} \qquad \Delta y = f'(a)\,\Delta x + \varepsilon\,\Delta x \qquad \text{where} \quad \varepsilon \to 0 \ \text{as} \ \Delta x \to 0$$

and ε is a continuous function of Δx. This property of differentiable functions is what enables us to prove the Chain Rule.

PROOF OF THE CHAIN RULE   Suppose $u = g(x)$ is differentiable at a and $y = f(u)$ is differentiable at $b = g(a)$. If Δx is an increment in x and Δu and Δy are the corresponding increments in u and y, then we can use Equation 5 to write

$$\boxed{6} \qquad \Delta u = g'(a)\,\Delta x + \varepsilon_1\,\Delta x = [g'(a) + \varepsilon_1]\,\Delta x$$

where $\varepsilon_1 \to 0$ as $\Delta x \to 0$. Similarly

$$\boxed{7} \qquad \Delta y = f'(b)\,\Delta u + \varepsilon_2\,\Delta u = [f'(b) + \varepsilon_2]\,\Delta u$$

where $\varepsilon_2 \to 0$ as $\Delta u \to 0$. If we now substitute the expression for Δu from Equation 6 into Equation 7, we get

$$\Delta y = [f'(b) + \varepsilon_2][g'(a) + \varepsilon_1]\,\Delta x$$

so
$$\frac{\Delta y}{\Delta x} = [f'(b) + \varepsilon_2][g'(a) + \varepsilon_1]$$

As $\Delta x \to 0$, Equation 6 shows that $\Delta u \to 0$. So both $\varepsilon_1 \to 0$ and $\varepsilon_2 \to 0$ as $\Delta x \to 0$. Therefore

$$\frac{dy}{dx} = \lim_{\Delta x \to 0} \frac{\Delta y}{\Delta x} = \lim_{\Delta x \to 0} [f'(b) + \varepsilon_2][g'(a) + \varepsilon_1]$$

$$= f'(b)\,g'(a) = f'(g(a))\,g'(a)$$

This proves the Chain Rule.   □

2.5 | EXERCISES

1–6 ▪ Write the composite function in the form $f(g(x))$. [Identify the inner function $u = g(x)$ and the outer function $y = f(u)$.] Then find the derivative dy/dx.

1. $y = \sqrt[3]{1 + 4x}$

2. $y = (2x^3 + 5)^4$

3. $y = \tan \pi x$

4. $y = \sin(\cot x)$

5. $y = \sqrt{\sin x}$

6. $y = \sin \sqrt{x}$

7–42 ▪ Find the derivative of the function.

7. $F(x) = (x^4 + 3x^2 - 2)^5$

8. $F(x) = (4x - x^2)^{100}$

9. $F(x) = \sqrt{1 - 2x}$

10. $f(x) = \dfrac{1}{(1 + \sec x)^2}$

11. $f(z) = \dfrac{1}{z^2 + 1}$

12. $f(t) = \sqrt[3]{1 + \tan t}$

13. $y = \cos(a^3 + x^3)$

14. $y = a^3 + \cos^3 x$

15. $y = x \sec kx$

16. $y = 3 \cot n\theta$

17. $f(x) = (2x - 3)^4(x^2 + x + 1)^5$

18. $g(x) = (x^2 + 1)^3(x^2 + 2)^6$

19. $h(t) = (t + 1)^{2/3}(2t^2 - 1)^3$

20. $F(t) = (3t - 1)^4(2t + 1)^{-3}$

21. $y = \left(\dfrac{x^2 + 1}{x^2 - 1}\right)^3$

22. $f(s) = \sqrt{\dfrac{s^2 + 1}{s^2 + 4}}$

23. $y = \sin(x \cos x)$

24. $f(x) = \dfrac{x}{\sqrt{7 - 3x}}$

25. $y = \dfrac{r}{\sqrt{r^2 + 1}}$

26. $G(y) = \dfrac{(y - 1)^4}{(y^2 + 2y)^5}$

27. $y = \sin\sqrt{1 + x^2}$

28. $F(v) = \left(\dfrac{v}{v^3 + 1}\right)^6$

29. $y = \sin(\tan 2x)$

30. $y = \sec^2(m\theta)$

31. $y = \sec^2 x + \tan^2 x$

32. $y = x \sin \dfrac{1}{x}$

33. $y = \left(\dfrac{1 - \cos 2x}{1 + \cos 2x} \right)^4$

34. $y = \left(ax + \sqrt{x^2 + b^2} \right)^{-2}$

35. $y = \cot^2(\sin \theta)$

36. $y = \sin(\sin(\sin x))$

37. $y = [x^2 + (1 - 3x)^5]^3$

38. $y = \sqrt{x + \sqrt{x + \sqrt{x}}}$

39. $g(x) = (2r \sin rx + n)^p$

40. $y = \cos^4(\sin^3 x)$

41. $y = \cos \sqrt{\sin(\tan \pi x)}$

42. $y = [x + (x + \sin^2 x)^3]^4$

43–46 ▪ Find the first and second derivatives of the function.

43. $y = \cos(x^2)$

44. $y = \cos^2 x$

45. $H(t) = \tan 3t$

46. $y = \dfrac{4x}{\sqrt{x + 1}}$

47–48 ▪ Find an equation of the tangent line to the curve at the given point.

47. $y = \sin(\sin x)$, $(\pi, 0)$

48. $y = \sqrt{1 + x^3}$, $(2, 3)$

---

49. (a) Find an equation of the tangent line to the curve $y = \tan(\pi x^2/4)$ at the point $(1, 1)$.

  (b) Illustrate part (a) by graphing the curve and the tangent line on the same screen.

50. (a) The curve $y = |x|/\sqrt{2 - x^2}$ is called a *bullet-nose curve*. Find an equation of the tangent line to this curve at the point $(1, 1)$.
  (b) Illustrate part (a) by graphing the curve and the tangent line on the same screen.

51. Find all points on the graph of the function $f(x) = 2 \sin x + \sin^2 x$ at which the tangent line is horizontal.

52. Find an equation of the tangent line to the curve $y = \sqrt{3 + x^2}$ that is parallel to the line $x - 2y = 1$.

53. If $F(x) = f(g(x))$, where $f(-2) = 8$, $f'(-2) = 4$, $f'(5) = 3$, $g(5) = -2$, and $g'(5) = 6$, find $F'(5)$.

54. If $h(x) = \sqrt{4 + 3f(x)}$, where $f(1) = 7$ and $f'(1) = 4$, find $h'(1)$.

55. A table of values for f, g, f', and g' is given.

x	$f(x)$	$g(x)$	$f'(x)$	$g'(x)$
1	3	2	4	6
2	1	8	5	7
3	7	2	7	9

(a) If $h(x) = f(g(x))$, find $h'(1)$.
(b) If $H(x) = g(f(x))$, find $H'(1)$.

56. Let f and g be the functions in Exercise 55.
(a) If $F(x) = f(f(x))$, find $F'(2)$.
(b) If $G(x) = g(g(x))$, find $G'(3)$.

57. If f and g are the functions whose graphs are shown, let $u(x) = f(g(x))$, $v(x) = g(f(x))$, and $w(x) = g(g(x))$. Find each derivative, if it exists. If it does not exist, explain why.
(a) $u'(1)$ (b) $v'(1)$ (c) $w'(1)$

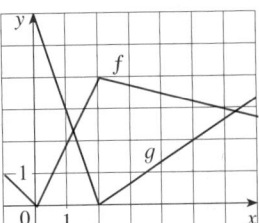

58. If f is the function whose graph is shown, let $h(x) = f(f(x))$ and $g(x) = f(x^2)$. Use the graph of f to estimate the value of each derivative.
(a) $h'(2)$ (b) $g'(2)$

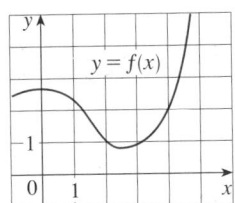

59. If $g(x) = \sqrt{f(x)}$, where the graph of f is shown, evaluate $g'(3)$.

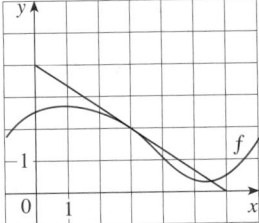

60. Suppose f is differentiable on $\mathbb{R}$ and α is a real number. Let $F(x) = f(x^\alpha)$ and $G(x) = [f(x)]^\alpha$. Find expressions for (a) $F'(x)$ and (b) $G'(x)$.

61. Let $r(x) = f(g(h(x)))$, where $h(1) = 2$, $g(2) = 3$, $h'(1) = 4$, $g'(2) = 5$, and $f'(3) = 6$. Find $r'(1)$.

62. If g is a twice differentiable function and $f(x) = xg(x^2)$, find f'' in terms of g, g', and g''.

63. Find the 50th derivative of $y = \cos 2x$.

64. If the equation of motion of a particle is given by $s = A \cos(\omega t + \delta)$, the particle is said to undergo *simple harmonic motion*.
(a) Find the velocity of the particle at time t.
(b) When is the velocity 0?

65. A Cepheid variable star is a star whose brightness alternately increases and decreases. The most easily visible such star is Delta Cephei, for which the interval between times of maximum brightness is 5.4 days. The average brightness of this star is 4.0 and its brightness changes by ± 0.35. In view of these data, the brightness of Delta Cephei at time t, where t is measured in days, has been modeled by the function

$$B(t) = 4.0 + 0.35 \sin(2\pi t/5.4)$$

(a) Find the rate of change of the brightness after t days.
(b) Find, correct to two decimal places, the rate of increase after one day.

66. A model for the length of daylight (in hours) in Philadelphia on the tth day of the year is given by the function

$$L(t) = 12 + 2.8 \sin\left[\frac{2\pi}{365}(t - 80)\right]$$

Use this model to compare how the number of hours of daylight is increasing in Philadelphia on March 21 and May 21.

67. The force F acting on a body with mass m and velocity v is the rate of change of momentum: $F = (d/dt)(mv)$. If m is constant, this becomes $F = ma$, where $a = dv/dt$ is the acceleration. But in the theory of relativity the mass of a particle varies with v as follows: $m = m_0/\sqrt{1 - v^2/c^2}$, where m_0 is the mass of the particle at rest and c is the speed of light. Show that

$$F = \frac{m_0 a}{(1 - v^2/c^2)^{3/2}}$$

68. Some of the highest tides in the world occur in the Bay of Fundy on the Atlantic Coast of Canada. At Hopewell Cape the water depth at low tide is about 2.0 m and at high tide it is about 12.0 m. The natural period of oscillation is a little more than 12 hours and on June 30, 2009, high tide occurred at 6:45 AM. This helps explain the following model for the water depth D (in meters) as a function of the time t (in hours after midnight) on that day:

$$D(t) = 7 + 5 \cos[0.503(t - 6.75)]$$

How fast was the tide rising (or falling) at the following times?
(a) 3:00 AM   (b) 6:00 AM
(c) 9:00 AM   (d) Noon

69. A particle moves along a straight line with displacement $s(t)$, velocity $v(t)$, and acceleration $a(t)$. Show that

$$a(t) = v(t)\frac{dv}{ds}$$

Explain the difference between the meanings of the derivatives dv/dt and dv/ds.

70. Air is being pumped into a spherical weather balloon. At any time t, the volume of the balloon is $V(t)$ and its radius is $r(t)$.
(a) What do the derivatives dV/dr and dV/dt represent?
(b) Express dV/dt in terms of dr/dt.

71. (a) If n is a positive integer, prove that

$$\frac{d}{dx}(\sin^n x \cos nx) = n \sin^{n-1} x \cos(n + 1)x$$

(b) Find a formula for the derivative of $y = \cos^n x \cos nx$ that is similar to the one in part (a).

72. Use the Chain Rule to prove the following.
(a) The derivative of an even function is an odd function.
(b) The derivative of an odd function is an even function.

73. Use the Chain Rule to show that if θ is measured in degrees, then

$$\frac{d}{d\theta}(\sin \theta) = \frac{\pi}{180}\cos \theta$$

(This gives one reason for the convention that radian measure is always used when dealing with trigonometric functions in calculus: The differentiation formulas would not be as simple if we used degree measure.)

74. Suppose $y = f(x)$ is a curve that always lies above the x-axis and never has a horizontal tangent, where f is differentiable everywhere. For what value of y is the rate of change of y^5 with respect to x eighty times the rate of change of y with respect to x?

75. If $F(x) = f(3f(4f(x)))$, where $f(0) = 0$ and $f'(0) = 2$, find $F'(0)$.

76. If $F(x) = f(xf(xf(x)))$, where $f(1) = 2, f(2) = 3$, $f'(1) = 4, f'(2) = 5$, and $f'(3) = 6$, find $F'(1)$.

77. If $y = f(u)$ and $u = g(x)$, where f and g are twice differentiable functions, show that

$$\frac{d^2y}{dx^2} = \frac{dy}{du}\frac{d^2u}{dx^2} + \frac{d^2y}{du^2}\left(\frac{du}{dx}\right)^2$$

78. (a) Write $|x| = \sqrt{x^2}$ and use the Chain Rule to show that

$$\frac{d}{dx}|x| = \frac{x}{|x|}$$

(b) If $f(x) = |\sin x|$, find $f'(x)$ and sketch the graphs of f and f'. Where is f not differentiable?
(c) If $g(x) = \sin|x|$, find $g'(x)$ and sketch the graphs of g and g'. Where is g not differentiable?

2.6 | IMPLICIT DIFFERENTIATION

The functions that we have met so far can be described by expressing one variable explicitly in terms of another variable—for example,

$$y = \sqrt{x^3 + 1} \qquad \text{or} \qquad y = x \sin x$$

or, in general, $y = f(x)$. Some functions, however, are defined implicitly by a relation between x and y such as

$$\boxed{1} \qquad\qquad\qquad x^2 + y^2 = 25$$

or

$$\boxed{2} \qquad\qquad\qquad x^3 + y^3 = 6xy$$

In some cases it is possible to solve such an equation for y as an explicit function (or several functions) of x. For instance, if we solve Equation 1 for y, we obtain $y = \pm\sqrt{25 - x^2}$, so two of the functions determined by the implicit Equation 1 are $f(x) = \sqrt{25 - x^2}$ and $g(x) = -\sqrt{25 - x^2}$. The graphs of f and g are the upper and lower semicircles of the circle $x^2 + y^2 = 25$. (See Figure 1.)

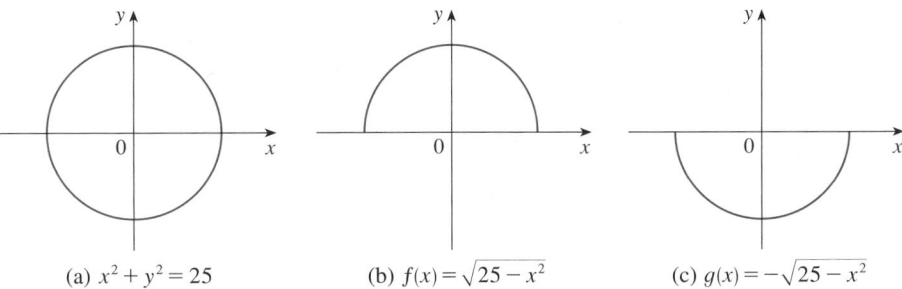

FIGURE 1   (a) $x^2 + y^2 = 25$ (b) $f(x) = \sqrt{25 - x^2}$ (c) $g(x) = -\sqrt{25 - x^2}$

It's not easy to solve Equation 2 for y explicitly as a function of x by hand. (A computer algebra system has no trouble, but the expressions it obtains are very complicated.) Nonetheless, $\boxed{2}$ is the equation of a curve called the **folium of Descartes** shown in Figure 2 and it implicitly defines y as several functions of x. The graphs of three such functions are shown in Figure 3. When we say that f is a function defined implicitly by Equation 2, we mean that the equation

$$x^3 + [f(x)]^3 = 6x f(x)$$

is true for all values of x in the domain of f.

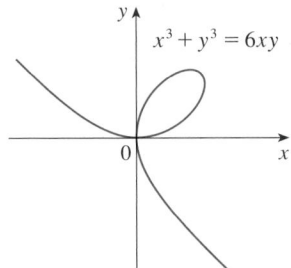

FIGURE 2   The folium of Descartes

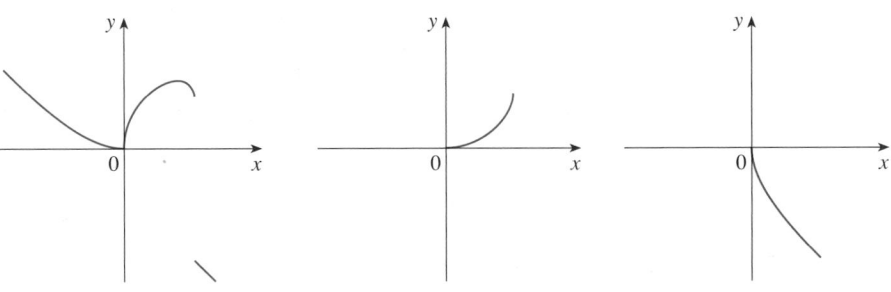

FIGURE 3   Graphs of three functions defined by the folium of Descartes

Fortunately, we don't need to solve an equation for y in terms of x in order to find the derivative of y. Instead we can use the method of **implicit differentiation**: This consists of differentiating both sides of the equation with respect to x and then solving the resulting equation for y'. In the examples and exercises of this section it is always assumed that the given equation determines y implicitly as a differentiable function of x so that the method of implicit differentiation can be applied.

▼ EXAMPLE 1

(a) If $x^2 + y^2 = 25$, find $\dfrac{dy}{dx}$.

(b) Find an equation of the tangent to the circle $x^2 + y^2 = 25$ at the point $(3, 4)$.

SOLUTION 1

(a) Differentiate both sides of the equation $x^2 + y^2 = 25$:

$$\frac{d}{dx}(x^2 + y^2) = \frac{d}{dx}(25)$$

$$\frac{d}{dx}(x^2) + \frac{d}{dx}(y^2) = 0$$

Remembering that y is a function of x and using the Chain Rule, we have

$$\frac{d}{dx}(y^2) = \frac{d}{dy}(y^2)\frac{dy}{dx} = 2y\frac{dy}{dx}$$

Thus
$$2x + 2y\frac{dy}{dx} = 0$$

Now we solve this equation for dy/dx:

$$\frac{dy}{dx} = -\frac{x}{y}$$

(b) At the point $(3, 4)$ we have $x = 3$ and $y = 4$, so

$$\frac{dy}{dx} = -\frac{3}{4}$$

An equation of the tangent to the circle at $(3, 4)$ is therefore

$$y - 4 = -\tfrac{3}{4}(x - 3) \qquad \text{or} \qquad 3x + 4y = 25$$

SOLUTION 2

(b) Solving the equation $x^2 + y^2 = 25$, we get $y = \pm\sqrt{25 - x^2}$. The point $(3, 4)$ lies on the upper semicircle $y = \sqrt{25 - x^2}$ and so we consider the function $f(x) = \sqrt{25 - x^2}$. Differentiating f using the Chain Rule, we have

$$f'(x) = \tfrac{1}{2}(25 - x^2)^{-1/2}\frac{d}{dx}(25 - x^2)$$

$$= \tfrac{1}{2}(25 - x^2)^{-1/2}(-2x) = -\frac{x}{\sqrt{25 - x^2}}$$

■ Example 1 illustrates that even when it is possible to solve an equation explicitly for y in terms of x, it may be easier to use implicit differentiation.

So
$$f'(3) = -\frac{3}{\sqrt{25 - 3^2}} = -\frac{3}{4}$$

and, as in Solution 1, an equation of the tangent is $3x + 4y = 25$. ■

V EXAMPLE 2
(a) Find y' if $x^3 + y^3 = 6xy$.
(b) Find the tangent to the folium of Descartes $x^3 + y^3 = 6xy$ at the point $(3, 3)$.
(c) At what point in the first quadrant is the tangent line horizontal?

SOLUTION
(a) Differentiating both sides of $x^3 + y^3 = 6xy$ with respect to x, regarding y as a function of x, and using the Chain Rule on the y^3 term and the Product Rule on the $6xy$ term, we get

$$3x^2 + 3y^2y' = 6y + 6xy'$$

or
$$x^2 + y^2y' = 2y + 2xy'$$

We now solve for y':
$$y^2y' - 2xy' = 2y - x^2$$

$$(y^2 - 2x)y' = 2y - x^2$$

$$y' = \frac{2y - x^2}{y^2 - 2x}$$

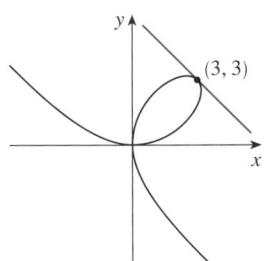

FIGURE 4

(b) When $x = y = 3$,

$$y' = \frac{2 \cdot 3 - 3^2}{3^2 - 2 \cdot 3} = -1$$

and a glance at Figure 4 confirms that this is a reasonable value for the slope at $(3, 3)$. So an equation of the tangent to the folium at $(3, 3)$ is

$$y - 3 = -1(x - 3) \qquad \text{or} \qquad x + y = 6$$

(c) The tangent line is horizontal if $y' = 0$. Using the expression for y' from part (a), we see that $y' = 0$ when $2y - x^2 = 0$ (provided that $y^2 - 2x \neq 0$). Substituting $y = \frac{1}{2}x^2$ in the equation of the curve, we get

$$x^3 + \left(\tfrac{1}{2}x^2\right)^3 = 6x\left(\tfrac{1}{2}x^2\right)$$

which simplifies to $x^6 = 16x^3$. Since $x \neq 0$ in the first quadrant, we have $x^3 = 16$. If $x = 16^{1/3} = 2^{4/3}$, then $y = \frac{1}{2}(2^{8/3}) = 2^{5/3}$. Thus the tangent is horizontal at $(2^{4/3}, 2^{5/3})$, which is approximately $(2.5198, 3.1748)$. Looking at Figure 5, we see that our answer is reasonable. ■

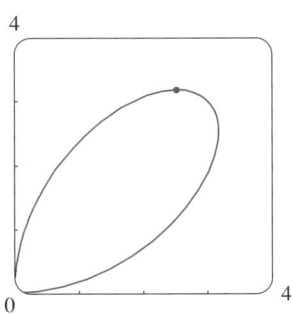

FIGURE 5

EXAMPLE 3 Find y' if $\sin(x + y) = y^2 \cos x$.

SOLUTION Differentiating implicitly with respect to x and remembering that y is a function of x, we get

$$\cos(x + y) \cdot (1 + y') = y^2(-\sin x) + (\cos x)(2yy')$$

(Note that we have used the Chain Rule on the left side and the Product Rule and

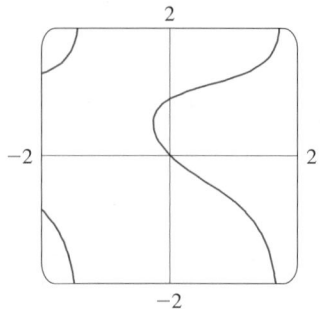

FIGURE 6

▪ www.stewartcalculus.com
See Additional Example A.

Chain Rule on the right side.) If we collect the terms that involve y', we get

$$\cos(x + y) + y^2 \sin x = (2y \cos x)y' - \cos(x + y) \cdot y'$$

So

$$y' = \frac{y^2 \sin x + \cos(x + y)}{2y \cos x - \cos(x + y)}$$

Figure 6, drawn with the implicit-plotting command of a computer algebra system, shows part of the curve $\sin(x + y) = y^2 \cos x$. As a check on our calculation, notice that $y' = -1$ when $x = y = 0$ and it appears from the graph that the slope is approximately -1 at the origin.   ■

EXAMPLE 4 Find y'' if $x^4 + y^4 = 16$.

SOLUTION Differentiating the equation implicitly with respect to x, we get

$$4x^3 + 4y^3 y' = 0$$

Solving for y' gives

$$\boxed{3} \qquad \qquad y' = -\frac{x^3}{y^3}$$

To find y'' we differentiate this expression for y' using the Quotient Rule and remembering that y is a function of x:

$$y'' = \frac{d}{dx}\left(-\frac{x^3}{y^3}\right)$$

$$= -\frac{y^3 \, (d/dx)(x^3) - x^3 \, (d/dx)(y^3)}{(y^3)^2}$$

$$= -\frac{y^3 \cdot 3x^2 - x^3(3y^2 y')}{y^6}$$

If we now substitute Equation 3 into this expression, we get

$$y'' = -\frac{3x^2 y^3 - 3x^3 y^2 \left(-\dfrac{x^3}{y^3}\right)}{y^6}$$

$$= -\frac{3(x^2 y^4 + x^6)}{y^7} = -\frac{3x^2(y^4 + x^4)}{y^7}$$

But the values of x and y must satisfy the original equation $x^4 + y^4 = 16$. So the answer simplifies to

$$y'' = -\frac{3x^2(16)}{y^7} = -48\,\frac{x^2}{y^7}$$ ■

▪ Figure 7 shows the graph of the curve $x^4 + y^4 = 16$ of Example 4. Notice that it's a stretched and flattened version of the circle $x^2 + y^2 = 4$. For this reason it's sometimes called a *fat circle*. It starts out very steep on the left but quickly becomes very flat. This can be seen from the expression

$$y' = -\frac{x^3}{y^3} = -\left(\frac{x}{y}\right)^3$$

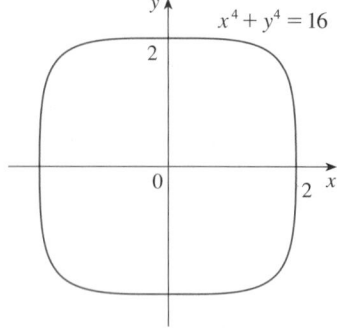

FIGURE 7

2.6 | EXERCISES

1–2 ∎
(a) Find y' by implicit differentiation.
(b) Solve the equation explicitly for y and differentiate to get y' in terms of x.
(c) Check that your solutions to parts (a) and (b) are consistent by substituting the expression for y into your solution for part (a).

1. $9x^2 - y^2 = 1$ **2.** $2x^2 + x + xy = 1$

3–16 ∎ Find dy/dx by implicit differentiation.

3. $x^3 + y^3 = 1$ **4.** $2x^3 + x^2y - xy^3 = 2$

5. $x^2 + xy - y^2 = 4$ **6.** $y^5 + x^2y^3 = 1 + x^4y$

7. $y \cos x = x^2 + y^2$ **8.** $\cos(xy) = 1 + \sin y$

9. $4 \cos x \sin y = 1$ **10.** $y \sin(x^2) = x \sin(y^2)$

11. $\tan(x/y) = x + y$ **12.** $\sqrt{x + y} = 1 + x^2y^2$

13. $\sqrt{xy} = 1 + x^2y$ **14.** $x \sin y + y \sin x = 1$

15. $y \cos x = 1 + \sin(xy)$ **16.** $\tan(x - y) = \dfrac{y}{1 + x^2}$

17. If $f(x) + x^2[f(x)]^3 = 10$ and $f(1) = 2$, find $f'(1)$.

18. If $g(x) + x \sin g(x) = x^2$, find $g'(0)$.

19–24 ∎ Use implicit differentiation to find an equation of the tangent line to the curve at the given point.

19. $x^2 + xy + y^2 = 3$, $(1, 1)$ (ellipse)

20. $x^2 + 2xy - y^2 + x = 2$, $(1, 2)$ (hyperbola)

21. $x^2 + y^2 = (2x^2 + 2y^2 - x)^2$ **22.** $x^{2/3} + y^{2/3} = 4$
$\left(0, \frac{1}{2}\right)$ (cardioid) $\left(-3\sqrt{3}, 1\right)$ (astroid)

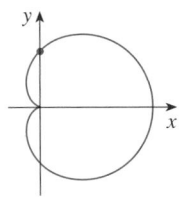   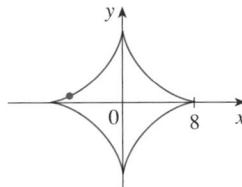

23. $2(x^2 + y^2)^2 = 25(x^2 - y^2)$ **24.** $y^2(y^2 - 4) = x^2(x^2 - 5)$
$(3, 1)$ (lemniscate) $(0, -2)$ (devil's curve)

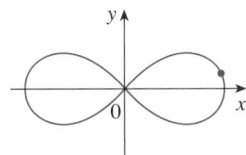   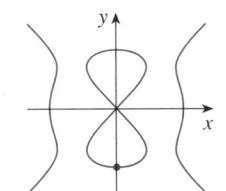

25–28 ∎ Find y'' by implicit differentiation.

25. $9x^2 + y^2 = 9$ **26.** $\sqrt{x} + \sqrt{y} = 1$

27. $x^3 + y^3 = 1$ **28.** $x^4 + y^4 = a^4$

29. (a) The curve with equation $y^2 = 5x^4 - x^2$ is called a **kampyle of Eudoxus**. Find an equation of the tangent line to this curve at the point $(1, 2)$.
(b) Illustrate part (a) by graphing the curve and the tangent line on a common screen. (If your graphing device will graph implicitly defined curves, then use that capability. If not, you can still graph this curve by graphing its upper and lower halves separately.)

30. (a) The curve with equation $y^2 = x^3 + 3x^2$ is called the **Tschirnhausen cubic**. Find an equation of the tangent line to this curve at the point $(1, -2)$.
(b) At what points does this curve have a horizontal tangent?
(c) Illustrate parts (a) and (b) by graphing the curve and the tangent lines on a common screen.

CAS **31.** Fanciful shapes can be created by using the implicit plotting capabilities of computer algebra systems.
(a) Graph the curve with equation
$$y(y^2 - 1)(y - 2) = x(x - 1)(x - 2)$$
At how many points does this curve have horizontal tangents? Estimate the x-coordinates of these points.
(b) Find equations of the tangent lines at the points $(0, 1)$ and $(0, 2)$.
(c) Find the exact x-coordinates of the points in part (a).
(d) Create even more fanciful curves by modifying the equation in part (a).

CAS **32.** (a) The curve with equation
$$2y^3 + y^2 - y^5 = x^4 - 2x^3 + x^2$$
has been likened to a bouncing wagon. Use a computer algebra system to graph this curve and discover why.
(b) At how many points does this curve have horizontal tangent lines? Find the x-coordinates of these points.

33. Find the points on the lemniscate in Exercise 23 where the tangent is horizontal.

34. Show by implicit differentiation that the tangent to the ellipse
$$\frac{x^2}{a^2} + \frac{y^2}{b^2} = 1$$
at the point (x_0, y_0) is
$$\frac{x_0 x}{a^2} + \frac{y_0 y}{b^2} = 1$$

35–38 ■ Two curves are **orthogonal** if their tangent lines are perpendicular at each point of intersection. Show that the given families of curves are **orthogonal trajectories** of each other, that is, every curve in one family is orthogonal to every curve in the other family. Sketch both families of curves on the same axes.

35. $x^2 + y^2 = r^2$, $\quad ax + by = 0$

36. $x^2 + y^2 = ax$, $\quad x^2 + y^2 = by$

37. $y = cx^2$, $\quad x^2 + 2y^2 = k$

38. $y = ax^3$, $\quad x^2 + 3y^2 = b$

39. (a) The *van der Waals equation* for n moles of a gas is

$$\left(P + \frac{n^2 a}{V^2}\right)(V - nb) = nRT$$

where P is the pressure, V is the volume, and T is the temperature of the gas. The constant R is the universal gas constant and a and b are positive constants that are characteristic of a particular gas. If T remains constant, use implicit differentiation to find dV/dP.

(b) Find the rate of change of volume with respect to pressure of 1 mole of carbon dioxide at a volume of $V = 10$ L and a pressure of $P = 2.5$ atm. Use $a = 3.592$ L²-atm/mole² and $b = 0.04267$ L/mole.

40. (a) Use implicit differentiation to find y' if $x^2 + xy + y^2 + 1 = 0$.

CAS (b) Plot the curve in part (a). What do you see? Prove that what you see is correct.

(c) In view of part (b), what can you say about the expression for y' that you found in part (a)?

41. Show, using implicit differentiation, that any tangent line at a point P to a circle with center O is perpendicular to the radius OP.

42. Show that the sum of the x- and y-intercepts of any tangent line to the curve $\sqrt{x} + \sqrt{y} = \sqrt{c}$ is equal to c.

43. The equation $x^2 - xy + y^2 = 3$ represents a "rotated ellipse," that is, an ellipse whose axes are not parallel to the coordinate axes. Find the points at which this ellipse crosses the x-axis and show that the tangent lines at these points are parallel.

44. (a) Where does the normal line to the ellipse $x^2 - xy + y^2 = 3$ at the point $(-1, 1)$ intersect the ellipse a second time?

(b) Illustrate part (a) by graphing the ellipse and the normal line.

45. Find all points on the curve $x^2 y^2 + xy = 2$ where the slope of the tangent line is -1.

46. Find equations of both the tangent lines to the ellipse $x^2 + 4y^2 = 36$ that pass through the point $(12, 3)$.

47. Show that the ellipse $x^2/a^2 + y^2/b^2 = 1$ and the hyperbola $x^2/A^2 - y^2/B^2 = 1$ are orthogonal trajectories if $A^2 < a^2$ and $a^2 - b^2 = A^2 + B^2$ (so the ellipse and hyperbola have the same foci).

48. Find the value of the number a such that the families of curves $y = (x + c)^{-1}$ and $y = a(x + k)^{1/3}$ are orthogonal trajectories.

49. The **Bessel function** of order 0, $y = J(x)$, satisfies the differential equation $xy'' + y' + xy = 0$ for all values of x and its value at 0 is $J(0) = 1$.

(a) Find $J'(0)$.

(b) Use implicit differentiation to find $J''(0)$.

50. The figure shows a lamp located three units to the right of the y-axis and a shadow created by the elliptical region $x^2 + 4y^2 \leq 5$. If the point $(-5, 0)$ is on the edge of the shadow, how far above the x-axis is the lamp located?

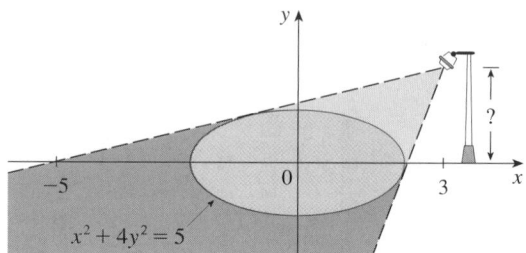

2.7 | RELATED RATES

If we are pumping air into a balloon, both the volume and the radius of the balloon are increasing and their rates of increase are related to each other. But it is much easier to measure directly the rate of increase of the volume than the rate of increase of the radius.

In a related rates problem the idea is to compute the rate of change of one quantity in terms of the rate of change of another quantity (which may be more easily measured). The procedure is to find an equation that relates the two quantities and then use the Chain Rule to differentiate both sides with respect to time.

▼ EXAMPLE 1 Air is being pumped into a spherical balloon so that its volume increases at a rate of 100 cm³/s. How fast is the radius of the balloon increasing when the diameter is 50 cm?

SOLUTION We start by identifying two things:

the *given information:*

> the rate of increase of the volume of air is 100 cm³/s

and the *unknown*:

> the rate of increase of the radius when the diameter is 50 cm

In order to express these quantities mathematically, we introduce some suggestive *notation*:

> Let V be the volume of the balloon and let r be its radius.

The key thing to remember is that rates of change are derivatives. In this problem, the volume and the radius are both functions of the time t. The rate of increase of the volume with respect to time is the derivative dV/dt, and the rate of increase of the radius is dr/dt. We can therefore restate the given and the unknown as follows:

$$\text{Given:} \qquad \frac{dV}{dt} = 100 \text{ cm}^3/\text{s}$$

$$\text{Unknown:} \qquad \frac{dr}{dt} \quad \text{when } r = 25 \text{ cm}$$

In order to connect dV/dt and dr/dt, we first relate V and r by the formula for the volume of a sphere:

$$V = \tfrac{4}{3}\pi r^3$$

In order to use the given information, we differentiate each side of this equation with respect to t. To differentiate the right side, we need to use the Chain Rule:

$$\frac{dV}{dt} = \frac{dV}{dr}\frac{dr}{dt} = 4\pi r^2 \frac{dr}{dt}$$

Now we solve for the unknown quantity:

$$\frac{dr}{dt} = \frac{1}{4\pi r^2}\frac{dV}{dt}$$

■ Notice that, although dV/dt is constant, dr/dt is *not* constant.

If we put $r = 25$ and $dV/dt = 100$ in this equation, we obtain

$$\frac{dr}{dt} = \frac{1}{4\pi(25)^2}100 = \frac{1}{25\pi}$$

The radius of the balloon is increasing at the rate of $1/(25\pi) \approx 0.0127$ cm/s. ■

EXAMPLE 2 A ladder 10 ft long rests against a vertical wall. If the bottom of the ladder slides away from the wall at a rate of 1 ft/s, how fast is the top of the ladder sliding down the wall when the bottom of the ladder is 6 ft from the wall?

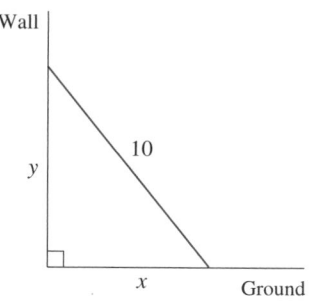

FIGURE 1

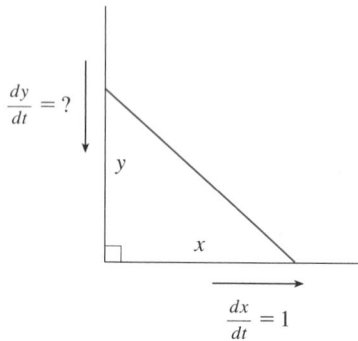

FIGURE 2

SOLUTION We first draw a diagram and label it as in Figure 1. Let x feet be the distance from the bottom of the ladder to the wall and y feet the distance from the top of the ladder to the ground. Note that x and y are both functions of t (time, measured in seconds).

We are given that $dx/dt = 1$ ft/s and we are asked to find dy/dt when $x = 6$ ft. (See Figure 2.) In this problem, the relationship between x and y is given by the Pythagorean Theorem:

$$x^2 + y^2 = 100$$

Differentiating each side with respect to t using the Chain Rule, we have

$$2x \frac{dx}{dt} + 2y \frac{dy}{dt} = 0$$

and solving this equation for the desired rate, we obtain

$$\frac{dy}{dt} = -\frac{x}{y} \frac{dx}{dt}$$

When $x = 6$, the Pythagorean Theorem gives $y = 8$ and so, substituting these values and $dx/dt = 1$, we have

$$\frac{dy}{dt} = -\frac{6}{8}(1) = -\frac{3}{4} \text{ ft/s}$$

The fact that dy/dt is negative means that the distance from the top of the ladder to the ground is *decreasing* at a rate of $\frac{3}{4}$ ft/s. In other words, the top of the ladder is sliding down the wall at a rate of $\frac{3}{4}$ ft/s. ∎

EXAMPLE 3 A water tank has the shape of an inverted circular cone with base radius 2 m and height 4 m. If water is being pumped into the tank at a rate of 2 m³/min, find the rate at which the water level is rising when the water is 3 m deep.

SOLUTION We first sketch the cone and label it as in Figure 3. Let V, r, and h be the volume of the water, the radius of the surface, and the height at time t, where t is measured in minutes.

We are given that $dV/dt = 2$ m³/min and we are asked to find dh/dt when h is 3 m. The quantities V and h are related by the equation

$$V = \tfrac{1}{3} \pi r^2 h$$

but it is very useful to express V as a function of h alone. In order to eliminate r, we use the similar triangles in Figure 3 to write

$$\frac{r}{h} = \frac{2}{4} \qquad r = \frac{h}{2}$$

and the expression for V becomes

$$V = \frac{1}{3} \pi \left(\frac{h}{2}\right)^2 h = \frac{\pi}{12} h^3$$

Now we can differentiate each side with respect to t:

$$\frac{dV}{dt} = \frac{\pi}{4} h^2 \frac{dh}{dt}$$

so

$$\frac{dh}{dt} = \frac{4}{\pi h^2} \frac{dV}{dt}$$

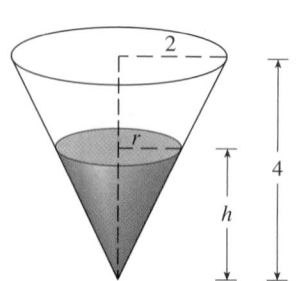

FIGURE 3

Substituting $h = 3$ m and $dV/dt = 2$ m³/min, we have

$$\frac{dh}{dt} = \frac{4}{\pi(3)^2} \cdot 2 = \frac{8}{9\pi}$$

The water level is rising at a rate of $8/(9\pi) \approx 0.28$ m/min. ∎

STRATEGY Examples 1–3 suggest the following steps in solving related rates problems:

1. Read the problem carefully.

2. Draw a diagram if possible.

3. Introduce notation. Assign symbols to all quantities that are functions of time.

4. Express the given information and the required rate in terms of derivatives.

5. Write an equation that relates the various quantities of the problem. If necessary, use the geometry of the situation to eliminate one of the variables by substitution (as in Example 3).

6. Use the Chain Rule to differentiate both sides of the equation with respect to t.

7. Substitute the given information into the resulting equation and solve for the unknown rate.

⊘ **WARNING** A common error is to substitute the given numerical information (for quantities that vary with time) too early. This should be done only *after* the differentiation. (Step 7 follows Step 6.) For instance, in Example 3 we dealt with general values of h until we finally substituted $h = 3$ at the last stage. (If we had put $h = 3$ earlier, we would have gotten $dV/dt = 0$, which is clearly wrong.)

The following examples are further illustrations of the strategy.

▼ **EXAMPLE 4** Car A is traveling west at 50 mi/h and car B is traveling north at 60 mi/h. Both are headed for the intersection of the two roads. At what rate are the cars approaching each other when car A is 0.3 mi and car B is 0.4 mi from the intersection?

SOLUTION We draw Figure 4, where C is the intersection of the roads. At a given time t, let x be the distance from car A to C, let y be the distance from car B to C, and let z be the distance between the cars, where x, y, and z are measured in miles.

We are given that $dx/dt = -50$ mi/h and $dy/dt = -60$ mi/h. (The derivatives are negative because x and y are decreasing.) We are asked to find dz/dt. The equation that relates x, y, and z is given by the Pythagorean Theorem:

$$z^2 = x^2 + y^2$$

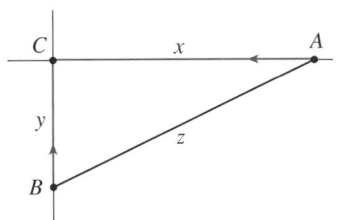

FIGURE 4

Differentiating each side with respect to t, we have

$$2z\frac{dz}{dt} = 2x\frac{dx}{dt} + 2y\frac{dy}{dt}$$

$$\frac{dz}{dt} = \frac{1}{z}\left(x\frac{dx}{dt} + y\frac{dy}{dt}\right)$$

When $x = 0.3$ mi and $y = 0.4$ mi, the Pythagorean Theorem gives $z = 0.5$ mi, so

$$\frac{dz}{dt} = \frac{1}{0.5}[0.3(-50) + 0.4(-60)]$$

$$= -78 \text{ mi/h}$$

The cars are approaching each other at a rate of 78 mi/h. ∎

▼ EXAMPLE 5 A man walks along a straight path at a speed of 4 ft/s. A searchlight is located on the ground 20 ft from the path and is kept focused on the man. At what rate is the searchlight rotating when the man is 15 ft from the point on the path closest to the searchlight?

SOLUTION We draw Figure 5 and let x be the distance from the man to the point on the path closest to the searchlight. We let θ be the angle between the beam of the searchlight and the perpendicular to the path.

We are given that $dx/dt = 4$ ft/s and are asked to find $d\theta/dt$ when $x = 15$. The equation that relates x and θ can be written from Figure 5:

$$\frac{x}{20} = \tan \theta \qquad x = 20 \tan \theta$$

Differentiating each side with respect to t, we get

$$\frac{dx}{dt} = 20 \sec^2\theta \, \frac{d\theta}{dt}$$

so

$$\frac{d\theta}{dt} = \tfrac{1}{20} \cos^2\theta \, \frac{dx}{dt} = \tfrac{1}{20} \cos^2\theta \, (4) = \tfrac{1}{5} \cos^2\theta$$

When $x = 15$ ft, the length of the beam is 25 ft, so $\cos \theta = \tfrac{4}{5}$ and

$$\frac{d\theta}{dt} = \frac{1}{5} \left(\frac{4}{5}\right)^2 = \frac{16}{125} = 0.128$$

The searchlight is rotating at a rate of 0.128 rad/s. ∎

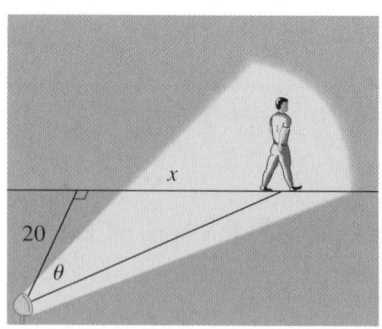

FIGURE 5

2.7 | EXERCISES

1. If V is the volume of a cube with edge length x and the cube expands as time passes, find dV/dt in terms of dx/dt.

2. (a) If A is the area of a circle with radius r and the circle expands as time passes, find dA/dt in terms of dr/dt.
   (b) Suppose oil spills from a ruptured tanker and spreads in a circular pattern. If the radius of the oil spill increases at a constant rate of 1 m/s, how fast is the area of the spill increasing when the radius is 30 m?

3. Each side of a square is increasing at a rate of 6 cm/s. At what rate is the area of the square increasing when the area of the square is 16 cm²?

4. The length of a rectangle is increasing at a rate of 8 cm/s and its width is increasing at a rate of 3 cm/s. When the length is 20 cm and the width is 10 cm, how fast is the area of the rectangle increasing?

5. A cylindrical tank with radius 5 m is being filled with water at a rate of 3 m³/min. How fast is the height of the water increasing?

6. The radius of a sphere is increasing at a rate of 4 mm/s. How fast is the volume increasing when the diameter is 80 mm?

7. Suppose $y = \sqrt{2x + 1}$, where x and y are functions of t.
   (a) If $dx/dt = 3$, find dy/dt when $x = 4$.
   (b) If $dy/dt = 5$, find dx/dt when $x = 12$.

8. Suppose $4x^2 + 9y^2 = 36$, where x and y are functions of t.
   (a) If $dy/dt = \tfrac{1}{3}$, find dx/dt when $x = 2$ and $y = \tfrac{2}{3}\sqrt{5}$.
   (b) If $dx/dt = 3$, find dy/dt when $x = -2$ and $y = \tfrac{2}{3}\sqrt{5}$.

9. If $x^2 + y^2 + z^2 = 9$, $dx/dt = 5$, and $dy/dt = 4$, find dz/dt when $(x, y, z) = (2, 2, 1)$.

10. A particle is moving along a hyperbola $xy = 8$. As it reaches the point $(4, 2)$, the y-coordinate is decreasing at a rate of 3 cm/s. How fast is the x-coordinate of the point changing at that instant?

11–14 ▪
(a) What quantities are given in the problem?
(b) What is the unknown?
(c) Draw a picture of the situation for any time t.

(d) Write an equation that relates the quantities.

(e) Finish solving the problem.

11. If a snowball melts so that its surface area decreases at a rate of 1 cm²/min, find the rate at which the diameter decreases when the diameter is 10 cm.

12. At noon, ship A is 150 km west of ship B. Ship A is sailing east at 35 km/h and ship B is sailing north at 25 km/h. How fast is the distance between the ships changing at 4:00 PM?

13. A plane flying horizontally at an altitude of 1 mi and a speed of 500 mi/h passes directly over a radar station. Find the rate at which the distance from the plane to the station is increasing when it is 2 mi away from the station.

14. A street light is mounted at the top of a 15-ft-tall pole. A man 6 ft tall walks away from the pole with a speed of 5 ft/s along a straight path. How fast is the tip of his shadow moving when he is 40 ft from the pole?

15. Two cars start moving from the same point. One travels south at 60 mi/h and the other travels west at 25 mi/h. At what rate is the distance between the cars increasing two hours later?

16. A spotlight on the ground shines on a wall 12 m away. If a man 2 m tall walks from the spotlight toward the building at a speed of 1.6 m/s, how fast is the length of his shadow on the building decreasing when he is 4 m from the building?

17. A man starts walking north at 4 ft/s from a point P. Five minutes later a woman starts walking south at 5 ft/s from a point 500 ft due east of P. At what rate are the people moving apart 15 minutes after the woman starts walking?

18. A baseball diamond is a square with side 90 ft. A batter hits the ball and runs toward first base with a speed of 24 ft/s.
(a) At what rate is his distance from second base decreasing when he is halfway to first base?
(b) At what rate is his distance from third base increasing at the same moment?

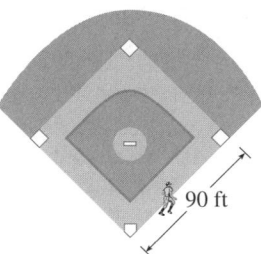

90 ft

19. The altitude of a triangle is increasing at a rate of 1 cm/min while the area of the triangle is increasing at a rate of 2 cm²/min. At what rate is the base of the triangle changing when the altitude is 10 cm and the area is 100 cm²?

20. A boat is pulled into a dock by a rope attached to the bow of the boat and passing through a pulley on the dock that is 1 m higher than the bow of the boat. If the rope is pulled in

at a rate of 1 m/s, how fast is the boat approaching the dock when it is 8 m from the dock?

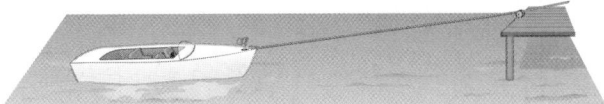

21. At noon, ship A is 100 km west of ship B. Ship A is sailing south at 35 km/h and ship B is sailing north at 25 km/h. How fast is the distance between the ships changing at 4:00 PM?

22. A particle moves along the curve $y = 2 \sin(\pi x/2)$. As the particle passes through the point $(\frac{1}{3}, 1)$, its x-coordinate increases at a rate of $\sqrt{10}$ cm/s. How fast is the distance from the particle to the origin changing at this instant?

23. Two carts, A and B, are connected by a rope 39 ft long that passes over a pulley P. The point Q is on the floor 12 ft directly beneath P and between the carts. Cart A is being pulled away from Q at a speed of 2 ft/s. How fast is cart B moving toward Q at the instant when cart A is 5 ft from Q?

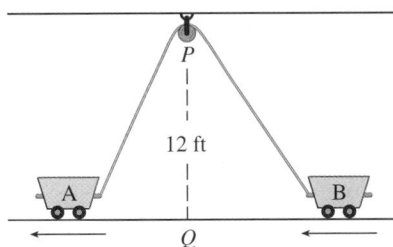

24. Water is leaking out of an inverted conical tank at a rate of 10,000 cm³/min at the same time that water is being pumped into the tank at a constant rate. The tank has height 6 m and the diameter at the top is 4 m. If the water level is rising at a rate of 20 cm/min when the height of the water is 2 m, find the rate at which water is being pumped into the tank.

25. A trough is 10 ft long and its ends have the shape of isosceles triangles that are 3 ft across at the top and have a height of 1 ft. If the trough is being filled with water at a rate of 12 ft³/min, how fast is the water level rising when the water is 6 inches deep?

26. A swimming pool is 20 ft wide, 40 ft long, 3 ft deep at the shallow end, and 9 ft deep at its deepest point. A cross-section is shown in the figure. If the pool is being filled at a rate of 0.8 ft³/min, how fast is the water level rising when the depth at the deepest point is 5 ft?

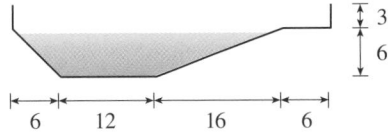

27. Gravel is being dumped from a conveyor belt at a rate of 30 ft³/min, and its coarseness is such that it forms a pile in the shape of a cone whose base diameter and height are always equal. How fast is the height of the pile increasing when the pile is 10 ft high?

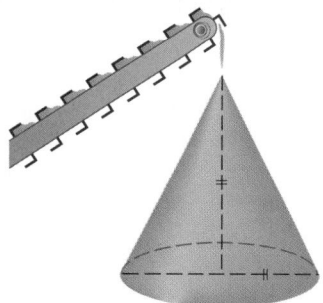

28. A kite 100 ft above the ground moves horizontally at a speed of 8 ft/s. At what rate is the angle between the string and the horizontal decreasing when 200 ft of string has been let out?

29. Two sides of a triangle are 4 m and 5 m in length and the angle between them is increasing at a rate of 0.06 rad/s. Find the rate at which the area of the triangle is increasing when the angle between the sides of fixed length is $\pi/3$.

30. Two sides of a triangle have lengths 12 m and 15 m. The angle between them is increasing at a rate of 2°/min. How fast is the length of the third side increasing when the angle between the sides of fixed length is 60°?

31. The top of a ladder slides down a vertical wall at a rate of 0.15 m/s. At the moment when the bottom of the ladder is 3 m from the wall, it slides away from the wall at a rate of 0.2 m/s. How long is the ladder?

32. A faucet is filling a hemispherical basin of diameter 60 cm with water at a rate of 2 L/min. Find the rate at which the water is rising in the basin when it is half full. [Use the following facts: 1 L is 1000 cm^3. The volume of the portion of a sphere with radius r from the bottom to a height h is $V = \pi\left(rh^2 - \frac{1}{3}h^3\right)$, as we will show in Chapter 7.]

33. Boyle's Law states that when a sample of gas is compressed at a constant temperature, the pressure P and volume V satisfy the equation $PV = C$, where C is a constant. Suppose that at a certain instant the volume is 600 cm^3, the pressure is 150 kPa, and the pressure is increasing at a rate of 20 kPa/min. At what rate is the volume decreasing at this instant?

34. When air expands adiabatically (without gaining or losing heat), its pressure P and volume V are related by the equation $PV^{1.4} = C$, where C is a constant. Suppose that at a certain instant the volume is 400 cm^3 and the pressure is 80 kPa and is decreasing at a rate of 10 kPa/min. At what rate is the volume increasing at this instant?

35. If two resistors with resistances R_1 and R_2 are connected in parallel, as in the figure, then the total resistance R,

measured in ohms (Ω), is given by

$$\frac{1}{R} = \frac{1}{R_1} + \frac{1}{R_2}$$

If R_1 and R_2 are increasing at rates of 0.3 Ω/s and 0.2 Ω/s, respectively, how fast is R changing when $R_1 = 80\ \Omega$ and $R_2 = 100\ \Omega$?

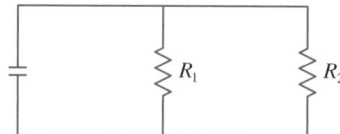

36. Brain weight B as a function of body weight W in fish has been modeled by the power function $B = 0.007W^{2/3}$, where B and W are measured in grams. A model for body weight as a function of body length L (measured in centimeters) is $W = 0.12L^{2.53}$. If, over 10 million years, the average length of a certain species of fish evolved from 15 cm to 20 cm at a constant rate, how fast was this species' brain growing when the average length was 18 cm?

37. A television camera is positioned 4000 ft from the base of a rocket launching pad. The angle of elevation of the camera has to change at the correct rate in order to keep the rocket in sight. Also, the mechanism for focusing the camera has to take into account the increasing distance from the camera to the rising rocket. Let's assume the rocket rises vertically and its speed is 600 ft/s when it has risen 3000 ft.
(a) How fast is the distance from the television camera to the rocket changing at that moment?
(b) If the television camera is always kept aimed at the rocket, how fast is the camera's angle of elevation changing at that same moment?

38. A lighthouse is located on a small island 3 km away from the nearest point P on a straight shoreline and its light makes four revolutions per minute. How fast is the beam of light moving along the shoreline when it is 1 km from P?

39. A plane flying with a constant speed of 300 km/h passes over a ground radar station at an altitude of 1 km and climbs at an angle of 30°. At what rate is the distance from the plane to the radar station increasing a minute later?

40. Two people start from the same point. One walks east at 3 mi/h and the other walks northeast at 2 mi/h. How fast is the distance between the people changing after 15 minutes?

41. A runner sprints around a circular track of radius 100 m at a constant speed of 7 m/s. The runner's friend is standing at a distance 200 m from the center of the track. How fast is the distance between the friends changing when the distance between them is 200 m?

42. The minute hand on a watch is 8 mm long and the hour hand is 4 mm long. How fast is the distance between the tips of the hands changing at one o'clock?

2.8 | LINEAR APPROXIMATIONS AND DIFFERENTIALS

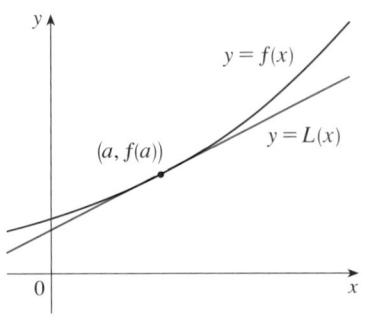

FIGURE 1

We have seen that a curve lies very close to its tangent line near the point of tangency. In fact, by zooming in toward a point on the graph of a differentiable function, we noticed that the graph looks more and more like its tangent line. (See Figure 4 in Section 2.1.) This observation is the basis for a method of finding approximate values of functions.

The idea is that it might be easy to calculate a value $f(a)$ of a function, but difficult (or even impossible) to compute nearby values of f. So we settle for the easily computed values of the linear function L whose graph is the tangent line of f at $(a, f(a))$. (See Figure 1.)

In other words, we use the tangent line at $(a, f(a))$ as an approximation to the curve $y = f(x)$ when x is near a. An equation of this tangent line is

$$y = f(a) + f'(a)(x - a)$$

and the approximation

$$\boxed{1} \qquad f(x) \approx f(a) + f'(a)(x - a)$$

is called the **linear approximation** or **tangent line approximation** of f at a. The linear function whose graph is this tangent line, that is,

$$\boxed{2} \qquad L(x) = f(a) + f'(a)(x - a)$$

is called the **linearization** of f at a.

▼ EXAMPLE 1 Find the linearization of the function $f(x) = \sqrt{x + 3}$ at $a = 1$ and use it to approximate the numbers $\sqrt{3.98}$ and $\sqrt{4.05}$. Are these approximations overestimates or underestimates?

SOLUTION The derivative of $f(x) = (x + 3)^{1/2}$ is

$$f'(x) = \tfrac{1}{2}(x + 3)^{-1/2} = \frac{1}{2\sqrt{x + 3}}$$

and so we have $f(1) = 2$ and $f'(1) = \tfrac{1}{4}$. Putting these values into Equation 2, we see that the linearization is

$$L(x) = f(1) + f'(1)(x - 1) = 2 + \tfrac{1}{4}(x - 1) = \frac{7}{4} + \frac{x}{4}$$

The corresponding linear approximation $\boxed{1}$ is

$$\sqrt{x + 3} \approx \frac{7}{4} + \frac{x}{4} \qquad \text{(when x is near 1)}$$

In particular, we have

$$\sqrt{3.98} \approx \tfrac{7}{4} + \tfrac{0.98}{4} = 1.995 \qquad \text{and} \qquad \sqrt{4.05} \approx \tfrac{7}{4} + \tfrac{1.05}{4} = 2.0125$$

The linear approximation is illustrated in Figure 2. We see that, indeed, the tangent line approximation is a good approximation to the given function when x is near 1. We also see that our approximations are overestimates because the tangent line lies above the curve.

Of course, a calculator could give us approximations for $\sqrt{3.98}$ and $\sqrt{4.05}$, but the linear approximation gives an approximation *over an entire interval*. ∎

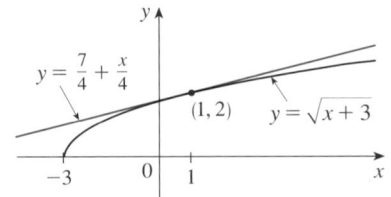

FIGURE 2

In the following table we compare the estimates from the linear approximation in Example 1 with the true values. Notice from this table, and also from Figure 2, that the tangent line approximation gives good estimates when x is close to 1 but the accuracy of the approximation deteriorates when x is farther away from 1.

	x	From $L(x)$	Actual value
$\sqrt{3.9}$	0.9	1.975	1.97484176...
$\sqrt{3.98}$	0.98	1.995	1.99499373...
$\sqrt{4}$	1	2	2.00000000...
$\sqrt{4.05}$	1.05	2.0125	2.01246117...
$\sqrt{4.1}$	1.1	2.025	2.02484567...
$\sqrt{5}$	2	2.25	2.23606797...
$\sqrt{6}$	3	2.5	2.44948974...

How good is the approximation that we obtained in Example 1? The next example shows that by using a graphing calculator or computer we can determine an interval throughout which a linear approximation provides a specified accuracy.

EXAMPLE 2 For what values of x is the linear approximation

$$\sqrt{x + 3} \approx \frac{7}{4} + \frac{x}{4}$$

accurate to within 0.5? What about accuracy to within 0.1?

SOLUTION Accuracy to within 0.5 means that the functions should differ by less than 0.5:

$$\left| \sqrt{x + 3} - \left(\frac{7}{4} + \frac{x}{4} \right) \right| < 0.5$$

Equivalently, we could write

$$\sqrt{x + 3} - 0.5 < \frac{7}{4} + \frac{x}{4} < \sqrt{x + 3} + 0.5$$

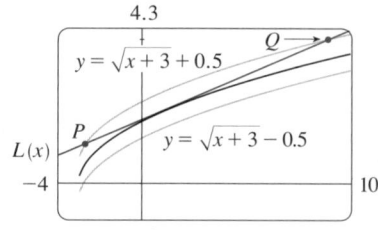

FIGURE 3

This says that the linear approximation should lie between the curves obtained by shifting the curve $y = \sqrt{x + 3}$ upward and downward by an amount 0.5. Figure 3 shows the tangent line $y = (7 + x)/4$ intersecting the upper curve $y = \sqrt{x + 3} + 0.5$ at P and Q. Zooming in and using the cursor, we estimate that the x-coordinate of P is about -2.66 and the x-coordinate of Q is about 8.66. Thus we see from the graph that the approximation

$$\sqrt{x + 3} \approx \frac{7}{4} + \frac{x}{4}$$

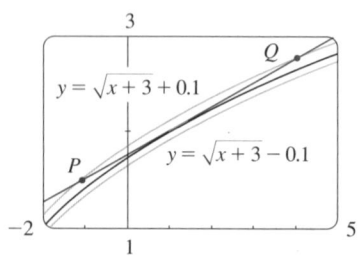

FIGURE 4

is accurate to within 0.5 when $-2.6 < x < 8.6$. (We have rounded to be safe.)

Similarly, from Figure 4 we see that the approximation is accurate to within 0.1 when $-1.1 < x < 3.9$. ∎

APPLICATIONS TO PHYSICS

Linear approximations are often used in physics. In analyzing the consequences of an equation, a physicist sometimes needs to simplify a function by replacing it with its linear approximation. For instance, in deriving a formula for the period of a pendulum, physics textbooks obtain the expression $a_T = -g \sin \theta$ for tangential accelera-tion and then replace $\sin \theta$ by θ with the remark that $\sin \theta$ is very close to θ if θ is not too large. [See, for example, *Physics: Calculus*, 2d ed., by Eugene Hecht (Pacific Grove, CA, 2000), p. 431.] You can verify that the linearization of the function $f(x) = \sin x$ at $a = 0$ is $L(x) = x$ and so the linear approximation at 0 is

$$\sin x \approx x$$

(see Exercise 28). So, in effect, the derivation of the formula for the period of a pen-dulum uses the tangent line approximation for the sine function.

Another example occurs in the theory of optics, where light rays that arrive at shallow angles relative to the optical axis are called *paraxial rays*. In paraxial (or Gaus-sian) optics, both $\sin \theta$ and $\cos \theta$ are replaced by their linearizations. In other words, the linear approximations

$$\sin \theta \approx \theta \qquad \text{and} \qquad \cos \theta \approx 1$$

are used because θ is close to 0. The results of calculations made with these approxi-mations became the basic theoretical tool used to design lenses. [See *Optics*, 4th ed., by Eugene Hecht (San Francisco, 2002), p. 154.]

In Section 8.8 we will present several other applications of the idea of linear approxi-mations to physics.

DIFFERENTIALS

The ideas behind linear approximations are sometimes formulated in the terminology and notation of *differentials*. If $y = f(x)$, where f is a differentiable function, then the **differential** dx is an independent variable; that is, dx can be given the value of any real number. The **differential** dy is then defined in terms of dx by the equation

$$\boxed{3} \qquad dy = f'(x)\,dx$$

■ If $dx \neq 0$, we can divide both sides of Equation 3 by dx to obtain

$$\frac{dy}{dx} = f'(x)$$

We have seen similar equations before, but now the left side can genuinely be interpreted as a ratio of differentials.

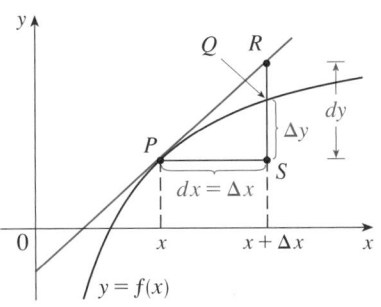

FIGURE 5

So dy is a dependent variable; it depends on the values of x and dx. If dx is given a specific value and x is taken to be some specific number in the domain of f, then the numerical value of dy is determined.

The geometric meaning of differentials is shown in Figure 5. Let $P(x, f(x))$ and $Q(x + \Delta x, f(x + \Delta x))$ be points on the graph of f and let $dx = \Delta x$. The correspond-ing change in y is

$$\Delta y = f(x + \Delta x) - f(x)$$

The slope of the tangent line PR is the derivative $f'(x)$. Thus the directed distance from S to R is $f'(x)\,dx = dy$. Therefore dy represents the amount that the tangent line rises or falls (the change in the linearization), whereas Δy represents the amount that the curve $y = f(x)$ rises or falls when x changes by an amount dx. Notice from Fig-ure 5 that the approximation $\Delta y \approx dy$ becomes better as Δx becomes smaller.

If we let $dx = x - a$, then $x = a + dx$ and we can rewrite the linear approxima-tion $\boxed{1}$ in the notation of differentials:

$$f(a + dx) \approx f(a) + dy$$

For instance, for the function $f(x) = \sqrt{x + 3}$ in Example 1, we have

$$dy = f'(x)\,dx = \frac{dx}{2\sqrt{x + 3}}$$

If $a = 1$ and $dx = \Delta x = 0.05$, then

$$dy = \frac{0.05}{2\sqrt{1 + 3}} = 0.0125$$

and

$$\sqrt{4.05} = f(1.05) \approx f(1) + dy = 2.0125$$

just as we found in Example 1.

Our final example illustrates the use of differentials in estimating the errors that occur because of approximate measurements.

V EXAMPLE 3 The radius of a sphere was measured and found to be 21 cm with a possible error in measurement of at most 0.05 cm. What is the maximum error in using this value of the radius to compute the volume of the sphere?

SOLUTION If the radius of the sphere is r, then its volume is $V = \frac{4}{3}\pi r^3$. If the error in the measured value of r is denoted by $dr = \Delta r$, then the corresponding error in the calculated value of V is ΔV, which can be approximated by the differential

$$dV = 4\pi r^2 \, dr$$

When $r = 21$ and $dr = 0.05$, this becomes

$$dV = 4\pi(21)^2 0.05 \approx 277$$

The maximum error in the calculated volume is about 277 cm³. ■

NOTE Although the possible error in Example 3 may appear to be rather large, a better picture of the error is given by the **relative error**, which is computed by dividing the error by the total volume:

$$\frac{\Delta V}{V} \approx \frac{dV}{V} = \frac{4\pi r^2 \, dr}{\frac{4}{3}\pi r^3} = 3\,\frac{dr}{r}$$

Therefore the relative error in the volume is approximately three times the relative error in the radius. In Example 3 the relative error in the radius is approximately $dr/r = 0.05/21 \approx 0.0024$ and it produces a relative error of about 0.007 in the volume. The errors could also be expressed as **percentage errors** of 0.24% in the radius and 0.7% in the volume.

2.8 | EXERCISES

1–4 ■ Find the linearization $L(x)$ of the function at a.

1. $f(x) = x^4 + 3x^2$, $a = -1$

2. $f(x) = \sin x$, $a = \pi/6$

3. $f(x) = \sqrt{x}$, $a = 4$ **4.** $f(x) = x^{3/4}$, $a = 16$

5. Find the linear approximation of the function $f(x) = \sqrt{1 - x}$ at $a = 0$ and use it to approximate the numbers $\sqrt{0.9}$ and $\sqrt{0.99}$. Illustrate by graphing f and the tangent line.

6. Find the linear approximation of the function $g(x) = \sqrt[3]{1 + x}$ at $a = 0$ and use it to approximate the numbers $\sqrt[3]{0.95}$ and $\sqrt[3]{1.1}$. Illustrate by graphing g and the tangent line.

7–10 ■ Verify the given linear approximation at $a = 0$. Then determine the values of x for which the linear approximation is accurate to within 0.1.

7. $\sqrt[4]{1 + 2x} \approx 1 + \frac{1}{2}x$ **8.** $\tan x \approx x$

9. $1/(1 + 2x)^4 \approx 1 - 8x$ **10.** $(1 + x)^{-3} \approx 1 - 3x$

11–14 ▪ Use a linear approximation (or differentials) to estimate the given number.

11. $(1.999)^4$ **12.** $\sqrt[3]{1001}$

13. $(8.06)^{2/3}$ **14.** $1/4.002$

15–16 ▪ Explain, in terms of linear approximations or differentials, why the approximation is reasonable.

15. $\sec 0.08 \approx 1$ **16.** $(1.01)^6 \approx 1.06$

17–18 ▪ Find the differential of each function.

17. (a) $y = \tan \sqrt{t}$ (b) $y = \dfrac{1 - v^2}{1 + v^2}$

18. (a) $y = s/(1 + 2s)$ (b) $y = u \cos u$

19. Let $y = \tan x$.
(a) Find the differential dy.
(b) Evaluate dy and Δy if $x = \pi/4$ and $dx = -0.1$.

20. Let $y = \sqrt{x}$.
(a) Find the differential dy.
(b) Evaluate dy and Δy if $x = 1$ and $dx = \Delta x = 1$.
(c) Sketch a diagram like Figure 5 showing the line segments with lengths dx, dy, and Δy.

21. The edge of a cube was found to be 30 cm with a possible error in measurement of 0.1 cm. Use differentials to estimate the maximum possible error, relative error, and percentage error in computing (a) the volume of the cube and (b) the surface area of the cube.

22. The radius of a circular disk is given as 24 cm with a maximum error in measurement of 0.2 cm.
(a) Use differentials to estimate the maximum error in the calculated area of the disk.
(b) What is the relative error? What is the percentage error?

23. The circumference of a sphere was measured to be 84 cm with a possible error of 0.5 cm.
(a) Use differentials to estimate the maximum error in the calculated surface area. What is the relative error?
(b) Use differentials to estimate the maximum error in the calculated volume. What is the relative error?

24. Use differentials to estimate the amount of paint needed to apply a coat of paint 0.05 cm thick to a hemispherical dome with diameter 50 m.

25. If a current I passes through a resistor with resistance R, Ohm's Law states that the voltage drop is $V = RI$. If V is constant and R is measured with a certain error, use differentials to show that the relative error in calculating I is approximately the same (in magnitude) as the relative error in R.

26. One side of a right triangle is known to be 20 cm long and the opposite angle is measured as 30°, with a possible error of $\pm1°$.
(a) Use differentials to estimate the error in computing the length of the hypotenuse.
(b) What is the percentage error?

27. When blood flows along a blood vessel, the flux F (the volume of blood per unit time that flows past a given point) is proportional to the fourth power of the radius R of the blood vessel:

$$F = kR^4$$

(This is known as Poiseuille's Law.) A partially clogged artery can be expanded by an operation called angioplasty, in which a balloon-tipped catheter is inflated inside the artery in order to widen it and restore the normal blood flow.
Show that the relative change in F is about four times the relative change in R. How will a 5% increase in the radius affect the flow of blood?

28. On page 431 of *Physics: Calculus*, 2d ed., by Eugene Hecht (Pacific Grove, CA, 2000), in the course of deriving the formula $T = 2\pi\sqrt{L/g}$ for the period of a pendulum of length L, the author obtains the equation $a_T = -g\sin\theta$ for the tangential acceleration of the bob of the pendulum. He then says, "for small angles, the value of θ in radians is very nearly the value of $\sin\theta$; they differ by less than 2% out to about 20°."
(a) Verify the linear approximation at 0 for the sine function:

$$\sin x \approx x$$

(b) Use a graphing device to determine the values of x for which $\sin x$ and x differ by less than 2%. Then verify Hecht's statement by converting from radians to degrees.

29. Suppose that the only information we have about a function f is that $f(1) = 5$ and the graph of its *derivative* is as shown.
(a) Use a linear approximation to estimate $f(0.9)$ and $f(1.1)$.
(b) Are your estimates in part (a) too large or too small? Explain.

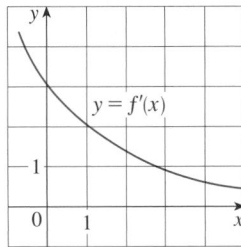

30. Suppose that we don't have a formula for $g(x)$ but we know that $g(2) = -4$ and $g'(x) = \sqrt{x^2 + 5}$ for all x.
(a) Use a linear approximation to estimate $g(1.95)$ and $g(2.05)$.
(b) Are your estimates in part (a) too large or too small? Explain.

CHAPTER 2   REVIEW

CONCEPT CHECK

1. Write an expression for the slope of the tangent line to the curve $y = f(x)$ at the point $(a, f(a))$.

2. Suppose an object moves along a straight line with position $f(t)$ at time t. Write an expression for the instantaneous velocity of the object at time $t = a$. How can you interpret this velocity in terms of the graph of f?

3. Define the derivative $f'(a)$. Discuss two ways of interpreting this number.

4. If $y = f(x)$ and x changes from x_1 to x_2, write expressions for the following.
  (a) The average rate of change of y with respect to x over the interval $[x_1, x_2]$.
  (b) The instantaneous rate of change of y with respect to x at $x = x_1$.

5. Define the second derivative of f. If $f(t)$ is the position function of a particle, how can you interpret the second derivative?

6. (a) What does it mean for f to be differentiable at a?
  (b) What is the relation between the differentiability and continuity of a function?

  (c) Sketch the graph of a function that is continuous but not differentiable at $a = 2$.

7. Describe several ways in which a function can fail to be differentiable. Illustrate with sketches.

8. State each differentiation rule both in symbols and in words.
  (a) The Power Rule   (b) The Constant Multiple Rule
  (c) The Sum Rule   (d) The Difference Rule
  (e) The Product Rule   (f) The Quotient Rule
  (g) The Chain Rule

9. State the derivative of each function.
  (a) $y = x^n$ (b) $y = \sin x$ (c) $y = \cos x$
  (d) $y = \tan x$ (e) $y = \csc x$ (f) $y = \sec x$
  (g) $y = \cot x$

10. Explain how implicit differentiation works.

11. (a) Write an expression for the linearization of f at a.
  (b) If $y = f(x)$, write an expression for the differential dy.
  (c) If $dx = \Delta x$, draw a picture showing the geometric meanings of Δy and dy.

TRUE-FALSE QUIZ

Determine whether the statement is true or false. If it is true, explain why. If it is false, explain why or give an example that disproves the statement.

1. If f is continuous at a, then f is differentiable at a.

2. If f and g are differentiable, then
$$\frac{d}{dx}[f(x) + g(x)] = f'(x) + g'(x)$$

3. If f and g are differentiable, then
$$\frac{d}{dx}[f(x)\, g(x)] = f'(x)\, g'(x)$$

4. If f and g are differentiable, then
$$\frac{d}{dx}[f(g(x))] = f'(g(x))\, g'(x)$$

5. If f is differentiable, then
$$\frac{d}{dx}\sqrt{f(x)} = \frac{f'(x)}{2\sqrt{f(x)}}$$

6. If f is differentiable, then
$$\frac{d}{dx} f(\sqrt{x}) = \frac{f'(x)}{2\sqrt{x}}$$

7. $\dfrac{d}{dx}\left| x^2 + x \right| = \left| 2x + 1 \right|$

8. If $f'(r)$ exists, then $\lim_{x \to r} f(x) = f(r)$.

9. If $g(x) = x^5$, then
$$\lim_{x \to 2} \frac{g(x) - g(2)}{x - 2} = 80$$

10. $\dfrac{d^2 y}{dx^2} = \left(\dfrac{dy}{dx} \right)^2$

11. An equation of the tangent line to the parabola $y = x^2$ at $(-2, 4)$ is $y - 4 = 2x(x + 2)$.

12. $\dfrac{d}{dx}(\tan^2 x) = \dfrac{d}{dx}(\sec^2 x)$

EXERCISES

1. For the function f whose graph is shown, arrange the following numbers in increasing order:

$$0 \quad 1 \quad f'(2) \quad f'(3) \quad f'(5) \quad f''(5)$$

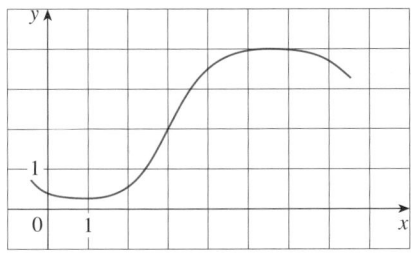

2. Find a function f and a number a such that

$$\lim_{h \to 0} \frac{(2+h)^6 - 64}{h} = f'(a)$$

3. The total cost of repaying a student loan at an interest rate of $r\%$ per year is $C = f(r)$.
 (a) What is the meaning of the derivative $f'(r)$? What are its units?
 (b) What does the statement $f'(10) = 1200$ mean?
 (c) Is $f'(r)$ always positive or does it change sign?

4–6 ▪ Trace or copy the graph of the function. Then sketch a graph of its derivative directly beneath.

4. 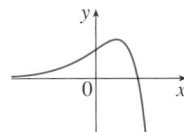   **5.** 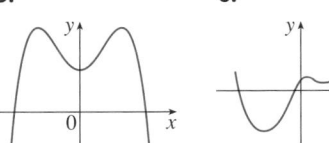   **6.**

7. The figure shows the graphs of f, f', and f''. Identify each curve, and explain your choices.

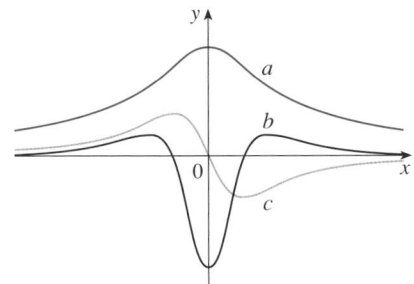

8. The *total fertility rate* at time t, denoted by $F(t)$, is an estimate of the average number of children born to each woman (assuming that current birth rates remain constant). The graph of the total fertility rate in the United States shows the fluctuations from 1940 to 1990.
 (a) Estimate the values of $F'(1950)$, $F'(1965)$, and $F'(1987)$.
 (b) What are the meanings of these derivatives?
 (c) Can you suggest reasons for the values of these derivatives?

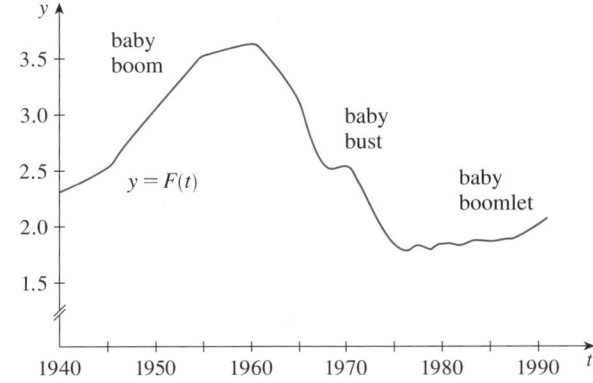

9. Let $C(t)$ be the total value of US currency (coins and banknotes) in circulation at time t. The table gives values of this function from 1980 to 2000, as of September 30, in billions of dollars. Interpret and estimate the value of $C'(1990)$.

t	1980	1985	1990	1995	2000
$C(t)$	129.9	187.3	271.9	409.3	568.6

10–11 ▪ Find $f'(x)$ from first principles, that is, directly from the definition of a derivative.

10. $f(x) = \dfrac{4-x}{3+x}$ **11.** $f(x) = x^3 + 5x + 4$

12. (a) If $f(x) = \sqrt{3 - 5x}$, use the definition of a derivative to find $f'(x)$.
 (b) Find the domains of f and f'.
 (c) Graph f and f' on a common screen. Compare the graphs to see whether your answer to part (a) is reasonable.

13–40 ▪ Calculate y'.

13. $y = (x^2 + x^3)^4$ **14.** $y = \dfrac{1}{\sqrt{x}} - \dfrac{1}{\sqrt[5]{x^3}}$

15. $y = \dfrac{x^2 - x + 2}{\sqrt{x}}$ **16.** $y = \dfrac{\tan x}{1 + \cos x}$

17. $y = x^2 \sin \pi x$ **18.** $y = \left(x + \dfrac{1}{x^2} \right)^{\sqrt{7}}$

19. $y = \dfrac{t^4 - 1}{t^4 + 1}$

20. $y = \sin(\cos x)$

21. $y = \tan \sqrt{1 - x}$

22. $y = \dfrac{1}{\sin(x - \sin x)}$

23. $xy^4 + x^2 y = x + 3y$

24. $y = \sec(1 + x^2)$

25. $y = \dfrac{\sec 2\theta}{1 + \tan 2\theta}$

26. $x^2 \cos y + \sin 2y = xy$

27. $y = (1 - x^{-1})^{-1}$

28. $y = 1/\sqrt[3]{x + \sqrt{x}}$

29. $\sin(xy) = x^2 - y$

30. $y = \sqrt{\sin \sqrt{x}}$

31. $y = \cot(3x^2 + 5)$

32. $y = \dfrac{(x + \lambda)^4}{x^4 + \lambda^4}$

33. $y = \sqrt{x} \cos \sqrt{x}$

34. $y = \dfrac{\sin mx}{x}$

35. $y = \tan^2(\sin \theta)$

36. $x \tan y = y - 1$

37. $y = \sqrt[5]{x \tan x}$

38. $y = \dfrac{(x - 1)(x - 4)}{(x - 2)(x - 3)}$

39. $y = \sin(\tan \sqrt{1 + x^3})$

40. $y = \sin^2(\cos \sqrt{\sin \pi x})$

41. If $f(t) = \sqrt{4t + 1}$, find $f''(2)$.

42. If $g(\theta) = \theta \sin \theta$, find $g''(\pi/6)$.

43. Find y'' if $x^6 + y^6 = 1$.

44. Find $f^{(n)}(x)$ if $f(x) = 1/(2 - x)$.

45–46 ▪ Find an equation of the tangent to the curve at the given point.

45. $y = 4 \sin^2 x$, $(\pi/6, 1)$

46. $y = \dfrac{x^2 - 1}{x^2 + 1}$, $(0, -1)$

47–48 ▪ Find equations of the tangent line and normal line to the curve at the given point.

47. $y = \sqrt{1 + 4 \sin x}$, $(0, 1)$

48. $x^2 + 4xy + y^2 = 13$, $(2, 1)$

49. At what points on the curve $y = \sin x + \cos x$, $0 \leq x \leq 2\pi$, is the tangent line horizontal?

50. Find the points on the ellipse $x^2 + 2y^2 = 1$ where the tangent line has slope 1.

51. Suppose that $h(x) = f(x) g(x)$ and $F(x) = f(g(x))$, where $f(2) = 3$, $g(2) = 5$, $g'(2) = 4$, $f'(2) = -2$, and $f'(5) = 11$. Find (a) $h'(2)$ and (b) $F'(2)$.

52. If f and g are the functions whose graphs are shown, let $P(x) = f(x) g(x)$, $Q(x) = f(x)/g(x)$, and $C(x) = f(g(x))$. Find (a) $P'(2)$, (b) $Q'(2)$, and (c) $C'(2)$.

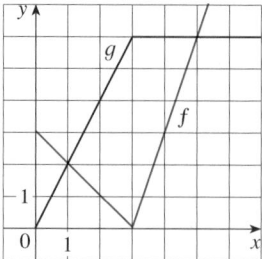

53–60 ▪ Find f' in terms of g'.

53. $f(x) = x^2 g(x)$

54. $f(x) = g(x^2)$

55. $f(x) = [g(x)]^2$

56. $f(x) = x^a g(x^b)$

57. $f(x) = g(g(x))$

58. $f(x) = \sin(g(x))$

59. $f(x) = g(\sin x)$

60. $f(x) = g(\tan \sqrt{x})$

61–62 ▪ Find h' in terms of f' and g'.

61. $h(x) = \dfrac{f(x) g(x)}{f(x) + g(x)}$

62. $h(x) = f(g(\sin 4x))$

63. The graph of f is shown. State, with reasons, the numbers at which f is not differentiable.

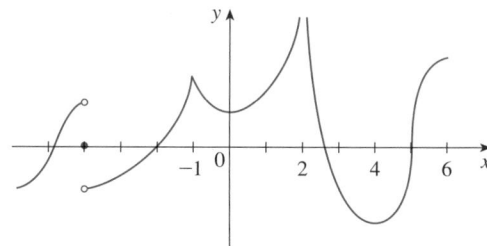

64. The volume of a right circular cone is $V = \frac{1}{3}\pi r^2 h$, where r is the radius of the base and h is the height.
(a) Find the rate of change of the volume with respect to the height if the radius is constant.
(b) Find the rate of change of the volume with respect to the radius if the height is constant.

65. A particle moves on a vertical line so that its coordinate at time t is $y = t^3 - 12t + 3$, $t \geq 0$.
(a) Find the velocity and acceleration functions.
(b) When is the particle moving upward and when is it moving downward?

(c) Find the distance that the particle travels in the time interval $0 \le t \le 3$.

(d) Graph the position, velocity, and acceleration functions for $0 \le t \le 3$.

66. The cost, in dollars, of producing x units of a certain commodity is

$$C(x) = 920 + 2x - 0.02x^2 + 0.00007x^3$$

(a) Find the marginal cost function.

(b) Find $C'(100)$ and explain its meaning.

(c) Compare $C'(100)$ with the cost of producing the 101st item.

67. The volume of a cube is increasing at a rate of 10 cm³/min. How fast is the surface area increasing when the length of an edge is 30 cm?

68. A paper cup has the shape of a cone with height 10 cm and radius 3 cm (at the top). If water is poured into the cup at a rate of 2 cm³/s, how fast is the water level rising when the water is 5 cm deep?

69. A balloon is rising at a constant speed of 5 ft/s. A boy is cycling along a straight road at a speed of 15 ft/s. When he passes under the balloon, it is 45 ft above him. How fast is the distance between the boy and the balloon increasing 3 s later?

70. A waterskier skis over the ramp shown in the figure at a speed of 30 ft/s. How fast is she rising as she leaves the ramp?

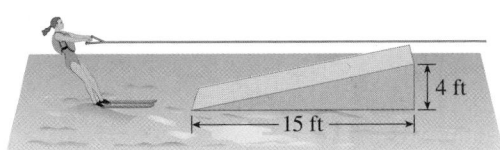

71. The angle of elevation of the sun is decreasing at a rate of 0.25 rad/h. How fast is the shadow cast by a 400-ft-tall building increasing when the angle of elevation of the sun is $\pi/6$?

72. (a) Find the linear approximation to $f(x) = \sqrt{25 - x^2}$ near 3.

(b) Illustrate part (a) by graphing f and the linear approximation.

(c) For what values of x is the linear approximation accurate to within 0.1?

73. (a) Find the linearization of $f(x) = \sqrt[3]{1 + 3x}$ at $a = 0$. State the corresponding linear approximation and use it to give an approximate value for $\sqrt[3]{1.03}$.

(b) Determine the values of x for which the linear approximation given in part (a) is accurate to within 0.1.

74. Evaluate dy if $y = x^3 - 2x^2 + 1$, $x = 2$, and $dx = 0.2$.

75. A window has the shape of a square surmounted by a semicircle. The base of the window is measured as having width 60 cm with a possible error in measurement of 0.1 cm. Use differentials to estimate the maximum error possible in computing the area of the window.

76–78 ■ Express the limit as a derivative and evaluate.

76. $\displaystyle \lim_{x \to 1} \frac{x^{17} - 1}{x - 1}$

77. $\displaystyle \lim_{h \to 0} \frac{\sqrt[4]{16 + h} - 2}{h}$

78. $\displaystyle \lim_{\theta \to \pi/3} \frac{\cos \theta - 0.5}{\theta - \pi/3}$

79. Evaluate $\displaystyle \lim_{x \to 0} \frac{\sqrt{1 + \tan x} - \sqrt{1 + \sin x}}{x^3}$.

80. Show that the length of the portion of any tangent line to the astroid $x^{2/3} + y^{2/3} = a^{2/3}$ cut off by the coordinate axes is constant.

81. If $f(x) = \displaystyle \lim_{t \to x} \frac{\sec t - \sec x}{t - x}$, find the value of $f'(\pi/4)$.

82. Find the values of the constants a and b such that

$$\lim_{x \to 0} \frac{\sqrt[3]{ax + b} - 2}{x} = \frac{5}{12}$$

3

INVERSE FUNCTIONS
Exponential, Logarithmic, and Inverse Trigonometric Functions

The common theme that links the functions of this chapter is that they occur as pairs of inverse functions. In particular, two of the most important functions that occur in mathematics and its applications are the exponential function $f(x) = a^x$ and its inverse function, the logarithmic function $g(x) = \log_a x$. Here we investigate their properties, compute their derivatives, and use them to describe exponential growth and decay in biology, physics, chemistry, and other sciences. We also study the inverses of the trigonometric and hyperbolic functions. Finally we look at a method (l'Hospital's Rule) for computing limits of such functions.

3.1 | EXPONENTIAL FUNCTIONS

The function $f(x) = 2^x$ is called an *exponential function* because the variable, x, is the exponent. It should not be confused with the power function $g(x) = x^2$, in which the variable is the base.

In general, an **exponential function** is a function of the form

$$f(x) = a^x$$

where a is a positive constant. Let's recall what this means.

If $x = n$, a positive integer, then

$$a^n = \underbrace{a \cdot a \cdot \cdots \cdot a}_{n \text{ factors}}$$

If $x = 0$, then $a^0 = 1$, and if $x = -n$, where n is a positive integer, then

$$a^{-n} = \frac{1}{a^n}$$

If x is a rational number, $x = p/q$, where p and q are integers and $q > 0$, then

$$a^x = a^{p/q} = \sqrt[q]{a^p} = \left(\sqrt[q]{a}\right)^p$$

But what is the meaning of a^x if x is an irrational number? For instance, what is meant by $2^{\sqrt{3}}$ or 5^π?

To help us answer this question we first look at the graph of the function $y = 2^x$, where x is rational. A representation of this graph is shown in Figure 1. We want to enlarge the domain of $y = 2^x$ to include both rational and irrational numbers.

There are holes in the graph in Figure 1 corresponding to irrational values of x. We want to fill in the holes by defining $f(x) = 2^x$, where $x \in \mathbb{R}$, so that f is an increasing continuous function. In particular, since the irrational number $\sqrt{3}$ satisfies

$$1.7 < \sqrt{3} < 1.8$$

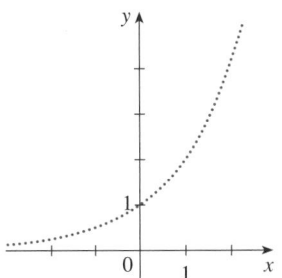

FIGURE 1
Representation of $y = 2^x$, x rational

we must have

$$2^{1.7} < 2^{\sqrt{3}} < 2^{1.8}$$

and we know what $2^{1.7}$ and $2^{1.8}$ mean because 1.7 and 1.8 are rational numbers. Similarly, if we use better approximations for $\sqrt{3}$, we obtain better approximations for $2^{\sqrt{3}}$:

$$1.73 < \sqrt{3} < 1.74 \quad\Rightarrow\quad 2^{1.73} < 2^{\sqrt{3}} < 2^{1.74}$$

$$1.732 < \sqrt{3} < 1.733 \quad\Rightarrow\quad 2^{1.732} < 2^{\sqrt{3}} < 2^{1.733}$$

$$1.7320 < \sqrt{3} < 1.7321 \quad\Rightarrow\quad 2^{1.7320} < 2^{\sqrt{3}} < 2^{1.7321}$$

$$1.73205 < \sqrt{3} < 1.73206 \quad\Rightarrow\quad 2^{1.73205} < 2^{\sqrt{3}} < 2^{1.73206}$$

$$\vdots \qquad \vdots \qquad\qquad \vdots \qquad \vdots$$

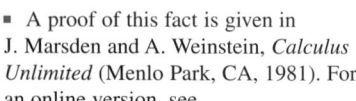

- A proof of this fact is given in J. Marsden and A. Weinstein, *Calculus Unlimited* (Menlo Park, CA, 1981). For an online version, see

www.cds.caltech.edu/~marsden/ volume/cu/CU.pdf

It can be shown that there is exactly one number that is greater than all of the numbers

$$2^{1.7}, \quad 2^{1.73}, \quad 2^{1.732}, \quad 2^{1.7320}, \quad 2^{1.73205}, \quad \ldots$$

and less than all of the numbers

$$2^{1.8}, \quad 2^{1.74}, \quad 2^{1.733}, \quad 2^{1.7321}, \quad 2^{1.73206}, \quad \ldots$$

We define $2^{\sqrt{3}}$ to be this number. Using the preceding approximation process, we can compute it correct to six decimal places:

$$2^{\sqrt{3}} \approx 3.321997$$

Similarly, we can define 2^x (or a^x, if $a > 0$) where x is any irrational number. Figure 2 shows how all the holes in Figure 1 have been filled to complete the graph of the function $f(x) = 2^x$, $x \in \mathbb{R}$.

In general, if a is any positive number, we define

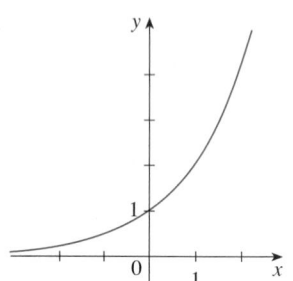

FIGURE 2
$y = 2^x$, x real

$\boxed{1}$
$$a^x = \lim_{r \to x} a^r \qquad r \text{ rational}$$

This definition makes sense because any irrational number can be approximated as closely as we like by a rational number. For instance, because $\sqrt{3}$ has the decimal representation $\sqrt{3} = 1.7320508\ldots$, Definition 1 says that $2^{\sqrt{3}}$ is the limit of the sequence of numbers

$$2^{1.7}, \quad 2^{1.73}, \quad 2^{1.732}, \quad 2^{1.7320}, \quad 2^{1.73205}, \quad 2^{1.732050}, \quad 2^{1.7320508}, \quad \ldots$$

Similarly, 5^{π} is the limit of the sequence of numbers

$$5^{3.1}, \quad 5^{3.14}, \quad 5^{3.141}, \quad 5^{3.1415}, \quad 5^{3.14159}, \quad 5^{3.141592}, \quad 5^{3.1415926}, \quad \ldots$$

It can be shown that Definition 1 uniquely specifies a^x and makes the function $f(x) = a^x$ continuous.

The graphs of members of the family of functions $y = a^x$ are shown in Figure 3 for various values of the base a. Notice that all of these graphs pass through the same point $(0, 1)$ because $a^0 = 1$ for $a \neq 0$. Notice also that as the base a gets larger, the exponential function grows more rapidly (for $x > 0$).

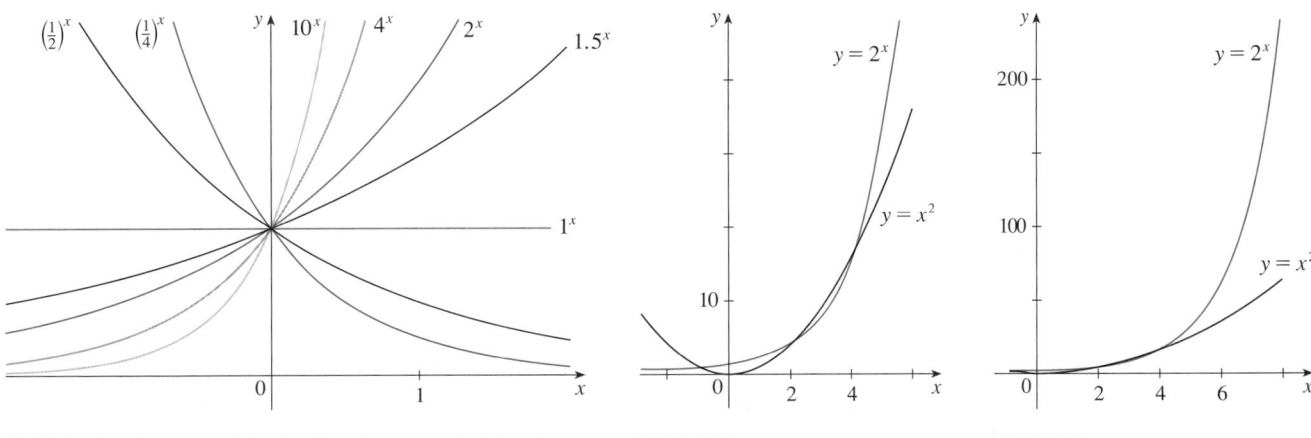

FIGURE 3 Members of the family of exponential functions   **FIGURE 4**   **FIGURE 5**

■ www.stewartcalculus.com
See Additional Example A.

Figure 4 shows how the exponential function $y = 2^x$ compares with the power function $y = x^2$. The graphs intersect three times, but ultimately the exponential curve $y = 2^x$ grows far more rapidly than the parabola $y = x^2$. (See also Figure 5.)

You can see from Figure 3 that there are basically three kinds of exponential functions $y = a^x$. If $0 < a < 1$, the exponential function decreases; if $a = 1$, it is a constant; and if $a > 1$, it increases. These three cases are illustrated in Figure 6. Since $(1/a)^x = 1/a^x = a^{-x}$, the graph of $y = (1/a)^x$ is just the reflection of the graph of $y = a^x$ about the y-axis.

FIGURE 6

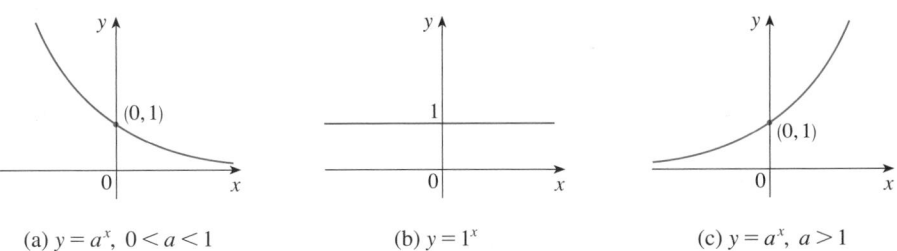

(a) $y = a^x$, $0 < a < 1$ (b) $y = 1^x$ (c) $y = a^x$, $a > 1$

The properties of the exponential function are summarized in the following theorem.

---

2 **THEOREM** If $a > 0$ and $a \neq 1$, then $f(x) = a^x$ is a continuous function with domain $\mathbb{R}$ and range $(0, \infty)$. In particular, $a^x > 0$ for all x. If $a, b > 0$ and $x, y \in \mathbb{R}$, then

1. $a^{x+y} = a^x a^y$ **2.** $a^{x-y} = \dfrac{a^x}{a^y}$ **3.** $(a^x)^y = a^{xy}$ **4.** $(ab)^x = a^x b^x$

The reason for the importance of the exponential function lies in properties 1–4, which are called the **Laws of Exponents**. If x and y are rational numbers, then these laws are well known from elementary algebra. For arbitrary real numbers x and y these laws can be deduced from the special case where the exponents are rational by using Equation 1.

■ In Appendix C we present a definition of the exponential function that enables us to give an easy proof of the Laws of Exponents.

The following limits can be read from the graphs shown in Figure 6 or proved from the definition of a limit at infinity. (See Exercise 79 in Section 3.2.)

$$\boxed{3} \qquad \text{If } a > 1, \text{ then} \qquad \lim_{x \to \infty} a^x = \infty \qquad \text{and} \qquad \lim_{x \to -\infty} a^x = 0$$

$$\text{If } 0 < a < 1, \text{ then} \qquad \lim_{x \to \infty} a^x = 0 \qquad \text{and} \qquad \lim_{x \to -\infty} a^x = \infty$$

In particular, if $a \neq 1$, then the x-axis is a horizontal asymptote of the graph of the exponential function $y = a^x$.

EXAMPLE 1
(a) Find $\lim_{x \to \infty} (2^{-x} - 1)$.
(b) Sketch the graph of the function $y = 2^{-x} - 1$.

SOLUTION
(a)
$$\lim_{x \to \infty} (2^{-x} - 1) = \lim_{x \to \infty} \left[\left(\tfrac{1}{2} \right)^x - 1 \right]$$
$$= 0 - 1 \qquad \left[\text{by } \boxed{3} \text{ with } a = \tfrac{1}{2} < 1 \right]$$
$$= -1$$

(b) We write $y = \left(\tfrac{1}{2} \right)^x - 1$ as in part (a). The graph of $y = \left(\tfrac{1}{2} \right)^x$ is shown in Figure 3, so we shift it down one unit to obtain the graph of $y = \left(\tfrac{1}{2} \right)^x - 1$ shown in Figure 7. (For a review of shifting graphs, see Section 1.2.) Part (a) shows that the line $y = -1$ is a horizontal asymptote. ■

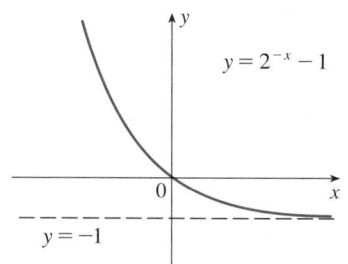

FIGURE 7

THE NUMBER e AND THE NATURAL EXPONENTIAL FUNCTION

Of all possible bases for an exponential function, there is one that is most convenient for the purposes of calculus. We will see in Section 3.3 that the differentiation formula for an exponential function is simplest when the base is chosen to be the number e, which is defined as follows:

$$\boxed{4} \qquad e = \lim_{x \to 0} (1 + x)^{1/x}$$

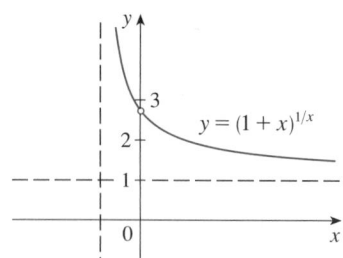

FIGURE 8

x	$(1 + x)^{1/x}$
0.1	2.59374246
0.01	2.70481383
0.001	2.71692393
0.0001	2.71814593
0.00001	2.71826824
0.000001	2.71828047
0.0000001	2.71828169
0.00000001	2.71828181

The graph of the function $y = (1 + x)^{1/x}$ is shown in Figure 8. It is not defined when $x = 0$, but its behavior when x is near 0 is indicated by the table of values correct to eight decimal places. These values suggest (but don't prove) that the limit in Definition 4 exists and that $e \approx 2.71828$. The existence of the limit is proved in Appendix D. The approximate value to 20 decimal places is

$$e \approx 2.71828182845904523536$$

The decimal expansion of e is nonrepeating because e is an irrational number. (See Exercise 30 in Section 8.8.) The notation e for this number was chosen by the Swiss mathematician Leonhard Euler in 1727, probably because it's the first letter of the word *exponential*.

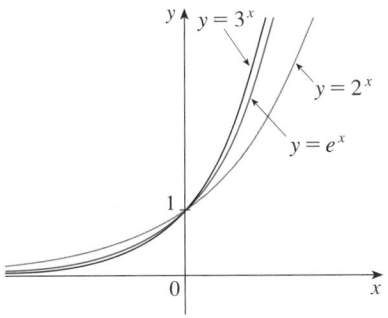

FIGURE 9

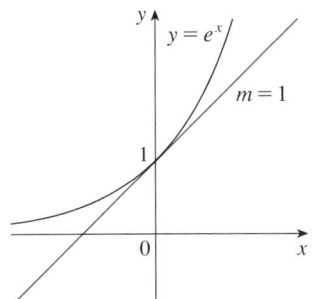

FIGURE 10
The natural exponential function
crosses the *y*-axis with a slope of 1.

The exponential function $y = e^x$ with base e is called the **natural exponential function**. Because e lies between 2 and 3, the graph of $y = e^x$ lies between the graphs of $y = 2^x$ and $y = 3^x$, as shown in Figure 9.

We will see in Section 3.3 that the natural exponential function is the exponential function whose graph crosses the *y*-axis with a slope of 1. (See Figure 10.) In fact, we can see why this might be true if we look at the limit in Definition 4. The slope of the tangent line to the graph of $f(x) = e^x$ at the point (0, 1) is

$$f'(0) = \lim_{h \to 0} \frac{f(0 + h) - f(0)}{h} = \lim_{h \to 0} \frac{e^h - 1}{h}$$

Replacing x by h in Definition 4, we see that, for small values of h,

$$e \approx (1 + h)^{1/h} \qquad \text{so} \qquad e^h \approx 1 + h \qquad \text{and} \qquad e^h - 1 \approx h$$

Thus if h is near 0, we have

$$\frac{e^h - 1}{h} \approx 1$$

and so it seems plausible that $f'(0) = 1$.

The exponential function $f(x) = e^x$ is one of the most frequently occurring functions in calculus and its applications, so it is important to be familiar with its graph (Figure 10) and properties. We summarize these properties as follows, using the fact that this function is just a special case of the exponential functions considered before but with base $a = e > 1$.

---

5 **PROPERTIES OF THE NATURAL EXPONENTIAL FUNCTION** The exponential function $f(x) = e^x$ is a continuous function with domain $\mathbb{R}$ and range $(0, \infty)$. Thus $e^x > 0$ for all x. Also

$$\lim_{x \to -\infty} e^x = 0 \qquad \lim_{x \to \infty} e^x = \infty$$

So the *x*-axis is a horizontal asymptote of $f(x) = e^x$.

■ www.stewartcalculus.com
See Additional Example B.

V EXAMPLE 2 Evaluate $\lim_{x \to 0^-} e^{1/x}$.

SOLUTION If we let $t = 1/x$, we know from Section 1.6 that $t \to -\infty$ as $x \to 0^-$. Therefore, by **5**,

$$\lim_{x \to 0^-} e^{1/x} = \lim_{t \to -\infty} e^t = 0 \qquad\qquad ■$$

EXAMPLE 3 Find $\lim_{x \to \infty} \frac{e^{2x}}{e^{2x} + 1}$.

SOLUTION We divide numerator and denominator by e^{2x}:

$$\lim_{x \to \infty} \frac{e^{2x}}{e^{2x} + 1} = \lim_{x \to \infty} \frac{1}{1 + e^{-2x}} = \frac{1}{1 + \lim_{x \to \infty} e^{-2x}} = \frac{1}{1 + 0} = 1$$

We have used the fact that $t = -2x \to -\infty$ as $x \to \infty$ and so

$$\lim_{x \to \infty} e^{-2x} = \lim_{t \to -\infty} e^t = 0 \qquad\qquad ■$$

3.1 | EXERCISES

1. (a) Write an equation that defines the exponential function with base $a > 0$.
   (b) What is the domain of this function?
   (c) If $a \neq 1$, what is the range of this function?
   (d) Sketch the general shape of the graph of the exponential function for each of the following cases.
      (i) $a > 1$ (ii) $a = 1$ (iii) $0 < a < 1$

2. (a) How is the number e defined?
   (b) What is an approximate value for e?
   (c) What is the natural exponential function?

 3–6 ▪ Graph the given functions on a common screen. How are these graphs related?

3. $y = 2^x$, $y = e^x$, $y = 5^x$, $y = 20^x$

4. $y = e^x$, $y = e^{-x}$, $y = 8^x$, $y = 8^{-x}$

5. $y = 3^x$, $y = 10^x$, $y = \left(\frac{1}{3}\right)^x$, $y = \left(\frac{1}{10}\right)^x$

6. $y = 0.9^x$, $y = 0.6^x$, $y = 0.3^x$, $y = 0.1^x$

7–12 ▪ Make a rough sketch of the graph of the function. Do not use a calculator. Just use the graphs given in Figures 3 and 9 and, if necessary, the transformations of Section 1.2.

7. $y = 10^{x+2}$ 8. $y = (0.5)^x - 2$

9. $y = -2^{-x}$ 10. $y = e^{|x|}$

11. $y = 1 - \frac{1}{2}e^{-x}$ 12. $y = 2(1 - e^x)$

13. Starting with the graph of $y = e^x$, write the equation of the graph that results from
   (a) shifting 2 units downward
   (b) shifting 2 units to the right
   (c) reflecting about the x-axis
   (d) reflecting about the y-axis
   (e) reflecting about the x-axis and then about the y-axis

14. Starting with the graph of $y = e^x$, find the equation of the graph that results from
   (a) reflecting about the line $y = 4$
   (b) reflecting about the line $x = 2$

15–16 ▪ Find the domain of each function.

15. (a) $f(x) = \dfrac{1 - e^{x^2}}{1 - e^{1-x^2}}$ (b) $f(x) = \dfrac{1+x}{e^{\cos x}}$

16. (a) $g(t) = \sin(e^{-t})$ (b) $g(t) = \sqrt{1 - 2^t}$

17–18 ▪ Find the exponential function $f(x) = Ca^x$ whose graph is given.

17.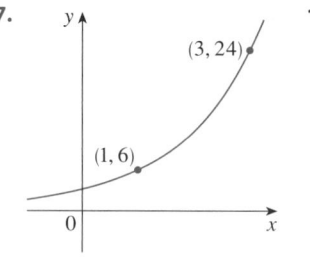
       18.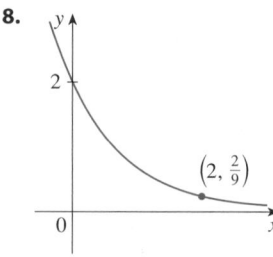

19. Suppose the graphs of $f(x) = x^2$ and $g(x) = 2^x$ are drawn on a coordinate grid where the unit of measurement is 1 inch. Show that, at a distance 2 ft to the right of the origin, the height of the graph of f is 48 ft but the height of the graph of g is about 265 mi.

 20. Compare the rates of growth of the functions $f(x) = x^5$ and $g(x) = 5^x$ by graphing both functions in several viewing rectangles. Find all points of intersection of the graphs correct to one decimal place.

 21. Compare the functions $f(x) = x^{10}$ and $g(x) = e^x$ by graphing both f and g in several viewing rectangles. When does the graph of g finally surpass the graph of f?

 22. Use a graph to estimate the values of x such that $e^x > 1{,}000{,}000{,}000$.

23–30 ▪ Find the limit.

23. $\lim\limits_{x \to \infty} (1.001)^x$ 24. $\lim\limits_{x \to \infty} e^{-x^2}$

25. $\lim\limits_{x \to \infty} \dfrac{e^{3x} - e^{-3x}}{e^{3x} + e^{-3x}}$ 26. $\lim\limits_{x \to \infty} \dfrac{2 + 10^x}{3 - 10^x}$

27. $\lim\limits_{x \to 2^+} e^{3/(2-x)}$ 28. $\lim\limits_{x \to 2^-} e^{3/(2-x)}$

29. $\lim\limits_{x \to \infty} (e^{-2x} \cos x)$ 30. $\lim\limits_{x \to (\pi/2)^+} e^{\tan x}$

 31. If you graph the function

$$f(x) = \frac{1 - e^{1/x}}{1 + e^{1/x}}$$

you'll see that f appears to be an odd function. Prove it.

 32. Graph several members of the family of functions

$$f(x) = \frac{1}{1 + ae^{bx}}$$

where $a > 0$. How does the graph change when b changes? How does it change when a changes?

 Graphing calculator or computer required   **CAS** Computer algebra system required   **1** Homework Hints at stewartcalculus.com

3.2 | INVERSE FUNCTIONS AND LOGARITHMS

Table 1 gives data from an experiment in which a bacteria culture started with 100 bacteria in a limited nutrient medium; the size of the bacteria population was recorded at hourly intervals. The number of bacteria N is a function of the time t: $N = f(t)$.

Suppose, however, that the biologist changes her point of view and becomes interested in the time required for the population to reach various levels. In other words, she is thinking of t as a function of N. This function is called the *inverse function* of f, denoted by f^{-1}, and read "f inverse." Thus $t = f^{-1}(N)$ is the time required for the population level to reach N. The values of f^{-1} can be found by reading Table 1 from right to left or by consulting Table 2. For instance, $f^{-1}(550) = 6$ because $f(6) = 550$.

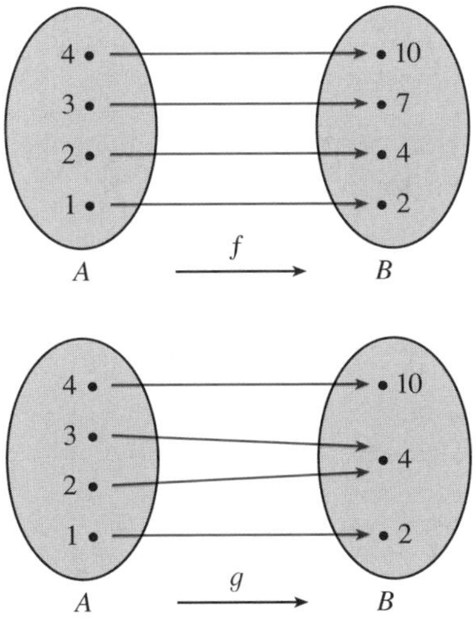

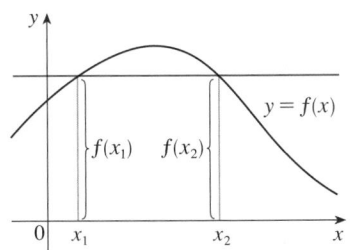

FIGURE 1

f is one-to-one; g is not.

TABLE 1 N as a function of t

t (hours)	$N = f(t)$ = population at time t
0	100
1	168
2	259
3	358
4	445
5	509
6	550
7	573
8	586

TABLE 2 t as a function of N

N	$t = f^{-1}(N)$ = time to reach N bacteria
100	0
168	1
259	2
358	3
445	4
509	5
550	6
573	7
586	8

Not all functions possess inverses. Let's compare the functions f and g whose arrow diagrams are shown in Figure 1. Note that f never takes on the same value twice (any two inputs in A have different outputs), whereas g does take on the same value twice (both 2 and 3 have the same output, 4). In symbols,

$$g(2) = g(3)$$

but
$$f(x_1) \neq f(x_2) \qquad \text{whenever } x_1 \neq x_2$$

Functions that share this property with f are called *one-to-one functions*.

■ In the language of inputs and outputs, Definition 1 says that f is one-to-one if each output corresponds to only one input.

> **1 DEFINITION** A function f is called a **one-to-one function** if it never takes on the same value twice; that is,
>
> $$f(x_1) \neq f(x_2) \qquad \text{whenever } x_1 \neq x_2$$

If a horizontal line intersects the graph of f in more than one point, then we see from Figure 2 that there are numbers x_1 and x_2 such that $f(x_1) = f(x_2)$. This means that f is not one-to-one. Therefore we have the following geometric method for determining whether a function is one-to-one.

FIGURE 2

This function is not one-to-one because $f(x_1) = f(x_2)$.

> **HORIZONTAL LINE TEST** A function is one-to-one if and only if no horizontal line intersects its graph more than once.

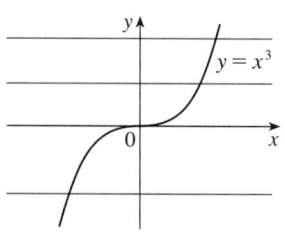

FIGURE 3
$f(x) = x^3$ is one-to-one.

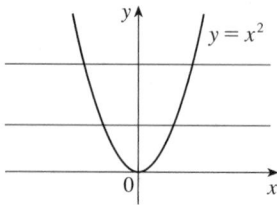

FIGURE 4
$g(x) = x^2$ is not one-to-one.

▼ EXAMPLE 1 Is the function $f(x) = x^3$ one-to-one?

SOLUTION 1 If $x_1 \neq x_2$, then $x_1^3 \neq x_2^3$ (two different numbers can't have the same cube). Therefore, by Definition 1, $f(x) = x^3$ is one-to-one.

SOLUTION 2 From Figure 3 we see that no horizontal line intersects the graph of $f(x) = x^3$ more than once. Therefore, by the Horizontal Line Test, f is one-to-one.   ■

▼ EXAMPLE 2 Is the function $g(x) = x^2$ one-to-one?

SOLUTION 1 This function is not one-to-one because, for instance,

$$g(1) = 1 = g(-1)$$

and so 1 and -1 have the same output.

SOLUTION 2 From Figure 4 we see that there are horizontal lines that intersect the graph of g more than once. Therefore, by the Horizontal Line Test, g is not one-to-one.   ■

One-to-one functions are important because they are precisely the functions that possess inverse functions according to the following definition.

---

2 DEFINITION Let f be a one-to-one function with domain A and range B. Then its **inverse function** f^{-1} has domain B and range A and is defined by

$$f^{-1}(y) = x \iff f(x) = y$$

for any y in B.

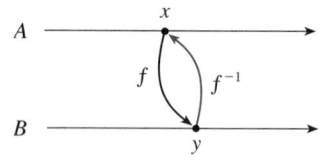

FIGURE 5

This definition says that if f maps x into y, then f^{-1} maps y back into x. (If f were not one-to-one, then f^{-1} would not be uniquely defined.) The arrow diagram in Figure 5 indicates that f^{-1} reverses the effect of f. Note that

$$\text{domain of } f^{-1} = \text{range of } f$$
$$\text{range of } f^{-1} = \text{domain of } f$$

For example, the inverse function of $f(x) = x^3$ is $f^{-1}(x) = x^{1/3}$ because if $y = x^3$, then

$$f^{-1}(y) = f^{-1}(x^3) = (x^3)^{1/3} = x$$

⊘ CAUTION Do not mistake the -1 in f^{-1} for an exponent. Thus

$$f^{-1}(x) \quad \text{does } not \text{ mean} \quad \frac{1}{f(x)}$$

The reciprocal $1/f(x)$ could, however, be written as $[f(x)]^{-1}$.

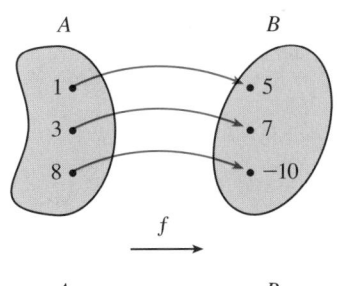

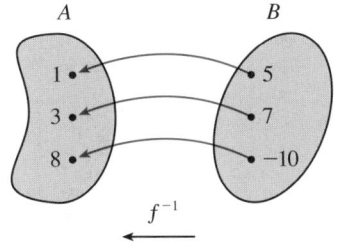

FIGURE 6

The inverse function reverses
inputs and outputs.

V EXAMPLE 3 If $f(1) = 5$, $f(3) = 7$, and $f(8) = -10$, find $f^{-1}(7)$, $f^{-1}(5)$, and $f^{-1}(-10)$.

SOLUTION From the definition of f^{-1} we have

$$f^{-1}(7) = 3 \qquad \text{because} \qquad f(3) = 7$$

$$f^{-1}(5) = 1 \qquad \text{because} \qquad f(1) = 5$$

$$f^{-1}(-10) = 8 \qquad \text{because} \qquad f(8) = -10$$

The diagram in Figure 6 makes it clear how f^{-1} reverses the effect of f in this case. ∎

The letter x is traditionally used as the independent variable, so when we concentrate on f^{-1} rather than on f, we usually reverse the roles of x and y in Definition 2 and write

$$\boxed{3} \qquad \boxed{\quad f^{-1}(x) = y \iff f(y) = x \quad}$$

By substituting for y in Definition 2 and substituting for x in $\boxed{3}$, we get the following **cancellation equations**:

$$\boxed{4} \qquad \boxed{\begin{array}{l} f^{-1}(f(x)) = x \quad \text{for every } x \text{ in } A \\ f(f^{-1}(x)) = x \quad \text{for every } x \text{ in } B \end{array}}$$

The first cancellation equation says that if we start with x, apply f, and then apply f^{-1}, we arrive back at x, where we started (see the machine diagram in Figure 7). Thus f^{-1} undoes what f does. The second equation says that f undoes what f^{-1} does.

FIGURE 7

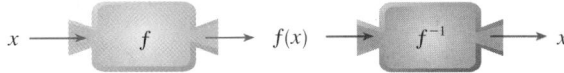

For example, if $f(x) = x^3$, then $f^{-1}(x) = x^{1/3}$ and so the cancellation equations become

$$f^{-1}(f(x)) = (x^3)^{1/3} = x$$

$$f(f^{-1}(x)) = (x^{1/3})^3 = x$$

These equations simply say that the cube function and the cube root function cancel each other when applied in succession.

Now let's see how to compute inverse functions. If we have a function $y = f(x)$ and are able to solve this equation for x in terms of y, then according to Definition 2 we must have $x = f^{-1}(y)$. If we want to call the independent variable x, we then interchange x and y and arrive at the equation $y = f^{-1}(x)$.

> **5** **HOW TO FIND THE INVERSE FUNCTION OF A ONE-TO-ONE FUNCTION** f
>
> **STEP 1** Write $y = f(x)$.
>
> **STEP 2** Solve this equation for x in terms of y (if possible).
>
> **STEP 3** To express f^{-1} as a function of x, interchange x and y. The resulting equation is $y = f^{-1}(x)$.

V **EXAMPLE 4** Find the inverse function of $f(x) = x^3 + 2$.

SOLUTION According to **5**, we first write

$$y = x^3 + 2$$

Then we solve this equation for x:

$$x^3 = y - 2$$

$$x = \sqrt[3]{y - 2}$$

Finally, we interchange x and y:

$$y = \sqrt[3]{x - 2}$$

■ In Example 4, notice how f^{-1} reverses the effect of f. The function f is the rule "Cube, then add 2"; f^{-1} is the rule "Subtract 2, then take the cube root."

Therefore the inverse function is $f^{-1}(x) = \sqrt[3]{x - 2}$. ■

The principle of interchanging x and y to find the inverse function also gives us the method for obtaining the graph of f^{-1} from the graph of f. Since $f(a) = b$ if and only if $f^{-1}(b) = a$, the point (a, b) is on the graph of f if and only if the point (b, a) is on the graph of f^{-1}. But we get the point (b, a) from (a, b) by reflecting about the line $y = x$. (See Figure 8.)

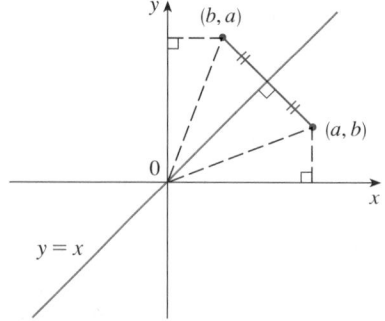

FIGURE 8

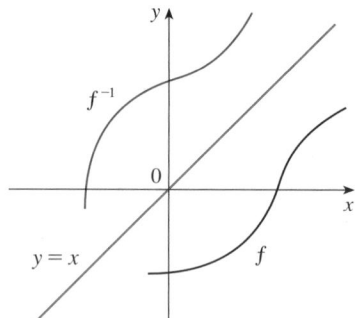

FIGURE 9

Therefore, as illustrated by Figure 9:

> The graph of f^{-1} is obtained by reflecting the graph of f about the line $y = x$.

EXAMPLE 5 Sketch the graphs of $f(x) = \sqrt{-1 - x}$ and its inverse function using the same coordinate axes.

SOLUTION First we sketch the curve $y = \sqrt{-1 - x}$ (the top half of the parabola $y^2 = -1 - x$, or $x = -y^2 - 1$) and then we reflect about the line $y = x$ to get the graph of f^{-1}. (See Figure 10.) As a check on our graph, notice that the expression for f^{-1} is $f^{-1}(x) = -x^2 - 1$, $x \geqslant 0$. So the graph of f^{-1} is the right half of the parabola $y = -x^2 - 1$ and this seems reasonable from Figure 10.

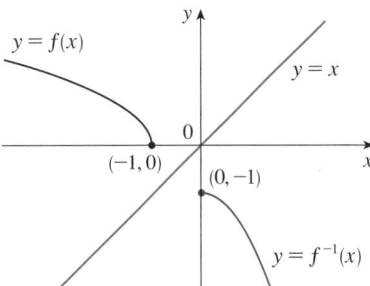

FIGURE 10

THE CALCULUS OF INVERSE FUNCTIONS

Now let's look at inverse functions from the point of view of calculus. Suppose that f is both one-to-one and continuous. We think of a continuous function as one whose graph has no break in it. (It consists of just one piece.) Since the graph of f^{-1} is obtained from the graph of f by reflecting about the line $y = x$, the graph of f^{-1} has no break in it either (see Figure 9). Thus we might expect that f^{-1} is also a continuous function.

This geometrical argument does not prove the following theorem but at least it makes the theorem plausible. A proof can be found in Appendix D.

> **6 THEOREM** If f is a one-to-one continuous function defined on an interval, then its inverse function f^{-1} is also continuous.

Now suppose that f is a one-to-one differentiable function. Geometrically we can think of a differentiable function as one whose graph has no corner or kink in it. We get the graph of f^{-1} by reflecting the graph of f about the line $y = x$, so the graph of f^{-1} has no corner or kink in it either. We therefore expect that f^{-1} is also differentiable (except where its tangents are vertical). In fact, we can predict the value of the derivative of f^{-1} at a given point by a geometric argument. In Figure 11 the graphs of f and its inverse f^{-1} are shown. If $f(b) = a$, then $f^{-1}(a) = b$ and $(f^{-1})'(a)$ is the slope of the tangent line L to the graph of f^{-1} at (a, b), which is $\Delta y / \Delta x$. Reflecting in the line $y = x$ has the effect of interchanging the x- and y-coordinates. So the slope of the reflected line ℓ [the tangent to the graph of f at (b, a)] is $\Delta x / \Delta y$. Thus the slope of L is the reciprocal of the slope of ℓ, that is,

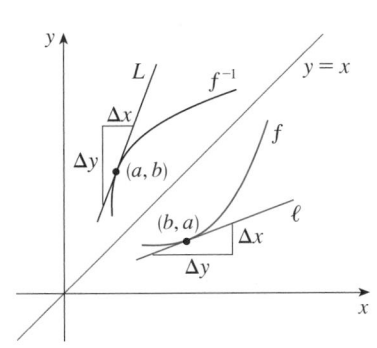

FIGURE 11

$$(f^{-1})'(a) = \frac{\Delta y}{\Delta x} = \frac{1}{\Delta x / \Delta y} = \frac{1}{f'(b)}$$

> **7** **THEOREM** If f is a one-to-one differentiable function with inverse function f^{-1} and $f'(f^{-1}(a)) \neq 0$, then the inverse function is differentiable at a and
> $$(f^{-1})'(a) = \frac{1}{f'(f^{-1}(a))}$$

PROOF Write the definition of derivative as in Equation 2.1.5:

$$(f^{-1})'(a) = \lim_{x \to a} \frac{f^{-1}(x) - f^{-1}(a)}{x - a}$$

If $f(b) = a$, then $f^{-1}(a) = b$. And if we let $y = f^{-1}(x)$, then $f(y) = x$. Since f is differentiable, it is continuous, so f^{-1} is continuous by Theorem 6. Thus if $x \to a$, then $f^{-1}(x) \to f^{-1}(a)$, that is, $y \to b$. Therefore

- Note that $x \neq a \Rightarrow f(y) \neq f(b)$ because f is one-to-one.

$$(f^{-1})'(a) = \lim_{x \to a} \frac{f^{-1}(x) - f^{-1}(a)}{x - a} = \lim_{y \to b} \frac{y - b}{f(y) - f(b)}$$

$$= \lim_{y \to b} \frac{1}{\dfrac{f(y) - f(b)}{y - b}} = \frac{1}{\displaystyle\lim_{y \to b} \dfrac{f(y) - f(b)}{y - b}}$$

$$= \frac{1}{f'(b)} = \frac{1}{f'(f^{-1}(a))} \qquad \square$$

NOTE 1 Replacing a by the general number x in the formula of Theorem 7, we get

8
$$(f^{-1})'(x) = \frac{1}{f'(f^{-1}(x))}$$

If we write $y = f^{-1}(x)$, then $f(y) = x$, so Equation 8, when expressed in Leibniz notation, becomes

$$\frac{dy}{dx} = \frac{1}{\dfrac{dx}{dy}}$$

NOTE 2 If it is known in advance that f^{-1} is differentiable, then its derivative can be computed more easily than in the proof of Theorem 7 by using implicit differentiation. If $y = f^{-1}(x)$, then $f(y) = x$. Differentiating the equation $f(y) = x$ implicitly with respect to x, remembering that y is a function of x, and using the Chain Rule, we get

$$f'(y) \frac{dy}{dx} = 1$$

Therefore
$$\frac{dy}{dx} = \frac{1}{f'(y)} = \frac{1}{\dfrac{dx}{dy}}$$

V EXAMPLE 6 If $f(x) = 2x + \cos x$, find $(f^{-1})'(1)$.

SOLUTION Notice that f is differentiable and one-to-one. (Its graph is shown in Figure 12.) To use Theorem 7 we need to know $f^{-1}(1)$ and we can find it by inspection:

$$f(0) = 1 \quad \Rightarrow \quad f^{-1}(1) = 0$$

Therefore
$$(f^{-1})'(1) = \frac{1}{f'(f^{-1}(1))} = \frac{1}{f'(0)} = \frac{1}{2 - \sin 0} = \frac{1}{2}$$

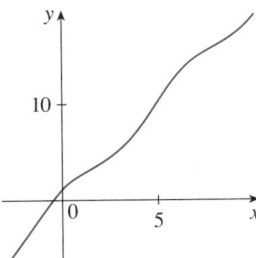

FIGURE 12

LOGARITHMIC FUNCTIONS

If $a > 0$ and $a \neq 1$, the exponential function $f(x) = a^x$ is either increasing or decreasing and so it is one-to-one by the Horizontal Line Test. It therefore has an inverse function f^{-1}, which is called the **logarithmic function with base a** and is denoted by $\log_a$. If we use the formulation of an inverse function given by $\boxed{3}$,

$$f^{-1}(x) = y \quad \Longleftrightarrow \quad f(y) = x$$

then we have

$\boxed{9}$
$$\log_a x = y \quad \Longleftrightarrow \quad a^y = x$$

Thus if $x > 0$, then $\log_a x$ is the exponent to which the base a must be raised to give x. For example, $\log_{10} 0.001 = -3$ because $10^{-3} = 0.001$.

The cancellation equations $\boxed{4}$, when applied to the functions $f(x) = a^x$ and $f^{-1}(x) = \log_a x$, become

$\boxed{10}$
$$\log_a(a^x) = x \quad \text{for every } x \in \mathbb{R}$$
$$a^{\log_a x} = x \quad \text{for every } x > 0$$

The logarithmic function $\log_a$ has domain $(0, \infty)$ and range $\mathbb{R}$ and is continuous since it is the inverse of a continuous function, namely, the exponential function. Its graph is the reflection of the graph of $y = a^x$ about the line $y = x$.

Figure 13 shows the case where $a > 1$. (The most important logarithmic functions have base $a > 1$.) The fact that $y = a^x$ is a very rapidly increasing function for $x > 0$ is reflected in the fact that $y = \log_a x$ is a very slowly increasing function for $x > 1$.

Figure 14 shows the graphs of $y = \log_a x$ with various values of the base $a > 1$. Since $\log_a 1 = 0$, the graphs of all logarithmic functions pass through the point $(1, 0)$.

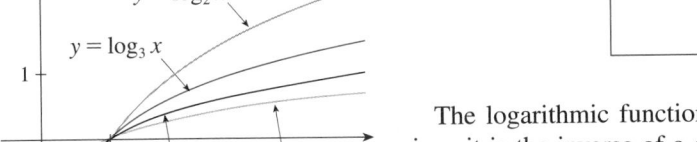

FIGURE 13

FIGURE 14

The following properties of logarithmic functions follow from the corresponding properties of exponential functions given in Section 3.1.

LAWS OF LOGARITHMS If x and y are positive numbers, then

1. $\log_a(xy) = \log_a x + \log_a y$

2. $\log_a\left(\dfrac{x}{y}\right) = \log_a x - \log_a y$

3. $\log_a(x^r) = r \log_a x$ (where r is any real number)

EXAMPLE 7 Use the laws of logarithms to evaluate $\log_2 80 - \log_2 5$.

SOLUTION Using Law 2, we have

$$\log_2 80 - \log_2 5 = \log_2\left(\frac{80}{5}\right) = \log_2 16 = 4$$

because $2^4 = 16$. ∎

The limits of exponential functions given in Section 3.1 are reflected in the following limits of logarithmic functions. (Compare with Figure 13.)

$\boxed{11}$ If $a > 1$, then

$$\lim_{x \to \infty} \log_a x = \infty \qquad \text{and} \qquad \lim_{x \to 0^+} \log_a x = -\infty$$

In particular, the y-axis is a vertical asymptote of the curve $y = \log_a x$.

EXAMPLE 8 Find $\lim_{x \to 0} \log_{10}(\tan^2 x)$.

SOLUTION As $x \to 0$, we know that $t = \tan^2 x \to \tan^2 0 = 0$ and the values of t are positive. So by $\boxed{11}$ with $a = 10 > 1$, we have

$$\lim_{x \to 0} \log_{10}(\tan^2 x) = \lim_{t \to 0^+} \log_{10} t = -\infty$$ ∎

NATURAL LOGARITHMS

■ **NOTATION FOR LOGARITHMS**

Most textbooks in calculus and the sciences, as well as calculators, use the notation $\ln x$ for the natural logarithm and $\log x$ for the "common logarithm," $\log_{10} x$. In the more advanced mathematical and scientific literature and in computer languages, however, the notation $\log x$ usually denotes the natural logarithm.

Of all possible bases a for logarithms, we will see in the next section that the most convenient choice of a base is the number e, which was defined in Section 3.1. The logarithm with base e is called the **natural logarithm** and has a special notation:

$$\log_e x = \ln x$$

If we put $a = e$ and replace $\log_e$ with "ln" in $\boxed{9}$ and $\boxed{10}$, then the defining properties of the natural logarithm function become

$\boxed{12}$

$$\ln x = y \iff e^y = x$$

$$\boxed{13} \qquad \boxed{\begin{array}{ll} \ln(e^x) = x & x \in \mathbb{R} \\[2mm] e^{\ln x} = x & x > 0 \end{array}}$$

In particular, if we set $x = 1$, we get

$$\boxed{\ln e = 1}$$

EXAMPLE 9 Find x if $\ln x = 5$.

SOLUTION 1 From $\boxed{12}$ we see that

$$\ln x = 5 \qquad \text{means} \qquad e^5 = x$$

Therefore $x = e^5$.

(If you have trouble working with the "ln" notation, just replace it by $\log_e$. Then the equation becomes $\log_e x = 5$; so, by the definition of logarithm, $e^5 = x$.)

SOLUTION 2 Start with the equation

$$\ln x = 5$$

and apply the exponential function to both sides of the equation:

$$e^{\ln x} = e^5$$

But the second cancellation equation in $\boxed{13}$ says that $e^{\ln x} = x$. Therefore $x = e^5$. ∎

V EXAMPLE 10 Solve the equation $e^{5-3x} = 10$.

SOLUTION We take natural logarithms of both sides of the equation and use $\boxed{13}$:

$$\ln(e^{5-3x}) = \ln 10$$

$$5 - 3x = \ln 10$$

$$3x = 5 - \ln 10$$

$$x = \tfrac{1}{3}(5 - \ln 10)$$

Since the natural logarithm is found on scientific calculators, we can approximate the solution to four decimal places: $x \approx 0.8991$. ∎

V EXAMPLE 11 Express $\ln a + \tfrac{1}{2} \ln b$ as a single logarithm.

SOLUTION Using Laws 3 and 1 of logarithms, we have

$$\ln a + \tfrac{1}{2} \ln b = \ln a + \ln b^{1/2} = \ln a + \ln \sqrt{b} = \ln\!\left(a \sqrt{b}\,\right)$$ ∎

The following formula shows that logarithms with any base can be expressed in terms of the natural logarithm.

14 **CHANGE OF BASE FORMULA** For any positive number a ($a \neq 1$), we have

$$\log_a x = \frac{\ln x}{\ln a}$$

PROOF Let $y = \log_a x$. Then, from **9**, we have $a^y = x$. Taking natural logarithms of both sides of this equation, we get $y \ln a = \ln x$. Therefore

$$y = \frac{\ln x}{\ln a}$$

Scientific calculators have a key for natural logarithms, so Formula 14 enables us to use a calculator to compute a logarithm with any base (as shown in the following example). Similarly, Formula 14 allows us to graph any logarithmic function on a graphing calculator or computer (see Exercises 55 and 56).

EXAMPLE 12 Evaluate $\log_8 5$ correct to six decimal places.

SOLUTION Formula 14 gives

$$\log_8 5 = \frac{\ln 5}{\ln 8} \approx 0.773976$$

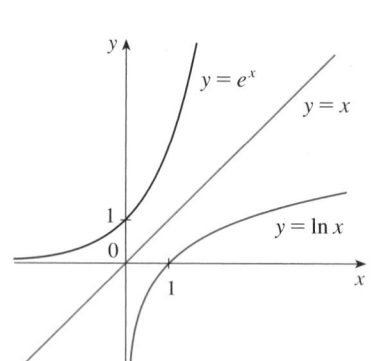

FIGURE 15

The graphs of the exponential function $y = e^x$ and its inverse function, the natural logarithm function, are shown in Figure 15. Because the curve $y = e^x$ crosses the y-axis with a slope of 1, it follows that the reflected curve $y = \ln x$ crosses the x-axis with a slope of 1.

In common with all other logarithmic functions with base greater than 1, the natural logarithm is a continuous, increasing function defined on $(0, \infty)$ and the y-axis is a vertical asymptote.

If we put $a = e$ in **11**, then we have the following limits:

15
$$\lim_{x \to \infty} \ln x = \infty \qquad \lim_{x \to 0^+} \ln x = -\infty$$

▼ EXAMPLE 13 Sketch the graph of the function $y = \ln(x - 2) - 1$.

SOLUTION We start with the graph of $y = \ln x$ as given in Figure 15. Using the transformations of Section 1.2, we shift it 2 units to the right to get the graph of $y = \ln(x - 2)$ and then we shift it 1 unit downward to get the graph of $y = \ln(x - 2) - 1$. (See Figure 16.) Notice that the line $x = 2$ is a vertical asymptote since

$$\lim_{x \to 2^+} [\ln(x - 2) - 1] = -\infty$$

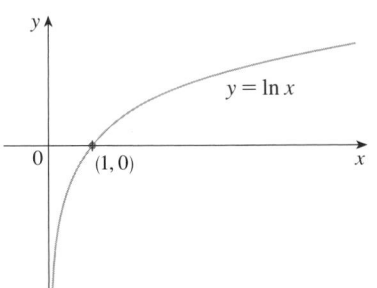
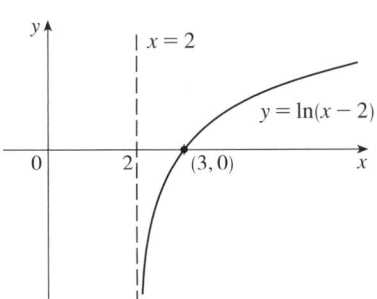
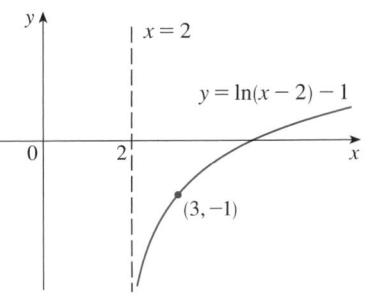

FIGURE 16

3.2 | EXERCISES

1. (a) What is a one-to-one function?
(b) How can you tell from the graph of a function whether it is one-to-one?

2. (a) Suppose f is a one-to-one function with domain A and range B. How is the inverse function f^{-1} defined? What is the domain of f^{-1}? What is the range of f^{-1}?
(b) If you are given a formula for f, how do you find a formula for f^{-1}?
(c) If you are given the graph of f, how do you find the graph of f^{-1}?

3–14 ■ A function is given by a table of values, a graph, a formula, or a verbal description. Determine whether it is one-to-one.

3.

x	1	2	3	4	5	6
$f(x)$	1.5	2.0	3.6	5.3	2.8	2.0

4.

x	1	2	3	4	5	6
$f(x)$	1.0	1.9	2.8	3.5	3.1	2.9

5.
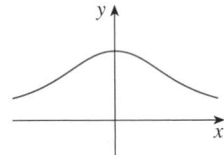

6.
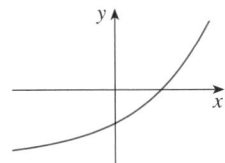

7.
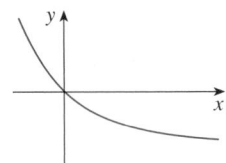

8.
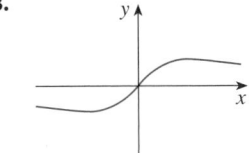

9. $f(x) = x^2 - 2x$

10. $f(x) = 10 - 3x$

11. $g(x) = 1/x$

12. $g(x) = \cos x$

13. $f(t)$ is the height of a football t seconds after kickoff.

14. $f(t)$ is your height at age t.

15. Assume that f is a one-to-one function.
(a) If $f(6) = 17$, what is $f^{-1}(17)$?
(b) If $f^{-1}(3) = 2$, what is $f(2)$?

16. If $f(x) = x^5 + x^3 + x$, find $f^{-1}(3)$ and $f\big(f^{-1}(2)\big)$.

17. If $g(x) = 3 + x + e^x$, find $g^{-1}(4)$.

18. The graph of f is given.
(a) Why is f one-to-one?
(b) What are the domain and range of f^{-1}?
(c) What is the value of $f^{-1}(2)$?
(d) Estimate the value of $f^{-1}(0)$.

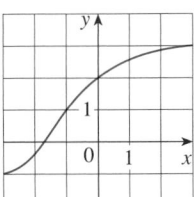

19. The formula $C = \frac{5}{9}(F - 32)$, where $F \geqslant -459.67$, expresses the Celsius temperature C as a function of the Fahrenheit temperature F. Find a formula for the inverse function and interpret it. What is the domain of the inverse function?

20. In the theory of relativity, the mass of a particle with speed v is
$$m = f(v) = \frac{m_0}{\sqrt{1 - v^2/c^2}}$$
where m_0 is the rest mass of the particle and c is the speed of light in a vacuum. Find the inverse function of f and explain its meaning.

21–26 ▪ Find a formula for the inverse of the function.

21. $f(x) = 1 + \sqrt{2 + 3x}$

22. $f(x) = \dfrac{4x - 1}{2x + 3}$

23. $f(x) = e^{2x-1}$

24. $y = x^2 - x, \quad x \geq \frac{1}{2}$

25. $y = \ln(x + 3)$

26. $y = \dfrac{e^x}{1 + 2e^x}$

 27–28 ▪ Find an explicit formula for f^{-1} and use it to graph f^{-1}, f, and the line $y = x$ on the same screen. To check your work, see whether the graphs of f and f^{-1} are reflections about the line.

27. $f(x) = x^4 + 1, \quad x \geq 0$

28. $f(x) = 2 - e^x$

29–30 ▪ Use the given graph of f to sketch the graph of f^{-1}.

29.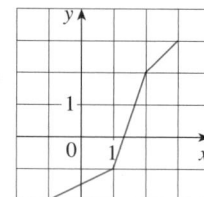

30.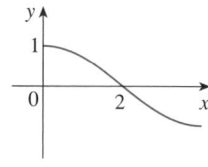

31–34 ▪
(a) Show that f is one-to-one.
(b) Use Theorem 7 to find $(f^{-1})'(a)$.
(c) Calculate $f^{-1}(x)$ and state the domain and range of f^{-1}.
(d) Calculate $(f^{-1})'(a)$ from the formula in part (c) and check that it agrees with the result of part (b).
(e) Sketch the graphs of f and f^{-1} on the same axes.

31. $f(x) = x^3, \quad a = 8$

32. $f(x) = \sqrt{x - 2}, \quad a = 2$

33. $f(x) = 9 - x^2, \quad 0 \leq x \leq 3, \quad a = 8$

34. $f(x) = 1/(x - 1), \quad x > 1, \quad a = 2$

35–38 ▪ Find $(f^{-1})'(a)$.

35. $f(x) = 2x^3 + 3x^2 + 7x + 4, \quad a = 4$

36. $f(x) = x^3 + 3\sin x + 2\cos x, \quad a = 2$

37. $f(x) = 3 + x^2 + \tan(\pi x/2), \quad -1 < x < 1, \quad a = 3$

38. $f(x) = \sqrt{x^3 + x^2 + x + 1}, \quad a = 2$

---

39. Suppose f^{-1} is the inverse function of a differentiable function f and $f(4) = 5$, $f'(4) = \frac{2}{3}$. Find $(f^{-1})'(5)$.

40. Suppose f^{-1} is the inverse function of a differentiable function f and let $G(x) = 1/f^{-1}(x)$. If $f(3) = 2$ and $f'(3) = \frac{1}{9}$, find $G'(2)$.

41. (a) How is the logarithmic function $y = \log_a x$ defined?
(b) What is the domain of this function?
(c) What is the range of this function?
(d) Sketch the general shape of the graph of the function $y = \log_a x$ if $a > 1$.

42. (a) What is the natural logarithm?
(b) What is the common logarithm?
(c) Sketch the graphs of the natural logarithm function and the natural exponential function with a common set of axes.

43–46 ▪ Find the exact value of each expression (without a calculator.)

43. (a) $\log_5 125$

(b) $\log_3\left(\frac{1}{27}\right)$

44. (a) $\ln(1/e)$

(b) $\log_{10} \sqrt{10}$

45. (a) $\log_2 6 - \log_2 15 + \log_2 20$
(b) $\log_3 100 - \log_3 18 - \log_3 50$

46. (a) $e^{-2\ln 5}$

(b) $\ln\left(\ln e^{e^{10}}\right)$

47–50 ▪ Use the properties of logarithms to expand the quantity.

47. $\ln \sqrt{ab}$

48. $\log_{10} \sqrt{\dfrac{x - 1}{x + 1}}$

49. $\ln \dfrac{x^2}{y^3 z^4}$

50. $\ln\left(s^4 \sqrt{t \sqrt{u}}\right)$

51–53 ▪ Express the given quantity as a single logarithm.

51. $\ln 5 + 5 \ln 3$

52. $\ln(a + b) + \ln(a - b) - 2\ln c$

53. $\frac{1}{3}\ln(x + 2)^3 + \frac{1}{2}\left[\ln x - \ln(x^2 + 3x + 2)^2\right]$

---

54. Use Formula 14 to evaluate each logarithm correct to six decimal places.
(a) $\log_{12} 10$

(b) $\log_2 8.4$

 55–56 ▪ Use Formula 14 to graph the given functions on a common screen. How are these graphs related?

55. $y = \log_{1.5} x, \quad y = \ln x, \quad y = \log_{10} x, \quad y = \log_{50} x$

56. $y = \ln x, \quad y = \log_{10} x, \quad y = e^x, \quad y = 10^x$

---

57. Suppose that the graph of $y = \log_2 x$ is drawn on a coordinate grid where the unit of measurement is an inch. How many miles to the right of the origin do we have to move before the height of the curve reaches 3 ft?

 58. Compare the functions $f(x) = x^{0.1}$ and $g(x) = \ln x$ by graphing both f and g in several viewing rectangles. When does the graph of f finally surpass the graph of g?

59–60 ▪ Make a rough sketch of the graph of each function. Do not use a calculator. Just use the graphs given in Figures 14 and 15 and, if necessary, the transformations of Section 1.2.

59. (a) $y = \log_{10}(x + 5)$ (b) $y = -\ln x$

60. (a) $y = \ln(-x)$ (b) $y = \ln|x|$

61–62 ▪
(a) What are the domain and range of f?
(b) What is the x-intercept of the graph of f?
(c) Sketch the graph of f.

61. $f(x) = \ln x + 2$ **62.** $f(x) = \ln(x - 1) - 1$

63–66 ▪ Solve each equation for x.

63. (a) $e^{7-4x} = 6$ (b) $\ln(3x - 10) = 2$

64. (a) $\ln(x^2 - 1) = 3$ (b) $e^{2x} - 3e^x + 2 = 0$

65. (a) $2^{x-5} = 3$ (b) $\ln x + \ln(x - 1) = 1$

66. (a) $\ln(\ln x) = 1$ (b) $e^{ax} = Ce^{bx}$, where $a \neq b$

67–68 ▪ Solve each inequality for x.

67. (a) $\ln x < 0$ (b) $e^x > 5$

68. (a) $1 < e^{3x-1} < 2$ (b) $1 - 2\ln x < 3$

69. (a) Find the domain of $f(x) = \ln(e^x - 3)$.
(b) Find f^{-1} and its domain.

70. (a) What are the values of $e^{\ln 300}$ and $\ln(e^{300})$?
(b) Use your calculator to evaluate $e^{\ln 300}$ and $\ln(e^{300})$. What do you notice? Can you explain why the calculator has trouble?

71–76 ▪ Find the limit.

71. $\lim\limits_{x \to 3^+} \ln(x^2 - 9)$ **72.** $\lim\limits_{x \to 2^-} \log_5(8x - x^4)$

73. $\lim\limits_{x \to 0} \ln(\cos x)$ **74.** $\lim\limits_{x \to 0^+} \ln(\sin x)$

75. $\lim\limits_{x \to \infty} [\ln(1 + x^2) - \ln(1 + x)]$

76. $\lim\limits_{x \to \infty} [\ln(2 + x) - \ln(1 + x)]$

CAS **77.** Graph the function $f(x) = \sqrt{x^3 + x^2 + x + 1}$ and explain why it is one-to-one. Then use a computer algebra system to find an explicit expression for $f^{-1}(x)$. (Your CAS will produce three possible expressions. Explain why two of them are irrelevant in this context.)

78. When a camera flash goes off, the batteries immediately begin to recharge the flash's capacitor, which stores electric charge given by

$$Q(t) = Q_0(1 - e^{-t/a})$$

(The maximum charge capacity is Q_0 and t is measured in seconds.)
(a) Find the inverse of this function and explain its meaning.
(b) How long does it take to recharge the capacitor to 90% of capacity if $a = 2$?

79. Let $a > 1$. Prove, using precise definitions, that
(a) $\lim\limits_{x \to -\infty} a^x = 0$ (b) $\lim\limits_{x \to \infty} a^x = \infty$

80. (a) If we shift a curve to the left, what happens to its reflection about the line $y = x$? In view of this geometric principle, find an expression for the inverse of $g(x) = f(x + c)$, where f is a one-to-one function.
(b) Find an expression for the inverse of $h(x) = f(cx)$, where $c \neq 0$.

3.3 | DERIVATIVES OF LOGARITHMIC AND EXPONENTIAL FUNCTIONS

In this section we find formulas for the derivatives of logarithmic functions and then use them to calculate the derivatives of exponential functions.

DERIVATIVES OF LOGARITHMIC FUNCTIONS

In using the definition of a derivative to differentiate the function $f(x) = \log_a x$, we use the fact that it is continuous, together with some of the laws of logarithms. We also need to recall the definition of e from Section 3.1:

$$e = \lim_{x \to 0} (1 + x)^{1/x}$$

> **1** **THEOREM** The function $f(x) = \log_a x$ is differentiable and
> $$f'(x) = \frac{1}{x}\log_a e$$

PROOF

$$f'(x) = \lim_{h \to 0} \frac{f(x+h) - f(x)}{h} = \lim_{h \to 0} \frac{\log_a(x+h) - \log_a x}{h}$$

$$= \lim_{h \to 0} \frac{\log_a\left(\dfrac{x+h}{x}\right)}{h} = \lim_{h \to 0} \frac{1}{h}\log_a\left(1 + \frac{h}{x}\right)$$

$$= \lim_{h \to 0} \frac{1}{x}\cdot\frac{x}{h}\log_a\left(1 + \frac{h}{x}\right)$$

$$= \frac{1}{x}\lim_{h \to 0} \frac{x}{h}\log_a\left(1 + \frac{h}{x}\right) \qquad \text{(by Limit Law 3)}$$

$$= \frac{1}{x}\lim_{h \to 0} \log_a\left(1 + \frac{h}{x}\right)^{x/h} \qquad \text{(by Law 3 of Logarithms)}$$

$$= \frac{1}{x}\log_a\left[\lim_{h \to 0}\left(1 + \frac{h}{x}\right)^{x/h}\right] \qquad \text{(since } \log_a \text{ is continuous)}$$

$$= \frac{1}{x}\log_a\left[\lim_{h \to 0}\left(1 + \frac{h}{x}\right)^{1/(h/x)}\right] = \frac{1}{x}\log_a e$$

The final step may be seen more clearly by making the change of variable $t = h/x$. As $h \to 0$, we also have $t \to 0$, so

$$\lim_{h \to 0}\left(1 + \frac{h}{x}\right)^{1/(h/x)} = \lim_{t \to 0}(1 + t)^{1/t} = e$$

by the definition of e. Thus

$$f'(x) = \frac{1}{x}\log_a e \qquad \square$$

NOTE We know from the Change of Base Formula (3.2.14) that

$$\log_a e = \frac{\ln e}{\ln a} = \frac{1}{\ln a}$$

and so the formula in Theorem 1 can be rewritten as follows:

2
$$\boxed{\frac{d}{dx}(\log_a x) = \frac{1}{x \ln a}}$$

EXAMPLE 1 Differentiate $f(x) = \log_{10}(2 + \sin x)$.

SOLUTION Using Formula 2 with $a = 10$, together with the Chain Rule, we have

$$f'(x) = \frac{d}{dx} \log_{10}(2 + \sin x) = \frac{1}{(2 + \sin x) \ln 10} \frac{d}{dx}(2 + \sin x)$$

$$= \frac{\cos x}{(2 + \sin x) \ln 10} \qquad\blacksquare$$

If we put $a = e$ in Formula 2, then the factor $\ln a$ on the right side becomes $\ln e = 1$ and we get the formula for the derivative of the natural logarithmic function $\log_e x = \ln x$:

---

3 **DERIVATIVE OF THE NATURAL LOGARITHMIC FUNCTION**

$$\frac{d}{dx}(\ln x) = \frac{1}{x}$$

By comparing Formulas 2 and 3, we see one of the main reasons that natural logarithms (logarithms with base e) are used in calculus: The differentiation formula is simplest when $a = e$ because $\ln e = 1$.

▼ EXAMPLE 2 Differentiate $y = \ln(x^3 + 1)$.

SOLUTION To use the Chain Rule, we let $u = x^3 + 1$. Then $y = \ln u$, so

$$\frac{dy}{dx} = \frac{dy}{du}\frac{du}{dx} = \frac{1}{u}\frac{du}{dx} = \frac{1}{x^3 + 1}(3x^2) = \frac{3x^2}{x^3 + 1} \qquad\blacksquare$$

In general, if we combine Formula 3 with the Chain Rule as in Example 2, we get

4
$$\frac{d}{dx}(\ln u) = \frac{1}{u}\frac{du}{dx} \qquad \text{or} \qquad \frac{d}{dx}[\ln g(x)] = \frac{g'(x)}{g(x)}$$

▼ EXAMPLE 3 Find $\dfrac{d}{dx}\ln(\sin x)$.

SOLUTION Using $\boxed{4}$, we have

$$\frac{d}{dx}\ln(\sin x) = \frac{1}{\sin x}\frac{d}{dx}(\sin x) = \frac{1}{\sin x}\cos x = \cot x \qquad\blacksquare$$

EXAMPLE 4 Differentiate $f(x) = \sqrt{\ln x}$.

SOLUTION This time the logarithm is the inner function, so the Chain Rule gives

$$f'(x) = \tfrac{1}{2}(\ln x)^{-1/2}\frac{d}{dx}(\ln x) = \frac{1}{2\sqrt{\ln x}} \cdot \frac{1}{x} = \frac{1}{2x\sqrt{\ln x}} \qquad\blacksquare$$

EXAMPLE 5 Find $\dfrac{d}{dx} \ln \dfrac{x+1}{\sqrt{x-2}}$.

SOLUTION 1

$$\frac{d}{dx} \ln \frac{x+1}{\sqrt{x-2}} = \frac{1}{\dfrac{x+1}{\sqrt{x-2}}} \, \frac{d}{dx} \frac{x+1}{\sqrt{x-2}}$$

$$= \frac{\sqrt{x-2}}{x+1} \, \frac{\sqrt{x-2} \cdot 1 - (x+1)(\tfrac{1}{2})(x-2)^{-1/2}}{x-2}$$

$$= \frac{x - 2 - \tfrac{1}{2}(x+1)}{(x+1)(x-2)} = \frac{x-5}{2(x+1)(x-2)}$$

SOLUTION 2 If we first simplify the given function using the laws of logarithms, then the differentiation becomes easier:

$$\frac{d}{dx} \ln \frac{x+1}{\sqrt{x-2}} = \frac{d}{dx} \left[\ln(x+1) - \tfrac{1}{2} \ln(x-2) \right]$$

$$= \frac{1}{x+1} - \frac{1}{2} \left(\frac{1}{x-2} \right)$$

(This answer can be left as written, but if we used a common denominator we would see that it gives the same answer as in Solution 1.) ∎

■ Figure 1 shows the graph of the function f of Example 5 together with the graph of its derivative. It gives a visual check on our calculation. Notice that $f'(x)$ is large negative when f is rapidly decreasing.

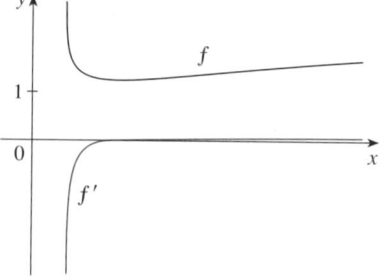

FIGURE 1

☑ EXAMPLE 6 Find $f'(x)$ if $f(x) = \ln |x|$.

SOLUTION Since

$$f(x) = \begin{cases} \ln x & \text{if } x > 0 \\ \ln(-x) & \text{if } x < 0 \end{cases}$$

it follows that

$$f'(x) = \begin{cases} \dfrac{1}{x} & \text{if } x > 0 \\[2mm] \dfrac{1}{-x}\,(-1) = \dfrac{1}{x} & \text{if } x < 0 \end{cases}$$

Thus $f'(x) = 1/x$ for all $x \neq 0$. ∎

■ Figure 2 shows the graph of the function $f(x) = \ln |x|$ in Example 6 and its derivative $f'(x) = 1/x$. Notice that when x is small, the graph of $y = \ln |x|$ is steep and so $f'(x)$ is large (positive or negative).

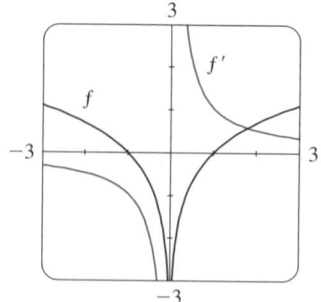

FIGURE 2

The result of Example 6 is worth remembering:

⑤

$$\frac{d}{dx} \ln |x| = \frac{1}{x}$$

LOGARITHMIC DIFFERENTIATION

The calculation of derivatives of complicated functions involving products, quotients, or powers can often be simplified by taking logarithms. The method used in the following example is called **logarithmic differentiation**.

▼ EXAMPLE 7 Differentiate $y = \dfrac{x^{3/4}\sqrt{x^2 + 1}}{(3x + 2)^5}$.

SOLUTION We take logarithms of both sides of the equation and use the Laws of Logarithms to simplify:

$$\ln y = \tfrac{3}{4} \ln x + \tfrac{1}{2} \ln(x^2 + 1) - 5 \ln(3x + 2)$$

Differentiating implicitly with respect to x gives

$$\frac{1}{y}\frac{dy}{dx} = \frac{3}{4} \cdot \frac{1}{x} + \frac{1}{2} \cdot \frac{2x}{x^2 + 1} - 5 \cdot \frac{3}{3x + 2}$$

Solving for dy/dx, we get

$$\frac{dy}{dx} = y\left(\frac{3}{4x} + \frac{x}{x^2 + 1} - \frac{15}{3x + 2}\right)$$

> ■ If we hadn't used logarithmic differentiation in Example 7, we would have had to use both the Quotient Rule and the Product Rule. The resulting calculation would have been horrendous.

Because we have an explicit expression for y, we can substitute and write

$$\frac{dy}{dx} = \frac{x^{3/4}\sqrt{x^2 + 1}}{(3x + 2)^5}\left(\frac{3}{4x} + \frac{x}{x^2 + 1} - \frac{15}{3x + 2}\right)$$ ■

STEPS IN LOGARITHMIC DIFFERENTIATION

1. Take natural logarithms of both sides of an equation $y = f(x)$ and use the Laws of Logarithms to simplify.
2. Differentiate implicitly with respect to x.
3. Solve the resulting equation for y'.

If $f(x) < 0$ for some values of x, then $\ln f(x)$ is not defined, but we can write $|y| = |f(x)|$ and use Equation 5. We illustrate this procedure by proving the general version of the Power Rule, as promised in Section 2.3.

THE POWER RULE If n is any real number and $f(x) = x^n$, then

$$f'(x) = nx^{n-1}$$

> ■ If $x = 0$, we can show that $f'(0) = 0$ for $n > 1$ directly from the definition of a derivative.

PROOF Let $y = x^n$ and use logarithmic differentiation:

$$\ln |y| = \ln |x|^n = n \ln |x| \qquad x \neq 0$$

Therefore

$$\frac{y'}{y} = \frac{n}{x}$$

Hence

$$y' = n\frac{y}{x} = n\frac{x^n}{x} = nx^{n-1}$$ □

DERIVATIVES OF EXPONENTIAL FUNCTIONS

To compute the derivative of the exponential function $y = a^x$ we use the fact that exponential and logarithmic functions are inverse functions.

---

6 THEOREM The exponential function $f(x) = a^x$, $a > 0$, is differentiable and

$$\frac{d}{dx}(a^x) = a^x \ln a$$

PROOF We know that the logarithmic function $y = \log_a x$ is differentiable (and its derivative is nonzero) by Theorem 1. So its inverse function $y = a^x$ is differentiable by Theorem 3.2.7.

If $y = a^x$, then $\log_a y = x$. Differentiating this equation implicitly with respect to x, we get

$$\frac{1}{y \ln a}\frac{dy}{dx} = 1$$

Thus
$$\frac{dy}{dx} = y \ln a = a^x \ln a$$ □

■ Another method for proving Theorem 6 is to use logarithmic differentiation.

EXAMPLE 8 Combining Formula 6 with the Chain Rule, we have

$$\frac{d}{dx}\left(10^{x^2}\right) = 10^{x^2}(\ln 10)\frac{d}{dx}(x^2) = (2 \ln 10)x10^{x^2}$$ ■

If we put $a = e$ in Theorem 6, the differentiation formula for exponential functions takes on a particularly simple form:

---

7 DERIVATIVE OF THE NATURAL EXPONENTIAL FUNCTION

$$\frac{d}{dx}(e^x) = e^x$$

This equation says that the exponential function $f(x) = e^x$ is its own derivative. Comparing Equations 6 and 7, we see that the simplest differentiation formula for an exponential function occurs when $a = e$. This is the reason that the natural exponential function is most often used in calculus.

TEC Visual 3.3 uses the slope-a-scope to illustrate this formula.

The geometric significance of Equation 7 is that the slope of a tangent to the curve $y = e^x$ at any point is equal to the y-coordinate of the point. In particular, if $f(x) = e^x$, then $f'(0) = e^0 = 1$. This means that of all the possible exponential functions $y = a^x$, $y = e^x$ is the one that crosses the y-axis with a slope of 1. (See Figure 3.)

EXAMPLE 9 Differentiate the function $y = e^{\tan x}$.

SOLUTION To use the Chain Rule, we let $u = \tan x$. Then we have $y = e^u$, so

$$\frac{dy}{dx} = \frac{dy}{du}\frac{du}{dx} = e^u\frac{du}{dx} = e^{\tan x}\sec^2 x$$ ■

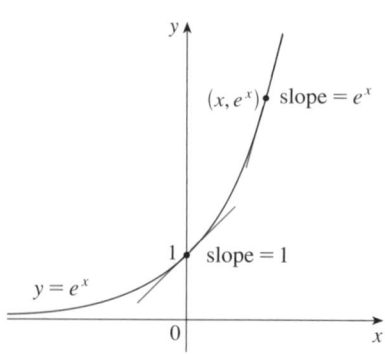

FIGURE 3

In general, if we combine Formula 7 with the Chain Rule, as in Example 9, we get

$$\boxed{8} \qquad \frac{d}{dx}(e^u) = e^u \frac{du}{dx}$$

EXAMPLE 10 Find y' if $y = e^{-4x} \sin 5x$.

SOLUTION Using Formula 8 and the Product Rule, we have

$$y' = e^{-4x}(\cos 5x)(5) + (\sin 5x)e^{-4x}(-4) = e^{-4x}(5\cos 5x - 4\sin 5x) \qquad \blacksquare$$

■ www.stewartcalculus.com
See Additional Examples A, B, C.

To differentiate a function of the form $y = [f(x)]^{g(x)}$, where both the base and the exponent are functions, logarithmic differentiation can be used as in the following example.

Ⅴ EXAMPLE 11 Differentiate $y = x^{\sqrt{x}}$.

SOLUTION 1 Using logarithmic differentiation, we have

$$\ln y = \ln x^{\sqrt{x}} = \sqrt{x}\, \ln x$$

$$\frac{y'}{y} = \sqrt{x} \cdot \frac{1}{x} + (\ln x)\frac{1}{2\sqrt{x}}$$

$$y' = y\left(\frac{1}{\sqrt{x}} + \frac{\ln x}{2\sqrt{x}}\right) = x^{\sqrt{x}}\left(\frac{2 + \ln x}{2\sqrt{x}}\right)$$

■ Figure 4 illustrates Example 11 by showing the graphs of $f(x) = x^{\sqrt{x}}$ and its derivative.

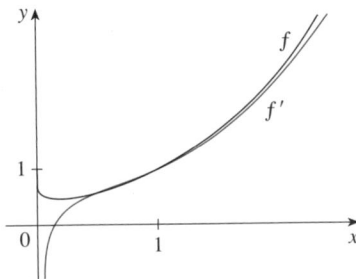

FIGURE 4

SOLUTION 2 Another method is to write $x^{\sqrt{x}} = (e^{\ln x})^{\sqrt{x}}$:

$$\frac{d}{dx}\left(x^{\sqrt{x}}\right) = \frac{d}{dx}\left(e^{\sqrt{x}\,\ln x}\right) = e^{\sqrt{x}\,\ln x}\frac{d}{dx}\left(\sqrt{x}\,\ln x\right)$$

$$= x^{\sqrt{x}}\left(\frac{2 + \ln x}{2\sqrt{x}}\right) \qquad \text{(as in Solution 1)} \qquad \blacksquare$$

3.3 | EXERCISES

1–40 ■ Differentiate the function.

1. $f(x) = \log_{10}(x^3 + 1)$

2. $f(x) = x \ln x - x$

3. $f(x) = \sin(\ln x)$

4. $f(x) = \ln(\sin^2 x)$

5. $f(x) = \ln\dfrac{1}{x}$

6. $y = \dfrac{1}{\ln x}$

7. $f(x) = \sin x \ln(5x)$

8. $f(x) = \log_5(xe^x)$

9. $g(x) = \ln\dfrac{a - x}{a + x}$

10. $f(u) = \dfrac{u}{1 + \ln u}$

11. $g(x) = \ln\left(x\sqrt{x^2 - 1}\right)$

12. $h(x) = \ln\left(x + \sqrt{x^2 - 1}\right)$

13. $G(y) = \ln\dfrac{(2y + 1)^5}{\sqrt{y^2 + 1}}$

14. $g(r) = r^2 \ln(2r + 1)$

15. $F(s) = \ln \ln s$

16. $y = \ln|\cos(\ln x)|$

17. $y = \tan[\ln(ax + b)]$

18. $H(z) = \ln\sqrt{\dfrac{a^2 - z^2}{a^2 + z^2}}$

19. $f(x) = (x^3 + 2x)e^x$

20. $g(x) = \sqrt{x}\, e^x$

21. $y = \dfrac{x}{e^x}$

22. $y = \dfrac{e^x}{1 - e^x}$

23. $y = \sqrt{1 + 2e^{3x}}$

24. $y = e^{-2t}\cos 4t$

25. $y = 5^{-1/x}$

26. $y = 10^{1-x^2}$

27. $F(t) = e^{t\sin 2t}$

28. $y = \dfrac{e^u - e^{-u}}{e^u + e^{-u}}$

29. $y = \ln|2 - x - 5x^2|$

30. $y = \sqrt{1 + xe^{-2x}}$

31. $f(t) = \tan(e^t) + e^{\tan t}$

32. $y = e^{k\tan\sqrt{x}}$

33. $y = \ln(e^{-x} + xe^{-x})$

34. $y = [\ln(1 + e^x)]^2$

35. $y = 2x\log_{10}\sqrt{x}$

36. $y = x^2 e^{-1/x}$

37. $f(t) = \sin^2(e^{\sin^2 t})$

38. $y = \log_2(e^{-x}\cos \pi x)$

39. $g(x) = (2ra^{rx} + n)^p$

40. $y = 2^{3^{x^2}}$

41–44 ▪ Find y' and y''.

41. $y = e^{\alpha x}\sin \beta x$

42. $y = \dfrac{\ln x}{x^2}$

43. $y = x\ln x$

44. $y = \ln(\sec x + \tan x)$

45–46 ▪ Find an equation of the tangent line to the curve at the given point.

45. $y = \ln(x^2 - 3x + 1)$, $(3, 0)$ **46.** $y = e^x/x$, $(1, e)$

47–48 ▪ Differentiate f and find the domain of f.

47. $f(x) = \dfrac{x}{1 - \ln(x - 1)}$

48. $f(x) = \ln\ln\ln x$

---

49. Let $f(x) = cx + \ln(\cos x)$. For what value of c is $f'(\pi/4) = 6$?

50. Let $f(x) = \log_a(3x^2 - 2)$. For what value of a is $f'(1) = 3$?

51–60 ▪ Use logarithmic differentiation or an alternative method to find the derivative of the function.

51. $y = (x^2 + 2)^2(x^4 + 4)^4$

52. $y = \dfrac{e^{-x}\cos^2 x}{x^2 + x + 1}$

53. $y = \sqrt{\dfrac{x - 1}{x^4 + 1}}$

54. $y = \sqrt{x}\,e^{x^2 - x}(x + 1)^{2/3}$

55. $y = x^x$

56. $y = x^{\cos x}$

57. $y = (\cos x)^x$

58. $y = \sqrt{x}^{\,x}$

59. $y = (\tan x)^{1/x}$

60. $y = (\sin x)^{\ln x}$

---

61. Find y' if $e^{x^2 y} = x + y$.

62. Find an equation of the tangent line to the curve $xe^y + ye^x = 1$ at the point $(0, 1)$.

63. Find y' if $y = \ln(x^2 + y^2)$.

64. Find y' if $x^y = y^x$.

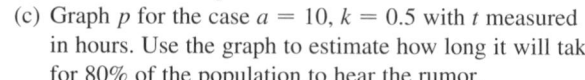 **65.** The motion of a spring that is subject to a frictional force or a damping force (such as a shock absorber in a car) is often modeled by the product of an exponential function and a sine or cosine function. Suppose the equation of motion of a point on such a spring is

$$s(t) = 2e^{-1.5t}\sin 2\pi t$$

where s is measured in centimeters and t in seconds. Find the velocity after t seconds and graph both the position and velocity functions for $0 \le t \le 2$.

66. Under certain circumstances a rumor spreads according to the equation

$$p(t) = \dfrac{1}{1 + ae^{-kt}}$$

where $p(t)$ is the proportion of the population that knows the rumor at time t and a and k are positive constants.
(a) Find $\lim_{t\to\infty} p(t)$.
(b) Find the rate of spread of the rumor.
(c) Graph p for the case $a = 10$, $k = 0.5$ with t measured in hours. Use the graph to estimate how long it will take for 80% of the population to hear the rumor.

67. Show that the function $y = Ae^{-x} + Bxe^{-x}$ satisfies the differential equation $y'' + 2y' + y = 0$.

68. For what values of r does the function $y = e^{rx}$ satisfy the equation $y'' + 5y' - 6y = 0$?

69. If $f(x) = e^{2x}$, find a formula for $f^{(n)}(x)$.

70. Find the thousandth derivative of $f(x) = xe^{-x}$.

71. Find a formula for $f^{(n)}(x)$ if $f(x) = \ln(x - 1)$.

72. Find $\dfrac{d^9}{dx^9}(x^8\ln x)$.

73. If $f(x) = 3 + x + e^x$, find $(f^{-1})'(4)$.

74. Evaluate $\lim\limits_{x\to\pi}\dfrac{e^{\sin x} - 1}{x - \pi}$.

3.4 | EXPONENTIAL GROWTH AND DECAY

In many natural phenomena, quantities grow or decay at a rate proportional to their size. For instance, if $y = f(t)$ is the number of individuals in a population of animals or bacteria at time t, then it seems reasonable to expect that the rate of growth $f'(t)$ is proportional to the population $f(t)$; that is, $f'(t) = kf(t)$ for some constant k. Indeed, under ideal conditions (unlimited environment, adequate nutrition, immunity to disease) the mathematical model given by the equation $f'(t) = kf(t)$ predicts what actually happens fairly accurately. Another example occurs in nuclear physics where the mass of a radioactive substance decays at a rate proportional to the mass. In chemistry, the rate of a unimolecular first-order reaction is proportional to the concentration of the substance. In finance, the value of a savings account with continuously compounded interest increases at a rate proportional to that value.

In general, if $y(t)$ is the value of a quantity y at time t and if the rate of change of y with respect to t is proportional to its size $y(t)$ at any time, then

$$\boxed{1} \qquad \boxed{\dfrac{dy}{dt} = ky}$$

where k is a constant. Equation 1 is sometimes called the **law of natural growth** (if $k > 0$) or the **law of natural decay** (if $k < 0$). It is called a **differential equation** because it involves an unknown function y and its derivative dy/dt.

It's not hard to think of a solution of Equation 1. This equation asks us to find a function whose derivative is a constant multiple of itself. We have met such functions in this chapter. Any exponential function of the form $y(t) = Ce^{kt}$, where C is a constant, satisfies

$$y'(t) = C(ke^{kt}) = k(Ce^{kt}) = ky(t)$$

We will see in Section 7.7 that *any* function that satsifies $dy/dt = ky$ must be of the form $y = Ce^{kt}$. To see the significance of the constant C, we observe that

$$y(0) = Ce^{k \cdot 0} = C$$

Therefore C is the initial value of the function.

$\boxed{2}$ **THEOREM** The only solutions of the differential equation $dy/dt = ky$ are the exponential functions

$$y(t) = y(0)e^{kt}$$

POPULATION GROWTH

What is the significance of the proportionality constant k? In the context of population growth, where $P(t)$ is the size of a population at time t, we can write

$$\boxed{3} \qquad \dfrac{dP}{dt} = kP \qquad \text{or} \qquad \dfrac{1}{P}\dfrac{dP}{dt} = k$$

The quantity

$$\frac{1}{P}\frac{dP}{dt}$$

is the growth rate divided by the population size; it is called the **relative growth rate**. According to ③, instead of saying "the growth rate is proportional to population size" we could say "the relative growth rate is constant." Then ② says that a population with constant relative growth rate must grow exponentially. Notice that the relative growth rate k appears as the coefficient of t in the exponential function Ce^{kt}. For instance, if

$$\frac{dP}{dt} = 0.02P$$

and t is measured in years, then the relative growth rate is $k = 0.02$ and the population grows at a relative rate of 2% per year. If the population at time 0 is P_0, then the expression for the population is

$$P(t) = P_0 e^{0.02t}$$

▼ EXAMPLE 1 Use the fact that the world population was 2560 million in 1950 and 3040 million in 1960 to model the population of the world in the second half of the 20th century. (Assume that the growth rate is proportional to the population size.) What is the relative growth rate? Use the model to estimate the world population in 1993 and to predict the population in the year 2020.

SOLUTION We measure the time t in years and let $t = 0$ in the year 1950. We measure the population $P(t)$ in millions of people. Then $P(0) = 2560$ and $P(10) = 3040$. Since we are assuming that $dP/dt = kP$, Theorem 2 gives

$$P(t) = P(0)e^{kt} = 2560e^{kt}$$

$$P(10) = 2560e^{10k} = 3040$$

$$k = \frac{1}{10}\ln\frac{3040}{2560} \approx 0.017185$$

The relative growth rate is about 1.7% per year and the model is

$$P(t) = 2560e^{0.017185t}$$

We estimate that the world population in 1993 was

$$P(43) = 2560e^{0.017185(43)} \approx 5360 \text{ million}$$

The model predicts that the population in 2020 will be

$$P(70) = 2560e^{0.017185(70)} \approx 8524 \text{ million}$$

■ www.stewartcalculus.com
See Additional Example A.

The graph in Figure 1 shows that the model is fairly accurate to date (the dots represent the actual population), so the estimate for 1993 is quite reliable. But the prediction for 2020 is riskier.

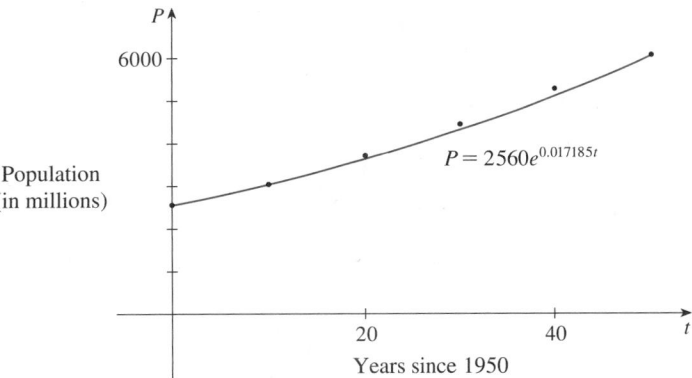

FIGURE 1

A model for world population growth in the second half of the 20th century

■

RADIOACTIVE DECAY

Radioactive substances decay by spontaneously emitting radiation. If $m(t)$ is the mass remaining from an initial mass m_0 of the substance after time t, then the relative decay rate

$$-\frac{1}{m}\frac{dm}{dt}$$

has been found experimentally to be constant. (Since dm/dt is negative, the relative decay rate is positive.) It follows that

$$\frac{dm}{dt} = km$$

where k is a negative constant. In other words, radioactive substances decay at a rate proportional to the remaining mass. This means that we can use $\boxed{2}$ to show that the mass decays exponentially:

$$m(t) = m_0 e^{kt}$$

Physicists express the rate of decay in terms of **half-life**, the time required for half of any given quantity to decay.

V EXAMPLE 2 The half-life of radium-226 ($^{226}_{88}$Ra) is 1590 years.
(a) A sample of radium-226 has a mass of 100 mg. Find a formula for the mass of $^{226}_{88}$Ra that remains after t years.
(b) Find the mass after 1000 years correct to the nearest milligram.
(c) When will the mass be reduced to 30 mg?

SOLUTION
(a) Let $m(t)$ be the mass of radium-226 (in milligrams) that remains after t years. Then $dm/dt = km$ and $y(0) = 100$, so $\boxed{2}$ gives

$$m(t) = m(0)e^{kt} = 100e^{kt}$$

In order to determine the value of k, we use the fact that $y(1590) = \frac{1}{2}(100)$. Thus

$$100e^{1590k} = 50 \qquad \text{so} \qquad e^{1590k} = \tfrac{1}{2}$$

and
$$1590k = \ln \tfrac{1}{2} = -\ln 2$$

$$k = -\frac{\ln 2}{1590}$$

Therefore
$$m(t) = 100e^{-(\ln 2)t/1590}$$

We could use the fact that $e^{\ln 2} = 2$ to write the expression for $m(t)$ in the alternative form

$$m(t) = 100 \times 2^{-t/1590}$$

(b) The mass after 1000 years is

$$m(1000) = 100e^{-(\ln 2)1000/1590} \approx 65 \text{ mg}$$

(c) We want to find the value of t such that $m(t) = 30$, that is,

$$100e^{-(\ln 2)t/1590} = 30 \qquad \text{or} \qquad e^{-(\ln 2)t/1590} = 0.3$$

We solve this equation for t by taking the natural logarithm of both sides:

$$-\frac{\ln 2}{1590}t = \ln 0.3$$

Thus
$$t = -1590\,\frac{\ln 0.3}{\ln 2} \approx 2762 \text{ years}$$

■

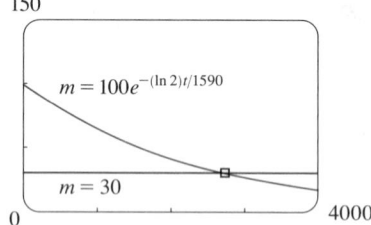

150

$m = 100e^{-(\ln 2)t/1590}$

$m = 30$

0   4000

FIGURE 2

As a check on our work in Example 2, we use a graphing device to draw the graph of $m(t)$ in Figure 2 together with the horizontal line $m = 30$. These curves intersect when $t \approx 2800$, and this agrees with the answer to part (c).

NEWTON'S LAW OF COOLING

Newton's Law of Cooling states that the rate of cooling of an object is proportional to the temperature difference between the object and its surroundings, provided that this difference is not too large. (This law also applies to warming.) If we let $T(t)$ be the temperature of the object at time t and T_s be the temperature of the surroundings, then we can formulate Newton's Law of Cooling as a differential equation:

$$\frac{dT}{dt} = k(T - T_s)$$

where k is a constant. This equation is not quite the same as Equation 1, so we make the change of variable $y(t) = T(t) - T_s$. Because T_s is constant, we have $y'(t) = T'(t)$ and so the equation becomes

$$\frac{dy}{dt} = ky$$

We can then use 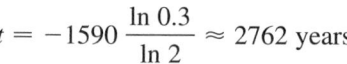 $\boxed{2}$ to find an expression for y, from which we can find T.

EXAMPLE 3 A bottle of soda pop at room temperature (72°F) is placed in a refrigerator where the temperature is 44°F. After half an hour the soda pop has cooled to 61°F.

(a) What is the temperature of the soda pop after another half hour?

(b) How long does it take for the soda pop to cool to 50°F?

SOLUTION

(a) Let $T(t)$ be the temperature of the soda after t minutes. The surrounding temperature is $T_s = 44°F$, so Newton's Law of Cooling states that

$$\frac{dT}{dt} = k(T - 44)$$

If we let $y = T - 44$, then $y(0) = T(0) - 44 = 72 - 44 = 28$, so y satisfies

$$\frac{dy}{dt} = ky \qquad y(0) = 28$$

and by $\boxed{2}$ we have

$$y(t) = y(0)e^{kt} = 28e^{kt}$$

We are given that $T(30) = 61$, so $y(30) = 61 - 44 = 17$ and

$$28e^{30k} = 17 \qquad e^{30k} = \tfrac{17}{28}$$

Taking logarithms, we have

$$k = \frac{\ln\left(\tfrac{17}{28}\right)}{30} \approx -0.01663$$

Thus

$$y(t) = 28e^{-0.01663t}$$

$$T(t) = 44 + 28e^{-0.01663t}$$

$$T(60) = 44 + 28e^{-0.01663(60)} \approx 54.3$$

So after another half hour the pop has cooled to about 54°F.

(b) We have $T(t) = 50$ when

$$44 + 28e^{-0.01663t} = 50$$

$$e^{-0.01663t} = \tfrac{6}{28}$$

$$t = \frac{\ln\left(\tfrac{6}{28}\right)}{-0.01663} \approx 92.6$$

The pop cools to 50°F after about 1 hour 33 minutes. ∎

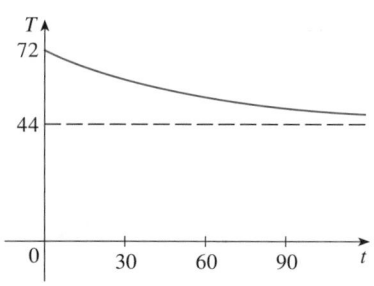

FIGURE 3

Notice that in Example 3, we have

$$\lim_{t \to \infty} T(t) = \lim_{t \to \infty} (44 + 28e^{-0.01663t}) = 44 + 28 \cdot 0 = 44$$

which is to be expected. The graph of the temperature function is shown in Figure 3.

CONTINUOUSLY COMPOUNDED INTEREST

EXAMPLE 4 If $1000 is invested at 6% interest, compounded annually, then after 1 year the investment is worth $1000(1.06) = $1060, after 2 years it's worth $[1000(1.06)]1.06 = $1123.60, and after t years it's worth $1000(1.06)^t$. In general, if an amount A_0 is invested at an interest rate r ($r = 0.06$ in this example), then after t years it's worth $A_0(1 + r)^t$. Usually, however, interest is compounded more frequently, say, n times a year. Then in each compounding period the interest rate is r/n and there are nt compounding periods in t years, so the value of the investment is

$$A_0 \left(1 + \frac{r}{n} \right)^{nt}$$

For instance, after 3 years at 6% interest a $1000 investment will be worth

$$\$1000(1.06)^3 = \$1191.02 \quad \text{with annual compounding}$$

$$\$1000(1.03)^6 = \$1194.05 \quad \text{with semiannual compounding}$$

$$\$1000(1.015)^{12} = \$1195.62 \quad \text{with quarterly compounding}$$

$$\$1000(1.005)^{36} = \$1196.68 \quad \text{with monthly compounding}$$

$$\$1000 \left(1 + \frac{0.06}{365} \right)^{365 \cdot 3} = \$1197.20 \quad \text{with daily compounding}$$

You can see that the interest paid increases as the number of compounding periods (n) increases. If we let $n \to \infty$, then we will be compounding the interest **continuously** and the value of the investment will be

$$A(t) = \lim_{n \to \infty} A_0 \left(1 + \frac{r}{n} \right)^{nt} = \lim_{n \to \infty} A_0 \left[\left(1 + \frac{r}{n} \right)^{n/r} \right]^{rt}$$

$$= A_0 \left[\lim_{n \to \infty} \left(1 + \frac{r}{n} \right)^{n/r} \right]^{rt}$$

$$= A_0 \left[\lim_{m \to \infty} \left(1 + \frac{1}{m} \right)^{m} \right]^{rt} \quad \text{(where } m = n/r\text{)}$$

■ Recall: $e = \lim_{x \to 0} (1 + x)^{1/x}$

If we put $n = 1/x$, then $n \to \infty$ as $x \to 0^+$ and so an alternative expression for e is

$$e = \lim_{n \to \infty} \left(1 + \frac{1}{n} \right)^{n}$$

But the limit in this expression is equal to the number e. So with continuous compounding of interest at interest rate r, the amount after t years is

$$A(t) = A_0 e^{rt}$$

If we differentiate this equation, we get

$$\frac{dA}{dt} = rA_0e^{rt} = rA(t)$$

which says that, with continuous compounding of interest, the rate of increase of an investment is proportional to its size.

Returning to the example of $1000 invested for 3 years at 6% interest, we see that with continuous compounding of interest the value of the investment will be

$$A(3) = \$1000e^{(0.06)3} = \$1197.22$$

Notice how close this is to the amount we calculated for daily compounding, $1197.20. But the amount is easier to compute if we use continuous compounding. ■

3.4 | EXERCISES

1. A population of protozoa develops with a constant relative growth rate of 0.7944 per member per day. On day zero the population consists of two members. Find the population size after six days.

2. A common inhabitant of human intestines is the bacterium *Escherichia coli*. A cell of this bacterium in a nutrient-broth medium divides into two cells every 20 minutes. The initial population of a culture is 60 cells.
(a) Find the relative growth rate.
(b) Find an expression for the number of cells after t hours.
(c) Find the number of cells after 8 hours.
(d) Find the rate of growth after 8 hours.
(e) When will the population reach 20,000 cells?

3. A bacteria culture initially contains 100 cells and grows at a rate proportional to its size. After an hour the population has increased to 420.
(a) Find an expression for the number of bacteria after t hours.
(b) Find the number of bacteria after 3 hours.
(c) Find the rate of growth after 3 hours.
(d) When will the population reach 10,000?

4. A bacteria culture grows with constant relative growth rate. The bacteria count was 400 after 2 hours and 25,600 after 6 hours.
(a) What is the relative growth rate? Express your answer as a percentage.
(b) What was the intitial size of the culture?
(c) Find an expression for the number of bacteria after t hours.
(d) Find the number of cells after 4.5 hours.
(e) Find the rate of growth after 4.5 hours.
(f) When will the population reach 50,000?

5. The table gives estimates of the world population, in millions, from 1750 to 2000:

Year	Population	Year	Population
1750	790	1900	1650
1800	980	1950	2560
1850	1260	2000	6080

(a) Use the exponential model and the population figures for 1750 and 1800 to predict the world population in 1900 and 1950. Compare with the actual figures.
(b) Use the exponential model and the population figures for 1850 and 1900 to predict the world population in 1950. Compare with the actual population.
(c) Use the exponential model and the population figures for 1900 and 1950 to predict the world population in 2000. Compare with the actual population and try to explain the discrepancy.

6. The table gives the population of India, in millions, for the second half of the 20th century.

Year	Population
1951	361
1961	439
1971	548
1981	683
1991	846
2001	1029

(a) Use the exponential model and the census figures for 1951 and 1961 to predict the population in 2001. Compare with the actual figure.

(b) Use the exponential model and the census figures for 1961 and 1981 to predict the population in 2001. Compare with the actual population. Then use this model to predict the population in the years 2010 and 2020.

(c) Graph both of the exponential functions in parts (a) and (b) together with a plot of the actual population. Are these models reasonable ones?

7. Experiments show that if the chemical reaction

$$N_2O_5 \rightarrow 2NO_2 + \tfrac{1}{2}O_2$$

takes place at 45°C, the rate of reaction of dinitrogen pentoxide is proportional to its concentration as follows:

$$-\frac{d[N_2O_5]}{dt} = 0.0005[N_2O_5]$$

(a) Find an expression for the concentration $[N_2O_5]$ after t seconds if the initial concentration is C.

(b) How long will the reaction take to reduce the concentration of N_2O_5 to 90% of its original value?

8. Strontium-90 has a half-life of 28 days.

(a) A sample has a mass of 50 mg initially. Find a formula for the mass remaining after t days.

(b) Find the mass remaining after 40 days.

(c) How long does it take the sample to decay to a mass of 2 mg?

(d) Sketch the graph of the mass function.

9. The half-life of cesium-137 is 30 years. Suppose we have a 100-mg sample.

(a) Find the mass that remains after t years.

(b) How much of the sample remains after 100 years?

(c) After how long will only 1 mg remain?

10. A sample of tritium-3 decayed to 94.5% of its original amount after a year.

(a) What is the half-life of tritium-3?

(b) How long would it take the sample to decay to 20% of its original amount?

11. Scientists can determine the age of ancient objects by the method of *radiocarbon dating*. The bombardment of the upper atmosphere by cosmic rays converts nitrogen to a radioactive isotope of carbon, ^{14}C, with a half-life of about 5730 years. Vegetation absorbs carbon dioxide through the atmosphere and animal life assimilates ^{14}C through food chains. When a plant or animal dies, it stops replacing its carbon and the amount of ^{14}C begins to decrease through radioactive decay. Therefore the level of radioactivity must also decay exponentially.

A parchment fragment was discovered that had about 74% as much ^{14}C radioactivity as does plant material on the earth today. Estimate the age of the parchment.

12. A curve passes through the point $(0, 5)$ and has the property that the slope of the curve at every point P is twice the y-coordinate of P. What is the equation of the curve?

13. A roast turkey is taken from an oven when its temperature has reached 185°F and is placed on a table in a room where the temperature is 75°F.

(a) If the temperature of the turkey is 150°F after half an hour, what is the temperature after 45 minutes?

(b) When will the turkey have cooled to 100°F?

14. In a murder investigation, the temperature of the corpse was 32.5°C at 1:30 PM and 30.3°C an hour later. Normal body temperature is 37.0°C and the temperature of the surroundings was 20.0°C. When did the murder take place?

15. When a cold drink is taken from a refrigerator, its temperature is 5°C. After 25 minutes in a 20°C room its temperature has increased to 10°C.

(a) What is the temperature of the drink after 50 minutes?

(b) When will its temperature be 15°C?

16. A freshly brewed cup of coffee has temperature 95°C in a 20°C room. When its temperature is 70°C, it is cooling at a rate of 1°C per minute. When does this occur?

17. The rate of change of atmospheric pressure P with respect to altitude h is proportional to P, provided that the temperature is constant. At 15°C the pressure is 101.3 kPa at sea level and 87.14 kPa at $h = 1000$ m.

(a) What is the pressure at an altitude of 3000 m?

(b) What is the pressure at the top of Mount McKinley, at an altitude of 6187 m?

18. (a) If $1000 is borrowed at 8% interest, find the amounts due at the end of 3 years if the interest is compounded (i) annually, (ii) quarterly, (iii) monthly, (iv) weekly, (v) daily, (vi) hourly, and (vii) continuously.

(b) Suppose $1000 is borrowed and the interest is compounded continuously. If $A(t)$ is the amount due after t years, where $0 \le t \le 3$, graph $A(t)$ for each of the interest rates 6%, 8%, and 10% on a common screen.

19. If $3000 is invested at 5% interest, find the value of the investment at the end of 5 years if the interest is compounded

(a) annually   (b) semiannually

(c) monthly   (d) weekly

(e) daily   (f) continuously

20. (a) How long will it take an investment to double in value if the interest rate is 6% compounded continuously?

(b) What is the equivalent annual interest rate?

3.5 | INVERSE TRIGONOMETRIC FUNCTIONS

Recall from Section 3.2 that the only functions that have inverse functions are one-to-one functions. Trigonometric functions, however, are not one-to-one and so they don't have inverse functions. But we can make them one-to-one by restricting their domains and we will see that the inverses of these restricted trigonometric functions play a major role in integral calculus.

You can see from Figure 1 that the sine function $y = \sin x$ is not one-to-one (use the Horizontal Line Test). But the function $f(x) = \sin x$, $-\pi/2 \le x \le \pi/2$ (see Figure 2), *is* one-to-one. The inverse function of this restricted sine function f exists and is denoted by $\sin^{-1}$ or arcsin. It is called the **inverse sine function** or the **arcsine function**.

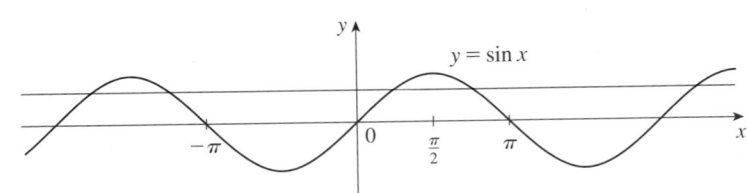

FIGURE 1

FIGURE 2 $y = \sin x$, $-\dfrac{\pi}{2} \le x \le \dfrac{\pi}{2}$

Since the definition of an inverse function says that

$$f^{-1}(x) = y \iff f(y) = x$$

we have

| 1 | $\sin^{-1}x = y \iff \sin y = x \quad \text{and} \quad -\dfrac{\pi}{2} \le y \le \dfrac{\pi}{2}$ |

$\oslash \quad \sin^{-1}x \ne \dfrac{1}{\sin x}$

Thus if $-1 \le x \le 1$, $\sin^{-1}x$ is the number between $-\pi/2$ and $\pi/2$ whose sine is x.

EXAMPLE 1 Evaluate (a) $\sin^{-1}\left(\frac{1}{2}\right)$ and (b) $\tan\left(\arcsin \frac{1}{3}\right)$.

SOLUTION
(a) We have

$$\sin^{-1}\left(\tfrac{1}{2}\right) = \dfrac{\pi}{6}$$

because $\sin(\pi/6) = \frac{1}{2}$ and $\pi/6$ lies between $-\pi/2$ and $\pi/2$.

(b) Let $\theta = \arcsin \frac{1}{3}$, so $\sin \theta = \frac{1}{3}$. Then we can draw a right triangle with angle θ as in Figure 3 and deduce from the Pythagorean Theorem that the third side has length $\sqrt{9 - 1} = 2\sqrt{2}$. This enables us to read from the triangle that

$$\tan\left(\arcsin \tfrac{1}{3}\right) = \tan \theta = \dfrac{1}{2\sqrt{2}}$$

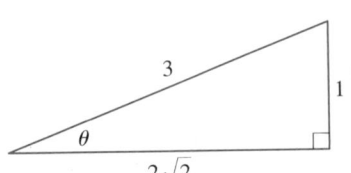

FIGURE 3

The cancellation equations for inverse functions become, in this case,

2

$$\sin^{-1}(\sin x) = x \quad \text{for } -\frac{\pi}{2} \le x \le \frac{\pi}{2}$$

$$\sin(\sin^{-1}x) = x \quad \text{for } -1 \le x \le 1$$

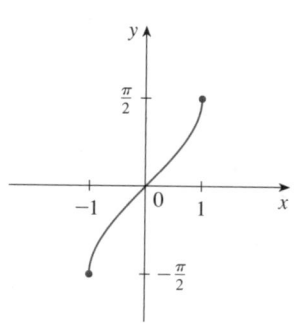

FIGURE 4 $y = \sin^{-1}x = \arcsin x$

■ www.stewartcalculus.com
See Additional Example A.

The inverse sine function, $\sin^{-1}$, has domain $[-1, 1]$ and range $[-\pi/2, \pi/2]$, and its graph, shown in Figure 4, is obtained from that of the restricted sine function (Figure 2) by reflection about the line $y = x$.

We know that the sine function f is continuous, so the inverse sine function is also continuous. We also know from Section 2.3 that the sine function is differentiable, so the inverse sine function is also differentiable. We could calculate the derivative of $\sin^{-1}$ by the formula in Theorem 3.2.7, but since we know that $\sin^{-1}$ is differentiable, we can just as easily calculate it by implicit differentiation as follows.

Let $y = \sin^{-1}x$. Then $\sin y = x$ and $-\pi/2 \le y \le \pi/2$. Differentiating $\sin y = x$ implicitly with respect to x, we obtain

$$\cos y \frac{dy}{dx} = 1$$

and

$$\frac{dy}{dx} = \frac{1}{\cos y}$$

Now $\cos y \ge 0$ since $-\pi/2 \le y \le \pi/2$, so

$$\cos y = \sqrt{1 - \sin^2 y} = \sqrt{1 - x^2}$$

Therefore

$$\frac{dy}{dx} = \frac{1}{\cos y} = \frac{1}{\sqrt{1 - x^2}}$$

3

$$\frac{d}{dx}(\sin^{-1}x) = \frac{1}{\sqrt{1 - x^2}} \qquad -1 < x < 1$$

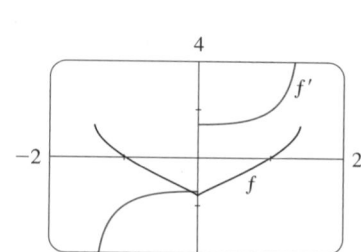

FIGURE 5

■ The graphs of the function f of Example 2 and its derivative are shown in Figure 5. Notice that f is not differentiable at 0 and this is consistent with the fact that the graph of f' makes a sudden jump at $x = 0$.

▼ EXAMPLE 2 If $f(x) = \sin^{-1}(x^2 - 1)$, find (a) the domain of f, (b) $f'(x)$, and (c) the domain of f'.

SOLUTION

(a) Since the domain of the inverse sine function is $[-1, 1]$, the domain of f is

$$\{x \mid -1 \le x^2 - 1 \le 1\} = \{x \mid 0 \le x^2 \le 2\}$$

$$= \{x \mid |x| \le \sqrt{2}\} = [-\sqrt{2}, \sqrt{2}]$$

(b) Combining Formula 3 with the Chain Rule, we have

$$f'(x) = \frac{1}{\sqrt{1 - (x^2 - 1)^2}} \frac{d}{dx}(x^2 - 1)$$

$$= \frac{1}{\sqrt{1 - (x^4 - 2x^2 + 1)}} 2x = \frac{2x}{\sqrt{2x^2 - x^4}}$$

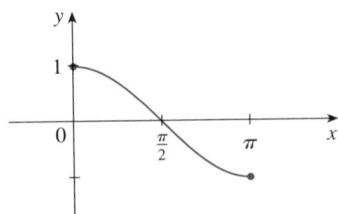

FIGURE 6 $y = \cos x, 0 \le x \le \pi$

(c) The domain of f' is

$$\{x \mid -1 < x^2 - 1 < 1\} = \{x \mid 0 < x^2 < 2\}$$
$$= \{x \mid 0 < |x| < \sqrt{2}\} = (-\sqrt{2}, 0) \cup (0, \sqrt{2}) \quad \blacksquare$$

The **inverse cosine function** is handled similarly. The restricted cosine function $f(x) = \cos x, 0 \le x \le \pi$, is one-to-one (see Figure 6) and so it has an inverse function denoted by $\cos^{-1}$ or arccos.

$$\boxed{4} \qquad \boxed{\cos^{-1} x = y \iff \cos y = x \quad \text{and} \quad 0 \le y \le \pi}$$

The cancellation equations are

$$\boxed{5} \qquad \boxed{\begin{array}{l} \cos^{-1}(\cos x) = x \quad \text{for } 0 \le x \le \pi \\[1mm] \cos(\cos^{-1} x) = x \quad \text{for } -1 \le x \le 1 \end{array}}$$

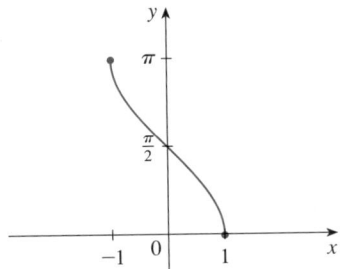

FIGURE 7 $y = \cos^{-1} x = \arccos x$

The inverse cosine function, $\cos^{-1}$, has domain $[-1, 1]$ and range $[0, \pi]$ and is a continuous function whose graph is shown in Figure 7. Its derivative is given by

$$\boxed{6} \qquad \boxed{\frac{d}{dx}(\cos^{-1} x) = -\frac{1}{\sqrt{1 - x^2}} \qquad -1 < x < 1}$$

Formula 6 can be proved by the same method as for Formula 3 and is left as Exercise 11.

The tangent function can be made one-to-one by restricting it to the interval $(-\pi/2, \pi/2)$. Thus the **inverse tangent function** is defined as the inverse of the function $f(x) = \tan x, -\pi/2 < x < \pi/2$. (See Figure 8.) It is denoted by $\tan^{-1}$ or arctan.

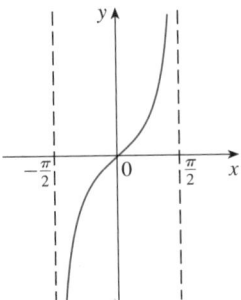

FIGURE 8
$y = \tan x, -\frac{\pi}{2} < x < \frac{\pi}{2}$

$$\boxed{7} \qquad \boxed{\tan^{-1} x = y \iff \tan y = x \quad \text{and} \quad -\frac{\pi}{2} < y < \frac{\pi}{2}}$$

EXAMPLE 3 Simplify the expression $\cos(\tan^{-1} x)$.

SOLUTION 1 Let $y = \tan^{-1} x$. Then $\tan y = x$ and $-\pi/2 < y < \pi/2$. We want to find $\cos y$ but, since $\tan y$ is known, it's easier to find $\sec y$ first:

$$\sec^2 y = 1 + \tan^2 y = 1 + x^2$$
$$\sec y = \sqrt{1 + x^2} \qquad \text{(since } \sec y > 0 \text{ for } -\pi/2 < y < \pi/2\text{)}$$

Thus
$$\cos(\tan^{-1} x) = \cos y = \frac{1}{\sec y} = \frac{1}{\sqrt{1 + x^2}}$$

SOLUTION 2 Instead of using trigonometric identities as in Solution 1, it is perhaps easier to use a diagram. If $y = \tan^{-1} x$, then $\tan y = x$, and we can read from

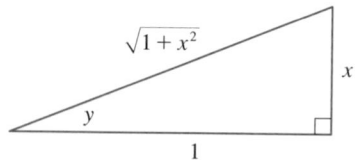

FIGURE 9

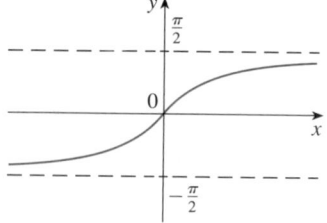

FIGURE 10

$y = \tan^{-1} x = \arctan x$

■ www.stewartcalculus.com
See Additional Example B.

Figure 9 (which illustrates the case $y > 0$) that

$$\cos(\tan^{-1}x) = \cos y = \frac{1}{\sqrt{1 + x^2}}$$ ■

The inverse tangent function, $\tan^{-1} = \arctan$, has domain $\mathbb{R}$ and its range is $(-\pi/2, \pi/2)$. Its graph is shown in Figure 10.

We know that

$$\lim_{x \to (\pi/2)^-} \tan x = \infty \quad \text{and} \quad \lim_{x \to -(\pi/2)^+} \tan x = -\infty$$

and so the lines $x = \pm\pi/2$ are vertical asymptotes of the graph of tan. Since the graph of $\tan^{-1}$ is obtained by reflecting the graph of the restricted tangent function about the line $y = x$, it follows that the lines $y = \pi/2$ and $y = -\pi/2$ are horizontal asymptotes of the graph of $\tan^{-1}$. This fact is expressed by the following limits:

8

$$\lim_{x \to \infty} \tan^{-1}x = \frac{\pi}{2} \qquad \lim_{x \to -\infty} \tan^{-1}x = -\frac{\pi}{2}$$

EXAMPLE 4 Evaluate $\displaystyle\lim_{x \to 2^+} \arctan\left(\frac{1}{x - 2}\right)$.

SOLUTION If we let $t = 1/(x - 2)$, we know that $t \to \infty$ as $x \to 2^+$. Therefore, by the second equation in **8**, we have

$$\lim_{x \to 2^+} \arctan\left(\frac{1}{x - 2}\right) = \lim_{t \to \infty} \arctan t = \frac{\pi}{2}$$ ■

Since tan is differentiable, $\tan^{-1}$ is also differentiable. To find its derivative, let $y = \tan^{-1}x$. Then $\tan y = x$. Differentiating this last equation implicitly with respect to x, we have

$$\sec^2 y \frac{dy}{dx} = 1$$

and so

$$\frac{dy}{dx} = \frac{1}{\sec^2 y} = \frac{1}{1 + \tan^2 y} = \frac{1}{1 + x^2}$$

9

$$\frac{d}{dx} (\tan^{-1}x) = \frac{1}{1 + x^2}$$

The remaining inverse trigonometric functions are not used as frequently and are summarized here.

10

$$y = \csc^{-1}x \ (|x| \geq 1) \iff \csc y = x \quad \text{and} \quad y \in (0, \pi/2] \cup (\pi, 3\pi/2]$$

$$y = \sec^{-1}x \ (|x| \geq 1) \iff \sec y = x \quad \text{and} \quad y \in [0, \pi/2) \cup [\pi, 3\pi/2)$$

$$y = \cot^{-1}x \ (x \in \mathbb{R}) \iff \cot y = x \quad \text{and} \quad y \in (0, \pi)$$

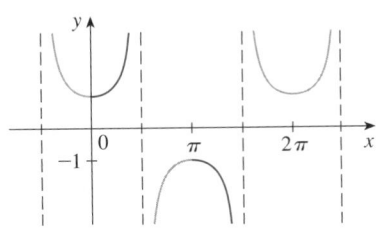

FIGURE 11

$y = \sec x$

The choice of intervals for y in the definitions of $\csc^{-1}$ and $\sec^{-1}$ is not universally agreed upon. For instance, some authors use $y \in [0, \pi/2) \cup (\pi/2, \pi]$ in the definition of $\sec^{-1}$. [You can see from the graph of the secant function in Figure 11 that both this choice and the one in $\boxed{10}$ will work.] The reason for the choice in $\boxed{10}$ is that the differentiation formulas are simpler (see Exercise 41).

We collect in Table 11 the differentiation formulas for all of the inverse trigonometric functions. The proofs of the formulas for the derivatives of $\csc^{-1}$, $\sec^{-1}$, and $\cot^{-1}$ are left as Exercises 13–15.

$\boxed{11}$ **TABLE OF DERIVATIVES OF INVERSE TRIGONOMETRIC FUNCTIONS**

$$\frac{d}{dx}(\sin^{-1}x) = \frac{1}{\sqrt{1-x^2}} \qquad \frac{d}{dx}(\csc^{-1}x) = -\frac{1}{x\sqrt{x^2-1}}$$

$$\frac{d}{dx}(\cos^{-1}x) = -\frac{1}{\sqrt{1-x^2}} \qquad \frac{d}{dx}(\sec^{-1}x) = \frac{1}{x\sqrt{x^2-1}}$$

$$\frac{d}{dx}(\tan^{-1}x) = \frac{1}{1+x^2} \qquad \frac{d}{dx}(\cot^{-1}x) = -\frac{1}{1+x^2}$$

Each of these formulas can be combined with the Chain Rule. For instance, if u is a differentiable function of x, then

$$\frac{d}{dx}(\sin^{-1}u) = \frac{1}{\sqrt{1-u^2}}\frac{du}{dx} \quad \text{and} \quad \frac{d}{dx}(\tan^{-1}u) = \frac{1}{1+u^2}\frac{du}{dx}$$

$\boxed{\text{V}}$ **EXAMPLE 5** Differentiate $f(x) = x\tan^{-1}\sqrt{x}$.

SOLUTION

$$f'(x) = x\frac{1}{1+(\sqrt{x})^2}\frac{1}{2}x^{-1/2} + \tan^{-1}\sqrt{x} = \frac{\sqrt{x}}{2(1+x)} + \tan^{-1}\sqrt{x} \qquad \blacksquare$$

3.5 | EXERCISES

1–6 ■ Find the exact value of each expression.

1. (a) $\sin^{-1}(\sqrt{3}/2)$ (b) $\cos^{-1}(-1)$

2. (a) $\tan^{-1}(1/\sqrt{3})$ (b) $\sec^{-1}2$

3. (a) $\arctan 1$ (b) $\sin^{-1}(1/\sqrt{2})$

4. (a) $\cot^{-1}(-\sqrt{3})$ (b) $\arccos(-\frac{1}{2})$

5. (a) $\tan(\arctan 10)$ (b) $\sin^{-1}(\sin(7\pi/3))$

6. (a) $\tan(\sec^{-1}4)$ (b) $\sin(2\sin^{-1}(\frac{3}{5}))$

7. Prove that $\cos(\sin^{-1}x) = \sqrt{1-x^2}$.

8–10 ■ Simplify the expression.

8. $\tan(\sin^{-1}x)$ **9.** $\sin(\tan^{-1}x)$

10. $\cos(2\tan^{-1}x)$

---

11. Prove Formula 6 for the derivative of $\cos^{-1}$ by the same method as for Formula 3.

12. (a) Prove that $\sin^{-1}x + \cos^{-1}x = \pi/2$.
    (b) Use part (a) to prove Formula 6.

13. Prove that $\dfrac{d}{dx}(\cot^{-1}x) = -\dfrac{1}{1+x^2}$.

14. Prove that $\dfrac{d}{dx}(\sec^{-1}x) = \dfrac{1}{x\sqrt{x^2-1}}$.

15. Prove that $\dfrac{d}{dx}(\csc^{-1}x) = -\dfrac{1}{x\sqrt{x^2-1}}$.

16–29 ▪ Find the derivative of the function. Simplify where possible.

16. $y = \tan^{-1}(x^2)$

17. $y = (\tan^{-1}x)^2$

18. $g(x) = \sqrt{x^2-1}\,\sec^{-1}x$

19. $y = \sin^{-1}(2x+1)$

20. $y = \tan^{-1}\!\left(x - \sqrt{1+x^2}\right)$

21. $G(x) = \sqrt{1-x^2}\,\arccos x$

22. $F(\theta) = \arcsin\sqrt{\sin\theta}$

23. $h(t) = \cot^{-1}(t) + \cot^{-1}(1/t)$

24. $y = \cos^{-1}(\sin^{-1}t)$

25. $y = \arctan(\cos\theta)$

26. $f(x) = x\ln(\arctan x)$

27. $y = x\sin^{-1}x + \sqrt{1-x^2}$

28. $y = \arctan\sqrt{\dfrac{1-x}{1+x}}$

29. $y = \arccos\!\left(\dfrac{b + a\cos x}{a + b\cos x}\right),\quad 0 \leqslant x \leqslant \pi,\ a > b > 0$

30–31 ▪ Find the derivative of the function. Find the domains of the function and its derivative.

30. $f(x) = \arcsin(e^x)$

31. $g(x) = \cos^{-1}(3-2x)$

32. Find y' if $\tan^{-1}(xy) = 1 + x^2y$.

33. If $g(x) = x\sin^{-1}(x/4) + \sqrt{16-x^2}$, find $g'(2)$.

34. Find an equation of the tangent line to the curve $y = 3\arccos(x/2)$ at the point $(1, \pi)$.

35–38 ▪ Find the limit.

35. $\displaystyle\lim_{x\to-1^+}\sin^{-1}x$

36. $\displaystyle\lim_{x\to\infty}\arccos\!\left(\dfrac{1+x^2}{1+2x^2}\right)$

37. $\displaystyle\lim_{x\to\infty}\arctan(e^x)$

38. $\displaystyle\lim_{x\to0^+}\tan^{-1}(\ln x)$

39. A ladder 10 ft long leans against a vertical wall. If the bottom of the ladder slides away from the base of the wall at a speed of 2 ft/s, how fast is the angle between the ladder and the wall changing when the bottom of the ladder is 6 ft from the base of the wall?

40. A lighthouse is located on a small island, 3 km away from the nearest point P on a straight shoreline, and its light makes four revolutions per minute. How fast is the beam of light moving along the shoreline when it is 1 km from P?

41. Some authors define $y = \sec^{-1}x \iff \sec y = x$ and $y \in [0, \pi/2) \cup (\pi/2, \pi]$. Show that with this definition, we have (instead of the formula given in Exercise 14)

$$\frac{d}{dx}(\sec^{-1}x) = \frac{1}{|x|\sqrt{x^2-1}} \qquad |x| > 1$$

42. (a) Sketch the graph of the function $f(x) = \sin(\sin^{-1}x)$.
(b) Sketch the graph of the function $g(x) = \sin^{-1}(\sin x)$, $x \in \mathbb{R}$.
(c) Show that $g'(x) = \dfrac{\cos x}{|\cos x|}$.
(d) Sketch the graph of $h(x) = \cos^{-1}(\sin x)$, $x \in \mathbb{R}$, and find its derivative.

3.6 | HYPERBOLIC FUNCTIONS

Certain even and odd combinations of the exponential functions e^x and e^{-x} arise so frequently in mathematics and its applications that they deserve to be given special names. In many ways they are analogous to the trigonometric functions, and they have the same relationship to the hyperbola that the trigonometric functions have to the circle. For this reason they are collectively called **hyperbolic functions** and individually called **hyperbolic sine**, **hyperbolic cosine**, and so on.

DEFINITION OF THE HYPERBOLIC FUNCTIONS

$$\sinh x = \frac{e^x - e^{-x}}{2} \qquad\qquad \operatorname{csch} x = \frac{1}{\sinh x}$$

$$\cosh x = \frac{e^x + e^{-x}}{2} \qquad\qquad \operatorname{sech} x = \frac{1}{\cosh x}$$

$$\tanh x = \frac{\sinh x}{\cosh x} \qquad\qquad \coth x = \frac{\cosh x}{\sinh x}$$

The graphs of hyperbolic sine and cosine can be sketched using graphical addition as in Figures 1 and 2.

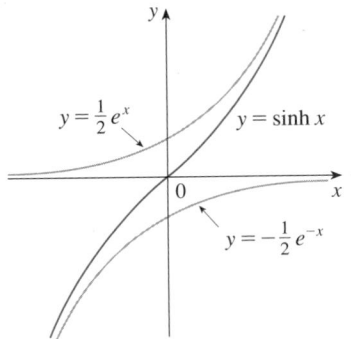

FIGURE 1
$y = \sinh x = \frac{1}{2}e^x - \frac{1}{2}e^{-x}$

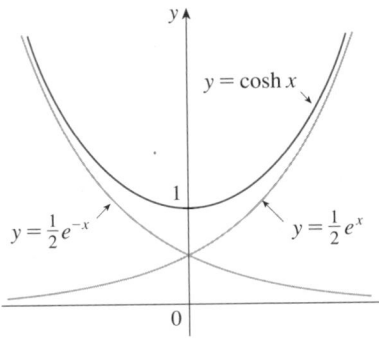

FIGURE 2
$y = \cosh x = \frac{1}{2}e^x + \frac{1}{2}e^{-x}$

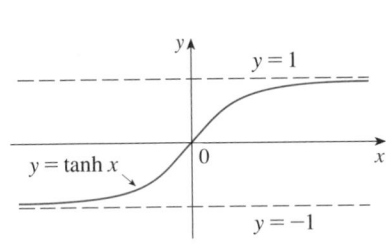

FIGURE 3

Note that sinh has domain $\mathbb{R}$ and range $\mathbb{R}$, while cosh has domain $\mathbb{R}$ and range $[1, \infty)$. The graph of tanh is shown in Figure 3. It has the horizontal asymptotes $y = \pm 1$. (See Exercise 19.)

Applications of hyperbolic functions to science and engineering occur whenever an entity such as light, velocity, electricity, or radioactivity is gradually absorbed or extinguished, for the decay can be represented by hyperbolic functions. The most famous application is the use of hyperbolic cosine to describe the shape of a hanging wire. It can be proved that if a heavy flexible cable (such as a telephone or power line) is suspended between two points at the same height, then it takes the shape of a curve with equation $y = c + a \cosh(x/a)$ called a *catenary* (see Figure 4). (The Latin word *catena* means "chain.")

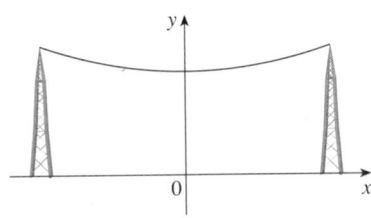

FIGURE 4
A catenary $y = c + a \cosh(x/a)$

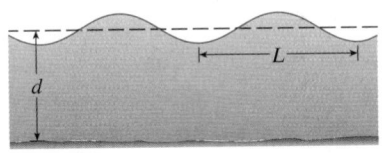

FIGURE 5
Idealized ocean wave

Another application of hyperbolic functions occurs in the description of ocean waves: The velocity of a water wave with length L moving across a body of water with depth d is modeled by the function

$$v = \sqrt{\frac{gL}{2\pi} \tanh\left(\frac{2\pi d}{L}\right)}$$

where g is the acceleration due to gravity. (See Figure 5 and Exercise 45.)

The hyperbolic functions satisfy a number of identities that are similar to well-known trigonometric identities. We list some of them here and leave most of the proofs to the exercises.

HYPERBOLIC IDENTITIES

$$\sinh(-x) = -\sinh x \qquad\qquad \cosh(-x) = \cosh x$$

$$\cosh^2 x - \sinh^2 x = 1 \qquad\qquad 1 - \tanh^2 x = \operatorname{sech}^2 x$$

$$\sinh(x + y) = \sinh x \cosh y + \cosh x \sinh y$$

$$\cosh(x + y) = \cosh x \cosh y + \sinh x \sinh y$$

EXAMPLE 1 Prove (a) $\cosh^2 x - \sinh^2 x = 1$ and (b) $1 - \tanh^2 x = \operatorname{sech}^2 x$.

SOLUTION

(a)
$$\cosh^2 x - \sinh^2 x = \left(\frac{e^x + e^{-x}}{2}\right)^2 - \left(\frac{e^x - e^{-x}}{2}\right)^2$$

$$= \frac{e^{2x} + 2 + e^{-2x}}{4} - \frac{e^{2x} - 2 + e^{-2x}}{4}$$

$$= \frac{4}{4} = 1$$

(b) We start with the identity proved in part (a):

$$\cosh^2 x - \sinh^2 x = 1$$

If we divide both sides by $\cosh^2 x$, we get

$$1 - \frac{\sinh^2 x}{\cosh^2 x} = \frac{1}{\cosh^2 x}$$

or
$$1 - \tanh^2 x = \operatorname{sech}^2 x \qquad \blacksquare$$

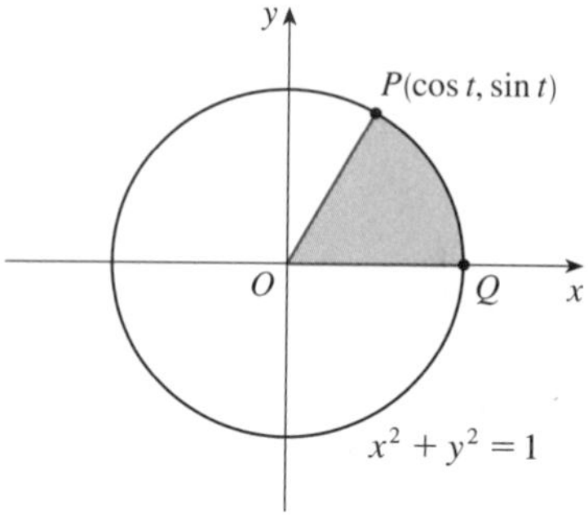

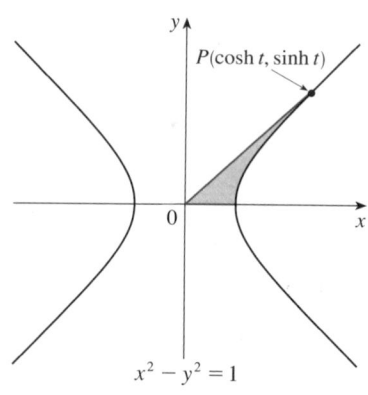

FIGURE 6

The identity proved in Example 1(a) gives a clue to the reason for the name "hyperbolic" functions:

If t is any real number, the point $P(\cos t, \sin t)$ lies on the unit circle $x^2 + y^2 = 1$ because $\cos^2 t + \sin^2 t = 1$. In fact, t can be interpreted as the radian measure of $\angle POQ$ in Figure 6. For this reason the trigonometric functions are sometimes called *circular* functions.

Likewise, if t is any real number, then the point $P(\cosh t, \sinh t)$ lies on the right branch of the hyperbola $x^2 - y^2 = 1$ because $\cosh^2 t - \sinh^2 t = 1$ and $\cosh t \geq 1$. This time, t does not represent the measure of an angle. However, it turns out that t represents twice the area of the shaded hyperbolic sector in Figure 7, just as in the trigonometric case t represents twice the area of the shaded circular sector in Figure 6.

The derivatives of the hyperbolic functions are easily computed. For example,

$$\frac{d}{dx}(\sinh x) = \frac{d}{dx}\left(\frac{e^x - e^{-x}}{2}\right) = \frac{e^x + e^{-x}}{2} = \cosh x$$

FIGURE 7

We list the differentiation formulas for the hyperbolic functions as Table 1. The remaining proofs are left as exercises. Note the analogy with the differentiation formulas for trigonometric functions, but beware that the signs are different in some cases.

---

1 DERIVATIVES OF HYPERBOLIC FUNCTIONS

$$\frac{d}{dx}(\sinh x) = \cosh x \qquad\qquad \frac{d}{dx}(\operatorname{csch} x) = -\operatorname{csch} x \coth x$$

$$\frac{d}{dx}(\cosh x) = \sinh x \qquad\qquad \frac{d}{dx}(\operatorname{sech} x) = -\operatorname{sech} x \tanh x$$

$$\frac{d}{dx}(\tanh x) = \operatorname{sech}^2 x \qquad\qquad \frac{d}{dx}(\coth x) = -\operatorname{csch}^2 x$$

▼ EXAMPLE 2 Any of these differentiation rules can be combined with the Chain Rule. For instance,

$$\frac{d}{dx}\left(\cosh\sqrt{x}\right) = \sinh\sqrt{x}\cdot\frac{d}{dx}\sqrt{x} = \frac{\sinh\sqrt{x}}{2\sqrt{x}}$$ ∎

INVERSE HYPERBOLIC FUNCTIONS

You can see from Figures 1 and 3 that sinh and tanh are one-to-one functions and so they have inverse functions denoted by $\sinh^{-1}$ and $\tanh^{-1}$. Figure 2 shows that cosh is not one-to-one, but when restricted to the domain $[0, \infty)$ it becomes one-to-one. The inverse hyperbolic cosine function is defined as the inverse of this restricted function.

$$\boxed{2}$$

$$
\begin{aligned}
y = \sinh^{-1}x &\iff \sinh y = x \\
y = \cosh^{-1}x &\iff \cosh y = x \quad \text{and} \quad y \geq 0 \\
y = \tanh^{-1}x &\iff \tanh y = x
\end{aligned}
$$

The remaining inverse hyperbolic functions are defined similarly (see Exercise 24).

We can sketch the graphs of $\sinh^{-1}$, $\cosh^{-1}$, and $\tanh^{-1}$ in Figures 8, 9, and 10 by using Figures 1, 2, and 3.

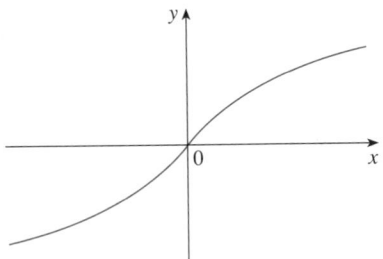

FIGURE 8 $y = \sinh^{-1}x$
domain $= \mathbb{R}$ range $= \mathbb{R}$

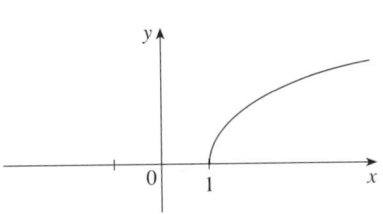

FIGURE 9 $y = \cosh^{-1}x$
domain $= [1, \infty)$ range $= [0, \infty)$

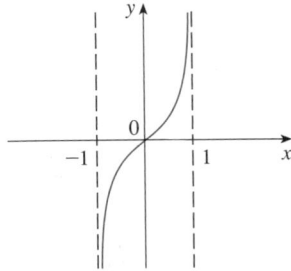

FIGURE 10 $y = \tanh^{-1}x$
domain $= (-1, 1)$ range $= \mathbb{R}$

Since the hyperbolic functions are defined in terms of exponential functions, it's not surprising to learn that the inverse hyperbolic functions can be expressed in terms of logarithms. In particular, we have:

■ Formula 3 is proved in Example 3. The proofs of Formulas 4 and 5 are requested in Exercises 22 and 23.

$$\boxed{3} \qquad \sinh^{-1}x = \ln\left(x + \sqrt{x^2 + 1}\right) \qquad x \in \mathbb{R}$$

$$\boxed{4} \qquad \cosh^{-1}x = \ln\left(x + \sqrt{x^2 - 1}\right) \qquad x \geq 1$$

$$\boxed{5} \qquad \tanh^{-1}x = \tfrac{1}{2}\ln\left(\frac{1 + x}{1 - x}\right) \qquad -1 < x < 1$$

EXAMPLE 3 Show that $\sinh^{-1}x = \ln\left(x + \sqrt{x^2 + 1}\right)$.

SOLUTION Let $y = \sinh^{-1}x$. Then

$$x = \sinh y = \frac{e^y - e^{-y}}{2}$$

so
$$e^y - 2x - e^{-y} = 0$$

or, multiplying by e^y,

$$e^{2y} - 2xe^y - 1 = 0$$

This is really a quadratic equation in e^y:

$$(e^y)^2 - 2x(e^y) - 1 = 0$$

Solving by the quadratic formula, we get

$$e^y = \frac{2x \pm \sqrt{4x^2 + 4}}{2} = x \pm \sqrt{x^2 + 1}$$

Note that $e^y > 0$, but $x - \sqrt{x^2 + 1} < 0$ $\left(\text{because } x < \sqrt{x^2 + 1}\right)$. Thus the minus sign is inadmissible and we have

$$e^y = x + \sqrt{x^2 + 1}$$

Therefore
$$y = \ln(e^y) = \ln\left(x + \sqrt{x^2 + 1}\right)$$

(See Exercise 21 for another method.) ∎

---

6 DERIVATIVES OF INVERSE HYPERBOLIC FUNCTIONS

$$\frac{d}{dx}(\sinh^{-1}x) = \frac{1}{\sqrt{1 + x^2}} \qquad \frac{d}{dx}(\operatorname{csch}^{-1}x) = -\frac{1}{|x|\sqrt{x^2 + 1}}$$

$$\frac{d}{dx}(\cosh^{-1}x) = \frac{1}{\sqrt{x^2 - 1}} \qquad \frac{d}{dx}(\operatorname{sech}^{-1}x) = -\frac{1}{x\sqrt{1 - x^2}}$$

$$\frac{d}{dx}(\tanh^{-1}x) = \frac{1}{1 - x^2} \qquad \frac{d}{dx}(\coth^{-1}x) = \frac{1}{1 - x^2}$$

■ Notice that the formulas for the derivatives of $\tanh^{-1}x$ and $\coth^{-1}x$ appear to be identical. But the domains of these functions have no numbers in common: $\tanh^{-1}x$ is defined for $|x| < 1$, whereas $\coth^{-1}x$ is defined for $|x| > 1$.

The inverse hyperbolic functions are all differentiable because the hyperbolic functions are differentiable. The formulas in Table 6 can be proved either by the method for inverse functions or by differentiating Formulas 3, 4, and 5.

Ⅴ EXAMPLE 4 Prove that $\dfrac{d}{dx}(\sinh^{-1}x) = \dfrac{1}{\sqrt{1 + x^2}}$.

SOLUTION Let $y = \sinh^{-1}x$. Then $\sinh y = x$. If we differentiate this equation implicitly with respect to x, we get

$$\cosh y \frac{dy}{dx} = 1$$

■ Another method for solving Example 4 is to differentiate Formula 3.

Since $\cosh^2 y - \sinh^2 y = 1$ and $\cosh y \geqslant 0$, we have $\cosh y = \sqrt{1 + \sinh^2 y}$, so

$$\frac{dy}{dx} = \frac{1}{\cosh y} = \frac{1}{\sqrt{1 + \sinh^2 y}} = \frac{1}{\sqrt{1 + x^2}}$$ ∎

V EXAMPLE 5 Find $\dfrac{d}{dx}[\tanh^{-1}(\sin x)]$.

SOLUTION Using Table 6 and the Chain Rule, we have

$$\frac{d}{dx}[\tanh^{-1}(\sin x)] = \frac{1}{1-(\sin x)^2}\frac{d}{dx}(\sin x)$$

$$= \frac{1}{1-\sin^2 x}\cos x = \frac{\cos x}{\cos^2 x} = \sec x \qquad \blacksquare$$

3.6 | EXERCISES

1–6 ▪ Find the numerical value of each expression.

1. (a) $\sinh 0$ (b) $\cosh 0$

2. (a) $\tanh 0$ (b) $\tanh 1$

3. (a) $\sinh(\ln 2)$ (b) $\sinh 2$

4. (a) $\cosh 3$ (b) $\cosh(\ln 3)$

5. (a) $\operatorname{sech} 0$ (b) $\cosh^{-1} 1$

6. (a) $\sinh 1$ (b) $\sinh^{-1} 1$

7–15 ▪ Prove the identity.

7. $\sinh(-x) = -\sinh x$
(This shows that sinh is an odd function.)

8. $\cosh(-x) = \cosh x$
(This shows that cosh is an even function.)

9. $\cosh x + \sinh x = e^x$

10. $\cosh x - \sinh x = e^{-x}$

11. $\sinh(x+y) = \sinh x \cosh y + \cosh x \sinh y$

12. $\cosh(x+y) = \cosh x \cosh y + \sinh x \sinh y$

13. $\sinh 2x = 2 \sinh x \cosh x$

14. $\dfrac{1+\tanh x}{1-\tanh x} = e^{2x}$

15. $(\cosh x + \sinh x)^n = \cosh nx + \sinh nx$
(n any real number)

16. If $\tanh x = \frac{12}{13}$, find the values of the other hyperbolic functions at x.

17. If $\cosh x = \frac{5}{3}$ and $x > 0$, find the values of the other hyperbolic functions at x.

18. (a) Use the graphs of sinh, cosh, and tanh in Figures 1–3 to draw the graphs of csch, sech, and coth.
(b) Check the graphs that you sketched in part (a) by using a graphing device to produce them.

19. Use the definitions of the hyperbolic functions to find each of the following limits.

(a) $\displaystyle\lim_{x\to\infty}\tanh x$ (b) $\displaystyle\lim_{x\to-\infty}\tanh x$

(c) $\displaystyle\lim_{x\to\infty}\sinh x$ (d) $\displaystyle\lim_{x\to-\infty}\sinh x$

(e) $\displaystyle\lim_{x\to\infty}\operatorname{sech} x$ (f) $\displaystyle\lim_{x\to\infty}\coth x$

(g) $\displaystyle\lim_{x\to 0^+}\coth x$ (h) $\displaystyle\lim_{x\to 0^-}\coth x$

(i) $\displaystyle\lim_{x\to-\infty}\operatorname{csch} x$

20. Prove the formulas given in Table 1 for the derivatives of the functions (a) cosh, (b) tanh, (c) csch, (d) sech, and (e) coth.

21. Give an alternative solution to Example 3 by letting $y = \sinh^{-1}x$ and then using Exercise 9 and Example 1(a) with x replaced by y.

22. Prove Equation 4.

23. Prove Formula 5 using (a) the method of Example 3 and (b) Exercise 14 with x replaced by y.

24. For each of the following functions (i) give a definition like those in $\boxed{2}$, (ii) sketch the graph, and (iii) find a formula similar to Formula 3.
(a) csch^{-1} (b) sech^{-1} (c) $\coth^{-1}$

25. Prove the formulas given in Table 6 for the derivatives of the following functions.
(a) $\cosh^{-1}$ (b) $\tanh^{-1}$ (c) sech^{-1}

26–41 ▪ Find the derivative. Simplify where possible.

26. $f(x) = \tanh(1+e^{2x})$ **27.** $f(x) = x \sinh x - \cosh x$

28. $g(x) = \cosh(\ln x)$ **29.** $h(x) = \ln(\cosh x)$

30. $y = x \coth(1+x^2)$ **31.** $y = e^{\cosh 3x}$

32. $f(t) = \operatorname{csch} t(1 - \ln \operatorname{csch} t)$

33. $f(t) = \operatorname{sech}^2(e^t)$

34. $y = \sinh(\cosh x)$ **35.** $G(x) = \dfrac{1-\cosh x}{1+\cosh x}$

36. $y = \sinh^{-1}(\tan x)$

37. $y = \cosh^{-1}\sqrt{x}$

38. $y = x \tanh^{-1}x + \ln \sqrt{1 - x^2}$

39. $y = x \sinh^{-1}(x/3) - \sqrt{9 + x^2}$

40. $y = \text{sech}^{-1}(e^{-x})$

41. $y = \coth^{-1}(\sec x)$

---

42. Show that $\dfrac{d}{dx} \sqrt[4]{\dfrac{1 + \tanh x}{1 - \tanh x}} = \frac{1}{2}e^{x/2}$.

43. Show that $\dfrac{d}{dx} \arctan(\tanh x) = \text{sech } 2x$.

44. The Gateway Arch in St. Louis was designed by Eero Saarinen and was constructed using the equation

$$y = 211.49 - 20.96 \cosh 0.03291765x$$

for the central curve of the arch, where x and y are measured in meters and $|x| \leqslant 91.20$.
  (a) Graph the central curve.
  (b) What is the height of the arch at its center?
  (c) At what points is the height 100 m?
  (d) What is the slope of the arch at the points in part (c)?

45. If a water wave with length L moves with velocity v in a body of water with depth d, then

$$v = \sqrt{\dfrac{gL}{2\pi} \tanh\left(\dfrac{2\pi d}{L}\right)}$$

where g is the acceleration due to gravity. (See Figure 5.) Explain why the approximation

$$v \approx \sqrt{\dfrac{gL}{2\pi}}$$

is appropriate in deep water.

46. A flexible cable always hangs in the shape of a catenary $y = c + a \cosh(x/a)$, where c and a are constants and $a > 0$ (see Figure 4 and Exercise 48). Graph several members of the family of functions $y = a \cosh(x/a)$. How does the graph change as a varies?

47. A telephone line hangs between two poles 14 m apart in the shape of the catenary $y = 20 \cosh(x/20) - 15$, where x and y are measured in meters.
  (a) Find the slope of this curve where it meets the right pole.

(b) Find the angle θ between the line and the pole.

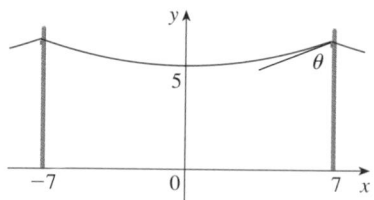

48. Using principles from physics it can be shown that when a cable is hung between two poles, it takes the shape of a curve $y = f(x)$ that satisfies the differential equation

$$\dfrac{d^2y}{dx^2} = \dfrac{\rho g}{T} \sqrt{1 + \left(\dfrac{dy}{dx}\right)^2}$$

where ρ is the linear density of the cable, g is the acceleration due to gravity, T is the tension in the cable at its lowest point, and the coordinate system is chosen appropriately. Verify that the function

$$y = f(x) = \dfrac{T}{\rho g} \cosh\left(\dfrac{\rho g x}{T}\right)$$

is a solution of this differential equation.

49. A cable with linear density $\rho = 2$ kg/m is strung from the tops of two poles that are 200 m apart.
  (a) Use Exercise 48 to find the tension T so that the cable is 60 m above the ground at its lowest point. How tall are the poles?
  (b) If the tension is doubled, what is the new low point of the cable? How tall are the poles now?

50. Evaluate $\displaystyle\lim_{x \to \infty} \dfrac{\sinh x}{e^x}$.

51. (a) Show that any function of the form

$$y = A \sinh mx + B \cosh mx$$

satisfies the differential equation $y'' = m^2 y$.
  (b) Find $y = y(x)$ such that $y'' = 9y$, $y(0) = -4$, and $y'(0) = 6$.

52. If $x = \ln(\sec \theta + \tan \theta)$, show that $\sec \theta = \cosh x$.

53. At what point of the curve $y = \cosh x$ does the tangent have slope 1?

54. Show that if $a \neq 0$ and $b \neq 0$, then there exist numbers α and β such that $ae^x + be^{-x}$ equals either $\alpha \sinh(x + \beta)$ or $\alpha \cosh(x + \beta)$. In other words, almost every function of the form $f(x) = ae^x + be^{-x}$ is a shifted and stretched hyperbolic sine or cosine function.

3.7 | INDETERMINATE FORMS AND L'HOSPITAL'S RULE

Suppose we are trying to analyze the behavior of the function

$$F(x) = \frac{\ln x}{x - 1}$$

Although F is not defined when $x = 1$, we need to know how F behaves *near* 1. In particular, we would like to know the value of the limit

$$\boxed{1} \qquad \lim_{x \to 1} \frac{\ln x}{x - 1}$$

In computing this limit we can't apply Law 5 of limits (the limit of a quotient is the quotient of the limits, see Section 1.4) because the limit of the denominator is 0. In fact, although the limit in $\boxed{1}$ exists, its value is not obvious because both numerator and denominator approach 0 and $\frac{0}{0}$ is not defined.

In general, if we have a limit of the form

$$\lim_{x \to a} \frac{f(x)}{g(x)}$$

where both $f(x) \to 0$ and $g(x) \to 0$ as $x \to a$, then this limit may or may not exist and is called an **indeterminate form of type $\frac{0}{0}$**. We met some limits of this type in Chapter 1. For rational functions, we can cancel common factors:

$$\lim_{x \to 1} \frac{x^2 - x}{x^2 - 1} = \lim_{x \to 1} \frac{x(x - 1)}{(x + 1)(x - 1)} = \lim_{x \to 1} \frac{x}{x + 1} = \frac{1}{2}$$

We used a geometric argument to show that

$$\lim_{x \to 0} \frac{\sin x}{x} = 1$$

But these methods do not work for limits such as $\boxed{1}$, so in this section we introduce a systematic method, known as *l'Hospital's Rule,* for the evaluation of indeterminate forms.

Another situation in which a limit is not obvious occurs when we look for a horizontal asymptote of F and need to evaluate the limit

$$\boxed{2} \qquad \lim_{x \to \infty} \frac{\ln x}{x - 1}$$

It isn't obvious how to evaluate this limit because both numerator and denominator become large as $x \to \infty$. There is a struggle between numerator and denominator. If the numerator wins, the limit will be ∞; if the denominator wins, the answer will be 0. Or there may be some compromise, in which case the answer may be some finite positive number.

In general, if we have a limit of the form

$$\lim_{x \to a} \frac{f(x)}{g(x)}$$

where both $f(x) \to \infty$ (or $-\infty$) and $g(x) \to \infty$ (or $-\infty$), then the limit may or may not exist and is called an **indeterminate form of type** ∞/∞. We saw in Section 1.6 that this type of limit can be evaluated for certain functions, including rational functions, by dividing numerator and denominator by the highest power of x that occurs in the denominator. For instance,

$$\lim_{x \to \infty} \frac{x^2 - 1}{2x^2 + 1} = \lim_{x \to \infty} \frac{1 - \dfrac{1}{x^2}}{2 + \dfrac{1}{x^2}} = \frac{1 - 0}{2 + 0} = \frac{1}{2}$$

This method does not work for limits such as $\boxed{2}$, but l'Hospital's Rule also applies to this type of indeterminate form.

- L'Hospital's Rule is named after a French nobleman, the Marquis de l'Hospital (1661–1704), but was discovered by a Swiss mathematician, John Bernoulli (1667–1748). See Exercise 49 for the example that the Marquis used to illustrate his rule.

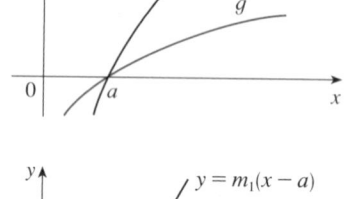

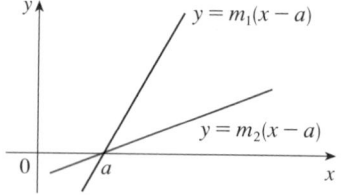

FIGURE 1

- Figure 1 suggests visually why l'Hospital's Rule might be true. The first graph shows two differentiable functions f and g, each of which approaches 0 as $x \to a$. If we were to zoom in toward the point $(a, 0)$, the graphs would start to look almost linear. But if the functions actually *were* linear, as in the second graph, then their ratio would be

$$\frac{m_1(x - a)}{m_2(x - a)} = \frac{m_1}{m_2}$$

which is the ratio of their derivatives. This suggests that

$$\lim_{x \to a} \frac{f(x)}{g(x)} = \lim_{x \to a} \frac{f'(x)}{g'(x)}$$

> **L'HOSPITAL'S RULE** Suppose f and g are differentiable and $g'(x) \neq 0$ near a (except possibly at a). Suppose that
>
> $$\lim_{x \to a} f(x) = 0 \qquad \text{and} \qquad \lim_{x \to a} g(x) = 0$$
>
> or that
>
> $$\lim_{x \to a} f(x) = \pm\infty \qquad \text{and} \qquad \lim_{x \to a} g(x) = \pm\infty$$
>
> (In other words, we have an indeterminate form of type $\frac{0}{0}$ or ∞/∞.) Then
>
> $$\lim_{x \to a} \frac{f(x)}{g(x)} = \lim_{x \to a} \frac{f'(x)}{g'(x)}$$
>
> if the limit on the right side exists (or is ∞ or $-\infty$).

NOTE 1 L'Hospital's Rule says that the limit of a quotient of functions is equal to the limit of the quotient of their derivatives, provided that the given conditions are satisfied. It is especially important to verify the conditions regarding the limits of f and g before using l'Hospital's Rule.

NOTE 2 L'Hospital's Rule is also valid for one-sided limits and for limits at infinity or negative infinity; that is, "$x \to a$" can be replaced by any of the symbols $x \to a^+$, $x \to a^-$, $x \to \infty$, or $x \to -\infty$.

NOTE 3 For the special case in which $f(a) = g(a) = 0$, f' and g' are continuous, and $g'(a) \neq 0$, it is easy to see why l'Hospital's Rule is true. In fact, using the alternative form of the definition of a derivative, we have

$$\lim_{x \to a} \frac{f'(x)}{g'(x)} = \frac{f'(a)}{g'(a)} = \frac{\displaystyle\lim_{x \to a} \frac{f(x) - f(a)}{x - a}}{\displaystyle\lim_{x \to a} \frac{g(x) - g(a)}{x - a}} = \lim_{x \to a} \frac{\dfrac{f(x) - f(a)}{x - a}}{\dfrac{g(x) - g(a)}{x - a}}$$

$$= \lim_{x \to a} \frac{f(x) - f(a)}{g(x) - g(a)} = \lim_{x \to a} \frac{f(x)}{g(x)}$$

The general version of l'Hospital's Rule is more difficult; its proof can be found in Appendix D.

V EXAMPLE 1 Find $\lim\limits_{x \to 1} \dfrac{\ln x}{x - 1}$.

SOLUTION Since

$$\lim_{x \to 1} \ln x = \ln 1 = 0 \qquad \text{and} \qquad \lim_{x \to 1} (x - 1) = 0$$

we can apply l'Hospital's Rule:

$$\lim_{x \to 1} \frac{\ln x}{x - 1} = \lim_{x \to 1} \frac{\dfrac{d}{dx}(\ln x)}{\dfrac{d}{dx}(x - 1)} = \lim_{x \to 1} \frac{1/x}{1} = \lim_{x \to 1} \frac{1}{x} = 1 \qquad \blacksquare$$

⊘ Notice that when using l'Hospital's Rule we differentiate the numerator and denominator *separately*. We do *not* use the Quotient Rule.

V EXAMPLE 2 Calculate $\lim\limits_{x \to \infty} \dfrac{e^x}{x^2}$.

SOLUTION We have $\lim_{x \to \infty} e^x = \infty$ and $\lim_{x \to \infty} x^2 = \infty$, so l'Hospital's Rule gives

$$\lim_{x \to \infty} \frac{e^x}{x^2} = \lim_{x \to \infty} \frac{\dfrac{d}{dx}(e^x)}{\dfrac{d}{dx}(x^2)} = \lim_{x \to \infty} \frac{e^x}{2x}$$

Since $e^x \to \infty$ and $2x \to \infty$ as $x \to \infty$, the limit on the right side is also indeterminate, but a second application of l'Hospital's Rule gives

$$\lim_{x \to \infty} \frac{e^x}{x^2} = \lim_{x \to \infty} \frac{e^x}{2x} = \lim_{x \to \infty} \frac{e^x}{2} = \infty \qquad \blacksquare$$

■ The graph of the function of Example 2 is shown in Figure 2. We have noticed previously that exponential functions grow far more rapidly than power functions, so the result of Example 2 is not unexpected. See also Exercise 41.

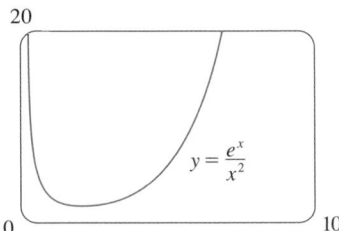

$y = \dfrac{e^x}{x^2}$

FIGURE 2

V EXAMPLE 3 Calculate $\lim\limits_{x \to \infty} \dfrac{\ln x}{\sqrt[3]{x}}$.

SOLUTION Since $\ln x \to \infty$ and $\sqrt[3]{x} \to \infty$ as $x \to \infty$, l'Hospital's Rule applies:

$$\lim_{x \to \infty} \frac{\ln x}{\sqrt[3]{x}} = \lim_{x \to \infty} \frac{1/x}{\frac{1}{3}x^{-2/3}}$$

Notice that the limit on the right side is now an indeterminate of type $\frac{0}{0}$. But instead of applying l'Hospital's Rule a second time as we did in Example 2, we simplify the expression and see that a second application is unnecessary:

$$\lim_{x \to \infty} \frac{\ln x}{\sqrt[3]{x}} = \lim_{x \to \infty} \frac{1/x}{\frac{1}{3}x^{-2/3}} = \lim_{x \to \infty} \frac{3}{\sqrt[3]{x}} = 0 \qquad \blacksquare$$

■ The graph of the function of Example 3 is shown in Figure 3. We have discussed previously the slow growth of logarithms, so it isn't surprising that this ratio approaches 0 as $x \to \infty$. See also Exercise 42.

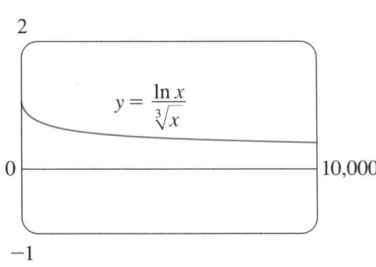

$y = \dfrac{\ln x}{\sqrt[3]{x}}$

FIGURE 3

EXAMPLE 4 Find $\lim\limits_{x \to 0} \dfrac{\tan x - x}{x^3}$. [See Exercise 22 in Section 1.3.]

SOLUTION Noting that both $\tan x - x \to 0$ and $x^3 \to 0$ as $x \to 0$, we use l'Hospital's Rule:

$$\lim_{x \to 0} \frac{\tan x - x}{x^3} = \lim_{x \to 0} \frac{\sec^2 x - 1}{3x^2}$$

■ The graph in Figure 4 gives visual confirmation of the result of Example 4. If we were to zoom in too far, however, we would get an inaccurate graph because tan x is close to x when x is small. See Exercise 22(d) in Section 1.3.

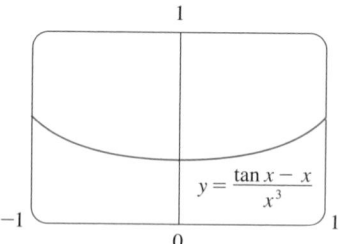

FIGURE 4

Since the limit on the right side is still indeterminate of type $\frac{0}{0}$, we apply l'Hospital's Rule again:

$$\lim_{x \to 0} \frac{\sec^2 x - 1}{3x^2} = \lim_{x \to 0} \frac{2 \sec^2 x \tan x}{6x}$$

Because $\lim_{x \to 0} \sec^2 x = 1$, we simplify the calculation by writing

$$\lim_{x \to 0} \frac{2 \sec^2 x \tan x}{6x} = \frac{1}{3} \lim_{x \to 0} \sec^2 x \cdot \lim_{x \to 0} \frac{\tan x}{x} = \frac{1}{3} \lim_{x \to 0} \frac{\tan x}{x}$$

We can evaluate this last limit either by using l'Hospital's Rule a third time or by writing $\tan x$ as $(\sin x)/(\cos x)$ and making use of our knowledge of trigonometric limits. Putting together all the steps, we get

$$\lim_{x \to 0} \frac{\tan x - x}{x^3} = \lim_{x \to 0} \frac{\sec^2 x - 1}{3x^2} = \lim_{x \to 0} \frac{2 \sec^2 x \tan x}{6x}$$

$$= \frac{1}{3} \lim_{x \to 0} \frac{\tan x}{x} = \frac{1}{3} \lim_{x \to 0} \frac{\sec^2 x}{1} = \frac{1}{3}$$ ■

V EXAMPLE 5 Find $\displaystyle \lim_{x \to \pi^-} \frac{\sin x}{1 - \cos x}$.

SOLUTION If we blindly attempted to use l'Hospital's Rule, we would get

$$\lim_{x \to \pi^-} \frac{\sin x}{1 - \cos x} = \lim_{x \to \pi^-} \frac{\cos x}{\sin x} = -\infty$$

This is **wrong!** Although the numerator $\sin x \to 0$ as $x \to \pi^-$, notice that the denominator $(1 - \cos x)$ does not approach 0, so l'Hospital's Rule can't be applied here.

The required limit is, in fact, easy to find because the function is continuous at π and the denominator is nonzero there:

$$\lim_{x \to \pi^-} \frac{\sin x}{1 - \cos x} = \frac{\sin \pi}{1 - \cos \pi} = \frac{0}{1 - (-1)} = 0$$ ■

Example 5 shows what can go wrong if you use l'Hospital's Rule without thinking. Other limits *can* be found using l'Hospital's Rule but are more easily found by other methods. (See Examples 2 and 4 in Section 1.4, Example 5 in Section 1.6, and the discussion at the beginning of this section.) So when evaluating any limit, you should consider other methods before using l'Hospital's Rule.

INDETERMINATE PRODUCTS

If $\lim_{x \to a} f(x) = 0$ and $\lim_{x \to a} g(x) = \infty$ (or $-\infty$), then it isn't clear what the value of $\lim_{x \to a} f(x) g(x)$, if any, will be. There is a struggle between f and g. If f wins, the answer will be 0; if g wins, the answer will be ∞ (or $-\infty$). Or there may be a compromise where the answer is a finite nonzero number. This kind of limit is called an **indeterminate form of type $0 \cdot \infty$.** We can deal with it by writing the product fg as a quotient:

$$fg = \frac{f}{1/g} \qquad \text{or} \qquad fg = \frac{g}{1/f}$$

This converts the given limit into an indeterminate form of type $\frac{0}{0}$ or ∞/∞ so that we can use l'Hospital's Rule.

☑ EXAMPLE 6 Evaluate $\lim\limits_{x \to 0^+} x \ln x$.

SOLUTION The given limit is indeterminate because, as $x \to 0^+$, the first factor (x) approaches 0 while the second factor $(\ln x)$ approaches $-\infty$. Writing $x = 1/(1/x)$, we have $1/x \to \infty$ as $x \to 0^+$, so l'Hospital's Rule gives

$$\lim_{x \to 0^+} x \ln x = \lim_{x \to 0^+} \frac{\ln x}{1/x} = \lim_{x \to 0^+} \frac{1/x}{-1/x^2} = \lim_{x \to 0^+} (-x) = 0 \qquad ■$$

NOTE In solving Example 6 another possible option would have been to write

$$\lim_{x \to 0^+} x \ln x = \lim_{x \to 0^+} \frac{x}{1/\ln x}$$

This gives an indeterminate form of the type 0/0, but if we apply l'Hospital's Rule we get a more complicated expression than the one we started with. In general, when we rewrite an indeterminate product, we try to choose the option that leads to the simpler limit.

- Figure 5 shows the graph of the function in Example 6. Notice that the function is undefined at $x = 0$; the graph approaches the origin but never quite reaches it.

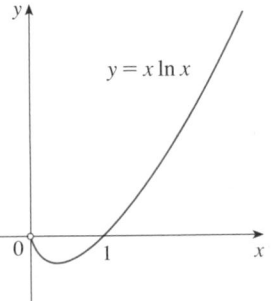

$y = x \ln x$

FIGURE 5

INDETERMINATE DIFFERENCES

If $\lim_{x \to a} f(x) = \infty$ and $\lim_{x \to a} g(x) = \infty$, then the limit

$$\lim_{x \to a} [f(x) - g(x)]$$

is called an **indeterminate form of type $\infty - \infty$**. Again there is a contest between f and g. Will the answer be ∞ (f wins) or will it be $-\infty$ (g wins) or will they compromise on a finite number? To find out, we try to convert the difference into a quotient (for instance, by using a common denominator, or rationalization, or factoring out a common factor) so that we have an indeterminate form of type $\frac{0}{0}$ or ∞/∞.

EXAMPLE 7 Compute $\lim\limits_{x \to (\pi/2)^-} (\sec x - \tan x)$.

SOLUTION First notice that $\sec x \to \infty$ and $\tan x \to \infty$ as $x \to (\pi/2)^-$, so the limit is indeterminate. Here we use a common denominator:

$$\lim_{x \to (\pi/2)^-} (\sec x - \tan x) = \lim_{x \to (\pi/2)^-} \left(\frac{1}{\cos x} - \frac{\sin x}{\cos x} \right)$$

$$= \lim_{x \to (\pi/2)^-} \frac{1 - \sin x}{\cos x} = \lim_{x \to (\pi/2)^-} \frac{-\cos x}{-\sin x} = 0$$

Note that the use of l'Hospital's Rule is justified because $1 - \sin x \to 0$ and $\cos x \to 0$ as $x \to (\pi/2)^-$. $\qquad ■$

INDETERMINATE POWERS

Several indeterminate forms arise from the limit

$$\lim_{x \to a} [f(x)]^{g(x)}$$

1. $\lim_{x \to a} f(x) = 0$ and $\lim_{x \to a} g(x) = 0$ type 0^0

2. $\lim_{x \to a} f(x) = \infty$ and $\lim_{x \to a} g(x) = 0$ type ∞^0

3. $\lim_{x \to a} f(x) = 1$ and $\lim_{x \to a} g(x) = \pm\infty$ type 1^∞

■ Although forms of the type 0^0, ∞^0, and 1^∞ are indeterminate, the form 0^∞ is not indeterminate. (See Exercise 52.)

Each of these three cases can be treated either by taking the natural logarithm:

$$\text{let} \quad y = [f(x)]^{g(x)}, \quad \text{then} \quad \ln y = g(x) \ln f(x)$$

or by writing the function as an exponential:

$$[f(x)]^{g(x)} = e^{g(x) \ln f(x)}$$

(Recall that both of these methods were used in differentiating such functions.) In either method we are led to the indeterminate product $g(x) \ln f(x)$, which is of type $0 \cdot \infty$.

EXAMPLE 8 Calculate $\lim_{x \to 0^+} (1 + \sin 4x)^{\cot x}$.

SOLUTION First notice that as $x \to 0^+$, we have $1 + \sin 4x \to 1$ and $\cot x \to \infty$, so the given limit is indeterminate. Let

$$y = (1 + \sin 4x)^{\cot x}$$

Then $$\ln y = \ln[(1 + \sin 4x)^{\cot x}] = \cot x \ln(1 + \sin 4x)$$

so l'Hospital's Rule gives

$$\lim_{x \to 0^+} \ln y = \lim_{x \to 0^+} \frac{\ln(1 + \sin 4x)}{\tan x} = \lim_{x \to 0^+} \frac{\dfrac{4 \cos 4x}{1 + \sin 4x}}{\sec^2 x} = 4$$

■ The graph of the function $y = x^x$, $x > 0$, is shown in Figure 6. Notice that although 0^0 is not defined, the values of the function approach 1 as $x \to 0^+$. This confirms the result of Example 9.

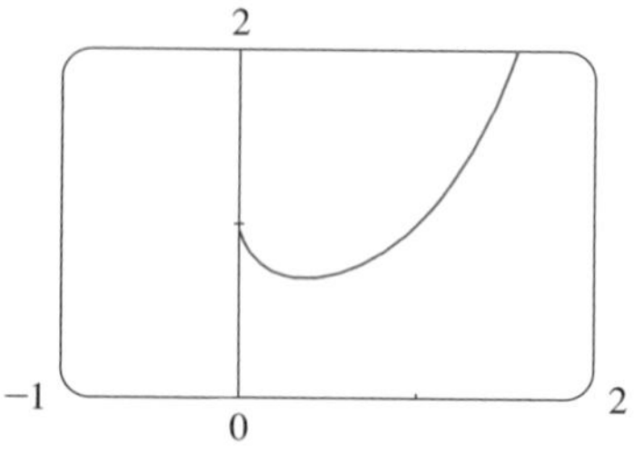

FIGURE 6

So far we have computed the limit of $\ln y$, but what we want is the limit of y. To find this we use the fact that $y = e^{\ln y}$:

$$\lim_{x \to 0^+} (1 + \sin 4x)^{\cot x} = \lim_{x \to 0^+} y = \lim_{x \to 0^+} e^{\ln y} = e^4$$

■

V EXAMPLE 9 Find $\lim_{x \to 0^+} x^x$.

SOLUTION Notice that this limit is indeterminate since $0^x = 0$ for any $x > 0$ but $x^0 = 1$ for any $x \neq 0$. We could proceed as in Example 8 or by writing the function as an exponential:

$$x^x = (e^{\ln x})^x = e^{x \ln x}$$

In Example 6 we used l'Hospital's Rule to show that

$$\lim_{x \to 0^+} x \ln x = 0$$

Therefore

$$\lim_{x \to 0^+} x^x = \lim_{x \to 0^+} e^{x \ln x} = e^0 = 1 \qquad \blacksquare$$

3.7 | EXERCISES

1–38 ■ Find the limit. Use l'Hospital's Rule where appropriate. If there is a more elementary method, consider using it. If l'Hospital's Rule doesn't apply, explain why.

1. $\displaystyle\lim_{x \to 1} \frac{x^2 - 1}{x^2 - x}$

2. $\displaystyle\lim_{x \to 2} \frac{x^2 + x - 6}{x - 2}$

3. $\displaystyle\lim_{x \to (\pi/2)^+} \frac{\cos x}{1 - \sin x}$

4. $\displaystyle\lim_{x \to 0} \frac{\sin 4x}{\tan 5x}$

5. $\displaystyle\lim_{t \to 0} \frac{e^{2t} - 1}{\sin t}$

6. $\displaystyle\lim_{x \to 0} \frac{x^2}{1 - \cos x}$

7. $\displaystyle\lim_{\theta \to \pi/2} \frac{1 - \sin \theta}{1 + \cos 2\theta}$

8. $\displaystyle\lim_{\theta \to \pi/2} \frac{1 - \sin \theta}{\csc \theta}$

9. $\displaystyle\lim_{x \to 0^+} \frac{\ln x}{x}$

10. $\displaystyle\lim_{x \to \infty} \frac{\ln \sqrt{x}}{x^2}$

11. $\displaystyle\lim_{t \to 1} \frac{t^8 - 1}{t^5 - 1}$

12. $\displaystyle\lim_{t \to 0} \frac{8^t - 5^t}{t}$

13. $\displaystyle\lim_{x \to 0} \frac{e^x - 1 - x}{x^2}$

14. $\displaystyle\lim_{u \to \infty} \frac{e^{u/10}}{u^3}$

15. $\displaystyle\lim_{x \to 0} \frac{x3^x}{3^x - 1}$

16. $\displaystyle\lim_{x \to 0} \frac{\cos mx - \cos nx}{x^2}$

17. $\displaystyle\lim_{x \to 1} \frac{1 - x + \ln x}{1 + \cos \pi x}$

18. $\displaystyle\lim_{x \to 0} \frac{x}{\tan^{-1}(4x)}$

19. $\displaystyle\lim_{x \to 1} \frac{x^a - ax + a - 1}{(x - 1)^2}$

20. $\displaystyle\lim_{x \to 0} \frac{e^x - e^{-x} - 2x}{x - \sin x}$

21. $\displaystyle\lim_{x \to 0} \frac{\cos x - 1 + \frac{1}{2}x^2}{x^4}$

22. $\displaystyle\lim_{x \to a^+} \frac{\cos x \ln(x - a)}{\ln(e^x - e^a)}$

23. $\displaystyle\lim_{x \to 0} \cot 2x \sin 6x$

24. $\displaystyle\lim_{x \to \infty} \sqrt{x}\, e^{-x/2}$

25. $\displaystyle\lim_{x \to \infty} x^3 e^{-x^2}$

26. $\displaystyle\lim_{x \to 0^+} \sin x \ln x$

27. $\displaystyle\lim_{x \to 1^+} \ln x \tan(\pi x/2)$

28. $\displaystyle\lim_{x \to \infty} x \tan(1/x)$

29. $\displaystyle\lim_{x \to 0^+} \left(\frac{1}{x} - \frac{1}{e^x - 1} \right)$

30. $\displaystyle\lim_{x \to 0} (\csc x - \cot x)$

31. $\displaystyle\lim_{x \to \infty} (x - \ln x)$

32. $\displaystyle\lim_{x \to 1^+} [\ln(x^7 - 1) - \ln(x^5 - 1)]$

33. $\displaystyle\lim_{x \to 0^+} x^{\sqrt{x}}$

34. $\displaystyle\lim_{x \to 0^+} (\tan 2x)^x$

35. $\displaystyle\lim_{x \to 0} (1 - 2x)^{1/x}$

36. $\displaystyle\lim_{x \to \infty} \left(1 + \frac{a}{x} \right)^{bx}$

37. $\displaystyle\lim_{x \to 1^+} x^{1/(1-x)}$

38. $\displaystyle\lim_{x \to \infty} (e^x + x)^{1/x}$

39–40 ■ Use a graph to estimate the value of the limit. Then use l'Hospital's Rule to find the exact value.

39. $\displaystyle\lim_{x \to \infty} \left(1 + \frac{2}{x} \right)^x$

40. $\displaystyle\lim_{x \to 0} \frac{5^x - 4^x}{3^x - 2^x}$

41. Prove that

$$\lim_{x \to \infty} \frac{e^x}{x^n} = \infty$$

for any positive integer n. This shows that the exponential function approaches infinity faster than any power of x.

42. Prove that

$$\lim_{x \to \infty} \frac{\ln x}{x^p} = 0$$

for any number $p > 0$. This shows that the logarithmic function approaches ∞ more slowly than any power of x.

43–44 ■ What happens if you try to use l'Hospital's Rule to find the limit? Evaluate the limit using another method.

43. $\displaystyle\lim_{x \to \infty} \frac{x}{\sqrt{x^2 + 1}}$

44. $\displaystyle\lim_{x \to (\pi/2)^-} \frac{\sec x}{\tan x}$

45. If an initial amount A_0 of money is invested at an interest rate r compounded n times a year, the value of the investment after t years is

$$A = A_0 \left(1 + \frac{r}{n} \right)^{nt}$$

If we let $n \to \infty$, we refer to the *continuous compounding* of interest. Use l'Hospital's Rule to show that if interest is compounded continuously, then the amount after t years is

$$A = A_0 e^{rt}$$

46. If an object with mass m is dropped from rest, one model for its speed v after t seconds, taking air resistance into account, is

$$v = \frac{mg}{c}(1 - e^{-ct/m})$$

where g is the acceleration due to gravity and c is a positive constant.

(a) Calculate $\lim_{t \to \infty} v$. What is the meaning of this limit?

(b) For fixed t, use l'Hospital's Rule to calculate $\lim_{c \to 0^+} v$. What can you conclude about the velocity of a falling object in a vacuum?

47. If an electrostatic field E acts on a liquid or a gaseous polar dielectric, the net dipole moment P per unit volume is

$$P(E) = \frac{e^E + e^{-E}}{e^E - e^{-E}} - \frac{1}{E}$$

Show that $\lim_{E \to 0^+} P(E) = 0$.

48. A metal cable has radius r and is covered by insulation, so that the distance from the center of the cable to the exterior of the insulation is R. The velocity v of an electrical impulse in the cable is

$$v = -c\left(\frac{r}{R}\right)^2 \ln\left(\frac{r}{R}\right)$$

where c is a positive constant. Find the following limits and interpret your answers.

(a) $\lim_{R \to r^+} v$ (b) $\lim_{r \to 0^+} v$

49. The first appearance in print of l'Hospital's Rule was in the book *Analyse des Infiniment Petits* published by the Marquis de l'Hospital in 1696. This was the first calculus *textbook* ever published and the example that the Marquis used in that book to illustrate his rule was to find the limit of the function

$$y = \frac{\sqrt{2a^3x - x^4} - a\sqrt[3]{aax}}{a - \sqrt[4]{ax^3}}$$

as x approaches a, where $a > 0$. (At that time it was common to write aa instead of a^2.) Solve this problem.

50. The figure shows a sector of a circle with central angle θ. Let $A(\theta)$ be the area of the segment between the chord PR and the arc PR. Let $B(\theta)$ be the area of the triangle PQR. Find $\lim_{\theta \to 0^+} A(\theta)/B(\theta)$.

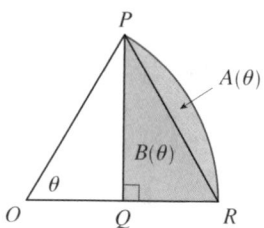

51. Evaluate $\lim_{x \to \infty}\left[x - x^2\ln\left(\frac{1 + x}{x}\right)\right]$.

52. Suppose f is a positive function. If $\lim_{x \to a} f(x) = 0$ and $\lim_{x \to a} g(x) = \infty$, show that

$$\lim_{x \to a}[f(x)]^{g(x)} = 0$$

This shows that 0^∞ is not an indeterminate form.

53. If f' is continuous, $f(2) = 0$, and $f'(2) = 7$, evaluate

$$\lim_{x \to 0} \frac{f(2 + 3x) + f(2 + 5x)}{x}$$

54. For what values of a and b is the following equation true?

$$\lim_{x \to 0}\left(\frac{\sin 2x}{x^3} + a + \frac{b}{x^2}\right) = 0$$

55. If f' is continuous, use l'Hospital's Rule to show that

$$\lim_{h \to 0} \frac{f(x + h) - f(x - h)}{2h} = f'(x)$$

Explain the meaning of this equation with the aid of a diagram.

56. If f'' is continuous, show that

$$\lim_{h \to 0} \frac{f(x + h) - 2f(x) + f(x - h)}{h^2} = f''(x)$$

57. Let

$$f(x) = \begin{cases} e^{-1/x^2} & \text{if } x \neq 0 \\ 0 & \text{if } x = 0 \end{cases}$$

(a) Use the definition of derivative to compute $f'(0)$.

(b) Show that f has derivatives of all orders that are defined on $\mathbb{R}$. [*Hint:* First show by induction that there is a polynomial $p_n(x)$ and a nonnegative integer k_n such that $f^{(n)}(x) = p_n(x)f(x)/x^{k_n}$ for $x \neq 0$.]

58. Let

$$f(x) = \begin{cases} |x|^x & \text{if } x \neq 0 \\ 1 & \text{if } x = 0 \end{cases}$$

(a) Show that f is continuous at 0.

(b) Investigate graphically whether f is differentiable at 0 by zooming in several times toward the point $(0, 1)$ on the graph of f.

(c) Show that f is not differentiable at 0. How can you reconcile this fact with the appearance of the graphs in part (b)?

CHAPTER 3   REVIEW

CONCEPT CHECK

1. (a) What is a one-to-one function? How can you tell if a function is one-to-one by looking at its graph?
  (b) If f is a one-to-one function, how is its inverse function f^{-1} defined? How do you obtain the graph of f^{-1} from the graph of f?
  (c) Suppose f is a one-to-one function. If $f'(f^{-1}(a)) \neq 0$, write a formula for $(f^{-1})'(a)$.

2. (a) Express e as a limit.
  (b) What is the value of e correct to five decimal places?
  (c) Why is the natural exponential function $y = e^x$ used more often in calculus than the other exponential functions $y = a^x$?
  (d) Why is the natural logarithmic function $y = \ln x$ used more often in calculus than the other logarithmic functions $y = \log_a x$?

3. (a) What are the domain and range of the natural exponential function $f(x) = e^x$?
  (b) What are the domain and range of the natural logarithmic function $f(x) = \ln x$?
  (c) How are the graphs of these functions related? Sketch these graphs, by hand, using the same axes.
  (d) If a is a positive number, $a \neq 1$, write an equation that expresses $\log_a x$ in terms of $\ln x$.

4. (a) How is the inverse sine function $f(x) = \sin^{-1}x$ defined? What are its domain and range?
  (b) How is the inverse cosine function $f(x) = \cos^{-1}x$ defined? What are its domain and range?

  (c) How is the inverse tangent function $f(x) = \tan^{-1}x$ defined? What are its domain and range? Sketch its graph.

5. Write the definitions of the hyperbolic functions $\sinh x$, $\cosh x$, and $\tanh x$.

6. State the derivative of each function.
  (a) $y = e^x$ (b) $y = a^x$ (c) $y = \ln x$
  (d) $y = \log_a x$ (e) $y = \sin^{-1}x$ (f) $y = \cos^{-1}x$
  (g) $y = \tan^{-1}x$ (h) $y = \sinh x$ (i) $y = \cosh x$
  (j) $y = \tanh x$ (k) $y = \sinh^{-1}x$ (l) $y = \cosh^{-1}x$
  (m) $y = \tanh^{-1}x$

7. (a) Write a differential equation that expresses the law of natural growth. What does it say in terms of relative growth rate?
  (b) Under what circumstances is this an appropriate model for population growth?
  (c) What are the solutions of this equation?

8. (a) What does l'Hospital's Rule say?
  (b) How can you use l'Hospital's Rule if you have a product $f(x)g(x)$ where $f(x) \to 0$ and $g(x) \to \infty$ as $x \to a$?
  (c) How can you use l'Hospital's Rule if you have a difference $f(x) - g(x)$ where $f(x) \to \infty$ and $g(x) \to \infty$ as $x \to a$?
  (d) How can you use l'Hospital's Rule if you have a power $[f(x)]^{g(x)}$ where $f(x) \to 0$ and $g(x) \to 0$ as $x \to a$?

TRUE-FALSE QUIZ

Determine whether the statement is true or false. If it is true, explain why. If it is false, explain why or give an example that disproves the statement.

1. If f is one-to-one, with domain $\mathbb{R}$, then $f^{-1}(f(6)) = 6$.

2. If f is one-to-one and differentiable, with domain $\mathbb{R}$, then $(f^{-1})'(6) = 1/f'(6)$.

3. The function $f(x) = \cos x$, $-\pi/2 \leqslant x \leqslant \pi/2$, is one-to-one.

4. $\tan^{-1}(-1) = 3\pi/4$

5. If $0 < a < b$, then $\ln a < \ln b$.

6. $\pi^{\sqrt{5}} = e^{\sqrt{5}\ln \pi}$

7. You can always divide by e^x.

8. If $a > 0$ and $b > 0$, then $\ln(a + b) = \ln a + \ln b$.

9. If $x > 0$, then $(\ln x)^6 = 6 \ln x$.

10. $\dfrac{d}{dx}(10^x) = x10^{x-1}$

11. $\dfrac{d}{dx}(\ln 10) = \dfrac{1}{10}$

12. The inverse function of $y = e^{3x}$ is $y = \frac{1}{3}\ln x$.

13. $\cos^{-1}x = \dfrac{1}{\cos x}$

14. $\tan^{-1}x = \dfrac{\sin^{-1}x}{\cos^{-1}x}$

15. $\cosh x \geqslant 1$ for all x

16. $\displaystyle\lim_{x \to \pi^-} \dfrac{\tan x}{1 - \cos x} = \lim_{x \to \pi^-} \dfrac{\sec^2x}{\sin x} = \infty$

EXERCISES

1. The graph of f is shown. Is f one-to-one? Explain.

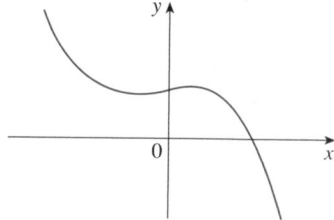

2. The graph of g is given.
(a) Why is g one-to-one?
(b) Estimate the value of $g^{-1}(2)$.
(c) Estimate the domain of g^{-1}.
(d) Sketch the graph of g^{-1}.

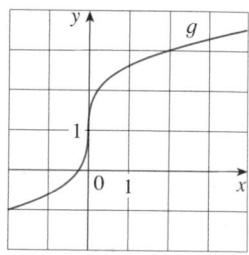

3. Suppose f is one-to-one, $f(7) = 3$, and $f'(7) = 8$. Find
(a) $f^{-1}(3)$ and (b) $(f^{-1})'(3)$.

4. Find the inverse function of $f(x) = \dfrac{x + 1}{2x + 1}$.

5–9 ▪ Sketch a rough graph of the function without using a calculator.

5. $y = 5^x - 1$
      6. $y = -e^{-x}$

7. $y = -\ln x$
      8. $y = \ln(x - 1)$

9. $y = 2 \arctan x$

10. Let $a > 1$. For large values of x, which of the functions $y = x^a$, $y = a^x$, and $y = \log_a x$ has the largest values and which has the smallest values?

11–12 ▪ Find the exact value of each expression.

11. (a) $e^{2 \ln 3}$
      (b) $\log_{10} 25 + \log_{10} 4$

12. (a) $\ln e^{\pi}$
      (b) $\tan\left(\arcsin \frac{1}{2}\right)$

13–16 ▪ Solve the equation for x.

13. (a) $e^x = 5$
      (b) $\ln x = 2$

14. (a) $e^{e^x} = 2$
      (b) $\tan^{-1}x = 1$

15. (a) $\ln(x + 1) + \ln(x - 1) = 1$
      (b) $\log_5(c^x) = d$

16. (a) $\ln(1 + e^{-x}) = 3$
      (b) $\sin x = 0.3$

17–43 ▪ Differentiate.

17. $y = \ln(x \ln x)$
      18. $y = e^{mx} \cos nx$

19. $y = \dfrac{e^{1/x}}{x^2}$
      20. $y = \ln \sec x$

21. $y = \sqrt{\arctan x}$
      22. $y = x \cos^{-1}x$

23. $f(t) = t^2 \ln t$
      24. $g(t) = \dfrac{e^t}{1 + e^t}$

25. $y = 3^{x \ln x}$
      26. $y = (\cos x)^x$

27. $y = x \sinh(x^2)$
      28. $xe^y = y \sin x$

29. $h(\theta) = e^{\tan 2\theta}$
      30. $y = (\arcsin 2x)^2$

31. $y = \ln \sin x - \frac{1}{2} \sin^2 x$
      32. $y = 10^{\tan \pi \theta}$

33. $y = \log_5(1 + 2x)$
      34. $y = e^{\cos x} + \cos(e^x)$

35. $y = \dfrac{\sqrt{x + 1}\,(2 - x)^5}{(x + 3)^7}$
      36. $y = \sin^{-1}(e^x)$

37. $y = x \tan^{-1}(4x)$
      38. $y = \dfrac{(x^2 + 1)^4}{(2x + 1)^3(3x - 1)^5}$

39. $y = \ln(\cosh 3x)$
      40. $y = \arctan(\arcsin \sqrt{x}\,)$

41. $y = \cosh^{-1}(\sinh x)$
      42. $y = x \tanh^{-1}\sqrt{x}$

43. $y = \cos(e^{\sqrt{\tan 3x}}\,)$

44. Show that
$$\frac{d}{dx}\left(\frac{1}{2} \tan^{-1}x + \frac{1}{4} \ln \frac{(x + 1)^2}{x^2 + 1}\right) = \frac{1}{(1 + x)(1 + x^2)}$$

45–48 ▪ Find f' in terms of g'.

45. $f(x) = e^{g(x)}$
      46. $f(x) = g(e^x)$

47. $f(x) = \ln |g(x)|$
      48. $f(x) = g(\ln x)$

49–50 ▪ Find $f^{(n)}(x)$.

49. $f(x) = 2^x$
      50. $f(x) = \ln(2x)$

51. Use mathematical induction to show that if $f(x) = xe^x$, then $f^{(n)}(x) = (x + n)e^x$.

52. Find an equation of the tangent to the curve $y = x \ln x$ at the point (e, e).

53. At what point on the curve $y = [\ln(x + 4)]^2$ is the tangent horizontal?

54. If $f(x) = xe^{\sin x}$, find $f'(x)$. Graph f and f' on the same screen and comment.

55. (a) Find an equation of the tangent to the curve $y = e^x$ that is parallel to the line $x - 4y = 1$.
(b) Find an equation of the tangent to the curve $y = e^x$ that passes through the origin.

56. The function $C(t) = K(e^{-at} - e^{-bt})$, where a, b, and K are positive constants and $b > a$, is used to model the concentration at time t of a drug injected into the bloodstream.
(a) Show that $\lim_{t \to \infty} C(t) = 0$.
(b) Find $C'(t)$, the rate at which the drug is cleared from circulation.
(c) When is this rate equal to 0?

57. A bacteria culture contains 200 cells initially and grows at a rate proportional to its size. After half an hour the population has increased to 360 cells.
(a) Find the number of bacteria after t hours.
(b) Find the number of bacteria after 4 hours.
(c) Find the rate of growth after 4 hours.
(d) When will the population reach 10,000?

58. Cobalt-60 has a half-life of 5.24 years.
(a) Find the mass that remains from a 100-mg sample after 20 years.
(b) How long would it take for the mass to decay to 1 mg?

59. Let $C(t)$ be the concentration of a drug in the bloodstream. As the body eliminates the drug, $C(t)$ decreases at a rate that is proportional to the amount of the drug that is present at the time. Thus $C'(t) = -kC(t)$, where k is a positive number called the *elimination constant* of the drug.
(a) If C_0 is the concentration at time $t = 0$, find the concentration at time t.
(b) If the body eliminates half the drug in 30 hours, how long does it take to eliminate 90% of the drug?

60. A cup of hot chocolate has temperature 80°C in a room kept at 20°C. After half an hour the hot chocolate cools to 60°C.
(a) What is the temperature of the chocolate after another half hour?
(b) When will the chocolate have cooled to 40°C?

61–76 ▪ Evaluate the limit.

61. $\lim_{x \to 0^+} \tan^{-1}(1/x)$

62. $\lim_{x \to \infty} e^{x - x^2}$

63. $\lim_{x \to 3^-} e^{2/(x-3)}$

64. $\lim_{x \to \infty} \arctan(x^3 - x)$

65. $\lim_{x \to 0^+} \ln(\sinh x)$

66. $\lim_{x \to \infty} e^{-x} \sin x$

67. $\lim_{x \to \infty} \dfrac{1 + 2^x}{1 - 2^x}$

68. $\lim_{x \to \infty} \left(1 + \dfrac{4}{x}\right)^x$

69. $\lim_{x \to 0} \dfrac{e^x - 1}{\tan x}$

70. $\lim_{x \to 0} \dfrac{\tan 4x}{x + \sin 2x}$

71. $\lim_{x \to 0} \dfrac{e^{4x} - 1 - 4x}{x^2}$

72. $\lim_{x \to \infty} \dfrac{e^{4x} - 1 - 4x}{x^2}$

73. $\lim_{x \to -\infty} (x^2 - x^3)e^{2x}$

74. $\lim_{x \to \pi^-} (x - \pi)\csc x$

75. $\lim_{x \to 1^+} \left(\dfrac{x}{x - 1} - \dfrac{1}{\ln x}\right)$

76. $\lim_{x \to (\pi/2)^-} (\tan x)^{\cos x}$

77. If $f(x) = \ln x + \tan^{-1}x$, find $(f^{-1})'(\pi/4)$.

78. Show that

$$\cos\{\arctan[\sin(\text{arccot } x)]\} = \sqrt{\dfrac{x^2 + 1}{x^2 + 2}}$$

4 | APPLICATIONS OF DIFFERENTIATION

We have already investigated some of the applications of derivatives, but now that we know the differentiation rules we are in a better position to pursue the applications of differentiation in greater depth. Here we learn how derivatives affect the shape of a graph of a function and, in particular, how they help us locate maximum and minimum values of functions. Many practical problems require us to minimize a cost or maximize an area or somehow find the best possible outcome of a situation. In particular, we will be able to investigate the optimal shape of a can and to explain the shape of cells in beehives.

4.1 | MAXIMUM AND MINIMUM VALUES

Some of the most important applications of differential calculus are *optimization problems*, in which we are required to find the optimal (best) way of doing something. Here are examples of such problems that we will solve in this chapter:

- What is the shape of a can that minimizes manufacturing costs?

- What is the maximum acceleration of a space shuttle? (This is an important question to the astronauts who have to withstand the effects of acceleration.)

- What is the radius of a contracted windpipe that expels air most rapidly during a cough?

- How far should you stand from a painting in an art gallery to get the best view of the painting?

These problems can be reduced to finding the maximum or minimum values of a function. Let's first explain exactly what we mean by maximum and minimum values.

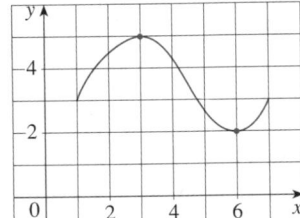

FIGURE 1

We see that the highest point on the graph of the function f shown in Figure 1 is the point $(3, 5)$. In other words, the largest value of f is $f(3) = 5$. Likewise, the smallest value is $f(6) = 2$. We say that $f(3) = 5$ is the *absolute maximum* of f and $f(6) = 2$ is the *absolute minimum*. In general, we use the following definition.

> **1 DEFINITION** Let c be a number in the domain D of a function f. Then $f(c)$ is the
>
> - **absolute maximum** value of f on D if $f(c) \geq f(x)$ for all x in D.
> - **absolute minimum** value of f on D if $f(c) \leq f(x)$ for all x in D.

An absolute maximum or minimum is sometimes called a **global** maximum or minimum. The maximum and minimum values of f are called **extreme values** of f.

Figure 2 shows the graph of a function f with absolute maximum at d and absolute minimum at a. Note that $(d, f(d))$ is the highest point on the graph and $(a, f(a))$ is the lowest point. In Figure 2, if we consider only values of x near b [for instance, if we restrict our attention to the interval (a, c)], then $f(b)$ is the largest of those values of $f(x)$ and is called a *local maximum value* of f. Likewise, $f(c)$ is called a *local minimum value* of f because $f(c) \leq f(x)$ for x near c [in the interval (b, d), for

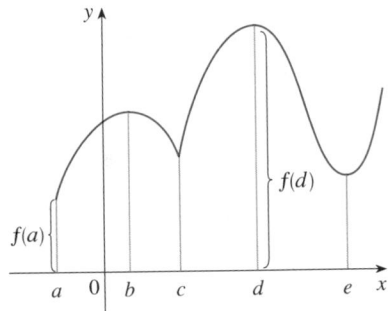

FIGURE 2
Abs min $f(a)$, abs max $f(d)$,
loc min $f(c)$, $f(e)$, loc max $f(b)$, $f(d)$

instance]. The function f also has a local minimum at e. In general, we have the following definition.

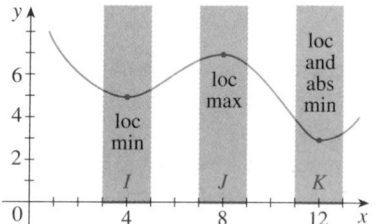

FIGURE 3

> **2** **DEFINITION** The number $f(c)$ is a
>
> - **local maximum** value of f if $f(c) \geq f(x)$ when x is near c.
> - **local minimum** value of f if $f(c) \leq f(x)$ when x is near c.

In Definition 2 (and elsewhere), if we say that something is true **near** c, we mean that it is true on some open interval containing c. For instance, in Figure 3 we see that $f(4) = 5$ is a local minimum because it's the smallest value of f on the interval I. It's not the absolute minimum because $f(x)$ takes smaller values when x is near 12 (in the interval K, for instance). In fact $f(12) = 3$ is both a local minimum and the absolute minimum. Similarly, $f(8) = 7$ is a local maximum, but not the absolute maximum because f takes larger values near 1.

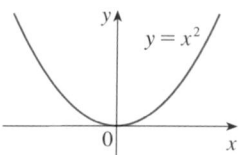

FIGURE 4

Minimum value 0, no maximum

EXAMPLE 1 The function $f(x) = \cos x$ takes on its (local and absolute) maximum value of 1 infinitely many times, since $\cos 2n\pi = 1$ for any integer n and $-1 \leq \cos x \leq 1$ for all x. Likewise, $\cos(2n + 1)\pi = -1$ is its minimum value, where n is any integer. ∎

EXAMPLE 2 If $f(x) = x^2$, then $f(x) \geq f(0)$ because $x^2 \geq 0$ for all x. Therefore $f(0) = 0$ is the absolute (and local) minimum value of f. This corresponds to the fact that the origin is the lowest point on the parabola $y = x^2$. (See Figure 4.) However, there is no highest point on the parabola and so this function has no maximum value. ∎

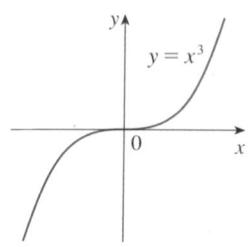

FIGURE 5

No minimum, no maximum

EXAMPLE 3 From the graph of the function $f(x) = x^3$, shown in Figure 5, we see that this function has neither an absolute maximum value nor an absolute minimum value. In fact, it has no local extreme values either. ∎

V **EXAMPLE 4** The graph of the function

$$f(x) = 3x^4 - 16x^3 + 18x^2 \qquad -1 \leq x \leq 4$$

is shown in Figure 6. You can see that $f(1) = 5$ is a local maximum, whereas the absolute maximum is $f(-1) = 37$. (This absolute maximum is not a local maximum because it occurs at an endpoint.) Also, $f(0) = 0$ is a local minimum and $f(3) = -27$ is both a local and an absolute minimum. Note that f has neither a local nor an absolute maximum at $x = 4$. ∎

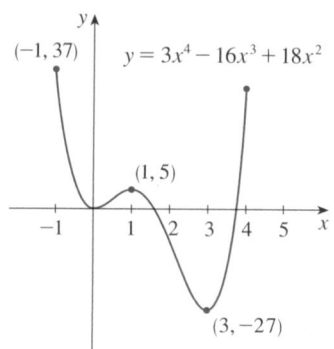

FIGURE 6

We have seen that some functions have extreme values, whereas others do not. The following theorem gives conditions under which a function is guaranteed to possess extreme values.

> **3** **THE EXTREME VALUE THEOREM** If f is continuous on a closed interval $[a, b]$, then f attains an absolute maximum value $f(c)$ and an absolute minimum value $f(d)$ at some numbers c and d in $[a, b]$.

The Extreme Value Theorem is illustrated in Figure 7. Note that an extreme value can be taken on more than once. Although the Extreme Value Theorem is intuitively very plausible, it is difficult to prove and so we omit the proof.

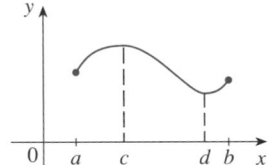 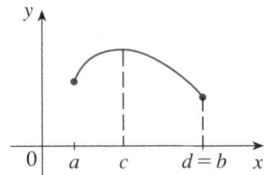 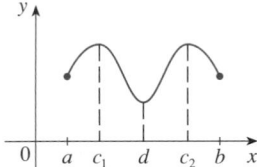

FIGURE 7

Figures 8 and 9 show that a function need not possess extreme values if either hypothesis (continuity or closed interval) is omitted from the Extreme Value Theorem.

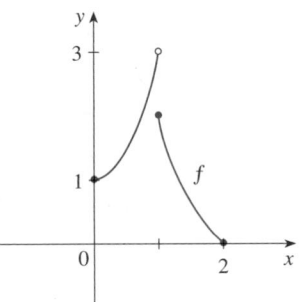

FIGURE 8
This function has minimum value $f(2) = 0$, but no maximum value.

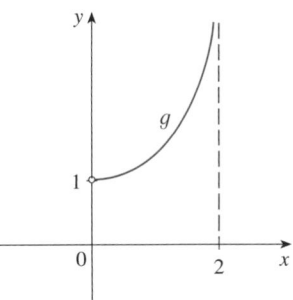

FIGURE 9
This continuous function g has no maximum or minimum.

The function f whose graph is shown in Figure 8 is defined on the closed interval [0, 2] but has no maximum value. [Notice that the range of f is [0, 3). The function takes on values arbitrarily close to 3, but never actually attains the value 3.] This does not contradict the Extreme Value Theorem because f is not continuous. [Nonetheless, a discontinuous function *could* have maximum and minimum values. See Exercise 13(b).]

The function g shown in Figure 9 is continuous on the open interval (0, 2) but has neither a maximum nor a minimum value. [The range of g is (1, ∞). The function takes on arbitrarily large values.] This does not contradict the Extreme Value Theorem because the interval (0, 2) is not closed.

The Extreme Value Theorem says that a continuous function on a closed interval has a maximum value and a minimum value, but it does not tell us how to find these extreme values. We start by looking for local extreme values.

Figure 10 shows the graph of a function f with a local maximum at c and a local minimum at d. It appears that at the maximum and minimum points the tangent lines are horizontal and therefore each has slope 0. We know that the derivative is the slope of the tangent line, so it appears that $f'(c) = 0$ and $f'(d) = 0$. The following theorem says that this is always true for differentiable functions.

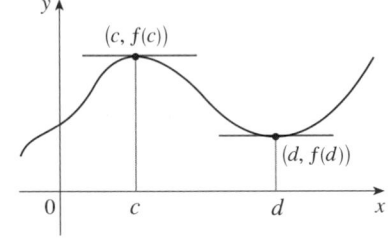

FIGURE 10

4 **FERMAT'S THEOREM** If f has a local maximum or minimum at c, and if $f'(c)$ exists, then $f'(c) = 0$.

■ **FERMAT**

Fermat's Theorem is named after Pierre Fermat (1601–1665), a French lawyer who took up mathematics as a hobby. Despite his amateur status, Fermat was one of the two inventors of analytic geometry (Descartes was the other). His methods for finding tangents to curves and maximum and minimum values (before the invention of limits and derivatives) made him a forerunner of Newton in the creation of differential calculus.

PROOF Suppose, for the sake of definiteness, that f has a local maximum at c. Then, according to Definition 2, $f(c) \geq f(x)$ if x is sufficiently close to c. This implies that if h is sufficiently close to 0, with h being positive or negative, then

$$f(c) \geq f(c + h)$$

and therefore

$$\boxed{5} \qquad f(c + h) - f(c) \leq 0$$

We can divide both sides of an inequality by a positive number. Thus if $h > 0$ and h is sufficiently small, we have

$$\frac{f(c + h) - f(c)}{h} \leq 0$$

Taking the right-hand limit of both sides of this inequality (using Theorem 1.4.3), we get

$$\lim_{h \to 0^+} \frac{f(c + h) - f(c)}{h} \leq \lim_{h \to 0^+} 0 = 0$$

But since $f'(c)$ exists, we have

$$f'(c) = \lim_{h \to 0} \frac{f(c + h) - f(c)}{h} = \lim_{h \to 0^+} \frac{f(c + h) - f(c)}{h}$$

and so we have shown that $f'(c) \leq 0$.

If $h < 0$, then the direction of the inequality $\boxed{5}$ is reversed when we divide by h:

$$\frac{f(c + h) - f(c)}{h} \geq 0 \qquad h < 0$$

So, taking the left-hand limit, we have

$$f'(c) = \lim_{h \to 0} \frac{f(c + h) - f(c)}{h} = \lim_{h \to 0^-} \frac{f(c + h) - f(c)}{h} \geq 0$$

We have shown that $f'(c) \geq 0$ and also that $f'(c) \leq 0$. Since both of these inequalities must be true, the only possibility is that $f'(c) = 0$.

We have proved Fermat's Theorem for the case of a local maximum. The case of a local minimum can be proved in a similar manner, or we could use Exercise 64 to deduce it from the case we have just proved (see Exercise 65). ☐

Although Fermat's Theorem is very useful, we have to guard against reading too much into it. If $f(x) = x^3$, then $f'(x) = 3x^2$, so $f'(0) = 0$. But f has no maximum or minimum at 0, as you can see from its graph in Figure 11. The fact that $f'(0) = 0$ simply means that the curve $y = x^3$ has a horizontal tangent at $(0, 0)$. Instead of having a maximum or minimum at $(0, 0)$, the curve crosses its horizontal tangent there.

Thus, when $f'(c) = 0$, f doesn't necessarily have a maximum or minimum at c. (In other words, the converse of Fermat's Theorem is false in general.)

We should bear in mind that there may be an extreme value where $f'(c)$ does not exist. For instance, the function $f(x) = |x|$ has its (local and absolute) minimum value at 0 (see Figure 12), but that value cannot be found by setting $f'(x) = 0$ because, as was shown in Example 5 in Section 2.2, $f'(0)$ does not exist.

Fermat's Theorem does suggest that we should at least *start* looking for extreme values of f at the numbers c where $f'(c) = 0$ or where $f'(c)$ does not exist. Such numbers are given a special name.

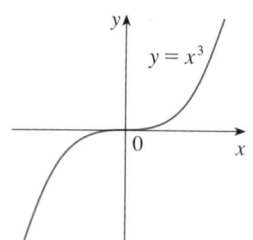

FIGURE 11
If $f(x) = x^3$, then $f'(0) = 0$ but f has no maximum or minimum.

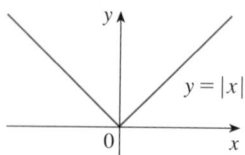

FIGURE 12
If $f(x) = |x|$, then $f(0) = 0$ is a minimum value, but $f'(0)$ does not exist.

> **6** **DEFINITION** A **critical number** of a function f is a number c in the domain of f such that either $f'(c) = 0$ or $f'(c)$ does not exist.

■ Figure 13 shows a graph of the function f in Example 5. It supports our answer because there is a horizontal tangent when $x = 1.5$ and a vertical tangent when $x = 0$.

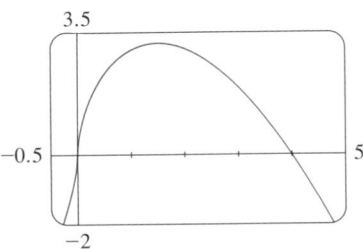

FIGURE 13

V EXAMPLE 5 Find the critical numbers of $f(x) = x^{3/5}(4 - x)$.

SOLUTION The Product Rule gives

$$f'(x) = x^{3/5}(-1) + \tfrac{3}{5}x^{-2/5}(4 - x) = -x^{3/5} + \frac{3(4 - x)}{5x^{2/5}}$$

$$= \frac{-5x + 3(4 - x)}{5x^{2/5}} = \frac{12 - 8x}{5x^{2/5}}$$

[The same result could be obtained by first writing $f(x) = 4x^{3/5} - x^{8/5}$.] Therefore $f'(x) = 0$ if $12 - 8x = 0$, that is, $x = \tfrac{3}{2}$, and $f'(x)$ does not exist when $x = 0$. Thus the critical numbers are $\tfrac{3}{2}$ and 0. ■

In terms of critical numbers, Fermat's Theorem can be rephrased as follows (compare Definition 6 with Theorem 4):

> **7** If f has a local maximum or minimum at c, then c is a critical number of f.

To find an absolute maximum or minimum of a continuous function on a closed interval, we note that either it is local [in which case it occurs at a critical number by **7**] or it occurs at an endpoint of the interval. Thus the following three-step procedure always works.

> **THE CLOSED INTERVAL METHOD** To find the *absolute* maximum and minimum values of a continuous function f on a closed interval $[a, b]$:
>
> **1.** Find the values of f at the critical numbers of f in (a, b).
>
> **2.** Find the values of f at the endpoints of the interval.
>
> **3.** The largest of the values from Steps 1 and 2 is the absolute maximum value; the smallest of these values is the absolute minimum value.

■ www.stewartcalculus.com
See Additional Example A.

V EXAMPLE 6 Find the absolute maximum and minimum values of the function

$$f(x) = x^3 - 3x^2 + 1 \qquad -\tfrac{1}{2} \leqslant x \leqslant 4$$

SOLUTION Since f is continuous on $\left[-\tfrac{1}{2}, 4\right]$, we can use the Closed Interval Method:

$$f(x) = x^3 - 3x^2 + 1$$

$$f'(x) = 3x^2 - 6x = 3x(x - 2)$$

Since $f'(x)$ exists for all x, the only critical numbers of f occur when $f'(x) = 0$, that is, $x = 0$ or $x = 2$. Notice that each of these critical numbers lies in the interval

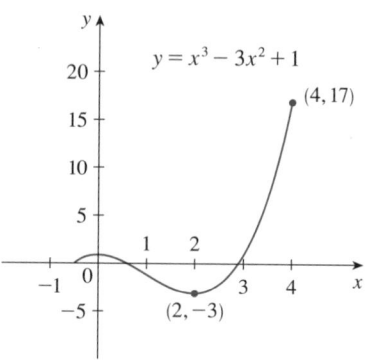

FIGURE 14

$\left(-\frac{1}{2}, 4\right)$. The values of f at these critical numbers are

$$f(0) = 1 \qquad f(2) = -3$$

The values of f at the endpoints of the interval are

$$f\left(-\frac{1}{2}\right) = \frac{1}{8} \qquad f(4) = 17$$

Comparing these four numbers, we see that the absolute maximum value is $f(4) = 17$ and the absolute minimum value is $f(2) = -3$.

Note that in this example the absolute maximum occurs at an endpoint, whereas the absolute minimum occurs at a critical number. The graph of f is sketched in Figure 14. ∎

4.1 | EXERCISES

1. Explain the difference between an absolute minimum and a local minimum.

2. Suppose f is a continuous function defined on a closed interval $[a, b]$.
(a) What theorem guarantees the existence of an absolute maximum value and an absolute minimum value for f?
(b) What steps would you take to find those maximum and minimum values?

3–4 ▪ For each of the numbers a, b, c, d, r, and s, state whether the function whose graph is shown has an absolute maximum or minimum, a local maximum or minimum, or neither a maximum nor a minimum.

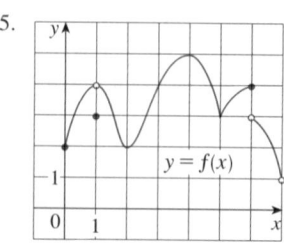
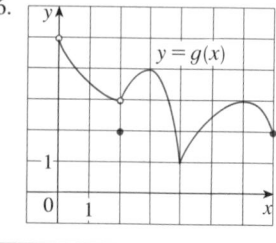

5–6 ▪ Use the graph to state the absolute and local maximum and minimum values of the function.

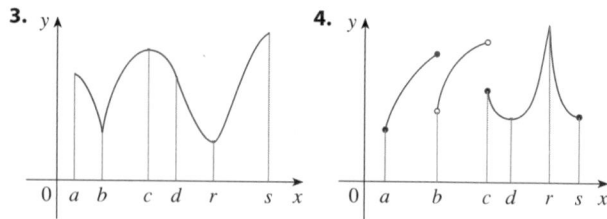

7–10 ▪ Sketch the graph of a function f that is continuous on [1, 5] and has the given properties.

7. Absolute minimum at 2, absolute maximum at 3, local minimum at 4

8. Absolute minimum at 1, absolute maximum at 5, local maximum at 2, local minimum at 4

9. Absolute maximum at 5, absolute minimum at 2, local maximum at 3, local minima at 2 and 4

10. f has no local maximum or minimum, but 2 and 4 are critical numbers

11. (a) Sketch the graph of a function that has a local maximum at 2 and is differentiable at 2.
(b) Sketch the graph of a function that has a local maximum at 2 and is continuous but not differentiable at 2.
(c) Sketch the graph of a function that has a local maximum at 2 and is not continuous at 2.

12. (a) Sketch the graph of a function on $[-1, 2]$ that has an absolute maximum but no local maximum.
(b) Sketch the graph of a function on $[-1, 2]$ that has a local maximum but no absolute maximum.

13. (a) Sketch the graph of a function on $[-1, 2]$ that has an absolute maximum but no absolute minimum.
(b) Sketch the graph of a function on $[-1, 2]$ that is discontinuous but has both an absolute maximum and an absolute minimum.

14. (a) Sketch the graph of a function that has two local maxima, one local minimum, and no absolute minimum.
(b) Sketch the graph of a function that has three local minima, two local maxima, and seven critical numbers.

⊞ Graphing calculator or computer required ‖ **CAS** Computer algebra system required ‖ **1.** Homework Hints at stewartcalculus.com

15–22 ▪ Sketch the graph of f by hand and use your sketch to find the absolute and local maximum and minimum values of f. (Use the graphs and transformations of Sections 1.2.)

15. $f(x) = \frac{1}{2}(3x - 1), \quad x \leq 3$

16. $f(x) = 2 - \frac{1}{3}x, \quad x \geq -2$

17. $f(x) = \sin x, \quad 0 \leq x < \pi/2$

18. $f(t) = \cos t, \quad -3\pi/2 \leq t \leq 3\pi/2$

19. $f(x) = \ln x, \quad 0 < x \leq 2$

20. $f(x) = 1/x, \quad 1 < x < 3$

21. $f(x) = 1 - \sqrt{x}$

22. $f(x) = \begin{cases} 4 - x^2 & \text{if } -2 \leq x < 0 \\ 2x - 1 & \text{if } 0 \leq x \leq 2 \end{cases}$

23–36 ▪ Find the critical numbers of the function.

23. $f(x) = 4 + \frac{1}{3}x - \frac{1}{2}x^2$

24. $f(x) = x^3 + 6x^2 - 15x$

25. $f(x) = 2x^3 - 3x^2 - 36x$

26. $f(x) = 2x^3 + x^2 + 2x$

27. $g(t) = t^4 + t^3 + t^2 + 1$

28. $g(t) = |3t - 4|$

29. $g(y) = \dfrac{y - 1}{y^2 - y + 1}$

30. $h(p) = \dfrac{p - 1}{p^2 + 4}$

31. $F(x) = x^{4/5}(x - 4)^2$

32. $g(x) = x^{1/3} - x^{-2/3}$

33. $f(\theta) = 2 \cos \theta + \sin^2\theta$

34. $g(\theta) = 4\theta - \tan \theta$

35. $f(x) = x^2 e^{-3x}$

36. $f(x) = x^{-2} \ln x$

37–50 ▪ Find the absolute maximum and absolute minimum values of f on the given interval.

37. $f(x) = 12 + 4x - x^2, \quad [0, 5]$

38. $f(x) = 5 + 54x - 2x^3, \quad [0, 4]$

39. $f(x) = 2x^3 - 3x^2 - 12x + 1, \quad [-2, 3]$

40. $f(x) = x^3 - 6x^2 + 5, \quad [-3, 5]$

41. $f(x) = 3x^4 - 4x^3 - 12x^2 + 1, \quad [-2, 3]$

42. $f(x) = (x^2 - 1)^3, \quad [-1, 2]$

43. $f(t) = t\sqrt{4 - t^2}, \quad [-1, 2]$

44. $f(x) = \dfrac{x}{x^2 - x + 1}, \quad [0, 3]$

45. $f(t) = 2\cos t + \sin 2t, \quad [0, \pi/2]$

46. $f(t) = t + \cot(\frac{1}{2}t), \quad [\pi/4, 7\pi/4]$

47. $f(x) = xe^{-x^2/8}, \quad [-1, 4]$

48. $f(x) = x - \ln x, \quad [\frac{1}{2}, 2]$

49. $f(x) = \ln(x^2 + x + 1), \quad [-1, 1]$

50. $f(x) = x - 2 \tan^{-1}x, \quad [0, 4]$

51. If a and b are positive numbers, find the maximum value of $f(x) = x^a(1 - x)^b, 0 \leq x \leq 1$.

52. Use a graph to estimate the critical numbers of $f(x) = |x^3 - 3x^2 + 2|$ correct to one decimal place.

53–56 ▪
(a) Use a graph to estimate the absolute maximum and minimum values of the function to two decimal places.
(b) Use calculus to find the exact maximum and minimum values.

53. $f(x) = x^5 - x^3 + 2, \quad -1 \leq x \leq 1$

54. $f(x) = e^x + e^{-2x}, \quad 0 \leq x \leq 1$

55. $f(x) = x\sqrt{x - x^2}$

56. $f(x) = x - 2 \cos x, \quad -2 \leq x \leq 0$

57. Between 0°C and 30°C, the volume V (in cubic centimeters) of 1 kg of water at a temperature T is given approximately by the formula

$$V = 999.87 - 0.06426T + 0.0085043T^2 - 0.0000679T^3$$

Find the temperature at which water has its maximum density.

58. An object with weight W is dragged along a horizontal plane by a force acting along a rope attached to the object. If the rope makes an angle θ with the plane, then the magnitude of the force is

$$F = \frac{\mu W}{\mu \sin \theta + \cos \theta}$$

where μ is a positive constant called the *coefficient of friction* and where $0 \leq \theta \leq \pi/2$. Show that F is minimized when $\tan \theta = \mu$.

59. A model for the US average price of a pound of white sugar from 1993 to 2003 is given by the function

$$S(t) = -0.00003237t^5 + 0.0009037t^4 - 0.008956t^3$$
$$+ 0.03629t^2 - 0.04458t + 0.4074$$

where t is measured in years since August of 1993. Estimate the times when sugar was cheapest and most expensive during the period 1993–2003.

60. The Hubble Space Telescope was deployed April 24, 1990, by the space shuttle *Discovery*. A model for the velocity of the shuttle during this mission, from liftoff at $t = 0$ until the solid rocket boosters were jettisoned at $t = 126$ s, is given by

$$v(t) = 0.001302t^3 - 0.09029t^2 + 23.61t - 3.083$$

(in feet per second). Using this model, estimate the absolute maximum and minimum values of the *acceleration* of the shuttle between liftoff and the jettisoning of the boosters.

61. When a foreign object lodged in the trachea (windpipe) forces a person to cough, the diaphragm thrusts upward causing an increase in pressure in the lungs. This is accompanied by a contraction of the trachea, making a narrower channel for the expelled air to flow through. For a given amount of air to escape in a fixed time, it must move faster through the narrower channel than the wider one. The greater the velocity of the airstream, the greater the force on the foreign object. X-rays show that the radius of the circular tracheal tube contracts to about two-thirds of its normal radius during a cough. According to a mathematical model of coughing, the velocity v of the airstream is related to the radius r of the trachea by the equation

$$v(r) = k(r_0 - r)r^2 \qquad \tfrac{1}{2}r_0 \leq r \leq r_0$$

where k is a constant and r_0 is the normal radius of the trachea. The restriction on r is due to the fact that the tracheal wall stiffens under pressure and a contraction greater than $\frac{1}{2}r_0$ is prevented (otherwise the person would suffocate).
(a) Determine the value of r in the interval $\left[\tfrac{1}{2}r_0, r_0\right]$ at which v has an absolute maximum. How does this compare with experimental evidence?
(b) What is the absolute maximum value of v on the interval?
(c) Sketch the graph of v on the interval $[0, r_0]$.

62. Show that 5 is a critical number of the function $g(x) = 2 + (x - 5)^3$ but g does not have a local extreme value at 5.

63. Prove that the function $f(x) = x^{101} + x^{51} + x + 1$ has neither a local maximum nor a local minimum.

64. If f has a local minimum value at c, show that the function $g(x) = -f(x)$ has a local maximum value at c.

65. Prove Fermat's Theorem for the case in which f has a local minimum at c.

66. A cubic function is a polynomial of degree 3; that is, it has the form $f(x) = ax^3 + bx^2 + cx + d$, where $a \neq 0$.
(a) Show that a cubic function can have two, one, or no critical number(s). Give examples and sketches to illustrate the three possibilities.
(b) How many local extreme values can a cubic function have?

4.2 | THE MEAN VALUE THEOREM

We will see that many of the results of this chapter depend on one central fact, which is called the Mean Value Theorem. But to arrive at the Mean Value Theorem we first need the following result.

■ ROLLE

Rolle's Theorem was first published in 1691 by the French mathematician Michel Rolle (1652–1719) in a book entitled *Méthode pour resoudre les Egalitez.* He was a vocal critic of the methods of his day and attacked calculus as being a "collection of ingenious fallacies." Later, however, he became convinced of the essential correctness of the methods of calculus.

ROLLE'S THEOREM Let f be a function that satisfies the following three hypotheses:

1. f is continuous on the closed interval $[a, b]$.

2. f is differentiable on the open interval (a, b).

3. $f(a) = f(b)$

Then there is a number c in (a, b) such that $f'(c) = 0$.

Before giving the proof let's take a look at the graphs of some typical functions that satisfy the three hypotheses. Figure 1 shows the graphs of four such functions. In each case it appears that there is at least one point $(c, f(c))$ on the graph where the tangent is horizontal and therefore $f'(c) = 0$. Thus Rolle's Theorem is plausible.

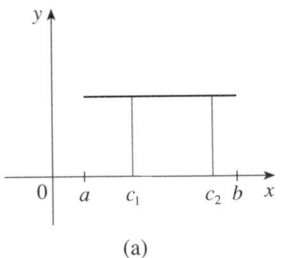

(a)

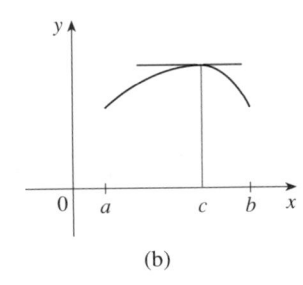

(b)

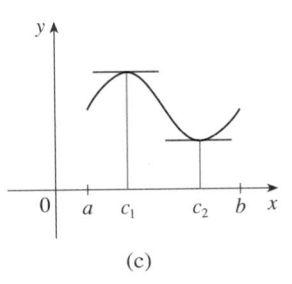

(c)

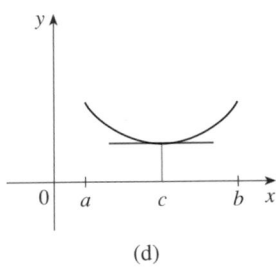

(d)

FIGURE 1

PROOF   There are three cases:

CASE I $f(x) = k$, **a constant**
Then $f'(x) = 0$, so the number c can be taken to be *any* number in (a, b).

CASE II $f(x) > f(a)$ **for some x in (a, b)**   [as in Figure 1(b) or (c)]
By the Extreme Value Theorem (which we can apply by hypothesis 1), f has a maximum value somewhere in $[a, b]$. Since $f(a) = f(b)$, it must attain this maximum value at a number c in the open interval (a, b). Then f has a *local* maximum at c and, by hypothesis 2, f is differentiable at c. Therefore $f'(c) = 0$ by Fermat's Theorem.

CASE III $f(x) < f(a)$ **for some x in (a, b)**   [as in Figure 1(c) or (d)]
By the Extreme Value Theorem, f has a minimum value in $[a, b]$ and, since $f(a) = f(b)$, it attains this minimum value at a number c in (a, b). Again $f'(c) = 0$ by Fermat's Theorem.   □

EXAMPLE 1   Let's apply Rolle's Theorem to the position function $s = f(t)$ of a moving object. If the object is in the same place at two different instants $t = a$ and $t = b$, then $f(a) = f(b)$. Rolle's Theorem says that there is some instant of time $t = c$ between a and b when $f'(c) = 0$; that is, the velocity is 0. (In particular, you can see that this is true when a ball is thrown directly upward.)   ∎

EXAMPLE 2   Prove that the equation $x^3 + x - 1 = 0$ has exactly one real root.

SOLUTION   First we use the Intermediate Value Theorem (1.5.9) to show that a root exists. Let $f(x) = x^3 + x - 1$. Then $f(0) = -1 < 0$ and $f(1) = 1 > 0$. Since f is a polynomial, it is continuous, so the Intermediate Value Theorem states that there is a number c between 0 and 1 such that $f(c) = 0$. Thus the given equation has a root.

To show that the equation has no other real root, we use Rolle's Theorem and argue by contradiction. Suppose that it had two roots a and b. Then $f(a) = 0 = f(b)$ and, since f is a polynomial, it is differentiable on (a, b) and continuous on $[a, b]$. Thus by Rolle's Theorem there is a number c between a and b such that $f'(c) = 0$. But

$$f'(x) = 3x^2 + 1 \geq 1 \qquad \text{for all } x$$

(since $x^2 \geq 0$) so $f'(x)$ can never be 0. This gives a contradiction. Therefore the equation can't have two real roots.   ∎

Our main use of Rolle's Theorem is in proving the following important theorem, which was first stated by another French mathematician, Joseph-Louis Lagrange.

■ Figure 2 shows a graph of the function $f(x) = x^3 + x - 1$ discussed in Example 2. Rolle's Theorem shows that, no matter how much we enlarge the viewing rectangle, we can never find a second x-intercept.

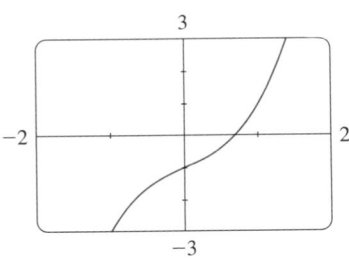

FIGURE 2

■ The Mean Value Theorem is an example of what is called an existence theorem. Like the Intermediate Value Theorem, the Extreme Value Theorem, and Rolle's Theorem, it guarantees that there *exists* a number with a certain property, but it doesn't tell us how to find the number.

THE MEAN VALUE THEOREM   Let f be a function that satisfies the following hypotheses:

1. f is continuous on the closed interval $[a, b]$.

2. f is differentiable on the open interval (a, b).

Then there is a number c in (a, b) such that

$$\boxed{1} \qquad f'(c) = \frac{f(b) - f(a)}{b - a}$$

or, equivalently,

$$\boxed{2} \qquad f(b) - f(a) = f'(c)(b - a)$$

Before proving this theorem, we can see that it is reasonable by interpreting it geometrically. Figures 3 and 4 show the points $A(a, f(a))$ and $B(b, f(b))$ on the graphs of two differentiable functions. The slope of the secant line AB is

$$\boxed{3} \qquad m_{AB} = \frac{f(b) - f(a)}{b - a}$$

which is the same expression as on the right side of Equation 1. Since $f'(c)$ is the slope of the tangent line at the point $(c, f(c))$, the Mean Value Theorem, in the form given by Equation 1, says that there is at least one point $P(c, f(c))$ on the graph where the slope of the tangent line is the same as the slope of the secant line AB. In other words, there is a point P where the tangent line is parallel to the secant line AB. (Imagine a line parallel to AB, starting far away and moving parallel to itself until it touches the graph for the first time.)

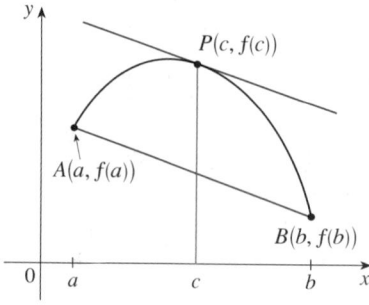

FIGURE 3

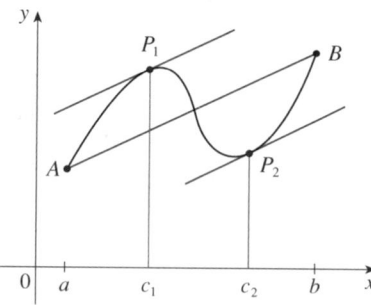

FIGURE 4

PROOF   We apply Rolle's Theorem to a new function h defined as the difference between f and the function whose graph is the secant line AB. Using Equation 3, we see that the equation of the line AB can be written as

$$y - f(a) = \frac{f(b) - f(a)}{b - a} (x - a)$$

or as

$$y = f(a) + \frac{f(b) - f(a)}{b - a} (x - a)$$

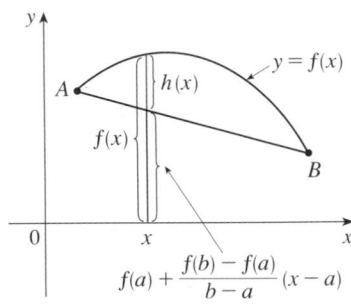

FIGURE 5

So, as shown in Figure 5,

$$\boxed{4} \qquad h(x) = f(x) - f(a) - \frac{f(b) - f(a)}{b - a}(x - a)$$

First we must verify that h satisfies the three hypotheses of Rolle's Theorem.

1. The function h is continuous on $[a, b]$ because it is the sum of f and a first-degree polynomial, both of which are continuous.

2. The function h is differentiable on (a, b) because both f and the first-degree polynomial are differentiable. In fact, we can compute h' directly from Equation 4:

$$h'(x) = f'(x) - \frac{f(b) - f(a)}{b - a}$$

(Note that $f(a)$ and $[f(b) - f(a)]/(b - a)$ are constants.)

3. $\qquad h(a) = f(a) - f(a) - \frac{f(b) - f(a)}{b - a}(a - a) = 0$

$$h(b) = f(b) - f(a) - \frac{f(b) - f(a)}{b - a}(b - a)$$

$$= f(b) - f(a) - [f(b) - f(a)] = 0$$

Therefore $h(a) = h(b)$.

Since h satisfies the hypotheses of Rolle's Theorem, that theorem says there is a number c in (a, b) such that $h'(c) = 0$. Therefore

$$0 = h'(c) = f'(c) - \frac{f(b) - f(a)}{b - a}$$

and so
$$f'(c) = \frac{f(b) - f(a)}{b - a} \qquad \square$$

▼ EXAMPLE 3   To illustrate the Mean Value Theorem with a specific function, let's consider $f(x) = x^3 - x$, $a = 0$, $b = 2$. Since f is a polynomial, it is continuous and differentiable for all x, so it is certainly continuous on $[0, 2]$ and differentiable on $(0, 2)$. Therefore, by the Mean Value Theorem, there is a number c in $(0, 2)$ such that

$$f(2) - f(0) = f'(c)(2 - 0)$$

Now $f(2) = 6$, $f(0) = 0$, and $f'(x) = 3x^2 - 1$, so this equation becomes

$$6 = (3c^2 - 1)2 = 6c^2 - 2$$

which gives $c^2 = \frac{4}{3}$, that is, $c = \pm 2/\sqrt{3}$. But c must lie in $(0, 2)$, so $c = 2/\sqrt{3}$. Figure 6 illustrates this calculation: The tangent line at this value of c is parallel to the secant line OB. ∎

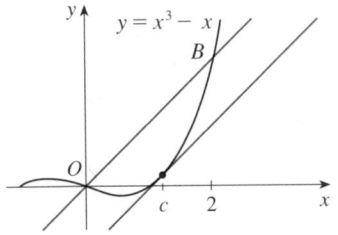

FIGURE 6

▼ EXAMPLE 4   If an object moves in a straight line with position function $s = f(t)$, then the average velocity between $t = a$ and $t = b$ is

$$\frac{f(b) - f(a)}{b - a}$$

and the velocity at $t = c$ is $f'(c)$. Thus the Mean Value Theorem (in the form of Equation 1) tells us that at some time $t = c$ between a and b the instantaneous velocity $f'(c)$ is equal to that average velocity. For instance, if a car traveled 180 km in 2 hours, then the speedometer must have read 90 km/h at least once.

In general, the Mean Value Theorem can be interpreted as saying that there is a number at which the instantaneous rate of change is equal to the average rate of change over an interval. ∎

The main significance of the Mean Value Theorem is that it enables us to obtain information about a function from information about its derivative. The next example provides an instance of this principle.

V EXAMPLE 5 Suppose that $f(0) = -3$ and $f'(x) \le 5$ for all values of x. How large can $f(2)$ possibly be?

SOLUTION We are given that f is differentiable (and therefore continuous) everywhere. In particular, we can apply the Mean Value Theorem on the interval $[0, 2]$. There exists a number c such that

$$f(2) - f(0) = f'(c)(2 - 0)$$

so

$$f(2) = f(0) + 2f'(c) = -3 + 2f'(c)$$

We are given that $f'(x) \le 5$ for all x, so in particular we know that $f'(c) \le 5$. Multiplying both sides of this inequality by 2, we have $2f'(c) \le 10$, so

$$f(2) = -3 + 2f'(c) \le -3 + 10 = 7$$

The largest possible value for $f(2)$ is 7. ∎

The Mean Value Theorem can be used to establish some of the basic facts of differential calculus. One of these basic facts is the following theorem. Others will be found in the following sections.

---

5 THEOREM If $f'(x) = 0$ for all x in an interval (a, b), then f is constant on (a, b).

PROOF Let x_1 and x_2 be any two numbers in (a, b) with $x_1 < x_2$. Since f is differentiable on (a, b), it must be differentiable on (x_1, x_2) and continuous on $[x_1, x_2]$. By applying the Mean Value Theorem to f on the interval $[x_1, x_2]$, we get a number c such that $x_1 < c < x_2$ and

$$\boxed{6} \qquad f(x_2) - f(x_1) = f'(c)(x_2 - x_1)$$

Since $f'(x) = 0$ for all x, we have $f'(c) = 0$, and so Equation 6 becomes

$$f(x_2) - f(x_1) = 0 \qquad \text{or} \qquad f(x_2) = f(x_1)$$

Therefore f has the same value at *any* two numbers x_1 and x_2 in (a, b). This means that f is constant on (a, b). ☐

> **7   COROLLARY**   If $f'(x) = g'(x)$ for all x in an interval (a, b), then $f - g$ is constant on (a, b); that is, $f(x) = g(x) + c$ where c is a constant.

PROOF   Let $F(x) = f(x) - g(x)$. Then

$$F'(x) = f'(x) - g'(x) = 0$$

for all x in (a, b). Thus, by Theorem 5, F is constant; that is, $f - g$ is constant.   □

NOTE   Care must be taken in applying Theorem 5. Let

$$f(x) = \frac{x}{|x|} = \begin{cases} 1 & \text{if } x > 0 \\ -1 & \text{if } x < 0 \end{cases}$$

The domain of f is $D = \{x \mid x \neq 0\}$ and $f'(x) = 0$ for all x in D. But f is obviously not a constant function. This does not contradict Theorem 5 because D is not an interval. Notice that f is constant on the interval $(0, \infty)$ and also on the interval $(-\infty, 0)$.

We will make extensive use of Theorem 5 and Corollary 7 when we study antiderivatives in Section 4.7.

■ www.stewartcalculus.com
See Additional Example A.

4.2 | EXERCISES

1–4 ■ Verify that the function satisfies the three hypotheses of Rolle's Theorem on the given interval. Then find all numbers c that satisfy the conclusion of Rolle's Theorem.

1. $f(x) = 5 - 12x + 3x^2$, $\quad [1, 3]$

2. $f(x) = x^3 - x^2 - 6x + 2$, $\quad [0, 3]$

3. $f(x) = \sqrt{x} - \frac{1}{3}x$, $\quad [0, 9]$

4. $f(x) = \cos 2x$, $\quad [\pi/8, 7\pi/8]$

5. Let $f(x) = 1 - x^{2/3}$. Show that $f(-1) = f(1)$ but there is no number c in $(-1, 1)$ such that $f'(c) = 0$. Why does this not contradict Rolle's Theorem?

6. Let $f(x) = \tan x$. Show that $f(0) = f(\pi)$ but there is no number c in $(0, \pi)$ such that $f'(c) = 0$. Why does this not contradict Rolle's Theorem?

7. Use the graph of f to estimate the values of c that satisfy the conclusion of the Mean Value Theorem for the interval $[0, 8]$.

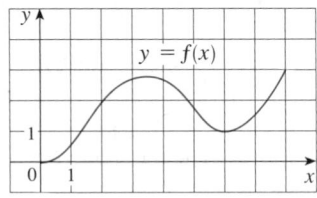

8. Use the graph of f given in Exercise 7 to estimate the values of c that satisfy the conclusion of the Mean Value Theorem for the interval $[1, 7]$.

9–12 ■ Verify that the function satisfies the hypotheses of the Mean Value Theorem on the given interval. Then find all numbers c that satisfy the conclusion of the Mean Value Theorem.

9. $f(x) = 2x^2 - 3x + 1$, $\quad [0, 2]$

10. $f(x) = x^3 - 3x + 2$, $\quad [-2, 2]$

11. $f(x) = \ln x$, $\quad [1, 4]$

12. $f(x) = 1/x$, $\quad [1, 3]$

13–14 ■ Find the number c that satisfies the conclusion of the Mean Value Theorem on the given interval. Graph the function, the secant line through the endpoints, and the tangent line at $(c, f(c))$. Are the secant line and the tangent line parallel?

13. $f(x) = \sqrt{x}$, $\quad [0, 4]$ **14.** $f(x) = e^{-x}$, $\quad [0, 2]$

15. Let $f(x) = (x - 3)^{-2}$. Show that there is no value of c in $(1, 4)$ such that $f(4) - f(1) = f'(c)(4 - 1)$. Why does this not contradict the Mean Value Theorem?

16. Let $f(x) = 2 - |2x - 1|$. Show that there is no value of c such that $f(3) - f(0) = f'(c)(3 - 0)$. Why does this not contradict the Mean Value Theorem?

17–18 ▪ Show that the equation has exactly one real root.

17. $2x + \cos x = 0$ **18.** $x^3 + e^x = 0$

19. Show that the equation $x^3 - 15x + c = 0$ has at most one root in the interval $[-2, 2]$.

20. Show that the equation $x^4 + 4x + c = 0$ has at most two real roots.

21. (a) Show that a polynomial of degree 3 has at most three real roots.
(b) Show that a polynomial of degree n has at most n real roots.

22. (a) Suppose that f is differentiable on $\mathbb{R}$ and has two roots. Show that f' has at least one root.
(b) Suppose f is twice differentiable on $\mathbb{R}$ and has three roots. Show that f'' has at least one real root.
(c) Can you generalize parts (a) and (b)?

23. If $f(1) = 10$ and $f'(x) \geqslant 2$ for $1 \leqslant x \leqslant 4$, how small can $f(4)$ possibly be?

24. Suppose that $3 \leqslant f'(x) \leqslant 5$ for all values of x. Show that $18 \leqslant f(8) - f(2) \leqslant 30$.

25. Does there exist a function f such that $f(0) = -1$, $f(2) = 4$, and $f'(x) \leqslant 2$ for all x?

26. Suppose that f and g are continuous on $[a, b]$ and differentiable on (a, b). Suppose also that $f(a) = g(a)$ and $f'(x) < g'(x)$ for $a < x < b$. Prove that $f(b) < g(b)$. [*Hint:* Apply the Mean Value Theorem to the function $h = f - g$.]

27. Show that $\sqrt{1 + x} < 1 + \frac{1}{2}x$ if $x > 0$.

28. Suppose f is an odd function and is differentiable everywhere. Prove that for every positive number b, there exists a number c in $(-b, b)$ such that $f'(c) = f(b)/b$.

29. Use the Mean Value Theorem to prove the inequality

$$|\sin a - \sin b| \leqslant |a - b| \qquad \text{for all } a \text{ and } b$$

30. If $f'(x) = c$ (c a constant) for all x, use Corollary 7 to show that $f(x) = cx + d$ for some constant d.

31. Let $f(x) = 1/x$ and

$$g(x) = \begin{cases} \dfrac{1}{x} & \text{if } x > 0 \\ 1 + \dfrac{1}{x} & \text{if } x < 0 \end{cases}$$

Show that $f'(x) = g'(x)$ for all x in their domains. Can we conclude from Corollary 7 that $f - g$ is constant?

32. Use Theorem 5 to prove the identity

$$2 \sin^{-1}x = \cos^{-1}(1 - 2x^2) \qquad x \geqslant 0$$

33. Prove the identity

$$\arcsin \frac{x - 1}{x + 1} = 2 \arctan \sqrt{x} - \frac{\pi}{2}$$

34. At 2:00 PM a car's speedometer reads 30 mi/h. At 2:10 PM it reads 50 mi/h. Show that at some time between 2:00 and 2:10 the acceleration is exactly 120 mi/h^2.

35. Two runners start a race at the same time and finish in a tie. Prove that at some time during the race they have the same speed. [*Hint:* Consider $f(t) = g(t) - h(t)$, where g and h are the position functions of the two runners.]

36. A number a is called a **fixed point** of a function f if $f(a) = a$. Prove that if $f'(x) \neq 1$ for all real numbers x, then f has at most one fixed point.

4.3 | DERIVATIVES AND THE SHAPES OF GRAPHS

Many of the applications of calculus depend on our ability to deduce facts about a function f from information concerning its derivatives. Because $f'(x)$ represents the slope of the curve $y = f(x)$ at the point $(x, f(x))$, it tells us the direction in which the curve proceeds at each point. So it is reasonable to expect that information about $f'(x)$ will provide us with information about $f(x)$.

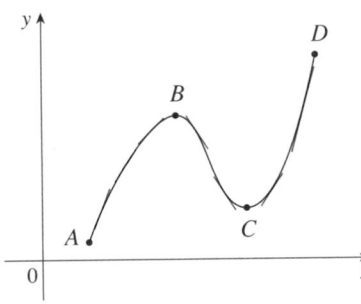

FIGURE 1

■ Let's abbreviate the name of this test to the I/D Test.

WHAT DOES f' SAY ABOUT f?

To see how the derivative of f can tell us where a function is increasing or decreasing, look at Figure 1. (Increasing functions and decreasing functions were defined in Section 1.1.) Between A and B and between C and D, the tangent lines have positive slopes and so $f'(x) > 0$. Between B and C, the tangent lines have negative slopes and so $f'(x) < 0$. Thus it appears that f increases when $f'(x)$ is positive and decreases when $f'(x)$ is negative. To prove that this is always the case, we use the Mean Value Theorem.

INCREASING/DECREASING TEST

(a) If $f'(x) > 0$ on an interval, then f is increasing on that interval.

(b) If $f'(x) < 0$ on an interval, then f is decreasing on that interval.

PROOF

(a) Let x_1 and x_2 be any two numbers in the interval with $x_1 < x_2$. According to the definition of an increasing function (page 7) we have to show that $f(x_1) < f(x_2)$.

Because we are given that $f'(x) > 0$, we know that f is differentiable on $[x_1, x_2]$. So by the Mean Value Theorem, there is a number c between x_1 and x_2 such that

$$\boxed{1} \qquad f(x_2) - f(x_1) = f'(c)(x_2 - x_1)$$

Now $f'(c) > 0$ by assumption and $x_2 - x_1 > 0$ because $x_1 < x_2$. Thus the right side of Equation 1 is positive, and so

$$f(x_2) - f(x_1) > 0 \qquad \text{or} \qquad f(x_1) < f(x_2)$$

This shows that f is increasing.

Part (b) is proved similarly. $\qquad\qquad\qquad\qquad\qquad\qquad\qquad\qquad\qquad\qquad$ □

▼ EXAMPLE 1 Find where the function $f(x) = 3x^4 - 4x^3 - 12x^2 + 5$ is increasing and where it is decreasing.

SOLUTION $\qquad f'(x) = 12x^3 - 12x^2 - 24x = 12x(x - 2)(x + 1)$

To use the I/D Test we have to know where $f'(x) > 0$ and where $f'(x) < 0$. This depends on the signs of the three factors of $f'(x)$, namely, $12x$, $x - 2$, and $x + 1$. We divide the real line into intervals whose endpoints are the critical numbers -1, 0, and 2 and arrange our work in a chart. A plus sign indicates that the given expression is positive, and a minus sign indicates that it is negative. The last column of the chart gives the conclusion based on the I/D Test. For instance, $f'(x) < 0$ for $0 < x < 2$, so f is decreasing on $(0, 2)$. (It would also be true to say that f is decreasing on the closed interval $[0, 2]$.)

Interval	$12x$	$x - 2$	$x + 1$	$f'(x)$	f
$x < -1$	$-$	$-$	$-$	$-$	decreasing on $(-\infty, -1)$
$-1 < x < 0$	$-$	$-$	$+$	$+$	increasing on $(-1, 0)$
$0 < x < 2$	$+$	$-$	$+$	$-$	decreasing on $(0, 2)$
$x > 2$	$+$	$+$	$+$	$+$	increasing on $(2, \infty)$

FIGURE 2

The graph of f shown in Figure 2 confirms the information in the chart. ■

Recall from Section 4.1 that if f has a local maximum or minimum at c, then c must be a critical number of f (by Fermat's Theorem), but not every critical number gives rise to a maximum or a minimum. We therefore need a test that will tell us whether or not f has a local maximum or minimum at a critical number.

You can see from Figure 2 that $f(0) = 5$ is a local maximum value of f because f increases on $(-1, 0)$ and decreases on $(0, 2)$. Or, in terms of derivatives, $f'(x) > 0$ for $-1 < x < 0$ and $f'(x) < 0$ for $0 < x < 2$. In other words, the sign of $f'(x)$ changes from positive to negative at 0. This observation is the basis of the following test.

THE FIRST DERIVATIVE TEST Suppose that c is a critical number of a continuous function f.

(a) If f' changes from positive to negative at c, then f has a local maximum at c.

(b) If f' changes from negative to positive at c, then f has a local minimum at c.

(c) If f' does not change sign at c (for example, if f' is positive on both sides of c or negative on both sides), then f has no local maximum or minimum at c.

The First Derivative Test is a consequence of the I/D Test. In part (a), for instance, since the sign of $f'(x)$ changes from positive to negative at c, f is increasing to the left of c and decreasing to the right of c. It follows that f has a local maximum at c.

It is easy to remember the First Derivative Test by visualizing diagrams such as those in Figure 3.

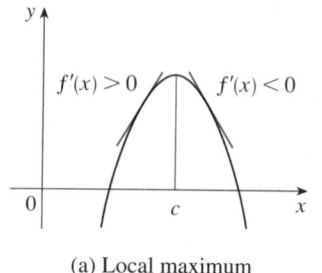

(a) Local maximum

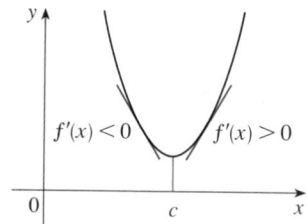

(b) Local minimum

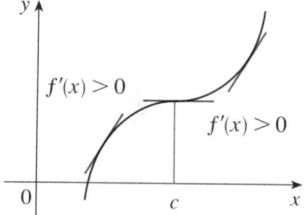

(c) No maximum or minimum

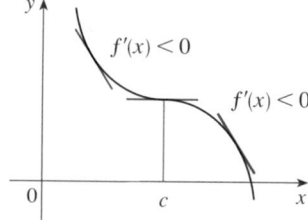

(d) No maximum or minimum

FIGURE 3

▼ **EXAMPLE 2** Find the local minimum and maximum values of the function f in Example 1.

SOLUTION From the chart in the solution to Example 1 we see that $f'(x)$ changes from negative to positive at -1, so $f(-1) = 0$ is a local minimum value by the First Derivative Test. Similarly, f' changes from negative to positive at 2, so $f(2) = -27$ is also a local minimum value. As previously noted, $f(0) = 5$ is a local maximum value because $f'(x)$ changes from positive to negative at 0. ∎

EXAMPLE 3 Find the local maximum and minimum values of the function

$$g(x) = x + 2 \sin x \qquad 0 \leqslant x \leqslant 2\pi$$

SOLUTION To find the critical numbers of g, we differentiate:

$$g'(x) = 1 + 2 \cos x$$

So $g'(x) = 0$ when $\cos x = -\frac{1}{2}$. The solutions of this equation are $2\pi/3$ and $4\pi/3$. Because g is differentiable everywhere, the only critical numbers are $2\pi/3$ and $4\pi/3$ and so we analyze g in the following table.

Interval	$g'(x) = 1 + 2\cos x$	g
$0 < x < 2\pi/3$	$+$	increasing on $(0, 2\pi/3)$
$2\pi/3 < x < 4\pi/3$	$-$	decreasing on $(2\pi/3, 4\pi/3)$
$4\pi/3 < x < 2\pi$	$+$	increasing on $(4\pi/3, 2\pi)$

■ The + signs in the table come from the fact that $g'(x) > 0$ when $\cos x > -\frac{1}{2}$. From the graph of $y = \cos x$, this is true in the indicated intervals.

Because $g'(x)$ changes from positive to negative at $2\pi/3$, the First Derivative Test tells us that there is a local maximum at $2\pi/3$ and the local maximum value is

$$g(2\pi/3) = \frac{2\pi}{3} + 2\sin\frac{2\pi}{3} = \frac{2\pi}{3} + 2\left(\frac{\sqrt{3}}{2}\right) = \frac{2\pi}{3} + \sqrt{3} \approx 3.83$$

Likewise, $g'(x)$ changes from negative to positive at $4\pi/3$ and so

$$g(4\pi/3) = \frac{4\pi}{3} + 2\sin\frac{4\pi}{3} = \frac{4\pi}{3} + 2\left(-\frac{\sqrt{3}}{2}\right) = \frac{4\pi}{3} - \sqrt{3} \approx 2.46$$

is a local minimum value. The graph of g in Figure 4 supports our conclusion.   ■

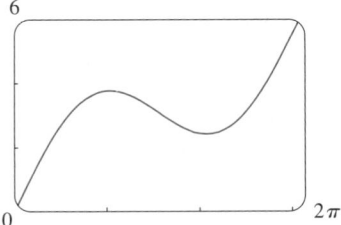

FIGURE 4
$y = x + 2\sin x$

WHAT DOES f'' SAY ABOUT f?

Figure 5 shows the graphs of two increasing functions on (a, b). Both graphs join point A to point B but they look different because they bend in different directions. How can we distinguish between these two types of behavior? In Figure 6 tangents to these curves have been drawn at several points. In (a) the curve lies above the tangents and f is called *concave upward* on (a, b). In (b) the curve lies below the tangents and g is called *concave downward* on (a, b).

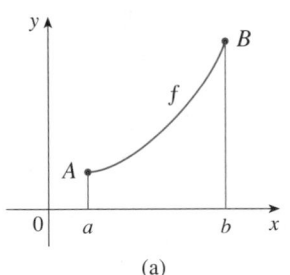
(a)

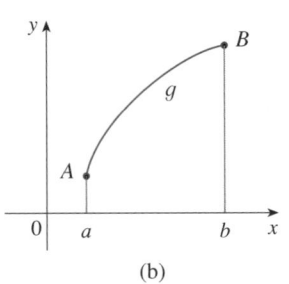
(b)

FIGURE 5

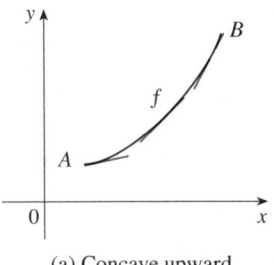
(a) Concave upward

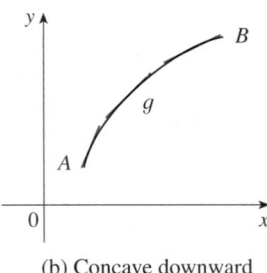
(b) Concave downward

FIGURE 6

DEFINITION If the graph of f lies above all of its tangents on an interval I, then it is called **concave upward** on I. If the graph of f lies below all of its tangents on I, it is called **concave downward** on I.

Figure 7 shows the graph of a function that is concave upward (abbreviated CU) on the intervals (b, c), (d, e), and (e, p) and concave downward (CD) on the intervals (a, b), (c, d), and (p, q).

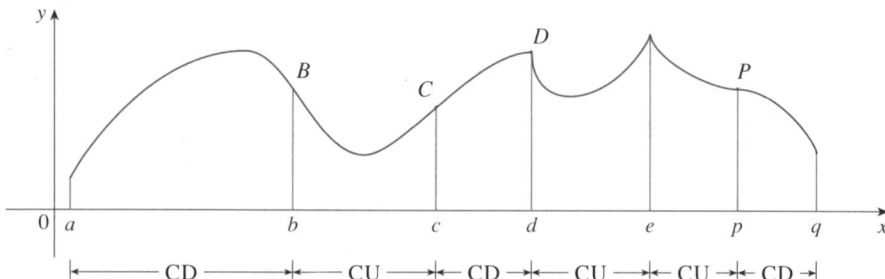

FIGURE 7

DEFINITION A point P on a curve $y = f(x)$ is called an **inflection point** if f is continuous there and the curve changes from concave upward to concave downward or from concave downward to concave upward at P.

For instance, in Figure 7, B, C, D, and P are the points of inflection. Notice that if a curve has a tangent at a point of inflection, then the curve crosses its tangent there.

■ www.stewartcalculus.com
See Additional Example A.

Let's see how the second derivative helps determine the intervals of concavity. Looking at Figure 6(a), you can see that, going from left to right, the slope of the tangent increases. This means that the derivative f' is an increasing function and therefore its derivative f'' is positive. Likewise, in Figure 6(b) the slope of the tangent decreases from left to right, so f' decreases and therefore f'' is negative. This reasoning can be reversed and suggests that the following theorem is true. A proof is given in Appendix D with the help of the Mean Value Theorem.

CONCAVITY TEST

(a) If $f''(x) > 0$ for all x in I, then the graph of f is concave upward on I.

(b) If $f''(x) < 0$ for all x in I, then the graph of f is concave downward on I.

In view of the Concavity Test, there is a point of inflection at any point where the second derivative changes sign.

☑ EXAMPLE 4 Sketch a possible graph of a function f that satisfies the following conditions:

(i) $f'(x) > 0$ on $(-\infty, 1)$, $f'(x) < 0$ on $(1, \infty)$

(ii) $f''(x) > 0$ on $(-\infty, -2)$ and $(2, \infty)$, $f''(x) < 0$ on $(-2, 2)$

(iii) $\lim\limits_{x \to -\infty} f(x) = -2$, $\lim\limits_{x \to \infty} f(x) = 0$

SOLUTION Condition (i) tells us that f is increasing on $(-\infty, 1)$ and decreasing on $(1, \infty)$. Condition (ii) says that f is concave upward on $(-\infty, -2)$ and $(2, \infty)$, and concave downward on $(-2, 2)$. From condition (iii) we know that the graph of f has two horizontal asymptotes: $y = -2$ and $y = 0$.

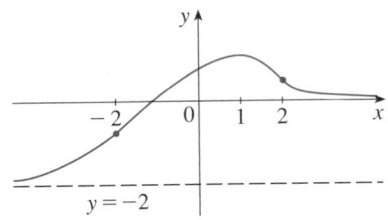

FIGURE 8

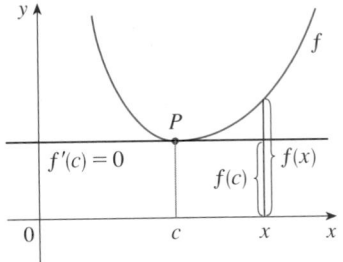

FIGURE 9
$f''(c) > 0$, f is concave upward

We first draw the horizontal asymptote $y = -2$ as a dashed line (see Figure 8). We then draw the graph of f approaching this asymptote at the far left, increasing to its maximum point at $x = 1$ and decreasing toward the x-axis at the far right. We also make sure that the graph has inflection points when $x = -2$ and 2. Notice that we made the curve bend upward for $x < -2$ and $x > 2$, and bend downward when x is between -2 and 2. ∎

Another application of the second derivative is the following test for maximum and minimum values. It is a consequence of the Concavity Test.

> **THE SECOND DERIVATIVE TEST** Suppose f'' is continuous near c.
> (a) If $f'(c) = 0$ and $f''(c) > 0$, then f has a local minimum at c.
> (b) If $f'(c) = 0$ and $f''(c) < 0$, then f has a local maximum at c.

For instance, part (a) is true because $f''(x) > 0$ near c and so f is concave upward near c. This means that the graph of f lies *above* its horizontal tangent at c and so f has a local minimum at c. (See Figure 9.)

☑ **EXAMPLE 5** Discuss the curve $y = x^4 - 4x^3$ with respect to concavity, points of inflection, and local maxima and minima. Use this information to sketch the curve.

SOLUTION If $f(x) = x^4 - 4x^3$, then

$$f'(x) = 4x^3 - 12x^2 = 4x^2(x - 3)$$
$$f''(x) = 12x^2 - 24x = 12x(x - 2)$$

To find the critical numbers we set $f'(x) = 0$ and obtain $x = 0$ and $x = 3$. To use the Second Derivative Test we evaluate f'' at these critical numbers:

$$f''(0) = 0 \qquad f''(3) = 36 > 0$$

Since $f'(3) = 0$ and $f''(3) > 0$, $f(3) = -27$ is a local minimum. Since $f''(0) = 0$, the Second Derivative Test gives no information about the critical number 0. But since $f'(x) < 0$ for $x < 0$ and also for $0 < x < 3$, the First Derivative Test tells us that f does not have a local maximum or minimum at 0. [In fact, the expression for $f'(x)$ shows that f decreases to the left of 3 and increases to the right of 3.]

Since $f''(x) = 0$ when $x = 0$ or 2, we divide the real line into intervals with these numbers as endpoints and complete the following chart.

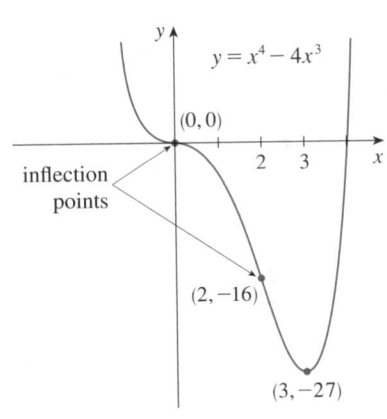

$y = x^4 - 4x^3$

(0, 0)

inflection points

(2, −16)

(3, −27)

FIGURE 10

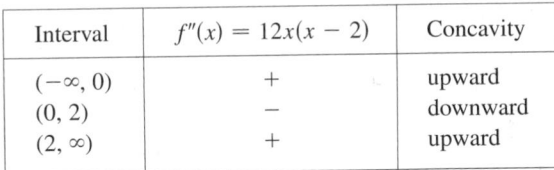

Interval	$f''(x) = 12x(x - 2)$	Concavity
$(-\infty, 0)$	$+$	upward
$(0, 2)$	$-$	downward
$(2, \infty)$	$+$	upward

The point $(0, 0)$ is an inflection point since the curve changes from concave upward to concave downward there. Also $(2, -16)$ is an inflection point since the curve changes from concave downward to concave upward there.

Using the local minimum, the intervals of concavity, and the inflection points, we sketch the curve in Figure 10. ∎

■ **www.stewartcalculus.com**
See Additional Examples B, C, D.

NOTE The Second Derivative Test is inconclusive when $f''(c) = 0$. In other words, at such a point there might be a maximum, there might be a minimum, or there might be neither (as in Example 5). This test also fails when $f''(c)$ does not exist. In such cases the First Derivative Test must be used. In fact, even when both tests apply, the First Derivative Test is often the easier one to use.

EXAMPLE 6 Sketch the graph of the function $f(x) = x^{2/3}(6 - x)^{1/3}$.

SOLUTION You can use the differentiation rules to check that the first two derivatives are

$$f'(x) = \frac{4 - x}{x^{1/3}(6 - x)^{2/3}} \qquad f''(x) = \frac{-8}{x^{4/3}(6 - x)^{5/3}}$$

Since $f'(x) = 0$ when $x = 4$ and $f'(x)$ does not exist when $x = 0$ or $x = 6$, the critical numbers are 0, 4, and 6.

■ Try reproducing the graph in Figure 11 with a graphing calculator or computer. Some machines produce the complete graph, some produce only the portion to the right of the y-axis, and some produce only the portion between $x = 0$ and $x = 6$. An equivalent expression that gives the correct graph is

$$y = (x^2)^{1/3} \cdot \frac{6 - x}{|6 - x|} |6 - x|^{1/3}$$

Interval	$4 - x$	$x^{1/3}$	$(6 - x)^{2/3}$	$f'(x)$	f
$x < 0$	$+$	$-$	$+$	$-$	decreasing on $(-\infty, 0)$
$0 < x < 4$	$+$	$+$	$+$	$+$	increasing on $(0, 4)$
$4 < x < 6$	$-$	$+$	$+$	$-$	decreasing on $(4, 6)$
$x > 6$	$-$	$+$	$+$	$-$	decreasing on $(6, \infty)$

To find the local extreme values we use the First Derivative Test. Since f' changes from negative to positive at 0, $f(0) = 0$ is a local minimum. Since f' changes from positive to negative at 4, $f(4) = 2^{5/3}$ is a local maximum. The sign of f' does not change at 6, so there is no minimum or maximum there. (The Second Derivative Test could be used at 4 but not at 0 or 6 since f'' does not exist at either of these numbers.)

Looking at the expression for $f''(x)$ and noting that $x^{4/3} \geq 0$ for all x, we have $f''(x) < 0$ for $x < 0$ and for $0 < x < 6$ and $f''(x) > 0$ for $x > 6$. So f is concave downward on $(-\infty, 0)$ and $(0, 6)$ and concave upward on $(6, \infty)$, and the only inflection point is $(6, 0)$. The graph is sketched in Figure 11. Note that the curve has vertical tangents at $(0, 0)$ and $(6, 0)$ because $|f'(x)| \to \infty$ as $x \to 0$ and as $x \to 6$. ■

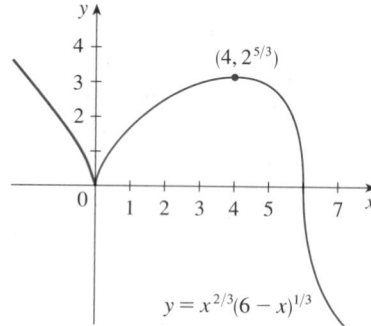

FIGURE 11

4.3 | EXERCISES

1–10 ■
(a) Find the intervals on which f is increasing or decreasing.
(b) Find the local maximum and minimum values of f.
(c) Find the intervals of concavity and the inflection points.

1. $f(x) = 2x^3 + 3x^2 - 36x$

2. $f(x) = 4x^3 + 3x^2 - 6x + 1$

3. $f(x) = x^4 - 2x^2 + 3$

4. $f(x) = \dfrac{x}{x^2 + 1}$

5. $f(x) = \sin x + \cos x, \quad 0 \leq x \leq 2\pi$

6. $f(x) = \cos^2 x - 2 \sin x, \quad 0 \leq x \leq 2\pi$

7. $f(x) = e^{2x} + e^{-x}$

8. $f(x) = x^2 \ln x$

9. $f(x) = x^2 - x - \ln x$

10. $f(x) = x^4 e^{-x}$

11–12 ■ Find the local maximum and minimum values of f using both the First and Second Derivative Tests. Which method do you prefer?

11. $f(x) = 1 + 3x^2 - 2x^3$

12. $f(x) = \dfrac{x^2}{x - 1}$

13. Suppose f'' is continuous on $(-\infty, \infty)$.
(a) If $f'(2) = 0$ and $f''(2) = -5$, what can you say about f?
(b) If $f'(6) = 0$ and $f''(6) = 0$, what can you say about f?

14. (a) Find the critical numbers of $f(x) = x^4(x - 1)^3$.
  (b) What does the Second Derivative Test tell you about the behavior of f at these critical numbers?
  (c) What does the First Derivative Test tell you?

15. In each part state the x-coordinates of the inflection points of f. Give reasons for your answers.
  (a) The curve is the graph of f.
  (b) The curve is the graph of f'.
  (c) The curve is the graph of f''.

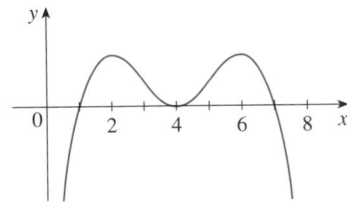

16. The graph of the first derivative f' of a function f is shown.
  (a) On what intervals is f increasing? Explain.
  (b) At what values of x does f have a local maximum or minimum? Explain.
  (c) On what intervals is f concave upward or concave downward? Explain.
  (d) What are the x-coordinates of the inflection points of f? Why?

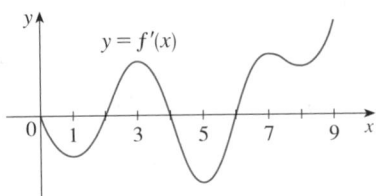

17–22 ▪ Sketch the graph of a function that satisfies all of the given conditions.

17. $f'(x)$ and $f''(x)$ are always negative.

18. Vertical asymptote $x = 0$, $f'(x) > 0$ if $x < -2$,
  $f'(x) < 0$ if $x > -2$ $(x \neq 0)$,
  $f''(x) < 0$ if $x < 0$, $f''(x) > 0$ if $x > 0$

19. $f'(0) = f'(2) = f'(4) = 0$,
  $f'(x) > 0$ if $x < 0$ or $2 < x < 4$,
  $f'(x) < 0$ if $0 < x < 2$ or $x > 4$,
  $f''(x) > 0$ if $1 < x < 3$, $f''(x) < 0$ if $x < 1$ or $x > 3$

20. $f'(1) = f'(-1) = 0$, $f'(x) < 0$ if $|x| < 1$,
  $f'(x) > 0$ if $1 < |x| < 2$, $f'(x) = -1$ if $|x| > 2$,
  $f''(x) < 0$ if $-2 < x < 0$,   inflection point $(0, 1)$

21. $f'(x) > 0$ if $|x| < 2$, $f'(x) < 0$ if $|x| > 2$,
  $f'(-2) = 0$, $\lim_{x \to 2} |f'(x)| = \infty$, $f''(x) > 0$ if $x \neq 2$

22. $f'(x) > 0$ if $|x| < 2$, $f'(x) < 0$ if $|x| > 2$,
  $f'(2) = 0$, $\lim_{x \to \infty} f(x) = 1$, $f(-x) = -f(x)$,
  $f''(x) < 0$ if $0 < x < 3$, $f''(x) > 0$ if $x > 3$

23–24 ▪ The graph of the derivative f' of a continuous function f is shown.
  (a) On what intervals is f increasing or decreasing?
  (b) At what values of x does f have a local maximum or minimum?
  (c) On what intervals is f concave upward or downward?
  (d) State the x-coordinate(s) of the point(s) of inflection.
  (e) Assuming that $f(0) = 0$, sketch a graph of f.

23.

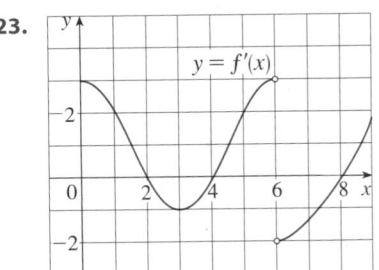

24.

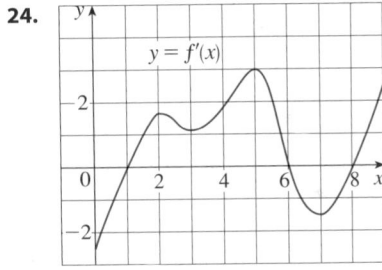

25–36 ▪
  (a) Find the intervals of increase or decrease.
  (b) Find the local maximum and minimum values.
  (c) Find the intervals of concavity and the inflection points.
  (d) Use the information from parts (a)–(c) to sketch the graph. Check your work with a graphing device if you have one.

25. $f(x) = x^3 - 12x + 2$ **26.** $f(x) = 36x + 3x^2 - 2x^3$

27. $f(x) = 2 + 2x^2 - x^4$ **28.** $g(x) = 200 + 8x^3 + x^4$

29. $h(x) = (x + 1)^5 - 5x - 2$ **30.** $h(x) = 5x^3 - 3x^5$

31. $F(x) = x\sqrt{6 - x}$ **32.** $G(x) = 5x^{2/3} - 2x^{5/3}$

33. $C(x) = x^{1/3}(x + 4)$ **34.** $f(x) = \ln(x^4 + 27)$

35. $f(\theta) = 2\cos\theta + \cos^2\theta$, $0 \leq \theta \leq 2\pi$

36. $S(x) = x - \sin x$, $0 \leq x \leq 4\pi$

37–44 ■

(a) Find the vertical and horizontal asymptotes.
(b) Find the intervals of increase or decrease.
(c) Find the local maximum and minimum values.
(d) Find the intervals of concavity and the inflection points.
(e) Use the information from parts (a)–(d) to sketch the graph of f.

37. $f(x) = 1 + \dfrac{1}{x} - \dfrac{1}{x^2}$

38. $f(x) = \dfrac{x^2 - 4}{x^2 + 4}$

39. $f(x) = \sqrt{x^2 + 1} - x$

40. $f(x) = \dfrac{e^x}{1 - e^x}$

41. $f(x) = e^{-x^2}$

42. $f(x) = x - \frac{1}{6}x^2 - \frac{2}{3}\ln x$

43. $f(x) = \ln(1 - \ln x)$

44. $f(x) = e^{\arctan x}$

---

45. Suppose the derivative of a function f is
$f'(x) = (x + 1)^2(x - 3)^5(x - 6)^4$. On what interval is f increasing?

46. Use the methods of this section to sketch the curve
$y = x^3 - 3a^2x + 2a^3$, where a is a positive constant. What do the members of this family of curves have in common? How do they differ from each other?

47–48 ■

(a) Use a graph of f to estimate the maximum and minimum values. Then find the exact values.
(b) Estimate the value of x at which f increases most rapidly. Then find the exact value.

47. $f(x) = \dfrac{x + 1}{\sqrt{x^2 + 1}}$

48. $f(x) = x^2 e^{-x}$

---

49. A *drug response curve* describes the level of medication in the bloodstream after a drug is administered. A surge function $S(t) = At^p e^{-kt}$ is often used to model the response curve, reflecting an initial surge in the drug level and then a more gradual decline. If, for a particular drug, $A = 0.01$, $p = 4$, $k = 0.07$, and t is measured in minutes, estimate the times corresponding to the inflection points and explain their significance. If you have a graphing device, use it to graph the drug response curve.

50. The family of bell-shaped curves

$$y = \dfrac{1}{\sigma\sqrt{2\pi}}\, e^{-(x-\mu)^2/(2\sigma^2)}$$

occurs in probability and statistics, where it is called the *normal density function*. The constant μ is called the *mean* and the positive constant σ is called the *standard deviation*. For simplicity, let's scale the function so as to remove the factor $1/(\sigma\sqrt{2\pi})$ and let's analyze the special case where $\mu = 0$. So we study the function

$$f(x) = e^{-x^2/(2\sigma^2)}$$

(a) Find the asymptote, maximum value, and inflection points of f.
(b) What role does σ play in the shape of the curve?
(c) Illustrate by graphing four members of this family on the same screen.

51. Find a cubic function $f(x) = ax^3 + bx^2 + cx + d$ that has a local maximum value of 3 at $x = -2$ and a local minimum value of 0 at $x = 1$.

52. For what values of the numbers a and b does the function

$$f(x) = axe^{bx^2}$$

have the maximum value $f(2) = 1$?

53. (a) If the function $f(x) = x^3 + ax^2 + bx$ has the local minimum value $-\frac{2}{9}\sqrt{3}$ at $x = 1/\sqrt{3}$, what are the values of a and b?
(b) Which of the tangent lines to the curve in part (a) has the smallest slope?

54. Show that the inflection points of the curve $y = x \sin x$ lie on the curve $y^2(x^2 + 4) = 4x^2$.

55. Show that the curve $y = (1 + x)/(1 + x^2)$ has three points of inflection and they all lie on one straight line.

56. Show that the curves $y = e^{-x}$ and $y = -e^{-x}$ touch the curve $y = e^{-x}\sin x$ at its inflection points.

57. Show that $\tan x > x$ for $0 < x < \pi/2$. [*Hint:* Show that $f(x) = \tan x - x$ is increasing on $(0, \pi/2)$.]

58. (a) Show that $e^x \geqslant 1 + x$ for $x \geqslant 0$.
(b) Deduce that $e^x \geqslant 1 + x + \frac{1}{2}x^2$ for $x \geqslant 0$.
(c) Use mathematical induction to prove that for $x \geqslant 0$ and any positive integer n,

$$e^x \geqslant 1 + x + \dfrac{x^2}{2!} + \cdots + \dfrac{x^n}{n!}$$

59. Show that a cubic function (a third-degree polynomial) always has exactly one point of inflection. If its graph has three x-intercepts x_1, x_2, and x_3, show that the x-coordinate of the inflection point is $(x_1 + x_2 + x_3)/3$.

60. For what values of c does the polynomial $P(x) = x^4 + cx^3 + x^2$ have two inflection points? One inflection point? None? Illustrate by graphing P for several values of c. How does the graph change as c decreases?

61. Prove that if $(c, f(c))$ is a point of inflection of the graph of f and f'' exists in an open interval that contains c, then $f''(c) = 0$. [*Hint:* Apply the First Derivative Test and Fermat's Theorem to the function $g = f'$.]

62. Show that if $f(x) = x^4$, then $f''(0) = 0$, but $(0, 0)$ is not an inflection point of the graph of f.

63. Show that the function $g(x) = x|x|$ has an inflection point at $(0, 0)$ but $g''(0)$ does not exist.

64. Suppose that f''' is continuous and $f'(c) = f''(c) = 0$, but $f'''(c) > 0$. Does f have a local maximum or minimum at c? Does f have a point of inflection at c?

65. Suppose f is differentiable on an interval I and $f'(x) > 0$ for all numbers x in I except for a single number c. Prove that f is increasing on the entire interval I.

66. For what values of c is the function

$$f(x) = cx + \frac{1}{x^2 + 3}$$

increasing on $(-\infty, \infty)$?

4.4 | CURVE SKETCHING

So far we have been concerned with some particular aspects of curve sketching: domain, range, symmetry, limits, continuity, and asymptotes in Chapter 1; derivatives and tangents in Chapters 2 and 3; l'Hospital's Rule in Chapter 3; and extreme values, intervals of increase and decrease, concavity, and points of inflection in this chapter. It's now time to put all of this information together to sketch graphs that reveal the important features of functions.

You may ask: Why don't we just use a graphing calculator or computer to graph a curve? Why do we need to use calculus?

It's true that modern technology is capable of producing very accurate graphs. But even the best graphing devices have to be used intelligently. The use of calculus enables us to discover the most interesting aspects of curves and to detect behavior that we might otherwise overlook. We will see in Example 4 how calculus helps us to avoid the pitfalls of technology.

GUIDELINES FOR SKETCHING A CURVE

The following checklist is intended as a guide to sketching a curve $y = f(x)$ by hand. Not every item is relevant to every function. (For instance, a given curve might not have an asymptote or possess symmetry.) But the guidelines provide all the information you need to make a sketch that displays the most important aspects of the function.

A. Domain It's often useful to start by determining the domain D of f, that is, the set of values of x for which $f(x)$ is defined.

B. Intercepts The y-intercept is $f(0)$ and this tells us where the curve intersects the y-axis. To find the x-intercepts, we set $y = 0$ and solve for x. (You can omit this step if the equation is difficult to solve.)

C. Symmetry

(i) If $f(-x) = f(x)$ for all x in D, that is, the equation of the curve is unchanged when x is replaced by $-x$, then f is an **even function** and the curve is symmetric about the y-axis. This means that our work is cut in half. If we know what the curve looks like for $x \geq 0$, then we need only reflect about the y-axis to obtain the complete curve [see Figure 1(a)]. Here are some examples: $y = x^2$, $y = x^4$, $y = |x|$, and $y = \cos x$.

(ii) If $f(-x) = -f(x)$ for all x in D, then f is an **odd function** and the curve is symmetric about the origin. Again we can obtain the complete curve if we know what it looks like for $x \geq 0$. [Rotate $180°$ about the origin; see Figure 1(b).] Some simple examples of odd functions are $y = x$, $y = x^3$, $y = x^5$, and $y = \sin x$.

(iii) If $f(x + p) = f(x)$ for all x in D, where p is a positive constant, then f is called a **periodic function** and the smallest such number p is called the **period**.

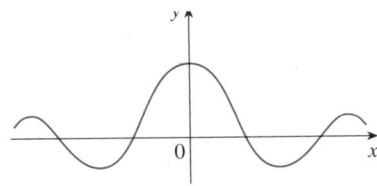

(a) Even function: reflectional symmetry

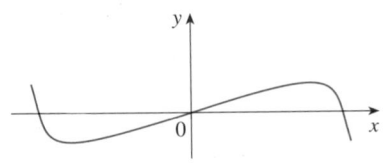

(b) Odd function: rotational symmetry

FIGURE 1

For instance, $y = \sin x$ has period 2π and $y = \tan x$ has period π. If we know what the graph looks like in an interval of length p, then we can use translation to sketch the entire graph (see Figure 2).

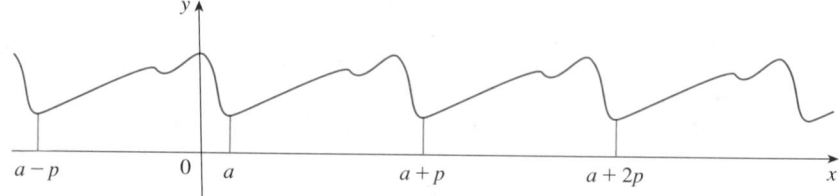

FIGURE 2

Periodic function: translational symmetry

D. Asymptotes

(i) *Horizontal Asymptotes.* Recall from Section 1.6 that if $\lim_{x \to \infty} f(x) = L$ or $\lim_{x \to -\infty} f(x) = L$, then the line $y = L$ is a horizontal asymptote of the curve $y = f(x)$. If it turns out that $\lim_{x \to \infty} f(x) = \infty$ (or $-\infty$), then we do not have an asymptote to the right, but that is still useful information for sketching the curve.

(ii) *Vertical Asymptotes.* Recall from Section 1.6 that the line $x = a$ is a vertical asymptote if at least one of the following statements is true:

$$\boxed{1} \qquad \lim_{x \to a^+} f(x) = \infty \qquad \lim_{x \to a^-} f(x) = \infty$$

$$\lim_{x \to a^+} f(x) = -\infty \qquad \lim_{x \to a^-} f(x) = -\infty$$

(For rational functions you can locate the vertical asymptotes by equating the denominator to 0 after canceling any common factors. But for other functions this method does not apply.) Furthermore, in sketching the curve it is very useful to know exactly which of the statements in $\boxed{1}$ is true. If $f(a)$ is not defined but a is an endpoint of the domain of f, then you should compute $\lim_{x \to a^-} f(x)$ or $\lim_{x \to a^+} f(x)$, whether or not this limit is infinite.

E. Intervals of Increase or Decrease
Use the I/D Test. Compute $f'(x)$ and find the intervals on which $f'(x)$ is positive (f is increasing) and the intervals on which $f'(x)$ is negative (f is decreasing).

F. Local Maximum and Minimum Values
Find the critical numbers of f [the numbers c where $f'(c) = 0$ or $f'(c)$ does not exist]. Then use the First Derivative Test. If f' changes from positive to negative at a critical number c, then $f(c)$ is a local maximum. If f' changes from negative to positive at c, then $f(c)$ is a local minimum. Although it is usually preferable to use the First Derivative Test, you can use the Second Derivative Test if $f'(c) = 0$ and $f''(c) \neq 0$. Then $f''(c) > 0$ implies that $f(c)$ is a local minimum, whereas $f''(c) < 0$ implies that $f(c)$ is a local maximum.

G. Concavity and Points of Inflection
Compute $f''(x)$ and use the Concavity Test. The curve is concave upward where $f''(x) > 0$ and concave downward where $f''(x) < 0$. Inflection points occur where the direction of concavity changes.

H. Sketch the Curve
Using the information in items A–G, draw the graph. Sketch the asymptotes as dashed lines. Plot the intercepts, maximum and minimum points, and inflection points. Then make the curve pass through these points, rising and falling according to E, with concavity according to G, and approaching the asymptotes. If additional accuracy is desired near any point, you can compute the value of the derivative there. The tangent indicates the direction in which the curve proceeds.

TEC In Module 4.4 you can practice using information about f' and f'' to determine the shape of the graph of f.

Ⅴ EXAMPLE 1 Use the guidelines to sketch the curve $y = \dfrac{2x^2}{x^2 - 1}$.

A. The domain is

$$\{x \mid x^2 - 1 \neq 0\} = \{x \mid x \neq \pm 1\} = (-\infty, -1) \cup (-1, 1) \cup (1, \infty)$$

B. The x- and y-intercepts are both 0.

C. Since $f(-x) = f(x)$, the function f is even. The curve is symmetric about the y-axis.

D.
$$\lim_{x \to \pm\infty} \frac{2x^2}{x^2 - 1} = \lim_{x \to \pm\infty} \frac{2}{1 - 1/x^2} = 2$$

Therefore the line $y = 2$ is a horizontal asymptote.

Since the denominator is 0 when $x = \pm 1$, we compute the following limits:

$$\lim_{x \to 1^+} \frac{2x^2}{x^2 - 1} = \infty \qquad \lim_{x \to 1^-} \frac{2x^2}{x^2 - 1} = -\infty$$

$$\lim_{x \to -1^+} \frac{2x^2}{x^2 - 1} = -\infty \qquad \lim_{x \to -1^-} \frac{2x^2}{x^2 - 1} = \infty$$

Therefore the lines $x = 1$ and $x = -1$ are vertical asymptotes. This information about limits and asymptotes enables us to draw the preliminary sketch in Figure 3, showing the parts of the curve near the asymptotes.

E.
$$f'(x) = \frac{4x(x^2 - 1) - 2x^2 \cdot 2x}{(x^2 - 1)^2} = \frac{-4x}{(x^2 - 1)^2}$$

Since $f'(x) > 0$ when $x < 0$ $(x \neq -1)$ and $f'(x) < 0$ when $x > 0$ $(x \neq 1)$, f is increasing on $(-\infty, -1)$ and $(-1, 0)$ and decreasing on $(0, 1)$ and $(1, \infty)$.

F. The only critical number is $x = 0$. Since f' changes from positive to negative at 0, $f(0) = 0$ is a local maximum by the First Derivative Test.

G.
$$f''(x) = \frac{-4(x^2 - 1)^2 + 4x \cdot 2(x^2 - 1)2x}{(x^2 - 1)^4} = \frac{12x^2 + 4}{(x^2 - 1)^3}$$

Since $12x^2 + 4 > 0$ for all x, we have

$$f''(x) > 0 \iff x^2 - 1 > 0 \iff |x| > 1$$

and $f''(x) < 0 \iff |x| < 1$. Thus the curve is concave upward on the intervals $(-\infty, -1)$ and $(1, \infty)$ and concave downward on $(-1, 1)$. It has no point of inflection since 1 and -1 are not in the domain of f.

H. Using the information in E–G, we finish the sketch in Figure 4. ∎

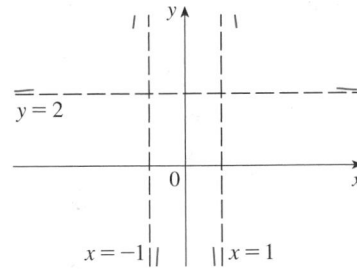

FIGURE 3

Preliminary sketch

■ We have shown the curve approaching its horizontal asymptote from above in Figure 3. This is confirmed by the intervals of increase and decrease.

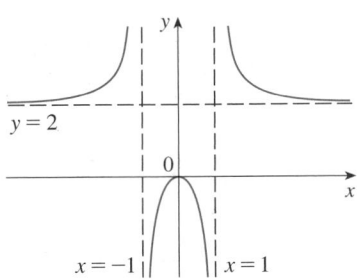

FIGURE 4

Finished sketch of $y = \dfrac{2x^2}{x^2 - 1}$

Ⅴ EXAMPLE 2 Sketch the graph of $f(x) = xe^x$.

A. The domain is $\mathbb{R}$.

B. The x- and y-intercepts are both 0.

C. Symmetry: None

D. Because both x and e^x become large as $x \to \infty$, we have $\lim_{x \to \infty} xe^x = \infty$. As $x \to -\infty$, however, $e^x \to 0$ and so we have an indeterminate product that requires the use of l'Hospital's Rule:

$$\lim_{x \to -\infty} xe^x = \lim_{x \to -\infty} \frac{x}{e^{-x}} = \lim_{x \to -\infty} \frac{1}{-e^{-x}} = \lim_{x \to -\infty} (-e^x) = 0$$

Thus the x-axis is a horizontal asymptote.

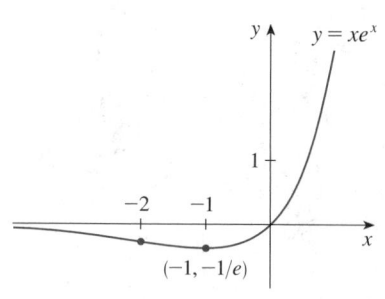

FIGURE 5

E.
$$f'(x) = xe^x + e^x = (x + 1)e^x$$

Since e^x is always positive, we see that $f'(x) > 0$ when $x + 1 > 0$, and $f'(x) < 0$ when $x + 1 < 0$. So f is increasing on $(-1, \infty)$ and decreasing on $(-\infty, -1)$.

F. Because $f'(-1) = 0$ and f' changes from negative to positive at $x = -1$, $f(-1) = -e^{-1}$ is a local (and absolute) minimum.

G.
$$f''(x) = (x + 1)e^x + e^x = (x + 2)e^x$$

Since $f''(x) > 0$ if $x > -2$ and $f''(x) < 0$ if $x < -2$, f is concave upward on $(-2, \infty)$ and concave downward on $(-\infty, -2)$. The inflection point is $(-2, -2e^{-2})$.

H. We use this information to sketch the curve in Figure 5.

EXAMPLE 3 Sketch the graph of $f(x) = \dfrac{\cos x}{2 + \sin x}$.

A. The domain is $\mathbb{R}$.

B. The y-intercept is $f(0) = \frac{1}{2}$. The x-intercepts occur when $\cos x = 0$, that is, $x = (2n + 1)\pi/2$, where n is an integer.

C. f is neither even nor odd, but $f(x + 2\pi) = f(x)$ for all x and so f is periodic and has period 2π. Thus in what follows we need to consider only $0 \le x \le 2\pi$ and then extend the curve by translation in part H.

D. Asymptotes: None

■ www.stewartcalculus.com
See Additional Examples A, B.

E.
$$f'(x) = \frac{(2 + \sin x)(-\sin x) - \cos x\,(\cos x)}{(2 + \sin x)^2} = -\frac{2 \sin x + 1}{(2 + \sin x)^2}$$

Thus $f'(x) > 0$ when $2 \sin x + 1 < 0 \iff \sin x < -\frac{1}{2} \iff 7\pi/6 < x < 11\pi/6$. So f is increasing on $(7\pi/6, 11\pi/6)$ and decreasing on $(0, 7\pi/6)$ and $(11\pi/6, 2\pi)$.

F. From part E and the First Derivative Test, we see that the local minimum value is $f(7\pi/6) = -1/\sqrt{3}$ and the local maximum value is $f(11\pi/6) = 1/\sqrt{3}$.

G. If we use the Quotient Rule again and simplify, we get
$$f''(x) = -\frac{2 \cos x\,(1 - \sin x)}{(2 + \sin x)^3}$$

Because $(2 + \sin x)^3 > 0$ and $1 - \sin x \ge 0$ for all x, we know that $f''(x) > 0$ when $\cos x < 0$, that is, $\pi/2 < x < 3\pi/2$. So f is concave upward on $(\pi/2, 3\pi/2)$ and concave downward on $(0, \pi/2)$ and $(3\pi/2, 2\pi)$. The inflection points are $(\pi/2, 0)$ and $(3\pi/2, 0)$.

H. The graph of the function restricted to $0 \le x \le 2\pi$ is shown in Figure 6. Then we extend it, using periodicity, to the complete graph in Figure 7.

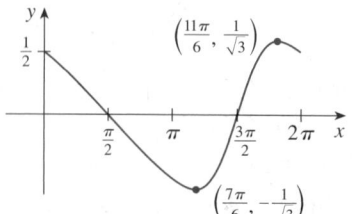

FIGURE 6

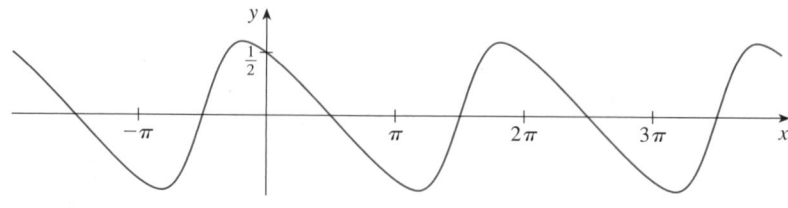

FIGURE 7

GRAPHING WITH TECHNOLOGY

When we use technology to graph a curve, our strategy is different from that in Examples 1–3. Here we *start* with a graph produced by a graphing calculator or computer and then we refine it. We use calculus to make sure that we reveal all the important features of the curve. And with the use of graphing devices we can tackle curves that would be far too complicated to consider without technology.

EXAMPLE 4 Graph the polynomial $f(x) = 2x^6 + 3x^5 + 3x^3 - 2x^2$. Use the graphs of f' and f'' to estimate all maximum and minimum points and intervals of concavity.

SOLUTION If we specify a domain but not a range, many graphing devices will deduce a suitable range from the values computed. Figure 8 shows the plot from one such device if we specify that $-5 \leq x \leq 5$. Although this viewing rectangle is useful for showing that the asymptotic behavior (or end behavior) is the same as for $y = 2x^6$, it is obviously hiding some finer detail. So we change to the viewing rectangle $[-3, 2]$ by $[-50, 100]$ shown in Figure 9.

From this graph it appears that there is an absolute minimum value of about -15.33 when $x \approx -1.62$ (by using the cursor) and f is decreasing on $(-\infty, -1.62)$ and increasing on $(-1.62, \infty)$. Also there appears to be a horizontal tangent at the origin and inflection points when $x = 0$ and when x is somewhere between -2 and -1.

Now let's try to confirm these impressions using calculus. We differentiate and get

$$f'(x) = 12x^5 + 15x^4 + 9x^2 - 4x$$

$$f''(x) = 60x^4 + 60x^3 + 18x - 4$$

When we graph f' in Figure 10 we see that $f'(x)$ changes from negative to positive when $x \approx -1.62$; this confirms (by the First Derivative Test) the minimum value that we found earlier. But, perhaps to our surprise, we also notice that $f'(x)$ changes from positive to negative when $x = 0$ and from negative to positive when $x \approx 0.35$. This means that f has a local maximum at 0 and a local minimum when $x \approx 0.35$, but these were hidden in Figure 9. Indeed, if we now zoom in toward the origin in Figure 11, we see what we missed before: a local maximum value of 0 when $x = 0$ and a local minimum value of about -0.1 when $x \approx 0.35$.

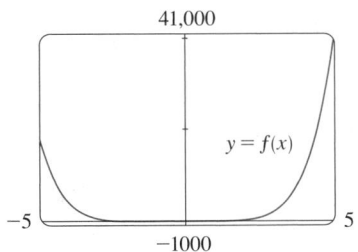

FIGURE 8

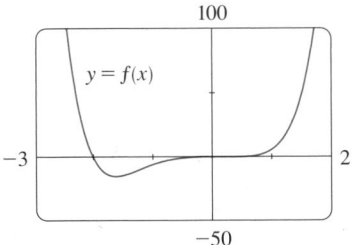

FIGURE 9

■ www.stewartcalculus.com
See Additional Examples C–F.

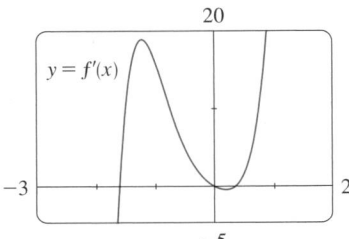

FIGURE 10

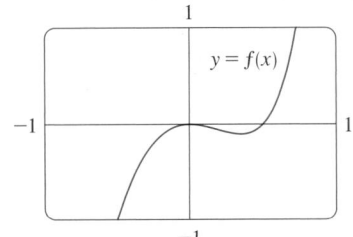

FIGURE 11

What about concavity and inflection points? From Figures 9 and 11 there appear to be inflection points when x is a little to the left of -1 and when x is a little to the

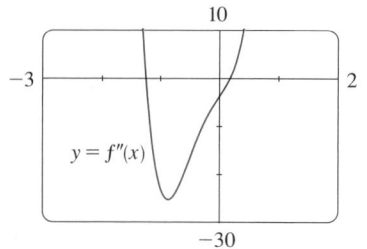

FIGURE 12

right of 0. But it's difficult to determine inflection points from the graph of f, so we graph the second derivative f'' in Figure 12. We see that f'' changes from positive to negative when $x \approx -1.23$ and from negative to positive when $x \approx 0.19$. So, correct to two decimal places, f is concave upward on $(-\infty, -1.23)$ and $(0.19, \infty)$ and concave downward on $(-1.23, 0.19)$. The inflection points are $(-1.23, -10.18)$ and $(0.19, -0.05)$.

We have discovered that no single graph reveals all the important features of this polynomial. But Figures 9 and 11, when taken together, do provide an accurate picture.   ■

4.4 | EXERCISES

1–44 ■ Use the guidelines of this section to sketch the curve.

1. $y = x^3 - 12x^2 + 36x$

2. $y = 2 + 3x^2 - x^3$

3. $y = x^4 - 4x$

4. $y = x^4 - 8x^2 + 8$

5. $y = x(x - 4)^3$

6. $y = x^5 - 5x$

7. $y = \frac{1}{5}x^5 - \frac{8}{3}x^3 + 16x$

8. $y = (4 - x^2)^5$

9. $y = \dfrac{x}{x - 1}$

10. $y = 1 + \dfrac{1}{x} + \dfrac{1}{x^2}$

11. $y = \dfrac{1}{x^2 - 9}$

12. $y = \dfrac{x}{x^2 - 9}$

13. $y = \dfrac{x}{x^2 + 9}$

14. $y = \dfrac{x^2}{x^2 + 9}$

15. $y = \dfrac{x - 1}{x^2}$

16. $y = \dfrac{x}{x^3 - 1}$

17. $y = x\sqrt{5 - x}$

18. $y = 2\sqrt{x} - x$

19. $y = \dfrac{x}{\sqrt{x^2 + 1}}$

20. $y = \sqrt{x^2 + x} - x$

21. $y = \dfrac{\sqrt{1 - x^2}}{x}$

22. $y = x\sqrt{2 - x^2}$

23. $y = x - 3x^{1/3}$

24. $y = x^{5/3} - 5x^{2/3}$

25. $y = \sqrt[3]{x^2 - 1}$

26. $y = \sqrt[3]{x^3 + 1}$

27. $y = \sin^3 x$

28. $y = x + \cos x$

29. $y = x \tan x, \quad -\pi/2 < x < \pi/2$

30. $y = 2x - \tan x, \quad -\pi/2 < x < \pi/2$

31. $y = \frac{1}{2}x - \sin x, \quad 0 < x < 3\pi$

32. $y = \sec x + \tan x, \quad 0 < x < \pi/2$

33. $y = \dfrac{\sin x}{1 + \cos x}$

34. $y = \dfrac{\sin x}{2 + \cos x}$

35. $y = 1/(1 + e^{-x})$

36. $y = e^{2x} - e^x$

37. $y = x \ln x$

38. $y = e^x/x^2$

39. $y = xe^{-x}$

40. $y = \ln(x^2 - 3x + 2)$

41. $y = \ln(\sin x)$

42. $y = e^x - 3e^{-x} - 4x$

43. $y = xe^{-1/x}$

44. $y = \tan^{-1}\left(\dfrac{x - 1}{x + 1}\right)$

45. In the theory of relativity, the mass of a particle is

$$m = \frac{m_0}{\sqrt{1 - v^2/c^2}}$$

where m_0 is the rest mass of the particle, m is the mass when the particle moves with speed v relative to the observer, and c is the speed of light. Sketch the graph of m as a function of v.

46. In the theory of relativity, the energy of a particle is

$$E = \sqrt{m_0^2 c^4 + h^2 c^2/\lambda^2}$$

where m_0 is the rest mass of the particle, λ is its wave length, and h is Planck's constant. Sketch the graph of E as a function of λ. What does the graph say about the energy?

47. The figure shows a beam of length L embedded in concrete walls. If a constant load W is distributed evenly along its length, the beam takes the shape of the deflection curve

$$y = -\frac{W}{24EI}x^4 + \frac{WL}{12EI}x^3 - \frac{WL^2}{24EI}x^2$$

where E and I are positive constants. (E is Young's modulus of elasticity and I is the moment of inertia of a cross-section of the beam.) Sketch the graph of the deflection curve.

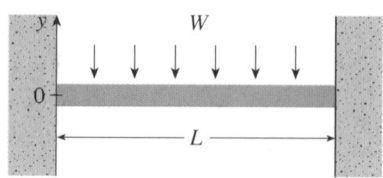

48. Coulomb's Law states that the force of attraction between two charged particles is directly proportional to the product of the charges and inversely proportional to the square of the distance between them. The figure shows particles with charge 1 located at positions 0 and 2 on a coordinate line and a particle with charge -1 at a position x between them. It follows from Coulomb's Law that the net force acting on the middle particle is

$$F(x) = -\frac{k}{x^2} + \frac{k}{(x-2)^2} \qquad 0 < x < 2$$

where k is a positive constant. Sketch the graph of the net force function. What does the graph say about the force?

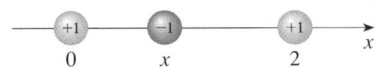

49–52 ▪ The line $y = mx + b$ is called a **slant asymptote** if $f(x) - (mx + b) \to 0$ as $x \to \infty$ or $x \to -\infty$ because the vertical distance between the curve $y = f(x)$ and the line $y = mx + b$ approaches 0 as x becomes large. Find an equation of the slant asymptote of the function and use it to help sketch the graph. [For rational functions, a slant asymptote occurs when the degree of the numerator is one more than the degree of the denominator. To find it, use long division to write $f(x) = mx + b + R(x)/Q(x)$.]

49. $y = \dfrac{x^2}{x-1}$

50. $y = \dfrac{1 + 5x - 2x^2}{x-2}$

51. $y = \dfrac{x^3 + 4}{x^2}$

52. $y = 1 - x + e^{1+x/3}$

53. Show that the curve $y = x - \tan^{-1}x$ has two slant asymptotes: $y = x + \pi/2$ and $y = x - \pi/2$. Use this fact to help sketch the curve.

54. Show that the curve $y = \sqrt{x^2 + 4x}$ has two slant asymptotes: $y = x + 2$ and $y = -x - 2$. Use this fact to help sketch the curve.

55–58 ▪ Produce graphs of f that reveal all the important aspects of the curve. In particular, you should use graphs of f' and f'' to estimate the intervals of increase and decrease, extreme values, intervals of concavity, and inflection points.

55. $f(x) = 4x^4 - 32x^3 + 89x^2 - 95x + 29$

56. $f(x) = x^6 - 15x^5 + 75x^4 - 125x^3 - x$

57. $f(x) = 6\sin x + \cot x, \quad -\pi \leqslant x \leqslant \pi$

58. $f(x) = e^x - 0.186x^4$

59–60 ▪ Produce graphs of f that reveal all the important aspects of the curve. Estimate the intervals of increase and decrease and intervals of concavity, and use calculus to find these intervals exactly.

59. $f(x) = 1 + \dfrac{1}{x} + \dfrac{8}{x^2} + \dfrac{1}{x^3}$ **60.** $f(x) = \dfrac{1}{x^8} - \dfrac{2 \times 10^8}{x^4}$

61–65 ▪ Describe how the graph of f varies as c varies. Graph several members of the family to illustrate the trends that you discover. In particular, you should investigate how maximum and minimum points and inflection points move when c changes. You should also identify any transitional values of c at which the basic shape of the curve changes.

61. $f(x) = \sqrt{x^4 + cx^2}$ **62.** $f(x) = x^3 + cx$

63. $f(x) = e^{-c/x^2}$ **64.** $f(x) = \ln(x^2 + c)$

65. $f(x) = cx + \sin x$

66. Investigate the family of curves given by the equation $f(x) = x^4 + cx^2 + x$. Start by determining the transitional value of c at which the number of inflection points changes. Then graph several members of the family to see what shapes are possible. There is another transitional value of c at which the number of critical numbers changes. Try to discover it graphically. Then prove what you have discovered.

4.5 | OPTIMIZATION PROBLEMS

The methods we have learned in this chapter for finding extreme values have practical applications in many areas of life. A businessperson wants to minimize costs and maximize profits. A traveler wants to minimize transportation time. Fermat's Principle in optics states that light follows the path that takes the least time. In this section and the next we solve such problems as maximizing areas, volumes, and profits and minimizing distances, times, and costs.

In solving such practical problems the greatest challenge is often to convert the word problem into a mathematical optimization problem by setting up the function that is to be maximized or minimized. The following steps may be useful.

STEPS IN SOLVING OPTIMIZATION PROBLEMS

1. **Understand the Problem**   The first step is to read the problem carefully until it is clearly understood. Ask yourself: What is the unknown? What are the given quantities? What are the given conditions?

2. **Draw a Diagram**   In most problems it is useful to draw a diagram and identify the given and required quantities on the diagram.

3. **Introduce Notation**   Assign a symbol to the quantity that is to be maximized or minimized (let's call it Q for now). Also select symbols $(a, b, c, \ldots, x, y)$ for other unknown quantities and label the diagram with these symbols. It may help to use initials as suggestive symbols—for example, A for area, h for height, t for time.

4. Express Q in terms of some of the other symbols from Step 3.

5. If Q has been expressed as a function of more than one variable in Step 4, use the given information to find relationships (in the form of equations) among these variables. Then use these equations to eliminate all but one of the variables in the expression for Q. Thus Q will be expressed as a function of *one* variable x, say, $Q = f(x)$. Write the domain of this function.

6. Use the methods of Sections 4.1 and 4.3 to find the *absolute* maximum or minimum value of f. In particular, if the domain of f is a closed interval, then the Closed Interval Method in Section 4.1 can be used.

EXAMPLE 1   A farmer has 2400 ft of fencing and wants to fence off a rectangular field that borders a straight river. He needs no fence along the river. What are the dimensions of the field that has the largest area?

SOLUTION   In order to get a feeling for what is happening in this problem, let's experiment with some special cases. Figure 1 (not to scale) shows three possible ways of laying out the 2400 ft of fencing. We see that when we try shallow, wide fields or deep, narrow fields, we get relatively small areas. It seems plausible that there is some intermediate configuration that produces the largest area.

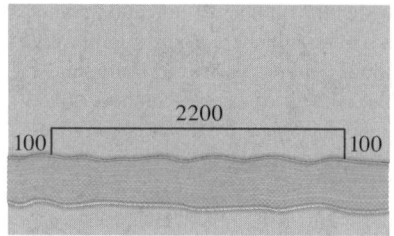

Area = 100 · 2200 = 220,000 ft²

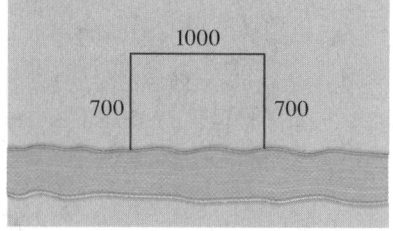

Area = 700 · 1000 = 700,000 ft²

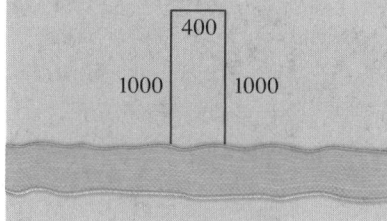

Area = 1000 · 400 = 400,000 ft²

FIGURE 1

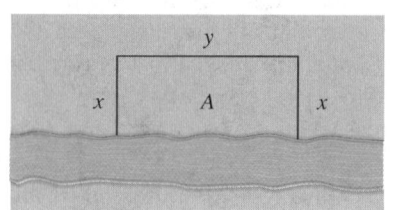

FIGURE 2

Figure 2 illustrates the general case. We wish to maximize the area A of the rectangle. Let x and y be the depth and width of the rectangle (in feet). Then we express A in terms of x and y:

$$A = xy$$

We want to express A as a function of just one variable, so we eliminate y by expressing it in terms of x. To do this we use the given information that the total length of the fencing is 2400 ft. Thus

$$2x + y = 2400$$

From this equation we have $y = 2400 - 2x$, which gives

$$A = x(2400 - 2x) = 2400x - 2x^2$$

Note that $x \geqslant 0$ and $x \leqslant 1200$ (otherwise $A < 0$). So the function that we wish to maximize is

$$A(x) = 2400x - 2x^2 \qquad 0 \leqslant x \leqslant 1200$$

The derivative is $A'(x) = 2400 - 4x$, so to find the critical numbers we solve the equation

$$2400 - 4x = 0$$

which gives $x = 600$. The maximum value of A must occur either at this critical number or at an endpoint of the interval. Since $A(0) = 0$, $A(600) = 720{,}000$, and $A(1200) = 0$, the Closed Interval Method gives the maximum value as $A(600) = 720{,}000$.

[Alternatively, we could have observed that $A''(x) = -4 < 0$ for all x, so A is always concave downward and the local maximum at $x = 600$ must be an absolute maximum.]

Thus the rectangular field should be 600 ft deep and 1200 ft wide. ■

V EXAMPLE 2 A cylindrical can is to be made to hold 1 L of oil. Find the dimensions that will minimize the cost of the metal to manufacture the can.

SOLUTION Draw the diagram as in Figure 3, where r is the radius and h the height (both in centimeters). In order to minimize the cost of the metal, we minimize the total surface area of the cylinder (top, bottom, and sides). From Figure 4 we see that the sides are made from a rectangular sheet with dimensions $2\pi r$ and h. So the surface area is

$$A = 2\pi r^2 + 2\pi rh$$

To eliminate h we use the fact that the volume is given as 1 L, which we take to be 1000 cm³. Thus

$$\pi r^2 h = 1000$$

which gives $h = 1000/(\pi r^2)$. Substitution of this into the expression for A gives

$$A = 2\pi r^2 + 2\pi r\left(\frac{1000}{\pi r^2}\right) = 2\pi r^2 + \frac{2000}{r}$$

Therefore the function that we want to minimize is

$$A(r) = 2\pi r^2 + \frac{2000}{r} \qquad r > 0$$

To find the critical numbers, we differentiate:

$$A'(r) = 4\pi r - \frac{2000}{r^2} = \frac{4(\pi r^3 - 500)}{r^2}$$

Then $A'(r) = 0$ when $\pi r^3 = 500$, so the only critical number is $r = \sqrt[3]{500/\pi}$.

Since the domain of A is $(0, \infty)$, we can't use the argument of Example 1 concerning endpoints. But we can observe that $A'(r) < 0$ for $r < \sqrt[3]{500/\pi}$ and $A'(r) > 0$ for $r > \sqrt[3]{500/\pi}$, so A is decreasing for *all* r to the left of the critical

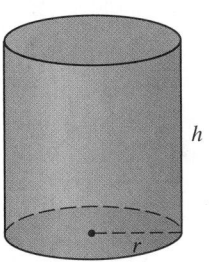

FIGURE 3

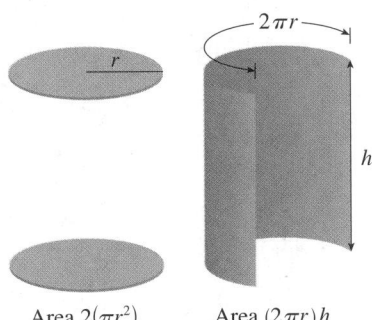

Area $2(\pi r^2)$ Area $(2\pi r)h$

FIGURE 4

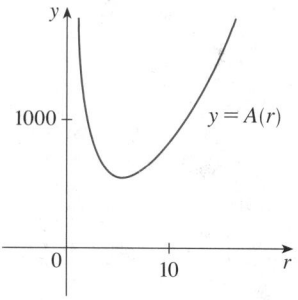

FIGURE 5

TEC Module 4.5 takes you through six additional optimization problems, including animations of the physical situations.

number and increasing for *all* r to the right. Thus $r = \sqrt[3]{500/\pi}$ must give rise to an *absolute* minimum.

[Alternatively, we could argue that $A(r) \rightarrow \infty$ as $r \rightarrow 0^+$ and $A(r) \rightarrow \infty$ as $r \rightarrow \infty$, so there must be a minimum value of $A(r)$, which must occur at the critical number. See Figure 5.]

The value of h corresponding to $r = \sqrt[3]{500/\pi}$ is

$$h = \frac{1000}{\pi r^2} = \frac{1000}{\pi(500/\pi)^{2/3}} = 2\sqrt[3]{\frac{500}{\pi}} = 2r$$

Thus to minimize the cost of the can, the radius should be $\sqrt[3]{500/\pi} \approx 5.42$ cm and the height should be equal to twice the radius, namely, the diameter. ∎

NOTE 1 The argument used in Example 2 to justify the absolute minimum is a variant of the First Derivative Test (which applies only to *local* maximum or minimum values) and is stated here for future reference.

FIRST DERIVATIVE TEST FOR ABSOLUTE EXTREME VALUES Suppose that c is a critical number of a continuous function f defined on an interval.

(a) If $f'(x) > 0$ for all $x < c$ and $f'(x) < 0$ for all $x > c$, then $f(c)$ is the absolute maximum value of f.

(b) If $f'(x) < 0$ for all $x < c$ and $f'(x) > 0$ for all $x > c$, then $f(c)$ is the absolute minimum value of f.

NOTE 2 An alternative method for solving optimization problems is to use implicit differentiation. Let's look at Example 2 again to illustrate the method. We work with the same equations

$$A = 2\pi r^2 + 2\pi rh \qquad \pi r^2 h = 100$$

but instead of eliminating h, we differentiate both equations implicitly with respect to r:

$$A' = 4\pi r + 2\pi rh' + 2\pi h \qquad \pi r^2 h' + 2\pi rh = 0$$

The minimum occurs at a critical number, so we set $A' = 0$, simplify, and arrive at the equations

$$2r + h + rh' = 0 \qquad 2h + rh' = 0$$

and subtraction gives $2r - h = 0$, or $h = 2r$.

▼ EXAMPLE 3 Find the point on the parabola $y^2 = 2x$ that is closest to the point $(1, 4)$.

SOLUTION The distance between the point $(1, 4)$ and the point (x, y) is

$$d = \sqrt{(x - 1)^2 + (y - 4)^2}$$

(See Figure 6.) But if (x, y) lies on the parabola, then $x = y^2/2$, so the expression for d becomes

$$d = \sqrt{\left(\tfrac{1}{2}y^2 - 1\right)^2 + (y - 4)^2}$$

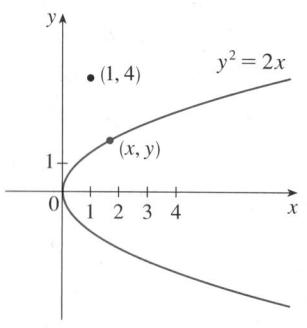

FIGURE 6

(Alternatively, we could have substituted $y = \sqrt{2x}$ to get d in terms of x alone.) Instead of minimizing d, we minimize its square:

$$d^2 = f(y) = \left(\tfrac{1}{2}y^2 - 1\right)^2 + (y - 4)^2$$

(You should convince yourself that the minimum of d occurs at the same point as the minimum of d^2, but d^2 is easier to work with.) Differentiating, we obtain

$$f'(y) = 2\left(\tfrac{1}{2}y^2 - 1\right)y + 2(y - 4) = y^3 - 8$$

so $f'(y) = 0$ when $y = 2$. Observe that $f'(y) < 0$ when $y < 2$ and $f'(y) > 0$ when $y > 2$, so by the First Derivative Test for Absolute Extreme Values, the absolute minimum occurs when $y = 2$. (Or we could simply say that because of the geometric nature of the problem, it's obvious that there is a closest point but not a farthest point.) The corresponding value of x is $x = y^2/2 = 2$. Thus the point on $y^2 = 2x$ closest to $(1, 4)$ is $(2, 2)$. ∎

EXAMPLE 4 A man launches his boat from point A on a bank of a straight river, 3 km wide, and wants to reach point B, 8 km downstream on the opposite bank, as quickly as possible (see Figure 7). He could row his boat directly across the river to point C and then run to B, or he could row directly to B, or he could row to some point D between C and B and then run to B. If he can row 6 km/h and run 8 km/h, where should he land to reach B as soon as possible? (We assume that the speed of the water is negligible compared with the speed at which the man rows.)

SOLUTION If we let x be the distance from C to D, then the running distance is $|DB| = 8 - x$ and the Pythagorean Theorem gives the rowing distance as $|AD| = \sqrt{x^2 + 9}$. We use the equation

$$\text{time} = \frac{\text{distance}}{\text{rate}}$$

Then the rowing time is $\sqrt{x^2 + 9}/6$ and the running time is $(8 - x)/8$, so the total time T as a function of x is

$$T(x) = \frac{\sqrt{x^2 + 9}}{6} + \frac{8 - x}{8}$$

The domain of this function T is $[0, 8]$. Notice that if $x = 0$ he rows to C and if $x = 8$ he rows directly to B. The derivative of T is

$$T'(x) = \frac{x}{6\sqrt{x^2 + 9}} - \frac{1}{8}$$

Thus, using the fact that $x \geqslant 0$, we have

$$T'(x) = 0 \iff \frac{x}{6\sqrt{x^2 + 9}} = \frac{1}{8} \iff 4x = 3\sqrt{x^2 + 9}$$

$$\iff 16x^2 = 9(x^2 + 9) \iff 7x^2 = 81$$

$$\iff x = \frac{9}{\sqrt{7}}$$

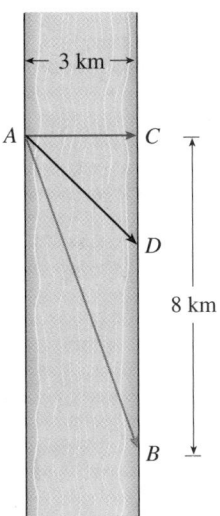

← 3 km →

A C

D

8 km

B

FIGURE 7

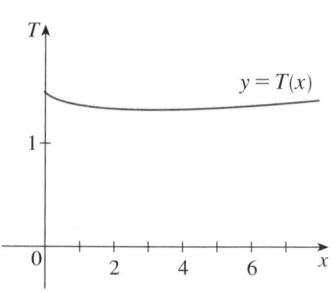

FIGURE 8

The only critical number is $x = 9/\sqrt{7}$. To see whether the minimum occurs at this critical number or at an endpoint of the domain $[0, 8]$, we evaluate T at all three points:

$$T(0) = 1.5 \qquad T\left(\frac{9}{\sqrt{7}}\right) = 1 + \frac{\sqrt{7}}{8} \approx 1.33 \qquad T(8) = \frac{\sqrt{73}}{6} \approx 1.42$$

Since the smallest of these values of T occurs when $x = 9/\sqrt{7}$, the absolute minimum value of T must occur there. Figure 8 illustrates this calculation by showing the graph of T.

Thus the man should land the boat at a point $9/\sqrt{7}$ km (≈ 3.4 km) downstream from his starting point. ∎

V EXAMPLE 5 Find the area of the largest rectangle that can be inscribed in a semicircle of radius r.

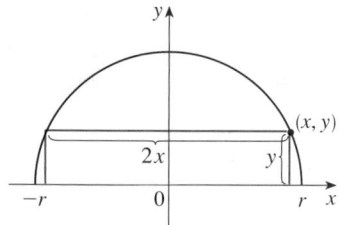

FIGURE 9

SOLUTION 1 Let's take the semicircle to be the upper half of the circle $x^2 + y^2 = r^2$ with center the origin. Then the word *inscribed* means that the rectangle has two vertices on the semicircle and two vertices on the x-axis as shown in Figure 9.

Let (x, y) be the vertex that lies in the first quadrant. Then the rectangle has sides of lengths $2x$ and y, so its area is

$$A = 2xy$$

To eliminate y we use the fact that (x, y) lies on the circle $x^2 + y^2 = r^2$ and so $y = \sqrt{r^2 - x^2}$. Thus

$$A = 2x\sqrt{r^2 - x^2}$$

The domain of this function is $0 \le x \le r$. Its derivative is

$$A' = -\frac{2x^2}{\sqrt{r^2 - x^2}} + 2\sqrt{r^2 - x^2} = \frac{2(r^2 - 2x^2)}{\sqrt{r^2 - x^2}}$$

which is 0 when $2x^2 = r^2$, that is, $x = r/\sqrt{2}$ (since $x \ge 0$). This value of x gives a maximum value of A since $A(0) = 0$ and $A(r) = 0$. Therefore the area of the largest inscribed rectangle is

$$A\left(\frac{r}{\sqrt{2}}\right) = 2\frac{r}{\sqrt{2}}\sqrt{r^2 - \frac{r^2}{2}} = r^2$$

SOLUTION 2 A simpler solution is possible if we think of using an angle as a variable. Let θ be the angle shown in Figure 10. Then the area of the rectangle is

$$A(\theta) = (2r\cos\theta)(r\sin\theta) = r^2(2\sin\theta\,\cos\theta) = r^2\sin 2\theta$$

We know that $\sin 2\theta$ has a maximum value of 1 and it occurs when $2\theta = \pi/2$. So $A(\theta)$ has a maximum value of r^2 and it occurs when $\theta = \pi/4$.

Notice that this trigonometric solution doesn't involve differentiation. In fact, we didn't need to use calculus at all. ∎

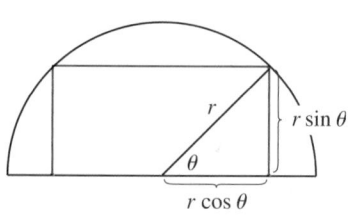

FIGURE 10

APPLICATIONS TO BUSINESS AND ECONOMICS

In Example 10 in Section 2.3 we introduced the idea of marginal cost. Recall that if $C(x)$, the **cost function**, is the cost of producing x units of a certain product, then the **marginal cost** is the rate of change of C with respect to x. In other words, the marginal cost function is the derivative, $C'(x)$, of the cost function.

Now let's consider marketing. Let $p(x)$ be the price per unit that the company can charge if it sells x units. Then p is called the **demand function** (or **price function**) and we would expect it to be a decreasing function of x. If x units are sold and the price per unit is $p(x)$, then the total revenue is

$$R(x) = xp(x)$$

and R is called the **revenue function**. The derivative R' of the revenue function is called the **marginal revenue function** and is the rate of change of revenue with respect to the number of units sold.

If x units are sold, then the total profit is

$$P(x) = R(x) - C(x)$$

and P is called the **profit function**. The **marginal profit function** is P', the derivative of the profit function. In Exercises 43–48 you are asked to use the marginal cost, revenue, and profit functions to minimize costs and maximize revenues and profits.

V EXAMPLE 6 A store has been selling 200 DVD burners a week at \$350 each. A market survey indicates that for each \$10 rebate offered to buyers, the number of units sold will increase by 20 a week. Find the demand function and the revenue function. How large a rebate should the store offer to maximize its revenue?

SOLUTION If x is the number of DVD burners sold per week, then the weekly increase in sales is $x - 200$. For each increase of 20 units sold, the price is decreased by \$10. So for each additional unit sold, the decrease in price will be $\frac{1}{20} \times 10$ and the demand function is

$$p(x) = 350 - \tfrac{10}{20}(x - 200) = 450 - \tfrac{1}{2}x$$

The revenue function is

$$R(x) = xp(x) = 450x - \tfrac{1}{2}x^2$$

Since $R'(x) = 450 - x$, we see that $R'(x) = 0$ when $x = 450$. This value of x gives an absolute maximum by the First Derivative Test (or simply by observing that the graph of R is a parabola that opens downward). The corresponding price is

$$p(450) = 450 - \tfrac{1}{2}(450) = 225$$

and the rebate is $350 - 225 = 125$. So to maximize revenue, the store should offer a rebate of \$125. ∎

4.5 | EXERCISES

1. Consider the following problem: Find two numbers whose sum is 23 and whose product is a maximum.
   (a) Make a table of values, like the following one, so that the sum of the numbers in the first two columns is always 23. On the basis of the evidence in your table, estimate the answer to the problem.

First number	Second number	Product
1	22	22
2	21	42
3	20	60
.	.	.
.	.	.
.	.	.

   (b) Use calculus to solve the problem and compare with your answer to part (a).

2. Find two numbers whose difference is 100 and whose product is a minimum.

3. Find two positive numbers whose product is 100 and whose sum is a minimum.

4. The sum of two positive numbers is 16. What is the smallest possible value of the sum of their squares?

5. What is the maximum vertical distance between the line $y = x + 2$ and the parabola $y = x^2$ for $-1 \leqslant x \leqslant 2$?

6. What is the minimum vertical distance between the parabolas $y = x^2 + 1$ and $y = x - x^2$?

7. Find the dimensions of a rectangle with perimeter 100 m whose area is as large as possible.

8. Find the dimensions of a rectangle with area 1000 m² whose perimeter is as small as possible.

9. Consider the following problem: A farmer with 750 ft of fencing wants to enclose a rectangular area and then divide it into four pens with fencing parallel to one side of the rectangle. What is the largest possible total area of the four pens?
   (a) Draw several diagrams illustrating the situation, some with shallow, wide pens and some with deep, narrow pens. Find the total areas of these configurations. Does it appear that there is a maximum area? If so, estimate it.
   (b) Draw a diagram illustrating the general situation. Introduce notation and label the diagram with your symbols.
   (c) Write an expression for the total area.
   (d) Use the given information to write an equation that relates the variables.
   (e) Use part (d) to write the total area as a function of one variable.
   (f) Finish solving the problem and compare the answer with your estimate in part (a).

10. Consider the following problem: A box with an open top is to be constructed from a square piece of cardboard, 3 ft wide, by cutting out a square from each of the four corners and bending up the sides. Find the largest volume that such a box can have.
    (a) Draw several diagrams to illustrate the situation, some short boxes with large bases and some tall boxes with small bases. Find the volumes of several such boxes. Does it appear that there is a maximum volume? If so, estimate it.
    (b) Draw a diagram illustrating the general situation. Introduce notation and label the diagram with your symbols.
    (c) Write an expression for the volume.
    (d) Use the given information to write an equation that relates the variables.
    (e) Use part (d) to write the volume as a function of one variable.
    (f) Finish solving the problem and compare the answer with your estimate in part (a).

11. If 1200 cm² of material is available to make a box with a square base and an open top, find the largest possible volume of the box.

12. A box with a square base and open top must have a volume of 32,000 cm³. Find the dimensions of the box that minimize the amount of material used.

13. (a) Show that of all the rectangles with a given area, the one with smallest perimeter is a square.
    (b) Show that of all the rectangles with a given perimeter, the one with greatest area is a square.

14. A rectangular storage container with an open top is to have a volume of 10 m³. The length of its base is twice the width. Material for the base costs $10 per square meter. Material for the sides costs $6 per square meter. Find the cost of materials for the cheapest such container.

15. Find the point on the line $y = 2x + 3$ that is closest to the origin.

16. Find the point on the curve $y = \sqrt{x}$ that is closest to the point $(3, 0)$.

17. Find the points on the ellipse $4x^2 + y^2 = 4$ that are farthest away from the point $(1, 0)$.

18. Find, correct to two decimal places, the coordinates of the point on the curve $y = \sin x$ that is closest to the point $(4, 2)$.

19. Find the dimensions of the rectangle of largest area that can be inscribed in an equilateral triangle of side L if one side of the rectangle lies on the base of the triangle.

20. Find the area of the largest trapezoid that can be inscribed in a circle of radius 1 and whose base is a diameter of the circle.

21. Find the dimensions of the isosceles triangle of largest area that can be inscribed in a circle of radius r.

22. Find the area of the largest rectangle that can be inscribed in a right triangle with legs of lengths 3 cm and 4 cm if two sides of the rectangle lie along the legs.

23. A right circular cylinder is inscribed in a sphere of radius r. Find the largest possible volume of such a cylinder.

24. Find the area of the largest rectangle that can be inscribed in the ellipse $x^2/a^2 + y^2/b^2 = 1$.

25. A Norman window has the shape of a rectangle surmounted by a semicircle. (Thus the diameter of the semicircle is equal to the width of the rectangle.) If the perimeter of the window is 30 ft, find the dimensions of the window so that the greatest possible amount of light is admitted.

26. A right circular cylinder is inscribed in a cone with height h and base radius r. Find the largest possible volume of such a cylinder.

27. A piece of wire 10 m long is cut into two pieces. One piece is bent into a square and the other is bent into an equilateral triangle. How should the wire be cut so that the total area enclosed is (a) a maximum? (b) A minimum?

28. A fence 8 ft tall runs parallel to a tall building at a distance of 4 ft from the building. What is the length of the shortest ladder that will reach from the ground over the fence to the wall of the building?

29. A cone-shaped drinking cup is made from a circular piece of paper of radius R by cutting out a sector and joining the edges CA and CB. Find the maximum capacity of such a cup.

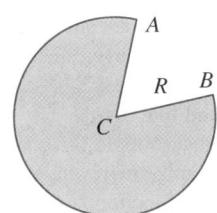

30. A cone-shaped paper drinking cup is to be made to hold 27 cm³ of water. Find the height and radius of the cup that will use the smallest amount of paper.

31. A cone with height h is inscribed in a larger cone with height H so that its vertex is at the center of the base of the larger cone. Show that the inner cone has maximum volume when $h = \frac{1}{3}H$.

32. The graph shows the fuel consumption c of a car (measured in gallons per hour) as a function of the speed v of the car. At very low speeds the engine runs inefficiently, so initially c decreases as the speed increases. But at high speeds the fuel consumption increases. You can see that $c(v)$ is minimized for this car when $v \approx 30$ mi/h. However, for fuel

efficiency, what must be minimized is not the consumption in gallons per hour but rather the fuel consumption in gallons *per mile*. Let's call this consumption G. Using the graph, estimate the speed at which G has its minimum value.

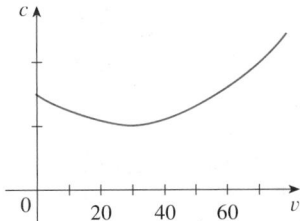

33. If a resistor of R ohms is connected across a battery of E volts with internal resistance r ohms, then the power (in watts) in the external resistor is

$$P = \frac{E^2R}{(R + r)^2}$$

If E and r are fixed but R varies, what is the maximum value of the power?

34. For a fish swimming at a speed v relative to the water, the energy expenditure per unit time is proportional to v^3. It is believed that migrating fish try to minimize the total energy required to swim a fixed distance. If the fish are swimming against a current u ($u < v$), then the time required to swim a distance L is $L/(v - u)$ and the total energy E required to swim the distance is given by

$$E(v) = av^3 \cdot \frac{L}{v - u}$$

where a is the proportionality constant.
(a) Determine the value of v that minimizes E.
(b) Sketch the graph of E.

Note: This result has been verified experimentally; migrating fish swim against a current at a speed 50% greater than the current speed.

35. In a beehive, each cell is a regular hexagonal prism, open at one end with a trihedral angle at the other end as in the figure. It is believed that bees form their cells in such a way

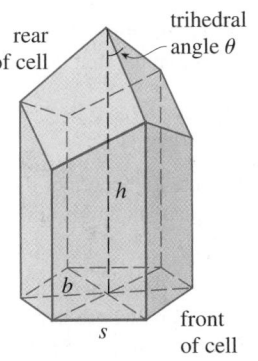

as to minimize the surface area, thus using the least amount of wax in cell construction. Examination of these cells has shown that the measure of the apex angle θ is amazingly consistent. Based on the geometry of the cell, it can be shown that the surface area S is given by

$$S = 6sh - \tfrac{3}{2}s^2 \cot \theta + \left(3s^2\sqrt{3}/2\right)\csc \theta$$

where s, the length of the sides of the hexagon, and h, the height, are constants.
(a) Calculate $dS/d\theta$.
(b) What angle should the bees prefer?
(c) Determine the minimum surface area of the cell (in terms of s and h).

Note: Actual measurements of the angle θ in beehives have been made, and the measures of these angles seldom differ from the calculated value by more than 2°.

36. A boat leaves a dock at 2:00 PM and travels due south at a speed of 20 km/h. Another boat has been heading due east at 15 km/h and reaches the same dock at 3:00 PM. At what time were the two boats closest together?

37. The illumination of an object by a light source is directly proportional to the strength of the source and inversely proportional to the square of the distance from the source. If two light sources, one three times as strong as the other, are placed 10 ft apart, where should an object be placed on the line between the sources so as to receive the least illumination?

38. A woman at a point A on the shore of a circular lake with radius 2 mi wants to arrive at the point C diametrically opposite A on the other side of the lake in the shortest possible time. She can walk at the rate of 4 mi/h and row a boat at 2 mi/h. How should she proceed?

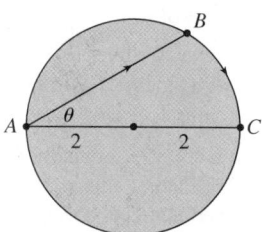

39. Find an equation of the line through the point $(3, 5)$ that cuts off the least area from the first quadrant.

40. At which points on the curve $y = 1 + 40x^3 - 3x^5$ does the tangent line have the largest slope?

41. What is the shortest possible length of the line segment that is cut off by the first quadrant and is tangent to the curve $y = 3/x$ at some point?

42. What is the smallest possible area of the triangle that is cut off by the first quadrant and whose hypotenuse is tangent to the parabola $y = 4 - x^2$ at some point?

43. (a) If $C(x)$ is the cost of producing x units of a commodity, then the **average cost** per unit is $c(x) = C(x)/x$. Show that if the average cost is a minimum, then the marginal cost equals the average cost.
(b) If $C(x) = 16{,}000 + 200x + 4x^{3/2}$, in dollars, find (i) the cost, average cost, and marginal cost at a production level of 1000 units; (ii) the production level that will minimize the average cost; and (iii) the minimum average cost.

44. (a) Show that if the profit $P(x)$ is a maximum, then the marginal revenue equals the marginal cost.
(b) If $C(x) = 16{,}000 + 500x - 1.6x^2 + 0.004x^3$ is the cost function and $p(x) = 1700 - 7x$ is the demand function, find the production level that will maximize profit.

45. A baseball team plays in a stadium that holds 55,000 spectators. With ticket prices at $10, the average attendance had been 27,000. When ticket prices were lowered to $8, the average attendance rose to 33,000.
(a) Find the demand function, assuming that it is linear.
(b) How should ticket prices be set to maximize revenue?

46. During the summer months Terry makes and sells necklaces on the beach. Last summer he sold the necklaces for $10 each and his sales averaged 20 per day. When he increased the price by $1, he found that the average decreased by two sales per day.
(a) Find the demand function, assuming that it is linear.
(b) If the material for each necklace costs Terry $6, what should the selling price be to maximize his profit?

47. A manufacturer has been selling 1000 flat-screen TVs a week at $450 each. A market survey indicates that for each $10 rebate offered to the buyer, the number of TVs sold will increase by 100 per week.
(a) Find the demand function.
(b) How large a rebate should the company offer the buyer in order to maximize its revenue?
(c) If its weekly cost function is $C(x) = 68{,}000 + 150x$, how should the manufacturer set the size of the rebate in order to maximize its profit?

48. The manager of a 100-unit apartment complex knows from experience that all units will be occupied if the rent is $800 per month. A market survey suggests that, on average, one additional unit will remain vacant for each $10 increase in rent. What rent should the manager charge to maximize revenue?

49. Let a and b be positive numbers. Find the length of the shortest line segment that is cut off by the first quadrant and passes through the point (a, b).

CAS **50.** The frame for a kite is to be made from six pieces of wood. The four exterior pieces have been cut with the lengths indicated in the figure. To maximize the area of the kite, how long should the diagonal pieces be?

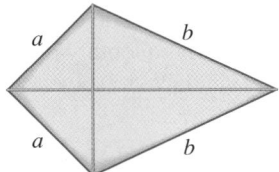

51. Let v_1 be the velocity of light in air and v_2 the velocity of light in water. According to Fermat's Principle, a ray of light will travel from a point A in the air to a point B in the water by a path ACB that minimizes the time taken. Show that

$$\frac{\sin \theta_1}{\sin \theta_2} = \frac{v_1}{v_2}$$

where θ_1 (the angle of incidence) and θ_2 (the angle of refraction) are as shown. This equation is known as Snell's Law.

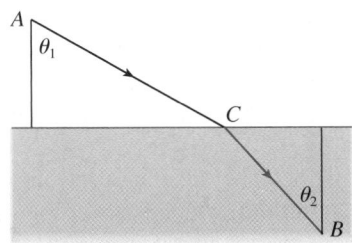

52. Two vertical poles PQ and ST are secured by a rope PRS going from the top of the first pole to a point R on the ground between the poles and then to the top of the second pole as in the figure. Show that the shortest length of such a rope occurs when $\theta_1 = \theta_2$.

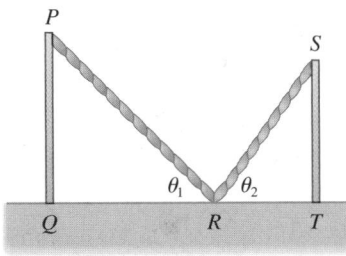

53. The upper right-hand corner of a piece of paper, 12 in. by 8 in., as in the figure, is folded over to the bottom edge. How

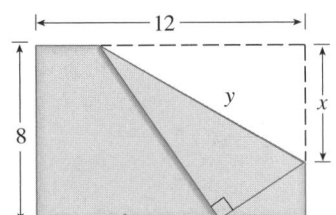

would you fold it so as to minimize the length of the fold? In other words, how would you choose x to minimize y?

54. A steel pipe is being carried down a hallway 9 ft wide. At the end of the hall there is a right-angled turn into a narrower hallway 6 ft wide. What is the length of the longest pipe that can be carried horizontally around the corner?

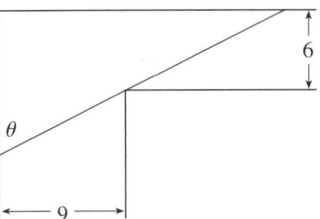

55. Find the maximum area of a rectangle that can be circumscribed about a given rectangle with length L and width W. [*Hint:* Express the area as a function of an angle θ.]

56. A rain gutter is to be constructed from a metal sheet of width 30 cm by bending up one-third of the sheet on each side through an angle θ. How should θ be chosen so that the gutter will carry the maximum amount of water?

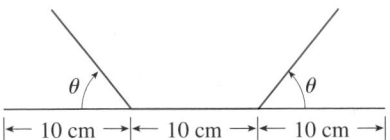

57. Where should the point P be chosen on the line segment AB so as to maximize the angle θ?

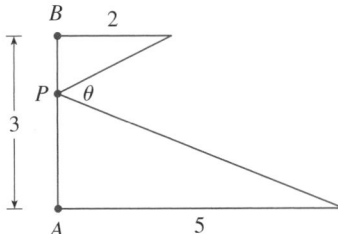

58. A painting in an art gallery has height h and is hung so that its lower edge is a distance d above the eye of an observer (as in the figure). How far from the wall should the observer stand to get the best view? (In other words, where should the observer stand so as to maximize the angle θ subtended at his eye by the painting?)

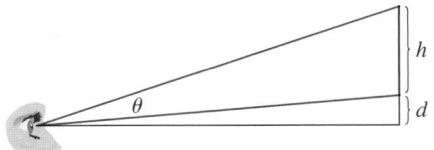

4.6 | NEWTON'S METHOD

Suppose that a car dealer offers to sell you a car for $18,000 or for payments of $375 per month for five years. You would like to know what monthly interest rate the dealer is, in effect, charging you. To find the answer, you have to solve the equation

$$\boxed{1} \qquad 48x(1 + x)^{60} - (1 + x)^{60} + 1 = 0$$

(The details are explained in Exercise 31.) How would you solve such an equation?

For a quadratic equation $ax^2 + bx + c = 0$ there is a well-known formula for the roots. For third- and fourth-degree equations there are also formulas for the roots, but they are extremely complicated. If f is a polynomial of degree 5 or higher, there is no such formula. Likewise, there is no formula that will enable us to find the exact roots of a transcendental equation such as $\cos x = x$.

We can find an *approximate* solution to Equation 1 by plotting the left side of the equation. Using a graphing device, and after experimenting with viewing rectangles, we produce the graph in Figure 1.

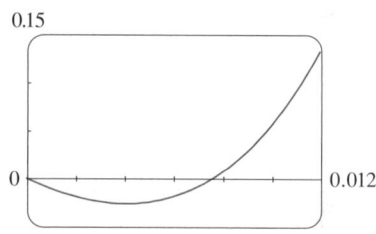

FIGURE 1

We see that in addition to the solution $x = 0$, which doesn't interest us, there is a solution between 0.007 and 0.008. Zooming in shows that the root is approximately 0.0076. If we need more accuracy we could zoom in repeatedly, but that becomes tiresome. A faster alternative is to use a numerical rootfinder on a calculator or computer algebra system. If we do so, we find that the root, correct to nine decimal places, is 0.007628603.

How do those numerical rootfinders work? They use a variety of methods, but most of them make some use of **Newton's method**, which is also called the **Newton-Raphson method**. We will explain how this method works, partly to show what happens inside a calculator or computer, and partly as an application of the idea of linear approximation.

The geometry behind Newton's method is shown in Figure 2, where the root that we are trying to find is labeled r. We start with a first approximation x_1, which is obtained by guessing, or from a rough sketch of the graph of f, or from a computer-generated graph of f. Consider the tangent line L to the curve $y = f(x)$ at the point $(x_1, f(x_1))$ and look at the x-intercept of L, labeled x_2. The idea behind Newton's method is that the tangent line is close to the curve and so its x-intercept, x_2, is close to the x-intercept of the curve (namely, the root r that we are seeking). Because the tangent is a line, we can easily find its x-intercept.

■ Try to solve Equation 1 using the numerical rootfinder on your calculator or computer. Some machines are not able to solve it. Others are successful but require you to specify a starting point for the search.

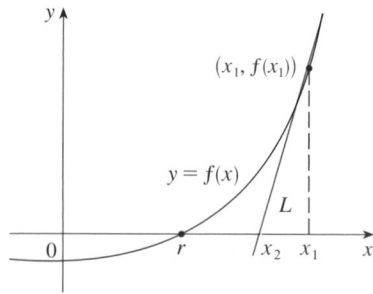

FIGURE 2

To find a formula for x_2 in terms of x_1 we use the fact that the slope of L is $f'(x_1)$, so its equation is

$$y - f(x_1) = f'(x_1)(x - x_1)$$

Since the x-intercept of L is x_2, we set $y = 0$ and obtain

$$0 - f(x_1) = f'(x_1)(x_2 - x_1)$$

If $f'(x_1) \neq 0$, we can solve this equation for x_2:

$$x_2 = x_1 - \frac{f(x_1)}{f'(x_1)}$$

We use x_2 as a second approximation to r.

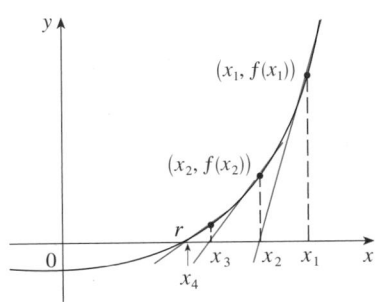

FIGURE 3

■ The convergence of infinite sequences is discussed in detail in Section 8.1.

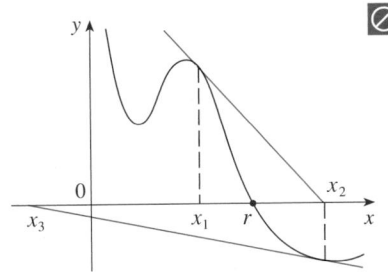

FIGURE 4

TEC In Module 4.6 you can investigate how Newton's method works for several functions and what happens when you change x_1.

■ Figure 5 shows the geometry behind the first step in Newton's method in Example 1. Since $f'(2) = 10$, the tangent line to $y = x^3 - 2x - 5$ at $(2, -1)$ has equation $y = 10x - 21$ and so its x-intercept is $x_2 = 2.1$.

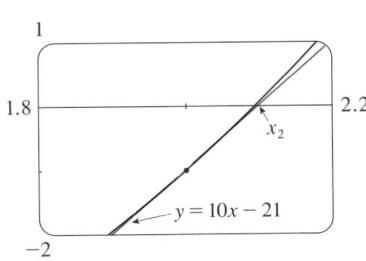

FIGURE 5

Unless otherwise noted, all content on this page is © Cengage Learning.

Next we repeat this procedure with x_1 replaced by x_2, using the tangent line at $(x_2, f(x_2))$. This gives a third approximation:

$$x_3 = x_2 - \frac{f(x_2)}{f'(x_2)}$$

If we keep repeating this process, we obtain a sequence of approximations x_1, x_2, x_3, x_4, ... as shown in Figure 3. In general, if the nth approximation is x_n and $f'(x_n) \neq 0$, then the next approximation is given by

[2]
$$x_{n+1} = x_n - \frac{f(x_n)}{f'(x_n)}$$

If the numbers x_n become closer and closer to r as n becomes large, then we say that the sequence *converges* to r and we write

$$\lim_{n \to \infty} x_n = r$$

⊘ Although the sequence of successive approximations converges to the desired root for functions of the type illustrated in Figure 3, in certain circumstances the sequence may not converge. For example, consider the situation shown in Figure 4. You can see that x_2 is a worse approximation than x_1. This is likely to be the case when $f'(x_1)$ is close to 0. It might even happen that an approximation (such as x_3 in Figure 4) falls outside the domain of f. Then Newton's method fails and a better initial approximation x_1 should be chosen. See Exercises 25–27 for specific examples in which Newton's method works very slowly or does not work at all.

▼ EXAMPLE 1 Starting with $x_1 = 2$, find the third approximation x_3 to the root of the equation $x^3 - 2x - 5 = 0$.

SOLUTION We apply Newton's method with

$$f(x) = x^3 - 2x - 5 \qquad \text{and} \qquad f'(x) = 3x^2 - 2$$

Newton himself used this equation to illustrate his method and he chose $x_1 = 2$ after some experimentation because $f(1) = -6$, $f(2) = -1$, and $f(3) = 16$. Equation 2 becomes

$$x_{n+1} = x_n - \frac{x_n^3 - 2x_n - 5}{3x_n^2 - 2}$$

With $n = 1$ we have

$$x_2 = x_1 - \frac{x_1^3 - 2x_1 - 5}{3x_1^2 - 2}$$

$$= 2 - \frac{2^3 - 2(2) - 5}{3(2)^2 - 2} = 2.1$$

Then with $n = 2$ we obtain

$$x_3 = x_2 - \frac{x_2^3 - 2x_2 - 5}{3x_2^2 - 2}$$

$$= 2.1 - \frac{(2.1)^3 - 2(2.1) - 5}{3(2.1)^2 - 2} \approx 2.0946$$

It turns out that this third approximation $x_3 \approx 2.0946$ is accurate to four decimal places. ∎

Suppose that we want to achieve a given accuracy, say to eight decimal places, using Newton's method. How do we know when to stop? The rule of thumb that is generally used is that we can stop when successive approximations x_n and x_{n+1} agree to eight decimal places. (A precise statement concerning accuracy in Newton's method will be given in Exercise 29 in Section 8.8.)

Notice that the procedure in going from n to $n + 1$ is the same for all values of n. (It is called an *iterative* process.) This means that Newton's method is particularly convenient for use with a programmable calculator or a computer.

V EXAMPLE 2 Use Newton's method to find $\sqrt[6]{2}$ correct to eight decimal places.

SOLUTION First we observe that finding $\sqrt[6]{2}$ is equivalent to finding the positive root of the equation

$$x^6 - 2 = 0$$

so we take $f(x) = x^6 - 2$. Then $f'(x) = 6x^5$ and Formula 2 (Newton's method) becomes

$$x_{n+1} = x_n - \frac{x_n^6 - 2}{6x_n^5}$$

If we choose $x_1 = 1$ as the initial approximation, then we obtain

$$x_2 \approx 1.16666667$$
$$x_3 \approx 1.12644368$$
$$x_4 \approx 1.12249707$$
$$x_5 \approx 1.12246205$$
$$x_6 \approx 1.12246205$$

Since x_5 and x_6 agree to eight decimal places, we conclude that

$$\sqrt[6]{2} \approx 1.12246205$$

to eight decimal places. ∎

V EXAMPLE 3 Find, correct to six decimal places, the root of the equation $\cos x = x$.

■ www.stewartcalculus.com
See Additional Example A.

SOLUTION We first rewrite the equation in standard form:

$$\cos x - x = 0$$

Therefore we let $f(x) = \cos x - x$. Then $f'(x) = -\sin x - 1$, so Formula 2 becomes

$$x_{n+1} = x_n - \frac{\cos x_n - x_n}{-\sin x_n - 1} = x_n + \frac{\cos x_n - x_n}{\sin x_n + 1}$$

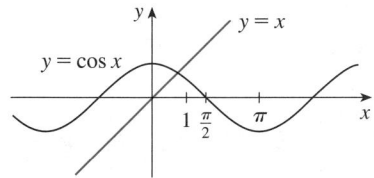

FIGURE 6

In order to guess a suitable value for x_1 we sketch the graphs of $y = \cos x$ and $y = x$ in Figure 6. It appears that they intersect at a point whose x-coordinate is somewhat less than 1, so let's take $x_1 = 1$ as a convenient first approximation. Then, remembering to put our calculator in radian mode, we get

$$x_2 \approx 0.75036387$$

$$x_3 \approx 0.73911289$$

$$x_4 \approx 0.73908513$$

$$x_5 \approx 0.73908513$$

Since x_4 and x_5 agree to six decimal places (eight, in fact), we conclude that the root of the equation, correct to six decimal places, is 0.739085. ∎

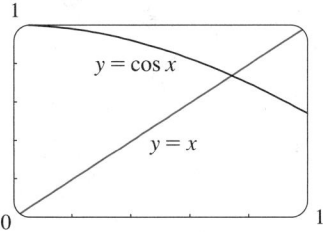

FIGURE 7

Instead of using the rough sketch in Figure 6 to get a starting approximation for Newton's method in Example 3, we could have used the more accurate graph that a calculator or computer provides. Figure 7 suggests that we use $x_1 = 0.75$ as the initial approximation. Then Newton's method gives

$$x_2 \approx 0.73911114 \qquad x_3 \approx 0.73908513 \qquad x_4 \approx 0.73908513$$

and so we obtain the same answer as before, but with one fewer step.

4.6 | EXERCISES

1. The figure shows the graph of a function f. Suppose that Newton's method is used to approximate the root r of the equation $f(x) = 0$ with initial approximation $x_1 = 1$.
  (a) Draw the tangent lines that are used to find x_2 and x_3, and estimate the numerical values of x_2 and x_3.
  (b) Would $x_1 = 5$ be a better first approximation? Explain.

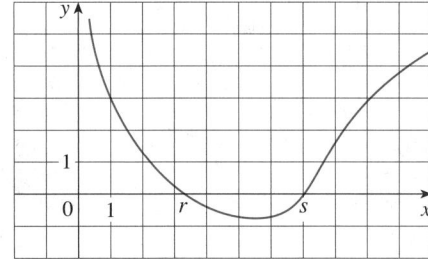

2. Follow the instructions for Exercise 1(a) but use $x_1 = 9$ as the starting approximation for finding the root s.

3. Suppose the tangent line to the curve $y = f(x)$ at the point $(2, 5)$ has the equation $y = 9 - 2x$. If Newton's method is used to locate a root of the equation $f(x) = 0$ and the initial approximation is $x_1 = 2$, find the second approximation x_2.

4. For each initial approximation, determine graphically what happens if Newton's method is used for the function whose graph is shown.
  (a) $x_1 = 0$ (b) $x_1 = 1$ (c) $x_1 = 3$
  (d) $x_1 = 4$ (e) $x_1 = 5$

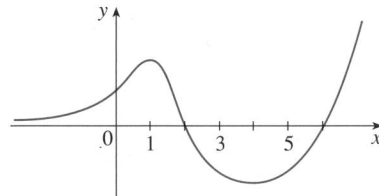

5. For which of the initial approximations $x_1 = a$, b, c, and d do you think Newton's method will work and lead to the root of the equation $f(x) = 0$?

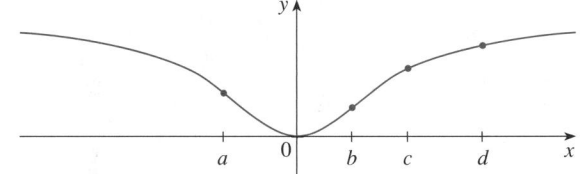

6–8 ■ Use Newton's method with the specified initial approximation x_1 to find x_3, the third approximation to the root of the given equation. (Give your answer to four decimal places.)

6. $\frac{1}{3}x^3 + \frac{1}{2}x^2 + 3 = 0, \quad x_1 = -3$

7. $x^5 - x - 1 = 0, \quad x_1 = 1$

8. $x^7 + 4 = 0, \quad x_1 = -1$

9. Use Newton's method with initial approximation $x_1 = -1$ to find x_2, the second approximation to the root of the equation $x^3 + x + 3 = 0$. Explain how the method works by first graphing the function and its tangent line at $(-1, 1)$.

10. Use Newton's method with initial approximation $x_1 = 1$ to find x_2, the second approximation to the root of the equation $x^4 - x - 1 = 0$. Explain how the method works by first graphing the function and its tangent line at $(1, -1)$.

11–12 ■ Use Newton's method to approximate the given number correct to eight decimal places.

11. $\sqrt[5]{20}$

12. $\sqrt[100]{100}$

13–14 ■ Use Newton's method to approximate the indicated root of the equation correct to six decimal places.

13. The root of $x^4 - 2x^3 + 5x^2 - 6 = 0$ in the interval $[1, 2]$

14. The positive root of $3 \sin x = x$

15–22 ■ Use Newton's method to find all the roots of the equation correct to eight decimal places. Start by drawing a graph to find initial approximations.

15. $x^6 - x^5 - 6x^4 - x^2 + x + 10 = 0$

16. $x^5 - 3x^4 + x^3 - x^2 - x + 6 = 0$

17. $e^{-x} = 2 + x$

18. $x^2(4 - x^2) = \dfrac{4}{x^2 + 1}$

19. $x^2\sqrt{2 - x - x^2} = 1$

20. $\cos(x^2 - x) = x^4$

21. $4e^{-x^2} \sin x = x^2 - x + 1$

22. $e^{\arctan x} = \sqrt{x^3 + 1}$

23. (a) Apply Newton's method to the equation $x^2 - a = 0$ to derive the following square-root algorithm (used by the ancient Babylonians to compute $\sqrt{a}$):

$$x_{n+1} = \frac{1}{2}\left(x_n + \frac{a}{x_n}\right)$$

(b) Use part (a) to compute $\sqrt{1000}$ correct to six decimal places.

24. (a) Apply Newton's method to the equation $1/x - a = 0$ to derive the following reciprocal algorithm:

$$x_{n+1} = 2x_n - ax_n^2$$

(This algorithm enables a computer to find reciprocals without actually dividing.)

(b) Use part (a) to compute $1/1.6984$ correct to six decimal places.

25. Explain why Newton's method doesn't work for finding the root of the equation $x^3 - 3x + 6 = 0$ if the initial approximation is chosen to be $x_1 = 1$.

26. (a) Use Newton's method with $x_1 = 1$ to find the root of the equation $x^3 - x = 1$ correct to six decimal places.
(b) Solve the equation in part (a) using $x_1 = 0.6$ as the initial approximation.
(c) Solve the equation in part (a) using $x_1 = 0.57$. (You definitely need a programmable calculator for this part.)
(d) Graph $f(x) = x^3 - x - 1$ and its tangent lines at $x_1 = 1$, 0.6, and 0.57 to explain why Newton's method is so sensitive to the value of the initial approximation.

27. Explain why Newton's method fails when applied to the equation $\sqrt[3]{x} = 0$ with any initial approximation $x_1 \neq 0$. Illustrate your explanation with a sketch.

28. Use Newton's method to find the absolute maximum value of the function $f(x) = x \cos x$, $0 \leqslant x \leqslant \pi$, correct to six decimal places.

29. Use Newton's method to find the coordinates of the inflection point of the curve $y = x^2 \sin x$, $0 \leqslant x \leqslant \pi$, correct to six decimal places.

30. Of the infinitely many lines that are tangent to the curve $y = -\sin x$ and pass through the origin, there is one that has the largest slope. Use Newton's method to find the slope of that line correct to six decimal places.

31. A car dealer sells a new car for $18,000. He also offers to sell the same car for payments of $375 per month for five years. What monthly interest rate is this dealer charging?

To solve this problem you will need to use the formula for the present value A of an annuity consisting of n equal payments of size R with interest rate i per time period:

$$A = \frac{R}{i}\left[1 - (1 + i)^{-n}\right]$$

Replacing i by x, show that

$$48x(1 + x)^{60} - (1 + x)^{60} + 1 = 0$$

Use Newton's method to solve this equation.

32. The figure shows the sun located at the origin and the earth at the point $(1, 0)$. (The unit here is the distance between the centers of the earth and the sun, called an *astronomical unit:* 1 AU $\approx 1.496 \times 10^8$ km.) There are five locations $L_1, L_2, L_3, L_4,$ and L_5 in this plane of rotation of the earth about the sun where a satellite remains motionless with respect to the earth because the forces acting on the satellite (including the gravitational attractions of the earth and the sun) balance each other. These locations are called *libration points.* (A solar research satellite has been placed at one of these libration points.) If m_1 is the mass of the sun, m_2 is the mass of the earth, and $r = m_2/(m_1 + m_2)$, it turns out that the x-coordinate of L_1 is the unique root of the fifth-degree equation

$$p(x) = x^5 - (2 + r)x^4 + (1 + 2r)x^3 - (1 - r)x^2$$
$$+ 2(1 - r)x + r - 1 = 0$$

and the x-coordinate of L_2 is the root of the equation

$$p(x) - 2rx^2 = 0$$

Using the value $r \approx 3.04042 \times 10^{-6}$, find the locations of the libration points (a) L_1 and (b) L_2.

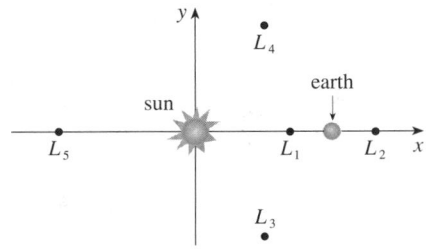

4.7 | ANTIDERIVATIVES

A physicist who knows the velocity of a particle might wish to know its position at a given time. An engineer who can measure the variable rate at which water is leaking from a tank wants to know the amount leaked over a certain time period. A biologist who knows the rate at which a bacteria population is increasing might want to deduce what the size of the population will be at some future time. In each case, the problem is to find a function F whose derivative is a known function f. If such a function F exists, it is called an *antiderivative* of f.

DEFINITION   A function F is called an **antiderivative** of f on an interval I if $F'(x) = f(x)$ for all x in I.

For instance, let $f(x) = x^2$. It isn't difficult to discover an antiderivative of f if we keep the Power Rule in mind. In fact, if $F(x) = \frac{1}{3}x^3$, then $F'(x) = x^2 = f(x)$. But the function $G(x) = \frac{1}{3}x^3 + 100$ also satisfies $G'(x) = x^2$. Therefore both F and G are antiderivatives of f. Indeed, any function of the form $H(x) = \frac{1}{3}x^3 + C$, where C is a constant, is an antiderivative of f. The question arises: Are there any others?

To answer this question, recall that in Section 4.2 we used the Mean Value Theorem to prove that if two functions have identical derivatives on an interval, then they must differ by a constant (Corollary 4.2.7). Thus if F and G are any two antiderivatives of f, then

$$F'(x) = f(x) = G'(x)$$

so $G(x) - F(x) = C$, where C is a constant. We can write this as $G(x) = F(x) + C$, so we have the following result.

---

1   **THEOREM**   If F is an antiderivative of f on an interval I, then the most general antiderivative of f on I is

$$F(x) + C$$

where C is an arbitrary constant.

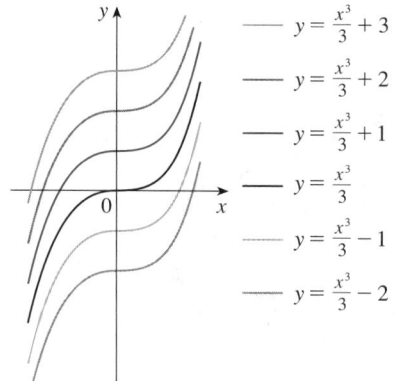

FIGURE 1

Members of the family of
antiderivatives of $f(x) = x^2$

Going back to the function $f(x) = x^2$, we see that the general antiderivative of f is $\frac{1}{3}x^3 + C$. By assigning specific values to the constant C, we obtain a family of functions whose graphs are vertical translates of one another (see Figure 1). This makes sense because each curve must have the same slope at any given value of x.

EXAMPLE 1 Find the most general antiderivative of each of the following functions.
(a) $f(x) = \sin x$ (b) $f(x) = 1/x$ (c) $f(x) = x^n, \quad n \neq -1$

SOLUTION

(a) If $F(x) = -\cos x$, then $F'(x) = \sin x$, so an antiderivative of $\sin x$ is $-\cos x$. By Theorem 1, the most general antiderivative is $G(x) = -\cos x + C$.

(b) Recall from Section 3.3 that

$$\frac{d}{dx}(\ln x) = \frac{1}{x}$$

So on the interval $(0, \infty)$ the general antiderivative of $1/x$ is $\ln x + C$. We also learned that

$$\frac{d}{dx}(\ln |x|) = \frac{1}{x}$$

for all $x \neq 0$. Theorem 1 then tells us that the general antiderivative of $f(x) = 1/x$ is $\ln |x| + C$ on any interval that doesn't contain 0. In particular, this is true on each of the intervals $(-\infty, 0)$ and $(0, \infty)$. So the general antiderivative of f is

$$F(x) = \begin{cases} \ln x + C_1 & \text{if } x > 0 \\ \ln(-x) + C_2 & \text{if } x < 0 \end{cases}$$

(c) We use the Power Rule to discover an antiderivative of x^n. In fact, if $n \neq -1$, then

$$\frac{d}{dx}\left(\frac{x^{n+1}}{n+1}\right) = \frac{(n+1)x^n}{n+1} = x^n$$

Thus the general antiderivative of $f(x) = x^n$ is

$$F(x) = \frac{x^{n+1}}{n+1} + C$$

This is valid for $n \geq 0$ since then $f(x) = x^n$ is defined on an interval. If n is negative (but $n \neq -1$), it is valid on any interval that doesn't contain 0. ∎

As in Example 1, every differentiation formula, when read from right to left, gives rise to an antidifferentiation formula. In Table 2 we list some particular antiderivatives. Each formula in the table is true because the derivative of the function in the right column appears in the left column. In particular, the first formula says that the antiderivative of a constant times a function is the constant times the antiderivative of the function. The second formula says that the antiderivative of a sum is the sum of the antiderivatives. (We use the notation $F' = f$, $G' = g$.)

2 **TABLE OF ANTIDIFFERENTIATION FORMULAS**

Function	Particular antiderivative	Function	Particular antiderivative
$cf(x)$	$cF(x)$	$\sec^2 x$	$\tan x$
$f(x) + g(x)$	$F(x) + G(x)$	$\sec x \tan x$	$\sec x$
$x^n \; (n \neq -1)$	$\dfrac{x^{n+1}}{n+1}$	$\dfrac{1}{\sqrt{1-x^2}}$	$\sin^{-1} x$
$\dfrac{1}{x}$	$\ln|x|$	$\dfrac{1}{1+x^2}$	$\tan^{-1} x$
e^x	e^x	$\cosh x$	$\sinh x$
$\cos x$	$\sin x$	$\sinh x$	$\cosh x$
$\sin x$	$-\cos x$		

■ To obtain the most general antiderivative from the particular ones in Table 2, we have to add a constant (or constants), as in Example 1.

EXAMPLE 2 Find all functions g such that

$$g'(x) = 4 \sin x + \frac{2x^5 - \sqrt{x}}{x}$$

SOLUTION We first rewrite the given function as follows:

$$g'(x) = 4 \sin x + \frac{2x^5}{x} - \frac{\sqrt{x}}{x} = 4 \sin x + 2x^4 - \frac{1}{\sqrt{x}}$$

Thus we want to find an antiderivative of

$$g'(x) = 4 \sin x + 2x^4 - x^{-1/2}$$

Using the formulas in Table 2 together with Theorem 1, we obtain

$$g(x) = 4(-\cos x) + 2\frac{x^5}{5} - \frac{x^{1/2}}{\frac{1}{2}} + C$$

$$= -4 \cos x + \tfrac{2}{5} x^5 - 2\sqrt{x} + C \qquad ■$$

In applications of calculus it is very common to have a situation as in Example 2, where it is required to find a function, given knowledge about its derivatives. An equation that involves the derivatives of a function is called a **differential equation**. These will be studied in some detail in Section 7.7, but for the present we can solve some elementary differential equations. The general solution of a differential equation involves an arbitrary constant (or constants) as in Example 2. However, there may be some extra conditions given that will determine the constants and therefore uniquely specify the solution.

EXAMPLE 3 Find f if $f'(x) = e^x + 20(1 + x^2)^{-1}$ and $f(0) = -2$.

SOLUTION The general antiderivative of

$$f'(x) = e^x + \frac{20}{1+x^2}$$

is $\qquad\qquad\qquad f(x) = e^x + 20 \tan^{-1} x + C$

■ Figure 2 shows the graphs of the function f' in Example 3 and its antiderivative f. Notice that $f'(x) > 0$, so f is always increasing. Also notice that when f' has a maximum or minimum, f appears to have an inflection point. So the graph serves as a check on our calculation.

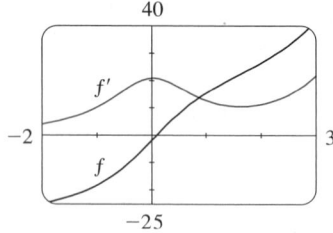

FIGURE 2

To determine C we use the fact that $f(0) = -2$:

$$f(0) = e^0 + 20 \tan^{-1} 0 + C = -2$$

Thus we have $C = -2 - 1 = -3$, so the particular solution is

$$f(x) = e^x + 20 \tan^{-1}x - 3$$ ■

V EXAMPLE 4 Find f if $f''(x) = 12x^2 + 6x - 4$, $f(0) = 4$, and $f(1) = 1$.

SOLUTION The general antiderivative of $f''(x) = 12x^2 + 6x - 4$ is

$$f'(x) = 12\frac{x^3}{3} + 6\frac{x^2}{2} - 4x + C = 4x^3 + 3x^2 - 4x + C$$

Using the antidifferentiation rules once more, we find that

$$f(x) = 4\frac{x^4}{4} + 3\frac{x^3}{3} - 4\frac{x^2}{2} + Cx + D = x^4 + x^3 - 2x^2 + Cx + D$$

To determine C and D we use the given conditions that $f(0) = 4$ and $f(1) = 1$. Since $f(0) = 0 + D = 4$, we have $D = 4$. Since

$$f(1) = 1 + 1 - 2 + C + 4 = 1$$

we have $C = -3$. Therefore the required function is

$$f(x) = x^4 + x^3 - 2x^2 - 3x + 4$$ ■

RECTILINEAR MOTION

Antidifferentiation is particularly useful in analyzing the motion of an object moving in a straight line. Recall that if the object has position function $s = f(t)$, then the velocity function is $v(t) = s'(t)$. This means that the position function is an antiderivative of the velocity function. Likewise, the acceleration function is $a(t) = v'(t)$, so the velocity function is an antiderivative of the acceleration. If the acceleration and the initial values $s(0)$ and $v(0)$ are known, then the position function can be found by antidifferentiating twice.

■ **www.stewartcalculus.com**
See Additional Example A.

V EXAMPLE 5 A particle moves in a straight line and has acceleration given by $a(t) = 6t + 4$. Its initial velocity is $v(0) = -6$ cm/s and its initial displacement is $s(0) = 9$ cm. Find its position function $s(t)$.

SOLUTION Since $v'(t) = a(t) = 6t + 4$, antidifferentiation gives

$$v(t) = 6\frac{t^2}{2} + 4t + C = 3t^2 + 4t + C$$

Note that $v(0) = C$. But we are given that $v(0) = -6$, so $C = -6$ and

$$v(t) = 3t^2 + 4t - 6$$

Since $v(t) = s'(t)$, s is the antiderivative of v:

$$s(t) = 3\frac{t^3}{3} + 4\frac{t^2}{2} - 6t + D = t^3 + 2t^2 - 6t + D$$

This gives $s(0) = D$. We are given that $s(0) = 9$, so $D = 9$ and the required position function is

$$s(t) = t^3 + 2t^2 - 6t + 9$$

An object near the surface of the earth is subject to a gravitational force that produces a downward acceleration denoted by g. For motion close to the ground we may assume that g is constant, its value being about 9.8 m/s^2 (or 32 ft/s^2).

EXAMPLE 6 A ball is thrown upward with a speed of 48 ft/s from the edge of a cliff 432 ft above the ground. Find its height above the ground t seconds later. When does it reach its maximum height? When does it hit the ground?

SOLUTION The motion is vertical and we choose the positive direction to be upward. At time t the distance above the ground is $s(t)$ and the velocity $v(t)$ is decreasing. Therefore the acceleration must be negative and we have

$$a(t) = \frac{dv}{dt} = -32$$

Taking antiderivatives, we have

$$v(t) = -32t + C$$

To determine C we use the given information that $v(0) = 48$. This gives $48 = 0 + C$, so

$$v(t) = -32t + 48$$

The maximum height is reached when $v(t) = 0$, that is, after 1.5 s. Since $s'(t) = v(t)$, we antidifferentiate again and obtain

$$s(t) = -16t^2 + 48t + D$$

Using the fact that $s(0) = 432$, we have $432 = 0 + D$ and so

$$s(t) = -16t^2 + 48t + 432$$

The expression for $s(t)$ is valid until the ball hits the ground. This happens when $s(t) = 0$, that is, when

$$-16t^2 + 48t + 432 = 0$$

or, equivalently,

$$t^2 - 3t - 27 = 0$$

Using the quadratic formula to solve this equation, we get

$$t = \frac{3 \pm 3\sqrt{13}}{2}$$

We reject the solution with the minus sign since it gives a negative value for t. Therefore the ball hits the ground after $3\left(1 + \sqrt{13}\right)/2 \approx 6.9$ s.

■ Figure 3 shows the position function of the ball in Example 6. The graph corroborates the conclusions we reached: The ball reaches its maximum height after 1.5 s and hits the ground after 6.9 s.

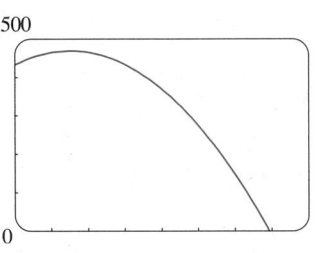

500

0   8

FIGURE 3

4.7 | EXERCISES

1–14 ▪ Find the most general antiderivative of the function. (Check your answer by differentiation.)

1. $f(x) = \frac{1}{2} + \frac{3}{4}x^2 - \frac{4}{5}x^3$

2. $f(x) = 8x^9 - 3x^6 + 12x^3$

3. $f(x) = 7x^{2/5} + 8x^{-4/5}$

4. $f(x) = \sqrt[3]{x^2} + x\sqrt{x}$

5. $f(x) = 3\sqrt{x} - 2\sqrt[3]{x}$

6. $f(t) = \dfrac{3t^4 - t^3 + 6t^2}{t^4}$

7. $g(t) = \dfrac{1 + t + t^2}{\sqrt{t}}$

8. $r(\theta) = \sec\theta\tan\theta - 2e^\theta$

9. $h(\theta) = 2\sin\theta - \sec^2\theta$

10. $f(t) = \sin t + 2\sinh t$

11. $f(x) = 5e^x - 3\cosh x$

12. $f(x) = 2\sqrt{x} + 6\cos x$

13. $f(x) = \dfrac{x^5 - x^3 + 2x}{x^4}$

14. $f(x) = \dfrac{2 + x^2}{1 + x^2}$

15–16 ▪ Find the antiderivative F of f that satisfies the given condition. Check your answer by comparing the graphs of f and F.

15. $f(x) = 5x^4 - 2x^5$, $F(0) = 4$

16. $f(x) = 4 - 3(1 + x^2)^{-1}$, $F(1) = 0$

17–34 ▪ Find f.

17. $f''(x) = 20x^3 - 12x^2 + 6x$

18. $f''(x) = x^6 - 4x^4 + x + 1$

19. $f''(x) = \frac{2}{3}x^{2/3}$

20. $f''(x) = 6x + \sin x$

21. $f'''(t) = \cos t$

22. $f'''(t) = e^t + t^{-4}$

23. $f'(x) = 1 + 3\sqrt{x}$, $f(4) = 25$

24. $f'(x) = 5x^4 - 3x^2 + 4$, $f(-1) = 2$

25. $f'(t) = 4/(1 + t^2)$, $f(1) = 0$

26. $f'(t) = t + 1/t^3$, $t > 0$, $f(1) = 6$

27. $f'(t) = 2\cos t + \sec^2 t$, $-\pi/2 < t < \pi/2$, $f(\pi/3) = 4$

28. $f'(x) = 4/\sqrt{1 - x^2}$, $f(\frac{1}{2}) = 1$

29. $f''(x) = -2 + 12x - 12x^2$, $f(0) = 4$, $f'(0) = 12$

30. $f''(x) = 8x^3 + 5$, $f(1) = 0$, $f'(1) = 8$

31. $f''(\theta) = \sin\theta + \cos\theta$, $f(0) = 3$, $f'(0) = 4$

32. $f''(t) = 3/\sqrt{t}$, $f(4) = 20$, $f'(4) = 7$

33. $f''(x) = x^{-2}$, $x > 0$, $f(1) = 0$, $f(2) = 0$

34. $f''(t) = 2e^t + 3\sin t$, $f(0) = 0$, $f(\pi) = 0$

35. Given that the graph of f passes through the point $(1, 6)$ and that the slope of its tangent line at $(x, f(x))$ is $2x + 1$, find $f(2)$.

36. Find a function f such that $f'(x) = x^3$ and the line $x + y = 0$ is tangent to the graph of f.

37–38 ▪ The graph of a function f is shown. Which graph is an antiderivative of f and why?

37.

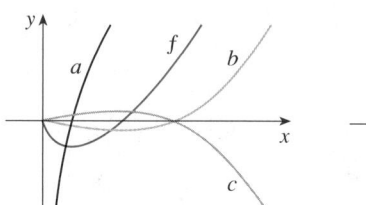

38.

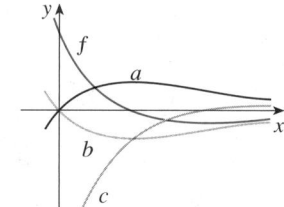

39–42 ▪ A particle is moving with the given data. Find the position of the particle.

39. $v(t) = \sin t - \cos t$, $s(0) = 0$

40. $v(t) = 1.5\sqrt{t}$, $s(4) = 10$

41. $a(t) = 10\sin t + 3\cos t$, $s(0) = 0$, $s(2\pi) = 12$

42. $a(t) = t^2 - 4t + 6$, $s(0) = 0$, $s(1) = 20$

43. A stone is dropped from the upper observation deck (the Space Deck) of the CN Tower, 450 m above the ground.
(a) Find the distance of the stone above ground level at time t.
(b) How long does it take the stone to reach the ground?
(c) With what velocity does it strike the ground?
(d) If the stone is thrown downward with a speed of 5 m/s, how long does it take to reach the ground?

44. Show that for motion in a straight line with constant acceleration a, initial velocity v_0, and initial displacement s_0, the displacement after time t is

$$s = \tfrac{1}{2}at^2 + v_0 t + s_0$$

45. An object is projected upward with initial velocity v_0 meters per second from a point s_0 meters above the ground. Show that

$$[v(t)]^2 = v_0^2 - 19.6[s(t) - s_0]$$

46. Two balls are thrown upward from the edge of the cliff in Example 6. The first is thrown with a speed of 48 ft/s and the other is thrown a second later with a speed of 24 ft/s. Do the balls ever pass each other?

47. A stone was dropped off a cliff and hit the ground with a speed of 120 ft/s. What is the height of the cliff?

48. If a diver of mass m stands at the end of a diving board with length L and linear density ρ, then the board takes on the shape of a curve $y = f(x)$, where

$$EIy'' = mg(L - x) + \tfrac{1}{2}\rho g(L - x)^2$$

E and I are positive constants that depend on the material of the board and $g \,(< 0)$ is the acceleration due to gravity.
(a) Find an expression for the shape of the curve.
(b) Use $f(L)$ to estimate the distance below the horizontal at the end of the board.

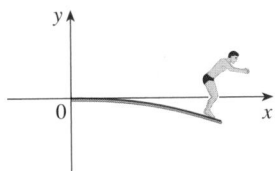

49. Since raindrops grow as they fall, their surface area increases and therefore the resistance to their falling increases. A raindrop has an initial downward velocity of 10 m/s and its downward acceleration is

$$a = \begin{cases} 9 - 0.9t & \text{if } 0 \le t \le 10 \\ 0 & \text{if } t > 10 \end{cases}$$

If the raindrop is initially 500 m above the ground, how long does it take to fall?

50. A car is traveling at 50 mi/h when the brakes are fully applied, producing a constant deceleration of 22 ft/s². What is the distance traveled before the car comes to a stop?

51. What constant acceleration is required to increase the speed of a car from 30 mi/h to 50 mi/h in 5 s?

52. A car braked with a constant deceleration of 16 ft/s², producing skid marks measuring 200 ft before coming to a stop. How fast was the car traveling when the brakes were first applied?

53. A car is traveling at 100 km/h when the driver sees an accident 80 m ahead and slams on the brakes. What constant deceleration is required to stop the car in time to avoid a pileup?

54. A model rocket is fired vertically upward from rest. Its acceleration for the first three seconds is $a(t) = 60t$, at which time the fuel is exhausted and it becomes a freely "falling" body. Fourteen seconds later, the rocket's parachute opens, and the (downward) velocity slows linearly to -18 ft/s in 5 s. The rocket then "floats" to the ground at that rate.
(a) Determine the position function s and the velocity function v (for all times t). Sketch the graphs of s and v.
(b) At what time does the rocket reach its maximum height, and what is that height?
(c) At what time does the rocket land?

55. A high-speed bullet train accelerates and decelerates at the rate of 4 ft/s². Its maximum cruising speed is 90 mi/h.
(a) What is the maximum distance the train can travel if it accelerates from rest until it reaches its cruising speed and then runs at that speed for 15 minutes?
(b) Suppose that the train starts from rest and must come to a complete stop in 15 minutes. What is the maximum distance it can travel under these conditions?
(c) Find the minimum time that the train takes to travel between two consecutive stations that are 45 miles apart.
(d) The trip from one station to the next takes 37.5 minutes. How far apart are the stations?

CHAPTER 4   REVIEW

CONCEPT CHECK

1. Explain the difference between an absolute maximum and a local maximum. Illustrate with a sketch.

2. (a) What does the Extreme Value Theorem say?
(b) Explain how the Closed Interval Method works.

3. (a) State Fermat's Theorem.
(b) Define a critical number of f.

4. (a) State Rolle's Theorem.
(b) State the Mean Value Theorem and give a geometric interpretation.

5. (a) State the Increasing/Decreasing Test.
(b) What does it mean to say that f is concave upward on an interval I?

(c) State the Concavity Test.
(d) What are inflection points? How do you find them?

6. (a) State the First Derivative Test.
(b) State the Second Derivative Test.
(c) What are the relative advantages and disadvantages of these tests?

7. If you have a graphing calculator or computer, why do you need calculus to graph a function?

8. (a) Given an initial approximation x_1 to a root of the equation $f(x) = 0$, explain geometrically, with a diagram, how the second approximation x_2 in Newton's method is obtained.

(b) Write an expression for x_2 in terms of x_1, $f(x_1)$, and $f'(x_1)$.

(c) Write an expression for x_{n+1} in terms of x_n, $f(x_n)$, and $f'(x_n)$.

(d) Under what circumstances is Newton's method likely to fail or to work very slowly?

9. (a) What is an antiderivative of a function f?

(b) Suppose F_1 and F_2 are both antiderivatives of f on an interval I. How are F_1 and F_2 related?

TRUE-FALSE QUIZ

Determine whether the statement is true or false. If it is true, explain why. If it is false, explain why or give an example that disproves the statement.

1. If $f'(c) = 0$, then f has a local maximum or minimum at c.

2. If f has an absolute minimum value at c, then $f'(c) = 0$.

3. If f is continuous on (a, b), then f attains an absolute maximum value $f(c)$ and an absolute minimum value $f(d)$ at some numbers c and d in (a, b).

4. If f is differentiable and $f(-1) = f(1)$, then there is a number c such that $|c| < 1$ and $f'(c) = 0$.

5. If $f'(x) < 0$ for $1 < x < 6$, then f is decreasing on $(1, 6)$.

6. If $f''(2) = 0$, then $(2, f(2))$ is an inflection point of the curve $y = f(x)$.

7. If $f'(x) = g'(x)$ for $0 < x < 1$, then $f(x) = g(x)$ for $0 < x < 1$.

8. There exists a function f such that $f(1) = -2$, $f(3) = 0$, and $f'(x) > 1$ for all x.

9. There exists a function f such that $f(x) > 0$, $f'(x) < 0$, and $f''(x) > 0$ for all x.

10. There exists a function f such that $f(x) < 0$, $f'(x) < 0$, and $f''(x) > 0$ for all x.

11. If f and g are increasing on an interval I, then $f + g$ is increasing on I.

12. If f and g are increasing on an interval I, then $f - g$ is increasing on I.

13. If f and g are increasing on an interval I, then fg is increasing on I.

14. If f and g are positive increasing functions on an interval I, then fg is increasing on I.

15. If f is increasing and $f(x) > 0$ on I, then $g(x) = 1/f(x)$ is decreasing on I.

16. If f is even, then f' is even.

17. If f is periodic, then f' is periodic.

18. The most general antiderivative of $f(x) = x^{-2}$ is

$$F(x) = -\frac{1}{x} + C$$

19. If $f'(x)$ exists and is nonzero for all x, then $f(1) \neq f(0)$.

EXERCISES

1–4 ▪ Find the local and absolute extreme values of the function on the given interval.

1. $f(x) = x^3 - 6x^2 + 9x + 1$, $[2, 4]$

2. $f(x) = x\sqrt{1 - x}$, $[-1, 1]$

3. $f(x) = \dfrac{3x - 4}{x^2 + 1}$, $[-2, 2]$ **4.** $f(x) = x^2 e^{-x}$, $[-1, 3]$

5–7 ▪ Sketch the graph of a function that satisfies the given conditions.

5. $f(0) = 0$, $f'(-2) = f'(1) = f'(9) = 0$,

$\lim_{x \to \infty} f(x) = 0$, $\lim_{x \to 6} f(x) = -\infty$,

$f'(x) < 0$ on $(-\infty, -2)$, $(1, 6)$, and $(9, \infty)$,

$f'(x) > 0$ on $(-2, 1)$ and $(6, 9)$,

$f''(x) > 0$ on $(-\infty, 0)$ and $(12, \infty)$,

$f''(x) < 0$ on $(0, 6)$ and $(6, 12)$

6. $f(0) = 0$, f is continuous and even,

$f'(x) = 2x$ if $0 < x < 1$, $f'(x) = -1$ if $1 < x < 3$,

$f'(x) = 1$ if $x > 3$

7. f is odd, $f'(x) < 0$ for $0 < x < 2$,

$f'(x) > 0$ for $x > 2$, $f''(x) > 0$ for $0 < x < 3$,

$f''(x) < 0$ for $x > 3$, $\lim_{x \to \infty} f(x) = -2$

8. The figure shows the graph of the *derivative* f' of a function f.

(a) On what intervals is f increasing or decreasing?

(b) For what values of x does f have a local maximum or minimum?

(c) Sketch the graph of f''.

(d) Sketch a possible graph of f.

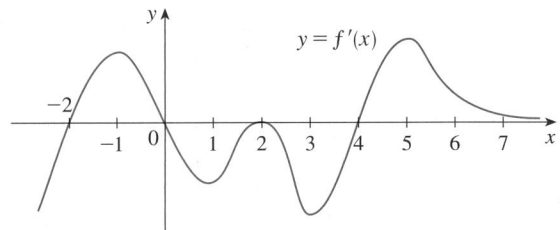

$y = f'(x)$

9–14 ■
(a) Find the vertical and horizontal asymptotes, if any.
(b) Find the intervals of increase or decrease.
(c) Find the local maximum and minimum values.
(d) Find the intervals of concavity and the inflection points.
(e) Use the information from parts (a)–(d) to sketch the graph of f. Check your work with a graphing device.

9. $f(x) = 2 - 2x - x^3$

10. $f(x) = \dfrac{1}{1 - x^2}$

11. $y = \sin^2 x - 2\cos x$

12. $y = e^{2x - x^2}$

13. $y = e^x + e^{-3x}$

14. $y = \ln(x^2 - 1)$

15–24 ■ Use the guidelines of Section 4.4 to sketch the curve.

15. $y = x^4 - 3x^3 + 3x^2 - x$

16. $y = x^3 - 6x^2 - 15x + 4$

17. $y = \dfrac{1}{x(x - 3)^2}$

18. $y = \dfrac{x}{1 - x^2}$

19. $y = x\sqrt{2 + x}$

20. $y = \sqrt{1 - x} + \sqrt{1 + x}$

21. $y = \sin^{-1}(1/x)$

22. $y = 4x - \tan x, \quad -\pi/2 < x < \pi/2$

23. $y = (x - 2)e^{-x}$

24. $y = x + \ln(x^2 + 1)$

25–28 ■ Produce graphs of f that reveal all the important aspects of the curve. Use graphs of f' and f'' to estimate the intervals of increase and decrease, extreme values, intervals of concavity, and inflection points. In Exercise 25 use calculus to find these quantities exactly.

25. $f(x) = \dfrac{x^2 - 1}{x^3}$

26. $f(x) = \dfrac{x^3 - x}{x^2 + x + 3}$

27. $f(x) = 3x^6 - 5x^5 + x^4 - 5x^3 - 2x^2 + 2$

28. $f(x) = x^2 + 6.5\sin x, \quad -5 \leq x \leq 5$

29. Graph $f(x) = e^{-1/x^2}$ in a viewing rectangle that shows all the main aspects of this function. Estimate the inflection points. Then use calculus to find them exactly.

CAS 30. (a) Graph the function $f(x) = 1/(1 + e^{1/x})$.
(b) Explain the shape of the graph by computing the limits of $f(x)$ as x approaches ∞, $-\infty$, 0^+, and 0^-.

(c) Use the graph of f to estimate the coordinates of the inflection points.
(d) Use your CAS to compute and graph f''.
(e) Use the graph in part (d) to estimate the inflection points more accurately.

CAS 31–32 ■ Use the graphs of f, f', and f'' to estimate the x-coordinates of the maximum and minimum points and inflection points of f.

31. $f(x) = \dfrac{\cos^2 x}{\sqrt{x^2 + x + 1}}, \quad -\pi \leq x \leq \pi$

32. $f(x) = e^{-0.1x}\ln(x^2 - 1)$

33. Investigate the family of functions $f(x) = \ln(\sin x + C)$. What features do the members of this family have in common? How do they differ? For which values of C is f continuous on $(-\infty, \infty)$? For which values of C does f have no graph at all? What happens as $C \to \infty$?

34. Investigate the family of functions $f(x) = cxe^{-cx^2}$. What happens to the maximum and minimum points and the inflection points as c changes? Illustrate your conclusions by graphing several members of the family.

35. Show that the equation $3x + 2\cos x + 5 = 0$ has exactly one real root.

36. Suppose that f is continuous on $[0, 4]$, $f(0) = 1$, and $2 \leq f'(x) \leq 5$ for all x in $(0, 4)$. Show that $9 \leq f(4) \leq 21$.

37. By applying the Mean Value Theorem to the function $f(x) = x^{1/5}$ on the interval $[32, 33]$, show that

$$2 < \sqrt[5]{33} < 2.0125$$

38. For what values of the constants a and b is $(1, 3)$ a point of inflection of the curve $y = ax^3 + bx^2$?

39. Find two positive integers such that the sum of the first number and four times the second number is 1000 and the product of the numbers is as large as possible.

40. Find the point on the hyperbola $xy = 8$ that is closest to the point $(3, 0)$.

41. Find the smallest possible area of an isosceles triangle that is circumscribed about a circle of radius r.

42. Find the volume of the largest circular cone that can be inscribed in a sphere of radius r.

43. In $\triangle ABC$, D lies on AB, $|CD| = 5$ cm, $|AD| = 4$ cm, $|BD| = 4$ cm, and $CD \perp AB$. Where should a point P be chosen on CD so that the sum $|PA| + |PB| + |PC|$ is a minimum? What if $|CD| = 2$ cm?

44. An observer stands at a point P, one unit away from a track. Two runners start at the point S in the figure and run along the track. One runner runs three times as fast as the other. Find the maximum value of the observer's angle of sight θ between the runners. [*Hint:* Maximize $\tan\theta$.]

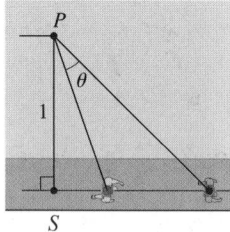

45. The velocity of a wave of length L in deep water is

$$v = K\sqrt{\frac{L}{C} + \frac{C}{L}}$$

where K and C are known positive constants. What is the length of the wave that gives the minimum velocity?

46. A metal storage tank with volume V is to be constructed in the shape of a right circular cylinder surmounted by a hemisphere. What dimensions will require the least amount of metal?

47. A hockey team plays in an arena with a seating capacity of 15,000 spectators. With the ticket price set at \$12, average attendance at a game has been 11,000. A market survey indicates that for each dollar the ticket price is lowered, average attendance will increase by 1000. How should the owners of the team set the ticket price to maximize their revenue from ticket sales?

48. Use Newton's method to find all roots of the equation $\sin x = x^2 - 3x + 1$ correct to six decimal places.

49. Use Newton's method to find the absolute maximum value of the function $f(t) = \cos t + t - t^2$ correct to eight decimal places.

50. Use the guidelines in Section 4.4 to sketch the curve $y = x \sin x$, $0 \le x \le 2\pi$. Use Newton's method when necessary.

51–52 ■ Find the most general antiderivative of the function.

51. $f(x) = e^x - \left(2/\sqrt{x}\right)$ **52.** $g(t) = (1 + t)/\sqrt{t}$

53–56 ■ Find $f(x)$.

53. $f'(x) = 2/(1 + x^2)$, $f(0) = -1$

54. $f'(u) = \dfrac{u^2 + \sqrt{u}}{u}$, $f(1) = 3$

55. $f''(x) = 1 - 6x + 48x^2$, $f(0) = 1$, $f'(0) = 2$

56. $f''(x) = 2x^3 + 3x^2 - 4x + 5$, $f(0) = 2$, $f(1) = 0$

57–58 ■ A particle is moving with the given data. Find the position of the particle.

57. $v(t) = 2t - 1/(1 + t^2)$, $s(0) = 1$

58. $a(t) = \sin t + 3 \cos t$, $s(0) = 0$, $v(0) = 2$

59. A canister is dropped from a helicopter 500 m above the ground. Its parachute does not open, but the canister has been designed to withstand an impact velocity of 100 m/s. Will it burst?

60. Investigate the family of curves given by

$$f(x) = x^4 + x^3 + cx^2$$

In particular you should determine the transitional value of c at which the number of critical numbers changes and the transitional value at which the number of inflection points changes. Illustrate the various possible shapes with graphs.

61. A rectangular beam will be cut from a cylindrical log of radius 10 inches.
  (a) Show that the beam of maximal cross-sectional area is a square.
  (b) Four rectangular planks will be cut from the four sections of the log that remain after cutting the square beam. Determine the dimensions of the planks that will have maximal cross-sectional area.
  (c) Suppose that the strength of a rectangular beam is proportional to the product of its width and the square of its depth. Find the dimensions of the strongest beam that can be cut from the cylindrical log.

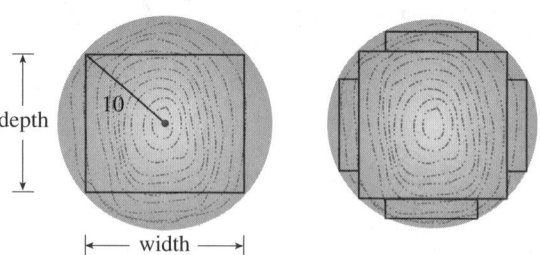

5

INTEGRALS

In Chapter 2 we used the tangent and velocity problems to introduce the derivative, which is the central idea in differential calculus. In much the same way, this chapter starts with the area and distance problems and uses them to formulate the idea of a definite integral, which is the basic concept of integral calculus. We will see in Chapter 7 how to use the integral to solve problems concerning volumes, lengths of curves, work, forces on a dam, and centers of mass, among many others.

There is a connection between integral calculus and differential calculus. The Fundamental Theorem of Calculus relates the integral to the derivative, and we will see in this chapter that it greatly simplifies the solution of many problems.

5.1 | AREAS AND DISTANCES

In this section we discover that in trying to find the area under a curve or the distance traveled by a car, we end up with the same special type of limit.

THE AREA PROBLEM

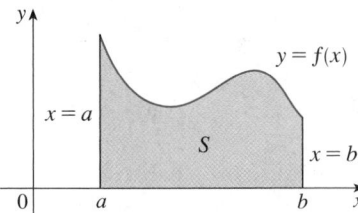

FIGURE 1
$S = \{(x, y) \mid a \leq x \leq b, 0 \leq y \leq f(x)\}$

We begin by attempting to solve the *area problem*: Find the area of the region S that lies under the curve $y = f(x)$ from a to b. This means that S, illustrated in Figure 1, is bounded by the graph of a continuous function f [where $f(x) \geq 0$], the vertical lines $x = a$ and $x = b$, and the x-axis.

In trying to solve the area problem we have to ask ourselves: What is the meaning of the word *area*? This question is easy to answer for regions with straight sides. For a rectangle, the area is defined as the product of the length and the width. The area of a triangle is half the base times the height. The area of a polygon is found by dividing it into triangles (as in Figure 2) and adding the areas of the triangles.

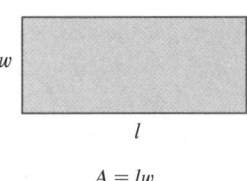

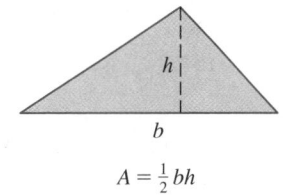

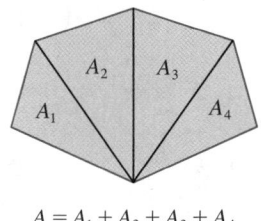

FIGURE 2 $A = lw$ $A = \frac{1}{2} bh$ $A = A_1 + A_2 + A_3 + A_4$

However, it isn't so easy to find the area of a region with curved sides. We all have an intuitive idea of what the area of a region is. But part of the area problem is to make this intuitive idea precise by giving an exact definition of area.

Recall that in defining a tangent we first approximated the slope of the tangent line by slopes of secant lines and then we took the limit of these approximations. We pursue a similar idea for areas. We first approximate the region S by rectangles and then we take the limit of the areas of these rectangles as we increase the number of rectangles. The following example illustrates the procedure.

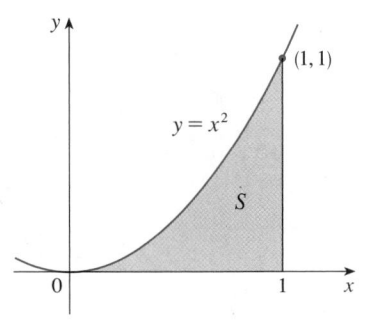

FIGURE 3

V EXAMPLE 1 Use rectangles to estimate the area under the parabola $y = x^2$ from 0 to 1 (the parabolic region S illustrated in Figure 3).

SOLUTION We first notice that the area of S must be somewhere between 0 and 1 because S is contained in a square with side length 1, but we can certainly do better than that. Suppose we divide S into four strips S_1, S_2, S_3, and S_4 by drawing the vertical lines $x = \frac{1}{4}$, $x = \frac{1}{2}$, and $x = \frac{3}{4}$ as in Figure 4(a).

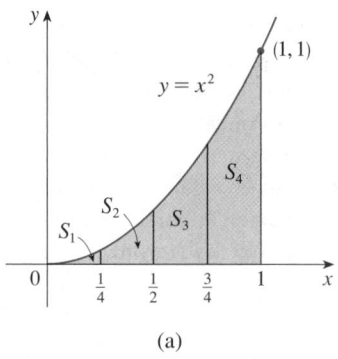
(a)

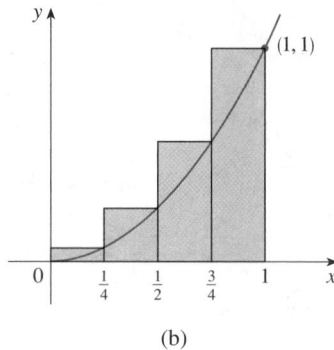
(b)

FIGURE 4

We can approximate each strip by a rectangle whose base is the same as the strip and whose height is the same as the right edge of the strip [see Figure 4(b)]. In other words, the heights of these rectangles are the values of the function $f(x) = x^2$ at the right endpoints of the subintervals $\left[0, \frac{1}{4}\right]$, $\left[\frac{1}{4}, \frac{1}{2}\right]$, $\left[\frac{1}{2}, \frac{3}{4}\right]$, and $\left[\frac{3}{4}, 1\right]$.

Each rectangle has width $\frac{1}{4}$ and the heights are $\left(\frac{1}{4}\right)^2$, $\left(\frac{1}{2}\right)^2$, $\left(\frac{3}{4}\right)^2$, and 1^2. If we let R_4 be the sum of the areas of these approximating rectangles, we get

$$R_4 = \frac{1}{4} \cdot \left(\frac{1}{4}\right)^2 + \frac{1}{4} \cdot \left(\frac{1}{2}\right)^2 + \frac{1}{4} \cdot \left(\frac{3}{4}\right)^2 + \frac{1}{4} \cdot 1^2 = \frac{15}{32} = 0.46875$$

From Figure 4(b) we see that the area A of S is less than R_4, so

$$A < 0.46875$$

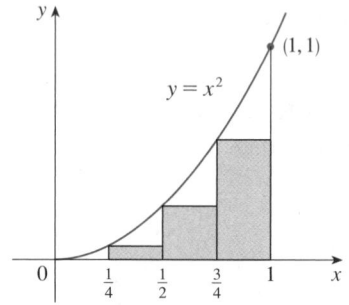

FIGURE 5

Instead of using the rectangles in Figure 4(b) we could use the smaller rectangles in Figure 5 whose heights are the values of f at the left endpoints of the subintervals. (The leftmost rectangle has collapsed because its height is 0.) The sum of the areas of these approximating rectangles is

$$L_4 = \frac{1}{4} \cdot 0^2 + \frac{1}{4} \cdot \left(\frac{1}{4}\right)^2 + \frac{1}{4} \cdot \left(\frac{1}{2}\right)^2 + \frac{1}{4} \cdot \left(\frac{3}{4}\right)^2 = \frac{7}{32} = 0.21875$$

We see that the area of S is larger than L_4, so we have lower and upper estimates for A:

$$0.21875 < A < 0.46875$$

We can repeat this procedure with a larger number of strips. Figure 6 shows what happens when we divide the region S into eight strips of equal width.

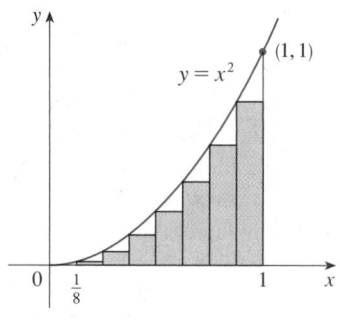
(a) Using left endpoints

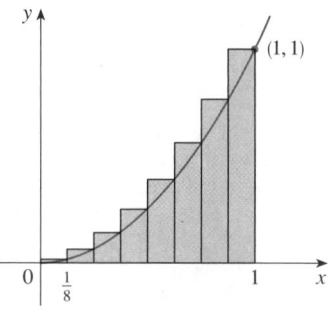
(b) Using right endpoints

FIGURE 6
Approximating S with eight rectangles

By computing the sum of the areas of the smaller rectangles (L_8) and the sum of the areas of the larger rectangles (R_8), we obtain better lower and upper estimates for A:

$$0.2734375 < A < 0.3984375$$

So one possible answer to the question is to say that the true area of S lies somewhere between 0.2734375 and 0.3984375.

We could obtain better estimates by increasing the number of strips. The table at the left shows the results of similar calculations (with a computer) using n rectangles whose heights are found with left endpoints (L_n) or right endpoints (R_n). In particular, we see by using 50 strips that the area lies between 0.3234 and 0.3434. With 1000 strips we narrow it down even more: A lies between 0.3328335 and 0.3338335. A good estimate is obtained by averaging these numbers: $A \approx 0.3333335$. ∎

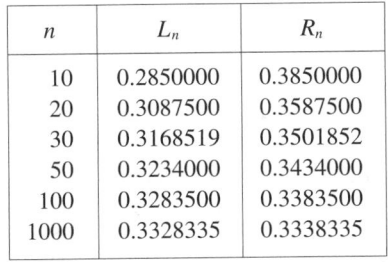

n	L_n	R_n
10	0.2850000	0.3850000
20	0.3087500	0.3587500
30	0.3168519	0.3501852
50	0.3234000	0.3434000
100	0.3283500	0.3383500
1000	0.3328335	0.3338335

From the values in the table in Example 1, it looks as if R_n is approaching $\frac{1}{3}$ as n increases. We confirm this in the next example.

▼ EXAMPLE 2   For the region S in Example 1, show that the sum of the areas of the upper approximating rectangles approaches $\frac{1}{3}$, that is,

$$\lim_{n \to \infty} R_n = \tfrac{1}{3}$$

SOLUTION R_n is the sum of the areas of the n rectangles in Figure 7. Each rectangle has width $1/n$ and the heights are the values of the function $f(x) = x^2$ at the points $1/n, 2/n, 3/n, \ldots, n/n$; that is, the heights are $(1/n)^2, (2/n)^2, (3/n)^2, \ldots, (n/n)^2$. Thus

$$R_n = \frac{1}{n}\left(\frac{1}{n}\right)^2 + \frac{1}{n}\left(\frac{2}{n}\right)^2 + \frac{1}{n}\left(\frac{3}{n}\right)^2 + \cdots + \frac{1}{n}\left(\frac{n}{n}\right)^2$$

$$= \frac{1}{n} \cdot \frac{1}{n^2}\left(1^2 + 2^2 + 3^2 + \cdots + n^2\right)$$

$$= \frac{1}{n^3}\left(1^2 + 2^2 + 3^2 + \cdots + n^2\right)$$

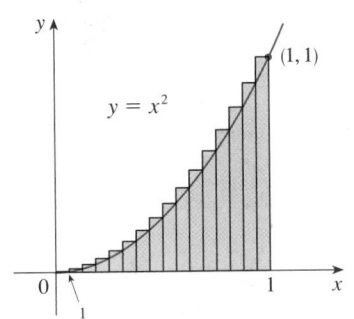

FIGURE 7

Here we need the formula for the sum of the squares of the first n positive integers:

| 1 |

$$1^2 + 2^2 + 3^2 + \cdots + n^2 = \frac{n(n+1)(2n+1)}{6}$$

Perhaps you have seen this formula before. It is proved in Example 5 in Appendix B. Putting Formula 1 into our expression for R_n, we get

$$R_n = \frac{1}{n^3} \cdot \frac{n(n+1)(2n+1)}{6} = \frac{(n+1)(2n+1)}{6n^2}$$

■ Here we are computing the limit of the sequence $\{R_n\}$. Sequences and their limits will be studied in detail in Section 8.1. The idea is very similar to a limit at infinity (Section 1.6) except that in writing $\lim_{n \to \infty}$ we restrict n to be a positive integer. In particular, we know that

$$\lim_{n \to \infty} \frac{1}{n} = 0$$

When we write $\lim_{n \to \infty} R_n = \frac{1}{3}$ we mean that we can make R_n as close to $\frac{1}{3}$ as we like by taking n sufficiently large.

Thus we have

$$\lim_{n \to \infty} R_n = \lim_{n \to \infty} \frac{(n + 1)(2n + 1)}{6n^2}$$

$$= \lim_{n \to \infty} \frac{1}{6} \left(\frac{n + 1}{n} \right) \left(\frac{2n + 1}{n} \right)$$

$$= \lim_{n \to \infty} \frac{1}{6} \left(1 + \frac{1}{n} \right) \left(2 + \frac{1}{n} \right)$$

$$= \frac{1}{6} \cdot 1 \cdot 2 = \frac{1}{3} \qquad ■$$

It can be shown that the lower approximating sums also approach $\frac{1}{3}$, that is,

$$\lim_{n \to \infty} L_n = \frac{1}{3}$$

From Figures 8 and 9 it appears that, as n increases, both L_n and R_n become better and better approximations to the area of S. Therefore we *define* the area A to be the limit of the sums of the areas of the approximating rectangles, that is,

$$A = \lim_{n \to \infty} R_n = \lim_{n \to \infty} L_n = \frac{1}{3}$$

TEC In Visual 5.1 you can create pictures like those in Figures 8 and 9 for other values of n.

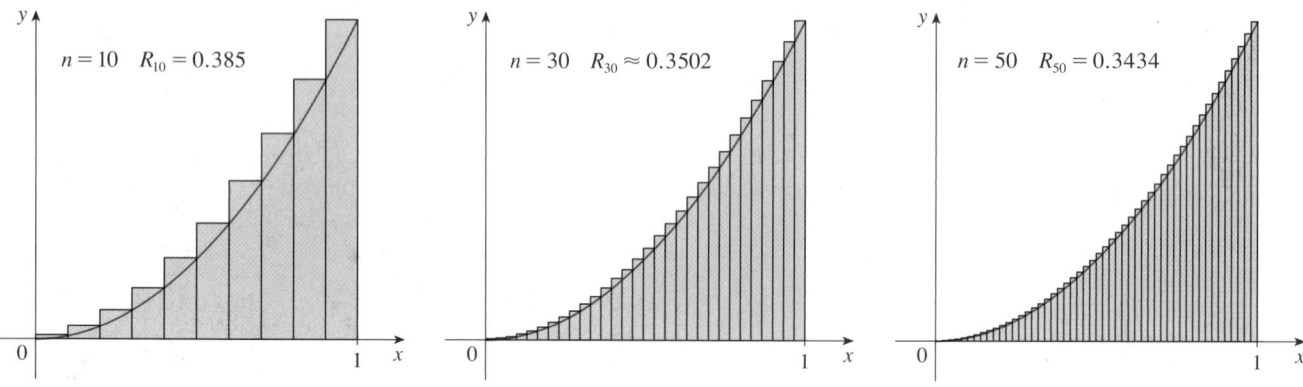

FIGURE 8 Right endpoints produce upper sums because $f(x) = x^2$ is increasing.

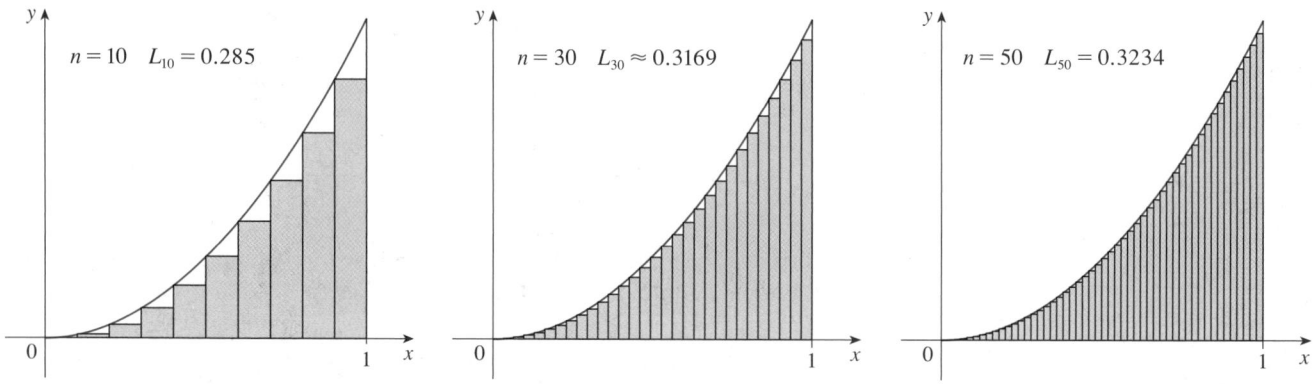

FIGURE 9 Left endpoints produce lower sums because $f(x) = x^2$ is increasing.

Let's apply the idea of Examples 1 and 2 to the more general region S of Figure 1. We start by subdividing S into n strips $S_1, S_2, \ldots, S_n$ of equal width as in Figure 10.

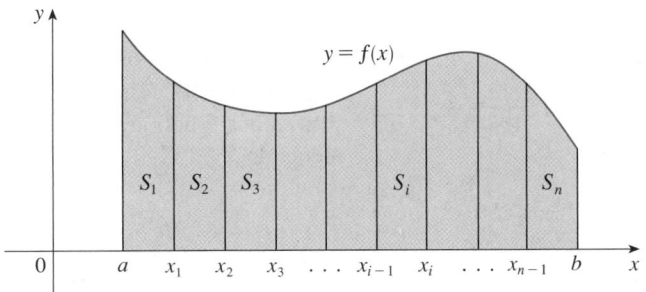

FIGURE 10

The width of the interval $[a, b]$ is $b - a$, so the width of each of the n strips is

$$\Delta x = \frac{b - a}{n}$$

These strips divide the interval $[a, b]$ into n subintervals

$$[x_0, x_1], \quad [x_1, x_2], \quad [x_2, x_3], \quad \ldots, \quad [x_{n-1}, x_n]$$

where $x_0 = a$ and $x_n = b$. The right endpoints of the subintervals are

$$x_1 = a + \Delta x,$$

$$x_2 = a + 2\,\Delta x,$$

$$x_3 = a + 3\,\Delta x,$$

$$\vdots$$

Let's approximate the ith strip S_i by a rectangle with width Δx and height $f(x_i)$, which is the value of f at the right endpoint (see Figure 11). Then the area of the ith rectangle is $f(x_i)\,\Delta x$. What we think of intuitively as the area of S is approximated by the sum of the areas of these rectangles, which is

$$R_n = f(x_1)\,\Delta x + f(x_2)\,\Delta x + \cdots + f(x_n)\,\Delta x$$

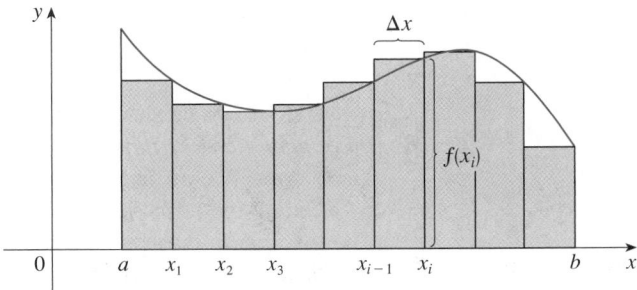

FIGURE 11

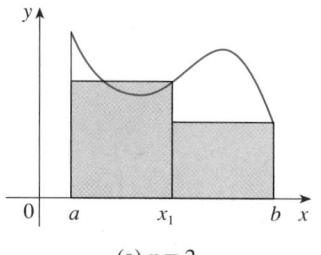

(a) $n = 2$

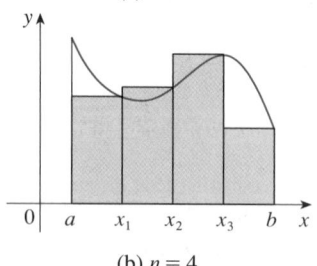

(b) $n = 4$

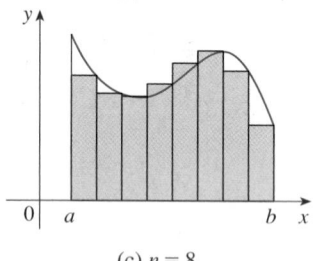

(c) $n = 8$

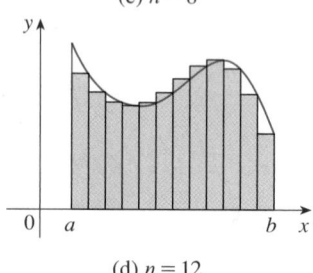

(d) $n = 12$

FIGURE 12

Figure 12 shows this approximation for $n = 2, 4, 8,$ and 12. Notice that this approximation appears to become better and better as the number of strips increases, that is, as $n \to \infty$. Therefore we define the area A of the region S in the following way.

---

2   **DEFINITION**   The **area** A of the region S that lies under the graph of the continuous function f is the limit of the sum of the areas of approximating rectangles:

$$A = \lim_{n \to \infty} R_n = \lim_{n \to \infty} \left[f(x_1) \, \Delta x + f(x_2) \, \Delta x + \cdots + f(x_n) \, \Delta x \right]$$

It can be proved that the limit in Definition 2 always exists, since we are assuming that f is continuous. It can also be shown that we get the same value if we use left endpoints:

$$\boxed{3} \qquad A = \lim_{n \to \infty} L_n = \lim_{n \to \infty} \left[f(x_0) \, \Delta x + f(x_1) \, \Delta x + \cdots + f(x_{n-1}) \, \Delta x \right]$$

In fact, instead of using left endpoints or right endpoints, we could take the height of the ith rectangle to be the value of f at *any* number x_i^* in the ith subinterval $[x_{i-1}, x_i]$. We call the numbers $x_1^*, x_2^*, \ldots, x_n^*$ the **sample points**. Figure 13 shows approximating rectangles when the sample points are not chosen to be endpoints. So a more general expression for the area of S is

$$\boxed{4} \qquad A = \lim_{n \to \infty} \left[f(x_1^*) \, \Delta x + f(x_2^*) \, \Delta x + \cdots + f(x_n^*) \, \Delta x \right]$$

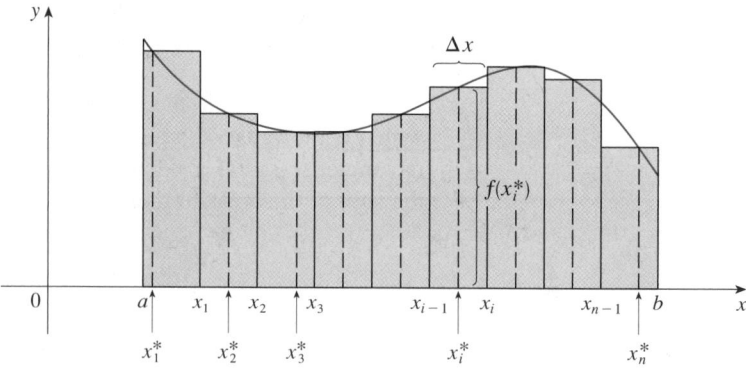

FIGURE 13

NOTE It can be shown that an equivalent definition of area is the following: *A is the unique number that is smaller than all the upper sums and bigger than all the lower sums.* We saw in Examples 1 and 2, for instance, that the area $\left(A = \frac{1}{3} \right)$ is trapped between all the left approximating sums L_n and all the right approximating sums R_n. The function in those examples, $f(x) = x^2$, happens to be increasing on $[0, 1]$ and so the lower sums arise from left endpoints and the upper sums from right endpoints. (See Figures 8 and 9.) In general, we form **lower** (and **upper**) **sums** by choosing the sample points x_i^* so that $f(x_i^*)$ is the minimum (and maximum) value of f on the ith subinterval. (See Figure 14 and Exercises 7–8.)

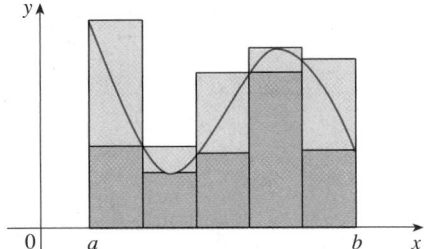

FIGURE 14

Lower sums (short rectangles) and upper sums (tall rectangles)

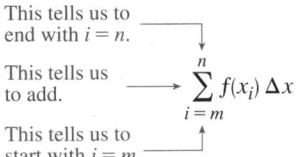

This tells us to end with $i = n$.

This tells us to add.

$$\sum_{i=m}^{n} f(x_i)\,\Delta x$$

This tells us to start with $i = m$.

■ If you need practice with sigma notation, look at the examples and try some of the exercises in Appendix B.

We often use **sigma notation** to write sums with many terms more compactly. For instance,

$$\sum_{i=1}^{n} f(x_i)\,\Delta x = f(x_1)\,\Delta x + f(x_2)\,\Delta x + \cdots + f(x_n)\,\Delta x$$

So the expressions for area in Equations 2, 3, and 4 can be written as follows:

$$A = \lim_{n\to\infty} \sum_{i=1}^{n} f(x_i)\,\Delta x$$

$$A = \lim_{n\to\infty} \sum_{i=1}^{n} f(x_{i-1})\,\Delta x$$

$$A = \lim_{n\to\infty} \sum_{i=1}^{n} f(x_i^*)\,\Delta x$$

We can also rewrite Formula 1 in the following way:

$$\sum_{i=1}^{n} i^2 = \frac{n(n+1)(2n+1)}{6}$$

EXAMPLE 3 Let A be the area of the region that lies under the graph of $f(x) = e^{-x}$ between $x = 0$ and $x = 2$.
(a) Using right endpoints, find an expression for A as a limit. Do not evaluate the limit.
(b) Estimate the area by taking the sample points to be midpoints and using four subintervals and then ten subintervals.

SOLUTION
(a) Since $a = 0$ and $b = 2$, the width of a subinterval is

$$\Delta x = \frac{2-0}{n} = \frac{2}{n}$$

So $x_1 = 2/n$, $x_2 = 4/n$, $x_3 = 6/n$, $x_i = 2i/n$, and $x_n = 2n/n$. The sum of the areas of the approximating rectangles is

$$R_n = f(x_1)\,\Delta x + f(x_2)\,\Delta x + \cdots + f(x_n)\,\Delta x$$

$$= e^{-x_1}\,\Delta x + e^{-x_2}\,\Delta x + \cdots + e^{-x_n}\,\Delta x$$

$$= e^{-2/n}\left(\frac{2}{n}\right) + e^{-4/n}\left(\frac{2}{n}\right) + \cdots + e^{-2n/n}\left(\frac{2}{n}\right)$$

According to Definition 2, the area is

$$A = \lim_{n \to \infty} R_n = \lim_{n \to \infty} \frac{2}{n}\left(e^{-2/n} + e^{-4/n} + e^{-6/n} + \cdots + e^{-2n/n}\right)$$

Using sigma notation we could write

$$A = \lim_{n \to \infty} \frac{2}{n} \sum_{i=1}^{n} e^{-2i/n}$$

It is difficult to evaluate this limit directly by hand, but with the aid of a computer algebra system it isn't hard (see Exercise 22). In Section 5.3 we will be able to find A more easily using a different method.

(b) With $n = 4$ the subintervals of equal width $\Delta x = 0.5$ are $[0, 0.5]$, $[0.5, 1]$, $[1, 1.5]$, and $[1.5, 2]$. The midpoints of these subintervals are $x_1^* = 0.25$, $x_2^* = 0.75$, $x_3^* = 1.25$, and $x_4^* = 1.75$, and the sum of the areas of the four approximating rectangles (see Figure 15) is

$$\begin{aligned}
M_4 &= \sum_{i=1}^{4} f(x_i^*)\,\Delta x \\
&= f(0.25)\,\Delta x + f(0.75)\,\Delta x + f(1.25)\,\Delta x + f(1.75)\,\Delta x \\
&= e^{-0.25}(0.5) + e^{-0.75}(0.5) + e^{-1.25}(0.5) + e^{-1.75}(0.5) \\
&= \tfrac{1}{2}\left(e^{-0.25} + e^{-0.75} + e^{-1.25} + e^{-1.75}\right) \approx 0.8557
\end{aligned}$$

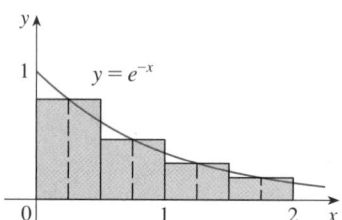

FIGURE 15

So an estimate for the area is

$$A \approx 0.8557$$

With $n = 10$ the subintervals are $[0, 0.2]$, $[0.2, 0.4]$, $\ldots$, $[1.8, 2]$ and the midpoints are $x_1^* = 0.1$, $x_2^* = 0.3$, $x_3^* = 0.5$, $\ldots$, $x_{10}^* = 1.9$. Thus

$$\begin{aligned}
A &\approx M_{10} = f(0.1)\,\Delta x + f(0.3)\,\Delta x + f(0.5)\,\Delta x + \cdots + f(1.9)\,\Delta x \\
&= 0.2\left(e^{-0.1} + e^{-0.3} + e^{-0.5} + \cdots + e^{-1.9}\right) \approx 0.8632
\end{aligned}$$

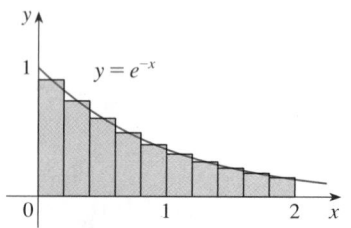

FIGURE 16

From Figure 16 it appears that this estimate is better than the estimate with $n = 4$. ∎

THE DISTANCE PROBLEM

Now let's consider the *distance problem*: Find the distance traveled by an object during a certain time period if the velocity of the object is known at all times. (In a sense this is the inverse problem of the velocity problem that we discussed in Section 2.1.) If the velocity remains constant, then the distance problem is easy to solve by means of the formula

$$\text{distance} = \text{velocity} \times \text{time}$$

But if the velocity varies, it's not so easy to find the distance traveled. We investigate the problem in the following example.

▼ **EXAMPLE 4** Suppose the odometer on our car is broken and we want to estimate the distance driven over a 30-second time interval. We take speedometer readings every five seconds and record them in the following table:

Time (s)	0	5	10	15	20	25	30
Velocity (mi/h)	17	21	24	29	32	31	28

In order to have the time and the velocity in consistent units, let's convert the velocity readings to feet per second (1 mi/h = 5280/3600 ft/s):

Time (s)	0	5	10	15	20	25	30
Velocity (ft/s)	25	31	35	43	47	46	41

During the first five seconds the velocity doesn't change very much, so we can estimate the distance traveled during that time by assuming that the velocity is constant. If we take the velocity during that time interval to be the initial velocity (25 ft/s), then we obtain the approximate distance traveled during the first five seconds:

$$25 \text{ ft/s} \times 5 \text{ s} = 125 \text{ ft}$$

Similarly, during the second time interval the velocity is approximately constant and we take it to be the velocity when $t = 5$ s. So our estimate for the distance traveled from $t = 5$ s to $t = 10$ s is

$$31 \text{ ft/s} \times 5 \text{ s} = 155 \text{ ft}$$

If we add similar estimates for the other time intervals, we obtain an estimate for the total distance traveled:

$$(25 \times 5) + (31 \times 5) + (35 \times 5) + (43 \times 5) + (47 \times 5) + (46 \times 5) = 1135 \text{ ft}$$

We could just as well have used the velocity at the *end* of each time period instead of the velocity at the beginning as our assumed constant velocity. Then our estimate becomes

$$(31 \times 5) + (35 \times 5) + (43 \times 5) + (47 \times 5) + (46 \times 5) + (41 \times 5) = 1215 \text{ ft}$$

If we had wanted a more accurate estimate, we could have taken velocity readings every two seconds, or even every second. ∎

Perhaps the calculations in Example 4 remind you of the sums we used earlier to estimate areas. The similarity is explained when we sketch a graph of the velocity function of the car in Figure 17 and draw rectangles whose heights are the initial velocities for each time interval. The area of the first rectangle is $25 \times 5 = 125$, which is also our estimate for the distance traveled in the first five seconds. In fact, the area of each rectangle can be interpreted as a distance because the height represents velocity and the width represents time. The sum of the areas of the rectangles in Figure 17 is $L_6 = 1135$, which is our initial estimate for the total distance traveled.

In general, suppose an object moves with velocity $v = f(t)$, where $a \le t \le b$ and $f(t) \ge 0$ (so the object always moves in the positive direction). We take velocity readings at times $t_0 \, (= a), t_1, t_2, \ldots, t_n \, (= b)$ so that the velocity is approximately constant on each subinterval. If these times are equally spaced, then the time between consecutive readings is $\Delta t = (b - a)/n$. During the first time interval the velocity is approximately $f(t_0)$ and so the distance traveled is approximately $f(t_0) \, \Delta t$. Similarly, the distance traveled during the second time interval is about $f(t_1) \, \Delta t$ and the total dis-

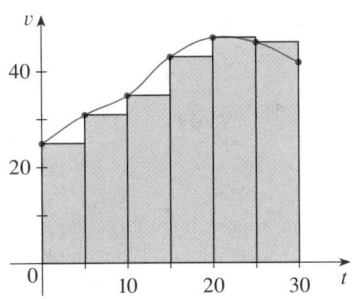

FIGURE 17

tance traveled during the time interval $[a, b]$ is approximately

$$f(t_0) \, \Delta t + f(t_1) \, \Delta t + \cdots + f(t_{n-1}) \, \Delta t = \sum_{i=1}^{n} f(t_{i-1}) \, \Delta t$$

If we use the velocity at right endpoints instead of left endpoints, our estimate for the total distance becomes

$$f(t_1) \, \Delta t + f(t_2) \, \Delta t + \cdots + f(t_n) \, \Delta t = \sum_{i=1}^{n} f(t_i) \, \Delta t$$

The more frequently we measure the velocity, the more accurate our estimates become, so it seems plausible that the *exact* distance d traveled is the *limit* of such expressions:

$$\boxed{5} \qquad d = \lim_{n \to \infty} \sum_{i=1}^{n} f(t_{i-1}) \, \Delta t = \lim_{n \to \infty} \sum_{i=1}^{n} f(t_i) \, \Delta t$$

We will see in Section 5.3 that this is indeed true.

Because Equation 5 has the same form as our expressions for area in Equations 2 and 3, it follows that the distance traveled is equal to the area under the graph of the velocity function. In Chapter 7 we will see that other quantities of interest in the natural and social sciences—such as the work done by a variable force—can also be interpreted as the area under a curve. So when we compute areas in this chapter, bear in mind that they can be interpreted in a variety of practical ways.

5.1 | EXERCISES

1. (a) By reading values from the given graph of f, use five rectangles to find a lower estimate and an upper estimate for the area under the given graph of f from $x = 0$ to $x = 10$. In each case sketch the rectangles that you use.
(b) Find new estimates using ten rectangles in each case.

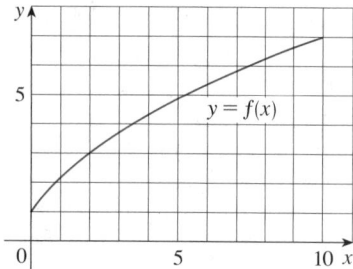

2. (a) Use six rectangles to find estimates of each type for the area under the given graph of f from $x = 0$ to $x = 12$.
  (i) L_6 (sample points are left endpoints)
  (ii) R_6 (sample points are right endpoints)
  (iii) M_6 (sample points are midpoints)
(b) Is L_6 an underestimate or overestimate of the true area?
(c) Is R_6 an underestimate or overestimate of the true area?

(d) Which of the numbers L_6, R_6, or M_6 gives the best estimate? Explain.

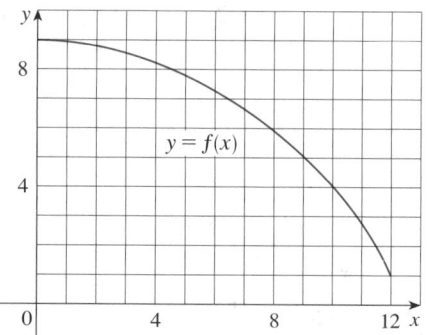

3. (a) Estimate the area under the graph of $f(x) = \cos x$ from $x = 0$ to $x = \pi/2$ using four approximating rectangles and right endpoints. Sketch the graph and the rectangles. Is your estimate an underestimate or an overestimate?
(b) Repeat part (a) using left endpoints.

4. (a) Estimate the area under the graph of $f(x) = \sqrt{x}$ from $x = 0$ to $x = 4$ using four approximating rectangles and right endpoints. Sketch the graph and the rectangles. Is your estimate an underestimate or an overestimate?

(b) Repeat part (a) using left endpoints.

5. (a) Estimate the area under the graph of $f(x) = 1 + x^2$ from $x = -1$ to $x = 2$ using three rectangles and right endpoints. Then improve your estimate by using six rectangles. Sketch the curve and the approximating rectangles.

(b) Repeat part (a) using left endpoints.

(c) Repeat part (a) using midpoints.

(d) From your sketches in parts (a)–(c), which appears to be the best estimate?

6. (a) Graph the function

$$f(x) = x - 2 \ln x \qquad 1 \le x \le 5$$

(b) Estimate the area under the graph of f using four approximating rectangles and taking the sample points to be (i) right endpoints and (ii) midpoints. In each case sketch the curve and the rectangles.

(c) Improve your estimates in part (b) by using eight rectangles.

7. Evaluate the upper and lower sums for $f(x) = 2 + \sin x$, $0 \le x \le \pi$, with $n = 2$, 4, and 8. Illustrate with diagrams like Figure 14.

8. Evaluate the upper and lower sums for $f(x) = 1 + x^2$, $-1 \le x \le 1$, with $n = 3$ and 4. Illustrate with diagrams like Figure 14.

9. The speed of a runner increased steadily during the first three seconds of a race. Her speed at half-second intervals is given in the table. Find lower and upper estimates for the distance that she traveled during these three seconds.

t (s)	0	0.5	1.0	1.5	2.0	2.5	3.0
v (ft/s)	0	6.2	10.8	14.9	18.1	19.4	20.2

10. Speedometer readings for a motorcycle at 12-second intervals are given in the table.

t (s)	0	12	24	36	48	60
v (ft/s)	30	28	25	22	24	27

(a) Estimate the distance traveled by the motorcycle during this time period using the velocities at the beginning of the time intervals.

(b) Give another estimate using the velocities at the end of the time periods.

(c) Are your estimates in parts (a) and (b) upper and lower estimates? Explain.

11. Oil leaked from a tank at a rate of $r(t)$ liters per hour. The rate decreased as time passed and values of the rate at two-hour time intervals are shown in the table. Find lower and upper estimates for the total amount of oil that leaked out.

t (h)	0	2	4	6	8	10
$r(t)$ (L/h)	8.7	7.6	6.8	6.2	5.7	5.3

12. When we estimate distances from velocity data, it is sometimes necessary to use times $t_0, t_1, t_2, t_3, \ldots$ that are not equally spaced. We can still estimate distances using the time periods $\Delta t_i = t_i - t_{i-1}$. For example, on May 7, 1992, the space shuttle *Endeavour* was launched on mission STS-49, the purpose of which was to install a new perigee kick motor in an Intelsat communications satellite. The table, provided by NASA, gives the velocity data for the shuttle between liftoff and the jettisoning of the solid rocket boosters. Use these data to estimate the height above the earth's surface of the space shuttle *Endeavour*, 62 seconds after liftoff.

Event	Time (s)	Velocity (ft/s)
Launch	0	0
Begin roll maneuver	10	185
End roll maneuver	15	319
Throttle to 89%	20	447
Throttle to 67%	32	742
Throttle to 104%	59	1325
Maximum dynamic pressure	62	1445
Solid rocket booster separation	125	4151

13. The velocity graph of a braking car is shown. Use it to estimate the distance traveled by the car while the brakes are applied.

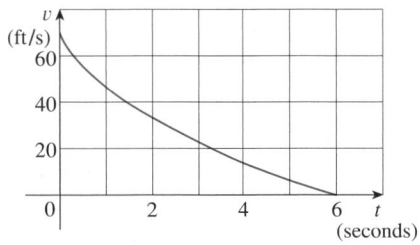

14. The velocity graph of a car accelerating from rest to a speed of 120 km/h over a period of 30 seconds is shown. Estimate the distance traveled during this period.

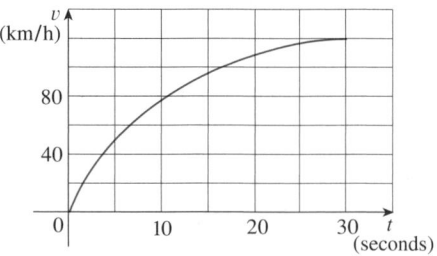

15–16 ▪ Use Definition 2 to find an expression for the area under the graph of f as a limit. Do not evaluate the limit.

15. $f(x) = \dfrac{2x}{x^2 + 1}$, $1 \le x \le 3$

16. $f(x) = x^2 + \sqrt{1 + 2x}$, $4 \le x \le 7$

17. Determine a region whose area is equal to

$$\lim_{n \to \infty} \sum_{i=1}^{n} \frac{\pi}{4n} \tan \frac{i\pi}{4n}$$

Do not evaluate the limit.

18. (a) Use Definition 2 to find an expression for the area under the curve $y = x^3$ from 0 to 1 as a limit.
  (b) The following formula for the sum of the cubes of the first n integers is proved in Appendix B. Use it to evaluate the limit in part (a).

$$1^3 + 2^3 + 3^3 + \cdots + n^3 = \left[\frac{n(n+1)}{2}\right]^2$$

19. Let A be the area under the graph of an increasing continuous function f from a to b, and let L_n and R_n be the approximations to A with n subintervals using left and right endpoints, respectively.
  (a) How are A, L_n, and R_n related?

(b) Show that

$$R_n - L_n = \frac{b - a}{n} [f(b) - f(a)]$$

Then draw a diagram to illustrate this equation by showing that the n rectangles representing $R_n - L_n$ can be reassembled to form a single rectangle whose area is the right side of the equation.
  (c) Deduce that

$$R_n - A < \frac{b - a}{n} [f(b) - f(a)]$$

20. If A is the area under the curve $y = e^x$ from 1 to 3, use Exercise 19 to find a value of n such that $R_n - A < 0.0001$.

CAS **21.** (a) Express the area under the curve $y = x^5$ from 0 to 2 as a limit.
  (b) Use a computer algebra system to find the sum in your expression from part (a).
  (c) Evaluate the limit in part (a).

CAS **22.** Find the exact area of the region under the graph of $y = e^{-x}$ from 0 to 2 by using a computer algebra system to evaluate the sum and then the limit in Example 3(a). Compare your answer with the estimate obtained in Example 3(b).

CAS **23.** Find the exact area under the cosine curve $y = \cos x$ from $x = 0$ to $x = b$, where $0 \le b \le \pi/2$. (Use a computer algebra system both to evaluate the sum and compute the limit.) In particular, what is the area if $b = \pi/2$?

24. (a) Let A_n be the area of a polygon with n equal sides inscribed in a circle with radius r. By dividing the polygon into n congruent triangles with central angle $2\pi/n$, show that

$$A_n = \tfrac{1}{2} n r^2 \sin\left(\frac{2\pi}{n}\right)$$

  (b) Show that $\lim_{n \to \infty} A_n = \pi r^2$. [*Hint:* Use Equation 1.4.6 on page 42.]

5.2 | THE DEFINITE INTEGRAL

We saw in Section 5.1 that a limit of the form

$$\boxed{1} \quad \lim_{n \to \infty} \sum_{i=1}^{n} f(x_i^*)\,\Delta x = \lim_{n \to \infty} [f(x_1^*)\,\Delta x + f(x_2^*)\,\Delta x + \cdots + f(x_n^*)\,\Delta x]$$

arises when we compute an area. We also saw that it arises when we try to find the distance traveled by an object. It turns out that this same type of limit occurs in a wide variety of situations even when f is not necessarily a positive function. Here we consider limits similar to $\boxed{1}$ but in which f need not be positive or continuous and the subintervals don't necessarily have the same length.

In general, we start with any function f defined on $[a, b]$ and we divide $[a, b]$ into n smaller subintervals by choosing partition points $x_0, x_1, x_2, \ldots, x_n$ so that

$$a = x_0 < x_1 < x_2 < \cdots < x_{n-1} < x_n = b$$

The resulting collection of subintervals

$$[x_0, x_1], \quad [x_1, x_2], \quad [x_2, x_3], \quad \ldots, \quad [x_{n-1}, x_n]$$

is called a **partition** P of $[a, b]$. We use the notation Δx_i for the length of the ith subinterval $[x_{i-1}, x_i]$. Thus

$$\Delta x_i = x_i - x_{i-1}$$

Then we choose **sample points** $x_1^*, x_2^*, \ldots, x_n^*$ in the subintervals with x_i^* in the ith subinterval $[x_{i-1}, x_i]$. These sample points could be left endpoints or right endpoints or any numbers between the endpoints. Figure 1 shows an example of a partition and sample points.

FIGURE 1
A partition of $[a, b]$ with sample points x_i^*

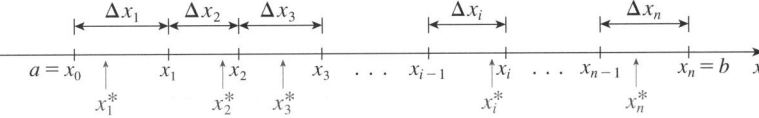

■ The Riemann sum is named after the German mathematician Bernhard Riemann (1826–1866). See the biographical note on page 270.

A **Riemann sum** associated with a partition P and a function f is constructed by evaluating f at the sample points, multiplying by the lengths of the corresponding subintervals, and adding:

$$\sum_{i=1}^{n} f(x_i^*)\, \Delta x_i = f(x_1^*)\, \Delta x_1 + f(x_2^*)\, \Delta x_2 + \cdots + f(x_n^*)\, \Delta x_n$$

The geometric interpretation of a Riemann sum is shown in Figure 2. Notice that if $f(x_i^*)$ is negative, then $f(x_i^*)\, \Delta x_i$ is negative and so we have to *subtract* the area of the corresponding rectangle.

FIGURE 2
A Riemann sum is the sum of the areas of the rectangles above the x-axis and the negatives of the areas of the rectangles below the x-axis.

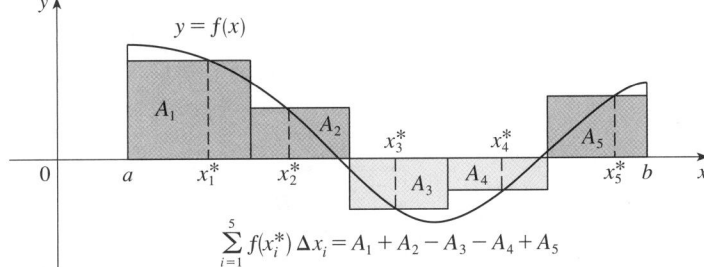

$$\sum_{i=1}^{5} f(x_i^*)\, \Delta x_i = A_1 + A_2 - A_3 - A_4 + A_5$$

If we imagine all possible partitions of $[a, b]$ and all possible choices of sample points, we can think of taking the limit of all possible Riemann sums as n becomes large by analogy with the definition of area. But because we are now allowing subintervals with different lengths, we need to ensure that all of these lengths Δx_i approach 0.

We can do that by insisting that the *largest* of these lengths, which we denote by max Δx_i, approaches 0. The result is called the *definite integral* of f from a to b.

$\boxed{2}$ **DEFINITION OF A DEFINITE INTEGRAL** If f is a function defined on $[a, b]$, the **definite integral of f from a to b** is the number

$$\int_a^b f(x)\,dx = \lim_{\max \Delta x_i \to 0} \sum_{i=1}^n f(x_i^*)\,\Delta x_i$$

provided that this limit exists. If it does exist, we say that f is **integrable** on $[a, b]$.

The precise meaning of the limit that defines the integral in Definition 2 is as follows: $\int_a^b f(x)\,dx = I$ means that for every $\varepsilon > 0$ there is a corresponding number $\delta > 0$ such that

$$\left| I - \sum_{i=1}^n f(x_i^*)\,\Delta x_i \right| < \varepsilon$$

for all partitions P of $[a, b]$ with max $\Delta x_i < \delta$ and for all possible choices of x_i^* in $[x_{i-1}, x_i]$.

This means that a definite integral can be approximated to within any desired degree of accuracy by a Riemann sum.

NOTE 1 The symbol $\int$ was introduced by Leibniz and is called an **integral sign**. It is an elongated S and was chosen because an integral is a limit of sums. In the notation $\int_a^b f(x)\,dx$, $f(x)$ is called the **integrand** and a and b are called the **limits of integration**; a is the **lower limit** and b is the **upper limit**. For now, the symbol dx has no official meaning by itself; $\int_a^b f(x)\,dx$ is all one symbol. The dx simply indicates that the independent variable is x. The procedure of calculating an integral is called **integration**.

NOTE 2 The definite integral $\int_a^b f(x)\,dx$ is a number; it does not depend on x. In fact, we could use any letter in place of x without changing the value of the integral:

$$\int_a^b f(x)\,dx = \int_a^b f(t)\,dt = \int_a^b f(r)\,dr$$

We have defined the definite integral for an integrable function, but not all functions are integrable. The following theorem shows that the most commonly occurring functions are in fact integrable. The theorem is proved in more advanced courses.

$\boxed{3}$ **THEOREM** If f is continuous on $[a, b]$, or if f has only a finite number of jump discontinuities, then f is integrable on $[a, b]$; that is, the definite integral $\int_a^b f(x)\,dx$ exists.

If f is integrable on $[a, b]$, then the Riemann sums in Definition 2 must approach $\int_a^b f(x)\,dx$ as max $\Delta x_i \to 0$ no matter how the partitions and sample points are chosen. So in calculating the value of an integral we are free to choose partitions P and sample points x_i^* to simplify the calculation. It's often convenient to take P to be a **regular partition**; that is, all the subintervals have the same length Δx. Then

$$\Delta x = \Delta x_1 = \Delta x_2 = \cdots = \Delta x_n = \frac{b - a}{n}$$

and $\qquad x_0 = a, \quad x_1 = a + \Delta x, \quad x_2 = a + 2\,\Delta x, \quad \ldots, \quad x_i = a + i\,\Delta x$

If we choose x_i^* to be the right endpoint of the ith subinterval, then

$$x_i^* = x_i = a + i\,\Delta x = a + i\frac{b - a}{n}$$

In this case, max $\Delta x_i = \Delta x = (b - a)/n \to 0$ as $n \to \infty$, so Definition 2 gives

$$\int_a^b f(x)\,dx = \lim_{\Delta x \to 0} \sum_{i=1}^n f(x_i)\,\Delta x = \lim_{n \to \infty} \sum_{i=1}^n f(x_i)\,\Delta x$$

---

4 **THEOREM**   If f is integrable on $[a, b]$, then

$$\int_a^b f(x)\,dx = \lim_{n \to \infty} \sum_{i=1}^n f(x_i)\,\Delta x$$

where $\qquad \Delta x = \dfrac{b - a}{n} \qquad$ and $\qquad x_i = a + i\,\Delta x$

In computing the value of an integral, Theorem 4 is much simpler to use than Definition 2.

EXAMPLE 1   Express

$$\lim_{n \to \infty} \sum_{i=1}^n (x_i^3 + x_i \sin x_i)\,\Delta x$$

as an integral on the interval $[0, \pi]$.

SOLUTION   Comparing the given limit with the limit in Theorem 4, we see that they will be identical if we choose $f(x) = x^3 + x \sin x$. We are given that $a = 0$ and $b = \pi$. Therefore, by Theorem 4, we have

$$\lim_{n \to \infty} \sum_{i=1}^n (x_i^3 + x_i \sin x_i)\,\Delta x = \int_0^\pi (x^3 + x \sin x)\,dx \qquad \blacksquare$$

Later, when we apply the definite integral to physical situations, it will be important to recognize limits of sums as integrals, as we did in Example 1. When Leibniz chose the notation for an integral, he chose the ingredients as reminders of the limiting process. In general, when we write

$$\lim_{n \to \infty} \sum_{i=1}^n f(x_i^*)\,\Delta x = \int_a^b f(x)\,dx$$

we replace lim Σ by $\int$, x_i^* by x, and Δx by dx.

NOTE 3 If f happens to be positive, then the Riemann sum can be interpreted as a sum of areas of approximating rectangles (see Figure 3). By comparing Theorem 4 with the definition of area in Section 5.1, we see that the definite integral $\int_a^b f(x)\,dx$ can be interpreted as the area under the curve $y = f(x)$ from a to b. (See Figure 4.)

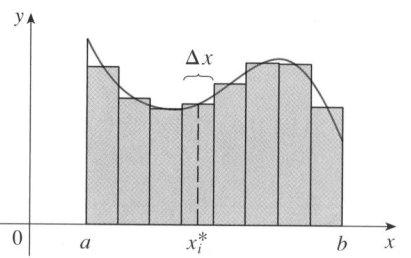

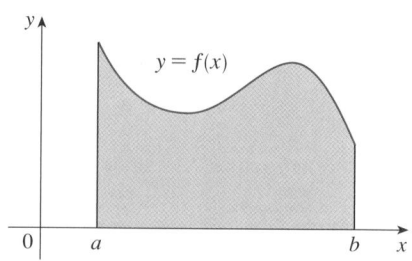

FIGURE 3
If $f(x) \geqslant 0$, the Riemann sum $\Sigma f(x_i^*)\,\Delta x$ is the sum of areas of rectangles.

FIGURE 4
If $f(x) \geqslant 0$, the integral $\int_a^b f(x)\,dx$ is the area under the curve $y = f(x)$ from a to b.

If f takes on both positive and negative values, as in Figure 5, then the Riemann sum is the sum of the areas of the rectangles that lie above the x-axis and the *negatives* of the areas of the rectangles that lie below the x-axis (the areas of the dark blue rectangles *minus* the areas of the light blue rectangles). When we take the limit of such Riemann sums, we get the situation illustrated in Figure 6. A definite integral can be interpreted as a **net area**, that is, a difference of areas:

$$\int_a^b f(x)\,dx = A_1 - A_2$$

where A_1 is the area of the region above the x-axis and below the graph of f, and A_2 is the area of the region below the x-axis and above the graph of f.

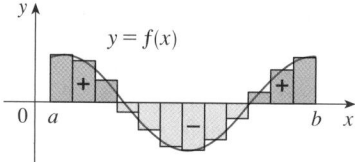

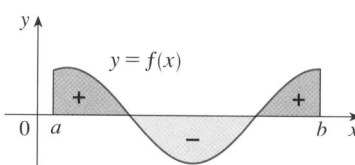

FIGURE 5
$\Sigma f(x_i^*)\,\Delta x$ is an approximation to the net area.

FIGURE 6
$\int_a^b f(x)\,dx$ is the net area.

EVALUATING INTEGRALS

When we use the definition or Theorem 4 to evaluate a definite integral, we need to know how to work with sums. The following three equations give formulas for sums of powers of positive integers. Equation 5 may be familiar to you from a course in algebra. Equations 6 and 7 were discussed in Section 5.1 and are proved in Appendix B.

5
$$\sum_{i=1}^{n} i = \frac{n(n+1)}{2}$$

$$\boxed{6} \qquad \sum_{i=1}^{n} i^2 = \frac{n(n+1)(2n+1)}{6}$$

$$\boxed{7} \qquad \sum_{i=1}^{n} i^3 = \left[\frac{n(n+1)}{2} \right]^2$$

The remaining formulas are simple rules for working with sigma notation:

- Formulas 8–11 are proved by writing out each side in expanded form. The left side of Equation 9 is

$$ca_1 + ca_2 + \cdots + ca_n$$

The right side is

$$c(a_1 + a_2 + \cdots + a_n)$$

These are equal by the distributive property. The other formulas are discussed in Appendix B.

$$\boxed{8} \qquad \sum_{i=1}^{n} c = nc$$

$$\boxed{9} \qquad \sum_{i=1}^{n} ca_i = c \sum_{i=1}^{n} a_i$$

$$\boxed{10} \qquad \sum_{i=1}^{n} (a_i + b_i) = \sum_{i=1}^{n} a_i + \sum_{i=1}^{n} b_i$$

$$\boxed{11} \qquad \sum_{i=1}^{n} (a_i - b_i) = \sum_{i=1}^{n} a_i - \sum_{i=1}^{n} b_i$$

EXAMPLE 2

(a) Evaluate the Riemann sum for $f(x) = x^3 - 6x$ taking the sample points to be right endpoints and $a = 0$, $b = 3$, and $n = 6$.

(b) Evaluate $\int_0^3 (x^3 - 6x)\, dx$.

SOLUTION

(a) With $n = 6$ the interval width is

$$\Delta x = \frac{b - a}{n} = \frac{3 - 0}{6} = \frac{1}{2}$$

and the right endpoints are $x_1 = 0.5$, $x_2 = 1.0$, $x_3 = 1.5$, $x_4 = 2.0$, $x_5 = 2.5$, and $x_6 = 3.0$. So the Riemann sum is

$$R_6 = \sum_{i=1}^{6} f(x_i)\, \Delta x$$

$$= f(0.5)\, \Delta x + f(1.0)\, \Delta x + f(1.5)\, \Delta x + f(2.0)\, \Delta x + f(2.5)\, \Delta x + f(3.0)\, \Delta x$$

$$= \tfrac{1}{2}(-2.875 - 5 - 5.625 - 4 + 0.625 + 9)$$

$$= -3.9375$$

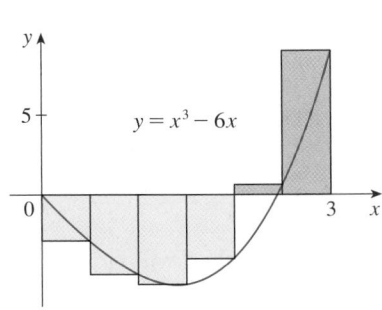

FIGURE 7

Notice that f is not a positive function and so the Riemann sum does not represent a sum of areas of rectangles. But it does represent the sum of the areas of the dark blue rectangles (above the x-axis) minus the sum of the areas of the light blue rectangles (below the x-axis) in Figure 7.

(b) With n subintervals we have

$$\Delta x = \frac{b - a}{n} = \frac{3}{n}$$

Thus $x_0 = 0$, $x_1 = 3/n$, $x_2 = 6/n$, $x_3 = 9/n$, and, in general, $x_i = 3i/n$. Since we are using right endpoints, we can use Theorem 4:

$$\int_0^3 (x^3 - 6x)\, dx = \lim_{n \to \infty} \sum_{i=1}^n f(x_i)\, \Delta x = \lim_{n \to \infty} \sum_{i=1}^n f\left(\frac{3i}{n}\right) \frac{3}{n}$$

- In the sum, n is a constant (unlike i), so we can move $3/n$ in front of the Σ sign.

$$= \lim_{n \to \infty} \frac{3}{n} \sum_{i=1}^n \left[\left(\frac{3i}{n}\right)^3 - 6\left(\frac{3i}{n}\right)\right] \qquad \text{(Equation 9 with } c = 3/n\text{)}$$

$$= \lim_{n \to \infty} \frac{3}{n} \sum_{i=1}^n \left[\frac{27}{n^3} i^3 - \frac{18}{n} i\right]$$

$$= \lim_{n \to \infty} \left[\frac{81}{n^4} \sum_{i=1}^n i^3 - \frac{54}{n^2} \sum_{i=1}^n i\right] \qquad \text{(Equations 11 and 9)}$$

$$= \lim_{n \to \infty} \left\{\frac{81}{n^4}\left[\frac{n(n+1)}{2}\right]^2 - \frac{54}{n^2}\frac{n(n+1)}{2}\right\} \qquad \text{(Equations 7 and 5)}$$

$$= \lim_{n \to \infty} \left[\frac{81}{4}\left(1 + \frac{1}{n}\right)^2 - 27\left(1 + \frac{1}{n}\right)\right]$$

$$= \frac{81}{4} - 27 = -\frac{27}{4} = -6.75$$

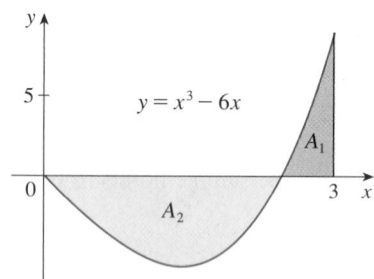

FIGURE 8
$\int_0^3 (x^3 - 6x)\, dx = A_1 - A_2 = -6.75$

This integral can't be interpreted as an area because f takes on both positive and negative values. But it can be interpreted as the difference of areas $A_1 - A_2$, where A_1 and A_2 are shown in Figure 8.

Figure 9 illustrates the calculation by showing the positive and negative terms in the right Riemann sum R_n for $n = 40$. The values in the table show the Riemann sums approaching the exact value of the integral, -6.75, as $n \to \infty$.

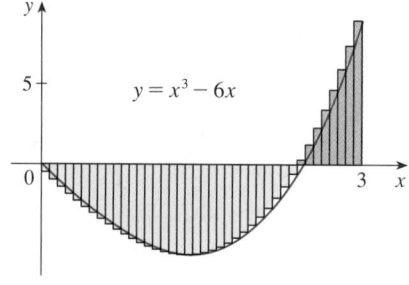

FIGURE 9
$R_{40} \approx -6.3998$

n	R_n
40	-6.3998
100	-6.6130
500	-6.7229
1000	-6.7365
5000	-6.7473

- www.stewartcalculus.com
See Additional Example A.

A much simpler method for evaluating the integral in Example 2 will be given in Section 5.3 after we have proved the Evaluation Theorem.

V EXAMPLE 3 Evaluate the following integrals by interpreting each in terms of areas.

(a) $\displaystyle\int_0^1 \sqrt{1 - x^2}\, dx$
(b) $\displaystyle\int_0^3 (x - 1)\, dx$

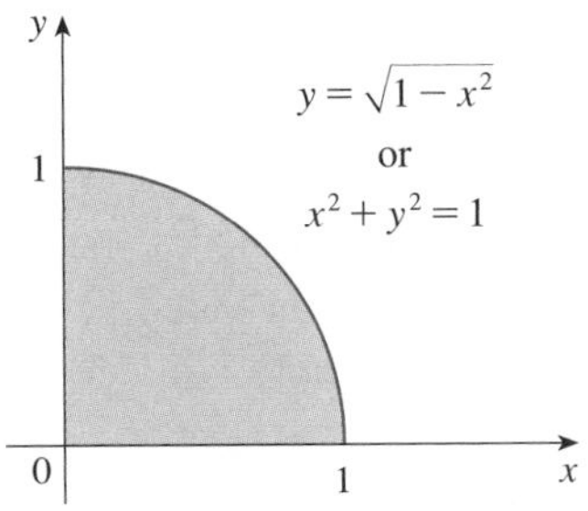

FIGURE 10

SOLUTION

(a) Since $f(x) = \sqrt{1 - x^2} \geqslant 0$, we can interpret this integral as the area under the curve $y = \sqrt{1 - x^2}$ from 0 to 1. But, since $y^2 = 1 - x^2$, we get $x^2 + y^2 = 1$, which shows that the graph of f is the quarter-circle with radius 1 in Figure 10. Therefore

$$\int_0^1 \sqrt{1 - x^2}\, dx = \tfrac{1}{4}\pi(1)^2 = \frac{\pi}{4}$$

(In Section 6.2 we will be able to *prove* that the area of a circle of radius r is πr^2.)

(b) The graph of $y = x - 1$ is the line with slope 1 shown in Figure 11. We compute the integral as the difference of the areas of the two triangles:

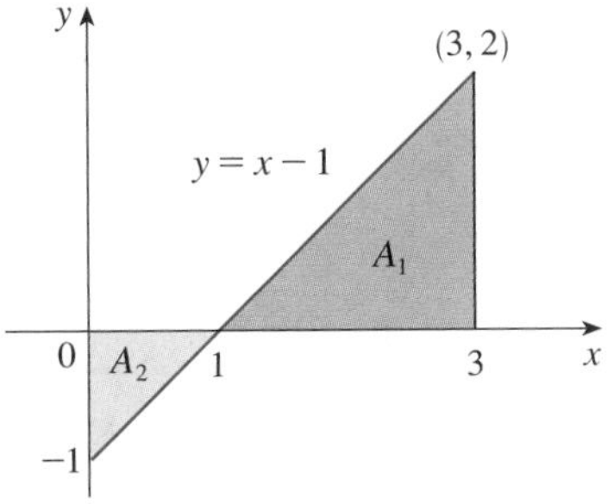

FIGURE 11

$$\int_0^3 (x - 1)\, dx = A_1 - A_2 = \tfrac{1}{2}(2 \cdot 2) - \tfrac{1}{2}(1 \cdot 1) = 1.5$$ ■

THE MIDPOINT RULE

We often choose the sample point x_i^* to be the right endpoint of the ith subinterval because it is convenient for computing the limit. But if the purpose is to find an *approximation* to an integral, it is usually better to choose x_i^* to be the midpoint of the interval, which we denote by $\bar{x}_i$. Any Riemann sum is an approximation to an integral, but if we use midpoints and a regular partition we get the following approximation.

TEC Module 5.2/6.5 shows how the Midpoint Rule estimates improve as n increases.

MIDPOINT RULE

$$\int_a^b f(x)\, dx \approx \sum_{i=1}^{n} f(\bar{x}_i)\, \Delta x = \Delta x\, [f(\bar{x}_1) + \cdots + f(\bar{x}_n)]$$

where $\quad \Delta x = \dfrac{b - a}{n}$

and $\quad \bar{x}_i = \tfrac{1}{2}(x_{i-1} + x_i) = \text{midpoint of } [x_{i-1}, x_i]$

V EXAMPLE 4 Use the Midpoint Rule with $n = 5$ to approximate $\displaystyle\int_1^2 \frac{1}{x}\, dx$.

SOLUTION The endpoints of the five subintervals are 1, 1.2, 1.4, 1.6, 1.8, and 2.0, so the midpoints are 1.1, 1.3, 1.5, 1.7, and 1.9. The width of the subintervals is $\Delta x = (2 - 1)/5 = \frac{1}{5}$, so the Midpoint Rule gives

$$\int_1^2 \frac{1}{x}\, dx \approx \Delta x\, [f(1.1) + f(1.3) + f(1.5) + f(1.7) + f(1.9)]$$

$$= \frac{1}{5}\left(\frac{1}{1.1} + \frac{1}{1.3} + \frac{1}{1.5} + \frac{1}{1.7} + \frac{1}{1.9}\right) \approx 0.691908$$

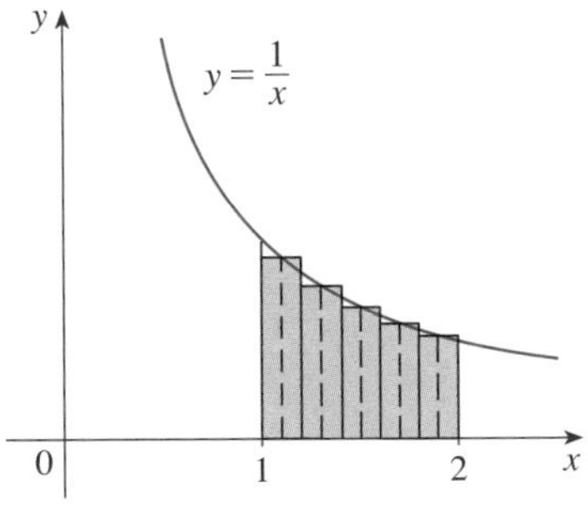

FIGURE 12

Since $f(x) = 1/x > 0$ for $1 \leqslant x \leqslant 2$, the integral represents an area, and the approximation given by the Midpoint Rule is the sum of the areas of the rectangles shown in Figure 12. ■

At the moment we don't know how accurate the approximation in Example 4 is, but in Section 6.5 we will learn a method for estimating the error involved in using the Midpoint Rule. At that time we will discuss other methods for approximating definite integrals.

If we apply the Midpoint Rule to the integral in Example 2, we get the picture in Figure 13. The approximation $M_{40} \approx -6.7563$ is much closer to the true value -6.75 than the right endpoint approximation, $R_{40} \approx -6.3998$, shown in Figure 9.

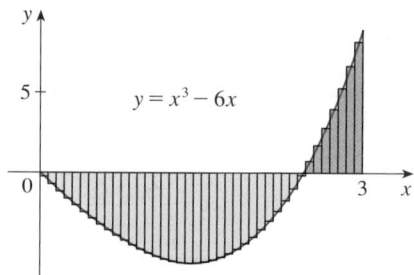

FIGURE 13
$M_{40} \approx -6.7563$

PROPERTIES OF THE DEFINITE INTEGRAL

We now develop some basic properties of integrals that will help us to evaluate integrals in a simple manner. We assume that f and g are integrable functions.

When we defined the definite integral $\int_a^b f(x)\,dx$, we implicitly assumed that $a < b$. But the definition as a limit of Riemann sums makes sense even if $a > b$. Notice that if we reverse a and b in Theorem 4, then Δx changes from $(b - a)/n$ to $(a - b)/n$. Therefore

$$\int_b^a f(x)\,dx = -\int_a^b f(x)\,dx$$

If $a = b$, then $\Delta x = 0$ and so

$$\int_a^a f(x)\,dx = 0$$

PROPERTIES OF THE INTEGRAL Suppose all the following integrals exist.

1. $\displaystyle\int_a^b c\,dx = c(b - a),$ where c is any constant

2. $\displaystyle\int_a^b [f(x) + g(x)]\,dx = \int_a^b f(x)\,dx + \int_a^b g(x)\,dx$

3. $\displaystyle\int_a^b cf(x)\,dx = c\int_a^b f(x)\,dx,$ where c is any constant

4. $\displaystyle\int_a^b [f(x) - g(x)]\,dx = \int_a^b f(x)\,dx - \int_a^b g(x)\,dx$

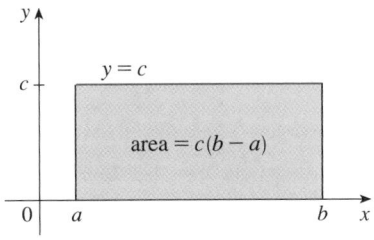

FIGURE 14

$$\int_a^b c \, dx = c(b - a)$$

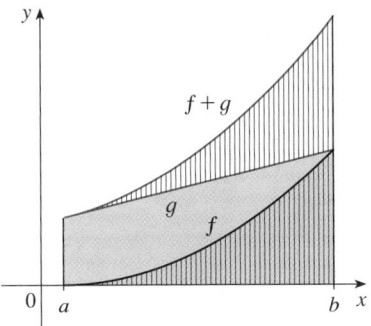

FIGURE 15

$$\int_a^b [f(x) + g(x)] \, dx =$$
$$\int_a^b f(x) \, dx + \int_a^b g(x) \, dx$$

■ Property 3 seems intuitively reasonable because we know that multiplying a function by a positive number c stretches or shrinks its graph vertically by a factor of c. So it stretches or shrinks each approximating rectangle by a factor c and therefore it has the effect of multiplying the area by c.

Property 1 says that the integral of a constant function $f(x) = c$ is the constant times the length of the interval. If $c > 0$ and $a < b$, this is to be expected because $c(b - a)$ is the area of the shaded rectangle in Figure 14.

Property 2 says that the integral of a sum is the sum of the integrals. For positive functions it says that the area under $f + g$ is the area under f plus the area under g. Figure 15 helps us understand why this is true: In view of how graphical addition works, the corresponding vertical line segments have equal height.

In general, Property 2 follows from Theorem 4 and the fact that the limit of a sum is the sum of the limits:

$$\int_a^b [f(x) + g(x)] \, dx = \lim_{n \to \infty} \sum_{i=1}^n [f(x_i) + g(x_i)] \Delta x$$

$$= \lim_{n \to \infty} \left[\sum_{i=1}^n f(x_i) \Delta x + \sum_{i=1}^n g(x_i) \Delta x \right]$$

$$= \lim_{n \to \infty} \sum_{i=1}^n f(x_i) \Delta x + \lim_{n \to \infty} \sum_{i=1}^n g(x_i) \Delta x$$

$$= \int_a^b f(x) \, dx + \int_a^b g(x) \, dx$$

Property 3 can be proved in a similar manner and says that the integral of a constant times a function is the constant times the integral of the function. In other words, a constant (but *only* a constant) can be taken in front of an integral sign. Property 4 is proved by writing $f - g = f + (-g)$ and using Properties 2 and 3 with $c = -1$.

EXAMPLE 5 Use the properties of integrals to evaluate $\int_0^1 (4 + 3x^2) \, dx$.

SOLUTION Using Properties 2 and 3 of integrals, we have

$$\int_0^1 (4 + 3x^2) \, dx = \int_0^1 4 \, dx + \int_0^1 3x^2 \, dx = \int_0^1 4 \, dx + 3 \int_0^1 x^2 \, dx$$

We know from Property 1 that

$$\int_0^1 4 \, dx = 4(1 - 0) = 4$$

and we found in Example 2 in Section 5.1 that $\int_0^1 x^2 \, dx = \frac{1}{3}$. So

$$\int_0^1 (4 + 3x^2) \, dx = \int_0^1 4 \, dx + 3 \int_0^1 x^2 \, dx$$

$$= 4 + 3 \cdot \tfrac{1}{3} = 5 \qquad ■$$

The next property tells us how to combine integrals of the same function over adjacent intervals:

5. $$\int_a^c f(x) \, dx + \int_c^b f(x) \, dx = \int_a^b f(x) \, dx$$

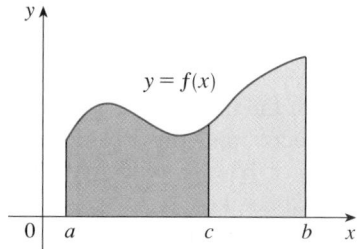

FIGURE 16

Property 5 is more complicated and is proved in Appendix D, but for the case where $f(x) \geqslant 0$ and $a < c < b$ it can be seen from the geometric interpretation in Figure 16: The area under $y = f(x)$ from a to c plus the area from c to b is equal to the total area from a to b.

▼ EXAMPLE 6 If it is known that $\int_0^{10} f(x)\,dx = 17$ and $\int_0^8 f(x)\,dx = 12$, find $\int_8^{10} f(x)\,dx$.

SOLUTION By Property 5, we have

$$\int_0^8 f(x)\,dx + \int_8^{10} f(x)\,dx = \int_0^{10} f(x)\,dx$$

so

$$\int_8^{10} f(x)\,dx = \int_0^{10} f(x)\,dx - \int_0^8 f(x)\,dx = 17 - 12 = 5 \qquad \blacksquare$$

Notice that Properties 1–5 are true whether $a < b$, $a = b$, or $a > b$. The following properties, in which we compare sizes of functions and sizes of integrals, are true only if $a \leqslant b$.

COMPARISON PROPERTIES OF THE INTEGRAL

6. If $f(x) \geqslant 0$ for $a \leqslant x \leqslant b$, then $\int_a^b f(x)\,dx \geqslant 0$.

7. If $f(x) \geqslant g(x)$ for $a \leqslant x \leqslant b$, then $\int_a^b f(x)\,dx \geqslant \int_a^b g(x)\,dx$.

8. If $m \leqslant f(x) \leqslant M$ for $a \leqslant x \leqslant b$, then

$$m(b - a) \leqslant \int_a^b f(x)\,dx \leqslant M(b - a)$$

If $f(x) \geqslant 0$, then $\int_a^b f(x)\,dx$ represents the area under the graph of f, so the geometric interpretation of Property 6 is simply that areas are positive. (It also follows directly from the definition because all the quantities involved are positive.). Property 7 says that a bigger function has a bigger integral. It follows from Properties 6 and 4 because $f - g \geqslant 0$.

Property 8 is illustrated by Figure 17 for the case where $f(x) \geqslant 0$. If f is continuous we could take m and M to be the absolute minimum and maximum values of f on the interval $[a, b]$. In this case Property 8 says that the area under the graph of f is greater than the area of the rectangle with height m and less than the area of the rectangle with height M.

In general, since $m \leqslant f(x) \leqslant M$, Property 7 gives

$$\int_a^b m\,dx \leqslant \int_a^b f(x)\,dx \leqslant \int_a^b M\,dx$$

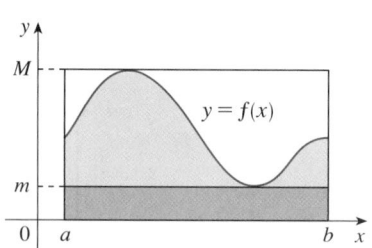

FIGURE 17

Using Property 1 to evaluate the integrals on the left- and right-hand sides, we obtain

$$m(b - a) \leqslant \int_a^b f(x)\,dx \leqslant M(b - a)$$

Property 8 is useful when all we want is a rough estimate of the size of an integral without going to the bother of using the Midpoint Rule.

EXAMPLE 7 Use Property 8 to estimate $\int_0^1 e^{-x^2}\, dx$.

SOLUTION Because $f(x) = e^{-x^2}$ is a decreasing function on $[0, 1]$, its absolute maximum value is $M = f(0) = 1$ and its absolute minimum value is $m = f(1) = e^{-1}$. Thus, by Property 8,

$$e^{-1}(1 - 0) \le \int_0^1 e^{-x^2}\, dx \le 1(1 - 0)$$

or

$$e^{-1} \le \int_0^1 e^{-x^2}\, dx \le 1$$

Since $e^{-1} \approx 0.3679$, we can write

$$0.367 \le \int_0^1 e^{-x^2}\, dx \le 1$$ ∎

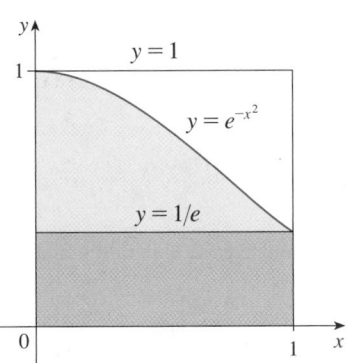

FIGURE 18

The result of Example 7 is illustrated in Figure 18. The integral is greater than the area of the lower rectangle and less than the area of the square.

5.2 | EXERCISES

1. Evaluate the Riemann sum for $f(x) = 3 - \frac{1}{2}x$, $2 \le x \le 14$, with six subintervals, taking the sample points to be left endpoints. Explain, with the aid of a diagram, what the Riemann sum represents.

2. If $f(x) = x^2 - 2x$, $0 \le x \le 3$, evaluate the Riemann sum with $n = 6$, taking the sample points to be right endpoints. What does the Riemann sum represent? Illustrate with a diagram.

3. If $f(x) = e^x - 2$, $0 \le x \le 2$, find the Riemann sum with $n = 4$ correct to six decimal places, taking the sample points to be midpoints. What does the Riemann sum represent? Illustrate with a diagram.

4. (a) Find the Riemann sum for $f(x) = \sin x$, $0 \le x \le 3\pi/2$, with six terms, taking the sample points to be right endpoints. (Give your answer correct to six decimal places.) Explain what the Riemann sum represents with the aid of a sketch.
(b) Repeat part (a) with midpoints as the sample points.

5. Find the Riemann sum for $f(x) = x^3$, $-1 \le x \le 1$, if the partition points are -1, -0.5, 0, 0.5, 1 and the sample points are -1, -0.4, 0.2, 1.

6. Find the Riemann sum for $f(x) = x + x^2$, $-2 \le x \le 0$, if the partition points are -2, -1.5, -1, -0.7, -0.4, 0 and the sample points are left endpoints. What is max Δx_i?

7. The graph of a function f is given. Estimate $\int_0^{10} f(x)\, dx$ using five subintervals with (a) right endpoints, (b) left endpoints, and (c) midpoints.

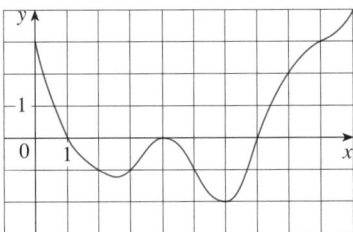

8. The graph of g is shown. Estimate $\int_{-2}^4 g(x)\, dx$ with six subintervals using (a) right endpoints, (b) left endpoints, and (c) midpoints.

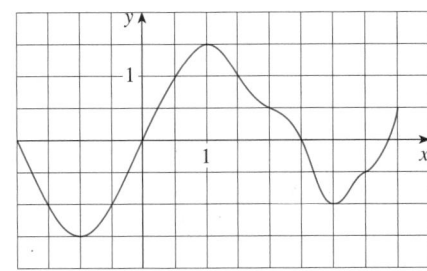

9. A table of values of an increasing function f is shown. Use the table to find lower and upper estimates for $\int_{10}^{30} f(x)\,dx$.

x	10	14	18	22	26	30
$f(x)$	-12	-6	-2	1	3	8

10. The table gives the values of a function obtained from an experiment. Use them to estimate $\int_3^9 f(x)\,dx$ using three equal subintervals with (a) right endpoints, (b) left endpoints, and (c) midpoints. If the function is known to be an increasing function, can you say whether your estimates are less than or greater than the exact value of the integral?

x	3	4	5	6	7	8	9
$f(x)$	-3.4	-2.1	-0.6	0.3	0.9	1.4	1.8

11–14 ▪ Use the Midpoint Rule with the given value of n to approximate the integral. Round the answer to four decimal places.

11. $\int_0^8 \sin\sqrt{x}\,dx, \quad n = 4$

12. $\int_0^{\pi/2} \cos^4 x\,dx, \quad n = 4$

13. $\int_0^2 \dfrac{x}{x+1}\,dx, \quad n = 5$

14. $\int_1^5 x^2 e^{-x}\,dx, \quad n = 4$

15–18 ▪ Express the limit as a definite integral on the given interval.

15. $\lim\limits_{n \to \infty} \sum\limits_{i=1}^{n} x_i \ln(1 + x_i^2)\,\Delta x, \quad [2, 6]$

16. $\lim\limits_{n \to \infty} \sum\limits_{i=1}^{n} \dfrac{\cos x_i}{x_i}\,\Delta x, \quad [\pi, 2\pi]$

17. $\lim\limits_{n \to \infty} \sum\limits_{i=1}^{n} [5(x_i^*)^3 - 4x_i^*]\,\Delta x, \quad [2, 7]$

18. $\lim\limits_{n \to \infty} \sum\limits_{i=1}^{n} \dfrac{x_i^*}{(x_i^*)^2 + 4}\,\Delta x, \quad [1, 3]$

19–23 ▪ Use the form of the definition of the integral given in Theorem 4 to evaluate the integral.

19. $\int_2^5 (4 - 2x)\,dx$

20. $\int_1^4 (x^2 - 4x + 2)\,dx$

21. $\int_{-2}^0 (x^2 + x)\,dx$

22. $\int_0^2 (2x - x^3)\,dx$

23. $\int_0^1 (x^3 - 3x^2)\,dx$

24. (a) Find an approximation to the integral $\int_0^4 (x^2 - 3x)\,dx$ using a Riemann sum with right endpoints and $n = 8$.
(b) Draw a diagram like Figure 2 to illustrate the approximation in part (a).

(c) Use Theorem 4 to evaluate $\int_0^4 (x^2 - 3x)\,dx$.
(d) Interpret the integral in part (c) as a difference of areas and illustrate with a diagram like Figure 6.

25–26 ▪ Express the integral as a limit of Riemann sums. Do not evaluate the limit.

25. $\int_2^6 \dfrac{x}{1 + x^5}\,dx$

26. $\int_1^{10} (x - 4 \ln x)\,dx$

CAS **27–28** ▪ Express the integral as a limit of sums. Then evaluate, using a computer algebra system to find both the sum and the limit.

27. $\int_0^\pi \sin 5x\,dx$

28. $\int_2^{10} x^6\,dx$

29. The graph of f is shown. Evaluate each integral by interpreting it in terms of areas.

(a) $\int_0^2 f(x)\,dx$ 　　　　(b) $\int_0^5 f(x)\,dx$

(c) $\int_5^7 f(x)\,dx$ 　　　　(d) $\int_0^9 f(x)\,dx$

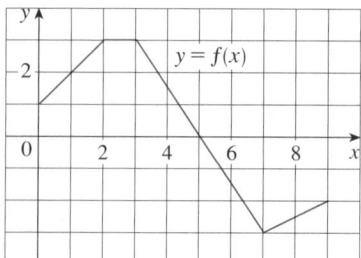

30. The graph of g consists of two straight lines and a semicircle. Use it to evaluate each integral.

(a) $\int_0^2 g(x)\,dx$ 　(b) $\int_2^6 g(x)\,dx$ 　(c) $\int_0^7 g(x)\,dx$

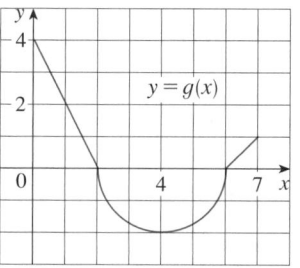

31–36 ▪ Evaluate the integral by interpreting it in terms of areas.

31. $\int_{-1}^2 (1 - x)\,dx$

32. $\int_0^9 \left(\tfrac{1}{3}x - 2\right)\,dx$

33. $\int_{-3}^0 \left(1 + \sqrt{9 - x^2}\right)\,dx$

34. $\int_{-5}^5 \left(x - \sqrt{25 - x^2}\right)\,dx$

35. $\int_{-1}^2 |x|\,dx$

36. $\int_0^{10} |x - 5|\,dx$

37. Evaluate $\int_{\pi}^{\pi} \sin^2 x \, \cos^4 x \, dx$.

38. Given that $\int_{0}^{1} 3x\sqrt{x^2 + 4} \, dx = 5\sqrt{5} - 8$, what is $\int_{1}^{0} 3u\sqrt{u^2 + 4} \, du$?

39. Write as a single integral in the form $\int_{a}^{b} f(x) \, dx$:

$$\int_{-2}^{2} f(x) \, dx + \int_{2}^{5} f(x) \, dx - \int_{-2}^{-1} f(x) \, dx$$

40. If $\int_{1}^{5} f(x) \, dx = 12$ and $\int_{4}^{5} f(x) \, dx = 3.6$, find $\int_{1}^{4} f(x) \, dx$.

41. If $\int_{0}^{9} f(x) \, dx = 37$ and $\int_{0}^{9} g(x) \, dx = 16$, find $\int_{0}^{9} [2f(x) + 3g(x)] \, dx$.

42. Find $\int_{0}^{5} f(x) \, dx$ if

$$f(x) = \begin{cases} 3 & \text{for } x < 3 \\ x & \text{for } x \geq 3 \end{cases}$$

43. In Example 2 in Section 5.1 we showed that $\int_{0}^{1} x^2 \, dx = \frac{1}{3}$. Use this fact and the properties of integrals to evaluate $\int_{0}^{1} (5 - 6x^2) \, dx$.

44. If $F(x) = \int_{2}^{x} f(t) \, dt$, where f is the function whose graph is given, which of the following values is largest?
(A) $F(0)$ (B) $F(1)$
(C) $F(2)$ (D) $F(3)$
(E) $F(4)$

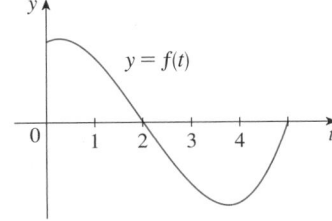

45. Each of the regions A, B, and C bounded by the graph of f and the x-axis has area 3. Find the value of

$$\int_{-4}^{2} [f(x) + 2x + 5] \, dx$$

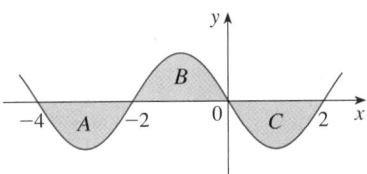

46. Suppose f has absolute minimum value m and absolute maximum value M. Between what two values must $\int_{0}^{2} f(x) \, dx$ lie? Which property of integrals allows you to make your conclusion?

47. Use the properties of integrals to verify that

$$0 \leq \int_{1}^{3} \ln x \, dx \leq 2 \ln 3$$

without evaluating the integral.

48–52 ▪ Use Property 8 to estimate the value of the integral.

48. $\int_{0}^{2} \sqrt{x^3 + 1} \, dx$ **49.** $\int_{1}^{2} \frac{1}{x} \, dx$

50. $\int_{0}^{2} (x^3 - 3x + 3) \, dx$ **51.** $\int_{\pi/4}^{\pi/3} \tan x \, dx$

52. $\int_{\pi/4}^{3\pi/4} \sin^2 x \, dx$

53. Express the following limit as a definite integral:

$$\lim_{n \to \infty} \sum_{i=1}^{n} \frac{i^4}{n^5}$$

5.3 | EVALUATING DEFINITE INTEGRALS

In Section 5.2 we computed integrals from the definition as a limit of Riemann sums and we saw that this procedure is sometimes long and difficult. Sir Isaac Newton discovered a much simpler method for evaluating integrals and a few years later Leibniz made the same discovery. They realized that they could calculate $\int_{a}^{b} f(x) \, dx$ if they happened to know an antiderivative F of f. Their discovery, called the Evaluation Theorem, is part of the Fundamental Theorem of Calculus, which is discussed in the next section.

> **EVALUATION THEOREM** If f is continuous on the interval $[a, b]$, then
>
> $$\int_a^b f(x)\,dx = F(b) - F(a)$$
>
> where F is any antiderivative of f, that is, $F' = f$.

This theorem states that if we know an antiderivative F of f, then we can evaluate $\int_a^b f(x)\,dx$ simply by subtracting the values of F at the endpoints of the interval $[a, b]$. It is very surprising that $\int_a^b f(x)\,dx$, which was defined by a complicated procedure involving all of the values of $f(x)$ for $a \le x \le b$, can be found by knowing the values of $F(x)$ at only two points, a and b.

For instance, we know from Section 4.7 that an antiderivative of $f(x) = x^2$ is $F(x) = \frac{1}{3}x^3$, so the Evaluation Theorem tells us that

$$\int_0^1 x^2\,dx = F(1) - F(0) = \frac{1}{3} \cdot 1^3 - \frac{1}{3} \cdot 0^3 = \frac{1}{3}$$

Comparing this method with the calculation in Example 2 in Section 5.1, where we found the area under the parabola $y = x^2$ from 0 to 1 by computing a limit of sums, we see that the Evaluation Theorem provides us with a simple and powerful method.

Although the Evaluation Theorem may be surprising at first glance, it becomes plausible if we interpret it in physical terms. If $v(t)$ is the velocity of an object and $s(t)$ is its position at time t, then $v(t) = s'(t)$, so s is an antiderivative of v. In Section 5.1 we considered an object that always moves in the positive direction and made the guess that the area under the velocity curve is equal to the distance traveled. In symbols:

$$\int_a^b v(t)\,dt = s(b) - s(a)$$

That is exactly what the Evaluation Theorem says in this context.

PROOF OF THE EVALUATION THEOREM We divide the interval $[a, b]$ into n subintervals with endpoints $x_0 \,(= a)$, x_1, x_2, ..., $x_n \,(= b)$ and with length $\Delta x = (b - a)/n$. Let F be any antiderivative of f. By subtracting and adding like terms, we can express the total difference in the F values as the sum of the differences over the subintervals:

$$F(b) - F(a) = F(x_n) - F(x_0)$$

$$= F(x_n) - F(x_{n-1}) + F(x_{n-1}) - F(x_{n-2}) + \cdots + F(x_2) - F(x_1) + F(x_1) - F(x_0)$$

$$= \sum_{i=1}^{n} [F(x_i) - F(x_{i-1})]$$

■ See Section 4.2 for The Mean Value Theorem.

Now F is continuous (because it's differentiable) and so we can apply the Mean Value Theorem to F on each subinterval $[x_{i-1}, x_i]$. Thus there exists a number x_i^* between x_{i-1} and x_i such that

$$F(x_i) - F(x_{i-1}) = F'(x_i^*)(x_i - x_{i-1}) = f(x_i^*)\,\Delta x$$

Therefore
$$F(b) - F(a) = \sum_{i=1}^{n} f(x_i^*)\,\Delta x$$

Now we take the limit of each side of this equation as $n \to \infty$. The left side is a constant and the right side is a Riemann sum for the function f, so

$$F(b) - F(a) = \lim_{n \to \infty} \sum_{i=1}^{n} f(x_i^*)\,\Delta x = \int_a^b f(x)\,dx \qquad \square$$

When applying the Evaluation Theorem we use the notation

$$F(x)\Big]_a^b = F(b) - F(a)$$

and so we can write

$$\int_a^b f(x)\,dx = F(x)\Big]_a^b \qquad \text{where} \qquad F' = f$$

Other common notations are $F(x)\big|_a^b$ and $[F(x)]_a^b$.

▼ EXAMPLE 1 Evaluate $\int_1^3 e^x\,dx$.

SOLUTION An antiderivative of $f(x) = e^x$ is $F(x) = e^x$, so we use the Evaluation Theorem as follows:

$$\int_1^3 e^x\,dx = e^x\Big]_1^3 = e^3 - e \qquad \blacksquare$$

EXAMPLE 2 Find the area under the cosine curve from 0 to b, where $0 \le b \le \pi/2$.

SOLUTION Since an antiderivative of $f(x) = \cos x$ is $F(x) = \sin x$, we have

$$A = \int_0^b \cos x\,dx = \sin x\Big]_0^b = \sin b - \sin 0 = \sin b$$

In particular, taking $b = \pi/2$, we have proved that the area under the cosine curve from 0 to $\pi/2$ is $\sin(\pi/2) = 1$. (See Figure 1.) $\qquad \blacksquare$

■ In applying the Evaluation Theorem we use a particular antiderivative F of f. It is not necessary to use the most general antiderivative ($e^x + C$).

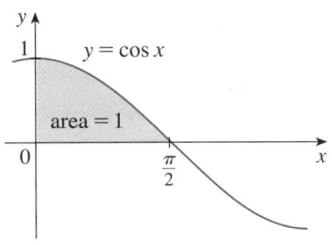

FIGURE 1

■ www.stewartcalculus.com
See Additional Example A.

When the French mathematician Gilles de Roberval first found the area under the sine and cosine curves in 1635, this was a very challenging problem that required a great deal of ingenuity. If we didn't have the benefit of the Evaluation Theorem, we would have to compute a difficult limit of sums using obscure trigonometric identities (or a computer algebra system as in Exercise 23 in Section 5.1). It was even more difficult for Roberval because the apparatus of limits had not been invented in 1635. But in the 1660s and 1670s, when the Evaluation Theorem was discovered by Newton and Leibniz, such problems became very easy, as you can see from Example 2.

INDEFINITE INTEGRALS

We need a convenient notation for antiderivatives that makes them easy to work with. Because of the relation given by the Evaluation Theorem between antiderivatives and integrals, the notation $\int f(x)\,dx$ is traditionally used for an antiderivative of f and is called an **indefinite integral**. Thus

$$\int f(x)\,dx = F(x) \qquad \text{means} \qquad F'(x) = f(x)$$

⊘ You should distinguish carefully between definite and indefinite integrals. A definite integral $\int_a^b f(x)\,dx$ is a *number*, whereas an indefinite integral $\int f(x)\,dx$ is a *function* (or family of functions). The connection between them is given by the Evaluation Theorem: If f is continuous on $[a, b]$, then

$$\int_a^b f(x)\,dx = \int f(x)\,dx \Bigg]_a^b$$

Recall from Section 4.7 that if F is an antiderivative of f on an interval I, then the most general antiderivative of f on I is $F(x) + C$, where C is an arbitrary constant. For instance, the formula

$$\int \frac{1}{x}\,dx = \ln|x| + C$$

is valid (on any interval that doesn't contain 0) because $(d/dx)\ln|x| = 1/x$. So an indefinite integral $\int f(x)\,dx$ can represent either a particular antiderivative of f or an entire *family* of antiderivatives (one for each value of the constant C).

The effectiveness of the Evaluation Theorem depends on having a supply of antiderivatives of functions. We therefore restate the Table of Antidifferentiation Formulas from Section 4.7, together with a few others, in the notation of indefinite integrals. Any formula can be verified by differentiating the function on the right side and obtaining the integrand. For instance,

$$\int \sec^2 x\,dx = \tan x + C \qquad \text{because} \qquad \frac{d}{dx}(\tan x + C) = \sec^2 x$$

- We adopt the convention that when a formula for a general indefinite integral is given, it is valid only on an interval.

---

1 TABLE OF INDEFINITE INTEGRALS

$$\int [f(x) + g(x)]\,dx = \int f(x)\,dx + \int g(x)\,dx \qquad \int cf(x)\,dx = c\int f(x)\,dx$$

$$\int x^n\,dx = \frac{x^{n+1}}{n+1} + C \quad (n \neq -1) \qquad\qquad \int \frac{1}{x}\,dx = \ln|x| + C$$

$$\int e^x\,dx = e^x + C \qquad\qquad\qquad\qquad \int a^x\,dx = \frac{a^x}{\ln a} + C$$

$$\int \sin x\,dx = -\cos x + C \qquad\qquad\qquad \int \cos x\,dx = \sin x + C$$

$$\int \sec^2 x\,dx = \tan x + C \qquad\qquad\qquad \int \csc^2 x\,dx = -\cot x + C$$

$$\int \sec x \tan x\,dx = \sec x + C \qquad\qquad \int \csc x \cot x\,dx = -\csc x + C$$

$$\int \frac{1}{x^2 + 1}\,dx = \tan^{-1}x + C \qquad\qquad \int \frac{1}{\sqrt{1 - x^2}}\,dx = \sin^{-1}x + C$$

$$\int \sinh x\,dx = \cosh x + C \qquad\qquad\qquad \int \cosh x\,dx = \sinh x + C$$

■ The indefinite integral in Example 3 is graphed in Figure 2 for several values of C. Here the value of C is the y-intercept.

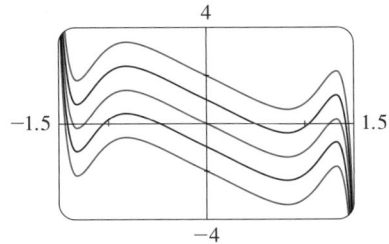

FIGURE 2

EXAMPLE 3 Find the general indefinite integral

$$\int (10x^4 - 2 \sec^2 x)\, dx$$

SOLUTION Using our convention and Table 1, we have

$$\int (10x^4 - 2 \sec^2 x)\, dx = 10 \int x^4\, dx - 2 \int \sec^2 x\, dx$$

$$= 10\, \frac{x^5}{5} - 2 \tan x + C$$

$$= 2x^5 - 2 \tan x + C$$

You should check this answer by differentiating it. ■

EXAMPLE 4 Evaluate $\int_0^3 (x^3 - 6x)\, dx$.

SOLUTION Using the Evaluation Theorem and Table 1, we have

$$\int_0^3 (x^3 - 6x)\, dx = \frac{x^4}{4} - 6\, \frac{x^2}{2} \Big]_0^3$$

$$= \left(\tfrac{1}{4} \cdot 3^4 - 3 \cdot 3^2\right) - \left(\tfrac{1}{4} \cdot 0^4 - 3 \cdot 0^2\right)$$

$$= \tfrac{81}{4} - 27 - 0 + 0 = -6.75$$

Compare this calculation with Example 2(b) in Section 5.2. ■

▼ **EXAMPLE 5** Find $\int_0^2 \left(2x^3 - 6x + \dfrac{3}{x^2 + 1}\right) dx$ and interpret the result in terms of areas.

SOLUTION The Evaluation Theorem gives

$$\int_0^2 \left(2x^3 - 6x + \frac{3}{x^2 + 1}\right) dx = 2\, \frac{x^4}{4} - 6\, \frac{x^2}{2} + 3 \tan^{-1} x \Big]_0^2$$

$$= \tfrac{1}{2} x^4 - 3x^2 + 3 \tan^{-1} x \Big]_0^2$$

$$= \tfrac{1}{2}(2^4) - 3(2^2) + 3 \tan^{-1} 2 - 0$$

$$= -4 + 3 \tan^{-1} 2$$

■ Figure 3 shows the graph of the integrand in Example 5. We know from Section 5.2 that the value of the integral can be interpreted as the sum of the areas labeled with a plus sign minus the area labeled with a minus sign.

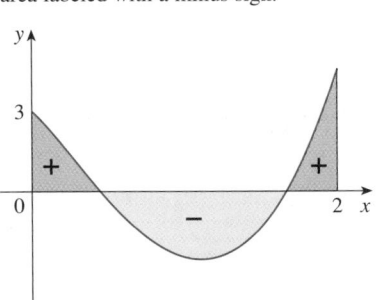

FIGURE 3

This is the exact value of the integral. If a decimal approximation is desired, we can use a calculator to approximate $\tan^{-1} 2$. Doing so, we get

$$\int_0^2 \left(2x^3 - 6x + \frac{3}{x^2 + 1}\right) dx \approx -0.67855$$ ■

EXAMPLE 6 Evaluate $\displaystyle\int_1^9 \frac{2t^2 + t^2\sqrt{t} - 1}{t^2}\, dt$.

SOLUTION First we need to write the integrand in a simpler form by carrying out the division:

■ www.stewartcalculus.com
See Additional Example B.

$$\int_1^9 \frac{2t^2 + t^2\sqrt{t} - 1}{t^2}\, dt = \int_1^9 (2 + t^{1/2} - t^{-2})\, dt$$

$$= 2t + \frac{t^{3/2}}{\frac{3}{2}} - \frac{t^{-1}}{-1}\Bigg]_1^9 = 2t + \tfrac{2}{3}t^{3/2} + \frac{1}{t}\Bigg]_1^9$$

$$= \left[2 \cdot 9 + \tfrac{2}{3}(9)^{3/2} + \tfrac{1}{9}\right] - \left(2 \cdot 1 + \tfrac{2}{3} \cdot 1^{3/2} + \tfrac{1}{1}\right)$$

$$= 18 + 18 + \tfrac{1}{9} - 2 - \tfrac{2}{3} - 1 = 32\tfrac{4}{9} \qquad ■$$

APPLICATIONS

The Evaluation Theorem says that if f is continuous on $[a, b]$, then

$$\int_a^b f(x)\, dx = F(b) - F(a)$$

where F is any antiderivative of f. This means that $F' = f$, so the equation can be rewritten as

$$\int_a^b F'(x)\, dx = F(b) - F(a)$$

We know that $F'(x)$ represents the rate of change of $y = F(x)$ with respect to x and $F(b) - F(a)$ is the change in y when x changes from a to b. [Note that y could, for instance, increase, then decrease, then increase again. Although y might change in both directions, $F(b) - F(a)$ represents the *net* change in y.] So we can reformulate the Evaluation Theorem in words as follows.

NET CHANGE THEOREM   The integral of a rate of change is the net change:

$$\int_a^b F'(x)\, dx = F(b) - F(a)$$

This principle can be applied to all of the rates of change in the natural and social sciences. Here are a few instances of this idea:

■ If $V(t)$ is the volume of water in a reservoir at time t, then its derivative $V'(t)$ is the rate at which water flows into the reservoir at time t. So

$$\int_{t_1}^{t_2} V'(t)\, dt = V(t_2) - V(t_1)$$

is the change in the amount of water in the reservoir between time t_1 and time t_2.

■ If $[C](t)$ is the concentration of the product of a chemical reaction at time t, then the rate of reaction is the derivative $d[C]/dt$. So

$$\int_{t_1}^{t_2} \frac{d[C]}{dt}\, dt = [C](t_2) - [C](t_1)$$

is the change in the concentration of C from time t_1 to time t_2.

- If the rate of growth of a population is dn/dt, then

$$\int_{t_1}^{t_2} \frac{dn}{dt} \, dt = n(t_2) - n(t_1)$$

is the net change in population during the time period from t_1 to t_2. (The population increases when births happen and decreases when deaths occur. The net change takes into account both births and deaths.)

- If an object moves along a straight line with position function $s(t)$, then its velocity is $v(t) = s'(t)$, so

$$\boxed{2} \qquad \int_{t_1}^{t_2} v(t) \, dt = s(t_2) - s(t_1)$$

is the net change of position, or *displacement,* of the particle during the time period from t_1 to t_2. In Section 5.1 we guessed that this was true for the case where the object moves in the positive direction, but now we have proved that it is always true.

- If we want to calculate the distance traveled during the time interval, we have to consider the intervals when $v(t) \geqslant 0$ (the particle moves to the right) and also the intervals when $v(t) \leqslant 0$ (the particle moves to the left). In both cases the distance is computed by integrating $|v(t)|$, the speed. Therefore

$$\boxed{3} \qquad \int_{t_1}^{t_2} |v(t)| \, dt = \text{total distance traveled}$$

Figure 4 shows how both displacement and distance traveled can be interpreted in terms of areas under a velocity curve.

- The acceleration of the object is $a(t) = v'(t)$, so

$$\int_{t_1}^{t_2} a(t) \, dt = v(t_2) - v(t_1)$$

is the change in velocity from time t_1 to time t_2.

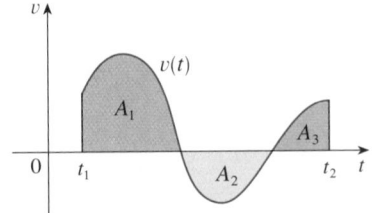

$$\text{displacement} = \int_{t_1}^{t_2} v(t) \, dt = A_1 - A_2 + A_3$$

$$\text{distance} = \int_{t_1}^{t_2} |v(t)| \, dt = A_1 + A_2 + A_3$$

FIGURE 4

V EXAMPLE 7 A particle moves along a line so that its velocity at time t is $v(t) = t^2 - t - 6$ (measured in meters per second).
(a) Find the displacement of the particle during the time period $1 \leqslant t \leqslant 4$.
(b) Find the distance traveled during this time period.

SOLUTION
(a) By Equation 2, the displacement is

$$s(4) - s(1) = \int_1^4 v(t) \, dt = \int_1^4 (t^2 - t - 6) \, dt$$

$$= \left[\frac{t^3}{3} - \frac{t^2}{2} - 6t \right]_1^4 = -\frac{9}{2}$$

This means that the particle's position at time $t = 4$ is 4.5 m to the left of its position at the start of the time period.

(b) Note that $v(t) = t^2 - t - 6 = (t - 3)(t + 2)$ and so $v(t) \leq 0$ on the interval $[1, 3]$ and $v(t) \geq 0$ on $[3, 4]$. Thus, from Equation 3, the distance traveled is

■ To integrate the absolute value of $v(t)$, we use Property 5 of integrals from Section 5.2 to split the integral into two parts, one where $v(t) \leq 0$ and one where $v(t) \geq 0$.

$$\int_1^4 |v(t)| \, dt = \int_1^3 [-v(t)] \, dt + \int_3^4 v(t) \, dt$$

$$= \int_1^3 (-t^2 + t + 6) \, dt + \int_3^4 (t^2 - t - 6) \, dt$$

$$= \left[-\frac{t^3}{3} + \frac{t^2}{2} + 6t \right]_1^3 + \left[\frac{t^3}{3} - \frac{t^2}{2} - 6t \right]_3^4$$

$$= \frac{61}{6} \approx 10.17 \text{ m}$$

■

V EXAMPLE 8 Figure 5 shows the power consumption in the city of San Francisco for a day in September (P is measured in megawatts; t is measured in hours starting at midnight). Estimate the energy used on that day.

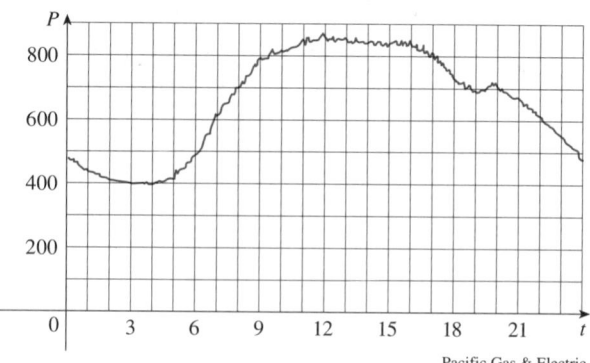

FIGURE 5

Pacific Gas & Electric

SOLUTION Power is the rate of change of energy: $P(t) = E'(t)$. So, by the Net Change Theorem,

$$\int_0^{24} P(t) \, dt = \int_0^{24} E'(t) \, dt = E(24) - E(0)$$

is the total amount of energy used on that day. We approximate the value of the integral using the Midpoint Rule with 12 subintervals and $\Delta t = 2$:

$$\int_0^{24} P(t) \, dt \approx [P(1) + P(3) + P(5) + \cdots + P(21) + P(23)] \, \Delta t$$

$$\approx (440 + 400 + 420 + 620 + 790 + 840 + 850$$

$$+ 840 + 810 + 690 + 670 + 550)(2)$$

$$= 15{,}840$$

The energy used was approximately 15,840 megawatt-hours. ■

■ A note on units

How did we know what units to use for energy in Example 8? The integral $\int_0^{24} P(t) \, dt$ is defined as the limit of sums of terms of the form $P(t_i^*) \, \Delta t$. Now $P(t_i^*)$ is measured in megawatts and Δt is measured in hours, so their product is measured in megawatt-hours. The same is true of the limit. In general, the unit of measurement for $\int_a^b f(x) \, dx$ is the product of the unit for $f(x)$ and the unit for x.

5.3 | EXERCISES

1–30 ▪ Evaluate the integral.

1. $\int_{-2}^{3} (x^2 - 3)\, dx$

2. $\int_{1}^{2} (4x^3 - 3x^2 + 2x)\, dx$

3. $\int_{-2}^{0} \left(\frac{1}{2}t^4 + \frac{1}{4}t^3 - t\right) dt$

4. $\int_{0}^{3} (1 + 6w^2 - 10w^4)\, dw$

5. $\int_{0}^{2} (2x - 3)(4x^2 + 1)\, dx$

6. $\int_{-1}^{1} t(1 - t)^2\, dt$

7. $\int_{0}^{\pi} (5e^x + 3 \sin x)\, dx$

8. $\int_{1}^{2} \left(\frac{1}{x^2} - \frac{4}{x^3}\right) dx$

9. $\int_{1}^{4} \left(\frac{4 + 6u}{\sqrt{u}}\right) du$

10. $\int_{0}^{4} (3\sqrt{t} - 2e^t)\, dt$

11. $\int_{0}^{1} x(\sqrt[3]{x} + \sqrt[4]{x})\, dx$

12. $\int_{1}^{4} \frac{\sqrt{y} - y}{y^2}\, dy$

13. $\int_{1}^{2} \left(\frac{x}{2} - \frac{2}{x}\right) dx$

14. $\int_{0}^{1} (5x - 5^x)\, dx$

15. $\int_{0}^{1} (x^{10} + 10^x)\, dx$

16. $\int_{\pi/4}^{\pi/3} \csc^2 \theta\, d\theta$

17. $\int_{0}^{\pi/4} \frac{1 + \cos^2 \theta}{\cos^2 \theta}\, d\theta$

18. $\int_{0}^{\pi/3} \frac{\sin \theta + \sin \theta \tan^2 \theta}{\sec^2 \theta}\, d\theta$

19. $\int_{0}^{1} \cosh t\, dt$

20. $\int_{-10}^{10} \frac{2e^x}{\sinh x + \cosh x}\, dx$

21. $\int_{0}^{\sqrt{3}/2} \frac{dr}{\sqrt{1 - r^2}}$

22. $\int_{1/\sqrt{3}}^{\sqrt{3}} \frac{8}{1 + x^2}\, dx$

23. $\int_{1}^{e} \frac{x^2 + x + 1}{x}\, dx$

24. $\int_{\pi/4}^{\pi/3} \sec \theta \tan \theta\, d\theta$

25. $\int_{-1}^{1} e^{u+1}\, du$

26. $\int_{1}^{2} \frac{(x - 1)^3}{x^2}\, dx$

27. $\int_{0}^{1/\sqrt{3}} \frac{t^2 - 1}{t^4 - 1}\, dt$

28. $\int_{0}^{2} |2x - 1|\, dx$

29. $\int_{-1}^{2} (x - 2|x|)\, dx$

30. $\int_{0}^{3\pi/2} |\sin x|\, dx$

31–32 ▪ What is wrong with the equation?

31. $\int_{-1}^{3} \frac{1}{x^2}\, dx = \left. \frac{x^{-1}}{-1} \right]_{-1}^{3} = -\frac{4}{3}$

32. $\int_{0}^{\pi} \sec^2 x\, dx = \tan x \big]_{0}^{\pi} = 0$

33–34 ▪ Calculate the area of the region that lies under the curve and above the x-axis.

33. $y = 1 - x^2$

34. $y = 2x - x^2$

35–36 ▪ Use a graph to give a rough estimate of the area of the region that lies beneath the given curve. Then find the exact area.

35. $y = \sin x$, $0 \leqslant x \leqslant \pi$

36. $y = \sec^2 x$, $0 \leqslant x \leqslant \pi/3$

37–38 ▪ Evaluate the integral and interpret it as a difference of areas. Illustrate with a sketch.

37. $\int_{-1}^{2} x^3\, dx$

38. $\int_{\pi/4}^{5\pi/2} \sin x\, dx$

39–40 ▪ Verify by differentiation that the formula is correct.

39. $\int \frac{x}{\sqrt{x^2 + 1}}\, dx = \sqrt{x^2 + 1} + C$

40. $\int \cos^2 x\, dx = \frac{1}{2}x + \frac{1}{4} \sin 2x + C$

41–42 ▪ Find the general indefinite integral. Illustrate by graphing several members of the family on the same screen.

41. $\int x\sqrt{x}\, dx$

42. $\int (\cos x - 2 \sin x)\, dx$

43–48 ▪ Find the general indefinite integral.

43. $\int (\sin x + \sinh x)\, dx$

44. $\int (\sqrt{x^3} + \sqrt[3]{x^2})\, dx$

45. $\int (u + 4)(2u + 1)\, du$

46. $\int v(v^2 + 2)^2\, dv$

47. $\int \frac{\sin x}{1 - \sin^2 x}\, dx$

48. $\int \frac{\sin 2x}{\sin x}\, dx$

49. The area of the region that lies to the right of the y-axis and to the left of the parabola $x = 2y - y^2$ (the shaded region in the figure) is given by the integral $\int_{0}^{2} (2y - y^2)\, dy$. (Turn

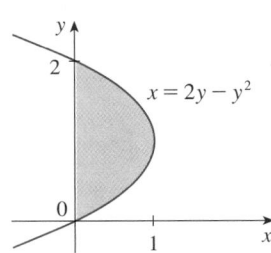

your head clockwise and think of the region as lying below the curve $x = 2y - y^2$ from $y = 0$ to $y = 2$.) Find the area of the region.

50. The boundaries of the shaded region are the y-axis, the line $y = 1$, and the curve $y = \sqrt[4]{x}$. Find the area of this region by writing x as a function of y and integrating with respect to y (as in Exercise 49).

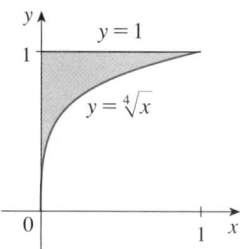

51. If $w'(t)$ is the rate of growth of a child in pounds per year, what does $\int_5^{10} w'(t)\,dt$ represent?

52. The current in a wire is defined as the derivative of the charge: $I(t) = Q'(t)$. What does $\int_a^b I(t)\,dt$ represent?

53. If oil leaks from a tank at a rate of $r(t)$ gallons per minute at time t, what does $\int_0^{120} r(t)\,dt$ represent?

54. A honeybee population starts with 100 bees and increases at a rate of $n'(t)$ bees per week. What does $100 + \int_0^{15} n'(t)\,dt$ represent?

55. In Section 4.5 we defined the marginal revenue function $R'(x)$ as the derivative of the revenue function $R(x)$, where x is the number of units sold. What does $\int_{1000}^{5000} R'(x)\,dx$ represent?

56. If $f(x)$ is the slope of a trail at a distance of x miles from the start of the trail, what does $\int_3^5 f(x)\,dx$ represent?

57. If x is measured in meters and $f(x)$ is measured in newtons, what are the units for $\int_0^{100} f(x)\,dx$?

58. If the units for x are feet and the units for $a(x)$ are pounds per foot, what are the units for da/dx? What units does $\int_2^8 a(x)\,dx$ have?

59–60 ▪ The velocity function (in meters per second) is given for a particle moving along a line. Find (a) the displacement and (b) the distance traveled by the particle during the given time interval.

59. $v(t) = 3t - 5, \quad 0 \le t \le 3$

60. $v(t) = t^2 - 2t - 8, \quad 1 \le t \le 6$

61–62 ▪ The acceleration function (in m/s²) and the initial velocity are given for a particle moving along a line. Find (a) the

velocity at time t and (b) the distance traveled during the given time interval.

61. $a(t) = t + 4, \quad v(0) = 5, \quad 0 \le t \le 10$

62. $a(t) = 2t + 3, \quad v(0) = -4, \quad 0 \le t \le 3$

63. The velocity of a car was read from its speedometer at 10-second intervals and recorded in the table. Use the Midpoint Rule to estimate the distance traveled by the car.

t (s)	v (mi/h)	t (s)	v (mi/h)
0	0	60	56
10	38	70	53
20	52	80	50
30	58	90	47
40	55	100	45
50	51		

64. Suppose that a volcano is erupting and readings of the rate $r(t)$ at which solid materials are spewed into the atmosphere are given in the table. The time t is measured in seconds and the units for $r(t)$ are tonnes (metric tons) per second.

t	0	1	2	3	4	5	6
$r(t)$	2	10	24	36	46	54	60

(a) Give upper and lower estimates for the total quantity $Q(6)$ of erupted materials after 6 seconds.
(b) Use the Midpoint Rule to estimate $Q(6)$.

65. Water flows from the bottom of a storage tank at a rate of $r(t) = 200 - 4t$ liters per minute, where $0 \le t \le 50$. Find the amount of water that flows from the tank during the first 10 minutes.

66. Water flows into and out of a storage tank. A graph of the rate of change $r(t)$ of the volume of water in the tank, in liters per day, is shown. If the amount of water in the tank at time $t = 0$ is 25,000 L, use the Midpoint Rule to estimate the amount of water four days later.

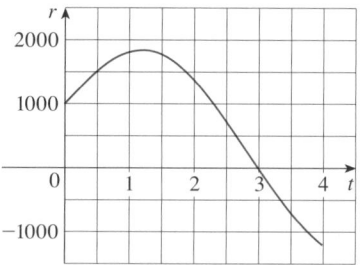

67. A bacteria population is 4000 at time $t = 0$ and its rate of growth is $1000 \cdot 2^t$ bacteria per hour after t hours. What is the population after one hour?

68. Shown is the graph of traffic on an Internet service provider's T1 data line from midnight to 8:00 A.M. D is the data throughput, measured in megabits per second. Use the Midpoint Rule to estimate the total amount of data transmitted during that time period.

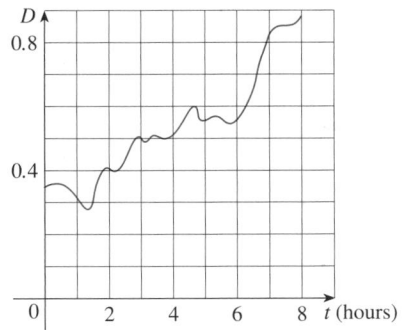

69. Suppose h is a function such that $h(1) = -2$, $h'(1) = 2$, $h''(1) = 3$, $h(2) = 6$, $h'(2) = 5$, $h''(2) = 13$, and h'' is continuous everywhere. Evaluate $\int_1^2 h''(u)\, du$.

70. The area labeled B is three times the area labeled A. Express b in terms of a.

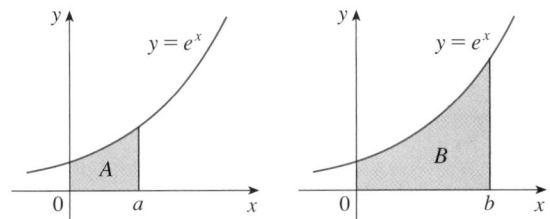

5.4 | THE FUNDAMENTAL THEOREM OF CALCULUS

The Fundamental Theorem of Calculus is appropriately named because it establishes a connection between the two branches of calculus: differential calculus and integral calculus. Differential calculus arose from the tangent problem, whereas integral calculus arose from a seemingly unrelated problem, the area problem. Newton's mentor at Cambridge, Isaac Barrow (1630–1677), discovered that these two problems are actually closely related. In fact, he realized that differentiation and integration are inverse processes. The Fundamental Theorem of Calculus gives the precise inverse relationship between the derivative and the integral. It was Newton and Leibniz who exploited this relationship and used it to develop calculus into a systematic mathematical method.

The first part of the Fundamental Theorem deals with functions defined by an equation of the form

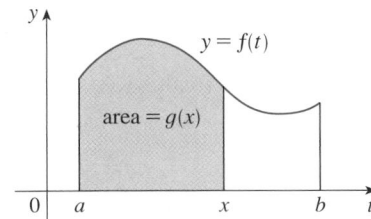

FIGURE 1

$$\boxed{1} \qquad g(x) = \int_a^x f(t)\, dt$$

where f is a continuous function on $[a, b]$ and x varies between a and b. Observe that g depends only on x, which appears as the variable upper limit in the integral. If x is a fixed number, then the integral $\int_a^x f(t)\, dt$ is a definite number. If we then let x vary, the number $\int_a^x f(t)\, dt$ also varies and defines a function of x denoted by $g(x)$.

If f happens to be a positive function, then $g(x)$ can be interpreted as the area under the graph of f from a to x, where x can vary from a to b. (Think of g as the "area so far" function; see Figure 1.)

◢ **EXAMPLE 1** If f is the function whose graph is shown in Figure 2 and $g(x) = \int_0^x f(t)\, dt$, find the values of $g(0)$, $g(1)$, $g(2)$, $g(3)$, $g(4)$, and $g(5)$. Then sketch a rough graph of g.

SOLUTION First we notice that $g(0) = \int_0^0 f(t)\, dt = 0$. From Figure 3 we see that $g(1)$ is the area of a triangle:

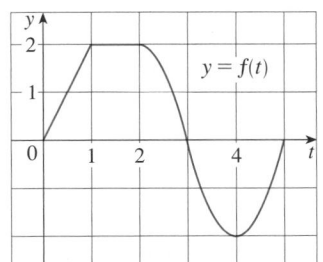

FIGURE 2

$$g(1) = \int_0^1 f(t)\, dt = \tfrac{1}{2}(1 \cdot 2) = 1$$

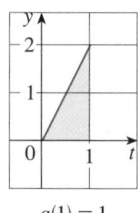
$g(1) = 1$

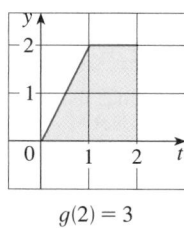
$g(2) = 3$

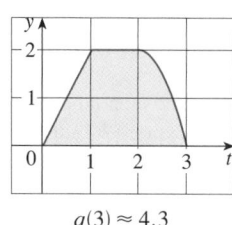
$g(3) \approx 4.3$

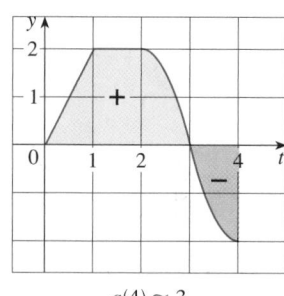
$g(4) \approx 3$

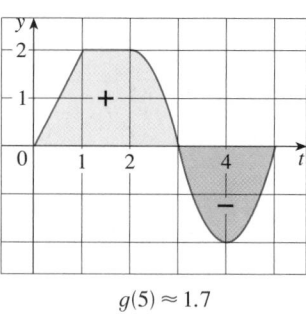
$g(5) \approx 1.7$

FIGURE 3

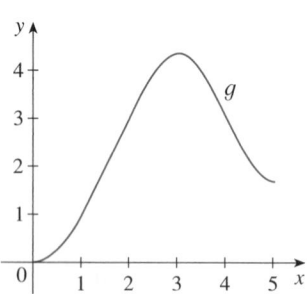

FIGURE 4
$g(x) = \int_a^x f(t)\, dt$

To find $g(2)$ we add to $g(1)$ the area of a rectangle:

$$g(2) = \int_0^2 f(t)\, dt = \int_0^1 f(t)\, dt + \int_1^2 f(t)\, dt = 1 + (1 \cdot 2) = 3$$

We estimate that the area under f from 2 to 3 is about 1.3, so

$$g(3) = g(2) + \int_2^3 f(t)\, dt \approx 3 + 1.3 = 4.3$$

For $t > 3$, $f(t)$ is negative and so we start subtracting areas:

$$g(4) = g(3) + \int_3^4 f(t)\, dt \approx 4.3 + (-1.3) = 3.0$$

$$g(5) = g(4) + \int_4^5 f(t)\, dt \approx 3 + (-1.3) = 1.7$$

We use these values to sketch the graph of g in Figure 4. Notice that, because $f(t)$ is positive for $t < 3$, we keep adding area for $t < 3$ and so g is increasing up to $x = 3$, where it attains a maximum value. For $x > 3$, g decreases because $f(t)$ is negative. ∎

EXAMPLE 2 If $g(x) = \int_a^x f(t)\, dt$, where $a = 1$ and $f(t) = t^2$, find a formula for $g(x)$ and calculate $g'(x)$.

SOLUTION In this case we can compute $g(x)$ explicitly using the Evaluation Theorem:

$$g(x) = \int_1^x t^2\, dt = \frac{t^3}{3}\Bigg]_1^x = \frac{x^3 - 1}{3}$$

Then

$$g'(x) = \frac{d}{dx}\left(\tfrac{1}{3}x^3 - \tfrac{1}{3}\right) = x^2 \qquad ∎$$

For the function in Example 2 notice that $g'(x) = x^2$, that is $g' = f$. In other words, if g is defined as the integral of f by Equation 1, then g turns out to be an antiderivative of f, at least in this case. And if we sketch the derivative of the function g shown in Figure 4 by estimating slopes of tangents, we get a graph like that of f in Figure 2. So we suspect that $g' = f$ in Example 1 too.

To see why this might be generally true we consider any continuous function f with $f(x) \geq 0$. Then $g(x) = \int_a^x f(t)\, dt$ can be interpreted as the area under the graph of f from a to x, as in Figure 1.

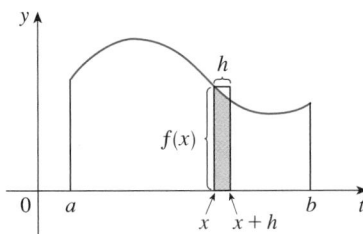

FIGURE 5

In order to compute $g'(x)$ from the definition of derivative we first observe that, for $h > 0$, $g(x + h) - g(x)$ is obtained by subtracting areas, so it is the area under the graph of f from x to $x + h$ (the shaded area in Figure 5). For small h you can see from the figure that this area is approximately equal to the area of the rectangle with height $f(x)$ and width h:

$$g(x + h) - g(x) \approx hf(x)$$

so

$$\frac{g(x + h) - g(x)}{h} \approx f(x)$$

Intuitively, we therefore expect that

$$g'(x) = \lim_{h \to 0} \frac{g(x + h) - g(x)}{h} = f(x)$$

The fact that this is true, even when f is not necessarily positive, is the first part of the Fundamental Theorem of Calculus.

■ We abbreviate the name of this theorem as FTC1. In words, it says that the derivative of a definite integral with respect to its upper limit is the integrand evaluated at the upper limit.

THE FUNDAMENTAL THEOREM OF CALCULUS, PART 1 If f is continuous on $[a, b]$, then the function g defined by

$$g(x) = \int_a^x f(t)\, dt \qquad a \le x \le b$$

is an antiderivative of f, that is, $g'(x) = f(x)$ for $a < x < b$.

TEC Module 5.4 provides visual evidence for FTC1.

PROOF If x and $x + h$ are in the open interval (a, b), then

$$g(x + h) - g(x) = \int_a^{x+h} f(t)\, dt - \int_a^x f(t)\, dt$$

$$= \left(\int_a^x f(t)\, dt + \int_x^{x+h} f(t)\, dt \right) - \int_a^x f(t)\, dt \quad \text{(by Property 5)}$$

$$= \int_x^{x+h} f(t)\, dt$$

and so, for $h \ne 0$,

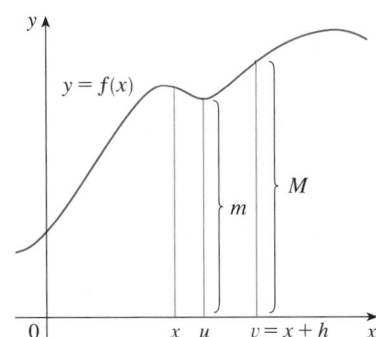

FIGURE 6

$$\boxed{2} \qquad \frac{g(x + h) - g(x)}{h} = \frac{1}{h} \int_x^{x+h} f(t)\, dt$$

For now let's assume that $h > 0$. Since f is continuous on $[x, x + h]$, the Extreme Value Theorem says that there are numbers u and v in $[x, x + h]$ such that $f(u) = m$ and $f(v) = M$, where m and M are the absolute minimum and maximum values of f on $[x, x + h]$. (See Figure 6.)

By Property 8 of integrals, we have

$$mh \le \int_x^{x+h} f(t)\, dt \le Mh$$

that is,

$$f(u)h \le \int_x^{x+h} f(t)\, dt \le f(v)h$$

Since $h > 0$, we can divide this inequality by h:

$$f(u) \leqslant \frac{1}{h} \int_x^{x+h} f(t)\, dt \leqslant f(v)$$

Now we use Equation 2 to replace the middle part of this inequality:

3 $$f(u) \leqslant \frac{g(x+h) - g(x)}{h} \leqslant f(v)$$

Inequality 3 can be proved in a similar manner for the case $h < 0$.

Now we let $h \to 0$. Then $u \to x$ and $v \to x$, since u and v lie between x and $x + h$. Thus

$$\lim_{h \to 0} f(u) = \lim_{u \to x} f(u) = f(x) \quad \text{and} \quad \lim_{h \to 0} f(v) = \lim_{v \to x} f(v) = f(x)$$

because f is continuous at x. We conclude, from 3 and the Squeeze Theorem, that

4 $$g'(x) = \lim_{h \to 0} \frac{g(x+h) - g(x)}{h} = f(x)$$

If $x = a$ or b, then Equation 4 can be interpreted as a one-sided limit. Then Theorem 2.2.4 (modified for one-sided limits) shows that g is continuous on $[a, b]$. $\square$

Using Leibniz notation for derivatives, we can write FTC1 as

$$\frac{d}{dx} \int_a^x f(t)\, dt = f(x)$$

when f is continuous. Roughly speaking, this equation says that if we first integrate f and then differentiate the result, we get back to the original function f.

☑ **EXAMPLE 3** Find the derivative of the function $g(x) = \int_0^x \sqrt{1 + t^2}\, dt$.

SOLUTION Since $f(t) = \sqrt{1 + t^2}$ is continuous, Part 1 of the Fundamental Theorem of Calculus gives

$$g'(x) = \sqrt{1 + x^2}$$ ∎

EXAMPLE 4 Although a formula of the form $g(x) = \int_a^x f(t)\, dt$ may seem like a strange way of defining a function, books on physics, chemistry, and statistics are full of such functions. For instance, the **Fresnel function**

$$S(x) = \int_0^x \sin(\pi t^2 / 2)\, dt$$

is named after the French physicist Augustin Fresnel (1788–1827), who is famous for his works in optics. This function first appeared in Fresnel's theory of the diffraction of light waves, but more recently it has been applied to the design of highways.

Part 1 of the Fundamental Theorem tells us how to differentiate the Fresnel function:

$$S'(x) = \sin(\pi x^2 / 2)$$

This means that we can apply all the methods of differential calculus to analyze S (see Exercise 29).

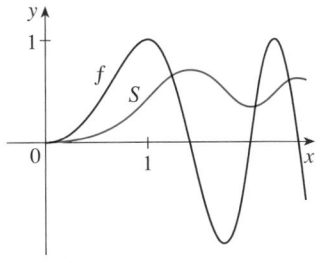

FIGURE 7

$f(x) = \sin(\pi x^2/2)$

$S(x) = \displaystyle\int_0^x \sin(\pi t^2/2)\,dt$

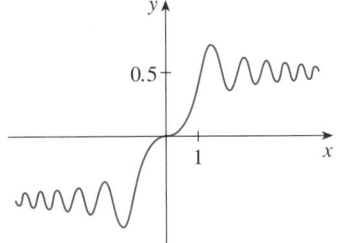

FIGURE 8

The Fresnel function

$S(x) = \displaystyle\int_0^x \sin(\pi t^2/2)\,dt$

Figure 7 shows the graphs of $f(x) = \sin(\pi x^2/2)$ and the Fresnel function $S(x) = \int_0^x f(t)\,dt$. A computer was used to graph S by computing the value of this integral for many values of x. It does indeed look as if $S(x)$ is the area under the graph of f from 0 to x [until $x \approx 1.4$, when $S(x)$ becomes a difference of areas]. Figure 8 shows a larger part of the graph of S.

If we now start with the graph of S in Figure 7 and think about what its derivative should look like, it seems reasonable that $S'(x) = f(x)$. [For instance, S is increasing when $f(x) > 0$ and decreasing when $f(x) < 0$.] So this gives a visual confirmation of Part 1 of the Fundamental Theorem of Calculus. ■

EXAMPLE 5 Find $\dfrac{d}{dx} \displaystyle\int_1^{x^4} \sec t\,dt$.

SOLUTION Here we have to be careful to use the Chain Rule in conjunction with Part 1 of the Fundamental Theorem. Let $u = x^4$. Then

$$\frac{d}{dx}\int_1^{x^4} \sec t\,dt = \frac{d}{dx}\int_1^{u} \sec t\,dt$$

$$= \frac{d}{du}\left[\int_1^{u} \sec t\,dt\right]\frac{du}{dx} \qquad \text{(by the Chain Rule)}$$

$$= \sec u\,\frac{du}{dx} \qquad \text{(by FTC1)}$$

$$= \sec(x^4)\cdot 4x^3 \qquad\qquad ■$$

DIFFERENTIATION AND INTEGRATION AS INVERSE PROCESSES

We now bring together the two parts of the Fundamental Theorem. We regard Part 1 as fundamental because it relates integration and differentiation. But the Evaluation Theorem from Section 5.3 also relates integrals and derivatives, so we rename it as Part 2 of the Fundamental Theorem.

> **THE FUNDAMENTAL THEOREM OF CALCULUS** Suppose f is continuous on $[a, b]$.
>
> **1.** If $g(x) = \int_a^x f(t)\,dt$, then $g'(x) = f(x)$.
>
> **2.** $\int_a^b f(x)\,dx = F(b) - F(a)$, where F is any antiderivative of f, that is, $F' = f$.

We noted that Part 1 can be rewritten as

$$\frac{d}{dx}\int_a^x f(t)\,dt = f(x)$$

which says that if f is integrated and the result is then differentiated, we arrive back at the original function f. In Section 5.3 we reformulated Part 2 as the Net Change Theorem:

$$\int_a^b F'(x)\,dx = F(b) - F(a)$$

This version says that if we take a function F, first differentiate it, and then integrate the result, we arrive back at the original function F, but in the form $F(b) - F(a)$. Taken together, the two parts of the Fundamental Theorem of Calculus say that differentiation and integration are inverse processes. Each undoes what the other does.

The Fundamental Theorem of Calculus is unquestionably the most important theorem in calculus and, indeed, it ranks as one of the great accomplishments of the human mind. Before it was discovered, from the time of Eudoxus and Archimedes to the time of Galileo and Fermat, problems of finding areas, volumes, and lengths of curves were so difficult that only a genius could meet the challenge. But now, armed with the systematic method that Newton and Leibniz fashioned out of the Fundamental Theorem, we will see in the chapters to come that these challenging problems are accessible to all of us.

AVERAGE VALUE OF A FUNCTION

It's easy to calculate the average value of finitely many numbers $y_1, y_2, \ldots, y_n$:

$$y_{\text{ave}} = \frac{y_1 + y_2 + \cdots + y_n}{n}$$

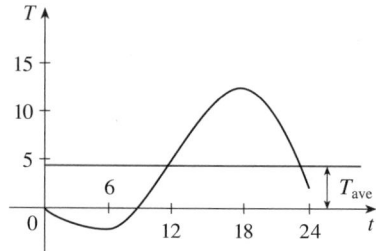

FIGURE 9

But how do we compute the average temperature during a day if infinitely many temperature readings are possible? Figure 9 shows the graph of a temperature function $T(t)$, where t is measured in hours and T in °C, and a guess at the average temperature, T_{ave}.

In general, let's try to compute the average value of a function $y = f(x)$, $a \leq x \leq b$. We start by dividing the interval $[a, b]$ into n equal subintervals, each with length $\Delta x = (b - a)/n$. Then we choose points $x_1^*, \ldots, x_n^*$ in successive subintervals and calculate the average of the numbers $f(x_1^*), \ldots, f(x_n^*)$:

$$\frac{f(x_1^*) + \cdots + f(x_n^*)}{n}$$

(For example, if f represents a temperature function and $n = 24$, this means that we take temperature readings every hour and then average them.) Since $\Delta x = (b - a)/n$, we can write $n = (b - a)/\Delta x$ and the average value becomes

$$\frac{f(x_1^*) + \cdots + f(x_n^*)}{\dfrac{b - a}{\Delta x}} = \frac{1}{b - a} \left[f(x_1^*)\,\Delta x + \cdots + f(x_n^*)\,\Delta x \right]$$

$$= \frac{1}{b - a} \sum_{i=1}^{n} f(x_i^*)\,\Delta x$$

If we let n increase, we would be computing the average value of a large number of closely spaced values. (For example, we would be averaging temperature readings taken every minute or even every second.) The limiting value is

$$\lim_{n \to \infty} \frac{1}{b - a} \sum_{i=1}^{n} f(x_i^*)\,\Delta x = \frac{1}{b - a} \int_a^b f(x)\,dx$$

by the definition of a definite integral.

Therefore, we define the **average value of f** on the interval $[a, b]$ as

- For a positive function, we can think of this definition as saying

$$\frac{\text{area}}{\text{width}} = \text{average height}$$

$$f_{\text{ave}} = \frac{1}{b - a} \int_a^b f(x)\, dx$$

■ **EXAMPLE 6** Find the average value of the function $f(x) = 1 + x^2$ on the interval $[-1, 2]$.

SOLUTION With $a = -1$ and $b = 2$ we have

$$f_{\text{ave}} = \frac{1}{b - a} \int_a^b f(x)\, dx = \frac{1}{2 - (-1)} \int_{-1}^2 (1 + x^2)\, dx = \frac{1}{3}\left[x + \frac{x^3}{3} \right]_{-1}^2 = 2 \quad ■$$

If $T(t)$ is the temperature at time t, we might wonder if there is a specific time when the temperature is the same as the average temperature. For the temperature function graphed in Figure 9, we see that there are two such times—just before noon and just before midnight. In general, is there a number c at which the value of a function f is exactly equal to the average value of the function, that is, $f(c) = f_{\text{ave}}$? The following theorem says that this is true for continuous functions.

THE MEAN VALUE THEOREM FOR INTEGRALS   If f is continuous on $[a, b]$, then there exists a number c in $[a, b]$ such that

$$f(c) = f_{\text{ave}} = \frac{1}{b - a} \int_a^b f(x)\, dx$$

that is,

$$\int_a^b f(x)\, dx = f(c)(b - a)$$

PROOF Let $F(x) = \int_a^x f(t)\, dt$ for $a \leq x \leq b$. By the Mean Value Theorem for derivatives, there is a number c between a and b such that

$$F(b) - F(a) = F'(c)(b - a)$$

But $F'(x) = f(x)$ by FTC1. Therefore

$$\int_a^b f(t)\, dt - 0 = f(c)(b - a) \qquad \square$$

The geometric interpretation of the Mean Value Theorem for Integrals is that, for *positive* functions f, there is a number c such that the rectangle with base $[a, b]$ and height $f(c)$ has the same area as the region under the graph of f from a to b. (See Figure 10 and the more picturesque interpretation in the margin note.)

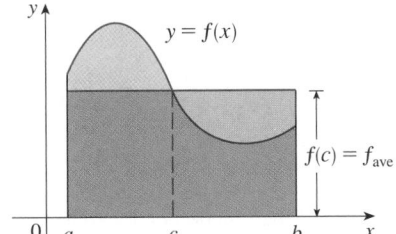

FIGURE 10

- You can always chop off the top of a (two-dimensional) mountain at a certain height and use it to fill in the valleys so that the mountaintop becomes completely flat.

■ **EXAMPLE 7** Since $f(x) = 1 + x^2$ is continuous on the interval $[-1, 2]$, the Mean Value Theorem for Integrals says there is a number c in $[-1, 2]$ such that

$$\int_{-1}^2 (1 + x^2)\, dx = f(c)[2 - (-1)]$$

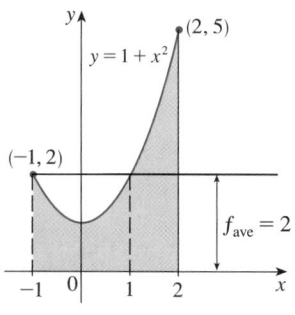

FIGURE 11

In this particular case we can find c explicitly. From Example 6 we know that $f_{ave} = 2$, so the value of c satisfies

$$f(c) = f_{ave} = 2$$

Therefore $1 + c^2 = 2$ so $c^2 = 1$

Thus in this case there happen to be two numbers $c = \pm 1$ in the interval $[-1, 2]$ that work in the Mean Value Theorem for Integrals.   ■

Examples 6 and 7 are illustrated by Figure 11.

5.4 | EXERCISES

1. Let $g(x) = \int_0^x f(t)\, dt$, where f is the function whose graph is shown.
   (a) Evaluate $g(0)$, $g(1)$, $g(2)$, $g(3)$, and $g(6)$.
   (b) On what interval is g increasing?
   (c) Where does g have a maximum value?
   (d) Sketch a rough graph of g.

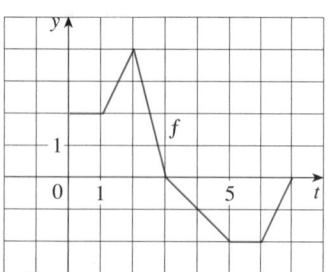

2. Let $g(x) = \int_0^x f(t)\, dt$, where f is the function whose graph is shown.
   (a) Evaluate $g(x)$ for $x = 0, 1, 2, 3, 4, 5,$ and 6.
   (b) Estimate $g(7)$.
   (c) Where does g have a maximum value? Where does it have a minimum value?
   (d) Sketch a rough graph of g.

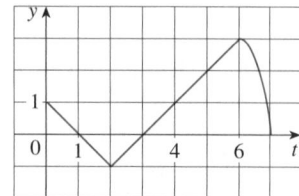

3–4 ▪ Sketch the area represented by $g(x)$. Then find $g'(x)$ in two ways: (a) by using Part 1 of the Fundamental Theorem and (b) by evaluating the integral using Part 2 and then differentiating.

3. $g(x) = \int_1^x t^2\, dt$

4. $g(x) = \int_0^x (2 + \sin t)\, dt$

5–14 ▪ Use Part 1 of the Fundamental Theorem of Calculus to find the derivative of the function.

5. $g(x) = \int_1^x \dfrac{1}{t^3 + 1}\, dt$

6. $g(x) = \int_3^x e^{t^2 - t}\, dt$

7. $g(s) = \int_5^s (t - t^2)^8\, dt$

8. $F(x) = \int_x^{10} \tan\theta\, d\theta$

$$\left[\text{Hint: } \int_x^{10} \tan\theta\, d\theta = -\int_{10}^x \tan\theta\, d\theta \right]$$

9. $h(x) = \int_2^{1/x} \arctan t\, dt$

10. $h(x) = \int_0^{x^2} \sqrt{1 + r^3}\, dr$

11. $y = \int_0^{\tan x} \sqrt{t + \sqrt{t}}\, dt$

12. $y = \int_{\sin x}^1 \sqrt{1 + t^2}\, dt$

13. $g(x) = \int_{2x}^{3x} \dfrac{u^2 - 1}{u^2 + 1}\, du$

$$\left[\text{Hint: } \int_{2x}^{3x} f(u)\, du = \int_{2x}^0 f(u)\, du + \int_0^{3x} f(u)\, du \right]$$

14. $y = \int_{\sin x}^{\cos x} (1 + v^2)^{10}\, dv$

15–18 ▪ Find the average value of the function on the given interval.

15. $g(x) = \sqrt[3]{x}$, $[1, 8]$

16. $f(x) = 1/x$, $[1, 4]$

17. $g(x) = \cos x$, $[0, \pi/2]$

18. $f(\theta) = \sec\theta\tan\theta$, $[0, \pi/4]$

19–20 ▪
   (a) Find the average value of f on the given interval.
   (b) Find c such that $f_{ave} = f(c)$.

(c) Sketch the graph of f and a rectangle whose area is the same as the area under the graph of f.

19. $f(x) = (x - 3)^2$, $[2, 5]$ **20.** $f(x) = \sqrt{x}$, $[0, 4]$

21. Find the average value of f on $[0, 8]$.

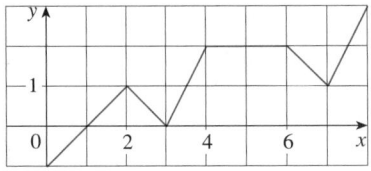

22. The velocity graph of an accelerating car is shown.
(a) Estimate the average velocity of the car during the first 12 seconds.
(b) At what time was the instantaneous velocity equal to the average velocity?

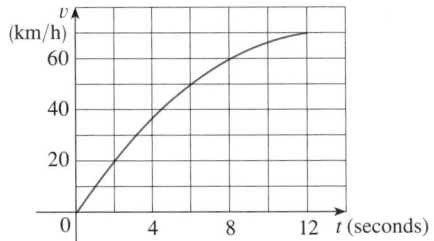

23. On what interval is the curve

$$y = \int_0^x \frac{t^2}{t^2 + t + 2} \, dt$$

concave downward?

24. If $f(x) = \int_0^{\sin x} \sqrt{1 + t^2} \, dt$ and $g(y) = \int_3^y f(x) \, dx$, find $g''(\pi/6)$.

25–26 ■ Let $g(x) = \int_0^x f(t) \, dt$, where f is the function whose graph is shown.
(a) At what values of x do the local maximum and minimum values of g occur?
(b) Where does g attain its absolute maximum value?
(c) On what intervals is g concave downward?
(d) Sketch the graph of g.

25.

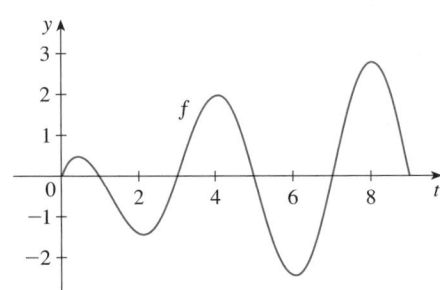

26.

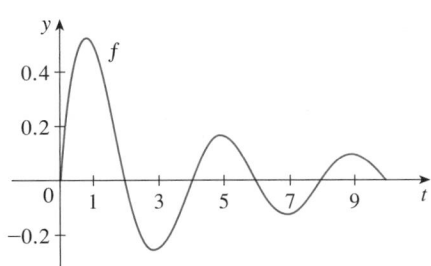

27. If $f(1) = 12$, f' is continuous, and $\int_1^4 f'(x) \, dx = 17$, what is the value of $f(4)$?

28. The **error function**

$$\text{erf}(x) = \frac{2}{\sqrt{\pi}} \int_0^x e^{-t^2} \, dt$$

is used in probability, statistics, and engineering.
(a) Show that $\int_a^b e^{-t^2} \, dt = \frac{1}{2} \sqrt{\pi} \, [\text{erf}(b) - \text{erf}(a)]$.
(b) Show that the function $y = e^{x^2} \text{erf}(x)$ satisfies the differential equation $y' = 2xy + 2/\sqrt{\pi}$.

29. The Fresnel function S was defined in Example 4 and graphed in Figures 7 and 8.
(a) At what values of x does this function have local maximum values?
(b) On what intervals is the function concave upward?
[CAS] (c) Use a graph to solve the following equation correct to two decimal places:

$$\int_0^x \sin(\pi t^2/2) \, dt = 0.2$$

[CAS] **30.** The **sine integral function**

$$\text{Si}(x) = \int_0^x \frac{\sin t}{t} \, dt$$

is important in electrical engineering. [The integrand $f(t) = (\sin t)/t$ is not defined when $t = 0$, but we know that its limit is 1 when $t \to 0$. So we define $f(0) = 1$ and this makes f a continuous function everywhere.]
(a) Draw the graph of Si.
(b) At what values of x does this function have local maximum values?
(c) Find the coordinates of the first inflection point to the right of the origin.
(d) Does this function have horizontal asymptotes?
(e) Solve the following equation correct to one decimal place:

$$\int_0^x \frac{\sin t}{t} \, dt = 1$$

31. Find a function f and a number a such that

$$6 + \int_a^x \frac{f(t)}{t^2} \, dt = 2\sqrt{x}$$

for all $x > 0$.

32. A high-tech company purchases a new computing system whose initial value is V. The system will depreciate at the rate $f = f(t)$ and will accumulate maintenance costs at the rate $g = g(t)$, where t is the time measured in months. The company wants to determine the optimal time to replace the system.

(a) Let

$$C(t) = \frac{1}{t} \int_0^t [f(s) + g(s)] \, ds$$

Show that the critical numbers of C occur at the numbers t where $C(t) = f(t) + g(t)$.

(b) Suppose that

$$f(t) = \begin{cases} \dfrac{V}{15} - \dfrac{V}{450}\, t & \text{if } 0 < t \le 30 \\ 0 & \text{if } t > 30 \end{cases}$$

and $\qquad g(t) = \dfrac{Vt^2}{12{,}900} \qquad t > 0$

Determine the length of time T for the total depreciation $D(t) = \int_0^t f(s) \, ds$ to equal the initial value V.

(c) Determine the absolute minimum of C on $(0, T]$.

(d) Sketch the graphs of C and $f + g$ in the same coordinate system, and verify the result in part (a) in this case.

33. A manufacturing company owns a major piece of equipment that depreciates at the (continuous) rate $f = f(t)$, where t is the time measured in months since its last overhaul. Because a fixed cost A is incurred each time the machine is overhauled, the company wants to determine the optimal time T (in months) between overhauls.

(a) Explain why $\int_0^t f(s) \, ds$ represents the loss in value of the machine over the period of time t since the last overhaul.

(b) Let $C = C(t)$ be given by

$$C(t) = \frac{1}{t}\left[A + \int_0^t f(s) \, ds \right]$$

What does C represent and why would the company want to minimize C?

(c) Show that C has a minimum value at the numbers $t = T$ where $C(T) = f(T)$.

5.5 | THE SUBSTITUTION RULE

Because of the Fundamental Theorem, it's important to be able to find antiderivatives. But our antidifferentiation formulas don't tell us how to evaluate integrals such as

$$\boxed{1} \qquad\qquad \int 2x \sqrt{1 + x^2} \, dx$$

To evaluate this integral our strategy is to simplify the integral by changing from the variable x to a new variable u. Suppose that we let u be the quantity under the root sign in $\boxed{1}$, $u = 1 + x^2$. Then the differential of u is $du = 2x \, dx$. Notice that if the dx in the notation for an integral were to be interpreted as a differential, then the differential $2x \, dx$ would occur in $\boxed{1}$ and so, formally, without justifying our calculation, we could write

■ Differentials were defined in Section 2.8. If $u = f(x)$, then

$$du = f'(x) \, dx$$

$$\boxed{2} \qquad \int 2x\sqrt{1 + x^2} \, dx = \int \sqrt{1 + x^2} \; 2x \, dx = \int \sqrt{u} \; du$$

$$= \tfrac{2}{3} u^{3/2} + C = \tfrac{2}{3}(x^2 + 1)^{3/2} + C$$

But now we can check that we have the correct answer by using the Chain Rule to differentiate the final function of Equation 2:

$$\frac{d}{dx}\left[\tfrac{2}{3}(x^2 + 1)^{3/2} + C \right] = \tfrac{2}{3} \cdot \tfrac{3}{2}(x^2 + 1)^{1/2} \cdot 2x = 2x\sqrt{x^2 + 1}$$

In general, this method works whenever we have an integral that we can write in the form $\int f(g(x))\,g'(x)\,dx$. Observe that if $F' = f$, then

$$\boxed{3} \qquad \int F'(g(x))\,g'(x)\,dx = F(g(x)) + C$$

because, by the Chain Rule,

$$\frac{d}{dx}\,[F(g(x))] = F'(g(x))\,g'(x)$$

If we make the "change of variable" or "substitution" $u = g(x)$, then from Equation 3 we have

$$\int F'(g(x))\,g'(x)\,dx = F(g(x)) + C = F(u) + C = \int F'(u)\,du$$

or, writing $F' = f$, we get

$$\int f(g(x))\,g'(x)\,dx = \int f(u)\,du$$

Thus we have proved the following rule.

$\boxed{4}$ **THE SUBSTITUTION RULE** If $u = g(x)$ is a differentiable function whose range is an interval I and f is continuous on I, then

$$\int f(g(x))\,g'(x)\,dx = \int f(u)\,du$$

Notice that the Substitution Rule for integration was proved using the Chain Rule for differentiation. Notice also that if $u = g(x)$, then $du = g'(x)\,dx$, so a way to remember the Substitution Rule is to think of dx and du in $\boxed{4}$ as differentials.

Thus the Substitution Rule says: **It is permissible to operate with dx and du after integral signs as if they were differentials.**

EXAMPLE 1 Find $\int x^3 \cos(x^4 + 2)\,dx$.

SOLUTION We make the substitution $u = x^4 + 2$ because its differential is $du = 4x^3\,dx$, which, apart from the constant factor 4, occurs in the integral. Thus, using $x^3\,dx = \frac{1}{4}\,du$ and the Substitution Rule, we have

$$\int x^3 \cos(x^4 + 2)\,dx = \int \cos u \cdot \tfrac{1}{4}\,du = \tfrac{1}{4}\int \cos u\,du$$

■ Check the answer by differentiating it.

$$= \tfrac{1}{4}\sin u + C$$

$$= \tfrac{1}{4}\sin(x^4 + 2) + C$$

Notice that at the final stage we had to return to the original variable x. ■

The idea behind the Substitution Rule is to replace a relatively complicated integral by a simpler integral. This is accomplished by changing from the original variable x to a new variable u that is a function of x. Thus in Example 1 we replaced the integral $\int x^3 \cos(x^4 + 2)\, dx$ by the simpler integral $\frac{1}{4} \int \cos u\, du$.

The main challenge in using the Substitution Rule is to think of an appropriate substitution. You should try to choose u to be some function in the integrand whose differential also occurs (except for a constant factor). This was the case in Example 1. If that is not possible, try choosing u to be some complicated part of the integrand (perhaps the inner function in a composite function). Finding the right substitution is a bit of an art. It's not unusual to guess wrong; if your first guess doesn't work, try another substitution.

EXAMPLE 2 Evaluate $\int \sqrt{2x + 1}\, dx$.

SOLUTION 1 Let $u = 2x + 1$. Then $du = 2\, dx$, so $dx = \frac{1}{2}\, du$. Thus the Substitution Rule gives

$$\int \sqrt{2x + 1}\, dx = \int \sqrt{u} \cdot \tfrac{1}{2}\, du = \tfrac{1}{2} \int u^{1/2}\, du$$

$$= \frac{1}{2} \cdot \frac{u^{3/2}}{3/2} + C = \tfrac{1}{3} u^{3/2} + C$$

$$= \tfrac{1}{3}(2x + 1)^{3/2} + C$$

SOLUTION 2 Another possible substitution is $u = \sqrt{2x + 1}$. Then

$$du = \frac{dx}{\sqrt{2x + 1}} \qquad \text{so} \qquad dx = \sqrt{2x + 1}\, du = u\, du$$

(Or observe that $u^2 = 2x + 1$, so $2u\, du = 2\, dx$.) Therefore

$$\int \sqrt{2x + 1}\, dx = \int u \cdot u\, du = \int u^2\, du$$

$$= \frac{u^3}{3} + C = \tfrac{1}{3}(2x + 1)^{3/2} + C \qquad \blacksquare$$

V **EXAMPLE 3** Find $\displaystyle\int \frac{x}{\sqrt{1 - 4x^2}}\, dx$.

SOLUTION Let $u = 1 - 4x^2$. Then $du = -8x\, dx$, so $x\, dx = -\frac{1}{8}\, du$ and

$$\int \frac{x}{\sqrt{1 - 4x^2}}\, dx = -\tfrac{1}{8} \int \frac{1}{\sqrt{u}}\, du = -\tfrac{1}{8} \int u^{-1/2}\, du$$

$$= -\tfrac{1}{8}\left(2\sqrt{u}\right) + C = -\tfrac{1}{4}\sqrt{1 - 4x^2} + C \qquad \blacksquare$$

The answer to Example 3 could be checked by differentiation, but instead let's check it with a graph. In Figure 1 we have used a computer to graph both the integrand $f(x) = x/\sqrt{1 - 4x^2}$ and its indefinite integral $g(x) = -\frac{1}{4}\sqrt{1 - 4x^2}$ (we take the case

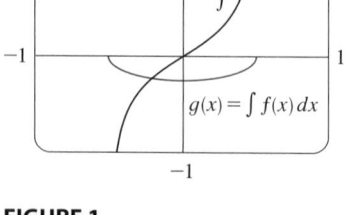

FIGURE 1

$$f(x) = \frac{x}{\sqrt{1 - 4x^2}}$$

$$g(x) = \int f(x)\, dx = -\tfrac{1}{4}\sqrt{1 - 4x^2}$$

$C = 0$). Notice that $g(x)$ decreases when $f(x)$ is negative, increases when $f(x)$ is positive, and has its minimum value when $f(x) = 0$. So it seems reasonable, from the graphical evidence, that g is an antiderivative of f.

EXAMPLE 4 Calculate $\displaystyle\int e^{5x}\,dx$.

SOLUTION If we let $u = 5x$, then $du = 5\,dx$, so $dx = \frac{1}{5}\,du$. Therefore

$$\int e^{5x}\,dx = \tfrac{1}{5}\int e^u\,du = \tfrac{1}{5}e^u + C = \tfrac{1}{5}e^{5x} + C$$ ∎

NOTE With some experience, you might be able to evaluate integrals like those in Examples 1–4 without going to the trouble of making an explicit substitution. By recognizing the pattern in Equation 3, where the integrand on the left side is the product of the derivative of an outer function and the derivative of the inner function, we could work Example 1 as follows:

$$\int x^3 \cos(x^4 + 2)\,dx = \int \cos(x^4 + 2) \cdot x^3\,dx = \tfrac{1}{4}\int \cos(x^4 + 2) \cdot (4x^3)\,dx$$

$$= \tfrac{1}{4}\int \cos(x^4 + 2) \cdot \frac{d}{dx}(x^4 + 2)\,dx = \tfrac{1}{4}\sin(x^4 + 2) + C$$

Similarly, the solution to Example 4 could be written like this:

$$\int e^{5x}\,dx = \tfrac{1}{5}\int 5e^{5x}\,dx = \tfrac{1}{5}\int \frac{d}{dx}(e^{5x})\,dx = \tfrac{1}{5}e^{5x} + C$$

The following example, however, is more complicated and so an explicit substitution is advisable.

Ⅴ EXAMPLE 5 Calculate $\displaystyle\int \tan x\,dx$.

SOLUTION First we write tangent in terms of sine and cosine:

$$\int \tan x\,dx = \int \frac{\sin x}{\cos x}\,dx$$

■ **www.stewartcalculus.com**
See Additional Example A.

This suggests that we should substitute $u = \cos x$, since then $du = -\sin x\,dx$ and so $\sin x\,dx = -du$:

$$\int \tan x\,dx = \int \frac{\sin x}{\cos x}\,dx = -\int \frac{1}{u}\,du$$

$$= -\ln|u| + C = -\ln|\cos x| + C$$ ∎

Since $-\ln|\cos x| = \ln(|\cos x|^{-1}) = \ln(1/|\cos x|) = \ln|\sec x|$, the result of Example 5 can also be written as

5

$$\int \tan x\,dx = \ln|\sec x| + C$$

DEFINITE INTEGRALS

When evaluating a *definite* integral by substitution, two methods are possible. One method is to evaluate the indefinite integral first and then use the Evaluation Theorem. For instance, using the result of Example 2, we have

$$\int_0^4 \sqrt{2x+1}\,dx = \int \sqrt{2x+1}\,dx\Big]_0^4 = \tfrac{1}{3}(2x+1)^{3/2}\Big]_0^4$$

$$= \tfrac{1}{3}(9)^{3/2} - \tfrac{1}{3}(1)^{3/2} = \tfrac{1}{3}(27-1) = \tfrac{26}{3}$$

Another method, which is usually preferable, is to change the limits of integration when the variable is changed.

■ This rule says that when using a substitution in a definite integral, we must put everything in terms of the new variable u, not only x and dx but also the limits of integration. The new limits of integration are the values of u that correspond to $x = a$ and $x = b$.

> **6 THE SUBSTITUTION RULE FOR DEFINITE INTEGRALS** If g' is continuous on $[a, b]$ and f is continuous on the range of $u = g(x)$, then
>
> $$\int_a^b f(g(x))\,g'(x)\,dx = \int_{g(a)}^{g(b)} f(u)\,du$$

■ www.stewartcalculus.com
See Additional Example B.

PROOF Let F be an antiderivative of f. Then, by $\boxed{3}$, $F(g(x))$ is an antiderivative of $f(g(x))\,g'(x)$ and so, by the Evaluation Theorem, we have

$$\int_a^b f(g(x))\,g'(x)\,dx = F(g(x))\Big]_a^b = F(g(b)) - F(g(a))$$

But, applying the Evaluation Theorem a second time, we also have

$$\int_{g(a)}^{g(b)} f(u)\,du = F(u)\Big]_{g(a)}^{g(b)} = F(g(b)) - F(g(a)) \qquad \square$$

■ The integral given in Example 6 is an abbreviation for

$$\int_1^2 \frac{1}{(3-5x)^2}\,dx$$

EXAMPLE 6 Evaluate $\displaystyle\int_1^2 \frac{dx}{(3-5x)^2}$.

SOLUTION Let $u = 3 - 5x$. Then $du = -5\,dx$, so $dx = -\tfrac{1}{5}\,du$. To find the new limits of integration we note that

when $x = 1$, $u = 3 - 5(1) = -2$ and   when $x = 2$, $u = 3 - 5(2) = -7$

Therefore

$$\int_1^2 \frac{dx}{(3-5x)^2} = -\frac{1}{5}\int_{-2}^{-7} \frac{du}{u^2} = -\frac{1}{5}\left[-\frac{1}{u}\right]_{-2}^{-7}$$

$$= \frac{1}{5u}\bigg]_{-2}^{-7} = \frac{1}{5}\left(-\frac{1}{7} + \frac{1}{2}\right) = \frac{1}{14}$$

Observe that when using $\boxed{6}$ we do not return to the variable x after integrating. We simply evaluate the expression in u between the appropriate values of u. ■

- Since the function $f(x) = (\ln x)/x$ in Example 7 is positive for $x > 1$, the integral represents the area of the shaded region in Figure 2.

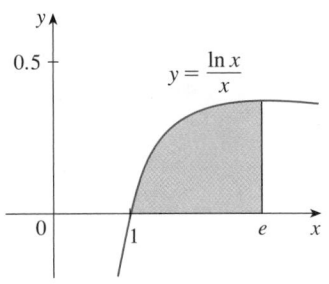

FIGURE 2

V EXAMPLE 7 Calculate $\int_1^e \dfrac{\ln x}{x}\,dx$.

SOLUTION We let $u = \ln x$ because its differential $du = dx/x$ occurs in the integral. When $x = 1$, $u = \ln 1 = 0$; when $x = e$, $u = \ln e = 1$. Thus

$$\int_1^e \frac{\ln x}{x}\,dx = \int_0^1 u\,du = \frac{u^2}{2}\bigg]_0^1 = \frac{1}{2}$$ ∎

SYMMETRY

The next theorem uses the Substitution Rule for Definite Integrals $\boxed{6}$ to simplify the calculation of integrals of functions that possess symmetry properties.

$\boxed{7}$ **INTEGRALS OF SYMMETRIC FUNCTIONS** Suppose f is continuous on $[-a, a]$.

(a) If f is even $[f(-x) = f(x)]$, then $\int_{-a}^a f(x)\,dx = 2\int_0^a f(x)\,dx$.

(b) If f is odd $[f(-x) = -f(x)]$, then $\int_{-a}^a f(x)\,dx = 0$.

PROOF We split the integral in two:

$$\boxed{8} \quad \int_{-a}^a f(x)\,dx = \int_{-a}^0 f(x)\,dx + \int_0^a f(x)\,dx = -\int_0^{-a} f(x)\,dx + \int_0^a f(x)\,dx$$

In the first integral on the far right side we make the substitution $u = -x$. Then $du = -dx$ and when $x = -a$, $u = a$. Therefore

$$-\int_0^{-a} f(x)\,dx = -\int_0^a f(-u)\,(-du) = \int_0^a f(-u)\,du$$

and so Equation 8 becomes

$$\boxed{9} \qquad\qquad \int_{-a}^a f(x)\,dx = \int_0^a f(-u)\,du + \int_0^a f(x)\,dx$$

(a) If f is even, then $f(-u) = f(u)$ so Equation 9 gives

$$\int_{-a}^a f(x)\,dx = \int_0^a f(u)\,du + \int_0^a f(x)\,dx = 2\int_0^a f(x)\,dx$$

(b) If f is odd, then $f(-u) = -f(u)$ and so Equation 9 gives

$$\int_{-a}^a f(x)\,dx = -\int_0^a f(u)\,du + \int_0^a f(x)\,dx = 0$$ ☐

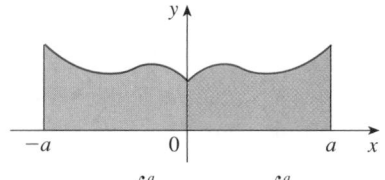

(a) f even, $\displaystyle\int_{-a}^{a} f(x)\,dx = 2\int_{0}^{a} f(x)\,dx$

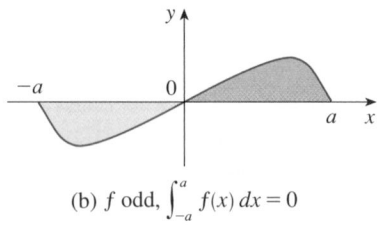

(b) f odd, $\displaystyle\int_{-a}^{a} f(x)\,dx = 0$

FIGURE 3

Theorem 7 is illustrated by Figure 3. For the case where f is positive and even, part (a) says that the area under $y = f(x)$ from $-a$ to a is twice the area from 0 to a because of symmetry. Recall that an integral $\int_{a}^{b} f(x)\,dx$ can be expressed as the area above the x-axis and below $y = f(x)$ minus the area below the axis and above the curve. Thus part (b) says the integral is 0 because the areas cancel.

▼ **EXAMPLE 8** Since $f(x) = x^6 + 1$ satisfies $f(-x) = f(x)$, it is even and so

$$\int_{-2}^{2} (x^6 + 1)\,dx = 2\int_{0}^{2} (x^6 + 1)\,dx$$
$$= 2\left[\tfrac{1}{7}x^7 + x\right]_{0}^{2} = 2\left(\tfrac{128}{7} + 2\right) = \tfrac{284}{7} \qquad ∎$$

EXAMPLE 9 Since $f(x) = (\tan x)/(1 + x^2 + x^4)$ satisfies $f(-x) = -f(x)$, it is odd and so

$$\int_{-1}^{1} \frac{\tan x}{1 + x^2 + x^4}\,dx = 0 \qquad ∎$$

5.5 | EXERCISES

1–6 ■ Evaluate the integral by making the given substitution.

1. $\displaystyle\int e^{-x}\,dx, \quad u = -x$

2. $\displaystyle\int x^3(2 + x^4)^5\,dx, \quad u = 2 + x^4$

3. $\displaystyle\int x^2\sqrt{x^3 + 1}\,dx, \quad u = x^3 + 1$

4. $\displaystyle\int \frac{dt}{(1 - 6t)^4}, \quad u = 1 - 6t$

5. $\displaystyle\int \cos^3\theta \sin\theta\,d\theta, \quad u = \cos\theta$

6. $\displaystyle\int \frac{\sec^2(1/x)}{x^2}\,dx, \quad u = 1/x$

7–36 ■ Evaluate the indefinite integral.

7. $\displaystyle\int x \sin(x^2)\,dx$

8. $\displaystyle\int x^2 e^{x^3}\,dx$

9. $\displaystyle\int (1 - 2x)^9\,dx$

10. $\displaystyle\int (3t + 2)^{2.4}\,dt$

11. $\displaystyle\int \frac{(\ln x)^2}{x}\,dx$

12. $\displaystyle\int \sec^2 2\theta\,d\theta$

13. $\displaystyle\int \frac{dx}{5 - 3x}$

14. $\displaystyle\int \frac{x}{(x^2 + 1)^2}\,dx$

15. $\displaystyle\int \frac{a + bx^2}{\sqrt{3ax + bx^3}}\,dx$

16. $\displaystyle\int u\sqrt{1 - u^2}\,du$

17. $\displaystyle\int \sin \pi t\,dt$

18. $\displaystyle\int \frac{\sin\sqrt{x}}{\sqrt{x}}\,dx$

19. $\displaystyle\int e^x\sqrt{1 + e^x}\,dx$

20. $\displaystyle\int \cos^4\theta \sin\theta\,d\theta$

21. $\displaystyle\int \sinh^2 x \cosh x\,dx$

22. $\displaystyle\int \frac{\sin(\ln x)}{x}\,dx$

23. $\displaystyle\int \sqrt{\cot x}\,\csc^2 x\,dx$

24. $\displaystyle\int \frac{\tan^{-1}x}{1 + x^2}\,dx$

25. $\displaystyle\int \frac{dx}{\sqrt{1 - x^2}\,\sin^{-1}x}$

26. $\displaystyle\int \frac{\cos(\pi/x)}{x^2}\,dx$

27. $\displaystyle\int \sec^3 x \tan x\,dx$

28. $\displaystyle\int \frac{2^t}{2^t + 3}\,dt$

29. $\displaystyle\int 5^t \sin(5^t)\,dt$

30. $\displaystyle\int \frac{dt}{\cos^2 t\sqrt{1 + \tan t}}$

31. $\displaystyle\int \frac{\sin 2x}{1 + \cos^2 x}\,dx$

32. $\displaystyle\int \frac{\sin x}{1 + \cos^2 x}\,dx$

33. $\displaystyle\int x(2x + 5)^8\,dx$

34. $\displaystyle\int \frac{x^3}{\sqrt{x^2 + 1}}\,dx$

35. $\displaystyle\int \frac{1 + x}{1 + x^2}\,dx$

36. $\displaystyle\int \frac{x}{1 + x^4}\,dx$

37–52 ▪ Evaluate the definite integral.

37. $\int_0^1 \cos(\pi t/2)\, dt$

38. $\int_0^1 (3t - 1)^{50}\, dt$

39. $\int_0^1 \sqrt[3]{1 + 7x}\, dx$

40. $\int_0^3 \dfrac{dx}{5x + 1}$

41. $\int_0^\pi \sec^2(t/4)\, dt$

42. $\int_{1/6}^{1/2} \csc \pi t \, \cot \pi t \, dt$

43. $\int_1^2 \dfrac{e^{1/x}}{x^2}\, dx$

44. $\int_0^{\pi/2} \cos x \, \sin(\sin x)\, dx$

45. $\int_1^2 x\sqrt{x - 1}\, dx$

46. $\int_{-\pi/2}^{\pi/2} \dfrac{x^2 \sin x}{1 + x^6}\, dx$

47. $\int_0^1 \dfrac{e^z + 1}{e^z + z}\, dz$

48. $\int_0^4 \dfrac{x}{\sqrt{1 + 2x}}\, dx$

49. $\int_{-\pi/4}^{\pi/4} (x^3 + x^4 \tan x)\, dx$

50. $\int_0^a x\sqrt{a^2 - x^2}\, dx$

51. $\int_e^{e^4} \dfrac{dx}{x\sqrt{\ln x}}$

52. $\int_0^{1/2} \dfrac{\sin^{-1}x}{\sqrt{1 - x^2}}\, dx$

53–56 ▪ Find the average value of the function on the given interval.

53. $f(t) = te^{-t^2}, \quad [0, 5]$

54. $f(x) = \sin 4x, \quad [-\pi, \pi]$

55. $h(x) = \cos^4 x \sin x, \quad [0, \pi]$

56. $h(r) = 3/(1 + r)^2, \quad [1, 6]$

---

57. Evaluate $\int_{-2}^2 (x + 3)\sqrt{4 - x^2}\, dx$ by writing it as a sum of two integrals and interpreting one of those integrals in terms of an area.

58. Evaluate $\int_0^1 x\sqrt{1 - x^4}\, dx$ by making a substitution and interpreting the resulting integral in terms of an area.

59. Which of the following areas are equal? Why?

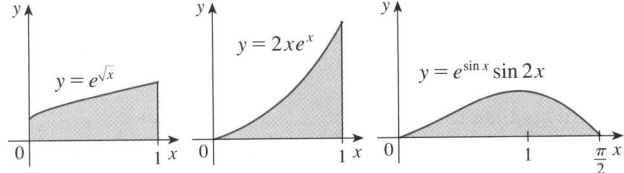

60. A model for the basal metabolism rate, in kcal/h, of a young man is $R(t) = 85 - 0.18 \cos(\pi t/12)$, where t is the time in hours measured from 5:00 AM. What is the total

basal metabolism of this man, $\int_0^{24} R(t)\, dt$, over a 24-hour time period?

61. An oil storage tank ruptures at time $t = 0$ and oil leaks from the tank at a rate of $r(t) = 100e^{-0.01t}$ liters per minute. How much oil leaks out during the first hour?

62. A bacteria population starts with 400 bacteria and grows at a rate of $r(t) = (450.268)e^{1.12567t}$ bacteria per hour. How many bacteria will there be after three hours?

63. Breathing is cyclic and a full respiratory cycle from the beginning of inhalation to the end of exhalation takes about 5 s. The maximum rate of air flow into the lungs is about 0.5 L/s. This explains, in part, why the function $f(t) = \frac{1}{2} \sin(2\pi t/5)$ has often been used to model the rate of air flow into the lungs. Use this model to find the volume of inhaled air in the lungs at time t.

64. Alabama Instruments Company has set up a production line to manufacture a new calculator. The rate of production of these calculators after t weeks is

$$\frac{dx}{dt} = 5000\left(1 - \frac{100}{(t + 10)^2}\right) \text{ calculators/week}$$

(Notice that production approaches 5000 per week as time goes on, but the initial production is lower because of the workers' unfamiliarity with the new techniques.) Find the number of calculators produced from the beginning of the third week to the end of the fourth week.

65. If f is continuous and $\int_0^4 f(x)\, dx = 10$, find $\int_0^2 f(2x)\, dx$.

66. If f is continuous and $\int_0^9 f(x)\, dx = 4$, find $\int_0^3 xf(x^2)\, dx$.

67. If f is continuous on $\mathbb{R}$, prove that

$$\int_a^b f(-x)\, dx = \int_{-b}^{-a} f(x)\, dx$$

For the case where $f(x) \geqslant 0$ and $0 < a < b$, draw a diagram to interpret this equation geometrically as an equality of areas.

68. If f is continuous on $\mathbb{R}$, prove that

$$\int_a^b f(x + c)\, dx = \int_{a+c}^{b+c} f(x)\, dx$$

For the case where $f(x) \geqslant 0$, draw a diagram to interpret this equation geometrically as an equality of areas.

69. If a and b are positive numbers, show that

$$\int_0^1 x^a(1 - x)^b\, dx = \int_0^1 x^b(1 - x)^a\, dx$$

CHAPTER 5　REVIEW

1. (a) Write an expression for a Riemann sum of a function f. Explain the meaning of the notation that you use.

(b) If $f(x) \geq 0$, what is the geometric interpretation of a Riemann sum? Illustrate with a diagram.

(c) If $f(x)$ takes on both positive and negative values, what is the geometric interpretation of a Riemann sum? Illustrate with a diagram.

2. (a) Write the definition of the definite integral of f from a to b. How does the definition simplify if you know that f is continuous and you use equal subintervals?

(b) What is the geometric interpretation of $\int_a^b f(x)\, dx$ if $f(x) \geq 0$?

(c) What is the geometric interpretation of $\int_a^b f(x)\, dx$ if $f(x)$ takes on both positive and negative values? Illustrate with a diagram.

3. State the Midpoint Rule.

4. (a) State the Evaluation Theorem.

(b) State the Net Change Theorem.

(c) If $r(t)$ is the rate at which water flows into a reservoir, what does $\int_{t_1}^{t_2} r(t)\, dt$ represent?

5. (a) Explain the meaning of the indefinite integral $\int f(x)\, dx$.

(b) What is the connection between the definite integral $\int_a^b f(x)\, dx$ and the indefinite integral $\int f(x)\, dx$?

6. State both parts of the Fundamental Theorem of Calculus.

7. Suppose a particle moves back and forth along a straight line with velocity $v(t)$, measured in feet per second, and acceleration $a(t)$.

(a) What is the meaning of $\int_{60}^{120} v(t)\, dt$?

(b) What is the meaning of $\int_{60}^{120} |v(t)|\, dt$?

(c) What is the meaning of $\int_{60}^{120} a(t)\, dt$?

8. (a) What is the average value of a function f on an interval $[a, b]$?

(b) What does the Mean Value Theorem for Integrals say? What is its geometric interpretation?

9. Explain exactly what is meant by the statement that "differentiation and integration are inverse processes."

10. State the Substitution Rule. In practice, how do you use it?

Determine whether the statement is true or false. If it is true, explain why. If it is false, explain why or give an example that disproves the statement.

1. If f and g are continuous on $[a, b]$, then

$$\int_a^b [f(x) + g(x)]\, dx = \int_a^b f(x)\, dx + \int_a^b g(x)\, dx$$

2. If f and g are continuous on $[a, b]$, then

$$\int_a^b [f(x)\, g(x)]\, dx = \left(\int_a^b f(x)\, dx\right)\left(\int_a^b g(x)\, dx\right)$$

3. If f is continuous on $[a, b]$, then

$$\int_a^b 5f(x)\, dx = 5\int_a^b f(x)\, dx$$

4. If f is continuous on $[a, b]$, then

$$\int_a^b xf(x)\, dx = x\int_a^b f(x)\, dx$$

5. If f is continuous on $[a, b]$ and $f(x) \geq 0$, then

$$\int_a^b \sqrt{f(x)}\, dx = \sqrt{\int_a^b f(x)\, dx}$$

6. If f' is continuous on $[1, 3]$, then $\int_1^3 f'(v)\, dv = f(3) - f(1)$.

7. If f and g are continuous and $f(x) \geq g(x)$ for $a \leq x \leq b$, then

$$\int_a^b f(x)\, dx \geq \int_a^b g(x)\, dx$$

8. If f and g are differentiable and $f(x) \geq g(x)$ for $a < x < b$, then $f'(x) \geq g'(x)$ for $a < x < b$.

9. $\int_{-1}^1 \left(x^5 - 6x^9 + \dfrac{\sin x}{(1 + x^4)^2}\right) dx = 0$

10. $\int_{-5}^5 (ax^2 + bx + c)\, dx = 2\int_0^5 (ax^2 + c)\, dx$

11. All continuous functions have derivatives.

12. All continuous functions have antiderivatives.

13. $\int_0^3 e^{x^2}\, dx = \int_0^5 e^{x^2}\, dx + \int_5^3 e^{x^2}\, dx$

14. If $\int_0^1 f(x)\, dx = 0$, then $f(x) = 0$ for $0 \leq x \leq 1$.

15. If f is continuous on $[a, b]$, then

$$\frac{d}{dx}\left(\int_a^b f(x)\,dx\right) = f(x)$$

16. $\int_0^2 (x - x^3)\,dx$ represents the area under the curve $y = x - x^3$ from 0 to 2.

17. $\int_{-2}^1 \dfrac{1}{x^4}\,dx = -\dfrac{3}{8}$

18. If f has a discontinuity at 0, then $\int_{-1}^1 f(x)\,dx$ does not exist.

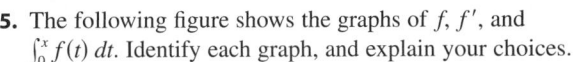

EXERCISES

1. Use the given graph of f to find the Riemann sum with six subintervals. Take the sample points to be (a) left endpoints and (b) midpoints. In each case draw a diagram and explain what the Riemann sum represents.

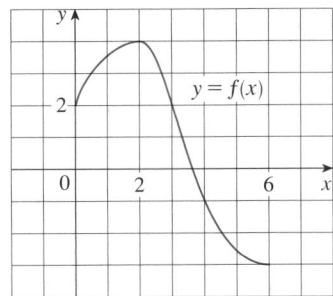

2. (a) Evaluate the Riemann sum for

$$f(x) = x^2 - x \qquad 0 \leqslant x \leqslant 2$$

with four subintervals, taking the sample points to be right endpoints. Explain, with the aid of a diagram, what the Riemann sum represents.

(b) Use the definition of a definite integral (with right endpoints) to calculate the value of the integral

$$\int_0^2 (x^2 - x)\,dx$$

(c) Use the Fundamental Theorem to check your answer to part (b).

(d) Draw a diagram to explain the geometric meaning of the integral in part (b).

3. Evaluate

$$\int_0^1 \left(x + \sqrt{1 - x^2}\right) dx$$

by interpreting it in terms of areas.

4. Express

$$\lim_{n \to \infty} \sum_{i=1}^n \sin x_i\, \Delta x$$

as a definite integral on the interval $[0, \pi]$ and then evaluate the integral.

5. The following figure shows the graphs of f, f', and $\int_0^x f(t)\,dt$. Identify each graph, and explain your choices.

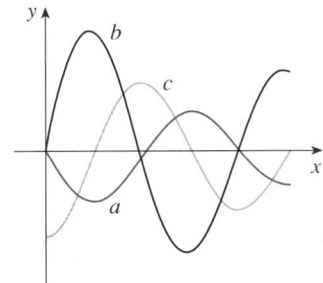

6. Evaluate:

(a) $\displaystyle \int_0^1 \frac{d}{dx}\left(e^{\arctan x}\right) dx$ (b) $\displaystyle \frac{d}{dx} \int_0^1 e^{\arctan x}\,dx$

(c) $\displaystyle \frac{d}{dx} \int_0^x e^{\arctan t}\,dt$

7–32 ▪ Evaluate the integral, if it exists.

7. $\displaystyle \int_1^2 (8x^3 + 3x^2)\,dx$ **8.** $\displaystyle \int_0^T (x^4 - 8x + 7)\,dx$

9. $\displaystyle \int_0^1 (1 - x^9)\,dx$ **10.** $\displaystyle \int_0^1 (1 - x)^9\,dx$

11. $\displaystyle \int_1^9 \frac{\sqrt{u} - 2u^2}{u}\,du$ **12.** $\displaystyle \int_0^1 \left(\sqrt[4]{u} + 1\right)^2 du$

13. $\displaystyle \int_0^1 y(y^2 + 1)^5\,dy$ **14.** $\displaystyle \int_0^2 y^2\sqrt{1 + y^3}\,dy$

15. $\displaystyle \int_0^1 v^2 \cos(v^3)\,dv$ **16.** $\displaystyle \int_0^1 \sin(3\pi t)\,dt$

17. $\displaystyle \int_{-\pi/4}^{\pi/4} \frac{t^4 \tan t}{2 + \cos t}\,dt$ **18.** $\displaystyle \int_{-1}^1 \frac{\sin x}{1 + x^2}\,dx$

19. $\displaystyle \int \left(\frac{1 - x}{x}\right)^2 dx$ **20.** $\displaystyle \int_0^1 \frac{e^x}{1 + e^{2x}}\,dx$

21. $\displaystyle \int \frac{x + 2}{\sqrt{x^2 + 4x}}\,dx$ **22.** $\displaystyle \int \frac{\csc^2 x}{1 + \cot x}\,dx$

23. $\int \sin \pi t \, \cos \pi t \, dt$

24. $\int \sin x \, \cos(\cos x) \, dx$

25. $\int \dfrac{e^{\sqrt{x}}}{\sqrt{x}} \, dx$

26. $\int \dfrac{\cos(\ln x)}{x} \, dx$

27. $\int \tan x \, \ln(\cos x) \, dx$

28. $\int \dfrac{x}{\sqrt{1 - x^4}} \, dx$

29. $\int \dfrac{x^3}{1 + x^4} \, dx$

30. $\int \sinh(1 + 4x) \, dx$

31. $\int \dfrac{\sec \theta \, \tan \theta}{1 + \sec \theta} \, d\theta$

32. $\int_0^{\pi/4} (1 + \tan t)^3 \sec^2 t \, dt$

---

33. Use a graph to give a rough estimate of the area of the region that lies under the curve $y = x\sqrt{x}$, $0 \le x \le 4$. Then find the exact area.

34. Graph the function $f(x) = \cos^2 x \, \sin x$ and use the graph to guess the value of the integral $\int_0^{2\pi} f(x) \, dx$. Then evaluate the integral to confirm your guess.

35–38 ▪ Find the derivative of the function.

35. $F(x) = \int_1^x \sqrt{1 + t^4} \, dt$

36. $g(x) = \int_1^{\cos x} \sqrt[3]{1 - t^2} \, dt$

37. $y = \int_{\sqrt{x}}^x \dfrac{e^t}{t} \, dt$

38. $y = \int_{2x}^{3x+1} \sin(t^4) \, dt$

---

39. Use Property 8 of integrals to estimate the value of

$$\int_1^3 \sqrt{x^2 + 3} \, dx$$

40. Use the properties of integrals to verify that

$$0 \le \int_0^1 x^4 \cos x \, dx \le 0.2$$

41. Use the Midpoint Rule with $n = 5$ to approximate $\int_0^1 \sqrt{1 + x^3} \, dx$.

42. A particle moves along a line with velocity function $v(t) = t^2 - t$, where v is measured in meters per second. Find (a) the displacement and (b) the distance traveled by the particle during the time interval $[0, 5]$.

43. Let $r(t)$ be the rate at which the world's oil is consumed, where t is measured in years starting at $t = 0$ on January 1, 2000, and $r(t)$ is measured in barrels per year. What does $\int_0^3 r(t) \, dt$ represent?

44. A radar gun was used to record the speed of a runner at the times given in the table. Use the Midpoint Rule to estimate the distance the runner covered during those 5 seconds.

t (s)	v (m/s)	t (s)	v (m/s)
0	0	3.0	10.51
0.5	4.67	3.5	10.67
1.0	7.34	4.0	10.76
1.5	8.86	4.5	10.81
2.0	9.73	5.0	10.81
2.5	10.22		

45. A population of honeybees increased at a rate of $r(t)$ bees per week, where the graph of r is as shown. Use the Midpoint Rule with six subintervals to estimate the increase in the bee population during the first 24 weeks.

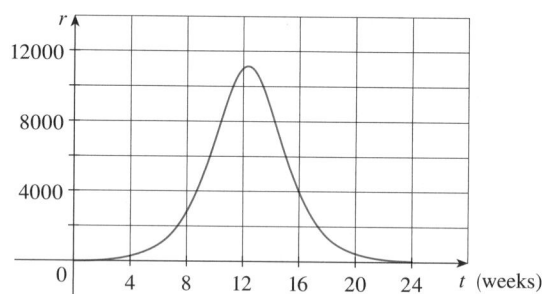

46. Find the average value of the function $f(x) = t \sin(t^2)$ on the interval $[0, 2]$.

47. If f is a continuous function, what is the limit as $h \to 0$ of the average value of f on the interval $[x, x + h]$?

48. Let

$$f(x) = \begin{cases} -x - 1 & \text{if } -3 \le x \le 0 \\ -\sqrt{1 - x^2} & \text{if } 0 \le x \le 1 \end{cases}$$

Evaluate $\int_{-3}^1 f(x) \, dx$ by interpreting the integral as a difference of areas.

49. Estimate the value of the number c such that the area under the curve $y = \sinh cx$ between $x = 0$ and $x = 1$ is equal to 1.

50. Evaluate

$$\lim_{n \to \infty} \frac{1}{n} \left[\left(\frac{1}{n} \right)^9 + \left(\frac{2}{n} \right)^9 + \left(\frac{3}{n} \right)^9 + \cdots + \left(\frac{n}{n} \right)^9 \right]$$

6

TECHNIQUES OF INTEGRATION

Because of the Fundamental Theorem of Calculus, we can integrate a function if we know an antiderivative, that is, an indefinite integral. We summarize here the most important integrals that we have learned so far.

$$\int x^n \, dx = \frac{x^{n+1}}{n+1} + C, \quad n \neq -1 \qquad \int \frac{1}{x} \, dx = \ln|x| + C$$

$$\int e^x \, dx = e^x + C \qquad \int a^x \, dx = \frac{a^x}{\ln a} + C$$

$$\int \sin x \, dx = -\cos x + C \qquad \int \cos x \, dx = \sin x + C$$

$$\int \sec^2 x \, dx = \tan x + C \qquad \int \csc^2 x \, dx = -\cot x + C$$

$$\int \sec x \tan x \, dx = \sec x + C \qquad \int \csc x \cot x \, dx = -\csc x + C$$

$$\int \sinh x \, dx = \cosh x + C \qquad \int \cosh x \, dx = \sinh x + C$$

$$\int \tan x \, dx = \ln|\sec x| + C \qquad \int \cot x \, dx = \ln|\sin x| + C$$

$$\int \frac{1}{x^2 + a^2} \, dx = \frac{1}{a} \tan^{-1}\left(\frac{x}{a}\right) + C \qquad \int \frac{1}{\sqrt{a^2 - x^2}} \, dx = \sin^{-1}\left(\frac{x}{a}\right) + C, \quad a > 0$$

In this chapter we develop techniques for using these basic integration formulas to obtain indefinite integrals of more complicated functions. We learned the most important method of integration, the Substitution Rule, in Section 5.5. The other general technique, integration by parts, is presented in Section 6.1. Then we learn methods that are special to particular classes of functions such as trigonometric functions and rational functions.

6.1 | INTEGRATION BY PARTS

Every differentiation rule has a corresponding integration rule. For instance, the Substitution Rule for integration corresponds to the Chain Rule for differentiation. The rule that corresponds to the Product Rule for differentiation is called the rule for *integration by parts*.

The Product Rule states that if f and g are differentiable functions, then

$$\frac{d}{dx}[f(x)g(x)] = f(x)g'(x) + g(x)f'(x)$$

In the notation for indefinite integrals this equation becomes

$$\int [f(x)g'(x) + g(x)f'(x)] \, dx = f(x)g(x)$$

or

$$\int f(x)g'(x) \, dx + \int g(x)f'(x) \, dx = f(x)g(x)$$

We can rearrange this equation as

☐1
$$\int f(x)g'(x)\,dx = f(x)g(x) - \int g(x)f'(x)\,dx$$

Formula 1 is called the **formula for integration by parts**. It is perhaps easier to remember in the following notation. Let $u = f(x)$ and $v = g(x)$. Then the differentials are $du = f'(x)\,dx$ and $dv = g'(x)\,dx$, so, by the Substitution Rule, the formula for integration by parts becomes

☐2
$$\int u\,dv = uv - \int v\,du$$

EXAMPLE 1 Find $\int x \sin x\,dx$.

SOLUTION USING FORMULA 1 Suppose we choose $f(x) = x$ and $g'(x) = \sin x$. Then $f'(x) = 1$ and $g(x) = -\cos x$. (For g we can choose *any* antiderivative of g'.) Thus, using Formula 1, we have

$$\int x \sin x\,dx = f(x)g(x) - \int g(x)f'(x)\,dx$$

$$= x(-\cos x) - \int (-\cos x)\,dx$$

$$= -x\cos x + \int \cos x\,dx$$

$$= -x\cos x + \sin x + C$$

It's wise to check the answer by differentiating it. If we do so, we get $x \sin x$, as expected.

SOLUTION USING FORMULA 2 Let

- It is helpful to use the pattern:
  $u = \square \qquad dv = \square$
  $du = \square \qquad v = \square$

$$u = x \qquad\qquad dv = \sin x\,dx$$

Then
$$du = dx \qquad\qquad v = -\cos x$$

and so

$$\int x \sin x\,dx = \int \overbrace{x}^{u}\ \overbrace{\sin x\,dx}^{dv} = \overbrace{x}^{u}\ \overbrace{(-\cos x)}^{v} - \int \overbrace{(-\cos x)}^{v}\ \overbrace{dx}^{du}$$

$$= -x\cos x + \int \cos x\,dx$$

$$= -x\cos x + \sin x + C$$

∎

NOTE Our aim in using integration by parts is to obtain a simpler integral than the one we started with. Thus in Example 1 we started with $\int x \sin x\,dx$ and expressed it in terms of the simpler integral $\int \cos x\,dx$. If we had chosen $u = \sin x$ and $dv = x\,dx$, then $du = \cos x \, dx$ and $v = x^2/2$, so integration by parts gives

$$\int x \sin x \, dx = (\sin x) \frac{x^2}{2} - \frac{1}{2} \int x^2 \cos x \, dx$$

Although this is true, $\int x^2 \cos x \, dx$ is a more difficult integral than the one we started with. In general, when deciding on a choice for u and dv, we usually try to choose $u = f(x)$ to be a function that becomes simpler when differentiated (or at least not more complicated) as long as $dv = g'(x) \, dx$ can be readily integrated to give v.

V EXAMPLE 2 Evaluate $\int \ln x \, dx$.

SOLUTION Here we don't have much choice for u and dv. Let

$$u = \ln x \qquad dv = dx$$

Then

$$du = \frac{1}{x} \, dx \qquad v = x$$

Integrating by parts, we get

$$\int \ln x \, dx = x \ln x - \int x \frac{dx}{x}$$

- It's customary to write $\int 1 \, dx$ as $\int dx$.

$$= x \ln x - \int dx$$

- Check the answer by differentiating it.

$$= x \ln x - x + C$$

Integration by parts is effective in this example because the derivative of the function $f(x) = \ln x$ is simpler than f. ∎

V EXAMPLE 3 Find $\int t^2 e^t \, dt$.

SOLUTION Notice that t^2 becomes simpler when differentiated (whereas e^t is unchanged when differentiated or integrated), so we choose

$$u = t^2 \qquad dv = e^t \, dt$$

Then

$$du = 2t \, dt \qquad v = e^t$$

Integration by parts gives

$$\boxed{3} \qquad \int t^2 e^t \, dt = t^2 e^t - 2 \int t e^t \, dt$$

The integral that we obtained, $\int t e^t \, dt$, is simpler than the original integral but is still not obvious. Therefore we use integration by parts a second time, this time with $u = t$ and $dv = e^t \, dt$. Then $du = dt$, $v = e^t$, and

$$\int t e^t \, dt = t e^t - \int e^t \, dt$$

$$= t e^t - e^t + C$$

Putting this in Equation 3, we get

$$\begin{aligned}\int t^2e^t\,dt &= t^2e^t - 2\int te^t\,dt\\ &= t^2e^t - 2(te^t - e^t + C)\\ &= t^2e^t - 2te^t + 2e^t + C_1 \qquad \text{where } C_1 = -2C\end{aligned}$$ ■

V EXAMPLE 4 Evaluate $\int e^x \sin x\,dx$.

■ **www.stewartcalculus.com** An easier method, using complex numbers, is given under *Additional Topics*. Click on *Complex Numbers* and see Exercise 50.

SOLUTION Neither e^x nor $\sin x$ becomes simpler when differentiated, but we try choosing $u = e^x$ and $dv = \sin x\,dx$ anyway. Then $du = e^x\,dx$ and $v = -\cos x$, so integration by parts gives

$$\boxed{4} \qquad \int e^x \sin x\,dx = -e^x \cos x + \int e^x \cos x\,dx$$

The integral that we have obtained, $\int e^x \cos x\,dx$, is no simpler than the original one, but at least it's no more difficult. Having had success in the preceding example integrating by parts twice, we persevere and integrate by parts again. This time we use $u = e^x$ and $dv = \cos x\,dx$. Then $du = e^x\,dx$, $v = \sin x$, and

$$\boxed{5} \qquad \int e^x \cos x\,dx = e^x \sin x - \int e^x \sin x\,dx$$

■ Figure 1 illustrates Example 4 by showing the graphs of $f(x) = e^x \sin x$ and $F(x) = \frac{1}{2}e^x(\sin x - \cos x)$. As a visual check on our work, notice that $f(x) = 0$ when F has a maximum or minimum.

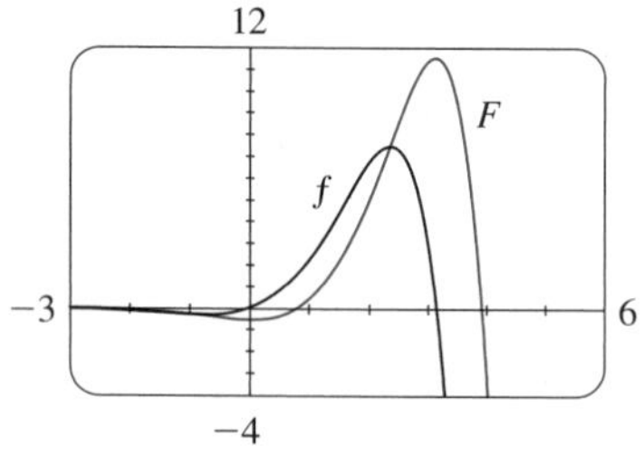

FIGURE 1

At first glance, it appears as if we have accomplished nothing because we have arrived at $\int e^x \sin x\,dx$, which is where we started. However, if we put the expression for $\int e^x \cos x\,dx$ from Equation 5 into Equation 4 we get

$$\int e^x \sin x\,dx = -e^x \cos x + e^x \sin x - \int e^x \sin x\,dx$$

This can be regarded as an equation to be solved for the unknown integral. Adding $\int e^x \sin x\,dx$ to both sides, we obtain

$$2\int e^x \sin x\,dx = -e^x \cos x + e^x \sin x$$

Dividing by 2 and adding the constant of integration, we get

$$\int e^x \sin x\,dx = \tfrac{1}{2}e^x(\sin x - \cos x) + C$$ ■

If we combine the formula for integration by parts with the Evaluation Theorem, we can evaluate definite integrals by parts. Evaluating both sides of Formula 1 between a and b, assuming f' and g' are continuous, and using the Evaluation Theorem, we obtain

$$\boxed{6} \qquad \boxed{\int_a^b f(x)g'(x)\,dx = f(x)g(x)\Big]_a^b - \int_a^b g(x)f'(x)\,dx}$$

EXAMPLE 5 Calculate $\int_0^1 \tan^{-1}x \, dx$.

SOLUTION Let

$$u = \tan^{-1}x \qquad dv = dx$$

Then

$$du = \frac{dx}{1 + x^2} \qquad v = x$$

So Formula 6 gives

$$\int_0^1 \tan^{-1}x \, dx = x \tan^{-1}x \Big]_0^1 - \int_0^1 \frac{x}{1 + x^2} \, dx$$

$$= 1 \cdot \tan^{-1} 1 - 0 \cdot \tan^{-1} 0 - \int_0^1 \frac{x}{1 + x^2} \, dx$$

$$= \frac{\pi}{4} - \int_0^1 \frac{x}{1 + x^2} \, dx$$

- Since $\tan^{-1}x \geq 0$ for $x \geq 0$, the integral in Example 5 can be interpreted as the area of the region shown in Figure 2.

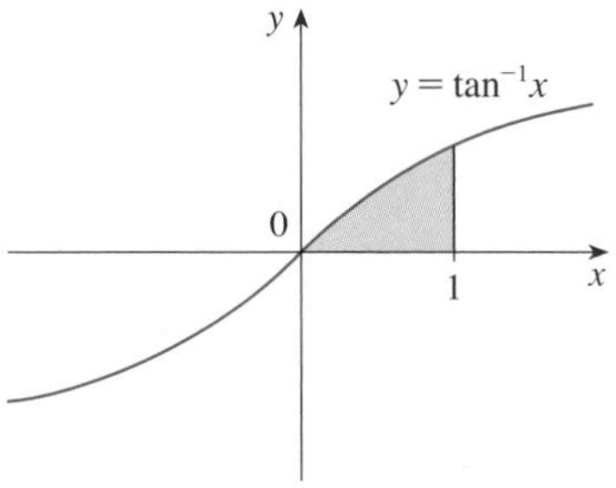

To evaluate this integral we use the substitution $t = 1 + x^2$ (since u has another meaning in this example). Then $dt = 2x \, dx$, so $x \, dx = \frac{1}{2} \, dt$. When $x = 0$, $t = 1$; when $x = 1$, $t = 2$; so

$$\int_0^1 \frac{x}{1 + x^2} \, dx = \frac{1}{2} \int_1^2 \frac{dt}{t} = \frac{1}{2} \ln |t| \Big]_1^2$$

$$= \frac{1}{2}(\ln 2 - \ln 1) = \frac{1}{2} \ln 2$$

FIGURE 2

Therefore

$$\int_0^1 \tan^{-1}x \, dx = \frac{\pi}{4} - \int_0^1 \frac{x}{1 + x^2} \, dx = \frac{\pi}{4} - \frac{\ln 2}{2} \qquad \blacksquare$$

EXAMPLE 6 Prove the reduction formula

- Equation 7 is called a *reduction formula* because the exponent n has been *reduced* to $n - 1$ and $n - 2$.

$$\boxed{7} \qquad \int \sin^n x \, dx = -\frac{1}{n} \cos x \sin^{n-1}x + \frac{n-1}{n} \int \sin^{n-2}x \, dx$$

where $n \geq 2$ is an integer.

SOLUTION Let

$$u = \sin^{n-1}x \qquad\qquad dv = \sin x \, dx$$

Then

$$du = (n-1)\sin^{n-2}x \cos x \, dx \qquad v = -\cos x$$

so integration by parts gives

$$\int \sin^n x \, dx = -\cos x \sin^{n-1}x + (n-1)\int \sin^{n-2}x \cos^2 x \, dx$$

Since $\cos^2 x = 1 - \sin^2 x$, we have

$$\int \sin^n x \, dx = -\cos x \sin^{n-1}x + (n-1)\int \sin^{n-2}x \, dx - (n-1)\int \sin^n x \, dx$$

As in Example 4, we solve this equation for the desired integral by taking the last term on the right side to the left side. Thus we have

$$n \int \sin^n x \, dx = -\cos x \sin^{n-1} x + (n-1) \int \sin^{n-2} x \, dx$$

or

$$\int \sin^n x \, dx = -\frac{1}{n} \cos x \sin^{n-1} x + \frac{n-1}{n} \int \sin^{n-2} x \, dx \qquad \blacksquare$$

The reduction formula $\boxed{7}$ is useful because by using it repeatedly we could eventually express $\int \sin^n x \, dx$ in terms of $\int \sin x \, dx$ (if n is odd) or $\int (\sin x)^0 \, dx = \int dx$ (if n is even).

6.1 | EXERCISES

1–2 ■ Evaluate the integral using integration by parts with the indicated choices of u and dv.

1. $\int x^2 \ln x \, dx; \quad u = \ln x, \ dv = x^2 \, dx$

2. $\int \theta \cos \theta \, d\theta; \quad u = \theta, \ dv = \cos \theta \, d\theta$

─────────────────────

3–26 ■ Evaluate the integral.

3. $\int x \cos 5x \, dx$

4. $\int y e^{0.2y} \, dy$

5. $\int t e^{-3t} \, dt$

6. $\int (x-1) \sin \pi x \, dx$

7. $\int (x^2 + 2x) \cos x \, dx$

8. $\int t^2 \sin \beta t \, dt$

9. $\int \ln(2x+1) \, dx$

10. $\int p^5 \ln p \, dp$

11. $\int \arctan 4t \, dt$

12. $\int \sin^{-1} x \, dx$

13. $\int e^{2\theta} \sin 3\theta \, d\theta$

14. $\int e^{-\theta} \cos 2\theta \, d\theta$

15. $\int \frac{x e^{2x}}{(1+2x)^2} \, dx$

16. $\int t^3 e^t \, dt$

17. $\int_0^{1/2} x \cos \pi x \, dx$

18. $\int_0^1 (x^2 + 1) e^{-x} \, dx$

19. $\int_1^3 r^3 \ln r \, dr$

20. $\int_4^9 \frac{\ln y}{\sqrt{y}} \, dy$

21. $\int_0^1 t \cosh t \, dt$

22. $\int_1^{\sqrt{3}} \arctan(1/x) \, dx$

23. $\int_0^{1/2} \cos^{-1} x \, dx$

24. $\int_0^1 \frac{r^3}{\sqrt{4+r^2}} \, dr$

25. $\int_1^2 (\ln x)^2 \, dx$

26. $\int_0^t e^s \sin(t-s) \, ds$

─────────────────────

27–30 ■ First make a substitution and then use integration by parts to evaluate the integral.

27. $\int \cos \sqrt{x} \, dx$

28. $\int t^3 e^{-t^2} \, dt$

29. $\int_{\sqrt{\pi/2}}^{\sqrt{\pi}} \theta^3 \cos(\theta^2) \, d\theta$

30. $\int_1^4 e^{\sqrt{x}} \, dx$

─────────────────────

31. (a) Use the reduction formula in Example 6 to show that

$$\int \sin^2 x \, dx = \frac{x}{2} - \frac{\sin 2x}{4} + C$$

(b) Use part (a) and the reduction formula to evaluate $\int \sin^4 x \, dx$.

32. (a) Prove the reduction formula

$$\int \cos^n x \, dx = \frac{1}{n} \cos^{n-1} x \sin x + \frac{n-1}{n} \int \cos^{n-2} x \, dx$$

(b) Use part (a) to evaluate $\int \cos^2 x \, dx$.
(c) Use parts (a) and (b) to evaluate $\int \cos^4 x \, dx$.

33. (a) Use the reduction formula in Example 6 to show that

$$\int_0^{\pi/2} \sin^n x \, dx = \frac{n-1}{n} \int_0^{\pi/2} \sin^{n-2} x \, dx$$

where $n \geqslant 2$ is an integer.

─────────────────────

⌨ Graphing calculator or computer required   CAS Computer algebra system required   **1.** Homework Hints at stewartcalculus.com

(b) Use part (a) to evaluate $\int_0^{\pi/2} \sin^3 x \, dx$ and $\int_0^{\pi/2} \sin^5 x \, dx$.

(c) Use part (a) to show that, for odd powers of sine,

$$\int_0^{\pi/2} \sin^{2n+1} x \, dx = \frac{2 \cdot 4 \cdot 6 \cdot \, \cdots \, \cdot 2n}{3 \cdot 5 \cdot 7 \cdot \, \cdots \, \cdot (2n + 1)}$$

34. Prove that, for even powers of sine,

$$\int_0^{\pi/2} \sin^{2n} x \, dx = \frac{1 \cdot 3 \cdot 5 \cdot \, \cdots \, \cdot (2n - 1)}{2 \cdot 4 \cdot 6 \cdot \, \cdots \, \cdot 2n} \frac{\pi}{2}$$

35–38 ▪ Use integration by parts to prove the reduction formula.

35. $\displaystyle\int (\ln x)^n \, dx = x(\ln x)^n - n \int (\ln x)^{n-1} \, dx$

36. $\displaystyle\int x^n e^x \, dx = x^n e^x - n \int x^{n-1} e^x \, dx$

37. $\displaystyle\int \tan^n x \, dx = \frac{\tan^{n-1} x}{n - 1} - \int \tan^{n-2} x \, dx \quad (n \neq 1)$

38. $\displaystyle\int \sec^n x \, dx = \frac{\tan x \sec^{n-2} x}{n - 1} + \frac{n - 2}{n - 1} \int \sec^{n-2} x \, dx$
$(n \neq 1)$

39. Use Exercise 35 to find $\int (\ln x)^3 \, dx$.

40. Use Exercise 36 to find $\int x^4 e^x \, dx$.

41. Calculate the average value of $f(x) = x \sec^2 x$ on the interval $[0, \pi/4]$.

42. A rocket accelerates by burning its onboard fuel, so its mass decreases with time. Suppose the initial mass of the rocket at liftoff (including its fuel) is m, the fuel is consumed at

rate r, and the exhaust gases are ejected with constant velocity v_e (relative to the rocket). A model for the velocity of the rocket at time t is given by the equation

$$v(t) = -gt - v_e \ln \frac{m - rt}{m}$$

where g is the acceleration due to gravity and t is not too large. If $g = 9.8$ m/s^2, $m = 30{,}000$ kg, $r = 160$ kg/s, and $v_e = 3000$ m/s, find the height of the rocket one minute after liftoff.

43. A particle that moves along a straight line has velocity $v(t) = t^2 e^{-t}$ meters per second after t seconds. How far will it travel during the first t seconds?

44. If $f(0) = g(0) = 0$ and f'' and g'' are continuous, show that

$$\int_0^a f(x) g''(x) \, dx = f(a) g'(a) - f'(a) g(a) + \int_0^a f''(x) g(x) \, dx$$

45. Suppose that $f(1) = 2$, $f(4) = 7$, $f'(1) = 5$, $f'(4) = 3$, and f'' is continuous. Find the value of $\int_1^4 x f''(x) \, dx$.

46. (a) Use integration by parts to show that

$$\int f(x) \, dx = x f(x) - \int x f'(x) \, dx$$

(b) If f and g are inverse functions and f' is continuous, prove that

$$\int_a^b f(x) \, dx = b f(b) - a f(a) - \int_{f(a)}^{f(b)} g(y) \, dy$$

   [*Hint:* Use part (a) and make the substitution $y = f(x)$.]

(c) In the case where f and g are positive functions and $b > a > 0$, draw a diagram to give a geometric interpretation of part (b).

(d) Use part (b) to evaluate $\int_1^e \ln x \, dx$.

6.2 | TRIGONOMETRIC INTEGRALS AND SUBSTITUTIONS

In this section we look at integrals involving trigonometric functions and integrals that can be transformed into trigonometric integrals by substitution.

TRIGONOMETRIC INTEGRALS

Here we use trigonometric identities to integrate certain combinations of trigonometric functions. We start with powers of sine and cosine.

EXAMPLE 1 Evaluate $\int \cos^3 x \, dx$.

SOLUTION Simply substituting $u = \cos x$ isn't helpful, since then $du = -\sin x \, dx$. In order to integrate powers of cosine, we would need an extra $\sin x$ factor. Similarly, a power of sine would require an extra $\cos x$ factor. Thus here we can separate one cosine factor and convert the remaining $\cos^2 x$ factor to an expression involving sine using the identity $\sin^2 x + \cos^2 x = 1$:

$$\cos^3 x = \cos^2 x \cdot \cos x = (1 - \sin^2 x) \cos x$$

We can then evaluate the integral by substituting $u = \sin x$, so $du = \cos x \, dx$ and

$$\int \cos^3 x \, dx = \int \cos^2 x \cdot \cos x \, dx = \int (1 - \sin^2 x) \cos x \, dx$$

$$= \int (1 - u^2) \, du = u - \tfrac{1}{3} u^3 + C = \sin x - \tfrac{1}{3} \sin^3 x + C \qquad \blacksquare$$

In general, we try to write an integrand involving powers of sine and cosine in a form where we have only one sine factor (and the remainder of the expression in terms of cosine) or only one cosine factor (and the remainder of the expression in terms of sine). The identity $\sin^2 x + \cos^2 x = 1$ enables us to convert back and forth between even powers of sine and cosine.

V EXAMPLE 2 Find $\displaystyle \int \sin^5 x \cos^2 x \, dx$.

SOLUTION We could convert $\cos^2 x$ to $1 - \sin^2 x$, but we would be left with an expression in terms of $\sin x$ with no extra $\cos x$ factor. Instead, we separate a single sine factor and rewrite the remaining $\sin^4 x$ factor in terms of $\cos x$:

$$\sin^5 x \cos^2 x = (\sin^2 x)^2 \cos^2 x \sin x = (1 - \cos^2 x)^2 \cos^2 x \sin x$$

Substituting $u = \cos x$, we have $du = -\sin x \, dx$ and so

$$\int \sin^5 x \cos^2 x \, dx = \int (\sin^2 x)^2 \cos^2 x \sin x \, dx = \int (1 - \cos^2 x)^2 \cos^2 x \sin x \, dx$$

$$= \int (1 - u^2)^2 u^2 (-du) = -\int (u^2 - 2u^4 + u^6) \, du$$

$$= -\left(\frac{u^3}{3} - 2\frac{u^5}{5} + \frac{u^7}{7} \right) + C$$

$$= -\tfrac{1}{3} \cos^3 x + \tfrac{2}{5} \cos^5 x - \tfrac{1}{7} \cos^7 x + C \qquad \blacksquare$$

In the preceding examples, an odd power of sine or cosine enabled us to separate a single factor and convert the remaining even power. If the integrand contains even powers of both sine and cosine, this strategy fails. In this case, we can take advantage of the following half-angle identities (see Equations 17b and 17a in Appendix A):

$$\sin^2 x = \tfrac{1}{2}(1 - \cos 2x) \qquad \text{and} \qquad \cos^2 x = \tfrac{1}{2}(1 + \cos 2x)$$

■ Figure 1 shows the graphs of the integrand $\sin^5 x \cos^2 x$ in Example 2 and its indefinite integral (with $C = 0$). Which is which?

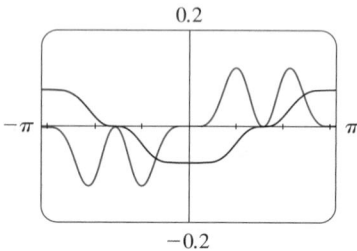

FIGURE 1

■ Example 3 shows that the area of the region shown in Figure 2 is $\pi/2$.

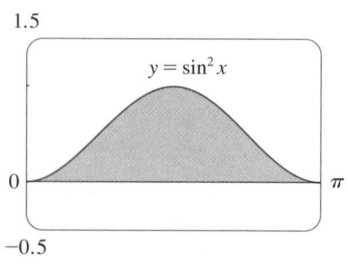

FIGURE 2

V EXAMPLE 3 Evaluate $\displaystyle \int_0^{\pi} \sin^2 x \, dx$.

SOLUTION If we write $\sin^2 x = 1 - \cos^2 x$, the integral is no simpler to evaluate. Using the half-angle formula for $\sin^2 x$, however, we have

$$\int_0^{\pi} \sin^2 x \, dx = \tfrac{1}{2} \int_0^{\pi} (1 - \cos 2x) \, dx = \left[\tfrac{1}{2}\left(x - \tfrac{1}{2} \sin 2x\right) \right]_0^{\pi}$$

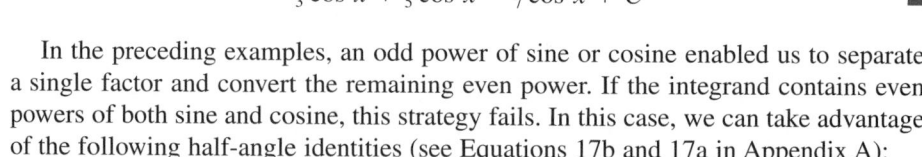

$$= \tfrac{1}{2}\left(\pi - \tfrac{1}{2} \sin 2\pi\right) - \tfrac{1}{2}\left(0 - \tfrac{1}{2} \sin 0\right) = \tfrac{1}{2}\pi$$

Notice that we mentally made the substitution $u = 2x$ when integrating $\cos 2x$. Another method for evaluating this integral was given in Exercise 31 in Section 6.1. $\qquad \blacksquare$

EXAMPLE 4 Find $\int \sin^4 x \, dx$.

SOLUTION We could evaluate this integral using the reduction formula for $\int \sin^n x \, dx$ (Equation 6.1.7) together with Example 3 (as in Exercise 31 in Section 6.1), but a better method is to write $\sin^4 x = (\sin^2 x)^2$ and use a half-angle formula:

$$\int \sin^4 x \, dx = \int (\sin^2 x)^2 \, dx$$

$$= \int \left(\frac{1 - \cos 2x}{2} \right)^2 dx$$

$$= \tfrac{1}{4} \int (1 - 2\cos 2x + \cos^2 2x) \, dx$$

Since $\cos^2 2x$ occurs, we must use another half-angle formula

$$\cos^2 2x = \tfrac{1}{2}(1 + \cos 4x)$$

This gives

$$\int \sin^4 x \, dx = \tfrac{1}{4} \int \left[1 - 2\cos 2x + \tfrac{1}{2}(1 + \cos 4x) \right] dx$$

$$= \tfrac{1}{4} \int \left(\tfrac{3}{2} - 2\cos 2x + \tfrac{1}{2}\cos 4x \right) dx$$

$$= \tfrac{1}{4} \left(\tfrac{3}{2}x - \sin 2x + \tfrac{1}{8}\sin 4x \right) + C \qquad \blacksquare$$

■ **HOW TO INTEGRATE POWERS OF $\sin x$ AND $\cos x$**
From Examples 1–4 we see that the following strategy works:

(i) If the power of $\cos x$ is odd, save one cosine factor and use $\cos^2 x = 1 - \sin^2 x$ to express the remaining factors in terms of $\sin x$. Then substitute $u = \sin x$.

(ii) If the power of $\sin x$ is odd, save one sine factor and use $\sin^2 x = 1 - \cos^2 x$ to express the remaining factors in terms of $\cos x$. Then substitute $u = \cos x$.

(iii) If the powers of both sine and cosine are even, use the half-angle identities:

$$\sin^2 x = \tfrac{1}{2}(1 - \cos 2x)$$
$$\cos^2 x = \tfrac{1}{2}(1 + \cos 2x)$$

It is sometimes helpful to use the identity

$$\sin x \cos x = \tfrac{1}{2}\sin 2x$$

We can use a similar strategy to evaluate integrals of the form $\int \tan^m x \, \sec^n x \, dx$. Since $(d/dx)\tan x = \sec^2 x$, we can separate a $\sec^2 x$ factor and convert the remaining (even) power of secant to an expression involving tangent using the identity $\sec^2 x = 1 + \tan^2 x$. Or, since $(d/dx)\sec x = \sec x \tan x$, we can separate a $\sec x \, \tan x$ factor and convert the remaining (even) power of tangent to secant.

V EXAMPLE 5 Evaluate $\int \tan^6 x \, \sec^4 x \, dx$.

SOLUTION If we separate one $\sec^2 x$ factor, we can express the remaining $\sec^2 x$ factor in terms of tangent using the identity $\sec^2 x = 1 + \tan^2 x$. We can then evaluate the integral by substituting $u = \tan x$ with $du = \sec^2 x \, dx$:

$$\int \tan^6 x \, \sec^4 x \, dx = \int \tan^6 x \, \sec^2 x \, \sec^2 x \, dx$$

$$= \int \tan^6 x \, (1 + \tan^2 x) \, \sec^2 x \, dx$$

$$= \int u^6 (1 + u^2) \, du = \int (u^6 + u^8) \, du$$

$$= \frac{u^7}{7} + \frac{u^9}{9} + C$$

$$= \tfrac{1}{7}\tan^7 x + \tfrac{1}{9}\tan^9 x + C \qquad \blacksquare$$

EXAMPLE 6 Find $\int \tan^5\theta \, \sec^7\theta \, d\theta$.

SOLUTION If we separate a $\sec^2\theta$ factor, as in the preceding example, we are left with a $\sec^5\theta$ factor, which isn't easily converted to tangent. However, if we separate a $\sec\theta \, \tan\theta$ factor, we can convert the remaining power of tangent to an expression involving only secant using the identity $\tan^2\theta = \sec^2\theta - 1$. We can then evaluate the integral by substituting $u = \sec\theta$, so $du = \sec\theta \, \tan\theta \, d\theta$:

$$\int \tan^5\theta \, \sec^7\theta \, d\theta = \int \tan^4\theta \, \sec^6\theta \, \sec\theta \, \tan\theta \, d\theta$$

$$= \int (\sec^2\theta - 1)^2 \sec^6\theta \, \sec\theta \, \tan\theta \, d\theta$$

$$= \int (u^2 - 1)^2 u^6 \, du = \int (u^{10} - 2u^8 + u^6) \, du$$

$$= \frac{u^{11}}{11} - 2\frac{u^9}{9} + \frac{u^7}{7} + C$$

$$= \tfrac{1}{11} \sec^{11}\theta - \tfrac{2}{9} \sec^9\theta + \tfrac{1}{7} \sec^7\theta + C \qquad \blacksquare$$

■ **HOW TO INTEGRATE POWERS OF $\tan x$ AND $\sec x$**

From Examples 5 and 6 we have a strategy for two cases:

(i) If the power of $\sec x$ is even, save a factor of $\sec^2 x$ and use $\sec^2 x = 1 + \tan^2 x$ to express the remaining factors in terms of $\tan x$. Then substitute $u = \tan x$.

(ii) If the power of $\tan x$ is odd, save a factor of $\sec x \, \tan x$ and use $\tan^2 x = \sec^2 x - 1$ to express the remaining factors in terms of $\sec x$. Then substitute $u = \sec x$.

For other cases, the guidelines are not as clear-cut. We may need to use identities, integration by parts, and occasionally a little ingenuity. We will sometimes need to be able to integrate $\tan x$ by using the formula established in (5.5.5):

$$\int \tan x \, dx = \ln |\sec x| + C$$

We will also need the indefinite integral of secant:

■ Formula 1 was discovered by James Gregory in 1668. (See his biography on page 115.) Gregory used this formula to solve a problem in constructing nautical tables.

$$\boxed{1} \qquad \boxed{\int \sec x \, dx = \ln |\sec x + \tan x| + C}$$

We could verify Formula 1 by differentiating the right side, or as follows. First we multiply numerator and denominator by $\sec x + \tan x$:

$$\int \sec x \, dx = \int \sec x \, \frac{\sec x + \tan x}{\sec x + \tan x} \, dx$$

$$= \int \frac{\sec^2 x + \sec x \, \tan x}{\sec x + \tan x} \, dx$$

If we substitute $u = \sec x + \tan x$, then $du = (\sec x \, \tan x + \sec^2 x) \, dx$, so the integral becomes $\int (1/u) \, du = \ln |u| + C$. Thus we have

$$\int \sec x \, dx = \ln |\sec x + \tan x| + C$$

EXAMPLE 7 Find $\int \tan^3 x \, dx$.

SOLUTION Here only $\tan x$ occurs, so we use $\tan^2 x = \sec^2 x - 1$ to rewrite a $\tan^2 x$ factor in terms of $\sec^2 x$:

$$\int \tan^3 x \, dx = \int \tan x \, \tan^2 x \, dx = \int \tan x \, (\sec^2 x - 1) \, dx$$

$$= \int \tan x \, \sec^2 x \, dx - \int \tan x \, dx$$

$$= \frac{\tan^2 x}{2} - \ln |\sec x| + C$$

In the first integral we mentally substituted $u = \tan x$ so that $du = \sec^2 x \, dx$. ∎

If an even power of tangent appears with an odd power of secant, it is helpful to express the integrand completely in terms of $\sec x$. Powers of $\sec x$ may require integration by parts, as shown in the following example.

EXAMPLE 8 Find $\int \sec^3 x \, dx$.

SOLUTION Here we integrate by parts with

$$u = \sec x \qquad\qquad dv = \sec^2 x \, dx$$

$$du = \sec x \, \tan x \, dx \qquad\qquad v = \tan x$$

Then

$$\int \sec^3 x \, dx = \sec x \, \tan x - \int \sec x \, \tan^2 x \, dx$$

$$= \sec x \, \tan x - \int \sec x \, (\sec^2 x - 1) \, dx$$

$$= \sec x \, \tan x - \int \sec^3 x \, dx + \int \sec x \, dx$$

Using Formula 1 and solving for the required integral, we get

$$\int \sec^3 x \, dx = \tfrac{1}{2}\big(\sec x \, \tan x + \ln |\sec x + \tan x|\big) + C \qquad\qquad ∎$$

Integrals such as the one in Example 8 may seem very special but they occur frequently in applications of integration, as we will see in Chapter 7. Integrals of the form $\int \cot^m x \, \csc^n x \, dx$ can be found by similar methods because of the identity $1 + \cot^2 x = \csc^2 x$.

TRIGONOMETRIC SUBSTITUTIONS

In finding the area of a circle or an ellipse, an integral of the form $\int \sqrt{a^2 - x^2} \, dx$ arises, where $a > 0$. If it were $\int x\sqrt{a^2 - x^2} \, dx$, the substitution $u = a^2 - x^2$ would be effective but, as it stands, $\int \sqrt{a^2 - x^2} \, dx$ is more difficult. If we change the variable from x to θ by the substitution $x = a \sin \theta$, then the identity $1 - \sin^2 \theta = \cos^2 \theta$ allows us to get rid of the root sign because

$$\sqrt{a^2 - x^2} = \sqrt{a^2 - a^2 \sin^2 \theta} = \sqrt{a^2(1 - \sin^2 \theta)} = \sqrt{a^2 \cos^2 \theta} = a |\cos \theta|$$

Notice the difference between the substitution $u = a^2 - x^2$ (in which the new variable is a function of the old one) and the substitution $x = a \sin \theta$ (the old variable is a function of the new one).

In general, we can make a substitution of the form $x = g(t)$ by using the Substitution Rule in reverse. To make our calculations simpler, we assume that g has an inverse function; that is, g is one-to-one. In this case, if we replace u by x and x by t in the Substitution Rule (Equation 5.5.4), we obtain

$$\int f(x)\, dx = \int f(g(t))g'(t)\, dt$$

This kind of substitution is called *inverse substitution*.

We can make the inverse substitution $x = a \sin \theta$ provided that it defines a one-to-one function. We accomplish this by restricting θ to lie in the interval $[-\pi/2, \pi/2]$.

In the following table we list trigonometric substitutions that are effective for the given radical expressions because of the specified trigonometric identities. In each case the restriction on θ is imposed to ensure that the function that defines the substitution is one-to-one. (These are the same intervals used in Section 3.5 in defining the inverse functions.)

TABLE OF TRIGONOMETRIC SUBSTITUTIONS

Expression	Substitution	Identity
$\sqrt{a^2 - x^2}$	$x = a \sin \theta, \quad -\dfrac{\pi}{2} \leqslant \theta \leqslant \dfrac{\pi}{2}$	$1 - \sin^2\theta = \cos^2\theta$
$\sqrt{a^2 + x^2}$	$x = a \tan \theta, \quad -\dfrac{\pi}{2} < \theta < \dfrac{\pi}{2}$	$1 + \tan^2\theta = \sec^2\theta$
$\sqrt{x^2 - a^2}$	$x = a \sec \theta, \quad 0 \leqslant \theta < \dfrac{\pi}{2} \text{ or } \pi \leqslant \theta < \dfrac{3\pi}{2}$	$\sec^2\theta - 1 = \tan^2\theta$

▼ EXAMPLE 9 Evaluate $\displaystyle\int \frac{\sqrt{9 - x^2}}{x^2}\, dx$.

SOLUTION Let $x = 3 \sin \theta$, where $-\pi/2 \leqslant \theta \leqslant \pi/2$. Then $dx = 3 \cos \theta\, d\theta$ and

$$\sqrt{9 - x^2} = \sqrt{9 - 9 \sin^2\theta} = \sqrt{9 \cos^2\theta} = 3\,|\cos \theta| = 3 \cos \theta$$

(Note that $\cos \theta \geqslant 0$ because $-\pi/2 \leqslant \theta \leqslant \pi/2$.) Thus, using inverse substitution, we get

$$\int \frac{\sqrt{9 - x^2}}{x^2}\, dx = \int \frac{3 \cos \theta}{9 \sin^2\theta}\, 3 \cos \theta\, d\theta = \int \frac{\cos^2\theta}{\sin^2\theta}\, d\theta$$

$$= \int \cot^2\theta\, d\theta = \int (\csc^2\theta - 1)\, d\theta = -\cot \theta - \theta + C$$

Since this is an indefinite integral, we must return to the original variable x. This can be done either by using trigonometric identities to express $\cot \theta$ in terms of $\sin \theta = x/3$ or by drawing a diagram, as in Figure 3, where θ is interpreted as an angle of a right triangle. Since $\sin \theta = x/3$, we label the opposite side and the hypotenuse as having lengths x and 3. Then the Pythagorean Theorem gives the length of

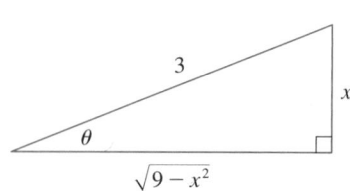

FIGURE 3

$\sin \theta = \dfrac{x}{3}$

the adjacent side as $\sqrt{9 - x^2}$, so we can simply read the value of $\cot \theta$ from the figure:

$$\cot \theta = \frac{\sqrt{9 - x^2}}{x}$$

(Although $\theta > 0$ in the diagram, this expression for $\cot \theta$ is valid even when $\theta < 0$.) Since $\sin \theta = x/3$, we have $\theta = \sin^{-1}(x/3)$ and so

$$\int \frac{\sqrt{9 - x^2}}{x^2}\, dx = -\frac{\sqrt{9 - x^2}}{x} - \sin^{-1}\left(\frac{x}{3}\right) + C \qquad\blacksquare$$

V EXAMPLE 10 Find the area enclosed by the ellipse

$$\frac{x^2}{a^2} + \frac{y^2}{b^2} = 1$$

SOLUTION Solving the equation of the ellipse for y, we get

$$\frac{y^2}{b^2} = 1 - \frac{x^2}{a^2} = \frac{a^2 - x^2}{a^2} \qquad \text{or} \qquad y = \pm\frac{b}{a}\sqrt{a^2 - x^2}$$

Because the ellipse is symmetric with respect to both axes, the total area A is four times the area in the first quadrant (see Figure 4). The part of the ellipse in the first quadrant is given by the function

$$y = \frac{b}{a}\sqrt{a^2 - x^2} \qquad 0 \le x \le a$$

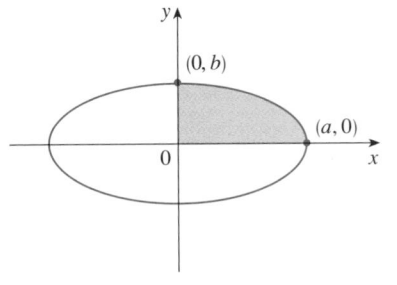

FIGURE 4

$$\frac{x^2}{a^2} + \frac{y^2}{b^2} = 1$$

and so

$$\tfrac{1}{4}A = \int_0^a \frac{b}{a}\sqrt{a^2 - x^2}\, dx$$

To evaluate this integral we substitute $x = a \sin \theta$. Then $dx = a \cos \theta\, d\theta$. To change the limits of integration we note that when $x = 0$, $\sin \theta = 0$, so $\theta = 0$; when $x = a$, $\sin \theta = 1$, so $\theta = \pi/2$. Also

$$\sqrt{a^2 - x^2} = \sqrt{a^2 - a^2 \sin^2\theta} = \sqrt{a^2 \cos^2\theta} = a\,|\cos \theta| = a \cos \theta$$

since $0 \le \theta \le \pi/2$. Therefore

$$A = 4\,\frac{b}{a}\int_0^a \sqrt{a^2 - x^2}\, dx = 4\,\frac{b}{a}\int_0^{\pi/2} a \cos \theta \cdot a \cos \theta\, d\theta$$

$$= 4ab\int_0^{\pi/2} \cos^2\theta\, d\theta = 4ab\int_0^{\pi/2} \tfrac{1}{2}(1 + \cos 2\theta)\, d\theta$$

$$= 2ab\left[\theta + \tfrac{1}{2}\sin 2\theta\right]_0^{\pi/2} = 2ab\left(\frac{\pi}{2} + 0 - 0\right) = \pi ab$$

We have shown that the area of an ellipse with semiaxes a and b is πab. In particular, taking $a = b = r$, we have proved the famous formula that the area of a circle with radius r is πr^2. $\qquad\blacksquare$

NOTE Since the integral in Example 10 was a definite integral, we changed the limits of integration and did not have to convert back to the original variable x.

V EXAMPLE 11 Find $\int \dfrac{1}{x^2\sqrt{x^2+4}} \, dx$.

SOLUTION Let $x = 2\tan\theta$, $-\pi/2 < \theta < \pi/2$. Then $dx = 2\sec^2\theta \, d\theta$ and

$$\sqrt{x^2+4} = \sqrt{4(\tan^2\theta+1)} = \sqrt{4\sec^2\theta} = 2\,|\sec\theta| = 2\sec\theta$$

Thus we have

$$\int \frac{dx}{x^2\sqrt{x^2+4}} = \int \frac{2\sec^2\theta \, d\theta}{4\tan^2\theta \cdot 2\sec\theta} = \frac{1}{4}\int \frac{\sec\theta}{\tan^2\theta} \, d\theta$$

To evaluate this trigonometric integral we put everything in terms of $\sin\theta$ and $\cos\theta$:

$$\frac{\sec\theta}{\tan^2\theta} = \frac{1}{\cos\theta} \cdot \frac{\cos^2\theta}{\sin^2\theta} = \frac{\cos\theta}{\sin^2\theta}$$

Therefore, making the substitution $u = \sin\theta$, we have

$$\int \frac{dx}{x^2\sqrt{x^2+4}} = \frac{1}{4}\int \frac{\cos\theta}{\sin^2\theta} \, d\theta = \frac{1}{4}\int \frac{du}{u^2}$$

$$= \frac{1}{4}\left(-\frac{1}{u}\right) + C = -\frac{1}{4\sin\theta} + C$$

$$= -\frac{\csc\theta}{4} + C$$

We use Figure 5 to determine that $\csc\theta = \sqrt{x^2+4}/x$ and so

$$\int \frac{dx}{x^2\sqrt{x^2+4}} = -\frac{\sqrt{x^2+4}}{4x} + C \qquad \blacksquare$$

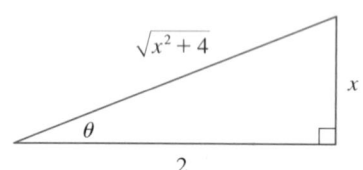

FIGURE 5

$\tan\theta = \dfrac{x}{2}$

EXAMPLE 12 Find $\int \dfrac{x}{\sqrt{x^2+4}} \, dx$.

SOLUTION It would be possible to use the trigonometric substitution $x = 2\tan\theta$ here (as in Example 11). But the direct substitution $u = x^2 + 4$ is simpler, because then $du = 2x \, dx$ and

$$\int \frac{x}{\sqrt{x^2+4}} \, dx = \frac{1}{2}\int \frac{du}{\sqrt{u}} = \sqrt{u} + C = \sqrt{x^2+4} + C \qquad \blacksquare$$

■ Example 12 illustrates the fact that even when trigonometric substitutions are possible, they may not give the easiest solution. You should look for a simpler method first.

EXAMPLE 13 Evaluate $\int \dfrac{dx}{\sqrt{x^2-a^2}}$, where $a > 0$.

SOLUTION We let $x = a\sec\theta$, where $0 < \theta < \pi/2$ or $\pi < \theta < 3\pi/2$. Then $dx = a\sec\theta\tan\theta \, d\theta$ and

$$\sqrt{x^2-a^2} = \sqrt{a^2(\sec^2\theta-1)} = \sqrt{a^2\tan^2\theta} = a\,|\tan\theta| = a\tan\theta$$

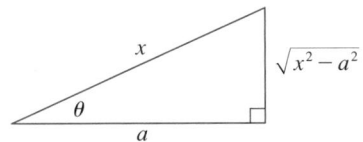

FIGURE 6

$\sec \theta = \dfrac{x}{a}$

Therefore

$$\int \frac{dx}{\sqrt{x^2 - a^2}} = \int \frac{a \sec \theta \tan \theta}{a \tan \theta} \, d\theta$$

$$= \int \sec \theta \, d\theta = \ln |\sec \theta + \tan \theta| + C$$

The triangle in Figure 6 gives $\tan \theta = \sqrt{x^2 - a^2}/a$, so we have

$$\int \frac{dx}{\sqrt{x^2 - a^2}} = \ln \left| \frac{x}{a} + \frac{\sqrt{x^2 - a^2}}{a} \right| + C$$

$$= \ln \left| x + \sqrt{x^2 - a^2} \right| - \ln a + C$$

Writing $C_1 = C - \ln a$, we have

$$\int \frac{dx}{\sqrt{x^2 - a^2}} = \ln \left| x + \sqrt{x^2 - a^2} \right| + C_1 \qquad \blacksquare$$

▪ As Example 14 shows, trigonometric substitution is sometimes a good idea when $(x^2 + a^2)^{n/2}$ occurs in an integral, where n is any integer. The same is true when $(a^2 - x^2)^{n/2}$ or $(x^2 - a^2)^{n/2}$ occur.

EXAMPLE 14 Find $\displaystyle\int_0^{3\sqrt{3}/2} \frac{x^3}{(4x^2 + 9)^{3/2}} \, dx$.

SOLUTION First we note that $(4x^2 + 9)^{3/2} = (\sqrt{4x^2 + 9})^3$ so trigonometric substitution is appropriate. Although $\sqrt{4x^2 + 9}$ is not quite one of the expressions in the table of trigonometric substitutions, it becomes one of them if we make the preliminary substitution $u = 2x$. When we combine this with the tangent substitution, we have $x = \frac{3}{2} \tan \theta$, which gives $dx = \frac{3}{2} \sec^2\theta \, d\theta$ and

$$\sqrt{4x^2 + 9} = \sqrt{9 \tan^2\theta + 9} = 3 \sec \theta$$

When $x = 0$, $\tan \theta = 0$, so $\theta = 0$; when $x = 3\sqrt{3}/2$, $\tan \theta = \sqrt{3}$, so $\theta = \pi/3$.

$$\int_0^{3\sqrt{3}/2} \frac{x^3}{(4x^2 + 9)^{3/2}} \, dx = \int_0^{\pi/3} \frac{\frac{27}{8} \tan^3\theta}{27 \sec^3\theta} \, \frac{3}{2} \sec^2\theta \, d\theta$$

$$= \frac{3}{16} \int_0^{\pi/3} \frac{\tan^3\theta}{\sec \theta} \, d\theta = \frac{3}{16} \int_0^{\pi/3} \frac{\sin^3\theta}{\cos^2\theta} \, d\theta$$

$$= \frac{3}{16} \int_0^{\pi/3} \frac{1 - \cos^2\theta}{\cos^2\theta} \sin \theta \, d\theta$$

▪ **www.stewartcalculus.com**
See Additional Example A.

Now we substitute $u = \cos \theta$ so that $du = -\sin \theta \, d\theta$. When $\theta = 0$, $u = 1$; when $\theta = \pi/3$, $u = \frac{1}{2}$. Therefore

$$\int_0^{3\sqrt{3}/2} \frac{x^3}{(4x^2 + 9)^{3/2}} \, dx = -\frac{3}{16} \int_1^{1/2} \frac{1 - u^2}{u^2} \, du = \frac{3}{16} \int_1^{1/2} (1 - u^{-2}) \, du$$

$$= \frac{3}{16} \left[u + \frac{1}{u} \right]_1^{1/2} = \frac{3}{16} \left[\left(\tfrac{1}{2} + 2 \right) - (1 + 1) \right] = \frac{3}{32} \qquad \blacksquare$$

6.2 | EXERCISES

1–36 ▪ Evaluate the integral.

1. $\int \sin^2 x \, \cos^3 x \, dx$

2. $\int \sin^3 \theta \, \cos^4 \theta \, d\theta$

3. $\int_0^{\pi/2} \sin^7 \theta \, \cos^5 \theta \, d\theta$

4. $\int_0^{\pi/2} \sin^5 x \, dx$

5. $\int_0^{\pi/2} \cos^2 \theta \, d\theta$

6. $\int_0^{2\pi} \sin^2 \left(\tfrac{1}{3} \theta \right) d\theta$

7. $\int_0^{\pi} \cos^4 (2t) \, dt$

8. $\int_0^{\pi/2} (2 - \sin \theta)^2 \, d\theta$

9. $\int_0^{\pi/2} \sin^2 x \, \cos^2 x \, dx$

10. $\int_0^{\pi} \cos^6 \theta \, d\theta$

11. $\int t \sin^2 t \, dt$

12. $\int x \cos^2 x \, dx$

13. $\int \cos^2 x \, \tan^3 x \, dx$

14. $\int \cot^5 \theta \, \sin^4 \theta \, d\theta$

15. $\int \dfrac{1 - \sin x}{\cos x} \, dx$

16. $\int \cos^2 x \, \sin 2x \, dx$

17. $\int \tan x \, \sec^3 x \, dx$

18. $\int \tan^2 \theta \, \sec^4 \theta \, d\theta$

19. $\int \tan^2 x \, dx$

20. $\int (\tan^2 x + \tan^4 x) \, dx$

21. $\int \tan^4 x \, \sec^6 x \, dx$

22. $\int_0^{\pi/4} \sec^4 \theta \, \tan^4 \theta \, d\theta$

23. $\int_0^{\pi/3} \tan^5 x \, \sec^4 x \, dx$

24. $\int \tan^5 x \, \sec^3 x \, dx$

25. $\int \tan^3 x \, \sec x \, dx$

26. $\int_0^{\pi/4} \tan^4 t \, dt$

27. $\int \tan^5 x \, dx$

28. $\int \tan^2 x \, \sec x \, dx$

29. $\int_{\pi/6}^{\pi/2} \cot^2 x \, dx$

30. $\int_{\pi/4}^{\pi/2} \cot^3 x \, dx$

31. $\int_{\pi/4}^{\pi/2} \cot^5 \phi \, \csc^3 \phi \, d\phi$

32. $\int \csc^4 x \, \cot^6 x \, dx$

33. $\int \csc x \, dx$

34. $\int \dfrac{1 - \tan^2 x}{\sec^2 x} \, dx$

35. $\int_0^{\pi/6} \sqrt{1 + \cos 2x} \, dx$

36. $\int \dfrac{dx}{\cos x - 1}$

37. (a) Use the formulas for $\cos(A + B)$ and $\cos(A - B)$ to show that

$$\sin A \, \sin B = \tfrac{1}{2}[\cos(A - B) - \cos(A + B)]$$

(b) Use part (a) to evaluate $\int \sin 5x \, \sin 2x \, dx$.

38. (a) Use the formulas for $\sin(A + B)$ and $\sin(A - B)$ to show that

$$\sin A \, \cos B = \tfrac{1}{2}[\sin(A - B) + \sin(A + B)]$$

(b) Use part (a) to evaluate $\int \sin 3x \, \cos x \, dx$.

39–41 ▪ Evaluate the integral using the indicated trigonometric substitution. Sketch and label the associated right triangle.

39. $\int \dfrac{dx}{x^2 \sqrt{4 - x^2}}$ $\qquad x = 2 \sin \theta$

40. $\int \dfrac{x^3}{\sqrt{x^2 + 4}} \, dx$ $\qquad x = 2 \tan \theta$

41. $\int \dfrac{\sqrt{x^2 - 4}}{x} \, dx$ $\qquad x = 2 \sec \theta$

42–60 Evaluate the integral.

42. $\int_0^1 x^3 \sqrt{1 - x^2} \, dx$

43. $\int_{\sqrt{2}}^2 \dfrac{1}{t^3 \sqrt{t^2 - 1}} \, dt$

44. $\int_0^2 x^3 \sqrt{x^2 + 4} \, dx$

45. $\int_0^a \dfrac{dx}{(a^2 + x^2)^{3/2}}, \quad a > 0$

46. $\int \dfrac{dt}{t^2 \sqrt{t^2 - 16}}$

47. $\int \dfrac{dx}{\sqrt{x^2 + 16}}$

48. $\int \dfrac{t^5}{\sqrt{t^2 + 2}} \, dt$

49. $\int \sqrt{1 - 4x^2} \, dx$

50. $\int \dfrac{du}{u \sqrt{5 - u^2}}$

51. $\int \dfrac{\sqrt{x^2 - 9}}{x^3} \, dx$

52. $\int \dfrac{x}{\sqrt{1 + x^2}} \, dx$

53. $\int_0^{0.6} \dfrac{x^2}{\sqrt{9 - 25x^2}} \, dx$

54. $\int \dfrac{dx}{[(ax)^2 - b^2]^{3/2}}$

55. $\int \dfrac{x}{\sqrt{x^2 - 7}} \, dx$

56. $\int_0^1 \sqrt{x^2 + 1} \, dx$

57. $\int \dfrac{\sqrt{1 + x^2}}{x} \, dx$

58. $\int_0^1 \dfrac{dx}{(x^2 + 1)^2}$

59. $\int x \sqrt{1 - x^4} \, dx$

60. $\int_0^{\pi/2} \dfrac{\cos t}{\sqrt{1 + \sin^2 t}} \, dt$

61. Evaluate the integral

$$\int \frac{1}{\sqrt{9x^2 + 6x - 8}}\, dx$$

by first completing the square and using the substitution $u = 3x + 1$.

62–64 ▪ Evaluate the integral by first completing the square.

62. $\displaystyle \int \frac{x^2}{(3 + 4x - 4x^2)^{3/2}}\, dx$

63. $\displaystyle \int \sqrt{5 + 4x - x^2}\, dx$

64. $\displaystyle \int \frac{x^2 + 1}{(x^2 - 2x + 2)^2}\, dx$

65. A particle moves on a straight line with velocity function $v(t) = \sin \omega t \, \cos^2 \omega t$. Find its position function $s = f(t)$ if $f(0) = 0$.

66. Household electricity is supplied in the form of alternating current that varies from 155 V to -155 V with a frequency of 60 cycles per second (Hz). The voltage is thus given by the equation

$$E(t) = 155 \sin(120\pi t)$$

where t is the time in seconds. Voltmeters read the RMS (root-mean-square) voltage, which is the square root of the average value of $[E(t)]^2$ over one cycle.
(a) Calculate the RMS voltage of household current.
(b) Many electric stoves require an RMS voltage of 220 V. Find the corresponding amplitude A needed for the voltage $E(t) = A \sin(120\pi t)$.

67. Find the average value of $f(x) = \sqrt{x^2 - 1}/x$, $1 \leqslant x \leqslant 7$.

68. Find the area of the region bounded by the hyperbola $9x^2 - 4y^2 = 36$ and the line $x = 3$.

69. Prove the formula $A = \frac{1}{2}r^2\theta$ for the area of a sector of a circle with radius r and central angle θ. [*Hint:* Assume $0 < \theta < \pi/2$ and place the center of the circle at the origin so it has the equation $x^2 + y^2 = r^2$. Then A is the sum of the area of the triangle POQ and the area of the region PQR in the figure.]

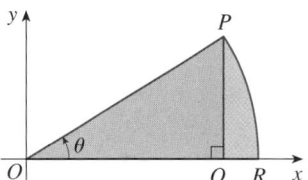

70. A charged rod of length L produces an electric field at point $P(a, b)$ given by

$$E(P) = \int_{-a}^{L-a} \frac{\lambda b}{4\pi\varepsilon_0(x^2 + b^2)^{3/2}}\, dx$$

where λ is the charge density per unit length on the rod and ε_0 is the free space permittivity (see the figure). Evaluate the integral to determine an expression for the electric field $E(P)$.

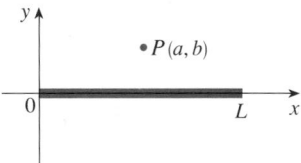

6.3 | PARTIAL FRACTIONS

In this section we show how to integrate any rational function (a ratio of polynomials) by expressing it as a sum of simpler fractions, called *partial fractions*, that we already know how to integrate. To illustrate the method, observe that by taking the fractions $2/(x - 1)$ and $1/(x + 2)$ to a common denominator we obtain

$$\frac{2}{x - 1} - \frac{1}{x + 2} = \frac{2(x + 2) - (x - 1)}{(x - 1)(x + 2)} = \frac{x + 5}{x^2 + x - 2}$$

If we now reverse the procedure, we see how to integrate the function on the right side of this equation:

$$\int \frac{x + 5}{x^2 + x - 2}\, dx = \int \left(\frac{2}{x - 1} - \frac{1}{x + 2} \right) dx$$

$$= 2 \ln |x - 1| - \ln |x + 2| + C$$

To see how the method of partial fractions works in general, let's consider a rational function

$$f(x) = \frac{P(x)}{Q(x)}$$

where P and Q are polynomials. It's possible to express f as a sum of simpler fractions provided that the degree of P is less than the degree of Q. Such a rational function is called *proper*. Recall that if

$$P(x) = a_n x^n + a_{n-1} x^{n-1} + \cdots + a_1 x + a_0$$

where $a_n \neq 0$, then the degree of P is n and we write $\deg(P) = n$.

If f is improper, that is, $\deg(P) \geq \deg(Q)$, then we must take the preliminary step of dividing Q into P (by long division) until a remainder $R(x)$ is obtained such that $\deg(R) < \deg(Q)$. The division statement is

$$\boxed{1} \qquad\qquad f(x) = \frac{P(x)}{Q(x)} = S(x) + \frac{R(x)}{Q(x)}$$

where S and R are also polynomials.

As the following example illustrates, sometimes this preliminary step is all that is required.

V EXAMPLE 1 Find $\displaystyle\int \frac{x^3 + x}{x - 1}\, dx$.

SOLUTION Since the degree of the numerator is greater than the degree of the denominator, we first perform the long division. This enables us to write

$$\int \frac{x^3 + x}{x - 1}\, dx = \int \left(x^2 + x + 2 + \frac{2}{x - 1} \right) dx$$

$$= \frac{x^3}{3} + \frac{x^2}{2} + 2x + 2 \ln|x - 1| + C \qquad \blacksquare$$

$$
\begin{array}{r}
x^2 + x + 2 \\
x - 1 \overline{)\, x^3 \phantom{{}+ x^2} + x } \\
\underline{x^3 - x^2 } \\
x^2 + x \\
\underline{x^2 - x } \\
2x \\
\underline{2x - 2} \\
2
\end{array}
$$

The next step is to factor the denominator $Q(x)$ as far as possible. It can be shown that any polynomial Q can be factored as a product of linear factors (of the form $ax + b$) and irreducible quadratic factors (of the form $ax^2 + bx + c$, where $b^2 - 4ac < 0$). For instance, if $Q(x) = x^4 - 16$, we could factor it as

$$Q(x) = (x^2 - 4)(x^2 + 4) = (x - 2)(x + 2)(x^2 + 4)$$

The third step is to express the proper rational function $R(x)/Q(x)$ (from Equation 1) as a sum of **partial fractions** of the form

$$\frac{A}{(ax + b)^i} \qquad \text{or} \qquad \frac{Ax + B}{(ax^2 + bx + c)^j}$$

A theorem in algebra guarantees that it is always possible to do this. We explain the details for the four cases that occur.

CASE I The denominator $Q(x)$ is a product of distinct linear factors.

This means that we can write

$$Q(x) = (a_1 x + b_1)(a_2 x + b_2) \cdots (a_k x + b_k)$$

where no factor is repeated (and no factor is a constant multiple of another). In this case the partial fraction theorem states that there exist constants $A_1, A_2, \ldots, A_k$ such that

$$\boxed{2} \qquad \frac{R(x)}{Q(x)} = \frac{A_1}{a_1 x + b_1} + \frac{A_2}{a_2 x + b_2} + \cdots + \frac{A_k}{a_k x + b_k}$$

These constants can be determined as in the following example.

▼ EXAMPLE 2 Evaluate $\displaystyle\int \frac{x^2 + 2x - 1}{2x^3 + 3x^2 - 2x}\, dx$.

SOLUTION Since the degree of the numerator is less than the degree of the denominator, we don't need to divide. We factor the denominator as

$$2x^3 + 3x^2 - 2x = x(2x^2 + 3x - 2) = x(2x - 1)(x + 2)$$

Since the denominator has three distinct linear factors, the partial fraction decomposition of the integrand $\boxed{2}$ has the form

$$\boxed{3} \qquad \frac{x^2 + 2x - 1}{x(2x - 1)(x + 2)} = \frac{A}{x} + \frac{B}{2x - 1} + \frac{C}{x + 2}$$

- Another method for finding A, B, and C is given in the note after this example.

To determine the values of A, B, and C, we multiply both sides of this equation by the product of the denominators, $x(2x - 1)(x + 2)$, obtaining

$$\boxed{4} \qquad x^2 + 2x - 1 = A(2x - 1)(x + 2) + Bx(x + 2) + Cx(2x - 1)$$

Expanding the right side of Equation 4 and writing it in the standard form for polynomials, we get

$$\boxed{5} \qquad x^2 + 2x - 1 = (2A + B + 2C)x^2 + (3A + 2B - C)x - 2A$$

The polynomials in Equation 5 are identical, so their coefficients must be equal. The coefficient of x^2 on the right side, $2A + B + 2C$, must equal the coefficient of x^2 on the left side—namely, 1. Likewise, the coefficients of x are equal and the constant terms are equal. This gives the following system of equations for A, B, and C:

$$\begin{aligned} 2A + \ \ B + 2C &= 1 \\ 3A + 2B - \ \ C &= 2 \\ -2A \qquad\qquad &= -1 \end{aligned}$$

- Figure 1 shows the graphs of the integrand in Example 2 and its indefinite integral (with $K = 0$). Which is which?

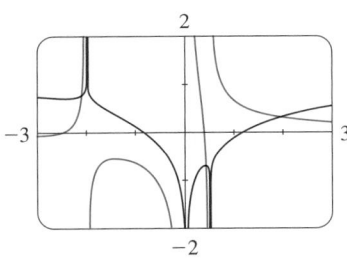

FIGURE 1

Solving, we get $A = \frac{1}{2}$, $B = \frac{1}{5}$, and $C = -\frac{1}{10}$, and so

$$\int \frac{x^2 + 2x - 1}{2x^3 + 3x^2 - 2x}\, dx = \int \left(\frac{1}{2}\frac{1}{x} + \frac{1}{5}\frac{1}{2x - 1} - \frac{1}{10}\frac{1}{x + 2} \right) dx$$

- We could check our work by taking the terms to a common denominator and adding them.

$$= \tfrac{1}{2} \ln |x| + \tfrac{1}{10} \ln |2x - 1| - \tfrac{1}{10} \ln |x + 2| + K$$

In integrating the middle term we have made the mental substitution $u = 2x - 1$, which gives $du = 2\, dx$ and $dx = du/2$. ∎

NOTE We can use an alternative method to find the coefficients A, B, and C in Example 2. Equation 4 is an identity; it is true for every value of x. Let's choose val-

■ **www.stewartcalculus.com**
See Additional Examples A, B.

ues of x that simplify the equation. If we put $x = 0$ in Equation 4, then the second and third terms on the right side vanish and the equation then becomes $-2A = -1$, or $A = \frac{1}{2}$. Likewise, $x = \frac{1}{2}$ gives $5B/4 = \frac{1}{4}$ and $x = -2$ gives $10C = -1$, so $B = \frac{1}{5}$ and $C = -\frac{1}{10}$. (You may object that Equation 3 is not valid for $x = 0, \frac{1}{2}$, or -2, so why should Equation 4 be valid for those values? In fact, Equation 4 is true for all values of x, even $x = 0, \frac{1}{2}$, and -2. See Exercise 45 for the reason.)

CASE II $Q(x)$ **is a product of linear factors, some of which are repeated.**

Suppose the first linear factor $(a_1x + b_1)$ is repeated r times; that is, $(a_1x + b_1)^r$ occurs in the factorization of $Q(x)$. Then instead of the single term $A_1/(a_1x + b_1)$ in Equation 2, we would use

$$\boxed{6} \qquad \frac{A_1}{a_1x + b_1} + \frac{A_2}{(a_1x + b_1)^2} + \cdots + \frac{A_r}{(a_1x + b_1)^r}$$

By way of illustration, we could write

$$\frac{x^3 - x + 1}{x^2(x - 1)^3} = \frac{A}{x} + \frac{B}{x^2} + \frac{C}{x - 1} + \frac{D}{(x - 1)^2} + \frac{E}{(x - 1)^3}$$

but we prefer to work out in detail a simpler example.

EXAMPLE 3 Find $\displaystyle\int \frac{x^4 - 2x^2 + 4x + 1}{x^3 - x^2 - x + 1}\,dx$.

SOLUTION The first step is to divide. The result of long division is

$$\frac{x^4 - 2x^2 + 4x + 1}{x^3 - x^2 - x + 1} = x + 1 + \frac{4x}{x^3 - x^2 - x + 1}$$

The second step is to factor the denominator $Q(x) = x^3 - x^2 - x + 1$. Since $Q(1) = 0$, we know that $x - 1$ is a factor and we obtain

$$x^3 - x^2 - x + 1 = (x - 1)(x^2 - 1) = (x - 1)(x - 1)(x + 1)$$
$$= (x - 1)^2(x + 1)$$

Since the linear factor $x - 1$ occurs twice, the partial fraction decomposition is

$$\frac{4x}{(x - 1)^2(x + 1)} = \frac{A}{x - 1} + \frac{B}{(x - 1)^2} + \frac{C}{x + 1}$$

Multiplying by the least common denominator, $(x - 1)^2(x + 1)$, we get

$$\boxed{7} \qquad 4x = A(x - 1)(x + 1) + B(x + 1) + C(x - 1)^2$$
$$= (A + C)x^2 + (B - 2C)x + (-A + B + C)$$

Now we equate coefficients:

$$A \qquad\quad + \;\; C = 0$$
$$B - 2C = 4$$
$$-A + B + \;\; C = 0$$

■ Another method for finding the coefficients:
Put $x = 1$ in $\boxed{7}$: $B = 2$.
Put $x = -1$: $C = -1$.
Put $x = 0$: $A = B + C = 1$.

Solving, we obtain $A = 1$, $B = 2$, and $C = -1$, so

$$\int \frac{x^4 - 2x^2 + 4x + 1}{x^3 - x^2 - x + 1} \, dx = \int \left[x + 1 + \frac{1}{x - 1} + \frac{2}{(x - 1)^2} - \frac{1}{x + 1} \right] dx$$

$$= \frac{x^2}{2} + x + \ln |x - 1| - \frac{2}{x - 1} - \ln |x + 1| + K$$

$$= \frac{x^2}{2} + x - \frac{2}{x - 1} + \ln \left| \frac{x - 1}{x + 1} \right| + K \qquad \blacksquare$$

- Here we use $\ln \dfrac{a}{b} = \ln a - \ln b$.

CASE III $Q(x)$ **contains irreducible quadratic factors, none of which is repeated.**

If $Q(x)$ has the factor $ax^2 + bx + c$, where $b^2 - 4ac < 0$, then, in addition to the partial fractions in Equations 2 and 6, the expression for $R(x)/Q(x)$ will have a term of the form

$$\boxed{8} \qquad \frac{Ax + B}{ax^2 + bx + c}$$

where A and B are constants to be determined. For instance, the function given by $f(x) = x/[(x - 2)(x^2 + 1)(x^2 + 4)]$ has a partial fraction decomposition of the form

$$\frac{x}{(x - 2)(x^2 + 1)(x^2 + 4)} = \frac{A}{x - 2} + \frac{Bx + C}{x^2 + 1} + \frac{Dx + E}{x^2 + 4}$$

The term in $\boxed{8}$ can be integrated by completing the square (if necessary) and using the formula

$$\boxed{9} \qquad \int \frac{dx}{x^2 + a^2} = \frac{1}{a} \tan^{-1} \left(\frac{x}{a} \right) + C$$

▼ EXAMPLE 4 Evaluate $\displaystyle\int \frac{2x^2 - x + 4}{x^3 + 4x} \, dx$.

SOLUTION Since $x^3 + 4x = x(x^2 + 4)$ can't be factored further, we write

$$\frac{2x^2 - x + 4}{x(x^2 + 4)} = \frac{A}{x} + \frac{Bx + C}{x^2 + 4}$$

Multiplying by $x(x^2 + 4)$, we have

$$2x^2 - x + 4 = A(x^2 + 4) + (Bx + C)x$$

$$= (A + B)x^2 + Cx + 4A$$

Equating coefficients, we obtain

$$A + B = 2 \qquad C = -1 \qquad 4A = 4$$

Thus $A = 1$, $B = 1$, and $C = -1$ and so

$$\int \frac{2x^2 - x + 4}{x^3 + 4x} \, dx = \int \left(\frac{1}{x} + \frac{x - 1}{x^2 + 4} \right) dx$$

In order to integrate the second term we split it into two parts:

$$\int \frac{x - 1}{x^2 + 4} \, dx = \int \frac{x}{x^2 + 4} \, dx - \int \frac{1}{x^2 + 4} \, dx$$

We make the substitution $u = x^2 + 4$ in the first of these integrals so that $du = 2x \, dx$. We evaluate the second integral by means of Formula 9 with $a = 2$:

$$\int \frac{2x^2 - x + 4}{x(x^2 + 4)} \, dx = \int \frac{1}{x} \, dx + \int \frac{x}{x^2 + 4} \, dx - \int \frac{1}{x^2 + 4} \, dx$$

$$= \ln|x| + \tfrac{1}{2}\ln(x^2 + 4) - \tfrac{1}{2}\tan^{-1}(x/2) + K \qquad \blacksquare$$

EXAMPLE 5 Evaluate $\displaystyle\int \frac{4x^2 - 3x + 2}{4x^2 - 4x + 3} \, dx.$

SOLUTION Since the degree of the numerator is *not less than* the degree of the denominator, we first divide and obtain

$$\frac{4x^2 - 3x + 2}{4x^2 - 4x + 3} = 1 + \frac{x - 1}{4x^2 - 4x + 3}$$

■ **www.stewartcalculus.com**
We have now learned several techniques for evaluating integrals. How do you know which technique to use on a given integral? For advice, click on *Additional Topics* and then on *Strategy for Integration*.

Notice that the quadratic $4x^2 - 4x + 3$ is irreducible because its discriminant is $b^2 - 4ac = -32 < 0$. This means it can't be factored, so we don't need to use the partial fraction technique.

To integrate the given function we complete the square in the denominator:

$$4x^2 - 4x + 3 = (2x - 1)^2 + 2$$

This suggests that we make the substitution $u = 2x - 1$. Then, $du = 2 \, dx$ and $x = (u + 1)/2$, so

$$\int \frac{4x^2 - 3x + 2}{4x^2 - 4x + 3} \, dx = \int \left(1 + \frac{x - 1}{4x^2 - 4x + 3} \right) dx$$

$$= x + \tfrac{1}{2}\int \frac{\tfrac{1}{2}(u + 1) - 1}{u^2 + 2} \, du = x + \tfrac{1}{4}\int \frac{u - 1}{u^2 + 2} \, du$$

$$= x + \tfrac{1}{4}\int \frac{u}{u^2 + 2} \, du - \tfrac{1}{4}\int \frac{1}{u^2 + 2} \, du$$

$$= x + \tfrac{1}{8}\ln(u^2 + 2) - \frac{1}{4} \cdot \frac{1}{\sqrt{2}} \tan^{-1}\left(\frac{u}{\sqrt{2}} \right) + C$$

$$= x + \tfrac{1}{8}\ln(4x^2 - 4x + 3) - \frac{1}{4\sqrt{2}} \tan^{-1}\left(\frac{2x - 1}{\sqrt{2}} \right) + C \qquad \blacksquare$$

NOTE Example 5 illustrates the general procedure for integrating a partial fraction of the form

$$\frac{Ax + B}{ax^2 + bx + c} \qquad \text{where } b^2 - 4ac < 0$$

We complete the square in the denominator and then make a substitution that brings the integral into the form

$$\int \frac{Cu + D}{u^2 + a^2}\, du = C \int \frac{u}{u^2 + a^2}\, du + D \int \frac{1}{u^2 + a^2}\, du$$

Then the first integral is a logarithm and the second is expressed in terms of $\tan^{-1}$.

CASE IV $Q(x)$ **contains a repeated irreducible quadratic factor.**

If $Q(x)$ has the factor $(ax^2 + bx + c)^r$, where $b^2 - 4ac < 0$, then instead of the single partial fraction $\boxed{8}$, the sum

$$\boxed{10} \qquad \frac{A_1x + B_1}{ax^2 + bx + c} + \frac{A_2x + B_2}{(ax^2 + bx + c)^2} + \cdots + \frac{A_rx + B_r}{(ax^2 + bx + c)^r}$$

occurs in the partial fraction decomposition of $R(x)/Q(x)$. Each of the terms in $\boxed{10}$ can be integrated by completing the square and using a substitution (if necessary).

■ It would be extremely tedious to work out by hand the numerical values of the coefficients in Example 6. Most computer algebra systems, however, can find the numerical values very quickly. For instance, the Maple command

`convert(f, parfrac, x)`

or the Mathematica command

`Apart[f]`

gives the following values:

$A = -1,\quad B = \frac{1}{8},\quad C = D = -1,$

$E = \frac{15}{8},\quad F = -\frac{1}{8},\quad G = H = \frac{3}{4}$

$I = -\frac{1}{2},\quad J = \frac{1}{2}$

EXAMPLE 6 Write out the form of the partial fraction decomposition of the function

$$\frac{x^3 + x^2 + 1}{x(x - 1)(x^2 + x + 1)(x^2 + 1)^3}$$

SOLUTION

$$\frac{x^3 + x^2 + 1}{x(x - 1)(x^2 + x + 1)(x^2 + 1)^3}$$

$$= \frac{A}{x} + \frac{B}{x - 1} + \frac{Cx + D}{x^2 + x + 1} + \frac{Ex + F}{x^2 + 1} + \frac{Gx + H}{(x^2 + 1)^2} + \frac{Ix + J}{(x^2 + 1)^3} \qquad ■$$

EXAMPLE 7 Evaluate $\displaystyle\int \frac{1 - x + 2x^2 - x^3}{x(x^2 + 1)^2}\, dx$.

SOLUTION The form of the partial fraction decomposition is

$$\frac{1 - x + 2x^2 - x^3}{x(x^2 + 1)^2} = \frac{A}{x} + \frac{Bx + C}{x^2 + 1} + \frac{Dx + E}{(x^2 + 1)^2}$$

Multiplying by $x(x^2 + 1)^2$, we have

$$-x^3 + 2x^2 - x + 1 = A(x^2 + 1)^2 + (Bx + C)x(x^2 + 1) + (Dx + E)x$$

$$= A(x^4 + 2x^2 + 1) + B(x^4 + x^2) + C(x^3 + x) + Dx^2 + Ex$$

$$= (A + B)x^4 + Cx^3 + (2A + B + D)x^2 + (C + E)x + A$$

If we equate coefficients, we get the system

$$A + B = 0 \qquad C = -1 \qquad 2A + B + D = 2 \qquad C + E = -1 \qquad A = 1$$

which has the solution $A = 1, B = -1, C = -1, D = 1$, and $E = 0$. Thus

$$\int \frac{1 - x + 2x^2 - x^3}{x(x^2 + 1)^2} \, dx = \int \left(\frac{1}{x} - \frac{x + 1}{x^2 + 1} + \frac{x}{(x^2 + 1)^2} \right) dx$$

$$= \int \frac{dx}{x} - \int \frac{x}{x^2 + 1} \, dx - \int \frac{dx}{x^2 + 1} + \int \frac{x \, dx}{(x^2 + 1)^2}$$

- In the second and fourth terms we made the mental substitution $u = x^2 + 1$.

$$= \ln |x| - \tfrac{1}{2} \ln(x^2 + 1) - \tan^{-1}x - \frac{1}{2(x^2 + 1)} + K \quad \blacksquare$$

6.3 | EXERCISES

1–6 ▪ Write out the form of the partial fraction decomposition of the function (as in Example 6). Do not determine the numerical values of the coefficients.

1. (a) $\dfrac{1 + 6x}{(4x - 3)(2x + 5)}$ 　(b) $\dfrac{10}{5x^2 - 2x^3}$

2. (a) $\dfrac{x}{x^2 + x - 2}$ 　(b) $\dfrac{x^2}{x^2 + x + 2}$

3. (a) $\dfrac{x^4 + 1}{x^5 + 4x^3}$ 　(b) $\dfrac{1}{(x^2 - 9)^2}$

4. (a) $\dfrac{x^4 - 2x^3 + x^2 + 2x - 1}{x^2 - 2x + 1}$ 　(b) $\dfrac{x^2 - 1}{x^3 + x^2 + x}$

5. (a) $\dfrac{x^6}{x^2 - 4}$ 　(b) $\dfrac{x^4}{(x^2 - x + 1)(x^2 + 2)^2}$

6. (a) $\dfrac{t^6 + 1}{t^6 + t^3}$ 　(b) $\dfrac{x^5 + 1}{(x^2 - x)(x^4 + 2x^2 + 1)}$

7–34 ▪ Evaluate the integral.

7. $\displaystyle\int \frac{x^4}{x - 1} \, dx$ 　**8.** $\displaystyle\int \frac{3t - 2}{t + 1} \, dt$

9. $\displaystyle\int \frac{5x + 1}{(2x + 1)(x - 1)} \, dx$ 　**10.** $\displaystyle\int \frac{y}{(y + 4)(2y - 1)} \, dy$

11. $\displaystyle\int_0^1 \frac{2}{2x^2 + 3x + 1} \, dx$ 　**12.** $\displaystyle\int_0^1 \frac{x - 4}{x^2 - 5x + 6} \, dx$

13. $\displaystyle\int \frac{ax}{x^2 - bx} \, dx$ 　**14.** $\displaystyle\int \frac{1}{(x + a)(x + b)} \, dx$

15. $\displaystyle\int_0^1 \frac{2x + 3}{(x + 1)^2} \, dx$ 　**16.** $\displaystyle\int_0^1 \frac{x^3 - 4x - 10}{x^2 - x - 6} \, dx$

17. $\displaystyle\int_1^2 \frac{4y^2 - 7y - 12}{y(y + 2)(y - 3)} \, dy$ 　**18.** $\displaystyle\int \frac{x^2 + 2x - 1}{x^3 - x} \, dx$

19. $\displaystyle\int \frac{x^2 + 1}{(x - 3)(x - 2)^2} \, dx$ 　**20.** $\displaystyle\int \frac{x^2 - 5x + 16}{(2x + 1)(x - 2)^2} \, dx$

21. $\displaystyle\int \frac{x^3 + 4}{x^2 + 4} \, dx$ 　**22.** $\displaystyle\int \frac{x^2 - x + 6}{x^3 + 3x} \, dx$

23. $\displaystyle\int \frac{10}{(x - 1)(x^2 + 9)} \, dx$ 　**24.** $\displaystyle\int \frac{x^2 - 2x - 1}{(x - 1)^2(x^2 + 1)} \, dx$

25. $\displaystyle\int \frac{x^3 + x^2 + 2x + 1}{(x^2 + 1)(x^2 + 2)} \, dx$ 　**26.** $\displaystyle\int \frac{x^3 - 2x^2 + x + 1}{x^4 + 5x^2 + 4} \, dx$

27. $\displaystyle\int \frac{x + 4}{x^2 + 2x + 5} \, dx$ 　**28.** $\displaystyle\int_0^1 \frac{x}{x^2 + 4x + 13} \, dx$

29. $\displaystyle\int \frac{1}{x^3 - 1} \, dx$ 　**30.** $\displaystyle\int \frac{x^5 + x - 1}{x^3 + 1} \, dx$

31. $\displaystyle\int \frac{dx}{x(x^2 + 4)^2}$ 　**32.** $\displaystyle\int \frac{x^4 + 3x^2 + 1}{x^5 + 5x^3 + 5x} \, dx$

33. $\displaystyle\int \frac{x - 3}{(x^2 + 2x + 4)^2} \, dx$ 　**34.** $\displaystyle\int \frac{x^4 + 1}{x(x^2 + 1)^2} \, dx$

35–40 ▪ Make a substitution to express the integrand as a rational function and then evaluate the integral.

35. $\displaystyle\int_9^{16} \frac{\sqrt{x}}{x - 4} \, dx$ 　$\left(\text{Let } u = \sqrt{x}.\right)$

36. $\displaystyle\int_0^1 \frac{1}{1 + \sqrt[3]{x}} \, dx$ 　$\left(\text{Let } u = \sqrt[3]{x}.\right)$

37. $\displaystyle\int \frac{x^3}{\sqrt[3]{x^2 + 1}} \, dx$ 　**38.** $\displaystyle\int_{1/3}^3 \frac{\sqrt{x}}{x^2 + x} \, dx$

39. $\displaystyle\int \frac{e^{2x}}{e^{2x} + 3e^x + 2} \, dx$ 　**40.** $\displaystyle\int \frac{\sin x}{\cos^2 x - 3 \cos x} \, dx$

41–42 ▪ Use integration by parts, together with the techniques of this section, to evaluate the integral.

41. $\int \ln(x^2 - x + 2)\, dx$ **42.** $\int x \tan^{-1}x\, dx$

---

43. One method of slowing the growth of an insect population without using pesticides is to introduce into the population a number of sterile males that mate with fertile females but produce no offspring. If P represents the number of female insects in a population, S the number of sterile males introduced each generation, and r the population's natural growth rate, then the female population is related to time t by

$$t = \int \frac{P + S}{P[(r-1)P - S]}\, dP$$

Suppose an insect population with 10,000 females grows at a rate of $r = 0.10$ and 900 sterile males are added. Evaluate the integral to give an equation relating the female population to time. (Note that the resulting equation can't be solved explicitly for P.)

44. Factor $x^4 + 1$ as a difference of squares by first adding and subtracting the same quantity. Use this factorization to evaluate $\int 1/(x^4 + 1)\, dx$.

45. Suppose that F, G, and Q are polynomials and

$$\frac{F(x)}{Q(x)} = \frac{G(x)}{Q(x)}$$

for all x except when $Q(x) = 0$. Prove that $F(x) = G(x)$ for all x. [*Hint:* Use continuity.]

46. If f is a quadratic function such that $f(0) = 1$ and

$$\int \frac{f(x)}{x^2(x+1)^3}\, dx$$

is a rational function, find the value of $f'(0)$.

47. If $a \neq 0$ and n is a positive integer, find the partial fraction decomposition of

$$f(x) = \frac{1}{x^n(x-a)}$$

[*Hint:* First find the coefficient of $1/(x - a)$. Then subtract the resulting term and simplify what is left.]

6.4 | INTEGRATION WITH TABLES AND COMPUTER ALGEBRA SYSTEMS

In this section we describe how to evaluate integrals using tables and computer algebra systems.

TABLES OF INTEGRALS

Tables of indefinite integrals are very useful when we are confronted by an integral that is difficult to evaluate by hand and we don't have access to a computer algebra system. A relatively brief table of 120 integrals, categorized by form, is provided on the Reference Pages at the back of the book. More extensive tables are available in *CRC Standard Mathematical Tables and Formulae*, 32nd ed, by Daniel Zwillinger (Boca Raton, FL, 2012) (709 entries) or in Gradshteyn and Ryzhik's *Table of Integrals, Series, and Products*, 7e, edited by A. Jefferey and D. Zwillinger (Boston, 2007), which contains hundreds of pages of integrals. It should be remembered, however, that integrals do not often occur in exactly the form listed in a table. Usually we need to use the Substitution Rule or algebraic manipulation to transform a given integral into one of the forms in the table.

▪ The Table of Integrals appears on Reference Pages 6–10 at the back of the book.

V EXAMPLE 1 Use the Table of Integrals to find $\displaystyle\int \frac{x^2}{\sqrt{5 - 4x^2}}\, dx$.

SOLUTION If we look at the section of the table entitled *Forms involving* $\sqrt{a^2 - u^2}$, we see that the closest entry is number 34:

$$\int \frac{u^2}{\sqrt{a^2 - u^2}}\, du = -\frac{u}{2}\sqrt{a^2 - u^2} + \frac{a^2}{2}\sin^{-1}\left(\frac{u}{a}\right) + C$$

This is not exactly what we have, but we will be able to use it if we first make the substitution $u = 2x$:

$$\int \frac{x^2}{\sqrt{5 - 4x^2}}\, dx = \int \frac{(u/2)^2}{\sqrt{5 - u^2}}\frac{du}{2} = \frac{1}{8}\int \frac{u^2}{\sqrt{5 - u^2}}\, du$$

Then we use Formula 34 with $a^2 = 5$ (so $a = \sqrt{5}$):

$$\int \frac{x^2}{\sqrt{5 - 4x^2}}\, dx = \frac{1}{8}\int \frac{u^2}{\sqrt{5 - u^2}}\, du = \frac{1}{8}\left[-\frac{u}{2}\sqrt{5 - u^2} + \frac{5}{2}\sin^{-1}\frac{u}{\sqrt{5}}\right] + C$$

$$= -\frac{x}{8}\sqrt{5 - 4x^2} + \frac{5}{16}\sin^{-1}\left(\frac{2x}{\sqrt{5}}\right) + C \qquad \blacksquare$$

EXAMPLE 2 Use the Table of Integrals to find $\int x^3 \sin x\, dx$.

SOLUTION If we look in the section called *Trigonometric Forms*, we see that none of the entries explicitly includes a u^3 factor. However, we can use the reduction formula in entry 84 with $n = 3$:

$$\int x^3 \sin x\, dx = -x^3 \cos x + 3\int x^2 \cos x\, dx$$

85. $\int u^n \cos u\, du$
$$= u^n \sin u - n\int u^{n-1}\sin u\, du$$

We now need to evaluate $\int x^2 \cos x\, dx$. We can use the reduction formula in entry 85 with $n = 2$, followed by entry 82:

$$\int x^2 \cos x\, dx = x^2 \sin x - 2\int x \sin x\, dx$$

$$= x^2 \sin x - 2(\sin x - x \cos x) + K$$

Combining these calculations, we get

$$\int x^3 \sin x\, dx = -x^3 \cos x + 3x^2 \sin x + 6x \cos x - 6 \sin x + C$$

where $C = 3K$. $\qquad \blacksquare$

▼ EXAMPLE 3 Use the Table of Integrals to find $\int x\sqrt{x^2 + 2x + 4}\, dx$.

SOLUTION Since the table gives forms involving $\sqrt{a^2 + x^2}$, $\sqrt{a^2 - x^2}$, and $\sqrt{x^2 - a^2}$, but not $\sqrt{ax^2 + bx + c}$, we first complete the square:

$$x^2 + 2x + 4 = (x + 1)^2 + 3$$

If we make the substitution $u = x + 1$ (so $x = u - 1$), the integrand will involve the pattern $\sqrt{a^2 + u^2}$:

$$\int x\sqrt{x^2 + 2x + 4} \; dx = \int (u - 1) \sqrt{u^2 + 3} \; du$$

$$= \int u\sqrt{u^2 + 3} \; du - \int \sqrt{u^2 + 3} \; du$$

The first integral is evaluated using the substitution $t = u^2 + 3$:

$$\int u\sqrt{u^2 + 3} \; du = \tfrac{1}{2} \int \sqrt{t} \; dt = \tfrac{1}{2} \cdot \tfrac{2}{3} t^{3/2} = \tfrac{1}{3}(u^2 + 3)^{3/2}$$

21. $\displaystyle\int \sqrt{a^2 + u^2} \; du = \frac{u}{2} \sqrt{a^2 + u^2}$

$\displaystyle + \frac{a^2}{2} \ln(u + \sqrt{a^2 + u^2}) + C$

For the second integral we use Formula 21 with $a = \sqrt{3}$:

$$\int \sqrt{u^2 + 3} \; du = \frac{u}{2} \sqrt{u^2 + 3} + \tfrac{3}{2} \ln(u + \sqrt{u^2 + 3})$$

Thus

$$\int x\sqrt{x^2 + 2x + 4} \; dx$$

$$= \tfrac{1}{3}(x^2 + 2x + 4)^{3/2} - \frac{x + 1}{2} \sqrt{x^2 + 2x + 4} - \tfrac{3}{2} \ln(x + 1 + \sqrt{x^2 + 2x + 4}) + C$$

∎

COMPUTER ALGEBRA SYSTEMS

We have seen that the use of tables involves matching the form of the given integrand with the forms of the integrands in the tables. Computers are particularly good at matching patterns. And just as we used substitutions in conjunction with tables, a CAS can perform substitutions that transform a given integral into one that occurs in its stored formulas. So it isn't surprising that computer algebra systems excel at integration. That doesn't mean that integration by hand is an obsolete skill. We will see that a hand computation sometimes produces an indefinite integral in a form that is more convenient than a machine answer.

To begin, let's see what happens when we ask a machine to integrate the relatively simple function $y = 1/(3x - 2)$. Using the substitution $u = 3x - 2$, an easy calculation by hand gives

$$\int \frac{1}{3x - 2} \; dx = \tfrac{1}{3} \ln|3x - 2| + C$$

whereas Derive, Mathematica, and Maple all return the answer

$$\tfrac{1}{3} \ln(3x - 2)$$

The first thing to notice is that computer algebra systems omit the constant of integration. In other words, they produce a *particular* antiderivative, not the most general one. Therefore, when making use of a machine integration, we might have to add a constant. Second, the absolute value signs are omitted in the machine answer. That is fine if our problem is concerned only with values of x greater than $\tfrac{2}{3}$. But if we are interested in other values of x, then we need to insert the absolute value symbol.

In the next example we reconsider the integral of Example 3, but this time we ask a machine for the answer.

EXAMPLE 4 Use a computer algebra system to find $\int x\sqrt{x^2 + 2x + 4}\, dx$.

SOLUTION Maple responds with the answer

$$\tfrac{1}{3}(x^2 + 2x + 4)^{3/2} - \tfrac{1}{4}(2x + 2)\sqrt{x^2 + 2x + 4} - \frac{3}{2}\operatorname{arcsinh}\frac{\sqrt{3}}{3}(1 + x)$$

This looks different from the answer we found in Example 3, but it is equivalent because the third term can be rewritten using the identity

$$\operatorname{arcsinh} x = \ln\!\left(x + \sqrt{x^2 + 1}\right)$$

■ This is Equation 3.6.3.

Thus

$$\operatorname{arcsinh}\frac{\sqrt{3}}{3}(1 + x) = \ln\!\left[\frac{\sqrt{3}}{3}(1 + x) + \sqrt{\tfrac{1}{3}(1 + x)^2 + 1}\right]$$

$$= \ln\frac{1}{\sqrt{3}}\left[1 + x + \sqrt{(1 + x)^2 + 3}\right]$$

$$= \ln\frac{1}{\sqrt{3}} + \ln\!\left(x + 1 + \sqrt{x^2 + 2x + 4}\right)$$

The resulting extra term $-\tfrac{3}{2}\ln\!\left(1/\sqrt{3}\right)$ can be absorbed into the constant of integration.

Mathematica gives the answer

$$\left(\frac{5}{6} + \frac{x}{6} + \frac{x^2}{3}\right)\sqrt{x^2 + 2x + 4} - \frac{3}{2}\operatorname{arcsinh}\!\left(\frac{1 + x}{\sqrt{3}}\right)$$

Mathematica combined the first two terms of Example 3 (and the Maple result) into a single term by factoring.

Derive gives the answer

$$\tfrac{1}{6}\sqrt{x^2 + 2x + 4}\,(2x^2 + x + 5) - \tfrac{3}{2}\ln\!\left(\sqrt{x^2 + 2x + 4} + x + 1\right)$$

The first term is like the first term in the Mathematica answer, and the second term is identical to the last term in Example 3. ■

EXAMPLE 5 Use a CAS to evaluate $\int x(x^2 + 5)^8\, dx$.

SOLUTION Maple and Mathematica give the same answer:

$$\tfrac{1}{18}x^{18} + \tfrac{5}{2}x^{16} + 50x^{14} + \tfrac{1750}{3}x^{12} + 4375x^{10} + 21875x^8 + \tfrac{218750}{3}x^6 + 156250x^4 + \tfrac{390625}{2}x^2$$

It's clear that both systems must have expanded $(x^2 + 5)^8$ by the Binomial Theorem and then integrated each term.

If we integrate by hand instead, using the substitution $u = x^2 + 5$, we get

■ Derive and the TI-89 and TI-92 also give this answer.

$$\int x(x^2 + 5)^8\, dx = \tfrac{1}{18}(x^2 + 5)^9 + C$$

For most purposes, this is a more convenient form of the answer. ■

EXAMPLE 6 Use a CAS to find $\int \sin^5 x \cos^2 x \, dx$.

SOLUTION In Example 2 in Section 6.2 we found that

$$\boxed{1} \qquad \int \sin^5 x \cos^2 x \, dx = -\tfrac{1}{3} \cos^3 x + \tfrac{2}{5} \cos^5 x - \tfrac{1}{7} \cos^7 x + C$$

Derive and Maple report the answer

$$-\tfrac{1}{7} \sin^4 x \cos^3 x - \tfrac{4}{35} \sin^2 x \cos^3 x - \tfrac{8}{105} \cos^3 x$$

whereas Mathematica produces

$$-\tfrac{5}{64} \cos x - \tfrac{1}{192} \cos 3x + \tfrac{3}{320} \cos 5x - \tfrac{1}{448} \cos 7x$$

We suspect that there are trigonometric identities which show these three answers are equivalent. Indeed, if we ask Derive, Maple, and Mathematica to simplify their expressions using trigonometric identities, they ultimately produce the same form of the answer as in Equation 1. ∎

CAN WE INTEGRATE ALL CONTINUOUS FUNCTIONS?

The question arises: Will our basic integration formulas, together with the Substitution Rule, integration by parts, tables of integrals, and computer algebra systems, enable us to find the integral of every continuous function? In particular, can we use it to evaluate $\int e^{x^2} \, dx$? The answer is No, at least not in terms of the functions that we are familiar with.

Most of the functions that we have been dealing with in this book are what are called **elementary functions**. These are the polynomials, rational functions, power functions (x^a), exponential functions (a^x), logarithmic functions, trigonometric and inverse trigonometric functions, and all functions that can be obtained from these by the five operations of addition, subtraction, multiplication, division, and composition. For instance, the function

$$f(x) = \sqrt{\frac{x^2 - 1}{x^3 + 2x - 1}} + \ln(\cos x) - xe^{\sin 2x}$$

is an elementary function.

If f is an elementary function, then f' is an elementary function but $\int f(x) \, dx$ need not be an elementary function. Consider $f(x) = e^{x^2}$. Since f is continuous, its integral exists, and if we define the function F by

$$F(x) = \int_0^x e^{t^2} \, dt$$

then we know from Part 1 of the Fundamental Theorem of Calculus that

$$F'(x) = e^{x^2}$$

Thus $f(x) = e^{x^2}$ has an antiderivative F, but it can be proved that F is not an elementary function. This means that no matter how hard we try, we will never succeed in evaluating $\int e^{x^2} \, dx$ in terms of the functions we know. (In Chapter 8, however, we will see how to express $\int e^{x^2} \, dx$ as an infinite series.) The same can be said of the following integrals:

$$\int \frac{e^x}{x}\, dx \qquad \int \sin(x^2)\, dx \qquad \int \cos(e^x)\, dx$$

$$\int \sqrt{x^3 + 1}\, dx \qquad \int \frac{1}{\ln x}\, dx \qquad \int \frac{\sin x}{x}\, dx$$

In fact, the majority of elementary functions don't have elementary antiderivatives.

6.4 | EXERCISES

1–22 ▪ Use the Table of Integrals on Reference Pages 6–10 to evaluate the integral.

1. $\displaystyle\int_0^{\pi/8} \arctan 2x\, dx$

2. $\displaystyle\int_0^2 x^2\sqrt{4 - x^2}\, dx$

3. $\displaystyle\int \frac{\cos x}{\sin^2 x - 9}\, dx$

4. $\displaystyle\int \frac{\ln\left(1 + \sqrt{x}\right)}{\sqrt{x}}\, dx$

5. $\displaystyle\int \frac{dx}{x^2\sqrt{4x^2 + 9}}$

6. $\displaystyle\int \frac{\sqrt{2y^2 - 3}}{y^2}\, dy$

7. $\displaystyle\int_0^\pi x^3 \sin x\, dx$

8. $\displaystyle\int \frac{dx}{2x^3 - 3x^2}$

9. $\displaystyle\int \frac{\tan^3(1/z)}{z^2}\, dz$

10. $\displaystyle\int \sin^{-1}\sqrt{x}\, dx$

11. $\displaystyle\int y\sqrt{6 + 4y - 4y^2}\, dy$

12. $\displaystyle\int x \sin(x^2)\, \cos(3x^2)\, dx$

13. $\displaystyle\int \sin^2 x \cos x \ln(\sin x)\, dx$

14. $\displaystyle\int \sin^6 2x\, dx$

15. $\displaystyle\int \frac{e^x}{3 - e^{2x}}\, dx$

16. $\displaystyle\int_0^2 x^3\sqrt{4x^2 - x^4}\, dx$

17. $\displaystyle\int \frac{x^4\, dx}{\sqrt{x^{10} - 2}}$

18. $\displaystyle\int_0^1 x^4 e^{-x}\, dx$

19. $\displaystyle\int \frac{\sqrt{4 + (\ln x)^2}}{x}\, dx$

20. $\displaystyle\int \frac{\sec^2\theta \, \tan^2\theta}{\sqrt{9 - \tan^2\theta}}\, d\theta$

21. $\displaystyle\int \sqrt{e^{2x} - 1}\, dx$

22. $\displaystyle\int e^t \sin(\alpha t - 3)\, dt$

23. Verify Formula 53 in the Table of Integrals (a) by differentiation and (b) by using the substitution $t = a + bu$.

24. Verify Formula 31 (a) by differentiation and (b) by substituting $u = a \sin \theta$.

CAS **25–32** ▪ Use a computer algebra system to evaluate the integral. Compare the answer with the result of using tables. If the answers are not the same, show that they are equivalent.

25. $\displaystyle\int \sec^4 x\, dx$

26. $\displaystyle\int \csc^5 x\, dx$

27. $\displaystyle\int x^2\sqrt{x^2 + 4}\, dx$

28. $\displaystyle\int \frac{dx}{e^x(3e^x + 2)}$

29. $\displaystyle\int \cos^4 x\, dx$

30. $\displaystyle\int x^2\sqrt{1 - x^2}\, dx$

31. $\displaystyle\int \tan^5 x\, dx$

32. $\displaystyle\int \frac{1}{\sqrt{1 + \sqrt[3]{x}}}\, dx$

CAS **33.** (a) Use the table of integrals to evaluate $F(x) = \int f(x)\, dx$, where

$$f(x) = \frac{1}{x\sqrt{1 - x^2}}$$

What is the domain of f and F?

(b) Use a CAS to evaluate $F(x)$. What is the domain of the function F that the CAS produces? Is there a discrepancy between this domain and the domain of the function F that you found in part (a)?

CAS **34.** Computer algebra systems sometimes need a helping hand from human beings. Try to evaluate

$$\int (1 + \ln x)\, \sqrt{1 + (x \ln x)^2}\, dx$$

with a computer algebra system. If it doesn't return an answer, make a substitution that changes the integral into one that the CAS *can* evaluate.

6.5 | APPROXIMATE INTEGRATION

There are two situations in which it is impossible to find the exact value of a definite integral.

The first situation arises from the fact that in order to find $\int_a^b f(x)\,dx$ using the Evaluation Theorem we need to know an antiderivative of f. Sometimes, however, it is difficult, or even impossible, to find an antiderivative (see Section 6.4). For example, it is impossible to evaluate the following integrals exactly:

$$\int_0^1 e^{x^2}\,dx \qquad \int_{-1}^1 \sqrt{1+x^3}\,dx$$

The second situation arises when the function is determined from a scientific experiment through instrument readings or collected data. There may be no formula for the function (see Example 5).

In both cases we need to find approximate values of definite integrals. We already know one such method. Recall that the definite integral is defined as a limit of Riemann sums, so any Riemann sum could be used as an approximation to the integral: If we divide $[a, b]$ into n subintervals of equal length $\Delta x = (b - a)/n$, then we have

$$\int_a^b f(x)\,dx \approx \sum_{i=1}^n f(x_i^*)\,\Delta x$$

where x_i^* is any point in the ith subinterval $[x_{i-1}, x_i]$. If x_i^* is chosen to be the left endpoint of the interval, then $x_i^* = x_{i-1}$ and we have

$$\boxed{1} \qquad \int_a^b f(x)\,dx \approx L_n = \sum_{i=1}^n f(x_{i-1})\,\Delta x$$

If $f(x) \geqslant 0$, then the integral represents an area and $\boxed{1}$ represents an approximation of this area by the rectangles shown in Figure 1(a). If we choose x_i^* to be the right endpoint, then $x_i^* = x_i$ and we have

$$\boxed{2} \qquad \int_a^b f(x)\,dx \approx R_n = \sum_{i=1}^n f(x_i)\,\Delta x$$

[See Figure 1(b).] The approximations L_n and R_n defined by Equations 1 and 2 are called the **left endpoint approximation** and **right endpoint approximation**, respectively.

In Section 5.2 we also considered the case where x_i^* is chosen to be the midpoint $\bar{x}_i$ of the subinterval $[x_{i-1}, x_i]$. Figure 1(c) shows the midpoint approximation M_n, which appears to be better than either L_n or R_n.

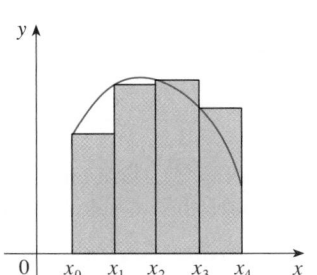

(a) Left endpoint approximation

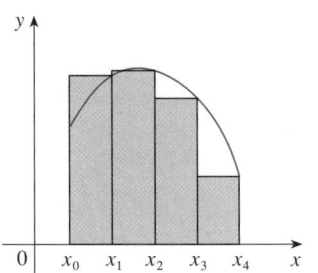

(b) Right endpoint approximation

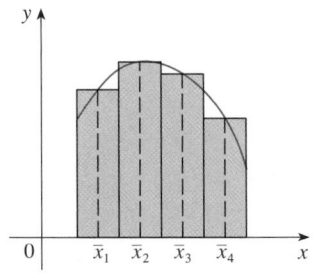

(c) Midpoint approximation

FIGURE 1

MIDPOINT RULE

$$\int_a^b f(x)\,dx \approx M_n = \Delta x\,[f(\bar{x}_1) + f(\bar{x}_2) + \cdots + f(\bar{x}_n)]$$

where

$$\Delta x = \frac{b - a}{n}$$

and

$$\bar{x}_i = \tfrac{1}{2}(x_{i-1} + x_i) = \text{midpoint of } [x_{i-1}, x_i]$$

Another approximation, called the Trapezoidal Rule, results from averaging the approximations in Equations 1 and 2:

$$\int_a^b f(x)\,dx \approx \frac{1}{2}\left[\sum_{i=1}^n f(x_{i-1})\,\Delta x + \sum_{i=1}^n f(x_i)\,\Delta x\right] = \frac{\Delta x}{2}\left[\sum_{i=1}^n \left(f(x_{i-1}) + f(x_i)\right)\right]$$

$$= \frac{\Delta x}{2}\left[\left(f(x_0) + f(x_1)\right) + \left(f(x_1) + f(x_2)\right) + \cdots + \left(f(x_{n-1}) + f(x_n)\right)\right]$$

$$= \frac{\Delta x}{2}\left[f(x_0) + 2f(x_1) + 2f(x_2) + \cdots + 2f(x_{n-1}) + f(x_n)\right]$$

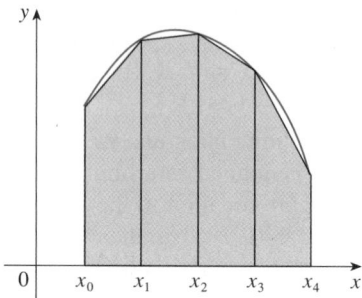

FIGURE 2

Trapezoidal approximation

TRAPEZOIDAL RULE

$$\int_a^b f(x)\,dx \approx T_n = \frac{\Delta x}{2}\left[f(x_0) + 2f(x_1) + 2f(x_2) + \cdots + 2f(x_{n-1}) + f(x_n)\right]$$

where $\Delta x = (b - a)/n$ and $x_i = a + i\,\Delta x$.

The reason for the name Trapezoidal Rule can be seen from Figure 2, which illustrates the case $f(x) \geq 0$. The area of the trapezoid that lies above the ith subinterval is

$$\Delta x\left(\frac{f(x_{i-1}) + f(x_i)}{2}\right) = \frac{\Delta x}{2}\left[f(x_{i-1}) + f(x_i)\right]$$

and if we add the areas of all these trapezoids, we get the right side of the Trapezoidal Rule.

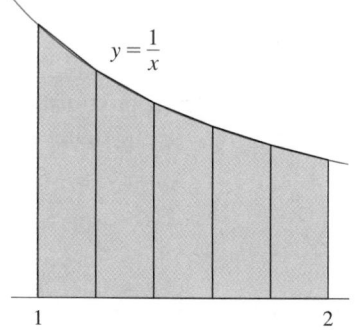

FIGURE 3

EXAMPLE 1 Use (a) the Trapezoidal Rule and (b) the Midpoint Rule with $n = 5$ to approximate the integral $\int_1^2 (1/x)\,dx$.

SOLUTION

(a) With $n = 5$, $a = 1$, and $b = 2$, we have $\Delta x = (2 - 1)/5 = 0.2$, and so the Trapezoidal Rule gives

$$\int_1^2 \frac{1}{x}\,dx \approx T_5 = \frac{0.2}{2}\left[f(1) + 2f(1.2) + 2f(1.4) + 2f(1.6) + 2f(1.8) + f(2)\right]$$

$$= 0.1\left(\frac{1}{1} + \frac{2}{1.2} + \frac{2}{1.4} + \frac{2}{1.6} + \frac{2}{1.8} + \frac{1}{2}\right)$$

$$\approx 0.695635$$

This approximation is illustrated in Figure 3.

(b) The midpoints of the five subintervals are 1.1, 1.3, 1.5, 1.7, and 1.9, so the Midpoint Rule gives

$$\int_1^2 \frac{1}{x}\,dx \approx \Delta x\left[f(1.1) + f(1.3) + f(1.5) + f(1.7) + f(1.9)\right]$$

$$= \frac{1}{5}\left(\frac{1}{1.1} + \frac{1}{1.3} + \frac{1}{1.5} + \frac{1}{1.7} + \frac{1}{1.9}\right)$$

$$\approx 0.691908$$

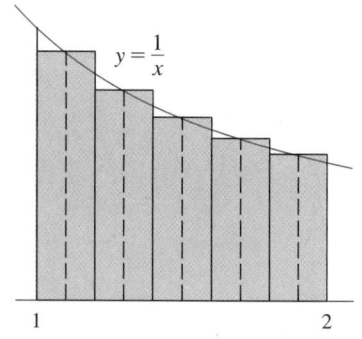

FIGURE 4

This approximation is illustrated in Figure 4. ∎

In Example 1 we deliberately chose an integral whose value can be computed explicitly so that we can see how accurate the Trapezoidal and Midpoint Rules are. By the Fundamental Theorem of Calculus,

$$\int_1^2 \frac{1}{x}\,dx = \ln x\Big]_1^2 = \ln 2 = 0.693147\ldots$$

$\int_a^b f(x)\,dx = \text{approximation} + \text{error}$

The **error** in using an approximation is defined to be the amount that needs to be added to the approximation to make it exact. From the values in Example 1 we see that the errors in the Trapezoidal and Midpoint Rule approximations for $n = 5$ are

$$E_T \approx -0.002488 \qquad \text{and} \qquad E_M \approx 0.001239$$

In general, we have

$$E_T = \int_a^b f(x)\,dx - T_n \qquad \text{and} \qquad E_M = \int_a^b f(x)\,dx - M_n$$

TEC Module 5.2/6.5 allows you to compare approximation methods.

The following tables show the results of calculations similar to those in Example 1, but for $n = 5$, 10, and 20 and for the left and right endpoint approximations as well as the Trapezoidal and Midpoint Rules.

Approximations to $\int_1^2 \frac{1}{x}\,dx$

n	L_n	R_n	T_n	M_n
5	0.745635	0.645635	0.695635	0.691908
10	0.718771	0.668771	0.693771	0.692835
20	0.705803	0.680803	0.693303	0.693069

Corresponding errors

n	E_L	E_R	E_T	E_M
5	−0.052488	0.047512	−0.002488	0.001239
10	−0.025624	0.024376	−0.000624	0.000312
20	−0.012656	0.012344	−0.000156	0.000078

We can make several observations from these tables:

1. In all of the methods we get more accurate approximations when we increase the value of n. (But very large values of n result in so many arithmetic operations that we have to beware of accumulated round-off error.)

2. The errors in the left and right endpoint approximations are opposite in sign and appear to decrease by a factor of about 2 when we double the value of n.

■ It turns out that these observations are true in most cases.

3. The Trapezoidal and Midpoint Rules are much more accurate than the endpoint approximations.

4. The errors in the Trapezoidal and Midpoint Rules are opposite in sign and appear to decrease by a factor of about 4 when we double the value of n.

5. The size of the error in the Midpoint Rule is about half the size of the error in the Trapezoidal Rule.

Figure 5 (on page 344) shows why we can usually expect the Midpoint Rule to be more accurate than the Trapezoidal Rule. The area of a typical rectangle in the Midpoint Rule is the same as the trapezoid $ABCD$ whose upper side is tangent to the graph

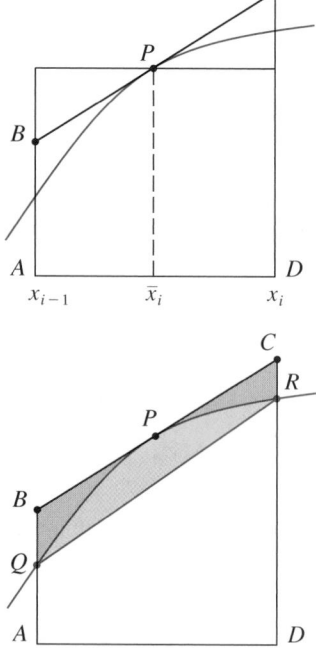

FIGURE 5

- K can be any number larger than all the values of $|f''(x)|$, but smaller values of K give better error bounds.

at P. The area of this trapezoid is closer to the area under the graph than is the area of the trapezoid $AQRD$ used in the Trapezoidal Rule. [The midpoint error (shaded gray) is smaller than the trapezoidal error (shaded blue).]

These observations are corroborated in the following error estimates, which are proved in books on numerical analysis. Notice that Observation 4 corresponds to the n^2 in each denominator because $(2n)^2 = 4n^2$. The fact that the estimates depend on the size of the second derivative is not surprising if you look at Figure 5, because $f''(x)$ measures how much the graph is curved. [Recall that $f''(x)$ measures how fast the slope of $y = f(x)$ changes.]

> **3 ERROR BOUNDS** Suppose $|f''(x)| \leq K$ for $a \leq x \leq b$. If E_T and E_M are the errors in the Trapezoidal and Midpoint Rules, then
>
> $$|E_T| \leq \frac{K(b-a)^3}{12n^2} \qquad \text{and} \qquad |E_M| \leq \frac{K(b-a)^3}{24n^2}$$

Let's apply this error estimate to the Trapezoidal Rule approximation in Example 1. If $f(x) = 1/x$, then $f'(x) = -1/x^2$ and $f''(x) = 2/x^3$. Since $1 \leq x \leq 2$, we have $1/x \leq 1$, so

$$|f''(x)| = \left|\frac{2}{x^3}\right| \leq \frac{2}{1^3} = 2$$

Therefore, taking $K = 2$, $a = 1$, $b = 2$, and $n = 5$ in the error estimate $\boxed{3}$, we see that

$$|E_T| \leq \frac{2(2-1)^3}{12(5)^2} = \frac{1}{150} \approx 0.006667$$

Comparing this error estimate of 0.006667 with the actual error of about 0.002488, we see that it can happen that the actual error is substantially less than the upper bound for the error given by $\boxed{3}$.

V EXAMPLE 2 How large should we take n in order to guarantee that the Trapezoidal and Midpoint Rule approximations for $\int_1^2 (1/x)\, dx$ are accurate to within 0.0001?

SOLUTION We saw in the preceding calculation that $|f''(x)| \leq 2$ for $1 \leq x \leq 2$, so we can take $K = 2$, $a = 1$, and $b = 2$ in $\boxed{3}$. Accuracy to within 0.0001 means that the size of the error should be less than 0.0001. Therefore we choose n so that

$$\frac{2(1)^3}{12n^2} < 0.0001$$

Solving the inequality for n, we get

$$n^2 > \frac{2}{12(0.0001)}$$

- It's quite possible that a lower value for n would suffice, but 41 is the smallest value for which the error bound formula can *guarantee* us accuracy to within 0.0001.

or

$$n > \frac{1}{\sqrt{0.0006}} \approx 40.8$$

Thus $n = 41$ will ensure the desired accuracy.

For the same accuracy with the Midpoint Rule we choose n so that

$$\frac{2(1)^3}{24n^2} < 0.0001$$

which gives

$$n > \frac{1}{\sqrt{0.0012}} \approx 29 \quad\blacksquare$$

▼ EXAMPLE 3
(a) Use the Midpoint Rule with $n = 10$ to approximate the integral $\int_0^1 e^{x^2}\,dx$.
(b) Give an upper bound for the error involved in this approximation.

SOLUTION
(a) Since $a = 0$, $b = 1$, and $n = 10$, the Midpoint Rule gives

$$\int_0^1 e^{x^2}\,dx \approx \Delta x\,[f(0.05) + f(0.15) + \cdots + f(0.85) + f(0.95)]$$

$$= 0.1[e^{0.0025} + e^{0.0225} + e^{0.0625} + e^{0.1225} + e^{0.2025} + e^{0.3025}$$

$$+ e^{0.4225} + e^{0.5625} + e^{0.7225} + e^{0.9025}]$$

$$\approx 1.460393$$

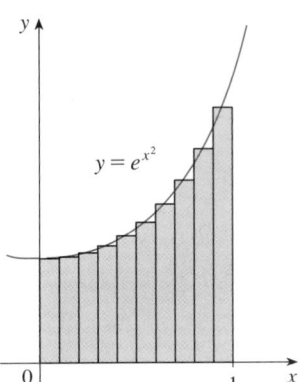

FIGURE 6

Figure 6 illustrates this approximation.

(b) Since $f(x) = e^{x^2}$, we have $f'(x) = 2xe^{x^2}$ and $f''(x) = (2 + 4x^2)e^{x^2}$. Also, since $0 \leqslant x \leqslant 1$, we have $x^2 \leqslant 1$ and so

$$0 \leqslant f''(x) = (2 + 4x^2)e^{x^2} \leqslant 6e$$

■ Error estimates are upper bounds for the error. They give theoretical, worst-case scenarios. The actual error in this case turns out to be about 0.0023.

Taking $K = 6e$, $a = 0$, $b = 1$, and $n = 10$ in the error estimate $\boxed{3}$, we see that an upper bound for the error is

$$\frac{6e(1)^3}{24(10)^2} = \frac{e}{400} \approx 0.007 \quad\blacksquare$$

SIMPSON'S RULE

Another rule for approximate integration results from using parabolas instead of straight line segments to approximate a curve. As before, we divide $[a, b]$ into n sub-intervals of equal length $h = \Delta x = (b - a)/n$, but this time we assume that n is an even number. Then on each consecutive pair of intervals we approximate the curve $y = f(x) \geqslant 0$ by a parabola as shown in Figure 7. If $y_i = f(x_i)$, then $P_i(x_i, y_i)$ is the

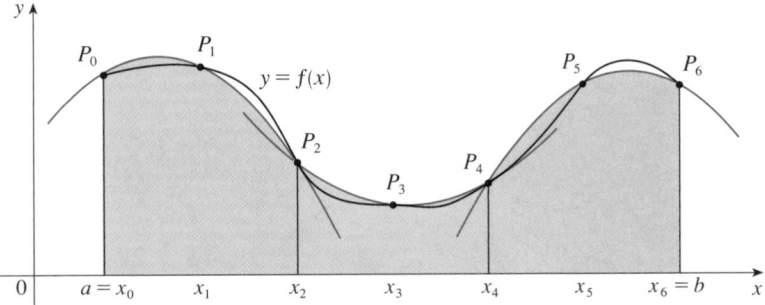

FIGURE 7

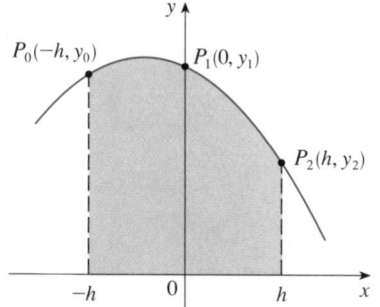

FIGURE 8

■ Here we have used Theorem 5.5.7. Notice that $Ax^2 + C$ is even and Bx is odd.

point on the curve lying above x_i. A typical parabola passes through three consecutive points P_i, P_{i+1}, and P_{i+2}.

To simplify our calculations, we first consider the case where $x_0 = -h$, $x_1 = 0$, and $x_2 = h$. (See Figure 8.) We know that the equation of the parabola through P_0, P_1, and P_2 is of the form $y = Ax^2 + Bx + C$ and so the area under the parabola from $x = -h$ to $x = h$ is

$$\int_{-h}^{h} (Ax^2 + Bx + C)\, dx = 2 \int_{0}^{h} (Ax^2 + C)\, dx$$

$$= 2 \left[A\frac{x^3}{3} + Cx \right]_0^h$$

$$= 2 \left(A\frac{h^3}{3} + Ch \right) = \frac{h}{3}(2Ah^2 + 6C)$$

But, since the parabola passes through $P_0(-h, y_0)$, $P_1(0, y_1)$, and $P_2(h, y_2)$, we have

$$y_0 = A(-h)^2 + B(-h) + C = Ah^2 - Bh + C$$

$$y_1 = C$$

$$y_2 = Ah^2 + Bh + C$$

and therefore
$$y_0 + 4y_1 + y_2 = 2Ah^2 + 6C$$

Thus we can rewrite the area under the parabola as

$$\frac{h}{3}(y_0 + 4y_1 + y_2)$$

By shifting this parabola horizontally we do not change the area under it. This means that the area under the parabola through P_0, P_1, and P_2 from $x = x_0$ to $x = x_2$ in Figure 7 is still

$$\frac{h}{3}(y_0 + 4y_1 + y_2)$$

Similarly, the area under the parabola through P_2, P_3, and P_4 from $x = x_2$ to $x = x_4$ is

$$\frac{h}{3}(y_2 + 4y_3 + y_4)$$

If we compute the areas under all the parabolas in this manner and add the results, we get

$$\int_a^b f(x)\, dx \approx \frac{h}{3}(y_0 + 4y_1 + y_2) + \frac{h}{3}(y_2 + 4y_3 + y_4) + \cdots + \frac{h}{3}(y_{n-2} + 4y_{n-1} + y_n)$$

$$= \frac{h}{3}(y_0 + 4y_1 + 2y_2 + 4y_3 + 2y_4 + \cdots + 2y_{n-2} + 4y_{n-1} + y_n)$$

Although we have derived this approximation for the case in which $f(x) \geqslant 0$, it is a reasonable approximation for any continuous function f and is called Simpson's Rule after the English mathematician Thomas Simpson (1710–1761). Note the pattern of coefficients: 1, 4, 2, 4, 2, 4, 2, ..., 4, 2, 4, 1

■ **SIMPSON**
Thomas Simpson was a weaver who taught himself mathematics and went on to become one of the best English mathematicians of the 18th century. What we call Simpson's Rule was actually known to Cavalieri and Gregory in the 17th century, but Simpson popularized it in his best-selling calculus textbook, entitled *A New Treatise of Fluxions*.

SIMPSON'S RULE

$$\int_a^b f(x)\, dx \approx S_n = \frac{\Delta x}{3}\left[f(x_0) + 4f(x_1) + 2f(x_2) + 4f(x_3) + \cdots \right.$$
$$\left. + 2f(x_{n-2}) + 4f(x_{n-1}) + f(x_n)\right]$$

where n is even and $\Delta x = (b - a)/n$.

EXAMPLE 4 Use Simpson's Rule with $n = 10$ to approximate $\int_1^2 (1/x)\, dx$.

SOLUTION Putting $f(x) = 1/x$, $n = 10$, and $\Delta x = 0.1$ in Simpson's Rule, we obtain

$$\int_1^2 \frac{1}{x}\, dx \approx S_{10}$$

$$= \frac{\Delta x}{3}\left[f(1) + 4f(1.1) + 2f(1.2) + 4f(1.3) + \cdots + 2f(1.8) + 4f(1.9) + f(2)\right]$$

$$= \frac{0.1}{3}\left(\frac{1}{1} + \frac{4}{1.1} + \frac{2}{1.2} + \frac{4}{1.3} + \frac{2}{1.4} + \frac{4}{1.5} + \frac{2}{1.6} + \frac{4}{1.7} + \frac{2}{1.8} + \frac{4}{1.9} + \frac{1}{2}\right)$$

$$\approx 0.693150 \qquad\qquad ■$$

Notice that, in Example 4, Simpson's Rule gives us a *much* better approximation ($S_{10} \approx 0.693150$) to the true value of the integral ($\ln 2 \approx 0.693147\ldots$) than does the Trapezoidal Rule ($T_{10} \approx 0.693771$) or the Midpoint Rule ($M_{10} \approx 0.692835$). It turns out (see Exercise 42) that the approximations in Simpson's Rule are weighted averages of those in the Trapezoidal and Midpoint Rules:

$$S_{2n} = \tfrac{1}{3}T_n + \tfrac{2}{3}M_n$$

(Recall that E_T and E_M usually have opposite signs and $|E_M|$ is about half the size of $|E_T|$.)

In many applications of calculus we need to evaluate an integral even if no explicit formula is known for y as a function of x. A function may be given graphically or as a table of values of collected data. If there is evidence that the values are not changing rapidly, then the Trapezoidal Rule or Simpson's Rule can still be used to find an approximate value for $\int_a^b y\, dx$, the integral of y with respect to x.

V EXAMPLE 5 Figure 9 shows data traffic on the link from the United States to SWITCH, the Swiss education and research network, on February 10, 1998. $D(t)$ is the data throughput, measured in megabits per second (Mb/s). Use Simpson's Rule to estimate the total amount of data transmitted on the link up to noon on that day.

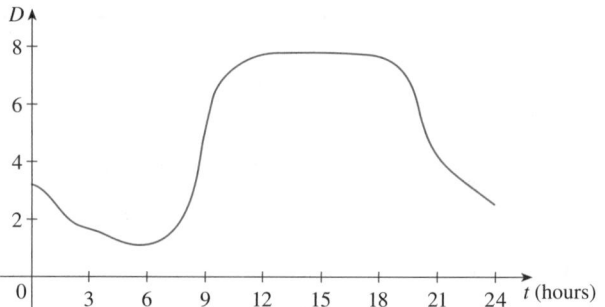

FIGURE 9

SOLUTION Because we want the units to be consistent and $D(t)$ is measured in megabits per second, we convert the units for t from hours to seconds. If we let $A(t)$ be the amount of data (in megabits) transmitted by time t, where t is measured in seconds, then $A'(t) = D(t)$. So, by the Net Change Theorem (see Section 5.3), the total amount of data transmitted by noon (when $t = 12 \times 60^2 = 43{,}200$) is

$$A(43{,}200) = \int_0^{43{,}200} D(t)\, dt$$

We estimate the values of $D(t)$ at hourly intervals from the graph and compile them in the table.

t (hours)	t (seconds)	$D(t)$	t (hours)	t (seconds)	$D(t)$
0	0	3.2	7	25,200	1.3
1	3,600	2.7	8	28,800	2.8
2	7,200	1.9	9	32,400	5.7
3	10,800	1.7	10	36,000	7.1
4	14,400	1.3	11	39,600	7.7
5	18,000	1.0	12	43,200	7.9
6	21,600	1.1			

Then we use Simpson's Rule with $n = 12$ and $\Delta t = 3600$ to estimate the integral:

$$\int_0^{43{,}200} A(t)\, dt \approx \frac{\Delta t}{3}\left[D(0) + 4D(3600) + 2D(7200) + \cdots + 4D(39{,}600) + D(43{,}200)\right]$$

$$\approx \frac{3600}{3}\,[3.2 + 4(2.7) + 2(1.9) + 4(1.7) + 2(1.3) + 4(1.0)$$

$$+\, 2(1.1) + 4(1.3) + 2(2.8) + 4(5.7) + 2(7.1) + 4(7.7) + 7.9]$$

$$= 143{,}880$$

Thus the total amount of data transmitted up to noon is about 144,000 megabits, or 144 gigabits. ∎

n	M_n	S_n
4	0.69121989	0.69315453
8	0.69266055	0.69314765
16	0.69302521	0.69314721

n	E_M	E_S
4	0.00192729	-0.00000735
8	0.00048663	-0.00000047
16	0.00012197	-0.00000003

The table in the margin shows how Simpson's Rule compares with the Midpoint Rule for the integral $\int_1^2 (1/x)\,dx$, whose value is about 0.69314718. The second table shows how the error E_S in Simpson's Rule decreases by a factor of about 16 when n is doubled. (In Exercise 24 you are asked to verify this for another integral.) That is consistent with the appearance of n^4 in the denominator of the following error estimate for Simpson's Rule. It is similar to the estimates given in $\boxed{3}$ for the Trapezoidal and Midpoint Rules, but it uses the fourth derivative of f.

$\boxed{4}$ **ERROR BOUND FOR SIMPSON'S RULE** Suppose that $|f^{(4)}(x)| \le K$ for $a \le x \le b$. If E_S is the error involved in using Simpson's Rule, then

$$|E_S| \le \frac{K(b-a)^5}{180n^4}$$

■ Many calculators and computer algebra systems have a built-in algorithm that computes an approximation of a definite integral. Some of these machines use Simpson's Rule; others use more sophisticated techniques such as *adaptive* numerical integration. This means that if a function fluctuates much more on a certain part of the interval than it does elsewhere, then that part gets divided into more subintervals. This strategy reduces the number of calculations required to achieve a prescribed accuracy.

EXAMPLE 6 How large should we take n in order to guarantee that the Simpson's Rule approximation for $\int_1^2 (1/x)\,dx$ is accurate to within 0.0001?

SOLUTION If $f(x) = 1/x$, then $f^{(4)}(x) = 24/x^5$. Since $x \ge 1$, we have $1/x \le 1$ and so

$$|f^{(4)}(x)| = \left|\frac{24}{x^5}\right| \le 24$$

Therefore we can take $K = 24$ in $\boxed{4}$. Thus for an error less than 0.0001 we should choose n so that

$$\frac{24(1)^5}{180n^4} < 0.0001$$

This gives

$$n^4 > \frac{24}{180(0.0001)}$$

or

$$n > \frac{1}{\sqrt[4]{0.00075}} \approx 6.04$$

Therefore $n = 8$ (n must be even) gives the desired accuracy. (Compare this with Example 2, where we obtained $n = 41$ for the Trapezoidal Rule and $n = 29$ for the Midpoint Rule.) ■

EXAMPLE 7
(a) Use Simpson's Rule with $n = 10$ to approximate the integral $\int_0^1 e^{x^2}\,dx$.
(b) Estimate the error involved in this approximation.

SOLUTION
(a) If $n = 10$, then $\Delta x = 0.1$ and Simpson's Rule gives

$$\int_0^1 e^{x^2}\,dx \approx \frac{\Delta x}{3}\left[f(0) + 4f(0.1) + 2f(0.2) + \cdots + 2f(0.8) + 4f(0.9) + f(1)\right]$$

$$= \frac{0.1}{3}\left[e^0 + 4e^{0.01} + 2e^{0.04} + 4e^{0.09} + 2e^{0.16} + 4e^{0.25} + 2e^{0.36} \right.$$

$$\left. + 4e^{0.49} + 2e^{0.64} + 4e^{0.81} + e^1\right]$$

■ Figure 10 illustrates the calculation in Example 7. Notice that the parabolic arcs are so close to the graph of $y = e^{x^2}$ that they are practically indistinguishable from it.

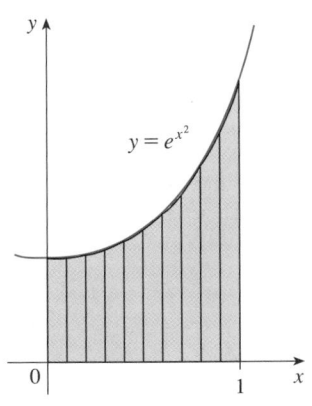

FIGURE 10

≈ 1.462681

(b) The fourth derivative of $f(x) = e^{x^2}$ is

$$f^{(4)}(x) = (12 + 48x^2 + 16x^4)e^{x^2}$$

and so, since $0 \le x \le 1$, we have

$$0 \le f^{(4)}(x) \le (12 + 48 + 16)e^1 = 76e$$

Therefore, putting $K = 76e$, $a = 0$, $b = 1$, and $n = 10$ in $\boxed{4}$, we see that the error is at most

$$\frac{76e(1)^5}{180(10)^4} \approx 0.000115$$

[Compare this with Example 3(b).] Thus, correct to three decimal places, we have

$$\int_0^1 e^{x^2}\,dx \approx 1.463$$ ∎

6.5 | EXERCISES

1. Let $I = \int_0^4 f(x)\,dx$, where f is the function whose graph is shown.
   (a) Use the graph to find L_2, R_2, and M_2.
   (b) Are these underestimates or overestimates of I?
   (c) Use the graph to find T_2. How does it compare with I?
   (d) For any value of n, list the numbers L_n, R_n, M_n, T_n, and I in increasing order.

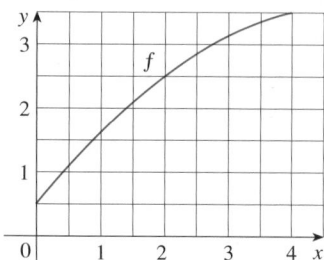

2. The left, right, Trapezoidal, and Midpoint Rule approximations were used to estimate $\int_0^2 f(x)\,dx$, where f is the function whose graph is shown. The estimates were 0.7811, 0.8675, 0.8632, and 0.9540, and the same number of subintervals were used in each case.
   (a) Which rule produced which estimate?
   (b) Between which two approximations does the true value of $\int_0^2 f(x)\,dx$ lie?

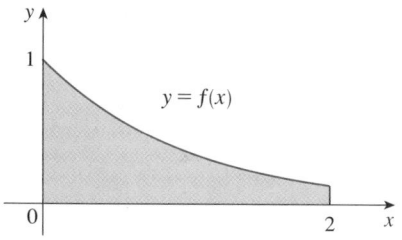

3. Estimate $\int_0^1 \cos(x^2)\,dx$ using (a) the Trapezoidal Rule and (b) the Midpoint Rule, each with $n = 4$. From a graph of the integrand, decide whether your answers are underestimates or overestimates. What can you conclude about the true value of the integral?

4. Draw the graph of $f(x) = \sin\left(\frac{1}{2}x^2\right)$ in the viewing rectangle $[0, 1]$ by $[0, 0.5]$ and let $I = \int_0^1 f(x)\,dx$.
   (a) Use the graph to decide whether L_2, R_2, M_2, and T_2 underestimate or overestimate I.
   (b) For any value of n, list the numbers L_n, R_n, M_n, T_n, and I in increasing order.
   (c) Compute L_5, R_5, M_5, and T_5. From the graph, which do you think gives the best estimate of I?

5–6 ▪ Use (a) the Midpoint Rule and (b) Simpson's Rule to approximate the given integral with the specified value of n. (Round your answers to six decimal places.) Compare your results to the actual value to determine the error in each approximation.

5. $\int_0^2 \dfrac{x}{1+x^2}\,dx$, $\quad n = 10$ 6. $\int_0^\pi x \cos x\,dx$, $\quad n = 4$

7–16 ▪ Use (a) the Trapezoidal Rule, (b) the Midpoint Rule, and (c) Simpson's Rule to approximate the given integral with the specified value of n. (Round your answers to six decimal places.)

7. $\int_1^2 \sqrt{x^3 - 1}\,dx$, $\quad n = 10$ 8. $\int_0^2 \dfrac{1}{1+x^6}\,dx$, $\quad n = 8$

9. $\int_0^2 \dfrac{e^x}{1+x^2}\,dx$, $\quad n = 10$ 10. $\int_0^{\pi/2} \sqrt[3]{1 + \cos x}$, $\quad n = 4$

11. $\int_0^4 e^{\sqrt{t}} \sin t \, dt, \quad n = 8$ **12.** $\int_0^1 \sin(x^3) \, dx, \quad n = 10$

13. $\int_1^5 \frac{\cos x}{x} \, dx, \quad n = 8$

14. $\int_4^6 \ln(x^3 + 2) \, dx, \quad n = 10$

15. $\int_0^3 \frac{1}{1 + y^5} \, dy, \quad n = 6$ **16.** $\int_0^1 \sqrt{z}\, e^{-z} \, dz, \quad n = 10$

17. (a) Find the approximations T_8 and M_8 for the integral $\int_0^1 \cos(x^2) \, dx$.
  (b) Estimate the errors in the approximations of part (a).
  (c) How large do we have to choose n so that the approximations T_n and M_n to the integral in part (a) are accurate to within 0.0001?

18. (a) Find the approximations T_{10} and M_{10} for $\int_1^2 e^{1/x} \, dx$.
  (b) Estimate the errors involved in the approximations of part (a).
  (c) How large do we have to choose n so that the approximations T_n and M_n to the integral in part (a) are accurate to within 0.0001?

19. (a) Find the approximations T_{10}, M_{10}, and S_{10} for $\int_0^\pi \sin x \, dx$ and the corresponding errors E_T, E_M, and E_S.
  (b) Compare the actual errors in part (a) with the error estimates given by $\boxed{3}$ and $\boxed{4}$.
  (c) How large do we have to choose n so that the approximations T_n, M_n, and S_n to the integral in part (a) are accurate to within 0.00001?

20. How large should n be to guarantee that the Simpson's Rule approximation to $\int_0^1 e^{x^2} \, dx$ is accurate to within 0.00001?

CAS 21. The trouble with the error estimates is that it is often very difficult to compute four derivatives and obtain a good upper bound K for $|f^{(4)}(x)|$ by hand. But computer algebra systems have no problem computing $f^{(4)}$ and graphing it, so we can easily find a value for K from a machine graph. This exercise deals with approximations to the integral $I = \int_0^{2\pi} f(x) \, dx$, where $f(x) = e^{\cos x}$.
  (a) Use a graph to get a good upper bound for $|f''(x)|$.
  (b) Use M_{10} to approximate I.
  (c) Use part (a) to estimate the error in part (b).
  (d) Use the built-in numerical integration capability of your CAS to approximate I.
  (e) How does the actual error compare with the error estimate in part (c)?
  (f) Use a graph to get a good upper bound for $|f^{(4)}(x)|$.
  (g) Use S_{10} to approximate I.
  (h) Use part (f) to estimate the error in part (g).
  (i) How does the actual error compare with the error estimate in part (h)?
  (j) How large should n be to guarantee that the size of the error in using S_n is less than 0.0001?

CAS 22. Repeat Exercise 21 for the integral $\int_{-1}^1 \sqrt{4 - x^3} \, dx$.

23. Find the approximations L_n, R_n, T_n, and M_n to the integral $\int_0^1 xe^x \, dx$ for $n = 5$, 10, and 20. Then compute the corresponding errors E_L, E_R, E_T, and E_M. (Round your answers to six decimal places. You may wish to use the sum command on a computer algebra system.) What observations can you make? In particular, what happens to the errors when n is doubled?

24. Find the approximations T_n, M_n, and S_n to the integral $\int_1^4 1/\sqrt{x} \, dx$ for $n = 6$ and 12. Then compute the corresponding errors E_T, E_M, and E_S. (Round your answers to six decimal places. You may wish to use the sum command on a computer algebra system.) What observations can you make? In particular, what happens to the errors when n is doubled?

25. Estimate the area under the graph in the figure by using (a) the Trapezoidal Rule, (b) the Midpoint Rule, and (c) Simpson's Rule, each with $n = 6$.

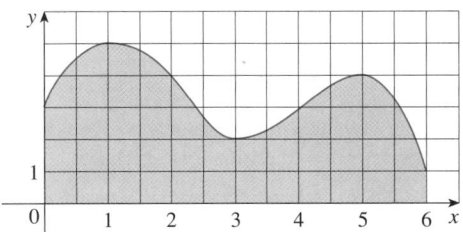

26. A radar gun was used to record the speed of a runner during the first 5 seconds of a race (see the table). Use Simpson's Rule to estimate the distance the runner covered during those 5 seconds.

t (s)	v (m/s)	t (s)	v (m/s)
0	0	3.0	10.51
0.5	4.67	3.5	10.67
1.0	7.34	4.0	10.76
1.5	8.86	4.5	10.81
2.0	9.73	5.0	10.81
2.5	10.22		

27. The graph of the acceleration $a(t)$ of a car measured in ft/s² is shown. Use Simpson's Rule to estimate the increase in the velocity of the car during the 6-second time interval.

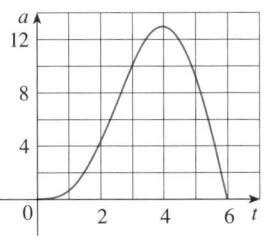

28. Water leaked from a tank at a rate of $r(t)$ liters per hour, where the graph of r is as shown. Use Simpson's Rule to estimate the total amount of water that leaked out during the first six hours.

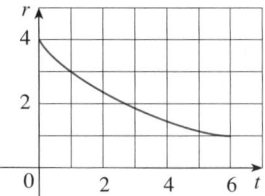

29. A graph of the temperature in New York City on September 19, 2009 is shown. Use Simpson's Rule with $n = 12$ to estimate the average temperature on that day.

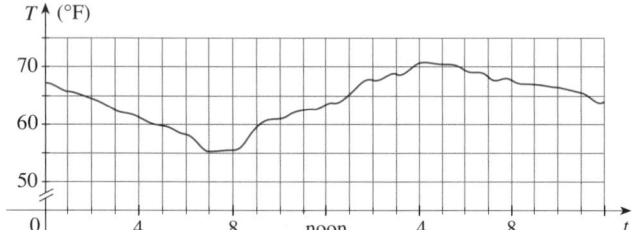

30. (a) A table of values of a function g is given. Use Simpson's Rule to estimate $\int_0^{1.6} g(x)\, dx$.

x	$g(x)$	x	$g(x)$
0.0	12.1	1.0	12.2
0.2	11.6	1.2	12.6
0.4	11.3	1.4	13.0
0.6	11.1	1.6	13.2
0.8	11.7		

(b) If $-5 \le g^{(4)}(x) \le 2$ for $0 \le x \le 1.6$, estimate the error involved in the approximation in part (a).

31. (a) Use the Midpoint Rule and the given data to estimate the value of the integral $\int_1^5 f(x)\, dx$.

x	$f(x)$	x	$f(x)$
1.0	2.4	3.5	4.0
1.5	2.9	4.0	4.1
2.0	3.3	4.5	3.9
2.5	3.6	5.0	3.5
3.0	3.8		

(b) If it is known that $-2 \le f''(x) \le 3$ for all x, estimate the error involved in the approximation in part (a).

32. Shown is the graph of traffic on an Internet service provider's T1 data line from midnight to 8:00 AM. D is the data throughput, measured in megabits per second. Use Simpson's Rule to estimate the total amount of data transmitted during that time period.

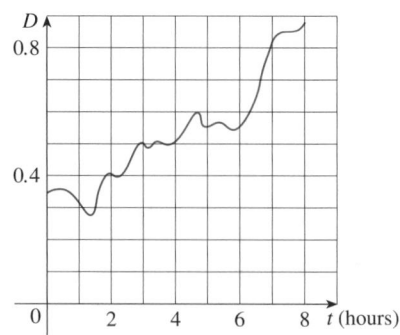

33. The table (supplied by San Diego Gas and Electric) gives the power consumption P in megawatts in San Diego County from midnight to 6:00 AM on a day in December. Use Simpson's Rule to estimate the energy used during that time period. (Use the fact that power is the derivative of energy.)

t	P	t	P
0:00	1814	3:30	1611
0:30	1735	4:00	1621
1:00	1686	4:30	1666
1:30	1646	5:00	1745
2:00	1637	5:30	1886
2:30	1609	6:00	2052
3:00	1604		

CAS **34.** The figure shows a pendulum with length L that makes a maximum angle θ_0 with the vertical. Using Newton's Second Law, it can be shown that the period T (the time for one complete swing) is given by

$$T = 4\sqrt{\frac{L}{g}} \int_0^{\pi/2} \frac{dx}{\sqrt{1 - k^2 \sin^2 x}}$$

where $k = \sin\left(\frac{1}{2}\theta_0\right)$ and g is the acceleration due to gravity. If $L = 1$ m and $\theta_0 = 42°$, use Simpson's Rule with $n = 10$ to find the period.

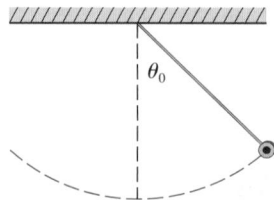

35. The intensity of light with wavelength λ traveling through a diffraction grating with N slits at an angle θ is given by $I(\theta) = N^2 \sin^2 k / k^2$, where $k = (\pi N d \sin \theta)/\lambda$ and d is the distance between adjacent slits. A helium-neon laser with wavelength $\lambda = 632.8 \times 10^{-9}$ m is emitting a narrow band of light, given by $-10^{-6} < \theta < 10^{-6}$, through a grating with 10,000 slits spaced 10^{-4} m apart. Use the Midpoint Rule with $n = 10$ to estimate the total light intensity $\int_{-10^{-6}}^{10^{-6}} I(\theta)\, d\theta$ emerging from the grating.

36. Sketch the graph of a continuous function on $[0, 2]$ for which the right endpoint approximation with $n = 2$ is more accurate than Simpson's Rule.

37. Sketch the graph of a continuous function on $[0, 2]$ for which the Trapezoidal Rule with $n = 2$ is more accurate than the Midpoint Rule.

38. Use the Trapezoidal Rule with $n = 10$ to approximate $\int_0^{20} \cos(\pi x)\, dx$. Compare your result to the actual value. Can you explain the discrepancy?

39. If f is a positive function and $f''(x) < 0$ for $a \le x \le b$, show that

$$T_n < \int_a^b f(x)\, dx < M_n$$

40. Show that if f is a polynomial of degree 3 or lower, then Simpson's Rule gives the exact value of $\int_a^b f(x)\, dx$.

41. Show that $\frac{1}{2}(T_n + M_n) = T_{2n}$.

42. Show that $\frac{1}{3}T_n + \frac{2}{3}M_n = S_{2n}$.

6.6 | IMPROPER INTEGRALS

In defining a definite integral $\int_a^b f(x)\, dx$ we dealt with a function f defined on a finite interval $[a, b]$. In this section we extend the concept of a definite integral to the case where the interval is infinite and also to the case where f has an infinite discontinuity in $[a, b]$. In either case the integral is called an *improper integral*.

TYPE 1: INFINITE INTERVALS

Consider the infinite region S that lies under the curve $y = 1/x^2$, above the x-axis, and to the right of the line $x = 1$. You might think that, since S is infinite in extent, its area must be infinite, but let's take a closer look. The area of the part of S that lies to the left of the line $x = t$ (shaded in Figure 1) is

$$A(t) = \int_1^t \frac{1}{x^2}\, dx = -\frac{1}{x}\Big]_1^t = 1 - \frac{1}{t}$$

FIGURE 1

Notice that $A(t) < 1$ no matter how large t is chosen.

We also observe that

$$\lim_{t \to \infty} A(t) = \lim_{t \to \infty} \left(1 - \frac{1}{t}\right) = 1$$

The area of the shaded region approaches 1 as $t \to \infty$ (see Figure 2), so we say that the area of the infinite region S is equal to 1 and we write

$$\int_1^\infty \frac{1}{x^2}\, dx = \lim_{t \to \infty} \int_1^t \frac{1}{x^2}\, dx = 1$$

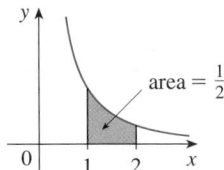
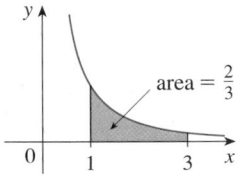
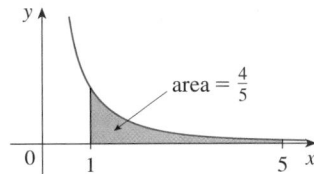
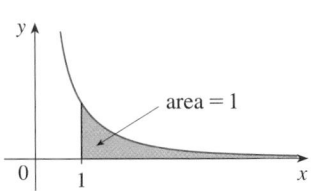

FIGURE 2

Using this example as a guide, we define the integral of f (not necessarily a positive function) over an infinite interval as the limit of integrals over finite intervals.

---

1 DEFINITION OF AN IMPROPER INTEGRAL OF TYPE 1

(a) If $\int_a^t f(x)\, dx$ exists for every number $t \geq a$, then

$$\int_a^\infty f(x)\, dx = \lim_{t \to \infty} \int_a^t f(x)\, dx$$

provided this limit exists (as a finite number).

(b) If $\int_t^b f(x)\, dx$ exists for every number $t \leq b$, then

$$\int_{-\infty}^b f(x)\, dx = \lim_{t \to -\infty} \int_t^b f(x)\, dx$$

provided this limit exists (as a finite number).

The improper integrals $\int_a^\infty f(x)\, dx$ and $\int_{-\infty}^b f(x)\, dx$ are called **convergent** if the corresponding limit exists and **divergent** if the limit does not exist.

(c) If both $\int_a^\infty f(x)\, dx$ and $\int_{-\infty}^a f(x)\, dx$ are convergent, then we define

$$\int_{-\infty}^\infty f(x)\, dx = \int_{-\infty}^a f(x)\, dx + \int_a^\infty f(x)\, dx$$

In part (c) any real number a can be used (see Exercise 52).

Any of the improper integrals in Definition 1 can be interpreted as an area provided that f is a positive function. For instance, in case (a) if $f(x) \geq 0$ and the integral $\int_a^\infty f(x)\, dx$ is convergent, then we define the area of the region

$$S = \{(x, y) \mid x \geq a,\ 0 \leq y \leq f(x)\}$$

in Figure 3 to be

$$A(S) = \int_a^\infty f(x)\, dx$$

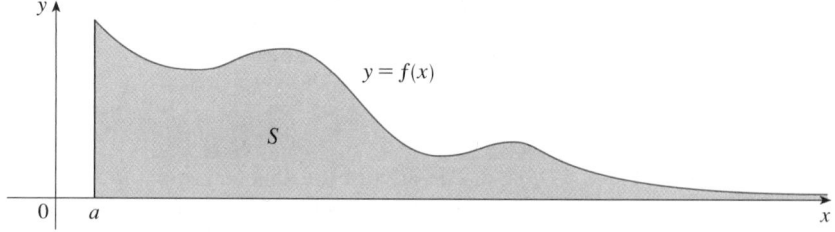

FIGURE 3

This is appropriate because $\int_a^\infty f(x)\, dx$ is the limit as $t \to \infty$ of the area under the graph of f from a to t.

V EXAMPLE 1 Determine whether the integral $\int_1^\infty (1/x)\, dx$ is convergent or divergent.

SOLUTION According to part (a) of Definition 1, we have

$$\int_1^\infty \frac{1}{x}\, dx = \lim_{t \to \infty} \int_1^t \frac{1}{x}\, dx = \lim_{t \to \infty} \ln|x|\,\big]_1^t$$

$$= \lim_{t \to \infty} (\ln t - \ln 1) = \lim_{t \to \infty} \ln t = \infty$$

The limit does not exist as a finite number and so the improper integral $\int_1^\infty (1/x)\,dx$ is divergent. ∎

Let's compare the result of Example 1 with the example given at the beginning of this section:

$$\int_1^\infty \frac{1}{x^2}\,dx \text{ converges} \qquad \int_1^\infty \frac{1}{x}\,dx \text{ diverges}$$

Geometrically, this says that although the curves $y = 1/x^2$ and $y = 1/x$ look very similar for $x > 0$, the region under $y = 1/x^2$ to the right of $x = 1$ (the shaded region in Figure 4) has finite area whereas the corresponding region under $y = 1/x$ (in Figure 5) has infinite area. Note that both $1/x^2$ and $1/x$ approach 0 as $x \to \infty$ but $1/x^2$ approaches 0 faster than $1/x$. The values of $1/x$ don't decrease fast enough for its integral to have a finite value.

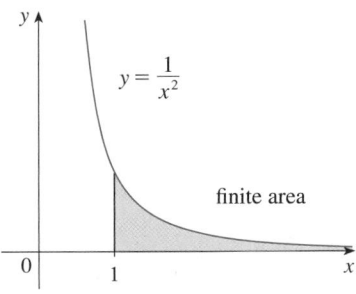

FIGURE 4 $\int_1^\infty (1/x^2)\,dx$ converges.

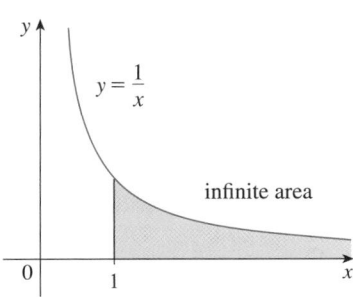

FIGURE 5 $\int_1^\infty (1/x)\,dx$ diverges.

EXAMPLE 2 Evaluate $\int_{-\infty}^0 xe^x\,dx$.

SOLUTION Using part (b) of Definition 1, we have

$$\int_{-\infty}^0 xe^x\,dx = \lim_{t \to -\infty} \int_t^0 xe^x\,dx$$

We integrate by parts with $u = x$, $dv = e^x\,dx$, so that $du = dx$, $v = e^x$:

$$\int_t^0 xe^x\,dx = xe^x\Big]_t^0 - \int_t^0 e^x\,dx = -te^t - 1 + e^t$$

We know that $e^t \to 0$ as $t \to -\infty$, and by l'Hospital's Rule we have

$$\lim_{t \to -\infty} te^t = \lim_{t \to -\infty} \frac{t}{e^{-t}} = \lim_{t \to -\infty} \frac{1}{-e^{-t}} = \lim_{t \to -\infty} (-e^t) = 0$$

Therefore

$$\int_{-\infty}^0 xe^x\,dx = \lim_{t \to -\infty} (-te^t - 1 + e^t) = -0 - 1 + 0 = -1 \qquad ∎$$

EXAMPLE 3 Evaluate $\int_{-\infty}^\infty \frac{1}{1 + x^2}\,dx$.

SOLUTION It's convenient to choose $a = 0$ in Definition 1(c):

$$\int_{-\infty}^\infty \frac{1}{1 + x^2}\,dx = \int_{-\infty}^0 \frac{1}{1 + x^2}\,dx + \int_0^\infty \frac{1}{1 + x^2}\,dx$$

We must now evaluate the integrals on the right side separately:

$$\int_0^\infty \frac{1}{1+x^2}\,dx = \lim_{t\to\infty} \int_0^t \frac{dx}{1+x^2} = \lim_{t\to\infty} \tan^{-1}x \Big]_0^t$$

$$= \lim_{t\to\infty}(\tan^{-1}t - \tan^{-1}0) = \lim_{t\to\infty}\tan^{-1}t = \frac{\pi}{2}$$

$$\int_{-\infty}^0 \frac{1}{1+x^2}\,dx = \lim_{t\to-\infty} \int_t^0 \frac{dx}{1+x^2} = \lim_{t\to-\infty} \tan^{-1}x \Big]_t^0$$

$$= \lim_{t\to-\infty}(\tan^{-1}0 - \tan^{-1}t)$$

$$= 0 - \left(-\frac{\pi}{2}\right) = \frac{\pi}{2}$$

Since both of these integrals are convergent, the given integral is convergent and

$$\int_{-\infty}^\infty \frac{1}{1+x^2}\,dx = \frac{\pi}{2} + \frac{\pi}{2} = \pi$$

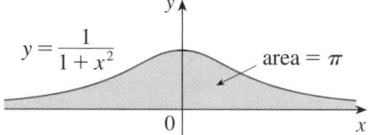

Since $1/(1+x^2) > 0$, the given improper integral can be interpreted as the area of the infinite region that lies under the curve $y = 1/(1+x^2)$ and above the x-axis (see Figure 6).

FIGURE 6

EXAMPLE 4 For what values of p is the following integral convergent?

$$\int_1^\infty \frac{1}{x^p}\,dx$$

SOLUTION We know from Example 1 that if $p = 1$, then the integral is divergent, so let's assume that $p \neq 1$. Then

$$\int_1^\infty \frac{1}{x^p}\,dx = \lim_{t\to\infty} \int_1^t x^{-p}\,dx = \lim_{t\to\infty} \frac{x^{-p+1}}{-p+1}\Bigg]_{x=1}^{x=t}$$

$$= \lim_{t\to\infty} \frac{1}{1-p}\left[\frac{1}{t^{p-1}} - 1\right]$$

If $p > 1$, then $p - 1 > 0$, so as $t \to \infty$, $t^{p-1} \to \infty$ and $1/t^{p-1} \to 0$. Therefore

$$\int_1^\infty \frac{1}{x^p}\,dx = \frac{1}{p-1} \qquad \text{if } p > 1$$

and so the integral converges. But if $p < 1$, then $p - 1 < 0$ and so

$$\frac{1}{t^{p-1}} = t^{1-p} \to \infty \qquad \text{as } t \to \infty$$

and the integral diverges.

We summarize the result of Example 4 for future reference:

2     $\displaystyle\int_1^\infty \frac{1}{x^p}\,dx$ is convergent if $p > 1$ and divergent if $p \le 1$.

TYPE 2: DISCONTINUOUS INTEGRANDS

Suppose that f is a positive continuous function defined on a finite interval $[a, b)$ but has a vertical asymptote at b. Let S be the unbounded region under the graph of f and above the x-axis between a and b. (For Type 1 integrals, the regions extended indefinitely in a horizontal direction. Here the region is infinite in a vertical direction.) The area of the part of S between a and t (the shaded region in Figure 7) is

$$A(t) = \int_a^t f(x)\,dx$$

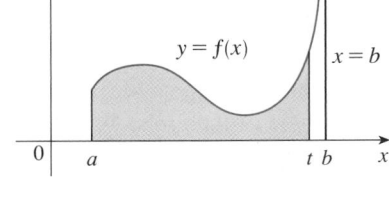

FIGURE 7

If it happens that $A(t)$ approaches a definite number A as $t \to b^-$, then we say that the area of the region S is A and we write

$$\int_a^b f(x)\,dx = \lim_{t \to b^-} \int_a^t f(x)\,dx$$

We use this equation to define an improper integral of Type 2 even when f is not a positive function, no matter what type of discontinuity f has at b.

---

3   **DEFINITION OF AN IMPROPER INTEGRAL OF TYPE 2**

(a) If f is continuous on $[a, b)$ and is discontinuous at b, then

$$\int_a^b f(x)\,dx = \lim_{t \to b^-} \int_a^t f(x)\,dx$$

    if this limit exists (as a finite number).

(b) If f is continuous on $(a, b]$ and is discontinuous at a, then

$$\int_a^b f(x)\,dx = \lim_{t \to a^+} \int_t^b f(x)\,dx$$

    if this limit exists (as a finite number).

The improper integral $\int_a^b f(x)\,dx$ is called **convergent** if the corresponding limit exists and **divergent** if the limit does not exist.

(c) If f has a discontinuity at c, where $a < c < b$, and both $\int_a^c f(x)\,dx$ and $\int_c^b f(x)\,dx$ are convergent, then we define

$$\int_a^b f(x)\,dx = \int_a^c f(x)\,dx + \int_c^b f(x)\,dx$$

▪ Parts (b) and (c) of Definition 3 are illustrated in Figures 8 and 9 for the case where $f(x) \ge 0$ and f has vertical asymptotes at a and c, respectively.

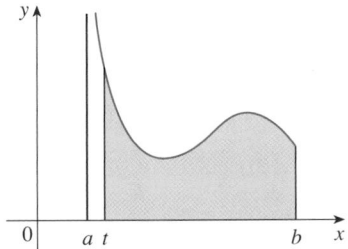

FIGURE 8

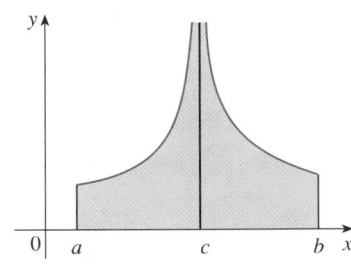

FIGURE 9

EXAMPLE 5 Find $\displaystyle\int_2^5 \frac{1}{\sqrt{x-2}}\,dx$.

SOLUTION We note first that the given integral is improper because $f(x) = 1/\sqrt{x-2}$ has the vertical asymptote $x = 2$. Since the infinite discontinuity

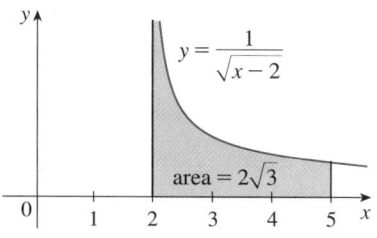

FIGURE 10

occurs at the left endpoint of $[2, 5]$, we use part (b) of Definition 3:

$$\int_2^5 \frac{dx}{\sqrt{x-2}} = \lim_{t \to 2^+} \int_t^5 \frac{dx}{\sqrt{x-2}} = \lim_{t \to 2^+} 2\sqrt{x-2} \Big]_t^5$$

$$= \lim_{t \to 2^+} 2\left(\sqrt{3} - \sqrt{t-2}\right) = 2\sqrt{3}$$

Thus the given improper integral is convergent and, since the integrand is positive, we can interpret the value of the integral as the area of the shaded region in Figure 10. ∎

V **EXAMPLE 6** Determine whether $\int_0^{\pi/2} \sec x \, dx$ converges or diverges.

SOLUTION Note that the given integral is improper because $\lim_{x \to (\pi/2)^-} \sec x = \infty$. Using part (a) of Definition 3 and Formula 14 from the Table of Integrals, we have

$$\int_0^{\pi/2} \sec x \, dx = \lim_{t \to (\pi/2)^-} \int_0^t \sec x \, dx$$

$$= \lim_{t \to (\pi/2)^-} \ln |\sec x + \tan x| \Big]_0^t$$

$$= \lim_{t \to (\pi/2)^-} [\ln(\sec t + \tan t) - \ln 1] = \infty$$

because $\sec t \to \infty$ and $\tan t \to \infty$ as $t \to (\pi/2)^-$. Thus the given improper integral is divergent. ∎

EXAMPLE 7 Evaluate $\int_0^3 \frac{dx}{x-1}$ if possible.

SOLUTION Observe that the line $x = 1$ is a vertical asymptote of the integrand. Since it occurs in the middle of the interval $[0, 3]$, we must use part (c) of Definition 3 with $c = 1$:

$$\int_0^3 \frac{dx}{x-1} = \int_0^1 \frac{dx}{x-1} + \int_1^3 \frac{dx}{x-1}$$

■ www.stewartcalculus.com
See Additional Example A.

where

$$\int_0^1 \frac{dx}{x-1} = \lim_{t \to 1^-} \int_0^t \frac{dx}{x-1} = \lim_{t \to 1^-} \ln |x-1| \Big]_0^t$$

$$= \lim_{t \to 1^-} \left(\ln |t-1| - \ln |-1|\right)$$

$$= \lim_{t \to 1^-} \ln(1-t) = -\infty$$

because $1 - t \to 0^+$ as $t \to 1^-$. Thus $\int_0^1 dx/(x-1)$ is divergent. This implies that $\int_0^3 dx/(x-1)$ is divergent. [We do not need to evaluate $\int_1^3 dx/(x-1)$.] ∎

⊘   **WARNING** If we had not noticed the asymptote $x = 1$ in Example 7 and had instead confused the integral with an ordinary integral, then we might have made the following erroneous calculation:

$$\int_0^3 \frac{dx}{x-1} = \ln |x-1| \Big]_0^3 = \ln 2 - \ln 1 = \ln 2$$

This is wrong because the integral is improper and must be calculated in terms of limits.

From now on, whenever you meet the symbol $\int_a^b f(x)\,dx$ you must decide, by look-ing at the function f on $[a, b]$, whether it is an ordinary definite integral or an improper integral.

A COMPARISON TEST FOR IMPROPER INTEGRALS

Sometimes it is impossible to find the exact value of an improper integral and yet it is important to know whether it is convergent or divergent. In such cases the follow-ing theorem is useful. Although we state it for Type 1 integrals, a similar theorem is true for Type 2 integrals.

COMPARISON THEOREM   Suppose that f and g are continuous functions with $f(x) \geq g(x) \geq 0$ for $x \geq a$.

(a) If $\int_a^\infty f(x)\,dx$ is convergent, then $\int_a^\infty g(x)\,dx$ is convergent.

(b) If $\int_a^\infty g(x)\,dx$ is divergent, then $\int_a^\infty f(x)\,dx$ is divergent.

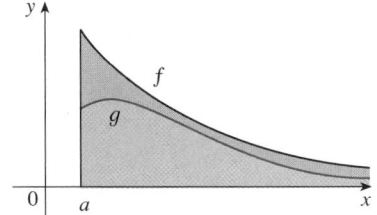

FIGURE 11

We omit the proof of the Comparison Theorem, but Figure 11 makes it seem plau-sible. If the area under the top curve $y = f(x)$ is finite, then so is the area under the bottom curve $y = g(x)$. And if the area under $y = g(x)$ is infinite, then so is the area under $y = f(x)$. [Note that the reverse is not necessarily true: If $\int_a^\infty g(x)\,dx$ is conver-gent, $\int_a^\infty f(x)\,dx$ may or may not be convergent, and if $\int_a^\infty f(x)\,dx$ is divergent, $\int_a^\infty g(x)\,dx$ may or may not be divergent.]

V EXAMPLE 8   Show that $\displaystyle\int_0^\infty e^{-x^2}dx$ is convergent.

SOLUTION   We can't evaluate the integral directly because the antiderivative of e^{-x^2} is not an elementary function (as explained in Section 6.4). We write

$$\int_0^\infty e^{-x^2}dx = \int_0^1 e^{-x^2}dx + \int_1^\infty e^{-x^2}dx$$

and observe that the first integral on the right-hand side is just an ordinary definite integral. In the second integral we use the fact that for $x \geq 1$ we have $x^2 \geq x$, so $-x^2 \leq -x$ and therefore $e^{-x^2} \leq e^{-x}$. (See Figure 12.) The integral of e^{-x} is easy to evaluate:

$$\int_1^\infty e^{-x}dx = \lim_{t\to\infty}\int_1^t e^{-x}dx = \lim_{t\to\infty}(e^{-1} - e^{-t}) = e^{-1}$$

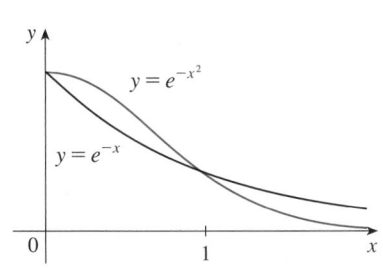

FIGURE 12

Thus, taking $f(x) = e^{-x}$ and $g(x) = e^{-x^2}$ in the Comparison Theorem, we see that $\int_1^\infty e^{-x^2}dx$ is convergent. It follows that $\int_0^\infty e^{-x^2}dx$ is convergent.   ∎

In Example 8 we showed that $\int_0^\infty e^{-x^2}dx$ is convergent without computing its value. In Exercise 58 we indicate how to show that its value is approximately 0.8862. In probability theory it is important to know the exact value of this improper integral; using the methods of multivariable calculus it can be shown that the exact value is $\sqrt{\pi}/2$. Table 1 illustrates the definition of an improper integral by showing how the (computer-generated) values of $\int_0^t e^{-x^2}dx$ approach $\sqrt{\pi}/2$ as t becomes large. In fact, these values converge quite quickly because $e^{-x^2} \to 0$ very rapidly as $x \to \infty$.

TABLE 1

t	$\int_0^t e^{-x^2}dx$
1	0.7468241328
2	0.8820813908
3	0.8862073483
4	0.8862269118
5	0.8862269255
6	0.8862269255

TABLE 2

t	$\int_1^t [(1 + e^{-x})/x]\, dx$
2	0.8636306042
5	1.8276735512
10	2.5219648704
100	4.8245541204
1000	7.1271392134
10000	9.4297243064

EXAMPLE 9 The integral $\displaystyle\int_1^\infty \frac{1 + e^{-x}}{x}\, dx$ is divergent by the Comparison Theorem because

$$\frac{1 + e^{-x}}{x} > \frac{1}{x}$$

and $\int_1^\infty (1/x)\, dx$ is divergent by Example 1 [or by $\boxed{2}$ with $p = 1$]. ∎

Table 2 illustrates the divergence of the integral in Example 9. It appears that the values are not approaching any fixed number.

6.6 | EXERCISES

1. Explain why each of the following integrals is improper.

(a) $\displaystyle\int_1^2 \frac{x}{x - 1}\, dx$ (b) $\displaystyle\int_0^\infty \frac{1}{1 + x^3}\, dx$

(c) $\displaystyle\int_{-\infty}^\infty x^2 e^{-x^2}\, dx$ (d) $\displaystyle\int_0^{\pi/4} \cot x\, dx$

2. Which of the following integrals are improper? Why?

(a) $\displaystyle\int_0^{\pi/4} \tan x\, dx$ (b) $\displaystyle\int_0^\pi \tan x\, dx$

(c) $\displaystyle\int_{-1}^1 \frac{dx}{x^2 - x - 2}$ (d) $\displaystyle\int_0^\infty e^{-x^3}\, dx$

3. Find the area under the curve $y = 1/x^3$ from $x = 1$ to $x = t$ and evaluate it for $t = 10$, 100, and 1000. Then find the total area under this curve for $x \geq 1$.

4. (a) Graph the functions $f(x) = 1/x^{1.1}$ and $g(x) = 1/x^{0.9}$ in the viewing rectangles $[0, 10]$ by $[0, 1]$ and $[0, 100]$ by $[0, 1]$.

(b) Find the areas under the graphs of f and g from $x = 1$ to $x = t$ and evaluate for $t = 10$, 100, 10^4, 10^6, 10^{10}, and 10^{20}.

(c) Find the total area under each curve for $x \geq 1$, if it exists.

5–32 ▪ Determine whether each integral is convergent or divergent. Evaluate those that are convergent.

5. $\displaystyle\int_3^\infty \frac{1}{(x - 2)^{3/2}}\, dx$ **6.** $\displaystyle\int_0^\infty \frac{1}{\sqrt[4]{1 + x}}\, dx$

7. $\displaystyle\int_{-\infty}^0 \frac{1}{3 - 4x}\, dx$ **8.** $\displaystyle\int_1^\infty \frac{1}{(2x + 1)^3}\, dx$

9. $\displaystyle\int_2^\infty e^{-5p}\, dp$ **10.** $\displaystyle\int_{-\infty}^0 2^r\, dr$

11. $\displaystyle\int_0^\infty \frac{x^2}{\sqrt{1 + x^3}}\, dx$ **12.** $\displaystyle\int_{-\infty}^\infty (y^3 - 3y^2)\, dy$

13. $\displaystyle\int_{-\infty}^\infty x e^{-x^2}\, dx$ **14.** $\displaystyle\int_{-\infty}^\infty x^2 e^{-x^3}\, dx$

15. $\displaystyle\int_{-\infty}^0 z e^{2z}\, dz$ **16.** $\displaystyle\int_{-\infty}^\infty \cos \pi t\, dt$

17. $\displaystyle\int_1^\infty \frac{\ln x}{x}\, dx$ **18.** $\displaystyle\int_{-\infty}^6 r e^{r/3}\, dr$

19. $\displaystyle\int_1^\infty \frac{1}{x^2 + x}\, dx$ **20.** $\displaystyle\int_1^\infty \frac{\ln x}{x^3}\, dx$

21. $\displaystyle\int_{-\infty}^\infty \frac{x^2}{9 + x^6}\, dx$ **22.** $\displaystyle\int_0^\infty \frac{e^x}{e^{2x} + 3}\, dx$

23. $\displaystyle\int_0^1 \frac{3}{x^5}\, dx$ **24.** $\displaystyle\int_2^3 \frac{1}{\sqrt{3 - x}}\, dx$

25. $\displaystyle\int_{-2}^{14} \frac{dx}{\sqrt[4]{x + 2}}$ **26.** $\displaystyle\int_6^8 \frac{4}{(x - 6)^3}\, dx$

27. $\displaystyle\int_0^9 \frac{1}{\sqrt[3]{x - 1}}\, dx$ **28.** $\displaystyle\int_0^5 \frac{w}{w - 2}\, dw$

29. $\displaystyle\int_{-1}^1 \frac{e^x}{e^x - 1}\, dx$ **30.** $\displaystyle\int_0^1 \frac{dx}{\sqrt{1 - x^2}}$

31. $\displaystyle\int_0^2 z^2 \ln z\, dz$ **32.** $\displaystyle\int_0^1 \frac{\ln x}{\sqrt{x}}\, dx$

33–38 ▪ Sketch the region and find its area (if the area is finite).

33. $S = \{(x, y) \mid x \geq 1,\ 0 \leq y \leq e^{-x}\}$

34. $S = \{(x, y) \mid x \leq 0,\ 0 \leq y \leq e^x\}$

35. $S = \{(x, y) \mid x \geq 1,\ 0 \leq y \leq 1/(x^3 + x)\}$

36. $S = \{(x, y) \mid x \geq 0,\ 0 \leq y \leq x e^{-x}\}$

37. $S = \{(x, y) \mid 0 \le x < \pi/2, \ 0 \le y \le \sec^2 x\}$

38. $S = \{(x, y) \mid -2 < x \le 0, \ 0 \le y \le 1/\sqrt{x + 2}\,\}$

39. (a) If $g(x) = (\sin^2 x)/x^2$, use your calculator or computer to make a table of approximate values of $\int_1^t g(x)\,dx$ for $t = 2, 5, 10, 100, 1000,$ and $10,000$. Does it appear that $\int_1^\infty g(x)\,dx$ is convergent?
(b) Use the Comparison Theorem with $f(x) = 1/x^2$ to show that $\int_1^\infty g(x)\,dx$ is convergent.
(c) Illustrate part (b) by graphing f and g on the same screen for $1 \le x \le 10$. Use your graph to explain intuitively why $\int_1^\infty g(x)\,dx$ is convergent.

40. (a) If $g(x) = 1/(\sqrt{x} - 1)$, use your calculator or computer to make a table of approximate values of $\int_2^t g(x)\,dx$ for $t = 5, 10, 100, 1000,$ and $10,000$. Does it appear that $\int_2^\infty g(x)\,dx$ is convergent or divergent?
(b) Use the Comparison Theorem with $f(x) = 1/\sqrt{x}$ to show that $\int_2^\infty g(x)\,dx$ is divergent.
(c) Illustrate part (b) by graphing f and g on the same screen for $2 \le x \le 20$. Use your graph to explain intuitively why $\int_2^\infty g(x)\,dx$ is divergent.

41–46 ▪ Use the Comparison Theorem to determine whether the integral is convergent or divergent.

41. $\displaystyle\int_0^\infty \frac{x}{x^3 + 1}\,dx$

42. $\displaystyle\int_1^\infty \frac{2 + e^{-x}}{x}\,dx$

43. $\displaystyle\int_1^\infty \frac{x + 1}{\sqrt{x^4 - x}}\,dx$

44. $\displaystyle\int_0^\infty \frac{\arctan x}{2 + e^x}\,dx$

45. $\displaystyle\int_0^1 \frac{\sec^2 x}{x\sqrt{x}}\,dx$

46. $\displaystyle\int_0^\pi \frac{\sin^2 x}{\sqrt{x}}\,dx$

47. The integral

$$\int_0^\infty \frac{1}{\sqrt{x}\,(1 + x)}\,dx$$

is improper for two reasons: The interval $[0, \infty)$ is infinite and the integrand has an infinite discontinuity at 0. Evaluate it by expressing it as a sum of improper integrals of Type 2 and Type 1 as follows:

$$\int_0^\infty \frac{1}{\sqrt{x}\,(1 + x)}\,dx = \int_0^1 \frac{1}{\sqrt{x}\,(1 + x)}\,dx + \int_1^\infty \frac{1}{\sqrt{x}\,(1 + x)}\,dx$$

48–49 ▪ Find the values of p for which the integral converges and evaluate the integral for those values of p.

48. $\displaystyle\int_e^\infty \frac{1}{x(\ln x)^p}\,dx$

49. $\displaystyle\int_0^1 \frac{1}{x^p}\,dx$

50. (a) Evaluate the integral $\int_0^\infty x^n e^{-x}\,dx$ for $n = 0, 1, 2,$ and 3.
(b) Guess the value of $\int_0^\infty x^n e^{-x}\,dx$ when n is an arbitrary positive integer.
(c) Prove your guess using mathematical induction.

51. (a) Show that $\int_{-\infty}^\infty x\,dx$ is divergent.
(b) Show that

$$\lim_{t \to \infty} \int_{-t}^t x\,dx = 0$$

This shows that we can't define

$$\int_{-\infty}^\infty f(x)\,dx = \lim_{t \to \infty} \int_{-t}^t f(x)\,dx$$

52. If $\int_{-\infty}^\infty f(x)\,dx$ is convergent and a and b are real numbers, show that

$$\int_{-\infty}^a f(x)\,dx + \int_a^\infty f(x)\,dx = \int_{-\infty}^b f(x)\,dx + \int_b^\infty f(x)\,dx$$

53. A manufacturer of lightbulbs wants to produce bulbs that last about 700 hours but, of course, some bulbs burn out faster than others. Let $F(t)$ be the fraction of the company's bulbs that burn out before t hours, so $F(t)$ always lies between 0 and 1.
(a) Make a rough sketch of what you think the graph of F might look like.
(b) What is the meaning of the derivative $r(t) = F'(t)$?
(c) What is the value of $\int_0^\infty r(t)\,dt$? Why?

54. The *average speed* of molecules in an ideal gas is

$$\bar{v} = \frac{4}{\sqrt{\pi}} \left(\frac{M}{2RT}\right)^{3/2} \int_0^\infty v^3 e^{-Mv^2/(2RT)}\,dv$$

where M is the molecular weight of the gas, R is the gas constant, T is the gas temperature, and v is the molecular speed. Show that

$$\bar{v} = \sqrt{\frac{8RT}{\pi M}}$$

55. As we saw in Section 3.4, a radioactive substance decays exponentially: The mass at time t is $m(t) = m(0)e^{kt}$, where $m(0)$ is the initial mass and k is a negative constant. The *mean life* M of an atom in the substance is

$$M = -k \int_0^\infty t e^{kt}\,dt$$

For the radioactive carbon isotope, ^{14}C, used in radiocarbon dating, the value of k is -0.000121. Find the mean life of a ^{14}C atom.

56. Astronomers use a technique called *stellar stereography* to determine the density of stars in a star cluster from the observed (two-dimensional) density that can be analyzed from a photograph. Suppose that in a spherical cluster of radius R the density of stars depends only on the distance r from the center of the cluster. If the perceived star density is given by $y(s)$, where s is the observed planar distance from the center of the cluster, and $x(r)$ is the actual density, it can be shown that

$$y(s) = \int_s^R \frac{2r}{\sqrt{r^2 - s^2}} \, x(r) \, dr$$

If the actual density of stars in a cluster is $x(r) = \frac{1}{2}(R - r)^2$, find the perceived density $y(s)$.

57. Determine how large the number a has to be so that

$$\int_a^\infty \frac{1}{x^2 + 1} \, dx < 0.001$$

58. Estimate the numerical value of $\int_0^\infty e^{-x^2} \, dx$ by writing it as the sum of $\int_0^4 e^{-x^2} \, dx$ and $\int_4^\infty e^{-x^2} \, dx$. Approximate the first integral by using Simpson's Rule with $n = 8$ and show that the second integral is smaller than $\int_4^\infty e^{-4x} \, dx$, which is less than 0.0000001.

59. Show that $\int_0^\infty x^2 e^{-x^2} \, dx = \frac{1}{2} \int_0^\infty e^{-x^2} \, dx$.

60. Show that $\int_0^\infty e^{-x^2} \, dx = \int_0^1 \sqrt{-\ln y} \, dy$ by interpreting the integrals as areas.

61. Find the value of the constant C for which the integral

$$\int_0^\infty \left(\frac{1}{\sqrt{x^2 + 4}} - \frac{C}{x + 2} \right) dx$$

converges. Evaluate the integral for this value of C.

62. Find the value of the constant C for which the integral

$$\int_0^\infty \left(\frac{x}{x^2 + 1} - \frac{C}{3x + 1} \right) dx$$

converges. Evaluate the integral for this value of C.

63. Suppose f is continuous on $[0, \infty)$ and $\lim_{x \to \infty} f(x) = 1$. Is it possible that $\int_0^\infty f(x) \, dx$ is convergent?

64. Show that if $a > -1$ and $b > a + 1$, then the following integral is convergent.

$$\int_0^\infty \frac{x^a}{1 + x^b} \, dx$$

CHAPTER 6 REVIEW

CONCEPT CHECK

1. State the rule for integration by parts. In practice, how do you use it?

2. How do you evaluate $\int \sin^m x \cos^n x \, dx$ if m is odd? What if n is odd? What if m and n are both even?

3. If the expression $\sqrt{a^2 - x^2}$ occurs in an integral, what substitution might you try? What if $\sqrt{a^2 + x^2}$ occurs? What if $\sqrt{x^2 - a^2}$ occurs?

4. What is the form of the partial fraction expansion of a rational function $P(x)/Q(x)$ if the degree of P is less than the degree of Q and $Q(x)$ has only distinct linear factors? What if a linear factor is repeated? What if $Q(x)$ has an irreducible quadratic factor (not repeated)? What if the quadratic factor is repeated?

5. State the rules for approximating the definite integral $\int_a^b f(x) \, dx$ with the Midpoint Rule, the Trapezoidal Rule, and Simpson's Rule. Which would you expect to give the best estimate? How do you approximate the error for each rule?

6. Define the following improper integrals.
   (a) $\int_a^\infty f(x) \, dx$ (b) $\int_{-\infty}^b f(x) \, dx$ (c) $\int_{-\infty}^\infty f(x) \, dx$

7. Define the improper integral $\int_a^b f(x) \, dx$ for each of the following cases.
   (a) f has an infinite discontinuity at a.
   (b) f has an infinite discontinuity at b.
   (c) f has an infinite discontinuity at c, where $a < c < b$.

8. State the Comparison Theorem for improper integrals.

TRUE-FALSE QUIZ

Determine whether the statement is true or false. If it is true, explain why. If it is false, explain why or give an example that disproves the statement.

1. $\dfrac{x(x^2 + 4)}{x^2 - 4}$ can be put in the form $\dfrac{A}{x + 2} + \dfrac{B}{x - 2}$.

2. $\dfrac{x^2 + 4}{x(x^2 - 4)}$ can be put in the form $\dfrac{A}{x} + \dfrac{B}{x + 2} + \dfrac{C}{x - 2}$.

3. $\dfrac{x^2 + 4}{x^2(x - 4)}$ can be put in the form $\dfrac{A}{x^2} + \dfrac{B}{x - 4}$.

4. $\dfrac{x^2 - 4}{x(x^2 + 4)}$ can be put in the form $\dfrac{A}{x} + \dfrac{B}{x^2 + 4}$.

5. $\int_0^4 \dfrac{x}{x^2 - 1} \, dx = \frac{1}{2} \ln 15$

6. $\int_1^\infty \dfrac{1}{x^{\sqrt{2}}}\,dx$ is convergent.

7. If f is continuous, then $\int_{-\infty}^\infty f(x)\,dx = \lim_{t\to\infty}\int_{-t}^t f(x)\,dx$.

8. The Midpoint Rule is always more accurate than the Trapezoidal Rule.

9. (a) Every elementary function has an elementary derivative.
(b) Every elementary function has an elementary antiderivative.

10. If f is continuous on $[0, \infty)$ and $\int_1^\infty f(x)\,dx$ is convergent, then $\int_0^\infty f(x)\,dx$ is convergent.

11. If f is a continuous, decreasing function on $[1, \infty)$ and $\lim_{x\to\infty} f(x) = 0$, then $\int_1^\infty f(x)\,dx$ is convergent.

12. If $\int_a^\infty f(x)\,dx$ and $\int_a^\infty g(x)\,dx$ are both convergent, then $\int_a^\infty [f(x) + g(x)]\,dx$ is convergent.

13. If $\int_a^\infty f(x)\,dx$ and $\int_a^\infty g(x)\,dx$ are both divergent, then $\int_a^\infty [f(x) + g(x)]\,dx$ is divergent.

14. If $f(x) \leqslant g(x)$ and $\int_0^\infty g(x)\,dx$ diverges, then $\int_0^\infty f(x)\,dx$ also diverges.

EXERCISES

1–40 ■ Evaluate the integral.

1. $\int_1^2 \dfrac{(x + 1)^2}{x}\,dx$

2. $\int_1^2 \dfrac{x}{(x + 1)^2}\,dx$

3. $\int_0^{\pi/2} \sin\theta\, e^{\cos\theta}\,d\theta$

4. $\int_0^{\pi/6} t\sin 2t\,dt$

5. $\int \dfrac{dt}{2t^2 + 3t + 1}$

6. $\int_1^2 x^5 \ln x\,dx$

7. $\int \dfrac{\sin(\ln t)}{t}\,dt$

8. $\int \dfrac{dx}{x^2\sqrt{1 + x^2}}$

9. $\int_1^4 x^{3/2}\ln x\,dx$

10. $\int_0^1 \dfrac{\sqrt{\arctan x}}{1 + x^2}\,dx$

11. $\int_1^2 \dfrac{\sqrt{x^2 - 1}}{x}\,dx$

12. $\int_{-\pi/4}^{\pi/4} \dfrac{\tan x}{4 + x^2}\,dx$

13. $\int \dfrac{dx}{x^3 + x}$

14. $\int \dfrac{x^2 + 2}{x + 2}\,dx$

15. $\int_0^{\pi/2} \sin^3\theta\,\cos^2\theta\,d\theta$

16. $\int \dfrac{\sec^6\theta}{\tan^2\theta}\,d\theta$

17. $\int x\sec x\tan x\,dx$

18. $\int \dfrac{x^2 + 8x - 3}{x^3 + 3x^2}\,dx$

19. $\int \dfrac{x + 1}{9x^2 + 6x + 5}\,dx$

20. $\int \dfrac{dt}{\sin^2 t + \cos 2t}$

21. $\int \dfrac{dx}{\sqrt{x^2 - 4x}}$

22. $\int \dfrac{x^3}{(x + 1)^{10}}\,dx$

23. $\int \csc^4 4x\,dx$

24. $\int e^x\cos x\,dx$

25. $\int \dfrac{3x^3 - x^2 + 6x - 4}{(x^2 + 1)(x^2 + 2)}\,dx$

26. $\int \tan^5\theta\,\sec^3\theta\,d\theta$

27. $\int_0^{\pi/2} \cos^3 x\,\sin 2x\,dx$

28. $\int \dfrac{\sqrt[3]{x} + 1}{\sqrt[3]{x} - 1}\,dx$

29. $\int_{-3}^3 \dfrac{x}{1 + |x|}\,dx$

30. $\int \dfrac{dx}{e^x\sqrt{1 - e^{-2x}}}$

31. $\int_0^{\ln 10} \dfrac{e^x\sqrt{e^x - 1}}{e^x + 8}\,dx$

32. $\int_0^{\pi/4} \dfrac{x\sin x}{\cos^3 x}\,dx$

33. $\int \dfrac{x^2}{(4 - x^2)^{3/2}}\,dx$

34. $\int (\arcsin x)^2\,dx$

35. $\int \dfrac{1}{\sqrt{x + x^{3/2}}}\,dx$

36. $\int \dfrac{1 - \tan\theta}{1 + \tan\theta}\,d\theta$

37. $\int (\cos x + \sin x)^2 \cos 2x\,dx$

38. $\int \dfrac{2^{\sqrt{x}}}{\sqrt{x}}\,dx$

39. $\int_0^{1/2} \dfrac{xe^{2x}}{(1 + 2x)^2}\,dx$

40. $\int_{\pi/4}^{\pi/3} \dfrac{\sqrt{\tan\theta}}{\sin 2\theta}\,d\theta$

41–50 ■ Evaluate the integral or show that it is divergent.

41. $\int_1^\infty \dfrac{1}{(2x + 1)^3}\,dx$

42. $\int_1^\infty \dfrac{\ln x}{x^4}\,dx$

43. $\int_2^\infty \dfrac{dx}{x\ln x}$

44. $\int_2^6 \dfrac{y}{\sqrt{y - 2}}\,dy$

45. $\int_0^4 \dfrac{\ln x}{\sqrt{x}}\,dx$

46. $\int_0^1 \dfrac{1}{2 - 3x}\,dx$

47. $\int_0^1 \dfrac{x - 1}{\sqrt{x}}\,dx$

48. $\int_{-1}^1 \dfrac{dx}{x^2 - 2x}$

49. $\int_{-\infty}^\infty \dfrac{dx}{4x^2 + 4x + 5}$

50. $\int_1^\infty \dfrac{\tan^{-1}x}{x^2}\,dx$

51–54 ■ Use the Table of Integrals on the Reference Pages to evaluate the integral.

51. $\int \sqrt{4x^2 - 4x - 3}\,dx$

52. $\int \csc^5 t\,dt$

53. $\displaystyle\int \cos x \sqrt{4 + \sin^2 x}\ dx$ **54.** $\displaystyle\int \frac{\cot x}{\sqrt{1 + 2 \sin x}}\ dx$

---

55. Is it possible to find a number n such that $\int_0^\infty x^n\ dx$ is convergent?

56. For what values of a is $\int_0^\infty e^{ax} \cos x\ dx$ convergent? Evaluate the integral for those values of a.

57–58 ■ Use (a) the Trapezoidal Rule, (b) the Midpoint Rule, and (c) Simpson's Rule with $n = 10$ to approximate the given integral. Round your answers to six decimal places.

57. $\displaystyle\int_2^4 \frac{1}{\ln x}\ dx$ **58.** $\displaystyle\int_1^4 \sqrt{x}\ \cos x\ dx$

---

59. Estimate the errors involved in Exercise 57, parts (a) and (b). How large should n be in each case to guarantee an error of less than 0.00001?

60. Use Simpson's Rule with $n = 6$ to estimate the area under the curve $y = e^x/x$ from $x = 1$ to $x = 4$.

61. The speedometer reading (v) on a car was observed at 1-minute intervals and recorded in the chart. Use Simpson's Rule to estimate the distance traveled by the car.

t (min)	v (mi/h)	t (min)	v (mi/h)
0	40	6	56
1	42	7	57
2	45	8	57
3	49	9	55
4	52	10	56
5	54		

62. A population of honeybees increased at a rate of $r(t)$ bees per week, where the graph of r is as shown. Use Simpson's Rule with six subintervals to estimate the increase in the bee population during the first 24 weeks.

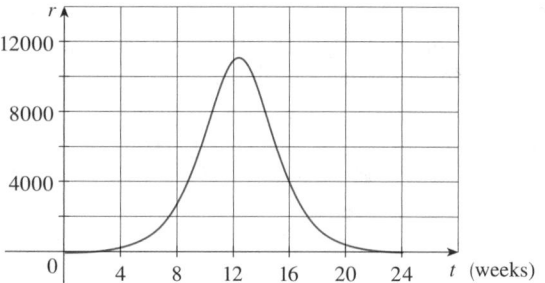

63. (a) If $f(x) = \sin(\sin x)$, use a graph to find an upper bound for $|f^{(4)}(x)|$.
  (b) Use Simpson's Rule with $n = 10$ to approximate $\int_0^\pi f(x)\ dx$ and use part (a) to estimate the error.
  (c) How large should n be to guarantee that the size of the error in using S_n is less than 0.00001?   CAS

64. Use the Comparison Theorem to determine whether the integral

$$\int_1^\infty \frac{x^3}{x^5 + 2}\ dx$$

is convergent or divergent.

65. If f' is continuous on $[0, \infty)$ and $\lim_{x\to\infty} f(x) = 0$, show that

$$\int_0^\infty f'(x)\ dx = -f(0)$$

7

APPLICATIONS OF INTEGRATION

In this chapter we explore some of the applications of the definite integral by using it to compute areas between curves, volumes of solids, lengths of curves, the work done by a varying force, the center of gravity of a plate, and the force on a dam. The common theme in most of these applications is the following general method, which is similar to the one we used to find areas under curves: We break up a quantity Q into a large number of small parts. We next approximate each small part by a quantity of the form $f(x_i^*) \, \Delta x$ and thus approximate Q by a Riemann sum. Then we take the limit and express Q as an integral. Finally we evaluate the integral by using the Evaluation Theorem, or Simpson's Rule, or technology.

In the final section we look at what is perhaps the most important of all the applications of integration: differential equations. When a scientist uses calculus, more often than not it is to solve a differential equation that has arisen in the description of some physical process.

7.1 | AREAS BETWEEN CURVES

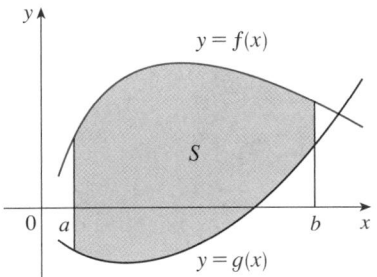

FIGURE 1
$S = \{(x, y) \mid a \leqslant x \leqslant b, g(x) \leqslant y \leqslant f(x)\}$

In Chapter 5 we defined and calculated areas of regions that lie under the graphs of functions. Here we use integrals to find areas of regions that lie between the graphs of two functions.

Consider the region S that lies between two curves $y = f(x)$ and $y = g(x)$ and between the vertical lines $x = a$ and $x = b$, where f and g are continuous functions and $f(x) \geqslant g(x)$ for all x in $[a, b]$. (See Figure 1.)

Just as we did for areas under curves in Section 5.1, we divide S into n strips of equal width and then we approximate the ith strip by a rectangle with base Δx and height $f(x_i^*) - g(x_i^*)$. (See Figure 2. If we like, we could take all of the sample points to be right endpoints, in which case $x_i^* = x_i$.) The Riemann sum

$$\sum_{i=1}^{n} [f(x_i^*) - g(x_i^*)] \, \Delta x$$

is therefore an approximation to what we intuitively think of as the area of S.

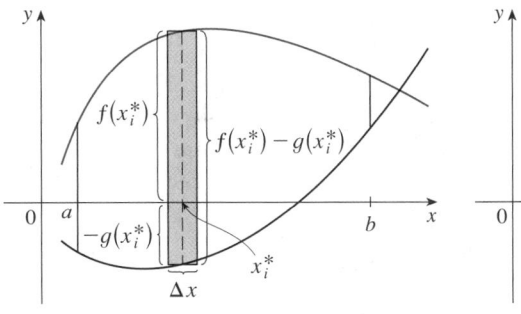

FIGURE 2
       (a) Typical rectangle   (b) Approximating rectangles

This approximation appears to become better and better as $n \to \infty$. Therefore we define the **area** A of S as the limiting value of the sum of the areas of these approximating rectangles.

$$\boxed{1} \qquad A = \lim_{n \to \infty} \sum_{i=1}^{n} \left[f(x_i^*) - g(x_i^*) \right] \Delta x$$

We recognize the limit in $\boxed{1}$ as the definite integral of $f - g$. Therefore we have the following formula for area.

> $\boxed{2}$ The area A of the region bounded by the curves $y = f(x)$, $y = g(x)$, and the lines $x = a$, $x = b$, where f and g are continuous and $f(x) \geqslant g(x)$ for all x in $[a, b]$, is
> $$A = \int_a^b \left[f(x) - g(x) \right] dx$$

Notice that in the special case where $g(x) = 0$, S is the region under the graph of f and our general definition of area $\boxed{1}$ reduces to our previous definition (Definition 5.1.2).

In the case where both f and g are positive, you can see from Figure 3 why $\boxed{2}$ is true:

$$A = [\text{area under } y = f(x)] - [\text{area under } y = g(x)]$$

$$= \int_a^b f(x)\,dx - \int_a^b g(x)\,dx = \int_a^b \left[f(x) - g(x) \right] dx$$

EXAMPLE 1 Find the area of the region bounded above by $y = e^x$, bounded below by $y = x$, and bounded on the sides by $x = 0$ and $x = 1$.

SOLUTION The region is shown in Figure 4. The upper boundary curve is $y = e^x$ and the lower boundary curve is $y = x$. So we use the area formula $\boxed{2}$ with $f(x) = e^x$, $g(x) = x$, $a = 0$, and $b = 1$:

$$A = \int_0^1 (e^x - x)\,dx = e^x - \tfrac{1}{2}x^2 \Big]_0^1 = e - \tfrac{1}{2} - 1 = e - 1.5 \qquad \blacksquare$$

In Figure 4 we drew a typical approximating rectangle with width Δx as a reminder of the procedure by which the area is defined in $\boxed{1}$. In general, when we set up an integral for an area, it's helpful to sketch the region to identify the top curve y_T, the bottom curve y_B, and a typical approximating rectangle as in Figure 5. Then the area of a typical rectangle is $(y_T - y_B)\,\Delta x$ and the equation

$$A = \lim_{n \to \infty} \sum_{i=1}^{n} (y_T - y_B)\,\Delta x = \int_a^b (y_T - y_B)\,dx$$

summarizes the procedure of adding (in a limiting sense) the areas of all the typical rectangles.

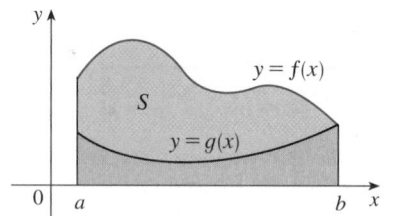

FIGURE 3

$A = \int_a^b f(x)\,dx - \int_a^b g(x)\,dx$

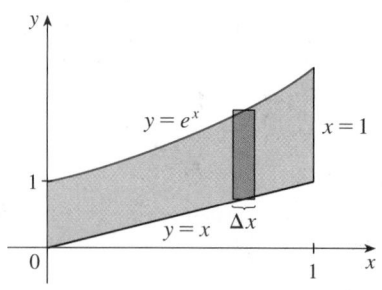

FIGURE 4

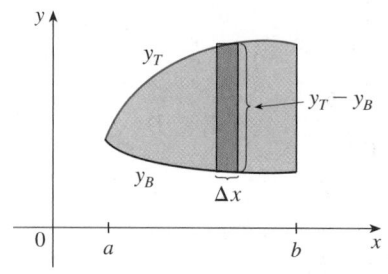

FIGURE 5

Notice that in Figure 5 the left-hand boundary reduces to a point, whereas in Figure 3 the right-hand boundary reduces to a point. In the next example both of the side boundaries reduce to a point, so the first step is to find a and b.

V EXAMPLE 2 Find the area of the region enclosed by the parabolas $y = x^2$ and $y = 2x - x^2$.

SOLUTION We first find the points of intersection of the parabolas by solving their equations simultaneously. This gives $x^2 = 2x - x^2$, or $2x^2 - 2x = 0$. Therefore $2x(x - 1) = 0$, so $x = 0$ or 1. The points of intersection are $(0, 0)$ and $(1, 1)$.

We see from Figure 6 that the top and bottom boundaries are

$$y_T = 2x - x^2 \qquad \text{and} \qquad y_B = x^2$$

The area of a typical rectangle is

$$(y_T - y_B)\,\Delta x = (2x - x^2 - x^2)\,\Delta x = (2x - 2x^2)\,\Delta x$$

and the region lies between $x = 0$ and $x = 1$. So the total area is

$$A = \int_0^1 (2x - 2x^2)\,dx = 2\int_0^1 (x - x^2)\,dx$$

$$= 2\left[\frac{x^2}{2} - \frac{x^3}{3}\right]_0^1 = 2\left(\frac{1}{2} - \frac{1}{3}\right) = \frac{1}{3} \qquad \blacksquare$$

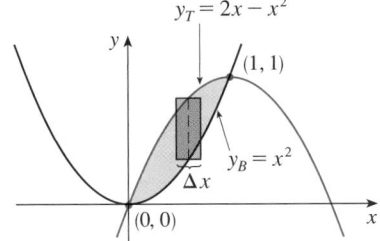

FIGURE 6

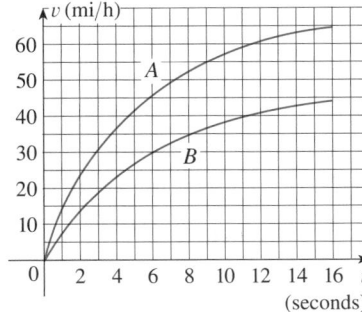

FIGURE 7

EXAMPLE 3 Figure 7 shows velocity curves for two cars, A and B, that start side by side and move along the same road. What does the area between the curves represent? Use Simpson's Rule to estimate it.

SOLUTION We know from Section 5.3 that the area under the velocity curve A represents the distance traveled by car A during the first 16 seconds. Similarly, the area under curve B is the distance traveled by car B during that time period. So the area between these curves, which is the difference of the areas under the curves, is the distance between the cars after 16 seconds. We read the velocities from the graph and convert them to feet per second ($1 \text{ mi/h} = \frac{5280}{3600} \text{ ft/s}$).

t	0	2	4	6	8	10	12	14	16
v_A	0	34	54	67	76	84	89	92	95
v_B	0	21	34	44	51	56	60	63	65
$v_A - v_B$	0	13	20	23	25	28	29	29	30

Using Simpson's Rule with $n = 8$ intervals, so that $\Delta t = 2$, we estimate the distance between the cars after 16 seconds:

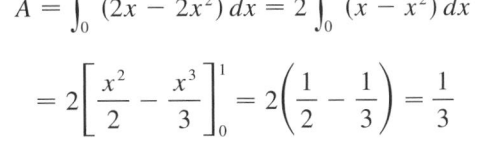

$$\int_0^{16} (v_A - v_B)\,dt$$

$$\approx \tfrac{2}{3}[0 + 4(13) + 2(20) + 4(23) + 2(25) + 4(28) + 2(29) + 4(29) + 30]$$

$$\approx 367 \text{ ft} \qquad \blacksquare$$

■ www.stewartcalculus.com
See Additional Examples A, B.

Some regions are best treated by regarding x as a function of y. If a region is bounded by curves with equations $x = f(y)$, $x = g(y)$, $y = c$, and $y = d$, where f and g are continuous and $f(y) \geq g(y)$ for $c \leq y \leq d$ (see Figure 8), then its area is

$$A = \int_c^d [f(y) - g(y)] \, dy$$

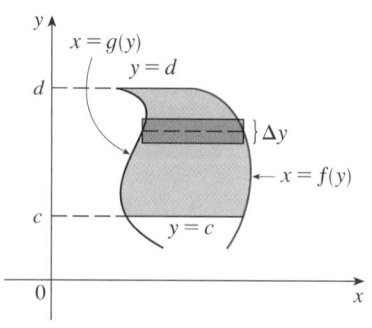

FIGURE 8

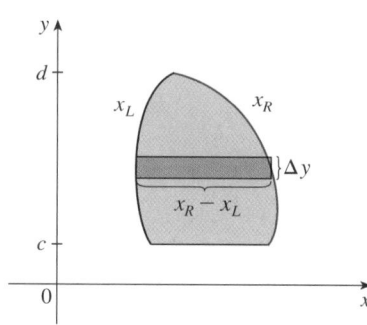

FIGURE 9

If we write x_R for the right boundary and x_L for the left boundary, then, as Figure 9 illustrates, we have

$$A = \int_c^d (x_R - x_L) \, dy$$

Here a typical approximating rectangle has dimensions $x_R - x_L$ and Δy.

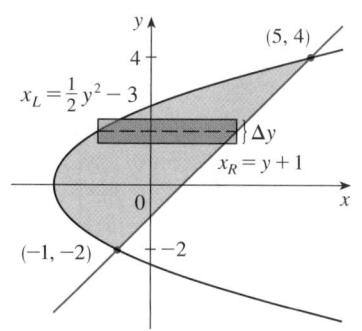

FIGURE 10

V EXAMPLE 4 Find the area enclosed by the line $y = x - 1$ and the parabola $y^2 = 2x + 6$.

SOLUTION By solving the two equations we find that the points of intersection are $(-1, -2)$ and $(5, 4)$. We solve the equation of the parabola for x and notice from Figure 10 that the left and right boundary curves are

$$x_L = \tfrac{1}{2}y^2 - 3 \qquad \text{and} \qquad x_R = y + 1$$

We must integrate between the appropriate y-values, $y = -2$ and $y = 4$. Thus

$$A = \int_{-2}^4 (x_R - x_L) \, dy = \int_{-2}^4 \left[(y + 1) - \left(\tfrac{1}{2}y^2 - 3 \right) \right] dy$$

$$= \int_{-2}^4 \left(-\tfrac{1}{2}y^2 + y + 4 \right) dy$$

$$= -\frac{1}{2} \left(\frac{y^3}{3} \right) + \frac{y^2}{2} + 4y \Bigg]_{-2}^4$$

$$= -\tfrac{1}{6}(64) + 8 + 16 - \left(\tfrac{4}{3} + 2 - 8 \right) = 18 \qquad \blacksquare$$

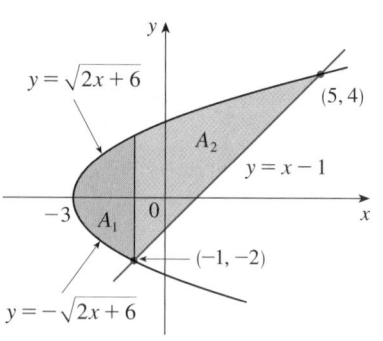

FIGURE 11

NOTE We could have found the area in Example 4 by integrating with respect to x instead of y, but the calculation is much more involved. It would have meant splitting the region in two and computing the areas labeled A_1 and A_2 in Figure 11. The method we used in Example 4 is *much* easier.

7.1 | EXERCISES

1–4 ■ Find the area of the shaded region.

1.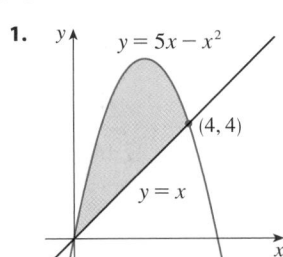

2.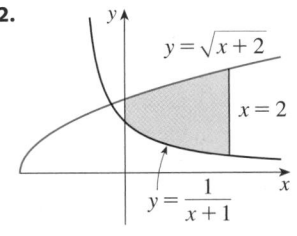

3.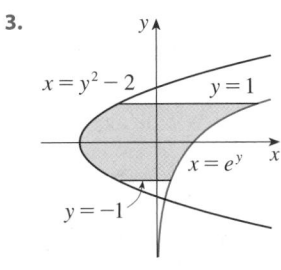

4.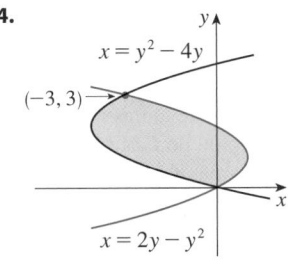

5–10 ■ Sketch the region enclosed by the given curves. Decide whether to integrate with respect to x or y. Draw a typical approximating rectangle and label its height and width. Then find the area of the region.

5. $y = e^x$, $y = x^2 - 1$, $x = -1$, $x = 1$

6. $y = \sin x$, $y = x$, $x = \pi/2$, $x = \pi$

7. $y = (x - 2)^2$, $y = x$

8. $y = \sin x$, $y = 2x/\pi$, $x \geqslant 0$

9. $x = 1 - y^2$, $x = y^2 - 1$

10. $4x + y^2 = 12$, $x = y$

11–20 ■ Sketch the region enclosed by the given curves and find its area.

11. $y = 12 - x^2$, $y = x^2 - 6$

12. $y = x^2$, $y = 4x - x^2$

13. $y = e^x$, $y = xe^x$, $x = 0$

14. $y = \cos x$, $y = 2 - \cos x$, $0 \leqslant x \leqslant 2\pi$

15. $x = 2y^2$, $x = 4 + y^2$

16. $x = y^4$, $y = \sqrt{2 - x}$, $y = 0$

17. $y = \cos \pi x$, $y = 4x^2 - 1$

18. $y = |x|$, $y = x^2 - 2$

19. $y = 1/x$, $y = x$, $y = \frac{1}{4}x$, $x > 0$

20. $y = \frac{1}{4}x^2$, $y = 2x^2$, $x + y = 3$, $x \geqslant 0$

21. Sketch the region that lies between the curves $y = \cos x$ and $y = \sin 2x$ and between $x = 0$ and $x = \pi/2$. Notice that the region consists of two separate parts. Find the area of this region.

22. Graph the curves $y = x^2 - x$ and $y = x^3 - 4x^2 + 3x$ on a common screen and observe that the region between them consists of two parts. Find the area of this region.

23–26 ■ Graph the region between the curves and use your calculator to compute the area correct to five decimal places.

23. $y = \dfrac{2}{1 + x^4}$, $y = x^2$ **24.** $y = e^{1-x^2}$, $y = x^4$

25. $y = \tan^2 x$, $y = \sqrt{x}$

26. $y = \cos x$, $y = x + 2\sin^4 x$

27. Racing cars driven by Chris and Kelly are side by side at the start of a race. The table shows the velocities of each car (in miles per hour) during the first ten seconds of the race. Use Simpson's Rule to estimate how much farther Kelly travels than Chris does during the first ten seconds.

t	v_C	v_K	t	v_C	v_K
0	0	0	6	69	80
1	20	22	7	75	86
2	32	37	8	81	93
3	46	52	9	86	98
4	54	61	10	90	102
5	62	71			

28. Two cars, A and B, start side by side and accelerate from rest. The figure shows the graphs of their velocity functions.
(a) Which car is ahead after one minute? Explain.
(b) What is the meaning of the area of the shaded region?
(c) Which car is ahead after two minutes? Explain.
(d) Estimate the time at which the cars are again side by side.

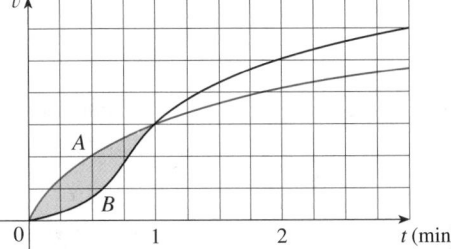

Graphing calculator or computer required   **CAS** Computer algebra system required   **1.** Homework Hints at stewartcalculus.com

29. The widths (in meters) of a kidney-shaped swimming pool were measured at 2-meter intervals as indicated in the figure. Use Simpson's Rule to estimate the area of the pool.

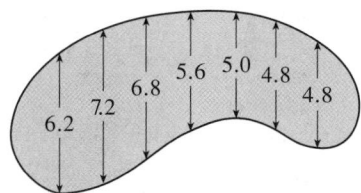

30. A cross-section of an airplane wing is shown. Measurements of the height of the wing, in centimeters, at 20-centimeter intervals are 5.8, 20.3, 26.7, 29.0, 27.6, 27.3, 23.8, 20.5, 15.1, 8.7, and 2.8. Use Simpson's Rule to estimate the area of the wing's cross-section.

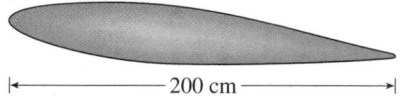

|← 200 cm →|

31. If the birth rate of a population is $b(t) = 2200e^{0.024t}$ people per year and the death rate is $d(t) = 1460e^{0.018t}$ people per year, find the area between these curves for $0 \le t \le 10$. What does this area represent?

32. A water storage tank has the shape of a cylinder with diameter 10 ft. It is mounted so that the circular cross-sections are vertical. If the depth of the water is 7 ft, what percentage of the total capacity is being used?

33. Find the area of the crescent-shaped region (called a *lune*) bounded by arcs of circles with radii r and R (see the figure).

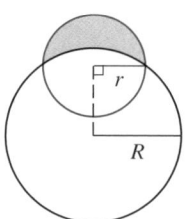

34. Sketch the region in the xy-plane defined by the inequalities $x - 2y^2 \ge 0$, $1 - x - |y| \ge 0$ and find its area.

35. Find the values of c such that the area of the region bounded by the parabolas $y = x^2 - c^2$ and $y = c^2 - x^2$ is 576.

36. Find the area of the region bounded by the parabola $y = x^2$, the tangent line to this parabola at $(1, 1)$, and the x-axis.

37. Find the number b such that the line $y = b$ divides the region bounded by the curves $y = x^2$ and $y = 4$ into two regions with equal area.

38. (a) Find the number a such that the line $x = a$ bisects the area under the curve $y = 1/x^2$, $1 \le x \le 4$.
(b) Find the number b such that the line $y = b$ bisects the area in part (a).

39. Find a positive continuous function f such that the area under the graph of f from 0 to t is $A(t) = t^3$ for all $t > 0$.

40. Suppose that $0 < c < \pi/2$. For what value of c is the area of the region enclosed by the curves $y = \cos x$, $y = \cos(x - c)$, and $x = 0$ equal to the area of the region enclosed by the curves $y = \cos(x - c)$, $x = \pi$, and $y = 0$?

41. For what values of m do the line $y = mx$ and the curve $y = x/(x^2 + 1)$ enclose a region? Find the area of the region.

7.2 | VOLUMES

In trying to find the volume of a solid we face the same type of problem as in finding areas. We have an intuitive idea of what volume means, but we must make this idea precise by using calculus to give an exact definition of volume.

We start with a simple type of solid called a **cylinder** (or, more precisely, a *right cylinder*). As illustrated in Figure 1(a), a cylinder is bounded by a plane region B_1, called the **base**, and a congruent region B_2 in a parallel plane. The cylinder consists of all points on line segments that are perpendicular to the base and join B_1 to B_2. If the area of the base is A and the height of the cylinder (the distance from B_1 to B_2) is h, then the volume V of the cylinder is defined as

$$V = Ah$$

In particular, if the base is a circle with radius r, then the cylinder is a circular cylinder with volume $V = \pi r^2 h$ [see Figure 1(b)], and if the base is a rectangle with length l and width w, then the cylinder is a rectangular box (also called a *rectangular parallelepiped*) with volume $V = lwh$ [see Figure 1(c)].

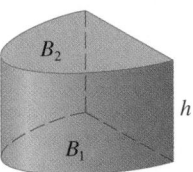 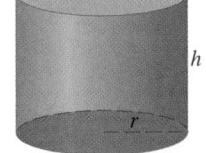 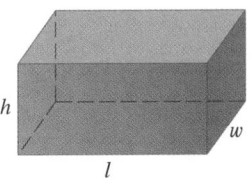

FIGURE 1   (a) Cylinder $V = Ah$ (b) Circular cylinder $V = \pi r^2 h$ (c) Rectangular box $V = lwh$

For a solid S that isn't a cylinder we first "cut" S into pieces and approximate each piece by a cylinder. We estimate the volume of S by adding the volumes of the cylinders. We arrive at the exact volume of S through a limiting process in which the number of pieces becomes large.

We start by intersecting S with a plane and obtaining a plane region that is called a **cross-section** of S. Let $A(x)$ be the area of the cross-section of S in a plane P_x perpendicular to the x-axis and passing through the point x, where $a \leqslant x \leqslant b$. (See Figure 2. Think of slicing S with a knife through x and computing the area of this slice.) The cross-sectional area $A(x)$ will vary as x increases from a to b.

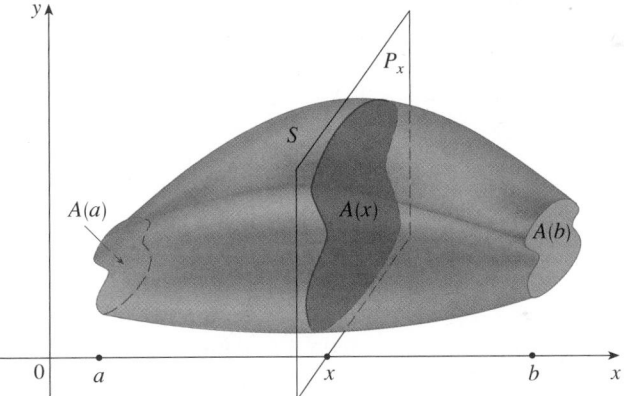

FIGURE 2

We consider a partition of the interval $[a, b]$ into n subintervals with partition points $x_0, x_1, x_2, \ldots, x_n$. We divide S into n "slabs" of width $\Delta x_i = x_i - x_{i-1}$ by using the planes $P_{x_1}, P_{x_2}, \ldots$ to slice the solid. (Think of slicing a loaf of bread.) If we choose sample points x_i^* in $[x_{i-1}, x_i]$, we can approximate the ith slab S_i (the part of S that lies between the planes $P_{x_{i-1}}$ and P_{x_i}) by a cylinder with base area $A(x_i^*)$ and "height" Δx_i. (See Figure 3.)

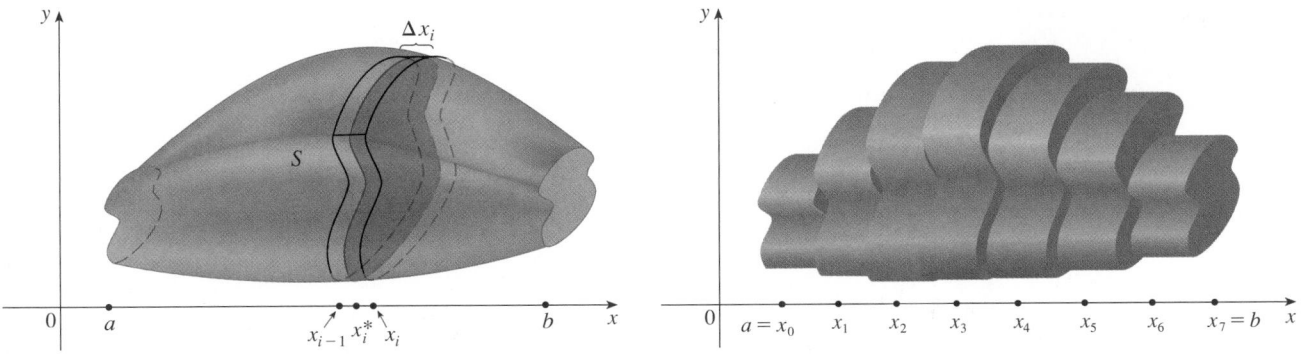

FIGURE 3

The volume of this cylinder is $A(x_i^*)\,\Delta x_i$, so an approximation to our intuitive conception of the volume of the ith slab S_i is

$$V(S_i) \approx A(x_i^*)\,\Delta x_i$$

Adding the volumes of these slabs, we get an approximation to the total volume (that is, what we think of intuitively as the volume):

$$V \approx \sum_{i=1}^{n} A(x_i^*)\,\Delta x_i$$

This approximation appears to become better and better as the slices become thinner and thinner. So we *define* the volume as the limit of these sums as max $\Delta x_i \to 0$. But we recognize the limit of Riemann sums as a definite integral and so we have the following definition.

■ It can be proved that this definition is independent of how S is situated with respect to the x-axis. In other words, no matter how we slice S with parallel planes, we always get the same answer for V.

DEFINITION OF VOLUME   Let S be a solid that lies between $x = a$ and $x = b$. If the cross-sectional area of S in the plane P_x, through x and perpendicular to the x-axis, is $A(x)$, where A is an integrable function, then the **volume** of S is

$$V = \lim_{\max \Delta x_i \to 0} \sum_{i=1}^{n} A(x_i^*)\,\Delta x_i = \int_a^b A(x)\,dx$$

When we use the volume formula $V = \int_a^b A(x)\,dx$ it is important to remember that $A(x)$ is the area of a moving cross-section obtained by slicing through x perpendicular to the x-axis.

Notice that, for a cylinder, the cross-sectional area is constant: $A(x) = A$ for all x. So our definition of volume gives $V = \int_a^b A\,dx = A(b - a)$; this agrees with the formula $V = Ah$.

EXAMPLE 1   Show that the volume of a sphere of radius r is $V = \frac{4}{3}\pi r^3$.

SOLUTION   If we place the sphere so that its center is at the origin (see Figure 4), then the plane P_x intersects the sphere in a circle whose radius (from the Pythagorean Theorem) is $y = \sqrt{r^2 - x^2}$. So the cross-sectional area is

$$A(x) = \pi y^2 = \pi(r^2 - x^2)$$

Using the definition of volume with $a = -r$ and $b = r$, we have

$$V = \int_{-r}^{r} A(x)\,dx = \int_{-r}^{r} \pi(r^2 - x^2)\,dx$$

$$= 2\pi \int_0^r (r^2 - x^2)\,dx \qquad \text{(The integrand is even.)}$$

$$= 2\pi \left[r^2 x - \frac{x^3}{3} \right]_0^r = 2\pi \left(r^3 - \frac{r^3}{3} \right)$$

$$= \tfrac{4}{3}\pi r^3 \qquad\qquad\qquad\qquad\qquad ■$$

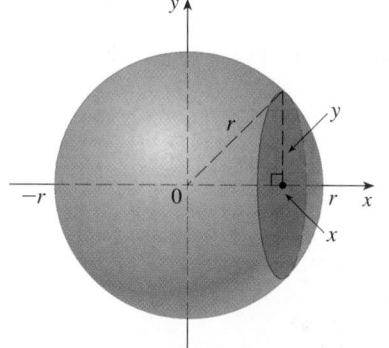

FIGURE 4

Figure 5 illustrates the definition of volume when the solid is a sphere with radius $r = 1$. From the result of Example 1, we know that the volume of the sphere is $\frac{4}{3}\pi \approx 4.18879$. Here the slabs are circular cylinders, or *disks*, and the three parts of

Figure 5 show the geometric interpretations of the Riemann sums

$$\sum_{i=1}^{n} A(\bar{x}_i)\,\Delta x = \sum_{i=1}^{n} \pi(1^2 - \bar{x}_i^2)\,\Delta x$$

when $n = 5$, 10, and 20 if we use regular partitions and choose the sample points x_i^* to be the midpoints $\bar{x}_i$. Notice that as we increase the number of approximating cylinders, the corresponding Riemann sums become closer to the true volume.

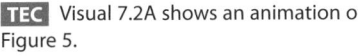

TEC Visual 7.2A shows an animation of Figure 5.

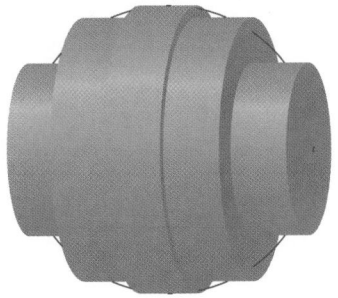

(a) Using 5 disks, $V \approx 4.2726$

(b) Using 10 disks, $V \approx 4.2097$

(c) Using 20 disks, $V \approx 4.1940$

FIGURE 5
Approximating the volume
of a sphere with radius 1

▼ EXAMPLE 2 Find the volume of the solid obtained by rotating about the x-axis the region under the curve $y = \sqrt{x}$ from 0 to 1. Illustrate the definition of volume by sketching a typical approximating cylinder.

SOLUTION The region is shown in Figure 6(a). If we rotate about the x-axis, we get the solid shown in Figure 6(b). When we slice through the point x, we get a disk with radius $\sqrt{x}$. The area of this cross-section is

$$A(x) = \pi\left(\sqrt{x}\,\right)^2 = \pi x$$

and the volume of the approximating cylinder (a disk with thickness Δx) is

$$A(x)\,\Delta x = \pi x\,\Delta x$$

The solid lies between $x = 0$ and $x = 1$, so its volume is

$$V = \int_0^1 A(x)\,dx = \int_0^1 \pi x\,dx = \pi\,\frac{x^2}{2}\bigg]_0^1 = \frac{\pi}{2}$$

■ Did we get a reasonable answer in Example 2? As a check on our work, let's replace the given region by a square with base $[0, 1]$ and height 1. If we rotate this square, we get a cylinder with radius 1, height 1, and volume $\pi \cdot 1^2 \cdot 1 = \pi$. We computed that the given solid has half this volume. That seems about right.

TEC Visual 7.2B shows how the solids of revolution in Examples 2–6 are formed.

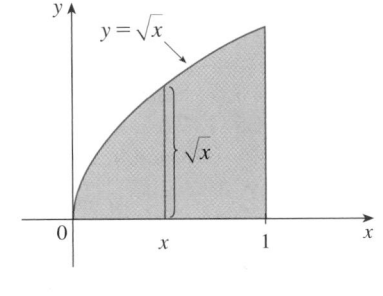

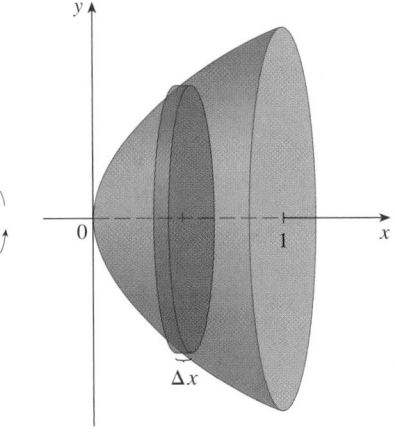

FIGURE 6

(a)

(b)

V EXAMPLE 3 Find the volume of the solid obtained by rotating the region bounded by $y = x^3$, $y = 8$, and $x = 0$ about the y-axis.

SOLUTION The region is shown in Figure 7(a) and the resulting solid is shown in Figure 7(b). Because the region is rotated about the y-axis, it makes sense to slice the solid perpendicular to the y-axis and therefore to integrate with respect to y. If we slice at height y, we get a circular disk with radius x, where $x = \sqrt[3]{y}$. So the area of a cross-section through y is

$$A(y) = \pi x^2 = \pi(\sqrt[3]{y})^2 = \pi y^{2/3}$$

and the volume of the approximating cylinder pictured in Figure 7(b) is

$$A(y)\,\Delta y = \pi y^{2/3}\,\Delta y$$

Since the solid lies between $y = 0$ and $y = 8$, its volume is

$$V = \int_0^8 A(y)\,dy = \int_0^8 \pi y^{2/3}\,dy = \pi \left[\tfrac{3}{5} y^{5/3} \right]_0^8 = \frac{96\pi}{5}$$

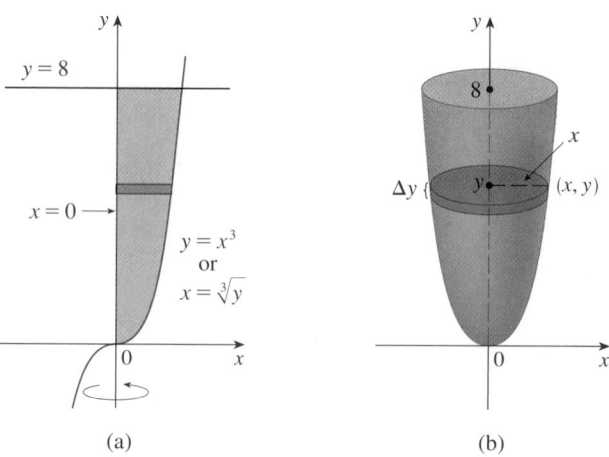

FIGURE 7 (a) (b)

EXAMPLE 4 The region $\mathcal{R}$ enclosed by the curves $y = x$ and $y = x^2$ is rotated about the x-axis. Find the volume of the resulting solid.

SOLUTION The curves $y = x$ and $y = x^2$ intersect at the points $(0, 0)$ and $(1, 1)$. The region between them, the solid of rotation, and a cross-section perpendicular to the x-axis are shown in Figure 8. A cross-section in the plane P_x has the shape of a *washer* (an annular ring) with inner radius x^2 and outer radius x, so we find the cross-sectional area by subtracting the area of the inner circle from the area of the outer circle:

$$A(x) = \pi x^2 - \pi(x^2)^2 = \pi(x^2 - x^4)$$

Therefore we have

$$V = \int_0^1 A(x)\,dx = \int_0^1 \pi(x^2 - x^4)\,dx$$

$$= \pi \left[\frac{x^3}{3} - \frac{x^5}{5} \right]_0^1 = \frac{2\pi}{15}$$

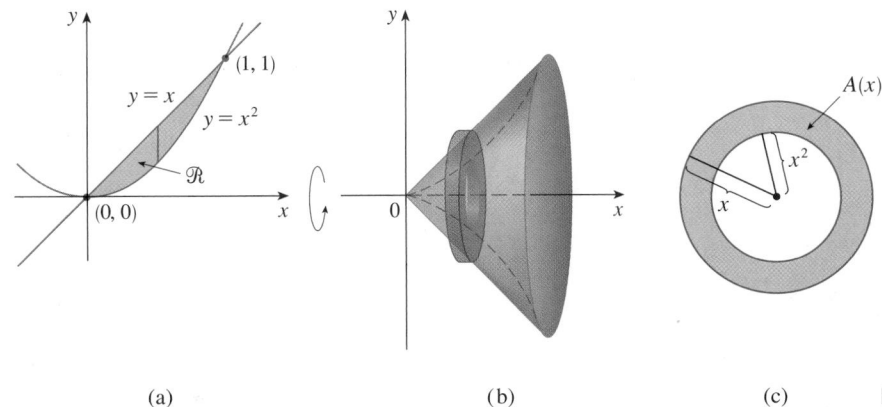

FIGURE 8
(a)
(b)
(c)

EXAMPLE 5 Find the volume of the solid obtained by rotating the region in Example 4 about the line $y = 2$.

SOLUTION The solid and a cross-section are shown in Figure 9. Again the cross-section is a washer, but this time the inner radius is $2 - x$ and the outer radius is $2 - x^2$. The cross-sectional area is

$$A(x) = \pi(2 - x^2)^2 - \pi(2 - x)^2$$

and so the volume of S is

$$V = \int_0^1 A(x)\, dx = \pi \int_0^1 \left[(2 - x^2)^2 - (2 - x)^2 \right] dx$$

$$= \pi \int_0^1 (x^4 - 5x^2 + 4x)\, dx$$

$$= \pi \left[\frac{x^5}{5} - 5\frac{x^3}{3} + 4\frac{x^2}{2} \right]_0^1 = \frac{8\pi}{15}$$

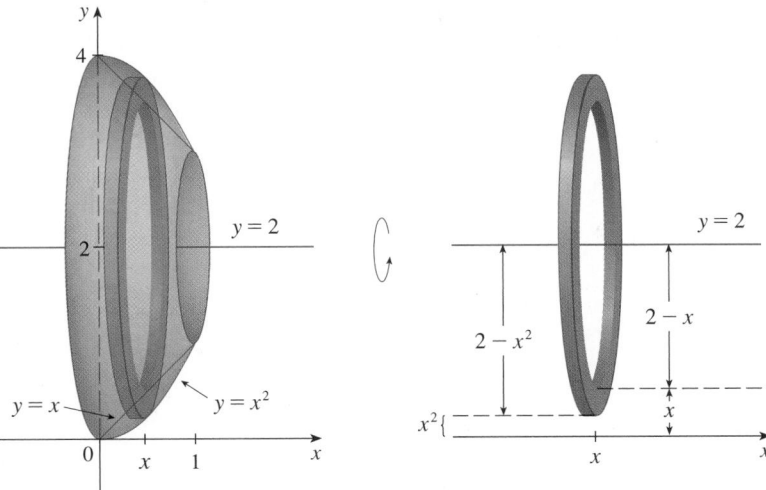

FIGURE 9

The solids in Examples 1–5 are all called **solids of revolution** because they are obtained by revolving a region about a line. In general, we calculate the volume of a solid of revolution by using the basic defining formula

$$V = \int_a^b A(x)\, dx \qquad \text{or} \qquad V = \int_c^d A(y)\, dy$$

and we find the cross-sectional area $A(x)$ or $A(y)$ in one of the following ways:

- If the cross-section is a disk (as in Examples 1–3), we find the radius of the disk (in terms of x or y) and use

$$A = \pi (\text{radius})^2$$

- If the cross-section is a washer (as in Examples 4 and 5), we find the inner radius r_{in} and outer radius r_{out} from a sketch (as in Figures 9 and 10) and compute the area of the washer by subtracting the area of the inner disk from the area of the outer disk:

$$A = \pi (\text{outer radius})^2 - \pi (\text{inner radius})^2$$

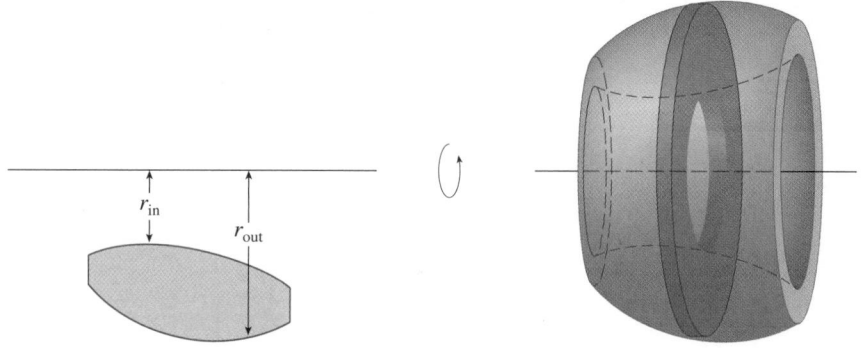

FIGURE 10

The next example gives a further illustration of the procedure.

EXAMPLE 6 Find the volume of the solid obtained by rotating the region in Example 4 about the line $x = -1$.

SOLUTION Figure 11 shows a horizontal cross-section. It is a washer with inner radius $1 + y$ and outer radius $1 + \sqrt{y}$, so the cross-sectional area is

$$A(y) = \pi (\text{outer radius})^2 - \pi (\text{inner radius})^2$$
$$= \pi (1 + \sqrt{y})^2 - \pi (1 + y)^2$$

The volume is

$$V = \int_0^1 A(y)\, dy = \pi \int_0^1 \left[(1 + \sqrt{y})^2 - (1 + y)^2 \right] dy$$

$$= \pi \int_0^1 \left(2\sqrt{y} - y - y^2 \right) dy$$

$$= \pi \left[\frac{4y^{3/2}}{3} - \frac{y^2}{2} - \frac{y^3}{3} \right]_0^1 = \frac{\pi}{2}$$

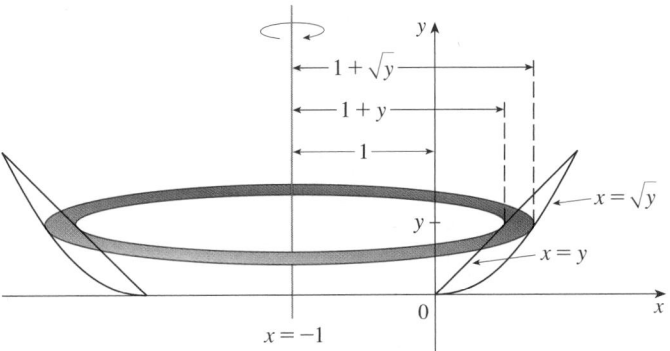

FIGURE 11 ■

TEC Visual 7.2C shows how the solid in Figure 12 is generated.

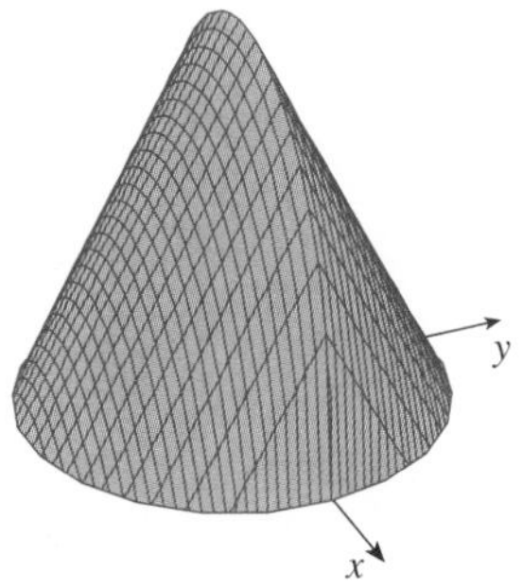

FIGURE 12
Computer-generated picture of the solid in Example 7

We now find the volumes of two solids that are *not* solids of revolution.

EXAMPLE 7 Figure 12 shows a solid with a circular base of radius 1. Parallel cross-sections perpendicular to the base are equilateral triangles. Find the volume of the solid.

SOLUTION Let's take the circle to be $x^2 + y^2 = 1$. The solid, its base, and a typical cross-section at a distance x from the origin are shown in Figure 13.

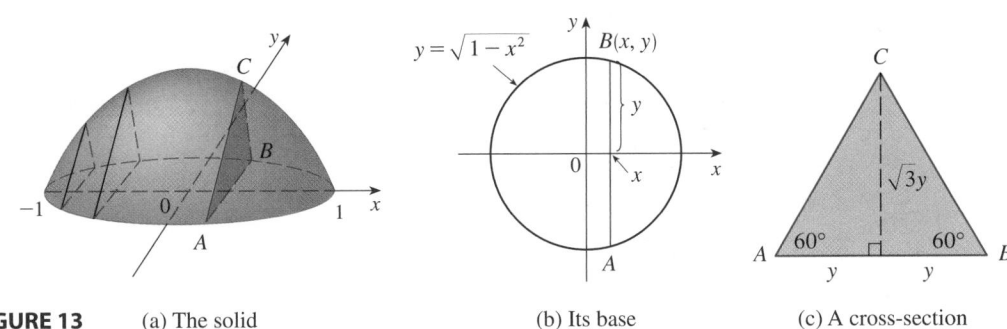

FIGURE 13 (a) The solid (b) Its base (c) A cross-section

Since B lies on the circle, we have $y = \sqrt{1 - x^2}$ and so the base of the triangle ABC is $|AB| = 2\sqrt{1 - x^2}$. Since the triangle is equilateral, we see from Figure 13(c) that its height is $\sqrt{3}\,y = \sqrt{3}\sqrt{1 - x^2}$. The cross-sectional area is therefore

$$A(x) = \tfrac{1}{2} \cdot 2\sqrt{1 - x^2} \cdot \sqrt{3}\sqrt{1 - x^2} = \sqrt{3}\,(1 - x^2)$$

and the volume of the solid is

$$V = \int_{-1}^{1} A(x)\,dx = \int_{-1}^{1} \sqrt{3}\,(1 - x^2)\,dx$$

$$= 2\int_{0}^{1} \sqrt{3}\,(1 - x^2)\,dx = 2\sqrt{3}\left[x - \frac{x^3}{3}\right]_0^1 = \frac{4\sqrt{3}}{3}$$ ■

V EXAMPLE 8 Find the volume of a pyramid whose base is a square with side L and whose height is h.

SOLUTION We place the origin O at the vertex of the pyramid and the x-axis along its central axis as in Figure 14. Any plane P_x that passes through x and is perpendicular to the x-axis intersects the pyramid in a square with side of length s, say. We can express s in terms of x by observing from the similar triangles in Figure 15 that

$$\frac{x}{h} = \frac{s/2}{L/2} = \frac{s}{L}$$

and so $s = Lx/h$. [Another method is to observe that the line OP has slope $L/(2h)$ and so its equation is $y = Lx/(2h)$.] Thus the cross-sectional area is

$$A(x) = s^2 = \frac{L^2}{h^2} x^2$$

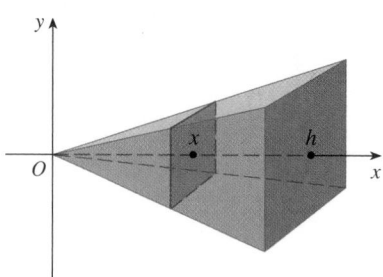

FIGURE 14   **FIGURE 15**

The pyramid lies between $x = 0$ and $x = h$, so its volume is

$$V = \int_0^h A(x)\,dx = \int_0^h \frac{L^2}{h^2} x^2\,dx = \frac{L^2}{h^2}\frac{x^3}{3}\Bigg]_0^h = \frac{L^2 h}{3} \qquad \blacksquare$$

NOTE We didn't need to place the vertex of the pyramid at the origin in Example 8. We did so merely to make the equations simple. If, instead, we had placed the center of the base at the origin and the vertex on the positive y-axis, as in Figure 16, you can verify that we would have obtained the integral

$$V = \int_0^h \frac{L^2}{h^2}(h - y)^2\,dy = \frac{L^2 h}{3}$$

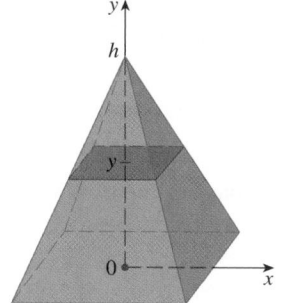

FIGURE 16

7.2 | EXERCISES

1–12 ▪ Find the volume of the solid obtained by rotating the region bounded by the given curves about the specified line. Sketch the region, the solid, and a typical disk or washer.

1. $y = 2 - \frac{1}{2}x$, $y = 0$, $x = 1$, $x = 2$; about the x-axis

2. $y = 1 - x^2$, $y = 0$; about the x-axis

3. $x = 2\sqrt{y}$, $x = 0$, $y = 9$; about the y-axis

4. $y = \ln x$, $y = 1$, $y = 2$, $x = 0$; about the y-axis

5. $y = x^3$, $y = x$, $x \geqslant 0$; about the x-axis

6. $y = \frac{1}{4}x^2$, $y = 5 - x^2$; about the x-axis

7. $y^2 = x$, $x = 2y$; about the y-axis

8. $y = \frac{1}{4}x^2$, $x = 2$, $y = 0$; about the y-axis

9. $y = x$, $y = \sqrt{x}$; about $y = 1$

10. $y = e^{-x}$, $y = 1$, $x = 2$; about $y = 2$

11. $y = 1 + \sec x$, $y = 3$; about $y = 1$

12. $y = x$, $y = \sqrt{x}$; about $x = 2$

13–18 ▪ The region enclosed by the given curves is rotated about the specified line. Find the volume of the resulting solid.

13. $y = 1/x$, $x = 1$, $x = 2$, $y = 0$; about the x-axis

14. $x = 2y - y^2$, $x = 0$; about the y-axis

15. $x - y = 1$, $y = x^2 - 4x + 3$; about $y = 3$

16. $x = y^2$, $x = 1$; about $x = 1$

17. $y = x^3$, $y = \sqrt{x}$; about $x = 1$

18. $y = x^3$, $y = \sqrt{x}$; about $y = 1$

19–22 ▪ Set up an integral for the volume of the solid obtained by rotating the region bounded by the given curves about the specified line. Then use your calculator to evaluate the integral correct to five decimal places.

19. $y = e^{-x^2}$, $y = 0$, $x = -1$, $x = 1$
(a) About the x-axis   (b) About $y = -1$

20. $y = 0$, $y = \cos^2 x$, $-\pi/2 \leqslant x \leqslant \pi/2$
(a) About the x-axis   (b) About $y = 1$

21. $x^2 + 4y^2 = 4$
(a) About $y = 2$ (b) About $x = 2$

22. $y = x^2$, $x^2 + y^2 = 1$, $y \geqslant 0$
(a) About the x-axis   (b) About the y-axis

⊞ **23–24** ▪ Use a graph to find approximate x-coordinates of the points of intersection of the given curves. Then use your calculator to find (approximately) the volume of the solid obtained by rotating about the x-axis the region bounded by these curves.

23. $y = 2 + x^2 \cos x$, $y = x^4 + x + 1$

24. $y = 3 \sin(x^2)$, $y = e^{x/2} + e^{-2x}$

CAS **25–26** ▪ Use a computer algebra system to find the exact volume of the solid obtained by rotating the region bounded by the given curves about the specified line.

25. $y = \sin^2 x$, $y = 0$, $0 \leqslant x \leqslant \pi$; about $y = -1$

26. $y = x$, $y = xe^{1-x/2}$; about $y = 3$

27–28 ▪ Each integral represents the volume of a solid. Describe the solid.

27. (a) $\pi \displaystyle\int_0^{\pi/2} \cos^2 x \, dx$ (b) $\pi \displaystyle\int_0^1 (y^4 - y^8) \, dy$

28. (a) $\pi \displaystyle\int_2^5 y \, dy$ (b) $\pi \displaystyle\int_0^{\pi/2} [(1 + \cos x)^2 - 1^2] \, dx$

---

29. A CAT scan produces equally spaced cross-sectional views of a human organ that provide information about the organ otherwise obtained only by surgery. Suppose that a CAT scan of a human liver shows cross-sections spaced 1.5 cm apart. The liver is 15 cm long and the cross-sectional areas, in square centimeters, are 0, 18, 58, 79, 94, 106, 117, 128, 63, 39, and 0. Use the Midpoint Rule to estimate the volume of the liver.

30. A log 10 m long is cut at 1-meter intervals and its cross-sectional areas A (at a distance x from the end of the log) are listed in the table. Use the Midpoint Rule with $n = 5$ to estimate the volume of the log.

x (m)	A (m^2)	x (m)	A (m^2)
0	0.68	6	0.53
1	0.65	7	0.55
2	0.64	8	0.52
3	0.61	9	0.50
4	0.58	10	0.48
5	0.59		

31–43 ▪ Find the volume of the described solid S.

31. A right circular cone with height h and base radius r

32. A frustum of a right circular cone with height h, lower base radius R, and top radius r

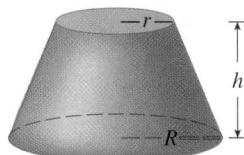

33. A cap of a sphere with radius r and height h

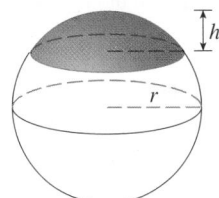

34. A frustum of a pyramid with square base of side b, square top of side a, and height h

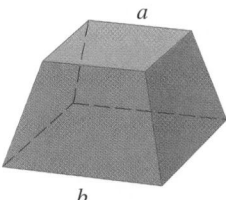

What happens if $a = b$? What happens if $a = 0$?

35. A pyramid with height h and rectangular base with dimensions b and $2b$

36. A pyramid with height h and base an equilateral triangle with side a (a tetrahedron)

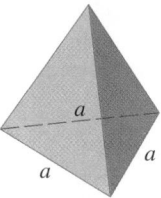

37. A tetrahedron with three mutually perpendicular faces and three mutually perpendicular edges with lengths 3 cm, 4 cm, and 5 cm

38. The base of S is a circular disk with radius r. Parallel cross-sections perpendicular to the base are squares.

39. The base of S is an elliptical region with boundary curve $9x^2 + 4y^2 = 36$. Cross-sections perpendicular to the x-axis are isosceles right triangles with hypotenuse in the base.

40. The base of S is the triangular region with vertices $(0, 0)$, $(1, 0)$, and $(0, 1)$. Cross-sections perpendicular to the y-axis are equilateral triangles.

41. The base of S is the same base as in Exercise 40, but cross-sections perpendicular to the x-axis are squares.

42. The base of S is the region enclosed by the parabola $y = 1 - x^2$ and the x-axis. Cross-sections perpendicular to the y-axis are squares.

43. The base of S is the same base as in Exercise 42, but cross-sections perpendicular to the x-axis are isosceles triangles with height equal to the base.

44. The base of S is a circular disk with radius r. Parallel cross-sections perpendicular to the base are isosceles triangles with height h and unequal side in the base.
(a) Set up an integral for the volume of S.
(b) By interpreting the integral as an area, find the volume of S.

45. Some of the pioneers of calculus, such as Kepler and Newton, were inspired by the problem of finding the volumes of wine barrels. (In fact Kepler published a book *Stereometria doliorum* in 1715 devoted to methods for finding the volumes of barrels.) They often approximated the shape of the sides by parabolas.
(a) A barrel with height h and maximum radius R is constructed by rotating about the x-axis the parabola $y = R - cx^2$, $-h/2 \le x \le h/2$, where c is a positive constant. Show that the radius of each end of the barrel is $r = R - d$, where $d = ch^2/4$.
(b) Show that the volume enclosed by the barrel is

$$V = \tfrac{1}{3}\pi h\left(2R^2 + r^2 - \tfrac{2}{5}d^2\right)$$

CAS **46.** (a) A model for the shape of a bird's egg is obtained by rotating about the x-axis the region under the graph of

$$f(x) = (ax^3 + bx^2 + cx + d)\sqrt{1 - x^2}$$

Use a CAS to find the volume of such an egg.
(b) For a Red-throated Loon, $a = -0.06$, $b = 0.04$, $c = 0.1$, and $d = 0.54$. Graph f and find the volume of an egg of this bird.

47. (a) Set up an integral for the volume of a solid *torus* (the donut-shaped solid shown in the figure) with radii r and R.
(b) By interpreting the integral as an area, find the volume of the torus.

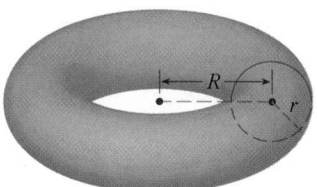

48. A wedge is cut out of a circular cylinder of radius 4 by two planes. One plane is perpendicular to the axis of the cylinder. The other intersects the first at an angle of 30° along a diameter of the cylinder. Find the volume of the wedge.

49. (a) Cavalieri's Principle states that if a family of parallel planes gives equal cross-sectional areas for two solids S_1 and S_2, then the volumes of S_1 and S_2 are equal. Prove this principle.
(b) Use Cavalieri's Principle to find the volume of the oblique cylinder shown in the figure.

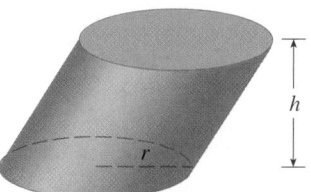

50. Find the volume common to two circular cylinders, each with radius r, if the axes of the cylinders intersect at right angles.

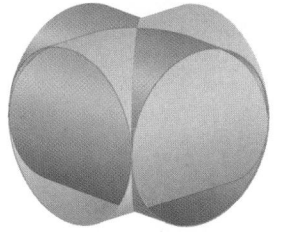 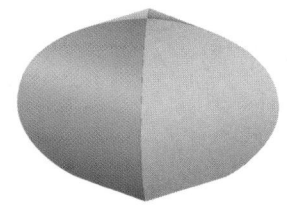

51. Find the volume common to two spheres, each with radius r, if the center of each sphere lies on the surface of the other sphere.

52. A bowl is shaped like a hemisphere with diameter 30 cm. A ball with diameter 10 cm is placed in the bowl and water is poured into the bowl to a depth of h centimeters. Find the volume of water in the bowl.

53. A hole of radius r is bored through a cylinder of radius $R > r$ at right angles to the axis of the cylinder. Set up, but do not evaluate, an integral for the volume cut out.

54. A hole of radius r is bored through the center of a sphere of radius $R > r$. Find the volume of the remaining portion of the sphere.

7.3 | VOLUMES BY CYLINDRICAL SHELLS

Some volume problems are very difficult to handle by the methods of the preceding section. For instance, let's consider the problem of finding the volume of the solid obtained by rotating about the y-axis the region bounded by $y = 2x^2 - x^3$ and $y = 0$. (See Figure 1.) If we slice perpendicular to the y-axis, we get a washer. But to compute the inner radius and the outer radius of the washer, we would have to solve the cubic equation $y = 2x^2 - x^3$ for x in terms of y; that's not easy.

Fortunately, there is a method, called the **method of cylindrical shells**, that is easier to use in such a case. Figure 2 shows a cylindrical shell with inner radius r_1, outer radius r_2, and height h. Its volume V is calculated by subtracting the volume V_1 of the inner cylinder from the volume V_2 of the outer cylinder:

$$V = V_2 - V_1$$
$$= \pi r_2^2 h - \pi r_1^2 h = \pi (r_2^2 - r_1^2)h$$
$$= \pi (r_2 + r_1)(r_2 - r_1)h$$
$$= 2\pi \frac{r_2 + r_1}{2} h(r_2 - r_1)$$

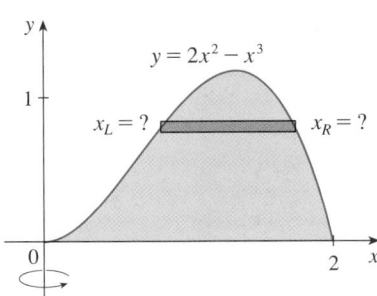

FIGURE 1

FIGURE 2

If we let $\Delta r = r_2 - r_1$ (the thickness of the shell) and $r = \frac{1}{2}(r_2 + r_1)$ (the average radius of the shell), then this formula for the volume of a cylindrical shell becomes

$$\boxed{1} \qquad \boxed{V = 2\pi r h \, \Delta r}$$

and it can be remembered as

$$V = [\text{circumference}][\text{height}][\text{thickness}]$$

Now let S be the solid obtained by rotating about the y-axis the region bounded by $y = f(x)$ [where f is continuous and $f(x) \geq 0$], $y = 0$, $x = a$, and $x = b$, where $b > a \geq 0$. (See Figure 3 on page 382.)

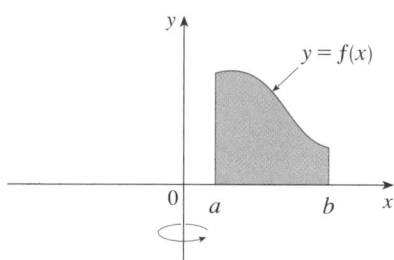
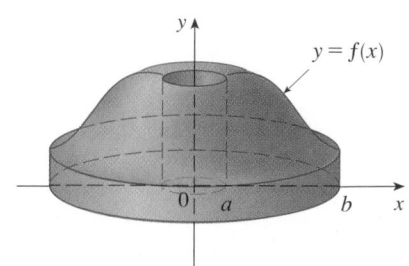

FIGURE 3

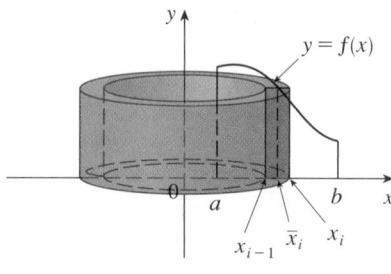

We divide the interval $[a, b]$ into n subintervals $[x_{i-1}, x_i]$ of equal width Δx and let $\bar{x}_i$ be the midpoint of the ith subinterval. If the rectangle with base $[x_{i-1}, x_i]$ and height $f(\bar{x}_i)$ is rotated about the y-axis, then the result is a cylindrical shell with average radius $\bar{x}_i$, height $f(\bar{x}_i)$, and thickness Δx (see Figure 4), so by Formula 1 its volume is

$$V_i = (2\pi\bar{x}_i)[f(\bar{x}_i)]\,\Delta x$$

Therefore an approximation to the volume V of S is given by the sum of the volumes of these shells:

$$V \approx \sum_{i=1}^{n} V_i = \sum_{i=1}^{n} 2\pi\bar{x}_i f(\bar{x}_i)\,\Delta x$$

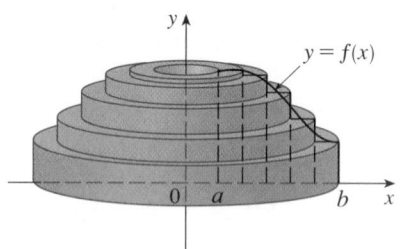

This approximation appears to become better as $n \to \infty$. But, from the definition of an integral, we know that

$$\lim_{n\to\infty} \sum_{i=1}^{n} 2\pi\bar{x}_i f(\bar{x}_i)\,\Delta x = \int_a^b 2\pi x f(x)\,dx$$

Thus the following appears plausible; a proof is outlined in Exercise 43.

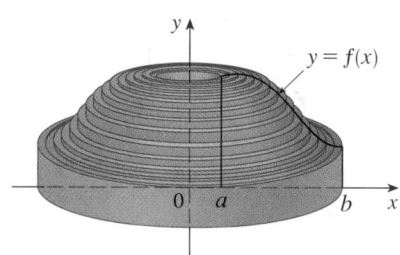

> **2**   The volume of the solid in Figure 3, obtained by rotating about the y-axis the region under the curve $y = f(x)$ from a to b, is
>
> $$V = \int_a^b 2\pi x f(x)\,dx \qquad \text{where } 0 \leq a < b$$

FIGURE 4

The best way to remember Formula 2 is to think of a typical shell, cut and flattened as in Figure 5, with radius x, circumference $2\pi x$, height $f(x)$, and thickness Δx or dx:

$$\int_a^b \underbrace{(2\pi x)}_{\text{circumference}}\ \underbrace{[f(x)]}_{\text{height}}\ dx$$

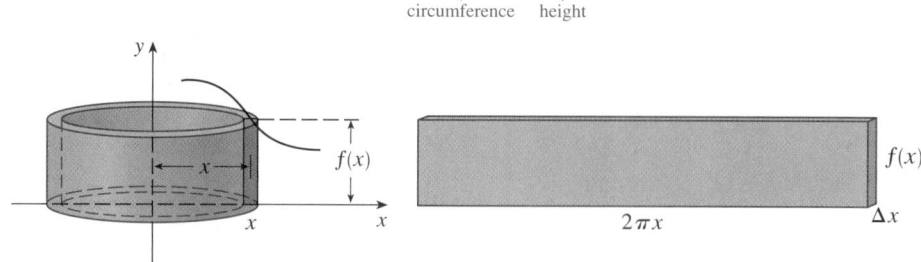

FIGURE 5

This type of reasoning will be helpful in other situations, such as when we rotate about lines other than the y-axis.

EXAMPLE 1 Find the volume of the solid obtained by rotating about the y-axis the region bounded by $y = 2x^2 - x^3$ and $y = 0$.

SOLUTION From the sketch in Figure 6 we see that a typical shell has radius x, circumference $2\pi x$, and height $f(x) = 2x^2 - x^3$. So, by the shell method, the volume is

$$V = \int_0^2 (2\pi x)(2x^2 - x^3)\,dx = 2\pi \int_0^2 (2x^3 - x^4)\,dx$$

$$= 2\pi\left[\tfrac{1}{2}x^4 - \tfrac{1}{5}x^5\right]_0^2 = 2\pi\left(8 - \tfrac{32}{5}\right) = \tfrac{16}{5}\pi$$

It can be verified that the shell method gives the same answer as slicing. ∎

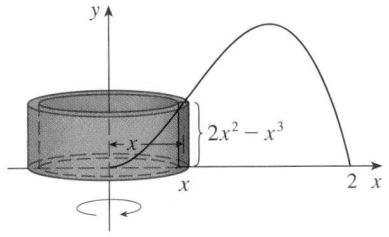

FIGURE 6

■ Figure 7 shows a computer-generated picture of the solid whose volume we computed in Example 1.

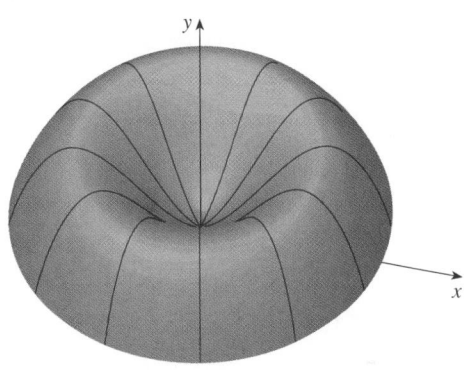

FIGURE 7

NOTE Comparing the solution of Example 1 with the remarks at the beginning of this section, we see that the method of cylindrical shells is much easier than the washer method for this problem. We did not have to find the coordinates of the local maximum and we did not have to solve the equation of the curve for x in terms of y. However, in other examples the methods of the preceding section may be easier.

V EXAMPLE 2 Find the volume of the solid obtained by rotating about the y-axis the region between $y = x$ and $y = x^2$.

SOLUTION The region and a typical shell are shown in Figure 8. We see that the shell has radius x, circumference $2\pi x$, and height $x - x^2$. So the volume is

$$V = \int_0^1 (2\pi x)(x - x^2)\,dx = 2\pi \int_0^1 (x^2 - x^3)\,dx$$

$$= 2\pi\left[\frac{x^3}{3} - \frac{x^4}{4}\right]_0^1 = \frac{\pi}{6}$$

∎

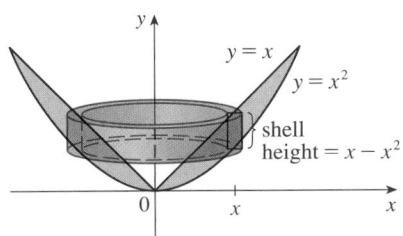

FIGURE 8

As the following example shows, the shell method works just as well if we rotate about the x-axis. We simply have to draw a diagram to identify the radius and height of a shell.

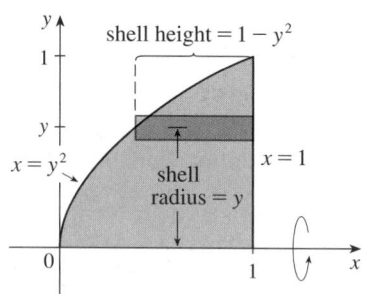

FIGURE 9

▼ **EXAMPLE 3** Use cylindrical shells to find the volume of the solid obtained by rotating about the *x*-axis the region under the curve $y = \sqrt{x}$ from 0 to 1.

SOLUTION This problem was solved using disks in Example 2 in Section 7.2. To use shells we relabel the curve $y = \sqrt{x}$ (in the figure in that example) as $x = y^2$ in Figure 9. For rotation about the *x*-axis we see that a typical shell has radius *y*, circumference $2\pi y$, and height $1 - y^2$. So the volume is

$$V = \int_0^1 (2\pi y)(1 - y^2)\, dy = 2\pi \int_0^1 (y - y^3)\, dy$$

$$= 2\pi \left[\frac{y^2}{2} - \frac{y^4}{4} \right]_0^1 = \frac{\pi}{2}$$

In this problem the disk method was simpler.   ∎

▼ **EXAMPLE 4** Find the volume of the solid obtained by rotating the region bounded by $y = x - x^2$ and $y = 0$ about the line $x = 2$.

SOLUTION Figure 10 shows the region and a cylindrical shell formed by rotation about the line $x = 2$. It has radius $2 - x$, circumference $2\pi(2 - x)$, and height $x - x^2$.

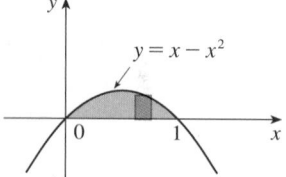

FIGURE 10

The volume of the given solid is

$$V = \int_0^1 2\pi(2 - x)(x - x^2)\, dx = 2\pi \int_0^1 (x^3 - 3x^2 + 2x)\, dx$$

$$= 2\pi \left[\frac{x^4}{4} - x^3 + x^2 \right]_0^1 = \frac{\pi}{2}$$ ∎

7.3 | EXERCISES

1. Let *S* be the solid obtained by rotating the region shown in the figure about the *y*-axis. Explain why it is awkward to use slicing to find the volume *V* of *S*. Sketch a typical approximating shell. What are its circumference and height? Use shells to find *V*.

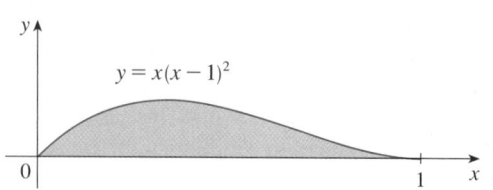

2. Let *S* be the solid obtained by rotating the region shown in the figure about the *y*-axis. Sketch a typical cylindrical shell and find its circumference and height. Use shells to find the volume of *S*. Do you think this method is preferable to slicing? Explain.

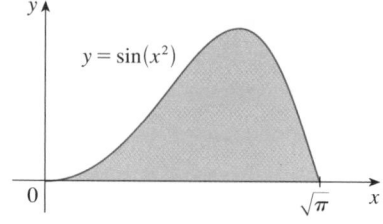

3–7 ▪ Use the method of cylindrical shells to find the volume generated by rotating the region bounded by the given curves about the *y*-axis.

3. $y = \sqrt[3]{x}$, $y = 0$, $x = 1$

4. $y = x^3$, $y = 0$, $x = 1$, $x = 2$

5. $y = e^{-x^2}$, $y = 0$, $x = 0$, $x = 1$

6. $y = 4x - x^2$, $y = x$

7. $y = x^2$, $y = 6x - 2x^2$

8. Let *V* be the volume of the solid obtained by rotating about the *y*-axis the region bounded by $y = \sqrt{x}$ and $y = x^2$. Find *V* both by slicing and by cylindrical shells. In both cases draw a diagram to explain your method.

9–14 ▪ Use the method of cylindrical shells to find the volume of the solid obtained by rotating the region bounded by the given curves about the *x*-axis.

9. $xy = 1$, $x = 0$, $y = 1$, $y = 3$

10. $y = \sqrt{x}$, $x = 0$, $y = 2$

11. $y = x^3$, $y = 8$, $x = 0$

12. $x = 4y^2 - y^3$, $x = 0$

13. $x = 1 + (y - 2)^2$, $x = 2$

14. $x + y = 3$, $x = 4 - (y - 1)^2$

15–20 ▪ Use the method of cylindrical shells to find the volume generated by rotating the region bounded by the given curves about the specified axis.

15. $y = x^4$, $y = 0$, $x = 1$;   about $x = 2$

16. $y = \sqrt{x}$, $y = 0$, $x = 1$;   about $x = -1$

17. $y = 4x - x^2$, $y = 3$;   about $x = 1$

18. $y = x^2$, $y = 2 - x^2$;   about $x = 1$

19. $y = x^3$, $y = 0$, $x = 1$;   about $y = 1$

20. $x = y^2 + 1$, $x = 2$;   about $y = -2$

21–26 ▪

(a) Set up an integral for the volume of the solid obtained by rotating the region bounded by the given curves about the specified axis.

(b) Use your calculator to evaluate the integral correct to five decimal places.

21. $y = xe^{-x}$, $y = 0$, $x = 2$;   about the *y*-axis

22. $y = \tan x$, $y = 0$, $x = \pi/4$;   about $x = \pi/2$

23. $y = \cos^4 x$, $y = -\cos^4 x$, $-\pi/2 \leqslant x \leqslant \pi/2$;   about $x = \pi$

24. $y = x$, $y = 2x/(1 + x^3)$;   about $x = -1$

25. $x = \sqrt{\sin y}$, $0 \leqslant y \leqslant \pi$, $x = 0$;   about $y = 4$

26. $x^2 - y^2 = 7$, $x = 4$;   about $y = 5$

27. Use the Midpoint Rule with $n = 4$ to estimate the volume obtained by rotating about the *y*-axis the region under the curve $y = \tan x$, $0 \leqslant x \leqslant \pi/4$.

28. (a) If the region shown in the figure is rotated about the *y*-axis to form a solid, use Simpson's Rule with $n = 8$ to estimate the volume of the solid.
(b) Estimate the volume if the region is rotated about the *x*-axis.

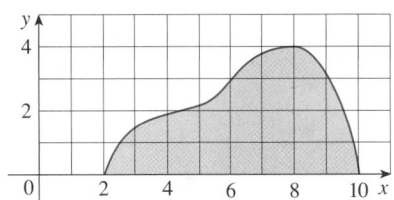

29–32 ▪ Each integral represents the volume of a solid. Describe the solid.

29. $\displaystyle\int_0^3 2\pi x^5 \, dx$

30. $\displaystyle 2\pi \int_0^2 \frac{y}{1 + y^2} \, dy$

31. $\displaystyle\int_0^1 2\pi (3 - y)(1 - y^2) \, dy$

32. $\displaystyle\int_0^{\pi/4} 2\pi (\pi - x)(\cos x - \sin x) \, dx$

33–38 ▪ The region bounded by the given curves is rotated about the specified axis. Find the volume of the resulting solid by any method.

33. $y = -x^2 + 6x - 8$, $y = 0$;   about the *y*-axis

34. $y = -x^2 + 6x - 8$, $y = 0$;   about the *x*-axis

35. $y^2 - x^2 = 1$, $y = 2$;   about the *x*-axis

36. $y^2 - x^2 = 1$, $y = 2$;   about the *y*-axis

37. $x^2 + (y - 1)^2 = 1$;   about the *y*-axis

38. $x = (y - 3)^2$, $x = 4$;   about $y = 1$

39–41 ▪ Use cylindrical shells to find the volume of the solid.

39. A sphere of radius *r*

40. The solid torus of Exercise 47 in Section 7.2

41. A right circular cone with height *h* and base radius *r*

42. Suppose you make napkin rings by drilling holes with different diameters through two wooden balls (which also have different diameters). You discover that both napkin rings have the same height h, as shown in the figure.

(a) Guess which ring has more wood in it.

(b) Check your guess: Use cylindrical shells to compute the volume of a napkin ring created by drilling a hole with radius r through the center of a sphere of radius R and express the answer in terms of h.

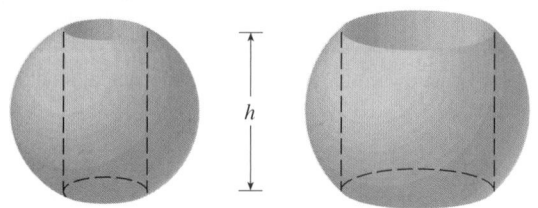

43. Use the following steps to prove Formula 2 for the case where f is one-to-one and therefore has an inverse function f^{-1}: Use the figure to show that

$$V = \pi b^2 d - \pi a^2 c - \int_c^d \pi [f^{-1}(y)]^2 \, dy$$

Make the substitution $y = f(x)$ and then use integration by parts on the resulting integral to prove that

$$V = \int_a^b 2\pi x f(x) \, dx$$

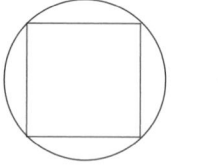

7.4 | ARC LENGTH

FIGURE 1

What is the length of this curve?

What do we mean by the length of a curve? We might think of fitting a piece of string to the curve in Figure 1 and then measuring the string against a ruler. But that might be difficult to do with much accuracy if we have a complicated curve. We need a precise definition for the length of an arc of a curve, in the same spirit as the definitions we developed for the concepts of area and volume.

If the curve is a polygon, we can easily find its length; we just add the lengths of the line segments that form the polygon. (We can use the distance formula to find the distance between the endpoints of each segment.) We are going to define the length of a general curve by first approximating it by a polygon and then taking a limit as the number of segments of the polygon is increased. This process is familiar for the case of a circle, where the circumference is the limit of lengths of inscribed polygons (see Figure 2).

TEC Visual 7.4 shows an animation of Figure 2.

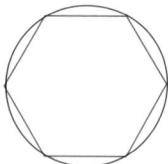  

FIGURE 2

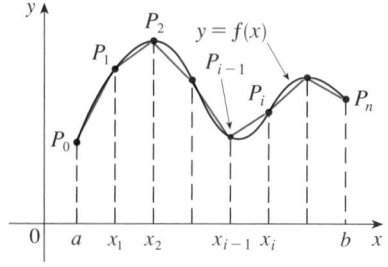

FIGURE 3

Now suppose that a curve C is defined by the equation $y = f(x)$, where f is continuous and $a \le x \le b$. We obtain a polygonal approximation to C by dividing the interval $[a, b]$ into n subintervals with endpoints $x_0, x_1, \ldots, x_n$ and equal width Δx. If $y_i = f(x_i)$, then the point $P_i(x_i, y_i)$ lies on C and the polygon with vertices $P_0, P_1, \ldots, P_n$, illustrated in Figure 3, is an approximation to C. The length L of C is approximately the length of this polygon and the approximation gets better as we let n increase. (See Figure 4, where the arc of the curve between P_{i-1} and P_i has been magnified and approximations with successively smaller values of Δx are shown.) Therefore we define the **length** L of the curve C with equation $y = f(x)$, $a \le x \le b$, as the limit of the lengths of these inscribed polygons (if the limit exists):

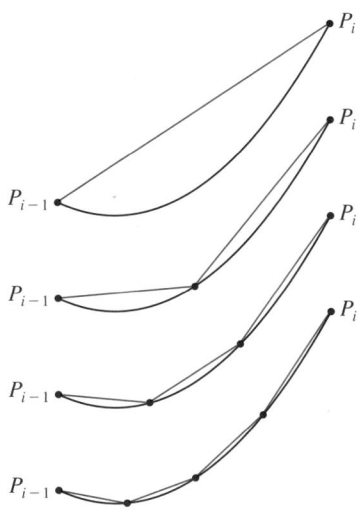

$$\boxed{1} \qquad L = \lim_{n \to \infty} \sum_{i=1}^{n} |P_{i-1}P_i|$$

Notice that the procedure for defining arc length is very similar to the procedure we used for defining area and volume: We divided the curve into a large number of small parts. We then found the approximate lengths of the small parts and added them. Finally, we took the limit as $n \to \infty$. (For the limit of a sequence, see the margin note on page 260 or Section 8.1.)

The definition of arc length given by Equation 1 is not very convenient for computational purposes, but we can derive an integral formula for L in the case where f has a continuous derivative. [Such a function f is called **smooth** because a small change in x produces a small change in $f'(x)$.]

If we let $\Delta y_i = y_i - y_{i-1}$, then

$$|P_{i-1}P_i| = \sqrt{(x_i - x_{i-1})^2 + (y_i - y_{i-1})^2} = \sqrt{(\Delta x)^2 + (\Delta y_i)^2}$$

By applying the Mean Value Theorem to f on the interval $[x_{i-1}, x_i]$, we find that there is a number x_i^* between x_{i-1} and x_i such that

$$f(x_i) - f(x_{i-1}) = f'(x_i^*)(x_i - x_{i-1})$$

that is, $\qquad\qquad\qquad\qquad \Delta y_i = f'(x_i^*)\,\Delta x$

Thus we have

$$|P_{i-1}P_i| = \sqrt{(\Delta x)^2 + (\Delta y_i)^2}$$
$$= \sqrt{(\Delta x)^2 + [f'(x_i^*)\,\Delta x]^2}$$
$$= \sqrt{1 + [f'(x_i^*)]^2}\,\sqrt{(\Delta x)^2}$$
$$= \sqrt{1 + [f'(x_i^*)]^2}\,\Delta x \qquad (\text{since } \Delta x > 0)$$

Therefore, by Definition 1,

$$L = \lim_{n \to \infty} \sum_{i=1}^{n} |P_{i-1}P_i| = \lim_{n \to \infty} \sum_{i=1}^{n} \sqrt{1 + [f'(x_i^*)]^2}\,\Delta x$$

We recognize this expression as being equal to

$$\int_a^b \sqrt{1 + [f'(x)]^2}\,dx$$

by the definition of a definite integral. This integral exists because the function $g(x) = \sqrt{1 + [f'(x)]^2}$ is continuous. Thus we have proved the following theorem:

> $\boxed{2}$ **THE ARC LENGTH FORMULA** If f' is continuous on $[a, b]$, then the length of the curve $y = f(x)$, $a \leqslant x \leqslant b$, is
>
> $$L = \int_a^b \sqrt{1 + [f'(x)]^2}\,dx$$

FIGURE 4

P_i
P_i
P_i
P_i
P_{i-1}
P_{i-1}
P_{i-1}
P_{i-1}

If we use Leibniz notation for derivatives, we can write the arc length formula as follows:

$$\boxed{3} \qquad L = \int_a^b \sqrt{1 + \left(\frac{dy}{dx}\right)^2}\, dx$$

EXAMPLE 1 Find the length of the arc of the semicubical parabola $y^2 = x^3$ between the points $(1, 1)$ and $(4, 8)$. (See Figure 5.)

SOLUTION For the top half of the curve we have

$$y = x^{3/2} \qquad \frac{dy}{dx} = \tfrac{3}{2} x^{1/2}$$

and so the arc length formula gives

$$L = \int_1^4 \sqrt{1 + \left(\frac{dy}{dx}\right)^2}\, dx = \int_1^4 \sqrt{1 + \tfrac{9}{4}x}\, dx$$

If we substitute $u = 1 + \tfrac{9}{4}x$, then $du = \tfrac{9}{4}\, dx$. When $x = 1$, $u = \tfrac{13}{4}$; when $x = 4$, $u = 10$. Therefore

$$L = \tfrac{4}{9} \int_{13/4}^{10} \sqrt{u}\, du = \tfrac{4}{9} \cdot \tfrac{2}{3} u^{3/2} \Big]_{13/4}^{10}$$

$$= \tfrac{8}{27}\left[10^{3/2} - \left(\tfrac{13}{4}\right)^{3/2}\right] = \tfrac{1}{27}\left(80\sqrt{10} - 13\sqrt{13}\right) \qquad \blacksquare$$

If a curve has the equation $x = g(y)$, $c \le y \le d$, and $g'(y)$ is continuous, then by interchanging the roles of x and y in Formula 2 or Equation 3, we obtain the following formula for its length:

$$\boxed{4} \qquad L = \int_c^d \sqrt{1 + [g'(y)]^2}\, dy = \int_c^d \sqrt{1 + \left(\frac{dx}{dy}\right)^2}\, dy$$

▼ **EXAMPLE 2** Find the length of the arc of the parabola $y^2 = x$ from $(0, 0)$ to $(1, 1)$.

SOLUTION Since $x = y^2$, we have $dx/dy = 2y$, and Formula 4 gives

$$L = \int_0^1 \sqrt{1 + \left(\frac{dx}{dy}\right)^2}\, dy = \int_0^1 \sqrt{1 + 4y^2}\, dy$$

We make the trigonometric substitution $y = \tfrac{1}{2} \tan\theta$, which gives $dy = \tfrac{1}{2} \sec^2\theta\, d\theta$ and $\sqrt{1 + 4y^2} = \sqrt{1 + \tan^2\theta} = \sec\theta$. When $y = 0$, $\tan\theta = 0$, so $\theta = 0$; when $y = 1$, $\tan\theta = 2$, so $\theta = \tan^{-1}2 = \alpha$, say. Thus

$$L = \int_0^\alpha \sec\theta \cdot \tfrac{1}{2}\sec^2\theta\, d\theta = \tfrac{1}{2}\int_0^\alpha \sec^3\theta\, d\theta$$

$$= \tfrac{1}{2} \cdot \tfrac{1}{2}\left[\sec\theta\,\tan\theta + \ln|\sec\theta + \tan\theta|\right]_0^\alpha \qquad \text{(from Example 8 in Section 6.2)}$$

$$= \tfrac{1}{4}\left(\sec\alpha\,\tan\alpha + \ln|\sec\alpha + \tan\alpha|\right)$$

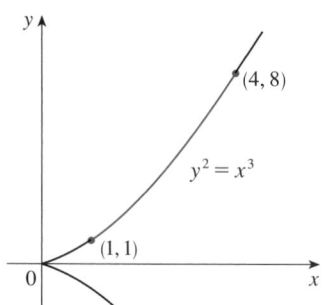

FIGURE 5

■ As a check on our answer to Example 1, notice from Figure 5 that it ought to be slightly larger than the distance from $(1, 1)$ to $(4, 8)$, which is

$$\sqrt{58} \approx 7.615773$$

According to our calculation in Example 1, we have

$$L = \tfrac{1}{27}\left(80\sqrt{10} - 13\sqrt{13}\right) \approx 7.633705$$

Sure enough, this is a bit greater than the length of the line segment.

(We could have used Formula 21 in the Table of Integrals.) Since $\tan \alpha = 2$, we have $\sec^2 \alpha = 1 + \tan^2 \alpha = 5$, so $\sec \alpha = \sqrt{5}$ and

$$L = \frac{\sqrt{5}}{2} + \frac{\ln(\sqrt{5} + 2)}{4} \qquad \blacksquare$$

■ Figure 6 shows the arc of the parabola whose length is computed in Example 2, together with polygonal approximations having $n = 1$ and $n = 2$ line segments, respectively. For $n = 1$ the approximate length is $L_1 = \sqrt{2}$, the diagonal of a square. The table shows the approximations L_n that we get by dividing $[0, 1]$ into n equal subintervals. Notice that each time we double the number of sides of the polygon, we get closer to the exact length, which is

$$L = \frac{\sqrt{5}}{2} + \frac{\ln(\sqrt{5} + 2)}{4} \approx 1.478943$$

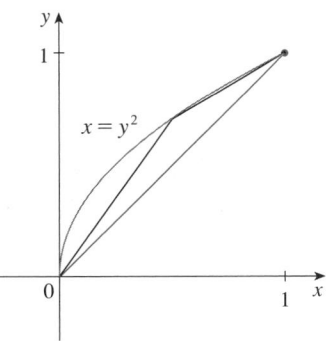

n	L_n
1	1.414
2	1.445
4	1.464
8	1.472
16	1.476
32	1.478
64	1.479

FIGURE 6

Because of the presence of the square root sign in Formulas 2 and 4, the calculation of an arc length often leads to an integral that is very difficult or even impossible to evaluate explicitly. Thus we sometimes have to be content with finding an approximation to the length of a curve as in the following example.

▾ EXAMPLE 3
(a) Set up an integral for the length of the arc of the hyperbola $xy = 1$ from the point $(1, 1)$ to the point $\left(2, \frac{1}{2}\right)$.
(b) Use Simpson's Rule with $n = 10$ to estimate the arc length.

SOLUTION
(a) We have

$$y = \frac{1}{x} \qquad \frac{dy}{dx} = -\frac{1}{x^2}$$

and so the arc length is

$$L = \int_1^2 \sqrt{1 + \left(\frac{dy}{dx}\right)^2} \; dx = \int_1^2 \sqrt{1 + \frac{1}{x^4}} \; dx = \int_1^2 \frac{\sqrt{x^4 + 1}}{x^2} \; dx$$

(b) Using Simpson's Rule (see Section 6.5) with $a = 1$, $b = 2$, $n = 10$, $\Delta x = 0.1$, and $f(x) = \sqrt{1 + 1/x^4}$, we have

■ Checking the value of the definite integral with a more accurate approximation produced by a computer algebra system, we see that the approximation using Simpson's Rule is accurate to four decimal places.

$$L = \int_1^2 \sqrt{1 + \frac{1}{x^4}} \; dx$$

$$\approx \frac{\Delta x}{3} [f(1) + 4f(1.1) + 2f(1.2) + 4f(1.3) + \cdots + 2f(1.8) + 4f(1.9) + f(2)]$$

$$\approx 1.1321 \qquad \blacksquare$$

THE ARC LENGTH FUNCTION

We will find it useful to have a function that measures the arc length of a curve from a particular starting point to any other point on the curve. Thus if a smooth curve C has the equation $y = f(x)$, $a \leq x \leq b$, let $s(x)$ be the distance along C from the initial point $P_0(a, f(a))$ to the point $Q(x, f(x))$. Then s is a function, called the **arc length function**, and, by Formula 2,

$$\boxed{5} \qquad s(x) = \int_a^x \sqrt{1 + [f'(t)]^2} \, dt$$

(We have replaced the variable of integration by t so that x does not have two meanings.) We can use Part 1 of the Fundamental Theorem of Calculus to differentiate Equation 5 (since the integrand is continuous):

$$\boxed{6} \qquad \frac{ds}{dx} = \sqrt{1 + [f'(x)]^2} = \sqrt{1 + \left(\frac{dy}{dx}\right)^2}$$

Equation 6 shows that the rate of change of s with respect to x is always at least 1 and is equal to 1 when $f'(x)$, the slope of the curve, is 0. The differential of arc length is

$$\boxed{7} \qquad ds = \sqrt{1 + \left(\frac{dy}{dx}\right)^2} \, dx$$

and this equation is sometimes written in the symmetric form

$$\boxed{8} \qquad (ds)^2 = (dx)^2 + (dy)^2$$

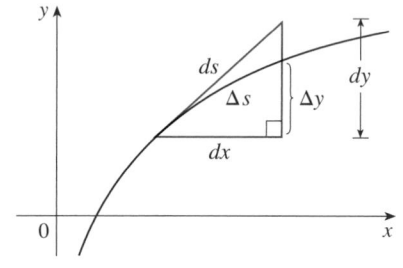

FIGURE 7

The geometric interpretation of Equation 8 is shown in Figure 7. It can be used as a mnemonic device for remembering both of the Formulas 3 and 4. If we write $L = \int ds$, then from Equation 8 either we can solve to get $\boxed{7}$, which gives $\boxed{3}$, or we can solve to get

$$ds = \sqrt{1 + \left(\frac{dx}{dy}\right)^2} \, dy$$

which gives $\boxed{4}$.

▼ **EXAMPLE 4** Find the arc length function for the curve $y = x^2 - \frac{1}{8} \ln x$ taking $P_0(1, 1)$ as the starting point.

SOLUTION If $f(x) = x^2 - \frac{1}{8} \ln x$, then

$$f'(x) = 2x - \frac{1}{8x}$$

$$1 + [f'(x)]^2 = 1 + \left(2x - \frac{1}{8x}\right)^2 = 1 + 4x^2 - \frac{1}{2} + \frac{1}{64x^2}$$

$$= 4x^2 + \frac{1}{2} + \frac{1}{64x^2} = \left(2x + \frac{1}{8x}\right)^2$$

$$\sqrt{1 + [f'(x)]^2} = 2x + \frac{1}{8x}$$

Thus the arc length function is given by

$$s(x) = \int_1^x \sqrt{1 + [f'(t)]^2}\, dt$$

$$= \int_1^x \left(2t + \frac{1}{8t}\right) dt = t^2 + \tfrac{1}{8}\ln t\Big]_1^x$$

$$= x^2 + \tfrac{1}{8}\ln x - 1$$

For instance, the arc length along the curve from $(1, 1)$ to $(3, f(3))$ is

$$s(3) = 3^2 + \tfrac{1}{8}\ln 3 - 1 = 8 + \frac{\ln 3}{8} \approx 8.1373$$ ∎

■ Figure 8 shows the interpretation of the arc length function in Example 4. Figure 9 shows the graph of this arc length function. Why is $s(x)$ negative when x is less than 1?

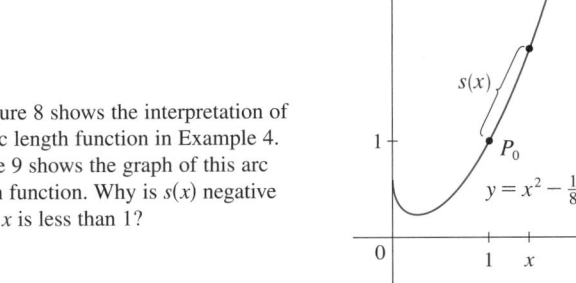

FIGURE 8

FIGURE 9

7.4 | EXERCISES

1. Use the arc length formula ⟨3⟩ to find the length of the curve $y = 2x - 5$, $-1 \leqslant x \leqslant 3$. Check your answer by noting that the curve is a line segment and calculating its length by the distance formula.

2. Use the arc length formula to find the length of the curve $y = \sqrt{2 - x^2}$, $0 \leqslant x \leqslant 1$. Check your answer by noting that the curve is part of a circle.

3–6 ■ Set up an integral that represents the length of the curve. Then use your calculator to find the length correct to four decimal places.

3. $y = \sin x$, $0 \leqslant x \leqslant \pi$

4. $y = xe^{-x}$, $0 \leqslant x \leqslant 2$

5. $x = \sqrt{y} - y$, $1 \leqslant y \leqslant 4$

6. $x = y^2 - 2y$, $0 \leqslant y \leqslant 2$

7–18 ■ Find the exact length of the curve.

7. $y = 1 + 6x^{3/2}$, $0 \leqslant x \leqslant 1$

8. $y^2 = 4(x + 4)^3$, $0 \leqslant x \leqslant 2$, $y > 0$

9. $y = \dfrac{x^3}{3} + \dfrac{1}{4x}$, $1 \leqslant x \leqslant 2$

10. $x = \dfrac{y^4}{8} + \dfrac{1}{4y^2}$, $1 \leqslant y \leqslant 2$

11. $x = \tfrac{1}{3}\sqrt{y}\,(y - 3)$, $1 \leqslant y \leqslant 9$

12. $y = \ln(\cos x)$, $0 \leqslant x \leqslant \pi/3$

13. $y = \ln(\sec x)$, $0 \leqslant x \leqslant \pi/4$

14. $y = 3 + \tfrac{1}{2}\cosh 2x$, $0 \leqslant x \leqslant 1$

15. $y = \tfrac{1}{4}x^2 - \tfrac{1}{2}\ln x$, $1 \leqslant x \leqslant 2$

16. $y = \sqrt{x - x^2} + \sin^{-1}(\sqrt{x}\,)$

17. $y = \ln(1 - x^2)$, $0 \leqslant x \leqslant \tfrac{1}{2}$

18. $y = 1 - e^{-x}$, $0 \leqslant x \leqslant 2$

19–20 ▪ Graph the curve and visually estimate its length. Then use your calculator to find the length correct to four decimal places.

19. $y = x^2 + x^3$, $1 \le x \le 2$

20. $y = x + \cos x$, $0 \le x \le \pi/2$

21–24 ▪ Use Simpson's Rule with $n = 10$ to estimate the arc length of the curve. Compare your answer with the value of the integral produced by your calculator.

21. $y = x \sin x$, $0 \le x \le 2\pi$

22. $y = \sqrt[3]{x}$, $1 \le x \le 6$

23. $y = \ln(1 + x^3)$, $0 \le x \le 5$

24. $y = e^{-x^2}$, $0 \le x \le 2$

CAS **25.** Use either a computer algebra system or a table of integrals to find the *exact* length of the arc of the curve $x = \ln(1 - y^2)$ that lies between the points $(0, 0)$ and $\left(\ln \frac{3}{4}, \frac{1}{2}\right)$.

CAS **26.** Use either a computer algebra system or a table of integrals to find the *exact* length of the arc of the curve $y = x^{4/3}$ that lies between the points $(0, 0)$ and $(1, 1)$. If your CAS has trouble evaluating the integral, make a substitution that changes the integral into one that the CAS can evaluate.

27. Sketch the curve with equation $x^{2/3} + y^{2/3} = 1$ and use symmetry to find its length.

28. (a) Sketch the curve $y^3 = x^2$.
(b) Use Formulas 3 and 4 to set up two integrals for the arc length from $(0, 0)$ to $(1, 1)$. Observe that one of these is an improper integral and evaluate both of them.
(c) Find the length of the arc of this curve from $(-1, 1)$ to $(8, 4)$.

29. Find the arc length function for the curve $y = 2x^{3/2}$ with starting point $P_0(1, 2)$.

30. (a) Find the arc length function for the curve $y = \ln(\sin x), 0 < x < \pi$, with starting point $(\pi/2, 0)$.
(b) Graph both the curve and its arc length function on the same screen.

31. Find the arc length function for the curve $y = \sin^{-1} x + \sqrt{1 - x^2}$ with starting point $(0, 1)$.

32. A steady wind blows a kite due west. The kite's height above ground from horizontal position $x = 0$ to $x = 80$ ft is given by $y = 150 - \frac{1}{40}(x - 50)^2$. Find the distance traveled by the kite.

33. A hawk flying at 15 m/s at an altitude of 180 m accidentally drops its prey. The parabolic trajectory of the falling prey is described by the equation

$$y = 180 - \frac{x^2}{45}$$

until it hits the ground, where y is its height above the ground and x is the horizontal distance traveled in meters. Calculate the distance traveled by the prey from the time it is dropped until the time it hits the ground. Express your answer correct to the nearest tenth of a meter.

34. (a) The figure shows a telephone wire hanging between two poles at $x = -b$ and $x = b$. It takes the shape of a catenary with equation $y = c + a \cosh(x/a)$. Find the length of the wire.
(b) Suppose two telephone poles are 50 ft apart and the length of the wire between the poles is 51 ft. If the lowest point of the wire must be 20 ft above the ground, how high up on each pole should the wire be attached?

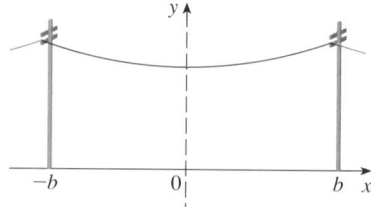

35. A manufacturer of corrugated metal roofing wants to produce panels that are 28 in. wide and 2 in. thick by processing flat sheets of metal as shown in the figure. The profile of the roofing takes the shape of a sine wave. Verify that the sine curve has equation $y = \sin(\pi x/7)$ and find the width w of a flat metal sheet that is needed to make a 28-inch panel. (Use your calculator to evaluate the integral correct to four significant digits.)

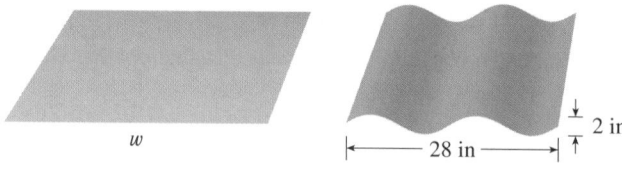

36. The curves with equations $x^n + y^n = 1, n = 4, 6, 8, \ldots$, are called **fat circles**. Graph the curves with $n = 2, 4, 6, 8$, and 10 to see why. Set up an integral for the length L_{2k} of the fat circle with $n = 2k$. Without attempting to evaluate this integral, state the value of $\lim_{k \to \infty} L_{2k}$.

7.5 | AREA OF A SURFACE OF REVOLUTION

A surface of revolution is formed when a curve is rotated about a line. Such a surface is the lateral boundary of a solid of revolution of the type discussed in Sections 7.2 and 7.3.

We want to define the area of a surface of revolution in such a way that it corresponds to our intuition. If the surface area is A, we can imagine that painting the surface would require the same amount of paint as does a flat region with area A.

Let's start with some simple surfaces. The lateral surface area of a circular cylinder with radius r and height h is taken to be $A = 2\pi rh$ because we can imagine cutting the cylinder and unrolling it (as in Figure 1) to obtain a rectangle with dimensions $2\pi r$ and h.

Likewise, we can take a circular cone with base radius r and slant height l, cut it along the dashed line in Figure 2, and flatten it to form a sector of a circle with radius l and central angle $\theta = 2\pi r/l$. We know that, in general, the area of a sector of a circle with radius l and angle θ is $\frac{1}{2}l^2\theta$ (see Exercise 69 in Section 6.2) and so in this case the area is

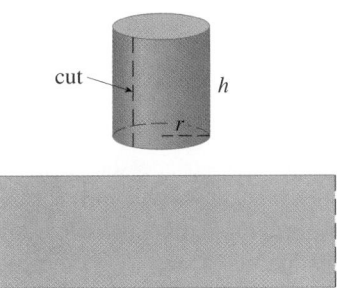

FIGURE 1

$$A = \tfrac{1}{2}l^2\theta = \tfrac{1}{2}l^2\left(\frac{2\pi r}{l}\right) = \pi rl$$

Therefore we define the lateral surface area of a cone to be $A = \pi rl$.

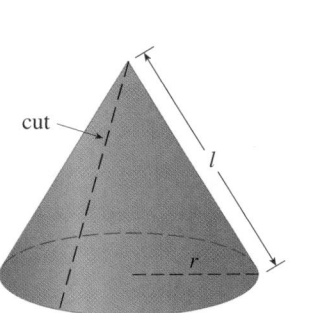
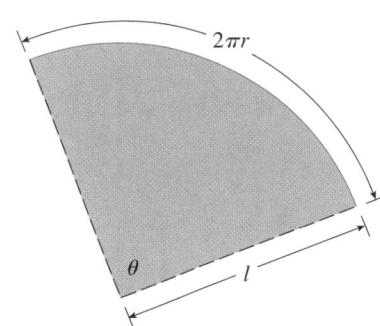

FIGURE 2

What about more complicated surfaces of revolution? If we follow the strategy we used with arc length, we can approximate the original curve by a polygon. When this polygon is rotated about an axis, it creates a simpler surface whose surface area approximates the actual surface area. By taking a limit, we can determine the exact surface area.

The approximating surface, then, consists of a number of *bands*, each formed by rotating a line segment about an axis. To find the surface area, each of these bands can be considered a portion of a circular cone, as shown in Figure 3. The area of the band (or frustum of a cone) with slant height l and upper and lower radii r_1 and r_2 is found by subtracting the areas of two cones:

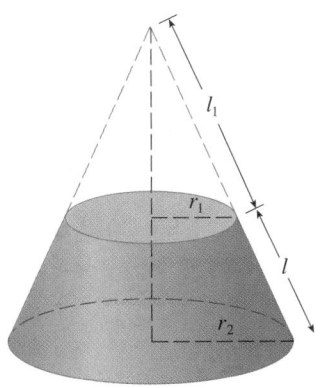

FIGURE 3

$$\boxed{1} \qquad A = \pi r_2(l_1 + l) - \pi r_1 l_1 = \pi[(r_2 - r_1)l_1 + r_2 l]$$

From similar triangles we have

$$\frac{l_1}{r_1} = \frac{l_1 + l}{r_2}$$

which gives

$$r_2 l_1 = r_1 l_1 + r_1 l \qquad \text{or} \qquad (r_2 - r_1)l_1 = r_1 l$$

Putting this in Equation 1, we get

$$A = \pi(r_1 l + r_2 l)$$

or

2

$$\boxed{A = 2\pi r l}$$

where $r = \frac{1}{2}(r_1 + r_2)$ is the average radius of the band.

Now we apply this formula to our strategy. Consider the surface shown in Figure 4, which is obtained by rotating the curve $y = f(x)$, $a \le x \le b$, about the x-axis, where f is positive and has a continuous derivative. In order to define its surface area, we divide the interval $[a, b]$ into n subintervals with endpoints $x_0, x_1, \ldots, x_n$ and equal width Δx, as we did in determining arc length. If $y_i = f(x_i)$, then the point $P_i(x_i, y_i)$ lies on the curve. The part of the surface between x_{i-1} and x_i is approximated by taking the line segment $P_{i-1}P_i$ and rotating it about the x-axis. The result is a band with slant height $l = |P_{i-1}P_i|$ and average radius $r = \frac{1}{2}(y_{i-1} + y_i)$ so, by Formula 2, its surface area is

$$2\pi \frac{y_{i-1} + y_i}{2} |P_{i-1}P_i|$$

As in the proof of Theorem 7.4.2, we have

$$|P_{i-1}P_i| = \sqrt{1 + [f'(x_i^*)]^2}\, \Delta x$$

where x_i^* is some number in $[x_{i-1}, x_i]$. When Δx is small, we have $y_i = f(x_i) \approx f(x_i^*)$ and also $y_{i-1} = f(x_{i-1}) \approx f(x_i^*)$, since f is continuous. Therefore

$$2\pi \frac{y_{i-1} + y_i}{2} |P_{i-1}P_i| \approx 2\pi f(x_i^*) \sqrt{1 + [f'(x_i^*)]^2}\, \Delta x$$

and so an approximation to what we think of as the area of the complete surface of revolution is

3

$$\sum_{i=1}^{n} 2\pi f(x_i^*) \sqrt{1 + [f'(x_i^*)]^2}\, \Delta x$$

This approximation appears to become better as $n \to \infty$ and, recognizing 3 as a Riemann sum for the function $g(x) = 2\pi f(x) \sqrt{1 + [f'(x)]^2}$, we have

$$\lim_{n \to \infty} \sum_{i=1}^{n} 2\pi f(x_i^*) \sqrt{1 + [f'(x_i^*)]^2}\, \Delta x = \int_a^b 2\pi f(x) \sqrt{1 + [f'(x)]^2}\, dx$$

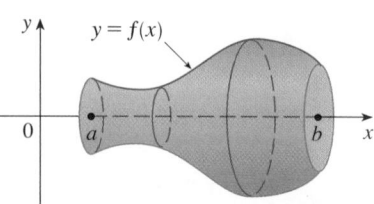

(a) Surface of revolution

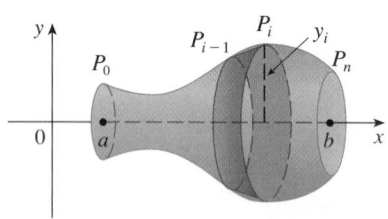

(b) Approximating band

FIGURE 4

Therefore, in the case where f is positive and has a continuous derivative, we define the **surface area** of the surface obtained by rotating the curve $y = f(x)$, $a \leq x \leq b$, about the x-axis as

$$\boxed{4} \qquad S = \int_a^b 2\pi f(x) \sqrt{1 + [f'(x)]^2} \, dx$$

With the Leibniz notation for derivatives, this formula becomes

$$\boxed{5} \qquad S = \int_a^b 2\pi y \sqrt{1 + \left(\frac{dy}{dx}\right)^2} \, dx$$

If the curve is described as $x = g(y)$, $c \leq y \leq d$, then the formula for surface area becomes

$$\boxed{6} \qquad S = \int_c^d 2\pi y \sqrt{1 + \left(\frac{dx}{dy}\right)^2} \, dy$$

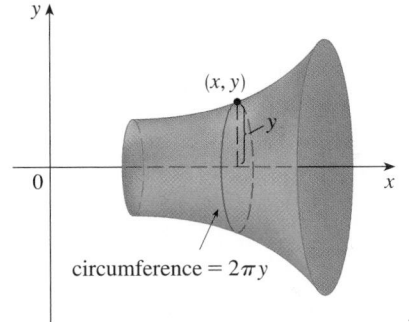

(a) Rotation about x-axis: $S = \int 2\pi y \, ds$

and both Formulas 5 and 6 can be summarized symbolically, using the notation for arc length given in Section 7.4, as

$$\boxed{7} \qquad S = \int 2\pi y \, ds$$

For rotation about the y-axis, the surface area formula becomes

$$\boxed{8} \qquad S = \int 2\pi x \, ds$$

where, as before, we can use either

$$ds = \sqrt{1 + \left(\frac{dy}{dx}\right)^2} \, dx \qquad \text{or} \qquad ds = \sqrt{1 + \left(\frac{dx}{dy}\right)^2} \, dy$$

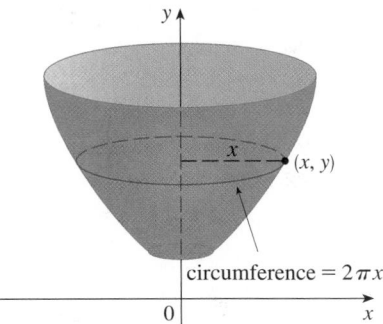

(b) Rotation about y-axis: $S = \int 2\pi x \, ds$

FIGURE 5

These formulas can be remembered by thinking of $2\pi y$ or $2\pi x$ as the circumference of a circle traced out by the point (x, y) on the curve as it is rotated about the x-axis or y-axis, respectively (see Figure 5).

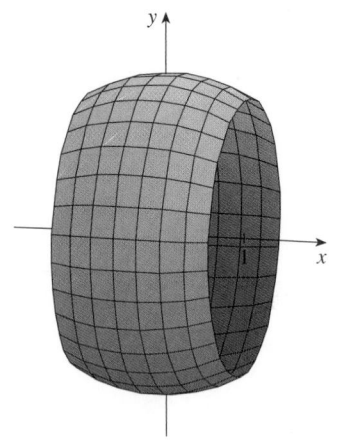

FIGURE 6

■ Figure 6 shows the portion of the sphere whose surface area is computed in Example 1.

■ Figure 7 shows the surface of revolution whose area is computed in Example 2.

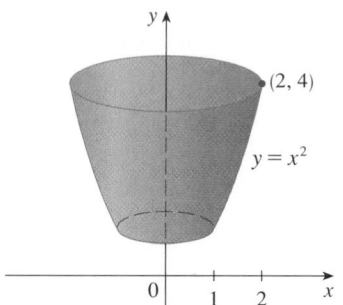

FIGURE 7

■ As a check on our answer to Example 2, notice from Figure 7 that the surface area should be close to that of a circular cylinder with the same height and radius halfway between the upper and lower radius of the surface: $2\pi(1.5)(3) \approx 28.27$. We computed that the surface area was

$$\frac{\pi}{6}\left(17\sqrt{17} - 5\sqrt{5}\right) \approx 30.85$$

which seems reasonable. Alternatively, the surface area should be slightly larger than the area of a frustum of a cone with the same top and bottom edges. From Equation 2, this is $2\pi(1.5)(\sqrt{10}) \approx 29.80$.

V EXAMPLE 1 The curve $y = \sqrt{4 - x^2}$, $-1 \leqslant x \leqslant 1$, is an arc of the circle $x^2 + y^2 = 4$. Find the area of the surface obtained by rotating this arc about the x-axis. (The surface is a portion of a sphere of radius 2. See Figure 6.)

SOLUTION We have

$$\frac{dy}{dx} = \tfrac{1}{2}(4 - x^2)^{-1/2}(-2x) = \frac{-x}{\sqrt{4 - x^2}}$$

and so, by Formula 5, the surface area is

$$S = \int_{-1}^{1} 2\pi y \sqrt{1 + \left(\frac{dy}{dx}\right)^2}\, dx$$

$$= 2\pi \int_{-1}^{1} \sqrt{4 - x^2} \sqrt{1 + \frac{x^2}{4 - x^2}}\, dx$$

$$= 2\pi \int_{-1}^{1} \sqrt{4 - x^2} \frac{2}{\sqrt{4 - x^2}}\, dx$$

$$= 4\pi \int_{-1}^{1} 1\, dx = 4\pi(2) = 8\pi$$ ■

V EXAMPLE 2 The arc of the parabola $y = x^2$ from $(1, 1)$ to $(2, 4)$ is rotated about the y-axis. Find the area of the resulting surface.

SOLUTION 1 Using

$$y = x^2 \qquad \text{and} \qquad \frac{dy}{dx} = 2x$$

we have, from Formula 8,

$$S = \int 2\pi x\, ds$$

$$= \int_{1}^{2} 2\pi x \sqrt{1 + \left(\frac{dy}{dx}\right)^2}\, dx$$

$$= 2\pi \int_{1}^{2} x \sqrt{1 + 4x^2}\, dx$$

Substituting $u = 1 + 4x^2$, we have $du = 8x\, dx$. Remembering to change the limits of integration, we have

$$S = \frac{\pi}{4} \int_{5}^{17} \sqrt{u}\, du = \frac{\pi}{4} \left[\tfrac{2}{3} u^{3/2}\right]_{5}^{17}$$

$$= \frac{\pi}{6}\left(17\sqrt{17} - 5\sqrt{5}\right)$$

SOLUTION 2 Using

$$x = \sqrt{y} \qquad \text{and} \qquad \frac{dx}{dy} = \frac{1}{2\sqrt{y}}$$

we have

$$S = \int 2\pi x \, ds = \int_1^4 2\pi x \sqrt{1 + \left(\frac{dx}{dy}\right)^2} \, dy$$

$$= 2\pi \int_1^4 \sqrt{y} \sqrt{1 + \frac{1}{4y}} \, dy = \pi \int_1^4 \sqrt{4y + 1} \, dy$$

$$= \frac{\pi}{4} \int_5^{17} \sqrt{u} \, du \quad \text{(where } u = 1 + 4y\text{)}$$

$$= \frac{\pi}{6}\left(17\sqrt{17} - 5\sqrt{5}\right) \quad \text{(as in Solution 1)} \qquad \blacksquare$$

▪ **www.stewartcalculus.com**
See Additional Example A.

7.5 | EXERCISES

1–4 ▪
(a) Set up an integral for the area of the surface obtained by rotating the curve about (i) the *x*-axis and (ii) the *y*-axis.
(b) Use the numerical integration capability of your calculator to evaluate the surface areas correct to four decimal places.

1. $y = \tan x, \quad 0 \leqslant x \leqslant \pi/3$

2. $y = x^{-2}, \quad 1 \leqslant x \leqslant 2$

3. $y = e^{-x^2}, \quad -1 \leqslant x \leqslant 1$

4. $x = \ln(2y + 1), \quad 0 \leqslant y \leqslant 1$

5–12 ▪ Find the exact area of the surface obtained by rotating the curve about the *x*-axis.

5. $y = x^3, \quad 0 \leqslant x \leqslant 2$

6. $9x = y^2 + 18, \quad 2 \leqslant x \leqslant 6$

7. $y = \sqrt{1 + 4x}, \quad 1 \leqslant x \leqslant 5$

8. $y = \sqrt{1 + e^x}, \quad 0 \leqslant x \leqslant 1$

9. $y = \sin \pi x, \quad 0 \leqslant x \leqslant 1$

10. $y = \dfrac{x^3}{6} + \dfrac{1}{2x}, \quad \tfrac{1}{2} \leqslant x \leqslant 1$

11. $x = \tfrac{1}{3}(y^2 + 2)^{3/2}, \quad 1 \leqslant y \leqslant 2$

12. $x = 1 + 2y^2, \quad 1 \leqslant y \leqslant 2$

13–16 ▪ The given curve is rotated about the *y*-axis. Find the area of the resulting surface.

13. $y = \sqrt[3]{x}, \quad 1 \leqslant y \leqslant 2$

14. $y = 1 - x^2, \quad 0 \leqslant x \leqslant 1$

15. $x = \sqrt{a^2 - y^2}, \quad 0 \leqslant y \leqslant a/2$

16. $y = \tfrac{1}{4}x^2 - \tfrac{1}{2}\ln x, \quad 1 \leqslant x \leqslant 2$

CAS **17–18** ▪ Use a CAS to find the exact area of the surface obtained by rotating the curve about the *y*-axis. If your CAS has trouble evaluating the integral, express the surface area as an integral in the other variable.

17. $y = x^3, \quad 0 \leqslant y \leqslant 1$

18. $y = \ln(x + 1), \quad 0 \leqslant x \leqslant 1$

19. (a) If $a > 0$, find the area of the surface generated by rotating the loop of the curve $3ay^2 = x(a - x)^2$ about the *x*-axis.
(b) Find the surface area if the loop is rotated about the *y*-axis.

20. A group of engineers is building a parabolic satellite dish whose shape will be formed by rotating the curve $y = ax^2$ about the *y*-axis. If the dish is to have a 10-ft diameter and a maximum depth of 2 ft, find the value of *a* and the surface area of the dish.

21. (a) The ellipse

$$\frac{x^2}{a^2} + \frac{y^2}{b^2} = 1 \qquad a > b$$

is rotated about the *x*-axis to form a surface called an *ellipsoid*, or *prolate spheroid*. Find the surface area of this ellipsoid.
(b) If the ellipse in part (a) is rotated about its minor axis (the *y*-axis), the resulting ellipsoid is called an *oblate spheroid*. Find the surface area of this ellipsoid.

22. Find the surface area of the torus in Exercise 47 in Section 7.2.

23. If the curve $y = f(x)$, $a \leqslant x \leqslant b$, is rotated about the horizontal line $y = c$, where $f(x) \leqslant c$, find a formula for the area of the resulting surface.

CAS **24.** Use the result of Exercise 23 to set up an integral to find the area of the surface generated by rotating the curve $y = \sqrt{x}$, $0 \leqslant x \leqslant 4$, about the line $y = 4$. Then use a CAS to evaluate the integral.

25. Find the area of the surface obtained by rotating the circle $x^2 + y^2 = r^2$ about the line $y = r$.

26. (a) Show that the surface area of a zone of a sphere that lies between two parallel planes is $S = 2\pi Rh$, where R is the radius of the sphere and h is the distance between the planes. (Notice that S depends only on the distance between the planes and not on their location, provided that both planes intersect the sphere.)

(b) Show that the surface area of a zone of a *cylinder* with radius R and height h is the same as the surface area of the zone of a *sphere* in part (a).

7.6   APPLICATIONS TO PHYSICS AND ENGINEERING

■ As a consequence of a calculation of work, you will be able to compute the velocity needed for a rocket to escape the earth's gravitational field. (See Exercise 23.)

Among the many applications of integral calculus to physics and engineering, we consider three: work, force due to water pressure, and centers of mass. As with our previous applications to geometry (areas, volumes, and lengths), our strategy is to break up the physical quantity into a large number of small parts, approximate each small part, add the results, take the limit, and evaluate the resulting integral.

WORK

The term *work* is used in everyday language to mean the total amount of effort required to perform a task. In physics it has a technical meaning that depends on the idea of a *force*. Intuitively, you can think of a force as describing a push or pull on an object—for example, a horizontal push of a book across a table or the downward pull of the earth's gravity on a ball. In general, if an object moves along a straight line with position function $s(t)$, then the **force** F on the object (in the same direction) is given by Newton's Second Law of Motion as the product of its mass m and its acceleration:

$$\boxed{1} \qquad\qquad F = m\,\frac{d^2s}{dt^2}$$

In the SI metric system, the mass is measured in kilograms (kg), the displacement in meters (m), the time in seconds (s), and the force in newtons (N = kg·m/s²). Thus a force of 1 N acting on a mass of 1 kg produces an acceleration of 1 m/s². In the US Customary system the fundamental unit is chosen to be the unit of force, which is the pound.

In the case of constant acceleration, the force F is also constant and the work done is defined to be the product of the force F and the distance d that the object moves:

$$\boxed{2} \qquad\qquad W = Fd \qquad \text{work} = \text{force} \times \text{distance}$$

If F is measured in newtons and d in meters, then the unit for W is a newton-meter, which is called a joule (J). If F is measured in pounds and d in feet, then the unit for W is a foot-pound (ft-lb), which is about 1.36 J.

For instance, suppose you lift a 1.2-kg book off the floor to put it on a desk that is 0.7 m high. The force you exert is equal and opposite to that exerted by gravity, so Equation 1 gives

$$F = mg = (1.2)(9.8) = 11.76 \text{ N}$$

and then Equation 2 gives the work done as

$$W = Fd = (11.76)(0.7) \approx 8.2 \text{ J}$$

But if a 20-lb weight is lifted 6 ft off the ground, then the force is given as $F = 20$ lb, so the work done is

$$W = Fd = 20 \cdot 6 = 120 \text{ ft-lb}$$

Here we didn't multiply by g because we were given the *weight* (a force) and not the mass.

Equation 2 defines work as long as the force is constant, but what happens if the force is variable? Let's suppose that the object moves along the x-axis in the positive direction, from $x = a$ to $x = b$, and at each point x between a and b a force $f(x)$ acts on the object, where f is a continuous function. We divide the interval $[a, b]$ into n subintervals with endpoints $x_0, x_1, \ldots, x_n$ and equal width Δx. We choose a sample point x_i^* in the ith subinterval $[x_{i-1}, x_i]$. Then the force at that point is $f(x_i^*)$. If n is large, then Δx is small, and since f is continuous, the values of f don't change very much over the interval $[x_{i-1}, x_i]$. In other words, f is almost constant on the interval and so the work W_i that is done in moving the particle from x_{i-1} to x_i is approximately given by Equation 2:

$$W_i \approx f(x_i^*) \, \Delta x$$

Thus we can approximate the total work by

$$\boxed{3} \qquad W \approx \sum_{i=1}^{n} f(x_i^*) \, \Delta x$$

It seems that this approximation becomes better as we make n larger. Therefore we define the **work done in moving the object from a to b** as the limit of this quantity as $n \to \infty$. Since the right side of $\boxed{3}$ is a Riemann sum, we recognize its limit as being a definite integral and so

$$\boxed{4} \qquad W = \lim_{n \to \infty} \sum_{i=1}^{n} f(x_i^*) \, \Delta x = \int_a^b f(x) \, dx$$

EXAMPLE 1 When a particle is located a distance x feet from the origin, a force of $x^2 + 2x$ pounds acts on it. How much work is done in moving it from $x = 1$ to $x = 3$?

SOLUTION $\qquad W = \int_1^3 (x^2 + 2x) \, dx = \dfrac{x^3}{3} + x^2 \Big]_1^3 = \dfrac{50}{3}$

The work done is $16\frac{2}{3}$ ft-lb. ■

In the next example we use a law from physics: **Hooke's Law** states that the force required to maintain a spring stretched x units beyond its natural length is proportional to x:

$$f(x) = kx$$

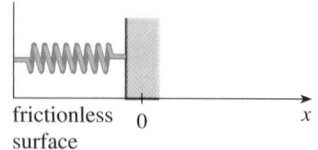

frictionless   0 x
surface

(a) Natural position of spring

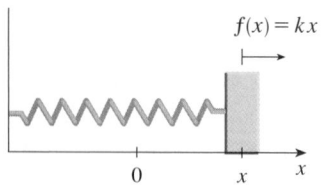

$f(x) = kx$

     0   x x

(b) Stretched position of spring

FIGURE 1

Hooke's Law

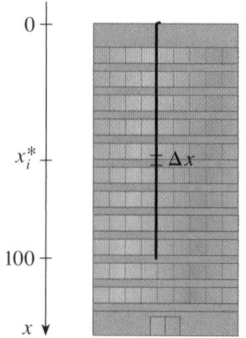

FIGURE 2

- If we had placed the origin at the bottom of the cable and the x-axis upward, we would have gotten

$$W = \int_0^{100} 2(100 - x)\, dx$$

which gives the same answer.

where k is a positive constant (called the **spring constant**). Hooke's Law holds provided that x is not too large (see Figure 1).

☑ EXAMPLE 2 A force of 40 N is required to hold a spring that has been stretched from its natural length of 10 cm to a length of 15 cm. How much work is done in stretching the spring from 15 cm to 18 cm?

SOLUTION According to Hooke's Law, the force required to hold the spring stretched x meters beyond its natural length is $f(x) = kx$. When the spring is stretched from 10 cm to 15 cm, the amount stretched is 5 cm $= 0.05$ m. This means that $f(0.05) = 40$, so

$$0.05k = 40 \qquad k = \tfrac{40}{0.05} = 800$$

Thus $f(x) = 800x$ and the work done in stretching the spring from 15 cm to 18 cm is

$$W = \int_{0.05}^{0.08} 800x\, dx = 800\, \frac{x^2}{2} \Bigg]_{0.05}^{0.08}$$

$$= 400[(0.08)^2 - (0.05)^2] = 1.56 \text{ J} \qquad \blacksquare$$

☑ EXAMPLE 3 A 200-lb cable is 100 ft long and hangs vertically from the top of a tall building. How much work is required to lift the cable to the top of the building?

SOLUTION Here we don't have a formula for the force function, but we can use an argument similar to the one that led to Definition 4.

Let's place the origin at the top of the building and the x-axis pointing downward as in Figure 2. We divide the cable into small parts with length Δx. If x_i^* is a point in the ith such interval, then all points in the interval are lifted by approximately the same amount, namely x_i^*. The cable weighs 2 pounds per foot, so the weight of the ith part is $2\Delta x$. Thus the work done on the ith part, in foot-pounds, is

$$\underbrace{(2\,\Delta x)}_{\text{force}} \cdot \underbrace{x_i^*}_{\text{distance}} = 2x_i^*\, \Delta x$$

We get the total work done by adding all these approximations and letting the number of parts become large (so $\Delta x \to 0$):

$$W = \lim_{n \to \infty} \sum_{i=1}^{n} 2x_i^*\, \Delta x = \int_0^{100} 2x\, dx$$

$$= x^2 \Big]_0^{100} = 10{,}000 \text{ ft-lb} \qquad \blacksquare$$

EXAMPLE 4 A tank has the shape of an inverted circular cone with height 10 m and base radius 4 m. It is filled with water to a height of 8 m. Find the work required to empty the tank by pumping all of the water to the top of the tank. (The density of water is 1000 kg/m^3.)

SOLUTION Let's measure depths from the top of the tank by introducing a vertical coordinate line as in Figure 3. The water extends from a depth of 2 m to a depth of 10 m and so we divide the interval $[2, 10]$ into n subintervals with endpoints $x_0, x_1, \ldots, x_n$ and choose x_i^* in the ith subinterval. This divides the water into n layers. The ith layer is approximated by a circular cylinder with radius r_i and

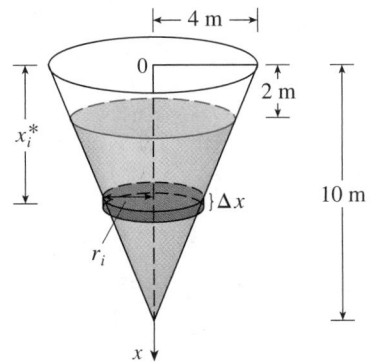

FIGURE 3

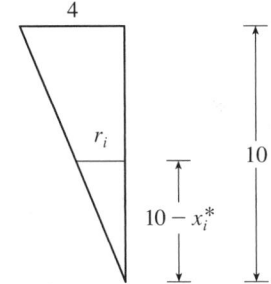

FIGURE 4

height Δx. We can compute r_i from similar triangles, using Figure 4, as follows:

$$\frac{r_i}{10 - x_i^*} = \frac{4}{10} \qquad r_i = \tfrac{2}{5}(10 - x_i^*)$$

Thus an approximation to the volume of the ith layer of water is

$$V_i \approx \pi r_i^2 \, \Delta x = \frac{4\pi}{25}(10 - x_i^*)^2 \, \Delta x$$

and so its mass is

$$m_i = \text{density} \times \text{volume}$$

$$\approx 1000 \cdot \frac{4\pi}{25}(10 - x_i^*)^2 \, \Delta x = 160\pi(10 - x_i^*)^2 \, \Delta x$$

The force required to raise this layer must overcome the force of gravity and so

$$F_i = m_i g \approx (9.8)160\pi(10 - x_i^*)^2 \, \Delta x \approx 1568\pi(10 - x_i^*)^2 \, \Delta x$$

Each particle in the layer must travel a distance of approximately x_i^*. The work W_i done to raise this layer to the top is approximately the product of the force F_i and the distance x_i^*:

$$W_i \approx F_i x_i^* \approx 1568\pi x_i^*(10 - x_i^*)^2 \, \Delta x$$

To find the total work done in emptying the entire tank, we add the contributions of each of the n layers and then take the limit as $n \to \infty$:

$$W = \lim_{n \to \infty} \sum_{i=1}^{n} 1568\pi x_i^*(10 - x_i^*)^2 \, \Delta x = \int_2^{10} 1568\pi x(10 - x)^2 \, dx$$

$$= 1568\pi \int_2^{10} (100x - 20x^2 + x^3) \, dx = 1568\pi \left[50x^2 - \frac{20x^3}{3} + \frac{x^4}{4} \right]_2^{10}$$

$$= 1568\pi \left(\tfrac{2048}{3} \right) \approx 3.4 \times 10^6 \text{ J} \qquad \blacksquare$$

HYDROSTATIC PRESSURE AND FORCE

Deep-sea divers realize that water pressure increases as they dive deeper. This is because the weight of the water above them increases.

In general, suppose that a thin horizontal plate with area A square meters is submerged in a fluid of density ρ kilograms per cubic meter at a depth d meters below the surface of the fluid as in Figure 5. The fluid directly above the plate has volume $V = Ad$, so its mass is $m = \rho V = \rho A d$. The force exerted by the fluid on the plate is therefore

$$F = mg = \rho g A d$$

where g is the acceleration due to gravity. The pressure P on the plate is defined to be the force per unit area:

$$P = \frac{F}{A} = \rho g d$$

surface of fluid

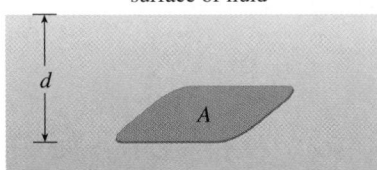

FIGURE 5

■ When using US Customary units, we write $P = \rho gd = \delta d$, where $\delta = \rho g$ is the *weight density* (as opposed to ρ, which is the *mass density*). For instance, the weight density of water is $\delta = 62.5$ lb/ft³.

The SI unit for measuring pressure is newtons per square meter, which is called a pascal (abbreviation: 1 N/m² $= 1$ Pa). Since this is a small unit, the kilopascal (kPa) is often used. For instance, because the density of water is $\rho = 1000$ kg/m³, the pressure at the bottom of a swimming pool 2 m deep is

$$P = \rho gd = 1000 \text{ kg/m}^3 \times 9.8 \text{ m/s}^2 \times 2 \text{ m}$$

$$= 19,600 \text{ Pa} = 19.6 \text{ kPa}$$

An important principle of fluid pressure is the experimentally verified fact that *at any point in a liquid the pressure is the same in all directions.* (A diver feels the same pressure on nose and both ears.) Thus the pressure in *any* direction at a depth d in a fluid with mass density ρ is given by

$$\boxed{5} \qquad\qquad P = \rho gd = \delta d$$

This helps us determine the hydrostatic force against a *vertical* plate or wall or dam in a fluid. This is not a straightforward problem because the pressure is not constant but increases as the depth increases.

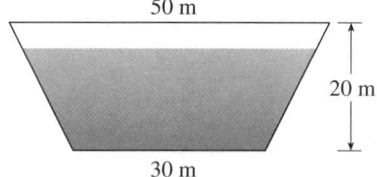

FIGURE 6

▼ **EXAMPLE 5** A dam has the shape of the trapezoid shown in Figure 6. The height is 20 m and the width is 50 m at the top and 30 m at the bottom. Find the force on the dam due to hydrostatic pressure if the water level is 4 m from the top of the dam.

SOLUTION We choose a vertical x-axis with origin at the surface of the water as in Figure 7(a). The depth of the water is 16 m, so we divide the interval $[0, 16]$ into subintervals with endpoints x_i and we choose $x_i^* \in [x_{i-1}, x_i]$. The ith horizontal strip of the dam is approximated by a rectangle with height Δx and width w_i, where, from similar triangles in Figure 7(b),

$$\frac{a}{16 - x_i^*} = \frac{10}{20} \qquad \text{or} \qquad a = \frac{16 - x_i^*}{2} = 8 - \frac{x_i^*}{2}$$

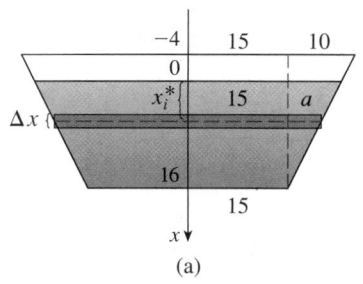

and so

$$w_i = 2(15 + a) = 2\left(15 + 8 - \tfrac{1}{2}x_i^*\right) = 46 - x_i^*$$

If A_i is the area of the ith strip, then

$$A_i \approx w_i \, \Delta x = (46 - x_i^*) \, \Delta x$$

If Δx is small, then the pressure P_i on the ith strip is almost constant and we can use Equation 5 to write

$$P_i \approx 1000 g x_i^*$$

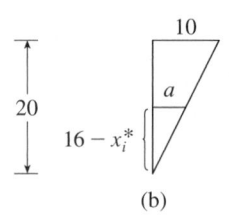

FIGURE 7

The hydrostatic force F_i acting on the ith strip is the product of the pressure and the area:

$$F_i = P_i A_i \approx 1000 g x_i^* (46 - x_i^*) \, \Delta x$$

Adding these forces and taking the limit as $n \to \infty$, we obtain the total hydrostatic force on the dam:

$$F = \lim_{n \to \infty} \sum_{i=1}^{n} 1000 g x_i^*(46 - x_i^*)\, \Delta x = \int_0^{16} 1000 g x(46 - x)\, dx$$

$$= 1000(9.8) \int_0^{16} (46x - x^2)\, dx = 9800 \left[23x^2 - \frac{x^3}{3} \right]_0^{16}$$

$$\approx 4.43 \times 10^7 \text{ N}$$

EXAMPLE 6 Find the hydrostatic force on one end of a cylindrical drum with radius 3 ft if the drum is submerged in water 10 ft deep.

SOLUTION In this example it is convenient to choose the axes as in Figure 8 so that the origin is placed at the center of the drum. Then the circle has a simple equation, $x^2 + y^2 = 9$. As in Example 5 we divide the circular region into horizontal strips of equal width. From the equation of the circle, we see that the length of the ith strip is $2\sqrt{9 - (y_i^*)^2}$ and so its area is

$$A_i = 2\sqrt{9 - (y_i^*)^2}\, \Delta y$$

The pressure on this strip is approximately

$$\delta d_i = 62.5(7 - y_i^*)$$

and so the force on the strip is approximately

$$\delta d_i A_i = 62.5(7 - y_i^*)2\sqrt{9 - (y_i^*)^2}\, \Delta y$$

The total force is obtained by adding the forces on all the strips and taking the limit:

$$F = \lim_{n \to \infty} \sum_{i=1}^{n} 62.5(7 - y_i^*)2\sqrt{9 - (y_i^*)^2}\, \Delta y$$

$$= 125 \int_{-3}^{3} (7 - y)\sqrt{9 - y^2}\, dy$$

$$= 125 \cdot 7 \int_{-3}^{3} \sqrt{9 - y^2}\, dy - 125 \int_{-3}^{3} y\sqrt{9 - y^2}\, dy$$

The second integral is 0 because the integrand is an odd function (see Theorem 5.5.7). The first integral can be evaluated using the trigonometric substitution $y = 3\sin\theta$, but it's simpler to observe that it is the area of a semicircular disk with radius 3. Thus

$$F = 875 \int_{-3}^{3} \sqrt{9 - y^2}\, dy = 875 \cdot \tfrac{1}{2}\pi(3)^2$$

$$= \frac{7875\pi}{2} \approx 12{,}370 \text{ lb}$$

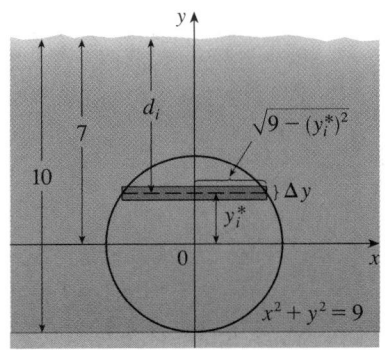

FIGURE 8

MOMENTS AND CENTERS OF MASS

Our main objective here is to find the point P on which a thin plate of any given shape balances horizontally as in Figure 9. This point is called the **center of mass** (or center of gravity) of the plate.

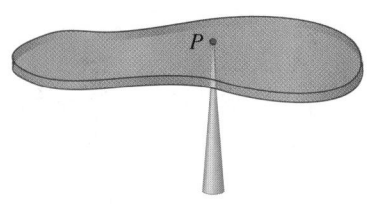

FIGURE 9

We first consider the simpler situation illustrated in Figure 10, where two masses m_1 and m_2 are attached to a rod of negligible mass on opposite sides of a fulcrum and at distances d_1 and d_2 from the fulcrum. The rod will balance if

$$\boxed{6} \qquad m_1 d_1 = m_2 d_2$$

This is an experimental fact discovered by Archimedes and called the Law of the Lever. (Think of a lighter person balancing a heavier one on a seesaw by sitting farther away from the center.)

Now suppose that the rod lies along the x-axis with m_1 at x_1 and m_2 at x_2 and the center of mass at $\bar{x}$. If we compare Figures 10 and 11, we see that $d_1 = \bar{x} - x_1$ and $d_2 = x_2 - \bar{x}$ and so Equation 6 gives

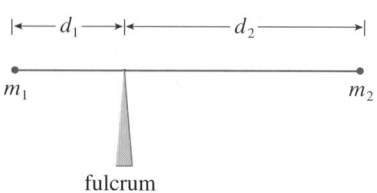

FIGURE 10

$$m_1(\bar{x} - x_1) = m_2(x_2 - \bar{x})$$

$$m_1\bar{x} + m_2\bar{x} = m_1 x_1 + m_2 x_2$$

$$\boxed{7} \qquad \bar{x} = \frac{m_1 x_1 + m_2 x_2}{m_1 + m_2}$$

The numbers $m_1 x_1$ and $m_2 x_2$ are called the **moments** of the masses m_1 and m_2 (with respect to the origin), and Equation 7 says that the center of mass $\bar{x}$ is obtained by adding the moments of the masses and dividing by the total mass $m = m_1 + m_2$.

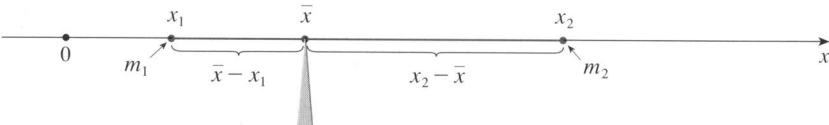

FIGURE 11

In general, if we have a system of n particles with masses $m_1, m_2, \ldots, m_n$ located at the points $x_1, x_2, \ldots, x_n$ on the x-axis, it can be shown similarly that the center of mass of the system is located at

$$\boxed{8} \qquad \bar{x} = \frac{\displaystyle\sum_{i=1}^{n} m_i x_i}{\displaystyle\sum_{i=1}^{n} m_i} = \frac{\displaystyle\sum_{i=1}^{n} m_i x_i}{m}$$

where $m = \sum m_i$ is the total mass of the system, and the sum of the individual moments

$$M = \sum_{i=1}^{n} m_i x_i$$

is called the **moment of the system about the origin**. Then Equation 8 could be rewritten as $m\bar{x} = M$, which says that if the total mass were considered as being concentrated at the center of mass $\bar{x}$, then its moment would be the same as the moment of the system.

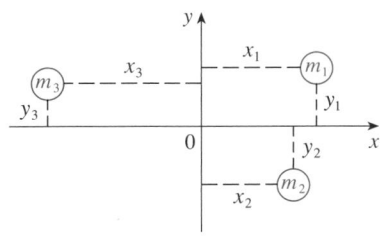

FIGURE 12

Now we consider a system of n particles with masses $m_1, m_2, \ldots, m_n$ located at the points (x_1, y_1), (x_2, y_2), $\ldots$, (x_n, y_n) in the xy-plane as shown in Figure 12. By analogy with the one-dimensional case, we define the **moment of the system about the y-axis** to be

$$\boxed{9} \qquad M_y = \sum_{i=1}^{n} m_i x_i$$

and the **moment of the system about the x-axis** as

$$\boxed{10} \qquad M_x = \sum_{i=1}^{n} m_i y_i$$

Then M_y measures the tendency of the system to rotate about the y-axis and M_x measures the tendency to rotate about the x-axis.

As in the one-dimensional case, the coordinates $(\bar{x}, \bar{y})$ of the center of mass are given in terms of the moments by the formulas

$$\boxed{11} \qquad \bar{x} = \frac{M_y}{m} \qquad \bar{y} = \frac{M_x}{m}$$

where $m = \Sigma\, m_i$ is the total mass. Since $m\bar{x} = M_y$ and $m\bar{y} = M_x$, the center of mass $(\bar{x}, \bar{y})$ is the point where a single particle of mass m would have the same moments as the system.

V EXAMPLE 7 Find the moments and center of mass of the system of objects that have masses 3, 4, and 8 at the points $(-1, 1)$, $(2, -1)$, and $(3, 2)$.

SOLUTION We use Equations 9 and 10 to compute the moments:

$$M_y = 3(-1) + 4(2) + 8(3) = 29$$

$$M_x = 3(1) + 4(-1) + 8(2) = 15$$

Since $m = 3 + 4 + 8 = 15$, we use Equations 11 to obtain

$$\bar{x} = \frac{M_y}{m} = \frac{29}{15} \qquad \bar{y} = \frac{M_x}{m} = \frac{15}{15} = 1$$

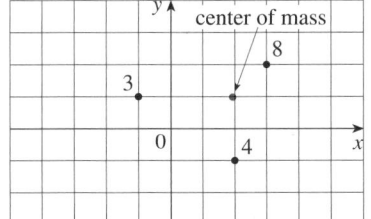

FIGURE 13

Thus the center of mass is $\left(1\frac{14}{15}, 1\right)$. (See Figure 13.) ∎

Next we consider a flat plate (called a *lamina*) with uniform density ρ that occupies a region $\mathcal{R}$ of the plane. We wish to locate the center of mass of the plate, which is called the **centroid** of $\mathcal{R}$. In doing so we use the following physical principles: The **symmetry principle** says that if $\mathcal{R}$ is symmetric about a line l, then the centroid of $\mathcal{R}$ lies on l. (If $\mathcal{R}$ is reflected about l, then $\mathcal{R}$ remains the same so its centroid remains fixed. But the only fixed points lie on l.) Thus the centroid of a rectangle is its center. Moments should be defined so that if the entire mass of a region is concentrated at the center of mass, then its moments remain unchanged. Also, the moment of the union of two nonoverlapping regions should be the sum of the moments of the individual regions.

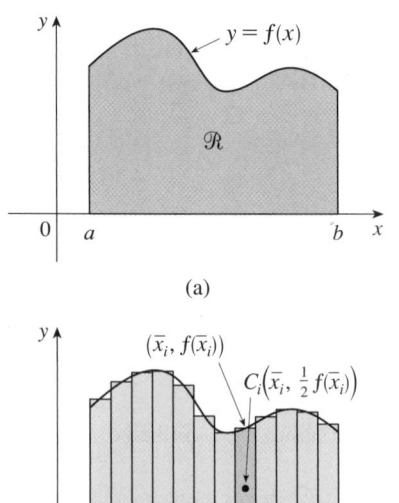

(a)

(b)

FIGURE 14

Suppose that the region $\mathcal{R}$ is of the type shown in Figure 14(a); that is, $\mathcal{R}$ lies between the lines $x = a$ and $x = b$, above the x-axis, and beneath the graph of f, where f is a continuous function. We divide the interval $[a, b]$ into n subintervals with endpoints $x_0, x_1, \ldots, x_n$ and equal width Δx. We choose the sample point x_i^* to be the midpoint $\bar{x}_i$ of the ith subinterval, that is, $\bar{x}_i = (x_{i-1} + x_i)/2$. This determines the polygonal approximation to $\mathcal{R}$ shown in Figure 14(b). The centroid of the ith approximating rectangle R_i is its center $C_i\left(\bar{x}_i, \frac{1}{2}f(\bar{x}_i)\right)$. Its area is $f(\bar{x}_i)\,\Delta x$, so its mass is

$$\rho f(\bar{x}_i)\,\Delta x$$

The moment of R_i about the y-axis is the product of its mass and the distance from C_i to the y-axis, which is $\bar{x}_i$. Thus

$$M_y(R_i) = [\rho f(\bar{x}_i)\,\Delta x]\,\bar{x}_i = \rho \bar{x}_i f(\bar{x}_i)\,\Delta x$$

Adding these moments, we obtain the moment of the polygonal approximation to $\mathcal{R}$, and then by taking the limit as $n \to \infty$ we obtain the moment of $\mathcal{R}$ itself about the y-axis:

$$M_y = \lim_{n \to \infty} \sum_{i=1}^{n} \rho \bar{x}_i f(\bar{x}_i)\,\Delta x = \rho \int_a^b x f(x)\,dx$$

In a similar fashion we compute the moment of R_i about the x-axis as the product of its mass and the distance from C_i to the x-axis:

$$M_x(R_i) = [\rho f(\bar{x}_i)\,\Delta x]\tfrac{1}{2}f(\bar{x}_i) = \rho \cdot \tfrac{1}{2}[f(\bar{x}_i)]^2\,\Delta x$$

Again we add these moments and take the limit to obtain the moment of $\mathcal{R}$ about the x-axis:

$$M_x = \lim_{n \to \infty} \sum_{i=1}^{n} \rho \cdot \tfrac{1}{2}[f(\bar{x}_i)]^2\,\Delta x = \rho \int_a^b \tfrac{1}{2}[f(x)]^2\,dx$$

Just as for systems of particles, the center of mass of the plate is defined so that $m\bar{x} = M_y$ and $m\bar{y} = M_x$. But the mass of the plate is the product of its density and its area:

$$m = \rho A = \rho \int_a^b f(x)\,dx$$

and so

$$\bar{x} = \frac{M_y}{m} = \frac{\rho \displaystyle\int_a^b x f(x)\,dx}{\rho \displaystyle\int_a^b f(x)\,dx} = \frac{\displaystyle\int_a^b x f(x)\,dx}{\displaystyle\int_a^b f(x)\,dx}$$

$$\bar{y} = \frac{M_x}{m} = \frac{\rho \displaystyle\int_a^b \tfrac{1}{2}[f(x)]^2\,dx}{\rho \displaystyle\int_a^b f(x)\,dx} = \frac{\displaystyle\int_a^b \tfrac{1}{2}[f(x)]^2\,dx}{\displaystyle\int_a^b f(x)\,dx}$$

Notice the cancellation of the ρ's. The location of the center of mass is independent of the density.

In summary, the center of mass of the plate (or the centroid of $\mathcal{R}$) is located at the point $(\bar{x}, \bar{y})$, where

$$\boxed{12} \qquad \bar{x} = \frac{1}{A} \int_a^b x f(x)\, dx \qquad \bar{y} = \frac{1}{A} \int_a^b \tfrac{1}{2}[f(x)]^2\, dx$$

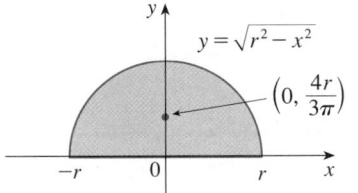

FIGURE 15

EXAMPLE 8 Find the center of mass of a semicircular plate of radius r.

SOLUTION In order to use $\boxed{12}$ we place the semicircle as in Figure 15 so that $f(x) = \sqrt{r^2 - x^2}$ and $a = -r$, $b = r$. Here there is no need to use the formula to calculate $\bar{x}$ because, by the symmetry principle, the center of mass must lie on the y-axis, so $\bar{x} = 0$. The area of the semicircle is $A = \tfrac{1}{2}\pi r^2$, so

$$\bar{y} = \frac{1}{A} \int_{-r}^r \tfrac{1}{2}[f(x)]^2\, dx = \frac{1}{\tfrac{1}{2}\pi r^2} \cdot \tfrac{1}{2} \int_{-r}^r \left(\sqrt{r^2 - x^2}\right)^2 dx$$

$$= \frac{2}{\pi r^2} \int_0^r (r^2 - x^2)\, dx = \frac{2}{\pi r^2}\left[r^2 x - \frac{x^3}{3} \right]_0^r$$

$$= \frac{2}{\pi r^2}\frac{2r^3}{3} = \frac{4r}{3\pi}$$

The center of mass is located at the point $(0, 4r/(3\pi))$. ∎

▪ www.stewartcalculus.com
See Additional Example A.

If the region $\mathcal{R}$ lies between two curves $y = f(x)$ and $y = g(x)$, where $f(x) \geq g(x)$, as illustrated in Figure 16, then the same sort of argument that led to Formulas 12 can be used to show that the centroid of $\mathcal{R}$ is $(\bar{x}, \bar{y})$, where

$$\boxed{13} \qquad \bar{x} = \frac{1}{A} \int_a^b x[f(x) - g(x)]\, dx \qquad \bar{y} = \frac{1}{A} \int_a^b \tfrac{1}{2}\{[f(x)]^2 - [g(x)]^2\}\, dx$$

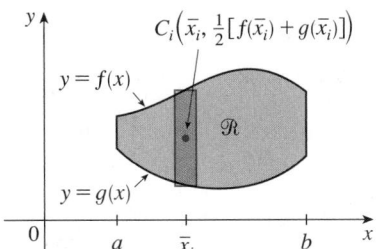

FIGURE 16

(See Exercise 55.)

EXAMPLE 9 Find the centroid of the region bounded by the line $y = x$ and the parabola $y = x^2$.

SOLUTION The region is sketched in Figure 17. We take $f(x) = x$, $g(x) = x^2$, $a = 0$, and $b = 1$ in Formulas 13. First we note that the area of the region is

$$A = \int_0^1 (x - x^2)\, dx = \frac{x^2}{2} - \frac{x^3}{3}\Big]_0^1 = \frac{1}{6}$$

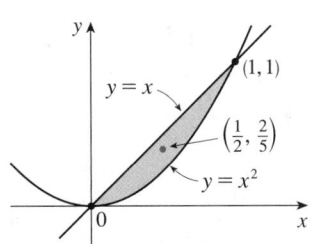

FIGURE 17

Therefore

$$\bar{x} = \frac{1}{A} \int_0^1 x[f(x) - g(x)]\, dx = \frac{1}{\tfrac{1}{6}} \int_0^1 x(x - x^2)\, dx$$

$$= 6 \int_0^1 (x^2 - x^3)\, dx = 6\left[\frac{x^3}{3} - \frac{x^4}{4} \right]_0^1 = \frac{1}{2}$$

$$\bar{y} = \frac{1}{A} \int_0^1 \frac{1}{2}\{[f(x)]^2 - [g(x)]^2\}\, dx = \frac{1}{\frac{1}{6}} \int_0^1 \frac{1}{2}(x^2 - x^4)\, dx$$

$$= 3\left[\frac{x^3}{3} - \frac{x^5}{5}\right]_0^1 = \frac{2}{5}$$

The centroid is $\left(\frac{1}{2}, \frac{2}{5}\right)$. ∎

We end this section by showing a surprising connection between centroids and volumes of revolution.

■ This theorem is named after the Greek mathematician Pappus of Alexandria, who lived in the fourth century A.D.

> **THEOREM OF PAPPUS** Let $\mathcal{R}$ be a plane region that lies entirely on one side of a line l in the plane. If $\mathcal{R}$ is rotated about l, then the volume of the resulting solid is the product of the area A of $\mathcal{R}$ and the distance d traveled by the centroid of $\mathcal{R}$.

PROOF We give the proof for the special case in which the region lies between $y = f(x)$ and $y = g(x)$ as in Figure 16 and the line l is the y-axis. Using the method of cylindrical shells (see Section 7.3), we have

$$V = \int_a^b 2\pi x[f(x) - g(x)]\, dx = 2\pi \int_a^b x[f(x) - g(x)]\, dx$$

$$= 2\pi(\bar{x}A) \qquad \text{(by Formulas 13)}$$

$$= (2\pi\bar{x})A = Ad$$

where $d = 2\pi\bar{x}$ is the distance traveled by the centroid during one rotation about the y-axis. □

▶ **EXAMPLE 10** A torus is formed by rotating a circle of radius r about a line in the plane of the circle that is a distance $R\ (> r)$ from the center of the circle. Find the volume of the torus.

SOLUTION The circle has area $A = \pi r^2$. By the symmetry principle, its centroid is its center and so the distance traveled by the centroid during a rotation is $d = 2\pi R$. Therefore, by the Theorem of Pappus, the volume of the torus is

$$V = Ad = (\pi r^2)(2\pi R) = 2\pi^2 r^2 R \qquad ∎$$

The method of Example 10 should be compared with the method of Exercise 47 in Section 7.2.

7.6 | EXERCISES

1. A variable force of $5x^{-2}$ pounds moves an object along a straight line when it is x feet from the origin. Calculate the work done in moving the object from $x = 1$ ft to $x = 10$ ft.

2. When a particle is located a distance x meters from the origin, a force of $\cos(\pi x/3)$ newtons acts on it. How much

work is done in moving the particle from $x = 1$ to $x = 2$? Interpret your answer by considering the work done from $x = 1$ to $x = 1.5$ and from $x = 1.5$ to $x = 2$.

3. Shown is the graph of a force function (in newtons) that increases to its maximum value and then remains constant.

How much work is done by the force in moving an object a distance of 8 m?

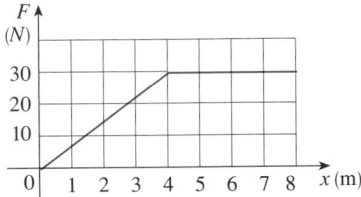

4. The table shows values of a force function $f(x)$, where x is measured in meters and $f(x)$ in newtons. Use the Midpoint Rule to estimate the work done by the force in moving an object from $x = 4$ to $x = 20$.

x	4	6	8	10	12	14	16	18	20
$f(x)$	5	5.8	7.0	8.8	9.6	8.2	6.7	5.2	4.1

5. A force of 10 lb is required to hold a spring stretched 4 in. beyond its natural length. How much work is done in stretching it from its natural length to 6 in. beyond its natural length?

6. A spring has a natural length of 20 cm. If a 25-N force is required to keep it stretched to a length of 30 cm, how much work is required to stretch it from 20 cm to 25 cm?

7. Suppose that 2 J of work is needed to stretch a spring from its natural length of 30 cm to a length of 42 cm.
   (a) How much work is needed to stretch the spring from 35 cm to 40 cm?
   (b) How far beyond its natural length will a force of 30 N keep the spring stretched?

8. If 6 J of work is needed to stretch a spring from 10 cm to 12 cm and another 10 J is needed to stretch it from 12 cm to 14 cm, what is the natural length of the spring?

9–16 ■ Show how to approximate the required work by a Riemann sum. Then express the work as an integral and evaluate it.

9. A heavy rope, 50 ft long, weighs 0.5 lb/ft and hangs over the edge of a building 120 ft high.
   (a) How much work is done in pulling the rope to the top of the building?
   (b) How much work is done in pulling half the rope to the top of the building?

10. A chain lying on the ground is 10 m long and its mass is 80 kg. How much work is required to raise one end of the chain to a height of 6 m?

11. A cable that weighs 2 lb/ft is used to lift 800 lb of coal up a mine shaft 500 ft deep. Find the work done.

12. A bucket that weighs 4 lb and a rope of negligible weight are used to draw water from a well that is 80 ft deep. The bucket is filled with 40 lb of water and is pulled up at a rate of 2 ft/s, but water leaks out of a hole in the bucket at a rate of 0.2 lb/s. Find the work done in pulling the bucket to the top of the well.

13. A leaky 10-kg bucket is lifted from the ground to a height of 12 m at a constant speed with a rope that weighs 0.8 kg/m. Initially the bucket contains 36 kg of water, but the water leaks at a constant rate and finishes draining just as the bucket reaches the 12-meter level. How much work is done?

14. A 10-ft chain weighs 25 lb and hangs from a ceiling. Find the work done in lifting the lower end of the chain to the ceiling so that it's level with the upper end.

15. An aquarium 2 m long, 1 m wide, and 1 m deep is full of water. Find the work needed to pump half of the water out of the aquarium. (Use the fact that the density of water is 1000 kg/m³.)

16. A circular swimming pool has a diameter of 24 ft, the sides are 5 ft high, and the depth of the water is 4 ft. How much work is required to pump all of the water out over the side? (Use the fact that water weighs 62.5 lb/ft³.)

17. The tank shown is full of water.
   (a) Find the work required to pump the water out of the spout.
   (b) Suppose that the pump breaks down after 4.7×10^5 J of work has been done. What is the depth of the water remaining in the tank?

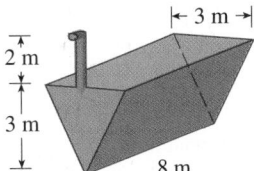

18. (a) The tank shown is full of water. Given that the water weighs 62.5 lb/ft³, find the work required to pump the water out of the spout.
   (b) What if the tank is half full of oil that has a density of 900 kg/m³?

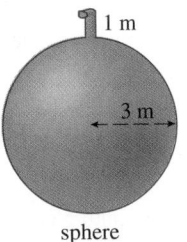

sphere

19. When gas expands in a cylinder with radius r, the pressure at any given time is a function of the volume: $P = P(V)$. The force exerted by the gas on the piston (see the figure) is the product of the pressure and the area: $F = \pi r^2 P$. Show that the work done by the gas when the volume expands from volume V_1 to volume V_2 is

$$W = \int_{V_1}^{V_2} P\, dV$$

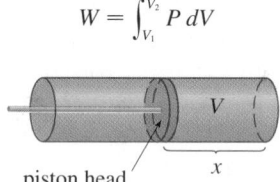

piston head

20. In a steam engine the pressure P and volume V of steam satisfy the equation $PV^{1.4} = k$, where k is a constant. (This is true for adiabatic expansion, that is, expansion in which there is no heat transfer between the cylinder and its surroundings.) Use Exercise 19 to calculate the work done by the engine during a cycle when the steam starts at a pressure of 160 lb/in² and a volume of 100 in³ and expands to a volume of 800 in³.

21. (a) Newton's Law of Gravitation states that two bodies with masses m_1 and m_2 attract each other with a force

$$F = G\frac{m_1 m_2}{r^2}$$

where r is the distance between the bodies and G is the gravitational constant. If one of the bodies is fixed, find the work needed to move the other from $r = a$ to $r = b$.

(b) Compute the work required to launch a 1000-kg satellite vertically to a height of 1000 km. You may assume that the earth's mass is 5.98×10^{24} kg and is concentrated at its center. Take the radius of the earth to be 6.37×10^6 m and $G = 6.67 \times 10^{-11}$ N·m²/kg².

22. Use an improper integral and information from Exercise 21 to find the work needed to propel a 1000-kg satellite out of the earth's gravitational field.

23. Find the *escape velocity* v_0 that is needed to propel a rocket of mass m out of the gravitational field of a planet with mass M and radius R. (Use the fact that the initial kinetic energy of $\frac{1}{2}mv_0^2$ supplies the needed work.)

24. The Great Pyramid of King Khufu was built of limestone in Egypt over a 20-year time period from 2580 BC to 2560 BC. Its base is a square with side length 756 ft and its height when built was 481 ft. (It was the tallest man-made structure in the world for more than 3800 years.) The density of the limestone is about 150 lb/ft³.

(a) Estimate the total work done in building the pyramid.

(b) If each laborer worked 10 hours a day for 20 years, for 340 days a year, and did 200 ft-lb/h of work in lifting the limestone blocks into place, about how many laborers were needed to construct the pyramid?

25–29 ▪ A vertical plate is submerged (or partially submerged) in water and has the indicated shape. Explain how to approximate the hydrostatic force against one side of the plate by a Riemann sum. Then express the force as an integral and evaluate it.

25.

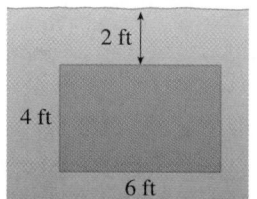

26.

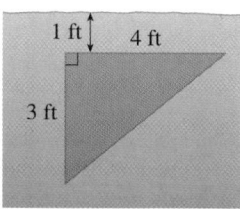

27.

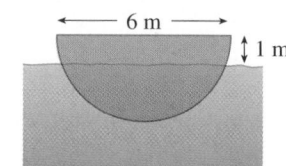

28.

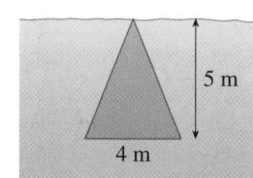

29.

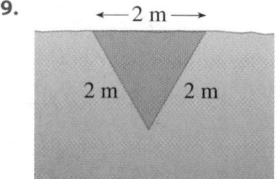

30. A milk truck carries milk with density 64.6 lb/ft³ in a horizontal cylindrical tank with diameter 6 ft.

(a) Find the force exerted by the milk on one end of the tank when the tank is full.

(b) What if the tank is half full?

31. A trough is filled with a liquid of density 840 kg/m³. The ends of the trough are equilateral triangles with sides 8 m long and vertex at the bottom. Find the hydrostatic force on one end of the trough.

32. A dam is inclined at an angle of 30° from the vertical and has the shape of an isosceles trapezoid 100 ft wide at the top and 50 ft wide at the bottom and with a slant height of 70 ft. Find the hydrostatic force on the dam when it is full of water.

33. A swimming pool is 20 ft wide and 40 ft long and its bottom is an inclined plane, the shallow end having a depth of 3 ft and the deep end, 9 ft. If the pool is full of water, estimate the hydrostatic force on (a) the shallow end, (b) the deep end, (c) one of the sides, and (d) the bottom of the pool.

34. A vertical dam has a semicircular gate as shown in the figure. Find the hydrostatic force against the gate.

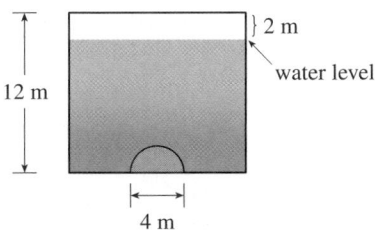

35. A metal plate was found submerged vertically in sea water, which has density 64 lb/ft^3. Measurements of the width of the plate were taken at the indicated depths. Use Simpson's Rule to estimate the force of the water against the plate.

Depth (m)	7.0	7.4	7.8	8.2	8.6	9.0	9.4
Plate width (m)	1.2	1.8	2.9	3.8	3.6	4.2	4.4

36. Point-masses m_i are located on the x-axis as shown. Find the moment M of the system about the origin and the center of mass $\bar{x}$.

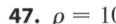

37–38 ■ The masses m_i are located at the points P_i. Find the moments M_x and M_y and the center of mass of the system.

37. $m_1 = 4,\ m_2 = 2,\ m_3 = 4;$
$P_1(2, -3),\ P_2(-3, 1),\ P_3(3, 5)$

38. $m_1 = 5,\ m_2 = 4,\ m_3 = 3,\ m_4 = 6;$
$P_1(-4, 2),\ P_2(0, 5),\ P_3(3, 2),\ P_4(1, -2)$

39–42 ■ Sketch the region bounded by the curves, and visually estimate the location of the centroid. Then find the exact coordinates of the centroid.

39. $y = 2x,\ y = 0,\ x = 1$

40. $y = \sqrt{x},\ y = 0,\ x = 4$

41. $y = e^x,\ y = 0,\ x = 0,\ x = 1$

42. $y = \sin x,\ y = 0,\ 0 \le x \le \pi$

43–46 ■ Find the centroid of the region bounded by the given curves.

43. $y = x^2,\ x = y^2$

44. $y = 2 - x^2,\ y = x$

45. $y = \sin x,\ y = \cos x,\ x = 0,\ x = \pi/4$

46. $y = x^3,\ x + y = 2,\ y = 0$

47–48 ■ Calculate the moments M_x and M_y and the center of mass of a lamina with the given density and shape.

47. $\rho = 10$

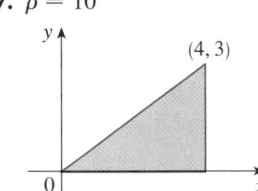

48. $\rho = 3$

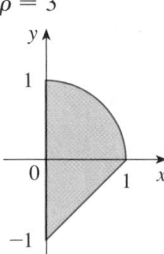

49. Prove that the centroid of any triangle is located at the point of intersection of the medians. [*Hints:* Place the axes so that the vertices are $(a, 0)$, $(0, b)$, and $(c, 0)$. Recall that a median is a line segment from a vertex to the midpoint of the opposite side. Recall also that the medians intersect at a point two-thirds of the way from each vertex (along the median) to the opposite side.]

50–51 ■ Find the centroid of the region shown, not by integration, but by locating the centroids of the rectangles and triangles (from Exercise 49) and using additivity of moments.

50.

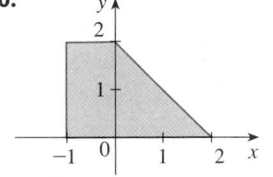

51.

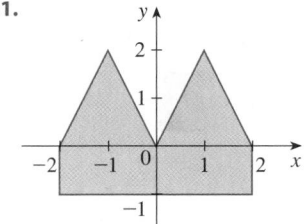

52–54 ■ Use the Theorem of Pappus to find the volume of the given solid.

52. A sphere of radius r (Use Example 8.)

53. A cone with height h and base radius r

54. The solid obtained by rotating the triangle with vertices $(2, 3)$, $(2, 5)$, and $(5, 4)$ about the x-axis

55. Prove Formulas 13.

56. Let $\mathcal{R}$ be the region that lies between the curves $y = x^m$ and $y = x^n$, $0 \le x \le 1$, where m and n are integers with $0 \le n < m$.
(a) Sketch the region $\mathcal{R}$.
(b) Find the coordinates of the centroid of $\mathcal{R}$.
(c) Try to find values of m and n such that the centroid lies *outside* $\mathcal{R}$.

7.7 | DIFFERENTIAL EQUATIONS

A **differential equation** is an equation that contains an unknown function and one or more of its derivatives. Here are some examples:

$$\boxed{1} \qquad\qquad y' = xy$$

$$\boxed{2} \qquad\qquad y'' + 2y' + y = 0$$

$$\boxed{3} \qquad\qquad \frac{d^3y}{dx^3} + x\frac{d^2y}{dx^2} + \frac{dy}{dx} - 2y = e^{-x}$$

In each of these differential equations y is an unknown function of x. The importance of differential equations lies in the fact that when a scientist or engineer formulates a physical law in mathematical terms, it frequently turns out to be a differential equation.

The **order** of a differential equation is the order of the highest derivative that occurs in the equation. Thus Equations 1, 2, and 3 are of order 1, 2, and 3, respectively.

A function f is called a **solution** of a differential equation if the equation is satisfied when $y = f(x)$ and its derivatives are substituted into the equation. Thus f is a solution of Equation 1 if

$$f'(x) = xf(x)$$

for all values of x in some interval.

When we are asked to *solve* a differential equation we are expected to find all possible solutions of the equation. We have already solved some particularly simple differential equations, namely, those of the form $y' = f(x)$. For instance, we know that the general solution of the differential equation $y' = x^3$ is given by $y = \frac{1}{4}x^4 + C$, where C is an arbitrary constant.

But, in general, solving a differential equation is not an easy matter. There is no systematic technique that enables us to solve all differential equations. In this section we learn how to solve a certain type of differential equation called a *separable equation*. At the end of the section, however, we will see how to sketch a rough graph of a solution of a first-order differential equation, even when it is impossible to find a formula for the solution.

SEPARABLE EQUATIONS

A **separable equation** is a first-order differential equation that can be written in the form

$$\frac{dy}{dx} = g(x)f(y)$$

The name *separable* comes from the fact that the expression on the right side can be "separated" into a function of x and a function of y. Equivalently, if $f(y) \neq 0$, we could write

$$\boxed{4} \qquad\qquad \frac{dy}{dx} = \frac{g(x)}{h(y)}$$

■ The technique for solving separable differential equations was first used by James Bernoulli (in 1690) in solving a problem about pendulums and by Leibniz (in a letter to Huygens in 1691). John Bernoulli explained the general method in a paper published in 1694.

where $h(y) = 1/f(y)$. To solve this equation we rewrite it in the differential form

$$h(y)\, dy = g(x)\, dx$$

so that all y's are on one side of the equation and all x's are on the other side. Then we integrate both sides of the equation:

⑤

$$\int h(y)\, dy = \int g(x)\, dx$$

Equation 5 defines y implicitly as a function of x. In some cases we may be able to solve for y in terms of x.

We use the Chain Rule to justify this procedure: If h and g satisfy ⑤, then

$$\frac{d}{dx}\left(\int h(y)\, dy\right) = \frac{d}{dx}\left(\int g(x)\, dx\right)$$

so

$$\frac{d}{dy}\left(\int h(y)\, dy\right)\frac{dy}{dx} = g(x)$$

and

$$h(y)\frac{dy}{dx} = g(x)$$

Thus Equation 4 is satisfied.

When applying differential equations, we are usually not as interested in finding a family of solutions (the *general solution*) as we are in finding a solution that satisfies some additional requirement. In many physical problems we need to find the particular solution that satisfies a condition of the form $y(x_0) = y_0$. This is called an **initial condition**, and the problem of finding a solution of the differential equation that satisfies the initial condition is called an **initial-value problem**.

EXAMPLE 1

(a) Solve the differential equation $\dfrac{dy}{dx} = \dfrac{x^2}{y^2}$.

(b) Find the solution of this equation that satisfies the initial condition $y(0) = 2$.

SOLUTION

(a) We write the equation in terms of differentials and integrate both sides:

$$y^2\, dy = x^2\, dx$$

$$\int y^2\, dy = \int x^2\, dx$$

$$\tfrac{1}{3}y^3 = \tfrac{1}{3}x^3 + C$$

■ Figure 1 shows graphs of several members of the family of solutions of the differential equation in Example 1. The solution of the initial-value problem in part (b) is shown in blue.

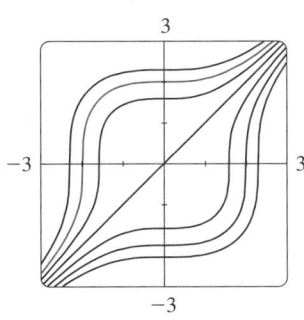

FIGURE 1

where C is an arbitrary constant. (We could have used a constant C_1 on the left side and another constant C_2 on the right side. But then we could combine these constants by writing $C = C_2 - C_1$.)

Solving for y, we get

$$y = \sqrt[3]{x^3 + 3C}$$

We could leave the solution like this or we could write it in the form

$$y = \sqrt[3]{x^3 + K}$$

where $K = 3C$. (Since C is an arbitrary constant, so is K.)

(b) If we put $x = 0$ in the general solution in part (a), we get $y(0) = \sqrt[3]{K}$. To satisfy the initial condition $y(0) = 2$, we must have $\sqrt[3]{K} = 2$ and so $K = 8$.

Thus the solution of the initial-value problem is

$$y = \sqrt[3]{x^3 + 8} \qquad ■$$

■ Some computer algebra systems can plot curves defined by implicit equations. Figure 2 shows the graphs of several members of the family of solutions of the differential equation in Example 2. As we look at the curves from left to right, the values of C are 3, 2, 1, 0, -1, -2, and -3.

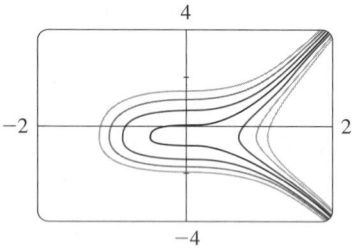

FIGURE 2

V EXAMPLE 2 Solve the differential equation $\dfrac{dy}{dx} = \dfrac{6x^2}{2y + \cos y}$.

SOLUTION Writing the equation in differential form and integrating both sides, we have

$$(2y + \cos y)\,dy = 6x^2\,dx$$

$$\int (2y + \cos y)\,dy = \int 6x^2\,dx$$

$$\boxed{6} \qquad y^2 + \sin y = 2x^3 + C$$

where C is a constant. Equation 6 gives the general solution implicitly. In this case it's impossible to solve the equation to express y explicitly as a function of x. ■

EXAMPLE 3 Solve the equation $y' = x^2 y$.

SOLUTION First we rewrite the equation using Leibniz notation:

$$\frac{dy}{dx} = x^2 y$$

■ If a solution y is a function that satisfies $y(x) \neq 0$ for some x, it follows from a uniqueness theorem for solutions of differential equations that $y(x) \neq 0$ for all x.

If $y \neq 0$, we can rewrite it in differential notation and integrate:

$$\frac{dy}{y} = x^2\,dx \qquad y \neq 0$$

$$\int \frac{dy}{y} = \int x^2\,dx$$

$$\ln |y| = \frac{x^3}{3} + C$$

■ Several solutions of the differential equation in Example 3 are graphed in Figure 3. The values of A are the same as the y-intercepts.

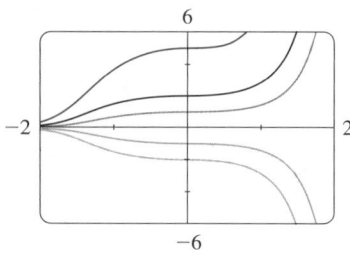

FIGURE 3

This equation defines y implicitly as a function of x. But in this case we can solve explicitly for y as follows:

$$|y| = e^{\ln |y|} = e^{(x^3/3) + C} = e^C e^{x^3/3}$$

so

$$y = \pm e^C e^{x^3/3}$$

We can easily verify that the function $y = 0$ is also a solution of the given differential equation. So we can write the general solution in the form

$$y = A e^{x^3/3}$$

where A is an arbitrary constant ($A = e^C$, or $A = -e^C$, or $A = 0$). ■

EXAMPLE 4 Solve the equation $\dfrac{dy}{dt} = ky$.

SOLUTION This differential equation was studied in Section 3.4, where it was called the law of natural growth (or decay). Since it is a separable equation, we can solve it by the methods of this section as follows:

$$\int \frac{dy}{y} = \int k\,dt \quad y \neq 0$$

$$\ln|y| = kt + C$$

$$|y| = e^{kt+C} = e^C e^{kt}$$

$$y = Ae^{kt}$$

where $A\ (= \pm e^C$ or 0) is an arbitrary constant. ∎

▪ www.stewartcalculus.com
See Additional Examples A, B.

LOGISTIC GROWTH

The differential equation of Example 4 is appropriate for modeling population growth ($y' = ky$ says that the rate of growth is proportional to the size of the population) under conditions of unlimited environment and food supply. However, in a restricted environment and with limited food supply, the population cannot exceed a maximal size M (called the **carrying capacity**) at which it consumes its entire food supply. If we make the assumption that the rate of growth of population is jointly proportional to the size of the population (y) and the amount by which y falls short of the maximal size ($M - y$), then we have the equation

7

$$\frac{dy}{dt} = ky(M - y)$$

where k is a constant. Equation 7 is called the **logistic differential equation** and was used by the Dutch mathematical biologist Pierre-François Verhulst in the 1840s to model world population growth.

The logistic equation is separable, so we write it in the form

$$\int \frac{dy}{y(M - y)} = \int k\,dt$$

Using partial fractions, we have

$$\frac{1}{y(M - y)} = \frac{1}{M}\left[\frac{1}{y} + \frac{1}{M - y}\right]$$

and so

$$\frac{1}{M}\left[\int \frac{dy}{y} + \int \frac{dy}{M - y}\right] = \int k\,dt = kt + C$$

$$\frac{1}{M}\left(\ln|y| - \ln|M - y|\right) = kt + C$$

But $|y| = y$ and $|M - y| = M - y$, since $0 < y < M$, so we have

$$\ln \frac{y}{M - y} = M(kt + C)$$

$$\frac{y}{M - y} = Ae^{kMt} \quad (A = e^{MC})$$

If the population at time $t = 0$ is $y(0) = y_0$, then $A = y_0/(M - y_0)$, so

$$\frac{y}{M - y} = \frac{y_0}{M - y_0} e^{kMt}$$

If we solve this equation for y, we get

$$\boxed{8} \qquad y = \frac{y_0 M e^{kMt}}{M - y_0 + y_0 e^{kMt}} = \frac{y_0 M}{y_0 + (M - y_0)e^{-kMt}}$$

Using the latter expression for y, we see that

$$\lim_{t \to \infty} y(t) = M$$

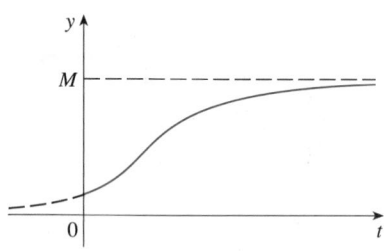

FIGURE 4
Logistic growth function

which is to be expected.

The graph of the logistic growth function is shown in Figure 4. At first the graph is concave upward and the growth curve appears to be almost exponential, but then it is concave downward and approaches the limiting population M.

MIXING PROBLEMS

A typical mixing problem involves a tank of fixed capacity filled with a thoroughly mixed solution of some substance, such as salt. A solution of a given concentration enters the tank at a fixed rate and the mixture, thoroughly stirred, leaves at a fixed rate, which may differ from the entering rate. If $y(t)$ denotes the amount of substance in the tank at time t, then $y'(t)$ is the rate at which the substance is being added minus the rate at which it is being removed. The mathematical description of this situation often leads to a first-order separable differential equation. We can use the same type of reasoning to model a variety of phenomena: chemical reactions, discharge of pollutants into a lake, injection of a drug into the bloodstream.

EXAMPLE 5 A tank contains 20 kg of salt dissolved in 5000 L of water. Brine that contains 0.03 kg of salt per liter of water enters the tank at a rate of 25 L/min. The solution is kept thoroughly mixed and drains from the tank at the same rate. How much salt remains in the tank after half an hour?

SOLUTION Let $y(t)$ be the amount of salt (in kilograms) after t minutes. We are given that $y(0) = 20$ and we want to find $y(30)$. We do this by finding a differential equation satisfied by $y(t)$. Note that dy/dt is the rate of change of the amount of salt, so

$$\boxed{9} \qquad \frac{dy}{dt} = (\text{rate in}) - (\text{rate out})$$

where (rate in) is the rate at which salt enters the tank and (rate out) is the rate at which salt leaves the tank. We have

$$\text{rate in} = \left(0.03 \frac{\text{kg}}{\text{L}}\right)\left(25 \frac{\text{L}}{\text{min}}\right) = 0.75 \frac{\text{kg}}{\text{min}}$$

The tank always contains 5000 L of liquid, so the concentration at time t is $y(t)/5000$ (measured in kilograms per liter). Since the brine flows out at a rate of 25 L/min, we have

$$\text{rate out} = \left(\frac{y(t)}{5000} \frac{\text{kg}}{\text{L}}\right)\left(25 \frac{\text{L}}{\text{min}}\right) = \frac{y(t)}{200} \frac{\text{kg}}{\text{min}}$$

Thus from Equation 9 we get

$$\frac{dy}{dt} = 0.75 - \frac{y(t)}{200} = \frac{150 - y(t)}{200}$$

Solving this separable differential equation, we obtain

$$\int \frac{dy}{150 - y} = \int \frac{dt}{200}$$

$$-\ln|150 - y| = \frac{t}{200} + C$$

Since $y(0) = 20$, we have $-\ln 130 = C$, so

$$-\ln|150 - y| = \frac{t}{200} - \ln 130$$

Therefore $\qquad |150 - y| = 130e^{-t/200}$

Since $y(t)$ is continuous and $y(0) = 20$ and the right side is never 0, we deduce that $150 - y(t)$ is always positive. Thus $|150 - y| = 150 - y$ and so

$$y(t) = 150 - 130e^{-t/200}$$

The amount of salt after 30 min is

$$y(30) = 150 - 130e^{-30/200} \approx 38.1 \text{ kg} \qquad \blacksquare$$

■ Figure 5 shows the graph of the function $y(t)$ of Example 5. Notice that, as time goes by, the amount of salt approaches 150 kg.

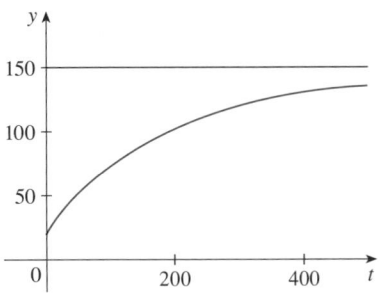

FIGURE 5

DIRECTION FIELDS

Suppose we are given a first-order differential equation of the form

$$y' = F(x, y)$$

where $F(x, y)$ is some expression in x and y. [Recall that a separable equation is the special case in which $F(x, y)$ can be factored as a function of x times a function of y.] Even if it is impossible to find a formula for the solution, we can still visualize the solution curves by means of a direction field. If a solution curve passes through a point (x, y), then its slope at that point is y', which is equal to $F(x, y)$. If we draw short line segments with slopes $F(x, y)$ at several points (x, y), the result is called a **direction field** (or **slope field**). These line segments indicate the direction in which a solution curve is heading, so the direction field helps us visualize the general shape of these curves.

▼ EXAMPLE 6

(a) Sketch the direction field for the differential equation $y' = x^2 + y^2 - 1$.
(b) Use part (a) to sketch the solution curve that passes through the origin.

SOLUTION

(a) We start by computing the slope at several points in the following chart:

x	-2	-1	0	1	2	-2	-1	0	1	2	$\ldots$
y	0	0	0	0	0	1	1	1	1	1	$\ldots$
$y' = x^2 + y^2 - 1$	3	0	-1	0	3	4	1	0	1	4	$\ldots$

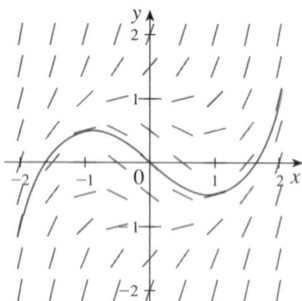

FIGURE 6

Now we draw short line segments with these slopes at these points. The result is the direction field shown in Figure 6.

(b) We start at the origin and move to the right in the direction of the line segment (which has slope -1). We continue to draw the solution curve so that it moves parallel to the nearby line segments. The resulting solution curve is shown in Figure 7. Returning to the origin, we draw the solution curve to the left as well. ■

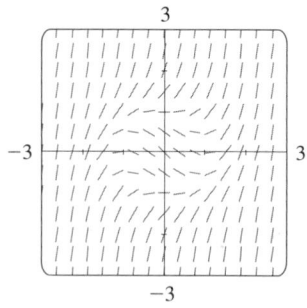

FIGURE 7

The more line segments we draw in a direction field, the clearer the picture becomes. Of course, it's tedious to compute slopes and draw line segments for a huge number of points by hand, but computers are well suited for this task. Figure 8 shows a more detailed, computer-drawn direction field for the differential equation in Example 6. It enables us to draw, with reasonable accuracy, the solution curves shown in Figure 9 with y-intercepts -2, -1, 0, 1, and 2.

TEC Module 7.7 shows direction fields and solution curves for a variety of differential equations.

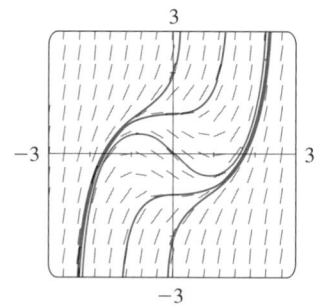

FIGURE 8   **FIGURE 9**

7.7 | EXERCISES

1–8 ▪ Solve the differential equation.

1. $\dfrac{dy}{dx} = xy^2$

2. $\dfrac{dy}{dx} = xe^{-y}$

3. $xy^2y' = x + 1$

4. $(y^2 + xy^2)y' = 1$

5. $(y + \sin y)y' = x + x^3$

6. $\dfrac{dy}{d\theta} = \dfrac{e^y \sin^2\theta}{y \sec \theta}$

7. $\dfrac{dp}{dt} = t^2 p - p + t^2 - 1$ **8.** $\dfrac{dz}{dt} + e^{t+z} = 0$

9–14 ▪ Find the solution of the differential equation that satisfies the given initial condition.

9. $\dfrac{dy}{dx} = \dfrac{x}{y}$, $y(0) = -3$

10. $\dfrac{dy}{dx} = \dfrac{\ln x}{xy}$, $y(1) = 2$

11. $\dfrac{du}{dt} = \dfrac{2t + \sec^2 t}{2u}$, $u(0) = -5$

12. $\dfrac{dP}{dt} = \sqrt{Pt}$, $P(1) = 2$

13. $y' \tan x = a + y,\ y(\pi/3) = a,\ \ 0 < x < \pi/2$

14. $\dfrac{dL}{dt} = kL^2 \ln t,\ L(1) = -1$

---

15. Find an equation of the curve that passes through the point $(0, 1)$ and whose slope at (x, y) is xy.

16. Find the function f such that $f'(x) = f(x)(1 - f(x))$ and $f(0) = \frac{1}{2}$.

17. (a) Solve the differential equation $y' = 2x\sqrt{1 - y^2}$.
    (b) Solve the initial-value problem $y' = 2x\sqrt{1 - y^2}$, $y(0) = 0$, and graph the solution.
    (c) Does the initial-value problem $y' = 2x\sqrt{1 - y^2}$, $y(0) = 2$, have a solution? Explain.

18. Solve the equation $e^{-y}y' + \cos x = 0$ and graph several members of the family of solutions. How does the solution curve change as the constant C varies?

CAS **19.** Solve the initial-value problem $y' = (\sin x)/\sin y$, $y(0) = \pi/2$, and graph the solution (if your CAS does implicit plots).

CAS **20.** Solve the equation $y' = x\sqrt{x^2 + 1}/(ye^y)$ and graph several members of the family of solutions (if your CAS does implicit plots). How does the solution curve change as the constant C varies?

21–24 ▪ Match the differential equation with its direction field (labeled I–IV). Give reasons for your answer.

21. $y' = 2 - y$ **22.** $y' = x(2 - y)$

23. $y' = x + y - 1$ **24.** $y' = \sin x \, \sin y$

I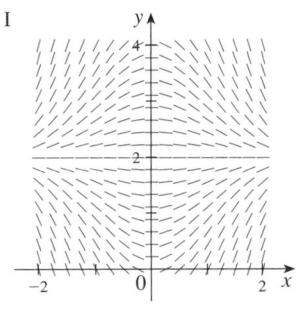

II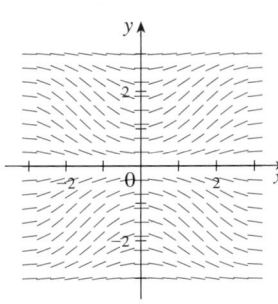

III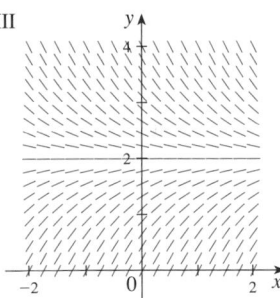

IV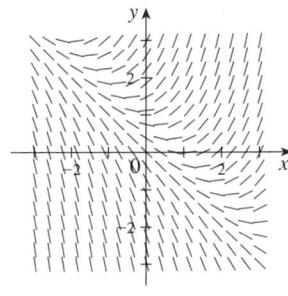

25–26 ▪ Refer to the direction fields in Exercises 21–24.

25. Use field II to sketch the graphs of the solutions that satisfy the given initial conditions.
    (a) $y(0) = 1$ (b) $y(0) = 2$ (c) $y(0) = -1$

26. Use field IV to sketch the graphs of the solutions that satisfy the given initial conditions.
    (a) $y(0) = -1$ (b) $y(0) = 0$ (c) $y(0) = 1$

27–28 ▪ Sketch a direction field for the differential equation. Then use it to sketch three solution curves.

27. $y' = \frac{1}{2}y$ **28.** $y' = x - y + 1$

29–32 ▪ Sketch the direction field of the differential equation. Then use it to sketch a solution curve that passes through the given point.

29. $y' = y - 2x$, $(1, 0)$ **30.** $y' = 1 - xy$, $(0, 0)$

31. $y' = y + xy$, $(0, 1)$ **32.** $y' = x - xy$, $(1, 0)$

---

33. Psychologists interested in learning theory study **learning curves**. A learning curve is the graph of a function $P(t)$, the performance of someone learning a skill as a function of the training time t. The derivative dP/dt represents the rate at which performance improves.
    (a) If M is the maximum level of performance of which the learner is capable, explain why the differential equation

$$\frac{dP}{dt} = k(M - P) \qquad k \text{ a positive constant}$$

is a reasonable model for learning.

(b) Solve the differential equation in part (a) to find an expression for $P(t)$. What is the limit of this expression?

34. A sphere with radius 1 m has temperature 15°C. It lies inside a concentric sphere with radius 2 m and temperature 25°C. The temperature $T(r)$ at a distance r from the common center of the spheres satisfies the differential equation

$$\frac{d^2T}{dr^2} + \frac{2}{r}\frac{dT}{dr} = 0$$

If we let $S = dT/dr$, then S satisfies a first-order differential equation. Solve it to find an expression for the temperature $T(r)$ between the spheres.

35. A glucose solution is administered intravenously into the bloodstream at a constant rate r. As the glucose is added, it is converted into other substances and removed from the bloodstream at a rate that is proportional to the concentration at that time. Thus a model for the concentration $C = C(t)$ of the glucose solution in the bloodstream is

$$\frac{dC}{dt} = r - kC$$

where k is a positive constant.
(a) Suppose that the concentration at time $t = 0$ is C_0. Determine the concentration at any time t by solving the differential equation.
(b) Assuming that $C_0 < r/k$, find $\lim_{t \to \infty} C(t)$ and interpret your answer.

36. A certain small country has $10 billion in paper currency in circulation, and each day $50 million comes into the country's banks. The government decides to introduce new currency by having the banks replace old bills with new ones whenever old currency comes into the banks. Let $x = x(t)$ denote the amount of new currency in circulation at time t, with $x(0) = 0$.
(a) Formulate a mathematical model in the form of an initial-value problem that represents the "flow" of the new currency into circulation.
(b) Solve the initial-value problem found in part (a).
(c) How long will it take for the new bills to account for 90% of the currency in circulation?

37. Write the solution of the logistic initial-value problem

$$\frac{dP}{dt} = 0.00008P(1000 - P) \qquad P(0) = 100$$

and use it to find the population sizes $P(40)$ and $P(80)$. At what time does the population reach 900?

38. The Pacific halibut fishery has been modeled by the differential equation

$$\frac{dy}{dt} = ky(M - y)$$

where $y(t)$ is the biomass (the total mass of the members of the population) in kilograms at time t (measured in years), the carrying capacity is estimated to be $M = 8 \times 10^7$ kg, and $k = 8.875 \times 10^{-9}$ per year.
(a) If $y(0) = 2 \times 10^7$ kg, find the biomass a year later.
(b) How long will it take for the biomass to reach 4×10^7 kg?

39. One model for the spread of a rumor is that the rate of spread is proportional to the product of the fraction y of the population who have heard the rumor and the fraction who have not heard the rumor.
(a) Write a differential equation that is satisfied by y.
(b) Solve the differential equation.
(c) A small town has 1000 inhabitants. At 8 AM, 80 people have heard a rumor. By noon half the town has heard it. At what time will 90% of the population have heard the rumor?

40. Biologists stocked a lake with 400 fish and estimated the carrying capacity (the maximal population for the fish of that species in that lake) to be 10,000. The number of fish tripled in the first year.
(a) Assuming that the size of the fish population satisfies the logistic equation, find an expression for the size of the population after t years.
(b) How long will it take for the population to increase to 5000?

41. (a) Show that if y satisfies the logistic equation $\boxed{7}$, then

$$\frac{d^2y}{dt^2} = k^2y(M - y)(M - 2y)$$

(b) Deduce that a population grows fastest when it reaches half its carrying capacity.

42. For a fixed value of M (say $M = 10$), the family of logistic functions given by Equation 8 depends on the initial value y_0 and the proportionality constant k. Graph several members of this family. How does the graph change when y_0 varies? How does it change when k varies?

43. A tank contains 1000 L of brine with 15 kg of dissolved salt. Pure water enters the tank at a rate of 10 L/min. The solution is kept thoroughly mixed and drains from the tank at the same rate. How much salt is in the tank (a) after t minutes and (b) after 20 minutes?

44. The air in a room with volume 180 m³ contains 0.15% carbon dioxide initially. Fresher air with only 0.05% carbon dioxide flows into the room at a rate of 2 m³/min and the mixed air flows out at the same rate. Find the percentage of carbon dioxide in the room as a function of time. What happens in the long run?

45. A vat with 500 gallons of beer contains 4% alcohol (by volume). Beer with 6% alcohol is pumped into the vat at a rate of 5 gal/min and the mixture is pumped out at the same rate. What is the percentage of alcohol after an hour?

46. A tank contains 1000 L of pure water. Brine that contains 0.05 kg of salt per liter of water enters the tank at a rate of 5 L/min. Brine that contains 0.04 kg of salt per liter of water enters the tank at a rate of 10 L/min. The solution is kept thoroughly mixed and drains from the tank at a rate of 15 L/min. How much salt is in the tank (a) after t minutes and (b) after one hour?

47. When a raindrop falls, it increases in size and so its mass at time t is a function of t, $m(t)$. The rate of growth of the mass is $km(t)$ for some positive constant k. When we apply Newton's Law of Motion to the raindrop, we get $(mv)' = gm$, where v is the velocity of the raindrop (directed downward) and g is the acceleration due to gravity. The *terminal velocity* of the raindrop is $\lim_{t \to \infty} v(t)$. Find an expression for the terminal velocity in terms of g and k.

48. An object of mass m is moving horizontally through a medium which resists the motion with a force that is a function of the velocity; that is,

$$m \frac{d^2 s}{dt^2} = m \frac{dv}{dt} = f(v)$$

where $v = v(t)$ and $s = s(t)$ represent the velocity and position of the object at time t, respectively. For example, think of a boat moving through the water.

(a) Suppose that the resisting force is proportional to the velocity, that is, $f(v) = -kv$, k a positive constant. (This model is appropriate for small values of v.) Let $v(0) = v_0$ and $s(0) = s_0$ be the initial values of v and s. Determine v and s at any time t. What is the total distance that the object travels from time $t = 0$?

(b) For larger values of v a better model is obtained by supposing that the resisting force is proportional to the square of the velocity, that is, $f(v) = -kv^2$, $k > 0$. (This model was first proposed by Newton.) Let v_0 and s_0 be the initial values of v and s. Determine v and s at any time t. What is the total distance that the object travels in this case?

49. Let $A(t)$ be the area of a tissue culture at time t and let M be the final area of the tissue when growth is complete. Most cell divisions occur on the periphery of the tissue and the number of cells on the periphery is proportional to $\sqrt{A(t)}$. So a reasonable model for the growth of tissue is obtained by assuming that the rate of growth of the area is jointly proportional to $\sqrt{A(t)}$ and $M - A(t)$.

(a) Formulate a differential equation and use it to show that the tissue grows fastest when $A(t) = \frac{1}{3}M$.

CAS

(b) Solve the differential equation to find an expression for $A(t)$. Use a computer algebra system to perform the integration.

50. According to Newton's Law of Universal Gravitation, the gravitational force on an object of mass m that has been projected vertically upward from the earth's surface is

$$F = \frac{mgR^2}{(x + R)^2}$$

where $x = x(t)$ is the object's distance above the surface at time t, R is the earth's radius, and g is the acceleration due to gravity. Also, by Newton's Second Law, $F = ma = m(dv/dt)$ and so

$$m \frac{dv}{dt} = -\frac{mgR^2}{(x + R)^2}$$

(a) Suppose a rocket is fired vertically upward with an initial velocity v_0. Let h be the maximum height above the surface reached by the object. Show that

$$v_0 = \sqrt{\frac{2gRh}{R + h}}$$

[*Hint:* By the Chain Rule, $m(dv/dt) = mv(dv/dx)$.]

(b) Calculate $v_e = \lim_{h \to \infty} v_0$. This limit is called the *escape velocity* for the earth.

(c) Use $R = 3960$ mi and $g = 32$ ft/s² to calculate v_e in feet per second and in miles per second.

CHAPTER 7   REVIEW

CONCEPT CHECK

1. (a) Draw two typical curves $y = f(x)$ and $y = g(x)$, where $f(x) \geqslant g(x)$ for $a \leqslant x \leqslant b$. Show how to approximate the area between these curves by a Riemann sum and sketch the corresponding approximating rectangles. Then write an expression for the exact area.

(b) Explain how the situation changes if the curves have equations $x = f(y)$ and $x = g(y)$, where $f(y) \geqslant g(y)$ for $c \leqslant y \leqslant d$.

2. Suppose that Sue runs faster than Kathy throughout a 1500-meter race. What is the physical meaning of the area between their velocity curves for the first minute of the race?

3. (a) Suppose S is a solid with known cross-sectional areas. Explain how to approximate the volume of S by a Riemann sum. Then write an expression for the exact volume.

(b) If S is a solid of revolution, how do you find the cross-sectional areas?

4. (a) What is the volume of a cylindrical shell?

(b) Explain how to use cylindrical shells to find the volume of a solid of revolution.

(c) Why might you want to use the shell method instead of slicing?

5. (a) How is the length of a curve defined?
  (b) Write an expression for the length of a smooth curve given by $y = f(x)$, $a \le x \le b$.
  (c) What if x is given as a function of y?

6. (a) Write an expression for the surface area of the surface obtained by rotating the curve $y = f(x)$, $a \le x \le b$, about the x-axis.
  (b) What if x is given as a function of y?
  (c) What if the curve is rotated about the y-axis?

7. Suppose that you push a book across a 6-meter-long table by exerting a force $f(x)$ at each point from $x = 0$ to $x = 6$. What does $\int_0^6 f(x)\,dx$ represent? If $f(x)$ is measured in newtons, what are the units for the integral?

8. Describe how we can find the hydrostatic force against a vertical wall submersed in a fluid.

9. (a) What is the physical significance of the center of mass of a thin plate?
  (b) If the plate lies between $y = f(x)$ and $y = 0$, where $a \le x \le b$, write expressions for the coordinates of the center of mass.

10. What does the Theorem of Pappus say?

11. (a) What is a differential equation?
  (b) What is the order of a differential equation?
  (c) What is an initial condition?

12. What is a direction field for the differential equation $y' = F(x, y)$?

13. What is a separable differential equation? How do you solve it?

EXERCISES

1–4 ■ Find the area of the region bounded by the given curves.

1. $y = x^2$, $y = 4x - x^2$

2. $y = 1/x$, $y = x^2$, $y = 0$, $x = e$

3. $y = 1 - 2x^2$, $y = |x|$

4. $x + y = 0$, $x = y^2 + 3y$

5–9 ■ Find the volume of the solid obtained by rotating the region bounded by the given curves about the specified axis.

5. $y = 2x$, $y = x^2$; about the x-axis

6. $x = 1 + y^2$, $y = x - 3$; about the y-axis

7. $x = 0$, $x = 9 - y^2$; about $x = -1$

8. $y = x^2 + 1$, $y = 9 - x^2$; about $y = -1$

9. $x^2 - y^2 = a^2$, $x = a + h$ (where $a > 0$, $h > 0$); about the y-axis

---

10–12 ■ Set up, but do not evaluate, an integral for the volume of the solid obtained by rotating the region bounded by the given curves about the specified axis.

10. $y = \tan x$, $y = x$, $x = \pi/3$; about the y-axis

11. $y = \cos^2 x$, $|x| \le \pi/2$, $y = \frac{1}{4}$; about $x = \pi/2$

12. $y = \sqrt{x}$, $y = x^2$; about $y = 2$

---

13. Find the volumes of the solids obtained by rotating the region bounded by the curves $y = x$ and $y = x^2$ about the following lines.
  (a) The x-axis   (b) The y-axis   (c) $y = 2$

14. Let $\mathcal{R}$ be the region in the first quadrant bounded by the curves $y = x^3$ and $y = 2x - x^2$. Calculate the following quantities.
  (a) The area of $\mathcal{R}$
  (b) The volume obtained by rotating $\mathcal{R}$ about the x-axis
  (c) The volume obtained by rotating $\mathcal{R}$ about the y-axis

15. Let $\mathcal{R}$ be the region bounded by the curves $y = \tan(x^2)$, $x = 1$, and $y = 0$. Use the Midpoint Rule with $n = 4$ to estimate the following quantities.
  (a) The area of $\mathcal{R}$
  (b) The volume obtained by rotating $\mathcal{R}$ about the x-axis

16. Let $\mathcal{R}$ be the region bounded by the curves $y = 1 - x^2$ and $y = x^6 - x + 1$. Estimate the following quantities.
  (a) The x-coordinates of the points of intersection of the curves
  (b) The area of $\mathcal{R}$
  (c) The volume generated when $\mathcal{R}$ is rotated about the x-axis
  (d) The volume generated when $\mathcal{R}$ is rotated about the y-axis

17–20 ■ Each integral represents the volume of a solid. Describe the solid.

17. $\int_0^{\pi/2} 2\pi x \cos x\,dx$ **18.** $\int_0^{\pi/2} 2\pi \cos^2 x\,dx$

19. $\int_0^{\pi} \pi(2 - \sin x)^2\,dx$

20. $\int_0^4 2\pi(6 - y)(4y - y^2)\,dy$

---

21. The base of a solid is a circular disk with radius 3. Find the volume of the solid if parallel cross-sections perpendicular to the base are isosceles right triangles with hypotenuse lying along the base.

22. The base of a solid is the region bounded by the parabolas $y = x^2$ and $y = 2 - x^2$. Find the volume of the solid if the cross-sections perpendicular to the x-axis are squares with one side lying along the base.

23. The height of a monument is 20 m. A horizontal cross-section at a distance x meters from the top is an equilateral triangle with side $\frac{1}{4}x$ meters. Find the volume of the monument.

24. (a) The base of a solid is a square with vertices located at $(1, 0)$, $(0, 1)$, $(-1, 0)$, and $(0, -1)$. Each cross-section perpendicular to the x-axis is a semicircle. Find the volume of the solid.
(b) Show that by cutting the solid of part (a), we can rearrange it to form a cone. Thus compute its volume more simply.

25–26 ▪ Find the length of the curve.

25. $y = \frac{1}{6}(x^2 + 4)^{3/2}$, $\quad 0 \leqslant x \leqslant 3$

26. $y = 2 \ln(\sin \frac{1}{2}x)$, $\quad \pi/3 \leqslant x \leqslant \pi$

27. (a) Find the length of the curve

$$y = \frac{x^4}{16} + \frac{1}{2x^2} \qquad 1 \leqslant x \leqslant 2$$

(b) Find the area of the surface obtained by rotating the curve in part (a) about the y-axis.

28. (a) The curve $y = x^2$, $0 \leqslant x \leqslant 1$, is rotated about the y-axis. Find the area of the resulting surface.
(b) Find the area of the surface obtained by rotating the curve in part (a) about the x-axis.

29. Use Simpson's Rule with $n = 10$ to estimate the length of the sine curve $y = \sin x$, $0 \leqslant x \leqslant \pi$.

30. Use Simpson's Rule with $n = 10$ to estimate the area of the surface obtained by rotating the sine curve in Exercise 29 about the x-axis.

31. Find the length of the curve

$$y = \int_1^x \sqrt{\sqrt{t} - 1} \; dt \qquad 1 \leqslant x \leqslant 16$$

32. Find the area of the surface obtained by rotating the curve in Exercise 31 about the y-axis.

33. A force of 30 N is required to maintain a spring stretched from its natural length of 12 cm to a length of 15 cm. How much work is done in stretching the spring from 12 cm to 20 cm?

34. A 1600-pound elevator is suspended by a 200-foot cable that weighs 10 lb/ft. How much work is required to raise the elevator from the basement to the third floor, a distance of 30 ft?

35. A tank full of water has the shape of a paraboloid of revolution as shown in the figure; that is, its shape is obtained by rotating a parabola about a vertical axis.
(a) If its height is 4 ft and the radius at the top is 4 ft, find the work required to pump the water out of the tank.
(b) After 4000 ft-lb of work has been done, what is the depth of the water remaining in the tank?

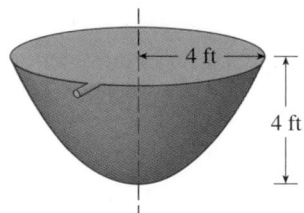

36. A trough is filled with water and its vertical ends have the shape of the parabolic region in the figure. Find the hydrostatic force on one end of the trough.

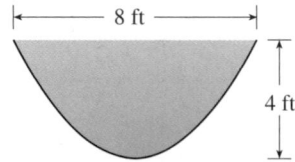

37. A gate in an irrigation canal is constructed in the form of a trapezoid 3 ft wide at the bottom, 5 ft wide at the top, and 2 ft high. It is placed vertically in the canal so that the water just covers the gate. Find the hydrostatic force on one side of the gate.

38. Find the centroid of the region shown.

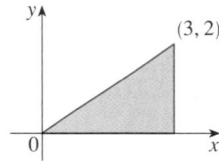

39–40 ▪ Find the centroid of the region bounded by the given curves.

39. $y = \frac{1}{2}x$, $\quad y = \sqrt{x}$

40. $y = \sin x$, $\quad y = 0$, $\quad x = \pi/4$, $\quad x = 3\pi/4$

41. Find the volume obtained when the circle of radius 1 with center $(1, 0)$ is rotated about the y-axis.

42. Use the Theorem of Pappus and the fact that the volume of a sphere of radius r is $\frac{4}{3}\pi r^3$ to find the centroid of the semi-circular region bounded by the curve $y = \sqrt{r^2 - x^2}$ and the x-axis.

43–44 ▪ Solve the differential equation.

43. $2ye^{y^2}y' = 2x + 3\sqrt{x}$

44. $\dfrac{dx}{dt} = 1 - t + x - tx$

45–46 ▪ Solve the initial-value problem.

45. $\dfrac{dr}{dt} + 2tr = r, \quad r(0) = 5$

46. $(1 + \cos x)y' = (1 + e^{-y})\sin x, \quad y(0) = 0$

---

47. (a) Sketch a direction field for the differential equation
$y' = x/y$. Then use it to sketch the four solutions that
satisfy the initial conditions $y(0) = 1$, $y(0) = -1$,
$y(2) = 1$, and $y(-2) = 1$.

(b) Check your work in part (a) by solving the differential
equation explicitly. What type of curve is each solution
curve?

48. Let $\mathcal{R}_1$ be the region bounded by $y = x^2$, $y = 0$, and $x = b$,
where $b > 0$. Let $\mathcal{R}_2$ be the region bounded by $y = x^2$,
$x = 0$, and $y = b^2$.
(a) Is there a value of b such that $\mathcal{R}_1$ and $\mathcal{R}_2$ have the same
area?
(b) Is there a value of b such that $\mathcal{R}_1$ sweeps out the same
volume when rotated about the x-axis and the y-axis?
(c) Is there a value of b such that $\mathcal{R}_1$ and $\mathcal{R}_2$ sweep out the
same volume when rotated about the x-axis?
(d) Is there a value of b such that $\mathcal{R}_1$ and $\mathcal{R}_2$ sweep out the
same volume when rotated about the y-axis?

8

SERIES

Infinite series are sums of infinitely many terms. (One of our aims in this chapter is to define exactly what is meant by an infinite sum.) Their importance in calculus stems from Newton's idea of representing functions as sums of infinite series. For instance, in finding areas he often integrated a function by first expressing it as a series and then integrating each term of the series. We will pursue his idea in Section 8.7 in order to integrate such functions as e^{-x^2}. (Recall that we have previously been unable to do this.) Many of the functions that arise in mathematical physics and chemistry, such as Bessel functions, are defined as sums of series, so it is important to be familiar with the basic concepts of convergence of infinite sequences and series.

Physicists also use series in another way, as we will see in Section 8.8. In studying fields as diverse as optics, special relativity, and electromagnetism, they analyze phenomena by replacing a function with the first few terms in the series that represents it.

8.1 | SEQUENCES

A **sequence** can be thought of as a list of numbers written in a definite order:

$$a_1, \ a_2, \ a_3, \ a_4, \ \ldots, \ a_n, \ldots$$

The number a_1 is called the *first term*, a_2 is the *second term*, and in general a_n is the *nth term*. We will deal exclusively with infinite sequences and so each term a_n will have a successor a_{n+1}.

Notice that for every positive integer n there is a corresponding number a_n and so a sequence can be defined as a function whose domain is the set of positive integers. But we usually write a_n instead of the function notation $f(n)$ for the value of the function at the number n.

NOTATION The sequence $\{a_1, a_2, a_3, \ldots\}$ is also denoted by

$$\{a_n\} \qquad \text{or} \qquad \{a_n\}_{n=1}^{\infty}$$

EXAMPLE 1 Some sequences can be defined by giving a formula for the nth term. In the following examples we give three descriptions of the sequence: one by using the preceding notation, another by using the defining formula, and a third by writing out the terms of the sequence. Notice that n doesn't have to start at 1.

(a) $\left\{ \dfrac{n}{n+1} \right\}_{n=1}^{\infty}$ $\qquad a_n = \dfrac{n}{n+1}$ $\qquad \left\{ \dfrac{1}{2}, \dfrac{2}{3}, \dfrac{3}{4}, \dfrac{4}{5}, \ldots, \dfrac{n}{n+1}, \ldots \right\}$

(b) $\left\{ \dfrac{(-1)^n(n+1)}{3^n} \right\}$ $\qquad a_n = \dfrac{(-1)^n(n+1)}{3^n}$ $\qquad \left\{ -\dfrac{2}{3}, \dfrac{3}{9}, -\dfrac{4}{27}, \dfrac{5}{81}, \ldots, \dfrac{(-1)^n(n+1)}{3^n}, \ldots \right\}$

(c) $\left\{ \sqrt{n-3} \right\}_{n=3}^{\infty}$ $\qquad a_n = \sqrt{n-3}, \ n \geq 3$ $\qquad \left\{ 0, 1, \sqrt{2}, \sqrt{3}, \ldots, \sqrt{n-3}, \ldots \right\}$

(d) $\left\{ \cos \dfrac{n\pi}{6} \right\}_{n=0}^{\infty}$ $\qquad a_n = \cos \dfrac{n\pi}{6}, \ n \geq 0$ $\qquad \left\{ 1, \dfrac{\sqrt{3}}{2}, \dfrac{1}{2}, 0, \ldots, \cos \dfrac{n\pi}{6}, \ldots \right\}$ ∎

▼ EXAMPLE 2 Find a formula for the general term a_n of the sequence

$$\left\{ \frac{3}{5}, -\frac{4}{25}, \frac{5}{125}, -\frac{6}{625}, \frac{7}{3125}, \ldots \right\}$$

assuming that the pattern of the first few terms continues.

SOLUTION We are given that

$$a_1 = \frac{3}{5} \qquad a_2 = -\frac{4}{25} \qquad a_3 = \frac{5}{125} \qquad a_4 = -\frac{6}{625} \qquad a_5 = \frac{7}{3125}$$

Notice that the numerators of these fractions start with 3 and increase by 1 whenever we go to the next term. The second term has numerator 4, the third term has numerator 5; in general, the nth term will have numerator $n + 2$. The denominators are the powers of 5, so a_n has denominator 5^n. The signs of the terms are alternately positive and negative, so we need to multiply by a power of -1. In Example 1(b) the factor $(-1)^n$ meant we started with a negative term. Here we want to start with a positive term and so we use $(-1)^{n-1}$ or $(-1)^{n+1}$. Therefore

$$a_n = (-1)^{n-1} \frac{n + 2}{5^n} \qquad\qquad ∎$$

EXAMPLE 3 Here are some sequences that don't have a simple defining equation.

(a) The sequence $\{p_n\}$, where p_n is the population of the world as of January 1 in the year n.

(b) If we let a_n be the digit in the nth decimal place of the number e, then $\{a_n\}$ is a well-defined sequence whose first few terms are

$$\{7, 1, 8, 2, 8, 1, 8, 2, 8, 4, 5, \ldots\}$$

(c) The **Fibonacci sequence** $\{f_n\}$ is defined recursively by the conditions

$$f_1 = 1 \qquad f_2 = 1 \qquad f_n = f_{n-1} + f_{n-2} \qquad n \geqslant 3$$

Each term is the sum of the two preceding terms. The first few terms are

$$\{1, 1, 2, 3, 5, 8, 13, 21, \ldots\}$$

This sequence arose when the 13th-century Italian mathematician known as Fibonacci solved a problem concerning the breeding of rabbits (see Exercise 45). ∎

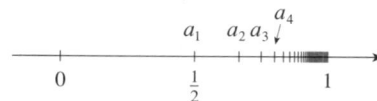

FIGURE 1

A sequence such as the one in Example 1(a), $a_n = n/(n + 1)$, can be pictured either by plotting its terms on a number line as in Figure 1 or by plotting its graph as in Figure 2. Note that, since a sequence is a function whose domain is the set of positive integers, its graph consists of isolated points with coordinates

$$(1, a_1) \qquad (2, a_2) \qquad (3, a_3) \qquad \ldots \qquad (n, a_n) \qquad \ldots$$

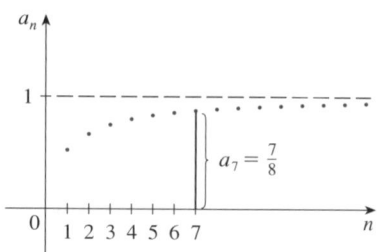

FIGURE 2

From Figure 1 or 2 it appears that the terms of the sequence $a_n = n/(n + 1)$ are approaching 1 as n becomes large. In fact, the difference

$$1 - \frac{n}{n + 1} = \frac{1}{n + 1}$$

can be made as small as we like by taking n sufficiently large. We indicate this by writing

$$\lim_{n \to \infty} \frac{n}{n+1} = 1$$

In general, the notation

$$\lim_{n \to \infty} a_n = L$$

means that the terms of the sequence $\{a_n\}$ approach L as n becomes large. Notice that the following definition of the limit of a sequence is very similar to the definition of a limit of a function at infinity given in Section 1.6.

---

1 DEFINITION A sequence $\{a_n\}$ has the **limit** L and we write

$$\lim_{n \to \infty} a_n = L \qquad \text{or} \qquad a_n \to L \text{ as } n \to \infty$$

if we can make the terms a_n as close to L as we like by taking n sufficiently large. If $\lim_{n \to \infty} a_n$ exists, we say the sequence **converges** (or is **convergent**). Otherwise, we say the sequence **diverges** (or is **divergent**).

Figure 3 illustrates Definition 1 by showing the graphs of two sequences that have the limit L.

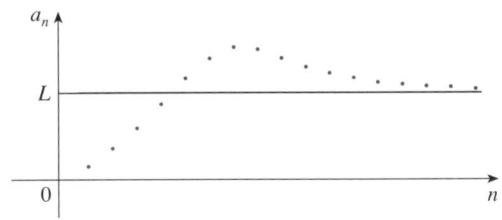
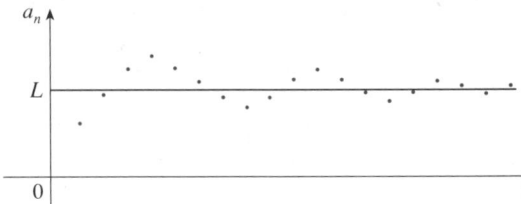

FIGURE 3
Graphs of two sequences with $\lim_{n \to \infty} a_n = L$

A more precise version of Definition 1 is as follows.

---

2 DEFINITION A sequence $\{a_n\}$ has the **limit** L and we write

$$\lim_{n \to \infty} a_n = L \qquad \text{or} \qquad a_n \to L \text{ as } n \to \infty$$

if for every $\varepsilon > 0$ there is a corresponding integer N such that

$$\text{if} \qquad n > N \qquad \text{then} \qquad |a_n - L| < \varepsilon$$

■ Compare this definition with Definition 1.6.7.

Definition 2 is illustrated by Figure 4, in which the terms $a_1, a_2, a_3, \ldots$ are plotted on a number line. No matter how small an interval $(L - \varepsilon, L + \varepsilon)$ is chosen, there exists an N such that all terms of the sequence from a_{N+1} onward must lie in that interval.

FIGURE 4

Another illustration of Definition 2 is given in Figure 5. The points on the graph of $\{a_n\}$ must lie between the horizontal lines $y = L + \varepsilon$ and $y = L - \varepsilon$ if $n > N$. This picture must be valid no matter how small ε is chosen, but usually a smaller ε requires a larger N.

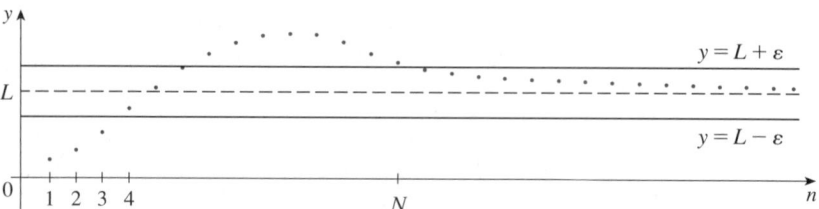

FIGURE 5

If you compare Definition 2 with Definition 1.6.7, you will see that the only difference between $\lim_{n \to \infty} a_n = L$ and $\lim_{x \to \infty} f(x) = L$ is that n is required to be an integer. Thus we have the following theorem, which is illustrated by Figure 6.

> **3** **THEOREM**   If $\lim_{x \to \infty} f(x) = L$ and $f(n) = a_n$ when n is an integer, then $\lim_{n \to \infty} a_n = L$.

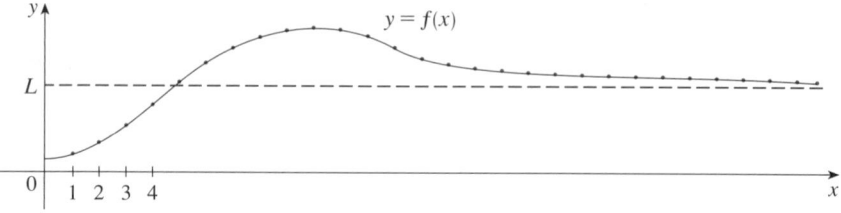

FIGURE 6

In particular, since we know that $\lim_{x \to \infty} (1/x^r) = 0$ when $r > 0$, we have

4
$$\lim_{n \to \infty} \frac{1}{n^r} = 0 \qquad \text{if } r > 0$$

If a_n becomes large as n becomes large, we use the notation $\lim_{n \to \infty} a_n = \infty$. The following precise definition is similar to Definition 1.6.8.

> **5** **DEFINITION** $\lim_{n \to \infty} a_n = \infty$ means that for every positive number M there is a positive integer N such that
>
> $$\text{if} \qquad n > N \qquad \text{then} \qquad a_n > M$$

If $\lim_{n \to \infty} a_n = \infty$, then the sequence $\{a_n\}$ is divergent but in a special way. We say that $\{a_n\}$ diverges to ∞.

The Limit Laws given in Section 1.4 also hold for the limits of sequences and their proofs are similar.

Limit Laws for Sequences

If $\{a_n\}$ and $\{b_n\}$ are convergent sequences and c is a constant, then

$$\lim_{n \to \infty} (a_n + b_n) = \lim_{n \to \infty} a_n + \lim_{n \to \infty} b_n$$

$$\lim_{n \to \infty} (a_n - b_n) = \lim_{n \to \infty} a_n - \lim_{n \to \infty} b_n$$

$$\lim_{n \to \infty} ca_n = c \lim_{n \to \infty} a_n \qquad\qquad \lim_{n \to \infty} c = c$$

$$\lim_{n \to \infty} (a_n b_n) = \lim_{n \to \infty} a_n \cdot \lim_{n \to \infty} b_n$$

$$\lim_{n \to \infty} \frac{a_n}{b_n} = \frac{\displaystyle\lim_{n \to \infty} a_n}{\displaystyle\lim_{n \to \infty} b_n} \quad \text{if } \lim_{n \to \infty} b_n \neq 0$$

$$\lim_{n \to \infty} a_n^p = \left[\lim_{n \to \infty} a_n \right]^p \quad \text{if } p > 0 \text{ and } a_n > 0$$

The Squeeze Theorem can also be adapted for sequences as follows (see Figure 7).

Squeeze Theorem for Sequences

If $a_n \leq b_n \leq c_n$ for $n \geq n_0$ and $\lim_{n \to \infty} a_n = \lim_{n \to \infty} c_n = L$, then $\lim_{n \to \infty} b_n = L$.

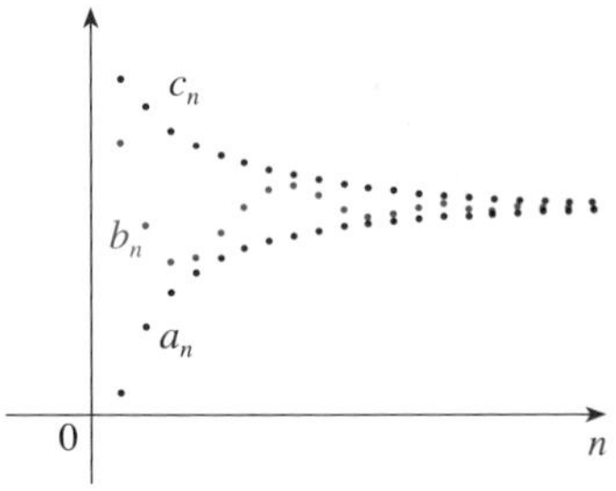

Another useful fact about limits of sequences is given by the following theorem, whose proof is left as Exercise 49.

6 THEOREM   If $\lim_{n \to \infty} |a_n| = 0$, then $\lim_{n \to \infty} a_n = 0$.

FIGURE 7

The sequence $\{b_n\}$ is squeezed between the sequences $\{a_n\}$ and $\{c_n\}$.

EXAMPLE 4 Find $\lim_{n \to \infty} \dfrac{n}{n + 1}$.

SOLUTION The method is similar to the one we used in Section 1.6: Divide numerator and denominator by the highest power of n that occurs in the denominator and then use the Limit Laws.

$$\lim_{n \to \infty} \frac{n}{n + 1} = \lim_{n \to \infty} \frac{1}{1 + \dfrac{1}{n}} = \frac{\displaystyle\lim_{n \to \infty} 1}{\displaystyle\lim_{n \to \infty} 1 + \lim_{n \to \infty} \dfrac{1}{n}}$$

$$= \frac{1}{1 + 0} = 1$$

■ This shows that the guess we made earlier from Figures 1 and 2 was correct.

Here we used Equation 4 with $r = 1$. ∎

■ www.stewartcalculus.com
See Additional Example A.

EXAMPLE 5 Calculate $\lim\limits_{n \to \infty} \dfrac{\ln n}{n}$.

SOLUTION Notice that both numerator and denominator approach infinity as $n \to \infty$. We can't apply l'Hospital's Rule directly because it applies not to sequences but to functions of a real variable. However, we can apply l'Hospital's Rule to the related function $f(x) = (\ln x)/x$ and obtain

$$\lim_{x \to \infty} \frac{\ln x}{x} = \lim_{x \to \infty} \frac{1/x}{1} = 0$$

Therefore, by Theorem 3, we have

$$\lim_{n \to \infty} \frac{\ln n}{n} = 0 \qquad ■$$

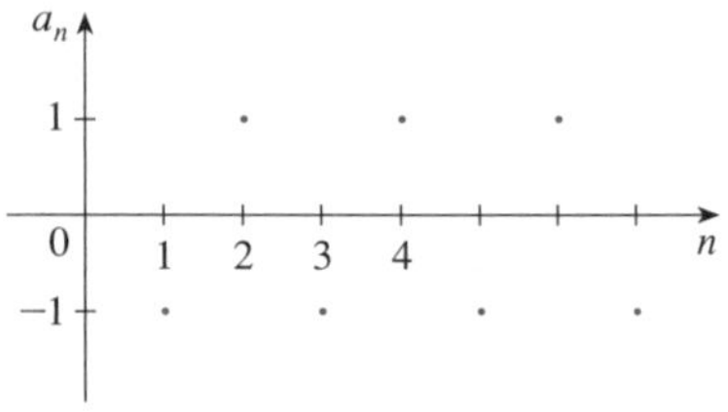

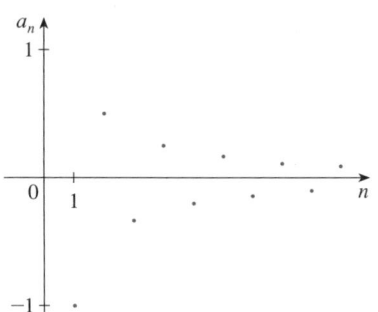

FIGURE 8

EXAMPLE 6 Determine whether the sequence $a_n = (-1)^n$ is convergent or divergent.

SOLUTION If we write out the terms of the sequence, we obtain

$$\{-1, 1, -1, 1, -1, 1, -1, \ldots\}$$

The graph of this sequence is shown in Figure 8. Since the terms oscillate between 1 and -1 infinitely often, a_n does not approach any number. Thus $\lim_{n \to \infty} (-1)^n$ does not exist; that is, the sequence $\{(-1)^n\}$ is divergent. ■

■ The graph of the sequence in Example 7 is shown in Figure 9 and supports the answer.

EXAMPLE 7 Evaluate $\lim\limits_{n \to \infty} \dfrac{(-1)^n}{n}$ if it exists.

SOLUTION

$$\lim_{n \to \infty} \left| \frac{(-1)^n}{n} \right| = \lim_{n \to \infty} \frac{1}{n} = 0$$

Therefore, by Theorem 6,

$$\lim_{n \to \infty} \frac{(-1)^n}{n} = 0 \qquad ■$$

FIGURE 9

The following theorem says that if we apply a continuous function to the terms of a convergent sequence, the result is also convergent. The proof is left as Exercise 50.

CONTINUITY AND CONVERGENCE THEOREM If $\lim\limits_{n \to \infty} a_n = L$ and the function f is continuous at L, then

$$\lim_{n \to \infty} f(a_n) = f(L)$$

EXAMPLE 8 Find $\lim\limits_{n \to \infty} \sin(\pi/n)$.

SOLUTION Because the sine function is continuous at 0, the Continuity and Convergence Theorem enables us to write

$$\lim_{n \to \infty} \sin(\pi/n) = \sin\left(\lim_{n \to \infty} (\pi/n) \right) = \sin 0 = 0 \qquad ■$$

☑ **EXAMPLE 9** Discuss the convergence of the sequence $a_n = n!/n^n$, where $n! = 1 \cdot 2 \cdot 3 \cdot \cdots \cdot n$.

SOLUTION Both numerator and denominator approach infinity as $n \to \infty$ but here we have no corresponding function for use with l'Hospital's Rule ($x!$ is not defined when x is not an integer). Let's write out a few terms to get a feeling for what happens to a_n as n gets large:

$$a_1 = 1 \qquad a_2 = \frac{1 \cdot 2}{2 \cdot 2} \qquad a_3 = \frac{1 \cdot 2 \cdot 3}{3 \cdot 3 \cdot 3}$$

$$\boxed{7} \qquad a_n = \frac{1 \cdot 2 \cdot 3 \cdot \cdots \cdot n}{n \cdot n \cdot n \cdot \cdots \cdot n}$$

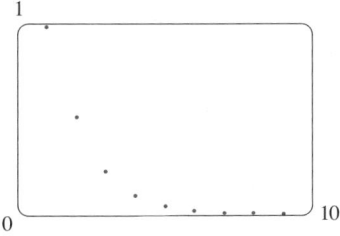

FIGURE 10

It appears from these expressions and the graph in Figure 10 that the terms are decreasing and perhaps approach 0. To confirm this, observe from Equation 7 that

$$a_n = \frac{1}{n} \left(\frac{2 \cdot 3 \cdot \cdots \cdot n}{n \cdot n \cdot \cdots \cdot n} \right)$$

Notice that the expression in parentheses is at most 1 because the numerator is less than (or equal to) the denominator. So

$$0 < a_n \le \frac{1}{n}$$

We know that $1/n \to 0$ as $n \to \infty$. Therefore $a_n \to 0$ as $n \to \infty$ by the Squeeze Theorem. ∎

☑ **EXAMPLE 10** For what values of r is the sequence $\{r^n\}$ convergent?

SOLUTION We know from Section 1.6 and the graphs of the exponential functions in Section 3.1 that $\lim_{x \to \infty} a^x = \infty$ for $a > 1$ and $\lim_{x \to \infty} a^x = 0$ for $0 < a < 1$. Therefore, putting $a = r$ and using Theorem 3, we have

$$\lim_{n \to \infty} r^n = \begin{cases} \infty & \text{if } r > 1 \\ 0 & \text{if } 0 < r < 1 \end{cases}$$

For the cases $r = 1$ and $r = 0$ we have

$$\lim_{n \to \infty} 1^n = \lim_{n \to \infty} 1 = 1 \qquad \text{and} \qquad \lim_{n \to \infty} 0^n = \lim_{n \to \infty} 0 = 0$$

If $-1 < r < 0$, then $0 < |r| < 1$, so

$$\lim_{n \to \infty} |r^n| = \lim_{n \to \infty} |r|^n = 0$$

and therefore $\lim_{n \to \infty} r^n = 0$ by Theorem 6. If $r \le -1$, then $\{r^n\}$ diverges as in

Example 6. Figure 11 shows the graphs for various values of r. (The case $r = -1$ is shown in Figure 8.)

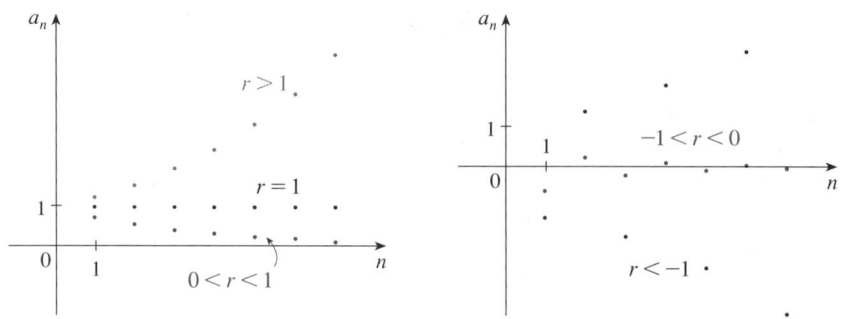

FIGURE 11
The sequence $a_n = r^n$

The results of Example 10 are summarized for future use as follows.

> **8** The sequence $\{r^n\}$ is convergent if $-1 < r \leq 1$ and divergent for all other values of r.
>
> $$\lim_{n \to \infty} r^n = \begin{cases} 0 & \text{if } -1 < r < 1 \\ 1 & \text{if } r = 1 \end{cases}$$

> **9** **DEFINITION** A sequence $\{a_n\}$ is called **increasing** if $a_n < a_{n+1}$ for all $n \geq 1$, that is, $a_1 < a_2 < a_3 < \cdots$. It is called **decreasing** if $a_n > a_{n+1}$ for all $n \geq 1$. A sequence is **monotonic** if it is either increasing or decreasing.

EXAMPLE 11 The sequence $\left\{ \dfrac{3}{n+5} \right\}$ is decreasing because

■ The right side is smaller because it has a larger denominator.

$$\frac{3}{n+5} > \frac{3}{(n+1)+5} = \frac{3}{n+6}$$

and so $a_n > a_{n+1}$ for all $n \geq 1$. ■

EXAMPLE 12 Show that the sequence $a_n = \dfrac{n}{n^2+1}$ is decreasing.

SOLUTION We must show that $a_{n+1} < a_n$, that is,

$$\frac{n+1}{(n+1)^2+1} < \frac{n}{n^2+1}$$

■ Another way to do Example 12 is to show that the function

$$f(x) = \frac{x}{x^2+1} \qquad x \geq 1$$

is decreasing because $f'(x) < 0$ for $x > 1$.

This inequality is equivalent to the one we get by cross-multiplication:

$$\frac{n+1}{(n+1)^2+1} < \frac{n}{n^2+1} \iff (n+1)(n^2+1) < n[(n+1)^2+1]$$

$$\iff n^3 + n^2 + n + 1 < n^3 + 2n^2 + 2n$$

$$\iff 1 < n^2 + n$$

Since $n \geqslant 1$, we know that the inequality $n^2 + n > 1$ is true. Therefore $a_{n+1} < a_n$ and so $\{a_n\}$ is decreasing. ∎

---

10 DEFINITION A sequence $\{a_n\}$ is **bounded above** if there is a number M such that

$$a_n \leqslant M \qquad \text{for all } n \geqslant 1$$

It is **bounded below** if there is a number m such that

$$m \leqslant a_n \qquad \text{for all } n \geqslant 1$$

If it is bounded above and below, then $\{a_n\}$ is a **bounded sequence**.

For instance, the sequence $a_n = n$ is bounded below ($a_n > 0$) but not above. The sequence $a_n = n/(n + 1)$ is bounded because $0 < a_n < 1$ for all n.

We know that not every bounded sequence is convergent [for instance, the sequence $a_n = (-1)^n$ satisfies $-1 \leqslant a_n \leqslant 1$ but is divergent from Example 6] and not every monotonic sequence is convergent ($a_n = n \to \infty$). But if a sequence is both bounded *and* monotonic, then it must be convergent. This fact is proved as Theorem 11, but intuitively you can understand why it is true by looking at Figure 12. If $\{a_n\}$ is increasing and $a_n \leqslant M$ for all n, then the terms are forced to crowd together and approach some number L.

The proof of Theorem 11 is based on the **Completeness Axiom** for the set $\mathbb{R}$ of real numbers, which says that if S is a nonempty set of real numbers that has an upper bound M ($x \leqslant M$ for all x in S), then S has a **least upper bound** b. (This means that b is an upper bound for S, but if M is any other upper bound, then $b \leqslant M$.) The Completeness Axiom is an expression of the fact that there is no gap or hole in the real number line.

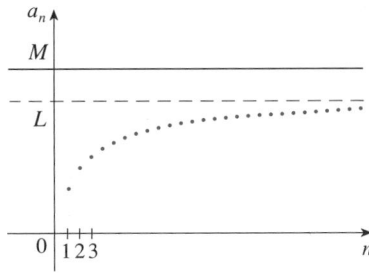

FIGURE 12

---

11 MONOTONIC SEQUENCE THEOREM Every bounded, monotonic sequence is convergent.

PROOF Suppose $\{a_n\}$ is an increasing sequence. Since $\{a_n\}$ is bounded, the set $S = \{a_n \mid n \geqslant 1\}$ has an upper bound. By the Completeness Axiom it has a least upper bound L. Given $\varepsilon > 0$, $L - \varepsilon$ is *not* an upper bound for S (since L is the *least* upper bound). Therefore

$$a_N > L - \varepsilon \qquad \text{for some integer } N$$

But the sequence is increasing so $a_n \geqslant a_N$ for every $n > N$. Thus if $n > N$ we have

$$a_n > L - \varepsilon$$

so

$$0 \leqslant L - a_n < \varepsilon$$

since $a_n \le L$. Thus

$$|L - a_n| < \varepsilon \qquad \text{whenever} \quad n > N$$

so $\lim_{n\to\infty} a_n = L$.

■ www.stewartcalculus.com
See Additional Example B.

A similar proof (using the greatest lower bound) works if $\{a_n\}$ is decreasing. ☐

The proof of Theorem 11 shows that a sequence that is increasing and bounded above is convergent. (Likewise, a decreasing sequence that is bounded below is convergent.) This fact is used many times in dealing with infinite series in Sections 8.2 and 8.3.

Another use of Theorem 11 is indicated in Exercises 42–44.

8.1 | EXERCISES

1. (a) What is a sequence?
(b) What does it mean to say that $\lim_{n\to\infty} a_n = 8$?
(c) What does it mean to say that $\lim_{n\to\infty} a_n = \infty$?

2. (a) What is a convergent sequence? Give two examples.
(b) What is a divergent sequence? Give two examples.

3. List the first six terms of the sequence defined by

$$a_n = \frac{n}{2n + 1}$$

Does the sequence appear to have a limit? If so, find it.

4. List the first nine terms of the sequence $\{\cos(n\pi/3)\}$. Does this sequence appear to have a limit? If so, find it. If not, explain why.

5–8 ■ Find a formula for the general term a_n of the sequence, assuming that the pattern of the first few terms continues.

5. $\left\{-3, 2, -\frac{4}{3}, \frac{8}{9}, -\frac{16}{27}, \ldots\right\}$

6. $\left\{1, -\frac{1}{3}, \frac{1}{9}, -\frac{1}{27}, \frac{1}{81}, \ldots\right\}$

7. $\left\{\frac{1}{2}, -\frac{4}{3}, \frac{9}{4}, -\frac{16}{5}, \frac{25}{6}, \ldots\right\}$

8. $\{5, 8, 11, 14, 17, \ldots\}$

9–32 ■ Determine whether the sequence converges or diverges. If it converges, find the limit.

9. $a_n = 1 - (0.2)^n$

10. $a_n = \dfrac{n^3}{n^3 + 1}$

11. $a_n = \dfrac{3 + 5n^2}{n + n^2}$

12. $a_n = \dfrac{n^3}{n + 1}$

13. $a_n = \tan\left(\dfrac{2n\pi}{1 + 8n}\right)$

14. $a_n = \dfrac{3^{n+2}}{5^n}$

15. $a_n = \dfrac{n^2}{\sqrt{n^3 + 4n}}$

16. $a_n = \sqrt{\dfrac{n + 1}{9n + 1}}$

17. $a_n = \dfrac{(-1)^n}{2\sqrt{n}}$

18. $a_n = \dfrac{(-1)^{n+1}n}{n + \sqrt{n}}$

19. $a_n = \cos(n/2)$

20. $a_n = \cos(2/n)$

21. $\left\{\dfrac{(2n - 1)!}{(2n + 1)!}\right\}$

22. $a_n = \dfrac{\tan^{-1}n}{n}$

23. $\{n^2 e^{-n}\}$

24. $a_n = \ln(n + 1) - \ln n$

25. $a_n = \dfrac{\cos^2 n}{2^n}$

26. $a_n = 2^{-n}\cos n\pi$

27. $a_n = \left(1 + \dfrac{2}{n}\right)^n$

28. $a_n = \dfrac{\sin 2n}{1 + \sqrt{n}}$

29. $\{0, 1, 0, 0, 1, 0, 0, 0, 1, \ldots\}$

30. $a_n = \dfrac{(\ln n)^2}{n}$

31. $a_n = \ln(2n^2 + 1) - \ln(n^2 + 1)$

32. $a_n = \dfrac{(-3)^n}{n!}$

33. If \$1000 is invested at 6% interest, compounded annually, then after n years the investment is worth $a_n = 1000(1.06)^n$ dollars.
(a) Find the first five terms of the sequence $\{a_n\}$.
(b) Is the sequence convergent or divergent? Explain.

⌨ Graphing calculator or computer required   [CAS] Computer algebra system required   **1.** Homework Hints at stewartcalculus.com

34. Find the first 40 terms of the sequence defined by

$$a_{n+1} = \begin{cases} \frac{1}{2}a_n & \text{if } a_n \text{ is an even number} \\ 3a_n + 1 & \text{if } a_n \text{ is an odd number} \end{cases}$$

and $a_1 = 11$. Do the same if $a_1 = 25$. Make a conjecture about this type of sequence.

35. Suppose you know that $\{a_n\}$ is a decreasing sequence and all its terms lie between the numbers 5 and 8. Explain why the sequence has a limit. What can you say about the value of the limit?

36. (a) If $\{a_n\}$ is convergent, show that

$$\lim_{n \to \infty} a_{n+1} = \lim_{n \to \infty} a_n$$

(b) A sequence $\{a_n\}$ is defined by $a_1 = 1$ and $a_{n+1} = 1/(1 + a_n)$ for $n \geq 1$. Assuming that $\{a_n\}$ is convergent, find its limit.

37–40 ■ Determine whether the sequence is increasing, decreasing, or not monotonic. Is the sequence bounded?

37. $a_n = \dfrac{1}{2n + 3}$

38. $a_n = \dfrac{2n - 3}{3n + 4}$

39. $a_n = n(-1)^n$

40. $a_n = n + \dfrac{1}{n}$

41. Find the limit of the sequence

$$\left\{ \sqrt{2}, \sqrt{2\sqrt{2}}, \sqrt{2\sqrt{2\sqrt{2}}}, \ldots \right\}$$

42. A sequence $\{a_n\}$ is given by $a_1 = \sqrt{2}$, $a_{n+1} = \sqrt{2 + a_n}$.
(a) By induction or otherwise, show that $\{a_n\}$ is increasing and bounded above by 3. Apply the Monotonic Sequence Theorem to show that $\lim_{n \to \infty} a_n$ exists.
(b) Find $\lim_{n \to \infty} a_n$.

43. Use induction to show that the sequence defined by $a_1 = 1$, $a_{n+1} = 3 - 1/a_n$ is increasing and $a_n < 3$ for all n. Deduce that $\{a_n\}$ is convergent and find its limit.

44. Show that the sequence defined by

$$a_1 = 2 \qquad a_{n+1} = \dfrac{1}{3 - a_n}$$

satisfies $0 < a_n \leq 2$ and is decreasing. Deduce that the sequence is convergent and find its limit.

45. (a) Fibonacci posed the following problem: Suppose that rabbits live forever and that every month each pair produces a new pair which becomes productive at age 2 months. If we start with one newborn pair, how many pairs of rabbits will we have in the nth month? Show

that the answer is f_n, where $\{f_n\}$ is the Fibonacci sequence defined in Example 3(c).
(b) Let $a_n = f_{n+1}/f_n$ and show that $a_{n-1} = 1 + 1/a_{n-2}$. Assuming that $\{a_n\}$ is convergent, find its limit.

46. (a) Let $a_1 = a$, $a_2 = f(a)$, $a_3 = f(a_2) = f(f(a))$, $\ldots$, $a_{n+1} = f(a_n)$, where f is a continuous function. If $\lim_{n \to \infty} a_n = L$, show that $f(L) = L$.
(b) Illustrate part (a) by taking $f(x) = \cos x$, $a = 1$, and estimating the value of L to five decimal places.

47. We know that $\lim_{n \to \infty} (0.8)^n = 0$ [from $\boxed{8}$ with $r = 0.8$]. Use logarithms to determine how large n has to be so that $(0.8)^n < 0.000001$.

48. Use Definition 2 directly to prove that $\lim_{n \to \infty} r^n = 0$ when $|r| < 1$.

49. Prove Theorem 6.
[*Hint:* Use either Definition 2 or the Squeeze Theorem.]

50. Prove the Continuity and Convergence Theorem.

51. Prove that if $\lim_{n \to \infty} a_n = 0$ and $\{b_n\}$ is bounded, then $\lim_{n \to \infty} (a_n b_n) = 0$.

52. (a) Show that if $\lim_{n \to \infty} a_{2n} = L$ and $\lim_{n \to \infty} a_{2n+1} = L$, then $\{a_n\}$ is convergent and $\lim_{n \to \infty} a_n = L$.
(b) If $a_1 = 1$ and

$$a_{n+1} = 1 + \dfrac{1}{1 + a_n}$$

find the first eight terms of the sequence $\{a_n\}$. Then use part (a) to show that $\lim_{n \to \infty} a_n = \sqrt{2}$. This gives the **continued fraction expansion**

$$\sqrt{2} = 1 + \cfrac{1}{2 + \cfrac{1}{2 + \cdots}}$$

53. The size of an undisturbed fish population has been modeled by the formula

$$p_{n+1} = \dfrac{bp_n}{a + p_n}$$

where p_n is the fish population after n years and a and b are positive constants that depend on the species and its environment. Suppose that the population in year 0 is $p_0 > 0$.
(a) Show that if $\{p_n\}$ is convergent, then the only possible values for its limit are 0 and $b - a$.
(b) Show that $p_{n+1} < (b/a)p_n$.
(c) Use part (b) to show that if $a > b$, then $\lim_{n \to \infty} p_n = 0$; in other words, the population dies out.
(d) Now assume that $a < b$. Show that if $p_0 < b - a$, then $\{p_n\}$ is increasing and $0 < p_n < b - a$. Show also that if $p_0 > b - a$, then $\{p_n\}$ is decreasing and $p_n > b - a$. Deduce that if $a < b$, then $\lim_{n \to \infty} p_n = b - a$.

8.2 | SERIES

■ The current record (2011) is that π has been computed to more than ten trillion decimal places by Shigeru Kondo and Alexander Yee.

What do we mean when we express a number as an infinite decimal? For instance, what does it mean to write

$$\pi = 3.14159\ 26535\ 89793\ 23846\ 26433\ 83279\ 50288 \ldots$$

The convention behind our decimal notation is that any number can be written as an infinite sum. Here it means that

$$\pi = 3 + \frac{1}{10} + \frac{4}{10^2} + \frac{1}{10^3} + \frac{5}{10^4} + \frac{9}{10^5} + \frac{2}{10^6} + \frac{6}{10^7} + \frac{5}{10^8} + \cdots$$

where the three dots ($\cdots$) indicate that the sum continues forever, and the more terms we add, the closer we get to the actual value of π.

In general, if we try to add the terms of an infinite sequence $\{a_n\}_{n=1}^{\infty}$ we get an expression of the form

$$\boxed{1} \qquad a_1 + a_2 + a_3 + \cdots + a_n + \cdots$$

which is called an **infinite series** (or just a **series**) and is denoted, for short, by the symbol

$$\sum_{n=1}^{\infty} a_n \qquad \text{or} \qquad \sum a_n$$

Does it make sense to talk about the sum of infinitely many terms?

It would be impossible to find a finite sum for the series

$$1 + 2 + 3 + 4 + 5 + \cdots + n + \cdots$$

because if we start adding the terms we get the cumulative sums 1, 3, 6, 10, 15, 21, . . . and, after the nth term, we get $n(n + 1)/2$, which becomes very large as n increases.

However, if we start to add the terms of the series

$$\frac{1}{2} + \frac{1}{4} + \frac{1}{8} + \frac{1}{16} + \frac{1}{32} + \frac{1}{64} + \cdots + \frac{1}{2^n} + \cdots$$

n	Sum of first n terms
1	0.50000000
2	0.75000000
3	0.87500000
4	0.93750000
5	0.96875000
6	0.98437500
7	0.99218750
10	0.99902344
15	0.99996948
20	0.99999905
25	0.99999997

we get $\frac{1}{2}, \frac{3}{4}, \frac{7}{8}, \frac{15}{16}, \frac{31}{32}, \frac{63}{64}, \ldots, 1 - 1/2^n, \ldots$. The table shows that as we add more and more terms, these *partial sums* become closer and closer to 1. In fact, by adding sufficiently many terms of the series we can make the partial sums as close as we like to 1. So it seems reasonable to say that the sum of this infinite series is 1 and to write

$$\sum_{n=1}^{\infty} \frac{1}{2^n} = \frac{1}{2} + \frac{1}{4} + \frac{1}{8} + \frac{1}{16} + \cdots + \frac{1}{2^n} + \cdots = 1$$

We use a similar idea to determine whether or not a general series $\boxed{1}$ has a sum. We consider the **partial sums**

$$s_1 = a_1$$
$$s_2 = a_1 + a_2$$
$$s_3 = a_1 + a_2 + a_3$$
$$s_4 = a_1 + a_2 + a_3 + a_4$$

and, in general,

$$s_n = a_1 + a_2 + a_3 + \cdots + a_n = \sum_{i=1}^{n} a_i$$

These partial sums form a new sequence $\{s_n\}$, which may or may not have a limit. If $\lim_{n \to \infty} s_n = s$ exists (as a finite number), then, as in the preceding example, we call it the sum of the infinite series $\Sigma\, a_n$.

---

2 DEFINITION Given a series $\sum_{n=1}^{\infty} a_n = a_1 + a_2 + a_3 + \cdots$, let s_n denote its nth partial sum:

$$s_n = \sum_{i=1}^{n} a_i = a_1 + a_2 + \cdots + a_n$$

If the sequence $\{s_n\}$ is convergent and $\lim_{n \to \infty} s_n = s$ exists as a real number, then the series $\Sigma\, a_n$ is called **convergent** and we write

$$a_1 + a_2 + \cdots + a_n + \cdots = s \qquad \text{or} \qquad \sum_{n=1}^{\infty} a_n = s$$

The number s is called the **sum** of the series. If the sequence $\{s_n\}$ is divergent, then the series is called **divergent**.

Thus the sum of a series is the limit of the sequence of partial sums. So when we write $\sum_{n=1}^{\infty} a_n = s$ we mean that by adding sufficiently many terms of the series we can get as close as we like to the number s. Notice that

$$\sum_{n=1}^{\infty} a_n = \lim_{n \to \infty} \sum_{i=1}^{n} a_i$$

■ Compare with the improper integral

$$\int_1^{\infty} f(x)\, dx = \lim_{t \to \infty} \int_1^t f(x)\, dx$$

To find this integral we integrate from 1 to t and then let $t \to \infty$. For a series, we sum from 1 to n and then let $n \to \infty$.

EXAMPLE 1 An important example of an infinite series is the **geometric series**

$$a + ar + ar^2 + ar^3 + \cdots + ar^{n-1} + \cdots = \sum_{n=1}^{\infty} ar^{n-1} \qquad a \neq 0$$

Each term is obtained from the preceding one by multiplying it by the **common ratio** r. $\left(\text{We have already considered the special case where } a = \tfrac{1}{2} \text{ and } r = \tfrac{1}{2} \text{ on page 436.}\right)$

If $r = 1$, then $s_n = a + a + \cdots + a = na \to \pm\infty$. Since $\lim_{n \to \infty} s_n$ doesn't exist, the geometric series diverges in this case.

If $r \neq 1$, we have

$$s_n = a + ar + ar^2 + \cdots + ar^{n-1}$$

and

$$rs_n = \qquad ar + ar^2 + \cdots + ar^{n-1} + ar^n$$

Subtracting these equations, we get

$$s_n - rs_n = a - ar^n$$

3
$$s_n = \frac{a(1 - r^n)}{1 - r}$$

- Figure 1 provides a geometric demonstration of the result in Example 1. If the triangles are constructed as shown and s is the sum of the series, then, by similar triangles,

$$\frac{s}{a} = \frac{a}{a - ar} \qquad \text{so} \qquad s = \frac{a}{1 - r}$$

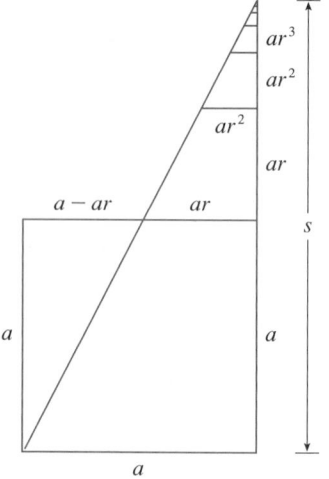

FIGURE 1

If $-1 < r < 1$, we know from (8.1.8) that $r^n \to 0$ as $n \to \infty$, so

$$\lim_{n \to \infty} s_n = \lim_{n \to \infty} \frac{a(1 - r^n)}{1 - r} = \frac{a}{1 - r} - \frac{a}{1 - r} \lim_{n \to \infty} r^n = \frac{a}{1 - r}$$

Thus when $|r| < 1$ the geometric series is convergent and its sum is $a/(1 - r)$.

If $r \leq -1$ or $r > 1$, the sequence $\{r^n\}$ is divergent by (8.1.8) and so, by Equation 3, $\lim_{n \to \infty} s_n$ does not exist. Therefore the geometric series diverges in those cases. ∎

We summarize the results of Example 1 as follows.

---

4 The geometric series

$$\sum_{n=1}^{\infty} ar^{n-1} = a + ar + ar^2 + \cdots$$

is convergent if $|r| < 1$ and its sum is

$$\sum_{n=1}^{\infty} ar^{n-1} = \frac{a}{1 - r} \qquad |r| < 1$$

If $|r| \geq 1$, the geometric series is divergent.

V EXAMPLE 2 Find the sum of the geometric series

$$5 - \tfrac{10}{3} + \tfrac{20}{9} - \tfrac{40}{27} + \cdots$$

SOLUTION The first term is $a = 5$ and the common ratio is $r = -\tfrac{2}{3}$. Since $|r| = \tfrac{2}{3} < 1$, the series is convergent by **4** and its sum is

$$5 - \frac{10}{3} + \frac{20}{9} - \frac{40}{27} + \cdots = \frac{5}{1 - \left(-\tfrac{2}{3}\right)} = \frac{5}{\tfrac{5}{3}} = 3 \qquad ∎$$

- What do we really mean when we say that the sum of the series in Example 2 is 3? Of course, we can't literally add an infinite number of terms, one by one. But, according to Definition 2, the total sum is the limit of the sequence of partial sums. So, by taking the sum of sufficiently many terms, we can get as close as we like to the number 3. The table shows the first ten partial sums s_n and the graph in Figure 2 shows how the sequence of partial sums approaches 3.

n	s_n
1	5.000000
2	1.666667
3	3.888889
4	2.407407
5	3.395062
6	2.736626
7	3.175583
8	2.882945
9	3.078037
10	2.947975

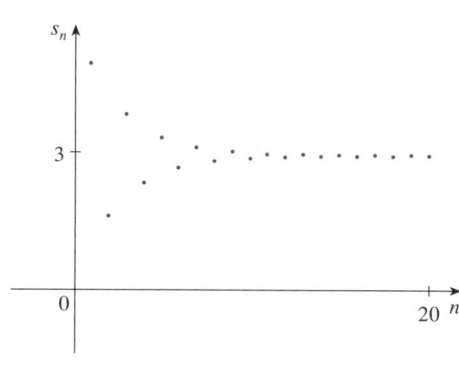

FIGURE 2

EXAMPLE 3 Is the series $\sum_{n=1}^{\infty} 2^{2n} 3^{1-n}$ convergent or divergent?

- Another way to identify a and r is to write out the first few terms:

$$4 + \tfrac{16}{3} + \tfrac{64}{9} + \cdots$$

SOLUTION Let's rewrite the nth term of the series in the form ar^{n-1}:

$$\sum_{n=1}^{\infty} 2^{2n} 3^{1-n} = \sum_{n=1}^{\infty} (2^2)^n 3^{-(n-1)} = \sum_{n=1}^{\infty} \frac{4^n}{3^{n-1}} = \sum_{n=1}^{\infty} 4\left(\tfrac{4}{3}\right)^{n-1}$$

We recognize this series as a geometric series with $a = 4$ and $r = \frac{4}{3}$. Since $r > 1$, the series diverges by $\boxed{4}$. ∎

V EXAMPLE 4 Write the number $2.3\overline{17} = 2.3171717\ldots$ as a ratio of integers.

SOLUTION

$$2.3171717\ldots = 2.3 + \frac{17}{10^3} + \frac{17}{10^5} + \frac{17}{10^7} + \cdots$$

After the first term we have a geometric series with $a = 17/10^3$ and $r = 1/10^2$. Therefore

$$2.3\overline{17} = 2.3 + \frac{\dfrac{17}{10^3}}{1 - \dfrac{1}{10^2}} = 2.3 + \frac{\dfrac{17}{1000}}{\dfrac{99}{100}}$$

$$= \frac{23}{10} + \frac{17}{990} = \frac{1147}{495}$$ ∎

EXAMPLE 5 Find the sum of the series $\displaystyle\sum_{n=0}^{\infty} x^n$, where $|x| < 1$.

SOLUTION Notice that this series starts with $n = 0$ and so the first term is $x^0 = 1$. (With series, we adopt the convention that $x^0 = 1$ even when $x = 0$.) Thus

$$\sum_{n=0}^{\infty} x^n = 1 + x + x^2 + x^3 + x^4 + \cdots$$

This is a geometric series with $a = 1$ and $r = x$. Since $|r| = |x| < 1$, it converges and $\boxed{4}$ gives

$$\boxed{5} \qquad \sum_{n=0}^{\infty} x^n = \frac{1}{1 - x}$$ ∎

EXAMPLE 6 Show that the series $\displaystyle\sum_{n=1}^{\infty} \frac{1}{n(n+1)}$ is convergent, and find its sum.

SOLUTION This is not a geometric series, so we go back to the definition of a convergent series and compute the partial sums.

$$s_n = \sum_{i=1}^{n} \frac{1}{i(i+1)} = \frac{1}{1 \cdot 2} + \frac{1}{2 \cdot 3} + \frac{1}{3 \cdot 4} + \cdots + \frac{1}{n(n+1)}$$

We can simplify this expression if we use the partial fraction decomposition

$$\frac{1}{i(i+1)} = \frac{1}{i} - \frac{1}{i+1}$$

(see Section 6.3). Thus we have

$$s_n = \sum_{i=1}^{n} \frac{1}{i(i+1)} = \sum_{i=1}^{n} \left(\frac{1}{i} - \frac{1}{i+1} \right)$$

$$= \left(1 - \frac{1}{2} \right) + \left(\frac{1}{2} - \frac{1}{3} \right) + \left(\frac{1}{3} - \frac{1}{4} \right) + \cdots + \left(\frac{1}{n} - \frac{1}{n+1} \right)$$

$$= 1 - \frac{1}{n+1}$$

■ **TEC** Module 8.2 explores a series that depends on an angle θ in a triangle and enables you to see how rapidly the series converges when θ varies.

■ Notice that the terms cancel in pairs. This is an example of a **telescoping sum**: Because of all the cancellations, the sum collapses (like a pirate's collapsing telescope) into just two terms.

■ Figure 3 illustrates Example 6 by showing the graphs of the sequence of terms $a_n = 1/[n(n + 1)]$ and the sequence $\{s_n\}$ of partial sums. Notice that $a_n \to 0$ and $s_n \to 1$. See Exercises 46 and 47 for two geometric interpretations of Example 6.

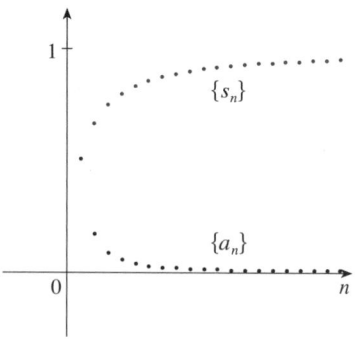

FIGURE 3

and so
$$\lim_{n \to \infty} s_n = \lim_{n \to \infty}\left(1 - \frac{1}{n + 1}\right) = 1 - 0 = 1$$

Therefore the given series is convergent and

$$\sum_{n=1}^{\infty} \frac{1}{n(n + 1)} = 1$$

■ **EXAMPLE 7** Show that the **harmonic series**

$$\sum_{n=1}^{\infty} \frac{1}{n} = 1 + \frac{1}{2} + \frac{1}{3} + \frac{1}{4} + \cdots$$

is divergent.

SOLUTION For this particular series it's convenient to consider the partial sums s_2, s_4, s_8, s_{16}, s_{32}, ... and show that they become large.

$$s_2 = 1 + \tfrac{1}{2}$$

$$s_4 = 1 + \tfrac{1}{2} + \left(\tfrac{1}{3} + \tfrac{1}{4}\right) > 1 + \tfrac{1}{2} + \left(\tfrac{1}{4} + \tfrac{1}{4}\right) = 1 + \tfrac{2}{2}$$

$$s_8 = 1 + \tfrac{1}{2} + \left(\tfrac{1}{3} + \tfrac{1}{4}\right) + \left(\tfrac{1}{5} + \tfrac{1}{6} + \tfrac{1}{7} + \tfrac{1}{8}\right)$$

$$> 1 + \tfrac{1}{2} + \left(\tfrac{1}{4} + \tfrac{1}{4}\right) + \left(\tfrac{1}{8} + \tfrac{1}{8} + \tfrac{1}{8} + \tfrac{1}{8}\right)$$

$$= 1 + \tfrac{1}{2} + \tfrac{1}{2} + \tfrac{1}{2} = 1 + \tfrac{3}{2}$$

$$s_{16} = 1 + \tfrac{1}{2} + \left(\tfrac{1}{3} + \tfrac{1}{4}\right) + \left(\tfrac{1}{5} + \cdots + \tfrac{1}{8}\right) + \left(\tfrac{1}{9} + \cdots + \tfrac{1}{16}\right)$$

$$> 1 + \tfrac{1}{2} + \left(\tfrac{1}{4} + \tfrac{1}{4}\right) + \left(\tfrac{1}{8} + \cdots + \tfrac{1}{8}\right) + \left(\tfrac{1}{16} + \cdots + \tfrac{1}{16}\right)$$

$$= 1 + \tfrac{1}{2} + \tfrac{1}{2} + \tfrac{1}{2} + \tfrac{1}{2} = 1 + \tfrac{4}{2}$$

Similarly, $s_{32} > 1 + \tfrac{5}{2}$, $s_{64} > 1 + \tfrac{6}{2}$, and in general

$$s_{2^n} > 1 + \frac{n}{2}$$

■ The method used in Example 7 for showing that the harmonic series diverges is due to the French scholar Nicole Oresme (1323–1382).

This shows that $s_{2^n} \to \infty$ as $n \to \infty$ and so $\{s_n\}$ is divergent. Therefore the harmonic series diverges. ■

> **6 THEOREM** If the series $\displaystyle\sum_{n=1}^{\infty} a_n$ is convergent, then $\displaystyle\lim_{n \to \infty} a_n = 0$.

PROOF Let $s_n = a_1 + a_2 + \cdots + a_n$. Then $a_n = s_n - s_{n-1}$. Since $\sum a_n$ is convergent, the sequence $\{s_n\}$ is convergent. Let $\lim_{n \to \infty} s_n = s$. Since $n - 1 \to \infty$ as $n \to \infty$, we also have $\lim_{n \to \infty} s_{n-1} = s$. Therefore

$$\lim_{n \to \infty} a_n = \lim_{n \to \infty}(s_n - s_{n-1}) = \lim_{n \to \infty} s_n - \lim_{n \to \infty} s_{n-1} = s - s = 0 \qquad \square$$

NOTE 1 With any *series* $\sum a_n$ we associate two *sequences:* the sequence $\{s_n\}$ of its partial sums and the sequence $\{a_n\}$ of its terms. If $\sum a_n$ is convergent, then the limit of the sequence $\{s_n\}$ is s (the sum of the series) and, as Theorem 6 asserts, the limit of the sequence $\{a_n\}$ is 0.

⊘ **NOTE 2** The converse of Theorem 6 is not true in general. If $\lim_{n\to\infty} a_n = 0$, we cannot conclude that $\Sigma\, a_n$ is convergent. Observe that for the harmonic series $\Sigma\, 1/n$ we have $a_n = 1/n \to 0$ as $n \to \infty$, but we showed in Example 7 that $\Sigma\, 1/n$ is divergent.

---

7 **TEST FOR DIVERGENCE** If $\lim\limits_{n\to\infty} a_n$ does not exist or if $\lim\limits_{n\to\infty} a_n \neq 0$, then the series $\sum\limits_{n=1}^{\infty} a_n$ is divergent.

The Test for Divergence follows from Theorem 6 because, if the series is not divergent, then it is convergent, and so $\lim_{n\to\infty} a_n = 0$.

EXAMPLE 8 Show that the series $\sum\limits_{n=1}^{\infty} \dfrac{n^2}{5n^2 + 4}$ diverges.

SOLUTION

$$\lim_{n\to\infty} a_n = \lim_{n\to\infty} \frac{n^2}{5n^2 + 4} = \lim_{n\to\infty} \frac{1}{5 + 4/n^2} = \frac{1}{5} \neq 0$$

So the series diverges by the Test for Divergence. ∎

NOTE 3 If we find that $\lim_{n\to\infty} a_n \neq 0$, we know that $\Sigma\, a_n$ is divergent. If we find that $\lim_{n\to\infty} a_n = 0$, we know *nothing* about the convergence or divergence of $\Sigma\, a_n$. Remember the warning in Note 2: If $\lim_{n\to\infty} a_n = 0$, the series $\Sigma\, a_n$ might converge or it might diverge.

---

8 **THEOREM** If $\Sigma\, a_n$ and $\Sigma\, b_n$ are convergent series, then so are the series $\Sigma\, ca_n$ (where c is a constant), $\Sigma\, (a_n + b_n)$, and $\Sigma\, (a_n - b_n)$, and

(i) $\sum\limits_{n=1}^{\infty} ca_n = c \sum\limits_{n=1}^{\infty} a_n$ (ii) $\sum\limits_{n=1}^{\infty} (a_n + b_n) = \sum\limits_{n=1}^{\infty} a_n + \sum\limits_{n=1}^{\infty} b_n$

(iii) $\sum\limits_{n=1}^{\infty} (a_n - b_n) = \sum\limits_{n=1}^{\infty} a_n - \sum\limits_{n=1}^{\infty} b_n$

These properties of convergent series follow from the corresponding Limit Laws for Sequences in Section 8.1. For instance, here is how part (ii) of Theorem 8 is proved:

Let

$$s_n = \sum_{i=1}^{n} a_i \qquad s = \sum_{n=1}^{\infty} a_n \qquad t_n = \sum_{i=1}^{n} b_i \qquad t = \sum_{n=1}^{\infty} b_n$$

The nth partial sum for the series $\Sigma\, (a_n + b_n)$ is

$$u_n = \sum_{i=1}^{n} (a_i + b_i)$$

and, using Equation 5.2.10, we have

$$\lim_{n \to \infty} u_n = \lim_{n \to \infty} \sum_{i=1}^{n} (a_i + b_i) = \lim_{n \to \infty} \left(\sum_{i=1}^{n} a_i + \sum_{i=1}^{n} b_i \right)$$

$$= \lim_{n \to \infty} \sum_{i=1}^{n} a_i + \lim_{n \to \infty} \sum_{i=1}^{n} b_i$$

$$= \lim_{n \to \infty} s_n + \lim_{n \to \infty} t_n = s + t$$

Therefore $\Sigma (a_n + b_n)$ is convergent and its sum is

$$\sum_{n=1}^{\infty} (a_n + b_n) = s + t = \sum_{n=1}^{\infty} a_n + \sum_{n=1}^{\infty} b_n \qquad \square$$

EXAMPLE 9 Find the sum of the series $\displaystyle\sum_{n=1}^{\infty} \left(\frac{3}{n(n + 1)} + \frac{1}{2^n} \right)$.

SOLUTION The series $\Sigma \, 1/2^n$ is a geometric series with $a = \frac{1}{2}$ and $r = \frac{1}{2}$, so

$$\sum_{n=1}^{\infty} \frac{1}{2^n} = \frac{\frac{1}{2}}{1 - \frac{1}{2}} = 1$$

In Example 6 we found that

$$\sum_{n=1}^{\infty} \frac{1}{n(n + 1)} = 1$$

So, by Theorem 8, the given series is convergent and

$$\sum_{n=1}^{\infty} \left(\frac{3}{n(n + 1)} + \frac{1}{2^n} \right) = 3 \sum_{n=1}^{\infty} \frac{1}{n(n + 1)} + \sum_{n=1}^{\infty} \frac{1}{2^n} = 3 \cdot 1 + 1 = 4 \qquad \blacksquare$$

NOTE 4 A finite number of terms doesn't affect the convergence or divergence of a series. For instance, suppose that we were able to show that the series

$$\sum_{n=4}^{\infty} \frac{n}{n^3 + 1}$$

is convergent. Since

$$\sum_{n=1}^{\infty} \frac{n}{n^3 + 1} = \frac{1}{2} + \frac{2}{9} + \frac{3}{28} + \sum_{n=4}^{\infty} \frac{n}{n^3 + 1}$$

it follows that the entire series $\sum_{n=1}^{\infty} n/(n^3 + 1)$ is convergent. Similarly, if it is known that the series $\sum_{n=N+1}^{\infty} a_n$ converges, then the full series

$$\sum_{n=1}^{\infty} a_n = \sum_{n=1}^{N} a_n + \sum_{n=N+1}^{\infty} a_n$$

is also convergent.

8.2 | EXERCISES

1. (a) What is the difference between a sequence and a series?
(b) What is a convergent series? What is a divergent series?

2. Explain what it means to say that $\sum_{n=1}^{\infty} a_n = 5$.

3–6 ■ Calculate the first eight terms of the sequence of partial sums correct to four decimal places. Does it appear that the series is convergent or divergent?

3. $\displaystyle\sum_{n=1}^{\infty} \frac{1}{n^3}$

4. $\displaystyle\sum_{n=1}^{\infty} \frac{1}{\ln(n+1)}$

5. $\displaystyle\sum_{n=1}^{\infty} \frac{n}{1+\sqrt{n}}$

6. $\displaystyle\sum_{n=1}^{\infty} \frac{(-1)^{n-1}}{n!}$

7–12 ■ Determine whether the geometric series is convergent or divergent. If it is convergent, find its sum.

7. $10 - 2 + 0.4 - 0.08 + \cdots$

8. $2 + 0.5 + 0.125 + 0.03125 + \cdots$

9. $\displaystyle\sum_{n=1}^{\infty} \frac{(-3)^{n-1}}{4^n}$

10. $\displaystyle\sum_{n=1}^{\infty} \frac{10^n}{(-9)^{n-1}}$

11. $\displaystyle\sum_{n=0}^{\infty} \frac{\pi^n}{3^{n+1}}$

12. $\displaystyle\sum_{n=0}^{\infty} \frac{1}{(\sqrt{2})^n}$

13–24 ■ Determine whether the series is convergent or divergent. If it is convergent, find its sum.

13. $\displaystyle\sum_{n=1}^{\infty} \frac{3^n}{e^{n-1}}$

14. $\displaystyle\sum_{k=1}^{\infty} \frac{k(k+2)}{(k+3)^2}$

15. $\displaystyle\sum_{n=1}^{\infty} \frac{n-1}{3n-1}$

16. $\displaystyle\sum_{n=1}^{\infty} \frac{1+3^n}{2^n}$

17. $\displaystyle\sum_{n=1}^{\infty} \frac{1+2^n}{3^n}$

18. $\displaystyle\sum_{n=1}^{\infty} \cos\frac{1}{n}$

19. $\displaystyle\sum_{n=1}^{\infty} \sqrt[n]{2}$

20. $\displaystyle\sum_{n=1}^{\infty} [(0.8)^{n-1} - (0.3)^n]$

21. $\displaystyle\sum_{n=1}^{\infty} \arctan n$

22. $\displaystyle\sum_{k=1}^{\infty} (\cos 1)^k$

23. $\dfrac{1}{3} + \dfrac{1}{6} + \dfrac{1}{9} + \dfrac{1}{12} + \dfrac{1}{15} + \cdots$

24. $\dfrac{1}{3} + \dfrac{2}{9} + \dfrac{1}{27} + \dfrac{2}{81} + \dfrac{1}{243} + \dfrac{2}{729} + \cdots$

25–28 ■ Determine whether the series is convergent or divergent by expressing s_n as a telescoping sum (as in Example 6). If it is convergent, find its sum.

25. $\displaystyle\sum_{n=2}^{\infty} \frac{2}{n^2-1}$

26. $\displaystyle\sum_{n=1}^{\infty} \ln\frac{n}{n+1}$

27. $\displaystyle\sum_{n=1}^{\infty} \frac{3}{n(n+3)}$

28. $\displaystyle\sum_{n=1}^{\infty} \left(e^{1/n} - e^{1/(n+1)}\right)$

29. Let $x = 0.99999\ldots.$
(a) Do you think that $x < 1$ or $x = 1$?
(b) Sum a geometric series to find the value of x.
(c) How many decimal representations does the number 1 have?
(d) Which numbers have more than one decimal representation?

30. A sequence of terms is defined by
$$a_1 = 1 \qquad a_n = (5-n)a_{n-1}$$
Calculate $\sum_{n=1}^{\infty} a_n$.

31–34 ■ Express the number as a ratio of integers.

31. $0.\overline{8} = 0.8888\ldots$

32. $0.\overline{46} = 0.46464646\ldots$

33. $2.\overline{516} = 2.516516516\ldots$

34. $10.1\overline{35} = 10.135353535\ldots$

35–37 ■ Find the values of x for which the series converges. Find the sum of the series for those values of x.

35. $\displaystyle\sum_{n=1}^{\infty} (-5)^n x^n$

36. $\displaystyle\sum_{n=0}^{\infty} (-4)^n (x-5)^n$

37. $\displaystyle\sum_{n=0}^{\infty} \frac{(x-2)^n}{3^n}$

38. We have seen that the harmonic series is a divergent series whose terms approach 0. Show that
$$\sum_{n=1}^{\infty} \ln\left(1 + \frac{1}{n}\right)$$
is another series with this property.

39. If the nth partial sum of a series $\sum_{n=1}^{\infty} a_n$ is
$$s_n = \frac{n-1}{n+1}$$
find a_n and $\sum_{n=1}^{\infty} a_n$.

40. If the nth partial sum of a series $\sum_{n=1}^{\infty} a_n$ is $s_n = 3 - n2^{-n}$, find a_n and $\sum_{n=1}^{\infty} a_n$.

41. A patient takes 150 mg of a drug at the same time every day. Just before each tablet is taken, 5% of the drug remains in the body.
  (a) What quantity of the drug is in the body after the third tablet? After the nth tablet?
  (b) What quantity of the drug remains in the body in the long run?

42. After injection of a dose D of insulin, the concentration of insulin in a patient's system decays exponentially and so it can be written as De^{-at}, where t represents time in hours and a is a positive constant.
  (a) If a dose D is injected every T hours, write an expression for the sum of the residual concentrations just before the $(n + 1)$st injection.
  (b) Determine the limiting pre-injection concentration.
  (c) If the concentration of insulin must always remain at or above a critical value C, determine a minimal dosage D in terms of C, a, and T.

43. When money is spent on goods and services, those who receive the money also spend some of it. The people receiving some of the twice-spent money will spend some of that, and so on. Economists call this chain reaction the *multiplier effect*. In a hypothetical isolated community, the local government begins the process by spending D dollars. Suppose that each recipient of spent money spends $100c\%$ and saves $100s\%$ of the money that he or she receives. The values c and s are called the *marginal propensity to consume* and the *marginal propensity to save* and, of course, $c + s = 1$.
  (a) Let S_n be the total spending that has been generated after n transactions. Find an equation for S_n.
  (b) Show that $\lim_{n\to\infty} S_n = kD$, where $k = 1/s$. The number k is called the *multiplier*. What is the multiplier if the marginal propensity to consume is 80%?

  Note: The federal government uses this principle to justify deficit spending. Banks use this principle to justify lending a large percentage of the money that they receive in deposits.

44. A certain ball has the property that each time it falls from a height h onto a hard, level surface, it rebounds to a height rh, where $0 < r < 1$. Suppose that the ball is dropped from an initial height of H meters.
  (a) Assuming that the ball continues to bounce indefinitely, find the total distance that it travels.
  (b) Calculate the total time that the ball travels. (Use the fact that the ball falls $\frac{1}{2}gt^2$ meters in t seconds.)
  (c) Suppose that each time the ball strikes the surface with velocity v it rebounds with velocity $-kv$, where $0 < k < 1$. How long will it take for the ball to come to rest?

45. Find the value of c if $\displaystyle\sum_{n=2}^{\infty} (1 + c)^{-n} = 2$.

46. Graph the curves $y = x^n$, $0 \le x \le 1$, for $n = 0, 1, 2, 3, 4, \ldots$ on a common screen. By finding the areas between successive curves, give a geometric demonstration of the fact, shown in Example 6, that

$$\sum_{n=1}^{\infty} \frac{1}{n(n + 1)} = 1$$

47. The figure shows two circles C and D of radius 1 that touch at P. T is a common tangent line; C_1 is the circle that touches C, D, and T; C_2 is the circle that touches C, D, and C_1; C_3 is the circle that touches C, D, and C_2. This procedure can be continued indefinitely and produces an infinite sequence of circles $\{C_n\}$. Find an expression for the diameter of C_n and thus provide another geometric demonstration of Example 6.

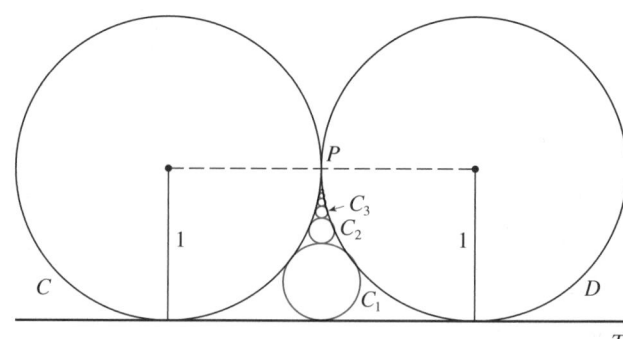

48. A right triangle ABC is given with $\angle A = \theta$ and $|AC| = b$. CD is drawn perpendicular to AB, DE is drawn perpendicular to BC, $EF \perp AB$, and this process is continued indefinitely as shown in the figure. Find the total length of all the perpendiculars

$$|CD| + |DE| + |EF| + |FG| + \cdots$$

in terms of b and θ.

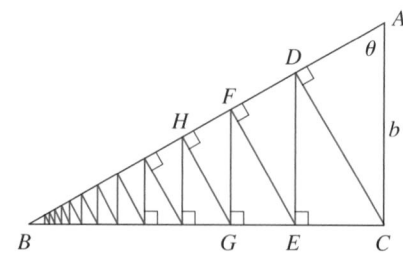

49. What is wrong with the following calculation?

$$0 = 0 + 0 + 0 + \cdots$$
$$= (1 - 1) + (1 - 1) + (1 - 1) + \cdots$$
$$= 1 - 1 + 1 - 1 + 1 - 1 + \cdots$$
$$= 1 + (-1 + 1) + (-1 + 1) + (-1 + 1) + \cdots$$
$$= 1 + 0 + 0 + 0 + \cdots = 1$$

(Guido Ubaldus thought that this proved the existence of God because "something has been created out of nothing.")

50. Suppose that $\sum_{n=1}^{\infty} a_n$ ($a_n \neq 0$) is known to be a convergent series. Prove that $\sum_{n=1}^{\infty} 1/a_n$ is a divergent series.

51. Prove part (i) of Theorem 8.

52. If $\sum a_n$ is divergent and $c \neq 0$, show that $\sum c a_n$ is divergent.

53. If $\sum a_n$ is convergent and $\sum b_n$ is divergent, show that the series $\sum (a_n + b_n)$ is divergent. [*Hint:* Argue by contradiction.]

54. If $\sum a_n$ and $\sum b_n$ are both divergent, is $\sum (a_n + b_n)$ necessarily divergent?

55. Suppose that a series $\sum a_n$ has positive terms and its partial sums s_n satisfy the inequality $s_n \leqslant 1000$ for all n. Explain why $\sum a_n$ must be convergent.

56. The Fibonacci sequence was defined in Section 8.1 by the equations

$$f_1 = 1, \quad f_2 = 1, \quad f_n = f_{n-1} + f_{n-2} \quad n \geqslant 3$$

Show that each of the following statements is true.

(a) $\dfrac{1}{f_{n-1} f_{n+1}} = \dfrac{1}{f_{n-1} f_n} - \dfrac{1}{f_n f_{n+1}}$ (b) $\displaystyle\sum_{n=2}^{\infty} \dfrac{1}{f_{n-1} f_{n+1}} = 1$

(c) $\displaystyle\sum_{n=2}^{\infty} \dfrac{f_n}{f_{n-1} f_{n+1}} = 2$

57. The **Cantor set**, named after the German mathematician Georg Cantor (1845–1918), is constructed as follows. We start with the closed interval $[0, 1]$ and remove the open interval $\left(\frac{1}{3}, \frac{2}{3}\right)$. That leaves the two intervals $\left[0, \frac{1}{3}\right]$ and $\left[\frac{2}{3}, 1\right]$ and we remove the open middle third of each. Four intervals remain and again we remove the open middle third of each of them. We continue this procedure indefinitely, at each step removing the open middle third of every interval that remains from the preceding step. The Cantor set consists of the numbers that remain in $[0, 1]$ after all those intervals have been removed.

(a) Show that the total length of all the intervals that are removed is 1. Despite that, the Cantor set contains infinitely many numbers. Give examples of some numbers in the Cantor set.

(b) The **Sierpinski carpet** is a two-dimensional counterpart of the Cantor set. It is constructed by removing the center one-ninth of a square of side 1, then removing the centers of the eight smaller remaining squares, and so on. (The figure shows the first three steps of the construction.) Show that the sum of the areas of the removed squares is 1. This implies that the Sierpinski carpet has area 0.

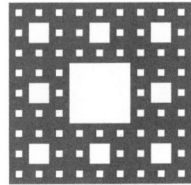

58. (a) A sequence $\{a_n\}$ is defined recursively by the equation $a_n = \frac{1}{2}(a_{n-1} + a_{n-2})$ for $n \geqslant 3$, where a_1 and a_2 can be any real numbers. Experiment with various values of a_1 and a_2 and use your calculator to guess the limit of the sequence.

(b) Find $\lim_{n \to \infty} a_n$ in terms of a_1 and a_2 by expressing $a_{n+1} - a_n$ in terms of $a_2 - a_1$ and summing a series.

59. Consider the series

$$\sum_{n=1}^{\infty} \frac{n}{(n + 1)!}$$

(a) Find the partial sums s_1, s_2, s_3, and s_4. Do you recognize the denominators? Use the pattern to guess a formula for s_n.

(b) Use mathematical induction to prove your guess.

(c) Show that the given infinite series is convergent, and find its sum.

60. In the figure there are infinitely many circles approaching the vertices of an equilateral triangle, each circle touching other circles and sides of the triangle. If the triangle has sides of length 1, find the total area occupied by the circles.

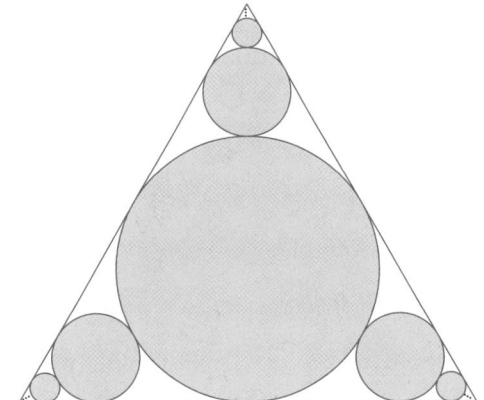

8.3 | THE INTEGRAL AND COMPARISON TESTS

In general, it is difficult to find the exact sum of a series. We were able to accomplish this for geometric series and the series $\Sigma\, 1/[n(n+1)]$ because in each of those cases we could find a simple formula for the nth partial sum s_n. But usually it isn't easy to compute $\lim_{n\to\infty} s_n$. Therefore, in this section and the next, we develop tests that enable us to determine whether a series is convergent or divergent without explicitly finding its sum.

In this section we deal only with series with positive terms, so the partial sums are increasing. In view of the Monotonic Sequence Theorem, to decide whether a series is convergent or divergent, we need to determine whether the partial sums are bounded or not.

TESTING WITH AN INTEGRAL

Let's investigate the series whose terms are the reciprocals of the squares of the positive integers:

$$\sum_{n=1}^{\infty} \frac{1}{n^2} = \frac{1}{1^2} + \frac{1}{2^2} + \frac{1}{3^2} + \frac{1}{4^2} + \frac{1}{5^2} + \cdots$$

n	$s_n = \displaystyle\sum_{i=1}^{n} \frac{1}{i^2}$
5	1.4636
10	1.5498
50	1.6251
100	1.6350
500	1.6429
1000	1.6439
5000	1.6447

There's no simple formula for the sum s_n of the first n terms, but the computer-generated table of values given in the margin suggests that the partial sums are approaching a number near 1.64 as $n \to \infty$ and so it looks as if the series is convergent.

We can confirm this impression with a geometric argument. Figure 1 shows the curve $y = 1/x^2$ and rectangles that lie below the curve. The base of each rectangle is an interval of length 1; the height is equal to the value of the function $y = 1/x^2$ at the right endpoint of the interval. So the sum of the areas of the rectangles is

$$\frac{1}{1^2} + \frac{1}{2^2} + \frac{1}{3^2} + \frac{1}{4^2} + \frac{1}{5^2} + \cdots = \sum_{n=1}^{\infty} \frac{1}{n^2}$$

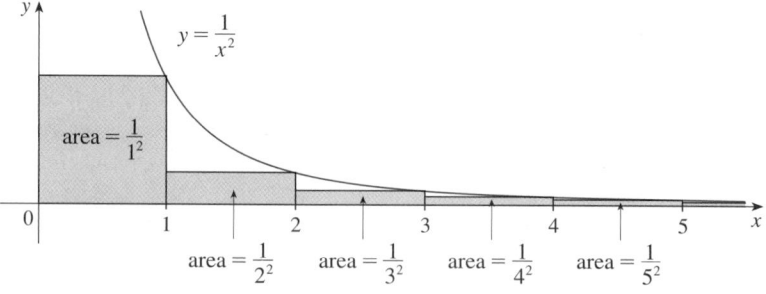

FIGURE 1

If we exclude the first rectangle, the total area of the remaining rectangles is smaller than the area under the curve $y = 1/x^2$ for $x \geq 1$, which is the value of the integral $\int_1^{\infty} (1/x^2)\, dx$. In Section 6.6 we discovered that this improper integral is convergent and has value 1. So the picture shows that all the partial sums are less than

$$\frac{1}{1^2} + \int_1^{\infty} \frac{1}{x^2}\, dx = 2$$

Thus the partial sums are bounded and the series converges. The sum of the series (the limit of the partial sums) is also less than 2:

$$\sum_{n=1}^{\infty} \frac{1}{n^2} = \frac{1}{1^2} + \frac{1}{2^2} + \frac{1}{3^2} + \frac{1}{4^2} + \cdots < 2$$

[The exact sum of this series was found by the Swiss mathematician Leonhard Euler (1707–1783) to be $\pi^2/6$, but the proof of this fact is beyond the scope of this book.]

Now let's look at the series

$$\sum_{n=1}^{\infty} \frac{1}{\sqrt{n}} = \frac{1}{\sqrt{1}} + \frac{1}{\sqrt{2}} + \frac{1}{\sqrt{3}} + \frac{1}{\sqrt{4}} + \frac{1}{\sqrt{5}} + \cdots$$

n	$s_n = \displaystyle\sum_{i=1}^{n} \frac{1}{\sqrt{i}}$
5	3.2317
10	5.0210
50	12.7524
100	18.5896
500	43.2834
1000	61.8010
5000	139.9681

The table of values of s_n suggests that the partial sums aren't approaching a finite number, so we suspect that the given series may be divergent. Again we use a picture for confirmation. Figure 2 shows the curve $y = 1/\sqrt{x}$, but this time we use rectangles whose tops lie *above* the curve.

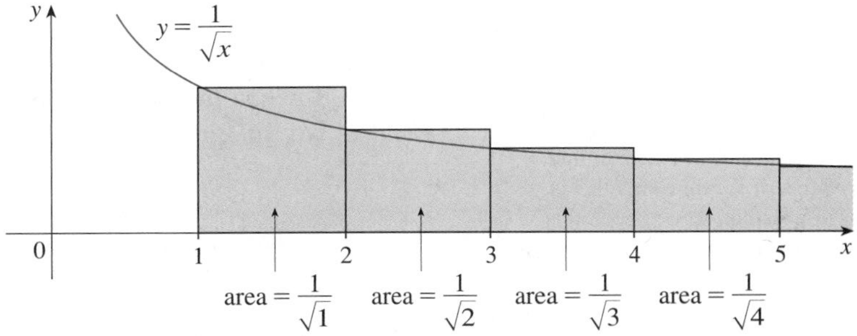

FIGURE 2

The base of each rectangle is an interval of length 1. The height is equal to the value of the function $y = 1/\sqrt{x}$ at the *left* endpoint of the interval. So the sum of the areas of all the rectangles is

$$\frac{1}{\sqrt{1}} + \frac{1}{\sqrt{2}} + \frac{1}{\sqrt{3}} + \frac{1}{\sqrt{4}} + \frac{1}{\sqrt{5}} + \cdots = \sum_{n=1}^{\infty} \frac{1}{\sqrt{n}}$$

This total area is greater than the area under the curve $y = 1/\sqrt{x}$ for $x \geqslant 1$, which is equal to the integral $\int_1^{\infty} \left(1/\sqrt{x}\right) dx$. But we know from Section 6.6 that this improper integral is divergent. In other words, the area under the curve is infinite. So the sum of the series must be infinite, that is, the series is divergent.

The same sort of geometric reasoning that we used for these two series can be used to prove the following test. (The proof is given at the end of this section.)

THE INTEGRAL TEST Suppose f is a continuous, positive, decreasing function on $[1, \infty)$ and let $a_n = f(n)$. Then the series $\sum_{n=1}^{\infty} a_n$ is convergent if and only if the improper integral $\int_1^{\infty} f(x) \, dx$ is convergent. In other words:

(i) If $\displaystyle\int_1^{\infty} f(x) \, dx$ is convergent, then $\displaystyle\sum_{n=1}^{\infty} a_n$ is convergent.

(ii) If $\displaystyle\int_1^{\infty} f(x) \, dx$ is divergent, then $\displaystyle\sum_{n=1}^{\infty} a_n$ is divergent.

■ www.stewartcalculus.com
See Additional Example A.

NOTE When we use the Integral Test it is not necessary to start the series or the integral at $n = 1$. For instance, in testing the series

$$\sum_{n=4}^{\infty} \frac{1}{(n-3)^2} \qquad \text{we use} \qquad \int_4^{\infty} \frac{1}{(x-3)^2} \, dx$$

Also, it is not necessary that f be always decreasing. What is important is that f be *ultimately* decreasing, that is, decreasing for x larger than some number N. Then $\sum_{n=N}^{\infty} a_n$ is convergent, so $\sum_{n=1}^{\infty} a_n$ is convergent by Note 4 of Section 8.2.

V EXAMPLE 1 Determine whether the series $\sum_{n=1}^{\infty} \frac{\ln n}{n}$ converges or diverges.

SOLUTION The function $f(x) = (\ln x)/x$ is positive and continuous for $x > 1$ because the logarithm function is positive and continuous there. But it is not obvious whether or not f is decreasing, so we compute its derivative:

$$f'(x) = \frac{x(1/x) - \ln x}{x^2} = \frac{1 - \ln x}{x^2}$$

■ In order to use the Integral Test we need to be able to evaluate $\int_1^{\infty} f(x) \, dx$ and therefore we have to be able to find an antiderivative of f. Frequently this is difficult or impossible, so we need other tests for convergence too.

Thus $f'(x) < 0$ when $\ln x > 1$, that is, $x > e$. It follows that f is decreasing when $x > e$ and so we can apply the Integral Test:

$$\int_1^{\infty} \frac{\ln x}{x} \, dx = \lim_{t \to \infty} \int_1^t \frac{\ln x}{x} \, dx = \lim_{t \to \infty} \frac{(\ln x)^2}{2} \Big]_1^t$$

$$= \lim_{t \to \infty} \frac{(\ln t)^2}{2} = \infty$$

Since this improper integral is divergent, the series $\sum (\ln n)/n$ is also divergent by the Integral Test. ∎

V EXAMPLE 2 For what values of p is the series $\sum_{n=1}^{\infty} \frac{1}{n^p}$ convergent?

SOLUTION If $p < 0$, then $\lim_{n \to \infty} (1/n^p) = \infty$. If $p = 0$, then $\lim_{n \to \infty} (1/n^p) = 1$. In either case $\lim_{n \to \infty} (1/n^p) \neq 0$, so the given series diverges by the Test for Divergence [see (8.2.7)].

If $p > 0$, then the function $f(x) = 1/x^p$ is clearly continuous, positive, and decreasing on $[1, \infty)$. We found in Chapter 6 [see (6.6.2)] that

■ Exercises 33–38 show how to estimate the sum of a series that is convergent by the Integral Test.

$$\int_1^{\infty} \frac{1}{x^p} \, dx \quad \text{converges if } p > 1 \text{ and diverges if } p \leq 1$$

It follows from the Integral Test that the series $\sum 1/n^p$ converges if $p > 1$ and diverges if $0 < p \leq 1$. (For $p = 1$, this series is the harmonic series discussed in Example 7 in Section 8.2.) ∎

The series in Example 2 is called the **p-series**. It is important in the rest of this chapter, so we summarize the results of Example 2 for future reference as follows.

---

1 The p-series $\sum_{n=1}^{\infty} \frac{1}{n^p}$ is convergent if $p > 1$ and divergent if $p \leq 1$.

For instance, the series

$$\sum_{n=1}^{\infty} \frac{1}{n^3} = \frac{1}{1^3} + \frac{1}{2^3} + \frac{1}{3^3} + \frac{1}{4^3} + \cdots$$

is convergent because it is a *p*-series with $p = 3 > 1$. But the series

$$\sum_{n=1}^{\infty} \frac{1}{n^{1/3}} = \sum_{n=1}^{\infty} \frac{1}{\sqrt[3]{n}} = 1 + \frac{1}{\sqrt[3]{2}} + \frac{1}{\sqrt[3]{3}} + \frac{1}{\sqrt[3]{4}} + \cdots$$

is divergent because it is a *p*-series with $p = \frac{1}{3} < 1$.

TESTING BY COMPARING

The series

$$\boxed{2} \qquad\qquad \sum_{n=1}^{\infty} \frac{1}{2^n + 1}$$

reminds us of the series $\sum_{n=1}^{\infty} 1/2^n$, which is a geometric series with $a = \frac{1}{2}$ and $r = \frac{1}{2}$ and is therefore convergent. Because the series $\boxed{2}$ is so similar to a convergent series, we have the feeling that it too must be convergent. Indeed, it is. The inequality

$$\frac{1}{2^n + 1} < \frac{1}{2^n}$$

shows that our given series $\boxed{2}$ has smaller terms than those of the geometric series and therefore all its partial sums are also smaller than 1 (the sum of the geometric series). This means that its partial sums form a bounded increasing sequence, which is convergent. It also follows that the sum of the series is less than the sum of the geometric series:

$$\sum_{n=1}^{\infty} \frac{1}{2^n + 1} < 1$$

Similar reasoning can be used to prove the following test, which applies only to series whose terms are positive. The first part says that if we have a series whose terms are *smaller* than those of a known *convergent* series, then our series is also convergent. The second part says that if we start with a series whose terms are *larger* than those of a known *divergent* series, then it too is divergent.

THE COMPARISON TEST Suppose that $\Sigma\, a_n$ and $\Sigma\, b_n$ are series with positive terms.

(i) If $\Sigma\, b_n$ is convergent and $a_n \leqslant b_n$ for all n, then $\Sigma\, a_n$ is also convergent.

(ii) If $\Sigma\, b_n$ is divergent and $a_n \geqslant b_n$ for all n, then $\Sigma\, a_n$ is also divergent.

■ It is important to keep in mind the distinction between a sequence and a series. A sequence is a list of numbers, whereas a series is a sum. With every series $\Sigma\, a_n$ there are associated two sequences: the sequence $\{a_n\}$ of terms and the sequence $\{s_n\}$ of partial sums.

PROOF

(i) Let

$$s_n = \sum_{i=1}^{n} a_i \qquad t_n = \sum_{i=1}^{n} b_i \qquad t = \sum_{n=1}^{\infty} b_n$$

Since both series have positive terms, the sequences $\{s_n\}$ and $\{t_n\}$ are increasing

($s_{n+1} = s_n + a_{n+1} \geqslant s_n$). Also $t_n \to t$, so $t_n \leqslant t$ for all n. Since $a_i \leqslant b_i$, we have $s_n \leqslant t_n$. Thus $s_n \leqslant t$ for all n. This means that $\{s_n\}$ is increasing and bounded above and therefore converges by the Monotonic Sequence Theorem. Thus $\Sigma\, a_n$ converges.

(ii) If $\Sigma\, b_n$ is divergent, then $t_n \to \infty$ (since $\{t_n\}$ is increasing). But $a_i \geqslant b_i$ so $s_n \geqslant t_n$. Thus $s_n \to \infty$. Therefore $\Sigma\, a_n$ diverges. $\qquad\square$

Standard Series for Use with the Comparison Test

In using the Comparison Test we must, of course, have some known series $\Sigma\, b_n$ for the purpose of comparison. Most of the time we use one of these series:

- A p-series $\left[\Sigma\, 1/n^p \text{ converges if } p > 1 \text{ and diverges if } p \leqslant 1; \text{ see } \boxed{1}\,\right]$

- A geometric series $\left[\Sigma\, ar^{n-1} \text{ converges if } |r| < 1 \text{ and diverges if } |r| \geqslant 1; \text{ see } (8.2.4)\right]$

$\blacktriangledown$ EXAMPLE 3 Determine whether the series $\displaystyle\sum_{n=1}^{\infty} \frac{5}{2n^2 + 4n + 3}$ converges or diverges.

SOLUTION For large n the dominant term in the denominator is $2n^2$, so we compare the given series with the series $\Sigma\, 5/(2n^2)$. Observe that

$$\frac{5}{2n^2 + 4n + 3} < \frac{5}{2n^2}$$

because the left side has a bigger denominator. (In the notation of the Comparison Test, a_n is the left side and b_n is the right side.) We know that

$$\sum_{n=1}^{\infty} \frac{5}{2n^2} = \frac{5}{2} \sum_{n=1}^{\infty} \frac{1}{n^2}$$

is convergent (p-series with $p = 2 > 1$). Therefore

$$\sum_{n=1}^{\infty} \frac{5}{2n^2 + 4n + 3}$$

is convergent by part (i) of the Comparison Test. $\qquad\blacksquare$

Although the condition $a_n \leqslant b_n$ or $a_n \geqslant b_n$ in the Comparison Test is given for all n, we need verify only that it holds for $n \geqslant N$, where N is some fixed integer, because the convergence of a series is not affected by a finite number of terms. This is illustrated in the next example.

$\blacktriangledown$ EXAMPLE 4 Test the series $\displaystyle\sum_{k=1}^{\infty} \frac{\ln k}{k}$ for convergence or divergence.

SOLUTION We used the Integral Test to test this series in Example 1, but we can also test it by comparing it with the harmonic series. Observe that $\ln k > 1$ for $k \geqslant 3$ and so

$$\frac{\ln k}{k} > \frac{1}{k} \qquad k \geqslant 3$$

We know that $\Sigma\, 1/k$ is divergent (p-series with $p = 1$). Thus the given series is divergent by the Comparison Test. $\qquad\blacksquare$

NOTE The terms of the series being tested must be smaller than those of a convergent series or larger than those of a divergent series. If the terms are larger than the terms of a convergent series or smaller than those of a divergent series, then the Comparison Test doesn't apply. Consider, for instance, the series

$$\sum_{n=1}^{\infty} \frac{1}{2^n - 1}$$

The inequality

$$\frac{1}{2^n - 1} > \frac{1}{2^n}$$

is useless as far as the Comparison Test is concerned because $\sum b_n = \sum \left(\frac{1}{2}\right)^n$ is convergent and $a_n > b_n$. Nonetheless, we have the feeling that $\sum 1/(2^n - 1)$ ought to be convergent because it is very similar to the convergent geometric series $\sum \left(\frac{1}{2}\right)^n$. In such cases the following test can be used.

▪ Exercises 42 and 43 deal with the cases $c = 0$ and $c = \infty$.

THE LIMIT COMPARISON TEST Suppose that $\sum a_n$ and $\sum b_n$ are series with positive terms. If

$$\lim_{n \to \infty} \frac{a_n}{b_n} = c$$

where c is a finite number and $c > 0$, then either both series converge or both diverge.

PROOF Let m and M be positive numbers such that $m < c < M$. Because a_n/b_n is close to c for large n, there is an integer N such that

$$m < \frac{a_n}{b_n} < M \qquad \text{when } n > N$$

and so

$$mb_n < a_n < Mb_n \qquad \text{when } n > N$$

If $\sum b_n$ converges, so does $\sum Mb_n$. Thus $\sum a_n$ converges by part (i) of the Comparison Test. If $\sum b_n$ diverges, so does $\sum mb_n$ and part (ii) of the Comparison Test shows that $\sum a_n$ diverges. ☐

EXAMPLE 5 Test the series $\sum_{n=1}^{\infty} \dfrac{1}{2^n - 1}$ for convergence or divergence.

SOLUTION We use the Limit Comparison Test with

$$a_n = \frac{1}{2^n - 1} \qquad b_n = \frac{1}{2^n}$$

and obtain

▪ **www.stewartcalculus.com**
See Additional Example B.

$$\lim_{n \to \infty} \frac{a_n}{b_n} = \lim_{n \to \infty} \frac{1/(2^n - 1)}{1/2^n} = \lim_{n \to \infty} \frac{2^n}{2^n - 1} = \lim_{n \to \infty} \frac{1}{1 - 1/2^n} = 1 > 0$$

Since this limit exists and $\sum 1/2^n$ is a convergent geometric series, the given series converges by the Limit Comparison Test. ■

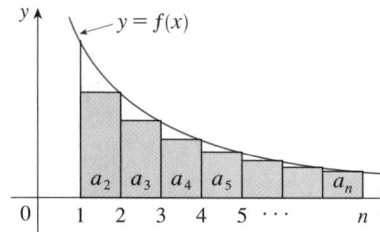

FIGURE 3

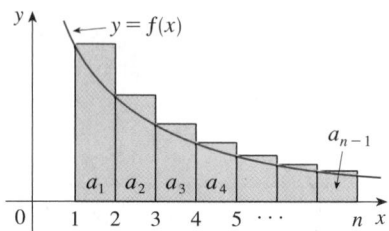

FIGURE 4

PROOF OF THE INTEGRAL TEST

We have already seen the basic idea behind the proof of the Integral Test in Figures 1 and 2 for the series $\Sigma \, 1/n^2$ and $\Sigma \, 1/\sqrt{n}$. For the general series $\Sigma \, a_n$ look at Figures 3 and 4. The area of the first shaded rectangle in Figure 3 is the value of f at the right endpoint of $[1, 2]$, that is, $f(2) = a_2$. So, comparing the areas of the shaded rectangles with the area under $y = f(x)$ from 1 to n, we see that

$$\boxed{3} \qquad a_2 + a_3 + \cdots + a_n \leqslant \int_1^n f(x)\, dx$$

(Notice that this inequality depends on the fact that f is decreasing.) Likewise, Figure 4 shows that

$$\boxed{4} \qquad \int_1^n f(x)\, dx \leqslant a_1 + a_2 + \cdots + a_{n-1}$$

(i) If $\int_1^\infty f(x)\, dx$ is convergent, then $\boxed{3}$ gives

$$\sum_{i=2}^n a_i \leqslant \int_1^n f(x)\, dx \leqslant \int_1^\infty f(x)\, dx$$

since $f(x) \geqslant 0$. Therefore

$$s_n = a_1 + \sum_{i=2}^n a_i \leqslant a_1 + \int_1^\infty f(x)\, dx = M$$

where M is a constant. Since $s_n \leqslant M$ for all n, the sequence $\{s_n\}$ is bounded above. Also

$$s_{n+1} = s_n + a_{n+1} \geqslant s_n$$

since $a_{n+1} = f(n + 1) \geqslant 0$. Thus $\{s_n\}$ is an increasing bounded sequence and so it is convergent by the Monotonic Sequence Theorem (8.1.11). This means that $\Sigma \, a_n$ is convergent.

(ii) If $\int_1^\infty f(x)\, dx$ is divergent, then $\int_1^n f(x)\, dx \to \infty$ as $n \to \infty$ because $f(x) \geqslant 0$. But $\boxed{4}$ gives

$$\int_1^n f(x)\, dx \leqslant \sum_{i=1}^{n-1} a_i = s_{n-1}$$

and so $s_{n-1} \to \infty$. This implies that $s_n \to \infty$ and so $\Sigma \, a_n$ diverges. $\qquad \square$

8.3 | EXERCISES

1. Draw a picture to show that

$$\sum_{n=2}^\infty \frac{1}{n^{1.3}} < \int_1^\infty \frac{1}{x^{1.3}}\, dx$$

What can you conclude about the series?

2. Suppose f is a continuous positive decreasing function for $x \geqslant 1$ and $a_n = f(n)$. By drawing a picture, rank the following three quantities in increasing order:

$$\int_1^6 f(x)\, dx \qquad \sum_{i=1}^5 a_i \qquad \sum_{i=2}^6 a_i$$

3. Suppose $\Sigma\, a_n$ and $\Sigma\, b_n$ are series with positive terms and $\Sigma\, b_n$ is known to be convergent.
(a) If $a_n > b_n$ for all n, what can you say about $\Sigma\, a_n$? Why?
(b) If $a_n < b_n$ for all n, what can you say about $\Sigma\, a_n$? Why?

4. Suppose $\Sigma\, a_n$ and $\Sigma\, b_n$ are series with positive terms and $\Sigma\, b_n$ is known to be divergent.
(a) If $a_n > b_n$ for all n, what can you say about $\Sigma\, a_n$? Why?
(b) If $a_n < b_n$ for all n, what can you say about $\Sigma\, a_n$? Why?

5. It is important to distinguish between

$$\sum_{n=1}^{\infty} n^b \quad \text{and} \quad \sum_{n=1}^{\infty} b^n$$

What name is given to the first series? To the second? For what values of b does the first series converge? For what values of b does the second series converge?

6–8 ▪ Use the Integral Test to determine whether the series is convergent or divergent.

6. $\displaystyle\sum_{n=1}^{\infty} \frac{1}{n^5}$

7. $\displaystyle\sum_{n=1}^{\infty} \frac{1}{\sqrt[5]{n}}$

8. $\displaystyle\sum_{n=1}^{\infty} \frac{1}{\sqrt{n+4}}$

9–10 ▪ Use the Comparison Test to determine whether the series is convergent or divergent.

9. $\displaystyle\sum_{n=1}^{\infty} \frac{n}{2n^3 + 1}$

10. $\displaystyle\sum_{n=2}^{\infty} \frac{n^3}{n^4 - 1}$

11–30 ▪ Determine whether the series is convergent or divergent.

11. $\displaystyle\sum_{n=1}^{\infty} \frac{2}{n^{0.85}}$

12. $\displaystyle\sum_{n=1}^{\infty} (n^{-1.4} + 3n^{-1.2})$

13. $1 + \dfrac{1}{8} + \dfrac{1}{27} + \dfrac{1}{64} + \dfrac{1}{125} + \cdots$

14. $1 + \dfrac{1}{2\sqrt{2}} + \dfrac{1}{3\sqrt{3}} + \dfrac{1}{4\sqrt{4}} + \dfrac{1}{5\sqrt{5}} + \cdots$

15. $1 + \dfrac{1}{3} + \dfrac{1}{5} + \dfrac{1}{7} + \dfrac{1}{9} + \cdots$

16. $\dfrac{1}{5} + \dfrac{1}{8} + \dfrac{1}{11} + \dfrac{1}{14} + \dfrac{1}{17} + \cdots$

17. $\displaystyle\sum_{n=1}^{\infty} n e^{-n}$

18. $\displaystyle\sum_{n=1}^{\infty} \frac{n^2}{n^3 + 1}$

19. $\displaystyle\sum_{n=2}^{\infty} \frac{1}{n \ln n}$

20. $\displaystyle\sum_{n=1}^{\infty} \frac{n^2 - 1}{3n^4 + 1}$

21. $\displaystyle\sum_{n=1}^{\infty} \frac{\cos^2 n}{n^2 + 1}$

22. $\displaystyle\sum_{n=1}^{\infty} \frac{4 + 3^n}{2^n}$

23. $\displaystyle\sum_{n=1}^{\infty} \frac{n - 1}{n 4^n}$

24. $\displaystyle\sum_{n=1}^{\infty} \frac{1}{\sqrt{n^3 + 1}}$

25. $\displaystyle\sum_{n=1}^{\infty} \frac{1 + 4^n}{1 + 3^n}$

26. $\displaystyle\sum_{n=1}^{\infty} \frac{1}{2n + 3}$

27. $\displaystyle\sum_{n=1}^{\infty} \frac{2 + (-1)^n}{n\sqrt{n}}$

28. $\displaystyle\sum_{n=0}^{\infty} \frac{1 + \sin n}{10^n}$

29. $\displaystyle\sum_{n=1}^{\infty} \sin\left(\frac{1}{n}\right)$

30. $\displaystyle\sum_{n=1}^{\infty} \frac{n + 5}{\sqrt[3]{n^7 + n^2}}$

31–32 ▪ Find the values of p for which the series is convergent.

31. $\displaystyle\sum_{n=2}^{\infty} \frac{1}{n(\ln n)^p}$

32. $\displaystyle\sum_{n=1}^{\infty} \frac{\ln n}{n^p}$

33. Let s be the sum of a series $\Sigma\, a_n$ that has been shown to be convergent by the Integral Test and let $f(x)$ be the function in that test. The remainder after n terms is

$$R_n = s - s_n = a_{n+1} + a_{n+2} + a_{n+3} + \cdots$$

Thus R_n is the error made when s_n, the sum of the first n terms, is used as an approximation to the total sum s.
(a) By comparing areas in a diagram like Figures 3 and 4 (but with $x \geqslant n$), show that

$$\int_{n+1}^{\infty} f(x)\, dx \leqslant R_n \leqslant \int_{n}^{\infty} f(x)\, dx$$

(b) Deduce from part (a) that

$$s_n + \int_{n+1}^{\infty} f(x)\, dx \leqslant s \leqslant s_n + \int_{n}^{\infty} f(x)\, dx$$

34. (a) Find the partial sum s_{10} of the series $\sum_{n=1}^{\infty} 1/n^4$. Use Exercise 33(a) to estimate the error in using s_{10} as an approximation to the sum of the series.
(b) Use Exercise 33(b) with $n = 10$ to give an improved estimate of the sum.
(c) Find a value of n so that s_n is within 0.00001 of the sum.

35. (a) Use the sum of the first 10 terms and Exercise 33(a) to estimate the sum of the series $\sum_{n=1}^{\infty} 1/n^2$. How good is this estimate?
(b) Improve this estimate using Exercise 33(b) with $n = 10$.
(c) Find a value of n that will ensure that the error in the approximation $s \approx s_n$ is less than 0.001.

36. Find the sum of the series $\sum_{n=1}^{\infty} 1/n^5$ correct to three decimal places.

37. (a) Use a graph of $y = 1/x$ to show that if s_n is the nth partial sum of the harmonic series, then

$$s_n \leqslant 1 + \ln n$$

(b) The harmonic series diverges, but very slowly. Use part (a) to show that the sum of the first million terms is less than 15 and the sum of the first billion terms is less than 22.

38. Show that if we want to approximate the sum of the series $\sum_{n=1}^{\infty} n^{-1.001}$ so that the error is less than 5 in the ninth decimal place, then we need to add more than $10^{11,301}$ terms!

39. The meaning of the decimal representation of a number $0.d_1d_2d_3 \ldots$ (where the digit d_i is one of the numbers 0, 1, 2, . . . , 9) is that

$$0.d_1d_2d_3d_4 \ldots = \frac{d_1}{10} + \frac{d_2}{10^2} + \frac{d_3}{10^3} + \frac{d_4}{10^4} + \cdots$$

Show that this series always converges.

40. Show that if $a_n > 0$ and $\sum a_n$ is convergent, then $\sum \ln(1 + a_n)$ is convergent.

41. If $\sum a_n$ is a convergent series with positive terms, is it true that $\sum \sin(a_n)$ is also convergent?

42. (a) Suppose that $\sum a_n$ and $\sum b_n$ are series with positive terms and $\sum b_n$ is convergent. Prove that if

$$\lim_{n \to \infty} \frac{a_n}{b_n} = 0$$

then $\sum a_n$ is also convergent.

(b) Use part (a) to show that the series converges.

(i) $\displaystyle\sum_{n=1}^{\infty} \frac{\ln n}{n^3}$　　(ii) $\displaystyle\sum_{n=1}^{\infty} \frac{\ln n}{\sqrt{n}\, e^n}$

43. (a) Suppose that $\sum a_n$ and $\sum b_n$ are series with positive terms and $\sum b_n$ is divergent. Prove that if

$$\lim_{n \to \infty} \frac{a_n}{b_n} = \infty$$

then $\sum a_n$ is also divergent.

(b) Use part (a) to show that the series diverges.

(i) $\displaystyle\sum_{n=2}^{\infty} \frac{1}{\ln n}$　　(ii) $\displaystyle\sum_{n=1}^{\infty} \frac{\ln n}{n}$

44. Give an example of a pair of series $\sum a_n$ and $\sum b_n$ with positive terms where $\lim_{n \to \infty} (a_n/b_n) = 0$ and $\sum b_n$ diverges, but $\sum a_n$ converges. [Compare with Exercise 42.]

45. Prove that if $a_n \geqslant 0$ and $\sum a_n$ converges, then $\sum a_n^2$ also converges.

46. Find all positive values of b for which the series $\sum_{n=1}^{\infty} b^{\ln n}$ converges.

47. Show that if $a_n > 0$ and $\lim_{n \to \infty} na_n \neq 0$, then $\sum a_n$ is divergent.

48. Find all values of c for which the following series converges.

$$\sum_{n=1}^{\infty} \left(\frac{c}{n} - \frac{1}{n+1} \right)$$

8.4 | OTHER CONVERGENCE TESTS

The convergence tests that we have looked at so far apply only to series with positive terms. In this section we learn how to deal with series whose terms are not necessarily positive.

ALTERNATING SERIES

An **alternating series** is a series whose terms are alternately positive and negative. Here are two examples:

$$1 - \frac{1}{2} + \frac{1}{3} - \frac{1}{4} + \frac{1}{5} - \frac{1}{6} + \cdots = \sum_{n=1}^{\infty} (-1)^{n-1} \frac{1}{n}$$

$$-\frac{1}{2} + \frac{2}{3} - \frac{3}{4} + \frac{4}{5} - \frac{5}{6} + \frac{6}{7} - \cdots = \sum_{n=1}^{\infty} (-1)^n \frac{n}{n+1}$$

We see from these examples that the nth term of an alternating series is of the form

$$a_n = (-1)^{n-1} b_n \qquad \text{or} \qquad a_n = (-1)^n b_n$$

where b_n is a positive number. $\big($In fact, $b_n = |a_n|.\big)$

The following test says that if the terms of an alternating series decrease to 0 in absolute value, then the series converges.

THE ALTERNATING SERIES TEST   If the alternating series

$$\sum_{n=1}^{\infty} (-1)^{n-1} b_n = b_1 - b_2 + b_3 - b_4 + b_5 - b_6 + \cdots \qquad b_n > 0$$

satisfies

(i) $b_{n+1} \leqslant b_n$ for all n

(ii) $\lim\limits_{n \to \infty} b_n = 0$

then the series is convergent.

Before giving the proof let's look at Figure 1, which gives a picture of the idea behind the proof. We first plot $s_1 = b_1$ on a number line. To find s_2 we subtract b_2, so s_2 is to the left of s_1. Then to find s_3 we add b_3, so s_3 is to the right of s_2. But, since $b_3 < b_2$, s_3 is to the left of s_1. Continuing in this manner, we see that the partial sums oscillate back and forth. Since $b_n \to 0$, the successive steps are becoming smaller and smaller. The even partial sums $s_2, s_4, s_6, \ldots$ are increasing and the odd partial sums s_1, $s_3, s_5, \ldots$ are decreasing. Thus it seems plausible that both are converging to some number s, which is the sum of the series. Therefore, in the following proof, we consider the even and odd partial sums separately.

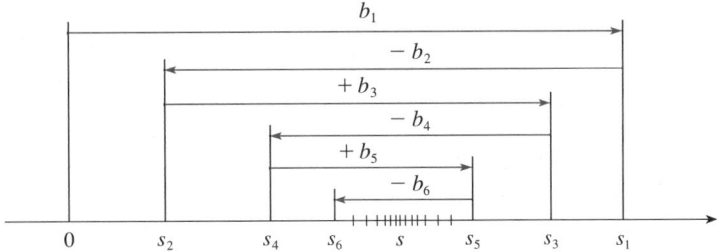

FIGURE 1

PROOF OF THE ALTERNATING SERIES TEST   We first consider the even partial sums:

$$s_2 = b_1 - b_2 \geqslant 0 \qquad \text{since } b_2 \leqslant b_1$$

$$s_4 = s_2 + (b_3 - b_4) \geqslant s_2 \qquad \text{since } b_4 \leqslant b_3$$

In general $s_{2n} = s_{2n-2} + (b_{2n-1} - b_{2n}) \geqslant s_{2n-2} \qquad \text{since } b_{2n} \leqslant b_{2n-1}$

Thus $0 \leqslant s_2 \leqslant s_4 \leqslant s_6 \leqslant \cdots \leqslant s_{2n} \leqslant \cdots$

But we can also write

$$s_{2n} = b_1 - (b_2 - b_3) - (b_4 - b_5) - \cdots - (b_{2n-2} - b_{2n-1}) - b_{2n}$$

Every term in brackets is positive, so $s_{2n} \leqslant b_1$ for all n. Therefore the sequence $\{s_{2n}\}$ of even partial sums is increasing and bounded above. So it is convergent by the Monotonic Sequence Theorem. Let's call its limit s, that is,

$$\lim_{n \to \infty} s_{2n} = s$$

Now we compute the limit of the odd partial sums:

$$\lim_{n\to\infty} s_{2n+1} = \lim_{n\to\infty}(s_{2n} + b_{2n+1})$$

$$= \lim_{n\to\infty} s_{2n} + \lim_{n\to\infty} b_{2n+1}$$

$$= s + 0 \qquad \text{[by condition (ii)]}$$

$$= s$$

Since both the even and odd partial sums converge to s, we have $\lim_{n\to\infty} s_n = s$ (see Exercise 52 in Section 8.1) and so the series is convergent. □

▼ **EXAMPLE 1** The alternating harmonic series

$$1 - \frac{1}{2} + \frac{1}{3} - \frac{1}{4} + \cdots = \sum_{n=1}^{\infty} \frac{(-1)^{n-1}}{n}$$

satisfies

(i) $b_{n+1} < b_n$ because $\dfrac{1}{n+1} < \dfrac{1}{n}$

(ii) $\lim_{n\to\infty} b_n = \lim_{n\to\infty} \dfrac{1}{n} = 0$

so the series is convergent by the Alternating Series Test. ∎

■ Figure 2 illustrates Example 1 by showing the graphs of the terms $a_n = (-1)^{n-1}/n$ and the partial sums s_n. Notice how the values of s_n zigzag across the limiting value, which appears to be about 0.7. In fact, it can be proved that the exact sum of the series is $\ln 2 \approx 0.693$.

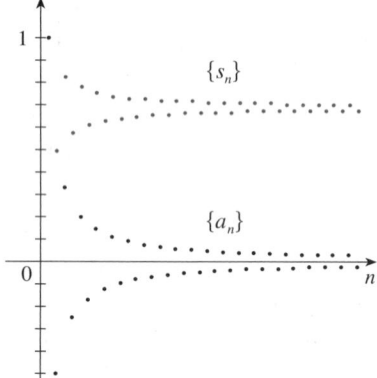

FIGURE 2

▼ **EXAMPLE 2** The series $\displaystyle\sum_{n=1}^{\infty} \frac{(-1)^n 3n}{4n-1}$ is alternating, but

$$\lim_{n\to\infty} b_n = \lim_{n\to\infty} \frac{3n}{4n-1} = \lim_{n\to\infty} \frac{3}{4 - \dfrac{1}{n}} = \frac{3}{4}$$

so condition (ii) is not satisfied. Instead, we look at the limit of the nth term of the series:

$$\lim_{n\to\infty} a_n = \lim_{n\to\infty} \frac{(-1)^n 3n}{4n-1}$$

This limit does not exist, so the series diverges by the Test for Divergence. ∎

EXAMPLE 3 Test the series $\displaystyle\sum_{n=1}^{\infty} (-1)^{n+1} \frac{n^2}{n^3+1}$ for convergence or divergence.

SOLUTION The given series is alternating so we try to verify conditions (i) and (ii) of the Alternating Series Test.

Unlike the situation in Example 1, it is not obvious that the sequence given by $b_n = n^2/(n^3+1)$ is decreasing. However, if we consider the related function $f(x) = x^2/(x^3+1)$, we find that

$$f'(x) = \frac{x(2-x^3)}{(x^3+1)^2}$$

■ Instead of verifying condition (i) of the Alternating Series Test by computing a derivative, we could verify that $b_{n+1} < b_n$ directly by using the technique of Example 12 in Section 8.1.

Since we are considering only positive x, we see that $f'(x) < 0$ if $2 - x^3 < 0$, that is, $x > \sqrt[3]{2}$. Thus f is decreasing on the interval $(\sqrt[3]{2}, \infty)$. This means that $f(n + 1) < f(n)$ and therefore $b_{n+1} < b_n$ when $n \geq 2$. (The inequality $b_2 < b_1$ can be verified directly but all that really matters is that the sequence $\{b_n\}$ is *eventually* decreasing.)

Condition (ii) is readily verified:

$$\lim_{n \to \infty} b_n = \lim_{n \to \infty} \frac{n^2}{n^3 + 1} = \lim_{n \to \infty} \frac{\dfrac{1}{n}}{1 + \dfrac{1}{n^3}} = 0$$

Thus the given series is convergent by the Alternating Series Test. ∎

A partial sum s_n of any convergent series can be used as an approximation to the total sum s, but this is not of much use unless we can estimate the accuracy of the approximation. The error involved in using $s \approx s_n$ is the remainder $R_n = s - s_n$. The next theorem says that for series that satisfy the conditions of the Alternating Series Test, the size of the error is smaller than b_{n+1}, which is the absolute value of the first neglected term.

■ You can see geometrically why the Alternating Series Estimation Theorem is true by looking at Figure 1 (on page 455). Notice that $s - s_4 < b_5$, $|s - s_5| < b_6$, and so on. Notice also that s lies between any two consecutive partial sums.

ALTERNATING SERIES ESTIMATION THEOREM If $s = \sum (-1)^{n-1} b_n$ is the sum of an alternating series that satisfies

> (i) $0 \leq b_{n+1} \leq b_n$ and (ii) $\lim_{n \to \infty} b_n = 0$

then

$$|R_n| = |s - s_n| \leq b_{n+1}$$

PROOF We know from the proof of the Alternating Series Test that s lies between any two consecutive partial sums s_n and s_{n+1}. (There we showed that s is larger than all the even partial sums. A similar argument shows that s is smaller than all the odd sums.) It follows that

$$|s - s_n| \leq |s_{n+1} - s_n| = b_{n+1} \qquad \square$$

▼ EXAMPLE 4 Find the sum of the series $\displaystyle\sum_{n=0}^{\infty} \frac{(-1)^n}{n!}$ correct to three decimal places. (By definition, $0! = 1$.)

SOLUTION We first observe that the series is convergent by the Alternating Series Test because

> (i) $b_{n+1} = \dfrac{1}{(n + 1)!} = \dfrac{1}{n!(n + 1)} < \dfrac{1}{n!} = b_n$

> (ii) $0 < \dfrac{1}{n!} < \dfrac{1}{n} \to 0$ so $b_n = \dfrac{1}{n!} \to 0$ as $n \to \infty$

To get a feel for how many terms we need to use in our approximation, let's write out the first few terms of the series:

$$s = \frac{1}{0!} - \frac{1}{1!} + \frac{1}{2!} - \frac{1}{3!} + \frac{1}{4!} - \frac{1}{5!} + \frac{1}{6!} - \frac{1}{7!} + \cdots$$

$$= 1 - 1 + \tfrac{1}{2} - \tfrac{1}{6} + \tfrac{1}{24} - \tfrac{1}{120} + \tfrac{1}{720} - \tfrac{1}{5040} + \cdots$$

Notice that
$$b_7 = \frac{1}{5040} < \frac{1}{5000} = 0.0002$$

and
$$s_6 = 1 - 1 + \frac{1}{2} - \frac{1}{6} + \frac{1}{24} - \frac{1}{120} + \frac{1}{720} \approx 0.368056$$

By the Alternating Series Estimation Theorem we know that
$$|s - s_6| \leq b_7 < 0.0002$$

This error of less than 0.0002 does not affect the third decimal place, so we have
$$s \approx 0.368$$

correct to three decimal places. ∎

⊘ **NOTE** The rule that the error (in using s_n to approximate s) is smaller than the first neglected term is, in general, valid only for alternating series that satisfy the conditions of the Alternating Series Estimation Theorem. The rule does not apply to other types of series.

ABSOLUTE CONVERGENCE

Given any series $\Sigma \, a_n$, we can consider the corresponding series
$$\sum_{n=1}^{\infty} |a_n| = |a_1| + |a_2| + |a_3| + \cdots$$

whose terms are the absolute values of the terms of the original series.

> **DEFINITION** A series $\Sigma \, a_n$ is called **absolutely convergent** if the series of absolute values $\Sigma \, |a_n|$ is convergent.

Notice that if $\Sigma \, a_n$ is a series with positive terms, then $|a_n| = a_n$ and so absolute convergence is the same as convergence.

EXAMPLE 5 The series
$$\sum_{n=1}^{\infty} \frac{(-1)^{n-1}}{n^2} = 1 - \frac{1}{2^2} + \frac{1}{3^2} - \frac{1}{4^2} + \cdots$$

is absolutely convergent because
$$\sum_{n=1}^{\infty} \left| \frac{(-1)^{n-1}}{n^2} \right| = \sum_{n=1}^{\infty} \frac{1}{n^2} = 1 + \frac{1}{2^2} + \frac{1}{3^2} + \frac{1}{4^2} + \cdots$$

is a convergent p-series ($p = 2$). ∎

EXAMPLE 6 We know that the alternating harmonic series
$$\sum_{n=1}^{\infty} \frac{(-1)^{n-1}}{n} = 1 - \frac{1}{2} + \frac{1}{3} - \frac{1}{4} + \cdots$$

■ In Section 8.7 we will prove that $e^x = \Sigma_{n=0}^{\infty} x^n/n!$ for all x, so what we have obtained in Example 4 is actually an approximation to the number e^{-1}.

■ We have convergence tests for series with positive terms and for alternating series. But what if the signs of the terms switch back and forth irregularly? We will see in Example 7 that the idea of absolute convergence sometimes helps in such cases.

is convergent (see Example 1), but it is not absolutely convergent because the corresponding series of absolute values is

$$\sum_{n=1}^{\infty} \left| \frac{(-1)^{n-1}}{n} \right| = \sum_{n=1}^{\infty} \frac{1}{n} = 1 + \frac{1}{2} + \frac{1}{3} + \frac{1}{4} + \cdots$$

which is the harmonic series (p-series with $p = 1$) and is therefore divergent. ■

DEFINITION   A series $\sum a_n$ is called **conditionally convergent** if it is convergent but not absolutely convergent.

■ It can be proved that if the terms of an absolutely convergent series are rearranged in a different order, then the sum is unchanged. But if a conditionally convergent series is rearranged, the sum could be different.

Example 6 shows that the alternating harmonic series is conditionally convergent. Thus it is possible for a series to be convergent but not absolutely convergent. However, the next theorem shows that absolute convergence implies convergence.

---

1 **THEOREM**   If a series $\sum a_n$ is absolutely convergent, then it is convergent.

PROOF   Observe that the inequality

$$0 \le a_n + |a_n| \le 2|a_n|$$

is true because $|a_n|$ is either a_n or $-a_n$. If $\sum a_n$ is absolutely convergent, then $\sum |a_n|$ is convergent, so $\sum 2|a_n|$ is convergent. Therefore, by the Comparison Test, $\sum (a_n + |a_n|)$ is convergent. Then

$$\sum a_n = \sum (a_n + |a_n|) - \sum |a_n|$$

is the difference of two convergent series and is therefore convergent.   □

■ Figure 3 shows the graphs of the terms a_n and partial sums s_n of the series in Example 7. Notice that the series is not alternating but has positive and negative terms.

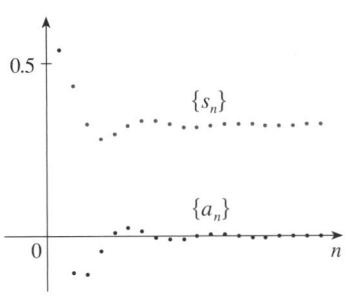

FIGURE 3

V **EXAMPLE 7**   Determine whether the series

$$\sum_{n=1}^{\infty} \frac{\cos n}{n^2} = \frac{\cos 1}{1^2} + \frac{\cos 2}{2^2} + \frac{\cos 3}{3^2} + \cdots$$

is convergent or divergent.

SOLUTION   This series has both positive and negative terms, but it is not alternating. (The first term is positive, the next three are negative, and the following three are positive. The signs change irregularly.) We can apply the Comparison Test to the series of absolute values

$$\sum_{n=1}^{\infty} \left| \frac{\cos n}{n^2} \right| = \sum_{n=1}^{\infty} \frac{|\cos n|}{n^2}$$

Since $|\cos n| \le 1$ for all n, we have

$$\frac{|\cos n|}{n^2} \le \frac{1}{n^2}$$

We know that $\sum 1/n^2$ is convergent (p-series with $p = 2$) and therefore $\sum |\cos n|/n^2$ is convergent by the Comparison Test. Thus the given series $\sum (\cos n)/n^2$ is absolutely convergent and therefore convergent by Theorem 1. ■

THE RATIO TEST

The following test is very useful in determining whether a given series is absolutely convergent.

THE RATIO TEST

(i) If $\lim\limits_{n \to \infty} \left| \dfrac{a_{n+1}}{a_n} \right| = L < 1$, then the series $\sum\limits_{n=1}^{\infty} a_n$ is absolutely convergent (and therefore convergent).

(ii) If $\lim\limits_{n \to \infty} \left| \dfrac{a_{n+1}}{a_n} \right| = L > 1$ or $\lim\limits_{n \to \infty} \left| \dfrac{a_{n+1}}{a_n} \right| = \infty$, then the series $\sum\limits_{n=1}^{\infty} a_n$ is divergent.

(iii) If $\lim\limits_{n \to \infty} \left| \dfrac{a_{n+1}}{a_n} \right| = 1$, the Ratio Test is inconclusive; that is, no conclusion can be drawn about the convergence or divergence of $\sum a_n$.

PROOF

(i) The idea is to compare the given series with a convergent geometric series. Since $L < 1$, we can choose a number r such that $L < r < 1$. Since

$$\lim_{n \to \infty} \left| \frac{a_{n+1}}{a_n} \right| = L \qquad \text{and} \qquad L < r$$

the ratio $\left| a_{n+1}/a_n \right|$ will eventually be less than r; that is, there exists an integer N such that

$$\left| \frac{a_{n+1}}{a_n} \right| < r \qquad \text{whenever } n \geqslant N$$

or, equivalently,

$$\boxed{2} \qquad \qquad |a_{n+1}| < |a_n|r \qquad \text{whenever } n \geqslant N$$

Putting n successively equal to N, $N + 1$, $N + 2$, . . . in $\boxed{2}$, we obtain

$$|a_{N+1}| < |a_N|r$$
$$|a_{N+2}| < |a_{N+1}|r < |a_N|r^2$$
$$|a_{N+3}| < |a_{N+2}|r < |a_N|r^3$$

and, in general,

$$\boxed{3} \qquad \qquad |a_{N+k}| < |a_N|r^k \qquad \text{for all } k \geqslant 1$$

Now the series

$$\sum_{k=1}^{\infty} |a_N|r^k = |a_N|r + |a_N|r^2 + |a_N|r^3 + \cdots$$

is convergent because it is a geometric series with $0 < r < 1$. So the inequality $\boxed{3}$, together with the Comparison Test, shows that the series

$$\sum_{n=N+1}^{\infty} |a_n| = \sum_{k=1}^{\infty} |a_{N+k}| = |a_{N+1}| + |a_{N+2}| + |a_{N+3}| + \cdots$$

is also convergent. It follows that the series $\sum_{n=1}^{\infty} |a_n|$ is convergent. (Recall that a finite number of terms doesn't affect convergence.) Therefore $\sum a_n$ is absolutely convergent.

(ii) If $|a_{n+1}/a_n| \to L > 1$ or $|a_{n+1}/a_n| \to \infty$, then the ratio $|a_{n+1}/a_n|$ will eventually be greater than 1; that is, there exists an integer N such that

$$\left| \frac{a_{n+1}}{a_n} \right| > 1 \qquad \text{whenever } n \geq N$$

This means that $|a_{n+1}| > |a_n|$ whenever $n \geq N$ and so

$$\lim_{n \to \infty} a_n \neq 0$$

Therefore $\sum a_n$ diverges by the Test for Divergence. ☐

NOTE Part (iii) of the Ratio Test says that if $\lim_{n \to \infty} |a_{n+1}/a_n| = 1$, the test gives no information. For instance, for the convergent series $\sum 1/n^2$ we have

$$\left| \frac{a_{n+1}}{a_n} \right| = \frac{\dfrac{1}{(n+1)^2}}{\dfrac{1}{n^2}} = \frac{n^2}{(n+1)^2} = \frac{1}{\left(1 + \dfrac{1}{n}\right)^2} \to 1 \qquad \text{as } n \to \infty$$

whereas for the divergent series $\sum 1/n$ we have

$$\left| \frac{a_{n+1}}{a_n} \right| = \frac{\dfrac{1}{n+1}}{\dfrac{1}{n}} = \frac{n}{n+1} = \frac{1}{1 + \dfrac{1}{n}} \to 1 \qquad \text{as } n \to \infty$$

Therefore, if $\lim_{n \to \infty} |a_{n+1}/a_n| = 1$, the series $\sum a_n$ might converge or it might diverge. In this case the Ratio Test fails and we must use some other test.

EXAMPLE 8 Test the series $\sum_{n=1}^{\infty} (-1)^n \dfrac{n^3}{3^n}$ for absolute convergence.

■ Series that involve factorials or other products (including a constant raised to the nth power) are often conveniently tested using the Ratio Test.

SOLUTION We use the Ratio Test with $a_n = (-1)^n n^3/3^n$:

$$\left| \frac{a_{n+1}}{a_n} \right| = \left| \frac{\dfrac{(-1)^{n+1}(n+1)^3}{3^{n+1}}}{\dfrac{(-1)^n n^3}{3^n}} \right| = \frac{(n+1)^3}{3^{n+1}} \cdot \frac{3^n}{n^3}$$

$$= \frac{1}{3} \left(\frac{n+1}{n} \right)^3 = \frac{1}{3} \left(1 + \frac{1}{n} \right)^3 \to \frac{1}{3} < 1$$

Thus by the Ratio Test the given series is absolutely convergent and therefore convergent. ■

V EXAMPLE 9 Test the convergence of the series $\sum_{n=1}^{\infty} \dfrac{n^n}{n!}$.

SOLUTION Since the terms $a_n = n^n/n!$ are positive, we don't need the absolute value signs.

$$\frac{a_{n+1}}{a_n} = \frac{(n + 1)^{n+1}}{(n + 1)!} \cdot \frac{n!}{n^n} = \frac{(n + 1)(n + 1)^n}{(n + 1)n!} \cdot \frac{n!}{n^n}$$

$$= \left(\frac{n + 1}{n} \right)^n = \left(1 + \frac{1}{n} \right)^n \to e \qquad \text{as } n \to \infty$$

- We know that
$$\lim_{x \to 0} (1 + x)^{1/x} = e$$
by the definition of e. If we let $n = 1/x$, then $n \to \infty$ as $x \to 0^+$ and so
$$\lim_{n \to \infty} (1 + 1/n)^n = e$$

Since $e > 1$, the given series is divergent by the Ratio Test. ∎

The following test is convenient to apply when nth powers occur. Its proof is similar to the proof of the Ratio Test and is left as Exercise 47.

THE ROOT TEST

(i) If $\lim\limits_{n \to \infty} \sqrt[n]{|a_n|} = L < 1$, then the series $\sum\limits_{n=1}^{\infty} a_n$ is absolutely convergent (and therefore convergent).

(ii) If $\lim\limits_{n \to \infty} \sqrt[n]{|a_n|} = L > 1$ or $\lim\limits_{n \to \infty} \sqrt[n]{|a_n|} = \infty$, then the series $\sum\limits_{n=1}^{\infty} a_n$ is divergent.

(iii) If $\lim\limits_{n \to \infty} \sqrt[n]{|a_n|} = 1$, the Root Test is inconclusive.

If $\lim_{n \to \infty} \sqrt[n]{|a_n|} = 1$, then part (iii) of the Root Test says that the test gives no information. The series $\sum a_n$ could converge or diverge. (If $L = 1$ in the Ratio Test, don't try the Root Test because L will again be 1. And if $L = 1$ in the Root Test, don't try the Ratio Test because it will fail too.)

V EXAMPLE 10 Test the convergence of the series $\sum_{n=1}^{\infty} \left(\dfrac{2n + 3}{3n + 2} \right)^n$.

SOLUTION

$$a_n = \left(\frac{2n + 3}{3n + 2} \right)^n$$

- **www.stewartcalculus.com**
We now have several tests for convergence of series. So, given a series, how do you know which test to use? For advice, click on *Additional Topics* and then on *Strategy for Testing Series*.

$$\sqrt[n]{|a_n|} = \frac{2n + 3}{3n + 2} = \frac{2 + \dfrac{3}{n}}{3 + \dfrac{2}{n}} \to \frac{2}{3} < 1$$

Thus the given series converges by the Root Test. ∎

8.4 | EXERCISES

1. (a) What is an alternating series?
   (b) Under what conditions does an alternating series converge?
   (c) If these conditions are satisfied, what can you say about the remainder after n terms?

2. What can you say about the series $\Sigma\, a_n$ in each of the following cases?

   (a) $\displaystyle\lim_{n\to\infty}\left|\frac{a_{n+1}}{a_n}\right| = 8$ (b) $\displaystyle\lim_{n\to\infty}\left|\frac{a_{n+1}}{a_n}\right| = 0.8$

   (c) $\displaystyle\lim_{n\to\infty}\left|\frac{a_{n+1}}{a_n}\right| = 1$

3–8 ▪ Test the series for convergence or divergence.

3. $\dfrac{4}{7} - \dfrac{4}{8} + \dfrac{4}{9} - \dfrac{4}{10} + \dfrac{4}{11} - \cdots$

4. $-\dfrac{3}{4} + \dfrac{5}{5} - \dfrac{7}{6} + \dfrac{9}{7} - \dfrac{11}{8} + \cdots$

5. $\displaystyle\sum_{n=1}^{\infty} \frac{(-1)^{n-1}}{2n+1}$ **6.** $\displaystyle\sum_{n=1}^{\infty} (-1)^n \frac{n}{\sqrt{n^3+2}}$

7. $\displaystyle\sum_{n=1}^{\infty} (-1)^n \frac{3n-1}{2n+1}$ **8.** $\displaystyle\sum_{n=1}^{\infty} (-1)^n \cos\!\left(\frac{\pi}{n}\right)$

9–12 ▪ Show that the series is convergent. How many terms of the series do we need to add in order to find the sum to the indicated accuracy?

9. $\displaystyle\sum_{n=1}^{\infty} \frac{(-1)^{n+1}}{n^6}$ $(\,|\,\text{error}\,| < 0.00005)$

10. $\displaystyle\sum_{n=1}^{\infty} \frac{(-1)^n}{n\,5^n}$ $(\,|\,\text{error}\,| < 0.0001)$

11. $\displaystyle\sum_{n=0}^{\infty} \frac{(-1)^n}{10^n n!}$ $(\,|\,\text{error}\,| < 0.000005)$

12. $\displaystyle\sum_{n=1}^{\infty} (-1)^{n-1} n e^{-n}$ $(\,|\,\text{error}\,| < 0.01)$

13–16 ▪ Approximate the sum of the series correct to four decimal places.

13. $\displaystyle\sum_{n=1}^{\infty} \frac{(-1)^n}{(2n)!}$ **14.** $\displaystyle\sum_{n=1}^{\infty} \frac{(-1)^{n+1}}{n^6}$

15. $\displaystyle\sum_{n=1}^{\infty} \frac{(-1)^{n-1} n^2}{10^n}$ **16.** $\displaystyle\sum_{n=1}^{\infty} \frac{(-1)^n}{3^n n!}$

17. Is the 50th partial sum s_{50} of the alternating series $\sum_{n=1}^{\infty} (-1)^{n-1}/n$ an overestimate or an underestimate of the total sum? Explain.

18. For what values of p is the following series convergent?

$$\sum_{n=1}^{\infty} \frac{(-1)^{n-1}}{n^p}$$

19–40 ▪ Determine whether the series is absolutely convergent, conditionally convergent, or divergent.

19. $\displaystyle\sum_{n=1}^{\infty} \frac{n}{5^n}$ **20.** $\displaystyle\sum_{n=1}^{\infty} \frac{(-2)^n}{n^2}$

21. $\displaystyle\sum_{n=0}^{\infty} \frac{(-10)^n}{n!}$ **22.** $\displaystyle\sum_{n=1}^{\infty} (-1)^{n-1} \frac{n}{n^2+4}$

23. $\displaystyle\sum_{n=0}^{\infty} \frac{(-1)^n}{5n+1}$ **24.** $\displaystyle\sum_{n=0}^{\infty} \frac{(-3)^n}{(2n+1)!}$

25. $\displaystyle\sum_{k=1}^{\infty} k\!\left(\frac{2}{3}\right)^k$ **26.** $\displaystyle\sum_{n=1}^{\infty} \frac{n!}{100^n}$

27. $\displaystyle\sum_{n=1}^{\infty} \frac{10^n}{(n+1)4^{2n+1}}$ **28.** $\displaystyle\sum_{n=1}^{\infty} \frac{\sin 4n}{4^n}$

29. $\displaystyle\sum_{n=1}^{\infty} \frac{\cos(n\pi/3)}{n!}$ **30.** $\displaystyle\sum_{n=2}^{\infty} \frac{(-1)^n}{n \ln n}$

31. $\displaystyle\sum_{n=1}^{\infty} \frac{(-1)^n \arctan n}{n^2}$ **32.** $\displaystyle\sum_{n=1}^{\infty} \frac{(-2)^n}{n^n}$

33. $\displaystyle\sum_{n=1}^{\infty} \left(\frac{n^2+1}{2n^2+1}\right)^n$ **34.** $\displaystyle\sum_{n=2}^{\infty} \left(\frac{-2n}{n+1}\right)^{5n}$

35. $\displaystyle\sum_{n=1}^{\infty} \left(1 + \frac{1}{n}\right)^{n^2}$ **36.** $\displaystyle\sum_{n=1}^{\infty} \frac{2^{n^2}}{n!}$

37. $1 - \dfrac{1\cdot 3}{3!} + \dfrac{1\cdot 3\cdot 5}{5!} - \dfrac{1\cdot 3\cdot 5\cdot 7}{7!} + \cdots$

$\quad + (-1)^{n-1} \dfrac{1\cdot 3\cdot 5\cdot\,\cdots\,\cdot(2n-1)}{(2n-1)!} + \cdots$

38. $\dfrac{2}{5} + \dfrac{2\cdot 6}{5\cdot 8} + \dfrac{2\cdot 6\cdot 10}{5\cdot 8\cdot 11} + \dfrac{2\cdot 6\cdot 10\cdot 14}{5\cdot 8\cdot 11\cdot 14} + \cdots$

39. $\displaystyle\sum_{n=1}^{\infty} \frac{2\cdot 4\cdot 6\cdot\,\cdots\,\cdot(2n)}{n!}$

40. $\displaystyle\sum_{n=1}^{\infty} (-1)^n \frac{2^n n!}{5\cdot 8\cdot 11\cdot\,\cdots\,\cdot(3n+2)}$

41–42 ■ Let $\{b_n\}$ be a sequence of positive numbers that converges to $\frac{1}{2}$. Determine whether the given series is absolutely convergent.

41. $\displaystyle\sum_{n=1}^{\infty} \frac{b_n^n \cos n\pi}{n}$

42. $\displaystyle\sum_{n=1}^{\infty} \frac{(-1)^n n!}{n^n b_1 b_2 b_3 \cdots b_n}$

43. For which of the following series is the Ratio Test inconclusive (that is, it fails to give a definite answer)?

(a) $\displaystyle\sum_{n=1}^{\infty} \frac{1}{n^3}$

(b) $\displaystyle\sum_{n=1}^{\infty} \frac{n}{2^n}$

(c) $\displaystyle\sum_{n=1}^{\infty} \frac{(-3)^{n-1}}{\sqrt{n}}$

(d) $\displaystyle\sum_{n=1}^{\infty} \frac{\sqrt{n}}{1 + n^2}$

44. For which positive integers k is the following series convergent?

$$\sum_{n=1}^{\infty} \frac{(n!)^2}{(kn)!}$$

45. (a) Show that $\sum_{n=0}^{\infty} x^n/n!$ converges for all x.
(b) Deduce that $\lim_{n\to\infty} x^n/n! = 0$ for all x.

46. Around 1910, the Indian mathematician Srinivasa Ramanujan discovered the formula

$$\frac{1}{\pi} = \frac{2\sqrt{2}}{9801} \sum_{n=0}^{\infty} \frac{(4n)!(1103 + 26390n)}{(n!)^4 396^{4n}}$$

William Gosper used this series in 1985 to compute the first 17 million digits of π.
(a) Verify that the series is convergent.
(b) How many correct decimal places of π do you get if you use just the first term of the series? What if you use two terms?

47. Prove the Root Test.   [*Hint for part (i):* Take any number r such that $L < r < 1$ and use the fact that there is an integer N such that $\sqrt[n]{|a_n|} < r$ whenever $n \geqslant N$.]

8.5 | POWER SERIES

A **power series** is a series of the form

$$\boxed{1} \qquad \sum_{n=0}^{\infty} c_n x^n = c_0 + c_1 x + c_2 x^2 + c_3 x^3 + \cdots$$

where x is a variable and the c_n's are constants called the **coefficients** of the series. For each fixed x, the series $\boxed{1}$ is a series of constants that we can test for convergence or divergence. A power series may converge for some values of x and diverge for other values of x. The sum of the series is a function

$$f(x) = c_0 + c_1 x + c_2 x^2 + \cdots + c_n x^n + \cdots$$

whose domain is the set of all x for which the series converges. Notice that f resembles a polynomial. The only difference is that f has infinitely many terms.

For instance, if we take $c_n = 1$ for all n, the power series becomes the geometric series

$$\sum_{n=0}^{\infty} x^n = 1 + x + x^2 + \cdots + x^n + \cdots$$

which converges when $-1 < x < 1$ and diverges when $|x| \geqslant 1$ (see Equation 8.2.5).

More generally, a series of the form

$$\boxed{2} \qquad \sum_{n=0}^{\infty} c_n(x - a)^n = c_0 + c_1(x - a) + c_2(x - a)^2 + \cdots$$

is called a **power series in** $(x - a)$ or a **power series centered at** a or a **power series about** a. Notice that in writing out the term corresponding to $n = 0$ in Equations 1 and 2 we have adopted the convention that $(x - a)^0 = 1$ even when $x = a$. Notice also that when $x = a$ all of the terms are 0 for $n \geqslant 1$ and so the power series $\boxed{2}$ always converges when $x = a$.

■ **TRIGONOMETRIC SERIES**

A power series is a series in which each term is a power function. A **trigonometric series**

$$\sum_{n=0}^{\infty} (a_n \cos nx + b_n \sin nx)$$

is a series whose terms are trigonometric functions. This type of series is discussed on the website

www.stewartcalculus.com

Click on *Additional Topics* and then on *Fourier Series*.

V EXAMPLE 1 For what values of x is the series $\sum_{n=0}^{\infty} n! x^n$ convergent?

SOLUTION We use the Ratio Test. If we let a_n, as usual, denote the nth term of the series, then $a_n = n! x^n$. If $x \neq 0$, we have

$$\lim_{n \to \infty} \left| \frac{a_{n+1}}{a_n} \right| = \lim_{n \to \infty} \left| \frac{(n+1)! x^{n+1}}{n! x^n} \right|$$

$$= \lim_{n \to \infty} (n+1) |x| = \infty$$

- Notice that
$$(n+1)! = (n+1)n(n-1) \cdot \cdots \cdot 3 \cdot 2 \cdot 1$$
$$= (n+1)n!$$

By the Ratio Test, the series diverges when $x \neq 0$. Thus the given series converges only when $x = 0$. ∎

V EXAMPLE 2 For what values of x does the series $\sum_{n=1}^{\infty} \frac{(x-3)^n}{n}$ converge?

SOLUTION Let $a_n = (x-3)^n / n$. Then

$$\left| \frac{a_{n+1}}{a_n} \right| = \left| \frac{(x-3)^{n+1}}{n+1} \cdot \frac{n}{(x-3)^n} \right|$$

$$= \frac{1}{1 + \dfrac{1}{n}} |x - 3| \to |x - 3| \quad \text{as } n \to \infty$$

By the Ratio Test, the given series is absolutely convergent, and therefore convergent, when $|x - 3| < 1$ and divergent when $|x - 3| > 1$. Now

$$|x - 3| < 1 \iff -1 < x - 3 < 1 \iff 2 < x < 4$$

so the series converges when $2 < x < 4$ and diverges when $x < 2$ or $x > 4$.

The Ratio Test gives no information when $|x - 3| = 1$ so we must consider $x = 2$ and $x = 4$ separately. If we put $x = 4$ in the series, it becomes $\Sigma\, 1/n$, the harmonic series, which is divergent. If $x = 2$, the series is $\Sigma\, (-1)^n / n$, which converges by the Alternating Series Test. Thus the given power series converges for $2 \leq x < 4$. ∎

We will see that the main use of a power series is that it provides a way to represent some of the most important functions that arise in mathematics, physics, and chemistry. In particular, the sum of the power series in the next example is called a **Bessel function**, after the German astronomer Friedrich Bessel (1784–1846), and the function given in Exercise 29 is another example of a Bessel function. In fact, these functions first arose when Bessel solved Kepler's equation for describing planetary motion. Since that time, these functions have been applied in many different physical situations, including the temperature distribution in a circular plate and the shape of a vibrating drumhead.

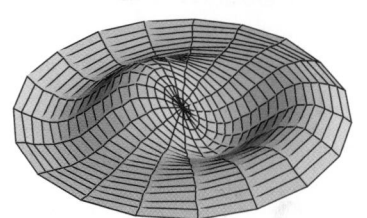

- Notice how closely the computer-generated model (which involves Bessel functions and cosine functions) matches the photograph of a vibrating rubber membrane.

EXAMPLE 3 Find the domain of the Bessel function of order 0 defined by

$$J_0(x) = \sum_{n=0}^{\infty} \frac{(-1)^n x^{2n}}{2^{2n}(n!)^2}$$

SOLUTION Let $a_n = (-1)^n x^{2n}/[2^{2n}(n!)^2]$. Then

$$\left| \frac{a_{n+1}}{a_n} \right| = \left| \frac{(-1)^{n+1} x^{2(n+1)}}{2^{2(n+1)}[(n+1)!]^2} \cdot \frac{2^{2n}(n!)^2}{(-1)^n x^{2n}} \right|$$

$$= \frac{x^{2n+2}}{2^{2n+2}(n+1)^2(n!)^2} \cdot \frac{2^{2n}(n!)^2}{x^{2n}}$$

$$= \frac{x^2}{4(n+1)^2} \to 0 < 1 \qquad \text{for all } x$$

Thus by the Ratio Test the given series converges for all values of x. In other words, the domain of the Bessel function J_0 is $(-\infty, \infty) = \mathbb{R}$. ∎

Recall that the sum of a series is equal to the limit of the sequence of partial sums. So when we define the Bessel function in Example 3 as the sum of a series we mean that, for every real number x,

$$J_0(x) = \lim_{n \to \infty} s_n(x) \qquad \text{where} \qquad s_n(x) = \sum_{i=0}^{n} \frac{(-1)^i x^{2i}}{2^{2i}(i!)^2}$$

The first few partial sums are

$$s_0(x) = 1 \qquad s_1(x) = 1 - \frac{x^2}{4} \qquad s_2(x) = 1 - \frac{x^2}{4} + \frac{x^4}{64}$$

$$s_3(x) = 1 - \frac{x^2}{4} + \frac{x^4}{64} - \frac{x^6}{2304} \qquad s_4(x) = 1 - \frac{x^2}{4} + \frac{x^4}{64} - \frac{x^6}{2304} + \frac{x^8}{147{,}456}$$

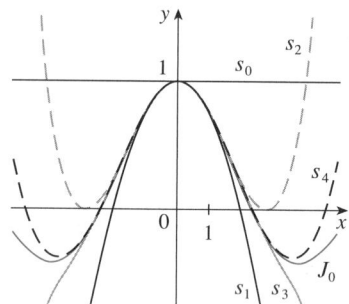

FIGURE 1

Partial sums of the Bessel function J_0

Figure 1 shows the graphs of these partial sums, which are polynomials. They are all approximations to the function J_0, but notice that the approximations become better when more terms are included. Figure 2 shows a more complete graph of the Bessel function.

For the power series that we have looked at so far, the set of values of x for which the series is convergent has always turned out to be an interval [a finite interval for the geometric series and the series in Example 2, the infinite interval $(-\infty, \infty)$ in Example 3, and a collapsed interval $[0, 0] = \{0\}$ in Example 1]. The following theorem, proved in Appendix D, says that this is true in general.

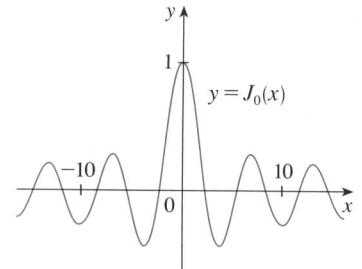

FIGURE 2

---

3 THEOREM For a given power series $\sum_{n=0}^{\infty} c_n(x - a)^n$ there are only three possibilities:

(i) The series converges only when $x = a$.

(ii) The series converges for all x.

(iii) There is a positive number R such that the series converges if $|x - a| < R$ and diverges if $|x - a| > R$.

The number R in case (iii) is called the **radius of convergence** of the power series. By convention, the radius of convergence is $R = 0$ in case (i) and $R = \infty$ in case (ii). The **interval of convergence** of a power series is the interval that consists of all values of x for which the series converges. In case (i) the interval consists of just a single point a. In case (ii) the interval is $(-\infty, \infty)$. In case (iii) note that the inequality $|x - a| < R$ can be rewritten as $a - R < x < a + R$. When x is an *endpoint* of the interval, that is, $x = a \pm R$, anything can happen—the series might converge at one or both endpoints or it might diverge at both endpoints. Thus in case (iii) there are four possibilities for the interval of convergence:

$$(a - R, a + R) \qquad (a - R, a + R] \qquad [a - R, a + R) \qquad [a - R, a + R]$$

The situation is illustrated in Figure 3.

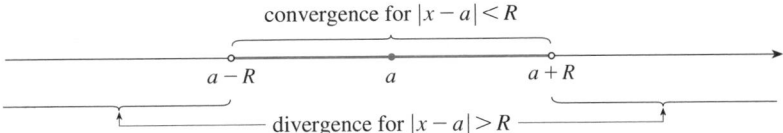

FIGURE 3

We summarize here the radius and interval of convergence for each of the examples already considered in this section.

	Series	Radius of convergence	Interval of convergence
Geometric series	$\sum\limits_{n=0}^{\infty} x^n$	$R = 1$	$(-1, 1)$
Example 1	$\sum\limits_{n=0}^{\infty} n!\, x^n$	$R = 0$	$\{0\}$
Example 2	$\sum\limits_{n=1}^{\infty} \dfrac{(x - 3)^n}{n}$	$R = 1$	$[2, 4)$
Example 3	$\sum\limits_{n=0}^{\infty} \dfrac{(-1)^n x^{2n}}{2^{2n}(n!)^2}$	$R = \infty$	$(-\infty, \infty)$

The Ratio Test (or sometimes the Root Test) should be used to determine the radius of convergence R in most cases. The Ratio and Root Tests always fail when x is an endpoint of the interval of convergence, so the endpoints must be checked with some other test.

EXAMPLE 4 Find the radius of convergence and interval of convergence of the series

$$\sum_{n=0}^{\infty} \frac{(-3)^n x^n}{\sqrt{n + 1}}$$

SOLUTION Let $a_n = (-3)^n x^n / \sqrt{n + 1}$. Then

$$\left| \frac{a_{n+1}}{a_n} \right| = \left| \frac{(-3)^{n+1} x^{n+1}}{\sqrt{n + 2}} \cdot \frac{\sqrt{n + 1}}{(-3)^n x^n} \right| = \left| -3x \sqrt{\frac{n + 1}{n + 2}} \right|$$

$$= 3 \sqrt{\frac{1 + (1/n)}{1 + (2/n)}} |x| \;\rightarrow\; 3|x| \qquad \text{as } n \rightarrow \infty$$

By the Ratio Test, the given series converges if $3|x| < 1$ and diverges if $3|x| > 1$. Thus it converges if $|x| < \frac{1}{3}$ and diverges if $|x| > \frac{1}{3}$. This means that the radius of convergence is $R = \frac{1}{3}$.

We know the series converges in the interval $\left(-\frac{1}{3}, \frac{1}{3}\right)$, but we must now test for convergence at the endpoints of this interval. If $x = -\frac{1}{3}$, the series becomes

$$\sum_{n=0}^{\infty} \frac{(-3)^n\left(-\frac{1}{3}\right)^n}{\sqrt{n+1}} = \sum_{n=0}^{\infty} \frac{1}{\sqrt{n+1}} = \frac{1}{\sqrt{1}} + \frac{1}{\sqrt{2}} + \frac{1}{\sqrt{3}} + \frac{1}{\sqrt{4}} + \cdots$$

which diverges. $\left(\text{Use the Integral Test or simply observe that it is a } p\text{-series with } p = \frac{1}{2} < 1.\right)$ If $x = \frac{1}{3}$, the series is

$$\sum_{n=0}^{\infty} \frac{(-3)^n\left(\frac{1}{3}\right)^n}{\sqrt{n+1}} = \sum_{n=0}^{\infty} \frac{(-1)^n}{\sqrt{n+1}}$$

which converges by the Alternating Series Test. Therefore the given power series converges when $-\frac{1}{3} < x \leq \frac{1}{3}$, so the interval of convergence is $\left(-\frac{1}{3}, \frac{1}{3}\right]$. ∎

V **EXAMPLE 5** Find the radius of convergence and interval of convergence of the series

$$\sum_{n=0}^{\infty} \frac{n(x+2)^n}{3^{n+1}}$$

SOLUTION If $a_n = n(x+2)^n/3^{n+1}$, then

$$\left| \frac{a_{n+1}}{a_n} \right| = \left| \frac{(n+1)(x+2)^{n+1}}{3^{n+2}} \cdot \frac{3^{n+1}}{n(x+2)^n} \right|$$

$$= \left(1 + \frac{1}{n} \right) \frac{|x+2|}{3} \rightarrow \frac{|x+2|}{3} \qquad \text{as } n \rightarrow \infty$$

Using the Ratio Test, we see that the series converges if $|x+2|/3 < 1$ and it diverges if $|x+2|/3 > 1$. So it converges if $|x+2| < 3$ and diverges if $|x+2| > 3$. Thus the radius of convergence is $R = 3$.

The inequality $|x+2| < 3$ can be written as $-5 < x < 1$, so we test the series at the endpoints -5 and 1. When $x = -5$, the series is

$$\sum_{n=0}^{\infty} \frac{n(-3)^n}{3^{n+1}} = \frac{1}{3} \sum_{n=0}^{\infty} (-1)^n n$$

which diverges by the Test for Divergence $[(-1)^n n$ doesn't converge to $0]$. When $x = 1$, the series is

$$\sum_{n=0}^{\infty} \frac{n(3)^n}{3^{n+1}} = \frac{1}{3} \sum_{n=0}^{\infty} n$$

which also diverges by the Test for Divergence. Thus the series converges only when $-5 < x < 1$, so the interval of convergence is $(-5, 1)$. ∎

8.5 | EXERCISES

1. What is a power series?

2. (a) What is the radius of convergence of a power series? How do you find it?
   (b) What is the interval of convergence of a power series? How do you find it?

3–22 ■ Find the radius of convergence and interval of convergence of the series.

3. $\displaystyle\sum_{n=1}^{\infty} (-1)^n n x^n$

4. $\displaystyle\sum_{n=1}^{\infty} \frac{(-1)^n x^n}{\sqrt[3]{n}}$

5. $\displaystyle\sum_{n=1}^{\infty} \frac{x^n}{2n-1}$

6. $\displaystyle\sum_{n=1}^{\infty} \frac{(-1)^n x^n}{n^2}$

7. $\displaystyle\sum_{n=0}^{\infty} \frac{x^n}{n!}$

8. $\displaystyle\sum_{n=1}^{\infty} \frac{x^n}{n3^n}$

9. $\displaystyle\sum_{n=1}^{\infty} (-1)^n \frac{n^2 x^n}{2^n}$

10. $\displaystyle\sum_{n=1}^{\infty} n^n x^n$

11. $\displaystyle\sum_{n=2}^{\infty} (-1)^n \frac{x^n}{4^n \ln n}$

12. $\displaystyle\sum_{n=1}^{\infty} \frac{x^n}{5^n n^5}$

13. $\displaystyle\sum_{n=1}^{\infty} \frac{(-3)^n}{n\sqrt{n}} x^n$

14. $\displaystyle\sum_{n=0}^{\infty} (-1)^n \frac{x^{2n+1}}{(2n+1)!}$

15. $\displaystyle\sum_{n=0}^{\infty} \frac{(x-2)^n}{n^2+1}$

16. $\displaystyle\sum_{n=0}^{\infty} (-1)^n \frac{(x-3)^n}{2n+1}$

17. $\displaystyle\sum_{n=1}^{\infty} \frac{n}{b^n} (x-a)^n, \quad b > 0$

18. $\displaystyle\sum_{n=1}^{\infty} \frac{n}{4^n} (x+1)^n$

19. $\displaystyle\sum_{n=1}^{\infty} n!(2x-1)^n$

20. $\displaystyle\sum_{n=1}^{\infty} \frac{(2x-1)^n}{5^n\sqrt{n}}$

21. $\displaystyle\sum_{n=1}^{\infty} \frac{x^n}{1\cdot 3\cdot 5\cdot\,\cdots\,\cdot(2n-1)}$

22. $\displaystyle\sum_{n=1}^{\infty} \frac{n^2 x^n}{2\cdot 4\cdot 6\cdot\,\cdots\,\cdot(2n)}$

23. If $\sum_{n=0}^{\infty} c_n 4^n$ is convergent, does it follow that the following series are convergent?

(a) $\displaystyle\sum_{n=0}^{\infty} c_n(-2)^n$　　(b) $\displaystyle\sum_{n=0}^{\infty} c_n(-4)^n$

24. Suppose that $\sum_{n=0}^{\infty} c_n x^n$ converges when $x = -4$ and diverges when $x = 6$. What can be said about the convergence or divergence of the following series?

(a) $\displaystyle\sum_{n=0}^{\infty} c_n$　　(b) $\displaystyle\sum_{n=0}^{\infty} c_n 8^n$

(c) $\displaystyle\sum_{n=0}^{\infty} c_n(-3)^n$　　(d) $\displaystyle\sum_{n=0}^{\infty} (-1)^n c_n 9^n$

25. If k is a positive integer, find the radius of convergence of the series

$$\sum_{n=0}^{\infty} \frac{(n!)^k}{(kn)!} x^n$$

26. Let p and q be real numbers with $p < q$. Find a power series whose interval of convergence is
(a) (p, q)　　　　(b) $(p, q]$
(c) $[p, q)$　　　　(d) $[p, q]$

27. Is it possible to find a power series whose interval of convergence is $[0, \infty)$? Explain.

28. Graph the first several partial sums $s_n(x)$ of the series $\sum_{n=0}^{\infty} x^n$, together with the sum function $f(x) = 1/(1-x)$, on a common screen. On what interval do these partial sums appear to be converging to $f(x)$?

29. The function J_1 defined by

$$J_1(x) = \sum_{n=0}^{\infty} \frac{(-1)^n x^{2n+1}}{n!(n+1)! 2^{2n+1}}$$

is called the *Bessel function of order 1*.
(a) Find its domain.
(b) Graph the first several partial sums on a common screen.
(c) If your CAS has built-in Bessel functions, graph J_1 on the same screen as the partial sums in part (b) and observe how the partial sums approximate J_1.

30. The function A defined by

$$A(x) = 1 + \frac{x^3}{2\cdot 3} + \frac{x^6}{2\cdot 3\cdot 5\cdot 6} + \frac{x^9}{2\cdot 3\cdot 5\cdot 6\cdot 8\cdot 9} + \cdots$$

is called an *Airy function* after the English mathematician and astronomer Sir George Airy (1801–1892).
(a) Find the domain of the Airy function.
(b) Graph the first several partial sums on a common screen.
(c) If your CAS has built-in Airy functions, graph A on the same screen as the partial sums in part (b) and observe how the partial sums approximate A.

31. A function f is defined by

$$f(x) = 1 + 2x + x^2 + 2x^3 + x^4 + \cdots$$

that is, its coefficients are $c_{2n} = 1$ and $c_{2n+1} = 2$ for all $n \geq 0$. Find the interval of convergence of the series and find an explicit formula for $f(x)$.

32. If $f(x) = \sum_{n=0}^{\infty} c_n x^n$, where $c_{n+4} = c_n$ for all $n \geq 0$, find the interval of convergence of the series and a formula for $f(x)$.

33. Show that if $\lim_{n\to\infty} \sqrt[n]{|c_n|} = c$, where $c \neq 0$, then the radius of convergence of the power series $\sum c_n x^n$ is $R = 1/c$.

34. Suppose that the power series $\sum c_n(x - a)^n$ satisfies $c_n \neq 0$ for all n. Show that if $\lim_{n\to\infty} |c_n/c_{n+1}|$ exists, then it is equal to the radius of convergence of the power series.

35. Suppose the series $\sum c_n x^n$ has radius of convergence 2 and the series $\sum d_n x^n$ has radius of convergence 3. What is the radius of convergence of the series $\sum (c_n + d_n)x^n$?

36. Suppose that the radius of convergence of the power series $\sum c_n x^n$ is R. What is the radius of convergence of the power series $\sum c_n x^{2n}$?

8.6 | REPRESENTING FUNCTIONS AS POWER SERIES

In this section we learn how to represent certain types of functions as sums of power series by manipulating geometric series or by differentiating or integrating such a series. You might wonder why we would ever want to express a known function as a sum of infinitely many terms. This strategy is useful for integrating functions that don't have elementary antiderivatives, for solving differential equations, and for approximating functions by polynomials. (Scientists do this to simplify the expressions they deal with; computer scientists do this to represent functions on calculators and computers.)

We start with an equation that we have seen before:

$$\boxed{\,\mathbf{1}\,}\qquad \frac{1}{1-x} = 1 + x + x^2 + x^3 + \cdots = \sum_{n=0}^{\infty} x^n \qquad |x| < 1$$

We first encountered this equation in Example 5 in Section 8.2, where we obtained it by observing that the series is a geometric series with $a = 1$ and $r = x$. But here our point of view is different. We now regard Equation 1 as expressing the function $f(x) = 1/(1 - x)$ as a sum of a power series.

■ A geometric illustration of Equation 1 is shown in Figure 1. Because the sum of a series is the limit of the sequence of partial sums, we have

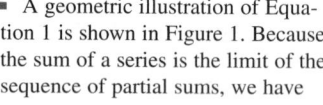

$$\frac{1}{1-x} = \lim_{n \to \infty} s_n(x)$$

where

$$s_n(x) = 1 + x + x^2 + \cdots + x^n$$

is the nth partial sum. Notice that as n increases, $s_n(x)$ becomes a better approximation to $f(x)$ for $-1 < x < 1$.

FIGURE 1

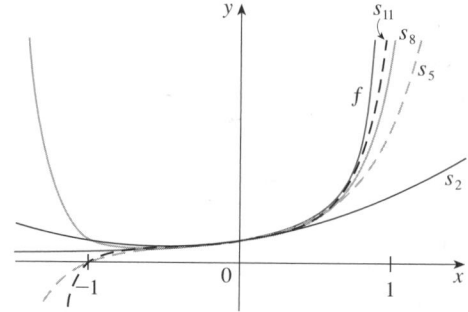

▼ EXAMPLE 1 Express $1/(1 + x^2)$ as the sum of a power series and find the interval of convergence.

SOLUTION Replacing x by $-x^2$ in Equation 1, we have

$$\frac{1}{1+x^2} = \frac{1}{1-(-x^2)} = \sum_{n=0}^{\infty} (-x^2)^n$$

$$= \sum_{n=0}^{\infty} (-1)^n x^{2n} = 1 - x^2 + x^4 - x^6 + x^8 - \cdots$$

Because this is a geometric series, it converges when $|-x^2| < 1$, that is, $x^2 < 1$, or $|x| < 1$. Therefore the interval of convergence is $(-1, 1)$. (Of course, we could have determined the radius of convergence by applying the Ratio Test, but that much work is unnecessary here.) ■

EXAMPLE 2 Find a power series representation for $1/(x + 2)$.

SOLUTION In order to put this function in the form of the left side of Equation 1 we first factor a 2 from the denominator:

$$\frac{1}{2+x} = \frac{1}{2\left(1+\dfrac{x}{2}\right)} = \frac{1}{2\left[1-\left(-\dfrac{x}{2}\right)\right]}$$

$$= \frac{1}{2} \sum_{n=0}^{\infty} \left(-\frac{x}{2}\right)^n = \sum_{n=0}^{\infty} \frac{(-1)^n}{2^{n+1}} x^n$$

This series converges when $\left|-x/2\right| < 1$, that is, $|x| < 2$. So the interval of convergence is $(-2, 2)$. ∎

EXAMPLE 3 Find a power series representation of $x^3/(x + 2)$.

SOLUTION Since this function is just x^3 times the function in Example 2, all we have to do is to multiply that series by x^3:

- It's legitimate to move x^3 across the sigma sign because it doesn't depend on n. [Use Theorem 8.2.8(i) with $c = x^3$.]

$$\frac{x^3}{x+2} = x^3 \cdot \frac{1}{x+2} = x^3 \sum_{n=0}^{\infty} \frac{(-1)^n}{2^{n+1}} x^n = \sum_{n=0}^{\infty} \frac{(-1)^n}{2^{n+1}} x^{n+3}$$

$$= \tfrac{1}{2} x^3 - \tfrac{1}{4} x^4 + \tfrac{1}{8} x^5 - \tfrac{1}{16} x^6 + \cdots$$

Another way of writing this series is as follows:

$$\frac{x^3}{x+2} = \sum_{n=3}^{\infty} \frac{(-1)^{n-1}}{2^{n-2}} x^n$$

As in Example 2, the interval of convergence is $(-2, 2)$. ∎

DIFFERENTIATION AND INTEGRATION OF POWER SERIES

The sum of a power series is a function $f(x) = \sum_{n=0}^{\infty} c_n(x - a)^n$ whose domain is the interval of convergence of the series. We would like to be able to differentiate and integrate such functions, and the following theorem (which we won't prove) says that we can do so by differentiating or integrating each individual term in the series, just as we would for a polynomial. This is called **term-by-term differentiation and integration**.

---

2 THEOREM If the power series $\sum c_n(x - a)^n$ has radius of convergence $R > 0$, then the function f defined by

$$f(x) = c_0 + c_1(x - a) + c_2(x - a)^2 + \cdots = \sum_{n=0}^{\infty} c_n(x - a)^n$$

is differentiable (and therefore continuous) on the interval $(a - R, a + R)$ and

(i) $f'(x) = c_1 + 2c_2(x - a) + 3c_3(x - a)^2 + \cdots = \sum_{n=1}^{\infty} nc_n(x - a)^{n-1}$

- In part (ii), $\int c_0 \, dx = c_0 x + C_1$ is written as $c_0(x - a) + C$, where $C = C_1 + ac_0$, so all the terms of the series have the same form.

(ii) $\displaystyle \int f(x) \, dx = C + c_0(x - a) + c_1 \frac{(x - a)^2}{2} + c_2 \frac{(x - a)^3}{3} + \cdots$

$$= C + \sum_{n=0}^{\infty} c_n \frac{(x - a)^{n+1}}{n + 1}$$

The radii of convergence of the power series in Equations (i) and (ii) are both R.

NOTE 1 Equations (i) and (ii) in Theorem 2 can be rewritten in the form

$$\text{(iii)}\quad \frac{d}{dx}\left[\sum_{n=0}^{\infty} c_n(x-a)^n\right] = \sum_{n=0}^{\infty} \frac{d}{dx}\left[c_n(x-a)^n\right]$$

$$\text{(iv)}\quad \int\left[\sum_{n=0}^{\infty} c_n(x-a)^n\right]dx = \sum_{n=0}^{\infty} \int c_n(x-a)^n\,dx$$

▪ **www.stewartcalculus.com**
The idea of differentiating a power series term by term is the basis for a powerful method for solving differential equations. Click on *Additional Topics* and then on *Using Series to Solve Differential Equations*.

We know that, for finite sums, the derivative of a sum is the sum of the derivatives and the integral of a sum is the sum of the integrals. Equations (iii) and (iv) assert that the same is true for infinite sums, provided we are dealing with *power series.* (For other types of series of functions the situation is not as simple; see Exercise 38.)

NOTE 2 Although Theorem 2 says that the radius of convergence remains the same when a power series is differentiated or integrated, this does not mean that the *interval* of convergence remains the same. It may happen that the original series converges at an endpoint, whereas the differentiated series diverges there. (See Exercise 39.)

EXAMPLE 4 In Example 3 in Section 8.5 we saw that the Bessel function

$$J_0(x) = \sum_{n=0}^{\infty} \frac{(-1)^n x^{2n}}{2^{2n}(n!)^2}$$

is defined for all x. Thus by Theorem 2, J_0 is differentiable for all x and its derivative is found by term-by-term differentiation as follows:

$$J_0'(x) = \sum_{n=0}^{\infty} \frac{d}{dx}\frac{(-1)^n x^{2n}}{2^{2n}(n!)^2} = \sum_{n=1}^{\infty} \frac{(-1)^n 2n x^{2n-1}}{2^{2n}(n!)^2} \qquad ∎$$

V EXAMPLE 5 Express $1/(1-x)^2$ as a power series by differentiating Equation 1. What is the radius of convergence?

SOLUTION Differentiating each side of the equation

$$\frac{1}{1-x} = 1 + x + x^2 + x^3 + \cdots = \sum_{n=0}^{\infty} x^n$$

we get

$$\frac{1}{(1-x)^2} = 1 + 2x + 3x^2 + \cdots = \sum_{n=1}^{\infty} nx^{n-1}$$

If we wish, we can replace n by $n+1$ and write the answer as

$$\frac{1}{(1-x)^2} = \sum_{n=0}^{\infty} (n+1)x^n$$

According to Theorem 2, the radius of convergence of the differentiated series is the same as the radius of convergence of the original series, namely, $R = 1$. $\qquad ∎$

EXAMPLE 6 Find a power series representation for $\ln(1+x)$ and its radius of convergence.

SOLUTION We notice that the derivative of this function is $1/(1+x)$. From Equation 1 we have

$$\frac{1}{1+x} = \frac{1}{1-(-x)} = 1 - x + x^2 - x^3 + \cdots \qquad |x| < 1$$

Integrating both sides of this equation, we get

$$\ln(1 + x) = \int \frac{1}{1 + x}\,dx = \int (1 - x + x^2 - x^3 + \cdots)\,dx$$

$$= x - \frac{x^2}{2} + \frac{x^3}{3} - \frac{x^4}{4} + \cdots + C$$

$$= \sum_{n=1}^{\infty} (-1)^{n-1} \frac{x^n}{n} + C \qquad |x| < 1$$

To determine the value of C we put $x = 0$ in this equation and obtain $\ln(1 + 0) = C$. Thus $C = 0$ and

$$\ln(1 + x) = x - \frac{x^2}{2} + \frac{x^3}{3} - \frac{x^4}{4} + \cdots = \sum_{n=1}^{\infty} (-1)^{n-1} \frac{x^n}{n} \qquad |x| < 1$$

The radius of convergence is the same as for the original series: $R = 1$. ∎

V EXAMPLE 7 Find a power series representation for $f(x) = \tan^{-1}x$.

SOLUTION We observe that $f'(x) = 1/(1 + x^2)$ and find the required series by integrating the power series for $1/(1 + x^2)$ found in Example 1.

$$\tan^{-1}x = \int \frac{1}{1 + x^2}\,dx = \int (1 - x^2 + x^4 - x^6 + \cdots)\,dx$$

$$= C + x - \frac{x^3}{3} + \frac{x^5}{5} - \frac{x^7}{7} + \cdots$$

To find C we put $x = 0$ and obtain $C = \tan^{-1} 0 = 0$. Therefore

$$\tan^{-1}x = x - \frac{x^3}{3} + \frac{x^5}{5} - \frac{x^7}{7} + \cdots = \sum_{n=0}^{\infty} (-1)^n \frac{x^{2n+1}}{2n + 1}$$

Since the radius of convergence of the series for $1/(1 + x^2)$ is 1, the radius of convergence of this series for $\tan^{-1}x$ is also 1. ∎

■ The power series for $\tan^{-1}x$ obtained in Example 7 is called *Gregory's series* after the Scottish mathematician James Gregory (1638–1675), who had anticipated some of Newton's discoveries. We have shown that Gregory's series is valid when $-1 < x < 1$, but it turns out (although it isn't easy to prove) that it is also valid when $x = \pm 1$. Notice that when $x = 1$ the series becomes

$$\frac{\pi}{4} = 1 - \frac{1}{3} + \frac{1}{5} - \frac{1}{7} + \cdots$$

This beautiful result is known as the Leibniz formula for π.

EXAMPLE 8 Evaluate $\int [1/(1 + x^7)]\,dx$ as a power series.

SOLUTION The first step is to express the integrand, $1/(1 + x^7)$, as the sum of a power series. As in Example 1, we start with Equation 1 and replace x by $-x^7$:

$$\frac{1}{1 + x^7} = \frac{1}{1 - (-x^7)} = \sum_{n=0}^{\infty} (-x^7)^n$$

$$= \sum_{n=0}^{\infty} (-1)^n x^{7n} = 1 - x^7 + x^{14} - \cdots$$

■ This example demonstrates one way in which power series representations are useful. Integrating $1/(1 + x^7)$ by hand is incredibly difficult. Different computer algebra systems return different forms of the answer, but they are all extremely complicated. (If you have a CAS, try it yourself.) The infinite series answer that we obtain in Example 8 is actually much easier to deal with than the finite answer provided by a CAS.

Now we integrate term by term:

$$\int \frac{1}{1 + x^7}\, dx = \int \sum_{n=0}^{\infty} (-1)^n x^{7n}\, dx = C + \sum_{n=0}^{\infty} (-1)^n \frac{x^{7n+1}}{7n + 1}$$

$$= C + x - \frac{x^8}{8} + \frac{x^{15}}{15} - \frac{x^{22}}{22} + \cdots$$

This series converges for $|-x^7| < 1$, that is, for $|x| < 1$. ∎

8.6 | EXERCISES

1. If the radius of convergence of the power series $\sum_{n=0}^{\infty} c_n x^n$ is 10, what is the radius of convergence of the series $\sum_{n=1}^{\infty} n c_n x^{n-1}$? Why?

2. Suppose you know that the series $\sum_{n=0}^{\infty} b_n x^n$ converges for $|x| < 2$. What can you say about the following series? Why?

$$\sum_{n=0}^{\infty} \frac{b_n}{n + 1} x^{n+1}$$

3–10 ■ Find a power series representation for the function and determine the interval of convergence.

3. $f(x) = \dfrac{1}{1 + x}$

4. $f(x) = \dfrac{5}{1 - 4x^2}$

5. $f(x) = \dfrac{2}{3 - x}$

6. $f(x) = \dfrac{1}{x + 10}$

7. $f(x) = \dfrac{x}{9 + x^2}$

8. $f(x) = \dfrac{x}{2x^2 + 1}$

9. $f(x) = \dfrac{1 + x}{1 - x}$

10. $f(x) = \dfrac{x^2}{a^3 - x^3}$

11–12 ■ Express the function as the sum of a power series by first using partial fractions. Find the interval of convergence.

11. $f(x) = \dfrac{3}{x^2 - x - 2}$

12. $f(x) = \dfrac{x + 2}{2x^2 - x - 1}$

13. (a) Use differentiation to find a power series representation for

$$f(x) = \frac{1}{(1 + x)^2}$$

What is the radius of convergence?

(b) Use part (a) to find a power series for

$$f(x) = \frac{1}{(1 + x)^3}$$

(c) Use part (b) to find a power series for

$$f(x) = \frac{x^2}{(1 + x)^3}$$

14. (a) Use Equation 1 to find a power series representation for $f(x) = \ln(1 - x)$. What is the radius of convergence?
(b) Use part (a) to find a power series for $f(x) = x \ln(1 - x)$.
(c) By putting $x = \frac{1}{2}$ in your result from part (a), express $\ln 2$ as the sum of an infinite series.

15–20 ■ Find a power series representation for the function and determine the radius of convergence.

15. $f(x) = \ln(5 - x)$

16. $f(x) = x^2 \tan^{-1}(x^3)$

17. $f(x) = \dfrac{x}{(1 + 4x)^2}$

18. $f(x) = \left(\dfrac{x}{2 - x}\right)^3$

19. $f(x) = \dfrac{1 + x}{(1 - x)^2}$

20. $f(x) = \dfrac{x^2 + x}{(1 - x)^3}$

21–24 ■ Find a power series representation for f, and graph f and several partial sums $s_n(x)$ on the same screen. What happens as n increases?

21. $f(x) = \dfrac{x}{x^2 + 16}$

22. $f(x) = \ln(x^2 + 4)$

23. $f(x) = \ln\left(\dfrac{1 + x}{1 - x}\right)$

24. $f(x) = \tan^{-1}(2x)$

25–28 ▪ Evaluate the indefinite integral as a power series. What is the radius of convergence?

25. $\displaystyle\int \frac{t}{1 - t^8}\, dt$

26. $\displaystyle\int \frac{t}{1 + t^3}\, dt$

27. $\displaystyle\int x^2 \ln(1 + x)\, dx$

28. $\displaystyle\int \frac{\tan^{-1}x}{x}\, dx$

29–32 ▪ Use a power series to approximate the definite integral to six decimal places.

29. $\displaystyle\int_0^{0.2} \frac{1}{1 + x^5}\, dx$

30. $\displaystyle\int_0^{0.4} \ln(1 + x^4)\, dx$

31. $\displaystyle\int_0^{0.1} x \arctan(3x)\, dx$

32. $\displaystyle\int_0^{0.3} \frac{x^2}{1 + x^4}\, dx$

---

33. Use the result of Example 7 to compute arctan 0.2 correct to five decimal places.

34. Show that the function

$$f(x) = \sum_{n=0}^{\infty} \frac{(-1)^n x^{2n}}{(2n)!}$$

is a solution of the differential equation

$$f''(x) + f(x) = 0$$

35. (a) Show that J_0 (the Bessel function of order 0 given in Example 4) satisfies the differential equation

$$x^2 J_0''(x) + x J_0'(x) + x^2 J_0(x) = 0$$

(b) Evaluate $\int_0^1 J_0(x)\, dx$ correct to three decimal places.

36. The Bessel function of order 1 is defined by

$$J_1(x) = \sum_{n=0}^{\infty} \frac{(-1)^n x^{2n+1}}{n!\,(n+1)!\,2^{2n+1}}$$

(a) Show that J_1 satisfies the differential equation

$$x^2 J_1''(x) + x J_1'(x) + (x^2 - 1) J_1(x) = 0$$

(b) Show that $J_0'(x) = -J_1(x)$.

37. (a) Show that the function

$$f(x) = \sum_{n=0}^{\infty} \frac{x^n}{n!}$$

is a solution of the differential equation

$$f'(x) = f(x)$$

(b) Show that $f(x) = e^x$.

38. Let $f_n(x) = (\sin nx)/n^2$. Show that the series $\Sigma f_n(x)$ converges for all values of x but the series of derivatives $\Sigma f_n'(x)$ diverges when $x = 2n\pi$, n an integer. For what values of x does the series $\Sigma f_n''(x)$ converge?

39. Let

$$f(x) = \sum_{n=1}^{\infty} \frac{x^n}{n^2}$$

Find the intervals of convergence for f, f', and f''.

40. (a) Starting with the geometric series $\Sigma_{n=0}^{\infty} x^n$, find the sum of the series

$$\sum_{n=1}^{\infty} n x^{n-1} \qquad |x| < 1$$

(b) Find the sum of each of the following series.

(i) $\displaystyle\sum_{n=1}^{\infty} n x^n, \quad |x| < 1$ (ii) $\displaystyle\sum_{n=1}^{\infty} \frac{n}{2^n}$

(c) Find the sum of each of the following series.

(i) $\displaystyle\sum_{n=2}^{\infty} n(n-1) x^n, \quad |x| < 1$

(ii) $\displaystyle\sum_{n=2}^{\infty} \frac{n^2 - n}{2^n}$ (iii) $\displaystyle\sum_{n=1}^{\infty} \frac{n^2}{2^n}$

41. Use the power series for $\tan^{-1}x$ to prove the following expression for π as the sum of an infinite series:

$$\pi = 2\sqrt{3} \sum_{n=0}^{\infty} \frac{(-1)^n}{(2n+1)3^n}$$

42. Find the sum of the series

$$\sum_{n=1}^{\infty} \frac{4^n}{n 5^n}$$

8.7 | TAYLOR AND MACLAURIN SERIES

In the preceding section we were able to find power series representations for a certain restricted class of functions. Here we investigate more general problems: Which functions have power series representations? How can we find such representations?

We start by supposing that f is any function that can be represented by a power series:

$$\boxed{1} \quad f(x) = c_0 + c_1(x - a) + c_2(x - a)^2 + c_3(x - a)^3 + c_4(x - a)^4 + \cdots$$
$$|x - a| < R$$

Let's try to determine what the coefficients c_n must be in terms of f. To begin, notice that if we put $x = a$ in Equation 1, then all terms after the first one are 0 and we get

$$f(a) = c_0$$

By Theorem 8.6.2, we can differentiate the series in Equation 1 term by term:

$$\boxed{2} \quad f'(x) = c_1 + 2c_2(x - a) + 3c_3(x - a)^2 + 4c_4(x - a)^3 + \cdots \qquad |x - a| < R$$

and substitution of $x = a$ in Equation 2 gives

$$f'(a) = c_1$$

Now we differentiate both sides of Equation 2 and obtain

$$\boxed{3} \quad f''(x) = 2c_2 + 2 \cdot 3c_3(x - a) + 3 \cdot 4c_4(x - a)^2 + \cdots \qquad |x - a| < R$$

Again we put $x = a$ in Equation 3. The result is

$$f''(a) = 2c_2$$

Let's apply the procedure one more time. Differentiation of the series in Equation 3 gives

$$\boxed{4} \quad f'''(x) = 2 \cdot 3c_3 + 2 \cdot 3 \cdot 4c_4(x - a) + 3 \cdot 4 \cdot 5c_5(x - a)^2 + \cdots \qquad |x - a| < R$$

and substitution of $x = a$ in Equation 4 gives

$$f'''(a) = 2 \cdot 3c_3 = 3!c_3$$

By now you can see the pattern. If we continue to differentiate and substitute $x = a$, we obtain

$$f^{(n)}(a) = 2 \cdot 3 \cdot 4 \cdot \cdots \cdot nc_n = n!c_n$$

Solving this equation for the nth coefficient c_n, we get

$$c_n = \frac{f^{(n)}(a)}{n!}$$

This formula remains valid even for $n = 0$ if we adopt the conventions that $0! = 1$ and $f^{(0)} = f$. Thus we have proved the following theorem.

---

5 THEOREM If f has a power series representation (expansion) at a, that is, if

$$f(x) = \sum_{n=0}^{\infty} c_n(x - a)^n \qquad |x - a| < R$$

then its coefficients are given by the formula

$$c_n = \frac{f^{(n)}(a)}{n!}$$

Substituting this formula for c_n back into the series, we see that *if f has a power series expansion at a, then it must be of the following form.*

---

6 $f(x) = \sum_{n=0}^{\infty} \frac{f^{(n)}(a)}{n!} (x - a)^n$

$$= f(a) + \frac{f'(a)}{1!} (x - a) + \frac{f''(a)}{2!} (x - a)^2 + \frac{f'''(a)}{3!} (x - a)^3 + \cdots$$

The series in Equation 6 is called the **Taylor series of the function f at a** (or **about a** or **centered at a**). For the special case $a = 0$ the Taylor series becomes

---

7 $f(x) = \sum_{n=0}^{\infty} \frac{f^{(n)}(0)}{n!} x^n = f(0) + \frac{f'(0)}{1!} x + \frac{f''(0)}{2!} x^2 + \cdots$

This case arises frequently enough that it is given the special name **Maclaurin series**.

NOTE We have shown that *if f can be represented as a power series about a, then f is equal to the sum of its Taylor series.* But there exist functions that are not equal to the sum of their Taylor series. An example of such a function is given in Exercise 70.

■ The Taylor series is named after the English mathematician Brook Taylor (1685–1731) and the Maclaurin series is named in honor of the Scottish mathematician Colin Maclaurin (1698–1746) despite the fact that the Maclaurin series is really just a special case of the Taylor series. But the idea of representing particular functions as sums of power series goes back to Newton, and the general Taylor series was known to the Scottish mathematician James Gregory in 1668 and to the Swiss mathematician John Bernoulli in the 1690s. Taylor was apparently unaware of the work of Gregory and Bernoulli when he published his discoveries on series in 1715 in his book *Methodus incrementorum directa et inversa*. Maclaurin series are named after Colin Maclaurin because he popularized them in his calculus textbook *Treatise of Fluxions* published in 1742.

V EXAMPLE 1 Find the Maclaurin series of the function $f(x) = e^x$ and its radius of convergence.

SOLUTION If $f(x) = e^x$, then $f^{(n)}(x) = e^x$, so $f^{(n)}(0) = e^0 = 1$ for all n. Therefore the Taylor series for f at 0 (that is, the Maclaurin series) is

$$\sum_{n=0}^{\infty} \frac{f^{(n)}(0)}{n!} x^n = \sum_{n=0}^{\infty} \frac{x^n}{n!} = 1 + \frac{x}{1!} + \frac{x^2}{2!} + \frac{x^3}{3!} + \cdots$$

To find the radius of convergence we let $a_n = x^n/n!$. Then

$$\left| \frac{a_{n+1}}{a_n} \right| = \left| \frac{x^{n+1}}{(n + 1)!} \cdot \frac{n!}{x^n} \right| = \frac{|x|}{n + 1} \rightarrow 0 < 1$$

so, by the Ratio Test, the series converges for all x and the radius of convergence is $R = \infty$. ∎

The conclusion we can draw from Theorem 5 and Example 1 is that *if e^x has a power series expansion at 0, then*

$$e^x = \sum_{n=0}^{\infty} \frac{x^n}{n!}$$

So how can we determine whether e^x *does* have a power series representation?

Let's investigate the more general question: Under what circumstances is a function equal to the sum of its Taylor series? In other words, if f has derivatives of all orders, when is it true that

$$f(x) = \sum_{n=0}^{\infty} \frac{f^{(n)}(a)}{n!}(x-a)^n$$

As with any convergent series, this means that $f(x)$ is the limit of the sequence of partial sums. In the case of the Taylor series, the partial sums are

$$T_n(x) = \sum_{i=0}^{n} \frac{f^{(i)}(a)}{i!}(x-a)^i$$

$$= f(a) + \frac{f'(a)}{1!}(x-a) + \frac{f''(a)}{2!}(x-a)^2 + \cdots + \frac{f^{(n)}(a)}{n!}(x-a)^n$$

Notice that T_n is a polynomial of degree n called the **nth-degree Taylor polynomial of f at a.** For instance, for the exponential function $f(x) = e^x$, the result of Example 1 shows that the Taylor polynomials at 0 (or Maclaurin polynomials) with $n = 1$, 2, and 3 are

$$T_1(x) = 1 + x \qquad T_2(x) = 1 + x + \frac{x^2}{2!} \qquad T_3(x) = 1 + x + \frac{x^2}{2!} + \frac{x^3}{3!}$$

The graphs of the exponential function and these three Taylor polynomials are drawn in Figure 1.

In general, $f(x)$ is the sum of its Taylor series if

$$f(x) = \lim_{n \to \infty} T_n(x)$$

If we let

$$R_n(x) = f(x) - T_n(x) \qquad \text{so that} \qquad f(x) = T_n(x) + R_n(x)$$

then $R_n(x)$ is called the **remainder** of the Taylor series. If we can somehow show that $\lim_{n \to \infty} R_n(x) = 0$, then it follows that

$$\lim_{n \to \infty} T_n(x) = \lim_{n \to \infty} [f(x) - R_n(x)] = f(x) - \lim_{n \to \infty} R_n(x) = f(x)$$

We have therefore proved the following.

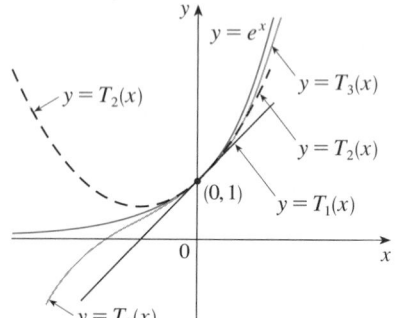

FIGURE 1

∎ As n increases, $T_n(x)$ appears to approach e^x in Figure 1. This suggests that e^x is equal to the sum of its Taylor series.

---

8 **THEOREM** If $f(x) = T_n(x) + R_n(x)$, where T_n is the nth-degree Taylor polynomial of f at a and

$$\lim_{n \to \infty} R_n(x) = 0$$

for $|x - a| < R$, then f is equal to the sum of its Taylor series on the interval $|x - a| < R$.

In trying to show that $\lim_{n \to \infty} R_n(x) = 0$ for a specific function f, we usually use the expression in the next theorem.

---

9 TAYLOR'S FORMULA If f has $n + 1$ derivatives in an interval I that contains the number a, then for x in I there is a number z strictly between x and a such that the remainder term in the Taylor series can be expressed as

$$R_n(x) = \frac{f^{(n+1)}(z)}{(n+1)!} (x - a)^{n+1}$$

NOTE 1 For the special case $n = 0$, if we put $x = b$ and $z = c$ in Taylor's Formula, we get $f(b) = f(a) + f'(c)(b - a)$, which is the Mean Value Theorem. In fact, Theorem 9 can be proved by a method similar to the proof of the Mean Value Theorem. The proof is given at the end of this section.

NOTE 2 Notice that the remainder term

$$\boxed{10} \qquad\qquad R_n(x) = \frac{f^{(n+1)}(z)}{(n+1)!} (x - a)^{n+1}$$

is very similar to the terms in the Taylor series except that $f^{(n+1)}$ is evaluated at z instead of at a. All we can say about the number z is that it lies somewhere between x and a. The expression for $R_n(x)$ in Equation 10 is known as **Lagrange's form of the remainder term**.

NOTE 3 In Section 8.8 we will explore the use of Taylor's Formula in approximating functions. Our immediate use of it is in conjunction with Theorem 8.

In applying Theorems 8 and 9 it is often helpful to make use of the following fact.

$$\boxed{11} \qquad\qquad \lim_{n \to \infty} \frac{x^n}{n!} = 0 \qquad \text{for every real number } x$$

This is true because we know from Example 1 that the series $\sum x^n/n!$ converges for all x and so its nth term approaches 0.

EXAMPLE 2 Prove that e^x is equal to the sum of its Taylor series.

SOLUTION If $f(x) = e^x$, then $f^{(n+1)}(x) = e^x$, so the remainder term in Taylor's Formula is

$$R_n(x) = \frac{e^z}{(n+1)!} x^{n+1}$$

where z lies between 0 and x. (Note, however, that z depends on n.) If $x > 0$, then $0 < z < x$, so $e^z < e^x$. Therefore

$$0 < R_n(x) = \frac{e^z}{(n+1)!} x^{n+1} < e^x \frac{x^{n+1}}{(n+1)!} \to 0$$

by Equation 11, so $R_n(x) \to 0$ as $n \to \infty$ by the Squeeze Theorem. If $x < 0$, then

$x < z < 0$, so $e^z < e^0 = 1$ and

$$|R_n(x)| < \frac{|x|^{n+1}}{(n+1)!} \to 0$$

Again $R_n(x) \to 0$. Thus, by Theorem 8, e^x is equal to the sum of its Taylor series, that is,

$$\boxed{12} \qquad \boxed{e^x = \sum_{n=0}^{\infty} \frac{x^n}{n!} \qquad \text{for all } x}$$ ∎

In particular, if we put $x = 1$ in Equation 12, we obtain the following expression for the number e as a sum of an infinite series:

■ In 1748 Leonard Euler used Equation 13 to find the value of e correct to 23 digits. In 2010 Shigeru Kondo, again using the series in $\boxed{13}$, computed e to more than a trillion decimal places!

$$\boxed{13} \qquad \boxed{e = \sum_{n=0}^{\infty} \frac{1}{n!} = 1 + \frac{1}{1!} + \frac{1}{2!} + \frac{1}{3!} + \cdots}$$

EXAMPLE 3 Find the Taylor series for $f(x) = e^x$ at $a = 2$.

SOLUTION We have $f^{(n)}(2) = e^2$ and so, putting $a = 2$ in the definition of a Taylor series $\boxed{6}$, we get

$$\sum_{n=0}^{\infty} \frac{f^{(n)}(2)}{n!} (x-2)^n = \sum_{n=0}^{\infty} \frac{e^2}{n!} (x-2)^n$$

Again it can be verified, as in Example 1, that the radius of convergence is $R = \infty$. As in Example 2 we can verify that $\lim_{n \to \infty} R_n(x) = 0$, so

$$\boxed{14} \qquad e^x = \sum_{n=0}^{\infty} \frac{e^2}{n!} (x-2)^n \qquad \text{for all } x$$ ∎

We have two power series expansions for e^x, the Maclaurin series in Equation 12 and the Taylor series in Equation 14. The first is better if we are interested in values of x near 0 and the second is better if x is near 2.

EXAMPLE 4 Find the Maclaurin series for $\sin x$ and prove that it represents $\sin x$ for all x.

SOLUTION We arrange our computation in two columns as follows:

$$f(x) = \sin x \qquad\qquad f(0) = 0$$
$$f'(x) = \cos x \qquad\qquad f'(0) = 1$$
$$f''(x) = -\sin x \qquad\qquad f''(0) = 0$$
$$f'''(x) = -\cos x \qquad\qquad f'''(0) = -1$$
$$f^{(4)}(x) = \sin x \qquad\qquad f^{(4)}(0) = 0$$

■ Figure 2 shows the graph of $\sin x$ together with its Taylor (or Maclaurin) polynomials

$$T_1(x) = x$$

$$T_3(x) = x - \frac{x^3}{3!}$$

$$T_5(x) = x - \frac{x^3}{3!} + \frac{x^5}{5!}$$

Notice that, as n increases, $T_n(x)$ becomes a better approximation to $\sin x$.

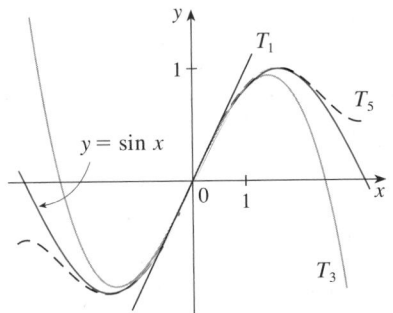

FIGURE 2

Since the derivatives repeat in a cycle of four, we can write the Maclaurin series as follows:

$$f(0) + \frac{f'(0)}{1!}x + \frac{f''(0)}{2!}x^2 + \frac{f'''(0)}{3!}x^3 + \cdots$$

$$= x - \frac{x^3}{3!} + \frac{x^5}{5!} - \frac{x^7}{7!} + \cdots = \sum_{n=0}^{\infty}(-1)^n \frac{x^{2n+1}}{(2n+1)!}$$

Using the remainder term $\boxed{10}$ with $a = 0$, we have

$$R_n(x) = \frac{f^{(n+1)}(z)}{(n+1)!}x^{n+1}$$

where $f(x) = \sin x$ and z lies between 0 and x. But $f^{(n+1)}(z)$ is $\pm\sin z$ or $\pm\cos z$. In any case, $|f^{(n+1)}(z)| \le 1$ and so

$$\boxed{15} \qquad 0 \le |R_n(x)| = \frac{|f^{(n+1)}(z)|}{(n+1)!}|x^{n+1}| \le \frac{|x|^{n+1}}{(n+1)!}$$

By Equation 11 the right side of this inequality approaches 0 as $n \to \infty$, so $|R_n(x)| \to 0$ by the Squeeze Theorem. It follows that $R_n(x) \to 0$ as $n \to \infty$, so $\sin x$ is equal to the sum of its Maclaurin series by Theorem 8. ■

We state the result of Example 4 for future reference.

$$\boxed{16} \qquad \sin x = x - \frac{x^3}{3!} + \frac{x^5}{5!} - \frac{x^7}{7!} + \cdots$$

$$= \sum_{n=0}^{\infty}(-1)^n \frac{x^{2n+1}}{(2n+1)!} \qquad \text{for all } x$$

EXAMPLE 5 Find the Maclaurin series for $\cos x$.

SOLUTION We could proceed directly as in Example 4 but it's easier to differentiate the Maclaurin series for $\sin x$ given by Equation 16:

$$\cos x = \frac{d}{dx}(\sin x) = \frac{d}{dx}\left(x - \frac{x^3}{3!} + \frac{x^5}{5!} - \frac{x^7}{7!} + \cdots\right)$$

$$= 1 - \frac{3x^2}{3!} + \frac{5x^4}{5!} - \frac{7x^6}{7!} + \cdots = 1 - \frac{x^2}{2!} + \frac{x^4}{4!} - \frac{x^6}{6!} + \cdots$$

■ The Maclaurin series for e^x, $\sin x$, and $\cos x$ that we found in Examples 2, 4, and 5 were discovered, using different methods, by Newton. These equations are remarkable because they say we know everything about each of these functions if we know all its derivatives at the single number 0.

Since the Maclaurin series for $\sin x$ converges for all x, Theorem 8.6.2 tells us that the differentiated series for $\cos x$ also converges for all x. Thus

$$\boxed{17} \qquad \cos x = 1 - \frac{x^2}{2!} + \frac{x^4}{4!} - \frac{x^6}{6!} + \cdots$$

$$= \sum_{n=0}^{\infty}(-1)^n \frac{x^{2n}}{(2n)!} \qquad \text{for all } x$$

■

EXAMPLE 6 Find the Maclaurin series for the function $f(x) = x \cos x$.

SOLUTION Instead of computing derivatives and substituting in Equation 7, it's easier to multiply the series for $\cos x$ (Equation 17) by x:

$$x \cos x = x \sum_{n=0}^{\infty} (-1)^n \frac{x^{2n}}{(2n)!} = \sum_{n=0}^{\infty} (-1)^n \frac{x^{2n+1}}{(2n)!} \qquad \blacksquare$$

■ **www.stewartcalculus.com**
See Additional Example A.

The power series that we obtained by indirect methods in Examples 5 and 6 and in Section 8.6 are indeed the Taylor or Maclaurin series of the given functions because Theorem 5 asserts that, no matter how we obtain a power series representation $f(x) = \sum c_n (x - a)^n$, it is always true that $c_n = f^{(n)}(a)/n!$. In other words, the coefficients are uniquely determined.

EXAMPLE 7 Find the Maclaurin series for $f(x) = (1 + x)^k$, where k is any real number.

SOLUTION Arranging our work in columns, we have

$$f(x) = (1 + x)^k \qquad\qquad\qquad f(0) = 1$$

$$f'(x) = k(1 + x)^{k-1} \qquad\qquad f'(0) = k$$

$$f''(x) = k(k - 1)(1 + x)^{k-2} \qquad f''(0) = k(k - 1)$$

$$f'''(x) = k(k - 1)(k - 2)(1 + x)^{k-3} \qquad f'''(0) = k(k - 1)(k - 2)$$

$$\vdots \qquad\qquad\qquad\qquad\qquad \vdots$$

$$f^{(n)}(x) = k(k - 1) \cdots (k - n + 1)(1 + x)^{k-n} \qquad f^{(n)}(0) = k(k - 1) \cdots (k - n + 1)$$

Therefore the Maclaurin series of $f(x) = (1 + x)^k$ is

$$\sum_{n=0}^{\infty} \frac{f^{(n)}(0)}{n!} x^n = \sum_{n=0}^{\infty} \frac{k(k - 1) \cdots (k - n + 1)}{n!} x^n$$

This series is called the **binomial series**. Notice that if k is a nonnegative integer, then the terms are eventually 0 and so the series is finite. For other values of k none of the terms is 0 and so we can try the Ratio Test. If the nth term is a_n, then

$$\left| \frac{a_{n+1}}{a_n} \right| = \left| \frac{k(k - 1) \cdots (k - n + 1)(k - n)x^{n+1}}{(n + 1)!} \cdot \frac{n!}{k(k - 1) \cdots (k - n + 1)x^n} \right|$$

$$= \frac{|k - n|}{n + 1} |x| = \frac{\left| 1 - \dfrac{k}{n} \right|}{1 + \dfrac{1}{n}} |x| \rightarrow |x| \qquad \text{as } n \rightarrow \infty$$

Thus by the Ratio Test the binomial series converges if $|x| < 1$ and diverges if $|x| > 1$.

$\blacksquare$

The traditional notation for the coefficients in the binomial series is

$$\binom{k}{n} = \frac{k(k-1)(k-2)\cdots(k-n+1)}{n!}$$

and these numbers are called the **binomial coefficients**.

The following theorem states that $(1+x)^k$ is equal to the sum of its Maclaurin series. It is possible to prove this by showing that the remainder term $R_n(x)$ approaches 0, but that turns out to be quite difficult. The proof outlined in Exercise 69 is much easier.

---

18 THE BINOMIAL SERIES If k is any real number and $|x| < 1$, then

$$(1+x)^k = \sum_{n=0}^{\infty} \binom{k}{n} x^n = 1 + kx + \frac{k(k-1)}{2!}x^2 + \frac{k(k-1)(k-2)}{3!}x^3 + \cdots$$

Although the binomial series always converges when $|x| < 1$, the question of whether or not it converges at the endpoints, ± 1, depends on the value of k. It turns out that the series converges at 1 if $-1 < k \leq 0$ and at both endpoints if $k \geq 0$. Notice that if k is a positive integer and $n > k$, then the expression for $\binom{k}{n}$ contains a factor $(k-k)$, so $\binom{k}{n} = 0$ for $n > k$. This means that the series terminates and reduces to the ordinary Binomial Theorem when k is a positive integer. (See Reference Page 1.)

■ www.stewartcalculus.com
See Additional Example B.

V EXAMPLE 8 Find the Maclaurin series for the function $f(x) = \dfrac{1}{\sqrt{4-x}}$ and its radius of convergence.

SOLUTION We write $f(x)$ in a form where we can use the binomial series:

$$\frac{1}{\sqrt{4-x}} = \frac{1}{\sqrt{4\left(1-\dfrac{x}{4}\right)}} = \frac{1}{2\sqrt{1-\dfrac{x}{4}}} = \frac{1}{2}\left(1-\frac{x}{4}\right)^{-1/2}$$

Using the binomial series with $k = -\frac{1}{2}$ and with x replaced by $-x/4$, we have

$$\frac{1}{\sqrt{4-x}} = \frac{1}{2}\left(1-\frac{x}{4}\right)^{-1/2} = \frac{1}{2}\sum_{n=0}^{\infty} \binom{-\frac{1}{2}}{n}\left(-\frac{x}{4}\right)^n$$

$$= \frac{1}{2}\left[1 + \left(-\frac{1}{2}\right)\left(-\frac{x}{4}\right) + \frac{\left(-\frac{1}{2}\right)\left(-\frac{3}{2}\right)}{2!}\left(-\frac{x}{4}\right)^2 + \frac{\left(-\frac{1}{2}\right)\left(-\frac{3}{2}\right)\left(-\frac{5}{2}\right)}{3!}\left(-\frac{x}{4}\right)^3 \right.$$

$$\left. + \cdots + \frac{\left(-\frac{1}{2}\right)\left(-\frac{3}{2}\right)\left(-\frac{5}{2}\right)\cdots\left(-\frac{1}{2}-n+1\right)}{n!}\left(-\frac{x}{4}\right)^n + \cdots\right]$$

$$= \frac{1}{2}\left[1 + \frac{1}{8}x + \frac{1\cdot 3}{2!8^2}x^2 + \frac{1\cdot 3\cdot 5}{3!8^3}x^3 + \cdots + \frac{1\cdot 3\cdot 5\cdot\cdots\cdot(2n-1)}{n!8^n}x^n + \cdots\right]$$

We know from 18 that this series converges when $|-x/4| < 1$, that is, $|x| < 4$, so the radius of convergence is $R = 4$.

We collect in the following table, for future reference, some important Maclaurin series that we have derived in this section and the preceding one.

TABLE 1

Important Maclaurin Series and Their Radii of Convergence

$$\frac{1}{1 - x} = \sum_{n=0}^{\infty} x^n = 1 + x + x^2 + x^3 + \cdots \qquad R = 1$$

$$e^x = \sum_{n=0}^{\infty} \frac{x^n}{n!} = 1 + \frac{x}{1!} + \frac{x^2}{2!} + \frac{x^3}{3!} + \cdots \qquad R = \infty$$

$$\sin x = \sum_{n=0}^{\infty} (-1)^n \frac{x^{2n+1}}{(2n + 1)!} = x - \frac{x^3}{3!} + \frac{x^5}{5!} - \frac{x^7}{7!} + \cdots \qquad R = \infty$$

$$\cos x = \sum_{n=0}^{\infty} (-1)^n \frac{x^{2n}}{(2n)!} = 1 - \frac{x^2}{2!} + \frac{x^4}{4!} - \frac{x^6}{6!} + \cdots \qquad R = \infty$$

$$\tan^{-1} x = \sum_{n=0}^{\infty} (-1)^n \frac{x^{2n+1}}{2n + 1} = x - \frac{x^3}{3} + \frac{x^5}{5} - \frac{x^7}{7} + \cdots \qquad R = 1$$

$$\ln(1 + x) = \sum_{n=1}^{\infty} (-1)^{n-1} \frac{x^n}{n} = x - \frac{x^2}{2} + \frac{x^3}{3} - \frac{x^4}{4} + \cdots \qquad R = 1$$

$$(1 + x)^k = \sum_{n=0}^{\infty} \binom{k}{n} x^n = 1 + kx + \frac{k(k - 1)}{2!} x^2 + \frac{k(k - 1)(k - 2)}{3!} x^3 + \cdots \quad R = 1$$

TEC Module 8.7/8.8 enables you to see how successive Taylor polynomials approach the original function.

One reason that Taylor series are important is that they enable us to integrate functions that we couldn't previously handle. In fact, in the introduction to this chapter we mentioned that Newton often integrated functions by first expressing them as power series and then integrating the series term by term. The function $f(x) = e^{-x^2}$ can't be integrated by techniques discussed so far because its antiderivative is not an elementary function (see Section 6.4). In the following example we use Newton's idea to integrate this function.

▼ EXAMPLE 9

(a) Evaluate $\int e^{-x^2}\, dx$ as an infinite series.

(b) Evaluate $\int_0^1 e^{-x^2}\, dx$ correct to within an error of 0.001.

SOLUTION

(a) First we find the Maclaurin series for $f(x) = e^{-x^2}$. Although it's possible to use the direct method, let's find it simply by replacing x with $-x^2$ in the series for e^x given in the table of Maclaurin series. Thus, for all values of x,

$$e^{-x^2} = \sum_{n=0}^{\infty} \frac{(-x^2)^n}{n!} = \sum_{n=0}^{\infty} (-1)^n \frac{x^{2n}}{n!} = 1 - \frac{x^2}{1!} + \frac{x^4}{2!} - \frac{x^6}{3!} + \cdots$$

Now we integrate term by term:

$$\int e^{-x^2}\, dx = \int \left(1 - \frac{x^2}{1!} + \frac{x^4}{2!} - \frac{x^6}{3!} + \cdots + (-1)^n \frac{x^{2n}}{n!} + \cdots\right) dx$$

$$= C + x - \frac{x^3}{3 \cdot 1!} + \frac{x^5}{5 \cdot 2!} - \frac{x^7}{7 \cdot 3!} + \cdots + (-1)^n \frac{x^{2n+1}}{(2n + 1)n!} + \cdots$$

This series converges for all x because the original series for e^{-x^2} converges for all x.

(b) The Evaluation Theorem gives

$$\int_0^1 e^{-x^2}\, dx = \left[x - \frac{x^3}{3 \cdot 1!} + \frac{x^5}{5 \cdot 2!} - \frac{x^7}{7 \cdot 3!} + \frac{x^9}{9 \cdot 4!} - \cdots \right]_0^1$$

$$= 1 - \tfrac{1}{3} + \tfrac{1}{10} - \tfrac{1}{42} + \tfrac{1}{216} - \cdots$$

$$\approx 1 - \tfrac{1}{3} + \tfrac{1}{10} - \tfrac{1}{42} + \tfrac{1}{216} \approx 0.7475$$

- We can take $C = 0$ in the antiderivative in part (a).

The Alternating Series Estimation Theorem shows that the error involved in this approximation is less than

$$\frac{1}{11 \cdot 5!} = \frac{1}{1320} < 0.001 \qquad \blacksquare$$

Another use of Taylor series is illustrated in the next example. The limit could be found with l'Hospital's Rule, but instead we use a series.

EXAMPLE 10 Evaluate $\displaystyle\lim_{x \to 0} \frac{e^x - 1 - x}{x^2}$.

SOLUTION Using the Maclaurin series for e^x, we have

- Some computer algebra systems compute limits in this way.

$$\lim_{x \to 0} \frac{e^x - 1 - x}{x^2} = \lim_{x \to 0} \frac{\left(1 + \dfrac{x}{1!} + \dfrac{x^2}{2!} + \dfrac{x^3}{3!} + \cdots\right) - 1 - x}{x^2}$$

$$= \lim_{x \to 0} \frac{\dfrac{x^2}{2!} + \dfrac{x^3}{3!} + \dfrac{x^4}{4!} + \cdots}{x^2}$$

$$= \lim_{x \to 0} \left(\frac{1}{2} + \frac{x}{3!} + \frac{x^2}{4!} + \frac{x^3}{5!} + \cdots\right) = \frac{1}{2}$$

because power series are continuous functions. $\qquad \blacksquare$

MULTIPLICATION AND DIVISION OF POWER SERIES

If power series are added or subtracted, they behave like polynomials (Theorem 8.2.8 shows this). In fact, as the following example illustrates, they can also be multiplied and divided like polynomials. We find only the first few terms because the calculations for the later terms become tedious and the initial terms are the most important ones.

EXAMPLE 11 Find the first three nonzero terms in the Maclaurin series for
(a) $e^x \sin x$ and (b) $\tan x$.

SOLUTION
(a) Using the Maclaurin series for e^x and $\sin x$ in Table 1, we have

$$e^x \sin x = \left(1 + \frac{x}{1!} + \frac{x^2}{2!} + \frac{x^3}{3!} + \cdots\right)\left(x - \frac{x^3}{3!} + \cdots\right)$$

We multiply these expressions, collecting like terms just as for polynomials:

$$1 + x + \tfrac{1}{2}x^2 + \tfrac{1}{6}x^3 + \cdots$$
$$\times \qquad x \qquad\quad - \tfrac{1}{6}x^3 + \cdots$$
$$\overline{\quad x + \quad x^2 + \tfrac{1}{2}x^3 + \tfrac{1}{6}x^4 + \cdots}$$
$$+ \qquad\qquad\quad\; - \tfrac{1}{6}x^3 - \tfrac{1}{6}x^4 - \cdots$$
$$\overline{\quad x + \quad x^2 + \tfrac{1}{3}x^3 + \cdots}$$

Thus
$$e^x \sin x = x + x^2 + \tfrac{1}{3}x^3 + \cdots$$

(b) Using the Maclaurin series in the table, we have

$$\tan x = \frac{\sin x}{\cos x} = \frac{x - \dfrac{x^3}{3!} + \dfrac{x^5}{5!} - \cdots}{1 - \dfrac{x^2}{2!} + \dfrac{x^4}{4!} - \cdots}$$

We use a procedure like long division:

$$
\begin{array}{r}
x + \tfrac{1}{3}x^3 + \tfrac{2}{15}x^5 + \cdots \\
1 - \tfrac{1}{2}x^2 + \tfrac{1}{24}x^4 - \cdots \overline{\big)\, x - \tfrac{1}{6}x^3 + \tfrac{1}{120}x^5 - \cdots} \\
\underline{x - \tfrac{1}{2}x^3 + \tfrac{1}{24}x^5 - \cdots} \\
\tfrac{1}{3}x^3 - \tfrac{1}{30}x^5 + \cdots \\
\underline{\tfrac{1}{3}x^3 - \tfrac{1}{6}x^5 + \cdots} \\
\tfrac{2}{15}x^5 + \cdots
\end{array}
$$

Thus
$$\tan x = x + \tfrac{1}{3}x^3 + \tfrac{2}{15}x^5 + \cdots \qquad\blacksquare$$

Although we have not attempted to justify the formal manipulations used in Example 11, they are legitimate. There is a theorem which states that if both $f(x) = \Sigma\, c_n x^n$ and $g(x) = \Sigma\, b_n x^n$ converge for $|x| < R$ and the series are multiplied as if they were polynomials, then the resulting series also converges for $|x| < R$ and represents $f(x)\,g(x)$. For division we require $b_0 \neq 0$; the resulting series converges for sufficiently small $|x|$.

PROOF OF TAYLOR'S FORMULA

We conclude this section by giving the promised proof of Theorem 9.

Let $R_n(x) = f(x) - T_n(x)$, where T_n is the nth-degree Taylor polynomial of f at a. The idea for the proof is the same as that for the Mean Value Theorem: We apply Rolle's Theorem to a specially constructed function. We think of x as a constant, $x \neq a$, and we define a function g on I by

$$g(t) = f(x) - f(t) - f'(t)(x - t) - \frac{f''(t)}{2!}(x - t)^2 - \cdots$$

$$- \frac{f^{(n)}(t)}{n!}(x - t)^n - R_n(x)\frac{(x - t)^{n+1}}{(x - a)^{n+1}}$$

Then

$$g(x) = f(x) - f(x) - 0 - \cdots - 0 = 0$$

$$g(a) = f(x) - [T_n(x) + R_n(x)] = f(x) - f(x) = 0$$

Thus, by Rolle's Theorem (applied to g on the interval from a to x), there is a number z between x and a such that $g'(z) = 0$. If we differentiate the expression for g, then most terms cancel. We leave it to you to verify that the expression for $g'(t)$ simplifies to

$$g'(t) = -\frac{f^{(n+1)}(t)}{n!}(x - t)^n + (n + 1)R_n(x)\frac{(x - t)^n}{(x - a)^{n+1}}$$

Thus we have

$$g'(z) = -\frac{f^{(n+1)}(z)}{n!}(x - z)^n + (n + 1)R_n(x)\frac{(x - z)^n}{(x - a)^{n+1}} = 0$$

and so

$$R_n(x) = \frac{f^{(n+1)}(z)}{(n + 1)!}(x - a)^{n+1}$$

8.7 | EXERCISES

1. If $f(x) = \sum_{n=0}^{\infty} b_n(x - 5)^n$ for all x, write a formula for b_8.

2. The graph of f is shown.

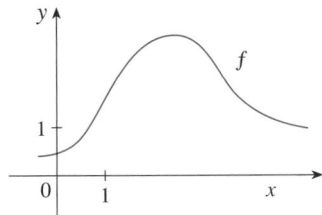

(a) Explain why the series

$$1.6 - 0.8(x - 1) + 0.4(x - 1)^2 - 0.1(x - 1)^3 + \cdots$$

is *not* the Taylor series of f centered at 1.
(b) Explain why the series

$$2.8 + 0.5(x - 2) + 1.5(x - 2)^2 - 0.1(x - 2)^3 + \cdots$$

is *not* the Taylor series of f centered at 2.

3. If $f^{(n)}(0) = (n + 1)!$ for $n = 0, 1, 2, \ldots$, find the Maclaurin series for f and its radius of convergence.

4. Find the Taylor series for f centered at 4 if

$$f^{(n)}(4) = \frac{(-1)^n n!}{3^n(n + 1)}$$

What is the radius of convergence of the Taylor series?

5–10 ▪ Find the Maclaurin series for $f(x)$ using the definition of a Maclaurin series. [Assume that f has a power series expansion. Do not show that $R_n(x) \to 0$.] Also find the associated radius of convergence.

5. $f(x) = (1 - x)^{-2}$ **6.** $f(x) = e^{-2x}$

7. $f(x) = \sin \pi x$ **8.** $f(x) = x \cos x$

9. $f(x) = \sinh x$ **10.** $f(x) = \cosh x$

11–18 ▪ Find the Taylor series for $f(x)$ centered at the given value of a. [Assume that f has a power series expansion. Do not show that $R_n(x) \to 0$.]

11. $f(x) = x^4 - 3x^2 + 1, \quad a = 1$

12. $f(x) = x - x^3, \quad a = -2$

13. $f(x) = \ln x, \quad a = 2$ **14.** $f(x) = 1/x, \quad a = -3$

15. $f(x) = e^{2x}, \quad a = 3$ **16.** $f(x) = \sin x, \quad a = \pi/2$

17. $f(x) = \cos x, \quad a = \pi$ **18.** $f(x) = \sqrt{x}, \quad a = 16$

19. Prove that the series obtained in Exercise 7 represents $\sin \pi x$ for all x.

20. Prove that the series obtained in Exercise 16 represents $\sin x$ for all x.

21. Prove that the series obtained in Exercise 9 represents $\sinh x$ for all x.

22. Prove that the series obtained in Exercise 10 represents $\cosh x$ for all x.

23–26 ■ Use the binomial series to expand the function as a power series. State the radius of convergence.

23. $\sqrt[4]{1 - x}$

24. $\sqrt[3]{8 + x}$

25. $\dfrac{1}{(2 + x)^3}$

26. $(1 - x)^{2/3}$

27–36 ■ Use a Maclaurin series in Table 1 to obtain the Maclaurin series for the given function.

27. $f(x) = \sin \pi x$

28. $f(x) = \cos(\pi x/2)$

29. $f(x) = e^x + e^{2x}$

30. $f(x) = e^x + 2e^{-x}$

31. $f(x) = x \cos(\frac{1}{2}x^2)$

32. $f(x) = x^2 \ln(1 + x^3)$

33. $f(x) = \dfrac{x}{\sqrt{4 + x^2}}$

34. $f(x) = \dfrac{x^2}{\sqrt{2 + x}}$

35. $f(x) = \sin^2 x$ $\left[\textit{Hint: } \text{Use } \sin^2 x = \frac{1}{2}(1 - \cos 2x).\right]$

36. $f(x) = \begin{cases} \dfrac{x - \sin x}{x^3} & \text{if } x \neq 0 \\ \dfrac{1}{6} & \text{if } x = 0 \end{cases}$

37–38 ■ Find the Maclaurin series of f (by any method) and its radius of convergence. Graph f and its first few Taylor polynomials on the same screen. What do you notice about the relationship between these polynomials and f?

37. $f(x) = \cos(x^2)$

38. $f(x) = e^{-x^2} + \cos x$

---

39. Use the Maclaurin series for $\cos x$ to compute $\cos 5°$ correct to five decimal places.

40. Use the Maclaurin series for e^x to calculate $1/\sqrt[10]{e}$ correct to five decimal places.

41. (a) Use the binomial series to expand $1/\sqrt{1 - x^2}$.
    (b) Use part (a) to find the Maclaurin series for $\sin^{-1} x$.

42. (a) Expand $1/\sqrt[4]{1 + x}$ as a power series.
    (b) Use part (a) to estimate $1/\sqrt[4]{1.1}$ correct to three decimal places.

43–46 ■ Evaluate the indefinite integral as an infinite series.

43. $\displaystyle\int x \cos(x^3)\,dx$

44. $\displaystyle\int \frac{e^x - 1}{x}\,dx$

45. $\displaystyle\int \frac{\cos x - 1}{x}\,dx$

46. $\displaystyle\int \arctan(x^2)\,dx$

47–50 ■ Use series to approximate the definite integral to within the indicated accuracy.

47. $\displaystyle\int_0^1 x \cos(x^3)\,dx$ (three decimal places)

48. $\displaystyle\int_0^1 \sin(x^4)\,dx$ (four decimal places)

49. $\displaystyle\int_0^{0.1} \frac{dx}{\sqrt{1 + x^3}}$ $\left(|\,\text{error}\,| < 10^{-8}\right)$

50. $\displaystyle\int_0^{0.5} x^2 e^{-x^2}\,dx$ $\left(|\,\text{error}\,| < 0.001\right)$

51–53 ■ Use series to evaluate the limit.

51. $\displaystyle\lim_{x \to 0} \frac{x - \ln(1 + x)}{x^2}$

52. $\displaystyle\lim_{x \to 0} \frac{1 - \cos x}{1 + x - e^x}$

53. $\displaystyle\lim_{x \to 0} \frac{\sin x - x + \frac{1}{6}x^3}{x^5}$

---

54. Use the series in Example 11(b) to evaluate

$$\lim_{x \to 0} \frac{\tan x - x}{x^3}$$

We found this limit in Example 4 in Section 3.7 using l'Hospital's Rule three times. Which method do you prefer?

55–58 ■ Use multiplication or division of power series to find the first three nonzero terms in the Maclaurin series for each function.

55. $y = e^{-x^2} \cos x$

56. $y = \sec x$

57. $y = \dfrac{x}{\sin x}$

58. $y = e^x \ln(1 + x)$

59–64 ■ Find the sum of the series.

59. $\displaystyle\sum_{n=0}^{\infty} (-1)^n \frac{x^{4n}}{n!}$

60. $\displaystyle\sum_{n=0}^{\infty} \frac{(-1)^n \pi^{2n}}{6^{2n}(2n)!}$

61. $\displaystyle\sum_{n=0}^{\infty} \frac{(-1)^n \pi^{2n+1}}{4^{2n+1}(2n + 1)!}$

62. $\displaystyle\sum_{n=0}^{\infty} \frac{3^n}{5^n n!}$

63. $3 + \dfrac{9}{2!} + \dfrac{27}{3!} + \dfrac{81}{4!} + \cdots$

64. $1 - \ln 2 + \dfrac{(\ln 2)^2}{2!} - \dfrac{(\ln 2)^3}{3!} + \cdots$

---

65. (a) Expand $f(x) = x/(1-x)^2$ as a power series.
(b) Use part (a) to find the sum of the series

$$\sum_{n=1}^{\infty} \frac{n}{2^n}$$

66. (a) Expand $f(x) = (x+x^2)/(1-x)^3$ as a power series.
(b) Use part (a) to find the sum of the series

$$\sum_{n=1}^{\infty} \frac{n^2}{2^n}$$

67. Show that if p is an nth-degree polynomial, then

$$p(x+1) = \sum_{i=0}^{n} \frac{p^{(i)}(x)}{i!}$$

68. If $f(x) = (1+x^3)^{30}$, what is $f^{(58)}(0)$?

69. Use the following steps to prove $\boxed{18}$.
(a) Let $g(x) = \sum_{n=0}^{\infty} \binom{k}{n} x^n$. Differentiate this series to show that

$$g'(x) = \frac{kg(x)}{1+x} \qquad -1 < x < 1$$

(b) Let $h(x) = (1+x)^{-k} g(x)$ and show that $h'(x) = 0$.
(c) Deduce that $g(x) = (1+x)^k$.

70. (a) Show that the function defined by

$$f(x) = \begin{cases} e^{-1/x^2} & \text{if } x \neq 0 \\ 0 & \text{if } x = 0 \end{cases}$$

is not equal to its Maclaurin series.
(b) Graph the function in part (a) and comment on its behavior near the origin.

8.8 | APPLICATIONS OF TAYLOR POLYNOMIALS

In this section we explore two types of applications of Taylor polynomials. First we look at how they are used to approximate functions—computer scientists like them because polynomials are the simplest of functions. Then we investigate how physicists and engineers use them in such fields as relativity, electric dipoles, the velocity of water waves, and building highways across a desert.

APPROXIMATING FUNCTIONS BY POLYNOMIALS

Suppose that $f(x)$ is equal to the sum of its Taylor series at a:

$$f(x) = \sum_{n=0}^{\infty} \frac{f^{(n)}(a)}{n!} (x-a)^n$$

In Section 8.7 we introduced the notation $T_n(x)$ for the nth partial sum of this series and called it the nth-degree Taylor polynomial of f at a. Thus

$$T_n(x) = \sum_{i=0}^{n} \frac{f^{(i)}(a)}{i!} (x-a)^i$$

$$= f(a) + \frac{f'(a)}{1!}(x-a) + \frac{f''(a)}{2!}(x-a)^2 + \cdots + \frac{f^{(n)}(a)}{n!}(x-a)^n$$

Since f is the sum of its Taylor series, we know that $T_n(x) \to f(x)$ as $n \to \infty$ and so T_n can be used as an approximation to f: $f(x) \approx T_n(x)$.

Notice that the first-degree Taylor polynomial

$$T_1(x) = f(a) + f'(a)(x-a)$$

is the same as the linearization of f at a that we discussed in Section 2.8. Notice also that T_1 and its derivative have the same values at a that f and f' have. In general, it can be shown that the derivatives of T_n at a agree with those of f up to and including derivatives of order n.

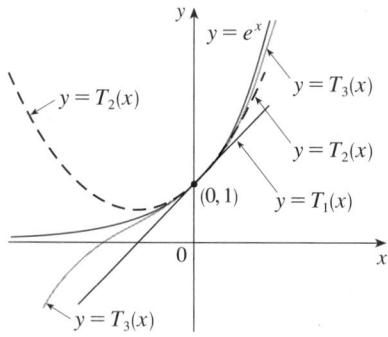

FIGURE 1

x	$x = 0.2$	$x = 3.0$
$T_2(x)$	1.220000	8.500000
$T_4(x)$	1.221400	16.375000
$T_6(x)$	1.221403	19.412500
$T_8(x)$	1.221403	20.009152
$T_{10}(x)$	1.221403	20.079665
e^x	1.221403	20.085537

To illustrate these ideas let's take another look at the graphs of $y = e^x$ and its first few Taylor polynomials, as shown in Figure 1. The graph of T_1 is the tangent line to $y = e^x$ at $(0, 1)$; this tangent line is the best linear approximation to e^x near $(0, 1)$. The graph of T_2 is the parabola $y = 1 + x + x^2/2$, and the graph of T_3 is the cubic curve $y = 1 + x + x^2/2 + x^3/6$, which is a closer fit to the exponential curve $y = e^x$ than T_2. The next Taylor polynomial T_4 would be an even better approximation, and so on.

The values in the table in the margin give a numerical demonstration of the convergence of the Taylor polynomials $T_n(x)$ to the function $y = e^x$. We see that when $x = 0.2$ the convergence is very rapid, but when $x = 3$ it is somewhat slower. In fact, the farther x is from 0, the more slowly $T_n(x)$ converges to e^x.

When using a Taylor polynomial T_n to approximate a function f, we have to ask the questions: How good an approximation is it? How large should we take n to be in order to achieve a desired accuracy? To answer these questions we need to look at the absolute value of the remainder:

$$|R_n(x)| = |f(x) - T_n(x)|$$

There are three possible methods for estimating the size of the error:

1. If a graphing device is available, we can use it to graph $|R_n(x)|$ and thereby estimate the error.

2. If the series happens to be an alternating series, we can use the Alternating Series Estimation Theorem.

3. In all cases we can use Taylor's Formula (8.7.9), which says that

$$R_n(x) = \frac{f^{(n+1)}(z)}{(n+1)!}(x - a)^{n+1}$$

where z is a number that lies between x and a.

V EXAMPLE 1
(a) Approximate the function $f(x) = \sqrt[3]{x}$ by a Taylor polynomial of degree 2 at $a = 8$.
(b) How accurate is this approximation when $7 \leq x \leq 9$?

SOLUTION
(a)
$$f(x) = \sqrt[3]{x} = x^{1/3} \qquad f(8) = 2$$
$$f'(x) = \tfrac{1}{3}x^{-2/3} \qquad f'(8) = \tfrac{1}{12}$$
$$f''(x) = -\tfrac{2}{9}x^{-5/3} \qquad f''(8) = -\tfrac{1}{144}$$
$$f'''(x) = \tfrac{10}{27}x^{-8/3}$$

Thus the second-degree Taylor polynomial is

$$T_2(x) = f(8) + \frac{f'(8)}{1!}(x - 8) + \frac{f''(8)}{2!}(x - 8)^2$$
$$= 2 + \tfrac{1}{12}(x - 8) - \tfrac{1}{288}(x - 8)^2$$

The desired approximation is

$$\sqrt[3]{x} \approx T_2(x) = 2 + \tfrac{1}{12}(x - 8) - \tfrac{1}{288}(x - 8)^2$$

(b) The Taylor series is not alternating when $x < 8$, so we can't use the Alternating Series Estimation Theorem in this example. But using Taylor's Formula we can write

$$R_2(x) = \frac{f'''(z)}{3!}(x-8)^3 = \tfrac{10}{27}z^{-8/3}\frac{(x-8)^3}{3!} = \frac{5(x-8)^3}{81z^{8/3}}$$

where z lies between 8 and x. In order to estimate the error we note that if $7 \le x \le 9$, then $-1 \le x - 8 \le 1$, so $|x - 8| \le 1$ and therefore $|x - 8|^3 \le 1$. Also, since $z > 7$, we have

$$z^{8/3} > 7^{8/3} > 179$$

and so

$$|R_2(x)| \le \frac{5|x-8|^3}{81z^{8/3}} < \frac{5 \cdot 1}{81 \cdot 179} < 0.0004$$

Thus if $7 \le x \le 9$, the approximation in part (a) is accurate to within 0.0004. ■

Let's use a graphing device to check the calculation in Example 1. Figure 2 shows that the graphs of $y = \sqrt[3]{x}$ and $y = T_2(x)$ are very close to each other when x is near 8. Figure 3 shows the graph of $|R_2(x)|$ computed from the expression

$$|R_2(x)| = |\sqrt[3]{x} - T_2(x)|$$

We see from this graph that

$$|R_2(x)| < 0.0003$$

when $7 \le x \le 9$. Thus the error estimate from graphical methods is slightly better than the error estimate from Taylor's Formula in this case.

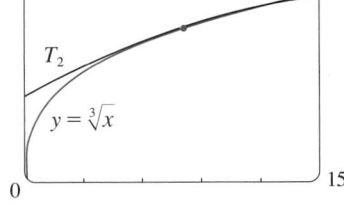

FIGURE 2

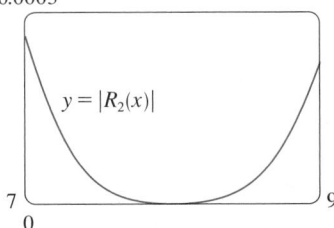

FIGURE 3

▼ EXAMPLE 2

(a) What is the maximum error possible in using the approximation

$$\sin x \approx x - \frac{x^3}{3!} + \frac{x^5}{5!}$$

when $-0.3 \le x \le 0.3$? Use this approximation to find $\sin 12°$ correct to six decimal places.

(b) For what values of x is this approximation accurate to within 0.00005?

SOLUTION

(a) Notice that the Maclaurin series

$$\sin x = x - \frac{x^3}{3!} + \frac{x^5}{5!} - \frac{x^7}{7!} + \cdots$$

is alternating for all nonzero values of x, and the successive terms decrease in size because $|x| < 1$, so we can use the Alternating Series Estimation Theorem. The error in approximating $\sin x$ by the first three terms of its Maclaurin series is at most

$$\left|\frac{x^7}{7!}\right| = \frac{|x|^7}{5040}$$

If $-0.3 \leqslant x \leqslant 0.3$, then $|x| \leqslant 0.3$, so the error is smaller than

$$\frac{(0.3)^7}{5040} \approx 4.3 \times 10^{-8}$$

To find $\sin 12°$ we first convert to radian measure.

$$\sin 12° = \sin\left(\frac{12\pi}{180}\right) = \sin\left(\frac{\pi}{15}\right)$$

$$\approx \frac{\pi}{15} - \left(\frac{\pi}{15}\right)^3 \frac{1}{3!} + \left(\frac{\pi}{15}\right)^5 \frac{1}{5!}$$

$$\approx 0.20791169$$

Thus, correct to six decimal places, $\sin 12° \approx 0.207912$.

(b) The error will be smaller than 0.00005 if

$$\frac{|x|^7}{5040} < 0.00005$$

Solving this inequality for x, we get

$$|x|^7 < 0.252 \qquad \text{or} \qquad |x| < (0.252)^{1/7} \approx 0.821$$

So the given approximation is accurate to within 0.00005 when $|x| < 0.82$. ∎

TEC Module 8.7/8.8 graphically shows the remainders in Taylor polynomial approximations.

What if we had used Taylor's Formula to solve Example 2? The remainder term is

$$R_6(x) = \frac{f^{(7)}(z)}{7!} x^7 = -\cos z \frac{x^7}{7!}$$

(Note that $T_5 = T_6$.) But $|-\cos z| \leqslant 1$, so $|R_6(x)| \leqslant |x|^7/7!$ and we get the same estimates as with the Alternating Series Estimation Theorem.

What about graphical methods? Figure 4 shows the graph of

$$|R_6(x)| = \left| \sin x - \left(x - \tfrac{1}{6}x^3 + \tfrac{1}{120}x^5\right) \right|$$

and we see from it that $|R_6(x)| < 4.3 \times 10^{-8}$ when $|x| \leqslant 0.3$. This is the same estimate that we obtained in Example 2. For part (b) we want $|R_6(x)| < 0.00005$, so we graph both $y = |R_6(x)|$ and $y = 0.00005$ in Figure 5. By placing the cursor on the right intersection point we find that the inequality is satisfied when $|x| < 0.82$. Again this is the same estimate that we obtained in the solution to Example 2.

If we had been asked to approximate $\sin 72°$ instead of $\sin 12°$ in Example 2, it would have been wise to use the Taylor polynomials at $a = \pi/3$ (instead of $a = 0$) because they are better approximations to $\sin x$ for values of x close to $\pi/3$. Notice that $72°$ is close to $60°$ (or $\pi/3$ radians) and the derivatives of $\sin x$ are easy to compute at $\pi/3$.

Figure 6 shows the graphs of the Maclaurin polynomial approximations

$$T_1(x) = x \qquad\qquad T_3(x) = x - \frac{x^3}{3!}$$

$$T_5(x) = x - \frac{x^3}{3!} + \frac{x^5}{5!} \qquad\qquad T_7(x) = x - \frac{x^3}{3!} + \frac{x^5}{5!} - \frac{x^7}{7!}$$

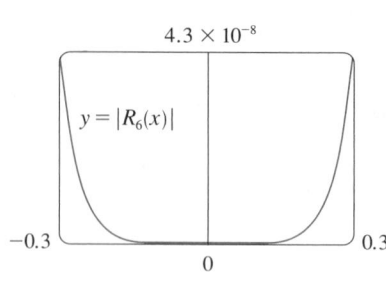

FIGURE 4

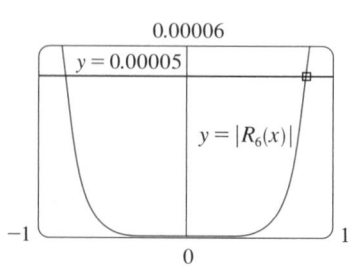

FIGURE 5

to the sine curve. You can see that as n increases, $T_n(x)$ is a good approximation to sin x on a larger and larger interval.

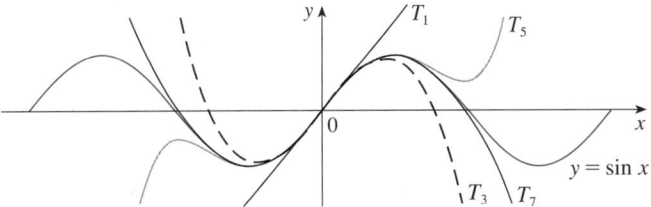

FIGURE 6

One use of the type of calculation done in Examples 1 and 2 occurs in calculators and computers. For instance, when you press the sin or e^x key on your calculator, or when a computer programmer uses a subroutine for a trigonometric or exponential or Bessel function, in many machines a polynomial approximation is calculated. The polynomial is often a Taylor polynomial that has been modified so that the error is spread more evenly throughout an interval.

APPLICATIONS TO PHYSICS

Taylor polynomials are also used frequently in physics. In order to gain insight into an equation, a physicist often simplifies a function by considering only the first two or three terms in its Taylor series. In other words, the physicist uses a Taylor polynomial as an approximation to the function. Taylor's Formula can then be used to gauge the accuracy of the approximation. The following example shows one way in which this idea is used in special relativity. Other applications are explored in Exercises 24–28.

▼ EXAMPLE 3 In Einstein's theory of special relativity the mass of an object moving with velocity v is

$$m = \frac{m_0}{\sqrt{1 - v^2/c^2}}$$

where m_0 is the mass of the object when at rest and c is the speed of light. The kinetic energy of the object is the difference between its total energy and its energy at rest:

$$K = mc^2 - m_0 c^2$$

(a) Show that when v is very small compared with c, this expression for K agrees with classical Newtonian physics: $K = \frac{1}{2} m_0 v^2$.
(b) Use Taylor's Formula to estimate the difference in these expressions for K when $|v| \leq 100$ m/s.

SOLUTION
(a) Using the expressions given for K and m, we get

$$K = mc^2 - m_0 c^2 = \frac{m_0 c^2}{\sqrt{1 - v^2/c^2}} - m_0 c^2$$

$$= m_0 c^2 \left[\left(1 - \frac{v^2}{c^2} \right)^{-1/2} - 1 \right]$$

With $x = -v^2/c^2$, the Maclaurin series for $(1 + x)^{-1/2}$ is most easily computed as a binomial series with $k = -\frac{1}{2}$. (Notice that $|x| < 1$ because $v < c$.) Therefore we

■ The upper curve in Figure 7 is the graph of the expression for the kinetic energy K of an object with velocity v in special relativity. The lower curve shows the function used for K in classical Newtonian physics. When v is much smaller than the speed of light, the curves are practically identical.

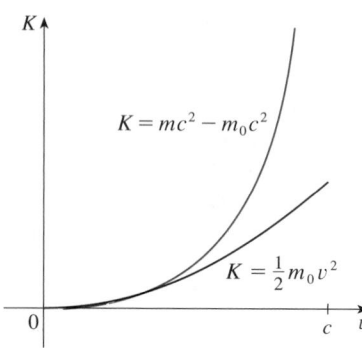

FIGURE 7

have

$$(1 + x)^{-1/2} = 1 - \tfrac{1}{2}x + \frac{\left(-\tfrac{1}{2}\right)\left(-\tfrac{3}{2}\right)}{2!} x^2 + \frac{\left(-\tfrac{1}{2}\right)\left(-\tfrac{3}{2}\right)\left(-\tfrac{5}{2}\right)}{3!} x^3 + \cdots$$

$$= 1 - \tfrac{1}{2}x + \tfrac{3}{8}x^2 - \tfrac{5}{16}x^3 + \cdots$$

and

$$K = m_0 c^2 \left[\left(1 + \frac{1}{2} \frac{v^2}{c^2} + \frac{3}{8} \frac{v^4}{c^4} + \frac{5}{16} \frac{v^6}{c^6} + \cdots \right) - 1 \right]$$

$$= m_0 c^2 \left(\frac{1}{2} \frac{v^2}{c^2} + \frac{3}{8} \frac{v^4}{c^4} + \frac{5}{16} \frac{v^6}{c^6} + \cdots \right)$$

If v is much smaller than c, then all terms after the first are very small when compared with the first term. If we omit them, we get

$$K \approx m_0 c^2 \left(\frac{1}{2} \frac{v^2}{c^2} \right) = \tfrac{1}{2} m_0 v^2$$

(b) By Taylor's Formula we can write the remainder term as

$$R_1(x) = \frac{f''(z)}{2!} x^2$$

where $f(x) = m_0 c^2 [(1 + x)^{-1/2} - 1]$ and $x = -v^2/c^2$. Since $f''(x) = \tfrac{3}{4} m_0 c^2 (1 + x)^{-5/2}$, we get

$$R_1(x) = \frac{3 m_0 c^2}{8(1 + z)^{5/2}} \cdot \frac{v^4}{c^4}$$

where z lies between 0 and $-v^2/c^2$. We have $c = 3 \times 10^8$ m/s and $|v| \leq 100$ m/s, so

$$R_1(x) \leq \frac{\tfrac{3}{8} m_0 (9 \times 10^{16})(100/c)^4}{(1 - 100^2/c^2)^{5/2}} < (4.17 \times 10^{-10}) m_0$$

Thus when $|v| \leq 100$ m/s, the magnitude of the error in using the Newtonian expression for kinetic energy is at most $(4.2 \times 10^{-10}) m_0$. ■

8.8 | EXERCISES

1. (a) Find the Taylor polynomials up to degree 6 for $f(x) = \cos x$ centered at $a = 0$. Graph f and these polynomials on a common screen.
  (b) Evaluate f and these polynomials at $x = \pi/4$, $\pi/2$, and π.
  (c) Comment on how the Taylor polynomials converge to $f(x)$.

2. (a) Find the Taylor polynomials up to degree 3 for $f(x) = 1/x$ centered at $a = 1$. Graph f and these polynomials on a common screen.
  (b) Evaluate f and these polynomials at $x = 0.9$ and 1.3.

(c) Comment on how the Taylor polynomials converge to $f(x)$.

3–8 ■ Find the Taylor polynomial $T_3(x)$ for the function f centered at the number a. Graph f and T_3 on the same screen.

3. $f(x) = 1/x$, $\quad a = 2$

4. $f(x) = e^{-x} \sin x$, $\quad a = 0$

5. $f(x) = \cos x$, $\quad a = \pi/2$

6. $f(x) = \dfrac{\ln x}{x}$, $\quad a = 1$

7. $f(x) = xe^{-2x}$, $a = 0$

8. $f(x) = \tan^{-1}x$, $a = 1$

9–16 ■
(a) Approximate f by a Taylor polynomial with degree n at the number a.
(b) Use Taylor's Formula to estimate the accuracy of the approximation $f(x) \approx T_n(x)$ when x lies in the given interval.
(c) Check your result in part (b) by graphing $|R_n(x)|$.

9. $f(x) = \sqrt{x}$, $a = 4$, $n = 2$, $4 \leq x \leq 4.2$

10. $f(x) = x^{-2}$, $a = 1$, $n = 2$, $0.9 \leq x \leq 1.1$

11. $f(x) = x^{2/3}$, $a = 1$, $n = 3$, $0.8 \leq x \leq 1.2$

12. $f(x) = \sin x$, $a = \pi/6$, $n = 4$, $0 \leq x \leq \pi/3$

13. $f(x) = e^{x^2}$, $a = 0$, $n = 3$, $0 \leq x \leq 0.1$

14. $f(x) = \ln(1 + 2x)$, $a = 1$, $n = 3$, $0.5 \leq x \leq 1.5$

15. $f(x) = x \sin x$, $a = 0$, $n = 4$, $-1 \leq x \leq 1$

16. $f(x) = x \ln x$, $a = 1$, $n = 3$, $0.5 \leq x \leq 1.5$

17. Use the information from Exercise 5 to estimate $\cos 80°$ correct to five decimal places.

18. Use the information from Exercise 12 to estimate $\sin 38°$ correct to five decimal places.

19. Use Taylor's Formula to determine the number of terms of the Maclaurin series for e^x that should be used to estimate $e^{0.1}$ to within 0.00001.

20. Suppose you know that

$$f^{(n)}(4) = \frac{(-1)^n n!}{3^n (n + 1)}$$

and the Taylor series of f centered at 4 converges to $f(x)$ for all x in the interval of convergence. Show that the fifth-degree Taylor polynomial approximates $f(5)$ with error less than 0.0002.

21–22 ■ Use the Alternating Series Estimation Theorem or Taylor's Formula to estimate the range of values of x for which the given approximation is accurate to within the stated error. Check your answer graphically.

21. $\sin x \approx x - \dfrac{x^3}{6}$ $(\,|\,\text{error}\,| < 0.01)$

22. $\cos x \approx 1 - \dfrac{x^2}{2} + \dfrac{x^4}{24}$ $(\,|\,\text{error}\,| < 0.005)$

23. A car is moving with speed 20 m/s and acceleration 2 m/s² at a given instant. Using a second-degree Taylor polynomial, estimate how far the car moves in the next second. Would it be reasonable to use this polynomial to estimate the distance traveled during the next minute?

24. The resistivity ρ of a conducting wire is the reciprocal of the conductivity and is measured in units of ohm-meters (Ω-m). The resistivity of a given metal depends on the temperature according to the equation

$$\rho(t) = \rho_{20} e^{\alpha(t-20)}$$

where t is the temperature in °C. There are tables that list the values of α (called the temperature coefficient) and ρ_{20} (the resistivity at 20°C) for various metals. Except at very low temperatures, the resistivity varies almost linearly with temperature and so it is common to approximate the expression for $\rho(t)$ by its first- or second-degree Taylor polynomial at $t = 20$.
(a) Find expressions for these linear and quadratic approximations.
 (b) For copper, the tables give $\alpha = 0.0039/°C$ and $\rho_{20} = 1.7 \times 10^{-8}$ Ω-m. Graph the resistivity of copper and the linear and quadratic approximations for $-250°C \leq t \leq 1000°C$.
 (c) For what values of t does the linear approximation agree with the exponential expression to within one percent?

25. An electric dipole consists of two electric charges of equal magnitude and opposite sign. If the charges are q and $-q$ and are located at a distance d from each other, then the electric field E at the point P in the figure is

$$E = \frac{q}{D^2} - \frac{q}{(D + d)^2}$$

By expanding this expression for E as a series in powers of d/D, show that E is approximately proportional to $1/D^3$ when P is far away from the dipole.

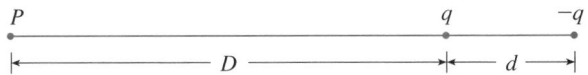

26. If a water wave with length L moves with velocity v across a body of water with depth d, as in the figure on page 496, then

$$v^2 = \frac{gL}{2\pi} \tanh \frac{2\pi d}{L}$$

(a) If the water is deep, show that $v \approx \sqrt{gL/(2\pi)}$.

(b) If the water is shallow, use the Maclaurin series for tanh to show that $v \approx \sqrt{gd}$. (Thus in shallow water the velocity of a wave tends to be independent of the length of the wave.)

(c) Use the Alternating Series Estimation Theorem to show that if $L > 10d$, then the estimate $v^2 \approx gd$ is accurate to within $0.014gL$.

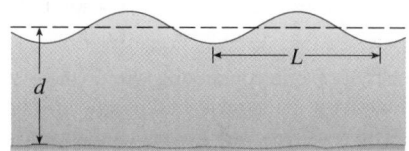

27. If a surveyor measures differences in elevation when making plans for a highway across a desert, corrections must be made for the curvature of the earth.

(a) If R is the radius of the earth and L is the length of the highway, show that the correction is

$$C = R \sec(L/R) - R$$

(b) Use a Taylor polynomial to show that

$$C \approx \frac{L^2}{2R} + \frac{5L^4}{24R^3}$$

(c) Compare the corrections given by the formulas in parts (a) and (b) for a highway that is 100 km long. (Take the radius of the earth to be 6370 km.)

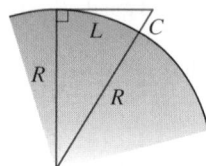

28. The period of a pendulum with length L that makes a maximum angle θ_0 with the vertical is

$$T = 4 \sqrt{\frac{L}{g}} \int_0^{\pi/2} \frac{dx}{\sqrt{1 - k^2 \sin^2 x}}$$

where $k = \sin\left(\frac{1}{2}\theta_0\right)$ and g is the acceleration due to gravity. (In Exercise 34 in Section 6.5 we approximated this integral using Simpson's Rule.)

(a) Expand the integrand as a binomial series and use the result of Exercise 34 in Section 6.1 to show that

$$T = 2\pi \sqrt{\frac{L}{g}} \left[1 + \frac{1^2}{2^2} k^2 + \frac{1^2 3^2}{2^2 4^2} k^4 + \frac{1^2 3^2 5^2}{2^2 4^2 6^2} k^6 + \cdots \right]$$

If θ_0 is not too large, the approximation $T \approx 2\pi\sqrt{L/g}$, obtained by using only the first term in the series, is

often used. A better approximation is obtained by using two terms:

$$T \approx 2\pi \sqrt{\frac{L}{g}} \left(1 + \tfrac{1}{4} k^2 \right)$$

(b) Notice that all the terms in the series after the first one have coefficients that are at most $\frac{1}{4}$. Use this fact to compare this series with a geometric series and show that

$$2\pi \sqrt{\frac{L}{g}} \left(1 + \tfrac{1}{4} k^2 \right) \leq T \leq 2\pi \sqrt{\frac{L}{g}} \frac{4 - 3k^2}{4 - 4k^2}$$

(c) Use the inequalities in part (b) to estimate the period of a pendulum with $L = 1$ meter and $\theta_0 = 10°$. How does it compare with the estimate $T \approx 2\pi\sqrt{L/g}$? What if $\theta_0 = 42°$?

29. In Section 4.6 we considered Newton's method for approximating a root r of the equation $f(x) = 0$, and from an initial approximation x_1 we obtained successive approximations $x_2, x_3, \ldots$, where

$$x_{n+1} = x_n - \frac{f(x_n)}{f'(x_n)}$$

Use Taylor's Formula with $n = 1$, $a = x_n$, and $x = r$ to show that if $f''(x)$ exists on an interval I containing r, x_n, and x_{n+1}, and $|f''(x)| \leq M$, $|f'(x)| \geq K$ for all $x \in I$, then

$$|x_{n+1} - r| \leq \frac{M}{2K} |x_n - r|^2$$

[This means that if x_n is accurate to d decimal places, then x_{n+1} is accurate to about $2d$ decimal places. More precisely, if the error at stage n is at most 10^{-m}, then the error at stage $n + 1$ is at most $(M/2K)10^{-2m}$.]

30. Use the following outline to prove that e is an irrational number.

(a) If e were rational, then it would be of the form $e = p/q$, where p and q are positive integers and $q > 2$. Use Taylor's Formula to write

$$\frac{p}{q} = e = 1 + \frac{1}{1!} + \frac{1}{2!} + \cdots + \frac{1}{q!} + \frac{e^z}{(q + 1)!}$$

$$= s_q + \frac{e^z}{(q + 1)!}$$

where $0 < z < 1$.

(b) Show that $q!(e - s_q)$ is an integer.

(c) Show that $q!(e - s_q) < 1$.

(d) Use parts (b) and (c) to deduce that e is irrational.

CHAPTER 8 REVIEW

CONCEPT CHECK

1. (a) What is a convergent sequence?
(b) What is a convergent series?
(c) What does $\lim_{n \to \infty} a_n = 3$ mean?
(d) What does $\sum_{n=1}^{\infty} a_n = 3$ mean?

2. (a) What is a bounded sequence?
(b) What is a monotonic sequence?
(c) What can you say about a bounded monotonic sequence?

3. (a) What is a geometric series? Under what circumstances is it convergent? What is its sum?
(b) What is a p-series? Under what circumstances is it convergent?

4. Suppose $\Sigma\, a_n = 3$ and s_n is the nth partial sum of the series. What is $\lim_{n \to \infty} a_n$? What is $\lim_{n \to \infty} s_n$?

5. State the following.
(a) The Test for Divergence
(b) The Integral Test
(c) The Comparison Test
(d) The Limit Comparison Test
(e) The Alternating Series Test
(f) The Ratio Test
(g) The Root Test

6. (a) What is an absolutely convergent series?
(b) What can you say about such a series?
(c) What is a conditionally convergent series?

7. If a series is convergent by the Alternating Series Test, how do you estimate its sum?

8. (a) Write the general form of a power series.
(b) What is the radius of convergence of a power series?
(c) What is the interval of convergence of a power series?

9. Suppose $f(x)$ is the sum of a power series with radius of convergence R.
(a) How do you differentiate f? What is the radius of convergence of the series for f'?
(b) How do you integrate f? What is the radius of convergence of the series for $\int f(x)\,dx$?

10. (a) Write an expression for the nth-degree Taylor polynomial of f centered at a.
(b) Write an expression for the Taylor series of f centered at a.
(c) Write an expression for the Maclaurin series of f.
(d) How do you show that $f(x)$ is equal to the sum of its Taylor series?
(e) State Taylor's Formula.

11. Write the Maclaurin series and the interval of convergence for each of the following functions.
(a) $1/(1-x)$ (b) e^x (c) $\sin x$
(d) $\cos x$ (e) $\tan^{-1}x$ (f) $\ln(1+x)$

12. Write the binomial series expansion of $(1+x)^k$. What is the radius of convergence of this series?

TRUE-FALSE QUIZ

Determine whether the statement is true or false. If it is true, explain why. If it is false, explain why or give an example that disproves the statement.

1. If $\lim_{n \to \infty} a_n = 0$, then $\Sigma\, a_n$ is convergent.

2. The series $\sum_{n=1}^{\infty} n^{-\sin 1}$ is convergent.

3. If $\lim_{n \to \infty} a_n = L$, then $\lim_{n \to \infty} a_{2n+1} = L$.

4. If $\Sigma\, c_n 6^n$ is convergent, then $\Sigma\, c_n(-2)^n$ is convergent.

5. If $\Sigma\, c_n 6^n$ is convergent, then $\Sigma\, c_n(-6)^n$ is convergent.

6. If $\Sigma\, c_n x^n$ diverges when $x = 6$, then it diverges when $x = 10$.

7. The Ratio Test can be used to determine whether $\Sigma\, 1/n^3$ converges.

8. The Ratio Test can be used to determine whether $\Sigma\, 1/n!$ converges.

9. If $0 \leq a_n \leq b_n$ and $\Sigma\, b_n$ diverges, then $\Sigma\, a_n$ diverges.

10. $\displaystyle\sum_{n=0}^{\infty} \frac{(-1)^n}{n!} = \frac{1}{e}$

11. If $-1 < \alpha < 1$, then $\lim_{n \to \infty} \alpha^n = 0$.

12. If $\Sigma\, a_n$ is divergent, then $\Sigma\, |a_n|$ is divergent.

13. If $f(x) = 2x - x^2 + \frac{1}{3}x^3 - \cdots$ converges for all x, then $f'''(0) = 2$.

14. If $\{a_n\}$ and $\{b_n\}$ are divergent, then $\{a_n + b_n\}$ is divergent.

15. If $\{a_n\}$ and $\{b_n\}$ are divergent, then $\{a_n b_n\}$ is divergent.

16. If $\{a_n\}$ is decreasing and $a_n > 0$ for all n, then $\{a_n\}$ is convergent.

17. If $a_n > 0$ and $\Sigma\, a_n$ converges, then $\Sigma\, (-1)^n a_n$ converges.

18. If $a_n > 0$ and $\lim_{n \to \infty} (a_{n+1}/a_n) < 1$, then $\lim_{n \to \infty} a_n = 0$.

19. $0.99999\ldots = 1$

20. If $\lim_{n \to \infty} a_n = 2$, then $\lim_{n \to \infty} (a_{n+3} - a_n) = 0$.

21. If a finite number of terms are added to a convergent series, then the new series is still convergent.

22. If $\sum_{n=1}^{\infty} a_n = A$ and $\sum_{n=1}^{\infty} b_n = B$, then $\sum_{n=1}^{\infty} a_n b_n = AB$.

EXERCISES

1–8 ▪ Determine whether the sequence is convergent or divergent. If it is convergent, find its limit.

1. $a_n = \dfrac{2 + n^3}{1 + 2n^3}$

2. $a_n = \dfrac{9^{n+1}}{10^n}$

3. $a_n = \dfrac{n^3}{1 + n^2}$

4. $a_n = \cos(n\pi/2)$

5. $a_n = \dfrac{n \sin n}{n^2 + 1}$

6. $a_n = \dfrac{\ln n}{\sqrt{n}}$

7. $\{(1 + 3/n)^{4n}\}$

8. $\{(-10)^n/n!\}$

9–20 ▪ Determine whether the series is convergent or divergent.

9. $\displaystyle\sum_{n=1}^{\infty} \dfrac{n}{n^3 + 1}$

10. $\displaystyle\sum_{n=1}^{\infty} \dfrac{n^2 + 1}{n^3 + 1}$

11. $\displaystyle\sum_{n=1}^{\infty} \dfrac{n^3}{5^n}$

12. $\displaystyle\sum_{n=1}^{\infty} \dfrac{(-1)^n}{\sqrt{n + 1}}$

13. $\displaystyle\sum_{n=2}^{\infty} \dfrac{1}{n\sqrt{\ln n}}$

14. $\displaystyle\sum_{n=1}^{\infty} \ln\!\left(\dfrac{n}{3n + 1}\right)$

15. $\displaystyle\sum_{n=1}^{\infty} \dfrac{\cos 3n}{1 + (1.2)^n}$

16. $\displaystyle\sum_{n=1}^{\infty} \dfrac{n^{2n}}{(1 + 2n^2)^n}$

17. $\displaystyle\sum_{n=1}^{\infty} \dfrac{1 \cdot 3 \cdot 5 \cdot \cdots \cdot (2n - 1)}{5^n n!}$

18. $\displaystyle\sum_{n=1}^{\infty} \dfrac{(-5)^{2n}}{n^2 9^n}$

19. $\displaystyle\sum_{n=1}^{\infty} (-1)^{n-1} \dfrac{\sqrt{n}}{n + 1}$

20. $\displaystyle\sum_{n=1}^{\infty} \dfrac{\sqrt{n + 1} - \sqrt{n - 1}}{n}$

21–24 ▪ Determine whether the series is conditionally convergent, absolutely convergent, or divergent.

21. $\displaystyle\sum_{n=1}^{\infty} (-1)^{n-1} n^{-1/3}$

22. $\displaystyle\sum_{n=1}^{\infty} (-1)^{n-1} n^{-3}$

23. $\displaystyle\sum_{n=1}^{\infty} \dfrac{(-1)^n (n + 1)3^n}{2^{2n+1}}$

24. $\displaystyle\sum_{n=2}^{\infty} \dfrac{(-1)^n \sqrt{n}}{\ln n}$

25–29 ▪ Find the sum of the series.

25. $\displaystyle\sum_{n=1}^{\infty} \dfrac{2^{2n+1}}{5^n}$

26. $\displaystyle\sum_{n=1}^{\infty} \dfrac{1}{n(n + 3)}$

27. $\displaystyle\sum_{n=1}^{\infty} [\tan^{-1}(n + 1) - \tan^{-1}n]$

28. $\displaystyle\sum_{n=0}^{\infty} \dfrac{(-1)^n \pi^n}{3^{2n}(2n)!}$

29. $1 - e + \dfrac{e^2}{2!} - \dfrac{e^3}{3!} + \dfrac{e^4}{4!} - \cdots$

---

30. Express the repeating decimal $4.17326326326\ldots$ as a fraction.

31. Show that $\cosh x \geqslant 1 + \frac{1}{2}x^2$ for all x.

32. For what values of x does the series $\sum_{n=1}^{\infty} (\ln x)^n$ converge?

33. Find the sum of the series $\displaystyle\sum_{n=1}^{\infty} \dfrac{(-1)^{n+1}}{n^5}$ correct to four decimal places.

34. (a) Show that the series $\displaystyle\sum_{n=1}^{\infty} \dfrac{n^n}{(2n)!}$ is convergent.

 (b) Deduce that $\displaystyle\lim_{n \to \infty} \dfrac{n^n}{(2n)!} = 0$.

35. Prove that if the series $\sum_{n=1}^{\infty} a_n$ is absolutely convergent, then the series

$$\sum_{n=1}^{\infty} \left(\dfrac{n + 1}{n}\right) a_n$$

is also absolutely convergent.

36–39 ▪ Find the radius of convergence and interval of convergence of the series.

36. $\displaystyle\sum_{n=1}^{\infty} (-1)^n \dfrac{x^n}{n^2 5^n}$

37. $\displaystyle\sum_{n=1}^{\infty} \dfrac{(x + 2)^n}{n \, 4^n}$

38. $\sum_{n=1}^{\infty} \dfrac{2^n(x-2)^n}{(n+2)!}$

39. $\sum_{n=0}^{\infty} \dfrac{2^n(x-3)^n}{\sqrt{n+3}}$

40. Find the radius of convergence of the series

$$\sum_{n=1}^{\infty} \frac{(2n)!}{(n!)^2} x^n$$

41. Find the Taylor series of $f(x) = \sin x$ at $a = \pi/6$.

42. Find the Taylor series of $f(x) = \cos x$ at $a = \pi/3$.

43–50 ▪ Find the Maclaurin series for f and its radius of convergence. You may use either the direct method (definition of a Maclaurin series) or known series such as geometric series, binomial series, or the Maclaurin series for e^x, $\sin x$, and $\tan^{-1} x$.

43. $f(x) = \dfrac{x^2}{1+x}$

44. $f(x) = \tan^{-1}(x^2)$

45. $f(x) = \ln(4-x)$

46. $f(x) = xe^{2x}$

47. $f(x) = \sin(x^4)$

48. $f(x) = 10^x$

49. $f(x) = 1/\sqrt[4]{16-x}$

50. $f(x) = (1-3x)^{-5}$

51. Evaluate $\displaystyle\int \frac{e^x}{x}\, dx$ as an infinite series.

52. Use series to approximate $\int_0^1 \sqrt{1+x^4}\, dx$ correct to two decimal places.

53–54 ▪
(a) Approximate f by a Taylor polynomial with degree n at the number a.
(b) Graph f and T_n on a common screen.
(c) Use Taylor's Formula to estimate the accuracy of the approximation $f(x) \approx T_n(x)$ when x lies in the given interval.
(d) Check your result in part (c) by graphing $|R_n(x)|$.

53. $f(x) = \sqrt{x}$, $\quad a = 1$, $\quad n = 3$, $\quad 0.9 \leqslant x \leqslant 1.1$

54. $f(x) = \sec x$, $\quad a = 0$, $\quad n = 2$, $\quad 0 \leqslant x \leqslant \pi/6$

55. Use series to evaluate the following limit.

$$\lim_{x \to 0} \frac{\sin x - x}{x^3}$$

56. The force due to gravity on an object with mass m at a height h above the surface of the earth is

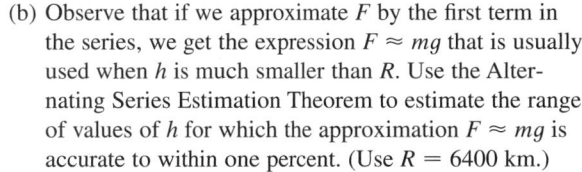

$$F = \frac{mgR^2}{(R+h)^2}$$

where R is the radius of the earth and g is the acceleration due to gravity.
(a) Express F as a series in powers of h/R.
(b) Observe that if we approximate F by the first term in the series, we get the expression $F \approx mg$ that is usually used when h is much smaller than R. Use the Alternating Series Estimation Theorem to estimate the range of values of h for which the approximation $F \approx mg$ is accurate to within one percent. (Use $R = 6400$ km.)

57. A sequence is defined recursively by the equations $a_1 = 1$, $a_{n+1} = \frac{1}{3}(a_n + 4)$. Show that $\{a_n\}$ is increasing and $a_n < 2$ for all n. Deduce that $\{a_n\}$ is convergent and find its limit.

58. Show that $\lim_{n \to \infty} n^4 e^{-n} = 0$ and use a graph to find the smallest value of N that corresponds to $\varepsilon = 0.1$ in the precise definition of a limit.

59. Suppose that $f(x) = \sum_{n=0}^{\infty} c_n x^n$ for all x.
(a) If f is an odd function, show that

$$c_0 = c_2 = c_4 = \cdots = 0$$

(b) If f is an even function, show that

$$c_1 = c_3 = c_5 = \cdots = 0$$

60. If $f(x) = e^{x^2}$, show that $f^{(2n)}(0) = \dfrac{(2n)!}{n!}$.

9

PARAMETRIC EQUATIONS AND POLAR COORDINATES

So far we have described plane curves by giving y as a function of x $[y = f(x)]$ or x as a function of y $[x = g(y)]$ or by giving a relation between x and y that defines y implicitly as a function of x $[f(x, y) = 0]$. In this chapter we discuss two new methods for describing curves.

Some curves, such as the cycloid, are best handled when both x and y are given in terms of a third variable t called a parameter $[x = f(t), y = g(t)]$. Other curves, such as the cardioid, have their most convenient description when we use a new coordinate system, called the polar coordinate system.

9.1 | PARAMETRIC CURVES

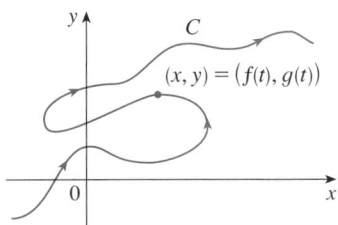

FIGURE 1

TEC Module 9.1A gives an animation of the relationship between motion along a parametric curve $x = f(t)$, $y = g(t)$ and motion along the graphs of f and g as functions of t.

Imagine that a particle moves along the curve C shown in Figure 1. It is impossible to describe C by an equation of the form $y = f(x)$ because C fails the Vertical Line Test. But the x- and y-coordinates of the particle are functions of time and so we can write $x = f(t)$ and $y = g(t)$. Such a pair of equations is often a convenient way of describing a curve and gives rise to the following definition.

Suppose that x and y are both given as functions of a third variable t (called a **parameter**) by the equations

$$x = f(t) \qquad y = g(t)$$

(called **parametric equations**). Each value of t determines a point (x, y), which we can plot in a coordinate plane. As t varies, the point $(x, y) = (f(t), g(t))$ varies and traces out a curve C, which we call a **parametric curve**. The parameter t does not necessarily represent time and, in fact, we could use a letter other than t for the parameter. But in many applications of parametric curves, t does denote time and therefore we can interpret $(x, y) = (f(t), g(t))$ as the position of a particle at time t.

EXAMPLE 1 Sketch and identify the curve defined by the parametric equations

$$x = t^2 - 2t \qquad y = t + 1$$

SOLUTION Each value of t gives a point on the curve, as shown in the table. For instance, if $t = 0$, then $x = 0$, $y = 1$ and so the corresponding point is $(0, 1)$. In Figure 2 we plot the points (x, y) determined by several values of the parameter t and we join them to produce a curve.

t	x	y
-2	8	-1
-1	3	0
0	0	1
1	-1	2
2	0	3
3	3	4
4	8	5

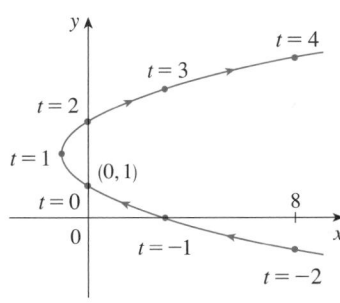

FIGURE 2

A particle whose position is given by the parametric equations moves along the curve in the direction of the arrows as t increases. Notice that the consecutive points marked on the curve appear at equal time intervals but not at equal distances. That is because the particle slows down and then speeds up as t increases.

It appears from Figure 2 that the curve traced out by the particle may be a parabola. This can be confirmed by eliminating the parameter t as follows. We obtain $t = y - 1$ from the second equation and substitute into the first equation. This gives

$$x = t^2 - 2t = (y - 1)^2 - 2(y - 1) = y^2 - 4y + 3$$

and so the curve represented by the given parametric equations is the parabola $x = y^2 - 4y + 3$. ∎

No restriction was placed on the parameter t in Example 1, so we assumed that t could be any real number. But sometimes we restrict t to lie in a finite interval. For instance, the parametric curve

$$x = t^2 - 2t \qquad y = t + 1 \qquad 0 \le t \le 4$$

shown in Figure 3 is the part of the parabola in Example 1 that starts at the point $(0, 1)$ and ends at the point $(8, 5)$. The arrowhead indicates the direction in which the curve is traced as t increases from 0 to 4.

In general, the curve with parametric equations

$$x = f(t) \qquad y = g(t) \qquad a \le t \le b$$

has **initial point** $(f(a), g(a))$ and **terminal point** $(f(b), g(b))$.

> ■ This equation in x and y describes *where* the particle has been, but it doesn't tell us *when* the particle was at a particular point. The parametric equations have an advantage—they tell us *when* the particle was at a point. They also indicate the *direction* of the motion.

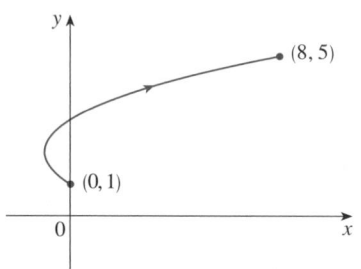

FIGURE 3

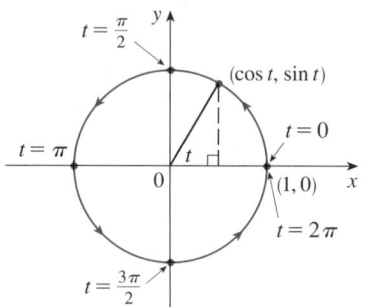

FIGURE 4

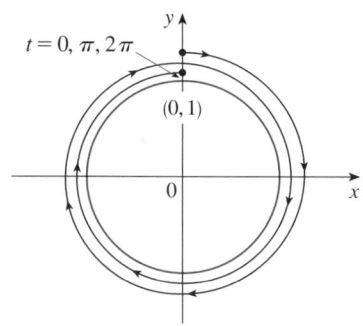

FIGURE 5

▼ EXAMPLE 2 What curve is represented by the following parametric equations?

$$x = \cos t \qquad y = \sin t \qquad 0 \le t \le 2\pi$$

SOLUTION If we plot points, it appears that the curve is a circle. We can confirm this impression by eliminating t. Observe that

$$x^2 + y^2 = \cos^2 t + \sin^2 t = 1$$

Thus the point (x, y) moves on the unit circle $x^2 + y^2 = 1$. Notice that in this example the parameter t can be interpreted as the angle (in radians) shown in Figure 4. As t increases from 0 to 2π, the point $(x, y) = (\cos t, \sin t)$ moves once around the circle in the counterclockwise direction starting from the point $(1, 0)$. ∎

EXAMPLE 3 What curve is represented by the given parametric equations?

$$x = \sin 2t \qquad y = \cos 2t \qquad 0 \le t \le 2\pi$$

SOLUTION Again we have

$$x^2 + y^2 = \sin^2 2t + \cos^2 2t = 1$$

so the parametric equations again represent the unit circle $x^2 + y^2 = 1$. But as t increases from 0 to 2π, the point $(x, y) = (\sin 2t, \cos 2t)$ starts at $(0, 1)$ and moves *twice* around the circle in the clockwise direction as indicated in Figure 5. ∎

Examples 2 and 3 show that different sets of parametric equations can represent the same curve. Thus we distinguish between a *curve*, which is a set of points, and a *parametric curve*, in which the points are traced in a particular way.

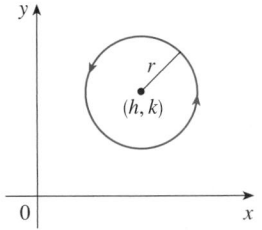

FIGURE 6
$x = h + r \cos t,\ y = k + r \sin t$

EXAMPLE 4 Find parametric equations for the circle with center (h, k) and radius r.

SOLUTION If we take the equations of the unit circle in Example 2 and multiply the expressions for x and y by r, we get $x = r \cos t$, $y = r \sin t$. You can verify that these equations represent a circle with radius r and center the origin traced counterclockwise. We now shift h units in the x-direction and k units in the y-direction and obtain parametric equations of the circle (Figure 6) with center (h, k) and radius r:

$$x = h + r \cos t \qquad y = k + r \sin t \qquad 0 \leqslant t \leqslant 2\pi$$ ∎

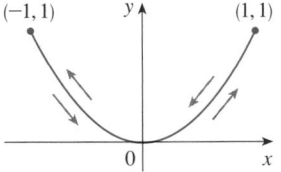

FIGURE 7

V EXAMPLE 5 Sketch the curve with parametric equations $x = \sin t$, $y = \sin^2 t$.

SOLUTION Observe that $y = (\sin t)^2 = x^2$ and so the point (x, y) moves on the parabola $y = x^2$. But note also that, since $-1 \leqslant \sin t \leqslant 1$, we have $-1 \leqslant x \leqslant 1$, so the parametric equations represent only the part of the parabola for which $-1 \leqslant x \leqslant 1$. Since $\sin t$ is periodic, the point $(x, y) = (\sin t, \sin^2 t)$ moves back and forth infinitely often along the parabola from $(-1, 1)$ to $(1, 1)$. (See Figure 7.) ∎

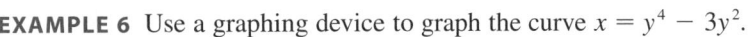

GRAPHING DEVICES

Most graphing calculators and computer graphing programs can be used to graph curves defined by parametric equations. In fact, it's instructive to watch a parametric curve being drawn by a graphing calculator because the points are plotted in order as the corresponding parameter values increase.

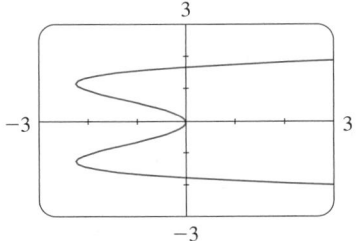

FIGURE 8

EXAMPLE 6 Use a graphing device to graph the curve $x = y^4 - 3y^2$.

SOLUTION If we let the parameter be $t = y$, then we have the equations

$$x = t^4 - 3t^2 \qquad y = t$$

Using these parametric equations to graph the curve, we obtain Figure 8. It would be possible to solve the given equation ($x = y^4 - 3y^2$) for y as four functions of x and graph them individually, but the parametric equations provide a much easier method. ∎

In general, if we need to graph an equation of the form $x = g(y)$, we can use the parametric equations

$$x = g(t) \qquad y = t$$

▪ **www.stewartcalculus.com**
See Additional Example A.

Notice also that curves with equations $y = f(x)$ (the ones we are most familiar with—graphs of functions) can also be regarded as curves with parametric equations

$$x = t \qquad y = f(t)$$

Graphing devices are particularly useful for sketching complicated curves. For instance, the curves shown in Figures 9, 10, and 11 would be virtually impossible to produce by hand.

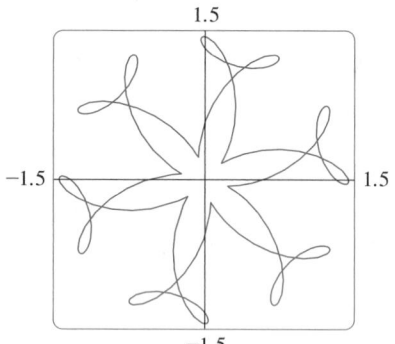

FIGURE 9
$x = \sin t + \frac{1}{2} \cos 5t + \frac{1}{4} \sin 13t$
$y = \cos t + \frac{1}{2} \sin 5t + \frac{1}{4} \cos 13t$

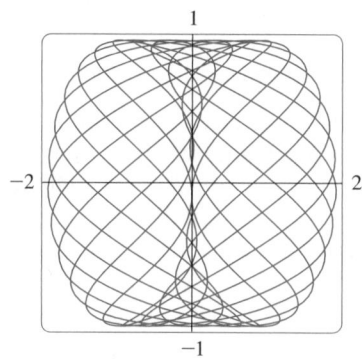

FIGURE 10
$x = \sin t - \sin 2.3t$
$y = \cos t$

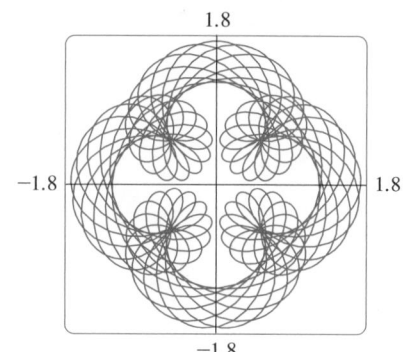

FIGURE 11
$x = \sin t + \frac{1}{2} \sin 5t + \frac{1}{4} \cos 2.3t$
$y = \cos t + \frac{1}{2} \cos 5t + \frac{1}{4} \sin 2.3t$

THE CYCLOID

 An animation in Module 9.1B shows how the cycloid is formed as the circle moves.

EXAMPLE 7 The curve traced out by a point P on the circumference of a circle as the circle rolls along a straight line is called a **cycloid** (see Figure 12). If the circle has radius r and rolls along the x-axis and if one position of P is the origin, find parametric equations for the cycloid.

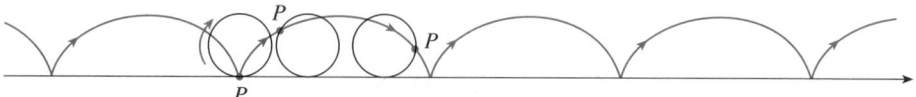

FIGURE 12

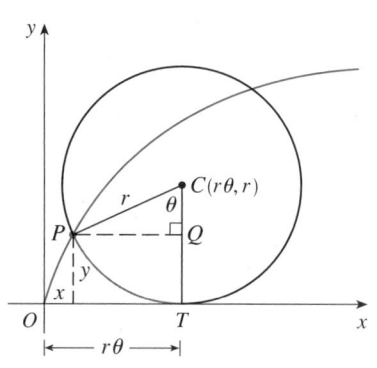

FIGURE 13

SOLUTION We choose as parameter the angle of rotation θ of the circle ($\theta = 0$ when P is at the origin). Suppose the circle has rotated through θ radians. Because the circle has been in contact with the line, we see from Figure 13 that the distance it has rolled from the origin is

$$|OT| = \text{arc } PT = r\theta$$

Therefore the center of the circle is $C(r\theta, r)$. Let the coordinates of P be (x, y). Then from Figure 13 we see that

$$x = |OT| - |PQ| = r\theta - r \sin \theta = r(\theta - \sin \theta)$$

$$y = |TC| - |QC| = r - r \cos \theta = r(1 - \cos \theta)$$

Therefore parametric equations of the cycloid are

$$\boxed{1} \qquad x = r(\theta - \sin \theta) \qquad y = r(1 - \cos \theta) \qquad \theta \in \mathbb{R}$$

One arch of the cycloid comes from one rotation of the circle and so is described by $0 \le \theta \le 2\pi$. Although Equations 1 were derived from Figure 13, which illustrates the case $0 < \theta < \pi/2$, it can be seen that these equations are still valid for other values of θ (see Exercise 33).

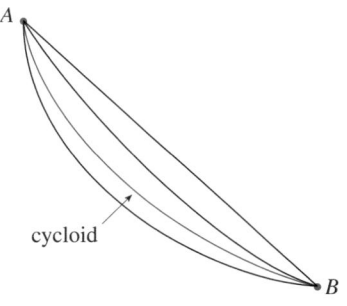

FIGURE 14

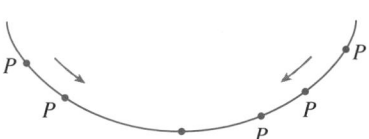

FIGURE 15

Although it is possible to eliminate the parameter θ from Equations 1, the resulting Cartesian equation in x and y is very complicated and not as convenient to work with as the parametric equations. ∎

One of the first people to study the cycloid was Galileo, who proposed that bridges be built in the shape of cycloids and who tried to find the area under one arch of a cycloid. Later this curve arose in connection with the *brachistochrone problem*: Find the curve along which a particle will slide in the shortest time (under the influence of gravity) from a point A to a lower point B not directly beneath A. The Swiss mathematician John Bernoulli, who posed this problem in 1696, showed that among all possible curves that join A to B, as in Figure 14, the particle will take the least time sliding from A to B if the curve is part of an inverted arch of a cycloid.

The Dutch physicist Huygens had already shown that the cycloid is also the solution to the *tautochrone problem*; that is, no matter where a particle P is placed on an inverted cycloid, it takes the same time to slide to the bottom (see Figure 15). Huygens proposed that pendulum clocks (which he invented) should swing in cycloidal arcs because then the pendulum takes the same time to make a complete oscillation whether it swings through a wide or a small arc.

9.1 | EXERCISES

1–4 ▪ Sketch the curve by using the parametric equations to plot points. Indicate with an arrow the direction in which the curve is traced as t increases.

1. $x = t^2 + t$, $\quad y = t^2 - t$, $\quad -2 \leq t \leq 2$

2. $x = t^2$, $\quad y = t^3 - 4t$, $\quad -3 \leq t \leq 3$

3. $x = \cos^2 t$, $\quad y = 1 - \sin t$, $\quad 0 \leq t \leq \pi/2$

4. $x = e^{-t} + t$, $\quad y = e^t - t$, $\quad -2 \leq t \leq 2$

5–8 ▪

(a) Sketch the curve by using the parametric equations to plot points. Indicate with an arrow the direction in which the curve is traced as t increases.

(b) Eliminate the parameter to find a Cartesian equation of the curve.

5. $x = 3 - 4t$, $\quad y = 2 - 3t$

6. $x = t - 1$, $\quad y = t^3 + 1$, $\quad -2 \leq t \leq 2$

7. $x = \sqrt{t}$, $\quad y = 1 - t$ $\qquad$ **8.** $x = t^2$, $\quad y = t^3$

9–14 ▪

(a) Eliminate the parameter to find a Cartesian equation of the curve.

(b) Sketch the curve and indicate with an arrow the direction in which the curve is traced as the parameter increases.

9. $x = \sin \frac{1}{2}\theta$, $\quad y = \cos \frac{1}{2}\theta$, $\quad -\pi \leq \theta \leq \pi$

10. $x = \frac{1}{2}\cos\theta$, $\quad y = 2\sin\theta$, $\quad 0 \leq \theta \leq \pi$

11. $x = \sin t$, $\quad y = \csc t$, $\quad 0 < t < \pi/2$

12. $x = e^t - 1$, $\quad y = e^{2t}$

13. $x = e^{2t}$, $\quad y = t + 1$

14. $y = \sqrt{t + 1}$, $\quad y = \sqrt{t - 1}$

15–18 ▪ Describe the motion of a particle with position (x, y) as t varies in the given interval.

15. $x = 3 + 2\cos t$, $\quad y = 1 + 2\sin t$, $\quad \pi/2 \leq t \leq 3\pi/2$

16. $x = 2\sin t$, $\quad y = 4 + \cos t$, $\quad 0 \leq t \leq 3\pi/2$

17. $x = 5\sin t$, $\quad y = 2\cos t$, $\quad -\pi \leq t \leq 5\pi$

18. $x = \sin t$, $\quad y = \cos^2 t$, $\quad -2\pi \leq t \leq 2\pi$

19–21 ▪ Use the graphs of $x = f(t)$ and $y = g(t)$ to sketch the parametric curve $x = f(t)$, $y = g(t)$. Indicate with arrows the direction in which the curve is traced as t increases.

19.

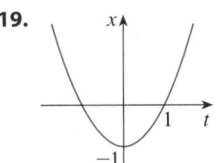
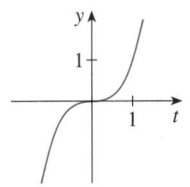

Graphing calculator or computer required $\qquad$ **CAS** Computer algebra system required $\qquad$ **1.** Homework Hints at stewartcalculus.com

20.

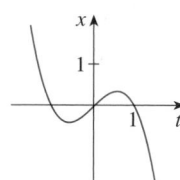

21.

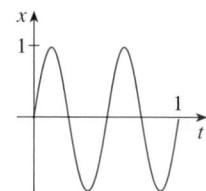
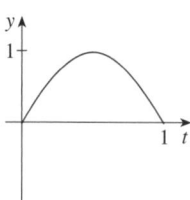

22. Match the parametric equations with the graphs labeled I–VI. Give reasons for your choices. (Do not use a graphing device.)

(a) $x = t^4 - t + 1, \quad y = t^2$

(b) $x = t^2 - 2t, \quad y = \sqrt{t}$

(c) $x = \sin 2t, \quad y = \sin(t + \sin 2t)$

(d) $x = \cos 5t, \quad y = \sin 2t$

(e) $x = t + \sin 4t, \quad y = t^2 + \cos 3t$

(f) $x = \dfrac{\sin 2t}{4 + t^2}, \quad y = \dfrac{\cos 2t}{4 + t^2}$

I

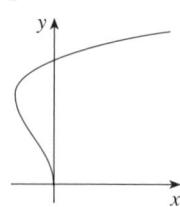

II

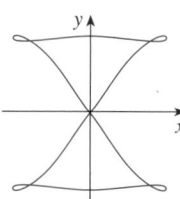

III

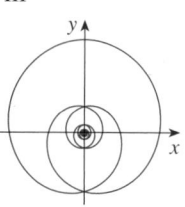

IV

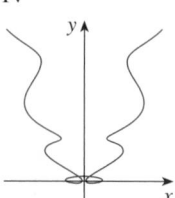

V

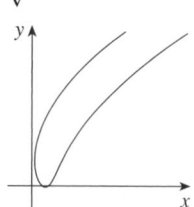

VI

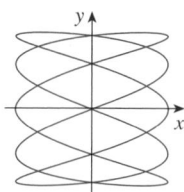

23. Graph the curve $x = y - 2 \sin \pi y$.

24. Graph the curves $y = x^3 - 4x$ and $x = y^3 - 4y$ and find their points of intersection correct to one decimal place.

25. (a) Show that the parametric equations

$$x = x_1 + (x_2 - x_1)t \qquad y = y_1 + (y_2 - y_1)t$$

where $0 \le t \le 1$, describe the line segment that joins the points $P_1(x_1, y_1)$ and $P_2(x_2, y_2)$.

(b) Find parametric equations to represent the line segment from $(-2, 7)$ to $(3, -1)$.

26. Use a graphing device and the result of Exercise 25(a) to draw the triangle with vertices $A(1, 1)$, $B(4, 2)$, and $C(1, 5)$.

27. Find parametric equations for the path of a particle that moves along the circle $x^2 + (y - 1)^2 = 4$ in the manner described.

(a) Once around clockwise, starting at $(2, 1)$

(b) Three times around counterclockwise, starting at $(2, 1)$

(c) Halfway around counterclockwise, starting at $(0, 3)$

28. (a) Find parametric equations for the ellipse $x^2/a^2 + y^2/b^2 = 1$. [*Hint:* Modify the equations of the circle in Example 2.]

(b) Use these parametric equations to graph the ellipse when $a = 3$ and $b = 1, 2, 4$, and 8.

(c) How does the shape of the ellipse change as b varies?

29–30 ▪ Use a graphing calculator or computer to reproduce the picture.

29.

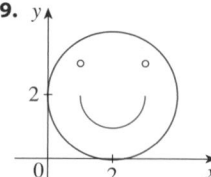

30.

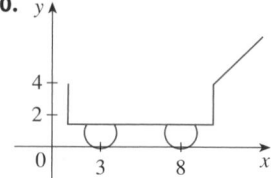

31–32 ▪ Compare the curves represented by the parametric equations. How do they differ?

31. (a) $x = t^3, \quad y = t^2$ (b) $x = t^6, \quad y = t^4$

(c) $x = e^{-3t}, \quad y = e^{-2t}$

32. (a) $x = t, \quad y = t^{-2}$ (b) $x = \cos t, \quad y = \sec^2 t$

(c) $x = e^t, \quad y = e^{-2t}$

33. Derive Equations 1 for the case $\pi/2 < \theta < \pi$.

34. Let P be a point at a distance d from the center of a circle of radius r. The curve traced out by P as the circle rolls along a straight line is called a **trochoid**. (Think of the motion of a point on a spoke of a bicycle wheel.) The cycloid is the special case of a trochoid with $d = r$. Using the same parameter θ as for the cycloid and, assuming the line is the x-axis and $\theta = 0$ when P is at one of its lowest points, show that parametric equations of the trochoid are

$$x = r\theta - d \sin \theta \qquad y = r - d \cos \theta$$

Sketch the trochoid for the cases $d < r$ and $d > r$.

35. If a and b are fixed numbers, find parametric equations for the curve that consists of all possible positions of the point P in the figure, using the angle θ as the parameter. Then eliminate the parameter and identify the curve.

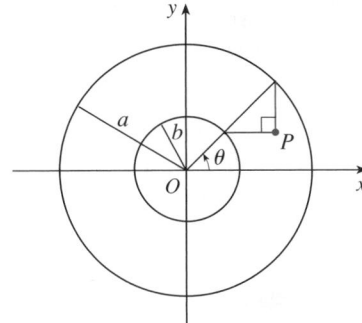

36. A curve, called a **witch of Maria Agnesi**, consists of all possible positions of the point P in the figure. Show that parametric equations for this curve can be written as

$$x = 2a \cot \theta \qquad y = 2a \sin^2\theta$$

Sketch the curve.

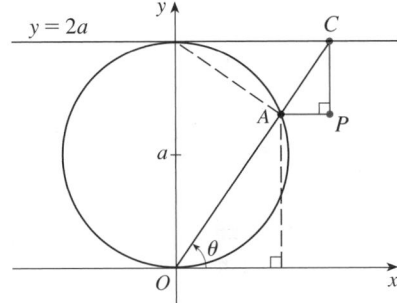

37. Suppose that the position of one particle at time t is given by

$$x_1 = 3 \sin t \qquad y_1 = 2 \cos t \qquad 0 \le t \le 2\pi$$

and the position of a second particle is given by

$$x_2 = -3 + \cos t \qquad y_2 = 1 + \sin t \qquad 0 \le t \le 2\pi$$

(a) Graph the paths of both particles. How many points of intersection are there?

(b) Are any of these points of intersection *collision points*? In other words, are the particles ever at the same place at the same time? If so, find the collision points.

(c) Describe what happens if the path of the second particle is given by

$$x_2 = 3 + \cos t \qquad y_2 = 1 + \sin t \qquad 0 \le t \le 2\pi$$

38. If a projectile is fired with an initial velocity of v_0 meters per second at an angle α above the horizontal and air resistance is assumed to be negligible, then its position after t seconds is given by the parametric equations

$$x = (v_0 \cos \alpha)t \qquad y = (v_0 \sin \alpha)t - \tfrac{1}{2}gt^2$$

where g is the acceleration due to gravity (9.8 m/s^2).

(a) If a gun is fired with $\alpha = 30°$ and $v_0 = 500$ m/s, when will the bullet hit the ground? How far from the gun will it hit the ground? What is the maximum height reached by the bullet?

(b) Use a graphing device to check your answers to part (a). Then graph the path of the projectile for several other values of the angle α to see where it hits the ground. Summarize your findings.

(c) Show that the path is parabolic by eliminating the parameter.

39. Investigate the family of curves defined by the parametric equations $x = t^2$, $y = t^3 - ct$. How does the shape change as c increases? Illustrate by graphing several members of the family.

40. The **swallowtail catastrophe curves** are defined by the parametric equations $x = 2ct - 4t^3$, $y = -ct^2 + 3t^4$. Graph several of these curves. What features do the curves have in common? How do they change when c increases?

41. Graph several members of the family of curves with parametric equations $x = t + a \cos t$, $y = t + a \sin t$, where $a > 0$. How does the shape change as a increases? For what values of a does the curve have a loop?

42. Graph several members of the family of curves $x = \sin t + \sin nt$, $y = \cos t + \cos nt$, where n is a positive integer. What features do the curves have in common? What happens as n increases?

43. The curves with equations $x = a \sin nt$, $y = b \cos t$ are called **Lissajous figures**. Investigate how these curves vary when a, b, and n vary. (Take n to be a positive integer.)

44. Investigate the family of curves defined by the parametric equations $x = \cos t$, $y = \sin t - \sin ct$, where $c > 0$. Start by letting c be a positive integer and see what happens to the shape as c increases. Then explore some of the possibilities that occur when c is a fraction.

9.2 | CALCULUS WITH PARAMETRIC CURVES

Having seen how to represent curves by parametric equations, we now apply the methods of calculus to these parametric curves. In particular, we solve problems involving tangents, areas, and arc length.

TANGENTS

Suppose f and g are differentiable functions and we want to find the tangent line at a point on the parametric curve $x = f(t)$, $y = g(t)$ where y is also a differentiable function of x. Then the Chain Rule gives

$$\frac{dy}{dt} = \frac{dy}{dx} \cdot \frac{dx}{dt}$$

If $dx/dt \neq 0$, we can solve for dy/dx:

■ If we think of the curve as being traced out by a moving particle, then dy/dt and dx/dt are the vertical and horizontal velocities of the particle and Formula 1 says that the slope of the tangent is the ratio of these velocities.

$$\boxed{1} \qquad \frac{dy}{dx} = \frac{\dfrac{dy}{dt}}{\dfrac{dx}{dt}} \qquad \text{if} \quad \frac{dx}{dt} \neq 0$$

Equation 1 (which you can remember by thinking of canceling the dt's) enables us to find the slope dy/dx of the tangent to a parametric curve without having to eliminate the parameter t. We see from $\boxed{1}$ that the curve has a horizontal tangent when $dy/dt = 0$ (provided that $dx/dt \neq 0$) and it has a vertical tangent when $dx/dt = 0$ (provided that $dy/dt \neq 0$). This information is useful for sketching parametric curves.

As we know from Chapter 4, it is also useful to consider d^2y/dx^2. This can be found by replacing y by dy/dx in Equation 1:

⊘ Note that $\dfrac{d^2y}{dx^2} \neq \dfrac{\dfrac{d^2y}{dt^2}}{\dfrac{d^2x}{dt^2}}$.

$$\frac{d^2y}{dx^2} = \frac{d}{dx}\left(\frac{dy}{dx}\right) = \frac{\dfrac{d}{dt}\left(\dfrac{dy}{dx}\right)}{\dfrac{dx}{dt}}$$

EXAMPLE 1 A curve C is defined by the parametric equations $x = t^2$, $y = t^3 - 3t$.
(a) Show that C has two tangents at the point $(3, 0)$ and find their equations.
(b) Find the points on C where the tangent is horizontal or vertical.
(c) Determine where the curve is concave upward or downward.
(d) Sketch the curve.

SOLUTION
(a) Notice that $y = t^3 - 3t = t(t^2 - 3) = 0$ when $t = 0$ or $t = \pm\sqrt{3}$. Therefore the point $(3, 0)$ on C arises from two values of the parameter, $t = \sqrt{3}$ and $t = -\sqrt{3}$. This indicates that C crosses itself at $(3, 0)$. Since

$$\frac{dy}{dx} = \frac{dy/dt}{dx/dt} = \frac{3t^2 - 3}{2t} = \frac{3}{2}\left(t - \frac{1}{t}\right)$$

■ **www.stewartcalculus.com**
See Additional Example A.

the slope of the tangent when $t = \pm\sqrt{3}$ is $dy/dx = \pm6/(2\sqrt{3}) = \pm\sqrt{3}$, so the equations of the tangents at $(3, 0)$ are

$$y = \sqrt{3}\,(x - 3) \qquad \text{and} \qquad y = -\sqrt{3}\,(x - 3)$$

(b) C has a horizontal tangent when $dy/dx = 0$, that is, when $dy/dt = 0$ and $dx/dt \neq 0$. Since $dy/dt = 3t^2 - 3$, this happens when $t^2 = 1$, that is, $t = \pm1$. The corresponding points on C are $(1, -2)$ and $(1, 2)$. C has a vertical tangent when $dx/dt = 2t = 0$, that is, $t = 0$. (Note that $dy/dt \neq 0$ there.) The corresponding point on C is $(0, 0)$.

(c) To determine concavity we calculate the second derivative:

$$\frac{d^2y}{dx^2} = \frac{\dfrac{d}{dt}\left(\dfrac{dy}{dx}\right)}{\dfrac{dx}{dt}} = \frac{\dfrac{3}{2}\left(1 + \dfrac{1}{t^2}\right)}{2t} = \frac{3(t^2 + 1)}{4t^3}$$

Thus the curve is concave upward when $t > 0$ and concave downward when $t < 0$.

(d) Using the information from parts (b) and (c), we sketch C in Figure 1. ∎

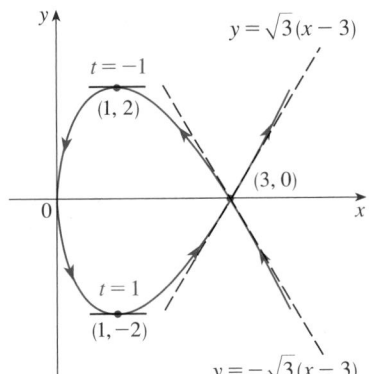

FIGURE 1

▶ EXAMPLE 2

(a) Find the tangent to the cycloid $x = r(\theta - \sin\theta)$, $y = r(1 - \cos\theta)$ at the point where $\theta = \pi/3$. (See Example 7 in Section 9.1.)

(b) At what points is the tangent horizontal? When is it vertical?

SOLUTION

(a) The slope of the tangent line is

$$\frac{dy}{dx} = \frac{dy/d\theta}{dx/d\theta} = \frac{r\sin\theta}{r(1 - \cos\theta)} = \frac{\sin\theta}{1 - \cos\theta}$$

When $\theta = \pi/3$, we have

$$x = r\left(\frac{\pi}{3} - \sin\frac{\pi}{3}\right) = r\left(\frac{\pi}{3} - \frac{\sqrt{3}}{2}\right) \qquad y = r\left(1 - \cos\frac{\pi}{3}\right) = \frac{r}{2}$$

and

$$\frac{dy}{dx} = \frac{\sin(\pi/3)}{1 - \cos(\pi/3)} = \frac{\sqrt{3}/2}{1 - \frac{1}{2}} = \sqrt{3}$$

Therefore the slope of the tangent is $\sqrt{3}$ and its equation is

$$y - \frac{r}{2} = \sqrt{3}\left(x - \frac{r\pi}{3} + \frac{r\sqrt{3}}{2}\right) \qquad \text{or} \qquad \sqrt{3}\,x - y = r\left(\frac{\pi}{\sqrt{3}} - 2\right)$$

The tangent is sketched in Figure 2.

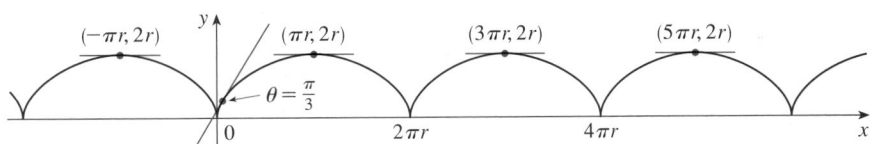

FIGURE 2

(b) The tangent is horizontal when $dy/dx = 0$, which occurs when $\sin \theta = 0$ and $1 - \cos \theta \neq 0$, that is, $\theta = (2n - 1)\pi$, n an integer. The corresponding point on the cycloid is $((2n - 1)\pi r, 2r)$.

When $\theta = 2n\pi$, both $dx/d\theta$ and $dy/d\theta$ are 0. It appears from the graph that there are vertical tangents at those points. We can verify this by using l'Hospital's Rule as follows:

$$\lim_{\theta \to 2n\pi^+} \frac{dy}{dx} = \lim_{\theta \to 2n\pi^+} \frac{\sin \theta}{1 - \cos \theta} = \lim_{\theta \to 2n\pi^+} \frac{\cos \theta}{\sin \theta} = \infty$$

A similar computation shows that $dy/dx \to -\infty$ as $\theta \to 2n\pi^-$, so indeed there are vertical tangents when $\theta = 2n\pi$, that is, when $x = 2n\pi r$. ■

AREAS

We know that the area under a curve $y = F(x)$ from a to b is $A = \int_a^b F(x)\,dx$, where $F(x) \geqslant 0$. If the curve is given by parametric equations $x = f(t)$, $y = g(t)$ and is traversed once as t increases from α to β, then we can adapt the earlier formula by using the Substitution Rule for Definite Integrals as follows:

$$A = \int_a^b y\,dx = \int_\alpha^\beta g(t)f'(t)\,dt$$

$$\left[\text{or} \quad \int_\beta^\alpha g(t)f'(t)\,dt \quad \text{if } \left(f(\beta), g(\beta)\right) \text{ is the leftmost endpoint} \right]$$

◪ **EXAMPLE 3** Find the area under one arch of the cycloid $x = r(\theta - \sin \theta)$, $y = r(1 - \cos \theta)$. (See Figure 3.)

SOLUTION One arch of the cycloid is given by $0 \leqslant \theta \leqslant 2\pi$. Using the Substitution Rule with $y = r(1 - \cos \theta)$ and $dx = r(1 - \cos \theta)\,d\theta$, we have

$$A = \int_0^{2\pi r} y\,dx = \int_0^{2\pi} r(1 - \cos \theta)\, r(1 - \cos \theta)\,d\theta$$

$$= r^2 \int_0^{2\pi} (1 - \cos \theta)^2\,d\theta = r^2 \int_0^{2\pi} (1 - 2\cos \theta + \cos^2\theta)\,d\theta$$

$$= r^2 \int_0^{2\pi} \left[1 - 2\cos \theta + \tfrac{1}{2}(1 + \cos 2\theta) \right] d\theta$$

$$= r^2 \left[\tfrac{3}{2}\theta - 2\sin \theta + \tfrac{1}{4}\sin 2\theta \right]_0^{2\pi} = r^2\left(\tfrac{3}{2} \cdot 2\pi \right) = 3\pi r^2 \qquad ■$$

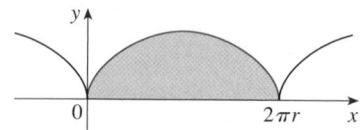

FIGURE 3

■ The result of Example 3 says that the area under one arch of the cycloid is three times the area of the rolling circle that generates the cycloid (see Example 7 in Section 9.1). Galileo guessed this result but it was first proved by the French mathematician Roberval and the Italian mathematician Torricelli.

ARC LENGTH

We already know how to find the length L of a curve C given in the form $y = F(x)$, $a \leqslant x \leqslant b$. Formula 7.4.3 says that if F' is continuous, then

$$\boxed{2} \qquad L = \int_a^b \sqrt{1 + \left(\frac{dy}{dx}\right)^2}\,dx$$

Suppose that C can also be described by the parametric equations $x = f(t)$, $y = g(t)$, $\alpha \leqslant t \leqslant \beta$, where $dx/dt = f'(t) > 0$. This means that C is traversed once, from left to right, as t increases from α to β and $f(\alpha) = a$, $f(\beta) = b$. Putting Formula 1 into Formula 2 and using the Substitution Rule, we obtain

$$L = \int_a^b \sqrt{1 + \left(\frac{dy}{dx}\right)^2}\, dx = \int_\alpha^\beta \sqrt{1 + \left(\frac{dy/dt}{dx/dt}\right)^2}\, \frac{dx}{dt}\, dt$$

Since $dx/dt > 0$, we have

$$\boxed{3} \qquad\qquad L = \int_\alpha^\beta \sqrt{\left(\frac{dx}{dt}\right)^2 + \left(\frac{dy}{dt}\right)^2}\, dt$$

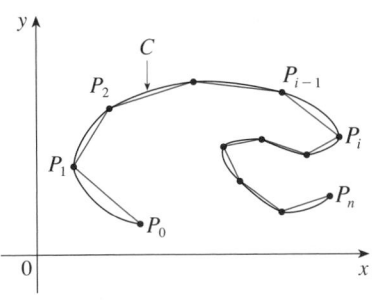

FIGURE 4

Even if C can't be expressed in the form $y = F(x)$, Formula 3 is still valid but we obtain it by polygonal approximations. We divide the parameter interval $[\alpha, \beta]$ into n subintervals of equal width Δt. If $t_0, t_1, t_2, \ldots, t_n$ are the endpoints of these subintervals, then $x_i = f(t_i)$ and $y_i = g(t_i)$ are the coordinates of points $P_i(x_i, y_i)$ that lie on C and the polygon with vertices $P_0, P_1, \ldots, P_n$ approximates C (see Figure 4).

As in Section 7.4, we define the length L of C to be the limit of the lengths of these approximating polygons as $n \to \infty$:

$$L = \lim_{n\to\infty} \sum_{i=1}^n |P_{i-1}P_i|$$

The Mean Value Theorem, when applied to f on the interval $[t_{i-1}, t_i]$, gives a number t_i^* in (t_{i-1}, t_i) such that

$$f(t_i) - f(t_{i-1}) = f'(t_i^*)(t_i - t_{i-1})$$

If we let $\Delta x_i = x_i - x_{i-1}$ and $\Delta y_i = y_i - y_{i-1}$, this equation becomes

$$\Delta x_i = f'(t_i^*)\, \Delta t$$

Similarly, when applied to g, the Mean Value Theorem gives a number t_i^{**} in (t_{i-1}, t_i) such that

$$\Delta y_i = g'(t_i^{**})\, \Delta t$$

Therefore

$$|P_{i-1}P_i| = \sqrt{(\Delta x_i)^2 + (\Delta y_i)^2} = \sqrt{[f'(t_i^*)\Delta t]^2 + [g'(t_i^{**})\Delta t]^2}$$
$$= \sqrt{[f'(t_i^*)]^2 + [g'(t_i^{**})]^2}\, \Delta t$$

and so

$$\boxed{4} \qquad\qquad L = \lim_{n\to\infty} \sum_{i=1}^n \sqrt{[f'(t_i^*)]^2 + [g'(t_i^{**})]^2}\, \Delta t$$

The sum in $\boxed{4}$ resembles a Riemann sum for the function $\sqrt{[f'(t)]^2 + [g'(t)]^2}$ but it is not exactly a Riemann sum because $t_i^* \neq t_i^{**}$ in general. Nevertheless, if f' and g' are continuous, it can be shown that the limit in $\boxed{4}$ is the same as if t_i^* and t_i^{**} were equal, namely,

$$L = \int_\alpha^\beta \sqrt{[f'(t)]^2 + [g'(t)]^2}\, dt$$

Thus, using Leibniz notation, we have the following result, which has the same form as $\boxed{3}$.

$\boxed{5}$ THEOREM If a curve C is described by the parametric equations $x = f(t)$, $y = g(t)$, $\alpha \leqslant t \leqslant \beta$, where f' and g' are continuous on $[\alpha, \beta]$ and C is traversed exactly once as t increases from α to β, then the length of C is

$$L = \int_{\alpha}^{\beta} \sqrt{\left(\frac{dx}{dt}\right)^2 + \left(\frac{dy}{dt}\right)^2}\, dt$$

Notice that the formula in Theorem 5 is consistent with the general formulas $L = \int ds$ and $(ds)^2 = (dx)^2 + (dy)^2$ of Section 7.4.

EXAMPLE 4 If we use the representation of the unit circle given in Example 2 in Section 9.1,

$$x = \cos t \qquad y = \sin t \qquad 0 \leqslant t \leqslant 2\pi$$

then $dx/dt = -\sin t$ and $dy/dt = \cos t$, so Theorem 5 gives

$$L = \int_0^{2\pi} \sqrt{\left(\frac{dx}{dt}\right)^2 + \left(\frac{dy}{dt}\right)^2}\, dt = \int_0^{2\pi} \sqrt{\sin^2 t + \cos^2 t}\, dt$$

$$= \int_0^{2\pi} dt = 2\pi$$

as expected. If, on the other hand, we use the representation given in Example 3 in Section 9.1,

$$x = \sin 2t \qquad y = \cos 2t \qquad 0 \leqslant t \leqslant 2\pi$$

then $dx/dt = 2\cos 2t$, $dy/dt = -2\sin 2t$, and the integral in Theorem 5 gives

$$\int_0^{2\pi} \sqrt{\left(\frac{dx}{dt}\right)^2 + \left(\frac{dy}{dt}\right)^2}\, dt = \int_0^{2\pi} \sqrt{4\cos^2(2t) + 4\sin^2(2t)}\, dt = \int_0^{2\pi} 2\, dt = 4\pi$$

⊘ Notice that the integral gives twice the arc length of the circle because as t increases from 0 to 2π, the point $(\sin 2t, \cos 2t)$ traverses the circle twice. In general, when finding the length of a curve C from a parametric representation, we have to be careful to ensure that C is traversed only once as t increases from α to β. ∎

$\boxed{V}$ EXAMPLE 5 Find the length of one arch of the cycloid $x = r(\theta - \sin\theta)$, $y = r(1 - \cos\theta)$.

SOLUTION From Example 3 we see that one arch is described by the parameter interval $0 \leqslant \theta \leqslant 2\pi$. Since

$$\frac{dx}{d\theta} = r(1 - \cos\theta) \qquad \text{and} \qquad \frac{dy}{d\theta} = r\sin\theta$$

we have

$$L = \int_0^{2\pi} \sqrt{\left(\frac{dx}{d\theta}\right)^2 + \left(\frac{dy}{d\theta}\right)^2} \, d\theta = \int_0^{2\pi} \sqrt{r^2(1 - \cos\theta)^2 + r^2 \sin^2\theta} \, d\theta$$

$$= \int_0^{2\pi} \sqrt{r^2(1 - 2\cos\theta + \cos^2\theta + \sin^2\theta)} \, d\theta = r \int_0^{2\pi} \sqrt{2(1 - \cos\theta)} \, d\theta$$

▪ The result of Example 5 says that the length of one arch of a cycloid is eight times the radius of the generating circle (see Figure 5). This was first proved in 1658 by Sir Christopher Wren, who later became the architect of St. Paul's Cathedral in London.

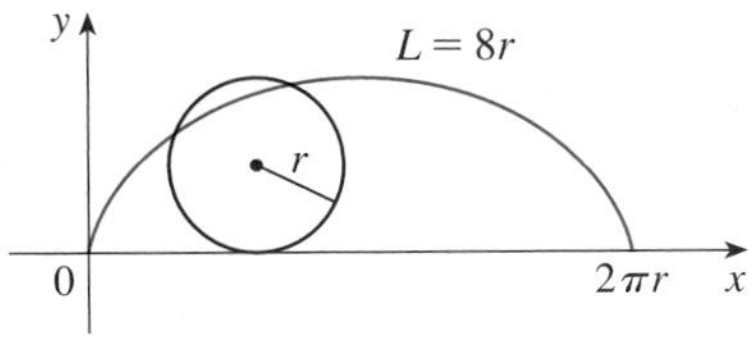

To evaluate this integral we use the identity $\sin^2 x = \frac{1}{2}(1 - \cos 2x)$ with $\theta = 2x$, which gives $1 - \cos\theta = 2\sin^2(\theta/2)$. Since $0 \le \theta \le 2\pi$, we have $0 \le \theta/2 \le \pi$ and so $\sin(\theta/2) \ge 0$. Therefore

$$\sqrt{2(1 - \cos\theta)} = \sqrt{4\sin^2(\theta/2)} = 2|\sin(\theta/2)| = 2\sin(\theta/2)$$

and so

$$L = 2r \int_0^{2\pi} \sin(\theta/2) \, d\theta = 2r\left[-2\cos(\theta/2)\right]_0^{2\pi}$$

$$= 2r[2 + 2] = 8r \qquad ■$$

FIGURE 5

9.2 | EXERCISES

1–2 ▪ Find dy/dx.

1. $x = t \sin t, \quad y = t^2 + t$

2. $x = 1/t, \quad y = \sqrt{t} \, e^{-t}$

3–6 ▪ Find an equation of the tangent to the curve at the point corresponding to the given value of the parameter.

3. $x = 1 + 4t - t^2, \quad y = 2 - t^3; \quad t = 1$

4. $x = t - t^{-1}, \quad y = 1 + t^2; \quad t = 1$

5. $x = t \cos t, \quad y = t \sin t; \quad t = \pi$

6. $x = \sin^3\theta, \quad y = \cos^3\theta; \quad \theta = \pi/6$

---

7. Find an equation of the tangent to the curve $x = 1 + \ln t$, $y = t^2 + 2$ at the point $(1, 3)$ by two methods: (a) without eliminating the parameter and (b) by first eliminating the parameter.

8. Find equations of the tangents to the curve $x = \sin t$, $y = \sin(t + \sin t)$ at the origin. Then graph the curve and the tangents.

9–12 ▪ Find dy/dx and d^2y/dx^2. For which values of t is the curve concave upward?

9. $x = t^2 + 1, \quad y = t^2 + t$

10. $x = t^3 + 1, \quad y = t^2 - t$

11. $x = e^t, \quad y = te^{-t}$

12. $x = \cos 2t, \quad y = \cos t, \quad 0 < t < \pi$

13–16 ▪ Find the points on the curve where the tangent is horizontal or vertical. If you have a graphing device, graph the curve to check your work.

13. $x = t^3 - 3t, \quad y = t^2 - 3$

14. $x = t^3 - 3t, \quad y = t^3 - 3t^2$

15. $x = 2\cos\theta, \quad y = \sin 2\theta$

16. $x = e^{\sin\theta}, \quad y = e^{\cos\theta}$

17. Use a graph to estimate the coordinates of the rightmost point on the curve $x = t - t^6, y = e^t$. Then use calculus to find the exact coordinates.

18. Use a graph to estimate the coordinates of the lowest point and the leftmost point on the curve $x = t^4 - 2t$, $y = t + t^4$. Then find the exact coordinates.

19–20 ▪ Graph the curve in a viewing rectangle that displays all the important aspects of the curve.

19. $x = t^4 - 2t^3 - 2t^2, \quad y = t^3 - t$

20. $x = t^4 + 4t^3 - 8t^2, \quad y = 2t^2 - t$

---

21. Show that the curve $x = \cos t, y = \sin t \cos t$ has two tangents at $(0, 0)$ and find their equations. Sketch the curve.

22. Graph the curve $x = \cos t + 2 \cos 2t$, $y = \sin t + 2 \sin 2t$ to discover where it crosses itself. Then find equations of both tangents at that point.

23. (a) Find the slope of the tangent line to the trochoid $x = r\theta - d \sin \theta$, $y = r - d \cos \theta$ in terms of θ. (See Exercise 34 in Section 9.1.)
(b) Show that if $d < r$, then the trochoid does not have a vertical tangent.

24. (a) Find the slope of the tangent to the astroid $x = a \cos^3\theta$, $y = a \sin^3\theta$ in terms of θ.
(b) At what points is the tangent horizontal or vertical?
(c) At what points does the tangent have slope 1 or -1?

25. At what points on the curve $x = 2t^3$, $y = 1 + 4t - t^2$ does the tangent line have slope 1?

26. Find equations of the tangents to the curve $x = 3t^2 + 1$, $y = 2t^3 + 1$ that pass through the point $(4, 3)$.

27. Use the parametric equations of an ellipse, $x = a \cos \theta$, $y = b \sin \theta$, $0 \le \theta \le 2\pi$, to find the area that it encloses.

28. Find the area enclosed by the curve $x = t^2 - 2t$, $y = \sqrt{t}$ and the y-axis.

29. Find the area enclosed by the x-axis and the curve $x = 1 + e^t$, $y = t - t^2$.

30. Find the area of the region enclosed by the astroid $x = a \cos^3\theta$, $y = a \sin^3\theta$.

31. Find the area under one arch of the trochoid of Exercise 34 in Section 9.1 for the case $d < r$.

32. Let $\mathcal{R}$ be the region enclosed by the loop of the curve in Example 1.
(a) Find the area of $\mathcal{R}$.
(b) If $\mathcal{R}$ is rotated about the x-axis, find the volume of the resulting solid.
(c) Find the centroid of $\mathcal{R}$.

33–36 ■ Set up an integral that represents the length of the curve. Then use your calculator to find the length correct to four decimal places.

33. $x = t + e^{-t}$, $y = t - e^{-t}$, $0 \le t \le 2$

34. $x = t^2 - t$, $y = t^4$, $1 \le t \le 4$

35. $x = t - 2 \sin t$, $y = 1 - 2 \cos t$, $0 \le t \le 4\pi$

36. $x = t + \sqrt{t}$, $y = t - \sqrt{t}$, $0 \le t \le 1$

37–40 ■ Find the exact length of the curve.

37. $x = 1 + 3t^2$, $y = 4 + 2t^3$, $0 \le t \le 1$

38. $x = e^t + e^{-t}$, $y = 5 - 2t$, $0 \le t \le 3$

39. $x = t \sin t$, $y = t \cos t$, $0 \le t \le 1$

40. $x = 3 \cos t - \cos 3t$, $y = 3 \sin t - \sin 3t$, $0 \le t \le \pi$

41–43 ■ Graph the curve and find its length.

41. $x = e^t \cos t$, $y = e^t \sin t$, $0 \le t \le \pi$

42. $x = \cos t + \ln(\tan \frac{1}{2}t)$, $y = \sin t$, $\pi/4 \le t \le 3\pi/4$

43. $x = e^t - t$, $y = 4e^{t/2}$, $-8 \le t \le 3$

44. Find the length of the loop of the curve $x = 3t - t^3$, $y = 3t^2$.

45. Use Simpson's Rule with $n = 6$ to estimate the length of the curve $x = t - e^t$, $y = t + e^t$, $-6 \le t \le 6$.

46. In Exercise 36 in Section 9.1 you were asked to derive the parametric equations $x = 2a \cot \theta$, $y = 2a \sin^2\theta$ for the curve called the witch of Maria Agnesi. Use Simpson's Rule with $n = 4$ to estimate the length of the arc of this curve given by $\pi/4 \le \theta \le \pi/2$.

47–48 ■ Find the distance traveled by a particle with position (x, y) as t varies in the given time interval. Compare with the length of the curve.

47. $x = \sin^2t$, $y = \cos^2t$, $0 \le t \le 3\pi$

48. $x = \cos^2t$, $y = \cos t$, $0 \le t \le 4\pi$

49. Show that the total length of the ellipse $x = a \sin \theta$, $y = b \cos \theta$, $a > b > 0$, is

$$L = 4a \int_0^{\pi/2} \sqrt{1 - e^2 \sin^2\theta} \; d\theta$$

where e is the eccentricity of the ellipse $(e = c/a$, where $c = \sqrt{a^2 - b^2}$).

50. Find the total length of the astroid $x = a \cos^3\theta$, $y = a \sin^3\theta$, where $a > 0$.

51. (a) Graph the **epitrochoid** with equations

$$x = 11 \cos t - 4 \cos(11t/2)$$

$$y = 11 \sin t - 4 \sin(11t/2)$$

What parameter interval gives the complete curve?
(b) Use your CAS to find the approximate length of this curve.

CAS **52.** A curve called **Cornu's spiral** is defined by the parametric equations

$$x = C(t) = \int_0^t \cos(\pi u^2/2)\, du$$

$$y = S(t) = \int_0^t \sin(\pi u^2/2)\, du$$

where C and S are the Fresnel functions that were introduced in Chapter 5.
(a) Graph this curve. What happens as $t \to \infty$ and as $t \to -\infty$?
(b) Find the length of Cornu's spiral from the origin to the point with parameter value t.

53. A string is wound around a circle and then unwound while being held taut. The curve traced by the point P at the end of the string is called the **involute** of the circle. If the circle has radius r and center O and the initial position of P is $(r, 0)$, and if the parameter θ is chosen as in the figure, show that parametric equations of the involute are

$$x = r(\cos\theta + \theta \sin\theta) \qquad y = r(\sin\theta - \theta \cos\theta)$$

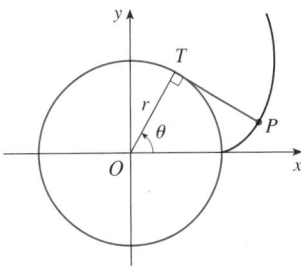

54. A cow is tied to a silo with radius r by a rope just long enough to reach the opposite side of the silo. Find the area available for grazing by the cow.

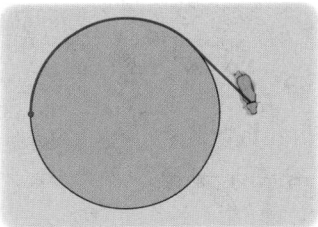

9.3 | POLAR COORDINATES

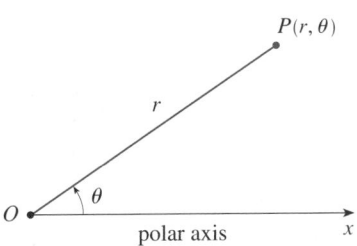

FIGURE 1

A coordinate system represents a point in the plane by an ordered pair of numbers called coordinates. Usually we use Cartesian coordinates, which are directed distances from two perpendicular axes. Here we describe a coordinate system introduced by Newton, called the **polar coordinate system**, which is more convenient for many purposes.

We choose a point in the plane that is called the **pole** (or origin) and is labeled O. Then we draw a ray (half-line) starting at O called the **polar axis**. This axis is usually drawn horizontally to the right and corresponds to the positive x-axis in Cartesian coordinates.

If P is any other point in the plane, let r be the distance from O to P and let θ be the angle (usually measured in radians) between the polar axis and the line OP as in Figure 1. Then the point P is represented by the ordered pair (r, θ) and r, θ are called **polar coordinates** of P. We use the convention that an angle is positive if measured in the counterclockwise direction from the polar axis and negative in the clockwise direction. If $P = O$, then $r = 0$ and we agree that $(0, \theta)$ represents the pole for any value of θ.

We extend the meaning of polar coordinates (r, θ) to the case in which r is negative by agreeing that, as in Figure 2, the points $(-r, \theta)$ and (r, θ) lie on the same line through O and at the same distance $|r|$ from O, but on opposite sides of O. If $r > 0$, the point (r, θ) lies in the same quadrant as θ; if $r < 0$, it lies in the quadrant on the opposite side of the pole. Notice that $(-r, \theta)$ represents the same point as $(r, \theta + \pi)$.

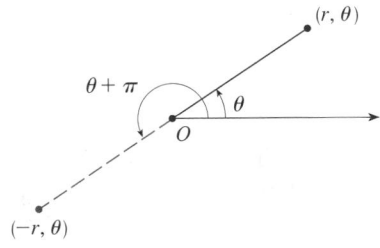

FIGURE 2

EXAMPLE 1 Plot the points whose polar coordinates are given.

(a) $(1, 5\pi/4)$ (b) $(2, 3\pi)$ (c) $(2, -2\pi/3)$ (d) $(-3, 3\pi/4)$

SOLUTION The points are plotted in Figure 3. In part (d) the point $(-3, 3\pi/4)$ is located three units from the pole in the fourth quadrant because the angle $3\pi/4$ is in the second quadrant and $r = -3$ is negative.

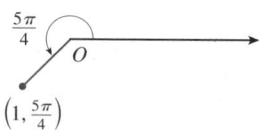 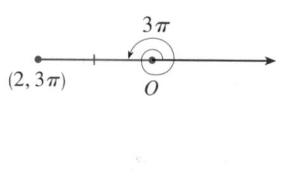 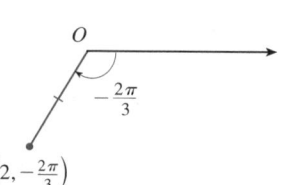 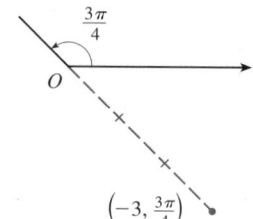

FIGURE 3

In the Cartesian coordinate system every point has only one representation, but in the polar coordinate system each point has many representations. For instance, the point $(1, 5\pi/4)$ in Example 1(a) could be written as $(1, -3\pi/4)$ or $(1, 13\pi/4)$ or $(-1, \pi/4)$. (See Figure 4.)

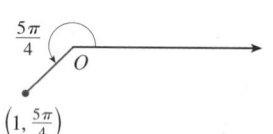 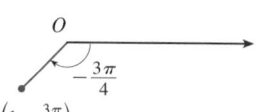 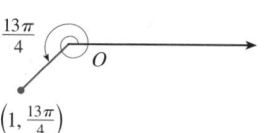 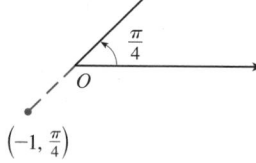

FIGURE 4

In fact, since a complete counterclockwise rotation is given by an angle 2π, the point represented by polar coordinates (r, θ) is also represented by

$$(r, \theta + 2n\pi) \qquad \text{and} \qquad (-r, \theta + (2n + 1)\pi)$$

where n is any integer.

The connection between polar and Cartesian coordinates can be seen from Figure 5, in which the pole corresponds to the origin and the polar axis coincides with the positive x-axis. If the point P has Cartesian coordinates (x, y) and polar coordinates (r, θ), then, from the figure, we have

$$\cos \theta = \frac{x}{r} \qquad \sin \theta = \frac{y}{r}$$

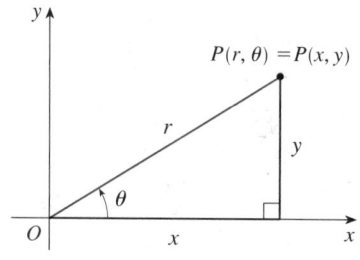

FIGURE 5

and so

1
$$x = r \cos \theta \qquad y = r \sin \theta$$

Although Equations 1 were deduced from Figure 5, which illustrates the case where $r > 0$ and $0 < \theta < \pi/2$, these equations are valid for all values of r and θ. (See the general definition of $\sin \theta$ and $\cos \theta$ in Appendix A.)

Equations 1 allow us to find the Cartesian coordinates of a point when the polar coordinates are known. To find r and θ when x and y are known, we use the equations

$$\boxed{2} \qquad r^2 = x^2 + y^2 \qquad \tan \theta = \frac{y}{x}$$

which can be deduced from Equations 1 or simply read from Figure 5.

EXAMPLE 2 Convert the point $(2, \pi/3)$ from polar to Cartesian coordinates.

SOLUTION Since $r = 2$ and $\theta = \pi/3$, Equations 1 give

$$x = r \cos \theta = 2 \cos \frac{\pi}{3} = 2 \cdot \frac{1}{2} = 1$$

$$y = r \sin \theta = 2 \sin \frac{\pi}{3} = 2 \cdot \frac{\sqrt{3}}{2} = \sqrt{3}$$

Therefore the point is $\left(1, \sqrt{3}\right)$ in Cartesian coordinates.   ■

EXAMPLE 3 Represent the point with Cartesian coordinates $(1, -1)$ in terms of polar coordinates.

SOLUTION If we choose r to be positive, then Equations 2 give

$$r = \sqrt{x^2 + y^2} = \sqrt{1^2 + (-1)^2} = \sqrt{2}$$

$$\tan \theta = \frac{y}{x} = -1$$

Since the point $(1, -1)$ lies in the fourth quadrant, we can choose $\theta = -\pi/4$ or $\theta = 7\pi/4$. Thus one possible answer is $\left(\sqrt{2}, -\pi/4\right)$; another is $(\sqrt{2}, 7\pi/4)$.   ■

NOTE Equations 2 do not uniquely determine θ when x and y are given because, as θ increases through the interval $0 \leqslant \theta < 2\pi$, each value of $\tan \theta$ occurs twice. Therefore, in converting from Cartesian to polar coordinates, it's not good enough just to find r and θ that satisfy Equations 2. As in Example 3, we must choose θ so that the point (r, θ) lies in the correct quadrant.

POLAR CURVES

The **graph of a polar equation** $r = f(\theta)$, or more generally $F(r, \theta) = 0$, consists of all points P that have at least one polar representation (r, θ) whose coordinates satisfy the equation.

▼ EXAMPLE 4 What curve is represented by the polar equation $r = 2$?

SOLUTION The curve consists of all points (r, θ) with $r = 2$. Since r represents the distance from the point to the pole, the curve $r = 2$ represents the circle with center O and radius 2. In general, the equation $r = a$ represents a circle with center O and radius $|a|$. (See Figure 6.)   ■

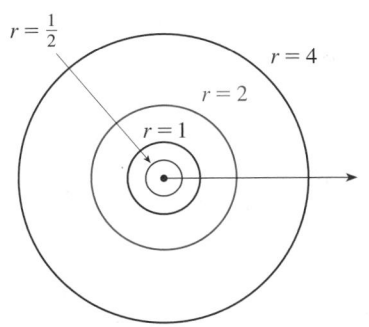

FIGURE 6

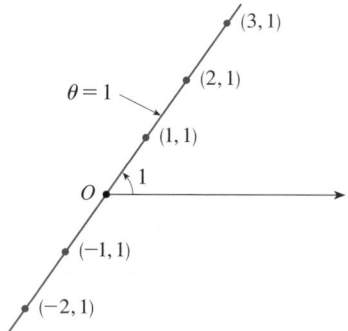

FIGURE 7

EXAMPLE 5 Sketch the polar curve $\theta = 1$.

SOLUTION This curve consists of all points (r, θ) such that the polar angle θ is 1 radian. It is the straight line that passes through O and makes an angle of 1 radian with the polar axis (see Figure 7). Notice that the points $(r, 1)$ on the line with $r > 0$ are in the first quadrant, whereas those with $r < 0$ are in the third quadrant. ■

EXAMPLE 6

(a) Sketch the curve with polar equation $r = 2 \cos \theta$.

(b) Find a Cartesian equation for this curve.

SOLUTION

(a) In Figure 8 we find the values of r for some convenient values of θ and plot the corresponding points (r, θ). Then we join these points to sketch the curve, which appears to be a circle. We have used only values of θ between 0 and π, since if we let θ increase beyond π, we obtain the same points again.

- The curve in Example 6 is symmetric about the polar axis because $\cos(-\theta) = \cos \theta$.

θ	$r = 2 \cos \theta$
0	2
$\pi/6$	$\sqrt{3}$
$\pi/4$	$\sqrt{2}$
$\pi/3$	1
$\pi/2$	0
$2\pi/3$	-1
$3\pi/4$	$-\sqrt{2}$
$5\pi/6$	$-\sqrt{3}$
π	-2

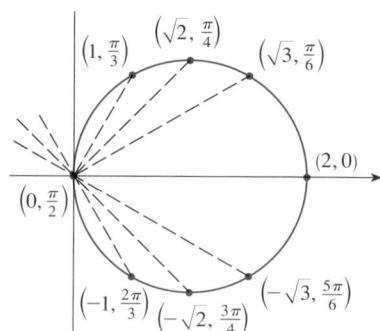

FIGURE 8

Table of values and graph of $r = 2 \cos \theta$

(b) To convert the given equation into a Cartesian equation we use Equations 1 and 2. From $x = r \cos \theta$ we have $\cos \theta = x/r$, so the equation $r = 2 \cos \theta$ becomes $r = 2x/r$, which gives

$$2x = r^2 = x^2 + y^2 \qquad \text{or} \qquad x^2 + y^2 - 2x = 0$$

Completing the square, we obtain

$$(x - 1)^2 + y^2 = 1$$

which is an equation of a circle with center $(1, 0)$ and radius 1. ■

- Figure 9 shows a geometrical illustration that the circle in Example 6 has the equation $r = 2 \cos \theta$. The angle OPQ is a right angle (Why?) and so $r/2 = \cos \theta$.

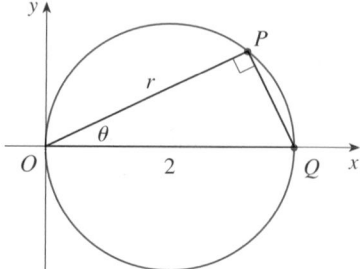

FIGURE 9

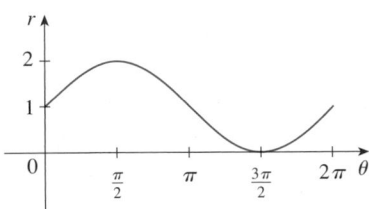

FIGURE 10
$r = 1 + \sin\theta$ in Cartesian coordinates, $0 \le \theta \le 2\pi$

V EXAMPLE 7 Sketch the curve $r = 1 + \sin\theta$.

SOLUTION Instead of plotting points as in Example 6, we first sketch the graph of $r = 1 + \sin\theta$ in *Cartesian* coordinates in Figure 10 by shifting the sine curve up one unit. This enables us to read at a glance the values of r that correspond to increasing values of θ. For instance, we see that as θ increases from 0 to $\pi/2$, r (the distance from O) increases from 1 to 2, so we sketch the corresponding part of the polar curve in Figure 11(a). As θ increases from $\pi/2$ to π, Figure 10 shows that r decreases from 2 to 1, so we sketch the next part of the curve as in Figure 11(b). As θ increases from π to $3\pi/2$, r decreases from 1 to 0 as shown in part (c). Finally, as θ increases from $3\pi/2$ to 2π, r increases from 0 to 1 as shown in part (d). If we let θ increase beyond 2π or decrease beyond 0, we would simply retrace our path. Putting together the parts of the curve from Figure 11(a)–(d), we sketch the complete curve in part (e). It is called a **cardioid** because it's shaped like a heart.

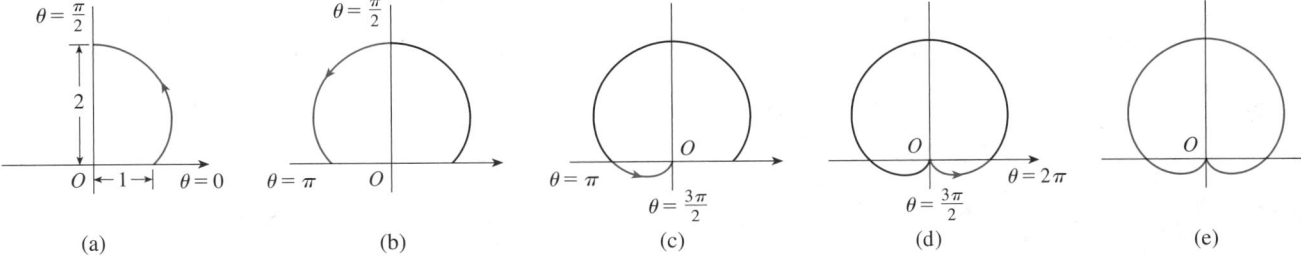

(a)   (b)   (c)   (d)   (e)

FIGURE 11   Stages in sketching the cardioid $r = 1 + \sin\theta$ ■

EXAMPLE 8 Sketch the curve $r = \cos 2\theta$.

SOLUTION As in Example 7, we first sketch $r = \cos 2\theta$, $0 \le \theta \le 2\pi$, in Cartesian coordinates in Figure 12. As θ increases from 0 to $\pi/4$, Figure 12 shows that r decreases from 1 to 0 and so we draw the corresponding portion of the polar curve in Figure 13 (indicated by ①). As θ increases from $\pi/4$ to $\pi/2$, r goes from 0 to -1. This means that the distance from O increases from 0 to 1, but instead of being in the first quadrant this portion of the polar curve (indicated by ②) lies on the opposite side of the pole in the third quadrant. The remainder of the curve is drawn in a similar fashion, with the arrows and numbers indicating the order in which the portions are traced out. The resulting curve has four loops and is called a **four-leaved rose**.

TEC Module 9.3 helps you see how polar curves are traced out by showing animations similar to Figures 10–13.

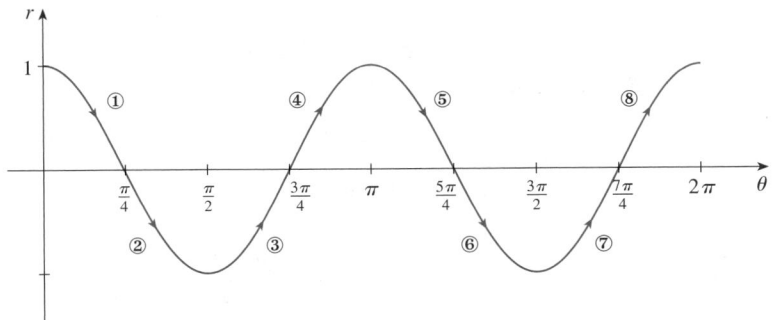

FIGURE 12
$r = \cos 2\theta$ in Cartesian coordinates

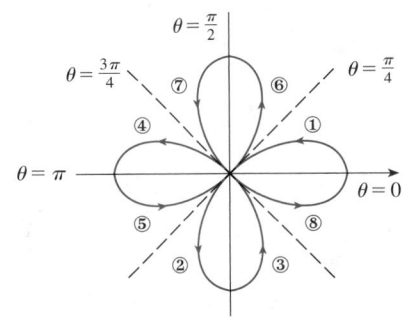

FIGURE 13
Four-leaved rose $r = \cos 2\theta$ ■

TANGENTS TO POLAR CURVES

To find a tangent line to a polar curve $r = f(\theta)$ we regard θ as a parameter and write its parametric equations as

$$x = r \cos \theta = f(\theta) \cos \theta \qquad y = r \sin \theta = f(\theta) \sin \theta$$

Then, using the method for finding slopes of parametric curves (Equation 9.2.1) and the Product Rule, we have

$$\boxed{3} \qquad \frac{dy}{dx} = \frac{\dfrac{dy}{d\theta}}{\dfrac{dx}{d\theta}} = \frac{\dfrac{dr}{d\theta} \sin \theta + r \cos \theta}{\dfrac{dr}{d\theta} \cos \theta - r \sin \theta}$$

We locate horizontal tangents by finding the points where $dy/d\theta = 0$ (provided that $dx/d\theta \neq 0$). Likewise, we locate vertical tangents at the points where $dx/d\theta = 0$ (provided that $dy/d\theta \neq 0$).

Notice that if we are looking for tangent lines at the pole, then $r = 0$ and Equation 3 simplifies to

$$\frac{dy}{dx} = \tan \theta \qquad \text{if} \quad \frac{dr}{d\theta} \neq 0$$

For instance, in Example 8 we found that $r = \cos 2\theta = 0$ when $\theta = \pi/4$ or $3\pi/4$. This means that the lines $\theta = \pi/4$ and $\theta = 3\pi/4$ (or $y = x$ and $y = -x$) are tangent lines to $r = \cos 2\theta$ at the origin.

EXAMPLE 9

(a) For the cardioid $r = 1 + \sin \theta$ of Example 7, find the slope of the tangent line when $\theta = \pi/3$.

(b) Find the points on the cardioid where the tangent line is horizontal or vertical.

SOLUTION Using Equation 3 with $r = 1 + \sin \theta$, we have

$$\frac{dy}{dx} = \frac{\dfrac{dr}{d\theta} \sin \theta + r \cos \theta}{\dfrac{dr}{d\theta} \cos \theta - r \sin \theta} = \frac{\cos \theta \sin \theta + (1 + \sin \theta) \cos \theta}{\cos \theta \cos \theta - (1 + \sin \theta) \sin \theta}$$

$$= \frac{\cos \theta \, (1 + 2 \sin \theta)}{1 - 2 \sin^2\theta - \sin \theta} = \frac{\cos \theta \, (1 + 2 \sin \theta)}{(1 + \sin \theta)(1 - 2 \sin \theta)}$$

(a) The slope of the tangent at the point where $\theta = \pi/3$ is

$$\left. \frac{dy}{dx} \right|_{\theta = \pi/3} = \frac{\cos(\pi/3)(1 + 2 \sin(\pi/3))}{(1 + \sin(\pi/3))(1 - 2 \sin(\pi/3))} = \frac{\frac{1}{2}\left(1 + \sqrt{3}\,\right)}{\left(1 + \sqrt{3}/2\right)\left(1 - \sqrt{3}\,\right)}$$

$$= \frac{1 + \sqrt{3}}{\left(2 + \sqrt{3}\,\right)\left(1 - \sqrt{3}\,\right)} = \frac{1 + \sqrt{3}}{-1 - \sqrt{3}} = -1$$

(b) Observe that

$$\frac{dy}{d\theta} = \cos \theta \, (1 + 2 \sin \theta) = 0 \qquad \text{when } \theta = \frac{\pi}{2}, \frac{3\pi}{2}, \frac{7\pi}{6}, \frac{11\pi}{6}$$

$$\frac{dx}{d\theta} = (1 + \sin \theta)(1 - 2 \sin \theta) = 0 \qquad \text{when } \theta = \frac{3\pi}{2}, \frac{\pi}{6}, \frac{5\pi}{6}$$

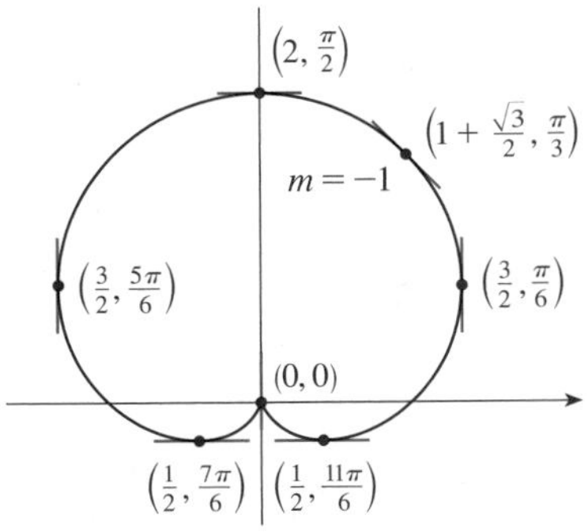

Therefore there are horizontal tangents at the points $(2, \pi/2)$, $\left(\frac{1}{2}, 7\pi/6\right)$, $\left(\frac{1}{2}, 11\pi/6\right)$ and vertical tangents at $\left(\frac{3}{2}, \pi/6\right)$ and $\left(\frac{3}{2}, 5\pi/6\right)$. When $\theta = 3\pi/2$, both $dy/d\theta$ and $dx/d\theta$ are 0, so we must be careful. Using l'Hospital's Rule, we have

$$\lim_{\theta \to (3\pi/2)^-} \frac{dy}{dx} = \left(\lim_{\theta \to (3\pi/2)^-} \frac{1 + 2 \sin \theta}{1 - 2 \sin \theta} \right) \left(\lim_{\theta \to (3\pi/2)^-} \frac{\cos \theta}{1 + \sin \theta} \right)$$

$$= -\frac{1}{3} \lim_{\theta \to (3\pi/2)^-} \frac{\cos \theta}{1 + \sin \theta}$$

$$= -\frac{1}{3} \lim_{\theta \to (3\pi/2)^-} \frac{-\sin \theta}{\cos \theta} = \infty$$

By symmetry,

$$\lim_{\theta \to (3\pi/2)^+} \frac{dy}{dx} = -\infty$$

Thus there is a vertical tangent line at the pole (see Figure 14). ■

FIGURE 14
Tangent lines for $r = 1 + \sin \theta$

(Figure 14 shows the curve with labeled points: $(2, \frac{\pi}{2})$, $(1 + \frac{\sqrt{3}}{2}, \frac{\pi}{3})$, $m = -1$, $(\frac{3}{2}, \frac{5\pi}{6})$, $(\frac{3}{2}, \frac{\pi}{6})$, $(0, 0)$, $(\frac{1}{2}, \frac{7\pi}{6})$, $(\frac{1}{2}, \frac{11\pi}{6})$.)

NOTE Instead of having to remember Equation 3, we could employ the method used to derive it. For instance, in Example 9 we could have written

$$x = r \cos \theta = (1 + \sin \theta) \cos \theta = \cos \theta + \tfrac{1}{2} \sin 2\theta$$

$$y = r \sin \theta = (1 + \sin \theta) \sin \theta = \sin \theta + \sin^2\theta$$

Then we have

$$\frac{dy}{dx} = \frac{dy/d\theta}{dx/d\theta} = \frac{\cos \theta + 2 \sin \theta \cos \theta}{-\sin \theta + \cos 2\theta} = \frac{\cos \theta + \sin 2\theta}{-\sin \theta + \cos 2\theta}$$

which is equivalent to our previous expression.

GRAPHING POLAR CURVES WITH GRAPHING DEVICES

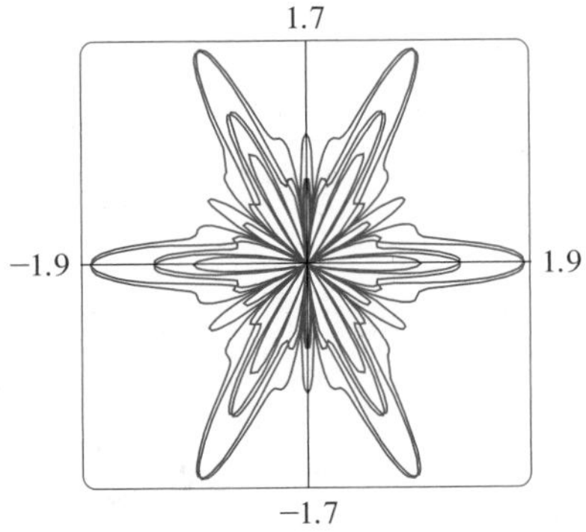

Although it's useful to be able to sketch simple polar curves by hand, we need to use a graphing calculator or computer when we are faced with a curve as complicated as the one shown in Figure 15.

Some graphing devices have commands that enable us to graph polar curves directly. With other machines we need to convert to parametric equations first. In this case we take the polar equation $r = f(\theta)$ and write its parametric equations as

$$x = r \cos \theta = f(\theta) \cos \theta \qquad y = r \sin \theta = f(\theta) \sin \theta$$

Some machines require that the parameter be called t rather than θ.

FIGURE 15
$r = \sin^2(1.2\theta) + \cos^3(6\theta)$

EXAMPLE 10 Graph the curve $r = \sin(8\theta/5)$.

■ **www.stewartcalculus.com**
See Additional Example A.

SOLUTION Let's assume that our graphing device doesn't have a built-in polar graphing command. In this case we need to work with the corresponding parametric equations, which are

$$x = r\cos\theta = \sin(8\theta/5)\cos\theta$$

$$y = r\sin\theta = \sin(8\theta/5)\sin\theta$$

In any case we need to determine the domain for θ. So we ask ourselves: How many complete rotations are required until the curve starts to repeat itself? If the answer is n, then

$$\sin\frac{8(\theta + 2n\pi)}{5} = \sin\left(\frac{8\theta}{5} + \frac{16n\pi}{5}\right) = \sin\frac{8\theta}{5}$$

and so we require that $16n\pi/5$ be an even multiple of π. This will first occur when $n = 5$. Therefore we will graph the entire curve if we specify that $0 \le \theta \le 10\pi$. Switching from θ to t, we have the equations

$$x = \sin(8t/5)\cos t \qquad y = \sin(8t/5)\sin t \qquad 0 \le t \le 10\pi$$

and Figure 16 shows the resulting curve. Notice that this rose has 16 loops. ■

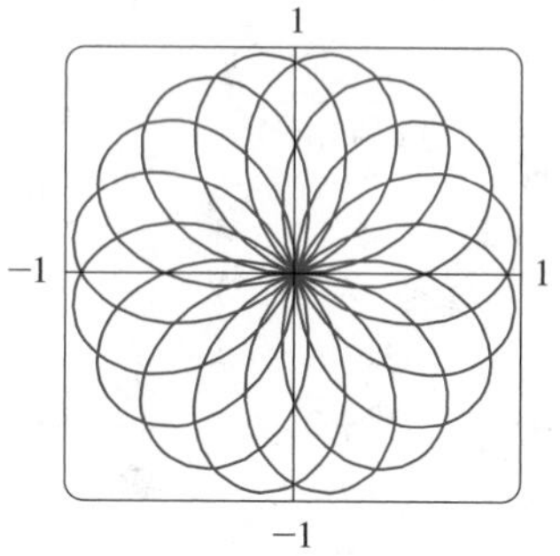

FIGURE 16
$r = \sin(8\theta/5)$

9.3 | EXERCISES

1–2 ■ Plot the point whose polar coordinates are given. Then find two other pairs of polar coordinates of this point, one with $r > 0$ and one with $r < 0$.

1. (a) $(2, \pi/3)$ (b) $(1, -3\pi/4)$ (c) $(-1, \pi/2)$

2. (a) $(1, 7\pi/4)$ (b) $(-3, \pi/6)$ (c) $(1, -1)$

3–4 ■ Plot the point whose polar coordinates are given. Then find the Cartesian coordinates of the point.

3. (a) $(1, \pi)$ (b) $\left(2, -2\pi/3\right)$ (c) $(-2, 3\pi/4)$

4. (a) $\left(-\sqrt{2}, 5\pi/4\right)$ (b) $(1, 5\pi/2)$ (c) $(2, -7\pi/6)$

5–6 ■ The Cartesian coordinates of a point are given.
(i) Find polar coordinates (r, θ) of the point, where $r > 0$ and $0 \le \theta < 2\pi$.
(ii) Find polar coordinates (r, θ) of the point, where $r < 0$ and $0 \le \theta < 2\pi$.

5. (a) $(2, -2)$ (b) $\left(-1, \sqrt{3}\right)$

6. (a) $\left(3\sqrt{3}, 3\right)$ (b) $(1, -2)$

7–12 ■ Sketch the region in the plane consisting of points whose polar coordinates satisfy the given conditions.

7. $1 \le r \le 2$

8. $0 \le r < 2, \quad \pi \le \theta \le 3\pi/2$

9. $r \ge 0, \quad \pi/4 \le \theta \le 3\pi/4$

10. $1 \le r \le 3, \quad \pi/6 < \theta < 5\pi/6$

11. $2 < r < 3, \quad 5\pi/3 \le \theta \le 7\pi/3$

12. $-1 \le r \le 1, \quad \pi/4 \le \theta \le 3\pi/4$

13–16 ■ Identify the curve by finding a Cartesian equation for the curve.

13. $r = 2\cos\theta$ **14.** $\theta = \pi/3$

15. $r^2\cos 2\theta = 1$ **16.** $r = \tan\theta \sec\theta$

17–20 ■ Find a polar equation for the curve represented by the given Cartesian equation.

17. $y = 1 + 3x$ **18.** $4y^2 = x$

19. $x^2 + y^2 = 2cx$ **20.** $xy = 4$

21–22 ▪ For each of the described curves, decide if the curve would be more easily given by a polar equation or a Cartesian equation. Then write an equation for the curve.

21. (a) A line through the origin that makes an angle of $\pi/6$ with the positive x-axis
  (b) A vertical line through the point $(3, 3)$

22. (a) A circle with radius 5 and center $(2, 3)$
  (b) A circle centered at the origin with radius 4

23–40 ▪ Sketch the curve with the given polar equation by first sketching the graph of r as a function of θ in Cartesian coordinates.

23. $r = -2 \sin \theta$ **24.** $r = 1 - \cos \theta$

25. $r = 2(1 + \cos \theta)$ **26.** $r = 1 + 2 \cos \theta$

27. $r = \theta, \ \theta \geqslant 0$ **28.** $r = \ln \theta, \ \theta \geqslant 1$

29. $r = 4 \sin 3\theta$ **30.** $r = \cos 5\theta$

31. $r = 2 \cos 4\theta$ **32.** $r = 3 \cos 6\theta$

33. $r = 1 - 2 \sin \theta$ **34.** $r = 2 + \sin \theta$

35. $r^2 = 9 \sin 2\theta$ **36.** $r^2 = \cos 4\theta$

37. $r = 2 + \sin 3\theta$ **38.** $r^2 \theta = 1$

39. $r = 1 + 2 \cos 2\theta$ **40.** $r = 3 + 4 \cos \theta$

41–42 ▪ The figure shows a graph of r as a function of θ in Cartesian coordinates. Use it to sketch the corresponding polar curve.

41. 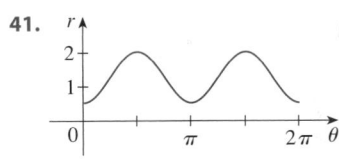 **42.**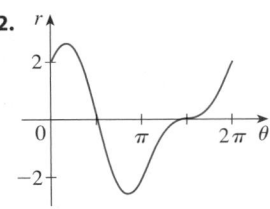

---

43. Show that the polar curve $r = 4 + 2 \sec \theta$ (called a **conchoid**) has the line $x = 2$ as a vertical asymptote by showing that $\lim_{r \to \pm\infty} x = 2$. Use this fact to help sketch the conchoid.

44. Sketch the curve $(x^2 + y^2)^3 = 4x^2y^2$.

45. Show that the curve $r = \sin \theta \tan \theta$ (called a **cissoid of Diocles**) has the line $x = 1$ as a vertical asymptote. Show also that the curve lies entirely within the vertical strip $0 \leqslant x < 1$. Use these facts to help sketch the cissoid.

46. Match the polar equations with the graphs labeled I–VI. Give reasons for your choices. (Don't use a graphing device.)
  (a) $r = \sqrt{\theta}, \ 0 \leqslant \theta \leqslant 16\pi$ (b) $r = \theta^2, \ 0 \leqslant \theta \leqslant 16\pi$
  (c) $r = \cos(\theta/3)$ (d) $r = 1 + 2\cos \theta$
  (e) $r = 2 + \sin 3\theta$ (f) $r = 1 + 2 \sin 3\theta$

I 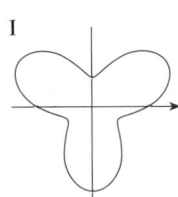 II 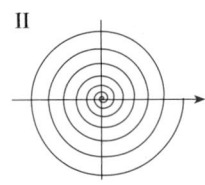 III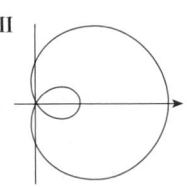

IV 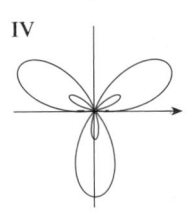 V 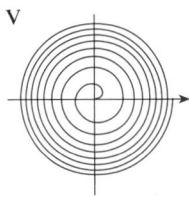 VI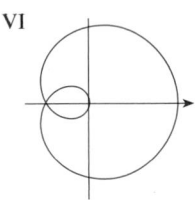

47–50 ▪ Find the slope of the tangent line to the given polar curve at the point specified by the value of θ.

47. $r = 2 \sin \theta, \quad \theta = \pi/6$ **48.** $r = 2 - \sin \theta, \quad \theta = \pi/3$

49. $r = 1/\theta, \quad \theta = \pi$ **50.** $r = \cos(\theta/3), \quad \theta = \pi$

51–54 ▪ Find the points on the given curve where the tangent line is horizontal or vertical.

51. $r = 3 \cos \theta$ **52.** $r = e^\theta$

53. $r = 1 + \cos \theta$ **54.** $r^2 = \sin 2\theta$

---

55. Show that the polar equation $r = a \sin \theta + b \cos \theta$, where $ab \neq 0$, represents a circle, and find its center and radius.

56. Show that the curves $r = a \sin \theta$ and $r = a \cos \theta$ intersect at right angles.

57–60 ▪ Use a graphing device to graph the polar curve. Choose the parameter interval to make sure that you produce the entire curve.

57. $r = e^{\sin \theta} - 2 \cos(4\theta)$ (butterfly curve)

58. $r = \sin^2(4\theta) + \cos(4\theta)$

59. $r = 1 + \cos^{999}\theta$ (PacMan curve)

60. $r = |\tan \theta|^{|\cot \theta|}$ (valentine curve)

61. How are the graphs of $r = 1 + \sin(\theta - \pi/6)$ and $r = 1 + \sin(\theta - \pi/3)$ related to the graph of $r = 1 + \sin \theta$? In general, how is the graph of $r = f(\theta - \alpha)$ related to the graph of $r = f(\theta)$?

62. Use a graph to estimate the y-coordinate of the highest points on the curve $r = \sin 2\theta$. Then use calculus to find the exact value.

63. (a) Investigate the family of curves defined by the polar equations $r = \sin n\theta$, where n is a positive integer. How is the number of loops related to n?
(b) What happens if the equation in part (a) is replaced by $r = |\sin n\theta|$?

64. A family of curves is given by the equations $r = 1 + c \sin n\theta$, where c is a real number and n is a positive integer. How does the graph change as n increases? How does it change as c changes? Illustrate by graphing enough members of the family to support your conclusions.

65. A family of curves has polar equations

$$r = \frac{1 - a \cos \theta}{1 + a \cos \theta}$$

Investigate how the graph changes as the number a changes. In particular, you should identify the transitional values of a for which the basic shape of the curve changes.

66. The astronomer Giovanni Cassini (1625–1712) studied the family of curves with polar equations

$$r^4 - 2c^2 r^2 \cos 2\theta + c^4 - a^4 = 0$$

where a and c are positive real numbers. These curves are called the **ovals of Cassini** even though they are oval

shaped only for certain values of a and c. (Cassini thought that these curves might represent planetary orbits better than Kepler's ellipses.) Investigate the variety of shapes that these curves may have. In particular, how are a and c related to each other when the curve splits into two parts?

67. Let P be any point (except the origin) on the curve $r = f(\theta)$. If ψ is the angle between the tangent line at P and the radial line OP, show that

$$\tan \psi = \frac{r}{dr/d\theta}$$

[*Hint:* Observe that $\psi = \phi - \theta$ in the figure.]

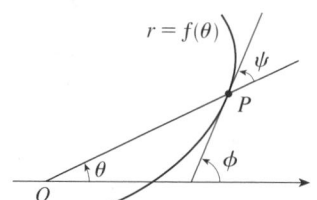

68. (a) Use Exercise 67 to show that the angle between the tangent line and the radial line is $\psi = \pi/4$ at every point on the curve $r = e^\theta$.
(b) Illustrate part (a) by graphing the curve and the tangent lines at the points where $\theta = 0$ and $\pi/2$.
(c) Prove that any polar curve $r = f(\theta)$ with the property that the angle ψ between the radial line and the tangent line is a constant must be of the form $r = Ce^{k\theta}$, where C and k are constants.

9.4 | AREAS AND LENGTHS IN POLAR COORDINATES

In this section we develop the formula for the area of a region whose boundary is given by a polar equation. We need to use the formula for the area of a sector of a circle

$$\boxed{1} \qquad\qquad A = \tfrac{1}{2} r^2 \theta$$

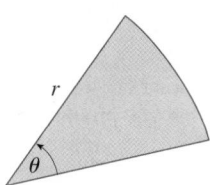

FIGURE 1

where, as in Figure 1, r is the radius and θ is the radian measure of the central angle. Formula 1 follows from the fact that the area of a sector is proportional to its central angle: $A = (\theta/2\pi)\pi r^2 = \tfrac{1}{2} r^2 \theta$. (See also Exercise 69 in Section 6.2.)

Let $\mathcal{R}$ be the region, illustrated in Figure 2, bounded by the polar curve $r = f(\theta)$ and by the rays $\theta = a$ and $\theta = b$, where f is a positive continuous function and where $0 < b - a \leq 2\pi$. We divide the interval $[a, b]$ into subintervals with endpoints θ_0, θ_1, θ_2, ..., θ_n and equal width $\Delta\theta$. The rays $\theta = \theta_i$ then divide $\mathcal{R}$ into n smaller regions with central angle $\Delta\theta = \theta_i - \theta_{i-1}$. If we choose θ_i^* in the ith subinterval $[\theta_{i-1}, \theta_i]$, then the area ΔA_i of the ith region is approximated by the area of the sector of a circle with central angle $\Delta\theta$ and radius $f(\theta_i^*)$. (See Figure 3.)

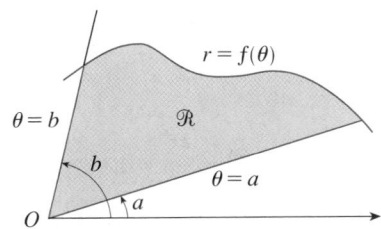

FIGURE 2

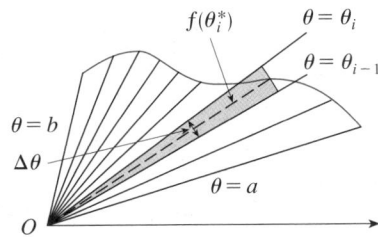

FIGURE 3

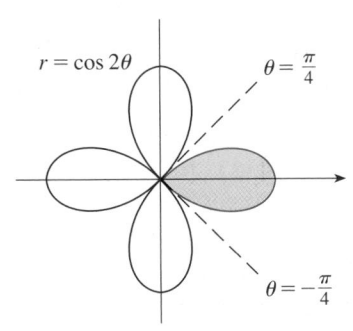

FIGURE 4

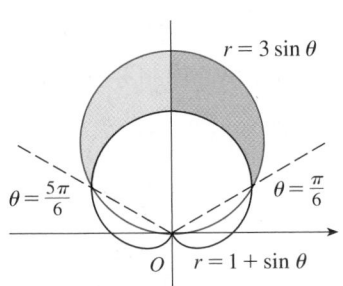

FIGURE 5

Thus from Formula 1 we have

$$\Delta A_i \approx \tfrac{1}{2}[f(\theta_i^*)]^2 \Delta\theta$$

and so an approximation to the total area A of $\mathcal{R}$ is

$$\boxed{2} \qquad A \approx \sum_{i=1}^{n} \tfrac{1}{2}[f(\theta_i^*)]^2 \Delta\theta$$

It appears from Figure 3 that the approximation in $\boxed{2}$ improves as $n \to \infty$. But the sums in $\boxed{2}$ are Riemann sums for the function $g(\theta) = \tfrac{1}{2}[f(\theta)]^2$, so

$$\lim_{n \to \infty} \sum_{i=1}^{n} \tfrac{1}{2}[f(\theta_i^*)]^2 \Delta\theta = \int_a^b \tfrac{1}{2}[f(\theta)]^2 \, d\theta$$

It therefore appears plausible (and can in fact be proved) that the formula for the area A of the polar region $\mathcal{R}$ is

$$\boxed{3} \qquad A = \int_a^b \tfrac{1}{2}[f(\theta)]^2 \, d\theta$$

Formula 3 is often written as

$$\boxed{4} \qquad A = \int_a^b \tfrac{1}{2}r^2 \, d\theta$$

with the understanding that $r = f(\theta)$. Note the similarity between Formulas 1 and 4.

When we apply Formula 3 or 4 it is helpful to think of the area as being swept out by a rotating ray through O that starts with angle a and ends with angle b.

▼ EXAMPLE 1 Find the area enclosed by one loop of the four-leaved rose $r = \cos 2\theta$.

SOLUTION The curve $r = \cos 2\theta$ was sketched in Example 8 in Section 9.3. Notice from Figure 4 that the region enclosed by the right loop is swept out by a ray that rotates from $\theta = -\pi/4$ to $\theta = \pi/4$. Therefore Formula 4 gives

$$A = \int_{-\pi/4}^{\pi/4} \tfrac{1}{2}r^2 \, d\theta = \tfrac{1}{2}\int_{-\pi/4}^{\pi/4} \cos^2 2\theta \, d\theta = \int_0^{\pi/4} \cos^2 2\theta \, d\theta$$

$$= \int_0^{\pi/4} \tfrac{1}{2}(1 + \cos 4\theta) \, d\theta = \tfrac{1}{2}\big[\theta + \tfrac{1}{4}\sin 4\theta\big]_0^{\pi/4} = \frac{\pi}{8} \qquad \blacksquare$$

▼ EXAMPLE 2 Find the area of the region that lies inside the circle $r = 3 \sin \theta$ and outside the cardioid $r = 1 + \sin \theta$.

SOLUTION The cardioid (see Example 7 in Section 9.3) and the circle are sketched in Figure 5 and the desired region is shaded. The values of a and b in Formula 4 are determined by finding the points of intersection of the two curves. They intersect when $3 \sin \theta = 1 + \sin \theta$, which gives $\sin \theta = \tfrac{1}{2}$, so $\theta = \pi/6, 5\pi/6$. The desired area can be found by subtracting the area inside the cardioid between $\theta = \pi/6$ and

$\theta = 5\pi/6$ from the area inside the circle from $\pi/6$ to $5\pi/6$. Thus

$$A = \tfrac{1}{2} \int_{\pi/6}^{5\pi/6} (3 \sin \theta)^2 \, d\theta - \tfrac{1}{2} \int_{\pi/6}^{5\pi/6} (1 + \sin \theta)^2 \, d\theta$$

Since the region is symmetric about the vertical axis $\theta = \pi/2$, we can write

$$A = 2 \left[\tfrac{1}{2} \int_{\pi/6}^{\pi/2} 9 \sin^2\theta \, d\theta - \tfrac{1}{2} \int_{\pi/6}^{\pi/2} (1 + 2 \sin \theta + \sin^2\theta) \, d\theta \right]$$

$$= \int_{\pi/6}^{\pi/2} (8 \sin^2\theta - 1 - 2 \sin \theta) \, d\theta$$

$$= \int_{\pi/6}^{\pi/2} (3 - 4 \cos 2\theta - 2 \sin \theta) \, d\theta \qquad \left[\text{because } \sin^2\theta = \tfrac{1}{2}(1 - \cos 2\theta) \right]$$

$$= 3\theta - 2 \sin 2\theta + 2 \cos \theta \Big]_{\pi/6}^{\pi/2} = \pi \qquad \blacksquare$$

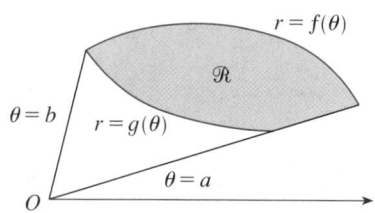

$r = f(\theta)$

$\mathcal{R}$

$\theta = b$ $r = g(\theta)$

$\theta = a$

O

FIGURE 6

Example 2 illustrates the procedure for finding the area of the region bounded by two polar curves. In general, let $\mathcal{R}$ be a region, as illustrated in Figure 6, that is bounded by curves with polar equations $r = f(\theta)$, $r = g(\theta)$, $\theta = a$, and $\theta = b$, where $f(\theta) \geqslant g(\theta) \geqslant 0$ and $0 < b - a \leqslant 2\pi$. The area A of $\mathcal{R}$ is found by subtracting the area inside $r = g(\theta)$ from the area inside $r = f(\theta)$, so using Formula 3 we have

$$A = \int_a^b \tfrac{1}{2}[f(\theta)]^2 \, d\theta - \int_a^b \tfrac{1}{2}[g(\theta)]^2 \, d\theta = \tfrac{1}{2} \int_a^b \left([f(\theta)]^2 - [g(\theta)]^2 \right) d\theta$$

⊘ **CAUTION** The fact that a single point has many representations in polar coordinates sometimes makes it difficult to find all the points of intersection of two polar curves. For instance, it is obvious from Figure 5 that the circle and the cardioid have three points of intersection; however, in Example 2 we solved the equations $r = 3 \sin \theta$ and $r = 1 + \sin \theta$ and found only two such points, $\left(\tfrac{3}{2}, \pi/6 \right)$ and $\left(\tfrac{3}{2}, 5\pi/6 \right)$. The origin is also a point of intersection, but we can't find it by solving the equations of the curves because the origin has no single representation in polar coordinates that satisfies both equations. Notice that, when represented as $(0, 0)$ or $(0, \pi)$, the origin satisfies $r = 3 \sin \theta$ and so it lies on the circle; when represented as $(0, 3\pi/2)$, it satisfies $r = 1 + \sin \theta$ and so it lies on the cardioid. Think of two points moving along the curves as the parameter value θ increases from 0 to 2π. On one curve the origin is reached at $\theta = 0$ and $\theta = \pi$; on the other curve it is reached at $\theta = 3\pi/2$. The points don't collide at the origin because they reach the origin at different times, but the curves intersect there nonetheless.

Thus, to find *all* points of intersection of two polar curves, it is recommended that you draw the graphs of both curves. It is especially convenient to use a graphing calculator or computer to help with this task.

EXAMPLE 3 Find all points of intersection of the curves $r = \cos 2\theta$ and $r = \tfrac{1}{2}$.

SOLUTION If we solve the equations $r = \cos 2\theta$ and $r = \tfrac{1}{2}$, we get $\cos 2\theta = \tfrac{1}{2}$ and therefore $2\theta = \pi/3, 5\pi/3, 7\pi/3, 11\pi/3$. Thus the values of θ between 0 and 2π that satisfy both equations are $\theta = \pi/6, 5\pi/6, 7\pi/6, 11\pi/6$. We have found four points of intersection: $\left(\tfrac{1}{2}, \pi/6 \right)$, $\left(\tfrac{1}{2}, 5\pi/6 \right)$, $\left(\tfrac{1}{2}, 7\pi/6 \right)$, and $\left(\tfrac{1}{2}, 11\pi/6 \right)$.

However, you can see from Figure 7 that the curves have four other points of intersection—namely, $\left(\tfrac{1}{2}, \pi/3 \right)$, $\left(\tfrac{1}{2}, 2\pi/3 \right)$, $\left(\tfrac{1}{2}, 4\pi/3 \right)$, and $\left(\tfrac{1}{2}, 5\pi/3 \right)$. These can be found using symmetry or by noticing that another equation of the circle is $r = -\tfrac{1}{2}$ and then solving the equations $r = \cos 2\theta$ and $r = -\tfrac{1}{2}$. $\blacksquare$

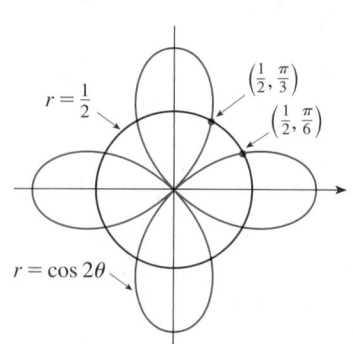

$r = \tfrac{1}{2}$

$\left(\tfrac{1}{2}, \tfrac{\pi}{3} \right)$

$\left(\tfrac{1}{2}, \tfrac{\pi}{6} \right)$

$r = \cos 2\theta$

FIGURE 7

ARC LENGTH

To find the length of a polar curve $r = f(\theta)$, $a \leq \theta \leq b$, we regard θ as a parameter and write the parametric equations of the curve as

$$x = r \cos \theta = f(\theta) \cos \theta \qquad y = r \sin \theta = f(\theta) \sin \theta$$

Using the Product Rule and differentiating with respect to θ, we obtain

$$\frac{dx}{d\theta} = \frac{dr}{d\theta} \cos \theta - r \sin \theta \qquad \frac{dy}{d\theta} = \frac{dr}{d\theta} \sin \theta + r \cos \theta$$

so, using $\cos^2\theta + \sin^2\theta = 1$, we have

$$\left(\frac{dx}{d\theta}\right)^2 + \left(\frac{dy}{d\theta}\right)^2 = \left(\frac{dr}{d\theta}\right)^2 \cos^2\theta - 2r \frac{dr}{d\theta} \cos \theta \sin \theta + r^2 \sin^2\theta$$

$$+ \left(\frac{dr}{d\theta}\right)^2 \sin^2\theta + 2r \frac{dr}{d\theta} \sin \theta \cos \theta + r^2 \cos^2\theta$$

$$= \left(\frac{dr}{d\theta}\right)^2 + r^2$$

Assuming that f' is continuous, we can use Formula 9.2.5 to write the arc length as

$$L = \int_a^b \sqrt{\left(\frac{dx}{d\theta}\right)^2 + \left(\frac{dy}{d\theta}\right)^2}\, d\theta$$

Therefore the length of a curve with polar equation $r = f(\theta)$, $a \leq \theta \leq b$, is

5
$$L = \int_a^b \sqrt{r^2 + \left(\frac{dr}{d\theta}\right)^2}\, d\theta$$

◢ EXAMPLE 4 Find the length of the cardioid $r = 1 + \sin \theta$.

SOLUTION The cardioid is shown in Figure 8. (We sketched it in Example 7 in Section 9.3.) Its full length is given by the parameter interval $0 \leq \theta \leq 2\pi$, so Formula 5 gives

$$L = \int_0^{2\pi} \sqrt{r^2 + \left(\frac{dr}{d\theta}\right)^2}\, d\theta = \int_0^{2\pi} \sqrt{(1 + \sin \theta)^2 + \cos^2\theta}\, d\theta$$

$$= \int_0^{2\pi} \sqrt{2 + 2 \sin \theta}\, d\theta$$

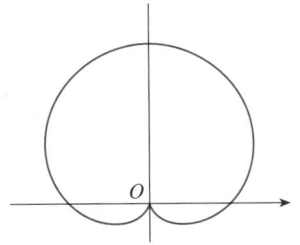

FIGURE 8
$r = 1 + \sin \theta$

We could evaluate this integral by multiplying and dividing the integrand by $\sqrt{2 - 2 \sin \theta}$, or we could use a computer algebra system. In any event, we find that the length of the cardioid is $L = 8$. ∎

9.4 | EXERCISES

1–4 ■ Find the area of the region that is bounded by the given curve and lies in the specified sector.

1. $r = e^{-\theta/4}$, $\pi/2 \leq \theta \leq \pi$

2. $r = \cos \theta$, $0 \leq \theta \leq \pi/6$

3. $r^2 = 9 \sin 2\theta$, $r \geq 0$, $0 \leq \theta \leq \pi/2$

4. $r = \tan \theta$, $\pi/6 \leq \theta \leq \pi/3$

5–8 ■ Find the area of the shaded region.

5.
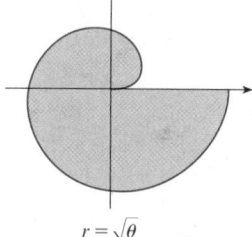
$r = \sqrt{\theta}$

6.
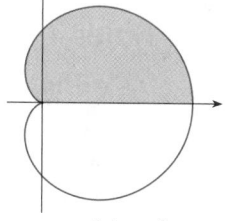
$r = 1 + \cos \theta$

7.
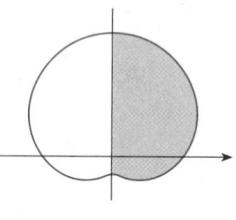
$r = 4 + 3 \sin \theta$

8.
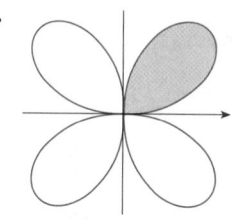
$r = \sin 2\theta$

9–12 ■ Sketch the curve and find the area that it encloses.

9. $r = 2 \sin \theta$

10. $r = 1 - \sin \theta$

11. $r = 3 + 2 \cos \theta$

12. $r = 4 + 3 \sin \theta$

13–14 ■ Graph the curve and find the area that it encloses.

13. $r = \sqrt{1 + \cos^2(5\theta)}$

14. $r = 3 - 2 \cos 4\theta$

15–18 ■ Find the area of the region enclosed by one loop of the curve.

15. $r = 4 \cos 3\theta$

16. $r^2 = \sin 2\theta$

17. $r = 1 + 2 \sin \theta$ (inner loop)

18. $r = 2 \cos \theta - \sec \theta$

19–22 ■ Find the area of the region that lies inside the first curve and outside the second curve.

19. $r = 2 \cos \theta$, $r = 1$

20. $r = 1 - \sin \theta$, $r = 1$

21. $r = 3 \cos \theta$, $r = 1 + \cos \theta$

22. $r = 2 + \sin \theta$, $r = 3 \sin \theta$

23–26 ■ Find the area of the region that lies inside both curves.

23. $r = \sqrt{3} \cos \theta$, $r = \sin \theta$

24. $r = 1 + \cos \theta$, $r = 1 - \cos \theta$

25. $r = \sin 2\theta$, $r = \cos 2\theta$

26. $r^2 = 2 \sin 2\theta$, $r = 1$

27. Find the area inside the larger loop and outside the smaller loop of the limaçon $r = \frac{1}{2} + \cos \theta$.

28. When recording live performances, sound engineers often use a microphone with a cardioid pickup pattern because it suppresses noise from the audience. Suppose the microphone is placed 4 m from the front of the stage (as in the figure) and the boundary of the optimal pickup region is given by the cardioid $r = 8 + 8 \sin \theta$, where r is measured in meters and the microphone is at the pole. The musicians want to know the area they will have on stage within the optimal pickup range of the microphone. Answer their question.

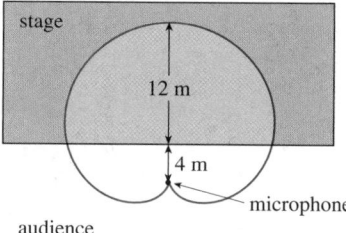

29–32 ■ Find all points of intersection of the given curves.

29. $r = 1 + \sin \theta$, $r = 3 \sin \theta$

30. $r = \cos 3\theta$, $r = \sin 3\theta$

31. $r = \sin \theta$, $r = \sin 2\theta$

32. $r^2 = \sin 2\theta$, $r^2 = \cos 2\theta$

33–36 ▪ Find the exact length of the polar curve.

33. $r = 3 \sin \theta, \quad 0 \leqslant \theta \leqslant \pi/3$

34. $r = e^{2\theta}, \quad 0 \leqslant \theta \leqslant 2\pi$

35. $r = \theta^2, \quad 0 \leqslant \theta \leqslant 2\pi$ **36.** $r = 2(1 + \cos \theta)$

37–38 ▪ Use a calculator to find the length of the curve correct to four decimal places. If necessary, graph the curve to determine the parameter interval.

37. $r = \sin(6 \sin \theta)$

38. $r = \sin(\theta/4)$

9.5 | CONIC SECTIONS IN POLAR COORDINATES

In your previous study of conic sections, parabolas were defined in terms of a focus and directrix whereas ellipses and hyperbolas were defined in terms of two foci. After reviewing those definitions and equations, we present a more unified treatment of all three types of conic sections in terms of a focus and directrix. Furthermore, if we place the focus at the origin, then a conic section has a simple polar equation. In Chapter 10 we will use the polar equation of an ellipse to derive Kepler's laws of planetary motion.

CONICS IN CARTESIAN COORDINATES

Here we provide a brief reminder of what you need to know about conic sections. A more thorough review can be found on the website www.stewartcalculus.com. (Click on *Review: Conic Sections.*)

Recall that a **parabola** is the set of points in a plane that are equidistant from a fixed point F (called the **focus**) and a fixed line (called the **directrix**). This definition is illustrated by Figure 1. Notice that the point halfway between the focus and the directrix lies on the parabola; it is called the **vertex**. The line through the focus perpendicular to the directrix is called the **axis** of the parabola.

A parabola has a very simple equation if its vertex is placed at the origin and its directrix is parallel to the x-axis or y-axis. If the focus is on the y-axis at the point $(0, p)$, then the directrix has the equation $y = -p$ and an equation of the parabola is $x^2 = 4py$. [See parts (a) and (b) of Figure 2.] If the focus is on the x-axis at $(p, 0)$, then the directrix is $x = -p$ and an equation is $y^2 = 4px$ as in parts (c) and (d).

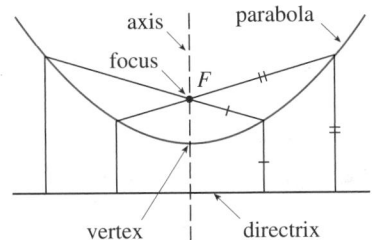

FIGURE 1

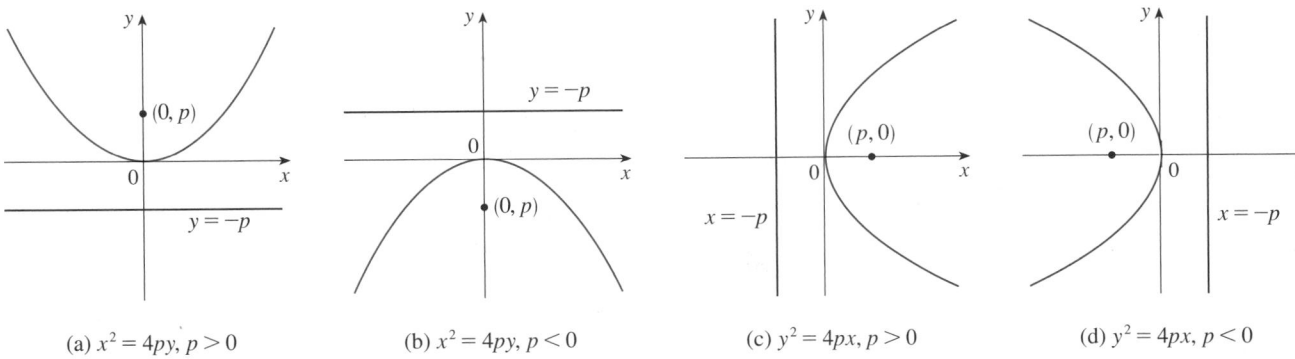

(a) $x^2 = 4py, \; p > 0$ (b) $x^2 = 4py, \; p < 0$ (c) $y^2 = 4px, \; p > 0$ (d) $y^2 = 4px, \; p < 0$

FIGURE 2

An **ellipse** is the set of points in a plane the sum of whose distances from two fixed points F_1 and F_2 is a constant. These two fixed points are called the **foci** (plural of **focus**).

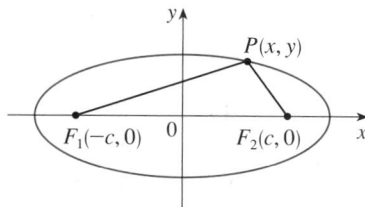

FIGURE 3

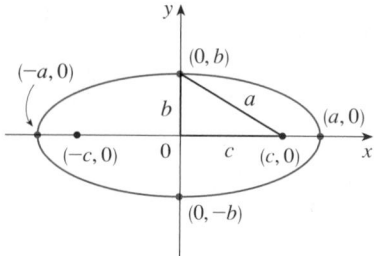

FIGURE 4
$$\frac{x^2}{a^2} + \frac{y^2}{b^2} = 1$$

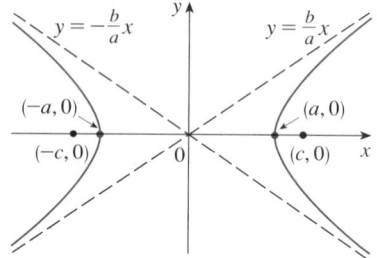

FIGURE 5
$$\frac{x^2}{a^2} - \frac{y^2}{b^2} = 1$$

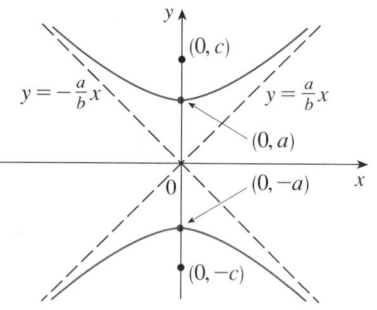

FIGURE 6
$$\frac{y^2}{a^2} - \frac{x^2}{b^2} = 1$$

An ellipse has a simple equation if we place the foci on the x-axis at the points $(-c, 0)$ and $(c, 0)$ as in Figure 3 so that the origin is halfway between the foci. If the sum of the distances from a point on the ellipse to the foci is $2a$, then the points $(a, 0)$ and $(-a, 0)$ where the ellipse meets the x-axis are called the **vertices**. The y-intercepts are $\pm b$, where $b^2 = a^2 - c^2$. (See Figure 4.)

> **1** The ellipse
> $$\frac{x^2}{a^2} + \frac{y^2}{b^2} = 1 \qquad a \geq b > 0$$
> has foci $(\pm c, 0)$, where $c^2 = a^2 - b^2$, and vertices $(\pm a, 0)$.

If the foci of an ellipse are located on the y-axis at $(0, \pm c)$, then we can find its equation by interchanging x and y in **1**.

A **hyperbola** is the set of all points in a plane the difference of whose distances from two fixed points F_1 and F_2 (the foci) is a constant. Notice that the definition of a hyperbola is similar to that of an ellipse; the only change is that the sum of distances has become a difference of distances. If the foci are on the x-axis at $(\pm c, 0)$ and the difference of distances is $\pm 2a$, then the equation of the hyperbola is $(x^2/a^2) - (y^2/b^2) = 1$, where $b^2 = c^2 - a^2$. The x-intercepts are $\pm a$ and the points $(a, 0)$ and $(-a, 0)$ are the **vertices** of the hyperbola. There is no y-intercept and the hyperbola consists of two parts, called its *branches*. (See Figure 5.)

When we draw a hyperbola it is useful to first draw its **asymptotes**, which are the dashed lines $y = (b/a)x$ and $y = -(b/a)x$ shown in Figure 5. Both branches of the hyperbola approach the asymptotes; that is, they come arbitrarily close to the asymptotes.

> **2** The hyperbola
> $$\frac{x^2}{a^2} - \frac{y^2}{b^2} = 1$$
> has foci $(\pm c, 0)$, where $c^2 = a^2 + b^2$, vertices $(\pm a, 0)$, and asymptotes $y = \pm(b/a)x$.

If the foci of a hyperbola are on the y-axis, then by reversing the roles of x and y we get the graph shown in Figure 6.

We have given the standard equations of the conic sections, but any of them can be shifted by replacing x by $x - h$ and y by $y - k$. For instance, an ellipse with center (h, k) has an equation of the form

$$\frac{(x - h)^2}{a^2} + \frac{(y - k)^2}{b^2} = 1$$

CONICS IN POLAR COORDINATES

In the following theorem we show how all three types of conic sections can be characterized in terms of a focus and directrix.

> **3** **THEOREM** Let F be a fixed point (called the **focus**) and l be a fixed line (called the **directrix**) in a plane. Let e be a fixed positive number (called the **eccentricity**). The set of all points P in the plane such that
>
> $$\frac{|PF|}{|Pl|} = e$$
>
> (that is, the ratio of the distance from F to the distance from l is the constant e) is a conic section. The conic is
>
> (a) an ellipse if $e < 1$
>
> (b) a parabola if $e = 1$
>
> (c) a hyperbola if $e > 1$

PROOF   Notice that if the eccentricity is $e = 1$, then $|PF| = |Pl|$ and so the given condition simply becomes the definition of a parabola.

Let us place the focus F at the origin and the directrix parallel to the y-axis and d units to the right. Thus the directrix has equation $x = d$ and is perpendicular to the polar axis. If the point P has polar coordinates (r, θ), we see from Figure 7 that

$$|PF| = r \qquad |Pl| = d - r\cos\theta$$

Thus the condition $|PF|/|Pl| = e$, or $|PF| = e|Pl|$, becomes

$$\boxed{4} \qquad\qquad r = e(d - r\cos\theta)$$

If we square both sides of this polar equation and convert to rectangular coordinates, we get

$$x^2 + y^2 = e^2(d - x)^2 = e^2(d^2 - 2dx + x^2)$$

or

$$(1 - e^2)x^2 + 2de^2x + y^2 = e^2d^2$$

After completing the square, we have

$$\boxed{5} \qquad\qquad \left(x + \frac{e^2d}{1 - e^2}\right)^2 + \frac{y^2}{1 - e^2} = \frac{e^2d^2}{(1 - e^2)^2}$$

If $e < 1$, we recognize Equation 5 as the equation of an ellipse. In fact, it is of the form

$$\frac{(x - h)^2}{a^2} + \frac{y^2}{b^2} = 1$$

where

$$\boxed{6} \qquad h = -\frac{e^2d}{1 - e^2} \qquad a^2 = \frac{e^2d^2}{(1 - e^2)^2} \qquad b^2 = \frac{e^2d^2}{1 - e^2}$$

We know that the foci of an ellipse are at a distance c from the center, where

$$\boxed{7} \qquad\qquad c^2 = a^2 - b^2 = \frac{e^4d^2}{(1 - e^2)^2}$$

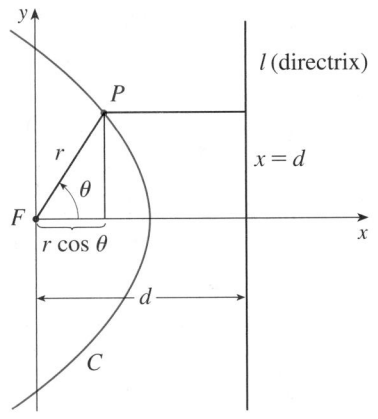

FIGURE 7

This shows that

$$c = \frac{e^2 d}{1 - e^2} = -h$$

and confirms that the focus as defined in Theorem 3 means the same as the focus defined earlier. It also follows from Equations 6 and 7 that the eccentricity is given by

$$e = \frac{c}{a}$$

If $e > 1$, then $1 - e^2 < 0$ and we see that Equation 5 represents a hyperbola. Just as we did before, we could rewrite Equation 5 in the form

$$\frac{(x - h)^2}{a^2} - \frac{y^2}{b^2} = 1$$

and see that

$$e = \frac{c}{a} \qquad \text{where} \quad c^2 = a^2 + b^2 \qquad \square$$

By solving Equation 4 for r, we see that the polar equation of the conic shown in Figure 7 can be written as

$$r = \frac{ed}{1 + e \cos \theta}$$

If the directrix is chosen to be to the left of the focus as $x = -d$, or if the directrix is chosen to be parallel to the polar axis as $y = \pm d$, then the polar equation of the conic is given by the following theorem, which is illustrated by Figure 8. (See Exercises 19–21.)

---

8 THEOREM A polar equation of the form

$$r = \frac{ed}{1 \pm e \cos \theta} \qquad \text{or} \qquad r = \frac{ed}{1 \pm e \sin \theta}$$

represents a conic section with eccentricity e. The conic is an ellipse if $e < 1$, a parabola if $e = 1$, or a hyperbola if $e > 1$.

V EXAMPLE 1 Find a polar equation for a parabola that has its focus at the origin and whose directrix is the line $y = -6$.

SOLUTION Using Theorem 8 with $e = 1$ and $d = 6$, and using part (d) of Figure 8, we see that the equation of the parabola is

$$r = \frac{6}{1 - \sin \theta} \qquad \blacksquare$$

V EXAMPLE 2 A conic is given by the polar equation

$$r = \frac{10}{3 - 2 \cos \theta}$$

Find the eccentricity, identify the conic, locate the directrix, and sketch the conic.

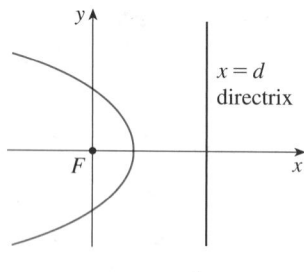

(a) $r = \dfrac{ed}{1 + e \cos \theta}$

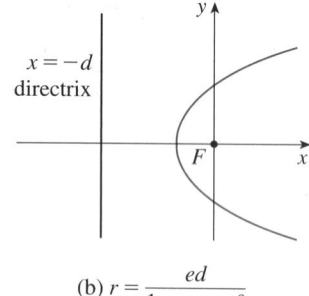

(b) $r = \dfrac{ed}{1 - e \cos \theta}$

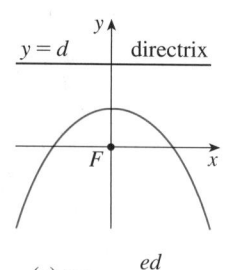

(c) $r = \dfrac{ed}{1 + e \sin \theta}$

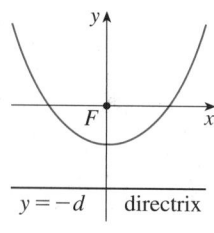

(d) $r = \dfrac{ed}{1 - e \sin \theta}$

FIGURE 8
Polar equations of conics

SOLUTION Dividing numerator and denominator by 3, we write the equation as

$$r = \frac{\frac{10}{3}}{1 - \frac{2}{3}\cos\theta}$$

From Theorem 8 we see that this represents an ellipse with $e = \frac{2}{3}$. Since $ed = \frac{10}{3}$, we have

$$d = \frac{\frac{10}{3}}{e} = \frac{\frac{10}{3}}{\frac{2}{3}} = 5$$

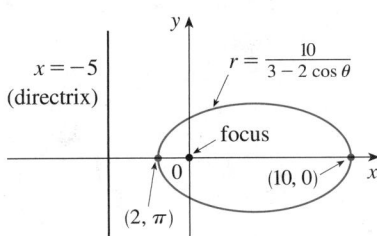

so the directrix has Cartesian equation $x = -5$. When $\theta = 0$, $r = 10$; when $\theta = \pi$, $r = 2$. So the vertices have polar coordinates $(10, 0)$ and $(2, \pi)$. The ellipse is sketched in Figure 9. ■

y
x
x = −5
(directrix)
r = 10/(3 − 2 cos θ)
focus
0
(10, 0)
(2, π)

FIGURE 9

EXAMPLE 3 Sketch the conic $r = \dfrac{12}{2 + 4\sin\theta}$.

SOLUTION Writing the equation in the form

$$r = \frac{6}{1 + 2\sin\theta}$$

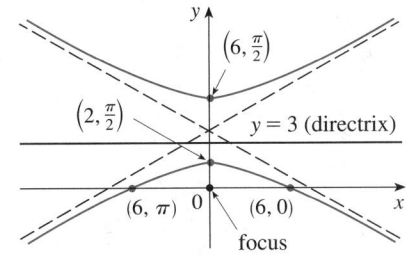

we see that the eccentricity is $e = 2$ and the equation therefore represents a hyperbola. Since $ed = 6$, $d = 3$ and the directrix has equation $y = 3$. The vertices occur when $\theta = \pi/2$ and $3\pi/2$, so they are $(2, \pi/2)$ and $(-6, 3\pi/2)$, which can be written as $(6, \pi/2)$. It is also useful to plot the x-intercepts. These occur when $\theta = 0, \pi$; in both cases $r = 6$. For additional accuracy we could draw the asymptotes. Note that $r \to \pm\infty$ when $2 + 4\sin\theta \to 0^+$ or 0^- and $2 + 4\sin\theta = 0$ when $\sin\theta = -\frac{1}{2}$. Thus the asymptotes are parallel to the rays $\theta = 7\pi/6$ and $\theta = 11\pi/6$. The hyperbola is sketched in Figure 10. ■

y
x
(6, π/2)
(2, π/2)
y = 3 (directrix)
(6, π)
0
(6, 0)
focus

FIGURE 10
$r = \dfrac{12}{2 + 4\sin\theta}$

In Figure 11 we use a computer to sketch a number of conics to demonstrate the effect of varying the eccentricity e. Notice that when e is close to 0 the ellipse is nearly circular, whereas it becomes more elongated as $e \to 1^-$. When $e = 1$, of course, the conic is a parabola.

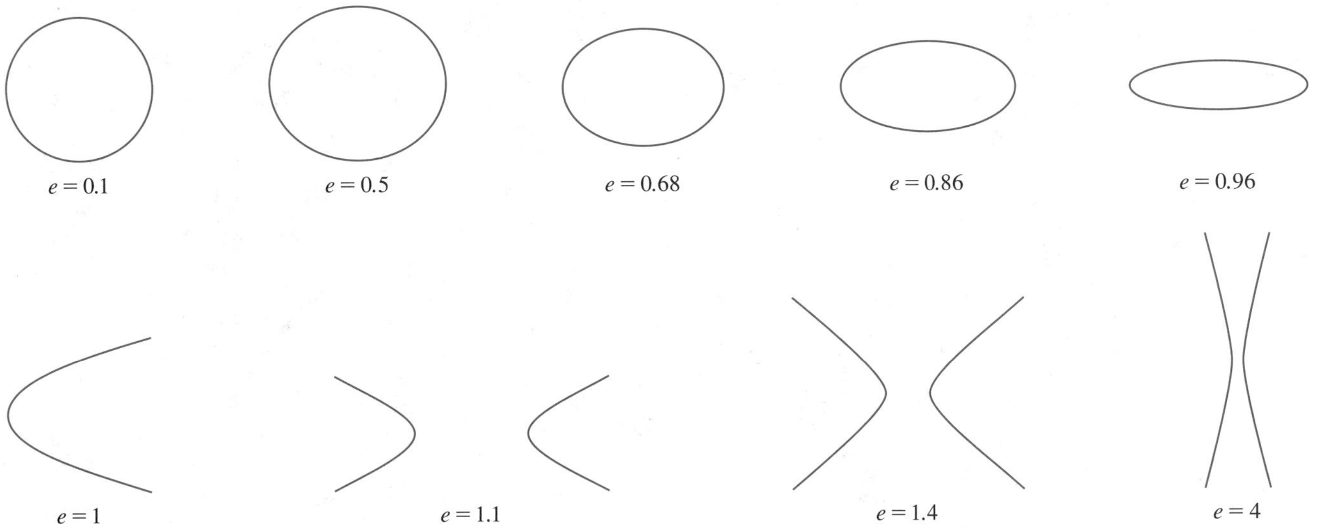

e = 0.1
e = 0.5
e = 0.68
e = 0.86
e = 0.96
e = 1
e = 1.1
e = 1.4
e = 4

FIGURE 11

9.5 | EXERCISES

1–8 ▪ Write a polar equation of a conic with the focus at the origin and the given data.

1. Ellipse, eccentricity $\frac{1}{2}$, directrix $x = 4$

2. Parabola, directrix $x = -3$

3. Hyperbola, eccentricity 1.5, directrix $y = 2$

4. Hyperbola, eccentricity 3, directrix $x = 3$

5. Parabola, vertex $(4, 3\pi/2)$

6. Ellipse, eccentricity 0.8, vertex $(1, \pi/2)$

7. Ellipse, eccentricity $\frac{1}{2}$, directrix $r = 4 \sec \theta$

8. Hyperbola, eccentricity 3, directrix $r = -6 \csc \theta$

9–16 ▪ (a) Find the eccentricity, (b) identify the conic, (c) give an equation of the directrix, and (d) sketch the conic.

9. $r = \dfrac{4}{5 - 4 \sin \theta}$

10. $r = \dfrac{12}{3 - 10 \cos \theta}$

11. $r = \dfrac{2}{3 + 3 \sin \theta}$

12. $r = \dfrac{3}{2 + 2 \cos \theta}$

13. $r = \dfrac{9}{6 + 2 \cos \theta}$

14. $r = \dfrac{5}{2 - 2 \sin \theta}$

15. $r = \dfrac{3}{4 - 8 \cos \theta}$

16. $r = \dfrac{4}{2 + \cos \theta}$

---

17. Graph the conics $r = e/(1 - e \cos \theta)$ with $e = 0.4, 0.6, 0.8$, and 1.0 on a common screen. How does the value of e affect the shape of the curve?

18. (a) Graph the conics $r = ed/(1 + e \sin \theta)$ for $e = 1$ and various values of d. How does the value of d affect the shape of the conic?
(b) Graph these conics for $d = 1$ and various values of e. How does the value of e affect the shape of the conic?

19. Show that a conic with focus at the origin, eccentricity e, and directrix $x = -d$ has polar equation

$$r = \frac{ed}{1 - e \cos \theta}$$

20. Show that a conic with focus at the origin, eccentricity e, and directrix $y = d$ has polar equation

$$r = \frac{ed}{1 + e \sin \theta}$$

21. Show that a conic with focus at the origin, eccentricity e, and directrix $y = -d$ has polar equation

$$r = \frac{ed}{1 - e \sin \theta}$$

22. Show that the parabolas $r = c/(1 + \cos \theta)$ and $r = d/(1 - \cos \theta)$ intersect at right angles.

23. (a) Show that the polar equation of an ellipse with directrix $x = -d$ can be written in the form

$$r = \frac{a(1 - e^2)}{1 - e \cos \theta}$$

(b) Find an approximate polar equation for the elliptical orbit of the earth around the sun (at one focus) given that the eccentricity is about 0.017 and the length of the major axis is about 2.99×10^8 km.

24. (a) The planets move around the sun in elliptical orbits with the sun at one focus. The positions of a planet that are closest to and farthest from the sun are called its *perihelion* and *aphelion,* respectively. Use Exercise 23(a) to show that the perihelion distance from a planet to the sun is $a(1 - e)$ and the aphelion distance is $a(1 + e)$.

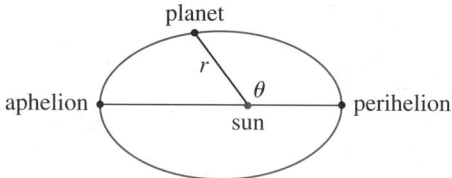

(b) Use the data of Exercise 23(b) to find the distances from the earth to the sun at perihelion and at aphelion.

25. The orbit of Halley's comet, last seen in 1986 and due to return in 2062, is an ellipse with eccentricity 0.97 and one focus at the sun. The length of its major axis is 36.18 AU. [An astronomical unit (AU) is the mean distance between the earth and the sun, about 93 million miles.] Find a polar equation for the orbit of Halley's comet. What is the maximum distance from the comet to the sun?

26. The Hale-Bopp comet, discovered in 1995, has an elliptical orbit with eccentricity 0.9951 and the length of the major axis is 356.5 AU. Find a polar equation for the orbit of this comet. How close to the sun does it come?

27. The planet Mercury travels in an elliptical orbit with eccentricity 0.206. Its minimum distance from the sun is 4.6×10^7 km. Use the results of Exercise 24(a) to find its maximum distance from the sun.

28. The distance from the planet Pluto to the sun is 4.43×10^9 km at perihelion and 7.37×10^9 km at aphelion. Use Exercise 24 to find the eccentricity of Pluto's orbit.

29. Using the data from Exercise 27, find the distance traveled by the planet Mercury during one complete orbit around the sun. (If your calculator or computer algebra system evaluates definite integrals, use it. Otherwise, use Simpson's Rule.)

CHAPTER 9 REVIEW

1. (a) What is a parametric curve?
(b) How do you sketch a parametric curve?

2. (a) How do you find the slope of a tangent to a parametric curve?
(b) How do you find the area under a parametric curve?

3. Write an expression for the length of a parametric curve.

4. (a) Use a diagram to explain the meaning of the polar coordinates (r, θ) of a point.
(b) Write equations that express the Cartesian coordinates (x, y) of a point in terms of the polar coordinates.

(c) What equations would you use to find the polar coordinates of a point if you knew the Cartesian coordinates?

5. (a) How do you find the slope of a tangent line to a polar curve?
(b) How do you find the area of a region bounded by a polar curve?
(c) How do you find the length of a polar curve?

6. (a) What is the eccentricity of a conic section?
(b) What can you say about the eccentricity if the conic section is an ellipse? A hyperbola? A parabola?
(c) Write a polar equation for a conic section with eccentricity e and directrix $x = d$. What if the directrix is $x = -d$? $y = d$? $y = -d$?

Determine whether the statement is true or false. If it is true, explain why. If it is false, explain why or give an example that disproves the statement.

1. If the parametric curve $x = f(t)$, $y = g(t)$ satisfies $g'(1) = 0$, then it has a horizontal tangent when $t = 1$.

2. If $x = f(t)$ and $y = g(t)$ are twice differentiable, then
$$\frac{d^2y}{dx^2} = \frac{d^2y/dt^2}{d^2x/dt^2}$$

3. The length of the curve $x = f(t)$, $y = g(t)$, $a \le t \le b$, is $\int_a^b \sqrt{[f'(t)]^2 + [g'(t)]^2}\, dt$.

4. If a point is represented by (x, y) in Cartesian coordinates (where $x \ne 0$) and (r, θ) in polar coordinates, then $\theta = \tan^{-1}(y/x)$.

5. The polar curves $r = 1 - \sin 2\theta$ and $r = \sin 2\theta - 1$ have the same graph.

6. The equations $r = 2$, $x^2 + y^2 = 4$, and $x = 2 \sin 3t$, $y = 2 \cos 3t$ $(0 \le t \le 2\pi)$ all have the same graph.

7. The parametric equations $x = t^2$, $y = t^4$ have the same graph as $x = t^3$, $y = t^6$.

8. A hyperbola never intersects its directrix.

1–4 ■ Sketch the parametric curve and eliminate the parameter to find the Cartesian equation of the curve.

1. $x = t^2 + 4t$, $\quad y = 2 - t$, $\quad -4 \le t \le 1$

2. $x = 1 + e^{2t}$, $\quad y = e^t$

3. $x = \cos\theta$, $\quad y = \sec\theta$, $\quad 0 \le \theta < \pi/2$

4. $x = 2\cos\theta$, $\quad y = 1 + \sin\theta$

5. Write three different sets of parametric equations for the curve $y = \sqrt{x}$.

6. Use the following graphs of $x = f(t)$ and $y = g(t)$ to sketch the parametric curve $x = f(t)$, $y = g(t)$. Indicate with arrows the direction in which the curve is traced as t increases.

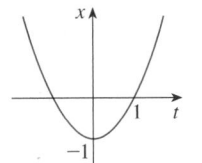 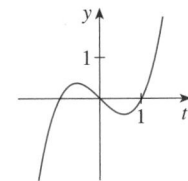

7. (a) Plot the point with polar coordinates $(4, 2\pi/3)$. Then find its Cartesian coordinates.
(b) The Cartesian coordinates of a point are $(-3, 3)$. Find two sets of polar coordinates for the point.

8. Sketch the region consisting of points whose polar coordinates satisfy $1 \le r < 2$ and $\pi/6 \le \theta \le 5\pi/6$.

9–16 ▪ Sketch the polar curve.

9. $r = 1 - \cos\theta$

10. $r = \sin 4\theta$

11. $r = \cos 3\theta$

12. $r = 3 + \cos 3\theta$

13. $r = 1 + \cos 2\theta$

14. $r = 2\cos(\theta/2)$

15. $r = \dfrac{3}{1 + 2\sin\theta}$

16. $r = \dfrac{3}{2 - 2\cos\theta}$

17–18 ▪ Find a polar equation for the curve represented by the given Cartesian equation.

17. $x + y = 2$

18. $x^2 + y^2 = 2$

19. The curve with polar equation $r = (\sin\theta)/\theta$ is called a **cochleoid**. Use a graph of r as a function of θ in Cartesian coordinates to sketch the cochleoid by hand. Then graph it with a machine to check your sketch.

20. Graph the ellipse $r = 2/(4 - 3\cos\theta)$ and its directrix. Also graph the ellipse obtained by rotation about the origin through an angle $2\pi/3$.

21–24 ▪ Find the slope of the tangent line to the given curve at the point corresponding to the specified value of the parameter.

21. $x = \ln t,\ y = 1 + t^2;\ \ t = 1$

22. $x = t^3 + 6t + 1,\ \ y = 2t - t^2;\ \ t = -1$

23. $r = e^{-\theta};\ \ \theta = \pi$

24. $r = 3 + \cos 3\theta;\ \ \theta = \pi/2$

25–26 ▪ Find dy/dx and d^2y/dx^2.

25. $x = t + \sin t,\ \ y = t - \cos t$

26. $x = 1 + t^2,\ \ y = t - t^3$

27. Use a graph to estimate the coordinates of the lowest point on the curve $x = t^3 - 3t,\ y = t^2 + t + 1$. Then use calculus to find the exact coordinates.

28. Find the area enclosed by the loop of the curve in Exercise 27.

29. At what points does the curve

$$x = 2a\cos t - a\cos 2t \qquad y = 2a\sin t - a\sin 2t$$

have vertical or horizontal tangents? Use this information to help sketch the curve.

30. Find the area enclosed by the curve in Exercise 29.

31. Find the area enclosed by the curve $r^2 = 9\cos 5\theta$.

32. Find the area enclosed by the inner loop of the curve $r = 1 - 3\sin\theta$.

33. Find the points of intersection of the curves $r = 2$ and $r = 4\cos\theta$.

34. Find the points of intersection of the curves $r = \cot\theta$ and $r = 2\cos\theta$.

35. Find the area of the region that lies inside both of the circles $r = 2\sin\theta$ and $r = \sin\theta + \cos\theta$.

36. Find the area of the region that lies inside the curve $r = 2 + \cos 2\theta$ but outside the curve $r = 2 + \sin\theta$.

37–40 ▪ Find the length of the curve.

37. $x = 3t^2,\ \ y = 2t^3,\ \ 0 \le t \le 2$

38. $x = 2 + 3t,\ \ y = \cosh 3t,\ \ 0 \le t \le 1$

39. $r = 1/\theta,\ \ \pi \le \theta \le 2\pi$

40. $r = \sin^3(\theta/3),\ \ 0 \le \theta \le \pi$

41. The curves defined by the parametric equations

$$x = \frac{t^2 - c}{t^2 + 1} \qquad y = \frac{t(t^2 - c)}{t^2 + 1}$$

are called **strophoids** (from a Greek word meaning "to turn or twist"). Investigate how these curves vary as c varies.

42. A family of curves has polar equations $r^a = |\sin 2\theta|$ where a is a positive number. Investigate how the curves change as a changes.

43. Find a polar equation for the ellipse with focus at the origin, eccentricity $\frac{1}{3}$, and directrix with equation $r = 4\sec\theta$.

44. Show that the angles between the polar axis and the asymptotes of the hyperbola $r = ed/(1 - e\cos\theta),\ e > 1$, are given by $\cos^{-1}(\pm 1/e)$.

45. In the figure the circle of radius a is stationary, and for every θ, the point P is the midpoint of the segment QR. The curve traced out by P for $0 < \theta < \pi$ is called the **longbow curve**. Find parametric equations for this curve.

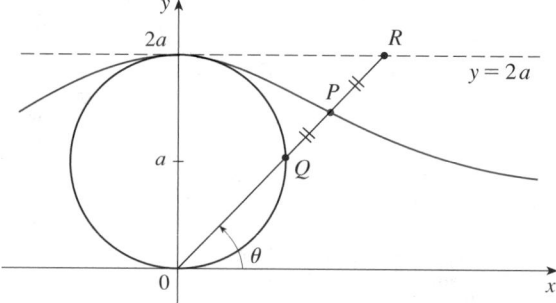

10 | VECTORS AND THE GEOMETRY OF SPACE

In this chapter we introduce vectors and coordinate systems for three-dimensional space. This will be the setting for the study of functions of two variables in Chapter 11 because the graph of such a function is a surface in space. In this chapter we will see that vectors provide particularly simple descriptions of lines, planes, and curves. We will also use vector-valued functions to describe the motion of objects through space. In particular, we will use them to derive Kepler's laws of planetary motion.

10.1 | THREE-DIMENSIONAL COORDINATE SYSTEMS

To locate a point in a plane, two numbers are necessary. We know that any point in the plane can be represented as an ordered pair (a, b) of real numbers, where a is the x-coordinate and b is the y-coordinate. For this reason, a plane is called two-dimensional. To locate a point in space, three numbers are required. We represent any point in space by an ordered triple (a, b, c) of real numbers.

In order to represent points in space, we first choose a fixed point O (the origin) and three directed lines through O that are perpendicular to each other, called the **coordinate axes** and labeled the x-axis, y-axis, and z-axis. Usually we think of the x- and y-axes as being horizontal and the z-axis as being vertical, and we draw the orientation of the axes as in Figure 1. The direction of the z-axis is determined by the **right-hand rule** as illustrated in Figure 2: If you curl the fingers of your right hand around the z-axis in the direction of a $90°$ counterclockwise rotation from the positive x-axis to the positive y-axis, then your thumb points in the positive direction of the z-axis.

The three coordinate axes determine the three **coordinate planes** illustrated in Figure 3(a). The xy-plane is the plane that contains the x- and y-axes; the yz-plane contains the y- and z-axes; the xz-plane contains the x- and z-axes. These three coordinate planes divide space into eight parts, called **octants**. The **first octant**, in the foreground, is determined by the positive axes.

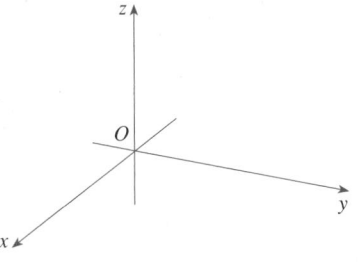

FIGURE 1
Coordinate axes

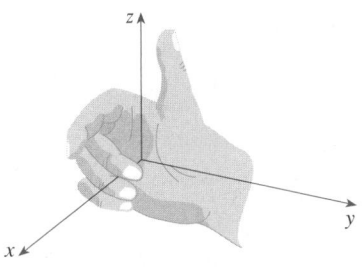

FIGURE 2
Right-hand rule

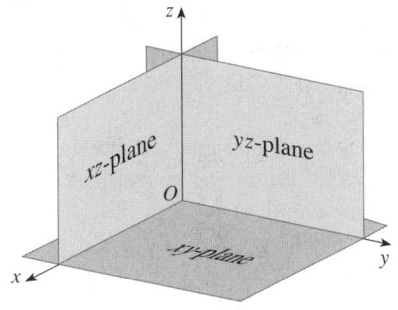
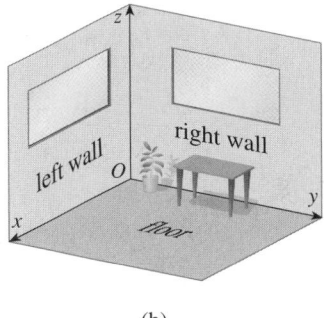

FIGURE 3

(a) Coordinate planes

(b)

Because many people have some difficulty visualizing diagrams of three-dimensional figures, you may find it helpful to do the following [see Figure 3(b)]. Look at any bottom corner of a room and call the corner the origin. The wall on your left is in the xz-plane, the wall on your right is in the yz-plane, and the floor is in the xy-plane. The x-axis runs along the intersection of the floor and the left wall. The y-axis runs along the intersection of the floor and the right wall. The z-axis runs up from the floor toward the ceiling along the intersection of the two walls. You are situated in the first octant, and you can now imagine seven other rooms situated in the other seven octants (three on the same floor and four on the floor below), all connected by the common corner point O.

Now if P is any point in space, let a be the (directed) distance from the yz-plane to P, let b be the distance from the xz-plane to P, and let c be the distance from the xy-plane to P. We represent the point P by the ordered triple (a, b, c) of real numbers and we call a, b, and c the **coordinates** of P; a is the x-coordinate, b is the y-coordinate, and c is the z-coordinate. Thus to locate the point (a, b, c) we can start at the origin O and move a units along the x-axis, then b units parallel to the y-axis, and then c units parallel to the z-axis as in Figure 4.

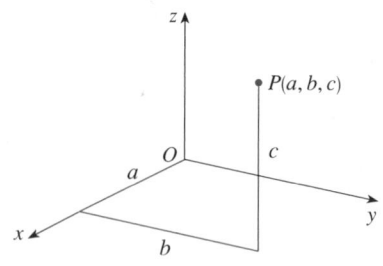

FIGURE 4

The point $P(a, b, c)$ determines a rectangular box as in Figure 5. If we drop a perpendicular from P to the xy-plane, we get a point Q with coordinates $(a, b, 0)$ called the **projection** of P on the xy-plane. Similarly, $R(0, b, c)$ and $S(a, 0, c)$ are the projections of P on the yz-plane and xz-plane, respectively.

As numerical illustrations, the points $(-4, 3, -5)$ and $(3, -2, -6)$ are plotted in Figure 6.

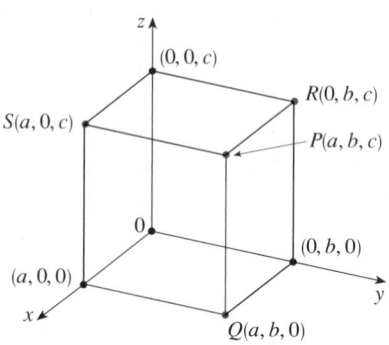

FIGURE 5

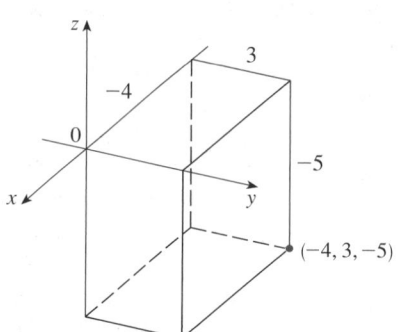

FIGURE 6

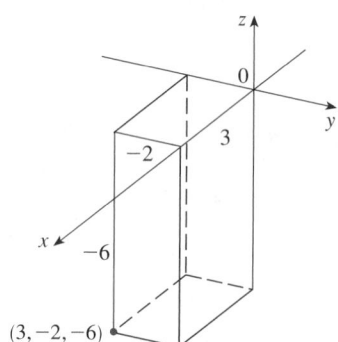

The Cartesian product $\mathbb{R} \times \mathbb{R} \times \mathbb{R} = \{(x, y, z) \mid x, y, z \in \mathbb{R}\}$ is the set of all ordered triples of real numbers and is denoted by $\mathbb{R}^3$. We have given a one-to-one correspondence between points P in space and ordered triples (a, b, c) in $\mathbb{R}^3$. It is called a **three-dimensional rectangular coordinate system**. Notice that, in terms of coordinates, the first octant can be described as the set of points whose coordinates are all positive.

In two-dimensional analytic geometry, the graph of an equation involving x and y is a curve in $\mathbb{R}^2$. In three-dimensional analytic geometry, an equation in x, y, and z represents a *surface* in $\mathbb{R}^3$.

▼ EXAMPLE 1 What surfaces in $\mathbb{R}^3$ are represented by the following equations?
(a) $z = 3$ (b) $y = 5$

SOLUTION
(a) The equation $z = 3$ represents the set $\{(x, y, z) \mid z = 3\}$, which is the set of all points in $\mathbb{R}^3$ whose z-coordinate is 3. This is the horizontal plane that is parallel to the xy-plane and three units above it as in Figure 7(a).

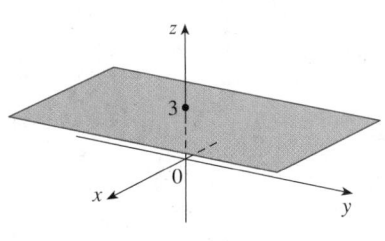
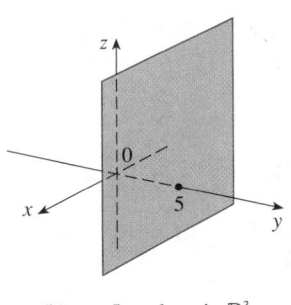
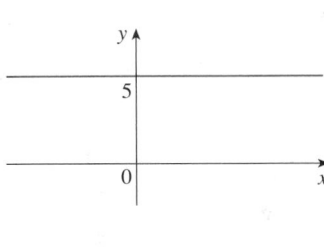

FIGURE 7   (a) $z = 3$, a plane in $\mathbb{R}^3$ (b) $y = 5$, a plane in $\mathbb{R}^3$ (c) $y = 5$, a line in $\mathbb{R}^2$

■ www.stewartcalculus.com
See Additional Example A.

(b) The equation $y = 5$ represents the set of all points in $\mathbb{R}^3$ whose y-coordinate is 5. This is the vertical plane that is parallel to the xz-plane and five units to the right of it as in Figure 7(b). ■

NOTE When an equation is given, we must understand from the context whether it represents a curve in $\mathbb{R}^2$ or a surface in $\mathbb{R}^3$. In Example 1, $y = 5$ represents a plane in $\mathbb{R}^3$, but of course $y = 5$ can also represent a line in $\mathbb{R}^2$ if we are dealing with two-dimensional analytic geometry. See Figure 7, parts (b) and (c).

In general, if k is a constant, then $x = k$ represents a plane parallel to the yz-plane, $y = k$ is a plane parallel to the xz-plane, and $z = k$ is a plane parallel to the xy-plane. In Figure 5, the faces of the rectangular box are formed by the three coordinate planes $x = 0$ (the yz-plane), $y = 0$ (the xz-plane), and $z = 0$ (the xy-plane), and the planes $x = a$, $y = b$, and $z = c$.

V EXAMPLE 2 Describe and sketch the surface in $\mathbb{R}^3$ represented by the equation $y = x$.

SOLUTION The equation represents the set of all points in $\mathbb{R}^3$ whose x- and y-coordinates are equal, that is, $\{(x, x, z) \mid x \in \mathbb{R}, z \in \mathbb{R}\}$. This is a vertical plane that intersects the xy-plane in the line $y = x$, $z = 0$. The portion of this plane that lies in the first octant is sketched in Figure 8. ■

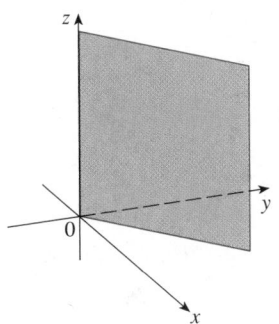

FIGURE 8
The plane $y = x$

The familiar formula for the distance between two points in a plane is easily extended to the following three-dimensional formula.

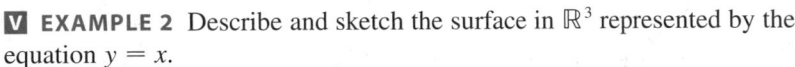

DISTANCE FORMULA IN THREE DIMENSIONS The distance $|P_1P_2|$ between the points $P_1(x_1, y_1, z_1)$ and $P_2(x_2, y_2, z_2)$ is

$$|P_1P_2| = \sqrt{(x_2 - x_1)^2 + (y_2 - y_1)^2 + (z_2 - z_1)^2}$$

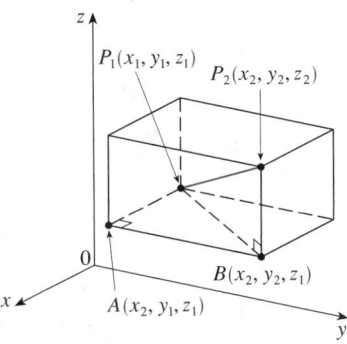

To see why this formula is true, we construct a rectangular box as in Figure 9, where P_1 and P_2 are opposite vertices and the faces of the box are parallel to the coordinate planes. If $A(x_2, y_1, z_1)$ and $B(x_2, y_2, z_1)$ are the vertices of the box indicated in the figure, then

$$|P_1A| = |x_2 - x_1| \qquad |AB| = |y_2 - y_1| \qquad |BP_2| = |z_2 - z_1|$$

FIGURE 9

Because triangles P_1BP_2 and P_1AB are both right-angled, two applications of the Pythagorean Theorem give

$$|P_1P_2|^2 = |P_1B|^2 + |BP_2|^2$$

and

$$|P_1B|^2 = |P_1A|^2 + |AB|^2$$

Combining these equations, we get

$$\begin{aligned}|P_1P_2|^2 &= |P_1A|^2 + |AB|^2 + |BP_2|^2 \\ &= |x_2 - x_1|^2 + |y_2 - y_1|^2 + |z_2 - z_1|^2 \\ &= (x_2 - x_1)^2 + (y_2 - y_1)^2 + (z_2 - z_1)^2\end{aligned}$$

Therefore $\qquad |P_1P_2| = \sqrt{(x_2 - x_1)^2 + (y_2 - y_1)^2 + (z_2 - z_1)^2}$

EXAMPLE 3 The distance from the point $P(2, -1, 7)$ to the point $Q(1, -3, 5)$ is

$$|PQ| = \sqrt{(1 - 2)^2 + (-3 + 1)^2 + (5 - 7)^2} = \sqrt{1 + 4 + 4} = 3 \qquad \blacksquare$$

V EXAMPLE 4 Find an equation of a sphere with radius r and center $C(h, k, l)$.

SOLUTION By definition, a sphere is the set of all points $P(x, y, z)$ whose distance from C is r. (See Figure 10.) Thus P is on the sphere if and only if $|PC| = r$. Squaring both sides, we have $|PC|^2 = r^2$ or

$$(x - h)^2 + (y - k)^2 + (z - l)^2 = r^2 \qquad \blacksquare$$

The result of Example 4 is worth remembering.

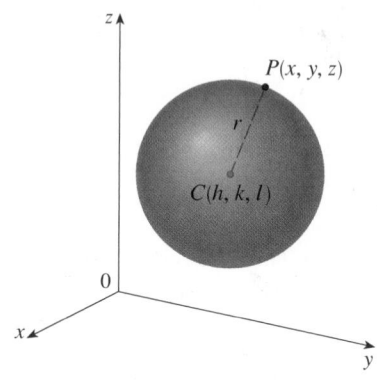

FIGURE 10

EQUATION OF A SPHERE An equation of a sphere with center $C(h, k, l)$ and radius r is

$$(x - h)^2 + (y - k)^2 + (z - l)^2 = r^2$$

In particular, if the center is the origin O, then an equation of the sphere is

$$x^2 + y^2 + z^2 = r^2$$

EXAMPLE 5 Show that $x^2 + y^2 + z^2 + 4x - 6y + 2z + 6 = 0$ is the equation of a sphere, and find its center and radius.

SOLUTION We can rewrite the given equation in the form of an equation of a sphere if we complete squares:

$$(x^2 + 4x + 4) + (y^2 - 6y + 9) + (z^2 + 2z + 1) = -6 + 4 + 9 + 1$$
$$(x + 2)^2 + (y - 3)^2 + (z + 1)^2 = 8$$

Comparing this equation with the standard form, we see that it is the equation of a sphere with center $(-2, 3, -1)$ and radius $\sqrt{8} = 2\sqrt{2}$. $\qquad \blacksquare$

EXAMPLE 6 What region in $\mathbb{R}^3$ is represented by the following inequalities?

$$1 \le x^2 + y^2 + z^2 \le 4 \qquad z \le 0$$

SOLUTION The inequalities

$$1 \le x^2 + y^2 + z^2 \le 4$$

can be rewritten as

$$1 \le \sqrt{x^2 + y^2 + z^2} \le 2$$

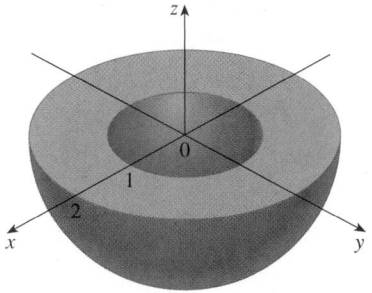

FIGURE 11

so they represent the points (x, y, z) whose distance from the origin is at least 1 and at most 2. But we are also given that $z \le 0$, so the points lie on or below the xy-plane. Thus the given inequalities represent the region that lies between (or on) the spheres $x^2 + y^2 + z^2 = 1$ and $x^2 + y^2 + z^2 = 4$ and beneath (or on) the xy-plane. It is sketched in Figure 11. ∎

10.1 | EXERCISES

1. Suppose you start at the origin, move along the x-axis a distance of 4 units in the positive direction, and then move downward a distance of 3 units. What are the coordinates of your position?

2. Sketch the points $(0, 5, 2)$, $(4, 0, -1)$, $(2, 4, 6)$, and $(1, -1, 2)$ on a single set of coordinate axes.

3. Which of the points $A(-4, 0, -1)$, $B(3, 1, -5)$, and $C(2, 4, 6)$ is closest to the yz-plane? Which point lies in the xz-plane?

4. What are the projections of the point $(2, 3, 5)$ on the xy-, yz-, and xz-planes? Draw a rectangular box with the origin and $(2, 3, 5)$ as opposite vertices and with its faces parallel to the coordinate planes. Label all vertices of the box. Find the length of the diagonal of the box.

5. Describe and sketch the surface in $\mathbb{R}^3$ represented by the equation $x + y = 2$.

6. (a) What does the equation $x = 4$ represent in $\mathbb{R}^2$? What does it represent in $\mathbb{R}^3$? Illustrate with sketches.
   (b) What does the equation $y = 3$ represent in $\mathbb{R}^3$? What does $z = 5$ represent? What does the pair of equations $y = 3$, $z = 5$ represent? In other words, describe the set of points (x, y, z) such that $y = 3$ and $z = 5$. Illustrate with a sketch.

7. Find the lengths of the sides of the triangle PQR. Is it a right triangle? Is it an isosceles triangle?
   (a) $P(3, -2, -3)$, $Q(7, 0, 1)$, $R(1, 2, 1)$
   (b) $P(2, -1, 0)$, $Q(4, 1, 1)$, $R(4, -5, 4)$

8. Find the distance from $(4, -2, 6)$ to each of the following.
   (a) The xy-plane   (b) The yz-plane
   (c) The xz-plane   (d) The x-axis
   (e) The y-axis   (f) The z-axis

9. Determine whether the points lie on a straight line.
   (a) $A(2, 4, 2)$, $B(3, 7, -2)$, $C(1, 3, 3)$
   (b) $D(0, -5, 5)$, $E(1, -2, 4)$, $F(3, 4, 2)$

10. Find an equation of the sphere with center $(2, -6, 4)$ and radius 5. Describe its intersection with each of the coordinate planes.

11. Find an equation of the sphere that passes through the point $(4, 3, -1)$ and has center $(3, 8, 1)$.

12. Find an equation of the sphere that passes through the origin and whose center is $(1, 2, 3)$.

13–16 ■ Show that the equation represents a sphere, and find its center and radius.

13. $x^2 + y^2 + z^2 - 2x - 4y + 8z = 15$

14. $x^2 + y^2 + z^2 + 8x - 6y + 2z + 17 = 0$

15. $2x^2 + 2y^2 + 2z^2 = 8x - 24z + 1$

16. $3x^2 + 3y^2 + 3z^2 = 10 + 6y + 12z$

17. (a) Prove that the midpoint of the line segment from $P_1(x_1, y_1, z_1)$ to $P_2(x_2, y_2, z_2)$ is

$$\left(\frac{x_1 + x_2}{2}, \frac{y_1 + y_2}{2}, \frac{z_1 + z_2}{2} \right)$$

(b) Find the lengths of the medians of the triangle with vertices $A(1, 2, 3)$, $B(-2, 0, 5)$, and $C(4, 1, 5)$.

18. Find an equation of a sphere if one of its diameters has endpoints $(2, 1, 4)$ and $(4, 3, 10)$.

19. Find equations of the spheres with center $(2, -3, 6)$ that touch (a) the xy-plane, (b) the yz-plane, (c) the xz-plane.

20. Find an equation of the largest sphere with center $(5, 4, 9)$ that is contained in the first octant.

21–30 ■ Describe in words the region of $\mathbb{R}^3$ represented by the equations or inequalities.

21. $x = 5$

22. $y = -2$

23. $y < 8$

24. $x \geqslant -3$

25. $0 \leqslant z \leqslant 6$

26. $z^2 = 1$

27. $x^2 + y^2 + z^2 \leqslant 3$

28. $x = z$

29. $x^2 + z^2 \leqslant 9$

30. $x^2 + y^2 + z^2 > 2z$

31–34 ■ Write inequalities to describe the region.

31. The region between the yz-plane and the vertical plane $x = 5$

32. The solid cylinder that lies on or below the plane $z = 8$ and on or above the disk in the xy-plane with center the origin and radius 2

33. The region consisting of all points between (but not on) the spheres of radius r and R centered at the origin, where $r < R$

34. The solid upper hemisphere of the sphere of radius 2 centered at the origin

35. Find an equation of the set of all points equidistant from the points $A(-1, 5, 3)$ and $B(6, 2, -2)$. Describe the set.

36. Find the volume of the solid that lies inside both of the spheres

$$x^2 + y^2 + z^2 + 4x - 2y + 4z + 5 = 0$$

and

$$x^2 + y^2 + z^2 = 4$$

37. Find the distance between the spheres $x^2 + y^2 + z^2 = 4$ and $x^2 + y^2 + z^2 = 4x + 4y + 4z - 11$.

38. Describe and sketch a solid with the following properties. When illuminated by rays parallel to the z-axis, its shadow is a circular disk. If the rays are parallel to the y-axis, its shadow is a square. If the rays are parallel to the x-axis, its shadow is an isosceles triangle.

10.2 | VECTORS

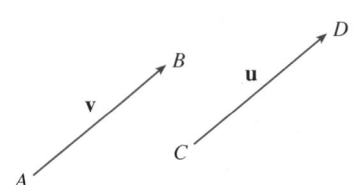

FIGURE 1
Equivalent vectors

The term **vector** is used by scientists to indicate a quantity (such as displacement or velocity or force) that has both magnitude and direction. A vector is often represented by an arrow or a directed line segment. The length of the arrow represents the magnitude of the vector and the arrow points in the direction of the vector. We denote a vector by printing a letter in boldface (**v**) or by putting an arrow above the letter ($\vec{v}$).

For instance, suppose a particle moves along a line segment from point A to point B. The corresponding **displacement vector v**, shown in Figure 1, has **initial point** A (the tail) and **terminal point** B (the tip) and we indicate this by writing $\mathbf{v} = \overrightarrow{AB}$. Notice that the vector $\mathbf{u} = \overrightarrow{CD}$ has the same length and the same direction as **v** even though it is in a different position. We say that **u** and **v** are **equivalent** (or **equal**) and we write $\mathbf{u} = \mathbf{v}$. The **zero vector**, denoted by **0**, has length 0. It is the only vector with no specific direction.

COMBINING VECTORS

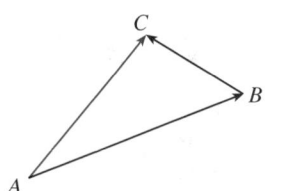

FIGURE 2

Suppose a particle moves from A to B, so its displacement vector is $\overrightarrow{AB}$. Then the particle changes direction and moves from B to C, with displacement vector $\overrightarrow{BC}$ as in Figure 2. The combined effect of these displacements is that the particle has moved from A to C. The resulting displacement vector $\overrightarrow{AC}$ is called the *sum* of $\overrightarrow{AB}$ and $\overrightarrow{BC}$ and we write

$$\overrightarrow{AC} = \overrightarrow{AB} + \overrightarrow{BC}$$

In general, if we start with vectors **u** and **v**, we first move **v** so that its tail coincides with the tip of **u** and define the sum of **u** and **v** as follows.

> **DEFINITION OF VECTOR ADDITION**   If **u** and **v** are vectors positioned so the initial point of **v** is at the terminal point of **u**, then the **sum u + v** is the vector from the initial point of **u** to the terminal point of **v**.

The definition of vector addition is illustrated in Figure 3. You can see why this definition is sometimes called the **Triangle Law**.

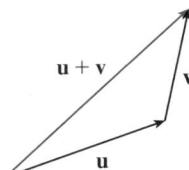

FIGURE 3   The Triangle Law

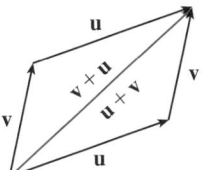

FIGURE 4   The Parallelogram Law

In Figure 4 we start with the same vectors **u** and **v** as in Figure 3 and draw another copy of **v** with the same initial point as **u**. Completing the parallelogram, we see that **u + v = v + u**. This also gives another way to construct the sum: If we place **u** and **v** so they start at the same point, then **u + v** lies along the diagonal of the parallelogram with **u** and **v** as sides. (This is called the **Parallelogram Law**.)

▼ EXAMPLE 1   Draw the sum of the vectors **a** and **b** shown in Figure 5.

SOLUTION   First we translate **b** and place its tail at the tip of **a**, being careful to draw a copy of **b** that has the same length and direction. Then we draw the vector **a + b** [see Figure 6(a)] starting at the initial point of **a** and ending at the terminal point of the copy of **b**.

Alternatively, we could place **b** so it starts where **a** starts and construct **a + b** by the Parallelogram Law as in Figure 6(b). ∎

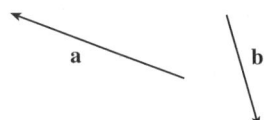

FIGURE 5

TEC   Visual 10.2 shows how the Triangle and Parallelogram Laws work for various vectors **u** and **v**.

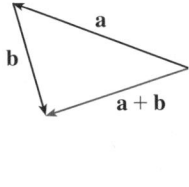 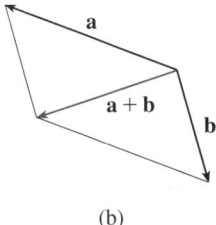

FIGURE 6   (a)   (b)

It is possible to multiply a vector by a real number c. (In this context we call the real number c a **scalar** to distinguish it from a vector.) For instance, we want $2\mathbf{v}$ to be the same vector as $\mathbf{v} + \mathbf{v}$, which has the same direction as **v** but is twice as long. In general, we multiply a vector by a scalar as follows.

> **DEFINITION OF SCALAR MULTIPLICATION**   If c is a scalar and **v** is a vector, then the **scalar multiple** $c\mathbf{v}$ is the vector whose length is $|c|$ times the length of **v** and whose direction is the same as **v** if $c > 0$ and is opposite to **v** if $c < 0$. If $c = 0$ or $\mathbf{v} = \mathbf{0}$, then $c\mathbf{v} = \mathbf{0}$.

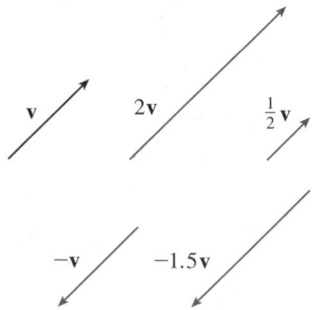

FIGURE 7

Scalar multiples of **v**

This definition is illustrated in Figure 7. We see that real numbers work like scaling factors here; that's why we call them scalars. Notice that two nonzero vectors are **parallel** if they are scalar multiples of one another. In particular, the vector $-\mathbf{v} = (-1)\mathbf{v}$ has the same length as **v** but points in the opposite direction. We call it the **negative** of **v**.

By the **difference** $\mathbf{u} - \mathbf{v}$ of two vectors we mean

$$\mathbf{u} - \mathbf{v} = \mathbf{u} + (-\mathbf{v})$$

So we can construct $\mathbf{u} - \mathbf{v}$ by first drawing the negative of **v**, $-\mathbf{v}$, and then adding it to **u** by the Parallelogram Law as in Figure 8(a). Alternatively, since $\mathbf{v} + (\mathbf{u} - \mathbf{v}) = \mathbf{u}$, the vector $\mathbf{u} - \mathbf{v}$, when added to **v**, gives **u**. So we could construct $\mathbf{u} - \mathbf{v}$ as in Figure 8(b) by means of the Triangle Law.

FIGURE 8

Drawing $\mathbf{u} - \mathbf{v}$

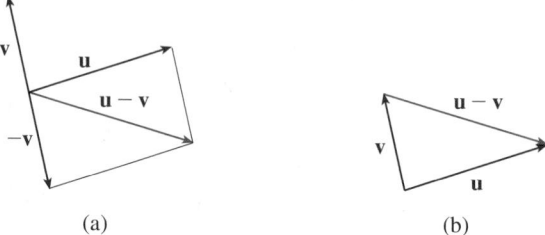

(a)   (b)

EXAMPLE 2 If **a** and **b** are the vectors shown in Figure 9, draw $\mathbf{a} - 2\mathbf{b}$.

SOLUTION We first draw the vector $-2\mathbf{b}$ pointing in the direction opposite to **b** and twice as long. We place it with its tail at the tip of **a** and then use the Triangle Law to draw $\mathbf{a} + (-2\mathbf{b})$ as in Figure 10.

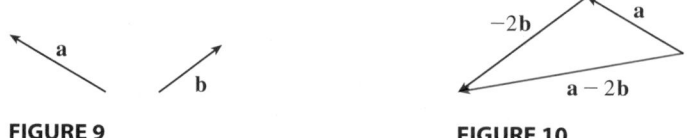

FIGURE 9   **FIGURE 10**   ■

COMPONENTS

For some purposes it's best to introduce a coordinate system and treat vectors algebraically. If we place the initial point of a vector **a** at the origin of a rectangular coordinate system, then the terminal point of **a** has coordinates of the form (a_1, a_2) or (a_1, a_2, a_3), depending on whether our coordinate system is two- or three-dimensional (see Figure 11). These coordinates are called the **components** of **a** and we write

$$\mathbf{a} = \langle a_1, a_2 \rangle \qquad \text{or} \qquad \mathbf{a} = \langle a_1, a_2, a_3 \rangle$$

We use the notation $\langle a_1, a_2 \rangle$ for the ordered pair that refers to a vector so as not to confuse it with the ordered pair (a_1, a_2) that refers to a point in the plane.

For instance, the vectors shown in Figure 12 are all equivalent to the vector $\overrightarrow{OP} = \langle 3, 2 \rangle$ whose terminal point is $P(3, 2)$. What they have in common is that the terminal point is reached from the initial point by a displacement of three units to the right and two upward. We can think of all these geometric vectors as **representations** of the algebraic vector $\mathbf{a} = \langle 3, 2 \rangle$. The particular representation $\overrightarrow{OP}$ from the origin to the point $P(3, 2)$ is called the **position vector** of the point P.

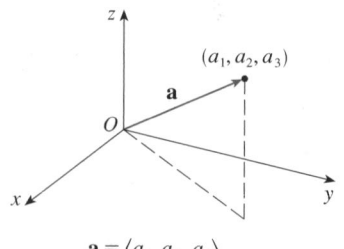

$$\mathbf{a} = \langle a_1, a_2 \rangle$$

$$\mathbf{a} = \langle a_1, a_2, a_3 \rangle$$

FIGURE 11

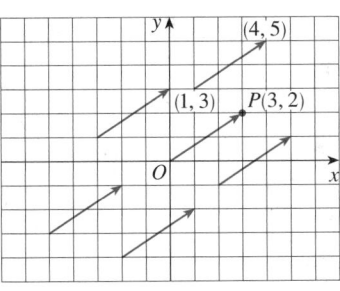

FIGURE 12
Representations of the vector $\mathbf{a} = \langle 3, 2 \rangle$

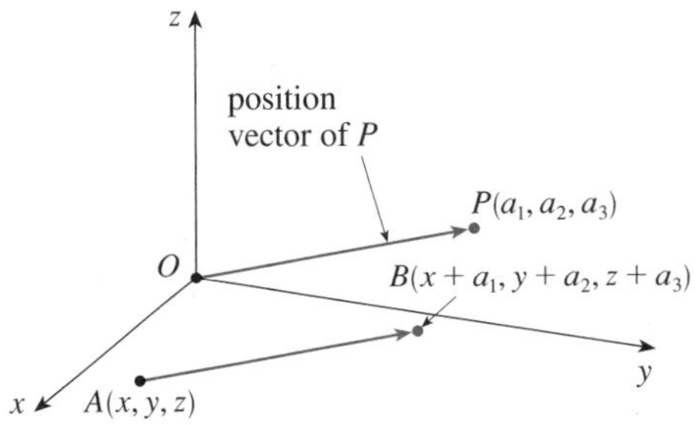

FIGURE 13
Representations of $\mathbf{a} = \langle a_1, a_2, a_3 \rangle$

In three dimensions, the vector $\mathbf{a} = \overrightarrow{OP} = \langle a_1, a_2, a_3 \rangle$ is the **position vector** of the point $P(a_1, a_2, a_3)$. (See Figure 13.) Let's consider any other representation $\overrightarrow{AB}$ of $\mathbf{a}$, where the initial point is $A(x_1, y_1, z_1)$ and the terminal point is $B(x_2, y_2, z_2)$. Then we must have $x_1 + a_1 = x_2$, $y_1 + a_2 = y_2$, and $z_1 + a_3 = z_2$ and so $a_1 = x_2 - x_1$, $a_2 = y_2 - y_1$, and $a_3 = z_2 - z_1$. Thus we have the following result.

> **1** Given the points $A(x_1, y_1, z_1)$ and $B(x_2, y_2, z_2)$, the vector $\mathbf{a}$ with representation $\overrightarrow{AB}$ is
> $$\mathbf{a} = \langle x_2 - x_1, y_2 - y_1, z_2 - z_1 \rangle$$

▼ EXAMPLE 3 Find the vector represented by the directed line segment with initial point $A(2, -3, 4)$ and terminal point $B(-2, 1, 1)$.

SOLUTION By $\boxed{1}$, the vector corresponding to $\overrightarrow{AB}$ is
$$\mathbf{a} = \langle -2 - 2, 1 - (-3), 1 - 4 \rangle = \langle -4, 4, -3 \rangle \qquad \blacksquare$$

The **magnitude** or **length** of the vector $\mathbf{v}$ is the length of any of its representations and is denoted by the symbol $|\mathbf{v}|$ or $\|\mathbf{v}\|$. By using the distance formula to compute the length of a segment OP, we obtain the following formulas.

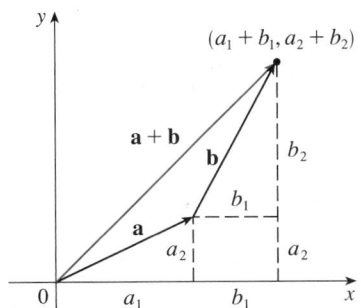

FIGURE 14

> The length of the two-dimensional vector $\mathbf{a} = \langle a_1, a_2 \rangle$ is
> $$|\mathbf{a}| = \sqrt{a_1^2 + a_2^2}$$
> The length of the three-dimensional vector $\mathbf{a} = \langle a_1, a_2, a_3 \rangle$ is
> $$|\mathbf{a}| = \sqrt{a_1^2 + a_2^2 + a_3^2}$$

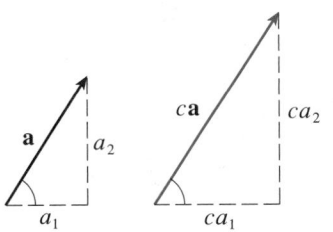

FIGURE 15

How do we add vectors algebraically? Figure 14 shows that if $\mathbf{a} = \langle a_1, a_2 \rangle$ and $\mathbf{b} = \langle b_1, b_2 \rangle$, then the sum is $\mathbf{a} + \mathbf{b} = \langle a_1 + b_1, a_2 + b_2 \rangle$, at least for the case where the components are positive. In other words, *to add algebraic vectors we add their components*. Similarly, *to subtract vectors we subtract components*. From the similar triangles in Figure 15 we see that the components of $c\mathbf{a}$ are ca_1 and ca_2. So *to multiply a vector by a scalar we multiply each component by that scalar*.

If $\mathbf{a} = \langle a_1, a_2 \rangle$ and $\mathbf{b} = \langle b_1, b_2 \rangle$, then

$$\mathbf{a} + \mathbf{b} = \langle a_1 + b_1, a_2 + b_2 \rangle \qquad \mathbf{a} - \mathbf{b} = \langle a_1 - b_1, a_2 - b_2 \rangle$$

$$c\mathbf{a} = \langle ca_1, ca_2 \rangle$$

Similarly, for three-dimensional vectors,

$$\langle a_1, a_2, a_3 \rangle + \langle b_1, b_2, b_3 \rangle = \langle a_1 + b_1, a_2 + b_2, a_3 + b_3 \rangle$$

$$\langle a_1, a_2, a_3 \rangle - \langle b_1, b_2, b_3 \rangle = \langle a_1 - b_1, a_2 - b_2, a_3 - b_3 \rangle$$

$$c\langle a_1, a_2, a_3 \rangle = \langle ca_1, ca_2, ca_3 \rangle$$

▼ EXAMPLE 4 If $\mathbf{a} = \langle 4, 0, 3 \rangle$ and $\mathbf{b} = \langle -2, 1, 5 \rangle$, find $|\mathbf{a}|$ and the vectors $\mathbf{a} + \mathbf{b}$, $\mathbf{a} - \mathbf{b}$, $3\mathbf{b}$, and $2\mathbf{a} + 5\mathbf{b}$.

SOLUTION $|\mathbf{a}| = \sqrt{4^2 + 0^2 + 3^2} = \sqrt{25} = 5$

$$\mathbf{a} + \mathbf{b} = \langle 4, 0, 3 \rangle + \langle -2, 1, 5 \rangle$$

$$= \langle 4 - 2, 0 + 1, 3 + 5 \rangle = \langle 2, 1, 8 \rangle$$

$$\mathbf{a} - \mathbf{b} = \langle 4, 0, 3 \rangle - \langle -2, 1, 5 \rangle$$

$$= \langle 4 - (-2), 0 - 1, 3 - 5 \rangle = \langle 6, -1, -2 \rangle$$

$$3\mathbf{b} = 3\langle -2, 1, 5 \rangle = \langle 3(-2), 3(1), 3(5) \rangle = \langle -6, 3, 15 \rangle$$

$$2\mathbf{a} + 5\mathbf{b} = 2\langle 4, 0, 3 \rangle + 5\langle -2, 1, 5 \rangle$$

$$= \langle 8, 0, 6 \rangle + \langle -10, 5, 25 \rangle = \langle -2, 5, 31 \rangle \qquad ■$$

■ Vectors in n dimensions are used to list various quantities in an organized way. For instance, the components of a six-dimensional vector

$$\mathbf{p} = \langle p_1, p_2, p_3, p_4, p_5, p_6 \rangle$$

might represent the prices of six different ingredients required to make a particular product. Four-dimensional vectors $\langle x, y, z, t \rangle$ are used in relativity theory, where the first three components specify a position in space and the fourth represents time.

We denote by V_2 the set of all two-dimensional vectors and by V_3 the set of all three-dimensional vectors. More generally, we will later need to consider the set V_n of all n-dimensional vectors. An n-dimensional vector is an ordered n-tuple:

$$\mathbf{a} = \langle a_1, a_2, \ldots, a_n \rangle$$

where $a_1, a_2, \ldots, a_n$ are real numbers that are called the components of $\mathbf{a}$. Addition and scalar multiplication are defined in terms of components just as for the cases $n = 2$ and $n = 3$.

PROPERTIES OF VECTORS   If $\mathbf{a}$, $\mathbf{b}$, and $\mathbf{c}$ are vectors in V_n and c and d are scalars, then

1. $\mathbf{a} + \mathbf{b} = \mathbf{b} + \mathbf{a}$
2. $\mathbf{a} + (\mathbf{b} + \mathbf{c}) = (\mathbf{a} + \mathbf{b}) + \mathbf{c}$
3. $\mathbf{a} + \mathbf{0} = \mathbf{a}$
4. $\mathbf{a} + (-\mathbf{a}) = \mathbf{0}$
5. $c(\mathbf{a} + \mathbf{b}) = c\mathbf{a} + c\mathbf{b}$
6. $(c + d)\mathbf{a} = c\mathbf{a} + d\mathbf{a}$
7. $(cd)\mathbf{a} = c(d\mathbf{a})$
8. $1\mathbf{a} = \mathbf{a}$

These eight properties of vectors can be readily verified either geometrically or algebraically. For instance, Property 1 can be seen from Figure 4 (it's equivalent to the

Parallelogram Law) or as follows for the case $n = 2$:

$$\mathbf{a} + \mathbf{b} = \langle a_1, a_2 \rangle + \langle b_1, b_2 \rangle = \langle a_1 + b_1, a_2 + b_2 \rangle$$

$$= \langle b_1 + a_1, b_2 + a_2 \rangle = \langle b_1, b_2 \rangle + \langle a_1, a_2 \rangle$$

$$= \mathbf{b} + \mathbf{a}$$

We can see why Property 2 (the associative law) is true by looking at Figure 16 and applying the Triangle Law several times: The vector $\overrightarrow{PQ}$ is obtained either by first constructing $\mathbf{a} + \mathbf{b}$ and then adding $\mathbf{c}$ or by adding $\mathbf{a}$ to the vector $\mathbf{b} + \mathbf{c}$.

Three vectors in V_3 play a special role. Let

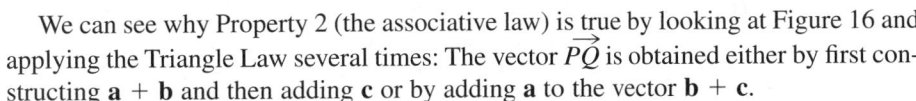

$$\mathbf{i} = \langle 1, 0, 0 \rangle \qquad \mathbf{j} = \langle 0, 1, 0 \rangle \qquad \mathbf{k} = \langle 0, 0, 1 \rangle$$

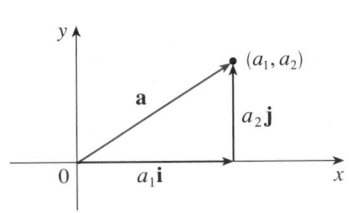

(a) $\mathbf{a} = a_1\mathbf{i} + a_2\mathbf{j}$

These vectors $\mathbf{i}$, $\mathbf{j}$, and $\mathbf{k}$ are called the **standard basis vectors**. They have length 1 and point in the directions of the positive x-, y-, and z-axes. Similarly, in two dimensions we define $\mathbf{i} = \langle 1, 0 \rangle$ and $\mathbf{j} = \langle 0, 1 \rangle$. (See Figure 17.)

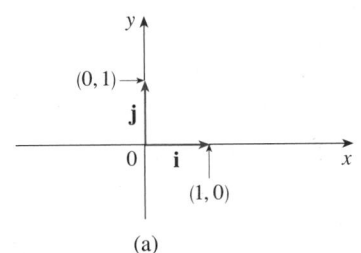

(a)

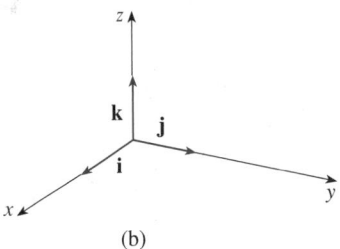

(b)

FIGURE 16

FIGURE 17

Standard basis vectors in V_2 and V_3

If $\mathbf{a} = \langle a_1, a_2, a_3 \rangle$, then we can write

$$\mathbf{a} = \langle a_1, a_2, a_3 \rangle = \langle a_1, 0, 0 \rangle + \langle 0, a_2, 0 \rangle + \langle 0, 0, a_3 \rangle$$

$$= a_1 \langle 1, 0, 0 \rangle + a_2 \langle 0, 1, 0 \rangle + a_3 \langle 0, 0, 1 \rangle$$

$$\boxed{2} \qquad \mathbf{a} = a_1\mathbf{i} + a_2\mathbf{j} + a_3\mathbf{k}$$

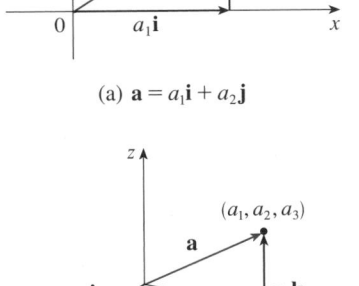

(b) $\mathbf{a} = a_1\mathbf{i} + a_2\mathbf{j} + a_3\mathbf{k}$

FIGURE 18

Thus any vector in V_3 can be expressed in terms of $\mathbf{i}$, $\mathbf{j}$, and $\mathbf{k}$. For instance,

$$\langle 1, -2, 6 \rangle = \mathbf{i} - 2\mathbf{j} + 6\mathbf{k}$$

Similarly, in two dimensions, we can write

$$\boxed{3} \qquad \mathbf{a} = \langle a_1, a_2 \rangle = a_1\mathbf{i} + a_2\mathbf{j}$$

See Figure 18 for the geometric interpretation of Equations 3 and 2 and compare with Figure 17.

EXAMPLE 5 If $\mathbf{a} = \mathbf{i} + 2\mathbf{j} - 3\mathbf{k}$ and $\mathbf{b} = 4\mathbf{i} + 7\mathbf{k}$, express the vector $2\mathbf{a} + 3\mathbf{b}$ in terms of $\mathbf{i}$, $\mathbf{j}$, and $\mathbf{k}$.

SOLUTION Using Properties 1, 2, 5, 6, and 7 of vectors, we have

$$2\mathbf{a} + 3\mathbf{b} = 2(\mathbf{i} + 2\mathbf{j} - 3\mathbf{k}) + 3(4\mathbf{i} + 7\mathbf{k})$$

$$= 2\mathbf{i} + 4\mathbf{j} - 6\mathbf{k} + 12\mathbf{i} + 21\mathbf{k} = 14\mathbf{i} + 4\mathbf{j} + 15\mathbf{k} \qquad ∎$$

A **unit vector** is a vector whose length is 1. For instance, **i**, **j**, and **k** are all unit vectors. In general, if $\mathbf{a} \neq \mathbf{0}$, then the unit vector that has the same direction as **a** is

$$\boxed{4} \qquad \mathbf{u} = \frac{1}{|\mathbf{a}|}\mathbf{a} = \frac{\mathbf{a}}{|\mathbf{a}|}$$

In order to verify this we let $c = 1/|\mathbf{a}|$. Then $\mathbf{u} = c\mathbf{a}$ and c is a positive scalar, so **u**
has the same direction as **a**. Also

$$|\mathbf{u}| = |c\mathbf{a}| = |c||\mathbf{a}| = \frac{1}{|\mathbf{a}|}|\mathbf{a}| = 1$$

EXAMPLE 6 Find the unit vector in the direction of the vector $2\mathbf{i} - \mathbf{j} - 2\mathbf{k}$.

SOLUTION The given vector has length

$$|2\mathbf{i} - \mathbf{j} - 2\mathbf{k}| = \sqrt{2^2 + (-1)^2 + (-2)^2} = \sqrt{9} = 3$$

so, by Equation 4, the unit vector with the same direction is

$$\tfrac{1}{3}(2\mathbf{i} - \mathbf{j} - 2\mathbf{k}) = \tfrac{2}{3}\mathbf{i} - \tfrac{1}{3}\mathbf{j} - \tfrac{2}{3}\mathbf{k} \qquad ■$$

APPLICATIONS

Vectors are useful in many aspects of physics and engineering. In Section 10.9 we will
see how they describe the velocity and acceleration of objects moving in space. Here
we look at forces.

A force is represented by a vector because it has both a magnitude (measured in
pounds or newtons) and a direction. If several forces are acting on an object, the **resultant force** experienced by the object is the vector sum of these forces.

EXAMPLE 7 A 100-lb weight hangs from two wires as shown in Figure 19. Find the
tensions (forces) $\mathbf{T}_1$ and $\mathbf{T}_2$ in both wires and their magnitudes.

SOLUTION We first express $\mathbf{T}_1$ and $\mathbf{T}_2$ in terms of their horizontal and vertical
components. From Figure 20 we see that

$$\boxed{5} \qquad \mathbf{T}_1 = -|\mathbf{T}_1|\cos 50° \,\mathbf{i} + |\mathbf{T}_1|\sin 50° \,\mathbf{j}$$

$$\boxed{6} \qquad \mathbf{T}_2 = |\mathbf{T}_2|\cos 32° \,\mathbf{i} + |\mathbf{T}_2|\sin 32° \,\mathbf{j}$$

The resultant $\mathbf{T}_1 + \mathbf{T}_2$ of the tensions counterbalances the weight **w** and so we must
have

$$\mathbf{T}_1 + \mathbf{T}_2 = -\mathbf{w} = 100\mathbf{j}$$

Thus

$$\left(-|\mathbf{T}_1|\cos 50° + |\mathbf{T}_2|\cos 32°\right)\mathbf{i} + \left(|\mathbf{T}_1|\sin 50° + |\mathbf{T}_2|\sin 32°\right)\mathbf{j} = 100\mathbf{j}$$

Equating components, we get

$$-|\mathbf{T}_1|\cos 50° + |\mathbf{T}_2|\cos 32° = 0$$

$$|\mathbf{T}_1|\sin 50° + |\mathbf{T}_2|\sin 32° = 100$$

Solving the first of these equations for $|\mathbf{T}_2|$ and substituting into the second, we get

$$|\mathbf{T}_1|\sin 50° + \frac{|\mathbf{T}_1|\cos 50°}{\cos 32°}\sin 32° = 100$$

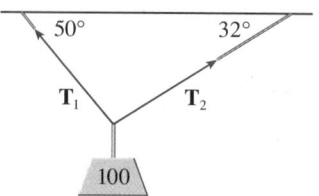

FIGURE 19

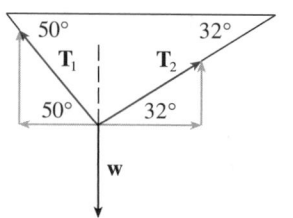

FIGURE 20

So the magnitudes of the tensions are

$$|\mathbf{T}_1| = \frac{100}{\sin 50° + \tan 32° \cos 50°} \approx 85.64 \text{ lb}$$

and

$$|\mathbf{T}_2| = \frac{|\mathbf{T}_1| \cos 50°}{\cos 32°} \approx 64.91 \text{ lb}$$

Substituting these values in $\boxed{5}$ and $\boxed{6}$, we obtain the tension vectors

$$\mathbf{T}_1 \approx -55.05\,\mathbf{i} + 65.60\,\mathbf{j} \qquad \mathbf{T}_2 \approx 55.05\,\mathbf{i} + 34.40\,\mathbf{j} \qquad \blacksquare$$

10.2 | EXERCISES

1. Name all the equal vectors in the parallelogram shown.

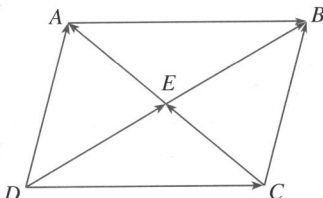

2. Write each combination of vectors as a single vector.

(a) $\overrightarrow{AB} + \overrightarrow{BC}$ (b) $\overrightarrow{CD} + \overrightarrow{DB}$

(c) $\overrightarrow{DB} - \overrightarrow{AB}$ (d) $\overrightarrow{DC} + \overrightarrow{CA} + \overrightarrow{AB}$

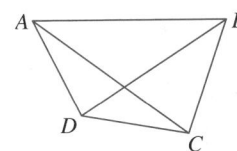

3. Copy the vectors in the figure and use them to draw the following vectors.

(a) $\mathbf{u} + \mathbf{v}$ (b) $\mathbf{u} + \mathbf{w}$

(c) $\mathbf{v} + \mathbf{w}$ (d) $\mathbf{u} - \mathbf{v}$

(e) $\mathbf{v} + \mathbf{u} + \mathbf{w}$ (f) $\mathbf{u} - \mathbf{w} - \mathbf{v}$

4. Copy the vectors in the figure and use them to draw the following vectors.

(a) $\mathbf{a} + \mathbf{b}$ (b) $\mathbf{a} - \mathbf{b}$

(c) $\frac{1}{2}\mathbf{a}$ (d) $-3\mathbf{b}$

(e) $\mathbf{a} + 2\mathbf{b}$ (f) $2\mathbf{b} - \mathbf{a}$

5–8 ▪ Find a vector $\mathbf{a}$ with representation given by the directed line segment $\overrightarrow{AB}$. Draw $\overrightarrow{AB}$ and the equivalent representation starting at the origin.

5. $A(-1, 1)$, $B(3, 2)$ **6.** $A(-4, -1)$, $B(1, 2)$

7. $A(0, 3, 1)$, $B(2, 3, -1)$ **8.** $A(4, 0, -2)$, $B(4, 2, 1)$

9–12 ▪ Find the sum of the given vectors and illustrate geometrically.

9. $\langle -1, 4 \rangle$, $\langle 6, -2 \rangle$ **10.** $\langle 3, -1 \rangle$, $\langle -1, 5 \rangle$

11. $\langle 3, 0, 1 \rangle$, $\langle 0, 8, 0 \rangle$ **12.** $\langle 1, 3, -2 \rangle$, $\langle 0, 0, 6 \rangle$

13–16 ▪ Find $\mathbf{a} + \mathbf{b}$, $2\mathbf{a} + 3\mathbf{b}$, $|\mathbf{a}|$, and $|\mathbf{a} - \mathbf{b}|$.

13. $\mathbf{a} = \langle 5, -12 \rangle$, $\mathbf{b} = \langle -3, -6 \rangle$

14. $\mathbf{a} = 4\mathbf{i} + \mathbf{j}$, $\mathbf{b} = \mathbf{i} - 2\mathbf{j}$

15. $\mathbf{a} = \mathbf{i} + 2\mathbf{j} - 3\mathbf{k}$, $\mathbf{b} = -2\mathbf{i} - \mathbf{j} + 5\mathbf{k}$

16. $\mathbf{a} = 2\mathbf{i} - 4\mathbf{j} + 4\mathbf{k}$, $\mathbf{b} = 2\mathbf{j} - \mathbf{k}$

17. Find a unit vector with the same direction as $8\mathbf{i} - \mathbf{j} + 4\mathbf{k}$.

18. Find a vector that has the same direction as $\langle -2, 4, 2 \rangle$ but has length 6.

19–20 ▪ What is the angle between the given vector and the positive direction of the x-axis?

19. $\mathbf{i} + \sqrt{3}\,\mathbf{j}$ **20.** $8\mathbf{i} + 6\mathbf{j}$

21. If $\mathbf{v}$ lies in the first quadrant and makes an angle $\pi/3$ with the positive x-axis and $|\mathbf{v}| = 4$, find $\mathbf{v}$ in component form.

22. If a child pulls a sled through the snow on a level path with a force of 50 N exerted at an angle of $38°$ above the horizontal, find the horizontal and vertical components of the force.

23. A quarterback throws a football with angle of elevation 40° and speed 60 ft/s. Find the horizontal and vertical components of the velocity vector.

24–25 ▪ Find the magnitude of the resultant force and the angle it makes with the positive *x*-axis.

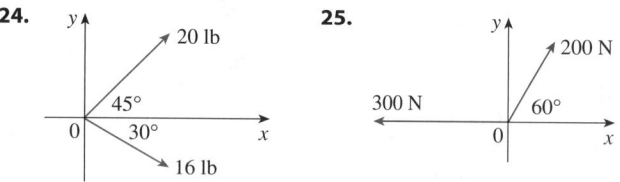

26. The magnitude of a velocity vector is called *speed.* Suppose that a wind is blowing from the direction N45°W at a speed of 50 km/h. (This means that the direction from which the wind blows is 45° west of the northerly direction.) A pilot is steering a plane in the direction N60°E at an airspeed (speed in still air) of 250 km/h. The *true course,* or *track,* of the plane is the direction of the resultant of the velocity vectors of the plane and the wind. The *ground speed* of the plane is the magnitude of the resultant. Find the true course and the ground speed of the plane.

27. A woman walks due west on the deck of a ship at 3 mi/h. The ship is moving north at a speed of 22 mi/h. Find the speed and direction of the woman relative to the surface of the water.

28. Ropes 3 m and 5 m in length are fastened to a holiday decoration that is suspended over a town square. The decoration has a mass of 5 kg. The ropes, fastened at different heights, make angles of 52° and 40° with the horizontal. Find the tension in each wire and the magnitude of each tension.

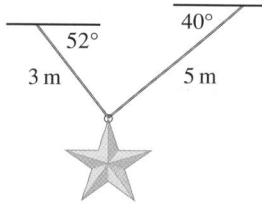

29. A clothesline is tied between two poles, 8 m apart. The line is quite taut and has negligible sag. When a wet shirt with a mass of 0.8 kg is hung at the middle of the line, the midpoint is pulled down 8 cm. Find the tension in each half of the clothesline.

30. The tension **T** at each end of the chain has magnitude 25 N (see the figure). What is the weight of the chain?

31. A boatman wants to cross a canal that is 3 km wide and wants to land at a point 2 km upstream from his starting point. The current in the canal flows at 3.5 km/h and the speed of his boat is 13 km/h.
(a) In what direction should he steer?
(b) How long will the trip take?

32. Three forces act on an object. Two of the forces are at an angle of 100° to each other and have magnitudes 25 N and 12 N. The third is perpendicular to the plane of these two forces and has magnitude 4 N. Calculate the magnitude of the force that would exactly counterbalance these three forces.

33. Find the unit vectors that are parallel to the tangent line to the parabola $y = x^2$ at the point $(2, 4)$.

34. (a) Find the unit vectors that are parallel to the tangent line to the curve $y = 2 \sin x$ at the point $(\pi/6, 1)$.
(b) Find the unit vectors that are perpendicular to the tangent line.
(c) Sketch the curve $y = 2 \sin x$ and the vectors in parts (a) and (b), all starting at $(\pi/6, 1)$.

35. (a) Draw the vectors $\mathbf{a} = \langle 3, 2 \rangle$, $\mathbf{b} = \langle 2, -1 \rangle$, and $\mathbf{c} = \langle 7, 1 \rangle$.
(b) Show, by means of a sketch, that there are scalars s and t such that $\mathbf{c} = s\mathbf{a} + t\mathbf{b}$.
(c) Use the sketch to estimate the values of s and t.
(d) Find the exact values of s and t.

36. Suppose that **a** and **b** are nonzero vectors that are not parallel and **c** is any vector in the plane determined by **a** and **b**. Give a geometric argument to show that **c** can be written as $\mathbf{c} = s\mathbf{a} + t\mathbf{b}$ for suitable scalars s and t. Then give an argument using components.

37. If $\mathbf{r} = \langle x, y, z \rangle$ and $\mathbf{r}_0 = \langle x_0, y_0, z_0 \rangle$, describe the set of all points (x, y, z) such that $|\mathbf{r} - \mathbf{r}_0| = 1$.

38. If $\mathbf{r} = \langle x, y \rangle$, $\mathbf{r}_1 = \langle x_1, y_1 \rangle$, and $\mathbf{r}_2 = \langle x_2, y_2 \rangle$, describe the set of all points (x, y) such that $|\mathbf{r} - \mathbf{r}_1| + |\mathbf{r} - \mathbf{r}_2| = k$, where $k > |\mathbf{r}_1 - \mathbf{r}_2|$.

39. Figure 16 gives a geometric demonstration of Property 2 of vectors. Use components to give an algebraic proof of this fact for the case $n = 2$.

40. Prove Property 5 of vectors algebraically for the case $n = 3$. Then use similar triangles to give a geometric proof.

41. Use vectors to prove that the line joining the midpoints of two sides of a triangle is parallel to the third side and half its length.

10.3 | THE DOT PRODUCT

So far we have added two vectors and multiplied a vector by a scalar. The question arises: Is it possible to multiply two vectors so that their product is a useful quantity? One such product is the dot product, whose definition follows. Another is the cross product, which is discussed in the next section.

---

1 DEFINITION If $\mathbf{a} = \langle a_1, a_2, a_3 \rangle$ and $\mathbf{b} = \langle b_1, b_2, b_3 \rangle$, then the **dot product** of $\mathbf{a}$ and $\mathbf{b}$ is the number $\mathbf{a} \cdot \mathbf{b}$ given by

$$\mathbf{a} \cdot \mathbf{b} = a_1 b_1 + a_2 b_2 + a_3 b_3$$

Thus to find the dot product of $\mathbf{a}$ and $\mathbf{b}$ we multiply corresponding components and add. The result is not a vector. It is a real number, that is, a scalar. For this reason, the dot product is sometimes called the **scalar product** (or **inner product**). Although Definition 1 is given for three-dimensional vectors, the dot product of two-dimensional vectors is defined in a similar fashion:

$$\langle a_1, a_2 \rangle \cdot \langle b_1, b_2 \rangle = a_1 b_1 + a_2 b_2$$

▼ EXAMPLE 1

$$\langle 2, 4 \rangle \cdot \langle 3, -1 \rangle = 2(3) + 4(-1) = 2$$

$$\langle -1, 7, 4 \rangle \cdot \left\langle 6, 2, -\tfrac{1}{2} \right\rangle = (-1)(6) + 7(2) + 4\left(-\tfrac{1}{2}\right) = 6$$

$$(\mathbf{i} + 2\mathbf{j} - 3\mathbf{k}) \cdot (2\mathbf{j} - \mathbf{k}) = 1(0) + 2(2) + (-3)(-1) = 7 \qquad \blacksquare$$

The dot product obeys many of the laws that hold for ordinary products of real numbers. These are stated in the following theorem.

---

2 PROPERTIES OF THE DOT PRODUCT If $\mathbf{a}$, $\mathbf{b}$, and $\mathbf{c}$ are vectors in V_3 and c is a scalar, then

1. $\mathbf{a} \cdot \mathbf{a} = |\mathbf{a}|^2$ 　　　　　　　　　**2.** $\mathbf{a} \cdot \mathbf{b} = \mathbf{b} \cdot \mathbf{a}$

3. $\mathbf{a} \cdot (\mathbf{b} + \mathbf{c}) = \mathbf{a} \cdot \mathbf{b} + \mathbf{a} \cdot \mathbf{c}$ 　　　**4.** $(c\mathbf{a}) \cdot \mathbf{b} = c(\mathbf{a} \cdot \mathbf{b}) = \mathbf{a} \cdot (c\mathbf{b})$

5. $\mathbf{0} \cdot \mathbf{a} = 0$

These properties are easily proved using Definition 1. For instance, here are the proofs of Properties 1 and 3:

1. $\mathbf{a} \cdot \mathbf{a} = a_1^2 + a_2^2 + a_3^2 = |\mathbf{a}|^2$

3. $\mathbf{a} \cdot (\mathbf{b} + \mathbf{c}) = \langle a_1, a_2, a_3 \rangle \cdot \langle b_1 + c_1, b_2 + c_2, b_3 + c_3 \rangle$

$$= a_1(b_1 + c_1) + a_2(b_2 + c_2) + a_3(b_3 + c_3)$$

$$= a_1 b_1 + a_1 c_1 + a_2 b_2 + a_2 c_2 + a_3 b_3 + a_3 c_3$$

$$= (a_1 b_1 + a_2 b_2 + a_3 b_3) + (a_1 c_1 + a_2 c_2 + a_3 c_3)$$

$$= \mathbf{a} \cdot \mathbf{b} + \mathbf{a} \cdot \mathbf{c}$$

The proofs of the remaining properties are left as exercises. 　　　　　　　　□

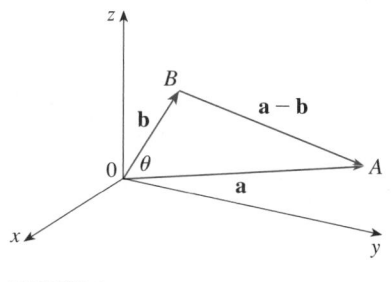

FIGURE 1

The dot product $\mathbf{a} \cdot \mathbf{b}$ can be given a geometric interpretation in terms of the **angle** θ **between a and b**, which is defined to be the angle between the representations of $\mathbf{a}$ and $\mathbf{b}$ that start at the origin, where $0 \leq \theta \leq \pi$. In other words, θ is the angle between the line segments $\overrightarrow{OA}$ and $\overrightarrow{OB}$ in Figure 1. Note that if $\mathbf{a}$ and $\mathbf{b}$ are parallel vectors, then $\theta = 0$ or $\theta = \pi$.

The formula in the following theorem is used by physicists as the *definition* of the dot product.

---

3 **THEOREM** If θ is the angle between the vectors $\mathbf{a}$ and $\mathbf{b}$, then

$$\mathbf{a} \cdot \mathbf{b} = |\mathbf{a}| |\mathbf{b}| \cos \theta$$

PROOF If we apply the Law of Cosines to triangle OAB in Figure 1, we get

$$\boxed{4} \qquad |AB|^2 = |OA|^2 + |OB|^2 - 2|OA| |OB| \cos \theta$$

(Observe that the Law of Cosines still applies in the limiting cases when $\theta = 0$ or π, or $\mathbf{a} = \mathbf{0}$ or $\mathbf{b} = \mathbf{0}$.) But $|OA| = |\mathbf{a}|$, $|OB| = |\mathbf{b}|$, and $|AB| = |\mathbf{a} - \mathbf{b}|$, so Equation 4 becomes

$$\boxed{5} \qquad |\mathbf{a} - \mathbf{b}|^2 = |\mathbf{a}|^2 + |\mathbf{b}|^2 - 2|\mathbf{a}| |\mathbf{b}| \cos \theta$$

Using Properties 1, 2, and 3 of the dot product, we can rewrite the left side of this equation as follows:

$$|\mathbf{a} - \mathbf{b}|^2 = (\mathbf{a} - \mathbf{b}) \cdot (\mathbf{a} - \mathbf{b}) = \mathbf{a} \cdot \mathbf{a} - \mathbf{a} \cdot \mathbf{b} - \mathbf{b} \cdot \mathbf{a} + \mathbf{b} \cdot \mathbf{b}$$

$$= |\mathbf{a}|^2 - 2\mathbf{a} \cdot \mathbf{b} + |\mathbf{b}|^2$$

Therefore Equation 5 gives

$$|\mathbf{a}|^2 - 2\mathbf{a} \cdot \mathbf{b} + |\mathbf{b}|^2 = |\mathbf{a}|^2 + |\mathbf{b}|^2 - 2|\mathbf{a}| |\mathbf{b}| \cos \theta$$

Thus
$$-2\mathbf{a} \cdot \mathbf{b} = -2|\mathbf{a}| |\mathbf{b}| \cos \theta$$

or
$$\mathbf{a} \cdot \mathbf{b} = |\mathbf{a}| |\mathbf{b}| \cos \theta \qquad \square$$

EXAMPLE 2 If the vectors $\mathbf{a}$ and $\mathbf{b}$ have lengths 4 and 6, and the angle between them is $\pi/3$, find $\mathbf{a} \cdot \mathbf{b}$.

SOLUTION Using Theorem 3, we have

$$\mathbf{a} \cdot \mathbf{b} = |\mathbf{a}| |\mathbf{b}| \cos(\pi/3) = 4 \cdot 6 \cdot \tfrac{1}{2} = 12 \qquad \blacksquare$$

The formula in Theorem 3 also enables us to find the angle between two vectors.

---

6 **COROLLARY** If θ is the angle between the nonzero vectors $\mathbf{a}$ and $\mathbf{b}$, then

$$\cos \theta = \frac{\mathbf{a} \cdot \mathbf{b}}{|\mathbf{a}| |\mathbf{b}|}$$

▼ EXAMPLE 3 Find the angle between the vectors $\mathbf{a} = \langle 2, 2, -1 \rangle$ and $\mathbf{b} = \langle 5, -3, 2 \rangle$.

SOLUTION Since

$$|\mathbf{a}| = \sqrt{2^2 + 2^2 + (-1)^2} = 3 \quad \text{and} \quad |\mathbf{b}| = \sqrt{5^2 + (-3)^2 + 2^2} = \sqrt{38}$$

and since

$$\mathbf{a} \cdot \mathbf{b} = 2(5) + 2(-3) + (-1)(2) = 2$$

we have, from Corollary 6,

$$\cos \theta = \frac{\mathbf{a} \cdot \mathbf{b}}{|\mathbf{a}||\mathbf{b}|} = \frac{2}{3\sqrt{38}}$$

So the angle between $\mathbf{a}$ and $\mathbf{b}$ is

$$\theta = \cos^{-1}\left(\frac{2}{3\sqrt{38}}\right) \approx 1.46 \quad (\text{or } 84°) \qquad \blacksquare$$

Two nonzero vectors $\mathbf{a}$ and $\mathbf{b}$ are called **perpendicular** or **orthogonal** if the angle between them is $\theta = \pi/2$. Then Theorem 3 gives

$$\mathbf{a} \cdot \mathbf{b} = |\mathbf{a}||\mathbf{b}| \cos(\pi/2) = 0$$

and conversely if $\mathbf{a} \cdot \mathbf{b} = 0$, then $\cos \theta = 0$, so $\theta = \pi/2$. The zero vector $\mathbf{0}$ is considered to be perpendicular to all vectors. Therefore we have the following method for determining whether two vectors are orthogonal.

> **7**   Two vectors $\mathbf{a}$ and $\mathbf{b}$ are orthogonal if and only if $\mathbf{a} \cdot \mathbf{b} = 0$.

EXAMPLE 4 Show that $2\mathbf{i} + 2\mathbf{j} - \mathbf{k}$ is perpendicular to $5\mathbf{i} - 4\mathbf{j} + 2\mathbf{k}$.

SOLUTION Since

$$(2\mathbf{i} + 2\mathbf{j} - \mathbf{k}) \cdot (5\mathbf{i} - 4\mathbf{j} + 2\mathbf{k}) = 2(5) + 2(-4) + (-1)(2) = 0$$

these vectors are perpendicular by $\boxed{7}$. $\qquad \blacksquare$

Because $\cos \theta > 0$ if $0 \leq \theta < \pi/2$ and $\cos \theta < 0$ if $\pi/2 < \theta \leq \pi$, we see that $\mathbf{a} \cdot \mathbf{b}$ is positive for $\theta < \pi/2$ and negative for $\theta > \pi/2$. We can think of $\mathbf{a} \cdot \mathbf{b}$ as measuring the extent to which $\mathbf{a}$ and $\mathbf{b}$ point in the same direction. The dot product $\mathbf{a} \cdot \mathbf{b}$ is positive if $\mathbf{a}$ and $\mathbf{b}$ point in the same general direction, 0 if they are perpendicular, and negative if they point in generally opposite directions (see Figure 2). In the extreme case where $\mathbf{a}$ and $\mathbf{b}$ point in exactly the same direction, we have $\theta = 0$, so $\cos \theta = 1$ and

$$\mathbf{a} \cdot \mathbf{b} = |\mathbf{a}||\mathbf{b}|$$

If $\mathbf{a}$ and $\mathbf{b}$ point in exactly opposite directions, then $\theta = \pi$ and so $\cos \theta = -1$ and $\mathbf{a} \cdot \mathbf{b} = -|\mathbf{a}||\mathbf{b}|$.

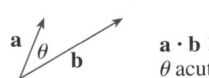
$\mathbf{a} \cdot \mathbf{b} > 0$
θ acute

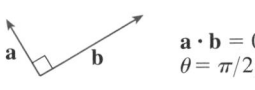
$\mathbf{a} \cdot \mathbf{b} = 0$
$\theta = \pi/2$

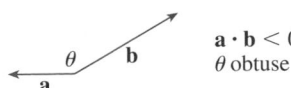
$\mathbf{a} \cdot \mathbf{b} < 0$
θ obtuse

FIGURE 2

TEC Visual 10.3A shows an animation of Figure 2.

PROJECTIONS

Figure 3 shows representations $\overrightarrow{PQ}$ and $\overrightarrow{PR}$ of two vectors $\mathbf{a}$ and $\mathbf{b}$ with the same initial point P. If S is the foot of the perpendicular from R to the line containing $\overrightarrow{PQ}$, then the vector with representation $\overrightarrow{PS}$ is called the **vector projection** of $\mathbf{b}$ onto $\mathbf{a}$ and is denoted by $\text{proj}_{\mathbf{a}}\, \mathbf{b}$. (You can think of it as a shadow of $\mathbf{b}$).

TEC Visual 10.3B shows how Figure 3 changes when we vary $\mathbf{a}$ and $\mathbf{b}$.

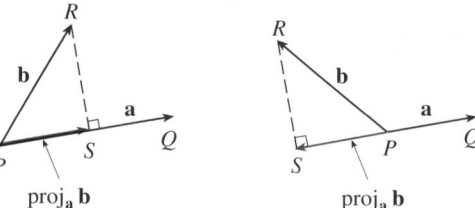

FIGURE 3

Vector projections

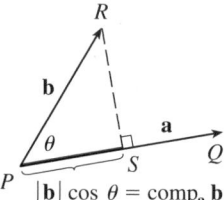

FIGURE 4

Scalar projection

The **scalar projection** of $\mathbf{b}$ onto $\mathbf{a}$ (also called the **component of b along a**) is defined to be the signed magnitude of the vector projection, which is the number $|\mathbf{b}| \cos \theta$, where θ is the angle between $\mathbf{a}$ and $\mathbf{b}$. (See Figure 4.) This is denoted by $\text{comp}_{\mathbf{a}}\, \mathbf{b}$. Observe that it is negative if $\pi/2 < \theta \le \pi$.

The equation

$$\mathbf{a} \cdot \mathbf{b} = |\mathbf{a}|\,|\mathbf{b}| \cos \theta = |\mathbf{a}|(|\mathbf{b}| \cos \theta)$$

shows that the dot product of $\mathbf{a}$ and $\mathbf{b}$ can be interpreted as the length of $\mathbf{a}$ times the scalar projection of $\mathbf{b}$ onto $\mathbf{a}$. Since

$$|\mathbf{b}| \cos \theta = \frac{\mathbf{a} \cdot \mathbf{b}}{|\mathbf{a}|} = \frac{\mathbf{a}}{|\mathbf{a}|} \cdot \mathbf{b}$$

the component of $\mathbf{b}$ along $\mathbf{a}$ can be computed by taking the dot product of $\mathbf{b}$ with the unit vector in the direction of $\mathbf{a}$. To summarize:

Scalar projection of $\mathbf{b}$ onto $\mathbf{a}$:	$\text{comp}_{\mathbf{a}}\, \mathbf{b} = \dfrac{\mathbf{a} \cdot \mathbf{b}}{|\mathbf{a}|}$
Vector projection of $\mathbf{b}$ onto $\mathbf{a}$:	$\text{proj}_{\mathbf{a}}\, \mathbf{b} = \left(\dfrac{\mathbf{a} \cdot \mathbf{b}}{|\mathbf{a}|} \right) \dfrac{\mathbf{a}}{|\mathbf{a}|} = \dfrac{\mathbf{a} \cdot \mathbf{b}}{|\mathbf{a}|^2}\, \mathbf{a}$

Notice that the vector projection is the scalar projection times the unit vector in the direction of $\mathbf{a}$.

V EXAMPLE 5 Find the scalar projection and vector projection of $\mathbf{b} = \langle 1, 1, 2 \rangle$ onto $\mathbf{a} = \langle -2, 3, 1 \rangle$.

SOLUTION Since $|\mathbf{a}| = \sqrt{(-2)^2 + 3^2 + 1^2} = \sqrt{14}$, the scalar projection of $\mathbf{b}$ onto $\mathbf{a}$ is

$$\text{comp}_{\mathbf{a}}\, \mathbf{b} = \frac{\mathbf{a} \cdot \mathbf{b}}{|\mathbf{a}|} = \frac{(-2)(1) + 3(1) + 1(2)}{\sqrt{14}} = \frac{3}{\sqrt{14}}$$

The vector projection is this scalar projection times the unit vector in the direction of **a**:

$$\text{proj}_{\mathbf{a}}\, \mathbf{b} = \frac{3}{\sqrt{14}}\,\frac{\mathbf{a}}{|\mathbf{a}|} = \frac{3}{14}\,\mathbf{a} = \left\langle -\frac{3}{7}, \frac{9}{14}, \frac{3}{14} \right\rangle$$ ∎

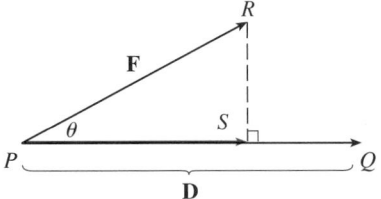

FIGURE 5

One use of projections occurs in physics in calculating work. In Section 7.6 we defined the work done by a constant force F in moving an object through a distance d as $W = Fd$, but this applies only when the force is directed along the line of motion of the object. Suppose, however, that the constant force is a vector $\mathbf{F} = \overrightarrow{PR}$ pointing in some other direction as in Figure 5. If the force moves the object from P to Q, then the **displacement vector** is $\mathbf{D} = \overrightarrow{PQ}$. The **work** done by this force is defined to be the product of the component of the force along **D** and the distance moved:

$$W = (|\mathbf{F}|\cos\theta)|\mathbf{D}|$$

But then, from Theorem 3, we have

$$\boxed{8} \qquad W = |\mathbf{F}||\mathbf{D}|\cos\theta = \mathbf{F}\cdot\mathbf{D}$$

Thus the work done by a constant force **F** is the dot product $\mathbf{F}\cdot\mathbf{D}$, where **D** is the displacement vector.

EXAMPLE 6 A wagon is pulled a distance of 100 m along a horizontal path by a constant force of 70 N. The handle of the wagon is held at an angle of 35° above the horizontal. Find the work done by the force.

SOLUTION If **F** and **D** are the force and displacement vectors, as pictured in Figure 6, then the work done is

$$W = \mathbf{F}\cdot\mathbf{D} = |\mathbf{F}||\mathbf{D}|\cos 35°$$

$$= (70)(100)\cos 35° \approx 5734\ \text{N·m} = 5734\ \text{J}$$ ∎

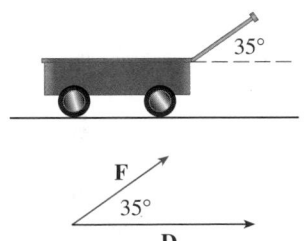

FIGURE 6

EXAMPLE 7 A force is given by a vector $\mathbf{F} = 3\mathbf{i} + 4\mathbf{j} + 5\mathbf{k}$ and moves a particle from the point $P(2, 1, 0)$ to the point $Q(4, 6, 2)$. Find the work done.

SOLUTION The displacement vector is $\mathbf{D} = \overrightarrow{PQ} = \langle 2, 5, 2\rangle$, so by Equation 8, the work done is

$$W = \mathbf{F}\cdot\mathbf{D} = \langle 3, 4, 5\rangle \cdot \langle 2, 5, 2\rangle$$

$$= 6 + 20 + 10 = 36$$

If the unit of length is meters and the magnitude of the force is measured in newtons, then the work done is 36 J. ∎

10.3 | EXERCISES

1. Which of the following expressions are meaningful? Which are meaningless? Explain.
(a) $(\mathbf{a} \cdot \mathbf{b}) \cdot \mathbf{c}$
(b) $(\mathbf{a} \cdot \mathbf{b})\mathbf{c}$
(c) $|\mathbf{a}|(\mathbf{b} \cdot \mathbf{c})$
(d) $\mathbf{a} \cdot (\mathbf{b} + \mathbf{c})$
(e) $\mathbf{a} \cdot \mathbf{b} + \mathbf{c}$
(f) $|\mathbf{a}| \cdot (\mathbf{b} + \mathbf{c})$

2–10 ■ Find $\mathbf{a} \cdot \mathbf{b}$.

2. $\mathbf{a} = \langle -2, 3 \rangle, \quad \mathbf{b} = \langle 0.7, 1.2 \rangle$

3. $\mathbf{a} = \langle -2, \frac{1}{3} \rangle, \quad \mathbf{b} = \langle -5, 12 \rangle$

4. $\mathbf{a} = \langle 6, -2, 3 \rangle, \quad \mathbf{b} = \langle 2, 5, -1 \rangle$

5. $\mathbf{a} = \langle 4, 1, \frac{1}{4} \rangle, \quad \mathbf{b} = \langle 6, -3, -8 \rangle$

6. $\mathbf{a} = \langle p, -p, 2p \rangle, \quad \mathbf{b} = \langle 2q, q, -q \rangle$

7. $\mathbf{a} = 2\mathbf{i} + \mathbf{j}, \quad \mathbf{b} = \mathbf{i} - \mathbf{j} + \mathbf{k}$

8. $\mathbf{a} = 3\mathbf{i} + 2\mathbf{j} - \mathbf{k}, \quad \mathbf{b} = 4\mathbf{i} + 5\mathbf{k}$

9. $|\mathbf{a}| = 6, \quad |\mathbf{b}| = 5,$ the angle between $\mathbf{a}$ and $\mathbf{b}$ is $2\pi/3$

10. $|\mathbf{a}| = 3, \quad |\mathbf{b}| = \sqrt{6},$ the angle between $\mathbf{a}$ and $\mathbf{b}$ is $45°$

11–12 ■ If $\mathbf{u}$ is a unit vector, find $\mathbf{u} \cdot \mathbf{v}$ and $\mathbf{u} \cdot \mathbf{w}$.

11.

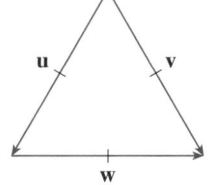

12.

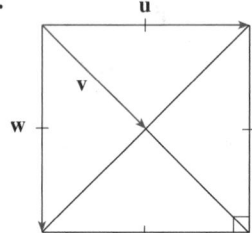

13. (a) Show that $\mathbf{i} \cdot \mathbf{j} = \mathbf{j} \cdot \mathbf{k} = \mathbf{k} \cdot \mathbf{i} = 0$.
(b) Show that $\mathbf{i} \cdot \mathbf{i} = \mathbf{j} \cdot \mathbf{j} = \mathbf{k} \cdot \mathbf{k} = 1$.

14. A street vendor sells a hamburgers, b hot dogs, and c soft drinks on a given day. He charges \$2 for a hamburger, \$1.50 for a hot dog, and \$1 for a soft drink. If $\mathbf{A} = \langle a, b, c \rangle$ and $\mathbf{P} = \langle 2, 1.5, 1 \rangle$, what is the meaning of the dot product $\mathbf{A} \cdot \mathbf{P}$?

15–17 ■ Find the angle between the vectors. (First find an exact expression and then approximate to the nearest degree.)

15. $\mathbf{a} = \langle 4, 3 \rangle, \quad \mathbf{b} = \langle 2, -1 \rangle$

16. $\mathbf{a} = \langle 4, 0, 2 \rangle, \quad \mathbf{b} = \langle 2, -1, 0 \rangle$

17. $\mathbf{a} = 4\mathbf{i} - 3\mathbf{j} + \mathbf{k}, \quad \mathbf{b} = 2\mathbf{i} - \mathbf{k}$

18. Find, correct to the nearest degree, the three angles of the triangle with vertices $A(1, 0, -1)$, $B(3, -2, 0)$, and $C(1, 3, 3)$.

19–20 ■ Determine whether the given vectors are orthogonal, parallel, or neither.

19. (a) $\mathbf{a} = \langle -5, 3, 7 \rangle, \quad \mathbf{b} = \langle 6, -8, 2 \rangle$
(b) $\mathbf{a} = \langle 4, 6 \rangle, \quad \mathbf{b} = \langle -3, 2 \rangle$
(c) $\mathbf{a} = -\mathbf{i} + 2\mathbf{j} + 5\mathbf{k}, \quad \mathbf{b} = 3\mathbf{i} + 4\mathbf{j} - \mathbf{k}$
(d) $\mathbf{a} = 2\mathbf{i} + 6\mathbf{j} - 4\mathbf{k}, \quad \mathbf{b} = -3\mathbf{i} - 9\mathbf{j} + 6\mathbf{k}$

20. (a) $\mathbf{u} = \langle -3, 9, 6 \rangle, \quad \mathbf{v} = \langle 4, -12, -8 \rangle$
(b) $\mathbf{u} = \mathbf{i} - \mathbf{j} + 2\mathbf{k}, \quad \mathbf{v} = 2\mathbf{i} - \mathbf{j} + \mathbf{k}$
(c) $\mathbf{u} = \langle a, b, c \rangle, \quad \mathbf{v} = \langle -b, a, 0 \rangle$

21. Use vectors to decide whether the triangle with vertices $P(1, -3, -2)$, $Q(2, 0, -4)$, and $R(6, -2, -5)$ is right-angled.

22. Find the values of x such that the angle between the vectors $\langle 2, 1, -1 \rangle$, and $\langle 1, x, 0 \rangle$ is $45°$.

23. Find a unit vector that is orthogonal to both $\mathbf{i} + \mathbf{j}$ and $\mathbf{i} + \mathbf{k}$.

24. Find two unit vectors that make an angle of $60°$ with $\mathbf{v} = \langle 3, 4 \rangle$.

25–26 ■ Find the acute angle between the lines.

25. $2x - y = 3, \quad 3x + y = 7$

26. $x + 2y = 7, \quad 5x - y = 2$

27–28 ■ Find the acute angles between the curves at their points of intersection. (The angle between two curves is the angle between their tangent lines at the point of intersection.)

27. $y = x^2, \quad y = x^3$

28. $y = \sin x, \quad y = \cos x, \quad 0 \leq x \leq \pi/2$

29–32 ■ Find the scalar and vector projections of $\mathbf{b}$ onto $\mathbf{a}$.

29. $\mathbf{a} = \langle -5, 12 \rangle, \quad \mathbf{b} = \langle 4, 6 \rangle$

30. $\mathbf{a} = \langle 1, 4 \rangle, \quad \mathbf{b} = \langle 2, 3 \rangle$

31. $\mathbf{a} = \langle 3, 6, -2 \rangle, \quad \mathbf{b} = \langle 1, 2, 3 \rangle$

32. $\mathbf{a} = \mathbf{i} + \mathbf{j} + \mathbf{k}, \quad \mathbf{b} = \mathbf{i} - \mathbf{j} + \mathbf{k}$

33. Show that the vector $\text{orth}_a\,\mathbf{b} = \mathbf{b} - \text{proj}_a\,\mathbf{b}$ is orthogonal to **a**. (It is called an **orthogonal projection** of **b**.)

34. For the vectors in Exercise 30, find $\text{orth}_a\,\mathbf{b}$ and illustrate by drawing the vectors **a**, **b**, $\text{proj}_a\,\mathbf{b}$, and $\text{orth}_a\,\mathbf{b}$.

35. If $\mathbf{a} = \langle 3, 0, -1 \rangle$, find a vector **b** such that $\text{comp}_a\,\mathbf{b} = 2$.

36. Suppose that **a** and **b** are nonzero vectors.
(a) Under what circumstances is $\text{comp}_a\,\mathbf{b} = \text{comp}_b\,\mathbf{a}$?
(b) Under what circumstances is $\text{proj}_a\,\mathbf{b} = \text{proj}_b\,\mathbf{a}$?

37. Find the work done by a force $\mathbf{F} = 8\,\mathbf{i} - 6\,\mathbf{j} + 9\,\mathbf{k}$ that moves an object from the point $(0, 10, 8)$ to the point $(6, 12, 20)$ along a straight line. The distance is measured in meters and the force in newtons.

38. A tow truck drags a stalled car along a road. The chain makes an angle of $30°$ with the road and the tension in the chain is 1500 N. How much work is done by the truck in pulling the car 1 km?

39. A sled is pulled along a level path through snow by a rope. A 30-lb force acting at an angle of $40°$ above the horizontal moves the sled 80 ft. Find the work done by the force.

40. A boat sails south with the help of a wind blowing in the direction S36°E with magnitude 400 lb. Find the work done by the wind as the boat moves 120 ft.

41. Use a scalar projection to show that the distance from a point $P_1(x_1, y_1)$ to the line $ax + by + c = 0$ is

$$\frac{|ax_1 + by_1 + c|}{\sqrt{a^2 + b^2}}$$

Use this formula to find the distance from the point $(-2, 3)$ to the line $3x - 4y + 5 = 0$.

42. If $\mathbf{r} = \langle x, y, z \rangle$, $\mathbf{a} = \langle a_1, a_2, a_3 \rangle$, and $\mathbf{b} = \langle b_1, b_2, b_3 \rangle$, show that the vector equation $(\mathbf{r} - \mathbf{a}) \cdot (\mathbf{r} - \mathbf{b}) = 0$ represents a sphere, and find its center and radius.

43. Find the angle between a diagonal of a cube and one of its edges.

44. Find the angle between a diagonal of a cube and a diagonal of one of its faces.

45. A molecule of methane, CH_4, is structured with the four hydrogen atoms at the vertices of a regular tetrahedron and the carbon atom at the centroid. The *bond angle* is the angle formed by the H—C—H combination; it is the angle between the lines that join the carbon atom to two of the hydrogen atoms. Show that the bond angle is about $109.5°$. [*Hint:* Take the vertices of the tetrahedron to be the points $(1, 0, 0)$, $(0, 1, 0)$, $(0, 0, 1)$, and $(1, 1, 1)$, as shown in the figure. Then the centroid is $\left(\frac{1}{2}, \frac{1}{2}, \frac{1}{2}\right)$.]

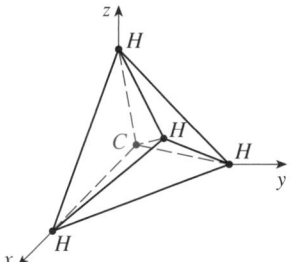

46. If $\mathbf{c} = |\mathbf{a}|\,\mathbf{b} + |\mathbf{b}|\,\mathbf{a}$, where **a**, **b**, and **c** are all nonzero vectors, show that **c** bisects the angle between **a** and **b**.

47. Prove Properties 2, 4, and 5 of the dot product (Theorem 2).

48. Suppose that all sides of a quadrilateral are equal in length and opposite sides are parallel. Use vector methods to show that the diagonals are perpendicular.

49. Use Theorem 3 to prove the Cauchy-Schwarz Inequality:

$$|\mathbf{a} \cdot \mathbf{b}| \leq |\mathbf{a}|\,|\mathbf{b}|$$

50. The Triangle Inequality for vectors is

$$|\mathbf{a} + \mathbf{b}| \leq |\mathbf{a}| + |\mathbf{b}|$$

(a) Give a geometric interpretation of the Triangle Inequality.
(b) Use the Cauchy-Schwarz Inequality from Exercise 49 to prove the Triangle Inequality. [*Hint:* Use the fact that $|\mathbf{a} + \mathbf{b}|^2 = (\mathbf{a} + \mathbf{b}) \cdot (\mathbf{a} + \mathbf{b})$ and use Property 3 of the dot product.]

51. The Parallelogram Law states that

$$|\mathbf{a} + \mathbf{b}|^2 + |\mathbf{a} - \mathbf{b}|^2 = 2|\mathbf{a}|^2 + 2|\mathbf{b}|^2$$

(a) Give a geometric interpretation of the Parallelogram Law.
(b) Prove the Parallelogram Law. (See the hint in Exercise 50.)

52. Show that if $\mathbf{u} + \mathbf{v}$ and $\mathbf{u} - \mathbf{v}$ are orthogonal, then the vectors **u** and **v** must have the same length.

10.4 | THE CROSS PRODUCT

The **cross product a** $\times$ **b** of two vectors **a** and **b**, unlike the dot product, is a vector. For this reason it is also called the **vector product**. Note that **a** $\times$ **b** is defined only when **a** and **b** are *three-dimensional* vectors.

> **1 DEFINITION** If $\mathbf{a} = \langle a_1, a_2, a_3 \rangle$ and $\mathbf{b} = \langle b_1, b_2, b_3 \rangle$, then the **cross product** of **a** and **b** is the vector
>
> $$\mathbf{a} \times \mathbf{b} = \langle a_2 b_3 - a_3 b_2, \, a_3 b_1 - a_1 b_3, \, a_1 b_2 - a_2 b_1 \rangle$$

■ HAMILTON

The cross product was invented by the Irish mathematician Sir William Rowan Hamilton (1805–1865), who had created a precursor of vectors, called quaternions. When he was five years old Hamilton could read Latin, Greek, and Hebrew. At age eight he added French and Italian and when ten he could read Arabic and Sanskrit. At the age of 21, while still an undergraduate at Trinity College in Dublin, Hamilton was appointed Professor of Astronomy at the university and Royal Astronomer of Ireland!

This may seem like a strange way of defining a product. The reason for the particular form of Definition 1 is that the cross product defined in this way has many useful properties, as we will soon see. In particular, we will show that the vector **a** $\times$ **b** is perpendicular to both **a** and **b**.

In order to make Definition 1 easier to remember, we use the notation of determinants. A **determinant of order 2** is defined by

$$\begin{vmatrix} a & b \\ c & d \end{vmatrix} = ad - bc$$

For example,
$$\begin{vmatrix} 2 & 1 \\ -6 & 4 \end{vmatrix} = 2(4) - 1(-6) = 14$$

A **determinant of order 3** can be defined in terms of second-order determinants as follows:

$$\boxed{2} \qquad \begin{vmatrix} a_1 & a_2 & a_3 \\ b_1 & b_2 & b_3 \\ c_1 & c_2 & c_3 \end{vmatrix} = a_1 \begin{vmatrix} b_2 & b_3 \\ c_2 & c_3 \end{vmatrix} - a_2 \begin{vmatrix} b_1 & b_3 \\ c_1 & c_3 \end{vmatrix} + a_3 \begin{vmatrix} b_1 & b_2 \\ c_1 & c_2 \end{vmatrix}$$

Observe that each term on the right side of Equation 2 involves a number a_i in the first row of the determinant, and a_i is multiplied by the second-order determinant obtained from the left side by deleting the row and column in which a_i appears. Notice also the minus sign in the second term. For example,

$$\begin{vmatrix} 1 & 2 & -1 \\ 3 & 0 & 1 \\ -5 & 4 & 2 \end{vmatrix} = 1 \begin{vmatrix} 0 & 1 \\ 4 & 2 \end{vmatrix} - 2 \begin{vmatrix} 3 & 1 \\ -5 & 2 \end{vmatrix} + (-1) \begin{vmatrix} 3 & 0 \\ -5 & 4 \end{vmatrix}$$

$$= 1(0 - 4) - 2(6 + 5) + (-1)(12 - 0) = -38$$

If we now rewrite Definition 1 using second-order determinants and the standard basis vectors **i**, **j**, and **k**, we see that the cross product of $\mathbf{a} = a_1 \mathbf{i} + a_2 \mathbf{j} + a_3 \mathbf{k}$ and

$\mathbf{b} = b_1\mathbf{i} + b_2\mathbf{j} + b_3\mathbf{k}$ is

$$\boxed{3} \qquad\qquad \mathbf{a} \times \mathbf{b} = \begin{vmatrix} a_2 & a_3 \\ b_2 & b_3 \end{vmatrix} \mathbf{i} - \begin{vmatrix} a_1 & a_3 \\ b_1 & b_3 \end{vmatrix} \mathbf{j} + \begin{vmatrix} a_1 & a_2 \\ b_1 & b_2 \end{vmatrix} \mathbf{k}$$

In view of the similarity between Equations 2 and 3, we often write

$$\boxed{4} \qquad\qquad \mathbf{a} \times \mathbf{b} = \begin{vmatrix} \mathbf{i} & \mathbf{j} & \mathbf{k} \\ a_1 & a_2 & a_3 \\ b_1 & b_2 & b_3 \end{vmatrix}$$

Although the first row of the symbolic determinant in Equation 4 consists of vectors, if we expand it as if it were an ordinary determinant using the rule in Equation 2, we obtain Equation 3. The symbolic formula in Equation 4 is probably the easiest way of remembering and computing cross products.

▼ EXAMPLE 1 If $\mathbf{a} = \langle 1, 3, 4 \rangle$ and $\mathbf{b} = \langle 2, 7, -5 \rangle$, then

$$\mathbf{a} \times \mathbf{b} = \begin{vmatrix} \mathbf{i} & \mathbf{j} & \mathbf{k} \\ 1 & 3 & 4 \\ 2 & 7 & -5 \end{vmatrix} = \begin{vmatrix} 3 & 4 \\ 7 & -5 \end{vmatrix} \mathbf{i} - \begin{vmatrix} 1 & 4 \\ 2 & -5 \end{vmatrix} \mathbf{j} + \begin{vmatrix} 1 & 3 \\ 2 & 7 \end{vmatrix} \mathbf{k}$$

$$= (-15 - 28)\,\mathbf{i} - (-5 - 8)\,\mathbf{j} + (7 - 6)\,\mathbf{k} = -43\mathbf{i} + 13\mathbf{j} + \mathbf{k} \qquad ■$$

▼ EXAMPLE 2 Show that $\mathbf{a} \times \mathbf{a} = \mathbf{0}$ for any vector $\mathbf{a}$ in V_3.

SOLUTION If $\mathbf{a} = \langle a_1, a_2, a_3 \rangle$, then

$$\mathbf{a} \times \mathbf{a} = \begin{vmatrix} \mathbf{i} & \mathbf{j} & \mathbf{k} \\ a_1 & a_2 & a_3 \\ a_1 & a_2 & a_3 \end{vmatrix}$$

$$= (a_2 a_3 - a_3 a_2)\,\mathbf{i} - (a_1 a_3 - a_3 a_1)\,\mathbf{j} + (a_1 a_2 - a_2 a_1)\,\mathbf{k}$$

$$= 0\mathbf{i} - 0\mathbf{j} + 0\mathbf{k} = \mathbf{0} \qquad ■$$

One of the most important properties of the cross product is given by the following theorem.

$\boxed{5}$ **THEOREM** The vector $\mathbf{a} \times \mathbf{b}$ is orthogonal to both $\mathbf{a}$ and $\mathbf{b}$.

PROOF In order to show that $\mathbf{a} \times \mathbf{b}$ is orthogonal to $\mathbf{a}$, we compute their dot product as follows:

$$(\mathbf{a} \times \mathbf{b}) \cdot \mathbf{a} = \begin{vmatrix} a_2 & a_3 \\ b_2 & b_3 \end{vmatrix} a_1 - \begin{vmatrix} a_1 & a_3 \\ b_1 & b_3 \end{vmatrix} a_2 + \begin{vmatrix} a_1 & a_2 \\ b_1 & b_2 \end{vmatrix} a_3$$

$$= a_1(a_2 b_3 - a_3 b_2) - a_2(a_1 b_3 - a_3 b_1) + a_3(a_1 b_2 - a_2 b_1)$$

$$= a_1 a_2 b_3 - a_1 b_2 a_3 - a_1 a_2 b_3 + b_1 a_2 a_3 + a_1 b_2 a_3 - b_1 a_2 a_3$$

$$= 0$$

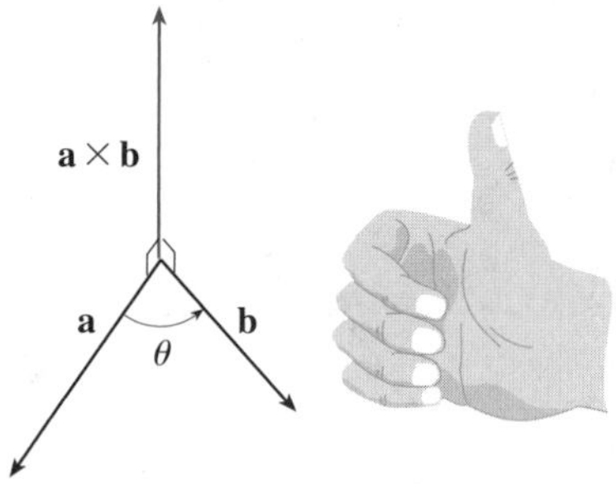

FIGURE 1
The right-hand rule gives
the direction of $\mathbf{a} \times \mathbf{b}$.

TEC Visual 10.4 shows how $\mathbf{a} \times \mathbf{b}$
changes as $\mathbf{b}$ changes.

A similar computation shows that $(\mathbf{a} \times \mathbf{b}) \cdot \mathbf{b} = 0$. Therefore $\mathbf{a} \times \mathbf{b}$ is orthogonal to both $\mathbf{a}$ and $\mathbf{b}$. □

If $\mathbf{a}$ and $\mathbf{b}$ are represented by directed line segments with the same initial point (as in Figure 1), then Theorem 5 says that the cross product $\mathbf{a} \times \mathbf{b}$ points in a direction perpendicular to the plane through $\mathbf{a}$ and $\mathbf{b}$. It turns out that the direction of $\mathbf{a} \times \mathbf{b}$ is given by the *right-hand rule:* If the fingers of your right hand curl in the direction of a rotation (through an angle less than $180°$) from $\mathbf{a}$ to $\mathbf{b}$, then your thumb points in the direction of $\mathbf{a} \times \mathbf{b}$.

Now that we know the direction of the vector $\mathbf{a} \times \mathbf{b}$, the remaining thing we need to complete its geometric description is its length $|\mathbf{a} \times \mathbf{b}|$. This is given by the following theorem.

---

6 **THEOREM** If θ is the angle between $\mathbf{a}$ and $\mathbf{b}$ (so $0 \leqslant \theta \leqslant \pi$), then

$$|\mathbf{a} \times \mathbf{b}| = |\mathbf{a}| |\mathbf{b}| \sin \theta$$

PROOF From the definitions of the cross product and length of a vector, we have

$$|\mathbf{a} \times \mathbf{b}|^2 = (a_2 b_3 - a_3 b_2)^2 + (a_3 b_1 - a_1 b_3)^2 + (a_1 b_2 - a_2 b_1)^2$$

$$= a_2^2 b_3^2 - 2 a_2 a_3 b_2 b_3 + a_3^2 b_2^2 + a_3^2 b_1^2 - 2 a_1 a_3 b_1 b_3 + a_1^2 b_3^2$$

$$\quad + a_1^2 b_2^2 - 2 a_1 a_2 b_1 b_2 + a_2^2 b_1^2$$

$$= (a_1^2 + a_2^2 + a_3^2)(b_1^2 + b_2^2 + b_3^2) - (a_1 b_1 + a_2 b_2 + a_3 b_3)^2$$

$$= |\mathbf{a}|^2 |\mathbf{b}|^2 - (\mathbf{a} \cdot \mathbf{b})^2$$

$$= |\mathbf{a}|^2 |\mathbf{b}|^2 - |\mathbf{a}|^2 |\mathbf{b}|^2 \cos^2 \theta \qquad \text{(by Theorem 10.3.3)}$$

$$= |\mathbf{a}|^2 |\mathbf{b}|^2 (1 - \cos^2 \theta)$$

$$= |\mathbf{a}|^2 |\mathbf{b}|^2 \sin^2 \theta$$

Taking square roots and observing that $\sqrt{\sin^2 \theta} = \sin \theta$ because $\sin \theta \geqslant 0$ when $0 \leqslant \theta \leqslant \pi$, we have

$$|\mathbf{a} \times \mathbf{b}| = |\mathbf{a}| |\mathbf{b}| \sin \theta \qquad \qquad □$$

Geometric characterization of $\mathbf{a} \times \mathbf{b}$

Since a vector is completely determined by its magnitude and direction, we can now say that $\mathbf{a} \times \mathbf{b}$ is the vector that is perpendicular to both $\mathbf{a}$ and $\mathbf{b}$, whose orientation is determined by the right-hand rule, and whose length is $|\mathbf{a}| |\mathbf{b}| \sin \theta$. In fact, that is exactly how physicists *define* $\mathbf{a} \times \mathbf{b}$.

---

7 **COROLLARY** Two nonzero vectors $\mathbf{a}$ and $\mathbf{b}$ are parallel if and only if

$$\mathbf{a} \times \mathbf{b} = \mathbf{0}$$

PROOF Two nonzero vectors $\mathbf{a}$ and $\mathbf{b}$ are parallel if and only if $\theta = 0$ or π. In either case $\sin \theta = 0$, so $|\mathbf{a} \times \mathbf{b}| = 0$ and therefore $\mathbf{a} \times \mathbf{b} = \mathbf{0}$. □

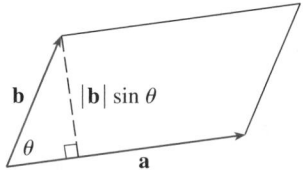

FIGURE 2

The geometric interpretation of Theorem 6 can be seen by looking at Figure 2. If **a** and **b** are represented by directed line segments with the same initial point, then they determine a parallelogram with base $|\mathbf{a}|$, altitude $|\mathbf{b}|\sin\theta$, and area

$$A = |\mathbf{a}|(|\mathbf{b}|\sin\theta) = |\mathbf{a} \times \mathbf{b}|$$

Thus we have the following way of interpreting the magnitude of a cross product.

> The length of the cross product $\mathbf{a} \times \mathbf{b}$ is equal to the area of the parallelogram determined by **a** and **b**.

EXAMPLE 3 Find a vector perpendicular to the plane that passes through the points $P(1, 4, 6)$, $Q(-2, 5, -1)$, and $R(1, -1, 1)$.

SOLUTION The vector $\overrightarrow{PQ} \times \overrightarrow{PR}$ is perpendicular to both $\overrightarrow{PQ}$ and $\overrightarrow{PR}$ and is therefore perpendicular to the plane through P, Q, and R. We know from (10.2.1) that

$$\overrightarrow{PQ} = (-2 - 1)\mathbf{i} + (5 - 4)\mathbf{j} + (-1 - 6)\mathbf{k} = -3\mathbf{i} + \mathbf{j} - 7\mathbf{k}$$

$$\overrightarrow{PR} = (1 - 1)\mathbf{i} + (-1 - 4)\mathbf{j} + (1 - 6)\mathbf{k} = -5\mathbf{j} - 5\mathbf{k}$$

We compute the cross product of these vectors:

$$\overrightarrow{PQ} \times \overrightarrow{PR} = \begin{vmatrix} \mathbf{i} & \mathbf{j} & \mathbf{k} \\ -3 & 1 & -7 \\ 0 & -5 & -5 \end{vmatrix}$$

$$= (-5 - 35)\mathbf{i} - (15 - 0)\mathbf{j} + (15 - 0)\mathbf{k} = -40\mathbf{i} - 15\mathbf{j} + 15\mathbf{k}$$

So the vector $\langle -40, -15, 15 \rangle$ is perpendicular to the given plane. Any nonzero scalar multiple of this vector, such as $\langle -8, -3, 3 \rangle$, is also perpendicular to the plane. ∎

EXAMPLE 4 Find the area of the triangle with vertices $P(1, 4, 6)$, $Q(-2, 5, -1)$, and $R(1, -1, 1)$.

SOLUTION In Example 3 we computed that $\overrightarrow{PQ} \times \overrightarrow{PR} = \langle -40, -15, 15 \rangle$. The area of the parallelogram with adjacent sides PQ and PR is the length of this cross product:

$$\left|\overrightarrow{PQ} \times \overrightarrow{PR}\right| = \sqrt{(-40)^2 + (-15)^2 + 15^2} = 5\sqrt{82}$$

The area A of the triangle PQR is half the area of this parallelogram, that is, $\frac{5}{2}\sqrt{82}$. ∎

If we apply Theorems 5 and 6 to the standard basis vectors **i**, **j**, and **k** using $\theta = \pi/2$, we obtain

$$\mathbf{i} \times \mathbf{j} = \mathbf{k} \qquad \mathbf{j} \times \mathbf{k} = \mathbf{i} \qquad \mathbf{k} \times \mathbf{i} = \mathbf{j}$$

$$\mathbf{j} \times \mathbf{i} = -\mathbf{k} \qquad \mathbf{k} \times \mathbf{j} = -\mathbf{i} \qquad \mathbf{i} \times \mathbf{k} = -\mathbf{j}$$

Observe that

$$\mathbf{i} \times \mathbf{j} \neq \mathbf{j} \times \mathbf{i}$$

Thus the cross product is not commutative. Also

$$\mathbf{i} \times (\mathbf{i} \times \mathbf{j}) = \mathbf{i} \times \mathbf{k} = -\mathbf{j}$$

whereas

$$(\mathbf{i} \times \mathbf{i}) \times \mathbf{j} = \mathbf{0} \times \mathbf{j} = \mathbf{0}$$

So the associative law for multiplication does not usually hold; that is, in general,

⊘

$$(\mathbf{a} \times \mathbf{b}) \times \mathbf{c} \neq \mathbf{a} \times (\mathbf{b} \times \mathbf{c})$$

However, some of the usual laws of algebra *do* hold for cross products. The following theorem summarizes the properties of vector products.

---

8 **THEOREM** If $\mathbf{a}$, $\mathbf{b}$, and $\mathbf{c}$ are vectors and c is a scalar, then

1. $\mathbf{a} \times \mathbf{b} = -\mathbf{b} \times \mathbf{a}$

2. $(c\mathbf{a}) \times \mathbf{b} = c(\mathbf{a} \times \mathbf{b}) = \mathbf{a} \times (c\mathbf{b})$

3. $\mathbf{a} \times (\mathbf{b} + \mathbf{c}) = \mathbf{a} \times \mathbf{b} + \mathbf{a} \times \mathbf{c}$

4. $(\mathbf{a} + \mathbf{b}) \times \mathbf{c} = \mathbf{a} \times \mathbf{c} + \mathbf{b} \times \mathbf{c}$

5. $\mathbf{a} \cdot (\mathbf{b} \times \mathbf{c}) = (\mathbf{a} \times \mathbf{b}) \cdot \mathbf{c}$

6. $\mathbf{a} \times (\mathbf{b} \times \mathbf{c}) = (\mathbf{a} \cdot \mathbf{c})\mathbf{b} - (\mathbf{a} \cdot \mathbf{b})\mathbf{c}$

These properties can be proved by writing the vectors in terms of their components and using the definition of a cross product. We give the proof of Property 5 and leave the remaining proofs as exercises.

PROOF OF PROPERTY 5 If $\mathbf{a} = \langle a_1, a_2, a_3 \rangle$, $\mathbf{b} = \langle b_1, b_2, b_3 \rangle$, and $\mathbf{c} = \langle c_1, c_2, c_3 \rangle$, then

9
$$\begin{aligned} \mathbf{a} \cdot (\mathbf{b} \times \mathbf{c}) &= a_1(b_2 c_3 - b_3 c_2) + a_2(b_3 c_1 - b_1 c_3) + a_3(b_1 c_2 - b_2 c_1) \\ &= a_1 b_2 c_3 - a_1 b_3 c_2 + a_2 b_3 c_1 - a_2 b_1 c_3 + a_3 b_1 c_2 - a_3 b_2 c_1 \\ &= (a_2 b_3 - a_3 b_2)c_1 + (a_3 b_1 - a_1 b_3)c_2 + (a_1 b_2 - a_2 b_1)c_3 \\ &= (\mathbf{a} \times \mathbf{b}) \cdot \mathbf{c} \end{aligned}$$
□

TRIPLE PRODUCTS

The product $\mathbf{a} \cdot (\mathbf{b} \times \mathbf{c})$ that occurs in Property 5 is called the **scalar triple product** of the vectors $\mathbf{a}$, $\mathbf{b}$, and $\mathbf{c}$. Notice from Equation 9 that we can write the scalar triple product as a determinant:

10
$$\mathbf{a} \cdot (\mathbf{b} \times \mathbf{c}) = \begin{vmatrix} a_1 & a_2 & a_3 \\ b_1 & b_2 & b_3 \\ c_1 & c_2 & c_3 \end{vmatrix}$$

The geometric significance of the scalar triple product can be seen by considering the parallelepiped determined by the vectors $\mathbf{a}$, $\mathbf{b}$, and $\mathbf{c}$. (See Figure 3.) The area of the base parallelogram is $A = |\mathbf{b} \times \mathbf{c}|$. If θ is the angle between $\mathbf{a}$ and $\mathbf{b} \times \mathbf{c}$, then the height h of the parallelepiped is $h = |\mathbf{a}| |\cos \theta|$. (We must use $|\cos \theta|$ instead of

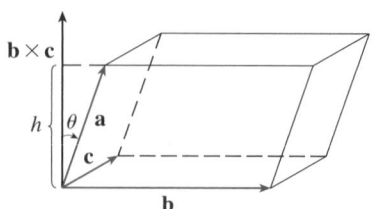

FIGURE 3

$\cos\theta$ in case $\theta > \pi/2$.) Therefore the volume of the parallelepiped is

$$V = Ah = |\mathbf{b} \times \mathbf{c}||\mathbf{a}||\cos\theta| = |\mathbf{a} \cdot (\mathbf{b} \times \mathbf{c})|$$

Thus we have proved the following formula.

$\boxed{11}$ The volume of the parallelepiped determined by the vectors $\mathbf{a}$, $\mathbf{b}$, and $\mathbf{c}$ is the magnitude of their scalar triple product:

$$V = |\mathbf{a} \cdot (\mathbf{b} \times \mathbf{c})|$$

If we use the formula in $\boxed{11}$ and discover that the volume of the parallelepiped determined by $\mathbf{a}$, $\mathbf{b}$, and $\mathbf{c}$ is 0, then the vectors must lie in the same plane; that is, they are **coplanar**.

▼ EXAMPLE 5 Use the scalar triple product to show that the vectors $\mathbf{a} = \langle 1, 4, -7 \rangle$, $\mathbf{b} = \langle 2, -1, 4 \rangle$, and $\mathbf{c} = \langle 0, -9, 18 \rangle$ are coplanar.

SOLUTION We use Equation 10 to compute their scalar triple product:

$$\mathbf{a} \cdot (\mathbf{b} \times \mathbf{c}) = \begin{vmatrix} 1 & 4 & -7 \\ 2 & -1 & 4 \\ 0 & -9 & 18 \end{vmatrix}$$

$$= 1 \begin{vmatrix} -1 & 4 \\ -9 & 18 \end{vmatrix} - 4 \begin{vmatrix} 2 & 4 \\ 0 & 18 \end{vmatrix} - 7 \begin{vmatrix} 2 & -1 \\ 0 & -9 \end{vmatrix}$$

$$= 1(18) - 4(36) - 7(-18) = 0$$

Therefore by $\boxed{11}$ the volume of the parallelepiped determined by $\mathbf{a}$, $\mathbf{b}$, and $\mathbf{c}$ is 0. This means that $\mathbf{a}$, $\mathbf{b}$, and $\mathbf{c}$ are coplanar. ■

The product $\mathbf{a} \times (\mathbf{b} \times \mathbf{c})$ that occurs in Property 6 is called the **vector triple product** of $\mathbf{a}$, $\mathbf{b}$, and $\mathbf{c}$. Property 6 will be used to derive Kepler's First Law of planetary motion in Section 10.9. Its proof is left as Exercise 50.

TORQUE

The idea of a cross product occurs often in physics. In particular, we consider a force $\mathbf{F}$ acting on a rigid body at a point given by a position vector $\mathbf{r}$. (For instance, if we tighten a bolt by applying a force to a wrench as in Figure 4, we produce a turning effect.) The **torque** $\boldsymbol{\tau}$ (relative to the origin) is defined to be the cross product of the position and force vectors

$$\boldsymbol{\tau} = \mathbf{r} \times \mathbf{F}$$

and measures the tendency of the body to rotate about the origin. The direction of the torque vector indicates the axis of rotation. According to Theorem 6, the magnitude of the torque vector is

$$|\boldsymbol{\tau}| = |\mathbf{r} \times \mathbf{F}| = |\mathbf{r}||\mathbf{F}|\sin\theta$$

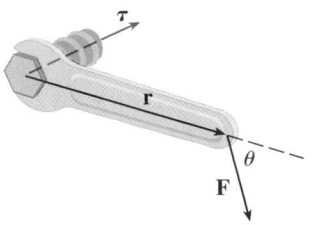

FIGURE 4

where θ is the angle between the position and force vectors. Observe that the only component of **F** that can cause a rotation is the one perpendicular to **r**, that is, $|\mathbf{F}| \sin \theta$. The magnitude of the torque is equal to the area of the parallelogram determined by **r** and **F**.

EXAMPLE 6 A bolt is tightened by applying a 40-N force to a 0.25-m wrench as shown in Figure 5. Find the magnitude of the torque about the center of the bolt.

SOLUTION The magnitude of the torque vector is

$$|\boldsymbol{\tau}| = |\mathbf{r} \times \mathbf{F}| = |\mathbf{r}||\mathbf{F}| \sin 75° = (0.25)(40) \sin 75°$$
$$= 10 \sin 75° \approx 9.66 \text{ N·m}$$

If the bolt is right-threaded, then the torque vector itself is

$$\boldsymbol{\tau} = |\boldsymbol{\tau}| \, \mathbf{n} \approx 9.66 \, \mathbf{n}$$

where **n** is a unit vector directed down into the page. ∎

FIGURE 5

10.4 | EXERCISES

1–7 ■ Find the cross product $\mathbf{a} \times \mathbf{b}$ and verify that it is orthogonal to both **a** and **b**.

1. $\mathbf{a} = \langle 6, 0, -2 \rangle$, $\quad \mathbf{b} = \langle 0, 8, 0 \rangle$

2. $\mathbf{a} = \langle 1, 1, -1 \rangle$, $\quad \mathbf{b} = \langle 2, 4, 6 \rangle$

3. $\mathbf{a} = \mathbf{i} + 3\mathbf{j} - 2\mathbf{k}$, $\quad \mathbf{b} = -\mathbf{i} + 5\mathbf{k}$

4. $\mathbf{a} = \mathbf{j} + 7\mathbf{k}$, $\quad \mathbf{b} = 2\mathbf{i} - \mathbf{j} + 4\mathbf{k}$

5. $\mathbf{a} = \mathbf{i} - \mathbf{j} - \mathbf{k}$, $\quad \mathbf{b} = \frac{1}{2}\mathbf{i} + \mathbf{j} + \frac{1}{2}\mathbf{k}$

6. $\mathbf{a} = t\mathbf{i} + \cos t\,\mathbf{j} + \sin t\,\mathbf{k}$, $\quad \mathbf{b} = \mathbf{i} - \sin t\,\mathbf{j} + \cos t\,\mathbf{k}$

7. $\mathbf{a} = \langle t, 1, 1/t \rangle$, $\quad \mathbf{b} = \langle t^2, t^2, 1 \rangle$

8. If $\mathbf{a} = \mathbf{i} - 2\mathbf{k}$ and $\mathbf{b} = \mathbf{j} + \mathbf{k}$, find $\mathbf{a} \times \mathbf{b}$. Sketch **a**, **b**, and $\mathbf{a} \times \mathbf{b}$ as vectors starting at the origin.

9–12 ■ Find the vector, not with determinants, but by using properties of cross products.

9. $(\mathbf{i} \times \mathbf{j}) \times \mathbf{k}$

10. $\mathbf{k} \times (\mathbf{i} - 2\mathbf{j})$

11. $(\mathbf{j} - \mathbf{k}) \times (\mathbf{k} - \mathbf{i})$

12. $(\mathbf{i} + \mathbf{j}) \times (\mathbf{i} - \mathbf{j})$

13. State whether each expression is meaningful. If not, explain why. If so, state whether it is a vector or a scalar.
(a) $\mathbf{a} \cdot (\mathbf{b} \times \mathbf{c})$
(b) $\mathbf{a} \times (\mathbf{b} \cdot \mathbf{c})$
(c) $\mathbf{a} \times (\mathbf{b} \times \mathbf{c})$
(d) $\mathbf{a} \cdot (\mathbf{b} \cdot \mathbf{c})$
(e) $(\mathbf{a} \cdot \mathbf{b}) \times (\mathbf{c} \cdot \mathbf{d})$
(f) $(\mathbf{a} \times \mathbf{b}) \cdot (\mathbf{c} \times \mathbf{d})$

14–15 ■ Find $|\mathbf{u} \times \mathbf{v}|$ and determine whether $\mathbf{u} \times \mathbf{v}$ is directed into the page or out of the page.

14.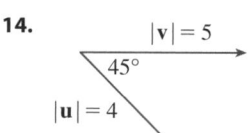

15.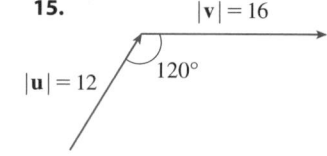

16. The figure shows a vector **a** in the xy-plane and a vector **b** in the direction of **k**. Their lengths are $|\mathbf{a}| = 3$ and $|\mathbf{b}| = 2$.
(a) Find $|\mathbf{a} \times \mathbf{b}|$.
(b) Use the right-hand rule to decide whether the components of $\mathbf{a} \times \mathbf{b}$ are positive, negative, or 0.

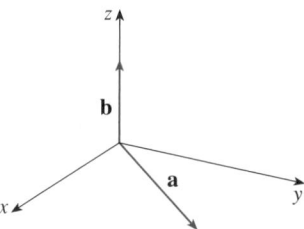

17. If $\mathbf{a} = \langle 2, -1, 3 \rangle$ and $\mathbf{b} = \langle 4, 2, 1 \rangle$, find $\mathbf{a} \times \mathbf{b}$ and $\mathbf{b} \times \mathbf{a}$.

18. If $\mathbf{a} = \langle 1, 0, 1 \rangle$, $\mathbf{b} = \langle 2, 1, -1 \rangle$, and $\mathbf{c} = \langle 0, 1, 3 \rangle$, show that $\mathbf{a} \times (\mathbf{b} \times \mathbf{c}) \neq (\mathbf{a} \times \mathbf{b}) \times \mathbf{c}$.

19. Find two unit vectors orthogonal to both $\langle 3, 2, 1 \rangle$ and $\langle -1, 1, 0 \rangle$.

20. Find two unit vectors orthogonal to both $\mathbf{j} - \mathbf{k}$ and $\mathbf{i} + \mathbf{j}$.

21. Show that $\mathbf{0} \times \mathbf{a} = \mathbf{0} = \mathbf{a} \times \mathbf{0}$ for any vector $\mathbf{a}$ in V_3.

22. Show that $(\mathbf{a} \times \mathbf{b}) \cdot \mathbf{b} = 0$ for all vectors $\mathbf{a}$ and $\mathbf{b}$ in V_3.

23. Prove Property 1 of Theorem 8.

24. Prove Property 2 of Theorem 8.

25. Prove Property 3 of Theorem 8.

26. Prove Property 4 of Theorem 8.

27. Find the area of the parallelogram with vertices $A(-2, 1)$, $B(0, 4)$, $C(4, 2)$, and $D(2, -1)$.

28. Find the area of the parallelogram with vertices $K(1, 2, 3)$, $L(1, 3, 6)$, $M(3, 8, 6)$, and $N(3, 7, 3)$.

29–32 ▪ (a) Find a nonzero vector orthogonal to the plane through the points P, Q, and R, and (b) find the area of triangle PQR.

29. $P(1, 0, 1)$, $Q(-2, 1, 3)$, $R(4, 2, 5)$

30. $P(0, 0, -3)$, $Q(4, 2, 0)$, $R(3, 3, 1)$

31. $P(0, -2, 0)$, $Q(4, 1, -2)$, $R(5, 3, 1)$

32. $P(-1, 3, 1)$, $Q(0, 5, 2)$, $R(4, 3, -1)$

33–34 ▪ Find the volume of the parallelepiped determined by the vectors $\mathbf{a}$, $\mathbf{b}$, and $\mathbf{c}$.

33. $\mathbf{a} = \langle 1, 2, 3 \rangle$, $\mathbf{b} = \langle -1, 1, 2 \rangle$, $\mathbf{c} = \langle 2, 1, 4 \rangle$

34. $\mathbf{a} = \mathbf{i} + \mathbf{j}$, $\mathbf{b} = \mathbf{j} + \mathbf{k}$, $\mathbf{c} = \mathbf{i} + \mathbf{j} + \mathbf{k}$

35–36 ▪ Find the volume of the parallelepiped with adjacent edges PQ, PR, and PS.

35. $P(-2, 1, 0)$, $Q(2, 3, 2)$, $R(1, 4, -1)$, $S(3, 6, 1)$

36. $P(3, 0, 1)$, $Q(-1, 2, 5)$, $R(5, 1, -1)$, $S(0, 4, 2)$

37. Use the scalar triple product to verify that the vectors $\mathbf{u} = \mathbf{i} + 5\mathbf{j} - 2\mathbf{k}$, $\mathbf{v} = 3\mathbf{i} - \mathbf{j}$, and $\mathbf{w} = 5\mathbf{i} + 9\mathbf{j} - 4\mathbf{k}$ are coplanar.

38. Use the scalar triple product to determine whether the points $A(1, 3, 2)$, $B(3, -1, 6)$, $C(5, 2, 0)$, and $D(3, 6, -4)$ lie in the same plane.

39. A bicycle pedal is pushed by a foot with a 60-N force as shown. The shaft of the pedal is 18 cm long. Find the magnitude of the torque about P.

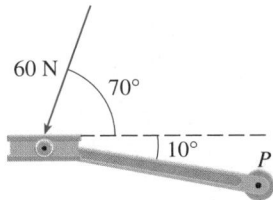

40. Find the magnitude of the torque about P if a 36-lb force is applied as shown.

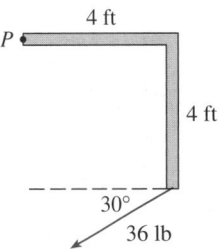

41. A wrench 30 cm long lies along the positive y-axis and grips a bolt at the origin. A force is applied in the direction $\langle 0, 3, -4 \rangle$ at the end of the wrench. Find the magnitude of the force needed to supply 100 N·m of torque to the bolt.

42. Let $\mathbf{v} = 5\mathbf{j}$ and let $\mathbf{u}$ be a vector with length 3 that starts at the origin and rotates in the xy-plane. Find the maximum and minimum values of the length of the vector $\mathbf{u} \times \mathbf{v}$. In what direction does $\mathbf{u} \times \mathbf{v}$ point?

43. If $\mathbf{a} \cdot \mathbf{b} = \sqrt{3}$ and $\mathbf{a} \times \mathbf{b} = \langle 1, 2, 2 \rangle$, find the angle between $\mathbf{a}$ and $\mathbf{b}$.

44. (a) Find all vectors $\mathbf{v}$ such that

$$\langle 1, 2, 1 \rangle \times \mathbf{v} = \langle 3, 1, -5 \rangle$$

(b) Explain why there is no vector $\mathbf{v}$ such that

$$\langle 1, 2, 1 \rangle \times \mathbf{v} = \langle 3, 1, 5 \rangle$$

45. (a) Let P be a point not on the line L that passes through the points Q and R. Show that the distance d from the point P to the line L is

$$d = \frac{|\mathbf{a} \times \mathbf{b}|}{|\mathbf{a}|}$$

where $\mathbf{a} = \overrightarrow{QR}$ and $\mathbf{b} = \overrightarrow{QP}$.
(b) Use the formula in part (a) to find the distance from the point $P(1, 1, 1)$ to the line through $Q(0, 6, 8)$ and $R(-1, 4, 7)$.

46. (a) Let P be a point not on the plane that passes through the points Q, R, and S. Show that the distance d from P to the plane is

$$d = \frac{|\mathbf{a} \cdot (\mathbf{b} \times \mathbf{c})|}{|\mathbf{a} \times \mathbf{b}|}$$

where $\mathbf{a} = \overrightarrow{QR}$, $\mathbf{b} = \overrightarrow{QS}$, and $\mathbf{c} = \overrightarrow{QP}$.
(b) Use the formula in part (a) to find the distance from the point $P(2, 1, 4)$ to the plane through the points $Q(1, 0, 0)$, $R(0, 2, 0)$, and $S(0, 0, 3)$.

47. Show that $|\mathbf{a} \times \mathbf{b}|^2 = |\mathbf{a}|^2 |\mathbf{b}|^2 - (\mathbf{a} \cdot \mathbf{b})^2$.

48. If $\mathbf{a} + \mathbf{b} + \mathbf{c} = \mathbf{0}$, show that

$$\mathbf{a} \times \mathbf{b} = \mathbf{b} \times \mathbf{c} = \mathbf{c} \times \mathbf{a}$$

49. Prove that $(\mathbf{a} - \mathbf{b}) \times (\mathbf{a} + \mathbf{b}) = 2(\mathbf{a} \times \mathbf{b})$.

50. Prove Property 6 of Theorem 8, that is,

$$\mathbf{a} \times (\mathbf{b} \times \mathbf{c}) = (\mathbf{a} \cdot \mathbf{c})\mathbf{b} - (\mathbf{a} \cdot \mathbf{b})\mathbf{c}$$

51. Use Exercise 50 to prove that

$$\mathbf{a} \times (\mathbf{b} \times \mathbf{c}) + \mathbf{b} \times (\mathbf{c} \times \mathbf{a}) + \mathbf{c} \times (\mathbf{a} \times \mathbf{b}) = \mathbf{0}$$

52. Prove that

$$(\mathbf{a} \times \mathbf{b}) \cdot (\mathbf{c} \times \mathbf{d}) = \begin{vmatrix} \mathbf{a} \cdot \mathbf{c} & \mathbf{b} \cdot \mathbf{c} \\ \mathbf{a} \cdot \mathbf{d} & \mathbf{b} \cdot \mathbf{d} \end{vmatrix}$$

53. Suppose that $\mathbf{a} \neq \mathbf{0}$.
 (a) If $\mathbf{a} \cdot \mathbf{b} = \mathbf{a} \cdot \mathbf{c}$, does it follow that $\mathbf{b} = \mathbf{c}$?
 (b) If $\mathbf{a} \times \mathbf{b} = \mathbf{a} \times \mathbf{c}$, does it follow that $\mathbf{b} = \mathbf{c}$?
 (c) If $\mathbf{a} \cdot \mathbf{b} = \mathbf{a} \cdot \mathbf{c}$ and $\mathbf{a} \times \mathbf{b} = \mathbf{a} \times \mathbf{c}$, does it follow that $\mathbf{b} = \mathbf{c}$?

54. If $\mathbf{v}_1$, $\mathbf{v}_2$, and $\mathbf{v}_3$ are noncoplanar vectors, let

$$\mathbf{k}_1 = \frac{\mathbf{v}_2 \times \mathbf{v}_3}{\mathbf{v}_1 \cdot (\mathbf{v}_2 \times \mathbf{v}_3)} \qquad \mathbf{k}_2 = \frac{\mathbf{v}_3 \times \mathbf{v}_1}{\mathbf{v}_1 \cdot (\mathbf{v}_2 \times \mathbf{v}_3)}$$

$$\mathbf{k}_3 = \frac{\mathbf{v}_1 \times \mathbf{v}_2}{\mathbf{v}_1 \cdot (\mathbf{v}_2 \times \mathbf{v}_3)}$$

(These vectors occur in the study of crystallography. Vectors of the form $n_1\mathbf{v}_1 + n_2\mathbf{v}_2 + n_3\mathbf{v}_3$, where each n_i is an integer, form a *lattice* for a crystal. Vectors written similarly in terms of $\mathbf{k}_1$, $\mathbf{k}_2$, and $\mathbf{k}_3$ form the *reciprocal lattice*.)
 (a) Show that $\mathbf{k}_i$ is perpendicular to $\mathbf{v}_j$ if $i \neq j$.
 (b) Show that $\mathbf{k}_i \cdot \mathbf{v}_i = 1$ for $i = 1, 2, 3$.
 (c) Show that $\mathbf{k}_1 \cdot (\mathbf{k}_2 \times \mathbf{k}_3) = \dfrac{1}{\mathbf{v}_1 \cdot (\mathbf{v}_2 \times \mathbf{v}_3)}$.

10.5 | EQUATIONS OF LINES AND PLANES

A line in the xy-plane is determined when a point on the line and the direction of the line (its slope or angle of inclination) are given. The equation of the line can then be written using the point-slope form.

Likewise, a line L in three-dimensional space is determined when we know a point $P_0(x_0, y_0, z_0)$ on L and the direction of L. In three dimensions the direction of a line is conveniently described by a vector, so we let $\mathbf{v}$ be a vector parallel to L. Let $P(x, y, z)$ be an arbitrary point on L and let $\mathbf{r}_0$ and $\mathbf{r}$ be the position vectors of P_0 and P (that is, they have representations $\overrightarrow{OP_0}$ and $\overrightarrow{OP}$). If $\mathbf{a}$ is the vector with representation $\overrightarrow{P_0P}$, as in Figure 1, then the Triangle Law for vector addition gives $\mathbf{r} = \mathbf{r}_0 + \mathbf{a}$. But, since $\mathbf{a}$ and $\mathbf{v}$ are parallel vectors, there is a scalar t such that $\mathbf{a} = t\mathbf{v}$. Thus

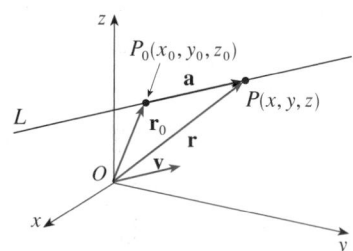

FIGURE 1

$$\boxed{1} \qquad \boxed{\mathbf{r} = \mathbf{r}_0 + t\mathbf{v}}$$

which is a **vector equation** of L. Each value of the **parameter** t gives the position vector $\mathbf{r}$ of a point on L. In other words, as t varies, the line is traced out by the tip of the vector $\mathbf{r}$. As Figure 2 indicates, positive values of t correspond to points on L that lie on one side of P_0, whereas negative values of t correspond to points that lie on the other side of P_0.

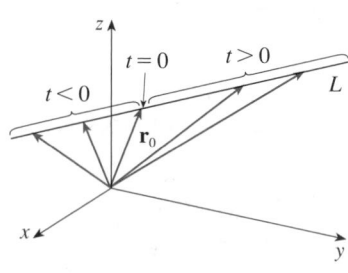

FIGURE 2

If the vector $\mathbf{v}$ that gives the direction of the line L is written in component form as $\mathbf{v} = \langle a, b, c \rangle$, then we have $t\mathbf{v} = \langle ta, tb, tc \rangle$. We can also write $\mathbf{r} = \langle x, y, z \rangle$ and $\mathbf{r}_0 = \langle x_0, y_0, z_0 \rangle$, so the vector equation $\boxed{1}$ becomes

$$\langle x, y, z \rangle = \langle x_0 + ta, y_0 + tb, z_0 + tc \rangle$$

Two vectors are equal if and only if corresponding components are equal. Therefore we have the three scalar equations:

$$\boxed{2} \qquad \boxed{x = x_0 + at \qquad y = y_0 + bt \qquad z = z_0 + ct}$$

where $t \in \mathbb{R}$. These equations are called **parametric equations** of the line L through the point $P_0(x_0, y_0, z_0)$ and parallel to the vector $\mathbf{v} = \langle a, b, c \rangle$. Each value of the parameter t gives a point (x, y, z) on L.

EXAMPLE 1

(a) Find a vector equation and parametric equations for the line that passes through the point $(5, 1, 3)$ and is parallel to the vector $\mathbf{i} + 4\mathbf{j} - 2\mathbf{k}$.

(b) Find two other points on the line.

SOLUTION

(a) Here $\mathbf{r}_0 = \langle 5, 1, 3 \rangle = 5\mathbf{i} + \mathbf{j} + 3\mathbf{k}$ and $\mathbf{v} = \mathbf{i} + 4\mathbf{j} - 2\mathbf{k}$, so the vector equation $\boxed{1}$ becomes

$$\mathbf{r} = (5\mathbf{i} + \mathbf{j} + 3\mathbf{k}) + t(\mathbf{i} + 4\mathbf{j} - 2\mathbf{k})$$

or $$\mathbf{r} = (5 + t)\,\mathbf{i} + (1 + 4t)\,\mathbf{j} + (3 - 2t)\,\mathbf{k}$$

Parametric equations are

$$x = 5 + t \qquad y = 1 + 4t \qquad z = 3 - 2t$$

(b) Choosing the parameter value $t = 1$ gives $x = 6$, $y = 5$, and $z = 1$, so $(6, 5, 1)$ is a point on the line. Similarly, $t = -1$ gives the point $(4, -3, 5)$. ∎

The vector equation and parametric equations of a line are not unique. If we change the point or the parameter or choose a different parallel vector, then the equations change. For instance, if, instead of $(5, 1, 3)$, we choose the point $(6, 5, 1)$ in Example 1, then the parametric equations of the line become

$$x = 6 + t \qquad y = 5 + 4t \qquad z = 1 - 2t$$

Or, if we stay with the point $(5, 1, 3)$ but choose the parallel vector $2\mathbf{i} + 8\mathbf{j} - 4\mathbf{k}$, we arrive at the equations

$$x = 5 + 2t \qquad y = 1 + 8t \qquad z = 3 - 4t$$

In general, if a vector $\mathbf{v} = \langle a, b, c \rangle$ is used to describe the direction of a line L, then the numbers a, b, and c are called **direction numbers** of L. Since any vector parallel to $\mathbf{v}$ could also be used, we see that any three numbers proportional to a, b, and c could also be used as a set of direction numbers for L.

Another way of describing a line L is to eliminate the parameter t from Equations 2. If none of a, b, or c is 0, we can solve each of these equations for t, equate the results, and obtain

$\boxed{3}$
$$\frac{x - x_0}{a} = \frac{y - y_0}{b} = \frac{z - z_0}{c}$$

These equations are called **symmetric equations** of L. Notice that the numbers a, b, and c that appear in the denominators of Equations 3 are direction numbers of L, that is, components of a vector parallel to L. If one of a, b, or c is 0, we can still elim-

■ Figure 3 shows the line L in Example 1 and its relation to the given point and to the vector that gives its direction.

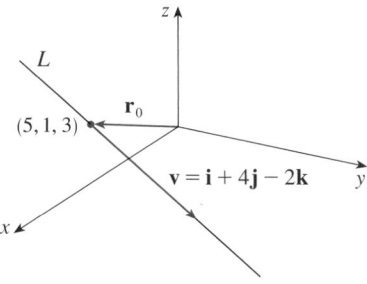

FIGURE 3

inate t. For instance, if $a = 0$, we could write the equations of L as

$$x = x_0 \qquad \frac{y - y_0}{b} = \frac{z - z_0}{c}$$

This means that L lies in the vertical plane $x = x_0$.

■ Figure 4 shows the line L in Example 2 and the point P where it intersects the xy-plane.

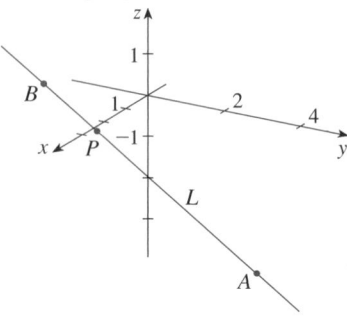

FIGURE 4

EXAMPLE 2

(a) Find parametric equations and symmetric equations of the line that passes through the points $A(2, 4, -3)$ and $B(3, -1, 1)$.
(b) At what point does this line intersect the xy-plane?

SOLUTION

(a) We are not explicitly given a vector parallel to the line, but observe that the vector $\mathbf{v}$ with representation $\overrightarrow{AB}$ is parallel to the line and

$$\mathbf{v} = \langle 3 - 2, -1 - 4, 1 - (-3) \rangle = \langle 1, -5, 4 \rangle$$

Thus direction numbers are $a = 1$, $b = -5$, and $c = 4$. Taking the point $(2, 4, -3)$ as P_0, we see that parametric equations $\boxed{2}$ are

$$x = 2 + t \qquad y = 4 - 5t \qquad z = -3 + 4t$$

and symmetric equations $\boxed{3}$ are

$$\frac{x - 2}{1} = \frac{y - 4}{-5} = \frac{z + 3}{4}$$

(b) The line intersects the xy-plane when $z = 0$, so we put $z = 0$ in the symmetric equations and obtain

$$\frac{x - 2}{1} = \frac{y - 4}{-5} = \frac{3}{4}$$

This gives $x = \frac{11}{4}$ and $y = \frac{1}{4}$, so the line intersects the xy-plane at the point $\left(\frac{11}{4}, \frac{1}{4}, 0\right)$. ■

In general, the procedure of Example 2 shows that direction numbers of the line L through the points $P_0(x_0, y_0, z_0)$ and $P_1(x_1, y_1, z_1)$ are $x_1 - x_0$, $y_1 - y_0$, and $z_1 - z_0$ and so symmetric equations of L are

$$\frac{x - x_0}{x_1 - x_0} = \frac{y - y_0}{y_1 - y_0} = \frac{z - z_0}{z_1 - z_0}$$

Often, we need a description, not of an entire line, but of just a line segment. How, for instance, could we describe the line segment AB in Example 2? If we put $t = 0$ in the parametric equations in Example 2(a), we get the point $(2, 4, -3)$ and if we put $t = 1$ we get $(3, -1, 1)$. So the line segment AB is described by the parametric equations

$$x = 2 + t \qquad y = 4 - 5t \qquad z = -3 + 4t \qquad 0 \le t \le 1$$

or by the corresponding vector equation

$$\mathbf{r}(t) = \langle 2 + t, 4 - 5t, -3 + 4t \rangle \qquad 0 \le t \le 1$$

In general, we know from Equation 1 that the vector equation of a line through the (tip of the) vector $\mathbf{r}_0$ in the direction of a vector $\mathbf{v}$ is $\mathbf{r} = \mathbf{r}_0 + t\mathbf{v}$. If the line also passes through (the tip of) $\mathbf{r}_1$, then we can take $\mathbf{v} = \mathbf{r}_1 - \mathbf{r}_0$ and so its vector equation is

$$\mathbf{r} = \mathbf{r}_0 + t(\mathbf{r}_1 - \mathbf{r}_0) = (1 - t)\mathbf{r}_0 + t\mathbf{r}_1$$

The line segment from $\mathbf{r}_0$ to $\mathbf{r}_1$ is given by the parameter interval $0 \leq t \leq 1$.

> **4** The line segment from $\mathbf{r}_0$ to $\mathbf{r}_1$ is given by the vector equation
>
> $$\mathbf{r}(t) = (1 - t)\mathbf{r}_0 + t\mathbf{r}_1 \qquad 0 \leq t \leq 1$$

- The lines L_1 and L_2 in Example 3, shown in Figure 5, are skew lines.

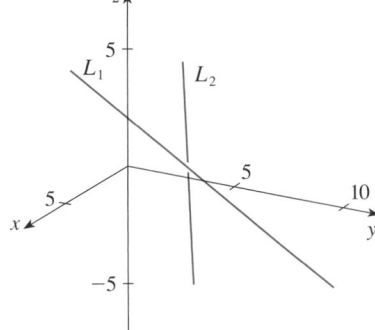

FIGURE 5

☑ **EXAMPLE 3** Show that the lines L_1 and L_2 with parametric equations

$$x = 1 + t \qquad y = -2 + 3t \qquad z = 4 - t$$
$$x = 2s \qquad\quad y = 3 + s \qquad\quad z = -3 + 4s$$

are **skew lines**; that is, they do not intersect and are not parallel (and therefore do not lie in the same plane).

SOLUTION The lines are not parallel because the corresponding vectors $\langle 1, 3, -1 \rangle$ and $\langle 2, 1, 4 \rangle$ are not parallel. (Their components are not proportional.) If L_1 and L_2 had a point of intersection, there would be values of t and s such that

$$1 + t = 2s$$
$$-2 + 3t = 3 + s$$
$$4 - t = -3 + 4s$$

But if we solve the first two equations, we get $t = \frac{11}{5}$ and $s = \frac{8}{5}$, and these values don't satisfy the third equation. Therefore there are no values of t and s that satisfy the three equations, so L_1 and L_2 do not intersect. Thus L_1 and L_2 are skew lines. ∎

PLANES

Although a line in space is determined by a point and a direction, a plane in space is more difficult to describe. A single vector parallel to a plane is not enough to convey the "direction" of the plane, but a vector perpendicular to the plane does completely specify its direction. Thus a plane in space is determined by a point $P_0(x_0, y_0, z_0)$ in the plane and a vector $\mathbf{n}$ that is orthogonal to the plane. This orthogonal vector $\mathbf{n}$ is called a **normal vector**. Let $P(x, y, z)$ be an arbitrary point in the plane, and let $\mathbf{r}_0$ and $\mathbf{r}$ be the position vectors of P_0 and P. Then the vector $\mathbf{r} - \mathbf{r}_0$ is represented by $\overrightarrow{P_0P}$. (See Figure 6.) The normal vector $\mathbf{n}$ is orthogonal to every vector in the given plane. In particular, $\mathbf{n}$ is orthogonal to $\mathbf{r} - \mathbf{r}_0$ and so we have

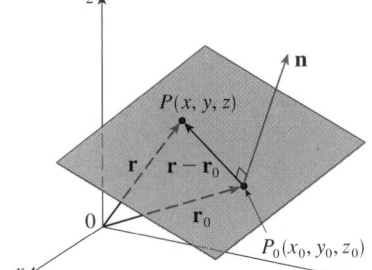

FIGURE 6

> **5** $$\mathbf{n} \cdot (\mathbf{r} - \mathbf{r}_0) = 0$$

which can be rewritten as

> **6** $$\mathbf{n} \cdot \mathbf{r} = \mathbf{n} \cdot \mathbf{r}_0$$

Either Equation 5 or Equation 6 is called a **vector equation of the plane**.

To obtain a scalar equation for the plane, we write $\mathbf{n} = \langle a, b, c \rangle$, $\mathbf{r} = \langle x, y, z \rangle$, and $\mathbf{r}_0 = \langle x_0, y_0, z_0 \rangle$. Then the vector equation $\boxed{5}$ becomes

$$\langle a, b, c \rangle \cdot \langle x - x_0, y - y_0, z - z_0 \rangle = 0$$

or

$\boxed{7}$
$$a(x - x_0) + b(y - y_0) + c(z - z_0) = 0$$

Equation 7 is the **scalar equation of the plane through** $P_0(x_0, y_0, z_0)$ **with normal vector** $\mathbf{n} = \langle a, b, c \rangle$.

▼ EXAMPLE 4 Find an equation of the plane through the point $(2, 4, -1)$ with normal vector $\mathbf{n} = \langle 2, 3, 4 \rangle$. Find the intercepts and sketch the plane.

SOLUTION Putting $a = 2$, $b = 3$, $c = 4$, $x_0 = 2$, $y_0 = 4$, and $z_0 = -1$ in Equation 7, we see that an equation of the plane is

$$2(x - 2) + 3(y - 4) + 4(z + 1) = 0$$

or
$$2x + 3y + 4z = 12$$

To find the x-intercept we set $y = z = 0$ in this equation and obtain $x = 6$. Similarly, the y-intercept is 4 and the z-intercept is 3. This enables us to sketch the portion of the plane that lies in the first octant (see Figure 7). ∎

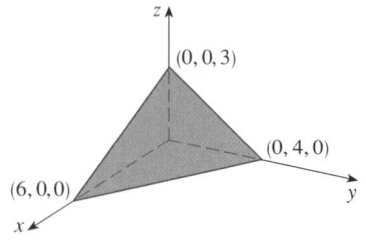

FIGURE 7

By collecting terms in Equation 7 as we did in Example 4, we can rewrite the equation of a plane as

$\boxed{8}$
$$ax + by + cz + d = 0$$

where $d = -(ax_0 + by_0 + cz_0)$. Equation 8 is called a **linear equation** in x, y, and z. Conversely, it can be shown that if a, b, and c are not all 0, then the linear equation $\boxed{8}$ represents a plane with normal vector $\langle a, b, c \rangle$. (See Exercise 59.)

EXAMPLE 5 Find an equation of the plane that passes through the points $P(1, 3, 2)$, $Q(3, -1, 6)$, and $R(5, 2, 0)$.

- Figure 8 shows the portion of the plane in Example 5 that is enclosed by triangle PQR.

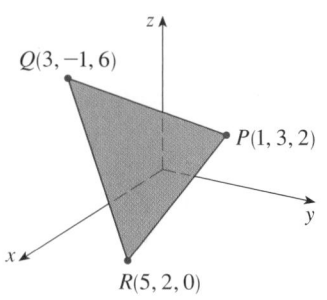

FIGURE 8

SOLUTION The vectors $\mathbf{a}$ and $\mathbf{b}$ corresponding to $\overrightarrow{PQ}$ and $\overrightarrow{PR}$ are

$$\mathbf{a} = \langle 2, -4, 4 \rangle \qquad \mathbf{b} = \langle 4, -1, -2 \rangle$$

Since both $\mathbf{a}$ and $\mathbf{b}$ lie in the plane, their cross product $\mathbf{a} \times \mathbf{b}$ is orthogonal to the plane and can be taken as the normal vector. Thus

$$\mathbf{n} = \mathbf{a} \times \mathbf{b} = \begin{vmatrix} \mathbf{i} & \mathbf{j} & \mathbf{k} \\ 2 & -4 & 4 \\ 4 & -1 & -2 \end{vmatrix} = 12\mathbf{i} + 20\mathbf{j} + 14\mathbf{k}$$

With the point $P(1, 3, 2)$ and the normal vector $\mathbf{n}$, an equation of the plane is

$$12(x - 1) + 20(y - 3) + 14(z - 2) = 0$$

or
$$6x + 10y + 7z = 50 \qquad ∎$$

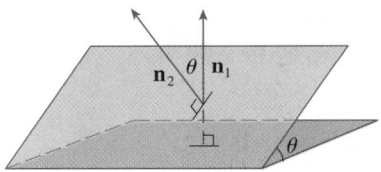

FIGURE 9

▪ Figure 10 shows the planes in Example 6 and their line of intersection L.

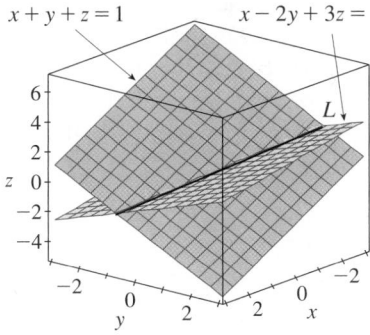

FIGURE 10

▪ Another way to find the line of intersection is to solve the equations of the planes for two of the variables in terms of the third, which can be taken as the parameter.

Two planes are **parallel** if their normal vectors are parallel. For instance, the planes $x + 2y - 3z = 4$ and $2x + 4y - 6z = 3$ are parallel because their normal vectors are $\mathbf{n}_1 = \langle 1, 2, -3 \rangle$ and $\mathbf{n}_2 = \langle 2, 4, -6 \rangle$ and $\mathbf{n}_2 = 2\mathbf{n}_1$. If two planes are not parallel, then they intersect in a straight line and the angle between the two planes is defined as the acute angle between their normal vectors (see angle θ in Figure 9).

▼ EXAMPLE 6

(a) Find the angle between the planes $x + y + z = 1$ and $x - 2y + 3z = 1$.
(b) Find symmetric equations for the line of intersection L of these two planes.

SOLUTION

(a) The normal vectors of these planes are

$$\mathbf{n}_1 = \langle 1, 1, 1 \rangle \qquad \mathbf{n}_2 = \langle 1, -2, 3 \rangle$$

and so, if θ is the angle between the planes,

$$\cos \theta = \frac{\mathbf{n}_1 \cdot \mathbf{n}_2}{|\mathbf{n}_1||\mathbf{n}_2|} = \frac{1(1) + 1(-2) + 1(3)}{\sqrt{1 + 1 + 1}\,\sqrt{1 + 4 + 9}} = \frac{2}{\sqrt{42}}$$

$$\theta = \cos^{-1}\left(\frac{2}{\sqrt{42}}\right) \approx 72°$$

(b) We first need to find a point on L. For instance, we can find the point where the line intersects the xy-plane by setting $z = 0$ in the equations of both planes. This gives the equations $x + y = 1$ and $x - 2y = 1$, whose solution is $x = 1$, $y = 0$. So the point $(1, 0, 0)$ lies on L.

Now we observe that, since L lies in both planes, it is perpendicular to both of the normal vectors. Thus a vector $\mathbf{v}$ parallel to L is given by the cross product

$$\mathbf{v} = \mathbf{n}_1 \times \mathbf{n}_2 = \begin{vmatrix} \mathbf{i} & \mathbf{j} & \mathbf{k} \\ 1 & 1 & 1 \\ 1 & -2 & 3 \end{vmatrix} = 5\mathbf{i} - 2\mathbf{j} - 3\mathbf{k}$$

and so the symmetric equations of L can be written as

$$\frac{x - 1}{5} = \frac{y}{-2} = \frac{z}{-3}$$

∎

EXAMPLE 7 Find a formula for the distance D from a point $P_1(x_1, y_1, z_1)$ to the plane $ax + by + cz + d = 0$.

SOLUTION Let $P_0(x_0, y_0, z_0)$ be any point in the given plane and let $\mathbf{b}$ be the vector corresponding to $\overrightarrow{P_0P_1}$. Then

$$\mathbf{b} = \langle x_1 - x_0, y_1 - y_0, z_1 - z_0 \rangle$$

From Figure 11 you can see that the distance D from P_1 to the plane is equal to the absolute value of the scalar projection of $\mathbf{b}$ onto the normal vector $\mathbf{n} = \langle a, b, c \rangle$. (See Section 10.3.) Thus

$$D = |\text{comp}_\mathbf{n}\, \mathbf{b}| = \frac{|\mathbf{n} \cdot \mathbf{b}|}{|\mathbf{n}|} = \frac{|a(x_1 - x_0) + b(y_1 - y_0) + c(z_1 - z_0)|}{\sqrt{a^2 + b^2 + c^2}}$$

$$= \frac{|(ax_1 + by_1 + cz_1) - (ax_0 + by_0 + cz_0)|}{\sqrt{a^2 + b^2 + c^2}}$$

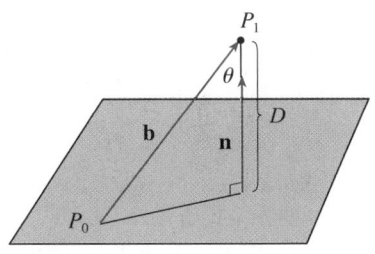

FIGURE 11

Since P_0 lies in the plane, its coordinates satisfy the equation of the plane and so we have $ax_0 + by_0 + cz_0 + d = 0$. Thus the formula for D can be written as

$$\boxed{9} \qquad \boxed{D = \frac{|ax_1 + by_1 + cz_1 + d|}{\sqrt{a^2 + b^2 + c^2}}} \qquad ■$$

EXAMPLE 8 Find the distance between the parallel planes $10x + 2y - 2z = 5$ and $5x + y - z = 1$.

SOLUTION First we note that the planes are parallel because their normal vectors $\langle 10, 2, -2 \rangle$ and $\langle 5, 1, -1 \rangle$ are parallel. To find the distance D between the planes, we choose any point on one plane and calculate its distance to the other plane. In particular, if we put $y = z = 0$ in the equation of the first plane, we get $10x = 5$ and so $\left(\frac{1}{2}, 0, 0\right)$ is a point in this plane. By Formula 9, the distance between $\left(\frac{1}{2}, 0, 0\right)$ and the plane $5x + y - z - 1 = 0$ is

■ www.stewartcalculus.com
See Additional Examples A, B.

$$D = \frac{\left|5\left(\frac{1}{2}\right) + 1(0) - 1(0) - 1\right|}{\sqrt{5^2 + 1^2 + (-1)^2}} = \frac{\frac{3}{2}}{3\sqrt{3}} = \frac{\sqrt{3}}{6}$$

So the distance between the planes is $\sqrt{3}/6$. ■

10.5 | EXERCISES

1. Determine whether each statement is true or false.
  (a) Two lines parallel to a third line are parallel.
  (b) Two lines perpendicular to a third line are parallel.
  (c) Two planes parallel to a third plane are parallel.
  (d) Two planes perpendicular to a third plane are parallel.
  (e) Two lines parallel to a plane are parallel.
  (f) Two lines perpendicular to a plane are parallel.
  (g) Two planes parallel to a line are parallel.
  (h) Two planes perpendicular to a line are parallel.
  (i) Two planes either intersect or are parallel.
  (j) Two lines either intersect or are parallel.
  (k) A plane and a line either intersect or are parallel.

2–5 ■ Find a vector equation and parametric equations for the line.

2. The line through the point $(6, -5, 2)$ and parallel to the vector $\left\langle 1, 3, -\frac{2}{3} \right\rangle$

3. The line through the point $(2, 2.4, 3.5)$ and parallel to the vector $3\mathbf{i} + 2\mathbf{j} - \mathbf{k}$

4. The line through the point $(0, 14, -10)$ and parallel to the line $x = -1 + 2t$, $y = 6 - 3t$, $z = 3 + 9t$

5. The line through the point $(1, 0, 6)$ and perpendicular to the plane $x + 3y + z = 5$

6–10 ■ Find parametric equations and symmetric equations for the line.

6. The line through the points $(1.0, 2.4, 4.6)$ and $(2.6, 1.2, 0.3)$

7. The line through the points $\left(0, \frac{1}{2}, 1\right)$ and $(2, 1, -3)$

8. The line through $(2, 1, 0)$ and perpendicular to both $\mathbf{i} + \mathbf{j}$ and $\mathbf{j} + \mathbf{k}$

9. The line through $(1, -1, 1)$ and parallel to the line $x + 2 = \frac{1}{2}y = z - 3$

10. The line of intersection of the planes $x + 2y + 3z = 1$ and $x - y + z = 1$

11. Is the line through $(-4, -6, 1)$ and $(-2, 0, -3)$ parallel to the line through $(10, 18, 4)$ and $(5, 3, 14)$?

12. Is the line through $(-2, 4, 0)$ and $(1, 1, 1)$ perpendicular to the line through $(2, 3, 4)$ and $(3, -1, -8)$?

13. (a) Find symmetric equations for the line that passes through the point $(1, -5, 6)$ and is parallel to the vector $\langle -1, 2, -3 \rangle$.
  (b) Find the points in which the required line in part (a) intersects the coordinate planes.

14. (a) Find parametric equations for the line through $(2, 4, 6)$ that is perpendicular to the plane $x - y + 3z = 7$.

(b) In what points does this line intersect the coordinate planes?

15. Find a vector equation for the line segment from $(2, -1, 4)$ to $(4, 6, 1)$.

16. Find parametric equations for the line segment from $(10, 3, 1)$ to $(5, 6, -3)$.

17–20 ▪ Determine whether the lines L_1 and L_2 are parallel, skew, or intersecting. If they intersect, find the point of intersection.

17. L_1: $x = 3 + 2t$, $y = 4 - t$, $z = 1 + 3t$
    L_2: $x = 1 + 4s$, $y = 3 - 2s$, $z = 4 + 5s$

18. L_1: $x = 5 - 12t$, $y = 3 + 9t$, $z = 1 - 3t$
    L_2: $x = 3 + 8s$, $y = -6s$, $z = 7 + 2s$

19. L_1: $\dfrac{x - 2}{1} = \dfrac{y - 3}{-2} = \dfrac{z - 1}{-3}$
    L_2: $\dfrac{x - 3}{1} = \dfrac{y + 4}{3} = \dfrac{z - 2}{-7}$

20. L_1: $\dfrac{x}{1} = \dfrac{y - 1}{-1} = \dfrac{z - 2}{3}$
    L_2: $\dfrac{x - 2}{2} = \dfrac{y - 3}{-2} = \dfrac{z}{7}$

21–32 ▪ Find an equation of the plane.

21. The plane through the point $\left(-1, \frac{1}{2}, 3\right)$ and with normal vector $\mathbf{i} + 4\mathbf{j} + \mathbf{k}$

22. The plane through the point $(2, 0, 1)$ and perpendicular to the line $x = 3t, y = 2 - t, z = 3 + 4t$

23. The plane through the point $(1, -1, -1)$ and parallel to the plane $5x - y - z = 6$

24. The plane that contains the line $x = 1 + t, y = 2 - t$, $z = 4 - 3t$ and is parallel to the plane $5x + 2y + z = 1$

25. The plane through the points $(0, 1, 1)$, $(1, 0, 1)$, and $(1, 1, 0)$

26. The plane through the origin and the points $(2, -4, 6)$ and $(5, 1, 3)$

27. The plane that passes through the point $(6, 0, -2)$ and contains the line $x = 4 - 2t, y = 3 + 5t, z = 7 + 4t$

28. The plane that passes through the point $(1, -1, 1)$ and contains the line with symmetric equations $x = 2y = 3z$

29. The plane that passes through the point $(-1, 2, 1)$ and contains the line of intersection of the planes $x + y - z = 2$ and $2x - y + 3z = 1$

30. The plane that passes through the points $(0, -2, 5)$ and $(-1, 3, 1)$ and is perpendicular to the plane $2z = 5x + 4y$

31. The plane that passes through the point $(1, 5, 1)$ and is perpendicular to the planes $2x + y - 2z = 2$ and $x + 3z = 4$

32. The plane that passes through the line of intersection of the planes $x - z = 1$ and $y + 2z = 3$ and is perpendicular to the plane $x + y - 2z = 1$

33. Find the point at which the line $x = 3 - t, y = 2 + t$, $z = 5t$ intersects the plane $x - y + 2z = 9$.

34. Where does the line through $(1, 0, 1)$ and $(4, -2, 2)$ intersect the plane $x + y + z = 6$?

35–38 ▪ Determine whether the planes are parallel, perpendicular, or neither. If neither, find the angle between them.

35. $x + y + z = 1$, $x - y + z = 1$

36. $2x - 3y + 4z = 5$, $x + 6y + 4z = 3$

37. $x = 4y - 2z$, $8y = 1 + 2x + 4z$

38. $x + 2y + 2z = 1$, $2x - y + 2z = 1$

39. (a) Find parametric equations for the line of intersection of the planes $x + y + z = 1$ and $x + 2y + 2z = 1$.
     (b) Find the angle between these planes.

40. Find an equation for the plane consisting of all points that are equidistant from the points $(2, 5, 5)$ and $(-6, 3, 1)$.

41. Find an equation of the plane with x-intercept a, y-intercept b, and z-intercept c.

42. (a) Find the point at which the given lines intersect:
$$\mathbf{r} = \langle 1, 1, 0 \rangle + t\langle 1, -1, 2 \rangle$$
$$\mathbf{r} = \langle 2, 0, 2 \rangle + s\langle -1, 1, 0 \rangle$$
     (b) Find an equation of the plane that contains these lines.

43. Find parametric equations for the line through the point $(0, 1, 2)$ that is parallel to the plane $x + y + z = 2$ and perpendicular to the line $x = 1 + t, y = 1 - t, z = 2t$.

44. Find parametric equations for the line through the point $(0, 1, 2)$ that is perpendicular to the line $x = 1 + t$, $y = 1 - t, z = 2t$ and intersects this line.

45. Which of the following four planes are parallel? Are any of them identical?

   P_1: $3x + 6y - 3z = 6$ P_2: $4x - 12y + 8z = 5$
   P_3: $9y = 1 + 3x + 6z$ P_4: $z = x + 2y - 2$

46. Which of the following four lines are parallel? Are any of them identical?

$$L_1: x = 1 + 6t, \quad y = 1 - 3t, \quad z = 12t + 5$$
$$L_2: x = 1 + 2t, \quad y = t, \quad z = 1 + 4t$$
$$L_3: 2x - 2 = 4 - 4y = z + 1$$
$$L_4: \mathbf{r} = \langle 3, 1, 5 \rangle + t\langle 4, 2, 8 \rangle$$

47–48 ▪ Use the formula in Exercise 45 in Section 10.4 to find the distance from the point to the given line.

47. $(4, 1, -2)$; $\quad x = 1 + t, \; y = 3 - 2t, \; z = 4 - 3t$

48. $(0, 1, 3)$; $\quad x = 2t, \; y = 6 - 2t, \; z = 3 + t$

49–50 ▪ Find the distance from the point to the given plane.

49. $(1, -2, 4)$, $\quad 3x + 2y + 6z = 5$

50. $(-6, 3, 5)$, $\quad x - 2y - 4z = 8$

51–52 ▪ Find the distance between the given parallel planes.

51. $2x - 3y + z = 4$, $\quad 4x - 6y + 2z = 3$

52. $6z = 4y - 2x$, $\quad 9z = 1 - 3x + 6y$

53. Show that the distance between the parallel planes $ax + by + cz + d_1 = 0$ and $ax + by + cz + d_2 = 0$ is

$$D = \frac{|d_1 - d_2|}{\sqrt{a^2 + b^2 + c^2}}$$

54. Find equations of the planes that are parallel to the plane $x + 2y - 2z = 1$ and two units away from it.

55. Show that the lines with symmetric equations $x = y = z$ and $x + 1 = y/2 = z/3$ are skew, and find the distance between these lines. [*Hint:* The skew lines lie in parallel planes.]

56. Find the distance between the skew lines with parametric equations $x = 1 + t$, $y = 1 + 6t$, $z = 2t$, and $x = 1 + 2s$, $y = 5 + 15s$, $z = -2 + 6s$.

57. Let L_1 be the line through the origin and the point $(2, 0, -1)$. Let L_2 be the line through the points $(1, -1, 1)$ and $(4, 1, 3)$. Find the distance between L_1 and L_2.

58. Let L_1 be the line through the points $(1, 2, 6)$ and $(2, 4, 8)$. Let L_2 be the line of intersection of the planes π_1 and π_2, where π_1 is the plane $x - y + 2z + 1 = 0$ and π_2 is the plane through the points $(3, 2, -1)$, $(0, 0, 1)$, and $(1, 2, 1)$. Calculate the distance between L_1 and L_2.

59. If a, b, and c are not all 0, show that the equation $ax + by + cz + d = 0$ represents a plane and $\langle a, b, c \rangle$ is a normal vector to the plane.
  Hint: Suppose $a \neq 0$ and rewrite the equation in the form

$$a\left(x + \frac{d}{a}\right) + b(y - 0) + c(z - 0) = 0$$

60. Give a geometric description of each family of planes.
  (a) $x + y + z = c$ $\qquad$ (b) $x + y + cz = 1$
  (c) $y \cos\theta + z \sin\theta = 1$

10.6 | CYLINDERS AND QUADRIC SURFACES

We have already looked at two special types of surfaces—planes (in Section 10.5) and spheres (in Section 10.1). Here we investigate two other types of surfaces—cylinders and quadric surfaces.

In order to sketch the graph of a surface, it is useful to determine the curves of intersection of the surface with planes parallel to the coordinate planes. These curves are called **traces** (or cross-sections) of the surface.

CYLINDERS

A **cylinder** is a surface that consists of all lines (called **rulings**) that are parallel to a given line and pass through a given plane curve.

▶ **EXAMPLE 1** Sketch the graph of the surface $z = x^2$.

SOLUTION Notice that the equation of the graph, $z = x^2$, doesn't involve y. This means that any vertical plane with equation $y = k$ (parallel to the xz-plane) intersects the graph in a curve with equation $z = x^2$. So these vertical traces are parabolas. Figure 1 shows how the graph is formed by taking the parabola $z = x^2$ in the xz-plane and moving it in the direction of the y-axis. The graph is a surface, called a

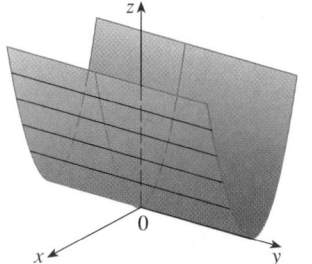

FIGURE 1
The surface $z = x^2$ is a parabolic cylinder.

parabolic cylinder, made up of infinitely many shifted copies of the same parabola. Here the rulings of the cylinder are parallel to the *y*-axis. ■

We noticed that the variable *y* is missing from the equation of the cylinder in Example 1. This is typical of a surface whose rulings are parallel to one of the coordinate axes. If one of the variables *x*, *y*, or *z* is missing from the equation of a surface, then the surface is a cylinder.

EXAMPLE 2 Identify and sketch the surfaces.
(a) $x^2 + y^2 = 1$ (b) $y^2 + z^2 = 1$

SOLUTION
(a) Since *z* is missing and the equations $x^2 + y^2 = 1$, $z = k$ represent a circle with radius 1 in the plane $z = k$, the surface $x^2 + y^2 = 1$ is a circular cylinder whose axis is the *z*-axis (see Figure 2). Here the rulings are vertical lines.

(b) In this case *x* is missing and the surface is a circular cylinder whose axis is the *x*-axis (see Figure 3). It is obtained by taking the circle $y^2 + z^2 = 1$, $x = 0$ in the *yz*-plane and moving it parallel to the *x*-axis.

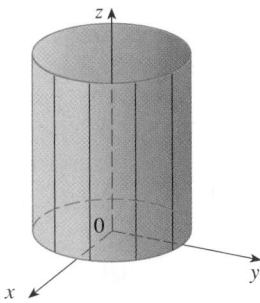

FIGURE 2 $x^2 + y^2 = 1$

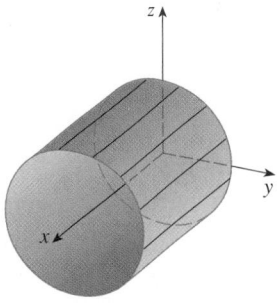

FIGURE 3 $y^2 + z^2 = 1$ ■

NOTE When you are dealing with surfaces, it is important to recognize that an equation like $x^2 + y^2 = 1$ represents a cylinder and not a circle. The trace of the cylinder $x^2 + y^2 = 1$ in the *xy*-plane is the circle with equations $x^2 + y^2 = 1$, $z = 0$.

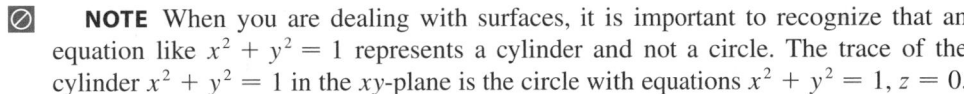

QUADRIC SURFACES

A **quadric surface** is the graph of a second-degree equation in three variables *x*, *y*, and *z*. The most general such equation is

$$Ax^2 + By^2 + Cz^2 + Dxy + Eyz + Fxz + Gx + Hy + Iz + J = 0$$

where $A, B, C, \ldots, J$ are constants, but by translation and rotation it can be brought into one of the two standard forms

$$Ax^2 + By^2 + Cz^2 + J = 0 \qquad \text{or} \qquad Ax^2 + By^2 + Iz = 0$$

Quadric surfaces are the counterparts in three dimensions of the conic sections in the plane. (See Section 9.5 for a review of conic sections.)

EXAMPLE 3 Use traces to sketch the quadric surface with equation

$$x^2 + \frac{y^2}{9} + \frac{z^2}{4} = 1$$

SOLUTION By substituting $z = 0$, we find that the trace in the xy-plane is $x^2 + y^2/9 = 1$, which we recognize as an equation of an ellipse. In general, the horizontal trace in the plane $z = k$ is

$$x^2 + \frac{y^2}{9} = 1 - \frac{k^2}{4} \qquad z = k$$

which is an ellipse, provided that $k^2 < 4$, that is, $-2 < k < 2$.

Similarly, the vertical traces are also ellipses:

$$\frac{y^2}{9} + \frac{z^2}{4} = 1 - k^2 \qquad x = k \qquad \text{(if } -1 < k < 1)$$

$$x^2 + \frac{z^2}{4} = 1 - \frac{k^2}{9} \qquad y = k \qquad \text{(if } -3 < k < 3)$$

Figure 4 shows how drawing some traces indicates the shape of the surface. It's called an **ellipsoid** because all of its traces are ellipses. Notice that it is symmetric with respect to each coordinate plane; this is a reflection of the fact that its equation involves only even powers of x, y, and z. ∎

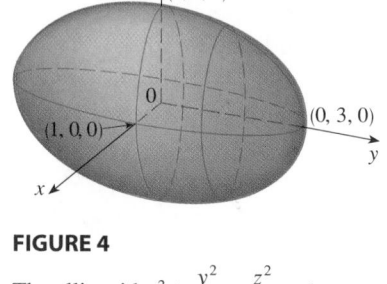

FIGURE 4

The ellipsoid $x^2 + \dfrac{y^2}{9} + \dfrac{z^2}{4} = 1$

EXAMPLE 4 Use traces to sketch the surface $z = 4x^2 + y^2$.

SOLUTION If we put $x = 0$, we get $z = y^2$, so the yz-plane intersects the surface in a parabola. If we put $x = k$ (a constant), we get $z = y^2 + 4k^2$. This means that if we slice the graph with any plane parallel to the yz-plane, we obtain a parabola that opens upward. Similarly, if $y = k$, the trace is $z = 4x^2 + k^2$, which is again a parabola that opens upward. If we put $z = k$, we get the horizontal traces $4x^2 + y^2 = k$, which we recognize as a family of ellipses. Knowing the shapes of the traces, we can sketch the graph in Figure 5. Because of the elliptical and parabolic traces, the quadric surface $z = 4x^2 + y^2$ is called an **elliptic paraboloid**.

FIGURE 5

The surface $z = 4x^2 + y^2$ is an elliptic paraboloid. Horizontal traces are ellipses; vertical traces are parabolas.

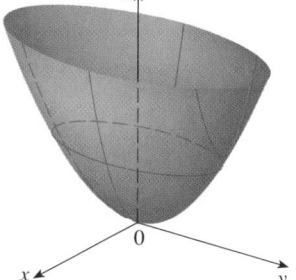

▼ EXAMPLE 5 Sketch the surface $z = y^2 - x^2$.

SOLUTION The traces in the vertical planes $x = k$ are the parabolas $z = y^2 - k^2$, which open upward. The traces in $y = k$ are the parabolas $z = -x^2 + k^2$, which open downward. The horizontal traces are $y^2 - x^2 = k$, a family of hyperbolas. We draw the families of traces in Figure 6, and we show how the traces appear when placed in their correct planes in Figure 7.

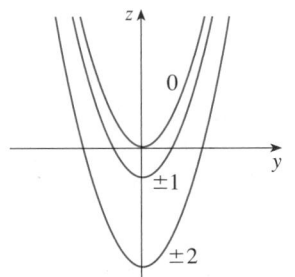 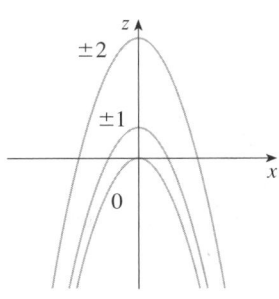 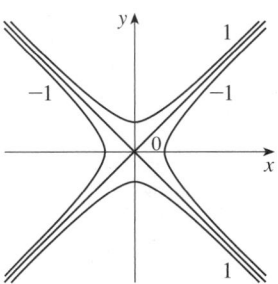

FIGURE 6

Vertical traces are parabolas; horizontal traces are hyperbolas. All traces are labeled with the value of k.

Traces in $x = k$ are $z = y^2 - k^2$ Traces in $y = k$ are $z = -x^2 + k^2$ Traces in $z = k$ are $y^2 - x^2 = k$

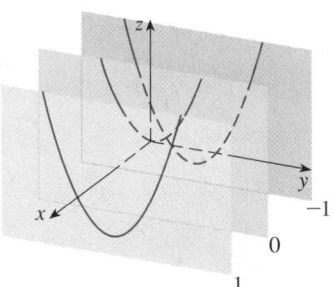 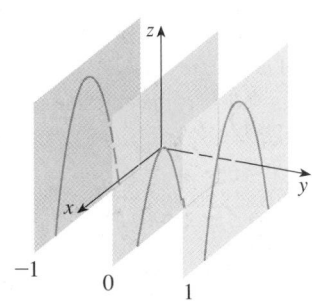 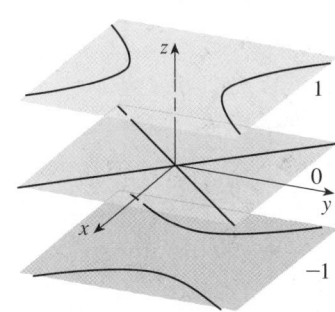

FIGURE 7

Traces moved to their correct planes

Traces in $x = k$ Traces in $y = k$ Traces in $z = k$

TEC In Module 10.6A you can investigate how traces determine the shape of a surface.

     In Figure 8 we fit together the traces from Figure 7 to form the surface $z = y^2 - x^2$, a **hyperbolic paraboloid**. Notice that the shape of the surface near the origin resembles that of a saddle. This surface will be investigated further in Section 11.7 when we discuss saddle points.

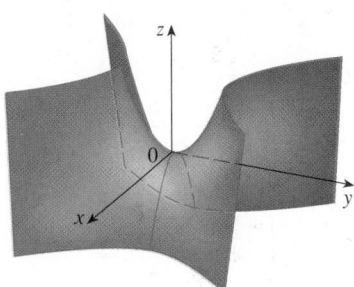

FIGURE 8

The surface $z = y^2 - x^2$ is a hyperbolic paraboloid.

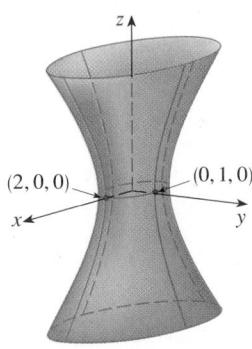

FIGURE 9

EXAMPLE 6 Sketch the surface $\dfrac{x^2}{4} + y^2 - \dfrac{z^2}{4} = 1$.

SOLUTION The trace in any horizontal plane $z = k$ is the ellipse

$$\frac{x^2}{4} + y^2 = 1 + \frac{k^2}{4} \qquad z = k$$

but the traces in the xz- and yz-planes are the hyperbolas

$$\frac{x^2}{4} - \frac{z^2}{4} = 1 \qquad y = 0 \qquad \text{and} \qquad y^2 - \frac{z^2}{4} = 1 \qquad x = 0$$

This surface is called a **hyperboloid of one sheet** and is sketched in Figure 9. ∎

The idea of using traces to draw a surface is employed in three-dimensional graphing software for computers. In most such software, traces in the vertical planes $x = k$ and $y = k$ are drawn for equally spaced values of k, and parts of the graph are eliminated using hidden line removal. Table 1 shows computer-drawn graphs of the six basic types of quadric surfaces in standard form. All surfaces are symmetric with respect to the z-axis. If a quadric surface is symmetric about a different axis, its equation changes accordingly.

TEC In Module 10.6B you can see how changing a, b, and c in Table 1 affects the shape of the quadric surface.

TABLE 1 Graphs of Quadric Surfaces

Surface	Equation	Surface	Equation
Ellipsoid	$$\frac{x^2}{a^2} + \frac{y^2}{b^2} + \frac{z^2}{c^2} = 1$$ All traces are ellipses. If $a = b = c$, the ellipsoid is a sphere.	Cone	$$\frac{z^2}{c^2} = \frac{x^2}{a^2} + \frac{y^2}{b^2}$$ Horizontal traces are ellipses. Vertical traces in the planes $x = k$ and $y = k$ are hyperbolas if $k \neq 0$ but are pairs of lines if $k = 0$.
Elliptic Paraboloid	$$\frac{z}{c} = \frac{x^2}{a^2} + \frac{y^2}{b^2}$$ Horizontal traces are ellipses. Vertical traces are parabolas. The variable raised to the first power indicates the axis of the paraboloid.	Hyperboloid of One Sheet	$$\frac{x^2}{a^2} + \frac{y^2}{b^2} - \frac{z^2}{c^2} = 1$$ Horizontal traces are ellipses. Vertical traces are hyperbolas. The axis of symmetry corresponds to the variable whose coefficient is negative.
Hyperbolic Paraboloid	$$\frac{z}{c} = \frac{x^2}{a^2} - \frac{y^2}{b^2}$$ Horizontal traces are hyperbolas. Vertical traces are parabolas. The case where $c < 0$ is illustrated.	Hyperboloid of Two Sheets	$$-\frac{x^2}{a^2} - \frac{y^2}{b^2} + \frac{z^2}{c^2} = 1$$ Horizontal traces in $z = k$ are ellipses if $k > c$ or $k < -c$. Vertical traces are hyperbolas. The two minus signs indicate two sheets.

EXAMPLE 7 Classify the quadric surface $x^2 + 2z^2 - 6x - y + 10 = 0$.

SOLUTION By completing the square we rewrite the equation as

$$y - 1 = (x - 3)^2 + 2z^2$$

Comparing this equation with Table 1, we see that it represents an elliptic parabo-

- **www.stewartcalculus.com**

See Additional Example A.

loid. Here, however, the axis of the paraboloid is parallel to the y-axis, and it has been shifted so that its vertex is the point $(3, 1, 0)$. The traces in the plane $y = k$ ($k > 1$) are the ellipses

$$(x - 3)^2 + 2z^2 = k - 1 \qquad y = k$$

The trace in the xy-plane is the parabola with equation $y = 1 + (x - 3)^2$, $z = 0$. The paraboloid is sketched in Figure 10.

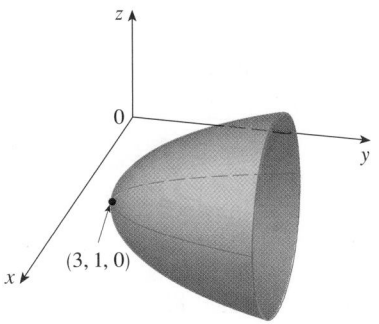

FIGURE 10
$x^2 + 2z^2 - 6x - y + 10 = 0$

10.6 | EXERCISES

1. (a) What does the equation $y = x^2$ represent as a curve in $\mathbb{R}^2$?
 (b) What does it represent as a surface in $\mathbb{R}^3$?
 (c) What does the equation $z = y^2$ represent?

2. (a) Sketch the graph of $y = e^x$ as a curve in $\mathbb{R}^2$.
 (b) Sketch the graph of $y = e^x$ as a surface in $\mathbb{R}^3$.
 (c) Describe and sketch the surface $z = e^y$.

3–8 ▪ Describe and sketch the surface.

3. $x^2 + z^2 = 1$
4. $4x^2 + y^2 = 4$
5. $z = 1 - y^2$
6. $y = z^2$
7. $xy = 1$
8. $z = \sin y$

---

9. (a) Find and identify the traces of the quadric surface $x^2 + y^2 - z^2 = 1$ and explain why the graph looks like the graph of the hyperboloid of one sheet in Table 1.
 (b) If we change the equation in part (a) to $x^2 - y^2 + z^2 = 1$, how is the graph affected?
 (c) What if we change the equation in part (a) to $x^2 + y^2 + 2y - z^2 = 0$?

10. (a) Find and identify the traces of the quadric surface $-x^2 - y^2 + z^2 = 1$ and explain why the graph looks like the graph of the hyperboloid of two sheets in Table 1.
 (b) If the equation in part (a) is changed to $x^2 - y^2 - z^2 = 1$, what happens to the graph? Sketch the new graph.

11–20 ▪ Use traces to sketch and identify the surface.

11. $x = y^2 + 4z^2$
12. $9x^2 - y^2 + z^2 = 0$
13. $x^2 = y^2 + 4z^2$
14. $25x^2 + 4y^2 + z^2 = 100$
15. $-x^2 + 4y^2 - z^2 = 4$
16. $4x^2 + 9y^2 + z = 0$
17. $36x^2 + y^2 + 36z^2 = 36$
18. $4x^2 - 16y^2 + z^2 = 16$
19. $y = z^2 - x^2$
20. $x = y^2 - z^2$

21–28 ▪ Reduce the equation to one of the standard forms, classify the surface, and sketch it.

21. $y^2 = x^2 + \frac{1}{9}z^2$

22. $4x^2 - y + 2z^2 = 0$

23. $x^2 + 2y - 2z^2 = 0$

24. $y^2 = x^2 + 4z^2 + 4$

25. $4x^2 + y^2 + 4z^2 - 4y - 24z + 36 = 0$

26. $4y^2 + z^2 - x - 16y - 4z + 20 = 0$

27. $x^2 - y^2 + z^2 - 4x - 2y - 2z + 4 = 0$

28. $x^2 - y^2 + z^2 - 2x + 2y + 4z + 2 = 0$

---

29. Sketch the region bounded by the surfaces $z = \sqrt{x^2 + y^2}$ and $x^2 + y^2 = 1$ for $1 \leqslant z \leqslant 2$.

30. Sketch the region bounded by the paraboloids $z = x^2 + y^2$ and $z = 2 - x^2 - y^2$.

31. Find an equation for the surface consisting of all points that are equidistant from the point $(-1, 0, 0)$ and the plane $x = 1$. Identify the surface.

32. Find an equation for the surface consisting of all points P for which the distance from P to the x-axis is twice the distance from P to the yz-plane. Identify the surface.

33. Graph the surfaces $z = x^2 + y^2$ and $z = 1 - y^2$ on a common screen using the domain $|x| \leq 1.2$, $|y| \leq 1.2$ and observe the curve of intersection of these surfaces. Show that the projection of this curve onto the xy-plane is an ellipse.

34. Show that the curve of intersection of the surfaces $x^2 + 2y^2 - z^2 + 3x = 1$ and $2x^2 + 4y^2 - 2z^2 - 5y = 0$ lies in a plane.

10.7 | VECTOR FUNCTIONS AND SPACE CURVES

In general, a function is a rule that assigns to each element in the domain an element in the range. A **vector-valued function**, or **vector function**, is simply a function whose domain is a set of real numbers and whose range is a set of vectors. We are most interested in vector functions $\mathbf{r}$ whose values are three-dimensional vectors. This means that for every number t in the domain of $\mathbf{r}$ there is a unique vector in V_3 denoted by $\mathbf{r}(t)$. If $f(t)$, $g(t)$, and $h(t)$ are the components of the vector $\mathbf{r}(t)$, then f, g, and h are real-valued functions called the **component functions** of $\mathbf{r}$ and we can write

$$\mathbf{r}(t) = \langle f(t), g(t), h(t) \rangle = f(t)\,\mathbf{i} + g(t)\,\mathbf{j} + h(t)\,\mathbf{k}$$

We use the letter t to denote the independent variable because it represents time in most applications of vector functions.

EXAMPLE 1 If

$$\mathbf{r}(t) = \left\langle t^3, \ln(3 - t), \sqrt{t} \right\rangle$$

then the component functions are

$$f(t) = t^3 \qquad g(t) = \ln(3 - t) \qquad h(t) = \sqrt{t}$$

By our usual convention, the domain of $\mathbf{r}$ consists of all values of t for which the expression for $\mathbf{r}(t)$ is defined. The expressions t^3, $\ln(3 - t)$, and $\sqrt{t}$ are all defined when $3 - t > 0$ and $t \geq 0$. Therefore the domain of $\mathbf{r}$ is the interval $[0, 3)$. ∎

The **limit** of a vector function $\mathbf{r}$ is defined by taking the limits of its component functions as follows.

▪ If $\lim_{t \to a} \mathbf{r}(t) = \mathbf{L}$, this definition is equivalent to saying that the length and direction of the vector $\mathbf{r}(t)$ approach the length and direction of the vector $\mathbf{L}$.

> **1** If $\mathbf{r}(t) = \langle f(t), g(t), h(t) \rangle$, then
>
> $$\lim_{t \to a} \mathbf{r}(t) = \left\langle \lim_{t \to a} f(t), \lim_{t \to a} g(t), \lim_{t \to a} h(t) \right\rangle$$
>
> provided the limits of the component functions exist.

Equivalently, we could have used an ε-δ definition (see Exercise 70). Limits of vector functions obey the same rules as limits of real-valued functions (see Exercise 69).

EXAMPLE 2 Find $\lim\limits_{t \to 0} \mathbf{r}(t)$, where $\mathbf{r}(t) = (1 + t^3)\,\mathbf{i} + te^{-t}\,\mathbf{j} + \dfrac{\sin t}{t}\,\mathbf{k}$.

SOLUTION According to Definition 1, the limit of $\mathbf{r}$ is the vector whose components are the limits of the component functions of $\mathbf{r}$:

$$\lim_{t \to 0} \mathbf{r}(t) = \left[\lim_{t \to 0}(1 + t^3)\right]\mathbf{i} + \left[\lim_{t \to 0} te^{-t}\right]\mathbf{j} + \left[\lim_{t \to 0}\frac{\sin t}{t}\right]\mathbf{k}$$

$$= \mathbf{i} + \mathbf{k} \qquad \text{(by Equation 1.4.6)} \qquad\blacksquare$$

A vector function $\mathbf{r}$ is **continuous at a** if

$$\lim_{t \to a} \mathbf{r}(t) = \mathbf{r}(a)$$

In view of Definition 1, we see that $\mathbf{r}$ is continuous at a if and only if its component functions f, g, and h are continuous at a.

There is a close connection between continuous vector functions and space curves. Suppose that f, g, and h are continuous real-valued functions on an interval I. Then the set C of all points (x, y, z) in space, where

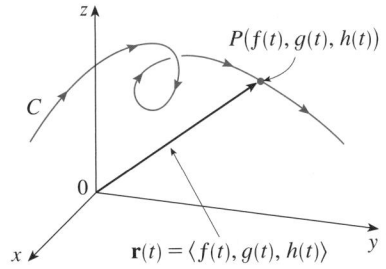

$$\boxed{2} \qquad\qquad x = f(t) \qquad y = g(t) \qquad z = h(t)$$

and t varies throughout the interval I, is called a **space curve**. The equations in $\boxed{2}$ are called **parametric equations of C** and t is called a **parameter**. We can think of C as being traced out by a moving particle whose position at time t is $\big(f(t), g(t), h(t)\big)$. If we now consider the vector function $\mathbf{r}(t) = \langle f(t), g(t), h(t) \rangle$, then $\mathbf{r}(t)$ is the position vector of the point $P\big(f(t), g(t), h(t)\big)$ on C. Thus any continuous vector function $\mathbf{r}$ defines a space curve C that is traced out by the tip of the moving vector $\mathbf{r}(t)$, as shown in Figure 1.

FIGURE 1
C is traced out by the tip of a moving position vector $\mathbf{r}(t)$.

TEC Visual 10.7A shows several curves being traced out by position vectors, including those in Figures 1 and 2.

$\boxed{\text{V}}$ **EXAMPLE 3** Describe the curve defined by the vector function

$$\mathbf{r}(t) = \langle 1 + t,\ 2 + 5t,\ -1 + 6t \rangle$$

SOLUTION The corresponding parametric equations are

$$x = 1 + t \qquad y = 2 + 5t \qquad z = -1 + 6t$$

which we recognize from Equations 10.5.2 as parametric equations of a line passing through the point $(1, 2, -1)$ and parallel to the vector $\langle 1, 5, 6 \rangle$. Alternatively, we could observe that the function can be written as $\mathbf{r} = \mathbf{r}_0 + t\mathbf{v}$, where $\mathbf{r}_0 = \langle 1, 2, -1 \rangle$ and $\mathbf{v} = \langle 1, 5, 6 \rangle$, and this is the vector equation of a line as given by Equation 10.5.1. $\qquad\blacksquare$

Plane curves can also be represented in vector notation. For instance, the curve given by the parametric equations $x = t^2 - 2t$ and $y = t + 1$ (see Example 1 in Section 9.1) could also be described by the vector equation

$$\mathbf{r}(t) = \langle t^2 - 2t,\ t + 1 \rangle = (t^2 - 2t)\,\mathbf{i} + (t + 1)\,\mathbf{j}$$

where $\mathbf{i} = \langle 1, 0 \rangle$ and $\mathbf{j} = \langle 0, 1 \rangle$.

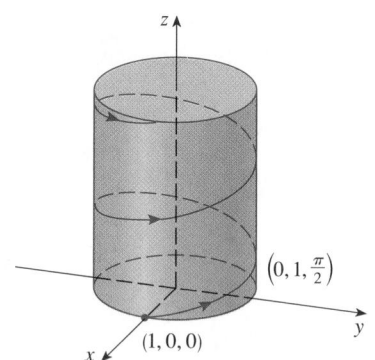

FIGURE 2

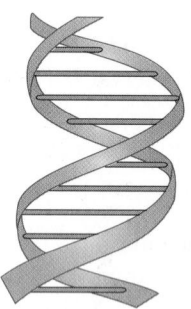

FIGURE 3

■ Figure 4 shows the line segment PQ in Example 5.

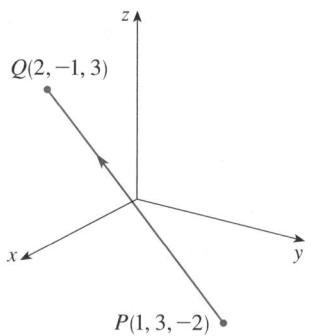

FIGURE 4

▼ **EXAMPLE 4** Sketch the curve whose vector equation is

$$\mathbf{r}(t) = \cos t\, \mathbf{i} + \sin t\, \mathbf{j} + t\, \mathbf{k}$$

SOLUTION The parametric equations for this curve are

$$x = \cos t \qquad y = \sin t \qquad z = t$$

Since $x^2 + y^2 = \cos^2 t + \sin^2 t = 1$, the curve must lie on the circular cylinder $x^2 + y^2 = 1$. The point (x, y, z) lies directly above the point $(x, y, 0)$, which moves counterclockwise around the circle $x^2 + y^2 = 1$ in the xy-plane. (The projection of the curve onto the xy-plane has vector equation $\mathbf{r}(t) = \langle \cos t, \sin t, 0 \rangle$. See Example 2 in Section 9.1.) Since $z = t$, the curve spirals upward around the cylinder as t increases. The curve, shown in Figure 2, is called a **helix**. ■

The corkscrew shape of the helix in Example 4 is familiar from its occurrence in coiled springs. It also occurs in the model of DNA (deoxyribonucleic acid, the genetic material of living cells). In 1953 James Watson and Francis Crick showed that the structure of the DNA molecule is that of two linked, parallel helixes that are intertwined as in Figure 3.

In Examples 3 and 4 we were given vector equations of curves and asked for a geometric description or sketch. In the next two examples we are given a geometric description of a curve and are asked to find parametric equations for the curve.

EXAMPLE 5 Find a vector equation and parametric equations for the line segment that joins the point $P(1, 3, -2)$ to the point $Q(2, -1, 3)$.

SOLUTION In Section 10.5 we found a vector equation for the line segment that joins the tip of the vector $\mathbf{r}_0$ to the tip of the vector $\mathbf{r}_1$:

$$\mathbf{r}(t) = (1 - t)\mathbf{r}_0 + t\mathbf{r}_1 \qquad 0 \leqslant t \leqslant 1$$

(See Equation 10.5.4.) Here we take $\mathbf{r}_0 = \langle 1, 3, -2 \rangle$ and $\mathbf{r}_1 = \langle 2, -1, 3 \rangle$ to obtain a vector equation of the line segment from P to Q:

$$\mathbf{r}(t) = (1 - t)\langle 1, 3, -2 \rangle + t\langle 2, -1, 3 \rangle \qquad 0 \leqslant t \leqslant 1$$

or $\qquad \mathbf{r}(t) = \langle 1 + t, 3 - 4t, -2 + 5t \rangle \qquad 0 \leqslant t \leqslant 1$

The corresponding parametric equations are

$$x = 1 + t \qquad y = 3 - 4t \qquad z = -2 + 5t \qquad 0 \leqslant t \leqslant 1 \qquad ■$$

▼ **EXAMPLE 6** Find a vector function that represents the curve of intersection of the cylinder $x^2 + y^2 = 1$ and the plane $y + z = 2$.

SOLUTION Figure 5 shows how the plane and the cylinder intersect, and Figure 6 shows the curve of intersection C, which is an ellipse.

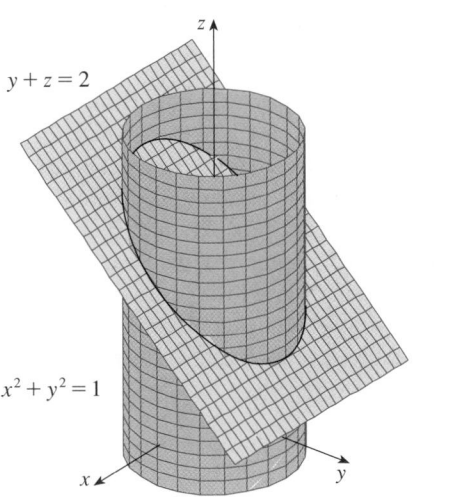

FIGURE 5

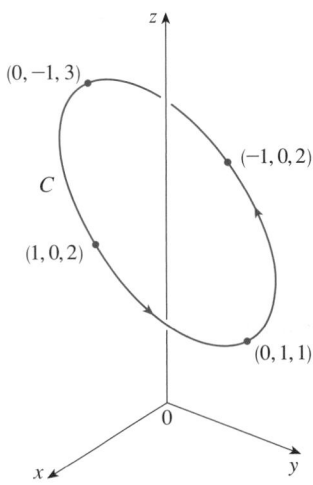

FIGURE 6

The projection of C onto the xy-plane is the circle $x^2 + y^2 = 1$, $z = 0$. So we know from Example 2 in Section 9.1 that we can write

$$x = \cos t \qquad y = \sin t \qquad 0 \leq t \leq 2\pi$$

From the equation of the plane, we have

$$z = 2 - y = 2 - \sin t$$

So we can write parametric equations for C as

$$x = \cos t \qquad y = \sin t \qquad z = 2 - \sin t \qquad 0 \leq t \leq 2\pi$$

The corresponding vector equation is

$$\mathbf{r}(t) = \cos t\, \mathbf{i} + \sin t\, \mathbf{j} + (2 - \sin t)\, \mathbf{k} \qquad 0 \leq t \leq 2\pi$$

This equation is called a *parametrization* of the curve C. The arrows in Figure 6 indicate the direction in which C is traced as the parameter t increases. ∎

USING COMPUTERS TO DRAW SPACE CURVES

Space curves are inherently more difficult to draw by hand than plane curves; for an accurate representation we need to use technology. For instance, Figure 7 shows a computer-generated graph of the curve with parametric equations

$$x = (4 + \sin 20t) \cos t \qquad y = (4 + \sin 20t) \sin t \qquad z = \cos 20t$$

It's called a **toroidal spiral** because it lies on a torus. Another interesting curve, the **trefoil knot**, with equations

$$x = (2 + \cos 1.5t) \cos t \qquad y = (2 + \cos 1.5t) \sin t \qquad z = \sin 1.5t$$

is graphed in Figure 8. It wouldn't be easy to plot either of these curves by hand.

Even when a computer is used to draw a space curve, optical illusions make it difficult to get a good impression of what the curve really looks like. (This is especially true in Figure 8.) The next example shows how to cope with this problem.

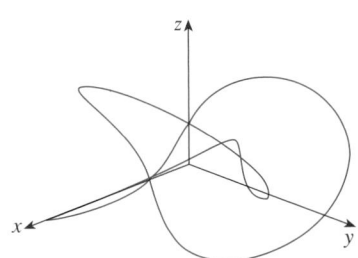

FIGURE 7   A toroidal spiral

FIGURE 8   A trefoil knot

EXAMPLE 7 Use a computer to draw the curve with vector equation $\mathbf{r}(t) = \langle t, t^2, t^3 \rangle$. This curve is called a **twisted cubic**.

SOLUTION We start by using the computer to plot the curve with parametric equations $x = t$, $y = t^2$, $z = t^3$ for $-2 \leq t \leq 2$. The result is shown in Figure 9(a), but it's hard to see the true nature of the curve from that graph alone. Most three-dimensional computer graphing programs allow the user to enclose a curve or surface in a box instead of displaying the coordinate axes. When we look at the same curve in a box in Figure 9(b), we have a much clearer picture of the curve. We can see that it climbs from a lower corner of the box to the upper corner nearest us, and it twists as it climbs. We get an even better idea of the curve when we view it from different vantage points. Part (c) shows the result of rotating the box to give another viewpoint.

TEC In Visual 10.7B you can rotate the box in Figure 9 to see the curve from any viewpoint.

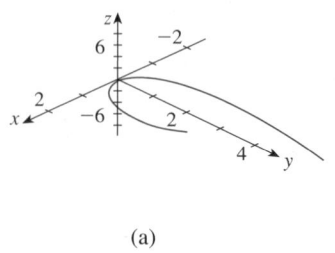

(a)

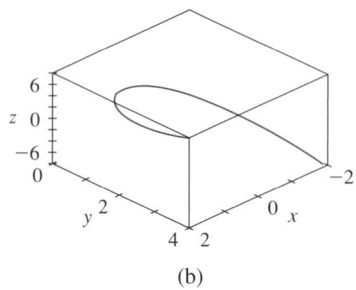

(b)

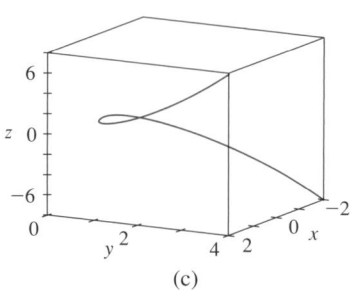

(c)

FIGURE 9
Views of the twisted cubic

DERIVATIVES

The **derivative $\mathbf{r}'$** of a vector function $\mathbf{r}$ is defined in much the same way as for real-valued functions:

$$\boxed{3} \qquad \frac{d\mathbf{r}}{dt} = \mathbf{r}'(t) = \lim_{h \to 0} \frac{\mathbf{r}(t + h) - \mathbf{r}(t)}{h}$$

if this limit exists. The geometric significance of this definition is shown in Figure 10.

TEC Visual 10.7C shows an animation of Figure 10.

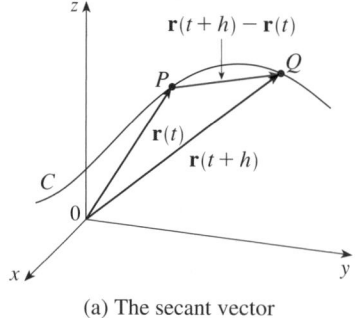

(a) The secant vector

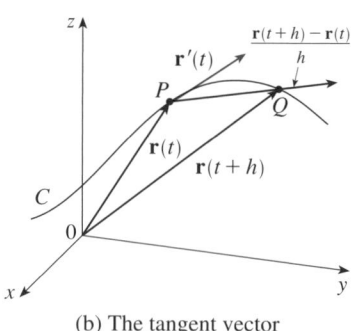

(b) The tangent vector

FIGURE 10

If the points P and Q have position vectors $\mathbf{r}(t)$ and $\mathbf{r}(t + h)$, then $\overrightarrow{PQ}$ represents the vector $\mathbf{r}(t + h) - \mathbf{r}(t)$, which can therefore be regarded as a secant vector. If $h > 0$, the scalar multiple $(1/h)(\mathbf{r}(t + h) - \mathbf{r}(t))$ has the same direction as $\mathbf{r}(t + h) - \mathbf{r}(t)$. As $h \rightarrow 0$, it appears that this vector approaches a vector that lies on the tangent line. For this reason, the vector $\mathbf{r}'(t)$ is called the **tangent vector** to the curve defined by $\mathbf{r}$ at the point P, provided that $\mathbf{r}'(t)$ exists and $\mathbf{r}'(t) \neq \mathbf{0}$. The **tangent line** to C at P is defined to be the line through P parallel to the tangent vector $\mathbf{r}'(t)$. We will also have occasion to consider the **unit tangent vector**, which is

$$\mathbf{T}(t) = \frac{\mathbf{r}'(t)}{|\mathbf{r}'(t)|}$$

The following theorem gives us a convenient method for computing the derivative of a vector function $\mathbf{r}$: just differentiate each component of $\mathbf{r}$.

---

4 **THEOREM** If $\mathbf{r}(t) = \langle f(t), g(t), h(t) \rangle = f(t)\,\mathbf{i} + g(t)\,\mathbf{j} + h(t)\,\mathbf{k}$, where f, g, and h are differentiable functions, then

$$\mathbf{r}'(t) = \langle f'(t), g'(t), h'(t) \rangle = f'(t)\,\mathbf{i} + g'(t)\,\mathbf{j} + h'(t)\,\mathbf{k}$$

PROOF

$$\mathbf{r}'(t) = \lim_{\Delta t \to 0} \frac{1}{\Delta t} [\mathbf{r}(t + \Delta t) - \mathbf{r}(t)]$$

$$= \lim_{\Delta t \to 0} \frac{1}{\Delta t} [\langle f(t + \Delta t), g(t + \Delta t), h(t + \Delta t) \rangle - \langle f(t), g(t), h(t) \rangle]$$

$$= \lim_{\Delta t \to 0} \left\langle \frac{f(t + \Delta t) - f(t)}{\Delta t}, \frac{g(t + \Delta t) - g(t)}{\Delta t}, \frac{h(t + \Delta t) - h(t)}{\Delta t} \right\rangle$$

$$= \left\langle \lim_{\Delta t \to 0} \frac{f(t + \Delta t) - f(t)}{\Delta t}, \lim_{\Delta t \to 0} \frac{g(t + \Delta t) - g(t)}{\Delta t}, \lim_{\Delta t \to 0} \frac{h(t + \Delta t) - h(t)}{\Delta t} \right\rangle$$

$$= \langle f'(t), g'(t), h'(t) \rangle$$

☐

▼ EXAMPLE 8
(a) Find the derivative of $\mathbf{r}(t) = (1 + t^3)\,\mathbf{i} + te^{-t}\,\mathbf{j} + \sin 2t\,\mathbf{k}$.
(b) Find the unit tangent vector at the point where $t = 0$.

SOLUTION
(a) According to Theorem 4, we differentiate each component of $\mathbf{r}$:

$$\mathbf{r}'(t) = 3t^2\,\mathbf{i} + (1 - t)e^{-t}\,\mathbf{j} + 2\cos 2t\,\mathbf{k}$$

(b) Since $\mathbf{r}(0) = \mathbf{i}$ and $\mathbf{r}'(0) = \mathbf{j} + 2\mathbf{k}$, the unit tangent vector at the point $(1, 0, 0)$ is

$$\mathbf{T}(0) = \frac{\mathbf{r}'(0)}{|\mathbf{r}'(0)|} = \frac{\mathbf{j} + 2\mathbf{k}}{\sqrt{1 + 4}} = \frac{1}{\sqrt{5}}\,\mathbf{j} + \frac{2}{\sqrt{5}}\,\mathbf{k}$$

■

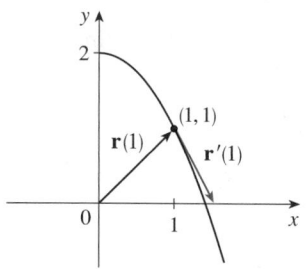

FIGURE 11

EXAMPLE 9 For the curve $\mathbf{r}(t) = \sqrt{t}\,\mathbf{i} + (2-t)\,\mathbf{j}$, find $\mathbf{r}'(t)$ and sketch the position vector $\mathbf{r}(1)$ and the tangent vector $\mathbf{r}'(1)$.

SOLUTION We have

$$\mathbf{r}'(t) = \frac{1}{2\sqrt{t}}\,\mathbf{i} - \mathbf{j} \qquad \text{and} \qquad \mathbf{r}'(1) = \frac{1}{2}\,\mathbf{i} - \mathbf{j}$$

The curve is a plane curve and elimination of the parameter from the equations $x = \sqrt{t}$, $y = 2 - t$ gives $y = 2 - x^2$, $x \geqslant 0$. In Figure 11 we draw the position vector $\mathbf{r}(1) = \mathbf{i} + \mathbf{j}$ starting at the origin and the tangent vector $\mathbf{r}'(1)$ starting at the corresponding point $(1, 1)$. ▪

▼ EXAMPLE 10 Find parametric equations for the tangent line to the helix with parametric equations

$$x = 2\cos t \qquad y = \sin t \qquad z = t$$

at the point $(0, 1, \pi/2)$.

SOLUTION The vector equation of the helix is $\mathbf{r}(t) = \langle 2\cos t, \sin t, t \rangle$, so

$$\mathbf{r}'(t) = \langle -2\sin t, \cos t, 1 \rangle$$

The parameter value corresponding to the point $(0, 1, \pi/2)$ is $t = \pi/2$, so the tangent vector there is $\mathbf{r}'(\pi/2) = \langle -2, 0, 1 \rangle$. The tangent line is the line through $(0, 1, \pi/2)$ parallel to the vector $\langle -2, 0, 1 \rangle$, so by Equations 10.5.2 its parametric equations are

$$x = -2t \qquad y = 1 \qquad z = \frac{\pi}{2} + t$$ ▪

■ The helix and the tangent line in Example 10 are shown in Figure 12.

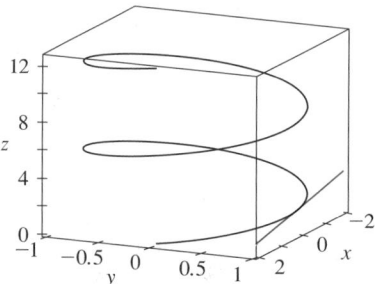

FIGURE 12

■ In Section 10.9 we will see how $\mathbf{r}'(t)$ and $\mathbf{r}''(t)$ can be interpreted as the velocity and acceleration vectors of a particle moving through space with position vector $\mathbf{r}(t)$ at time t.

Just as for real-valued functions, the **second derivative** of a vector function $\mathbf{r}$ is the derivative of $\mathbf{r}'$, that is, $\mathbf{r}'' = (\mathbf{r}')'$. For instance, the second derivative of the function in Example 10 is

$$\mathbf{r}''(t) = \langle -2\cos t, -\sin t, 0 \rangle$$

DIFFERENTIATION RULES

The next theorem shows that the differentiation formulas for real-valued functions have their counterparts for vector-valued functions.

5 **THEOREM** Suppose $\mathbf{u}$ and $\mathbf{v}$ are differentiable vector functions, c is a scalar, and f is a real-valued function. Then

1. $\dfrac{d}{dt}[\mathbf{u}(t) + \mathbf{v}(t)] = \mathbf{u}'(t) + \mathbf{v}'(t)$

2. $\dfrac{d}{dt}[c\mathbf{u}(t)] = c\mathbf{u}'(t)$

3. $\dfrac{d}{dt}[f(t)\,\mathbf{u}(t)] = f'(t)\,\mathbf{u}(t) + f(t)\,\mathbf{u}'(t)$

4. $\dfrac{d}{dt}[\mathbf{u}(t) \cdot \mathbf{v}(t)] = \mathbf{u}'(t) \cdot \mathbf{v}(t) + \mathbf{u}(t) \cdot \mathbf{v}'(t)$

5. $\dfrac{d}{dt}[\mathbf{u}(t) \times \mathbf{v}(t)] = \mathbf{u}'(t) \times \mathbf{v}(t) + \mathbf{u}(t) \times \mathbf{v}'(t)$

6. $\dfrac{d}{dt}[\mathbf{u}(f(t))] = f'(t)\,\mathbf{u}'(f(t))$ (Chain Rule)

This theorem can be proved either directly from Definition 3 or by using Theorem 4 and the corresponding differentiation formulas for real-valued functions. The proof of Formula 4 follows; the remaining proofs are left as exercises.

PROOF OF FORMULA 4 Let

$$\mathbf{u}(t) = \langle f_1(t), f_2(t), f_3(t) \rangle \qquad \mathbf{v}(t) = \langle g_1(t), g_2(t), g_3(t) \rangle$$

Then $\mathbf{u}(t) \cdot \mathbf{v}(t) = f_1(t)g_1(t) + f_2(t)g_2(t) + f_3(t)g_3(t) = \displaystyle\sum_{i=1}^{3} f_i(t)\,g_i(t)$

so the Product Rule for scalar functions gives

$$\frac{d}{dt}[\mathbf{u}(t) \cdot \mathbf{v}(t)] = \frac{d}{dt}\sum_{i=1}^{3} f_i(t)\,g_i(t) = \sum_{i=1}^{3} \frac{d}{dt}[f_i(t)\,g_i(t)]$$

$$= \sum_{i=1}^{3} [f_i'(t)\,g_i(t) + f_i(t)\,g_i'(t)]$$

$$= \sum_{i=1}^{3} f_i'(t)\,g_i(t) + \sum_{i=1}^{3} f_i(t)\,g_i'(t)$$

$$= \mathbf{u}'(t) \cdot \mathbf{v}(t) + \mathbf{u}(t) \cdot \mathbf{v}'(t) \qquad \square$$

▼ EXAMPLE 11 Show that if $|\mathbf{r}(t)| = c$ (a constant), then $\mathbf{r}'(t)$ is orthogonal to $\mathbf{r}(t)$ for all t.

SOLUTION Since

$$\mathbf{r}(t) \cdot \mathbf{r}(t) = |\mathbf{r}(t)|^2 = c^2$$

and c^2 is a constant, Formula 4 of Theorem 5 gives

$$0 = \frac{d}{dt}[\mathbf{r}(t) \cdot \mathbf{r}(t)] = \mathbf{r}'(t) \cdot \mathbf{r}(t) + \mathbf{r}(t) \cdot \mathbf{r}'(t) = 2\mathbf{r}'(t) \cdot \mathbf{r}(t)$$

Thus $\mathbf{r}'(t) \cdot \mathbf{r}(t) = 0$, which says that $\mathbf{r}'(t)$ is orthogonal to $\mathbf{r}(t)$.

Geometrically, this result says that if a curve lies on a sphere with center the origin, then the tangent vector $\mathbf{r}'(t)$ is always perpendicular to the position vector $\mathbf{r}(t)$. ■

INTEGRALS

The **definite integral** of a continuous vector function $\mathbf{r}(t)$ can be defined in much the same way as for real-valued functions except that the integral is a vector. But then we can express the integral of $\mathbf{r}$ in terms of the integrals of its component functions f, g, and h as follows. (We use the notation of Chapter 5.)

$$\int_a^b \mathbf{r}(t)\, dt = \lim_{n \to \infty} \sum_{i=1}^n \mathbf{r}(t_i^*)\, \Delta t$$

$$= \lim_{n \to \infty} \left[\left(\sum_{i=1}^n f(t_i^*)\, \Delta t \right) \mathbf{i} + \left(\sum_{i=1}^n g(t_i^*)\, \Delta t \right) \mathbf{j} + \left(\sum_{i=1}^n h(t_i^*)\, \Delta t \right) \mathbf{k} \right]$$

and so

$$\int_a^b \mathbf{r}(t)\, dt = \left(\int_a^b f(t)\, dt \right) \mathbf{i} + \left(\int_a^b g(t)\, dt \right) \mathbf{j} + \left(\int_a^b h(t)\, dt \right) \mathbf{k}$$

This means that we can evaluate an integral of a vector function by integrating each component function.

We can extend the Fundamental Theorem of Calculus to continuous vector functions as follows:

$$\int_a^b \mathbf{r}(t)\, dt = \mathbf{R}(t) \Big]_a^b = \mathbf{R}(b) - \mathbf{R}(a)$$

where $\mathbf{R}$ is an antiderivative of $\mathbf{r}$, that is, $\mathbf{R}'(t) = \mathbf{r}(t)$. We use the notation $\int \mathbf{r}(t)\, dt$ for indefinite integrals (antiderivatives).

EXAMPLE 12 If $\mathbf{r}(t) = 2\cos t\,\mathbf{i} + \sin t\,\mathbf{j} + 2t\,\mathbf{k}$, then

$$\int \mathbf{r}(t)\, dt = \left(\int 2\cos t\, dt \right) \mathbf{i} + \left(\int \sin t\, dt \right) \mathbf{j} + \left(\int 2t\, dt \right) \mathbf{k}$$

$$= 2\sin t\,\mathbf{i} - \cos t\,\mathbf{j} + t^2\,\mathbf{k} + \mathbf{C}$$

where $\mathbf{C}$ is a vector constant of integration, and

$$\int_0^{\pi/2} \mathbf{r}(t)\, dt = \left[2\sin t\,\mathbf{i} - \cos t\,\mathbf{j} + t^2\,\mathbf{k} \right]_0^{\pi/2} = 2\mathbf{i} + \mathbf{j} + \frac{\pi^2}{4}\,\mathbf{k}$$ ■

10.7 | EXERCISES

1–2 ▪ Find the domain of the vector function.

1. $\mathbf{r}(t) = \left\langle \sqrt{4 - t^2}, e^{-3t}, \ln(t + 1) \right\rangle$

2. $\mathbf{r}(t) = \dfrac{t - 2}{t + 2}\mathbf{i} + \sin t\,\mathbf{j} + \ln(9 - t^2)\mathbf{k}$

3–4 ▪ Find the limit.

3. $\displaystyle \lim_{t \to 0} \left(e^{-3t}\mathbf{i} + \dfrac{t^2}{\sin^2 t}\mathbf{j} + \cos 2t\,\mathbf{k} \right)$

4. $\displaystyle \lim_{t \to 1} \left(\dfrac{t^2 - t}{t - 1}\mathbf{i} + \sqrt{t + 8}\,\mathbf{j} + \dfrac{\sin \pi t}{\ln t}\mathbf{k} \right)$

5–12 ▪ Sketch the curve with the given vector equation. Indicate with an arrow the direction in which t increases.

5. $\mathbf{r}(t) = \langle \sin t, t \rangle$

6. $\mathbf{r}(t) = \langle t^3, t^2 \rangle$

7. $\mathbf{r}(t) = \langle t, 2 - t, 2t \rangle$

8. $\mathbf{r}(t) = \langle \sin \pi t, t, \cos \pi t \rangle$

9. $\mathbf{r}(t) = \langle 1, \cos t, 2 \sin t \rangle$

10. $\mathbf{r}(t) = t^2\mathbf{i} + t\mathbf{j} + 2\mathbf{k}$

11. $\mathbf{r}(t) = t^2\mathbf{i} + t^4\mathbf{j} + t^6\mathbf{k}$

12. $\mathbf{r}(t) = \cos t\,\mathbf{i} - \cos t\,\mathbf{j} + \sin t\,\mathbf{k}$

13–16 ▪ Find a vector equation and parametric equations for the line segment that joins P to Q.

13. $P(2, 0, 0), \quad Q(6, 2, -2)$

14. $P(-1, 2, -2), \quad Q(-3, 5, 1)$

15. $P(0, -1, 1), \quad Q\left(\frac{1}{2}, \frac{1}{3}, \frac{1}{4}\right)$

16. $P(a, b, c), \quad Q(u, v, w)$

17–22 ▪ Match the parametric equations with the graphs (labeled I–VI). Give reasons for your choices.

I
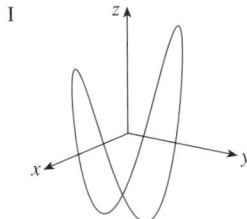

II
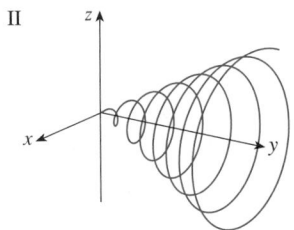

III
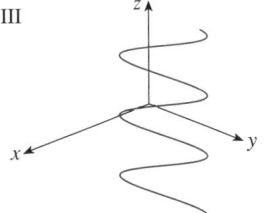

IV
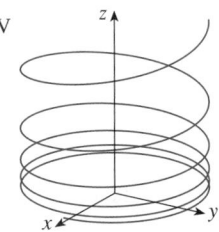

V
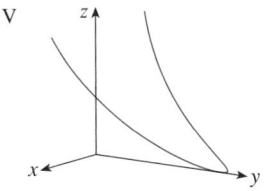

VI
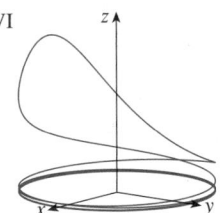

17. $x = t \cos t, \quad y = t, \quad z = t \sin t, \quad t \geq 0$

18. $x = \cos t, \quad y = \sin t, \quad z = 1/(1 + t^2)$

19. $x = t, \quad y = 1/(1 + t^2), \quad z = t^2$

20. $x = \cos t, \quad y = \sin t, \quad z = \cos 2t$

21. $x = \cos 8t, \quad y = \sin 8t, \quad z = e^{0.8t}, \quad t \geq 0$

22. $x = \cos^2 t, \quad y = \sin^2 t, \quad z = t$

---

23. Show that the curve with parametric equations $x = t \cos t$, $y = t \sin t$, $z = t$ lies on the cone $z^2 = x^2 + y^2$, and use this fact to help sketch the curve.

24. Show that the curve with parametric equations $x = \sin t$, $y = \cos t$, $z = \sin^2 t$ is the curve of intersection of the surfaces $z = x^2$ and $x^2 + y^2 = 1$. Use this fact to help sketch the curve.

25. At what points does the curve $\mathbf{r}(t) = t\,\mathbf{i} + (2t - t^2)\mathbf{k}$ intersect the paraboloid $z = x^2 + y^2$?

26. Graph the curve with parametric equations

$$x = \sqrt{1 - 0.25 \cos^2(10t)}\,\cos t$$
$$y = \sqrt{1 - 0.25 \cos^2(10t)}\,\sin t$$
$$z = 0.5 \cos(10t)$$

Explain the appearance of the graph by showing that it lies on a sphere.

27. Show that the curve with parametric equations $x = t^2$, $y = 1 - 3t$, $z = 1 + t^3$ passes through the points $(1, 4, 0)$ and $(9, -8, 28)$ but not through the point $(4, 7, -6)$.

28–30 ■ Find a vector function that represents the curve of intersection of the two surfaces.

28. The cylinder $x^2 + y^2 = 4$ and the surface $z = xy$

29. The cone $z = \sqrt{x^2 + y^2}$ and the plane $z = 1 + y$

30. The semiellipsoid $x^2 + y^2 + 4z^2 = 4$, $y \geqslant 0$, and the cylinder $x^2 + z^2 = 1$

31. Try to sketch by hand the curve of intersection of the circular cylinder $x^2 + y^2 = 4$ and the parabolic cylinder $z = x^2$. Then find parametric equations for this curve and use these equations and a computer to graph the curve.

32. Try to sketch by hand the curve of intersection of the parabolic cylinder $y = x^2$ and the top half of the ellipsoid $x^2 + 4y^2 + 4z^2 = 16$. Then find parametric equations for this curve and use these equations and a computer to graph the curve.

33–38 ■
(a) Sketch the plane curve with the given vector equation.
(b) Find $\mathbf{r}'(t)$.
(c) Sketch the position vector $\mathbf{r}(t)$ and the tangent vector $\mathbf{r}'(t)$ for the given value of t.

33. $\mathbf{r}(t) = \langle t - 2, t^2 + 1 \rangle$, $\quad t = -1$

34. $\mathbf{r}(t) = \langle t^2, t^3 \rangle$, $\quad t = 1$

35. $\mathbf{r}(t) = \sin t\,\mathbf{i} + 2 \cos t\,\mathbf{j}$, $\quad t = \pi/4$

36. $\mathbf{r}(t) = e^t\,\mathbf{i} + e^{-t}\,\mathbf{j}$, $\quad t = 0$

37. $\mathbf{r}(t) = e^{2t}\,\mathbf{i} + e^t\,\mathbf{j}$, $\quad t = 0$

38. $\mathbf{r}(t) = (1 + \cos t)\,\mathbf{i} + (2 + \sin t)\,\mathbf{j}$, $\quad t = \pi/6$

39–44 ■ Find the derivative of the vector function.

39. $\mathbf{r}(t) = \langle t \sin t, t^2, t \cos 2t \rangle$

40. $\mathbf{r}(t) = \langle \tan t, \sec t, 1/t^2 \rangle$

41. $\mathbf{r}(t) = e^{t^2}\,\mathbf{i} - \mathbf{j} + \ln(1 + 3t)\,\mathbf{k}$

42. $\mathbf{r}(t) = at \cos 3t\,\mathbf{i} + b \sin^3 t\,\mathbf{j} + c \cos^3 t\,\mathbf{k}$

43. $\mathbf{r}(t) = \mathbf{a} + t\,\mathbf{b} + t^2\,\mathbf{c}$

44. $\mathbf{r}(t) = t\,\mathbf{a} \times (\mathbf{b} + t\,\mathbf{c})$

45–46 ■ Find the unit tangent vector $\mathbf{T}(t)$ at the point with the given value of the parameter t.

45. $\mathbf{r}(t) = \cos t\,\mathbf{i} + 3t\,\mathbf{j} + 2 \sin 2t\,\mathbf{k}$, $\quad t = 0$

46. $\mathbf{r}(t) = \langle t^3 + 3t, t^2 + 1, 3t + 4 \rangle$, $\quad t = 1$

47. If $\mathbf{r}(t) = \langle t, t^2, t^3 \rangle$, find $\mathbf{r}'(t)$, $\mathbf{T}(1)$, $\mathbf{r}''(t)$, and $\mathbf{r}'(t) \times \mathbf{r}''(t)$.

48. If $\mathbf{r}(t) = \langle e^{2t}, e^{-2t}, te^{2t} \rangle$, find $\mathbf{T}(0)$, $\mathbf{r}''(0)$, and $\mathbf{r}'(t) \cdot \mathbf{r}''(t)$.

49–52 ■ Find parametric equations for the tangent line to the curve with the given parametric equations at the specified point.

49. $x = 1 + 2\sqrt{t}$, $\quad y = t^3 - t$, $\quad z = t^3 + t$; $\quad (3, 0, 2)$

50. $x = e^t$, $\quad y = te^t$, $\quad z = te^{t^2}$; $\quad (1, 0, 0)$

51. $x = e^{-t} \cos t$, $\quad y = e^{-t} \sin t$, $\quad z = e^{-t}$; $\quad (1, 0, 1)$

52. $x = \sqrt{t^2 + 3}$, $\quad y = \ln(t^2 + 3)$, $\quad z = t$; $\quad (2, \ln 4, 1)$

53. Find a vector equation for the tangent line to the curve of intersection of the cylinders $x^2 + y^2 = 25$ and $y^2 + z^2 = 20$ at the point $(3, 4, 2)$.

54. Find the point on the curve $\mathbf{r}(t) = \langle 2 \cos t, 2 \sin t, e^t \rangle$, $0 \leqslant t \leqslant \pi$, where the tangent line is parallel to the plane $\sqrt{3}\,x + y = 1$.

CAS **55.** Find parametric equations for the tangent line to the curve $x = t \cos t$, $y = t$, $z = t \sin t$ at the point $(-\pi, \pi, 0)$. Illustrate by graphing both the curve and the tangent line on a common screen.

56. (a) Find the point of intersection of the tangent lines to the curve $\mathbf{r}(t) = \langle \sin \pi t, 2 \sin \pi t, \cos \pi t \rangle$ at the points where $t = 0$ and $t = 0.5$.
(b) Illustrate by graphing the curve and both tangent lines.

57. The curves $\mathbf{r}_1(t) = \langle t, t^2, t^3 \rangle$ and $\mathbf{r}_2(t) = \langle \sin t, \sin 2t, t \rangle$ intersect at the origin. Find their angle of intersection correct to the nearest degree.

58. At what point do the curves $\mathbf{r}_1(t) = \langle t, 1 - t, 3 + t^2 \rangle$ and $\mathbf{r}_2(s) = \langle 3 - s, s - 2, s^2 \rangle$ intersect? Find their angle of intersection correct to the nearest degree.

59–64 ■ Evaluate the integral.

59. $\displaystyle\int_0^2 (t\,\mathbf{i} - t^3\,\mathbf{j} + 3t^5\,\mathbf{k})\,dt$

60. $\displaystyle\int_0^1 \left(\frac{4}{1 + t^2}\,\mathbf{j} + \frac{2t}{1 + t^2}\,\mathbf{k} \right) dt$

61. $\displaystyle\int_0^{\pi/2} (3 \sin^2 t \cos t\,\mathbf{i} + 3 \sin t \cos^2 t\,\mathbf{j} + 2 \sin t \cos t\,\mathbf{k})\,dt$

62. $\displaystyle\int_1^2 \left(t^2\,\mathbf{i} + t\sqrt{t - 1}\,\mathbf{j} + t \sin \pi t\,\mathbf{k} \right) dt$

63. $\displaystyle\int (\sec^2 t\,\mathbf{i} + t(t^2 + 1)^3\,\mathbf{j} + t^2 \ln t\,\mathbf{k})\,dt$

64. $\displaystyle\int \left(te^{2t}\,\mathbf{i} + \frac{t}{1 - t}\,\mathbf{j} + \frac{1}{\sqrt{1 - t^2}}\,\mathbf{k} \right) dt$

65. Find $\mathbf{r}(t)$ if $\mathbf{r}'(t) = 2t\,\mathbf{i} + 3t^2\,\mathbf{j} + \sqrt{t}\,\mathbf{k}$ and $\mathbf{r}(1) = \mathbf{i} + \mathbf{j}$.

66. Find $\mathbf{r}(t)$ if $\mathbf{r}'(t) = t\,\mathbf{i} + e^t\,\mathbf{j} + te^t\,\mathbf{k}$ and $\mathbf{r}(0) = \mathbf{i} + \mathbf{j} + \mathbf{k}$.

67. If two objects travel through space along two different curves, it's often important to know whether they will collide. (Will a missile hit its moving target? Will two aircraft collide?) The curves might intersect, but we need to know whether the objects are in the same position *at the same time*. Suppose the trajectories of two particles are given by the vector functions

$$\mathbf{r}_1(t) = \langle t^2, 7t - 12, t^2 \rangle \qquad \mathbf{r}_2(t) = \langle 4t - 3, t^2, 5t - 6 \rangle$$

for $t \geqslant 0$. Do the particles collide?

68. Two particles travel along the space curves

$$\mathbf{r}_1(t) = \langle t, t^2, t^3 \rangle \qquad \mathbf{r}_2(t) = \langle 1 + 2t, 1 + 6t, 1 + 14t \rangle$$

Do the particles collide? Do their paths intersect?

69. Suppose $\mathbf{u}$ and $\mathbf{v}$ are vector functions that possess limits as $t \to a$ and let c be a constant. Prove the following properties of limits.

(a) $\lim_{t \to a} [\mathbf{u}(t) + \mathbf{v}(t)] = \lim_{t \to a} \mathbf{u}(t) + \lim_{t \to a} \mathbf{v}(t)$

(b) $\lim_{t \to a} c\mathbf{u}(t) = c \lim_{t \to a} \mathbf{u}(t)$

(c) $\lim_{t \to a} [\mathbf{u}(t) \cdot \mathbf{v}(t)] = \lim_{t \to a} \mathbf{u}(t) \cdot \lim_{t \to a} \mathbf{v}(t)$

(d) $\lim_{t \to a} [\mathbf{u}(t) \times \mathbf{v}(t)] = \lim_{t \to a} \mathbf{u}(t) \times \lim_{t \to a} \mathbf{v}(t)$

70. Show that $\lim_{t \to a} \mathbf{r}(t) = \mathbf{b}$ if and only if for every $\varepsilon > 0$ there is a number $\delta > 0$ such that $|\mathbf{r}(t) - \mathbf{b}| < \varepsilon$ whenever $0 < |t - a| < \delta$.

71. Prove Formula 1 of Theorem 5.

72. Prove Formula 3 of Theorem 5.

73. Prove Formula 5 of Theorem 5.

74. Prove Formula 6 of Theorem 5.

75. If $\mathbf{u}(t) = \langle \sin t, \cos t, t \rangle$ and $\mathbf{v}(t) = \langle t, \cos t, \sin t \rangle$, use Formula 4 of Theorem 5 to find

$$\frac{d}{dt} [\mathbf{u}(t) \cdot \mathbf{v}(t)]$$

76. If $\mathbf{u}$ and $\mathbf{v}$ are the vector functions in Exercise 75, use Formula 5 of Theorem 5 to find

$$\frac{d}{dt} [\mathbf{u}(t) \times \mathbf{v}(t)]$$

77. Find $f'(2)$, where $f(t) = \mathbf{u}(t) \cdot \mathbf{v}(t)$, $\mathbf{u}(2) = \langle 1, 2, -1 \rangle$, $\mathbf{u}'(2) = \langle 3, 0, 4 \rangle$, and $\mathbf{v}(t) = \langle t, t^2, t^3 \rangle$.

78. If $\mathbf{r}(t) = \mathbf{u}(t) \times \mathbf{v}(t)$, where $\mathbf{u}$ and $\mathbf{v}$ are the vector functions in Exercise 77, find $\mathbf{r}'(2)$.

79. Show that if $\mathbf{r}$ is a vector function such that $\mathbf{r}''$ exists, then

$$\frac{d}{dt} [\mathbf{r}(t) \times \mathbf{r}'(t)] = \mathbf{r}(t) \times \mathbf{r}''(t)$$

80. Find an expression for $\dfrac{d}{dt} [\mathbf{u}(t) \cdot (\mathbf{v}(t) \times \mathbf{w}(t))]$.

81. If $\mathbf{r}(t) \neq \mathbf{0}$, show that $\dfrac{d}{dt} |\mathbf{r}(t)| = \dfrac{1}{|\mathbf{r}(t)|} \mathbf{r}(t) \cdot \mathbf{r}'(t)$.

$\left[\textit{Hint:} \; |\mathbf{r}(t)|^2 = \mathbf{r}(t) \cdot \mathbf{r}(t) \right]$

82. If a curve has the property that the position vector $\mathbf{r}(t)$ is always perpendicular to the tangent vector $\mathbf{r}'(t)$, show that the curve lies on a sphere with center the origin.

83. If $\mathbf{u}(t) = \mathbf{r}(t) \cdot [\mathbf{r}'(t) \times \mathbf{r}''(t)]$, show that

$$\mathbf{u}'(t) = \mathbf{r}(t) \cdot [\mathbf{r}'(t) \times \mathbf{r}'''(t)]$$

10.8 | ARC LENGTH AND CURVATURE

In Section 9.2 we defined the length of a plane curve with parametric equations $x = f(t)$, $y = g(t)$, $a \leqslant t \leqslant b$, as the limit of lengths of inscribed polygons and, for the case where f' and g' are continuous, we arrived at the formula

$$\boxed{1} \qquad L = \int_a^b \sqrt{[f'(t)]^2 + [g'(t)]^2} \, dt$$

$$= \int_a^b \sqrt{\left(\frac{dx}{dt}\right)^2 + \left(\frac{dy}{dt}\right)^2} \, dt$$

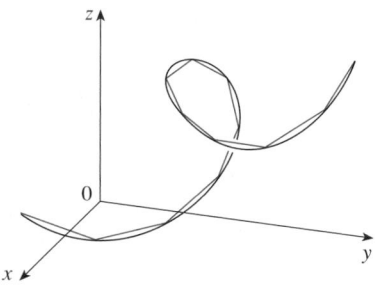

FIGURE 1
The length of a space curve is the limit
of lengths of inscribed polygons.

The length of a space curve is defined in exactly the same way (see Figure 1). Suppose that the curve has the vector equation $\mathbf{r}(t) = \langle f(t), g(t), h(t) \rangle$, $a \leq t \leq b$, or, equivalently, the parametric equations $x = f(t)$, $y = g(t)$, $z = h(t)$, where f', g', and h' are continuous. If the curve is traversed exactly once as t increases from a to b, then it can be shown that its length is

$$\boxed{2} \qquad \begin{aligned} L &= \int_a^b \sqrt{[f'(t)]^2 + [g'(t)]^2 + [h'(t)]^2}\, dt \\ &= \int_a^b \sqrt{\left(\frac{dx}{dt}\right)^2 + \left(\frac{dy}{dt}\right)^2 + \left(\frac{dz}{dt}\right)^2}\, dt \end{aligned}$$

Notice that both of the arc length formulas $\boxed{1}$ and $\boxed{2}$ can be put into the more compact form

$$\boxed{3} \qquad L = \int_a^b |\mathbf{r}'(t)|\, dt$$

because, for plane curves $\mathbf{r}(t) = f(t)\,\mathbf{i} + g(t)\,\mathbf{j}$,

$$|\mathbf{r}'(t)| = |f'(t)\,\mathbf{i} + g'(t)\,\mathbf{j}| = \sqrt{[f'(t)]^2 + [g'(t)]^2}$$

whereas, for space curves $\mathbf{r}(t) = f(t)\,\mathbf{i} + g(t)\,\mathbf{j} + h(t)\,\mathbf{k}$,

$$|\mathbf{r}'(t)| = |f'(t)\,\mathbf{i} + g'(t)\,\mathbf{j} + h'(t)\,\mathbf{k}| = \sqrt{[f'(t)]^2 + [g'(t)]^2 + [h'(t)]^2}$$

■ Figure 2 shows the arc of the helix whose length is computed in Example 1.

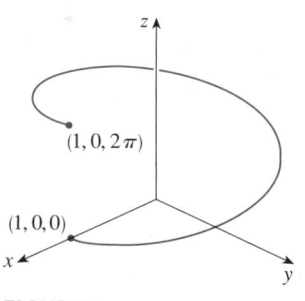

FIGURE 2

✔ EXAMPLE 1 Find the length of the arc of the circular helix with vector equation $\mathbf{r}(t) = \cos t\,\mathbf{i} + \sin t\,\mathbf{j} + t\,\mathbf{k}$ from the point $(1, 0, 0)$ to the point $(1, 0, 2\pi)$.

SOLUTION Since $\mathbf{r}'(t) = -\sin t\,\mathbf{i} + \cos t\,\mathbf{j} + \mathbf{k}$, we have

$$|\mathbf{r}'(t)| = \sqrt{(-\sin t)^2 + \cos^2 t + 1} = \sqrt{2}$$

The arc from $(1, 0, 0)$ to $(1, 0, 2\pi)$ is described by the parameter interval $0 \leq t \leq 2\pi$ and so, from Formula 3, we have

$$L = \int_0^{2\pi} |\mathbf{r}'(t)|\, dt = \int_0^{2\pi} \sqrt{2}\, dt = 2\sqrt{2}\, \pi \qquad ■$$

A single curve C can be represented by more than one vector function. For instance, the twisted cubic

$$\boxed{4} \qquad \mathbf{r}_1(t) = \langle t, t^2, t^3 \rangle \qquad 1 \leq t \leq 2$$

could also be represented by the function

$$\boxed{5} \qquad \mathbf{r}_2(u) = \langle e^u, e^{2u}, e^{3u} \rangle \qquad 0 \leq u \leq \ln 2$$

where the connection between the parameters t and u is given by $t = e^u$. We say that Equations 4 and 5 are **parametrizations** of the curve C. If we were to use Equation 3 to compute the length of C using Equations 4 and 5, we would get the same answer.

In general, it can be shown that when Equation 3 is used to compute arc length, the answer is independent of the parametrization that is used.

Now we suppose that C is a curve given by a vector function

$$\mathbf{r}(t) = f(t)\mathbf{i} + g(t)\mathbf{j} + h(t)\mathbf{k} \qquad a \le t \le b$$

where $\mathbf{r}'$ is continuous and C is traversed exactly once as t increases from a to b. We define its **arc length function** s by

$$\boxed{6} \qquad s(t) = \int_a^t |\mathbf{r}'(u)|\, du = \int_a^t \sqrt{\left(\frac{dx}{du}\right)^2 + \left(\frac{dy}{du}\right)^2 + \left(\frac{dz}{du}\right)^2}\, du$$

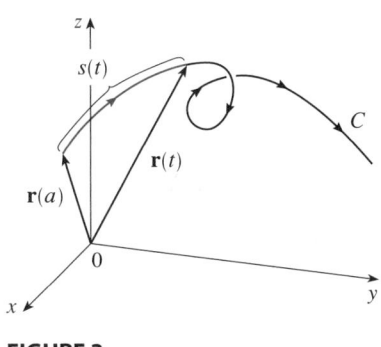

FIGURE 3

Thus $s(t)$ is the length of the part of C between $\mathbf{r}(a)$ and $\mathbf{r}(t)$. (See Figure 3.) If we differentiate both sides of Equation 6 using Part 1 of the Fundamental Theorem of Calculus, we obtain

$$\boxed{7} \qquad \frac{ds}{dt} = |\mathbf{r}'(t)|$$

It is often useful to **parametrize a curve with respect to arc length** because arc length arises naturally from the shape of the curve and does not depend on a particular coordinate system. If a curve $\mathbf{r}(t)$ is already given in terms of a parameter t and $s(t)$ is the arc length function given by Equation 6, then we may be able to solve for t as a function of s: $t = t(s)$. Then the curve can be reparametrized in terms of s by substituting for t: $\mathbf{r} = \mathbf{r}(t(s))$. Thus if $s = 3$ for instance, $\mathbf{r}(t(3))$ is the position vector of the point 3 units of length along the curve from its starting point.

EXAMPLE 2 Reparametrize the helix $\mathbf{r}(t) = \cos t\, \mathbf{i} + \sin t\, \mathbf{j} + t\, \mathbf{k}$ with respect to arc length measured from $(1, 0, 0)$ in the direction of increasing t.

SOLUTION The initial point $(1, 0, 0)$ corresponds to the parameter value $t = 0$. From Example 1 we have

$$\frac{ds}{dt} = |\mathbf{r}'(t)| = \sqrt{2}$$

and so

$$s = s(t) = \int_0^t |\mathbf{r}'(u)|\, du = \int_0^t \sqrt{2}\, du = \sqrt{2}\, t$$

Therefore $t = s/\sqrt{2}$ and the required reparametrization is obtained by substituting for t:

$$\mathbf{r}(t(s)) = \cos(s/\sqrt{2})\, \mathbf{i} + \sin(s/\sqrt{2})\, \mathbf{j} + (s/\sqrt{2})\, \mathbf{k} \qquad ■$$

CURVATURE

A parametrization $\mathbf{r}(t)$ is called **smooth** on an interval I if $\mathbf{r}'$ is continuous and $\mathbf{r}'(t) \ne \mathbf{0}$ on I. A curve is called **smooth** if it has a smooth parametrization. A smooth curve has no sharp corners or cusps; when the tangent vector turns, it does so continuously.

If C is a smooth curve defined by the vector function $\mathbf{r}$, recall that the unit tangent vector $\mathbf{T}(t)$ is given by

$$\mathbf{T}(t) = \frac{\mathbf{r}'(t)}{|\mathbf{r}'(t)|}$$

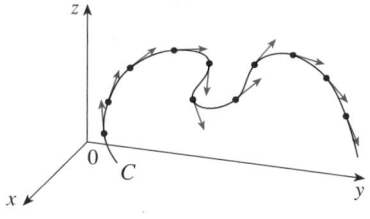

FIGURE 4

Unit tangent vectors at equally spaced points on C

TEC Visual 10.8A shows animated unit tangent vectors, like those in Figure 4, for a variety of plane curves and space curves.

and indicates the direction of the curve. From Figure 4 you can see that $\mathbf{T}(t)$ changes direction very slowly when C is fairly straight, but it changes direction more quickly when C bends or twists more sharply.

The curvature of C at a given point is a measure of how quickly the curve changes direction at that point. Specifically, we define it to be the magnitude of the rate of change of the unit tangent vector with respect to arc length. (We use arc length so that the curvature will be independent of the parametrization.)

> **8** **DEFINITION** The **curvature** of a curve is
>
> $$\kappa = \left| \frac{d\mathbf{T}}{ds} \right|$$
>
> where $\mathbf{T}$ is the unit tangent vector.

The curvature is easier to compute if it is expressed in terms of the parameter t instead of s, so we use the Chain Rule (Theorem 10.7.5, Formula 6) to write

$$\frac{d\mathbf{T}}{dt} = \frac{d\mathbf{T}}{ds}\frac{ds}{dt} \qquad \text{and} \qquad \kappa = \left| \frac{d\mathbf{T}}{ds} \right| = \left| \frac{d\mathbf{T}/dt}{ds/dt} \right|$$

But $ds/dt = |\mathbf{r}'(t)|$ from Equation 7, so

$$\boxed{\; \kappa(t) = \frac{|\mathbf{T}'(t)|}{|\mathbf{r}'(t)|} \;} \qquad \textbf{9}$$

V EXAMPLE 3 Show that the curvature of a circle of radius a is $1/a$.

SOLUTION We can take the circle to have center the origin, and then a parametrization is

$$\mathbf{r}(t) = a \cos t \, \mathbf{i} + a \sin t \, \mathbf{j}$$

Therefore

$$\mathbf{r}'(t) = -a \sin t \, \mathbf{i} + a \cos t \, \mathbf{j} \qquad \text{and} \qquad |\mathbf{r}'(t)| = a$$

so

$$\mathbf{T}(t) = \frac{\mathbf{r}'(t)}{|\mathbf{r}'(t)|} = -\sin t \, \mathbf{i} + \cos t \, \mathbf{j}$$

and

$$\mathbf{T}'(t) = -\cos t \, \mathbf{i} - \sin t \, \mathbf{j}$$

This gives $|\mathbf{T}'(t)| = 1$, so using Equation 9, we have

$$\kappa(t) = \frac{|\mathbf{T}'(t)|}{|\mathbf{r}'(t)|} = \frac{1}{a}$$

The result of Example 3 shows that small circles have large curvature and large circles have small curvature, in accordance with our intuition. We can see directly from the definition of curvature that the curvature of a straight line is always 0 because the tangent vector is constant.

Although Formula 9 can be used in all cases to compute the curvature, the formula given by the following theorem is often more convenient to apply.

---

10 THEOREM The curvature of the curve given by the vector function **r** is

$$\kappa(t) = \frac{|\mathbf{r}'(t) \times \mathbf{r}''(t)|}{|\mathbf{r}'(t)|^3}$$

PROOF Since $\mathbf{T} = \mathbf{r}'/|\mathbf{r}'|$ and $|\mathbf{r}'| = ds/dt$, we have

$$\mathbf{r}' = |\mathbf{r}'|\mathbf{T} = \frac{ds}{dt}\mathbf{T}$$

so the Product Rule (Theorem 10.7.5, Formula 3) gives

$$\mathbf{r}'' = \frac{d^2s}{dt^2}\mathbf{T} + \frac{ds}{dt}\mathbf{T}'$$

Using the fact that $\mathbf{T} \times \mathbf{T} = \mathbf{0}$ (see Example 2 in Section 10.4), we have

$$\mathbf{r}' \times \mathbf{r}'' = \left(\frac{ds}{dt}\right)^2 (\mathbf{T} \times \mathbf{T}')$$

Now $|\mathbf{T}(t)| = 1$ for all t, so **T** and **T**′ are orthogonal by Example 11 in Section 10.7. Therefore, by Theorem 10.4.6,

$$|\mathbf{r}' \times \mathbf{r}''| = \left(\frac{ds}{dt}\right)^2 |\mathbf{T} \times \mathbf{T}'| = \left(\frac{ds}{dt}\right)^2 |\mathbf{T}||\mathbf{T}'| = \left(\frac{ds}{dt}\right)^2 |\mathbf{T}'|$$

Thus

$$|\mathbf{T}'| = \frac{|\mathbf{r}' \times \mathbf{r}''|}{(ds/dt)^2} = \frac{|\mathbf{r}' \times \mathbf{r}''|}{|\mathbf{r}'|^2}$$

and

$$\kappa = \frac{|\mathbf{T}'|}{|\mathbf{r}'|} = \frac{|\mathbf{r}' \times \mathbf{r}''|}{|\mathbf{r}'|^3}$$

$\square$

EXAMPLE 4 Find the curvature of the twisted cubic $\mathbf{r}(t) = \langle t, t^2, t^3 \rangle$ at a general point and at $(0, 0, 0)$.

SOLUTION We first compute the required ingredients:

$$\mathbf{r}'(t) = \langle 1, 2t, 3t^2 \rangle \qquad \mathbf{r}''(t) = \langle 0, 2, 6t \rangle$$

$$|\mathbf{r}'(t)| = \sqrt{1 + 4t^2 + 9t^4}$$

$$\mathbf{r}'(t) \times \mathbf{r}''(t) = \begin{vmatrix} \mathbf{i} & \mathbf{j} & \mathbf{k} \\ 1 & 2t & 3t^2 \\ 0 & 2 & 6t \end{vmatrix} = 6t^2\mathbf{i} - 6t\mathbf{j} + 2\mathbf{k}$$

$$|\mathbf{r}'(t) \times \mathbf{r}''(t)| = \sqrt{36t^4 + 36t^2 + 4} = 2\sqrt{9t^4 + 9t^2 + 1}$$

Theorem 10 then gives

$$\kappa(t) = \frac{|\mathbf{r}'(t) \times \mathbf{r}''(t)|}{|\mathbf{r}'(t)|^3} = \frac{2\sqrt{1 + 9t^2 + 9t^4}}{(1 + 4t^2 + 9t^4)^{3/2}}$$

At the origin, where $t = 0$, the curvature is $\kappa(0) = 2$.   ■

For the special case of a plane curve with equation $y = f(x)$, we choose x as the parameter and write $\mathbf{r}(x) = x\,\mathbf{i} + f(x)\,\mathbf{j}$. Then $\mathbf{r}'(x) = \mathbf{i} + f'(x)\,\mathbf{j}$ and $\mathbf{r}''(x) = f''(x)\,\mathbf{j}$. Since $\mathbf{i} \times \mathbf{j} = \mathbf{k}$ and $\mathbf{j} \times \mathbf{j} = \mathbf{0}$, we have $\mathbf{r}'(x) \times \mathbf{r}''(x) = f''(x)\,\mathbf{k}$. We also have $|\mathbf{r}'(x)| = \sqrt{1 + [f'(x)]^2}$ and so, by Theorem 10,

$$\boxed{11} \qquad \boxed{\kappa(x) = \frac{|f''(x)|}{[1 + (f'(x))^2]^{3/2}}}$$

EXAMPLE 5 Find the curvature of the parabola $y = x^2$ at the points $(0, 0)$, $(1, 1)$, and $(2, 4)$.

SOLUTION Since $y' = 2x$ and $y'' = 2$, Formula 11 gives

$$\kappa(x) = \frac{|y''|}{[1 + (y')^2]^{3/2}} = \frac{2}{(1 + 4x^2)^{3/2}}$$

The curvature at $(0, 0)$ is $\kappa(0) = 2$. At $(1, 1)$ it is $\kappa(1) = 2/5^{3/2} \approx 0.18$. At $(2, 4)$ it is $\kappa(2) = 2/17^{3/2} \approx 0.03$. Observe from the expression for $\kappa(x)$ or the graph of κ in Figure 5 that $\kappa(x) \to 0$ as $x \to \pm\infty$. This corresponds to the fact that the parabola appears to become flatter as $x \to \pm\infty$.   ■

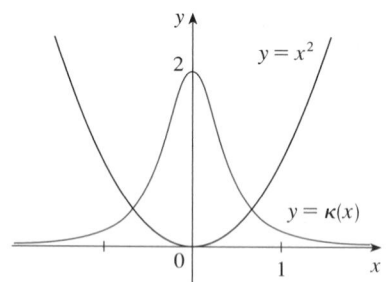

FIGURE 5

The parabola $y = x^2$ and its curvature function $y = \kappa(x)$

THE NORMAL AND BINORMAL VECTORS

At a given point on a smooth space curve $\mathbf{r}(t)$, there are many vectors that are orthogonal to the unit tangent vector $\mathbf{T}(t)$. We single out one by observing that, because $|\mathbf{T}(t)| = 1$ for all t, we have $\mathbf{T}(t) \cdot \mathbf{T}'(t) = 0$ by Example 11 in Section 10.7, so $\mathbf{T}'(t)$ is orthogonal to $\mathbf{T}(t)$. Note that $\mathbf{T}'(t)$ is itself not a unit vector. But at any point where $\kappa \neq 0$ we can define the **principal unit normal vector** $\mathbf{N}(t)$ (or simply **unit normal**) as

$$\mathbf{N}(t) = \frac{\mathbf{T}'(t)}{|\mathbf{T}'(t)|}$$

■ We can think of the normal vector as indicating the direction in which the curve is turning at each point.

The vector $\mathbf{B}(t) = \mathbf{T}(t) \times \mathbf{N}(t)$ is called the **binormal vector**. It is perpendicular to both $\mathbf{T}$ and $\mathbf{N}$ and is also a unit vector. (See Figure 6.)

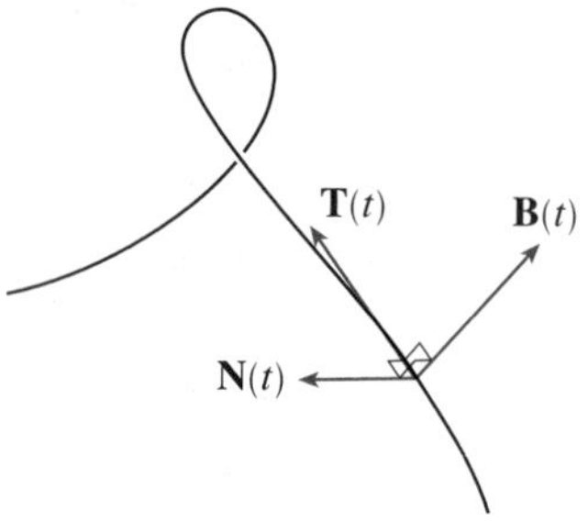

EXAMPLE 6 Find the unit normal and binormal vectors for the circular helix

$$\mathbf{r}(t) = \cos t\,\mathbf{i} + \sin t\,\mathbf{j} + t\,\mathbf{k}$$

FIGURE 6

■ Figure 7 illustrates Example 6 by showing the vectors **T**, **N**, and **B** at two locations on the helix. In general, the vectors **T**, **N**, and **B**, starting at the various points on a curve, form a set of orthogonal vectors, called the **TNB** frame, that moves along the curve as t varies. This **TNB** frame plays an important role in the branch of mathematics known as differential geometry and in its applications to the motion of spacecraft.

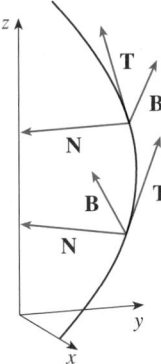

FIGURE 7

TEC Visual 10.8B shows how the TNB frame moves along several curves.

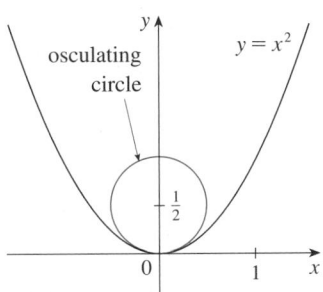

FIGURE 8

TEC Visual 10.8C shows how the osculating circle changes as a point moves along a curve.

■ www.stewartcalculus.com
See Additional Examples A, B.

SOLUTION We first compute the ingredients needed for the unit normal vector:

$$\mathbf{r}'(t) = -\sin t \, \mathbf{i} + \cos t \, \mathbf{j} + \mathbf{k}$$

$$|\mathbf{r}'(t)| = \sqrt{2}$$

$$\mathbf{T}(t) = \frac{\mathbf{r}'(t)}{|\mathbf{r}'(t)|} = \frac{1}{\sqrt{2}}(-\sin t \, \mathbf{i} + \cos t \, \mathbf{j} + \mathbf{k})$$

$$\mathbf{T}'(t) = \frac{1}{\sqrt{2}}(-\cos t \, \mathbf{i} - \sin t \, \mathbf{j}) \qquad |\mathbf{T}'(t)| = \frac{1}{\sqrt{2}}$$

$$\mathbf{N}(t) = \frac{\mathbf{T}'(t)}{|\mathbf{T}'(t)|} = -\cos t \, \mathbf{i} - \sin t \, \mathbf{j} = \langle -\cos t, -\sin t, 0 \rangle$$

This shows that the normal vector at a point on the helix is horizontal and points toward the z-axis. The binormal vector is

$$\mathbf{B}(t) = \mathbf{T}(t) \times \mathbf{N}(t) = \frac{1}{\sqrt{2}} \begin{bmatrix} \mathbf{i} & \mathbf{j} & \mathbf{k} \\ -\sin t & \cos t & 1 \\ -\cos t & -\sin t & 0 \end{bmatrix}$$

$$= \frac{1}{\sqrt{2}} \langle \sin t, -\cos t, 1 \rangle \qquad \blacksquare$$

The plane determined by the normal and binormal vectors **N** and **B** at a point P on a curve C is called the **normal plane** of C at P. It consists of all lines that are orthogonal to the tangent vector **T**. The plane determined by the vectors **T** and **N** is called the **osculating plane** of C at P. The name comes from the Latin *osculum*, meaning "kiss." It is the plane that comes closest to containing the part of the curve near P. (For a plane curve, the osculating plane is simply the plane that contains the curve.)

The circle that lies in the osculating plane of C at P, has the same tangent as C at P, lies on the concave side of C (toward which **N** points), and has radius $\rho = 1/\kappa$ (the reciprocal of the curvature) is called the **osculating circle** (or the **circle of curvature**) of C at P. It is the circle that best describes how C behaves near P; it shares the same tangent, normal, and curvature at P. (See Figure 8.)

We summarize here the formulas for unit tangent, unit normal and binormal vectors, and curvature.

$$\mathbf{T}(t) = \frac{\mathbf{r}'(t)}{|\mathbf{r}'(t)|} \qquad \mathbf{N}(t) = \frac{\mathbf{T}'(t)}{|\mathbf{T}'(t)|} \qquad \mathbf{B}(t) = \mathbf{T}(t) \times \mathbf{N}(t)$$

$$\kappa = \left| \frac{d\mathbf{T}}{ds} \right| = \frac{|\mathbf{T}'(t)|}{|\mathbf{r}'(t)|} = \frac{|\mathbf{r}'(t) \times \mathbf{r}''(t)|}{|\mathbf{r}'(t)|^3}$$

10.8 | EXERCISES

1–4 ■ Find the length of the curve.

1. $\mathbf{r}(t) = \langle t, 3\cos t, 3\sin t \rangle, \quad -5 \leq t \leq 5$

2. $\mathbf{r}(t) = \cos t\,\mathbf{i} + \sin t\,\mathbf{j} + \ln\cos t\,\mathbf{k}, \quad 0 \leq t \leq \pi/4$

3. $\mathbf{r}(t) = \mathbf{i} + t^2\,\mathbf{j} + t^3\,\mathbf{k}, \quad 0 \leq t \leq 1$

4. $\mathbf{r}(t) = 12t\,\mathbf{i} + 8t^{3/2}\,\mathbf{j} + 3t^2\,\mathbf{k}, \quad 0 \leq t \leq 1$

5–6 ■ Find the length of the curve correct to four decimal places. (Use your calculator to approximate the integral.)

5. $\mathbf{r}(t) = \langle t^2, t^3, t^4 \rangle, \quad 0 \leq t \leq 2$

6. $\mathbf{r}(t) = \langle t, e^{-t}, te^{-t} \rangle, \quad 1 \leq t \leq 3$

7. Let C be the curve of intersection of the parabolic cylinder $x^2 = 2y$ and the surface $3z = xy$. Find the exact length of C from the origin to the point $(6, 18, 36)$.

8. Graph the curve with parametric equations $x = \cos t$, $y = \sin 3t$, $z = \sin t$. Find the total length of this curve correct to four decimal places.

9–10 ■ Reparametrize the curve with respect to arc length measured from the point where $t = 0$ in the direction of increasing t.

9. $\mathbf{r}(t) = 2t\,\mathbf{i} + (1 - 3t)\,\mathbf{j} + (5 + 4t)\,\mathbf{k}$

10. $\mathbf{r}(t) = e^{2t}\cos 2t\,\mathbf{i} + 2\,\mathbf{j} + e^{2t}\sin 2t\,\mathbf{k}$

11. Suppose you start at the point $(0, 0, 3)$ and move 5 units along the curve $x = 3\sin t$, $y = 4t$, $z = 3\cos t$ in the positive direction. Where are you now?

12. Reparametrize the curve

$$\mathbf{r}(t) = \left(\frac{2}{t^2 + 1} - 1\right)\mathbf{i} + \frac{2t}{t^2 + 1}\,\mathbf{j}$$

with respect to arc length measured from the point $(1, 0)$ in the direction of increasing t. Express the reparametrization in its simplest form. What can you conclude about the curve?

13–16 ■
(a) Find the unit tangent and unit normal vectors $\mathbf{T}(t)$ and $\mathbf{N}(t)$.
(b) Use Formula 9 to find the curvature.

13. $\mathbf{r}(t) = \langle t, 3\cos t, 3\sin t \rangle$

14. $\mathbf{r}(t) = \langle t^2, \sin t - t\cos t, \cos t + t\sin t \rangle, \quad t > 0$

15. $\mathbf{r}(t) = \langle \sqrt{2}\,t, e^t, e^{-t} \rangle$

16. $\mathbf{r}(t) = \langle t, \tfrac{1}{2}t^2, t^2 \rangle$

17–19 ■ Use Theorem 10 to find the curvature.

17. $\mathbf{r}(t) = t^3\,\mathbf{j} + t^2\,\mathbf{k}$

18. $\mathbf{r}(t) = t\,\mathbf{i} + t^2\,\mathbf{j} + e^t\,\mathbf{k}$

19. $\mathbf{r}(t) = 3t\,\mathbf{i} + 4\sin t\,\mathbf{j} + 4\cos t\,\mathbf{k}$

20. Find the curvature of $\mathbf{r}(t) = \langle t^2, \ln t, t\ln t \rangle$ at the point $(1, 0, 0)$.

21. Find the curvature of $\mathbf{r}(t) = \langle t, t^2, t^3 \rangle$ at the point $(1, 1, 1)$.

22. Graph the curve with parametric equations $x = \cos t$, $y = \sin t$, $z = \sin 5t$ and find the curvature at the point $(1, 0, 0)$.

23–25 ■ Use Formula 11 to find the curvature.

23. $y = x^4$ **24.** $y = \tan x$ **25.** $y = xe^x$

26–27 ■ At what point does the curve have maximum curvature? What happens to the curvature as $x \to \infty$?

26. $y = \ln x$ **27.** $y = e^x$

28. Find an equation of a parabola that has curvature 4 at the origin.

29. (a) Is the curvature of the curve C shown in the figure greater at P or at Q? Explain.
(b) Estimate the curvature at P and at Q by sketching the osculating circles at those points.

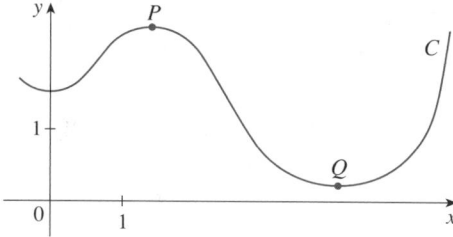

30–31 ■ Use a graphing calculator or computer to graph both the curve and its curvature function $\kappa(x)$ on the same screen. Is the graph of κ what you would expect?

30. $y = x^4 - 2x^2$ **31.** $y = x^{-2}$

CAS **32–33** ▪ Plot the space curve and its curvature function $\kappa(t)$. Comment on how the curvature reflects the shape of the curve.

32. $\mathbf{r}(t) = \langle t - \sin t, 1 - \cos t, 4 \cos(t/2) \rangle, \quad 0 \le t \le 8\pi$

33. $\mathbf{r}(t) = \langle te^t, e^{-t}, \sqrt{2}\,t \rangle, \quad -5 \le t \le 5$

34–35 ▪ Two graphs, a and b, are shown. One is a curve $y = f(x)$ and the other is the graph of its curvature function $y = \kappa(x)$. Identify each curve and explain your choices.

34.   **35.**

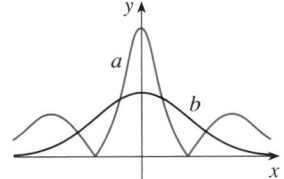

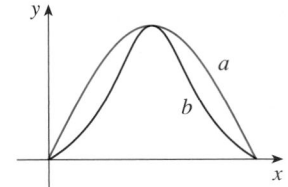

36. Use Theorem 10 to show that the curvature of a plane parametric curve $x = f(t)$, $y = g(t)$ is

$$\kappa = \frac{|\dot{x}\ddot{y} - \dot{y}\ddot{x}|}{[\dot{x}^2 + \dot{y}^2]^{3/2}}$$

where the dots indicate derivatives with respect to t.

37–38 ▪ Use the formula in Exercise 36 to find the curvature.

37. $x = e^t \cos t, \quad y = e^t \sin t$

38. $x = a \cos \omega t, \quad y = b \sin \omega t$

39–40 ▪ Find the vectors $\mathbf{T}$, $\mathbf{N}$, and $\mathbf{B}$ at the given point.

39. $\mathbf{r}(t) = \langle t^2, \frac{2}{3}t^3, t \rangle, \quad \left(1, \frac{2}{3}, 1\right)$

40. $\mathbf{r}(t) = \langle \cos t, \sin t, \ln \cos t \rangle, \quad (1, 0, 0)$

41–42 ▪ Find equations of the normal plane and osculating plane of the curve at the given point.

41. $x = 2 \sin 3t, \quad y = t, \quad z = 2 \cos 3t; \quad (0, \pi, -2)$

42. $x = t, \quad y = t^2, \quad z = t^3; \quad (1, 1, 1)$

43. Find equations of the osculating circles of the ellipse $9x^2 + 4y^2 = 36$ at the points $(2, 0)$ and $(0, 3)$. Use a graphing calculator or computer to graph the ellipse and both osculating circles on the same screen.

44. Find equations of the osculating circles of the parabola $y = \frac{1}{2}x^2$ at the points $(0, 0)$ and $\left(1, \frac{1}{2}\right)$. Graph both osculating circles and the parabola on the same screen.

45. At what point on the curve $x = t^3$, $y = 3t$, $z = t^4$ is the normal plane parallel to the plane $6x + 6y - 8z = 1$?

CAS **46.** Is there a point on the curve in Exercise 45 where the osculating plane is parallel to the plane $x + y + z = 1$? [*Note:* You will need a CAS for differentiating, for simplifying, and for computing a cross product.]

47. Show that the curvature κ is related to the tangent and normal vectors by the equation

$$\frac{d\mathbf{T}}{ds} = \kappa\mathbf{N}$$

48. Show that the curvature of a plane curve is $\kappa = |d\phi/ds|$, where ϕ is the angle between $\mathbf{T}$ and $\mathbf{i}$; that is, ϕ is the angle of inclination of the tangent line.

49. (a) Show that $d\mathbf{B}/ds$ is perpendicular to $\mathbf{B}$.
 (b) Show that $d\mathbf{B}/ds$ is perpendicular to $\mathbf{T}$.
 (c) Deduce from parts (a) and (b) that $d\mathbf{B}/ds = -\tau(s)\mathbf{N}$ for some number $\tau(s)$ called the **torsion** of the curve. (The torsion measures the degree of twisting of a curve.)
 (d) Show that for a plane curve the torsion is $\tau(s) = 0$.

50. The following formulas, called the **Frenet-Serret formulas**, are of fundamental importance in differential geometry:

   1. $d\mathbf{T}/ds = \kappa\mathbf{N}$

   2. $d\mathbf{N}/ds = -\kappa\mathbf{T} + \tau\mathbf{B}$

   3. $d\mathbf{B}/ds = -\tau\mathbf{N}$

(Formula 1 comes from Exercise 47 and Formula 3 comes from Exercise 49.) Use the fact that $\mathbf{N} = \mathbf{B} \times \mathbf{T}$ to deduce Formula 2 from Formulas 1 and 3.

51. Use the Frenet-Serret formulas to prove each of the following. (Primes denote derivatives with respect to t. Start as in the proof of Theorem 10.)
 (a) $\mathbf{r}'' = s''\mathbf{T} + \kappa(s')^2\mathbf{N}$
 (b) $\mathbf{r}' \times \mathbf{r}'' = \kappa(s')^3\mathbf{B}$
 (c) $\mathbf{r}''' = [s''' - \kappa^2(s')^3]\mathbf{T} + [3\kappa s' s'' + \kappa'(s')^2]\mathbf{N} + \kappa\tau(s')^3\mathbf{B}$
 (d) $\tau = \dfrac{(\mathbf{r}' \times \mathbf{r}'') \cdot \mathbf{r}'''}{|\mathbf{r}' \times \mathbf{r}''|^2}$

52. Show that the circular helix $\mathbf{r}(t) = \langle a \cos t, a \sin t, bt \rangle$, where a and b are positive constants, has constant curvature and constant torsion. [Use the result of Exercise 51(d).]

53. The DNA molecule has the shape of a double helix (see Figure 3 on page 582). The radius of each helix is about 10 angstroms (1 Å $= 10^{-8}$ cm). Each helix rises about 34 Å during each complete turn, and there are about 2.9×10^8 complete turns. Estimate the length of each helix.

54. Let's consider the problem of designing a railroad track to make a smooth transition between sections of straight track. Existing track along the negative x-axis is to be joined smoothly to a track along the line $y = 1$ for $x \geq 1$.

(a) Find a polynomial $P = P(x)$ of degree 5 such that the function F defined by

$$F(x) = \begin{cases} 0 & \text{if } x \leq 0 \\ P(x) & \text{if } 0 < x < 1 \\ 1 & \text{if } x \geq 1 \end{cases}$$

is continuous and has continuous slope and continuous curvature.

(b) Use a graphing calculator or computer to draw the graph of F.

10.9 | MOTION IN SPACE: VELOCITY AND ACCELERATION

In this section we show how the ideas of tangent and normal vectors and curvature can be used in physics to study the motion of an object, including its velocity and acceleration, along a space curve. In particular, we follow in the footsteps of Newton by using these methods to derive Kepler's First Law of planetary motion.

Suppose a particle moves through space so that its position vector at time t is $\mathbf{r}(t)$. Notice from Figure 1 that, for small values of h, the vector

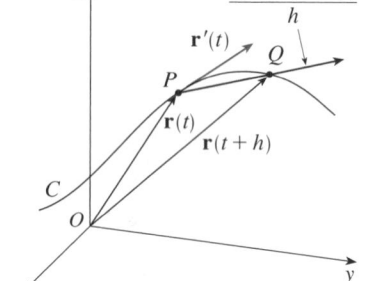

FIGURE 1

$$\boxed{1} \qquad \frac{\mathbf{r}(t + h) - \mathbf{r}(t)}{h}$$

approximates the direction of the particle moving along the curve $\mathbf{r}(t)$. Its magnitude measures the size of the displacement vector per unit time. The vector $\boxed{1}$ gives the average velocity over a time interval of length h and its limit is the **velocity vector** $\mathbf{v}(t)$ at time t:

$$\boxed{2} \qquad \mathbf{v}(t) = \lim_{h \to 0} \frac{\mathbf{r}(t + h) - \mathbf{r}(t)}{h} = \mathbf{r}'(t)$$

Thus the velocity vector is also the tangent vector and points in the direction of the tangent line.

The **speed** of the particle at time t is the magnitude of the velocity vector, that is, $|\mathbf{v}(t)|$. This is appropriate because, from $\boxed{2}$ and from Equation 10.8.7, we have

$$|\mathbf{v}(t)| = |\mathbf{r}'(t)| = \frac{ds}{dt} = \text{rate of change of distance with respect to time}$$

As in the case of one-dimensional motion, the **acceleration** of the particle is defined as the derivative of the velocity:

$$\mathbf{a}(t) = \mathbf{v}'(t) = \mathbf{r}''(t)$$

EXAMPLE 1 The position vector of an object moving in a plane is given by $\mathbf{r}(t) = t^3\,\mathbf{i} + t^2\,\mathbf{j}$. Find its velocity, speed, and acceleration when $t = 1$ and illustrate geometrically.

SOLUTION The velocity and acceleration at time t are

$$\mathbf{v}(t) = \mathbf{r}'(t) = 3t^2\,\mathbf{i} + 2t\,\mathbf{j}$$

$$\mathbf{a}(t) = \mathbf{r}''(t) = 6t\,\mathbf{i} + 2\,\mathbf{j}$$

and the speed is

$$\left|\mathbf{v}(t)\right| = \sqrt{(3t^2)^2 + (2t)^2} = \sqrt{9t^4 + 4t^2}$$

When $t = 1$, we have

$$\mathbf{v}(1) = 3\,\mathbf{i} + 2\,\mathbf{j} \qquad \mathbf{a}(1) = 6\,\mathbf{i} + 2\,\mathbf{j} \qquad \left|\mathbf{v}(1)\right| = \sqrt{13}$$

These velocity and acceleration vectors are shown in Figure 2. ∎

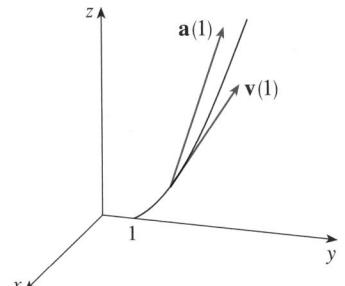

FIGURE 2

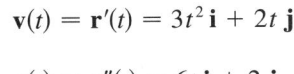 Visual 10.9 shows animated velocity and acceleration vectors for objects moving along various curves.

▪ Figure 3 shows the path of the particle in Example 2 with the velocity and acceleration vectors when $t = 1$.

FIGURE 3

EXAMPLE 2 Find the velocity, acceleration, and speed of a particle with position vector $\mathbf{r}(t) = \langle t^2,\, e^t,\, te^t \rangle$.

SOLUTION

$$\mathbf{v}(t) = \mathbf{r}'(t) = \langle 2t,\, e^t,\, (1 + t)e^t \rangle$$

$$\mathbf{a}(t) = \mathbf{v}'(t) = \langle 2,\, e^t,\, (2 + t)e^t \rangle$$

$$\left|\mathbf{v}(t)\right| = \sqrt{4t^2 + e^{2t} + (1 + t)^2 e^{2t}}$$

∎

The vector integrals that were introduced in Section 10.7 can be used to find position vectors when velocity or acceleration vectors are known, as in the next example.

▼ **EXAMPLE 3** A moving particle starts at an initial position $\mathbf{r}(0) = \langle 1, 0, 0 \rangle$ with initial velocity $\mathbf{v}(0) = \mathbf{i} - \mathbf{j} + \mathbf{k}$. Its acceleration is $\mathbf{a}(t) = 4t\,\mathbf{i} + 6t\,\mathbf{j} + \mathbf{k}$. Find its velocity and position at time t.

SOLUTION Since $\mathbf{a}(t) = \mathbf{v}'(t)$, we have

$$\mathbf{v}(t) = \int \mathbf{a}(t)\,dt = \int (4t\,\mathbf{i} + 6t\,\mathbf{j} + \mathbf{k})\,dt$$

$$= 2t^2\,\mathbf{i} + 3t^2\,\mathbf{j} + t\,\mathbf{k} + \mathbf{C}$$

To determine the value of the constant vector $\mathbf{C}$, we use the fact that $\mathbf{v}(0) = \mathbf{i} - \mathbf{j} + \mathbf{k}$. The preceding equation gives $\mathbf{v}(0) = \mathbf{C}$, so $\mathbf{C} = \mathbf{i} - \mathbf{j} + \mathbf{k}$ and

$$\mathbf{v}(t) = 2t^2\,\mathbf{i} + 3t^2\,\mathbf{j} + t\,\mathbf{k} + \mathbf{i} - \mathbf{j} + \mathbf{k}$$

$$= (2t^2 + 1)\,\mathbf{i} + (3t^2 - 1)\,\mathbf{j} + (t + 1)\,\mathbf{k}$$

■ The expression for $\mathbf{r}(t)$ that we obtained in Example 3 was used to plot the path of the particle in Figure 4 for $0 \leqslant t \leqslant 3$.

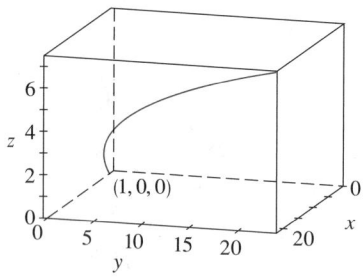

FIGURE 4

Since $\mathbf{v}(t) = \mathbf{r}'(t)$, we have

$$\mathbf{r}(t) = \int \mathbf{v}(t)\, dt$$

$$= \int \left[(2t^2 + 1)\,\mathbf{i} + (3t^2 - 1)\,\mathbf{j} + (t + 1)\,\mathbf{k}\right] dt$$

$$= \left(\tfrac{2}{3}t^3 + t\right)\mathbf{i} + (t^3 - t)\,\mathbf{j} + \left(\tfrac{1}{2}t^2 + t\right)\mathbf{k} + \mathbf{D}$$

Putting $t = 0$, we find that $\mathbf{D} = \mathbf{r}(0) = \mathbf{i}$, so

$$\mathbf{r}(t) = \left(\tfrac{2}{3}t^3 + t + 1\right)\mathbf{i} + (t^3 - t)\,\mathbf{j} + \left(\tfrac{1}{2}t^2 + t\right)\mathbf{k} \qquad ■$$

In general, vector integrals allow us to recover velocity when acceleration is known and position when velocity is known:

$$\mathbf{v}(t) = \mathbf{v}(t_0) + \int_{t_0}^{t} \mathbf{a}(u)\, du \qquad\qquad \mathbf{r}(t) = \mathbf{r}(t_0) + \int_{t_0}^{t} \mathbf{v}(u)\, du$$

If the force that acts on a particle is known, then the acceleration can be found from **Newton's Second Law of Motion**. The vector version of this law states that if, at any time t, a force $\mathbf{F}(t)$ acts on an object of mass m producing an acceleration $\mathbf{a}(t)$, then

$$\mathbf{F}(t) = m\mathbf{a}(t)$$

■ The angular speed of the object moving with position P is $\omega = d\theta/dt$, where θ is the angle shown in Figure 5.

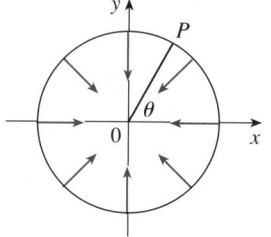

FIGURE 5

EXAMPLE 4 An object with mass m that moves in a circular path with constant angular speed ω has position vector $\mathbf{r}(t) = a \cos \omega t\,\mathbf{i} + a \sin \omega t\,\mathbf{j}$. Find the force acting on the object and show that it is directed toward the origin.

SOLUTION $\qquad \mathbf{v}(t) = \mathbf{r}'(t) = -a\omega \sin \omega t\,\mathbf{i} + a\omega \cos \omega t\,\mathbf{j}$

$$\mathbf{a}(t) = \mathbf{v}'(t) = -a\omega^2 \cos \omega t\,\mathbf{i} - a\omega^2 \sin \omega t\,\mathbf{j}$$

Therefore Newton's Second Law gives the force as

$$\mathbf{F}(t) = m\mathbf{a}(t) = -m\omega^2(a \cos \omega t\,\mathbf{i} + a \sin \omega t\,\mathbf{j})$$

Notice that $\mathbf{F}(t) = -m\omega^2\mathbf{r}(t)$. This shows that the force acts in the direction opposite to the radius vector $\mathbf{r}(t)$ and therefore points toward the origin (see Figure 5). Such a force is called a *centripetal* (center-seeking) force. ■

☑ EXAMPLE 5 A projectile is fired with angle of elevation α and initial velocity $\mathbf{v}_0$. (See Figure 6.) Assuming that air resistance is negligible and the only external force is due to gravity, find the position function $\mathbf{r}(t)$ of the projectile. What value of α maximizes the range (the horizontal distance traveled)?

SOLUTION We set up the axes so that the projectile starts at the origin. Since the force due to gravity acts downward, we have

$$\mathbf{F} = m\mathbf{a} = -mg\,\mathbf{j}$$

where $g = |\mathbf{a}| \approx 9.8 \text{ m/s}^2$. Thus

$$\mathbf{a} = -g\,\mathbf{j}$$

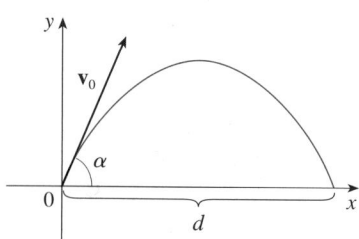

FIGURE 6

Since $\mathbf{v}'(t) = \mathbf{a}$, we have

$$\mathbf{v}(t) = -gt\,\mathbf{j} + \mathbf{C}$$

where $\mathbf{C} = \mathbf{v}(0) = \mathbf{v}_0$. Therefore

$$\mathbf{r}'(t) = \mathbf{v}(t) = -gt\,\mathbf{j} + \mathbf{v}_0$$

Integrating again, we obtain

$$\mathbf{r}(t) = -\tfrac{1}{2}gt^2\,\mathbf{j} + t\,\mathbf{v}_0 + \mathbf{D}$$

But $\mathbf{D} = \mathbf{r}(0) = \mathbf{0}$, so the position vector of the projectile is given by

$$\boxed{3} \qquad\qquad \mathbf{r}(t) = -\tfrac{1}{2}gt^2\,\mathbf{j} + t\,\mathbf{v}_0$$

If we write $|\mathbf{v}_0| = v_0$ (the initial speed of the projectile), then

$$\mathbf{v}_0 = v_0\cos\alpha\,\mathbf{i} + v_0\sin\alpha\,\mathbf{j}$$

and Equation 3 becomes

$$\mathbf{r}(t) = (v_0\cos\alpha)t\,\mathbf{i} + \left[(v_0\sin\alpha)t - \tfrac{1}{2}gt^2\right]\mathbf{j}$$

The parametric equations of the trajectory are therefore

■ If you eliminate t from Equations 4, you will see that y is a quadratic function of x. So the path of the projectile is part of a parabola.

$$\boxed{4} \qquad\qquad x = (v_0\cos\alpha)t \qquad y = (v_0\sin\alpha)t - \tfrac{1}{2}gt^2$$

The horizontal distance d is the value of x when $y = 0$. Setting $y = 0$, we obtain $t = 0$ or $t = (2v_0\sin\alpha)/g$. The latter value of t then gives

$$d = x = (v_0\cos\alpha)\,\frac{2v_0\sin\alpha}{g} = \frac{v_0^2(2\sin\alpha\cos\alpha)}{g} = \frac{v_0^2\sin 2\alpha}{g}$$

Clearly, d has its maximum value when $\sin 2\alpha = 1$, that is, $\alpha = \pi/4$.   ■

V EXAMPLE 6 A projectile is fired with muzzle speed 150 m/s and angle of elevation 45° from a position 10 m above ground level. Where does the projectile hit the ground, and with what speed?

SOLUTION If we place the origin at ground level, then the initial position of the projectile is $(0, 10)$ and so we need to adjust Equations 4 by adding 10 to the expression for y. With $v_0 = 150$ m/s, $\alpha = 45°$, and $g = 9.8$ m/s^2, we have

$$x = 150\cos(\pi/4)t = 75\sqrt{2}\,t$$

$$y = 10 + 150\sin(\pi/4)t - \tfrac{1}{2}(9.8)t^2 = 10 + 75\sqrt{2}\,t - 4.9t^2$$

Impact occurs when $y = 0$, that is, $4.9t^2 - 75\sqrt{2}t - 10 = 0$. Solving this quadratic equation (and using only the positive value of t), we get

$$t = \frac{75\sqrt{2} + \sqrt{11{,}250 + 196}}{9.8} \approx 21.74$$

Then $x \approx 75\sqrt{2}\,(21.74) \approx 2306$, so the projectile hits the ground about 2306 m away.

The velocity of the projectile is

$$\mathbf{v}(t) = \mathbf{r}'(t) = 75\sqrt{2}\,\mathbf{i} + \left(75\sqrt{2} - 9.8t\right)\mathbf{j}$$

So its speed at impact is

$$\left|\mathbf{v}(21.74)\right| = \sqrt{\left(75\sqrt{2}\right)^2 + \left(75\sqrt{2} - 9.8 \cdot 21.74\right)^2} \approx 151 \text{ m/s} \qquad \blacksquare$$

TANGENTIAL AND NORMAL COMPONENTS OF ACCELERATION

When we study the motion of a particle, it is often useful to resolve the acceleration into two components, one in the direction of the tangent and the other in the direction of the normal. If we write $v = |\mathbf{v}|$ for the speed of the particle, then

$$\mathbf{T}(t) = \frac{\mathbf{r}'(t)}{|\mathbf{r}'(t)|} = \frac{\mathbf{v}(t)}{|\mathbf{v}(t)|} = \frac{\mathbf{v}}{v}$$

and so

$$\mathbf{v} = v\mathbf{T}$$

If we differentiate both sides of this equation with respect to t, we get

$$\boxed{5} \qquad \mathbf{a} = \mathbf{v}' = v'\mathbf{T} + v\mathbf{T}'$$

If we use the expression for the curvature given by Equation 10.8.9, then we have

$$\boxed{6} \qquad \kappa = \frac{|\mathbf{T}'|}{|\mathbf{r}'|} = \frac{|\mathbf{T}'|}{v} \qquad \text{so} \qquad |\mathbf{T}'| = \kappa v$$

The unit normal vector was defined in the preceding section as $\mathbf{N} = \mathbf{T}'/|\mathbf{T}'|$, so $\boxed{6}$ gives

$$\mathbf{T}' = |\mathbf{T}'|\mathbf{N} = \kappa v\mathbf{N}$$

and Equation 5 becomes

$$\boxed{7} \qquad \boxed{\mathbf{a} = v'\mathbf{T} + \kappa v^2 \mathbf{N}}$$

Writing a_T and a_N for the tangential and normal components of acceleration, we have

$$\mathbf{a} = a_T\mathbf{T} + a_N\mathbf{N}$$

where

$$\boxed{8} \qquad a_T = v' \qquad \text{and} \qquad a_N = \kappa v^2$$

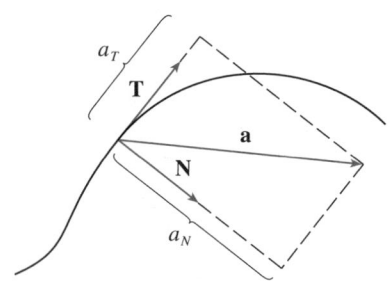

FIGURE 7

This resolution is illustrated in Figure 7.

Let's look at what Formula 7 says. The first thing to notice is that the binormal vector $\mathbf{B}$ is absent. No matter how an object moves through space, its acceleration always lies in the plane of $\mathbf{T}$ and $\mathbf{N}$ (the osculating plane). (Recall that $\mathbf{T}$ gives the direction of motion and $\mathbf{N}$ points in the direction the curve is turning.) Next we notice that the tangential component of acceleration is v', the rate of change of speed, and the normal component of acceleration is κv^2, the curvature times the square of the speed. This makes sense if we think of a passenger in a car—a sharp turn in a road means a large value of the curvature κ, so the component of the acceleration perpendicular to the motion is large and the passenger is thrown against a car door. High speed around the turn has the same effect; in fact, if you double your speed, a_N is increased by a factor of 4.

Although we have expressions for the tangential and normal components of acceleration in Equations 8, it's desirable to have expressions that depend only on $\mathbf{r}$, $\mathbf{r}'$, and $\mathbf{r}''$. To this end we take the dot product of $\mathbf{v} = v\mathbf{T}$ with $\mathbf{a}$ as given by Equation 7:

$$\mathbf{v} \cdot \mathbf{a} = v\mathbf{T} \cdot (v'\mathbf{T} + \kappa v^2\mathbf{N})$$

$$= vv'\mathbf{T} \cdot \mathbf{T} + \kappa v^3\mathbf{T} \cdot \mathbf{N}$$

$$= vv' \qquad \text{(since } \mathbf{T} \cdot \mathbf{T} = 1 \text{ and } \mathbf{T} \cdot \mathbf{N} = 0\text{)}$$

Therefore

$$\boxed{9} \qquad a_T = v' = \frac{\mathbf{v} \cdot \mathbf{a}}{v} = \frac{\mathbf{r}'(t) \cdot \mathbf{r}''(t)}{|\mathbf{r}'(t)|}$$

Using the formula for curvature given by Theorem 10.8.10, we have

$$\boxed{10} \qquad a_N = \kappa v^2 = \frac{|\mathbf{r}'(t) \times \mathbf{r}''(t)|}{|\mathbf{r}'(t)|^3}|\mathbf{r}'(t)|^2 = \frac{|\mathbf{r}'(t) \times \mathbf{r}''(t)|}{|\mathbf{r}'(t)|}$$

EXAMPLE 7 A particle moves with position function $\mathbf{r}(t) = \langle t^2, t^2, t^3 \rangle$. Find the tangential and normal components of acceleration.

SOLUTION

$$\mathbf{r}(t) = t^2\mathbf{i} + t^2\mathbf{j} + t^3\mathbf{k}$$

$$\mathbf{r}'(t) = 2t\mathbf{i} + 2t\mathbf{j} + 3t^2\mathbf{k}$$

$$\mathbf{r}''(t) = 2\mathbf{i} + 2\mathbf{j} + 6t\mathbf{k}$$

$$|\mathbf{r}'(t)| = \sqrt{8t^2 + 9t^4}$$

Therefore Equation 9 gives the tangential component as

$$a_T = \frac{\mathbf{r}'(t) \cdot \mathbf{r}''(t)}{|\mathbf{r}'(t)|} = \frac{8t + 18t^3}{\sqrt{8t^2 + 9t^4}}$$

Since

$$\mathbf{r}'(t) \times \mathbf{r}''(t) = \begin{vmatrix} \mathbf{i} & \mathbf{j} & \mathbf{k} \\ 2t & 2t & 3t^2 \\ 2 & 2 & 6t \end{vmatrix} = 6t^2\mathbf{i} - 6t^2\mathbf{j}$$

Equation 10 gives the normal component as

$$a_N = \frac{|\mathbf{r}'(t) \times \mathbf{r}''(t)|}{|\mathbf{r}'(t)|} = \frac{6\sqrt{2}\,t^2}{\sqrt{8t^2 + 9t^4}}$$

∎

KEPLER'S LAWS OF PLANETARY MOTION

We now describe one of the great accomplishments of calculus by showing how the material of this chapter can be used to prove Kepler's laws of planetary motion. After 20 years of studying the astronomical observations of the Danish astronomer Tycho Brahe, the German mathematician and astronomer Johannes Kepler (1571–1630) formulated the following three laws.

KEPLER'S LAWS

1. A planet revolves around the sun in an elliptical orbit with the sun at one focus.

2. The line joining the sun to a planet sweeps out equal areas in equal times.

3. The square of the period of revolution of a planet is proportional to the cube of the length of the major axis of its orbit.

In his book *Principia Mathematica* of 1687, Sir Isaac Newton was able to show that these three laws are consequences of two of his own laws, the Second Law of Motion and the Law of Universal Gravitation. In what follows we prove Kepler's First Law. The remaining laws are proved as a Web Project (with hints).

Since the gravitational force of the sun on a planet is so much larger than the forces exerted by other celestial bodies, we can safely ignore all bodies in the universe except the sun and one planet revolving about it. We use a coordinate system with the sun at the origin and we let $\mathbf{r} = \mathbf{r}(t)$ be the position vector of the planet. (Equally well, $\mathbf{r}$ could be the position vector of the moon or a satellite moving around the earth or a comet moving around a star.) The velocity vector is $\mathbf{v} = \mathbf{r}'$ and the acceleration vector is $\mathbf{a} = \mathbf{r}''$. We use the following laws of Newton:

$$\text{Second Law of Motion:} \quad \mathbf{F} = m\mathbf{a}$$

$$\text{Law of Gravitation:} \quad \mathbf{F} = -\frac{GMm}{r^3}\,\mathbf{r} = -\frac{GMm}{r^2}\,\mathbf{u}$$

where $\mathbf{F}$ is the gravitational force on the planet, m and M are the masses of the planet and the sun, G is the gravitational constant, $r = |\mathbf{r}|$, and $\mathbf{u} = (1/r)\mathbf{r}$ is the unit vector in the direction of $\mathbf{r}$.

We first show that the planet moves in one plane. By equating the expressions for $\mathbf{F}$ in Newton's two laws, we find that

$$\mathbf{a} = -\frac{GM}{r^3}\,\mathbf{r}$$

and so $\mathbf{a}$ is parallel to $\mathbf{r}$. It follows that $\mathbf{r} \times \mathbf{a} = \mathbf{0}$. We use Formula 5 in Theorem 10.7.5 to write

$$\frac{d}{dt}(\mathbf{r} \times \mathbf{v}) = \mathbf{r}' \times \mathbf{v} + \mathbf{r} \times \mathbf{v}'$$

$$= \mathbf{v} \times \mathbf{v} + \mathbf{r} \times \mathbf{a} = \mathbf{0} + \mathbf{0} = \mathbf{0}$$

Therefore $\qquad\qquad\qquad \mathbf{r} \times \mathbf{v} = \mathbf{h}$

where **h** is a constant vector. (We may assume that $\mathbf{h} \neq \mathbf{0}$; that is, **r** and **v** are not parallel.) This means that the vector $\mathbf{r} = \mathbf{r}(t)$ is perpendicular to **h** for all values of t, so the planet always lies in the plane through the origin perpendicular to **h**. Thus the orbit of the planet is a plane curve.

To prove Kepler's First Law we rewrite the vector **h** as follows:

$$\mathbf{h} = \mathbf{r} \times \mathbf{v} = \mathbf{r} \times \mathbf{r}' = r\mathbf{u} \times (r\mathbf{u})'$$
$$= r\mathbf{u} \times (r\mathbf{u}' + r'\mathbf{u}) = r^2(\mathbf{u} \times \mathbf{u}') + rr'(\mathbf{u} \times \mathbf{u})$$
$$= r^2(\mathbf{u} \times \mathbf{u}')$$

Then

$$\mathbf{a} \times \mathbf{h} = \frac{-GM}{r^2} \mathbf{u} \times (r^2\mathbf{u} \times \mathbf{u}') = -GM\,\mathbf{u} \times (\mathbf{u} \times \mathbf{u}')$$
$$= -GM[(\mathbf{u} \cdot \mathbf{u}')\mathbf{u} - (\mathbf{u} \cdot \mathbf{u})\mathbf{u}'] \quad \text{(by Theorem 10.4.8, Property 6)}$$

But $\mathbf{u} \cdot \mathbf{u} = |\mathbf{u}|^2 = 1$ and, since $|\mathbf{u}(t)| = 1$, it follows from Example 11 in Section 10.7 that $\mathbf{u} \cdot \mathbf{u}' = 0$. Therefore

$$\mathbf{a} \times \mathbf{h} = GM\,\mathbf{u}'$$

and so

$$(\mathbf{v} \times \mathbf{h})' = \mathbf{v}' \times \mathbf{h} = \mathbf{a} \times \mathbf{h} = GM\,\mathbf{u}'$$

Integrating both sides of this equation, we get

$$\boxed{11} \qquad \mathbf{v} \times \mathbf{h} = GM\,\mathbf{u} + \mathbf{c}$$

where **c** is a constant vector.

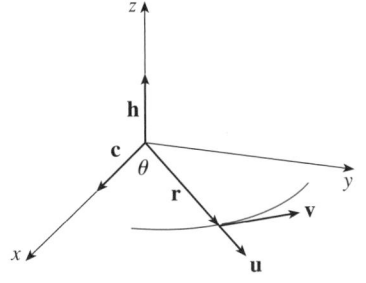

FIGURE 8

At this point it is convenient to choose the coordinate axes so that the standard basis vector **k** points in the direction of the vector **h**. Then the planet moves in the xy-plane. Since both $\mathbf{v} \times \mathbf{h}$ and **u** are perpendicular to **h**, Equation 11 shows that **c** lies in the xy-plane. This means that we can choose the x- and y-axes so that the vector **i** lies in the direction of **c**, as shown in Figure 8.

If θ is the angle between **c** and **r**, then (r, θ) are polar coordinates of the planet. From Equation 11 we have

$$\mathbf{r} \cdot (\mathbf{v} \times \mathbf{h}) = \mathbf{r} \cdot (GM\,\mathbf{u} + \mathbf{c}) = GM\,\mathbf{r} \cdot \mathbf{u} + \mathbf{r} \cdot \mathbf{c}$$
$$= GMr\,\mathbf{u} \cdot \mathbf{u} + |\mathbf{r}||\mathbf{c}|\cos\theta = GMr + rc\cos\theta$$

where $c = |\mathbf{c}|$. Then

$$r = \frac{\mathbf{r} \cdot (\mathbf{v} \times \mathbf{h})}{GM + c\cos\theta} = \frac{1}{GM}\frac{\mathbf{r} \cdot (\mathbf{v} \times \mathbf{h})}{1 + e\cos\theta}$$

where $e = c/(GM)$. But

$$\mathbf{r} \cdot (\mathbf{v} \times \mathbf{h}) = (\mathbf{r} \times \mathbf{v}) \cdot \mathbf{h} = \mathbf{h} \cdot \mathbf{h} = |\mathbf{h}|^2 = h^2$$

where $h = |\mathbf{h}|$. So

$$r = \frac{h^2/(GM)}{1 + e \cos \theta} = \frac{eh^2/c}{1 + e \cos \theta}$$

Writing $d = h^2/c$, we obtain the equation

$$\boxed{12} \qquad r = \frac{ed}{1 + e \cos \theta}$$

Comparing with Theorem 9.5.8, we see that Equation 12 is the polar equation of a conic section with focus at the origin and eccentricity e. We know that the orbit of a planet is a closed curve and so the conic must be an ellipse.

This completes the derivation of Kepler's First Law. A Web Project will guide you through the derivation of the Second and Third Laws. The proofs of these three laws show that the methods of this chapter provide a powerful tool for describing some of the laws of nature.

■ **www.stewartcalculus.com**
Click on *Projects* and select
Applied Project: Kepler's Laws.

10.9 | EXERCISES

1–6 ■ Find the velocity, acceleration, and speed of a particle with the given position function. Sketch the path of the particle and draw the velocity and acceleration vectors for the specified value of t.

1. $\mathbf{r}(t) = \left\langle -\frac{1}{2}t^2, t \right\rangle$, $\quad t = 2$

2. $\mathbf{r}(t) = \left\langle 2 - t, 4\sqrt{t} \right\rangle$, $\quad t = 1$

3. $\mathbf{r}(t) = 3 \cos t \, \mathbf{i} + 2 \sin t \, \mathbf{j}$, $\quad t = \pi/3$

4. $\mathbf{r}(t) = e^t \mathbf{i} + e^{2t} \mathbf{j}$, $\quad t = 0$

5. $\mathbf{r}(t) = t \mathbf{i} + t^2 \mathbf{j} + 2 \mathbf{k}$, $\quad t = 1$

6. $\mathbf{r}(t) = t \mathbf{i} + 2 \cos t \, \mathbf{j} + \sin t \, \mathbf{k}$, $\quad t = 0$

7–10 ■ Find the velocity, acceleration, and speed of a particle with the given position function.

7. $\mathbf{r}(t) = \left\langle t^2 + t, t^2 - t, t^3 \right\rangle$

8. $\mathbf{r}(t) = \left\langle 2 \cos t, 3t, 2 \sin t \right\rangle$

9. $\mathbf{r}(t) = \sqrt{2}\, t \, \mathbf{i} + e^t \mathbf{j} + e^{-t} \mathbf{k}$

10. $\mathbf{r}(t) = t^2 \mathbf{i} + 2t \, \mathbf{j} + \ln t \, \mathbf{k}$

11–12 ■ Find the velocity and position vectors of a particle that has the given acceleration and the given initial velocity and position.

11. $\mathbf{a}(t) = \mathbf{i} + 2 \mathbf{j}$, $\quad \mathbf{v}(0) = \mathbf{k}$, $\quad \mathbf{r}(0) = \mathbf{i}$

12. $\mathbf{a}(t) = 2 \mathbf{i} + 6t \, \mathbf{j} + 12t^2 \mathbf{k}$, $\quad \mathbf{v}(0) = \mathbf{i}$, $\quad \mathbf{r}(0) = \mathbf{j} - \mathbf{k}$

13–14 ■

(a) Find the position vector of a particle that has the given acceleration and the specified initial velocity and position.

(b) Use a computer to graph the path of the particle.

13. $\mathbf{a}(t) = 2t \, \mathbf{i} + \sin t \, \mathbf{j} + \cos 2t \, \mathbf{k}$, $\quad \mathbf{v}(0) = \mathbf{i}$, $\quad \mathbf{r}(0) = \mathbf{j}$

14. $\mathbf{a}(t) = t \, \mathbf{i} + e^t \mathbf{j} + e^{-t} \mathbf{k}$, $\quad \mathbf{v}(0) = \mathbf{k}$, $\quad \mathbf{r}(0) = \mathbf{j} + \mathbf{k}$

15. The position function of a particle is given by $\mathbf{r}(t) = \left\langle t^2, 5t, t^2 - 16t \right\rangle$. When is the speed a minimum?

16. What force is required so that a particle of mass m has the position function $\mathbf{r}(t) = t^3 \mathbf{i} + t^2 \mathbf{j} + t^3 \mathbf{k}$?

17. A force with magnitude 20 N acts directly upward from the xy-plane on an object with mass 4 kg. The object starts at the origin with initial velocity $\mathbf{v}(0) = \mathbf{i} - \mathbf{j}$. Find its position function and its speed at time t.

18. Show that if a particle moves with constant speed, then the velocity and acceleration vectors are orthogonal.

19. A projectile is fired with an initial speed of 200 m/s and angle of elevation 60°. Find (a) the range of the projectile, (b) the maximum height reached, and (c) the speed at impact.

20. Rework Exercise 19 if the projectile is fired from a position 100 m above the ground.

21. A ball is thrown at an angle of 45° to the ground. If the ball lands 90 m away, what was the initial speed of the ball?

22. A gun is fired with angle of elevation 30°. What is the muzzle speed if the maximum height of the shell is 500 m?

23. A gun has muzzle speed 150 m/s. Find two angles of elevation that can be used to hit a target 800 m away.

24. A batter hits a baseball 3 ft above the ground toward the center field fence, which is 10 ft high and 400 ft from home plate. The ball leaves the bat with speed 115 ft/s at an angle 50° above the horizontal. Is it a home run? (In other words, does the ball clear the fence?)

25. A medieval city has the shape of a square and is protected by walls with length 500 m and height 15 m. You are the commander of an attacking army and the closest you can get to the wall is 100 m. Your plan is to set fire to the city by catapulting heated rocks over the wall (with an initial speed of 80 m/s). At what range of angles should you tell your men to set the catapult? (Assume the path of the rocks is perpendicular to the wall.)

26. Show that a projectile reaches three-quarters of its maximum height in half the time needed to reach its maximum height.

27. A ball is thrown eastward into the air from the origin (in the direction of the positive x-axis). The initial velocity is $50\,\mathbf{i} + 80\,\mathbf{k}$, with speed measured in feet per second. The spin of the ball results in a southward acceleration of 4 ft/s^2, so the acceleration vector is $\mathbf{a} = -4\,\mathbf{j} - 32\,\mathbf{k}$. Where does the ball land and with what speed?

28. A ball with mass 0.8 kg is thrown southward into the air with a speed of 30 m/s at an angle of 30° to the ground. A west wind applies a steady force of 4 N to the ball in an easterly direction. Where does the ball land and with what speed?

29. Water traveling along a straight portion of a river normally flows fastest in the middle, and the speed slows to almost zero at the banks. Consider a long straight stretch of river flowing north, with parallel banks 40 m apart. If the maximum water speed is 3 m/s, we can use a quadratic function as a basic model for the rate of water flow x units from the west bank: $f(x) = \frac{3}{400}x(40 - x)$.
(a) A boat proceeds at a constant speed of 5 m/s from a point A on the west bank while maintaining a heading perpendicular to the bank. How far down the river on the opposite bank will the boat touch shore? Graph the path of the boat.
(b) Suppose we would like to pilot the boat to land at the point B on the east bank directly opposite A. If we maintain a constant speed of 5 m/s and a constant heading, find the angle at which the boat should head. Then graph the actual path the boat follows. Does the path seem realistic?

30–33 ▪ Find the tangential and normal components of the acceleration vector.

30. $\mathbf{r}(t) = (1 + t)\,\mathbf{i} + (t^2 - 2t)\,\mathbf{j}$

31. $\mathbf{r}(t) = \cos t\,\mathbf{i} + \sin t\,\mathbf{j} + t\,\mathbf{k}$

32. $\mathbf{r}(t) = t\,\mathbf{i} + \cos^2 t\,\mathbf{j} + \sin^2 t\,\mathbf{k}$

33. $\mathbf{r}(t) = (3t - t^3)\,\mathbf{i} + 3t^2\,\mathbf{j}$

---

34. If a particle with mass m moves with position vector $\mathbf{r}(t)$, then its **angular momentum** is defined as $\mathbf{L}(t) = m\mathbf{r}(t) \times \mathbf{v}(t)$ and its **torque** as $\boldsymbol{\tau}(t) = m\mathbf{r}(t) \times \mathbf{a}(t)$. Show that $\mathbf{L}'(t) = \boldsymbol{\tau}(t)$. Deduce that if $\boldsymbol{\tau}(t) = \mathbf{0}$ for all t, then $\mathbf{L}(t)$ is constant. (This is the *law of conservation of angular momentum*.)

35. The position function of a spaceship is

$$\mathbf{r}(t) = (3 + t)\,\mathbf{i} + (2 + \ln t)\,\mathbf{j} + \left(7 - \frac{4}{t^2 + 1}\right)\mathbf{k}$$

and the coordinates of a space station are (6, 4, 9). The captain wants the spaceship to coast into the space station. When should the engines be turned off?

36. A rocket burning its onboard fuel while moving through space has velocity $\mathbf{v}(t)$ and mass $m(t)$ at time t. If the exhaust gases escape with velocity $\mathbf{v}_e$ relative to the rocket, it can be deduced from Newton's Second Law of Motion that

$$m\frac{d\mathbf{v}}{dt} = \frac{dm}{dt}\mathbf{v}_e$$

(a) Show that $\mathbf{v}(t) = \mathbf{v}(0) - \ln\dfrac{m(0)}{m(t)}\mathbf{v}_e$.
(b) For the rocket to accelerate in a straight line from rest to twice the speed of its own exhaust gases, what fraction of its initial mass would the rocket have to burn as fuel?

CHAPTER 10   REVIEW

1. What is the difference between a vector and a scalar?

2. How do you add two vectors geometrically? How do you add them algebraically?

3. If **a** is a vector and c is a scalar, how is c**a** related to **a** geometrically? How do you find c**a** algebraically?

4. How do you find the vector from one point to another?

5. How do you find the dot product **a** · **b** of two vectors if you know their lengths and the angle between them? What if you know their components?

6. How are dot products useful?

7. Write expressions for the scalar and vector projections of **b** onto **a**. Illustrate with diagrams.

8. How do you find the cross product **a** × **b** of two vectors if you know their lengths and the angle between them? What if you know their components?

9. How are cross products useful?

10. (a) How do you find the area of the parallelogram determined by **a** and **b**?
(b) How do you find the volume of the parallelepiped determined by **a**, **b**, and **c**?

11. How do you find a vector perpendicular to a plane?

12. How do you find the angle between two intersecting planes?

13. Write a vector equation, parametric equations, and symmetric equations for a line.

14. Write a vector equation and a scalar equation for a plane.

15. (a) How do you tell if two vectors are parallel?
(b) How do you tell if two vectors are perpendicular?
(c) How do you tell if two planes are parallel?

16. (a) Describe a method for determining whether three points P, Q, and R lie on the same line.
(b) Describe a method for determining whether four points P, Q, R, and S lie in the same plane.

17. (a) How do you find the distance from a point to a line?
(b) How do you find the distance from a point to a plane?

18. What are the traces of a surface? How do you find them?

19. Write equations in standard form of the six types of quadric surfaces.

20. What is a vector function? How do you find its derivative and its integral?

21. What is the connection between vector functions and space curves?

22. How do you find the tangent vector to a smooth curve at a point? How do you find the tangent line? The unit tangent vector?

23. If **u** and **v** are differentiable vector functions, c is a scalar, and f is a real-valued function, write the rules for differentiating the following vector functions.
(a) $\mathbf{u}(t) + \mathbf{v}(t)$ 　　(b) $c\mathbf{u}(t)$ 　　(c) $f(t)\,\mathbf{u}(t)$
(d) $\mathbf{u}(t) \cdot \mathbf{v}(t)$ 　　(e) $\mathbf{u}(t) \times \mathbf{v}(t)$ 　　(f) $\mathbf{u}(f(t))$

24. How do you find the length of a space curve given by a vector function $\mathbf{r}(t)$?

25. (a) What is the definition of curvature?
(b) Write a formula for curvature in terms of $\mathbf{r}'(t)$ and $\mathbf{T}'(t)$.
(c) Write a formula for curvature in terms of $\mathbf{r}'(t)$ and $\mathbf{r}''(t)$.
(d) Write a formula for the curvature of a plane curve with equation $y = f(x)$.

26. Write formulas for the unit normal and binormal vectors of a smooth space curve $\mathbf{r}(t)$.

27. (a) How do you find the velocity, speed, and acceleration of a particle that moves along a space curve?
(b) Write the acceleration in terms of its tangential and normal components.

28. State Kepler's Laws.

Determine whether the statement is true or false. If it is true, explain why. If it is false, explain why or give an example that disproves the statement.

1. If $\mathbf{u} = \langle u_1, u_2 \rangle$ and $\mathbf{v} = \langle v_1, v_2 \rangle$, then $\mathbf{u} \cdot \mathbf{v} = \langle u_1 v_1, u_2 v_2 \rangle$.

2. For any vectors $\mathbf{u}$ and $\mathbf{v}$ in V_3, $|\mathbf{u} + \mathbf{v}| = |\mathbf{u}| + |\mathbf{v}|$.

3. For any vectors $\mathbf{u}$ and $\mathbf{v}$ in V_3, $|\mathbf{u} \cdot \mathbf{v}| = |\mathbf{u}||\mathbf{v}|$.

4. For any vectors $\mathbf{u}$ and $\mathbf{v}$ in V_3, $|\mathbf{u} \times \mathbf{v}| = |\mathbf{u}||\mathbf{v}|$.

5. For any vectors $\mathbf{u}$ and $\mathbf{v}$ in V_3, $\mathbf{u} \cdot \mathbf{v} = \mathbf{v} \cdot \mathbf{u}$.

6. For any vectors $\mathbf{u}$ and $\mathbf{v}$ in V_3, $\mathbf{u} \times \mathbf{v} = \mathbf{v} \times \mathbf{u}$.

7. For any vectors $\mathbf{u}$ and $\mathbf{v}$ in V_3, $|\mathbf{u} \times \mathbf{v}| = |\mathbf{v} \times \mathbf{u}|$.

8. For any vectors $\mathbf{u}$ and $\mathbf{v}$ in V_3 and any scalar k, $k(\mathbf{u} \cdot \mathbf{v}) = (k\mathbf{u}) \cdot \mathbf{v}$.

9. For any vectors $\mathbf{u}$ and $\mathbf{v}$ in V_3 and any scalar k, $k(\mathbf{u} \times \mathbf{v}) = (k\mathbf{u}) \times \mathbf{v}$.

10. For any vectors $\mathbf{u}$, $\mathbf{v}$, and $\mathbf{w}$ in V_3, $(\mathbf{u} + \mathbf{v}) \times \mathbf{w} = \mathbf{u} \times \mathbf{w} + \mathbf{v} \times \mathbf{w}$.

11. For any vectors $\mathbf{u}$, $\mathbf{v}$, and $\mathbf{w}$ in V_3, $\mathbf{u} \cdot (\mathbf{v} \times \mathbf{w}) = (\mathbf{u} \times \mathbf{v}) \cdot \mathbf{w}$.

12. For any vectors $\mathbf{u}$, $\mathbf{v}$, and $\mathbf{w}$ in V_3, $\mathbf{u} \times (\mathbf{v} \times \mathbf{w}) = (\mathbf{u} \times \mathbf{v}) \times \mathbf{w}$.

13. For any vectors $\mathbf{u}$ and $\mathbf{v}$ in V_3, $(\mathbf{u} \times \mathbf{v}) \cdot \mathbf{u} = 0$.

14. For any vectors $\mathbf{u}$ and $\mathbf{v}$ in V_3, $(\mathbf{u} + \mathbf{v}) \times \mathbf{v} = \mathbf{u} \times \mathbf{v}$.

15. The vector $\langle 3, -1, 2 \rangle$ is parallel to the plane $6x - 2y + 4z = 1$.

16. A linear equation $Ax + By + Cz + D = 0$ represents a line in space.

17. The set of points $\{(x, y, z) \mid x^2 + y^2 = 1\}$ is a circle.

18. In $\mathbb{R}^3$ the graph of $y = x^2$ is a paraboloid.

19. If $\mathbf{u} \cdot \mathbf{v} = 0$, then $\mathbf{u} = \mathbf{0}$ or $\mathbf{v} = \mathbf{0}$.

20. If $\mathbf{u} \times \mathbf{v} = \mathbf{0}$, then $\mathbf{u} = \mathbf{0}$ or $\mathbf{v} = \mathbf{0}$.

21. If $\mathbf{u} \cdot \mathbf{v} = 0$ and $\mathbf{u} \times \mathbf{v} = \mathbf{0}$, then $\mathbf{u} = \mathbf{0}$ or $\mathbf{v} = \mathbf{0}$.

22. If $\mathbf{u}$ and $\mathbf{v}$ are in V_3, then $|\mathbf{u} \cdot \mathbf{v}| \leq |\mathbf{u}||\mathbf{v}|$.

23. The curve with vector equation $\mathbf{r}(t) = t^3\mathbf{i} + 2t^3\mathbf{j} + 3t^3\mathbf{k}$ is a line.

24. The curve $\mathbf{r}(t) = \langle 0, t^2, 4t \rangle$ is a parabola.

25. The curve $\mathbf{r}(t) = \langle 2t, 3 - t, 0 \rangle$ is a line that passes through the origin.

26. The derivative of a vector function is obtained by differentiating each component function.

27. If $\mathbf{u}(t)$ and $\mathbf{v}(t)$ are differentiable vector functions, then
$$\frac{d}{dt}[\mathbf{u}(t) \times \mathbf{v}(t)] = \mathbf{u}'(t) \times \mathbf{v}'(t)$$

28. If $\mathbf{r}(t)$ is a differentiable vector function, then
$$\frac{d}{dt}|\mathbf{r}(t)| = |\mathbf{r}'(t)|$$

29. If $\mathbf{T}(t)$ is the unit tangent vector of a smooth curve, then the curvature is $\kappa = |d\mathbf{T}/dt|$.

30. The binormal vector is $\mathbf{B}(t) = \mathbf{N}(t) \times \mathbf{T}(t)$.

31. Suppose f is twice continuously differentiable. At an inflection point of the curve $y = f(x)$, the curvature is 0.

32. If $\kappa(t) = 0$ for all t, the curve is a straight line.

33. If $|\mathbf{r}(t)| = 1$ for all t, then $|\mathbf{r}'(t)|$ is a constant.

34. If $|\mathbf{r}(t)| = 1$ for all t, then $\mathbf{r}'(t)$ is orthogonal to $\mathbf{r}(t)$ for all t.

EXERCISES

1. (a) Find an equation of the sphere that passes through the point $(6, -2, 3)$ and has center $(-1, 2, 1)$.
   (b) Find the curve in which this sphere intersects the yz-plane.
   (c) Find the center and radius of the sphere

$$x^2 + y^2 + z^2 - 8x + 2y + 6z + 1 = 0$$

2. Copy the vectors in the figure and use them to draw each of the following vectors.
   (a) $\mathbf{a} + \mathbf{b}$ (b) $\mathbf{a} - \mathbf{b}$ (c) $-\frac{1}{2}\mathbf{a}$ (d) $2\mathbf{a} + \mathbf{b}$

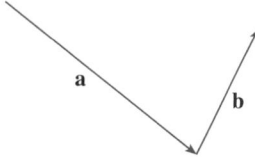

3. If $\mathbf{u}$ and $\mathbf{v}$ are the vectors shown in the figure, find $\mathbf{u} \cdot \mathbf{v}$ and $|\mathbf{u} \times \mathbf{v}|$. Is $\mathbf{u} \times \mathbf{v}$ directed into the page or out of it?

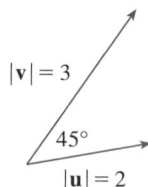

4. Calculate the given quantity if

$$\mathbf{a} = \mathbf{i} + \mathbf{j} - 2\mathbf{k}$$
$$\mathbf{b} = 3\mathbf{i} - 2\mathbf{j} + \mathbf{k}$$
$$\mathbf{c} = \mathbf{j} - 5\mathbf{k}$$

   (a) $2\mathbf{a} + 3\mathbf{b}$ (b) $|\mathbf{b}|$
   (c) $\mathbf{a} \cdot \mathbf{b}$ (d) $\mathbf{a} \times \mathbf{b}$
   (e) $|\mathbf{b} \times \mathbf{c}|$ (f) $\mathbf{a} \cdot (\mathbf{b} \times \mathbf{c})$
   (g) $\mathbf{c} \times \mathbf{c}$ (h) $\mathbf{a} \times (\mathbf{b} \times \mathbf{c})$
   (i) $\text{comp}_{\mathbf{a}}\, \mathbf{b}$ (j) $\text{proj}_{\mathbf{a}}\, \mathbf{b}$
   (k) The angle between $\mathbf{a}$ and $\mathbf{b}$ (correct to the nearest degree)

5. Find the values of x such that the vectors $\langle 3, 2, x \rangle$ and $\langle 2x, 4, x \rangle$ are orthogonal.

6. Find two unit vectors that are orthogonal to both $\mathbf{j} + 2\mathbf{k}$ and $\mathbf{i} - 2\mathbf{j} + 3\mathbf{k}$.

7. Suppose that $\mathbf{u} \cdot (\mathbf{v} \times \mathbf{w}) = 2$. Find
   (a) $(\mathbf{u} \times \mathbf{v}) \cdot \mathbf{w}$ (b) $\mathbf{u} \cdot (\mathbf{w} \times \mathbf{v})$
   (c) $\mathbf{v} \cdot (\mathbf{u} \times \mathbf{w})$ (d) $(\mathbf{u} \times \mathbf{v}) \cdot \mathbf{v}$

8. Show that if $\mathbf{a}$, $\mathbf{b}$, and $\mathbf{c}$ are in V_3, then

$$(\mathbf{a} \times \mathbf{b}) \cdot [(\mathbf{b} \times \mathbf{c}) \times (\mathbf{c} \times \mathbf{a})] = [\mathbf{a} \cdot (\mathbf{b} \times \mathbf{c})]^2$$

9. Find the acute angle between two diagonals of a cube.

10. Given the points $A(1, 0, 1)$, $B(2, 3, 0)$, $C(-1, 1, 4)$, and $D(0, 3, 2)$, find the volume of the parallelepiped with adjacent edges AB, AC, and AD.

11. (a) Find a vector perpendicular to the plane through the points $A(1, 0, 0)$, $B(2, 0, -1)$, and $C(1, 4, 3)$.
   (b) Find the area of triangle ABC.

12. A constant force $\mathbf{F} = 3\mathbf{i} + 5\mathbf{j} + 10\mathbf{k}$ moves an object along the line segment from $(1, 0, 2)$ to $(5, 3, 8)$. Find the work done if the distance is measured in meters and the force in newtons.

13. A boat is pulled onto shore using two ropes, as shown in the diagram. If a force of 255 N is needed, find the magnitude of the force in each rope.

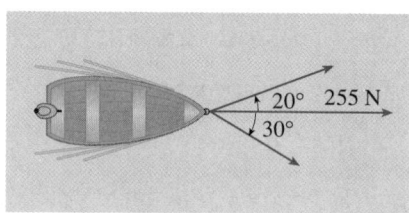

14. Find the magnitude of the torque about P if a 50-N force is applied as shown.

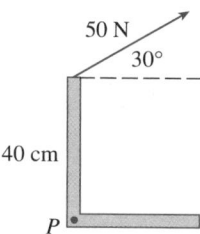

15–17 ▪ Find parametric equations for the line.

15. The line through $(4, -1, 2)$ and $(1, 1, 5)$

16. The line through $(1, 0, -1)$ and parallel to the line $\frac{1}{3}(x - 4) = \frac{1}{2}y = z + 2$

17. The line through $(-2, 2, 4)$ and perpendicular to the plane $2x - y + 5z = 12$

18–20 ▪ Find an equation of the plane.

18. The plane through $(2, 1, 0)$ and parallel to $x + 4y - 3z = 1$

19. The plane through $(3, -1, 1)$, $(4, 0, 2)$, and $(6, 3, 1)$

20. The plane through $(1, 2, -2)$ that contains the line $x = 2t$, $y = 3 - t$, $z = 1 + 3t$

21. Find the point in which the line with parametric equations $x = 2 - t$, $y = 1 + 3t$, $z = 4t$ intersects the plane $2x - y + z = 2$.

22. Find the distance from the origin to the line $x = 1 + t$, $y = 2 - t$, $z = -1 + 2t$.

23. Determine whether the lines given by the symmetric equations

$$\frac{x - 1}{2} = \frac{y - 2}{3} = \frac{z - 3}{4}$$

and

$$\frac{x + 1}{6} = \frac{y - 3}{-1} = \frac{z + 5}{2}$$

are parallel, skew, or intersecting.

24. (a) Show that the planes $x + y - z = 1$ and $2x - 3y + 4z = 5$ are neither parallel nor perpendicular.
(b) Find, correct to the nearest degree, the angle between these planes.

25. Find an equation of the plane through the line of intersection of the planes $x - z = 1$ and $y + 2z = 3$ and perpendicular to the plane $x + y - 2z = 1$.

26. (a) Find an equation of the plane that passes through the points $A(2, 1, 1)$, $B(-1, -1, 10)$, and $C(1, 3, -4)$.
(b) Find symmetric equations for the line through B that is perpendicular to the plane in part (a).
(c) A second plane passes through $(2, 0, 4)$ and has normal vector $\langle 2, -4, -3 \rangle$. Show that the acute angle between the planes is approximately $43°$.
(d) Find parametric equations for the line of intersection of the two planes.

27. Find the distance between the planes $3x + y - 4z = 2$ and $3x + y - 4z = 24$.

28–36 ▪ Identify and sketch the graph of each surface.

28. $x = 3$

29. $x = z$

30. $y = z^2$

31. $x^2 = y^2 + 4z^2$

32. $4x - y + 2z = 4$

33. $-4x^2 + y^2 - 4z^2 = 4$

34. $y^2 + z^2 = 1 + x^2$

35. $4x^2 + 4y^2 - 8y + z^2 = 0$

36. $x = y^2 + z^2 - 2y - 4z + 5$

37. An ellipsoid is created by rotating the ellipse $4x^2 + y^2 = 16$ about the x-axis. Find an equation of the ellipsoid.

38. A surface consists of all points P such that the distance from P to the plane $y = 1$ is twice the distance from P to the point $(0, -1, 0)$. Find an equation for this surface and identify it.

39. (a) Sketch the curve with vector function

$$\mathbf{r}(t) = t\,\mathbf{i} + \cos \pi t\,\mathbf{j} + \sin \pi t\,\mathbf{k} \qquad t \geqslant 0$$

(b) Find $\mathbf{r}'(t)$ and $\mathbf{r}''(t)$.

40. Let $\mathbf{r}(t) = \langle \sqrt{2 - t}, (e^t - 1)/t, \ln(t + 1) \rangle$.
(a) Find the domain of $\mathbf{r}$.   (b) Find $\lim_{t \to 0} \mathbf{r}(t)$.
(c) Find $\mathbf{r}'(t)$.

41. Find a vector function that represents the curve of intersection of the cylinder $x^2 + y^2 = 16$ and the plane $x + z = 5$.

42. Find parametric equations for the tangent line to the curve $x = 2 \sin t$, $y = 2 \sin 2t$, $z = 2 \sin 3t$ at the point $(1, \sqrt{3}, 2)$. Graph the curve and the tangent line on a common screen.

43. If $\mathbf{r}(t) = t^2\,\mathbf{i} + t \cos \pi t\,\mathbf{j} + \sin \pi t\,\mathbf{k}$, evaluate $\int_0^1 \mathbf{r}(t)\,dt$.

44. Let C be the curve with equations $x = 2 - t^3$, $y = 2t - 1$, $z = \ln t$. Find (a) the point where C intersects the xz-plane, (b) parametric equations of the tangent line at $(1, 1, 0)$, and (c) an equation of the normal plane to C at $(1, 1, 0)$.

45. Use Simpson's Rule with $n = 6$ to estimate the length of the arc of the curve with equations $x = t^2$, $y = t^3$, $z = t^4$, $0 \leqslant t \leqslant 3$.

46. Find the length of the curve $\mathbf{r}(t) = \langle 2t^{3/2}, \cos 2t, \sin 2t \rangle$, $0 \leqslant t \leqslant 1$.

47. The helix $\mathbf{r}_1(t) = \cos t\,\mathbf{i} + \sin t\,\mathbf{j} + t\,\mathbf{k}$ intersects the curve $\mathbf{r}_2(t) = (1 + t)\mathbf{i} + t^2\,\mathbf{j} + t^3\,\mathbf{k}$ at the point $(1, 0, 0)$. Find the angle of intersection of these curves.

48. Reparametrize the curve $\mathbf{r}(t) = e^t\,\mathbf{i} + e^t \sin t\,\mathbf{j} + e^t \cos t\,\mathbf{k}$ with respect to arc length measured from the point $(1, 0, 1)$ in the direction of increasing t.

49. For the curve given by $\mathbf{r}(t) = \langle \frac{1}{3}t^3, \frac{1}{2}t^2, t \rangle$, find (a) the unit tangent vector, (b) the unit normal vector, and (c) the curvature.

50. Find the curvature of the ellipse $x = 3 \cos t$, $y = 4 \sin t$ at the points $(3, 0)$ and $(0, 4)$.

51. Find the curvature of the curve $y = x^4$ at the point $(1, 1)$.

52. Find an equation of the osculating circle of the curve $y = x^4 - x^2$ at the origin. Graph both the curve and its osculating circle.

53. A particle moves with position function

$$\mathbf{r}(t) = t \ln t \, \mathbf{i} + t \, \mathbf{j} + e^{-t} \, \mathbf{k}$$

Find the velocity, speed, and acceleration of the particle.

54. A particle starts at the origin with initial velocity $\mathbf{i} - \mathbf{j} + 3\mathbf{k}$. Its acceleration is

$$\mathbf{a}(t) = 6t \, \mathbf{i} + 12t^2 \, \mathbf{j} - 6t \, \mathbf{k}$$

Find its position function.

55. An athlete throws a shot at an angle of $45°$ to the horizontal at an initial speed of 43 ft/s. It leaves his hand 7 ft above the ground.
(a) Where is the shot 2 seconds later?
(b) How high does the shot go?
(c) Where does the shot land?

56. Find the tangential and normal components of the acceleration vector of a particle with position function

$$\mathbf{r}(t) = t \, \mathbf{i} + 2t \, \mathbf{j} + t^2 \, \mathbf{k}$$

57. Find the curvature of the curve with parametric equations

$$x = \int_0^t \sin\left(\tfrac{1}{2} \pi \theta^2\right) d\theta \qquad y = \int_0^t \cos\left(\tfrac{1}{2} \pi \theta^2\right) d\theta$$

11 PARTIAL DERIVATIVES

So far we have dealt with the calculus of functions of a single variable. But, in the real world, physical quantities often depend on two or more variables, so in this chapter we turn our attention to functions of several variables and extend the basic ideas of differential calculus to such functions.

11.1 FUNCTIONS OF SEVERAL VARIABLES

The temperature T at a point on the surface of the earth at any given time depends on the longitude x and latitude y of the point. We can think of T as being a function of the two variables x and y, or as a function of the pair (x, y). We indicate this functional dependence by writing $T = f(x, y)$.

The volume V of a circular cylinder depends on its radius r and its height h. In fact, we know that $V = \pi r^2 h$. We say that V is a function of r and h, and we write $V(r, h) = \pi r^2 h$.

DEFINITION A **function f of two variables** is a rule that assigns to each ordered pair of real numbers (x, y) in a set D a unique real number denoted by $f(x, y)$. The set D is the **domain** of f and its **range** is the set of values that f takes on, that is, $\{f(x, y) \mid (x, y) \in D\}$.

We often write $z = f(x, y)$ to make explicit the value taken on by f at the general point (x, y). The variables x and y are **independent variables** and z is the **dependent variable**. [Compare this with the notation $y = f(x)$ for functions of a single variable.]

A function of two variables is just a function whose domain is a subset of $\mathbb{R}^2$ and whose range is a subset of $\mathbb{R}$. One way of visualizing such a function is by means of an arrow diagram (see Figure 1), where the domain D is represented as a subset of the xy-plane and the range is a set of numbers on a real line, shown as a z-axis. For instance, if $f(x, y)$ represents the temperature at a point (x, y) in a flat metal plate with the shape of D, we can think of the z-axis as a thermometer displaying the recorded temperatures.

If a function f is given by a formula and no domain is specified, then the domain of f is understood to be the set of all pairs (x, y) for which the given expression is a well-defined real number.

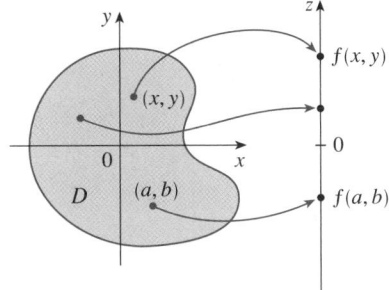

FIGURE 1

EXAMPLE 1 Find the domains of the following functions and evaluate $f(3, 2)$.

(a) $f(x, y) = \dfrac{\sqrt{x + y + 1}}{x - 1}$ 　　　　　　(b) $f(x, y) = x \ln(y^2 - x)$

SOLUTION

(a) 　　　　　$f(3, 2) = \dfrac{\sqrt{3 + 2 + 1}}{3 - 1} = \dfrac{\sqrt{6}}{2}$

The expression for f makes sense if the denominator is not 0 and the quantity under

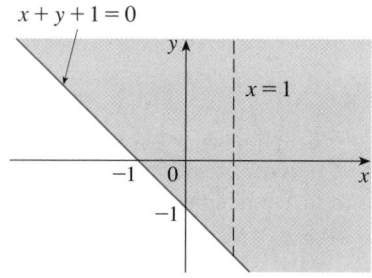

FIGURE 2

Domain of $f(x, y) = \dfrac{\sqrt{x + y + 1}}{x - 1}$

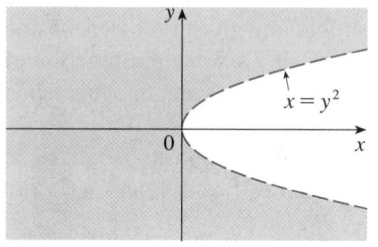

FIGURE 3

Domain of $f(x, y) = x \ln(y^2 - x)$

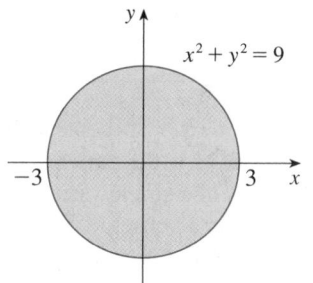

FIGURE 4

Domain of $g(x, y) = \sqrt{9 - x^2 - y^2}$

the square root sign is nonnegative. So the domain of f is

$$D = \{(x, y) \mid x + y + 1 \geqslant 0, \ x \neq 1\}$$

The inequality $x + y + 1 \geqslant 0$, or $y \geqslant -x - 1$, describes the points that lie on or above the line $y = -x - 1$, while $x \neq 1$ means that the points on the line $x = 1$ must be excluded from the domain (see Figure 2).

(b) $$f(3, 2) = 3 \ln(2^2 - 3) = 3 \ln 1 = 0$$

Since $\ln(y^2 - x)$ is defined only when $y^2 - x > 0$, that is, $x < y^2$, the domain of f is $D = \{(x, y) \mid x < y^2\}$. This is the set of points to the left of the parabola $x = y^2$. (See Figure 3.) ∎

EXAMPLE 2 Find the domain and range of $g(x, y) = \sqrt{9 - x^2 - y^2}$.

SOLUTION The domain of g is

$$D = \{(x, y) \mid 9 - x^2 - y^2 \geqslant 0\} = \{(x, y) \mid x^2 + y^2 \leqslant 9\}$$

which is the disk with center $(0, 0)$ and radius 3. (See Figure 4.) The range of g is

$$\left\{z \mid z = \sqrt{9 - x^2 - y^2}, \ (x, y) \in D\right\}$$

Since z is a positive square root, $z \geqslant 0$. Also

$$9 - x^2 - y^2 \leqslant 9 \quad \Rightarrow \quad \sqrt{9 - x^2 - y^2} \leqslant 3$$

So the range is

$$\{z \mid 0 \leqslant z \leqslant 3\} = [0, 3]$$ ∎

GRAPHS

Another way of visualizing the behavior of a function of two variables is to consider its graph.

> **DEFINITION** If f is a function of two variables with domain D, then the **graph** of f is the set of all points (x, y, z) in $\mathbb{R}^3$ such that $z = f(x, y)$ and (x, y) is in D.

Just as the graph of a function f of one variable is a curve C with equation $y = f(x)$, so the graph of a function f of two variables is a surface S with equation

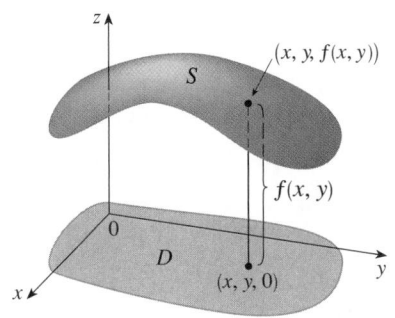

FIGURE 5

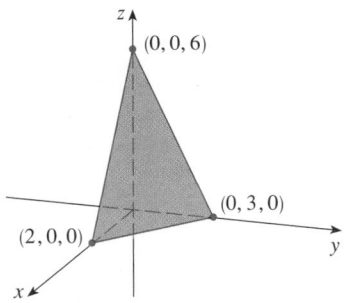

FIGURE 6
Graph of $f(x, y) = 6 - 3x - 2y$

FIGURE 7
Graph of $g(x, y) = \sqrt{9 - x^2 - y^2}$

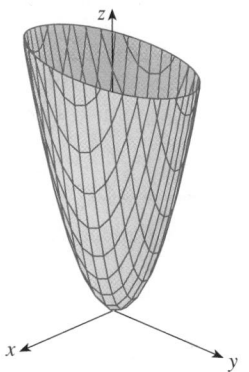

FIGURE 8
Graph of $h(x, y) = 4x^2 + y^2$

$z = f(x, y)$. We can visualize the graph S of f as lying directly above or below its domain D in the xy-plane (see Figure 5).

EXAMPLE 3 Sketch the graph of the function $f(x, y) = 6 - 3x - 2y$.

SOLUTION The graph of f has the equation $z = 6 - 3x - 2y$, or $3x + 2y + z = 6$, which represents a plane. To graph the plane we first find the intercepts. Putting $y = z = 0$ in the equation, we get $x = 2$ as the x-intercept. Similarly, the y-intercept is 3 and the z-intercept is 6. This helps us sketch the portion of the graph that lies in the first octant (Figure 6). ∎

The function in Example 3 is a special case of the function

$$f(x, y) = ax + by + c$$

which is called a **linear function**. The graph of such a function has the equation $z = ax + by + c$, or $ax + by - z + c = 0$, so it is a plane. In much the same way that linear functions of one variable are important in single-variable calculus, we will see that linear functions of two variables play a central role in multivariable calculus.

V EXAMPLE 4 Sketch the graph of $g(x, y) = \sqrt{9 - x^2 - y^2}$.

SOLUTION The graph has equation $z = \sqrt{9 - x^2 - y^2}$. We square both sides of this equation to obtain $z^2 = 9 - x^2 - y^2$, or $x^2 + y^2 + z^2 = 9$, which we recognize as an equation of the sphere with center the origin and radius 3. But, since $z \geq 0$, the graph of g is just the top half of this sphere (see Figure 7).

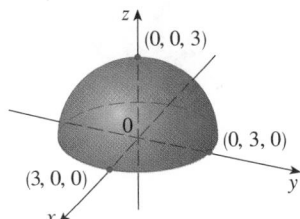

V EXAMPLE 5 Find the domain and range and sketch the graph of

$$h(x, y) = 4x^2 + y^2$$

SOLUTION Notice that $h(x, y)$ is defined for all possible ordered pairs of real numbers (x, y), so the domain is $\mathbb{R}^2$, the entire xy-plane. The range of h is the set $[0, \infty)$ of all nonnegative real numbers. [Notice that $x^2 \geq 0$ and $y^2 \geq 0$, so $h(x, y) \geq 0$ for all x and y.]

The graph of h has the equation $z = 4x^2 + y^2$, which is the elliptic paraboloid that we sketched in Example 4 in Section 10.6. Horizontal traces are ellipses and vertical traces are parabolas (see Figure 8). ∎

Computer programs are readily available for graphing functions of two variables. In most such programs, traces in the vertical planes $x = k$ and $y = k$ are drawn for equally spaced values of k and parts of the graph are eliminated using hidden line removal.

Figure 9 shows computer-generated graphs of several functions. Notice that we get an especially good picture of a function when rotation is used to give views from different vantage points. In parts (a) and (b) the graph of f is very flat and close to the xy-plane except near the origin; this is because $e^{-x^2-y^2}$ is very small when x or y is large.

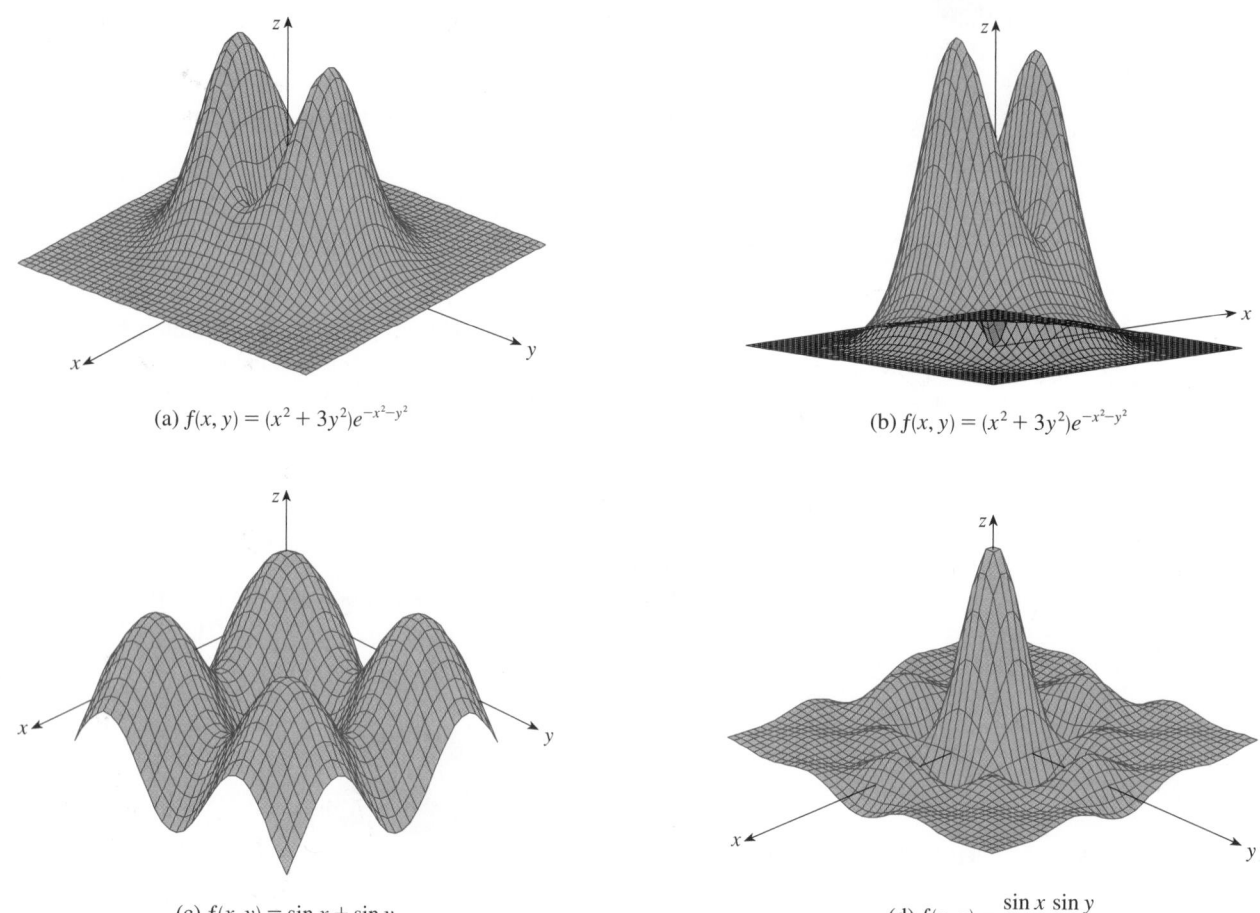

(a) $f(x, y) = (x^2 + 3y^2)e^{-x^2-y^2}$

(b) $f(x, y) = (x^2 + 3y^2)e^{-x^2-y^2}$

(c) $f(x, y) = \sin x + \sin y$

(d) $f(x, y) = \dfrac{\sin x \sin y}{xy}$

FIGURE 9

LEVEL CURVES

So far we have two methods for visualizing functions: arrow diagrams and graphs. A third method, borrowed from mapmakers, is a contour map on which points of constant elevation are joined to form *contour curves,* or *level curves.*

> **DEFINITION** The **level curves** of a function f of two variables are the curves with equations $f(x, y) = k$, where k is a constant (in the range of f).

A level curve $f(x, y) = k$ is the set of all points in the domain of f at which f takes on a given value k. In other words, it shows where the graph of f has height k.

You can see from Figure 10 the relation between level curves and horizontal traces. The level curves $f(x, y) = k$ are just the traces of the graph of f in the horizontal plane $z = k$ projected down to the xy-plane. So if you draw the level curves of a function

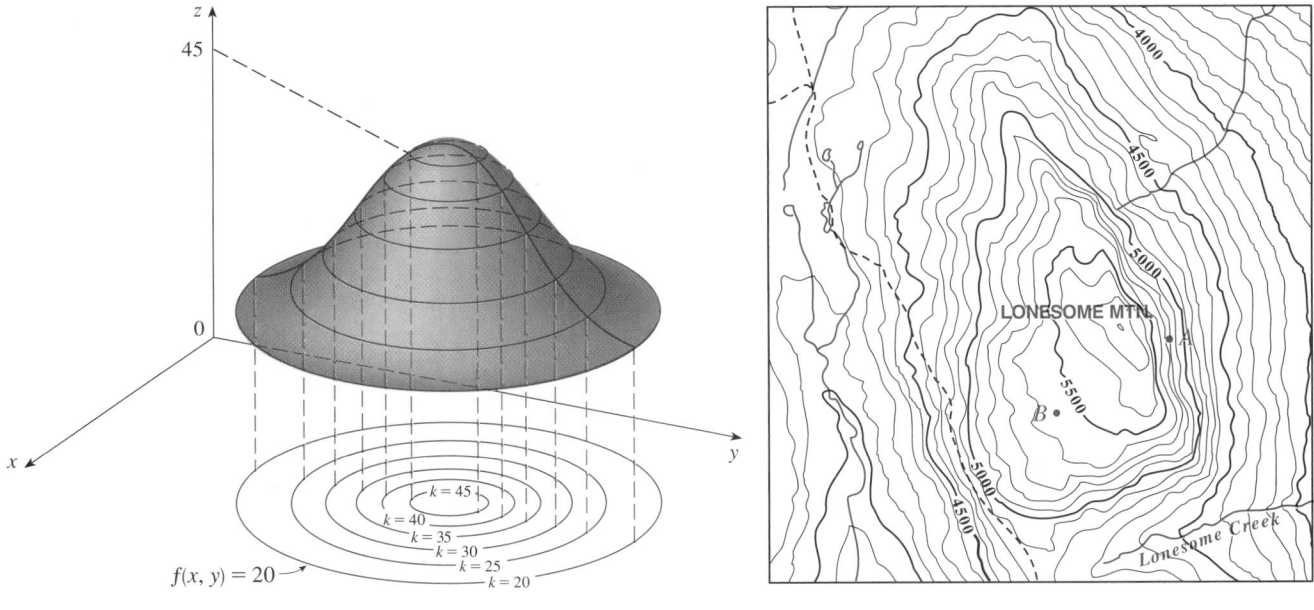

FIGURE 10

FIGURE 11

TEC Visual 11.1A animates Figure 10 by showing level curves being lifted up to graphs of functions.

and visualize them being lifted up to the surface at the indicated height, then you can mentally piece together a picture of the graph. The surface is steep where the level curves are close together. It is somewhat flatter where they are farther apart.

One common example of level curves occurs in topographic maps of mountainous regions, such as the map in Figure 11. The level curves are curves of constant elevation above sea level. If you walk along one of these contour lines you neither ascend nor descend. Another common example is the temperature function introduced in the opening paragraph of this section. Here the level curves are called **isothermals** and join locations with the same temperature. Figure 12 shows a weather map of the world indicating the average January temperatures. The isothermals are the curves that separate the shaded bands.

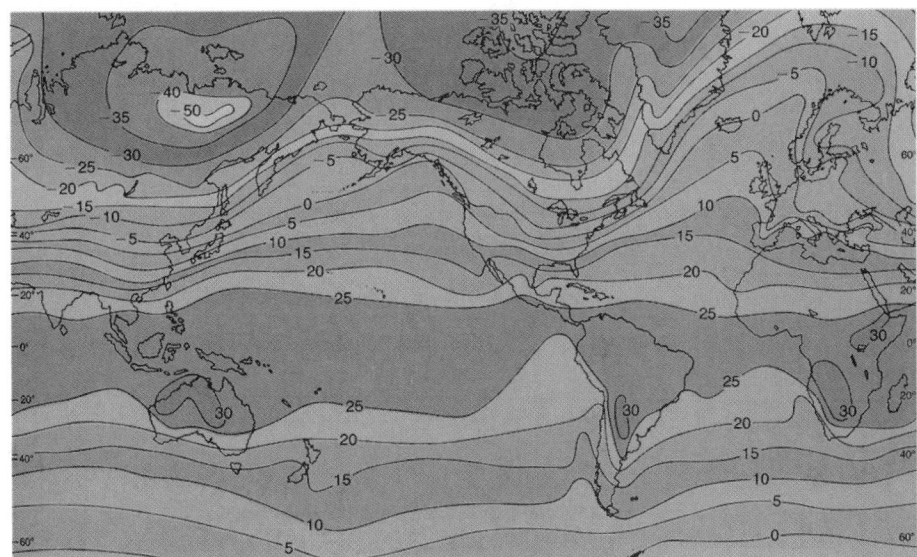

FIGURE 12

World mean sea-level temperatures in January in degrees Celsius

LUTGENS, FREDERICK K.; TARBUCK, EDWARD J.; TASA, DENNIS, *ATMOSPHERE, THE: AN INTRODUCTION TO METEOROLOGY,* 11th ed., © 2010. Printed and electronically reproduced by permission of Pearson Education, Inc., Upper Saddle River, NJ

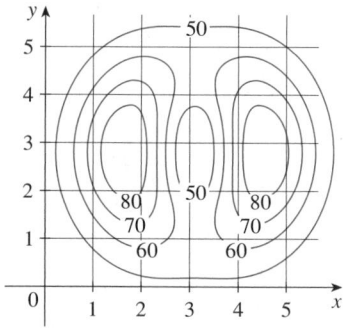

FIGURE 13

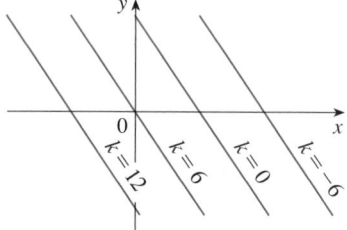

FIGURE 14
Contour map of
$f(x, y) = 6 - 3x - 2y$

EXAMPLE 6 A contour map for a function f is shown in Figure 13. Use it to estimate the values of $f(1, 3)$ and $f(4, 5)$.

SOLUTION The point $(1, 3)$ lies partway between the level curves with z-values 70 and 80. We estimate that

$$f(1, 3) \approx 73$$

Similarly, we estimate that

$$f(4, 5) \approx 56$$

EXAMPLE 7 Sketch the level curves of the function $f(x, y) = 6 - 3x - 2y$ for the values $k = -6, 0, 6, 12$.

SOLUTION The level curves are

$$6 - 3x - 2y = k \qquad \text{or} \qquad 3x + 2y + (k - 6) = 0$$

This is a family of lines with slope $-\frac{3}{2}$. The four particular level curves with $k = -6, 0, 6,$ and 12 are $3x + 2y - 12 = 0$, $3x + 2y - 6 = 0$, $3x + 2y = 0$, and $3x + 2y + 6 = 0$. They are sketched in Figure 14. The level curves are equally spaced parallel lines because the graph of f is a plane (see Figure 6).

▼ **EXAMPLE 8** Sketch the level curves of the function

$$g(x, y) = \sqrt{9 - x^2 - y^2} \qquad \text{for} \quad k = 0, 1, 2, 3$$

SOLUTION The level curves are

$$\sqrt{9 - x^2 - y^2} = k \qquad \text{or} \qquad x^2 + y^2 = 9 - k^2$$

This is a family of concentric circles with center $(0, 0)$ and radius $\sqrt{9 - k^2}$. The cases $k = 0, 1, 2, 3$ are shown in Figure 15. Try to visualize these level curves lifted up to form a surface and compare with the graph of g (a hemisphere) in Figure 7. (See TEC Visual 11.1A.)

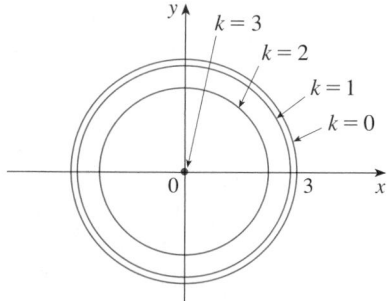

FIGURE 15
Contour map of $g(x, y) = \sqrt{9 - x^2 - y^2}$

EXAMPLE 9 Sketch some level curves of the function $h(x, y) = 4x^2 + y^2 + 1$.

SOLUTION The level curves are

$$4x^2 + y^2 + 1 = k \qquad \text{or} \qquad \frac{x^2}{\frac{1}{4}(k - 1)} + \frac{y^2}{k - 1} = 1$$

which, for $k > 1$, describes a family of ellipses with semiaxes $\frac{1}{2}\sqrt{k - 1}$ and $\sqrt{k - 1}$. Figure 16(a) shows a contour map of h drawn by a computer. Figure 16(b) shows these level curves lifted up to the graph of h (an elliptic paraboloid) where they become horizontal traces. We see from Figure 16 how the graph of h is put together from the level curves.

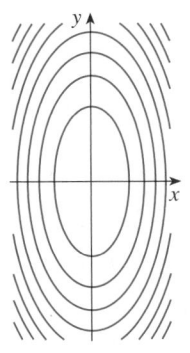

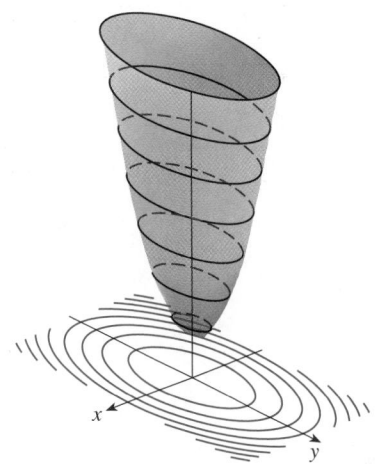

FIGURE 16

The graph of $h(x, y) = 4x^2 + y^2 + 1$ is formed by lifting the level curves.

(a) Contour map

(b) Horizontal traces are raised level curves

Figure 17 shows some computer-generated level curves together with the corresponding computer-generated graphs. Notice that the level curves in part (c) crowd together near the origin. That corresponds to the fact that the graph in part (d) is very steep near the origin.

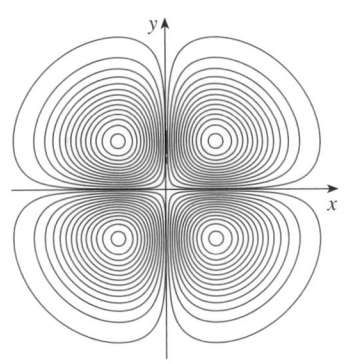

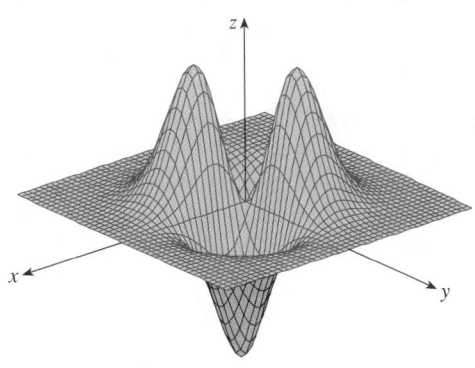

(a) Level curves of $f(x, y) = -xye^{-x^2-y^2}$

(b) Two views of $f(x, y) = -xye^{-x^2-y^2}$

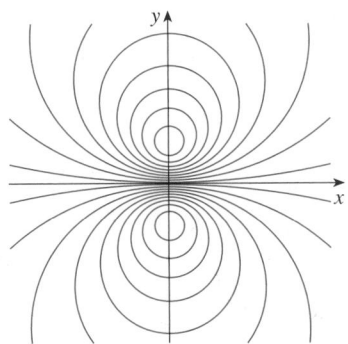

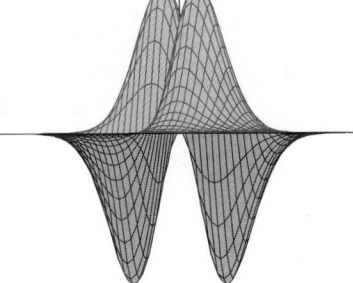

FIGURE 17

(c) Level curves of $f(x, y) = \dfrac{-3y}{x^2 + y^2 + 1}$

(d) $f(x, y) = \dfrac{-3y}{x^2 + y^2 + 1}$

FUNCTIONS OF THREE OR MORE VARIABLES

A **function of three variables**, f, is a rule that assigns to each ordered triple (x, y, z) in a domain $D \subset \mathbb{R}^3$ a unique real number denoted by $f(x, y, z)$. For instance, the temperature T at a point on the surface of the earth depends on the longitude x and latitude y of the point and on the time t, so we could write $T = f(x, y, t)$.

EXAMPLE 10 Find the domain of f if $f(x, y, z) = \ln(z - y) + xy \sin z$.

SOLUTION The expression for $f(x, y, z)$ is defined as long as $z - y > 0$, so the domain of f is

$$D = \{(x, y, z) \in \mathbb{R}^3 \mid z > y\}$$

This is a **half-space** consisting of all points that lie above the plane $z = y$. ∎

It's very difficult to visualize a function f of three variables by its graph, since that would lie in a four-dimensional space. However, we do gain some insight into f by examining its **level surfaces**, which are the surfaces with equations $f(x, y, z) = k$, where k is a constant. If the point (x, y, z) moves along a level surface, the value of $f(x, y, z)$ remains fixed.

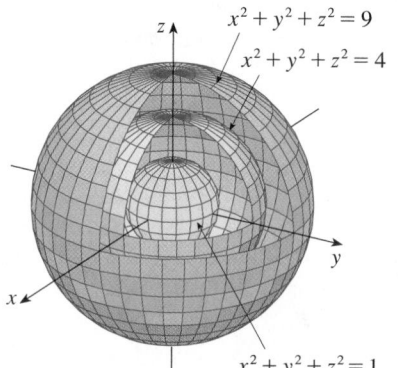

$x^2 + y^2 + z^2 = 9$
$x^2 + y^2 + z^2 = 4$
$x^2 + y^2 + z^2 = 1$

FIGURE 18

EXAMPLE 11 Find the level surfaces of the function $f(x, y, z) = x^2 + y^2 + z^2$.

SOLUTION The level surfaces are $x^2 + y^2 + z^2 = k$, where $k \geq 0$. These form a family of concentric spheres with radius $\sqrt{k}$. (See Figure 18.) Thus as (x, y, z) varies over any sphere with center O, the value of $f(x, y, z)$ remains fixed. ∎

Functions of any number of variables can be considered. A **function of n variables** is a rule that assigns a number $z = f(x_1, x_2, \ldots, x_n)$ to an n-tuple $(x_1, x_2, \ldots, x_n)$ of real numbers. We denote by $\mathbb{R}^n$ the set of all such n-tuples. For example, if a company uses n different ingredients in making a food product, c_i is the cost per unit of the ith ingredient, and x_i units of the ith ingredient are used, then the total cost C of the ingredients is a function of the n variables $x_1, x_2, \ldots, x_n$:

$$\boxed{1} \qquad C = f(x_1, x_2, \ldots, x_n) = c_1 x_1 + c_2 x_2 + \cdots + c_n x_n$$

The function f is a real-valued function whose domain is a subset of $\mathbb{R}^n$. Sometimes we will use vector notation in order to write such functions more compactly: If $\mathbf{x} = \langle x_1, x_2, \ldots, x_n \rangle$, we often write $f(\mathbf{x})$ in place of $f(x_1, x_2, \ldots, x_n)$. With this notation we can rewrite the function defined in Equation 1 as

$$f(\mathbf{x}) = \mathbf{c} \cdot \mathbf{x}$$

where $\mathbf{c} = \langle c_1, c_2, \ldots, c_n \rangle$ and $\mathbf{c} \cdot \mathbf{x}$ denotes the dot product of the vectors $\mathbf{c}$ and $\mathbf{x}$ in V_n.

In view of the one-to-one correspondence between points $(x_1, x_2, \ldots, x_n)$ in $\mathbb{R}^n$ and their position vectors $\mathbf{x} = \langle x_1, x_2, \ldots, x_n \rangle$ in V_n, we have three ways of looking at a function f defined on a subset of $\mathbb{R}^n$:

1. As a function of n real variables $x_1, x_2, \ldots, x_n$

2. As a function of a single point variable $(x_1, x_2, \ldots, x_n)$

3. As a function of a single vector variable $\mathbf{x} = \langle x_1, x_2, \ldots, x_n \rangle$

We will see that all three points of view are useful.

11.1 | EXERCISES

1. Let $g(x, y) = \cos(x + 2y)$.
(a) Evaluate $g(2, -1)$.
(b) Find the domain of g.
(c) Find the range of g.

2. Let $F(x, y) = 1 + \sqrt{4 - y^2}$.
(a) Evaluate $F(3, 1)$.
(b) Find and sketch the domain of F.
(c) Find the range of F.

3. Let $f(x, y, z) = \sqrt{x} + \sqrt{y} + \sqrt{z} + \ln(4 - x^2 - y^2 - z^2)$.
(a) Evaluate $f(1, 1, 1)$.
(b) Find and describe the domain of f.

4. Let $g(x, y, z) = x^3 y^2 z \sqrt{10 - x - y - z}$.
(a) Evaluate $g(1, 2, 3)$.
(b) Find and describe the domain of g.

5–12 ▪ Find and sketch the domain of the function.

5. $f(x, y) = \sqrt{2x - y}$

6. $f(x, y) = \sqrt{xy}$

7. $f(x, y) = \ln(9 - x^2 - 9y^2)$

8. $f(x, y) = \sqrt{y} + \sqrt{25 - x^2 - y^2}$

9. $f(x, y) = \dfrac{\sqrt{y - x^2}}{1 - x^2}$

10. $f(x, y) = \arcsin(x^2 + y^2 - 2)$

11. $f(x, y, z) = \sqrt{1 - x^2 - y^2 - z^2}$

12. $f(x, y, z) = \ln(16 - 4x^2 - 4y^2 - z^2)$

13–20 ▪ Sketch the graph of the function.

13. $f(x, y) = 10 - 4x - 5y$

14. $f(x, y) = 2 - x$

15. $f(x, y) = y^2 + 1$

16. $f(x, y) = e^{-y}$

17. $f(x, y) = 9 - x^2 - 9y^2$

18. $f(x, y) = 1 + 2x^2 + 2y^2$

19. $f(x, y) = \sqrt{4 - 4x^2 - y^2}$

20. $f(x, y) = \sqrt{4x^2 + y^2}$

21. A contour map for a function f is shown. Use it to estimate the values of $f(-3, 3)$ and $f(3, -2)$. What can you say about the shape of the graph?

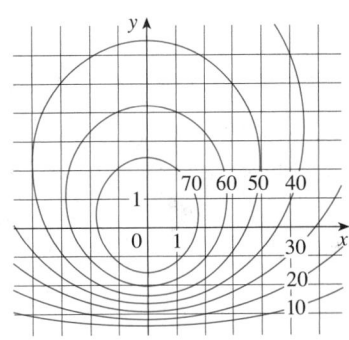

22. Two contour maps are shown. One is for a function f whose graph is a cone. The other is for a function g whose graph is a paraboloid. Which is which, and why?

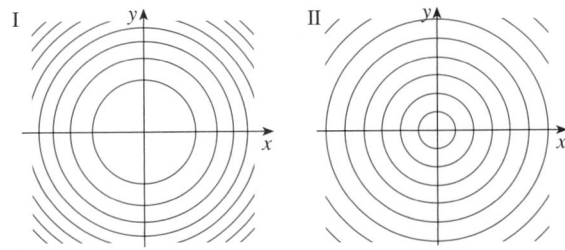

23. Locate the points A and B on the map of Lonesome Mountain (Figure 11). How would you describe the terrain near A? Near B?

24. Make a rough sketch of a contour map for the function whose graph is shown.

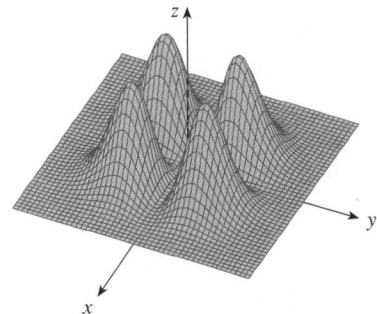

25–32 ▪ Draw a contour map of the function showing several level curves.

25. $f(x, y) = (y - 2x)^2$

26. $f(x, y) = x^3 - y$

27. $f(x, y) = \sqrt{x} + y$

28. $f(x, y) = \ln(x^2 + 4y^2)$

29. $f(x, y) = ye^x$

30. $f(x, y) = y \sec x$

31. $f(x, y) = \sqrt{y^2 - x^2}$

32. $f(x, y) = y/(x^2 + y^2)$

▨ Graphing calculator or computer required   CAS Computer algebra system required   **1.** Homework Hints at stewartcalculus.com

33–34 ▪ Sketch both a contour map and a graph of the function and compare them.

33. $f(x, y) = x^2 + 9y^2$

34. $f(x, y) = \sqrt{36 - 9x^2 - 4y^2}$

35. A thin metal plate, located in the xy-plane, has temperature $T(x, y)$ at the point (x, y). The level curves of T are called *isothermals* because at all points on such a curve the temperature is the same. Sketch some isothermals if the temperature function is given by

$$T(x, y) = \frac{100}{1 + x^2 + 2y^2}$$

36. If $V(x, y)$ is the electric potential at a point (x, y) in the xy-plane, then the level curves of V are called *equipotential curves* because at all points on such a curve the electric potential is the same. Sketch some equipotential curves if $V(x, y) = c/\sqrt{r^2 - x^2 - y^2}$, where c is a positive constant.

37–40 ▪ Use a computer to graph the function using various domains and viewpoints. Get a printout of one that, in your opinion, gives a good view. If your software also produces level curves, then plot some contour lines of the same function and compare with the graph.

37. $f(x, y) = xy^2 - x^3$ (monkey saddle)

38. $f(x, y) = xy^3 - yx^3$ (dog saddle)

39. $f(x, y) = e^{-(x^2+y^2)/3}(\sin(x^2) + \cos(y^2))$

40. $f(x, y) = \cos x \cos y$

41–46 ▪ Match the function (a) with its graph (labeled A–F on page 625) and (b) with its contour map (labeled I–VI). Give reasons for your choices.

41. $z = \sin(xy)$

42. $z = e^x \cos y$

43. $z = \sin(x - y)$

44. $z = \sin x - \sin y$

45. $z = (1 - x^2)(1 - y^2)$

46. $z = \dfrac{x - y}{1 + x^2 + y^2}$

47–50 ▪ Describe the level surfaces of the function.

47. $f(x, y, z) = x + 3y + 5z$

48. $f(x, y, z) = x^2 + 3y^2 + 5z^2$

49. $f(x, y, z) = y^2 + z^2$

50. $f(x, y, z) = x^2 - y^2 - z^2$

51–52 ▪ Describe how the graph of g is obtained from the graph of f.

51. (a) $g(x, y) = f(x, y) + 2$
(b) $g(x, y) = 2f(x, y)$
(c) $g(x, y) = -f(x, y)$
(d) $g(x, y) = 2 - f(x, y)$

52. (a) $g(x, y) = f(x - 2, y)$
(b) $g(x, y) = f(x, y + 2)$
(c) $g(x, y) = f(x + 3, y - 4)$

53. Use a computer to graph the function

$$f(x, y) = \frac{x + y}{x^2 + y^2}$$

using various domains and viewpoints. Comment on the limiting behavior of the function. What happens as both x and y become large? What happens as (x, y) approaches the origin?

54. Use a computer to investigate the family of surfaces

$$z = (ax^2 + by^2)e^{-x^2-y^2}$$

How does the shape of the graph depend on the numbers a and b?

55. Use a computer to investigate the family of functions $f(x, y) = e^{cx^2+y^2}$. How does the shape of the graph depend on c?

56. Graph the functions

$$f(x, y) = \sqrt{x^2 + y^2}$$

$$f(x, y) = e^{\sqrt{x^2+y^2}}$$

$$f(x, y) = \ln\sqrt{x^2 + y^2}$$

$$f(x, y) = \sin(\sqrt{x^2 + y^2})$$

and

$$f(x, y) = \frac{1}{\sqrt{x^2 + y^2}}$$

In general, if g is a function of one variable, how is the graph of

$$f(x, y) = g(\sqrt{x^2 + y^2})$$

obtained from the graph of g?

Graphs and Contour Maps for Exercises 41–46

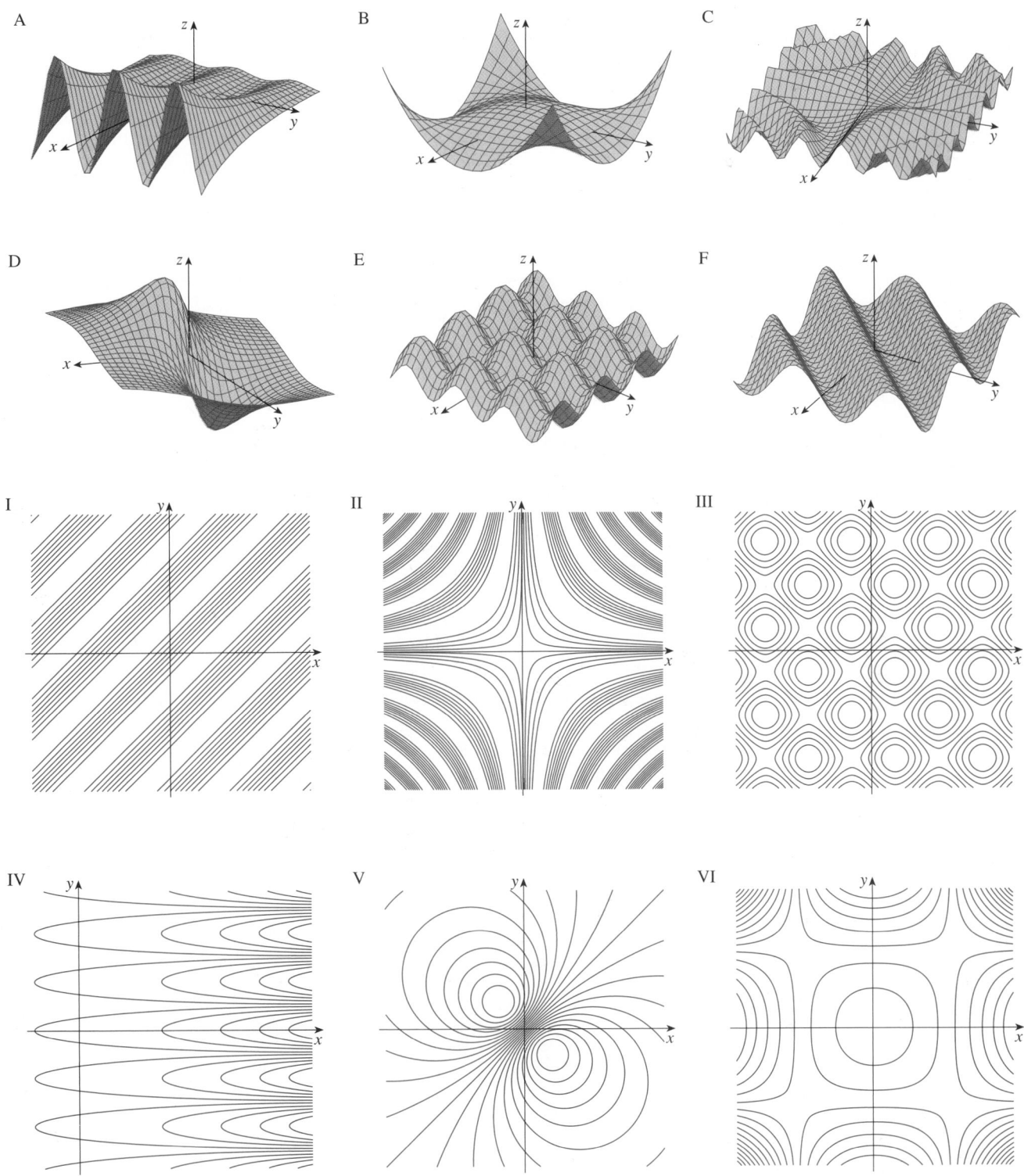

11.2 | LIMITS AND CONTINUITY

The limit of a function of two or more variables is similar to the limit of a function of a single variable. We use the notation

$$\lim_{(x, y) \to (a, b)} f(x, y) = L$$

to indicate that the values of $f(x, y)$ approach the number L as the point (x, y) approaches the point (a, b) along any path that stays within the domain of f. In other words, we can make the values of $f(x, y)$ as close to L as we like by taking the point (x, y) sufficiently close to the point (a, b), but not equal to (a, b). A more precise definition follows.

---

1 DEFINITION Let f be a function of two variables whose domain D includes points arbitrarily close to (a, b). Then we say that the **limit of $f(x, y)$ as (x, y) approaches (a, b)** is L and we write

$$\lim_{(x, y) \to (a, b)} f(x, y) = L$$

if for every number $\varepsilon > 0$ there is a corresponding number $\delta > 0$ such that

if $(x, y) \in D$ and $0 < \sqrt{(x - a)^2 + (y - b)^2} < \delta$ then $|f(x, y) - L| < \varepsilon$

Other notations for the limit in Definition 1 are

$$\lim_{\substack{x \to a \\ y \to b}} f(x, y) = L \qquad \text{and} \qquad f(x, y) \to L \text{ as } (x, y) \to (a, b)$$

Notice that $|f(x, y) - L|$ is the distance between the numbers $f(x, y)$ and L, and $\sqrt{(x - a)^2 + (y - b)^2}$ is the distance between the point (x, y) and the point (a, b). Thus Definition 1 says that the distance between $f(x, y)$ and L can be made arbitrarily small by making the distance from (x, y) to (a, b) sufficiently small (but not 0). Figure 1 illustrates Definition 1 by means of an arrow diagram. If any small interval $(L - \varepsilon, L + \varepsilon)$ is given around L, then we can find a disk D_δ with center (a, b) and radius $\delta > 0$ such that f maps all the points in D_δ [except possibly (a, b)] into the interval $(L - \varepsilon, L + \varepsilon)$.

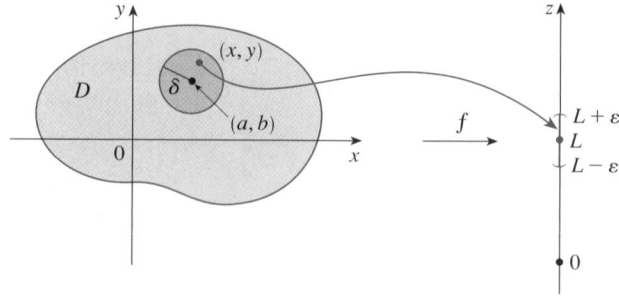

FIGURE 1

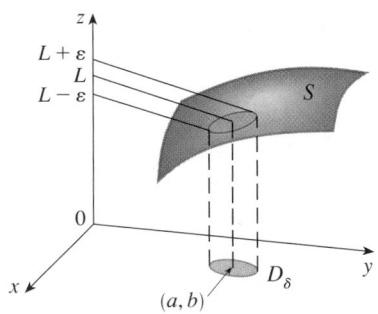

FIGURE 2

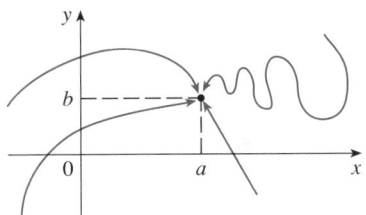

FIGURE 3

Another illustration of Definition 1 is given in Figure 2 where the surface S is the graph of f. If $\varepsilon > 0$ is given, we can find $\delta > 0$ such that if (x, y) is restricted to lie in the disk D_δ and $(x, y) \neq (a, b)$, then the corresponding part of S lies between the horizontal planes $z = L - \varepsilon$ and $z = L + \varepsilon$.

For functions of a single variable, when we let x approach a, there are only two possible directions of approach, from the left or from the right. We recall from Chapter 1 that if $\lim_{x \to a^-} f(x) \neq \lim_{x \to a^+} f(x)$, then $\lim_{x \to a} f(x)$ does not exist.

For functions of two variables the situation is not as simple because we can let (x, y) approach (a, b) from an infinite number of directions in any manner whatsoever (see Figure 3) as long as (x, y) stays within the domain of f.

Definition 1 says that the distance between $f(x, y)$ and L can be made arbitrarily small by making the distance from (x, y) to (a, b) sufficiently small (but not 0). The definition refers only to the *distance* between (x, y) and (a, b). It does not refer to the direction of approach. Therefore, if the limit exists, then $f(x, y)$ must approach the same limit no matter how (x, y) approaches (a, b). Thus if we can find two different paths of approach along which the function $f(x, y)$ has different limits, then it follows that $\lim_{(x, y) \to (a, b)} f(x, y)$ does not exist.

If $f(x, y) \to L_1$ as $(x, y) \to (a, b)$ along a path C_1 and $f(x, y) \to L_2$ as $(x, y) \to (a, b)$ along a path C_2, where $L_1 \neq L_2$, then $\lim_{(x, y) \to (a, b)} f(x, y)$ does not exist.

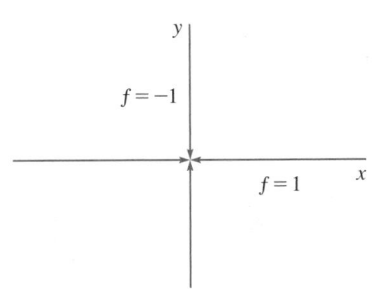

FIGURE 4

V EXAMPLE 1 Show that $\displaystyle\lim_{(x, y) \to (0, 0)} \frac{x^2 - y^2}{x^2 + y^2}$ does not exist.

SOLUTION Let $f(x, y) = (x^2 - y^2)/(x^2 + y^2)$. First let's approach $(0, 0)$ along the x-axis. Then $y = 0$ gives $f(x, 0) = x^2/x^2 = 1$ for all $x \neq 0$, so

$$f(x, y) \to 1 \qquad \text{as} \qquad (x, y) \to (0, 0) \text{ along the } x\text{-axis}$$

We now approach along the y-axis by putting $x = 0$. Then $f(0, y) = \dfrac{-y^2}{y^2} = -1$ for all $y \neq 0$, so

$$f(x, y) \to -1 \qquad \text{as} \qquad (x, y) \to (0, 0) \text{ along the } y\text{-axis}$$

(See Figure 4.) Since f has two different limits along two different lines, the given limit does not exist. ∎

EXAMPLE 2 If $f(x, y) = xy/(x^2 + y^2)$, does $\displaystyle\lim_{(x, y) \to (0, 0)} f(x, y)$ exist?

SOLUTION If $y = 0$, then $f(x, 0) = 0/x^2 = 0$. Therefore

$$f(x, y) \to 0 \qquad \text{as} \qquad (x, y) \to (0, 0) \text{ along the } x\text{-axis}$$

If $x = 0$, then $f(0, y) = 0/y^2 = 0$, so

$$f(x, y) \to 0 \qquad \text{as} \qquad (x, y) \to (0, 0) \text{ along the } y\text{-axis}$$

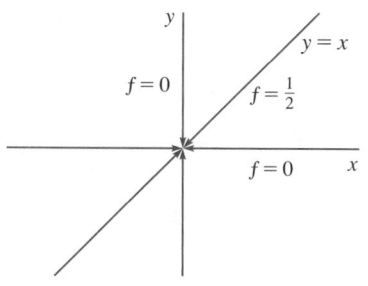

FIGURE 5

In Visual 11.2 a rotating line on the surface in Figure 6 shows different limits at the origin from different directions.

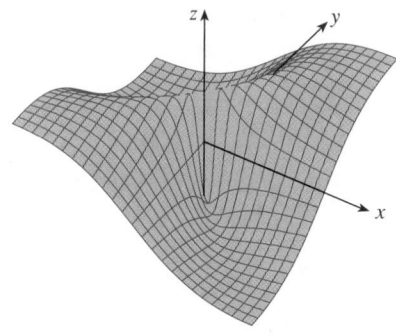

FIGURE 6 $f(x, y) = \dfrac{xy}{x^2 + y^2}$

■ Figure 7 shows the graph of the function in Example 3. Notice the ridge above the parabola $x = y^2$.

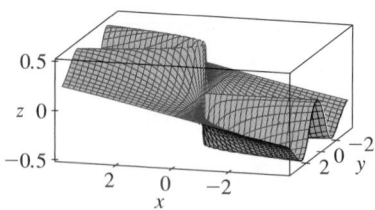

FIGURE 7

Although we have obtained identical limits along the axes, that does not show that the given limit is 0. Let's now approach $(0, 0)$ along another line, say $y = x$. For all $x \neq 0$,

$$f(x, x) = \frac{x^2}{x^2 + x^2} = \frac{1}{2}$$

Therefore $f(x, y) \to \frac{1}{2}$ as $(x, y) \to (0, 0)$ along $y = x$

(See Figure 5.) Since we have obtained different limits along different paths, the given limit does not exist. ■

Figure 6 sheds some light on Example 2. The ridge that occurs above the line $y = x$ corresponds to the fact that $f(x, y) = \frac{1}{2}$ for all points (x, y) on that line except the origin.

V EXAMPLE 3 If $f(x, y) = \dfrac{xy^2}{x^2 + y^4}$, does $\lim\limits_{(x, y) \to (0, 0)} f(x, y)$ exist?

SOLUTION With the solution of Example 2 in mind, let's try to save time by letting $(x, y) \to (0, 0)$ along any nonvertical line through the origin. Then $y = mx$, where m is the slope, and

$$f(x, y) = f(x, mx) = \frac{x(mx)^2}{x^2 + (mx)^4} = \frac{m^2 x^3}{x^2 + m^4 x^4} = \frac{m^2 x}{1 + m^4 x^2}$$

So $f(x, y) \to 0$ as $(x, y) \to (0, 0)$ along $y = mx$

Thus f has the same limiting value along every nonvertical line through the origin. But that does not show that the given limit is 0, for if we now let $(x, y) \to (0, 0)$ along the parabola $x = y^2$, we have

$$f(x, y) = f(y^2, y) = \frac{y^2 \cdot y^2}{(y^2)^2 + y^4} = \frac{y^4}{2y^4} = \frac{1}{2}$$

so $f(x, y) \to \frac{1}{2}$ as $(x, y) \to (0, 0)$ along $x = y^2$

Since different paths lead to different limiting values, the given limit does not exist. ■

Now let's look at limits that *do* exist. Just as for functions of one variable, the calculation of limits for functions of two variables can be greatly simplified by the use of properties of limits. The Limit Laws listed in Section 1.4 can be extended to functions of two variables: The limit of a sum is the sum of the limits, the limit of a product is the product of the limits, and so on. In particular, the following equations are true.

$$\boxed{2} \qquad \lim_{(x, y) \to (a, b)} x = a \qquad \lim_{(x, y) \to (a, b)} y = b \qquad \lim_{(x, y) \to (a, b)} c = c$$

The Squeeze Theorem also holds.

EXAMPLE 4 Find $\lim\limits_{(x, y) \to (0, 0)} \dfrac{3x^2 y}{x^2 + y^2}$ if it exists.

SOLUTION As in Example 3, we could show that the limit along any line through the origin is 0. This doesn't prove that the given limit is 0, but the limits along the parabolas $y = x^2$ and $x = y^2$ also turn out to be 0, so we begin to suspect that the limit does exist and is equal to 0.

Let $\varepsilon > 0$. We want to find $\delta > 0$ such that

$$\text{if} \quad 0 < \sqrt{x^2 + y^2} < \delta \quad \text{then} \quad \left| \frac{3x^2 y}{x^2 + y^2} - 0 \right| < \varepsilon$$

that is,

$$\text{if} \quad 0 < \sqrt{x^2 + y^2} < \delta \quad \text{then} \quad \frac{3x^2 |y|}{x^2 + y^2} < \varepsilon$$

But $x^2 \leq x^2 + y^2$ since $y^2 \geq 0$, so $x^2/(x^2 + y^2) \leq 1$ and therefore

$$\boxed{3} \qquad\qquad \frac{3x^2 |y|}{x^2 + y^2} \leq 3|y| = 3\sqrt{y^2} \leq 3\sqrt{x^2 + y^2}$$

Thus if we choose $\delta = \varepsilon/3$ and let $0 < \sqrt{x^2 + y^2} < \delta$, then

$$\left| \frac{3x^2 y}{x^2 + y^2} - 0 \right| \leq 3\sqrt{x^2 + y^2} < 3\delta = 3\left(\frac{\varepsilon}{3}\right) = \varepsilon$$

■ Another way to do Example 4 is to use the Squeeze Theorem instead of Definition 1. From $\boxed{2}$ it follows that

$$\lim_{(x, y) \to (0, 0)} 3|y| = 0$$

and so the first inequality in $\boxed{3}$ shows that the given limit is 0.

Hence, by Definition 1,

$$\lim_{(x, y) \to (0, 0)} \frac{3x^2 y}{x^2 + y^2} = 0 \qquad\qquad ■$$

CONTINUITY

Recall that evaluating limits of *continuous* functions of a single variable is easy. It can be accomplished by direct substitution because the defining property of a continuous function is $\lim_{x \to a} f(x) = f(a)$. Continuous functions of two variables are also defined by the direct substitution property.

> $\boxed{4}$ **DEFINITION** A function f of two variables is called **continuous at** (a, b) if
>
> $$\lim_{(x, y) \to (a, b)} f(x, y) = f(a, b)$$
>
> We say f is **continuous on** D if f is continuous at every point (a, b) in D.

The intuitive meaning of continuity is that if the point (x, y) changes by a small amount, then the value of $f(x, y)$ changes by a small amount. This means that a surface that is the graph of a continuous function has no hole or break.

Using the properties of limits, you can see that sums, differences, products, and quotients of continuous functions are continuous on their domains. Let's use this fact to give examples of continuous functions.

A **polynomial function of two variables** (or polynomial, for short) is a sum of terms of the form $cx^m y^n$, where c is a constant and m and n are nonnegative integers. A **rational function** is a ratio of polynomials. For instance,

$$f(x, y) = x^4 + 5x^3 y^2 + 6xy^4 - 7y + 6$$

is a polynomial, whereas

$$g(x, y) = \frac{2xy + 1}{x^2 + y^2}$$

is a rational function.

The limits in $\boxed{2}$ show that the functions $f(x, y) = x$, $g(x, y) = y$, and $h(x, y) = c$ are continuous. Since any polynomial can be built up out of the simple functions f, g, and h by multiplication and addition, it follows that *all polynomials are continuous on* $\mathbb{R}^2$. Likewise, any rational function is continuous on its domain because it is a quotient of continuous functions.

V EXAMPLE 5 Evaluate $\lim\limits_{(x, y) \to (1, 2)} (x^2y^3 - x^3y^2 + 3x + 2y)$.

SOLUTION Since $f(x, y) = x^2y^3 - x^3y^2 + 3x + 2y$ is a polynomial, it is continuous everywhere, so we can find the limit by direct substitution:

$$\lim\limits_{(x, y) \to (1, 2)} (x^2y^3 - x^3y^2 + 3x + 2y) = 1^2 \cdot 2^3 - 1^3 \cdot 2^2 + 3 \cdot 1 + 2 \cdot 2 = 11 \quad \blacksquare$$

EXAMPLE 6 Where is the function $f(x, y) = \dfrac{x^2 - y^2}{x^2 + y^2}$ continuous?

SOLUTION The function f is discontinuous at $(0, 0)$ because it is not defined there. Since f is a rational function, it is continuous on its domain, which is the set $D = \{(x, y) \mid (x, y) \neq (0, 0)\}$. $\quad \blacksquare$

EXAMPLE 7 Let

$$g(x, y) = \begin{cases} \dfrac{x^2 - y^2}{x^2 + y^2} & \text{if } (x, y) \neq (0, 0) \\ 0 & \text{if } (x, y) = (0, 0) \end{cases}$$

Here g is defined at $(0, 0)$ but g is still discontinuous there because $\lim\limits_{(x, y) \to (0, 0)} g(x, y)$ does not exist (see Example 1). $\quad \blacksquare$

EXAMPLE 8 Let

$$f(x, y) = \begin{cases} \dfrac{3x^2y}{x^2 + y^2} & \text{if } (x, y) \neq (0, 0) \\ 0 & \text{if } (x, y) = (0, 0) \end{cases}$$

We know f is continuous for $(x, y) \neq (0, 0)$ since it is equal to a rational function there. Also, from Example 4 we have

$$\lim\limits_{(x, y) \to (0, 0)} f(x, y) = \lim\limits_{(x, y) \to (0, 0)} \frac{3x^2y}{x^2 + y^2} = 0 = f(0, 0)$$

Therefore f is continuous at $(0, 0)$, and so it is continuous on $\mathbb{R}^2$. $\quad \blacksquare$

Just as for functions of one variable, composition is another way of combining two continuous functions to get a third. In fact, it can be shown that if f is a continuous function of two variables and g is a continuous function of a single variable that is defined on the range of f, then the composite function $h = g \circ f$ defined by $h(x, y) = g(f(x, y))$ is also a continuous function.

■ Figure 8 shows the graph of the continuous function in Example 8.

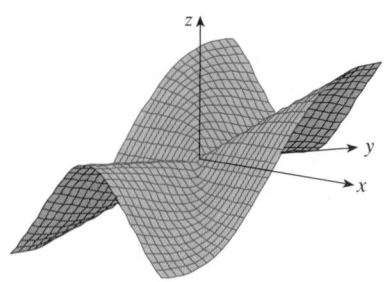

FIGURE 8

EXAMPLE 9 Where is the function $h(x, y) = \arctan(y/x)$ continuous?

SOLUTION The function $f(x, y) = y/x$ is a rational function and therefore continuous except on the line $x = 0$. The function $g(t) = \arctan t$ is continuous everywhere. So the composite function

$$g(f(x, y)) = \arctan(y/x) = h(x, y)$$

is continuous except where $x = 0$. The graph in Figure 9 shows the break in the graph of h above the y-axis. ∎

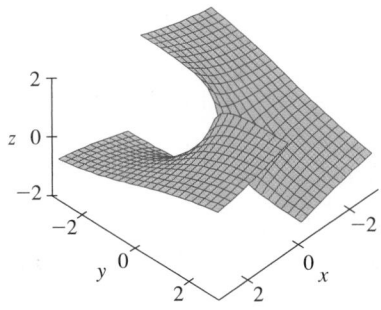

FIGURE 9
The function $h(x, y) = \arctan(y/x)$ is discontinuous where $x = 0$.

FUNCTIONS OF THREE OR MORE VARIABLES

Everything that we have done in this section can be extended to functions of three or more variables. The notation

$$\lim_{(x, y, z) \to (a, b, c)} f(x, y, z) = L$$

means that the values of $f(x, y, z)$ approach the number L as the point (x, y, z) approaches the point (a, b, c) along any path in the domain of f. Because the distance between two points (x, y, z) and (a, b, c) is $\sqrt{(x - a)^2 + (y - b)^2 + (z - c)^2}$, we can write the precise definition as follows: For every number $\varepsilon > 0$ there is a corresponding number $\delta > 0$ such that

if (x, y, z) is in the domain of f and $0 < \sqrt{(x - a)^2 + (y - b)^2 + (z - c)^2} < \delta$

then $|f(x, y, z) - L| < \varepsilon$

The function f is **continuous** at (a, b, c) if

$$\lim_{(x, y, z) \to (a, b, c)} f(x, y, z) = f(a, b, c)$$

For instance, the function

$$f(x, y, z) = \frac{1}{x^2 + y^2 + z^2 - 1}$$

is a rational function of three variables and so is continuous at every point in $\mathbb{R}^3$ except where $x^2 + y^2 + z^2 = 1$. In other words, it is discontinuous on the sphere with center the origin and radius 1.

If we use the vector notation introduced at the end of Section 11.1, then we can write the definitions of a limit for functions of two or three variables in a single compact form as follows.

---

5 If f is defined on a subset D of $\mathbb{R}^n$, then $\lim_{\mathbf{x} \to \mathbf{a}} f(\mathbf{x}) = L$ means that for every number $\varepsilon > 0$ there is a corresponding number $\delta > 0$ such that

if $\mathbf{x} \in D$ and $0 < |\mathbf{x} - \mathbf{a}| < \delta$ then $|f(\mathbf{x}) - L| < \varepsilon$

Notice that if $n = 1$, then $\mathbf{x} = x$ and $\mathbf{a} = a$, and **5** is just the definition of a limit for functions of a single variable. For the case $n = 2$, we have $\mathbf{x} = \langle x, y \rangle$, $\mathbf{a} = \langle a, b \rangle$, and $|\mathbf{x} - \mathbf{a}| = \sqrt{(x - a)^2 + (y - b)^2}$, so **5** becomes Definition 1. If $n = 3$, then $\mathbf{x} = \langle x, y, z \rangle$, $\mathbf{a} = \langle a, b, c \rangle$, and **5** becomes the definition of a limit of a function of three variables. In each case the definition of continuity can be written as

$$\lim_{\mathbf{x} \to \mathbf{a}} f(\mathbf{x}) = f(\mathbf{a})$$

11.2 | EXERCISES

1. Suppose that $\lim_{(x, y) \to (3, 1)} f(x, y) = 6$. What can you say about the value of $f(3, 1)$? What if f is continuous?

2. Explain why each function is continuous or discontinuous.
(a) The outdoor temperature as a function of longitude, latitude, and time
(b) Elevation (height above sea level) as a function of longitude, latitude, and time
(c) The cost of a taxi ride as a function of distance traveled and time

3–16 ▪ Find the limit, if it exists, or show that the limit does not exist.

3. $\lim_{(x, y) \to (1, 2)} (5x^3 - x^2 y^2)$

4. $\lim_{(x, y) \to (1, -1)} e^{-xy} \cos(x + y)$

5. $\lim_{(x, y) \to (0, 0)} \dfrac{x^4 - 4y^2}{x^2 + 2y^2}$

6. $\lim_{(x, y) \to (0, 0)} \dfrac{5y^4 \cos^2 x}{x^4 + y^4}$

7. $\lim_{(x, y) \to (0, 0)} \dfrac{y^2 \sin^2 x}{x^4 + y^4}$

8. $\lim_{(x, y) \to (1, 0)} \dfrac{xy - y}{(x - 1)^2 + y^2}$

9. $\lim_{(x, y) \to (0, 0)} \dfrac{xy}{\sqrt{x^2 + y^2}}$

10. $\lim_{(x, y) \to (0, 0)} \dfrac{x^2 \sin^2 y}{x^2 + 2y^2}$

11. $\lim_{(x, y) \to (0, 0)} \dfrac{x^2 y e^y}{x^4 + 4y^2}$

12. $\lim_{(x, y) \to (0, 0)} \dfrac{xy^4}{x^2 + y^8}$

13. $\lim_{(x, y) \to (0, 0)} \dfrac{x^2 + y^2}{\sqrt{x^2 + y^2 + 1} - 1}$

14. $\lim_{(x, y) \to (0, 0)} \dfrac{x^4 - y^4}{x^2 + y^2}$

15. $\lim_{(x, y, z) \to (0, 0, 0)} \dfrac{xy + yz^2 + xz^2}{x^2 + y^2 + z^4}$

16. $\lim_{(x, y, z) \to (0, 0, 0)} \dfrac{yz}{x^2 + 4y^2 + 9z^2}$

17–18 ▪ Use a computer graph of the function to explain why the limit does not exist.

17. $\lim_{(x, y) \to (0, 0)} \dfrac{2x^2 + 3xy + 4y^2}{3x^2 + 5y^2}$

18. $\lim_{(x, y) \to (0, 0)} \dfrac{xy^3}{x^2 + y^6}$

19–20 ▪ Find $h(x, y) = g(f(x, y))$ and the set on which h is continuous.

19. $g(t) = t^2 + \sqrt{t}$, $f(x, y) = 2x + 3y - 6$

20. $g(t) = t + \ln t$, $f(x, y) = \dfrac{1 - xy}{1 + x^2 y^2}$

21–28 ▪ Determine the set of points at which the function is continuous.

21. $F(x, y) = \dfrac{1 + x^2 + y^2}{1 - x^2 - y^2}$

22. $F(x, y) = \cos \sqrt{1 + x - y}$

23. $G(x, y) = \ln(x^2 + y^2 - 4)$

24. $H(x, y) = \dfrac{e^x + e^y}{e^{xy} - 1}$

25. $f(x, y, z) = \arcsin(x^2 + y^2 + z^2)$

26. $f(x, y, z) = \sqrt{y - x^2} \, \ln z$

27. $f(x, y) = \begin{cases} \dfrac{x^2 y^3}{2x^2 + y^2} & \text{if } (x, y) \neq (0, 0) \\ 1 & \text{if } (x, y) = (0, 0) \end{cases}$

28. $f(x, y) = \begin{cases} \dfrac{xy}{x^2 + xy + y^2} & \text{if } (x, y) \neq (0, 0) \\ 0 & \text{if } (x, y) = (0, 0) \end{cases}$

29–31 ▪ Use polar coordinates to find the limit. [If (r, θ) are polar coordinates of the point (x, y) with $r \geq 0$, note that $r \to 0^+$ as $(x, y) \to (0, 0)$.]

29. $\lim_{(x, y) \to (0, 0)} \dfrac{x^3 + y^3}{x^2 + y^2}$

30. $\lim_{(x, y) \to (0, 0)} (x^2 + y^2) \ln(x^2 + y^2)$

31. $\lim_{(x, y) \to (0, 0)} \dfrac{e^{-x^2 - y^2} - 1}{x^2 + y^2}$

32. Graph and discuss the continuity of the function

$$f(x, y) = \begin{cases} \dfrac{\sin xy}{xy} & \text{if } xy \neq 0 \\ 1 & \text{if } xy = 0 \end{cases}$$

33. Show that the function f given by $f(\mathbf{x}) = |\mathbf{x}|$ is continuous on $\mathbb{R}^n$. [*Hint*: Consider $|\mathbf{x} - \mathbf{a}|^2 = (\mathbf{x} - \mathbf{a}) \cdot (\mathbf{x} - \mathbf{a})$.]

34. If $\mathbf{c} \in V_n$, show that the function f given by $f(\mathbf{x}) = \mathbf{c} \cdot \mathbf{x}$ is continuous on $\mathbb{R}^n$.

11.3 | PARTIAL DERIVATIVES

If f is a function of two variables x and y, suppose we let only x vary while keeping y fixed, say $y = b$, where b is a constant. Then we are really considering a function of a single variable x, namely, $g(x) = f(x, b)$. If g has a derivative at a, then we call it the **partial derivative of f with respect to x at (a, b)** and denote it by $f_x(a, b)$. Thus

$$\boxed{1} \qquad f_x(a, b) = g'(a) \qquad \text{where} \qquad g(x) = f(x, b)$$

By the definition of a derivative, we have

$$g'(a) = \lim_{h \to 0} \frac{g(a + h) - g(a)}{h}$$

and so Equation 1 becomes

$$\boxed{2} \qquad f_x(a, b) = \lim_{h \to 0} \frac{f(a + h, b) - f(a, b)}{h}$$

Similarly, the **partial derivative of f with respect to y at (a, b)**, denoted by $f_y(a, b)$, is obtained by keeping x fixed ($x = a$) and finding the ordinary derivative at b of the function $G(y) = f(a, y)$:

$$\boxed{3} \qquad f_y(a, b) = \lim_{h \to 0} \frac{f(a, b + h) - f(a, b)}{h}$$

If we now let the point (a, b) vary in Equations 2 and 3, f_x and f_y become functions of two variables.

> $\boxed{4}$ If f is a function of two variables, its **partial derivatives** are the functions f_x and f_y defined by
>
> $$f_x(x, y) = \lim_{h \to 0} \frac{f(x + h, y) - f(x, y)}{h}$$
>
> $$f_y(x, y) = \lim_{h \to 0} \frac{f(x, y + h) - f(x, y)}{h}$$

There are many alternative notations for partial derivatives. For instance, instead of f_x we can write f_1 or $D_1 f$ (to indicate differentiation with respect to the *first* variable) or $\partial f/\partial x$. But here $\partial f/\partial x$ can't be interpreted as a ratio of differentials.

> **NOTATIONS FOR PARTIAL DERIVATIVES** If $z = f(x, y)$, we write
>
> $$f_x(x, y) = f_x = \frac{\partial f}{\partial x} = \frac{\partial}{\partial x} f(x, y) = \frac{\partial z}{\partial x} = f_1 = D_1 f = D_x f$$
>
> $$f_y(x, y) = f_y = \frac{\partial f}{\partial y} = \frac{\partial}{\partial y} f(x, y) = \frac{\partial z}{\partial y} = f_2 = D_2 f = D_y f$$

To compute partial derivatives, all we have to do is remember from Equation 1 that the partial derivative with respect to x is just the *ordinary* derivative of the function g of a single variable that we get by keeping y fixed. Thus we have the following rule.

> **RULE FOR FINDING PARTIAL DERIVATIVES OF $z = f(x, y)$**
>
> **1.** To find f_x, regard y as a constant and differentiate $f(x, y)$ with respect to x.
>
> **2.** To find f_y, regard x as a constant and differentiate $f(x, y)$ with respect to y.

EXAMPLE 1 If $f(x, y) = x^3 + x^2 y^3 - 2y^2$, find $f_x(2, 1)$ and $f_y(2, 1)$.

SOLUTION Holding y constant and differentiating with respect to x, we get

$$f_x(x, y) = 3x^2 + 2xy^3$$

and so

$$f_x(2, 1) = 3 \cdot 2^2 + 2 \cdot 2 \cdot 1^3 = 16$$

Holding x constant and differentiating with respect to y, we get

$$f_y(x, y) = 3x^2 y^2 - 4y$$

$$f_y(2, 1) = 3 \cdot 2^2 \cdot 1^2 - 4 \cdot 1 = 8$$ ∎

INTERPRETATIONS OF PARTIAL DERIVATIVES

To give a geometric interpretation of partial derivatives, we recall that the equation $z = f(x, y)$ represents a surface S (the graph of f). If $f(a, b) = c$, then the point $P(a, b, c)$ lies on S. By fixing $y = b$, we are restricting our attention to the curve C_1 in which the vertical plane $y = b$ intersects S. (In other words, C_1 is the trace of S in the plane $y = b$.) Likewise, the vertical plane $x = a$ intersects S in a curve C_2. Both of the curves C_1 and C_2 pass through the point P. (See Figure 1.)

Notice that the curve C_1 is the graph of the function $g(x) = f(x, b)$, so the slope of its tangent T_1 at P is $g'(a) = f_x(a, b)$. The curve C_2 is the graph of the function $G(y) = f(a, y)$, so the slope of its tangent T_2 at P is $G'(b) = f_y(a, b)$.

Thus the partial derivatives $f_x(a, b)$ and $f_y(a, b)$ can be interpreted geometrically as the slopes of the tangent lines at $P(a, b, c)$ to the traces C_1 and C_2 of S in the planes $y = b$ and $x = a$.

Partial derivatives can also be interpreted as *rates of change*. If $z = f(x, y)$, then $\partial z / \partial x$ represents the rate of change of z with respect to x when y is fixed. Similarly, $\partial z / \partial y$ represents the rate of change of z with respect to y when x is fixed.

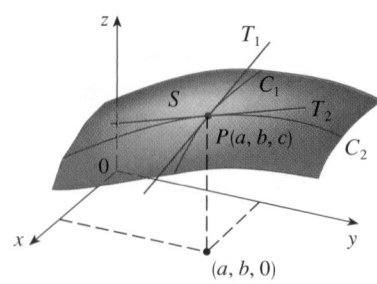

FIGURE 1

The partial derivatives of f at (a, b) are the slopes of the tangents to C_1 and C_2.

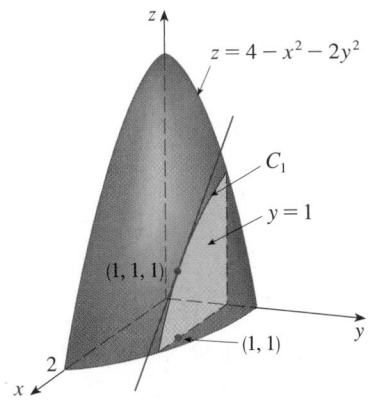

$z = 4 - x^2 - 2y^2$

C_1

$y = 1$

$(1, 1, 1)$

$(1, 1)$

FIGURE 2

EXAMPLE 2 If $f(x, y) = 4 - x^2 - 2y^2$, find $f_x(1, 1)$ and $f_y(1, 1)$ and interpret these numbers as slopes.

SOLUTION We have

$$f_x(x, y) = -2x \qquad f_y(x, y) = -4y$$

$$f_x(1, 1) = -2 \qquad f_y(1, 1) = -4$$

The graph of f is the paraboloid $z = 4 - x^2 - 2y^2$ and the vertical plane $y = 1$ intersects it in the parabola $z = 2 - x^2$, $y = 1$. (As in the preceding discussion, we label it C_1 in Figure 2.) The slope of the tangent line to this parabola at the point $(1, 1, 1)$ is $f_x(1, 1) = -2$. Similarly, the curve C_2 in which the plane $x = 1$ intersects the paraboloid is the parabola $z = 3 - 2y^2$, $x = 1$, and the slope of the tangent line at $(1, 1, 1)$ is $f_y(1, 1) = -4$. (See Figure 3.) ∎

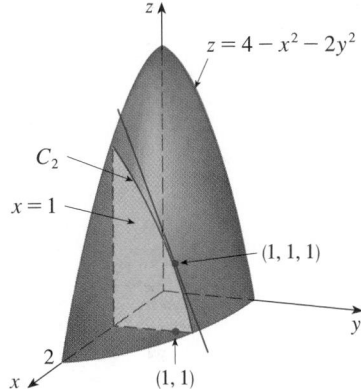

$z = 4 - x^2 - 2y^2$

C_2

$x = 1$

$(1, 1, 1)$

$(1, 1)$

FIGURE 3

▼ EXAMPLE 3 If $f(x, y) = \sin\left(\dfrac{x}{1 + y}\right)$, calculate $\dfrac{\partial f}{\partial x}$ and $\dfrac{\partial f}{\partial y}$.

SOLUTION Using the Chain Rule for functions of one variable, we have

$$\frac{\partial f}{\partial x} = \cos\left(\frac{x}{1 + y}\right) \cdot \frac{\partial}{\partial x}\left(\frac{x}{1 + y}\right) = \cos\left(\frac{x}{1 + y}\right) \cdot \frac{1}{1 + y}$$

$$\frac{\partial f}{\partial y} = \cos\left(\frac{x}{1 + y}\right) \cdot \frac{\partial}{\partial y}\left(\frac{x}{1 + y}\right) = -\cos\left(\frac{x}{1 + y}\right) \cdot \frac{x}{(1 + y)^2} \qquad ∎$$

▼ EXAMPLE 4 Find $\partial z/\partial x$ and $\partial z/\partial y$ if z is defined implicitly as a function of x and y by the equation

$$x^3 + y^3 + z^3 + 6xyz = 1$$

SOLUTION To find $\partial z/\partial x$, we differentiate implicitly with respect to x, being careful to treat y as a constant:

$$3x^2 + 3z^2\frac{\partial z}{\partial x} + 6yz + 6xy\frac{\partial z}{\partial x} = 0$$

■ Some computer algebra systems can plot surfaces defined by implicit equations in three variables. Figure 4 shows such a plot of the surface defined by the equation in Example 4.

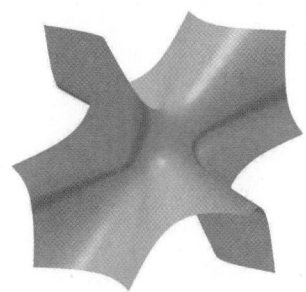

FIGURE 4

Solving this equation for $\partial z/\partial x$, we obtain

$$\frac{\partial z}{\partial x} = -\frac{x^2 + 2yz}{z^2 + 2xy}$$

Similarly, implicit differentiation with respect to y gives

$$\frac{\partial z}{\partial y} = -\frac{y^2 + 2xz}{z^2 + 2xy} \qquad ∎$$

FUNCTIONS OF MORE THAN TWO VARIABLES

Partial derivatives can also be defined for functions of three or more variables. For example, if f is a function of three variables x, y, and z, then its partial derivative with respect to x is defined as

$$f_x(x, y, z) = \lim_{h \to 0} \frac{f(x + h, y, z) - f(x, y, z)}{h}$$

and it is found by regarding y and z as constants and differentiating $f(x, y, z)$ with respect to x. If $w = f(x, y, z)$, then $f_x = \partial w / \partial x$ can be interpreted as the rate of change of w with respect to x when y and z are held fixed. But we can't interpret it geometrically because the graph of f lies in four-dimensional space.

In general, if u is a function of n variables, $u = f(x_1, x_2, \ldots, x_n)$, its partial derivative with respect to the ith variable x_i is

$$\frac{\partial u}{\partial x_i} = \lim_{h \to 0} \frac{f(x_1, \ldots, x_{i-1}, x_i + h, x_{i+1}, \ldots, x_n) - f(x_1, \ldots, x_i, \ldots, x_n)}{h}$$

and we also write

$$\frac{\partial u}{\partial x_i} = \frac{\partial f}{\partial x_i} = f_{x_i} = f_i = D_i f$$

EXAMPLE 5 Find f_x, f_y, and f_z if $f(x, y, z) = e^{xy} \ln z$.

SOLUTION Holding y and z constant and differentiating with respect to x, we have

$$f_x = y e^{xy} \ln z$$

Similarly, $\qquad f_y = x e^{xy} \ln z \qquad$ and $\qquad f_z = \dfrac{e^{xy}}{z}$ ∎

HIGHER DERIVATIVES

If f is a function of two variables, then its partial derivatives f_x and f_y are also functions of two variables, so we can consider their partial derivatives $(f_x)_x$, $(f_x)_y$, $(f_y)_x$, and $(f_y)_y$, which are called the **second partial derivatives** of f. If $z = f(x, y)$, we use the following notation:

$$(f_x)_x = f_{xx} = f_{11} = \frac{\partial}{\partial x}\left(\frac{\partial f}{\partial x}\right) = \frac{\partial^2 f}{\partial x^2} = \frac{\partial^2 z}{\partial x^2}$$

$$(f_x)_y = f_{xy} = f_{12} = \frac{\partial}{\partial y}\left(\frac{\partial f}{\partial x}\right) = \frac{\partial^2 f}{\partial y\, \partial x} = \frac{\partial^2 z}{\partial y\, \partial x}$$

$$(f_y)_x = f_{yx} = f_{21} = \frac{\partial}{\partial x}\left(\frac{\partial f}{\partial y}\right) = \frac{\partial^2 f}{\partial x\, \partial y} = \frac{\partial^2 z}{\partial x\, \partial y}$$

$$(f_y)_y = f_{yy} = f_{22} = \frac{\partial}{\partial y}\left(\frac{\partial f}{\partial y}\right) = \frac{\partial^2 f}{\partial y^2} = \frac{\partial^2 z}{\partial y^2}$$

Thus the notation f_{xy} (or $\partial^2 f / \partial y\, \partial x$) means that we first differentiate with respect to x and then with respect to y, whereas in computing f_{yx} the order is reversed.

EXAMPLE 6 Find the second partial derivatives of

$$f(x, y) = x^3 + x^2y^3 - 2y^2$$

SOLUTION In Example 1 we found that

$$f_x(x, y) = 3x^2 + 2xy^3 \qquad f_y(x, y) = 3x^2y^2 - 4y$$

Therefore

$$f_{xx} = \frac{\partial}{\partial x}(3x^2 + 2xy^3) = 6x + 2y^3 \qquad f_{xy} = \frac{\partial}{\partial y}(3x^2 + 2xy^3) = 6xy^2$$

$$f_{yx} = \frac{\partial}{\partial x}(3x^2y^2 - 4y) = 6xy^2 \qquad f_{yy} = \frac{\partial}{\partial y}(3x^2y^2 - 4y) = 6x^2y - 4 \quad \blacksquare$$

Notice that $f_{xy} = f_{yx}$ in Example 6. This is not just a coincidence. It turns out that the mixed partial derivatives f_{xy} and f_{yx} are equal for most functions that one meets in practice. The following theorem, which was discovered by the French mathematician Alexis Clairaut (1713–1765), gives conditions under which we can assert that $f_{xy} = f_{yx}$. The proof is given in Appendix D.

> **CLAIRAUT'S THEOREM** Suppose f is defined on a disk D that contains the point (a, b). If the functions f_{xy} and f_{yx} are both continuous on D, then
>
> $$f_{xy}(a, b) = f_{yx}(a, b)$$

Partial derivatives of order 3 or higher can also be defined. For instance,

$$f_{xyy} = (f_{xy})_y = \frac{\partial}{\partial y}\left(\frac{\partial^2 f}{\partial y \, \partial x}\right) = \frac{\partial^3 f}{\partial y^2 \, \partial x}$$

and using Clairaut's Theorem it can be shown that $f_{xyy} = f_{yxy} = f_{yyx}$ if these functions are continuous.

▼ EXAMPLE 7 Calculate f_{xxyz} if $f(x, y, z) = \sin(3x + yz)$.

SOLUTION
$$f_x = 3\cos(3x + yz)$$
$$f_{xx} = -9\sin(3x + yz)$$
$$f_{xxy} = -9z\cos(3x + yz)$$
$$f_{xxyz} = -9\cos(3x + yz) + 9yz\sin(3x + yz) \quad \blacksquare$$

■ Alexis Clairaut was a child prodigy in mathematics, having read l'Hospital's textbook on calculus when he was ten and presented a paper on geometry to the French Academy of Sciences when he was 13. At the age of 18, Clairaut published *Recherches sur les courbes à double courbure*, which was the first systematic treatise on three-dimensional analytic geometry and included the calculus of space curves.

PARTIAL DIFFERENTIAL EQUATIONS

Partial derivatives occur in *partial differential equations* that express certain physical laws. For instance, the partial differential equation

$$\frac{\partial^2 u}{\partial x^2} + \frac{\partial^2 u}{\partial y^2} = 0$$

is called **Laplace's equation** after Pierre Laplace (1749–1827). Solutions of this equation are called **harmonic functions** and play a role in problems of heat conduction, fluid flow, and electric potential.

EXAMPLE 8 Show that the function $u(x, y) = e^x \sin y$ is a solution of Laplace's equation.

SOLUTION

$$u_x = e^x \sin y \qquad\qquad u_y = e^x \cos y$$

$$u_{xx} = e^x \sin y \qquad\qquad u_{yy} = -e^x \sin y$$

$$u_{xx} + u_{yy} = e^x \sin y - e^x \sin y = 0$$

Therefore u satisfies Laplace's equation. ∎

The **wave equation**

$$\frac{\partial^2 u}{\partial t^2} = a^2 \frac{\partial^2 u}{\partial x^2}$$

describes the motion of a waveform, which could be an ocean wave, a sound wave, a light wave, or a wave traveling along a vibrating string. For instance, if $u(x, t)$ represents the displacement of a vibrating violin string at time t and at a distance x from one end of the string (as in Figure 5), then $u(x, t)$ satisfies the wave equation. Here the constant a depends on the density of the string and on the tension in the string.

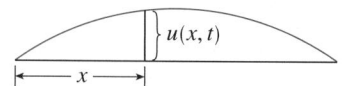

FIGURE 5

EXAMPLE 9 Verify that the function $u(x, t) = \sin(x - at)$ satisfies the wave equation.

SOLUTION

$$u_x = \cos(x - at) \qquad\qquad u_{xx} = -\sin(x - at)$$

$$u_t = -a \cos(x - at) \qquad\qquad u_{tt} = -a^2 \sin(x - at) = a^2 u_{xx}$$

So u satisfies the wave equation. ∎

11.3 | EXERCISES

1. The temperature T at a location in the Northern Hemisphere depends on the longitude x, latitude y, and time t, so we can write $T = f(x, y, t)$. Let's measure time in hours from the beginning of January.
  (a) What are the meanings of the partial derivatives $\partial T/\partial x$, $\partial T/\partial y$, and $\partial T/\partial t$?
  (b) Honolulu has longitude 158° W and latitude 21° N. Suppose that at 9:00 AM on January 1 the wind is blowing hot air to the northeast, so the air to the west and south is warm and the air to the north and east is cooler. Would you expect $f_x(158, 21, 9)$, $f_y(158, 21, 9)$, and $f_t(158, 21, 9)$ to be positive or negative? Explain.

2. A contour map is given for a function f. Use it to estimate $f_x(2, 1)$ and $f_y(2, 1)$.

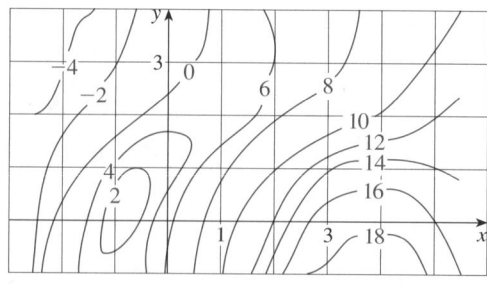

3–4 ▪ Determine the signs of the partial derivatives for the function f whose graph is shown.

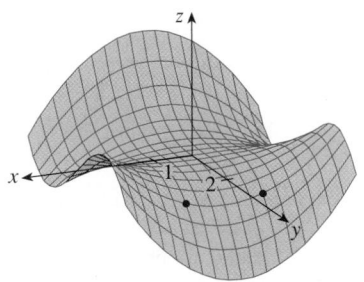

3. (a) $f_x(1, 2)$ (b) $f_y(1, 2)$

4. (a) $f_x(-1, 2)$ (b) $f_y(-1, 2)$
    (c) $f_{xx}(-1, 2)$ (d) $f_{yy}(-1, 2)$

5. If $f(x, y) = 16 - 4x^2 - y^2$, find $f_x(1, 2)$ and $f_y(1, 2)$ and interpret these numbers as slopes. Illustrate with either hand-drawn sketches or computer plots.

6. If $f(x, y) = \sqrt{4 - x^2 - 4y^2}$, find $f_x(1, 0)$ and $f_y(1, 0)$ and interpret these numbers as slopes. Illustrate with either hand-drawn sketches or computer plots.

7–30 ▪ Find the first partial derivatives of the function.

7. $f(x, y) = y^5 - 3xy$ **8.** $f(x, y) = x^4y^3 + 8x^2y$

9. $f(x, t) = e^{-t}\cos \pi x$ **10.** $f(x, t) = \sqrt{x}\ln t$

11. $f(x, y) = \dfrac{x}{y}$ **12.** $f(x, y) = \dfrac{x}{(x + y)^2}$

13. $f(x, y) = \dfrac{ax + by}{cx + dy}$ **14.** $w = \dfrac{e^v}{u + v^2}$

15. $g(u, v) = (u^2v - v^3)^5$ **16.** $u(r, \theta) = \sin(r\cos\theta)$

17. $R(p, q) = \tan^{-1}(pq^2)$ **18.** $f(x, y) = x^y$

19. $F(x, y) = \displaystyle\int_y^x \cos(e^t)\, dt$ **20.** $F(\alpha, \beta) = \displaystyle\int_\alpha^\beta \sqrt{t^3 + 1}\, dt$

21. $f(x, y, z) = xz - 5x^2y^3z^4$ **22.** $f(x, y, z) = x\sin(y - z)$

23. $w = \ln(x + 2y + 3z)$ **24.** $w = ze^{xyz}$

25. $u = xy\sin^{-1}(yz)$ **26.** $u = x^{y/z}$

27. $h(x, y, z, t) = x^2y\cos(z/t)$ **28.** $\phi(x, y, z, t) = \dfrac{\alpha x + \beta y^2}{\gamma z + \delta t^2}$

29. $u = \sqrt{x_1^2 + x_2^2 + \cdots + x_n^2}$

30. $u = \sin(x_1 + 2x_2 + \cdots + nx_n)$

31–34 ▪ Find the indicated partial derivative.

31. $f(x, y) = \ln(x + \sqrt{x^2 + y^2})$; $f_x(3, 4)$

32. $f(x, y) = \arctan(y/x)$; $f_x(2, 3)$

33. $f(x, y, z) = \dfrac{y}{x + y + z}$; $f_y(2, 1, -1)$

34. $f(x, y, z) = \sqrt{\sin^2x + \sin^2y + \sin^2z}$; $f_z(0, 0, \pi/4)$

35–36 ▪ Use the definition of partial derivatives as limits $\boxed{4}$ to find $f_x(x, y)$ and $f_y(x, y)$.

35. $f(x, y) = xy^2 - x^3y$ **36.** $f(x, y) = \dfrac{x}{x + y^2}$

37–38 ▪ Find f_x and f_y and graph f, f_x, and f_y with domains and viewpoints that enable you to see the relationships between them.

37. $f(x, y) = x^2y^3$ **38.** $f(x, y) = \dfrac{y}{1 + x^2y^2}$

39–42 ▪ Use implicit differentiation to find $\partial z/\partial x$ and $\partial z/\partial y$.

39. $x^2 + 2y^2 + 3z^2 = 1$ **40.** $x^2 - y^2 + z^2 - 2z = 4$

41. $e^z = xyz$ **42.** $yz + x\ln y = z^2$

43–44 ▪ Find $\partial z/\partial x$ and $\partial z/\partial y$.

43. (a) $z = f(x) + g(y)$ (b) $z = f(x + y)$

44. (a) $z = f(x)g(y)$ (b) $z = f(xy)$
    (c) $z = f(x/y)$

45–50 ▪ Find all the second partial derivatives.

45. $f(x, y) = x^3y^5 + 2x^4y$ **46.** $f(x, y) = \sin^2(mx + ny)$

47. $w = \sqrt{u^2 + v^2}$ **48.** $v = \dfrac{xy}{x - y}$

49. $z = \arctan\dfrac{x + y}{1 - xy}$ **50.** $v = e^{xe^y}$

51–52 ▪ Verify that the conclusion of Clairaut's Theorem holds, that is, $u_{xy} = u_{yx}$.

51. $u = x^4y^3 - y^4$ **52.** $u = e^{xy}\sin y$

53–58 ▪ Find the indicated partial derivative(s).

53. $f(x, y) = x^4y^2 - x^3y$; f_{xxx}, f_{xyx}

54. $f(x, y) = \sin(2x + 5y)$; f_{yxy}

55. $f(x, y, z) = e^{xyz^2}$; f_{xyz}

56. $g(r, s, t) = e^r\sin(st)$; g_{rst}

57. $u = e^{r\theta}\sin\theta$; $\dfrac{\partial^3 u}{\partial r^2\, \partial\theta}$ **58.** $u = x^ay^bz^c$; $\dfrac{\partial^6 u}{\partial x\, \partial y^2\, \partial z^3}$

59. If $f(x, y, z) = xy^2z^3 + \arcsin(x\sqrt{z})$, find f_{xzy}. [*Hint:* Which order of differentiation is easiest?]

60. If $g(x, y, z) = \sqrt{1 + xz} + \sqrt{1 - xy}$, find g_{xyz}. [*Hint:* Use a different order of differentiation for each term.]

61. Verify that the function $u = e^{-\alpha^2 k^2 t} \sin kx$ is a solution of the *heat conduction equation* $u_t = \alpha^2 u_{xx}$.

62. Determine whether each of the following functions is a solution of Laplace's equation $u_{xx} + u_{yy} = 0$.
(a) $u = x^2 + y^2$ (b) $u = x^2 - y^2$
(c) $u = x^3 + 3xy^2$ (d) $u = \ln\sqrt{x^2 + y^2}$
(e) $u = \sin x \cosh y + \cos x \sinh y$
(f) $u = e^{-x} \cos y - e^{-y} \cos x$

63. Verify that the function $u = 1/\sqrt{x^2 + y^2 + z^2}$ is a solution of the three-dimensional Laplace equation $u_{xx} + u_{yy} + u_{zz} = 0$.

64. Show that each of the following functions is a solution of the wave equation $u_{tt} = a^2 u_{xx}$.
(a) $u = \sin(kx)\sin(akt)$ (b) $u = t/(a^2t^2 - x^2)$
(c) $u = (x - at)^6 + (x + at)^6$
(d) $u = \sin(x - at) + \ln(x + at)$

65. If f and g are twice differentiable functions of a single variable, show that the function
$$u(x, t) = f(x + at) + g(x - at)$$
is a solution of the wave equation given in Exercise 64.

66. If $u = e^{a_1x_1 + a_2x_2 + \cdots + a_nx_n}$, where $a_1^2 + a_2^2 + \cdots + a_n^2 = 1$, show that
$$\frac{\partial^2 u}{\partial x_1^2} + \frac{\partial^2 u}{\partial x_2^2} + \cdots + \frac{\partial^2 u}{\partial x_n^2} = u$$

67. Show that the function $z = xe^y + ye^x$ is a solution of the equation
$$\frac{\partial^3 z}{\partial x^3} + \frac{\partial^3 z}{\partial y^3} = x\frac{\partial^3 z}{\partial x\, \partial y^2} + y\frac{\partial^3 z}{\partial x^2\, \partial y}$$

68. The temperature at a point (x, y) on a flat metal plate is given by $T(x, y) = 60/(1 + x^2 + y^2)$, where T is measured in °C and x, y in meters. Find the rate of change of temperature with respect to distance at the point $(2, 1)$ in (a) the x-direction and (b) the y-direction.

69. The total resistance R produced by three conductors with resistances R_1, R_2, R_3 connected in a parallel electrical circuit is given by the formula
$$\frac{1}{R} = \frac{1}{R_1} + \frac{1}{R_2} + \frac{1}{R_3}$$
Find $\partial R/\partial R_1$.

70. (a) The gas law for a fixed mass m of an ideal gas at absolute temperature T, pressure P, and volume V is $PV = mRT$, where R is the gas constant. Show that
$$\frac{\partial P}{\partial V}\frac{\partial V}{\partial T}\frac{\partial T}{\partial P} = -1$$

(b) For the ideal gas of part (a), show that
$$T\frac{\partial P}{\partial T}\frac{\partial V}{\partial T} = mR$$

71. The *van der Waals equation* for n moles of a gas is
$$\left(P + \frac{n^2a}{V^2}\right)(V - nb) = nRT$$
where P is the pressure, V is the volume, and T is the temperature of the gas. The constant R is the universal gas constant and a and b are positive constants that are characteristic of a particular gas. Calculate $\partial T/\partial P$ and $\partial P/\partial V$.

72. The *wind-chill index* is a measure of how cold it feels in windy weather. It is modeled by the function
$$W = 13.12 + 0.6215T - 11.37v^{0.16} + 0.3965Tv^{0.16}$$
where T is the temperature (in °C) and v is the wind speed (in km/h). When $T = -15$°C and $v = 30$ km/h, by how much would you expect the apparent temperature W to drop if the actual temperature decreases by 1°C? What if the wind speed increases by 1 km/h?

73. The kinetic energy of a body with mass m and velocity v is $K = \frac{1}{2}mv^2$. Show that
$$\frac{\partial K}{\partial m}\frac{\partial^2 K}{\partial v^2} = K$$

74. If a, b, c are the sides of a triangle and A, B, C are the opposite angles, find $\partial A/\partial a$, $\partial A/\partial b$, $\partial A/\partial c$ by implicit differentiation of the Law of Cosines.

75. You are told that there is a function f whose partial derivatives are $f_x(x, y) = x + 4y$ and $f_y(x, y) = 3x - y$. Should you believe it?

76. The paraboloid $z = 6 - x - x^2 - 2y^2$ intersects the plane $x = 1$ in a parabola. Find parametric equations for the tangent line to this parabola at the point $(1, 2, -4)$. Use a computer to graph the paraboloid, the parabola, and the tangent line on the same screen.

77. The ellipsoid $4x^2 + 2y^2 + z^2 = 16$ intersects the plane $y = 2$ in an ellipse. Find parametric equations for the tangent line to this ellipse at the point $(1, 2, 2)$.

78. In a study of frost penetration it was found that the temperature T at time t (measured in days) at a depth x (measured in feet) can be modeled by the function
$$T(x, t) = T_0 + T_1 e^{-\lambda x} \sin(\omega t - \lambda x)$$
where $\omega = 2\pi/365$ and λ is a positive constant.
(a) Find $\partial T/\partial x$. What is its physical significance?
(b) Find $\partial T/\partial t$. What is its physical significance?

(c) Show that T satisfies the heat equation $T_t = kT_{xx}$ for a certain constant k.

(d) If $\lambda = 0.2$, $T_0 = 0$, and $T_1 = 10$, use a computer to graph $T(x, t)$.

(e) What is the physical significance of the term $-\lambda x$ in the expression $\sin(\omega t - \lambda x)$?

79. Use Clairaut's Theorem to show that if the third-order partial derivatives of f are continuous, then

$$f_{xyy} = f_{yxy} = f_{yyx}$$

80. (a) How many nth-order partial derivatives does a function of two variables have?

(b) If these partial derivatives are all continuous, how many of them can be distinct?

(c) Answer the question in part (a) for a function of three variables.

81. If $f(x, y) = x(x^2 + y^2)^{-3/2} e^{\sin(x^2 y)}$, find $f_x(1, 0)$. [*Hint:* Instead of finding $f_x(x, y)$ first, note that it's easier to use Equation 1 or Equation 2.]

82. If $f(x, y) = \sqrt[3]{x^3 + y^3}$, find $f_x(0, 0)$.

83. Let

$$f(x, y) = \begin{cases} \dfrac{x^3 y - xy^3}{x^2 + y^2} & \text{if } (x, y) \neq (0, 0) \\ 0 & \text{if } (x, y) = (0, 0) \end{cases}$$

(a) Use a computer to graph f.

(b) Find $f_x(x, y)$ and $f_y(x, y)$ when $(x, y) \neq (0, 0)$.

(c) Find $f_x(0, 0)$ and $f_y(0, 0)$ using Equations 2 and 3.

(d) Show that $f_{xy}(0, 0) = -1$ and $f_{yx}(0, 0) = 1$.

(e) Does the result of part (d) contradict Clairaut's Theorem? Use graphs of f_{xy} and f_{yx} to illustrate your answer.

11.4 | TANGENT PLANES AND LINEAR APPROXIMATIONS

One of the most important ideas in single-variable calculus is that as we zoom in toward a point on the graph of a differentiable function, the graph becomes indistinguishable from its tangent line and we can approximate the function by a linear function. (See Section 2.8.) Here we develop similar ideas in three dimensions. As we zoom in toward a point on a surface that is the graph of a differentiable function of two variables, the surface looks more and more like a plane (its tangent plane) and we can approximate the function by a linear function of two variables. We also extend the idea of a differential to functions of two or more variables.

TANGENT PLANES

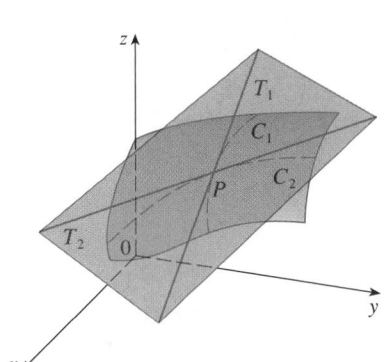

FIGURE 1

The tangent plane contains the tangent lines T_1 and T_2.

Suppose a surface S has equation $z = f(x, y)$, where f has continuous first partial derivatives, and let $P(x_0, y_0, z_0)$ be a point on S. As in the preceding section, let C_1 and C_2 be the curves obtained by intersecting the vertical planes $y = y_0$ and $x = x_0$ with the surface S. Then the point P lies on both C_1 and C_2. Let T_1 and T_2 be the tangent lines to the curves C_1 and C_2 at the point P. Then the **tangent plane** to the surface S at the point P is defined to be the plane that contains both tangent lines T_1 and T_2. (See Figure 1.)

We will see in Section 11.6 that if C is any other curve that lies on the surface S and passes through P, then its tangent line at P also lies in the tangent plane. Therefore you can think of the tangent plane to S at P as consisting of all possible tangent lines at P to curves that lie on S and pass through P. The tangent plane at P is the plane that most closely approximates the surface S near the point P.

We know from Equation 10.5.7 that any plane passing through the point $P(x_0, y_0, z_0)$ has an equation of the form

$$A(x - x_0) + B(y - y_0) + C(z - z_0) = 0$$

By dividing this equation by C and letting $a = -A/C$ and $b = -B/C$, we can write it in the form

$$\boxed{1} \qquad z - z_0 = a(x - x_0) + b(y - y_0)$$

If Equation 1 represents the tangent plane at P, then its intersection with the plane $y = y_0$ must be the tangent line T_1. Setting $y = y_0$ in Equation 1 gives

$$z - z_0 = a(x - x_0) \qquad y = y_0$$

and we recognize these as the equations (in point-slope form) of a line with slope a. But from Section 11.3 we know that the slope of the tangent T_1 is $f_x(x_0, y_0)$. Therefore $a = f_x(x_0, y_0)$.

Similarly, putting $x = x_0$ in Equation 1, we get $z - z_0 = b(y - y_0)$, which must represent the tangent line T_2, so $b = f_y(x_0, y_0)$.

■ Note the similarity between the equation of a tangent plane and the equation of a tangent line:

$$y - y_0 = f'(x_0)(x - x_0)$$

> $\boxed{2}$ Suppose f has continuous partial derivatives. An equation of the tangent plane to the surface $z = f(x, y)$ at the point $P(x_0, y_0, z_0)$ is
>
> $$z - z_0 = f_x(x_0, y_0)(x - x_0) + f_y(x_0, y_0)(y - y_0)$$

V EXAMPLE 1 Find the tangent plane to the elliptic paraboloid $z = 2x^2 + y^2$ at the point $(1, 1, 3)$.

SOLUTION Let $f(x, y) = 2x^2 + y^2$. Then

$$f_x(x, y) = 4x \qquad f_y(x, y) = 2y$$

$$f_x(1, 1) = 4 \qquad f_y(1, 1) = 2$$

Then $\boxed{2}$ gives the equation of the tangent plane at $(1, 1, 3)$ as

$$z - 3 = 4(x - 1) + 2(y - 1)$$

or $\qquad z = 4x + 2y - 3$ ∎

TEC Visual 11.4 shows an animation of Figure 2.

Figure 2(a) shows the elliptic paraboloid and its tangent plane at $(1, 1, 3)$ that we found in Example 1. In parts (b) and (c) we zoom in toward the point $(1, 1, 3)$ by restricting the domain of the function $f(x, y) = 2x^2 + y^2$. Notice that the more we zoom in, the flatter the graph appears and the more it resembles its tangent plane.

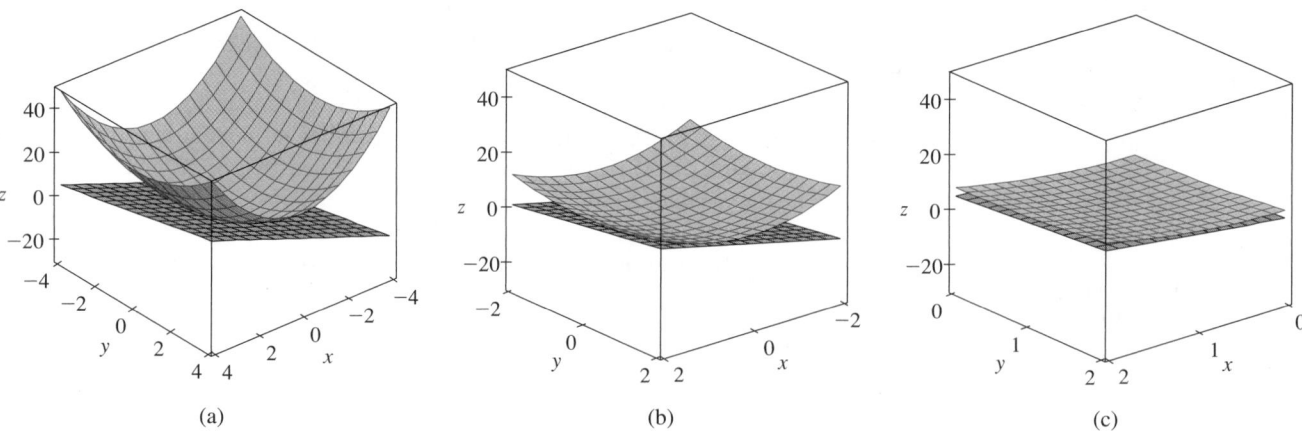

| (a) | (b) | (c) |

FIGURE 2 The elliptic paraboloid $z = 2x^2 + y^2$ appears to coincide with its tangent plane as we zoom in toward $(1, 1, 3)$.

In Figure 3 we corroborate this impression by zooming in toward the point $(1, 1)$ on a contour map of the function $f(x, y) = 2x^2 + y^2$. Notice that the more we zoom in, the more the level curves look like equally spaced parallel lines, which is characteristic of a plane.

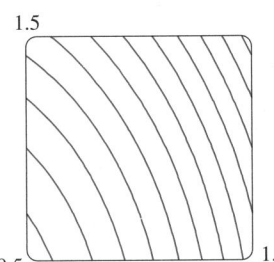 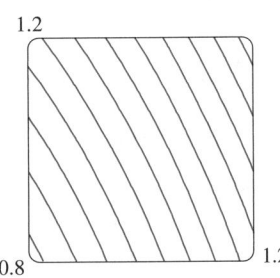 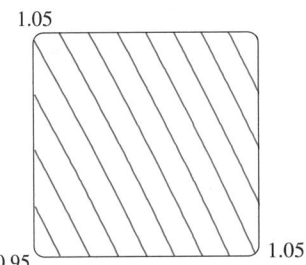

FIGURE 3

Zooming in toward $(1, 1)$
on a contour map of
$f(x, y) = 2x^2 + y^2$

LINEAR APPROXIMATIONS

In Example 1 we found that an equation of the tangent plane to the graph of the function $f(x, y) = 2x^2 + y^2$ at the point $(1, 1, 3)$ is $z = 4x + 2y - 3$. Therefore, in view of the visual evidence in Figures 2 and 3, the linear function of two variables

$$L(x, y) = 4x + 2y - 3$$

is a good approximation to $f(x, y)$ when (x, y) is near $(1, 1)$. The function L is called the *linearization* of f at $(1, 1)$ and the approximation

$$f(x, y) \approx 4x + 2y - 3$$

is called the *linear approximation* or *tangent plane approximation* of f at $(1, 1)$.

For instance, at the point $(1.1, 0.95)$ the linear approximation gives

$$f(1.1, 0.95) \approx 4(1.1) + 2(0.95) - 3 = 3.3$$

which is quite close to the true value of $f(1.1, 0.95) = 2(1.1)^2 + (0.95)^2 = 3.3225$. But if we take a point farther away from $(1, 1)$, such as $(2, 3)$, we no longer get a good approximation. In fact, $L(2, 3) = 11$ whereas $f(2, 3) = 17$.

In general, we know from $\boxed{2}$ that an equation of the tangent plane to the graph of a function f of two variables at the point $(a, b, f(a, b))$ is

$$z = f(a, b) + f_x(a, b)(x - a) + f_y(a, b)(y - b)$$

if f_x and f_y are continuous. The linear function whose graph is this tangent plane, namely

$$\boxed{3} \qquad L(x, y) = f(a, b) + f_x(a, b)(x - a) + f_y(a, b)(y - b)$$

is called the **linearization** of f at (a, b) and the approximation

$$\boxed{4} \qquad f(x, y) \approx f(a, b) + f_x(a, b)(x - a) + f_y(a, b)(y - b)$$

is called the **linear approximation** or the **tangent plane approximation** of f at (a, b).

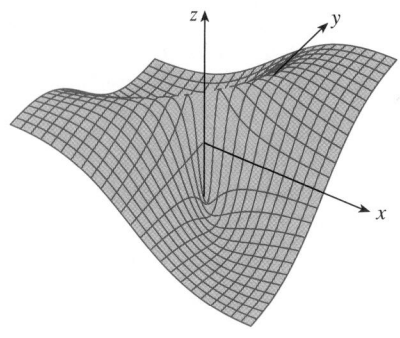

FIGURE 4

$f(x, y) = \dfrac{xy}{x^2 + y^2}$ if $(x, y) \neq (0, 0)$,

$f(0, 0) = 0$

- This is Equation 2.5.5.

- Theorem 8 is proved in Appendix D.

We have defined tangent planes for surfaces $z = f(x, y)$, where f has continuous first partial derivatives. What happens if f_x and f_y are not continuous? Figure 4 pictures such a function; its equation is

$$f(x, y) = \begin{cases} \dfrac{xy}{x^2 + y^2} & \text{if } (x, y) \neq (0, 0) \\ 0 & \text{if } (x, y) = (0, 0) \end{cases}$$

You can verify (see Exercise 38) that its partial derivatives exist at the origin and, in fact, $f_x(0, 0) = 0$ and $f_y(0, 0) = 0$, but f_x and f_y are not continuous. The linear approximation would be $f(x, y) \approx 0$, but $f(x, y) = \frac{1}{2}$ at all points on the line $y = x$ (other than the origin). So a function of two variables can behave badly even though both of its partial derivatives exist. To rule out such behavior, we formulate the idea of a differentiable function of two variables.

Recall that for a function of one variable, $y = f(x)$, if x changes from a to $a + \Delta x$, we defined the increment of y as

$$\Delta y = f(a + \Delta x) - f(a)$$

In Chapter 2 we showed that if f is differentiable at a, then

$$\boxed{5} \qquad \Delta y = f'(a)\,\Delta x + \varepsilon\,\Delta x \qquad \text{where } \varepsilon \to 0 \text{ as } \Delta x \to 0$$

Now consider a function of two variables, $z = f(x, y)$, and suppose x changes from a to $a + \Delta x$ and y changes from b to $b + \Delta y$. Then the corresponding **increment** of z is

$$\boxed{6} \qquad \Delta z = f(a + \Delta x, b + \Delta y) - f(a, b)$$

Thus the increment Δz represents the change in the value of f when (x, y) changes from (a, b) to $(a + \Delta x, b + \Delta y)$. By analogy with $\boxed{5}$ we define the differentiability of a function of two variables as follows.

$\boxed{7}$ **DEFINITION**   If $z = f(x, y)$, then f is **differentiable** at (a, b) if Δz can be expressed in the form

$$\Delta z = f_x(a, b)\,\Delta x + f_y(a, b)\,\Delta y + \varepsilon_1\,\Delta x + \varepsilon_2\,\Delta y$$

where ε_1 and $\varepsilon_2 \to 0$ as $(\Delta x, \Delta y) \to (0, 0)$.

Definition 7 says that a differentiable function is one for which the linear approximation $\boxed{4}$ is a good approximation when (x, y) is near (a, b). In other words, the tangent plane approximates the graph of f well near the point of tangency.

It's sometimes hard to use Definition 7 directly to check the differentiability of a function, but the following theorem provides a convenient sufficient condition for differentiability.

$\boxed{8}$ **THEOREM**   If the partial derivatives f_x and f_y exist near (a, b) and are continuous at (a, b), then f is differentiable at (a, b).

V EXAMPLE 2 Show that $f(x, y) = xe^{xy}$ is differentiable at $(1, 0)$ and find its linearization there. Then use it to approximate $f(1.1, -0.1)$.

SOLUTION The partial derivatives are

$$f_x(x, y) = e^{xy} + xye^{xy} \qquad f_y(x, y) = x^2 e^{xy}$$

$$f_x(1, 0) = 1 \qquad\qquad\qquad f_y(1, 0) = 1$$

Both f_x and f_y are continuous functions, so f is differentiable by Theorem 8. The linearization is

$$L(x, y) = f(1, 0) + f_x(1, 0)(x - 1) + f_y(1, 0)(y - 0)$$

$$= 1 + 1(x - 1) + 1 \cdot y = x + y$$

The corresponding linear approximation is

$$xe^{xy} \approx x + y$$

so

$$f(1.1, -0.1) \approx 1.1 - 0.1 = 1$$

Compare this with the actual value of $f(1.1, -0.1) = 1.1e^{-0.11} \approx 0.98542$. ∎

■ Figure 5 shows the graphs of the function f and its linearization L in Example 2.

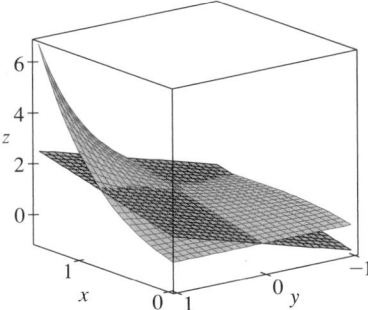

FIGURE 5

DIFFERENTIALS

For a differentiable function of one variable, $y = f(x)$, we define the differential dx to be an independent variable; that is, dx can be given the value of any real number. The differential of y is then defined as

$$\boxed{9} \qquad\qquad dy = f'(x)\, dx$$

(See Section 2.8.) Figure 6 shows the relationship between the increment Δy and the differential dy: Δy represents the change in height of the curve $y = f(x)$ and dy represents the change in height of the tangent line when x changes by an amount $dx = \Delta x$.

For a differentiable function of two variables, $z = f(x, y)$, we define the **differentials** dx and dy to be independent variables; that is, they can be given any values. Then the **differential** dz, also called the **total differential**, is defined by

$$\boxed{10} \qquad dz = f_x(x, y)\, dx + f_y(x, y)\, dy = \frac{\partial z}{\partial x}\, dx + \frac{\partial z}{\partial y}\, dy$$

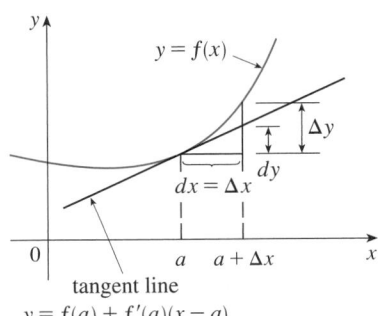

FIGURE 6

(Compare with Equation 9.) Sometimes the notation df is used in place of dz.

If we take $dx = \Delta x = x - a$ and $dy = \Delta y = y - b$ in Equation 10, then the differential of z is

$$dz = f_x(a, b)(x - a) + f_y(a, b)(y - b)$$

So, in the notation of differentials, the linear approximation $\boxed{4}$ can be written as

$$f(x, y) \approx f(a, b) + dz$$

Figure 7 is the three-dimensional counterpart of Figure 6 and shows the geometric interpretation of the differential dz and the increment Δz: dz represents the change in height of the tangent plane, whereas Δz represents the change in height of the surface $z = f(x, y)$ when (x, y) changes from (a, b) to $(a + \Delta x, b + \Delta y)$.

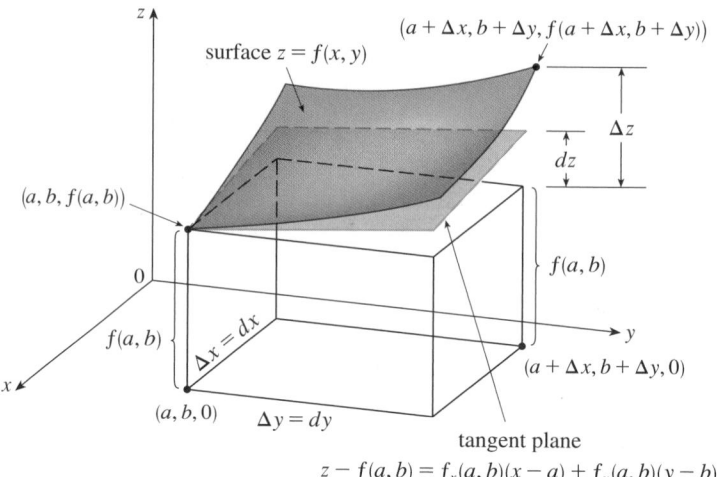

FIGURE 7

$$z - f(a, b) = f_x(a, b)(x - a) + f_y(a, b)(y - b)$$

V EXAMPLE 3

(a) If $z = f(x, y) = x^2 + 3xy - y^2$, find the differential dz.

(b) If x changes from 2 to 2.05 and y changes from 3 to 2.96, compare the values of Δz and dz.

SOLUTION

(a) Definition 10 gives

$$dz = \frac{\partial z}{\partial x}\, dx + \frac{\partial z}{\partial y}\, dy = (2x + 3y)\, dx + (3x - 2y)\, dy$$

■ In Example 3, dz is close to Δz because the tangent plane is a good approximation to the surface $z = x^2 + 3xy - y^2$ near $(2, 3, 13)$. (See Figure 8.)

(b) Putting $x = 2$, $dx = \Delta x = 0.05$, $y = 3$, and $dy = \Delta y = -0.04$, we get

$$dz = [2(2) + 3(3)]0.05 + [3(2) - 2(3)](-0.04)$$
$$= 0.65$$

The increment of z is

$$\Delta z = f(2.05, 2.96) - f(2, 3)$$
$$= [(2.05)^2 + 3(2.05)(2.96) - (2.96)^2] - [2^2 + 3(2)(3) - 3^2]$$
$$= 0.6449$$

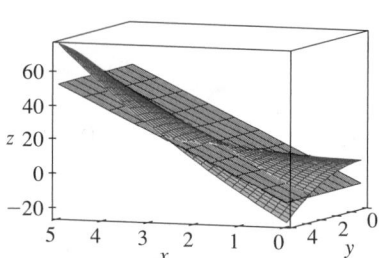

FIGURE 8

Notice that $\Delta z \approx dz$ but dz is easier to compute. ∎

EXAMPLE 4 The base radius and height of a right circular cone are measured as 10 cm and 25 cm, respectively, with a possible error in measurement of as much as 0.1 cm in each. Use differentials to estimate the maximum error in the calculated volume of the cone.

SOLUTION The volume V of a cone with base radius r and height h is $V = \pi r^2 h/3$. So the differential of V is

$$dV = \frac{\partial V}{\partial r}\, dr + \frac{\partial V}{\partial h}\, dh = \frac{2\pi rh}{3}\, dr + \frac{\pi r^2}{3}\, dh$$

Since each error is at most 0.1 cm, we have $|\Delta r| \leq 0.1$, $|\Delta h| \leq 0.1$. To estimate the largest error in the volume we take the largest error in the measurement of r and of h. Therefore we take $dr = 0.1$ and $dh = 0.1$ along with $r = 10$, $h = 25$. This gives

$$dV = \frac{500\pi}{3}\, (0.1) + \frac{100\pi}{3}\, (0.1) = 20\pi$$

Thus the maximum error in the calculated volume is about 20π cm$^3 \approx 63$ cm^3. ■

FUNCTIONS OF THREE OR MORE VARIABLES

Linear approximations, differentiability, and differentials can be defined in a similar manner for functions of more than two variables. A differentiable function is defined by an expression similar to the one in Definition 7. For such functions the **linear approximation** is

$$f(x, y, z) \approx f(a, b, c) + f_x(a, b, c)(x - a) + f_y(a, b, c)(y - b) + f_z(a, b, c)(z - c)$$

and the linearization $L(x, y, z)$ is the right side of this expression.

If $w = f(x, y, z)$, then the **increment** of w is

$$\Delta w = f(x + \Delta x, y + \Delta y, z + \Delta z) - f(x, y, z)$$

The **differential** dw is defined in terms of the differentials dx, dy, and dz of the independent variables by

$$dw = \frac{\partial w}{\partial x}\, dx + \frac{\partial w}{\partial y}\, dy + \frac{\partial w}{\partial z}\, dz$$

EXAMPLE 5 The dimensions of a rectangular box are measured to be 75 cm, 60 cm, and 40 cm, and each measurement is correct to within 0.2 cm. Use differentials to estimate the largest possible error when the volume of the box is calculated from these measurements.

SOLUTION If the dimensions of the box are x, y, and z, its volume is $V = xyz$ and so

$$dV = \frac{\partial V}{\partial x}\, dx + \frac{\partial V}{\partial y}\, dy + \frac{\partial V}{\partial z}\, dz = yz\, dx + xz\, dy + xy\, dz$$

We are given that $|\Delta x| \leq 0.2$, $|\Delta y| \leq 0.2$, and $|\Delta z| \leq 0.2$. To estimate the largest error in the volume, we therefore use $dx = 0.2$, $dy = 0.2$, and $dz = 0.2$ together with $x = 75$, $y = 60$, and $z = 40$:

$$\Delta V \approx dV = (60)(40)(0.2) + (75)(40)(0.2) + (75)(60)(0.2) = 1980$$

Thus an error of only 0.2 cm in measuring each dimension could lead to an error of approximately 1980 cm^3 in the calculated volume! This may seem like a large error, but it's only about 1% of the volume of the box. ■

11.4 | EXERCISES

1–6 ▪ Find an equation of the tangent plane to the given surface at the specified point.

1. $z = 3y^2 - 2x^2 + x$, $(2, -1, -3)$

2. $z = 3(x - 1)^2 + 2(y + 3)^2 + 7$, $(2, -2, 12)$

3. $z = \sqrt{xy}$, $(1, 1, 1)$

4. $z = xe^{xy}$, $(2, 0, 2)$

5. $z = x \sin(x + y)$, $(-1, 1, 0)$

6. $z = \ln(x - 2y)$, $(3, 1, 0)$

7–8 ▪ Graph the surface and the tangent plane at the given point. (Choose the domain and viewpoint so that you get a good view of both the surface and the tangent plane.) Then zoom in until the surface and the tangent plane become indistinguishable.

7. $z = x^2 + xy + 3y^2$, $(1, 1, 5)$

8. $z = \arctan(xy^2)$, $(1, 1, \pi/4)$

CAS **9–10** ▪ Draw the graph of f and its tangent plane at the given point. (Use your computer algebra system both to compute the partial derivatives and to graph the surface and its tangent plane.) Then zoom in until the surface and the tangent plane become indistinguishable.

9. $f(x, y) = \dfrac{xy \sin(x - y)}{1 + x^2 + y^2}$, $(1, 1, 0)$

10. $f(x, y) = e^{-xy/10}\left(\sqrt{x} + \sqrt{y} + \sqrt{xy}\right)$, $(1, 1, 3e^{-0.1})$

11–14 ▪ Explain why the function is differentiable at the given point. Then find the linearization $L(x, y)$ of the function at that point.

11. $f(x, y) = 1 + x \ln(xy - 5)$, $(2, 3)$

12. $f(x, y) = y + \sin(x/y)$, $(0, 3)$

13. $f(x, y) = e^{-xy} \cos y$, $(\pi, 0)$

14. $f(x, y) = \sqrt{x + e^{4y}}$, $(3, 0)$

15–16 ▪ Verify the linear approximation at $(0, 0)$.

15. $\dfrac{2x + 3}{4y + 1} \approx 3 + 2x - 12y$

16. $\sqrt{y + \cos^2 x} \approx 1 + \tfrac{1}{2}y$

---

17. Given that f is a differentiable function with $f(2, 5) = 6$, $f_x(2, 5) = 1$, and $f_y(2, 5) = -1$, use a linear approximation to estimate $f(2.2, 4.9)$.

18. Find the linear approximation of the function $f(x, y) = 1 - xy \cos \pi y$ at $(1, 1)$ and use it to approximate $f(1.02, 0.97)$. Illustrate by graphing f and the tangent plane.

19. Find the linear approximation of the function $f(x, y, z) = \sqrt{x^2 + y^2 + z^2}$ at $(3, 2, 6)$ and use it to approximate the number $\sqrt{(3.02)^2 + (1.97)^2 + (5.99)^2}$.

20–24 ▪ Find the differential of the function.

20. $u = \sqrt{x^2 + 3y^2}$

21. $m = p^5 q^3$

22. $T = \dfrac{v}{1 + uvw}$

23. $R = \alpha \beta^2 \cos \gamma$

24. $L = xze^{-y^2 - z^2}$

---

25. If $z = 5x^2 + y^2$ and (x, y) changes from $(1, 2)$ to $(1.05, 2.1)$, compare the values of Δz and dz.

26. If $z = x^2 - xy + 3y^2$ and (x, y) changes from $(3, -1)$ to $(2.96, -0.95)$, compare the values of Δz and dz.

27. The length and width of a rectangle are measured as 30 cm and 24 cm, respectively, with an error in measurement of at most 0.1 cm in each. Use differentials to estimate the maximum error in the calculated area of the rectangle.

28. Use differentials to estimate the amount of metal in a closed cylindrical can that is 10 cm high and 4 cm in diameter if the metal in the top and bottom is 0.1 cm thick and the metal in the sides is 0.05 cm thick.

29. Use differentials to estimate the amount of tin in a closed tin can with diameter 8 cm and height 12 cm if the tin is 0.04 cm thick.

30. The pressure, volume, and temperature of a mole of an ideal gas are related by the equation $PV = 8.31T$, where P is measured in kilopascals, V in liters, and T in kelvins. Use differentials to find the approximate change in the pressure if the volume increases from 12 L to 12.3 L and the temperature decreases from 310 K to 305 K.

31. A model for the surface area of a human body is given by $S = 0.1091w^{0.425}h^{0.725}$, where w is the weight (in pounds), h is the height (in inches), and S is measured in square feet. If the errors in measurement of w and h are at most 2%, use differentials to estimate the maximum percentage error in the calculated surface area.

32. The wind-chill index is modeled by the function

$$W = 13.12 + 0.6215T - 11.37v^{0.16} + 0.3965Tv^{0.16}$$

where T is the temperature (in °C) and v is the wind speed (in km/h). The wind speed is measured as 26 km/h, with a possible error of ± 2 km/h, and the temperature is measured as $-11°C$, with a possible error of $\pm 1°C$. Use differentials to estimate the maximum error in the calculated value of W due to the measurement errors in T and v.

33. If R is the total resistance of three resistors, connected in parallel, with resistances R_1, R_2, R_3, then

$$\frac{1}{R} = \frac{1}{R_1} + \frac{1}{R_2} + \frac{1}{R_3}$$

If the resistances are measured in ohms as $R_1 = 25\ \Omega$, $R_2 = 40\ \Omega$, and $R_3 = 50\ \Omega$, with a possible error of 0.5% in each case, estimate the maximum error in the calculated value of R.

34. Suppose you need to know an equation of the tangent plane to a surface S at the point $P(2, 1, 3)$. You don't have an equation for S but you know that the curves

$$\mathbf{r}_1(t) = \langle 2 + 3t,\ 1 - t^2,\ 3 - 4t + t^2 \rangle$$

$$\mathbf{r}_2(u) = \langle 1 + u^2,\ 2u^3 - 1,\ 2u + 1 \rangle$$

both lie on S. Find an equation of the tangent plane at P.

35–36 ■ Show that the function is differentiable by finding values of ε_1 and ε_2 that satisfy Definition 7.

35. $f(x, y) = x^2 + y^2$

36. $f(x, y) = xy - 5y^2$

37. Prove that if f is a function of two variables that is differentiable at (a, b), then f is continuous at (a, b).
   Hint: Show that

$$\lim_{(\Delta x,\, \Delta y) \to (0,\, 0)} f(a + \Delta x, b + \Delta y) = f(a, b)$$

38. (a) The function

$$f(x, y) = \begin{cases} \dfrac{xy}{x^2 + y^2} & \text{if } (x, y) \neq (0, 0) \\ 0 & \text{if } (x, y) = (0, 0) \end{cases}$$

was graphed in Figure 4. Show that $f_x(0, 0)$ and $f_y(0, 0)$ both exist but f is not differentiable at $(0, 0)$. [*Hint:* Use the result of Exercise 37.]
   (b) Explain why f_x and f_y are not continuous at $(0, 0)$.

11.5 | THE CHAIN RULE

Recall that the Chain Rule for functions of a single variable gives the rule for differentiating a composite function: If $y = f(x)$ and $x = g(t)$, where f and g are differentiable functions, then y is indirectly a differentiable function of t and

$$\boxed{1} \qquad \frac{dy}{dt} = \frac{dy}{dx} \frac{dx}{dt}$$

For functions of more than one variable, the Chain Rule has several versions, each of them giving a rule for differentiating a composite function. The first version (Theorem 2) deals with the case where $z = f(x, y)$ and each of the variables x and y is, in turn, a function of a variable t. This means that z is indirectly a function of t, $z = f(g(t), h(t))$, and the Chain Rule gives a formula for differentiating z as a function of t. We assume that f is differentiable (Definition 11.4.7). Recall that this is the case when f_x and f_y are continuous (Theorem 11.4.8).

> **2** **THE CHAIN RULE (CASE 1)** Suppose that $z = f(x, y)$ is a differentiable function of x and y, where $x = g(t)$ and $y = h(t)$ are both differentiable functions of t. Then z is a differentiable function of t and
>
> $$\frac{dz}{dt} = \frac{\partial f}{\partial x} \frac{dx}{dt} + \frac{\partial f}{\partial y} \frac{dy}{dt}$$

PROOF A change of Δt in t produces changes of Δx in x and Δy in y. These, in turn, produce a change of Δz in z, and from Definition 11.4.7 we have

$$\Delta z = \frac{\partial f}{\partial x} \Delta x + \frac{\partial f}{\partial y} \Delta y + \varepsilon_1 \Delta x + \varepsilon_2 \Delta y$$

where $\varepsilon_1 \to 0$ and $\varepsilon_2 \to 0$ as $(\Delta x, \Delta y) \to (0, 0)$. [If the functions ε_1 and ε_2 are not defined at $(0, 0)$, we can define them to be 0 there.] Dividing both sides of this equation by Δt, we have

$$\frac{\Delta z}{\Delta t} = \frac{\partial f}{\partial x} \frac{\Delta x}{\Delta t} + \frac{\partial f}{\partial y} \frac{\Delta y}{\Delta t} + \varepsilon_1 \frac{\Delta x}{\Delta t} + \varepsilon_2 \frac{\Delta y}{\Delta t}$$

If we now let $\Delta t \to 0$, then $\Delta x = g(t + \Delta t) - g(t) \to 0$ because g is differentiable and therefore continuous. Similarly, $\Delta y \to 0$. This, in turn, means that $\varepsilon_1 \to 0$ and $\varepsilon_2 \to 0$, so

$$\frac{dz}{dt} = \lim_{\Delta t \to 0} \frac{\Delta z}{\Delta t}$$

$$= \frac{\partial f}{\partial x} \lim_{\Delta t \to 0} \frac{\Delta x}{\Delta t} + \frac{\partial f}{\partial y} \lim_{\Delta t \to 0} \frac{\Delta y}{\Delta t} + \lim_{\Delta t \to 0} \varepsilon_1 \lim_{\Delta t \to 0} \frac{\Delta x}{\Delta t} + \lim_{\Delta t \to 0} \varepsilon_2 \lim_{\Delta t \to 0} \frac{\Delta y}{\Delta t}$$

$$= \frac{\partial f}{\partial x} \frac{dx}{dt} + \frac{\partial f}{\partial y} \frac{dy}{dt} + 0 \cdot \frac{dx}{dt} + 0 \cdot \frac{dy}{dt}$$

$$= \frac{\partial f}{\partial x} \frac{dx}{dt} + \frac{\partial f}{\partial y} \frac{dy}{dt}$$

Since we often write $\partial z/\partial x$ in place of $\partial f/\partial x$, we can rewrite the Chain Rule in the form

$$\boxed{\frac{dz}{dt} = \frac{\partial z}{\partial x} \frac{dx}{dt} + \frac{\partial z}{\partial y} \frac{dy}{dt}}$$

▪ Notice the similarity to the definition of the differential:

$$dz = \frac{\partial z}{\partial x} dx + \frac{\partial z}{\partial y} dy$$

EXAMPLE 1 If $z = x^2 y + 3xy^4$, where $x = \sin 2t$ and $y = \cos t$, find dz/dt when $t = 0$.

SOLUTION The Chain Rule gives

$$\frac{dz}{dt} = \frac{\partial z}{\partial x} \frac{dx}{dt} + \frac{\partial z}{\partial y} \frac{dy}{dt}$$

$$= (2xy + 3y^4)(2 \cos 2t) + (x^2 + 12xy^3)(-\sin t)$$

It's not necessary to substitute the expressions for x and y in terms of t. We simply observe that when $t = 0$ we have $x = \sin 0 = 0$ and $y = \cos 0 = 1$. Therefore

$$\frac{dz}{dt}\bigg|_{t=0} = (0 + 3)(2 \cos 0) + (0 + 0)(-\sin 0) = 6$$

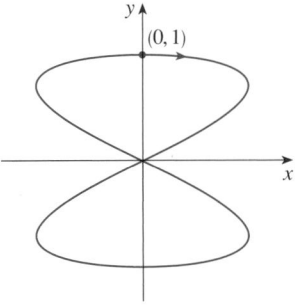

FIGURE 1
The curve $x = \sin 2t$, $y = \cos t$

The derivative in Example 1 can be interpreted as the rate of change of z with respect to t as the point (x, y) moves along the curve C with parametric equations $x = \sin 2t$, $y = \cos t$. (See Figure 1.) In particular, when $t = 0$, the point (x, y) is $(0, 1)$ and $dz/dt = 6$ is the rate of increase as we move along the curve C through $(0, 1)$. If, for instance, $z = T(x, y) = x^2y + 3xy^4$ represents the temperature at the point (x, y), then the composite function $z = T(\sin 2t, \cos t)$ represents the temperature at points on C and the derivative dz/dt represents the rate at which the temperature changes along C.

▼ EXAMPLE 2 The pressure P (in kilopascals), volume V (in liters), and temperature T (in kelvins) of a mole of an ideal gas are related by the equation $PV = 8.31T$. Find the rate at which the pressure is changing when the temperature is 300 K and increasing at a rate of 0.1 K/s and the volume is 100 L and increasing at a rate of 0.2 L/s.

SOLUTION If t represents the time elapsed in seconds, then at the given instant we have $T = 300$, $dT/dt = 0.1$, $V = 100$, $dV/dt = 0.2$. Since

$$P = 8.31\frac{T}{V}$$

the Chain Rule gives

$$\frac{dP}{dt} = \frac{\partial P}{\partial T}\frac{dT}{dt} + \frac{\partial P}{\partial V}\frac{dV}{dt} = \frac{8.31}{V}\frac{dT}{dt} - \frac{8.31T}{V^2}\frac{dV}{dt}$$

$$= \frac{8.31}{100}(0.1) - \frac{8.31(300)}{100^2}(0.2) = -0.04155$$

The pressure is decreasing at a rate of about 0.042 kPa/s. ∎

We now consider the situation where $z = f(x, y)$ but each of x and y is a function of two variables s and t: $x = g(s, t)$, $y = h(s, t)$. Then z is indirectly a function of s and t and we wish to find $\partial z/\partial s$ and $\partial z/\partial t$. Recall that in computing $\partial z/\partial t$ we hold s fixed and compute the ordinary derivative of z with respect to t. Therefore we can apply Theorem 2 to obtain

$$\frac{\partial z}{\partial t} = \frac{\partial z}{\partial x}\frac{\partial x}{\partial t} + \frac{\partial z}{\partial y}\frac{\partial y}{\partial t}$$

A similar argument holds for $\partial z/\partial s$ and so we have proved the following version of the Chain Rule.

---

3 THE CHAIN RULE (CASE 2) Suppose that $z = f(x, y)$ is a differentiable function of x and y, where $x = g(s, t)$ and $y = h(s, t)$ are differentiable functions of s and t. Then

$$\frac{\partial z}{\partial s} = \frac{\partial z}{\partial x}\frac{\partial x}{\partial s} + \frac{\partial z}{\partial y}\frac{\partial y}{\partial s} \qquad \frac{\partial z}{\partial t} = \frac{\partial z}{\partial x}\frac{\partial x}{\partial t} + \frac{\partial z}{\partial y}\frac{\partial y}{\partial t}$$

EXAMPLE 3 If $z = e^x \sin y$, where $x = st^2$ and $y = s^2t$, find $\partial z/\partial s$ and $\partial z/\partial t$.

SOLUTION Applying Case 2 of the Chain Rule, we get

$$\frac{\partial z}{\partial s} = \frac{\partial z}{\partial x}\frac{\partial x}{\partial s} + \frac{\partial z}{\partial y}\frac{\partial y}{\partial s} = (e^x \sin y)(t^2) + (e^x \cos y)(2st)$$

$$= t^2 e^{st^2} \sin(s^2t) + 2st e^{st^2} \cos(s^2t)$$

$$\frac{\partial z}{\partial t} = \frac{\partial z}{\partial x}\frac{\partial x}{\partial t} + \frac{\partial z}{\partial y}\frac{\partial y}{\partial t} = (e^x \sin y)(2st) + (e^x \cos y)(s^2)$$

$$= 2st e^{st^2} \sin(s^2t) + s^2 e^{st^2} \cos(s^2t) \qquad\blacksquare$$

Case 2 of the Chain Rule contains three types of variables: s and t are **independent** variables, x and y are called **intermediate** variables, and z is the **dependent** variable. Notice that Theorem 3 has one term for each intermediate variable and each of these terms resembles the one-dimensional Chain Rule in Equation 1.

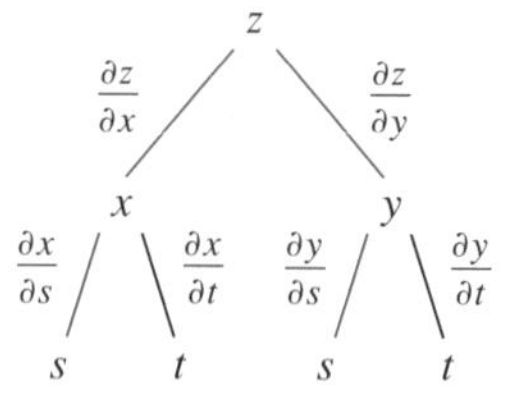

To remember the Chain Rule it's helpful to draw the **tree diagram** in Figure 2. We draw branches from the dependent variable z to the intermediate variables x and y to indicate that z is a function of x and y. Then we draw branches from x and y to the independent variables s and t. On each branch we write the corresponding partial derivative. To find $\partial z/\partial s$ we find the product of the partial derivatives along each path from z to s and then add these products:

$$\frac{\partial z}{\partial s} = \frac{\partial z}{\partial x}\frac{\partial x}{\partial s} + \frac{\partial z}{\partial y}\frac{\partial y}{\partial s}$$

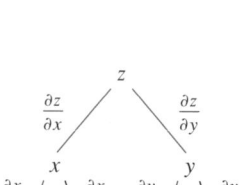

FIGURE 2

Similarly, we find $\partial z/\partial t$ by using the paths from z to t.

Now we consider the general situation in which a dependent variable u is a function of n intermediate variables $x_1, \ldots, x_n$, each of which is, in turn, a function of m independent variables $t_1, \ldots, t_m$. Notice that there are n terms, one for each intermediate variable. The proof is similar to that of Case 1.

---

4 **THE CHAIN RULE (GENERAL VERSION)** Suppose that u is a differentiable function of the n variables $x_1, x_2, \ldots, x_n$ and each x_j is a differentiable function of the m variables $t_1, t_2, \ldots, t_m$. Then u is a function of $t_1, t_2, \ldots, t_m$ and

$$\frac{\partial u}{\partial t_i} = \frac{\partial u}{\partial x_1}\frac{\partial x_1}{\partial t_i} + \frac{\partial u}{\partial x_2}\frac{\partial x_2}{\partial t_i} + \cdots + \frac{\partial u}{\partial x_n}\frac{\partial x_n}{\partial t_i}$$

for each $i = 1, 2, \ldots, m$.

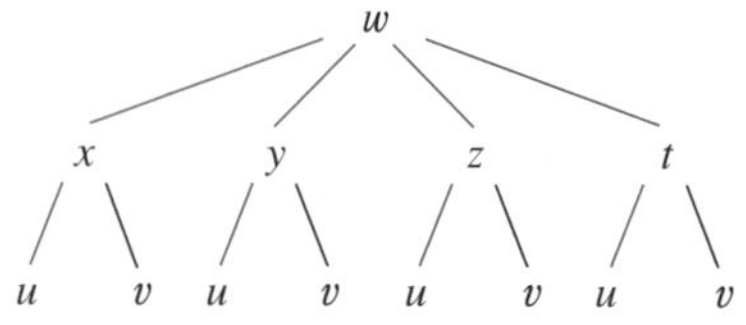

▼ EXAMPLE 4 Write out the Chain Rule for the case where $w = f(x, y, z, t)$ and $x = x(u, v)$, $y = y(u, v)$, $z = z(u, v)$, and $t = t(u, v)$.

SOLUTION We apply Theorem 4 with $n = 4$ and $m = 2$. Figure 3 shows the tree diagram. Although we haven't written the derivatives on the branches, it's understood that if a branch leads from y to u, then the partial derivative for that branch is

FIGURE 3

$\partial y/\partial u$. With the aid of the tree diagram we can now write the required expressions:

$$\frac{\partial w}{\partial u} = \frac{\partial w}{\partial x}\frac{\partial x}{\partial u} + \frac{\partial w}{\partial y}\frac{\partial y}{\partial u} + \frac{\partial w}{\partial z}\frac{\partial z}{\partial u} + \frac{\partial w}{\partial t}\frac{\partial t}{\partial u}$$

$$\frac{\partial w}{\partial v} = \frac{\partial w}{\partial x}\frac{\partial x}{\partial v} + \frac{\partial w}{\partial y}\frac{\partial y}{\partial v} + \frac{\partial w}{\partial z}\frac{\partial z}{\partial v} + \frac{\partial w}{\partial t}\frac{\partial t}{\partial v} \qquad \blacksquare$$

V EXAMPLE 5 If $u = x^4 y + y^2 z^3$, where $x = rse^t$, $y = rs^2 e^{-t}$, and $z = r^2 s \sin t$, find the value of $\partial u/\partial s$ when $r = 2$, $s = 1$, $t = 0$.

SOLUTION With the help of the tree diagram in Figure 4, we have

$$\frac{\partial u}{\partial s} = \frac{\partial u}{\partial x}\frac{\partial x}{\partial s} + \frac{\partial u}{\partial y}\frac{\partial y}{\partial s} + \frac{\partial u}{\partial z}\frac{\partial z}{\partial s}$$

$$= (4x^3 y)(re^t) + (x^4 + 2yz^3)(2rse^{-t}) + (3y^2 z^2)(r^2 \sin t)$$

When $r = 2$, $s = 1$, and $t = 0$, we have $x = 2$, $y = 2$, and $z = 0$, so

$$\frac{\partial u}{\partial s} = (64)(2) + (16)(4) + (0)(0) = 192 \qquad \blacksquare$$

EXAMPLE 6 If $g(s, t) = f(s^2 - t^2, t^2 - s^2)$ and f is differentiable, show that g satisfies the equation

$$t\frac{\partial g}{\partial s} + s\frac{\partial g}{\partial t} = 0$$

SOLUTION Let $x = s^2 - t^2$ and $y = t^2 - s^2$. Then $g(s, t) = f(x, y)$ and the Chain Rule gives

$$\frac{\partial g}{\partial s} = \frac{\partial f}{\partial x}\frac{\partial x}{\partial s} + \frac{\partial f}{\partial y}\frac{\partial y}{\partial s} = \frac{\partial f}{\partial x}(2s) + \frac{\partial f}{\partial y}(-2s)$$

$$\frac{\partial g}{\partial t} = \frac{\partial f}{\partial x}\frac{\partial x}{\partial t} + \frac{\partial f}{\partial y}\frac{\partial y}{\partial t} = \frac{\partial f}{\partial x}(-2t) + \frac{\partial f}{\partial y}(2t)$$

Therefore

$$t\frac{\partial g}{\partial s} + s\frac{\partial g}{\partial t} = \left(2st\frac{\partial f}{\partial x} - 2st\frac{\partial f}{\partial y}\right) + \left(-2st\frac{\partial f}{\partial x} + 2st\frac{\partial f}{\partial y}\right) = 0 \qquad \blacksquare$$

EXAMPLE 7 If $z = f(x, y)$ has continuous second-order partial derivatives and $x = r^2 + s^2$ and $y = 2rs$, find (a) $\partial z/\partial r$ and (b) $\partial^2 z/\partial r^2$.

SOLUTION
(a) The Chain Rule gives

$$\frac{\partial z}{\partial r} = \frac{\partial z}{\partial x}\frac{\partial x}{\partial r} + \frac{\partial z}{\partial y}\frac{\partial y}{\partial r} = \frac{\partial z}{\partial x}(2r) + \frac{\partial z}{\partial y}(2s)$$

(b) Applying the Product Rule to the expression in part (a), we get

$$\boxed{5} \qquad \frac{\partial^2 z}{\partial r^2} = \frac{\partial}{\partial r}\left(2r\frac{\partial z}{\partial x} + 2s\frac{\partial z}{\partial y}\right)$$

$$= 2\frac{\partial z}{\partial x} + 2r\frac{\partial}{\partial r}\left(\frac{\partial z}{\partial x}\right) + 2s\frac{\partial}{\partial r}\left(\frac{\partial z}{\partial y}\right)$$

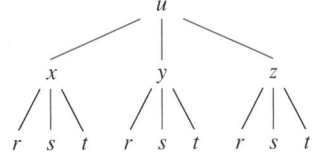

FIGURE 4

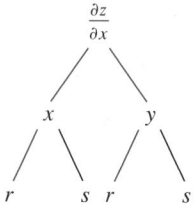

$\dfrac{\partial z}{\partial x}$

$x \qquad y$

$r \quad s \quad r \quad s$

FIGURE 5

But, using the Chain Rule again (see Figure 5), we have

$$\frac{\partial}{\partial r}\left(\frac{\partial z}{\partial x}\right) = \frac{\partial}{\partial x}\left(\frac{\partial z}{\partial x}\right)\frac{\partial x}{\partial r} + \frac{\partial}{\partial y}\left(\frac{\partial z}{\partial x}\right)\frac{\partial y}{\partial r} = \frac{\partial^2 z}{\partial x^2}\,(2r) + \frac{\partial^2 z}{\partial y\,\partial x}\,(2s)$$

$$\frac{\partial}{\partial r}\left(\frac{\partial z}{\partial y}\right) = \frac{\partial}{\partial x}\left(\frac{\partial z}{\partial y}\right)\frac{\partial x}{\partial r} + \frac{\partial}{\partial y}\left(\frac{\partial z}{\partial y}\right)\frac{\partial y}{\partial r} = \frac{\partial^2 z}{\partial x\,\partial y}\,(2r) + \frac{\partial^2 z}{\partial y^2}\,(2s)$$

Putting these expressions into Equation 5 and using the equality of the mixed second-order derivatives, we obtain

$$\frac{\partial^2 z}{\partial r^2} = 2\,\frac{\partial z}{\partial x} + 2r\left(2r\,\frac{\partial^2 z}{\partial x^2} + 2s\,\frac{\partial^2 z}{\partial y\,\partial x}\right) + 2s\left(2r\,\frac{\partial^2 z}{\partial x\,\partial y} + 2s\,\frac{\partial^2 z}{\partial y^2}\right)$$

$$= 2\,\frac{\partial z}{\partial x} + 4r^2\,\frac{\partial^2 z}{\partial x^2} + 8rs\,\frac{\partial^2 z}{\partial x\,\partial y} + 4s^2\,\frac{\partial^2 z}{\partial y^2} \qquad \blacksquare$$

IMPLICIT DIFFERENTIATION

The Chain Rule can be used to give a more complete description of the process of implicit differentiation that was introduced in Sections 2.6 and 11.3. We suppose that an equation of the form $F(x, y) = 0$ defines y implicitly as a differentiable function of x, that is, $y = f(x)$, where $F(x, f(x)) = 0$ for all x in the domain of f. If F is differentiable, we can apply Case 1 of the Chain Rule to differentiate both sides of the equation $F(x, y) = 0$ with respect to x. Since both x and y are functions of x, we obtain

$$\frac{\partial F}{\partial x}\frac{dx}{dx} + \frac{\partial F}{\partial y}\frac{dy}{dx} = 0$$

But $dx/dx = 1$, so if $\partial F/\partial y \neq 0$ we solve for dy/dx and obtain

$$\boxed{6} \qquad \boxed{\; \frac{dy}{dx} = -\frac{\dfrac{\partial F}{\partial x}}{\dfrac{\partial F}{\partial y}} = -\frac{F_x}{F_y} \;}$$

To derive this equation we assumed that $F(x, y) = 0$ defines y implicitly as a function of x. The **Implicit Function Theorem**, proved in advanced calculus, gives conditions under which this assumption is valid. It states that if F is defined on a disk containing (a, b), where $F(a, b) = 0$, $F_y(a, b) \neq 0$, and F_x and F_y are continuous on the disk, then the equation $F(x, y) = 0$ defines y as a function of x near the point (a, b) and the derivative of this function is given by Equation 6.

EXAMPLE 8 Find y' if $x^3 + y^3 = 6xy$.

SOLUTION The given equation can be written as

$$F(x, y) = x^3 + y^3 - 6xy = 0$$

so Equation 6 gives

■ The solution to Example 8 should be compared to the one in Example 2 in Section 2.6.

$$\frac{dy}{dx} = -\frac{F_x}{F_y} = -\frac{3x^2 - 6y}{3y^2 - 6x} = -\frac{x^2 - 2y}{y^2 - 2x}$$

■

Now we suppose that z is given implicitly as a function $z = f(x, y)$ by an equation of the form $F(x, y, z) = 0$. This means that $F(x, y, f(x, y)) = 0$ for all (x, y) in the domain of f. If F and f are differentiable, then we can use the Chain Rule to differentiate the equation $F(x, y, z) = 0$ as follows:

$$\frac{\partial F}{\partial x}\frac{\partial x}{\partial x} + \frac{\partial F}{\partial y}\frac{\partial y}{\partial x} + \frac{\partial F}{\partial z}\frac{\partial z}{\partial x} = 0$$

But

$$\frac{\partial}{\partial x}(x) = 1 \quad \text{and} \quad \frac{\partial}{\partial x}(y) = 0$$

so this equation becomes

$$\frac{\partial F}{\partial x} + \frac{\partial F}{\partial z}\frac{\partial z}{\partial x} = 0$$

If $\partial F/\partial z \neq 0$, we solve for $\partial z/\partial x$ and obtain the first formula in Equations 7. The formula for $\partial z/\partial y$ is obtained in a similar manner.

$$\boxed{7} \qquad \frac{\partial z}{\partial x} = -\frac{\dfrac{\partial F}{\partial x}}{\dfrac{\partial F}{\partial z}} \qquad \frac{\partial z}{\partial y} = -\frac{\dfrac{\partial F}{\partial y}}{\dfrac{\partial F}{\partial z}}$$

Again, a version of the **Implicit Function Theorem** gives conditions under which our assumption is valid. If F is defined within a sphere containing (a, b, c), where $F(a, b, c) = 0$, $F_z(a, b, c) \neq 0$, and F_x, F_y, and F_z are continuous inside the sphere, then the equation $F(x, y, z) = 0$ defines z as a function of x and y near the point (a, b, c) and this function is differentiable, with partial derivatives given by $\boxed{7}$.

EXAMPLE 9 Find $\dfrac{\partial z}{\partial x}$ and $\dfrac{\partial z}{\partial y}$ if $x^3 + y^3 + z^3 + 6xyz = 1$.

SOLUTION Let $F(x, y, z) = x^3 + y^3 + z^3 + 6xyz - 1$. Then, from Equations 7, we have

■ The solution to Example 9 should be compared to the one in Example 4 in Section 11.3.

$$\frac{\partial z}{\partial x} = -\frac{F_x}{F_z} = -\frac{3x^2 + 6yz}{3z^2 + 6xy} = -\frac{x^2 + 2yz}{z^2 + 2xy}$$

$$\frac{\partial z}{\partial y} = -\frac{F_y}{F_z} = -\frac{3y^2 + 6xz}{3z^2 + 6xy} = -\frac{y^2 + 2xz}{z^2 + 2xy}$$

■

11.5 | EXERCISES

1–4 ▪ Use the Chain Rule to find dz/dt or dw/dt.

1. $z = x^2 + y^2 + xy$, $x = \sin t$, $y = e^t$

2. $z = \cos(x + 4y)$, $x = 5t^4$, $y = 1/t$

3. $w = xe^{y/z}$, $x = t^2$, $y = 1 - t$, $z = 1 + 2t$

4. $w = \ln\sqrt{x^2 + y^2 + z^2}$, $x = \sin t$, $y = \cos t$, $z = \tan t$

5–8 ▪ Use the Chain Rule to find $\partial z/\partial s$ and $\partial z/\partial t$.

5. $z = x^2y^3$, $x = s \cos t$, $y = s \sin t$

6. $z = \arcsin(x - y)$, $x = s^2 + t^2$, $y = 1 - 2st$

7. $z = e^r \cos \theta$, $r = st$, $\theta = \sqrt{s^2 + t^2}$

8. $z = \tan(u/v)$, $u = 2s + 3t$, $v = 3s - 2t$

9. If $z = f(x, y)$, where f is differentiable, and

$$x = g(t) \qquad\qquad y = h(t)$$
$$g(3) = 2 \qquad\qquad h(3) = 7$$
$$g'(3) = 5 \qquad\qquad h'(3) = -4$$
$$f_x(2, 7) = 6 \qquad\qquad f_y(2, 7) = -8$$

find dz/dt when $t = 3$.

10. Let $W(s, t) = F(u(s, t), v(s, t))$, where F, u, and v are differentiable, and

$$u(1, 0) = 2 \qquad\qquad v(1, 0) = 3$$
$$u_s(1, 0) = -2 \qquad\qquad v_s(1, 0) = 5$$
$$u_t(1, 0) = 6 \qquad\qquad v_t(1, 0) = 4$$
$$F_u(2, 3) = -1 \qquad\qquad F_v(2, 3) = 10$$

Find $W_s(1, 0)$ and $W_t(1, 0)$.

11. Suppose f is a differentiable function of x and y, and $g(u, v) = f(e^u + \sin v, e^u + \cos v)$. Use the table of values to calculate $g_u(0, 0)$ and $g_v(0, 0)$.

	f	g	f_x	f_y
$(0, 0)$	3	6	4	8
$(1, 2)$	6	3	2	5

12. Suppose f is a differentiable function of x and y, and $g(r, s) = f(2r - s, s^2 - 4r)$. Use the table of values in Exercise 11 to calculate $g_r(1, 2)$ and $g_s(1, 2)$.

13–16 ▪ Use a tree diagram to write out the Chain Rule for the given case. Assume all functions are differentiable.

13. $u = f(x, y)$,   where $x = x(r, s, t)$, $y = y(r, s, t)$

14. $R = f(x, y, z, t)$,   where $x = x(u, v, w)$, $y = y(u, v, w)$, $z = z(u, v, w)$, $t = t(u, v, w)$

15. $w = f(r, s, t)$,   where $r = r(x, y)$, $s = s(x, y)$, $t = t(x, y)$

16. $t = f(u, v, w)$,   where $u = u(p, q, r, s)$, $v = v(p, q, r, s)$, $w = w(p, q, r, s)$

17–21 ▪ Use the Chain Rule to find the indicated partial derivatives.

17. $z = x^4 + x^2y$, $x = s + 2t - u$, $y = stu^2$;

$\dfrac{\partial z}{\partial s}, \dfrac{\partial z}{\partial t}, \dfrac{\partial z}{\partial u}$ when $s = 4, t = 2, u = 1$

18. $T = \dfrac{v}{2u + v}$, $u = pq\sqrt{r}$, $v = p\sqrt{q}\,r$;

$\dfrac{\partial T}{\partial p}, \dfrac{\partial T}{\partial q}, \dfrac{\partial T}{\partial r}$ when $p = 2, q = 1, r = 4$

19. $w = xy + yz + zx$, $x = r \cos \theta$, $y = r \sin \theta$, $z = r\theta$;

$\dfrac{\partial w}{\partial r}, \dfrac{\partial w}{\partial \theta}$ when $r = 2, \theta = \pi/2$

20. $P = \sqrt{u^2 + v^2 + w^2}$, $u = xe^y$, $v = ye^x$, $w = e^{xy}$;

$\dfrac{\partial P}{\partial x}, \dfrac{\partial P}{\partial y}$ when $x = 0, y = 2$

21. $N = \dfrac{p + q}{p + r}$, $p = u + vw$, $q = v + uw$, $r = w + uv$;

$\dfrac{\partial N}{\partial u}, \dfrac{\partial N}{\partial v}, \dfrac{\partial N}{\partial w}$ when $u = 2, v = 3, w = 4$

22–24 ▪ Use Equation 6 to find dy/dx.

22. $\cos(xy) = 1 + \sin y$

23. $\tan^{-1}(x^2y) = x + xy^2$ **24.** $e^y \sin x = x + xy$

25–28 ▪ Use Equations 7 to find $\partial z/\partial x$ and $\partial z/\partial y$.

25. $x^2 + 2y^2 + 3z^2 = 1$ **26.** $xyz = \cos(x + y + z)$

27. $e^z = xyz$ **28.** $yz + x \ln y = z^2$

29. The temperature at a point (x, y) is $T(x, y)$, measured in degrees Celsius. A bug crawls so that its position after t seconds is given by $x = \sqrt{1 + t}$, $y = 2 + \frac{1}{3}t$, where x and y are measured in centimeters. The temperature function satisfies $T_x(2, 3) = 4$ and $T_y(2, 3) = 3$. How fast is the temperature rising on the bug's path after 3 seconds?

30. Wheat production W in a given year depends on the average temperature T and the annual rainfall R. Scientists estimate that the average temperature is rising at a rate of $0.15°C/\text{year}$ and rainfall is decreasing at a rate of

0.1 cm/year. They also estimate that, at current production levels, $\partial W/\partial T = -2$ and $\partial W/\partial R = 8$.
(a) What is the significance of the signs of these partial derivatives?
(b) Estimate the current rate of change of wheat production, dW/dt.

31. The speed of sound traveling through ocean water with salinity 35 parts per thousand has been modeled by the equation

$$C = 1449.2 + 4.6T - 0.055T^2 + 0.00029T^3 + 0.016D$$

where C is the speed of sound (in meters per second), T is the temperature (in degrees Celsius), and D is the depth below the ocean surface (in meters). A scuba diver began a leisurely dive into the ocean water; the diver's depth and the surrounding water temperature over time are recorded in the following graphs. Estimate the rate of change (with respect to time) of the speed of sound through the ocean water experienced by the diver 20 minutes into the dive. What are the units?

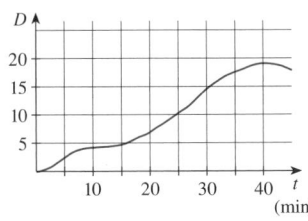
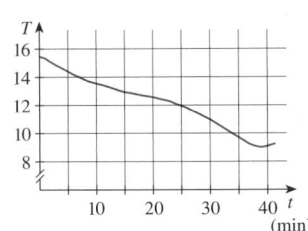

32. The radius of a right circular cone is increasing at a rate of 1.8 in/s while its height is decreasing at a rate of 2.5 in/s. At what rate is the volume of the cone changing when the radius is 120 in. and the height is 140 in.?

33. The length ℓ, width w, and height h of a box change with time. At a certain instant the dimensions are $\ell = 1$ m and $w = h = 2$ m, and ℓ and w are increasing at a rate of 2 m/s while h is decreasing at a rate of 3 m/s. At that instant find the rates at which the following quantities are changing.
(a) The volume
(b) The surface area
(c) The length of a diagonal

34. The voltage V in a simple electrical circuit is slowly decreasing as the battery wears out. The resistance R is slowly increasing as the resistor heats up. Use Ohm's Law, $V = IR$, to find how the current I is changing at the moment when $R = 400\ \Omega$, $I = 0.08$ A, $dV/dt = -0.01$ V/s, and $dR/dt = 0.03\ \Omega/s$.

35. The pressure of 1 mole of an ideal gas is increasing at a rate of 0.05 kPa/s and the temperature is increasing at a rate of 0.15 K/s. Use the equation in Example 2 to find the rate of change of the volume when the pressure is 20 kPa and the temperature is 320 K.

36. If a sound with frequency f_s is produced by a source traveling along a line with speed v_s and an observer is traveling with speed v_o along the same line from the opposite direction toward the source, then the frequency of the sound heard by the observer is

$$f_o = \left(\frac{c + v_o}{c - v_s}\right)f_s$$

where c is the speed of sound, about 332 m/s. (This is the **Doppler effect.**) Suppose that, at a particular moment, you are in a train traveling at 34 m/s and accelerating at 1.2 m/s². A train is approaching you from the opposite direction on the other track at 40 m/s, accelerating at 1.4 m/s², and sounds its whistle, which has a frequency of 460 Hz. At that instant, what is the perceived frequency that you hear and how fast is it changing?

37–40 ▪ Assume that all the given functions are differentiable.

37. If $z = f(x, y)$, where $x = r \cos \theta$ and $y = r \sin \theta$, (a) find $\partial z/\partial r$ and $\partial z/\partial \theta$ and (b) show that

$$\left(\frac{\partial z}{\partial x}\right)^2 + \left(\frac{\partial z}{\partial y}\right)^2 = \left(\frac{\partial z}{\partial r}\right)^2 + \frac{1}{r^2}\left(\frac{\partial z}{\partial \theta}\right)^2$$

38. If $u = f(x, y)$, where $x = e^s \cos t$ and $y = e^s \sin t$, show that

$$\left(\frac{\partial u}{\partial x}\right)^2 + \left(\frac{\partial u}{\partial y}\right)^2 = e^{-2s}\left[\left(\frac{\partial u}{\partial s}\right)^2 + \left(\frac{\partial u}{\partial t}\right)^2\right]$$

39. If $z = f(x - y)$, show that $\dfrac{\partial z}{\partial x} + \dfrac{\partial z}{\partial y} = 0$.

40. If $z = f(x, y)$, where $x = s + t$ and $y = s - t$, show that

$$\left(\frac{\partial z}{\partial x}\right)^2 - \left(\frac{\partial z}{\partial y}\right)^2 = \frac{\partial z}{\partial s}\frac{\partial z}{\partial t}$$

41–46 ▪ Assume that all the given functions have continuous second-order partial derivatives.

41. Show that any function of the form

$$z = f(x + at) + g(x - at)$$

is a solution of the wave equation

$$\frac{\partial^2 z}{\partial t^2} = a^2 \frac{\partial^2 z}{\partial x^2}$$

[*Hint:* Let $u = x + at$, $v = x - at$.]

42. If $u = f(x, y)$, where $x = e^s \cos t$ and $y = e^s \sin t$, show that

$$\frac{\partial^2 u}{\partial x^2} + \frac{\partial^2 u}{\partial y^2} = e^{-2s}\left[\frac{\partial^2 u}{\partial s^2} + \frac{\partial^2 u}{\partial t^2}\right]$$

43. If $z = f(x, y)$, where $x = r^2 + s^2$ and $y = 2rs$, find $\partial^2 z/\partial r\ \partial s$. (Compare with Example 7.)

44. If $z = f(x, y)$, where $x = r \cos \theta$ and $y = r \sin \theta$, find
(a) $\partial z / \partial r$, (b) $\partial z / \partial \theta$, and (c) $\partial^2 z / \partial r \, \partial \theta$.

45. If $z = f(x, y)$, where $x = r \cos \theta$ and $y = r \sin \theta$, show that

$$\frac{\partial^2 z}{\partial x^2} + \frac{\partial^2 z}{\partial y^2} = \frac{\partial^2 z}{\partial r^2} + \frac{1}{r^2} \frac{\partial^2 z}{\partial \theta^2} + \frac{1}{r} \frac{\partial z}{\partial r}$$

46. Suppose $z = f(x, y)$, where $x = g(s, t)$ and $y = h(s, t)$.
(a) Show that

$$\frac{\partial^2 z}{\partial t^2} = \frac{\partial^2 z}{\partial x^2} \left(\frac{\partial x}{\partial t} \right)^2 + 2 \frac{\partial^2 z}{\partial x \, \partial y} \frac{\partial x}{\partial t} \frac{\partial y}{\partial t} + \frac{\partial^2 z}{\partial y^2} \left(\frac{\partial y}{\partial t} \right)^2$$

$$+ \frac{\partial z}{\partial x} \frac{\partial^2 x}{\partial t^2} + \frac{\partial z}{\partial y} \frac{\partial^2 y}{\partial t^2}$$

(b) Find a similar formula for $\partial^2 z / \partial s \, \partial t$.

47. Suppose that the equation $F(x, y, z) = 0$ implicitly defines each of the three variables x, y, and z as functions of the other two: $z = f(x, y)$, $y = g(x, z)$, $x = h(y, z)$. If F is differentiable and F_x, F_y, and F_z are all nonzero, show that

$$\frac{\partial z}{\partial x} \frac{\partial x}{\partial y} \frac{\partial y}{\partial z} = -1$$

48. Equation 6 is a formula for the derivative dy/dx of a function defined implicitly by an equation $F(x, y) = 0$, provided that F is differentiable and $F_y \neq 0$. Prove that if F has continuous second derivatives, then a formula for the second derivative of y is

$$\frac{d^2 y}{dx^2} = -\frac{F_{xx} F_y^2 - 2 F_{xy} F_x F_y + F_{yy} F_x^2}{F_y^3}$$

11.6 | DIRECTIONAL DERIVATIVES AND THE GRADIENT VECTOR

Recall that if $z = f(x, y)$, then the partial derivatives f_x and f_y are defined as

1
$$f_x(x_0, y_0) = \lim_{h \to 0} \frac{f(x_0 + h, y_0) - f(x_0, y_0)}{h}$$

$$f_y(x_0, y_0) = \lim_{h \to 0} \frac{f(x_0, y_0 + h) - f(x_0, y_0)}{h}$$

and represent the rates of change of z in the x- and y-directions, that is, in the directions of the unit vectors **i** and **j**.

Suppose that we now wish to find the rate of change of z at (x_0, y_0) in the direction of an arbitrary unit vector $\mathbf{u} = \langle a, b \rangle$. (See Figure 1.) To do this we consider the surface S with equation $z = f(x, y)$ (the graph of f) and we let $z_0 = f(x_0, y_0)$. Then the point $P(x_0, y_0, z_0)$ lies on S. The vertical plane that passes through P in the direction of **u** intersects S in a curve C. (See Figure 2.) The slope of the tangent line T to C at the point P is the rate of change of z in the direction of **u**.

If $Q(x, y, z)$ is another point on C and P', Q' are the projections of P, Q on the xy-plane, then the vector $\overrightarrow{P'Q'}$ is parallel to **u** and so

$$\overrightarrow{P'Q'} = h\mathbf{u} = \langle ha, hb \rangle$$

for some scalar h. Therefore $x - x_0 = ha$, $y - y_0 = hb$, so $x = x_0 + ha$, $y = y_0 + hb$, and

$$\frac{\Delta z}{h} = \frac{z - z_0}{h} = \frac{f(x_0 + ha, y_0 + hb) - f(x_0, y_0)}{h}$$

If we take the limit as $h \to 0$, we obtain the rate of change of z (with respect to distance) in the direction of **u**, which is called the *directional derivative* of f in the direction of **u**.

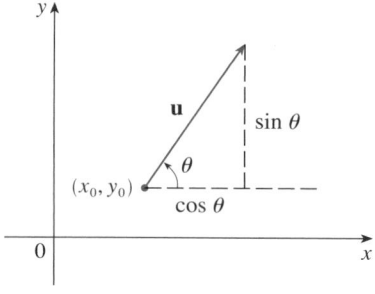

FIGURE 1
A unit vector $\mathbf{u} = \langle a, b \rangle = \langle \cos \theta, \sin \theta \rangle$

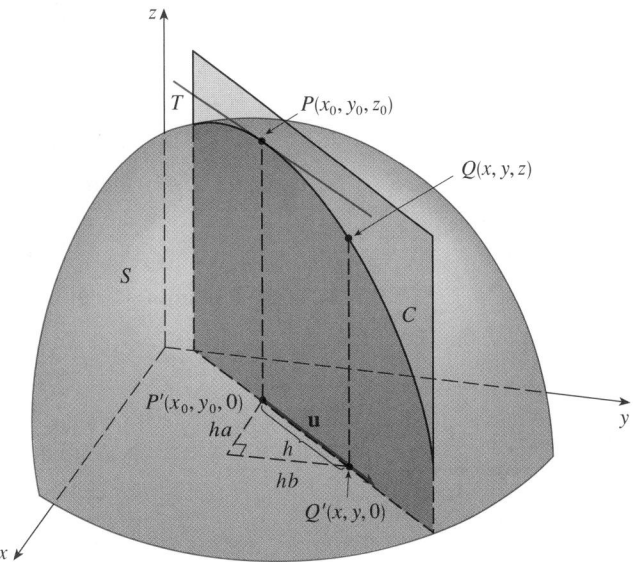

TEC Visual 11.6A animates Figure 2 by rotating **u** and therefore T.

FIGURE 2

---

2 DEFINITION The **directional derivative** of f at (x_0, y_0) in the direction of a unit vector $\mathbf{u} = \langle a, b \rangle$ is

$$D_{\mathbf{u}} f(x_0, y_0) = \lim_{h \to 0} \frac{f(x_0 + ha, y_0 + hb) - f(x_0, y_0)}{h}$$

if this limit exists.

By comparing Definition 2 with Equations 1, we see that if $\mathbf{u} = \mathbf{i} = \langle 1, 0 \rangle$, then $D_{\mathbf{i}} f = f_x$ and if $\mathbf{u} = \mathbf{j} = \langle 0, 1 \rangle$, then $D_{\mathbf{j}} f = f_y$. In other words, the partial derivatives of f with respect to x and y are just special cases of the directional derivative.

When we compute the directional derivative of a function defined by a formula, we generally use the following theorem.

---

3 THEOREM If f is a differentiable function of x and y, then f has a directional derivative in the direction of any unit vector $\mathbf{u} = \langle a, b \rangle$ and

$$D_{\mathbf{u}} f(x, y) = f_x(x, y)a + f_y(x, y)b$$

PROOF If we define a function g of the single variable h by

$$g(h) = f(x_0 + ha, y_0 + hb)$$

then by the definition of a derivative we have

$$
\boxed{4} \quad g'(0) = \lim_{h \to 0} \frac{g(h) - g(0)}{h} = \lim_{h \to 0} \frac{f(x_0 + ha, y_0 + hb) - f(x_0, y_0)}{h}
$$

$$= D_{\mathbf{u}} f(x_0, y_0)$$

On the other hand, we can write $g(h) = f(x, y)$, where $x = x_0 + ha$, $y = y_0 + hb$, so the Chain Rule (Theorem 11.5.2) gives

$$g'(h) = \frac{\partial f}{\partial x}\frac{dx}{dh} + \frac{\partial f}{\partial y}\frac{dy}{dh} = f_x(x, y)a + f_y(x, y)b$$

If we now put $h = 0$, then $x = x_0$, $y = y_0$, and

$$\boxed{5} \qquad g'(0) = f_x(x_0, y_0)a + f_y(x_0, y_0)b$$

Comparing Equations 4 and 5, we see that

$$D_{\mathbf{u}} f(x_0, y_0) = f_x(x_0, y_0)a + f_y(x_0, y_0)b \qquad \square$$

If the unit vector $\mathbf{u}$ makes an angle θ with the positive x-axis (as in Figure 1), then we can write $\mathbf{u} = \langle \cos\theta, \sin\theta \rangle$ and the formula in Theorem 3 becomes

$$\boxed{6} \qquad D_{\mathbf{u}} f(x, y) = f_x(x, y)\cos\theta + f_y(x, y)\sin\theta$$

■ The directional derivative $D_{\mathbf{u}} f(1, 2)$ in Example 1 represents the rate of change of z in the direction of $\mathbf{u}$. This is the slope of the tangent line to the curve of intersection of the surface $z = x^3 - 3xy + 4y^2$ and the vertical plane through $(1, 2, 0)$ in the direction of $\mathbf{u}$ shown in Figure 3.

EXAMPLE 1 Find the directional derivative $D_{\mathbf{u}} f(x, y)$ if $f(x, y) = x^3 - 3xy + 4y^2$ and $\mathbf{u}$ is the unit vector given by angle $\theta = \pi/6$. What is $D_{\mathbf{u}} f(1, 2)$?

SOLUTION Formula 6 gives

$$D_{\mathbf{u}} f(x, y) = f_x(x, y)\cos\frac{\pi}{6} + f_y(x, y)\sin\frac{\pi}{6} = (3x^2 - 3y)\frac{\sqrt{3}}{2} + (-3x + 8y)\tfrac{1}{2}$$
$$= \tfrac{1}{2}\left[3\sqrt{3}\,x^2 - 3x + \left(8 - 3\sqrt{3}\right)y\right]$$

Therefore

$$D_{\mathbf{u}} f(1, 2) = \tfrac{1}{2}\left[3\sqrt{3}\,(1)^2 - 3(1) + \left(8 - 3\sqrt{3}\right)(2)\right] = \frac{13 - 3\sqrt{3}}{2} \qquad \blacksquare$$

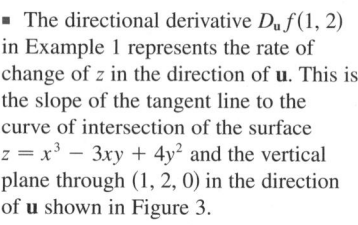

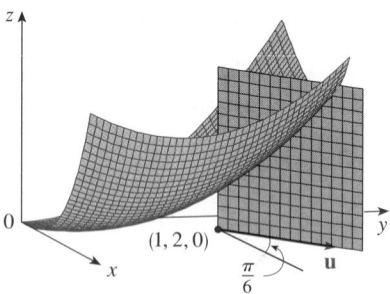

FIGURE 3

THE GRADIENT VECTOR

Notice from Theorem 3 that the directional derivative can be written as the dot product of two vectors:

$$\boxed{7} \qquad D_{\mathbf{u}} f(x, y) = f_x(x, y)a + f_y(x, y)b$$
$$= \langle f_x(x, y), f_y(x, y) \rangle \cdot \langle a, b \rangle$$
$$= \langle f_x(x, y), f_y(x, y) \rangle \cdot \mathbf{u}$$

The first vector in this dot product occurs not only in computing directional derivatives but in many other contexts as well. So we give it a special name (the *gradient* of f) and a special notation (**grad** f or ∇f, which is read "del f").

$\boxed{8}$ **DEFINITION** If f is a function of two variables x and y, then the **gradient** of f is the vector function ∇f defined by

$$\nabla f(x, y) = \langle f_x(x, y), f_y(x, y) \rangle = \frac{\partial f}{\partial x}\mathbf{i} + \frac{\partial f}{\partial y}\mathbf{j}$$

EXAMPLE 2 If $f(x, y) = \sin x + e^{xy}$, then

$$\nabla f(x, y) = \langle f_x, f_y \rangle = \langle \cos x + y e^{xy}, x e^{xy} \rangle$$

and
$$\nabla f(0, 1) = \langle 2, 0 \rangle \qquad \blacksquare$$

With this notation for the gradient vector, we can rewrite the expression $\boxed{7}$ for the directional derivative as

$$\boxed{9} \qquad \boxed{\quad D_{\mathbf{u}} f(x, y) = \nabla f(x, y) \cdot \mathbf{u} \quad}$$

This expresses the directional derivative in the direction of $\mathbf{u}$ as the scalar projection of the gradient vector onto $\mathbf{u}$.

$\boxed{\text{V}}$ **EXAMPLE 3** Find the directional derivative of the function $f(x, y) = x^2 y^3 - 4y$ at the point $(2, -1)$ in the direction of the vector $\mathbf{v} = 2\mathbf{i} + 5\mathbf{j}$.

SOLUTION We first compute the gradient vector at $(2, -1)$:

$$\nabla f(x, y) = 2xy^3 \mathbf{i} + (3x^2 y^2 - 4)\mathbf{j}$$

$$\nabla f(2, -1) = -4\mathbf{i} + 8\mathbf{j}$$

Note that $\mathbf{v}$ is not a unit vector, but since $|\mathbf{v}| = \sqrt{29}$, the unit vector in the direction of $\mathbf{v}$ is

$$\mathbf{u} = \frac{\mathbf{v}}{|\mathbf{v}|} = \frac{2}{\sqrt{29}}\mathbf{i} + \frac{5}{\sqrt{29}}\mathbf{j}$$

Therefore, by Equation 9, we have

$$D_{\mathbf{u}} f(2, -1) = \nabla f(2, -1) \cdot \mathbf{u} = (-4\mathbf{i} + 8\mathbf{j}) \cdot \left(\frac{2}{\sqrt{29}}\mathbf{i} + \frac{5}{\sqrt{29}}\mathbf{j} \right)$$

$$= \frac{-4 \cdot 2 + 8 \cdot 5}{\sqrt{29}} = \frac{32}{\sqrt{29}} \qquad \blacksquare$$

■ The gradient vector $\nabla f(2, -1)$ in Example 3 is shown in Figure 4 with initial point $(2, -1)$. Also shown is the vector $\mathbf{v}$ that gives the direction of the directional derivative. Both of these vectors are superimposed on a contour plot of the graph of f.

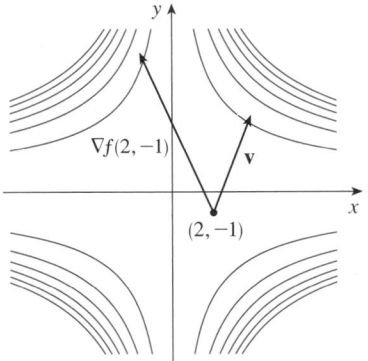

FIGURE 4

FUNCTIONS OF THREE VARIABLES

For functions of three variables we can define directional derivatives in a similar manner. Again $D_{\mathbf{u}} f(x, y, z)$ can be interpreted as the rate of change of the function in the direction of a unit vector $\mathbf{u}$.

$\boxed{10}$ **DEFINITION** The **directional derivative** of f at (x_0, y_0, z_0) in the direction of a unit vector $\mathbf{u} = \langle a, b, c \rangle$ is

$$D_{\mathbf{u}} f(x_0, y_0, z_0) = \lim_{h \to 0} \frac{f(x_0 + ha, y_0 + hb, z_0 + hc) - f(x_0, y_0, z_0)}{h}$$

if this limit exists.

If we use vector notation, then we can write both definitions (2 and 10) of the directional derivative in the compact form

$$\boxed{\, \text{11} \qquad D_{\mathbf{u}} f(\mathbf{x}_0) = \lim_{h \to 0} \frac{f(\mathbf{x}_0 + h\mathbf{u}) - f(\mathbf{x}_0)}{h} \,}$$

where $\mathbf{x}_0 = \langle x_0, y_0 \rangle$ if $n = 2$ and $\mathbf{x}_0 = \langle x_0, y_0, z_0 \rangle$ if $n = 3$. This is reasonable since the vector equation of the line through $\mathbf{x}_0$ in the direction of the vector $\mathbf{u}$ is given by $\mathbf{x} = \mathbf{x}_0 + t\mathbf{u}$ (Equation 10.5.1) and so $f(\mathbf{x}_0 + h\mathbf{u})$ represents the value of f at a point on this line.

If $f(x, y, z)$ is differentiable and $\mathbf{u} = \langle a, b, c \rangle$, then the same method that was used to prove Theorem 3 can be used to show that

$$\text{12} \qquad D_{\mathbf{u}} f(x, y, z) = f_x(x, y, z)a + f_y(x, y, z)b + f_z(x, y, z)c$$

For a function f of three variables, the **gradient vector**, denoted by ∇f or **grad** f, is

$$\nabla f(x, y, z) = \langle f_x(x, y, z), f_y(x, y, z), f_z(x, y, z) \rangle$$

or, for short,

$$\boxed{\, \text{13} \qquad \nabla f = \langle f_x, f_y, f_z \rangle = \frac{\partial f}{\partial x} \mathbf{i} + \frac{\partial f}{\partial y} \mathbf{j} + \frac{\partial f}{\partial z} \mathbf{k} \,}$$

Then, just as with functions of two variables, Formula 12 for the directional derivative can be rewritten as

$$\boxed{\, \text{14} \qquad D_{\mathbf{u}} f(x, y, z) = \nabla f(x, y, z) \cdot \mathbf{u} \,}$$

▼ EXAMPLE 4 If $f(x, y, z) = x \sin yz$, (a) find the gradient of f and (b) find the directional derivative of f at $(1, 3, 0)$ in the direction of $\mathbf{v} = \mathbf{i} + 2\mathbf{j} - \mathbf{k}$.

SOLUTION
(a) The gradient of f is

$$\nabla f(x, y, z) = \langle f_x(x, y, z), f_y(x, y, z), f_z(x, y, z) \rangle$$

$$= \langle \sin yz, \, xz \cos yz, \, xy \cos yz \rangle$$

(b) At $(1, 3, 0)$ we have $\nabla f(1, 3, 0) = \langle 0, 0, 3 \rangle$. The unit vector in the direction of $\mathbf{v} = \mathbf{i} + 2\mathbf{j} - \mathbf{k}$ is

$$\mathbf{u} = \frac{1}{\sqrt{6}} \mathbf{i} + \frac{2}{\sqrt{6}} \mathbf{j} - \frac{1}{\sqrt{6}} \mathbf{k}$$

Therefore Equation 14 gives

$$D_{\mathbf{u}} f(1, 3, 0) = \nabla f(1, 3, 0) \cdot \mathbf{u}$$

$$= 3\mathbf{k} \cdot \left(\frac{1}{\sqrt{6}} \mathbf{i} + \frac{2}{\sqrt{6}} \mathbf{j} - \frac{1}{\sqrt{6}} \mathbf{k} \right)$$

$$= 3\left(-\frac{1}{\sqrt{6}} \right) = -\sqrt{\frac{3}{2}}$$

MAXIMIZING THE DIRECTIONAL DERIVATIVE

Suppose we have a function f of two or three variables and we consider all possible directional derivatives of f at a given point. These give the rates of change of f in all possible directions. We can then ask the questions: In which of these directions does f change fastest and what is the maximum rate of change? The answers are provided by the following theorem.

TEC Visual 11.6B provides visual confirmation of Theorem 15.

> **15   THEOREM**   Suppose f is a differentiable function of two or three variables. The maximum value of the directional derivative $D_{\mathbf{u}} f(\mathbf{x})$ is $|\nabla f(\mathbf{x})|$ and it occurs when $\mathbf{u}$ has the same direction as the gradient vector $\nabla f(\mathbf{x})$.

PROOF   From Equation 9 or 14 we have

$$D_{\mathbf{u}} f = \nabla f \cdot \mathbf{u} = |\nabla f| |\mathbf{u}| \cos \theta = |\nabla f| \cos \theta$$

where θ is the angle between ∇f and $\mathbf{u}$. The maximum value of $\cos \theta$ is 1 and this occurs when $\theta = 0$. Therefore the maximum value of $D_{\mathbf{u}} f$ is $|\nabla f|$ and it occurs when $\theta = 0$, that is, when $\mathbf{u}$ has the same direction as ∇f.

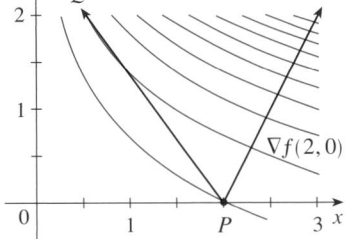

FIGURE 5

■ At $(2, 0)$ the function in Example 5 increases fastest in the direction of the gradient vector $\nabla f(2, 0) = \langle 1, 2 \rangle$. Notice from Figure 5 that this vector appears to be perpendicular to the level curve through $(2, 0)$. Figure 6 shows the graph of f and the gradient vector.

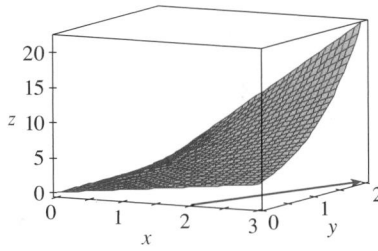

FIGURE 6

EXAMPLE 5
(a) If $f(x, y) = xe^y$, find the rate of change of f at the point $P(2, 0)$ in the direction from P to $Q(\frac{1}{2}, 2)$.
(b) In what direction does f have the maximum rate of change? What is this maximum rate of change?

SOLUTION
(a) We first compute the gradient vector:

$$\nabla f(x, y) = \langle f_x, f_y \rangle = \langle e^y, xe^y \rangle$$

$$\nabla f(2, 0) = \langle 1, 2 \rangle$$

The unit vector in the direction of $\overrightarrow{PQ} = \langle -1.5, 2 \rangle$ is $\mathbf{u} = \langle -\frac{3}{5}, \frac{4}{5} \rangle$, so the rate of change of f in the direction from P to Q is

$$D_{\mathbf{u}} f(2, 0) = \nabla f(2, 0) \cdot \mathbf{u} = \langle 1, 2 \rangle \cdot \langle -\frac{3}{5}, \frac{4}{5} \rangle$$

$$= 1\left(-\frac{3}{5} \right) + 2\left(\frac{4}{5} \right) = 1$$

(b) According to Theorem 15, f increases fastest in the direction of the gradient vector $\nabla f(2, 0) = \langle 1, 2 \rangle$. The maximum rate of change is

$$|\nabla f(2, 0)| = |\langle 1, 2 \rangle| = \sqrt{5}$$

EXAMPLE 6 Suppose that the temperature at a point (x, y, z) in space is given by $T(x, y, z) = 80/(1 + x^2 + 2y^2 + 3z^2)$, where T is measured in degrees Celsius and x, y, z in meters. In which direction does the temperature increase fastest at the point $(1, 1, -2)$? What is the maximum rate of increase?

SOLUTION The gradient of T is

$$\nabla T = \frac{\partial T}{\partial x}\, \mathbf{i} + \frac{\partial T}{\partial y}\, \mathbf{j} + \frac{\partial T}{\partial z}\, \mathbf{k}$$

$$= -\frac{160x}{(1 + x^2 + 2y^2 + 3z^2)^2}\, \mathbf{i} - \frac{320y}{(1 + x^2 + 2y^2 + 3z^2)^2}\, \mathbf{j} - \frac{480z}{(1 + x^2 + 2y^2 + 3z^2)^2}\, \mathbf{k}$$

$$= \frac{160}{(1 + x^2 + 2y^2 + 3z^2)^2}\, (-x\, \mathbf{i} - 2y\, \mathbf{j} - 3z\, \mathbf{k})$$

At the point $(1, 1, -2)$ the gradient vector is

$$\nabla T(1, 1, -2) = \tfrac{160}{256}(-\mathbf{i} - 2\mathbf{j} + 6\mathbf{k}) = \tfrac{5}{8}(-\mathbf{i} - 2\mathbf{j} + 6\mathbf{k})$$

By Theorem 15 the temperature increases fastest in the direction of the gradient vector $\nabla T(1, 1, -2) = \tfrac{5}{8}(-\mathbf{i} - 2\mathbf{j} + 6\mathbf{k})$ or, equivalently, in the direction of $-\mathbf{i} - 2\mathbf{j} + 6\mathbf{k}$ or the unit vector $(-\mathbf{i} - 2\mathbf{j} + 6\mathbf{k})/\sqrt{41}$. The maximum rate of increase is the length of the gradient vector:

$$|\nabla T(1, 1, -2)| = \tfrac{5}{8}|-\mathbf{i} - 2\mathbf{j} + 6\mathbf{k}| = \frac{5\sqrt{41}}{8}$$

Therefore the maximum rate of increase of temperature is $5\sqrt{41}/8 \approx 4°C/m$. ∎

TANGENT PLANES TO LEVEL SURFACES

Suppose S is a surface with equation $F(x, y, z) = k$, that is, it is a level surface of a function F of three variables, and let $P(x_0, y_0, z_0)$ be a point on S. Let C be any curve that lies on the surface S and passes through the point P. Recall from Section 10.7 that the curve C is described by a continuous vector function $\mathbf{r}(t) = \langle x(t), y(t), z(t) \rangle$. Let t_0 be the parameter value corresponding to P; that is, $\mathbf{r}(t_0) = \langle x_0, y_0, z_0 \rangle$. Since C lies on S, any point $(x(t), y(t), z(t))$ must satisfy the equation of S, that is,

$$\boxed{16} \qquad\qquad F\big(x(t), y(t), z(t)\big) = k$$

If x, y, and z are differentiable functions of t and F is also differentiable, then we can use the Chain Rule to differentiate both sides of Equation 16 as follows:

$$\boxed{17} \qquad\qquad \frac{\partial F}{\partial x}\frac{dx}{dt} + \frac{\partial F}{\partial y}\frac{dy}{dt} + \frac{\partial F}{\partial z}\frac{dz}{dt} = 0$$

But, since $\nabla F = \langle F_x, F_y, F_z \rangle$ and $\mathbf{r}'(t) = \langle x'(t), y'(t), z'(t) \rangle$, Equation 17 can be written in terms of a dot product as

$$\nabla F \cdot \mathbf{r}'(t) = 0$$

In particular, when $t = t_0$ we have $\mathbf{r}(t_0) = \langle x_0, y_0, z_0 \rangle$, so

$$\boxed{18} \qquad \nabla F(x_0, y_0, z_0) \cdot \mathbf{r}'(t_0) = 0$$

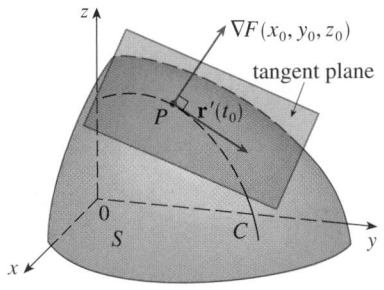

FIGURE 7

Equation 18 says that *the gradient vector at P, $\nabla F(x_0, y_0, z_0)$, is perpendicular to the tangent vector $\mathbf{r}'(t_0)$ to any curve C on S that passes through P.* (See Figure 7.) If $\nabla F(x_0, y_0, z_0) \neq \mathbf{0}$, it is therefore natural to define the **tangent plane to the level surface** $F(x, y, z) = k$ **at** $P(x_0, y_0, z_0)$ as the plane that passes through P and has normal vector $\nabla F(x_0, y_0, z_0)$. Using the standard equation of a plane (Equation 10.5.7), we can write the equation of this tangent plane as

$$\boxed{19} \quad F_x(x_0, y_0, z_0)(x - x_0) + F_y(x_0, y_0, z_0)(y - y_0) + F_z(x_0, y_0, z_0)(z - z_0) = 0$$

The **normal line** to S at P is the line passing through P and perpendicular to the tangent plane. The direction of the normal line is therefore given by the gradient vector $\nabla F(x_0, y_0, z_0)$ and so, by Equation 10.5.3, its symmetric equations are

$$\boxed{20} \qquad \frac{x - x_0}{F_x(x_0, y_0, z_0)} = \frac{y - y_0}{F_y(x_0, y_0, z_0)} = \frac{z - z_0}{F_z(x_0, y_0, z_0)}$$

In the special case in which the equation of a surface S is of the form $z = f(x, y)$ (that is, S is the graph of a function f of two variables), we can rewrite the equation as

$$F(x, y, z) = f(x, y) - z = 0$$

and regard S as a level surface (with $k = 0$) of F. Then

$$F_x(x_0, y_0, z_0) = f_x(x_0, y_0)$$

$$F_y(x_0, y_0, z_0) = f_y(x_0, y_0)$$

$$F_z(x_0, y_0, z_0) = -1$$

so Equation 19 becomes

$$f_x(x_0, y_0)(x - x_0) + f_y(x_0, y_0)(y - y_0) - (z - z_0) = 0$$

which is equivalent to Equation 11.4.2. Thus our new, more general, definition of a tangent plane is consistent with the definition that was given for the special case of Section 11.4.

▼ EXAMPLE 7 Find the equations of the tangent plane and normal line at the point $(-2, 1, -3)$ to the ellipsoid

$$\frac{x^2}{4} + y^2 + \frac{z^2}{9} = 3$$

SOLUTION The ellipsoid is the level surface (with $k = 3$) of the function

$$F(x, y, z) = \frac{x^2}{4} + y^2 + \frac{z^2}{9}$$

■ Figure 8 shows the ellipsoid, tangent plane, and normal line in Example 7.

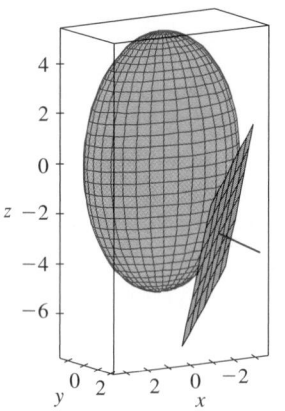

FIGURE 8

Therefore we have

$$F_x(x, y, z) = \frac{x}{2} \qquad F_y(x, y, z) = 2y \qquad F_z(x, y, z) = \frac{2z}{9}$$

$$F_x(-2, 1, -3) = -1 \qquad F_y(-2, 1, -3) = 2 \qquad F_z(-2, 1, -3) = -\tfrac{2}{3}$$

Then Equation 19 gives the equation of the tangent plane at $(-2, 1, -3)$ as

$$-1(x + 2) + 2(y - 1) - \tfrac{2}{3}(z + 3) = 0$$

which simplifies to $3x - 6y + 2z + 18 = 0$.

By Equation 20, symmetric equations of the normal line are

$$\frac{x + 2}{-1} = \frac{y - 1}{2} = \frac{z + 3}{-\tfrac{2}{3}}$$ ■

SIGNIFICANCE OF THE GRADIENT VECTOR

We now summarize the ways in which the gradient vector is significant. We first consider a function f of three variables and a point $P(x_0, y_0, z_0)$ in its domain. On the one hand, we know from Theorem 15 that the gradient vector $\nabla f(x_0, y_0, z_0)$ gives the direction of fastest increase of f. On the other hand, we know that $\nabla f(x_0, y_0, z_0)$ is orthogonal to the level surface S of f through P. (Refer to Figure 7.) These two properties are quite compatible intuitively because as we move away from P on the level surface S, the value of f does not change at all. So it seems reasonable that if we move in the perpendicular direction, we get the maximum increase.

In like manner we consider a function f of two variables and a point $P(x_0, y_0)$ in its domain. Again the gradient vector $\nabla f(x_0, y_0)$ gives the direction of fastest increase of f. Also, by considerations similar to our discussion of tangent planes, it can be shown that $\nabla f(x_0, y_0)$ is perpendicular to the level curve $f(x, y) = k$ that passes through P. Again this is intuitively plausible because the values of f remain constant as we move along the curve. (See Figure 9.)

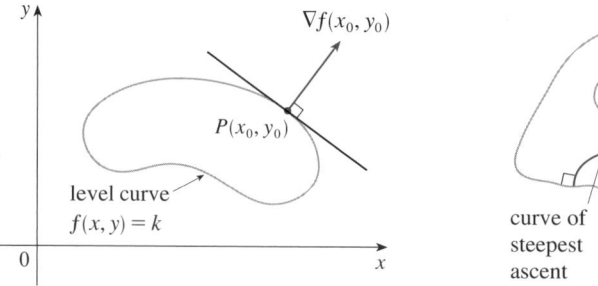

FIGURE 9

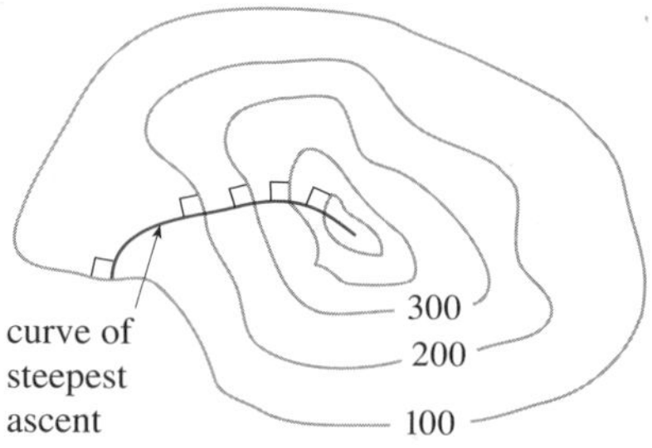

FIGURE 10

If we consider a topographical map of a hill and let $f(x, y)$ represent the height above sea level at a point with coordinates (x, y), then a curve of steepest ascent can be drawn as in Figure 10 by making it perpendicular to all of the contour lines. This phenomenon can also be noticed in Figure 11 in Section 11.1, where Lonesome Creek follows a curve of steepest descent.

11.6 | EXERCISES

1–2 ▪ Find the directional derivative of f at the given point in the direction indicated by the angle θ.

1. $f(x, y) = ye^{-x}$, $(0, 4)$, $\theta = 2\pi/3$

2. $f(x, y) = x^3y^4 + x^4y^3$, $(1, 1)$, $\theta = \pi/6$

3–6 ▪
(a) Find the gradient of f.
(b) Evaluate the gradient at the point P.
(c) Find the rate of change of f at P in the direction of the vector $\mathbf{u}$.

3. $f(x, y) = \sin(2x + 3y)$, $P(-6, 4)$, $\mathbf{u} = \frac{1}{2}(\sqrt{3}\,\mathbf{i} - \mathbf{j})$

4. $f(x, y) = y^2/x$, $P(1, 2)$, $\mathbf{u} = \frac{1}{3}(2\mathbf{i} + \sqrt{5}\,\mathbf{j})$

5. $f(x, y, z) = x^2yz - xyz^3$, $P(2, -1, 1)$, $\mathbf{u} = \langle 0, \frac{4}{5}, -\frac{3}{5} \rangle$

6. $f(x, y, z) = y^2e^{xyz}$, $P(0, 1, -1)$, $\mathbf{u} = \langle \frac{3}{13}, \frac{4}{13}, \frac{12}{13} \rangle$

7–11 ▪ Find the directional derivative of the function at the given point in the direction of the vector $\mathbf{v}$.

7. $f(x, y) = e^x \sin y$, $(0, \pi/3)$, $\mathbf{v} = \langle -6, 8 \rangle$

8. $f(x, y) = \dfrac{x}{x^2 + y^2}$, $(1, 2)$, $\mathbf{v} = \langle 3, 5 \rangle$

9. $g(p, q) = p^4 - p^2q^3$, $(2, 1)$, $\mathbf{v} = \mathbf{i} + 3\mathbf{j}$

10. $g(r, s) = \tan^{-1}(rs)$, $(1, 2)$, $\mathbf{v} = 5\mathbf{i} + 10\mathbf{j}$

11. $f(x, y, z) = xe^y + ye^z + ze^x$, $(0, 0, 0)$, $\mathbf{v} = \langle 5, 1, -2 \rangle$

12. Use the figure to estimate $D_{\mathbf{u}}f(2, 2)$.

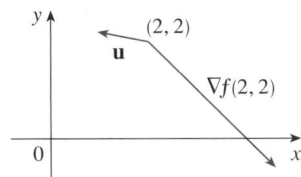

13. Find the directional derivative of $f(x, y) = \sqrt{xy}$ at $P(2, 8)$ in the direction of $Q(5, 4)$.

14. Find the directional derivative of $f(x, y, z) = xy + yz + zx$ at $P(1, -1, 3)$ in the direction of $Q(2, 4, 5)$.

15–18 ▪ Find the maximum rate of change of f at the given point and the direction in which it occurs.

15. $f(x, y) = \sin(xy)$, $(1, 0)$

16. $f(s, t) = te^{st}$, $(0, 2)$

17. $f(x, y, z) = \sqrt{x^2 + y^2 + z^2}$, $(3, 6, -2)$

18. $f(p, q, r) = \arctan(pqr)$, $(1, 2, 1)$

19. (a) Show that a differentiable function f decreases most rapidly at $\mathbf{x}$ in the direction opposite to the gradient vector, that is, in the direction of $-\nabla f(\mathbf{x})$.
(b) Use the result of part (a) to find the direction in which the function $f(x, y) = x^4y - x^2y^3$ decreases fastest at the point $(2, -3)$.

20. Find the directions in which the directional derivative of $f(x, y) = x^2 + \sin xy$ at the point $(1, 0)$ has the value 1.

21. Find all points at which the direction of fastest change of the function $f(x, y) = x^2 + y^2 - 2x - 4y$ is $\mathbf{i} + \mathbf{j}$.

22. Near a buoy, the depth of a lake at the point with coordinates (x, y) is $z = 200 + 0.02x^2 - 0.001y^3$, where x, y, and z are measured in meters. A fisherman in a small boat starts at the point $(80, 60)$ and moves toward the buoy, which is located at $(0, 0)$. Is the water under the boat getting deeper or shallower when he departs? Explain.

23. The temperature T in a metal ball is inversely proportional to the distance from the center of the ball, which we take to be the origin. The temperature at the point $(1, 2, 2)$ is $120°$.
(a) Find the rate of change of T at $(1, 2, 2)$ in the direction toward the point $(2, 1, 3)$.
(b) Show that at any point in the ball the direction of greatest increase in temperature is given by a vector that points toward the origin.

24. The temperature at a point (x, y, z) is given by

$$T(x, y, z) = 200e^{-x^2 - 3y^2 - 9z^2}$$

where T is measured in $°C$ and x, y, z in meters.
(a) Find the rate of change of temperature at the point $P(2, -1, 2)$ in the direction toward the point $(3, -3, 3)$.
(b) In which direction does the temperature increase fastest at P?
(c) Find the maximum rate of increase at P.

25. Suppose that over a certain region of space the electrical potential V is given by $V(x, y, z) = 5x^2 - 3xy + xyz$.
(a) Find the rate of change of the potential at $P(3, 4, 5)$ in the direction of the vector $\mathbf{v} = \mathbf{i} + \mathbf{j} - \mathbf{k}$.
(b) In which direction does V change most rapidly at P?
(c) What is the maximum rate of change at P?

26. Suppose you are climbing a hill whose shape is given by the equation $z = 1000 - 0.005x^2 - 0.01y^2$, where x, y, and z are measured in meters, and you are standing at a point with coordinates $(60, 40, 966)$. The positive x-axis points east and the positive y-axis points north.
(a) If you walk due south, will you start to ascend or descend? At what rate?

(b) If you walk northwest, will you start to ascend or descend? At what rate?

(c) In which direction is the slope largest? What is the rate of ascent in that direction? At what angle above the horizontal does the path in that direction begin?

27. Let f be a function of two variables that has continuous partial derivatives and consider the points $A(1, 3)$, $B(3, 3)$, $C(1, 7)$, and $D(6, 15)$. The directional derivative of f at A in the direction of the vector $\overrightarrow{AB}$ is 3 and the directional derivative at A in the direction of $\overrightarrow{AC}$ is 26. Find the directional derivative of f at A in the direction of the vector $\overrightarrow{AD}$.

28. Shown is a topographic map of Blue River Pine Provincial Park in British Columbia. Draw curves of steepest descent from point A (descending to Mud Lake) and from point B.

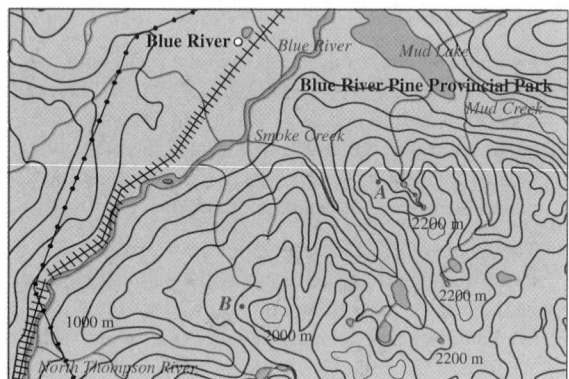

29. Show that the operation of taking the gradient of a function has the given property. Assume that u and v are differentiable functions of x and y and that a, b are constants.

(a) $\nabla(au + bv) = a\nabla u + b\nabla v$

(b) $\nabla(uv) = u\nabla v + v\nabla u$

(c) $\nabla\left(\dfrac{u}{v}\right) = \dfrac{v\nabla u - u\nabla v}{v^2}$ (d) $\nabla u^n = nu^{n-1}\nabla u$

30. Sketch the gradient vector $\nabla f(4, 6)$ for the function f whose level curves are shown. Explain how you chose the direction and length of this vector.

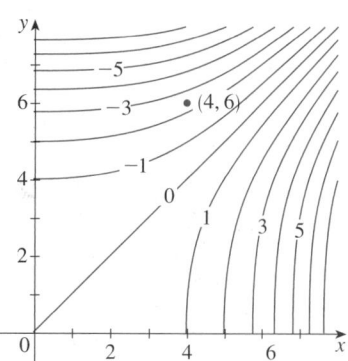

31–36 ■ Find equations of (a) the tangent plane and (b) the normal line to the given surface at the specified point.

31. $2(x - 2)^2 + (y - 1)^2 + (z - 3)^2 = 10$, $(3, 3, 5)$

32. $y = x^2 - z^2$, $(4, 7, 3)$

33. $xyz^2 = 6$, $(3, 2, 1)$

34. $xy + yz + zx = 5$, $(1, 2, 1)$

35. $x + y + z = e^{xyz}$, $(0, 0, 1)$

36. $x^4 + y^4 + z^4 = 3x^2y^2z^2$, $(1, 1, 1)$

37–38 ■ Use a computer to graph the surface, the tangent plane, and the normal line on the same screen. Choose the domain carefully so that you avoid extraneous vertical planes. Choose the viewpoint so that you get a good view of all three objects.

37. $xy + yz + zx = 3$, $(1, 1, 1)$ **38.** $xyz = 6$, $(1, 2, 3)$

39. If $f(x, y) = xy$, find the gradient vector $\nabla f(3, 2)$ and use it to find the tangent line to the level curve $f(x, y) = 6$ at the point $(3, 2)$. Sketch the level curve, the tangent line, and the gradient vector.

40. If $g(x, y) = x^2 + y^2 - 4x$, find the gradient vector $\nabla g(1, 2)$ and use it to find the tangent line to the level curve $g(x, y) = 1$ at the point $(1, 2)$. Sketch the level curve, the tangent line, and the gradient vector.

41. Show that the equation of the tangent plane to the ellipsoid $x^2/a^2 + y^2/b^2 + z^2/c^2 = 1$ at the point (x_0, y_0, z_0) can be written as

$$\frac{xx_0}{a^2} + \frac{yy_0}{b^2} + \frac{zz_0}{c^2} = 1$$

42. At what point on the paraboloid $y = x^2 + z^2$ is the tangent plane parallel to the plane $x + 2y + 3z = 1$?

43. Are there any points on the hyperboloid $x^2 - y^2 - z^2 = 1$ where the tangent plane is parallel to the plane $z = x + y$?

44. Show that the ellipsoid $3x^2 + 2y^2 + z^2 = 9$ and the sphere $x^2 + y^2 + z^2 - 8x - 6y - 8z + 24 = 0$ are tangent to each other at the point $(1, 1, 2)$. (This means that they have a common tangent plane at the point.)

45. Where does the normal line to the paraboloid $z = x^2 + y^2$ at the point $(1, 1, 2)$ intersect the paraboloid a second time?

46. At what points does the normal line through the point $(1, 2, 1)$ on the ellipsoid $4x^2 + y^2 + 4z^2 = 12$ intersect the sphere $x^2 + y^2 + z^2 = 102$?

47. Show that the sum of the x-, y-, and z-intercepts of any tangent plane to the surface $\sqrt{x} + \sqrt{y} + \sqrt{z} = \sqrt{c}$ is a constant.

48. Show that every normal line to the sphere $x^2 + y^2 + z^2 = r^2$ passes through the center of the sphere.

49. Find parametric equations for the tangent line to the curve of intersection of the paraboloid $z = x^2 + y^2$ and the ellipsoid $4x^2 + y^2 + z^2 = 9$ at the point $(-1, 1, 2)$.

50. (a) The plane $y + z = 3$ intersects the cylinder $x^2 + y^2 = 5$ in an ellipse. Find parametric equations for the tangent line to this ellipse at the point $(1, 2, 1)$.

(b) Graph the cylinder, the plane, and the tangent line on the same screen.

51. (a) Two surfaces are called **orthogonal** at a point of intersection if their normal lines are perpendicular at that point. Show that surfaces with equations $F(x, y, z) = 0$ and $G(x, y, z) = 0$ are orthogonal at a point P where $\nabla F \neq \mathbf{0}$ and $\nabla G \neq \mathbf{0}$ if and only if

$$F_x G_x + F_y G_y + F_z G_z = 0 \quad \text{at } P$$

(b) Use part (a) to show that the surfaces $z^2 = x^2 + y^2$ and $x^2 + y^2 + z^2 = r^2$ are orthogonal at every point of intersection. Can you see why this is true without using calculus?

52. (a) Show that the function $f(x, y) = \sqrt[3]{xy}$ is continuous and the partial derivatives f_x and f_y exist at the origin but the directional derivatives in all other directions do not exist.

(b) Graph f near the origin and comment on how the graph confirms part (a).

53. Suppose that the directional derivatives of $f(x, y)$ are known at a given point in two nonparallel directions given by unit vectors $\mathbf{u}$ and $\mathbf{v}$. Is it possible to find ∇f at this point? If so, how would you do it?

54. Show that if $z = f(x, y)$ is differentiable at $\mathbf{x}_0 = \langle x_0, y_0 \rangle$, then

$$\lim_{\mathbf{x} \to \mathbf{x}_0} \frac{f(\mathbf{x}) - f(\mathbf{x}_0) - \nabla f(\mathbf{x}_0) \cdot (\mathbf{x} - \mathbf{x}_0)}{|\mathbf{x} - \mathbf{x}_0|} = 0$$

[*Hint:* Use Definition 11.4.7 directly.]

11.7 | MAXIMUM AND MINIMUM VALUES

As we saw in Chapter 4, one of the main uses of ordinary derivatives is in finding maximum and minimum values. In this section we see how to use partial derivatives to locate maxima and minima of functions of two variables. In particular, in Example 5 we will see how to maximize the volume of a box without a lid if we have a fixed amount of cardboard to work with.

Look at the hills and valleys in the graph of f shown in Figure 1. There are two points (a, b) where f has a *local maximum*, that is, where $f(a, b)$ is larger than nearby values of $f(x, y)$. The larger of these two values is the *absolute maximum*. Likewise, f has two *local minima*, where $f(a, b)$ is smaller than nearby values. The smaller of these two values is the *absolute minimum*.

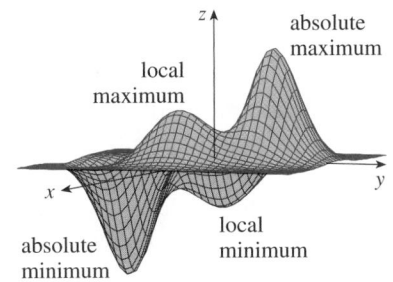

FIGURE 1

$\boxed{1}$ **DEFINITION**   A function of two variables has a **local maximum** at (a, b) if $f(x, y) \leqslant f(a, b)$ when (x, y) is near (a, b). [This means that $f(x, y) \leqslant f(a, b)$ for all points (x, y) in some disk with center (a, b).] The number $f(a, b)$ is called a **local maximum value**. If $f(x, y) \geqslant f(a, b)$ when (x, y) is near (a, b), then $f(a, b)$ is a **local minimum value**.

If the inequalities in Definition 1 hold for *all* points (x, y) in the domain of f, then f has an **absolute maximum** (or **absolute minimum**) at (a, b).

■ Notice that the conclusion of Theorem 2 can be stated in the notation of gradient vectors as $\nabla f(a, b) = \mathbf{0}$.

$\boxed{2}$ **THEOREM**   If f has a local maximum or minimum at (a, b) and the first-order partial derivatives of f exist there, then $f_x(a, b) = 0$ and $f_y(a, b) = 0$.

PROOF Let $g(x) = f(x, b)$. If f has a local maximum (or minimum) at (a, b), then g has a local maximum (or minimum) at a, so $g'(a) = 0$ by Fermat's Theorem (see Theorem 4.1.4). But $g'(a) = f_x(a, b)$ (see Equation 11.3.1) and so $f_x(a, b) = 0$. Similarly, by applying Fermat's Theorem to the function $G(y) = f(a, y)$, we obtain $f_y(a, b) = 0$. $\square$

If we put $f_x(a, b) = 0$ and $f_y(a, b) = 0$ in the equation of a tangent plane (Equation 11.4.2), we get $z = z_0$. Thus the geometric interpretation of Theorem 2 is that if the graph of f has a tangent plane at a local maximum or minimum, then the tangent plane must be horizontal.

A point (a, b) is called a **critical point** (or *stationary point*) of f if $f_x(a, b) = 0$ and $f_y(a, b) = 0$, or if one of these partial derivatives does not exist. Theorem 2 says that if f has a local maximum or minimum at (a, b), then (a, b) is a critical point of f. However, as in single-variable calculus, not all critical points give rise to maxima or minima. At a critical point, a function could have a local maximum or a local minimum or neither.

EXAMPLE 1 Let $f(x, y) = x^2 + y^2 - 2x - 6y + 14$. Then

$$f_x(x, y) = 2x - 2 \qquad f_y(x, y) = 2y - 6$$

These partial derivatives are equal to 0 when $x = 1$ and $y = 3$, so the only critical point is $(1, 3)$. By completing the square, we find that

$$f(x, y) = 4 + (x - 1)^2 + (y - 3)^2$$

Since $(x - 1)^2 \geqslant 0$ and $(y - 3)^2 \geqslant 0$, we have $f(x, y) \geqslant 4$ for all values of x and y. Therefore $f(1, 3) = 4$ is a local minimum, and in fact it is the absolute minimum of f. This can be confirmed geometrically from the graph of f, which is the elliptic paraboloid with vertex $(1, 3, 4)$ shown in Figure 2. ∎

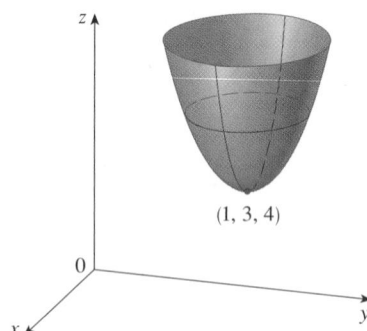

FIGURE 2
$z = x^2 + y^2 - 2x - 6y + 14$

EXAMPLE 2 Find the extreme values of $f(x, y) = y^2 - x^2$.

SOLUTION Since $f_x = -2x$ and $f_y = 2y$, the only critical point is $(0, 0)$. Notice that for points on the x-axis we have $y = 0$, so $f(x, y) = -x^2 < 0$ (if $x \neq 0$). However, for points on the y-axis we have $x = 0$, so $f(x, y) = y^2 > 0$ (if $y \neq 0$). Thus every disk with center $(0, 0)$ contains points where f takes positive values as well as points where f takes negative values. Therefore $f(0, 0) = 0$ can't be an extreme value for f, so f has no extreme value. ∎

Example 2 illustrates the fact that a function need not have a maximum or minimum value at a critical point. Figure 3 shows how this is possible. The graph of f is the hyperbolic paraboloid $z = y^2 - x^2$, which has a horizontal tangent plane ($z = 0$) at the origin. You can see that $f(0, 0) = 0$ is a maximum in the direction of the x-axis but a minimum in the direction of the y-axis. Near the origin the graph has the shape of a saddle and so $(0, 0)$ is called a *saddle point* of f.

We need to be able to determine whether or not a function has an extreme value at a critical point. The following test, which is proved in Appendix D, is analogous to the Second Derivative Test for functions of one variable.

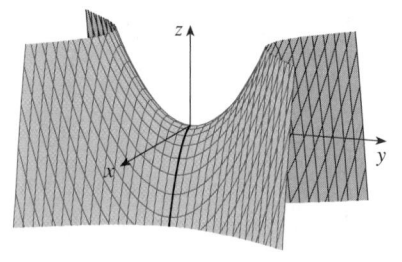

FIGURE 3
$z = y^2 - x^2$

> **3** **SECOND DERIVATIVES TEST** Suppose the second partial derivatives of f are continuous on a disk with center (a, b), and suppose that $f_x(a, b) = 0$ and $f_y(a, b) = 0$ [that is, (a, b) is a critical point of f]. Let
>
> $$D = D(a, b) = f_{xx}(a, b) f_{yy}(a, b) - [f_{xy}(a, b)]^2$$
>
> (a) If $D > 0$ and $f_{xx}(a, b) > 0$, then $f(a, b)$ is a local minimum.
>
> (b) If $D > 0$ and $f_{xx}(a, b) < 0$, then $f(a, b)$ is a local maximum.
>
> (c) If $D < 0$, then $f(a, b)$ is not a local maximum or minimum.

NOTE 1 In case (c) the point (a, b) is called a **saddle point** of f and the graph of f crosses its tangent plane at (a, b).

NOTE 2 If $D = 0$, the test gives no information: f could have a local maximum or local minimum at (a, b), or (a, b) could be a saddle point of f.

NOTE 3 To remember the formula for D it's helpful to write it as a determinant:

$$D = \begin{vmatrix} f_{xx} & f_{xy} \\ f_{yx} & f_{yy} \end{vmatrix} = f_{xx} f_{yy} - (f_{xy})^2$$

V EXAMPLE 3 Find the local maximum and minimum values and saddle points of $f(x, y) = x^4 + y^4 - 4xy + 1$.

SOLUTION We first locate the critical points:

$$f_x = 4x^3 - 4y \qquad f_y = 4y^3 - 4x$$

Setting these partial derivatives equal to 0, we obtain the equations

$$x^3 - y = 0 \qquad \text{and} \qquad y^3 - x = 0$$

To solve these equations we substitute $y = x^3$ from the first equation into the second one. This gives

$$0 = x^9 - x = x(x^8 - 1) = x(x^4 - 1)(x^4 + 1) = x(x^2 - 1)(x^2 + 1)(x^4 + 1)$$

so there are three real roots: $x = 0, 1, -1$. The three critical points are $(0, 0)$, $(1, 1)$, and $(-1, -1)$.

Next we calculate the second partial derivatives and $D(x, y)$:

$$f_{xx} = 12x^2 \qquad f_{xy} = -4 \qquad f_{yy} = 12y^2$$

$$D(x, y) = f_{xx} f_{yy} - (f_{xy})^2 = 144x^2 y^2 - 16$$

Since $D(0, 0) = -16 < 0$, it follows from case (c) of the Second Derivatives Test that the origin is a saddle point; that is, f has no local maximum or minimum at $(0, 0)$. Since $D(1, 1) = 128 > 0$ and $f_{xx}(1, 1) = 12 > 0$, we see from case (a) of the test that $f(1, 1) = -1$ is a local minimum. Similarly, we have $D(-1, -1) = 128 > 0$ and $f_{xx}(-1, -1) = 12 > 0$, so $f(-1, -1) = -1$ is also a local minimum.

The graph of f is shown in Figure 4. ∎

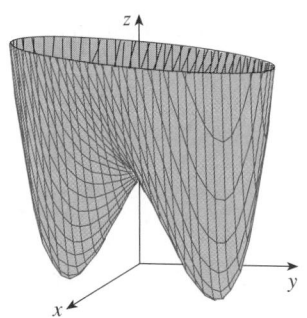

FIGURE 4
$z = x^4 + y^4 - 4xy + 1$

▪ A contour map of the function *f* in Example 3 is shown in Figure 5. The level curves near $(1, 1)$ and $(-1, -1)$ are oval in shape and indicate that as we move away from $(1, 1)$ or $(-1, -1)$ in any direction the values of *f* are increasing. The level curves near $(0, 0)$, on the other hand, resemble hyperbolas. They reveal that as we move away from the origin (where the value of *f* is 1), the values of *f* decrease in some directions but increase in other directions. Thus the contour map suggests the presence of the minima and saddle point that we found in Example 3.

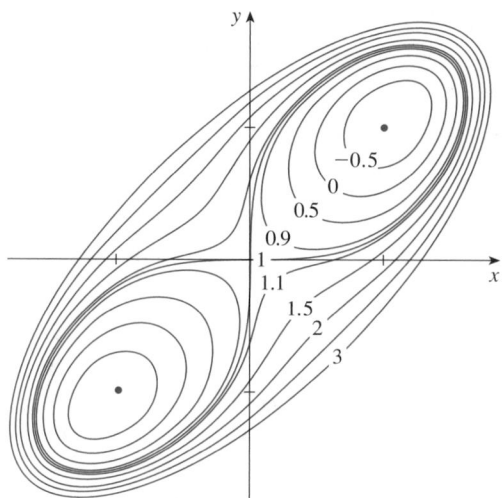

FIGURE 5

TEC In Module 11.7 you can use contour maps to estimate the locations of critical points.

▪ www.stewartcalculus.com
See Additional Example A.

▼ EXAMPLE 4 Find the shortest distance from the point $(1, 0, -2)$ to the plane $x + 2y + z = 4$.

SOLUTION The distance from any point (x, y, z) to the point $(1, 0, -2)$ is

$$d = \sqrt{(x - 1)^2 + y^2 + (z + 2)^2}$$

but if (x, y, z) lies on the plane $x + 2y + z = 4$, then $z = 4 - x - 2y$ and so we have $d = \sqrt{(x - 1)^2 + y^2 + (6 - x - 2y)^2}$. We can minimize d by minimizing the simpler expression

$$d^2 = f(x, y) = (x - 1)^2 + y^2 + (6 - x - 2y)^2$$

By solving the equations

$$f_x = 2(x - 1) - 2(6 - x - 2y) = 4x + 4y - 14 = 0$$

$$f_y = 2y - 4(6 - x - 2y) = 4x + 10y - 24 = 0$$

▪ Example 4 could also be solved using vectors. Compare with the methods of Section 10.5.

we find that the only critical point is $\left(\frac{11}{6}, \frac{5}{3}\right)$. Since $f_{xx} = 4$, $f_{xy} = 4$, and $f_{yy} = 10$, we have $D(x, y) = f_{xx}f_{yy} - (f_{xy})^2 = 24 > 0$ and $f_{xx} > 0$, so by the Second Derivatives Test f has a local minimum at $\left(\frac{11}{6}, \frac{5}{3}\right)$. Intuitively, we can see that this local minimum is actually an absolute minimum because there must be a point on the given plane that is closest to $(1, 0, -2)$. If $x = \frac{11}{6}$ and $y = \frac{5}{3}$, then

$$d = \sqrt{(x - 1)^2 + y^2 + (6 - x - 2y)^2} = \sqrt{\left(\tfrac{5}{6}\right)^2 + \left(\tfrac{5}{3}\right)^2 + \left(\tfrac{5}{6}\right)^2} = \frac{5\sqrt{6}}{6}$$

The shortest distance from $(1, 0, -2)$ to the plane $x + 2y + z = 4$ is $\frac{5}{6}\sqrt{6}$.   ■

▼ EXAMPLE 5 A rectangular box without a lid is to be made from 12 m² of cardboard. Find the maximum volume of such a box.

SOLUTION Let the length, width, and height of the box (in meters) be x, y, and z, as shown in Figure 6. Then the volume of the box is

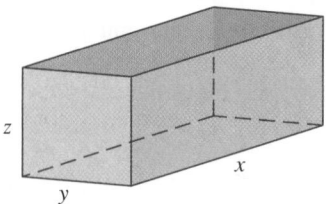

FIGURE 6

$$V = xyz$$

We can express V as a function of just two variables x and y by using the fact that the area of the four sides and the bottom of the box is

$$2xz + 2yz + xy = 12$$

Solving this equation for z, we get $z = (12 - xy)/[2(x + y)]$, so the expression for V becomes

$$V = xy \frac{12 - xy}{2(x + y)} = \frac{12xy - x^2y^2}{2(x + y)}$$

We compute the partial derivatives:

$$\frac{\partial V}{\partial x} = \frac{y^2(12 - 2xy - x^2)}{2(x + y)^2} \qquad \frac{\partial V}{\partial y} = \frac{x^2(12 - 2xy - y^2)}{2(x + y)^2}$$

If V is a maximum, then $\partial V/\partial x = \partial V/\partial y = 0$, but $x = 0$ or $y = 0$ gives $V = 0$, so we must solve the equations

$$12 - 2xy - x^2 = 0 \qquad 12 - 2xy - y^2 = 0$$

These imply that $x^2 = y^2$ and so $x = y$. (Note that x and y must both be positive in this problem.) If we put $x = y$ in either equation we get $12 - 3x^2 = 0$, which gives $x = 2$, $y = 2$, and $z = (12 - 2 \cdot 2)/[2(2 + 2)] = 1$.

We could use the Second Derivatives Test to show that this gives a local maximum of V, or we could simply argue from the physical nature of this problem that there must be an absolute maximum volume, which has to occur at a critical point of V, so it must occur when $x = 2$, $y = 2$, $z = 1$. Then $V = 2 \cdot 2 \cdot 1 = 4$, so the maximum volume of the box is 4 m³. ∎

ABSOLUTE MAXIMUM AND MINIMUM VALUES

For a function f of one variable the Extreme Value Theorem says that if f is continuous on a closed interval $[a, b]$, then f has an absolute minimum value and an absolute maximum value. According to the Closed Interval Method in Section 4.1, we found these by evaluating f not only at the critical numbers but also at the endpoints a and b.

There is a similar situation for functions of two variables. Just as a closed interval contains its endpoints, a **closed set** in $\mathbb{R}^2$ is one that contains all its boundary points. [A boundary point of D is a point (a, b) such that every disk with center (a, b) contains points in D and also points not in D.] For instance, the disk

$$D = \{(x, y) \mid x^2 + y^2 \leqslant 1\}$$

which consists of all points on and inside the circle $x^2 + y^2 = 1$, is a closed set because it contains all of its boundary points (which are the points on the circle $x^2 + y^2 = 1$). But if even one point on the boundary curve were omitted, the set would not be closed. (See Figure 7.)

A **bounded set** in $\mathbb{R}^2$ is one that is contained within some disk. In other words, it is finite in extent. Then, in terms of closed and bounded sets, we can state the following counterpart of the Extreme Value Theorem in two dimensions.

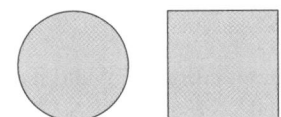

(a) Closed sets

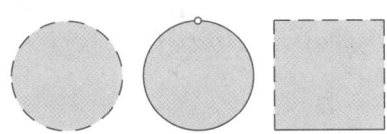

(b) Sets that are not closed

FIGURE 7

> **4** **EXTREME VALUE THEOREM FOR FUNCTIONS OF TWO VARIABLES** If f is continuous on a closed, bounded set D in $\mathbb{R}^2$, then f attains an absolute maximum value $f(x_1, y_1)$ and an absolute minimum value $f(x_2, y_2)$ at some points (x_1, y_1) and (x_2, y_2) in D.

To find the extreme values guaranteed by Theorem 4, we note that, by Theorem 2, if f has an extreme value at (x_1, y_1), then (x_1, y_1) is either a critical point of f or a boundary point of D. Thus we have the following extension of the Closed Interval Method.

> **5** To find the absolute maximum and minimum values of a continuous function f on a closed, bounded set D:
>
> **1.** Find the values of f at the critical points of f in D.
>
> **2.** Find the extreme values of f on the boundary of D.
>
> **3.** The largest of the values from steps 1 and 2 is the absolute maximum value; the smallest of these values is the absolute minimum value.

EXAMPLE 6 Find the absolute maximum and minimum values of the function $f(x, y) = x^2 - 2xy + 2y$ on the rectangle $D = \{(x, y) \mid 0 \le x \le 3, 0 \le y \le 2\}$.

SOLUTION Since f is a polynomial, it is continuous on the closed, bounded rectangle D, so Theorem 4 tells us there is both an absolute maximum and an absolute minimum. According to step 1 in $\boxed{5}$, we first find the critical points. These occur when

$$f_x = 2x - 2y = 0 \qquad f_y = -2x + 2 = 0$$

so the only critical point is $(1, 1)$, and the value of f there is $f(1, 1) = 1$.

In step 2 we look at the values of f on the boundary of D, which consists of the four line segments L_1, L_2, L_3, L_4 shown in Figure 8. On L_1 we have $y = 0$ and

$$f(x, 0) = x^2 \qquad 0 \le x \le 3$$

This is an increasing function of x, so its minimum value is $f(0, 0) = 0$ and its maximum value is $f(3, 0) = 9$. On L_2 we have $x = 3$ and

$$f(3, y) = 9 - 4y \qquad 0 \le y \le 2$$

This is a decreasing function of y, so its maximum value is $f(3, 0) = 9$ and its minimum value is $f(3, 2) = 1$. On L_3 we have $y = 2$ and

$$f(x, 2) = x^2 - 4x + 4 \qquad 0 \le x \le 3$$

By the methods of Chapter 4, or simply by observing that $f(x, 2) = (x - 2)^2$, we see that the minimum value of this function is $f(2, 2) = 0$ and the maximum value is $f(0, 2) = 4$. Finally, on L_4 we have $x = 0$ and

$$f(0, y) = 2y \qquad 0 \le y \le 2$$

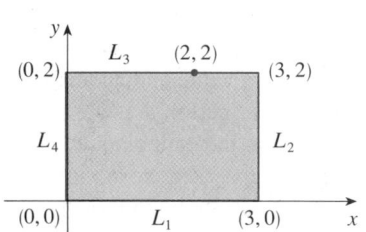

FIGURE 8

with maximum value $f(0, 2) = 4$ and minimum value $f(0, 0) = 0$. Thus, on the boundary, the minimum value of f is 0 and the maximum is 9.

In step 3 we compare these values with the value $f(1, 1) = 1$ at the critical point and conclude that the absolute maximum value of f on D is $f(3, 0) = 9$ and the absolute minimum value is $f(0, 0) = f(2, 2) = 0$. Figure 9 shows the graph of f.

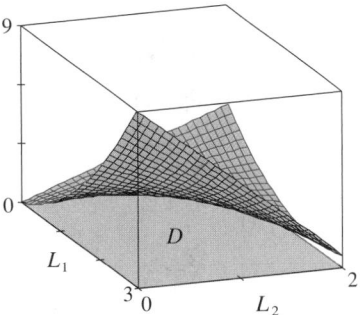

FIGURE 9

$f(x, y) = x^2 - 2xy + 2y$

11.7 | EXERCISES

1. Suppose $(1, 1)$ is a critical point of a function f with continuous second derivatives. In each case, what can you say about f?

  (a) $f_{xx}(1, 1) = 4$, $f_{xy}(1, 1) = 1$, $f_{yy}(1, 1) = 2$

  (b) $f_{xx}(1, 1) = 4$, $f_{xy}(1, 1) = 3$, $f_{yy}(1, 1) = 2$

2. Use the level curves in the figure to predict the location of the critical points of $f(x, y) = 3x - x^3 - 2y^2 + y^4$ and whether f has a saddle point or a local maximum or minimum at each of those points. Explain your reasoning. Then use the Second Derivatives Test to confirm your predictions.

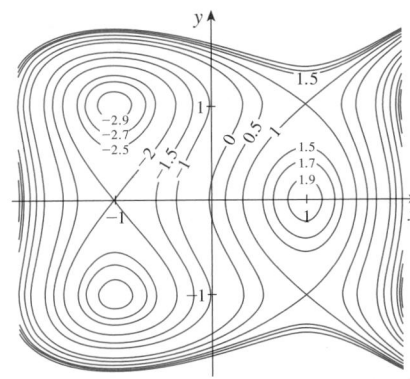

3–14 ▪ Find the local maximum and minimum values and saddle point(s) of the function. If you have three-dimensional graphing software, graph the function with a domain and viewpoint that reveal all the important aspects of the function.

3. $f(x, y) = x^2 + xy + y^2 + y$

4. $f(x, y) = xy - 2x - 2y - x^2 - y^2$

5. $f(x, y) = y^3 + 3x^2y - 6x^2 - 6y^2 + 2$

6. $f(x, y) = xe^{-2x^2-2y^2}$

7. $f(x, y) = x^3 - 12xy + 8y^3$

8. $f(x, y) = xy(1 - x - y)$

9. $f(x, y) = e^x \cos y$

10. $f(x, y) = xy + \dfrac{1}{x} + \dfrac{1}{y}$

11. $f(x, y) = (x^2 + y^2)e^{y^2-x^2}$

12. $f(x, y) = e^y(y^2 - x^2)$

13. $f(x, y) = y^2 - 2y \cos x$, $-1 \leq x \leq 7$

14. $f(x, y) = \sin x \sin y$, $-\pi < x < \pi$, $-\pi < y < \pi$

15–18 ▪ Use a graph or level curves or both to estimate the local maximum and minimum values and saddle point(s) of the function. Then use calculus to find these values precisely.

15. $f(x, y) = 3x^2y + y^3 - 3x^2 - 3y^2 + 2$

16. $f(x, y) = xye^{-x^2-y^2}$

17. $f(x, y) = \sin x + \sin y + \sin(x + y)$,

  $0 \leq x \leq 2\pi$, $0 \leq y \leq 2\pi$

18. $f(x, y) = \sin x + \sin y + \cos(x + y)$,

  $0 \leq x \leq \pi/4$, $0 \leq y \leq \pi/4$

19–22 ▪ Use a graphing device (or Newton's method or a rootfinder) to find the critical points of f correct to three decimal places. Then classify the critical points and find the highest or lowest points on the graph, if any.

19. $f(x, y) = x^4 + y^4 - 4x^2y + 2y$

20. $f(x, y) = y^6 - 2y^4 + x^2 - y^2 + y$

21. $f(x, y) = x^4 + y^3 - 3x^2 + y^2 + x - 2y + 1$

22. $f(x, y) = 20e^{-x^2-y^2}\sin 3x \cos 3y, \quad |x| \leq 1, \quad |y| \leq 1$

23–28 ▪ Find the absolute maximum and minimum values of f on the set D.

23. $f(x, y) = x^2 + y^2 - 2x$, D is the closed triangular region with vertices $(2, 0)$, $(0, 2)$, and $(0, -2)$

24. $f(x, y) = x + y - xy$, D is the closed triangular region with vertices $(0, 0)$, $(0, 2)$, and $(4, 0)$

25. $f(x, y) = x^2 + y^2 + x^2y + 4$,
$D = \{(x, y) \mid |x| \leq 1, |y| \leq 1\}$

26. $f(x, y) = 4x + 6y - x^2 - y^2$,
$D = \{(x, y) \mid 0 \leq x \leq 4, 0 \leq y \leq 5\}$

27. $f(x, y) = x^4 + y^4 - 4xy + 2$,
$D = \{(x, y) \mid 0 \leq x \leq 3, 0 \leq y \leq 2\}$

28. $f(x, y) = xy^2$, $D = \{(x, y) \mid x \geq 0, y \geq 0, x^2 + y^2 \leq 3\}$

---

29. For functions of one variable it is impossible for a continuous function to have two local maxima and no local minimum. But for functions of two variables such functions exist. Show that the function
$$f(x, y) = -(x^2 - 1)^2 - (x^2y - x - 1)^2$$
has only two critical points, but has local maxima at both of them. Then use a computer to produce a graph with a carefully chosen domain and viewpoint to see how this is possible.

30. If a function of one variable is continuous on an interval and has only one critical number, then a local maximum has to be an absolute maximum. But this is not true for functions of two variables. Show that the function
$$f(x, y) = 3xe^y - x^3 - e^{3y}$$
has exactly one critical point, and that f has a local maximum there that is not an absolute maximum. Then use a computer to produce a graph with a carefully chosen domain and viewpoint to see how this is possible.

31. Find the shortest distance from the point $(2, 0, -3)$ to the plane $x + y + z = 1$.

32. Find the point on the plane $x - 2y + 3z = 6$ that is closest to the point $(0, 1, 1)$.

33. Find the points on the cone $z^2 = x^2 + y^2$ that are closest to the point $(4, 2, 0)$.

34. Find the points on the surface $y^2 = 9 + xz$ that are closest to the origin.

35. Find three positive numbers whose sum is 100 and whose product is a maximum.

36. Find three positive numbers whose sum is 12 and the sum of whose squares is as small as possible.

37. Find the maximum volume of a rectangular box that is inscribed in a sphere of radius r.

38. Find the dimensions of the box with volume 1000 cm³ that has minimal surface area.

39. Find the volume of the largest rectangular box in the first octant with three faces in the coordinate planes and one vertex in the plane $x + 2y + 3z = 6$.

40. Find the dimensions of the rectangular box with largest volume if the total surface area is given as 64 cm².

41. Find the dimensions of a rectangular box of maximum volume such that the sum of the lengths of its 12 edges is a constant c.

42. The base of an aquarium with given volume V is made of slate and the sides are made of glass. If slate costs five times as much (per unit area) as glass, find the dimensions of the aquarium that minimize the cost of the materials.

43. A cardboard box without a lid is to have a volume of 32,000 cm³. Find the dimensions that minimize the amount of cardboard used.

44. A rectangular building is being designed to minimize heat loss. The east and west walls lose heat at a rate of 10 units/m² per day, the north and south walls at a rate of 8 units/m² per day, the floor at a rate of 1 unit/m² per day, and the roof at a rate of 5 units/m² per day. Each wall must be at least 30 m long, the height must be at least 4 m, and the volume must be exactly 4000 m³.
(a) Find and sketch the domain of the heat loss as a function of the lengths of the sides.
(b) Find the dimensions that minimize heat loss. (Check both the critical points and the points on the boundary of the domain.)
(c) Could you design a building with even less heat loss if the restrictions on the lengths of the walls were removed?

45. If the length of the diagonal of a rectangular box must be L, what is the largest possible volume?

46. Three alleles (alternative versions of a gene) A, B, and O determine the four blood types A (AA or AO), B (BB or BO), O (OO), and AB. The Hardy-Weinberg Law states that the proportion of individuals in a population who carry two different alleles is
$$P = 2pq + 2pr + 2rq$$
where p, q, and r are the proportions of A, B, and O in the population. Use the fact that $p + q + r = 1$ to show that P is at most $\frac{2}{3}$.

47. Suppose that a scientist has reason to believe that two quantities x and y are related linearly, that is, $y = mx + b$, at least approximately, for some values of m and b. The scientist performs an experiment and collects data in the form of points $(x_1, y_1), (x_2, y_2), \ldots, (x_n, y_n)$, and then plots these points. The points don't lie exactly on a straight line, so the scientist wants to find constants m and b so that the line $y = mx + b$ "fits" the points as well as possible (see the figure).

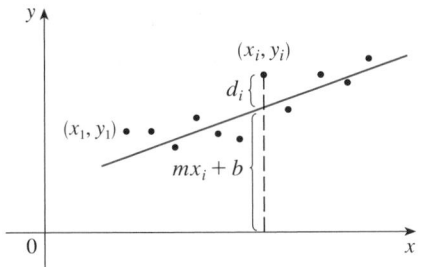

Let $d_i = y_i - (mx_i + b)$ be the vertical deviation of the point (x_i, y_i) from the line. The **method of least squares** determines m and b so as to minimize $\sum_{i=1}^{n} d_i^2$, the sum of the squares of these deviations. Show that, according to this method, the line of best fit is obtained when

$$m \sum_{i=1}^{n} x_i + bn = \sum_{i=1}^{n} y_i$$

$$m \sum_{i=1}^{n} x_i^2 + b \sum_{i=1}^{n} x_i = \sum_{i=1}^{n} x_i y_i$$

Thus the line is found by solving these two equations in the two unknowns m and b.

48. Find an equation of the plane that passes through the point $(1, 2, 3)$ and cuts off the smallest volume in the first octant.

11.8 | LAGRANGE MULTIPLIERS

In Example 5 in Section 11.7 we maximized a volume function $V = xyz$ subject to the constraint $2xz + 2yz + xy = 12$, which expressed the side condition that the surface area was 12 m². In this section we present Lagrange's method for maximizing or minimizing a general function $f(x, y, z)$ subject to a constraint (or side condition) of the form $g(x, y, z) = k$.

It's easier to explain the geometric basis of Lagrange's method for functions of two variables. So we start by trying to find the extreme values of $f(x, y)$ subject to a constraint of the form $g(x, y) = k$. In other words, we seek the extreme values of $f(x, y)$ when the point (x, y) is restricted to lie on the level curve $g(x, y) = k$. Figure 1 shows this curve together with several level curves of f. These have the equations $f(x, y) = c$, where $c = 7, 8, 9, 10, 11$. To maximize $f(x, y)$ subject to $g(x, y) = k$ is to find the largest value of c such that the level curve $f(x, y) = c$ intersects $g(x, y) = k$. It appears from Figure 1 that this happens when these curves just touch each other, that is, when they have a common tangent line. (Otherwise, the value of c could be increased further.) This means that the normal lines at the point (x_0, y_0) where they touch are identical. So the gradient vectors are parallel; that is, $\nabla f(x_0, y_0) = \lambda \nabla g(x_0, y_0)$ for some scalar λ.

This kind of argument also applies to the problem of finding the extreme values of $f(x, y, z)$ subject to the constraint $g(x, y, z) = k$. Thus the point (x, y, z) is restricted to lie on the level surface S with equation $g(x, y, z) = k$. Instead of the level curves in Figure 1, we consider the level surfaces $f(x, y, z) = c$ and argue that if the maximum value of f is $f(x_0, y_0, z_0) = c$, then the level surface $f(x, y, z) = c$ is tangent to the level surface $g(x, y, z) = k$ and so the corresponding gradient vectors are parallel.

This intuitive argument can be made precise as follows. Suppose that a function f has an extreme value at a point $P(x_0, y_0, z_0)$ on the surface S and let C be a curve with vector equation $\mathbf{r}(t) = \langle x(t), y(t), z(t) \rangle$ that lies on S and passes through P. If t_0 is the parameter value corresponding to the point P, then $\mathbf{r}(t_0) = \langle x_0, y_0, z_0 \rangle$. The composite function $h(t) = f(x(t), y(t), z(t))$ represents the values that f takes on the curve C. Since f has an extreme value at (x_0, y_0, z_0), it follows that h has an extreme value at

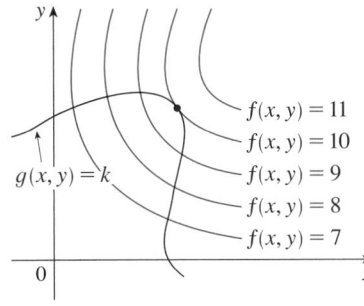

FIGURE 1

TEC Visual 11.8 animates Figure 1 for both level curves and level surfaces.

t_0, so $h'(t_0) = 0$. But if f is differentiable, we can use the Chain Rule to write

$$0 = h'(t_0)$$
$$= f_x(x_0, y_0, z_0)x'(t_0) + f_y(x_0, y_0, z_0)y'(t_0) + f_z(x_0, y_0, z_0)z'(t_0)$$
$$= \nabla f(x_0, y_0, z_0) \cdot \mathbf{r}'(t_0)$$

This shows that the gradient vector $\nabla f(x_0, y_0, z_0)$ is orthogonal to the tangent vector $\mathbf{r}'(t_0)$ to every such curve C. But we already know from Section 11.6 that the gradient vector of g, $\nabla g(x_0, y_0, z_0)$, is also orthogonal to $\mathbf{r}'(t_0)$. (See Equation 11.6.18.) This means that the gradient vectors $\nabla f(x_0, y_0, z_0)$ and $\nabla g(x_0, y_0, z_0)$ must be parallel. Therefore if $\nabla g(x_0, y_0, z_0) \neq \mathbf{0}$, there is a number λ such that

■ Lagrange multipliers are named after the French-Italian mathematician Joseph-Louis Lagrange (1736–1813). See page 213 for a biographical sketch of Lagrange.

$$\boxed{1} \qquad \boxed{\nabla f(x_0, y_0, z_0) = \lambda \, \nabla g(x_0, y_0, z_0)}$$

The number λ in Equation 1 is called a **Lagrange multiplier**. The procedure based on Equation 1 is as follows.

METHOD OF LAGRANGE MULTIPLIERS To find the maximum and minimum values of $f(x, y, z)$ subject to the constraint $g(x, y, z) = k$ [assuming that these extreme values exist and $\nabla g \neq \mathbf{0}$ on the surface $g(x, y, z) = k$]:

(a) Find all values of x, y, z, and λ such that

$$\nabla f(x, y, z) = \lambda \, \nabla g(x, y, z)$$

and $\qquad\qquad g(x, y, z) = k$

(b) Evaluate f at all the points (x, y, z) that result from step (a). The largest of these values is the maximum value of f; the smallest is the minimum value of f.

■ In deriving Lagrange's method we assumed that $\nabla g \neq \mathbf{0}$. In each of our examples you can check that $\nabla g \neq \mathbf{0}$ at all points where $g(x, y, z) = k$. See Exercise 23 for what can go wrong if $\nabla g = \mathbf{0}$.

If we write the vector equation $\nabla f = \lambda \, \nabla g$ in terms of its components, then the equations in step (a) become

$$f_x = \lambda g_x \qquad f_y = \lambda g_y \qquad f_z = \lambda g_z \qquad g(x, y, z) = k$$

This is a system of four equations in the four unknowns x, y, z, and λ, but it is not necessary to find explicit values for λ.

For functions of two variables the method of Lagrange multipliers is similar to the method just described. To find the extreme values of $f(x, y)$ subject to the constraint $g(x, y) = k$, we look for values of x, y, and λ such that

$$\nabla f(x, y) = \lambda \, \nabla g(x, y) \qquad \text{and} \qquad g(x, y) = k$$

This amounts to solving three equations in three unknowns:

$$f_x = \lambda g_x \qquad f_y = \lambda g_y \qquad g(x, y) = k$$

Our first illustration of Lagrange's method is to reconsider the problem given in Example 5 in Section 11.7.

▼ EXAMPLE 1 A rectangular box without a lid is to be made from 12 m² of cardboard. Find the maximum volume of such a box.

SOLUTION As in Example 5 in Section 11.7 we let x, y, and z be the length, width, and height, respectively, of the box, in meters. Then we wish to maximize

$$V = xyz$$

subject to the constraint

$$g(x, y, z) = 2xz + 2yz + xy = 12$$

Using the method of Lagrange multipliers, we look for values of x, y, z, and λ such that $\nabla V = \lambda \nabla g$ and $g(x, y, z) = 12$. This gives the equations

$$V_x = \lambda g_x \qquad V_y = \lambda g_y \qquad V_z = \lambda g_z \qquad 2xz + 2yz + xy = 12$$

which become

$$\boxed{2} \qquad\qquad\qquad yz = \lambda(2z + y)$$

$$\boxed{3} \qquad\qquad\qquad xz = \lambda(2z + x)$$

$$\boxed{4} \qquad\qquad\qquad xy = \lambda(2x + 2y)$$

$$\boxed{5} \qquad\qquad\qquad 2xz + 2yz + xy = 12$$

There are no general rules for solving systems of equations. Sometimes some ingenuity is required. In the present example you might notice that if we multiply $\boxed{2}$ by x, $\boxed{3}$ by y, and $\boxed{4}$ by z, then the left sides of these equations will be identical. Doing this, we have

$$\boxed{6} \qquad\qquad\qquad xyz = \lambda(2xz + xy)$$

$$\boxed{7} \qquad\qquad\qquad xyz = \lambda(2yz + xy)$$

$$\boxed{8} \qquad\qquad\qquad xyz = \lambda(2xz + 2yz)$$

■ Another method for solving the system of equations (2–5) is to solve each of Equations 2, 3, and 4 for λ and then to equate the resulting expressions.

We observe that $\lambda \neq 0$ because $\lambda = 0$ would imply $yz = xz = xy = 0$ from $\boxed{2}$, $\boxed{3}$, and $\boxed{4}$ and this would contradict $\boxed{5}$. Therefore from $\boxed{6}$ and $\boxed{7}$ we have

$$2xz + xy = 2yz + xy$$

which gives $xz = yz$. But $z \neq 0$ (since $z = 0$ would give $V = 0$), so $x = y$. From $\boxed{7}$ and $\boxed{8}$ we have

$$2yz + xy = 2xz + 2yz$$

which gives $2xz = xy$ and so (since $x \neq 0$) $y = 2z$. If we now put $x = y = 2z$ in $\boxed{5}$, we get

$$4z^2 + 4z^2 + 4z^2 = 12$$

Since x, y, and z are all positive, we therefore have $z = 1$, $x = 2$, and $y = 2$ as before. ■

■ In geometric terms, Example 2 asks for the highest and lowest points on the curve C in Figure 2 that lies on the paraboloid $z = x^2 + 2y^2$ and directly above the constraint circle $x^2 + y^2 = 1$.

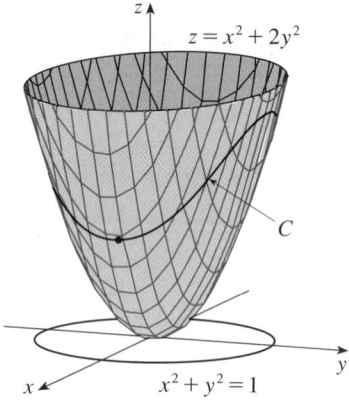

FIGURE 2

■ The geometry behind the use of Lagrange multipliers in Example 2 is shown in Figure 3. The extreme values of $f(x, y) = x^2 + 2y^2$ correspond to the level curves that touch the circle $x^2 + y^2 = 1$.

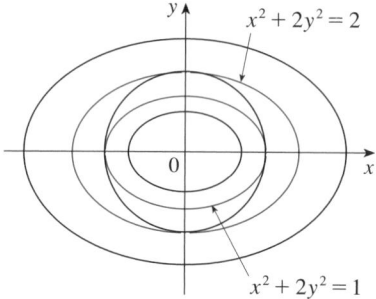

FIGURE 3

Ⅴ EXAMPLE 2 Find the extreme values of the function $f(x, y) = x^2 + 2y^2$ on the circle $x^2 + y^2 = 1$.

SOLUTION We are asked for the extreme values of f subject to the constraint $g(x, y) = x^2 + y^2 = 1$. Using Lagrange multipliers, we solve the equations $\nabla f = \lambda \nabla g$, $g(x, y) = 1$, which can be written as

$$f_x = \lambda g_x \qquad f_y = \lambda g_y \qquad g(x, y) = 1$$

or as

$$\boxed{9} \qquad\qquad 2x = 2x\lambda$$

$$\boxed{10} \qquad\qquad 4y = 2y\lambda$$

$$\boxed{11} \qquad\qquad x^2 + y^2 = 1$$

From $\boxed{9}$ we have $x = 0$ or $\lambda = 1$. If $x = 0$, then $\boxed{11}$ gives $y = \pm 1$. If $\lambda = 1$, then $y = 0$ from $\boxed{10}$, so then $\boxed{11}$ gives $x = \pm 1$. Therefore f has possible extreme values at the points $(0, 1)$, $(0, -1)$, $(1, 0)$, and $(-1, 0)$. Evaluating f at these four points, we find that

$$f(0, 1) = 2 \qquad f(0, -1) = 2 \qquad f(1, 0) = 1 \qquad f(-1, 0) = 1$$

Therefore the maximum value of f on the circle $x^2 + y^2 = 1$ is $f(0, \pm 1) = 2$ and the minimum value is $f(\pm 1, 0) = 1$. Checking with Figure 2, we see that these values look reasonable. ∎

EXAMPLE 3 Find the extreme values of $f(x, y) = x^2 + 2y^2$ on the disk $x^2 + y^2 \le 1$.

SOLUTION According to the procedure in (11.7.5), we compare the values of f at the critical points with values at the points on the boundary. Since $f_x = 2x$ and $f_y = 4y$, the only critical point is $(0, 0)$. We compare the value of f at that point with the extreme values on the boundary from Example 2:

$$f(0, 0) = 0 \qquad f(\pm 1, 0) = 1 \qquad f(0, \pm 1) = 2$$

Therefore the maximum value of f on the disk $x^2 + y^2 \le 1$ is $f(0, \pm 1) = 2$ and the minimum value is $f(0, 0) = 0$. ∎

EXAMPLE 4 Find the points on the sphere $x^2 + y^2 + z^2 = 4$ that are closest to and farthest from the point $(3, 1, -1)$.

SOLUTION The distance from a point (x, y, z) to the point $(3, 1, -1)$ is

$$d = \sqrt{(x - 3)^2 + (y - 1)^2 + (z + 1)^2}$$

but the algebra is simpler if we instead maximize and minimize the square of the distance:

$$d^2 = f(x, y, z) = (x - 3)^2 + (y - 1)^2 + (z + 1)^2$$

The constraint is that the point (x, y, z) lies on the sphere, that is,

$$g(x, y, z) = x^2 + y^2 + z^2 = 4$$

According to the method of Lagrange multipliers, we solve $\nabla f = \lambda \nabla g$, $g = 4$. This gives

$$\boxed{12} \qquad 2(x - 3) = 2x\lambda$$

$$\boxed{13} \qquad 2(y - 1) = 2y\lambda$$

$$\boxed{14} \qquad 2(z + 1) = 2z\lambda$$

$$\boxed{15} \qquad x^2 + y^2 + z^2 = 4$$

The simplest way to solve these equations is to solve for x, y, and z in terms of λ from $\boxed{12}$, $\boxed{13}$, and $\boxed{14}$, and then substitute these values into $\boxed{15}$. From $\boxed{12}$ we have

$$x - 3 = x\lambda \quad \text{or} \quad x(1 - \lambda) = 3 \quad \text{or} \quad x = \frac{3}{1 - \lambda}$$

[Note that $1 - \lambda \neq 0$ because $\lambda = 1$ is impossible from $\boxed{12}$.] Similarly, $\boxed{13}$ and $\boxed{14}$ give

$$y = \frac{1}{1 - \lambda} \qquad z = -\frac{1}{1 - \lambda}$$

Therefore from $\boxed{15}$ we have

$$\frac{3^2}{(1 - \lambda)^2} + \frac{1^2}{(1 - \lambda)^2} + \frac{(-1)^2}{(1 - \lambda)^2} = 4$$

which gives $(1 - \lambda)^2 = \frac{11}{4}$, $1 - \lambda = \pm\sqrt{11}/2$, so

$$\lambda = 1 \pm \frac{\sqrt{11}}{2}$$

These values of λ then give the corresponding points (x, y, z):

$$\left(\frac{6}{\sqrt{11}}, \frac{2}{\sqrt{11}}, -\frac{2}{\sqrt{11}} \right) \quad \text{and} \quad \left(-\frac{6}{\sqrt{11}}, -\frac{2}{\sqrt{11}}, \frac{2}{\sqrt{11}} \right)$$

It's easy to see that f has a smaller value at the first of these points, so the closest point is $\left(6/\sqrt{11}, 2/\sqrt{11}, -2/\sqrt{11} \right)$ and the farthest is $\left(-6/\sqrt{11}, -2/\sqrt{11}, 2/\sqrt{11} \right)$. ∎

■ Figure 4 shows the sphere and the nearest point P in Example 4. Can you see how to find the coordinates of P without using calculus?

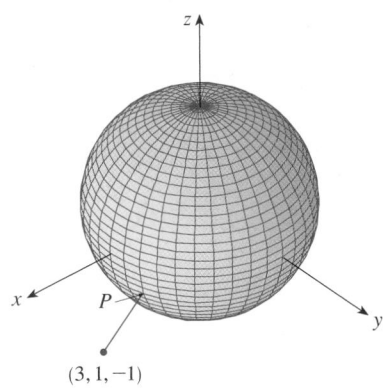

FIGURE 4

TWO CONSTRAINTS

Suppose now that we want to find the maximum and minimum values of a function $f(x, y, z)$ subject to two constraints (side conditions) of the form $g(x, y, z) = k$ and $h(x, y, z) = c$. Geometrically, this means that we are looking for the extreme values of f when (x, y, z) is restricted to lie on the curve of intersection C of the level surfaces $g(x, y, z) = k$ and $h(x, y, z) = c$. (See Figure 5.) Suppose f has such an extreme value at a point $P(x_0, y_0, z_0)$. We know from the beginning of this section that ∇f is orthogonal to C there. But we also know that ∇g is orthogonal to $g(x, y, z) = k$ and ∇h is orthogonal to $h(x, y, z) = c$, so ∇g and ∇h are both orthogonal to C. This means that the gradient vector $\nabla f(x_0, y_0, z_0)$ is in the plane determined by $\nabla g(x_0, y_0, z_0)$ and $\nabla h(x_0, y_0, z_0)$. (We assume that these gradient vectors are not zero and not parallel.)

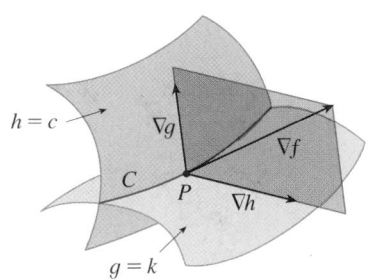

FIGURE 5

So there are numbers λ and μ (called Lagrange multipliers) such that

$$\boxed{16} \qquad \nabla f(x_0, y_0, z_0) = \lambda \, \nabla g(x_0, y_0, z_0) + \mu \, \nabla h(x_0, y_0, z_0)$$

In this case Lagrange's method is to look for extreme values by solving five equations in the five unknowns x, y, z, λ, and μ. These equations are obtained by writing Equation 16 in terms of its components and using the constraint equations:

$$f_x = \lambda g_x + \mu h_x$$

$$f_y = \lambda g_y + \mu h_y$$

$$f_z = \lambda g_z + \mu h_z$$

$$g(x, y, z) = k$$

$$h(x, y, z) = c$$

▼ EXAMPLE 5 Find the maximum value of the function $f(x, y, z) = x + 2y + 3z$ on the curve of intersection of the plane $x - y + z = 1$ and the cylinder $x^2 + y^2 = 1$.

SOLUTION We maximize the function $f(x, y, z) = x + 2y + 3z$ subject to the constraints $g(x, y, z) = x - y + z = 1$ and $h(x, y, z) = x^2 + y^2 = 1$. The Lagrange condition is $\nabla f = \lambda \, \nabla g + \mu \, \nabla h$, so we solve the equations

$$\boxed{17} \qquad\qquad 1 = \lambda + 2x\mu$$

$$\boxed{18} \qquad\qquad 2 = -\lambda + 2y\mu$$

$$\boxed{19} \qquad\qquad 3 = \lambda$$

$$\boxed{20} \qquad\qquad x - y + z = 1$$

$$\boxed{21} \qquad\qquad x^2 + y^2 = 1$$

Putting $\lambda = 3$ [from $\boxed{19}$] in $\boxed{17}$, we get $2x\mu = -2$, so $x = -1/\mu$. Similarly, $\boxed{18}$ gives $y = 5/(2\mu)$. Substitution in $\boxed{21}$ then gives

$$\frac{1}{\mu^2} + \frac{25}{4\mu^2} = 1$$

and so $\mu^2 = \frac{29}{4}$, $\mu = \pm\sqrt{29}/2$. Then $x = \mp 2/\sqrt{29}$, $y = \pm 5/\sqrt{29}$, and, from $\boxed{20}$, $z = 1 - x + y = 1 \pm 7/\sqrt{29}$. The corresponding values of f are

$$\mp\frac{2}{\sqrt{29}} + 2\left(\pm\frac{5}{\sqrt{29}}\right) + 3\left(1 \pm \frac{7}{\sqrt{29}}\right) = 3 \pm \sqrt{29}$$

Therefore the maximum value of f on the given curve is $3 + \sqrt{29}$.   ■

■ The cylinder $x^2 + y^2 = 1$ intersects the plane $x - y + z = 1$ in an ellipse (Figure 6). Example 5 asks for the maximum value of f when (x, y, z) is restricted to lie on the ellipse.

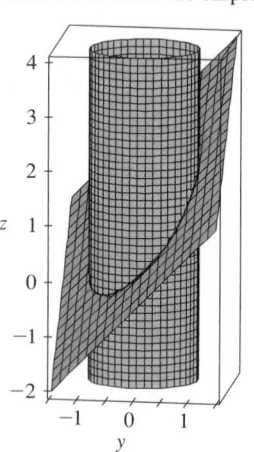

FIGURE 6

11.8 | EXERCISES

1–12 ▪ Use Lagrange multipliers to find the maximum and minimum values of the function subject to the given constraint.

1. $f(x, y) = x^2 + y^2; \quad xy = 1$

2. $f(x, y) = 3x + y; \quad x^2 + y^2 = 10$

3. $f(x, y) = y^2 - x^2; \quad \frac{1}{4}x^2 + y^2 = 1$

4. $f(x, y) = e^{xy}; \quad x^3 + y^3 = 16$

5. $f(x, y, z) = 2x + 2y + z; \quad x^2 + y^2 + z^2 = 9$

6. $f(x, y, z) = x^2 + y^2 + z^2; \quad x + y + z = 12$

7. $f(x, y, z) = xyz; \quad x^2 + 2y^2 + 3z^2 = 6$

8. $f(x, y, z) = x^2 y^2 z^2; \quad x^2 + y^2 + z^2 = 1$

9. $f(x, y, z) = x^2 + y^2 + z^2; \quad x^4 + y^4 + z^4 = 1$

10. $f(x, y, z) = x^4 + y^4 + z^4; \quad x^2 + y^2 + z^2 = 1$

11. $f(x, y, z, t) = x + y + z + t; \quad x^2 + y^2 + z^2 + t^2 = 1$

12. $f(x_1, x_2, \ldots, x_n) = x_1 + x_2 + \cdots + x_n;$
$x_1^2 + x_2^2 + \cdots + x_n^2 = 1$

13–16 ▪ Find the extreme values of f subject to both constraints.

13. $f(x, y, z) = x + 2y; \quad x + y + z = 1, \quad y^2 + z^2 = 4$

14. $f(x, y, z) = 3x - y - 3z;$
$x + y - z = 0, \quad x^2 + 2z^2 = 1$

15. $f(x, y, z) = yz + xy; \quad xy = 1, \quad y^2 + z^2 = 1$

16. $f(x, y, z) = x^2 + y^2 + z^2; \quad x - y = 1, \quad y^2 - z^2 = 1$

17–19 ▪ Find the extreme values of f on the region described by the inequality.

17. $f(x, y) = x^2 + y^2 + 4x - 4y, \quad x^2 + y^2 \leqslant 9$

18. $f(x, y) = 2x^2 + 3y^2 - 4x - 5, \quad x^2 + y^2 \leqslant 16$

19. $f(x, y) = e^{-xy}, \quad x^2 + 4y^2 \leqslant 1$

---

20. (a) Use a graphing calculator or computer to graph the circle $x^2 + y^2 = 1$. On the same screen, graph several curves of the form $x^2 + y = c$ until you find two that just touch the circle. What is the significance of the values of c for these two curves?
(b) Use Lagrange multipliers to find the extreme values of $f(x, y) = x^2 + y$ subject to the constraint $x^2 + y^2 = 1$. Compare your answers with those in part (a).

21. Pictured are a contour map of f and a curve with equation $g(x, y) = 8$. Estimate the maximum and minimum values of f subject to the constraint that $g(x, y) = 8$. Explain your reasoning.

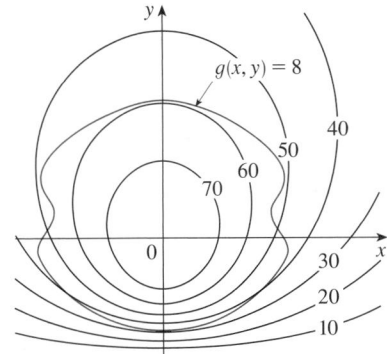

22. Consider the problem of maximizing the function $f(x, y) = 2x + 3y$ subject to the constraint $\sqrt{x} + \sqrt{y} = 5$.
(a) Try using Lagrange multipliers to solve the problem.
(b) Does $f(25, 0)$ give a larger value than the one in part (a)?
(c) Solve the problem by graphing the constraint equation and several level curves of f.
(d) Explain why the method of Lagrange multipliers fails to solve the problem.
(e) What is the significance of $f(9, 4)$?

23. Consider the problem of minimizing the function $f(x, y) = x$ on the curve $y^2 + x^4 - x^3 = 0$ (a piriform).
(a) Try using Lagrange multipliers to solve the problem.
(b) Show that the minimum value is $f(0, 0) = 0$ but the Lagrange condition $\nabla f(0, 0) = \lambda \nabla g(0, 0)$ is not satisfied for any value of λ.
(c) Explain why Lagrange multipliers fail to find the minimum value in this case.

CAS 24. (a) If your computer algebra system plots implicitly defined curves, use it to estimate the minimum and maximum values of $f(x, y) = x^3 + y^3 + 3xy$ subject to the constraint $(x - 3)^2 + (y - 3)^2 = 9$ by graphical methods.
(b) Solve the problem in part (a) with the aid of Lagrange multipliers. Use your CAS to solve the equations numerically. Compare your answers with those in part (a).

25. The total production P of a certain product depends on the amount L of labor used and the amount K of capital investment. The Cobb-Douglas model for the production function is $P = bL^\alpha K^{1-\alpha}$, where b and α are positive constants and $\alpha < 1$. If the cost of a unit of labor is m and the cost of a unit of capital is n, and the company can spend only p dollars as its total budget, then maximizing the production P is subject to the constraint $mL + nK = p$. Show that the maximum production occurs when

$$L = \frac{\alpha p}{m} \quad \text{and} \quad K = \frac{(1 - \alpha)p}{n}$$

26. Referring to Exercise 25, we now suppose that the production is fixed at $bL^\alpha K^{1-\alpha} = Q$, where Q is a constant. What values of L and K minimize the cost function $C(L, K) = mL + nK$?

27. Use Lagrange multipliers to prove that the rectangle with maximum area that has a given perimeter p is a square.

28. Use Lagrange multipliers to prove that the triangle with maximum area that has a given perimeter p is equilateral. *Hint:* Use Heron's formula for the area:

$$A = \sqrt{s(s - x)(s - y)(s - z)}$$

where $s = p/2$ and x, y, z are the lengths of the sides.

29–41 ▪ Use Lagrange multipliers to give an alternate solution to the indicated exercise in Section 11.7.

29. Exercise 31   **30.** Exercise 32

31. Exercise 33   **32.** Exercise 34

33. Exercise 35   **34.** Exercise 36

35. Exercise 37   **36.** Exercise 38

37. Exercise 39   **38.** Exercise 40

39. Exercise 41   **40.** Exercise 42

41. Exercise 45

42. Find the maximum and minimum volumes of a rectangular box whose surface area is 1500 cm² and whose total edge length is 200 cm.

43. The plane $x + y + 2z = 2$ intersects the paraboloid $z = x^2 + y^2$ in an ellipse. Find the points on this ellipse that are nearest to and farthest from the origin.

44. The plane $4x - 3y + 8z = 5$ intersects the cone $z^2 = x^2 + y^2$ in an ellipse.

    (a) Graph the cone and the plane, and observe the resulting ellipse.

    (b) Use Lagrange multipliers to find the highest and lowest points on the ellipse.

CAS **45–46** ▪ Find the maximum and minimum values of f subject to the given constraints. Use a computer algebra system to solve the system of equations that arises in using Lagrange multipliers. (If your CAS finds only one solution, you may need to use additional commands.)

45. $f(x, y, z) = ye^{x-z}$;
    $9x^2 + 4y^2 + 36z^2 = 36, \;\; xy + yz = 1$

46. $f(x, y, z) = x + y + z; \;\; x^2 - y^2 = z, \;\; x^2 + z^2 = 4$

47. (a) Find the maximum value of

$$f(x_1, x_2, \ldots, x_n) = \sqrt[n]{x_1 x_2 \cdots x_n}$$

    given that $x_1, x_2, \ldots, x_n$ are positive numbers and $x_1 + x_2 + \cdots + x_n = c$, where c is a constant.

    (b) Deduce from part (a) that if $x_1, x_2, \ldots, x_n$ are positive numbers, then

$$\sqrt[n]{x_1 x_2 \cdots x_n} \leqslant \frac{x_1 + x_2 + \cdots + x_n}{n}$$

    This inequality says that the geometric mean of n numbers is no larger than the arithmetic mean of the numbers. Under what circumstances are these two means equal?

48. (a) Maximize $\sum_{i=1}^{n} x_i y_i$ subject to the constraints $\sum_{i=1}^{n} x_i^2 = 1$ and $\sum_{i=1}^{n} y_i^2 = 1$.

    (b) Put

$$x_i = \frac{a_i}{\sqrt{\sum a_j^2}} \quad \text{and} \quad y_i = \frac{b_i}{\sqrt{\sum b_j^2}}$$

    to show that

$$\sum a_i b_i \leqslant \sqrt{\sum a_j^2} \sqrt{\sum b_j^2}$$

    for any numbers $a_1, \ldots, a_n, b_1, \ldots, b_n$. This inequality is known as the Cauchy-Schwarz Inequality.

CHAPTER 11 REVIEW

1. (a) What is a function of two variables?
(b) Describe three methods for visualizing a function of two variables.

2. What is a function of three variables? How can you visualize such a function?

3. What does
$$\lim_{(x,\,y)\to(a,\,b)} f(x,\,y) = L$$
mean? How can you show that such a limit does not exist?

4. (a) What does it mean to say that f is continuous at (a, b)?
(b) If f is continuous on $\mathbb{R}^2$, what can you say about its graph?

5. (a) Write expressions for the partial derivatives $f_x(a, b)$ and $f_y(a, b)$ as limits.
(b) How do you interpret $f_x(a, b)$ and $f_y(a, b)$ geometrically? How do you interpret them as rates of change?
(c) If $f(x, y)$ is given by a formula, how do you calculate f_x and f_y?

6. What does Clairaut's Theorem say?

7. How do you find a tangent plane to each of the following types of surfaces?
(a) A graph of a function of two variables, $z = f(x, y)$
(b) A level surface of a function of three variables, $F(x, y, z) = k$

8. Define the linearization of f at (a, b). What is the corresponding linear approximation? What is the geometric interpretation of the linear approximation?

9. (a) What does it mean to say that f is differentiable at (a, b)?
(b) How do you usually verify that f is differentiable?

10. If $z = f(x, y)$, what are the differentials dx, dy, and dz?

11. State the Chain Rule for the case where $z = f(x, y)$ and x and y are functions of one variable. What if x and y are functions of two variables?

12. If z is defined implicitly as a function of x and y by an equation of the form $F(x, y, z) = 0$, how do you find $\partial z/\partial x$ and $\partial z/\partial y$?

13. (a) Write an expression as a limit for the directional derivative of f at (x_0, y_0) in the direction of a unit vector $\mathbf{u} = \langle a, b \rangle$. How do you interpret it as a rate? How do you interpret it geometrically?
(b) If f is differentiable, write an expression for $D_{\mathbf{u}} f(x_0, y_0)$ in terms of f_x and f_y.

14. (a) Define the gradient vector ∇f for a function f of two or three variables.
(b) Express $D_{\mathbf{u}} f$ in terms of ∇f.
(c) Explain the geometric significance of the gradient.

15. What do the following statements mean?
(a) f has a local maximum at (a, b).
(b) f has an absolute maximum at (a, b).
(c) f has a local minimum at (a, b).
(d) f has an absolute minimum at (a, b).
(e) f has a saddle point at (a, b).

16. (a) If f has a local maximum at (a, b), what can you say about its partial derivatives at (a, b)?
(b) What is a critical point of f?

17. State the Second Derivatives Test.

18. (a) What is a closed set in $\mathbb{R}^2$? What is a bounded set?
(b) State the Extreme Value Theorem for functions of two variables.
(c) How do you find the values that the Extreme Value Theorem guarantees?

19. Explain how the method of Lagrange multipliers works in finding the extreme values of $f(x, y, z)$ subject to the constraint $g(x, y, z) = k$. What if there is a second constraint $h(x, y, z) = c$?

Determine whether the statement is true or false. If it is true, explain why. If it is false, explain why or give an example that disproves the statement.

1. $f_y(a, b) = \lim\limits_{y\to b} \dfrac{f(a, y) - f(a, b)}{y - b}$

2. There exists a function f with continuous second-order partial derivatives such that $f_x(x, y) = x + y^2$ and $f_y(x, y) = x - y^2$.

3. $f_{xy} = \dfrac{\partial^2 f}{\partial x\, \partial y}$

4. $D_{\mathbf{k}} f(x, y, z) = f_z(x, y, z)$

5. If $f(x, y) \to L$ as $(x, y) \to (a, b)$ along every straight line through (a, b), then $\lim_{(x,\,y)\to(a,\,b)} f(x, y) = L$.

6. If $f_x(a, b)$ and $f_y(a, b)$ both exist, then f is differentiable at (a, b).

7. If f has a local minimum at (a, b) and f is differentiable at (a, b), then $\nabla f(a, b) = \mathbf{0}$.

8. If f is a function, then

$$\lim_{(x, y) \to (2, 5)} f(x, y) = f(2, 5)$$

9. If $f(x, y) = \ln y$, then $\nabla f(x, y) = 1/y$.

10. If $(2, 1)$ is a critical point of f and

$$f_{xx}(2, 1) f_{yy}(2, 1) < [f_{xy}(2, 1)]^2$$

then f has a saddle point at $(2, 1)$.

11. If $f(x, y) = \sin x + \sin y$, then $-\sqrt{2} \leqslant D_{\mathbf{u}} f(x, y) \leqslant \sqrt{2}$.

12. If $f(x, y)$ has two local maxima, then f must have a local minimum.

EXERCISES

1–2 ▪ Find and sketch the domain of the function.

1. $f(x, y) = \ln(x + y + 1)$

2. $f(x, y) = \sqrt{4 - x^2 - y^2} + \sqrt{1 - x^2}$

3–4 ▪ Sketch the graph of the function.

3. $f(x, y) = 1 - y^2$

4. $f(x, y) = x^2 + (y - 2)^2$

5–6 ▪ Sketch several level curves of the function.

5. $f(x, y) = \sqrt{4x^2 + y^2}$ **6.** $f(x, y) = e^x + y$

---

7. Make a rough sketch of a contour map for the function whose graph is shown.

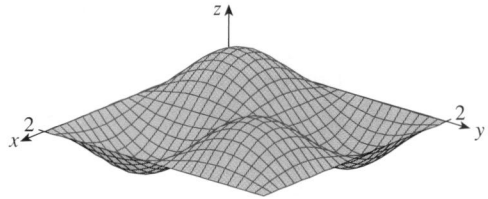

8. A contour map of a function f is shown. Use it to make a rough sketch of the graph of f.

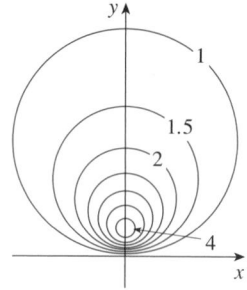

9–10 ▪ Evaluate the limit or show that it does not exist.

9. $\displaystyle\lim_{(x, y) \to (1, 1)} \frac{2xy}{x^2 + 2y^2}$ **10.** $\displaystyle\lim_{(x, y) \to (0, 0)} \frac{2xy}{x^2 + 2y^2}$

11–15 ▪ Find the first partial derivatives.

11. $f(x, y) = (5y^3 + 2x^2y)^8$ **12.** $g(u, v) = \dfrac{u + 2v}{u^2 + v^2}$

13. $F(\alpha, \beta) = \alpha^2 \ln(\alpha^2 + \beta^2)$ **14.** $G(x, y, z) = e^{xz} \sin(y/z)$

15. $S(u, v, w) = u \arctan(v\sqrt{w})$

---

16. The speed of sound traveling through ocean water is a function of temperature, salinity, and pressure. It has been modeled by the function

$$C = 1449.2 + 4.6T - 0.055T^2 + 0.00029T^3$$
$$+ (1.34 - 0.01T)(S - 35) + 0.016D$$

where C is the speed of sound (in meters per second), T is the temperature (in degrees Celsius), S is the salinity (the concentration of salts in parts per thousand, which means the number of grams of dissolved solids per 1000 g of water), and D is the depth below the ocean surface (in meters). Compute $\partial C/\partial T$, $\partial C/\partial S$, and $\partial C/\partial D$ when $T = 10°C$, $S = 35$ parts per thousand, and $D = 100$ m. Explain the physical significance of these partial derivatives.

17–20 ▪ Find all second partial derivatives of f.

17. $f(x, y) = 4x^3 - xy^2$ **18.** $z = xe^{-2y}$

19. $f(x, y, z) = x^k y^l z^m$ **20.** $v = r \cos(s + 2t)$

---

21. If $z = xy + xe^{y/x}$, show that $x\dfrac{\partial z}{\partial x} + y\dfrac{\partial z}{\partial y} = xy + z$.

22. If $z = \sin(x + \sin t)$, show that

$$\frac{\partial z}{\partial x} \frac{\partial^2 z}{\partial x \, \partial t} = \frac{\partial z}{\partial t} \frac{\partial^2 z}{\partial x^2}$$

23–27 ▪ Find equations of (a) the tangent plane and (b) the normal line to the given surface at the specified point.

23. $z = 3x^2 - y^2 + 2x$, $(1, -2, 1)$

24. $z = e^x \cos y$, $(0, 0, 1)$

25. $x^2 + 2y^2 - 3z^2 = 3$, $(2, -1, 1)$

26. $xy + yz + zx = 3$, $(1, 1, 1)$

27. $\sin(xyz) = x + 2y + 3z$, $(2, -1, 0)$

---

28. Use a computer to graph the surface $z = x^2 + y^4$ and its tangent plane and normal line at $(1, 1, 2)$ on the same screen. Choose the domain and viewpoint so that you get a good view of all three objects.

29. Find the points on the hyperboloid $x^2 + 4y^2 - z^2 = 4$ where the tangent plane is parallel to the plane $2x + 2y + z = 5$.

30. Find du if $u = \ln(1 + se^{2t})$.

31. Find the linear approximation of the function $f(x, y, z) = x^3\sqrt{y^2 + z^2}$ at the point $(2, 3, 4)$ and use it to estimate the number $(1.98)^3\sqrt{(3.01)^2 + (3.97)^2}$.

32. The two legs of a right triangle are measured as 5 m and 12 m with a possible error in measurement of at most 0.2 cm in each. Use differentials to estimate the maximum error in the calculated value of (a) the area of the triangle and (b) the length of the hypotenuse.

33. If $u = x^2y^3 + z^4$, where $x = p + 3p^2$, $y = pe^p$, and $z = p \sin p$, use the Chain Rule to find du/dp.

34. If $v = x^2 \sin y + ye^{xy}$, where $x = s + 2t$ and $y = st$, use the Chain Rule to find $\partial v/\partial s$ and $\partial v/\partial t$ when $s = 0$ and $t = 1$.

35. Suppose $z = f(x, y)$, where $x = g(s, t)$, $y = h(s, t)$, $g(1, 2) = 3$, $g_s(1, 2) = -1$, $g_t(1, 2) = 4$, $h(1, 2) = 6$, $h_s(1, 2) = -5$, $h_t(1, 2) = 10$, $f_x(3, 6) = 7$, and $f_y(3, 6) = 8$. Find $\partial z/\partial s$ and $\partial z/\partial t$ when $s = 1$ and $t = 2$.

36. Use a tree diagram to write out the Chain Rule for the case where $w = f(t, u, v)$, $t = t(p, q, r, s)$, $u = u(p, q, r, s)$, and $v = v(p, q, r, s)$ are all differentiable functions.

37. If $z = y + f(x^2 - y^2)$, where f is differentiable, show that

$$y \frac{\partial z}{\partial x} + x \frac{\partial z}{\partial y} = x$$

38. The length x of a side of a triangle is increasing at a rate of 3 in/s, the length y of another side is decreasing at a rate of 2 in/s, and the contained angle θ is increasing at a rate of 0.05 radian/s. How fast is the area of the triangle changing when $x = 40$ in, $y = 50$ in, and $\theta = \pi/6$?

39. If $z = f(u, v)$, where $u = xy$, $v = y/x$, and f has continuous second partial derivatives, show that

$$x^2 \frac{\partial^2 z}{\partial x^2} - y^2 \frac{\partial^2 z}{\partial y^2} = -4uv \frac{\partial^2 z}{\partial u \, \partial v} + 2v \frac{\partial z}{\partial v}$$

40. If $\cos(xyz) = 1 + x^2y^2 + z^2$, find $\dfrac{\partial z}{\partial x}$ and $\dfrac{\partial z}{\partial y}$.

41. Find the gradient of the function $f(x, y, z) = x^2e^{yz^2}$.

42. (a) When is the directional derivative of f a maximum?
(b) When is it a minimum?
(c) When is it 0?
(d) When is it half of its maximum value?

43–44 ▪ Find the directional derivative of f at the given point in the indicated direction.

43. $f(x, y) = x^2e^{-y}$, $(-2, 0)$,
in the direction toward the point $(2, -3)$

44. $f(x, y, z) = x^2y + x\sqrt{1 + z}$, $(1, 2, 3)$,
in the direction of $\mathbf{v} = 2\mathbf{i} + \mathbf{j} - 2\mathbf{k}$

---

45. Find the maximum rate of change of $f(x, y) = x^2y + \sqrt{y}$ at the point $(2, 1)$. In which direction does it occur?

46. Find parametric equations of the tangent line at the point $(-2, 2, 4)$ to the curve of intersection of the surface $z = 2x^2 - y^2$ and the plane $z = 4$.

47–50 ▪ Find the local maximum and minimum values and saddle points of the function. If you have three-dimensional graphing software, graph the function with a domain and viewpoint that reveal all the important aspects of the function.

47. $f(x, y) = x^2 - xy + y^2 + 9x - 6y + 10$

48. $f(x, y) = x^3 - 6xy + 8y^3$

49. $f(x, y) = 3xy - x^2y - xy^2$

50. $f(x, y) = (x^2 + y)e^{y/2}$

51–52 ▪ Find the absolute maximum and minimum values of f on the set D.

51. $f(x, y) = 4xy^2 - x^2y^2 - xy^3$; D is the closed triangular region in the xy-plane with vertices $(0, 0)$, $(0, 6)$, and $(6, 0)$

52. $f(x, y) = e^{-x^2-y^2}(x^2 + 2y^2)$; D is the disk $x^2 + y^2 \le 4$

53. Use a graph or level curves or both to estimate the local maximum and minimum values and saddle points of $f(x, y) = x^3 - 3x + y^4 - 2y^2$. Then use calculus to find these values precisely.

54. Use a graphing calculator or computer (or Newton's method or a computer algebra system) to find the critical points of $f(x, y) = 12 + 10y - 2x^2 - 8xy - y^4$ correct to three decimal places. Then classify the critical points and find the highest point on the graph.

55–58 ▪ Use Lagrange multipliers to find the maximum and minimum values of f subject to the given constraint(s).

55. $f(x, y) = x^2 y;$ $x^2 + y^2 = 1$

56. $f(x, y) = \dfrac{1}{x} + \dfrac{1}{y};$ $\dfrac{1}{x^2} + \dfrac{1}{y^2} = 1$

57. $f(x, y, z) = xyz;$ $x^2 + y^2 + z^2 = 3$

58. $f(x, y, z) = x^2 + 2y^2 + 3z^2;$
$x + y + z = 1,$ $x - y + 2z = 2$

---

59. Find the points on the surface $xy^2 z^3 = 2$ that are closest to the origin.

60. A package in the shape of a rectangular box can be mailed by the US Postal Service if the sum of its length and girth (the perimeter of a cross-section perpendicular to the length) is at most 108 in. Find the dimensions of the package with largest volume that can be mailed.

61. A pentagon is formed by placing an isosceles triangle on a rectangle, as shown in the figure. If the pentagon has fixed perimeter P, find the lengths of the sides of the pentagon that maximize the area of the pentagon.

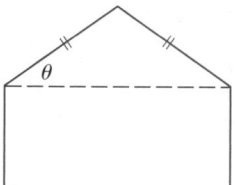

62. A particle of mass m moves on the surface $z = f(x, y)$. Let $x = x(t)$ and $y = y(t)$ be the x- and y-coordinates of the particle at time t.
  (a) Find the velocity vector $\mathbf{v}$ and the kinetic energy $K = \frac{1}{2} m |\mathbf{v}|^2$ of the particle.
  (b) Determine the acceleration vector $\mathbf{a}$.
  (c) Let $z = x^2 + y^2$ and $x(t) = t \cos t$, $y(t) = t \sin t$. Find the velocity vector, the kinetic energy, and the acceleration vector.

12 | MULTIPLE INTEGRALS

In this chapter we extend the idea of a definite integral to double and triple integrals of functions of two or three variables. These ideas are then used to compute volumes, masses, and centroids of more general regions than we were able to consider in Chapter 7. We will see that polar coordinates are useful in computing double integrals over some types of regions. In a similar way, we will introduce two new coordinate systems in three-dimensional space—cylindrical coordinates and spherical coordinates—that greatly simplify the computation of triple integrals over certain commonly occurring solid regions.

12.1 | DOUBLE INTEGRALS OVER RECTANGLES

In much the same way that our attempt to solve the area problem led to the definition of a definite integral, we now seek to find the volume of a solid and in the process we arrive at the definition of a double integral.

REVIEW OF THE DEFINITE INTEGRAL

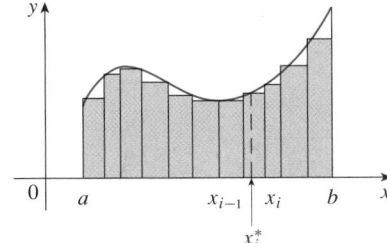

FIGURE 1

First let's recall the basic facts concerning definite integrals of functions of a single variable. If $f(x)$ is defined for $a \leq x \leq b$, we start by dividing the interval $[a, b]$ into n subintervals $[x_{i-1}, x_i]$ with length $\Delta x_i = x_i - x_{i-1}$ and we choose sample points x_i^* in these subintervals as in Figure 1. Then we form the Riemann sum

$$\boxed{1} \qquad \sum_{i=1}^{n} f(x_i^*) \, \Delta x_i$$

and take the limit of such sums as the largest of the lengths approaches 0 to obtain the definite integral of f from a to b:

$$\boxed{2} \qquad \int_a^b f(x) \, dx = \lim_{\max \Delta x_i \to 0} \sum_{i=1}^{n} f(x_i^*) \, \Delta x_i$$

In the special case where $f(x) \geq 0$, the Riemann sum can be interpreted as the sum of the areas of the approximating rectangles in Figure 1, and $\int_a^b f(x) \, dx$ represents the area under the curve $y = f(x)$ from a to b.

VOLUMES AND DOUBLE INTEGRALS

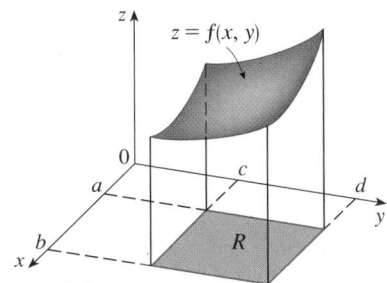

FIGURE 2

In a similar manner we consider a function f of two variables defined on a closed rectangle

$$R = [a, b] \times [c, d] = \left\{ (x, y) \in \mathbb{R}^2 \mid a \leq x \leq b, \ c \leq y \leq d \right\}$$

and we first suppose that $f(x, y) \geq 0$. The graph of f is a surface with equation $z = f(x, y)$. Let S be the solid that lies above R and under the graph of f, that is,

$$S = \left\{ (x, y, z) \in \mathbb{R}^3 \mid 0 \leq z \leq f(x, y), \ (x, y) \in R \right\}$$

(See Figure 2.) Our goal is to find the volume of S.

The first step is to take a partition P of R into subrectangles. This is accomplished by dividing the intervals $[a, b]$ and $[c, d]$ as follows:

$$a = x_0 < x_1 < \cdots < x_{i-1} < x_i < \cdots < x_m = b$$

$$c = y_0 < y_1 < \cdots < y_{j-1} < y_j < \cdots < y_n = d$$

By drawing lines parallel to the coordinate axes through these partition points as in Figure 3, we form the subrectangles

$$R_{ij} = [x_{i-1}, x_i] \times [y_{j-1}, y_j] = \{(x, y) \mid x_{i-1} \leqslant x \leqslant x_i,\ y_{j-1} \leqslant y \leqslant y_j\}$$

for $i = 1, \ldots, m$ and $j = 1, \ldots, n$. There are mn of these subrectangles, and they cover R. If we let $\Delta x_i = x_i - x_{i-1}$ and $\Delta y_j = y_j - y_{j-1}$, then the area of R_{ij} is

$$\Delta A_{ij} = \Delta x_i \, \Delta y_j$$

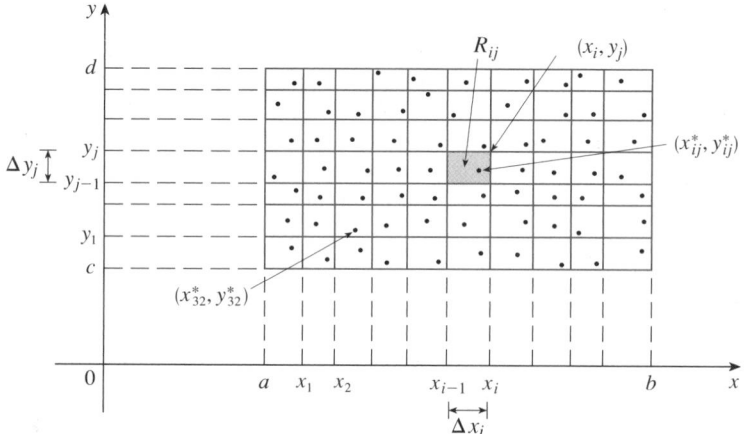

FIGURE 3

Partition of a rectangle

If we choose a **sample point** (x_{ij}^*, y_{ij}^*) in each R_{ij}, then we can approximate the part of S that lies above each R_{ij} by a thin rectangular box (or "column") with base R_{ij} and height $f(x_{ij}^*, y_{ij}^*)$ as shown in Figure 4. (Compare with Figure 1.) The volume of this box is the height of the box times the area of the base rectangle:

$$f(x_{ij}^*, y_{ij}^*) \, \Delta A_{ij}$$

If we follow this procedure for all the rectangles and add the volumes of the corresponding boxes, we get an approximation to the total volume of S:

$$\boxed{3} \qquad\qquad V \approx \sum_{i=1}^{m} \sum_{j=1}^{n} f(x_{ij}^*, y_{ij}^*) \, \Delta A_{ij}$$

(See Figure 5.) This double Riemann sum means that for each subrectangle we evaluate f at the chosen point and multiply by the area of the subrectangle, and then we add the results.

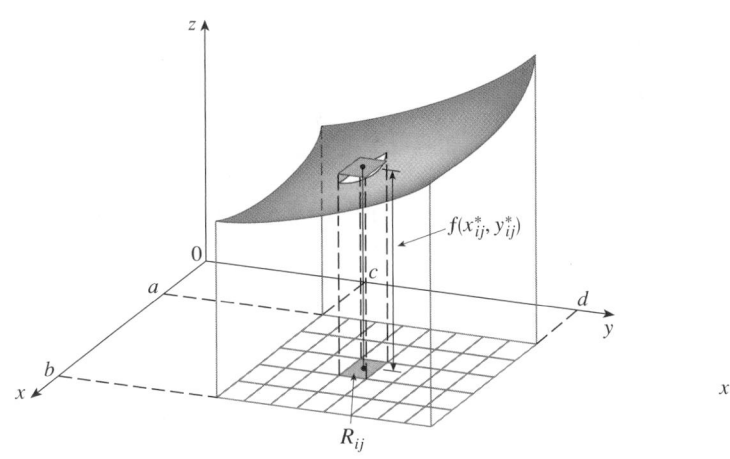

FIGURE 4

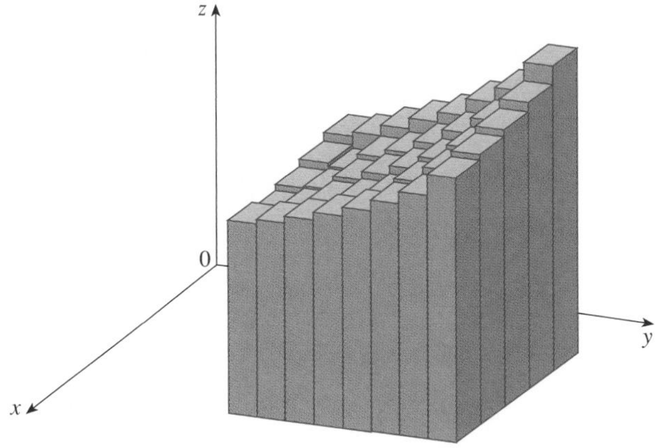

FIGURE 5

■ The meaning of the double limit in Equation 4 is that we can make the double sum as close as we like to the number V [for any choice of (x_{ij}^*, y_{ij}^*) in R_{ij}] by making the subrectangles sufficiently small.

Our intuition tells us that the approximation given in $\boxed{3}$ becomes better as the subrectangles become smaller. So if we denote by max $\Delta x_i, \Delta y_j$ the largest of the lengths of all the subintervals, we would expect that

$$\boxed{4} \qquad V = \lim_{\max \Delta x_i, \Delta y_j \to 0} \sum_{i=1}^{m} \sum_{j=1}^{n} f(x_{ij}^*, y_{ij}^*) \, \Delta A_{ij}$$

We use the expression in Equation 4 to define the **volume** of the solid S that lies under the graph of f and above the rectangle R. (It can be shown that this definition is consistent with our formula for volume in Section 7.2.)

Limits of the type that appear in Equation 4 occur frequently, not just in finding volumes but in a variety of other situations as well—as we will see in Section 12.4— even when f is not a positive function. So we make the following definition.

■ Notice the similarity between Definition 5 and the definition of a single integral in Equation 2.

$\boxed{5}$ **DEFINITION** The **double integral** of f over the rectangle R is

$$\iint\limits_R f(x, y) \, dA = \lim_{\max \Delta x_i, \Delta y_j \to 0} \sum_{i=1}^{m} \sum_{j=1}^{n} f(x_{ij}^*, y_{ij}^*) \, \Delta A_{ij}$$

if this limit exists.

The precise meaning of the limit in Definition 5 is that for every number $\varepsilon > 0$ there is a corresponding number δ such that

$$\left| \iint\limits_R f(x, y) \, dA - \sum_{i=1}^{m} \sum_{j=1}^{n} f(x_{ij}^*, y_{ij}^*) \, \Delta A_{ij} \right| < \varepsilon$$

for all partitions P of R whose subinterval lengths are less than δ, and for any choice of sample points (x_{ij}^*, y_{ij}^*) in R_{ij}.

A function f is called **integrable** if the limit in Definition 5 exists. It is shown in courses on advanced calculus that all continuous functions are integrable. In fact, the double integral of f exists provided that f is "not too discontinuous." In particular, if f is bounded [that is, there is a constant M such that $|f(x, y)| \leqslant M$ for all (x, y) in R], and f is continuous there, except on a finite number of smooth curves, then f is integrable over R.

If we know that f is integrable, we can choose the partitions P to be **regular**, that is, all the subrectangles R_{ij} have the same dimensions and therefore the same area: $\Delta A = \Delta x \, \Delta y$. In this case we can simply let $m \to \infty$ and $n \to \infty$. In addition, the sample point (x_{ij}^*, y_{ij}^*) can be chosen to be any point in the subrectangle R_{ij}, but if we choose it to be the upper right-hand corner of R_{ij} [namely (x_i, y_j), see Figure 3], then the expression for the double integral looks simpler:

$$\boxed{6} \qquad \iint\limits_R f(x, y) \, dA = \lim_{m, n \to \infty} \sum_{i=1}^{m} \sum_{j=1}^{n} f(x_i, y_j) \, \Delta A$$

By comparing Definitions 4 and 5, we see that a volume can be written as a double integral:

> If $f(x, y) \geqslant 0$, then the volume V of the solid that lies above the rectangle R and below the surface $z = f(x, y)$ is
>
> $$V = \iint\limits_R f(x, y) \, dA$$

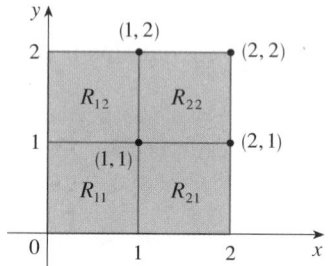

FIGURE 6

The sum in Definition 5,

$$\sum_{i=1}^{m} \sum_{j=1}^{n} f(x_{ij}^*, y_{ij}^*) \, \Delta A_{ij}$$

is called a **double Riemann sum** and is used as an approximation to the value of the double integral. [Notice how similar it is to the Riemann sum in $\boxed{1}$ for a function of a single variable.] If f happens to be a *positive* function, then the double Riemann sum represents the sum of volumes of columns, as in Figure 5, and is an approximation to the volume under the graph of f.

V EXAMPLE 1 Estimate the volume of the solid that lies above the square $R = [0, 2] \times [0, 2]$ and below the elliptic paraboloid $z = 16 - x^2 - 2y^2$. Divide R into four equal squares and choose the sample point to be the upper right corner of each square R_{ij}. Sketch the solid and the approximating rectangular boxes.

SOLUTION The squares are shown in Figure 6. The paraboloid is the graph of $f(x, y) = 16 - x^2 - 2y^2$ and the area of each square is 1. Approximating the volume by the Riemann sum with $m = n = 2$, we have

$$V \approx \sum_{i=1}^{2} \sum_{j=1}^{2} f(x_i, y_j) \, \Delta A$$

$$= f(1, 1) \, \Delta A + f(1, 2) \, \Delta A + f(2, 1) \, \Delta A + f(2, 2) \, \Delta A$$

$$= 13(1) + 7(1) + 10(1) + 4(1) = 34$$

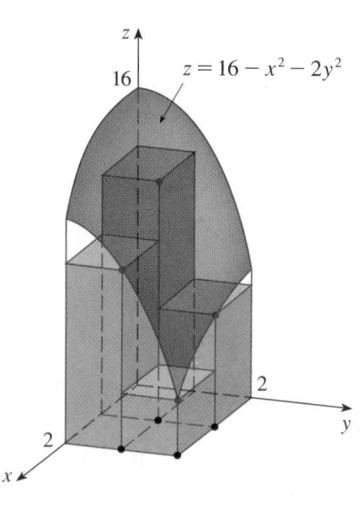

FIGURE 7

This is the volume of the approximating rectangular boxes shown in Figure 7. ∎

We get better approximations to the volume in Example 1 if we increase the number of squares. Figure 8 shows how the columns start to look more like the actual solid and the corresponding approximations become more accurate when we use 16, 64, and 256 squares. In Example 7 we will be able to show that the exact volume is 48.

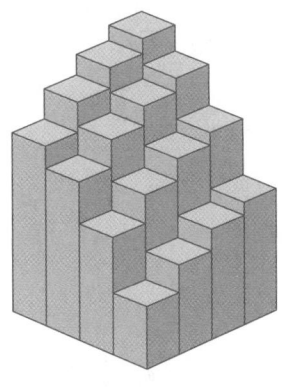 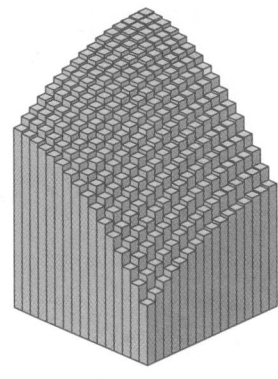

(a) $m = n = 4$, $V \approx 41.5$ (b) $m = n = 8$, $V \approx 44.875$ (c) $m = n = 16$, $V \approx 46.46875$

FIGURE 8

The Riemann sum approximations to the volume under $z = 16 - x^2 - 2y^2$ become more accurate as m and n increase.

▼ EXAMPLE 2 If $R = \{(x, y) \mid -1 \le x \le 1, -2 \le y \le 2\}$, evaluate the integral

$$\iint\limits_{R} \sqrt{1 - x^2} \, dA$$

SOLUTION It would be very difficult to evaluate this integral directly from Definition 5 but, because $\sqrt{1 - x^2} \ge 0$, we can compute the integral by interpreting it as a volume. If $z = \sqrt{1 - x^2}$, then $x^2 + z^2 = 1$ and $z \ge 0$, so the given double integral represents the volume of the solid S that lies below the circular cylinder $x^2 + z^2 = 1$ and above the rectangle R. (See Figure 9.) The volume of S is the area of a semicircle with radius 1 times the length of the cylinder. Thus

$$\iint\limits_{R} \sqrt{1 - x^2} \, dA = \tfrac{1}{2} \pi (1)^2 \times 4 = 2\pi \qquad \blacksquare$$

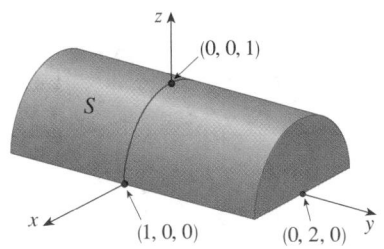

FIGURE 9

THE MIDPOINT RULE

The methods that we used for approximating single integrals (the Midpoint Rule, the Trapezoidal Rule, Simpson's Rule) all have counterparts for double integrals. Here we consider only the Midpoint Rule for double integrals. This means that we use a double Riemann sum with a regular partition to approximate the double integral, where all the subrectangles have area ΔA and the sample point (x_{ij}^*, y_{ij}^*) in R_{ij} is chosen to be the center $(\overline{x}_i, \overline{y}_j)$ of R_{ij}. In other words, $\overline{x}_i$ is the midpoint of $[x_{i-1}, x_i]$ and $\overline{y}_j$ is the midpoint of $[y_{j-1}, y_j]$.

■ **www.stewartcalculus.com**
See Additional Example A.

MIDPOINT RULE FOR DOUBLE INTEGRALS

$$\iint\limits_{R} f(x, y) \, dA \approx \sum_{i=1}^{m} \sum_{j=1}^{n} f(\overline{x}_i, \overline{y}_j) \, \Delta A$$

where $\overline{x}_i$ is the midpoint of $[x_{i-1}, x_i]$ and $\overline{y}_j$ is the midpoint of $[y_{j-1}, y_j]$.

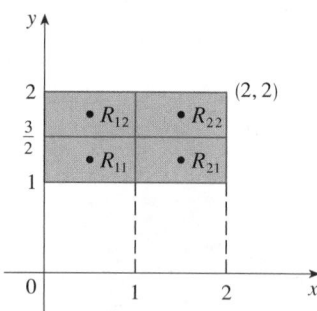

FIGURE 10

Number of subrectangles	Midpoint Rule approximations
1	-11.5000
4	-11.8750
16	-11.9687
64	-11.9922
256	-11.9980
1024	-11.9995

▼ **EXAMPLE 3** Use the Midpoint Rule with $m = n = 2$ to estimate the value of the integral $\iint_R (x - 3y^2)\, dA$, where $R = \{(x, y) \mid 0 \leqslant x \leqslant 2,\, 1 \leqslant y \leqslant 2\}$.

SOLUTION In using the Midpoint Rule with $m = n = 2$, we evaluate $f(x, y) = x - 3y^2$ at the centers of the four subrectangles shown in Figure 10. So $\bar{x}_1 = \frac{1}{2}$, $\bar{x}_2 = \frac{3}{2}$, $\bar{y}_1 = \frac{5}{4}$, and $\bar{y}_2 = \frac{7}{4}$. The area of each subrectangle is $\Delta A = \frac{1}{2}$. Thus

$$
\begin{aligned}
\iint_R (x - 3y^2)\, dA &\approx \sum_{i=1}^{2} \sum_{j=1}^{2} f(\bar{x}_i, \bar{y}_j)\, \Delta A \\
&= f(\bar{x}_1, \bar{y}_1)\, \Delta A + f(\bar{x}_1, \bar{y}_2)\, \Delta A + f(\bar{x}_2, \bar{y}_1)\, \Delta A + f(\bar{x}_2, \bar{y}_2)\, \Delta A \\
&= f\big(\tfrac{1}{2}, \tfrac{5}{4}\big)\, \Delta A + f\big(\tfrac{1}{2}, \tfrac{7}{4}\big)\, \Delta A + f\big(\tfrac{3}{2}, \tfrac{5}{4}\big)\, \Delta A + f\big(\tfrac{3}{2}, \tfrac{7}{4}\big)\, \Delta A \\
&= \big(-\tfrac{67}{16}\big)\tfrac{1}{2} + \big(-\tfrac{139}{16}\big)\tfrac{1}{2} + \big(-\tfrac{51}{16}\big)\tfrac{1}{2} + \big(-\tfrac{123}{16}\big)\tfrac{1}{2} \\
&= -\tfrac{95}{8} = -11.875
\end{aligned}
$$

Thus we have
$$
\iint_R (x - 3y^2)\, dA \approx -11.875
$$
∎

NOTE In Example 5 we will see that the exact value of the double integral in Example 3 is -12. (Remember that the interpretation of a double integral as a volume is valid only when the integrand f is a *positive* function. The integrand in Example 3 is not a positive function, so its integral is not a volume. In Examples 5 and 6 we will discuss how to interpret integrals of functions that are not always positive in terms of volumes.) If we keep dividing each subrectangle in Figure 10 into four smaller ones with similar shape, we get the Midpoint Rule approximations displayed in the chart in the margin. Notice how these approximations approach the exact value of the double integral, -12.

ITERATED INTEGRALS

Recall that it is usually difficult to evaluate single integrals directly from the definition of an integral, but the Evaluation Theorem (Part 2 of the Fundamental Theorem of Calculus) provides a much easier method. The evaluation of double integrals from first principles is even more difficult, but here we see how to express a double integral as an *iterated integral*, which can then be evaluated by calculating two single integrals.

Suppose that f is a function of two variables that is continuous on the rectangle $R = [a, b] \times [c, d]$. We use the notation $\int_c^d f(x, y)\, dy$ to mean that x is held fixed and $f(x, y)$ is integrated with respect to y from $y = c$ to $y = d$. This procedure is called *partial integration with respect to y*. (Notice its similarity to partial differentiation.) Now $\int_c^d f(x, y)\, dy$ is a number that depends on the value of x, so it defines a function of x:

$$
A(x) = \int_c^d f(x, y)\, dy
$$

If we now integrate the function A with respect to x from $x = a$ to $x = b$, we get

$$
\boxed{7} \qquad \int_a^b A(x)\, dx = \int_a^b \left[\int_c^d f(x, y)\, dy \right] dx
$$

The integral on the right side of Equation 7 is called an **iterated integral**. Usually the brackets are omitted. Thus

$$\boxed{8} \qquad \int_a^b \int_c^d f(x, y)\, dy\, dx = \int_a^b \left[\int_c^d f(x, y)\, dy \right] dx$$

means that we first integrate with respect to y from c to d and then with respect to x from a to b.

Similarly, the iterated integral

$$\boxed{9} \qquad \int_c^d \int_a^b f(x, y)\, dx\, dy = \int_c^d \left[\int_a^b f(x, y)\, dx \right] dy$$

means that we first integrate with respect to x (holding y fixed) from $x = a$ to $x = b$ and then we integrate the resulting function of y with respect to y from $y = c$ to $y = d$. Notice that in both Equations 8 and 9 we work *from the inside out*.

EXAMPLE 4 Evaluate the iterated integrals.

(a) $\displaystyle \int_0^3 \int_1^2 x^2 y\, dy\, dx$ \qquad\qquad (b) $\displaystyle \int_1^2 \int_0^3 x^2 y\, dx\, dy$

SOLUTION

(a) Regarding x as a constant, we obtain

$$\int_1^2 x^2 y\, dy = \left[x^2 \frac{y^2}{2} \right]_{y=1}^{y=2} = x^2 \left(\frac{2^2}{2} \right) - x^2 \left(\frac{1^2}{2} \right) = \tfrac{3}{2} x^2$$

Thus the function A in the preceding discussion is given by $A(x) = \tfrac{3}{2} x^2$ in this example. We now integrate this function of x from 0 to 3:

$$\int_0^3 \int_1^2 x^2 y\, dy\, dx = \int_0^3 \left[\int_1^2 x^2 y\, dy \right] dx$$

$$= \int_0^3 \tfrac{3}{2} x^2\, dx = \frac{x^3}{2} \bigg]_0^3 = \frac{27}{2}$$

(b) Here we first integrate with respect to x:

$$\int_1^2 \int_0^3 x^2 y\, dx\, dy = \int_1^2 \left[\int_0^3 x^2 y\, dx \right] dy = \int_1^2 \left[\frac{x^3}{3} y \right]_{x=0}^{x=3} dy$$

$$= \int_1^2 9y\, dy = 9 \frac{y^2}{2} \bigg]_1^2 = \frac{27}{2} \qquad\blacksquare$$

Notice that in Example 4 we obtained the same answer whether we integrated with respect to y or x first. In general, it turns out (see Theorem 10) that the two iterated integrals in Equations 8 and 9 are always equal; that is, the order of integration does not matter. (This is similar to Clairaut's Theorem on the equality of the mixed partial derivatives.)

The following theorem gives a practical method for evaluating a double integral by expressing it as an iterated integral (in either order).

■ Theorem 10 is named after the Italian mathematician Guido Fubini (1879–1943), who proved a very general version of this theorem in 1907. But the version for continuous functions was known to the French mathematician Augustin-Louis Cauchy almost a century earlier.

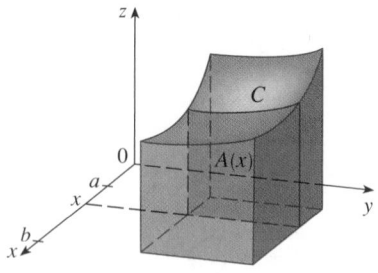

FIGURE 11

TEC Visual 12.1 illustrates Fubini's Theorem by showing an animation of Figures 11 and 12.

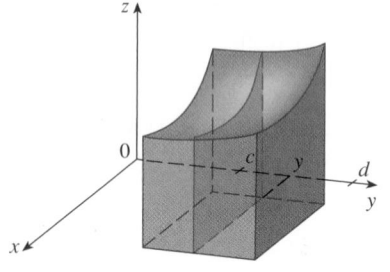

FIGURE 12

■ Notice the negative answer in Example 5; nothing is wrong with that. The function f in that example is not a positive function, so its integral doesn't represent a volume. From Figure 13 we see that f is always negative on R, so the value of the integral is the *negative* of the volume that lies *above* the graph of f and *below* R.

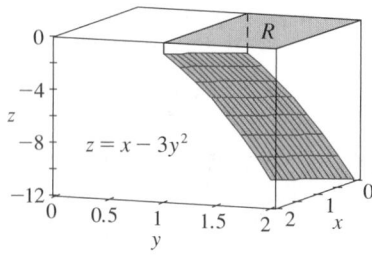

FIGURE 13

---

10 **FUBINI'S THEOREM** If f is continuous on the rectangle $R = \{(x, y) \mid a \leq x \leq b, c \leq y \leq d\}$, then

$$\iint_R f(x, y)\, dA = \int_a^b \int_c^d f(x, y)\, dy\, dx = \int_c^d \int_a^b f(x, y)\, dx\, dy$$

More generally, this is true if we assume that f is bounded on R, f is discontinuous only on a finite number of smooth curves, and the iterated integrals exist.

The proof of Fubini's Theorem is too difficult to include in this book, but we can at least give an intuitive indication of why it is true for the case where $f(x, y) \geq 0$. Recall that if f is positive, then we can interpret the double integral $\iint_R f(x, y)\, dA$ as the volume V of the solid S that lies above R and under the surface $z = f(x, y)$. But we have another formula that we used for volume in Chapter 7, namely,

$$V = \int_a^b A(x)\, dx$$

where $A(x)$ is the area of a cross-section of S in the plane through x perpendicular to the x-axis. From Figure 11 you can see that $A(x)$ is the area under the curve C whose equation is $z = f(x, y)$, where x is held constant and $c \leq y \leq d$. Therefore

$$A(x) = \int_c^d f(x, y)\, dy$$

and we have

$$\iint_R f(x, y)\, dA = V = \int_a^b A(x)\, dx = \int_a^b \int_c^d f(x, y)\, dy\, dx$$

A similar argument, using cross-sections perpendicular to the y-axis as in Figure 12, shows that

$$\iint_R f(x, y)\, dA = \int_c^d \int_a^b f(x, y)\, dx\, dy$$

V EXAMPLE 5 Evaluate the double integral $\iint_R (x - 3y^2)\, dA$, where $R = \{(x, y) \mid 0 \leq x \leq 2, 1 \leq y \leq 2\}$. (Compare with Example 3.)

SOLUTION 1 Fubini's Theorem gives

$$\iint_R (x - 3y^2)\, dA = \int_0^2 \int_1^2 (x - 3y^2)\, dy\, dx = \int_0^2 \left[xy - y^3 \right]_{y=1}^{y=2} dx$$

$$= \int_0^2 (x - 7)\, dx = \frac{x^2}{2} - 7x \Big]_0^2 = -12$$

SOLUTION 2 Again applying Fubini's Theorem, but this time integrating with respect to x first, we have

$$\iint_R (x - 3y^2)\, dA = \int_1^2 \int_0^2 (x - 3y^2)\, dx\, dy = \int_1^2 \left[\frac{x^2}{2} - 3xy^2 \right]_{x=0}^{x=2} dy$$

$$= \int_1^2 (2 - 6y^2)\, dy = 2y - 2y^3 \Big]_1^2 = -12 \qquad ■$$

■ For a function f that takes on both positive and negative values, $\iint_R f(x, y) \, dA$ is a difference of volumes: $V_1 - V_2$, where V_1 is the volume above R and below the graph of f and V_2 is the volume below R and above the graph. The fact that the integral in Example 6 is 0 means that these two volumes V_1 and V_2 are equal. (See Figure 14.)

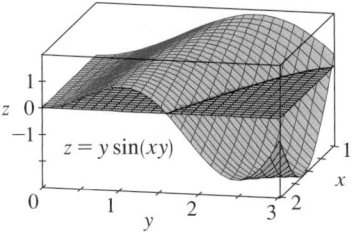

FIGURE 14

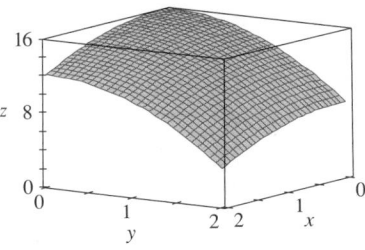

FIGURE 15

☑ EXAMPLE 6 Evaluate $\iint_R y \sin(xy) \, dA$, where $R = [1, 2] \times [0, \pi]$.

SOLUTION If we first integrate with respect to x, we get

$$\iint_R y \sin(xy) \, dA = \int_0^\pi \int_1^2 y \sin(xy) \, dx \, dy = \int_0^\pi \Big[-\cos(xy) \Big]_{x=1}^{x=2} dy$$

$$= \int_0^\pi (-\cos 2y + \cos y) \, dy = -\tfrac{1}{2} \sin 2y + \sin y \Big]_0^\pi = 0 \quad ■$$

NOTE If we first integrate with respect to y in Example 6, we get

$$\iint_R y \sin(xy) \, dA = \int_1^2 \int_0^\pi y \sin(xy) \, dy \, dx$$

but this order of integration is much more difficult than the method given in the example because it involves integration by parts twice. Therefore, when we evaluate double integrals, it is wise to choose the order of integration that gives simpler integrals.

☑ EXAMPLE 7 Find the volume of the solid S that is bounded by the elliptic paraboloid $x^2 + 2y^2 + z = 16$, the planes $x = 2$ and $y = 2$, and the three coordinate planes.

SOLUTION We first observe that S is the solid that lies under the surface $z = 16 - x^2 - 2y^2$ and above the square $R = [0, 2] \times [0, 2]$. (See Figure 15.) This solid was considered in Example 1, but we are now in a position to evaluate the double integral using Fubini's Theorem. Therefore

$$V = \iint_R (16 - x^2 - 2y^2) \, dA = \int_0^2 \int_0^2 (16 - x^2 - 2y^2) \, dx \, dy$$

$$= \int_0^2 \Big[16x - \tfrac{1}{3}x^3 - 2y^2x \Big]_{x=0}^{x=2} dy = \int_0^2 \left(\tfrac{88}{3} - 4y^2 \right) dy = \Big[\tfrac{88}{3}y - \tfrac{4}{3}y^3 \Big]_0^2 = 48 \quad ■$$

In the special case where $f(x, y)$ can be factored as the product of a function of x only and a function of y only, the double integral of f can be written in a particularly simple form. To be specific, suppose that $f(x, y) = g(x) h(y)$ and $R = [a, b] \times [c, d]$. Then Fubini's Theorem gives

$$\iint_R f(x, y) \, dA = \int_c^d \int_a^b g(x) h(y) \, dx \, dy = \int_c^d \left[\int_a^b g(x) h(y) \, dx \right] dy$$

In the inner integral y is a constant, so $h(y)$ is a constant and we can write

$$\int_c^d \left[\int_a^b g(x) h(y) \, dx \right] dy = \int_c^d \left[h(y) \left(\int_a^b g(x) \, dx \right) \right] dy = \int_a^b g(x) \, dx \int_c^d h(y) \, dy$$

since $\int_a^b g(x) \, dx$ is a constant. Therefore, in this case, the double integral of f can be written as the product of two single integrals:

$$\boxed{11} \quad \iint_R g(x) h(y) \, dA = \int_a^b g(x) \, dx \int_c^d h(y) \, dy \qquad \text{where } R = [a, b] \times [c, d]$$

■ The function $f(x, y) = \sin x \cos y$ in Example 8 is positive on R, so the integral represents the volume of the solid that lies above R and below the graph of f shown in Figure 16.

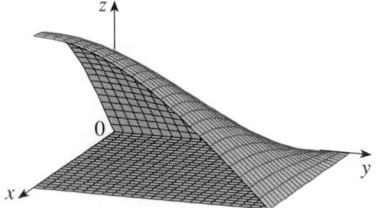

FIGURE 16

■ Double integrals behave this way because the double sums that define them behave this way.

EXAMPLE 8 If $R = [0, \pi/2] \times [0, \pi/2]$, then, by Equation 11,

$$\iint\limits_R \sin x \cos y \, dA = \int_0^{\pi/2} \sin x \, dx \int_0^{\pi/2} \cos y \, dy$$

$$= \left[-\cos x\right]_0^{\pi/2} \left[\sin y\right]_0^{\pi/2} = 1 \cdot 1 = 1 \qquad ■$$

PROPERTIES OF DOUBLE INTEGRALS

We list here three properties of double integrals that can be proved in the same manner as in Section 5.2. We assume that all of the integrals exist. Properties 12 and 13 are referred to as the *linearity* of the integral.

$$\boxed{12} \qquad \iint\limits_R [f(x, y) + g(x, y)] \, dA = \iint\limits_R f(x, y) \, dA + \iint\limits_R g(x, y) \, dA$$

$$\boxed{13} \qquad \iint\limits_R cf(x, y) \, dA = c \iint\limits_R f(x, y) \, dA \qquad \text{where } c \text{ is a constant}$$

If $f(x, y) \geq g(x, y)$ for all (x, y) in R, then

$$\boxed{14} \qquad \iint\limits_R f(x, y) \, dA \geq \iint\limits_R g(x, y) \, dA$$

12.1 | EXERCISES

1. (a) Estimate the volume of the solid that lies below the surface $z = xy$ and above the rectangle

$$R = \{(x, y) \mid 0 \leq x \leq 6, 0 \leq y \leq 4\}$$

Use a Riemann sum with $m = 3$, $n = 2$, and a regular partition, and take the sample point to be the upper right corner of each square.

(b) Use the Midpoint Rule to estimate the volume of the solid in part (a).

2. If $R = [0, 4] \times [-1, 2]$, use a Riemann sum with $m = 2$, $n = 3$ to estimate the value of $\iint_R (1 - xy^2) \, dA$. Take the sample points to be (a) the lower right corners and (b) the upper left corners of the rectangles.

3. (a) Use a Riemann sum with $m = n = 2$ to estimate the value of $\iint_R xe^{-xy} \, dA$, where $R = [0, 2] \times [0, 1]$. Take the sample points to be upper right corners.

(b) Use the Midpoint Rule to estimate the integral in part (a).

4. (a) Estimate the volume of the solid that lies below the surface $z = x + 2y^2$ and above the rectangle $R = [0, 2] \times [0, 4]$. Use a Riemann sum with

$m = n = 2$ and choose the sample points to be lower right corners.

(b) Use the Midpoint Rule to estimate the volume in part (a).

(c) Evaluate the double integral and compare your answer with the estimates in parts (a) and (b).

5. A contour map is shown for a function f on the square $R = [0, 4] \times [0, 4]$. Use the Midpoint Rule with $m = n = 2$ to estimate the value of $\iint_R f(x, y) \, dA$.

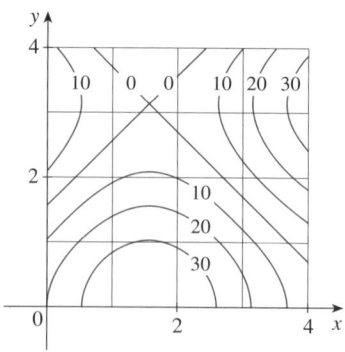

⊞ Graphing calculator or computer required   **CAS** Computer algebra system required   **1.** Homework Hints at stewartcalculus.com

6. A 20-ft-by-30-ft swimming pool is filled with water. The depth is measured at 5-foot intervals, starting at one corner of the pool, and the values are recorded in the table. Estimate the volume of water in the pool.

	0	5	10	15	20	25	30
0	2	3	4	6	7	8	8
5	2	3	4	7	8	10	8
10	2	4	6	8	10	12	10
15	2	3	4	5	6	8	7
20	2	2	2	2	3	4	4

7–9 ■ Evaluate the double integral by first identifying it as the volume of a solid.

7. $\iint_R 3 \, dA$, $R = \{(x, y) \mid -2 \leqslant x \leqslant 2, 1 \leqslant y \leqslant 6\}$

8. $\iint_R (5 - x) \, dA$, $R = \{(x, y) \mid 0 \leqslant x \leqslant 5, 0 \leqslant y \leqslant 3\}$

9. $\iint_R (4 - 2y) \, dA$, $R = [0, 1] \times [0, 1]$

10. The integral $\iint_R \sqrt{9 - y^2} \, dA$, where $R = [0, 4] \times [0, 2]$, represents the volume of a solid. Sketch the solid.

11–20 ■ Calculate the iterated integral.

11. $\int_1^4 \int_0^2 (6x^2y - 2x) \, dy \, dx$

12. $\int_0^1 \int_1^2 (4x^3 - 9x^2y^2) \, dy \, dx$

13. $\int_0^2 \int_0^4 y^3 e^{2x} \, dy \, dx$

14. $\int_1^3 \int_1^5 \frac{\ln y}{xy} \, dy \, dx$

15. $\int_{-3}^3 \int_0^{\pi/2} (y + y^2 \cos x) \, dx \, dy$

16. $\int_0^1 \int_0^3 e^{x+3y} \, dx \, dy$

17. $\int_1^4 \int_1^2 \left(\frac{x}{y} + \frac{y}{x} \right) dy \, dx$

18. $\int_0^1 \int_0^1 \sqrt{s + t} \, ds \, dt$

19. $\int_0^1 \int_0^1 v(u + v^2)^4 \, du \, dv$

20. $\int_0^1 \int_0^1 xy\sqrt{x^2 + y^2} \, dy \, dx$

21–26 ■ Calculate the double integral.

21. $\iint_R \frac{xy^2}{x^2 + 1} \, dA$, $R = \{(x, y) \mid 0 \leqslant x \leqslant 1, -3 \leqslant y \leqslant 3\}$

22. $\iint_R (y + xy^{-2}) \, dA$, $R = \{(x, y) \mid 0 \leqslant x \leqslant 2, 1 \leqslant y \leqslant 2\}$

23. $\iint_R x \sin(x + y) \, dA$, $R = [0, \pi/6] \times [0, \pi/3]$

24. $\iint_R \frac{1 + x^2}{1 + y^2} \, dA$, $R = \{(x, y) \mid 0 \leqslant x \leqslant 1, 0 \leqslant y \leqslant 1\}$

25. $\iint_R ye^{-xy} \, dA$, $R = [0, 2] \times [0, 3]$

26. $\iint_R \frac{x}{1 + xy} \, dA$, $R = [0, 1] \times [0, 1]$

27–28 ■ Sketch the solid whose volume is given by the iterated integral.

27. $\int_0^1 \int_0^1 (4 - x - 2y) \, dx \, dy$

28. $\int_0^1 \int_0^1 (2 - x^2 - y^2) \, dy \, dx$

29. Find the volume of the solid that lies under the plane $4x + 6y - 2z + 15 = 0$ and above the rectangle $R = \{(x, y) \mid -1 \leqslant x \leqslant 2, -1 \leqslant y \leqslant 1\}$.

30. Find the volume of the solid that lies under the hyperbolic paraboloid $z = 3y^2 - x^2 + 2$ and above the rectangle $R = [-1, 1] \times [1, 2]$.

31. Find the volume of the solid lying under the elliptic paraboloid $x^2/4 + y^2/9 + z = 1$ and above the rectangle $R = [-1, 1] \times [-2, 2]$.

32. Find the volume of the solid enclosed by the surface $z = 1 + e^x \sin y$ and the planes $x = \pm 1$, $y = 0$, $y = \pi$, and $z = 0$.

33. Find the volume of the solid enclosed by the surface $z = x \sec^2 y$ and the planes $z = 0$, $x = 0$, $x = 2$, $y = 0$, and $y = \pi/4$.

34. Find the volume of the solid in the first octant bounded by the cylinder $z = 16 - x^2$ and the plane $y = 5$.

35. Find the volume of the solid enclosed by the paraboloid $z = 2 + x^2 + (y - 2)^2$ and the planes $z = 1$, $x = 1$, $x = -1$, $y = 0$, and $y = 4$.

36. Graph the solid that lies between the surface $z = 2xy/(x^2 + 1)$ and the plane $z = x + 2y$ and is bounded by the planes $x = 0$, $x = 2$, $y = 0$, and $y = 4$. Then find its volume.

CAS **37.** Use a computer algebra system to find the exact value of the integral $\iint_R x^5 y^3 e^{xy} \, dA$, where $R = [0, 1] \times [0, 1]$. Then use the CAS to draw the solid whose volume is given by the integral.

CAS **38.** Graph the solid that lies between the surfaces $z = e^{-x^2} \cos(x^2 + y^2)$ and $z = 2 - x^2 - y^2$ for $|x| \leqslant 1$, $|y| \leqslant 1$. Use a computer algebra system to approximate the volume of this solid correct to four decimal places.

39–40 ▪ The **average value** of a function $f(x, y)$ over a rectangle R is defined to be

$$f_{ave} = \frac{1}{A(R)} \iint_R f(x, y) \, dA$$

(Compare with the definition for functions of one variable in Section 5.4.) Find the average value of f over the given rectangle.

39. $f(x, y) = x^2 y$,
R has vertices $(-1, 0)$, $(-1, 5)$, $(1, 5)$, $(1, 0)$

40. $f(x, y) = e^y \sqrt{x + e^y}$, $\quad R = [0, 4] \times [0, 1]$

41. If f is a constant function, $f(x, y) = k$, and $R = [a, b] \times [c, d]$, show that

$$\iint_R k \, dA = k(b - a)(d - c)$$

42. Use the result of Exercise 41 to show that

$$0 \le \iint_R \sin \pi x \cos \pi y \, dA \le \frac{1}{32}$$

where $R = \left[0, \frac{1}{4}\right] \times \left[\frac{1}{4}, \frac{1}{2}\right]$.

43–44 ▪ Use symmetry to evaluate the double integral.

43. $\displaystyle\iint_R \frac{xy}{1 + x^4} \, dA, \quad R = \{(x, y) \mid -1 \le x \le 1, 0 \le y \le 1\}$

44. $\displaystyle\iint_R (1 + x^2 \sin y + y^2 \sin x) \, dA, \quad R = [-\pi, \pi] \times [-\pi, \pi]$

CAS **45.** Use your CAS to compute the iterated integrals

$$\int_0^1 \int_0^1 \frac{x - y}{(x + y)^3} \, dy \, dx \quad \text{and} \quad \int_0^1 \int_0^1 \frac{x - y}{(x + y)^3} \, dx \, dy$$

Do the answers contradict Fubini's Theorem? Explain what is happening.

46. (a) In what way are the theorems of Fubini and Clairaut similar?
(b) If $f(x, y)$ is continuous on $[a, b] \times [c, d]$ and

$$g(x, y) = \int_a^x \int_c^y f(s, t) \, dt \, ds$$

for $a < x < b$, $c < y < d$, show that
$g_{xy} = g_{yx} = f(x, y)$.

12.2 | DOUBLE INTEGRALS OVER GENERAL REGIONS

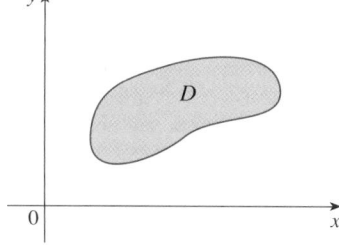

FIGURE 1

For single integrals, the region over which we integrate is always an interval. But for double integrals, we want to be able to integrate a function f not just over rectangles but also over regions D of more general shape, such as the one illustrated in Figure 1. We suppose that D is a bounded region, which means that D can be enclosed in a rectangular region R as in Figure 2. Then we define a new function F with domain R by

$$\boxed{1} \qquad F(x, y) = \begin{cases} f(x, y) & \text{if } (x, y) \text{ is in } D \\ 0 & \text{if } (x, y) \text{ is in } R \text{ but not in } D \end{cases}$$

If the double integral of F exists over R, then we define the **double integral of f over D** by

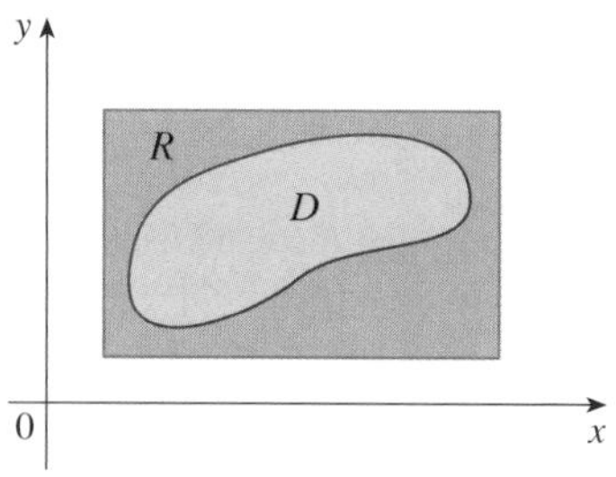

$$\boxed{2} \qquad \iint_D f(x, y) \, dA = \iint_R F(x, y) \, dA \qquad \text{where } F \text{ is given by Equation 1}$$

FIGURE 2

Definition 2 makes sense because R is a rectangle and so $\iint_R F(x, y) \, dA$ has been previously defined in Section 12.1. The procedure that we have used is reasonable because the values of $F(x, y)$ are 0 when (x, y) lies outside D and so they contribute nothing to the integral. This means that it doesn't matter what rectangle R we use as long as it contains D.

In the case where $f(x, y) \geqslant 0$ we can still interpret $\iint_D f(x, y) \, dA$ as the volume of the solid that lies above D and under the surface $z = f(x, y)$ (the graph of f). You can see that this is reasonable by comparing the graphs of f and F in Figures 3 and 4 and remembering that $\iint_R F(x, y) \, dA$ is the volume under the graph of F.

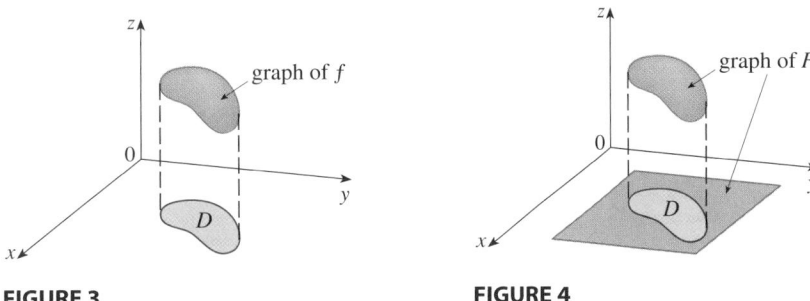

FIGURE 3

FIGURE 4

Figure 4 also shows that F is likely to have discontinuities at the boundary points of D. Nonetheless, if f is continuous on D and the boundary curve of D is "well behaved" (in a sense outside the scope of this book), then it can be shown that $\iint_R F(x, y) \, dA$ exists and therefore $\iint_D f(x, y) \, dA$ exists. In particular, this is the case for the following types of regions.

A plane region D is said to be of **type I** if it lies between the graphs of two continuous functions of x, that is,

$$D = \left\{ (x, y) \mid a \leqslant x \leqslant b, \ g_1(x) \leqslant y \leqslant g_2(x) \right\}$$

where g_1 and g_2 are continuous on $[a, b]$. Some examples of type I regions are shown in Figure 5.

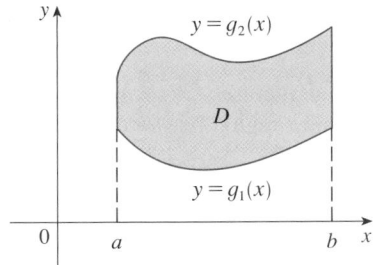

FIGURE 5 Some type I regions

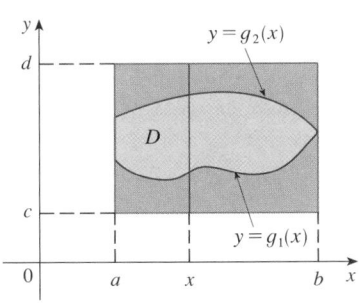

FIGURE 6

In order to evaluate $\iint_D f(x, y) \, dA$ when D is a region of type I, we choose a rectangle $R = [a, b] \times [c, d]$ that contains D, as in Figure 6, and we let F be the function given by Equation 1; that is, F agrees with f on D and F is 0 outside D. Then, by Fubini's Theorem,

$$\iint_D f(x, y) \, dA = \iint_R F(x, y) \, dA = \int_a^b \int_c^d F(x, y) \, dy \, dx$$

Observe that $F(x, y) = 0$ if $y < g_1(x)$ or $y > g_2(x)$ because (x, y) then lies outside D. Therefore

$$\int_c^d F(x, y) \, dy = \int_{g_1(x)}^{g_2(x)} F(x, y) \, dy = \int_{g_1(x)}^{g_2(x)} f(x, y) \, dy$$

because $F(x, y) = f(x, y)$ when $g_1(x) \leq y \leq g_2(x)$. Thus we have the following formula that enables us to evaluate the double integral as an iterated integral.

---

3 If f is continuous on a type I region D such that

$$D = \{(x, y) \mid a \leq x \leq b, \ g_1(x) \leq y \leq g_2(x)\}$$

then

$$\iint\limits_D f(x, y) \, dA = \int_a^b \int_{g_1(x)}^{g_2(x)} f(x, y) \, dy \, dx$$

The integral on the right side of **3** is an iterated integral that is similar to the ones we considered in the preceding section, except that in the inner integral we regard x as being constant not only in $f(x, y)$ but also in the limits of integration, $g_1(x)$ and $g_2(x)$.

We also consider plane regions of **type II**, which can be expressed as

4
$$D = \{(x, y) \mid c \leq y \leq d, \ h_1(y) \leq x \leq h_2(y)\}$$

where h_1 and h_2 are continuous. Two such regions are illustrated in Figure 7.

Using the same methods that were used in establishing **3**, we can show that

---

5
$$\iint\limits_D f(x, y) \, dA = \int_c^d \int_{h_1(y)}^{h_2(y)} f(x, y) \, dx \, dy$$

where D is a type II region given by Equation 4.

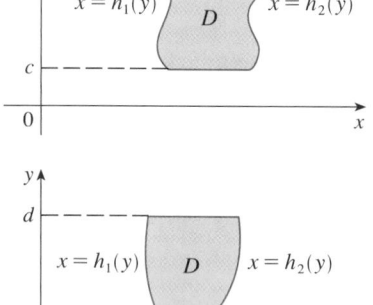

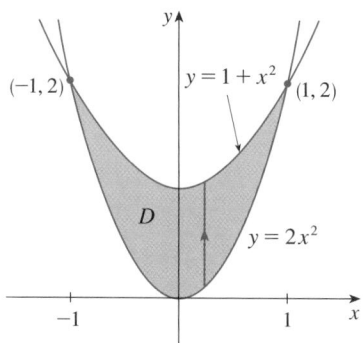

FIGURE 7
Some type II regions

V EXAMPLE 1 Evaluate $\iint_D (x + 2y) \, dA$, where D is the region bounded by the parabolas $y = 2x^2$ and $y = 1 + x^2$.

SOLUTION The parabolas intersect when $2x^2 = 1 + x^2$, that is, $x^2 = 1$, so $x = \pm 1$. We note that the region D, sketched in Figure 8, is a type I region but not a type II region and we can write

$$D = \{(x, y) \mid -1 \leq x \leq 1, \ 2x^2 \leq y \leq 1 + x^2\}$$

Since the lower boundary is $y = 2x^2$ and the upper boundary is $y = 1 + x^2$, Equation 3 gives

$$\iint\limits_D (x + 2y) \, dA = \int_{-1}^1 \int_{2x^2}^{1+x^2} (x + 2y) \, dy \, dx = \int_{-1}^1 \left[xy + y^2 \right]_{y=2x^2}^{y=1+x^2} dx$$

$$= \int_{-1}^1 \left[x(1 + x^2) + (1 + x^2)^2 - x(2x^2) - (2x^2)^2 \right] dx$$

$$= \int_{-1}^1 (-3x^4 - x^3 + 2x^2 + x + 1) \, dx$$

$$= -3\frac{x^5}{5} - \frac{x^4}{4} + 2\frac{x^3}{3} + \frac{x^2}{2} + x \bigg]_{-1}^1 = \frac{32}{15}$$ ∎

FIGURE 8

NOTE When we set up a double integral as in Example 1, it is essential to draw a diagram. Often it is helpful to draw a vertical arrow as in Figure 8. Then the limits of integration for the *inner* integral can be read from the diagram as follows: The arrow starts at the lower boundary $y = g_1(x)$, which gives the lower limit in the integral, and the arrow ends at the upper boundary $y = g_2(x)$, which gives the upper limit of integration. For a type II region the arrow is drawn horizontally from the left boundary to the right boundary.

EXAMPLE 2 Find the volume of the solid that lies under the paraboloid $z = x^2 + y^2$ and above the region D in the xy-plane bounded by the line $y = 2x$ and the parabola $y = x^2$.

SOLUTION 1 From Figure 9 we see that D is a type I region and

$$D = \left\{(x, y) \mid 0 \le x \le 2, \ x^2 \le y \le 2x\right\}$$

Therefore the volume under $z = x^2 + y^2$ and above D is

$$V = \iint_D (x^2 + y^2) \, dA = \int_0^2 \int_{x^2}^{2x} (x^2 + y^2) \, dy \, dx$$

$$= \int_0^2 \left[x^2 y + \frac{y^3}{3} \right]_{y=x^2}^{y=2x} dx$$

$$= \int_0^2 \left[x^2(2x) + \frac{(2x)^3}{3} - x^2 x^2 - \frac{(x^2)^3}{3} \right] dx$$

$$= \int_0^2 \left(-\frac{x^6}{3} - x^4 + \frac{14x^3}{3} \right) dx = -\frac{x^7}{21} - \frac{x^5}{5} + \frac{7x^4}{6} \bigg]_0^2 = \frac{216}{35}$$

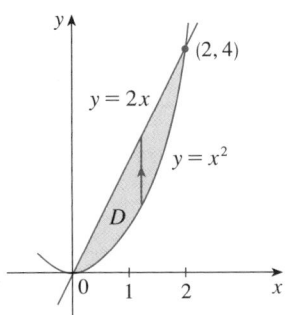

FIGURE 9
D as a type I region

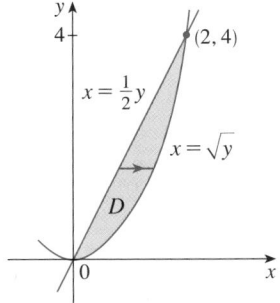

FIGURE 10
D as a type II region

SOLUTION 2 From Figure 10 we see that D can also be written as a type II region:

$$D = \left\{(x, y) \mid 0 \le y \le 4, \ \tfrac{1}{2}y \le x \le \sqrt{y}\right\}$$

Therefore another expression for V is

$$V = \iint_D (x^2 + y^2) \, dA = \int_0^4 \int_{\frac{1}{2}y}^{\sqrt{y}} (x^2 + y^2) \, dx \, dy$$

$$= \int_0^4 \left[\frac{x^3}{3} + y^2 x \right]_{x=\frac{1}{2}y}^{x=\sqrt{y}} dy = \int_0^4 \left(\frac{y^{3/2}}{3} + y^{5/2} - \frac{y^3}{24} - \frac{y^3}{2} \right) dy$$

$$= \tfrac{2}{15} y^{5/2} + \tfrac{2}{7} y^{7/2} - \tfrac{13}{96} y^4 \bigg]_0^4 = \frac{216}{35} \qquad \blacksquare$$

■ Figure 11 shows the solid whose volume is calculated in Example 2. It lies above the xy-plane, below the paraboloid $z = x^2 + y^2$, and between the plane $y = 2x$ and the parabolic cylinder $y = x^2$.

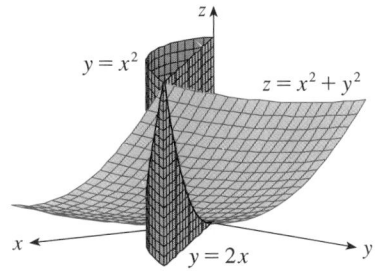

FIGURE 11

▼ **EXAMPLE 3** Evaluate $\iint_D xy \, dA$, where D is the region bounded by the line $y = x - 1$ and the parabola $y^2 = 2x + 6$.

SOLUTION The region D is shown in Figure 12. Again D is both type I and type II, but the description of D as a type I region is more complicated because the lower boundary consists of two parts. Therefore we prefer to express D as a type II region:

$$D = \left\{(x, y) \mid -2 \leqslant y \leqslant 4, \ \tfrac{1}{2}y^2 - 3 \leqslant x \leqslant y + 1\right\}$$

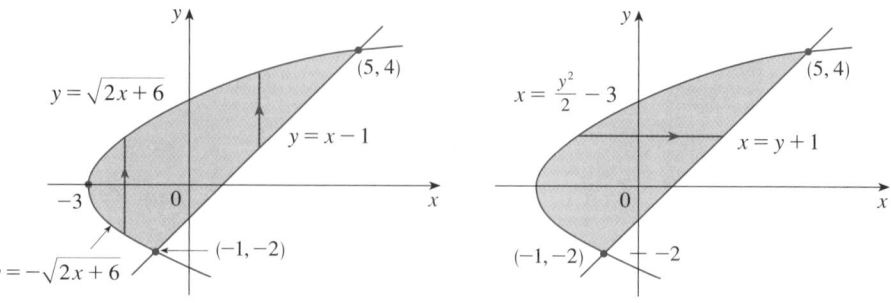

FIGURE 12   (a) D as a type I region   (b) D as a type II region

Then $\boxed{5}$ gives

$$\iint_D xy \, dA = \int_{-2}^4 \int_{\frac{1}{2}y^2-3}^{y+1} xy \, dx \, dy = \int_{-2}^4 \left[\frac{x^2}{2}\, y\right]_{x=\frac{1}{2}y^2-3}^{x=y+1} dy$$

$$= \tfrac{1}{2}\int_{-2}^4 y\left[(y + 1)^2 - \left(\tfrac{1}{2}y^2 - 3\right)^2\right] dy$$

$$= \tfrac{1}{2}\int_{-2}^4 \left(-\frac{y^5}{4} + 4y^3 + 2y^2 - 8y\right) dy$$

$$= \frac{1}{2}\left[-\frac{y^6}{24} + y^4 + 2\,\frac{y^3}{3} - 4y^2\right]_{-2}^4 = 36$$

If we had expressed D as a type I region using Figure 12(a), then we would have obtained

$$\iint_D xy \, dA = \int_{-3}^{-1} \int_{-\sqrt{2x+6}}^{\sqrt{2x+6}} xy \, dy \, dx + \int_{-1}^5 \int_{x-1}^{\sqrt{2x+6}} xy \, dy \, dx$$

but this would have involved more work than the other method. ∎

EXAMPLE 4 Find the volume of the tetrahedron bounded by the planes $x + 2y + z = 2$, $x = 2y$, $x = 0$, and $z = 0$.

SOLUTION In a question such as this, it's wise to draw two diagrams: one of the three-dimensional solid and another of the plane region D over which it lies. Figure 13 shows the tetrahedron T bounded by the coordinate planes $x = 0$, $z = 0$, the vertical plane $x = 2y$, and the plane $x + 2y + z = 2$. Since the plane $x + 2y + z = 2$ intersects the xy-plane (whose equation is $z = 0$) in the line $x + 2y = 2$, we see that T lies above the triangular region D in the xy-plane bounded by the lines $x = 2y$, $x + 2y = 2$, and $x = 0$. (See Figure 14.)

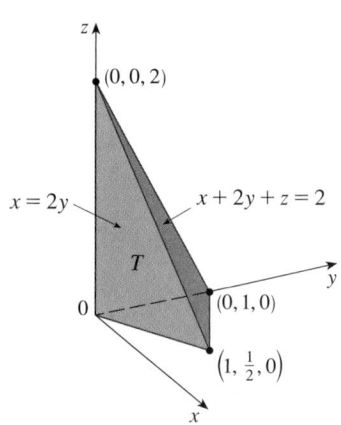

FIGURE 13

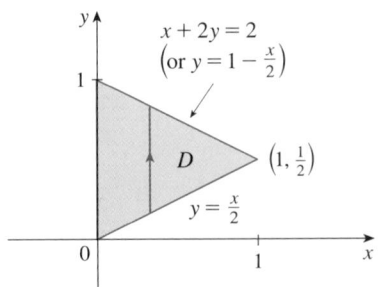

FIGURE 14

The plane $x + 2y + z = 2$ can be written as $z = 2 - x - 2y$, so the required volume lies under the graph of the function $z = 2 - x - 2y$ and above

$$D = \{(x, y) \mid 0 \leq x \leq 1, \ x/2 \leq y \leq 1 - x/2\}$$

Therefore

$$V = \iint\limits_{D} (2 - x - 2y) \, dA = \int_0^1 \int_{x/2}^{1-x/2} (2 - x - 2y) \, dy \, dx$$

$$= \int_0^1 \left[2y - xy - y^2 \right]_{y=x/2}^{y=1-x/2} dx$$

$$= \int_0^1 \left[2 - x - x\left(1 - \frac{x}{2}\right) - \left(1 - \frac{x}{2}\right)^2 - x + \frac{x^2}{2} + \frac{x^2}{4} \right] dx$$

$$= \int_0^1 (x^2 - 2x + 1) \, dx = \frac{x^3}{3} - x^2 + x \Big]_0^1 = \frac{1}{3} \qquad \blacksquare$$

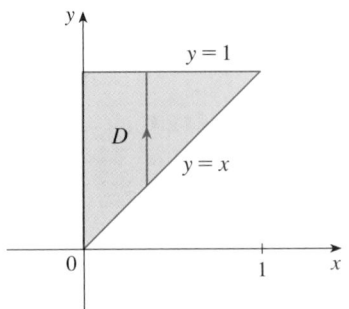

FIGURE 15

D as a type I region

◢ EXAMPLE 5 Evaluate the iterated integral $\int_0^1 \int_x^1 \sin(y^2) \, dy \, dx$.

SOLUTION If we try to evaluate the integral as it stands, we are faced with the task of first evaluating $\int \sin(y^2) \, dy$. But it's impossible to do so in finite terms since $\int \sin(y^2) \, dy$ is not an elementary function. (See the end of Section 6.4.) So we must change the order of integration. This is accomplished by first expressing the given iterated integral as a double integral. Using $\boxed{3}$ backward, we have

$$\int_0^1 \int_x^1 \sin(y^2) \, dy \, dx = \iint\limits_{D} \sin(y^2) \, dA$$

where

$$D = \{(x, y) \mid 0 \leq x \leq 1, \ x \leq y \leq 1\}$$

We sketch this region D in Figure 15. Then from Figure 16 we see that an alternative description of D is

$$D = \{(x, y) \mid 0 \leq y \leq 1, \ 0 \leq x \leq y\}$$

This enables us to use $\boxed{5}$ to express the double integral as an iterated integral in the reverse order:

$$\int_0^1 \int_x^1 \sin(y^2) \, dy \, dx = \iint\limits_{D} \sin(y^2) \, dA$$

$$= \int_0^1 \int_0^y \sin(y^2) \, dx \, dy = \int_0^1 \left[x \sin(y^2) \right]_{x=0}^{x=y} dy$$

$$= \int_0^1 y \sin(y^2) \, dy = -\tfrac{1}{2} \cos(y^2) \Big]_0^1$$

$$= \tfrac{1}{2}(1 - \cos 1) \qquad \blacksquare$$

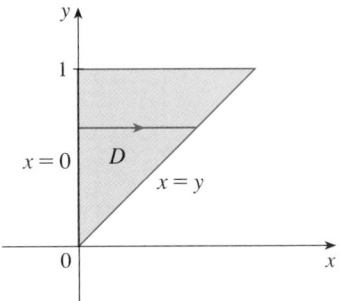

FIGURE 16

D as a type II region

PROPERTIES OF DOUBLE INTEGRALS

We assume that all of the following integrals exist. The first three properties of double integrals over a region D follow immediately from Definition 2 and Properties 12, 13, and 14 in Section 12.1.

$$\boxed{6} \qquad \iint\limits_D [f(x, y) + g(x, y)] \, dA = \iint\limits_D f(x, y) \, dA + \iint\limits_D g(x, y) \, dA$$

$$\boxed{7} \qquad \iint\limits_D c f(x, y) \, dA = c \iint\limits_D f(x, y) \, dA$$

If $f(x, y) \geq g(x, y)$ for all (x, y) in D, then

$$\boxed{8} \qquad \iint\limits_D f(x, y) \, dA \geq \iint\limits_D g(x, y) \, dA$$

The next property of double integrals is similar to the property of single integrals given by the equation $\int_a^b f(x) \, dx = \int_a^c f(x) \, dx + \int_c^b f(x) \, dx$.

If $D = D_1 \cup D_2$, where D_1 and D_2 don't overlap except perhaps on their boundaries (see Figure 17), then

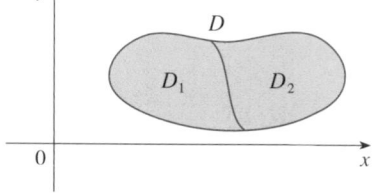

FIGURE 17

$$\boxed{9} \qquad \iint\limits_D f(x, y) \, dA = \iint\limits_{D_1} f(x, y) \, dA + \iint\limits_{D_2} f(x, y) \, dA$$

Property 9 can be used to evaluate double integrals over regions D that are neither type I nor type II but can be expressed as a union of regions of type I or type II. Figure 18 illustrates this procedure. (See Exercises 49 and 50.)

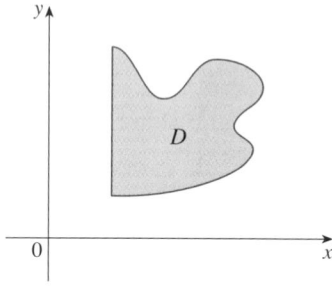

(a) D is neither type I nor type II.

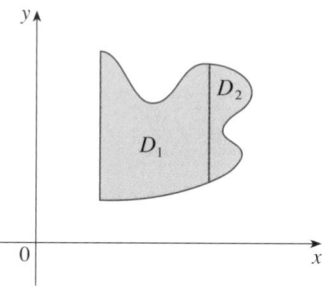

(b) $D = D_1 \cup D_2$, D_1 is type I, D_2 is type II.

FIGURE 18

The next property of integrals says that if we integrate the constant function $f(x, y) = 1$ over a region D, we get the area of D:

$$\boxed{10} \qquad \iint\limits_D 1 \, dA = A(D)$$

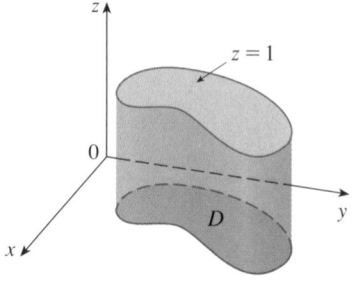

FIGURE 19
Cylinder with base D and height 1

Figure 19 illustrates why Equation 10 is true: A solid cylinder whose base is D and whose height is 1 has volume $A(D) \cdot 1 = A(D)$, but we know that we can also write its volume as $\iint_D 1 \, dA$.

Finally, we can combine Properties 7, 8, and 10 to prove the following property. (See Exercise 53.)

11 If $m \leqslant f(x, y) \leqslant M$ for all (x, y) in D, then

$$mA(D) \leqslant \iint\limits_{D} f(x, y) \, dA \leqslant MA(D)$$

EXAMPLE 6 Use Property 11 to estimate the integral $\iint_D e^{\sin x \cos y} \, dA$, where D is the disk with center the origin and radius 2.

SOLUTION Since $-1 \leqslant \sin x \leqslant 1$ and $-1 \leqslant \cos y \leqslant 1$, we have $-1 \leqslant \sin x \cos y \leqslant 1$ and therefore

$$e^{-1} \leqslant e^{\sin x \cos y} \leqslant e^{1} = e$$

Thus, using $m = e^{-1} = 1/e$, $M = e$, and $A(D) = \pi(2)^2$ in Property 11, we obtain

$$\frac{4\pi}{e} \leqslant \iint\limits_{D} e^{\sin x \cos y} \, dA \leqslant 4\pi e \qquad \blacksquare$$

12.2 | EXERCISES

1–6 ▪ Evaluate the iterated integral.

1. $\int_0^4 \int_0^{\sqrt{y}} xy^2 \, dx \, dy$

2. $\int_0^1 \int_{2x}^2 (x - y) \, dy \, dx$

3. $\int_0^1 \int_{x^2}^x (1 + 2y) \, dy \, dx$

4. $\int_0^2 \int_y^{2y} xy \, dx \, dy$

5. $\int_0^1 \int_0^{s^2} \cos(s^3) \, dt \, ds$

6. $\int_0^1 \int_0^{e^v} \sqrt{1 + e^v} \, dw \, dv$

7–10 ▪ Evaluate the double integral.

7. $\iint\limits_{D} y^2 \, dA, \quad D = \{(x, y) \mid -1 \leqslant y \leqslant 1, \; -y - 2 \leqslant x \leqslant y\}$

8. $\iint\limits_{D} \frac{y}{x^5 + 1} \, dA, \quad D = \{(x, y) \mid 0 \leqslant x \leqslant 1, \; 0 \leqslant y \leqslant x^2\}$

9. $\iint\limits_{D} x \, dA, \quad D = \{(x, y) \mid 0 \leqslant x \leqslant \pi, \; 0 \leqslant y \leqslant \sin x\}$

10. $\iint\limits_{D} x^3 \, dA, \quad D = \{(x, y) \mid 1 \leqslant x \leqslant e, \; 0 \leqslant y \leqslant \ln x\}$

11–12 ▪ Express D as a region of type I and also as a region of type II. Then evaluate the double integral in two ways.

11. $\iint\limits_{D} x \, dA, \quad D$ is enclosed by the lines $y = x$, $y = 0$, $x = 1$

12. $\iint\limits_{D} xy \, dA, \quad D$ is enclosed by the curves $y = x^2$, $y = 3x$

13–14 ▪ Set up iterated integrals for both orders of integration. Then evaluate the double integral using the easier order and explain why it's easier.

13. $\iint\limits_{D} y \, dA, \quad D$ is bounded by $y = x - 2$, $x = y^2$

14. $\iint\limits_{D} y^2 e^{xy} \, dA, \quad D$ is bounded by $y = x$, $y = 4$, $x = 0$

15–20 ▪ Evaluate the double integral.

15. $\iint\limits_{D} x \cos y \, dA, \quad D$ is bounded by $y = 0$, $y = x^2$, $x = 1$

16. $\iint\limits_{D} (x^2 + 2y) \, dA, \quad D$ is bounded by $y = x$, $y = x^3$, $x \geqslant 0$

17. $\iint\limits_{D} y^2 \, dA,$
D is the triangular region with vertices $(0, 1)$, $(1, 2)$, $(4, 1)$

18. $\iint\limits_{D} xy^2 \, dA, \quad D$ is enclosed by $x = 0$ and $x = \sqrt{1 - y^2}$

19. $\iint\limits_{D} (2x - y)\, dA$,

D is bounded by the circle with center the origin and radius 2

20. $\iint\limits_{D} 2xy\, dA$, D is the triangular region with vertices $(0, 0)$, $(1, 2)$, and $(0, 3)$

21–30 ▪ Find the volume of the given solid.

21. Under the plane $x - 2y + z = 1$ and above the region bounded by $x + y = 1$ and $x^2 + y = 1$

22. Under the surface $z = 1 + x^2 y^2$ and above the region enclosed by $x = y^2$ and $x = 4$

23. Under the surface $z = xy$ and above the triangle with vertices $(1, 1)$, $(4, 1)$, and $(1, 2)$

24. Enclosed by the paraboloid $z = x^2 + 3y^2$ and the planes $x = 0$, $y = 1$, $y = x$, $z = 0$

25. Bounded by the coordinate planes and the plane $3x + 2y + z = 6$

26. Bounded by the planes $z = x$, $y = x$, $x + y = 2$, and $z = 0$

27. Enclosed by the cylinders $z = x^2$, $y = x^2$ and the planes $z = 0$, $y = 4$

28. Bounded by the cylinder $y^2 + z^2 = 4$ and the planes $x = 2y$, $x = 0$, $z = 0$ in the first octant

29. Bounded by the cylinder $x^2 + y^2 = 1$ and the planes $y = z$, $x = 0$, $z = 0$ in the first octant

30. Bounded by the cylinders $x^2 + y^2 = r^2$ and $y^2 + z^2 = r^2$

31–32 ▪ Find the volume of the solid by subtracting two volumes.

31. The solid enclosed by the parabolic cylinders $y = 1 - x^2$, $y = x^2 - 1$ and the planes $x + y + z = 2$, $2x + 2y - z + 10 = 0$

32. The solid enclosed by the parabolic cylinder $y = x^2$ and the planes $z = 3y$, $z = 2 + y$

33–34 ▪ Sketch the solid whose volume is given by the iterated integral.

33. $\int_{0}^{1} \int_{0}^{1-x} (1 - x - y)\, dy\, dx$ **34.** $\int_{0}^{1} \int_{0}^{1-x^2} (1 - x)\, dy\, dx$

CAS **35–36** ▪ Use a computer algebra system to find the exact volume of the solid.

35. Enclosed by $z = 1 - x^2 - y^2$ and $z = 0$

36. Enclosed by $z = x^2 + y^2$ and $z = 2y$

37–42 ▪ Sketch the region of integration and change the order of integration.

37. $\int_{0}^{1} \int_{0}^{y} f(x, y)\, dx\, dy$ **38.** $\int_{0}^{2} \int_{x^2}^{4} f(x, y)\, dy\, dx$

39. $\int_{0}^{\pi/2} \int_{0}^{\cos x} f(x, y)\, dy\, dx$ **40.** $\int_{-2}^{2} \int_{0}^{\sqrt{4-y^2}} f(x, y)\, dx\, dy$

41. $\int_{1}^{2} \int_{0}^{\ln x} f(x, y)\, dy\, dx$ **42.** $\int_{0}^{1} \int_{\arctan x}^{\pi/4} f(x, y)\, dy\, dx$

43–48 ▪ Evaluate the integral by reversing the order of integration.

43. $\int_{0}^{1} \int_{3y}^{3} e^{x^2}\, dx\, dy$ **44.** $\int_{0}^{\sqrt{\pi}} \int_{y}^{\sqrt{\pi}} \cos(x^2)\, dx\, dy$

45. $\int_{0}^{4} \int_{\sqrt{x}}^{2} \frac{1}{y^3 + 1}\, dy\, dx$ **46.** $\int_{0}^{1} \int_{x}^{1} e^{x/y}\, dy\, dx$

47. $\int_{0}^{1} \int_{\arcsin y}^{\pi/2} \cos x \sqrt{1 + \cos^2 x}\, dx\, dy$

48. $\int_{0}^{8} \int_{\sqrt[3]{y}}^{2} e^{x^4}\, dx\, dy$

49–50 ▪ Express D as a union of regions of type I or type II and evaluate the integral.

49. $\iint\limits_{D} x^2\, dA$ **50.** $\iint\limits_{D} y\, dA$

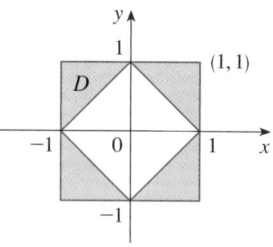   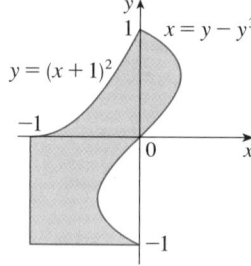

51–52 ▪ Use Property 11 to estimate the value of the integral.

51. $\iint\limits_{D} \sqrt{x^3 + y^3}\, dA$, $D = [0, 1] \times [0, 1]$

52. $\iint\limits_{D} e^{x^2+y^2}\, dA$,

D is the disk with center the origin and radius $\frac{1}{2}$

---

53. Prove Property 11.

54. In evaluating a double integral over a region D, a sum of iterated integrals was obtained as follows:

$$\iint\limits_{D} f(x, y)\, dA = \int_0^1 \int_0^{2y} f(x, y)\, dx\, dy + \int_1^3 \int_0^{3-y} f(x, y)\, dx\, dy$$

Sketch the region D and express the double integral as an iterated integral with reversed order of integration.

55–59 ▪ Use geometry or symmetry, or both, to evaluate the double integral.

55. $\iint\limits_{D} (x + 2)\, dA$, $D = \{(x, y) \mid 0 \leqslant y \leqslant \sqrt{9 - x^2}\}$

56. $\iint\limits_{D} \sqrt{R^2 - x^2 - y^2}\, dA$,

D is the disk with center the origin and radius R

57. $\iint\limits_{D} (2x + 3y)\, dA$,

D is the rectangle $0 \leqslant x \leqslant a,\ 0 \leqslant y \leqslant b$

58. $\iint\limits_{D} (2 + x^2 y^3 - y^2 \sin x)\, dA$,

$D = \{(x, y) \mid |x| + |y| \leqslant 1\}$

59. $\iint\limits_{D} \left(ax^3 + by^3 + \sqrt{a^2 - x^2}\right) dA$,

$D = [-a, a] \times [-b, b]$

[CAS] **60.** Graph the solid bounded by the plane $x + y + z = 1$ and the paraboloid $z = 4 - x^2 - y^2$ and find its exact volume. (Use your CAS to do the graphing, to find the equations of the boundary curves of the region of integration, and to evaluate the double integral.)

12.3 | DOUBLE INTEGRALS IN POLAR COORDINATES

Suppose that we want to evaluate a double integral $\iint_R f(x, y)\, dA$, where R is one of the regions shown in Figure 1. In either case the description of R in terms of rectangular coordinates is rather complicated but R is easily described using polar coordinates.

▪ Polar coordinates were introduced in Section 9.3.

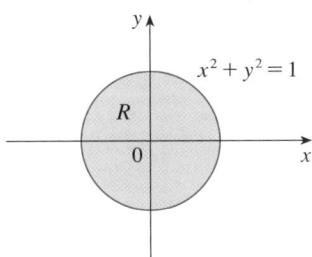

FIGURE 1

(a) $R = \{(r, \theta) \mid 0 \leqslant r \leqslant 1, 0 \leqslant \theta \leqslant 2\pi\}$

(b) $R = \{(r, \theta) \mid 1 \leqslant r \leqslant 2, 0 \leqslant \theta \leqslant \pi\}$

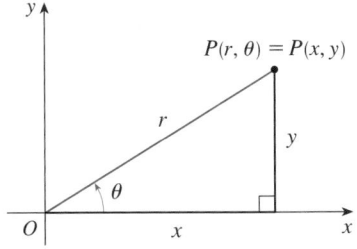

FIGURE 2

Recall from Figure 2 that the polar coordinates (r, θ) of a point are related to the rectangular coordinates (x, y) by the equations

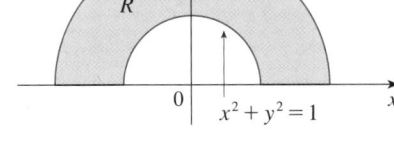

$$r^2 = x^2 + y^2 \qquad x = r \cos \theta \qquad y = r \sin \theta$$

The regions in Figure 1 are special cases of a **polar rectangle**

$$R = \{(r, \theta) \mid a \leqslant r \leqslant b, \alpha \leqslant \theta \leqslant \beta\}$$

which is shown in Figure 3. In order to compute the double integral $\iint_R f(x, y)\, dA$, where R is a polar rectangle, we divide the interval $[a, b]$ into m subintervals $[r_{i-1}, r_i]$ with lengths $\Delta r_i = r_i - r_{i-1}$ and we divide the interval $[\alpha, \beta]$ into n subintervals $[\theta_{j-1}, \theta_j]$ with lengths $\Delta\theta_j = \theta_j - \theta_{j-1}$. Then the circles $r = r_i$ and the rays $\theta = \theta_j$ divide the polar rectangle R into the small polar rectangles shown in Figure 4.

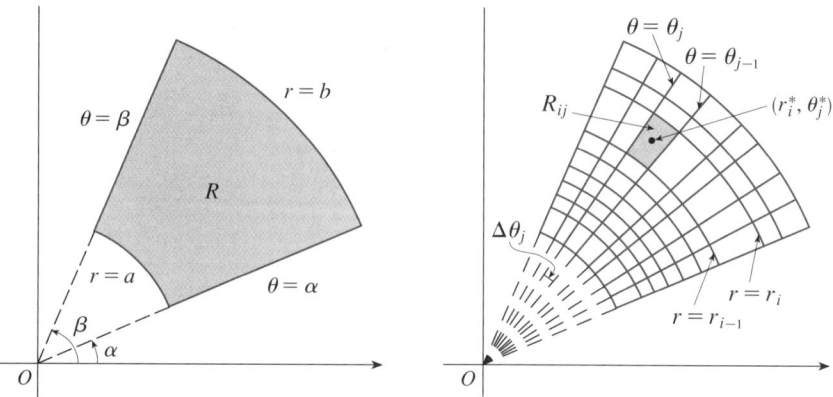

FIGURE 3 Polar rectangle   **FIGURE 4** Dividing R into polar subrectangles

The "center" of the polar subrectangle

$$R_{ij} = \left\{ (r, \theta) \mid r_{i-1} \leqslant r \leqslant r_i,\ \theta_{j-1} \leqslant \theta \leqslant \theta_j \right\}$$

has polar coordinates

$$r_i^* = \tfrac{1}{2}(r_{i-1} + r_i) \qquad \theta_j^* = \tfrac{1}{2}(\theta_{j-1} + \theta_j)$$

We compute the area of R_{ij} using the fact that the area of a sector of a circle with radius r and central angle θ is $\tfrac{1}{2}r^2\theta$. Subtracting the areas of two such sectors, each of which has central angle $\Delta\theta_j$, we find that the area of R_{ij} is

$$\Delta A_{ij} = \tfrac{1}{2}r_i^2\,\Delta\theta_j - \tfrac{1}{2}r_{i-1}^2\,\Delta\theta_j = \tfrac{1}{2}(r_i^2 - r_{i-1}^2)\,\Delta\theta_j$$

$$= \tfrac{1}{2}(r_i + r_{i-1})(r_i - r_{i-1})\,\Delta\theta_j = r_i^*\,\Delta r_i\,\Delta\theta_j$$

Although we have defined the double integral $\iint_R f(x, y)\, dA$ in terms of ordinary rectangles, it can be shown that, for continuous functions f, we always obtain the same answer using polar rectangles. The rectangular coordinates of the center of R_{ij} are $(r_i^* \cos\theta_j^*,\ r_i^* \sin\theta_j^*)$, so a typical Riemann sum is

$$\boxed{1}\quad \sum_{i=1}^{m}\sum_{j=1}^{n} f(r_i^* \cos\theta_j^*,\ r_i^* \sin\theta_j^*)\,\Delta A_{ij} = \sum_{i=1}^{m}\sum_{j=1}^{n} f(r_i^* \cos\theta_j^*,\ r_i^* \sin\theta_j^*)\,r_i^*\,\Delta r_i\,\Delta\theta_j$$

If we write $g(r, \theta) = r f(r\cos\theta,\ r\sin\theta)$, then the Riemann sum in Equation 1 can be written as

$$\sum_{i=1}^{m}\sum_{j=1}^{n} g(r_i^*, \theta_j^*)\,\Delta r_i\,\Delta\theta_j$$

which is a Riemann sum for the double integral

$$\int_\alpha^\beta \int_a^b g(r, \theta) \, dr \, d\theta$$

Therefore we have

$$\iint_R f(x, y) \, dA = \lim_{\max \Delta r_i, \Delta \theta_j \to 0} \sum_{i=1}^m \sum_{j=1}^n f(r_i^* \cos \theta_j^*, \, r_i^* \sin \theta_j^*) \, \Delta A_{ij}$$

$$= \lim_{\max \Delta r_i, \Delta \theta_j \to 0} \sum_{i=1}^m \sum_{j=1}^n g(r_i^*, \theta_j^*) \, \Delta r_i \, \Delta \theta_j = \int_\alpha^\beta \int_a^b g(r, \theta) \, dr \, d\theta$$

$$= \int_\alpha^\beta \int_a^b f(r \cos \theta, r \sin \theta) \, r \, dr \, d\theta$$

---

2   **CHANGE TO POLAR COORDINATES IN A DOUBLE INTEGRAL**   If f is continuous on a polar rectangle R given by $0 \le a \le r \le b$, $\alpha \le \theta \le \beta$, where $0 \le \beta - \alpha \le 2\pi$, then

$$\iint_R f(x, y) \, dA = \int_\alpha^\beta \int_a^b f(r \cos \theta, r \sin \theta) \, r \, dr \, d\theta$$

The formula in **2** says that we convert from rectangular to polar coordinates in a double integral by writing $x = r \cos \theta$ and $y = r \sin \theta$, using the appropriate limits of integration for r and θ, and replacing dA by $r \, dr \, d\theta$. Be careful not to forget the additional factor r on the right side of Formula 2. A classical method for remembering this is shown in Figure 5, where the "infinitesimal" polar rectangle can be thought of as an ordinary rectangle with dimensions $r \, d\theta$ and dr and therefore has "area" $dA = r \, dr \, d\theta$.

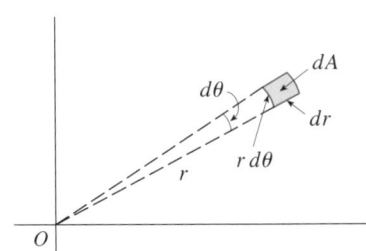

FIGURE 5

EXAMPLE 1   Evaluate $\iint_R (3x + 4y^2) \, dA$, where R is the region in the upper half-plane bounded by the circles $x^2 + y^2 = 1$ and $x^2 + y^2 = 4$.

SOLUTION   The region R can be described as

$$R = \{(x, y) \mid y \ge 0, \ 1 \le x^2 + y^2 \le 4\}$$

It is the half-ring shown in Figure 1(b), and in polar coordinates it is given by $1 \le r \le 2$, $0 \le \theta \le \pi$. Therefore, by Formula 2,

$$\iint_R (3x + 4y^2) \, dA = \int_0^\pi \int_1^2 (3r \cos \theta + 4r^2 \sin^2\theta) \, r \, dr \, d\theta$$

$$= \int_0^\pi \int_1^2 (3r^2 \cos \theta + 4r^3 \sin^2\theta) \, dr \, d\theta$$

$$= \int_0^\pi \left[r^3 \cos \theta + r^4 \sin^2\theta \right]_{r=1}^{r=2} d\theta = \int_0^\pi (7 \cos \theta + 15 \sin^2\theta) \, d\theta$$

- Here we use the trigonometric identity

$$\sin^2\theta = \tfrac{1}{2}(1 - \cos 2\theta)$$

as discussed in Section 6.2.

$$= \int_0^\pi \left[7 \cos \theta + \tfrac{15}{2}(1 - \cos 2\theta) \right] d\theta$$

$$= 7 \sin \theta + \frac{15\theta}{2} - \frac{15}{4} \sin 2\theta \bigg]_0^\pi = \frac{15\pi}{2} \quad \blacksquare$$

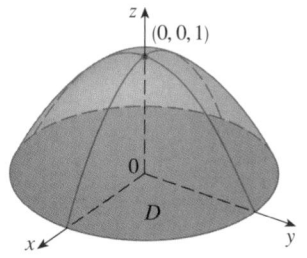

FIGURE 6

▼ EXAMPLE 2 Find the volume of the solid bounded by the plane $z = 0$ and the paraboloid $z = 1 - x^2 - y^2$.

SOLUTION If we put $z = 0$ in the equation of the paraboloid, we get $x^2 + y^2 = 1$. This means that the plane intersects the paraboloid in the circle $x^2 + y^2 = 1$, so the solid lies under the paraboloid and above the circular disk D given by $x^2 + y^2 \leq 1$ [see Figures 6 and 1(a)]. In polar coordinates D is given by $0 \leq r \leq 1$, $0 \leq \theta \leq 2\pi$. Since $1 - x^2 - y^2 = 1 - r^2$, the volume is

$$V = \iint\limits_{D} (1 - x^2 - y^2)\, dA = \int_0^{2\pi} \int_0^1 (1 - r^2)\, r\, dr\, d\theta$$

$$= \int_0^{2\pi} d\theta \int_0^1 (r - r^3)\, dr = 2\pi \left[\frac{r^2}{2} - \frac{r^4}{4} \right]_0^1 = \frac{\pi}{2}$$

If we had used rectangular coordinates instead of polar coordinates, then we would have obtained

$$V = \iint\limits_{D} (1 - x^2 - y^2)\, dA = \int_{-1}^{1} \int_{-\sqrt{1-x^2}}^{\sqrt{1-x^2}} (1 - x^2 - y^2)\, dy\, dx$$

which is not easy to evaluate because it involves finding $\int (1 - x^2)^{3/2}\, dx$. ∎

What we have done so far can be extended to the more complicated type of region shown in Figure 7. It's similar to the type II rectangular regions considered in Section 12.2. In fact, by combining Formula 2 in this section with Formula 12.2.5, we obtain the following formula.

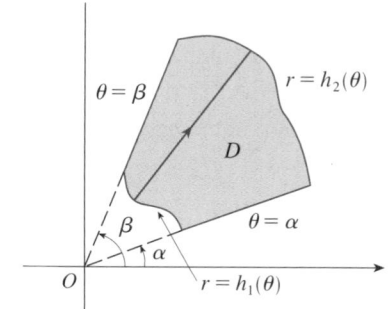

FIGURE 7
$D = \{(r, \theta) \mid \alpha \leq \theta \leq \beta, h_1(\theta) \leq r \leq h_2(\theta)\}$

---

3 If f is continuous on a polar region of the form

$$D = \left\{ (r, \theta) \mid \alpha \leq \theta \leq \beta,\ h_1(\theta) \leq r \leq h_2(\theta) \right\}$$

then

$$\iint\limits_{D} f(x, y)\, dA = \int_{\alpha}^{\beta} \int_{h_1(\theta)}^{h_2(\theta)} f(r\cos\theta, r\sin\theta)\, r\, dr\, d\theta$$

In particular, taking $f(x, y) = 1$, $h_1(\theta) = 0$, and $h_2(\theta) = h(\theta)$ in this formula, we see that the area of the region D bounded by $\theta = \alpha$, $\theta = \beta$, and $r = h(\theta)$ is

$$A(D) = \iint\limits_{D} 1\, dA = \int_{\alpha}^{\beta} \int_0^{h(\theta)} r\, dr\, d\theta$$

$$= \int_{\alpha}^{\beta} \left[\frac{r^2}{2} \right]_0^{h(\theta)} d\theta = \int_{\alpha}^{\beta} \tfrac{1}{2} [h(\theta)]^2\, d\theta$$

■ www.stewartcalculus.com
See Additional Example A.

and this agrees with Formula 9.4.3.

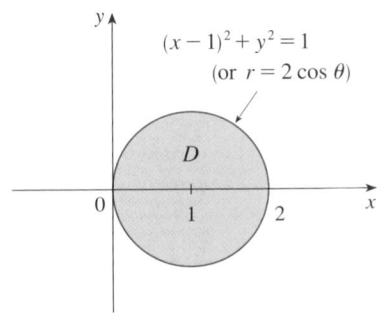

FIGURE 8

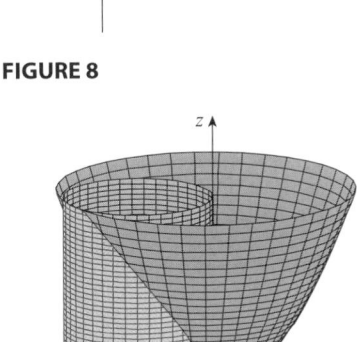

FIGURE 9

V EXAMPLE 3 Find the volume of the solid that lies under the paraboloid $z = x^2 + y^2$, above the xy-plane, and inside the cylinder $x^2 + y^2 = 2x$.

SOLUTION The solid lies above the disk D whose boundary circle has equation $x^2 + y^2 = 2x$ or, after completing the square,

$$(x - 1)^2 + y^2 = 1$$

(See Figures 8 and 9.) In polar coordinates we have $x^2 + y^2 = r^2$ and $x = r\cos\theta$, so the boundary circle becomes $r^2 = 2r\cos\theta$, or $r = 2\cos\theta$. Thus the disk D is given by

$$D = \{(r, \theta) \mid -\pi/2 \le \theta \le \pi/2, \ 0 \le r \le 2\cos\theta\}$$

and, by Formula 3, we have

$$V = \iint_D (x^2 + y^2)\, dA = \int_{-\pi/2}^{\pi/2} \int_0^{2\cos\theta} r^2\, r\, dr\, d\theta = \int_{-\pi/2}^{\pi/2} \left[\frac{r^4}{4}\right]_0^{2\cos\theta} d\theta$$

$$= 4\int_{-\pi/2}^{\pi/2} \cos^4\theta\, d\theta = 8\int_0^{\pi/2} \cos^4\theta\, d\theta = 8\int_0^{\pi/2} \left(\frac{1 + \cos 2\theta}{2}\right)^2 d\theta$$

$$= 2\int_0^{\pi/2} \left[1 + 2\cos 2\theta + \tfrac{1}{2}(1 + \cos 4\theta)\right] d\theta$$

$$= 2\left[\tfrac{3}{2}\theta + \sin 2\theta + \tfrac{1}{8}\sin 4\theta\right]_0^{\pi/2} = 2\left(\frac{3}{2}\right)\left(\frac{\pi}{2}\right) = \frac{3\pi}{2} \qquad \blacksquare$$

12.3 | EXERCISES

1–4 ■ A region R is shown. Decide whether to use polar coordinates or rectangular coordinates and write $\iint_R f(x, y)\, dA$ as an iterated integral, where f is an arbitrary continuous function on R.

1.

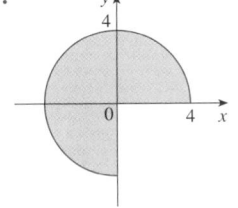

2.

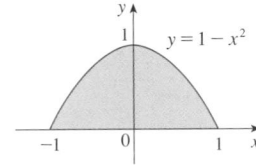

3.

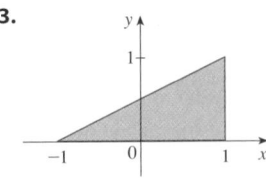

4.

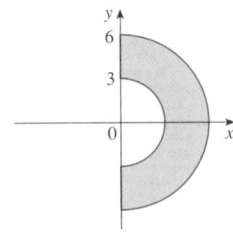

5–6 ■ Sketch the region whose area is given by the integral and evaluate the integral.

5. $\displaystyle\int_{\pi/4}^{3\pi/4} \int_1^2 r\, dr\, d\theta$

6. $\displaystyle\int_{\pi/2}^{\pi} \int_0^{2\sin\theta} r\, dr\, d\theta$

7–12 ■ Evaluate the given integral by changing to polar coordinates.

7. $\iint_D x^2 y\, dA$, where D is the top half of the disk with center the origin and radius 5

8. $\iint_R (2x - y)\, dA$, where R is the region in the first quadrant enclosed by the circle $x^2 + y^2 = 4$ and the lines $x = 0$ and $y = x$

9. $\iint_R \sin(x^2 + y^2)\, dA$, where R is the region in the first quadrant between the circles with center the origin and radii 1 and 3

10. $\iint_R \dfrac{y^2}{x^2 + y^2}\, dA$, where R is the region that lies between the circles $x^2 + y^2 = a^2$ and $x^2 + y^2 = b^2$ with $0 < a < b$

11. $\iint_R \arctan(y/x)\, dA$,
where $R = \{(x, y) \mid 1 \leqslant x^2 + y^2 \leqslant 4,\ 0 \leqslant y \leqslant x\}$

12. $\iint_D \cos\sqrt{x^2 + y^2}\, dA$, where D is the disk with center the origin and radius 2

13–19 ■ Use polar coordinates to find the volume of the given solid.

13. Under the cone $z = \sqrt{x^2 + y^2}$ and above the disk $x^2 + y^2 \leqslant 4$

14. Below the paraboloid $z = 18 - 2x^2 - 2y^2$ and above the xy-plane

15. A sphere of radius a

16. Inside the sphere $x^2 + y^2 + z^2 = 16$ and outside the cylinder $x^2 + y^2 = 4$

17. Above the cone $z = \sqrt{x^2 + y^2}$ and below the sphere $x^2 + y^2 + z^2 = 1$

18. Bounded by the paraboloids $z = 3x^2 + 3y^2$ and $z = 4 - x^2 - y^2$

19. Inside both the cylinder $x^2 + y^2 = 4$ and the ellipsoid $4x^2 + 4y^2 + z^2 = 64$

---

20. (a) A cylindrical drill with radius r_1 is used to bore a hole through the center of a sphere of radius r_2. Find the volume of the ring-shaped solid that remains.
(b) Express the volume in part (a) in terms of the height h of the ring. Notice that the volume depends only on h, not on r_1 or r_2.

21–22 ■ Use a double integral to find the area of the region.

21. One loop of the rose $r = \cos 3\theta$

22. The region enclosed by both of the cardioids $r = 1 + \cos\theta$ and $r = 1 - \cos\theta$

23–26 ■ Evaluate the iterated integral by converting to polar coordinates.

23. $\int_{-3}^{3} \int_{0}^{\sqrt{9-x^2}} \sin(x^2 + y^2)\, dy\, dx$ **24.** $\int_{0}^{a} \int_{-\sqrt{a^2-y^2}}^{0} x^2 y\, dx\, dy$

25. $\int_{0}^{1} \int_{y}^{\sqrt{2-y^2}} (x + y)\, dx\, dy$

26. $\int_{0}^{2} \int_{0}^{\sqrt{2x-x^2}} \sqrt{x^2 + y^2}\, dy\, dx$

---

27. A swimming pool is circular with a 40-ft diameter. The depth is constant along east-west lines and increases

linearly from 2 ft at the south end to 7 ft at the north end. Find the volume of water in the pool.

28. An agricultural sprinkler distributes water in a circular pattern of radius 100 ft. It supplies water to a depth of e^{-r} feet per hour at a distance of r feet from the sprinkler.
(a) If $0 < R \leqslant 100$, what is the total amount of water supplied per hour to the region inside the circle of radius R centered at the sprinkler?
(b) Determine an expression for the average amount of water per hour per square foot supplied to the region inside the circle of radius R.

29. Use polar coordinates to combine the sum

$$\int_{1/\sqrt{2}}^{1} \int_{\sqrt{1-x^2}}^{x} xy\, dy\, dx + \int_{1}^{\sqrt{2}} \int_{0}^{x} xy\, dy\, dx + \int_{\sqrt{2}}^{2} \int_{0}^{\sqrt{4-x^2}} xy\, dy\, dx$$

into one double integral. Then evaluate the double integral.

30. (a) We define the improper integral (over the entire plane $\mathbb{R}^2$)

$$I = \iint_{\mathbb{R}^2} e^{-(x^2+y^2)}\, dA = \int_{-\infty}^{\infty} \int_{-\infty}^{\infty} e^{-(x^2+y^2)}\, dy\, dx$$
$$= \lim_{a \to \infty} \iint_{D_a} e^{-(x^2+y^2)}\, dA$$

where D_a is the disk with radius a and center the origin. Show that

$$\int_{-\infty}^{\infty} \int_{-\infty}^{\infty} e^{-(x^2+y^2)}\, dA = \pi$$

(b) An equivalent definition of the improper integral in part (a) is

$$\iint_{\mathbb{R}^2} e^{-(x^2+y^2)}\, dA = \lim_{a \to \infty} \iint_{S_a} e^{-(x^2+y^2)}\, dA$$

where S_a is the square with vertices $(\pm a, \pm a)$. Use this to show that

$$\int_{-\infty}^{\infty} e^{-x^2}\, dx \int_{-\infty}^{\infty} e^{-y^2}\, dy = \pi$$

(c) Deduce that

$$\int_{-\infty}^{\infty} e^{-x^2}\, dx = \sqrt{\pi}$$

(d) By making the change of variable $t = \sqrt{2}\, x$, show that

$$\int_{-\infty}^{\infty} e^{-x^2/2}\, dx = \sqrt{2\pi}$$

(This is a fundamental result for probability and statistics.)

31. Use the result of Exercise 30 part (c) to evaluate the following integrals.

(a) $\int_{0}^{\infty} x^2 e^{-x^2}\, dx$ (b) $\int_{0}^{\infty} \sqrt{x}\, e^{-x}\, dx$

12.4 | **APPLICATIONS OF DOUBLE INTEGRALS**

We have already seen one application of double integrals: computing volumes. Another geometric application is finding areas of surfaces and this will be done in the next chapter. In this section we explore physical applications such as computing mass, electric charge, center of mass, and moment of inertia.

DENSITY AND MASS

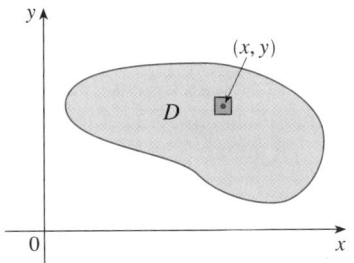

FIGURE 1

In Chapter 7 we were able to use single integrals to compute moments and the center of mass of a thin plate or lamina with constant density. But now, equipped with the double integral, we can consider a lamina with variable density. Suppose the lamina occupies a region D of the xy-plane and its **density** (in units of mass per unit area) at a point (x, y) in D is given by $\rho(x, y)$, where ρ is a continuous function on D. This means that

$$\rho(x, y) = \lim \frac{\Delta m}{\Delta A}$$

where Δm and ΔA are the mass and area of a small rectangle that contains (x, y) and the limit is taken as the dimensions of the rectangle approach 0. (See Figure 1.)

To find the total mass m of the lamina we divide a rectangle R containing D into subrectangles R_{ij} (as in Figure 2) and consider $\rho(x, y)$ to be 0 outside D. If we choose a point (x_{ij}^*, y_{ij}^*) in R_{ij}, then the mass of the part of the lamina that occupies R_{ij} is approximately $\rho(x_{ij}^*, y_{ij}^*)\,\Delta A_{ij}$, where ΔA_{ij} is the area of R_{ij}. If we add all such masses, we get an approximation to the total mass:

$$m \approx \sum_{i=1}^{k} \sum_{j=1}^{l} \rho(x_{ij}^*, y_{ij}^*)\,\Delta A_{ij}$$

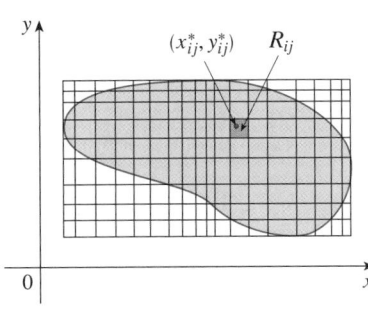

FIGURE 2

If we now take finer partitions by using smaller subrectangles, we obtain the total mass m of the lamina as the limiting value of the approximations:

$$\boxed{1} \qquad m = \lim_{\max \Delta x_i, \Delta y_j \to 0} \sum_{i=1}^{k} \sum_{j=1}^{l} \rho(x_{ij}^*, y_{ij}^*)\,\Delta A_{ij} = \iint\limits_{D} \rho(x, y)\, dA$$

Physicists also consider other types of density that can be treated in the same manner. For example, if an electric charge is distributed over a region D and the charge density (in units of charge per unit area) is given by $\sigma(x, y)$ at a point (x, y) in D, then the total charge Q is given by

$$\boxed{2} \qquad Q = \iint\limits_{D} \sigma(x, y)\, dA$$

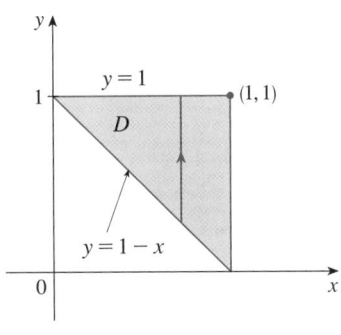

FIGURE 3

EXAMPLE 1 Charge is distributed over the triangular region D in Figure 3 so that the charge density at (x, y) is $\sigma(x, y) = xy$, measured in coulombs per square meter (C/m²). Find the total charge.

SOLUTION From Equation 2 and Figure 3 we have

$$Q = \iint_D \sigma(x, y) \, dA = \int_0^1 \int_{1-x}^1 xy \, dy \, dx$$

$$= \int_0^1 \left[x \frac{y^2}{2} \right]_{y=1-x}^{y=1} dx = \int_0^1 \frac{x}{2} [1^2 - (1-x)^2] \, dx$$

$$= \tfrac{1}{2} \int_0^1 (2x^2 - x^3) \, dx = \frac{1}{2} \left[\frac{2x^3}{3} - \frac{x^4}{4} \right]_0^1 = \frac{5}{24}$$

Thus the total charge is $\frac{5}{24}$ C.   ■

MOMENTS AND CENTERS OF MASS

In Section 7.6 we found the center of mass of a lamina with constant density; here we consider a lamina with variable density. Suppose the lamina occupies a region D and has density function $\rho(x, y)$. Recall from Chapter 7 that we defined the moment of a particle about an axis as the product of its mass and its directed distance from the axis. We divide D into small rectangles as in Figure 2. Then the mass of R_{ij} is approximately $\rho(x_{ij}^*, y_{ij}^*) \, \Delta A_{ij}$, so we can approximate the moment of R_{ij} with respect to the x-axis by

$$[\rho(x_{ij}^*, y_{ij}^*) \, \Delta A_{ij}] y_{ij}^*$$

If we now add these quantities and take the limit as the subrectangles become smaller, we obtain the **moment** of the entire lamina **about the x-axis**:

$$\boxed{3} \qquad M_x = \lim_{\max \Delta x_i, \Delta y_j \to 0} \sum_{i=1}^m \sum_{j=1}^n y_{ij}^* \, \rho(x_{ij}^*, y_{ij}^*) \, \Delta A_{ij} = \iint_D y\rho(x, y) \, dA$$

Similarly, the **moment about the y-axis** is

$$\boxed{4} \qquad M_y = \lim_{\max \Delta x_i, \Delta y_j \to 0} \sum_{i=1}^m \sum_{j=1}^n x_{ij}^* \, \rho(x_{ij}^*, y_{ij}^*) \, \Delta A_{ij} = \iint_D x\rho(x, y) \, dA$$

As before, we define the center of mass $(\bar{x}, \bar{y})$ so that $m\bar{x} = M_y$ and $m\bar{y} = M_x$. The physical significance is that the lamina behaves as if its entire mass is concentrated at its center of mass. Thus the lamina balances horizontally when supported at its center of mass (see Figure 4).

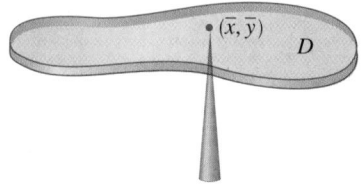

FIGURE 4

$\boxed{5}$ The coordinates $(\bar{x}, \bar{y})$ of the center of mass of a lamina occupying the region D and having density function $\rho(x, y)$ are

$$\bar{x} = \frac{M_y}{m} = \frac{1}{m} \iint_D x\rho(x, y) \, dA \qquad \bar{y} = \frac{M_x}{m} = \frac{1}{m} \iint_D y\rho(x, y) \, dA$$

where the mass m is given by

$$m = \iint_D \rho(x, y) \, dA$$

V **EXAMPLE 2** Find the mass and center of mass of a triangular lamina with vertices $(0, 0)$, $(1, 0)$, and $(0, 2)$ if the density function is $\rho(x, y) = 1 + 3x + y$.

SOLUTION The triangle is shown in Figure 5. (Note that the equation of the upper boundary is $y = 2 - 2x$.) The mass of the lamina is

$$m = \iint\limits_{D} \rho(x, y)\, dA = \int_{0}^{1} \int_{0}^{2-2x} (1 + 3x + y)\, dy\, dx$$

$$= \int_{0}^{1} \left[y + 3xy + \frac{y^2}{2} \right]_{y=0}^{y=2-2x} dx = 4 \int_{0}^{1} (1 - x^2)\, dx = 4 \left[x - \frac{x^3}{3} \right]_{0}^{1} = \frac{8}{3}$$

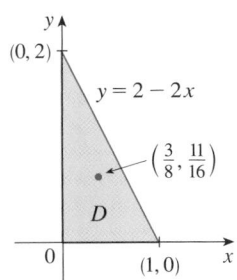

FIGURE 5

Then the formulas in $\boxed{5}$ give

$$\bar{x} = \frac{1}{m} \iint\limits_{D} x\rho(x, y)\, dA = \frac{3}{8} \int_{0}^{1} \int_{0}^{2-2x} (x + 3x^2 + xy)\, dy\, dx$$

$$= \frac{3}{8} \int_{0}^{1} \left[xy + 3x^2 y + x\frac{y^2}{2} \right]_{y=0}^{y=2-2x} dx = \frac{3}{2} \int_{0}^{1} (x - x^3)\, dx$$

$$= \frac{3}{2} \left[\frac{x^2}{2} - \frac{x^4}{4} \right]_{0}^{1} = \frac{3}{8}$$

$$\bar{y} = \frac{1}{m} \iint\limits_{D} y\rho(x, y)\, dA = \frac{3}{8} \int_{0}^{1} \int_{0}^{2-2x} (y + 3xy + y^2)\, dy\, dx$$

$$= \frac{3}{8} \int_{0}^{1} \left[\frac{y^2}{2} + 3x\frac{y^2}{2} + \frac{y^3}{3} \right]_{y=0}^{y=2-2x} dx = \frac{1}{4} \int_{0}^{1} (7 - 9x - 3x^2 + 5x^3)\, dx$$

$$= \frac{1}{4} \left[7x - 9\frac{x^2}{2} - x^3 + 5\frac{x^4}{4} \right]_{0}^{1} = \frac{11}{16}$$

The center of mass is at the point $\left(\frac{3}{8}, \frac{11}{16} \right)$. ∎

V **EXAMPLE 3** The density at any point on a semicircular lamina is proportional to the distance from the center of the circle. Find the center of mass of the lamina.

SOLUTION Let's place the lamina as the upper half of the circle $x^2 + y^2 = a^2$. (See Figure 6.) Then the distance from a point (x, y) to the center of the circle (the origin) is $\sqrt{x^2 + y^2}$. Therefore the density function is

$$\rho(x, y) = K\sqrt{x^2 + y^2}$$

where K is some constant. Both the density function and the shape of the lamina suggest that we convert to polar coordinates. Then $\sqrt{x^2 + y^2} = r$ and the region D is given by $0 \leq r \leq a$, $0 \leq \theta \leq \pi$. Thus the mass of the lamina is

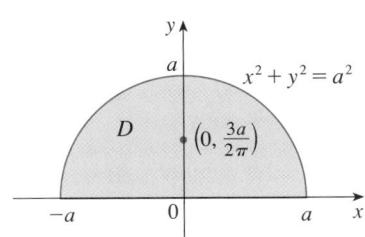

FIGURE 6

$$m = \iint\limits_{D} \rho(x, y)\, dA = \iint\limits_{D} K\sqrt{x^2 + y^2}\, dA = \int_{0}^{\pi} \int_{0}^{a} (Kr)\, r\, dr\, d\theta$$

$$= K \int_{0}^{\pi} d\theta \int_{0}^{a} r^2\, dr = K\pi \frac{r^3}{3} \Big]_{0}^{a} = \frac{K\pi a^3}{3}$$

Both the lamina and the density function are symmetric with respect to the y-axis, so

the center of mass must lie on the y-axis, that is, $\bar{x} = 0$. The y-coordinate is given by

$$\bar{y} = \frac{1}{m} \iint_D y\rho(x, y)\, dA = \frac{3}{K\pi a^3} \int_0^\pi \int_0^a r \sin\theta \,(Kr)\, r\, dr\, d\theta$$

$$= \frac{3}{\pi a^3} \int_0^\pi \sin\theta\, d\theta \int_0^a r^3\, dr = \frac{3}{\pi a^3} \Big[-\cos\theta\Big]_0^\pi \left[\frac{r^4}{4}\right]_0^a$$

$$= \frac{3}{\pi a^3} \frac{2a^4}{4} = \frac{3a}{2\pi}$$

- Compare the location of the center of mass in Example 3 with Example 8 in Section 7.6 where we found that the center of mass of a lamina with the same shape but uniform density is located at the point $(0, 4a/(3\pi))$.

Therefore the center of mass is located at the point $(0, 3a/(2\pi))$. ∎

MOMENT OF INERTIA

The **moment of inertia** (also called the **second moment**) of a particle of mass m about an axis is defined to be mr^2, where r is the distance from the particle to the axis. We extend this concept to a lamina with density function $\rho(x, y)$ and occupying a region D by proceeding as we did for ordinary moments. We divide D into small rectangles, approximate the moment of inertia of each subrectangle about the x-axis, and take the limit of the sum as the subrectangles become smaller. The result is the **moment of inertia** of the lamina **about the x-axis**:

$$\boxed{6} \qquad I_x = \lim_{\max \Delta x_i, \Delta y_j \to 0} \sum_{i=1}^m \sum_{j=1}^n (y_{ij}^*)^2 \rho(x_{ij}^*, y_{ij}^*)\, \Delta A_{ij} = \iint_D y^2 \rho(x, y)\, dA$$

Similarly, the **moment of inertia about the y-axis** is

$$\boxed{7} \qquad I_y = \lim_{\max \Delta x_i, \Delta y_j \to 0} \sum_{i=1}^m \sum_{j=1}^n (x_{ij}^*)^2 \rho(x_{ij}^*, y_{ij}^*)\, \Delta A_{ij} = \iint_D x^2 \rho(x, y)\, dA$$

It is also of interest to consider the **moment of inertia about the origin**, also called the **polar moment of inertia**:

$$\boxed{8} \qquad I_0 = \lim_{\max \Delta x_i, \Delta y_j \to 0} \sum_{i=1}^m \sum_{j=1}^n \left[(x_{ij}^*)^2 + (y_{ij}^*)^2\right] \rho(x_{ij}^*, y_{ij}^*)\, \Delta A_{ij}$$

$$= \iint_D (x^2 + y^2)\rho(x, y)\, dA$$

Note that $I_0 = I_x + I_y$.

V EXAMPLE 4 Find the moments of inertia I_x, I_y, and I_0 of a homogeneous disk D with density $\rho(x, y) = \rho$, center the origin, and radius a.

SOLUTION The boundary of D is the circle $x^2 + y^2 = a^2$ and in polar coordinates

D is described by $0 \leqslant \theta \leqslant 2\pi$, $0 \leqslant r \leqslant a$. Let's compute I_0 first:

$$I_0 = \iint\limits_{D} (x^2 + y^2)\rho \, dA = \rho \int_0^{2\pi} \int_0^a r^2 r \, dr \, d\theta$$

$$= \rho \int_0^{2\pi} d\theta \int_0^a r^3 \, dr = 2\pi\rho \left[\frac{r^4}{4}\right]_0^a = \frac{\pi\rho a^4}{2}$$

Instead of computing I_x and I_y directly, we use the facts that $I_x + I_y = I_0$ and $I_x = I_y$ (from the symmetry of the problem). Thus

$$I_x = I_y = \frac{I_0}{2} = \frac{\pi\rho a^4}{4}$$ ∎

In Example 4 notice that the mass of the disk is

$$m = \text{density} \times \text{area} = \rho(\pi a^2)$$

so the moment of inertia of the disk about the origin (like a wheel about its axle) can be written as

$$I_0 = \frac{\pi\rho a^4}{2} = \tfrac{1}{2}(\rho\pi a^2)a^2 = \tfrac{1}{2}ma^2$$

Thus if we increase the mass or the radius of the disk, we thereby increase the moment of inertia. In general, the moment of inertia plays much the same role in rotational motion that mass plays in linear motion. The moment of inertia of a wheel is what makes it difficult to start or stop the rotation of the wheel, just as the mass of a car is what makes it difficult to start or stop the motion of the car.

The **radius of gyration of a lamina about an axis** is the number R such that

[9] $$mR^2 = I$$

where m is the mass of the lamina and I is the moment of inertia about the given axis. Equation 9 says that if the mass of the lamina were concentrated at a distance R from the axis, then the moment of inertia of this "point mass" would be the same as the moment of inertia of the lamina.

In particular, the radius of gyration $\bar{\bar{y}}$ with respect to the x-axis and the radius of gyration $\bar{\bar{x}}$ with respect to the y-axis are given by the equations

[10] $$m\bar{\bar{y}}^2 = I_x \qquad m\bar{\bar{x}}^2 = I_y$$

Thus $(\bar{\bar{x}}, \bar{\bar{y}})$ is the point at which the mass of the lamina can be concentrated without changing the moments of inertia with respect to the coordinate axes. (Note the analogy with the center of mass.)

▼ **EXAMPLE 5** Find the radius of gyration about the x-axis of the disk in Example 4.

SOLUTION As noted, the mass of the disk is $m = \rho\pi a^2$, so from Equations 10 we have

$$\bar{\bar{y}}^2 = \frac{I_x}{m} = \frac{\tfrac{1}{4}\pi\rho a^4}{\rho\pi a^2} = \frac{a^2}{4}$$

Therefore the radius of gyration about the x-axis is $\bar{\bar{y}} = \tfrac{1}{2}a$, half the radius of the disk. ∎

12.4 | EXERCISES

1. Electric charge is distributed over the rectangle $0 \leq x \leq 5$, $2 \leq y \leq 5$ so that the charge density at (x, y) is $\sigma(x, y) = 2x + 4y$ (measured in coulombs per square meter). Find the total charge on the rectangle.

2. Electric charge is distributed over the disk $x^2 + y^2 \leq 1$ so that the charge density at (x, y) is $\sigma(x, y) = \sqrt{x^2 + y^2}$ (measured in coulombs per square meter). Find the total charge on the disk.

3–10 ▪ Find the mass and center of mass of the lamina that occupies the region D and has the given density function ρ.

3. $D = \{(x, y) \mid 1 \leq x \leq 3, 1 \leq y \leq 4\}$; $\rho(x, y) = ky^2$

4. $D = \{(x, y) \mid 0 \leq x \leq a, 0 \leq y \leq b\}$;
   $\rho(x, y) = 1 + x^2 + y^2$

5. D is the triangular region with vertices $(0, 0)$, $(2, 1)$, $(0, 3)$; $\rho(x, y) = x + y$

6. D is the triangular region enclosed by the lines $x = 0$, $y = x$, and $2x + y = 6$; $\rho(x, y) = x^2$

7. D is bounded by $y = 1 - x^2$ and $y = 0$; $\rho(x, y) = ky$

8. D is bounded by $y = x^2$ and $y = x + 2$; $\rho(x, y) = kx$

9. $D = \{(x, y) \mid 0 \leq y \leq \sin(\pi x/L), 0 \leq x \leq L\}$; $\rho(x, y) = y$

10. D is bounded by the parabolas $y = x^2$ and $x = y^2$;
    $\rho(x, y) = \sqrt{x}$

11. A lamina occupies the part of the disk $x^2 + y^2 \leq 1$ in the first quadrant. Find its center of mass if the density at any point is proportional to its distance from the x-axis.

12. Find the center of mass of the lamina in Exercise 11 if the density at any point is proportional to the square of its distance from the origin.

13. The boundary of a lamina consists of the semicircles $y = \sqrt{1 - x^2}$ and $y = \sqrt{4 - x^2}$ together with the portions of the x-axis that join them. Find the center of mass of the

lamina if the density at any point is proportional to its distance from the origin.

14. Find the center of mass of the lamina in Exercise 13 if the density at any point is inversely proportional to its distance from the origin.

15. Find the center of mass of a lamina in the shape of an isosceles right triangle with equal sides of length a if the density at any point is proportional to the square of the distance from the vertex opposite the hypotenuse.

16. A lamina occupies the region inside the circle $x^2 + y^2 = 2y$ but outside the circle $x^2 + y^2 = 1$. Find the center of mass if the density at any point is inversely proportional to its distance from the origin.

17. Find the moments of inertia I_x, I_y, I_0 for the lamina of Exercise 7.

18. Find the moments of inertia I_x, I_y, I_0 for the lamina of Exercise 12.

19. Find the moments of inertia I_x, I_y, I_0 for the lamina of Exercise 15.

20. Consider a square fan blade with sides of length 2 and the lower left corner placed at the origin. If the density of the blade is $\rho(x, y) = 1 + 0.1x$, is it more difficult to rotate the blade about the x-axis or the y-axis?

[CAS] **21–22** ▪ Use a computer algebra system to find the mass, center of mass, and moments of inertia of the lamina that occupies the region D and has the given density function.

21. D is enclosed by the right loop of the four-leaved rose $r = \cos 2\theta$; $\rho(x, y) = x^2 + y^2$

22. $D = \{(x, y) \mid 0 \leq y \leq xe^{-x}, 0 \leq x \leq 2\}$; $\rho(x, y) = x^2y^2$

23–24 ▪ A lamina with constant density $\rho(x, y) = \rho$ occupies the given region. Find the moments of inertia I_x and I_y and the radii of gyration $\bar{\bar{x}}$ and $\bar{\bar{y}}$.

23. The rectangle $0 \leq x \leq b$, $0 \leq y \leq h$

24. The region under the curve $y = \sin x$ from $x = 0$ to $x = \pi$

12.5 | TRIPLE INTEGRALS

Just as we defined single integrals for functions of one variable and double integrals for functions of two variables, so we can define triple integrals for functions of three variables. Let's first deal with the simplest case where f is defined on a rectangular box:

$$\boxed{1} \qquad B = \{(x, y, z) \mid a \leq x \leq b, \ c \leq y \leq d, \ r \leq z \leq s\}$$

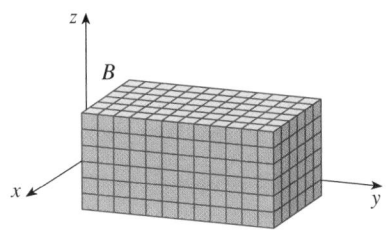

The first step is to divide B into sub-boxes. We do this by dividing the interval $[a, b]$ into l subintervals $[x_{i-1}, x_i]$ with lengths $\Delta x_i = x_i - x_{i-1}$, dividing $[c, d]$ into m subintervals with lengths $\Delta y_j = y_j - y_{j-1}$, and dividing $[r, s]$ into n subintervals with lengths $\Delta z_k = z_k - z_{k-1}$. The planes through the endpoints of these subintervals parallel to the coordinate planes divide the box B into lmn sub-boxes

$$B_{ijk} = [x_{i-1}, x_i] \times [y_{j-1}, y_j] \times [z_{k-1}, z_k]$$

which are shown in Figure 1. The sub-box B_{ijk} has volume $\Delta V_{ijk} = \Delta x_i \, \Delta y_j \, \Delta z_k$.

Then we form the **triple Riemann sum**

$$\boxed{2} \qquad \sum_{i=1}^{l} \sum_{j=1}^{m} \sum_{k=1}^{n} f(x_{ijk}^*, y_{ijk}^*, z_{ijk}^*) \, \Delta V_{ijk}$$

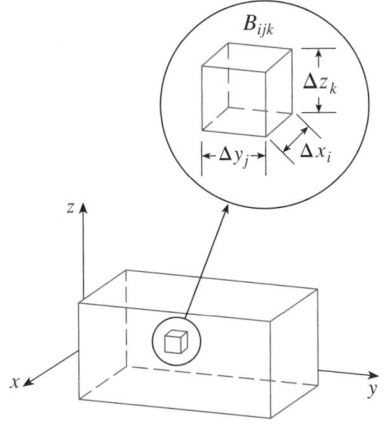

where the sample point $(x_{ijk}^*, y_{ijk}^*, z_{ijk}^*)$ is in B_{ijk}. By analogy with the definition of a double integral (12.1.5), we define the triple integral as the limit of the triple Riemann sums in $\boxed{2}$ as the sub-boxes shrink.

FIGURE 1

$\boxed{3}$ **DEFINITION** The **triple integral** of f over the box B is

$$\iiint_B f(x, y, z) \, dV = \lim_{\max \Delta x_i, \Delta y_j, \Delta z_k \to 0} \sum_{i=1}^{l} \sum_{j=1}^{m} \sum_{k=1}^{n} f(x_{ijk}^*, y_{ijk}^*, z_{ijk}^*) \, \Delta V_{ijk}$$

if this limit exists.

Again, the triple integral always exists if f is continuous. We can choose the sample point to be any point in the sub-box, but if we choose it to be the point (x_i, y_j, z_k), and if we choose sub-boxes with the same dimensions, so that $\Delta V_{ijk} = \Delta V$, we get a simpler-looking expression for the triple integral:

$$\iiint_B f(x, y, z) \, dV = \lim_{l, m, n \to \infty} \sum_{i=1}^{l} \sum_{j=1}^{m} \sum_{k=1}^{n} f(x_i, y_j, z_k) \, \Delta V$$

Just as for double integrals, the practical method for evaluating triple integrals is to express them as iterated integrals as follows.

$\boxed{4}$ **FUBINI'S THEOREM FOR TRIPLE INTEGRALS** If f is continuous on the rectangular box $B = [a, b] \times [c, d] \times [r, s]$, then

$$\iiint_B f(x, y, z) \, dV = \int_r^s \int_c^d \int_a^b f(x, y, z) \, dx \, dy \, dz$$

The iterated integral on the right side of Fubini's Theorem means that we integrate first with respect to x (keeping y and z fixed), then we integrate with respect to y (keeping z fixed), and finally we integrate with respect to z. There are five other possible orders in which we can integrate, all of which give the same value. For instance, if we integrate with respect to y, then z, and then x, we have

$$\iiint_B f(x, y, z)\, dV = \int_a^b \int_r^s \int_c^d f(x, y, z)\, dy\, dz\, dx$$

V EXAMPLE 1 Evaluate the triple integral $\iiint_B xyz^2\, dV$, where B is the rectangular box given by

$$B = \left\{ (x, y, z) \mid 0 \leqslant x \leqslant 1,\ -1 \leqslant y \leqslant 2,\ 0 \leqslant z \leqslant 3 \right\}$$

SOLUTION We could use any of the six possible orders of integration. If we choose to integrate with respect to x, then y, and then z, we obtain

$$\iiint_B xyz^2\, dV = \int_0^3 \int_{-1}^2 \int_0^1 xyz^2\, dx\, dy\, dz = \int_0^3 \int_{-1}^2 \left[\frac{x^2 yz^2}{2} \right]_{x=0}^{x=1} dy\, dz$$

$$= \int_0^3 \int_{-1}^2 \frac{yz^2}{2}\, dy\, dz = \int_0^3 \left[\frac{y^2 z^2}{4} \right]_{y=-1}^{y=2} dz = \int_0^3 \frac{3z^2}{4}\, dz = \frac{z^3}{4} \Big]_0^3 = \frac{27}{4} \qquad \blacksquare$$

Now we define the **triple integral over a general bounded region** E in three-dimensional space (a solid) by much the same procedure that we used for double integrals (12.2.2). We enclose E in a box B of the type given by Equation 1. Then we define a function F so that it agrees with f on E but is 0 for points in B that are outside E. By definition,

$$\iiint_E f(x, y, z)\, dV = \iiint_B F(x, y, z)\, dV$$

This integral exists if f is continuous and the boundary of E is "reasonably smooth." The triple integral has essentially the same properties as the double integral (Properties 6–9 in Section 12.2).

We restrict our attention to continuous functions f and to certain simple types of regions. A solid region E is said to be of **type 1** if it lies between the graphs of two continuous functions of x and y, that is,

$$\boxed{5} \qquad E = \left\{ (x, y, z) \mid (x, y) \in D,\ u_1(x, y) \leqslant z \leqslant u_2(x, y) \right\}$$

where D is the projection of E onto the xy-plane as shown in Figure 2. Notice that the upper boundary of the solid E is the surface with equation $z = u_2(x, y)$, while the lower boundary is the surface $z = u_1(x, y)$.

By the same sort of argument that led to (12.2.3), it can be shown that if E is a type 1 region given by Equation 5, then

$$\boxed{6} \qquad \iiint_E f(x, y, z)\, dV = \iint_D \left[\int_{u_1(x, y)}^{u_2(x, y)} f(x, y, z)\, dz \right] dA$$

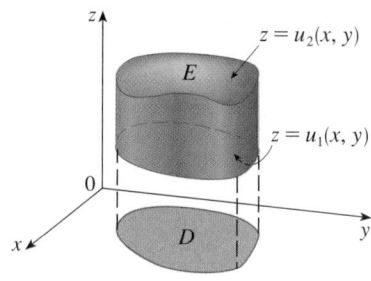

FIGURE 2
A type 1 solid region

The meaning of the inner integral on the right side of Equation 6 is that x and y are held fixed, and therefore $u_1(x, y)$ and $u_2(x, y)$ are regarded as constants, while $f(x, y, z)$ is integrated with respect to z.

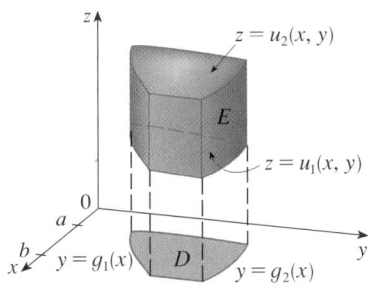

FIGURE 3

A type 1 solid region where the projection D is a type I plane region

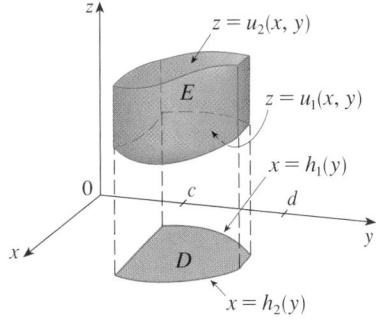

FIGURE 4

Another type 1 solid region with a type II projection

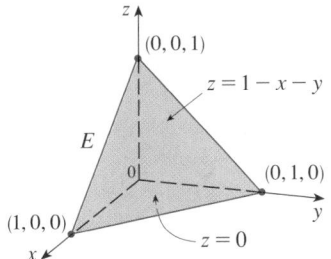

FIGURE 5

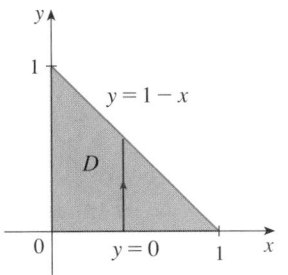

FIGURE 6

In particular, if the projection D of E onto the xy-plane is a type I plane region (as in Figure 3), then

$$E = \{(x, y, z) \mid a \leqslant x \leqslant b, \ g_1(x) \leqslant y \leqslant g_2(x), \ u_1(x, y) \leqslant z \leqslant u_2(x, y)\}$$

and Equation 6 becomes

$$\boxed{7} \qquad \iiint_E f(x, y, z) \, dV = \int_a^b \int_{g_1(x)}^{g_2(x)} \int_{u_1(x, y)}^{u_2(x, y)} f(x, y, z) \, dz \, dy \, dx$$

If, on the other hand, D is a type II plane region (as in Figure 4), then

$$E = \{(x, y, z) \mid c \leqslant y \leqslant d, \ h_1(y) \leqslant x \leqslant h_2(y), \ u_1(x, y) \leqslant z \leqslant u_2(x, y)\}$$

and Equation 6 becomes

$$\boxed{8} \qquad \iiint_E f(x, y, z) \, dV = \int_c^d \int_{h_1(y)}^{h_2(y)} \int_{u_1(x, y)}^{u_2(x, y)} f(x, y, z) \, dz \, dx \, dy$$

EXAMPLE 2 Evaluate $\iiint_E z \, dV$, where E is the solid tetrahedron bounded by the four planes $x = 0$, $y = 0$, $z = 0$, and $x + y + z = 1$.

SOLUTION When we set up a triple integral it's wise to draw *two* diagrams: one of the solid region E (see Figure 5) and one of its projection D on the xy-plane (see Figure 6). The lower boundary of the tetrahedron is the plane $z = 0$ and the upper boundary is the plane $x + y + z = 1$ (or $z = 1 - x - y$), so we use $u_1(x, y) = 0$ and $u_2(x, y) = 1 - x - y$ in Formula 7. Notice that the planes $x + y + z = 1$ and $z = 0$ intersect in the line $x + y = 1$ (or $y = 1 - x$) in the xy-plane. So the projection of E is the triangular region shown in Figure 6, and we have

$$\boxed{9} \qquad E = \{(x, y, z) \mid 0 \leqslant x \leqslant 1, \ 0 \leqslant y \leqslant 1 - x, \ 0 \leqslant z \leqslant 1 - x - y\}$$

This description of E as a type 1 region enables us to evaluate the integral as follows:

$$\iiint_E z \, dV = \int_0^1 \int_0^{1-x} \int_0^{1-x-y} z \, dz \, dy \, dx$$

$$= \int_0^1 \int_0^{1-x} \left[\frac{z^2}{2} \right]_{z=0}^{z=1-x-y} dy \, dx$$

$$= \frac{1}{2} \int_0^1 \int_0^{1-x} (1 - x - y)^2 \, dy \, dx$$

$$= \frac{1}{2} \int_0^1 \left[-\frac{(1 - x - y)^3}{3} \right]_{y=0}^{y=1-x} dx$$

$$= \frac{1}{6} \int_0^1 (1 - x)^3 \, dx = \frac{1}{6} \left[-\frac{(1 - x)^4}{4} \right]_0^1 = \frac{1}{24} \qquad \blacksquare$$

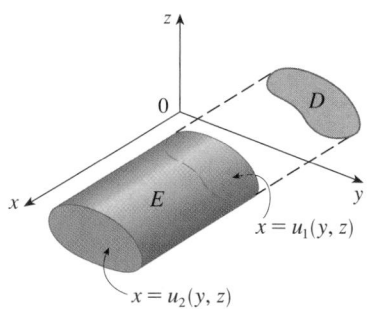

FIGURE 7
A type 2 region

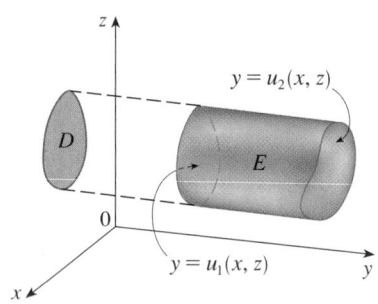

FIGURE 8
A type 3 region

A solid region E is of **type 2** if it is of the form

$$E = \{(x, y, z) \mid (y, z) \in D, \ u_1(y, z) \leq x \leq u_2(y, z)\}$$

where, this time, D is the projection of E onto the yz-plane (see Figure 7). The back surface is $x = u_1(y, z)$, the front surface is $x = u_2(y, z)$, and we have

$$\boxed{10} \qquad \iiint_E f(x, y, z) \, dV = \iint_D \left[\int_{u_1(y, z)}^{u_2(y, z)} f(x, y, z) \, dx \right] dA$$

Finally, a **type 3** region is of the form

$$E = \{(x, y, z) \mid (x, z) \in D, \ u_1(x, z) \leq y \leq u_2(x, z)\}$$

where D is the projection of E onto the xz-plane, $y = u_1(x, z)$ is the left surface, and $y = u_2(x, z)$ is the right surface (see Figure 8). For this type of region we have

$$\boxed{11} \qquad \iiint_E f(x, y, z) \, dV = \iint_D \left[\int_{u_1(x, z)}^{u_2(x, z)} f(x, y, z) \, dy \right] dA$$

In each of Equations 10 and 11 there may be two possible expressions for the integral depending on whether D is a type I or type II plane region (and corresponding to Equations 7 and 8).

▼ **EXAMPLE 3** Evaluate $\iiint_E \sqrt{x^2 + z^2} \, dV$, where E is the region bounded by the paraboloid $y = x^2 + z^2$ and the plane $y = 4$.

SOLUTION The solid E is shown in Figure 9. If we regard it as a type 1 region, then we need to consider its projection D_1 onto the xy-plane, which is the parabolic region in Figure 10. (The trace of $y = x^2 + z^2$ in the plane $z = 0$ is the parabola $y = x^2$.)

TEC Visual 12.5 illustrates how solid regions (including the one in Figure 9) project onto coordinate planes.

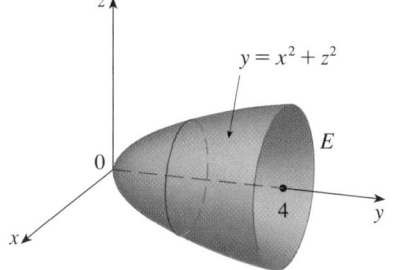

FIGURE 9
Region of integration

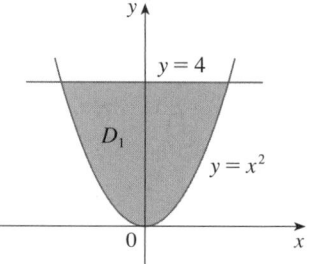

FIGURE 10
Projection onto xy-plane

From $y = x^2 + z^2$ we obtain $z = \pm\sqrt{y - x^2}$, so the lower boundary surface of E is $z = -\sqrt{y - x^2}$ and the upper surface is $z = \sqrt{y - x^2}$. Therefore the description of E as a type 1 region is

$$E = \left\{(x, y, z) \mid -2 \leq x \leq 2, \ x^2 \leq y \leq 4, \ -\sqrt{y - x^2} \leq z \leq \sqrt{y - x^2}\right\}$$

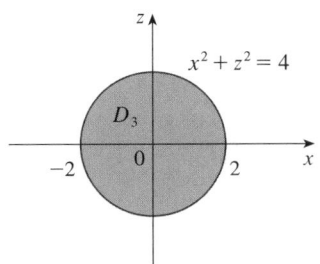

FIGURE 11
Projection onto xz-plane

and so we obtain

$$\iiint_E \sqrt{x^2 + z^2} \, dV = \int_{-2}^{2} \int_{x^2}^{4} \int_{-\sqrt{y-x^2}}^{\sqrt{y-x^2}} \sqrt{x^2 + z^2} \, dz \, dy \, dx$$

Although this expression is correct, it is extremely difficult to evaluate. So let's instead consider E as a type 3 region. As such, its projection D_3 onto the xz-plane is the disk $x^2 + z^2 \le 4$ shown in Figure 11.

Then the left boundary of E is the paraboloid $y = x^2 + z^2$ and the right boundary is the plane $y = 4$, so taking $u_1(x, z) = x^2 + z^2$ and $u_2(x, z) = 4$ in Equation 11, we have

$$\iiint_E \sqrt{x^2 + z^2} \, dV = \iint_{D_3} \left[\int_{x^2 + z^2}^{4} \sqrt{x^2 + z^2} \, dy \right] dA$$

$$= \iint_{D_3} (4 - x^2 - z^2)\sqrt{x^2 + z^2} \, dA$$

⊘ The most difficult step in evaluating a triple integral is setting up an expression for the region of integration (such as Equation 9 in Example 2). Remember that the limits of integration in the inner integral contain at most two variables, the limits of integration in the middle integral contain at most one variable, and the limits of integration in the outer integral must be constants.

Although this integral could be written as

$$\int_{-2}^{2} \int_{-\sqrt{4-x^2}}^{\sqrt{4-x^2}} (4 - x^2 - z^2)\sqrt{x^2 + z^2} \, dz \, dx$$

it's easier to convert to polar coordinates in the xz-plane: $x = r\cos\theta$, $z = r\sin\theta$. This gives

$$\iiint_E \sqrt{x^2 + z^2} \, dV = \iint_{D_3} (4 - x^2 - z^2)\sqrt{x^2 + z^2} \, dA = \int_{0}^{2\pi} \int_{0}^{2} (4 - r^2) r \, r \, dr \, d\theta$$

$$= \int_{0}^{2\pi} d\theta \int_{0}^{2} (4r^2 - r^4) \, dr = 2\pi \left[\frac{4r^3}{3} - \frac{r^5}{5} \right]_{0}^{2} = \frac{128\pi}{15} \qquad ∎$$

EXAMPLE 4 Express the iterated integral $\int_0^1 \int_0^{x^2} \int_0^y f(x, y, z) \, dz \, dy \, dx$ as a triple integral and then rewrite it as an iterated integral in a different order, integrating first with respect to x, then z, and then y.

SOLUTION We can write

$$\int_0^1 \int_0^{x^2} \int_0^y f(x, y, z) \, dz \, dy \, dx = \iiint_E f(x, y, z) \, dV$$

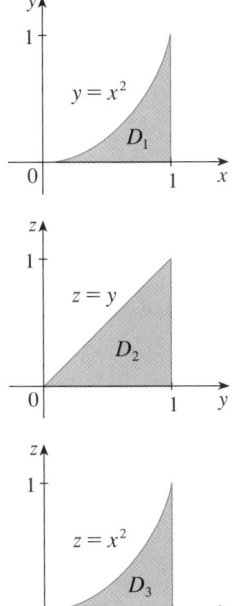

FIGURE 12
Projections of E

where $E = \{(x, y, z) \mid 0 \le x \le 1, 0 \le y \le x^2, 0 \le z \le y\}$. This description of E enables us to write projections onto the three coordinate planes as follows:

on the xy-plane: $D_1 = \{(x, y) \mid 0 \le x \le 1, 0 \le y \le x^2\}$
$= \{(x, y) \mid 0 \le y \le 1, \sqrt{y} \le x \le 1\}$

on the yz-plane: $D_2 = \{(x, y) \mid 0 \le y \le 1, 0 \le z \le y\}$

on the xz-plane: $D_3 = \{(x, y) \mid 0 \le x \le 1, 0 \le z \le x^2\}$

From the resulting sketches of the projections in Figure 12 we sketch the solid E in

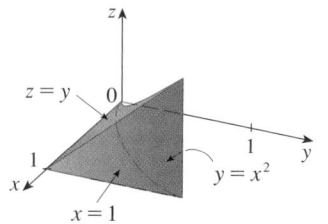

FIGURE 13
The solid E

Figure 13. We see that it is the solid enclosed by the planes $z = 0$, $x = 1$, $y = z$ and the parabolic cylinder $y = x^2$ $\left(\text{or } x = \sqrt{y}\,\right)$.

If we integrate first with respect to x, then z, and then y, we use an alternate description of E:

$$E = \left\{(x, y, z) \mid 0 \leqslant x \leqslant 1, 0 \leqslant z \leqslant y, \sqrt{y} \leqslant x \leqslant 1\right\}$$

Thus

$$\iiint_E f(x, y, z)\, dV = \int_0^1 \int_0^y \int_{\sqrt{y}}^1 f(x, y, z)\, dx\, dz\, dy \qquad \blacksquare$$

APPLICATIONS OF TRIPLE INTEGRALS

Recall that if $f(x) \geqslant 0$, then the single integral $\int_a^b f(x)\, dx$ represents the area under the curve $y = f(x)$ from a to b, and if $f(x, y) \geqslant 0$, then the double integral $\iint_D f(x, y)\, dA$ represents the volume under the surface $z = f(x, y)$ and above D. The corresponding interpretation of a triple integral $\iiint_E f(x, y, z)\, dV$, where $f(x, y, z) \geqslant 0$, is not very useful because it would be the "hypervolume" of a four-dimensional object and, of course, that is very difficult to visualize. (Remember that E is just the *domain* of the function f; the graph of f lies in four-dimensional space.) Nonetheless, the triple integral $\iiint_E f(x, y, z)\, dV$ can be interpreted in different ways in different physical situations, depending on the physical interpretations of x, y, z and $f(x, y, z)$.

Let's begin with the special case where $f(x, y, z) = 1$ for all points in E. Then the triple integral does represent the volume of E:

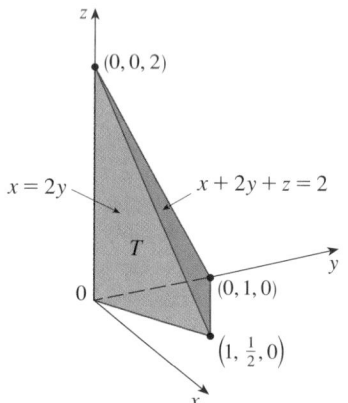

FIGURE 14

$$\boxed{12} \qquad \boxed{\quad V(E) = \iiint_E dV \quad}$$

For example, you can see this in the case of a type 1 region by putting $f(x, y, z) = 1$ in Formula 6:

$$\iiint_E 1\, dV = \iint_D \left[\int_{u_1(x, y)}^{u_2(x, y)} dz\right] dA = \iint_D [u_2(x, y) - u_1(x, y)]\, dA$$

and from Section 12.2 we know this represents the volume that lies between the surfaces $z = u_1(x, y)$ and $z = u_2(x, y)$.

EXAMPLE 5 Use a triple integral to find the volume of the tetrahedron T bounded by the planes $x + 2y + z = 2$, $x = 2y$, $x = 0$, and $z = 0$.

SOLUTION The tetrahedron T and its projection D on the xy-plane are shown in Figures 14 and 15. The lower boundary of T is the plane $z = 0$ and the upper boundary is the plane $x + 2y + z = 2$, that is, $z = 2 - x - 2y$. Therefore we have

$$V(T) = \iiint_T dV = \int_0^1 \int_{x/2}^{1-x/2} \int_0^{2-x-2y} dz\, dy\, dx$$

$$= \int_0^1 \int_{x/2}^{1-x/2} (2 - x - 2y)\, dy\, dx = \tfrac{1}{3}$$

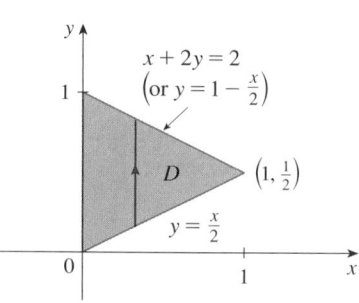

FIGURE 15

by the same calculation as in Example 4 in Section 12.2.

(Notice that it is not necessary to use triple integrals to compute volumes. They simply give an alternative method for setting up the calculation.) ∎

All the applications of double integrals in Section 12.4 can be immediately extended to triple integrals. For example, if the density function of a solid object that occupies the region E is $\rho(x, y, z)$, in units of mass per unit volume, at any given point (x, y, z), then its **mass** is

$$\boxed{13} \qquad m = \iiint_E \rho(x, y, z)\, dV$$

and its **moments** about the three coordinate planes are

$$\boxed{14} \qquad M_{yz} = \iiint_E x\rho(x, y, z)\, dV \qquad M_{xz} = \iiint_E y\rho(x, y, z)\, dV$$

$$M_{xy} = \iiint_E z\rho(x, y, z)\, dV$$

The **center of mass** is located at the point $(\bar{x}, \bar{y}, \bar{z})$, where

$$\boxed{15} \qquad \bar{x} = \frac{M_{yz}}{m} \qquad \bar{y} = \frac{M_{xz}}{m} \qquad \bar{z} = \frac{M_{xy}}{m}$$

If the density is constant, the center of mass of the solid is called the **centroid** of E. The **moments of inertia** about the three coordinate axes are

$$\boxed{16} \quad I_x = \iiint_E (y^2 + z^2)\rho(x, y, z)\, dV \qquad I_y = \iiint_E (x^2 + z^2)\rho(x, y, z)\, dV$$

$$I_z = \iiint_E (x^2 + y^2)\rho(x, y, z)\, dV$$

As in Section 12.4, the total **electric charge** on a solid object occupying a region E and having charge density $\sigma(x, y, z)$ is

$$Q = \iiint_E \sigma(x, y, z)\, dV$$

▼ EXAMPLE 6 Find the center of mass of a solid of constant density that is bounded by the parabolic cylinder $x = y^2$ and the planes $x = z$, $z = 0$, and $x = 1$.

SOLUTION The solid E and its projection onto the xy-plane are shown in Figure 16. The lower and upper surfaces of E are the planes $z = 0$ and $z = x$, so we describe E as a type 1 region:

$$E = \left\{(x, y, z) \mid -1 \leqslant y \leqslant 1,\ y^2 \leqslant x \leqslant 1,\ 0 \leqslant z \leqslant x\right\}$$

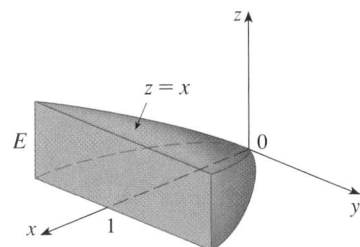

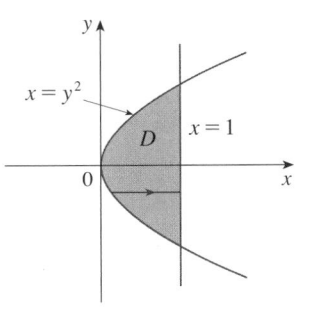

FIGURE 16

Then, if the density is $\rho(x, y, z) = \rho$, the mass is

$$m = \iiint_E \rho \, dV = \int_{-1}^1 \int_{y^2}^1 \int_0^x \rho \, dz \, dx \, dy = \rho \int_{-1}^1 \int_{y^2}^1 x \, dx \, dy$$

$$= \rho \int_{-1}^1 \left[\frac{x^2}{2} \right]_{x=y^2}^{x=1} dy = \frac{\rho}{2} \int_{-1}^1 (1 - y^4) \, dy$$

$$= \rho \int_0^1 (1 - y^4) \, dy = \rho \left[y - \frac{y^5}{5} \right]_0^1 = \frac{4\rho}{5}$$

Because of the symmetry of E and ρ about the xz-plane, we can immediately say that $M_{xz} = 0$ and therefore $\bar{y} = 0$. The other moments are

$$M_{yz} = \iiint_E x\rho \, dV = \int_{-1}^1 \int_{y^2}^1 \int_0^x x\rho \, dz \, dx \, dy = \rho \int_{-1}^1 \int_{y^2}^1 x^2 \, dx \, dy$$

$$= \rho \int_{-1}^1 \left[\frac{x^3}{3} \right]_{x=y^2}^{x=1} dy = \frac{2\rho}{3} \int_0^1 (1 - y^6) \, dy = \frac{2\rho}{3} \left[y - \frac{y^7}{7} \right]_0^1 = \frac{4\rho}{7}$$

$$M_{xy} = \iiint_E z\rho \, dV = \int_{-1}^1 \int_{y^2}^1 \int_0^x z\rho \, dz \, dx \, dy = \rho \int_{-1}^1 \int_{y^2}^1 \left[\frac{z^2}{2} \right]_{z=0}^{z=x} dx \, dy$$

$$= \frac{\rho}{2} \int_{-1}^1 \int_{y^2}^1 x^2 \, dx \, dy = \frac{\rho}{3} \int_0^1 (1 - y^6) \, dy = \frac{2\rho}{7}$$

Therefore the center of mass is

$$(\bar{x}, \bar{y}, \bar{z}) = \left(\frac{M_{yz}}{m}, \frac{M_{xz}}{m}, \frac{M_{xy}}{m} \right) = \left(\tfrac{5}{7}, 0, \tfrac{5}{14} \right)$$ ∎

12.5 | EXERCISES

1. Evaluate the integral in Example 1, integrating first with respect to z, then x, and then y.

2. Evaluate the integral $\iiint_E (xy + z^2) \, dV$, where

$$E = \{(x, y, z) \mid 0 \le x \le 2, 0 \le y \le 1, 0 \le z \le 3\}$$

using three different orders of integration.

3–6 ▪ Evaluate the iterated integral.

3. $\int_0^2 \int_0^{z^2} \int_0^{y-z} (2x - y) \, dx \, dy \, dz$

4. $\int_0^1 \int_0^1 \int_0^{\sqrt{1-z^2}} \frac{z}{y+1} \, dx \, dz \, dy$

5. $\int_0^{\pi/2} \int_0^y \int_0^x \cos(x + y + z) \, dz \, dx \, dy$

6. $\int_0^{\sqrt{\pi}} \int_0^x \int_0^{xz} x^2 \sin y \, dy \, dz \, dx$

7–16 ▪ Evaluate the triple integral.

7. $\iiint_E y \, dV$, where

$$E = \{(x, y, z) \mid 0 \le x \le 3, 0 \le y \le x, x - y \le z \le x + y\}$$

8. $\iiint_E e^{z/y} \, dV$, where

$$E = \{(x, y, z) \mid 0 \le y \le 1, y \le x \le 1, 0 \le z \le xy\}$$

9. $\iiint_E \frac{z}{x^2 + z^2} \, dV$, where

$$E = \{(x, y, z) \mid 1 \le y \le 4, y \le z \le 4, 0 \le x \le z\}$$

10. $\iiint_E \sin y \, dV$, where E lies below the plane $z = x$ and above the triangular region with vertices $(0, 0, 0)$, $(\pi, 0, 0)$, and $(0, \pi, 0)$

11. $\iiint_E 6xy \, dV$, where E lies under the plane $z = 1 + x + y$ and above the region in the xy-plane bounded by the curves $y = \sqrt{x}$, $y = 0$, and $x = 1$

12. $\iiint_E xy \, dV$, where E is bounded by the parabolic cylinders $y = x^2$ and $x = y^2$ and the planes $z = 0$ and $z = x + y$

13. $\iiint_T x^2 \, dV$, where T is the solid tetrahedron with vertices $(0, 0, 0)$, $(1, 0, 0)$, $(0, 1, 0)$, and $(0, 0, 1)$

14. $\iiint_T xyz \, dV$, where T is the solid tetrahedron with vertices $(0, 0, 0)$, $(1, 0, 0)$, $(1, 1, 0)$, and $(1, 0, 1)$

15. $\iiint_E x \, dV$, where E is bounded by the paraboloid $x = 4y^2 + 4z^2$ and the plane $x = 4$

16. $\iiint_E z \, dV$, where E is bounded by the cylinder $y^2 + z^2 = 9$ and the planes $x = 0$, $y = 3x$, and $z = 0$ in the first octant

---

17–20 ▪ Use a triple integral to find the volume of the given solid.

17. The tetrahedron enclosed by the coordinate planes and the plane $2x + y + z = 4$

18. The solid enclosed by the paraboloids $y = x^2 + z^2$ and $y = 8 - x^2 - z^2$

19. The solid enclosed by the cylinder $y = x^2$ and the planes $z = 0$ and $y + z = 1$

20. The solid enclosed by the cylinder $x^2 + z^2 = 4$ and the planes $y = -1$ and $y + z = 4$

---

21. (a) Express the volume of the wedge in the first octant that is cut from the cylinder $y^2 + z^2 = 1$ by the planes $y = x$ and $x = 1$ as a triple integral.

CAS   (b) Use either the Table of Integrals (on Reference Pages 6–10) or a computer algebra system to find the exact value of the triple integral in part (a).

22. (a) In the **Midpoint Rule for triple integrals** we use a triple Riemann sum to approximate a triple integral over a box B, where $f(x, y, z)$ is evaluated at the center $(\bar{x}_i, \bar{y}_j, \bar{z}_k)$ of the box B_{ijk}. Use the Midpoint Rule to estimate $\iiint_B \sqrt{x^2 + y^2 + z^2} \, dV$, where B is the cube defined by $0 \leq x \leq 4$, $0 \leq y \leq 4$, $0 \leq z \leq 4$. Divide B into eight cubes of equal size.

CAS   (b) Use a computer algebra system to approximate the integral in part (a) correct to the nearest integer. Compare with the answer to part (a).

23–24 ▪ Use the Midpoint Rule for triple integrals (Exercise 22) to estimate the value of the integral. Divide B into eight sub-boxes of equal size.

23. $\iiint_B \cos(xyz) \, dV$, where
$$B = \{(x, y, z) \mid 0 \leq x \leq 1, \, 0 \leq y \leq 1, \, 0 \leq z \leq 1\}$$

24. $\iiint_B \sqrt{x} \, e^{xyz} \, dV$, where
$$B = \{(x, y, z) \mid 0 \leq x \leq 4, \, 0 \leq y \leq 1, \, 0 \leq z \leq 2\}$$

25–26 ▪ Sketch the solid whose volume is given by the iterated integral.

25. $\int_0^1 \int_0^{1-x} \int_0^{2-2z} dy \, dz \, dx$ **26.** $\int_0^2 \int_0^{2-y} \int_0^{4-y^2} dx \, dz \, dy$

27–30 ▪ Express the integral $\iiint_E f(x, y, z) \, dV$ as an iterated integral in six different ways, where E is the solid bounded by the given surfaces.

27. $y = 4 - x^2 - 4z^2$, $y = 0$

28. $y^2 + z^2 = 9$, $x = -2$, $x = 2$

29. $y = x^2$, $z = 0$, $y + 2z = 4$

30. $x = 2$, $y = 2$, $z = 0$, $x + y - 2z = 2$

---

31. The figure shows the region of integration for the integral
$$\int_0^1 \int_{\sqrt{x}}^1 \int_0^{1-y} f(x, y, z) \, dz \, dy \, dx$$

Rewrite this integral as an equivalent iterated integral in the five other orders.

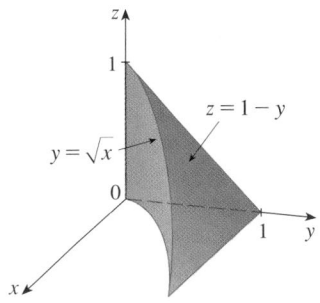

32. The figure shows the region of integration for the integral

$$\int_0^1 \int_0^{1-x^2} \int_0^{1-x} f(x, y, z)\, dy\, dz\, dx$$

Rewrite this integral as an equivalent iterated integral in the five other orders.

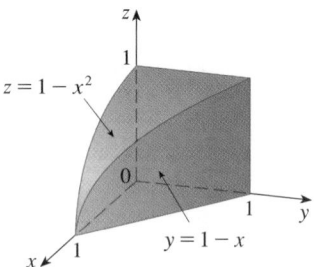

33–34 ▪ Write five other iterated integrals that are equal to the given iterated integral.

33. $\int_0^1 \int_y^1 \int_0^y f(x, y, z)\, dz\, dx\, dy$ **34.** $\int_0^1 \int_y^1 \int_0^z f(x, y, z)\, dx\, dz\, dy$

35–36 ▪ Evaluate the triple integral using only geometric interpretation and symmetry.

35. $\iiint_C (4 + 5x^2yz^2)\, dV$, where C is the cylindrical region $x^2 + y^2 \leq 4$, $-2 \leq z \leq 2$

36. $\iiint_B (z^3 + \sin y + 3)\, dV$, where B is the unit ball $x^2 + y^2 + z^2 \leq 1$

37–40 ▪ Find the mass and center of mass of the solid E with the given density function ρ.

37. E is the solid of Exercise 11; $\rho(x, y, z) = 2$

38. E is bounded by the parabolic cylinder $z = 1 - y^2$ and the planes $x + z = 1$, $x = 0$, and $z = 0$; $\rho(x, y, z) = 4$

39. E is the cube given by $0 \leq x \leq a$, $0 \leq y \leq a$, $0 \leq z \leq a$; $\rho(x, y, z) = x^2 + y^2 + z^2$

40. E is the tetrahedron bounded by the planes $x = 0$, $y = 0$, $z = 0$, $x + y + z = 1$; $\rho(x, y, z) = y$

41–42 ▪ Set up, but do not evaluate, integral expressions for (a) the mass, (b) the center of mass, and (c) the moment of inertia about the z-axis.

41. The solid of Exercise 19; $\rho(x, y, z) = \sqrt{x^2 + y^2}$

42. The hemisphere $x^2 + y^2 + z^2 \leq 1$, $z \geq 0$; $\rho(x, y, z) = \sqrt{x^2 + y^2 + z^2}$

CAS 43. Let E be the solid in the first octant bounded by the cylinder $x^2 + y^2 = 1$ and the planes $y = z$, $x = 0$, and $z = 0$ with the density function $\rho(x, y, z) = 1 + x + y + z$. Use a computer algebra system to find the exact values of the following quantities for E.
(a) The mass
(b) The center of mass
(c) The moment of inertia about the z-axis

CAS 44. If E is the solid of Exercise 16 with density function $\rho(x, y, z) = x^2 + y^2$, find the following quantities, correct to three decimal places.
(a) The mass
(b) The center of mass
(c) The moment of inertia about the z-axis

45. Find the moments of inertia for a cube of constant density k and side length L if one vertex is located at the origin and three edges lie along the coordinate axes.

46. Find the moments of inertia for a rectangular brick with dimensions a, b, and c, mass M, and constant density if the center of the brick is situated at the origin and the edges are parallel to the coordinate axes.

47. Find the moment of inertia about the z-axis of the solid cylinder $x^2 + y^2 \leq a^2$, $0 \leq z \leq h$.

48. Find the moment of inertia about the z-axis of the solid cone $\sqrt{x^2 + y^2} \leq z \leq h$.

49–50 ▪ The **average value** of a function $f(x, y, z)$ over a solid region E is defined to be

$$f_{ave} = \frac{1}{V(E)} \iiint_E f(x, y, z)\, dV$$

where $V(E)$ is the volume of E. For instance, if ρ is a density function, then ρ_{ave} is the average density of E.

49. Find the average value of the function $f(x, y, z) = xyz$ over the cube with side length L that lies in the first octant with one vertex at the origin and edges parallel to the coordinate axes.

50. Find the average value of the function $f(x, y, z) = x^2z + y^2z$ over the region enclosed by the paraboloid $z = 1 - x^2 - y^2$ and the plane $z = 0$.

51. (a) Find the region E for which the triple integral

$$\iiint_E (1 - x^2 - 2y^2 - 3z^2)\, dV$$

is a maximum.
CAS (b) Use a computer algebra system to calculate the exact maximum value of the triple integral in part (a).

12.6 | **TRIPLE INTEGRALS IN CYLINDRICAL COORDINATES**

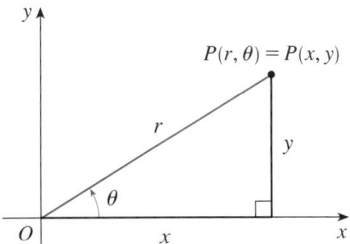

FIGURE 1

In plane geometry the polar coordinate system is used to give a convenient description of certain curves and regions. (See Section 9.3.) Figure 1 enables us to recall the connection between polar and Cartesian coordinates. If the point P has Cartesian coordinates (x, y) and polar coordinates (r, θ), then, from the figure,

$$x = r \cos \theta \qquad y = r \sin \theta$$

$$r^2 = x^2 + y^2 \qquad \tan \theta = \frac{y}{x}$$

In three dimensions there is a coordinate system, called *cylindrical coordinates*, that is similar to polar coordinates and gives convenient descriptions of some commonly occurring surfaces and solids. As we will see, some triple integrals are much easier to evaluate in cylindrical coordinates.

CYLINDRICAL COORDINATES

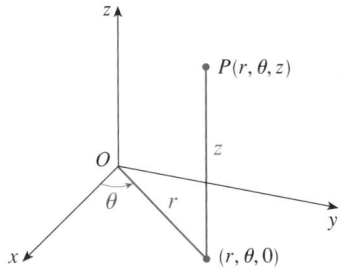

FIGURE 2
The cylindrical coordinates of a point

In the **cylindrical coordinate system**, a point P in three-dimensional space is represented by the ordered triple (r, θ, z), where r and θ are polar coordinates of the projection of P onto the xy-plane and z is the directed distance from the xy-plane to P. (See Figure 2.)

To convert from cylindrical to rectangular coordinates, we use the equations

$$\boxed{1} \qquad \boxed{\quad x = r \cos \theta \qquad y = r \sin \theta \qquad z = z \quad}$$

whereas to convert from rectangular to cylindrical coordinates, we use

$$\boxed{2} \qquad \boxed{\quad r^2 = x^2 + y^2 \qquad \tan \theta = \frac{y}{x} \qquad z = z \quad}$$

EXAMPLE 1
(a) Plot the point with cylindrical coordinates $(2, 2\pi/3, 1)$ and find its rectangular coordinates.
(b) Find cylindrical coordinates of the point with rectangular coordinates $(3, -3, -7)$.

SOLUTION
(a) The point with cylindrical coordinates $(2, 2\pi/3, 1)$ is plotted in Figure 3. From Equations 1, its rectangular coordinates are

$$x = 2 \cos \frac{2\pi}{3} = 2\left(-\frac{1}{2}\right) = -1$$

$$y = 2 \sin \frac{2\pi}{3} = 2\left(\frac{\sqrt{3}}{2}\right) = \sqrt{3}$$

$$z = 1$$

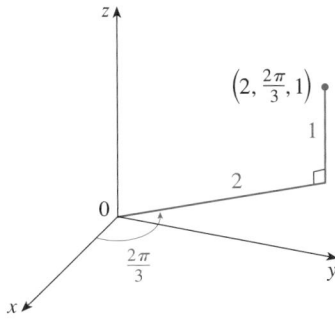

FIGURE 3

Thus the point is $\left(-1, \sqrt{3}, 1\right)$ in rectangular coordinates.

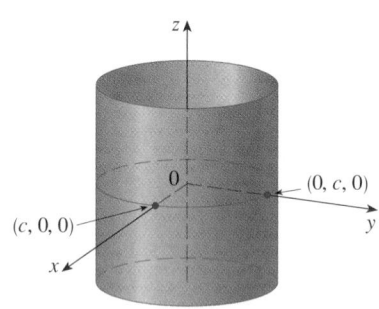

FIGURE 4
$r = c$, a cylinder

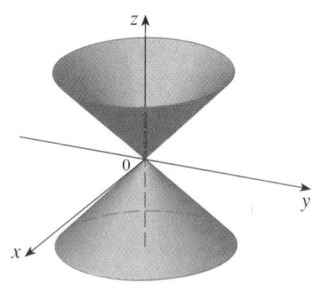

FIGURE 5
$z = r$, a cone

■ **www.stewartcalculus.com**
See Additional Example A.

(b) From Equations 2 we have

$$r = \sqrt{3^2 + (-3)^2} = 3\sqrt{2}$$

$$\tan \theta = \frac{-3}{3} = -1 \qquad \text{so} \qquad \theta = \frac{7\pi}{4} + 2n\pi$$

$$z = -7$$

Therefore one set of cylindrical coordinates is $\left(3\sqrt{2},\, 7\pi/4,\, -7\right)$. Another is $\left(3\sqrt{2},\, -\pi/4,\, -7\right)$. As with polar coordinates, there are infinitely many choices. ■

Cylindrical coordinates are useful in problems that involve symmetry about an axis, and the z-axis is chosen to coincide with this axis of symmetry. For instance, the axis of the circular cylinder with Cartesian equation $x^2 + y^2 = c^2$ is the z-axis. In cylindrical coordinates this cylinder has the very simple equation $r = c$. (See Figure 4.) This is the reason for the name "cylindrical" coordinates.

✔ EXAMPLE 2 Describe the surface whose equation in cylindrical coordinates is $z = r$.

SOLUTION The equation says that the z-value, or height, of each point on the surface is the same as r, the distance from the point to the z-axis. Because θ doesn't appear, it can vary. So any horizontal trace in the plane $z = k$ ($k > 0$) is a circle of radius k. These traces suggest that the surface is a cone. This prediction can be confirmed by converting the equation into rectangular coordinates. From the first equation in ⟨2⟩ we have

$$z^2 = r^2 = x^2 + y^2$$

We recognize the equation $z^2 = x^2 + y^2$ (by comparison with Table 1 in Section 10.6) as being a circular cone whose axis is the z-axis (see Figure 5). ■

EVALUATING TRIPLE INTEGRALS WITH CYLINDRICAL COORDINATES

Suppose that E is a type 1 region whose projection D onto the xy-plane is conveniently described in polar coordinates (see Figure 6). In particular, suppose that f is continuous and

$$E = \left\{ (x, y, z) \mid (x, y) \in D,\ u_1(x, y) \leqslant z \leqslant u_2(x, y) \right\}$$

where D is given in polar coordinates by

$$D = \left\{ (r, \theta) \mid \alpha \leqslant \theta \leqslant \beta,\ h_1(\theta) \leqslant r \leqslant h_2(\theta) \right\}$$

We know from Equation 12.5.6 that

$$\boxed{3} \qquad \iiint\limits_{E} f(x, y, z)\, dV = \iint\limits_{D} \left[\int_{u_1(x, y)}^{u_2(x, y)} f(x, y, z)\, dz \right] dA$$

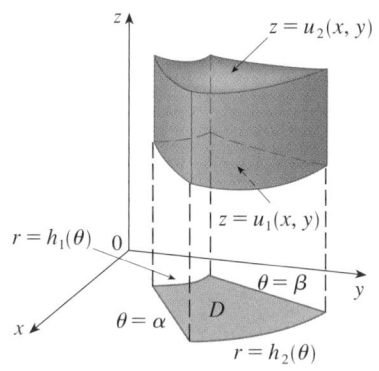

FIGURE 6

But we also know how to evaluate double integrals in polar coordinates. In fact, combining Equation 3 with Equation 12.3.3, we obtain

$$\boxed{4} \quad \iiint_E f(x, y, z)\, dV = \int_\alpha^\beta \int_{h_1(\theta)}^{h_2(\theta)} \int_{u_1(r\cos\theta,\, r\sin\theta)}^{u_2(r\cos\theta,\, r\sin\theta)} f(r\cos\theta, r\sin\theta, z)\, r\, dz\, dr\, d\theta$$

Formula 4 is the **formula for triple integration in cylindrical coordinates**. It says that we convert a triple integral from rectangular to cylindrical coordinates by writing $x = r\cos\theta$, $y = r\sin\theta$, leaving z as it is, using the appropriate limits of integration for z, r, and θ, and replacing dV by $r\, dz\, dr\, d\theta$. (Figure 7 shows how to remember this.) It is worthwhile to use this formula when E is a solid region easily described in cylindrical coordinates, and especially when the function $f(x, y, z)$ involves the expression $x^2 + y^2$.

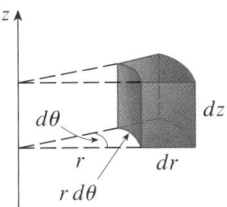

FIGURE 7
Volume element in cylindrical coordinates: $dV = r\, dz\, dr\, d\theta$

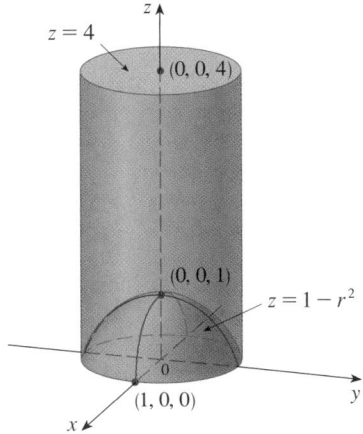

FIGURE 8

▼ EXAMPLE 3   A solid E lies within the cylinder $x^2 + y^2 = 1$, below the plane $z = 4$, and above the paraboloid $z = 1 - x^2 - y^2$. (See Figure 8.) The density at any point is proportional to its distance from the axis of the cylinder. Find the mass of E.

SOLUTION   In cylindrical coordinates the cylinder is $r = 1$ and the paraboloid is $z = 1 - r^2$, so we can write

$$E = \{(r, \theta, z) \mid 0 \le \theta \le 2\pi,\ 0 \le r \le 1,\ 1 - r^2 \le z \le 4\}$$

Since the density at (x, y, z) is proportional to the distance from the z-axis, the density function is

$$f(x, y, z) = K\sqrt{x^2 + y^2} = Kr$$

where K is the proportionality constant. Therefore, from Formula 12.5.13, the mass of E is

$$m = \iiint_E K\sqrt{x^2 + y^2}\, dV = \int_0^{2\pi} \int_0^1 \int_{1-r^2}^4 (Kr)\, r\, dz\, dr\, d\theta$$

$$= \int_0^{2\pi} \int_0^1 Kr^2[4 - (1 - r^2)]\, dr\, d\theta = K \int_0^{2\pi} d\theta \int_0^1 (3r^2 + r^4)\, dr$$

$$= 2\pi K \left[r^3 + \frac{r^5}{5} \right]_0^1 = \frac{12\pi K}{5} \qquad ∎$$

EXAMPLE 4   Evaluate $\int_{-2}^2 \int_{-\sqrt{4-x^2}}^{\sqrt{4-x^2}} \int_{\sqrt{x^2+y^2}}^2 (x^2 + y^2)\, dz\, dy\, dx.$

SOLUTION   This iterated integral is a triple integral over the solid region

$$E = \{(x, y, z) \mid -2 \le x \le 2,\ -\sqrt{4 - x^2} \le y \le \sqrt{4 - x^2},\ \sqrt{x^2 + y^2} \le z \le 2\}$$

and the projection of E onto the xy-plane is the disk $x^2 + y^2 \le 4$. The lower surface of E is the cone $z = \sqrt{x^2 + y^2}$ and its upper surface is the plane $z = 2$. (See

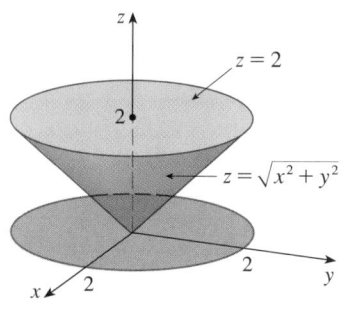

FIGURE 9

Figure 9.) This region has a much simpler description in cylindrical coordinates:

$$E = \left\{(r, \theta, z) \mid 0 \leqslant \theta \leqslant 2\pi, \ 0 \leqslant r \leqslant 2, \ r \leqslant z \leqslant 2\right\}$$

Therefore we have

$$\int_{-2}^{2} \int_{-\sqrt{4-x^2}}^{\sqrt{4-x^2}} \int_{\sqrt{x^2+y^2}}^{2} (x^2 + y^2)\, dz\, dy\, dx = \iiint_E (x^2 + y^2)\, dV$$

$$= \int_{0}^{2\pi} \int_{0}^{2} \int_{r}^{2} r^2\, r\, dz\, dr\, d\theta$$

$$= \int_{0}^{2\pi} d\theta \int_{0}^{2} r^3(2 - r)\, dr$$

$$= 2\pi \left[\tfrac{1}{2}r^4 - \tfrac{1}{5}r^5\right]_{0}^{2} = \tfrac{16}{5}\pi \quad \blacksquare$$

12.6 | EXERCISES

1–2 ▪ Plot the point whose cylindrical coordinates are given. Then find the rectangular coordinates of the point.

1. (a) $(4, \pi/3, -2)$ (b) $(2, -\pi/2, 1)$

2. (a) $\left(\sqrt{2}, 3\pi/4, 2\right)$ (b) $(1, 1, 1)$

3–4 ▪ Change from rectangular to cylindrical coordinates.

3. (a) $(-1, 1, 1)$ (b) $\left(-2, 2\sqrt{3}, 3\right)$

4. (a) $\left(2\sqrt{3}, 2, -1\right)$ (b) $(4, -3, 2)$

5–6 ▪ Describe in words the surface whose equation is given.

5. $\theta = \pi/4$ **6.** $r = 5$

7–8 ▪ Identify the surface whose equation is given.

7. $z = 4 - r^2$ **8.** $2r^2 + z^2 = 1$

9–10 ▪ Write the equations in cylindrical coordinates.

9. (a) $x^2 - x + y^2 + z^2 = 1$ (b) $z = x^2 - y^2$

10. (a) $3x + 2y + z = 6$ (b) $-x^2 - y^2 + z^2 = 1$

11–12 ▪ Sketch the solid described by the given inequalities.

11. $0 \leqslant r \leqslant 2, \quad -\pi/2 \leqslant \theta \leqslant \pi/2, \quad 0 \leqslant z \leqslant 1$

12. $0 \leqslant \theta \leqslant \pi/2, \quad r \leqslant z \leqslant 2$

---

13. A cylindrical shell is 20 cm long, with inner radius 6 cm and outer radius 7 cm. Write inequalities that describe the shell in an appropriate coordinate system. Explain how you have positioned the coordinate system with respect to the shell.

14. Use a graphing device to draw the solid enclosed by the paraboloids $z = x^2 + y^2$ and $z = 5 - x^2 - y^2$.

15–16 ▪ Sketch the solid whose volume is given by the integral and evaluate the integral.

15. $\int_{-\pi/2}^{\pi/2} \int_{0}^{2} \int_{0}^{r^2} r\, dz\, dr\, d\theta$ **16.** $\int_{0}^{2} \int_{0}^{2\pi} \int_{0}^{r} r\, dz\, d\theta\, dr$

17–28 ▪ Use cylindrical coordinates.

17. Evaluate $\iiint_E \sqrt{x^2 + y^2}\, dV$, where E is the region that lies inside the cylinder $x^2 + y^2 = 16$ and between the planes $z = -5$ and $z = 4$.

18. Evaluate $\iiint_E z\, dV$, where E is enclosed by the paraboloid $z = x^2 + y^2$ and the plane $z = 4$.

19. Evaluate $\iiint_E (x + y + z)\, dV$, where E is the solid in the first octant that lies under the paraboloid $z = 4 - x^2 - y^2$.

20. Evaluate $\iiint_E x\, dV$, where E is enclosed by the planes $z = 0$ and $z = x + y + 5$ and by the cylinders $x^2 + y^2 = 4$ and $x^2 + y^2 = 9$.

21. Evaluate $\iiint_E x^2 \, dV$, where E is the solid that lies within the cylinder $x^2 + y^2 = 1$, above the plane $z = 0$, and below the cone $z^2 = 4x^2 + 4y^2$.

22. Find the volume of the solid that lies within both the cylinder $x^2 + y^2 = 1$ and the sphere $x^2 + y^2 + z^2 = 4$.

23. Find the volume of the solid that is enclosed by the cone $z = \sqrt{x^2 + y^2}$ and the sphere $x^2 + y^2 + z^2 = 2$.

24. Find the volume of the solid that lies between the paraboloid $z = x^2 + y^2$ and the sphere $x^2 + y^2 + z^2 = 2$.

25. (a) Find the volume of the region E bounded by the paraboloids $z = x^2 + y^2$ and $z = 36 - 3x^2 - 3y^2$.
     (b) Find the centroid of E (the center of mass in the case where the density is constant).

26. (a) Find the volume of the solid that the cylinder $r = a \cos \theta$ cuts out of the sphere of radius a centered at the origin.
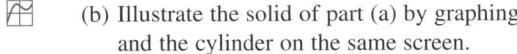
     (b) Illustrate the solid of part (a) by graphing the sphere and the cylinder on the same screen.

27. Find the mass and center of mass of the solid S bounded by the paraboloid $z = 4x^2 + 4y^2$ and the plane $z = a$ $(a > 0)$ if S has constant density K.

28. Find the mass of a ball B given by $x^2 + y^2 + z^2 \leqslant a^2$ if the density at any point is proportional to its distance from the z-axis.

29–30 ■ Evaluate the integral by changing to cylindrical coordinates.

29. $\displaystyle\int_{-2}^{2} \int_{-\sqrt{4-y^2}}^{\sqrt{4-y^2}} \int_{\sqrt{x^2+y^2}}^{2} xz \, dz \, dx \, dy$

30. $\displaystyle\int_{-3}^{3} \int_{0}^{\sqrt{9-x^2}} \int_{0}^{9-x^2-y^2} \sqrt{x^2 + y^2} \, dz \, dy \, dx$

31. When studying the formation of mountain ranges, geologists estimate the amount of work required to lift a mountain from sea level. Consider a mountain that is essentially in the shape of a right circular cone. Suppose that the weight density of the material in the vicinity of a point P is $g(P)$ and the height is $h(P)$.
     (a) Find a definite integral that represents the total work done in forming the mountain.
     (b) Assume that Mount Fuji in Japan is in the shape of a right circular cone with radius 62,000 ft, height 12,400 ft, and density a constant 200 lb/ft³. How much work was done in forming Mount Fuji if the land was initially at sea level?

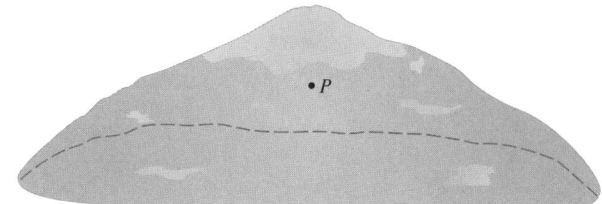

12.7 | TRIPLE INTEGRALS IN SPHERICAL COORDINATES

Another useful coordinate system in three dimensions is the *spherical coordinate system*. It simplifies the evaluation of triple integrals over regions bounded by spheres or cones.

SPHERICAL COORDINATES

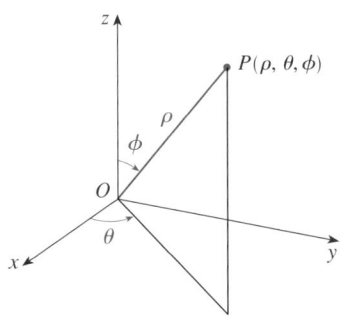

FIGURE 1
The spherical coordinates of a point

The **spherical coordinates** (ρ, θ, ϕ) of a point P in space are shown in Figure 1, where $\rho = |OP|$ is the distance from the origin to P, θ is the same angle as in cylindrical coordinates, and ϕ is the angle between the positive z-axis and the line segment OP. Note that

$$\rho \geqslant 0 \qquad 0 \leqslant \phi \leqslant \pi$$

The spherical coordinate system is especially useful in problems where there is symmetry about a point, and the origin is placed at this point. For example, the sphere with center the origin and radius c has the simple equation $\rho = c$ (see Figure 2); this is the reason for the name "spherical" coordinates. The graph of the equation $\theta = c$ is a vertical half-plane (see Figure 3), and the equation $\phi = c$ represents a half-cone with the z-axis as its axis (see Figure 4).

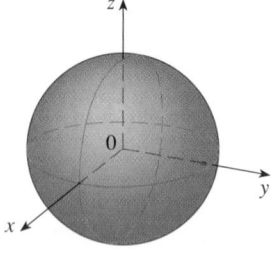

FIGURE 2 $\rho = c$, a sphere

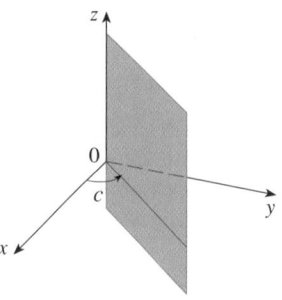

FIGURE 3 $\theta = c$, a half-plane

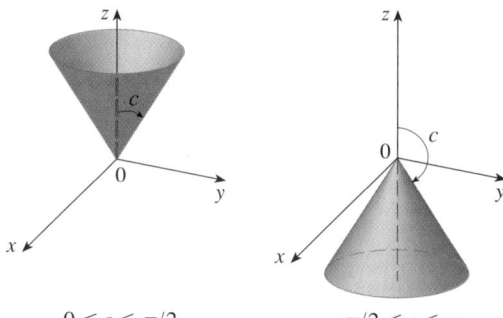

$0 < c < \pi/2$ $\pi/2 < c < \pi$

FIGURE 4 $\phi = c$, a half-cone

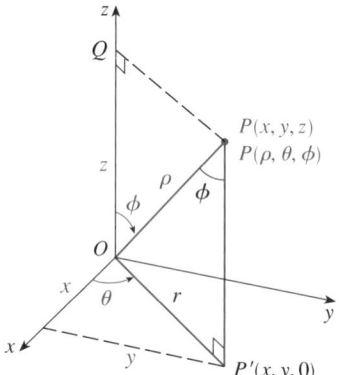

FIGURE 5

The relationship between rectangular and spherical coordinates can be seen from Figure 5. From triangles OPQ and OPP' we have

$$z = \rho \cos \phi \qquad r = \rho \sin \phi$$

But $x = r \cos \theta$ and $y = r \sin \theta$, so to convert from spherical to rectangular coordinates, we use the equations

$\boxed{1}$
$$x = \rho \sin \phi \cos \theta \qquad y = \rho \sin \phi \sin \theta \qquad z = \rho \cos \phi$$

Also, the distance formula shows that

$\boxed{2}$
$$\rho^2 = x^2 + y^2 + z^2$$

We use this equation in converting from rectangular to spherical coordinates.

⊘ **WARNING** There is not universal agreement on the notation for spherical coordinates. Most books on physics reverse the meanings of θ and ϕ and use r in place of ρ.

▼ **EXAMPLE 1** The point $(2, \pi/4, \pi/3)$ is given in spherical coordinates. Plot the point and find its rectangular coordinates.

SOLUTION We plot the point in Figure 6. From Equations 1 we have

$$x = \rho \sin \phi \cos \theta = 2 \sin \frac{\pi}{3} \cos \frac{\pi}{4} = 2 \left(\frac{\sqrt{3}}{2} \right) \left(\frac{1}{\sqrt{2}} \right) = \sqrt{\frac{3}{2}}$$

$$y = \rho \sin \phi \sin \theta = 2 \sin \frac{\pi}{3} \sin \frac{\pi}{4} = 2 \left(\frac{\sqrt{3}}{2} \right) \left(\frac{1}{\sqrt{2}} \right) = \sqrt{\frac{3}{2}}$$

$$z = \rho \cos \phi = 2 \cos \frac{\pi}{3} = 2 \left(\tfrac{1}{2} \right) = 1$$

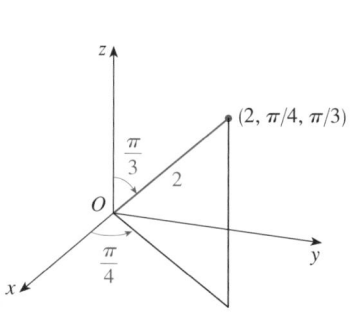

FIGURE 6

Thus the point $(2, \pi/4, \pi/3)$ is $\left(\sqrt{3/2}, \sqrt{3/2}, 1 \right)$ in rectangular coordinates. ■

■ www.stewartcalculus.com
See Additional Examples A–C.

▼ EXAMPLE 2 The point $\left(0, 2\sqrt{3}, -2\right)$ is given in rectangular coordinates. Find spherical coordinates for this point.

SOLUTION From Equation 2 we have

$$\rho = \sqrt{x^2 + y^2 + z^2} = \sqrt{0 + 12 + 4} = 4$$

and so Equations 1 give

$$\cos\phi = \frac{z}{\rho} = \frac{-2}{4} = -\frac{1}{2} \qquad \phi = \frac{2\pi}{3}$$

$$\cos\theta = \frac{x}{\rho \sin\phi} = 0 \qquad \theta = \frac{\pi}{2}$$

TEC In Module 12.7 you can investigate families of surfaces in cylindrical and spherical coordinates.

(Note that $\theta \neq 3\pi/2$ because $y = 2\sqrt{3} > 0$.) Therefore spherical coordinates of the given point are $\left(4, \pi/2, 2\pi/3\right)$. ∎

EVALUATING TRIPLE INTEGRALS WITH SPHERICAL COORDINATES

In the spherical coordinate system the counterpart of a rectangular box is a **spherical wedge**

$$E = \{(\rho, \theta, \phi) \mid a \leqslant \rho \leqslant b, \ \alpha \leqslant \theta \leqslant \beta, \ c \leqslant \phi \leqslant d\}$$

where $a \geqslant 0$ and $\beta - \alpha \leqslant 2\pi$, and $d - c \leqslant 2\pi$. Although we defined triple integrals by dividing solids into small boxes, it can be shown that dividing a solid into small spherical wedges always gives the same result. So we divide E into smaller spherical wedges E_{ijk} by means of spheres $\rho = \rho_i$, half-planes $\theta = \theta_j$, and half-cones $\phi = \phi_k$. Figure 7 shows that E_{ijk} is approximately a rectangular box with dimensions $\Delta\rho_i$, $\rho_i \Delta\phi_k$ (arc of a circle with radius ρ_i, angle $\Delta\phi_k$), and $\rho_i \sin\phi_k \Delta\theta_j$ (arc of a circle with radius $\rho_i \sin\phi_k$, angle $\Delta\theta_j$). So an approximation to the volume of E_{ijk} is given by

$$\Delta V_{ijk} \approx (\Delta\rho_i)(\rho_i \Delta\phi_k)(\rho_i \sin\phi_k \Delta\theta_j) = \rho_i^2 \sin\phi_k \Delta\rho_i \Delta\theta_j \Delta\phi_k$$

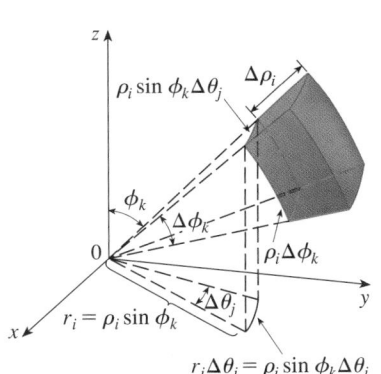

FIGURE 7

In fact, it can be shown, with the aid of the Mean Value Theorem (Exercise 45), that the volume of E_{ijk} is given exactly by

$$\Delta V_{ijk} = \tilde{\rho}_i^2 \sin\tilde{\phi}_k \Delta\rho_i \Delta\theta_j \Delta\phi_k$$

where $(\tilde{\rho}_i, \tilde{\theta}_j, \tilde{\phi}_k)$ is some point in E_{ijk}. Let $(x_{ijk}^*, y_{ijk}^*, z_{ijk}^*)$ be the rectangular coordinates of this point. Then

$$\iiint\limits_E f(x, y, z)\, dV = \lim_{\max \Delta\rho_i, \Delta\theta_j, \Delta\phi_k \to 0} \sum_{i=1}^{l} \sum_{j=1}^{m} \sum_{k=1}^{n} f(x_{ijk}^*, y_{ijk}^*, z_{ijk}^*)\, \Delta V_{ijk}$$

$$= \lim_{\max \Delta\rho_i, \Delta\theta_j, \Delta\phi_k \to 0} \sum_{i=1}^{l} \sum_{j=1}^{m} \sum_{k=1}^{n} f(\tilde{\rho}_i \sin\tilde{\phi}_k \cos\tilde{\theta}_j, \tilde{\rho}_i \sin\tilde{\phi}_k \sin\tilde{\theta}_j, \tilde{\rho}_i \cos\tilde{\phi}_k)\, \tilde{\rho}_i^2 \sin\tilde{\phi}_k \Delta\rho_i \Delta\theta_j \Delta\phi_k$$

But this sum is a Riemann sum for the function

$$F(\rho, \theta, \phi) = \rho^2 \sin\phi \, f(\rho \sin\phi \cos\theta, \rho \sin\phi \sin\theta, \rho \cos\phi)$$

Consequently, we have arrived at the following **formula for triple integration in spherical coordinates**.

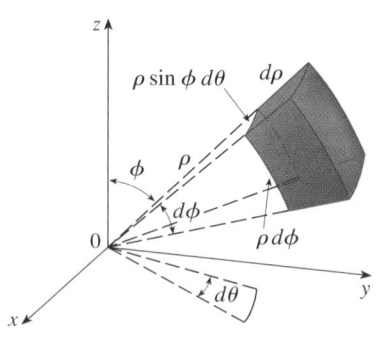

FIGURE 8

Volume element in spherical coordinates: $dV = \rho^2 \sin \phi \, d\rho \, d\theta \, d\phi$

$\boxed{3}$ $\displaystyle\iiint\limits_{E} f(x, y, z) \, dV$

$$= \int_c^d \int_\alpha^\beta \int_a^b f(\rho \sin \phi \cos \theta, \rho \sin \phi \sin \theta, \rho \cos \phi) \, \rho^2 \sin \phi \, d\rho \, d\theta \, d\phi$$

where E is a spherical wedge given by

$$E = \{(\rho, \theta, \phi) \mid a \le \rho \le b, \ \alpha \le \theta \le \beta, \ c \le \phi \le d\}$$

Formula 3 says that we convert a triple integral from rectangular coordinates to spherical coordinates by writing

$$x = \rho \sin \phi \cos \theta \qquad y = \rho \sin \phi \sin \theta \qquad z = \rho \cos \phi$$

using the appropriate limits of integration, and replacing dV by $\rho^2 \sin \phi \, d\rho \, d\theta \, d\phi$. This is illustrated in Figure 8.

This formula can be extended to include more general spherical regions such as

$$E = \{(\rho, \theta, \phi) \mid \alpha \le \theta \le \beta, \ c \le \phi \le d, \ g_1(\theta, \phi) \le \rho \le g_2(\theta, \phi)\}$$

In this case the formula is the same as in $\boxed{3}$ except that the limits of integration for ρ are $g_1(\theta, \phi)$ and $g_2(\theta, \phi)$.

Usually, spherical coordinates are used in triple integrals when surfaces such as cones and spheres form the boundary of the region of integration.

◤ EXAMPLE 3 Evaluate $\iiint_B e^{(x^2+y^2+z^2)^{3/2}} \, dV$, where B is the unit ball:

$$B = \{(x, y, z) \mid x^2 + y^2 + z^2 \le 1\}$$

SOLUTION Since the boundary of B is a sphere, we use spherical coordinates:

$$B = \{(\rho, \theta, \phi) \mid 0 \le \rho \le 1, \ 0 \le \theta \le 2\pi, \ 0 \le \phi \le \pi\}$$

In addition, spherical coordinates are appropriate because

$$x^2 + y^2 + z^2 = \rho^2$$

Thus $\boxed{3}$ gives

$$\iiint\limits_{B} e^{(x^2+y^2+z^2)^{3/2}} \, dV = \int_0^\pi \int_0^{2\pi} \int_0^1 e^{(\rho^2)^{3/2}} \rho^2 \sin \phi \, d\rho \, d\theta \, d\phi$$

$$= \int_0^\pi \sin \phi \, d\phi \int_0^{2\pi} d\theta \int_0^1 \rho^2 e^{\rho^3} \, d\rho$$

$$= \left[-\cos \phi \right]_0^\pi (2\pi) \left[\tfrac{1}{3} e^{\rho^3} \right]_0^1 = \tfrac{4}{3} \pi (e - 1) \qquad ■$$

NOTE It would have been extremely awkward to evaluate the integral in Example 3 without spherical coordinates. In rectangular coordinates the iterated integral would have been

$$\int_{-1}^1 \int_{-\sqrt{1-x^2}}^{\sqrt{1-x^2}} \int_{-\sqrt{1-x^2-y^2}}^{\sqrt{1-x^2-y^2}} e^{(x^2+y^2+z^2)^{3/2}} \, dz \, dy \, dx$$

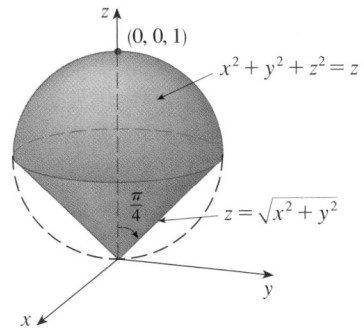

FIGURE 9

■ Figure 10 gives another look (this time drawn by Maple) at the solid of Example 4.

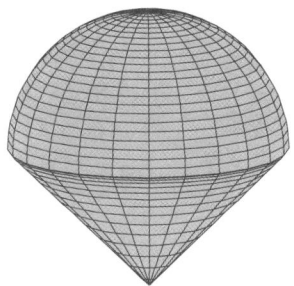

FIGURE 10

TEC Visual 12.7 shows an animation of Figure 11.

▼ EXAMPLE 4 Use spherical coordinates to find the volume of the solid that lies above the cone $z = \sqrt{x^2 + y^2}$ and below the sphere $x^2 + y^2 + z^2 = z$. (See Figure 9.)

SOLUTION Notice that the sphere passes through the origin and has center $\left(0, 0, \frac{1}{2}\right)$. We write the equation of the sphere in spherical coordinates as

$$\rho^2 = \rho \cos \phi \qquad \text{or} \qquad \rho = \cos \phi$$

The equation of the cone can be written as

$$\rho \cos \phi = \sqrt{\rho^2 \sin^2\phi \, \cos^2\theta + \rho^2 \sin^2\phi \, \sin^2\theta} = \rho \sin \phi$$

This gives $\sin \phi = \cos \phi$, or $\phi = \pi/4$. Therefore the description of the solid E in spherical coordinates is

$$E = \left\{(\rho, \theta, \phi) \mid 0 \leq \theta \leq 2\pi, \ 0 \leq \phi \leq \pi/4, \ 0 \leq \rho \leq \cos \phi\right\}$$

Figure 11 shows how E is swept out if we integrate first with respect to ρ, then ϕ, and then θ. The volume of E is

$$V(E) = \iiint\limits_{E} dV = \int_0^{2\pi} \int_0^{\pi/4} \int_0^{\cos \phi} \rho^2 \sin \phi \, d\rho \, d\phi \, d\theta$$

$$= \int_0^{2\pi} d\theta \int_0^{\pi/4} \sin \phi \left[\frac{\rho^3}{3}\right]_{\rho=0}^{\rho=\cos \phi} d\phi$$

$$= \frac{2\pi}{3} \int_0^{\pi/4} \sin \phi \, \cos^3\phi \, d\phi = \frac{2\pi}{3}\left[-\frac{\cos^4\phi}{4}\right]_0^{\pi/4} = \frac{\pi}{8}$$

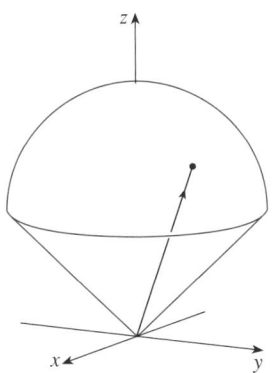

ρ varies from 0 to $\cos \phi$ while ϕ and θ are constant.

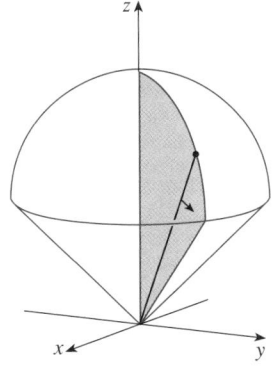

ϕ varies from 0 to $\pi/4$ while θ is constant.

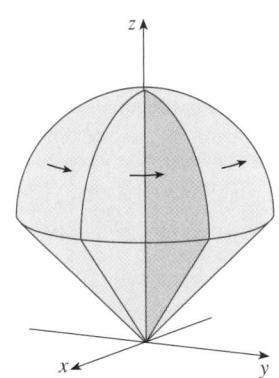

θ varies from 0 to 2π.

FIGURE 11

■

12.7 | EXERCISES

1–2 ▪ Plot the point whose spherical coordinates are given. Then find the rectangular coordinates of the point.

1. (a) $(6, \pi/3, \pi/6)$ (b) $(3, \pi/2, 3\pi/4)$

2. (a) $(2, \pi/2, \pi/2)$ (b) $(4, -\pi/4, \pi/3)$

3–4 ▪ Change from rectangular to spherical coordinates.

3. (a) $(0, -2, 0)$ (b) $\left(-1, 1, -\sqrt{2}\right)$

4. (a) $\left(1, 0, \sqrt{3}\right)$ (b) $\left(\sqrt{3}, -1, 2\sqrt{3}\right)$

5–6 ▪ Describe in words the surface whose equation is given.

5. $\phi = \pi/3$ **6.** $\rho = 3$

7–8 ▪ Identify the surface whose equation is given.

7. $\rho = \sin\theta \sin\phi$ **8.** $\rho = 2\cos\phi$

9–10 ▪ Write the equation in spherical coordinates.

9. (a) $z^2 = x^2 + y^2$ (b) $x^2 + z^2 = 9$

10. (a) $x^2 - 2x + y^2 + z^2 = 0$ (b) $x + 2y + 3z = 1$

11–14 ▪ Sketch the solid described by the given inequalities.

11. $2 \le \rho \le 4$, $\quad 0 \le \phi \le \pi/3$, $\quad 0 \le \theta \le \pi$

12. $1 \le \rho \le 2$, $\quad 0 \le \phi \le \pi/2$, $\quad \pi/2 \le \theta \le 3\pi/2$

13. $\rho \le 1$, $\quad 3\pi/4 \le \phi \le \pi$

14. $\rho \le 2$, $\quad \rho \le \csc\phi$

15. A solid lies above the cone $z = \sqrt{x^2 + y^2}$ and below the sphere $x^2 + y^2 + z^2 = z$. Write a description of the solid in terms of inequalities involving spherical coordinates.

16. (a) Find inequalities that describe a hollow ball with diameter 30 cm and thickness 0.5 cm. Explain how you have positioned the coordinate system that you have chosen.
(b) Suppose the ball is cut in half. Write inequalities that describe one of the halves.

17–18 ▪ Sketch the solid whose volume is given by the integral and evaluate the integral.

17. $\int_0^{\pi/6} \int_0^{\pi/2} \int_0^3 \rho^2 \sin\phi \, d\rho \, d\theta \, d\phi$

18. $\int_0^{2\pi} \int_{\pi/2}^{\pi} \int_1^2 \rho^2 \sin\phi \, d\rho \, d\phi \, d\theta$

19–20 ▪ Set up the triple integral of an arbitrary continuous function $f(x, y, z)$ in cylindrical or spherical coordinates over the solid shown.

19.
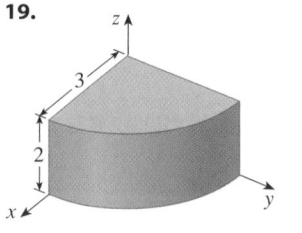

20.
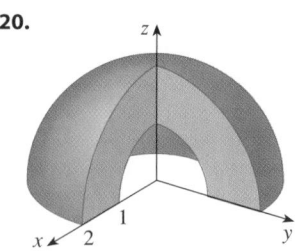

21–32 ▪ Use spherical coordinates.

21. Evaluate $\iiint_B (x^2 + y^2 + z^2)^2 \, dV$, where B is the ball with center the origin and radius 5.

22. Evaluate $\iiint_H (9 - x^2 - y^2) \, dV$, where H is the solid hemisphere $x^2 + y^2 + z^2 \le 9$, $z \ge 0$.

23. Evaluate $\iiint_E (x^2 + y^2) \, dV$, where E lies between the spheres $x^2 + y^2 + z^2 = 4$ and $x^2 + y^2 + z^2 = 9$.

24. Evaluate $\iiint_E xyz \, dV$, where E lies between the spheres $\rho = 2$ and $\rho = 4$ and above the cone $\phi = \pi/3$.

25. Evaluate $\iiint_E xe^{x^2+y^2+z^2} \, dV$, where E is the portion of the unit ball $x^2 + y^2 + z^2 \le 1$ that lies in the first octant.

26. Find the average distance from a point in a ball of radius a to its center.

27. (a) Find the volume of the solid that lies above the cone $\phi = \pi/3$ and below the sphere $\rho = 4\cos\phi$.
(b) Find the centroid of the solid in part (a).

28. Find the volume of the solid that lies within the sphere $x^2 + y^2 + z^2 = 4$, above the xy-plane, and below the cone $z = \sqrt{x^2 + y^2}$.

29. (a) Find the centroid of the solid in Example 4.
(b) Find the moment of inertia about the z-axis for this solid.

30. Let H be a solid hemisphere of radius a whose density at any point is proportional to its distance from the center of the base.
(a) Find the mass of H.
(b) Find the center of mass of H.
(c) Find the moment of inertia of H about its axis.

31. (a) Find the centroid of a solid homogeneous hemisphere of radius a.
(b) Find the moment of inertia of the solid in part (a) about a diameter of its base.

32. Find the mass and center of mass of a solid hemisphere of radius a if the density at any point is proportional to its distance from the base.

33–36 ▪ Use cylindrical or spherical coordinates, whichever seems more appropriate.

33. Find the volume and centroid of the solid E that lies above the cone $z = \sqrt{x^2 + y^2}$ and below the sphere $x^2 + y^2 + z^2 = 1$.

34. Find the volume of the smaller wedge cut from a sphere of radius a by two planes that intersect along a diameter at an angle of $\pi/6$.

CAS **35.** Evaluate $\iiint_E z \, dV$, where E lies above the paraboloid $z = x^2 + y^2$ and below the plane $z = 2y$. Use either the Table of Integrals (on Reference Pages 6–10) or a computer algebra system to evaluate the integral.

CAS **36.** (a) Find the volume enclosed by the torus $\rho = \sin \phi$.
(b) Use a computer to draw the torus.

37–39 ▪ Evaluate the integral by changing to spherical coordinates.

37. $\displaystyle\int_0^1 \int_0^{\sqrt{1-x^2}} \int_{\sqrt{x^2+y^2}}^{\sqrt{2-x^2-y^2}} xy \, dz \, dy \, dx$

38. $\displaystyle\int_{-a}^{a} \int_{-\sqrt{a^2-y^2}}^{\sqrt{a^2-y^2}} \int_{-\sqrt{a^2-x^2-y^2}}^{\sqrt{a^2-x^2-y^2}} (x^2 z + y^2 z + z^3) \, dz \, dx \, dy$

39. $\displaystyle\int_{-2}^{2} \int_{-\sqrt{4-x^2}}^{\sqrt{4-x^2}} \int_{2-\sqrt{4-x^2-y^2}}^{2+\sqrt{4-x^2-y^2}} (x^2 + y^2 + z^2)^{3/2} \, dz \, dy \, dx$

40. A model for the density δ of the earth's atmosphere near its surface is

$$\delta = 619.09 - 0.000097\rho$$

where ρ (the distance from the center of the earth) is measured in meters and δ is measured in kilograms per cubic meter. If we take the surface of the earth to be a sphere with radius 6370 km, then this model is a reasonable one for $6.370 \times 10^6 \le \rho \le 6.375 \times 10^6$. Use this model to estimate the mass of the atmosphere between the ground and an altitude of 5 km.

41. Use a graphing device to draw a silo consisting of a cylinder with radius 3 and height 10 surmounted by a hemisphere.

42. The latitude and longitude of a point P in the Northern Hemisphere are related to spherical coordinates ρ, θ, ϕ as follows. We take the origin to be the center of the earth and the positive z-axis to pass through the North Pole. The positive x-axis passes through the point where the prime meridian (the meridian through Greenwich, England) intersects the equator. Then the latitude of P is $\alpha = 90° - \phi°$ and the longitude is $\beta = 360° - \theta°$. Find the great-circle distance from Los Angeles (lat. 34.06° N, long. 118.25° W) to Montréal (lat. 45.50° N, long. 73.60° W). Take the radius of the earth to be 3960 mi. (A *great circle* is the circle of intersection of a sphere and a plane through the center of the sphere.)

CAS **43.** The surfaces $\rho = 1 + \frac{1}{5} \sin m\theta \sin n\phi$ have been used as models for tumors. The "bumpy sphere" with $m = 6$ and $n = 5$ is shown. Use a computer algebra system to find the volume it encloses.

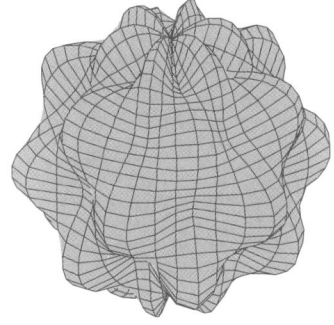

44. Show that

$$\int_{-\infty}^{\infty} \int_{-\infty}^{\infty} \int_{-\infty}^{\infty} \sqrt{x^2 + y^2 + z^2} \, e^{-(x^2+y^2+z^2)} \, dx \, dy \, dz = 2\pi$$

(The improper triple integral is defined as the limit of a triple integral over a solid sphere as the radius of the sphere increases indefinitely.)

45. (a) Use cylindrical coordinates to show that the volume of the solid bounded above by the sphere $r^2 + z^2 = a^2$ and below by the cone $z = r \cot \phi_0$ (or $\phi = \phi_0$), where $0 < \phi_0 < \pi/2$, is

$$V = \frac{2\pi a^3}{3} (1 - \cos \phi_0)$$

(b) Deduce that the volume of the spherical wedge given by $\rho_1 \le \rho \le \rho_2$, $\theta_1 \le \theta \le \theta_2$, $\phi_1 \le \phi \le \phi_2$ is

$$\Delta V = \frac{\rho_2^3 - \rho_1^3}{3} (\cos \phi_1 - \cos \phi_2)(\theta_2 - \theta_1)$$

(c) Use the Mean Value Theorem to show that the volume in part (b) can be written as

$$\Delta V = \bar{\rho}^2 \sin \tilde{\phi} \, \Delta\rho \, \Delta\theta \, \Delta\phi$$

where $\bar{\rho}$ lies between ρ_1 and ρ_2, $\tilde{\phi}$ lies between ϕ_1 and ϕ_2, $\Delta\rho = \rho_2 - \rho_1$, $\Delta\theta = \theta_2 - \theta_1$, and $\Delta\phi = \phi_2 - \phi_1$.

12.8 | CHANGE OF VARIABLES IN MULTIPLE INTEGRALS

In one-dimensional calculus we often use a change of variable (a substitution) to simplify an integral. By reversing the roles of x and u, we can write the Substitution Rule (5.5.6) as

$$\boxed{1} \qquad \int_a^b f(x)\,dx = \int_c^d f(g(u))g'(u)\,du$$

where $x = g(u)$ and $a = g(c)$, $b = g(d)$. Another way of writing Formula 1 is as follows:

$$\boxed{2} \qquad \int_a^b f(x)\,dx = \int_c^d f(x(u))\,\frac{dx}{du}\,du$$

A change of variables can also be useful in double integrals. We have already seen one example of this: conversion to polar coordinates. The new variables r and θ are related to the old variables x and y by the equations

$$x = r\cos\theta \qquad y = r\sin\theta$$

and the change of variables formula (12.3.2) can be written as

$$\iint\limits_R f(x,y)\,dA = \iint\limits_S f(r\cos\theta,\, r\sin\theta)\, r\,dr\,d\theta$$

where S is the region in the $r\theta$-plane that corresponds to the region R in the xy-plane.

More generally, we consider a change of variables that is given by a **transformation** T from the uv-plane to the xy-plane:

$$T(u, v) = (x, y)$$

where x and y are related to u and v by the equations

$$\boxed{3} \qquad x = g(u, v) \qquad y = h(u, v)$$

or, as we sometimes write,

$$x = x(u, v) \qquad y = y(u, v)$$

We usually assume that T is a C^1 **transformation**, which means that g and h have continuous first-order partial derivatives.

A transformation T is really just a function whose domain and range are both subsets of $\mathbb{R}^2$. If $T(u_1, v_1) = (x_1, y_1)$, then the point (x_1, y_1) is called the **image** of the point (u_1, v_1). If no two points have the same image, T is called **one-to-one**. Figure 1 shows the effect of a transformation T on a region S in the uv-plane. T transforms S into a region R in the xy-plane called the **image of S**, consisting of the images of all points in S.

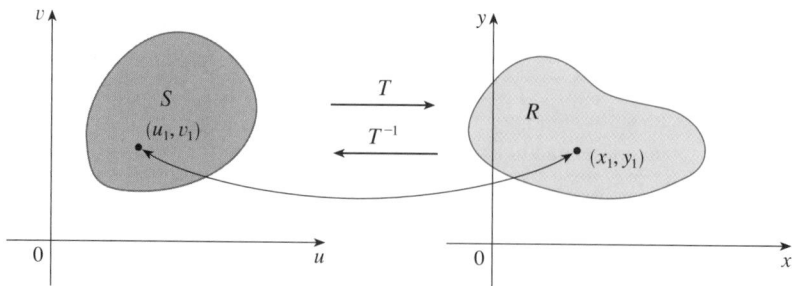

FIGURE 1

If T is a one-to-one transformation, then it has an **inverse transformation** T^{-1} from the xy-plane to the uv-plane and it may be possible to solve Equations 3 for u and v in terms of x and y:

$$u = G(x, y) \qquad v = H(x, y)$$

V EXAMPLE 1 A transformation is defined by the equations

$$x = u^2 - v^2 \qquad y = 2uv$$

Find the image of the square $S = \{(u, v) \mid 0 \le u \le 1, \ 0 \le v \le 1\}$.

SOLUTION The transformation maps the boundary of S into the boundary of the image. So we begin by finding the images of the sides of S. The first side, S_1, is given by $v = 0$ ($0 \le u \le 1$). (See Figure 2.) From the given equations we have $x = u^2$, $y = 0$, and so $0 \le x \le 1$. Thus S_1 is mapped into the line segment from $(0, 0)$ to $(1, 0)$ in the xy-plane. The second side, S_2, is $u = 1$ ($0 \le v \le 1$) and, putting $u = 1$ in the given equations, we get

$$x = 1 - v^2 \qquad y = 2v$$

Eliminating v, we obtain

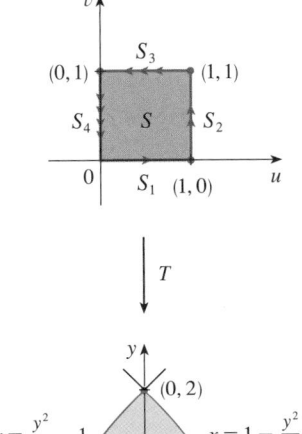

FIGURE 2

$$\boxed{4} \qquad\qquad x = 1 - \frac{y^2}{4} \qquad 0 \le x \le 1$$

which is part of a parabola. Similarly, S_3 is given by $v = 1$ ($0 \le u \le 1$), whose image is the parabolic arc

$$\boxed{5} \qquad\qquad x = \frac{y^2}{4} - 1 \qquad -1 \le x \le 0$$

Finally, S_4 is given by $u = 0$ ($0 \le v \le 1$) whose image is $x = -v^2$, $y = 0$, that is, $-1 \le x \le 0$. (Notice that as we move around the square in the counterclockwise direction, we also move around the parabolic region in the counterclockwise direction.) The image of S is the region R (shown in Figure 2) bounded by the x-axis and the parabolas given by Equations 4 and 5. ∎

Now let's see how a change of variables affects a double integral. We start with a small rectangle S in the uv-plane whose lower left corner is the point (u_0, v_0) and whose dimensions are Δu and Δv. (See Figure 3.)

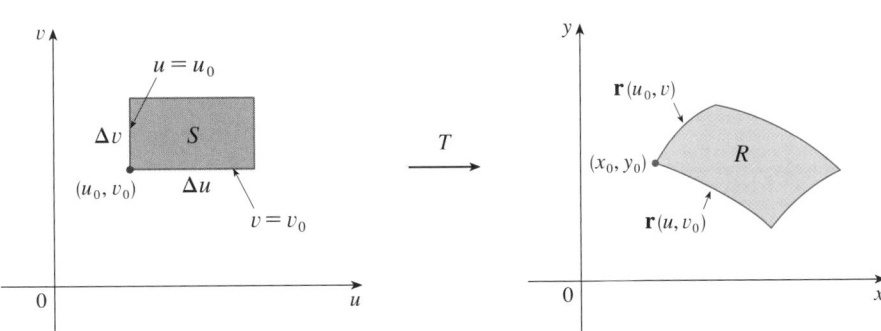

FIGURE 3

The image of S is a region R in the xy-plane, one of whose boundary points is $(x_0, y_0) = T(u_0, v_0)$. The vector

$$\mathbf{r}(u, v) = g(u, v)\,\mathbf{i} + h(u, v)\,\mathbf{j}$$

is the position vector of the image of the point (u, v). The equation of the lower side of S is $v = v_0$, whose image curve is given by the vector function $\mathbf{r}(u, v_0)$. The tangent vector at (x_0, y_0) to this image curve is

$$\mathbf{r}_u = g_u(u_0, v_0)\,\mathbf{i} + h_u(u_0, v_0)\,\mathbf{j} = \frac{\partial x}{\partial u}\,\mathbf{i} + \frac{\partial y}{\partial u}\,\mathbf{j}$$

Similarly, the tangent vector at (x_0, y_0) to the image curve of the left side of S (namely, $u = u_0$) is

$$\mathbf{r}_v = g_v(u_0, v_0)\,\mathbf{i} + h_v(u_0, v_0)\,\mathbf{j} = \frac{\partial x}{\partial v}\,\mathbf{i} + \frac{\partial y}{\partial v}\,\mathbf{j}$$

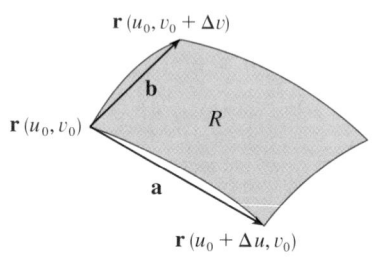

FIGURE 4

We can approximate the image region $R = T(S)$ by a parallelogram determined by the secant vectors

$$\mathbf{a} = \mathbf{r}(u_0 + \Delta u, v_0) - \mathbf{r}(u_0, v_0) \qquad \mathbf{b} = \mathbf{r}(u_0, v_0 + \Delta v) - \mathbf{r}(u_0, v_0)$$

shown in Figure 4. But

$$\mathbf{r}_u = \lim_{\Delta u \to 0} \frac{\mathbf{r}(u_0 + \Delta u, v_0) - \mathbf{r}(u_0, v_0)}{\Delta u}$$

and so

$$\mathbf{r}(u_0 + \Delta u, v_0) - \mathbf{r}(u_0, v_0) \approx \Delta u\,\mathbf{r}_u$$

Similarly

$$\mathbf{r}(u_0, v_0 + \Delta v) - \mathbf{r}(u_0, v_0) \approx \Delta v\,\mathbf{r}_v$$

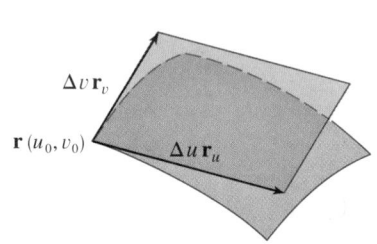

FIGURE 5

This means that we can approximate R by a parallelogram determined by the vectors $\Delta u\,\mathbf{r}_u$ and $\Delta v\,\mathbf{r}_v$. (See Figure 5.) Therefore we can approximate the area of R by the area of this parallelogram, which, from Section 10.4, is

$$\boxed{6} \qquad |(\Delta u\,\mathbf{r}_u) \times (\Delta v\,\mathbf{r}_v)| = |\mathbf{r}_u \times \mathbf{r}_v|\,\Delta u\,\Delta v$$

Computing the cross product, we obtain

$$\mathbf{r}_u \times \mathbf{r}_v = \begin{vmatrix} \mathbf{i} & \mathbf{j} & \mathbf{k} \\ \dfrac{\partial x}{\partial u} & \dfrac{\partial y}{\partial u} & 0 \\ \dfrac{\partial x}{\partial v} & \dfrac{\partial y}{\partial v} & 0 \end{vmatrix} = \begin{vmatrix} \dfrac{\partial x}{\partial u} & \dfrac{\partial y}{\partial u} \\ \dfrac{\partial x}{\partial v} & \dfrac{\partial y}{\partial v} \end{vmatrix} \mathbf{k} = \begin{vmatrix} \dfrac{\partial x}{\partial u} & \dfrac{\partial x}{\partial v} \\ \dfrac{\partial y}{\partial u} & \dfrac{\partial y}{\partial v} \end{vmatrix} \mathbf{k}$$

The determinant that arises in this calculation is called the *Jacobian* of the transformation and is given a special notation.

■ The Jacobian is named after the German mathematician Carl Gustav Jacob Jacobi (1804–1851). Although the French mathematician Cauchy first used these special determinants involving partial derivatives, Jacobi developed them into a method for evaluating multiple integrals.

> **7 DEFINITION** The **Jacobian** of the transformation T given by $x = g(u, v)$ and $y = h(u, v)$ is
>
> $$\frac{\partial(x, y)}{\partial(u, v)} = \begin{vmatrix} \dfrac{\partial x}{\partial u} & \dfrac{\partial x}{\partial v} \\ \dfrac{\partial y}{\partial u} & \dfrac{\partial y}{\partial v} \end{vmatrix} = \frac{\partial x}{\partial u}\frac{\partial y}{\partial v} - \frac{\partial x}{\partial v}\frac{\partial y}{\partial u}$$

With this notation we can use Equation 6 to give an approximation to the area ΔA of R:

$$\boxed{8} \qquad \Delta A \approx \left| \frac{\partial(x, y)}{\partial(u, v)} \right| \Delta u\, \Delta v$$

where the Jacobian is evaluated at (u_0, v_0).

Next we divide a region S in the uv-plane into rectangles S_{ij} and call their images in the xy-plane R_{ij}. (See Figure 6.)

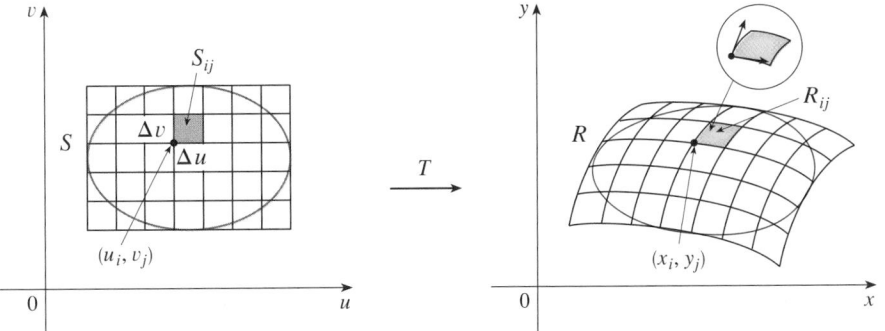

FIGURE 6

Applying the approximation $\boxed{8}$ to each R_{ij}, we approximate the double integral of f over R as follows:

$$\iint\limits_{R} f(x, y)\, dA \approx \sum_{i=1}^{m} \sum_{j=1}^{n} f(x_i, y_j)\, \Delta A$$

$$\approx \sum_{i=1}^{m} \sum_{j=1}^{n} f\big(g(u_i, v_j), h(u_i, v_j)\big) \left| \frac{\partial(x, y)}{\partial(u, v)} \right| \Delta u\, \Delta v$$

where the Jacobian is evaluated at (u_i, v_j). Notice that this double sum is a Riemann sum for the integral

$$\iint\limits_{S} f\big(g(u, v), h(u, v)\big) \left| \frac{\partial(x, y)}{\partial(u, v)} \right| du\, dv$$

The foregoing argument suggests that the following theorem is true. (A full proof is given in books on advanced calculus.)

> **9** **CHANGE OF VARIABLES IN A DOUBLE INTEGRAL** Suppose that T is a C^1 transformation whose Jacobian is nonzero and that maps a region S in the uv-plane onto a region R in the xy-plane. Suppose that f is continuous on R and that R and S are type I or type II plane regions. Suppose also that T is one-to-one, except perhaps on the boundary of S. Then
>
> $$\iint_R f(x, y)\, dA = \iint_S f\big(x(u, v), y(u, v)\big) \left| \frac{\partial(x, y)}{\partial(u, v)} \right| du\, dv$$

Theorem 9 says that we change from an integral in x and y to an integral in u and v by expressing x and y in terms of u and v and writing

$$dA = \left| \frac{\partial(x, y)}{\partial(u, v)} \right| du\, dv$$

Notice the similarity between Theorem 9 and the one-dimensional formula in Equation 2. Instead of the derivative dx/du, we have the absolute value of the Jacobian, that is, $|\partial(x, y)/\partial(u, v)|$.

As a first illustration of Theorem 9, we show that the formula for integration in polar coordinates is just a special case. Here the transformation T from the $r\theta$-plane to the xy-plane is given by

$$x = g(r, \theta) = r \cos\theta \qquad y = h(r, \theta) = r \sin\theta$$

and the geometry of the transformation is shown in Figure 7. T maps an ordinary rectangle in the $r\theta$-plane to a polar rectangle in the xy-plane. The Jacobian of T is

$$\frac{\partial(x, y)}{\partial(r, \theta)} = \begin{vmatrix} \dfrac{\partial x}{\partial r} & \dfrac{\partial x}{\partial \theta} \\[2mm] \dfrac{\partial y}{\partial r} & \dfrac{\partial y}{\partial \theta} \end{vmatrix} = \begin{vmatrix} \cos\theta & -r\sin\theta \\ \sin\theta & r\cos\theta \end{vmatrix} = r\cos^2\theta + r\sin^2\theta = r > 0$$

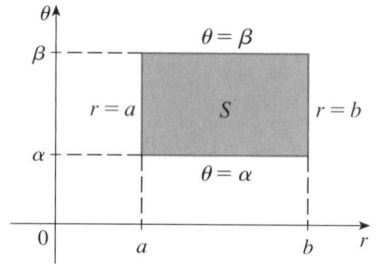

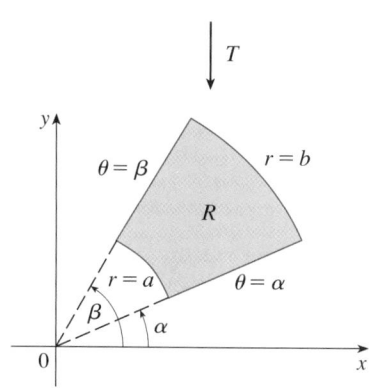

FIGURE 7

The polar coordinate transformation

Thus Theorem 9 gives

$$\iint_R f(x, y)\, dx\, dy = \iint_S f(r\cos\theta,\, r\sin\theta) \left| \frac{\partial(x, y)}{\partial(r, \theta)} \right| dr\, d\theta$$

$$= \int_\alpha^\beta \int_a^b f(r\cos\theta,\, r\sin\theta)\, r\, dr\, d\theta$$

which is the same as Formula 12.3.2.

☑ **EXAMPLE 2** Use the change of variables $x = u^2 - v^2$, $y = 2uv$ to evaluate the integral $\iint_R y\, dA$, where R is the region bounded by the x-axis and the parabolas $y^2 = 4 - 4x$ and $y^2 = 4 + 4x$, $y \geqslant 0$.

SOLUTION The region R is pictured in Figure 2 (on page 743). In Example 1 we discovered that $T(S) = R$, where S is the square $[0, 1] \times [0, 1]$. Indeed, the reason for making the change of variables to evaluate the integral is that S is a much simpler region than R. First we need to compute the Jacobian:

$$\frac{\partial(x, y)}{\partial(u, v)} = \begin{vmatrix} \dfrac{\partial x}{\partial u} & \dfrac{\partial x}{\partial v} \\[2mm] \dfrac{\partial y}{\partial u} & \dfrac{\partial y}{\partial v} \end{vmatrix} = \begin{vmatrix} 2u & -2v \\ 2v & 2u \end{vmatrix} = 4u^2 + 4v^2 > 0$$

Therefore, by Theorem 9,

$$\iint_R y \, dA = \iint_S 2uv \left| \frac{\partial(x, y)}{\partial(u, v)} \right| dA = \int_0^1 \int_0^1 (2uv)4(u^2 + v^2) \, du \, dv$$

$$= 8 \int_0^1 \int_0^1 (u^3 v + uv^3) \, du \, dv = 8 \int_0^1 \left[\tfrac{1}{4} u^4 v + \tfrac{1}{2} u^2 v^3 \right]_{u=0}^{u=1} dv$$

$$= \int_0^1 (2v + 4v^3) \, dv = \left[v^2 + v^4 \right]_0^1 = 2 \qquad \blacksquare$$

NOTE Example 2 was not a very difficult problem to solve because we were given a suitable change of variables. If we are not supplied with a transformation, then the first step is to think of an appropriate change of variables. If $f(x, y)$ is difficult to integrate, then the form of $f(x, y)$ may suggest a transformation. If the region of integration R is awkward, then the transformation should be chosen so that the corresponding region S in the uv-plane has a convenient description.

EXAMPLE 3 Evaluate the integral $\iint_R e^{(x+y)/(x-y)} \, dA$, where R is the trapezoidal region with vertices $(1, 0)$, $(2, 0)$, $(0, -2)$, and $(0, -1)$.

SOLUTION Since it isn't easy to integrate $e^{(x+y)/(x-y)}$, we make a change of variables suggested by the form of this function:

$$\boxed{10} \qquad\qquad u = x + y \qquad v = x - y$$

These equations define a transformation T^{-1} from the xy-plane to the uv-plane. Theorem 9 talks about a transformation T from the uv-plane to the xy-plane. It is obtained by solving Equations 10 for x and y:

$$\boxed{11} \qquad\qquad x = \tfrac{1}{2}(u + v) \qquad y = \tfrac{1}{2}(u - v)$$

The Jacobian of T is

$$\frac{\partial(x, y)}{\partial(u, v)} = \begin{vmatrix} \dfrac{\partial x}{\partial u} & \dfrac{\partial x}{\partial v} \\[2mm] \dfrac{\partial y}{\partial u} & \dfrac{\partial y}{\partial v} \end{vmatrix} = \begin{vmatrix} \tfrac{1}{2} & \tfrac{1}{2} \\[1mm] \tfrac{1}{2} & -\tfrac{1}{2} \end{vmatrix} = -\tfrac{1}{2}$$

To find the region S in the uv-plane corresponding to R, we note that the sides of R lie on the lines

$$y = 0 \qquad x - y = 2 \qquad x = 0 \qquad x - y = 1$$

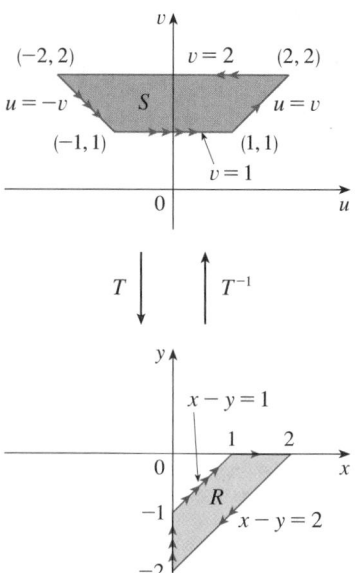

FIGURE 8

and, from either Equations 10 or Equations 11, the image lines in the uv-plane are

$$u = v \qquad v = 2 \qquad u = -v \qquad v = 1$$

Thus the region S is the trapezoidal region with vertices $(1, 1)$, $(2, 2)$, $(-2, 2)$, and $(-1, 1)$ shown in Figure 8. Since

$$S = \{(u, v) \mid 1 \leq v \leq 2, \ -v \leq u \leq v\}$$

Theorem 9 gives

$$\iint_R e^{(x+y)/(x-y)} \, dA = \iint_S e^{u/v} \left| \frac{\partial(x, y)}{\partial(u, v)} \right| du \, dv$$

$$= \int_1^2 \int_{-v}^v e^{u/v} \left(\tfrac{1}{2} \right) du \, dv = \tfrac{1}{2} \int_1^2 \left[v e^{u/v} \right]_{u=-v}^{u=v} \, dv$$

$$= \tfrac{1}{2} \int_1^2 (e - e^{-1}) v \, dv = \tfrac{3}{4}(e - e^{-1}) \qquad \blacksquare$$

TRIPLE INTEGRALS

There is a similar change of variables formula for triple integrals. Let T be a transformation that maps a region S in uvw-space onto a region R in xyz-space by means of the equations

$$x = g(u, v, w) \qquad y = h(u, v, w) \qquad z = k(u, v, w)$$

The **Jacobian** of T is the following 3×3 determinant:

$$\boxed{12} \qquad \frac{\partial(x, y, z)}{\partial(u, v, w)} = \begin{vmatrix} \dfrac{\partial x}{\partial u} & \dfrac{\partial x}{\partial v} & \dfrac{\partial x}{\partial w} \\[2mm] \dfrac{\partial y}{\partial u} & \dfrac{\partial y}{\partial v} & \dfrac{\partial y}{\partial w} \\[2mm] \dfrac{\partial z}{\partial u} & \dfrac{\partial z}{\partial v} & \dfrac{\partial z}{\partial w} \end{vmatrix}$$

Under hypotheses similar to those in Theorem 9, we have the following formula for triple integrals:

$$\boxed{13} \quad \iiint_R f(x, y, z) \, dV$$

$$= \iiint_S f\big(x(u, v, w), y(u, v, w), z(u, v, w)\big) \left| \frac{\partial(x, y, z)}{\partial(u, v, w)} \right| du \, dv \, dw$$

V EXAMPLE 4 Use Formula 13 to derive the formula for triple integration in spherical coordinates.

SOLUTION Here the change of variables is given by

$$x = \rho \sin \phi \cos \theta \qquad y = \rho \sin \phi \sin \theta \qquad z = \rho \cos \phi$$

We compute the Jacobian as follows:

$$\frac{\partial(x, y, z)}{\partial(\rho, \theta, \phi)} = \begin{vmatrix} \sin \phi \cos \theta & -\rho \sin \phi \sin \theta & \rho \cos \phi \cos \theta \\ \sin \phi \sin \theta & \rho \sin \phi \cos \theta & \rho \cos \phi \sin \theta \\ \cos \phi & 0 & -\rho \sin \phi \end{vmatrix}$$

$$= \cos \phi \begin{vmatrix} -\rho \sin \phi \sin \theta & \rho \cos \phi \cos \theta \\ \rho \sin \phi \cos \theta & \rho \cos \phi \sin \theta \end{vmatrix} - \rho \sin \phi \begin{vmatrix} \sin \phi \cos \theta & -\rho \sin \phi \sin \theta \\ \sin \phi \sin \theta & \rho \sin \phi \cos \theta \end{vmatrix}$$

$$= \cos \phi \, (-\rho^2 \sin \phi \cos \phi \sin^2\theta - \rho^2 \sin \phi \cos \phi \cos^2\theta)$$

$$\quad - \rho \sin \phi \, (\rho \sin^2\phi \cos^2\theta + \rho \sin^2\phi \sin^2\theta)$$

$$= -\rho^2 \sin \phi \cos^2\phi - \rho^2 \sin \phi \sin^2\phi = -\rho^2 \sin \phi$$

Since $0 \leqslant \phi \leqslant \pi$, we have $\sin \phi \geqslant 0$. Therefore

$$\left| \frac{\partial(x, y, z)}{\partial(\rho, \theta, \phi)} \right| = |-\rho^2 \sin \phi| = \rho^2 \sin \phi$$

and Formula 13 gives

$$\iiint_R f(x, y, z) \, dV = \iiint_S f(\rho \sin \phi \cos \theta, \rho \sin \phi \sin \theta, \rho \cos \phi) \, \rho^2 \sin \phi \, d\rho \, d\theta \, d\phi$$

which is equivalent to Formula 12.7.3.   ∎

12.8 | EXERCISES

1–6 ■ Find the Jacobian of the transformation.

1. $x = 5u - v, \quad y = u + 3v$

2. $x = uv, \quad y = u/v$

3. $x = e^{-r} \sin \theta, \quad y = e^r \cos \theta$

4. $x = e^{s+t}, \quad y = e^{s-t}$

5. $x = u/v, \quad y = v/w, \quad z = w/u$

6. $x = v + w^2, \quad y = w + u^2, \quad z = u + v^2$

7–10 ■ Find the image of the set S under the given transformation.

7. $S = \{(u, v) \mid 0 \leqslant u \leqslant 3, \ 0 \leqslant v \leqslant 2\};$
$x = 2u + 3v, \ y = u - v$

8. S is the square bounded by the lines $u = 0, u = 1, v = 0,$
$v = 1; \quad x = v, \ y = u(1 + v^2)$

9. S is the triangular region with vertices $(0, 0), (1, 1), (0, 1);$
$x = u^2, \ y = v$

10. S is the disk given by $u^2 + v^2 \leqslant 1; \quad x = au, \ y = bv$

11–14 ■ A region R in the xy-plane is given. Find equations for a transformation T that maps a rectangular region S in the uv-plane onto R, where the sides of S are parallel to the u- and v-axes.

11. R is bounded by $y = 2x - 1, y = 2x + 1, y = 1 - x,$
$y = 3 - x$

12. R is the parallelogram with vertices $(0, 0)$, $(4, 3)$, $(2, 4)$, $(-2, 1)$

13. R lies between the circles $x^2 + y^2 = 1$ and $x^2 + y^2 = 2$ in the first quadrant

14. R is bounded by the hyperbolas $y = 1/x$, $y = 4/x$ and the lines $y = x$, $y = 4x$ in the first quadrant

---

15–20 ▪ Use the given transformation to evaluate the integral.

15. $\iint_R (x - 3y)\, dA$, where R is the triangular region with vertices $(0, 0)$, $(2, 1)$, and $(1, 2)$; $x = 2u + v$, $y = u + 2v$

16. $\iint_R (4x + 8y)\, dA$, where R is the parallelogram with vertices $(-1, 3)$, $(1, -3)$, $(3, -1)$, and $(1, 5)$; $x = \frac{1}{4}(u + v)$, $y = \frac{1}{4}(v - 3u)$

17. $\iint_R x^2\, dA$, where R is the region bounded by the ellipse $9x^2 + 4y^2 = 36$; $x = 2u$, $y = 3v$

18. $\iint_R (x^2 - xy + y^2)\, dA$, where R is the region bounded by the ellipse $x^2 - xy + y^2 = 2$; $x = \sqrt{2}\, u - \sqrt{2/3}\, v$, $y = \sqrt{2}\, u + \sqrt{2/3}\, v$

19. $\iint_R xy\, dA$, where R is the region in the first quadrant bounded by the lines $y = x$ and $y = 3x$ and the hyperbolas $xy = 1$, $xy = 3$; $x = u/v$, $y = v$

20. $\iint_R y^2\, dA$, where R is the region bounded by the curves $xy = 1$, $xy = 2$, $xy^2 = 1$, $xy^2 = 2$; $u = xy$, $v = xy^2$. Illustrate by using a graphing calculator or computer to draw R.

---

21. (a) Evaluate $\iiint_E dV$, where E is the solid enclosed by the ellipsoid $x^2/a^2 + y^2/b^2 + z^2/c^2 = 1$. Use the transformation $x = au$, $y = bv$, $z = cw$.
(b) The earth is not a perfect sphere; rotation has resulted in flattening at the poles. So the shape can be approximated by an ellipsoid with $a = b = 6378$ km and

$c = 6356$ km. Use part (a) to estimate the volume of the earth.
(c) If the solid of part (a) has constant density k, find its moment of inertia about the z-axis.

22. An important problem in thermodynamics is to find the work done by an ideal Carnot engine. A cycle consists of alternating expansion and compression of gas in a piston. The work done by the engine is equal to the area of the region R enclosed by two isothermal curves $xy = a$, $xy = b$ and two adiabatic curves $xy^{1.4} = c$, $xy^{1.4} = d$, where $0 < a < b$ and $0 < c < d$. Compute the work done by determining the area of R.

23–27 ▪ Evaluate the integral by making an appropriate change of variables.

23. $\iint_R \dfrac{x - 2y}{3x - y}\, dA$, where R is the parallelogram enclosed by the lines $x - 2y = 0$, $x - 2y = 4$, $3x - y = 1$, and $3x - y = 8$

24. $\iint_R (x + y)e^{x^2 - y^2}\, dA$, where R is the rectangle enclosed by the lines $x - y = 0$, $x - y = 2$, $x + y = 0$, and $x + y = 3$

25. $\iint_R \cos\left(\dfrac{y - x}{y + x}\right) dA$, where R is the trapezoidal region with vertices $(1, 0)$, $(2, 0)$, $(0, 2)$, and $(0, 1)$

26. $\iint_R \sin(9x^2 + 4y^2)\, dA$, where R is the region in the first quadrant bounded by the ellipse $9x^2 + 4y^2 = 1$

27. $\iint_R e^{x+y}\, dA$, where R is given by the inequality $|x| + |y| \le 1$

---

28. Let f be continuous on $[0, 1]$ and let R be the triangular region with vertices $(0, 0)$, $(1, 0)$, and $(0, 1)$. Show that

$$\iint_R f(x + y)\, dA = \int_0^1 uf(u)\, du$$

CHAPTER 12 REVIEW

1. Suppose f is a continuous function defined on a rectangle $R = [a, b] \times [c, d]$.
(a) Write an expression for a double Riemann sum of f. If $f(x, y) \geq 0$, what does the sum represent?
(b) Write the definition of $\iint_R f(x, y) \, dA$ as a limit.
(c) What is the geometric interpretation of $\iint_R f(x, y) \, dA$ if $f(x, y) \geq 0$? What if f takes on both positive and negative values?
(d) How do you evaluate $\iint_R f(x, y) \, dA$?
(e) What does the Midpoint Rule for double integrals say?

2. (a) How do you define $\iint_D f(x, y) \, dA$ if D is a bounded region that is not a rectangle?
(b) What is a type I region? How do you evaluate $\iint_D f(x, y) \, dA$ if D is a type I region?
(c) What is a type II region? How do you evaluate $\iint_D f(x, y) \, dA$ if D is a type II region?
(d) What properties do double integrals have?

3. How do you change from rectangular coordinates to polar coordinates in a double integral? Why would you want to make the change?

4. If a lamina occupies a plane region D and has density function $\rho(x, y)$, write expressions for each of the following in terms of double integrals.
(a) The mass
(b) The moments about the axes
(c) The center of mass
(d) The moments of inertia about the axes and the origin

5. (a) Write the definition of the triple integral of f over a rectangular box B.
(b) How do you evaluate $\iiint_B f(x, y, z) \, dV$?

(c) How do you define $\iiint_E f(x, y, z) \, dV$ if E is a bounded solid region that is not a box?
(d) What is a type 1 solid region? How do you evaluate $\iiint_E f(x, y, z) \, dV$ if E is such a region?
(e) What is a type 2 solid region? How do you evaluate $\iiint_E f(x, y, z) \, dV$ if E is such a region?
(f) What is a type 3 solid region? How do you evaluate $\iiint_E f(x, y, z) \, dV$ if E is such a region?

6. Suppose a solid object occupies the region E and has density function $\rho(x, y, z)$. Write expressions for each of the following.
(a) The mass
(b) The moments about the coordinate planes
(c) The coordinates of the center of mass
(d) The moments of inertia about the axes

7. (a) Write the equations for converting from cylindrical to rectangular coordinates. In what situation would you use cylindrical coordinates?
(b) Write the equations for converting from spherical to rectangular coordinates. In what situation would you use spherical coordinates?

8. (a) How do you change from rectangular coordinates to cylindrical coordinates in a triple integral?
(b) How do you change from rectangular coordinates to spherical coordinates in a triple integral?
(c) In what situations would you change to cylindrical or spherical coordinates?

9. (a) If a transformation T is given by $x = g(u, v)$, $y = h(u, v)$, what is the Jacobian of T?
(b) How do you change variables in a double integral?
(c) How do you change variables in a triple integral?

Determine whether the statement is true or false. If it is true, explain why. If it is false, explain why or give an example that disproves the statement.

1. $\int_{-1}^{2} \int_{0}^{6} x^2 \sin(x - y) \, dx \, dy = \int_{0}^{6} \int_{-1}^{2} x^2 \sin(x - y) \, dy \, dx$

2. $\int_{0}^{1} \int_{0}^{x} \sqrt{x + y^2} \, dy \, dx = \int_{0}^{x} \int_{0}^{1} \sqrt{x + y^2} \, dx \, dy$

3. $\int_{1}^{2} \int_{3}^{4} x^2 e^y \, dy \, dx = \int_{1}^{2} x^2 \, dx \int_{3}^{4} e^y \, dy$

4. $\int_{-1}^{1} \int_{0}^{1} e^{x^2 + y^2} \sin y \, dx \, dy = 0$

5. If f is continuous on $[0, 1]$, then

$$\int_{0}^{1} \int_{0}^{1} f(x) f(y) \, dy \, dx = \left[\int_{0}^{1} f(x) \, dx \right]^2$$

6. $\int_{1}^{4} \int_{0}^{1} \left(x^2 + \sqrt{y} \right) \sin(x^2 y^2) \, dx \, dy \leq 9$

7. If D is the disk given by $x^2 + y^2 \leq 4$, then

$$\iint_D \sqrt{4 - x^2 - y^2} \, dA = \tfrac{16}{3} \pi$$

8. The integral

$$\iiint_E kr^3 \, dz \, dr \, d\theta$$

represents the moment of inertia about the z-axis of a solid E with constant density k.

9. The integral

$$\int_0^{2\pi} \int_0^2 \int_r^2 dz \, dr \, d\theta$$

represents the volume enclosed by the cone $z = \sqrt{x^2 + y^2}$ and the plane $z = 2$.

EXERCISES

1. A contour map is shown for a function f on the square $R = [0, 3] \times [0, 3]$. Use a Riemann sum with nine terms to estimate the value of $\iint_R f(x, y) \, dA$. Take the sample points to be the upper right corners of the squares.

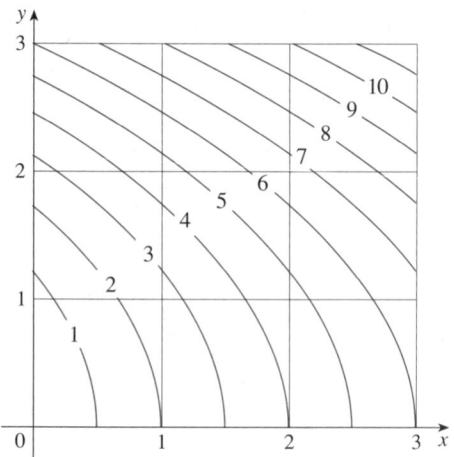

2. Use the Midpoint Rule to estimate the integral in Exercise 1.

3–8 ▪ Calculate the iterated integral.

3. $\displaystyle\int_1^2 \int_0^2 (y + 2xe^y) \, dx \, dy$

4. $\displaystyle\int_0^1 \int_0^1 ye^{xy} \, dx \, dy$

5. $\displaystyle\int_0^1 \int_0^x \cos(x^2) \, dy \, dx$

6. $\displaystyle\int_0^1 \int_x^{e^x} 3xy^2 \, dy \, dx$

7. $\displaystyle\int_0^{\pi} \int_0^1 \int_0^{\sqrt{1-y^2}} y \sin x \, dz \, dy \, dx$

8. $\displaystyle\int_0^1 \int_0^y \int_x^1 6xyz \, dz \, dx \, dy$

9–10 ▪ Write $\iint_R f(x, y) \, dA$ as an iterated integral, where R is the region shown and f is an arbitrary continuous function on R.

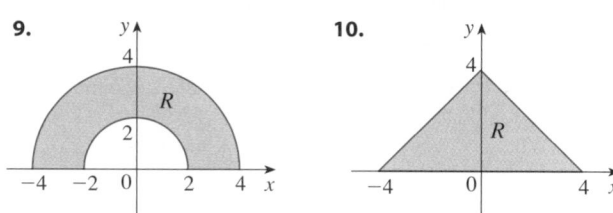

11. Describe the region whose area is given by the integral

$$\int_0^{\pi/2} \int_0^{\sin 2\theta} r \, dr \, d\theta$$

12. Describe the solid whose volume is given by the integral

$$\int_0^{\pi/2} \int_0^{\pi/2} \int_1^2 \rho^2 \sin \phi \, d\rho \, d\phi \, d\theta$$

and evaluate the integral.

13–14 ▪ Calculate the iterated integral by first reversing the order of integration.

13. $\displaystyle\int_0^1 \int_x^1 \cos(y^2) \, dy \, dx$

14. $\displaystyle\int_0^1 \int_{\sqrt{y}}^1 \frac{ye^{x^2}}{x^3} \, dx \, dy$

15–28 ▪ Calculate the value of the multiple integral.

15. $\iint_R ye^{xy} \, dA$, where $R = \{(x, y) \mid 0 \le x \le 2, \ 0 \le y \le 3\}$

16. $\iint_D xy \, dA$,
where $D = \{(x, y) \mid 0 \le y \le 1, \ y^2 \le x \le y + 2\}$

17. $\displaystyle\iint_D \frac{y}{1 + x^2} \, dA$,
where D is bounded by $y = \sqrt{x}$, $y = 0$, $x = 1$

18. $\displaystyle\iint_D \frac{1}{1 + x^2} \, dA$, where D is the triangular region with vertices $(0, 0)$, $(1, 1)$, and $(0, 1)$

19. $\iint_D y \, dA$, where D is the region in the first quadrant bounded by the parabolas $x = y^2$ and $x = 8 - y^2$

20. $\iint_D y \, dA$, where D is the region in the first quadrant that lies above the hyperbola $xy = 1$ and the line $y = x$ and below the line $y = 2$

21. $\iint_D (x^2 + y^2)^{3/2} \, dA$, where D is the region in the first quadrant bounded by the lines $y = 0$ and $y = \sqrt{3} \, x$ and the circle $x^2 + y^2 = 9$

22. $\iint_D x \, dA$, where D is the region in the first quadrant that lies between the circles $x^2 + y^2 = 1$ and $x^2 + y^2 = 2$

23. $\iiint_E xy \, dV$, where
$E = \{(x, y, z) \mid 0 \leq x \leq 3, \ 0 \leq y \leq x, \ 0 \leq z \leq x + y\}$

24. $\iiint_T xy \, dV$, where T is the solid tetrahedron with vertices $(0, 0, 0)$, $(\frac{1}{3}, 0, 0)$, $(0, 1, 0)$, and $(0, 0, 1)$

25. $\iiint_E y^2 z^2 \, dV$, where E is bounded by the paraboloid $x = 1 - y^2 - z^2$ and the plane $x = 0$

26. $\iiint_E z \, dV$, where E is bounded by the planes $y = 0$, $z = 0$, $x + y = 2$ and the cylinder $y^2 + z^2 = 1$ in the first octant

27. $\iiint_E yz \, dV$, where E lies above the plane $z = 0$, below the plane $z = y$, and inside the cylinder $x^2 + y^2 = 4$

28. $\iiint_H z^3 \sqrt{x^2 + y^2 + z^2} \, dV$, where H is the solid hemisphere that lies above the xy-plane and has center the origin and radius 1

29–34 ▪ Find the volume of the given solid.

29. Under the paraboloid $z = x^2 + 4y^2$ and above the rectangle $R = [0, 2] \times [1, 4]$

30. Under the surface $z = x^2 y$ and above the triangle in the xy-plane with vertices $(1, 0)$, $(2, 1)$, and $(4, 0)$

31. The solid tetrahedron with vertices $(0, 0, 0)$, $(0, 0, 1)$, $(0, 2, 0)$, and $(2, 2, 0)$

32. Bounded by the cylinder $x^2 + y^2 = 4$ and the planes $z = 0$ and $y + z = 3$

33. One of the wedges cut from the cylinder $x^2 + 9y^2 = a^2$ by the planes $z = 0$ and $z = mx$

34. Above the paraboloid $z = x^2 + y^2$ and below the half-cone $z = \sqrt{x^2 + y^2}$

35. Consider a lamina that occupies the region D bounded by the parabola $x = 1 - y^2$ and the coordinate axes in the first quadrant with density function $\rho(x, y) = y$.
(a) Find the mass of the lamina.
(b) Find the center of mass.
(c) Find the moments of inertia and radii of gyration about the x- and y-axes.

36. A lamina occupies the part of the disk $x^2 + y^2 \leq a^2$ that lies in the first quadrant.
(a) Find the centroid of the lamina.
(b) Find the center of mass of the lamina if the density function is $\rho(x, y) = xy^2$.

37. (a) Find the centroid of a right circular cone with height h and base radius a. (Place the cone so that its base is in the xy-plane with center the origin and its axis along the positive z-axis.)
(b) Find the moment of inertia of the cone about its axis (the z-axis).

CAS **38.** Find the center of mass of the solid tetrahedron with vertices $(0, 0, 0)$, $(1, 0, 0)$, $(0, 2, 0)$, $(0, 0, 3)$ and density function $\rho(x, y, z) = x^2 + y^2 + z^2$.

39. The cylindrical coordinates of a point are $(2\sqrt{3}, \pi/3, 2)$. Find the rectangular and spherical coordinates of the point.

40. The rectangular coordinates of a point are $(2, 2, -1)$. Find the cylindrical and spherical coordinates of the point.

41. The spherical coordinates of a point are $(8, \pi/4, \pi/6)$. Find the rectangular and cylindrical coordinates of the point.

42. Identify the surfaces whose equations are given.
(a) $\theta = \pi/4$ (b) $\phi = \pi/4$

43. Write the equation $x^2 + y^2 + z^2 = 4$ in cylindrical coordinates and in spherical coordinates.

44. Sketch the solid consisting of all points with spherical coordinates (ρ, θ, ϕ) such that $0 \leq \theta \leq \pi/2$, $0 \leq \phi \leq \pi/6$, and $0 \leq \rho \leq 2 \cos \phi$.

45. Use polar coordinates to evaluate
$$\int_0^3 \int_{-\sqrt{9-x^2}}^{\sqrt{9-x^2}} (x^3 + xy^2) \, dy \, dx$$

46. Use spherical coordinates to evaluate
$$\int_{-2}^2 \int_0^{\sqrt{4-y^2}} \int_{-\sqrt{4-x^2-y^2}}^{\sqrt{4-x^2-y^2}} y^2 \sqrt{x^2 + y^2 + z^2} \, dz \, dx \, dy$$

47. Rewrite the integral
$$\int_{-1}^1 \int_{x^2}^1 \int_0^{1-y} f(x, y, z) \, dz \, dy \, dx$$
as an iterated integral in the order $dx \, dy \, dz$.

48. Give five other iterated integrals that are equal to

$$\int_0^2 \int_0^{y^3} \int_0^{y^2} f(x, y, z) \, dz \, dx \, dy$$

49. Use the transformation $u = x - y$, $v = x + y$ to evaluate $\iint_R (x - y)/(x + y) \, dA$, where R is the square with vertices $(0, 2)$, $(1, 1)$, $(2, 2)$, and $(1, 3)$.

50. Use the transformation $x = u^2$, $y = v^2$, $z = w^2$ to find the volume of the region bounded by the surface $\sqrt{x} + \sqrt{y} + \sqrt{z} = 1$ and the coordinate planes.

51. Use the change of variables formula and an appropriate transformation to evaluate $\iint_R xy \, dA$, where R is the square with vertices $(0, 0)$, $(1, 1)$, $(2, 0)$, and $(1, -1)$.

52. (a) Evaluate $\iint_D \dfrac{1}{(x^2 + y^2)^{n/2}} \, dA$, where n is an integer and D is the region bounded by the circles with center the origin and radii r and R, $0 < r < R$.

(b) For what values of n does the integral in part (a) have a limit as $r \to 0^+$?

(c) Find $\iiint_E \dfrac{1}{(x^2 + y^2 + z^2)^{n/2}} \, dV$, where E is the region bounded by the spheres with center the origin and radii r and R, $0 < r < R$.

(d) For what values of n does the integral in part (c) have a limit as $r \to 0^+$?

13 VECTOR CALCULUS

In this chapter we study the calculus of vector fields. (These are functions that assign vectors to points in space.) In particular we define line integrals (which can be used to find the work done by a force field in moving an object along a curve). Then we define surface integrals (which can be used to find the rate of fluid flow across a surface). The connections between these new types of integrals and the single, double, and triple integrals that we have already met are given by the higher-dimensional versions of the Fundamental Theorem of Calculus: Green's Theorem, Stokes' Theorem, and the Divergence Theorem.

13.1 | VECTOR FIELDS

The vectors in Figure 1(a) are air velocity vectors that indicate the wind speed and direction at points 10 m above the surface elevation in the San Francisco Bay area at 6:00 PM on March 1, 2010. We see at a glance from the largest arrows that the greatest wind speeds at that time occurred as the winds entered the bay across the Golden Gate Bridge. Associated with every point in the air we can imagine a wind velocity vector. This is an example of a *velocity vector field*. Another example of a velocity vector field is illustrated in Figure 1(b).

(a) San Francisco Bay wind patterns

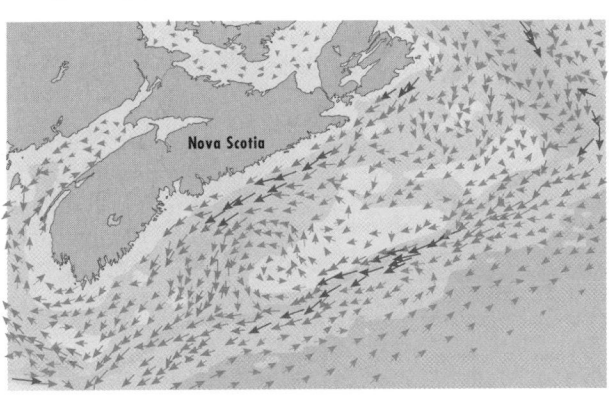

(b) Ocean currents off the coast of Nova Scotia

FIGURE 1 Velocity vector fields

Another type of vector field, called a *force field,* associates a force vector with each point in a region. An example is the gravitational force field that we will look at in Example 4.

In general, a vector field is a function whose domain is a set of points in $\mathbb{R}^2$ (or $\mathbb{R}^3$) and whose range is a set of vectors in V_2 (or V_3).

> **1 DEFINITION** Let D be a set in $\mathbb{R}^2$ (a plane region). A **vector field on** $\mathbb{R}^2$ is a function $\mathbf{F}$ that assigns to each point (x, y) in D a two-dimensional vector $\mathbf{F}(x, y)$.

The best way to picture a vector field is to draw the arrow representing the vector $\mathbf{F}(x, y)$ starting at the point (x, y). Of course, it's impossible to do this for all points (x, y), but we can gain a reasonable impression of $\mathbf{F}$ by doing it for a few representative points in D as in Figure 2. Since $\mathbf{F}(x, y)$ is a two-dimensional vector, we can write

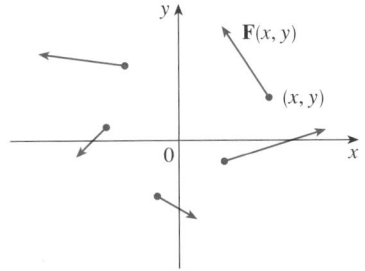

FIGURE 2
Vector field on $\mathbb{R}^2$

it in terms of its **component functions** P and Q as follows:

$$\mathbf{F}(x, y) = P(x, y)\,\mathbf{i} + Q(x, y)\,\mathbf{j} = \langle P(x, y), Q(x, y) \rangle$$

or, for short,
$$\mathbf{F} = P\,\mathbf{i} + Q\,\mathbf{j}$$

Notice that P and Q are scalar functions of two variables and are sometimes called **scalar fields** to distinguish them from vector fields.

> **2** **DEFINITION** Let E be a subset of $\mathbb{R}^3$. A **vector field on** $\mathbb{R}^3$ is a function $\mathbf{F}$ that assigns to each point (x, y, z) in E a three-dimensional vector $\mathbf{F}(x, y, z)$.

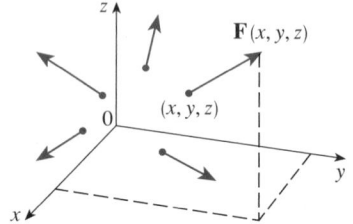

FIGURE 3
Vector field on $\mathbb{R}^3$

A vector field $\mathbf{F}$ on $\mathbb{R}^3$ is pictured in Figure 3. We can express it in terms of its component functions P, Q, and R as

$$\mathbf{F}(x, y, z) = P(x, y, z)\,\mathbf{i} + Q(x, y, z)\,\mathbf{j} + R(x, y, z)\,\mathbf{k}$$

As with the vector functions in Section 10.7, we can define continuity of vector fields and show that $\mathbf{F}$ is continuous if and only if its component functions P, Q, and R are continuous.

We sometimes identify a point (x, y, z) with its position vector $\mathbf{x} = \langle x, y, z \rangle$ and write $\mathbf{F}(\mathbf{x})$ instead of $\mathbf{F}(x, y, z)$. Then $\mathbf{F}$ becomes a function that assigns a vector $\mathbf{F}(\mathbf{x})$ to a vector $\mathbf{x}$.

▼ EXAMPLE 1 A vector field on $\mathbb{R}^2$ is defined by $\mathbf{F}(x, y) = -y\,\mathbf{i} + x\,\mathbf{j}$. Describe $\mathbf{F}$ by sketching some of the vectors $\mathbf{F}(x, y)$ as in Figure 2.

SOLUTION Since $\mathbf{F}(1, 0) = \mathbf{j}$, we draw the vector $\mathbf{j} = \langle 0, 1 \rangle$ starting at the point $(1, 0)$ in Figure 4. Since $\mathbf{F}(0, 1) = -\mathbf{i}$, we draw the vector $\langle -1, 0 \rangle$ with starting point $(0, 1)$. Continuing in this way, we calculate several other representative values of $\mathbf{F}(x, y)$ in the table and draw the corresponding vectors to represent the vector field in Figure 4.

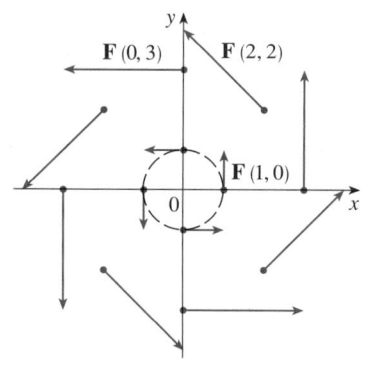

FIGURE 4
$\mathbf{F}(x, y) = -y\,\mathbf{i} + x\,\mathbf{j}$

(x, y)	$\mathbf{F}(x, y)$	(x, y)	$\mathbf{F}(x, y)$
$(1, 0)$	$\langle 0, 1 \rangle$	$(-1, 0)$	$\langle 0, -1 \rangle$
$(2, 2)$	$\langle -2, 2 \rangle$	$(-2, -2)$	$\langle 2, -2 \rangle$
$(3, 0)$	$\langle 0, 3 \rangle$	$(-3, 0)$	$\langle 0, -3 \rangle$
$(0, 1)$	$\langle -1, 0 \rangle$	$(0, -1)$	$\langle 1, 0 \rangle$
$(-2, 2)$	$\langle -2, -2 \rangle$	$(2, -2)$	$\langle 2, 2 \rangle$
$(0, 3)$	$\langle -3, 0 \rangle$	$(0, -3)$	$\langle 3, 0 \rangle$

It appears from Figure 4 that each arrow is tangent to a circle with center the origin. To confirm this, we take the dot product of the position vector $\mathbf{x} = x\,\mathbf{i} + y\,\mathbf{j}$ with the vector $\mathbf{F}(\mathbf{x}) = \mathbf{F}(x, y)$:

$$\mathbf{x} \cdot \mathbf{F}(\mathbf{x}) = (x\,\mathbf{i} + y\,\mathbf{j}) \cdot (-y\,\mathbf{i} + x\,\mathbf{j}) = -xy + yx = 0$$

This shows that $\mathbf{F}(x, y)$ is perpendicular to the position vector $\langle x, y \rangle$ and is therefore tangent to a circle with center the origin and radius $|\mathbf{x}| = \sqrt{x^2 + y^2}$. Notice also that

$$|\mathbf{F}(x, y)| = \sqrt{(-y)^2 + x^2} = \sqrt{x^2 + y^2} = |\mathbf{x}|$$

so the magnitude of the vector $\mathbf{F}(x, y)$ is equal to the radius of the circle. ∎

Some computer algebra systems are capable of plotting vector fields in two or three dimensions. They give a better impression of the vector field than is possible by hand because the computer can plot a large number of representative vectors. Figure 5 shows a computer plot of the vector field in Example 1; Figures 6 and 7 show two other vector fields. Notice that the computer scales the lengths of the vectors so they are not too long and yet are proportional to their true lengths.

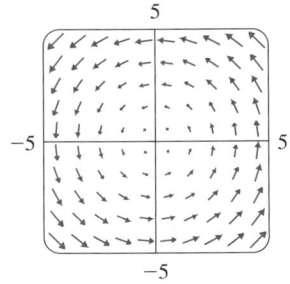

FIGURE 5
$\mathbf{F}(x, y) = \langle -y, x \rangle$

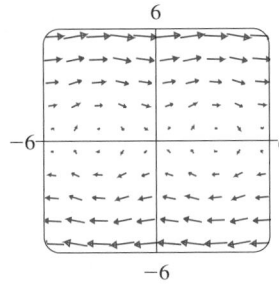

FIGURE 6
$\mathbf{F}(x, y) = \langle y, \sin x \rangle$

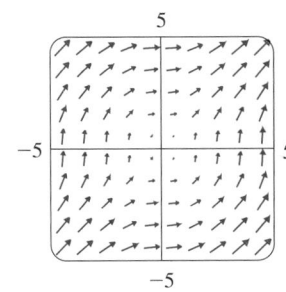

FIGURE 7
$\mathbf{F}(x, y) = \langle \ln(1 + y^2), \ln(1 + x^2) \rangle$

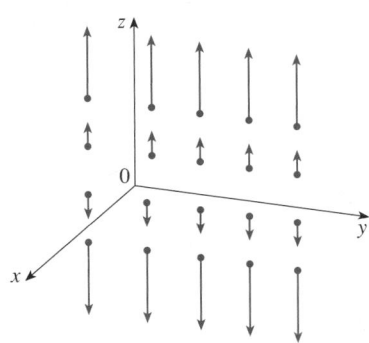

FIGURE 8
$\mathbf{F}(x, y, z) = z\,\mathbf{k}$

TEC In Visual 13.1 you can rotate the vector fields in Figures 9–11 as well as additional fields.

V **EXAMPLE 2** Sketch the vector field on $\mathbb{R}^3$ given by $\mathbf{F}(x, y, z) = z\,\mathbf{k}$.

SOLUTION The sketch is shown in Figure 8. Notice that all vectors are vertical and point upward above the xy-plane or downward below it. The magnitude increases with the distance from the xy-plane. ■

We were able to draw the vector field in Example 2 by hand because of its particularly simple formula. Most three-dimensional vector fields, however, are virtually impossible to sketch by hand and so we need to resort to a computer. Examples are shown in Figures 9, 10, and 11. Notice that the vector fields in Figures 9 and 10 have similar formulas, but all the vectors in Figure 10 point in the general direction of the negative y-axis because their y-components are all -2. If the vector field in Figure 11 represents a velocity field, then a particle would be swept upward and would spiral around the z-axis in the clockwise direction as viewed from above.

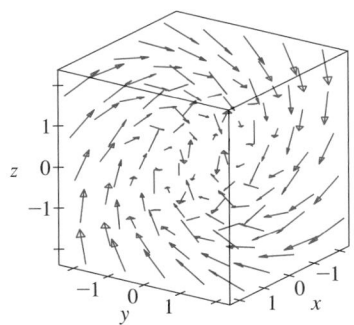

FIGURE 9
$\mathbf{F}(x, y, z) = y\,\mathbf{i} + z\,\mathbf{j} + x\,\mathbf{k}$

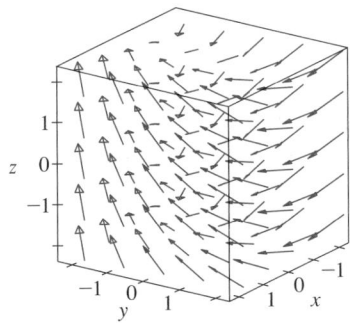

FIGURE 10
$\mathbf{F}(x, y, z) = y\,\mathbf{i} - 2\,\mathbf{j} + x\,\mathbf{k}$

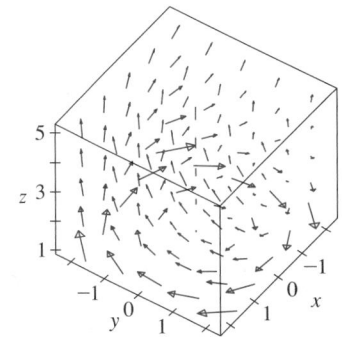

FIGURE 11
$\mathbf{F}(x, y, z) = \dfrac{y}{z}\,\mathbf{i} - \dfrac{x}{z}\,\mathbf{j} + \dfrac{z}{4}\,\mathbf{k}$

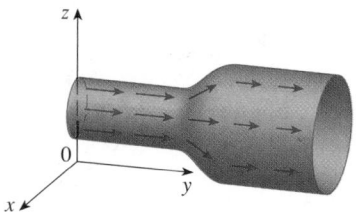

FIGURE 12
Velocity field in fluid flow

EXAMPLE 3 Imagine a fluid flowing steadily along a pipe and let $\mathbf{V}(x, y, z)$ be the velocity vector at a point (x, y, z). Then $\mathbf{V}$ assigns a vector to each point (x, y, z) in a certain domain E (the interior of the pipe) and so $\mathbf{V}$ is a vector field on $\mathbb{R}^3$ called a **velocity field**. A possible velocity field is illustrated in Figure 12. The speed at any given point is indicated by the length of the arrow.

Velocity fields also occur in other areas of physics. For instance, the vector field in Example 1 could be used as the velocity field describing the counterclockwise rotation of a wheel. We have seen other examples of velocity fields in Figure 1. ∎

EXAMPLE 4 Newton's Law of Gravitation states that the magnitude of the gravitational force between two objects with masses m and M is

$$|\mathbf{F}| = \frac{mMG}{r^2}$$

where r is the distance between the objects and G is the gravitational constant. (This is an example of an inverse square law.) Let's assume that the object with mass M is located at the origin in $\mathbb{R}^3$. (For instance, M could be the mass of the earth and the origin would be at its center.) Let the position vector of the object with mass m be $\mathbf{x} = \langle x, y, z \rangle$. Then $r = |\mathbf{x}|$, so $r^2 = |\mathbf{x}|^2$. The gravitational force exerted on this second object acts toward the origin, and the unit vector in this direction is

$$-\frac{\mathbf{x}}{|\mathbf{x}|}$$

Therefore the gravitational force acting on the object at $\mathbf{x} = \langle x, y, z \rangle$ is

$$\boxed{3} \qquad \mathbf{F}(\mathbf{x}) = -\frac{mMG}{|\mathbf{x}|^3}\,\mathbf{x}$$

[Physicists often use the notation $\mathbf{r}$ instead of $\mathbf{x}$ for the position vector, so you may see Formula 3 written in the form $\mathbf{F} = -(mMG/r^3)\mathbf{r}$.] The function given by Equation 3 is an example of a vector field, called the **gravitational field**, because it associates a vector [the force $\mathbf{F}(\mathbf{x})$] with every point $\mathbf{x}$ in space (except for the origin).

Formula 3 is a compact way of writing the gravitational field, but we can also write it in terms of its component functions by using the facts that $\mathbf{x} = x\,\mathbf{i} + y\,\mathbf{j} + z\,\mathbf{k}$ and $|\mathbf{x}| = \sqrt{x^2 + y^2 + z^2}$:

$$\mathbf{F}(x, y, z) = \frac{-mMGx}{(x^2 + y^2 + z^2)^{3/2}}\,\mathbf{i} + \frac{-mMGy}{(x^2 + y^2 + z^2)^{3/2}}\,\mathbf{j} + \frac{-mMGz}{(x^2 + y^2 + z^2)^{3/2}}\,\mathbf{k}$$

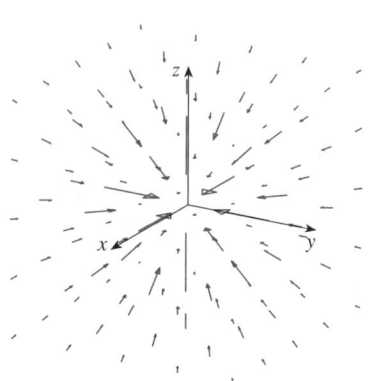

FIGURE 13
Gravitational force field

The gravitational field $\mathbf{F}$ is pictured in Figure 13. ∎

EXAMPLE 5 Suppose an electric charge Q is located at the origin. According to Coulomb's Law, the electric force $\mathbf{F}(\mathbf{x})$ exerted by this charge on a charge q located at a point (x, y, z) with position vector $\mathbf{x} = \langle x, y, z \rangle$ is

$$\boxed{4} \qquad \mathbf{F}(\mathbf{x}) = \frac{\varepsilon qQ}{|\mathbf{x}|^3}\,\mathbf{x}$$

where ε is a constant (that depends on the units used). For like charges, we have $qQ > 0$ and the force is repulsive; for unlike charges, we have $qQ < 0$ and the force is attractive. Notice the similarity between Formulas 3 and 4. Both vector fields are examples of **force fields**.

Instead of considering the electric force **F**, physicists often consider the force per unit charge:

$$\mathbf{E}(\mathbf{x}) = \frac{1}{q}\,\mathbf{F}(\mathbf{x}) = \frac{\varepsilon Q}{|\mathbf{x}|^3}\,\mathbf{x}$$

Then **E** is a vector field on $\mathbb{R}^3$ called the **electric field** of Q. ∎

GRADIENT FIELDS

If f is a scalar function of two variables, recall from Section 11.6 that its gradient ∇f (or grad f) is defined by

$$\nabla f(x, y) = f_x(x, y)\,\mathbf{i} + f_y(x, y)\,\mathbf{j}$$

Therefore ∇f is really a vector field on $\mathbb{R}^2$ and is called a **gradient vector field**. Likewise, if f is a scalar function of three variables, its gradient is a vector field on $\mathbb{R}^3$ given by

$$\nabla f(x, y, z) = f_x(x, y, z)\,\mathbf{i} + f_y(x, y, z)\,\mathbf{j} + f_z(x, y, z)\,\mathbf{k}$$

Ⅴ EXAMPLE 6 Find the gradient vector field of $f(x, y) = x^2 y - y^3$. Plot the gradient vector field together with a contour map of f. How are they related?

SOLUTION The gradient vector field is given by

$$\nabla f(x, y) = \frac{\partial f}{\partial x}\,\mathbf{i} + \frac{\partial f}{\partial y}\,\mathbf{j} = 2xy\,\mathbf{i} + (x^2 - 3y^2)\,\mathbf{j}$$

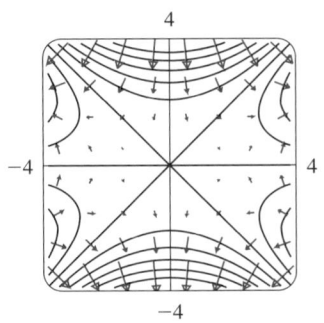

FIGURE 14

Figure 14 shows a contour map of f with the gradient vector field. Notice that the gradient vectors are perpendicular to the level curves, as we would expect from Section 11.6. Notice also that the gradient vectors are long where the level curves are close to each other and short where they are farther apart. That's because the length of the gradient vector is the value of the directional derivative of f and closely spaced level curves indicate a steep graph. ∎

A vector field **F** is called a **conservative vector field** if it is the gradient of some scalar function, that is, if there exists a function f such that $\mathbf{F} = \nabla f$. In this situation f is called a **potential function** for **F**.

Not all vector fields are conservative, but such fields do arise frequently in physics. For example, the gravitational field **F** in Example 4 is conservative because if we define

$$f(x, y, z) = \frac{mMG}{\sqrt{x^2 + y^2 + z^2}}$$

then

$$\nabla f(x, y, z) = \frac{\partial f}{\partial x}\,\mathbf{i} + \frac{\partial f}{\partial y}\,\mathbf{j} + \frac{\partial f}{\partial z}\,\mathbf{k}$$

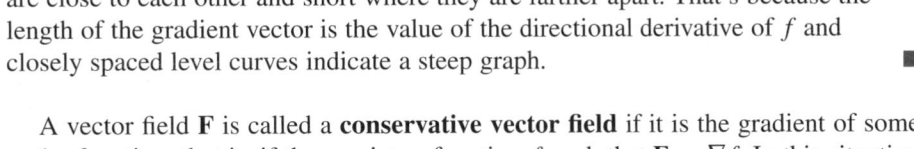

$$= \frac{-mMGx}{(x^2 + y^2 + z^2)^{3/2}}\,\mathbf{i} + \frac{-mMGy}{(x^2 + y^2 + z^2)^{3/2}}\,\mathbf{j} + \frac{-mMGz}{(x^2 + y^2 + z^2)^{3/2}}\,\mathbf{k}$$

$$= \mathbf{F}(x, y, z)$$

In Sections 13.3 and 13.5 we will learn how to tell whether or not a given vector field is conservative.

13.1 | EXERCISES

1–10 ▪ Sketch the vector field **F** by drawing a diagram like Figure 4 or Figure 8.

1. $\mathbf{F}(x, y) = 0.3\,\mathbf{i} - 0.4\,\mathbf{j}$ **2.** $\mathbf{F}(x, y) = \frac{1}{2}x\,\mathbf{i} + y\,\mathbf{j}$

3. $\mathbf{F}(x, y) = -\frac{1}{2}\,\mathbf{i} + (y - x)\,\mathbf{j}$

4. $\mathbf{F}(x, y) = y\,\mathbf{i} + (x + y)\,\mathbf{j}$

5. $\mathbf{F}(x, y) = \dfrac{y\,\mathbf{i} + x\,\mathbf{j}}{\sqrt{x^2 + y^2}}$

6. $\mathbf{F}(x, y) = \dfrac{y\,\mathbf{i} - x\,\mathbf{j}}{\sqrt{x^2 + y^2}}$

7. $\mathbf{F}(x, y, z) = \mathbf{k}$ **8.** $\mathbf{F}(x, y, z) = -y\,\mathbf{k}$

9. $\mathbf{F}(x, y, z) = x\,\mathbf{k}$

10. $\mathbf{F}(x, y, z) = \mathbf{j} - \mathbf{i}$

11–14 ▪ Match the vector fields **F** with the plots labeled I–IV. Give reasons for your choices.

11. $\mathbf{F}(x, y) = \langle x, -y \rangle$

12. $\mathbf{F}(x, y) = \langle y, x - y \rangle$

13. $\mathbf{F}(x, y) = \langle y, y + 2 \rangle$

14. $\mathbf{F}(x, y) = \langle \cos(x + y), x \rangle$

I
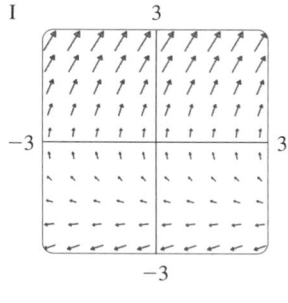

II
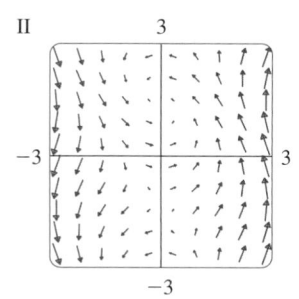

III
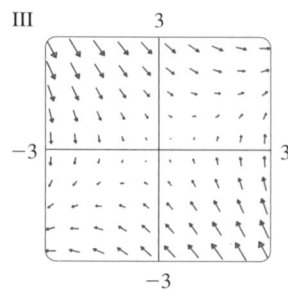

IV
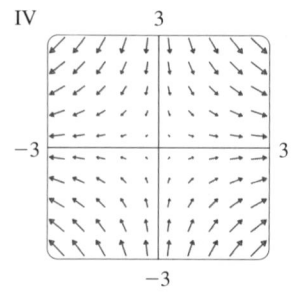

15–18 ▪ Match the vector fields **F** on $\mathbb{R}^3$ with the plots labeled I–IV. Give reasons for your choices.

15. $\mathbf{F}(x, y, z) = \mathbf{i} + 2\,\mathbf{j} + 3\,\mathbf{k}$

16. $\mathbf{F}(x, y, z) = \mathbf{i} + 2\,\mathbf{j} + z\,\mathbf{k}$

17. $\mathbf{F}(x, y, z) = x\,\mathbf{i} + y\,\mathbf{j} + 3\,\mathbf{k}$

18. $\mathbf{F}(x, y, z) = x\,\mathbf{i} + y\,\mathbf{j} + z\,\mathbf{k}$

I
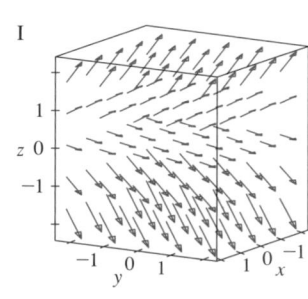

II
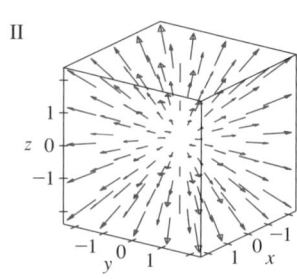

III
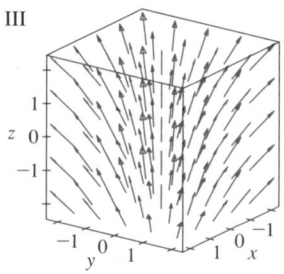

IV
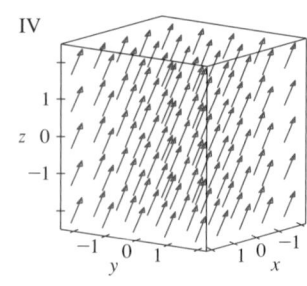

CAS **19.** If you have a CAS that plots vector fields (the command is `fieldplot` in Maple and `PlotVectorField` or `VectorPlot` in Mathematica), use it to plot

$$\mathbf{F}(x, y) = (y^2 - 2xy)\,\mathbf{i} + (3xy - 6x^2)\,\mathbf{j}$$

Explain the appearance by finding the set of points (x, y) such that $\mathbf{F}(x, y) = \mathbf{0}$.

CAS **20.** Let $\mathbf{F}(\mathbf{x}) = (r^2 - 2r)\mathbf{x}$, where $\mathbf{x} = \langle x, y \rangle$ and $r = |\mathbf{x}|$. Use a CAS to plot this vector field in various domains until you can see what is happening. Describe the appearance of the plot and explain it by finding the points where $\mathbf{F}(\mathbf{x}) = \mathbf{0}$.

21–24 ▪ Find the gradient vector field of f.

21. $f(x, y) = xe^{xy}$ **22.** $f(x, y) = \tan(3x - 4y)$

23. $f(x, y, z) = \sqrt{x^2 + y^2 + z^2}$

24. $f(x, y, z) = x \ln(y - 2z)$

⊞ Graphing calculator or computer required   **CAS** Computer algebra system required   **1.** Homework Hints at stewartcalculus.com

25–26 ▪ Find the gradient vector field ∇f of f and sketch it.

25. $f(x, y) = x^2 - y$ **26.** $f(x, y) = \sqrt{x^2 + y^2}$

CAS **27–28** ▪ Plot the gradient vector field of f together with a contour map of f. Explain how they are related to each other.

27. $f(x, y) = \ln(1 + x^2 + 2y^2)$

28. $f(x, y) = \cos x - 2 \sin y$

---

29. A particle moves in a velocity field $\mathbf{V}(x, y) = \langle x^2, x + y^2 \rangle$. If it is at position $(2, 1)$ at time $t = 3$, estimate its location at time $t = 3.01$.

30. At time $t = 1$, a particle is located at position $(1, 3)$. If it moves in a velocity field

$$\mathbf{F}(x, y) = \langle xy - 2, y^2 - 10 \rangle$$

find its approximate location at time $t = 1.05$.

31. The **flow lines** (or **streamlines**) of a vector field are the paths followed by a particle whose velocity field is the given vector field. Thus the vectors in a vector field are tangent to the flow lines.

(a) Use a sketch of the vector field $\mathbf{F}(x, y) = x\,\mathbf{i} - y\,\mathbf{j}$ to draw some flow lines. From your sketches, can you guess the equations of the flow lines?

(b) If parametric equations of a flow line are $x = x(t)$, $y = y(t)$, explain why these functions satisfy the differential equations $dx/dt = x$ and $dy/dt = -y$. Then solve the differential equations to find an equation of the flow line that passes through the point $(1, 1)$.

32. (a) Sketch the vector field $\mathbf{F}(x, y) = \mathbf{i} + x\,\mathbf{j}$ and then sketch some flow lines. What shape do these flow lines appear to have?

(b) If parametric equations of the flow lines are $x = x(t)$, $y = y(t)$, what differential equations do these functions satisfy? Deduce that $dy/dx = x$.

(c) If a particle starts at the origin in the velocity field given by $\mathbf{F}$, find an equation of the path it follows.

13.2 | LINE INTEGRALS

In this section we define an integral that is similar to a single integral except that instead of integrating over an interval $[a, b]$, we integrate over a curve C. Such integrals are called *line integrals*, although "curve integrals" would be better terminology. They were invented in the early 19th century to solve problems involving fluid flow, forces, electricity, and magnetism.

We start with a plane curve C given by the parametric equations

$$\boxed{1} \qquad\qquad x = x(t) \qquad y = y(t) \qquad a \le t \le b$$

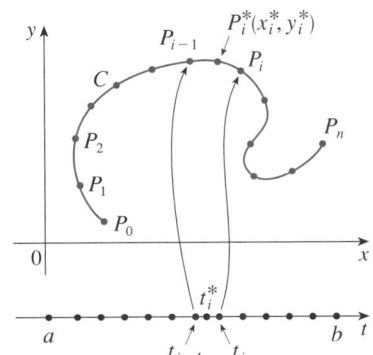

FIGURE 1

or, equivalently, by the vector equation $\mathbf{r}(t) = x(t)\,\mathbf{i} + y(t)\,\mathbf{j}$, and we assume that C is a smooth curve. [This means that $\mathbf{r}'$ is continuous and $\mathbf{r}'(t) \ne \mathbf{0}$. See Section 10.8.] If we divide the parameter interval $[a, b]$ into n subintervals $[t_{i-1}, t_i]$ and we let $x_i = x(t_i)$ and $y_i = y(t_i)$, then the corresponding points $P_i(x_i, y_i)$ divide C into n subarcs with lengths $\Delta s_1, \Delta s_2, \ldots, \Delta s_n$. (See Figure 1.) We choose any point $P_i^*(x_i^*, y_i^*)$ in the ith subarc. (This corresponds to a point t_i^* in $[t_{i-1}, t_i]$.) Now if f is any function of two variables whose domain includes the curve C, we evaluate f at the point (x_i^*, y_i^*), multiply by the length Δs_i of the subarc, and form the sum

$$\sum_{i=1}^{n} f(x_i^*, y_i^*)\,\Delta s_i$$

which is similar to a Riemann sum. Then we take the limit of these sums and make the following definition by analogy with a single integral.

> **2 DEFINITION** If f is defined on a smooth curve C given by Equations 1, then the **line integral of f along C** is
>
> $$\int_C f(x, y) \, ds = \lim_{\max \Delta s_i \to 0} \sum_{i=1}^{n} f(x_i^*, y_i^*) \, \Delta s_i$$
>
> if this limit exists.

In Section 9.2 we found that the length of C is

$$L = \int_a^b \sqrt{\left(\frac{dx}{dt}\right)^2 + \left(\frac{dy}{dt}\right)^2} \, dt$$

A similar type of argument can be used to show that if f is a continuous function, then the limit in Definition 2 always exists and the following formula can be used to evaluate the line integral:

> **3** $$\int_C f(x, y) \, ds = \int_a^b f\big(x(t), y(t)\big) \sqrt{\left(\frac{dx}{dt}\right)^2 + \left(\frac{dy}{dt}\right)^2} \, dt$$

The value of the line integral does not depend on the parametrization of the curve, provided that the curve is traversed exactly once as t increases from a to b.

If $s(t)$ is the length of C between $\mathbf{r}(a)$ and $\mathbf{r}(t)$, then

$$\frac{ds}{dt} = \sqrt{\left(\frac{dx}{dt}\right)^2 + \left(\frac{dy}{dt}\right)^2}$$

■ The arc length function s is discussed in Section 10.8.

So the way to remember Formula 3 is to express everything in terms of the parameter t: Use the parametric equations to express x and y in terms of t and write ds as

$$ds = \sqrt{\left(\frac{dx}{dt}\right)^2 + \left(\frac{dy}{dt}\right)^2} \, dt$$

In the special case where C is the line segment that joins $(a, 0)$ to $(b, 0)$, using x as the parameter, we can write the parametric equations of C as follows: $x = x$, $y = 0$, $a \le x \le b$. Formula 3 then becomes

$$\int_C f(x, y) \, ds = \int_a^b f(x, 0) \, dx$$

and so the line integral reduces to an ordinary single integral in this case.

Just as for an ordinary single integral, we can interpret the line integral of a *positive* function as an area. In fact, if $f(x, y) \ge 0$, $\int_C f(x, y) \, ds$ represents the area of one side of the "fence" or "curtain" in Figure 2, whose base is C and whose height above the point (x, y) is $f(x, y)$.

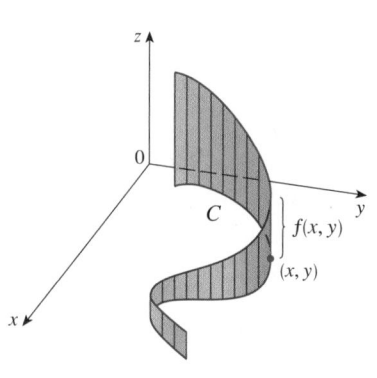

FIGURE 2

EXAMPLE 1 Evaluate $\int_C (2 + x^2 y) \, ds$, where C is the upper half of the unit circle $x^2 + y^2 = 1$.

SOLUTION In order to use Formula 3 we first need parametric equations to represent C. Recall that the unit circle can be parametrized by means of the equations

$$x = \cos t \qquad y = \sin t$$

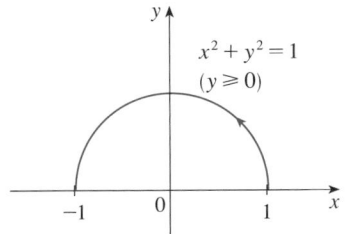

FIGURE 3

and the upper half of the circle is described by the parameter interval $0 \leqslant t \leqslant \pi$. (See Figure 3.) Therefore Formula 3 gives

$$\int_C (2 + x^2 y) \, ds = \int_0^\pi (2 + \cos^2 t \sin t) \sqrt{\left(\frac{dx}{dt}\right)^2 + \left(\frac{dy}{dt}\right)^2} \, dt$$

$$= \int_0^\pi (2 + \cos^2 t \sin t) \sqrt{\sin^2 t + \cos^2 t} \, dt$$

$$= \int_0^\pi (2 + \cos^2 t \sin t) \, dt = \left[2t - \frac{\cos^3 t}{3} \right]_0^\pi$$

$$= 2\pi + \tfrac{2}{3} \qquad \blacksquare$$

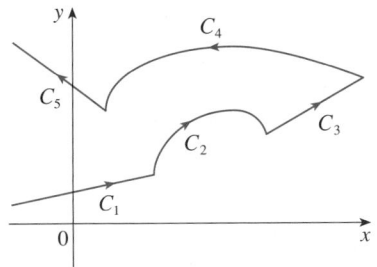

FIGURE 4

A piecewise-smooth curve

Suppose now that C is a **piecewise-smooth curve**; that is, C is a union of a finite number of smooth curves $C_1, C_2, \ldots, C_n$, where, as illustrated in Figure 4, the initial point of C_{i+1} is the terminal point of C_i. Then we define the integral of f along C as the sum of the integrals of f along each of the smooth pieces of C:

$$\int_C f(x, y) \, ds = \int_{C_1} f(x, y) \, ds + \int_{C_2} f(x, y) \, ds + \cdots + \int_{C_n} f(x, y) \, ds$$

EXAMPLE 2 Evaluate $\int_C 2x \, ds$, where C consists of the arc C_1 of the parabola $y = x^2$ from $(0, 0)$ to $(1, 1)$ followed by the vertical line segment C_2 from $(1, 1)$ to $(1, 2)$.

SOLUTION The curve C is shown in Figure 5. C_1 is the graph of a function of x, so we can choose x as the parameter and the equations for C_1 become

$$x = x \qquad y = x^2 \qquad 0 \leqslant x \leqslant 1$$

Therefore

$$\int_{C_1} 2x \, ds = \int_0^1 2x \sqrt{\left(\frac{dx}{dx}\right)^2 + \left(\frac{dy}{dx}\right)^2} \, dx = \int_0^1 2x \sqrt{1 + 4x^2} \, dx$$

$$= \tfrac{1}{4} \cdot \tfrac{2}{3} (1 + 4x^2)^{3/2} \big]_0^1 = \frac{5\sqrt{5} - 1}{6}$$

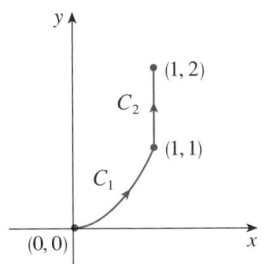

FIGURE 5

$C = C_1 \cup C_2$

On C_2 we choose y as the parameter, so the equations of C_2 are

$$x = 1 \qquad y = y \qquad 1 \leqslant y \leqslant 2$$

and

$$\int_{C_2} 2x \, ds = \int_1^2 2(1) \sqrt{\left(\frac{dx}{dy}\right)^2 + \left(\frac{dy}{dy}\right)^2} \, dy = \int_1^2 2 \, dy = 2$$

Thus

$$\int_C 2x \, ds = \int_{C_1} 2x \, ds + \int_{C_2} 2x \, ds = \frac{5\sqrt{5} - 1}{6} + 2 \qquad \blacksquare$$

Any physical interpretation of a line integral $\int_C f(x, y) \, ds$ depends on the physical interpretation of the function f. Suppose that $\rho(x, y)$ represents the linear density at a point (x, y) of a thin wire shaped like a curve C. Then the mass of the part of the wire from P_{i-1} to P_i in Figure 1 is approximately $\rho(x_i^*, y_i^*) \, \Delta s_i$ and so the total mass of the wire is approximately $\Sigma \, \rho(x_i^*, y_i^*) \, \Delta s_i$. By taking more and more points on the curve, we obtain the **mass** m of the wire as the limiting value of these approximations:

$$m = \lim_{\max \Delta s_i \to 0} \sum_{i=1}^{n} \rho(x_i^*, y_i^*) \, \Delta s_i = \int_C \rho(x, y) \, ds$$

[For example, if $f(x, y) = 2 + x^2 y$ represents the density of a semicircular wire, then the integral in Example 1 would represent the mass of the wire.] The **center of mass** of the wire with density function ρ is located at the point $(\bar{x}, \bar{y})$, where

$$\boxed{4} \qquad \bar{x} = \frac{1}{m} \int_C x \rho(x, y) \, ds \qquad \bar{y} = \frac{1}{m} \int_C y \rho(x, y) \, ds$$

Other physical interpretations of line integrals will be discussed later in this chapter.

▼ **EXAMPLE 3** A wire takes the shape of the semicircle $x^2 + y^2 = 1$, $y \geqslant 0$, and is thicker near its base than near the top. Find the center of mass of the wire if the linear density at any point is proportional to its distance from the line $y = 1$.

SOLUTION As in Example 1 we use the parametrization $x = \cos t$, $y = \sin t$, $0 \leqslant t \leqslant \pi$, and find that $ds = dt$. The linear density is

$$\rho(x, y) = k(1 - y)$$

where k is a constant, and so the mass of the wire is

$$m = \int_C k(1 - y) \, ds = \int_0^{\pi} k(1 - \sin t) \, dt = k\big[t + \cos t\big]_0^{\pi} = k(\pi - 2)$$

From Equations 4 we have

$$\bar{y} = \frac{1}{m} \int_C y \rho(x, y) \, ds = \frac{1}{k(\pi - 2)} \int_C y k(1 - y) \, ds = \frac{1}{\pi - 2} \int_0^{\pi} (\sin t - \sin^2 t) \, dt$$

$$= \frac{1}{\pi - 2} \Big[-\cos t - \tfrac{1}{2} t + \tfrac{1}{4} \sin 2t \Big]_0^{\pi} = \frac{4 - \pi}{2(\pi - 2)}$$

By symmetry we see that $\bar{x} = 0$, so the center of mass is

$$\left(0, \, \frac{4 - \pi}{2(\pi - 2)} \right) \approx (0, 0.38)$$

See Figure 6.   ■

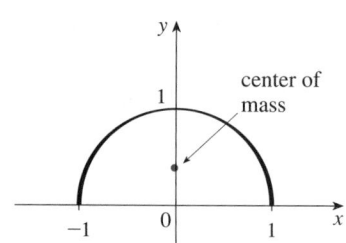

FIGURE 6

Two other line integrals are obtained by replacing Δs_i by either $\Delta x_i = x_i - x_{i-1}$ or $\Delta y_i = y_i - y_{i-1}$ in Definition 2. They are called the **line integrals of f along C with respect to x and y**:

$$\boxed{5} \qquad \int_C f(x, y) \, dx = \lim_{\max \Delta x_i \to 0} \sum_{i=1}^{n} f(x_i^*, y_i^*) \, \Delta x_i$$

$$\boxed{6} \qquad \int_C f(x, y) \, dy = \lim_{\max \Delta y_i \to 0} \sum_{i=1}^{n} f(x_i^*, y_i^*) \, \Delta y_i$$

When we want to distinguish the original line integral $\int_C f(x, y) \, ds$ from those in Equations 5 and 6, we call it the **line integral with respect to arc length**.

The following formulas say that line integrals with respect to x and y can also be evaluated by expressing everything in terms of t: $x = x(t)$, $y = y(t)$, $dx = x'(t)\,dt$, $dy = y'(t)\,dt$.

7

$$\int_C f(x, y)\,dx = \int_a^b f\big(x(t), y(t)\big)\,x'(t)\,dt$$

$$\int_C f(x, y)\,dy = \int_a^b f\big(x(t), y(t)\big)\,y'(t)\,dt$$

It frequently happens that line integrals with respect to x and y occur together. When this happens, it's customary to abbreviate by writing

$$\int_C P(x, y)\,dx + \int_C Q(x, y)\,dy = \int_C P(x, y)\,dx + Q(x, y)\,dy$$

When we are setting up a line integral, sometimes the most difficult thing is to think of a parametric representation for a curve whose geometric description is given. In particular, we often need to parametrize a line segment, so it's useful to remember that a vector representation of the line segment that starts at $\mathbf{r}_0$ and ends at $\mathbf{r}_1$ is given by

8

$$\mathbf{r}(t) = (1 - t)\mathbf{r}_0 + t\,\mathbf{r}_1 \qquad 0 \leqslant t \leqslant 1$$

(See Equation 10.5.4.)

▼ **EXAMPLE 4** Evaluate $\int_C y^2\,dx + x\,dy$, where (a) $C = C_1$ is the line segment from $(-5, -3)$ to $(0, 2)$ and (b) $C = C_2$ is the arc of the parabola $x = 4 - y^2$ from $(-5, -3)$ to $(0, 2)$. (See Figure 7.)

SOLUTION
(a) A parametric representation for the line segment is

$$x = 5t - 5 \qquad y = 5t - 3 \qquad 0 \leqslant t \leqslant 1$$

(Use Equation 8 with $\mathbf{r}_0 = \langle -5, -3 \rangle$ and $\mathbf{r}_1 = \langle 0, 2 \rangle$.) Then $dx = 5\,dt$, $dy = 5\,dt$, and Formulas 7 give

$$\int_{C_1} y^2\,dx + x\,dy = \int_0^1 (5t - 3)^2(5\,dt) + (5t - 5)(5\,dt)$$

$$= 5\int_0^1 (25t^2 - 25t + 4)\,dt$$

$$= 5\left[\frac{25t^3}{3} - \frac{25t^2}{2} + 4t\right]_0^1 = -\frac{5}{6}$$

(b) Since the parabola is given as a function of y, let's take y as the parameter and write C_2 as

$$x = 4 - y^2 \qquad y = y \qquad -3 \leqslant y \leqslant 2$$

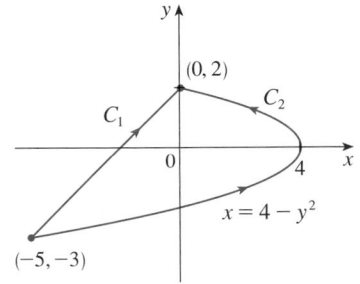

$x = 4 - y^2$

FIGURE 7

Then $dx = -2y\,dy$ and by Formulas 7 we have

$$\int_{C_2} y^2\,dx + x\,dy = \int_{-3}^{2} y^2(-2y)\,dy + (4 - y^2)\,dy = \int_{-3}^{2} (-2y^3 - y^2 + 4)\,dy$$

$$= \left[-\frac{y^4}{2} - \frac{y^3}{3} + 4y \right]_{-3}^{2} = 40\tfrac{5}{6} \qquad \blacksquare$$

Notice that we got different answers in parts (a) and (b) of Example 4 even though the two curves had the same endpoints. Thus, in general, the value of a line integral depends not just on the endpoints of the curve but also on the path. (But see Section 13.3 for conditions under which the integral is independent of the path.)

Notice also that the answers in Example 4 depend on the direction, or orientation, of the curve. If $-C_1$ denotes the line segment from $(0, 2)$ to $(-5, -3)$, then using the parametrization

$$x = -5t \qquad y = 2 - 5t \qquad 0 \le t \le 1$$

you can verify that

$$\int_{-C_1} y^2\,dx + x\,dy = \tfrac{5}{6}$$

In general, a given parametrization $x = x(t)$, $y = y(t)$, $a \le t \le b$, determines an **orientation** of a curve C, with the positive direction corresponding to increasing values of the parameter t. (See Figure 8, where the initial point A corresponds to the parameter value a and the terminal point B corresponds to $t = b$.)

If $-C$ denotes the curve consisting of the same points as C but with the opposite orientation (from initial point B to terminal point A in Figure 8), then we have

$$\int_{-C} f(x, y)\,dx = -\int_{C} f(x, y)\,dx \qquad \int_{-C} f(x, y)\,dy = -\int_{C} f(x, y)\,dy$$

But if we integrate with respect to arc length, the value of the line integral does *not* change when we reverse the orientation of the curve:

$$\int_{-C} f(x, y)\,ds = \int_{C} f(x, y)\,ds$$

This is because Δs_i is always positive, whereas Δx_i and Δy_i change sign when we reverse the orientation of C.

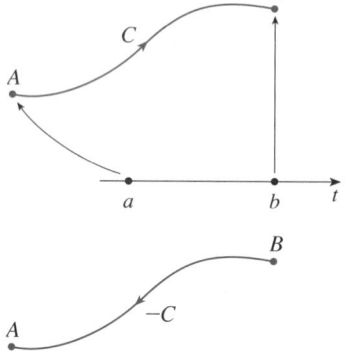

FIGURE 8

LINE INTEGRALS IN SPACE

We now suppose that C is a smooth space curve given by the parametric equations

$$x = x(t) \qquad y = y(t) \qquad z = z(t) \qquad a \le t \le b$$

or by a vector equation $\mathbf{r}(t) = x(t)\,\mathbf{i} + y(t)\,\mathbf{j} + z(t)\,\mathbf{k}$. If f is a function of three variables that is continuous on some region containing C, then we define the **line integral of f along C** (with respect to arc length) in a manner similar to that for plane curves:

$$\int_{C} f(x, y, z)\,ds = \lim_{\max \Delta s_i \to 0} \sum_{i=1}^{n} f(x_i^*, y_i^*, z_i^*)\,\Delta s_i$$

We evaluate it using a formula similar to Formula 3:

$$\boxed{9} \quad \int_C f(x, y, z)\, ds = \int_a^b f\big(x(t), y(t), z(t)\big) \sqrt{\left(\frac{dx}{dt}\right)^2 + \left(\frac{dy}{dt}\right)^2 + \left(\frac{dz}{dt}\right)^2}\, dt$$

Observe that the integrals in both Formulas 3 and 9 can be written in the more compact vector notation

$$\int_a^b f(\mathbf{r}(t))\, |\,\mathbf{r}'(t)\,|\, dt$$

For the special case $f(x, y, z) = 1$, we get

$$\int_C ds = \int_a^b |\,\mathbf{r}'(t)\,|\, dt = L$$

where L is the length of the curve C (see Formula 10.8.3).

Line integrals along C with respect to x, y, and z can also be defined. For example,

$$\int_C f(x, y, z)\, dz = \lim_{\max \Delta z_i \to 0} \sum_{i=1}^n f(x_i^*, y_i^*, z_i^*)\, \Delta z_i = \int_a^b f\big(x(t), y(t), z(t)\big) z'(t)\, dt$$

Therefore, as with line integrals in the plane, we evaluate integrals of the form

$$\boxed{10} \qquad \int_C P(x, y, z)\, dx + Q(x, y, z)\, dy + R(x, y, z)\, dz$$

by expressing everything (x, y, z, dx, dy, dz) in terms of the parameter t.

▼ EXAMPLE 5 Evaluate $\int_C y \sin z\, ds$, where C is the circular helix given by the equations $x = \cos t$, $y = \sin t$, $z = t$, $0 \leqslant t \leqslant 2\pi$. (See Figure 9.)

SOLUTION Formula 9 gives

$$\int_C y \sin z\, ds = \int_0^{2\pi} (\sin t) \sin t \sqrt{\left(\frac{dx}{dt}\right)^2 + \left(\frac{dy}{dt}\right)^2 + \left(\frac{dz}{dt}\right)^2}\, dt$$

$$= \int_0^{2\pi} \sin^2 t \sqrt{\sin^2 t + \cos^2 t + 1}\, dt = \sqrt{2} \int_0^{2\pi} \tfrac{1}{2}(1 - \cos 2t)\, dt$$

$$= \frac{\sqrt{2}}{2}\left[t - \tfrac{1}{2} \sin 2t \right]_0^{2\pi} = \sqrt{2}\,\pi \qquad\blacksquare$$

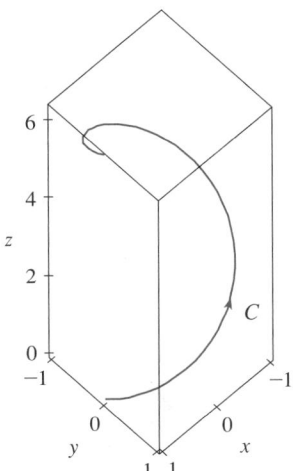

FIGURE 9

EXAMPLE 6 Evaluate $\int_C y\, dx + z\, dy + x\, dz$, where C consists of the line segment C_1 from $(2, 0, 0)$ to $(3, 4, 5)$ followed by the vertical line segment C_2 from $(3, 4, 5)$ to $(3, 4, 0)$.

SOLUTION The curve C is shown in Figure 10. Using Equation 8, we write C_1 as

$$\mathbf{r}(t) = (1 - t)\langle 2, 0, 0 \rangle + t\langle 3, 4, 5 \rangle = \langle 2 + t, 4t, 5t \rangle$$

or, in parametric form, as

$$x = 2 + t \qquad y = 4t \qquad z = 5t \qquad 0 \leqslant t \leqslant 1$$

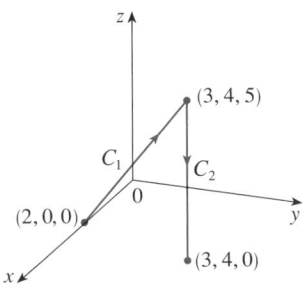

FIGURE 10

Thus

$$\int_{C_1} y\,dx + z\,dy + x\,dz = \int_0^1 (4t)\,dt + (5t)4\,dt + (2+t)5\,dt$$

$$= \int_0^1 (10 + 29t)\,dt = 10t + 29\,\frac{t^2}{2}\Big]_0^1 = 24.5$$

Likewise, C_2 can be written in the form

$$\mathbf{r}(t) = (1-t)\langle 3, 4, 5\rangle + t\langle 3, 4, 0\rangle = \langle 3, 4, 5-5t\rangle$$

or $\qquad\qquad x = 3 \qquad y = 4 \qquad z = 5 - 5t \qquad 0 \leqslant t \leqslant 1$

Then $dx = 0 = dy$, so

$$\int_{C_2} y\,dx + z\,dy + x\,dz = \int_0^1 3(-5)\,dt = -15$$

Adding the values of these integrals, we obtain

$$\int_C y\,dx + z\,dy + x\,dz = 24.5 - 15 = 9.5 \qquad\blacksquare$$

LINE INTEGRALS OF VECTOR FIELDS

Recall from Section 7.6 that the work done by a variable force $f(x)$ in moving a particle from a to b along the x-axis is $W = \int_a^b f(x)\,dx$. Then in Section 10.3 we found that the work done by a constant force $\mathbf{F}$ in moving an object from a point P to another point Q in space is $W = \mathbf{F} \cdot \mathbf{D}$, where $\mathbf{D} = \overrightarrow{PQ}$ is the displacement vector.

Now suppose that $\mathbf{F} = P\,\mathbf{i} + Q\,\mathbf{j} + R\,\mathbf{k}$ is a continuous force field on $\mathbb{R}^3$, such as the gravitational field of Example 4 in Section 13.1 or the electric force field of Example 5 in Section 13.1. (A force field on $\mathbb{R}^2$ could be regarded as a special case where $R = 0$ and P and Q depend only on x and y.) We wish to compute the work done by this force in moving a particle along a smooth curve C.

We divide C into subarcs $P_{i-1}P_i$ with lengths Δs_i by dividing the parameter interval $[a, b]$ into subintervals. (See Figure 1 for the two-dimensional case or Figure 11 for the three-dimensional case.) Choose a point $P_i^*(x_i^*, y_i^*, z_i^*)$ on the ith subarc corresponding to the parameter value t_i^*. If Δs_i is small, then as the particle moves from P_{i-1} to P_i along the curve, it proceeds approximately in the direction of $\mathbf{T}(t_i^*)$, the unit tangent vector at P_i^*. Thus the work done by the force $\mathbf{F}$ in moving the particle from P_{i-1} to P_i is approximately

$$\mathbf{F}(x_i^*, y_i^*, z_i^*) \cdot [\Delta s_i\,\mathbf{T}(t_i^*)] = [\mathbf{F}(x_i^*, y_i^*, z_i^*) \cdot \mathbf{T}(t_i^*)]\,\Delta s_i$$

and the total work done in moving the particle along C is approximately

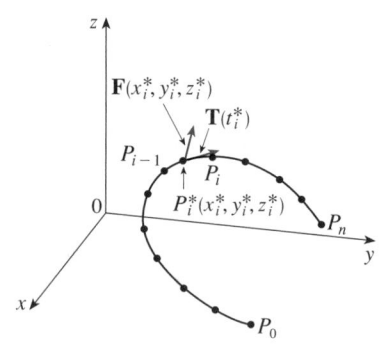

FIGURE 11

$$\boxed{11} \qquad \sum_{i=1}^n [\mathbf{F}(x_i^*, y_i^*, z_i^*) \cdot \mathbf{T}(x_i^*, y_i^*, z_i^*)]\,\Delta s_i$$

where $\mathbf{T}(x, y, z)$ is the unit tangent vector at the point (x, y, z) on C. Intuitively, we see that these approximations ought to become better as the subarcs become shorter. Therefore we define the **work** W done by the force field $\mathbf{F}$ as the limit of the Riemann sums in $\boxed{11}$, namely,

$$\boxed{12} \qquad W = \int_C \mathbf{F}(x, y, z) \cdot \mathbf{T}(x, y, z)\, ds = \int_C \mathbf{F} \cdot \mathbf{T}\, ds$$

Equation 12 says that *work is the line integral with respect to arc length of the tangential component of the force.*

If the curve C is given by the vector equation $\mathbf{r}(t) = x(t)\,\mathbf{i} + y(t)\,\mathbf{j} + z(t)\,\mathbf{k}$, then $\mathbf{T}(t) = \mathbf{r}'(t)/|\mathbf{r}'(t)|$, so using Equation 9 we can rewrite Equation 12 in the form

$$W = \int_a^b \left[\mathbf{F}(\mathbf{r}(t)) \cdot \frac{\mathbf{r}'(t)}{|\mathbf{r}'(t)|} \right] |\mathbf{r}'(t)|\, dt = \int_a^b \mathbf{F}(\mathbf{r}(t)) \cdot \mathbf{r}'(t)\, dt$$

This integral is often abbreviated as $\int_C \mathbf{F} \cdot d\mathbf{r}$ and occurs in other areas of physics as well. Therefore we make the following definition for the line integral of *any* continuous vector field.

$\boxed{13}$ **DEFINITION** Let $\mathbf{F}$ be a continuous vector field defined on a smooth curve C given by a vector function $\mathbf{r}(t)$, $a \leqslant t \leqslant b$. Then the **line integral of $\mathbf{F}$ along C** is

$$\int_C \mathbf{F} \cdot d\mathbf{r} = \int_a^b \mathbf{F}(\mathbf{r}(t)) \cdot \mathbf{r}'(t)\, dt = \int_C \mathbf{F} \cdot \mathbf{T}\, ds$$

When using Definition 13, remember that $\mathbf{F}(\mathbf{r}(t))$ is just an abbreviation for $\mathbf{F}(x(t), y(t), z(t))$, so we evaluate $\mathbf{F}(\mathbf{r}(t))$ simply by putting $x = x(t)$, $y = y(t)$, and $z = z(t)$ in the expression for $\mathbf{F}(x, y, z)$. Notice also that we can formally write $d\mathbf{r} = \mathbf{r}'(t)\, dt$.

■ Figure 12 shows the force field and the curve in Example 7. The work done is negative because the field impedes movement along the curve.

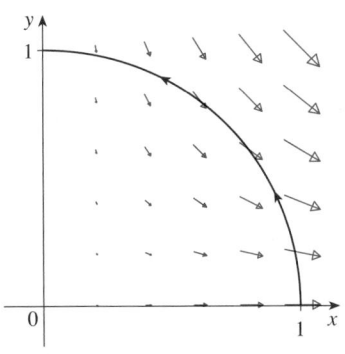

FIGURE 12

EXAMPLE 7 Find the work done by the force field $\mathbf{F}(x, y) = x^2\,\mathbf{i} - xy\,\mathbf{j}$ in moving a particle along the quarter-circle $\mathbf{r}(t) = \cos t\,\mathbf{i} + \sin t\,\mathbf{j}$, $0 \leqslant t \leqslant \pi/2$.

SOLUTION Since $x = \cos t$ and $y = \sin t$, we have

$$\mathbf{F}(\mathbf{r}(t)) = \cos^2 t\,\mathbf{i} - \cos t \sin t\,\mathbf{j}$$

and

$$\mathbf{r}'(t) = -\sin t\,\mathbf{i} + \cos t\,\mathbf{j}$$

Therefore the work done is

$$\int_C \mathbf{F} \cdot d\mathbf{r} = \int_0^{\pi/2} \mathbf{F}(\mathbf{r}(t)) \cdot \mathbf{r}'(t)\, dt = \int_0^{\pi/2} (-2\cos^2 t \sin t)\, dt$$

$$= 2\,\frac{\cos^3 t}{3} \bigg]_0^{\pi/2} = -\frac{2}{3} \qquad ■$$

NOTE Even though $\int_C \mathbf{F} \cdot d\mathbf{r} = \int_C \mathbf{F} \cdot \mathbf{T}\, ds$ and integrals with respect to arc length are unchanged when orientation is reversed, it is still true that

$$\int_{-C} \mathbf{F} \cdot d\mathbf{r} = -\int_C \mathbf{F} \cdot d\mathbf{r}$$

because the unit tangent vector $\mathbf{T}$ is replaced by its negative when C is replaced by $-C$.

■ Figure 13 shows the twisted cubic C in Example 8 and some typical vectors acting at three points on C.

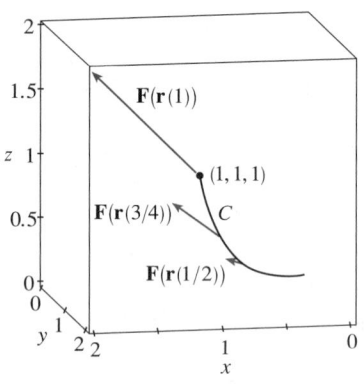

FIGURE 13

EXAMPLE 8 Evaluate $\int_C \mathbf{F} \cdot d\mathbf{r}$, where $\mathbf{F}(x, y, z) = xy\,\mathbf{i} + yz\,\mathbf{j} + zx\,\mathbf{k}$ and C is the twisted cubic given by

$$x = t \qquad y = t^2 \qquad z = t^3 \qquad 0 \le t \le 1$$

SOLUTION We have

$$\mathbf{r}(t) = t\,\mathbf{i} + t^2\,\mathbf{j} + t^3\,\mathbf{k}$$

$$\mathbf{r}'(t) = \mathbf{i} + 2t\,\mathbf{j} + 3t^2\,\mathbf{k}$$

$$\mathbf{F}(\mathbf{r}(t)) = t^3\,\mathbf{i} + t^5\,\mathbf{j} + t^4\,\mathbf{k}$$

Thus

$$\int_C \mathbf{F} \cdot d\mathbf{r} = \int_0^1 \mathbf{F}(\mathbf{r}(t)) \cdot \mathbf{r}'(t)\,dt$$

$$= \int_0^1 (t^3 + 5t^6)\,dt = \frac{t^4}{4} + \frac{5t^7}{7} \Big]_0^1 = \frac{27}{28} \qquad ■$$

Finally, we note the connection between line integrals of vector fields and line integrals of scalar fields. Suppose the vector field $\mathbf{F}$ on $\mathbb{R}^3$ is given in component form by the equation $\mathbf{F} = P\,\mathbf{i} + Q\,\mathbf{j} + R\,\mathbf{k}$. We use Definition 13 to compute its line integral along C:

$$\int_C \mathbf{F} \cdot d\mathbf{r} = \int_a^b \mathbf{F}(\mathbf{r}(t)) \cdot \mathbf{r}'(t)\,dt$$

$$= \int_a^b (P\,\mathbf{i} + Q\,\mathbf{j} + R\,\mathbf{k}) \cdot (x'(t)\,\mathbf{i} + y'(t)\,\mathbf{j} + z'(t)\,\mathbf{k})\,dt$$

$$= \int_a^b \left[P\big(x(t), y(t), z(t)\big)x'(t) + Q\big(x(t), y(t), z(t)\big)y'(t) + R\big(x(t), y(t), z(t)\big)z'(t) \right] dt$$

But this last integral is precisely the line integral in $\boxed{10}$. Therefore we have

$$\int_C \mathbf{F} \cdot d\mathbf{r} = \int_C P\,dx + Q\,dy + R\,dz \qquad \text{where } \mathbf{F} = P\,\mathbf{i} + Q\,\mathbf{j} + R\,\mathbf{k}$$

For example, the integral $\int_C y\,dx + z\,dy + x\,dz$ in Example 6 could be expressed as $\int_C \mathbf{F} \cdot d\mathbf{r}$ where

$$\mathbf{F}(x, y, z) = y\,\mathbf{i} + z\,\mathbf{j} + x\,\mathbf{k}$$

13.2 | EXERCISES

1–16 ■ Evaluate the line integral, where C is the given curve.

1. $\int_C y^3\,ds,\quad C: x = t^3,\ y = t,\ 0 \le t \le 2$

2. $\int_C xy\,ds,\quad C: x = t^2,\ y = 2t,\ 0 \le t \le 1$

3. $\int_C xy^4\,ds,\quad C$ is the right half of the circle $x^2 + y^2 = 16$

4. $\int_C x \sin y\,ds,\quad C$ is the line segment from $(0, 3)$ to $(4, 6)$

5. $\int_C \left(x^2 y^3 - \sqrt{x} \right) dy,$
C is the arc of the curve $y = \sqrt{x}$ from $(1, 1)$ to $(4, 2)$

6. $\int_C e^x\,dx,$
C is the arc of the curve $x = y^3$ from $(-1, -1)$ to $(1, 1)$

7. $\int_C (x + 2y)\, dx + x^2\, dy$, C consists of line segments from $(0, 0)$ to $(2, 1)$ and from $(2, 1)$ to $(3, 0)$

8. $\int_C x^2\, dx + y^2\, dy$, C consists of the arc of the circle $x^2 + y^2 = 4$ from $(2, 0)$ to $(0, 2)$ followed by the line segment from $(0, 2)$ to $(4, 3)$

9. $\int_C xyz\, ds$,
$C: x = 2 \sin t,\ y = t,\ z = -2 \cos t,\ 0 \le t \le \pi$

10. $\int_C xyz^2\, ds$,
C is the line segment from $(-1, 5, 0)$ to $(1, 6, 4)$

11. $\int_C xe^{yz}\, ds$,
C is the line segment from $(0, 0, 0)$ to $(1, 2, 3)$

12. $\int_C (x^2 + y^2 + z^2)\, ds$,
$C: x = t,\ y = \cos 2t,\ z = \sin 2t,\ 0 \le t \le 2\pi$

13. $\int_C xye^{yz}\, dy$,
$C: x = t,\ y = t^2,\ z = t^3,\ 0 \le t \le 1$

14. $\int_C y\, dx + z\, dy + x\, dz$,
$C: x = \sqrt{t},\ y = t,\ z = t^2,\ 1 \le t \le 4$

15. $\int_C z^2\, dx + x^2\, dy + y^2\, dz$,
C is the line segment from $(1, 0, 0)$ to $(4, 1, 2)$

16. $\int_C (y + z)\, dx + (x + z)\, dy + (x + y)\, dz$,
C consists of line segments from $(0, 0, 0)$ to $(1, 0, 1)$ and from $(1, 0, 1)$ to $(0, 1, 2)$

17. Let $\mathbf{F}$ be the vector field shown in the figure.
 (a) If C_1 is the vertical line segment from $(-3, -3)$ to $(-3, 3)$, determine whether $\int_{C_1} \mathbf{F} \cdot d\mathbf{r}$ is positive, negative, or zero.
 (b) If C_2 is the counterclockwise-oriented circle with radius 3 and center the origin, determine whether $\int_{C_2} \mathbf{F} \cdot d\mathbf{r}$ is positive, negative, or zero.

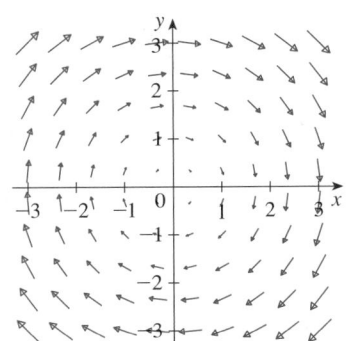

18. The figure shows a vector field $\mathbf{F}$ and two curves C_1 and C_2. Are the line integrals of $\mathbf{F}$ over C_1 and C_2 positive, negative, or zero? Explain.

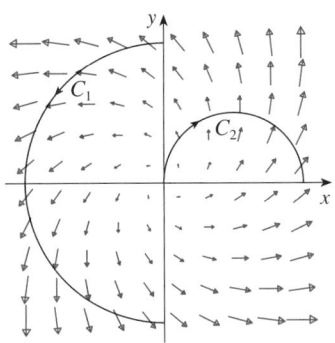

19–22 ■ Evaluate the line integral $\int_C \mathbf{F} \cdot d\mathbf{r}$, where C is given by the vector function $\mathbf{r}(t)$.

19. $\mathbf{F}(x, y) = xy\, \mathbf{i} + 3y^2\, \mathbf{j}$,
$\mathbf{r}(t) = 11t^4\, \mathbf{i} + t^3\, \mathbf{j},\quad 0 \le t \le 1$

20. $\mathbf{F}(x, y, z) = (x + y)\, \mathbf{i} + (y - z)\, \mathbf{j} + z^2\, \mathbf{k}$,
$\mathbf{r}(t) = t^2\, \mathbf{i} + t^3\, \mathbf{j} + t^2\, \mathbf{k},\quad 0 \le t \le 1$

21. $\mathbf{F}(x, y, z) = \sin x\, \mathbf{i} + \cos y\, \mathbf{j} + xz\, \mathbf{k}$,
$\mathbf{r}(t) = t^3\, \mathbf{i} - t^2\, \mathbf{j} + t\, \mathbf{k},\quad 0 \le t \le 1$

22. $\mathbf{F}(x, y, z) = x\, \mathbf{i} + y\, \mathbf{j} + xy\, \mathbf{k}$,
$\mathbf{r}(t) = \cos t\, \mathbf{i} + \sin t\, \mathbf{j} + t\, \mathbf{k},\quad 0 \le t \le \pi$

23–24 ■ Use a calculator or CAS to evaluate the line integral correct to four decimal places.

23. $\int_C \mathbf{F} \cdot d\mathbf{r}$, where $\mathbf{F}(x, y) = xy\, \mathbf{i} + \sin y\, \mathbf{j}$ and
$\mathbf{r}(t) = e^t\, \mathbf{i} + e^{-t^2}\, \mathbf{j},\ 1 \le t \le 2$

24. $\int_C ze^{-xy}\, ds$, where C has parametric equations $x = t$, $y = t^2,\ z = e^{-t},\ 0 \le t \le 1$

CAS 25–26 ■ Use a graph of the vector field $\mathbf{F}$ and the curve C to guess whether the line integral of $\mathbf{F}$ over C is positive, negative, or zero. Then evaluate the line integral.

25. $\mathbf{F}(x, y) = (x - y)\, \mathbf{i} + xy\, \mathbf{j}$,
C is the arc of the circle $x^2 + y^2 = 4$ traversed counterclockwise from $(2, 0)$ to $(0, -2)$

26. $\mathbf{F}(x, y) = \dfrac{x}{\sqrt{x^2 + y^2}}\, \mathbf{i} + \dfrac{y}{\sqrt{x^2 + y^2}}\, \mathbf{j}$,
C is the parabola $y = 1 + x^2$ from $(-1, 2)$ to $(1, 2)$

27. (a) Evaluate the line integral $\int_C \mathbf{F} \cdot d\mathbf{r}$, where $\mathbf{F}(x, y) = e^{x-1}\, \mathbf{i} + xy\, \mathbf{j}$ and C is given by $\mathbf{r}(t) = t^2\, \mathbf{i} + t^3\, \mathbf{j},\ 0 \le t \le 1$.

(b) Illustrate part (a) by using a graphing calculator or computer to graph C and the vectors from the vector field corresponding to $t = 0$, $1/\sqrt{2}$, and 1 (as in Figure 13).

28. (a) Evaluate the line integral $\int_C \mathbf{F} \cdot d\mathbf{r}$, where $\mathbf{F}(x, y, z) = x\,\mathbf{i} - z\,\mathbf{j} + y\,\mathbf{k}$ and C is given by $\mathbf{r}(t) = 2t\,\mathbf{i} + 3t\,\mathbf{j} - t^2\,\mathbf{k}$, $-1 \le t \le 1$.

(b) Illustrate part (a) by using a computer to graph C and the vectors from the vector field corresponding to $t = \pm 1$ and $\pm\frac{1}{2}$ (as in Figure 13).

CAS **29.** Find the exact value of $\int_C x^3 y^5\,ds$, where C is the part of the astroid $x = \cos^3 t$, $y = \sin^3 t$ in the first quadrant.

30. (a) Find the work done by the force field $\mathbf{F}(x, y) = x^2\,\mathbf{i} + xy\,\mathbf{j}$ on a particle that moves once around the circle $x^2 + y^2 = 4$ oriented in the counterclockwise direction.

CAS (b) Use a computer algebra system to graph the force field and circle on the same screen. Use the graph to explain your answer to part (a).

31. A thin wire is bent into the shape of a semicircle $x^2 + y^2 = 4$, $x \ge 0$. If the linear density is a constant k, find the mass and center of mass of the wire.

32. A thin wire has the shape of the first-quadrant part of the circle with center the origin and radius a. If the density function is $\rho(x, y) = kxy$, find the mass and center of mass of the wire.

33. (a) Write the formulas similar to Equations 4 for the center of mass $(\bar{x}, \bar{y}, \bar{z})$ of a thin wire in the shape of a space curve C if the wire has density function $\rho(x, y, z)$.

(b) Find the center of mass of a wire in the shape of the helix $x = 2\sin t$, $y = 2\cos t$, $z = 3t$, $0 \le t \le 2\pi$, if the density is a constant k.

34. Find the mass and center of mass of a wire in the shape of the helix $x = t$, $y = \cos t$, $z = \sin t$, $0 \le t \le 2\pi$, if the density at any point is equal to the square of the distance from the origin.

35. If a wire with linear density $\rho(x, y)$ lies along a plane curve C, its **moments of inertia** about the x- and y-axes are defined as

$$I_x = \int_C y^2 \rho(x, y)\,ds \qquad I_y = \int_C x^2 \rho(x, y)\,ds$$

Find the moments of inertia for the wire in Example 3.

36. If a wire with linear density $\rho(x, y, z)$ lies along a space curve C, its **moments of inertia** about the x-, y-, and z-axes are defined as

$$I_x = \int_C (y^2 + z^2)\rho(x, y, z)\,ds$$

$$I_y = \int_C (x^2 + z^2)\rho(x, y, z)\,ds$$

$$I_z = \int_C (x^2 + y^2)\rho(x, y, z)\,ds$$

Find the moments of inertia for the wire in Exercise 33.

37. Find the work done by the force field $\mathbf{F}(x, y) = x\,\mathbf{i} + (y + 2)\,\mathbf{j}$ in moving an object along an arch of the cycloid $\mathbf{r}(t) = (t - \sin t)\,\mathbf{i} + (1 - \cos t)\,\mathbf{j}$, $0 \le t \le 2\pi$.

38. Find the work done by the force field $\mathbf{F}(x, y) = x^2\,\mathbf{i} + ye^x\,\mathbf{j}$ on a particle that moves along the parabola $x = y^2 + 1$ from $(1, 0)$ to $(2, 1)$.

39. Find the work done by the force field $\mathbf{F}(x, y, z) = \langle x - y^2, y - z^2, z - x^2 \rangle$ on a particle that moves along the line segment from $(0, 0, 1)$ to $(2, 1, 0)$.

40. The force exerted by an electric charge at the origin on a charged particle at a point (x, y, z) with position vector $\mathbf{r} = \langle x, y, z \rangle$ is $\mathbf{F}(\mathbf{r}) = K\mathbf{r}/|\mathbf{r}|^3$ where K is a constant. (See Example 5 in Section 13.1.) Find the work done as the particle moves along a straight line from $(2, 0, 0)$ to $(2, 1, 5)$.

41. The position of an object with mass m at time t is $\mathbf{r}(t) = at^2\,\mathbf{i} + bt^3\,\mathbf{j}$, $0 \le t \le 1$.

(a) What is the force acting on the object at time t?

(b) What is the work done by the force during the time interval $0 \le t \le 1$?

42. An object with mass m moves with position function

$$\mathbf{r}(t) = a\sin t\,\mathbf{i} + b\cos t\,\mathbf{j} + ct\,\mathbf{k} \quad 0 \le t \le \pi/2$$

Find the work done on the object during this time period.

43. A 160-lb man carries a 25-lb can of paint up a helical staircase that encircles a silo with a radius of 20 ft. If the silo is 90 ft high and the man makes exactly three complete revolutions climbing to the top, how much work is done by the man against gravity?

44. Suppose there is a hole in the can of paint in Exercise 43 and 9 lb of paint leaks steadily out of the can during the man's ascent. How much work is done?

45. If C is a smooth curve given by a vector function $\mathbf{r}(t)$, $a \le t \le b$, and $\mathbf{v}$ is a constant vector, show that

$$\int_C \mathbf{v} \cdot d\mathbf{r} = \mathbf{v} \cdot [\mathbf{r}(b) - \mathbf{r}(a)]$$

46. If C is a smooth curve given by a vector function $\mathbf{r}(t)$, $a \le t \le b$, show that

$$\int_C \mathbf{r} \cdot d\mathbf{r} = \tfrac{1}{2}\big[|\mathbf{r}(b)|^2 - |\mathbf{r}(a)|^2\big]$$

47. (a) Show that a constant force field does zero work on a particle that moves once uniformly around the circle $x^2 + y^2 = 1$.

(b) Is this also true for a force field $\mathbf{F}(\mathbf{x}) = k\mathbf{x}$, where k is a constant and $\mathbf{x} = \langle x, y \rangle$?

48. Experiments show that a steady current I in a long wire produces a magnetic field $\mathbf{B}$ that is tangent to any circle that lies in the plane perpendicular to the wire and whose center is the axis of the wire (as in the figure at the right). *Ampère's Law* relates the electric current to its magnetic effects and states that

$$\int_C \mathbf{B} \cdot d\mathbf{r} = \mu_0 I$$

where I is the net current that passes through any surface bounded by a closed curve C, and μ_0 is a constant called the permeability of free space. By taking C to be a circle with radius r, show that the magnitude $B = |\mathbf{B}|$ of the magnetic field at a distance r from the center of the wire is

$$B = \frac{\mu_0 I}{2\pi r}$$

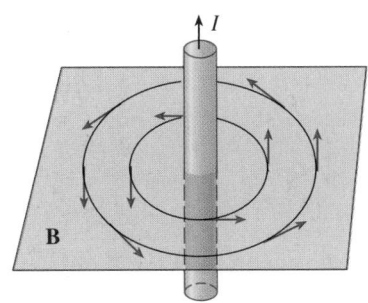

13.3 | THE FUNDAMENTAL THEOREM FOR LINE INTEGRALS

Recall from Section 5.4 that Part 2 of the Fundamental Theorem of Calculus can be written as

$$\boxed{1} \qquad \int_a^b F'(x)\, dx = F(b) - F(a)$$

where F' is continuous on $[a, b]$. We also called Equation 1 the Net Change Theorem: The integral of a rate of change is the net change.

If we think of the gradient vector ∇f of a function f of two or three variables as a sort of derivative of f, then the following theorem can be regarded as a version of the Fundamental Theorem for line integrals.

$\boxed{2}$ **THEOREM**   Let C be a smooth curve given by the vector function $\mathbf{r}(t)$, $a \leqslant t \leqslant b$. Let f be a differentiable function of two or three variables whose gradient vector ∇f is continuous on C. Then

$$\int_C \nabla f \cdot d\mathbf{r} = f(\mathbf{r}(b)) - f(\mathbf{r}(a))$$

NOTE   Theorem 2 says that we can evaluate the line integral of a conservative vector field (the gradient vector field of the potential function f) simply by knowing the value of f at the endpoints of C. In fact, Theorem 2 says that the line integral of ∇f is the net change in f. If f is a function of two variables and C is a plane curve with initial point $A(x_1, y_1)$ and terminal point $B(x_2, y_2)$, as in Figure 1, then Theorem 2 becomes

$$\int_C \nabla f \cdot d\mathbf{r} = f(x_2, y_2) - f(x_1, y_1)$$

If f is a function of three variables and C is a space curve joining the point $A(x_1, y_1, z_1)$ to the point $B(x_2, y_2, z_2)$, then we have

$$\int_C \nabla f \cdot d\mathbf{r} = f(x_2, y_2, z_2) - f(x_1, y_1, z_1)$$

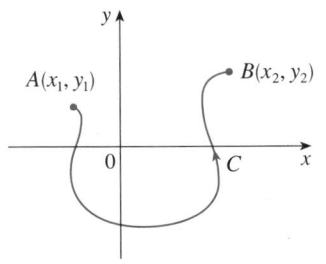

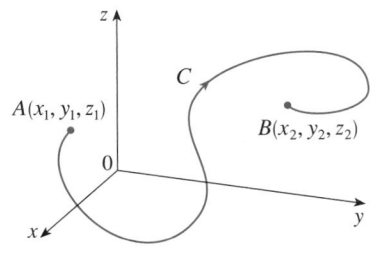

FIGURE 1

Let's prove Theorem 2 for this case.

PROOF OF THEOREM 2 Using Definition 13.2.13, we have

$$\int_C \nabla f \cdot d\mathbf{r} = \int_a^b \nabla f(\mathbf{r}(t)) \cdot \mathbf{r}'(t) \, dt$$

$$= \int_a^b \left(\frac{\partial f}{\partial x} \frac{dx}{dt} + \frac{\partial f}{\partial y} \frac{dy}{dt} + \frac{\partial f}{\partial z} \frac{dz}{dt} \right) dt$$

$$= \int_a^b \frac{d}{dt} f(\mathbf{r}(t)) \, dt \quad \text{(by the Chain Rule)}$$

$$= f(\mathbf{r}(b)) - f(\mathbf{r}(a))$$

The last step follows from the Fundamental Theorem of Calculus (Equation 1). $\square$

Although we have proved Theorem 2 for smooth curves, it is also true for piecewise-smooth curves. This can be seen by subdividing C into a finite number of smooth curves and adding the resulting integrals.

EXAMPLE 1 Find the work done by the gravitational field

$$\mathbf{F}(\mathbf{x}) = -\frac{mMG}{|\mathbf{x}|^3} \mathbf{x}$$

in moving a particle with mass m from the point $(3, 4, 12)$ to the point $(2, 2, 0)$ along a piecewise-smooth curve C. (See Example 4 in Section 13.1.)

SOLUTION From Section 13.1 we know that $\mathbf{F}$ is a conservative vector field and, in fact, $\mathbf{F} = \nabla f$, where

$$f(x, y, z) = \frac{mMG}{\sqrt{x^2 + y^2 + z^2}}$$

Therefore, by Theorem 2, the work done is

$$W = \int_C \mathbf{F} \cdot d\mathbf{r} = \int_C \nabla f \cdot d\mathbf{r}$$

$$= f(2, 2, 0) - f(3, 4, 12)$$

$$= \frac{mMG}{\sqrt{2^2 + 2^2}} - \frac{mMG}{\sqrt{3^2 + 4^2 + 12^2}} = mMG\left(\frac{1}{2\sqrt{2}} - \frac{1}{13} \right) \quad \blacksquare$$

INDEPENDENCE OF PATH

Suppose C_1 and C_2 are two piecewise-smooth curves (which are called **paths**) that have the same initial point A and terminal point B. We know from Example 4 in Section 13.2 that, in general, $\int_{C_1} \mathbf{F} \cdot d\mathbf{r} \neq \int_{C_2} \mathbf{F} \cdot d\mathbf{r}$. But one implication of Theorem 2 is that

$$\int_{C_1} \nabla f \cdot d\mathbf{r} = \int_{C_2} \nabla f \cdot d\mathbf{r}$$

whenever ∇f is continuous. In other words, the line integral of a *conservative* vector field depends only on the initial point and terminal point of a curve.

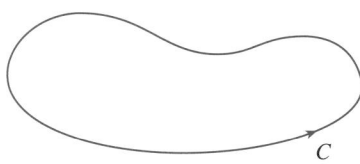

FIGURE 2
A closed curve

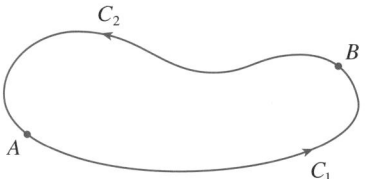

FIGURE 3

In general, if **F** is a continuous vector field with domain D, we say that the line integral $\int_C \mathbf{F} \cdot d\mathbf{r}$ is **independent of path** if $\int_{C_1} \mathbf{F} \cdot d\mathbf{r} = \int_{C_2} \mathbf{F} \cdot d\mathbf{r}$ for any two paths C_1 and C_2 in D that have the same initial and terminal points. With this terminology we can say that *line integrals of conservative vector fields are independent of path.*

A curve is called **closed** if its terminal point coincides with its initial point, that is, $\mathbf{r}(b) = \mathbf{r}(a)$. (See Figure 2.) If $\int_C \mathbf{F} \cdot d\mathbf{r}$ is independent of path in D and C is any closed path in D, we can choose any two points A and B on C and regard C as being composed of the path C_1 from A to B followed by the path C_2 from B to A. (See Figure 3.) Then

$$\int_C \mathbf{F} \cdot d\mathbf{r} = \int_{C_1} \mathbf{F} \cdot d\mathbf{r} + \int_{C_2} \mathbf{F} \cdot d\mathbf{r} = \int_{C_1} \mathbf{F} \cdot d\mathbf{r} - \int_{-C_2} \mathbf{F} \cdot d\mathbf{r} = 0$$

since C_1 and $-C_2$ have the same initial and terminal points.

Conversely, if it is true that $\int_C \mathbf{F} \cdot d\mathbf{r} = 0$ whenever C is a closed path in D, then we demonstrate independence of path as follows. Take any two paths C_1 and C_2 from A to B in D and define C to be the curve consisting of C_1 followed by $-C_2$. Then

$$0 = \int_C \mathbf{F} \cdot d\mathbf{r} = \int_{C_1} \mathbf{F} \cdot d\mathbf{r} + \int_{-C_2} \mathbf{F} \cdot d\mathbf{r} = \int_{C_1} \mathbf{F} \cdot d\mathbf{r} - \int_{C_2} \mathbf{F} \cdot d\mathbf{r}$$

and so $\int_{C_1} \mathbf{F} \cdot d\mathbf{r} = \int_{C_2} \mathbf{F} \cdot d\mathbf{r}$. Thus we have proved the following theorem.

---

3   THEOREM $\int_C \mathbf{F} \cdot d\mathbf{r}$ is independent of path in D if and only if $\int_C \mathbf{F} \cdot d\mathbf{r} = 0$ for every closed path C in D.

Since we know that the line integral of any conservative vector field **F** is independent of path, it follows that $\int_C \mathbf{F} \cdot d\mathbf{r} = 0$ for any closed path. The physical interpretation is that the work done by a conservative force field (such as the gravitational or electric field in Section 13.1) as it moves an object around a closed path is 0.

The following theorem says that the *only* vector fields that are independent of path are conservative. It is stated and proved for plane curves, but there is a similar version for space curves. We assume that D is **open**, which means that for every point P in D there is a disk with center P that lies entirely in D. (So D doesn't contain any of its boundary points.) In addition, we assume that D is **connected**. This means that any two points in D can be joined by a path that lies in D.

---

4   THEOREM   Suppose **F** is a vector field that is continuous on an open connected region D. If $\int_C \mathbf{F} \cdot d\mathbf{r}$ is independent of path in D, then **F** is a conservative vector field on D; that is, there exists a function f such that $\nabla f = \mathbf{F}$.

PROOF   Let $A(a, b)$ be a fixed point in D. We construct the desired potential function f by defining

$$f(x, y) = \int_{(a, b)}^{(x, y)} \mathbf{F} \cdot d\mathbf{r}$$

for any point (x, y) in D. Since $\int_C \mathbf{F} \cdot d\mathbf{r}$ is independent of path, it does not matter which path C from (a, b) to (x, y) is used to evaluate $f(x, y)$. Since D is open, there

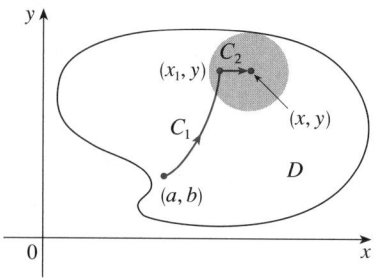

FIGURE 4

exists a disk contained in D with center (x, y). Choose any point (x_1, y) in the disk with $x_1 < x$ and let C consist of any path C_1 from (a, b) to (x_1, y) followed by the horizontal line segment C_2 from (x_1, y) to (x, y). (See Figure 4.) Then

$$f(x, y) = \int_{C_1} \mathbf{F} \cdot d\mathbf{r} + \int_{C_2} \mathbf{F} \cdot d\mathbf{r} = \int_{(a, b)}^{(x_1, y)} \mathbf{F} \cdot d\mathbf{r} + \int_{C_2} \mathbf{F} \cdot d\mathbf{r}$$

Notice that the first of these integrals does not depend on x, so

$$\frac{\partial}{\partial x} f(x, y) = 0 + \frac{\partial}{\partial x} \int_{C_2} \mathbf{F} \cdot d\mathbf{r}$$

If we write $\mathbf{F} = P\,\mathbf{i} + Q\,\mathbf{j}$, then

$$\int_{C_2} \mathbf{F} \cdot d\mathbf{r} = \int_{C_2} P\,dx + Q\,dy$$

On C_2, y is constant, so $dy = 0$. Using t as the parameter, where $x_1 \leq t \leq x$, we have

$$\frac{\partial}{\partial x} f(x, y) = \frac{\partial}{\partial x} \int_{C_2} P\,dx + Q\,dy$$

$$= \frac{\partial}{\partial x} \int_{x_1}^{x} P(t, y)\,dt = P(x, y)$$

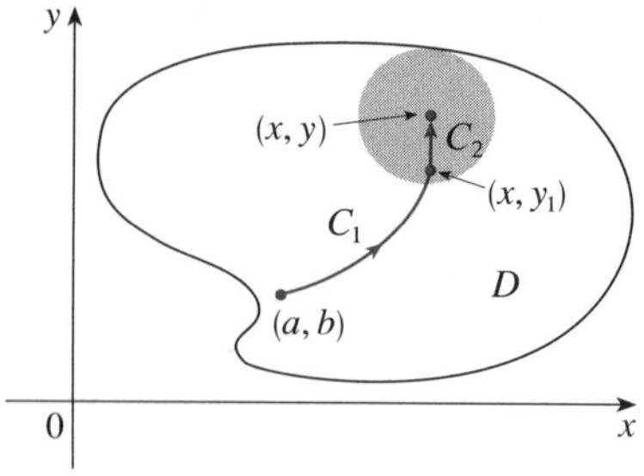

by Part 1 of the Fundamental Theorem of Calculus (see Section 5.4). A similar argument, using a vertical line segment (see Figure 5), shows that

$$\frac{\partial}{\partial y} f(x, y) = \frac{\partial}{\partial y} \int_{C_2} P\,dx + Q\,dy = \frac{\partial}{\partial y} \int_{y_1}^{y} Q(x, t)\,dt = Q(x, y)$$

Thus
$$\mathbf{F} = P\,\mathbf{i} + Q\,\mathbf{j} = \frac{\partial f}{\partial x}\,\mathbf{i} + \frac{\partial f}{\partial y}\,\mathbf{j} = \nabla f$$

FIGURE 5

which says that $\mathbf{F}$ is conservative. ☐

The question remains: How is it possible to determine whether or not a vector field $\mathbf{F}$ is conservative? Suppose it is known that $\mathbf{F} = P\,\mathbf{i} + Q\,\mathbf{j}$ is conservative, where P and Q have continuous first-order partial derivatives. Then there is a function f such that $\mathbf{F} = \nabla f$, that is,

$$P = \frac{\partial f}{\partial x} \qquad \text{and} \qquad Q = \frac{\partial f}{\partial y}$$

Therefore, by Clairaut's Theorem,

$$\frac{\partial P}{\partial y} = \frac{\partial^2 f}{\partial y\,\partial x} = \frac{\partial^2 f}{\partial x\,\partial y} = \frac{\partial Q}{\partial x}$$

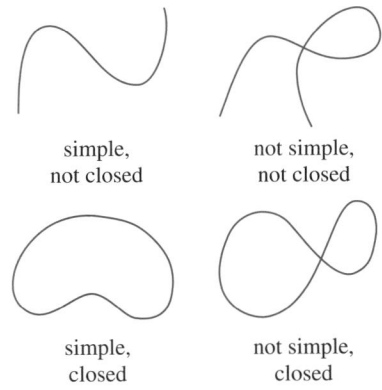

simple,
not closed

not simple,
not closed

simple,
closed

not simple,
closed

FIGURE 6

Types of curves

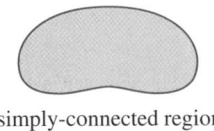

simply-connected region

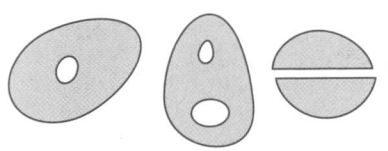

regions that are not simply-connected

FIGURE 7

5 **THEOREM**   If $\mathbf{F}(x, y) = P(x, y)\,\mathbf{i} + Q(x, y)\,\mathbf{j}$ is a conservative vector field, where P and Q have continuous first-order partial derivatives on a domain D, then throughout D we have

$$\frac{\partial P}{\partial y} = \frac{\partial Q}{\partial x}$$

The converse of Theorem 5 is true only for a special type of region. To explain this, we first need the concept of a **simple curve**, which is a curve that doesn't intersect itself anywhere between its endpoints. [See Figure 6; $\mathbf{r}(a) = \mathbf{r}(b)$ for a simple closed curve, but $\mathbf{r}(t_1) \neq \mathbf{r}(t_2)$ when $a < t_1 < t_2 < b$.]

In Theorem 4 we needed an open connected region. For the next theorem we need a stronger condition. A **simply-connected region** in the plane is a connected region D such that every simple closed curve in D encloses only points that are in D. Notice from Figure 7 that, intuitively speaking, a simply-connected region contains no hole and can't consist of two separate pieces.

In terms of simply-connected regions we can now state a partial converse to Theorem 5 that gives a convenient method for verifying that a vector field on $\mathbb{R}^2$ is conservative. The proof will be sketched in the next section as a consequence of Green's Theorem.

6 **THEOREM**   Let $\mathbf{F} = P\,\mathbf{i} + Q\,\mathbf{j}$ be a vector field on an open simply-connected region D. Suppose that P and Q have continuous first-order derivatives and

$$\frac{\partial P}{\partial y} = \frac{\partial Q}{\partial x} \qquad \text{throughout } D$$

Then $\mathbf{F}$ is conservative.

■ Figure 8 shows the vector field in Example 2. The vectors that start on the closed curve C all appear to point in roughly the same direction as C. So it looks as if $\int_C \mathbf{F} \cdot d\mathbf{r} > 0$ and therefore $\mathbf{F}$ is not conservative. The calculation in Example 2 confirms this impression.

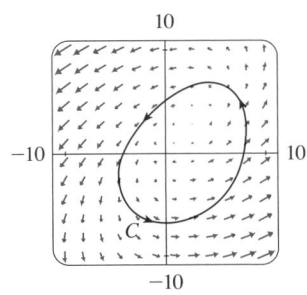

FIGURE 8

V **EXAMPLE 2**   Determine whether or not the vector field

$$\mathbf{F}(x, y) = (x - y)\,\mathbf{i} + (x - 2)\,\mathbf{j}$$

is conservative.

SOLUTION   Let $P(x, y) = x - y$ and $Q(x, y) = x - 2$. Then

$$\frac{\partial P}{\partial y} = -1 \qquad \frac{\partial Q}{\partial x} = 1$$

Since $\partial P/\partial y \neq \partial Q/\partial x$, $\mathbf{F}$ is not conservative by Theorem 5. ∎

V **EXAMPLE 3**   Determine whether or not the vector field

$$\mathbf{F}(x, y) = (3 + 2xy)\,\mathbf{i} + (x^2 - 3y^2)\,\mathbf{j}$$

is conservative.

SOLUTION   Let $P(x, y) = 3 + 2xy$ and $Q(x, y) = x^2 - 3y^2$. Then

$$\frac{\partial P}{\partial y} = 2x = \frac{\partial Q}{\partial x}$$

■ Figure 9 shows the vector field in Example 3. Some of the vectors near the curves C_1 and C_2 point in approximately the same direction as the curves, whereas others point in the opposite direction. So it appears plausible that line integrals around all closed paths are 0. Example 3 shows that **F** is indeed conservative.

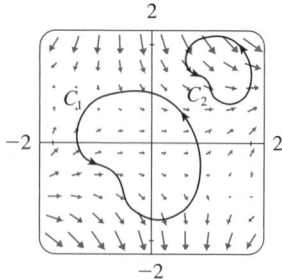

FIGURE 9

Also, the domain of **F** is the entire plane ($D = \mathbb{R}^2$), which is open and simply-connected. Therefore we can apply Theorem 6 and conclude that **F** is conservative. ■

In Example 3, Theorem 6 told us that **F** is conservative, but it did not tell us how to find the (potential) function f such that $\mathbf{F} = \nabla f$. The proof of Theorem 4 gives us a clue as to how to find f. We use "partial integration" as in the following example.

EXAMPLE 4

(a) If $\mathbf{F}(x, y) = (3 + 2xy)\,\mathbf{i} + (x^2 - 3y^2)\,\mathbf{j}$, find a function f such that $\mathbf{F} = \nabla f$.

(b) Evaluate the line integral $\int_C \mathbf{F} \cdot d\mathbf{r}$, where C is the curve given by $\mathbf{r}(t) = e^t \sin t\,\mathbf{i} + e^t \cos t\,\mathbf{j}$, $0 \le t \le \pi$.

SOLUTION

(a) From Example 3 we know that **F** is conservative and so there exists a function f with $\nabla f = \mathbf{F}$, that is,

$$\boxed{7} \qquad f_x(x, y) = 3 + 2xy$$

$$\boxed{8} \qquad f_y(x, y) = x^2 - 3y^2$$

Integrating $\boxed{7}$ with respect to x, we obtain

$$\boxed{9} \qquad f(x, y) = 3x + x^2 y + g(y)$$

Notice that the constant of integration is a constant with respect to x, that is, a function of y, which we have called $g(y)$. Next we differentiate both sides of $\boxed{9}$ with respect to y:

$$\boxed{10} \qquad f_y(x, y) = x^2 + g'(y)$$

Comparing $\boxed{8}$ and $\boxed{10}$, we see that

$$g'(y) = -3y^2$$

Integrating with respect to y, we have

$$g(y) = -y^3 + K$$

where K is a constant. Putting this in $\boxed{9}$, we have

$$f(x, y) = 3x + x^2 y - y^3 + K$$

as the desired potential function.

(b) To use Theorem 2 all we have to know are the initial and terminal points of C, namely, $\mathbf{r}(0) = (0, 1)$ and $\mathbf{r}(\pi) = (0, -e^\pi)$. In the expression for $f(x, y)$ in part (a), any value of the constant K will do, so let's choose $K = 0$. Then we have

$$\int_C \mathbf{F} \cdot d\mathbf{r} = \int_C \nabla f \cdot d\mathbf{r} = f(0, -e^\pi) - f(0, 1) = e^{3\pi} - (-1) = e^{3\pi} + 1$$

This method is much shorter than the straightforward method for evaluating line integrals that we learned in Section 13.2. ■

A criterion for determining whether or not a vector field **F** on $\mathbb{R}^3$ is conservative is given in Section 13.5. Meanwhile, the next example shows that the technique for finding the potential function is much the same as for vector fields on $\mathbb{R}^2$.

V EXAMPLE 5 If $\mathbf{F}(x, y, z) = y^2\,\mathbf{i} + (2xy + e^{3z})\,\mathbf{j} + 3ye^{3z}\,\mathbf{k}$, find a function f such that $\nabla f = \mathbf{F}$.

SOLUTION If there is such a function f, then

$$\boxed{11} \qquad\qquad f_x(x, y, z) = y^2$$

$$\boxed{12} \qquad\qquad f_y(x, y, z) = 2xy + e^{3z}$$

$$\boxed{13} \qquad\qquad f_z(x, y, z) = 3ye^{3z}$$

Integrating $\boxed{11}$ with respect to x, we get

$$\boxed{14} \qquad\qquad f(x, y, z) = xy^2 + g(y, z)$$

where $g(y, z)$ is a constant with respect to x. Then differentiating $\boxed{14}$ with respect to y, we have

$$f_y(x, y, z) = 2xy + g_y(y, z)$$

and comparison with $\boxed{12}$ gives

$$g_y(y, z) = e^{3z}$$

Thus $g(y, z) = ye^{3z} + h(z)$ and we rewrite $\boxed{14}$ as

$$f(x, y, z) = xy^2 + ye^{3z} + h(z)$$

Finally, differentiating with respect to z and comparing with $\boxed{13}$, we obtain $h'(z) = 0$ and therefore $h(z) = K$, a constant. The desired function is

$$f(x, y, z) = xy^2 + ye^{3z} + K$$

It is easily verified that $\nabla f = \mathbf{F}$. ∎

CONSERVATION OF ENERGY

Let's apply the ideas of this chapter to a continuous force field $\mathbf{F}$ that moves an object along a path C given by $\mathbf{r}(t)$, $a \leqslant t \leqslant b$, where $\mathbf{r}(a) = A$ is the initial point and $\mathbf{r}(b) = B$ is the terminal point of C. According to Newton's Second Law of Motion (see Section 10.9), the force $\mathbf{F}(\mathbf{r}(t))$ at a point on C is related to the acceleration $\mathbf{a}(t) = \mathbf{r}''(t)$ by the equation

$$\mathbf{F}(\mathbf{r}(t)) = m\mathbf{r}''(t)$$

So the work done by the force on the object is

$$W = \int_C \mathbf{F} \cdot d\mathbf{r} = \int_a^b \mathbf{F}(\mathbf{r}(t)) \cdot \mathbf{r}'(t)\,dt = \int_a^b m\mathbf{r}''(t) \cdot \mathbf{r}'(t)\,dt$$

$$= \frac{m}{2} \int_a^b \frac{d}{dt}\left[\mathbf{r}'(t) \cdot \mathbf{r}'(t)\right]dt \qquad\qquad \text{(Theorem 10.7.5, Formula 4)}$$

$$= \frac{m}{2} \int_a^b \frac{d}{dt}\,|\mathbf{r}'(t)|^2\,dt = \frac{m}{2}\left[|\mathbf{r}'(t)|^2\right]_a^b \qquad \text{(Fundamental Theorem of Calculus)}$$

$$= \frac{m}{2}\left(|\mathbf{r}'(b)|^2 - |\mathbf{r}'(a)|^2\right)$$

Therefore

$$15 \qquad W = \tfrac{1}{2} m \, | \, \mathbf{v}(b) \, |^2 - \tfrac{1}{2} m \, | \, \mathbf{v}(a) \, |^2$$

where $\mathbf{v} = \mathbf{r}'$ is the velocity.

The quantity $\tfrac{1}{2} m \, | \, \mathbf{v}(t) \, |^2$, that is, half the mass times the square of the speed, is called the **kinetic energy** of the object. Therefore we can rewrite Equation 15 as

$$16 \qquad W = K(B) - K(A)$$

which says that the work done by the force field along C is equal to the change in kinetic energy at the endpoints of C.

Now let's further assume that $\mathbf{F}$ is a conservative force field; that is, we can write $\mathbf{F} = \nabla f$. In physics, the **potential energy** of an object at the point (x, y, z) is defined as $P(x, y, z) = -f(x, y, z)$, so we have $\mathbf{F} = -\nabla P$. Then by Theorem 2 we have

$$W = \int_C \mathbf{F} \cdot d\mathbf{r} = -\int_C \nabla P \cdot d\mathbf{r} = -[P(\mathbf{r}(b)) - P(\mathbf{r}(a))] = P(A) - P(B)$$

Comparing this equation with Equation 16, we see that

$$P(A) + K(A) = P(B) + K(B)$$

which says that if an object moves from one point A to another point B under the influence of a conservative force field, then the sum of its potential energy and its kinetic energy remains constant. This is called the **Law of Conservation of Energy** and it is the reason the vector field is called *conservative*.

13.3 | EXERCISES

1. The figure shows a curve C and a contour map of a function f whose gradient is continuous. Find $\int_C \nabla f \cdot d\mathbf{r}$.

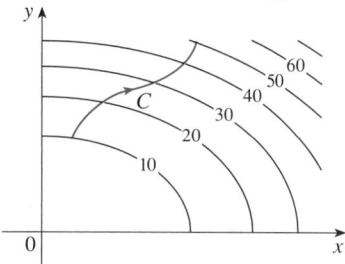

2. A table of values of a function f with continuous gradient is given. Find $\int_C \nabla f \cdot d\mathbf{r}$, where C has parametric equations

$$x = t^2 + 1 \qquad y = t^3 + t \qquad 0 \leqslant t \leqslant 1$$

x \ y	0	1	2
0	1	6	4
1	3	5	7
2	8	2	9

3–10 ■ Determine whether or not $\mathbf{F}$ is a conservative vector field. If it is, find a function f such that $\mathbf{F} = \nabla f$.

3. $\mathbf{F}(x, y) = (2x - 3y) \, \mathbf{i} + (-3x + 4y - 8) \, \mathbf{j}$

4. $\mathbf{F}(x, y) = e^x \sin y \, \mathbf{i} + e^x \cos y \, \mathbf{j}$

5. $\mathbf{F}(x, y) = e^x \cos y \, \mathbf{i} + e^x \sin y \, \mathbf{j}$

6. $\mathbf{F}(x, y) = (3x^2 - 2y^2) \, \mathbf{i} + (4xy + 3) \, \mathbf{j}$

7. $\mathbf{F}(x, y) = (ye^x + \sin y) \, \mathbf{i} + (e^x + x \cos y) \, \mathbf{j}$

8. $\mathbf{F}(x, y) = (2xy + y^{-2}) \, \mathbf{i} + (x^2 - 2xy^{-3}) \, \mathbf{j}, \quad y > 0$

9. $\mathbf{F}(x, y) = (\ln y + 2xy^3) \, \mathbf{i} + (3x^2y^2 + x/y) \, \mathbf{j}$

10. $\mathbf{F}(x, y) = (xy \cosh xy + \sinh xy) \, \mathbf{i} + (x^2 \cosh xy) \, \mathbf{j}$

11–16 ■ (a) Find a function f such that $\mathbf{F} = \nabla f$ and (b) use part (a) to evaluate $\int_C \mathbf{F} \cdot d\mathbf{r}$ along the given curve C.

11. $\mathbf{F}(x, y) = xy^2 \, \mathbf{i} + x^2y \, \mathbf{j}$,
    $C: \mathbf{r}(t) = \langle t + \sin \tfrac{1}{2}\pi t, \, t + \cos \tfrac{1}{2}\pi t \rangle, \quad 0 \leqslant t \leqslant 1$

12. $\mathbf{F}(x, y) = (1 + xy)e^{xy} \, \mathbf{i} + x^2 e^{xy} \, \mathbf{j}$,
    $C: \mathbf{r}(t) = \cos t \, \mathbf{i} + 2 \sin t \, \mathbf{j}, \quad 0 \leqslant t \leqslant \pi/2$

13. $\mathbf{F}(x, y, z) = yz\,\mathbf{i} + xz\,\mathbf{j} + (xy + 2z)\,\mathbf{k}$,
C is the line segment from $(1, 0, -2)$ to $(4, 6, 3)$

14. $\mathbf{F}(x, y, z) = (y^2z + 2xz^2)\,\mathbf{i} + 2xyz\,\mathbf{j} + (xy^2 + 2x^2z)\,\mathbf{k}$,
$C:\ x = \sqrt{t},\ y = t + 1,\ z = t^2,\quad 0 \le t \le 1$

15. $\mathbf{F}(x, y, z) = yze^{xz}\,\mathbf{i} + e^{xz}\,\mathbf{j} + xye^{xz}\,\mathbf{k}$,
$C:\ \mathbf{r}(t) = (t^2 + 1)\,\mathbf{i} + (t^2 - 1)\,\mathbf{j} + (t^2 - 2t)\,\mathbf{k}$,
$0 \le t \le 2$

16. $\mathbf{F}(x, y, z) = \sin y\,\mathbf{i} + (x\cos y + \cos z)\,\mathbf{j} - y\sin z\,\mathbf{k}$,
$C:\ \mathbf{r}(t) = \sin t\,\mathbf{i} + t\,\mathbf{j} + 2t\,\mathbf{k},\quad 0 \le t \le \pi/2$

17–18 ■ Show that the line integral is independent of path and evaluate the integral.

17. $\int_C 2xe^{-y}\,dx + (2y - x^2e^{-y})\,dy$,
C is any path from $(1, 0)$ to $(2, 1)$

18. $\int_C \sin y\,dx + (x\cos y - \sin y)\,dy$,
C is any path from $(2, 0)$ to $(1, \pi)$

19–20 ■ Find the work done by the force field $\mathbf{F}$ in moving an object from P to Q.

19. $\mathbf{F}(x, y) = 2y^{3/2}\,\mathbf{i} + 3x\sqrt{y}\,\mathbf{j};\quad P(1, 1),\ Q(2, 4)$

20. $\mathbf{F}(x, y) = e^{-y}\,\mathbf{i} - xe^{-y}\,\mathbf{j};\quad P(0, 1),\ Q(2, 0)$

21–22 ■ Is the vector field shown in the figure conservative? Explain.

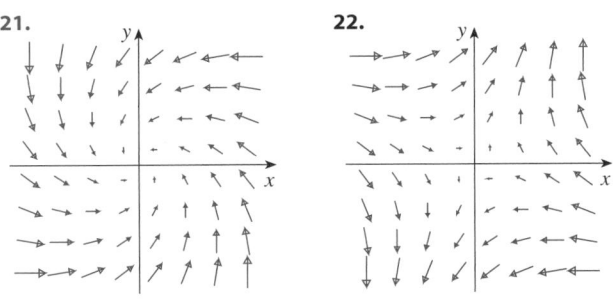

21.

22.

CAS 23. If $\mathbf{F}(x, y) = \sin y\,\mathbf{i} + (1 + x\cos y)\,\mathbf{j}$, use a plot to guess whether $\mathbf{F}$ is conservative. Then determine whether your guess is correct.

24. Let $\mathbf{F} = \nabla f$, where $f(x, y) = \sin(x - 2y)$. Find curves C_1 and C_2 that are not closed and satisfy the equation.

(a) $\int_{C_1} \mathbf{F} \cdot d\mathbf{r} = 0$ \qquad (b) $\int_{C_2} \mathbf{F} \cdot d\mathbf{r} = 1$

25. Show that if the vector field $\mathbf{F} = P\,\mathbf{i} + Q\,\mathbf{j} + R\,\mathbf{k}$ is conservative and P, Q, R have continuous first-order partial derivatives, then

$$\frac{\partial P}{\partial y} = \frac{\partial Q}{\partial x} \qquad \frac{\partial P}{\partial z} = \frac{\partial R}{\partial x} \qquad \frac{\partial Q}{\partial z} = \frac{\partial R}{\partial y}$$

26. Use Exercise 25 to show that the line integral
$\int_C y\,dx + x\,dy + xyz\,dz$ is not independent of path.

27–30 ■ Determine whether or not the given set is (a) open, (b) connected, and (c) simply-connected.

27. $\{(x, y) \mid 0 < y < 3\}$ \qquad **28.** $\{(x, y) \mid 1 < |x| < 2\}$

29. $\{(x, y) \mid 1 \le x^2 + y^2 \le 4, y \ge 0\}$

30. $\{(x, y) \mid (x, y) \ne (2, 3)\}$

---

31. Let $\mathbf{F}(x, y) = \dfrac{-y\,\mathbf{i} + x\,\mathbf{j}}{x^2 + y^2}$.

(a) Show that $\partial P/\partial y = \partial Q/\partial x$.

(b) Show that $\int_C \mathbf{F} \cdot d\mathbf{r}$ is not independent of path.
[*Hint:* Compute $\int_{C_1} \mathbf{F} \cdot d\mathbf{r}$ and $\int_{C_2} \mathbf{F} \cdot d\mathbf{r}$, where C_1 and C_2 are the upper and lower halves of the circle $x^2 + y^2 = 1$ from $(1, 0)$ to $(-1, 0)$.] Does this contradict Theorem 6?

32. (a) Suppose that $\mathbf{F}$ is an inverse square force field, that is,

$$\mathbf{F}(\mathbf{r}) = \frac{c\mathbf{r}}{|\mathbf{r}|^3}$$

for some constant c, where $\mathbf{r} = x\,\mathbf{i} + y\,\mathbf{j} + z\,\mathbf{k}$. Find the work done by $\mathbf{F}$ in moving an object from a point P_1 along a path to a point P_2 in terms of the distances d_1 and d_2 from these points to the origin.

(b) An example of an inverse square field is the gravitational field $\mathbf{F} = -(mMG)\mathbf{r}/|\mathbf{r}|^3$ discussed in Example 4 in Section 13.1. Use part (a) to find the work done by the gravitational field when the earth moves from aphelion (at a maximum distance of 1.52×10^8 km from the sun) to perihelion (at a minimum distance of 1.47×10^8 km). (Use the values $m = 5.97 \times 10^{24}$ kg, $M = 1.99 \times 10^{30}$ kg, and $G = 6.67 \times 10^{-11}$ N·m²/kg².)

(c) Another example of an inverse square field is the electric force field $\mathbf{F} = \varepsilon qQ\mathbf{r}/|\mathbf{r}|^3$ discussed in Example 5 in Section 13.1. Suppose that an electron with a charge of -1.6×10^{-19} C is located at the origin. A positive unit charge is positioned a distance 10^{-12} m from the electron and moves to a position half that distance from the electron. Use part (a) to find the work done by the electric force field. (Use the value $\varepsilon = 8.985 \times 10^9$.)

13.4 | GREEN'S THEOREM

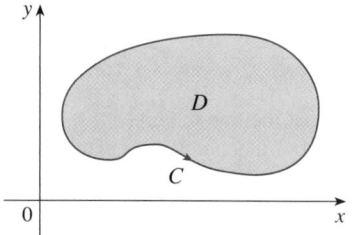

FIGURE 1

Green's Theorem gives the relationship between a line integral around a simple closed curve C and a double integral over the plane region D bounded by C. (See Figure 1. We assume that D consists of all points inside C as well as all points on C.) In stating Green's Theorem we use the convention that the **positive orientation** of a simple closed curve C refers to a single *counterclockwise* traversal of C. Thus if C is given by the vector function $\mathbf{r}(t)$, $a \leqslant t \leqslant b$, then the region D is always on the left as the point $\mathbf{r}(t)$ traverses C. (See Figure 2.)

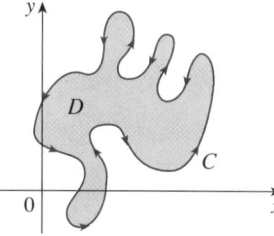

(a) Positive orientation

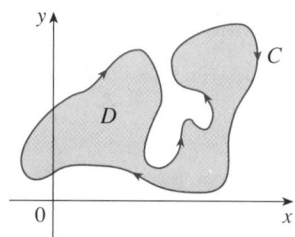

(b) Negative orientation

FIGURE 2

> **GREEN'S THEOREM** Let C be a positively oriented, piecewise-smooth, simple closed curve in the plane and let D be the region bounded by C. If P and Q have continuous partial derivatives on an open region that contains D, then
>
> $$\int_C P \, dx + Q \, dy = \iint_D \left(\frac{\partial Q}{\partial x} - \frac{\partial P}{\partial y} \right) dA$$

■ Recall that the left side of this equation is another way of writing $\int_C \mathbf{F} \cdot d\mathbf{r}$, where $\mathbf{F} = P\,\mathbf{i} + Q\,\mathbf{j}$.

NOTE The notation

$$\oint_C P \, dx + Q \, dy \qquad \text{or} \qquad \oint_C P \, dx + Q \, dy$$

is sometimes used to indicate that the line integral is calculated using the positive orientation of the closed curve C. Another notation for the positively oriented boundary curve of D is ∂D, so the equation in Green's Theorem can be written as

$$\boxed{1} \qquad \iint_D \left(\frac{\partial Q}{\partial x} - \frac{\partial P}{\partial y} \right) dA = \int_{\partial D} P \, dx + Q \, dy$$

Green's Theorem should be regarded as the counterpart of the Fundamental Theorem of Calculus for double integrals. Compare Equation 1 with the statement of the Fundamental Theorem of Calculus, Part 2, in the following equation:

$$\int_a^b F'(x) \, dx = F(b) - F(a)$$

In both cases there is an integral involving derivatives (F', $\partial Q/\partial x$, and $\partial P/\partial y$) on the left side of the equation. And in both cases the right side involves the values of the original functions (F, Q, and P) only on the *boundary* of the domain. (In the one-

Green's Theorem is named after the self-taught English scientist George Green (1793–1841). He worked full-time in his father's bakery from the age of nine and taught himself mathematics from library books. In 1828 he published privately *An Essay on the Application of Mathematical Analysis to the Theories of Electricity and Magnetism*, but only 100 copies were printed and most of those went to his friends. This pamphlet contained a theorem that is equivalent to what we know as Green's Theorem, but it didn't become widely known at that time. Finally, at age 40, Green entered Cambridge University as an undergraduate but died four years after graduation. In 1846 William Thomson (Lord Kelvin) located a copy of Green's essay, realized its significance, and had it reprinted. Green was the first person to try to formulate a mathematical theory of electricity and magnetism. His work was the basis for the subsequent electromagnetic theories of Thomson, Stokes, Rayleigh, and Maxwell.

FIGURE 3

dimensional case, the domain is an interval $[a, b]$ whose boundary consists of just two points, a and b.)

Green's Theorem is not easy to prove in general, but we can give a proof for the special case where the region is both of type I and of type II (see Section 12.2). Let's call such regions **simple regions**.

PROOF OF GREEN'S THEOREM FOR THE CASE IN WHICH D IS A SIMPLE REGION
Notice that Green's Theorem will be proved if we can show that

$$\boxed{2} \qquad \int_C P \, dx = -\iint_D \frac{\partial P}{\partial y} \, dA$$

and

$$\boxed{3} \qquad \int_C Q \, dy = \iint_D \frac{\partial Q}{\partial x} \, dA$$

We prove Equation 2 by expressing D as a type I region:

$$D = \{(x, y) \mid a \leqslant x \leqslant b, \ g_1(x) \leqslant y \leqslant g_2(x)\}$$

where g_1 and g_2 are continuous functions. This enables us to compute the double integral on the right side of Equation 2 as follows:

$$\boxed{4} \qquad \iint_D \frac{\partial P}{\partial y} \, dA = \int_a^b \int_{g_1(x)}^{g_2(x)} \frac{\partial P}{\partial y}(x, y) \, dy \, dx = \int_a^b [P(x, g_2(x)) - P(x, g_1(x))] \, dx$$

where the last step follows from the Fundamental Theorem of Calculus.

Now we compute the left side of Equation 2 by breaking up C as the union of the four curves C_1, C_2, C_3, and C_4 shown in Figure 3. On C_1 we take x as the parameter and write the parametric equations as $x = x$, $y = g_1(x)$, $a \leqslant x \leqslant b$. Thus

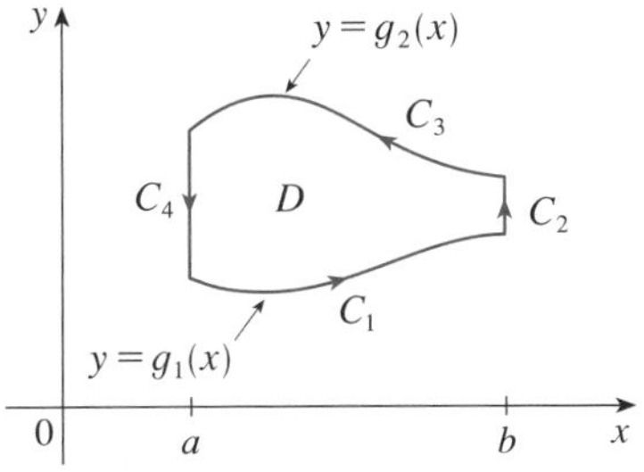

$$\int_{C_1} P(x, y) \, dx = \int_a^b P(x, g_1(x)) \, dx$$

Observe that C_3 goes from right to left but $-C_3$ goes from left to right, so we can write the parametric equations of $-C_3$ as $x = x$, $y = g_2(x)$, $a \leqslant x \leqslant b$. Therefore

$$\int_{C_3} P(x, y) \, dx = -\int_{-C_3} P(x, y) \, dx = -\int_a^b P(x, g_2(x)) \, dx$$

On C_2 or C_4 (either of which might reduce to just a single point), x is constant, so $dx = 0$ and

$$\int_{C_2} P(x, y) \, dx = 0 = \int_{C_4} P(x, y) \, dx$$

Hence

$$\int_C P(x, y) \, dx = \int_{C_1} P(x, y) \, dx + \int_{C_2} P(x, y) \, dx + \int_{C_3} P(x, y) \, dx + \int_{C_4} P(x, y) \, dx$$

$$= \int_a^b P(x, g_1(x)) \, dx - \int_a^b P(x, g_2(x)) \, dx$$

Comparing this expression with the one in Equation 4, we see that

$$\int_C P(x, y)\, dx = -\iint_D \frac{\partial P}{\partial y}\, dA$$

Equation 3 can be proved in much the same way by expressing D as a type II region (see Exercise 30). Then, by adding Equations 2 and 3, we obtain Green's Theorem. ☐

EXAMPLE 1   Evaluate $\int_C x^4\, dx + xy\, dy$, where C is the triangular curve consisting of the line segments from $(0, 0)$ to $(1, 0)$, from $(1, 0)$ to $(0, 1)$, and from $(0, 1)$ to $(0, 0)$.

SOLUTION   Although the given line integral could be evaluated as usual by the methods of Section 13.2, that would involve setting up three separate integrals along the three sides of the triangle, so let's use Green's Theorem instead. Notice that the region D enclosed by C is simple and C has positive orientation (see Figure 4). If we let $P(x, y) = x^4$ and $Q(x, y) = xy$, then we have

$$\int_C x^4\, dx + xy\, dy = \iint_D \left(\frac{\partial Q}{\partial x} - \frac{\partial P}{\partial y} \right) dA = \int_0^1 \int_0^{1-x} (y - 0)\, dy\, dx$$

$$= \int_0^1 \left[\tfrac{1}{2} y^2 \right]_{y=0}^{y=1-x}\, dx = \tfrac{1}{2} \int_0^1 (1 - x)^2\, dx = -\tfrac{1}{6}(1 - x)^3 \Big]_0^1 = \tfrac{1}{6}$$ ■

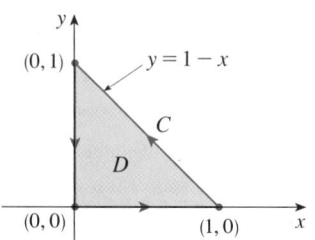

(0, 1) $y = 1 - x$

C

D

(0, 0)   (1, 0)

FIGURE 4

V EXAMPLE 2   Evaluate $\oint_C \left(3y - e^{\sin x}\right) dx + \left(7x + \sqrt{y^4 + 1}\right) dy$, where C is the circle $x^2 + y^2 = 9$.

SOLUTION   The region D bounded by C is the disk $x^2 + y^2 \leq 9$, so let's change to polar coordinates after applying Green's Theorem:

$$\oint_C \left(3y - e^{\sin x}\right) dx + \left(7x + \sqrt{y^4 + 1}\right) dy$$

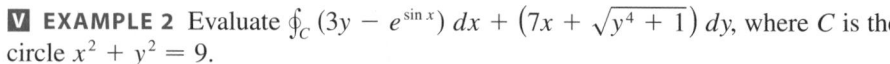

$$= \iint_D \left[\frac{\partial}{\partial x}\left(7x + \sqrt{y^4 + 1}\right) - \frac{\partial}{\partial y}\left(3y - e^{\sin x}\right) \right] dA$$

$$= \int_0^{2\pi} \int_0^3 (7 - 3)\, r\, dr\, d\theta = 4 \int_0^{2\pi} d\theta \int_0^3 r\, dr = 36\pi$$ ■

■ Instead of using polar coordinates, we could simply use the fact that D is a disk of radius 3 and write

$$\iint_D 4\, dA = 4 \cdot \pi(3)^2 = 36\pi$$

In Examples 1 and 2 we found that the double integral was easier to evaluate than the line integral. (Try setting up the line integral in Example 2 and you'll soon be convinced!) But sometimes it's easier to evaluate the line integral, and Green's Theorem is used in the reverse direction. For instance, if it is known that $P(x, y) = Q(x, y) = 0$ on the curve C, then Green's Theorem gives

$$\iint_D \left(\frac{\partial Q}{\partial x} - \frac{\partial P}{\partial y} \right) dA = \int_C P\, dx + Q\, dy = 0$$

no matter what values P and Q assume in the region D.

Another application of the reverse direction of Green's Theorem is in computing areas. Since the area of D is $\iint_D 1\, dA$, we wish to choose P and Q so that

$$\frac{\partial Q}{\partial x} - \frac{\partial P}{\partial y} = 1$$

There are several possibilities:

$$P(x, y) = 0 \qquad P(x, y) = -y \qquad P(x, y) = -\tfrac{1}{2}y$$

$$Q(x, y) = x \qquad Q(x, y) = 0 \qquad Q(x, y) = \tfrac{1}{2}x$$

Then Green's Theorem gives the following formulas for the area of D:

$$\boxed{5} \qquad \boxed{A = \oint_C x \, dy = -\oint_C y \, dx = \tfrac{1}{2} \oint_C x \, dy - y \, dx}$$

EXAMPLE 3 Find the area enclosed by the ellipse $\dfrac{x^2}{a^2} + \dfrac{y^2}{b^2} = 1$.

SOLUTION The ellipse has parametric equations $x = a \cos t$ and $y = b \sin t$, where $0 \leqslant t \leqslant 2\pi$. Using the third formula in Equation 5, we have

$$A = \tfrac{1}{2} \int_C x \, dy - y \, dx = \tfrac{1}{2} \int_0^{2\pi} (a \cos t)(b \cos t) \, dt - (b \sin t)(-a \sin t) \, dt$$

$$= \frac{ab}{2} \int_0^{2\pi} dt = \pi ab \qquad \blacksquare$$

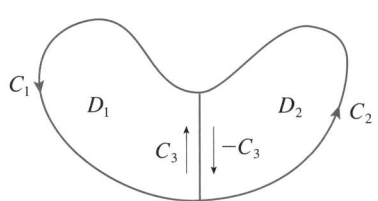

FIGURE 5

Although we have proved Green's Theorem only for the case where D is simple, we can now extend it to the case where D is a finite union of simple regions. For example, if D is the region shown in Figure 5, then we can write $D = D_1 \cup D_2$, where D_1 and D_2 are both simple. The boundary of D_1 is $C_1 \cup C_3$ and the boundary of D_2 is $C_2 \cup (-C_3)$ so, applying Green's Theorem to D_1 and D_2 separately, we get

$$\int_{C_1 \cup C_3} P \, dx + Q \, dy = \iint_{D_1} \left(\frac{\partial Q}{\partial x} - \frac{\partial P}{\partial y} \right) dA$$

$$\int_{C_2 \cup (-C_3)} P \, dx + Q \, dy = \iint_{D_2} \left(\frac{\partial Q}{\partial x} - \frac{\partial P}{\partial y} \right) dA$$

If we add these two equations, the line integrals along C_3 and $-C_3$ cancel, so we get

$$\int_{C_1 \cup C_2} P \, dx + Q \, dy = \iint_{D} \left(\frac{\partial Q}{\partial x} - \frac{\partial P}{\partial y} \right) dA$$

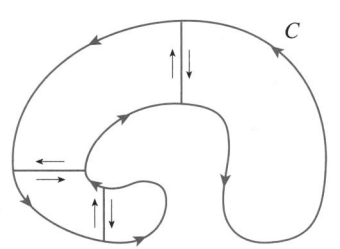

FIGURE 6

which is Green's Theorem for $D = D_1 \cup D_2$, since its boundary is $C = C_1 \cup C_2$.

The same sort of argument allows us to establish Green's Theorem for any finite union of nonoverlapping simple regions (see Figure 6).

◢ **EXAMPLE 4** Evaluate $\oint_C y^2 \, dx + 3xy \, dy$, where C is the boundary of the semi-annular region D in the upper half-plane between the circles $x^2 + y^2 = 1$ and $x^2 + y^2 = 4$.

SOLUTION Notice that although D is not simple, the y-axis divides it into two simple regions (see Figure 7). In polar coordinates we can write

$$D = \left\{ (r, \theta) \mid 1 \leqslant r \leqslant 2, \; 0 \leqslant \theta \leqslant \pi \right\}$$

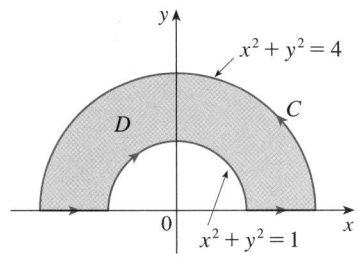

FIGURE 7

Therefore Green's Theorem gives

$$\oint_C y^2\,dx + 3xy\,dy = \iint_D \left[\frac{\partial}{\partial x}(3xy) - \frac{\partial}{\partial y}(y^2)\right] dA = \iint_D y\,dA = \int_0^\pi \int_1^2 (r\sin\theta)\,r\,dr\,d\theta$$

$$= \int_0^\pi \sin\theta\,d\theta \int_1^2 r^2\,dr = \big[-\cos\theta\big]_0^\pi \big[\tfrac{1}{3}r^3\big]_1^2 = \frac{14}{3}$$ ■

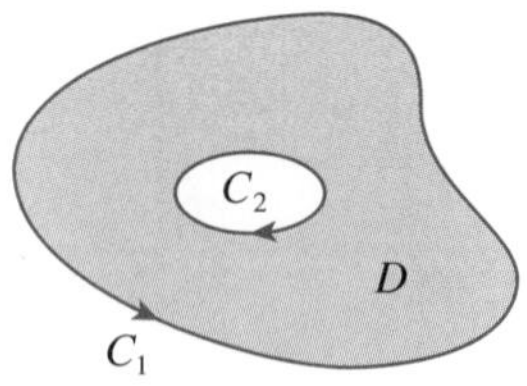

FIGURE 8

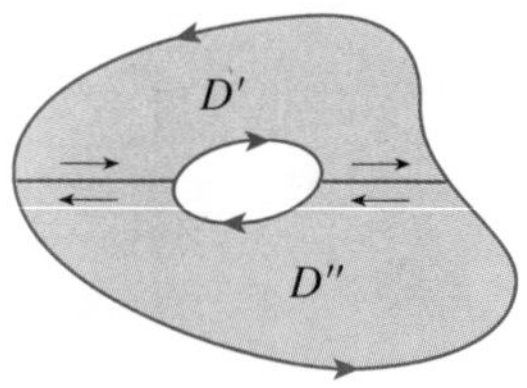

FIGURE 9

Green's Theorem can be extended to apply to regions with holes, that is, regions that are not simply-connected. Observe that the boundary C of the region D in Figure 8 consists of two simple closed curves C_1 and C_2. We assume that these boundary curves are oriented so that the region D is always on the left as the curve C is traversed. Thus the positive direction is counterclockwise for the outer curve C_1 but clockwise for the inner curve C_2. If we divide D into two regions D' and D'' by means of the lines shown in Figure 9 and then apply Green's Theorem to each of D' and D'', we get

$$\iint_D \left(\frac{\partial Q}{\partial x} - \frac{\partial P}{\partial y}\right) dA = \iint_{D'} \left(\frac{\partial Q}{\partial x} - \frac{\partial P}{\partial y}\right) dA + \iint_{D''} \left(\frac{\partial Q}{\partial x} - \frac{\partial P}{\partial y}\right) dA$$

$$= \int_{\partial D'} P\,dx + Q\,dy + \int_{\partial D''} P\,dx + Q\,dy$$

Since the line integrals along the common boundary lines are in opposite directions, they cancel and we get

$$\iint_D \left(\frac{\partial Q}{\partial x} - \frac{\partial P}{\partial y}\right) dA = \int_{C_1} P\,dx + Q\,dy + \int_{C_2} P\,dx + Q\,dy = \int_C P\,dx + Q\,dy$$

which is Green's Theorem for the region D.

V EXAMPLE 5 If $\mathbf{F}(x, y) = (-y\,\mathbf{i} + x\,\mathbf{j})/(x^2 + y^2)$, show that $\int_C \mathbf{F} \cdot d\mathbf{r} = 2\pi$ for every positively oriented simple closed path that encloses the origin.

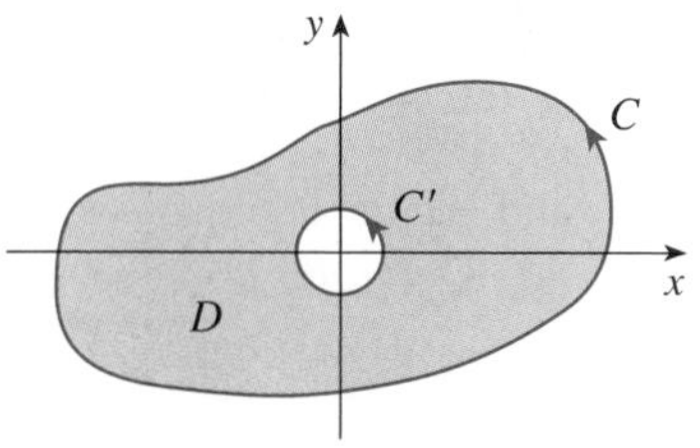

FIGURE 10

SOLUTION Since C is an *arbitrary* closed path that encloses the origin, it's difficult to compute the given integral directly. So let's consider a counterclockwise-oriented circle C' with center the origin and radius a, where a is chosen to be small enough that C' lies inside C. (See Figure 10.) Let D be the region bounded by C and C'. Then its positively oriented boundary is $C \cup (-C')$ and so the general version of Green's Theorem gives

$$\int_C P\,dx + Q\,dy + \int_{-C'} P\,dx + Q\,dy = \iint_D \left(\frac{\partial Q}{\partial x} - \frac{\partial P}{\partial y}\right) dA$$

$$= \iint_D \left[\frac{y^2 - x^2}{(x^2 + y^2)^2} - \frac{y^2 - x^2}{(x^2 + y^2)^2}\right] dA$$

$$= 0$$

Therefore $$\int_C P\,dx + Q\,dy = \int_{C'} P\,dx + Q\,dy$$

that is, $$\int_C \mathbf{F} \cdot d\mathbf{r} = \int_{C'} \mathbf{F} \cdot d\mathbf{r}$$

We now easily compute this last integral using the parametrization given by $\mathbf{r}(t) = a \cos t\,\mathbf{i} + a \sin t\,\mathbf{j}$, $0 \leqslant t \leqslant 2\pi$. Thus

$$\int_C \mathbf{F} \cdot d\mathbf{r} = \int_{C'} \mathbf{F} \cdot d\mathbf{r} = \int_0^{2\pi} \mathbf{F}(\mathbf{r}(t)) \cdot \mathbf{r}'(t)\,dt$$

$$= \int_0^{2\pi} \frac{(-a \sin t)(-a \sin t) + (a \cos t)(a \cos t)}{a^2 \cos^2 t + a^2 \sin^2 t}\,dt = \int_0^{2\pi} dt = 2\pi \qquad \blacksquare$$

We end this section by using Green's Theorem to discuss a result that was stated in the preceding section.

SKETCH OF PROOF OF THEOREM 13.3.6 We're assuming that $\mathbf{F} = P\,\mathbf{i} + Q\,\mathbf{j}$ is a vector field on an open simply-connected region D, that P and Q have continuous first-order partial derivatives, and that

$$\frac{\partial P}{\partial y} = \frac{\partial Q}{\partial x} \qquad \text{throughout } D$$

If C is any simple closed path in D and R is the region that C encloses, then Green's Theorem gives

$$\oint_C \mathbf{F} \cdot d\mathbf{r} = \oint_C P\,dx + Q\,dy = \iint_R \left(\frac{\partial Q}{\partial x} - \frac{\partial P}{\partial y} \right)\,dA = \iint_R 0\,dA = 0$$

A curve that is not simple crosses itself at one or more points and can be broken up into a number of simple curves. We have shown that the line integrals of $\mathbf{F}$ around these simple curves are all 0 and, adding these integrals, we see that $\int_C \mathbf{F} \cdot d\mathbf{r} = 0$ for any closed curve C. Therefore $\int_C \mathbf{F} \cdot d\mathbf{r}$ is independent of path in D by Theorem 13.3.3. It follows that $\mathbf{F}$ is a conservative vector field. $\qquad \square$

13.4 | EXERCISES

1–4 ▪ Evaluate the line integral by two methods: (a) directly and (b) using Green's Theorem.

1. $\oint_C (x - y)\,dx + (x + y)\,dy$,
C is the circle with center the origin and radius 2

2. $\oint_C xy\,dx + x^2\,dy$,
C is the rectangle with vertices $(0, 0)$, $(3, 0)$, $(3, 1)$, and $(0, 1)$

3. $\oint_C xy\,dx + x^2 y^3\,dy$,
C is the triangle with vertices $(0, 0)$, $(1, 0)$, and $(1, 2)$

4. $\oint_C x^2 y^2\,dx + xy\,dy$, C consists of the arc of the parabola $y = x^2$ from $(0, 0)$ to $(1, 1)$ and the line segments from $(1, 1)$ to $(0, 1)$ and from $(0, 1)$ to $(0, 0)$

5–10 ▪ Use Green's Theorem to evaluate the line integral along the given positively oriented curve.

5. $\int_C xy^2\,dx + 2x^2 y\,dy$,
C is the triangle with vertices $(0, 0)$, $(2, 2)$, and $(2, 4)$

6. $\int_C \cos y\,dx + x^2 \sin y\,dy$,
C is the rectangle with vertices $(0, 0)$, $(5, 0)$, $(5, 2)$, and $(0, 2)$

7. $\int_C \left(y + e^{\sqrt{x}} \right) dx + (2x + \cos y^2)\,dy$,
C is the boundary of the region enclosed by the parabolas $y = x^2$ and $x = y^2$

8. $\int_C y^4\,dx + 2xy^3\,dy$, C is the ellipse $x^2 + 2y^2 = 2$

9. $\int_C y^3 \, dx - x^3 \, dy$, C is the circle $x^2 + y^2 = 4$

10. $\int_C (1 - y^3) \, dx + (x^3 + e^{y^2}) \, dy$, C is the boundary of the region between the circles $x^2 + y^2 = 4$ and $x^2 + y^2 = 9$

11–14 ■ Use Green's Theorem to evaluate $\int_C \mathbf{F} \cdot d\mathbf{r}$. (Check the orientation of the curve before applying the theorem.)

11. $\mathbf{F}(x, y) = \langle y \cos x - xy \sin x, \, xy + x \cos x \rangle$, C is the triangle from $(0, 0)$ to $(0, 4)$ to $(2, 0)$ to $(0, 0)$

12. $\mathbf{F}(x, y) = \langle e^{-x} + y^2, \, e^{-y} + x^2 \rangle$, C consists of the arc of the curve $y = \cos x$ from $(-\pi/2, 0)$ to $(\pi/2, 0)$ and the line segment from $(\pi/2, 0)$ to $(-\pi/2, 0)$

13. $\mathbf{F}(x, y) = \langle y - \cos y, \, x \sin y \rangle$, C is the circle $(x - 3)^2 + (y + 4)^2 = 4$ oriented clockwise

14. $\mathbf{F}(x, y) = \langle \sqrt{x^2 + 1}, \, \tan^{-1} x \rangle$, C is the triangle from $(0, 0)$ to $(1, 1)$ to $(0, 1)$ to $(0, 0)$

CAS **15–16** ■ Verify Green's Theorem by using a computer algebra system to evaluate both the line integral and the double integral.

15. $P(x, y) = y^2 e^x$, $Q(x, y) = x^2 e^y$, C consists of the line segment from $(-1, 1)$ to $(1, 1)$ followed by the arc of the parabola $y = 2 - x^2$ from $(1, 1)$ to $(-1, 1)$

16. $P(x, y) = 2x - x^3 y^5$, $Q(x, y) = x^3 y^8$, C is the ellipse $4x^2 + y^2 = 4$

---

17. Use Green's Theorem to find the work done by the force $\mathbf{F}(x, y) = x(x + y)\, \mathbf{i} + xy^2 \, \mathbf{j}$ in moving a particle from the origin along the x-axis to $(1, 0)$, then along the line segment to $(0, 1)$, and then back to the origin along the y-axis.

18. A particle starts at the point $(-2, 0)$, moves along the x-axis to $(2, 0)$, and then along the semicircle $y = \sqrt{4 - x^2}$ to the starting point. Use Green's Theorem to find the work done on this particle by the force field $\mathbf{F}(x, y) = \langle x, x^3 + 3xy^2 \rangle$.

19. Use one of the formulas in $\boxed{5}$ to find the area under one arch of the cycloid $x = t - \sin t$, $y = 1 - \cos t$.

20. If a circle C with radius 1 rolls along the outside of the circle $x^2 + y^2 = 16$, a fixed point P on C traces out a curve called an *epicycloid*, with parametric equations $x = 5 \cos t - \cos 5t$, $y = 5 \sin t - \sin 5t$. Graph the epicycloid and use $\boxed{5}$ to find the area it encloses.

21. (a) If C is the line segment connecting the point (x_1, y_1) to the point (x_2, y_2), show that
$$\int_C x \, dy - y \, dx = x_1 y_2 - x_2 y_1$$

(b) If the vertices of a polygon, in counterclockwise order, are (x_1, y_1), (x_2, y_2), . . . , (x_n, y_n), show that the area of the polygon is
$$A = \tfrac{1}{2}[(x_1 y_2 - x_2 y_1) + (x_2 y_3 - x_3 y_2) + \cdots$$
$$+ (x_{n-1} y_n - x_n y_{n-1}) + (x_n y_1 - x_1 y_n)]$$

(c) Find the area of the pentagon with vertices $(0, 0)$, $(2, 1)$, $(1, 3)$, $(0, 2)$, and $(-1, 1)$.

22. Let D be a region bounded by a simple closed path C in the xy-plane. Use Green's Theorem to prove that the coordinates of the centroid $(\bar{x}, \bar{y})$ of D are
$$\bar{x} = \frac{1}{2A} \oint_C x^2 \, dy \qquad \bar{y} = -\frac{1}{2A} \oint_C y^2 \, dx$$
where A is the area of D.

23. Use Exercise 22 to find the centroid of a quarter-circular region of radius a.

24. Use Exercise 22 to find the centroid of the triangle with vertices $(0, 0)$, $(a, 0)$, and (a, b), where $a > 0$ and $b > 0$.

25. A plane lamina with constant density $\rho(x, y) = \rho$ occupies a region in the xy-plane bounded by a simple closed path C. Show that its moments of inertia about the axes are
$$I_x = -\frac{\rho}{3} \oint_C y^3 \, dx \qquad I_y = \frac{\rho}{3} \oint_C x^3 \, dy$$

26. Use Exercise 25 to find the moment of inertia of a circular disk of radius a with constant density ρ about a diameter. (Compare with Example 4 in Section 12.4.)

27. Use the method of Example 5 to calculate $\int_C \mathbf{F} \cdot d\mathbf{r}$, where
$$\mathbf{F}(x, y) = \frac{2xy \, \mathbf{i} + (y^2 - x^2)\, \mathbf{j}}{(x^2 + y^2)^2}$$
and C is any positively oriented simple closed curve that encloses the origin.

28. Calculate $\int_C \mathbf{F} \cdot d\mathbf{r}$, where $\mathbf{F}(x, y) = \langle x^2 + y, \, 3x - y^2 \rangle$ and C is the positively oriented boundary curve of a region D that has area 6.

29. If $\mathbf{F}$ is the vector field of Example 5, show that $\int_C \mathbf{F} \cdot d\mathbf{r} = 0$ for every simple closed path that does not pass through or enclose the origin.

30. Complete the proof of the special case of Green's Theorem by proving Equation 3.

31. Use Green's Theorem to prove the change of variables formula for a double integral (Formula 12.8.9) for the case where $f(x, y) = 1$:

$$\iint\limits_{R} dx\, dy = \iint\limits_{S} \left| \frac{\partial(x, y)}{\partial(u, v)} \right| du\, dv$$

Here R is the region in the xy-plane that corresponds to the region S in the uv-plane under the transformation given by $x = g(u, v)$, $y = h(u, v)$.

[*Hint:* Note that the left side is $A(R)$ and apply the first part of Equation 5. Convert the line integral over ∂R to a line integral over ∂S and apply Green's Theorem in the uv-plane.]

13.5 | CURL AND DIVERGENCE

In this section we define two operations that can be performed on vector fields and that play a basic role in the applications of vector calculus to fluid flow and electricity and magnetism. Each operation resembles differentiation, but one produces a vector field whereas the other produces a scalar field.

CURL

If $\mathbf{F} = P\,\mathbf{i} + Q\,\mathbf{j} + R\,\mathbf{k}$ is a vector field on $\mathbb{R}^3$ and the partial derivatives of P, Q, and R all exist, then the **curl** of $\mathbf{F}$ is the vector field on $\mathbb{R}^3$ defined by

$$\boxed{1} \qquad \operatorname{curl} \mathbf{F} = \left(\frac{\partial R}{\partial y} - \frac{\partial Q}{\partial z} \right) \mathbf{i} + \left(\frac{\partial P}{\partial z} - \frac{\partial R}{\partial x} \right) \mathbf{j} + \left(\frac{\partial Q}{\partial x} - \frac{\partial P}{\partial y} \right) \mathbf{k}$$

As an aid to our memory, let's rewrite Equation 1 using operator notation. We introduce the vector differential operator ∇ ("del") as

$$\nabla = \mathbf{i}\,\frac{\partial}{\partial x} + \mathbf{j}\,\frac{\partial}{\partial y} + \mathbf{k}\,\frac{\partial}{\partial z}$$

It has meaning when it operates on a scalar function to produce the gradient of f:

$$\nabla f = \mathbf{i}\,\frac{\partial f}{\partial x} + \mathbf{j}\,\frac{\partial f}{\partial y} + \mathbf{k}\,\frac{\partial f}{\partial z} = \frac{\partial f}{\partial x}\,\mathbf{i} + \frac{\partial f}{\partial y}\,\mathbf{j} + \frac{\partial f}{\partial z}\,\mathbf{k}$$

If we think of ∇ as a vector with components $\partial/\partial x$, $\partial/\partial y$, and $\partial/\partial z$, we can also consider the formal cross product of ∇ with the vector field $\mathbf{F}$ as follows:

$$\nabla \times \mathbf{F} = \begin{vmatrix} \mathbf{i} & \mathbf{j} & \mathbf{k} \\ \dfrac{\partial}{\partial x} & \dfrac{\partial}{\partial y} & \dfrac{\partial}{\partial z} \\ P & Q & R \end{vmatrix} = \left(\frac{\partial R}{\partial y} - \frac{\partial Q}{\partial z} \right) \mathbf{i} + \left(\frac{\partial P}{\partial z} - \frac{\partial R}{\partial x} \right) \mathbf{j} + \left(\frac{\partial Q}{\partial x} - \frac{\partial P}{\partial y} \right) \mathbf{k}$$

$$= \operatorname{curl} \mathbf{F}$$

Thus the easiest way to remember Definition 1 is by means of the symbolic expression

$$\boxed{2} \qquad \operatorname{curl} \mathbf{F} = \nabla \times \mathbf{F}$$

EXAMPLE 1 If $\mathbf{F}(x, y, z) = xz\,\mathbf{i} + xyz\,\mathbf{j} - y^2\,\mathbf{k}$, find curl $\mathbf{F}$.

SOLUTION Using Equation 2, we have

$$\operatorname{curl}\mathbf{F} = \nabla \times \mathbf{F} = \begin{vmatrix} \mathbf{i} & \mathbf{j} & \mathbf{k} \\ \dfrac{\partial}{\partial x} & \dfrac{\partial}{\partial y} & \dfrac{\partial}{\partial z} \\ xz & xyz & -y^2 \end{vmatrix}$$

$$= \left[\frac{\partial}{\partial y}(-y^2) - \frac{\partial}{\partial z}(xyz)\right]\mathbf{i} - \left[\frac{\partial}{\partial x}(-y^2) - \frac{\partial}{\partial z}(xz)\right]\mathbf{j} + \left[\frac{\partial}{\partial x}(xyz) - \frac{\partial}{\partial y}(xz)\right]\mathbf{k}$$

$$= (-2y - xy)\,\mathbf{i} - (0 - x)\,\mathbf{j} + (yz - 0)\,\mathbf{k} = -y(2 + x)\,\mathbf{i} + x\,\mathbf{j} + yz\,\mathbf{k} \qquad \blacksquare$$

- Most computer algebra systems have commands that compute the curl and divergence of vector fields. If you have access to a CAS, use these commands to check the answers to the examples and exercises in this section.

Recall that the gradient of a function f of three variables is a vector field on $\mathbb{R}^3$ and so we can compute its curl. The following theorem says that the curl of a gradient vector field is $\mathbf{0}$.

---

3 THEOREM If f is a function of three variables that has continuous second-order partial derivatives, then

$$\operatorname{curl}(\nabla f) = \mathbf{0}$$

PROOF We have

- Notice the similarity to what we know from Section 10.4: $\mathbf{a} \times \mathbf{a} = \mathbf{0}$ for every three-dimensional vector $\mathbf{a}$.

$$\operatorname{curl}(\nabla f) = \nabla \times (\nabla f) = \begin{vmatrix} \mathbf{i} & \mathbf{j} & \mathbf{k} \\ \dfrac{\partial}{\partial x} & \dfrac{\partial}{\partial y} & \dfrac{\partial}{\partial z} \\ \dfrac{\partial f}{\partial x} & \dfrac{\partial f}{\partial y} & \dfrac{\partial f}{\partial z} \end{vmatrix}$$

$$= \left(\frac{\partial^2 f}{\partial y\,\partial z} - \frac{\partial^2 f}{\partial z\,\partial y}\right)\mathbf{i} + \left(\frac{\partial^2 f}{\partial z\,\partial x} - \frac{\partial^2 f}{\partial x\,\partial z}\right)\mathbf{j} + \left(\frac{\partial^2 f}{\partial x\,\partial y} - \frac{\partial^2 f}{\partial y\,\partial x}\right)\mathbf{k}$$

$$= 0\,\mathbf{i} + 0\,\mathbf{j} + 0\,\mathbf{k} = \mathbf{0}$$

by Clairaut's Theorem. $\qquad \square$

- Compare this with Exercise 25 in Section 13.3.

Since a conservative vector field is one for which $\mathbf{F} = \nabla f$, Theorem 3 can be rephrased as follows:

If $\mathbf{F}$ is conservative, then curl $\mathbf{F} = \mathbf{0}$.

This gives us a way of verifying that a vector field is not conservative.

▼ EXAMPLE 2 Show that the vector field $\mathbf{F}(x, y, z) = xz\,\mathbf{i} + xyz\,\mathbf{j} - y^2\,\mathbf{k}$ is not conservative.

SOLUTION In Example 1 we showed that

$$\operatorname{curl}\mathbf{F} = -y(2 + x)\,\mathbf{i} + x\,\mathbf{j} + yz\,\mathbf{k}$$

This shows that curl $\mathbf{F} \neq \mathbf{0}$ and so, by Theorem 3, $\mathbf{F}$ is not conservative. $\qquad \blacksquare$

The converse of Theorem 3 is not true in general, but the following theorem says the converse is true if **F** is defined everywhere. (More generally it is true if the domain is simply-connected, that is, "has no hole.") Theorem 4 is the three-dimensional version of Theorem 13.3.6. Its proof requires Stokes' Theorem and is sketched at the end of Section 13.8.

---

4 **THEOREM**   If **F** is a vector field defined on all of $\mathbb{R}^3$ whose component functions have continuous partial derivatives and curl **F** = **0**, then **F** is a conservative vector field.

V **EXAMPLE 3**

(a) Show that
$$\mathbf{F}(x, y, z) = y^2z^3\,\mathbf{i} + 2xyz^3\,\mathbf{j} + 3xy^2z^2\,\mathbf{k}$$

is a conservative vector field.

(b) Find a function f such that $\mathbf{F} = \nabla f$.

SOLUTION

(a) We compute the curl of **F**:

$$\text{curl }\mathbf{F} = \nabla \times \mathbf{F} = \begin{vmatrix} \mathbf{i} & \mathbf{j} & \mathbf{k} \\ \dfrac{\partial}{\partial x} & \dfrac{\partial}{\partial y} & \dfrac{\partial}{\partial z} \\ y^2z^3 & 2xyz^3 & 3xy^2z^2 \end{vmatrix}$$

$$= (6xyz^2 - 6xyz^2)\,\mathbf{i} - (3y^2z^2 - 3y^2z^2)\,\mathbf{j} + (2yz^3 - 2yz^3)\,\mathbf{k} = \mathbf{0}$$

Since curl **F** = **0** and the domain of **F** is $\mathbb{R}^3$, **F** is a conservative vector field by Theorem 4.

(b) The technique for finding f was given in Section 13.3. We have

$$\boxed{5} \qquad\qquad f_x(x, y, z) = y^2z^3$$

$$\boxed{6} \qquad\qquad f_y(x, y, z) = 2xyz^3$$

$$\boxed{7} \qquad\qquad f_z(x, y, z) = 3xy^2z^2$$

Integrating $\boxed{5}$ with respect to x, we obtain

$$\boxed{8} \qquad\qquad f(x, y, z) = xy^2z^3 + g(y, z)$$

Differentiating $\boxed{8}$ with respect to y, we get $f_y(x, y, z) = 2xyz^3 + g_y(y, z)$, so comparison with $\boxed{6}$ gives $g_y(y, z) = 0$. Thus $g(y, z) = h(z)$ and

$$f_z(x, y, z) = 3xy^2z^2 + h'(z)$$

Then $\boxed{7}$ gives $h'(z) = 0$. Therefore

$$f(x, y, z) = xy^2z^3 + K \qquad\qquad\blacksquare$$

The reason for the name *curl* is that the curl vector is associated with rotations. One connection is explained in Exercise 35. Another occurs when **F** represents the velocity field in fluid flow (see Example 3 in Section 13.1). Particles near (x, y, z) in the fluid tend to rotate about the axis that points in the direction of curl $\mathbf{F}(x, y, z)$ and the length

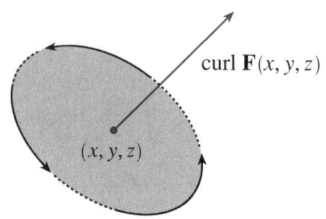

FIGURE 1

of this curl vector is a measure of how quickly the particles move around the axis (see Figure 1). If curl $\mathbf{F} = \mathbf{0}$ at a point P, then the fluid is free from rotations at P and $\mathbf{F}$ is called **irrotational** at P. In other words, there is no whirlpool or eddy at P. If curl $\mathbf{F} = \mathbf{0}$, then a tiny paddle wheel moves with the fluid but doesn't rotate about its axis. If curl $\mathbf{F} \neq \mathbf{0}$, the paddle wheel rotates about its axis. We give a more detailed explanation in Section 13.8 as a consequence of Stokes' Theorem.

DIVERGENCE

If $\mathbf{F} = P\,\mathbf{i} + Q\,\mathbf{j} + R\,\mathbf{k}$ is a vector field on $\mathbb{R}^3$ and $\partial P/\partial x$, $\partial Q/\partial y$, and $\partial R/\partial z$ exist, then the **divergence of F** is the function of three variables defined by

$$\boxed{9} \qquad \operatorname{div} \mathbf{F} = \frac{\partial P}{\partial x} + \frac{\partial Q}{\partial y} + \frac{\partial R}{\partial z}$$

Observe that curl $\mathbf{F}$ is a vector field but div $\mathbf{F}$ is a scalar field. In terms of the gradient operator $\nabla = (\partial/\partial x)\,\mathbf{i} + (\partial/\partial y)\,\mathbf{j} + (\partial/\partial z)\,\mathbf{k}$, the divergence of $\mathbf{F}$ can be written symbolically as the dot product of ∇ and $\mathbf{F}$:

$$\boxed{10} \qquad \operatorname{div} \mathbf{F} = \nabla \cdot \mathbf{F}$$

EXAMPLE 4 If $\mathbf{F}(x, y, z) = xz\,\mathbf{i} + xyz\,\mathbf{j} - y^2\,\mathbf{k}$, find div $\mathbf{F}$.

SOLUTION By the definition of divergence (Equation 9 or 10) we have

$$\operatorname{div} \mathbf{F} = \nabla \cdot \mathbf{F} = \frac{\partial}{\partial x}\,(xz) + \frac{\partial}{\partial y}\,(xyz) + \frac{\partial}{\partial z}\,(-y^2) = z + xz \qquad\blacksquare$$

If $\mathbf{F}$ is a vector field on $\mathbb{R}^3$, then curl $\mathbf{F}$ is also a vector field on $\mathbb{R}^3$. As such, we can compute its divergence. The next theorem shows that the result is 0.

---

11 THEOREM If $\mathbf{F} = P\,\mathbf{i} + Q\,\mathbf{j} + R\,\mathbf{k}$ is a vector field on $\mathbb{R}^3$ and P, Q, and R have continuous second-order partial derivatives, then

$$\operatorname{div} \operatorname{curl} \mathbf{F} = 0$$

■ Note the analogy with the scalar triple product: $\mathbf{a} \cdot (\mathbf{a} \times \mathbf{b}) = 0$.

PROOF Using the definitions of divergence and curl, we have

$$\operatorname{div} \operatorname{curl} \mathbf{F} = \nabla \cdot (\nabla \times \mathbf{F})$$

$$= \frac{\partial}{\partial x}\left(\frac{\partial R}{\partial y} - \frac{\partial Q}{\partial z}\right) + \frac{\partial}{\partial y}\left(\frac{\partial P}{\partial z} - \frac{\partial R}{\partial x}\right) + \frac{\partial}{\partial z}\left(\frac{\partial Q}{\partial x} - \frac{\partial P}{\partial y}\right)$$

$$= \frac{\partial^2 R}{\partial x\,\partial y} - \frac{\partial^2 Q}{\partial x\,\partial z} + \frac{\partial^2 P}{\partial y\,\partial z} - \frac{\partial^2 R}{\partial y\,\partial x} + \frac{\partial^2 Q}{\partial z\,\partial x} - \frac{\partial^2 P}{\partial z\,\partial y}$$

$$= 0$$

because the terms cancel in pairs by Clairaut's Theorem. □

V **EXAMPLE 5** Show that the vector field $\mathbf{F}(x, y, z) = xz\,\mathbf{i} + xyz\,\mathbf{j} - y^2\,\mathbf{k}$ can't be written as the curl of another vector field, that is, $\mathbf{F} \neq \operatorname{curl} \mathbf{G}$.

SOLUTION In Example 4 we showed that

$$\operatorname{div} \mathbf{F} = z + xz$$

and therefore $\operatorname{div} \mathbf{F} \neq 0$. If it were true that $\mathbf{F} = \operatorname{curl} \mathbf{G}$, then Theorem 11 would give

$$\operatorname{div} \mathbf{F} = \operatorname{div} \operatorname{curl} \mathbf{G} = 0$$

which contradicts $\operatorname{div} \mathbf{F} \neq 0$. Therefore $\mathbf{F}$ is not the curl of another vector field. ∎

■ The reason for this interpretation of div **F** will be explained at the end of Section 13.9 as a consequence of the Divergence Theorem.

Again, the reason for the name *divergence* can be understood in the context of fluid flow. If $\mathbf{F}(x, y, z)$ is the velocity of a fluid (or gas), then $\operatorname{div} \mathbf{F}(x, y, z)$ represents the net rate of change (with respect to time) of the mass of fluid (or gas) flowing from the point (x, y, z) per unit volume. In other words, $\operatorname{div} \mathbf{F}(x, y, z)$ measures the tendency of the fluid to diverge from the point (x, y, z). If $\operatorname{div} \mathbf{F} = 0$, then $\mathbf{F}$ is said to be **incompressible**.

Another differential operator occurs when we compute the divergence of a gradient vector field ∇f. If f is a function of three variables, we have

$$\operatorname{div}(\nabla f) = \nabla \cdot (\nabla f) = \frac{\partial^2 f}{\partial x^2} + \frac{\partial^2 f}{\partial y^2} + \frac{\partial^2 f}{\partial z^2}$$

and this expression occurs so often that we abbreviate it as $\nabla^2 f$. The operator

$$\nabla^2 = \nabla \cdot \nabla$$

is called the **Laplace operator** because of its relation to **Laplace's equation**

$$\nabla^2 f = \frac{\partial^2 f}{\partial x^2} + \frac{\partial^2 f}{\partial y^2} + \frac{\partial^2 f}{\partial z^2} = 0$$

We can also apply the Laplace operator ∇^2 to a vector field

$$\mathbf{F} = P\,\mathbf{i} + Q\,\mathbf{j} + R\,\mathbf{k}$$

in terms of its components:

$$\nabla^2 \mathbf{F} = \nabla^2 P\,\mathbf{i} + \nabla^2 Q\,\mathbf{j} + \nabla^2 R\,\mathbf{k}$$

VECTOR FORMS OF GREEN'S THEOREM

The curl and divergence operators allow us to rewrite Green's Theorem in versions that will be useful in our later work. We suppose that the plane region D, its boundary curve C, and the functions P and Q satisfy the hypotheses of Green's Theorem. Then we consider the vector field $\mathbf{F} = P\,\mathbf{i} + Q\,\mathbf{j}$. Its line integral is

$$\oint_C \mathbf{F} \cdot d\mathbf{r} = \oint_C P\,dx + Q\,dy$$

and, regarding **F** as a vector field on $\mathbb{R}^3$ with third component 0, we have

$$\text{curl } \mathbf{F} = \begin{vmatrix} \mathbf{i} & \mathbf{j} & \mathbf{k} \\ \dfrac{\partial}{\partial x} & \dfrac{\partial}{\partial y} & \dfrac{\partial}{\partial z} \\ P(x, y) & Q(x, y) & 0 \end{vmatrix} = \left(\dfrac{\partial Q}{\partial x} - \dfrac{\partial P}{\partial y} \right) \mathbf{k}$$

Therefore

$$(\text{curl } \mathbf{F}) \cdot \mathbf{k} = \left(\dfrac{\partial Q}{\partial x} - \dfrac{\partial P}{\partial y} \right) \mathbf{k} \cdot \mathbf{k} = \dfrac{\partial Q}{\partial x} - \dfrac{\partial P}{\partial y}$$

and we can now rewrite the equation in Green's Theorem in the vector form

12
$$\oint_C \mathbf{F} \cdot d\mathbf{r} = \iint_D (\text{curl } \mathbf{F}) \cdot \mathbf{k} \, dA$$

Equation 12 expresses the line integral of the tangential component of **F** along C as the double integral of the vertical component of curl **F** over the region D enclosed by C. We now derive a similar formula involving the *normal* component of **F**.

If C is given by the vector equation

$$\mathbf{r}(t) = x(t)\,\mathbf{i} + y(t)\,\mathbf{j} \qquad a \leqslant t \leqslant b$$

then the unit tangent vector (see Section 10.7) is

$$\mathbf{T}(t) = \dfrac{x'(t)}{|\mathbf{r}'(t)|}\,\mathbf{i} + \dfrac{y'(t)}{|\mathbf{r}'(t)|}\,\mathbf{j}$$

You can verify that the outward unit normal vector to C is given by

$$\mathbf{n}(t) = \dfrac{y'(t)}{|\mathbf{r}'(t)|}\,\mathbf{i} - \dfrac{x'(t)}{|\mathbf{r}'(t)|}\,\mathbf{j}$$

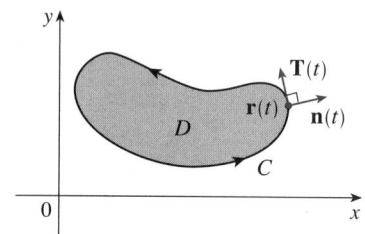

FIGURE 2

(See Figure 2.) Then, from Equation 13.2.3, we have

$$\oint_C \mathbf{F} \cdot \mathbf{n} \, ds = \int_a^b (\mathbf{F} \cdot \mathbf{n})(t)\,|\mathbf{r}'(t)|\,dt$$

$$= \int_a^b \left[\dfrac{P(x(t), y(t))\,y'(t)}{|\mathbf{r}'(t)|} - \dfrac{Q(x(t), y(t))\,x'(t)}{|\mathbf{r}'(t)|} \right] |\mathbf{r}'(t)|\,dt$$

$$= \int_a^b P(x(t), y(t))\,y'(t)\,dt - Q(x(t), y(t))\,x'(t)\,dt$$

$$= \int_C P\,dy - Q\,dx = \iint_D \left(\dfrac{\partial P}{\partial x} + \dfrac{\partial Q}{\partial y} \right) dA$$

by Green's Theorem. But the integrand in this double integral is just the divergence of **F**. So we have a second vector form of Green's Theorem.

$$\boxed{13} \qquad \oint_C \mathbf{F} \cdot \mathbf{n} \, ds = \iint_D \operatorname{div} \mathbf{F}(x, y) \, dA$$

This version says that the line integral of the normal component of $\mathbf{F}$ along C is equal to the double integral of the divergence of $\mathbf{F}$ over the region D enclosed by C.

13.5 | EXERCISES

1–7 ■ Find (a) the curl and (b) the divergence of the vector field.

1. $\mathbf{F}(x, y, z) = (x + yz)\,\mathbf{i} + (y + xz)\,\mathbf{j} + (z + xy)\,\mathbf{k}$

2. $\mathbf{F}(x, y, z) = xy^2z^3\,\mathbf{i} + x^3yz^2\,\mathbf{j} + x^2y^3z\,\mathbf{k}$

3. $\mathbf{F}(x, y, z) = xye^z\,\mathbf{i} + yze^x\,\mathbf{k}$

4. $\mathbf{F}(x, y, z) = \sin yz\,\mathbf{i} + \sin zx\,\mathbf{j} + \sin xy\,\mathbf{k}$

5. $\mathbf{F}(x, y, z) = \dfrac{1}{\sqrt{x^2 + y^2 + z^2}}\,(x\,\mathbf{i} + y\,\mathbf{j} + z\,\mathbf{k})$

6. $\mathbf{F}(x, y, z) = e^{xy}\sin z\,\mathbf{j} + y\tan^{-1}(x/z)\,\mathbf{k}$

7. $\mathbf{F}(x, y, z) = \langle e^x \sin y, \, e^y \sin z, \, e^z \sin x \rangle$

8–9 ■ The vector field $\mathbf{F}$ is shown in the xy-plane and looks the same in all other horizontal planes. (In other words, $\mathbf{F}$ is independent of z and its z-component is 0.)
(a) Is div $\mathbf{F}$ positive, negative, or zero? Explain.
(b) Determine whether curl $\mathbf{F} = \mathbf{0}$. If not, in which direction does curl $\mathbf{F}$ point?

8. 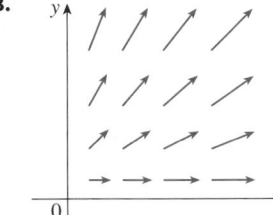   **9.**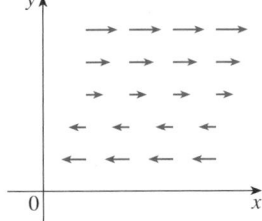

10. Let f be a scalar field and $\mathbf{F}$ a vector field. State whether each expression is meaningful. If not, explain why. If so, state whether it is a scalar field or a vector field.

(a) curl f (b) grad f

(c) div $\mathbf{F}$ (d) curl(grad f)

(e) grad $\mathbf{F}$ (f) grad(div $\mathbf{F}$)

(g) div(grad f)   (h) grad(div f)

(i) curl(curl $\mathbf{F}$)   (j) div(div $\mathbf{F}$)

(k) (grad f) $\times$ (div $\mathbf{F}$)   (l) div(curl(grad f))

11–16 ■ Determine whether or not the vector field is conservative. If it is conservative, find a function f such that $\mathbf{F} = \nabla f$.

11. $\mathbf{F}(x, y, z) = y^2z^3\,\mathbf{i} + 2xyz^3\,\mathbf{j} + 3xy^2z^2\,\mathbf{k}$

12. $\mathbf{F}(x, y, z) = xyz^2\,\mathbf{i} + x^2yz^2\,\mathbf{j} + x^2y^2z\,\mathbf{k}$

13. $\mathbf{F}(x, y, z) = 3xy^2z^2\,\mathbf{i} + 2x^2yz^3\,\mathbf{j} + 3x^2y^2z^2\,\mathbf{k}$

14. $\mathbf{F}(x, y, z) = \mathbf{i} + \sin z\,\mathbf{j} + y\cos z\,\mathbf{k}$

15. $\mathbf{F}(x, y, z) = e^{yz}\,\mathbf{i} + xze^{yz}\,\mathbf{j} + xye^{yz}\,\mathbf{k}$

16. $\mathbf{F}(x, y, z) = e^x \sin yz\,\mathbf{i} + ze^x \cos yz\,\mathbf{j} + ye^x \cos yz\,\mathbf{k}$

17. Is there a vector field $\mathbf{G}$ on $\mathbb{R}^3$ such that curl $\mathbf{G} = \langle x \sin y, \cos y, z - xy \rangle$? Explain.

18. Is there a vector field $\mathbf{G}$ on $\mathbb{R}^3$ such that curl $\mathbf{G} = \langle xyz, -y^2z, yz^2 \rangle$? Explain.

19. Show that any vector field of the form

$$\mathbf{F}(x, y, z) = f(x)\,\mathbf{i} + g(y)\,\mathbf{j} + h(z)\,\mathbf{k}$$

where f, g, h are differentiable functions, is irrotational.

20. Show that any vector field of the form

$$\mathbf{F}(x, y, z) = f(y, z)\,\mathbf{i} + g(x, z)\,\mathbf{j} + h(x, y)\,\mathbf{k}$$

is incompressible.

21–27 ■ Prove the identity, assuming that the appropriate partial derivatives exist and are continuous. If f is a scalar field and $\mathbf{F}$, $\mathbf{G}$ are vector fields, then $f\mathbf{F}$, $\mathbf{F} \cdot \mathbf{G}$, and $\mathbf{F} \times \mathbf{G}$ are defined by

$$(f\mathbf{F})(x, y, z) = f(x, y, z)\,\mathbf{F}(x, y, z)$$

$$(\mathbf{F} \cdot \mathbf{G})(x, y, z) = \mathbf{F}(x, y, z) \cdot \mathbf{G}(x, y, z)$$

$$(\mathbf{F} \times \mathbf{G})(x, y, z) = \mathbf{F}(x, y, z) \times \mathbf{G}(x, y, z)$$

21. $\operatorname{div}(\mathbf{F} + \mathbf{G}) = \operatorname{div} \mathbf{F} + \operatorname{div} \mathbf{G}$

22. $\operatorname{curl}(\mathbf{F} + \mathbf{G}) = \operatorname{curl} \mathbf{F} + \operatorname{curl} \mathbf{G}$

23. $\operatorname{div}(f\,\mathbf{F}) = f\,\operatorname{div}\mathbf{F} + \mathbf{F}\cdot\nabla f$

24. $\operatorname{curl}(f\,\mathbf{F}) = f\,\operatorname{curl}\mathbf{F} + (\nabla f)\times\mathbf{F}$

25. $\operatorname{div}(\mathbf{F}\times\mathbf{G}) = \mathbf{G}\cdot\operatorname{curl}\mathbf{F} - \mathbf{F}\cdot\operatorname{curl}\mathbf{G}$

26. $\operatorname{div}(\nabla f\times\nabla g) = 0$

27. $\operatorname{curl}(\operatorname{curl}\mathbf{F}) = \operatorname{grad}(\operatorname{div}\mathbf{F}) - \nabla^2\mathbf{F}$

28–30 ▪ Let $\mathbf{r} = x\,\mathbf{i} + y\,\mathbf{j} + z\,\mathbf{k}$ and $r = |\mathbf{r}|$.

28. Verify each identity.
  (a) $\nabla\cdot\mathbf{r} = 3$ (b) $\nabla\cdot(r\mathbf{r}) = 4r$
  (c) $\nabla^2 r^3 = 12r$

29. Verify each identity.
  (a) $\nabla r = \mathbf{r}/r$ (b) $\nabla\times\mathbf{r} = \mathbf{0}$
  (c) $\nabla(1/r) = -\mathbf{r}/r^3$ (d) $\nabla\ln r = \mathbf{r}/r^2$

30. If $\mathbf{F} = \mathbf{r}/r^p$, find div $\mathbf{F}$. Is there a value of p for which div $\mathbf{F} = 0$?

---

31. Use Green's Theorem in the form of Equation 13 to prove **Green's first identity**:

$$\iint_D f\nabla^2 g\,dA = \oint_C f(\nabla g)\cdot\mathbf{n}\,ds - \iint_D \nabla f\cdot\nabla g\,dA$$

where D and C satisfy the hypotheses of Green's Theorem and the appropriate partial derivatives of f and g exist and are continuous. (The quantity $\nabla g\cdot\mathbf{n} = D_{\mathbf{n}}g$ occurs in the line integral. This is the directional derivative in the direction of the normal vector $\mathbf{n}$ and is called the **normal derivative** of g.)

32. Use Green's first identity (Exercise 31) to prove **Green's second identity**:

$$\iint_D (f\nabla^2 g - g\nabla^2 f)\,dA = \oint_C (f\nabla g - g\nabla f)\cdot\mathbf{n}\,ds$$

where D and C satisfy the hypotheses of Green's Theorem and the appropriate partial derivatives of f and g exist and are continuous.

33. Recall from Section 11.3 that a function g is called *harmonic* on D if it satisfies Laplace's equation, that is, $\nabla^2 g = 0$ on D. Use Green's first identity (with the same hypotheses as in Exercise 31) to show that if g is harmonic on D, then $\oint_C D_{\mathbf{n}}g\,ds = 0$. Here $D_{\mathbf{n}}g$ is the normal derivative of g defined in Exercise 31.

34. Use Green's first identity to show that if f is harmonic on D, and if $f(x, y) = 0$ on the boundary curve C, then $\iint_D |\nabla f|^2\,dA = 0$. (Assume the same hypotheses as in Exercise 31.)

35. This exercise demonstrates a connection between the curl vector and rotations. Let B be a rigid body rotating about the z-axis. The rotation can be described by the vector $\mathbf{w} = \omega\mathbf{k}$, where ω is the angular speed of B, that is, the tangential speed of any point P in B divided by the distance d from the axis of rotation. Let $\mathbf{r} = \langle x, y, z\rangle$ be the position vector of P.
  (a) By considering the angle θ in the figure, show that the velocity field of B is given by $\mathbf{v} = \mathbf{w}\times\mathbf{r}$.
  (b) Show that $\mathbf{v} = -\omega y\,\mathbf{i} + \omega x\,\mathbf{j}$.
  (c) Show that curl $\mathbf{v} = 2\mathbf{w}$.

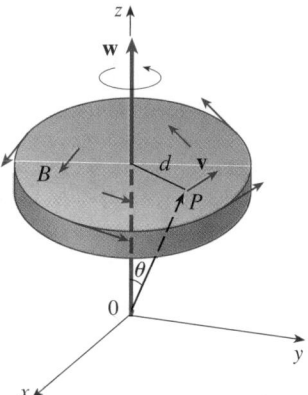

36. Maxwell's equations relating the electric field $\mathbf{E}$ and magnetic field $\mathbf{H}$ as they vary with time in a region containing no charge and no current can be stated as follows:

$$\operatorname{div}\mathbf{E} = 0 \qquad\qquad \operatorname{div}\mathbf{H} = 0$$

$$\operatorname{curl}\mathbf{E} = -\frac{1}{c}\frac{\partial\mathbf{H}}{\partial t} \qquad \operatorname{curl}\mathbf{H} = \frac{1}{c}\frac{\partial\mathbf{E}}{\partial t}$$

where c is the speed of light. Use these equations to prove the following:

  (a) $\nabla\times(\nabla\times\mathbf{E}) = -\dfrac{1}{c^2}\dfrac{\partial^2\mathbf{E}}{\partial t^2}$

  (b) $\nabla\times(\nabla\times\mathbf{H}) = -\dfrac{1}{c^2}\dfrac{\partial^2\mathbf{H}}{\partial t^2}$

  (c) $\nabla^2\mathbf{E} = \dfrac{1}{c^2}\dfrac{\partial^2\mathbf{E}}{\partial t^2}$ [*Hint:* Use Exercise 27.]

  (d) $\nabla^2\mathbf{H} = \dfrac{1}{c^2}\dfrac{\partial^2\mathbf{H}}{\partial t^2}$

13.6 | PARAMETRIC SURFACES AND THEIR AREAS

So far we have considered special types of surfaces: cylinders, quadric surfaces, graphs of functions of two variables, and level surfaces of functions of three variables. Here we use vector functions to describe more general surfaces, called *parametric surfaces*, and compute their areas. Then we take the general surface area formula and see how it applies to special surfaces.

PARAMETRIC SURFACES

In much the same way that we describe a space curve by a vector function $\mathbf{r}(t)$ of a single parameter t, we can describe a surface by a vector function $\mathbf{r}(u, v)$ of two parameters u and v. We suppose that

$$\boxed{1} \qquad \mathbf{r}(u, v) = x(u, v)\,\mathbf{i} + y(u, v)\,\mathbf{j} + z(u, v)\,\mathbf{k}$$

is a vector-valued function defined on a region D in the uv-plane. So x, y, and z, the component functions of $\mathbf{r}$, are functions of the two variables u and v with domain D. The set of all points (x, y, z) in $\mathbb{R}^3$ such that

$$\boxed{2} \qquad x = x(u, v) \qquad y = y(u, v) \qquad z = z(u, v)$$

and (u, v) varies throughout D, is called a **parametric surface** S and Equations 2 are called **parametric equations** of S. Each choice of u and v gives a point on S; by making all choices, we get all of S. In other words, the surface S is traced out by the tip of the position vector $\mathbf{r}(u, v)$ as (u, v) moves throughout the region D. (See Figure 1.)

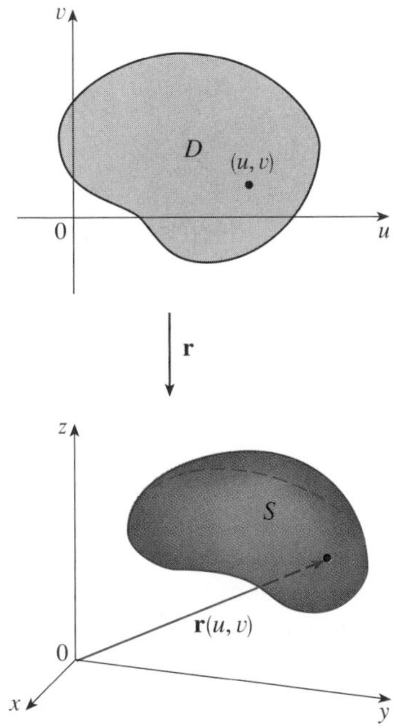

FIGURE 1
A parametric surface

EXAMPLE 1 Identify and sketch the surface with vector equation

$$\mathbf{r}(u, v) = 2 \cos u\,\mathbf{i} + v\,\mathbf{j} + 2 \sin u\,\mathbf{k}$$

SOLUTION The parametric equations for this surface are

$$x = 2 \cos u \qquad y = v \qquad z = 2 \sin u$$

So for any point (x, y, z) on the surface, we have

$$x^2 + z^2 = 4 \cos^2 u + 4 \sin^2 u = 4$$

This means that vertical cross-sections parallel to the xz-plane (that is, with y constant) are all circles with radius 2. Since $y = v$ and no restriction is placed on v, the surface is a circular cylinder with radius 2 whose axis is the y-axis (see Figure 2). ∎

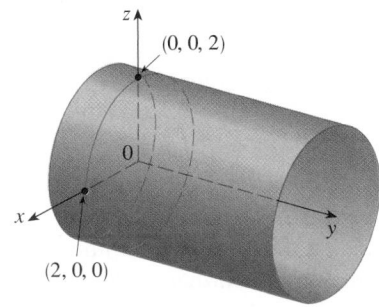

FIGURE 2

In Example 1 we placed no restrictions on the parameters u and v and so we got the entire cylinder. If, for instance, we restrict u and v by writing the parameter domain as

$$0 \le u \le \pi/2 \qquad 0 \le v \le 3$$

then $x \ge 0$, $z \ge 0$, $0 \le y \le 3$, and we get the quarter-cylinder with length 3 illustrated in Figure 3.

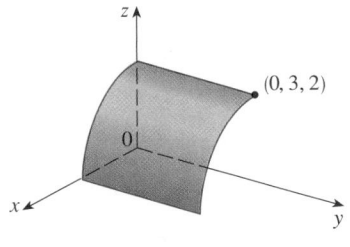

FIGURE 3

If a parametric surface S is given by a vector function $\mathbf{r}(u, v)$, then there are two useful families of curves that lie on S, one family with u constant and the other with v constant. These families correspond to vertical and horizontal lines in the uv-plane. If we keep u constant by putting $u = u_0$, then $\mathbf{r}(u_0, v)$ becomes a vector function of the single parameter v and defines a curve C_1 lying on S. (See Figure 4.)

TEC Visual 13.6 shows animated versions of Figures 4 and 5, with moving grid curves, for several parametric surfaces.

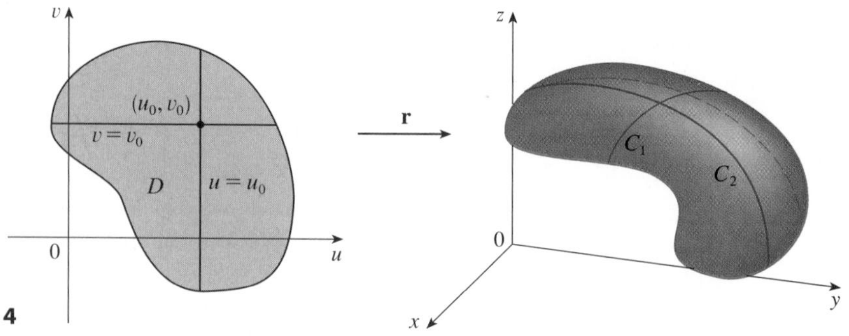

FIGURE 4

Similarly, if we keep v constant by putting $v = v_0$, we get a curve C_2 given by $\mathbf{r}(u, v_0)$ that lies on S. We call these curves **grid curves**. (In Example 1, for instance, the grid curves obtained by letting u be constant are horizontal lines whereas the grid curves with v constant are circles.) In fact, when a computer graphs a parametric surface, it usually depicts the surface by plotting these grid curves, as we see in the following example.

EXAMPLE 2 Use a computer algebra system to graph the surface

$$\mathbf{r}(u, v) = \langle (2 + \sin v) \cos u, (2 + \sin v) \sin u, u + \cos v \rangle$$

Which grid curves have u constant? Which have v constant?

SOLUTION We graph the portion of the surface with parameter domain $0 \le u \le 4\pi$, $0 \le v \le 2\pi$ in Figure 5. It has the appearance of a spiral tube. To identify the grid curves, we write the corresponding parametric equations:

$$x = (2 + \sin v) \cos u \qquad y = (2 + \sin v) \sin u \qquad z = u + \cos v$$

If v is constant, then $\sin v$ and $\cos v$ are constant, so the parametric equations resemble those of the helix in Example 4 in Section 10.7. So the grid curves with v constant are the spiral curves in Figure 5. We deduce that the grid curves with u constant must be the curves that look like circles in the figure. Further evidence for this assertion is that if u is kept constant, $u = u_0$, then the equation $z = u_0 + \cos v$ shows that the z-values vary from $u_0 - 1$ to $u_0 + 1$. ∎

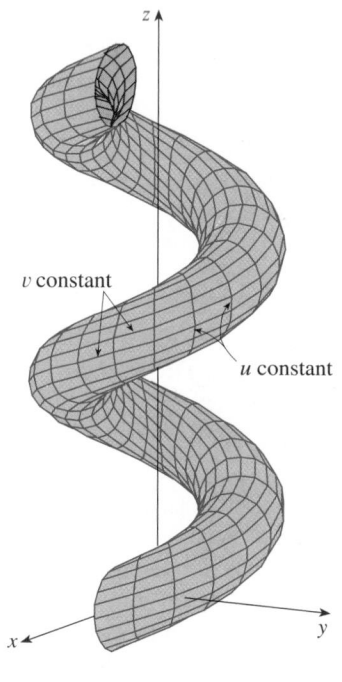

v constant

u constant

FIGURE 5

In Examples 1 and 2 we were given a vector equation and asked to graph the corresponding parametric surface. In the following examples, however, we are given the more challenging problem of finding a vector function to represent a given surface. In the rest of this chapter we will often need to do exactly that.

EXAMPLE 3 Find a vector function that represents the plane that passes through the point P_0 with position vector $\mathbf{r}_0$ and that contains two nonparallel vectors $\mathbf{a}$ and $\mathbf{b}$.

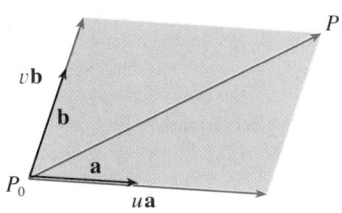

FIGURE 6

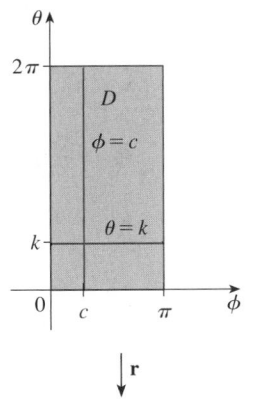

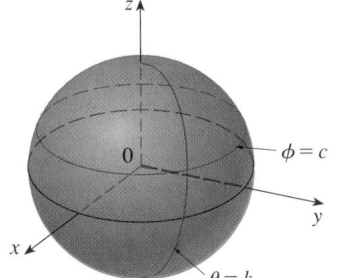

FIGURE 7

■ One of the uses of parametric surfaces is in computer graphics. Figure 8 shows the result of trying to graph the sphere $x^2 + y^2 + z^2 = 1$ by solving the equation for z and graphing the top and bottom hemispheres separately. Part of the sphere appears to be missing because of the rectangular grid system used by the computer. The much better picture in Figure 9 was produced by a computer using the parametric equations found in Example 4.

SOLUTION If P is any point in the plane, we can get from P_0 to P by moving a certain distance in the direction of $\mathbf{a}$ and another distance in the direction of $\mathbf{b}$. So there are scalars u and v such that $\overrightarrow{P_0P} = u\mathbf{a} + v\mathbf{b}$. (Figure 6 illustrates how this works, by means of the Parallelogram Law, for the case where u and v are positive. See also Exercise 36 in Section 10.2.) If $\mathbf{r}$ is the position vector of P, then

$$\mathbf{r} = \overrightarrow{OP_0} + \overrightarrow{P_0P} = \mathbf{r}_0 + u\mathbf{a} + v\mathbf{b}$$

So the vector equation of the plane can be written as

$$\mathbf{r}(u, v) = \mathbf{r}_0 + u\mathbf{a} + v\mathbf{b}$$

where u and v are real numbers.

If we write $\mathbf{r} = \langle x, y, z \rangle$, $\mathbf{r}_0 = \langle x_0, y_0, z_0 \rangle$, $\mathbf{a} = \langle a_1, a_2, a_3 \rangle$, and $\mathbf{b} = \langle b_1, b_2, b_3 \rangle$, then we can write the parametric equations of the plane through the point (x_0, y_0, z_0) as follows:

$$x = x_0 + ua_1 + vb_1 \qquad y = y_0 + ua_2 + vb_2 \qquad z = z_0 + ua_3 + vb_3 \qquad ■$$

▼ EXAMPLE 4 Find a parametric representation of the sphere $x^2 + y^2 + z^2 = a^2$.

SOLUTION The sphere has a simple representation $\rho = a$ in spherical coordinates, so let's choose the angles ϕ and θ in spherical coordinates as the parameters (see Section 12.7). Then, putting $\rho = a$ in the equations for conversion from spherical to rectangular coordinates (Equations 12.7.1), we obtain

$$x = a \sin \phi \cos \theta \qquad y = a \sin \phi \sin \theta \qquad z = a \cos \phi$$

as the parametric equations of the sphere. The corresponding vector equation is

$$\mathbf{r}(\phi, \theta) = a \sin \phi \cos \theta \, \mathbf{i} + a \sin \phi \sin \theta \, \mathbf{j} + a \cos \phi \, \mathbf{k}$$

We have $0 \le \phi \le \pi$ and $0 \le \theta \le 2\pi$, so the parameter domain is the rectangle $D = [0, \pi] \times [0, 2\pi]$. The grid curves with ϕ constant are the circles of constant latitude (including the equator). The grid curves with θ constant are the meridians (semicircles), which connect the north and south poles (see Figure 7). ■

NOTE We saw in Example 4 that the grid curves for a sphere are curves of constant latitude and longitude. For a general parametric surface we are really making a map and the grid curves are similar to lines of latitude and longitude. Describing a point on a parametric surface (like the one in Figure 5) by giving specific values of u and v is like giving the latitude and longitude of a point.

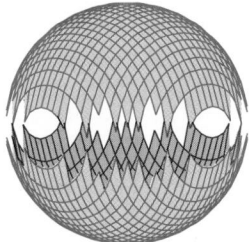

FIGURE 8

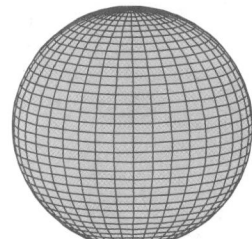

FIGURE 9

EXAMPLE 5 Find a parametric representation for the cylinder

$$x^2 + y^2 = 4 \qquad 0 \leqslant z \leqslant 1$$

SOLUTION The cylinder has a simple representation $r = 2$ in cylindrical coordinates, so we choose as parameters θ and z in cylindrical coordinates. Then the parametric equations of the cylinder are

$$x = 2\cos\theta \qquad y = 2\sin\theta \qquad z = z$$

where $0 \leqslant \theta \leqslant 2\pi$ and $0 \leqslant z \leqslant 1$. ∎

TEC In Module 13.6 you can investigate several families of parametric surfaces.

Parametric representations (also called parametrizations) of surfaces are not unique. The next example shows two ways to parametrize a cone.

EXAMPLE 6 Find a parametric representation for the surface $z = 2\sqrt{x^2 + y^2}$, that is, the top half of the cone $z^2 = 4x^2 + 4y^2$.

SOLUTION 1 If we regard x and y as parameters, then the parametric equations are simply

$$x = x \qquad y = y \qquad z = 2\sqrt{x^2 + y^2}$$

and the vector equation is

$$\mathbf{r}(x, y) = x\,\mathbf{i} + y\,\mathbf{j} + 2\sqrt{x^2 + y^2}\,\mathbf{k}$$

■ For some purposes the parametric representations in Solutions 1 and 2 are equally good, but Solution 2 might be preferable in certain situations. If we are interested only in the part of the cone that lies below the plane $z = 1$, for instance, all we have to do in Solution 2 is change the parameter domain to

$$0 \leqslant r \leqslant \tfrac{1}{2} \qquad 0 \leqslant \theta \leqslant 2\pi$$

SOLUTION 2 Another representation results from choosing as parameters the polar coordinates r and θ. A point (x, y, z) on the cone satisfies $x = r\cos\theta$, $y = r\sin\theta$, and $z = 2\sqrt{x^2 + y^2} = 2r$. So a vector equation for the cone is

$$\mathbf{r}(r, \theta) = r\cos\theta\,\mathbf{i} + r\sin\theta\,\mathbf{j} + 2r\,\mathbf{k}$$

where $r \geqslant 0$ and $0 \leqslant \theta \leqslant 2\pi$. ∎

As in the first solution of Example 6, a general surface given as the graph of a function of x and y, that is, with an equation of the form $z = f(x, y)$, can always be regarded as a parametric surface by taking x and y as parameters and writing the parametric equations as

$$x = x \qquad y = y \qquad z = f(x, y)$$

SURFACES OF REVOLUTION

Surfaces of revolution can be represented parametrically and thus graphed using a computer. For instance, let's consider the surface S obtained by rotating the curve $y = f(x)$, $a \leqslant x \leqslant b$, about the x-axis, where $f(x) \geqslant 0$. Let θ be the angle of rotation as shown in Figure 10. If (x, y, z) is a point on S, then

$$\boxed{3} \qquad x = x \qquad y = f(x)\cos\theta \qquad z = f(x)\sin\theta$$

Therefore we take x and θ as parameters and regard Equations 3 as parametric equations of S. The parameter domain is given by $a \leqslant x \leqslant b$, $0 \leqslant \theta \leqslant 2\pi$.

EXAMPLE 7 Find parametric equations for the surface generated by rotating the curve $y = \sin x$, $0 \leqslant x \leqslant 2\pi$, about the x-axis. Use these equations to graph the surface of revolution.

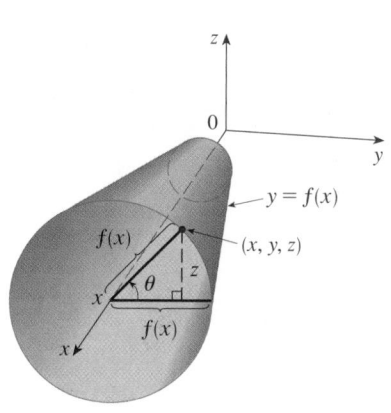

FIGURE 10

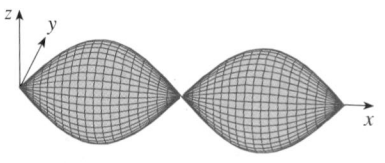

FIGURE 11

SOLUTION From Equations 3, the parametric equations are

$$x = x \qquad y = \sin x \cos \theta \qquad z = \sin x \sin \theta$$

and the parameter domain is $0 \le x \le 2\pi$, $0 \le \theta \le 2\pi$. Using a computer to plot these equations and rotate the image, we obtain the graph in Figure 11. ∎

We can adapt Equations 3 to represent a surface obtained through revolution about the y- or z-axis. (See Exercise 26.)

TANGENT PLANES

We now find the tangent plane to a parametric surface S traced out by a vector function

$$\mathbf{r}(u, v) = x(u, v)\,\mathbf{i} + y(u, v)\,\mathbf{j} + z(u, v)\,\mathbf{k}$$

at a point P_0 with position vector $\mathbf{r}(u_0, v_0)$. If we keep u constant by putting $u = u_0$, then $\mathbf{r}(u_0, v)$ becomes a vector function of the single parameter v and defines a grid curve C_1 lying on S. (See Figure 12.) The tangent vector to C_1 at P_0 is obtained by taking the partial derivative of $\mathbf{r}$ with respect to v:

$$\boxed{4} \qquad \mathbf{r}_v = \frac{\partial x}{\partial v}(u_0, v_0)\,\mathbf{i} + \frac{\partial y}{\partial v}(u_0, v_0)\,\mathbf{j} + \frac{\partial z}{\partial v}(u_0, v_0)\,\mathbf{k}$$

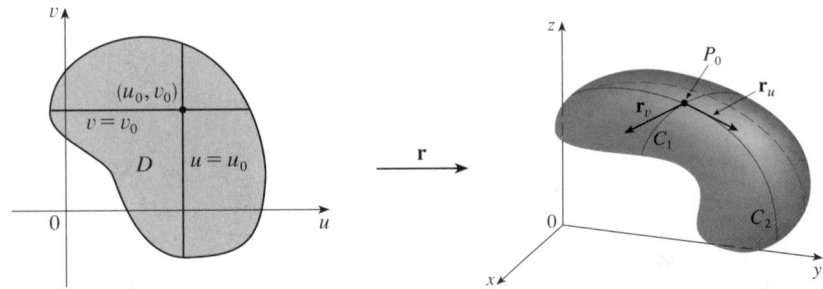

FIGURE 12

Similarly, if we keep v constant by putting $v = v_0$, we get a grid curve C_2 given by $\mathbf{r}(u, v_0)$ that lies on S, and its tangent vector at P_0 is

$$\boxed{5} \qquad \mathbf{r}_u = \frac{\partial x}{\partial u}(u_0, v_0)\,\mathbf{i} + \frac{\partial y}{\partial u}(u_0, v_0)\,\mathbf{j} + \frac{\partial z}{\partial u}(u_0, v_0)\,\mathbf{k}$$

If $\mathbf{r}_u \times \mathbf{r}_v$ is not $\mathbf{0}$, then the surface S is called **smooth** (it has no "corners"). For a smooth surface, the **tangent plane** is the plane that contains the tangent vectors $\mathbf{r}_u$ and $\mathbf{r}_v$, and the vector $\mathbf{r}_u \times \mathbf{r}_v$ is a normal vector to the tangent plane.

▼ EXAMPLE 8 Find the tangent plane to the surface with parametric equations $x = u^2$, $y = v^2$, $z = u + 2v$ at the point $(1, 1, 3)$.

SOLUTION We first compute the tangent vectors:

$$\mathbf{r}_u = \frac{\partial x}{\partial u}\,\mathbf{i} + \frac{\partial y}{\partial u}\,\mathbf{j} + \frac{\partial z}{\partial u}\,\mathbf{k} = 2u\,\mathbf{i} + \mathbf{k}$$

$$\mathbf{r}_v = \frac{\partial x}{\partial v}\,\mathbf{i} + \frac{\partial y}{\partial v}\,\mathbf{j} + \frac{\partial z}{\partial v}\,\mathbf{k} = 2v\,\mathbf{j} + 2\,\mathbf{k}$$

■ Figure 13 shows the self-intersecting surface in Example 8 and its tangent plane at (1, 1, 3).

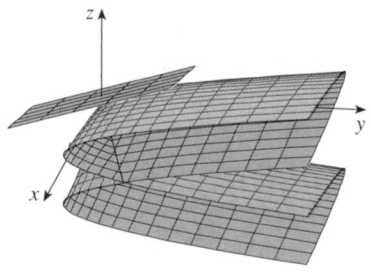

FIGURE 13

Thus a normal vector to the tangent plane is

$$\mathbf{r}_u \times \mathbf{r}_v = \begin{vmatrix} \mathbf{i} & \mathbf{j} & \mathbf{k} \\ 2u & 0 & 1 \\ 0 & 2v & 2 \end{vmatrix} = -2v\,\mathbf{i} - 4u\,\mathbf{j} + 4uv\,\mathbf{k}$$

Notice that the point (1, 1, 3) corresponds to the parameter values $u = 1$ and $v = 1$, so the normal vector there is

$$-2\,\mathbf{i} - 4\,\mathbf{j} + 4\,\mathbf{k}$$

Therefore an equation of the tangent plane at (1, 1, 3) is

$$-2(x - 1) - 4(y - 1) + 4(z - 3) = 0$$

or

$$x + 2y - 2z + 3 = 0 \qquad\blacksquare$$

SURFACE AREA

Now we define the surface area of a general parametric surface given by Equation 1. For simplicity we start by considering a surface whose parameter domain D is a rectangle, and we divide it into subrectangles R_{ij}. Let's choose (u_i^*, v_j^*) to be the lower left corner of R_{ij}. (See Figure 14.)

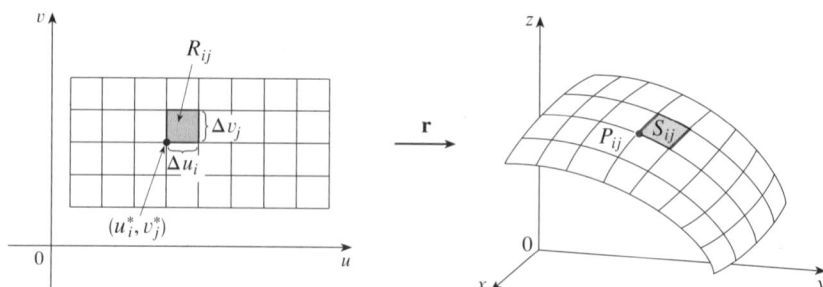

FIGURE 14

The image of the subrectangle R_{ij} is the patch S_{ij}.

The part S_{ij} of the surface S that corresponds to R_{ij} is called a *patch* and has the point P_{ij} with position vector $\mathbf{r}(u_i^*, v_j^*)$ as one of its corners. Let

$$\mathbf{r}_u^* = \mathbf{r}_u(u_i^*, v_j^*) \qquad \text{and} \qquad \mathbf{r}_v^* = \mathbf{r}_v(u_i^*, v_j^*)$$

be the tangent vectors at P_{ij} as given by Equations 5 and 4.

Figure 15(a) shows how the two edges of the patch that meet at P_{ij} can be approximated by vectors. These vectors, in turn, can be approximated by the vectors $\Delta u_i\,\mathbf{r}_u^*$ and $\Delta v_j\,\mathbf{r}_v^*$ because partial derivatives can be approximated by difference quotients. So we approximate S_{ij} by the parallelogram determined by the vectors $\Delta u_i\,\mathbf{r}_u^*$ and $\Delta v_j\,\mathbf{r}_v^*$. This parallelogram is shown in Figure 15(b) and lies in the tangent plane to S at P_{ij}. The area of this parallelogram is

$$\left| (\Delta u_i\,\mathbf{r}_u^*) \times (\Delta v_j\,\mathbf{r}_v^*) \right| = \left| \mathbf{r}_u^* \times \mathbf{r}_v^* \right| \Delta u_i\,\Delta v_j$$

and so an approximation to the area of S is

$$\sum_{i=1}^{m} \sum_{j=1}^{n} \left| \mathbf{r}_u^* \times \mathbf{r}_v^* \right| \Delta u_i\,\Delta v_j$$

Our intuition tells us that this approximation gets better as we increase the number of subrectangles and their dimensions decrease, and we recognize the double sum as a Riemann sum for the double integral $\iint_D \left| \mathbf{r}_u \times \mathbf{r}_v \right| du\,dv$. This motivates the following definition.

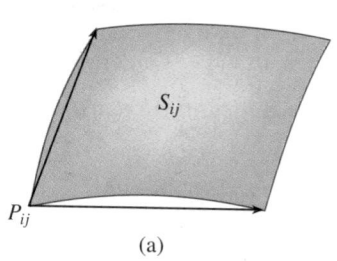

(a)

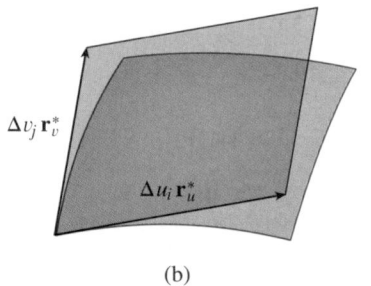

(b)

FIGURE 15

Approximating a patch by a parallelogram

---

6 **DEFINITION** If a smooth parametric surface S is given by the equation

$$\mathbf{r}(u, v) = x(u, v)\,\mathbf{i} + y(u, v)\,\mathbf{j} + z(u, v)\,\mathbf{k} \qquad (u, v) \in D$$

and S is covered just once as (u, v) ranges throughout the parameter domain D, then the **surface area** of S is

$$A(S) = \iint\limits_{D} |\mathbf{r}_u \times \mathbf{r}_v|\, dA$$

where $\qquad \mathbf{r}_u = \dfrac{\partial x}{\partial u}\,\mathbf{i} + \dfrac{\partial y}{\partial u}\,\mathbf{j} + \dfrac{\partial z}{\partial u}\,\mathbf{k} \qquad \mathbf{r}_v = \dfrac{\partial x}{\partial v}\,\mathbf{i} + \dfrac{\partial y}{\partial v}\,\mathbf{j} + \dfrac{\partial z}{\partial v}\,\mathbf{k}$

EXAMPLE 9 Find the surface area of a sphere of radius a.

SOLUTION In Example 4 we found the parametric representation

$$x = a \sin \phi \, \cos \theta \qquad y = a \sin \phi \, \sin \theta \qquad z = a \cos \phi$$

where the parameter domain is

$$D = \big\{ (\phi, \theta) \mid 0 \le \phi \le \pi, \ 0 \le \theta \le 2\pi \big\}$$

We first compute the cross product of the tangent vectors:

$$\mathbf{r}_\phi \times \mathbf{r}_\theta = \begin{vmatrix} \mathbf{i} & \mathbf{j} & \mathbf{k} \\[4pt] \dfrac{\partial x}{\partial \phi} & \dfrac{\partial y}{\partial \phi} & \dfrac{\partial z}{\partial \phi} \\[8pt] \dfrac{\partial x}{\partial \theta} & \dfrac{\partial y}{\partial \theta} & \dfrac{\partial z}{\partial \theta} \end{vmatrix} = \begin{vmatrix} \mathbf{i} & \mathbf{j} & \mathbf{k} \\[4pt] a \cos \phi \, \cos \theta & a \cos \phi \, \sin \theta & -a \sin \phi \\[4pt] -a \sin \phi \, \sin \theta & a \sin \phi \, \cos \theta & 0 \end{vmatrix}$$

$$= a^2 \sin^2\!\phi \, \cos \theta \, \mathbf{i} + a^2 \sin^2\!\phi \, \sin \theta \, \mathbf{j} + a^2 \sin \phi \, \cos \phi \, \mathbf{k}$$

Thus

$$|\mathbf{r}_\phi \times \mathbf{r}_\theta| = \sqrt{a^4 \sin^4\!\phi \, \cos^2\theta + a^4 \sin^4\!\phi \, \sin^2\theta + a^4 \sin^2\!\phi \, \cos^2\phi}$$

$$= \sqrt{a^4 \sin^4\!\phi + a^4 \sin^2\!\phi \, \cos^2\phi} = a^2 \sqrt{\sin^2\!\phi} = a^2 \sin \phi$$

since $\sin \phi \ge 0$ for $0 \le \phi \le \pi$. Therefore, by Definition 6, the area of the sphere is

$$A = \iint\limits_{D} |\mathbf{r}_\phi \times \mathbf{r}_\theta|\, dA = \int_0^{2\pi} \int_0^{\pi} a^2 \sin \phi \, d\phi \, d\theta$$

$$= a^2 \int_0^{2\pi} d\theta \int_0^{\pi} \sin \phi \, d\phi = a^2(2\pi)2 = 4\pi a^2 \qquad \blacksquare$$

SURFACE AREA OF THE GRAPH OF A FUNCTION

For the special case of a surface S with equation $z = f(x, y)$, where (x, y) lies in D and f has continuous partial derivatives, we take x and y as parameters. The parametric equations are

$$x = x \qquad y = y \qquad z = f(x, y)$$

so $$\mathbf{r}_x = \mathbf{i} + \left(\frac{\partial f}{\partial x}\right)\mathbf{k} \qquad \mathbf{r}_y = \mathbf{j} + \left(\frac{\partial f}{\partial y}\right)\mathbf{k}$$

and

$$\boxed{7} \qquad \mathbf{r}_x \times \mathbf{r}_y = \begin{vmatrix} \mathbf{i} & \mathbf{j} & \mathbf{k} \\ 1 & 0 & \dfrac{\partial f}{\partial x} \\ 0 & 1 & \dfrac{\partial f}{\partial y} \end{vmatrix} = -\frac{\partial f}{\partial x}\mathbf{i} - \frac{\partial f}{\partial y}\mathbf{j} + \mathbf{k}$$

Thus we have

$$\boxed{8} \qquad |\mathbf{r}_x \times \mathbf{r}_y| = \sqrt{\left(\frac{\partial f}{\partial x}\right)^2 + \left(\frac{\partial f}{\partial y}\right)^2 + 1} = \sqrt{1 + \left(\frac{\partial z}{\partial x}\right)^2 + \left(\frac{\partial z}{\partial y}\right)^2}$$

and the surface area formula in Definition 6 becomes

$$\boxed{9} \qquad A(S) = \iint_D \sqrt{1 + \left(\frac{\partial z}{\partial x}\right)^2 + \left(\frac{\partial z}{\partial y}\right)^2}\, dA$$

■ Notice the similarity between the surface area formula in Equation 9 and the arc length formula

$$L = \int_a^b \sqrt{1 + \left(\frac{dy}{dx}\right)^2}\, dx$$

from Section 7.4.

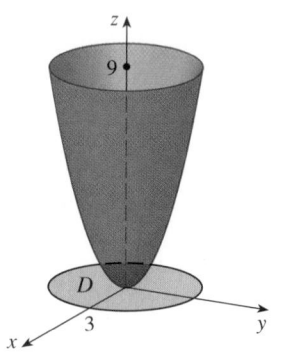

FIGURE 16

V EXAMPLE 10 Find the area of the part of the paraboloid $z = x^2 + y^2$ that lies under the plane $z = 9$.

SOLUTION The plane intersects the paraboloid in the circle $x^2 + y^2 = 9$, $z = 9$. Therefore the given surface lies above the disk D with center the origin and radius 3. (See Figure 16.) Using Formula 9, we have

$$A = \iint_D \sqrt{1 + \left(\frac{\partial z}{\partial x}\right)^2 + \left(\frac{\partial z}{\partial y}\right)^2}\, dA = \iint_D \sqrt{1 + (2x)^2 + (2y)^2}\, dA$$

$$= \iint_D \sqrt{1 + 4(x^2 + y^2)}\, dA$$

Converting to polar coordinates, we obtain

$$A = \int_0^{2\pi} \int_0^3 \sqrt{1 + 4r^2}\; r\, dr\, d\theta = \int_0^{2\pi} d\theta \int_0^3 r\sqrt{1 + 4r^2}\, dr$$

$$= 2\pi\left(\tfrac{1}{8}\right)\tfrac{2}{3}(1 + 4r^2)^{3/2}\Big]_0^3 = \frac{\pi}{6}\left(37\sqrt{37} - 1\right) \qquad \blacksquare$$

In Exercise 57 you are asked to show that the area of a surface of revolution given by $\boxed{3}$ is

$$\boxed{10} \qquad A = 2\pi \int_a^b f(x)\sqrt{1 + [f'(x)]^2}\, dx$$

This means that our definition of surface area $\boxed{6}$ is consistent with the surface area formula from single variable calculus (Formula 7.5.4).

13.6 | EXERCISES

1–4 ▪ Identify the surface with the given vector equation.

1. $\mathbf{r}(u, v) = (u + v)\,\mathbf{i} + (3 - v)\,\mathbf{j} + (1 + 4u + 5v)\,\mathbf{k}$

2. $\mathbf{r}(u, v) = 2 \sin u\,\mathbf{i} + 3 \cos u\,\mathbf{j} + v\,\mathbf{k}, \quad 0 \leqslant v \leqslant 2$

3. $\mathbf{r}(s, t) = \langle s, t, t^2 - s^2 \rangle$

4. $\mathbf{r}(s, t) = \langle s \sin 2t, s^2, s \cos 2t \rangle$

5–10 ▪ Use a computer to graph the parametric surface. Get a printout and indicate on it which grid curves have u constant and which have v constant.

5. $\mathbf{r}(u, v) = \langle u^2, v^2, u + v \rangle$,
$-1 \leqslant u \leqslant 1, \ -1 \leqslant v \leqslant 1$

6. $\mathbf{r}(u, v) = \langle u, v^3, -v \rangle$,
$-2 \leqslant u \leqslant 2, \ -2 \leqslant v \leqslant 2$

7. $\mathbf{r}(u, v) = \langle u \cos v, u \sin v, u^5 \rangle$,
$-1 \leqslant u \leqslant 1, \ 0 \leqslant v \leqslant 2\pi$

8. $\mathbf{r}(u, v) = \langle u, \sin(u + v), \sin v \rangle$,
$-\pi \leqslant u \leqslant \pi, \ -\pi \leqslant v \leqslant \pi$

9. $x = \sin v, \quad y = \cos u \sin 4v, \quad z = \sin 2u \sin 4v$,
$0 \leqslant u \leqslant 2\pi, \ -\pi/2 \leqslant v \leqslant \pi/2$

10. $x = \sin u, \quad y = \cos u \sin v, \quad z = \sin v$,
$0 \leqslant u \leqslant 2\pi, \ 0 \leqslant v \leqslant 2\pi$

11–14 ▪ Match the equations with the graphs labeled I–IV and give reasons for your answers. Determine which families of grid curves have u constant and which have v constant.

11. $\mathbf{r}(u, v) = u \cos v\,\mathbf{i} + u \sin v\,\mathbf{j} + v\,\mathbf{k}$

12. $\mathbf{r}(u, v) = u \cos v\,\mathbf{i} + u \sin v\,\mathbf{j} + \sin u\,\mathbf{k}, \quad -\pi \leqslant u \leqslant \pi$

13. $\mathbf{r}(u, v) = \sin v\,\mathbf{i} + \cos u \sin 2v\,\mathbf{j} + \sin u \sin 2v\,\mathbf{k}$

14. $x = (1 - u)(3 + \cos v) \cos 4\pi u$,
$y = (1 - u)(3 + \cos v) \sin 4\pi u$,
$z = 3u + (1 - u) \sin v$

I 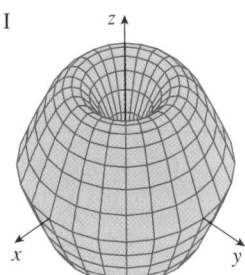 **II**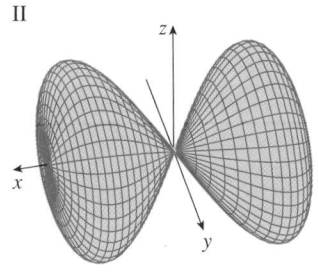

III 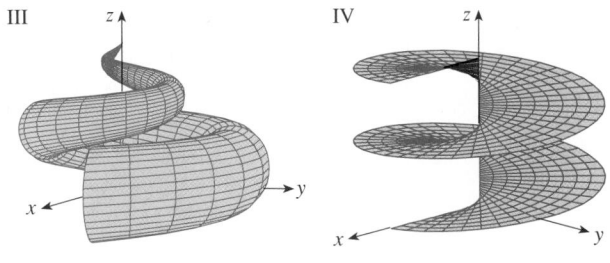 **IV**

15–22 ▪ Find a parametric representation for the surface.

15. The plane through the origin that contains the vectors $\mathbf{i} - \mathbf{j}$ and $\mathbf{j} - \mathbf{k}$

16. The plane that passes through the point $(0, -1, 5)$ and contains the vectors $\langle 2, 1, 4 \rangle$ and $\langle -3, 2, 5 \rangle$

17. The part of the hyperboloid $4x^2 - 4y^2 - z^2 = 4$ that lies in front of the yz-plane

18. The part of the ellipsoid $x^2 + 2y^2 + 3z^2 = 1$ that lies to the left of the xz-plane

19. The part of the sphere $x^2 + y^2 + z^2 = 4$ that lies above the cone $z = \sqrt{x^2 + y^2}$

20. The part of the sphere $x^2 + y^2 + z^2 = 16$ that lies between the planes $z = -2$ and $z = 2$

21. The part of the cylinder $y^2 + z^2 = 16$ that lies between the planes $x = 0$ and $x = 5$

22. The part of the plane $z = x + 3$ that lies inside the cylinder $x^2 + y^2 = 1$

CAS 23–24 ▪ Use a computer algebra system to produce a graph that looks like the given one.

23. 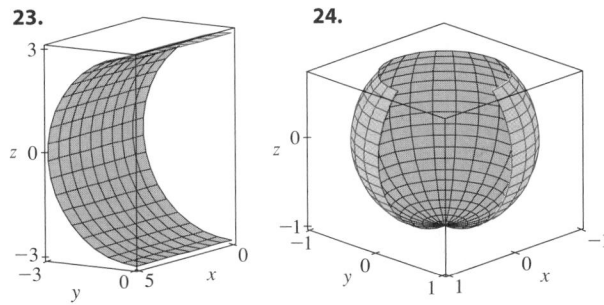 **24.**

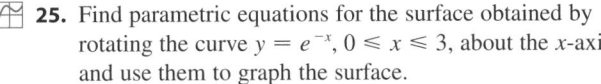

25. Find parametric equations for the surface obtained by rotating the curve $y = e^{-x}, 0 \leqslant x \leqslant 3$, about the x-axis and use them to graph the surface.

26. Find parametric equations for the surface obtained by rotating the curve $x = 4y^2 - y^4$, $-2 \leqslant y \leqslant 2$, about the y-axis and use them to graph the surface.

27. (a) What happens to the spiral tube in Example 2 (see Figure 5) if we replace $\cos u$ by $\sin u$ and $\sin u$ by $\cos u$?
(b) What happens if we replace $\cos u$ by $\cos 2u$ and $\sin u$ by $\sin 2u$?

28. The surface with parametric equations

$$x = 2\cos\theta + r\cos(\theta/2)$$

$$y = 2\sin\theta + r\cos(\theta/2)$$

$$z = r\sin(\theta/2)$$

where $-\frac{1}{2} \leqslant r \leqslant \frac{1}{2}$ and $0 \leqslant \theta \leqslant 2\pi$, is called a **Möbius strip**. Graph this surface with several viewpoints. What is unusual about it?

29–32 ▪ Find an equation of the tangent plane to the given parametric surface at the specified point. If you have software that graphs parametric surfaces, use a computer to graph the surface and the tangent plane.

29. $x = u + v$, $y = 3u^2$, $z = u - v$; $(2, 3, 0)$

30. $x = u^2 + 1$, $y = v^3 + 1$, $z = u + v$; $(5, 2, 3)$

31. $\mathbf{r}(u, v) = u\cos v\,\mathbf{i} + u\sin v\,\mathbf{j} + v\,\mathbf{k}$; $u = 1$, $v = \pi/3$

32. $\mathbf{r}(u, v) = \sin u\,\mathbf{i} + \cos u\sin v\,\mathbf{j} + \sin v\,\mathbf{k}$; $u = \pi/6$, $v = \pi/6$

33–44 ▪ Find the area of the surface.

33. The part of the plane $3x + 2y + z = 6$ that lies in the first octant

34. The part of the plane with vector equation $\mathbf{r}(u, v) = \langle u + v, 2 - 3u, 1 + u - v \rangle$ that is given by $0 \leqslant u \leqslant 2$, $-1 \leqslant v \leqslant 1$

35. The part of the plane $x + 2y + 3z = 1$ that lies inside the cylinder $x^2 + y^2 = 3$

36. The part of the cone $z = \sqrt{x^2 + y^2}$ that lies between the plane $y = x$ and the cylinder $y = x^2$

37. The surface $z = \frac{2}{3}(x^{3/2} + y^{3/2})$, $0 \leqslant x \leqslant 1$, $0 \leqslant y \leqslant 1$

38. The part of the surface $z = 1 + 3x + 2y^2$ that lies above the triangle with vertices $(0, 0)$, $(0, 1)$, and $(2, 1)$

39. The part of the surface $z = xy$ that lies within the cylinder $x^2 + y^2 = 1$

40. The part of the paraboloid $x = y^2 + z^2$ that lies inside the cylinder $y^2 + z^2 = 9$

41. The part of the surface $y = 4x + z^2$ that lies between the planes $x = 0$, $x = 1$, $z = 0$, and $z = 1$

42. The helicoid (or spiral ramp) with vector equation $\mathbf{r}(u, v) = u\cos v\,\mathbf{i} + u\sin v\,\mathbf{j} + v\,\mathbf{k}$, $0 \leqslant u \leqslant 1$, $0 \leqslant v \leqslant \pi$

43. The surface with parametric equations $x = uv$, $y = u + v$, $z = u - v$, $u^2 + v^2 \leqslant 1$

44. The part of the sphere $x^2 + y^2 + z^2 = b^2$ that lies inside the cylinder $x^2 + y^2 = a^2$, where $0 < a < b$

45. If the equation of a surface S is $z = f(x, y)$, where $x^2 + y^2 \leqslant R^2$, and you know that $|f_x| \leqslant 1$ and $|f_y| \leqslant 1$, what can you say about $A(S)$?

46–47 ▪ Find the area of the surface correct to four decimal places by expressing the area in terms of a single integral and using your calculator to estimate the integral.

46. The part of the surface $z = \cos(x^2 + y^2)$ that lies inside the cylinder $x^2 + y^2 = 1$

47. The part of the surface $z = e^{-x^2 - y^2}$ that lies above the disk $x^2 + y^2 \leqslant 4$

48. Find, to four decimal places, the area of the part of the surface $z = (1 + x^2)/(1 + y^2)$ that lies above the square $|x| + |y| \leqslant 1$. Illustrate by graphing this part of the surface.

49. (a) Use the Midpoint Rule for double integrals (see Section 12.1) with six squares to estimate the area of the surface $z = 1/(1 + x^2 + y^2)$, $0 \leqslant x \leqslant 6$, $0 \leqslant y \leqslant 4$.
(b) Use a computer algebra system to approximate the surface area in part (a) to four decimal places. Compare with the answer to part (a).

50. Find the area of the surface with vector equation $\mathbf{r}(u, v) = \langle \cos^3 u \cos^3 v, \sin^3 u \cos^3 v, \sin^3 v \rangle$, $0 \leqslant u \leqslant \pi$, $0 \leqslant v \leqslant 2\pi$. State your answer correct to four decimal places.

51. Find the exact area of the surface $z = 1 + 2x + 3y + 4y^2$, $1 \leqslant x \leqslant 4$, $0 \leqslant y \leqslant 1$.

52. (a) Set up, but do not evaluate, a double integral for the area of the surface with parametric equations $x = au\cos v$, $y = bu\sin v$, $z = u^2$, $0 \leqslant u \leqslant 2$, $0 \leqslant v \leqslant 2\pi$.
(b) Eliminate the parameters to show that the surface is an elliptic paraboloid and set up another double integral for the surface area.
(c) Use the parametric equations in part (a) with $a = 2$ and $b = 3$ to graph the surface.
(d) For the case $a = 2$, $b = 3$, use a computer algebra system to find the surface area correct to four decimal places.

53. (a) Show that the parametric equations $x = a\sin u\cos v$, $y = b\sin u\sin v$, $z = c\cos u$, $0 \leqslant u \leqslant \pi$, $0 \leqslant v \leqslant 2\pi$, represent an ellipsoid.

(b) Use the parametric equations in part (a) to graph the ellipsoid for the case $a = 1$, $b = 2$, $c = 3$.
(c) Set up, but do not evaluate, a double integral for the surface area of the ellipsoid in part (b).

54. (a) Show that the parametric equations $x = a \cosh u \cos v$, $y = b \cosh u \sin v$, $z = c \sinh u$, represent a hyperboloid of one sheet.
 (b) Use the parametric equations in part (a) to graph the hyperboloid for the case $a = 1$, $b = 2$, $c = 3$.
(c) Set up, but do not evaluate, a double integral for the surface area of the part of the hyperboloid in part (b) that lies between the planes $z = -3$ and $z = 3$.

55. Find the area of the part of the sphere $x^2 + y^2 + z^2 = 4z$ that lies inside the paraboloid $z = x^2 + y^2$.

56. The figure shows the surface created when the cylinder $y^2 + z^2 = 1$ intersects the cylinder $x^2 + z^2 = 1$. Find the area of this surface.

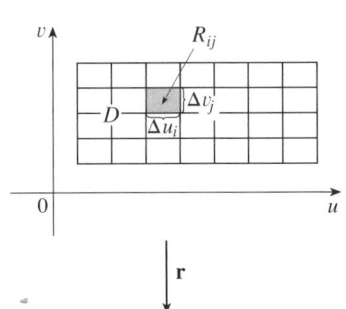

57. Use Definition 6 and the parametric equations for a surface of revolution ③ to verify Formula 10.

58–59 ▪ Use Formula 10 to find the area of the surface obtained by rotating the given curve about the x-axis.

58. $y = x^3$, $0 \le x \le 2$ **59.** $y = \sqrt{x}$, $4 \le x \le 9$

60. (a) Find a parametric representation for the torus obtained by rotating about the z-axis the circle in the xz-plane with center $(b, 0, 0)$ and radius $a < b$. [*Hint:* Take as parameters the angles θ and α shown in the figure.]
 (b) Use the parametric equations found in part (a) to graph the torus for several values of a and b.
(c) Use the parametric representation from part (a) to find the surface area of the torus.

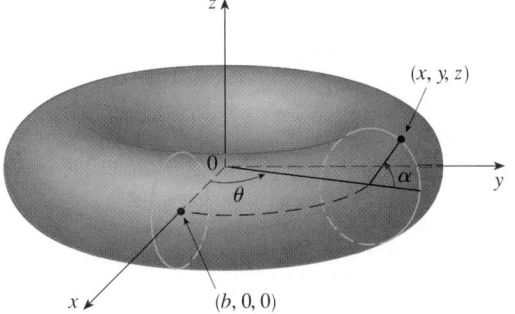

13.7 | **SURFACE INTEGRALS**

The relationship between surface integrals and surface area is much the same as the relationship between line integrals and arc length. Suppose f is a function of three variables whose domain includes a surface S. We will define the surface integral of f over S in such a way that, in the case where $f(x, y, z) = 1$, the value of the surface integral is equal to the surface area of S. We start with parametric surfaces and then deal with the special case where S is the graph of a function of two variables.

PARAMETRIC SURFACES

Suppose that a surface S has a vector equation

$$\mathbf{r}(u, v) = x(u, v)\,\mathbf{i} + y(u, v)\,\mathbf{j} + z(u, v)\,\mathbf{k} \qquad (u, v) \in D$$

We first assume that the parameter domain D is a rectangle and we divide it into subrectangles R_{ij} with dimensions Δu_i and Δv_j. Then the surface S is divided into corresponding patches S_{ij} as in Figure 1. We evaluate f at a point P_{ij}^* in each patch, multiply by the area ΔS_{ij} of the patch, and form the Riemann sum

$$\sum_{i=1}^{m} \sum_{j=1}^{n} f(P_{ij}^*)\, \Delta S_{ij}$$

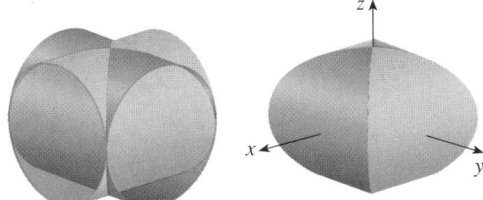

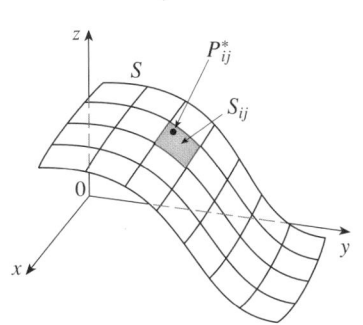

FIGURE 1

Then we take the limit as the number of patches increases (and their dimensions decrease) and define the **surface integral of f over the surface S** as

$$\boxed{1} \qquad \iint_S f(x, y, z)\, dS = \lim_{\max \Delta u_i,\, \Delta v_j \to 0} \sum_{i=1}^{m} \sum_{j=1}^{n} f(P_{ij}^*)\, \Delta S_{ij}$$

Notice the analogy with the definition of a line integral (13.2.2) and also the analogy with the definition of a double integral (12.1.5).

To evaluate the surface integral in Equation 1 we approximate the patch area ΔS_{ij} by the area of an approximating parallelogram in the tangent plane. In our discussion of surface area in Section 13.6 we made the approximation

$$\Delta S_{ij} \approx |\mathbf{r}_u \times \mathbf{r}_v|\, \Delta u_i\, \Delta v_j$$

where $\qquad \mathbf{r}_u = \dfrac{\partial x}{\partial u}\,\mathbf{i} + \dfrac{\partial y}{\partial u}\,\mathbf{j} + \dfrac{\partial z}{\partial u}\,\mathbf{k} \qquad \mathbf{r}_v = \dfrac{\partial x}{\partial v}\,\mathbf{i} + \dfrac{\partial y}{\partial v}\,\mathbf{j} + \dfrac{\partial z}{\partial v}\,\mathbf{k}$

are the tangent vectors at a corner of S_{ij}. If the components are continuous and $\mathbf{r}_u$ and $\mathbf{r}_v$ are nonzero and nonparallel in the interior of D, it can be shown from Definition 1, even when D is not a rectangle, that

- We assume that the surface is covered only once as (u, v) ranges throughout D. The value of the surface integral does not depend on the parametrization that is used.

$$\boxed{2} \qquad \iint_S f(x, y, z)\, dS = \iint_D f(\mathbf{r}(u, v))\, |\mathbf{r}_u \times \mathbf{r}_v|\, dA$$

This should be compared with the formula for a line integral:

$$\int_C f(x, y, z)\, ds = \int_a^b f(\mathbf{r}(t))\, |\mathbf{r}'(t)|\, dt$$

Observe also that

$$\iint_S 1\, dS = \iint_D |\mathbf{r}_u \times \mathbf{r}_v|\, dA = A(S)$$

Formula 2 allows us to compute a surface integral by converting it into a double integral over the parameter domain D. When using this formula, remember that $f(\mathbf{r}(u, v))$ is evaluated by writing $x = x(u, v)$, $y = y(u, v)$, and $z = z(u, v)$ in the formula for $f(x, y, z)$.

EXAMPLE 1 Compute the surface integral $\iint_S x^2\, dS$, where S is the unit sphere $x^2 + y^2 + z^2 = 1$.

SOLUTION As in Example 4 in Section 13.6, we use the parametric representation

$$x = \sin \phi \cos \theta \quad y = \sin \phi \sin \theta \quad z = \cos \phi \quad 0 \le \phi \le \pi \quad 0 \le \theta \le 2\pi$$

that is, $\qquad \mathbf{r}(\phi, \theta) = \sin \phi \cos \theta\, \mathbf{i} + \sin \phi \sin \theta\, \mathbf{j} + \cos \phi\, \mathbf{k}$

As in Example 9 in Section 13.6, we can compute that

$$|\mathbf{r}_\phi \times \mathbf{r}_\theta| = \sin \phi$$

Therefore, by Formula 2,

$$\iint\limits_{S} x^2 \, dS = \iint\limits_{D} (\sin\phi \, \cos\theta)^2 \, |\mathbf{r}_\phi \times \mathbf{r}_\theta| \, dA = \int_0^{2\pi} \int_0^{\pi} \sin^2\phi \, \cos^2\theta \, \sin\phi \, d\phi \, d\theta$$

$$= \int_0^{2\pi} \cos^2\theta \, d\theta \int_0^{\pi} \sin^3\phi \, d\phi = \int_0^{2\pi} \tfrac{1}{2}(1 + \cos 2\theta) \, d\theta \int_0^{\pi} (\sin\phi - \sin\phi \, \cos^2\phi) \, d\phi$$

$$= \tfrac{1}{2}\Big[\theta + \tfrac{1}{2}\sin 2\theta\Big]_0^{2\pi} \Big[-\cos\phi + \tfrac{1}{3}\cos^3\phi\Big]_0^{\pi} = \frac{4\pi}{3} \qquad\blacksquare$$

Surface integrals have applications similar to those for the integrals we have previously considered. For example, if a thin sheet (say, of aluminum foil) has the shape of a surface S and the density (mass per unit area) at the point (x, y, z) is $\rho(x, y, z)$, then the total **mass** of the sheet is

$$m = \iint\limits_{S} \rho(x, y, z) \, dS$$

and the **center of mass** is $(\bar{x}, \bar{y}, \bar{z})$, where

$$\bar{x} = \frac{1}{m} \iint\limits_{S} x\rho(x, y, z) \, dS \qquad \bar{y} = \frac{1}{m} \iint\limits_{S} y\rho(x, y, z) \, dS \qquad \bar{z} = \frac{1}{m} \iint\limits_{S} z\rho(x, y, z) \, dS$$

Moments of inertia can also be defined as before (see Exercise 39).

GRAPHS

Any surface S with equation $z = g(x, y)$ can be regarded as a parametric surface with parametric equations

$$x = x \qquad y = y \qquad z = g(x, y)$$

and so we have

$$\mathbf{r}_x = \mathbf{i} + \left(\frac{\partial g}{\partial x}\right)\mathbf{k} \qquad \mathbf{r}_y = \mathbf{j} + \left(\frac{\partial g}{\partial y}\right)\mathbf{k}$$

Thus

$$\boxed{3} \qquad\qquad \mathbf{r}_x \times \mathbf{r}_y = -\frac{\partial g}{\partial x}\mathbf{i} - \frac{\partial g}{\partial y}\mathbf{j} + \mathbf{k}$$

and

$$|\mathbf{r}_x \times \mathbf{r}_y| = \sqrt{\left(\frac{\partial z}{\partial x}\right)^2 + \left(\frac{\partial z}{\partial y}\right)^2 + 1}$$

Therefore, in this case, Formula 2 becomes

$$\boxed{4} \qquad \iint\limits_{S} f(x, y, z) \, dS = \iint\limits_{D} f(x, y, g(x, y)) \sqrt{\left(\frac{\partial z}{\partial x}\right)^2 + \left(\frac{\partial z}{\partial y}\right)^2 + 1} \, dA$$

Similar formulas apply when it is more convenient to project S onto the yz-plane or xz-plane. For instance, if S is a surface with equation $y = h(x, z)$ and D is its projection on the *xz*-plane, then

$$\iint\limits_{S} f(x, y, z)\, dS = \iint\limits_{D} f\big(x, h(x, z), z\big) \sqrt{\left(\frac{\partial y}{\partial x}\right)^2 + \left(\frac{\partial y}{\partial z}\right)^2 + 1}\, dA$$

EXAMPLE 2 Evaluate $\iint_S y\, dS$, where S is the surface $z = x + y^2$, $0 \leqslant x \leqslant 1$, $0 \leqslant y \leqslant 2$. (See Figure 2.)

SOLUTION Since $\qquad \dfrac{\partial z}{\partial x} = 1 \qquad$ and $\qquad \dfrac{\partial z}{\partial y} = 2y$

Formula 4 gives

$$\iint\limits_{S} y\, dS = \iint\limits_{D} y \sqrt{1 + \left(\frac{\partial z}{\partial x}\right)^2 + \left(\frac{\partial z}{\partial y}\right)^2}\, dA$$

$$= \int_0^1 \int_0^2 y\sqrt{1 + 1 + 4y^2}\, dy\, dx$$

$$= \int_0^1 dx\, \sqrt{2} \int_0^2 y\sqrt{1 + 2y^2}\, dy$$

$$= \sqrt{2}\, \big(\tfrac{1}{4}\big)\tfrac{2}{3}(1 + 2y^2)^{3/2}\big]_0^2 = \frac{13\sqrt{2}}{3} \qquad \blacksquare$$

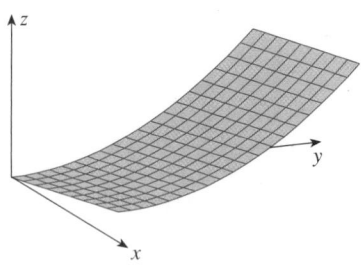

FIGURE 2

If S is a piecewise-smooth surface, that is, a finite union of smooth surfaces S_1, $S_2, \ldots, S_n$ that intersect only along their boundaries, then the surface integral of f over S is defined by

$$\iint\limits_{S} f(x, y, z)\, dS = \iint\limits_{S_1} f(x, y, z)\, dS + \cdots + \iint\limits_{S_n} f(x, y, z)\, dS$$

 EXAMPLE 3 Evaluate $\iint_S z\, dS$, where S is the surface whose sides S_1 are given by the cylinder $x^2 + y^2 = 1$, whose bottom S_2 is the disk $x^2 + y^2 \leqslant 1$ in the plane $z = 0$, and whose top S_3 is the part of the plane $z = 1 + x$ that lies above S_2.

SOLUTION The surface S is shown in Figure 3. (We have changed the usual position of the axes to get a better look at S.) For S_1 we use θ and z as parameters (see Example 5 in Section 13.6) and write its parametric equations as

$$x = \cos \theta \qquad y = \sin \theta \qquad z = z$$

where

$$0 \leqslant \theta \leqslant 2\pi \qquad \text{and} \qquad 0 \leqslant z \leqslant 1 + x = 1 + \cos \theta$$

Therefore

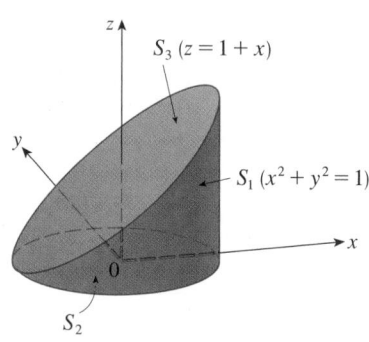

FIGURE 3

$$\mathbf{r}_\theta \times \mathbf{r}_z = \begin{vmatrix} \mathbf{i} & \mathbf{j} & \mathbf{k} \\ -\sin \theta & \cos \theta & 0 \\ 0 & 0 & 1 \end{vmatrix} = \cos \theta\, \mathbf{i} + \sin \theta\, \mathbf{j}$$

and

$$|\mathbf{r}_\theta \times \mathbf{r}_z| = \sqrt{\cos^2\theta + \sin^2\theta} = 1$$

Thus the surface integral over S_1 is

$$\iint_{S_1} z \, dS = \iint_D z \, |\mathbf{r}_\theta \times \mathbf{r}_z| \, dA = \int_0^{2\pi} \int_0^{1+\cos\theta} z \, dz \, d\theta = \int_0^{2\pi} \tfrac{1}{2}(1 + \cos\theta)^2 \, d\theta$$

$$= \tfrac{1}{2} \int_0^{2\pi} \left[1 + 2\cos\theta + \tfrac{1}{2}(1 + \cos 2\theta)\right] d\theta = \tfrac{1}{2}\left[\tfrac{3}{2}\theta + 2\sin\theta + \tfrac{1}{4}\sin 2\theta\right]_0^{2\pi} = \frac{3\pi}{2}$$

Since S_2 lies in the plane $z = 0$, we have

$$\iint_{S_2} z \, dS = \iint_{S_2} 0 \, dS = 0$$

The top surface S_3 lies above the unit disk D and is part of the plane $z = 1 + x$. So, taking $g(x, y) = 1 + x$ in Formula 4 and converting to polar coordinates, we have

$$\iint_{S_3} z \, dS = \iint_D (1 + x)\sqrt{1 + \left(\frac{\partial z}{\partial x}\right)^2 + \left(\frac{\partial z}{\partial y}\right)^2} \, dA$$

$$= \int_0^{2\pi} \int_0^1 (1 + r\cos\theta)\sqrt{1 + 1 + 0} \, r \, dr \, d\theta$$

$$= \sqrt{2} \int_0^{2\pi} \int_0^1 (r + r^2\cos\theta) \, dr \, d\theta$$

$$= \sqrt{2} \int_0^{2\pi} \left(\tfrac{1}{2} + \tfrac{1}{3}\cos\theta\right) d\theta = \sqrt{2}\left[\frac{\theta}{2} + \frac{\sin\theta}{3}\right]_0^{2\pi} = \sqrt{2}\,\pi$$

Therefore

$$\iint_S z \, dS = \iint_{S_1} z \, dS + \iint_{S_2} z \, dS + \iint_{S_3} z \, dS$$

$$= \frac{3\pi}{2} + 0 + \sqrt{2}\,\pi = \left(\tfrac{3}{2} + \sqrt{2}\right)\pi \qquad\blacksquare$$

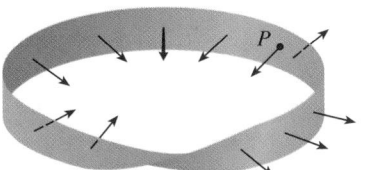

FIGURE 4

A Möbius strip

TEC Visual 13.7 shows a Möbius strip with a normal vector that can be moved along the surface.

ORIENTED SURFACES

To define surface integrals of vector fields, we need to rule out nonorientable surfaces such as the Möbius strip shown in Figure 4. [It is named after the German geometer August Möbius (1790–1868).] You can construct one for yourself by taking a long rectangular strip of paper, giving it a half-twist, and taping the short edges together as in Figure 5. If an ant were to crawl along the Möbius strip starting at a point P, it would end up on the "other side" of the strip (that is, with its upper side pointing in the opposite direction). Then, if the ant continued to crawl in the same direction, it would end up back at the same point P without ever having crossed an edge. (If you have constructed a Möbius strip, try drawing a pencil line down the middle.) Therefore a Möbius strip really has only one side. You can graph the Möbius strip using the parametric equations in Exercise 28 in Section 13.6.

FIGURE 5

Constructing a Möbius strip

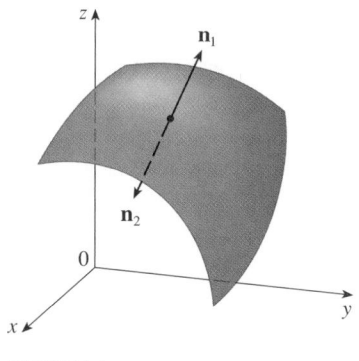

FIGURE 6

From now on we consider only orientable (two-sided) surfaces. We start with a surface S that has a tangent plane at every point (x, y, z) on S (except at any boundary point). There are two unit normal vectors $\mathbf{n}_1$ and $\mathbf{n}_2 = -\mathbf{n}_1$ at (x, y, z). (See Figure 6.)

If it is possible to choose a unit normal vector $\mathbf{n}$ at every such point (x, y, z) so that $\mathbf{n}$ varies continuously over S, then S is called an **oriented surface** and the given choice of $\mathbf{n}$ provides S with an **orientation**. There are two possible orientations for any orientable surface (see Figure 7).

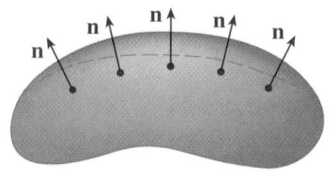 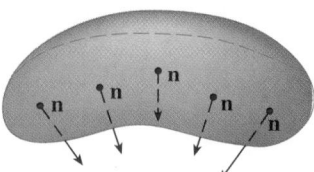

FIGURE 7

The two orientations
of an orientable surface

For a surface $z = g(x, y)$ given as the graph of g, we use Equation 3 to associate with the surface a natural orientation given by the unit normal vector

$$\boxed{5} \qquad \mathbf{n} = \frac{-\dfrac{\partial g}{\partial x}\,\mathbf{i} - \dfrac{\partial g}{\partial y}\,\mathbf{j} + \mathbf{k}}{\sqrt{1 + \left(\dfrac{\partial g}{\partial x}\right)^2 + \left(\dfrac{\partial g}{\partial y}\right)^2}}$$

Since the $\mathbf{k}$-component is positive, this gives the *upward* orientation of the surface.

If S is a smooth orientable surface given in parametric form by a vector function $\mathbf{r}(u, v)$, then it is automatically supplied with the orientation of the unit normal vector

$$\boxed{6} \qquad \mathbf{n} = \frac{\mathbf{r}_u \times \mathbf{r}_v}{|\mathbf{r}_u \times \mathbf{r}_v|}$$

and the opposite orientation is given by $-\mathbf{n}$. For instance, in Example 4 in Section 13.6 we found the parametric representation

$$\mathbf{r}(\phi, \theta) = a \sin \phi \cos \theta\, \mathbf{i} + a \sin \phi \sin \theta\, \mathbf{j} + a \cos \phi\, \mathbf{k}$$

for the sphere $x^2 + y^2 + z^2 = a^2$. Then in Example 9 in Section 13.6 we found that

$$\mathbf{r}_\phi \times \mathbf{r}_\theta = a^2 \sin^2\phi \cos \theta\, \mathbf{i} + a^2 \sin^2\phi \sin \theta\, \mathbf{j} + a^2 \sin \phi \cos \phi\, \mathbf{k}$$

and $$|\mathbf{r}_\phi \times \mathbf{r}_\theta| = a^2 \sin \phi$$

So the orientation induced by $\mathbf{r}(\phi, \theta)$ is defined by the unit normal vector

$$\mathbf{n} = \frac{\mathbf{r}_\phi \times \mathbf{r}_\theta}{|\mathbf{r}_\phi \times \mathbf{r}_\theta|} = \sin \phi \cos \theta\, \mathbf{i} + \sin \phi \sin \theta\, \mathbf{j} + \cos \phi\, \mathbf{k} = \frac{1}{a}\,\mathbf{r}(\phi, \theta)$$

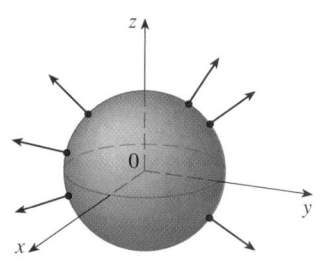

FIGURE 8
Positive orientation

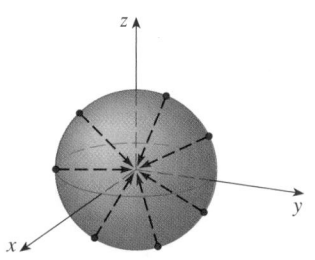

FIGURE 9
Negative orientation

Observe that $\mathbf{n}$ points in the same direction as the position vector, that is, outward from the sphere (see Figure 8). The opposite (inward) orientation would have been obtained (see Figure 9) if we had reversed the order of the parameters because $\mathbf{r}_\theta \times \mathbf{r}_\phi = -\mathbf{r}_\phi \times \mathbf{r}_\theta$.

For a **closed surface**, that is, a surface that is the boundary of a solid region E, the convention is that the **positive orientation** is the one for which the normal vectors point *outward* from E, and inward-pointing normals give the negative orientation (see Figures 8 and 9).

SURFACE INTEGRALS OF VECTOR FIELDS

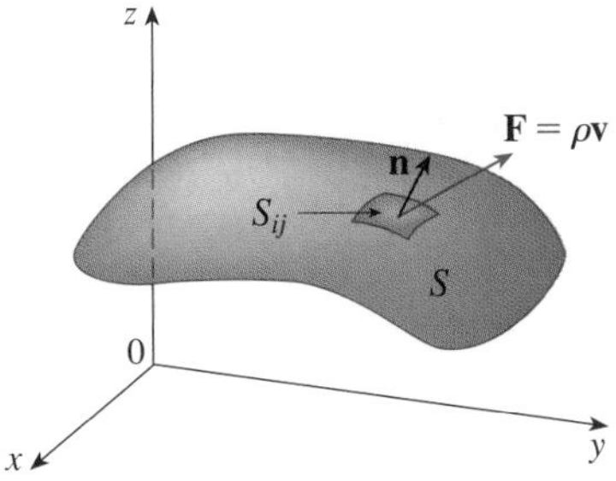

FIGURE 10

Suppose that S is an oriented surface with unit normal vector $\mathbf{n}$, and imagine a fluid with density $\rho(x, y, z)$ and velocity field $\mathbf{v}(x, y, z)$ flowing through S. (Think of S as an imaginary surface that doesn't impede the fluid flow, like a fishing net across a stream.) Then the rate of flow (mass per unit time) per unit area is $\rho\mathbf{v}$. If we divide S into small patches S_{ij}, as in Figure 10 (compare with Figure 1), then S_{ij} is nearly planar and so we can approximate the mass of fluid crossing S_{ij} in the direction of the normal $\mathbf{n}$ per unit time by the quantity

$$(\rho\mathbf{v} \cdot \mathbf{n})A(S_{ij})$$

where ρ, $\mathbf{v}$, and $\mathbf{n}$ are evaluated at some point on S_{ij}. (Recall that the component of the vector $\rho\mathbf{v}$ in the direction of the unit vector $\mathbf{n}$ is $\rho\mathbf{v} \cdot \mathbf{n}$.) By summing these quantities and taking the limit we get, according to Definition 1, the surface integral of the function $\rho\mathbf{v} \cdot \mathbf{n}$ over S:

$$\boxed{7} \qquad \iint_S \rho\mathbf{v} \cdot \mathbf{n} \, dS = \iint_S \rho(x, y, z)\mathbf{v}(x, y, z) \cdot \mathbf{n}(x, y, z) \, dS$$

and this is interpreted physically as the rate of flow through S.

If we write $\mathbf{F} = \rho\mathbf{v}$, then $\mathbf{F}$ is also a vector field on $\mathbb{R}^3$ and the integral in Equation 7 becomes

$$\iint_S \mathbf{F} \cdot \mathbf{n} \, dS$$

A surface integral of this form occurs frequently in physics, even when $\mathbf{F}$ is not $\rho\mathbf{v}$, and is called the *surface integral* (or *flux integral*) of $\mathbf{F}$ over S.

$\boxed{8}$ **DEFINITION** If $\mathbf{F}$ is a continuous vector field defined on an oriented surface S with unit normal vector $\mathbf{n}$, then the **surface integral of F over S** is

$$\iint_S \mathbf{F} \cdot d\mathbf{S} = \iint_S \mathbf{F} \cdot \mathbf{n} \, dS$$

This integral is also called the **flux** of $\mathbf{F}$ across S.

In words, Definition 8 says that the surface integral of a vector field over S is equal to the surface integral of its normal component over S (as previously defined).

If S is given by a vector function $\mathbf{r}(u, v)$, then $\mathbf{n}$ is given by Equation 6, and from Definition 8 and Equation 2 we have

$$\iint_S \mathbf{F} \cdot d\mathbf{S} = \iint_S \mathbf{F} \cdot \frac{\mathbf{r}_u \times \mathbf{r}_v}{|\mathbf{r}_u \times \mathbf{r}_v|} \, dS = \iint_D \left[\mathbf{F}(\mathbf{r}(u, v)) \cdot \frac{\mathbf{r}_u \times \mathbf{r}_v}{|\mathbf{r}_u \times \mathbf{r}_v|} \right] |\mathbf{r}_u \times \mathbf{r}_v| \, dA$$

where D is the parameter domain. Thus we have

■ Compare Equation 9 to the similar expression for evaluating line integrals of vector fields in Definition 13.2.13:

$$\int_C \mathbf{F} \cdot d\mathbf{r} = \int_a^b \mathbf{F}(\mathbf{r}(t)) \cdot \mathbf{r}'(t) \, dt$$

$$\boxed{9} \qquad \iint_S \mathbf{F} \cdot d\mathbf{S} = \iint_D \mathbf{F} \cdot (\mathbf{r}_u \times \mathbf{r}_v) \, dA$$

■ Figure 11 shows the vector field **F** in Example 4 at points on the unit sphere.

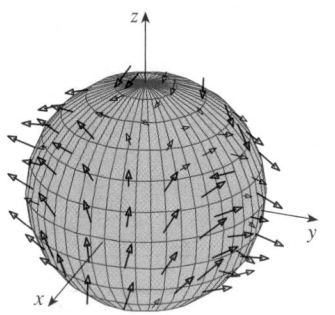

FIGURE 11

EXAMPLE 4 Find the flux of the vector field $\mathbf{F}(x, y, z) = z\,\mathbf{i} + y\,\mathbf{j} + x\,\mathbf{k}$ across the unit sphere $x^2 + y^2 + z^2 = 1$.

SOLUTION As in Example 1, we use the parametric representation

$$\mathbf{r}(\phi, \theta) = \sin\phi\cos\theta\,\mathbf{i} + \sin\phi\sin\theta\,\mathbf{j} + \cos\phi\,\mathbf{k} \qquad 0 \le \phi \le \pi \qquad 0 \le \theta \le 2\pi$$

Then
$$\mathbf{F}(\mathbf{r}(\phi, \theta)) = \cos\phi\,\mathbf{i} + \sin\phi\sin\theta\,\mathbf{j} + \sin\phi\cos\theta\,\mathbf{k}$$

and, from Example 9 in Section 13.6,

$$\mathbf{r}_\phi \times \mathbf{r}_\theta = \sin^2\phi\cos\theta\,\mathbf{i} + \sin^2\phi\sin\theta\,\mathbf{j} + \sin\phi\cos\phi\,\mathbf{k}$$

Therefore

$$\mathbf{F}(\mathbf{r}(\phi, \theta)) \cdot (\mathbf{r}_\phi \times \mathbf{r}_\theta) = \cos\phi\sin^2\phi\cos\theta + \sin^3\phi\sin^2\theta + \sin^2\phi\cos\phi\cos\theta$$

and, by Formula 9, the flux is

$$\iint_S \mathbf{F} \cdot d\mathbf{S} = \iint_D \mathbf{F} \cdot (\mathbf{r}_\phi \times \mathbf{r}_\theta)\, dA = \int_0^{2\pi} \int_0^\pi (2\sin^2\phi\cos\phi\cos\theta + \sin^3\phi\sin^2\theta)\, d\phi\, d\theta$$

$$= 2\int_0^\pi \sin^2\phi\cos\phi\, d\phi \int_0^{2\pi} \cos\theta\, d\theta + \int_0^\pi \sin^3\phi\, d\phi \int_0^{2\pi} \sin^2\theta\, d\theta$$

$$= 0 + \int_0^\pi \sin^3\phi\, d\phi \int_0^{2\pi} \sin^2\theta\, d\theta \qquad \left(\text{since } \int_0^{2\pi} \cos\theta\, d\theta = 0\right)$$

$$= \frac{4\pi}{3}$$

by the same calculation as in Example 1. ∎

If, for instance, the vector field in Example 4 is a velocity field describing the flow of a fluid with density 1, then the answer, $4\pi/3$, represents the rate of flow through the unit sphere in units of mass per unit time.

In the case of a surface S given by a graph $z = g(x, y)$, we can think of x and y as parameters and use Equation 3 to write

$$\mathbf{F} \cdot (\mathbf{r}_x \times \mathbf{r}_y) = (P\,\mathbf{i} + Q\,\mathbf{j} + R\,\mathbf{k}) \cdot \left(-\frac{\partial g}{\partial x}\,\mathbf{i} - \frac{\partial g}{\partial y}\,\mathbf{j} + \mathbf{k}\right)$$

Thus Formula 9 becomes

$$\boxed{10 \qquad \iint_S \mathbf{F} \cdot d\mathbf{S} = \iint_D \left(-P\frac{\partial g}{\partial x} - Q\frac{\partial g}{\partial y} + R\right) dA}$$

This formula assumes the upward orientation of S; for a downward orientation we multiply by -1. Similar formulas can be worked out if S is given by $y = h(x, z)$ or $x = k(y, z)$. (See Exercises 35 and 36.)

▼ EXAMPLE 5 Evaluate $\iint_S \mathbf{F} \cdot d\mathbf{S}$, where $\mathbf{F}(x, y, z) = y\,\mathbf{i} + x\,\mathbf{j} + z\,\mathbf{k}$ and S is the boundary of the solid region E enclosed by the paraboloid $z = 1 - x^2 - y^2$ and the plane $z = 0$.

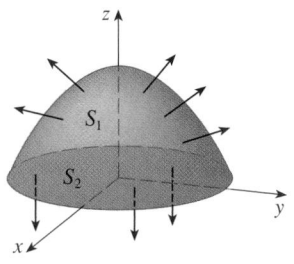

FIGURE 12

SOLUTION S consists of a parabolic top surface S_1 and a circular bottom surface S_2. (See Figure 12.) Since S is a closed surface, we use the convention of positive (outward) orientation. This means that S_1 is oriented upward and we can use Equation 10 with D being the projection of S_1 on the xy-plane, namely, the disk $x^2 + y^2 \leq 1$. Since

$$P(x, y, z) = y \qquad Q(x, y, z) = x \qquad R(x, y, z) = z = 1 - x^2 - y^2$$

on S_1 and

$$\frac{\partial g}{\partial x} = -2x \qquad \frac{\partial g}{\partial y} = -2y$$

we have

$$\iint\limits_{S_1} \mathbf{F} \cdot d\mathbf{S} = \iint\limits_{D} \left(-P\frac{\partial g}{\partial x} - Q\frac{\partial g}{\partial y} + R \right) dA = \iint\limits_{D} [-y(-2x) - x(-2y) + 1 - x^2 - y^2]\, dA$$

$$= \iint\limits_{D} (1 + 4xy - x^2 - y^2)\, dA = \int_0^{2\pi} \int_0^1 (1 + 4r^2 \cos\theta \sin\theta - r^2)\, r\, dr\, d\theta$$

$$= \int_0^{2\pi} \int_0^1 (r - r^3 + 4r^3 \cos\theta \sin\theta)\, dr\, d\theta = \int_0^{2\pi} \left(\tfrac{1}{4} + \cos\theta \sin\theta \right) d\theta = \tfrac{1}{4}(2\pi) + 0 = \frac{\pi}{2}$$

The disk S_2 is oriented downward, so its unit normal vector is $\mathbf{n} = -\mathbf{k}$ and we have

$$\iint\limits_{S_2} \mathbf{F} \cdot d\mathbf{S} = \iint\limits_{S_2} \mathbf{F} \cdot (-\mathbf{k})\, dS = \iint\limits_{D} (-z)\, dA = \iint\limits_{D} 0\, dA = 0$$

since $z = 0$ on S_2. Finally, we compute, by definition, $\iint_S \mathbf{F} \cdot d\mathbf{S}$ as the sum of the surface integrals of $\mathbf{F}$ over the pieces S_1 and S_2:

$$\iint\limits_{S} \mathbf{F} \cdot d\mathbf{S} = \iint\limits_{S_1} \mathbf{F} \cdot d\mathbf{S} + \iint\limits_{S_2} \mathbf{F} \cdot d\mathbf{S} = \frac{\pi}{2} + 0 = \frac{\pi}{2} \qquad \blacksquare$$

Although we motivated the surface integral of a vector field using the example of fluid flow, this concept also arises in other physical situations. For instance, if $\mathbf{E}$ is an electric field (see Example 5 in Section 13.1), then the surface integral

$$\iint\limits_{S} \mathbf{E} \cdot d\mathbf{S}$$

is called the **electric flux** of $\mathbf{E}$ through the surface S. One of the important laws of electrostatics is **Gauss's Law**, which says that the net charge enclosed by a closed surface S is

$$\boxed{11} \qquad Q = \varepsilon_0 \iint\limits_{S} \mathbf{E} \cdot d\mathbf{S}$$

where ε_0 is a constant (called the permittivity of free space) that depends on the units used. (In the SI system, $\varepsilon_0 \approx 8.8542 \times 10^{-12}\ \mathrm{C^2/N \cdot m^2}$.) Therefore, if the vector field $\mathbf{F}$ in Example 4 represents an electric field, we can conclude that the charge enclosed by S is $Q = 4\pi\varepsilon_0/3$.

Another application of surface integrals occurs in the study of heat flow. Suppose the temperature at a point (x, y, z) in a body is $u(x, y, z)$. Then the **heat flow** is defined as the vector field

$$\mathbf{F} = -K \, \nabla u$$

where K is an experimentally determined constant called the **conductivity** of the substance. The rate of heat flow across the surface S in the body is then given by the surface integral

$$\iint_S \mathbf{F} \cdot d\mathbf{S} = -K \iint_S \nabla u \cdot d\mathbf{S}$$

V **EXAMPLE 6** The temperature u in a metal ball is proportional to the square of the distance from the center of the ball. Find the rate of heat flow across a sphere S of radius a with center at the center of the ball.

SOLUTION Taking the center of the ball to be at the origin, we have

$$u(x, y, z) = C(x^2 + y^2 + z^2)$$

where C is the proportionality constant. Then the heat flow is

$$\mathbf{F}(x, y, z) = -K \, \nabla u = -KC(2x \, \mathbf{i} + 2y \, \mathbf{j} + 2z \, \mathbf{k})$$

where K is the conductivity of the metal. Instead of using the usual parametrization of the sphere as in Example 4, we observe that the outward unit normal to the sphere $x^2 + y^2 + z^2 = a^2$ at the point (x, y, z) is

$$\mathbf{n} = \frac{1}{a}(x \, \mathbf{i} + y \, \mathbf{j} + z \, \mathbf{k})$$

and so

$$\mathbf{F} \cdot \mathbf{n} = -\frac{2KC}{a}(x^2 + y^2 + z^2)$$

But on S we have $x^2 + y^2 + z^2 = a^2$, so $\mathbf{F} \cdot \mathbf{n} = -2aKC$. Therefore the rate of heat flow across S is

$$\iint_S \mathbf{F} \cdot d\mathbf{S} = \iint_S \mathbf{F} \cdot \mathbf{n} \, dS = -2aKC \iint_S dS$$

$$= -2aKCA(S) = -2aKC(4\pi a^2) = -8KC\pi a^3 \qquad \blacksquare$$

13.7 | EXERCISES

1. Let S be the boundary surface of the box enclosed by the planes $x = 0$, $x = 2$, $y = 0$, $y = 4$, $z = 0$, and $z = 6$. Approximate $\iint_S e^{-0.1(x+y+z)} \, dS$ by using a Riemann sum as in Definition 1, taking the patches S_{ij} to be the rectangles that are the faces of the box S and the points P_{ij}^* to be the centers of the rectangles.

2. A surface S consists of the cylinder $x^2 + y^2 = 1$, $-1 \leq z \leq 1$, together with its top and bottom disks. Suppose you know that f is a continuous function with

$$f(\pm 1, 0, 0) = 2 \qquad f(0, \pm 1, 0) = 3 \qquad f(0, 0, \pm 1) = 4$$

Estimate the value of $\iint_S f(x, y, z) \, dS$ by using a Riemann sum, taking the patches S_{ij} to be four quarter-cylinders and the top and bottom disks.

3. Let H be the hemisphere $x^2 + y^2 + z^2 = 50$, $z \geq 0$, and suppose f is a continuous function with $f(3, 4, 5) = 7$, $f(3, -4, 5) = 8$, $f(-3, 4, 5) = 9$, and $f(-3, -4, 5) = 12$. By dividing H into four patches, estimate the value of $\iint_H f(x, y, z) \, dS$.

4. Suppose that $f(x, y, z) = g(\sqrt{x^2 + y^2 + z^2})$, where g is a function of one variable such that $g(2) = -5$. Evaluate $\iint_S f(x, y, z) \, dS$, where S is the sphere $x^2 + y^2 + z^2 = 4$.

5–20 ▪ Evaluate the surface integral.

5. $\iint_S (x + y + z) \, dS$,
S is the parallelogram with parametric equations $x = u + v$, $y = u - v$, $z = 1 + 2u + v$, $0 \leq u \leq 2$, $0 \leq v \leq 1$

6. $\iint_S xyz \, dS$,
   S is the cone with parametric equations $x = u \cos v$,
   $y = u \sin v, z = u, 0 \leqslant u \leqslant 1, 0 \leqslant v \leqslant \pi/2$

7. $\iint_S y \, dS$, S is the helicoid with vector equation
   $\mathbf{r}(u, v) = \langle u \cos v, u \sin v, v \rangle, 0 \leqslant u \leqslant 1, 0 \leqslant v \leqslant \pi$

8. $\iint_S (x^2 + y^2) \, dS$,
   S is the surface with vector equation
   $\mathbf{r}(u, v) = \langle 2uv, u^2 - v^2, u^2 + v^2 \rangle, u^2 + v^2 \leqslant 1$

9. $\iint_S x^2 yz \, dS$,
   S is the part of the plane $z = 1 + 2x + 3y$ that lies above
   the rectangle $[0, 3] \times [0, 2]$

10. $\iint_S xz \, dS$,
   S is the part of the plane $2x + 2y + z = 4$ that lies in the
   first octant

11. $\iint_S x \, dS$,
   S is the triangular region with vertices $(1, 0, 0)$, $(0, -2, 0)$,
   and $(0, 0, 4)$

12. $\iint_S y \, dS$,
   S is the surface $z = \frac{2}{3}(x^{3/2} + y^{3/2}), 0 \leqslant x \leqslant 1, 0 \leqslant y \leqslant 1$

13. $\iint_S x^2 z^2 \, dS$,
   S is the part of the cone $z^2 = x^2 + y^2$ that lies between the
   planes $z = 1$ and $z = 3$

14. $\iint_S z \, dS$,
   S is the surface $x = y + 2z^2, 0 \leqslant y \leqslant 1, 0 \leqslant z \leqslant 1$

15. $\iint_S y \, dS$,
   S is the part of the paraboloid $y = x^2 + z^2$ that lies inside
   the cylinder $x^2 + z^2 = 4$

16. $\iint_S y^2 \, dS$,
   S is the part of the sphere $x^2 + y^2 + z^2 = 4$ that lies
   inside the cylinder $x^2 + y^2 = 1$ and above the xy-plane

17. $\iint_S (x^2 z + y^2 z) \, dS$,
   S is the hemisphere $x^2 + y^2 + z^2 = 4, z \geqslant 0$

18. $\iint_S xz \, dS$,
   S is the boundary of the region enclosed by the cylinder
   $y^2 + z^2 = 9$ and the planes $x = 0$ and $x + y = 5$

19. $\iint_S (z + x^2 y) \, dS$,
   S is the part of the cylinder $y^2 + z^2 = 1$ that lies between
   the planes $x = 0$ and $x = 3$ in the first octant

20. $\iint_S (x^2 + y^2 + z^2) \, dS$,
   S is the part of the cylinder $x^2 + y^2 = 9$ between the planes
   $z = 0$ and $z = 2$, together with its top and bottom disks

---

21–31 ■ Evaluate the surface integral $\iint_S \mathbf{F} \cdot d\mathbf{S}$ for the given
vector field $\mathbf{F}$ and the oriented surface S. In other words, find
the flux of $\mathbf{F}$ across S. For closed surfaces, use the positive
(outward) orientation.

21. $\mathbf{F}(x, y, z) = ze^{xy} \mathbf{i} - 3ze^{xy} \mathbf{j} + xy \mathbf{k}$,
   S is the parallelogram of Exercise 5 with upward orientation

22. $\mathbf{F}(x, y, z) = z \mathbf{i} + y \mathbf{j} + x \mathbf{k}$,
   S is the helicoid of Exercise 7 with upward orientation

23. $\mathbf{F}(x, y, z) = xy \mathbf{i} + yz \mathbf{j} + zx \mathbf{k}$, S is the part of the
   paraboloid $z = 4 - x^2 - y^2$ that lies above the square
   $0 \leqslant x \leqslant 1, 0 \leqslant y \leqslant 1$, and has upward orientation

24. $\mathbf{F}(x, y, z) = -x \mathbf{i} - y \mathbf{j} + z^3 \mathbf{k}$,
   S is the part of the cone $z = \sqrt{x^2 + y^2}$ between the planes
   $z = 1$ and $z = 3$ with downward orientation

25. $\mathbf{F}(x, y, z) = x \mathbf{i} - z \mathbf{j} + y \mathbf{k}$,
   S is the part of the sphere $x^2 + y^2 + z^2 = 4$ in the first
   octant, with orientation toward the origin

26. $\mathbf{F}(x, y, z) = xz \mathbf{i} + x \mathbf{j} + y \mathbf{k}$,
   S is the hemisphere $x^2 + y^2 + z^2 = 25, y \geqslant 0$, oriented in
   the direction of the positive y-axis

27. $\mathbf{F}(x, y, z) = y \mathbf{j} - z \mathbf{k}$,
   S consists of the paraboloid $y = x^2 + z^2, 0 \leqslant y \leqslant 1$,
   and the disk $x^2 + z^2 \leqslant 1, y = 1$

28. $\mathbf{F}(x, y, z) = xy \mathbf{i} + 4x^2 \mathbf{j} + yz \mathbf{k}$, S is the surface $z = xe^y$,
   $0 \leqslant x \leqslant 1, 0 \leqslant y \leqslant 1$, with upward orientation

29. $\mathbf{F}(x, y, z) = x \mathbf{i} + 2y \mathbf{j} + 3z \mathbf{k}$,
   S is the cube with vertices $(\pm 1, \pm 1, \pm 1)$

30. $\mathbf{F}(x, y, z) = y \mathbf{i} + (z - y) \mathbf{j} + x \mathbf{k}$,
   S is the surface of the tetrahedron with vertices $(0, 0, 0)$,
   $(1, 0, 0), (0, 1, 0)$, and $(0, 0, 1)$

31. $\mathbf{F}(x, y, z) = x^2 \mathbf{i} + y^2 \mathbf{j} + z^2 \mathbf{k}$, S is the boundary of the
   solid half-cylinder $0 \leqslant z \leqslant \sqrt{1 - y^2}, 0 \leqslant x \leqslant 2$

CAS **32.** Find the exact value of $\iint_S x^2 yz \, dS$, where S is the surface
   $z = xy, 0 \leqslant x \leqslant 1, 0 \leqslant y \leqslant 1$.

CAS **33.** Find the value of $\iint_S x^2 y^2 z^2 \, dS$ correct to four decimal
   places, where S is the part of the paraboloid
   $z = 3 - 2x^2 - y^2$ that lies above the xy-plane.

CAS **34.** Find the flux of

$$\mathbf{F}(x, y, z) = \sin(xyz) \mathbf{i} + x^2 y \mathbf{j} + z^2 e^{x/5} \mathbf{k}$$

   across the part of the cylinder $4y^2 + z^2 = 4$ that lies above
   the xy-plane and between the planes $x = -2$ and $x = 2$
   with upward orientation. Illustrate by using a computer
   algebra system to draw the cylinder and the vector field on
   the same screen.

35. Find a formula for $\iint_S \mathbf{F} \cdot d\mathbf{S}$ similar to Formula 10 for the
   case where S is given by $y = h(x, z)$ and $\mathbf{n}$ is the unit normal that points toward the left.

36. Find a formula for $\iint_S \mathbf{F} \cdot d\mathbf{S}$ similar to Formula 10 for the
   case where S is given by $x = k(y, z)$ and $\mathbf{n}$ is the unit normal that points forward (that is, toward the viewer when
   the axes are drawn in the usual way).

37. Find the center of mass of the hemisphere $x^2 + y^2 + z^2 = a^2$, $z \geqslant 0$, if it has constant density.

38. Find the mass of a thin funnel in the shape of a cone $z = \sqrt{x^2 + y^2}$, $1 \leqslant z \leqslant 4$, if its density function is $\rho(x, y, z) = 10 - z$.

39. (a) Give an integral expression for the moment of inertia I_z about the z-axis of a thin sheet in the shape of a surface S if the density function is ρ.
  (b) Find the moment of inertia about the z-axis of the funnel in Exercise 38.

40. Let S be the part of the sphere $x^2 + y^2 + z^2 = 25$ that lies above the plane $z = 4$. If S has constant density k, find (a) the center of mass and (b) the moment of inertia about the z-axis.

41. A fluid has density 870 kg/m^3 and flows with velocity $\mathbf{v} = z\,\mathbf{i} + y^2\,\mathbf{j} + x^2\,\mathbf{k}$, where x, y, and z are measured in meters and the components of $\mathbf{v}$ in meters per second. Find the rate of flow outward through the cylinder $x^2 + y^2 = 4$, $0 \leqslant z \leqslant 1$.

42. Seawater has density 1025 kg/m^3 and flows in a velocity field $\mathbf{v} = y\,\mathbf{i} + x\,\mathbf{j}$, where x, y, and z are measured in meters and the components of $\mathbf{v}$ in meters per second. Find the rate

of flow outward through the hemisphere $x^2 + y^2 + z^2 = 9$, $z \geqslant 0$.

43. Use Gauss's Law to find the charge contained in the solid hemisphere $x^2 + y^2 + z^2 \leqslant a^2$, $z \geqslant 0$, if the electric field is
$$\mathbf{E}(x, y, z) = x\,\mathbf{i} + y\,\mathbf{j} + 2z\,\mathbf{k}$$

44. Use Gauss's Law to find the charge enclosed by the cube with vertices $(\pm 1, \pm 1, \pm 1)$ if the electric field is
$$\mathbf{E}(x, y, z) = x\,\mathbf{i} + y\,\mathbf{j} + z\,\mathbf{k}$$

45. The temperature at the point (x, y, z) in a substance with conductivity $K = 6.5$ is $u(x, y, z) = 2y^2 + 2z^2$. Find the rate of heat flow inward across the cylindrical surface $y^2 + z^2 = 6$, $0 \leqslant x \leqslant 4$.

46. The temperature at a point in a ball with conductivity K is inversely proportional to the distance from the center of the ball. Find the rate of heat flow across a sphere S of radius a with center at the center of the ball.

47. Let $\mathbf{F}$ be an inverse square field, that is, $\mathbf{F}(\mathbf{r}) = c\mathbf{r}/|\mathbf{r}|^3$ for some constant c, where $\mathbf{r} = x\,\mathbf{i} + y\,\mathbf{j} + z\,\mathbf{k}$. Show that the flux of $\mathbf{F}$ across a sphere S with center the origin is independent of the radius of S.

13.8 | STOKES' THEOREM

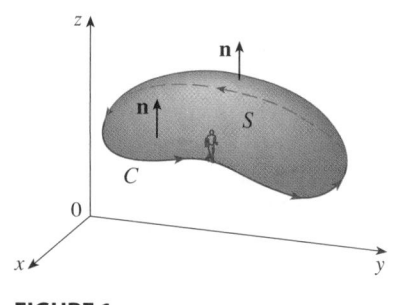

FIGURE 1

Stokes' Theorem can be regarded as a higher-dimensional version of Green's Theorem. Whereas Green's Theorem relates a double integral over a plane region D to a line integral around its plane boundary curve, Stokes' Theorem relates a surface integral over a surface S to a line integral around the boundary curve of S (which is a space curve). Figure 1 shows an oriented surface with unit normal vector $\mathbf{n}$. The orientation of S induces the **positive orientation of the boundary curve C** shown in the figure. This means that if you walk in the positive direction around C with your head pointing in the direction of $\mathbf{n}$, then the surface will always be on your left.

> **STOKES' THEOREM**   Let S be an oriented piecewise-smooth surface that is bounded by a simple, closed, piecewise-smooth boundary curve C with positive orientation. Let $\mathbf{F}$ be a vector field whose components have continuous partial derivatives on an open region in $\mathbb{R}^3$ that contains S. Then
> $$\int_C \mathbf{F} \cdot d\mathbf{r} = \iint_S \text{curl } \mathbf{F} \cdot d\mathbf{S}$$

Since

$$\int_C \mathbf{F} \cdot d\mathbf{r} = \int_C \mathbf{F} \cdot \mathbf{T} \, ds \quad \text{and} \quad \iint_S \text{curl } \mathbf{F} \cdot d\mathbf{S} = \iint_S \text{curl } \mathbf{F} \cdot \mathbf{n} \, dS$$

Stokes' Theorem says that the line integral around the boundary curve of S of the tangential component of $\mathbf{F}$ is equal to the surface integral of the normal component of the curl of $\mathbf{F}$.

The positively oriented boundary curve of the oriented surface S is often written as ∂S, so Stokes' Theorem can be expressed as

$$\boxed{1} \quad \iint_S \text{curl } \mathbf{F} \cdot d\mathbf{S} = \int_{\partial S} \mathbf{F} \cdot d\mathbf{r}$$

There is an analogy among Stokes' Theorem, Green's Theorem, and the Fundamental Theorem of Calculus. As before, there is an integral involving derivatives on the left side of Equation 1 (recall that curl $\mathbf{F}$ is a sort of derivative of $\mathbf{F}$) and the right side involves the values of $\mathbf{F}$ only on the *boundary* of S.

In fact, in the special case where the surface S is flat and lies in the xy-plane with upward orientation, the unit normal is $\mathbf{k}$, the surface integral becomes a double integral, and Stokes' Theorem becomes

$$\int_C \mathbf{F} \cdot d\mathbf{r} = \iint_S \text{curl } \mathbf{F} \cdot d\mathbf{S} = \iint_S (\text{curl } \mathbf{F}) \cdot \mathbf{k} \, dA$$

This is precisely the vector form of Green's Theorem given in Equation 13.5.12. Thus we see that Green's Theorem is really a special case of Stokes' Theorem.

Although Stokes' Theorem is too difficult for us to prove in its full generality, we can give a proof when S is a graph and $\mathbf{F}$, S, and C are well behaved.

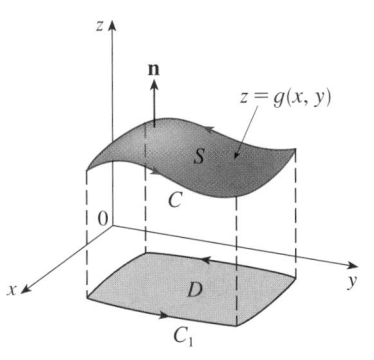

FIGURE 2

PROOF OF A SPECIAL CASE OF STOKES' THEOREM We assume that the equation of S is $z = g(x, y)$, $(x, y) \in D$, where g has continuous second-order partial derivatives and D is a simple plane region whose boundary curve C_1 corresponds to C. If the orientation of S is upward, then the positive orientation of C corresponds to the positive orientation of C_1. (See Figure 2.) We are also given that $\mathbf{F} = P\,\mathbf{i} + Q\,\mathbf{j} + R\,\mathbf{k}$, where the partial derivatives of P, Q, and R are continuous.

Since S is a graph of a function, we can apply Formula 13.7.10 with $\mathbf{F}$ replaced by curl $\mathbf{F}$. The result is

$$\boxed{2} \quad \iint_S \text{curl } \mathbf{F} \cdot d\mathbf{S}$$

$$= \iint_D \left[-\left(\frac{\partial R}{\partial y} - \frac{\partial Q}{\partial z}\right)\frac{\partial z}{\partial x} - \left(\frac{\partial P}{\partial z} - \frac{\partial R}{\partial x}\right)\frac{\partial z}{\partial y} + \left(\frac{\partial Q}{\partial x} - \frac{\partial P}{\partial y}\right) \right] dA$$

where the partial derivatives of P, Q, and R are evaluated at $(x, y, g(x, y))$. If

$$x = x(t) \qquad y = y(t) \qquad a \leqslant t \leqslant b$$

is a parametric representation of C_1, then a parametric representation of C is

$$x = x(t) \qquad y = y(t) \qquad z = g(x(t), y(t)) \qquad a \leqslant t \leqslant b$$

This allows us, with the aid of the Chain Rule, to evaluate the line integral as follows:

$$\int_C \mathbf{F} \cdot d\mathbf{r} = \int_a^b \left(P\frac{dx}{dt} + Q\frac{dy}{dt} + R\frac{dz}{dt} \right) dt = \int_a^b \left[P\frac{dx}{dt} + Q\frac{dy}{dt} + R\left(\frac{\partial z}{\partial x}\frac{dx}{dt} + \frac{\partial z}{\partial y}\frac{dy}{dt} \right) \right] dt$$

$$= \int_a^b \left[\left(P + R\frac{\partial z}{\partial x} \right)\frac{dx}{dt} + \left(Q + R\frac{\partial z}{\partial y} \right)\frac{dy}{dt} \right] dt = \int_{C_1} \left(P + R\frac{\partial z}{\partial x} \right) dx + \left(Q + R\frac{\partial z}{\partial y} \right) dy$$

$$= \iint_D \left[\frac{\partial}{\partial x}\left(Q + R\frac{\partial z}{\partial y} \right) - \frac{\partial}{\partial y}\left(P + R\frac{\partial z}{\partial x} \right) \right] dA$$

where we have used Green's Theorem in the last step. Then, using the Chain Rule again and remembering that P, Q, and R are functions of x, y, and z and that z is itself a function of x and y, we get

$$\int_C \mathbf{F} \cdot d\mathbf{r} = \iint_D \left[\left(\frac{\partial Q}{\partial x} + \frac{\partial Q}{\partial z}\frac{\partial z}{\partial x} + \frac{\partial R}{\partial x}\frac{\partial z}{\partial y} + \frac{\partial R}{\partial z}\frac{\partial z}{\partial x}\frac{\partial z}{\partial y} + R\frac{\partial^2 z}{\partial x\,\partial y} \right) \right.$$
$$\left. - \left(\frac{\partial P}{\partial y} + \frac{\partial P}{\partial z}\frac{\partial z}{\partial y} + \frac{\partial R}{\partial y}\frac{\partial z}{\partial x} + \frac{\partial R}{\partial z}\frac{\partial z}{\partial y}\frac{\partial z}{\partial x} + R\frac{\partial^2 z}{\partial y\,\partial x} \right) \right] dA$$

Four of the terms in this double integral cancel and the remaining six terms can be arranged to coincide with the right side of Equation 2. Therefore

$$\int_C \mathbf{F} \cdot d\mathbf{r} = \iint_S \text{curl } \mathbf{F} \cdot d\mathbf{S} \qquad \square$$

◢ **EXAMPLE 1** Evaluate $\int_C \mathbf{F} \cdot d\mathbf{r}$, where $\mathbf{F}(x, y, z) = -y^2\,\mathbf{i} + x\,\mathbf{j} + z^2\,\mathbf{k}$ and C is the curve of intersection of the plane $y + z = 2$ and the cylinder $x^2 + y^2 = 1$. (Orient C to be counterclockwise when viewed from above.)

SOLUTION The curve C (an ellipse) is shown in Figure 3. Although $\int_C \mathbf{F} \cdot d\mathbf{r}$ could be evaluated directly, it's easier to use Stokes' Theorem. We first compute

$$\text{curl } \mathbf{F} = \begin{vmatrix} \mathbf{i} & \mathbf{j} & \mathbf{k} \\ \dfrac{\partial}{\partial x} & \dfrac{\partial}{\partial y} & \dfrac{\partial}{\partial z} \\ -y^2 & x & z^2 \end{vmatrix} = (1 + 2y)\,\mathbf{k}$$

Although there are many surfaces with boundary C, the most convenient choice is the elliptical region S in the plane $y + z = 2$ that is bounded by C. If we orient S upward, then C has the induced positive orientation. The projection D of S on the xy-plane is the disk $x^2 + y^2 \leqslant 1$ and so using Equation 13.7.10 with $z = g(x, y) = 2 - y$, we have

$$\int_C \mathbf{F} \cdot d\mathbf{r} = \iint_S \text{curl } \mathbf{F} \cdot d\mathbf{S} = \iint_D (1 + 2y)\,dA = \int_0^{2\pi} \int_0^1 (1 + 2r\sin\theta)\,r\,dr\,d\theta$$

$$= \int_0^{2\pi} \left[\frac{r^2}{2} + 2\frac{r^3}{3}\sin\theta \right]_0^1 d\theta = \int_0^{2\pi} \left(\tfrac{1}{2} + \tfrac{2}{3}\sin\theta \right) d\theta = \tfrac{1}{2}(2\pi) + 0 = \pi \qquad \blacksquare$$

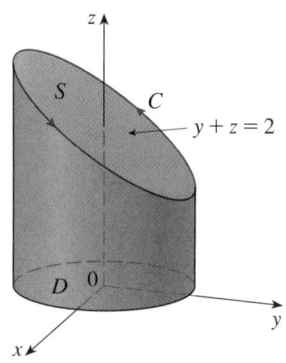

z

S

C

$y + z = 2$

D 0

y

x

FIGURE 3

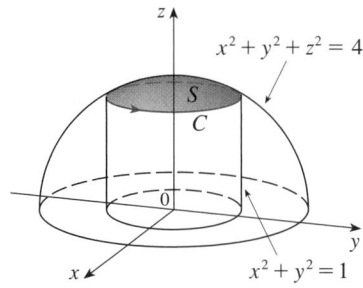

$x^2 + y^2 + z^2 = 4$

S

C

$x^2 + y^2 = 1$

FIGURE 4

V **EXAMPLE 2** Use Stokes' Theorem to compute the integral $\iint_S \text{curl } \mathbf{F} \cdot d\mathbf{S}$, where $\mathbf{F}(x, y, z) = xz\,\mathbf{i} + yz\,\mathbf{j} + xy\,\mathbf{k}$ and S is the part of the sphere $x^2 + y^2 + z^2 = 4$ that lies inside the cylinder $x^2 + y^2 = 1$ and above the xy-plane. (See Figure 4.)

SOLUTION To find the boundary curve C we solve the equations $x^2 + y^2 + z^2 = 4$ and $x^2 + y^2 = 1$. Subtracting, we get $z^2 = 3$ and so $z = \sqrt{3}$ (since $z > 0$). Thus C is the circle given by the equations $x^2 + y^2 = 1$, $z = \sqrt{3}$. A vector equation of C is

$$\mathbf{r}(t) = \cos t\,\mathbf{i} + \sin t\,\mathbf{j} + \sqrt{3}\,\mathbf{k} \qquad 0 \le t \le 2\pi$$

so

$$\mathbf{r}'(t) = -\sin t\,\mathbf{i} + \cos t\,\mathbf{j}$$

Also, we have

$$\mathbf{F}(\mathbf{r}(t)) = \sqrt{3}\,\cos t\,\mathbf{i} + \sqrt{3}\,\sin t\,\mathbf{j} + \cos t \sin t\,\mathbf{k}$$

Therefore, by Stokes' Theorem,

$$\iint_S \text{curl } \mathbf{F} \cdot d\mathbf{S} = \int_C \mathbf{F} \cdot d\mathbf{r} = \int_0^{2\pi} \mathbf{F}(\mathbf{r}(t)) \cdot \mathbf{r}'(t)\, dt$$

$$= \int_0^{2\pi} \left(-\sqrt{3}\,\cos t \sin t + \sqrt{3}\,\sin t \cos t \right) dt$$

$$= \sqrt{3} \int_0^{2\pi} 0\, dt = 0 \qquad \blacksquare$$

Note that in Example 2 we computed a surface integral simply by knowing the values of $\mathbf{F}$ on the boundary curve C. This means that if we have another oriented surface with the same boundary curve C, then we get exactly the same value for the surface integral!

In general, if S_1 and S_2 are oriented surfaces with the same oriented boundary curve C and both satisfy the hypotheses of Stokes' Theorem, then

$$\boxed{3} \qquad \iint_{S_1} \text{curl } \mathbf{F} \cdot d\mathbf{S} = \int_C \mathbf{F} \cdot d\mathbf{r} = \iint_{S_2} \text{curl } \mathbf{F} \cdot d\mathbf{S}$$

This fact is useful when it is difficult to integrate over one surface but easy to integrate over the other.

We now use Stokes' Theorem to throw some light on the meaning of the curl vector. Suppose that C is an oriented closed curve and $\mathbf{v}$ represents the velocity field in fluid flow. Consider the line integral

$$\int_C \mathbf{v} \cdot d\mathbf{r} = \int_C \mathbf{v} \cdot \mathbf{T}\, ds$$

and recall that $\mathbf{v} \cdot \mathbf{T}$ is the component of $\mathbf{v}$ in the direction of the unit tangent vector $\mathbf{T}$. This means that the closer the direction of $\mathbf{v}$ is to the direction of $\mathbf{T}$, the larger the value of $\mathbf{v} \cdot \mathbf{T}$. Thus $\int_C \mathbf{v} \cdot d\mathbf{r}$ is a measure of the tendency of the fluid to move around C and is called the **circulation** of $\mathbf{v}$ around C. (See Figure 5.)

Now let $P_0(x_0, y_0, z_0)$ be a point in the fluid and let S_a be a small disk with radius a and center P_0. Then $(\text{curl } \mathbf{F})(P) \approx (\text{curl } \mathbf{F})(P_0)$ for all points P on S_a because curl $\mathbf{F}$ is continuous. Thus, by Stokes' Theorem, we get the following approximation to the cir-

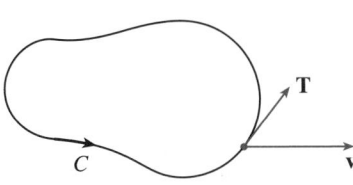

(a) $\int_C \mathbf{v} \cdot d\mathbf{r} > 0$, positive circulation

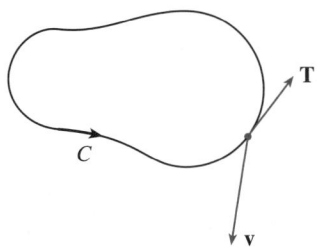

(b) $\int_C \mathbf{v} \cdot d\mathbf{r} < 0$, negative circulation

FIGURE 5

culation around the boundary circle C_a:

$$\int_{C_a} \mathbf{v} \cdot d\mathbf{r} = \iint_{S_a} \text{curl } \mathbf{v} \cdot d\mathbf{S} = \iint_{S_a} \text{curl } \mathbf{v} \cdot \mathbf{n} \, dS$$

$$\approx \iint_{S_a} \text{curl } \mathbf{v}(P_0) \cdot \mathbf{n}(P_0) \, dS = \text{curl } \mathbf{v}(P_0) \cdot \mathbf{n}(P_0) \pi a^2$$

■ Imagine a tiny paddle wheel placed in the fluid at a point P, as in Figure 6; the paddle wheel rotates fastest when its axis is parallel to curl $\mathbf{v}$.

This approximation becomes better as $a \to 0$ and we have

$$\boxed{4} \qquad \text{curl } \mathbf{v}(P_0) \cdot \mathbf{n}(P_0) = \lim_{a \to 0} \frac{1}{\pi a^2} \int_{C_a} \mathbf{v} \cdot d\mathbf{r}$$

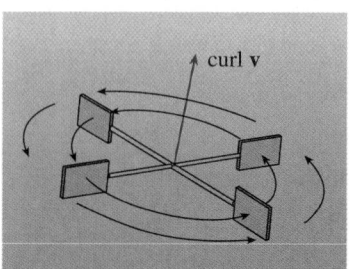

FIGURE 6

Equation 4 gives the relationship between the curl and the circulation. It shows that curl $\mathbf{v} \cdot \mathbf{n}$ is a measure of the rotating effect of the fluid about the axis $\mathbf{n}$. The curling effect is greatest about the axis parallel to curl $\mathbf{v}$.

Finally, we mention that Stokes' Theorem can be used to prove Theorem 13.5.4 (which states that if curl $\mathbf{F} = \mathbf{0}$ on all of $\mathbb{R}^3$, then $\mathbf{F}$ is conservative). From our previous work (Theorems 13.3.3 and 13.3.4), we know that $\mathbf{F}$ is conservative if $\int_C \mathbf{F} \cdot d\mathbf{r} = 0$ for every closed path C. Given C, suppose we can find an orientable surface S whose boundary is C. (This can be done, but the proof requires advanced techniques.) Then Stokes' Theorem gives

$$\int_C \mathbf{F} \cdot d\mathbf{r} = \iint_S \text{curl } \mathbf{F} \cdot d\mathbf{S} = \iint_S \mathbf{0} \cdot d\mathbf{S} = 0$$

A curve that is not simple can be broken into a number of simple curves, and the integrals around these simple curves are all 0. Adding these integrals, we obtain $\int_C \mathbf{F} \cdot d\mathbf{r} = 0$ for any closed curve C.

13.8 | EXERCISES

1–4 ■ Use Stokes' Theorem to evaluate $\iint_S \text{curl } \mathbf{F} \cdot d\mathbf{S}$.

1. $\mathbf{F}(x, y, z) = x^2 z^2 \, \mathbf{i} + y^2 z^2 \, \mathbf{j} + xyz \, \mathbf{k}$,
S is the part of the paraboloid $z = x^2 + y^2$ that lies inside the cylinder $x^2 + y^2 = 4$, oriented upward

2. $\mathbf{F}(x, y, z) = 2y \cos z \, \mathbf{i} + e^x \sin z \, \mathbf{j} + x e^y \, \mathbf{k}$,
S is the hemisphere $x^2 + y^2 + z^2 = 9$, $z \geq 0$, oriented upward

3. $\mathbf{F}(x, y, z) = xyz \, \mathbf{i} + xy \, \mathbf{j} + x^2 yz \, \mathbf{k}$,
S consists of the top and the four sides (but not the bottom) of the cube with vertices $(\pm 1, \pm 1, \pm 1)$, oriented outward

4. $\mathbf{F}(x, y, z) = \tan^{-1}(x^2 y z^2) \, \mathbf{i} + x^2 y \, \mathbf{j} + x^2 z^2 \, \mathbf{k}$,
S is the cone $x = \sqrt{y^2 + z^2}$, $0 \leq x \leq 2$, oriented in the direction of the positive x-axis

5–8 ■ Use Stokes' Theorem to evaluate $\int_C \mathbf{F} \cdot d\mathbf{r}$. In each case C is oriented counterclockwise as viewed from above.

5. $\mathbf{F}(x, y, z) = (x + y^2) \, \mathbf{i} + (y + z^2) \, \mathbf{j} + (z + x^2) \, \mathbf{k}$,
C is the triangle with vertices $(1, 0, 0)$, $(0, 1, 0)$, and $(0, 0, 1)$

6. $\mathbf{F}(x, y, z) = \mathbf{i} + (x + yz) \, \mathbf{j} + \left(xy - \sqrt{z}\right) \mathbf{k}$,
C is the boundary of the part of the plane $3x + 2y + z = 1$ in the first octant

7. $\mathbf{F}(x, y, z) = yz \, \mathbf{i} + 2xz \, \mathbf{j} + e^{xy} \, \mathbf{k}$,
C is the circle $x^2 + y^2 = 16$, $z = 5$

8. $\mathbf{F}(x, y, z) = xy \, \mathbf{i} + 2z \, \mathbf{j} + 3y \, \mathbf{k}$, C is the curve of intersection of the plane $x + z = 5$ and the cylinder $x^2 + y^2 = 9$

9. (a) Use Stokes' Theorem to evaluate $\int_C \mathbf{F} \cdot d\mathbf{r}$, where

$$\mathbf{F}(x, y, z) = x^2 z\,\mathbf{i} + xy^2\,\mathbf{j} + z^2\,\mathbf{k}$$

and C is the curve of intersection of the plane $x + y + z = 1$ and the cylinder $x^2 + y^2 = 9$ oriented counterclockwise as viewed from above.

 (b) Graph both the plane and the cylinder with domains chosen so that you can see the curve C and the surface that you used in part (a).

 (c) Find parametric equations for C and use them to graph C.

10. (a) Use Stokes' Theorem to evaluate $\int_C \mathbf{F} \cdot d\mathbf{r}$, where $\mathbf{F}(x, y, z) = x^2 y\,\mathbf{i} + \frac{1}{3}x^3\,\mathbf{j} + xy\,\mathbf{k}$ and C is the curve of intersection of the hyperbolic paraboloid $z = y^2 - x^2$ and the cylinder $x^2 + y^2 = 1$ oriented counterclockwise as viewed from above.

(b) Graph both the hyperbolic paraboloid and the cylinder with domains chosen so that you can see the curve C and the surface that you used in part (a).

(c) Find parametric equations for C and use them to graph C.

11–13 ▪ Verify that Stokes' Theorem is true for the given vector field $\mathbf{F}$ and surface S.

11. $\mathbf{F}(x, y, z) = -y\,\mathbf{i} + x\,\mathbf{j} - 2\,\mathbf{k}$,
S is the cone $z^2 = x^2 + y^2$, $0 \le z \le 4$, oriented downward

12. $\mathbf{F}(x, y, z) = -2yz\,\mathbf{i} + y\,\mathbf{j} + 3x\,\mathbf{k}$,
S is the part of the paraboloid $z = 5 - x^2 - y^2$ that lies above the plane $z = 1$, oriented upward

13. $\mathbf{F}(x, y, z) = y\,\mathbf{i} + z\,\mathbf{j} + x\,\mathbf{k}$,
S is the hemisphere $x^2 + y^2 + z^2 = 1$, $y \ge 0$, oriented in the direction of the positive y-axis

14. Let C be a simple closed smooth curve that lies in the plane $x + y + z = 1$. Show that the line integral

$$\int_C z\,dx - 2x\,dy + 3y\,dz$$

depends only on the area of the region enclosed by C and not on the shape of C or its location in the plane.

15. A particle moves along line segments from the origin to the points $(1, 0, 0)$, $(1, 2, 1)$, $(0, 2, 1)$, and back to the origin under the influence of the force field

$$\mathbf{F}(x, y, z) = z^2\,\mathbf{i} + 2xy\,\mathbf{j} + 4y^2\,\mathbf{k}$$

Find the work done.

16. Evaluate

$$\int_C (y + \sin x)\,dx + (z^2 + \cos y)\,dy + x^3\,dz$$

where C is the curve $\mathbf{r}(t) = \langle \sin t, \cos t, \sin 2t \rangle$, $0 \le t \le 2\pi$. [*Hint:* Observe that C lies on the surface $z = 2xy$.]

17. If S is a sphere and $\mathbf{F}$ satisfies the hypotheses of Stokes' Theorem, show that $\iint_S \text{curl } \mathbf{F} \cdot d\mathbf{S} = 0$.

18. Suppose S and C satisfy the hypotheses of Stokes' Theorem and f, g have continuous second-order partial derivatives. Use Exercises 22 and 24 in Section 13.5 to show the following.

(a) $\int_C (f\,\nabla g) \cdot d\mathbf{r} = \iint_S (\nabla f \times \nabla g) \cdot d\mathbf{S}$

(b) $\int_C (f\,\nabla f) \cdot d\mathbf{r} = 0$

(c) $\int_C (f\,\nabla g + g\,\nabla f) \cdot d\mathbf{r} = 0$

13.9 | THE DIVERGENCE THEOREM

In Section 13.5 we rewrote Green's Theorem in a vector version as

$$\int_C \mathbf{F} \cdot \mathbf{n}\,ds = \iint_D \text{div } \mathbf{F}(x, y)\,dA$$

where C is the positively oriented boundary curve of the plane region D. If we were seeking to extend this theorem to vector fields on $\mathbb{R}^3$, we might make the guess that

$$\boxed{1} \qquad \iint_S \mathbf{F} \cdot \mathbf{n}\,dS = \iiint_E \text{div } \mathbf{F}(x, y, z)\,dV$$

where S is the boundary surface of the solid region E. It turns out that Equation 1 is true, under appropriate hypotheses, and is called the Divergence Theorem. Notice its similarity to Green's Theorem and Stokes' Theorem in that it relates the integral of a derivative of a function (div $\mathbf{F}$ in this case) over a region to the integral of the original function $\mathbf{F}$ over the boundary of the region.

At this stage you may wish to review the various types of regions over which we were able to evaluate triple integrals in Section 12.5. We state and prove the Divergence Theorem for regions E that are simultaneously of types 1, 2, and 3 and we call such regions **simple solid regions**. (For instance, regions bounded by ellipsoids or rectangular boxes are simple solid regions.) The boundary of E is a closed surface, and we use the convention, introduced in Section 13.7, that the positive orientation is outward; that is, the unit normal vector $\mathbf{n}$ is directed outward from E.

■ The Divergence Theorem is sometimes called Gauss's Theorem after the great German mathematician Karl Friedrich Gauss (1777–1855), who discovered this theorem during his investigation of electrostatics. In Eastern Europe the Divergence Theorem is known as Ostrogradsky's Theorem after the Russian mathematician Mikhail Ostrogradsky (1801–1862), who published this result in 1826.

THE DIVERGENCE THEOREM Let E be a simple solid region and let S be the boundary surface of E, given with positive (outward) orientation. Let $\mathbf{F}$ be a vector field whose component functions have continuous partial derivatives on an open region that contains E. Then

$$\iint_S \mathbf{F} \cdot d\mathbf{S} = \iiint_E \text{div } \mathbf{F} \, dV$$

Thus the Divergence Theorem states that, under the given conditions, the flux of $\mathbf{F}$ across the boundary surface of E is equal to the triple integral of the divergence of $\mathbf{F}$ over E.

PROOF Let $\mathbf{F} = P\,\mathbf{i} + Q\,\mathbf{j} + R\,\mathbf{k}$. Then

$$\text{div } \mathbf{F} = \frac{\partial P}{\partial x} + \frac{\partial Q}{\partial y} + \frac{\partial R}{\partial z}$$

so

$$\iiint_E \text{div } \mathbf{F} \, dV = \iiint_E \frac{\partial P}{\partial x} \, dV + \iiint_E \frac{\partial Q}{\partial y} \, dV + \iiint_E \frac{\partial R}{\partial z} \, dV$$

If $\mathbf{n}$ is the unit outward normal of S, then the surface integral on the left side of the Divergence Theorem is

$$\iint_S \mathbf{F} \cdot d\mathbf{S} = \iint_S \mathbf{F} \cdot \mathbf{n} \, dS = \iint_S (P\,\mathbf{i} + Q\,\mathbf{j} + R\,\mathbf{k}) \cdot \mathbf{n} \, dS$$

$$= \iint_S P\,\mathbf{i} \cdot \mathbf{n} \, dS + \iint_S Q\,\mathbf{j} \cdot \mathbf{n} \, dS + \iint_S R\,\mathbf{k} \cdot \mathbf{n} \, dS$$

Therefore, to prove the Divergence Theorem, it suffices to prove the following three equations:

$$\boxed{2} \qquad \iint_S P\,\mathbf{i} \cdot \mathbf{n} \, dS = \iiint_E \frac{\partial P}{\partial x} \, dV$$

$$\boxed{3} \qquad \iint_S Q\,\mathbf{j} \cdot \mathbf{n} \, dS = \iiint_E \frac{\partial Q}{\partial y} \, dV$$

$$\boxed{4} \qquad \iint_S R\,\mathbf{k} \cdot \mathbf{n} \, dS = \iiint_E \frac{\partial R}{\partial z} \, dV$$

To prove Equation 4 we use the fact that E is a type 1 region:

$$E = \{(x, y, z) \mid (x, y) \in D, u_1(x, y) \leqslant z \leqslant u_2(x, y)\}$$

where D is the projection of E onto the xy-plane. By Equation 12.5.6, we have

$$\iiint_E \frac{\partial R}{\partial z}\, dV = \iint_D \left[\int_{u_1(x, y)}^{u_2(x, y)} \frac{\partial R}{\partial z} (x, y, z)\, dz \right] dA$$

and therefore, by the Fundamental Theorem of Calculus,

$$\boxed{5} \qquad \iiint_E \frac{\partial R}{\partial z}\, dV = \iint_D [R(x, y, u_2(x, y)) - R(x, y, u_1(x, y))]\, dA$$

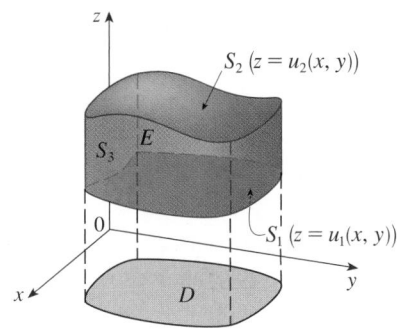

FIGURE 1

The boundary surface S consists of three pieces: the bottom surface S_1, the top surface S_2, and possibly a vertical surface S_3, which lies above the boundary curve of D. (See Figure 1. It might happen that S_3 doesn't appear, as in the case of a sphere.) Notice that on S_3 we have $\mathbf{k} \cdot \mathbf{n} = 0$, because $\mathbf{k}$ is vertical and $\mathbf{n}$ is horizontal, and so

$$\iint_{S_3} R\, \mathbf{k} \cdot \mathbf{n}\, dS = \iint_{S_3} 0\, dS = 0$$

Thus, regardless of whether there is a vertical surface, we can write

$$\boxed{6} \qquad \iint_S R\, \mathbf{k} \cdot \mathbf{n}\, dS = \iint_{S_1} R\, \mathbf{k} \cdot \mathbf{n}\, dS + \iint_{S_2} R\, \mathbf{k} \cdot \mathbf{n}\, dS$$

The equation of S_2 is $z = u_2(x, y)$, $(x, y) \in D$, and the outward normal $\mathbf{n}$ points upward, so from Equation 13.7.10 (with $\mathbf{F}$ replaced by $R\, \mathbf{k}$) we have

$$\iint_{S_2} R\, \mathbf{k} \cdot \mathbf{n}\, dS = \iint_D R(x, y, u_2(x, y))\, dA$$

On S_1 we have $z = u_1(x, y)$, but here the outward normal $\mathbf{n}$ points downward, so we multiply by -1:

$$\iint_{S_1} R\, \mathbf{k} \cdot \mathbf{n}\, dS = -\iint_D R(x, y, u_1(x, y))\, dA$$

Therefore Equation 6 gives

$$\iint_S R\, \mathbf{k} \cdot \mathbf{n}\, dS = \iint_D [R(x, y, u_2(x, y)) - R(x, y, u_1(x, y))]\, dA$$

Comparison with Equation 5 shows that

$$\iint_S R\, \mathbf{k} \cdot \mathbf{n}\, dS = \iiint_E \frac{\partial R}{\partial z}\, dV$$

■ Notice that the method of proof of the Divergence Theorem is very similar to that of Green's Theorem.

Equations 2 and 3 are proved in a similar manner using the expressions for E as a type 2 or type 3 region, respectively.

V **EXAMPLE 1** Find the flux of the vector field $\mathbf{F}(x, y, z) = z\,\mathbf{i} + y\,\mathbf{j} + x\,\mathbf{k}$ over the unit sphere $x^2 + y^2 + z^2 = 1$.

SOLUTION First we compute the divergence of $\mathbf{F}$:

$$\operatorname{div} \mathbf{F} = \frac{\partial}{\partial x}(z) + \frac{\partial}{\partial y}(y) + \frac{\partial}{\partial z}(x) = 1$$

The unit sphere S is the boundary of the unit ball B given by $x^2 + y^2 + z^2 \leqslant 1$. Thus the Divergence Theorem gives the flux as

$$\iint_S \mathbf{F} \cdot d\mathbf{S} = \iiint_B \operatorname{div} \mathbf{F}\, dV = \iiint_B 1\, dV = V(B) = \tfrac{4}{3}\pi(1)^3 = \frac{4\pi}{3}\qquad\blacksquare$$

■ The solution in Example 1 should be compared with the solution in Example 4 in Section 13.7.

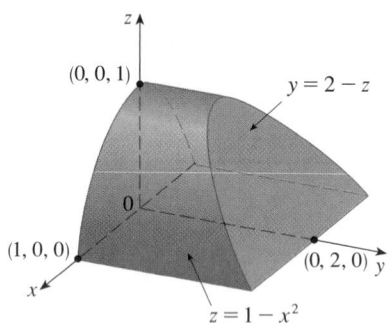

FIGURE 2

V **EXAMPLE 2** Evaluate $\displaystyle\iint_S \mathbf{F} \cdot d\mathbf{S}$, where

$$\mathbf{F}(x, y, z) = xy\,\mathbf{i} + \left(y^2 + e^{xz^2}\right)\mathbf{j} + \sin(xy)\,\mathbf{k}$$

and S is the surface of the region E bounded by the parabolic cylinder $z = 1 - x^2$ and the planes $z = 0$, $y = 0$, and $y + z = 2$. (See Figure 2.)

SOLUTION It would be extremely difficult to evaluate the given surface integral directly. (We would have to evaluate four surface integrals corresponding to the four pieces of S.) Furthermore, the divergence of $\mathbf{F}$ is much less complicated than $\mathbf{F}$ itself:

$$\operatorname{div} \mathbf{F} = \frac{\partial}{\partial x}(xy) + \frac{\partial}{\partial y}\left(y^2 + e^{xz^2}\right) + \frac{\partial}{\partial z}(\sin xy) = y + 2y = 3y$$

Therefore we use the Divergence Theorem to transform the given surface integral into a triple integral. The easiest way to evaluate the triple integral is to express E as a type 3 region:

$$E = \left\{(x, y, z) \mid -1 \leqslant x \leqslant 1,\ 0 \leqslant z \leqslant 1 - x^2,\ 0 \leqslant y \leqslant 2 - z\right\}$$

Then we have

$$\iint_S \mathbf{F} \cdot d\mathbf{S} = \iiint_E \operatorname{div} \mathbf{F}\, dV = \iiint_E 3y\, dV = 3\int_{-1}^{1}\int_{0}^{1-x^2}\int_{0}^{2-z} y\, dy\, dz\, dx$$

$$= 3\int_{-1}^{1}\int_{0}^{1-x^2} \frac{(2-z)^2}{2}\, dz\, dx = \frac{3}{2}\int_{-1}^{1}\left[-\frac{(2-z)^3}{3}\right]_{0}^{1-x^2} dx$$

$$= -\tfrac{1}{2}\int_{-1}^{1}\left[(x^2 + 1)^3 - 8\right] dx = -\int_{0}^{1}(x^6 + 3x^4 + 3x^2 - 7)\, dx = \tfrac{184}{35}\qquad\blacksquare$$

Although we have proved the Divergence Theorem only for simple solid regions, it can be proved for regions that are finite unions of simple solid regions. (The procedure is similar to the one we used in Section 13.4 to extend Green's Theorem.)

For example, let's consider the region E that lies between the closed surfaces S_1 and S_2, where S_1 lies inside S_2. Let $\mathbf{n}_1$ and $\mathbf{n}_2$ be outward normals of S_1 and S_2. Then the boundary surface of E is $S = S_1 \cup S_2$ and its normal $\mathbf{n}$ is given by $\mathbf{n} = -\mathbf{n}_1$ on S_1 and

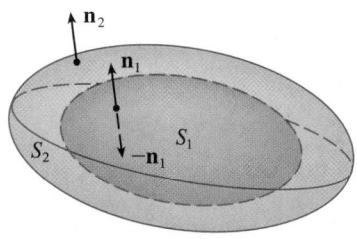

FIGURE 3

$\mathbf{n} = \mathbf{n}_2$ on S_2. (See Figure 3.) Applying the Divergence Theorem to S, we get

$$\boxed{7} \qquad \iiint_E \operatorname{div} \mathbf{F}\, dV = \iint_S \mathbf{F} \cdot d\mathbf{S} = \iint_S \mathbf{F} \cdot \mathbf{n}\, dS$$

$$= \iint_{S_1} \mathbf{F} \cdot (-\mathbf{n}_1)\, dS + \iint_{S_2} \mathbf{F} \cdot \mathbf{n}_2\, dS$$

$$= -\iint_{S_1} \mathbf{F} \cdot d\mathbf{S} + \iint_{S_2} \mathbf{F} \cdot d\mathbf{S}$$

EXAMPLE 3 In Example 5 in Section 13.1 we considered the electric field

$$\mathbf{E}(\mathbf{x}) = \frac{\varepsilon Q}{|\mathbf{x}|^3}\, \mathbf{x}$$

where the electric charge Q is located at the origin and $\mathbf{x} = \langle x, y, z \rangle$ is a position vector. Use the Divergence Theorem to show that the electric flux of $\mathbf{E}$ through any closed surface S_2 that encloses the origin is

$$\iint_{S_2} \mathbf{E} \cdot d\mathbf{S} = 4\pi\varepsilon Q$$

SOLUTION The difficulty is that we don't have an explicit equation for S_2 because it is *any* closed surface enclosing the origin. The simplest such surface would be a sphere, so we let S_1 be a small sphere with radius a and center the origin. You can verify that div $\mathbf{E} = 0$. (See Exercise 23.) Therefore Equation 7 gives

$$\iint_{S_2} \mathbf{E} \cdot d\mathbf{S} = \iint_{S_1} \mathbf{E} \cdot d\mathbf{S} + \iiint_E \operatorname{div} \mathbf{E}\, dV = \iint_{S_1} \mathbf{E} \cdot d\mathbf{S} = \iint_{S_1} \mathbf{E} \cdot \mathbf{n}\, dS$$

The point of this calculation is that we can compute the surface integral over S_1 because S_1 is a sphere. The normal vector at $\mathbf{x}$ is $\mathbf{x}/|\mathbf{x}|$. Therefore

$$\mathbf{E} \cdot \mathbf{n} = \frac{\varepsilon Q}{|\mathbf{x}|^3}\, \mathbf{x} \cdot \left(\frac{\mathbf{x}}{|\mathbf{x}|} \right) = \frac{\varepsilon Q}{|\mathbf{x}|^4}\, \mathbf{x} \cdot \mathbf{x} = \frac{\varepsilon Q}{|\mathbf{x}|^2} = \frac{\varepsilon Q}{a^2}$$

since the equation of S_1 is $|\mathbf{x}| = a$. Thus we have

$$\iint_{S_2} \mathbf{E} \cdot d\mathbf{S} = \iint_{S_1} \mathbf{E} \cdot \mathbf{n}\, dS = \frac{\varepsilon Q}{a^2} \iint_{S_1} dS = \frac{\varepsilon Q}{a^2}\, A(S_1) = \frac{\varepsilon Q}{a^2}\, 4\pi a^2 = 4\pi\varepsilon Q$$

This shows that the electric flux of $\mathbf{E}$ is $4\pi\varepsilon Q$ through *any* closed surface S_2 that contains the origin. [This is a special case of Gauss's Law (Equation 13.7.11) for a single charge. The relationship between ε and ε_0 is $\varepsilon = 1/(4\pi\varepsilon_0)$.] ∎

Another application of the Divergence Theorem occurs in fluid flow. Let $\mathbf{v}(x, y, z)$ be the velocity field of a fluid with constant density ρ. Then $\mathbf{F} = \rho\mathbf{v}$ is the rate of flow per unit area. If $P_0(x_0, y_0, z_0)$ is a point in the fluid and B_a is a ball with center P_0 and very small radius a, then div $\mathbf{F}(P) \approx$ div $\mathbf{F}(P_0)$ for all points in B_a since div $\mathbf{F}$ is continuous. We approximate the flux over the boundary sphere S_a as follows:

$$\iint_{S_a} \mathbf{F} \cdot d\mathbf{S} = \iiint_{B_a} \operatorname{div} \mathbf{F}\, dV \approx \iiint_{B_a} \operatorname{div} \mathbf{F}(P_0)\, dV = \operatorname{div} \mathbf{F}(P_0) V(B_a)$$

This approximation becomes better as $a \to 0$ and suggests that

$$\boxed{8} \qquad \operatorname{div} \mathbf{F}(P_0) = \lim_{a \to 0} \frac{1}{V(B_a)} \iint_{S_a} \mathbf{F} \cdot d\mathbf{S}$$

Equation 8 says that $\operatorname{div} \mathbf{F}(P_0)$ is the net rate of outward flux per unit volume at P_0. (This is the reason for the name *divergence*.) If $\operatorname{div} \mathbf{F}(P) > 0$, the net flow is outward near P and P is called a **source**. If $\operatorname{div} \mathbf{F}(P) < 0$, the net flow is inward near P and P is called a **sink**.

For the vector field in Figure 4, it appears that the vectors that end near P_1 are shorter than the vectors that start near P_1. Thus the net flow is outward near P_1, so $\operatorname{div} \mathbf{F}(P_1) > 0$ and P_1 is a source. Near P_2, on the other hand, the incoming arrows are longer than the outgoing arrows. Here the net flow is inward, so $\operatorname{div} \mathbf{F}(P_2) < 0$ and P_2 is a sink. We can use the formula for $\mathbf{F}$ to confirm this impression. Since $\mathbf{F} = x^2 \mathbf{i} + y^2 \mathbf{j}$, we have $\operatorname{div} \mathbf{F} = 2x + 2y$, which is positive when $y > -x$. So the points above the line $y = -x$ are sources and those below are sinks.

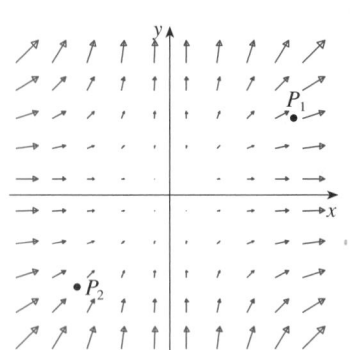

FIGURE 4

The vector field $\mathbf{F} = x^2 \mathbf{i} + y^2 \mathbf{j}$

SUMMARY

The main results of this chapter are all higher-dimensional versions of the Fundamental Theorem of Calculus. To help you remember them, we collect them together here (without hypotheses) so that you can see more easily their essential similarity. Notice that in each case we have an integral of a "derivative" over a region on the left side, and the right side involves the values of the original function only on the *boundary* of the region.

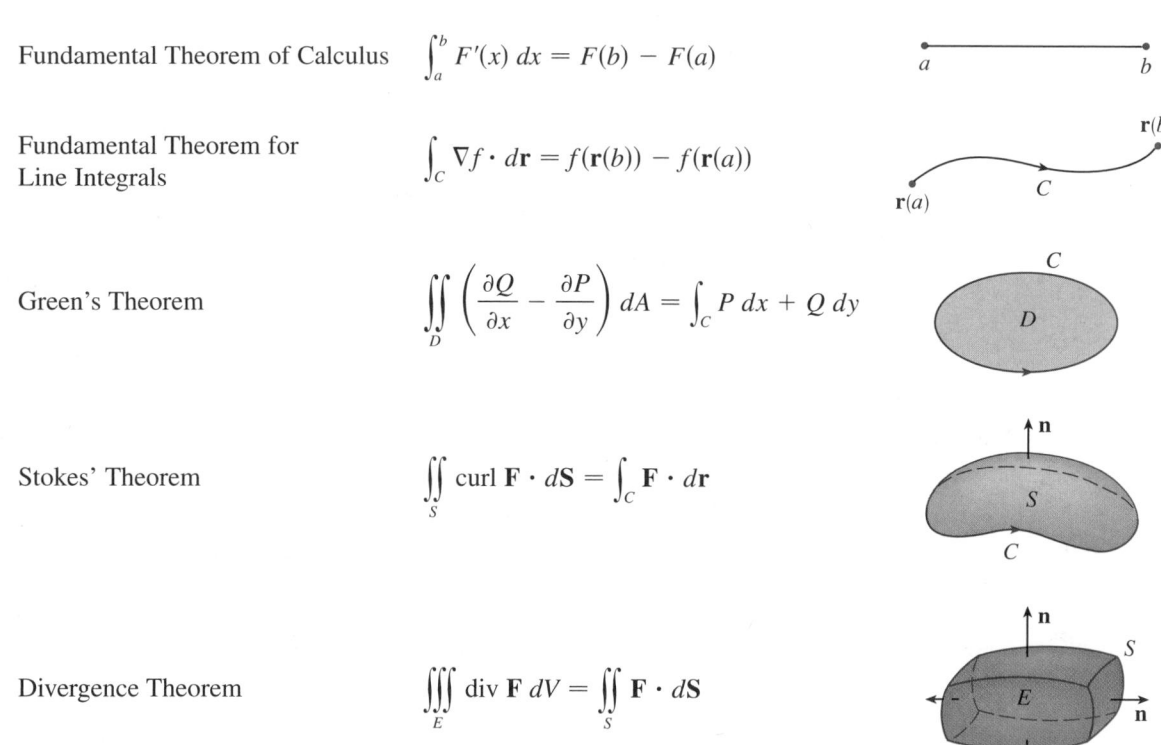

Fundamental Theorem of Calculus $\displaystyle \int_a^b F'(x)\, dx = F(b) - F(a)$

Fundamental Theorem for Line Integrals $\displaystyle \int_C \nabla f \cdot d\mathbf{r} = f(\mathbf{r}(b)) - f(\mathbf{r}(a))$

Green's Theorem $\displaystyle \iint_D \left(\frac{\partial Q}{\partial x} - \frac{\partial P}{\partial y} \right) dA = \int_C P\, dx + Q\, dy$

Stokes' Theorem $\displaystyle \iint_S \operatorname{curl} \mathbf{F} \cdot d\mathbf{S} = \int_C \mathbf{F} \cdot d\mathbf{r}$

Divergence Theorem $\displaystyle \iiint_E \operatorname{div} \mathbf{F}\, dV = \iint_S \mathbf{F} \cdot d\mathbf{S}$

13.9 | EXERCISES

1–4 ▪ Verify that the Divergence Theorem is true for the vector field **F** on the region E.

1. $\mathbf{F}(x, y, z) = 3x\,\mathbf{i} + xy\,\mathbf{j} + 2xz\,\mathbf{k}$,
E is the cube bounded by the planes $x = 0$, $x = 1$, $y = 0$, $y = 1$, $z = 0$, and $z = 1$

2. $\mathbf{F}(x, y, z) = x^2\,\mathbf{i} + xy\,\mathbf{j} + z\,\mathbf{k}$,
E is the solid bounded by the paraboloid $z = 4 - x^2 - y^2$ and the xy-plane

3. $\mathbf{F}(x, y, z) = \langle z, y, x \rangle$,
E is the solid ball $x^2 + y^2 + z^2 \le 16$

4. $\mathbf{F}(x, y, z) = \langle x^2, -y, z \rangle$,
E is the solid cylinder $y^2 + z^2 \le 9$, $0 \le x \le 2$

5–15 ▪ Use the Divergence Theorem to calculate the surface integral $\iint_S \mathbf{F} \cdot d\mathbf{S}$; that is, calculate the flux of **F** across S.

5. $\mathbf{F}(x, y, z) = xye^z\,\mathbf{i} + xy^2z^3\,\mathbf{j} - ye^z\,\mathbf{k}$,
S is the surface of the box bounded by the coordinate planes and the planes $x = 3$, $y = 2$, and $z = 1$

6. $\mathbf{F}(x, y, z) = x^2yz\,\mathbf{i} + xy^2z\,\mathbf{j} + xyz^2\,\mathbf{k}$,
S is the surface of the box enclosed by the planes $x = 0$, $x = a$, $y = 0$, $y = b$, $z = 0$, and $z = c$, where a, b, and c are positive numbers

7. $\mathbf{F}(x, y, z) = 3xy^2\,\mathbf{i} + xe^z\,\mathbf{j} + z^3\,\mathbf{k}$,
S is the surface of the solid bounded by the cylinder $y^2 + z^2 = 1$ and the planes $x = -1$ and $x = 2$

8. $\mathbf{F}(x, y, z) = (x^3 + y^3)\,\mathbf{i} + (y^3 + z^3)\,\mathbf{j} + (z^3 + x^3)\,\mathbf{k}$,
S is the sphere with center the origin and radius 2

9. $\mathbf{F}(x, y, z) = x^2\sin y\,\mathbf{i} + x\cos y\,\mathbf{j} - xz\sin y\,\mathbf{k}$,
S is the "fat sphere" $x^8 + y^8 + z^8 = 8$

10. $\mathbf{F}(x, y, z) = z\,\mathbf{i} + y\,\mathbf{j} + zx\,\mathbf{k}$,
S is the surface of the tetrahedron enclosed by the coordinate planes and the plane

$$\frac{x}{a} + \frac{y}{b} + \frac{z}{c} = 1$$

where a, b, and c are positive numbers

11. $\mathbf{F}(x, y, z) = (\cos z + xy^2)\,\mathbf{i} + xe^{-z}\,\mathbf{j} + (\sin y + x^2z)\,\mathbf{k}$,
S is the surface of the solid bounded by the paraboloid $z = x^2 + y^2$ and the plane $z = 4$

12. $\mathbf{F}(x, y, z) = x^4\,\mathbf{i} - x^3z^2\,\mathbf{j} + 4xy^2z\,\mathbf{k}$,
S is the surface of the solid bounded by the cylinder $x^2 + y^2 = 1$ and the planes $z = x + 2$ and $z = 0$

13. $\mathbf{F} = |\mathbf{r}|\,\mathbf{r}$, where $\mathbf{r} = x\,\mathbf{i} + y\,\mathbf{j} + z\,\mathbf{k}$,
S consists of the hemisphere $z = \sqrt{1 - x^2 - y^2}$ and the disk $x^2 + y^2 \le 1$ in the xy-plane

14. $\mathbf{F} = |\mathbf{r}|^2\mathbf{r}$, where $\mathbf{r} = x\,\mathbf{i} + y\,\mathbf{j} + z\,\mathbf{k}$,
S is the sphere with radius R and center the origin

CAS 15. $\mathbf{F}(x, y, z) = e^y\tan z\,\mathbf{i} + y\sqrt{3 - x^2}\,\mathbf{j} + x\sin y\,\mathbf{k}$,
S is the surface of the solid that lies above the xy-plane and below the surface $z = 2 - x^4 - y^4$, $-1 \le x \le 1$, $-1 \le y \le 1$

CAS 16. Use a computer algebra system to plot the vector field $\mathbf{F}(x, y, z) = \sin x\cos^2 y\,\mathbf{i} + \sin^3 y\cos^4 z\,\mathbf{j} + \sin^5 z\cos^6 x\,\mathbf{k}$ in the cube cut from the first octant by the planes $x = \pi/2$, $y = \pi/2$, and $z = \pi/2$. Then compute the flux across the surface of the cube.

17. Use the Divergence Theorem to evaluate $\iint_S \mathbf{F} \cdot d\mathbf{S}$, where $\mathbf{F}(x, y, z) = z^2x\,\mathbf{i} + \left(\frac{1}{3}y^3 + \tan z\right)\mathbf{j} + (x^2z + y^2)\,\mathbf{k}$ and S is the top half of the sphere $x^2 + y^2 + z^2 = 1$. [*Hint:* Note that S is not a closed surface. First compute integrals over S_1 and S_2, where S_1 is the disk $x^2 + y^2 \le 1$, oriented downward, and $S_2 = S \cup S_1$.]

18. Let $\mathbf{F}(x, y, z) = z\tan^{-1}(y^2)\,\mathbf{i} + z^3\ln(x^2 + 1)\,\mathbf{j} + z\,\mathbf{k}$. Find the flux of **F** across the part of the paraboloid $x^2 + y^2 + z = 2$ that lies above the plane $z = 1$ and is oriented upward.

19. A vector field **F** is shown. Use the interpretation of divergence derived in this section to determine whether div **F** is positive or negative at P_1 and at P_2.

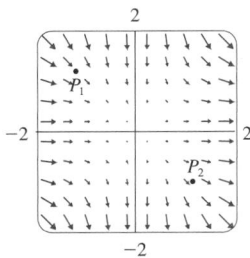

20. (a) Are the points P_1 and P_2 sources or sinks for the vector field **F** shown in the figure? Give an explanation based solely on the picture.
(b) Given that $\mathbf{F}(x, y) = \langle x, y^2 \rangle$, use the definition of divergence to verify your answer to part (a).

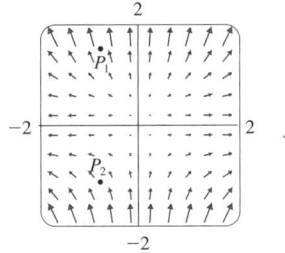

CAS **21–22** ▪ Plot the vector field and guess where div $\mathbf{F} > 0$ and where div $\mathbf{F} < 0$. Then calculate div $\mathbf{F}$ to check your guess.

21. $\mathbf{F}(x, y) = \langle xy, x + y^2 \rangle$ **22.** $\mathbf{F}(x, y) = \langle x^2, y^2 \rangle$

23. Verify that div $\mathbf{E} = 0$ for the electric field $\mathbf{E}(\mathbf{x}) = \dfrac{\varepsilon Q}{|\mathbf{x}|^3}\,\mathbf{x}$.

24. Use the Divergence Theorem to evaluate

$$\iint_S (2x + 2y + z^2)\, dS$$

where S is the sphere $x^2 + y^2 + z^2 = 1$.

25–30 ▪ Prove each identity, assuming that S and E satisfy the conditions of the Divergence Theorem and the scalar functions and components of the vector fields have continuous second-order partial derivatives.

25. $\displaystyle\iint_S \mathbf{a} \cdot \mathbf{n}\, dS = 0$, where $\mathbf{a}$ is a constant vector

26. $V(E) = \dfrac{1}{3}\displaystyle\iint_S \mathbf{F} \cdot d\mathbf{S}$, where $\mathbf{F}(x, y, z) = x\,\mathbf{i} + y\,\mathbf{j} + z\,\mathbf{k}$

27. $\displaystyle\iint_S \text{curl } \mathbf{F} \cdot d\mathbf{S} = 0$ **28.** $\displaystyle\iint_S D_{\mathbf{n}} f\, dS = \iiint_E \nabla^2 f\, dV$

29. $\displaystyle\iint_S (f\,\nabla g) \cdot \mathbf{n}\, dS = \iiint_E (f\,\nabla^2 g + \nabla f \cdot \nabla g)\, dV$

30. $\displaystyle\iint_S (f\,\nabla g - g\,\nabla f) \cdot \mathbf{n}\, dS = \iiint_E (f\,\nabla^2 g - g\,\nabla^2 f)\, dV$

CHAPTER 13 REVIEW

CONCEPT CHECK

1. What is a vector field? Give three examples that have physical meaning.

2. (a) What is a conservative vector field?
    (b) What is a potential function?

3. (a) Write the definition of the line integral of a scalar function f along a smooth curve C with respect to arc length.
    (b) How do you evaluate such a line integral?
    (c) Write expressions for the mass and center of mass of a thin wire shaped like a curve C if the wire has linear density function $\rho(x, y)$.
    (d) Write the definitions of the line integrals along C of a scalar function f with respect to x, y, and z.
    (e) How do you evaluate these line integrals?

4. (a) Define the line integral of a vector field $\mathbf{F}$ along a smooth curve C given by a vector function $\mathbf{r}(t)$.
    (b) If $\mathbf{F}$ is a force field, what does this line integral represent?
    (c) If $\mathbf{F} = \langle P, Q, R \rangle$, what is the connection between the line integral of $\mathbf{F}$ and the line integrals of the component functions P, Q, and R?

5. State the Fundamental Theorem for Line Integrals.

6. (a) What does it mean to say that $\int_C \mathbf{F} \cdot d\mathbf{r}$ is independent of path?
    (b) If you know that $\int_C \mathbf{F} \cdot d\mathbf{r}$ is independent of path, what can you say about $\mathbf{F}$?

7. State Green's Theorem.

8. Write expressions for the area enclosed by a curve C in terms of line integrals around C.

9. Suppose $\mathbf{F}$ is a vector field on $\mathbb{R}^3$.
    (a) Define curl $\mathbf{F}$.
    (b) Define div $\mathbf{F}$.
    (c) If $\mathbf{F}$ is a velocity field in fluid flow, what are the physical interpretations of curl $\mathbf{F}$ and div $\mathbf{F}$?

10. If $\mathbf{F} = P\,\mathbf{i} + Q\,\mathbf{j}$, how do you test to determine whether $\mathbf{F}$ is conservative? What if $\mathbf{F}$ is a vector field on $\mathbb{R}^3$?

11. (a) What is a parametric surface? What are its grid curves?
    (b) Write an expression for the area of a parametric surface.
    (c) What is the area of a surface given by an equation $z = g(x, y)$?

12. (a) Write the definition of the surface integral of a scalar function f over a surface S.
    (b) How do you evaluate such an integral if S is a parametric surface given by a vector function $\mathbf{r}(u, v)$?
    (c) What if S is given by an equation $z = g(x, y)$?
    (d) If a thin sheet has the shape of a surface S, and the density at (x, y, z) is $\rho(x, y, z)$, write expressions for the mass and center of mass of the sheet.

13. (a) What is an oriented surface? Give an example of a non-orientable surface.
    (b) Define the surface integral (or flux) of a vector field $\mathbf{F}$ over an oriented surface S with unit normal vector $\mathbf{n}$.

(c) How do you evaluate such an integral if S is a parametric surface given by a vector function $\mathbf{r}(u, v)$?

(d) What if S is given by an equation $z = g(x, y)$?

14. State Stokes' Theorem.

15. State the Divergence Theorem.

16. In what ways are the Fundamental Theorem for Line
Integrals, Green's Theorem, Stokes' Theorem, and the
Divergence Theorem similar?

TRUE-FALSE QUIZ

Determine whether the statement is true or false. If it is true, explain
why. If it is false, explain why or give an example that disproves the
statement.

1. If $\mathbf{F}$ is a vector field, then div $\mathbf{F}$ is a vector field.

2. If $\mathbf{F}$ is a vector field, then curl $\mathbf{F}$ is a vector field.

3. If f has continuous partial derivatives of all orders on $\mathbb{R}^3$,
then $\text{div}(\text{curl } \nabla f) = 0$.

4. If f has continuous partial derivatives on $\mathbb{R}^3$ and C is any
circle, then $\int_C \nabla f \cdot d\mathbf{r} = 0$.

5. If $\mathbf{F} = P\,\mathbf{i} + Q\,\mathbf{j}$ and $P_y = Q_x$ in an open region D, then $\mathbf{F}$
is conservative.

6. $\int_{-C} f(x, y)\, ds = -\int_C f(x, y)\, ds$

7. If $\mathbf{F}$ and $\mathbf{G}$ are vector fields and div $\mathbf{F}$ = div $\mathbf{G}$, then $\mathbf{F} = \mathbf{G}$.

8. The work done by a conservative force field in moving a
particle around a closed path is zero.

9. If $\mathbf{F}$ and $\mathbf{G}$ are vector fields, then
$$\text{curl}(\mathbf{F} + \mathbf{G}) = \text{curl } \mathbf{F} + \text{curl } \mathbf{G}$$

10. If $\mathbf{F}$ and $\mathbf{G}$ are vector fields, then
$$\text{curl}(\mathbf{F} \cdot \mathbf{G}) = \text{curl } \mathbf{F} \cdot \text{curl } \mathbf{G}$$

11. If S is a sphere and $\mathbf{F}$ is a constant vector field, then
$$\iint_S \mathbf{F} \cdot d\mathbf{S} = 0$$

12. There is a vector field $\mathbf{F}$ such that
$$\text{curl } \mathbf{F} = x\,\mathbf{i} + y\,\mathbf{j} + z\,\mathbf{k}$$

EXERCISES

1. A vector field $\mathbf{F}$, a curve C, and a point P are shown.

(a) Is $\int_C \mathbf{F} \cdot d\mathbf{r}$ positive, negative, or zero? Explain.

(b) Is div $\mathbf{F}(P)$ positive, negative, or zero? Explain.

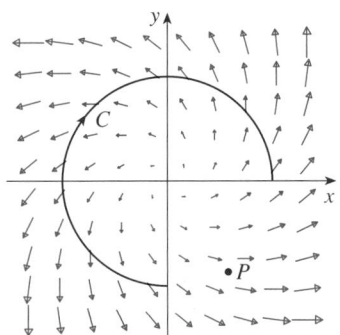

2–9 ■ Evaluate the line integral.

2. $\int_C x\, ds$,
C is the arc of the parabola $y = x^2$ from $(0, 0)$ to $(1, 1)$

3. $\int_C yz \cos x\, ds$,
C: $x = t$, $y = 3 \cos t$, $z = 3 \sin t$, $0 \leqslant t \leqslant \pi$

4. $\int_C y\, dx + (x + y^2)\, dy$, C is the ellipse $4x^2 + 9y^2 = 36$
with counterclockwise orientation

5. $\int_C y^3\, dx + x^2\, dy$, C is the arc of the parabola $x = 1 - y^2$
from $(0, -1)$ to $(0, 1)$

6. $\int_C \sqrt{xy}\, dx + e^y\, dy + xz\, dz$,
C is given by $\mathbf{r}(t) = t^4\,\mathbf{i} + t^2\,\mathbf{j} + t^3\,\mathbf{k}$, $0 \leqslant t \leqslant 1$

7. $\int_C xy\, dx + y^2\, dy + yz\, dz$,
C is the line segment from $(1, 0, -1)$, to $(3, 4, 2)$

8. $\int_C \mathbf{F} \cdot d\mathbf{r}$, where $\mathbf{F}(x, y) = xy\,\mathbf{i} + x^2\,\mathbf{j}$ and C is given by
$\mathbf{r}(t) = \sin t\,\mathbf{i} + (1 + t)\,\mathbf{j}$, $0 \leqslant t \leqslant \pi$

9. $\int_C \mathbf{F} \cdot d\mathbf{r}$, where $\mathbf{F}(x, y, z) = e^z\,\mathbf{i} + xz\,\mathbf{j} + (x + y)\,\mathbf{k}$ and
C is given by $\mathbf{r}(t) = t^2\,\mathbf{i} + t^3\,\mathbf{j} - t\,\mathbf{k}$, $0 \leqslant t \leqslant 1$

---

10. Find the work done by the force field
$$\mathbf{F}(x, y, z) = z\,\mathbf{i} + x\,\mathbf{j} + y\,\mathbf{k}$$
in moving a particle from the point $(3, 0, 0)$ to the point
$(0, \pi/2, 3)$ along

(a) a straight line

(b) the helix $x = 3 \cos t$, $y = t$, $z = 3 \sin t$

11–12 ▪ Show that $\mathbf{F}$ is a conservative vector field. Then find a function f such that $\mathbf{F} = \nabla f$.

11. $\mathbf{F}(x, y) = (1 + xy)e^{xy}\, \mathbf{i} + (e^y + x^2 e^{xy})\, \mathbf{j}$

12. $\mathbf{F}(x, y, z) = \sin y\, \mathbf{i} + x \cos y\, \mathbf{j} - \sin z\, \mathbf{k}$

13–14 ▪ Show that $\mathbf{F}$ is conservative and use this fact to evaluate $\int_C \mathbf{F} \cdot d\mathbf{r}$ along the given curve.

13. $\mathbf{F}(x, y) = (4x^3 y^2 - 2xy^3)\, \mathbf{i} + (2x^4 y - 3x^2 y^2 + 4y^3)\, \mathbf{j}$,
$C: \mathbf{r}(t) = (t + \sin \pi t)\, \mathbf{i} + (2t + \cos \pi t)\, \mathbf{j}, \quad 0 \leqslant t \leqslant 1$

14. $\mathbf{F}(x, y, z) = e^y\, \mathbf{i} + (xe^y + e^z)\, \mathbf{j} + ye^z\, \mathbf{k}$,
C is the line segment from $(0, 2, 0)$ to $(4, 0, 3)$

---

15. Verify that Green's Theorem is true for the line integral $\int_C xy^2\, dx - x^2 y\, dy$, where C consists of the parabola $y = x^2$ from $(-1, 1)$ to $(1, 1)$ and the line segment from $(1, 1)$ to $(-1, 1)$.

16. Use Green's Theorem to evaluate

$$\int_C \sqrt{1 + x^3}\, dx + 2xy\, dy$$

where C is the triangle with vertices $(0, 0)$, $(1, 0)$, and $(1, 3)$.

17. Use Green's Theorem to evaluate $\int_C x^2 y\, dx - xy^2\, dy$, where C is the circle $x^2 + y^2 = 4$ with counterclockwise orientation.

18. Find curl $\mathbf{F}$ and div $\mathbf{F}$ if

$$\mathbf{F}(x, y, z) = e^{-x} \sin y\, \mathbf{i} + e^{-y} \sin z\, \mathbf{j} + e^{-z} \sin x\, \mathbf{k}$$

19. Show that there is no vector field $\mathbf{G}$ such that

$$\text{curl } \mathbf{G} = 2x\, \mathbf{i} + 3yz\, \mathbf{j} - xz^2\, \mathbf{k}$$

20. Show that, under conditions to be stated on the vector fields $\mathbf{F}$ and $\mathbf{G}$,

$$\text{curl}(\mathbf{F} \times \mathbf{G}) = \mathbf{F}\, \text{div } \mathbf{G} - \mathbf{G}\, \text{div } \mathbf{F} + (\mathbf{G} \cdot \nabla)\mathbf{F} - (\mathbf{F} \cdot \nabla)\mathbf{G}$$

21. If C is any piecewise-smooth simple closed plane curve and f and g are differentiable functions, show that $\int_C f(x)\, dx + g(y)\, dy = 0$.

22. If f and g are twice differentiable functions, show that

$$\nabla^2(fg) = f\nabla^2 g + g\nabla^2 f + 2\nabla f \cdot \nabla g$$

23. If f is a harmonic function, that is, $\nabla^2 f = 0$, show that the line integral $\int f_y\, dx - f_x\, dy$ is independent of path in any simple region D.

24. (a) Sketch the curve C with parametric equations

$$x = \cos t \qquad y = \sin t \qquad z = \sin t \qquad 0 \leqslant t \leqslant 2\pi$$

(b) Find $\int_C 2xe^{2y}\, dx + (2x^2 e^{2y} + 2y \cot z)\, dy - y^2 \csc^2 z\, dz$.

25. Find the area of the part of the surface $z = x^2 + 2y$ that lies above the triangle with vertices $(0, 0)$, $(1, 0)$, and $(1, 2)$.

26. (a) Find an equation of the tangent plane at the point $(4, -2, 1)$ to the parametric surface S given by

$$\mathbf{r}(u, v) = v^2\, \mathbf{i} - uv\, \mathbf{j} + u^2\, \mathbf{k} \qquad 0 \leqslant u \leqslant 3, \, -3 \leqslant v \leqslant 3$$

▦ (b) Use a computer to graph the surface S and the tangent plane found in part (a).

(c) Set up, but do not evaluate, an integral for the surface area of S.

[CAS] (d) If

$$\mathbf{F}(x, y, z) = \frac{z^2}{1 + x^2}\, \mathbf{i} + \frac{x^2}{1 + y^2}\, \mathbf{j} + \frac{y^2}{1 + z^2}\, \mathbf{k}$$

find $\iint_S \mathbf{F} \cdot d\mathbf{S}$ correct to four decimal places.

27–30 ▪ Evaluate the surface integral.

27. $\iint_S z\, dS$, where S is the part of the paraboloid $z = x^2 + y^2$ that lies under the plane $z = 4$

28. $\iint_S (x^2 z + y^2 z)\, dS$, where S is the part of the plane $z = 4 + x + y$ that lies inside the cylinder $x^2 + y^2 = 4$

29. $\iint_S \mathbf{F} \cdot d\mathbf{S}$, where $\mathbf{F}(x, y, z) = xz\, \mathbf{i} - 2y\, \mathbf{j} + 3x\, \mathbf{k}$ and S is the sphere $x^2 + y^2 + z^2 = 4$ with outward orientation

30. $\iint_S \mathbf{F} \cdot d\mathbf{S}$, where $\mathbf{F}(x, y, z) = x^2\, \mathbf{i} + xy\, \mathbf{j} + z\, \mathbf{k}$ and S is the part of the paraboloid $z = x^2 + y^2$ below the plane $z = 1$ with upward orientation

---

31. Verify that Stokes' Theorem is true for the vector field $\mathbf{F}(x, y, z) = x^2\, \mathbf{i} + y^2\, \mathbf{j} + z^2\, \mathbf{k}$, where S is the part of the paraboloid $z = 1 - x^2 - y^2$ that lies above the xy-plane, and S has upward orientation.

32. Use Stokes' Theorem to evaluate $\iint_S \text{curl } \mathbf{F} \cdot d\mathbf{S}$, where $\mathbf{F}(x, y, z) = x^2 yz\, \mathbf{i} + yz^2\, \mathbf{j} + z^3 e^{xy}\, \mathbf{k}$, S is the part of the sphere $x^2 + y^2 + z^2 = 5$ that lies above the plane $z = 1$, and S is oriented upward.

33. Use Stokes' Theorem to evaluate $\int_C \mathbf{F} \cdot d\mathbf{r}$, where $\mathbf{F}(x, y, z) = xy\, \mathbf{i} + yz\, \mathbf{j} + zx\, \mathbf{k}$, and C is the triangle with vertices $(1, 0, 0)$, $(0, 1, 0)$, and $(0, 0, 1)$, oriented counterclockwise as viewed from above.

34. Use the Divergence Theorem to calculate the surface integral $\iint_S \mathbf{F} \cdot d\mathbf{S}$, where $\mathbf{F}(x, y, z) = x^3\, \mathbf{i} + y^3\, \mathbf{j} + z^3\, \mathbf{k}$ and S is the surface of the solid bounded by the cylinder $x^2 + y^2 = 1$ and the planes $z = 0$ and $z = 2$.

35. Verify that the Divergence Theorem is true for the vector field $\mathbf{F}(x, y, z) = x\,\mathbf{i} + y\,\mathbf{j} + z\,\mathbf{k}$, where E is the unit ball $x^2 + y^2 + z^2 \leq 1$.

36. Compute the outward flux of

$$\mathbf{F}(x, y, z) = \frac{x\,\mathbf{i} + y\,\mathbf{j} + z\,\mathbf{k}}{(x^2 + y^2 + z^2)^{3/2}}$$

through the ellipsoid $4x^2 + 9y^2 + 6z^2 = 36$.

37. Find $\iint_S \mathbf{F} \cdot \mathbf{n}\,dS$, where $\mathbf{F}(x, y, z) = x\,\mathbf{i} + y\,\mathbf{j} + z\,\mathbf{k}$ and S is the outwardly oriented surface shown in the figure (the boundary surface of a cube with a unit corner cube removed).

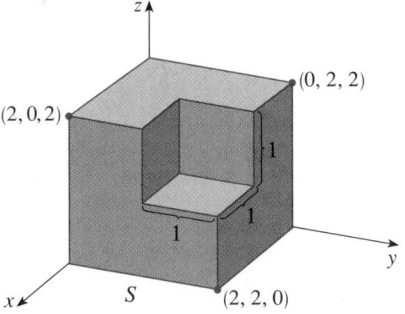

38. Let

$$\mathbf{F}(x, y) = \frac{(2x^3 + 2xy^2 - 2y)\,\mathbf{i} + (2y^3 + 2x^2y + 2x)\,\mathbf{j}}{x^2 + y^2}$$

Evaluate $\oint_C \mathbf{F} \cdot d\mathbf{r}$, where C is shown in the figure.

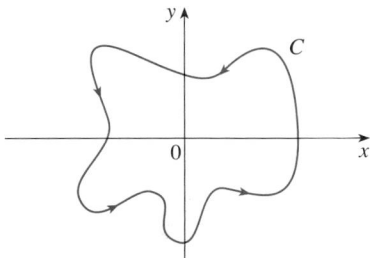

39. If $\mathbf{a}$ is a constant vector, $\mathbf{r} = x\,\mathbf{i} + y\,\mathbf{j} + z\,\mathbf{k}$, and S is an oriented, smooth surface with a simple, closed, smooth, positively oriented boundary curve C, show that

$$\iint_S 2\mathbf{a} \cdot d\mathbf{S} = \int_C (\mathbf{a} \times \mathbf{r}) \cdot d\mathbf{r}$$

40. If the components of $\mathbf{F}$ have continuous second partial derivatives and S is the boundary surface of a simple solid region, show that $\iint_S \operatorname{curl} \mathbf{F} \cdot d\mathbf{S} = 0$.

A | TRIGONOMETRY

ANGLES

Angles can be measured in degrees or in radians (abbreviated as rad). The angle given by a complete revolution contains 360°, which is the same as 2π rad. Therefore

$$\boxed{1} \qquad \boxed{\pi \text{ rad} = 180°}$$

and

$$\boxed{2} \qquad 1 \text{ rad} = \left(\frac{180}{\pi}\right)° \approx 57.3° \qquad 1° = \frac{\pi}{180} \text{ rad} \approx 0.017 \text{ rad}$$

EXAMPLE 1
(a) Find the radian measure of 60°.   (b) Express $5\pi/4$ rad in degrees.

SOLUTION
(a) From Equation 1 or 2 we see that to convert from degrees to radians we multiply by $\pi/180$. Therefore

$$60° = 60\left(\frac{\pi}{180}\right) = \frac{\pi}{3} \text{ rad}$$

(b) To convert from radians to degrees we multiply by $180/\pi$. Thus

$$\frac{5\pi}{4} \text{ rad} = \frac{5\pi}{4}\left(\frac{180}{\pi}\right) = 225° \qquad\blacksquare$$

In calculus we use radians to measure angles except when otherwise indicated. The following table gives the correspondence between degree and radian measures of some common angles.

Degrees	0°	30°	45°	60°	90°	120°	135°	150°	180°	270°	360°
Radians	0	$\frac{\pi}{6}$	$\frac{\pi}{4}$	$\frac{\pi}{3}$	$\frac{\pi}{2}$	$\frac{2\pi}{3}$	$\frac{3\pi}{4}$	$\frac{5\pi}{6}$	π	$\frac{3\pi}{2}$	2π

Figure 1 shows a sector of a circle with central angle θ and radius r subtending an arc with length a. Since the length of the arc is proportional to the size of the angle, and since the entire circle has circumference $2\pi r$ and central angle 2π, we have

$$\frac{\theta}{2\pi} = \frac{a}{2\pi r}$$

Solving this equation for θ and for a, we obtain

$$\boxed{3} \qquad \boxed{\theta = \frac{a}{r}} \qquad\qquad \boxed{a = r\theta}$$

Remember that Equations 3 are valid only when θ is measured in radians.

In particular, putting $a = r$ in Equation 3, we see that an angle of 1 rad is the angle subtended at the center of a circle by an arc equal in length to the radius of the circle (see Figure 2).

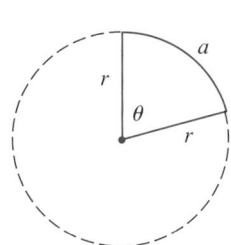

FIGURE 1

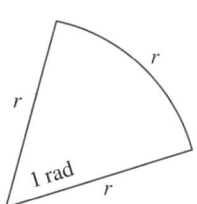

FIGURE 2

EXAMPLE 2
(a) If the radius of a circle is 5 cm, what angle is subtended by an arc of 6 cm?
(b) If a circle has radius 3 cm, what is the length of an arc subtended by a central angle of $3\pi/8$ rad?

SOLUTION
(a) Using Equation 3 with $a = 6$ and $r = 5$, we see that the angle is

$$\theta = \tfrac{6}{5} = 1.2 \text{ rad}$$

(b) With $r = 3$ cm and $\theta = 3\pi/8$ rad, the arc length is

$$a = r\theta = 3\left(\frac{3\pi}{8}\right) = \frac{9\pi}{8} \text{ cm}$$

The **standard position** of an angle occurs when we place its vertex at the origin of a coordinate system and its initial side on the positive x-axis as in Figure 3. A **positive** angle is obtained by rotating the initial side counterclockwise until it coincides with the terminal side. Likewise, **negative** angles are obtained by clockwise rotation as in Figure 4.

Figure 5 shows several examples of angles in standard position. Notice that different angles can have the same terminal side. For instance, the angles $3\pi/4$, $-5\pi/4$, and $11\pi/4$ have the same initial and terminal sides because

$$\frac{3\pi}{4} - 2\pi = -\frac{5\pi}{4} \qquad \frac{3\pi}{4} + 2\pi = \frac{11\pi}{4}$$

and 2π rad represents a complete revolution.

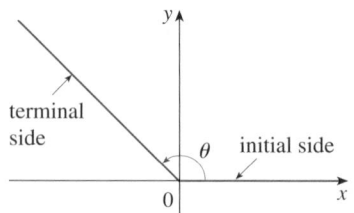

FIGURE 3 $\theta \geqslant 0$

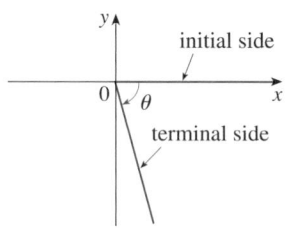

FIGURE 4 $\theta < 0$

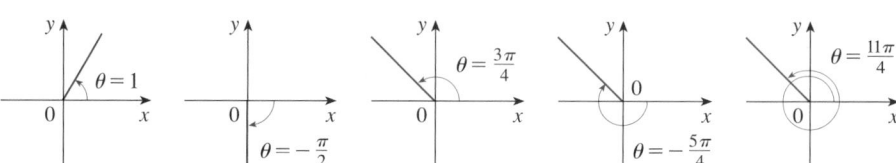

FIGURE 5
Angles in standard position

THE TRIGONOMETRIC FUNCTIONS

For an acute angle θ the six trigonometric functions are defined as ratios of lengths of sides of a right triangle as follows (see Figure 6).

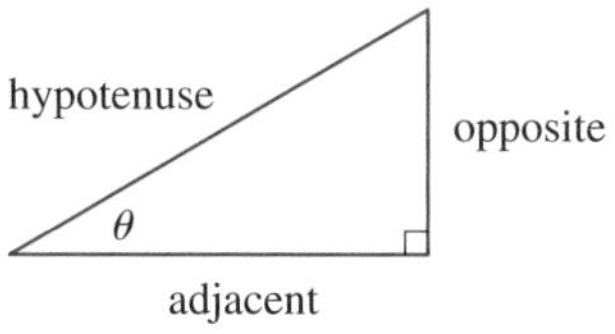

FIGURE 6

$$\boxed{4} \qquad \begin{array}{ll} \sin\theta = \dfrac{\text{opp}}{\text{hyp}} & \csc\theta = \dfrac{\text{hyp}}{\text{opp}} \\[2ex] \cos\theta = \dfrac{\text{adj}}{\text{hyp}} & \sec\theta = \dfrac{\text{hyp}}{\text{adj}} \\[2ex] \tan\theta = \dfrac{\text{opp}}{\text{adj}} & \cot\theta = \dfrac{\text{adj}}{\text{opp}} \end{array}$$

This definition doesn't apply to obtuse or negative angles, so for a general angle θ in standard position we let $P(x, y)$ be any point on the terminal side of θ and we let r

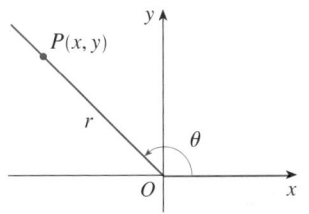

FIGURE 7

be the distance $|OP|$ as in Figure 7. Then we define

$\boxed{5}$

$$\sin \theta = \frac{y}{r} \qquad\qquad \csc \theta = \frac{r}{y}$$

$$\cos \theta = \frac{x}{r} \qquad\qquad \sec \theta = \frac{r}{x}$$

$$\tan \theta = \frac{y}{x} \qquad\qquad \cot \theta = \frac{x}{y}$$

Since division by 0 is not defined, $\tan \theta$ and $\sec \theta$ are undefined when $x = 0$ and $\csc \theta$ and $\cot \theta$ are undefined when $y = 0$. Notice that the definitions in $\boxed{4}$ and $\boxed{5}$ are consistent when θ is an acute angle.

If θ is a number, the convention is that $\sin \theta$ means the sine of the angle whose *radian* measure is θ. For example, the expression $\sin 3$ implies that we are dealing with an angle of 3 rad. When finding a calculator approximation to this number we must remember to set our calculator in radian mode, and then we obtain

$$\sin 3 \approx 0.14112$$

If we want to know the sine of the angle $3°$ we would write $\sin 3°$ and, with our calculator in degree mode, we find that

$$\sin 3° \approx 0.05234$$

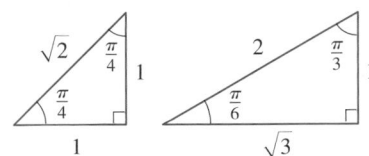

FIGURE 8

The exact trigonometric ratios for certain angles can be read from the triangles in Figure 8. For instance,

$$\sin \frac{\pi}{4} = \frac{1}{\sqrt{2}} \qquad \sin \frac{\pi}{6} = \frac{1}{2} \qquad \sin \frac{\pi}{3} = \frac{\sqrt{3}}{2}$$

$$\cos \frac{\pi}{4} = \frac{1}{\sqrt{2}} \qquad \cos \frac{\pi}{6} = \frac{\sqrt{3}}{2} \qquad \cos \frac{\pi}{3} = \frac{1}{2}$$

$$\tan \frac{\pi}{4} = 1 \qquad \tan \frac{\pi}{6} = \frac{1}{\sqrt{3}} \qquad \tan \frac{\pi}{3} = \sqrt{3}$$

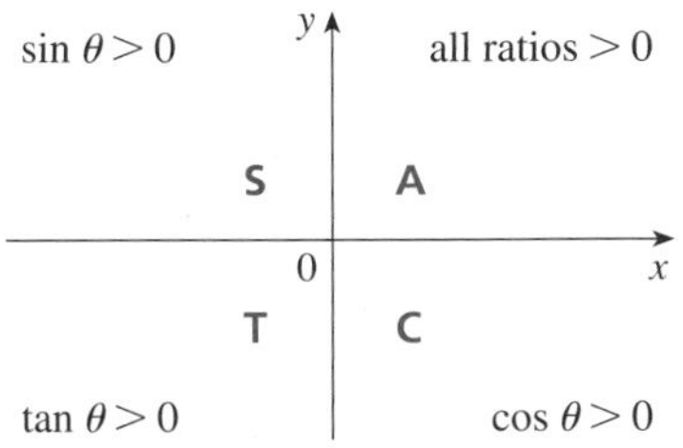

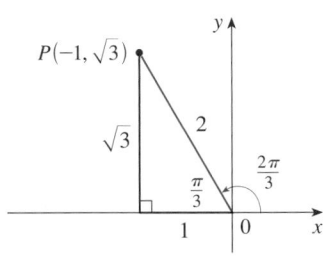

FIGURE 9

The signs of the trigonometric functions for angles in each of the four quadrants can be remembered by means of the rule "**A**ll **S**tudents **T**ake **C**alculus" shown in Figure 9.

EXAMPLE 3 Find the exact trigonometric ratios for $\theta = 2\pi/3$.

SOLUTION From Figure 10 we see that a point on the terminal line for $\theta = 2\pi/3$ is $P(-1, \sqrt{3})$. Therefore, taking

$$x = -1 \qquad y = \sqrt{3} \qquad r = 2$$

in the definitions of the trigonometric ratios, we have

$$\sin \frac{2\pi}{3} = \frac{\sqrt{3}}{2} \qquad \cos \frac{2\pi}{3} = -\frac{1}{2} \qquad \tan \frac{2\pi}{3} = -\sqrt{3}$$

$$\csc \frac{2\pi}{3} = \frac{2}{\sqrt{3}} \qquad \sec \frac{2\pi}{3} = -2 \qquad \cot \frac{2\pi}{3} = -\frac{1}{\sqrt{3}}$$

■

FIGURE 10

The following table gives some values of $\sin \theta$ and $\cos \theta$ found by the method of Example 3.

θ	0	$\dfrac{\pi}{6}$	$\dfrac{\pi}{4}$	$\dfrac{\pi}{3}$	$\dfrac{\pi}{2}$	$\dfrac{2\pi}{3}$	$\dfrac{3\pi}{4}$	$\dfrac{5\pi}{6}$	π	$\dfrac{3\pi}{2}$	2π
$\sin \theta$	0	$\dfrac{1}{2}$	$\dfrac{1}{\sqrt{2}}$	$\dfrac{\sqrt{3}}{2}$	1	$\dfrac{\sqrt{3}}{2}$	$\dfrac{1}{\sqrt{2}}$	$\dfrac{1}{2}$	0	-1	0
$\cos \theta$	1	$\dfrac{\sqrt{3}}{2}$	$\dfrac{1}{\sqrt{2}}$	$\dfrac{1}{2}$	0	$-\dfrac{1}{2}$	$-\dfrac{1}{\sqrt{2}}$	$-\dfrac{\sqrt{3}}{2}$	-1	0	1

EXAMPLE 4 If $\cos \theta = \frac{2}{5}$ and $0 < \theta < \pi/2$, find the other five trigonometric functions of θ.

SOLUTION Since $\cos \theta = \frac{2}{5}$, we can label the hypotenuse as having length 5 and the adjacent side as having length 2 in Figure 11. If the opposite side has length x, then the Pythagorean Theorem gives $x^2 + 4 = 25$ and so $x^2 = 21$, or $x = \sqrt{21}$. We can now use the diagram to write the other five trigonometric functions:

$$\sin \theta = \frac{\sqrt{21}}{5} \qquad \tan \theta = \frac{\sqrt{21}}{2}$$

$$\csc \theta = \frac{5}{\sqrt{21}} \qquad \sec \theta = \frac{5}{2} \qquad \cot \theta = \frac{2}{\sqrt{21}} \qquad ∎$$

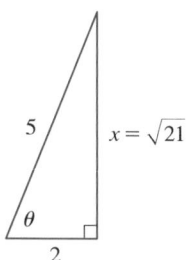

FIGURE 11

EXAMPLE 5 Use a calculator to approximate the value of x in Figure 12.

SOLUTION From the diagram we see that

$$\tan 40° = \frac{16}{x}$$

Therefore
$$x = \frac{16}{\tan 40°} \approx 19.07 \qquad ∎$$

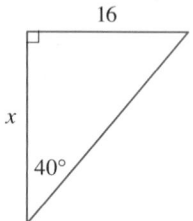

FIGURE 12

TRIGONOMETRIC IDENTITIES

A trigonometric identity is a relationship among the trigonometric functions. The most elementary are the following, which are immediate consequences of the definitions of the trigonometric functions.

$$\boxed{6} \quad \begin{array}{ccc} \csc \theta = \dfrac{1}{\sin \theta} & \sec \theta = \dfrac{1}{\cos \theta} & \cot \theta = \dfrac{1}{\tan \theta} \\[4mm] \tan \theta = \dfrac{\sin \theta}{\cos \theta} & & \cot \theta = \dfrac{\cos \theta}{\sin \theta} \end{array}$$

For the next identity we refer back to Figure 7. The distance formula (or, equivalently, the Pythagorean Theorem) tells us that $x^2 + y^2 = r^2$. Therefore

$$\sin^2\theta + \cos^2\theta = \frac{y^2}{r^2} + \frac{x^2}{r^2} = \frac{x^2 + y^2}{r^2} = \frac{r^2}{r^2} = 1$$

We have therefore proved one of the most useful of all trigonometric identities:

$$\boxed{\sin^2\theta + \cos^2\theta = 1}$$ **7**

If we now divide both sides of Equation 7 by $\cos^2\theta$ and use Equations 6, we get

$$\boxed{\tan^2\theta + 1 = \sec^2\theta}$$ **8**

Similarly, if we divide both sides of Equation 7 by $\sin^2\theta$, we get

$$\boxed{1 + \cot^2\theta = \csc^2\theta}$$ **9**

The identities

10a
10b
$$\boxed{\begin{aligned}\sin(-\theta) &= -\sin\theta \\ \cos(-\theta) &= \cos\theta\end{aligned}}$$

■ Odd functions and even functions are discussed in Section 1.1.

show that sin is an odd function and cos is an even function. They are easily proved by drawing a diagram showing θ and $-\theta$ in standard position (see Exercise 39).

Since the angles θ and $\theta + 2\pi$ have the same terminal side, we have

11
$$\boxed{\sin(\theta + 2\pi) = \sin\theta \qquad \cos(\theta + 2\pi) = \cos\theta}$$

These identities show that the sine and cosine functions are periodic with period 2π.

The remaining trigonometric identities are all consequences of two basic identities called the **addition formulas**:

12a
12b
$$\boxed{\begin{aligned}\sin(x + y) &= \sin x \cos y + \cos x \sin y \\ \cos(x + y) &= \cos x \cos y - \sin x \sin y\end{aligned}}$$

The proofs of these addition formulas are outlined in Exercises 85, 86, and 87.

By substituting $-y$ for y in Equations 12a and 12b and using Equations 10a and 10b, we obtain the following **subtraction formulas**:

13a
13b
$$\boxed{\begin{aligned}\sin(x - y) &= \sin x \cos y - \cos x \sin y \\ \cos(x - y) &= \cos x \cos y + \sin x \sin y\end{aligned}}$$

Then, by dividing the formulas in Equations 12 or Equations 13, we obtain the corresponding formulas for $\tan(x \pm y)$:

14a
14b
$$\boxed{\begin{aligned}\tan(x + y) &= \frac{\tan x + \tan y}{1 - \tan x \tan y} \\[2mm] \tan(x - y) &= \frac{\tan x - \tan y}{1 + \tan x \tan y}\end{aligned}}$$

If we put $y = x$ in the addition formulas $\boxed{12}$, we get the **double-angle formulas**:

$\boxed{15a}$

$$\sin 2x = 2 \sin x \cos x$$

$\boxed{15b}$

$$\cos 2x = \cos^2 x - \sin^2 x$$

Then, by using the identity $\sin^2 x + \cos^2 x = 1$, we obtain the following alternate forms of the double-angle formulas for $\cos 2x$:

$\boxed{16a}$

$$\cos 2x = 2 \cos^2 x - 1$$

$\boxed{16b}$

$$\cos 2x = 1 - 2 \sin^2 x$$

If we now solve these equations for $\cos^2 x$ and $\sin^2 x$, we get the following **half-angle formulas**, which are useful in integral calculus:

$\boxed{17a}$

$$\cos^2 x = \frac{1 + \cos 2x}{2}$$

$\boxed{17b}$

$$\sin^2 x = \frac{1 - \cos 2x}{2}$$

Finally, we state the **product formulas**, which can be deduced from Equations 12 and 13:

$\boxed{18a}$

$$\sin x \cos y = \tfrac{1}{2}[\sin(x + y) + \sin(x - y)]$$

$\boxed{18b}$

$$\cos x \cos y = \tfrac{1}{2}[\cos(x + y) + \cos(x - y)]$$

$\boxed{18c}$

$$\sin x \sin y = \tfrac{1}{2}[\cos(x - y) - \cos(x + y)]$$

There are many other trigonometric identities, but those we have stated are the ones used most often in calculus. If you forget any of them, remember that they can all be deduced from Equations 12a and 12b.

EXAMPLE 6 Find all values of x in the interval $[0, 2\pi]$ such that $\sin x = \sin 2x$.

SOLUTION Using the double-angle formula $\boxed{15a}$, we rewrite the given equation as

$$\sin x = 2 \sin x \cos x \qquad \text{or} \qquad \sin x \, (1 - 2 \cos x) = 0$$

Therefore there are two possibilities:

$$\sin x = 0 \qquad \text{or} \qquad 1 - 2 \cos x = 0$$

$$x = 0, \, \pi, \, 2\pi \qquad\qquad \cos x = \tfrac{1}{2}$$

$$x = \frac{\pi}{3}, \frac{5\pi}{3}$$

The given equation has five solutions: 0, $\pi/3$, π, $5\pi/3$, and 2π. ∎

GRAPHS OF TRIGONOMETRIC FUNCTIONS

The graph of the function $f(x) = \sin x$, shown in Figure 13(a), is obtained by plotting points for $0 \le x \le 2\pi$ and then using the periodic nature of the function (from Equation 11) to complete the graph. Notice that the zeros of the sine function occur at the integer multiples of π, that is,

$$\sin x = 0 \qquad \text{whenever } x = n\pi, \quad n \text{ an integer}$$

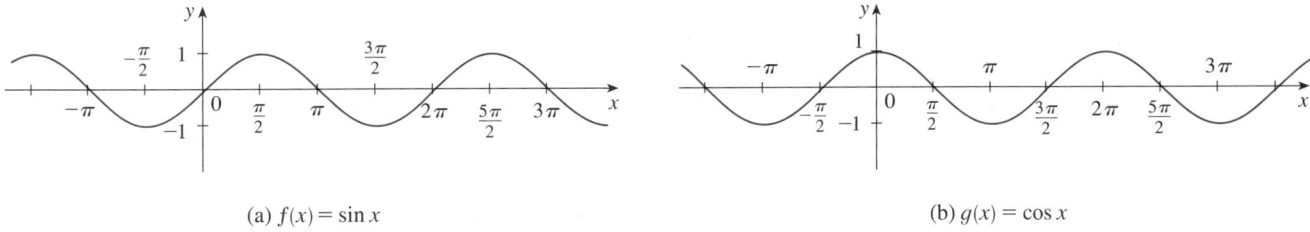

(a) $f(x) = \sin x$

(b) $g(x) = \cos x$

FIGURE 13

Because of the identity

$$\cos x = \sin\left(x + \frac{\pi}{2}\right)$$

(which can be verified using Equation 12a), the graph of cosine is obtained by shifting the graph of sine by an amount $\pi/2$ to the left [see Figure 13(b)]. Note that for both the sine and cosine functions the domain is $(-\infty, \infty)$ and the range is the closed interval $[-1, 1]$. Thus, for all values of x, we have

$$-1 \le \sin x \le 1 \qquad -1 \le \cos x \le 1$$

The graphs of the remaining four trigonometric functions are shown in Figure 14 and their domains are indicated there. Notice that tangent and cotangent have range $(-\infty, \infty)$, whereas cosecant and secant have range $(-\infty, -1] \cup [1, \infty)$. All four functions are periodic: tangent and cotangent have period π, whereas cosecant and secant have period 2π.

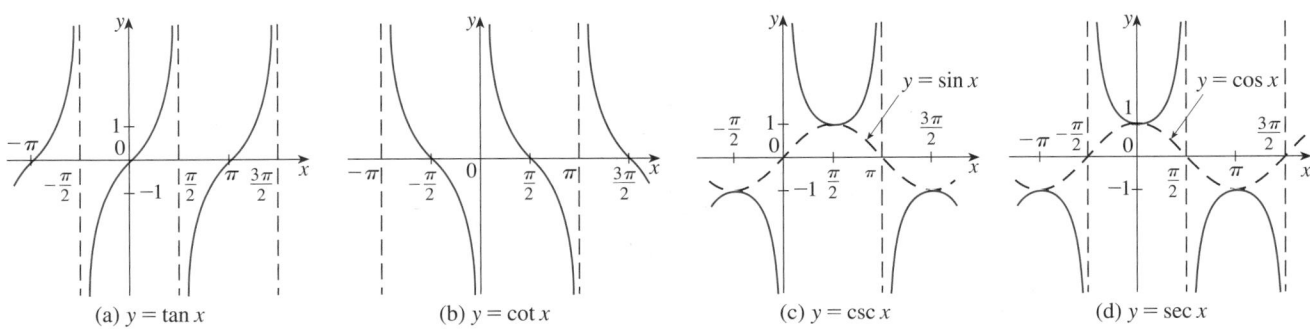

(a) $y = \tan x$

(b) $y = \cot x$

(c) $y = \csc x$

(d) $y = \sec x$

FIGURE 14

A | EXERCISES

1–6 ▪ Convert from degrees to radians.

1. $210°$ **2.** $300°$ **3.** $9°$

4. $-315°$ **5.** $900°$ **6.** $36°$

7–12 ▪ Convert from radians to degrees.

7. 4π **8.** $-\dfrac{7\pi}{2}$ **9.** $\dfrac{5\pi}{12}$

10. $\dfrac{8\pi}{3}$ **11.** $-\dfrac{3\pi}{8}$ **12.** 5

13. Find the length of a circular arc subtended by an angle of $\pi/12$ rad if the radius of the circle is 36 cm.

14. If a circle has radius 10 cm, find the length of the arc subtended by a central angle of $72°$.

15. A circle has radius 1.5 m. What angle is subtended at the center of the circle by an arc 1 m long?

16. Find the radius of a circular sector with angle $3\pi/4$ and arc length 6 cm.

17–22 ▪ Draw, in standard position, the angle whose measure is given.

17. $315°$ **18.** $-150°$ **19.** $-\dfrac{3\pi}{4}$ rad

20. $\dfrac{7\pi}{3}$ rad   **21.** 2 rad   **22.** -3 rad

23–28 ▪ Find the exact trigonometric ratios for the angle whose radian measure is given.

23. $\dfrac{3\pi}{4}$ **24.** $\dfrac{4\pi}{3}$ **25.** $\dfrac{9\pi}{2}$

26. -5π **27.** $\dfrac{5\pi}{6}$ **28.** $\dfrac{11\pi}{4}$

29–34 ▪ Find the remaining trigonometric ratios.

29. $\sin\theta = \dfrac{3}{5}$, $0 < \theta < \dfrac{\pi}{2}$

30. $\tan\alpha = 2$, $0 < \alpha < \dfrac{\pi}{2}$

31. $\sec\phi = -1.5$, $\dfrac{\pi}{2} < \phi < \pi$

32. $\cos x = -\dfrac{1}{3}$, $\pi < x < \dfrac{3\pi}{2}$

33. $\cot\beta = 3$, $\pi < \beta < 2\pi$

34. $\csc\theta = -\dfrac{4}{3}$, $\dfrac{3\pi}{2} < \theta < 2\pi$

35–38 ▪ Find, correct to five decimal places, the length of the side labeled x.

35.
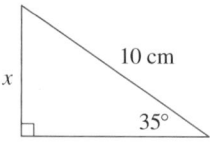

36.
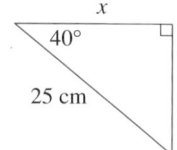

37.
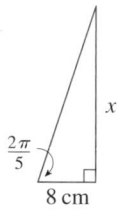

38.
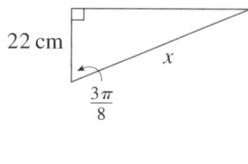

39–41 ▪ Prove each equation.

39. (a) Equation 10a   (b) Equation 10b

40. (a) Equation 14a   (b) Equation 14b

41. (a) Equation 18a   (b) Equation 18b
(c) Equation 18c

42–58 ▪ Prove the identity.

42. $\cos\left(\dfrac{\pi}{2} - x\right) = \sin x$

43. $\sin\left(\dfrac{\pi}{2} + x\right) = \cos x$ **44.** $\sin(\pi - x) = \sin x$

45. $\sin\theta \cot\theta = \cos\theta$

46. $(\sin x + \cos x)^2 = 1 + \sin 2x$

47. $\sec y - \cos y = \tan y \sin y$

48. $\tan^2\alpha - \sin^2\alpha = \tan^2\alpha \sin^2\alpha$

49. $\cot^2\theta + \sec^2\theta = \tan^2\theta + \csc^2\theta$

50. $2\csc 2t = \sec t \csc t$

51. $\tan 2\theta = \dfrac{2\tan\theta}{1 - \tan^2\theta}$

52. $\dfrac{1}{1 - \sin\theta} + \dfrac{1}{1 + \sin\theta} = 2\sec^2\theta$

▦ Graphing calculator or computer required   [CAS] Computer algebra system required   **1.** Homework Hints at stewartcalculus.com

53. $\sin x \sin 2x + \cos x \cos 2x = \cos x$

54. $\sin^2 x - \sin^2 y = \sin(x + y) \sin(x - y)$

55. $\dfrac{\sin \phi}{1 - \cos \phi} = \csc \phi + \cot \phi$

56. $\tan x + \tan y = \dfrac{\sin(x + y)}{\cos x \cos y}$

57. $\sin 3\theta + \sin \theta = 2 \sin 2\theta \cos \theta$

58. $\cos 3\theta = 4 \cos^3\theta - 3 \cos \theta$

59–64 ▪ If $\sin x = \frac{1}{3}$ and $\sec y = \frac{5}{4}$, where x and y lie between 0 and $\pi/2$, evaluate the expression.

59. $\sin(x + y)$ **60.** $\cos(x + y)$

61. $\cos(x - y)$ **62.** $\sin(x - y)$

63. $\sin 2y$ **64.** $\cos 2y$

65–72 ▪ Find all values of x in the interval $[0, 2\pi]$ that satisfy the equation.

65. $2 \cos x - 1 = 0$ **66.** $3 \cot^2 x = 1$

67. $2 \sin^2 x = 1$ **68.** $|\tan x| = 1$

69. $\sin 2x = \cos x$ **70.** $2 \cos x + \sin 2x = 0$

71. $\sin x = \tan x$ **72.** $2 + \cos 2x = 3 \cos x$

73–76 ▪ Find all values of x in the interval $[0, 2\pi]$ that satisfy the inequality.

73. $\sin x \leq \frac{1}{2}$ **74.** $2 \cos x + 1 > 0$

75. $-1 < \tan x < 1$ **76.** $\sin x > \cos x$

77–82 ▪ Graph the function by starting with the graphs in Figures 13 and 14 and applying the transformations of Section 1.3 where appropriate.

77. $y = \cos\left(x - \dfrac{\pi}{3}\right)$ **78.** $y = \tan 2x$

79. $y = \dfrac{1}{3} \tan\left(x - \dfrac{\pi}{2}\right)$ **80.** $y = 1 + \sec x$

81. $y = |\sin x|$ **82.** $y = 2 + \sin\left(x + \dfrac{\pi}{4}\right)$

83. Prove the **Law of Cosines**: If a triangle has sides with lengths a, b, and c, and θ is the angle between the sides with lengths a and b, then

$$c^2 = a^2 + b^2 - 2ab \cos \theta$$

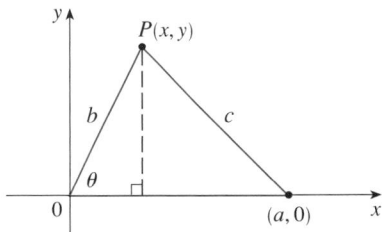

[*Hint:* Introduce a coordinate system so that θ is in standard position as in the figure. Express x and y in terms of θ and then use the distance formula to compute c.]

84. In order to find the distance $|AB|$ across a small inlet, a point C is located as in the figure and the following measurements were recorded:

$$\angle C = 103° \qquad |AC| = 820 \text{ m} \qquad |BC| = 910 \text{ m}$$

Use the Law of Cosines from Exercise 83 to find the required distance.

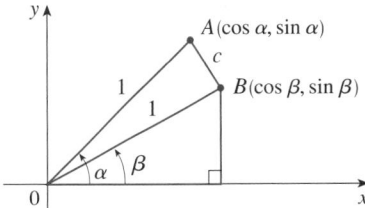

85. Use the figure to prove the subtraction formula

$$\cos(\alpha - \beta) = \cos \alpha \cos \beta + \sin \alpha \sin \beta$$

[*Hint:* Compute c^2 in two ways (using the Law of Cosines from Exercise 83 and also using the distance formula) and compare the two expressions.]

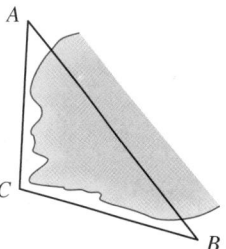

86. Use the formula in Exercise 85 to prove the addition formula for cosine 12b.

87. Use the addition formula for cosine and the identities

$$\cos\left(\dfrac{\pi}{2} - \theta\right) = \sin \theta \qquad \sin\left(\dfrac{\pi}{2} - \theta\right) = \cos \theta$$

to prove the subtraction formula for the sine function.

88. Show that the area of a triangle with sides of lengths a and b and with included angle θ is

$$A = \tfrac{1}{2}ab \sin \theta$$

89. Find the area of triangle ABC, correct to five decimal places, if

$$|AB| = 10 \text{ cm} \qquad |BC| = 3 \text{ cm} \qquad \angle ABC = 107°$$

B | SIGMA NOTATION

A convenient way of writing sums uses the Greek letter Σ (capital sigma, corresponding to our letter S) and is called **sigma notation**.

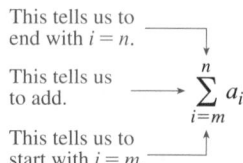
This tells us to end with $i = n$.
This tells us to add.
This tells us to start with $i = m$.

> **1 DEFINITION** If $a_m, a_{m+1}, \ldots, a_n$ are real numbers and m and n are integers such that $m \leq n$, then
>
> $$\sum_{i=m}^{n} a_i = a_m + a_{m+1} + a_{m+2} + \cdots + a_{n-1} + a_n$$

With function notation, Definition 1 can be written as

$$\sum_{i=m}^{n} f(i) = f(m) + f(m + 1) + f(m + 2) + \cdots + f(n - 1) + f(n)$$

Thus the symbol $\sum_{i=m}^{n}$ indicates a summation in which the letter i (called the **index of summation**) takes on consecutive integer values beginning with m and ending with n, that is, $m, m + 1, \ldots, n$. Other letters can also be used as the index of summation.

EXAMPLE 1

(a) $\displaystyle\sum_{i=1}^{4} i^2 = 1^2 + 2^2 + 3^2 + 4^2 = 30$

(b) $\displaystyle\sum_{i=3}^{n} i = 3 + 4 + 5 + \cdots + (n - 1) + n$

(c) $\displaystyle\sum_{j=0}^{5} 2^j = 2^0 + 2^1 + 2^2 + 2^3 + 2^4 + 2^5 = 63$

(d) $\displaystyle\sum_{k=1}^{n} \frac{1}{k} = 1 + \frac{1}{2} + \frac{1}{3} + \cdots + \frac{1}{n}$

(e) $\displaystyle\sum_{i=1}^{3} \frac{i - 1}{i^2 + 3} = \frac{1 - 1}{1^2 + 3} + \frac{2 - 1}{2^2 + 3} + \frac{3 - 1}{3^2 + 3} = 0 + \frac{1}{7} + \frac{1}{6} = \frac{13}{42}$

(f) $\displaystyle\sum_{i=1}^{4} 2 = 2 + 2 + 2 + 2 = 8$ ∎

EXAMPLE 2 Write the sum $2^3 + 3^3 + \cdots + n^3$ in sigma notation.

SOLUTION There is no unique way of writing a sum in sigma notation. We could write

$$2^3 + 3^3 + \cdots + n^3 = \sum_{i=2}^{n} i^3$$

or
$$2^3 + 3^3 + \cdots + n^3 = \sum_{j=1}^{n-1} (j + 1)^3$$

or
$$2^3 + 3^3 + \cdots + n^3 = \sum_{k=0}^{n-2} (k + 2)^3$$ ∎

The following theorem gives three simple rules for working with sigma notation.

> **2** **THEOREM** If c is any constant (that is, it does not depend on i), then
>
> (a) $\displaystyle\sum_{i=m}^{n} ca_i = c \sum_{i=m}^{n} a_i$ (b) $\displaystyle\sum_{i=m}^{n} (a_i + b_i) = \sum_{i=m}^{n} a_i + \sum_{i=m}^{n} b_i$
>
> (c) $\displaystyle\sum_{i=m}^{n} (a_i - b_i) = \sum_{i=m}^{n} a_i - \sum_{i=m}^{n} b_i$

PROOF To see why these rules are true, all we have to do is write both sides in expanded form. Rule (a) is just the distributive property of real numbers:

$$ca_m + ca_{m+1} + \cdots + ca_n = c(a_m + a_{m+1} + \cdots + a_n)$$

Rule (b) follows from the associative and commutative properties:

$$(a_m + b_m) + (a_{m+1} + b_{m+1}) + \cdots + (a_n + b_n)$$
$$= (a_m + a_{m+1} + \cdots + a_n) + (b_m + b_{m+1} + \cdots + b_n)$$

Rule (c) is proved similarly. ☐

EXAMPLE 3 Find $\displaystyle\sum_{i=1}^{n} 1$.

SOLUTION
$$\sum_{i=1}^{n} 1 = \underbrace{1 + 1 + \cdots + 1}_{n \text{ terms}} = n$$ ∎

EXAMPLE 4 Prove the formula for the sum of the first n positive integers:

$$\sum_{i=1}^{n} i = 1 + 2 + 3 + \cdots + n = \frac{n(n + 1)}{2}$$

SOLUTION This formula can be proved by mathematical induction (see page A12) or by the following method used by the German mathematician Karl Friedrich Gauss (1777–1855) when he was ten years old.

Write the sum S twice, once in the usual order and once in reverse order:

$$S = 1 + \quad 2 \quad + \quad 3 \quad + \cdots + (n - 1) + n$$
$$S = n + (n - 1) + (n - 2) + \cdots + \quad 2 \quad + 1$$

Adding all columns vertically, we get

$$2S = (n + 1) + (n + 1) + (n + 1) + \cdots + (n + 1) + (n + 1)$$

On the right side there are n terms, each of which is $n + 1$, so

$$2S = n(n + 1) \qquad \text{or} \qquad S = \frac{n(n + 1)}{2}$$

∎

EXAMPLE 5 Prove the formula for the sum of the squares of the first n positive integers:

$$\sum_{i=1}^{n} i^2 = 1^2 + 2^2 + 3^2 + \cdots + n^2 = \frac{n(n + 1)(2n + 1)}{6}$$

SOLUTION 1 Let S be the desired sum. We start with the *telescoping sum* (or collapsing sum):

Most terms cancel in pairs.

$$\sum_{i=1}^{n} [(1 + i)^3 - i^3] = (2^3 - 1^3) + (3^3 - 2^3) + (4^3 - 3^3) + \cdots + [(n + 1)^3 - n^3]$$

$$= (n + 1)^3 - 1^3 = n^3 + 3n^2 + 3n$$

On the other hand, using Theorem 2 and Examples 3 and 4, we have

$$\sum_{i=1}^{n} [(1 + i)^3 - i^3] = \sum_{i=1}^{n} [3i^2 + 3i + 1] = 3 \sum_{i=1}^{n} i^2 + 3 \sum_{i=1}^{n} i + \sum_{i=1}^{n} 1$$

$$= 3S + 3\frac{n(n + 1)}{2} + n = 3S + \tfrac{3}{2}n^2 + \tfrac{5}{2}n$$

Thus we have

$$n^3 + 3n^2 + 3n = 3S + \tfrac{3}{2}n^2 + \tfrac{5}{2}n$$

Solving this equation for S, we obtain

$$3S = n^3 + \tfrac{3}{2}n^2 + \tfrac{1}{2}n$$

or

$$S = \frac{2n^3 + 3n^2 + n}{6} = \frac{n(n + 1)(2n + 1)}{6}$$

■ **PRINCIPLE OF MATHEMATICAL INDUCTION**
Let S_n be a statement involving the positive integer n. Suppose that
1. S_1 is true.
2. If S_k is true, then S_{k+1} is true.
Then S_n is true for all positive integers n.

SOLUTION 2 Let S_n be the given formula.

1. S_1 is true because $\qquad 1^2 = \dfrac{1(1 + 1)(2 \cdot 1 + 1)}{6}$

2. Assume that S_k is true; that is,

$$1^2 + 2^2 + 3^2 + \cdots + k^2 = \frac{k(k + 1)(2k + 1)}{6}$$

Then

$$1^2 + 2^2 + 3^2 + \cdots + (k + 1)^2 = (1^2 + 2^2 + 3^2 + \cdots + k^2) + (k + 1)^2$$

$$= \frac{k(k + 1)(2k + 1)}{6} + (k + 1)^2 = (k + 1)\frac{k(2k + 1) + 6(k + 1)}{6}$$

$$= (k + 1)\frac{2k^2 + 7k + 6}{6} = \frac{(k + 1)(k + 2)(2k + 3)}{6}$$

$$= \frac{(k + 1)[(k + 1) + 1][2(k + 1) + 1]}{6}$$

So S_{k+1} is true.

By the Principle of Mathematical Induction, S_n is true for all n. ∎

We list the results of Examples 3, 4, and 5 together with a similar result for cubes (see Exercises 37–40) as Theorem 3. These formulas are needed for finding areas and evaluating integrals in Chapter 5.

> **3** **THEOREM** Let c be a constant and n a positive integer. Then
>
> (a) $\displaystyle\sum_{i=1}^{n} 1 = n$ $\qquad\qquad\qquad$ (b) $\displaystyle\sum_{i=1}^{n} c = nc$
>
> (c) $\displaystyle\sum_{i=1}^{n} i = \frac{n(n+1)}{2}$ $\qquad\qquad$ (d) $\displaystyle\sum_{i=1}^{n} i^2 = \frac{n(n+1)(2n+1)}{6}$
>
> (e) $\displaystyle\sum_{i=1}^{n} i^3 = \left[\frac{n(n+1)}{2}\right]^2$

EXAMPLE 6 Evaluate $\displaystyle\sum_{i=1}^{n} i(4i^2 - 3)$.

SOLUTION Using Theorems 2 and 3, we have

$$\sum_{i=1}^{n} i(4i^2 - 3) = \sum_{i=1}^{n} (4i^3 - 3i) = 4\sum_{i=1}^{n} i^3 - 3\sum_{i=1}^{n} i$$

$$= 4\left[\frac{n(n+1)}{2}\right]^2 - 3\,\frac{n(n+1)}{2}$$

$$= \frac{n(n+1)[2n(n+1) - 3]}{2}$$

$$= \frac{n(n+1)(2n^2 + 2n - 3)}{2}$$ ■

■ The type of calculation in Example 7 arises in Chapter 5 when we compute areas.

EXAMPLE 7 Find $\displaystyle\lim_{n\to\infty} \sum_{i=1}^{n} \frac{3}{n}\left[\left(\frac{i}{n}\right)^2 + 1\right]$.

SOLUTION

$$\lim_{n\to\infty} \sum_{i=1}^{n} \frac{3}{n}\left[\left(\frac{i}{n}\right)^2 + 1\right] = \lim_{n\to\infty} \sum_{i=1}^{n} \left[\frac{3}{n^3}i^2 + \frac{3}{n}\right]$$

$$= \lim_{n\to\infty} \left[\frac{3}{n^3}\sum_{i=1}^{n} i^2 + \frac{3}{n}\sum_{i=1}^{n} 1\right]$$

$$= \lim_{n\to\infty} \left[\frac{3}{n^3}\,\frac{n(n+1)(2n+1)}{6} + \frac{3}{n}\cdot n\right]$$

$$= \lim_{n\to\infty} \left[\frac{1}{2}\cdot\frac{n}{n}\cdot\left(\frac{n+1}{n}\right)\left(\frac{2n+1}{n}\right) + 3\right]$$

$$= \lim_{n\to\infty} \left[\frac{1}{2}\cdot 1\left(1 + \frac{1}{n}\right)\left(2 + \frac{1}{n}\right) + 3\right]$$

$$= \tfrac{1}{2}\cdot 1\cdot 1\cdot 2 + 3 = 4$$ ■

B | EXERCISES

1–10 ▪ Write the sum in expanded form.

1. $\displaystyle\sum_{i=1}^{5} \sqrt{i}$

2. $\displaystyle\sum_{i=1}^{6} \frac{1}{i+1}$

3. $\displaystyle\sum_{i=4}^{6} 3^i$

4. $\displaystyle\sum_{i=4}^{6} i^3$

5. $\displaystyle\sum_{k=0}^{4} \frac{2k-1}{2k+1}$

6. $\displaystyle\sum_{k=5}^{8} x^k$

7. $\displaystyle\sum_{i=1}^{n} i^{10}$

8. $\displaystyle\sum_{j=n}^{n+3} j^2$

9. $\displaystyle\sum_{j=0}^{n-1} (-1)^j$

10. $\displaystyle\sum_{i=1}^{n} f(x_i)\,\Delta x_i$

11–20 ▪ Write the sum in sigma notation.

11. $1 + 2 + 3 + 4 + \cdots + 10$

12. $\sqrt{3} + \sqrt{4} + \sqrt{5} + \sqrt{6} + \sqrt{7}$

13. $\frac{1}{2} + \frac{2}{3} + \frac{3}{4} + \frac{4}{5} + \cdots + \frac{19}{20}$

14. $\frac{3}{7} + \frac{4}{8} + \frac{5}{9} + \frac{6}{10} + \cdots + \frac{23}{27}$

15. $2 + 4 + 6 + 8 + \cdots + 2n$

16. $1 + 3 + 5 + 7 + \cdots + (2n-1)$

17. $1 + 2 + 4 + 8 + 16 + 32$

18. $\frac{1}{1} + \frac{1}{4} + \frac{1}{9} + \frac{1}{16} + \frac{1}{25} + \frac{1}{36}$

19. $x + x^2 + x^3 + \cdots + x^n$

20. $1 - x + x^2 - x^3 + \cdots + (-1)^n x^n$

21–35 ▪ Find the value of the sum.

21. $\displaystyle\sum_{i=4}^{8} (3i - 2)$

22. $\displaystyle\sum_{i=3}^{6} i(i+2)$

23. $\displaystyle\sum_{j=1}^{6} 3^{j+1}$

24. $\displaystyle\sum_{k=0}^{8} \cos k\pi$

25. $\displaystyle\sum_{n=1}^{20} (-1)^n$

26. $\displaystyle\sum_{i=1}^{100} 4$

27. $\displaystyle\sum_{i=0}^{4} (2^i + i^2)$

28. $\displaystyle\sum_{i=-2}^{4} 2^{3-i}$

29. $\displaystyle\sum_{i=1}^{n} 2i$

30. $\displaystyle\sum_{i=1}^{n} (2 - 5i)$

31. $\displaystyle\sum_{i=1}^{n} (i^2 + 3i + 4)$

32. $\displaystyle\sum_{i=1}^{n} (3 + 2i)^2$

33. $\displaystyle\sum_{i=1}^{n} (i+1)(i+2)$

34. $\displaystyle\sum_{i=1}^{n} i(i+1)(i+2)$

35. $\displaystyle\sum_{i=1}^{n} (i^3 - i - 2)$

---

36. Find the number n such that $\displaystyle\sum_{i=1}^{n} i = 78$.

37. Prove formula (b) of Theorem 3.

38. Prove formula (e) of Theorem 3 using mathematical induction.

39. Prove formula (e) of Theorem 3 using a method similar to that of Example 5, Solution 1 [start with $(1 + i)^4 - i^4$].

40. Prove formula (e) of Theorem 3 using the following method published by Abu Bekr Mohammed ibn Alhusain Alkarchi in about AD 1010. The figure shows a square $ABCD$ in which sides AB and AD have been divided into segments of lengths 1, 2, 3, . . . , n. Thus the side of the square has length $n(n + 1)/2$ so the area is $[n(n + 1)/2]^2$. But the area is also the sum of the areas of the n "gnomons" G_1, G_2, . . . , G_n shown in the figure. Show that the area of G_i is i^3 and conclude that formula (e) is true.

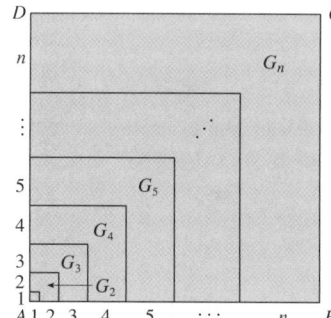

41. Evaluate each telescoping sum.

(a) $\displaystyle\sum_{i=1}^{n} [i^4 - (i-1)^4]$

(b) $\displaystyle\sum_{i=1}^{100} (5^i - 5^{i-1})$

(c) $\displaystyle\sum_{i=3}^{99} \left(\frac{1}{i} - \frac{1}{i+1} \right)$

(d) $\displaystyle\sum_{i=1}^{n} (a_i - a_{i-1})$

42. Prove the generalized triangle inequality:

$$\left| \sum_{i=1}^{n} a_i \right| \leq \sum_{i=1}^{n} |a_i|$$

43–46 ■ Find the limit.

43. $\lim\limits_{n \to \infty} \sum\limits_{i=1}^{n} \dfrac{1}{n} \left(\dfrac{i}{n} \right)^2$

44. $\lim\limits_{n \to \infty} \sum\limits_{i=1}^{n} \dfrac{1}{n} \left[\left(\dfrac{i}{n} \right)^3 + 1 \right]$

45. $\lim\limits_{n \to \infty} \sum\limits_{i=1}^{n} \dfrac{2}{n} \left[\left(\dfrac{2i}{n} \right)^3 + 5\left(\dfrac{2i}{n} \right) \right]$

46. $\lim\limits_{n \to \infty} \sum\limits_{i=1}^{n} \dfrac{3}{n} \left[\left(1 + \dfrac{3i}{n} \right)^3 - 2\left(1 + \dfrac{3i}{n} \right) \right]$

47. Prove the formula for the sum of a finite geometric series with first term a and common ratio $r \neq 1$:

$$\sum_{i=1}^{n} ar^{i-1} = a + ar + ar^2 + \cdots + ar^{n-1} = \frac{a(r^n - 1)}{r - 1}$$

48. Evaluate $\sum\limits_{i=1}^{n} \dfrac{3}{2^{i-1}}$.

49. Evaluate $\sum\limits_{i=1}^{n} (2i + 2^i)$.

50. Evaluate $\sum\limits_{i=1}^{m} \left[\sum\limits_{j=1}^{n} (i + j) \right]$.

C | THE LOGARITHM DEFINED AS AN INTEGRAL

The treatment of exponential and logarithmic functions presented in Chapter 3 relied on our intuition, which is based on numerical and visual evidence. Here we use the Fundamental Theorem of Calculus to give an alternative treatment that provides a surer footing for these functions.

Instead of starting with a^x and defining $\log_a x$ as its inverse, this time we start by defining $\ln x$ as an integral and then define the exponential function as its inverse. In this section you should bear in mind that we do not use any of our previous definitions and results concerning exponential and logarithmic functions.

THE NATURAL LOGARITHM

We first define $\ln x$ as an integral.

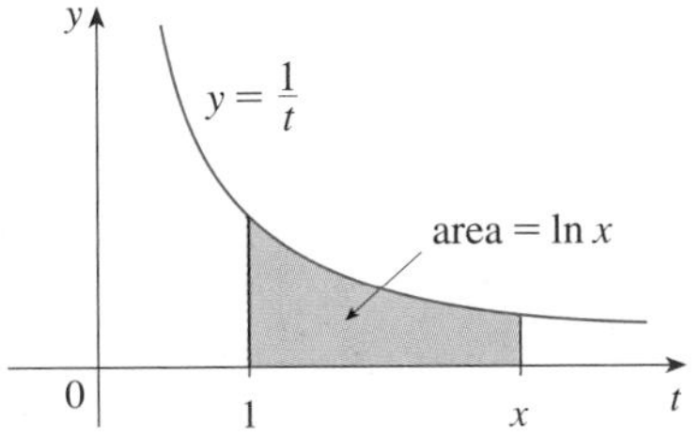

FIGURE 1

1 DEFINITION   The **natural logarithmic function** is the function defined by

$$\ln x = \int_1^x \frac{1}{t}\, dt \qquad x > 0$$

The existence of this function depends on the fact that the integral of a continuous function always exists. If $x > 1$, then $\ln x$ can be interpreted geometrically as the area under the hyperbola $y = 1/t$ from $t = 1$ to $t = x$. (See Figure 1.) For $x = 1$, we have

$$\ln 1 = \int_1^1 \frac{1}{t}\, dt = 0$$

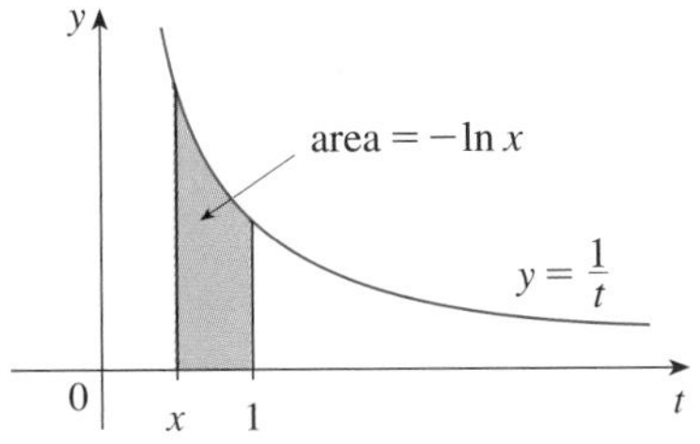

For $0 < x < 1$,

$$\ln x = \int_1^x \frac{1}{t}\, dt = -\int_x^1 \frac{1}{t}\, dt < 0$$

FIGURE 2

and so $\ln x$ is the negative of the area shown in Figure 2.

▼ EXAMPLE 1

(a) By comparing areas, show that $\frac{1}{2} < \ln 2 < \frac{3}{4}$.

(b) Use the Midpoint Rule with $n = 10$ to estimate the value of $\ln 2$.

SOLUTION

(a) We can interpret $\ln 2$ as the area under the curve $y = 1/t$ from 1 to 2. From Figure 3 we see that this area is larger than the area of rectangle $BCDE$ and smaller than the area of trapezoid $ABCD$. Thus we have

$$\tfrac{1}{2} \cdot 1 < \ln 2 < 1 \cdot \tfrac{1}{2}\left(1 + \tfrac{1}{2}\right)$$

$$\tfrac{1}{2} < \ln 2 < \tfrac{3}{4}$$

(b) If we use the Midpoint Rule with $f(t) = 1/t$, $n = 10$, and $\Delta t = 0.1$, we get

$$\ln 2 = \int_1^2 \frac{1}{t}\, dt \approx (0.1)[\,f(1.05) + f(1.15) + \cdots + f(1.95)\,]$$

$$= (0.1)\left(\frac{1}{1.05} + \frac{1}{1.15} + \cdots + \frac{1}{1.95}\right) \approx 0.693 \qquad ■$$

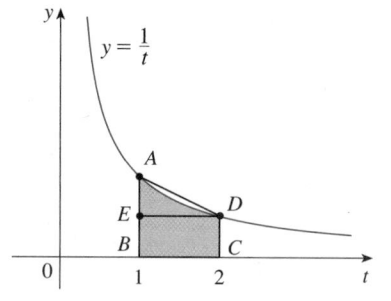

FIGURE 3

Notice that the integral that defines $\ln x$ is exactly the type of integral discussed in Part 1 of the Fundamental Theorem of Calculus (see Section 5.4). In fact, using that theorem, we have

$$\frac{d}{dx} \int_1^x \frac{1}{t}\, dt = \frac{1}{x}$$

and so

2

$$\frac{d}{dx}(\ln x) = \frac{1}{x}$$

We now use this differentiation rule to prove the following properties of the logarithm function.

---

3 **LAWS OF LOGARITHMS** If x and y are positive numbers and r is a rational number, then

1. $\ln(xy) = \ln x + \ln y$ **2.** $\ln\!\left(\dfrac{x}{y}\right) = \ln x - \ln y$ **3.** $\ln(x^r) = r \ln x$

PROOF

1. Let $f(x) = \ln(ax)$, where a is a positive constant. Then, using Equation 2 and the Chain Rule, we have

$$f'(x) = \frac{1}{ax}\frac{d}{dx}(ax) = \frac{1}{ax} \cdot a = \frac{1}{x}$$

Therefore $f(x)$ and $\ln x$ have the same derivative and so they must differ by a constant:

$$\ln(ax) = \ln x + C$$

Putting $x = 1$ in this equation, we get $\ln a = \ln 1 + C = 0 + C = C$. Thus

$$\ln(ax) = \ln x + \ln a$$

If we now replace the constant a by any number y, we have

$$\ln(xy) = \ln x + \ln y$$

2. Using Law 1 with $x = 1/y$, we have

$$\ln \frac{1}{y} + \ln y = \ln\left(\frac{1}{y} \cdot y\right) = \ln 1 = 0$$

and so

$$\ln \frac{1}{y} = -\ln y$$

Using Law 1 again, we have

$$\ln\left(\frac{x}{y}\right) = \ln\left(x \cdot \frac{1}{y}\right) = \ln x + \ln \frac{1}{y} = \ln x - \ln y$$

The proof of Law 3 is left as an exercise. □

In order to graph $y = \ln x$, we first determine its limits:

$$\boxed{4} \qquad \boxed{\text{(a)} \ \lim_{x \to \infty} \ln x = \infty \qquad \text{(b)} \ \lim_{x \to 0^+} \ln x = -\infty}$$

PROOF

(a) Using Law 3 with $x = 2$ and $r = n$ (where n is any positive integer), we have $\ln(2^n) = n \ln 2$. Now $\ln 2 > 0$, so this shows that $\ln(2^n) \to \infty$ as $n \to \infty$. But $\ln x$ is an increasing function since its derivative $1/x > 0$. Therefore $\ln x \to \infty$ as $x \to \infty$.

(b) If we let $t = 1/x$, then $t \to \infty$ as $x \to 0^+$. Thus, using (a), we have

$$\lim_{x \to 0^+} \ln x = \lim_{t \to \infty} \ln\left(\frac{1}{t}\right) = \lim_{t \to \infty} (-\ln t) = -\infty \qquad □$$

If $y = \ln x$, $x > 0$, then

$$\frac{dy}{dx} = \frac{1}{x} > 0 \qquad \text{and} \qquad \frac{d^2y}{dx^2} = -\frac{1}{x^2} < 0$$

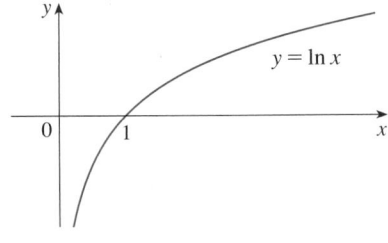

FIGURE 4

which shows that $\ln x$ is increasing and concave downward on $(0, \infty)$. Putting this information together with $\boxed{4}$, we draw the graph of $y = \ln x$ in Figure 4.

Since $\ln 1 = 0$ and $\ln x$ is an increasing continuous function that takes on arbitrarily large values, the Intermediate Value Theorem shows that there is a number where $\ln x$ takes on the value 1. (See Figure 5.) This important number is denoted by e.

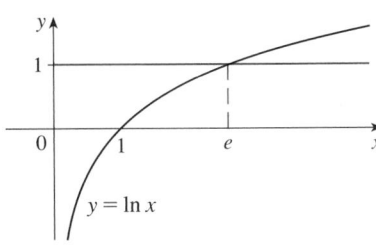

FIGURE 5

$\boxed{5}$ **DEFINITION** e is the number such that $\ln e = 1$.

We will show (in Theorem 19) that this definition is consistent with our previous definition of e.

THE NATURAL EXPONENTIAL FUNCTION

Since ln is an increasing function, it is one-to-one and therefore has an inverse function, which we denote by exp. Thus, according to the definition of an inverse function,

$$f^{-1}(x) = y \iff f(y) = x$$

| 6 | $\exp(x) = y \iff \ln y = x$ |

and the cancellation equations are

$$f^{-1}(f(x)) = x$$
$$f(f^{-1}(x)) = x$$

| 7 | $\exp(\ln x) = x \quad$ and $\quad \ln(\exp x) = x$ |

In particular, we have

$$\exp(0) = 1 \quad \text{since} \quad \ln 1 = 0$$

$$\exp(1) = e \quad \text{since} \quad \ln e = 1$$

We obtain the graph of $y = \exp x$ by reflecting the graph of $y = \ln x$ about the line $y = x$. (See Figure 6.) The domain of exp is the range of ln, that is, $(-\infty, \infty)$; the range of exp is the domain of ln, that is, $(0, \infty)$.

If r is any rational number, then the third law of logarithms gives

$$\ln(e^r) = r \ln e = r$$

Therefore, by 6, $\qquad \exp(r) = e^r$

Thus $\exp(x) = e^x$ whenever x is a rational number. This leads us to define e^x, even for irrational values of x, by the equation

$$e^x = \exp(x)$$

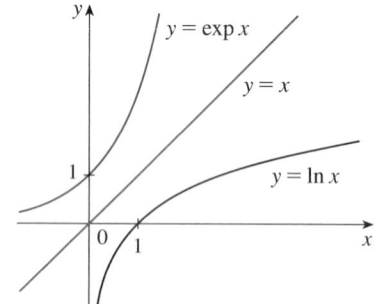

FIGURE 6

In other words, for the reasons given, we define e^x to be the inverse of the function $\ln x$. In this notation 6 becomes

| 8 | $e^x = y \iff \ln y = x$ |

and the cancellation equations 7 become

| 9 | $e^{\ln x} = x \qquad x > 0$ |

| 10 | $\ln(e^x) = x \qquad \text{for all } x$ |

The natural exponential function $f(x) = e^x$ is one of the most frequently occurring functions in calculus and its applications, so it is important to be familiar with its graph (Figure 7) and its properties (which follow from the fact that it is the inverse of the natural logarithmic function).

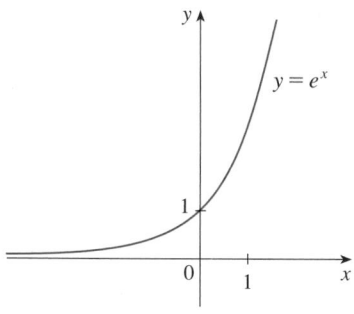

FIGURE 7
The natural exponential function

PROPERTIES OF THE EXPONENTIAL FUNCTION The exponential function $f(x) = e^x$ is an increasing continuous function with domain $\mathbb{R}$ and range $(0, \infty)$. Thus $e^x > 0$ for all x. Also

$$\lim_{x \to -\infty} e^x = 0 \qquad \lim_{x \to \infty} e^x = \infty$$

So the x-axis is a horizontal asymptote of $f(x) = e^x$.

We now verify that f has the other properties expected of an exponential function.

11 **LAWS OF EXPONENTS** If x and y are real numbers and r is rational, then

1. $e^{x+y} = e^x e^y$ **2.** $e^{x-y} = \dfrac{e^x}{e^y}$ **3.** $(e^x)^r = e^{rx}$

PROOF OF LAW 1 Using the first law of logarithms and Equation 10, we have

$$\ln(e^x e^y) = \ln(e^x) + \ln(e^y) = x + y = \ln(e^{x+y})$$

Since ln is a one-to-one function, it follows that $e^x e^y = e^{x+y}$.

Laws 2 and 3 are proved similarly (see Exercises 6 and 7). As we will soon see, Law 3 actually holds when r is any real number. ☐

We now prove the differentiation formula for e^x.

12
$$\frac{d}{dx}(e^x) = e^x$$

PROOF The function $y = e^x$ is differentiable because it is the inverse function of $y = \ln x$, which we know is differentiable with nonzero derivative. To find its derivative, we use the inverse function method. Let $y = e^x$. Then $\ln y = x$ and, differentiating this latter equation implicitly with respect to x, we get

$$\frac{1}{y}\frac{dy}{dx} = 1$$

$$\frac{dy}{dx} = y = e^x$$ ☐

GENERAL EXPONENTIAL FUNCTIONS

If $a > 0$ and r is any rational number, then by **9** and **11**,

$$a^r = (e^{\ln a})^r = e^{r \ln a}$$

Therefore, even for irrational numbers x, we *define*

13
$$a^x = e^{x \ln a}$$

Thus, for instance,

$$2^{\sqrt{3}} = e^{\sqrt{3}\ln 2} \approx e^{1.20} \approx 3.32$$

The function $f(x) = a^x$ is called the **exponential function with base a**. Notice that a^x is positive for all x because e^x is positive for all x.

Definition 13 allows us to extend one of the laws of logarithms. We already know that $\ln(a^r) = r \ln a$ when r is rational. But if we now let r be *any* real number we have, from Definition 13,

$$\ln a^r = \ln(e^{r \ln a}) = r \ln a$$

Thus

$$\boxed{14} \qquad \ln a^r = r \ln a \qquad \text{for any real number } r$$

The general laws of exponents follow from Definition 13 together with the laws of exponents for e^x.

$\boxed{15}$ **LAWS OF EXPONENTS** If x and y are real numbers and $a, b > 0$, then

1. $a^{x+y} = a^x a^y$ **2.** $a^{x-y} = a^x/a^y$ **3.** $(a^x)^y = a^{xy}$ **4.** $(ab)^x = a^x b^x$

PROOF

1. Using Definition 13 and the laws of exponents for e^x, we have

$$a^{x+y} = e^{(x+y)\ln a} = e^{x \ln a + y \ln a} = e^{x \ln a} e^{y \ln a} = a^x a^y$$

3. Using Equation 14 we obtain

$$(a^x)^y = e^{y \ln(a^x)} = e^{yx \ln a} = e^{xy \ln a} = a^{xy}$$

The remaining proofs are left as exercises. $\qquad \square$

The differentiation formula for exponential functions is also a consequence of Definition 13:

$$\boxed{16} \qquad \frac{d}{dx}(a^x) = a^x \ln a$$

PROOF $\qquad \dfrac{d}{dx}(a^x) = \dfrac{d}{dx}(e^{x \ln a}) = e^{x \ln a}\dfrac{d}{dx}(x \ln a) = a^x \ln a \qquad \square$

If $a > 1$, then $\ln a > 0$, so $(d/dx)\, a^x = a^x \ln a > 0$, which shows that $y = a^x$ is increasing (see Figure 8). If $0 < a < 1$, then $\ln a < 0$ and so $y = a^x$ is decreasing (see Figure 9).

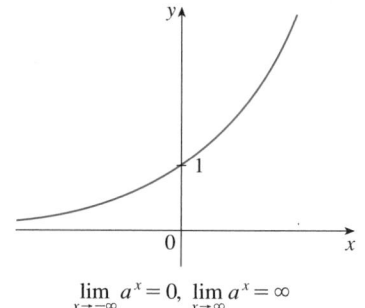

$$\lim_{x \to -\infty} a^x = 0, \ \lim_{x \to \infty} a^x = \infty$$

FIGURE 8
$y = a^x, \ a > 1$

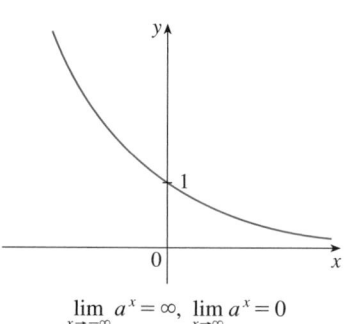

$$\lim_{x \to -\infty} a^x = \infty, \ \lim_{x \to \infty} a^x = 0$$

FIGURE 9
$y = a^x, \ 0 < a < 1$

GENERAL LOGARITHMIC FUNCTIONS

If $a > 0$ and $a \neq 1$, then $f(x) = a^x$ is a one-to-one function. Its inverse function is called the **logarithmic function with base a** and is denoted by $\log_a$. Thus

$$\boxed{\log_a x = y \iff a^y = x}$$

[17]

In particular, we see that

$$\log_e x = \ln x$$

The laws of logarithms are similar to those for the natural logarithm and can be deduced from the laws of exponents (see Exercise 10).

To differentiate $y = \log_a x$, we write the equation as $a^y = x$. From Equation 14 we have $y \ln a = \ln x$, so

$$\log_a x = y = \frac{\ln x}{\ln a}$$

Since $\ln a$ is a constant, we can differentiate as follows:

$$\frac{d}{dx} (\log_a x) = \frac{d}{dx} \frac{\ln x}{\ln a} = \frac{1}{\ln a} \frac{d}{dx} (\ln x) = \frac{1}{x \ln a}$$

[18]
$$\boxed{\frac{d}{dx} (\log_a x) = \frac{1}{x \ln a}}$$

THE NUMBER e EXPRESSED AS A LIMIT

In this section we defined e as the number such that $\ln e = 1$. The next theorem shows that this is the same as the number e defined in Section 3.1.

[19]
$$\boxed{e = \lim_{x \to 0} (1 + x)^{1/x}}$$

PROOF Let $f(x) = \ln x$. Then $f'(x) = 1/x$, so $f'(1) = 1$. But, by the definition of derivative,

$$f'(1) = \lim_{h \to 0} \frac{f(1 + h) - f(1)}{h} = \lim_{x \to 0} \frac{f(1 + x) - f(1)}{x}$$

$$= \lim_{x \to 0} \frac{\ln(1 + x) - \ln 1}{x} = \lim_{x \to 0} \frac{1}{x} \ln(1 + x)$$

$$= \lim_{x \to 0} \ln(1 + x)^{1/x}$$

Because $f'(1) = 1$, we have

$$\lim_{x \to 0} \ln(1 + x)^{1/x} = 1$$

Then, by Theorem 1.5.7 and the continuity of the exponential function, we have

$$e = e^1 = e^{\lim_{x \to 0} \ln(1+x)^{1/x}} = \lim_{x \to 0} e^{\ln(1+x)^{1/x}} = \lim_{x \to 0} (1 + x)^{1/x} \qquad \square$$

C | EXERCISES

1. (a) By comparing areas, show that

$$\tfrac{1}{3} < \ln 1.5 < \tfrac{5}{12}$$

   (b) Use the Midpoint Rule with $n = 10$ to estimate $\ln 1.5$.

2. Refer to Example 1.
   (a) Find the equation of the tangent line to the curve $y = 1/t$ that is parallel to the secant line AD.
   (b) Use part (a) to show that $\ln 2 > 0.66$.

3. By comparing areas, show that

$$\frac{1}{2} + \frac{1}{3} + \cdots + \frac{1}{n} < \ln n < 1 + \frac{1}{2} + \frac{1}{3} + \cdots + \frac{1}{n-1}$$

4. (a) By comparing areas, show that $\ln 2 < 1 < \ln 3$.
   (b) Deduce that $2 < e < 3$.

5. Prove the third law of logarithms. [*Hint:* Start by showing that both sides of the equation have the same derivative.]

6. Prove the second law of exponents for e^x [see [11]].

7. Prove the third law of exponents for e^x [see [11]].

8. Prove the second law of exponents [see [15]].

9. Prove the fourth law of exponents [see [15]].

10. Deduce the following laws of logarithms from [15]:
   (a) $\log_a(xy) = \log_a x + \log_a y$
   (b) $\log_a(x/y) = \log_a x - \log_a y$
   (c) $\log_a(x^y) = y \log_a x$

D | PROOFS

In this appendix we present proofs of several theorems that are stated in the main body of the text. The sections in which they occur are indicated in the margin.

SECTION 1.4

We start by proving the Triangle Inequality, which is an important property of absolute value.

THE TRIANGLE INEQUALITY If a and b are any real numbers, then

$$|a + b| \leq |a| + |b|$$

Observe that if the numbers a and b are both positive or both negative, then the two sides in the Triangle Inequality are actually equal. But if a and b have opposite signs, the left side involves a subtraction and the right side does not. This makes the Triangle Inequality seem reasonable, but we can prove it as follows.

PROOF Notice that

$$-|a| \leq a \leq |a|$$

is always true because a equals either $|a|$ or $-|a|$. The corresponding statement for b is

$$-|b| \leq b \leq |b|$$

Adding these inequalities, we get

$$-(|a| + |b|) \leq a + b \leq |a| + |b|$$

If we now use the fact that $|x| \leq a \iff -a \leq x \leq a$ (with x replaced by $a + b$ and a by $|a| + |b|$), we obtain

$$|a + b| \leq |a| + |b|$$

which is what we wanted to show. ☐

LIMIT LAWS Suppose that c is a constant and the limits

$$\lim_{x \to a} f(x) = L \qquad \text{and} \qquad \lim_{x \to a} g(x) = M$$

exist. Then

1. $\lim_{x \to a} [f(x) + g(x)] = L + M$ **2.** $\lim_{x \to a} [f(x) - g(x)] = L - M$

3. $\lim_{x \to a} [cf(x)] = cL$ **4.** $\lim_{x \to a} [f(x)g(x)] = LM$

5. $\lim_{x \to a} \dfrac{f(x)}{g(x)} = \dfrac{L}{M}$ if $M \neq 0$

PROOF OF LAW 1 Let $\varepsilon > 0$ be given. We must find $\delta > 0$ such that

$$\text{if} \qquad 0 < |x - a| < \delta \qquad \text{then} \qquad |f(x) + g(x) - (L + M)| < \varepsilon$$

Using the Triangle Inequality we can write

$$\boxed{1} \qquad |f(x) + g(x) - (L + M)| = |(f(x) - L) + (g(x) - M)|$$
$$\leq |f(x) - L| + |g(x) - M|$$

We will make $|f(x) + g(x) - (L + M)|$ less than ε by making each of the terms $|f(x) - L|$ and $|g(x) - M|$ less than $\varepsilon/2$.

Since $\varepsilon/2 > 0$ and $\lim_{x \to a} f(x) = L$, there exists a number $\delta_1 > 0$ such that

$$\text{if} \qquad 0 < |x - a| < \delta_1 \qquad \text{then} \qquad |f(x) - L| < \frac{\varepsilon}{2}$$

Similarly, since $\lim_{x \to a} g(x) = M$, there exists a number $\delta_2 > 0$ such that

$$\text{if} \qquad 0 < |x - a| < \delta_2 \qquad \text{then} \qquad |g(x) - M| < \frac{\varepsilon}{2}$$

Let $\delta = \min\{\delta_1, \delta_2\}$, the smaller of the numbers δ_1 and δ_2. Notice that

$$\text{if} \quad 0 < |x - a| < \delta \quad \text{then} \quad 0 < |x - a| < \delta_1 \quad \text{and} \quad 0 < |x - a| < \delta_2$$

and so

$$|f(x) - L| < \frac{\varepsilon}{2} \qquad \text{and} \qquad |g(x) - M| < \frac{\varepsilon}{2}$$

Therefore, by $\boxed{1}$,

$$|f(x) + g(x) - (L + M)| \leqslant |f(x) - L| + |g(x) - M| < \frac{\varepsilon}{2} + \frac{\varepsilon}{2} = \varepsilon$$

To summarize,

if $\quad 0 < |x - a| < \delta \quad$ then $\quad |f(x) + g(x) - (L + M)| < \varepsilon$

Thus, by the definition of a limit,

$$\lim_{x \to a} [f(x) + g(x)] = L + M \qquad \square$$

PROOF OF LAW 4 Let $\varepsilon > 0$ be given. We want to find $\delta > 0$ such that

if $\quad 0 < |x - a| < \delta \quad$ then $\quad |f(x)g(x) - LM| < \varepsilon$

In order to get terms that contain $|f(x) - L|$ and $|g(x) - M|$, we add and subtract $Lg(x)$ as follows:

$$
\begin{aligned}
|f(x)g(x) - LM| &= |f(x)g(x) - Lg(x) + Lg(x) - LM| \\
&= |[f(x) - L]g(x) + L[g(x) - M]| \\
&\leqslant |[f(x) - L]g(x)| + |L[g(x) - M]| \qquad \text{(Triangle Inequality)} \\
&= |f(x) - L||g(x)| + |L||g(x) - M|
\end{aligned}
$$

We want to make each of these terms less than $\varepsilon/2$.

Since $\lim_{x \to a} g(x) = M$, there is a number $\delta_1 > 0$ such that

if $\quad 0 < |x - a| < \delta_1 \quad$ then $\quad |g(x) - M| < \dfrac{\varepsilon}{2(1 + |L|)}$

Also, there is a number $\delta_2 > 0$ such that if $0 < |x - a| < \delta_2$, then

$$|g(x) - M| < 1$$

and therefore

$$|g(x)| = |g(x) - M + M| \leqslant |g(x) - M| + |M| < 1 + |M|$$

Since $\lim_{x \to a} f(x) = L$, there is a number $\delta_3 > 0$ such that

if $\quad 0 < |x - a| < \delta_3 \quad$ then $\quad |f(x) - L| < \dfrac{\varepsilon}{2(1 + |M|)}$

Let $\delta = \min\{\delta_1, \delta_2, \delta_3\}$. If $0 < |x - a| < \delta$, then we have $0 < |x - a| < \delta_1$, $0 < |x - a| < \delta_2$, and $0 < |x - a| < \delta_3$, so we can combine the inequalities to obtain

$$
\begin{aligned}
|f(x)g(x) - LM| &\leqslant |f(x) - L||g(x)| + |L||g(x) - M| \\
&< \frac{\varepsilon}{2(1 + |M|)} (1 + |M|) + |L| \frac{\varepsilon}{2(1 + |L|)} \\
&< \frac{\varepsilon}{2} + \frac{\varepsilon}{2} = \varepsilon
\end{aligned}
$$

This shows that $\lim_{x \to a} f(x)g(x) = LM$. $\qquad \square$

PROOF OF LAW 3 If we take $g(x) = c$ in Law 4, we get

$$\lim_{x \to a} [cf(x)] = \lim_{x \to a} [g(x)f(x)] = \lim_{x \to a} g(x) \cdot \lim_{x \to a} f(x)$$

$$= \lim_{x \to a} c \cdot \lim_{x \to a} f(x)$$

$$= c \lim_{x \to a} f(x) \qquad \text{(by Law 7)} \qquad \square$$

PROOF OF LAW 2 Using Law 1 and Law 3 with $c = -1$, we have

$$\lim_{x \to a} [f(x) - g(x)] = \lim_{x \to a} [f(x) + (-1)g(x)] = \lim_{x \to a} f(x) + \lim_{x \to a} (-1)g(x)$$

$$= \lim_{x \to a} f(x) + (-1) \lim_{x \to a} g(x) = \lim_{x \to a} f(x) - \lim_{x \to a} g(x) \qquad \square$$

PROOF OF LAW 5 First let us show that

$$\lim_{x \to a} \frac{1}{g(x)} = \frac{1}{M}$$

To do this we must show that, given $\varepsilon > 0$, there exists $\delta > 0$ such that

$$\text{if} \qquad 0 < |x - a| < \delta \qquad \text{then} \qquad \left| \frac{1}{g(x)} - \frac{1}{M} \right| < \varepsilon$$

Observe that

$$\left| \frac{1}{g(x)} - \frac{1}{M} \right| = \frac{|M - g(x)|}{|Mg(x)|}$$

We know that we can make the numerator small. But we also need to know that the denominator is not small when x is near a. Since $\lim_{x \to a} g(x) = M$, there is a number $\delta_1 > 0$ such that, whenever $0 < |x - a| < \delta_1$, we have

$$|g(x) - M| < \frac{|M|}{2}$$

and therefore

$$|M| = |M - g(x) + g(x)| \le |M - g(x)| + |g(x)| < \frac{|M|}{2} + |g(x)|$$

This shows that

$$\text{if} \qquad 0 < |x - a| < \delta_1 \qquad \text{then} \qquad |g(x)| > \frac{|M|}{2}$$

and so, for these values of x,

$$\frac{1}{|Mg(x)|} = \frac{1}{|M||g(x)|} < \frac{1}{|M|} \cdot \frac{2}{|M|} = \frac{2}{M^2}$$

Also, there exists $\delta_2 > 0$ such that

$$\text{if} \qquad 0 < |x - a| < \delta_2 \qquad \text{then} \qquad |g(x) - M| < \frac{M^2}{2} \varepsilon$$

Let $\delta = \min\{\delta_1, \delta_2\}$. Then, for $0 < |x - a| < \delta$, we have

$$\left| \frac{1}{g(x)} - \frac{1}{M} \right| = \frac{|M - g(x)|}{|Mg(x)|} < \frac{2}{M^2} \frac{M^2}{2} \varepsilon = \varepsilon$$

It follows that $\lim_{x \to a} 1/g(x) = 1/M$. Finally, using Law 4, we obtain

$$\lim_{x \to a} \frac{f(x)}{g(x)} = \lim_{x \to a} f(x)\left(\frac{1}{g(x)}\right) = \lim_{x \to a} f(x) \lim_{x \to a} \frac{1}{g(x)} = L \cdot \frac{1}{M} = \frac{L}{M} \qquad \square$$

---

3 **THEOREM** If $f(x) \leq g(x)$ for all x in an open interval that contains a (except possibly at a) and

$$\lim_{x \to a} f(x) = L \qquad \text{and} \qquad \lim_{x \to a} g(x) = M$$

then $L \leq M$.

PROOF We use the method of proof by contradiction. Suppose, if possible, that $L > M$. Law 2 of limits says that

$$\lim_{x \to a} [g(x) - f(x)] = M - L$$

Therefore, for any $\varepsilon > 0$, there exists $\delta > 0$ such that

$$\text{if} \qquad 0 < |x - a| < \delta \qquad \text{then} \qquad |[g(x) - f(x)] - (M - L)| < \varepsilon$$

In particular, taking $\varepsilon = L - M$ (noting that $L - M > 0$ by hypothesis), we have a number $\delta > 0$ such that

$$\text{if} \qquad 0 < |x - a| < \delta \qquad \text{then} \qquad |[g(x) - f(x)] - (M - L)| < L - M$$

Since $a \leq |a|$ for any number a, we have

$$\text{if} \qquad 0 < |x - a| < \delta \qquad \text{then} \qquad [g(x) - f(x)] - (M - L) < L - M$$

which simplifies to

$$\text{if} \qquad 0 < |x - a| < \delta \qquad \text{then} \qquad g(x) < f(x)$$

But this contradicts $f(x) \leq g(x)$. Thus the inequality $L > M$ must be false. Therefore $L \leq M$. $\qquad \square$

---

4 **THE SQUEEZE THEOREM** If $f(x) \leq g(x) \leq h(x)$ for all x in an open interval that contains a (except possibly at a) and

$$\lim_{x \to a} f(x) = \lim_{x \to a} h(x) = L$$

then

$$\lim_{x \to a} g(x) = L$$

PROOF Let $\varepsilon > 0$ be given. Since $\lim_{x \to a} f(x) = L$, there is a number $\delta_1 > 0$ such that

$$\text{if} \qquad 0 < |x - a| < \delta_1 \qquad \text{then} \qquad |f(x) - L| < \varepsilon$$

that is,

$$\text{if} \qquad 0 < |x - a| < \delta_1 \qquad \text{then} \qquad L - \varepsilon < f(x) < L + \varepsilon$$

Since $\lim_{x \to a} h(x) = L$, there is a number $\delta_2 > 0$ such that

$$\text{if} \qquad 0 < |x - a| < \delta_2 \qquad \text{then} \qquad |h(x) - L| < \varepsilon$$

that is,

$$\text{if} \qquad 0 < |x - a| < \delta_2 \qquad \text{then} \qquad L - \varepsilon < h(x) < L + \varepsilon$$

Let $\delta = \min\{\delta_1, \delta_2\}$. If $0 < |x - a| < \delta$, then $0 < |x - a| < \delta_1$ and $0 < |x - a| < \delta_2$, so

$$L - \varepsilon < f(x) \leqslant g(x) \leqslant h(x) < L + \varepsilon$$

In particular, $\qquad\qquad\qquad L - \varepsilon < g(x) < L + \varepsilon$

and so $|g(x) - L| < \varepsilon$. Therefore $\lim_{x \to a} g(x) = L$. $\qquad\qquad\square$

The proof of the following result was promised when we proved that $\lim_{\theta \to 0} \dfrac{\sin \theta}{\theta} = 1$.

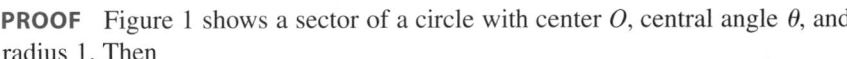

THEOREM   If $0 < \theta < \pi/2$, then $\theta \leqslant \tan \theta$.

PROOF   Figure 1 shows a sector of a circle with center O, central angle θ, and radius 1. Then

$$|AD| = |OA| \tan \theta = \tan \theta$$

We approximate the arc AB by an inscribed polygon consisting of n equal line segments and we look at a typical segment PQ. We extend the lines OP and OQ to meet AD in the points R and S. Then we draw $RT \parallel PQ$ as in Figure 1. Observe that

$$\angle RTO = \angle PQO < 90°$$

and so $\angle RTS > 90°$. Therefore we have

$$|PQ| < |RT| < |RS|$$

If we add n such inequalities, we get

$$L_n < |AD| = \tan \theta$$

where L_n is the length of the inscribed polygon. Thus, by Theorem 1.4.3, we have

$$\lim_{n \to \infty} L_n \leqslant \tan \theta$$

But the arc length is defined in Equation 7.4.1 as the limit of the lengths of inscribed polygons, so

$$\theta = \lim_{n \to \infty} L_n \leqslant \tan \theta \qquad\qquad\qquad\square$$

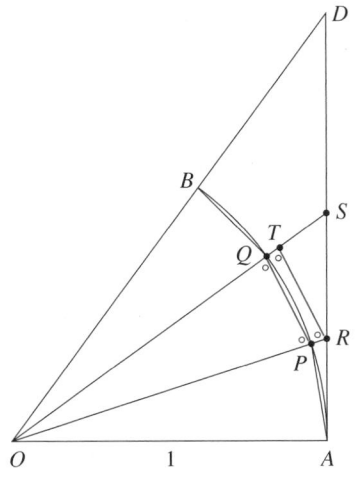

FIGURE 1

SECTION 1.5

> $\boxed{7}$ **THEOREM** If f is continuous at b and $\lim_{x \to a} g(x) = b$, then
>
> $$\lim_{x \to a} f(g(x)) = f(b)$$

PROOF Let $\varepsilon > 0$ be given. We want to find a number $\delta > 0$ such that

$$\text{if} \quad 0 < |x - a| < \delta \quad \text{then} \quad |f(g(x)) - f(b)| < \varepsilon$$

Since f is continuous at b, we have

$$\lim_{y \to b} f(y) = f(b)$$

and so there exists $\delta_1 > 0$ such that

$$\text{if} \quad 0 < |y - b| < \delta_1 \quad \text{then} \quad |f(y) - f(b)| < \varepsilon$$

Since $\lim_{x \to a} g(x) = b$, there exists $\delta > 0$ such that

$$\text{if} \quad 0 < |x - a| < \delta \quad \text{then} \quad |g(x) - b| < \delta_1$$

Combining these two statements, we see that whenever $0 < |x - a| < \delta$ we have $|g(x) - b| < \delta_1$, which implies that $|f(g(x)) - f(b)| < \varepsilon$. Therefore we have proved that $\lim_{x \to a} f(g(x)) = f(b)$. $\qquad \square$

SECTION 3.1

We are going to use the Monotonic Sequence Theorem from Section 8.1 to prove the existence of the limit $\lim_{x \to 0}(1 + x)^{1/x}$ that we used to define the number e. We will need the following result.

> **LEMMA** If $0 \leqslant a < b$ and n is a positive integer, then
>
> $$b^n[(n + 1)a - nb] < a^{n+1}$$

PROOF We begin by factoring $b^{n+1} - a^{n+1}$. Since $a < b$, we have

$$b^{n+1} - a^{n+1} = (b - a)(b^n + b^{n-1}a + b^{n-2}a^2 + \cdots + ba^{n-1} + a^n)$$

$$< (b - a)(b^n + b^{n-1}b + b^{n-2}b^2 + \cdots + bb^{n-1} + b^n)$$

$$= (b - a)(b^n + b^n + b^n + \cdots + b^n + b^n) = (b - a)(n + 1)b^n$$

We have shown that

$$b^{n+1} - a^{n+1} < (b - a)(n + 1)b^n$$

so $\qquad\qquad b^{n+1} - (n + 1)b^n(b - a) < a^{n+1}$

Factoring b^n from the left side of this inequality and simplifying, we get

$$b^n[(n + 1)a - nb] < a^{n+1} \qquad \square$$

THEOREM The limit $\lim\limits_{x \to 0} (1 + x)^{1/x}$ exists.

PROOF Let

$$a_n = \left(1 + \frac{1}{n}\right)^n$$

To show that $\{a_n\}$ is an increasing sequence, we put $a = 1 + 1/(n + 1)$ and $b = 1 + 1/n$ in the lemma:

$$\left(1 + \frac{1}{n}\right)^n \left[(n + 1)\left(1 + \frac{1}{n + 1}\right) - n\left(1 + \frac{1}{n}\right)\right] < \left(1 + \frac{1}{n + 1}\right)^{n+1}$$

Simplifying the left side of this inequality, we get

$$\left(1 + \frac{1}{n}\right)^n < \left(1 + \frac{1}{n + 1}\right)^{n+1}$$

This says that $a_n < a_{n+1}$, that is, $\{a_n\}$ is an increasing sequence.

Next we show that $\{a_n\}$ is a bounded sequence. If we let $a = 1$ and $b = 1 + 1/(2n)$ in the lemma, we get

$$\left(1 + \frac{1}{2n}\right)^n \left[n + 1 - n - \frac{1}{2}\right] < 1$$

so

$$\left(1 + \frac{1}{2n}\right)^n < 2 \qquad \text{and} \qquad \left(1 + \frac{1}{2n}\right)^{2n} < 4$$

This says that $a_{2n} < 4$. Since $\{a_n\}$ is increasing, we have $a_n < a_{2n}$ and so

$$0 < a_n < 4 \qquad \text{for all } n$$

Thus $\{a_n\}$ is a bounded sequence. It follows froom the Monotonic Sequence Theorem that $\{a_n\}$ is convergent. We denote its limit by e.

Now if x is any real number that satisfies

$$\frac{1}{n + 1} < x < \frac{1}{n}$$

then

$$\boxed{1} \qquad \left(1 + \frac{1}{n + 1}\right)^n < (1 + x)^{1/x} < \left(1 + \frac{1}{n}\right)^{n+1}$$

For instance, the right side of $\boxed{1}$ is true because $1 + x < 1 + 1/n$ and

$$x > \frac{1}{n + 1} \qquad \Rightarrow \qquad \frac{1}{x} < n + 1 \qquad \Rightarrow \qquad (1 + x)^{1/x} < \left(1 + \frac{1}{n}\right)^{n+1}$$

But

$$\lim_{n \to \infty} \left(1 + \frac{1}{n}\right)^{n+1} = \lim_{n \to \infty} \left[\left(1 + \frac{1}{n}\right)^{n}\left(1 + \frac{1}{n}\right)\right]$$

$$= \lim_{n \to \infty} \left(1 + \frac{1}{n}\right)^{n} \cdot \lim_{n \to \infty} \left(1 + \frac{1}{n}\right) = e \cdot 1 = e$$

Similarly

$$\lim_{n \to \infty} \left(1 + \frac{1}{n+1}\right)^{n} = e$$

It follows from the pair of inequalities in $\boxed{1}$ that

$$\lim_{x \to 0^+} (1 + x)^{1/x} \text{ exists}$$

Finally, if $x < 0$, let $t = -x$. Then

$$(1 + x)^{1/x} = (1 - t)^{-1/t} = \left(\frac{1}{1 - t}\right)^{1/t}$$

$$= \left(1 + \frac{t}{1 - t}\right)^{1/t} = \left(1 + \frac{t}{1 - t}\right)^{(1-t)/t} \left(1 + \frac{t}{1 - t}\right)$$

As $t \to 0^-$, we have $t/(1 - t) \to 0^+$ and

$$\lim_{x \to 0^-} (1 + x)^{1/x} = \lim_{t \to 0^+} \left(1 + \frac{t}{1 - t}\right)^{(1-t)/t} \left(1 + \frac{t}{1 - t}\right) = e \cdot 1 = e$$

We have therefore shown that

$$\lim_{x \to 0} (1 + x)^{1/x} \text{ exists}$$ $\square$

SECTION 3.2

> $\boxed{6}$ **THEOREM** If f is a one-to-one continuous function defined on an interval (a, b), then its inverse function f^{-1} is also continuous.

PROOF First we show that if f is both one-to-one and continuous on (a, b), then it must be either increasing or decreasing on (a, b). If it were neither increasing nor decreasing, then there would exist numbers x_1, x_2, and x_3 in (a, b) with $x_1 < x_2 < x_3$ such that $f(x_2)$ does not lie between $f(x_1)$ and $f(x_3)$. There are two possibilities: either (1) $f(x_3)$ lies between $f(x_1)$ and $f(x_2)$ or (2) $f(x_1)$ lies between $f(x_2)$ and $f(x_3)$. (Draw a picture.) In case (1) we apply the Intermediate Value Theorem to the continuous function f to get a number c between x_1 and x_2 such that $f(c) = f(x_3)$. In case (2) the Intermediate Value Theorem gives a number c between x_2 and x_3 such that $f(c) = f(x_1)$. In either case we have contradicted the fact that f is one-to-one.

Let us assume, for the sake of definiteness, that f is increasing on (a, b). We take any number y_0 in the domain of f^{-1} and we let $f^{-1}(y_0) = x_0$; that is, x_0 is the number in (a, b) such that $f(x_0) = y_0$. To show that f^{-1} is continuous at y_0 we take any $\varepsilon > 0$ such that the interval $(x_0 - \varepsilon, x_0 + \varepsilon)$ is contained in the interval (a, b). Since f is increasing, it maps the numbers in the interval $(x_0 - \varepsilon, x_0 + \varepsilon)$ onto the numbers in the interval $(f(x_0 - \varepsilon), f(x_0 + \varepsilon))$ and f^{-1} reverses the correspondence. If we let δ denote the smaller of the numbers $\delta_1 = y_0 - f(x_0 - \varepsilon)$ and $\delta_2 = f(x_0 + \varepsilon) - y_0$, then the interval $(y_0 - \delta, y_0 + \delta)$ is contained in the interval $(f(x_0 - \varepsilon), f(x_0 + \varepsilon))$ and so is mapped into the interval $(x_0 - \varepsilon, x_0 + \varepsilon)$ by f^{-1}.

(See the arrow diagram in Figure 2.) We have therefore found a number $\delta > 0$ such that

$$\text{if} \qquad |y - y_0| < \delta \qquad \text{then} \qquad |f^{-1}(y) - f^{-1}(y_0)| < \varepsilon$$

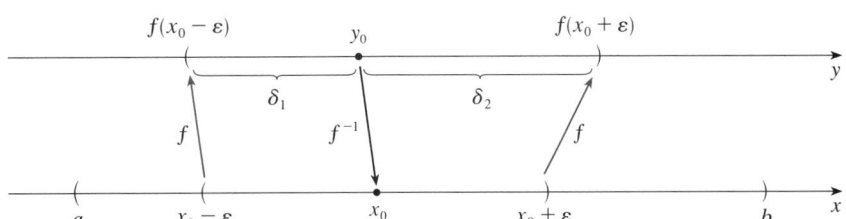

FIGURE 2

This shows that $\lim_{y \to y_0} f^{-1}(y) = f^{-1}(y_0)$ and so f^{-1} is continuous at any number y_0 in its domain. ☐

SECTION 3.7

In order to give the promised proof of l'Hospital's Rule we first need a generalization of the Mean Value Theorem. The following theorem is named after the French mathematician, Augustin-Louis Cauchy (1789–1857).

CAUCHY'S MEAN VALUE THEOREM Suppose that the functions f and g are continuous on $[a, b]$ and differentiable on (a, b), and $g'(x) \neq 0$ for all x in (a, b). Then there is a number c in (a, b) such that

$$\frac{f'(c)}{g'(c)} = \frac{f(b) - f(a)}{g(b) - g(a)}$$

Notice that if we take the special case in which $g(x) = x$, then $g'(c) = 1$ and Cauchy's Mean Value Theorem is just the ordinary Mean Value Theorem. Furthermore, it can be proved in a similar manner. You can verify that all we have to do is change the function h given by Equation 4.2.4 to the function

$$h(x) = f(x) - f(a) - \frac{f(b) - f(a)}{g(b) - g(a)} [g(x) - g(a)]$$

and apply Rolle's Theorem as before.

L'HOSPITAL'S RULE Suppose f and g are differentiable and $g'(x) \neq 0$ on an open interval I that contains a (except possibly at a). Suppose that

$$\lim_{x \to a} f(x) = 0 \qquad \text{and} \qquad \lim_{x \to a} g(x) = 0$$

or that

$$\lim_{x \to a} f(x) = \pm\infty \qquad \text{and} \qquad \lim_{x \to a} g(x) = \pm\infty$$

(In other words, we have an indeterminate form of type $\frac{0}{0}$ or ∞/∞.) Then

$$\lim_{x \to a} \frac{f(x)}{g(x)} = \lim_{x \to a} \frac{f'(x)}{g'(x)}$$

if the limit on the right side exists (or is ∞ or $-\infty$).

PROOF OF L'HOSPITAL'S RULE   We are assuming that $\lim_{x \to a} f(x) = 0$ and $\lim_{x \to a} g(x) = 0$. Let

$$L = \lim_{x \to a} \frac{f'(x)}{g'(x)}$$

We must show that $\lim_{x \to a} f(x)/g(x) = L$. Define

$$F(x) = \begin{cases} f(x) & \text{if } x \ne a \\ 0 & \text{if } x = a \end{cases} \qquad G(x) = \begin{cases} g(x) & \text{if } x \ne a \\ 0 & \text{if } x = a \end{cases}$$

Then F is continuous on I since f is continuous on $\{x \in I \mid x \ne a\}$ and

$$\lim_{x \to a} F(x) = \lim_{x \to a} f(x) = 0 = F(a)$$

Likewise, G is continuous on I. Let $x \in I$ and $x > a$. Then F and G are continuous on $[a, x]$ and differentiable on (a, x) and $G' \ne 0$ there (since $F' = f'$ and $G' = g'$). Therefore, by Cauchy's Mean Value Theorem, there is a number y such that $a < y < x$ and

$$\frac{F'(y)}{G'(y)} = \frac{F(x) - F(a)}{G(x) - G(a)} = \frac{F(x)}{G(x)}$$

Here we have used the fact that, by definition, $F(a) = 0$ and $G(a) = 0$. Now, if we let $x \to a^+$, then $y \to a^+$ (since $a < y < x$), so

$$\lim_{x \to a^+} \frac{f(x)}{g(x)} = \lim_{x \to a^+} \frac{F(x)}{G(x)} = \lim_{y \to a^+} \frac{F'(y)}{G'(y)} = \lim_{y \to a^+} \frac{f'(y)}{g'(y)} = L$$

A similar argument shows that the left-hand limit is also L. Therefore

$$\lim_{x \to a} \frac{f(x)}{g(x)} = L$$

This proves l'Hospital's Rule for the case where a is finite.

If a is infinite, we let $t = 1/x$. Then $t \to 0^+$ as $x \to \infty$, so we have

$$\lim_{x \to \infty} \frac{f(x)}{g(x)} = \lim_{t \to 0^+} \frac{f(1/t)}{g(1/t)}$$

$$= \lim_{t \to 0^+} \frac{f'(1/t)(-1/t^2)}{g'(1/t)(-1/t^2)} \qquad \text{(by l'Hospital's Rule for finite } a\text{)}$$

$$= \lim_{t \to 0^+} \frac{f'(1/t)}{g'(1/t)} = \lim_{x \to \infty} \frac{f'(x)}{g'(x)}$$

SECTION 4.3

CONCAVITY TEST

(a) If $f''(x) > 0$ for all x in I, then the graph of f is concave upward on I.

(b) If $f''(x) < 0$ for all x in I, then the graph of f is concave downward on I.

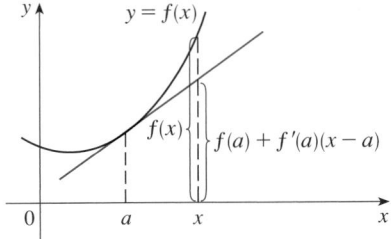

FIGURE 3

PROOF OF (a)   Let a be any number in I. We must show that the curve $y = f(x)$ lies above the tangent line at the point $(a, f(a))$. The equation of this tangent is

$$y = f(a) + f'(a)(x - a)$$

So we must show that

$$f(x) > f(a) + f'(a)(x - a)$$

whenever $x \in I$ $(x \neq a)$. (See Figure 3.)

First let us take the case where $x > a$. Applying the Mean Value Theorem to f on the interval $[a, x]$, we get a number c, with $a < c < x$, such that

$$\boxed{1} \qquad\qquad f(x) - f(a) = f'(c)(x - a)$$

Since $f'' > 0$ on I we know from the Increasing/Decreasing Test that f' is increasing on I. Thus, since $a < c$, we have

$$f'(a) < f'(c)$$

and so, multiplying this inequality by the positive number $x - a$, we get

$$\boxed{2} \qquad\qquad f'(a)(x - a) < f'(c)(x - a)$$

Now we add $f(a)$ to both sides of this inequality:

$$f(a) + f'(a)(x - a) < f(a) + f'(c)(x - a)$$

But from Equation 1 we have $f(x) = f(a) + f'(c)(x - a)$. So this inequality becomes

$$\boxed{3} \qquad\qquad f(x) > f(a) + f'(a)(x - a)$$

which is what we wanted to prove.

For the case where $x < a$ we have $f'(c) < f'(a)$, but multiplication by the negative number $x - a$ reverses the inequality, so we get $\boxed{2}$ and $\boxed{3}$ as before.   □

SECTION 5.2

PROPERTY 5 OF INTEGRALS

$$\int_a^b f(x)\, dx = \int_a^c f(x)\, dx + \int_c^b f(x)\, dx$$

if all of these integrals exist.

PROOF   We first assume that $a < c < b$. Since we are assuming that $\int_a^b f(x)\, dx$ exists, we can compute it as a limit of Riemann sums using only partitions P that include c as one of the partition points. If P is such a partition, let P_1 be the corresponding partition of $[a, c]$ determined by those partition points of P that lie in $[a, c]$. Similarly, P_2 will denote the corresponding partition of $[c, b]$.

We introduce the notation $\|P\|$ for the length of the longest subinterval in P, that is,

$$\|P\| = \max \Delta x_i = \max\{\Delta x_1, \Delta x_2, \ldots, \Delta x_n\}$$

Note that $\|P_1\| \leqslant \|P\|$ and $\|P_2\| \leqslant \|P\|$. Thus, if $\|P\| \to 0$, it follows that $\|P_1\| \to 0$ and $\|P_2\| \to 0$. If $\{x_i \mid 1 \leqslant i \leqslant n\}$ is the set of partition points for P and $n = k + m$, where k is the number of subintervals in $[a, c]$ and m is the number of subintervals in $[c, b]$, then $\{x_i \mid 1 \leqslant i \leqslant k\}$ is the set of partition points for P_1. If we write $t_j = x_{k+j}$ for the partition points to the right of c, then $\{t_j \mid 1 \leqslant j \leqslant m\}$ is the set of partition points for P_2. Thus we have

$$a = x_0 < x_1 < \cdots < x_k < x_{k+1} < \cdots < x_n = b$$

$$c < t_1 < \cdots < t_m = b$$

Choosing $x_i^* = x_i$ and letting $\Delta t_j = t_j - t_{j-1}$, we compute $\int_a^b f(x)\,dx$ as follows:

$$\int_a^b f(x)\,dx = \lim_{\|P\| \to 0} \sum_{i=1}^n f(x_i)\,\Delta x_i = \lim_{\|P\| \to 0} \left[\sum_{i=1}^k f(x_i)\,\Delta x_i + \sum_{i=k+1}^n f(x_i)\,\Delta x_i \right]$$

$$= \lim_{\|P\| \to 0} \left[\sum_{i=1}^k f(x_i)\,\Delta x_i + \sum_{j=1}^m f(t_j)\,\Delta t_j \right]$$

$$= \lim_{\|P_1\| \to 0} \sum_{i=1}^k f(x_i)\,\Delta x_i + \lim_{\|P_2\| \to 0} \sum_{j=1}^m f(t_j)\,\Delta t_j = \int_a^c f(x)\,dx + \int_c^b f(t)\,dt$$

Now suppose that $c < a < b$. By what we have already proved, we have

$$\int_c^b f(x)\,dx = \int_c^a f(x)\,dx + \int_a^b f(x)\,dx$$

Therefore

$$\int_a^b f(x)\,dx = -\int_c^a f(x)\,dx + \int_c^b f(x)\,dx = \int_a^c f(x)\,dx + \int_c^b f(x)\,dx$$

The proofs are similar for the remaining four orderings of a, b, and c. ☐

In order to prove Theorem 8.5.3 we first need the following results.

THEOREM

1. If a power series $\sum c_n x^n$ converges when $x = b$ (where $b \neq 0$), then it converges whenever $|x| < |b|$.

2. If a power series $\sum c_n x^n$ diverges when $x = d$ (where $d \neq 0$), then it diverges whenever $|x| > |d|$.

PROOF OF 1   Suppose that $\sum c_n b^n$ converges. Then, by Theorem 8.2.6, we have $\lim_{n \to \infty} c_n b^n = 0$. According to Definition 8.1.2 with $\varepsilon = 1$, there is a positive integer N such that $|c_n b^n| < 1$ whenever $n \geqslant N$. Thus for $n \geqslant N$ we have

$$|c_n x^n| = \left| \frac{c_n b^n x^n}{b^n} \right| = |c_n b^n| \left| \frac{x}{b} \right|^n < \left| \frac{x}{b} \right|^n$$

If $|x| < |b|$, then $|x/b| < 1$, so $\Sigma |x/b|^n$ is a convergent geometric series. Therefore, by the Comparison Test, the series $\sum_{n=N}^{\infty} |c_n x^n|$ is convergent. Thus the series $\Sigma \, c_n x^n$ is absolutely convergent and therefore convergent. □

PROOF OF 2 Suppose that $\Sigma \, c_n d^n$ diverges. If x is any number such that $|x| > |d|$, then $\Sigma \, c_n x^n$ cannot converge because, by part 1, the convergence of $\Sigma \, c_n x^n$ would imply the convergence of $\Sigma \, c_n d^n$. Therefore $\Sigma \, c_n x^n$ diverges whenever $|x| > |d|$. □

THEOREM For a power series $\Sigma \, c_n x^n$ there are only three possibilities:

1. The series converges only when $x = 0$.

2. The series converges for all x.

3. There is a positive number R such that the series converges if $|x| < R$ and diverges if $|x| > R$.

PROOF Suppose that neither case 1 nor case 2 is true. Then there are nonzero numbers b and d such that $\Sigma \, c_n x^n$ converges for $x = b$ and diverges for $x = d$. Therefore the set $S = \{x \mid \Sigma \, c_n x^n \text{ converges}\}$ is not empty. By the preceding theorem, the series diverges if $|x| > |d|$, so $|x| \le |d|$ for all $x \in S$. This says that $|d|$ is an upper bound for the set S. Thus, by the Completeness Axiom (see Section 8.1), S has a least upper bound R. If $|x| > R$, then $x \notin S$, so $\Sigma \, c_n x^n$ diverges. If $|x| < R$, then $|x|$ is not an upper bound for S and so there exists $b \in S$ such that $b > |x|$. Since $b \in S$, $\Sigma \, c_n b^n$ converges, so by the preceding theorem $\Sigma \, c_n x^n$ converges. □

---

3 THEOREM For a power series $\Sigma \, c_n (x - a)^n$ there are only three possibilities:

1. The series converges only when $x = a$.

2. The series converges for all x.

3. There is a positive number R such that the series converges if $|x - a| < R$ and diverges if $|x - a| > R$.

PROOF If we make the change of variable $u = x - a$, then the power series becomes $\Sigma \, c_n u^n$ and we can apply the preceding theorem to this series. In case 3 we have convergence for $|u| < R$ and divergence for $|u| > R$. Thus we have convergence for $|x - a| < R$ and divergence for $|x - a| > R$. □

SECTION 11.3

CLAIRAUT'S THEOREM Suppose f is defined on a disk D that contains the point (a, b). If the functions f_{xy} and f_{yx} are both continuous on D, then $f_{xy}(a, b) = f_{yx}(a, b)$.

PROOF For small values of h, $h \neq 0$, consider the difference

$$\Delta(h) = [f(a + h, b + h) - f(a + h, b)] - [f(a, b + h) - f(a, b)]$$

Notice that if we let $g(x) = f(x, b + h) - f(x, b)$, then

$$\Delta(h) = g(a + h) - g(a)$$

By the Mean Value Theorem, there is a number c between a and $a + h$ such that

$$g(a + h) - g(a) = g'(c)h = h[f_x(c, b + h) - f_x(c, b)]$$

Applying the Mean Value Theorem again, this time to f_x, we get a number d between b and $b + h$ such that

$$f_x(c, b + h) - f_x(c, b) = f_{xy}(c, d)h$$

Combining these equations, we obtain

$$\Delta(h) = h^2 f_{xy}(c, d)$$

If $h \rightarrow 0$, then $(c, d) \rightarrow (a, b)$, so the continuity of f_{xy} at (a, b) gives

$$\lim_{h \to 0} \frac{\Delta(h)}{h^2} = \lim_{(c, d) \to (a, b)} f_{xy}(c, d) = f_{xy}(a, b)$$

Similarly, by writing

$$\Delta(h) = [f(a + h, b + h) - f(a, b + h)] - [f(a + h, b) - f(a, b)]$$

and using the Mean Value Theorem twice and the continuity of f_{yx} at (a, b), we obtain

$$\lim_{h \to 0} \frac{\Delta(h)}{h^2} = f_{yx}(a, b)$$

It follows that $f_{xy}(a, b) = f_{yx}(a, b)$. $\qquad\square$

SECTION 11.4

> **8** **THEOREM**   If the partial derivatives f_x and f_y exist near (a, b) and are continuous at (a, b), then f is differentiable at (a, b).

PROOF Let

$$\Delta z = f(a + \Delta x, b + \Delta y) - f(a, b)$$

According to (11.4.7), to prove that f is differentiable at (a, b) we have to show that we can write Δz in the form

$$\Delta z = f_x(a, b)\, \Delta x + f_y(a, b)\, \Delta y + \varepsilon_1\, \Delta x + \varepsilon_2\, \Delta y$$

where ε_1 and $\varepsilon_2 \rightarrow 0$ as $(\Delta x, \Delta y) \rightarrow (0, 0)$.

Referring to Figure 4, we write

$$\boxed{1} \quad \Delta z = [f(a + \Delta x, b + \Delta y) - f(a, b + \Delta y)] + [f(a, b + \Delta y) - f(a, b)]$$

Observe that the function of a single variable

$$g(x) = f(x, b + \Delta y)$$

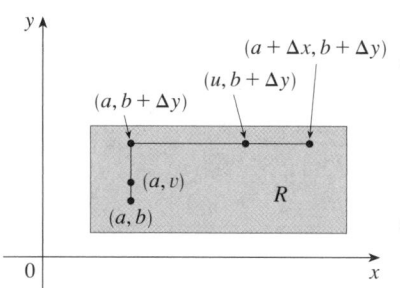

FIGURE 4

is defined on the interval $[a, a + \Delta x]$ and $g'(x) = f_x(x, b + \Delta y)$. If we apply the Mean Value Theorem to g, we get

$$g(a + \Delta x) - g(a) = g'(u)\, \Delta x$$

where u is some number between a and $a + \Delta x$. In terms of f, this equation becomes

$$f(a + \Delta x, b + \Delta y) - f(a, b + \Delta y) = f_x(u, b + \Delta y)\,\Delta x$$

This gives us an expression for the first part of the right side of Equation 1. For the second part we let $h(y) = f(a, y)$. Then h is a function of a single variable defined on the interval $[b, b + \Delta y]$ and $h'(y) = f_y(a, y)$. A second application of the Mean Value Theorem then gives

$$h(b + \Delta y) - h(b) = h'(v)\,\Delta y$$

where v is some number between b and $b + \Delta y$. In terms of f, this becomes

$$f(a, b + \Delta y) - f(a, b) = f_y(a, v)\,\Delta y$$

We now substitute these expressions into Equation 1 and obtain

$$\Delta z = f_x(u, b + \Delta y)\,\Delta x + f_y(a, v)\,\Delta y$$

$$= f_x(a, b)\,\Delta x + [f_x(u, b + \Delta y) - f_x(a, b)]\,\Delta x + f_y(a, b)\,\Delta y$$

$$+ [f_y(a, v) - f_y(a, b)]\,\Delta y$$

$$= f_x(a, b)\,\Delta x + f_y(a, b)\,\Delta y + \varepsilon_1\,\Delta x + \varepsilon_2\,\Delta y$$

where

$$\varepsilon_1 = f_x(u, b + \Delta y) - f_x(a, b)$$

$$\varepsilon_2 = f_y(a, v) - f_y(a, b)$$

Since $(u, b + \Delta y) \to (a, b)$ and $(a, v) \to (a, b)$ as $(\Delta x, \Delta y) \to (0, 0)$ and since f_x and f_y are continuous at (a, b), we see that $\varepsilon_1 \to 0$ and $\varepsilon_2 \to 0$ as $(\Delta x, \Delta y) \to (0, 0)$.

Therefore f is differentiable at (a, b). $\qquad\square$

SECTION 11.7

SECOND DERIVATIVES TEST Suppose the second partial derivatives of f are continuous on a disk with center (a, b), and suppose that $f_x(a, b) = 0$ and $f_y(a, b) = 0$ [that is, (a, b) is a critical point of f]. Let

$$D = D(a, b) = f_{xx}(a, b)f_{yy}(a, b) - [f_{xy}(a, b)]^2$$

(a) If $D > 0$ and $f_{xx}(a, b) > 0$, then $f(a, b)$ is a local minimum.

(b) If $D > 0$ and $f_{xx}(a, b) < 0$, then $f(a, b)$ is a local maximum.

(c) If $D < 0$, then $f(a, b)$ is not a local maximum or minimum.

PROOF OF PART (a) We compute the second-order directional derivative of f in the direction of $\mathbf{u} = \langle h, k \rangle$. The first-order derivative is given by Theorem 11.6.3:

$$D_{\mathbf{u}}f = f_x h + f_y k$$

Applying this theorem a second time, we have

$$D_{\mathbf{u}}^2 f = D_{\mathbf{u}}(D_{\mathbf{u}} f) = \frac{\partial}{\partial x}(D_{\mathbf{u}} f)h + \frac{\partial}{\partial y}(D_{\mathbf{u}} f)k$$

$$= (f_{xx}h + f_{yx}k)h + (f_{xy}h + f_{yy}k)k$$

$$= f_{xx}h^2 + 2f_{xy}hk + f_{yy}k^2 \qquad \text{(by Clairaut's Theorem)}$$

If we complete the square in this expression, we obtain

$$\boxed{1} \qquad D_{\mathbf{u}}^2 f = f_{xx}\left(h + \frac{f_{xy}}{f_{xx}}k\right)^2 + \frac{k^2}{f_{xx}}(f_{xx}f_{yy} - f_{xy}^2)$$

We are given that $f_{xx}(a, b) > 0$ and $D(a, b) > 0$. But f_{xx} and $D = f_{xx}f_{yy} - f_{xy}^2$ are continuous functions, so there is a disk B with center (a, b) and radius $\delta > 0$ such that $f_{xx}(x, y) > 0$ and $D(x, y) > 0$ whenever (x, y) is in B. Therefore, by looking at Equation 1, we see that $D_{\mathbf{u}}^2 f(x, y) > 0$ whenever (x, y) is in B. This means that if C is the curve obtained by intersecting the graph of f with the vertical plane through $P(a, b, f(a, b))$ in the direction of $\mathbf{u}$, then C is concave upward on an interval of length 2δ. This is true in the direction of every vector $\mathbf{u}$, so if we restrict (x, y) to lie in B, the graph of f lies above its horizontal tangent plane at P. Thus $f(x, y) \geq f(a, b)$ whenever (x, y) is in B. This shows that $f(a, b)$ is a local minimum.

□

E | ANSWERS TO ODD-NUMBERED EXERCISES

CHAPTER 1

EXERCISES 1.1 ■ PAGE 8

1. Yes
3. (a) 3   (b) −0.2   (c) 0, 3   (d) −0.8
(e) $[-2, 4], [-1, 3]$ (f) $[-2, 1]$
5. No   **7.** Yes, $[-3, 2], [-3, -2) \cup [-1, 3]$
9. Diet, exercise, or illness
11.

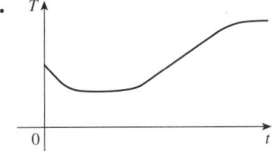

13.

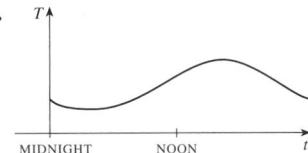

15.

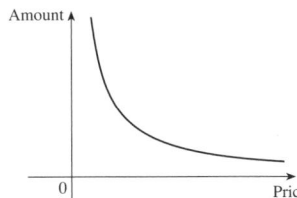

17.

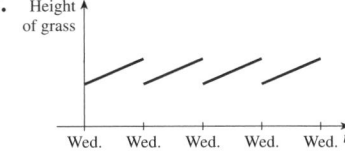

19. $12, 16, 3a^2 - a + 2, 3a^2 + a + 2, 3a^2 + 5a + 4,$
$6a^2 - 2a + 4, 12a^2 - 2a + 2, 3a^4 - a^2 + 2,$
$9a^4 - 6a^3 + 13a^2 - 4a + 4, 3a^2 + 6ah + 3h^2 - a - h + 2$
21. $-3 - h$ **23.** $-1/(ax)$
25. $(-\infty, -3) \cup (-3, 3) \cup (3, \infty)$ **27.** $[0, 4]$
29. $(-\infty, 0) \cup (5, \infty)$
31. $(-\infty, \infty)$

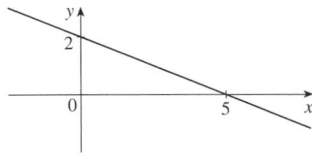

33. $(-\infty, \infty)$

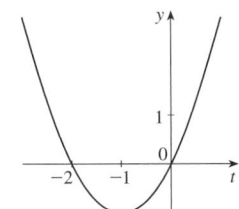

35. $[5, \infty)$

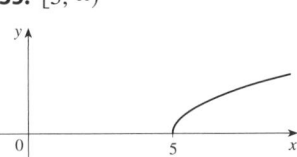

37. $(-\infty, 0) \cup (0, \infty)$

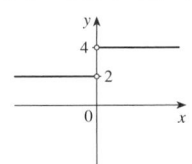

39. $(-\infty, \infty)$

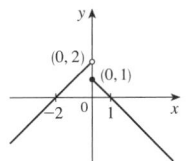

41. $(-\infty, \infty)$

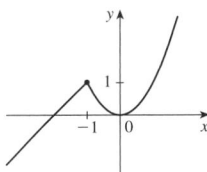

43. $f(x) = \frac{5}{2}x - \frac{11}{2}, 1 \le x \le 5$ **45.** $f(x) = 1 - \sqrt{-x}$
47. $A(L) = 10L - L^2, 0 < L < 10$
49. $A(x) = \sqrt{3}x^2/4, x > 0$ **51.** $S(x) = x^2 + (8/x), x > 0$
53. (a)

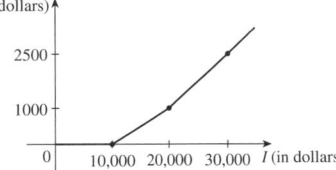

(b) $400, 1900
(c)

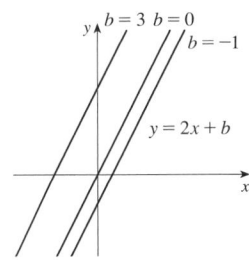

55. f is odd, g is even
57. (a) $(-5, 3)$ (b) $(-5, -3)$
59. Odd   **61.** Neither   **63.** Even
65. Even; odd; neither (unless $f = 0$ or $g = 0$)

EXERCISES 1.2 ■ PAGE 21

1. (a) $y = 2x + b$,
where b is the y-intercept

(b) $y = mx + 1 - 2m$, where m is the slope. See graph at right.
(c) $y = 2x - 3$

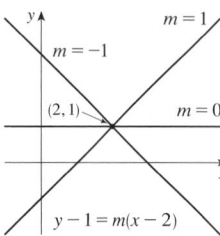

3. Their graphs have slope -1.

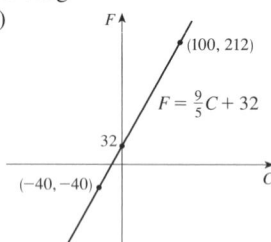

5. $f(x) = -3x(x + 1)(x - 2)$
7. (a) 8.34, change in mg for every 1 year change
(b) 8.34 mg
9. (a)

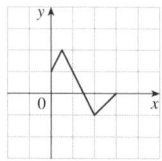

(b) $\frac{9}{5}$, change in °F for every 1°C change; 32, Fahrenheit temperature corresponding to 0°C

11. (a) $T = \frac{1}{6}N + \frac{307}{6}$ (b) $\frac{1}{6}$, change in °F for every chirp per minute change (c) 76°F
13. (a) $P = 0.434d + 15$ (b) 196 ft
15. Four times as bright
17. (a) $y = f(x) + 3$ (b) $y = f(x) - 3$ (c) $y = f(x - 3)$
(d) $y = f(x + 3)$ (e) $y = -f(x)$ (f) $y = f(-x)$
(g) $y = 3f(x)$ (h) $y = \frac{1}{3}f(x)$
19. (a) 3 (b) 1 (c) 4 (d) 5 (e) 2
21. (a)

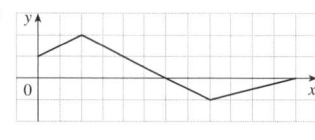

(c) (d)

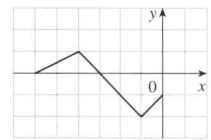

23.

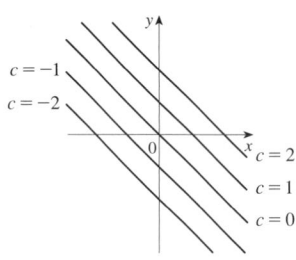

25.

27.

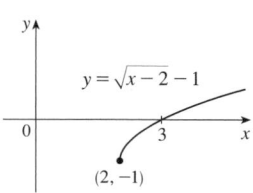

29.

31.

33. **35.**

37. (a) $(f + g)(x) = x^3 + 5x^2 - 1$, $(-\infty, \infty)$
(b) $(f - g)(x) = x^3 - x^2 + 1$, $(-\infty, \infty)$
(c) $(fg)(x) = 3x^5 + 6x^4 - x^3 - 2x^2$, $(-\infty, \infty)$
(d) $(f/g)(x) = (x^3 + 2x^2)/(3x^2 - 1)$, $\{x \mid x \neq \pm 1/\sqrt{3}\}$
39. (a) $(f \circ g)(x) = 4x^2 + 4x$, $(-\infty, \infty)$
(b) $(g \circ f)(x) = 2x^2 - 1$, $(-\infty, \infty)$
(c) $(f \circ f)(x) = x^4 - 2x^2$, $(-\infty, \infty)$
(d) $(g \circ g)(x) = 4x + 3$, $(-\infty, \infty)$
41. (a) $(f \circ g)(x) = 1 - 3 \cos x$, $(-\infty, \infty)$
(b) $(g \circ f)(x) = \cos(1 - 3x)$, $(-\infty, \infty)$
(c) $(f \circ f)(x) = 9x - 2$, $(-\infty, \infty)$
(d) $(g \circ g)(x) = \cos(\cos x)$, $(-\infty, \infty)$
43. (a) $(f \circ g)(x) = \dfrac{2x^2 + 6x + 5}{(x + 2)(x + 1)}$, $\{x \mid x \neq -2, -1\}$

(b) $(g \circ f)(x) = \dfrac{x^2 + x + 1}{(x + 1)^2}$, $\{x \mid x \neq -1, 0\}$

(c) $(f \circ f)(x) = \dfrac{x^4 + 3x^2 + 1}{x(x^2 + 1)}$, $\{x \mid x \neq 0\}$

(d) $(g \circ g)(x) = \dfrac{2x + 3}{3x + 5}$, $\{x \mid x \neq -2, -\frac{5}{3}\}$

45. $(f \circ g \circ h)(x) = \sqrt{x^6 + 4x^3 + 1}$
47. $g(x) = 2x + x^2$, $f(x) = x^4$
49. $g(t) = t^2$, $f(t) = \sec t \tan t$
51. $h(x) = \sqrt{x}$, $g(x) = x - 1$, $f(x) = \sqrt{x}$

53. $h(x) = \sqrt{x}$, $g(x) = \sec x$, $f(x) = x^4$
55. (a) 4 (b) 3 (c) 0 (d) Does not exist; $f(6) = 6$ is not in the domain of g. (e) 4 (f) -2
57. (a) $r(t) = 60t$ (b) $(A \circ r)(t) = 3600\pi t^2$; the area of the circle as a function of time
59. (a)

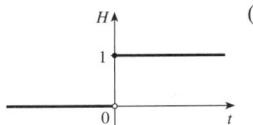

(b)

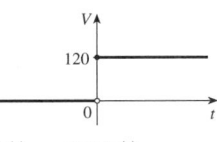

$$V(t) = 120H(t)$$

(c)

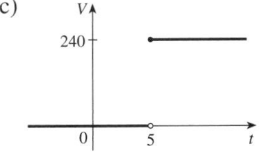

$$V(t) = 240H(t - 5)$$

61. Yes; $m_1 m_2$
63. (a) $f(x) = x^2 + 6$ (b) $g(x) = x^2 + x - 1$
65. Yes

EXERCISES 1.3 ■ PAGE 33

1. (a) (i) -32 ft/s (ii) -25.6 ft/s (iii) -24.8 ft/s
(iv) -24.16 ft/s (b) -24 ft/s
3. (a) 2 (b) 1 (c) 4 (d) Does not exist (e) 3
5. (a) -1 (b) -2 (c) Does not exist (d) 2 (e) 0
(f) Does not exist (g) 1 (h) 3
7.

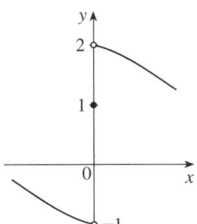

9.

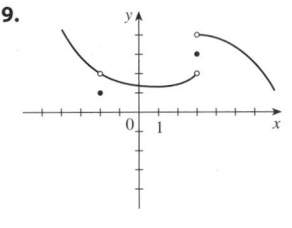

11. $\frac{2}{3}$ **13.** $\frac{1}{2}$ **15.** $\frac{1}{4}$ **17.** $\frac{3}{5}$ **19.** (a) -1.5
21. (a) 0.998000, 0.638259, 0.358484, 0.158680, 0.038851, 0.008928, 0.001465; 0
(b) 0.000572, -0.000614, -0.000907, -0.000978, -0.000993, -0.001000; -0.001
23. 1.44 (or any smaller positive number)
25. 0.0906 (or any smaller positive number)
27. (a) $\sqrt{1000/\pi}$ cm (b) Within approximately 0.0445 cm
(c) Radius; area; $\sqrt{1000/\pi}$; 1000; 5; ≈ 0.0445
45. (a) 0.093 (b) $\delta = (B^{2/3} - 12)/(6B^{1/3}) - 1$, where $B = 216 + 108\varepsilon + 12\sqrt{336 + 324\varepsilon + 81\varepsilon^2}$

EXERCISES 1.4 ■ PAGE 43

1. (a) -6 (b) -8 (c) 2 (d) -6
(e) Does not exist (f) 0
3. 105 **5.** $\frac{7}{8}$ **7.** 390 **9.** $\pi/2$ **11.** 4
13. Does not exist **15.** $\frac{6}{5}$ **17.** -10 **19.** $\frac{1}{12}$

21. $\frac{1}{6}$ **23.** $\frac{1}{128}$ **25.** $-\frac{1}{16}$ **27.** $3x^2$ **29.** (a), (b) $\frac{2}{3}$
33. 7 **37.** 6 **39.** -4 **41.** Does not exist
43. (a) (i) 5 (ii) -5 (b) Does not exist
(c)

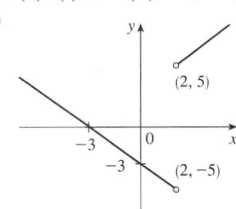

45. (a) (i) -2 (ii) Does not exist (iii) -3
(b) (i) $n - 1$ (ii) n (c) a is not an integer.
49. 3 **51.** 3 **53.** $-\frac{3}{4}$ **55.** $\frac{1}{2}$ **61.** 8
65. 15; -1

EXERCISES 1.5 ■ PAGE 54

1. $\lim_{x \to 4} f(x) = f(4)$
3. (a) $f(-4)$ is not defined and $\lim_{x \to a} f(x)$ [for $a = -2, 2,$ and 4] does not exist
(b) -4, neither; -2, left; 2, right; 4, right
5.

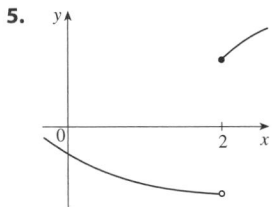

7.

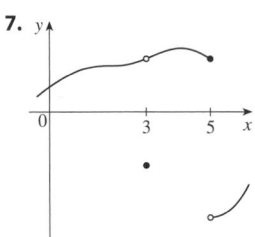

9. (a)

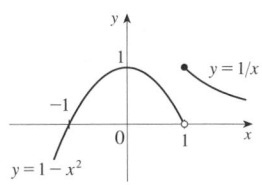

11. 4

15. $f(-2)$ is undefined.

17. $\lim_{x \to 1} f(x)$ does not exist

19. $(-\infty, \infty)$ **21.** $\left(-\infty, \sqrt[3]{2}\right) \cup \left(\sqrt[3]{2}, \infty\right)$
23. $(-\infty, -1] \cup (0, \infty)$

25. $x = (-\pi/2) + 2n\pi$, n an integer

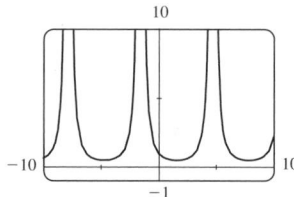

27. $\frac{7}{3}$ **31.** 0, right; 1, left

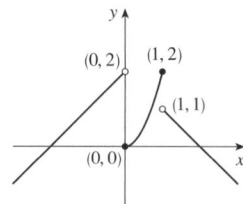

33. $\frac{2}{3}$

35. (a) $g(x) = x^3 + x^2 + x + 1$ (b) $g(x) = x^2 + x$

43. (b) $(0.86, 0.87)$

45. (b) 1.434   **47.** Yes

EXERCISES 1.6 ■ PAGE 67

1. (a) -2 (b) 2   (c) ∞ (d) $-\infty$
(e) $x = 1$, $x = 3$, $y = -2$, $y = 2$

3. 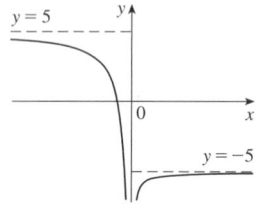   **5.**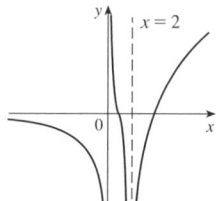

7.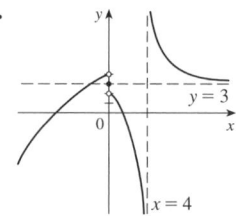

9. 0   **11.** $x \approx -1.62$, $x \approx 0.62$, $x = 1$; $y = 1$ **13.** $-\infty$
15. ∞ **17.** $-\infty$ **19.** $\frac{3}{2}$ **21.** -1 **23.** 4
25. $\frac{1}{6}$ **27.** ∞ **29.** Does not exist   **31.** ∞ **33.** $-\infty$
35. $y = 2$; $x = -2$, $x = 1$ **37.** (a), (b) $-\frac{1}{2}$
39. $y = 3$ **41.** $f(x) = \dfrac{2 - x}{x^2(x - 3)}$
43. (a) $\frac{5}{4}$ (b) 5
45. (a) 0   (b) $\pm\infty$
47. 4
49. (b) It approaches the concentration of the brine being pumped into the tank.
51. Within 0.1   **55.** $N \geqslant 15$ **57.** (a) $x > 100$

CHAPTER 1 REVIEW ■ PAGE 70

True-False Quiz

1. False   **3.** False   **5.** True   **7.** False   **9.** True
11. False   **13.** True   **15.** True   **17.** False
19. False   **21.** True   **23.** True   **25.** True

Exercises

1. (a) 2.7   (b) 2.3, 5.6   (c) $[-6, 6]$ (d) $[-4, 4]$
(e) $[-4, 4]$ (f) Odd; its graph is symmetric about the origin.
3. $\left(-\infty, \frac{1}{3}\right) \cup \left(\frac{1}{3}, \infty\right)$, $(-\infty, 0) \cup (0, \infty)$ **5.** $\mathbb{R}$, $[0, 2]$
7. (a) Shift the graph 8 units upward.
(b) Shift the graph 8 units to the left.
(c) Stretch the graph vertically by a factor of 2, then shift it 1 unit upward.
(d) Shift the graph 2 units to the right and 2 units downward.
(e) Reflect the graph about the x-axis.
(f) Reflect the graph about the x-axis, then shift it 3 units upward.

9.
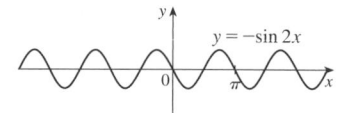

11.
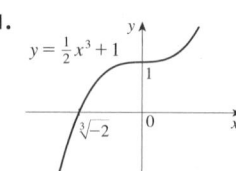

13.
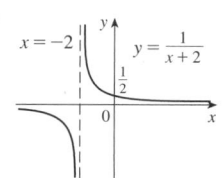

15. (a) Neither   (b) Odd   (c) Even   (d) Neither
17. (a) $(f \circ g)(x) = \sqrt{\sin x}$,
   $\{x \mid x \in [2n\pi, \pi + 2n\pi], n$ an integer$\}$
(b) $(g \circ f)(x) = \sin \sqrt{x}$, $[0, \infty)$ (c) $(f \circ f)(x) = \sqrt[4]{x}$, $[0, \infty)$
(d) $(g \circ g)(x) = \sin(\sin x)$, $\mathbb{R}$
19. (a) (i) 3   (ii) 0   (iii) Does not exist   (iv) 2
(v) ∞ (vi) $-\infty$ (vii) 4   (viii) -1
(b) $y = 4$, $y = -1$ (c) $x = 0$, $x = 2$ (d) $-3, 0, 2, 4$
21. 1   **23.** $\frac{3}{2}$ **25.** 3   **27.** ∞ **29.** $-\frac{1}{8}$ **31.** $-\frac{1}{2}$
33. 2   **35.** $\frac{1}{2}$ **37.** $x = 0$, $y = 0$ **39.** 1
45. (a) (i) 3   (ii) 0   (iii) Does not exist   (iv) 0   (v) 0
(vi) 0   (b) At 0 and 3   (c)

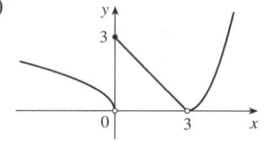

CHAPTER 2

EXERCISES 2.1 ▪ PAGE 80

1. (a) 2   (b) $y = 2x + 1$ (c)

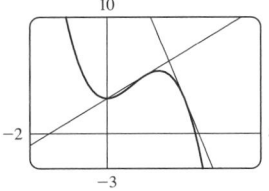

3. $y = -8x + 12$ **5.** $y = \frac{1}{2}x + \frac{1}{2}$

7. (a) $8a - 6a^2$ (b) $y = 2x + 3$, $y = -8x + 19$

(c)

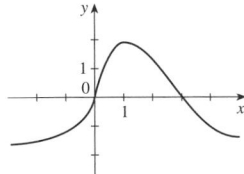

9. (a) 0   (b) C (c) Speeding up, slowing down, neither

(d) The car did not move.

11. -24 ft/s

13. $-2/a^3$ m/s; -2 m/s; $-\frac{1}{4}$ m/s; $-\frac{2}{27}$ m/s

15. $g'(0)$, 0, $g'(4)$, $g'(2)$, $g'(-2)$ **17.** $f(2) = 3$; $f'(2) = 4$

19.

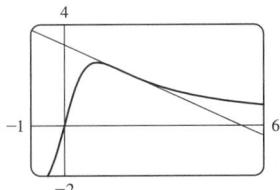

21. $y = 3x - 1$

23. (a) $-\frac{3}{5}$; $y = -\frac{3}{5}x + \frac{16}{5}$ (b)

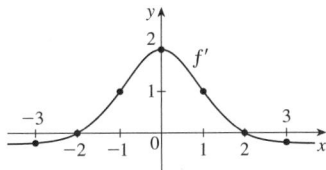

25. $6a - 4$ **27.** $\dfrac{5}{(a + 3)^2}$ **29.** $-\dfrac{1}{\sqrt{1 - 2a}}$

31. $f(x) = x^{10}$, $a = 1$ or $f(x) = (1 + x)^{10}$, $a = 0$

33. $f(x) = 2^x$, $a = 5$

35. $f(x) = \cos x$, $a = \pi$ or $f(x) = \cos(\pi + x)$, $a = 0$

37.   Greater (in magnitude)

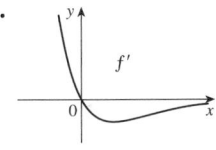

39. (a) (i) 23 million/year   (ii) 20.5 million/year

(iii) 16 million/year

(b) 18.25 million/year   (c) 17 million/year

41. (a) (i) $20.25/unit   (ii) $20.05/unit   (b) $20/unit

43. (a) The rate at which the cost is changing per ounce of gold produced; dollars per ounce

(b) When the 800th ounce of gold is produced, the cost of production is $17/oz.

(c) Decrease in the short term; increase in the long term

45. The rate at which the temperature is changing at 8:00 AM; 3.75°F/h

47. (a) The rate at which the oxygen solubility changes with respect to the water temperature; (mg/L)/°C

(b) $S'(16) \approx -0.25$; as the temperature increases past 16°C, the oxygen solubility is decreasing at a rate of 0.25 (mg/L)/°C.

49. Does not exist

EXERCISES 2.2 ▪ PAGE 92

1. (a) -0.2 (b) 0   (c) 1   (d) 2

(e) 1   (f) 0   (g) -0.2

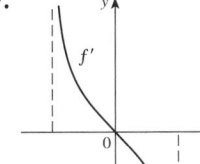

3. (a) II   (b) IV   (c) I   (d) III

5.   **7.**

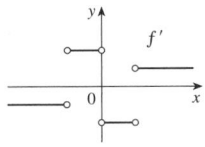

9.   **11.**

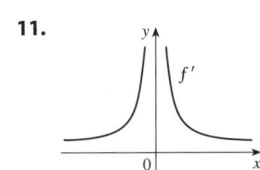

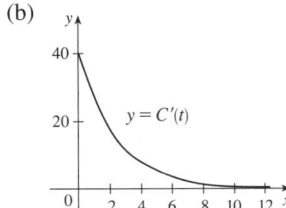

13. (a) The instantaneous rate of change of percentage of full capacity with respect to elapsed time in hours

(b)   The rate of change of percentage of full capacity is decreasing and approaching 0.

15.

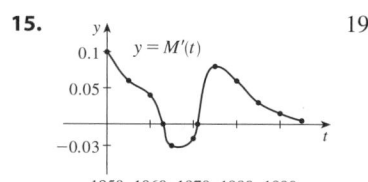

1963 to 1971

17. (a) 0, 1, 2, 4   (b) −1, −2, −4   (c) $f'(x) = 2x$
19. $f'(x) = \frac{1}{2}$, $\mathbb{R}$, $\mathbb{R}$ **21.** $f'(x) = 2x - 6x^2$, $\mathbb{R}$, $\mathbb{R}$

23. $g'(x) = -\dfrac{1}{2\sqrt{9-x}}$, $(-\infty, 9]$, $(-\infty, 9)$

25. $G'(t) = \dfrac{-7}{(3+t)^2}$, $(-\infty, -3) \cup (-3, \infty)$,
$(-\infty, -3) \cup (-3, \infty)$

27. $f'(x) = 4x^3$, $\mathbb{R}$, $\mathbb{R}$ **29.** (a) $f'(x) = 4x^3 + 2$

31. (a) The rate at which the unemployment rate is changing, in percent unemployed per year

(b)

t	$U'(t)$	t	$U'(t)$
1999	−0.2	2004	−0.45
2000	0.25	2005	−0.45
2001	0.9	2006	−0.25
2002	0.65	2007	0.6
2003	−0.15	2008	1.2

33. −4 (corner); 0 (discontinuity)
35. −1 (vertical tangent); 4 (corner)
37.

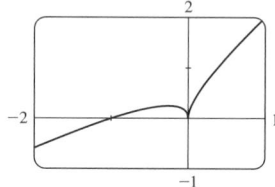

Differentiable at −1; not differentiable at 0

39. $a = f$, $b = f'$, $c = f''$
41. a = acceleration, b = velocity, c = position
43. $6x + 2$; 6

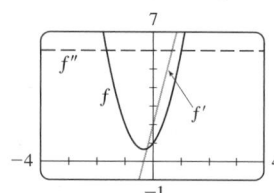

45. (a) $\frac{1}{3}a^{-2/3}$

47. $f'(x) = \begin{cases} -1 & \text{if } x < 6 \\ 1 & \text{if } x > 6 \end{cases}$

or $f'(x) = \dfrac{x-6}{|x-6|}$

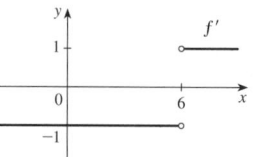

51. 63°

EXERCISES 2.3 ■ PAGE 105

1. $f'(x) = 0$ **3.** $f'(t) = -\frac{2}{3}$ **5.** $f'(x) = 3x^2 - 4$
7. $f'(x) = 6x + 2\sin x$ **9.** $g'(x) = 2x - 6x^2$
11. $g'(t) = -\frac{3}{2}t^{-7/4}$ **13.** $A'(s) = 60/s^6$
15. $R'(a) = 18a + 6$ **17.** $S'(p) = \frac{1}{2}p^{-1/2} - 1$
19. $y' = \frac{3}{2}\sqrt{x} + \dfrac{2}{\sqrt{x}} - \dfrac{3}{2x\sqrt{x}}$ **21.** $v' = 2t + \dfrac{3}{4t\sqrt[4]{t^3}}$
23. $z' = -10A/y^{11} - B\sin y$
25. $H'(x) = 3x^2 + 3 - 3x^{-2} - 3x^{-4}$
27. $y = -3\sqrt{3}x + 3 + \pi\sqrt{3}$, $y = \dfrac{x}{3\sqrt{3}} + 3 - \dfrac{\pi}{9\sqrt{3}}$
29. $y = 3x - 1$
31. $f'(x) = 4x^3 - 9x^2 + 16$, $f''(x) = 12x^2 - 18x$
33. $g'(t) = -2\sin t - 3\cos t$, $g''(t) = -2\cos t + 3\sin t$
35. $-\cos x$ **37.** $(2n+1)\pi \pm \frac{1}{3}\pi$, n an integer
41. $y = \frac{1}{3}x - \frac{1}{3}$
43. (a) $v(t) = 3t^2 - 3$, $a(t) = 6t$ (b) 12 m/s²
(c) $a(1) = 6$ m/s²
45. (a) $3t^2 - 24t + 36$
(b) −9 ft/s   (c) $t = 2, 6$
(d) $0 \leqslant t < 2$, $t > 6$
(e) 96 ft
(f) See graph at right.
(g) $6t - 24$; −6 ft/s²
(h) 40

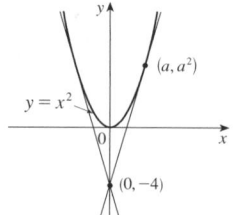

47. (a) $t = 4$ s
(b) $t = 1.5$ s; the velocity has a minimum.
49. (a) 7.56 m/s   (b) 6.24 m/s; −6.24 m/s
51. (a) $C'(x) = 12 - 0.2x + 0.0015x^2$
(b) \$32/yard; the cost of producing the 201st yard
(c) \$32.20
53. (a) 8π ft²/ft   (b) 16π ft²/ft   (c) 24π ft²/ft
The rate increases as the radius increases.
55. (a) $V = 5.3/P$
(b) −0.00212; instantaneous rate of change of the volume with respect to the pressure at 25°C; m³/kPa
59. $A = -\frac{3}{10}$, $B = -\frac{1}{10}$
61. $(\pm 2, 4)$

63. $a = -\frac{1}{2}$, $b = 2$ **65.** $y = \frac{3}{16}x^3 - \frac{9}{4}x + 3$
67. $y = 2x^2 - x$ **69.** 1000   **71.** 3; 1

EXERCISES 2.4 ▪ PAGE 112

1. $1 - 2x + 6x^2 - 8x^3$ **3.** $g'(t) = 3t^2 \cos t - t^3 \sin t$
5. $F'(y) = 5 + 14/y^2 + 9/y^4$ **7.** $f'(x) = \cos x - \frac{1}{2}\csc^2 x$
9. $h'(\theta) = \csc \theta - \theta \csc \theta \cot \theta + \csc^2 \theta$
11. $g'(x) = \dfrac{10}{(3 - 4x)^2}$ **13.** $y' = \dfrac{x^2(3 - x^2)}{(1 - x^2)^2}$
15. $y' = 2v - 1/\sqrt{v}$ **17.** $f'(t) = \dfrac{4 + t^{1/2}}{(2 + \sqrt{t}\,)^2}$
19. $y' = \dfrac{2 - \tan x + x \sec^2 x}{(2 - \tan x)^2}$ **21.** $f'(\theta) = \dfrac{\sec \theta \tan \theta}{(1 + \sec \theta)^2}$
23. $y' = \dfrac{(t^2 + t)\cos t + \sin t}{(1 + t)^2}$ **25.** $f'(x) = 2cx/(x^2 + c)^2$
27. $y = \frac{2}{3}x - \frac{2}{3}$ **29.** $y = x - \pi - 1$
31. (a) $y = \frac{1}{2}x + 1$ (b)

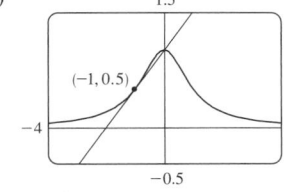

33. $\frac{1}{4}$ **35.** $\theta \cos \theta + \sin \theta$; $2 \cos \theta - \theta \sin \theta$
41. (a) -16 (b) $-\frac{20}{9}$ (c) 20 **43.** (a) 0 (b) $-\frac{2}{3}$
45. (a) $y' = xg'(x) + g(x)$ (b) $y' = [g(x) - xg'(x)]/[g(x)]^2$
(c) $y' = [xg'(x) - g(x)]/x^2$
47. Two, $\left(-2 \pm \sqrt{3}, \frac{1}{2}(1 \mp \sqrt{3})\right)$ **49.** 1
51. (a) $v(t) = 8 \cos t$, $a(t) = -8 \sin t$
(b) $4\sqrt{3}, -4, -4\sqrt{3}$; to the left; speeding up
53. -0.2436 K/min
55. (b) $y' = \sin x \cos x + x \cos^2 x - x \sin^2 x$
57. (b) $y' = -2x(2x^2 + 1)/(x^4 + x^2 + 1)^2$

EXERCISES 2.5 ▪ PAGE 120

1. $\dfrac{4}{3\sqrt[3]{(1 + 4x)^2}}$ **3.** $\pi \sec^2 \pi x$ **5.** $\dfrac{\cos x}{2\sqrt{\sin x}}$
7. $F'(x) = 10x(x^4 + 3x^2 - 2)^4(2x^2 + 3)$
9. $F'(x) = -\dfrac{1}{\sqrt{1 - 2x}}$ **11.** $f'(z) = -\dfrac{2z}{(z^2 + 1)^2}$
13. $y' = -3x^2 \sin(a^3 + x^3)$ **15.** $y' = \sec kx(kx \tan kx + 1)$
17. $f'(x) = (2x - 3)^3(x^2 + x + 1)^4(28x^2 - 12x - 7)$
19. $h'(t) = \frac{2}{3}(t + 1)^{-1/3}(2t^2 - 1)^2(20t^2 + 18t - 1)$
21. $y' = \dfrac{-12x(x^2 + 1)^2}{(x^2 - 1)^4}$
23. $y' = (\cos x - x \sin x)\cos(x \cos x)$
25. $y' = (r^2 + 1)^{-3/2}$
27. $y' = \left(x \cos\sqrt{1 + x^2}\right)/\sqrt{1 + x^2}$
29. $y' = 2 \cos(\tan 2x) \sec^2(2x)$
31. $y' = 4 \sec^2 x \tan x$
33. $y' = \dfrac{16 \sin 2x(1 - \cos 2x)^3}{(1 + \cos 2x)^5}$
35. $y' = -2 \cos \theta \cot(\sin \theta) \csc^2(\sin \theta)$

37. $y' = 3[x^2 + (1 - 3x)^5]^2[2x - 15(1 - 3x)^4]$
39. $g'(x) = p(2r \sin rx + n)^{p-1}(2r^2 \cos rx)$
41. $y' = \dfrac{-\pi \cos(\tan \pi x)\sec^2(\pi x)\sin\sqrt{\sin(\tan \pi x)}}{2\sqrt{\sin(\tan \pi x)}}$
43. $y' = -2x \sin(x^2)$; $y'' = -4x^2\cos(x^2) - 2 \sin(x^2)$
45. $H'(t) = 3 \sec^2 3t$, $H''(t) = 18 \sec^2 3t \tan 3t$
47. $y = -x + \pi$
49. (a) $y = \pi x - \pi + 1$ (b)

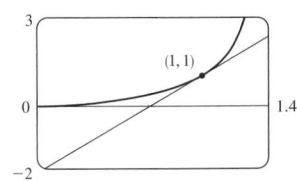

51. $((\pi/2) + 2n\pi, 3), ((3\pi/2) + 2n\pi, -1)$, n an integer
53. 24 **55.** (a) 30 (b) 36
57. (a) $\frac{3}{4}$ (b) Does not exist   (c) -2
59. $-\frac{1}{6}\sqrt{2}$ **61.** 120 **63.** $-2^{50} \cos 2x$
65. (a) $dB/dt = \frac{7}{54}\pi \cos(2\pi t/5.4)$ (b) 0.16
69. dv/dt is the rate of change of velocity with respect to time; dv/ds is the rate of change of velocity with respect to displacement
71. (b) $-n \cos^{n-1}x \sin[(n + 1)x]$ **75.** 96

EXERCISES 2.6 ▪ PAGE 127

1. (a) $y' = 9x/y$ (b) $y = \pm\sqrt{9x^2 - 1}$, $y' = \pm 9x/\sqrt{9x^2 - 1}$
3. $y' = -\dfrac{x^2}{y^2}$ **5.** $y' = \dfrac{2x + y}{2y - x}$ **7.** $y' = \dfrac{2x + y \sin x}{\cos x - 2y}$
9. $y' = \tan x \tan y$ **11.** $y' = \dfrac{y \sec^2(x/y) - y^2}{y^2 + x \sec^2(x/y)}$
13. $y' = \dfrac{4xy\sqrt{xy} - y}{x - 2x^2\sqrt{xy}}$ **15.** $y' = \dfrac{y \sin x + y \cos(xy)}{\cos x - x \cos(xy)}$
17. $-\frac{16}{13}$ **19.** $y = -x + 2$ **21.** $y = x + \frac{1}{2}$
23. $y = -\frac{9}{13}x + \frac{40}{13}$ **25.** $-81/y^3$ **27.** $-2x/y^5$
29. (a) $y = \frac{9}{2}x - \frac{5}{2}$ (b)

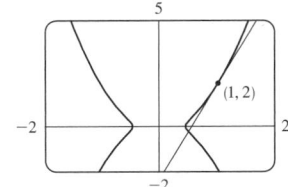

31. (a)

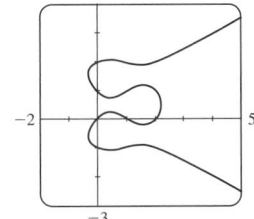

Eight; $x \approx 0.42, 1.58$

(b) $y = -x + 1$, $y = \frac{1}{3}x + 2$ (c) $1 \mp \sqrt{3}/3$
33. $\left(\pm\frac{5}{4}\sqrt{3}, \pm\frac{5}{4}\right)$

35. 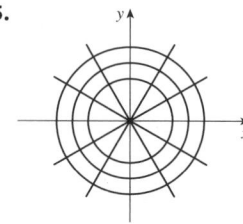   **37.**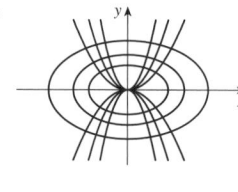

39. (a) $\dfrac{V^3(nb - V)}{PV^3 - n^2aV + 2n^3ab}$ (b) -4.04 L/atm

43. $\left(\pm\sqrt{3}, 0\right)$ **45.** $(-1, -1), (1, 1)$ **49.** (a) 0 (b) $-\frac{1}{2}$

EXERCISES 2.7 ▪ PAGE 132

1. $dV/dt = 3x^2\, dx/dt$ **3.** 48 cm²/s   **5.** $3/(25\pi)$ m/min

7. (a) 1 (b) 25 **9.** -18

11. (a) The rate of decrease of the surface area is 1 cm²/min.
(b) The rate of decrease of the diameter when the diameter is
10 cm
(c) 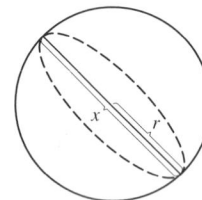   (d) $S = \pi x^2$
(e) $1/(20\pi)$ cm/min

13. (a) The plane's altitude is 1 mi and its speed is 500 mi/h.
(b) The rate at which the distance from the plane to the station
is increasing when the plane is 2 mi from the station
(c) 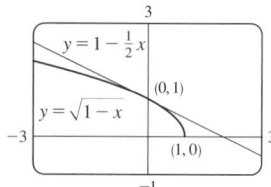   (d) $y^2 = x^2 + 1$
(e) $250\sqrt{3}$ mi/h

15. 65 mi/h   **17.** $837/\sqrt{8674} \approx 8.99$ ft/s

19. -1.6 cm/min   **21.** $\frac{720}{13} \approx 55.4$ km/h

23. $10/\sqrt{133} \approx 0.87$ ft/s   **25.** $\frac{4}{5}$ ft/min

27. $6/(5\pi) \approx 0.38$ ft/min   **29.** 0.3 m²/s

31. 5 m   **33.** 80 cm³/min   **35.** $\frac{107}{810} \approx 0.132$ Ω/s

37. (a) 360 ft/s   (b) 0.096 rad/s

39. $1650/\sqrt{31} \approx 296$ km/h   **41.** $\frac{7}{4}\sqrt{15} \approx 6.78$ m/s

EXERCISES 2.8 ▪ PAGE 138

1. $L(x) = -10x - 6$ **3.** $L(x) = \frac{1}{4}x + 1$

5. $\sqrt{1 - x} \approx 1 - \frac{1}{2}x$;
$\sqrt{0.9} \approx 0.95$,
$\sqrt{0.99} \approx 0.995$
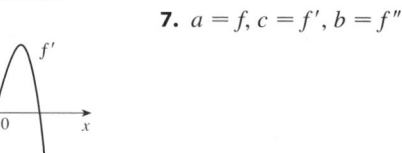

7. $-0.368 < x < 0.677$ **9.** $-0.045 < x < 0.055$

11. 15.968 **13.** 4.02

17. (a) $dy = \dfrac{\sec^2\sqrt{t}}{2\sqrt{t}}\, dt$ (b) $dy = \dfrac{-4v}{(1 + v^2)^2}\, dv$

19. (a) $dy = \sec^2 x\, dx$ (b) $dy = -0.2,\ \Delta y = -0.18237$

21. (a) 270 cm³, 0.01, 1% (b) 36 cm², 0.006, 0.6%

23. (a) $84/\pi \approx 27$ cm²; $\frac{1}{84} \approx 0.012$
(b) $1764/\pi^2 \approx 179$ cm³; $\frac{1}{56} \approx 0.018$

27. A 5% increase in the radius corresponds to a 20% increase
in blood flow.

29. (a) $4.8, 5.2$ (b) Too large

CHAPTER 2 REVIEW ▪ PAGE 140

True-False Quiz

1. False   **3.** False   **5.** True   **7.** False

9. True   **11.** False

Exercises

1. $f''(5), 0, f'(5), f'(2), 1, f'(3)$

3. (a) The rate at which the cost changes with respect to the
interest rate; dollars/(percent per year)
(b) As the interest rate increases past 10%, the cost is increasing
at a rate of $\$1200$/(percent per year).
(c) Always positive

5.   **7.** $a = f, c = f', b = f''$

9. The rate at which the total value of US currency in
circulation is changing in billions of dollars per year;
$\$22.2$ billion/year

11. $f'(x) = 3x^2 + 5$ **13.** $4x^7(x + 1)^3(3x + 2)$

15. $\frac{3}{2}\sqrt{x} - \dfrac{1}{2\sqrt{x}} - \dfrac{1}{\sqrt{x^3}}$ **17.** $x(\pi x \cos \pi x + 2 \sin \pi x)$

19. $\dfrac{8t^3}{(t^4 + 1)^2}$ **21.** $-\dfrac{\sec^2\sqrt{1 - x}}{2\sqrt{1 - x}}$ **23.** $\dfrac{1 - y^4 - 2xy}{4xy^3 + x^2 - 3}$

25. $\dfrac{2 \sec 2\theta\, (\tan 2\theta - 1)}{(1 + \tan 2\theta)^2}$ **27.** $-(x - 1)^{-2}$

29. $\dfrac{2x - y \cos(xy)}{x \cos(xy) + 1}$ **31.** $-6x \csc^2(3x^2 + 5)$

33. $\dfrac{\cos\sqrt{x} - \sqrt{x} \sin\sqrt{x}}{2\sqrt{x}}$ **35.** $2 \cos\theta \tan(\sin\theta) \sec^2(\sin\theta)$

37. $\frac{1}{5}(x \tan x)^{-4/5}(\tan x + x \sec^2 x)$

39. $\cos\left(\tan\sqrt{1 + x^3}\right)\left(\sec^2\sqrt{1 + x^3}\right)\dfrac{3x^2}{2\sqrt{1 + x^3}}$

41. $-\frac{4}{27}$

43. $-5x^4/y^{11}$ **45.** $y = 2\sqrt{3}x + 1 - \pi\sqrt{3}/3$

47. $y = 2x + 1$; $y = -\frac{1}{2}x + 1$

49. $\left(\pi/4, \sqrt{2}\right), \left(5\pi/4, -\sqrt{2}\right)$

51. (a) 2   (b) 44   **53.** $f'(x) = 2xg(x) + x^2g'(x)$
55. $f'(x) = 2g(x)g'(x)$ **57.** $f'(x) = g'(g(x))g'(x)$
59. $f'(x) = g'(\sin x) \cdot \cos x$
61. $h'(x) = \dfrac{f'(x)[g(x)]^2 + g'(x)[f(x)]^2}{[f(x) + g(x)]^2}$
63. -4 (discontinuity), -1 (corner), 2 (discontinuity), 5 (vertical tangent)
65. (a) $v(t) = 3t^2 - 12$, $a(t) = 6t$
(b) Upward when $t > 2$, downward when $0 \le t < 2$ (c) 23
(d)

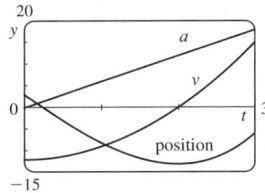

67. $\frac{4}{3}$ cm²/min   **69.** 13 ft/s   **71.** 400 ft/h
73. (a) $L(x) = 1 + x$; $\sqrt[3]{1 + 3x} \approx 1 + x$; $\sqrt[3]{1.03} \approx 1.01$
(b) $-0.23 < x < 0.40$
75. $12 + 3\pi/2 \approx 16.7$ cm²   **77.** $\frac{1}{32}$ **79.** $\frac{1}{4}$ **81.** $3\sqrt{2}$

CHAPTER 3

EXERCISES 3.1 ■ **PAGE 150**

1. (a) $f(x) = a^x, a > 0$ (b) $\mathbb{R}$ (c) $(0, \infty)$
(d) See Figures 6(c), 6(b), and 6(a), respectively.
3.

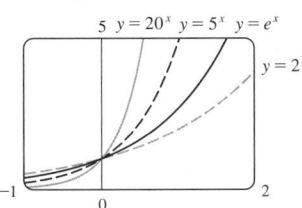

All approach 0 as $x \to -\infty$, all pass through $(0, 1)$, and all are increasing. The larger the base, the faster the rate of increase.

5.

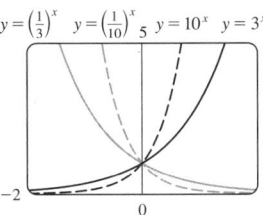

The functions with base greater than 1 are increasing and those with base less than 1 are decreasing. The latter are reflections of the former about the y-axis.

7.

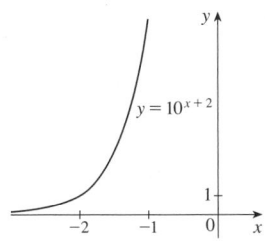

9.

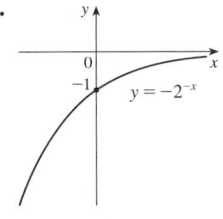

11.

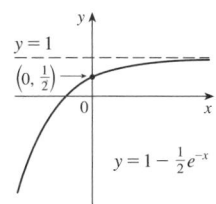

13. (a) $y = e^x - 2$ (b) $y = e^{x-2}$ (c) $y = -e^x$
(d) $y = e^{-x}$ (e) $y = -e^{-x}$
15. (a) $(-\infty, -1) \cup (-1, 1) \cup (1, \infty)$ (b) $(-\infty, \infty)$
17. $f(x) = 3 \cdot 2^x$ **21.** At $x \approx 35.8$
23. ∞ **25.** 1   **27.** 0   **29.** 0

EXERCISES 3.2 ■ **PAGE 161**

1. (a) See Definition 1.
(b) It must pass the Horizontal Line Test.
3. No   **5.** No   **7.** Yes   **9.** No   **11.** Yes   **13.** No
15. (a) 6   (b) 3   **17.** 0
19. $F = \frac{9}{5}C + 32$; the Fahrenheit temperature as a function of the Celsius temperature; $[-273.15, \infty)$
21. $y = \frac{1}{3}(x - 1)^2 - \frac{2}{3}, x \ge 1$
23. $y = \frac{1}{2}(1 + \ln x)$ **25.** $y = e^x - 3$
27. $f^{-1}(x) = \sqrt[4]{x - 1}$ **29.**

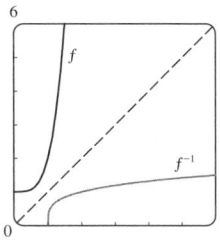
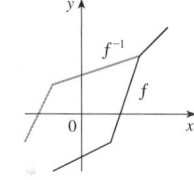

31. (b) $\frac{1}{12}$
(c) $f^{-1}(x) = \sqrt[3]{x}$,
domain $= \mathbb{R} =$ range
(e)

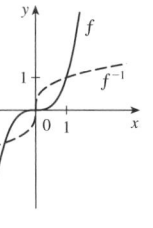

33. (b) $-\frac{1}{2}$
(c) $f^{-1}(x) = \sqrt{9 - x}$,
domain $= [0, 9]$,
range $= [0, 3]$
(e)

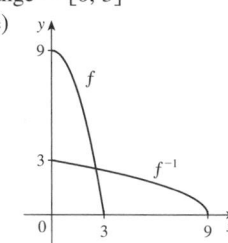

35. $\frac{1}{7}$ **37.** $2/\pi$ **39.** $\frac{3}{2}$
41. (a) It's defined as the inverse of the exponential function with base a, that is, $\log_a x = y \Longleftrightarrow a^y = x$.
(b) $(0, \infty)$ (c) $\mathbb{R}$ (d) See Figure 13.
43. (a) 3   (b) -3 **45.** (a) 3   (b) -2
47. $\frac{1}{2}\ln a + \frac{1}{2}\ln b$ **49.** $2 \ln x - 3 \ln y - 4 \ln z$
51. $\ln 1215$ **53.** $\ln \dfrac{\sqrt{x}}{x + 1}$

55.

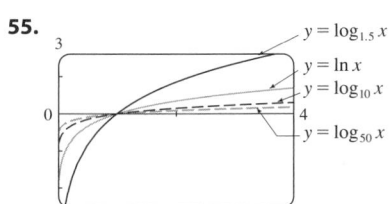

All graphs approach $-\infty$ as $x \to 0^+$, all pass through $(1, 0)$, and all are increasing. The larger the base, the slower the rate of increase.

57. About 1,084,588 mi

59. (a)

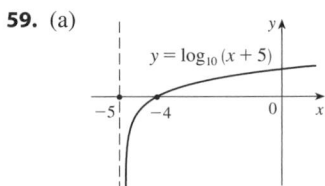

(b)

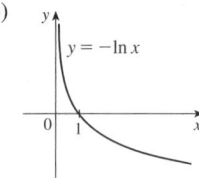

61. (a) $(0, \infty)$; $(-\infty, \infty)$ (b) e^{-2}

(c)

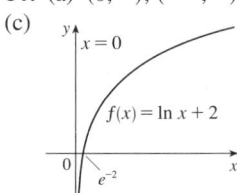

63. (a) $\frac{1}{4}(7 - \ln 6)$ (b) $\frac{1}{3}(e^2 + 10)$

65. (a) $5 + \log_2 3$ or $5 + (\ln 3)/\ln 2$ (b) $\frac{1}{2}\left(1 + \sqrt{1 + 4e}\right)$

67. (a) $0 < x < 1$ (b) $x > \ln 5$

69. (a) $(\ln 3, \infty)$ (b) $f^{-1}(x) = \ln(e^x + 3)$; $\mathbb{R}$

71. $-\infty$ **73.** 0   **75.** ∞

77.

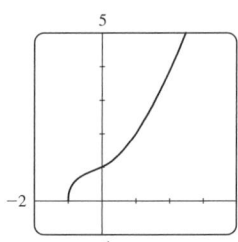

The graph passes the Horizontal Line Test.

$f^{-1}(x) = -\dfrac{\sqrt[3]{4}}{6}\left(\sqrt[3]{D - 27x^2 + 20} - \sqrt[3]{D + 27x^2 - 20} + \sqrt[3]{2}\right)$, where $D = 3\sqrt{3}\sqrt{27x^4 - 40x^2 + 16}$; two of the expressions are complex.

EXERCISES 3.3 ▪ PAGE 169

1. $f'(x) = \dfrac{3x^2}{(x^3 + 1)\ln 10}$ **3.** $f'(x) = \dfrac{\cos(\ln x)}{x}$

5. $f'(x) = -\dfrac{1}{x}$ **7.** $f'(x) = \dfrac{\sin x}{x} + \cos x \ln(5x)$

9. $g'(x) = -\dfrac{2a}{a^2 - x^2}$ **11.** $g'(x) = \dfrac{2x^2 - 1}{x(x^2 - 1)}$

13. $G'(y) = \dfrac{10}{2y + 1} - \dfrac{y}{y^2 + 1}$ **15.** $F'(s) = \dfrac{1}{s \ln s}$

17. $y' = \sec^2(\ln(ax + b))\dfrac{a}{ax + b}$

19. $f'(x) = e^x(x^3 + 3x^2 + 2x + 2)$ **21.** $y' = \dfrac{1 - x}{e^x}$

23. $y' = \dfrac{3e^{3x}}{\sqrt{1 + 2e^{3x}}}$ **25.** $y' = 5^{-1/x}(\ln 5)/x^2$

27. $F'(t) = e^{t \sin 2t}(2t \cos 2t + \sin 2t)$ **29.** $y' = \dfrac{10x + 1}{5x^2 + x - 2}$

31. $f'(t) = \sec^2(e^t)e^t + e^{\tan t}\sec^2 t$ **33.** $y' = -x/(1 + x)$

35. $y' = \dfrac{1}{\ln 10} + \log_{10} x$

37. $f'(t) = 4 \sin(e^{\sin^2 t}) \cos(e^{\sin^2 t}) e^{\sin^2 t} \sin t \cos t$

39. $g'(x) = 2r^2 p(\ln a)(2ra^{rx} + n)^{p-1}a^{rx}$

41. $y' = e^{\alpha x}(\beta \cos \beta x + \alpha \sin \beta x)$; $y'' = e^{\alpha x}[(\alpha^2 - \beta^2) \sin \beta x + 2\alpha\beta \cos \beta x]$

43. $y' = 1 + \ln x$, $y'' = 1/x$ **45.** $y = 3x - 9$

47. $f'(x) = \dfrac{2x - 1 - (x - 1)\ln(x - 1)}{(x - 1)[1 - \ln(x - 1)]^2}$; $(1, 1 + e) \cup (1 + e, \infty)$

49. 7   **51.** $y' = (x^2 + 2)^2(x^4 + 4)^4\left(\dfrac{4x}{x^2 + 2} + \dfrac{16x^3}{x^4 + 4}\right)$

53. $y' = \sqrt{\dfrac{x - 1}{x^4 + 1}}\left(\dfrac{1}{2x - 2} - \dfrac{2x^3}{x^4 + 1}\right)$

55. $y' = x^x(1 + \ln x)$

57. $y' = (\cos x)^x(-x \tan x + \ln \cos x)$

59. $y' = (\tan x)^{1/x}\left(\dfrac{\sec^2 x}{x \tan x} - \dfrac{\ln \tan x}{x^2}\right)$

61. $y' = \dfrac{1 - 2xye^{x^2 y}}{x^2 e^{x^2 y} - 1}$ **63.** $y' = \dfrac{2x}{x^2 + y^2 - 2y}$

65. $v(t) = 2e^{-1.5t}(2\pi \cos 2\pi t - 1.5 \sin 2\pi t)$

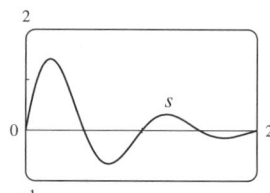 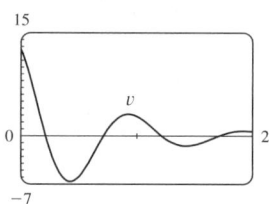

69. $f^{(n)}(x) = 2^n e^{2x}$ **71.** $f^{(n)}(x) = (-1)^{n-1}(n - 1)!/(x - 1)^n$

73. $\frac{1}{2}$

EXERCISES 3.4 ▪ PAGE 177

1. About 235

3. (a) $100(4.2)^t$ (b) ≈ 7409 (c) $\approx 10{,}632$ bacteria/h (d) $(\ln 100)/(\ln 4.2) \approx 3.2$ h

5. (a) 1508 million, 1871 million   (b) 2161 million (c) 3972 million; wars in the first half of century, increased life expectancy in second half

7. (a) $Ce^{-0.0005t}$ (b) $-2000 \ln 0.9 \approx 211$ s

9. (a) $100 \times 2^{-t/30}$ mg   (b) ≈ 9.92 mg   (c) ≈ 199.3 years

11. ≈ 2500 years   **13.** (a) $\approx 137°$F   (b) ≈ 116 min

15. (a) $13.3°$C   (b) ≈ 67.74 min

17. (a) ≈ 64.5 kPa   (b) ≈ 39.9 kPa

19. (a) \$3828.84 (b) \$3840.25 (c) \$3850.08
(d) \$3851.61 (e) \$3852.01 (f) \$3852.08

EXERCISES 3.5 ▪ PAGE 183

1. (a) $\pi/3$ (b) π **3.** (a) $\pi/4$ (b) $\pi/4$
5. (a) 10 (b) $\pi/3$ **9.** $x/\sqrt{1+x^2}$

17. $y' = \dfrac{2\tan^{-1}x}{1+x^2}$ **19.** $y' = \dfrac{1}{\sqrt{-x^2-x}}$

21. $G'(x) = -1 - \dfrac{x\arccos x}{\sqrt{1-x^2}}$ **23.** $h'(t) = 0$

25. $y' = -\dfrac{\sin\theta}{1+\cos^2\theta}$ **27.** $y' = \sin^{-1}x$

29. $y' = \dfrac{\sqrt{a^2-b^2}}{a+b\cos x}$

31. $g'(x) = \dfrac{2}{\sqrt{1-(3-2x)^2}}$; $[1,2]$, $(1,2)$

33. $\pi/6$ **35.** $-\pi/2$ **37.** $\pi/2$ **39.** $\frac{1}{4}$ rad/s

EXERCISES 3.6 ▪ PAGE 189

1. (a) 0 (b) 1 **3.** (a) $\frac{3}{4}$ (b) $\frac{1}{2}(e^2 - e^{-2}) \approx 3.62686$
5. (a) 1 (b) 0
17. $\operatorname{sech} x = \frac{3}{5}$, $\sinh x = \frac{4}{3}$, $\operatorname{csch} x = \frac{3}{4}$, $\tanh x = \frac{4}{5}$, $\coth x = \frac{5}{4}$
19. (a) 1 (b) -1 (c) ∞ (d) $-\infty$ (e) 0 (f) 1
(g) ∞ (h) $-\infty$ (i) 0
27. $f'(x) = x\cosh x$ **29.** $h'(x) = \tanh x$
31. $y' = 3e^{\cosh 3x}\sinh 3x$
33. $f'(t) = -2e^t\operatorname{sech}^2(e^t)\tanh(e^t)$

35. $G'(x) = \dfrac{-2\sinh x}{(1+\cosh x)^2}$ **37.** $y' = \dfrac{1}{2\sqrt{x(x-1)}}$

39. $y' = \sinh^{-1}(x/3)$ **41.** $y' = -\csc x$
47. (a) 0.3572 (b) 70.34°
49. (a) 164.50 m (b) 120 m; 164.13 m
51. (b) $y = 2\sinh 3x - 4\cosh 3x$
53. $\left(\ln(1+\sqrt{2}),\ \sqrt{2}\right)$

EXERCISES 3.7 ▪ PAGE 197

1. 2 **3.** $-\infty$ **5.** 2 **7.** $\frac{1}{4}$ **9.** $-\infty$ **11.** $\frac{8}{5}$
13. $\frac{1}{2}$ **15.** $1/\ln 3$ **17.** $-1/\pi^2$ **19.** $\frac{1}{2}a(a-1)$
21. $\frac{1}{24}$ **23.** 3 **25.** 0 **27.** $-2/\pi$ **29.** $\frac{1}{2}$ **31.** ∞
33. 1 **35.** e^{-2} **37.** $1/e$ **39.** e^2 **43.** 1 **49.** $\frac{16}{9}a$
51. $\frac{1}{2}$ **53.** 56 **57.** (a) 0

CHAPTER 3 REVIEW ▪ PAGE 199

True-False Quiz

1. True **3.** False **5.** True **7.** True **9.** False
11. False **13.** False **15.** True

Exercises

1. No **3.** (a) 7 (b) $\frac{1}{8}$

5.

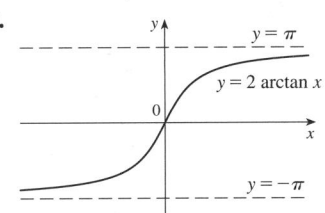

7.

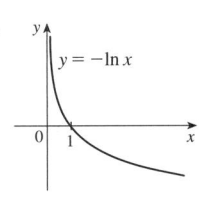

9.

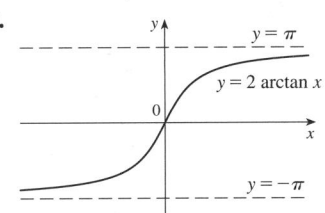

11. (a) 9 (b) 2 **13.** (a) $\ln 5$ (b) e^2

15. (a) $\sqrt{1+e}$ (b) $\dfrac{d\ln 5}{\ln c}$

17. $y' = \dfrac{1+\ln x}{x\ln x}$ **19.** $y' = -\dfrac{e^{1/x}(1+2x)}{x^4}$

21. $y' = \dfrac{1}{2\sqrt{\arctan x}\,(1+x^2)}$ **23.** $f'(t) = t + 2t\ln t$

25. $y' = 3^{x\ln x}(\ln 3)(1+\ln x)$
27. $y' = 2x^2\cosh(x^2) + \sinh(x^2)$
29. $h'(\theta) = 2\sec^2(2\theta)e^{\tan 2\theta}$ **31.** $y' = \cot x - \sin x\cos x$

33. $y' = \dfrac{2}{(1+2x)\ln 5}$

35. $y' = \dfrac{(x-2)^4(3x^2 - 55x - 52)}{2\sqrt{x+1}\,(x+3)^8}$

37. $y' = \dfrac{4x}{1+16x^2} + \tan^{-1}(4x)$ **39.** $y' = 3\tanh 3x$

41. $y' = \dfrac{\cosh x}{\sqrt{\sinh^2 x - 1}}$

43. $y' = \dfrac{-3\sin\!\left(e^{\sqrt{\tan 3x}}\right)e^{\sqrt{\tan 3x}}\sec^2(3x)}{2\sqrt{\tan 3x}}$

45. $f'(x) = g'(x)e^{g(x)}$ **47.** $f'(x) = g'(x)/g(x)$
49. $2^x(\ln 2)^n$ **53.** $(-3, 0)$
55. (a) $y = \frac{1}{4}x + \frac{1}{4}(\ln 4 + 1)$ (b) $y = ex$
57. (a) $200(3.24)^t$ (b) $\approx 22{,}040$
(c) $\approx 25{,}910$ bacteria/h (d) $(\ln 50)/(\ln 3.24) \approx 3.33$ h
59. (a) $C_0 e^{-kt}$ (b) ≈ 100 h **61.** $\pi/2$ **63.** 0
65. $-\infty$ **67.** -1 **69.** 1 **71.** 8
73. 0 **75.** $\frac{1}{2}$ **77.** $\frac{2}{3}$

CHAPTER 4

EXERCISES 4.1 ▪ PAGE 208

Abbreviations: abs, absolute; loc, local; max, maximum;
min, minimum

1. Abs min: smallest function value on the entire domain of the
function; loc min at c: smallest function value when x is near c
3. Abs max at s, abs min at r, loc max at c, loc min at b and r,
neither a max nor a min at a and d

5. Abs max $f(4) = 5$, loc max $f(4) = 5$ and $f(6) = 4$, loc min $f(2) = 2$ and $f(1) = f(5) = 3$

7. 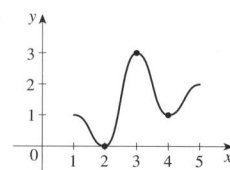 **9.**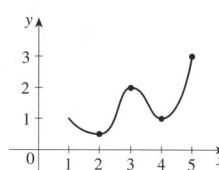

11. (a) 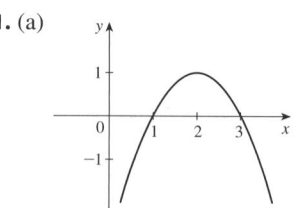 (b)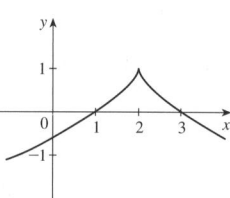

(c)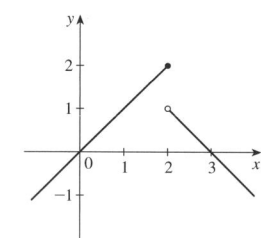

13. (a) (b)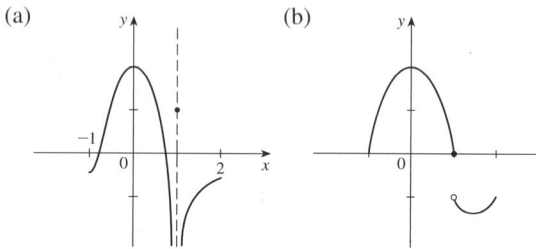

15. Abs max $f(3) = 4$ **17.** Abs min $f(0) = 0$
19. Abs max $f(2) = \ln 2$ **21.** Abs max $f(0) = 1$
23. $\frac{1}{3}$ **25.** $-2, 3$ **27.** 0 **29.** $0, 2$ **31.** $0, \frac{8}{7}, 4$
33. $n\pi$ (n an integer) **35.** $0, \frac{2}{3}$ **37.** $f(2) = 16, f(5) = 7$
39. $f(-1) = 8$, $f(2) = -19$ **41.** $f(-2) = 33, f(2) = -31$
43. $f(\sqrt{2}) = 2$, $f(-1) = -\sqrt{3}$
45. $f(\pi/6) = \frac{3}{2}\sqrt{3}$, $f(\pi/2) = 0$
47. $f(2) = 2/\sqrt{e}$, $f(-1) = -1/\sqrt[8]{e}$
49. $f(1) = \ln 3$, $f(-\frac{1}{2}) = \ln \frac{3}{4}$
51. $f\left(\dfrac{a}{a+b}\right) = \dfrac{a^a b^b}{(a+b)^{a+b}}$
53. (a) $2.19, 1.81$ (b) $\frac{6}{25}\sqrt{\frac{3}{5}} + 2, -\frac{6}{25}\sqrt{\frac{3}{5}} + 2$
55. (a) $0.32, 0.00$ (b) $\frac{3}{16}\sqrt{3}, 0$ **57.** $\approx 3.9665°C$
59. Cheapest, $t \approx 0.855$ (June 1994); most expensive, $t \approx 4.618$ (March 1998)
61. (a) $r = \frac{2}{3}r_0$ (b) $v = \frac{4}{27}kr_0^3$

(c)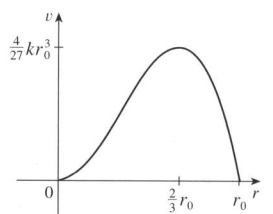

EXERCISES 4.2 ▪ PAGE 215

1. 2 **3.** $\frac{9}{4}$ **5.** f is not differentiable on $(-1, 1)$
7. $0.3, 3, 6.3$ **9.** 1 **11.** $3/\ln 4$ **13.** 1
15. f is not continous at 3. **23.** 16 **25.** No **31.** No

EXERCISES 4.3 ▪ PAGE 222

Abbreviations: inc, increasing; dec, decreasing; CD, concave downward; CU, concave upward; HA, horizontal asymptote; VA, vertical asymptote; IP, inflection point(s)

1. (a) Inc on $(-\infty, -3)$, $(2, \infty)$; dec on $(-3, 2)$
(b) Loc max $f(-3) = 81$; loc min $f(2) = -44$
(c) CU on $(-\frac{1}{2}, \infty)$; CD on $(-\infty, -\frac{1}{2})$; IP $(-\frac{1}{2}, \frac{37}{2})$

3. (a) Inc on $(-1, 0)$, $(1, \infty)$; dec on $(-\infty, -1)$, $(0, 1)$
(b) Loc max $f(0) = 3$; loc min $f(\pm 1) = 2$
(c) CU on $(-\infty, -\sqrt{3}/3)$, $(\sqrt{3}/3, \infty)$;
CD on $(-\sqrt{3}/3, \sqrt{3}/3)$; IP $(\pm\sqrt{3}/3, \frac{22}{9})$

5. (a) Inc on $(0, \pi/4)$, $(5\pi/4, 2\pi)$; dec on $(\pi/4, 5\pi/4)$
(b) Loc max $f(\pi/4) = \sqrt{2}$; loc min $f(5\pi/4) = -\sqrt{2}$
(c) CU on $(3\pi/4, 7\pi/4)$; CD on $(0, 3\pi/4)$, $(7\pi/4, 2\pi)$;
IP $(3\pi/4, 0)$, $(7\pi/4, 0)$

7. (a) Inc on $(-\frac{1}{3}\ln 2, \infty)$; dec on $(-\infty, -\frac{1}{3}\ln 2)$
(b) Loc min $f(-\frac{1}{3}\ln 2) = 2^{-2/3} + 2^{1/3}$ (c) CU on $(-\infty, \infty)$

9. (a) Inc on $(1, \infty)$; dec on $(0, 1)$ (b) Loc min $f(1) = 0$
(c) CU on $(0, \infty)$

11. Loc max $f(1) = 2$; loc min $f(0) = 1$

13. (a) f has a local maximum at 2.
(b) f has a horizontal tangent at 6.

15. (a) $3, 5$ (b) $2, 4, 6$ (c) $1, 7$
17. 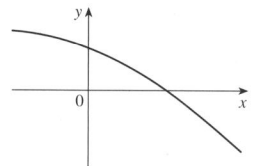 **19.**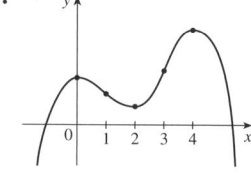

21.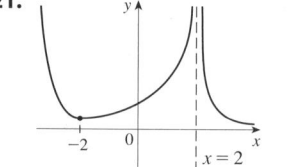

23. (a) Inc on $(0, 2)$, $(4, 6)$, $(8, \infty)$;
dec on $(2, 4)$, $(6, 8)$
(b) Loc max at $x = 2, 6$;
loc min at $x = 4, 8$
(c) CU on $(3, 6)$, $(6, \infty)$;
CD on $(0, 3)$
(d) 3
(e) See graph at right.

25. (a) Inc on $(-\infty, -2)$, $(2, \infty)$; dec on $(-2, 2)$
(b) Loc max $f(-2) = 18$; loc min $f(2) = -14$
(c) CU on $(0, \infty)$, CD on $(-\infty, 0)$; IP $(0, 2)$
(d)

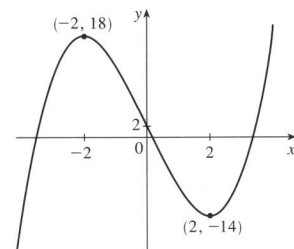

27. (a) Inc on $(-\infty, -1)$, $(0, 1)$;
dec on $(-1, 0)$, $(1, \infty)$
(b) Loc max $f(-1) = 3, f(1) = 3$;
loc min $f(0) = 2$
(c) CU on $\left(-1/\sqrt{3}, 1/\sqrt{3}\right)$;
CD on $\left(-\infty, -1/\sqrt{3}\right)$, $\left(1/\sqrt{3}, \infty\right)$;
IP $\left(\pm 1/\sqrt{3}, \frac{23}{9}\right)$
(d) See graph at right.

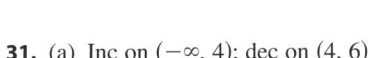

29. (a) Inc on $(-\infty, -2)$, $(0, \infty)$;
dec on $(-2, 0)$
(b) Loc max $h(-2) = 7$;
loc min $h(0) = -1$
(c) CU on $(-1, \infty)$;
CD on $(-\infty, -1)$; IP $(-1, 3)$
(d) See graph at right.

31. (a) Inc on $(-\infty, 4)$; dec on $(4, 6)$
(b) Loc max $f(4) = 4\sqrt{2}$
(c) CD on $(-\infty, 6)$
(d) See graph at right.

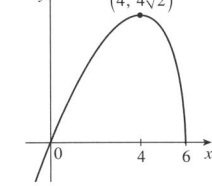

33. (a) Inc on $(-1, \infty)$;
dec on $(-\infty, -1)$
(b) Loc min $C(-1) = -3$
(c) CU on $(-\infty, 0)$, $(2, \infty)$;
CD on $(0, 2)$;
IP $(0, 0)$, $\left(2, 6\sqrt[3]{2}\right)$
(d) See graph at right.

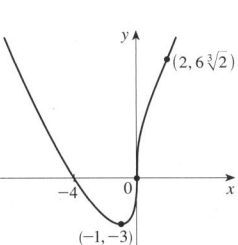

35. (a) Inc on $(\pi, 2\pi)$;
dec on $(0, \pi)$
(b) Loc min $f(\pi) = -1$
(c) CU on $(\pi/3, 5\pi/3)$;
CD on $(0, \pi/3)$, $(5\pi/3, 2\pi)$;
IP $\left(\pi/3, \frac{5}{4}\right)$, $\left(5\pi/3, \frac{5}{4}\right)$

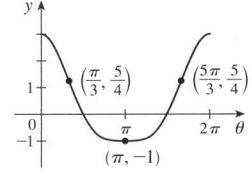

37. (a) VA $x = 0$; HA $y = 1$
(b) Inc on $(0, 2)$;
dec on $(-\infty, 0)$, $(2, \infty)$
(c) Loc max $f(2) = \frac{5}{4}$
(d) CU on $(3, \infty)$;
CD on $(-\infty, 0)$, $(0, 3)$; IP $\left(3, \frac{11}{9}\right)$
(e) See graph at right.

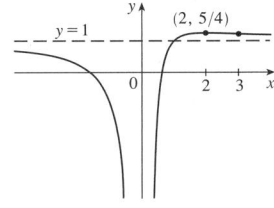

39. (a) HA $y = 0$
(b) Dec on $(-\infty, \infty)$
(c) None
(d) CU on $(-\infty, \infty)$
(e) See graph at right.

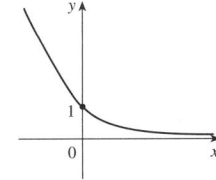

41. (a) HA $y = 0$
(b) Inc on $(-\infty, 0)$, dec on $(0, \infty)$
(c) Loc max $f(0) = 1$
(d) CU on $\left(-\infty, -1\sqrt{2}\right)$,
$\left(1/\sqrt{2}, \infty\right)$;
CD on $\left(-1/\sqrt{2}, 1/\sqrt{2}\right)$;
IP $\left(\pm 1/\sqrt{2}, e^{-1/2}\right)$
(e) See graph at right.

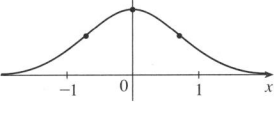

43. (a) VA $x = 0$, $x = e$
(b) Dec on $(0, e)$
(c) None
(d) CU on $(0, 1)$; CD on $(1, e)$;
IP $(1, 0)$
(e) See graph at right.

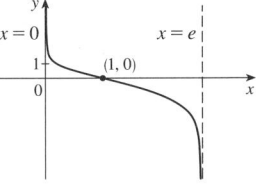

45. $(3, \infty)$ **47.** (a) Loc and abs max $f(1) = \sqrt{2}$, no min
(b) $\frac{1}{4}\left(3 - \sqrt{17}\right)$
49. 28.57 min, when the rate of increase of drug level in the bloodstream is greatest; 85.71 min, when rate of decrease is greatest
51. $f(x) = \frac{1}{9}(2x^3 + 3x^2 - 12x + 7)$
53. (a) $a = 0$, $b = -1$ (b) $y = -x$ at $(0, 0)$

EXERCISES 4.4 ■ PAGE 230

Abbreviations: int, intercept; SA, slant asymptote

1. A. $\mathbb{R}$ B. y-int 0; x-int 0, 6
C. None D. None
E. Inc on $(-\infty, 2)$, $(6, \infty)$;
dec on $(2, 6)$
F. Loc max $f(2) = 32$;
loc min $f(6) = 0$
G. CU on $(4, \infty)$; CD on $(-\infty, 4)$;
IP $(4, 16)$
H. See graph at right.

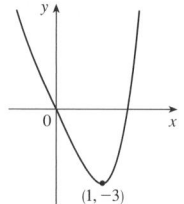

3. A. $\mathbb{R}$ B. y-int 0; x-int 0, $\sqrt[3]{4}$
C. None D. None
E. Inc on $(1, \infty)$; dec on $(-\infty, 1)$
F. Loc min $f(1) = -3$
G. CU on $(-\infty, \infty)$
H. See graph at right.

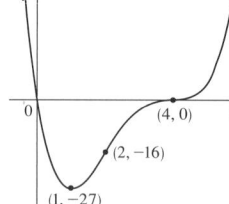

5. A. $\mathbb{R}$ B. y-int 0; x-int 0, 4
C. None D. None
E. Inc on $(1, \infty)$; dec on $(-\infty, 1)$
F. Loc min $f(1) = -27$
G. CU on $(-\infty, 2)$, $(4, \infty)$;
CD on $(2, 4)$;
IP $(2, -16)$, $(4, 0)$
H. See graph at right.

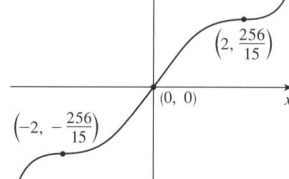

7. A. $\mathbb{R}$ B. y-int 0; x-int 0
C. About $(0, 0)$ D. None
E. Inc on $(-\infty, \infty)$
F. None
G. CU on $(-2, 0)$, $(2, \infty)$;
CD on $(-\infty, -2)$, $(0, 2)$;
IP $\left(-2, -\frac{256}{15}\right)$, $(0, 0)$, $\left(2, \frac{256}{15}\right)$
H. See graph at right.

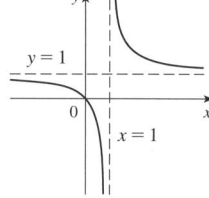

9. A. $\{x \mid x \neq 1\}$ B. y-int 0; x-int 0
C. None D. VA $x = 1$, HA $y = 1$
E. Dec on $(-\infty, 1)$, $(1, \infty)$
F. None
G. CU on $(1, \infty)$; CD on $(-\infty, 1)$
H. See graph at right.

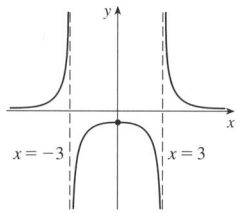

11. A. $\{x \mid x \neq \pm 3\}$ B. y-int $-\frac{1}{9}$
C. About y-axis D. VA $x = \pm 3$, HA $y = 0$
E. Inc on $(-\infty, -3)$, $(-3, 0)$;
dec on $(0, 3)$, $(3, \infty)$
F. Loc max $f(0) = -\frac{1}{9}$
G. CU on $(-\infty, -3)$, $(3, \infty)$;
CD on $(-3, 3)$
H. See graph at right.

13. A. $\mathbb{R}$ B. y-int 0; x-int 0
C. About the origin
D. HA $y = 0$
E. Inc on $(-3, 3)$;
dec on $(-\infty, -3)$, $(3, \infty)$
F. Loc min $f(-3) = -\frac{1}{6}$;
loc max $f(3) = \frac{1}{6}$;
G. CU on $\left(-3\sqrt{3}, 0\right)$, $\left(3\sqrt{3}, \infty\right)$;
CD on $\left(-\infty, -3\sqrt{3}\right)$, $\left(0, 3\sqrt{3}\right)$;
IP $(0, 0)$, $\left(\pm 3\sqrt{3}, \pm\sqrt{3}/12\right)$
H. See graph at right.

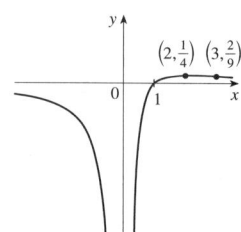

15. A. $(-\infty, 0) \cup (0, \infty)$ B. x-int 1
C. None D. HA $y = 0$; VA $x = 0$
E. Inc on $(0, 2)$;
dec on $(-\infty, 0)$, $(2, \infty)$
F. Loc max $f(2) = \frac{1}{4}$
G. CU on $(3, \infty)$;
CD on $(-\infty, 0)$, $(0, 3)$; IP $\left(3, \frac{2}{9}\right)$
H. See graph at right.

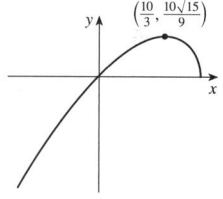

17. A. $(-\infty, 5]$ B. y-int. 0; x-int. 0, 5
C. None D. None
E. Inc. on $\left(-\infty, \frac{10}{3}\right)$; dec. on $\left(\frac{10}{3}, 5\right)$
F. Loc. max. $f\left(\frac{10}{3}\right) = \frac{10}{9}\sqrt{15}$
G. CD on $(-\infty, 5)$
H. See graph at right.

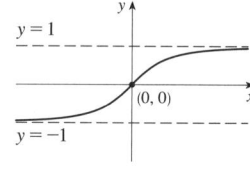

19. A. $\mathbb{R}$ B. y-int 0; x-int 0
C. About the origin
D. HA $y = \pm 1$
E. Inc on $(-\infty, \infty)$ F. None
G. CU on $(-\infty, 0)$;
CD on $(0, \infty)$; IP $(0, 0)$
H. See graph at right.

21. A. $\{x \mid |x| \leq 1, x \neq 0\} = [-1, 0) \cup (0, 1]$
B. x-int ± 1 C. About the origin
D. VA $x = 0$
E. Dec on $(-1, 0)$, $(0, 1)$
F. None
G. CU on $\left(-1, -\sqrt{2/3}\right)$, $\left(0, \sqrt{2/3}\right)$;
CD on $\left(-\sqrt{2/3}, 0\right)$, $\left(\sqrt{2/3}, 1\right)$;
IP $\left(\pm\sqrt{2/3}, \pm 1/\sqrt{2}\right)$
H. See graph at right.

23. A. $\mathbb{R}$ B. y-int 0; x-int 0, $\pm 3\sqrt{3}$ C. About the origin
D. None E. Inc on $(-\infty, -1)$, $(1, \infty)$; dec on $(-1, 1)$
F. Loc max $f(-1) = 2$;
loc min $f(1) = -2$
G. CU on $(0, \infty)$; CD on $(-\infty, 0)$;
IP $(0, 0)$
H. See graph at right.

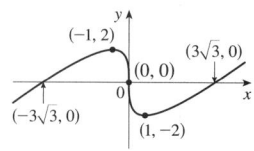

25. A. $\mathbb{R}$ **B.** y-int -1; x-int ± 1
C. About y-axis   **D.** None
E. Inc on $(0, \infty)$; dec on $(-\infty, 0)$
F. Loc min $f(0) = -1$
G. CU on $(-1, 1)$;
CD on $(-\infty, -1)$, $(1, \infty)$;
IP $(\pm 1, 0)$
H. See graph at right.

27. A. $\mathbb{R}$ **B.** y-int 0; x-int $n\pi$ (n an integer)
C. About the origin, period 2π **D.** None
E–G answers for $0 \le x \le \pi$:
E. Inc on $(0, \pi/2)$; dec on $(\pi/2, \pi)$ **F.** Loc max $f(\pi/2) = 1$
G. Let $\alpha = \sin^{-1}\sqrt{2/3}$; CU on $(0, \alpha)$, $(\pi - \alpha, \pi)$;
CD on $(\alpha, \pi - \alpha)$; IP at $x = 0, \pi, \alpha, \pi - \alpha$
H.

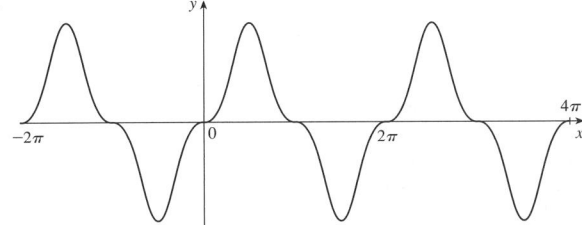

29. A. $(-\pi/2, \pi/2)$ **B.** y-int 0; x-int 0   **C.** About y-axis
D. VA $x = \pm \pi/2$
E. Inc on $(0, \pi/2)$;
dec on $(-\pi/2, 0)$
F. Loc min $f(0) = 0$
G. CU on $(-\pi/2, \pi/2)$
H. See graph at right.

31. A. $(0, 3\pi)$ **C.** None   **D.** None
E. Inc on $(\pi/3, 5\pi/3)$, $(7\pi/3, 3\pi)$;
dec on $(0, \pi/3)$, $(5\pi/3, 7\pi/3)$
F. Loc min $f(\pi/3) = (\pi/6) - \frac{1}{2}\sqrt{3}$, $f(7\pi/3) = (7\pi/6) - \frac{1}{2}\sqrt{3}$;
loc max $f(5\pi/3) = (5\pi/6) + \frac{1}{2}\sqrt{3}$
G. CU on $(0, \pi)$, $(2\pi, 3\pi)$;
CD on $(\pi, 2\pi)$;
IP $(\pi, \pi/2)$, $(2\pi, \pi)$
H. See graph at right.

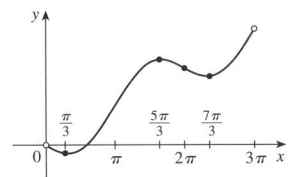

33. A. All reals except $(2n + 1)\pi$ (n an integer)
B. y-int 0; x-int $2n\pi$
C. About the origin, period 2π
D. VA $x = (2n + 1)\pi$
E. Inc on $((2n - 1)\pi, (2n + 1)\pi)$ **F.** None
G. CU on $(2n\pi, (2n + 1)\pi)$; CD on $((2n - 1)\pi, 2n\pi)$;
IP $(2n\pi, 0)$
H.

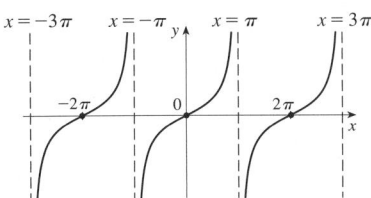

35. A. $\mathbb{R}$ **B.** y-int. $\frac{1}{2}$ **C.** None
D. HA $y = 0$, $y = 1$
E. Inc on $\mathbb{R}$ **F.** None
G. CU on $(-\infty, 0)$; CD on $(0, \infty)$;
IP $\left(0, \frac{1}{2}\right)$ **H.** See graph at right.

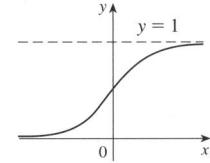

37. A. $(0, \infty)$ **B.** x-int. 1
C. None   **D.** None
E. Inc. on $(1/e, \infty)$; dec. on $(0, 1/e)$
F. Loc. min. $f(1/e) = -1/e$
G. CU on $(0, \infty)$
H. See graph at right.

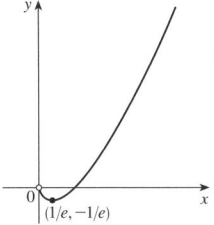

39. A. $\mathbb{R}$ **B.** y-int. 0; x-int. 0
C. None   **D.** HA $y = 0$
E. Inc. on $(-\infty, 1)$; dec. on $(1, \infty)$
F. Loc. max. $f(1) = 1/e$
G. CU on $(2, \infty)$; CD on $(-\infty, 2)$;
IP $(2, 2/e^2)$ **H.** See graph at right.

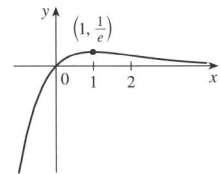

41. A. All x in $(2n\pi, (2n + 1)\pi)$ (n an integer)
B. x-int $\pi/2 + 2n\pi$ **C.** Period 2π **D.** VA $x = n\pi$
E. Inc on $(2n\pi, \pi/2 + 2n\pi)$; dec on $(\pi/2 + 2n\pi, (2n + 1)\pi)$
F. Loc max $f(\pi/2 + 2n\pi) = 0$ **G.** CD on $(2n\pi, (2n + 1)\pi)$
H.

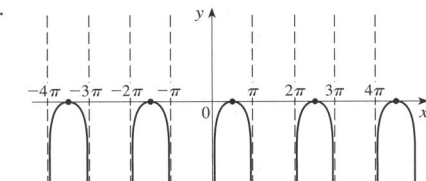

43. A. $(-\infty, 0) \cup (0, \infty)$
B. None   **C.** None   **D.** VA $x = 0$
E. Inc on $(-\infty, -1)$, $(0, \infty)$;
dec on $(-1, 0)$
F. Loc max $f(-1) = -e$
G. CU on $(0, \infty)$; CD on $(-\infty, 0)$
H. See graph at right.

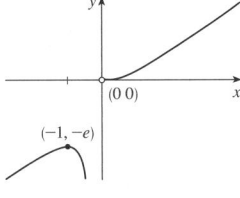

45.

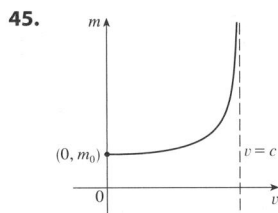

47.

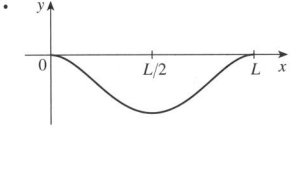

49. A. $(-\infty, 1) \cup (1, \infty)$
B. y-int 0; x-int 0   C. None
D. VA $x = 1$; SA $y = x + 1$
E. Inc on $(-\infty, 0)$, $(2, \infty)$;
dec on $(0, 1)$, $(1, 2)$
F. Loc max $f(0) = 0$;
loc min $f(2) = 4$
G. CU on $(1, \infty)$; CD on $(-\infty, 1)$
H. See graph at right.

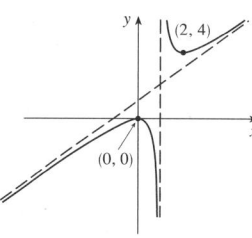

51. A. $(-\infty, 0) \cup (0, \infty)$
B. x-int $-\sqrt[3]{4}$ C. None
D. VA $x = 0$; SA $y = x$
E. Inc. on $(-\infty, 0)$, $(2, \infty)$;
dec on $(0, 2)$
F. Loc min $f(2) = 3$
G. CU on $(-\infty, 0)$, $(0, \infty)$
H. See graph at right.

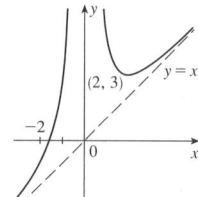

53.

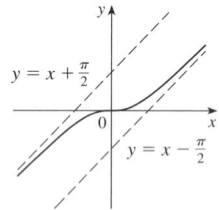

55. Inc on $(0.92, 2.5)$, $(2.58, \infty)$; dec on $(-\infty, 0.92)$, $(2.5, 2.58)$;
loc max $f(2.5) = 4$; loc min $f(0.92) \approx -5.12$, $f(2.58) \approx 3.998$;
CU on $(-\infty, 1.46)$, $(2.54, \infty)$;
CD on $(1.46, 2.54)$; IP $(1.46, -1.40)$, $(2.54, 3.999)$

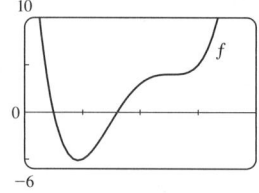 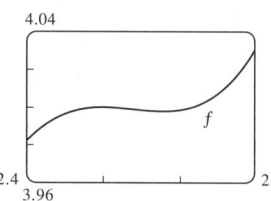

57. Inc on $(-1.40, -0.44)$, $(0.44, 1.40)$; dec on $(-\pi, -1.40)$,
$(-0.44, 0)$, $(0, 0.44)$, $(1.40, \pi)$; loc max $f(-0.44) \approx -4.68$,
$f(1.40) \approx 6.09$; loc min $f(-1.40) \approx -6.09$, $f(0.44) \approx 5.22$;
CU on $(-\pi, -0.77)$, $(0, 0.77)$; CD on $(-0.77, 0)$, $(0.77, \pi)$;
IP $(-0.77, -5.22)$, $(0.77, 5.22)$

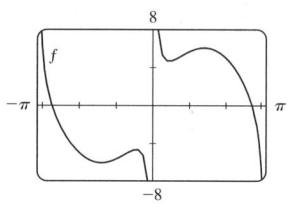

59. Inc on $\left(-8 - \sqrt{61}, -8 + \sqrt{61}\right)$; dec on $\left(-\infty, -8 - \sqrt{61}\right)$,
$\left(-8 + \sqrt{61}, 0\right)$, $(0, \infty)$; CU on $\left(-12 - \sqrt{138}, -12 + \sqrt{138}\right)$,
$(0, \infty)$; CD on $\left(-\infty, -12 - \sqrt{138}\right)$, $\left(-12 + \sqrt{138}, 0\right)$

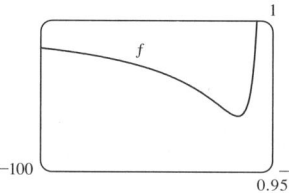 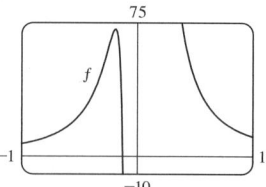

61. For $c \geq 0$, there is an absolute minimum at the origin. There
are no other maxima or minima. The more negative c becomes,
the farther the two IPs move from the origin. $c = 0$ is a transitional value.

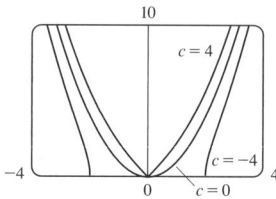

63. There is no maximum or minimum, regardless of the
value of c. For $c < 0$, there is a vertical asymptote at $x = 0$,
$\lim_{x \to 0} f(x) = \infty$, and $\lim_{x \to \pm\infty} f(x) = 1$.
$c = 0$ is a transitional value at which $f(x) = 1$ for $x \neq 0$.
For $c > 0$, $\lim_{x \to 0} f(x) = 0$, $\lim_{x \to \pm\infty} f(x) = 1$, and there are
two IPs, which move away from the y-axis as $c \to \infty$.

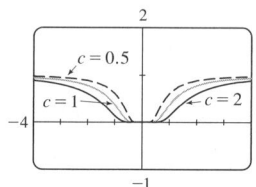 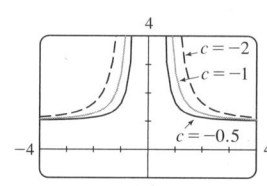

65. For $|c| < 1$, the graph has loc max and min values; for
$|c| \geq 1$ it does not. The function increases for $c \geq 1$ and
decreases for $c \leq -1$. As c changes, the IPs move vertically but
not horizontally.

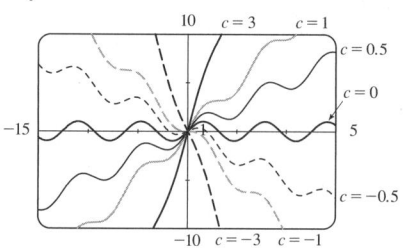

EXERCISES 4.5 ▪ PAGE 238

1. (a) 11, 12   (b) 11.5, 11.5   **3.** 10, 10   **5.** $\frac{9}{4}$
7. 25 m by 25 m

9. (a)

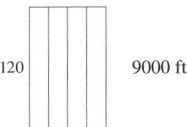

(b)

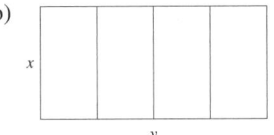

(c) $A = xy$ (d) $5x + 2y = 750$ (e) $A(x) = 375x - \frac{5}{2}x^2$
(f) $14{,}062.5$ ft^2
11. 4000 cm^3 **15.** $\left(-\frac{6}{5}, \frac{3}{5}\right)$ **17.** $\left(-\frac{1}{3}, \pm\frac{4}{3}\sqrt{2}\right)$
19. $L/2,\ \sqrt{3}\,L/4$ **21.** Base $\sqrt{3}\,r$, height $3r/2$
23. $4\pi r^3/(3\sqrt{3})$
25. Width $60/(4 + \pi)$ ft; rectangle height $30/(4 + \pi)$ ft
27. (a) Use all of the wire for the square
(b) $40\sqrt{3}/(9 + 4\sqrt{3})$ m for the square
29. $V = 2\pi R^3/(9\sqrt{3})$ **33.** $E^2/(4r)$
35. (a) $\frac{3}{2}s^2 \csc\theta\,(\csc\theta - \sqrt{3}\cot\theta)$ (b) $\cos^{-1}(1/\sqrt{3}) \approx 55°$
(c) $6s\left[h + s/(2\sqrt{2})\right]$
37. $10\sqrt[3]{3}/(1 + \sqrt[3]{3})$ ft from the stronger source
39. $y = -\frac{5}{3}x + 10$ **41.** $2\sqrt{6}$
43. (b) (i) $342{,}491$; 342/unit; 390/unit
(ii) 400 (iii) 320/unit
45. (a) $p(x) = 19 - \frac{1}{3000}x$ (b) 9.50
47. (a) $p(x) = 550 - \frac{1}{10}x$ (b) 175 (c) 100
49. $(a^{2/3} + b^{2/3})^{3/2}$ **53.** $x = 6$ in.
55. $\frac{1}{2}(L + W)^2$ **57.** At a distance $5 - 2\sqrt{5}$ from A

EXERCISES 4.6 ▪ PAGE 245

1. (a) $x_2 \approx 2.3,\ x_3 \approx 3$ (b) No
3. $\frac{9}{2}$ **5.** a, b, c **7.** 1.1785 **9.** -1.25
11. 1.82056420 **13.** 1.217562
15. $-1.93822883,\ -1.21997997,\ 1.13929375,\ 2.98984102$
17. -0.44285440 **19.** $-1.97806681,\ -0.82646233$
21. $0.21916368,\ 1.08422462$
23. (b) 31.622777 **29.** $(1.519855, 2.306964)$
31. 0.76286%

EXERCISES 4.7 ▪ PAGE 252

1. $F(x) = \frac{1}{2}x + \frac{1}{4}x^3 - \frac{1}{5}x^4 + C$
3. $F(x) = 5x^{7/5} + 40x^{1/5} + C$
5. $F(x) = 2x^{3/2} - \frac{3}{2}x^{4/3} + C$
7. $G(t) = 2t^{1/2} + \frac{2}{3}t^{3/2} + \frac{2}{5}t^{5/2} + C$
9. $H(\theta) = -2\cos\theta - \tan\theta + C_n$ on $(n\pi - \pi/2, n\pi + \pi/2)$,
n an integer
11. $F(x) = 5e^x - 3\sinh x + C$
13. $F(x) = \frac{1}{2}x^2 - \ln|x| - 1/x^2 + C$

15. $F(x) = x^5 - \frac{1}{3}x^6 + 4$
17. $f(x) = x^5 - x^4 + x^3 + Cx + D$
19. $f(x) = \frac{3}{20}x^{8/3} + Cx + D$
21. $f(t) = -\sin t + Ct^2 + Dt + E$
23. $f(x) = x + 2x^{3/2} + 5$ **25.** $f(t) = 4\arctan t - \pi$
27. $f(t) = 2\sin t + \tan t + 4 - 2\sqrt{3}$
29. $f(x) = -x^2 + 2x^3 - x^4 + 12x + 4$
31. $f(\theta) = -\sin\theta - \cos\theta + 5\theta + 4$
33. $f(x) = -\ln x + (\ln 2)x - \ln 2$ **35.** 10 **37.** b
39. $s(t) = 1 - \cos t - \sin t$
41. $s(t) = -10\sin t - 3\cos t + (6/\pi)t + 3$
43. (a) $s(t) = 450 - 4.9t^2$ (b) $\sqrt{450/4.9} \approx 9.58$ s
(c) $-9.8\sqrt{450/4.9} \approx -93.9$ m/s   (d) About 9.09 s
47. 225 ft   **49.** $\frac{130}{11} \approx 11.8$ s
51. $\frac{88}{15} \approx 5.87$ ft/s^2 **53.** $62{,}500$ km/h$^2 \approx 4.82$ m/s^2
55. (a) 22.9125 mi   (b) 21.675 mi   (c) 30 min 33 s
(d) 55.425 mi

CHAPTER 4 REVIEW ▪ PAGE 254

True-False Quiz

1. False   **3.** False   **5.** True   **7.** False   **9.** True
11. True   **13.** False   **15.** True   **17.** True
19. True

Exercises

1. Abs max $f(4) = 5$, abs and loc min $f(3) = 1$
3. Abs max $f(2) = \frac{2}{5}$, abs and loc min $f\left(-\frac{1}{3}\right) = -\frac{9}{2}$
5.

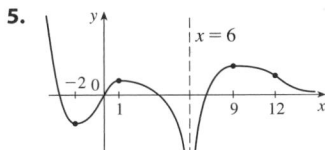

7.

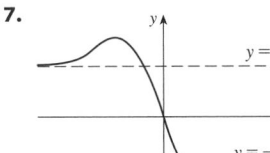

9. (a) None
(b) Dec. on $(-\infty, \infty)$
(c) None
(d) CU on $(-\infty, 0)$; CD on $(0, \infty)$;
IP $(0, 2)$
(e) See graph at right.

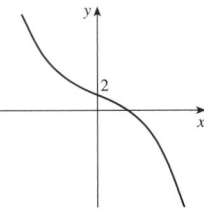

11. (a) None

(b) Inc on $(2n\pi, (2n + 1)\pi)$, n an integer;
dec on $((2n + 1)\pi, (2n + 2)\pi)$

(c) Loc max $f((2n + 1)\pi) = 2$; loc min $f(2n\pi) = -2$

(d) CU on $(2n\pi - (\pi/3), 2n\pi + (\pi/3))$;
CD on $(2n\pi + (\pi/3), 2n\pi + (5\pi/3))$; IPs $\left(2n\pi \pm (\pi/3), -\frac{1}{4}\right)$

(e)

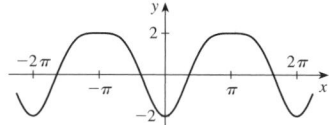

13. (a) None

(b) Inc on $\left(\frac{1}{4}\ln 3, \infty\right)$,
dec on $\left(-\infty, \frac{1}{4}\ln 3\right)$

(c) Loc min
$f\left(\frac{1}{4}\ln 3\right) = 3^{1/4} + 3^{-3/4}$

(d) CU on $(-\infty, \infty)$

(e) See graph at right.

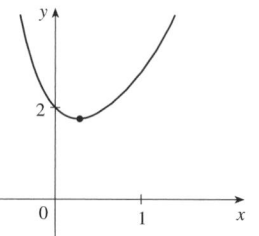

15. A. $\mathbb{R}$ B. y-int 0; x-int 0, 1

C. None   D. None

E. Inc on $\left(\frac{1}{4}, \infty\right)$; dec on $\left(-\infty, \frac{1}{4}\right)$

F. Loc min $f\left(\frac{1}{4}\right) = -\frac{27}{256}$

G. CU on $\left(-\infty, \frac{1}{2}\right), (1, \infty)$;
CD on $\left(\frac{1}{2}, 1\right)$; IP $\left(\frac{1}{2}, -\frac{1}{16}\right), (1, 0)$

H. See graph at right.

17. A. $\{x \mid x \neq 0, 3\}$

B. None   C. None

D. HA $y = 0$; VAs $x = 0$, $x = 3$

E. Inc on $(1, 3)$;
dec on $(-\infty, 0), (0, 1), (3, \infty)$

F. Loc min $f(1) = \frac{1}{4}$

G. CU on $(0, 3), (3, \infty)$; CD on $(-\infty, 0)$

H. See graph at right.

19. A. $[-2, \infty)$

B. y-int 0; x-int -2, 0

C. None   D. None

E. Inc on $\left(-\frac{4}{3}, \infty\right)$, dec on $\left(-2, -\frac{4}{3}\right)$

F. Loc min $f\left(-\frac{4}{3}\right) = -\frac{4}{9}\sqrt{6}$

G. CU on $(-2, \infty)$

H. See graph at right.

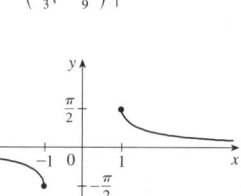

21. A. $\left\{x \mid |x| \geq 1\right\}$

B. None

C. About the origin

D. HA $y = 0$

E. Dec on $(-\infty, -1), (1, \infty)$

F. None

G. CU on $(1, \infty)$; CD on $(-\infty, -1)$

H. See graph at right.

23. A. $\mathbb{R}$

B. y-int -2; x-int 2

C. None   D. HA $y = 0$

E. Inc on $(-\infty, 3)$; dec on $(3, \infty)$

F. Loc max $f(3) = e^{-3}$

G. CU on $(4, \infty)$; CD on $(-\infty, 4)$;
IP $(4, 2e^{-4})$

H. See graph at right.

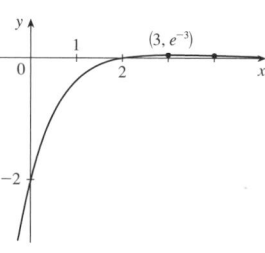

25. Inc on $\left(-\sqrt{3}, 0\right), \left(0, \sqrt{3}\right)$;
dec on $\left(-\infty, -\sqrt{3}\right), \left(\sqrt{3}, \infty\right)$;
loc max $f\left(\sqrt{3}\right) = \frac{2}{9}\sqrt{3}$,
loc min $f\left(-\sqrt{3}\right) = -\frac{2}{9}\sqrt{3}$;
CU on $\left(-\sqrt{6}, 0\right), \left(\sqrt{6}, \infty\right)$;
CD on $\left(-\infty, -\sqrt{6}\right), \left(0, \sqrt{6}\right)$;
IP $\left(\sqrt{6}, \frac{5}{36}\sqrt{6}\right), \left(-\sqrt{6}, -\frac{5}{36}\sqrt{6}\right)$

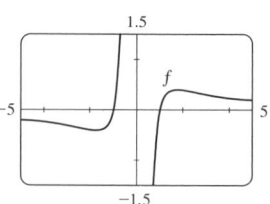

27. Inc on $(-0.23, 0), (1.62, \infty)$; dec on $(-\infty, -0.23), (0, 1.62)$;
loc max $f(0) = 2$; loc min $f(-0.23) \approx 1.96$, $f(1.62) \approx -19.2$;
CU on $(-\infty, -0.12), (1.24, \infty)$; CD on $(-0.12, 1.24)$;
IP $(-0.12, 1.98), (1.24, -12.1)$

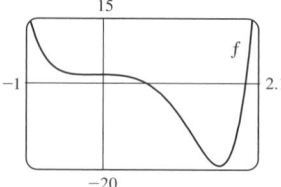
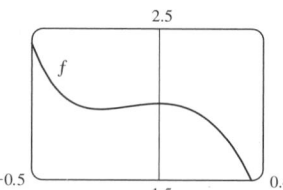

29. $(\pm 0.82, 0.22)$; $\left(\pm\sqrt{2/3}, e^{-3/2}\right)$

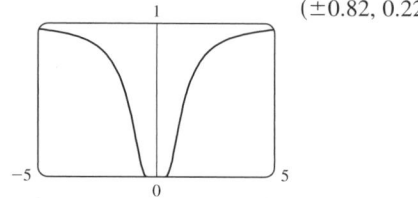

31. $-2.96, -0.18, 3.01$; $-1.57, 1.57$; $-2.16, -0.75, 0.46, 2.21$

33. For $C > -1$, f is periodic with period 2π and has local maxima at $2n\pi + \pi/2$, n an integer. For $C \leq -1$, f has no graph. For $-1 < C \leq 1$, f has vertical asymptotes. For $C > 1$, f is continuous on $\mathbb{R}$. As C increases, f moves upward and its oscillations become less pronounced.

39. 500, 125   **41.** $3\sqrt{3}\, r^2$ **43.** $4/\sqrt{3}$ cm from D; at C

45. $L = C$ **47.** \$11.50   **49.** 1.16718557

51. $F(x) = e^x - 4\sqrt{x} + C$ **53.** $2\arctan x - 1$

55. $\frac{1}{2}x^2 - x^3 + 4x^4 + 2x + 1$ **57.** $s(t) = t^2 - \tan^{-1}t + 1$

59. No

61. (b) About 8.5 in. by 2 in.   (c) $20/\sqrt{3}$ in. by $20\sqrt{2/3}$ in.

CHAPTER 5

EXERCISES 5.1 ■ PAGE 266

1. (a) 40, 52

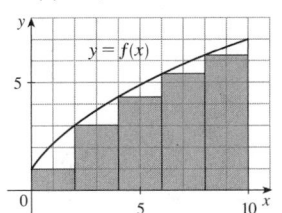 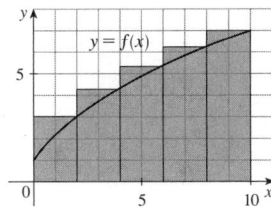

(b) 43.2, 49.2

3. (a) 0.7908, underestimate (b) 1.1835, overestimate

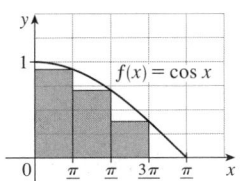 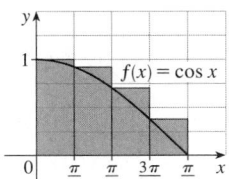

5. (a) 8, 6.875 (b) 5, 5.375

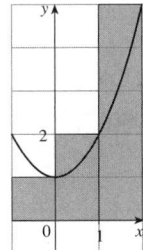 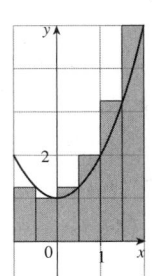 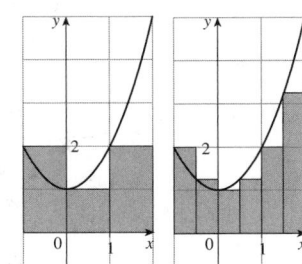

(c) 5.75, 5.9375

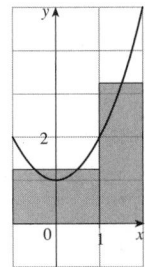 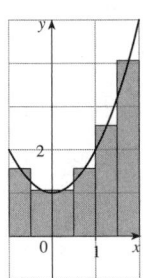

(d) M_6

7. $n = 2$: upper $= 3\pi \approx 9.42$, lower $= 2\pi \approx 6.28$

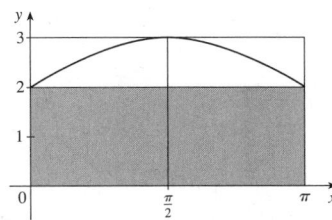

$n = 4$: upper $= \left(10 + \sqrt{2}\,\right)(\pi/4) \approx 8.96$,
lower $= \left(8 + \sqrt{2}\,\right)(\pi/4) \approx 7.39$

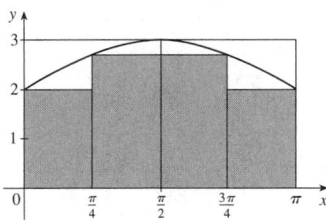

$n = 8$: upper ≈ 8.65, lower ≈ 7.86

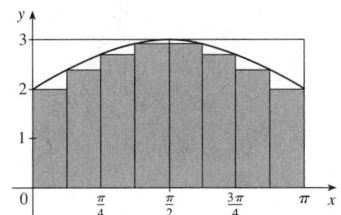

9. 34.7 ft, 44.8 ft **11.** 63.2 L, 70 L **13.** 155 ft

15. $\displaystyle \lim_{n\to\infty} \sum_{i=1}^{n} \frac{2(1 + 2i/n)}{(1 + 2i/n)^2 + 1} \cdot \frac{2}{n}$

17. The region under the graph of $y = \tan x$ from 0 to $\pi/4$

19. (a) $L_n < A < R_n$

21. (a) $\displaystyle \lim_{n\to\infty} \frac{64}{n^6} \sum_{i=1}^{n} i^5$ (b) $\dfrac{n^2(n + 1)^2(2n^2 + 2n - 1)}{12}$ (c) $\frac{32}{3}$

23. $\sin b$, 1

EXERCISES 5.2 ■ PAGE 279

1. -6
The Riemann sum represents
the sum of the areas of the two
rectangles above the x-axis
minus the sum of the areas of
the three rectangles below the
x-axis; that is, the net area of the
rectangles with respect to the
x-axis.

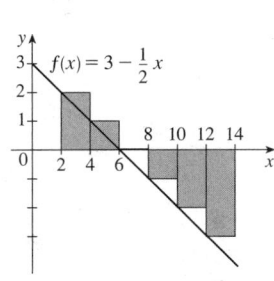

3. 2.322986
The Riemann sum represents the sum
of the areas of the three rectangles
above the x-axis minus the area of
the rectangle below the x-axis.

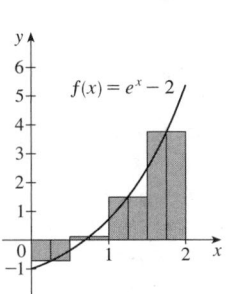

5. -0.028
7. (a) 6 (b) 4 (c) 2
9. Lower, $L_5 = -64$; upper, $R_5 = 16$
11. 6.1820 **13.** 0.9071

15. $\int_2^6 x \ln(1 + x^2) \, dx$ **17.** $\int_2^7 (5x^3 - 4x) \, dx$

19. -9 **21.** $\frac{2}{3}$ **23.** $-\frac{3}{4}$

25. $\displaystyle\lim_{n\to\infty} \sum_{i=1}^{n} \frac{2 + 4i/n}{1 + (2 + 4i/n)^5} \cdot \frac{4}{n}$

27. $\displaystyle\lim_{n\to\infty} \sum_{i=1}^{n} \left(\sin \frac{5\pi i}{n} \right) \frac{\pi}{n} = \frac{2}{5}$

29. (a) 4 (b) 10 (c) -3 (d) 2

31. $\frac{3}{2}$ **33.** $3 + \frac{9}{4}\pi$ **35.** $\frac{5}{2}$ **37.** 0

39. $\int_{-1}^{5} f(x) \, dx$ **41.** 122 **43.** 3 **45.** 15

49. $\frac{1}{2} \le \int_1^2 dx/x \le 1$ **51.** $\frac{\pi}{12} \le \int_{\pi/4}^{\pi/3} \tan x \, dx \le \frac{\pi}{12}\sqrt{3}$

53. $\int_0^1 x^4 \, dx$

EXERCISES 5.3 ▪ PAGE 289

1. $-\frac{10}{3}$ **3.** $\frac{21}{5}$ **5.** -2 **7.** $5e^\pi + 1$ **9.** 36

11. $\frac{55}{63}$ **13.** $\frac{3}{4} - 2\ln 2$ **15.** $\frac{1}{11} + \frac{9}{\ln 10}$

17. $1 + \pi/4$ **19.** $\frac{1}{2}(e - e^{-1})$ **21.** $\pi/3$

23. $\frac{1}{2}e^2 + e - \frac{1}{2}$ **25.** $e^2 - 1$ **27.** $\pi/6$ **29.** -3.5

31. The function $f(x) = 1/x^2$ is not continuous on the interval $[-1, 3]$, so the Evaluation Theorem cannot be applied.

33. $\frac{4}{3}$ **35.** 2

37. 3.75

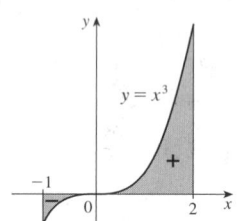

41. $\frac{2}{5}x^{5/2} + C$

43. $-\cos x + \cosh x + C$ **45.** $\frac{2}{3}u^3 + \frac{9}{2}u^2 + 4u + C$

47. $\sec x + C$ **49.** $\frac{4}{3}$

51. Increase in the child's weight (in pounds) between the ages of 5 and 10

53. Number of gallons of oil leaked in the first 2 hours

55. Increase in revenue when production is increased from 1000 to 5000 units

57. Newton-meters **59.** (a) $-\frac{3}{2}$ m (b) $\frac{41}{6}$ m

61. (a) $v(t) = \frac{1}{2}t^2 + 4t + 5$ m/s (b) $416\frac{2}{3}$ m

63. 1.4 mi **65.** 1800 L **67.** 5443 bacteria **69.** 3

EXERCISES 5.4 ▪ PAGE 298

1. (a) 0, 2, 5, 7, 3
(b) (0, 3)
(c) $x = 3$

(d)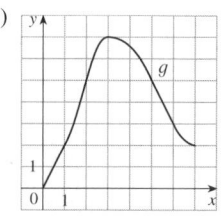

3.

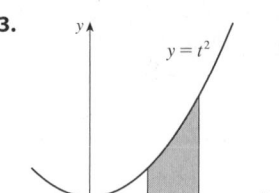

(a), (b) x^2

5. $g'(x) = 1/(x^3 + 1)$ **7.** $g'(s) = (s - s^2)^8$

9. $h'(x) = -\dfrac{\arctan(1/x)}{x^2}$ **11.** $y' = \sqrt{\tan x + \sqrt{\tan x}} \sec^2 x$

13. $g'(x) = \dfrac{-2(4x^2 - 1)}{4x^2 + 1} + \dfrac{3(9x^2 - 1)}{9x^2 + 1}$ **15.** $\frac{45}{28}$

17. $2/\pi$

19. (a) 1 (b) 2, 4 (c)

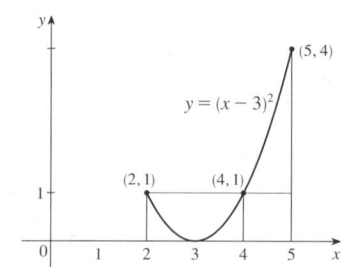

21. $\frac{9}{8}$ **23.** $(-4, 0)$

25. (a) Loc max at 1 and 5; loc min at 3 and 7
(b) $x = 9$
(c) $\left(\frac{1}{2}, 2\right)$, (4, 6), (8, 9)
(d) See graph at right.

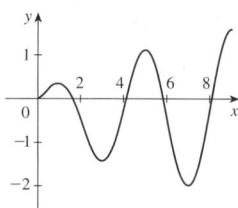

27. 29

29. (a) $-2\sqrt{n}$, $\sqrt{4n - 2}$, n an integer > 0
(b) (0, 1), $\left(-\sqrt{4n - 1}, -\sqrt{4n - 3}\right)$, and $\left(\sqrt{4n - 1}, \sqrt{4n + 1}\right)$, n an integer > 0 (c) ≈ 0.74

31. $f(x) = x^{3/2}$, $a = 9$

33. (b) Average expenditure over $[0, t]$; minimize average expenditure

EXERCISES 5.5 ▪ PAGE 306

1. $-e^{-x} + C$ **3.** $\frac{2}{9}(x^3 + 1)^{3/2} + C$ **5.** $-\frac{1}{4}\cos^4\theta + C$

7. $-\frac{1}{2}\cos(x^2) + C$ **9.** $-\frac{1}{20}(1 - 2x)^{10} + C$

11. $\frac{1}{3}(\ln x)^3 + C$ **13.** $-\frac{1}{3}\ln|5 - 3x| + C$

15. $\frac{2}{3}\sqrt{3ax + bx^3} + C$ **17.** $-(1/\pi)\cos \pi t + C$

19. $\frac{2}{3}(1 + e^x)^{3/2} + C$ **21.** $\frac{1}{3}\sinh^3 x + C$

23. $-\frac{2}{3}(\cot x)^{3/2} + C$ **25.** $\ln|\sin^{-1}x| + C$

27. $\frac{1}{3}\sec^3 x + C$ **29.** $-\dfrac{1}{\ln 5}\cos(5^t) + C$

31. $-\ln(1 + \cos^2 x) + C$

33. $\frac{1}{40}(2x + 5)^{10} - \frac{5}{36}(2x + 5)^9 + C$

35. $\tan^{-1}x + \frac{1}{2}\ln(1 + x^2) + C$

37. $2/\pi$ **39.** $\frac{45}{28}$ **41.** 4 **43.** $e - \sqrt{e}$

45. $\frac{16}{15}$ **47.** $\ln(e + 1)$ **49.** 0

51. 2 **53.** $\frac{1}{10}(1 - e^{-25})$ **55.** $2/(5\pi)$

57. 6π **59.** All three areas are equal.   **61.** ≈ 4512 L

63. $\dfrac{5}{4\pi}\left(1 - \cos\dfrac{2\pi t}{5}\right)$ L   **65.** 5

CHAPTER 5 REVIEW ▪ PAGE 308

True-False Quiz

1. True   **3.** True   **5.** False   **7.** True   **9.** True
11. False   **13.** True   **15.** False   **17.** False

Exercises

1. (a) 8   (b) 5.7

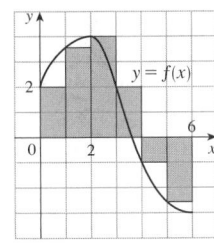 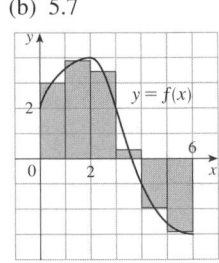

3. $\frac{1}{2} + \pi/4$ **5.** $f = c,\ f' = b,\ \int_0^x f(t)\,dt = a$

7. 37 **9.** $\frac{9}{10}$ **11.** -76 **13.** $\frac{21}{4}$ **15.** $\frac{1}{3}\sin 1$

17. 0 **19.** $-(1/x) - 2\ln|x| + x + C$

21. $\sqrt{x^2 + 4x} + C$ **23.** $\dfrac{1}{2\pi}\sin^2 \pi t + C$

25. $2e^{\sqrt{x}} + C$ **27.** $-\frac{1}{2}[\ln(\cos x)]^2 + C$

29. $\frac{1}{4}\ln(1 + x^4) + C$ **31.** $\ln|1 + \sec\theta| + C$

33. $\frac{64}{5}$ **35.** $F'(x) = \sqrt{1 + x^4}$

37. $y' = \left(2e^x - e^{\sqrt{x}}\right)/(2x)$

39. $4 \leqslant \int_1^3 \sqrt{x^2 + 3}\,dx \leqslant 4\sqrt{3}$ **41.** 1.11

43. Number of barrels of oil consumed from Jan. 1, 2000, through Jan. 1, 2003

45. $72{,}400$ **47.** $f(x)$ **49.** $c \approx 1.62$

CHAPTER 6

EXERCISES 6.1 ▪ PAGE 316

1. $\frac{1}{3}x^3 \ln x - \frac{1}{9}x^3 + C$ **3.** $\frac{1}{5}x\sin 5x + \frac{1}{25}\cos 5x + C$

5. $-\frac{1}{3}te^{-3t} - \frac{1}{9}e^{-3t} + C$

7. $(x^2 + 2x)\sin x + (2x + 2)\cos x - 2\sin x + C$

9. $\frac{1}{2}(2x + 1)\ln(2x + 1) - x + C$

11. $t\arctan 4t - \frac{1}{8}\ln(1 + 16t^2) + C$

13. $\frac{1}{13}e^{2\theta}(2\sin 3\theta - 3\cos 3\theta) + C$

15. $\dfrac{e^{2x}}{4(2x + 1)} + C$ **17.** $\dfrac{\pi - 2}{2\pi^2}$ **19.** $\frac{81}{4}\ln 3 - 5$

21. $1 - 1/e$ **23.** $\frac{1}{6}(\pi + 6 - 3\sqrt{3})$

25. $2(\ln 2)^2 - 4\ln 2 + 2$

27. $2\sqrt{x}\sin\sqrt{x} + 2\cos\sqrt{x} + C$ **29.** $-\frac{1}{2} - \pi/4$

31. (b) $-\frac{1}{4}\cos x \sin^3 x + \frac{3}{8}x - \frac{3}{16}\sin 2x + C$

33. (b) $\frac{2}{3},\ \frac{8}{15}$

39. $x[(\ln x)^3 - 3(\ln x)^2 + 6\ln x - 6] + C$

41. $1 - (2/\pi)\ln 2$

43. $2 - e^{-t}(t^2 + 2t + 2)$ meters   **45.** 2

EXERCISES 6.2 ▪ PAGE 326

1. $\frac{1}{3}\sin^3 x - \frac{1}{5}\sin^5 x + C$ **3.** $\frac{1}{120}$

5. $\pi/4$ **7.** $3\pi/8$ **9.** $\pi/16$

11. $\frac{1}{4}t^2 - \frac{1}{4}t\sin 2t - \frac{1}{8}\cos 2t + C$

13. $\frac{1}{2}\cos^2 x - \ln|\cos x| + C$

15. $\ln(1 + \sin x) + C$

17. $\frac{1}{3}\sec^3 x + C$ **19.** $\tan x - x + C$

21. $\frac{1}{9}\tan^9 x + \frac{2}{7}\tan^7 x + \frac{1}{5}\tan^5 x + C$ **23.** $\frac{117}{8}$

25. $\frac{1}{3}\sec^3 x - \sec x + C$

27. $\frac{1}{4}\sec^4 x - \tan^2 x + \ln|\sec x| + C$

29. $\sqrt{3} - \frac{1}{3}\pi$ **31.** $\frac{22}{105}\sqrt{2} - \frac{8}{105}$

33. $\ln|\csc x - \cot x| + C$ **35.** $\frac{1}{2}\sqrt{2}$

37. (b) $\frac{1}{6}\sin 3x - \frac{1}{14}\sin 7x + C$

39. $-\dfrac{\sqrt{4 - x^2}}{4x} + C$ **41.** $\sqrt{x^2 - 4} - 2\sec^{-1}\left(\dfrac{x}{2}\right) + C$

43. $\dfrac{\pi}{24} + \dfrac{\sqrt{3}}{8} - \dfrac{1}{4}$ **45.** $\dfrac{1}{\sqrt{2}a^2}$

47. $\ln(\sqrt{x^2 + 16} + x) + C$

49. $\frac{1}{4}\sin^{-1}(2x) + \frac{1}{2}x\sqrt{1 - 4x^2} + C$

51. $\frac{1}{6}\sec^{-1}(x/3) - \sqrt{x^2 - 9}/(2x^2) + C$ **53.** $\frac{9}{500}\pi$

55. $\sqrt{x^2 - 7} + C$

57. $\ln\left|(\sqrt{1 + x^2} - 1)/x\right| + \sqrt{1 + x^2} + C$

59. $\frac{1}{4}\sin^{-1}(x^2) + \frac{1}{4}x^2\sqrt{1 - x^4} + C$

61. $\frac{1}{3}\ln|3x + 1 + \sqrt{9x^2 + 6x - 8}| + C$

63. $\frac{9}{2}\sin^{-1}((x - 2)/3) + \frac{1}{2}(x - 2)\sqrt{5 + 4x - x^2} + C$

65. $s = (1 - \cos^3\omega t)/(3\omega)$ **67.** $\frac{1}{6}(\sqrt{48} - \sec^{-1}7)$

EXERCISES 6.3 ▪ PAGE 334

1. (a) $\dfrac{A}{4x-3}+\dfrac{B}{2x+5}$ (b) $\dfrac{A}{x}+\dfrac{B}{x^2}+\dfrac{C}{5-2x}$

3. (a) $\dfrac{A}{x}+\dfrac{B}{x^2}+\dfrac{C}{x^3}+\dfrac{Dx+E}{x^2+4}$

(b) $\dfrac{A}{x+3}+\dfrac{B}{(x+3)^2}+\dfrac{C}{x-3}+\dfrac{D}{(x-3)^2}$

5. (a) $x^4+4x^2+16+\dfrac{A}{x+2}+\dfrac{B}{x-2}$

(b) $\dfrac{Ax+B}{x^2-x+1}+\dfrac{Cx+D}{x^2+2}+\dfrac{Ex+F}{(x^2+2)^2}$

7. $\frac{1}{4}x^4+\frac{1}{3}x^3+\frac{1}{2}x^2+x+\ln|x-1|+C$

9. $\frac{1}{2}\ln|2x+1|+2\ln|x-1|+C$ **11.** $2\ln\frac{3}{2}$

13. $a\ln|x-b|+C$ **15.** $2\ln 2+\frac{1}{2}$

17. $\frac{27}{5}\ln 2-\frac{9}{5}\ln 3$ (or $\frac{9}{5}\ln\frac{8}{3}$)

19. $10\ln|x-3|-9\ln|x-2|+\dfrac{5}{x-2}+C$

21. $\frac{1}{2}x^2-2\ln(x^2+4)+2\tan^{-1}(x/2)+C$

23. $\ln|x-1|-\frac{1}{2}\ln(x^2+9)-\frac{1}{3}\tan^{-1}(x/3)+C$

25. $\frac{1}{2}\ln(x^2+1)+\left(1/\sqrt{2}\right)\tan^{-1}\!\left(x/\sqrt{2}\right)+C$

27. $\frac{1}{2}\ln(x^2+2x+5)+\frac{3}{2}\tan^{-1}\!\left(\dfrac{x+1}{2}\right)+C$

29. $\frac{1}{3}\ln|x-1|-\frac{1}{6}\ln(x^2+x+1)-\dfrac{1}{\sqrt{3}}\tan^{-1}\dfrac{2x+1}{\sqrt{3}}+C$

31. $\frac{1}{16}\ln|x|-\frac{1}{32}\ln(x^2+4)+\dfrac{1}{8(x^2+4)}+C$

33. $\dfrac{-1}{2(x^2+2x+4)}-\dfrac{2\sqrt{3}}{9}\tan^{-1}\!\left(\dfrac{x+1}{\sqrt{3}}\right)-\dfrac{2(x+1)}{3(x^2+2x+4)}+C$

35. $2+\ln\frac{25}{9}$ **37.** $\frac{3}{10}(x^2+1)^{5/3}-\frac{3}{4}(x^2+1)^{2/3}+C$

39. $\ln\left[\dfrac{(e^x+2)^2}{e^x+1}\right]+C$

41. $\left(x-\frac{1}{2}\right)\ln(x^2-x+2)-2x+\sqrt{7}\,\tan^{-1}\!\left(\dfrac{2x-1}{\sqrt{7}}\right)+C$

43. $t=-\ln P-\frac{1}{9}\ln(0.9P+900)+C$, where $C\approx 10.23$

47. $\dfrac{1}{a^n(x-a)}-\dfrac{1}{a^n x}-\dfrac{1}{a^{n-1}x^2}-\cdots-\dfrac{1}{ax^n}$

EXERCISES 6.4 ▪ PAGE 340

1. $\dfrac{\pi}{8}\arctan\dfrac{\pi}{4}-\frac{1}{4}\ln\!\left(1+\frac{1}{16}\pi^2\right)$ **3.** $\frac{1}{6}\ln\left|\dfrac{\sin x-3}{\sin x+3}\right|+C$

5. $-\sqrt{4x^2+9}/(9x)+C$ **7.** $\pi^3-6\pi$

9. $-\frac{1}{2}\tan^2(1/z)-\ln|\cos(1/z)|+C$

11. $\dfrac{2y-1}{8}\sqrt{6+4y-4y^2}+\frac{7}{8}\sin^{-1}\!\left(\dfrac{2y-1}{\sqrt{7}}\right)$
$-\frac{1}{12}(6+4y-4y^2)^{3/2}+C$

13. $\frac{1}{9}\sin^3 x\,[3\ln(\sin x)-1]+C$

15. $\dfrac{1}{2\sqrt{3}}\ln\left|\dfrac{e^x+\sqrt{3}}{e^x-\sqrt{3}}\right|+C$

17. $\frac{1}{5}\ln\left|x^5+\sqrt{x^{10}-2}\right|+C$

19. $\frac{1}{2}(\ln x)\sqrt{4+(\ln x)^2}+2\ln\!\left[\ln x+\sqrt{4+(\ln x)^2}\right]+C$

21. $\sqrt{e^{2x}-1}-\cos^{-1}(e^{-x})+C$

25. $\frac{1}{3}\tan x\sec^2 x+\frac{2}{3}\tan x+C$

27. $\frac{1}{4}x(x^2+2)\sqrt{x^2+4}-2\ln\!\left(\sqrt{x^2+4}+x\right)+C$

29. $\frac{1}{4}\cos^3 x\sin x+\frac{3}{8}x+\frac{3}{8}\sin x\cos x+C$

31. $\frac{1}{4}\tan^4 x-\frac{1}{2}\tan^2 x-\ln|\cos x|+C$

33. (a) $-\ln\left|\dfrac{1+\sqrt{1-x^2}}{x}\right|+C$;

both have domain $(-1,0)\cup(0,1)$

EXERCISES 6.5 ▪ PAGE 350

1. (a) $L_2=6,\ R_2=12,\ M_2\approx 9.6$
(b) L_2 is an underestimate, R_2 and M_2 are overestimates.
(c) $T_2=9<I$ (d) $L_n<T_n<I<M_n<R_n$

3. (a) $T_4\approx 0.895759$ (underestimate)
(b) $M_4\approx 0.908907$ (overestimate)
$T_4<I<M_4$

5. (a) $M_{10}\approx 0.806598,\ E_M\approx -0.001879$
(b) $S_{10}\approx 0.804779,\ E_S\approx -0.000060$

7. (a) 1.506361 (b) 1.518362 (c) 1.511519

9. (a) 2.660833 (b) 2.664377 (c) 2.663244

11. (a) 4.513618 (b) 4.748256 (c) 4.675111

13. (a) -0.495333 (b) -0.543321 (c) -0.526123

15. (a) 1.064275 (b) 1.067416 (c) 1.074915

17. (a) $T_8\approx 0.902333,\ M_8\approx 0.905620$
(b) $|E_T|\le 0.0078,\ |E_M|\le 0.0039$
(c) $n=71$ for T_n, $n=50$ for M_n

19. (a) $T_{10}\approx 1.983524,\ E_T\approx 0.016476$;
$M_{10}\approx 2.008248,\ E_M\approx -0.008248$;
$S_{10}\approx 2.000110,\ E_S\approx -0.000110$
(b) $|E_T|\le 0.025839,\ |E_M|\le 0.012919,\ |E_S|\le 0.000170$
(c) $n=509$ for T_n, $n=360$ for M_n, $n=22$ for S_n

21. (a) 2.8 (b) 7.954926518 (c) 0.2894
(d) 7.954926521 (e) The actual error is much smaller.
(f) 10.9 (g) 7.953789422 (h) 0.0593
(i) The actual error is smaller. (j) $n\ge 50$

23.

n	L_n	R_n	T_n	M_n
5	0.742943	1.286599	1.014771	0.992621
10	0.867782	1.139610	1.003696	0.998152
20	0.932967	1.068881	1.000924	0.999538

n	E_L	E_R	E_T	E_M
5	0.257057	-0.286599	-0.014771	0.007379
10	0.132218	-0.139610	-0.003696	0.001848
20	0.067033	-0.068881	-0.000924	0.000462

Observations are the same as after Example 1.

25. (a) 19.8 (b) 20.6 (c) $20.5\overline{3}$

27. $37.7\overline{3}$ ft/s **29.** $64.4°$F

31. (a) 14.4 (b) $\frac{1}{2}$
33. 10,177 megawatt-hours   **35.** 59.4
37.

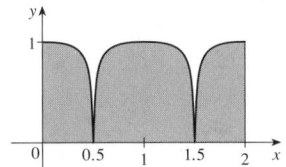

EXERCISES 6.6 ▪ PAGE 360

Abbreviations: C, convergent; D, divergent

1. (a), (d) Infinite discontinuity   (b), (c) Infinite interval
3. $\frac{1}{2} - 1/(2t^2)$; 0.495, 0.49995, 0.4999995; 0.5
5. 2   **7.** D   **9.** $\frac{1}{5}e^{-10}$ **11.** D   **13.** 0   **15.** $-\frac{1}{4}$
17. D   **19.** ln 2   **21.** $\pi/9$ **23.** D   **25.** $\frac{32}{3}$
27. $\frac{9}{2}$ **29.** D   **31.** $\frac{8}{3}\ln 2 - \frac{8}{9}$
33. $1/e$ **43.** $\frac{1}{2}\ln 2$

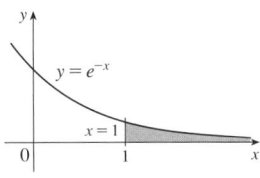

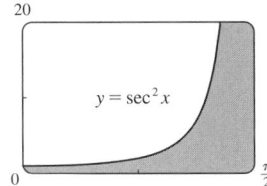

37. Infinite area

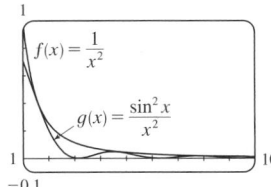

39. (a)

t	$\int_1^t [(\sin^2 x)/x^2]\,dx$
2	0.447453
5	0.577101
10	0.621306
100	0.668479
1,000	0.672957
10,000	0.673407

It appears that the integral is convergent.
(c)

41. C   **43.** D   **45.** D   **47.** π
49. $p < 1,\ 1/(1-p)$

53. (a)

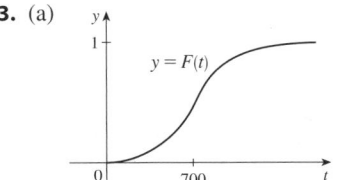

(b) The rate at which the fraction $F(t)$ increases as t increases
(c) 1; all bulbs burn out eventually
55. 8264.5 years   **57.** 1000   **61.** $C = 1$; ln 2   **63.** No

CHAPTER 6 REVIEW ▪ PAGE 362

True-False Quiz

1. False   **3.** False   **5.** False   **7.** False
9. (a) True   (b) False   **11.** False   **13.** False

Exercises

1. $\frac{7}{2} + \ln 2$ **3.** $e - 1$ **5.** $\ln|2t + 1| - \ln|t + 1| + C$
7. $-\cos(\ln t) + C$ **9.** $\frac{64}{5}\ln 4 - \frac{124}{25}$ **11.** $\sqrt{3} - (\pi/3)$
13. $\ln|x| - \frac{1}{2}\ln(x^2 + 1) + C$ **15.** $\frac{2}{15}$
17. $x \sec x - \ln|\sec x + \tan x| + C$
19. $\frac{1}{18}\ln(9x^2 + 6x + 5) + \frac{1}{9}\tan^{-1}\left(\frac{1}{2}(3x + 1)\right) + C$
21. $\ln\left|x - 2 + \sqrt{x^2 - 4x}\right| + C$
23. $-\frac{1}{12}(\cot^3 4x + 3\cot 4x) + C$
25. $\frac{3}{2}\ln(x^2 + 1) - 3\tan^{-1}x + \sqrt{2}\tan^{-1}\left(x/\sqrt{2}\right) + C$
27. $\frac{2}{5}$ **29.** 0   **31.** $6 - \frac{3}{2}\pi$
33. $\dfrac{x}{\sqrt{4 - x^2}} - \sin^{-1}\left(\dfrac{x}{2}\right) + C$
35. $4\sqrt{1 + \sqrt{x}} + C$ **37.** $\frac{1}{2}\sin 2x - \frac{1}{8}\cos 4x + C$
39. $\frac{1}{8}e - \frac{1}{4}$ **41.** $\frac{1}{36}$ **43.** D
45. $4\ln 4 - 8$ **47.** $-\frac{4}{3}$ **49.** $\pi/4$
51. $\frac{1}{4}(2x - 1)\sqrt{4x^2 - 4x - 3} -$
$\qquad\qquad \ln\left|2x - 1 + \sqrt{4x^2 - 4x - 3}\right| + C$
53. $\frac{1}{2}\sin x\sqrt{4 + \sin^2 x} + 2\ln\left(\sin x + \sqrt{4 + \sin^2 x}\right) + C$
55. No
57. (a) 1.925444   (b) 1.920915   (c) 1.922470
59. (a) 0.01348, $n \geqslant 368$ (b) 0.00674, $n \geqslant 260$
61. 8.6 mi
63. (a) 3.8   (b) 1.7867, 0.000646   (c) $n \geqslant 30$

CHAPTER 7

EXERCISES 7.1 ▪ PAGE 369

1. $\frac{32}{3}$ **3.** $e - (1/e) + \frac{10}{3}$ **5.** $e - (1/e) + \frac{4}{3}$ **7.** $\frac{9}{2}$
9. $\frac{8}{3}$ **11.** 72   **13.** $e - 2$ **15.** $\frac{32}{3}$ **17.** $2/\pi + \frac{2}{3}$
19. ln 2

21. $\frac{1}{2}$

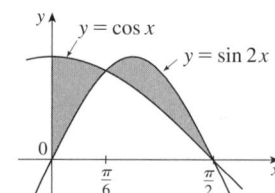

23. 2.80123 **25.** 0.25142 **27.** 118 ft **29.** 84 m²
31. 8868; increase in population over a 10-year period
33. $r\sqrt{R^2 - r^2} + \pi r^2/2 - R^2 \arcsin(r/R)$ **35.** ±6
37. $4^{2/3}$ **39.** $f(t) = 3t^2$ **41.** $0 < m < 1; m - \ln m - 1$

EXERCISES 7.2 ▪ PAGE 378

1. $19\pi/12$

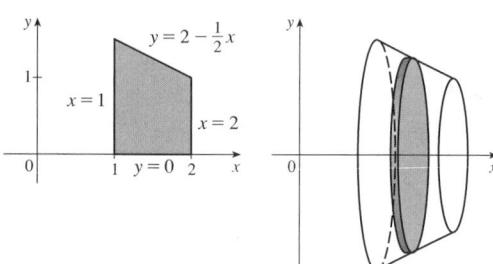

3. 162π

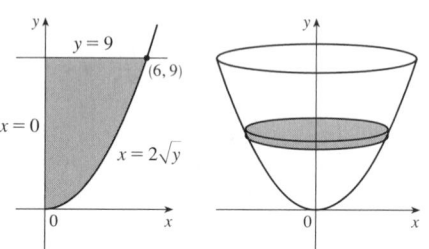

5. $4\pi/21$

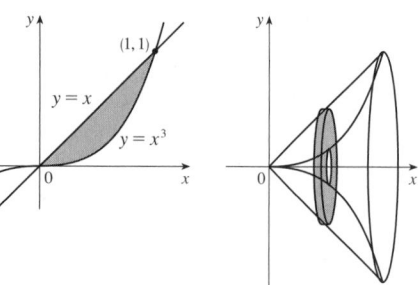

7. $64\pi/15$

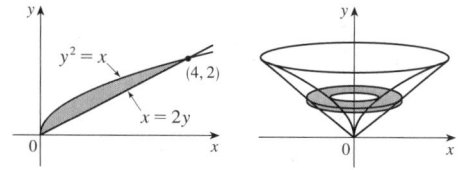

9. $\pi/6$

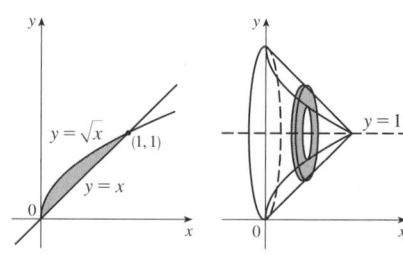

11. $2\pi\left(\frac{4}{3}\pi - \sqrt{3}\right)$

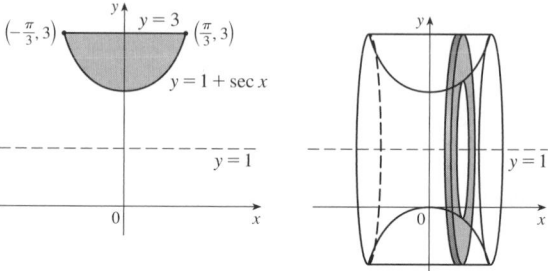

13. $\pi/2$ **15.** $108\pi/5$ **17.** $13\pi/30$
19. (a) $2\pi \int_0^1 e^{-2x^2} dx \approx 3.75825$
(b) $2\pi \int_0^1 \left(e^{-2x^2} + 2e^{-x^2}\right) dx \approx 13.14312$
21. (a) $2\pi \int_0^2 8\sqrt{1 - x^2/4} \, dx \approx 78.95684$
(b) $2\pi \int_0^1 8\sqrt{4 - 4y^2} \, dy \approx 78.95684$
23. $-1.288, 0.884; 23.780$ **25.** $\frac{11}{8}\pi^2$
27. (a) Solid obtained by rotating the region $0 \le y \le \cos x$,
$0 \le x \le \pi/2$ about the x-axis (b) Solid obtained by rotating
the region $y^4 \le x \le y^2$, $0 \le y \le 1$ about the y-axis
29. 1110 cm³ **31.** $\frac{1}{3}\pi r^2 h$ **33.** $\pi h^2(r - \frac{1}{3}h)$
35. $\frac{2}{3}b^2 h$ **37.** 10 cm³ **39.** 24 **41.** $\frac{1}{3}$ **43.** $\frac{8}{15}$
47. (a) $8\pi R \int_0^r \sqrt{r^2 - y^2} \, dy$ (b) $2\pi^2 r^2 R$ **49.** (b) $\pi r^2 h$
51. $\frac{5}{12}\pi r^3$ **53.** $8 \int_0^r \sqrt{R^2 - y^2}\sqrt{r^2 - y^2} \, dy$

EXERCISES 7.3 ▪ PAGE 384

1. Circumference $= 2\pi x$, height $= x(x - 1)^2$; $\pi/15$

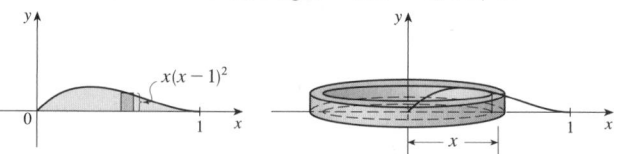

3. $6\pi/7$ **5.** $\pi(1 - 1/e)$ **7.** 8π **9.** 4π
11. $768\pi/7$ **13.** $16\pi/3$ **15.** $7\pi/15$ **17.** $8\pi/3$
19. $5\pi/14$ **21.** (a) $2\pi \int_0^2 x^2 e^{-x} dx$ (b) 4.06300
23. (a) $4\pi \int_{-\pi/2}^{\pi/2} (\pi - x) \cos^4 x \, dx$ (b) 46.50942
25. (a) $\int_0^\pi 2\pi(4 - y)\sqrt{\sin y} \, dy$ (b) 36.57476
27. 1.142
29. Solid obtained by rotating the region $0 \le y \le x^4$, $0 \le x \le 3$
about the y-axis

31. Solid obtained by rotating the region bounded by
(i) $x = 1 - y^2$, $x = 0$, and $y = 0$, or (ii) $x = y^2$, $x = 1$, and
$y = 0$ about the line $y = 3$
33. 8π **35.** $4\sqrt{3}\,\pi$ **37.** $4\pi/3$
39. $\frac{4}{3}\pi r^3$ **41.** $\frac{1}{3}\pi r^2 h$

EXERCISES 7.4 ▪ PAGE 391

1. $4\sqrt{5}$ **3.** 3.8202 **5.** 3.6095
7. $\frac{2}{243}\big(82\sqrt{82} - 1\big)$ **9.** $\frac{59}{24}$ **11.** $\frac{32}{3}$
13. $\ln(\sqrt{2} + 1)$ **15.** $\frac{3}{4} + \frac{1}{2}\ln 2$ **17.** $\ln 3 - \frac{1}{2}$
19. 10.0556 **21.** 15.49805; 15.374568
23. 7.094570; 7.118819 **25.** $\ln 3 - \frac{1}{2}$
27. 6

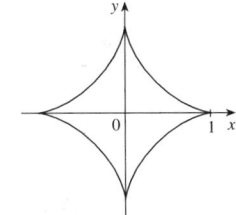

29. $s(x) = \frac{2}{27}\big[(1 + 9x)^{3/2} - 10\sqrt{10}\big]$
31. $2\sqrt{2}\big(\sqrt{1 + x} - 1\big)$
33. 209.1 m   **35.** 29.36 in.

EXERCISES 7.5 ▪ PAGE 397

1. (a) (i) $\int_0^{\pi/3} 2\pi \tan x\sqrt{1 + \sec^4 x}\; dx$
(ii) $\int_0^{\pi/3} 2\pi x\sqrt{1 + \sec^4 x}\; dx$ (b) (i) 10.5017 (ii) 7.9353
3. (a) (i) $\int_{-1}^{1} 2\pi e^{-x^2}\sqrt{1 + 4x^2 e^{-2x^2}}\; dx$
(ii) $\int_0^1 2\pi x\sqrt{1 + 4x^2 e^{-2x^2}}\; dx$ (b) (i) 11.0753 (ii) 3.9603
5. $\frac{1}{27}\pi\big(145\sqrt{145} - 1\big)$ **7.** $\frac{98}{3}\pi$
9. $2\sqrt{1 + \pi^2} + (2/\pi)\ln\big(\pi + \sqrt{1 + \pi^2}\big)$ **11.** $\frac{21}{2}\pi$
13. $\frac{1}{27}\pi\big(145\sqrt{145} - 10\sqrt{10}\big)$ **15.** πa^2
17. $\frac{1}{6}\pi\big[\ln(\sqrt{10} + 3) + 3\sqrt{10}\big]$
19. (a) $\frac{1}{3}\pi a^2$ (b) $\frac{56}{45}\pi\sqrt{3}\,a^2$
21. (a) $2\pi\left[b^2 + \dfrac{a^2 b \sin^{-1}\big(\sqrt{a^2 - b^2}/a\big)}{\sqrt{a^2 - b^2}}\right]$
(b) $2\pi a^2 + \dfrac{2\pi a b^2}{\sqrt{a^2 - b^2}}\ln\dfrac{a + \sqrt{a^2 - b^2}}{b}$
23. $\int_a^b 2\pi[c - f(x)]\sqrt{1 + [f'(x)]^2}\; dx$ **25.** $4\pi^2 r^2$

EXERCISES 7.6 ▪ PAGE 408

1. 4.5 ft-lb   **3.** 180 J   **5.** $\frac{15}{4}$ ft-lb
7. (a) $\frac{25}{24} \approx 1.04$ J   (b) 10.8 cm
9. (a) 625 ft-lb   (b) $\frac{1875}{4}$ ft-lb
11. $650{,}000$ ft-lb
13. 3857 J   **15.** 2450 J
17. (a) $\approx 1.06 \times 10^6$ J   (b) 2.0 m

21. (a) $Gm_1 m_2\left(\dfrac{1}{a} - \dfrac{1}{b}\right)$ (b) $\approx 8.50 \times 10^9$ J
23. $\sqrt{2GM/R}$ **25.** 6000 lb   **27.** 6.7×10^4 N
29. 9.8×10^3 N   **31.** 5.27×10^5 N
33. (a) 5.63×10^3 lb   (b) 5.06×10^4 lb
(c) 4.88×10^4 lb   (d) 3.03×10^5 lb
35. 4148 lb   **37.** 10; 14; $(1.4, 1)$ **39.** $\big(\frac{2}{3}, \frac{2}{3}\big)$
41. $\left(\dfrac{1}{e - 1}, \dfrac{e + 1}{4}\right)$ **43.** $\big(\frac{9}{20}, \frac{9}{20}\big)$
45. $\left(\dfrac{\pi\sqrt{2} - 4}{4(\sqrt{2} - 1)}, \dfrac{1}{4(\sqrt{2} - 1)}\right)$ **47.** 60; 160; $\big(\frac{8}{3}, 1\big)$
51. $\big(0, \frac{1}{12}\big)$ **53.** $\frac{1}{3}\pi r^2 h$

EXERCISES 7.7 ▪ PAGE 418

1. $y = \dfrac{2}{K - x^2}$, $y = 0$ **3.** $y = \sqrt[3]{3x + 3\ln|x| + K}$
5. $\frac{1}{2}y^2 - \cos y = \frac{1}{2}x^2 + \frac{1}{4}x^4 + C$ **7.** $p = Ke^{t^3/3 - t} - 1$
9. $y = -\sqrt{x^2 + 9}$ **11.** $u = -\sqrt{t^2 + \tan t + 25}$
13. $y = \dfrac{4a}{\sqrt{3}}\sin x - a$ **15.** $y = e^{x^2/2}$
17. (a) $\sin^{-1} y = x^2 + C$
(b) $y = \sin(x^2)$, $-\sqrt{\pi/2} \le x \le \sqrt{\pi/2}$ (c) No

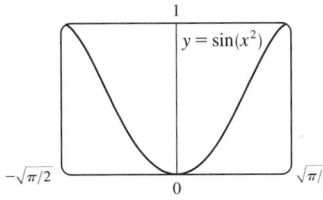

19. $\cos y = \cos x - 1$

21. III   **23.** IV

25.

27.

29. 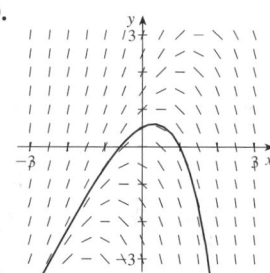   **31.**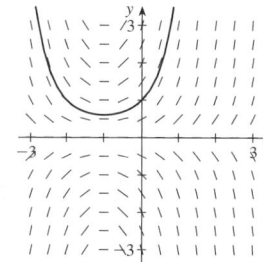

33. (b) $P(t) = M - Me^{-kt}$; M

35. (a) $C(t) = (C_0 - r/k)e^{-kt} + r/k$

(b) r/k; the concentration approaches r/k regardless of the value of C_0

37. $P(40) \approx 732$, $P(80) \approx 985$; $t \approx 55$

39. (a) $dy/dt = ky(1 - y)$

(b) $y = \dfrac{y_0}{y_0 + (1 - y_0)e^{-kt}}$ (c) 3:36 PM

43. (a) $15e^{-t/100}$ kg   (b) $15e^{-0.2} \approx 12.3$ kg

45. About 4.9%   **47.** g/k

49. (a) $dA/dt = k\sqrt{A}\,(M - A)$

(b) $A(t) = M\left(\dfrac{Ce^{\sqrt{M}kt} - 1}{Ce^{\sqrt{M}kt} + 1}\right)^2$, where $C = \dfrac{\sqrt{M} + \sqrt{A_0}}{\sqrt{M} - \sqrt{A_0}}$

and $A_0 = A(0)$

CHAPTER 7 REVIEW ■ **PAGE 422**

Exercises

1. $\frac{8}{3}$ **3.** $\frac{7}{12}$ **5.** $64\pi/15$ **7.** $1656\pi/5$

9. $\frac{4}{3}\pi(2ah + h^2)^{3/2}$ **11.** $\int_{-\pi/3}^{\pi/3} 2\pi(\pi/2 - x)(\cos^2 x - \frac{1}{4})\,dx$

13. (a) $2\pi/15$ (b) $\pi/6$ (c) $8\pi/15$

15. (a) 0.38   (b) 0.87

17. Solid obtained by rotating the region $0 \le y \le \cos x$, $0 \le x \le \pi/2$ about the y-axis

19. Solid obtained by rotating the region $0 \le x \le \pi$, $0 \le y \le 2 - \sin x$ about the x-axis

21. 36   **23.** $\frac{125}{3}\sqrt{3}$ m³   **25.** $\frac{15}{2}$ **27.** (a) $\frac{21}{16}$ (b) $\frac{41}{10}\pi$

29. 3.8202   **31.** $\frac{124}{5}$ **33.** 3.2 J

35. (a) $8000\pi/3 \approx 8378$ ft-lb   (b) 2.1 ft

37. ≈ 458 lb   **39.** $\left(\frac{8}{5}, 1\right)$ **41.** $2\pi^2$

43. $y = \pm\sqrt{\ln(x^2 + 2x^{3/2} + C)}$ **45.** $r(t) = 5e^{t-t^2}$

47. (a)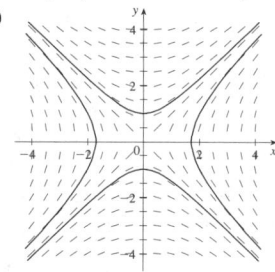

(b) The pair of lines $y = \pm x$, for $C = 0$; the hyperbola $x^2 - y^2 = -C$ for $C \ne 0$.

CHAPTER 8

EXERCISES 8.1 ■ PAGE 434

Abbreviations: C, convergent; D, divergent

1. (a) A sequence is an ordered list of numbers. It can also be defined as a function whose domain is the set of positive integers.

(b) The terms a_n approach 8 as n becomes large.

(c) The terms a_n become large as n becomes large.

3. $\frac{1}{3}, \frac{2}{5}, \frac{3}{7}, \frac{4}{9}, \frac{5}{11}, \frac{6}{13}$; yes; $\frac{1}{2}$ **5.** $a_n = -3\left(-\frac{2}{3}\right)^{n-1}$

7. $a_n = (-1)^{n+1}\dfrac{n^2}{n+1}$ **9.** 1   **11.** 5   **13.** 1

15. D   **17.** 0   **19.** D   **21.** 0   **23.** 0   **25.** 0

27. e^2 **29.** D   **31.** ln 2

33. (a) 1060, 1123.60, 1191.02, 1262.48, 1338.23   (b) D

35. Convergent by the Monotonic Sequence Theorem; $5 \le L < 8$

37. Decreasing; yes   **39.** Not monotonic; no   **41.** 2

43. $(3 + \sqrt{5})/2$ **45.** (b) $(1 + \sqrt{5})/2$ **47.** 62

EXERCISES 8.2 ■ PAGE 443

1. (a) A sequence is an ordered list of numbers whereas a series is the *sum* of a list of numbers.

(b) A series is convergent if the sequence of partial sums is a convergent sequence. A series is divergent if it is not convergent.

3. 1, 1.125, 1.1620, 1.1777, 1.1857, 1.1903, 1.1932, 1.1952; C

5. 0.5, 1.3284, 2.4265, 3.7598, 5.3049, 7.0443, 8.9644, 11.0540; D

7. $\frac{25}{3}$ **9.** $\frac{1}{7}$ **11.** D   **13.** D   **15.** D   **17.** $\frac{5}{2}$

19. D   **21.** D   **23.** D   **25.** $\frac{3}{2}$ **27.** $\frac{11}{6}$

29. (b) 1   (c) 2   (d) All rational numbers with a terminating decimal representation, except 0.

31. $\frac{8}{9}$ **33.** $\frac{838}{333}$

35. $-\dfrac{1}{5} < x < \dfrac{1}{5}$; $\dfrac{-5x}{1 + 5x}$

37. $-1 < x < 5$; $\dfrac{3}{5 - x}$

39. $a_1 = 0$, $a_n = 2/[n(n + 1)]$ for $n > 1$, sum $= 1$

41. (a) 157.875 mg; $\frac{3000}{19}(1 - 0.05^n)$ (b) 157.895 mg

43. (a) $S_n = \dfrac{D(1 - c^n)}{1 - c}$ (b) 5

45. $\frac{1}{2}(\sqrt{3} - 1)$ **47.** $1/[n(n + 1)]$

49. The series is divergent.

55. $\{s_n\}$ is bounded and increasing.

57. (a) $0, \frac{1}{9}, \frac{2}{9}, \frac{1}{3}, \frac{2}{3}, \frac{7}{9}, \frac{8}{9}, 1$

59. (a) $\frac{1}{2}, \frac{5}{6}, \frac{23}{24}, \frac{119}{120}$; $[(n + 1)! - 1]/(n + 1)!$ (c) 1

EXERCISES 8.3 ▪ PAGE 452

1. C

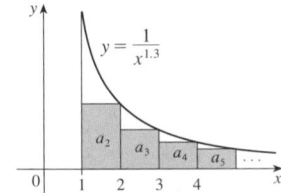

3. (a) Nothing (b) C
5. p-series; geometric series; $b < -1$; $-1 < b < 1$
7. D **9.** C **11.** D **13.** C **15.** D
17. C **19.** D **21.** C **23.** C **25.** D
27. C **29.** D **31.** $p > 1$
35. (a) 1.54977, error ≤ 0.1 (b) 1.64522, error ≤ 0.005
(c) $n > 1000$
41. Yes

EXERCISES 8.4 ▪ PAGE 463

Abbreviations: AC, absolutely convergent;
CC, conditionally convergent
1. (a) A series whose terms are alternately positive and
negative
(b) $0 < b_{n+1} \le b_n$ and $\lim_{n \to \infty} b_n = 0$, where $b_n = |a_n|$
(c) $|R_n| \le b_{n+1}$
3. C **5.** C **7.** D **9.** 5 **11.** 4
13. -0.4597 **15.** 0.0676 **17.** An underestimate
19. AC **21.** AC **23.** CC **25.** AC **27.** AC
29. AC **31.** AC **33.** AC **35.** D
37. AC **39.** D **41.** AC **43.** (a) and (d)

EXERCISES 8.5 ▪ PAGE 468

1. A series of the form $\sum_{n=0}^{\infty} c_n(x - a)^n$, where x is a variable and
a and the c_n's are constants
3. $1, (-1, 1)$ **5.** $1, [-1, 1)$ **7.** $\infty, (-\infty, \infty)$
9. $2, (-2, 2)$ **11.** $4, (-4, 4]$ **13.** $\frac{1}{3}, \left[-\frac{1}{3}, \frac{1}{3}\right]$
15. $1, [1, 3]$ **17.** $b, (a - b, a + b)$ **19.** $0, \left\{\frac{1}{2}\right\}$
21. $\infty, (-\infty, \infty)$ **23.** (a) Yes (b) No
25. k^k **27.** No
29. (a) $(-\infty, \infty)$
(b), (c)

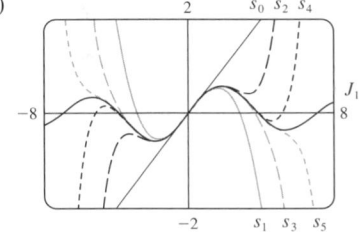

31. $(-1, 1), f(x) = (1 + 2x)/(1 - x^2)$
35. 2

EXERCISES 8.6 ▪ PAGE 474

1. 10 **3.** $\sum_{n=0}^{\infty} (-1)^n x^n, (-1, 1)$ **5.** $2 \sum_{n=0}^{\infty} \frac{1}{3^{n+1}} x^n, (-3, 3)$
7. $\sum_{n=0}^{\infty} (-1)^n \frac{1}{9^{n+1}} x^{2n+1}, (-3, 3)$ **9.** $1 + 2 \sum_{n=1}^{\infty} x^n, (-1, 1)$
11. $\sum_{n=0}^{\infty} \left[(-1)^{n+1} - \frac{1}{2^{n+1}}\right] x^n, (-1, 1)$
13. (a) $\sum_{n=0}^{\infty} (-1)^n (n + 1) x^n, R = 1$

(b) $\frac{1}{2} \sum_{n=0}^{\infty} (-1)^n (n + 2)(n + 1) x^n, R = 1$

(c) $\frac{1}{2} \sum_{n=2}^{\infty} (-1)^n n(n - 1) x^n, R = 1$

15. $\ln 5 - \sum_{n=1}^{\infty} \frac{x^n}{n 5^n}, R = 5$

17. $\sum_{n=0}^{\infty} (-1)^n 4^n (n + 1) x^{n+1}, R = \frac{1}{4}$

19. $\sum_{n=0}^{\infty} (2n + 1) x^n, R = 1$

21. $\sum_{n=0}^{\infty} (-1)^n \frac{1}{16^{n+1}} x^{2n+1}, R = 4$

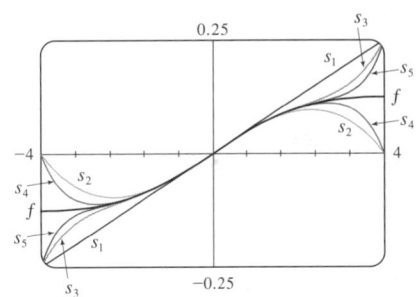

23. $\sum_{n=0}^{\infty} \frac{2x^{2n+1}}{2n + 1}, R = 1$

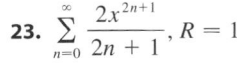
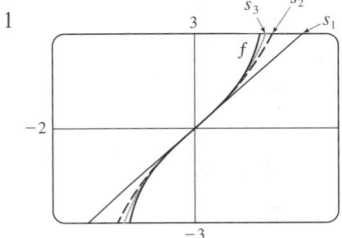

25. $C + \sum_{n=0}^{\infty} \frac{t^{8n+2}}{8n + 2}, R = 1$

27. $C + \sum_{n=1}^{\infty} (-1)^n \frac{x^{n+3}}{n(n + 3)}, R = 1$

29. 0.199989 **31.** 0.000983 **33.** 0.19740
35. (b) 0.920 **39.** $[-1, 1], [-1, 1), (-1, 1)$

EXERCISES 8.7 ▪ PAGE 487

1. $b_8 = f^{(8)}(5)/8!$ **3.** $\sum_{n=0}^{\infty} (n+1)x^n, R = 1$

5. $\sum_{n=0}^{\infty} (n+1)x^n, R = 1$

7. $\sum_{n=0}^{\infty} (-1)^n \dfrac{\pi^{2n+1}}{(2n+1)!} x^{2n+1}, R = \infty$

9. $\sum_{n=0}^{\infty} \dfrac{x^{2n+1}}{(2n+1)!}, R = \infty$

11. $-1 - 2(x-1) + 3(x-1)^2 + 4(x-1)^3 + (x-1)^4,$
$R = \infty$

13. $\ln 2 + \sum_{n=1}^{\infty} (-1)^{n+1} \dfrac{1}{n\,2^n}(x-2)^n, R = 2$

15. $\sum_{n=0}^{\infty} \dfrac{2^n e^6}{n!}(x-3)^n, R = \infty$

17. $\sum_{n=0}^{\infty} (-1)^{n+1} \dfrac{1}{(2n)!}(x-\pi)^{2n}, R = \infty$

23. $1 - \dfrac{1}{4}x - \sum_{n=2}^{\infty} \dfrac{3\cdot 7 \cdot \cdots \cdot (4n-5)}{4^n \cdot n!} x^n, R = 1$

25. $\sum_{n=0}^{\infty} (-1)^n \dfrac{(n+1)(n+2)}{2^{n+4}} x^n, R = 2$

27. $\sum_{n=0}^{\infty} (-1)^n \dfrac{\pi^{2n+1}}{(2n+1)!} x^{2n+1}, R = \infty$

29. $\sum_{n=0}^{\infty} \dfrac{2^n + 1}{n!} x^n, R = \infty$

31. $\sum_{n=0}^{\infty} (-1)^n \dfrac{1}{2^{2n}(2n)!} x^{4n+1}, R = \infty$

33. $\dfrac{1}{2}x + \sum_{n=1}^{\infty} (-1)^n \dfrac{1 \cdot 3 \cdot 5 \cdot \cdots \cdot (2n-1)}{n!\,2^{3n+1}} x^{2n+1}, R = 2$

35. $\sum_{n=1}^{\infty} (-1)^{n+1} \dfrac{2^{2n-1}}{(2n)!} x^{2n}, R = \infty$

37. $\sum_{n=0}^{\infty} (-1)^n \dfrac{1}{(2n)!} x^{4n}, R = \infty$

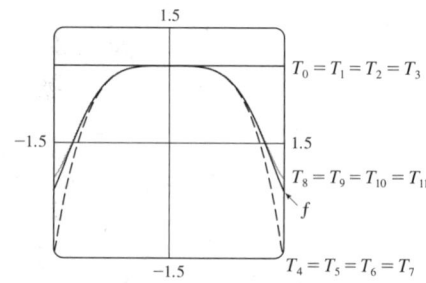

39. 0.99619

41. (a) $1 + \sum_{n=1}^{\infty} \dfrac{1 \cdot 3 \cdot 5 \cdot \cdots \cdot (2n-1)}{2^n n!} x^{2n}$

(b) $x + \sum_{n=1}^{\infty} \dfrac{1 \cdot 3 \cdot 5 \cdot \cdots \cdot (2n-1)}{(2n+1)2^n n!} x^{2n+1}$

43. $C + \sum_{n=0}^{\infty} (-1)^n \dfrac{x^{6n+2}}{(6n+2)(2n)!}, R = \infty$

45. $C + \sum_{n=1}^{\infty} (-1)^n \dfrac{1}{2n\,(2n)!} x^{2n}, R = \infty$

47. 0.440 **49.** 0.09998750 **51.** $\frac{1}{2}$ **53.** $\frac{1}{120}$

55. $1 - \frac{3}{2}x^2 + \frac{25}{24}x^4$ **57.** $1 + \frac{1}{6}x^2 + \frac{7}{360}x^4$ **59.** e^{-x^4}

61. $1/\sqrt{2}$ **63.** $e^3 - 1$

65. (a) $\sum_{n=1}^{\infty} nx^n$ (b) 2

EXERCISES 8.8 ▪ PAGE 494

1. (a) $T_0(x) = 1 = T_1(x)$, $T_2(x) = 1 - \frac{1}{2}x^2 = T_3(x)$,
$T_4(x) = 1 - \frac{1}{2}x^2 + \frac{1}{24}x^4 = T_5(x)$,
$T_6(x) = 1 - \frac{1}{2}x^2 + \frac{1}{24}x^4 - \frac{1}{720}x^6$

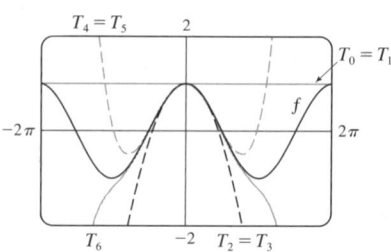

3. $\frac{1}{2} - \frac{1}{4}(x-2) + \frac{1}{8}(x-2)^2 - \frac{1}{16}(x-2)^3$

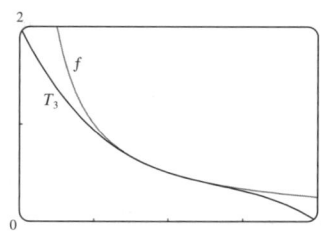

5. $-\left(x - \dfrac{\pi}{2}\right) + \dfrac{1}{6}\left(x - \dfrac{\pi}{2}\right)^3$

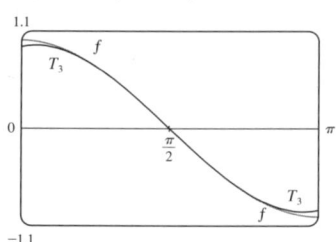

7. $x - 2x^2 + 2x^3$

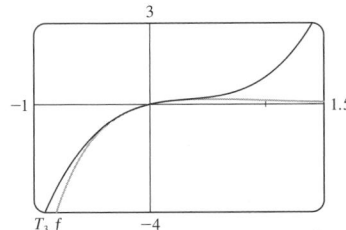

9. (a) $2 + \frac{1}{4}(x - 4) - \frac{1}{64}(x - 4)^2$ (b) 0.000015625

11. (a) $1 + \frac{2}{3}(x - 1) - \frac{1}{9}(x - 1)^2 + \frac{4}{81}(x - 1)^3$

(b) 0.000097

13. (a) $1 + x^2$ (b) 0.00006

15. (a) $x^2 - \frac{1}{6}x^4$ (b) 0.042 **17.** 0.17365 **19.** Four

21. $-1.037 < x < 1.037$ **23.** 21 m, no

27. (c) They differ by about 8×10^{-9} km.

CHAPTER 8 REVIEW ■ **PAGE 497**

True-False Quiz

1. False **3.** True **5.** False **7.** False **9.** False
11. True **13.** True **15.** False **17.** True
19. True **21.** True

Exercises

1. $\frac{1}{2}$ **3.** D **5.** 0 **7.** e^{12} **9.** C **11.** C
13. D **15.** C **17.** C **19.** C **21.** CC **23.** AC
25. 8 **27.** $\pi/4$ **29.** e^{-e} **33.** 0.9721
37. $4, [-6, 2)$ **39.** $0.5, [2.5, 3.5]$

41. $\frac{1}{2} \sum_{n=0}^{\infty} (-1)^n \left[\frac{1}{(2n)!} \left(x - \frac{\pi}{6} \right)^{2n} + \frac{\sqrt{3}}{(2n + 1)!} \left(x - \frac{\pi}{6} \right)^{2n+1} \right]$

43. $\sum_{n=0}^{\infty} (-1)^n x^{n+2}, R = 1$ **45.** $\ln 4 - \sum_{n=1}^{\infty} \frac{x^n}{n4^n}, R = 4$

47. $\sum_{n=0}^{\infty} (-1)^n \frac{x^{8n+4}}{(2n + 1)!}, R = \infty$

49. $\frac{1}{2} + \sum_{n=1}^{\infty} \frac{1 \cdot 5 \cdot 9 \cdot \cdots \cdot (4n - 3)}{n! 2^{6n+1}} x^n, R = 16$

51. $C + \ln |x| + \sum_{n=1}^{\infty} \frac{x^n}{n \cdot n!}$

53. (a) $1 + \frac{1}{2}(x - 1) - \frac{1}{8}(x - 1)^2 + \frac{1}{16}(x - 1)^3$

(b) (c) 0.000006

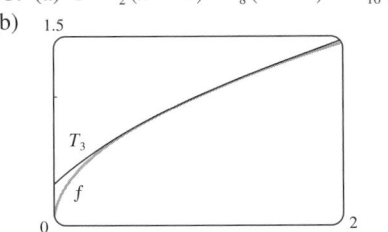

55. $-\frac{1}{6}$ **57.** 2

CHAPTER 9

EXERCISES 9.1 ■ **PAGE 505**

1. 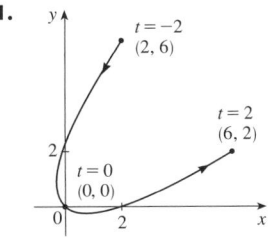 **3.**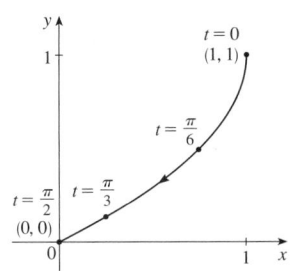

5. (a) 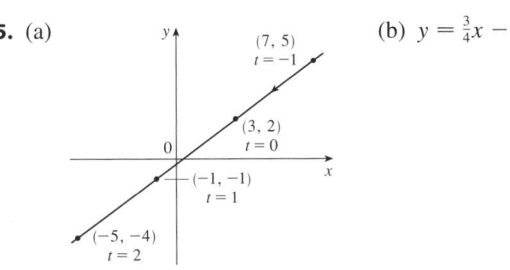 (b) $y = \frac{3}{4}x - \frac{1}{4}$

7. (a) 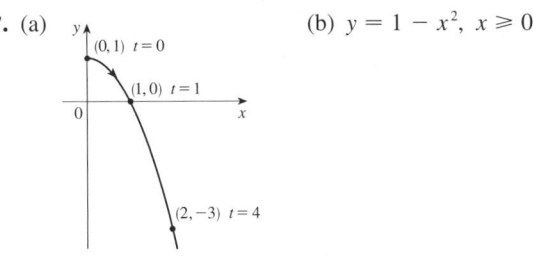 (b) $y = 1 - x^2, x \geq 0$

9. (a) $x^2 + y^2 = 1, y \geq 0$ (b)
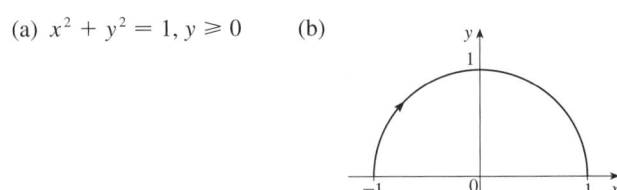

11. (a) $y = 1/x, y > 1$ (b)
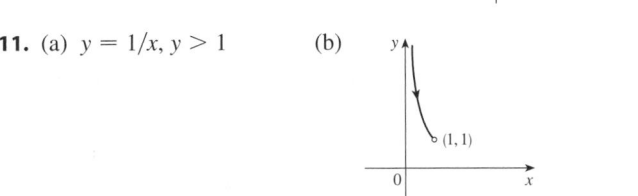

13. (a) $y = \frac{1}{2} \ln x + 1$ (b)
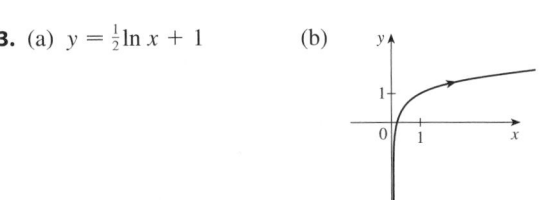

15. Moves counterclockwise along the circle
$(x - 3)^2 + (y - 1)^2 = 4$ from $(3, 3)$ to $(3, -1)$
17. Moves 3 times clockwise around the ellipse
$(x^2/25) + (y^2/4) = 1$, starting and ending at $(0, -2)$
19.

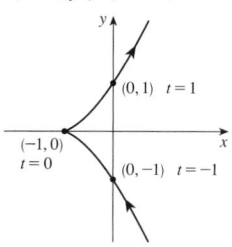

21.

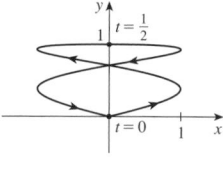

23.

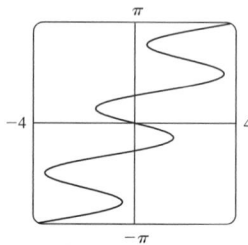

25. (b) $x = -2 + 5t$, $y = 7 - 8t$, $0 \le t \le 1$
27. (a) $x = 2 \cos t$, $y = 1 - 2 \sin t$, $0 \le t \le 2\pi$
(b) $x = 2 \cos t$, $y = 1 + 2 \sin t$, $0 \le t \le 6\pi$
(c) $x = 2 \cos t$, $y = 1 + 2 \sin t$, $\pi/2 \le t \le 3\pi/2$
31. The curve $y = x^{2/3}$ is generated in (a). In (b), only the portion with $x \ge 0$ is generated, and in (c) we get only the portion with $x > 0$.
35. $x = a \cos \theta$, $y = b \sin \theta$; $(x^2/a^2) + (y^2/b^2) = 1$, ellipse
37. (a) Two points of intersection

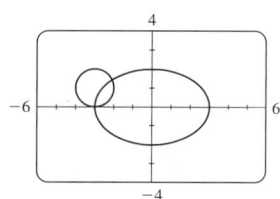

(b) One collision point at $(-3, 0)$ when $t = 3\pi/2$
(c) There are still two intersection points, but no collision point.
39. For $c = 0$, there is a cusp; for $c > 0$, there is a loop whose size increases as c increases.

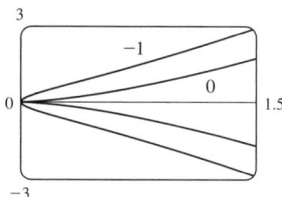

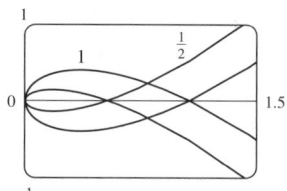

41. The curves roughly follow the line $y = x$, and they start having loops when a is between 1.4 and 1.6 (more precisely, when $a > \sqrt{2}$). The loops increase in size as a increases.
43. As n increases, the number of oscillations increases; a and b determine the width and height.

EXERCISES 9.2 ■ PAGE 513

1. $\dfrac{2t + 1}{t \cos t + \sin t}$ **3.** $y = -\dfrac{3}{2}x + 7$ **5.** $y = \pi x + \pi^2$
7. $y = 2x + 1$
9. $\dfrac{2t + 1}{2t}$, $-\dfrac{1}{4t^3}$, $t < 0$ **11.** $e^{-2t}(1 - t)$, $e^{-3t}(2t - 3)$, $t > \dfrac{3}{2}$
13. Horizontal at $(0, -3)$, vertical at $(\pm 2, -2)$
15. Horizontal at $(\pm\sqrt{2}, \pm 1)$ (four points), vertical at $(\pm 2, 0)$
17. $(0.6, 2)$; $\left(5 \cdot 6^{-6/5}, e^{6^{-1/5}}\right)$
19.

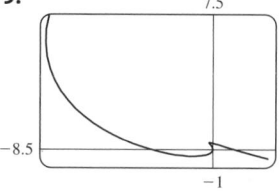

21. $y = x$, $y = -x$

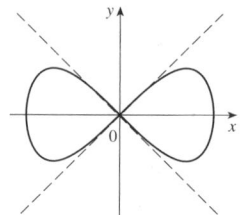

23. (a) $d \sin \theta/(r - d \cos \theta)$ **25.** $\left(\frac{16}{27}, \frac{29}{9}\right)$, $(-2, -4)$
27. πab **29.** $3 - e$ **31.** $2\pi r^2 + \pi d^2$
33. $\int_0^2 \sqrt{2 + 2e^{-2t}}\, dt \approx 3.1416$
35. $\int_0^{4\pi} \sqrt{5 - 4 \cos t}\, dt \approx 26.7298$ **37.** $4\sqrt{2} - 2$
39. $\frac{1}{2}\sqrt{2} + \frac{1}{2}\ln(1 + \sqrt{2})$
41. $\sqrt{2}\,(e^\pi - 1)$

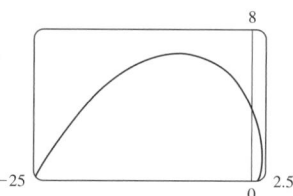

43. $e^3 + 11 - e^{-8}$

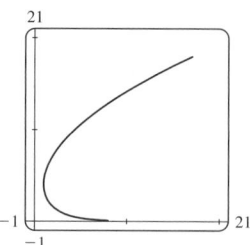

45. 612.3053   **47.** $6\sqrt{2}$, $\sqrt{2}$
51. (a) $t \in [0, 4\pi]$

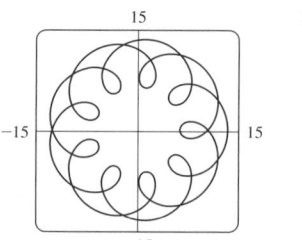

(b) 294

EXERCISES 9.3 ▪ **PAGE 522**

1. (a)

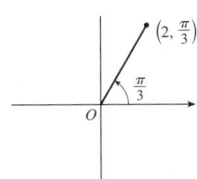

(b)

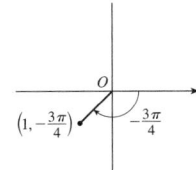

$(2, 7\pi/3)$, $(-2, 4\pi/3)$

$(1, 5\pi/4)$, $(-1, \pi/4)$

(c)

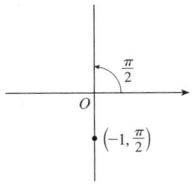

$(1, 3\pi/2)$, $(-1, 5\pi/2)$

3. (a)

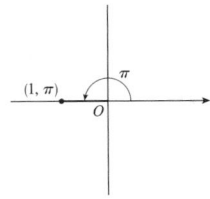

(b)

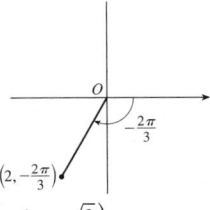

$(-1, 0)$

$(-1, -\sqrt{3})$

(c)

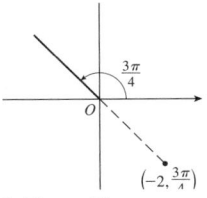

$(\sqrt{2}, -\sqrt{2})$

5. (a) (i) $(2\sqrt{2}, 7\pi/4)$ (ii) $(-2\sqrt{2}, 3\pi/4)$
(b) (i) $(2, 2\pi/3)$ (ii) $(-2, 5\pi/3)$

7.

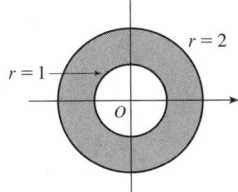

9.

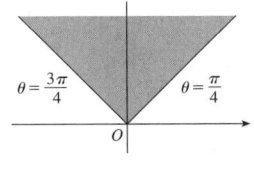

11.

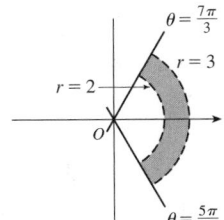

13. Circle, center $(1, 0)$, radius 1
15. Hyperbola, center O, foci on x-axis

17. $r = 1/(\sin\theta - 3\cos\theta)$
19. $r = 2c\cos\theta$ **21.** (a) $\theta = \pi/6$ (b) $x = 3$

23.

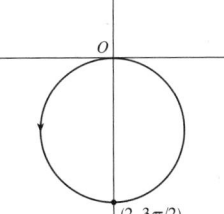

25.

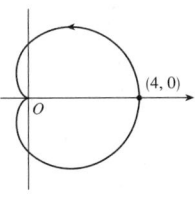

27.

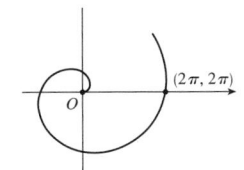

29.

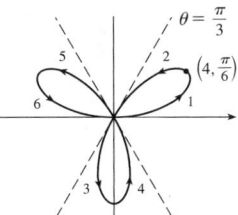

31.

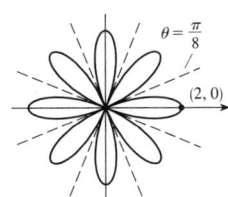

33.

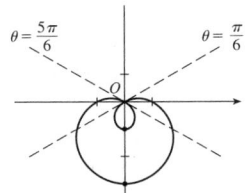

35.

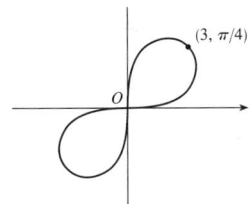

37.

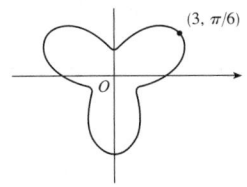

39.

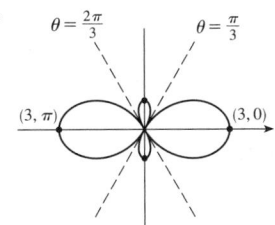

41.

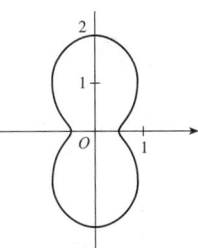

43.

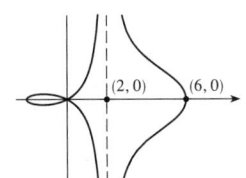

45.

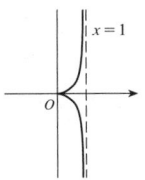

47. $\sqrt{3}$ **49.** $-\pi$

51. Horizontal at $(3/\sqrt{2},\ \pi/4)$, $(-3/\sqrt{2},\ 3\pi/4)$; vertical at $(3,\ 0)$, $(0,\ \pi/2)$

53. Horizontal at $\left(\frac{3}{2},\ \pi/3\right)$, $(0,\ \pi)$ [the pole], and $\left(\frac{3}{2},\ 5\pi/3\right)$; vertical at $(2,\ 0)$, $\left(\frac{1}{2},\ 2\pi/3\right)$, $\left(\frac{1}{2},\ 4\pi/3\right)$

55. Center $(b/2,\ a/2)$, radius $\sqrt{a^2 + b^2}/2$

57.

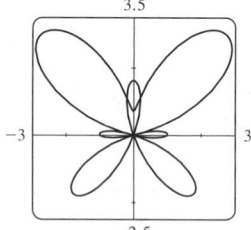

59.

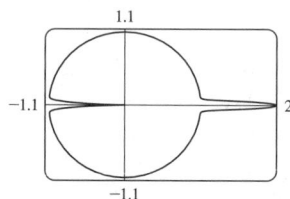

61. By counterclockwise rotation through angle $\pi/6$, $\pi/3$, or α about the origin

63. (a) A rose with n loops if n is odd and $2n$ loops if n is even (b) Number of loops is always $2n$

65. For $0 < a < 1$, the curve is an oval, which develops a dimple as $a \to 1^-$. When $a > 1$, the curve splits into two parts, one of which has a loop.

EXERCISES 9.4 ■ PAGE 528

1. $e^{-\pi/4} - e^{-\pi/2}$ **3.** $\frac{9}{2}$ **5.** π^2 **7.** $\frac{41}{4}\pi$

9. π **11.** 11π

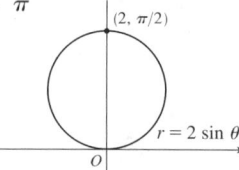

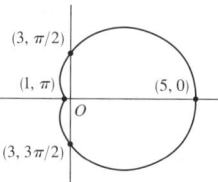

13. $\frac{3}{2}\pi$

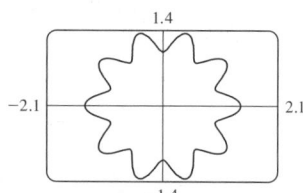

15. $\frac{4}{3}\pi$ **17.** $\pi - \frac{3}{2}\sqrt{3}$ **19.** $\frac{1}{3}\pi + \frac{1}{2}\sqrt{3}$ **21.** π

23. $\frac{5}{24}\pi - \frac{1}{4}\sqrt{3}$ **25.** $\frac{1}{2}\pi - 1$ **27.** $\frac{1}{4}(\pi + 3\sqrt{3})$

29. $\left(\frac{3}{2},\ \pi/6\right)$, $\left(\frac{3}{2},\ 5\pi/6\right)$, and the pole

31. $\left(\frac{1}{2}\sqrt{3},\ \pi/3\right)$, $\left(\frac{1}{2}\sqrt{3},\ 2\pi/3\right)$, and the pole

33. π **35.** $\frac{8}{3}\left[(\pi^2 + 1)^{3/2} - 1\right]$ **37.** 8.0091

EXERCISES 9.5 ■ PAGE 534

1. $r = \dfrac{4}{2 + \cos\theta}$ **3.** $r = \dfrac{6}{2 + 3\sin\theta}$

5. $r = \dfrac{8}{1 - \sin\theta}$ **7.** $r = \dfrac{4}{2 + \cos\theta}$

9. (a) $\frac{4}{5}$ (b) Ellipse (c) $y = -1$
(d)

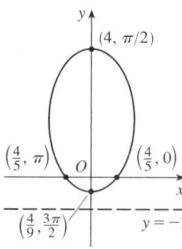

11. (a) 1 (b) Parabola (c) $y = \frac{2}{3}$
(d)

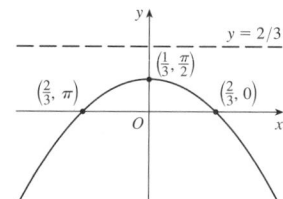

13. (a) $\frac{1}{3}$ (b) Ellipse (c) $x = \frac{9}{2}$
(d)

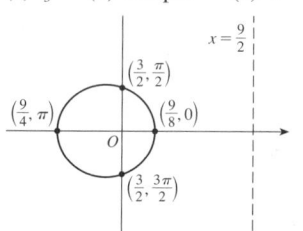

15. (a) 2 (b) Hyperbola (c) $x = -\frac{3}{8}$
(d)

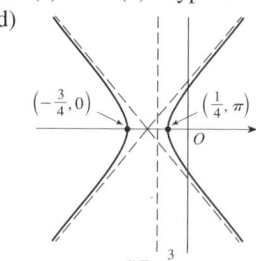

17. The ellipse is nearly circular when e is close to 0 and becomes more elongated as $e \to 1^-$. At $e = 1$, the curve becomes a parabola.

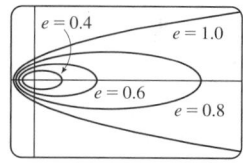

23. (b) $r = \dfrac{1.49 \times 10^8}{1 - 0.017 \cos \theta}$

25. 35.64 AU   **27.** 7.0×10^7 km   **29.** 3.6×10^8 km

CHAPTER 9 REVIEW ▪ PAGE 535

True-False Quiz

1. False   **3.** False   **5.** True   **7.** False

Exercises

1. $x = y^2 - 8y + 12$

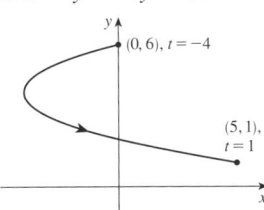

3. $y = 1/x$

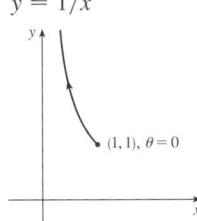

5. $x = t$, $y = \sqrt{t}$; $x = t^4$, $y = t^2$;
$x = \tan^2 t$, $y = \tan t$, $0 \leq t < \pi/2$

7. (a)

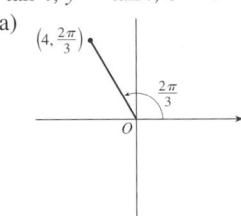

$\left(-2, 2\sqrt{3}\right)$

(b) $\left(3\sqrt{2}, 3\pi/4\right)$,
$\left(-3\sqrt{2}, 7\pi/4\right)$

9.

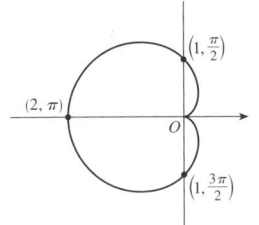

11.

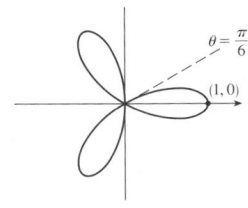

13.

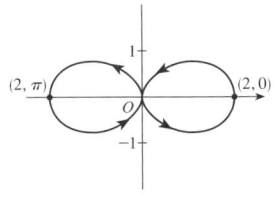

15.

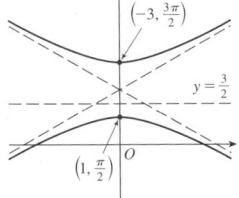

17. $r = \dfrac{2}{\cos \theta + \sin \theta}$

19.

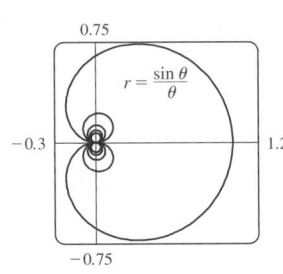

$r = \dfrac{\sin \theta}{\theta}$

21. 2   **23.** -1

25. $\dfrac{1 + \sin t}{1 + \cos t}$, $\dfrac{1 + \cos t + \sin t}{(1 + \cos t)^3}$ **27.** $\left(\frac{11}{8}, \frac{3}{4}\right)$

29. Vertical tangent at
$\left(\frac{3}{2}a, \pm\frac{1}{2}\sqrt{3}\,a\right)$, $(-3a, 0)$;
horizontal tangent at
$(a, 0)$, $\left(-\frac{1}{2}a, \pm\frac{3}{2}\sqrt{3}\,a\right)$

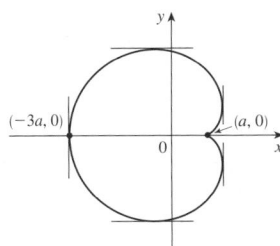

31. 18   **33.** $(2, \pm\pi/3)$ **35.** $\frac{1}{2}(\pi - 1)$

37. $2\left(5\sqrt{5} - 1\right)$

39. $\dfrac{2\sqrt{\pi^2 + 1} - \sqrt{4\pi^2 + 1}}{2\pi} + \ln\left(\dfrac{2\pi + \sqrt{4\pi^2 + 1}}{\pi + \sqrt{\pi^2 + 1}}\right)$

41. All curves have the vertical asymptote $x = 1$. For $c < -1$, the curve bulges to the right. At $c = -1$, the curve is the line $x = 1$. For $-1 < c < 0$, it bulges to the left. At $c = 0$ there is a cusp at $(0, 0)$. For $c > 0$, there is a loop.

43. $r = 4/(3 + \cos \theta)$

45. $x = a(\cot \theta + \sin \theta \cos \theta)$, $y = a(1 + \sin^2\theta)$

CHAPTER 10

EXERCISES 10.1 ▪ PAGE 541

1. $(4, 0, -3)$ **3.** $C; A$

5. A vertical plane that intersects the xy-plane in the line $y = 2 - x$, $z = 0$

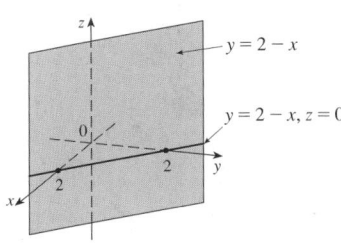

7. (a) $|PQ| = 6$, $|QR| = 2\sqrt{10}$, $|RP| = 6$; isosceles triangle
(b) $|PQ| = 3$, $|QR| = 3\sqrt{5}$, $|RP| = 6$; right triangle

9. (a) No   (b) Yes

11. $(x - 3)^2 + (y - 8)^2 + (z - 1)^2 = 30$

13. $(1, 2, -4)$, 6   **15.** $(2, 0, -6)$, $9/\sqrt{2}$

17. (b) $\frac{5}{2}$, $\frac{1}{2}\sqrt{94}$, $\frac{1}{2}\sqrt{85}$

19. (a) $(x - 2)^2 + (y + 3)^2 + (z - 6)^2 = 36$
(b) $(x - 2)^2 + (y + 3)^2 + (z - 6)^2 = 4$
(c) $(x - 2)^2 + (y + 3)^2 + (z - 6)^2 = 9$

21. A plane parallel to the yz-plane and 5 units in front of it

23. A half-space consisting of all points to the left of the plane $y = 8$

25. All points on or between the horizontal planes $z = 0$ and $z = 6$

27. All points on or inside a sphere with radius $\sqrt{3}$ and center O

29. All points on or inside a circular cylinder of radius 3 with axis the y-axis

31. $0 < x < 5$ **33.** $r^2 < x^2 + y^2 + z^2 < R^2$

35. $14x - 6y - 10z = 9$, a plane perpendicular to AB

37. $2\sqrt{3} - 3$

EXERCISES 10.2 ▪ PAGE 549

1. $\overrightarrow{AB} = \overrightarrow{DC}, \overrightarrow{DA} = \overrightarrow{CB}, \overrightarrow{DE} = \overrightarrow{EB}, \overrightarrow{EA} = \overrightarrow{CE}$

3. (a) 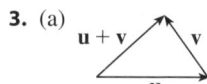 (b)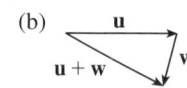

(c) 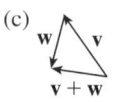 (d)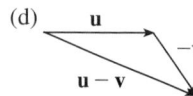

(e) 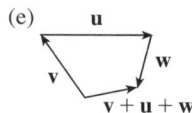 (f)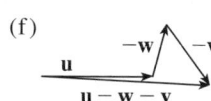

5. $\mathbf{a} = \langle 4, 1 \rangle$ **7.** $\mathbf{a} = \langle 2, 0, -2 \rangle$

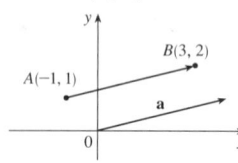 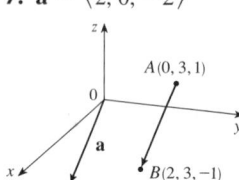

9. $\langle 5, 2 \rangle$ **11.** $\langle 3, 8, 1 \rangle$

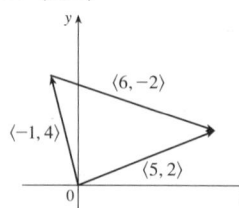 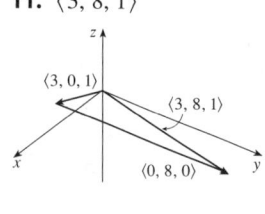

13. $\langle 2, -18 \rangle, \langle 1, -42 \rangle, 13, 10$

15. $-\mathbf{i} + \mathbf{j} + 2\mathbf{k}, -4\mathbf{i} + \mathbf{j} + 9\mathbf{k}, \sqrt{14}, \sqrt{82}$

17. $\frac{8}{9}\mathbf{i} - \frac{1}{9}\mathbf{j} + \frac{4}{9}\mathbf{k}$ **19.** $60°$

21. $\langle 2, 2\sqrt{3} \rangle$ **23.** ≈ 45.96 ft/s, ≈ 38.57 ft/s

25. $100\sqrt{7} \approx 264.6$ N, $\approx 139.1°$

27. $\sqrt{493} \approx 22.2$ mi/h, N8°W

29. $\mathbf{T}_1 = -196\,\mathbf{i} + 3.92\,\mathbf{j}, \mathbf{T}_2 = 196\,\mathbf{i} + 3.92\,\mathbf{j}$

31. (a) At an angle of 43.4° from the bank, toward upstream
(b) 20.2 min

33. $\pm(\mathbf{i} + 4\mathbf{j})/\sqrt{17}$

35. (a), (b) 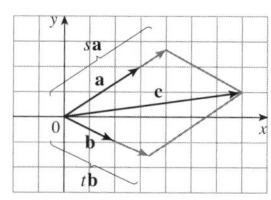 (d) $s = \frac{9}{7}, t = \frac{11}{7}$

37. A sphere with radius 1, centered at (x_0, y_0, z_0)

EXERCISES 10.3 ▪ PAGE 556

1. (b), (c), (d) are meaningful **3.** 14 **5.** 19 **7.** 1

9. -15 **11.** $\mathbf{u} \cdot \mathbf{v} = \frac{1}{2}, \mathbf{u} \cdot \mathbf{w} = -\frac{1}{2}$

15. $\cos^{-1}\left(\dfrac{1}{\sqrt{5}}\right) \approx 63°$ **17.** $\cos^{-1}\left(\dfrac{7}{\sqrt{130}}\right) \approx 52°$

19. (a) Neither (b) Orthogonal
(c) Orthogonal (d) Parallel

21. Yes **23.** $(\mathbf{i} - \mathbf{j} - \mathbf{k})/\sqrt{3}$ $\left[\text{or } (-\mathbf{i} + \mathbf{j} + \mathbf{k})/\sqrt{3}\right]$

25. $45°$ **27.** $0°$ at $(0, 0)$, $8.1°$ at $(1, 1)$

29. $4, \left\langle -\frac{20}{13}, \frac{48}{13} \right\rangle$ **31.** $\frac{9}{7}, \left\langle \frac{27}{49}, \frac{54}{49}, -\frac{18}{49} \right\rangle$

35. $\left\langle 0, 0, -2\sqrt{10} \right\rangle$ or any vector of the form $\left\langle s, t, 3s - 2\sqrt{10} \right\rangle, s, t \in \mathbb{R}$

37. 144 J **39.** $2400 \cos(40°) \approx 1839$ ft-lb

41. $\frac{13}{5}$ **43.** $\cos^{-1}(1/\sqrt{3}) \approx 55°$

EXERCISES 10.4 ▪ PAGE 564

1. $16\,\mathbf{i} + 48\,\mathbf{k}$ **3.** $15\,\mathbf{i} - 3\,\mathbf{j} + 3\,\mathbf{k}$ **5.** $\frac{1}{2}\mathbf{i} - \mathbf{j} + \frac{3}{2}\mathbf{k}$

7. $(1 - t)\,\mathbf{i} + (t^3 - t^2)\,\mathbf{k}$ **9.** 0 **11.** $\mathbf{i} + \mathbf{j} + \mathbf{k}$

13. (a) Scalar (b) Meaningless (c) Vector
(d) Meaningless (e) Meaningless (f) Scalar

15. $96\sqrt{3}$; into the page **17.** $\langle -7, 10, 8 \rangle, \langle 7, -10, -8 \rangle$

19. $\left\langle -\dfrac{1}{3\sqrt{3}}, -\dfrac{1}{3\sqrt{3}}, \dfrac{5}{3\sqrt{3}} \right\rangle, \left\langle \dfrac{1}{3\sqrt{3}}, \dfrac{1}{3\sqrt{3}}, -\dfrac{5}{3\sqrt{3}} \right\rangle$

27. 16 **29.** (a) $\langle 0, 18, -9 \rangle$ (b) $\frac{9}{2}\sqrt{5}$

31. (a) $\langle 13, -14, 5 \rangle$ (b) $\frac{1}{2}\sqrt{390}$

33. 9 **35.** 16 **39.** $10.8 \sin 80° \approx 10.6$ N·m

41. ≈ 417 N **43.** $60°$

45. (b) $\sqrt{97/3}$ **53.** (a) No (b) No (c) Yes

EXERCISES 10.5 ▪ PAGE 572

1. (a) True (b) False (c) True (d) False (e) False
(f) True (g) False (h) True (i) True (j) False
(k) True

3. $\mathbf{r} = (2\,\mathbf{i} + 2.4\,\mathbf{j} + 3.5\,\mathbf{k}) + t(3\,\mathbf{i} + 2\,\mathbf{j} - \mathbf{k})$;
$x = 2 + 3t, y = 2.4 + 2t, z = 3.5 - t$

5. $\mathbf{r} = (\mathbf{i} + 6\mathbf{k}) + t(\mathbf{i} + 3\mathbf{j} + \mathbf{k})$;
$x = 1 + t, y = 3t, z = 6 + t$

7. $x = 2 + 2t, y = 1 + \frac{1}{2}t, z = -3 - 4t$;
$(x - 2)/2 = 2y - 2 = (z + 3)/(-4)$

9. $x = 1 + t, y = -1 + 2t, z = 1 + t$;
$x - 1 = (y + 1)/2 = z - 1$

11. Yes
13. (a) $(x - 1)/(-1) = (y + 5)/2 = (z - 6)/(-3)$
(b) $(-1, -1, 0), \left(-\frac{3}{2}, 0, -\frac{3}{2}\right), (0, -3, 3)$
15. $\mathbf{r}(t) = (2\mathbf{i} - \mathbf{j} + 4\mathbf{k}) + t(2\mathbf{i} + 7\mathbf{j} - 3\mathbf{k}), 0 \leq t \leq 1$
17. Skew   **19.** $(4, -1, -5)$ **21.** $x + 4y + z = 4$
23. $5x - y - z = 7$ **25.** $x + y + z = 2$
27. $33x + 10y + 4z = 190$ **29.** $x - 2y + 4z = -1$
31. $3x - 8y - z = -38$ **33.** $(2, 3, 5)$
35. Neither, $\cos^{-1}\left(\frac{1}{3}\right) \approx 70.5°$ **37.** Parallel
39. (a) $x = 1, y = -t, z = t$ (b) $\cos^{-1}\left(\dfrac{5}{3\sqrt{3}}\right) \approx 15.8°$
41. $(x/a) + (y/b) + (z/c) = 1$
43. $x = 3t, y = 1 - t, z = 2 - 2t$
45. P_2 and P_3 are parallel, P_1 and P_4 are identical
47. $\sqrt{61/14}$ **49.** $\frac{18}{7}$ **51.** $5/(2\sqrt{14})$ **55.** $1/\sqrt{6}$
57. $13/\sqrt{69}$

EXERCISES 10.6 ▪ PAGE 579

1. (a) Parabola
(b) Parabolic cylinder with rulings parallel to the z-axis
(c) Parabolic cylinder with rulings parallel to the x-axis

3. Circular cylinder   **5.** Parabolic cylinder

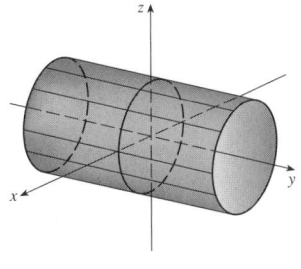   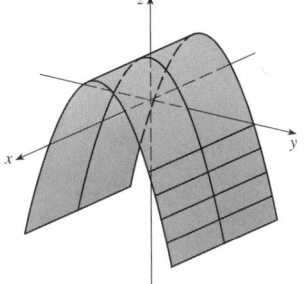

7. Hyperbolic cylinder

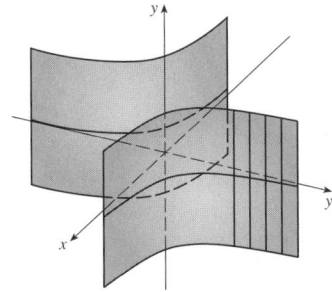

9. (a) $x = k, y^2 - z^2 = 1 - k^2$, hyperbola $(k \neq \pm 1)$;
$y = k, x^2 - z^2 = 1 - k^2$, hyperbola $(k \neq \pm 1)$;
$z = k, x^2 + y^2 = 1 + k^2$, circle
(b) The hyperboloid is rotated so that it has axis the y-axis
(c) The hyperboloid is shifted one unit in the negative y-direction

11. Elliptic paraboloid with axis the x-axis

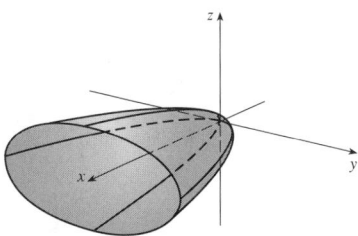

13. Elliptic cone with axis the x-axis

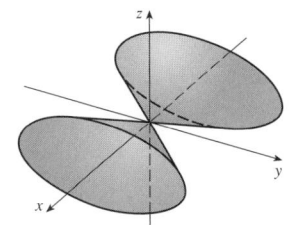

15. Hyperboloid of two sheets

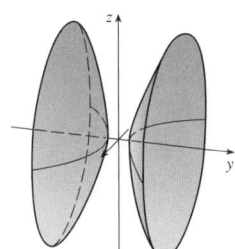

17. Ellipsoid

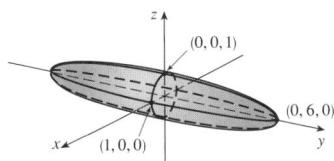

19. Hyperbolic paraboloid

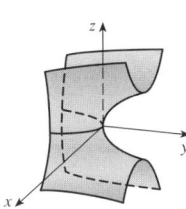

21. $y^2 = x^2 + \dfrac{z^2}{9}$

Elliptic cone with axis the

y-axis

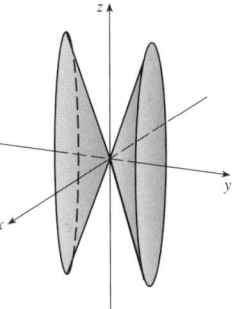

23. $y = z^2 - \dfrac{x^2}{2}$

Hyperbolic paraboloid

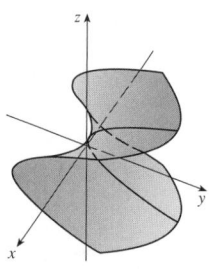

25. $x^2 + \dfrac{(y-2)^2}{4} + (z-3)^2 = 1$

Ellipsoid with center $(0, 2, 3)$

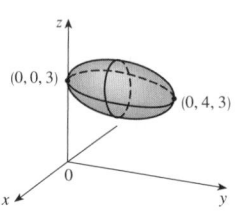

27. $(y+1)^2 = (x-2)^2 + (z-1)^2$
Circular cone with vertex $(2, -1, 1)$
and axis parallel to the y-axis

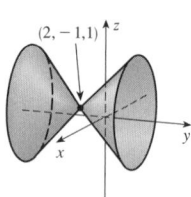

29.

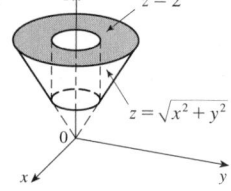

31. $-4x = y^2 + z^2$, paraboloid

33.

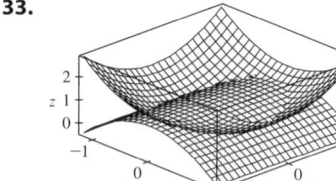

1. $(-1, 2]$ **3.** $\mathbf{i} + \mathbf{j} + \mathbf{k}$

5.

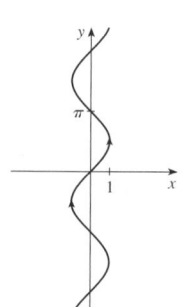

7.

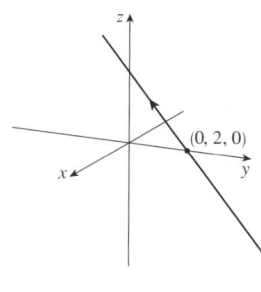

9.

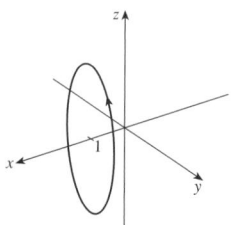

11.

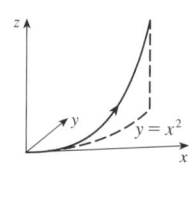

13. $\mathbf{r}(t) = \langle 2 + 4t, 2t, -2t \rangle$, $0 \le t \le 1$;
$x = 2 + 4t$, $y = 2t$, $z = -2t$, $0 \le t \le 1$
15. $\mathbf{r}(t) = \langle \frac{1}{2}t, -1 + \frac{4}{3}t, 1 - \frac{3}{4}t \rangle$, $0 \le t \le 1$;
$x = \frac{1}{2}t$, $y = -1 + \frac{4}{3}t$, $z = 1 - \frac{3}{4}t$, $0 \le t \le 1$
17. II   **19.** V   **21.** IV
23.   **25.** $(0, 0, 0)$, $(1, 0, 1)$

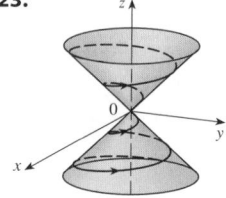

29. $\mathbf{r}(t) = t\,\mathbf{i} + \frac{1}{2}(t^2 - 1)\,\mathbf{j} + \frac{1}{2}(t^2 + 1)\,\mathbf{k}$
31. $x = 2\cos t$, $y = 2\sin t$, $z = 4\cos^2 t$
33. (a), (c)   (b) $\mathbf{r}'(t) = \langle 1, 2t \rangle$

35. (a), (c)   (b) $\mathbf{r}'(t) = \cos t\,\mathbf{i} - 2\sin t\,\mathbf{j}$

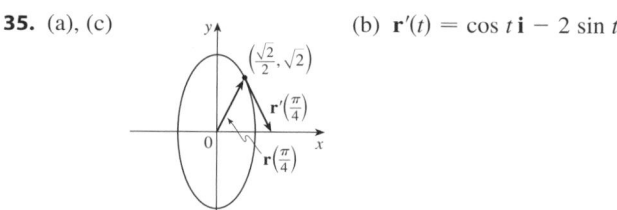

37. (a), (c) 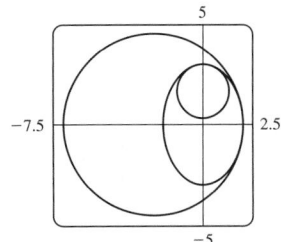   (b) $\mathbf{r}'(t) = 2e^{2t}\,\mathbf{i} + e^t\,\mathbf{j}$

39. $\mathbf{r}'(t) = \langle t \cos t + \sin t, 2t, \cos 2t - 2t \sin 2t \rangle$

41. $\mathbf{r}'(t) = 2te^{t^2}\,\mathbf{i} + [3/(1 + 3t)]\,\mathbf{k}$ **43.** $\mathbf{r}'(t) = \mathbf{b} + 2t\mathbf{c}$

45. $\frac{3}{5}\,\mathbf{j} + \frac{4}{5}\,\mathbf{k}$

47. $\langle 1, 2t, 3t^2 \rangle$, $\langle 1/\sqrt{14}, 2/\sqrt{14}, 3/\sqrt{14} \rangle$, $\langle 0, 2, 6t \rangle$, $\langle 6t^2, -6t, 2 \rangle$

49. $x = 3 + t, y = 2t, z = 2 + 4t$

51. $x = 1 - t, y = t, z = 1 - t$

53. $\mathbf{r}(t) = (3 - 4t)\,\mathbf{i} + (4 + 3t)\,\mathbf{j} + (2 - 6t)\,\mathbf{k}$

55. $x = -\pi - t, y = \pi + t, z = -\pi t$

57. $66°$ **59.** $2\,\mathbf{i} - 4\,\mathbf{j} + 32\,\mathbf{k}$ **61.** $\mathbf{i} + \mathbf{j} + \mathbf{k}$

63. $\tan t\,\mathbf{i} + \frac{1}{8}(t^2 + 1)^4\,\mathbf{j} + \left(\frac{1}{3}t^3 \ln t - \frac{1}{9}t^3\right)\mathbf{k} + \mathbf{C}$

65. $t^2\,\mathbf{i} + t^3\,\mathbf{j} + \left(\frac{2}{3}t^{3/2} - \frac{2}{3}\right)\mathbf{k}$

67. Yes

75. $2t \cos t + 2 \sin t - 2 \cos t \sin t$ **77.** 35

EXERCISES 10.8 ■ PAGE 598

1. $10\sqrt{10}$ **3.** $\frac{1}{27}(13^{3/2} - 8)$ **5.** 18.6833 **7.** 42

9. $\mathbf{r}(t(s)) = \dfrac{2}{\sqrt{29}}s\,\mathbf{i} + \left(1 - \dfrac{3}{\sqrt{29}}s\right)\mathbf{j} + \left(5 + \dfrac{4}{\sqrt{29}}s\right)\mathbf{k}$

11. $(3 \sin 1, 4, 3 \cos 1)$

13. (a) $\langle 1/\sqrt{10}, (-3/\sqrt{10})\sin t, (3/\sqrt{10})\cos t \rangle$, $\langle 0, -\cos t, -\sin t \rangle$ (b) $\frac{3}{10}$

15. (a) $\dfrac{1}{e^{2t} + 1}\langle \sqrt{2}e^t, e^{2t}, -1 \rangle$, $\dfrac{1}{e^{2t} + 1}\langle 1 - e^{2t}, \sqrt{2}e^t, \sqrt{2}e^t \rangle$

(b) $\sqrt{2}\,e^{2t}/(e^{2t} + 1)^2$

17. $6t^2/(9t^4 + 4t^2)^{3/2}$

19. $\frac{4}{25}$ **21.** $\frac{1}{7}\sqrt{\frac{19}{14}}$

23. $12x^2/(1 + 16x^6)^{3/2}$ **25.** $e^x|x + 2|/[1 + (xe^x + e^x)^2]^{3/2}$

27. $\left(-\frac{1}{2}\ln 2, 1/\sqrt{2}\right)$; approaches 0 **29.** (a) P (b) $1.3, 0.7$

31.

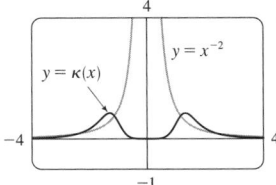

33.

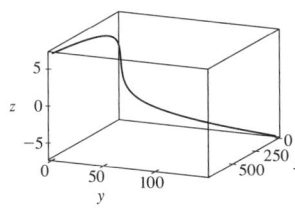 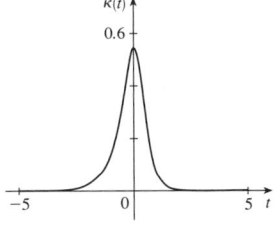

35. a is $y = f(x)$, b is $y = \kappa(x)$

37. $1/(\sqrt{2}e^t)$ **39.** $\langle \frac{2}{3}, \frac{2}{3}, \frac{1}{3} \rangle, \langle -\frac{1}{3}, \frac{2}{3}, -\frac{2}{3} \rangle, \langle -\frac{2}{3}, \frac{1}{3}, \frac{2}{3} \rangle$

41. $y = 6x + \pi, x + 6y = 6\pi$

43. $\left(x + \frac{5}{2}\right)^2 + y^2 = \frac{81}{4}, x^2 + \left(y - \frac{5}{3}\right)^2 = \frac{16}{9}$

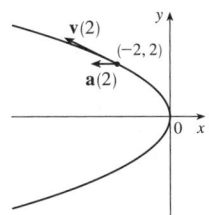

45. $(-1, -3, 1)$ **53.** 2.07×10^{10} Å ≈ 2 m

EXERCISES 10.9 ■ PAGE 608

1. $\mathbf{v}(t) = \langle -t, 1 \rangle$
$\mathbf{a}(t) = \langle -1, 0 \rangle$
$|\mathbf{v}(t)| = \sqrt{t^2 + 1}$

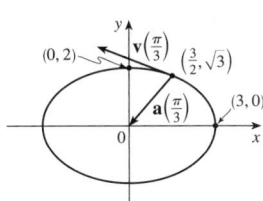

3. $\mathbf{v}(t) = -3 \sin t\,\mathbf{i} + 2 \cos t\,\mathbf{j}$
$\mathbf{a}(t) = -3 \cos t\,\mathbf{i} - 2 \sin t\,\mathbf{j}$
$|\mathbf{v}(t)| = \sqrt{5 \sin^2 t + 4}$

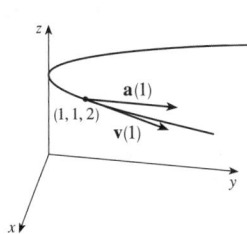

5. $\mathbf{v}(t) = \mathbf{i} + 2t\,\mathbf{j}$
$\mathbf{a}(t) = 2\,\mathbf{j}$
$|\mathbf{v}(t)| = \sqrt{1 + 4t^2}$

7. $\langle 2t + 1, 2t - 1, 3t^2 \rangle$, $\langle 2, 2, 6t \rangle$, $\sqrt{9t^4 + 8t^2 + 2}$

9. $\sqrt{2}\,\mathbf{i} + e^t\,\mathbf{j} - e^{-t}\,\mathbf{k}$, $e^t\,\mathbf{j} + e^{-t}\,\mathbf{k}$, $e^t + e^{-t}$

11. $\mathbf{v}(t) = t\,\mathbf{i} + 2t\,\mathbf{j} + \mathbf{k}$, $\mathbf{r}(t) = \left(\frac{1}{2}t^2 + 1\right)\mathbf{i} + t^2\,\mathbf{j} + t\,\mathbf{k}$

13. (a) $\mathbf{r}(t) = \left(\frac{1}{3}t^3 + t\right)\mathbf{i} + (t - \sin t + 1)\,\mathbf{j} + \left(\frac{1}{4} - \frac{1}{4}\cos 2t\right)\mathbf{k}$

(b)

15. $t = 4$

17. $\mathbf{r}(t) = t\,\mathbf{i} - t\,\mathbf{j} + \frac{5}{2}t^2\,\mathbf{k}$, $|\mathbf{v}(t)| = \sqrt{25t^2 + 2}$

19. (a) ≈ 3535 m (b) ≈ 1531 m (c) 200 m/s
21. 30 m/s **23.** $\approx 10.2°$, $\approx 79.8°$
25. $13.0° < \theta < 36.0°$, $55.4° < \theta < 85.5°$
27. $(250, -50, 0)$; $10\sqrt{93} \approx 96.4$ ft/s
29. (a) 16 m (b) $\approx 23.6°$ upstream

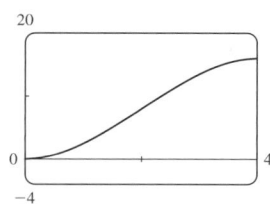

31. 0, 1 **33.** $6t, 6$ **35.** $t = 1$

CHAPTER 10 REVIEW ▪ **PAGE 610**

True-False Quiz

1. False **3.** False **5.** True **7.** True **9.** True
11. True **13.** True **15.** False **17.** False
19. False **21.** True **23.** True **25.** False
27. False **29.** False **31.** True **33.** False

Exercises

1. (a) $(x + 1)^2 + (y - 2)^2 + (z - 1)^2 = 69$
(b) $(y - 2)^2 + (z - 1)^2 = 68$, $x = 0$
(c) Center $(4, -1, -3)$, radius 5
3. $\mathbf{u} \cdot \mathbf{v} = 3\sqrt{2}$; $|\mathbf{u} \times \mathbf{v}| = 3\sqrt{2}$; out of the page
5. $-2, -4$ **7.** (a) 2 (b) -2 (c) -2 (d) 0
9. $\cos^{-1}(\frac{1}{3}) \approx 71°$ **11.** (a) $\langle 4, -3, 4 \rangle$ (b) $\sqrt{41}/2$
13. 166 N, 114 N
15. $x = 4 - 3t, y = -1 + 2t, z = 2 + 3t$
17. $x = -2 + 2t, y = 2 - t, z = 4 + 5t$
19. $-4x + 3y + z = -14$ **21.** $(1, 4, 4)$ **23.** Skew
25. $x + y + z = 4$ **27.** $22/\sqrt{26}$
29. Plane **31.** Cone

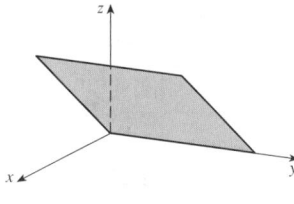

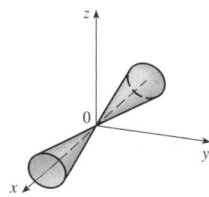

33. Hyperboloid of two sheets **35.** Ellipsoid

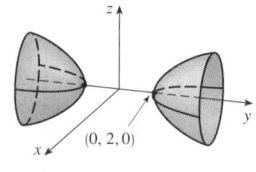

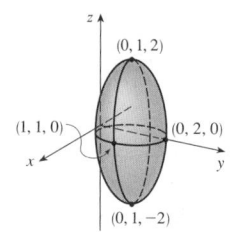

37. $4x^2 + y^2 + z^2 = 16$
39. (a)

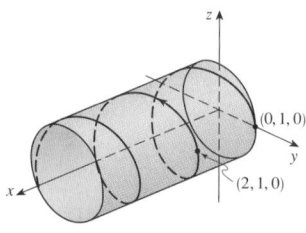

(b) $\mathbf{r}'(t) = \mathbf{i} - \pi \sin \pi t \, \mathbf{j} + \pi \cos \pi t \, \mathbf{k}$,
$\mathbf{r}''(t) = -\pi^2 \cos \pi t \, \mathbf{j} - \pi^2 \sin \pi t \, \mathbf{k}$
41. $\mathbf{r}(t) = 4 \cos t \, \mathbf{i} + 4 \sin t \, \mathbf{j} + (5 - 4 \cos t)\mathbf{k}$, $0 \le t \le 2\pi$
43. $\frac{1}{3}\mathbf{i} - (2/\pi^2)\mathbf{j} + (2/\pi)\mathbf{k}$ **45.** 86.631 **47.** $\pi/2$
49. (a) $\langle t^2, t, 1 \rangle/\sqrt{t^4 + t^2 + 1}$
(b) $\langle t^3 + 2t, 1 - t^4, -2t^3 - t \rangle/\sqrt{t^8 + 5t^6 + 6t^4 + 5t^2 + 1}$
(c) $\sqrt{t^8 + 5t^6 + 6t^4 + 5t^2 + 1}/(t^4 + t^2 + 1)^2$
51. $12/17^{3/2}$
53. $\mathbf{v}(t) = (1 + \ln t)\mathbf{i} + \mathbf{j} - e^{-t}\mathbf{k}$,
$|\mathbf{v}(t)| = \sqrt{2 + 2 \ln t + (\ln t)^2 + e^{-2t}}$, $\mathbf{a}(t) = (1/t)\mathbf{i} + e^{-t}\mathbf{k}$
55. (a) About 3.8 ft above the ground, 60.8 ft from the athlete
(b) ≈ 21.4 ft (c) ≈ 64.2 ft from the athlete
57. $\pi |t|$

CHAPTER 11

EXERCISES 11.1 ▪ **PAGE 623**

1. (a) 1 (b) $\mathbb{R}^2$ (c) $[-1, 1]$
3. (a) 3
(b) $\{(x, y, z) \mid x^2 + y^2 + z^2 < 4, x \ge 0, y \ge 0, z \ge 0\}$,
interior of a sphere of radius 2, center the origin, in the first octant
5. $\{(x, y) \mid y \le 2x\}$

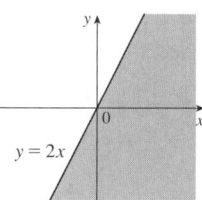

7. $\{(x, y) \mid \frac{1}{9}x^2 + y^2 < 1\}$

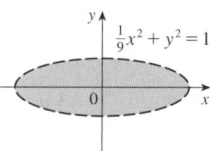

9. $\{(x, y) \mid y \ge x^2, x \ne \pm 1\}$

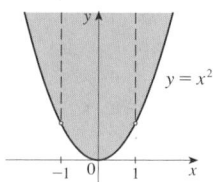

11. $\{(x, y, z) \mid x^2 + y^2 + z^2 \le 1\}$

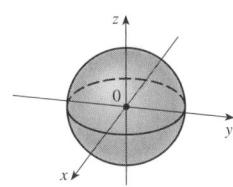

13. $4x + 5y + z = 10$, plane

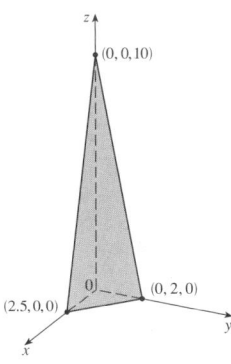

15. $z = y^2 + 1$, parabolic cylinder

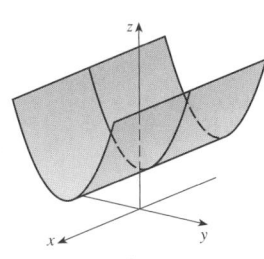

17. $z = 9 - x^2 - 9y^2$,
elliptic paraboloid

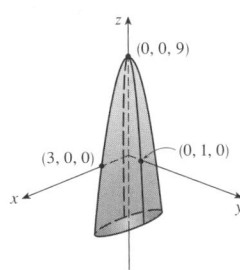

19. $z = \sqrt{4 - 4x^2 - y^2}$,
top half of ellipsoid

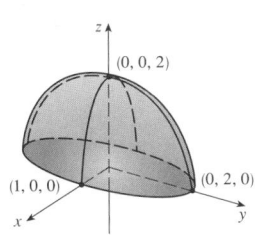

21. $\approx 56, \approx 35$ **23.** Steep; nearly flat

25. $(y - 2x)^2 = k$

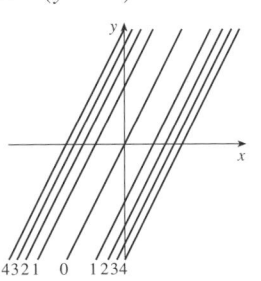

27. $y = -\sqrt{x} + k$

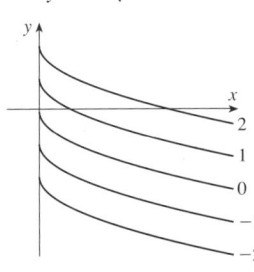

29. $y = ke^{-x}$

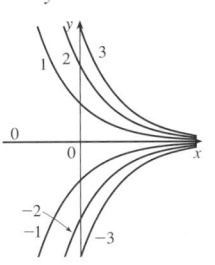

31. $y^2 - x^2 = k^2$

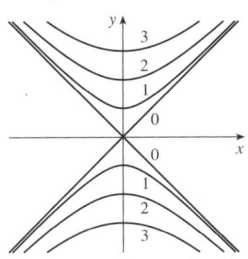

33. $x^2 + 9y^2 = k$

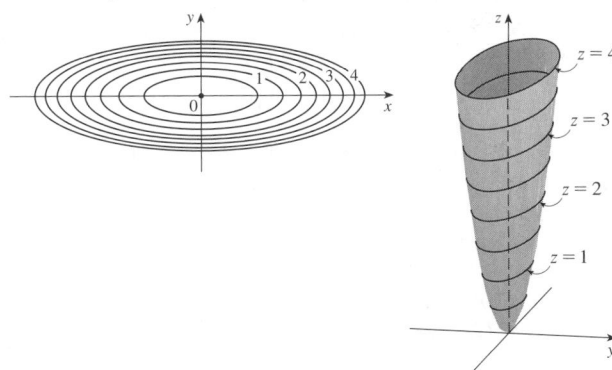

35.

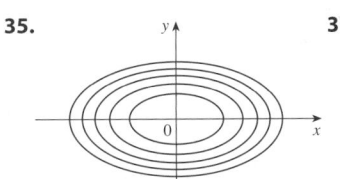

37.

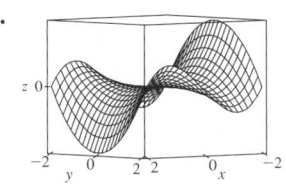

39.

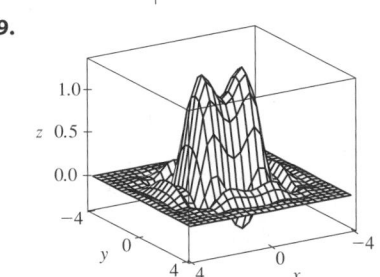

41. (a) C   (b) II   **43.** (a) F   (b) I
45. (a) B   (b) VI   **47.** Family of parallel planes
49. Family of circular cylinders with axis the x-axis $(k > 0)$

51. (a) Shift the graph of f upward 2 units
(b) Stretch the graph of f vertically by a factor of 2
(c) Reflect the graph of f about the xy-plane
(d) Reflect the graph of f about the xy-plane and then shift it upward 2 units

53.

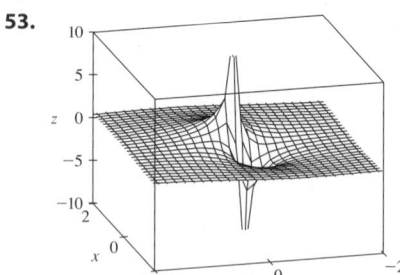

The function values approach 0 as x, y become large; as (x, y) approaches the origin, f approaches $\pm\infty$ or 0, depending on the direction of approach.

55. If $c = 0$, the graph is a cylindrical surface. For $c > 0$, the level curves are ellipses. The graph curves upward as we leave the origin, and the steepness increases as c increases. For $c < 0$, the level curves are hyperbolas. The graph curves upward in the y-direction and downward, approaching the xy-plane, in the x-direction giving a saddle-shaped appearance near $(0, 0, 1)$.

EXERCISES 11.2 ■ PAGE 632

1. Nothing; if f is continuous, $f(3, 1) = 6$ **3.** 1
5. Does not exist **7.** Does not exist
9. 0 **11.** Does not exist **13.** 2
15. Does not exist
17. The graph shows that the function approaches different numbers along different lines.
19. $h(x, y) = (2x + 3y - 6)^2 + \sqrt{2x + 3y - 6}$;
$\{(x, y) \mid 2x + 3y \geq 6\}$
21. $\{(x, y) \mid x^2 + y^2 \neq 1\}$
23. $\{(x, y) \mid x^2 + y^2 > 4\}$ **25.** $\{(x, y, z) \mid x^2 + y^2 + z^2 \leq 1\}$
27. $\{(x, y) \mid (x, y) \neq (0, 0)\}$ **29.** 0 **31.** -1

EXERCISES 11.3 ■ PAGE 638

1. (a) The rate of change of temperature as longitude varies, with latitude and time fixed; the rate of change as only latitude varies; the rate of change as only time varies.
(b) Positive, negative, positive
3. (a) Positive (b) Negative
5. $f_x(1, 2) = -8 =$ slope of C_1, $f_y(1, 2) = -4 =$ slope of C_2

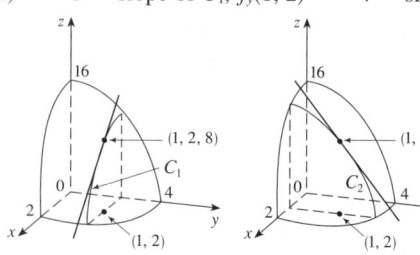

7. $f_x(x, y) = -3y$, $f_y(x, y) = 5y^4 - 3x$
9. $f_x(x, t) = -\pi e^{-t}\sin \pi x$, $f_t(x, t) = -e^{-t}\cos \pi x$
11. $f_x(x, y) = 1/y$, $f_y(x, y) = -x/y^2$
13. $f_x(x, y) = \dfrac{(ad - bc)y}{(cx + dy)^2}$, $f_y(x, y) = \dfrac{(bc - ad)x}{(cx + dy)^2}$
15. $g_u(u, v) = 10uv(u^2v - v^3)^4$,
$g_v(u, v) = 5(u^2 - 3v^2)(u^2v - v^3)^4$
17. $R_p(p, q) = \dfrac{q^2}{1 + p^2q^4}$, $R_q(p, q) = \dfrac{2pq}{1 + p^2q^4}$
19. $F_x(x, y) = \cos(e^x)$, $F_y(x, y) = -\cos(e^y)$
21. $f_x = z - 10xy^3z^4$, $f_y = -15x^2y^2z^4$, $f_z = x - 20x^2y^3z^3$
23. $\partial w/\partial x = 1/(x + 2y + 3z)$, $\partial w/\partial y = 2/(x + 2y + 3z)$, $\partial w/\partial z = 3/(x + 2y + 3z)$
25. $\partial u/\partial x = y \sin^{-1}(yz)$,
$\partial u/\partial y = x \sin^{-1}(yz) + xyz/\sqrt{1 - y^2z^2}$, $\partial u/\partial z = xy^2/\sqrt{1 - y^2z^2}$
27. $h_x = 2xy \cos(z/t)$, $h_y = x^2 \cos(z/t)$,
$h_z = (-x^2y/t) \sin(z/t)$, $h_t = (x^2yz/t^2) \sin(z/t)$
29. $\partial u/\partial x_i = x_i/\sqrt{x_1^2 + x_2^2 + \cdots + x_n^2}$ **31.** $\frac{1}{5}$ **33.** $\frac{1}{4}$
35. $f_x(x, y) = y^2 - 3x^2y$, $f_y(x, y) = 2xy - x^3$
37.

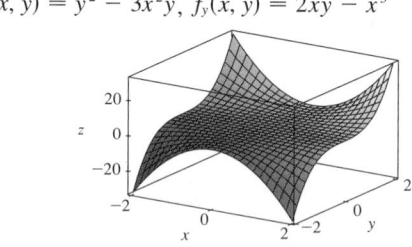

$f(x, y) = x^2y^3$

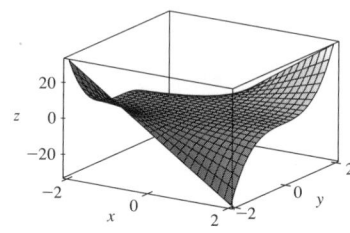

$f_x(x, y) = 2xy^3$

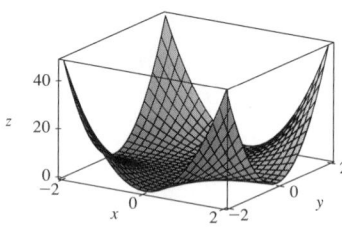

$f_y(x, y) = 3x^2y^2$

39. $\dfrac{\partial z}{\partial x} = -\dfrac{x}{3z}$, $\dfrac{\partial z}{\partial y} = -\dfrac{2y}{3z}$
41. $\dfrac{\partial z}{\partial x} = \dfrac{yz}{e^z - xy}$, $\dfrac{\partial z}{\partial y} = \dfrac{xz}{e^z - xy}$
43. (a) $f'(x)$, $g'(y)$ (b) $f'(x + y)$, $f'(x + y)$
45. $f_{xx} = 6xy^5 + 24x^2y$, $f_{xy} = 15x^2y^4 + 8x^3 = f_{yx}$, $f_{yy} = 20x^3y^3$

47. $w_{uu} = v^2/(u^2 + v^2)^{3/2}$, $w_{uv} = -uv/(u^2 + v^2)^{3/2} = w_{vu}$, $w_{vv} = u^2/(u^2 + v^2)^{3/2}$

49. $z_{xx} = -2x/(1 + x^2)^2$, $z_{xy} = 0 = z_{yx}$, $z_{yy} = -2y/(1 + y^2)^2$

53. $24xy^2 - 6y$, $24x^2y - 6x$

55. $(2x^2y^2z^5 + 6xyz^3 + 2z)e^{xyz^2}$

57. $\theta e^{r\theta}(2\sin\theta + \theta\cos\theta + r\theta\sin\theta)$ **59.** $6yz^2$

69. R^2/R_1^2

71. $\dfrac{\partial T}{\partial P} = \dfrac{V - nb}{nR}$, $\dfrac{\partial P}{\partial V} = \dfrac{2n^2a}{V^3} - \dfrac{nRT}{(V - nb)^2}$

75. No   **77.** $x = 1 + t, y = 2, z = 2 - 2t$ **81.** -2

83. (a)

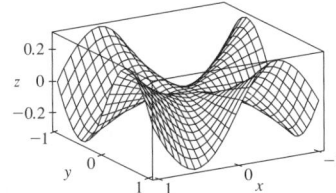

(b) $f_x(x, y) = \dfrac{x^4y + 4x^2y^3 - y^5}{(x^2 + y^2)^2}$, $f_y(x, y) = \dfrac{x^5 - 4x^3y^2 - xy^4}{(x^2 + y^2)^2}$

(c) $0, 0$ (e) No, since f_{xy} and f_{yx} are not continuous.

EXERCISES 11.4 ▪ PAGE 648

1. $z = -7x - 6y + 5$ **3.** $x + y - 2z = 0$

5. $x + y + z = 0$

7. 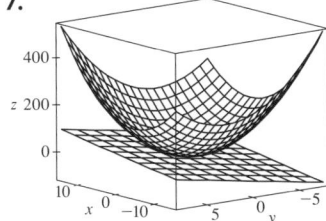   **9.**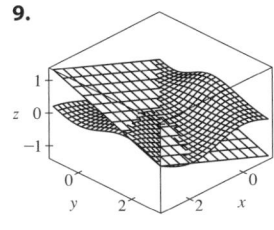

11. $6x + 4y - 23$ **13.** $1 - \pi y$ **17.** 6.3

19. $\frac{3}{7}x + \frac{2}{7}y + \frac{6}{7}z$; 6.9914

21. $dm = 5p^4q^3\,dp + 3p^5q^2\,dq$

23. $dR = \beta^2\cos\gamma\,d\alpha + 2\alpha\beta\cos\gamma\,d\beta - \alpha\beta^2\sin\gamma\,d\gamma$

25. $\Delta z = 0.9225$, $dz = 0.9$ **27.** 5.4 cm^2 **29.** 16 cm^3

31. 2.3% **33.** $\frac{1}{17} \approx 0.059$ Ω **35.** $\varepsilon_1 = \Delta x$, $\varepsilon_2 = \Delta y$

EXERCISES 11.5 ▪ PAGE 656

1. $(2x + y)\cos t + (2y + x)e^t$

3. $e^{y/z}[2t - (x/z) - (2xy/z^2)]$

5. $\partial z/\partial s = 2xy^3\cos t + 3x^2y^2\sin t$, $\partial z/\partial t = -2sxy^3\sin t + 3sx^2y^2\cos t$

7. $\dfrac{\partial z}{\partial s} = e^r\left(t\cos\theta - \dfrac{s}{\sqrt{s^2 + t^2}}\sin\theta\right)$, $\dfrac{\partial z}{\partial t} = e^r\left(s\cos\theta - \dfrac{t}{\sqrt{s^2 + t^2}}\sin\theta\right)$

9. 62 **11.** $7, 2$

13. $\dfrac{\partial u}{\partial r} = \dfrac{\partial u}{\partial x}\dfrac{\partial x}{\partial r} + \dfrac{\partial u}{\partial y}\dfrac{\partial y}{\partial r}$, $\dfrac{\partial u}{\partial s} = \dfrac{\partial u}{\partial x}\dfrac{\partial x}{\partial s} + \dfrac{\partial u}{\partial y}\dfrac{\partial y}{\partial s}$, $\dfrac{\partial u}{\partial t} = \dfrac{\partial u}{\partial x}\dfrac{\partial x}{\partial t} + \dfrac{\partial u}{\partial y}\dfrac{\partial y}{\partial t}$

15. $\dfrac{\partial w}{\partial x} = \dfrac{\partial w}{\partial r}\dfrac{\partial r}{\partial x} + \dfrac{\partial w}{\partial s}\dfrac{\partial s}{\partial x} + \dfrac{\partial w}{\partial t}\dfrac{\partial t}{\partial x}$, $\dfrac{\partial w}{\partial y} = \dfrac{\partial w}{\partial r}\dfrac{\partial r}{\partial y} + \dfrac{\partial w}{\partial s}\dfrac{\partial s}{\partial y} + \dfrac{\partial w}{\partial t}\dfrac{\partial t}{\partial y}$

17. $1582, 3164, -700$ **19.** $2\pi, -2\pi$

21. $\frac{5}{144}, -\frac{5}{96}, \frac{5}{144}$ **23.** $\dfrac{1 + x^4y^2 + y^2 + x^4y^4 - 2xy}{x^2 - 2xy - 2x^5y^3}$

25. $-\dfrac{x}{3z}, -\dfrac{2y}{3z}$ **27.** $\dfrac{yz}{e^z - xy}, \dfrac{xz}{e^z - xy}$

29. $2°$C/s   **31.** ≈ -0.33 m/s per minute

33. (a) 6 m^3/s   (b) 10 m^2/s   (c) 0 m/s

35. ≈ -0.27 L/s

37. (a) $\partial z/\partial r = (\partial z/\partial x)\cos\theta + (\partial z/\partial y)\sin\theta$, $\partial z/\partial\theta = -(\partial z/\partial x)r\sin\theta + (\partial z/\partial y)r\cos\theta$

43. $4rs\,\partial^2z/\partial x^2 + (4r^2 + 4s^2)\partial^2z/\partial x\,\partial y + 4rs\,\partial^2z/\partial y^2 + 2\,\partial z/\partial y$

EXERCISES 11.6 ▪ PAGE 667

1. $2 + \sqrt{3}/2$

3. (a) $\nabla f(x, y) = \langle 2\cos(2x + 3y), 3\cos(2x + 3y)\rangle$
(b) $\langle 2, 3\rangle$ (c) $\sqrt{3} - \frac{3}{2}$

5. (a) $\langle 2xyz - yz^3, x^2z - xz^3, x^2y - 3xyz^2\rangle$
(b) $\langle -3, 2, 2\rangle$ (c) $\frac{2}{5}$

7. $\dfrac{4 - 3\sqrt{3}}{10}$ **9.** $-8/\sqrt{10}$ **11.** $4/\sqrt{30}$ **13.** $2/5$

15. $1, \langle 0, 1\rangle$ **17.** $1, \langle 3, 6, -2\rangle$ **19.** (b) $\langle -12, 92\rangle$

21. All points on the line $y = x + 1$ **23.** (a) $-40/(3\sqrt{3})$

25. (a) $32/\sqrt{3}$ (b) $\langle 38, 6, 12\rangle$ (c) $2\sqrt{406}$

27. $\frac{327}{13}$

31. (a) $x + y + z = 11$ (b) $x - 3 = y - 3 = z - 5$

33. (a) $2x + 3y + 12z = 24$ (b) $\dfrac{x - 3}{2} = \dfrac{y - 2}{3} = \dfrac{z - 1}{12}$

35. (a) $x + y + z = 1$ (b) $x = y = z - 1$

37. 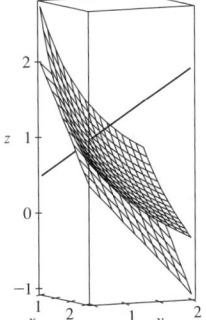   **39.** $\langle 2, 3\rangle$, $2x + 3y = 12$

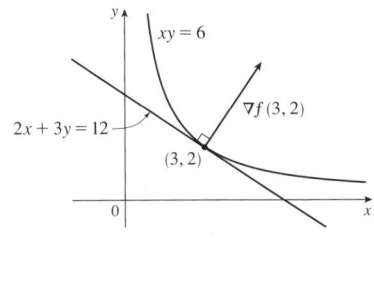

43. No   **45.** $\left(-\frac{5}{4}, -\frac{5}{4}, \frac{25}{8}\right)$

49. $x = -1 - 10t, y = 1 - 16t, z = 2 - 12t$

53. If $\mathbf{u} = \langle a, b\rangle$ and $\mathbf{v} = \langle c, d\rangle$, then $af_x + bf_y$ and $cf_x + df_y$ are known, so we solve linear equations for f_x and f_y.

EXERCISES 11.7 ■ **PAGE 675**

1. (a) f has a local minimum at $(1, 1)$.
(b) f has a saddle point at $(1, 1)$.
3. Minimum $f\left(\frac{1}{3}, -\frac{2}{3}\right) = -\frac{1}{3}$
5. Maximum $f(0, 0) = 2$, minimum $f(0, 4) = -30$,
saddle points at $(2, 2)$, $(-2, 2)$
7. Minimum $f(2, 1) = -8$, saddle point at $(0, 0)$
9. None **11.** Minimum $f(0, 0) = 0$, saddle points at $(\pm 1, 0)$
13. Minima $f(0, 1) = f(\pi, -1) = f(2\pi, 1) = -1$,
saddle points at $(\pi/2, 0)$, $(3\pi/2, 0)$
15. Maximum $f(0, 0) = 2$, minimum $f(0, 2) = -2$,
saddle points $(\pm 1, 1)$
17. Maximum $f(\pi/3, \pi/3) = 3\sqrt{3}/2$,
minimum $f(5\pi/3, 5\pi/3) = -3\sqrt{3}/2$, saddle point at (π, π)
19. Minima $f(0, -0.794) \approx -1.191$,
$f(\pm 1.592, 1.267) \approx -1.310$, saddle points $(\pm 0.720, 0.259)$,
lowest points $(\pm 1.592, 1.267, -1.310)$
21. Maximum $f(0.170, -1.215) \approx 3.197$,
minima $f(-1.301, 0.549) \approx -3.145$, $f(1.131, 0.549) \approx -0.701$,
saddle points $(-1.301, -1.215)$, $(0.170, 0.549)$, $(1.131, -1.215)$,
no highest or lowest point
23. Maximum $f(0, \pm 2) = 4$, minimum $f(1, 0) = -1$
25. Maximum $f(\pm 1, 1) = 7$, minimum $f(0, 0) = 4$
27. Maximum $f(3, 0) = 83$, minimum $f(1, 1) = 0$
29.

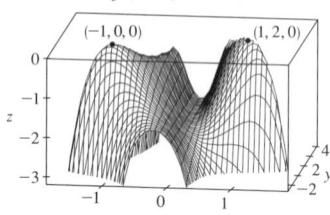

31. $2/\sqrt{3}$ **33.** $\left(2, 1, \sqrt{5}\right), \left(2, 1, -\sqrt{5}\right)$ **35.** $\frac{100}{3}, \frac{100}{3}, \frac{100}{3}$
37. $8r^3/(3\sqrt{3})$ **39.** $\frac{4}{3}$ **41.** Cube, edge length $c/12$
43. Square base of side 40 cm, height 20 cm **45.** $L^3/(3\sqrt{3})$

EXERCISES 11.8 ■ **PAGE 683**

1. No maximum, minimum $f(1, 1) = f(-1, -1) = 2$
3. Maximum $f(0, \pm 1) = 1$, minimum $f(\pm 2, 0) = -4$
5. Maximum $f(2, 2, 1) = 9$, minimum $f(-2, -2, -1) = -9$
7. Maximum $2/\sqrt{3}$, minimum $-2/\sqrt{3}$
9. Maximum $\sqrt{3}$, minimum 1
11. Maximum $f\left(\frac{1}{2}, \frac{1}{2}, \frac{1}{2}, \frac{1}{2}\right) = 2$,
minimum $f\left(-\frac{1}{2}, -\frac{1}{2}, -\frac{1}{2}, -\frac{1}{2}\right) = -2$
13. Maximum $f\left(1, \sqrt{2}, -\sqrt{2}\right) = 1 + 2\sqrt{2}$,
minimum $f\left(1, -\sqrt{2}, \sqrt{2}\right) = 1 - 2\sqrt{2}$
15. Maximum $\frac{3}{2}$, minimum $\frac{1}{2}$
17. Maximum $f\left(3/\sqrt{2}, -3/\sqrt{2}\right) = 9 + 12\sqrt{2}$,
minimum $f(-2, 2) = -8$
19. Maximum $f\left(\pm 1/\sqrt{2}, \mp 1/(2\sqrt{2})\right) = e^{1/4}$,
minimum $f\left(\pm 1/\sqrt{2}, \pm 1/(2\sqrt{2})\right) = e^{-1/4}$
21. $\approx 59, 30$
29–39. See Exercises 31–41 in Section 11.7.
41. $L^3/(3\sqrt{3})$

43. Nearest $\left(\frac{1}{2}, \frac{1}{2}, \frac{1}{2}\right)$, farthest $(-1, -1, 2)$
45. Maximum ≈ 9.7938, minimum ≈ -5.3506
47. (a) c/n (b) When $x_1 = x_2 = \cdots = x_n$

CHAPTER 11 REVIEW ■ **PAGE 685**

True-False Quiz

1. True **3.** False **5.** False **7.** True **9.** False
11. True

Exercises

1. $\{(x, y) \mid y > -x - 1\}$ **3.**

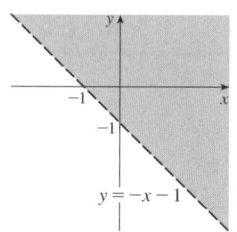

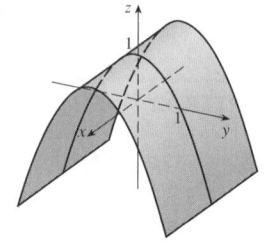

5.   **7.**

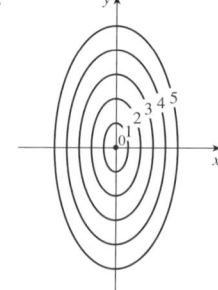

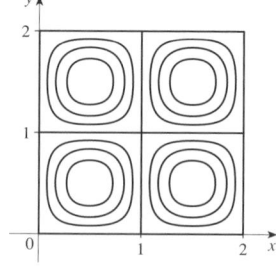

9. $\frac{2}{3}$
11. $f_x = 32xy(5y^3 + 2x^2y)^7, f_y = (16x^2 + 120y^2)(5y^3 + 2x^2y)^7$
13. $F_\alpha = \dfrac{2\alpha^3}{\alpha^2 + \beta^2} + 2\alpha \ln(\alpha^2 + \beta^2), F_\beta = \dfrac{2\alpha^2\beta}{\alpha^2 + \beta^2}$
15. $S_u = \arctan\left(v\sqrt{w}\right), S_v = \dfrac{u\sqrt{w}}{1 + v^2w}, S_w = \dfrac{uv}{2\sqrt{w}(1 + v^2w)}$
17. $f_{xx} = 24x, f_{xy} = -2y = f_{yx}, f_{yy} = -2x$
19. $f_{xx} = k(k - 1)x^{k-2}y^lz^m, f_{xy} = klx^{k-1}y^{l-1}z^m = f_{yx}$,
$f_{xz} = kmx^{k-1}y^lz^{m-1} = f_{zx}, f_{yy} = l(l - 1)x^ky^{l-2}z^m$,
$f_{yz} = lmx^ky^{l-1}z^{m-1} = f_{zy}, f_{zz} = m(m - 1)x^ky^lz^{m-2}$
23. (a) $z = 8x + 4y + 1$ (b) $\dfrac{x - 1}{8} = \dfrac{y + 2}{4} = \dfrac{z - 1}{-1}$
25. (a) $2x - 2y - 3z = 3$ (b) $\dfrac{x - 2}{4} = \dfrac{y + 1}{-4} = \dfrac{z - 1}{-6}$
27. (a) $x + 2y + 5z = 0$ (b) $x - 2 = \dfrac{y + 1}{2} = \dfrac{z}{5}$
29. $\left(2, \frac{1}{2}, -1\right), \left(-2, -\frac{1}{2}, 1\right)$
31. $60x + \frac{24}{5}y + \frac{32}{5}z - 120$; 38.656
33. $2xy^3(1 + 6p) + 3x^2y^2(pe^p + e^p) + 4z^3(p \cos p + \sin p)$
35. $-47, 108$ **41.** $\left\langle 2xe^{yz^2}, x^2z^2e^{yz^2}, 2x^2yze^{yz^2}\right\rangle$
43. $-\frac{4}{5}$ **45.** $\sqrt{145}/2, \left\langle 4, \frac{9}{2}\right\rangle$

47. Minimum $f(-4, 1) = -11$
49. Maximum $f(1, 1) = 1$; saddle points $(0, 0)$, $(0, 3)$, $(3, 0)$
51. Maximum $f(1, 2) = 4$, minimum $f(2, 4) = -64$
53. Maximum $f(-1, 0) = 2$, minima $f(1, \pm 1) = -3$,
saddle points $(-1, \pm 1)$, $(1, 0)$
55. Maximum $f\left(\pm\sqrt{2/3}, 1/\sqrt{3}\right) = 2/\left(3\sqrt{3}\right)$,
minimum $f\left(\pm\sqrt{2/3}, -1/\sqrt{3}\right) = -2/\left(3\sqrt{3}\right)$
57. Maximum 1, minimum -1
59. $\left(\pm 3^{-1/4}, 3^{-1/4}\sqrt{2}, \pm 3^{1/4}\right), \left(\pm 3^{-1/4}, -3^{-1/4}\sqrt{2}, \pm 3^{1/4}\right)$
61. $P\left(2 - \sqrt{3}\right), P\left(3 - \sqrt{3}\right)/6, P\left(2\sqrt{3} - 3\right)/3$

CHAPTER 12

EXERCISES 12.1 ▪ PAGE 698

1. (a) 288   (b) 144   **3.** (a) 0.990   (b) 1.151
5. 248   **7.** 60   **9.** 3   **11.** 222   **13.** $32(e^4 - 1)$
15. 18   **17.** $\frac{21}{2} \ln 2$ **19.** $\frac{31}{30}$ **21.** 9 ln 2
23. $\frac{1}{2}\left(\sqrt{3} - 1\right) - \frac{1}{12}\pi$ **25.** $\frac{1}{2}e^{-6} + \frac{5}{2}$
27.

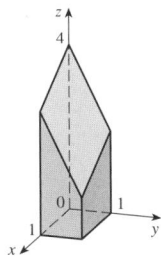

29. 51   **31.** $\frac{166}{27}$ **33.** 2   **35.** $\frac{64}{3}$
37. $21e - 57$

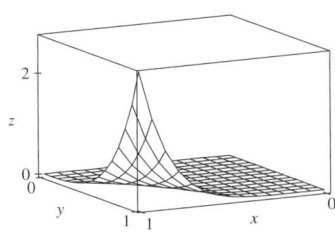

39. $\frac{5}{6}$ **43.** 0
45. Fubini's Theorem does not apply. The integrand has an infinite discontinuity at the origin.

EXERCISES 12.2 ▪ PAGE 707

1. 32   **3.** $\frac{3}{10}$ **5.** $\frac{1}{3} \sin 1$ **7.** $\frac{4}{3}$ **9.** π
11. Type I: $D = \{(x, y) \mid 0 \le x \le 1, 0 \le y \le x\}$,
type II: $D = \{(x, y) \mid 0 \le y \le 1, y \le x \le 1\}$; $\frac{1}{3}$
13. $\int_0^1 \int_{-\sqrt{x}}^{\sqrt{x}} y \, dy \, dx + \int_1^4 \int_{x-2}^{\sqrt{x}} y \, dy \, dx = \int_{-1}^2 \int_{y^2}^{y+2} y \, dx \, dy = \frac{9}{4}$
15. $\frac{1}{2}(1 - \cos 1)$ **17.** $\frac{11}{3}$ **19.** 0   **21.** $\frac{17}{60}$ **23.** $\frac{31}{8}$
25. 6   **27.** $\frac{128}{15}$ **29.** $\frac{1}{3}$ **31.** $\frac{64}{3}$

33.

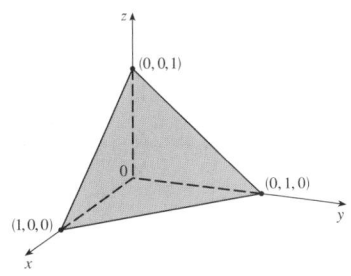

35. $\pi/2$
37. $\int_0^1 \int_x^1 f(x, y) \, dy \, dx$

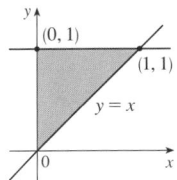

39. $\int_0^1 \int_0^{\cos^{-1} y} f(x, y) \, dx \, dy$

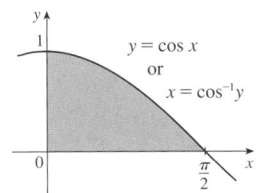

41. $\int_0^{\ln 2} \int_{e^y}^2 f(x, y) \, dx \, dy$

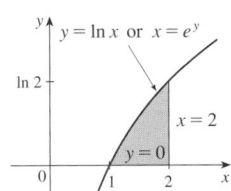

43. $\frac{1}{6}(e^9 - 1)$ **45.** $\frac{1}{3} \ln 9$ **47.** $\frac{1}{3}\left(2\sqrt{2} - 1\right)$ **49.** 1
51. $0 \le \iint_D \sqrt{x^3 + y^3} \, dA \le \sqrt{2}$ **55.** 9π
57. $a^2 b + \frac{3}{2} ab^2$ **59.** $\pi a^2 b$

EXERCISES 12.3 ▪ PAGE 713

1. $\int_0^{3\pi/2} \int_0^4 f(r \cos \theta, r \sin \theta) r \, dr \, d\theta$ **3.** $\int_{-1}^1 \int_0^{(x+1)/2} f(x, y) \, dy \, dx$
5.

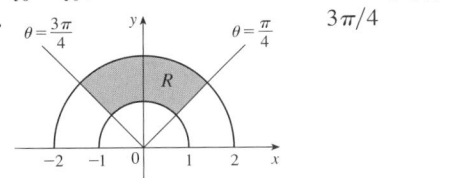

$3\pi/4$

7. $\frac{1250}{3}$ **9.** $(\pi/4)(\cos 1 - \cos 9)$ **11.** $\frac{3}{64} \pi^2$ **13.** $\frac{16}{3} \pi$
15. $\frac{4}{3} \pi a^3$ **17.** $(2\pi/3)\left[1 - \left(1/\sqrt{2}\right)\right]$
19. $(8\pi/3)\left(64 - 24\sqrt{3}\right)$ **21.** $\pi/12$ **23.** $\frac{1}{2}\pi(1 - \cos 9)$
25. $2\sqrt{2}/3$ **27.** $1800\pi \text{ ft}^3$ **29.** $\frac{15}{16}$
31. (a) $\sqrt{\pi}/4$ (b) $\sqrt{\pi}/2$

EXERCISES 12.4 ▪ PAGE 720

1. 285 C   **3.** $42k, \left(2, \frac{85}{28}\right)$ **5.** $6, \left(\frac{3}{4}, \frac{3}{2}\right)$ **7.** $\frac{8}{15}k, \left(0, \frac{4}{7}\right)$

9. $L/4, (L/2, 16/(9\pi))$ **11.** $\left(\frac{3}{8}, 3\pi/16\right)$

13. $(0, 45/(14\pi))$

15. $(2a/5, 2a/5)$ if vertex is $(0, 0)$ and sides are along positive axes

17. $\frac{64}{315}k, \frac{8}{105}k, \frac{88}{315}k$

19. $7ka^6/180, 7ka^6/180, 7ka^6/90$ if vertex is $(0, 0)$ and sides are along positive axes

21. $m = 3\pi/64, (\bar{x}, \bar{y}) = \left(\dfrac{16384\sqrt{2}}{10395\pi}, 0\right)$,

$I_x = \dfrac{5\pi}{384} - \dfrac{4}{105}, I_y = \dfrac{5\pi}{384} + \dfrac{4}{105}, I_0 = \dfrac{5\pi}{192}$

23. $\rho bh^3/3, \rho b^3h/3; b/\sqrt{3}, h/\sqrt{3}$

EXERCISES 12.5 ▪ PAGE 728

3. $\frac{16}{15}$ **5.** $-\frac{1}{3}$ **7.** $\frac{27}{2}$ **9.** $9\pi/8$

11. $\frac{65}{28}$ **13.** $\frac{1}{60}$ **15.** $16\pi/3$ **17.** $\frac{16}{3}$ **19.** $\frac{8}{15}$

21. (a) $\int_0^1 \int_0^x \int_0^{\sqrt{1-y^2}} dz\,dy\,dx$ (b) $\frac{1}{4}\pi - \frac{1}{3}$

23. 0.985

25.

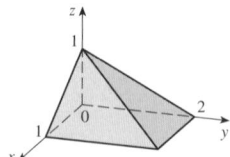

27. $\int_{-2}^2 \int_0^{4-x^2} \int_{-\sqrt{4-x^2-y}/2}^{\sqrt{4-x^2-y}/2} f(x, y, z)\,dz\,dy\,dx$

$= \int_0^4 \int_{-\sqrt{4-y}}^{\sqrt{4-y}} \int_{-\sqrt{4-x^2-y}/2}^{\sqrt{4-x^2-y}/2} f(x, y, z)\,dz\,dx\,dy$

$= \int_{-1}^1 \int_0^{4-4z^2} \int_{-\sqrt{4-y-4z^2}}^{\sqrt{4-y-4z^2}} f(x, y, z)\,dx\,dy\,dz$

$= \int_0^4 \int_{-\sqrt{4-y}/2}^{\sqrt{4-y}/2} \int_{-\sqrt{4-y-4z^2}}^{\sqrt{4-y-4z^2}} f(x, y, z)\,dx\,dz\,dy$

$= \int_{-2}^2 \int_{-\sqrt{4-x^2}/2}^{\sqrt{4-x^2}/2} \int_0^{4-x^2-4z^2} f(x, y, z)\,dy\,dz\,dx$

$= \int_{-1}^1 \int_{-\sqrt{4-4z^2}}^{\sqrt{4-4z^2}} \int_0^{4-x^2-4z^2} f(x, y, z)\,dy\,dx\,dz$

29. $\int_{-2}^2 \int_{x^2}^4 \int_0^{2-y/2} f(x, y, z)\,dz\,dy\,dx$

$= \int_0^4 \int_{-\sqrt{y}}^{\sqrt{y}} \int_0^{2-y/2} f(x, y, z)\,dz\,dx\,dy$

$= \int_0^2 \int_0^{4-2z} \int_{-\sqrt{y}}^{\sqrt{y}} f(x, y, z)\,dx\,dy\,dz$

$= \int_0^4 \int_0^{2-y/2} \int_{-\sqrt{y}}^{\sqrt{y}} f(x, y, z)\,dx\,dz\,dy$

$= \int_{-2}^2 \int_0^{2-x^2/2} \int_{x^2}^{4-2z} f(x, y, z)\,dy\,dz\,dx$

$= \int_0^2 \int_{-\sqrt{4-2z}}^{\sqrt{4-2z}} \int_{x^2}^{4-2z} f(x, y, z)\,dy\,dx\,dz$

31. $\int_0^1 \int_{\sqrt{x}}^1 \int_0^{1-y} f(x, y, z)\,dz\,dy\,dx = \int_0^1 \int_0^{y^2} \int_0^{1-y} f(x, y, z)\,dz\,dx\,dy$

$= \int_0^1 \int_0^{1-z} \int_0^{y^2} f(x, y, z)\,dx\,dy\,dz = \int_0^1 \int_0^{1-y} \int_0^{y^2} f(x, y, z)\,dx\,dz\,dy$

$= \int_0^1 \int_0^{1-\sqrt{x}} \int_{\sqrt{x}}^{1-z} f(x, y, z)\,dy\,dz\,dx$

$= \int_0^1 \int_0^{(1-z)^2} \int_{\sqrt{x}}^{1-z} f(x, y, z)\,dy\,dx\,dz$

33. $\int_0^1 \int_y^1 \int_0^y f(x, y, z)\,dz\,dx\,dy = \int_0^1 \int_0^x \int_0^y f(x, y, z)\,dz\,dy\,dx$

$= \int_0^1 \int_z^1 \int_y^1 f(x, y, z)\,dx\,dy\,dz = \int_0^1 \int_0^y \int_y^1 f(x, y, z)\,dx\,dz\,dy$

$= \int_0^1 \int_0^x \int_z^x f(x, y, z)\,dy\,dz\,dx = \int_0^1 \int_z^1 \int_z^x f(x, y, z)\,dy\,dx\,dz$

35. 64π **37.** $\frac{79}{30}, \left(\frac{358}{553}, \frac{33}{79}, \frac{571}{553}\right)$

39. $a^5, (7a/12, 7a/12, 7a/12)$

41. (a) $m = \int_{-1}^1 \int_{x^2}^1 \int_0^{1-y} \sqrt{x^2 + y^2}\,dz\,dy\,dx$

(b) $(\bar{x}, \bar{y}, \bar{z})$, where

$\bar{x} = (1/m) \int_{-1}^1 \int_{x^2}^1 \int_0^{1-y} x\sqrt{x^2 + y^2}\,dz\,dy\,dx$

$\bar{y} = (1/m) \int_{-1}^1 \int_{x^2}^1 \int_0^{1-y} y\sqrt{x^2 + y^2}\,dz\,dy\,dx$

$\bar{z} = (1/m) \int_{-1}^1 \int_{x^2}^1 \int_0^{1-y} z\sqrt{x^2 + y^2}\,dz\,dy\,dx$

(c) $\int_{-1}^1 \int_{x^2}^1 \int_0^{1-y} (x^2 + y^2)^{3/2}\,dz\,dy\,dx$

43. (a) $\frac{3}{32}\pi + \frac{11}{24}$

(b) $\left(\dfrac{28}{9\pi + 44}, \dfrac{30\pi + 128}{45\pi + 220}, \dfrac{45\pi + 208}{135\pi + 660}\right)$

(c) $\frac{1}{240}(68 + 15\pi)$

45. $I_x = I_y = I_z = \frac{2}{3}kL^5$ **47.** $\frac{1}{2}\pi kha^4$ **49.** $L^3/8$

51. (a) The region bounded by the ellipsoid $x^2 + 2y^2 + 3z^2 = 1$

(b) $4\sqrt{6}\pi/45$

EXERCISES 12.6 ▪ PAGE 734

1. (a)   (b)

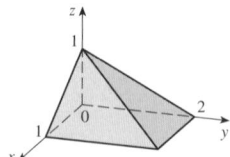

$(2, 2\sqrt{3}, -2)$ $(0, -2, 1)$

3. (a) $\left(\sqrt{2}, 3\pi/4, 1\right)$ (b) $(4, 2\pi/3, 3)$

5. Vertical half-plane through the z-axis

7. Circular paraboloid

9. (a) $z^2 = 1 + r\cos\theta - r^2$ (b) $z = r^2 \cos 2\theta$

11.

13. Cylindrical coordinates: $6 \le r \le 7, 0 \le \theta \le 2\pi,$
$0 \le z \le 20$

15. 4π

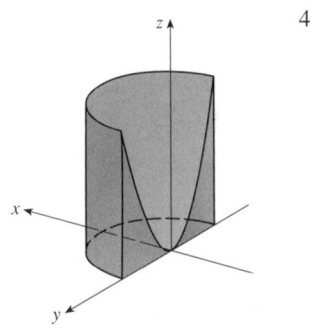

17. 384π **19.** $\frac{8}{3}\pi + \frac{128}{15}$ **21.** $2\pi/5$ **23.** $\frac{4}{3}\pi(\sqrt{2}-1)$
25. (a) 162π (b) $(0, 0, 15)$
27. $\pi Ka^2/8$, $(0, 0, 2a/3)$ **29.** 0
31. (a) $\iiint_C h(P)g(P)\, dV$, where C is the cone
(b) $\approx 3.1 \times 10^{19}$ ft-lb

EXERCISES 12.7 ▪ PAGE 740

1. (a)   (b)

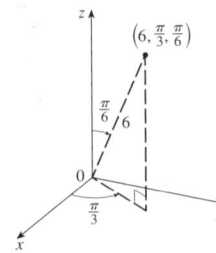 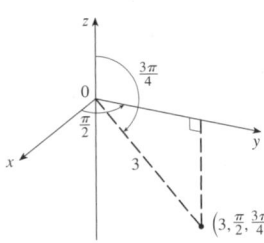

$\left(\dfrac{3}{2}, \dfrac{3\sqrt{3}}{2}, 3\sqrt{3}\right)$ $\left(0, \dfrac{3\sqrt{2}}{2}, -\dfrac{3\sqrt{2}}{2}\right)$

3. (a) $(2, 3\pi/2, \pi/2)$ (b) $(2, 3\pi/4, 3\pi/4)$
5. Half-cone **7.** Sphere, radius $\frac{1}{2}$, center $(0, \frac{1}{2}, 0)$
9. (a) $\cos^2\phi = \sin^2\phi$ (b) $\rho^2(\sin^2\phi\cos^2\theta + \cos^2\phi) = 9$
11.

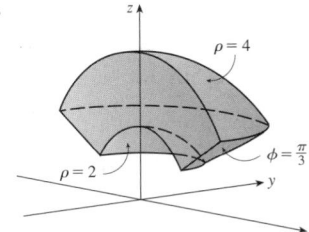

13.

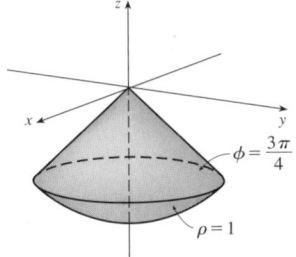

15. $0 \le \phi \le \pi/4$, $0 \le \rho \le \cos\phi$
17. $(9\pi/4)(2 - \sqrt{3})$

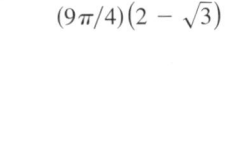

19. $\int_0^{\pi/2} \int_0^3 \int_0^2 f(r\cos\theta, r\sin\theta, z)\, r\, dz\, dr\, d\theta$
21. $312{,}500\,\pi/7$ **23.** $1688\pi/15$ **25.** $\pi/8$
27. (a) 10π (b) $(0, 0, 2.1)$
29. (a) $\left(0, 0, \frac{7}{12}\right)$ (b) $11K\pi/960$
31. (a) $\left(0, 0, \frac{3}{8}a\right)$ (b) $4K\pi a^5/15$
33. $\frac{1}{3}\pi(2 - \sqrt{2})$, $\left(0, 0, \dfrac{3}{8(2 - \sqrt{2})}\right)$ **35.** $5\pi/6$
37. $(4\sqrt{2} - 5)/15$ **39.** $4096\pi/21$
41.   **43.** $136\pi/99$

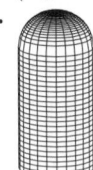

EXERCISES 12.8 ▪ PAGE 749

1. 16 **3.** $\sin^2\theta - \cos^2\theta$ **5.** 0
7. The parallelogram with vertices $(0, 0)$, $(6, 3)$, $(12, 1)$, $(6, -2)$
9. The region bounded by the line $y = 1$, the y-axis, and $y = \sqrt{x}$
11. $x = \frac{1}{3}(v - u)$, $y = \frac{1}{3}(u + 2v)$ is one possible transformation, where $S = \{(u, v) \mid -1 \le u \le 1, 1 \le v \le 3\}$
13. $x = u\cos v$, $y = u\sin v$ is one possible transformation, where $S = \{(u, v) \mid 1 \le u \le \sqrt{2}, 0 \le v \le \pi/2\}$
15. -3 **17.** 6π **19.** $2\ln 3$
21. (a) $\frac{4}{3}\pi abc$ (b) 1.083×10^{12} km^3
(c) $\frac{4}{15}\pi(a^2 + b^2)abck$
23. $\frac{8}{5}\ln 8$ **25.** $\frac{3}{2}\sin 1$ **27.** $e - e^{-1}$

CHAPTER 12 REVIEW ▪ PAGE 751

True-False Quiz

1. True **3.** True **5.** True **7.** True **9.** False

Exercises

1. 64.0 **3.** $4e^2 - 4e + 3$ **5.** $\frac{1}{2}\sin 1$ **7.** $\frac{2}{3}$
9. $\int_0^\pi \int_2^4 f(r\cos\theta, r\sin\theta)\, r\, dr\, d\theta$
11. The region inside the loop of the four-leaved rose $r = \sin 2\theta$ in the first quadrant
13. $\frac{1}{2}\sin 1$ **15.** $\frac{1}{2}e^6 - \frac{7}{2}$ **17.** $\frac{1}{4}\ln 2$ **19.** 8
21. $81\pi/5$ **23.** $\frac{81}{2}$ **25.** $\pi/96$ **27.** $\frac{64}{15}$
29. 176 **31.** $\frac{2}{3}$ **33.** $2ma^3/9$
35. (a) $\frac{1}{4}$ (b) $\left(\frac{1}{3}, \frac{8}{15}\right)$
(c) $I_x = \frac{1}{12}$, $I_y = \frac{1}{24}$; $\bar{\bar{y}} = 1/\sqrt{3}$, $\bar{\bar{x}} = 1/\sqrt{6}$

37. (a) $(0, 0, h/4)$ (b) $\pi a^4 h/10$

39. $(\sqrt{3}, 3, 2), (4, \pi/3, \pi/3)$

41. $(2\sqrt{2}, 2\sqrt{2}, 4\sqrt{3}), (4, \pi/4, 4\sqrt{3})$

43. $r^2 + z^2 = 4, \rho = 2$ **45.** 97.2

47. $\int_0^1 \int_0^{1-z} \int_{-\sqrt{y}}^{\sqrt{y}} f(x, y, z)\, dx\, dy\, dz$ **49.** $-\ln 2$ **51.** 0

CHAPTER 13

EXERCISES 13.1 ▪ PAGE 760

1.

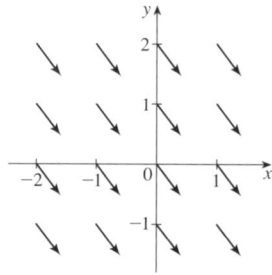

3.

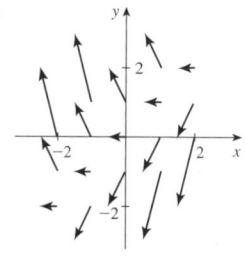

5.

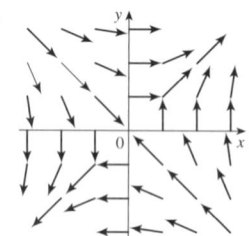

7.

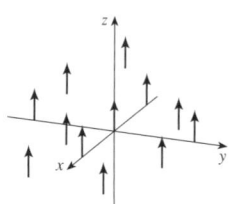

9.

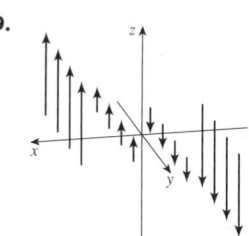

11. IV   **13.** I   **15.** IV   **17.** III

19.

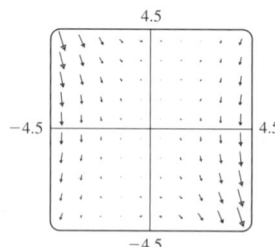

The line $y = 2x$

21. $\nabla f(x, y) = (xy + 1)e^{xy}\,\mathbf{i} + x^2 e^{xy}\,\mathbf{j}$

23. $\nabla f(x, y, z) = \dfrac{x}{\sqrt{x^2 + y^2 + z^2}}\,\mathbf{i}$
$+ \dfrac{y}{\sqrt{x^2 + y^2 + z^2}}\,\mathbf{j} + \dfrac{z}{\sqrt{x^2 + y^2 + z^2}}\,\mathbf{k}$

25. $\nabla f(x, y) = 2x\,\mathbf{i} - \mathbf{j}$

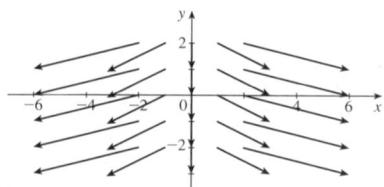

27.

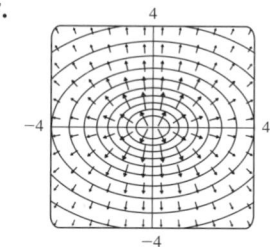

29. $(2.04, 1.03)$

31. (a)   (b) $y = 1/x, x > 0$

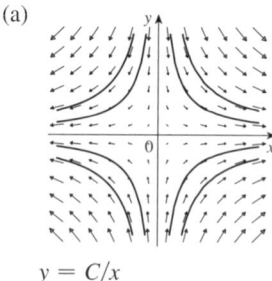

$y = C/x$

EXERCISES 13.2 ▪ PAGE 770

1. $\frac{1}{54}(145^{3/2} - 1)$ **3.** 1638.4   **5.** $\frac{243}{8}$ **7.** $\frac{5}{2}$

9. $\sqrt{5}\,\pi$ **11.** $\frac{1}{12}\sqrt{14}(e^6 - 1)$ **13.** $\frac{2}{5}(e - 1)$ **15.** $\frac{35}{3}$

17. (a) Positive   (b) Negative   **19.** 45

21. $\frac{6}{5} - \cos 1 - \sin 1$ **23.** 1.9633

25. $3\pi + \frac{2}{3}$

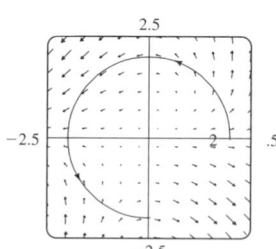

27. (a) $\frac{11}{8} - 1/e$ (b)

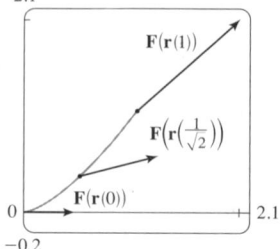

29. $\frac{945}{16,777,216}\pi$ **31.** $2\pi k, (4/\pi, 0)$

33. (a) $\bar{x} = (1/m)\int_C x\rho(x, y, z)\,ds,$

$\bar{y} = (1/m)\int_C y\rho(x, y, z)\,ds,$

$\bar{z} = (1/m)\int_C z\rho(x, y, z)\,ds,$ where $m = \int_C \rho(x, y, z)\,ds$

(b) $(0, 0, 3\pi)$

35. $I_x = k(\frac{1}{2}\pi - \frac{4}{3}), I_y = k(\frac{1}{2}\pi - \frac{2}{3})$ **37.** $2\pi^2$ **39.** $\frac{7}{3}$

41. (a) $2ma\,\mathbf{i} + 6mbt\,\mathbf{j}$ (b) $2ma^2 + \frac{9}{2}mb^2$

43. $\approx 1.67 \times 10^4$ ft-lb   **47.** (b) Yes

EXERCISES 13.3 ▪ PAGE 780

1. 40   **3.** $f(x, y) = x^2 - 3xy + 2y^2 - 8y + K$

5. Not conservative   **7.** $f(x, y) = ye^x + x\sin y + K$

9. $f(x, y) = x\ln y + x^2y^3 + K$

11. (a) $f(x, y) = \frac{1}{2}x^2y^2$ (b) 2

13. (a) $f(x, y, z) = xyz + z^2$ (b) 77

15. (a) $f(x, y, z) = ye^{xz}$ (b) 4   **17.** $4/e$

19. 30   **21.** No   **23.** Conservative

27. (a) Yes   (b) Yes   (c) Yes

29. (a) No   (b) Yes   (c) Yes

EXERCISES 13.4 ▪ PAGE 787

1. 8π **3.** $\frac{2}{3}$ **5.** 12   **7.** $\frac{1}{3}$ **9.** -24π **11.** $-\frac{16}{3}$

13. 4π **15.** $-8e + 48e^{-1}$ **17.** $-\frac{1}{12}$ **19.** 3π

21. (c) $\frac{9}{2}$

23. $(4a/3\pi, 4a/3\pi)$ if the region is the portion of the disk $x^2 + y^2 = a^2$ in the first quadrant

27. 0

EXERCISES 13.5 ▪ PAGE 795

1. (a) $\mathbf{0}$ (b) 3

3. (a) $ze^x\,\mathbf{i} + (xye^z - yze^x)\,\mathbf{j} - xe^z\,\mathbf{k}$ (b) $y(e^z + e^x)$

5. (a) $\mathbf{0}$ (b) $2/\sqrt{x^2 + y^2 + z^2}$

7. (a) $\langle -e^y\cos z, -e^z\cos x, -e^x\cos y\rangle$

(b) $e^x\sin y + e^y\sin z + e^z\sin x$

9. (a) Zero   (b) curl $\mathbf{F}$ points in the negative z-direction

11. $f(x, y, z) = xy^2z^3 + K$ **13.** Not conservative

15. $f(x, y, z) = xe^{yz} + K$ **17.** No

EXERCISES 13.6 ▪ PAGE 805

1. Plane through $(0, 3, 1)$ containing vectors $\langle 1, 0, 4\rangle$, $\langle 1, -1, 5\rangle$

3. Hyperbolic paraboloid

5.

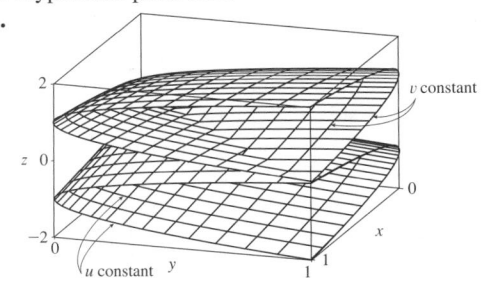

7.

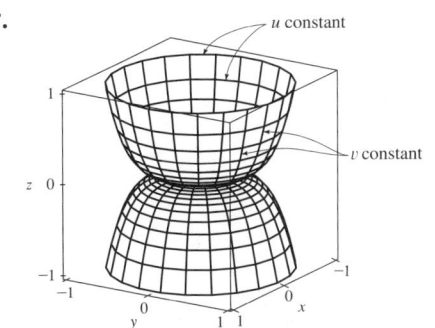

9.

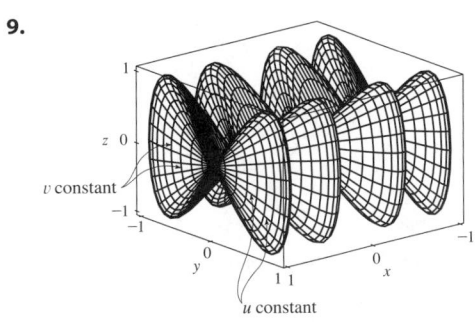

11. IV   **13.** II   **15.** $x = u, y = v - u, z = -v$

17. $y = y, z = z, x = \sqrt{1 + y^2 + \frac{1}{4}z^2}$

19. $x = 2\sin\phi\cos\theta, y = 2\sin\phi\sin\theta,$

$z = 2\cos\phi, 0 \le \phi \le \pi/4, 0 \le \theta \le 2\pi$

$\left[\text{or } x = x, y = y, z = \sqrt{4 - x^2 - y^2}, x^2 + y^2 \le 2\right]$

21. $x = x, y = 4\cos\theta, z = 4\sin\theta, 0 \le x \le 5, 0 \le \theta \le 2\pi$

25. $x = x, y = e^{-x}\cos\theta,$

$z = e^{-x}\sin\theta, 0 \le x \le 3,$

$0 \le \theta \le 2\pi$

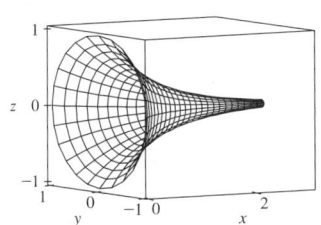

27. (a) Direction reverses   (b) Number of coils doubles

29. $3x - y + 3z = 3$ **31.** $\frac{\sqrt{3}}{2}x - \frac{1}{2}y + z = \frac{\pi}{3}$

33. $3\sqrt{14}$ **35.** $\sqrt{14}\,\pi$

37. $\frac{4}{15}(3^{5/2} - 2^{7/2} + 1)$ **39.** $(2\pi/3)(2\sqrt{2} - 1)$

41. $\frac{1}{2}\sqrt{21} + \frac{17}{4}[\ln(2 + \sqrt{21}) - \ln\sqrt{17}]$ **43.** $\pi(2\sqrt{6} - \frac{8}{3})$

45. $\pi R^2 \le A(S) \le \sqrt{3}\,\pi R^2$ **47.** 13.9783

49. (a) 24.2055   (b) 24.2476

51. $\frac{45}{8}\sqrt{14} + \frac{15}{16}\ln\left(\dfrac{11\sqrt{5} + 3\sqrt{70}}{3\sqrt{5} + \sqrt{70}}\right)$

53. (b)

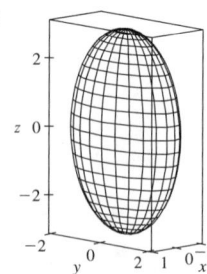

(c) $\int_0^{2\pi} \int_0^{\pi} \sqrt{36 \sin^4 u \cos^2 v + 9 \sin^4 u \sin^2 v + 4 \cos^2 u \sin^2 u} \, du \, dv$

55. 4π **59.** $(\pi/6)(37\sqrt{37} - 17\sqrt{17})$

EXERCISES 13.7 ▪ PAGE 816

1. 49.09 **3.** 900π **5.** $11\sqrt{14}$ **7.** $\frac{2}{3}(2\sqrt{2} - 1)$
9. $171\sqrt{14}$ **11.** $\sqrt{21}/3$ **13.** $364\sqrt{2}\,\pi/3$
15. $(\pi/60)(391\sqrt{17} + 1)$ **17.** 16π **19.** 12 **21.** 4
23. $\frac{713}{180}$ **25.** $-\frac{4}{3}\pi$ **27.** 0 **29.** 48 **31.** $2\pi + \frac{8}{3}$
33. 3.4895
35. $\iint_S \mathbf{F} \cdot d\mathbf{S} = \iint_D [P(\partial h/\partial x) - Q + R(\partial h/\partial z)] \, dA$,
where D = projection of S onto the xz-plane
37. $(0, 0, a/2)$
39. (a) $I_z = \iint_S (x^2 + y^2)\rho(x, y, z) \, dS$ (b) $4329\sqrt{2}\,\pi/5$
41. 0 kg/s **43.** $\frac{8}{3}\pi a^3 \varepsilon_0$ **45.** 1248π

EXERCISES 13.8 ▪ PAGE 822

1. 0 **3.** 0 **5.** -1 **7.** 80π
9. (a) $81\pi/2$ (b)

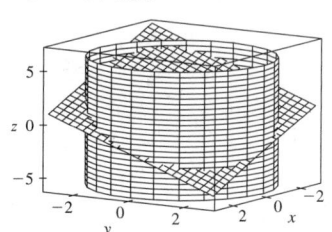

(c) $x = 3 \cos t$, $y = 3 \sin t$,
$z = 1 - 3(\cos t + \sin t)$,
$0 \leq t \leq 2\pi$

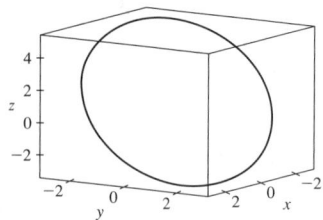

15. 3

EXERCISES 13.9 ▪ PAGE 829

5. $\frac{9}{2}$ **7.** $9\pi/2$ **9.** 0 **11.** $32\pi/3$ **13.** 2π
15. $341\sqrt{2}/60 + \frac{81}{20}\arcsin(\sqrt{3}/3)$
17. $13\pi/20$ **19.** Negative at P_1, positive at P_2
21. div $\mathbf{F} > 0$ in quadrants I, II; div $\mathbf{F} < 0$ in quadrants III, IV

CHAPTER 13 REVIEW ▪ PAGE 831

True-False Quiz

1. False **3.** True **5.** False
7. False **9.** True **11.** True

Exercises

1. (a) Negative (b) Positive **3.** $6\sqrt{10}$ **5.** $\frac{4}{15}$
7. $\frac{110}{3}$ **9.** $\frac{11}{12} - 4/e$ **11.** $f(x, y) = e^y + xe^{xy}$ **13.** 0
17. -8π **25.** $\frac{1}{6}(27 - 5\sqrt{5})$ **27.** $(\pi/60)(391\sqrt{17} + 1)$
29. $-64\pi/3$ **33.** $-\frac{1}{2}$ **37.** 21

APPENDIXES

EXERCISES A ▪ PAGE A8

1. $7\pi/6$ **3.** $\pi/20$ **5.** 5π **7.** $720°$ **9.** $75°$
11. $-67.5°$ **13.** 3π cm **15.** $\frac{2}{3}$ rad $= (120/\pi)°$
17. **19.**

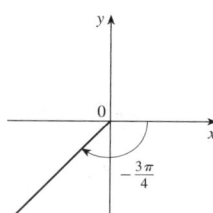

21.

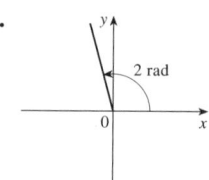

23. $\sin(3\pi/4) = 1/\sqrt{2}$, $\cos(3\pi/4) = -1/\sqrt{2}$, $\tan(3\pi/4) = -1$,
$\csc(3\pi/4) = \sqrt{2}$, $\sec(3\pi/4) = -\sqrt{2}$, $\cot(3\pi/4) = -1$
25. $\sin(9\pi/2) = 1$, $\cos(9\pi/2) = 0$, $\csc(9\pi/2) = 1$,
$\cot(9\pi/2) = 0$, $\tan(9\pi/2)$ and $\sec(9\pi/2)$ undefined
27. $\sin(5\pi/6) = \frac{1}{2}$, $\cos(5\pi/6) = -\sqrt{3}/2$, $\tan(5\pi/6) = -1/\sqrt{3}$,
$\csc(5\pi/6) = 2$, $\sec(5\pi/6) = -2/\sqrt{3}$, $\cot(5\pi/6) = -\sqrt{3}$
29. $\cos \theta = \frac{4}{5}$, $\tan \theta = \frac{3}{4}$, $\csc \theta = \frac{5}{3}$, $\sec \theta = \frac{5}{4}$, $\cot \theta = \frac{4}{3}$
31. $\sin \phi = \sqrt{5}/3$, $\cos \phi = -\frac{2}{3}$, $\tan \phi = -\sqrt{5}/2$,
$\csc \phi = 3/\sqrt{5}$, $\cot \phi = -2/\sqrt{5}$
33. $\sin \beta = -1/\sqrt{10}$, $\cos \beta = -3/\sqrt{10}$, $\tan \beta = \frac{1}{3}$,
$\csc \beta = -\sqrt{10}$, $\sec \beta = -\sqrt{10}/3$
35. 5.73576 cm **37.** 24.62147 cm
59. $(4 + 6\sqrt{2})/15$ **61.** $(3 + 8\sqrt{2})/15$
63. $\frac{24}{25}$ **65.** $\pi/3, 5\pi/3$ **67.** $\pi/4, 3\pi/4, 5\pi/4, 7\pi/4$
69. $\pi/6, \pi/2, 5\pi/6, 3\pi/2$ **71.** $0, \pi, 2\pi$
73. $0 \leq x \leq \pi/6$ and $5\pi/6 \leq x \leq 2\pi$
75. $0 \leq x < \pi/4, 3\pi/4 < x < 5\pi/4, 7\pi/4 < x \leq 2\pi$

77.

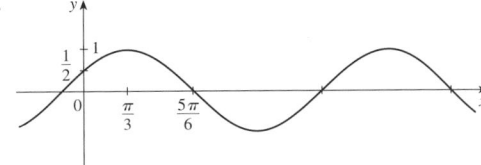

79.

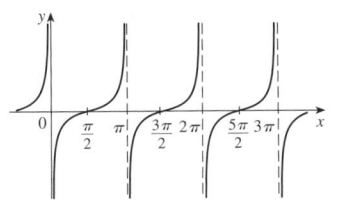

81.

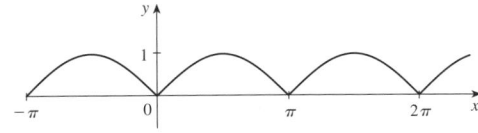

89. 14.34457 cm^2

EXERCISES B ▪ **PAGE A14**

1. $\sqrt{1} + \sqrt{2} + \sqrt{3} + \sqrt{4} + \sqrt{5}$ **3.** $3^4 + 3^5 + 3^6$

5. $-1 + \frac{1}{3} + \frac{3}{5} + \frac{5}{7} + \frac{7}{9}$ **7.** $1^{10} + 2^{10} + 3^{10} + \cdots + n^{10}$

9. $1 - 1 + 1 - 1 + \cdots + (-1)^{n-1}$ **11.** $\sum_{i=1}^{10} i$

13. $\sum_{i=1}^{19} \frac{i}{i+1}$ **15.** $\sum_{i=1}^{n} 2i$ **17.** $\sum_{i=0}^{5} 2^i$ **19.** $\sum_{i=1}^{n} x^i$

21. 80 **23.** 3276 **25.** 0 **27.** 61 **29.** $n(n+1)$

31. $n(n^2 + 6n + 17)/3$ **33.** $n(n^2 + 6n + 11)/3$

35. $n(n^3 + 2n^2 - n - 10)/4$

41. (a) n^4 (b) $5^{100} - 1$ (c) $\frac{97}{300}$ (d) $a_n - a_0$

43. $\frac{1}{3}$ **45.** 14 **49.** $2^{n+1} + n^2 + n - 2$

EXERCISES C ▪ **PAGE A22**

1. (b) 0.405

INDEX

RP denotes Reference Page numbers.

ALGEBRA

ARITHMETIC OPERATIONS

$$a(b + c) = ab + ac$$

$$\frac{a}{b} + \frac{c}{d} = \frac{ad + bc}{bd}$$

$$\frac{a + c}{b} = \frac{a}{b} + \frac{c}{b}$$

$$\frac{\dfrac{a}{b}}{\dfrac{c}{d}} = \frac{a}{b} \times \frac{d}{c} = \frac{ad}{bc}$$

EXPONENTS AND RADICALS

$$x^m x^n = x^{m+n}$$

$$\frac{x^m}{x^n} = x^{m-n}$$

$$(x^m)^n = x^{mn}$$

$$x^{-n} = \frac{1}{x^n}$$

$$(xy)^n = x^n y^n$$

$$\left(\frac{x}{y}\right)^n = \frac{x^n}{y^n}$$

$$x^{1/n} = \sqrt[n]{x}$$

$$x^{m/n} = \sqrt[n]{x^m} = \left(\sqrt[n]{x}\right)^m$$

$$\sqrt[n]{xy} = \sqrt[n]{x}\,\sqrt[n]{y}$$

$$\sqrt[n]{\frac{x}{y}} = \frac{\sqrt[n]{x}}{\sqrt[n]{y}}$$

FACTORING SPECIAL POLYNOMIALS

$$x^2 - y^2 = (x + y)(x - y)$$

$$x^3 + y^3 = (x + y)(x^2 - xy + y^2)$$

$$x^3 - y^3 = (x - y)(x^2 + xy + y^2)$$

BINOMIAL THEOREM

$$(x + y)^2 = x^2 + 2xy + y^2 \qquad (x - y)^2 = x^2 - 2xy + y^2$$

$$(x + y)^3 = x^3 + 3x^2y + 3xy^2 + y^3$$

$$(x - y)^3 = x^3 - 3x^2y + 3xy^2 - y^3$$

$$(x + y)^n = x^n + nx^{n-1}y + \frac{n(n - 1)}{2}x^{n-2}y^2$$

$$+ \cdots + \binom{n}{k}x^{n-k}y^k + \cdots + nxy^{n-1} + y^n$$

where $\displaystyle \binom{n}{k} = \frac{n(n - 1) \cdots (n - k + 1)}{1 \cdot 2 \cdot 3 \cdot \cdots \cdot k}$

QUADRATIC FORMULA

If $ax^2 + bx + c = 0$, then $x = \dfrac{-b \pm \sqrt{b^2 - 4ac}}{2a}$.

INEQUALITIES AND ABSOLUTE VALUE

If $a < b$ and $b < c$, then $a < c$.

If $a < b$, then $a + c < b + c$.

If $a < b$ and $c > 0$, then $ca < cb$.

If $a < b$ and $c < 0$, then $ca > cb$.

If $a > 0$, then

$$|x| = a \quad \text{means} \quad x = a \quad \text{or} \quad x = -a$$

$$|x| < a \quad \text{means} \quad -a < x < a$$

$$|x| > a \quad \text{means} \quad x > a \quad \text{or} \quad x < -a$$

GEOMETRY

GEOMETRIC FORMULAS

Formulas for area A, circumference C, and volume V:

Triangle
$$A = \tfrac{1}{2}bh$$
$$= \tfrac{1}{2}ab \sin \theta$$

Circle
$$A = \pi r^2$$
$$C = 2\pi r$$

Sector of Circle
$$A = \tfrac{1}{2}r^2\theta$$
$$s = r\theta \ (\theta \text{ in radians})$$

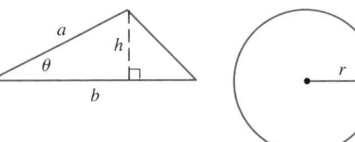 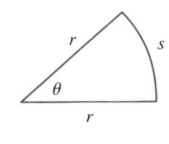

Sphere
$$V = \tfrac{4}{3}\pi r^3$$
$$A = 4\pi r^2$$

Cylinder
$$V = \pi r^2 h$$

Cone
$$V = \tfrac{1}{3}\pi r^2 h$$

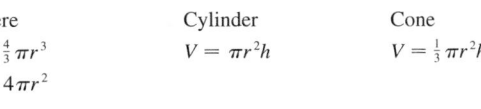

DISTANCE AND MIDPOINT FORMULAS

Distance between $P_1(x_1, y_1)$ and $P_2(x_2, y_2)$:

$$d = \sqrt{(x_2 - x_1)^2 + (y_2 - y_1)^2}$$

Midpoint of $\overline{P_1P_2}$: $\left(\dfrac{x_1 + x_2}{2}, \dfrac{y_1 + y_2}{2}\right)$

LINES

Slope of line through $P_1(x_1, y_1)$ and $P_2(x_2, y_2)$:

$$m = \frac{y_2 - y_1}{x_2 - x_1}$$

Point-slope equation of line through $P_1(x_1, y_1)$ with slope m:

$$y - y_1 = m(x - x_1)$$

Slope-intercept equation of line with slope m and y-intercept b:

$$y = mx + b$$

CIRCLES

Equation of the circle with center (h, k) and radius r:

$$(x - h)^2 + (y - k)^2 = r^2$$

TRIGONOMETRY

ANGLE MEASUREMENT

π radians $= 180°$

$1° = \dfrac{\pi}{180}$ rad $\qquad$ 1 rad $= \dfrac{180°}{\pi}$

$s = r\theta$

(θ in radians)

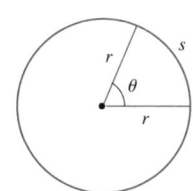

RIGHT ANGLE TRIGONOMETRY

$\sin\theta = \dfrac{\text{opp}}{\text{hyp}} \qquad \csc\theta = \dfrac{\text{hyp}}{\text{opp}}$

$\cos\theta = \dfrac{\text{adj}}{\text{hyp}} \qquad \sec\theta = \dfrac{\text{hyp}}{\text{adj}}$

$\tan\theta = \dfrac{\text{opp}}{\text{adj}} \qquad \cot\theta = \dfrac{\text{adj}}{\text{opp}}$

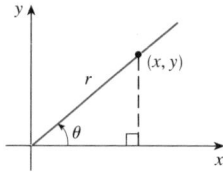

TRIGONOMETRIC FUNCTIONS

$\sin\theta = \dfrac{y}{r} \qquad \csc\theta = \dfrac{r}{y}$

$\cos\theta = \dfrac{x}{r} \qquad \sec\theta = \dfrac{r}{x}$

$\tan\theta = \dfrac{y}{x} \qquad \cot\theta = \dfrac{x}{y}$

GRAPHS OF THE TRIGONOMETRIC FUNCTIONS

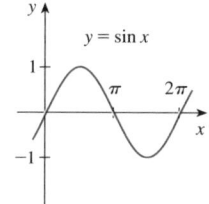

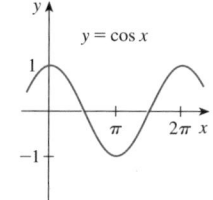

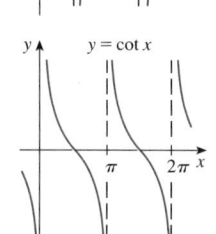

TRIGONOMETRIC FUNCTIONS OF IMPORTANT ANGLES

θ	radians	$\sin\theta$	$\cos\theta$	$\tan\theta$
0°	0	0	1	0
30°	$\pi/6$	$1/2$	$\sqrt{3}/2$	$\sqrt{3}/3$
45°	$\pi/4$	$\sqrt{2}/2$	$\sqrt{2}/2$	1
60°	$\pi/3$	$\sqrt{3}/2$	$1/2$	$\sqrt{3}$
90°	$\pi/2$	1	0	—

FUNDAMENTAL IDENTITIES

$\csc\theta = \dfrac{1}{\sin\theta} \qquad\qquad \sec\theta = \dfrac{1}{\cos\theta}$

$\tan\theta = \dfrac{\sin\theta}{\cos\theta} \qquad\qquad \cot\theta = \dfrac{\cos\theta}{\sin\theta}$

$\cot\theta = \dfrac{1}{\tan\theta} \qquad\qquad \sin^2\theta + \cos^2\theta = 1$

$1 + \tan^2\theta = \sec^2\theta \qquad\qquad 1 + \cot^2\theta = \csc^2\theta$

$\sin(-\theta) = -\sin\theta \qquad\qquad \cos(-\theta) = \cos\theta$

$\tan(-\theta) = -\tan\theta \qquad\qquad \sin\left(\dfrac{\pi}{2} - \theta\right) = \cos\theta$

$\cos\left(\dfrac{\pi}{2} - \theta\right) = \sin\theta \qquad\qquad \tan\left(\dfrac{\pi}{2} - \theta\right) = \cot\theta$

THE LAW OF SINES

$\dfrac{\sin A}{a} = \dfrac{\sin B}{b} = \dfrac{\sin C}{c}$

THE LAW OF COSINES

$a^2 = b^2 + c^2 - 2bc\cos A$

$b^2 = a^2 + c^2 - 2ac\cos B$

$c^2 = a^2 + b^2 - 2ab\cos C$

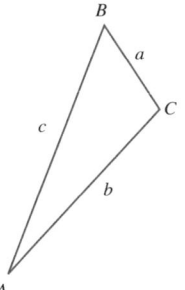

ADDITION AND SUBTRACTION FORMULAS

$\sin(x + y) = \sin x\cos y + \cos x\sin y$

$\sin(x - y) = \sin x\cos y - \cos x\sin y$

$\cos(x + y) = \cos x\cos y - \sin x\sin y$

$\cos(x - y) = \cos x\cos y + \sin x\sin y$

$\tan(x + y) = \dfrac{\tan x + \tan y}{1 - \tan x\tan y}$

$\tan(x - y) = \dfrac{\tan x - \tan y}{1 + \tan x\tan y}$

DOUBLE-ANGLE FORMULAS

$\sin 2x = 2\sin x\cos x$

$\cos 2x = \cos^2 x - \sin^2 x = 2\cos^2 x - 1 = 1 - 2\sin^2 x$

$\tan 2x = \dfrac{2\tan x}{1 - \tan^2 x}$

HALF-ANGLE FORMULAS

$\sin^2 x = \dfrac{1 - \cos 2x}{2} \qquad \cos^2 x = \dfrac{1 + \cos 2x}{2}$

TABLE OF INTERVALS

Notation	Set description	Picture		Notation	Set description	Picture
(a, b)	$\{x \mid a < x < b\}$			(a, ∞)	$\{x \mid x > a\}$	
$[a, b]$	$\{x \mid a \leq x \leq b\}$			$[a, \infty)$	$\{x \mid x \geq a\}$	
$[a, b)$	$\{x \mid a \leq x < b\}$			$(-\infty, b)$	$\{x \mid x < b\}$	
				$(-\infty, b]$	$\{x \mid x \leq b\}$	
$(a, b]$	$\{x \mid a < x \leq b\}$			$(-\infty, \infty)$	$\mathbb{R}$ (set of all real numbers)	

SPECIAL FUNCTIONS

POWER FUNCTIONS $f(x) = x^a$

(i) $f(x) = x^n$, n a positive integer

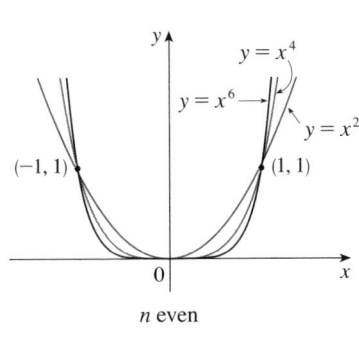

n even

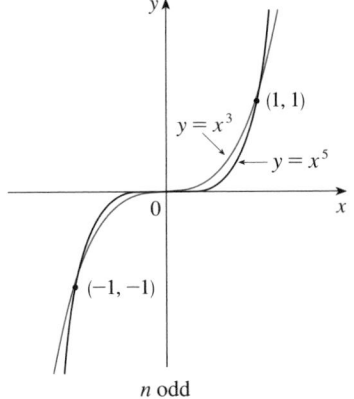

n odd

(ii) $f(x) = x^{1/n} = \sqrt[n]{x}$, n a positive integer

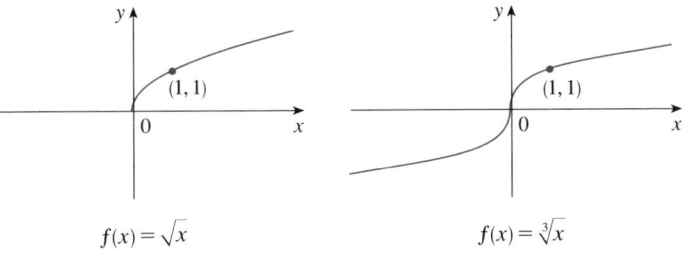

$f(x) = \sqrt{x}$ $f(x) = \sqrt[3]{x}$

(iii) $f(x) = x^{-1} = \dfrac{1}{x}$

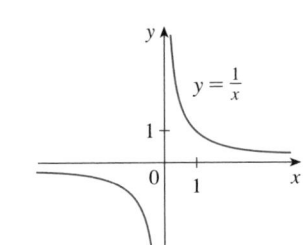

INVERSE TRIGONOMETRIC FUNCTIONS

$\arcsin x = \sin^{-1}x = y \iff \sin y = x \quad \text{and} \quad -\dfrac{\pi}{2} \leq y \leq \dfrac{\pi}{2}$

$\arccos x = \cos^{-1}x = y \iff \cos y = x \quad \text{and} \quad 0 \leq y \leq \pi$

$\arctan x = \tan^{-1}x = y \iff \tan y = x \quad \text{and} \quad -\dfrac{\pi}{2} < y < \dfrac{\pi}{2}$

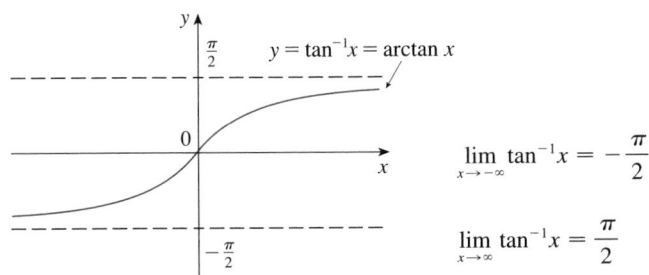

$\displaystyle\lim_{x \to -\infty} \tan^{-1}x = -\dfrac{\pi}{2}$

$\displaystyle\lim_{x \to \infty} \tan^{-1}x = \dfrac{\pi}{2}$

SPECIAL FUNCTIONS

EXPONENTIAL AND LOGARITHMIC FUNCTIONS

$\log_a x = y \iff a^y = x$

$\ln x = \log_e x, \quad \text{where} \quad \ln e = 1$

$\ln x = y \iff e^y = x$

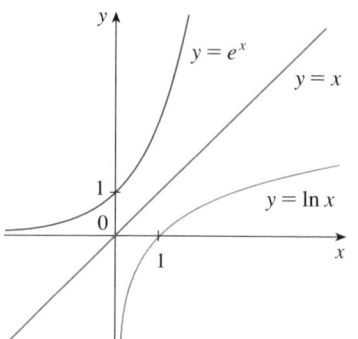

Cancellation Equations

$\log_a(a^x) = x \qquad a^{\log_a x} = x$

$\ln(e^x) = x \qquad e^{\ln x} = x$

Laws of Logarithms

1. $\log_a(xy) = \log_a x + \log_a y$

2. $\log_a\left(\dfrac{x}{y}\right) = \log_a x - \log_a y$

3. $\log_a(x^r) = r \log_a x$

$\displaystyle\lim_{x \to -\infty} e^x = 0 \qquad \lim_{x \to \infty} e^x = \infty$

$\displaystyle\lim_{x \to 0^+} \ln x = -\infty \qquad \lim_{x \to \infty} \ln x = \infty$

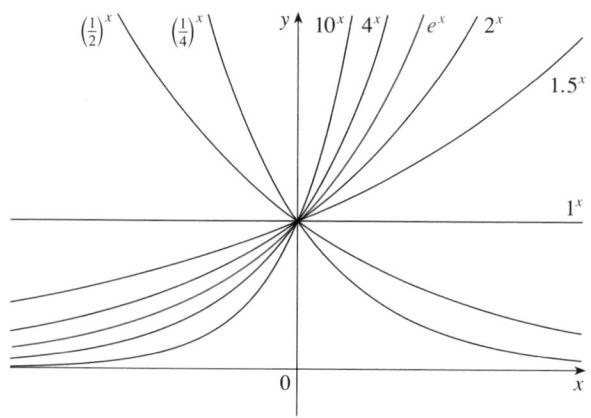

Exponential functions

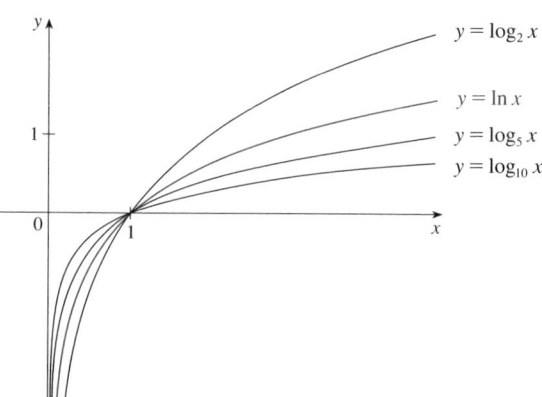

Logarithmic functions

HYPERBOLIC FUNCTIONS

$\sinh x = \dfrac{e^x - e^{-x}}{2}$

$\operatorname{csch} x = \dfrac{1}{\sinh x}$

$\cosh x = \dfrac{e^x + e^{-x}}{2}$

$\operatorname{sech} x = \dfrac{1}{\cosh x}$

$\tanh x = \dfrac{\sinh x}{\cosh x}$

$\coth x = \dfrac{\cosh x}{\sinh x}$

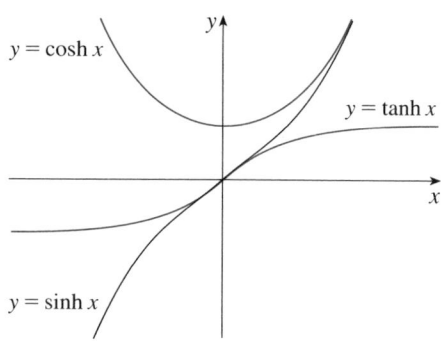

INVERSE HYPERBOLIC FUNCTIONS

$y = \sinh^{-1} x \iff \sinh y = x$

$\sinh^{-1} x = \ln\left(x + \sqrt{x^2 + 1}\right)$

$y = \cosh^{-1} x \iff \cosh y = x \quad \text{and} \quad y \geq 0$

$\cosh^{-1} x = \ln\left(x + \sqrt{x^2 - 1}\right)$

$y = \tanh^{-1} x \iff \tanh y = x$

$\tanh^{-1} x = \tfrac{1}{2} \ln\left(\dfrac{1 + x}{1 - x}\right)$

DIFFERENTIATION RULES

GENERAL FORMULAS

1. $\dfrac{d}{dx}(c) = 0$

2. $\dfrac{d}{dx}[cf(x)] = cf'(x)$

3. $\dfrac{d}{dx}[f(x) + g(x)] = f'(x) + g'(x)$

4. $\dfrac{d}{dx}[f(x) - g(x)] = f'(x) - g'(x)$

5. $\dfrac{d}{dx}[f(x)g(x)] = f(x)g'(x) + g(x)f'(x)$ (Product Rule)

6. $\dfrac{d}{dx}\left[\dfrac{f(x)}{g(x)}\right] = \dfrac{g(x)f'(x) - f(x)g'(x)}{[g(x)]^2}$ (Quotient Rule)

7. $\dfrac{d}{dx}f(g(x)) = f'(g(x))g'(x)$ (Chain Rule)

8. $\dfrac{d}{dx}(x^n) = nx^{n-1}$ (Power Rule)

EXPONENTIAL AND LOGARITHMIC FUNCTIONS

9. $\dfrac{d}{dx}(e^x) = e^x$

10. $\dfrac{d}{dx}(a^x) = a^x \ln a$

11. $\dfrac{d}{dx}\ln|x| = \dfrac{1}{x}$

12. $\dfrac{d}{dx}(\log_a x) = \dfrac{1}{x \ln a}$

TRIGONOMETRIC FUNCTIONS

13. $\dfrac{d}{dx}(\sin x) = \cos x$

14. $\dfrac{d}{dx}(\cos x) = -\sin x$

15. $\dfrac{d}{dx}(\tan x) = \sec^2 x$

16. $\dfrac{d}{dx}(\csc x) = -\csc x \cot x$

17. $\dfrac{d}{dx}(\sec x) = \sec x \tan x$

18. $\dfrac{d}{dx}(\cot x) = -\csc^2 x$

INVERSE TRIGONOMETRIC FUNCTIONS

19. $\dfrac{d}{dx}(\sin^{-1} x) = \dfrac{1}{\sqrt{1 - x^2}}$

20. $\dfrac{d}{dx}(\cos^{-1} x) = -\dfrac{1}{\sqrt{1 - x^2}}$

21. $\dfrac{d}{dx}(\tan^{-1} x) = \dfrac{1}{1 + x^2}$

22. $\dfrac{d}{dx}(\csc^{-1} x) = -\dfrac{1}{x\sqrt{x^2 - 1}}$

23. $\dfrac{d}{dx}(\sec^{-1} x) = \dfrac{1}{x\sqrt{x^2 - 1}}$

24. $\dfrac{d}{dx}(\cot^{-1} x) = -\dfrac{1}{1 + x^2}$

HYPERBOLIC FUNCTIONS

25. $\dfrac{d}{dx}(\sinh x) = \cosh x$

26. $\dfrac{d}{dx}(\cosh x) = \sinh x$

27. $\dfrac{d}{dx}(\tanh x) = \operatorname{sech}^2 x$

28. $\dfrac{d}{dx}(\operatorname{csch} x) = -\operatorname{csch} x \coth x$

29. $\dfrac{d}{dx}(\operatorname{sech} x) = -\operatorname{sech} x \tanh x$

30. $\dfrac{d}{dx}(\coth x) = -\operatorname{csch}^2 x$

INVERSE HYPERBOLIC FUNCTIONS

31. $\dfrac{d}{dx}(\sinh^{-1} x) = \dfrac{1}{\sqrt{1 + x^2}}$

32. $\dfrac{d}{dx}(\cosh^{-1} x) = \dfrac{1}{\sqrt{x^2 - 1}}$

33. $\dfrac{d}{dx}(\tanh^{-1} x) = \dfrac{1}{1 - x^2}$

34. $\dfrac{d}{dx}(\operatorname{csch}^{-1} x) = -\dfrac{1}{|x|\sqrt{x^2 + 1}}$

35. $\dfrac{d}{dx}(\operatorname{sech}^{-1} x) = -\dfrac{1}{x\sqrt{1 - x^2}}$

36. $\dfrac{d}{dx}(\coth^{-1} x) = \dfrac{1}{1 - x^2}$

Cut here and keep for reference

TABLE OF INTEGRALS

BASIC FORMS

1. $\int u \, dv = uv - \int v \, du$

2. $\int u^n \, du = \dfrac{u^{n+1}}{n+1} + C, \quad n \neq -1$

3. $\int \dfrac{du}{u} = \ln |u| + C$

4. $\int e^u \, du = e^u + C$

5. $\int a^u \, du = \dfrac{a^u}{\ln a} + C$

6. $\int \sin u \, du = -\cos u + C$

7. $\int \cos u \, du = \sin u + C$

8. $\int \sec^2 u \, du = \tan u + C$

9. $\int \csc^2 u \, du = -\cot u + C$

10. $\int \sec u \, \tan u \, du = \sec u + C$

11. $\int \csc u \, \cot u \, du = -\csc u + C$

12. $\int \tan u \, du = \ln |\sec u| + C$

13. $\int \cot u \, du = \ln |\sin u| + C$

14. $\int \sec u \, du = \ln |\sec u + \tan u| + C$

15. $\int \csc u \, du = \ln |\csc u - \cot u| + C$

16. $\int \dfrac{du}{\sqrt{a^2 - u^2}} = \sin^{-1} \dfrac{u}{a} + C, \quad a > 0$

17. $\int \dfrac{du}{a^2 + u^2} = \dfrac{1}{a} \tan^{-1} \dfrac{u}{a} + C$

18. $\int \dfrac{du}{u\sqrt{u^2 - a^2}} = \dfrac{1}{a} \sec^{-1} \dfrac{u}{a} + C$

19. $\int \dfrac{du}{a^2 - u^2} = \dfrac{1}{2a} \ln \left| \dfrac{u+a}{u-a} \right| + C$

20. $\int \dfrac{du}{u^2 - a^2} = \dfrac{1}{2a} \ln \left| \dfrac{u-a}{u+a} \right| + C$

FORMS INVOLVING $\sqrt{a^2 + u^2}$, $a > 0$

21. $\int \sqrt{a^2 + u^2} \, du = \dfrac{u}{2} \sqrt{a^2 + u^2} + \dfrac{a^2}{2} \ln\!\left(u + \sqrt{a^2 + u^2}\right) + C$

22. $\int u^2 \sqrt{a^2 + u^2} \, du = \dfrac{u}{8}(a^2 + 2u^2)\sqrt{a^2 + u^2} - \dfrac{a^4}{8} \ln\!\left(u + \sqrt{a^2 + u^2}\right) + C$

23. $\int \dfrac{\sqrt{a^2 + u^2}}{u} \, du = \sqrt{a^2 + u^2} - a \ln \left| \dfrac{a + \sqrt{a^2 + u^2}}{u} \right| + C$

24. $\int \dfrac{\sqrt{a^2 + u^2}}{u^2} \, du = -\dfrac{\sqrt{a^2 + u^2}}{u} + \ln\!\left(u + \sqrt{a^2 + u^2}\right) + C$

25. $\int \dfrac{du}{\sqrt{a^2 + u^2}} = \ln\!\left(u + \sqrt{a^2 + u^2}\right) + C$

26. $\int \dfrac{u^2 \, du}{\sqrt{a^2 + u^2}} = \dfrac{u}{2} \sqrt{a^2 + u^2} - \dfrac{a^2}{2} \ln\!\left(u + \sqrt{a^2 + u^2}\right) + C$

27. $\int \dfrac{du}{u\sqrt{a^2 + u^2}} = -\dfrac{1}{a} \ln \left| \dfrac{\sqrt{a^2 + u^2} + a}{u} \right| + C$

28. $\int \dfrac{du}{u^2 \sqrt{a^2 + u^2}} = -\dfrac{\sqrt{a^2 + u^2}}{a^2 u} + C$

29. $\int \dfrac{du}{(a^2 + u^2)^{3/2}} = \dfrac{u}{a^2 \sqrt{a^2 + u^2}} + C$

TABLE OF INTEGRALS

FORMS INVOLVING $\sqrt{a^2 - u^2}$, $a > 0$

30. $\displaystyle\int \sqrt{a^2 - u^2}\, du = \frac{u}{2}\sqrt{a^2 - u^2} + \frac{a^2}{2}\sin^{-1}\frac{u}{a} + C$

31. $\displaystyle\int u^2\sqrt{a^2 - u^2}\, du = \frac{u}{8}(2u^2 - a^2)\sqrt{a^2 - u^2} + \frac{a^4}{8}\sin^{-1}\frac{u}{a} + C$

32. $\displaystyle\int \frac{\sqrt{a^2 - u^2}}{u}\, du = \sqrt{a^2 - u^2} - a\ln\left|\frac{a + \sqrt{a^2 - u^2}}{u}\right| + C$

33. $\displaystyle\int \frac{\sqrt{a^2 - u^2}}{u^2}\, du = -\frac{1}{u}\sqrt{a^2 - u^2} - \sin^{-1}\frac{u}{a} + C$

34. $\displaystyle\int \frac{u^2\, du}{\sqrt{a^2 - u^2}} = -\frac{u}{2}\sqrt{a^2 - u^2} + \frac{a^2}{2}\sin^{-1}\frac{u}{a} + C$

35. $\displaystyle\int \frac{du}{u\sqrt{a^2 - u^2}} = -\frac{1}{a}\ln\left|\frac{a + \sqrt{a^2 - u^2}}{u}\right| + C$

36. $\displaystyle\int \frac{du}{u^2\sqrt{a^2 - u^2}} = -\frac{1}{a^2 u}\sqrt{a^2 - u^2} + C$

37. $\displaystyle\int (a^2 - u^2)^{3/2}\, du = -\frac{u}{8}(2u^2 - 5a^2)\sqrt{a^2 - u^2} + \frac{3a^4}{8}\sin^{-1}\frac{u}{a} + C$

38. $\displaystyle\int \frac{du}{(a^2 - u^2)^{3/2}} = \frac{u}{a^2\sqrt{a^2 - u^2}} + C$

FORMS INVOLVING $\sqrt{u^2 - a^2}$, $a > 0$

39. $\displaystyle\int \sqrt{u^2 - a^2}\, du = \frac{u}{2}\sqrt{u^2 - a^2} - \frac{a^2}{2}\ln\left|u + \sqrt{u^2 - a^2}\right| + C$

40. $\displaystyle\int u^2\sqrt{u^2 - a^2}\, du = \frac{u}{8}(2u^2 - a^2)\sqrt{u^2 - a^2} - \frac{a^4}{8}\ln\left|u + \sqrt{u^2 - a^2}\right| + C$

41. $\displaystyle\int \frac{\sqrt{u^2 - a^2}}{u}\, du = \sqrt{u^2 - a^2} - a\cos^{-1}\frac{a}{|u|} + C$

42. $\displaystyle\int \frac{\sqrt{u^2 - a^2}}{u^2}\, du = -\frac{\sqrt{u^2 - a^2}}{u} + \ln\left|u + \sqrt{u^2 - a^2}\right| + C$

43. $\displaystyle\int \frac{du}{\sqrt{u^2 - a^2}} = \ln\left|u + \sqrt{u^2 - a^2}\right| + C$

44. $\displaystyle\int \frac{u^2\, du}{\sqrt{u^2 - a^2}} = \frac{u}{2}\sqrt{u^2 - a^2} + \frac{a^2}{2}\ln\left|u + \sqrt{u^2 - a^2}\right| + C$

45. $\displaystyle\int \frac{du}{u^2\sqrt{u^2 - a^2}} = \frac{\sqrt{u^2 - a^2}}{a^2 u} + C$

46. $\displaystyle\int \frac{du}{(u^2 - a^2)^{3/2}} = -\frac{u}{a^2\sqrt{u^2 - a^2}} + C$

TABLE OF INTEGRALS

FORMS INVOLVING $a + bu$

47. $\displaystyle\int \frac{u\,du}{a + bu} = \frac{1}{b^2}\left(a + bu - a\ln|a + bu|\right) + C$

48. $\displaystyle\int \frac{u^2\,du}{a + bu} = \frac{1}{2b^3}\left[(a + bu)^2 - 4a(a + bu) + 2a^2\ln|a + bu|\right] + C$

49. $\displaystyle\int \frac{du}{u(a + bu)} = \frac{1}{a}\ln\left|\frac{u}{a + bu}\right| + C$

50. $\displaystyle\int \frac{du}{u^2(a + bu)} = -\frac{1}{au} + \frac{b}{a^2}\ln\left|\frac{a + bu}{u}\right| + C$

51. $\displaystyle\int \frac{u\,du}{(a + bu)^2} = \frac{a}{b^2(a + bu)} + \frac{1}{b^2}\ln|a + bu| + C$

52. $\displaystyle\int \frac{du}{u(a + bu)^2} = \frac{1}{a(a + bu)} - \frac{1}{a^2}\ln\left|\frac{a + bu}{u}\right| + C$

53. $\displaystyle\int \frac{u^2\,du}{(a + bu)^2} = \frac{1}{b^3}\left(a + bu - \frac{a^2}{a + bu} - 2a\ln|a + bu|\right) + C$

54. $\displaystyle\int u\sqrt{a + bu}\,du = \frac{2}{15b^2}(3bu - 2a)(a + bu)^{3/2} + C$

55. $\displaystyle\int \frac{u\,du}{\sqrt{a + bu}} = \frac{2}{3b^2}(bu - 2a)\sqrt{a + bu} + C$

56. $\displaystyle\int \frac{u^2\,du}{\sqrt{a + bu}} = \frac{2}{15b^3}(8a^2 + 3b^2u^2 - 4abu)\sqrt{a + bu} + C$

57. $\displaystyle\int \frac{du}{u\sqrt{a + bu}} = \frac{1}{\sqrt{a}}\ln\left|\frac{\sqrt{a + bu} - \sqrt{a}}{\sqrt{a + bu} + \sqrt{a}}\right| + C, \quad \text{if } a > 0$

$\displaystyle\qquad\qquad\qquad = \frac{2}{\sqrt{-a}}\tan^{-1}\sqrt{\frac{a + bu}{-a}} + C, \qquad \text{if } a < 0$

58. $\displaystyle\int \frac{\sqrt{a + bu}}{u}\,du = 2\sqrt{a + bu} + a\int \frac{du}{u\sqrt{a + bu}}$

59. $\displaystyle\int \frac{\sqrt{a + bu}}{u^2}\,du = -\frac{\sqrt{a + bu}}{u} + \frac{b}{2}\int \frac{du}{u\sqrt{a + bu}}$

60. $\displaystyle\int u^n\sqrt{a + bu}\,du = \frac{2}{b(2n + 3)}\left[u^n(a + bu)^{3/2} - na\int u^{n-1}\sqrt{a + bu}\,du\right]$

61. $\displaystyle\int \frac{u^n\,du}{\sqrt{a + bu}} = \frac{2u^n\sqrt{a + bu}}{b(2n + 1)} - \frac{2na}{b(2n + 1)}\int \frac{u^{n-1}\,du}{\sqrt{a + bu}}$

62. $\displaystyle\int \frac{du}{u^n\sqrt{a + bu}} = -\frac{\sqrt{a + bu}}{a(n - 1)u^{n-1}} - \frac{b(2n - 3)}{2a(n - 1)}\int \frac{du}{u^{n-1}\sqrt{a + bu}}$

TABLE OF INTEGRALS

TRIGONOMETRIC FORMS

63. $\displaystyle\int \sin^2 u\, du = \tfrac{1}{2}u - \tfrac{1}{4}\sin 2u + C$

64. $\displaystyle\int \cos^2 u\, du = \tfrac{1}{2}u + \tfrac{1}{4}\sin 2u + C$

65. $\displaystyle\int \tan^2 u\, du = \tan u - u + C$

66. $\displaystyle\int \cot^2 u\, du = -\cot u - u + C$

67. $\displaystyle\int \sin^3 u\, du = -\tfrac{1}{3}(2 + \sin^2 u)\cos u + C$

68. $\displaystyle\int \cos^3 u\, du = \tfrac{1}{3}(2 + \cos^2 u)\sin u + C$

69. $\displaystyle\int \tan^3 u\, du = \tfrac{1}{2}\tan^2 u + \ln|\cos u| + C$

70. $\displaystyle\int \cot^3 u\, du = -\tfrac{1}{2}\cot^2 u - \ln|\sin u| + C$

71. $\displaystyle\int \sec^3 u\, du = \tfrac{1}{2}\sec u \tan u + \tfrac{1}{2}\ln|\sec u + \tan u| + C$

72. $\displaystyle\int \csc^3 u\, du = -\tfrac{1}{2}\csc u \cot u + \tfrac{1}{2}\ln|\csc u - \cot u| + C$

73. $\displaystyle\int \sin^n u\, du = -\frac{1}{n}\sin^{n-1}u \cos u + \frac{n-1}{n}\int \sin^{n-2}u\, du$

74. $\displaystyle\int \cos^n u\, du = \frac{1}{n}\cos^{n-1}u \sin u + \frac{n-1}{n}\int \cos^{n-2}u\, du$

75. $\displaystyle\int \tan^n u\, du = \frac{1}{n-1}\tan^{n-1}u - \int \tan^{n-2}u\, du$

76. $\displaystyle\int \cot^n u\, du = \frac{-1}{n-1}\cot^{n-1}u - \int \cot^{n-2}u\, du$

77. $\displaystyle\int \sec^n u\, du = \frac{1}{n-1}\tan u \sec^{n-2}u + \frac{n-2}{n-1}\int \sec^{n-2}u\, du$

78. $\displaystyle\int \csc^n u\, du = \frac{-1}{n-1}\cot u \csc^{n-2}u + \frac{n-2}{n-1}\int \csc^{n-2}u\, du$

79. $\displaystyle\int \sin au \sin bu\, du = \frac{\sin(a-b)u}{2(a-b)} - \frac{\sin(a+b)u}{2(a+b)} + C$

80. $\displaystyle\int \cos au \cos bu\, du = \frac{\sin(a-b)u}{2(a-b)} + \frac{\sin(a+b)u}{2(a+b)} + C$

81. $\displaystyle\int \sin au \cos bu\, du = -\frac{\cos(a-b)u}{2(a-b)} - \frac{\cos(a+b)u}{2(a+b)} + C$

82. $\displaystyle\int u \sin u\, du = \sin u - u \cos u + C$

83. $\displaystyle\int u \cos u\, du = \cos u + u \sin u + C$

84. $\displaystyle\int u^n \sin u\, du = -u^n \cos u + n \int u^{n-1}\cos u\, du$

85. $\displaystyle\int u^n \cos u\, du = u^n \sin u - n \int u^{n-1}\sin u\, du$

86. $\displaystyle\int \sin^n u \cos^m u\, du = -\frac{\sin^{n-1}u \cos^{m+1}u}{n+m} + \frac{n-1}{n+m}\int \sin^{n-2}u \cos^m u\, du$

$\displaystyle\qquad = \frac{\sin^{n+1}u \cos^{m-1}u}{n+m} + \frac{m-1}{n+m}\int \sin^n u \cos^{m-2}u\, du$

INVERSE TRIGONOMETRIC FORMS

87. $\displaystyle\int \sin^{-1}u\, du = u \sin^{-1}u + \sqrt{1-u^2} + C$

88. $\displaystyle\int \cos^{-1}u\, du = u \cos^{-1}u - \sqrt{1-u^2} + C$

89. $\displaystyle\int \tan^{-1}u\, du = u \tan^{-1}u - \tfrac{1}{2}\ln(1+u^2) + C$

90. $\displaystyle\int u \sin^{-1}u\, du = \frac{2u^2-1}{4}\sin^{-1}u + \frac{u\sqrt{1-u^2}}{4} + C$

91. $\displaystyle\int u \cos^{-1}u\, du = \frac{2u^2-1}{4}\cos^{-1}u - \frac{u\sqrt{1-u^2}}{4} + C$

92. $\displaystyle\int u \tan^{-1}u\, du = \frac{u^2+1}{2}\tan^{-1}u - \frac{u}{2} + C$

93. $\displaystyle\int u^n \sin^{-1}u\, du = \frac{1}{n+1}\left[u^{n+1}\sin^{-1}u - \int \frac{u^{n+1}\, du}{\sqrt{1-u^2}} \right], \quad n \neq -1$

94. $\displaystyle\int u^n \cos^{-1}u\, du = \frac{1}{n+1}\left[u^{n+1}\cos^{-1}u + \int \frac{u^{n+1}\, du}{\sqrt{1-u^2}} \right], \quad n \neq -1$

95. $\displaystyle\int u^n \tan^{-1}u\, du = \frac{1}{n+1}\left[u^{n+1}\tan^{-1}u - \int \frac{u^{n+1}\, du}{1+u^2} \right], \quad n \neq -1$

TABLE OF INTEGRALS

EXPONENTIAL AND LOGARITHMIC FORMS

96. $\displaystyle\int ue^{au}\,du = \frac{1}{a^2}(au-1)e^{au} + C$

97. $\displaystyle\int u^n e^{au}\,du = \frac{1}{a}u^n e^{au} - \frac{n}{a}\int u^{n-1}e^{au}\,du$

98. $\displaystyle\int e^{au}\sin bu\,du = \frac{e^{au}}{a^2+b^2}(a\sin bu - b\cos bu) + C$

99. $\displaystyle\int e^{au}\cos bu\,du = \frac{e^{au}}{a^2+b^2}(a\cos bu + b\sin bu) + C$

100. $\displaystyle\int \ln u\,du = u\ln u - u + C$

101. $\displaystyle\int u^n \ln u\,du = \frac{u^{n+1}}{(n+1)^2}\big[(n+1)\ln u - 1\big] + C$

102. $\displaystyle\int \frac{1}{u\ln u}\,du = \ln|\ln u| + C$

HYPERBOLIC FORMS

103. $\displaystyle\int \sinh u\,du = \cosh u + C$

104. $\displaystyle\int \cosh u\,du = \sinh u + C$

105. $\displaystyle\int \tanh u\,du = \ln\cosh u + C$

106. $\displaystyle\int \coth u\,du = \ln|\sinh u| + C$

107. $\displaystyle\int \operatorname{sech} u\,du = \tan^{-1}|\sinh u| + C$

108. $\displaystyle\int \operatorname{csch} u\,du = \ln\left|\tanh\tfrac{1}{2}u\right| + C$

109. $\displaystyle\int \operatorname{sech}^2 u\,du = \tanh u + C$

110. $\displaystyle\int \operatorname{csch}^2 u\,du = -\coth u + C$

111. $\displaystyle\int \operatorname{sech} u\tanh u\,du = -\operatorname{sech} u + C$

112. $\displaystyle\int \operatorname{csch} u\coth u\,du = -\operatorname{csch} u + C$

FORMS INVOLVING $\sqrt{2au - u^2}$, $a > 0$

113. $\displaystyle\int \sqrt{2au-u^2}\,du = \frac{u-a}{2}\sqrt{2au-u^2} + \frac{a^2}{2}\cos^{-1}\!\left(\frac{a-u}{a}\right) + C$

114. $\displaystyle\int u\sqrt{2au-u^2}\,du = \frac{2u^2-au-3a^2}{6}\sqrt{2au-u^2} + \frac{a^3}{2}\cos^{-1}\!\left(\frac{a-u}{a}\right) + C$

115. $\displaystyle\int \frac{\sqrt{2au-u^2}}{u}\,du = \sqrt{2au-u^2} + a\cos^{-1}\!\left(\frac{a-u}{a}\right) + C$

116. $\displaystyle\int \frac{\sqrt{2au-u^2}}{u^2}\,du = -\frac{2\sqrt{2au-u^2}}{u} - \cos^{-1}\!\left(\frac{a-u}{a}\right) + C$

117. $\displaystyle\int \frac{du}{\sqrt{2au-u^2}} = \cos^{-1}\!\left(\frac{a-u}{a}\right) + C$

118. $\displaystyle\int \frac{u\,du}{\sqrt{2au-u^2}} = -\sqrt{2au-u^2} + a\cos^{-1}\!\left(\frac{a-u}{a}\right) + C$

119. $\displaystyle\int \frac{u^2\,du}{\sqrt{2au-u^2}} = -\frac{(u+3a)}{2}\sqrt{2au-u^2} + \frac{3a^2}{2}\cos^{-1}\!\left(\frac{a-u}{a}\right) + C$

120. $\displaystyle\int \frac{du}{u\sqrt{2au-u^2}} = -\frac{\sqrt{2au-u^2}}{au} + C$